T0398369

ENCYCLOPEDIA
AND HANDBOOK OF

MATERIALS,
PARTS,
and
FINISHES

THIRD EDITION

ENCYCLOPEDIA AND HANDBOOK OF

MATERIALS, PARTS, and FINISHES

THIRD EDITION

MEL SCHWARTZ

CRC Press
Taylor & Francis Group
Boca Raton London New York

CRC Press is an imprint of the
Taylor & Francis Group, an **informa** business

CRC Press
Taylor & Francis Group
6000 Broken Sound Parkway NW, Suite 300
Boca Raton, FL 33487-2742

© 2016 by Taylor & Francis Group, LLC
CRC Press is an imprint of Taylor & Francis Group, an Informa business

No claim to original U.S. Government works

Printed on acid-free paper
Version Date: 20160504

International Standard Book Number-13: 978-1-4665-6747-4 (Hardback)

This book contains information obtained from authentic and highly regarded sources. Reasonable efforts have been made to publish reliable data and information, but the author and publisher cannot assume responsibility for the validity of all materials or the consequences of their use. The authors and publishers have attempted to trace the copyright holders of all material reproduced in this publication and apologize to copyright holders if permission to publish in this form has not been obtained. If any copyright material has not been acknowledged please write and let us know so we may rectify in any future reprint. Printed in Canada.

Except as permitted under U.S. Copyright Law, no part of this book may be reprinted, reproduced, transmitted, or utilized in any form by any electronic, mechanical, or other means, now known or hereafter invented, including photocopying, microfilming, and recording, or in any information storage or retrieval system, without written permission from the publishers.

For permission to photocopy or use material electronically from this work, please access www.copyright.com (http://www.copyright.com/) or contact the Copyright Clearance Center, Inc. (CCC), 222 Rosewood Drive, Danvers, MA 01923, 978-750-8400. CCC is a not-for-profit organization that provides licenses and registration for a variety of users. For organizations that have been granted a photocopy license by the CCC, a separate system of payment has been arranged.

Trademark Notice: Product or corporate names may be trademarks or registered trademarks, and are used only for identification and explanation without intent to infringe.

Visit the Taylor & Francis Web site at
http://www.taylorandfrancis.com

and the CRC Press Web site at
http://www.crcpress.com

*This book is dedicated to three special family members, David Christopher Cumming,
Morton Irving Schwartz, who recently passed away, and Dr. Paul R. Kunk,
who is a wonderful and knowledgeable addition to our family.*

Contents

Preface...ix

A..1

B..73

C..121

D..245

E..283

F..321

G..379

H..407

I..437

J..475

K..479

L..485

M..515

N..581

O..609

P..629

Q..747

R..751

S..801

T..929

U..983

V..995

W 1015

X 1045

Y 1049

Z 1051

Preface

The encyclopedia and handbook represents an update of existing materials, parts, finishes (coatings), systems, and processes and includes new materials that have been invented or changed or modified either by new processes or by additions or innovative techniques—the book covers basic materials from "A" to "Z."

This encyclopedia is the culmination of over 70 years of various encyclopedias and material handbooks. With the advent of a steady increase in the number of materials and processes being developed over these past years, it is hoped that a one-volume encyclopedia would intelligently describe the important characteristics of commercially available materials without going into the details—an encyclopedia/handbook that would meet the job needs of managers and executives, purchasing and manufacturing managers, supervisors, engineers, metallurgists, chemists, students, technologists, teachers, and others.

The encyclopedia reflects the phenomenal proliferation in the number and variety of materials, processes, parts, coatings, and systems. More than 16,000 different materials are described. Despite this manifold increase over the years, it is virtually impossible to include all commercially available materials in a one-volume work of this kind. Nevertheless, the most important and most widely used of the thousands of materials introduced every year are included in this updated edition.

The diverse technologies that make up the field of materials and structures are at varying stages of commercialization. For example, piezoelectric and electrostrictive ceramics, piezoelectric polymers, and fiber-optic sensor systems are well-established commercial technologies, whereas microelectromechanical systems (MEMS), magnetostrictive materials, shape memory alloys (SMA) and polymers, conductive polymers, engineering-grade thermoplastics, nanomaterials, additive manufacturing (AM), and static-dissipative ABS plastics are, in some cases, in their developmental stages of commercialization.

There has been a tremendous increase in the variety of materials and processes, especially in the medical field, coupled with the rise of new and more severe service requirements, rigid testing, and qualification requirements, as well as a demand for lower cost. This has brought about many changes in the way these materials are utilized. Traditionally, users fit the design of a product to the properties of the material it was made of. This attitude has changed. The major concern now is finding and applying a material or materials with the right combination of properties to meet design and service conditions. Today, material selection is a complex process that operates throughout the entire span of a product's evolution. Thus, the new attitude is end-service oriented. The next logical step, of course, is tailor-made materials.

Analytical procedures have been developed to deal with the complex interaction between requirements and performance properties. And materials engineering departments and specialists who devote their full time to material selection problems are now a common adjunct to engineering or manufacturing departments.

As we will see, scientists, researchers, and technologists no longer accept the atomic arrangements nature gives us. Such is the case with, for example, high polymers, aggregates of giant, chain-like molecules. High polymers that nature gives us, such as wood, leather, and glue, have been used in engineering materials for centuries. But only during the past couple of decades have we acquired sufficient understanding of their molecular structure to improve upon nature. Now, by varying the chain length and degree of branching or cross-linking, materials can be produced with combinations of properties to meet specific application requirements.

The past century was dominated by the influence of emerging materials that were shaped and transformed into new mass-produced products that defined the twentieth century. Plastics, composites, aluminum, and advanced ceramics facilitated innovations that influenced culture and let us form products and machinery that made our lives easier—brightly colored plastics that pushed through labor-saving devices, materials engineered to make cars more comfortable, and touchscreen interactivity that simplified complex technology.

The next phase of material innovation is going to be a little more subtle and inconspicuous—not driven by materials and products that we can see, but rather by materials whose performance is more under the radar. These innovations are likely to manifest themselves through materials that intuitively make decisions for us. There is an increasing number of materials being developed as a response to how we live now and how we will live in the future. These new materials are often incorporating nanotechnology in some way to enhance performance or adapt, over time, to sensors and with the intention of improving our day-to-day life. Some materials that encompass this area are as follows:

- Nanocoated fabric, which repels stains
- Phase-change fabric
- Noise-absorbing material
- Self-healing plastic film
- Biodegradable additive for plastics
- Mineral plaster, which removes odor
- Foam crash mats, which keep you safe on the slopes

With regard to alloy names—metals, plastics, ceramics, and other material types—the choice between inclusion and exclusion is inevitably arbitrary. At one extreme, it would be ridiculous to offer an encyclopedia that excluded terms such as brass and solder or even admiralty brass or sterling silver. At the other extreme, it would be impossible, even assuming if it was desirable, to include all of the immense number of names introduced, and often discarded, at the whim of a manufacturer. The aim, therefore, will be to include all generic names, such as those mentioned here, together with a small number of trade or proprietary names, such as Dural and Nimonic, that are commonly used.

The aim of this book is to reflect the broadest usage of materials and processing technology that is current, and, furthermore, limited reference will be made to some terms once popular, but now fallen from favor. Consequently, a very large number of books, reports, and papers, and even conversations, have been absorbed in an attempt to provide a consensus and comprehensive view. Thus, rather than attempting the invidious task of identifying individual sources or influences, I prefer to offer thankful knowledge with gratitude to the many material engineers, metallurgists, technologists, scientists, chemists, and others whose work has directly or indirectly been utilized or described in this encyclopedia.

Mel Schwartz

A

A

The symbol for a repeating unit in a polymer chain.

Å

Angstrom, a unit of measure, 10^{-10} m.

A, A$_1$, A$_c$, etc.

Designations of critical temperatures in the transformation of steel.

ABA Copolymers

Block copolymers with three sequences but only two domains.

A-Basis

The "A" mechanical property value is the value above which at least 99% of the population of values is expected to fall, with a confidence of 95%. Also called A-allowable. See also *B-basis*, *S-basis*, and *typical-basis*.

Aberration

In microscopy, any error that results in image degradation. Such errors may be chromatic, spherical, astigmatic, chromatic, distortion, or curvature of field and can result from design or execution, or both.

Abhesive

A material that resists adhesion. A film or coating applied to surfaces to prevent sticking, heat sealing, and so on, such as a parting agent or mold release agent.

Ablative Plastic

A material that absorbs heat (with a low material loss and char rate) through a decomposition process (pyrolysis) that takes place at or near the surface exposed to the heat. This mechanism essentially provides thermal protection (insulation) of the subsurface materials and components by sacrificing the surface layer. Ablation is an exothermic process.

Ablatives

Materials used for the outward dissipation of extremely high heats by mass removal. Their most common use is as an external heat shield to protect supersonic aerospace vehicles from an excessive buildup of heat caused by air friction at the surface. The ablative material must have a low thermal conductivity in order that the heat may remain concentrated in the thin surface layer. As the surface of the ablator melts or sublimes, it is wiped away by the frictional forces that simultaneously heat newly exposed surfaces. The heat is carried off with the material removed. The less material that is lost, the more efficient is the ablative material. The ablative material also should have a high thermal capacity in the solid, liquid, and gaseous states, a high heat of fusion and evaporation, and a high heat of dissociation of its vapors. The ablative agent, or ablator, is usually a carbonaceous organic compound, such as a phenolic plastic. As the dissociation products are lost as liquid or vapor, the char is held in place by the refractivity reinforcing fibers, still giving a measure of heat resistance. The effective life of an ablative is short, calculated in seconds per millimeter of thickness for the distance traveled in the atmosphere.

Single ablative materials seldom have all of the desirable factors, and thus composites are used. Phenolic or epoxy resins are reinforced with asbestos fabric, carbonized cloth, or refractory fibers, such as asbestos, fused silica, and glass. The refractory fibers not only are incorporated for mechanical strength but also have a function in the ablative process, and surface-active agents may be added to speed the rate of evaporation. Ablative paint, for protecting woodwork, may be organic silicones, which convert to silica at temperatures above 2000°F (1093°C).

Abnormal Steel

A steel with variable hardenability leading to local soft spots when components are hardened. In the annealed or normalized conditions, these areas comprise abnormally coarse and irregular pearlite. The effect is associated with segregation of alloy elements or variation and aluminum (deoxidizer) content.

Abrasion

(1) A process in which hard particles or protuberances are forced against and moved along a solid surface. (2) A roughening or scratching of a surface due to abrasive wear. (3) The process of grinding or wearing away through the use of abrasives. The mechanism is predominantly a mechanical cutting action as opposed to the repeated welding/tearing action involved in adhesive wear.

Abrasion Artifact

A false structure introduced during an abrasion stage of a surface-preparation sequence.

Abrasion Fluid

A liquid added to an abrasion system. The liquid may act as a lubricant, as a coolant, or as a means of flushing abrasion debris from the abrasion track.

Abrasion Process

An abrasive machining procedure in which the surface of the workpiece is rubbed against a 2D array of abrasive particles under approximately constant load.

Abrasion Rate

The rate at which material is removed from a surface during abrasion. It is usually expressed in terms of the thickness removed per unit of time or distance traversed.

Abrasion Resistance

The ability of the material to resist surface wear.

Abrasion Soldering

A soldering process variation in which the faying surface of the base metal is mechanically abraded during soldering.

Abrasive

(1) A hard substance used for grinding, honing, lapping, superfinishing, polishing, pressure blasting, or barrel finishing. Abrasives in common use are natural and artificial. The natural abrasives include the diamond, emery, corundum, sand, crushed garnet and quartz, tripoli, and pumice. Artificial abrasives, or manufactured abrasives, are generally superior in uniformity to natural abrasives and are mostly silicon carbide, aluminum oxide, boron carbide, or boron nitride, marketed under trade names. Artificial diamonds are also now being produced. The massive natural abrasives, such as sandstone, are cut into grinding wheels from the natural block, but most abrasive materials are used as grains or built into artificial shapes.

For industrial grinding, artificial abrasives are preferred to natural abrasives because of their greater uniformity. Grading is important because uniform grinding requires grains of the same size. The abrasive grains are used as a grinding powder, are made into wheels, blocks, or stones, or are bonded to paper or cloth. An abrasive belt is a coated abrasive product in the form of a belt, used in production grinding and polishing.

Abrasive Blasting

A process for cleaning or finishing by means of an abrasive directed at high velocity against the workpiece.

Abrasive Cloth

It is made of cotton jean or drills to close tolerances of yarns and weaves, and the grains are attached with glue or resin.

Abrasive Disk

(1) A grinding wheel that is mounted on a steel plate, with the exposed flat side being used for grinding. (2) A disk-shaped, coated abrasive product.

Abrasive Erosion

Erosive wear caused by the relative motion of solid particles, which are entrained in a fluid, moving nearly parallel to a solid surface.

Abrasive Flow Machining

Removal of material by a viscous, abrasive media flowing under pressure through or across a workpiece.

Abrasive Jet Machining

Material removal from a workpiece, by impingement of fine abrasive particles, which are entrained in a focused, high-velocity gas stream.

Abrasive Machining

A machining process in which the points of abrasive particles are used as machining tools. Grinding is a typical abrasive machining process.

Abrasive Paper

Has the grains, usually aluminum oxide or silicon carbide, glued to one side of 40–130 lb kraft paper. The usual grain sizes are No. 16 to No. 500.

Abrasive Powder

Usually graded in sizes from 8 to 240 mesh. Coarse grain is to 24 mesh; fine-grain is 150–240. Levigated abrasives are fine powders for final burnishing of metals or for metallographic polishing, usually processed to make them chemically neutral. Green rouge is levigated chromic oxide, and mild polish may be levigated tin oxide; both are used for burnishing soft metals. Polishing powder may be aluminum oxide or metal oxide powders of ultrafine particle size down to 600 mesh.

Abrasive Sand

Any sand used for abrasive and grinding purposes, but the term does not include the sharp grains obtained by crushing quartz and used for sandpaper. The chief types of abrasive sand include sandblast sand, glass-grinding sand, and stone-cutting sand. Sand for stone sawing and for marble and glass grinding is usually ungraded, with no preparation other than screening, but it must have tough, uniform grains.

Abrasive Tumbling

See *barrel finishing*.

Abrasive Wear

The removal of material from a surface when hard particles slide or roll across the surface under pressure. The particles may be loose or may be part of another surface in contact with the surface being abraded. Compare with adhesive wear.

Abrasive Wheel

A grinding wheel composed of an abrasive grit and a bonding agent. Grinding of abrasive wheels is by grit size number from No. 10 to No. 600, which is 600 mesh; by grade of wheel, or strength of the bond, which is by letter designation, increasing in hardness from A to Z; and by grain spacing or structure number. The ideal condition is with a bond strong enough to hold the grains to accomplish the desired result and then release them before they become too dull. Essential qualities in the abrasive grain are penetration hardness, body strength sufficient to resist fracture until the points dull and then break to present a new edge, and an attrition resistance suitable to the work. Some wheels are made with a porous honeycombed structure to give free cutting and cooler operation on some types of metal grinding. Some diamond wheels are made with aluminum powder mixed with a thermosetting resin, and the diamond abrasive mix is hot pressed around this core wheel. Norton diamond wheels are of three types: metal bonded by powder metallurgy, resinoid bonded, and vitrified bonded.

Aluminum oxide wheels are used for grinding materials of high tensile strength. Silicon carbide is harder but is not as strong as aluminum oxide. It is used for grinding metals that have dense grain structure and for stone. Vitrified wheels are made by molding under heat and pressure. They are used for general and precision grinding where the wheel does not exceed a speed of 6500 surface ft/min (33 m/s). The rigidity gives high precision, and the porosity and strength of bond permit high stock removal. Silicate wheels have a silicate binder and are baked. The silicate bond releases the grains more easily than the vitrified and is used for grinding edge tools to reduce burning of the tool. Synthetic resins are used for bonding where greater strength is required and is obtained with the silicate, but less openness than with the vitrified. Resinoid bonds are used up to 16,000 surface ft/min (81 m/s) and are used especially for thread grinding and cut off wheels. Shellac binder is used for light work and for high finishing. Rubber is used for precision grinding and for centerless-feed machines.

Mild abrasives, uses silver polishes and window-cleaning compounds, such as chalk and talc, have a hardness of 1–2 Mohs. The milder abrasives for dental pastes and powders may be precipitated calcium carbonate, tricalcium phosphate, or combinations of sodium metaphosphate and tricalcium phosphate. Abrasives for metal polishes may also be pumice, diatomite, silica flour, tripoli, whiting, kaolin, tin oxide, or fuller's earth. This type of fine abrasive must be of very uniform grain in order to prevent scratching.

Abrasivity

The extent to which a surface, particle, or collection of particles will tend to cause abrasive wear when forced against a solid surface under relative motion and under prescribed conditions.

ABS Plastics

The letters ABS identify the family of acrylonitrile–butadiene–styrene resins. Common trade names for these materials are Cycolac, Magnum, and Lustran. ABS resins belong to a very versatile family of engineering thermoplastics, produced by combining three monomers—acrylonitrile, butadiene, and styrene—by a variety of methods involving polymerization, graft polymerization, physical mixing, and combinations thereof. Each monomer is an important component of ABS. Acrylonitrile contributes heat resistance, chemical resistance, and surface hardness to the system. The styrene component contributes processibility, rigidity, and strength. Butadiene contributes toughness and impact resistance. This family of plastics generally are opaque and distinguished by a good balance of properties, including high impact strength, rigidity, and hardness over a temperature range of −40°F to 230°F (−40°C to 110°C). Compared to other structural engineering plastics, they are generally considered to fall at the lower end of the scale. Medium impact grades are hard, rigid, and tough and are used for appearance parts

that require high strength, good fatigue resistance, and surface hardness and gloss. High impact grades are formulated for similar products where additional impact strength is gained at some sacrifice in rigidity and hardness. See Table A.1.

ABS plastics are two-phase systems. Styrene–acrylonitrile (SAN) forms the continuous matrix phase. The second phase is composed of dispersed polybutadiene particles, which have a layer of SAN grafted onto their surface. The layer of SAN at the interface makes the two phases compatible.

ABS plastics are readily processed by extrusion, injection molding, blow molding, calendering, and vacuum forming. Resins have been developed especially for cold forming or stamping from extruded sheet. Typical applications are helmets, refrigerator liners, luggage tote trays, housings, grills for hot air systems, and pump impellers. Extruded shapes include tubing and pipe. ABS-plated parts are now in wide use, replacing metal parts in the automotive and appliance field.

Absolute Density

See *density* and *absolute*.

Absolute Humidity

The weight of water vapor present in a unit volume of air, such as grams per cubic foot or grams per cubic meter. The amount of water vapor is also reported in terms of weight per unit weight of dry air, such as grams per pound of dry air, but this value differs from values calculated on a volume basis and should not be referred to as absolute humidity. It is designated as humidity ratio, specific humidity, or moisture content.

Absolute Impact Velocity

See *impact velocity*.

Absolute Pore Size

The maximum pore opening of a porous material, such as a filter, through which no large particle will pass.

Absolute Temperature

The temperature related to absolute zero. On the Celsius (centigrade, C) scale, 0° Kelvin (K) is −273.15°C and 1 Kelvin degree interval equals 1 Celsius degree interval. On the Fahrenheit (F) scale, 0° Rankine (R) is −459.67° Fahrenheit and 1 Rankine degree interval equals 1 Fahrenheit degree interval.

Absolute Viscosity

See *viscosity*.

Absorbance (A)

The logarithm to the base 10 of the reciprocal of the transmittance. The preferred term for photography is optical density.

Absorber

(1) A material capable of physically assimilating others. In a nuclear context, an absorber is a material capable of accepting and retaining

A

TABLE A.1
Properties of Acrylonitrile–Butadiene–Styrene and Styrene–Acrylonitrile

ASTM or UL Test	Property	Standard Acrylonitrile–Butadiene–Styrene Grades			Special-Purpose Acrylonitrile–Butadiene–Styrene Grades					Styrene–Acrylonitrile Grades
		High Impact	Superhigh Impact	Medium Impact	High Heat	Flame Retardant	Clear	Expandable	Plating	
Physical										
D792	Specific gravity	1.01–1.05	1.02–1.05	1.04–1.06	1.04–1.06	1.19–1.22	1.05	0.55–0.85	1.05–1.07	1.07–1.08
D792	Specific volume (in.^{3}/lb)	27	27	28	28	—	26	—	26	26
Mechanical										
D638	Tensile strength (psi)	6,000	5,000–6,300	6,000–7,500	6,000–7,500	5,500–10,000	5,800–6,300	3,000–4,000	5,500–6,600	9,000–12,000
D638	Elongation (%)	5–20	5–70	5–25	3–20	5–25	25–75	—	—	1–4
D638	Tensile modulus (10^5 psi)	3.3	2.0–3.4	3.6–3.8	3.0–4.0	3.2–3.7	3.0–3.3	1.0–2.5	3–3.8	4.5–5.6
D790	Flexural strength (psi)	10,500	6,000–11,500	11,500	10,000–13,000	9,000–12,250	10,500	3,000–8,000	8,700–11,500	14,000–17,000
D790	Flexural modulus (10^5 psi)	3.4	2.0–3.5	3.6–4.0	3.1–4.0	3.0–3.4	3.4–3.9	1.4–2.8	3.0–3.8	5.5
D256	Impact strength, Izod (ft lb/in. of notch)	6.5	7.0–8.0	4.0–5.5	2.3–6.0	4.0–13.0	2.5–4.0		5.0–7.0	0.35–0.50
D785	Hardness, Rockwell R	103	69–105	107	111	90–117	100–105	60–70[a]	103–109	M85
Thermal										
D696	Coefficient of thermal expansion (10^{-5}) in./in. °F	5.3	5.6	4.6	3.9–5.1	3.7–4.6	4.6	4.9	4.7–5.3	3.0
D648	Deflection temperature[b] (°F)									
	At 264 psi	188	192	184	220–240	180–220	168	160	189	210
	At 66 psi	203	208	201	230–245	198–238	180–185	185	214	—
UL94	Flammability rating	HB	HB	HB	HB	V-0 to V-1	HB	HB-V-0	HB	HB
Electrical										
D149	Dielectric strength (V/mils) Short time, 1/8 in. thk	400	350–500	350–500	350–500	400+	400	—	—	—
D495	Arc resistance (s)	89	50–85	50–85	50–85	20–60	120–130	—	—	—

Source: Mach. Design Basics Eng. Design, 674, June 1993. With permission.
ASTM, American Society for Testing and Materials; UL, Underwriters' Laboratories.
[a] Density has a marked effect.
[b] Unannealed.
[c] 0.060 in.—thick samples.

a large number of neutrons without fission. (2) A material or component intended to absorb kinetic energy as in the shock absorber of a vehicle.

Absorption

(1) The taking up of a liquid or gas by capillary, osmotic, or solvent action. (2) The capacity of a solid to receive and retain a substance, usually a liquid or gas, with the formation of an apparently homogeneous mixture. (3) Transformation of radiant energy to a different form of energy by interaction with matter. (4) The process by which a liquid is drawn into and tends to fill permeable pores in a porous solid body; also, the increase in mass of a porous solid body resulting from the penetration of a liquid into its permeable pores. See also *adsorption*.

Unless otherwise stated, it is usually implicit that the interaction is physical, there is no chemical reaction or alloying.

Absorption Contrast

In transmission electron microscopy, image contrast caused by differences in absorption within a sample due to regions of different mass density and thickness.

Absorption Edge

The wavelength or energy corresponding to a discontinuity in the plot of absorption coefficient versus wavelength for a specific medium.

Absorption Spectroscopy

The branch of spectroscopy treating the theory, interpretation, and application of spectra originating in the absorption of electromagnetic radiation by atoms, ions, radicals, and molecules.

Absorptive Lens (Eye Protection)

A filter lens whose physical properties are designed to attenuate the effects of glare and reflected and stray light. See also *filter plate*.

Absorptivity

A measure of radiant energy from an incident beam as it traverses an absorbing medium, equal to the absorbance of the medium divided by the product of the concentration of the substance and the sample half-life. Also known as *absorption coefficient*.

AC

Alternating current.

AC

See *acetal copolymers*, *acetal homopolymers*, and *acetal resins*.

AC Noncapacitive Arc

A high-voltage electrical discharge used in spectrochemical analysis to vaporize the sample material. See also *DC intermittent noncapacitive arc*.

A.C.P.D.

Alternating current, potential drop; a technique for measuring crack propagation, see *potential drop*.

Accelerated Aging

A process by which the effects of aging are accelerated under extreme and/or cycling temperature and humidity conditions. The process is meant to duplicate long-time environmental conditions in a relatively short space of time.

Accelerated Corrosion Test

Method designed to approximate, in a short time, the deteriorating effect under normal long-term service conditions.

Accelerated Life Test

A method designed to approximate, in a short time, the deteriorating effect obtained under normal long-term service conditions. See also *artificial aging*.

Accelerated Testing

A test performed on materials or assemblies that are meant to produce failures caused by the same failure mechanism as expected in field operation but in significantly shorter time. The failure mechanism is accelerated by changing one or more of the controlling test parameters.

Accelerating Potential

(1) A relatively high voltage applied between the cathode and anode of an electron gun to accelerate electrons. (2) The potential in electron beam welding that imparts the velocity to the electronics, thus giving them energy.

Accelerating Voltage

In various electron beam instruments and x-ray generators, the difference in potential between the filament (cathode) and the anode, causing acceleration of the electrons by 2–30 keV. See also *depth of penetration* and *resolution*.

Acceleration Period

In cavitation and liquid impingement erosion, the stage following the incubation period, during which the erosion rate increases from near zero to a maximum value.

Accelerator

A material that, when mixed with a catalyst or a resin, speeds up the chemical reaction between the catalyst and the resin (usually in the polymerizing of resins or vulcanization of rubbers). Also called promoter.

Acceptable Quality Level

(1) The lowest-quality level a supplier is permitted to present continually for acceptance. (2) The maximum percentage of defects or number of defective parts considered to be an acceptable average for a given process or technique.

Acceptable Weld

A weld that meets all the requirements and the acceptance criteria prescribed by the welding specifications.

A

Acceptance Test

A test, or series of tests, conducted by the procuring agency, or an agent thereof, upon receipt, to determine whether an individual lot of materials conforms to the purchase order or contract or to determine the degree of uniformity of the material supplied by the vendor or both. Compare to preproduction test and qualification test.

Accepted Reference Value

A value that serves as an agreed-on reference for comparison and which is derived as: (1) a theoretical or established value, based on scientific principles, (2) an assigned value, based on experimental work of some national or international standards organization, or (3) a consensus value, based on collaborative experimental work under the auspices of a scientific or engineering group. When the accepted reference value is the theoretical value, it is sometimes referred to as the "true" value.

Acceptor (in Semiconductor)

See *semiconductor*.

Access Hole

A hole or series of holes in successive layers of a multilayer printed circuit board that provide access to the surface of the land in one of the layers of the board.

Accessory Seal

On various types of engines, a seal that is employed for sealing an accessory shaft in the gearbox, such as a shaft for operating an oil pump, a fuel pump, a generator, a starter, or a deoiler.

Accordion

A type of printed circuit connector contact in which the spring is given a Z shape to permit high deflection without causing overstress.

Accumulation Period

See preferred term acceleration period.

Accumulator

An auxiliary cylinder and piston (plunger) mounted on injection molding or blowing machines and used to provide faster molding cycles. In blow molding, the accumulator cylinder is filled (during the time between parison deliveries, or "shots") with melted plastic coming from the main (primary) extruder. The plastic melt is stored, or *accumulated*, in this auxiliary cylinder until the next shot or parison is required. At that time, the piston in the accumulator cylinder forces the molten plastic into the dies that form the parison.

1. A reversible electrolytic cell in which electricity is stored as chemical energy.
2. A vessel serving as a reservoir for a fluid power source. A gas, often air, would be stored and compressed while a liquid, often water, might be pressurized or not.

Accumulator Metal Alloys

Based on lead with small quantities of tin and other elements used for electrical storage batteries.

Acetal (AC) Copolymers

A family of highly crystalline thermoplastics prepared by copolymerizing trioxane with small amounts of a comonomer that randomly distributes carbon–carbon bonds in the polymer chain. These bonds, as well as hydroxyethyl terminal units, give the acetal copolymers a high degree of thermal stability and resistance to strong alkaline environments.

The copolymers have an excellent balance of properties and processing characteristics. Melt temperature can range from 182°C to 232°C with little effect on part strength. UV-resistant grades (also available in colors), glass-reinforced grades, low-wear grades, and impact-modified grades are standard. Also available are electroplatable and dimensionally stable, low-warpage grades.

Properties

Acetal copolymers have high tensile flexural strength, fatigue resistance, and hardness. Lubricity is excellent. They retain much of their toughness through a broad temperature range and are among the most creep resistant of the crystalline thermoplastics. Moisture absorption is low, permitting molded parts to serve reliably and environments involving humidity changes.

Good electrical properties, combined with high mechanical strength and an Underwriters' Laboratories electrical rating of 100°C, qualify these materials for electrical applications requiring long-term stability.

Acetal copolymers have excellent resistance to chemicals and solvents. For example, specimens immersed for 12 months at room temperature in various inorganic solutions were unaffected except by strong mineral acids—sulfuric, nitric, and hydrochloric. Continuous contact is not recommended with strong oxidizing agents such as aqueous solutions containing high concentrations of hypochlorite ions. Solutions of 10% ammonium hydroxide and 10% sodium chloride discolor samples in prolonged immersion, but physical mechanical properties are not significantly changed. Most organic reagents tested have no effect, nor do mineral oil, motor oil, or brake fluids. Resistance to strong alkalies is essentially good; specimens immersed in boiling 50% sodium hydroxide solution and other strong bases for many months showed no property changes.

Strength of acetal copolymer is only slightly reduced after aging for 1 year in air at 116°C. Impact strength holds constant the first 6 months and falls off about one-third during the next 6-month period. Aging in air at 82°C for 2 years has little or no effect on properties, and immersion for 1 year in 82°C water leaves most properties virtually unchanged. Samples tested in boiling water retain nearly original tensile strength after 9 months.

The Creek-modulus curve of acetal copolymer under load shows a linear decrease on a log–log scale, typical of many plastics. Acetal springs lose over 50% of spring force after 1000 h and 60% in 10,000 h. The same spring loses 66% of its force after 100,000 h (about 11 years) under load.

Plastic springs are best used in applications where they generate a force at a specified deflection for limited time but otherwise remain relaxed. Ideally, spring should undergo occasional deflections where they have time to recover, at less than 50% design strain. Recovery time should be at least equal to time under load.

Application

Industrial and automotive applications of acetal copolymer include gears, cams, bushings, clips, lugs, door handles, window cranks, housings, and seat-belt components. Plumbing products such as valves, valve stems, pumps, faucets, and impellers utilize the lubricity and corrosion and hot water resistance of the copolymer. Mechanical components that require dimensional stability, such as watch gears, conveyor links, aerosols, and mechanical pen and pencil parts, or other uses applications for the FDA-approved grades include milk pumps, coffee spigots, filter housings, and food conveyors. Parts that require greater load-bearing stability at elevated temperatures, such as cams, gears, television tuner arms, and automotive underhood components, are molded from glass fiber–reinforced grades.

More costly acetal copolymer has excellent load-bearing characteristics for long-lasting plastic springs. To boost resin performance, engineers use fillers, reinforcing fibers, and additives. Although there are automotive uses for large fiber-reinforced composite leaf springs, unfilled resins are the better candidates for small springs. Glass fibers increase stiffness and strength, but they also limit deflection. And impact modifiers reduce modulus and make plastics more flexible but decrease creep resistance.

Acetal (AC) Homopolymers

Highly crystalline linear polymers formed by polymerizing formaldehyde and capping it with acetate and groups.

The homopolymers are available in several viscosity ranges that meet a variety of processing and end-use needs. The higher-viscosity materials are generally used for extrusions and for molded parts requiring maximum toughness; the lower-viscosity grades are used for injection molding. Elastomer-modified grades offer greatly improved toughness.

Properties

Acetal homopolymer resins have high tensile strength, stiffness, resilience, printing endurance, and moderate toughness under repeated impact. Some tough grades can deliver up to seven times greater toughness than unmodified acetal in Izod impact tests and up to 30 times greater toughness as measured by Gardner impact tests (Table A.2).

Homopolymer acetals have high resistance to organic solvents, excellent dimensional stability, a low coefficient of friction, and outstanding abrasion resistance among thermoplastics. The general-purpose resins can be used over a wide range of environmental conditions; special, UV-stabilized grades are recommended for applications requiring long-term exposure to weathering. However, prolonged exposure to strong acids and bases outside the range of pH 4–9 is not recommended.

Acetal homopolymer has the highest fatigue endurance of any unfilled commercial thermoplastic. Under completely reversed tensile and compressive stress, and with 100% relative humidity (at 73°F), fatigue endurance limit is 30.9 MPa at 10^6 cycles. Resistance to creep is excellent. Moisture, lubricants, and solvents including gasoline and gasohol have little effect on this property, which is important in parts incorporating self-threading screws or interference fits.

The low friction and good wear resistance of acetals against metals make these resins suitable for use in cams and gears having internal bearings. The coefficient of friction (nonlubricated) on steel in a

TABLE A.2
Properties of Acetals

ASTM or UL Test	Property	Copolymer	Homopolymer
Physical			
D792	Specific gravity	1.41	1.42
D792	Specific volume (in.³/lb)	19.7	19.5
D570	Water absorption, 24 h, 1/8 in. thk (%)	0.22	0.25
Mechanical			
D638	Tensile strength (psi)[a]		
	At 73°F	8,800	10,000
	At 160°F	5,000	6,900
D638	Elongation (%)[a]	60	40
D638	Tensile modulus (10^5 psi)[a]	4.1	5.2
D790	Flexural strength (psi)	13,000	14,100
D790	Flexural modulus (10^5 psi)		
	At 73°F	3.75	4.10
	At 160°F	1.80	2.30
D256	Impact strength, Izod (ft lb/in.)		
	Notched	1.3	1.4
	Unnotched	20	24
D671	Fatigue endurance limit, 10^7 cycles (psi)	3,300	4,300
D785	Hardness, Rockwell M	80	94
Thermal			
C177	Thermal conductivity		
	(10^{-4} cal cm/s cm² °C)	5.5	8.9
	(Btu in./h ft² °F)	1.6	2.6
D696	Coefficient of thermal expansion −40°C to +185°C (10^{-5} in./in. °C)	8.5	10.0
D648	Deflection temp (°F)		
	At 264 psi	230	277
	At 66 psi	316	342
UL94	Flammability rating	HB	HB
Electrical			
D149	Dielectric strength		
	Short time (V/mils)		
	5 mils	2,100	3,000
	20 mils	—	2,000
	90 mils	500	500
D150	Dielectric constant		
	At 1 kHz	3.7	3.7
	At 1 MHz	3.7	3.7
D150	Dissipation factor		
	At 1 kHz	0.001	0.001
	At 1 MHz	0.006	0.005
D257	Volume resistivity (Ω cm) At 73°F, 50% RH	10^{14}	10^{15}
D495	Arc resistance (s) 120 mils	240 (burns)	220 (burns, no tracking)
Frictional			
	Coefficient of friction		
	Self	0.35	0.3
	Against steel	0.15	0.15

Source: Mach. Design Basics Eng. Design, 676, June 1993. With permission.
[a] At 0.2 in./min loading rate.

rotating thrust washer test is 0.1–0.3, depending on pressure; little variation occurs from 22.8°C to 121°C. For even lower friction and wear, polytetrafluoroethylene fiber-filled and chemically lubricated formulations are available.

Properties of low moisture absorption, excellent creep resistance, and height deflection temperature suit acetal homopolymer or close-tolerance, high-performance parts.

Applications

Automotive applications of acetal homopolymer resins include fuel-system and seat-belt components, steering columns, window-support brackets, and handles. Typical plumbing applications that have replaced brass or zinc components are showerheads, ball cocks, faucet cartridges, and various fittings. Consumer items include quality toys, garden sprayers, stereo cassette parts, butane lighter bodies, zippers, and telephone components. Industrial applications of acetal homopolymer include couplings, pump impellers, conveyor plates, gears, sprockets, and springs.

Acetal (AC) Resins

Thermoplastics (polyformaldehyde and polyoxymethylene resins) produced by the addition polymerization of aldehydes by means of the carbonyl function, yielding unbranched polyoxymethylene chains of great length. The acetal resins, among the strongest and stiffest of all thermoplastics, are also characterized by good fatigue life, resilience, low moisture sensitivity, high solvent and chemical resistance, and good electrical properties. They may be processed by conventional injection molding and extrusion techniques and fabricated by welding methods used for other plastics.

Processing Acetals

Acetal resin can be molded in standard injection molding equipment at conventional production rates. The processing temperature is around 204°C. Satisfactory performance has been demonstrated in full-automatic injection machines using multicavity molds. Successful commercial moldings point up the ability of the material to be molded to form large-area parts with thin sections, heavy parts with thick sections, parts requiring glossy surfaces or different surface textures, parts requiring close tolerances, parts with undercuts for snap fits, parts requiring metal inserts, and parts requiring no flash. It can also be extruded as rod, tubing, sheeting, jacketing, wire coating, or shapes on standard commercial equipment. Extrusion temperatures are in the range of 199°C–204°C.

Generally, the same equipment and techniques for blow molding other thermoplastics work with acetal resin. Both thin-wall and thick-wall containers (aerosol type) can be produced in many shapes and surface textures.

Various sheet-forming techniques including vacuum, pressure, and matched mold have been successfully used with acetal resins.

Fabrication

Acetal resin is easy to machine (equal to or better than free-cutting brass) on standard production machine shop equipment. It can be sawed, drilled, turned, milled, shaped, reamed, threaded and tapped, blanked and punched, filed, sanded, and polished.

The material is easy to join and offers wide latitude in the choice of fast, economical methods of assembly. Integral bonds of acetal-to-acetal can be performed by welding with a heated metal surface, hot gas, hot wire, or spin-welding techniques. High-strength joints result from standard mechanical joining methods such as snap fits, interference or press fits, rivets, nailing, heading, threads, or self-tapping screws. Where low joint strengths are acceptable, several commercial adhesives can be used for bonding acetal to itself and other substrates.

Acetyl resin can be painted successfully with certain commercial paints and lacquers, using ordinary spraying equipment and a special surface treatment or followed by a baked topcoat. Successful first surface metalizing has been accomplished with conventional equipment and standard techniques for application of such coatings. Direct printing, process printing, and roll-leaf stamping (hot stamping) can be used for printing on acetyl resin. Baking at elevated temperatures is required for good adhesion of the ink in direct and screen-process printing. In hot stamping, the heated die provides the elevated temperature. Printing produced by these processes resists abrasion and listing by cellophane adhesive tape.

Acetal Plastics

Acetals are independent structural units or a part of certain biological and commercial polymers, and acetal resins are highly crystallized plastics based on formaldehyde polymerization technology. These engineering resins are strong, rigid, and have good moisture, heat, and solvent resistance.

Acetals were specially developed to compete with zinc and aluminum castings. The natural acetal resin is translucent white and can be readily colored with a high sparkle and brilliance. There are two basic types—homopolymer (Delrin) and copolymer (Celcon). In general, the homopolymers are harder, more rigid, have a higher tensile flexural and fatigue strength, but lower elongation; however, they have a higher melting point. Some high-molecular-weight homopolymer grades are extremely tough and have higher elongation than the copolymers. Homopolymer grades are available that are modified for improved hydrolysis resistance to 82°C, similar to copolymer materials.

The copolymers remain stable and long-term, high-temperature service and offer exceptional resistance to the effects of immersion in water at high temperatures. Neither of the types resists strong acids, and the copolymer is virtually unaffected by strong bases. Both types are available in a wide range of melt-flow grades, but the copolymers process more easily and faster than the conventional homopolymer grades.

Both the homopolymers and copolymers are available in several unmodified and glass fiber–reinforced injection molding grades. Both are available in polytetrafluoroethylene or silicone-filled grades, and the homopolymer is available in chemically lubricated low-friction formulations.

The acetals are also available in extruded rod and slab form for machine parts. Property data listed in Table A.2 applied to the general-purpose injection molding and extrusion grade of Delrin 500 and to Celcon M00.

Acetals are among the strongest and stiffest of the thermoplastics. Their tensile strength range is from 54.4 to 92.5 MPa, tensile modulus is about 3400 MPa, and fatigue strength at room temperature is about 34 MPa. Acetals are also among the best in creep resistance. This combined with low moisture absorption (less than 0.4%) gives them excellent dimensional stability. They are useful for continuous service up to about 104°C.

Injection-molding powders and extrusion powders are the most frequently used forms of the material. Sheets, rods, tubes, and pipe are also available. Colorability is excellent.

The range of desirable design properties and processing techniques provides outstanding design freedom in the areas (1) style (color, shape, surface texture, and decoration), (2) weight reduction, (3) assembly techniques, and (4) one-piece multifunctional parts (e.g., combined gear, cam, bearing, and shaft).

Acetic Acid

Also known as ethanoic acid. A colorless, corrosive liquid of pungent odor and composition is $CH_3 \cdot COOH$ having a wide variety of industrial uses as a reagent, solvent, and esterifier. A carboxylic acid, it is employed as a weak acid for etching and for soldering, in stain removers and bleaches; as a preservative, in photographic chemicals, for the manufacture of cellulose acetate and vinyl acetate; as a solvent for essential oils, resins, and gums; and as a precipitant for latex, in tanning leather, and in making artificial flavors. Acetic acid is found in the juices of many fruits and, in combination, in the stems or woody parts of plants. It is the active principle in vinegar, giving it the characteristic sour taste, acid flavor, and pungent odor. It is made commercially by oxidation of acetaldehyde (in the presence of manganese, cobalt, or copper acetate), butane, or naphtha. Its specific gravity is 1.049, its boiling point is 118°C, and it becomes a colorless solid below 16.6°C.

Acetone

An important industrial solvent, used in the manufacture of lacquers, plastics, smokeless powder, for dewaxing lubricating oils, for dissolving acetylene for storage, for dyeing cotton with aniline black, and as a raw material in the manufacture of other chemicals. It is a colorless, flammable liquid with a mint-like odor and is soluble in water and in ether. The composition is $CH_3 \cdot CO \cdot CH_3$, specific gravity 0.790, boiling point 56°C, and solidification point −94°C. Acetone is mainly produced as a by-product in the cleavage of cumene hydroperoxide into phenol. A secondary route is by catalytic dehydrogenation of isopropyl alcohol.

Synthetic methyl acetone is a mixture of about 50% acetone, 30 methyl acetate, and 20 methanol, used in lacquers, paint removers, and for coagulating latex. Dihydroxyacetone, a colorless crystalline solid produced from glycerin by sorbose bacteria reaction, is used in cosmetics, and in preparing foodstuff emulsions, plasticizers, and alkyd resins. It is soluble in water and alcohol.

Acetylene

A fuel gas, C_2H_2, colorless, a flammable gas with a garlic-like odor. Under compressed conditions, it is highly explosive.

Acetylene cannot be compressed to more than about 2 atm, so it is commonly stored in cylinders, which are packed with a suitable porous material such as carbon or kapok and then filled with acetone. The acetone is capable of absorbing about 25 times its own volume of acetylene for every atmosphere increase in pressure.

With its intense heat and controllability, the oxyacetylene flame can be used for many different welding and cutting operations including hard facing, brazing, beveling, gouging, and scarfing. The heating capability of acetylene also can be utilized in the bending, straightening, forming, hardening, softening, and strengthening of metals.

Achromatic

Free of color. A lens or objective is achromatic when corrected for longitudinal chromatic aberration for two colors. See also *achromatic objective*.

Achromatic Lens

A lens that is corrected for chromatic aberration so that its tendency to refract light differently as a function of wavelength is minimized. See also *achromatic* and *apochromatic lens*.

Achromatic Objective

Objectives are achromatic when corrected chromatically for two colors, generally red and green, and spherically for light of one color, usually in the yellow-green portion of the spectrum.

Acicular Alpha (Titanium)

A product of nucleation and growth from β to the lower temperature allotrope α phase. It may have a needlelike appearance and may have needle, lenticular, or flattened bar morphology in three dimensions.

Acicular Ferrite

A highly substructured nonequiaxed ferrite formed upon continuous cooling by a mix diffusion and shear mode of transformation that begins at a temperature slightly higher than the transformation temperature range for upper bainite. It is distinguished from bainite in that it has a limited amount of carbon available; thus, there is only a small amount of carbide present.

Acicular Ferrite Steels

Ultralow carbon (<0.08%) steels having a microstructure consisting of either acicular ferrite (low-carbon bainite) or a mixture of acicular and equiaxed ferrite.

Acicular Powder

A powder composed of needle or sliver-like particles.

Acid

A chemical substance that yields hydrogen ions (H^+) when dissolved in water. If it produces hydroxyl ions, (OH^-), it is termed "alkaline." Compare with base and pH. In a metallurgical context, the term indicates processes in which the environmental conditions such as the furnace refractory lining and slags, which contain a high percentage of silica and minerals, are acidic as opposed to basic. Siliceous refractories, for example, are acidic.

Acid Acceptor

A compound that acts as a stabilizer by chemically combining with acid that may be initially present in minute quantities in a plastic or that may be formed by the decomposition of the resin.

Acid Bottom and Lining

The inner bottom and lining of a melting furnace, consisting of materials like sand, siliceous rock, or silica brick that give an acid reaction at the operating temperature.

Acid Cleaning

The use of acid to dissolve undesirable materials, such as scale, from a metal surface. Dissolution of the underlying metal is intended to be minimal. Also termed "pickling" although this term also includes the use of alkaline solvents.

Acid Copper

(1) Copper electrodeposited from an acid solution of a copper salt, usually copper sulfate. (2) The solution referred to in (1).

Acid Core Solder

See *cored solder.*

Acid Dip

The use of acid solutions to clean components. The term tends to imply an intent to remove slight tarnishing rather than severe scale, that is, a less aggressive attack than implied by acid cleaning.

Acid Embrittlement

Any embrittlement resulting from contact with acid, for example, during pickling. It is a form of hydrogen embrittlement caused when hydrogen, released by the chemical reaction with the acid, enters the metal in quantities large enough to cause damage such as cracking or a tendency toward a nonductile failure during forming operations.

Acid Etching

The use of acid to dissolve, within controlled limits, the surface of the metal. It is used in metallography to reveal details of the microstructure of a metal, printing to provide a release effect or in industry to prepare a metal surface for treatments such as painting or electroplating.

Acid Extraction

Removal of phases by dissolution of the matrix metal in an acid. See also *extraction.*

Acid Lining

A refractory furnace lining, usually silica, which produces an acid reaction with the charge.

Acid Process

A steelmaking method using an acid refractory-lined furnace. Neither sulfur nor phosphorous is removed.

Acid Rain

Atmospheric precipitation with a pH below 5.6–5.7. Burning of fossil fuels for heat and power is the major factor in the generation of oxides of nitrogen and sulfur, which are converted into nitric and sulfuric acids washed down in the rain. See also *atmospheric corrosion.*

Acid Refractory

Siliceous ceramic materials of a high melting temperature, such as silica brick, used for metallurgical furnace linings. Compare with basic refractories. See also *refractories.*

Acid Steel

Steel melted in a furnace with an acid bottom and lining and under a slag containing an excess of an acid substance such as silica.

A_{cm}, A_1, A_3, A_4

Same as Ae_{cm}, Ae_1, Ae_3, Ae_4.

Acoustic Emission

A measure of integrity of a material, as determined by sound emission on a material, is stressed. Ideally, emissions can be correlated with defects and/or incipient failure.

Acoustic Fatigue

Fatigue induced by atmospheric vibration associated with noise, air turbulence, or vortex shedding. Also called sonic fatigue.

Acrylic Plastics

Acrylic plastics comprise a broad array of polymers and copolymers in which the major monomeric constituents belong to two families of esters-acrylates and methacrylates. These are used singly or in combination. Hard, clear acrylic sheet is made from methyl methacrylate, whereas molding and extrusion pellets are made from methyl methacrylate copolymerized with small percentages of other acrylates or methacrylates.

The resins produced may be in the form of casting syrups (for cast sheet) or pellets for molding and extrusion. The latter are made either in bulk (continuous solution polymerization), followed by extrusion and pelletizing, or continuously by polymerization in an extruder in which unconverted monomer under reduced pressure and recovered for recycling.

The most widely used acrylic plastics are based on polymers of methyl methacrylate. This primary constituent may be modified by copolymerizing or blending with other acrylic monomers or modifiers to obtain a variety of properties. Although acrylic polymers based on monomers other than methyl methacrylate have been investigated, they are not as important as commercial plastics and are generally confined to uses in fibers, rubbers, motor oil additives, and other special products.

Standard Acrylics

Poly(methyl methacrylate), the polymerized methyl ester of methacrylic acid, is thermoplastic. The method of polymerization may be varied to achieve specific physical properties, or the monomer may be combined with other components. Sheet materials may be prepared by casting the monomer in bulk. Suspension polymerization of the monomeric ester may be used to prepare molding powders.

Conventional poly(methyl methacrylate) is amorphous; however, reports have been published of methyl methacrylate polymers of regular configuration, which are susceptible to crystallization. Both the amorphous and crystalline forms of such crystallization-susceptible polymers possess physical properties that are different from those of the conventional polymer and suggest new applications.

Service Properties

Acrylic thermoplastics are known for their outstanding weatherability. They are available in cast sheet, rod, and tube, extruded sheet and film, and compounds for injection molding and extrusion. They are also characterized by good impact strength, formability, and excellent resistance to sunlight, weather, and most chemicals. Maximum service temperature of heat-resistant grades is about 200°F. Standard grades are rated as slow burning, but a special self-extinguishing grade of sheet is available. Although acrylic plastic weighs less than half as much as glass, it has many times greater impact resistance. As a thermal insulator, it is approximately 20% better than glass. It is tasteless and odorless.

When poly(methyl methacrylate) is manufactured with scrupulous care, excellent optical properties are obtained. Light transmission is 92%; colorants produce a full spectrum of transparent, translucent, or opaque colors. Most colors can be formulated for long-term outdoor durability. Acrylics are normally formulated to filter UV energy and the 360 nm and lower band. Because of its excellent transparency and favorable index of refraction, acrylic plastic is often used in the manufacture of optical lenses. Superior dimensional stability makes it practical to produce precision lenses by injection molding techniques.

In chemical resistance, poly(methyl methacrylate) is virtually unaffected by water, alkalies, weak acids, most inorganic solutions, mineral and animal oils, and low concentrations of alcohol. Oxidizing acids affect the material only at high concentrations. It is also virtually unaffected by paraffinic and olefinic hydrocarbons, amines, alkyl monohalides, and esters containing more than 10 carbon atoms. Lower esters, aromatic hydrocarbons, phenols, aryl halides, aliphatic acids, and alkyl polyhalides usually have a solvent action. Acrylic sheet and moldings are attacked, however, by chlorinated and aromatic hydrocarbons, esters, and ketones.

Mechanical properties of acrylics are high for short-term loading. However, for long-term service, tensile stresses must be limited to 1500 psi to avoid crazing or surface cracking.

The moderate impact resistance of standard formulations is maintained even under conditions of extreme cold. High-impact grades have considerably higher impact strength than standard grades at room temperature, but impact strength decreases as temperature drops. Special formulations ensure compliance with Underwriters' Laboratories standards for bullet resistance.

Although acrylic plastics are among the most scratch resistant of the thermoplastics, normal maintenance and cleaning operations scratch and abrade them. Special abrasive-resistant sheet is available that has the same optical and impact properties as standard grades.

Toughness of acrylic sheet, as measured by resistance to crack propagation, can be improved several folds by inducing molecular orientation during forming. Jet aircraft cabin windows, for example, are made from oriented acrylic sheet.

Transparency, gloss, and dimensional stability of acrylics are virtually unaffected by years of exposure to the elements, salt spray, or corrosive atmospheres. These materials withstand exposure to light from fluorescent lamps without darkening or deteriorating. They ultimately discolor, however, when exposed to high-intensity UV light below 265 nm. Special formulations resist UV emission from light sources such as mercury-vapor and sodium-vapor lamps.

Product Forms

Cell-cast sheet is produced in several sizes and thicknesses. The larger sheets available are 120 × 1 44 in., thicknesses from 0.0302 to 4.25 in. Continuous-cast material is supplied as flat sheet to 1/2 in. thick, in widths to 9 ft. Acrylic sheet cast by the continuous process (between stainless steel belts) is more uniform in thickness than cell-cast sheet. Cell-cast sheet, on the other hand, which is cast between glass plates, has superior optical properties and surface quality. Also, cell-cast sheet is available in a greater variety of colors

and compositions. Cast acrylic sheet is supplied in general-purpose grades and in UV-absorbing, mirrored, super-thermoformable, and cementable grades, and with various surface finishes. Sheets are available in transparent, translucent, and opaque colors.

Acrylic film is available in 2, 3, and 6 mil thicknesses, in clear form and in colors. It is supplied in rolls to 60 in. wide, principally for use as a protective laminated cover over other plastic materials.

Injection molding and extrusion compounds are available in both standard and high-molecular-weight grades. Property differences between the two formulations are principally in flow and heat resistance. Higher-molecular-weight resins have lower melt-flow rates and greater hot strength joint processing. Lower-molecular-weight grades flow more readily and are designed for making complex parts and hard to fill; see Table A.3.

Fabrication Characteristics

When heated to a pliable state, acrylic sheet can be formed to almost any shape. The forming operation is usually carried out at about 292°F–340°F. Aircraft canopies, for example, are usually made by differential air pressure, either with or without balls. Such canopies have been made from (1) monolithic sheet stock; (2) laminates of two layers of acrylic, bonded by a layer of polyvinyl butyral; and (3) stretched monolithic sheet. Irregular shapes, such as sign faces, lighting fixtures, or boxes, can be made by positive pressure forming, using molds.

Residual strains caused by forming are minimized by annealing, which also brings cemented joints to full strength. Cementing can be readily accomplished by using either solvent or polymerizable cements.

Acrylic plastic can be sawed, drilled, and machined like wood or soft metals. Saws should be hollow ground or have set teeth. Slow feed and coolant will prevent overheating. Drilling can be done with conventional metal-cutting drills. Routing requires high-speed cutters to prevent chipping. Finished parts can be sanded, and sanded surfaces can be polished with a high-speed buffing wheel. Cleaning should be by soap or detergent and water, not by solvent-type cleaners.

Acrylic molding powder may be used for injection, extrusion, or compression molding. The material is available on several grades, with a varying balance of flow characteristics and heat resistance. Acrylics give molded parts of excellent dimensional stability. Precise contours and sharp angles, important in such applications as lenses, are achieved without difficulty, and this accuracy of molding can be maintained throughout large production runs.

Since dirt, lint, and dust detract from the excellent clarity of acrylics, careful handling and storage of the molding powder are extremely important.

Applications

In merchandising, acrylic sheet has become a major sign material for internally lighted faces and letters, particularly for outdoor use where resistance to sunlight and weathering is important. In addition, acrylics are used for counter dividers, display fixtures and cases, transparent demonstration models of household appliances and industrial machines, and vending machine cases.

The ability of acrylics to resist breakage and corrosion and to transmit and diffuse light efficiently has led to many industrial and architectural applications. Industrial window glazing, safety shields, inspection windows, machine covers, and pump components are some of the uses commonly found in plants and factories. Acrylics are employed to good advantage as the diffusing medium in lighting fixtures and large luminous ceiling areas. Dome skylights form

TABLE A.3

Properties of Acrylics

ASTM Test	Property	Molding Grade		
		Cast Sheet	Standard	High Impact
Physical				
D792	Specific gravity	1.19	1.19	1.15–1.17
D792	Specific volume (in.^{3}/lb)	23.3	23.3	24.1
D570	Water absorption, 24 h, 1/8 in. thk (%)	0.2	0.3	0.3
Mechanical				
D638	Tensile strength (psi)	10,500	10,500	5,400–7,000
D638	Elongation (%)	5	5	50
D638	Tensile modulus (10^5 psi)	4.5	4.3	2.2–8.2
D790	Flexural strength (psi)	16,500	16,000	7,000–10,500
D790	Flexural modulus (10^5 psi)	4.5	4.5	0.65–2.5
D256	Impact strength, Izod (ft lb/in. of notch)	0.4	0.4	0.6–1.2
D785	Hardness, Rockwell	M100–102	M95	R99–M68
Thermal				
D696	Coefficient of thermal expansion (10^{-5}) in./in. °C	3.9	3.6	3.8
D648	Deflection temperature (°F)			
	At 264 psi	200–215	198	170–190
	At 66 psi	225	214	187
Electrical				
D149	Dielectric strength (V/mils) Short time, 1/8 in. thk	500	500	383–450
D150	Dielectric constant			
	At 1 kHz	3.3	3.3	3.9
	At 1 MHz	2.5	2.3	2.5–3.0
D150	Dissipation factor			
	At 1 kHz	0.04	0.04	—
	At 1 MHz	0.02–0.03	0.02–0.03	0.01–0.02
D257	Volume resistivity (Ω cm) at 73°F, 50% RH	>10^{17}	>10^{17}	>10^{15}
D495	Arc resistance(s)	No track	No track	No track
Optical				
D542	Refractive index	1.49	1.49	1.49
D1003	Transmittance (%)	92	92	90

Source: Mach. Design Basics Eng. Design, 678, June 1993. With permission.

from acrylic sheet are an increasingly popular means of admitting daylight to industrial, commercial, and public buildings even to private homes.

Shower enclosures and deeply formed components such as top-shower units, which are subsequently backed with glass fiber–reinforced polyester and decorated partitions, are other typical applications. A large volume of the material is used for curved and flat windshields or pleasure boats, both in board and outboard types.

Acrylic sheet is the standard transparent material for aircraft canopies, windows, instrument panels, and searchlight and landing light covers. To meet the increasingly severe service requirements of pressurized jet aircraft, new grades of acrylic have been developed that have improved resistance to heat and crazing. The stretching technique has made possible enhanced resistance to both crazing and shattering. Large sheets, edge lighted, are used as radar plotting boards and shipboard and ground-control stations.

In molded form, acrylics are used extensively for automotive parts, such as taillight and stoplight lenses, medallions, dials, instrument panels, and signal lights. The beauty and durability of molded acrylic products have led to their wide use for nameplates, control knobs, dials, and handles on all types of appliances. Acrylic molding powder is also used for the manufacture of pen and pencil barrels, hairbrush backs, watch and jewelry cases, and other accessories. Large-section moldings, such as covers for fluorescent streetlights, coin-operated phonograph panels, and fruit juice dispenser bowls, are being molded from acrylic powder. The extrusion of acrylic sheet from molding powder is particularly effective in the production of thin sheeting for use in such applications as signs, lighting, glazing, and partitions.

The transparency, strength, lightweight, and edge-lighting characteristics of acrylics have led to applications in the fields of hospital equipment, medical examination instruments, and orthopedic devices. The use of acrylic polymers in the preparation of dentures is an established practice. Contact lenses are also made of acrylics. The embedment of normal and pathological tissues in acrylic for preservation and instructional use is an accepted technique. This has been extended to include embedment of industrial machine parts, and sales aids, and the preparation of various types of home decorative articles.

High-Impact Acrylics

High-impact acrylic molding powder is used in large-volume, general use. It is used where toughness greater than that found in the standard acrylics is desired. Other advantages in include resistance to staining, high surface gloss, dimensional stability, chemical resistance, and stiffness, and they provide the same transparency and weatherability as the conventional acrylics.

High-impact acrylic is off-white and nearly opaque in its natural state and can be produced in a wide range of opaque colors. Several grades are available to meet requirements for different combinations of properties. Various members of the family have Izod impact strength from about 0.5 to as high as 4 ft lb/in. notch. Other mechanical properties are similar to those of conventional acrylics.

High-impact acrylics are used for hard service applications, such as women's thin-style shoe heels and housings, ranging from electric razors to outboard motors, piano and organ keys, and beverage vending machine housings and canisters—in short, applications where toughness, chemical resistance, dimensional stability, stiffness, resistance to staining, lack of unpleasant odor or taste, and high surface gloss are required.

Acrylic Resins

Colorless, highly transparent, thermoplastic, synthetic resins made by the polymerization of acrylic derivatives, chiefly from the esters of acrylic acid, $CH_2:CH \cdot COOH$, and methacrylic acid, $CH_2:C(CH_3) \cdot COOH$, ethyl acrylate, and methyl acrylate.

The resins vary from soft, sticky semisolids to hard, brittle solids, depending upon the constitution of the monomers and upon the polymerization. They are used for adhesives, protective coatings, finishes, laminated glass, transparent structural sheet, and molded products. Acrylic resins or acrylate resins are stable and resistant to chemicals. They do not cloud or fade in light when used as laminating material in glass and are used as air-curing adhesives to seal glass to metals or wood. Water-based acrylics are used for the formulation of calks and sealants. They have better adhesion and weather resistance than butyl rubbers and dry more quickly. The sealants usually contain about 80% solids.

Lucite is methyl methacrylate of DuPont, marketed as molding powder and in rods, tubes, and cast and molded sheets. Plexiglas, of Rohm and Haas Company, is transparent methyl methacrylate in sheets and rods. This plastic is used for aircraft windows. Plexiglas V is for injection molding, while Plexiglas VM is a molding powder to resist heat distortion to 174°F (79°C). Korad films, from Polymer Extruded Products, are weatherable, wood grain, acrylic laminating films for outdoor window and door profiles and for adhesive-free bonding to polypropylene sheet for thermoformed products. Acumer 3000, a water-treatment acrylic polymer from Rohm and Haas Company, controls silica and prevents formation of magnesium–silicate scale.

The Acryloid resins, of Rohm and Haas Company, are acrylic copolymer solid resins, and the Acrysol resins are solutions for coatings. Plexene M, of the same company, is a styrene–acrylic resin for injection molding.

Acrylonitrile

A monomer with the structure $(CH_2:CHCN)$. It is most useful in copolymers. Its copolymer with butadiene is nitrile rubber; acrylonitrile–butadiene copolymers with styrene are tougher than polystyrene. Acrylonitrile is also used as a synthetic fiber and as a chemical intermediate. Also called vinyl cyanide and propene nitrile.

It is used as a liquid in insecticides and for producing plastics and other chemicals. It is made by the addition of hydrocyanic acid to acetylene, by using propylene as the starter and reacting with ammonia or from petroleum. Acrylonitrile fiber, originally developed in Germany as a textile staple fiber and as a monofilament for screens in weaving, and known as Redon, has good dimensional stability and high dielectric strength and is resistant to water and solvents.

Orlon, of DuPont, is a polymerized acrylonitrile fiber. It is nearly as strong as nylon and has a softer feel. It can be crimped to facilitate spending with wool. It is used for clothing textiles and for filter fabrics. *Dynel*, of Union Carbide Corp., is an acrylonitrile–vinyl chloride copolymer staple fiber. It produces textiles with a warmth and feel like those of wool. It has good strength, is resilient, dyes easily, and is mothproof. *Verel*, of Eastman Chemical Company, is a similar acrylic fiber produced from acrylonitrile and vinylidene chloride, and *Creslan*, of American Cyanamid Company, called *Exlan* in Japan, is an acrylic fiber. *Acrilan*, of Monsanto, is a similar textile fiber and is an acrylonitrile–vinyl acetate copolymer. Acrylonitrile–styrene is a copolymer for injection molding and extruding that produces rigid thermoplastic parts of a higher tensile strength than those of the methacrylates and has good dimensional stability and scratch resistance. *Saran F-120*, of Dow Chemical Company, is a similar material.

Acrylonitrile–Butadiene–Styrene (ABS)

Acrylonitrile-butadiene-styrene (ABS) are three thermoplastic monomers.

Actinic

Of light, characterized by radiation that causes chemical changes, for example, the effect of light on photographic emulsions. Blue and ultraviolet are the most actinic regions of the spectrum.

Actinide Metals

The group of radioactive elements of atomic numbers 89–103 of the periodic system—namely, thorium, protactinium, uranium, neptunium, plutonium, americium, curium, berkelium, californium, einsteinium, fermium, mendelevium, nobelium, and lawrencium.

Actinium

A radioactive metallic element with no significant commercial application.

Activated

Any material treated in some way to enhance its response to its environment.

Activated Alumina

Partly dehydrated aluminum hydroxide. It is used as a desiccant and catalyst.

Activated Charcoal

A nearly chemically pure amorphous carbon made by carbonizing and treating dense material such as coconut shells, peach pits, or hardwood. When made from coal, or in the chemical industry, it is more usually called activated carbon or filter carbon.

This carbon-rich material is produced by heating carbonaceous material in the absence of oxygen. It is used to absorb contaminants. Furthermore, it is used as an adsorbent material for gas masks, for cigarette filters, and for purifying acids, recovering solvents, and decolorizing liquids. Activated carbon woven into garments protects members of the armed forces from chemical warfare. Garments with superactivated carbon are lighter in weight and much more absorbent.

Activated Rosin Flux

A rosin- or resin-base flux containing an additive that increases wetting by the solder.

Activated Sintering

The use of additives, such as chemical additions to the powder or additions to the sintering atmosphere, to improve the densification rate. See *sintering*.

Activating

A treatment that renders nonconductive material receptive to electroless deposition. Also called seeding, catalyzing, and sensitizing, all of which are not preferred terms.

Activation

(1) The changing of a passive surface of a metal to a chemically active state. In contrast with *passivation*. (2) The (usually) chemical process of making a surface more receptive to bonding with a coating or an encapsulating material.

Activation Analysis

A method of chemical analysis based on the detection of characteristic radionuclides following nuclear bombardment. See also *neutron activation analysis*.

Activation Energy

The additional energy required for initiating a metallurgical reaction or physical process—for example, plastic flow, diffusion, chemical reaction. The activation energy may be calculated from the slope of the line obtained by plotting the natural log of the reaction rate versus the reciprocal of the absolute temperature.

Activator

The additive used in activated sintering, also called a dopant.

Active

The negative direction of *electrode potential*. Also used to describe corrosion and its associated potential range when an electrode potential is more negative than an adjacent depressed corrosion rate (passive) range.

Active Area

In electronic packaging, the internal area of a package bottom, usually a cavity, is used for substrate attachment. The term is applied preferably to package cases of all-metal construction (as opposed to glass or ceramic).

Active Components

Electronic components, such as transistors, diodes, electron tubes, and thyristors, which can operate applied electrical signal in such a way as to change its basic characteristics, for example, rectification, amplification, and switching.

Active Devices

Parts of a circuit that are capable of amplification, usually silicon semiconductor devices. Transistors, for instance, are active devices. Components that cannot amplify are passive—for example, resistors and capacitors.

Active Mass

The molecular concentration.

Active Material

Reacting, or capable of reacting, with its environment.

Actuators

Actuators and materials play a key role in developing advanced precision engineering. The breakthroughs in this field are closely related to the development of various types of actuators and related materials. The successes of piezoelectric ceramics and ceramic actuators have, for instance, the propagating-wave type ultrasonic motor that produces precise rotational displacements has been used in autofocusing movie cameras and VCRs. Multimorph ceramic actuators prepared from electrostrictive $Pb(Mg_{1/3}Nb_{2/3})O_3$ ceramics are used as deformable mirrors to correct image distortions from atmospheric effects. The likelihood that the range of applications and demand for actuators will grow actively has stimulated intensive research on the piezoelectric ceramics.

Each kind of actuator has its own advantages and drawbacks, so their selection and optimization should be determined by the requirements of the application.

Piezoelectric/electrostrictive ceramics are widely used in many different types of sensing–actuating devices. This is particularly true for the whole family of micro- and macropiezoelectric ceramic actuators. Various types of ceramic actuators have been developed for different applications. From a structural point of view, ceramic actuators are classified as unimorph, bimorph, moonie, cymbal, and rainbow monomorph benders.

Another device developed to increase the force–displacement performance of a piezoelectric actuator is the rainbow, a monolithic monomorph that is produced from the conventional, high-lead piezoelectric ceramic disk that has one surface reduced to a nonpiezoelectric phase by a high-temperature, chemical reduction reaction.

The photostrictive actuator is another type of bimorph application. The photostrictive behavior results from a combined photovoltaic effect in a piezoelectric effect.

There is also increasing interest in electrostrictive ceramic actuators because electrostrictive ceramics do not contain ferroelectric domains, so that they can return to their original dimensions immediately, when the external electric field is reduced to zero. Therefore, the advantages of an electrostrictive actuator are the near absence of hysteresis and lack of aging behavior.

Piezoelectric/electrostrictive ceramics, magnetostrictive materials, and ferroelastic shape memory alloys are all used for actuators. However, different classes of ceramic actuators require somewhat different materials. In general, an ideal actuating material should exhibit a large stroke, high recovery force, and superior dynamic response.

Addition Agent

(1) A substance added to a solution for the purpose of altering or controlling a process or modifying its characteristics. Examples are wetting agents in acid pickles; brighteners or antipitting agents in plating solutions, to improve luster; or materials added to pickling baths to inhibit pitting. (2) Any material added to a charge of molten metal in a bath or ladle to bring the alloy to specifications.

Addition Polymerization

A chemical reaction in which simple molecules (monomers) are linked to each other to form long-chain molecules (polymers) by chain reaction.

Additive

(1) In lubrication, a material added to a lubricant for the purpose of imparting new properties or of enhancing existing properties. Main classes of additives include *anticorrosive, antifoam, antioxidant, antiwear, detergent, dispersant, extreme pressure,* and *viscosity index improver additives.*

(2) In polymer engineering, a substance added to another substance, usually to improve properties, such as plasticizers, initiators, light stabilizers, and flame retardants. See also *filler.*

Additive Process

A process for obtaining conductive patterns by the selective deposition of conductive material on clad or unclad base material. See also *semiadditive process, semiattractive process,* and *fully additive process.*

Adhere

To cause two surfaces to be held together by adhesion.

Adherence

In tribology, the physical attachment of material to a surface (either by adhesion or by other means of attachment) that results from the contact of two solid surfaces undergoing relative motion. Adhesive bonding is not a requirement for adherence because mechanism such as mechanical interlocking of asperities can also provide a means for adherence. See also *adhesion (adhesive force)* and *mechanical adhesion.*

Adherend

A body held to another body by an adhesive. See also *substrate.*

Adherend Preparation

See *surface preparation.*

Adhesion

(1) In frictional contacts, the attractive force between adjacent surfaces. In physical chemistry, adhesion denotes the attraction between a solid surface and a second (liquid or solid) phase. This definition is based on the assumption of a reversible equilibrium. In mechanical technology, adhesion is generally irreversible. In railway engineering, adhesion often means friction. (2) Force of attraction between the molecules (or atoms) of two different phases. In contrast with *cohesion.* (3) The state in which two surfaces are held together by interfacial forces, which may consist of valence forces, interlocking action, or both. (4) Bonding of components by means of some glue-like substance in the interface—galling and seizure are appropriate. See also *mechanical adhesion* and *specific adhesion.*

Adhesion Coefficient

See *coefficient of adhesion.*

Adhesion Promoter

A coating applied to a substrate, before it is coated with an adhesive, to improve the adhesion of the substrate. Also called primer.

Adhesion Promotion

The chemical process of preparing a surface to provide for a uniform, well-bonded interface.

Adhesive

A substance capable of holding materials together by surface attachment. Adhesive is a general term and includes, among others, cement, glue, mucilage, paste, and mastic. These terms are loosely used interchangeably. Various descriptive adjectives are applied to the term adhesive to indicate certain physical characteristics: *hot-melt adhesives, pressure-sensitive adhesives, structural adhesives, ultraviolet/electron beam-cured adhesives,* and *water-based adhesives.* The characteristics of these five groups of adhesives are summarized in Table A.4. Table A.5 lists the advantages and limitations of the five groups.

Adhesives are characterized by degree of tack, or stickiness, by strength of bond after setting or drying, by rapidity of bonding, and by durability. The strength of bond is inherent in the character of the adhesive itself, particularly in its ability to adhere intimately to the surface to be bonded. Adhesives prepared from organic products are in general subject to disintegration on exposure. The life of an adhesive usually depends upon the stability of the ingredient that gives the holding power, although otherwise good cements of synthetic materials may disintegrate by the oxidation of fillers or materials used to increase tack. Plasticizers usually reduce adhesion. Some fillers such as mineral fibers or walnut-shell flour increase the thixotropy and the strength, while some such as starch increase the tack but also increase the tendency to disintegrate.

TABLE A.4
Characteristics of Various Types of Adhesives

Structural	Hot Melt	Pressure Sensitive	Water Base	Ultraviolet/Electron Beam Cured
Bonds can be stressed to a high proportion of maximum failure load under service environments.	100% solid thermoplastics.	Hold substrates together upon brief application of pressure at room temperature.	Includes adhesives dissolved or dispersed (latex) in water.	100% reactive liquids cured to solids.
Most are thermosets.	Melt sharply to a low-viscosity liquid, which is applied to surface.	Available as organic solvent-base, water-base, or hot-melt systems.	On porous substrates, water is absorbed or evaporated in order to bond.	One substrate must be transparent for UV cure, except when dual-curing adhesives are used (see below).
One- or two-component systems.	Rapid setting, no cure.	Some require extensive compounding (rubber base) to achieve tackiness, whereas others (polyacrylates) do not.	On nonporous substrates, water must be removed prior to bonding.	Some UV-curable formulations are dual curing; a second cure mechanism introduces heat or moisture or eliminates oxygen (anaerobics).
Room- or elevated-temperature cures.	Melt viscosity is an important property.	Available supported (most) or unsupported on a substrate.	Some are bonded following reactivation of dried adhesive film under heat and pressure.	In electron beam (EB) curing, density of material affects penetration.
Wide range of costs.	Nonpressure sensitive and pressure sensitive.	Primarily used in tapes and labels.	Many are based on natural (vegetable or animal) adhesives.	UV/EB-curable formulations have laminating and pressure-sensitive adhesive (PSA) applications.
Various chemical families with varying strengths and flexibilities.	Compounded with additives for tack and wettability.		Nonpressure-sensitive (most) or pressure-sensitive applications.	UV-curable formulations have laminating, PSA, and structural adhesive applications.

TABLE A.5
Advantages and Limitations of Various Types of Adhesives

Structural	Hot Melt	Pressure Sensitive	Water Base	Ultraviolet/Electron Beam Cured
Advantages				
High strength.	100% solids, no solvents.	Labels and tapes have uniform thickness.	Low cost, nonflammable, nonhazardous solvent.	Fast cure (some in 2–60 s).
Capable of resisting loads.	Can bond impervious surfaces.	Permanent tack at room temperature.	Long shelf life.	One-component liquid no mixing, no solvents.
Good elevated-temperature resistance (cross-linked).	Rapid bond formation.	No activation required by heat, water, or solvents.	Easy to apply.	Heat-sensitive substrates can be bonded; cure is "cool."
Good solvent resistance.	Good gap-filling capability.	Cross-linking of some formulations possible.	Good solvent resistance.	Many are optically clear.
Good creep resistance.	Rigid to flexible bonds.	Soft or firm tapes and labels.	Cross-linking of some formulations possible.	High production rates.
Some available in film form.	Good barrier properties.	Easy to apply.	High-molecular-weight dispersions at high solids content with low viscosity.	Good tensile strength.
Limitations				
Two-component systems require careful proportioning and mixing.	Thermoplastics have limited elevated-temperature resistance.	Many are based on rubbers, requiring compounding.	Poor water resistance.	Equipment expensive.
Some have poor peel strength.	Poor creep resistance.	Poor gap fillers.	Slow drying.	High material cost.
Some are difficult to remove and repair.	Little penetration due to fast viscosity increase upon cooling.	Limited heat resistance.	Tendency to freeze.	UV cures only through transparent materials (or secondary cure required).
Some require heat to cure.	Limited toughness at usable viscosities.		Low strength under loads.	Difficult curing on parts with complex shapes.
Some yield by-products upon cure (condensation polymers).			Poor creep resistance.	Many UV cures have poor weatherability because they continue to absorb UV rays.
			Limited heat resistance.	
			Shrinkage of certain substrates in supported films and tapes.	

From an industrial manufacturing standpoint, the advent of the stealth aircraft and all the structural adhesive bonding it entails has drawn widespread attention to the real capabilities of adhesives. Structural bonding uses adhesives to join load-bearing assemblies. Most often, the assemblies are also subject to severe service conditions. Such adhesives, regardless of chemistry, generally have the following properties:

- Tensile strengths in the 1500–4500 psi range
- Very high impact and peel strength
- Service temperature ranges of about −65°F to 3500°F

If these types of working conditions are expected, then one should give special consideration to proper adhesive selection and durability testing.

Parameters

Innumerable adhesives and adhesive formulations are available today. The selection of the proper type for a specific application can only be made after a complete evaluation of the design, the service requirements, production feasibility, and cost considerations. Usually, such selection is best left to adhesive suppliers. Once they have been given the complete details of the application, they are in the best position to select both the type and specific adhesive formulation.

Types and Forms

Adhesives have been in use since ancient times and are even mentioned in the Bible. The first adhesives were of natural origin, for example, bitumen, fish oil, and tree resins. In more modern times, adhesives were still derived from natural products but were processed before use. These modern natural adhesives include animal-derived (such as blood, gelatin, and casein), vegetable-derived (such as soybean oil and wheat flour), and forest-derived (pine resins and cellulose derivatives) products.

Forms include liquid, paste, powder, and dry film. The commercial adhesives include pastes, glues, pyroxylin cements, rubber cements, latex cement, special cements of chlorinated rubber, synthetic rubbers or synthetic resins, and the natural mucilages.

Joint Design

Bonded joints have four main loading modes:

- Peel loads produced by out-of-plane loads acting on thin substrates.
- Shear stresses produced by tensile, torsional, or pure shear loads on substrates.
- Tensile stresses produced by out-of-plane tensile loads.
- Cleavage loads produced by out-of-plane tensile loads acting on the rigid and thick substrates at the end of the joint.
- Good joint design for substrates and structural adhesives maximizes tension, compression and shear, minimizes peel and cleavage, and increases the bond area—wider is better.

The aim is to design a joint that maintains the adhesive in a state of sheer or compression, since bonded joints are strongest under these conditions. Avoid or at least minimize tension, cleavage, and

peel forces. The presence of stresses compromises joint strength and fatigue resistance. It is important to ensure that the adhesive is not the weakest link in the bond. Structural adhesives are known for their relatively poor resistance to peel stresses, and to obtain a maximum bond efficiency joints must be designed to minimize tensile stresses. Potential failure modes for structural adhesives bonding metallic or composite joints include

- Tensile, compressive, or shear failure of the adherends
- Shear or peel in the adhesive
- Shear or peel in a composite near surface plies
- Shear or peel in the resin-rich layer on the composite's surface
- Adhesive failure at the metal/adhesive or composite/adhesive interface

Adhesive failure is a rupture of the adhesive bond, separating the adhesive from the adherend (substrate). Adhesive failure can also be called interfacial failure. Cohesive failure of the adhesive occurs when the load exceeds the inherent adhesive strength. Cohesive failure of the adherend occurs when the load exceeds the adherend strength.

Good bond joint design is critical to the successful use of structural adhesives. Joint designs can include lap/overlap, double lap, butt joint, scarf joint, strap or double strap joint, and a cylindrical joint. The joint design should increase bond strength. Butt joints are best used when stresses are concentrated along the bottom line and perpendicular forces are minimal. Scarf joints are useful for close-fitting parts that allow a large contact area. Lap and offset lap joints are well suited for bonding thin, cross-sectional, rigid parts. In lap joints, the bonded parts are slightly offset. See Figure A.1.

When working with liquid adhesives, a smooth and consistent bead is important to ensure a good bond. Too much or too little adhesive introduces inconsistencies and potential problems. With tape systems, using the right amount of pressure is critical to ensure that the tape adheres completely and fills in any gaps between materials, especially between machine materials.

FIGURE A.1   Good joint design for substrates and structural adhesives maximizes tension, compression, and shear, minimizes peel and cleavage, and increases the bond area—wider is better.

Current and Future Market

Structural adhesives can now be found in a range of industries, including transportation, construction, product assembly, medical device manufacturing, electrical and electronics, and defense. Within these industries, applications can include flanges for doors and enclosures, window/glass attachment, attachment of components, bonding of dissimilar substrates, body/rigid panel attachment, and engine components where vibration absorption is critical.

Classification

Adhesives can recruit into five classifications based on chemical composition. These are summarized in Table A.6.

Natural

These include vegetable- and animal-based adhesives and natural gums. They are inexpensive, easy to apply, and have a long shelf life. They develop tack quickly but provide only low-strength joints. Most are water soluble. They are supplied as liquids or as dry powders to be mixed with water.

Casein–latex type is an exception. It consists of combinations of casein with either natural or synthetic rubber latex. It is used to bond metal to wood for panel construction and to join laminated plastics and linoleum to wood and metal. Except for this type, most natural adhesives are used for bonding paper, cardboard, foil, and light wood.

Synthetic Polymer

The greatest growth in the development and use of organic compound-based adhesives came with the application of synthetically

TABLE A.6
Adhesives Classified by Chemical Composition

	Natural	Thermoplastic	Thermosetting	Elastomeric	Alloys[a]
Types within group	Casein, blood albumin, hide, bone, fish, starch (plain and modified); rosin, shellac, asphalt; inorganic (sodium silicate, litharge–glycerin)	Polyvinyl acetate, polyvinyl alcohol, acrylic, cellulose nitrate, asphalt, oleoresin	Phenolic, resorcinol, phenol-resorcinol, epoxy, epoxyphenolic, urea, melamine, alkyd	Natural rubber, reclaim rubber, grade of butadiene-styrene (GR-S), neoprene, acrylonitrile-butadiene (Buna-N), silicone	Phenolic-polyvinyl butyral, phenolic–polyvinyl formal, phenolic–neoprene rubber, phenolic–nitrile rubber, modified epoxy
Most used form	Liquid, powder	Liquid, some dry film	Liquid, but all forms common	Liquid, some film	Liquid, paste, film
Common further classifications	By vehicle (water emulsion is most common but many types are solvent dispersions)	By vehicle (most are solvent dispersions or water emulsions)	By cure requirements (heat and/or pressure most common but some are catalyst types)	By cure requirements (all are common); also by vehicle (most are solvent dispersions or water emulsions)	By cure requirements (usually heat and pressure except some epoxy types); by vehicle (most are solvent dispersions or 100% solids); and by type of adherends or end-service conditions
Bond characteristics	Wide range, but generally low strength; good resistance to heat, chemicals; generally poor moisture resistance	Good to 150°F–200°F; poor creep strength; fair peel strength	Good to 200°F–500°F; good creep strength; fair peel strength	Good to 150°F–400°F; never melt completely; low strength; high flexibility	Balanced combination of properties of other chemical groups depending on formulation; generally higher strength over wider temp range
Major type of use[b]	Household, general purpose, quick set, long shelf life	Unstressed joints; designs with caps, overlaps, stiffeners	Stressed joints at slightly elevated temp	Unstressed joints on lightweight materials; joints in flexure	Where highest and strictest end-service conditions must be met; sometimes regardless of cost, as military uses
Materials most commonly bonded	Wood (furniture), paper, cork, liners, packaging (food), textiles, some metals and plastics; industrial uses giving way to other groups	Formulation range covers all materials, but with emphasis on nonmetallics— including wood, leather, cork, and paper.	Epoxy-phenolics for structural uses of most materials; others mainly for wood; alkyds for laminations; most epoxies are modified (alloys)	Few used "straight" for rubber, fabric, foil, paper, leather, plastics, films; also as tapes; most modified with synthetic resins	Metals, ceramics, glass, thermosetting plastics; nature of aherends often not as vital as design or end-service conditions (i.e., high strength, temp)

Source: Schwartz, M., *Encyclopedia of Materials, Parts, and Finishes*, 2nd edn., CRC Press, p. 15.

[a] "Alloy," as used here, refers to formulations containing resins from two or more *different* chemical groups. There are also formulations that benefit from compounding two resin types from the same chemical group (e.g., epoxy-phenolic).

[b] Although some uses of the "nonalloyed" adhesives absorb a large percentage of the quantity of adhesives sold, the uses are narrow in scope; from the standpoint of diversified applications, by far the most important use of any group is the forming of adhesive alloys.

derived organic polymers. Broadly, these materials can be divided into two types, thermoplastics and thermosets. Thermoplastic adhesives become softer liquid upon heating and are also soluble. Thermoset adhesives cure upon heating and then become solid and insoluble. Those adhesives that cure under ambient conditions by appropriate choice of chemistry are also considered thermosets.

An example of a thermoplastic adhesive is a hot-melt adhesive. A well-known hot-melt adhesive used since the Middle Ages is sealing wax. Modern hot melts are heavily compounded with wax and other materials. Another widely used thermoplastic adhesive is polyvinyl acetate, which is applied from an emulsion.

Hot-Melt Adhesives

Hot-melt adhesives are 100% solid thermoplastics that are applied in a molten state and form a bond after cooling to a solid state. In contrast to other adhesives, which achieve the solid state through evaporation of solvents or chemical cure, hot-melt adhesives achieve a solid state and resultant strength by cooling. In general, hot-melt adhesives are solid at temperatures below 79°C (175°F). Ideally, as the temperatures increased beyond this point, the material rapidly melts to a low-viscosity fluid that can be easily applied. Upon cooling, the adhesive sets rapidly. Because these adhesives are thermoplastics, the melting-resolidification process is repeatable with the addition and removal of the required amount of heat. Typical application temperatures of hot-melt adhesives are 150°C–290°C (300°F–550°F). See Table A.7.

Materials that are primarily used as hot-melt adhesives include ethylene and vinyl acetate copolymers, polyvinyl acetates, polyethylene, amorphous polypropylene, thermoplastic elastomers such as polyurethane, polyether amide, and block copolymers (e.g., styrene–butadiene–styrene, styrene–isoprene–styrene, and styrene–olefin–styrene), polyamides, and polyesters. Hot-melt adhesives can also be divided into nonpressure-sensitive and pressure-sensitive types. Nonpressure-sensitive adhesives include those for direct bonding and heat sealing. Pressure-sensitive hot-melt adhesives are tacky to the touch and can be bonded by the application of pressure alone at room temperature.

Hot-melt adhesives are used to bond all types of substrates, including metals, glass, ceramics, rubbers, and wood. Primary areas of application include packaging, bookbinding, assembly bonding (such as air filters and footwear), and industrial bonding (such as carpet tape and backings).

Thermoplastic Adhesives

They can be softened or melted by heating and hardened by cooling. They are based on thermoplastic resins (including asphalt oleoresin adhesives) dissolved in solvent or emulsified in water. Most of them become brittle at subzero temperatures and may not be used under stress at temperatures much above 150°F. As they are relatively soft materials, thermoplastic adhesives have poor creep strength. Although lower in strength than all but natural adhesive and suitable only for noncritical service, they are also lower in costs than most adhesives. They are also odorless and tasteless and can be made fungus resistant.

TABLE A.7
Typical Properties of Hot-Melt Adhesives

Property	Ethyl Vinyl Acetate/ Polyolefin Homopolymers and Copolymers	Polyvinyl Acetate	Polyurethane	Polyamides	Polyamide Copolymer	Aromatic Polyamide
Brookfield viscosity (Pa·s)	1–30	1.6–10	2	0.5–7.5	11	2.2
Viscosity test temperature, °C (°F)	204 (400)	121 (250)	104 (220)	204 (400)	230 (446)	204 (400)
Softening temperature, °C (°F)	99–139 (211–282)	—	—	93–154 (200–310)	—	129–140 (265–285)
Application temperature, °C (°F)	—	121–177 (250–350)	—	—	—	—
Service temperature range, °C (°F)	−34 to 80 (−30 to 176)	−1 to 120 (30–248)	—	−40 to 185 (−40 to 365)	—	—
Relative cost[a]	Lowest	Low to medium	Medium to high	High	High	High
Bonding substrates	Paper, wood, selected thermoplastics, selected metals, selected glasses	Paper, wood, leather, glass, selected plastics, selected metals	Plastics	Wood, leather, selected plastics, selected metals	Selected metals, selected plastics	Selected metals, selected plastics
Applications	Bookbinding, packaging, toys, automotive, furniture, electronics	Tray forming, packaging, binding, sealing cases and cartons, bottle labels, cans, jars	Laminates	Packaging, electronics, furniture, footwear	Packaging, electronics, binding	Electronics, packaging, binding

[a] Relative to other hot-melt adhesives.

Pressure Sensitivity

Pressure-sensitive adhesives (PSAs) are mostly thermoplastic in nature and exhibit an important property known as tack. That is, PSAs exhibit a measurable adhesive strength with only a mildly applied pressure. PSAs are derived from elastomeric materials, such as polybutadiene or polyisoprene.

PSAs are capable of holding substrates together when they are brought into contact under brief pressure at room temperature. PSAs are either unsupported or are supported by various carriers, including paper, cellophane, plastic films, cloth, and metal foil. Both single- and double-sided tapes and films are included. Most of the adhesives are based on rubbers compounded with additives, including tackifiers. PSA materials include, in order of decreasing volume and increasing price, natural rubber, styrene–butadiene rubber, reclaimed rubber, butyl rubber, butadiene–acrylonitrile rubber, thermoplastic elastomers, polyacrylates, polyvinyl alkyl ethers, and silicones.

Materials used as PSAs are usually available as solvent systems or hot melts. Substrates are coated with one of these two types of adhesives, and usually no cure of the material is involved upon its application. Adhesive-coated substrates, in their dry state, are permanently tacky at room temperature and do not require activation by water, solvents, or heat. There are two major classes of PSAs: adhesives that are compounded to form PSAs and adhesives that are inherently pressure sensitive and require little or no compounding. Included in the former category are elastomers, and in the latter category are polyacrylates and polyvinyl alkyl ethers.

Tapes, the largest area of application of PSAs, can be classified by construction, function, application, or texture. The construction category includes fabric tapes, paper tapes, foil tapes, film tapes, nonwoven fabric tapes, reinforced tapes, foam tapes, two-faced tapes, and transfer tapes. The backing, rather than the adhesive, is a distinguishing feature of this tape classification. The function category includes masking, holding, ceiling, reinforcing, protecting, bundling, stenciling, splicing, identifying, insulating, packaging, and mounting. The application category includes hospital and first aid tapes, office and graphic art tapes, building industry tapes, packaging and surface protection tapes, electrical tapes, automotive industry tapes, shoe industry tapes, and corrosion protection tapes. The texture category includes floor tiles, wall coverings, automobile wood-grained films, and decorative sheets.

Labels are the second largest application for PSAs. Specific characteristics that are important, and which differentiate labels from tapes, include backing material printability, flatness, ease of die cutting, and release paper properties. The adhesive is applied to the release paper and allowed to dry. These materials are laminated to the label stock.

Thermosetting Adhesives

Based on thermosetting resins, they soften with heat only long enough for the cure to initiate. Once cured, they become relatively infusible up to their decomposition temperature. Although most such adhesives do not decompose at temperatures below 500°F, some are useful only to about 150°F. Different chemical types have different curing requirements. Some are supplied as two-part adhesives and mixed for use at room temperature; some require heat or pressure to bond.

As a group, these adhesives provide stronger bonds than natural, thermoplastic, or elastomeric adhesives. Creep strength is good and peel strength is fair. Generally, bonds are brittle and have little resilience and low impact strength.

Elastomeric Adhesives

Based on natural and synthetic rubbers, elastomeric adhesives are available as solvent dispersions, latexes, or water dispersions. They are primarily used as compounds that have been modified with resins to form some of the adhesive *alloys* discussed in the succeeding text. They are similar to thermoplastics in that they soften with heat but never melt completely. They generally provide high flexibility and low strength and, without resin modifiers, are used to bond paper and similar materials.

Alloy Adhesives

This term refers to adhesives compounded from resins of two or more different chemical families, for example, thermosetting and thermoplastic, or thermosetting and elastomeric. In such adhesives, the performance benefits of two or more types of resins can be combined. For example, thermosetting resins are plasticized by a second resin, resulting in improved toughness, flexibility, and impact resistance.

Structural Adhesives

Structural adhesives are, in general, of the alloy or thermosetting type and have the property of fastening adherends that are structural materials (such as metals and wood) for long periods of time even when the adhesive joint is under load. Phenolic-based structural adhesives were among the first structural adhesives to be developed and used.

Structural adhesives constitute about 35% of the total estimated sales of all adhesives and sealants. Their primary areas of application include automotive, aerospace, appliances, biomedical/dental construction, consumer electronics, fabric, furniture, industrial machines, and marine and sports equipment.

The most common type of structural adhesive is classified as a chemically reactive adhesive. The most widely used materials included in this classification are of epoxies, polyurethanes, modified acrylics, cyanoacrylates, and anaerobics. Chemically reactive adhesives can be subdivided into two groups: one-component systems, which include moisture cure and heat-activated cure categories, and two-component systems, which are subdivided into mix-in and no-mix systems. One-component formulations that cure by moisture from the surrounding air or by adsorbed moisture from the surface of a substrate include polyurethanes, cyanoacrylates, and silicones. A one-component heat-activated system usually consists of two components that are premixed. Chemical families in this group include epoxies and epoxy-nylons, polyurethanes, polyimides, polybenzimidazoles, and phenolics.

The most widely used structural adhesives are based on epoxy resins. Epoxy resin structural adhesives will cure at ambient or elevated temperatures, depending on the type of curative. Urethanes, generated by isocyanate–diol reactions, are also used as structural adhesives. Acrylic monomers have also been utilized as structural adhesives. These acrylic adhesives used an ambient-temperature surface-activated free radical cure. A special type of acrylic adhesive, based on cyanoacrylates (so-called superglue), is a structural adhesive that utilizes an anionic polymerization for its cure. Acrylic adhesives are known for their high strength and extremely rapid cure. Structural adhesives with resistance to high temperatures (in excess of 390°F, or 200°C) for long times can be generated from latter monomers such as polyimides and polyphenol quinoxalines.

Three of the most commonly used adhesives are the modified epoxies, neoprene-phenolics, in final formal phenolics. Modified epoxy adhesives are thermosetting and may be of either the

A

room-temperature-curing type, which cure by addition of a chemical activator, or the heat-curing type. They have high strength and resist temperature up to nearly 500°F (260°C).

The primary advantage of the epoxies is that they are 100% solids, and there is no problem of solvent evaporation after joining impervious surfaces. Other advantages include high shear strengths, rigidity, excellent self-filleting characteristics, and excellent wetting of metal and glass surfaces. Disadvantages include low peel strength, lack of flexibility, and inability to withstand high impact.

Neoprene-phenolic adhesives are alloys characterized by excellent peel strength but lower shear strength than modified epoxies. They are moderately priced, offer good flexibility and vibration absorption, and have good adhesion to most metals and plastics.

Neoprene-phenolics are solvent types, but special two-part chemically curing types are sometimes used to obtain specific properties.

Vinyl formal-phenolic adhesives are alloys whose properties fall between those of modified epoxies and the thermoset-elastomer types. Vinyl formal-phenolics have good shear, peel, fatigue, and creep strengths and good resistance to heat, although they soften somewhat at elevated temperatures.

They are supplied as solvent dispersions in solution or in film form. In the film form, the adhesive is coated on both sides of a reinforcing fabric. Sometimes, it is prepared by mixing a liquid phenolic resin with vinyl formal powder just prior to use.

Other Adhesives/Cements

Paste adhesives are usually water solutions of starches or dextrins, sometimes mixed with gums, resins, or glue to add strength, and containing antioxidants. They are the cheapest of the adhesives but deteriorate on exposure unless made with chemically altered starches. They are widely employed for the adhesion of paper and paperboard. Much of the so-called vegetable glue is tapioca paste. It is used for the cheaper plywoods, postage stamps, envelopes, and labeling. It has a quick tack and is valued for pastes for automatic box-making machines. Latex paste of the rub-off type is used for such purposes as photographic mounting as they do not shrink the paper as do the starch pastes. Glues are usually water solutions of animal gelatin, and the only difference between animal glues and edible gelatin is in the degree of purity. Hide and bone glues are marketed as dry flake, but fish glue is liquid. Mucilages are light vegetable glues, generally from water-soluble gums.

Rubber cements for paper bonding are simple solutions of rubber in a chemical solvent. They are like the latex pastes in that the excess can be rubbed off the paper. Stronger rubber cements are usually compounded with resins, gums, or synthetics. An infinite variety of these cements is possible, and they are all waterproof with good-in initial bond, but they are subject to deterioration on exposure, as a rubber is uncured. This type of cement is also made from synthetic rubbers that are self-curing. Curing cements are rubber compounds to be cured by heat and pressure or by chemical curing agents. When cured, they are stronger, give better adhesion to metal surfaces, and have longer life. Latex cements are solvent solutions of rubber latex. They provide excellent tack and give strong bonds to paper, leather, and fabric, but they are subject to rapid disintegration unless cured.

In general, natural rubber has the highest cohesive strength of the rubbers, with rapid initial tack and high bond strength. It also is odorless. Neoprene has the highest cohesive strength of the synthetic rubbers, but it requires tackifiers. Graphite–sulfur rubber (styrene–butadiene) is high-end specific adhesion for quick bonding but has low strength. Reclaimed rubber may be used in cements, but it has low initial tack and needs tackifiers.

Pyroxylin cements may be merely solutions of nitrocellulose and chemical solvents, or they may be compounded with resins or plasticized with gums or synthetics. They dry by the evaporation of the solvent and have little initial tack, but because of their ability to adhere to almost any type of surface, they are called household cements. Cellulose acetate may also be used. These cements are used for bonding the soles of women's shoes. The bonding strength is about 10 lb/in.^{2} (0.07 MPa) or equivalent to the adhesive strength of the outer fibers of the leather to be bonded. For hot-press lamination of wood, the plastic cement is sometimes marketed in the form of thin sheet.

Polyvinyl acetate–crotonic acid copolymer resin is used as a hot-dip adhesive for book and magazine binding. It is soluble in alkali solutions, and thus the trim is reusable. Polyvinyl alcohol, with fillers of clay and starch, is used for paperboard containers. Vinyl emulsions are much used as adhesives for laminates.

Epoxy resin cements give good adhesion to almost any material and are heat resistant to about 400°F (204°C). An epoxy resin will give a steel-to-steel bond of 3100 lb/in.^{2} (22 MPa) and aluminum-to-aluminum bond to 3800 lb/in.^{2} (26 MPa).

Some pressure-sensitive adhesives are mixtures of a phenolic resin and a nitrile rubber in a solvent, but adhesive tapes are made with a wide variety of rubber or resin compounds.

Furan cements, usually made with furfural alcohol resins, are strong and highly resistant to chemicals. They are valued for bonding acid-resistant brick and tile.

Acrylic adhesives are solutions of rubber-based polymers in methacrylate monomers. They are two-component systems and have characteristics similar to those of epoxy and urethane adhesives. They bond rapidly at room temperature, and adhesion is not greatly affected by oily or poorly prepared surfaces. Other advantages are low shrinkage during cure, high peel and shear strength, excellent impact resistance, and good elevated temperature properties. They can be used to bond a great variety of materials, such as wood, glass, aluminum, brass, copper, steel, most plastics, and dissimilar metals.

Ultraviolet cure adhesives are anaerobic structural adhesives formulated specifically for glass-bonding applications. The adhesive remains liquid after application until ultraviolet light triggers the curing mechanism.

A ceramic adhesive developed by the Air Force for bonding stainless steel to resist to 1500°F (816°C) is made with a porcelain enamel frit, iron oxide, and stainless steel powder. It is applied to both parts and fired at 1750°F (954°C), giving a shear strength of 1500 lb/in.^{2} (10 MPa) in the bond. But ceramic cements that require firing are generally classed with ordinary adhesives. Wash-away adhesives are used for holding lenses, electronic crystal wafers, or other small parts for grinding and polishing operations. They are based on acrylic or other low-melting thermoplastic resins. They can be removed with a solvent or by heating.

Electrically conductive adhesives are made by adding metallic fillers, such as gold, silver, nickel, copper, or carbon powder. Most conductive adhesives are epoxy-based systems, because of their excellent adhesion to metallic and nonmetallic surfaces. Silicones and polyimides are also frequently the base in adhesives used in bonding conductive gaskets to housings for electromagnetic and radio frequency interference applications.

Properties

An important property for a structural adhesive is resistance to fracture (toughness). Thermoplastics, because they are not cured, can deform under load and exhibit resistance to fracture. As a class, thermosets are quite brittle, and thermoset adhesives are modified by elastomers to increase their resistance to fracture. See Table A.8.

TABLE A.8
Properties of Diallyl Phthalate Molding Compounds

ASTM Test	Property	Filler			
		Polyester	Long Glass	Short Glass	Arc-Track Resistant
Physical					
D792	Specific gravity	1.39–1.42	1.70–1.90	1.6–1.8	1.87
D792	Specific volume (in.³/lb)	19.96–19.54	17.90–16.32	17.34–15.42	14.84
D570	Water absorption, 24 h, 1/8 in. thk (%)	0.2	0.05–0.2	0.05–0.2	0.14
Mechanical					
D638	Tensile strength (psi)	5,000	9,000	7,000	7,000–10,000
D790	Flexural strength (psi)	11,500–12,500	18,000	16,000	24,000
D790	Flexural modulus (10^5 psi)	6.4	16	17	19
D256	Impact strength, Izod (ft lb/in. of notch)	4.5–12	6.0	0.8	3.6
D785	Hardness, Rockwell M	108	105–110	105–110	112
Thermal					
C177	Thermal conductivity (10^{-4} cal cm/s cm² °C)	—	14–16	14–15	15–17
D696	Coefficient of thermal expansion (10^{-5} in./in. °C)	—	2.0–3.0	2.0–3.0	23–27
D648	Deflection temperature (°F) At 264 psi	290	450	420	>572
Electrical					
D149	Dielectric strength, (V/mils) Step by step, 1/8 in. thk	400	385	400	400
D150	Dielectric constant At 1 kHz	0.008	0.004–0.006	0.006	0.003–0.008
D150	Dissipation factor At 1 kHz	3.6	4.2	4.4	4.1–4.5
D257	Volume resistivity (Ω cm) At 73°F, 50% RH	2–3×10^{15}	2–3×10^{15}	2–3×10^{15}	10^{16}
D495	Arc resistance(s)	125	140	135	125–180
Frictional					
—	Coefficient of friction			Stat/Dyn	
	Self	—	—	0.14/0.13	—
	Against steel	—	—	0.20/0.19	—

Source: Mach. Design Basics Eng. Design, 680, 1993 June. With permission.

Applications

Hot-melt adhesives are used for the manufacture of corrugated paper, in packaging, in carpeting, in bookbinding, and in shoe manufacture. Pressure-sensitive adhesives are most widely used in the form of coatings on tapes. These pressure-sensitive adhesive tapes have numerous applications, from electrical tape to surgical tape. Structural adhesives are applied in the form of liquids, pastes, or 100% adhesive films. Epoxy liquids and pastes are very widely used adhesive materials, having application in many assembly operations ranging from general industrial to automotive to aerospace vehicle construction. Solid-film structural adhesives are used widely in aircraft construction. Acrylic adhesives are used in thread-locking operations and in small-assembly operations such as electronics manufacture, which require rapid cure times. The largest-volume use of adhesives is in plywood and other timber products manufacture. Adhesives for wood bonding range from the natural products (such as blood or casein) to the very durable phenolic-based adhesives.

Trends and Advantages of Adhesives

Design engineers have many reasons to replace mechanical fasteners with adhesives, including

- Higher energy costs pushing weight reduction in fuel efficiency
- Manufacturers wanting to reduce costs, reduce the number of components in an assembly, and improve productivity

- Manufacturers looking to improve product quality by eliminating process variables
- Design engineers working with dissimilar materials such as plastics, composites, and new metal alloys
- Components becoming smaller and lighter with less space in the design for mechanical fasteners and less space on the manufacturing floor for mechanical fastening processes
- Manufacturers automating or processes, with structural adhesives more adaptable to automated processes

Structural adhesives are generally defined as "load-bearing" adhesives. Depending on the application, they can withstand heavy loads to 4000 psi or even higher. Regardless of the application, a structural adhesive won't fail when a bonded joint is stressed to its yield point. Structural adhesives are part of the structure of their assemblies; they are designed to meet the performance specifications of the entire assembly.

Structural adhesives can include acrylics, epoxies, and urethanes. Acrylics are formulated to provide good adhesion with metal, composites, and thermoplastics. In general, they are fast setting and curing and offer significant product gains. Epoxies provide excellent adhesion to metals and rigid substrates that include thermoset plastics and composites. Epoxies are known for durability, chemical and temperature resistance, and low shrinkage. Urethanes and polyurethanes are suitable for use with thermoplastics; because of their flexibility, they work well with softer, less rigid materials.

Using structural adhesives instead of mechanical fasteners lets you assemble dissimilar materials with higher weight loads and distribute stress evenly across the bond. Structural adhesives won't negatively affect the integrity and strength of the materials—there are no holes, rivets, or other mechanical fasteners to weaken the structure of the joined materials. Structural adhesive also results in a more aesthetically pleasing design, an important criteria in many industries, including automotive, medical equipment, and appliance. In addition, structural adhesives can fill large gaps between substrates and seal joints against corrosion and environmental extremes. They are also easily integrated into automated manufacturing processes, letting you explore new designs and maximize the use of the materials.

In one application, a trade-show display manufacturer was looking for a lightweight, durable solution to bond aluminum hinges to polypropylene panels. The bonds needed to stand up to extreme shipping and handling conditions and heavy use. It was suggested and tested structural adhesives that offered advantages and bond strength and allowed for joining of dissimilar materials—requirements that traditional mechanical joining methods are not able to accommodate.

Perhaps the greatest advantage of structural adhesives is cost. These adhesives are lower in cost than labor-intensive mechanical fastening technologies like rivets, screws, or nuts and bolts.

Of course, structural adhesives are not appropriate for every application. The required time to cure and fixture, application can be messy, they must be properly handled, and they can sometimes be difficult to assemble.

As the use of structural adhesives grows, manufacturing engineers and engineers are looking for help who can best select the best structural adhesives and materials for their applications and manufacturing process. They need to

- Analyze the application for materials and adhesive requirements
- Recommend materials and adhesives and suppliers
- Test the materials and adhesives
- Develop prototypes for customer testing
- Provide design for manufacturability

In addition, the outside help if not available in-house should be able to qualify materials and adhesives based on the following in-house laboratory testing:

- Temperature resistance
- Performance at upper temperature limits
- Shear, tensile, and peel strength
- Outgassing
- Dielectric strength and electrical conductivity
- Thermal conductivity
- Slitting widths and tolerances

For structural pressure-sensitive and other tapes, one must be knowledgeable in precision die cutting using rotary or laser technology, water jet cutting, and dieless cutting. In addition, precision slitting and rewinding, laminating, and printing are also critical capabilities.

Adhesive Assembly

A group of materials or parts, including adhesive, that are placed together for bonding or that have been bonded together. See also *assembly adhesive*.

Adhesive Bond

Attractive forces, generally physical and character, between that adhesive in the base materials. Two principal interactions that contribute to the adhesion are van der Waals bonds and dipole bonds. See also *van der Waals bond*.

Adhesive Bonding

A material-joining process in which an adhesive, placed between the faying surfaces (adherends), solidifies to produce an adhesive bond.

Adhesive Dispersion

A two-phase system in which one phase is suspended in a liquid. Compare with *emulsion*.

Adhesive Failure

Rupture of an adhesive bond such that the separation appears to be at the adhesive–adherend interface. Sometimes termed failure in adhesion. Compare with *cohesive failure*.

Adhesive Film

A synthetic resin adhesive, with or without a film carrier fabric, usually of the thermosetting type, in the form of a thin film of resin, used under heat and pressure as an interleaf in the production of bonded structures.

Adhesive Joint

Location at which two adherends are held together with a layer of adhesive. See also *bond*.

Adhesive Strength

The strength of the bond between an adhesive and an adherend.

Adhesive System

An integrated engineering process that analyzes the total environment of a potential bonded assembly to select the most suitable adhesive, application method, and dispensing equipment.

Adhesive Wear

(1) Wear by transference of material from one surface to another during relative motion due to a process of solid-phase welding. Particles that are removed from one surface are either permanently or temporarily attached to the other surface. (2) Wear due to localized bonding between contacting solid surfaces leading to material transfer between the two surfaces or loss from either surface. (3) Removal of material from a metal surface by a repeated welding and tearing action. When two surfaces move relative to each other in unlubricated contact, the high points on the two surfaces can weld together as a result of local high temperatures and disruption of surface films. Continuing movement immediately breaks the weld but the resultant surface damage becomes a site for further welding and tearing cycles. As the damage increases, it may be referred to, progressively, as scuffing, galling, and finally, seizure. The damage may also be described as pitting, scoring, or gouging, but these terms are also used for damage by other mechanisms. Compare with *abrasive wear*.

Adhesive, Cold-Setting

See *cold-setting adhesive*.

Adhesive, Contact

See *contact adhesive*.

Adhesive, Gap-Filling

See *gap-filling adhesive*.

Adhesive, Heat-Activated

See *heat-activated adhesive*.

Adhesive, Heat-Sealing

See *heat-sealing adhesive*.

Adhesive, Hot-Melt

See *hot-melt adhesive*.

Adhesive, Hot-Setting

See *hot-setting adhesive*.

Adhesive, Intermediate-Temperature-Setting

See *intermediate-temperature-setting adhesive*.

Adhesive, Pressure-Sensitive

See *pressure-sensitive adhesive*.

Adhesive, Structural

See *structural adhesive*.

Adiabatic

Occurring with no addition or loss of heat from the system under consideration.

Adjustable Bed

Bed of a press designed so that the die space height can be varied conveniently.

Admiralty Metal

There are copper alloys noted for good resistance in low-velocity freshwater and seawater and thus are used for condenser, distiller, and heat-exchanger tubing and related equipment in these environments. Also known as inhibited admiralty metal and admiralty brass, they comprise nominally 71.5% copper, 28% zinc, plus small amounts of lead (0.07 maximum) and iron (0.06 maximum) and either 0.02–0.06 arsenic (arsenical admiralty metal), 0.02–0.10 antimony (antimonial admiralty metal), or 0.02–0.10 phosphorous (phosphorized admiralty metal). Admiralty brass, an alloy of 70% copper,

29% zinc, and 1% tin, has good corrosion resistance particularly in seawater, hence its use for condenser tubing an steam-driven vessels. Admiralty gunmetal, an alloy of 88% copper, 10% tin, and 2% zinc, has good strength, corrosion resistance and good casting characteristics, and historically used for naval guns. Though admiralty metal is available in the annealed and cold-work tempers, the annealed condition is often preferred because of the alloy's susceptibility to stress corrosion. Typical tensile properties in the annealed condition are 45,000–53,000 lb/in.^{2} (310–365 MPa) ultimate strength, 13,002–22,000 lb/in.^{2} (89–152 MPa) yield strength, and 65%–70% elongation. Cold working appreciably increases strength.

Admixture

(1) The addition and homogeneous dispersion of discrete components, before cure of a polymer. (2) A material other than water, aggregates, hydraulic cement, and fiber reinforcement used as an ingredient of concrete or mortar and added to the batch immediately before or during its mixing. (3) Material added to (cement) mortars as a water-repellent or coloring agent or to retard or hasten setting.

Adsorbent

A material used to remove odor, taste, haze, and color from oils, foods, pharmaceuticals, or chemicals by selected adsorption of the impurities. Such materials are also called adsorbates. The common adsorbates are activated carbon, or activated clays, alumina, magnesium silicate, or silica gel. The noncarbonaceous adsorbents are used for decolorizing vegetable, animal, or mineral oils, but activated carbon may also be used in conjunction with clays to adsorb color bodies not removed by the clay.

Absorbents called molecular sieves are used to separate chemicals of different molecular diameters without regard to their boiling points. The adsorbents used in vacuum tubes to adsorb or combine with residual gases are called getters. Bulk getters are sheets or wires of zirconium, tantalum, or columbium mounted on the hot electrode to trap gases at temperatures of 900°F–2200°F (482°C–1204°C). Thorium or thorium–misch metal may be used as getters for high temperatures by a coating sintered on the tube anode.

Adsorption

The adhesion of the molecules of gases, dissolved substances, or liquids in more or less concentrated form, to the surfaces of solids or liquids with which they are in contact. The concentration of a substance at a surface or interface of another substance forming a fairly loosely bonded molecular layer.

Adsorption Chromatography

Chromatography based on differing degrees of adsorption of sample compounds onto a polar stationary phase. See also *liquid–solid chromatography*.

Advanced Ceramics

Ceramic materials that exhibit superior mechanical properties, corrosion/oxidation resistance, or electrical, optical, and/or magnetic properties. This term includes many monolithic ceramics as well as particulate-, whisker-, and fiber-reinforced glass, glass-ceramics, and ceramic-matrix composites. Also known as engineering, fine, or technical ceramics. In contrast with *traditional ceramics*.

Advanced Composites

Composite materials that are reinforced with continuous fibers having a modulus higher than that of fiberglass fibers. The term includes metal-matrix and ceramic-matrix composites, as well as carbon–carbon composites.

Ae_{cm}, Ae_1, Ae_3, Ae_4

Defined under *transformation temperature*.

Aerate

To fluff up molding sand to reduce its density.

Aerated Bath Nitriding

A type of liquid nitriding in which air is pumped through the molten bath creating agitation and increased chemical activity.

Aeration

(1) Exposing to the action of air. (2) Causing air to bubble through. (3) Introducing air into a solution by spraying, stirring, or a similar method. (4) Supplying or infusing with air, as in sand or soil.

Aeration Cell (Oxygen Cell)

See *differential aeration cell*.

Aerodynamic Lubrication

See *gas lubrication*.

Aerogel

A highly porous, sometimes more than 98%, extremely lightweight, and nearly transparent solid formed from a gel by replacing the liquid with a gas with little change in volume. The fine, airy cell structure can support more than 1000 times its weight and is considered a superior thermal and perhaps acoustic insulator. Silica, the most common aerogel, is typically made by dissolving silicon alkoxides in a solution, curing it to form a gel, and drying the gel with carbon dioxide at about 1150 lb/in.^{2} (7.9 MPa) and a high temperature.

AFS 50-70 Test Sand

A rounded quartz sand specified for use as an abrasive in the dry sand-rubber wheel abrasive wear test (ASTM G 65).

Aerosol

A dispersion of particles in air, particularly the chemical dispensing of a liquid or a finely divided powder substance by a gas propellant under pressure. The common aerosol can system was developed during World War II for dispensing insecticides. Substances commonly dispensed by the aerosol process include resins, paints, waxes, and cosmetics. Chlorofluorocarbons, hydrocarbons, and carbon dioxide have been used as propellants. The main propellants now are liquefied hydrocarbons, carbon dioxide, and nitrogen. Aerothene MM, of Dow Chemical Company, is a methylene chloride that has properties needed for the efficient functioning of carbon dioxide and hydrocarbon propellants in cosmetic aerosol applications. In noncosmetic aerosol applications, such as paints and insecticides, hydrocarbons claimed 45% of the market. The principal objection to their use for cosmetic aerosols has been their flammability.

Aerostatic Lubrication

See *pressurized gas lubrication*.

Afterbake

See *post cure*.

Afterblow

In the basic Bessemer steelmaking process, the main flow produces a strong flame resulting from the carbon removal. Continuation of the blow after carbon removal, termed the afterblow, produces a lesser flame associated with phosphorus removal.

Agate

A natural mixture of crystalline and colloidal silica, but consisting mainly of the mineral chalcedony. It usually occurs in irregular banded layers of various colors derived from mineral salts, and when polished, it has a waxy luster. Agate is used for knife edges and bearings of instruments, for pestles and mortars, for textile rollers, and for ornamental articles, and the finer specimens are employed as gemstones.

Age

Any process taking place over a period of time. Examples include an increase in hardness, as in strain age hardening, or an increase in the hysteresis losses in magnetic steels and the term is a common contraction of age harden.

Age Hardening

Hardening by *aging* (heat treatment) usually after rapid cooling or cold working.

An increase in hardness occurring spontaneously over a period of time, particularly that observed in alloys that have been solution treated. See *precipitation hardening* and *strain age hardening*.

Age Hardening (of Grease)

The increasing consistency of a lubricating grease with time of storage.

Age Softening

Spontaneous decrease of strength and hardness that takes place at room temperature in certain strain hardened alloys, especially those of aluminum.

Agglomerate

This clustering together of a few or many particles, whiskers, or fibers, or a combination thereof, into a larger solid mass.

Aggregate

(1) A dense mass of particles held together by strong intermolecular or atomic cohesive forces. (2) Granular material, such as sand, gravel, crushed stone, or iron blast-furnace slag, used with a cementing medium to form hydraulic-cement concrete or mortar. (3) A hard, coarse material usually of mineral origin used with an epoxy binder (or other resin) in plastic tools. Also used in flooring or as a surface medium.

Aggressive Tack

Synonym for *dry tack*.

Aging

(1) The effect on materials of exposure to an environment for a prolonged interval of time. (2) The process of exposing materials to an environment for a prolonged interval of time in order to predict in-service lifetime. (3) Generally, the degradation of properties or function with time.

Aging (Heat Treatment)

A change in the properties of certain metals and alloys that occurs at ambient or moderately elevated temperatures after hot working or a heat treatment (quench aging in ferrous alloys, natural or artificial aging in ferrous and nonferrous alloys) or after a cold working operation (strain aging). The change in properties is often, but not always, due to a phase change (precipitation) but never involves a change in chemical composition of the metal or alloy. See also *age hardening*, *artificial aging*, *interrupted aging*, *natural aging*, *overaging*, *precipitation hardening*, *precipitation heat treatment*, *progressive aging*, *quench aging*, *step aging*, and *strain aging*.

Agitator

A device to intensify mixing. Example is a high-speed stirrer or paddle in a blender or drum of a mill.

Agricultural Steel

A largely obsolete term implying plain carbon steels as used for simple farming implements such as plows, spades possibly with a hard face deposit.

Air

The gas atmosphere in which we live. It comprises about 80% nitrogen, 20% oxygen, small quantities of carbon dioxide, rare gases such as argon, and pollutants such as sulfur dioxide.

Air Acetylene Welding

A fuel gas welding process in which coalescence is produced by heating with a gas flame or flames obtained from the combustion of acetylene with air, without the application of pressure, and with or without the use of filler metal.

Air Arc Cutting

Cutting processes in which the material to be cut is melted by an electric arc struck between an electrode and the workpiece and the molten materials ejected by at high-velocity air jet. Terms such as air carbon arc cutting and air metal arc cutting indicate the material of the electrode.

Air Bearing

A bearing for a shaft or sliding interface where air is induced to keep the faces separate and act as a lubricant.

Air Bend Die

Angle forming dies in which the metal is formed without striking the bottom of the die. Metal contact is made at only three points in the cross section; the nose of the male die and the two edges of a V-shape die opening.

Air Bending

Bending in an *air bend die*.

Air Blasting

See *blasting* or *blast cleaning*.

Air Cap (Thermal Spraying)

A device for forming, shaping, and directing an air pattern for the atomization of wire or ceramic rod.

Air Carbon Arc Cutting (AAC)

An arc cutting process in which metals to be cut are melted by the heat of a carbon arc, and the molten metal is removed by a blast of air.

Air Channel

A groove or hole that carries the vent from a core to the outside of a mold.

Air Circulation Furnace

An air furnace that has some system to circulate air to ensure even heating of the contents.

Air Classification

The separation of metal powder into particle-size fractions by means of an air stream of controlled velocity and application of the principle of *elutriation*.

Air Dried

Refers to the air drying of a casting core or mold without the application of heat.

Air Dried Strength

Strength (compressive, shear, or tensile) of the refractory (sand) mixture after being air dried at room temperature.

Air Feed

A *thermal spraying process* variation in which an air stream carries the powdered material to be sprayed through the gun in and to the heat source.

Air Furnace

Reverberatory-type furnace in which metal is melted by heat from fuel burning at one end of the hearth, passing over the bath toward the stack at the other end. It is also reflected from the roof and side walls. The atmosphere is untreated air as opposed to fuel combustion products, some gas deliberately introduced, or a vacuum. See also *reverberatory furnace*.

Air Gap

In extrusion coating, the distance from the die opening to the nip formed by the pressure roll and the chill roll.

Air Gap (in Welding)

The distance between the electrode and the workpiece over which an electric arc is struck.

Air Gap (in an Electromagnetic Context)

A gap crossed by the magnetic flux to complete the magnetic circuit. In a generator, it is the radial gap between rotor and stator, and even when the machine is hydrogen cooled and the gap is filled with hydrogen, it is still referred to as the air gap.

Air Hammer

In lubrication, a type of instability, basically a resonance, that occurs in externally pressured gas bearings.

Air Hole

A hole in a casting caused by air or gas trapped in the metal drawing solidification.

Air Setting

The characteristic of some materials, such as refractory cements, core pastes, binders, and plastics, to take permanent set at normal air temperatures.

Air Vent

A small outlet to prevent entrapment of gases in a molding or tooling fixture.

Air-Assist Forming

A method of thermoforming in which air flow or air pressure is employed to preform plastic sheet partially just before the final pull-down onto the mold using vacuum.

Air-Bubble Void

Air entrapment within a molded item or between the plies of reinforcement or within a bond line or encapsulated area; localized, noninterconnected, and spherical in shape.

Air-Hardening Steel

A steel containing sufficient carbon and other alloying elements to harden fully during cooling in air or other gaseous media from a temperature above its transformation range. The term should be restricted to steels that are capable of being hardened by cooling in air in fairly large sections, about 2 in. (50 mm) or more in diameter. Same as self-hardening steel.

Air-Lift Hammer

A type of gravity-drop hammer in which the ram is raised for each stroke by an air cylinder. Because length of stroke can be controlled, ram velocity and therefore the energy delivered to the workpiece can be varied. See also *drop hammer* and *gravity hammer*.

Air-Slip Forming

A variation of vacuum snap-back *thermoforming* in which the male mold is enclosed in a box such that when the mold moves forward toward the hot plastic, air is trapped between the mold in the plastic sheet. As the mold advances, the plastic is kept away from it by this air cushion until the full travel of the mold is completed, at which point, a vacuum is applied, destroying the cushion and forming the part against the plug.

Alabaster

The naturally occurring crystalline form of gypsum.

Albumin

The water-soluble and alcohol-soluble protein obtained from blood, eggs, or milk and used in adhesives, textile and paper finishes, leather coatings, varnishes, as a clarifying agent for tannins, and in oil emulsions. Crude blood albumin is a brown amorphous lumpy material obtained by clotting slaughterhouse blood and dissolving out the albumin. The remaining dark-red material is made into ground blood and marketed as a fertilizer. Blood albumin is sold as clear, pale, amber, and colored powders. Blood albumin from human blood is a stable, dry, white powder. It is used in water solution for treatment of shock.

The material of egg white is sometimes spelled albumen. Egg white is a complex mixture of at least eight proteins, with sugar and inorganic salts. More than half of the total is the protein oval albumin, a strong coagulating agent, and another large percentage consists of conalbumin, which forms metal complexes and unites with iron in the human system. Two of the proteins not so desirable in the human body are ovomucoid, which inhibits the action of the digestive enzyme trypsin, and avidin, which combines with and destroys the action of the growth vitamin biotin. Egg albumin is prepared from the dried egg white and is marketed in yellowish amorphous lumps or powdered.

Alclad

Composite wrought product comprised of an aluminum alloy core having one or both surfaces a metallurgically bonded aluminum or aluminum alloy coating that is anodic to the core and thus electrochemically protects the core against corrosion.

Alcohol

The common name for ethyl alcohol, but the term properly applies to a large group of organic compounds that have important uses in industry, especially in solvents and in the preparation of other materials. A characteristic of all alcohols is the monovalent $-OH$ group. In the primary alcohols, there is always a $\cdot CH_2OH$ group in the molecule. The secondary alcohols have a $\cdot CHOH$ group, and the tertiary alcohols have a distinctive $:COH$ group. Alcohols with one OH group are called monohydroxy alcohols or polyhydric alcohols. Another method of classification is by the terms "saturated" and "unsaturated." The common alcohols used in industry are ethyl, methyl, amyl, butyl, isopropyl, and octyl. The alcohols vary in consistency. Methyl alcohol is like water, amyl alcohol is oily, and melissyl alcohol is a solid. Many of the alcohols are most easily made by fermentation; others are produced from natural gas or from petroleum hydrocarbons. Much of the production of ethyl alcohol is from blackstrap molasses. Alcohols, generally colorless, are similar to water in some ways and are neither alkaline nor acid in reaction.

Methyl alcohol, commonly known as wood alcohol, has the chemical name methanol. It is also referred to as carbinol. A colorless, poisonous liquid of composition CH_3OH, it was originally made by the distillation of hardwoods. It is now produced chiefly by catalytic reduction of carbon monoxide and dioxide by hydrogen.

Fuel use, either directly or for making methyl tert-butyl ether (MTBE), is growing. MTBE makes up as much as 15% of reformulated gasoline. Methanol is used as a solvent in lacquers, varnishes, and shellac. On oxidation, it yields formaldehyde, and it is used in making the latter product for synthetic molding materials.

Solidified alcohol, marketed in tins and used as a fuel in small stoves, is a jellylike solution of nitrocellulose in methyl alcohol. It burns with a hot flame. Sterno is this material, while Trioxane, employed for the same purpose, is an anhydrous formaldehyde trimer but has the disadvantage of being water soluble.

Butyl alcohol is a colorless liquid used as a solvent for paints and for varnishes and in the manufacture of dyes, plastics, and many chemicals. There are four forms of this alcohol, with the normal or primary butyl alcohol beings the most important. Normal butyl alcohol, $CH_3 (CH_2)_2 \cdot OH$, in this form, known as butanol, has strong solvent power and is valued where a low evaporation rate is desired, such as in latexes and nitrocellulose lacquer.

Fluoroalcohols are alcohols in which fluorine is substituted for hydrogen in the nonalcohol branch. They have the general composition $H(CF_2CF_2)_2 CH_2OH$, and as solvents, they dissolve some synthetic resins that resist common solvents. Some of the esters are used as lubricants for temperatures to 500°F (260°C). Acetylenic alcohols are methyl butanol and used as a solvent, in metal pickling, in plating, and in vitamin manufacture, and methyl pentynol. It is a powerful solvent. It has hypnotic qualities and is also used for tranquilizing fish in transport.

Fatty acid alcohols, made from fatty acids synthetically, have the general formula $CH_3 (CH_2)_2OH$, ranging from the C_8 of octyl alcohol to the C_{18} of stearyl alcohol. They are easily esterified, oxidized or ethoxilated, and used for making cosmetics, detergents, emulsifiers, and other chemicals. Polyols are polyhydric alcohols containing many hydroxyl, $-OH$, radicals. They react easily with isocyanates to form urethane.

Alcomax

A series of alloys with strong permanent magnet characteristics. Iron based with about 20% cobalt, 15% nickel, 10% aluminum, and small quantities of other elements such as titanium, niobium, and copper. They are hard and brittle and can only be formed by casting or powder metal processes. Alnico alloys are similar.

Aldehydes

A group name for substances made by the dehydrogenation or oxidation of alcohols, such as formaldehyde from methyl alcohol. By further oxidation, the aldehydes formed corresponding acids, as formic acid. The aldehydes have the radical group $-CHO$ in the molecule, and because of their ease of oxidation, they are important reducing agents. They are also used in the manufacture of synthetic resins and many other chemicals. Aldehydes occur in animal tissues and in the odorous parts of plants.

Acetaldehyde is a water-white flammable liquid with an aromatic penetrating odor, used as a reducing agent, preservative, and for silvering mirrors, and in the manufacture of synthetic resins, dyestuffs, and explosives. Also called ethanal, it has the composition of $CH_3 \cdot CHO$ and is made by the direct liquid-phase oxidation of ethylene.

Aerolein is acrylic aldehyde, $CH_2 : CH \cdot CHO$, a colorless volatile liquid whose vapor is irritating to the eyes and nose, and the unpleasant effect of scorching fat is due to the acrolein formed. Acrolein is made by oxidation of propylene with a catalyst. It polymerizes easily and can be copolymerized with ethylene, styrene, epoxies, and other resins to form various types of plastics. Its reactive double bond and carbonyl group make it a useful material for chemical synthesis. It is used as an antimicrobial agent for controlling algae, microbes, mollusks, and aquatic weeds.

Algorithm

A procedure for solving a mathematical problem by a series of operations. In *computed tomography*, the mathematical process used to convert the transmission measurements into a cross-sectional image. Also known as reconstruction algorithm.

Aligning Bearing

A bearing with an external spherical seat surface that provides a compensation for shaft or housing deflection and misalignment. Compare with *self-aligning bearing*.

Alignment

In mechanical or electrical adjustment of the components of an optical device so that the path of the radiating beam coincides with the optical axis or other predetermined path in the system. See also *mechanical alignment* and *voltage alignment*.

Aliphatic Hydrocarbons

Saturated hydrocarbons having an open-chain structure, for example, gasoline and propane.

Aliphatic Polyketones

Semicrystalline thermoplastics developed by Shell Chemicals, which discontinued production in the year 2000. Called Carilon polymers, these polyketones consist of a perfectly alternating linear structure of ethylene and carbon monoxide with a minor amount of propylene for excellent chemical resistance and dimensional stability in harsh environments and good mechanical properties. They are especially resistant to salt solutions, hydrocarbons, oil field chemicals, weak acids and bases, and soaps and detergents, resist hydrolysis, dissolution, and plasticization in a broad range of chemicals; exhibit minimal swelling in harsh hydrocarbon environments; and provide good barrier properties, or permeation resistance, to methane, carbon dioxide, and hydrogen sulfide and has a melting temperature of 428°F (220°C).

Aliquot

A representative sample of a larger quantity.

Alkali

A caustic hydroxide characterized by its ability to neutralize acids and form soluble soaps with fatty acids. Fundamentally, alkalies are inorganic alcohols, with the monovalent hydroxyl group –OH in the molecule, but in the alkalies, this group is in combination with a metal or an ammonia group, and alkalies have none of the characteristics of alcohols. All alkalies are basic and have a pH value from 7 to 14. They neutralize acids to form a salt and water. The common alkalies are sodium hydroxide and potassium hydroxide, which are used in making soaps, soluble oils, and cutting compounds, and cleaning solutions, and for etching aluminum. All the alkalies have a brackish taste and a soapy feel; most corrode animal and vegetable tissues.

Alkali Metals

A metal in group IA of the periodic system—namely, lithium sodium, potassium, rubidium, cesium, and francium, as well as calcium, and barium because of the basic reaction of their oxides, hydroxides, and carbonates. Carbonates of these metals are called fixed alkalies. The metals show a gradation in properties and increase in chemical activity with increase in atomic weight.

All are silvery white and very soft. They tarnish rapidly in air and decompose water at ordinary temperatures. In the alkali metals, the electron bonding is so weak that even the impact of light rays knocks electrons free. All have remarkable affinity for oxygen. Rubidium and cesium ignite spontaneously in dry oxygen. Calcium, strontium, and barium are also called earth metals. Thin films of the alkali metals are transparent to ultraviolet light but opaque to visible light.

Alkaline

(1) Having properties of an alkali. (2) Having a pH greater than 7.

Alkaline Cleaner

A material blended from alkali hydroxides in such alkaline salts as borates, carbonates, phosphates, or silicates. The cleaning action may be enhanced by the addition of surface-active agents and special solvents.

Alkaline Earth Metal

A metal in group IIA of the periodic system—namely, beryllium, magnesium, calcium, strontium, barium, and radium—so-called because the oxides or "earths" of calcium, strontium, and barium were found by the early chemists to be alkaline in reaction.

Alkaline Sodium Picrate

An etchant for steel. Typically, 25 g of sodium hydroxide plus 2 g of picric acid dissolved in 100 cm^3 of hot water. When used hot, it will slowly darken cementite.

Alkyd Coatings

They are used for such diverse applications as air-drying water emulsion wall paints and baked enamels for automobiles and appliances. The properties of oil-modified alkyd coatings depend upon the specific oil used as well as the percentage of oil in the composition. In general, they are comparatively low-cost and have excellent color retention, heat resistance, and salt spray resistance. The low-modified alkyds can be further modified with other resins to produce resin-modified alkyds. The resin is sometimes added during manufacture of the alkyd and becomes an integral part of the alkyd, or the modifying resin is blended with the alkyd when the paint is formulated. When mixed with urea formaldehyde or melamine resin, harder and more resistant baked enamels are produced. Alkyds blended with ethyl cellulose are used as tough flexible coatings for electric cable. Other resins blended with alkyds to produce special or improved properties include phenolic, resin, vinyl, and silicone.

Alkyd Plastics

These are molding compounds composed of a polyester resin and usually a diallyl phthalate monomer plus various inorganic fillers, depending on the desired properties. The raw material is produced in three forms—granular, putty, and glass fiber reinforced. As a class, the alkyds have excellent heat resistance up to about 300°F (149°C), high stiffness, and moderate tensile and impact strength. Their low moisture absorption combined with good dielectric strength makes them particularly suitable for electronic and electrical hardware, such as switchgears, insulators, and parts for motor controllers and automotive ignition systems. They are easily molded at low pressures and cure rapidly.

These thermoset plastics are based on resins composed principally of polymeric esters, in which the recurring ester groups are an integral part of the main polymer chain, and in which ester groups occur in most cross-links that may be present between chains.

Alkyd Resins

A group of thermosetting synthetic resins known chemically as hydroxycarboxylic resins, of which the one produced from phthalic anhydride and glycerol is representative. They are made by the esterification of a polybasic acid with a polyhydric alcohol and have the characteristics of homogeneity and solubility that make them especially suitable for coatings and finishes, plastic molding compounds, caulking compounds, adhesives, and plasticizers for other resins. The resins have high adhesion to metals; are transparent, easily colored, tough, flexible, heat and chemical resistant; and have good dielectric strength. They vary greatly with the raw materials used in with varying percentage compositions, from soft rubbery gums to hard, brittle solids. Phthalic anhydride imparts hardness and stability. Maleic acid makes a higher–melting point resin. Azelaic acid gives a softer and less brittle resin. The long-chain dibasic acids, such as adipic acid, give resins of great toughness and flexibility. In place of glycerol, the glycols yield soft resins, and sometimes, the glycerol is modified with a proportion of glycol. The resins are reacted with oils, fatty acids, or other resins, such as urea or melamine, to make them compatible with drying oils and to import special characteristics.

Since alkyd resins are basically esterification products of innumerable polybasic acids and polyhydric alcohols and can be modified with many types of oils and resins; the actual number of different alkyd resins is unlimited, and the users' specification is normally by service requirements rather than composition.

Alkyds

Alkyds are part of the group of materials that includes bulk-molding compounds and sheet-molding compounds. They are processed by compression, transfer, or injection molding. Faster molding cycles at low pressure make alkyds easier to mold like many other thermosets. They represent the introduction to the thermosetting plastics industry of the concept of low-pressure, high-speed molding.

Alkyds are furnished in granular compounds, extruded ropes or logs, bulk-molding compound, flake, and puttylike sheets. Except for the putty grades, which may be used for encapsulation, these compounds contain fibrous reinforcement. Generally, the fiber reinforcement in rope and logs is longer than that in granular compounds and shorter than that in flake compounds. Thus, strength of these materials is between those of granular and flake compounds. Because fillers which are opaque and the resins are amber, translucent colors are not possible. Opaque, light shades can be produced in most colors, however.

Molded alkyd parts resist weak acids, organic solvents, and hydrocarbons such as alcohol and fatty acids; they are attacked by alkalies.

Depending on the properties desired in the finished compound, the fillers used are clay, asbestos, fibrous glass, or combinations of these materials. The resulting alkyd compounds are characterized in their molding behavior by the following significant features: (1) no liberation of volatiles during the cure, (2) extremely soft flow, and (3) fast cure at molding temperatures.

Although the general characteristics of fast cure at low-pressure requirements are common to all alkyd compounds, they may be divided into three different groups that are easily discernible by the physical form in which they are manufactured.

1. Granular types, which have mineral or modified mineral filters, providing superior dielectric properties and heat resistance
2. Putty types, which are quite soft and particularly well-suited for low-pressure molding
3. Glass fiber–reinforced types, which have superior mechanical strengths

Granular Types

The physical form of materials in this group is that of a free-flowing powder. Thus, these materials readily lend themselves to conventional molding practices such as volumetric loading, preforming, and high-speed automatic operations. The outstanding properties of parts molded from this group of compounds are high dielectric strength at elevated temperatures, high arc resistance, excellent dimensional stability, and high heat resistance. Compounds are available within this group that are self-extinguishing and certain recently developed types display exceptional retention of insulating properties under high humidity conditions.

These materials have found extensive use as high-grade electrical insulation, especially in the electronics field. One of the major electronic applications for alkyd compounds is in the construction of vacuum tube bases, where the high dry insulation resistance of the material is particularly useful in keeping the electrical leakage between pins to a minimum. In the television industry, tuner segments are frequently molded from granular alkyd compound since electrical and dimensional stability is necessary to prevent calibration shift in the tuner circuits. Also, the granular alkyds have received considerable usage in automotive ignition systems where retention of good dielectric characteristics at elevated temperatures is vitally important.

Putty Types

This group contains materials that are furnished in soft, putty-like sheets. They are characterized by very low-pressure molding requirements (less than 800 psi) and are used in molding around delicate inserts and solving special loading problems. Molders customarily extrude these materials into a ribbon of a specific size, which is then cut into preforms before molding. Whereas granular alkyds are rather diversified in their various applications, putty has found widespread use in one major application: molded encapsulation of small electronic components, such as mica, polyester film, and paper capacitors, deposited carbon resistors, small coils, and transformers.

The purpose here is to insulate the components electrically, as well as to seal out moisture. Use of alkyds has become especially popular because of their excellent electrical and thermal properties, which resulted in high functional efficiency of the unit in a minimum space, coupled with low-pressure molding requirements, which prevent distortion of the subassembly during molding.

Glass Fiber–Reinforced Types

This type of alkyd-molding compound is used in a large number of applications requiring high mechanical strength as well as electrical insulating properties. Glass fiber–reinforced alkyds can be either compression or plunger molded permitting a wide variety of types of applications, ranging from large circuit breaker housings to extremely delicate electronic components.

Other Types of Alkyd-Molding Compounds

Halogen- and/or phosphorous-bearing alkyd-molding compounds with antimony trioxide added provide improved flame resistance. Other flame-resistant compounds are available that do not contain halogenated resins. Many grades are Underwriters' Laboratories rated at 94 V-0 in sections under 1/16 in. Flammability ratings depend on specific formulations, however, and can vary from 94 HB to V-0. Flammability ratings also vary with section thickness.

Glass- and asbestos-filled compounds have better heat resistance than the cellulose-modified types. Depending on type, alkyds can be used continuously to 350°F and, for short periods, to 450°F.

Alkyd molding compounds retain their dimensional stability and electrical mechanical properties over a wide temperature range.

Alkylation

(1) A chemical process in which an alkyl radical is introduced into an organic compound by substitution or addition. (2) A refinery process for chemically combining isoparaffin with olefin hydrocarbons.

Alligator Skin

See *orange peel.*

Alligatoring

(1) Pronounced wide cracking over the entire surface of a coating having the appearance of alligator hide. (2) The longitudinal splitting of flat slabs in a plane parallel to the rolled surface. Also called fish mouthing.

All-Metal Package

A hybrid circuit package made solely of metal, excluding glass or ceramic. Its main applications are with microwave modules and large plug-ins.

Allomeric

Different substances having the same crystallographic structure.

Allomorphous

Having different crystalline forms for a given composition.

Allophanate

Reactive product of an isocyanate and the hydrogen atoms in a urethane.

Alloprene

Chlorinated rubber.

Allotriomorphic Crystal

A crystal whose lattice structure is normal but whose external surfaces are not bounded by regular crystal faces; rather, the external surfaces are impressed by contact with other crystals or another surface such as a mold wall or are irregularly shaped because of nonuniform growth. Compare with *idiomorphic crystal.*

Allotropic

Occurring in two or more solid forms having differing physical characteristics and where the change is reversible. If the change is not reversible, polymorphism is the usual term.

Allotropy

(1) A near synonym for *polymorphism.* Allotropy is generally restricted to describing polymorphic behavior and elements,

terminal phases, and alloys whose behavior *closely parallels* that of the predominant constituent element. (2) The existence of a substance, especially an element, and two or more physical states (e.g., crystals). See also *graphite.*

Allowance

(1) The specified difference in limiting sizes (minimum clearance or maximum interference) between mating parts, as computed arithmetically from the specified dimensions and tolerances of each part. (2) In a foundry, the specified clearance. See also *tolerance.*

Allowed (Energy) Bands

The band of energy in which the valence electrons of a metal crystal are allowed to exist. Individual bands may be empty or partially or completely filled, and they are separated by forbidden bands in which the electrons cannot normally exist other than to jump across. See also *band theory* and *semiconductors.*

Alloy

(1) A substance having metallic properties and being composed of two or more chemical elements of which at least one is a *metal.* (2) To make or melt an alloy. (3) In plastics, a blend of polymers or copolymers with other polymers or elastomers under selective conditions, for example, styrene–acrylonitrile. Also called polymer blend.

The term implies that the additional element has been introduced deliberately with the intention of improving some characteristic of the material. Brass, bronze, steel, and sterling silver are examples of alloys. Steel was an alloy of iron with a small but vital carbon content so the term "alloy steel" implies the addition of further elements such as nickel, chromium, and molybdenum to improve specific properties. The term is used colloquially, as in "alloy wheels" when referring to aluminum probably as a corruption of the long-established casual conversational use in some industries of "alley" referring to aluminum in either its pure or alloy form. The potential for confusion is obvious.

Alloys are used because they have specific properties or production characteristics that are most attractive than those of the pure, elemental metals. For example, some alloys possess high strength, others have low melting points, others are refractory with high melting temperatures, some are especially resistant to corrosion, and others have desirable magnetic, thermal, or electrical properties. These characteristics arise from both the internal and the electronic structure of the alloy. In recent years, the term *plastic alloy* also has been applied to plastics.

Metal alloys are more specifically described with reference to the major element by weight, which is also called the base metal or parent metal. Thus, the terms "aluminum alloy," "copper alloy," etc. Elements present in lesser quantities are called alloying elements. When one or more alloying elements are present in substantial quantity or, regardless of their amount, have a pronounced effect on the alloy, they, too, may be reflected in generic designations.

Metal alloys are also often designated by trade names or by trade association or society designations. Among the more common of the latter are the three-digit designations for the major families of stainless steels and the four-digit ones for aluminum alloys.

Structurally, there are two kinds of metal alloys—single phase and multiphase. Single-phase alloys are composed of crystals with the same type of structure. They are formed by "dissolving" together different elements to produce a solid solution. The crystal structure of a solid solution is normally that of the base metal.

In contrast to single-phase alloys, multiphase alloys are mixtures rather than solid solutions. They are composed of aggregates of two or more different phases. The individual phases making up the alloy are different from one another in their composition or structure. Solder, in which the metals lead and tin are present as a mechanical mixture of two separate phases, is an example of the simplest kind of multiphase alloy. In contrast, steel is a complex alloy composed of different phases, some of which are solid solutions. Multiphase alloys far outnumber single-phase alloys in the industrial material field, chiefly because they provide greater property flexibility. Thus, properties of multiphase alloys are dependent upon many factors, including the composition of the individual phases, the relative amounts of the different phases, and the positions of the various phases relative to one another.

When two different thermoplastic resins are plastic, alloy is obtained. Alloying permits resin polymers to be blended that cannot be polymerized. Not all plastics are amenable to alloying. Only resins that are compatible with each other—those that have similar melt traits—can be successfully blended.

Types of Alloys

Bearing Alloys

These alloys are used for metals that encounter sliding contact under pressure with another surface; the steel of a rotating shaft is a common example. Most bearing alloys contain particles of a hard intermetallic compound that resists wear. These particles, however, are embedded in a matrix of softer material that adjusts to the hard particles so that the shaft is uniformly loaded over the total surface. The most familiar bearing alloy is Babbitt metal, which contains 83%–91% tin (Sn); the remainder is made up of equal parts of antimony (Sb) and copper (Cu), which form hard particles of the compounds SbSn and CuSn in a soft tin matrix. Other bearing alloys are based on cadmium (Cd), copper, or silver (Ag). For example, an alloy of 70% copper and 30% lead (Pb) is used extensively for heavily loaded bearings. Bearings made by powder metallurgy techniques are widely used. These techniques are valuable because they permit the combination of materials that are incompatible as liquids, for example, bronze and graphite. Powder techniques also permit controlled ferocity within the bearing so that they can be saturated with oil before being used, the so-called oilless bearings.

Corrosion-Resisting Alloys

Certain alloys resist corrosion because they are noble metals. Among these alloys are the precious metal alloys, which will be discussed separately. Other alloys resist corrosion because a protective film develops on the metal surface. This passive film is an oxide that separates the metal from the corrosive environment. Stainless steels and aluminum alloys exemplify metals with this type of protection. Stainless steels, which are iron alloys containing more than 12% chromium (Cr) and 8% nickel (Ni), are the best known and possess a high degree of resistance to many corrosive environments. Aluminum (Al) alloys gain their corrosion-deterring characteristics by the formation of a very thin surface layer of aluminum oxide (Al_2O_3), which is inert to many environmental liquids. This layer is intentionally thickened by a commercial anodizing processes to give a more permanent Al_2O_3 coating. Monel, an alloy of approximately 70% nickel and 30% copper, is a well-known corrosion-resisting alloy that also has high strength. Another nickel-base alloy is Inconel, which contains 14% chromium and 6% iron (Fe). The bronzes, alloys of copper and tin, also may be considered to be corrosion resistant.

Dental Alloys

Amalgams are predominantly alloys of silver and mercury, but they may contain minor amounts of tin, copper, and zinc for hardening purposes, for example, 33% silver, 52% mercury, 12% tin, 2% copper, and less than 1% zinc. Liquid mercury is added to a powder of a precursor alloy of the other metals. After compaction, the mercury diffuses into the silver-base metal to give a completely solid alloy. Gold-base dental alloys are preferred over pure gold because gold is relatively soft. The most common dental gold alloy contains gold (80%–90%), silver (3%–12%), and copper (2%–4%). For higher strengths and hardnesses, palladium and platinum (up to 3%) are added, and the copper and silver are increased so that the gold content drops 60%–70%. Vitallium, an alloy of cobalt (65%), chromium (5%), and molybdenum (3%), and nickel (3%), and other corrosion-resistant alloys are used for bridge work and special applications.

Die-Casting Alloys

These alloys have melting temperatures low enough so that in the liquid form they can be injected under pressure into steel dies. Such castings are used for automotive parts and for office and household appliances that have moderately complex shapes. This processing procedure eliminates the need for expensive machining and forming operations. Most die castings are made from zinc-base or aluminum-base alloys. Magnesium-base alloys also find some application when weight reduction is paramount. Low-melting alloys of lead and tin are not common because they lack the necessary strength for the aforementioned applications. A common zinc-base alloy contains approximately 4% aluminum and up to 1% copper. These additions provide a second phase in the metal to give added strength. The alloy must be free of even minor amounts (less than 100 ppm) of impurities such as lead, cadmium, or tin, because impurities increase the rate of corrosion. Common alumina-base alloys contain 5%–12% silicon, which introduces hard-silicon particles into the tough alumina matrix. Unlike zinc-base alloys, aluminum-base alloys cannot be electroplated; however, they may be burnished or coated with enamel or lacquer.

Advances in high-temperature die-mold materials have focused attention on the die casting of copper-base and iron-base alloys. However, the high casting temperatures introduce costly production requirements, which must be justified on the basis of reduced machining costs.

Eutectic Alloys

In certain alloy systems, a liquid of a fixed composition freezes to form a mixture of two basically different solids or phases. An alloy that undergoes this type of solidification process is called a eutectic alloy. A typical eutectic alloy is formed by combining 28.1% of copper with 71.9% of silver. A homogeneous liquid of this composition on slow cooling freezes to form a mixture of particles of nearly pure copper embedded in a matrix (background) of nearly pure silver.

The advantageous mechanical properties inherent in composite materials such as plywood composed of sheets or lamellae of wood bonded together and fiberglass in which glass fibers are used to reinforce a plastic matrix have been known for many years. Attention is being given to eutectic alloys because they are basically natural composite materials. This is particularly true when they are directionally solidified to yield structures with parallel plates of the two phases (lamellar structure) or long fibers of one phase embedded in the other phase (fibrous structure). Directionally solidified eutectic alloys are being given serious consideration for use in fabricating jet engine turbine blades. For this purpose, eutectic alloys that freeze to form tantalum carbide (TaC) fibers in a matrix of a cobalt-rich alloy have been heavily studied.

Fusible Alloys

These alloys generally have melting temperatures below that of tin (450°F or 232°C) and in some cases as well as 120°F (50°C). Using eutectic compositions of metals such as lead, cadmium, bismuth, tin, antimony, and indium achieves these low melting temperatures. These alloys are used for many purposes, for example, in fusible elements in automatic sprinklers, forming and stretching dies, filler for thin-walled tubing that is being bent, and anchoring dies, punches, and parts being machined. Alloys rich in bismuth were formally used for type metal because these low-melting metals exhibited a slight expansion on solidification, thus replicating the font perfectly for printing and publication.

High-Temperature Alloys

Energy conversion is more efficient at high temperatures than at low; thus the need in power-generating plants, jet engines, and gas turbines for metals that have high strengths at high temperatures is obvious. In addition to having strength, these alloys must resist oxidation by fuel–air mixtures and steam vapor. At temperatures up to about 1380°F (750°C), the austenitic stainless steels (18% Cr–8% Ni) serve well. An additional 180°F (100°C) may be realized if the steels also contain 3% molybdenum. Both nickel-base and copper-base alloys, commonly categorized as superalloys, may serve useful functions up to 2000°F (1100°C). Nichrome, a nickel-base alloy containing 12%–15% chromium and 25% iron, is a fairly simple superalloy. More sophisticated alloys invariably contain five, six, or more components, for example, an alloy called Rene, 41 contains approximately 9% Cr, 1.5% Al, 3% Ti, 11% Co, 10% Mo, 3% Fe, 0.1% C, 0.005% B, and the balance Ni. Other alloys are equally complex. The major contributor to strengthen these alloys is the solution–precipitate phase of Ni_3 (TiAl). It provides strength because it is coherent with the nickel-rich phase. Cobalt-base superalloys maybe even more complex and generally contain carbon, which combines with the tungsten (W) and chromium to produce carbides that serve as the strengthening agent. In general, the cobalt-base superalloys are more resistant to oxidation than the nickel-based alloys are, but they are not as strong. Molybdenum-base alloys have exceptionally high strength at high temperatures, but their brittleness at lower temperatures and their poor oxidation resistance at high temperatures have limited their use. However, coatings permit the use of such alloys in an oxidizing atmosphere, and they are finding increased application. A group of materials called cermets, which are mixtures of metals and compounds such as oxides and carbides, have high strength at high temperatures, and although their ductility is low, they have been found to be usable. One of the better-known cermets consists of a mixture of TiC and nickel, the nickel acting as a binder or cement for the carbide.

Joining Alloys

Metals are bonded by three principal procedures: welding, brazing, and soldering. Welded joints melt the contact region of the adjacent metal; thus, the filler material is chosen to approximate the composition of the parts being joined. Brazing and soldering alloys are chosen to provide filler metal with an appreciably lower melting point than that of the joined parts. Typically, brazing alloys melt above 750°F (400°C), whereas solders melt at lower temperatures. A 57% Cu–42% Zn–1% Sn brass is a general-purpose alloy for brazing steel and many nonferrous metals. A Si–Al eutectic alloy is used for brazing aluminum, and an aluminum-containing magnesium eutectic alloy brazes magnesium parts. The most common solders are based on Pb–Sn alloys. The prevalent 60% Sn–40% Pb alloy has a range of solidification and is thus preferred as a wiping solder by plumbers.

Light-Metal Alloys

Aluminum and magnesium, with densities of 2.7 and 1.75 g/cm^3, respectively, are the basis for most of the light-metal alloys. Titanium (4.5 g/cm^3) may also be regarded as a light-metal alloy if comparisons are made with metal such as steel and copper. Aluminum and magnesium must be hardened to receive extensive application. Age-hardening processes are used for this purpose. Typical alloys are 90% Al–10% Mg, 95% Al–5% Cu, and 90% Mg–10% Al. Ternary (three element) and more complex alloys are very important light-metal alloys because of their better properties. The Al–Zn–Mg system of alloys, used extensively in aircraft applications, is a prime example of one such alloys system.

Low-Expansion Alloys

This group of alloys includes Invar (64% Fe–36% Ni), the dimensions of which do not vary over the atmospheric temperature range. It has special applications in watches and other temperature-sensitive devices. Glass-to-metal seals for electronic and related devices require a matching of the thermal-expansion characteristics of the two materials. Kovar (54% Fe–29% Ni–17% Co) is widely used because its expansion is low enough to match that of glass.

Magnetic Alloys

Soft and hard magnetic materials involve two distinct categories of alloys. The former consists of materials used for magnetic cores of transformers fan motors and must be magnetized and demagnetized easily. For AC applications, silicon ferrite is commonly used. This is an alloy of iron containing as much as 5% silicon. The silicon has little influence on the magnetic properties of the iron, but it increases the electric resistance appreciably and thereby decreases the core loss by induced currents. A higher magnetic permeability, and therefore greater transformer efficiency, is achieved if these silicon steels are grain oriented so that the crystal axes are closely aligned with the magnetic field. Permalloy (78.5% Ni–21.5% Fe) and some comparable cobalt-base alloys have very high permeabilities at low field strengths and thus are used in the communications industry. Ceramic ferrites, although not strictly alloys, are widely used in high-frequency applications because of their low electrical conductivity and negligible induced energy losses in the magnetic field. Permanent or hard magnets may be made from steels that are mechanically hardened, either by deformation or by quenching. Some precipitation-hardening, iron-base alloys are widely used for magnets. Typical of these are the Alnicos, for example, Alnico-4 (55% Fe–28% Ni–12% Al–5% Co). Since these alloys cannot be forged, they must be produced in the form of castings. Hard magnets are being produced from alloys of cobalt and the rare earth type of metals. The compound RCo_5, where R is samarium (Sm), lanthanum (La), cerium (Ce), and so on, has extremely high coercivity.

Precious-Metal Alloys

In addition to their use in coins and jewelry, precious metals such as silver, gold, and the heavier platinum (Pt) metals are used extensively in electrical devices in which contact resistances must remain low, in catalytic applications to aid chemical reactions, and in temperature-measuring devices such as resistance thermometers and thermocouples. The unit of alloy impurity is commonly expressed in karats, when each karat is 1/24 of the part. The most common precious-metal alloy is sterling silver (92.5% Ag, with the remainder being unspecified, but usually copper). Copper is very beneficial in that it makes the alloy harder and stronger than pure silver. Yellow gold is an Au–Ag–Cu alloy with an approximately 2:1:1 ratio. White gold is an alloy that ranges from 10 to 18 karats, the remainder being additions of nickel, silver, or zinc, with change of color from yellow to white. The alloy 87% platinum–13% rhodium (Rh), when joined with pure platinum, provides a widely used thermocouple for temperature measurements in the 1832°F–3000°F (1002°C–1650°C) temperature range.

Shape Memory Alloys

These alloys have a very interesting and desirable property. In a typical case, a metallic object of a given shape is cooled from a given temperature T_1, two or lower temperature T_2, where it is deformed to change its shape. Upon reheating from T_2 to T_1, the shape change accomplished at T_2 is recovered so that the object returns to its original configuration. This thermoelastic property of the shape memory alloys is associated with the fact that they undergo a martensitic phase transformation (i.e., a reversible change in crystal structure that does not involve diffusion) when they are cooled or heated between T_1 and T_2.

For a number of years, the shape memory materials were essentially scientific curiosities. Among the first alloys shown to possess these properties was one of gold alloyed with 47.5% cadmium. Considerable attention has been given to an alloy of nickel and titanium known as *nitinol*. The interest in shape memory alloys has increased because it has been realized that these alloys are capable of being employed in a number of useful applications. One example is for thermostats; another is for couplings on hydraulic lines or electrical circuits. The thermoelastic properties can also be used, at least in principle, to construct heat engines that will operate over a small temperature differential and will thus be of interest in the area of energy conversion.

Thermocouple Alloys

These include *chromel*, containing 90% Ni and 10% Cr, and *alumel*, containing 94% Ni, 2% Al, 3% Cr, and 1% Si. These two alloys together form the widely used *chromel-alumel* thermocouple, which can measure temperatures up to 2200°F (1204°C). Another common thermocouple alloy is *constantan*, consisting of 45% Ni and 55% Cu. It is used to form iron–constantan and copper–constantan couples, used at lower temperatures. For precise temperature measurements and for measuring temperatures up to 3000°F (1650°C), thermocouples are used in which one metal is platinum and the other metal is platinum plus either 10% or 13% rhodium.

Prosthetic Alloys

Prosthetic alloys are alloys used in internal prostheses, that is, surgical implants such as artificial hips and knees. External prostheses are devices that are worn by patients outside the body; alloy selection criteria are different from those for internal prostheses. In the United States, surgeons use about 250,000 artificial hips and knees and about 30,000 dental implants per year.

Alloy selection criteria for surgical implants can be stringent primarily because of biochemical and chemical aspects of the service environment. Mechanically, the properties and shape of an implant must meet anticipated functional demands, for example, hip joint replacements are routinely subjected to cyclic forces that can be several times body weight. Therefore, intrinsic mechanical properties of an alloy, for example, elastic modulus, yield strength, fatigue strength, ultimate tensile strength, and wear resistance, must all be considered. Similarly, because the pH and ionic conditions within a living organism define a relatively hostile corrosion environment for metals, corrosion properties are an important consideration. Corrosion must be avoided not only because of alloy deterioration but also because of the possible physiological effects of harmful or even cytotoxic corrosion products that may be released into the body. (Study of the biological effects of biomaterials is a broad subject in itself, often referred to as biocompatibility.) The corrosion

resistance of all modern alloys stems primarily from strongly adherent and passivating surface oxides, such as TiO_2 on titanium-base alloys and Cr_2O_3 on cobalt-based alloys.

The most widely used prosthetic alloys therefore include high strength, corrosion-resistant ferrous, cobalt-base, or titanium-base alloys. Examples include cold-worked stainless steel; cast Vitallium, a wrought alloy of cobalt, nickel, chromium, molybdenum, and titanium; titanium alloyed with aluminum and vanadium; and commercial-purity titanium. Specifications for nominal alloy compositions are designated by the American Society for Testing and Materials (ASTM).

Prosthetic alloys have a range of properties. Some are easier than others to fabricate into the complicated shapes dictated by anatomical constraints. Fabrication techniques include investment casting (solidifying molten metal in a mold), forging (forming metal by deformation), machining (forming by machine shop processes, including computer-aided design and manufacturing), and hot isostatic pressing (compacting fine powders of alloy into desired shapes under heat and pressure). Cobalt-base alloys are difficult to machine and are therefore usually made by casting or hot isostatic pressing. Some newer implant designs are porous coated, that is, they are made from the standard ASTM alloys but are coated with alloy beads or mesh applied to the surface by centering or other methods. The rationale for such coatings is implant fixation by bone ingrowth.

Some alloys are modified by nitriding or ion implantation of surface layers of enhanced service properties. A key point is that prosthetic alloys of identical composition can differ substantially in terms of structure and properties, depending on fabrication history. For example, the fatigue strength approximately triples for hot isostatically pressing versus as-cast Co–Cr–Mo alloy, primarily because of a much smaller grain size in the microstructure of the former.

No single alloy is vastly superior to all others, existing prosthetic alloys have all been used in successful and, indeed, unsuccessful implant designs. Alloy selection is only one determinant of performance of the implanted device.

Superconducting Alloys

Superconductors are materials that have zero resistance to the flow of electric current at low temperatures. There are more than 6000 elements, alloys, and compounds that are known superconductors. This remarkable property of zero resistance offers unprecedented technological advances such as the generation of intense magnetic fields. Realization of these new technologies requires development of specifically designed superconducting alloys and composite conductors. An alloy of niobium and titanium (NbTi) has a great number of applications and superconductivity; it becomes superconducting at 9.5 K (critical superconducting temperature, T_c). This alloy is preferred because of its ductility and its ability to carry large amounts of current at high magnetic fields, represented by $J_c(H)$ (where J_c is the critical current and H is a given magnetic field), and still retains its superconducting properties. Brittle compounds with intrinsically superior superconducting properties are also being developed for magnet applications. The most promising of these are compounds of niobium and strontium (Nb_3Sn), vanadium and gallium (V_3Ga), niobium and germanium (Nb_3Ge), and niobium and aluminum (Nb_3Al), which have higher T_c (15–23 K) and higher $J_c(H)$ than NbTi.

Superconducting materials possess other unique properties such as magnetic flux quantization and magnetic field–modulated supercurrent flow between two slightly separated superconductors.

These properties form the basis for electronic applications of superconductivity such as high-speed computers or ultrasensitive magnetometers. Development of these applications began using lead or niobium in bulk form, but the emphasis then was transferred to materials deposited in the thin-film form. PbIn and PbAu alloys are more desirable than pure lead films, as they are more stable. Improved vacuum deposition systems eventually led to the use of pure niobium films as they, in turn, were more stable that lead alloy films. Advances in thin-film synthesis techniques led to the use of the refractory compound niobium nitride (NbN) in electronic applications. This compound is very stable and possesses a higher T_c (15 K) than either lead or niobium.

Novel high-temperature superconducting materials have revolutionary impact on superconductivity and its applications. These materials are ceramic, copper oxide–based materials that contain at least four and as many as six elements. Typical examples are yttrium–barium–copper–oxygen (T_c [93 K]), bismuth–strontium–calcium–copper–oxygen (T_c [110 K]), and thallium–barium–calcium–copper (T_c [125 K]). These materials become superconducting at such high temperatures that refrigeration is simpler, more dependable, and less expensive. Much research and development has been done to improve the technologically important properties such as J_c (H), chemical and mechanical stability, and device-compatible processing procedures. It is anticipated that the new compounds will have a significant impact in the growing field of superconductivity.

Alloy Plating

The codeposition of two or more metallic elements. The electrodeposition of a coating comprising two metals such as copper and zinc to produce a brass plate. The two metals may be supplied from a single alloy anode or from individual anodes.

Alloy Powder; Alloyed Powder

A metal powder consisting of at least two constituents that are partially or completely alloyed with each other.

Alloy Steel

Steel containing specified quantities of alloying elements (other than carbon and the commonly accepted amounts of manganese, copper, silicon, sulfur, and phosphorus) within the limits recognized for constructional alloy steels, added to affect changes in mechanical or physical properties.

In general, the term applies to all steels exceeding the limits of manganese, silicon, and copper of carbon steels or which contain other alloying ingredients. Alloy steels often take the name of the alloying element or elements having the greatest influence on their performance characteristics or the name of the key characteristic, processing mechanism, or application. Thus, the prevalence of such terms as nickel steels; stainless, or corrosion-resistant, steels; maraging steels; precipitation-hardening steels; tool steels; valve steels; etc. Usually, however, the term excludes high-alloy steels and refers instead to the standard alloy steels of the American Iron and Steel Institute (AISI) and Society of Automotive Engineers (SAE) International, which contain low to moderate amounts of alloying elements, usually less than 5% total. The AISI or SAE designations of these steels are usually noted by four numerals—13XX–91XX. The first two numerals pertain to a specific alloying element or elements, and the last two numerals indicate carbon content in hundredths of 1%. Sometimes, three numerals are used to denote carbon content, and a letter, such as B for boron and L for lead, follows the first two numerals to indicate an alloying element not indicative of the first two numerals. A letter prefix is used occasionally to designate special furnace practice used to make the steel, and the suffix H is used to designate steels made to specific hardenability requirements. A three-numeral system, 9XX, is commonly used to designate high-strength, low-alloy (HSLA) steels, some of which are also called microalloyed steels because of the small amount of alloying elements, with the last two numerals indicating minimum tensile yield strength in 1000 lb/in.² (6.895 MPa). Although most alloy steels are heat treated by users and extremely high levels of strength and toughness can be achieved, HSLA steels are typically supplied to specific strength levels and are not heat treated by users.

Alloy Structures

Metals in actual commercial use are almost exclusively alloys, and not pure metals since it is possible for the designer to realize an extensive variety of physical properties in the product by varying the metallic composition of the alloy. As a case in point, commercially pure or cast iron is very brittle because of the small amount of carbon impurity always present, whereas the steels are much more ductile, with greater strength and better corrosion properties. In general, the highly purified single crystal of a metal is very soft and malleable, with high electrical conductivity, whereas the alloy is usually harder and may have a much lower conductivity. The conductivity will vary with the degree of order of the alloy, and the hardness will vary with the particular heat treatment used.

The basic knowledge of structural properties of alloys is still in large part empirical, and indeed, it will probably never be possible to derive formulas that will predict which metals to mix in a certain proportion and with a certain heat treatment to yield a specified property or set of properties. However, a set of rules exists that describes the qualitative behavior of certain groups of alloys. These rules are statements concerning the relative sizes of constituent atoms, for alloy formation, and concerning what kinds of phases to expect in terms of the valence of the constituent atoms. The rules were discovered in a strictly empirical way, and for the most part, the present theoretical understanding of alloys consists of rudimentary theories that describe how the rules arise from the basic principles of physics. These rules were proposed by W. Hume-Rothery concerning the binary substitutional alloys and phase diagrams.

Alloy System

A complete series of compositions produced by mixing in all proportions any group of two or more components, at least one of which is a metal.

Alloying Element

An element added to and remaining in a metal that changes structure and properties.

All-Position Electrode

In arc welding, a filler-metal electrode for depositing weld metal in the flat, horizontal, overhead, and vertical positions.

Alluvial Tin

This or other materials deposited on the beds of rivers and areas of slow flow, having been transported by the stream from locations where it was eroded from the rocks.

All-Weld-Metal Test Specimen

A test specimen wherein the portion being tested is composed wholly of weld metal.

Allylics (Diallyl Phthalate Plastics)

Allylics are thermosetting materials developed since World War II. The most important of these are diallyl phthalate (DAP) and diallyl isophthalate (DAIP), which are currently available in the form of monomers and prepolymers (resins). Both DAP and DAIP are readily converted to thermoset-molding compounds and resins for preimpregnated glass cloth and paper. Allyls are also used in crosslinking agents for unsaturated polyesters.

DAP resin is the first all-allylic polymer commercially available as a dry, free-flowing white powder. Chemically, DAP is a relatively linear partially polymerized resin that softens and flows under heat and pressure (as in molding and laminating) and cross-links to a 3D insoluble thermoset resin during curing. This family of thermoset resins made by *addition polymerization* of compounds containing the group CH_2: $CH–CH_2$, such as esters of allyl alcohol and dibasic acids. They are available as monomers, partially polymerized prepolymers, or molding compounds. Other members of the family besides the two mentioned earlier are diallyl maleate and diallyl chlorendate.

Properties

In preparing the resin, DAP is polymerized to a point where almost all the change in specific gravity has taken place. Final cure, therefore, produces very little additional shrinkage. In fact, DAP is cured by polymerization without water formation.

Allylic resins enjoy certain specific advantages over other plastics, which make them of interest in various special applications. Allylics exhibit superior electrical properties under severe temperature and humidity conditions. These good electrical properties (insulation resistance, low loss factor, arc resistance, etc.) are retained despite repeated exposure to high heat and humidity. DAP resin is resistant to 155°C–180°C temperatures, and the DAIP resin is good for continuous exposures up to 206°C–232°C temperatures. Allylic resins exhibit excellent postmold dimensional stability, low moisture absorption, good resistance to solvents, acids, alkalies, weathering, and wet and dry abrasion. They are chemically stable, have good surface finish, mold well around metal inserts, and can be formulated in pastel colors with excellent color retention at high temperatures.

DAP resin currently finds major use in (1) molding and (2) industrial and decorative laminates. Both applications utilize the desirable combination of low shrinkage, absence of volatiles, and superior electrical and physical properties common to DAP.

Molding Compounds

Compounds based on allyl prepolymers are reinforced with fibers (glass, polyester, or acrylic) and filled with particulate materials to improve properties. Glass fiber imparts maximum mechanical properties, acrylic fiber provides the best electrical properties, and polyester fiber improves impact resistance and strength in thin sections. Compounds can be made in a wide range of colors because the resin is essentially colorless. See Table A.9.

Prepregs (preimpregnated glass cloth) based on allyl prepolymers can be formulated for short cure cycles. They contain no toxic

TABLE A.9
Properties of DAP Molding Compounds

| ASTM Test | Property | Filler | | | |
		Polyester	Long Glass	Short Glass	Arc-Track Resistant
Physical					
D792	Specific gravity	1.39–1.42	1.70–1.90	1.6–1.8	1.87
D792	Specific volume (in.³/lb)	19.96–19.54	17.90–16.32	17.34–15.42	14.84
D570	Water absorption, 24 h, 1/8-in. thk (%)	0.2	0.05–0.2	0.05–0.2	0.14
Mechanical					
D638	Tensile strength (psi)	5,000	9,000	7,000	7,000–10,000
D790	Flexural strength (psi)	11,500–12,500	18,000	16,000	24,000
D790	Flexural modulus (10^5 psi)	6.4	16	17	19
D256	Impact strength, Izod (ft-lb/in. of notch)	4.5–12	6.0	0.8	3.6
D785	Hardness, Rockwell M	108	105–110	105–110	112
Thermal					
C177	Thermal conductivity (10^{-4} cal-cm/s-cm²-°C)	—	14–16	14–15	15–17
D696	Coefficient of thermal expansion (10^{-5} in./in.-°C)	—	2.0–3.0	2.0–3.0	23–27
D648	Deflection temperature (°F) At 264 psi	290	450	420	>572
Electrical					
D149	Dielectric strength, (V/mil) Step by step, 1/8-in. thk	400	385	400	400
D150	Dielectric constant At 1 kHz	0.008	0.004–0.006	0.006	0.003–0.008
D150	Dissipation factor At 1 kHz	3.6	4.2	4.4	4.1–4.5
D257	Volume resistivity (ohm-cm) At 73°F, 50% RH	$2–3 \times 10^{15}$	$2–3 \times 10^{15}$	$2–3 \times 10^{15}$	10^{16}
D495	Arc resistance (s)	125	140	135	125–180
Frictional					
—	Coefficient of friction			Stat/Dyn	
	Self	—	—	0.14/0.13	—
	Against steel	—	—	0.20/0.19	—

Source: Mach. Design Basics Eng. Design, June, p. 680, 1993. With permission.

additives, and they offer long storage stability and ease of handling and fabrication. Properties such as flame resistance can be incorporated. The allyl prepolymers contributed excellent chemical resistance and good electrical properties.

Other molding powders are compounded of DAP resin, DAP monomer, and various fillers like asbestos, Orlon, Dacron, cellulose, glass, and other fibers. Inert fillers used include ground quartz and clays, calcium carbonate, and talc.

Allyl moldings have low-mold shrinkage and postmold shrinkage—attributed to their nearly complete addition reaction in the mold—and have excellent stability under prolonged or cyclic heat exposure. Advantages of allyl systems over polyesters are freedom from styrene odor low toxicity, low evaporation losses during evacuation cycles, though subsequent oozing or bleed out, and long-term retention of electrical insulation characteristics.

Applications

Uses of such DAP-molding compounds are largely for electrical and electronic parts, connectors, resistors, panels, switches, and insulators. Other applications for molding compounds include appliance handles, control knobs, dinnerware, and cooking equipment.

Decorative laminates containing DAP resin can be made from glass cloth (or other woven and nonwoven materials), glass mat, or paper. Such laminates may be bonded directly to a variety of rigid surfaces at lower pressures (50–300 psi) than generally required for other plastic laminates. A short hot-to-hot cycle is employed, and press platens are always held at curing temperatures. DAP laminates can, therefore, be used to give a permanent finish to high-grade wood veneers (with a clear overlay sheet) or to upgrade low-cost core materials (by means of a patterned sheet).

Allyl prepolymers are particularly suited for critical electronic components that serve in severe environmental conditions. Chemical inertness qualifies the resins for molded pump impellers and other chemical-processing equipment. Their ability to withstand steam environments permits uses in sterilizing and hot-water equipment. Because of their excellent flow characteristics, DAP compounds are used for parts requiring extreme dimensional accuracy. Modified resin systems are used for encapsulation of electronic devices such as semiconductors and sealants for metal castings.

A major application area for allyl compounds is electrical connectors, used in communications, computer, and aerospace systems. The high thermal resistance of these materials permits their use in vapor-phase soldering operations. Uses for prepolymers include arc-track-resistant compounds for switchgear and television components. Other representative uses are for insulators, encapsulating shells, potentiometer components, circuit boards, junction boxes, and housings.

DAP and DAIP prepregs are used to make lightweight, intricate parts such as radomes, printed circuit boards, tubing, ducting, and aircraft parts. Another use is in copper-clad laminates for high-performance printed circuit boards.

Alnico

See Alcomax.

Alpha

The various phases occurring in our systems are designated by Greek letters, α, β, γ, etc., respectively, alpha, beta, gamma, etc. Alpha usually refers to the primary solid solution in any system. See also *Greek*.

Alpha (α) Cellulose

A very pure cellulose prepared by special chemical treatment.

Alpha (α) Loss Peak

In dynamic mechanical or dielectric measurement, the first peak in the damping curve below the melt, in order of decreasing temperature or increasing frequency.

Alpha Brass

A solid-solution phase of one or more alloying elements and copper having the same crystal lattice as copper.

Alpha Brasses

Brasses containing up to about 37% zinc. They are single, alpha phase and are readily cold worked following initial hard working.

Alpha Case

The oxygen-, nitrogen-, or carbon-enriched α-stabilized surface in titanium resulting from elevated temperature exposure. See also *alpha stabilizer*.

Alpha Double Prime (α'') (Orthorhombic Martinsite)

A supersaturated, nonequilibrium orthorhombic phase formed by a diffusionless transformation of the β phase in certain titanium alloys. It is often difficult to distinguish from a acicular α, although the latter is usually less well defined and frequently has curved, instead of straight sides.

Alpha Ferrite

See *ferrite*.

Alpha Iron

The body-centered cubic form of pure iron, stable below 910°C (1670°F).

Alpha Model 1

A type of wear-testing machine consisting of a conforming or flat-faced block pressed vertically downward by a deadweight loading arrangement against the circumference of a hardened steel ring that is rotating on a shaft.

Alpha Particles

Subatomic particles comprising two protons and two neutrons, that is, the helium nucleus, and hence positively charged. They have relatively low penetrating power.

Alpha Process

A *shell molding* and core making method in which a thin resin-bonded shell is baked with the less expensive, highly permeable material.

Alpha Stabilizer

An alloying element in titanium that dissolves preferentially in the α and α – β transformation temperature.

Alpha Transus

The temperature that designates the phase boundary between the α and α + β fields in titanium alloys.

Alpha–Beta Brasses

Brasses that contain about 40% zinc. They are duplex, alpha plus beta, and are readily hot worked but will accept only a small amount of cold work.

Alpha–Beta Structure

A titanium microstructure containing α and β as the principal phases at a specific temperature. See also *beta*.

Alsifer

A deoxidizer (20Al, 40Si, 40Fe) used for steel.

Alternate Immersion Test

A corrosion test in which the specimens are intermittently exposed to a liquid medium at definite time intervals.

Alternate Polarity Operation

A resistance welding process variation in which succeeding welds are made with pulses of alternating polarity.

Alternating Copolymer

A copolymer in which each repeating unit is joined to another repeating unit in the polymer chain (–A–B–A–B).

Alternating Current Resistance

The resistance offered by any circuit to the flow of alternating current.

Alternating Stress Amplitude

A test parameter of a dynamic fatigue test; one-half the algebraic difference between the maximum and minimum stress in one cycle.

Alum

A colorless to white crystalline potassium aluminum sulfate, $KAl(SO_4)_2 \cdot 12H_2O$, or $\{KAl(H_2O)_6\}SO_4 \cdot 6H_2O$, occurring naturally as the mineral kalunite, or kalinite, and in combination as the mineral alunite. It is also called potash alum to distinguish it from other forms. It has a sweetish taste and is very astringent. It is used as an additive in the leather and textile industries, in sizing paper, as a mordant in dyeing, in medicines as an astringent, and baking powder. It is made commercially by reacting bauxite with sulfuric acid and then potassium sulfate. It is an important water-purifying agent. From a water solution, it crystallizes out, forming positively charged particles, which attract the negatively charged organic impurities, thus purifying the water as they settle out.

Alumel

An alloy of nickel with about 2.5% manganese, 2% aluminum, and 1% silicon widely used in conjunction with chromel for pyrometric thermocouples.

Alumina

The oxide of aluminum is Al_2O_3. The natural crystalline mineral is called corundum, but the synthetic crystals used for abrasives are designated usually as aluminum oxide or marketed under trade names. For other uses and as a powder, it is generally called alumina. It is widely distributed in nature in combination with silica and other minerals and is an important constituent of the clays for making porcelain, bricks, pottery, and refractories.

The crushed and graded crystals of alumina when pure are nearly colorless, but the fine powder is white. Off colors are due to impurities. American aluminum oxide used for abrasives is at least 99.5% pure, in nearly colorless crystals melting at 2050°C. The chief uses for alumina are for the production of aluminum metal and for abrasives, but it is also used for ceramics, refractories, pigments, catalyst carriers, and in chemicals.

Aluminum oxide crystals are normally hexagonal and are minute in size. For abrasives, the grain sizes are usually from 100 to 600 mesh. The larger grain sizes are made up of many crystals, unlike the single-crystal large grains of SiC. The specific gravity is about 3.95, and a hardness is up to 2000 Knoop.

There are two kinds of ultrafine alumina abrasive powder. Type A is alpha alumina with hexagonal crystals with particle size of 0.3 μm, density 4.0, and hardness 9 Mohs, and type B is gamma alumina with cubic crystals with particle size under 0.1 μm, specific gravity of 3.6, and a hardness 8. Type A cuts faster, but type B gives a finer finish. At high temperatures, gamma alumina transforms to the alpha crystal. The aluminum oxide most frequently used for refractories is the beta alumina and hexagonal crystals heat-stabilized with sodium.

Activated alumina is partly dehydrated alumina trihydrate, which has a strong affinity for moisture or gases and is used for dehydrating organic solvents, and hydrated alumina is alumina trihydrate. $Al_2O_3 \cdot 3H_2O$ is used as a catalyst carrier.

Activated alumina F-1 is a porous form of alumina, Al_2O_3, used for drying gases or liquids and is also used as a catalyst for many chemical processes.

Alumina ceramics are the most widely used oxide-type ceramic, chiefly because Al_2O_3 is plentiful, relatively low in cost, and equal to or better than most oxides and mechanical properties. Density can be varied over a wide range, as purity—down to about 90% Al_2O_3—to meet specific application requirements. Al_2O_3 ceramics are the hardest, strongest, and stiffest of the oxides. They are also outstanding in electrical resistivity and dielectric strength, are resistant to a wide variety of chemicals, and are unaffected by air, water vapor, and sulfurous atmospheres. However, with a melting point of only 2037°C, they are relatively low in refractoriness, and at 1371°C retain only about 10% of room-temperature strength. Besides wide use as electrical insulators and chemical and aerospace applications, the high hardness and close dimensional tolerance capability of alumina make this ceramic suitable for such abrasion-resistant parts as textile guides, pump plungers, chute linings, discharge orifices, dies, and bearings.

Alumina Al-200, which is used for high-frequency insulators, gives a molded product with the tensile strength of 172 MPa, compressive strength of 2000 MPa, and specific gravity of 3.36. The coefficient of thermal expansion is half that of steel and the hardness about that of sapphire. Alumina AD-995 is a dense vacuum-tight ceramic for high-temperature electronic use. It is 99.5% Al_2O_3 with no SiO_2. The hardness is Rockwell N 80, and dielectric constant is 9.27. The maximum working temperature is 1760°C, and at 1093°C, it has a flexural strength of 200 MPa.

Other alumina products have found their way in the casting of hollow jet engine cores. These cores are then incorporated in molds into which eutectic superalloys are poured to form the turbine blades.

Alumina balls are available in sizes from 0.6 to 1.9 cm for reactor and catalytic beds. They are usually 99% alumina, with high resistance to heat and chemicals. Alumina fibers in the form of short linear crystals, called sapphire whiskers, have high strength up to 1375 MPa for use as a filler in plastics to increase heat resistance and dielectric properties. Continuous single-crystal sapphire (alumina filaments) has unusual physical properties: high tensile strength (over 2069 MPa) and modulus of elasticity of 448.2–482.7 GPa. The filaments are especially needed for use in metal composites at elevated temperatures and in highly corrosive environments. An unusual method for producing single-crystal fibers in lieu of a crystal-growing machine is the floating zone fiber-drawing process. The fibers are produced directly from a molten ceramic without using a crucible.

FP, a polycrystalline alumina (Al_2O_3) fiber, has been developed. The material has greater than 99% purity, and a melting point of 2045°C, which makes it attractive for use with high-temperature metal-matrix composite processing techniques. Thanks to a mechanism, currently not explainable by the developer of FP fibers (DuPont), a silica coating results in an increase in the tensile strength of the filaments to 1896 MPa even though the coating is approximately 0.25 μm thick and the modulus does not change. Fiber FP has been demonstrated as a reinforcement for magnesium, aluminum, lead, copper, and zinc, with emphasis to date on aluminum and magnesium materials.

Fumed alumina powder of submicrometer size is made by flame reduction of aluminum chloride. It is used in coatings and for plastics reinforcement and in the production of ferrite ceramic magnets.

Aluminum oxide film, or alumina film, used as a supporting material in ionizing tubes, is a strong, transparent sheet made by oxidizing aluminum foil, rubbing off the oxide on one side, and dissolving the foil in an acid solution to leave the oxide film on the other side. It is transparent to electrons. Alumina bubble brick is a lightweight refractory brick for kiln lining, made by passing molten alumina in front of an air jet, producing small hollow bubbles, which are then pressed into bricks and shapes.

The foam has a density of 448.5 kg/m³ and porosity of 85%. The thermal conductivity at 1093°C is 0.002 W/(cm²)(°C).

Aluminides

True metals include the alkali and alkaline earth metals, beryllium, magnesium, copper, silver, gold, and the transition elements. These metals exhibit those characteristics generally associated with the metallic state.

The B subgroups comprise the remaining metallic elements. These elements exhibit complex structures and significant departures from typically metallic properties. Aluminum, although considered under the B subgroup metals, is somewhat anomalous in that it exhibits many characteristics of a true metal.

The alloys of a true metal and a B subgroup element are very complex because their components differ electrochemically. This difference gives rise to a stronger tendency toward definite chemical combination of solid solution. Discreet geometrically ordered structures usually result. Such alloys are also termed "electron compounds." The aluminides are phases in such alloys or compounds. A substantial number of beta, gamma, and epsilon phases have been observed in electron compounds, but few have been isolated and evaluated.

The development of intermetallic alloys into useful and practical structural materials remains, despite recent successes, a major scientific and engineering challenge. As with many new and advanced materials, hope and the promise of major breakthroughs in the near future have kept a very active and resilient fraction of the metallurgical community focused on intermetallic alloys.

Compared to conventional aerospace materials, aluminides of titanium, nickel, iron, niobium, etc., with various compositions offer attractive properties for potential structural applications. The combination of good high-temperature strength and creep capability, improved high-temperature environmental resistance, and relatively low density makes this general class of materials good candidates to replace more conventional titanium alloys and, in some instances, nickel-base superalloys. Moreover, titanium aluminide matrix composites appear to have the potential to surpass the monolithic titanium aluminides in a number of important property areas, and fabrication into composites form may be a partial solution to some of the current shortcomings attributed to monolithic titanium aluminides.

The material classes include monolithic and continuous fiber composite materials based on the intermetallic compositions Ti_3Al (α_2-phase) and monolithic alloys based on the intermetallic composition TiAl (γ-phase). In their monolithic form, and as a matrix material for continuous fiber composites, titanium aluminides are important candidates to fill a need in the intermediate-temperature regime of 600°C–1000°C. Before these materials can become flightworthy, however, they must demonstrate reliable mechanical behavior over the range of anticipated surface conditions.

The β and γ phases that are found to exist in the Mo–Al alloy are generally considered to correspond to the compositions $MoAl_3$ and $MoAl_2$, respectively.

Powder metallurgy techniques have proved feasible for the production of alloys of molybdenum and aluminum, provided care is taken to employ raw materials of high purity (99%+). As the temperature of the compact is raised, a strong exothermic reaction occurs at about 640°C causing a rapid rise in temperature to above 960°C in a matter of seconds. Bloating occurs, transforming the compact into a porous mass. Complete alloying, however, is accomplished. This porous, friable mass can be subsequently finally comminuted, repressed, and sintered (or hot pressed) to form a useful body with quite uniform in composition. Vacuum sintering at 1300°C for 1 h at 0.04 μm produces clean, oxide-free metal throughout. Wet comminution prevents caking of the powder, and a pyrophoric powder can be produced by prolonged milling.

Hot pressing is a highly successful means of forming bodies of molybdenum and aluminum previously reacted as mentioned earlier. Graphite dies are employed to which resistance heating techniques are applicable. A parting compound is required since aluminum is highly reactive with carbon causing sticking to the die walls.

Hot-pressed small bars exhibit modulus of rupture strengths ranging from 40,000 to 50,000 psi at room temperature, decreasing to 38,000 to 40,000 psi at 1040°C. Room temperature resistant to fuming nitric acid is excellent.

As has been recognized for some time, ordered intermetallic compounds have a number of properties that make them intrinsically more appealing than other metallic systems for high-temperature use. The primary requirements for high-temperature structural intermetallics, as with any high-temperature structural material, are that they (1) have a high melting point, (2) possess some degree of resistance to environmental degradation, (3) maintain structural and chemical stability at high temperatures, and (4) retain high specific mechanical properties at elevated temperatures whether they are intended as monolithic components or as reinforcing fibers or matrix in composite structures.

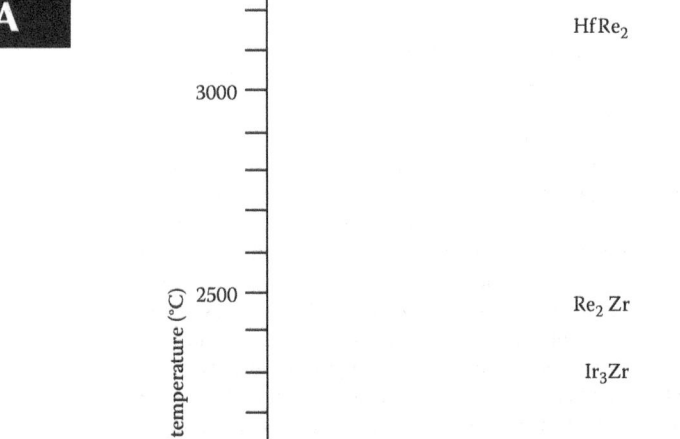

FIGURE A.2 Melting points of various intermetallic compounds relative to superalloys. (From Schwartz, M., *Emerging Technology*, Technomics Publishing Co., Lancaster, PA, 1996. With permission.)

Melting point is a useful first approximation of the high-temperature performance of a material, as various high-temperature mechanical properties (e.g., strengthen creep resistance) are limited by thermally assisted or diffusional processes and thus tend to scale with a melting point of the material. Therefore, the intermetallic can be crudely ranked in terms of their melting points to indicate their future applicability as high-temperature structural materials.

As may be seen in Figure A.2, metallic materials (intermetallics or otherwise) that are currently in use are being studied that melt at temperatures much lower than 1650°C. If these materials are discounted from consideration, the remaining intermetallics in Figure A.2 may be roughly divided into two groups; those that fall in the temperature range just above 1650°C and those whose melting points extend to much higher temperatures.

This second group of intermetallic compounds belongs to a group of intermetallics that are predicted on the basis of the Engel–Brewer phase stability theory.

There are several techniques that have been developed and used to improve the toughness of intermetallics as well as intermetallic compounds:

- Crystal structure modification (macroalloying)
- Microalloying
- Control of grain size or shape
- Reinforcement by ductile fibers or particles
- Control of substructure

Table A.10 includes the aforementioned major categories; however, the use of hydrostatic pressure and suppression of environment should also be cited.

TABLE A.10

Toughness and Ductility Improvements

Microalloying
    B in Ni_3Al, Ni_3Si, PdIn
    Be in Ni_3Al
    Fe, Mo, Ga in NiAl
    Ag in Ni_3Al
Macroalloying
    Fe in Co_3V
    Mn, V, Cr in TiAl
    Nb in Ti_3Al
    Mn, Cr in Al_3Ti
    Pd in Ni_3Al
Grain size refinement
    NiAl
Hydrostatic pressure
    Ni_3Al
Martensite transformation
    Fe in NiAl
Composites (fibers or tubes)
    NiAl/304SS
    Al_3Ta/Al_2O_3
    $MoSi_2$/Nb–1Zr
Composites (ductile particles)
    Nb in TiAl
    Fe, Mn in Ni_3Al
    Nb in $MoSi_2$

Source: Schwartz, M., *Emerging Technology*, Technomics Publishing Co., Lancaster, PA, 1996. With permission.

Additions of chromium and manganese have induced appreciable compressive ductility and modest improvements in bend ductility of Al_3Ti, but significant tensile ductility remains unattainable.

The interest in aluminides has covered the high-melting-point phases in metallic systems with aluminum.

Ordered intermetallics constitute a unique class of metallic materials that form long-range-ordered crystal structures (Figure A.3) below a critical temperature that is generally referred to as the critical ordering temperature (T_c). These intermetallics usually exist in relatively narrow compositional ranges around simple stoichiometric ratios.

The search for new high-temperature structural materials has stimulated much interest in ordered intermetallics. Recent interest has been focused on aluminides of nickel, iron, and titanium. These aluminides possess many attributes that make them attractive for high-temperature structural applications. They contain enough aluminum to form, in oxidizing environments, thin films of aluminide oxides that often are compact and protective. They have low densities, relatively high melting points, and good high-temperature strength properties. For example, N_3Al shows an increase rather than a decrease in yield strength with increasing temperatures. The aluminides of interest are described in Table A.11.

In the range of 14%–34%, aluminum by weight occurs in two intermetallic phases, Ni_3Al and NiAl. The alloys are prepared by powder metallurgy and by casting techniques. Compacts of NiAl + 5% Ni, produced by powder techniques, exhibit room-temperature modulus of rupture values of 144,000 psi, and heat shock resistance is considered excellent. They have good resistance to red and white fuming nitric acids.

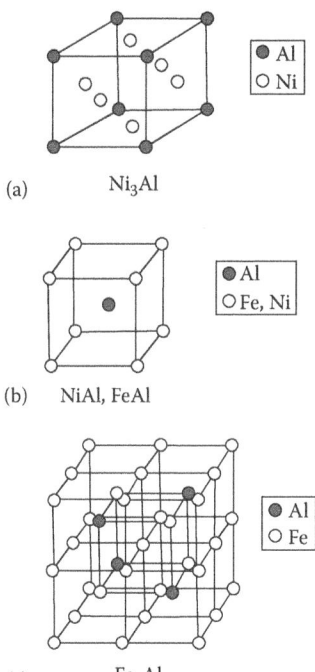

FIGURE A.3 The crystal structure of nickel and iron aluminides (a) LI₂, (b) B₂, and (c) DO₃. (From Schwartz, M., *Emerging Technology,* Technomics Publishing Co., Lancaster, PA, 1996. With permission.)

TABLE A.11
Properties of Nickel and Iron Aluminides

Alloy	Crystal Structure[a]	Critical Ordering Temp. (°C)	Melting Point (°C)	Material Density (g/cm³)	Young's Modulus (GPa)
Ni₃Al	LI₂ (ordered fcc)	1390	1390	7.50	178
NiAl	B₂ (ordered bcc)	1640	1640	5.86	294
Fe₃Al	DO₃ (ordered bcc)	540	1540	6.72	141
	B₂ (ordered bcc)	760	1540		
FeAl	B₂ (ordered bcc)	1250	1250	5.56	260

Source: Schwartz, M., *Emerging Technology*, Technomics Publishing Co., Lancaster, PA, 1996. With permission.

[a] fcc, face-centered cubic; bcc, body-centered cubic.

The cast alloys of nickel and aluminum exhibit an increasing exothermic character with increasing aluminum content. Alloys with the exception of the 34% aluminum alloy and NiAl (31.5% Al) produce sound castings free of excessive porosity. A room-temperature tensile strength of approximately 49,000 psi is exhibited by the Ni₃Al compound with about 5% elongation. The 25% Al (NiAl) alloy has a room temperature tensile strength of 24,000 psi.

The 17.5% aluminum alloy that contained a mixture of the phases Ni₃Al and NiAl exhibits the best strength properties. Room-temperature tensile strength is approximately 80,000 psi with about 2% elongation, while at 815°C, the tensile strength is about 50,000 psi. This alloy can be rolled at 1315°C and possesses good thermal shock and oxidation resistance. Impact resistance is fair. The 100 h stress-to-rupture strength at 734°C is 14,000 psi, but it should be noted creep rates are high.

Ni₃Al is an intermetallic phase that forms at the nickel-rich end of the Ni–Al system and has a crystal structure (Figure A.3). Ni₃Al is the most important strengthening constituent of nickel-base superalloys. This is because the aluminide possesses an excellent elevated-temperature strength in addition to good oxidation resistance. Unlike conventional materials, Ni₃Al and its alloy show a yield anomaly, that is, its yield strength increases rather than decreases with increasing temperature.

Boron has been found to be most effective in improving the tensile ductility of Ni₃Al (<25% aluminum) tested in air at room temperature. Other tests indicated foreign-doped Ni₃Al is not sensitive to test environment at room temperature.

Boron-free Ni₃Al (24% Al) has shown a ductility of only 8.2% in dry O₂ while boron-doped Ni₃Al exhibited 42.8%. This suggests that boron segregation also enhances the grain-boundary cohesion in Ni₃Al. In addition to boron, iron, chromium, and zirconium appear to enhance the grain-boundary properties and improve the tensile ductility of Ni₃Al at room temperature.

The mechanical properties of Ni₃Al strongly depend on deviation from alloy's stoichiometry (i.e., aluminum/nickel ratio) and ternary alloying additions. The aluminide is capable of dissolving substantial alloying additions without forming second phases; as a result, its deformation and fracture behavior can be strongly altered by alloying.

Mechanical and metallurgical properties of Ni₃Al can be improved by alloying additions. Recent alloy design efforts have led to the development of Ni₃Al-base alloys with the following composition range (in atomic percent) for structural use at elevated temperatures in hostile environments: Ni–(14%–18%) Al–(6%–9%) Cr–(1%–4%) Mo–(0.01%–1.5%) Zr/Hf–(0.01%–0.20%) B.

NiAl containing more than 41 at% forms a single-phase ordered B₂ structure based on the body centered cubic lattice (Figure A.3). In terms of thermophysical properties, B2 NiAl offers more potential for high-temperature structural applications than L1 NiAl. It has a higher melting point (1638°C), a substantially lower density (5.86 g/cm²), and a distinctly higher thermal conductivity 76 W/(m K), at ambient temperatures. In addition, NiAl has excellent oxidation resistance at high temperatures.

The structural use of NiAl suffers from two major drawbacks: poor fracture resistance at ambient temperatures and low strength and creep resistance at elevated temperatures. Single crystals of NiAl are quite ductile and compression, but both single-crystal and polycrystalline NiAl appear to be brittle in tension at ambient temperatures.

Because of the potential use of NiAl at high temperatures, considerable effort has been devoted to understanding brittle fracture and improving mechanical properties of NiAl during the past years.

The room-temperature tensile ductility of NiAl is as high as 6% by adding about 0.2% at% iron. The ductility decreases sharply when the iron content is beyond 0.4 at%. A similar effect has been observed for molybdenum and gallium.

Gamma alloys may be processed by

- Casting
- Ingot metallurgy (IM)
- Powder metallurgy
- Sheet forming

The XD™ process for producing gamma alloys containing TiB₂ₚ can be classified separately, although the XD alloys are produced by either the casting process or the IM process. Each process can involve one or more routes.

The conventional method of producing foils from brittle materials was chemical milling. The approach involves suspending hot-rolled and surface-ground 0.050 mm thick plates of the material in a pickling bath and dissolving the metal until the plates are about 0.015 mm thick. For γ-TiAl, the chemical milling technique yields a poor surface finish on the foil.

Engineers knew that the γ-TiAl, while brittle in tension, accepts high compression strain without failing. So they designed a new "isobaric" rolling mill to maintain triaxial compression on the material as it passes through the rolls. This proprietary process increases yield from the original ingot to around 50% and provides thinner (0.03–0.05 mm) foils with smaller grains. As a result, isobaric cold rolling is a breakthrough technology. A cold-rolled and annealed γ-Ti-48-2-2 strip has a hardness of 260 H_v (Vickers).

Other processing methods include the self-propagating high-temperature synthesis (SHS) process, which starts with a structure of alternating layers of pure metal sheets or foils (e.g., iron, nickel, or titanium) and aluminum sheets or foils. The ductile metal layers are formed into the desired shape, either before or during the SHS reaction. The SHS reaction is initiated by heating in a vacuum. When the system reaches the initiation temperature (typically less than 660°C, the melting point of aluminum) an SHS reaction begins at the metal interface, that is, between the aluminum and the iron, nickel, or titanium metals. The heat generated from the reaction melts the aluminum allowing rapid reaction with the adjacent metal layers. That is, the SHS reaction occurs between the elemental metals producing well-bonded layers of unreacted elemental metal and metal–aluminum intermetallics. In this technique, the aluminum completely reacts, forming a titanium–aluminide–titanium, a nickel–aluminide–nickel, or an iron–aluminide–iron layered composite structure.

A second technique takes elemental metal powder mixtures that are placed between layers of an elemental metal foil worksheet, after which the entire layout is heated under pressure in a vacuum. Starting powders have been iron, nickel, or titanium mixed with aluminum. Starting elemental metal sheets have been titanium, nickel, or iron.

IM Developments

A mixture of titanium and 34 wt.% aluminum, that is, 48 at%, was prepared by blending commercially available elemental powders with a maximum grain size of 100 mm compacted at room temperature in a uniaxial pressing machine to billets of 50 mm in diameter.

Titanium aluminide foils were produced using a combination of a standard Powder metallurgy (P/M) technique (extrusion), aluminum technology (rolling), and advanced titanium technology (Superplastic forming device). Claims have been made that since these techniques were available on a large scale, it should be possible to produce titanium aluminide foils in production.

A series of low-pressure turbine blades fabricated from titanium aluminide was successfully tested and this development could eventually result in reducing the weight of aviation gas turbines by hundreds of pounds. Blades of 49% Ti–47% Al–2% Cr–2% Nb, which have about half the weight of comparable components made from conventional nickel-base alloys, were run in the fifth low-pressure turbine stage of a CF6-80C2 power plant, and during the tests, the blades went through 1000 flight cycles. Compared with traditional nickel-base alloys, the material has half the density and is comparable in strength up to about 700°C. The titanium aluminide also is about 50% stiffer than conventional titanium alloys. If used in the low-pressure section of new aircraft engines as blade material, the titanium aluminide could cut engine weight by more than 136.01 kg.

The 98 plates used in the tests weighed 217 g. Comparable nickel alloy components weigh 383 g. In an ideal situation, the lighter blades would also allow turbine wheels to be lighter and less robust because the reduced-weight blades create lower stresses during operation.

The titanium aluminide alloys used to fabricate the blades were solid airfoils and were cast using conventional foundry techniques.

New titanium aluminum alloys, stronger and tougher than conventional α_2(Ti$_3$Al) materials, are based on the ordered orthorhombic-phase Ti$_3$AlNb (Ti–25 at% Al–25 at% Nb).

The materials are potential weight-saving alternatives to nickel-base superalloys and relatively low temperature (650°C) aircraft gas-turbine engine applications. Several examples include exhaust-nozzle structures, compressor casings, and various compressor components; while the density of Ti$_3$AlNb-base alloys (and other titanium aluminides) is less than two-thirds that of nickel-base superalloys, such as alloy 718, the lower weight can be translated into higher thrust-to-weight ratios or gains in fuel efficiency.

Of two new alloys, the Ti–22% Al–27% Nb has the best combination of high-temperature strength and room-temperature ductility and fracture toughness. The alloy is much stronger than conventional α_2 titanium aluminides such as Ti–24% Al–11% Nb and Ti–25% Al–10% Nb–3% V–1% Mo. It also has a higher strength-to-weight ratio, even though it is slightly denser. For example, Ti–22% Al–27% Nb as a factor-of-two advantage in strength-to-weight ratio over Ti–25% Al–10% Nb–3% V–1% Mo, an α_2-alloy, from room temperature to 650°C. Fracture toughness (K_{1c}) of Ti–22% Al–27% Nb averages about 28 MPa m$^{1/2}$, which is higher than those of the α_2-alloys Ti–25% Al–10% Nb–3% V–1% Mo (K_{1c} = 14 MPa m$^{1/2}$) and Ti–24% Al–17% Nb–1% Mo (19 MPa m$^{1/2}$). Creep behavior of the Ti$_3$AlNb-base alloy is competitive with lower toughness titanium aluminides that have been optimized for creep resistance.

Applications

Coatings

NiAl is the basis of a family of oxidation- and corrosion-resistant coatings that have been used on nickel-base and cobalt-base superalloys over the past 40 years. NiAl coatings 25–100 mm thick are typically applied by a pack cementation process.

Both inward diffusion and outward diffusion coatings have been applied. The mechanisms of coating degradation include a depletion due to spallation and reaction with the substrate. A variety of alloying additions, most notably chromium and yttrium, have been applied to improve performance and thermal cycling and hot corrosion.

Turbine Blades and Vanes

The advantages associated with NiAl as a candidate structural material in advanced gas turbine engines include a 30% reduction in density over current nickel-base superalloys, good intrinsic environmental resistance, a thermal conductivity three to eight times larger than nickel-base superalloys (for improved cooling efficiency), and a melting temperature that is approximately 128°C higher than superalloys (for higher operating temperatures).

The processing, fabrication, and engineering design of NiAl, along with an understanding and improvement in impact properties, will also provide significant technical challenges that must be overcome before NiAl can be successfully employed as a structural material that may compete with current nickel-base superalloys. It is important to point out that a single-crystal approach is crucial to obtaining a balance of both room-temperature ductility and high-temperature strength. Although NiAl-base composites may provide an alternative avenue to balanced properties, this approach is still in its infancy.

Electronic Applications

NiAl is an attractive material for metallization on III–V semiconductors and is also a critical component in complex *metal*/III–V semiconductor heterostructures, such as enhanced-barrier Schottky contacts and semiconductor-clad metallic quantum wells. The characteristics that make NiAl an attractive candidate include a good lattice parameter match with several semiconductors (for epitaxy), chemical stability, good electrical conductivity, and a stable native oxide that allows patterning in air without destroying electrical continuity. An NiAl layer 3.3 mm thick is currently the thinnest independently contacted electrode in any solid-state device.

High-Temperature Applications

Among the alloys currently being considered as replacements for superalloys in high-temperature applications, some of the most promising are the intermetallic compounds in the Nb–Al system. The three compounds $NbAl_3$, $NbAl_2$, and Nb_3Al have melting temperatures in excess of many candidate replacement materials and densities that are superior to or competitive with those materials. Furthermore, the crystal structures, although complex, are reasonably well understood. In addition to these physical properties, there is a distinct processing consideration that leads to a heightened interest in this system. The compounds of the Nb–Al system represent the highest melting temperature alloys that can be processed using currently active superalloy techniques, that is, the various vacuum and hearth melting techniques.

The intermetallic compounds of the Nb–Al system, when systematically viewed as a whole, demonstrate both an orderly progression of high-temperature properties and a great deal of promise for future applications. Vickers microhardness as a function of temperature proceeds from $NbAl_3$, which is inferior to Ni_3Al, to Nb_3Al, which is the highest-melting-temperature compound with a temperature superiority of about 600°C over Ni_3Al.

Iron aluminides based on Fe_3Al and FeAl (Figure A.3) have excellent oxidation and corrosion resistance because they are capable of forming protective oxide scales at elevated temperatures in hostile environments. These aluminides exhibit corrosion rates lower than those of promising iron-based alloys (including coating material) by a couple orders of magnitude when tested in a severe sulfidizing environment at 800°C. The major drawbacks of the aluminides are their poor ductility and brittle fracture at ambient temperatures and poor strength and creep resistance above 600°C.

Iron aluminides have been known to be brittle at room temperature for more than 60 years; however, the major cause of their low ductility and brittle fracture was not identified until recently. A study of the environmental effects on tensile properties indicates that the poor ductility commonly observed in air test is caused mainly by an extrinsic effect—environmental embrittlement.

The understanding of the cause of brittleness and iron aluminides has led to new directions in the design of ductile iron aluminide alloys. The schemes used to improve the ductility of Fe_3Al–FeAl alloys include control of surface conditions, reduction of hydrogen solubility and diffusion by alloying additions, refinement of grain structure by thermomechanical treatment, enhancement of grain-boundary cohesion by microalloying, and control of grain shape and recrystallization condition. The ductile Fe_3Al alloys show a distinctly high yield strength and a high ductility of 13%–60% when tested in air at room temperature. Similarly, FeAl (35.8 at% Al) alloys show an increase in room-temperature ductility from 2% to 11% as a result of alloying additions (zirconium, molybdenum, and boron) and grain-structure refinement. This ductility further increased to 14% by formation of protective oxide scales on specimen surfaces through preoxidation at 700°C. The room-temperature ductility of Fe_3Al and FeAl can be significantly improved by refining grain structure via powder extrusion with or without second-phase particles (such as TiB_2). The strength of Fe_3Al and FeAl alloys

at elevated temperatures can be improved by alloying with molybdenum, zirconium, niobium, titanium, and TiB_2.

Gamma alloys of titanium aluminide can be grouped into single-phase (γ)-alloys and two-phase ($\alpha_2 + \gamma$)-alloys. Single-phase alloys attracted attention initially because of their excellent resistance to environmental attack (e.g., oxidation of hydrogen absorption). In spite of a lack of progress toward overcoming their poor ductility and fracture toughness, interest in these engineering alloys has not diminished. Gamma titanium aluminide alloys of importance are thus two-phase alloys based on Ti–(45%–49%)Al with appropriate combinations of alloying elements.

Aluminized Steel

Standard type I aluminized steel consists of a hot coating of 92% aluminum and 8% silicon on cold-rolled sheet steel for oxidation and corrosion resistance. The coating provides good oxidation resistance in continuous service to 1250°F (675°C) and is used for auto exhaust-system components and other applications. The thicker coating provides the better environmental resistance, but the thinner one is preferred for parts requiring severe forming because of its better adhesion. Type 2 aluminized steel provides still better environmental resistance but is limited to parts requiring still less severe forming.

Galvalume

A cold-rolled steel, having a thin hot-dip coating of 55% aluminum, 43.4% zinc, and 1.6% silicon on both sides, was developed by Bethlehem Steel as an alternative to type 2 aluminized steel. It is substantially stronger and correspondingly less ductile than the aluminized steel, scale resistant to 1250°F (675°C), and superior to conventional galvanized steel and corrosion resistance and rural, industrial, and marine atmospheres. Applications include roofing for pre engineered buildings, auto exhaust-system parts, agricultural equipment, and appliance parts.

Aluminizing

Any treatment that causes aluminum to diffuse into the surface of a component (usually steel) to improve corrosion resistance, particularly resistance to oxidation at temperatures up to about 900°C. Processes include dipping in molten metal, metal spraying followed by a diffusion heat treatment, or exposure of the steel at elevated temperature to a gaseous or liquid environment capable of releasing aluminum into the metal surface. The common feature is that the diffusion and the formation of an intermetallic alloy layer ensures good adhesion in contrast to metal spraying, which merely applies a mechanically adherent coating.

Aluminothermic Welding

Same as thermite welding.

Aluminum

Called aluminium in England, aluminum is a white metal with a bluish tinge (symbol Al, atomic weight 26.97), obtained chiefly from bauxite. It is the most widely distributed of the elements next to O_2 and silicon, occurring in all common clays. Aluminum metal is produced by first extracting alumina (aluminum oxide) from the bauxite

A

by a chemical process. The alumina is then dissolved in a molten electrolyte, and an electric current is passed through it, causing the metallic aluminum to be deposited on the cathode. The metal was discovered in 1727 but was obtained only in small amounts until it was reduced electrolytically in 1885.

Pure (99.99%) aluminum has a specific gravity of 2.70 or a density of 2685 kg/m³, a melting point of 660°C, electrical and thermal conductivities about two-thirds that of copper, and a tensile modulus of elasticity of 62,000 MPa. The metal is nonmagnetic, highly reflective, and has a face-centered cubic crystal structure. Soft and ductile in the annealed condition, it is readily cold worked to moderate strength. It resists corrosion in many environments as a result of the presence of a thin aluminum oxide film.

Iron, silicon, and copper are the principal impurities in commercially pure aluminum, wrought products containing at least 99% alumina and a foil product containing at least 99.99%. Such unalloyed aluminum is available in a wide variety of mill forms and constitutes the aluminum 1XXX series in the designation system for wrought aluminum and aluminum alloys. Annealed sheet is quite ductile—35%–45% tensile elongation—but weak, having a tensile yield strength of 27–35 MPa. Cold reduction of 75% increases yield strength to 124–166 MPa. Thus, the unalloyed metal is used far more for its electrical, thermal, corrosion-resistant, and cosmetic characteristics than for its mechanical properties. Applications include electrical and thermal conductors, capacitor electrodes, heat exchangers, chemical equipment, packaging foil, heat and light reflectors, and decorative trim.

Aluminum flake enhances the reflectance and durability of paints. Aluminum powder is used for powder-metal parts. Aluminum and alumina paste are used in catalysts, soaps, explosives, fuels, and thermite welding. Aluminum shot is used to deoxidize steel, and aluminum foam, made by foaming the metal with zirconium hydride or other hydrides, is an effective core material for lightweight structures.

Anodized aluminum is aluminum with a hard aluminum oxide surface imparted electrolytically using the metal as the anode. The coating, which is much thicker than the naturally formed aluminum oxide film, provides additional corrosion and weathering resistance in certain environments but is generally no more protective in acidic or alkaline solutions outside the 4–8.5 pH range. The coating is nonconductive, wear resistant, and can be produced in various colors, thus enhancing decorative appeal. Whereas the naturally formed oxide film is less than a millionth of an inch thick, anodized coatings range from 0.005 to 0.008 mm for bright auto trim to as much as 0.030 mm for architectural applications.

Aluminum Alloys

Alloying aluminum with various elements markedly improves mechanical properties, strength primarily, at only a slight sacrifice in density, thus increasing specific strength, or strength-to-weight ratio. Traditionally, wrought alloys have been produced by thermomechanically processing cast ingot into mill products such as billet, bar, plate, sheet, extrusions, and wire. For some alloys, however, such mill products are now made by similarly processing "ingot" consolidated from powder. Such alloys are called powder metal (PM) wrought alloys or simply PM alloys. To distinguish the traditional type from these, they are now sometimes referred to as ingot-metallurgy (IM) alloys or ingot-cast alloys. Another class of PM alloys is used to make PM parts by pressing and sintering the powder to near-net shape. There are also many cast alloys. All told, there are about 100 commercial aluminum alloys.

There are two principal kinds of wrought alloys: (1) heat-treatable alloys—those strengthened primarily by solution heat treatment or solution heat treatment and artificially aging (precipitation hardening)—and (2) non-heat-treatable alloys—those that depend primarily on cold work for strengthening. Alloy designations are a continuation of the four-digit system noted for aluminum, followed with a letter to designate the temper or condition of the alloy: F (as-fabricated condition), O (annealed), H (strain hardened), W (solution heat treated and unstable, that is, the alloy is prone to natural aging in air at room temperature), and T (heat treated to a stable condition). Numerals following T and H designations further distinguish between tempers or conditions. T3, for example, refers to alloys that have been solution heat treated, cold worked, and naturally aged to a substantially stable condition. T6 denotes alloys that have been solution heat treated and artificially aged. H designations are followed by two or three digits. The first (1, 2, or 3) indicates a specific sequence of operations applied. The second (1–8) refers to the degree of strain hardening (the higher the number, the greater the amount of strain or hardening; 8 corresponds to the amount induced by a cold reduction of about 75%). The third (1–9) further distinguishes between middle treatments (Table A.12).

The aluminum alloy 2XXX series is characterized by copper (2.3%–6.3%) as the principal alloying element. Most of these alloys also contain lesser amounts of magnesium and manganese, and some may contain small amounts of other ingredients, such as iron, nickel, titanium, vanadium, zinc, and zirconium. The 2XXX alloys are strengthened mainly by solution heat treatment, sometimes by solution heat treatment and artificial aging. Among the more common, especially for structural aircraft applications, are aluminum alloys 2014, 2024, and 2219, which can be heat treated to tensile yield strengths in the range of 276–414 MPa. A recently developed alloy for auto body panels is aluminum alloy 2036, which in the T4 temper provides a tensile yield strength of about 195 MPa. Aluminum alloy 2XXX series is not as corrosion resistant as other aluminum alloys and thus is often clad with a thin layer of essentially pure aluminum or a more corrosion-resistant aluminum alloy, especially for aircraft applications.

Manganese (0.5%–1.2%) is a distinguishing alloying element in the aluminum alloy 3XXX series. Aluminum alloy 3003 also contains 0.12% copper. Aluminum alloys 3004 and 3105 also contain 1% and 0.5% magnesium, respectively. Strengthened by strain hardening, 3XXX alloys provide maximum tensile yield strengths in the range of 186–248 MPa and are used for chemical equipment, storage tanks, cooking utensils, furniture, builders' hardware, and residential siding.

The aluminum alloy 4XXX series is characterized by the addition of silicon: about 12%, for example, an aluminum alloy 4032 and 5% aluminum alloy 4043. Aluminum alloy 4032, which also contains 1% magnesium and almost as much copper and nickel, is heat treatable, providing a tensile yield strength of about 315 MPa in the T6 temper. This, combined with its high wear resistance and low thermal expansivity, has made it popular for forged engine pistons. Aluminum alloy 4043, which is alloyed only with silicon, is a strain-hardenable alloy used for welding rod and wire. Some Al–Si alloys are also used for brazing, others for architectural applications. Their appeal for architectural use stems from the dark-gray color they develop and anodizing.

The principal alloying element in the aluminum alloy 5XXX series is magnesium, which may range from about 1% to 5%, and is often combined with lesser amounts of manganese and/or chromium. Like 3XXX alloys, the 5XXX are hardenable only by strain hardening, and all are available in a wide variety of H tempers.

A

TABLE A.12
Aluminum Wrought Alloys

Designation	1100		2011		2014		2024		2036	2219		3003	
Temper	O	H18	T3	T8	O	T6	O	T4	T4	O	T62	O	H18
Yield strength (10^3 psi)	5	22	43	45	14	60	11	47	28	11	42	6	27
Tensile strength (10^3 psi)	13	24	55	59	27	70	27	68	49	25	60	16	29
Shear strength (10^3 psi)	9	13	32	35	18	42	18	41	30	—	—	11	16
Fatigue limit (10^3 psi)	5	9	18	18	13	18	13	20	18	—	15	7	10
Elongation in 2 in. (%)	45	15	15	12	18	13	22	19	24	—	—	40	10
Modulus of elasticity (10^6 psi)	10.0	10.0	10.2	10.2	10.6	10.6	10.6	10.3	10.6	10.6	—	10.0	10.0
Melting temperature (°F)	1190–1215	1190–1215	1005–1190	1005–1190	945–1080	945–1080	935–1180	935–1180	1030–1200	1010–1190	1010–1190	1190–1210	1190–1210
Coefficient of thermal expansion (10^{-6} in./in. °F)	13.1	13.1	12.7	12.7	12.8	12.8	12.9	12.9	13.0	12.4	12.4	12.9	12.9
Thermal conductivity (Btu in./h ft^3 °F)	1540	1510	1050	1190	1340	1070	1340	840	1100	1190	840	1340	1070
Electrical resistivity (Ω-cir mils/ft)	18	18	27	23	21	26	21	35	25	24	35	21	26

Designation	3004		5052		5083		6061		6066		6262	7050	7075
Temper	O	H38	O	H38	O	H321	O	T6	O	T6	T9	T74	T6
Yield strength (10^3 psi)	10	36	13	37	21	33	8	40	12	52	55	70	73
Tensile strength (10^3 psi)	26	41	28	42	42	46	18	45	22	57	58	81	83
Shear strength (10^3 psi)	16	21	18	24	25	—	12	30	14	34	35	45	48
Fatigue limit (10^3 psi)	14	16	16	20	—	23	9	14	—	16	13	21	23
Elongation in 2 in. (%)	25	6	30	8	22	16	30	17	18	12	10	10	11
Modulus of elasticity (10^6 psi)	10.0	10.0	10.2	10.2	10.3	10.3	10.0	10.0	10.0	10.0	10.0	10.3	10.4
Melting temperature (°F)	1165–1210	1165–1210	1125–1200	1125–1200	1095–1180	1095–1180	1080–1205	1080–1205	1045–1195	1045–1195	1080–1205	910–1165	890–1175
Coefficient of thermal expansion (10^{-6} in./in. °F)	13.3	13.3	13.2	13.2	13.2	13.2	13.1	13.1	12.9	12.9	13.0	12.8	13.1
Thermal conductivity (Btu in./h ft^3 °F)	1130	1130	960	960	810	—	1250	1160	1070	1020	1190	1092	900
Electrical resistivity (Ω-cir mils/ft)	25	25	30	30	36	—	22	24	26	28	24	35	31

Source: Mach. Design Basics Eng. Design, 54–55, June 1993. With permission.

Tensile yield strength range from less than 69 MPa in the annealed condition to more than 276 MPa in highly strained conditions. Aluminum alloys 5083, 5154, 5454, and 5456 are widely used for welded structures, pressure vessels, and storage tanks, others for more general applications, such as appliances, cooking utensils, builders' hardware, residential siding, auto panels and trim, cable sheathing, and hydraulic tubing.

The aluminum alloy 6XXX series is characterized by modest additions (0.4%–1.4%) of silicon and magnesium and can be strengthened by heat treatment. Except for auto sheet aluminum alloys 6009 and 6010, which are typically supplied and used in the T4 temper, the alloys are strengthened by solution heat treatment and artificial aging. As a class, these alloys are intermediate in strength to aluminum alloy 2XXX and aluminum alloy 7XXX but provide good overall fabricability. The auto sheet alloys are relatively new. More traditional among the dozen or so alloys of this kind are aluminum alloy 6061, which is used for truck, marine, and railroad-car structures, pipelines, and furniture, and aluminum alloy 6063 furniture, railings, and architectural applications. Other applications include complex forgings, high-strength conductors, and screw-machine products.

Zinc is the major alloying element in the aluminum alloy 7XXX series, and it is usually combined with magnesium for strengthening by heat treatment. An exception is aluminum alloy 7072, which is alloyed only with 1% zinc, is hardenable by strain hardening and is used for fin stock or as a clad for other aluminum alloys. The other alloys, such as aluminum alloy 7005, 7049, 7050, 7072, 7075, 7175, 7178, and 7475, contain 4.5%–7.6% Zn, 1.4%–2.7% Mg, and, in some cases, also include copper, magnesium, silicon, titanium, or zirconium. These heat-treatable alloys are the strongest of aluminum alloys, with tensile yield strengths of some exceeding 483 MPa in the T6 temper. They are widely used for high-strength structures, primarily in aircraft.

Cast Aluminum Alloys

Of some 40 or so standard casting alloys, 10 are aluminum die-casting alloys. The others are aluminum sand-casting alloys and/or aluminum permanent-mold-casting alloys. Some of the latter also are used for plaster-mold casting, investment casting, and centrifugal casting. Although the die-casting alloys are not normally heat treated, those for sand and/or permanent-mold casting often are, usually by solution heat treatment and artificial aging. Both heat-treatable and non-heat-treatable alloys are available. The heat-treatable alloys, which can be solution heat treated and aged similar to wrought heat-treatable grades, carry the temper designations T2, T4, T5, T6, or T7. Die castings are seldom solution heat-treated because of the danger of blistering.

The major alloying addition of several important casting alloys is silicon, which improves the castability of aluminum and provides moderate strength. Other elements are added primarily to increase the tensile strength. Most die castings are made of alloy 413.0 or 380.0. Alloy 443.0 has been very popular in architectural work, and 355.0 and 356.0 are the principal alloys for sand casting. Number 390.0 is employed for die-cast automotive engine cylinder blocks, and alloy F332.0 is used for pistons for internal combustion engines.

Casting alloys are significant users of secondary metal (recovered from scrap for reuse). Thus, casting alloys usually contain minor amounts of a variety of elements; these do no harm as long as they are kept within certain limits. The use of secondary metal is also increasing importance in wrought alloy manufacturing as producers take steps to reduce the energy that is required is producing fabricated aluminum products.

Aluminum alloy 380.0 and its modifications constitute the bulk of die-casting applications. Containing 8.5% Si, 3.5% Cu, and as much as 2% Fe, it provides a tensile yield strength of about 165 MPa, good corrosion resistance, and is quite fluid and free from hot shortness, and thus is relatively castable. Engine cylinder heads, typewriter frames, and various housings are among its applications. Aluminum alloy 390.0, a high-silicon (17%) Cu–Mg–Zn alloy, is the strongest of the die-casting alloys, providing a tensile yield strength of about 240 MPa as cast and 260 MPa in the T5 temper. Like high silicon in wrought alloys, it also features low thermal expansivity and excellent wear resistance. Typical applications include auto engine cylinder blocks, breakthroughs, compressors, and pumps requiring abrasion resistance. Aluminum alloys 383.0, 384.0, 413.0, and A413.0 provide the best die-filling capacity and have excellent resistance to hot cracking and die sticking. Aluminum alloy 518.0 provides the best corrosion resistance, machinability, and polishability but is more susceptible to hot cracking and die sticking than the other alloys.

Sand and/or permanent-mold casting alloys are encompassed in each of the aluminum alloy 2XX.X–8XX.X series. As-cast, tensile yield strengths range from about 97 MPa for 208.0–200 MPa or aluminum alloy A 390.0. In various solution-treated and aged conditions, several alloys can provide tensile yield strengths exceeding 280 MPa. Strongest of the alloys—415 MPa in the T7 temper and 435 MPa in the T6—is aluminum alloy 201.0, which contains 0.7% Ag and 0.25% Ti in addition to 4.6% Cu and small amounts of magnesium and manganese. The T7 temper is suggested for applications requiring stress-corrosion resistance. Applications for 201.0 include aircraft and ordinance fittings and housings, engine cylinder heads, and pistons, pumps, and impellers. Other more commonly used high-strength alloys are aluminum alloys 354.0, 355.0, and 356.0, which are alloyed primarily with silicon, copper, magnesium, manganese, iron, and zinc, and which are used for auto and aircraft components. Premium-quality aluminum alloy castings are those guaranteed to meet minimum tensile properties throughout the casting or in specifically designated areas.

Aluminum–Lithium Alloys

Aluminum–lithium (Al–Li) alloys are a significant recent development in high-strength wrought aluminum alloys. Lithium is the lightest metal in existence. For each weight percent of lithium added to aluminum, there is a corresponding decrease of 3% in the weight of the alloy. Therefore, Al–Li alloys are for the attractive property of low density. Because of the very low density of lithium, every 1% of this alkali metal can provide a 3% reduction in density and a 10% increase in stiffness-to-density ratio relative to conventional 2XXX and 7XXX alloys. Thus, these alloys, such as aluminum alloys 2090 and 2091 (there are several others, including proprietary ones), are extremely promising for future aircraft applications, see Table A.13. These alloys are comparable in strength with some of the strongest traditional alloys. Some are ingot-cast products, others PM-wrought alloys.

TABLE A.13
Aluminum–Lithium Alloys

Product Form	Alloy	Temper of End Use Condition and Definition	Property Category
Sheet	AA8090	T81—solution treated, quenched, stretched, and precipitation heat treated to an underaged condition	Damage tolerant: low strength but enhanced toughness and impact resistance.
	AA8090	T621—resolution treated by the user, quenched, and precipitation heat treated to an underaged condition (same aging time and temperature as for T81)	Resolution treatment by the user to promote enhanced formability.
	AA8090	T8—solution treated, quenched, stretched, and precipitation heat treated to a near-peak aged condition	Medium strength where some reduction in toughness and impact resistance compared to T81 and T621 can be tolerated.
Extruded sections (solid and hollow)	AA8090	T8511—solution treated, quenched, controlled stretched, and precipitation heat treated to a near-peak aged condition	Medium/high strength.
Die forgings	AA8090	T852—solution treated, quenched, cold compressed, and precipitation heat treated to a near-peak aged condition	Medium/high strength.
	AA5091	H112—as forged	Medium/high strength but where cold compression is impractical or insufficient to achieve properties if made from AA8090.

Source: Adv. Mater. Proc., 41–43, October 1992. With permission.

A second beneficial effect of lithium additions is the increase in elastic modulus (stiffness). Also, as the amount of lithium is increased, there is a corresponding increase in strength due to the presence of very small precipitates that act as strengthening agents to the aluminum. As the precipitates grow during heat treatment, the strength increases to a limit, then begins to decrease. Al–Li alloys therefore come under the classification of precipitation-strengthening alloys.

With the exception of the PM alloy AA 5091, all the current commercially available Al–Li alloys are produced by direct-chill casting and require a precipitation-aging heat treatment to achieve the required properties. In Al–Li alloys containing greater than 1.3% (by weight) of lithium, the intermetallic phase δ-Al₃Li precipitates upon natural or artificial aging, but the associated strengthening effect is insufficient to meet the medium or high strength levels usually required (the damage tolerant temper in AA 8090 is an exception).

Cold-working operation is a prime requirement for IM alloys containing relatively high levels of lithium, such as AA 8090 or AA 2090, if optimum medium-strength or high-strength properties are to be achieved. Cold working is readily applied to sheet, plate, and extrusions by stretching, although it must be recognized that in most cases, this can only be carried out by the metal manufacturer.

Alloy AA 5091 is made by the PM process of mechanical alloying (MA). This technology avoids the limitations arising from the need to cold-compress alloys AA 8090 and AA 2090 as forgings, because it is dispersion strengthened during manufacture. Through careful selection of the Al–Li–Mg base composition, a single-phase solid solution is achieved, and the need for heat treatment is eliminated.

The MA process is inherently more expensive than direct-chill-casting and is not a practical route for producing large rectangular-section ingots for sheet or plate rolling. However, it is useful for certain parts in which the configuration of the forged components would not allow adequate cold compression to be achieved in alloys AA 8090 or AA 2090.

Powder Metallurgy Alloys

Elements such as iron, cobalt, and nickel have been added to aluminum in large quantities to produce alloys that have relatively stable structures at elevated temperatures and higher elastic moduli than conventional aluminum alloys.

The PM process is a rapid solidification process that involves the transformation of finely divided liquid either in the form of droplets or thin sheets that become solid at high solidification rates (on the order of 10^4 K/s). Solidification occurs as heat is removed from the molten metal.

Rapid-solidification-process aluminum alloys containing iron and cerium appear to result in alloys that have refined particle-size distribution and improved high-temperature stability. These PM alloys appear competitive with titanium up to 191°C.

Aluminum PM parts are made mainly from aluminum alloys 201 AB and 601 AB. Both are copper, silicon, and magnesium compositions that can be heat treated to the T4 or T6 tempers after sintering. The 201 AB is the stronger, having 324 MPa tensile yield strength for 95% dense material in the T6 temper as compared with about 241 MPa for 601 AB for these conditions.

Mechanical Alloying

This process circumvents the limitations of conventional ingot casting. Blends of powders are mixed in a ball mill. A drum is mounted horizontally and half filled with steel balls and blends of elemental metal powders. As the drum rotates, the balls drop on the metal powder. The degree of homogeneity of the powder mixture is determined by the size of the particles in the mixture.

Aluminum–Beryllium Alloys (AlBeMet)

AlBeMetAE is an alloy of aluminum and beryllium. The Al–62Be alloy has the modulus of steel at 200 GPa and density of 2.1 g/cm³, approximately 20% that of steel, which gives a specific modulus of 950×10^6 cm.

Powder processing for AlBeMet is well established. First, inert-gas atomization produces prealloyed powder. The powder is cold isostatically pressed to a density of about 80% of theoretical, then consolidated by extrusion.

The alloy varies from 60% to 65% Be and 35% to 40% Al. The alloy is formed by melting spherical beryllium and aluminum powder and then is atomized, and the metals are mixed. As the metal mixture cools, an alloy with a matrix-like structure is formed. The resulting materials extruded and can be rolled or formed with hot isostatic pressing. AlBeMet also can be dip brazed or welded.

The extruded bar may be fabricated directly into parts, or cut for cross rolling into sheet. Sheet and plate thicknesses of 7.6 mm are produced by rolling, followed by grinding to final gauge. This sheet has essentially isotropic properties in the sheet plane.

A range of applications is made economically possible by the extrusion process, which has produced extrusions with diameters up to 25 cm. For example, profiled AlBeMet extrusions have been fabricated into computer hard disk drive arms, with over one million parts machined from extruded bar stock.

AlBeMet is also produced as hot isostatically pressed block. This is quite useful for prototype parts, and for components in which strength and ductility are of secondary importance to stiffness, density, modulus, thermal conductivity, and thermal expansion. The input powder for hot isostatic pressing is the same as that for wrought product manufacture. The processes carried out after degassing to remove adsorbed species such as water vapor.

The physical properties of AlBeMet are largely independent of processing route. However, they are critically dependent on the beryllium content of the alloy. Although AlBeMet 162 with 62 wt% Be is the most common composition, others can be quite useful. The most striking physical properties are modulus and density. The modulus is substantially higher, and the density is significantly lower than conventional materials such as Al–Li, which is typically 2.6 g/cm³.

Unlike its physical properties, the mechanical properties of AlBeMet 162 are strongly dependent upon the processing route.

To meet many of the economic requirements imposed by modern designs, that shape or near-net shape processes are required. Investment casting has long been recognized as a process capable of producing parts to near net shape. However, only recently has the capability of investment casting been developed for Al–Be alloy parts. AlBeCast 910 is the first of a family of Al–Be investment casting alloys.

The properties of AlBeCast 910 include 0.2% yield strength of 124 MPa, ultimate tensile strength of 180 MPa, and elongation of 3%.

In addition to conventional processes such as casting and mechanical working, significant progress has been made in adapting new technologies to AlBeMet. The semisolid processing of Al–Be alloys enables the net shape and high production capability of permanent mold die casting for a material whose high melting point and reactivity preclude conventional die casting. As a prototype, a sabot from a 30 mm projectile was semisolid formed, and the uniformity and repeatability of the process was proved out in a semiproduction run of approximately 2500 parts.

The high modulus-to-density ratio, 3.8 times that of aluminum or steel, minimizes flexure and reduces the chance of mechanically induced failure. Therefore, AlBeMet plays a dual role in its avionics applications.

The stiffness, low density, and thermal characteristics of Al–Be alloys combine to produce the highest-performance brake components for the demanding Formula I race car circuit, specifically the brake caliper from extrusion. It is the stiffness that is particularly important in such parts, because excessive bending of the caliper leads to reduced braking efficiency. In addition, the thermal management characteristics allow higher and more consistent braking performance than possible with conventional materials.

As a structural material, AlBeMet 162 sheet is being qualified as the rudder material for advanced versions of a U.S. Air Force tactical fighter. In addition, as an essentially isotropic material, it allows simplified design, reduced weight, and increased performance.

Space structures also benefit from AlBeMet, including the OrbComm vehicle. On this satellite, the face sheets for the honeycomb bus (the circular portion of the structure), the brackets that hold the bus to the payload, and the long boom are all fabricated AlBeMet structures.

Aluminum–Scandium Alloys (AlSc)

An AlSc base alloy, Sc 7000, has been developed for use as bicycle frame tubing, shock absorbers, and handlebars. The alloy is 50% stronger than aluminum bicycle alloys and reduces the weight of the bicycle frame by 10%–12%.

New and Special Alloys

Aluminum alloy 7033-T6 is 50% stronger than AA 6061-T6 and provides higher fatigue performance. Developed to provide improved fracture toughness for forgings, the alloy offers corrosion and stress-corrosion properties nearly equal to those of AA 6061-T6 and superior to those of AA 2014-T6.

There are some new Al–Li ingot-cast products, and other P/M wrought alloys. Alcoa's *aluminum alloy C155* is 7% more lightweight and has greater fracture toughness and resistance to fatigue-crack growth than 7075-T651 and 2024-T351. It is also 11% stronger than the latter alloys. Reynolds Metals' *Weldalite* 2195 is 5% more lightweight and 30% stronger than 2219 alloy and has good fracture toughness at cryogenic temperatures. *VacLite* aluminum–lithium alloys, of Comalco Aluminium Limited of Australia, are vacuum-refined alloys having low amounts of hydrogen and alkali metal impurities. Their high purity is believed to account for improved toughness, greater resistance to stress corrosion, and superior weldability, or less tendency for weld cracking than other aluminum–lithium alloys. One of the alloys, with 3.3% lithium, has a tensile yield strength of 504 MPa and a tensile modulus of 83,400 MPa. Aluminum–lithium alloys were developed primarily for aircraft applications. They are used in the horizontal stabilizer of two airbus models and in the cargo floor of the C-17 military transport.

Aluminum Powder

Aluminum powder is produced by drawing a stream of the molten metal through an atomizing nozzle and impinging that stream with compressed air or inert gas, solidifying and disintegrating the metal into small particles, which are then drawn into a collection system

and screened, graded, and packaged. Particle sizes range from –325 mesh (fine) to +200 mesh (granules). In the process, the metal reacts with oxygen in the air and moisture, causing a thin film of Al_2O_3 to form on the surface of the particles. Oxide content, which increases with decreasing particle size, ranges from 0.1% to 1.0% by weight.

First used to produce aluminum flake by baul milling for paint pigments, aluminum powder now finds many other uses: ferrous and nonferrous metals production, P/M parts, P/M-wrought aluminum alloy mill products, coatings for steel, asphalt-roof products, spray coatings, and vacuum metalizing. Other applications include rocket fuels and explosives, incendiary bombs, pyrotechnics, signal flares, and heat and magnetic shields. It is also used in permanent magnets, high-temperature lubricants, industrial cements, chalking compounds, printing inks, and cosmetic and medical products.

Aluminum–Silicon Alloys

Osprey aluminum alloys include a family of machine-readable, lightweight, aluminum–silicon alloys, featuring controlled thermal expansion, high thermal conductivity, electromagnetic or radio-frequency shielding capability, and thermal mechanical stability up to 932°F (500°C). Applications include radio frequency (RF) or microwave housings, electrooptical housings, integrated-circuit power devices, and waveguide- and microwave-filter components. Alloys include *aluminum alloy CE7, CE9, CE11*, and *CE13*. The numbers in these designations indicate the approximate coefficient of thermal expansion in ppm/°C. Compositions can be customized to thermal expansion requirements. For the common alloys, CE7 contains 7% by weight silicon, CE9 60%, CE11 50%, and CE13 58%.

Supral and DS Alloys

Wrought superplastic alloys, such as Superform Metals' *Supral* alloys and fine-grain *aluminum alloy 7475* or modifications thereof, permit far greater tensile elongation than conventional alloys at low strain rates and temperatures in the 700°F–980°F (371°C–537°C) range, permitting extremely complex shapes to be formed at low gas pressure (0.69–2.1 MPa).

For structural aircraft applications, high-strength alloys such as 2024 and 7075 are generally limited to temperatures below 400°F (204°C). Developmental *dispersion-strengthened aluminum alloys* may extend the range to 600°F (315°C). Key elements of these PM wrought systems include iron and cerium; iron and molybdenum; iron, vanadium, and silicon; or chromium, zirconium, and manganese. All provide tensile yield strengths greater than 270 MPa at 600°F (315°C), the strongest being an aluminum–iron–vanadium–silicon alloy. At room temperature, this alloy has an ultimate tensile strength of 625 MPa, a yield of 568 MPa, and 6% elongation. At 600°F, these values are 310, 300 MPa, and 7%, respectively.

Thixalloy 540, from Salzburger Aluminum AG of Austria, is intended for casting with partially molten alloy. This reduces thermal stress on tooling and solidification shrinkage. The aluminum–magnesium alloy contains 5% silicon, 2% manganese, and some chromium and provides a yield strength of 186 MPa and 18% elongation.

Aluminum Metal Matrix Composites

Aluminum alloys also constitute a major class of *metal-matrix composites*. *Duralcan composites*, from Duralcan USA of Alcan Aluminum Limited, comprise *alumina particulates* in a wrought matrix alloy and *silicon carbide particulates* in a cast matrix alloy. The amount of particulate pertains to the volume percent. The composites are made by mixing the particulate in a molten alloy, which is then poured into foundry ingot, extrusion billet, rolling bloom, or rolling ingot. These forms are then processed conventionally by casting, extruding, rolling, or forging. Matrix alloys are similar in composition to standard wrought and cast alloys, and all the composites are heat treatable. The composites, however, are stronger, more rigid, harder, and far more wear and abrasion resistant. On the other hand, their ductility and fracture toughness are less, and machining typically requires the use of polycrystalline diamond cutting tools for efficient production. In general, alumina decreases electrical and thermal conductivity and the coefficient of thermal expansion, whereas silicon carbide increases thermal conductivity while decreasing electrical conductivity and thermal expansivity.

Here are some typical properties of extruded bar and rod in the T6 temper; with 15% Al_2O_3, *aluminum alloy composite 2014 (Duralcan W2A)* has a tensile yield strength of 500 MPa, 2% elongation, and a tensile modulus of 93,800 MPa. Although yield strength is only 6% greater than that of 2014-T6 at room temperature, it is 16%–67% greater in the 200°F–700°F (93°C–371°C) range. *Aluminum alloy composite 2618 (Duralcan W2F)*, with 20% Al_2O_3, has a yield strength of 420 MPa, 2% elongation, and a modulus of 104,000 MPa. At 400°F (204°C), the yield strength is 320 MPa. With this alumina content, Duralcan W2F has a density of 0.109 lb/in.^{3} (3017 kg/m^3).

In the T4 temper, 15% SiC *aluminum alloy composite X 2080* extrusions have a density of 0.102 lb/in.^{3} (2823 kg/m^3), and ultimate tensile strength of 483 MPa, a yield strength of 365 MPa, 7.5% elongation, and a tensile modulus of 99,978 MPa. With 20% SiC, the density is 1.03 lb/in.^{3} (2851 kg/m^3) and the other properties 517 MPa, 393 MPa, and 6% and 110,320 MPa. With 25% SiC, *aluminum alloy composite 6113* extrusions in the T6 temper have similar density, tensile ultimate and yield strengths of 496 and 438 MPa, respectively, 3% elongation, and a tensile modulus of 120,662 MPa.

Cast Duralcan alloys are based on matrix alloys 339 and 3594 gravity casting and 360 and 380 alloys for die casting. The 359 is for general room-temperature applications, and the 339 is for elevated-temperature applications. Alloy 380 is also for general use, and 360 is for uses requiring additional corrosion resistance. In the T6 temper, *aluminum alloy composite 359 (Duralcan F3S)* with 20% SiC has a tensile yield strength of 340 MPa, 0.4% elongation, and a modulus of 98,600 MPa. Rockwell B hardness is 77, compressive yield strength is also 340 MPa, and the modulus and compression is 101,000 MPa. *Aluminum alloy composite 339 (Duralcan F3K)* with 10% SiC has a yield strength of 360 MPa and a modulus of 87,500 MPa. In the T5 temper, Rockwell B hardness is 63, and the tensile yield strength is 248 MPa at room temperature, 138 MPa at 400°F (204°C), and 69 MPa at 600°F (316°C). In the T5 temper, *aluminum alloy composite 380 (Duralcan F3D)* with 10% SiC has a Rockwell B hardness of 84, a tensile yield strength of 330 MPa, 0.7% elongation, and a tensile modulus of 93,800 MPa. It has a density of 0.100 lb/in.^{3} (2768 kg/m^3).

Boralyn, from Alyn Corp., is *boron carbide–reinforced aluminum alloy* made into billet by isostatic pressing and vacuum sintering powder of conventional wrought alloys and boron carbide (B_4C) particulate followed by extrusion, rolling, forging, or casting. Solution-treated and aged extrusion of *aluminum alloy composite Boralyn H-10*, 10% (B_4 C) 6092 alloy, has a density of 0.097 lb/in.^{3} (2685 kg/m^3), a tensile modulus of 84,119 MPa, and ultimate tensile strength of 414 MPa, a tensile yield strength of 345 MPa, and a thermal expansion of 11.7 ppm/°F (21.1 ppm/°C). *Aluminum alloy composite Boralyn H-30*, 30% B_4C 6092, has a density of 0.096 lb/in.^{3}

(2657 kg/m^3), a tensile modulus of 120,660 MPa, tensile strengths again of 60,000 lb/in.^{2} ultimate and 50,000 lb/in.^{2} yield, and a thermal expansion of 8.3 ppm/°F (14.9 ppm/°C). *Aluminum alloy composite Boralyn E-10*, 10% B_4C 7083 alloy, has a density of 0.103 (2851 kg/m^3), a tensile modulus of 86188 MPa, tensile strengths of 621 MPa ultimate and 586 MPa yield, and a thermal expansion of 11.7 ppm/°F (21.1 ppm/°C).

Composite castings also are made by casting alloys about selectively placed *fiber preforms*. Honda has cast auto engine blocks with alumina and graphite preforms along the cylinder-bore surfaces, eliminating the need for cast iron cylinder liners. *Silicon carbide whiskers* and preforms and have potential for squeeze-cast aluminum composites, especially for military vehicles. *ARALL*, of Alcoa, is a laminate composite comprising thin layers of aramid-reinforced epoxy and wrought high-strength aluminum alloy. Besides being more lightweight than aluminum alloys, the laminate has high fatigue resistance. Variants include *GLARE*, comprising layers of glass-reinforced epoxy and high-strength aluminum alloy, and *CARE*, layers of carbon-reinforced epoxy and high-strength aluminum alloy.

Aluminum Bomb

A bomb-shaped container used in determining the oxygen content in liquid steel.

Aluminum Brass

A wrought copper (76%–79%) and zinc alloy containing 1.8%–2.5% aluminum and 0.022%–0.06% arsenic. The aluminum edition markedly improves the alloy's resistance to impingement attack in high-velocity seawater, and the arsenic inhibits the dezincification. *Aluminum brass C 68700* is commonly used for condensers and heat exchangers and marine applications.

Aluminum Bronze

A series of wrought and *cast copper–aluminum alloys*, most of which also contain several other elements, principally iron. Aluminum content ranges from 4% to 15%, depending on the alloy, and iron, if present, from less than 1%–6%. Alloy designations for *wrought aluminum bronzes* are C60XXX, C61XXX, C62XXX, C63XXX, and C64XXX; cast alloys are C95XXX. The alloys are also referred to by their nominal aluminum content, such as 5% aluminum bronze, or by other distinguishing alloying ingredients, such as *nickel aluminum bronze*. Aluminum bronzes are noted for their moderate to high strength, and corrosion resistance and potable, brackish, and sea waters, industrial and marine atmospheres, and various chemicals. They are attacked, however, by oxidizing acids. Because of their pale gold look and amenability to lustrous finishing, some of the alloys are also used for their decorative appeal. Aluminum bronzes are available in a full range of wrought products, and the cast alloys are amenable to various casting methods. Typical applications include condenser and heat-exchanger tubing and tube sheets, fasteners, shafts, cams, gears, bearings, bushings, valves and valve seats, deep-drawn parts, wear plates, pickling equipment, impellers and agitators, nonsparking hardware, and decorative ware.

Single-phase (alpha) alloys containing only copper and aluminum can only be strengthened by strain hardening during cold working. The 4%–7% *aluminum bronze C60600* alloy in the form of 0.5 in. (13 mm) plate has a tensile yield strength of 115 MPa annealed and 165 MPa in the H04, or "hard," cold-work temper.

In these same conditions, 1 in. (25 mm) diameter rod of *aluminum (6%–8.5%) bronze C61000* has a tensile yield strength of 205 and 380 MPa. Two-phase aluminum bronzes, such as those alloying with iron or iron and nickel, are typically strengthened by quenching and then are tempered. One such casting alloy, *aluminum bronze C95500* (nominally 81% copper, 11 aluminum, 4 iron, 4 nickel), as a tensile yield strength of 415 MPa in the quenched and tempered (TQ50) condition, 50% greater than as cast.

Aluminum Nitride (AlN)

A high–thermal conductivity ceramic used as an electronic substrate. Also a key component in the production of *sialons*.

Aluminum Oxide (Al₂O₃)

See *alumina*.

Amalgam

A combination of a metal with mercury. The amalgams have the characteristics that when slightly heated, they are soft and easily workable, and they become very hard when set. They are used for filling where it is not possible to employ high temperatures. Although native amalgams are chemical combinations of the metals, some of the artificial amalgams are alloys and others are compounds. *Dental amalgams* are prepared by mixing mercury with finely divided alloys composed of varying proportions of silver, tin, and copper. A *silver–tin alloy*, developed at the National Institute of Standards and Technology (NIST), is mercury-free and less brittle than standard amalgams. Also promising is NIST's development of fine silver powder that can be compacted and consolidated with conventional dental tools.

Amalgamation Process

A process for the initial extraction of gold and silver. The crushed core is mixed with liquid mercury, which amalgamates with the metals leaving behind the impurities. The amalgam is then heated to drive off the mercury leaving impure gold and silver.

Ambient

The environment that surrounds and contacts a system or component.

American Wire Gauge

A standard used in the determination of the physical size of a conductor determined by its circular-mil area.

Amine

An amine as a member of a group of organic compounds that can be considered as derived from ammonia by replacement of one or more hydrogens by organic radicals. Generally, amines are bases of widely varying strengths, but a few that are acidic are known.

Amines constitute one of the most important classes of organic compounds. The lone pair of electrons on the amine nitrogen enables amines to participate in a large variety of reactions as a base or a nucleophile. Amines play prominent roles in vitamins, for example, epinephrine (adrenaline), thiamine, or vitamin B_{12} and Novocaine.

Amines are used to manufacture many medicinal chemicals, such as sulfa drugs and anesthetics. The important fiber nylon is an amine derivative.

Aramids are synthetic fibers produced from long-chain polyamides (nylons) in which 85% of the amine linkages are attached directly to two aromatic rings. The fibers are exceptionally stable and have good strength, toughness, and stiffness, which are retained well above 150°C. Two aramids are *Nomex* and *Kevlar* from DuPont Co., and *Twaron* from AkzoNobel NV. They have high strength, intermediate stiffness, and are suitable for cables, ropes, webbings, and tapes. *Kevlar 49*, with high strength and stiffness, is used for reinforcing plastics. *Nomex*, best known for its excellent flame and abrasion resistance, is used for protective clothing, air filtration bags, and electrical insulation.

Amino

Melamine and urea are the principal commercial thermosetting polymers called aminos. The amino resins are formed by an addition reaction of formaldehyde and compounds containing NH_3 amino groups. They are supplied as liquid or dry resins and filled molding compounds. Applying heat in the presence of a catalyst converts the materials into strong, hard products. Aminos are used as molding compounds, laminating resins, wood adhesives, coatings, wet-strength paper resins, and textile-treating resins.

Moldings made from amino compounds are hard, rigid, abrasion resistant, and have high resistance to deformation under load. They have excellent electrical insulation characteristics. Melamines are superior to ureas in resistance to acids, alkalies, heat, boiling water, and for applications involving wet/dry cycling.

Melamines and ureas are not resistant to strong oxidizing acids or strong alkalies, but they can be used safely with conventional household chemicals such as naphtha and detergents. They are unaffected by organic solvents such as acetone, carbon tetrachloride, ethyl alcohol, heptane, and isopropyl alcohol. Petroleum, paraffin hydrocarbons, gasoline, kerosene, motor oil, aromatic hydrocarbons, and fluorinated hydrocarbons (Freon) have no apparent effect on urea and melamine moldings. Dimensional stability is good, but moldings do swell and shrink slightly in varying moisture conditions.

Amino Resins

Resins made by the polycondensation of a compound containing amino groups, such as urea or melamine, within aldehyde, such as formaldehyde, or an aldehyde-yielding material. Melamine formaldehyde and urea formaldehyde resins are the most important family members. The resins can be dispersed in water to form colorless syrups. With appropriate catalysts, they can be cured at elevated temperatures.

Ammeter

An instrument for measuring the magnitude of electric current flow.

Ammonia

A gas of the formula NH_3, originally called *alkaline air* and *volatile alkali*. Ammonia is predominantly made by direct synthesis at high temperatures and pressures of hydrogen and nitrogen, derived by steam reforming hydrocarbons. It is a major building block for making fertilizers, nitric acid, plastics, and explosives. Ammonia is readily absorbed by water, which at 60°F (16°C) takes up 683 times its own volume of the gas, forming the liquid commonly called ammonia, but which is *ammonium hydroxide*, a colorless, strongly alkaline, and pungent liquid of composition NH_4OH with a boiling point of 100°F (38°C). At 80°F (27°C), it contains 29.4% ammonia in stable solution. It is also known as *ammonium hydrate* and *aqua ammonia*, and it is used for the saponification of fats and oils, as a deodorant, for cleaning and bleaching, for etching aluminum, and in chemical processing.

Ammonia gas is used in refrigeration, in nitriding steels, and in the manufacture of chemicals. Chlorine unites with it to form *chloramines*, which are used as solvents, chlorinating agents, and disinfectants. The gas does not burn in air, but a mixture of ammonia and oxygen explodes when ignited. *Anhydrous ammonia* is the purified gas liquefied under pressure and marketed in cylinders. The anhydrous ammonia used for controlled atmospheres for nitriding steel, bright annealing, and sintering metals contains 90% NH_3 and is oxygen-free.

Smelling salts, sometimes referred to as *ammonia*, and in solution as *aromatic spirits of ammonia*, are *ammonium carbonate*, forming in colorless or white crystals. *Ammonium bicarbonate*, or acid ammonium carbonate, is a water-soluble, white, crystalline powder used as a source of pure ammonia and carbon dioxide and to decrease density inorganic materials by creating voids, such as for making foamed rubber and in the food-baking industry. *Ammonium gluconate* is a water-soluble, white, crystalline powder used as an emulsifier for cheese and mayonnaise and as a catalyst and textile printing.

The ammonium radical, $-NH_4$, has the chemical reaction of an alkali metal and forms many important chemicals. *Ammonium nitrate* is made by the action of nitric acid on ammonium hydroxide. It is a colorless to white, crystalline, hygroscopic powder of composition NH_4NO_3, soluble in water, alcohol, and alkalies. It is also used in fertilizers, pyrotechnics, dental gas, insecticides, freezing mixtures, and explosives. For use as a slow-burning propellant for missiles, it is mixed with a burning-rate catalyst and a synthetic-rubber binder and pressed into blocks. *Riv*, a vapor-phase rust inhibitor, is ammonium nitrate. The explosive *amatol* is a mixture of ammonium nitrate and TNT, which explodes violently on detonation. The 50–50 mixture can be melted and poured, while the 80–20 mixture is like sugar and was used for filling large shells. *ANFO*, a mixture of ammonium nitrate and fuel oil formed into porous prills, is one of the largest-selling explosives. *Macite*, for tree trunk blasting, is ammonium nitrate coated with TNT, with a catalyst to make it more sensitive. Akremite is ammonium nitrate and carbon black, used as an explosive in strip mining.

Ammonium perchlorate is another explosive made by the action of perchloric acid on ammonium hydroxide. It is a white crystalline substance of composition NH_4ClO_4, is soluble in water, and decomposes on heating. *Nitrogen trichloride*, NCl_3, which forms in reactions of chlorine and ammonia when there is an excess of chlorine, is a highly explosive yellow oil. *Ammonium sulfate*, $(NH_4)_2SO_4$, is a gray crystalline, water-soluble material used as a fertilizer and for fireproofing. It is made by reacting sulfuric acid with ammonia by-product from coal distillation, and, increasingly, from sulfuric acid and synthetic ammonia, from caprolactam production, or from gypsum.

Ammonium chloride, or *sal ammoniac*, NH_4Cl, is a white crystalline powder used in electric batteries, in textile printing, as a soldering flux, and in making other compounds. Many salts and metallic soaps are also formed in the same manner as with the alkali metals.

Amorphous

Not having a crystal structure; noncrystalline, that is, the atoms are not arranged in any regular pattern.

Amorphous Metals

Also known as metallic glasses, amorphous metals are produced by rapid quenching of molten metal–metalloid alloys, resulting in a noncrystalline grain-free structure in the form of ribbon or narrow strips. They are extremely strong and hard yet reasonably ductile and quite corrosion resistant. Perhaps of greatest interest is the magnetic performance of several complex iron-based compositions, called *Metglas*. Because of their very low hysteresis and power losses, they can markedly reduce the size of transformer cores traditionally made of silicon steels. Several nickel-alloy compositions also are available for use as brazing foils. One such foil, *Metglas MBF60A*, contains 11% phosphorus and 0.1% maximum carbon and meets *BNi-6 brazing filler* specification of the American Welding Society. For aircraft turbines, it is used as a preform to braze *fiber metal* to 430 stainless steel for abradable seals.

A family of amorphous *zirconium–beryllium alloys* with substantial amounts of titanium, copper, and nickel can be cast into bulk shapes. The alloy family was developed by researchers at the California Institute of Technology and Amorphous Technologies International. The alloy 41Zr–14Ti–12Cu–10Ni has a hardness of 585 H_v. One alloy has exhibited the unusual behavior of giving off intense sparks when struck and broken by impact in air and giving off infrared emissions when struck and broken in a nitrogen atmosphere. This behavior is attributed to the intense heat generated by the highly localized deformation when metallic glasses break. Also, the alloy is not corrosion proof.

Amorcor hardfacing alloys, of Amorphous Metals Technologies, are iron–chromium alloys. There are three basic alloys: alloy M for general corrosion resistance plus resistance to carburization, oxidation, and sulfonation at temperatures up to 1700°F (927°C); alloy C, for resistance to bleach, chlorides, and medium-strength sulfuric acid and to wear, with a maximum use temperature of 1600°F (871°C); and alloy T for especially severe wear applications. The alloys have been used on down-hole drilling pipe and drill bits, engine valve guides, and steelmaking, coal mining, agricultural, and earth-moving equipment.

Amorphous Plastic

A plastic that has no crystalline component, no known order or pattern of molecule distribution, and no sharp melting point.

Amorphous Powder

A powder that consists of particles that are substantially noncrystalline in character.

Amorphous Silicon

A noncrystalline form of silicon or silicon hydride. It is formed as a thin film by chemical vapor deposition and is used as a semiconductor.

Amorphous Solid

A rigid material whose structure lacks crystalline periodicity, that is, the pattern of its constituent atoms or molecules does not repeat periodically in three dimensions. See also *metallic glass*.

Ampere

The unit used for measuring the quantity of an electric current flow. One ampere represents a flow of one coulomb per second.

Ampere Turns

The product of the number of turns in an electromagnetic coil and the current and amperes passing through the coil.

Amperometry

Chemical analysis by methods that involve measurements of electric currents.

Amphoteric

A term applied to oxides and hydroxides, which can act basic toward strong acids and acidic toward strong alkalies. Substances that can dissociate electrolytically to produce hydrogen or hydroxyl ions according to conditions.

Amplifier

A negative lens used instead of an eyepiece to project under magnification the image formed by an objective. The amplifier is designed for flatness of field and should be used with an apochromatic objective.

Amsler Wear Machine

A wear and traction-testing machine consisting of two disk-shaped specimens oriented such that their axes are parallel and whose circumferential, cylindrical surfaces are caused to roll or roll and slide against one another. The rotation rates of each disk may be varied so as to produce varying degrees of sliding and rolling motion.

Amylaceous

Pertaining to, or of the nature of, starch; starchy.

Anaerobic Adhesive

An adhesive that cures only in the absence of air after being confined between assembled parts.

Anaerobic Environments

Containing little or no oxygen.

Analog Circuits

Circuits that provide a continuous (versus discontinuous) relationship between the input and output.

Analog Computer

A computer that processes data that are continuously variable in nature.

Analog-to-Digital Converter

A device that converts a continuously variable electrical signal into discrete signals suitable for analysis by a digital computer.

Analysis

The ascertainment of the identity or concentration, or both, of the constituents or components of a sample. See also *determination*.

Analyte

In any analysis, the substance (element, ion, compound, and so on) being identified or determined.

Analytical Chemistry

The science of chemical characterization and measurement. Qualitative analysis is concerned with the description of chemical composition in terms of elements, compounds, or structural units; quantitative analysis is concerned with the precise measurement of amount. A variety of physical measurements are used, including methods based on spectroscopic, electrochemical, radiochemical, chromatographic, and nuclear principles.

Analytical Curve

The graphical representation of very relationship between (1) the intensity of the response to measurement (e.g., a mission, absorbance, and conductivity) and (2) the concentration or mass of the substance being measured. The curve is generated by measuring the responses for standards of known concentration. Also termed standard curve or working curve.

Analytical Electron Microscopy

The technique of materials analysis in the transmission electron microscope equipped to detect and quantify many different signals from the specimen. The technique usually involves a combination of imaging, chemical analysis, and crystallographic analysis by diffraction at high spatial resolution.

Analytical Gap

The region between two electrodes in which the sample is excited in the sources used for emission spectroscopy and spark source mass spectrometry.

Analytical Line

In spectroscopy, the particular spectral line of an element used in the identification or determination of the concentration of that element.

Analytical Wavelength

In spectroscopy, the particular wavelength used for the identification or determination of the concentration of an element or compound.

Analyzer

An optical device, capable of producing plane polarized light, used for detecting the state of polarization.

Anatomical Materials

Metals and other materials used for long-term insertion into the human body as prostheses, bone reinforcement, etc.

Anchor Pattern

A pattern made by blast cleaning abrasives on an adherend surface in preparation for adhesive application prior to bonding. Pattern is examined in profile.

Anchorage

Part of an insert that is molded inside the plastic and held fast by the shrinkage of the plastic.

Anchorite

A zinc–iron phosphate coating for iron and steel.

Andalusite

A mineral of composition Al_2O_3–SiO_2 used in the production of aluminosilicate bricks for use in blast furnaces, steel ladles, and torpedo ladles.

Anelastic Deformation

Any portion of the total deformation of a body that occurs as a function of time when the load is applied and which disappears completely after a period of time when the load is removed.

Anelasticity

The property of solids by virtue of which strain is not a single-value function of stress in the low-stress range where no permanent set occurs.

Angle of Attack

In tribology, the angle between the direction of motion of an impinging liquid or solid particle and the tangent to the surface at the point of impact.

Angle of Bend

The angle between the two legs of the specimen after bending is completed. It is measured before release of the bending force, unless otherwise specified.

Angle of Bevel

See preferred term *bevel angle*.

Angle of Bite

In the rolling of metals, the location where all of the force is transmitted through the rolls; the maximum attainable angle between the role radius at the first contact and the line of role centers. Operating angles less than the angle of bite are termed contact angles or rolling angles.

Angle of Contact

In a ball race, the angle between a diametral plane perpendicular to a ball-bearing axis and a line drawn between points of tangency of the balls to the inner and outer rings.

Angle of Incidence

(1) In tribology, the angle between the direction of motion of an impinging liquid or solid particle and the normal to the surface at the point of impact. (2) In materials characterization, the angle between an incident radiant beam and a perpendicular to the interface between two media.

Angle of Nip

In rolling, the *angle of bite*. In roll, jaw, or gyratory crushing, the entrance angle formed by the tangents at the two points of contact between the working surfaces and the (assumed) spherical particles to be crushed.

Angle of Reflection

(1) Reflection: The angle between the reflected beam and the normal to the reflecting surface. See also *normal*. (2) Diffraction: The angle between the diffracted beam and the diffracted planes.

Angle of Repose

The angular contour that a powder pile assumes.

Angle Press

A hydraulic molding press equipped with horizontal and vertical rams, specially designed for the production of complex plastic moldings having deep undercuts.

Angle Wrap

Tape fabric that is wrapped on a starter dam mandrel at an angle to the centerline.

Angle-Ply Laminate

A laminate having fibers of adjacent plies, oriented at alternating angles.

Angstrom Unit (Å)

A unit of linear measure equal to 10^{-10} m, or 0.1 nm. Although not an accepted SI unit, it is occasionally used for small distances, such as interatomic distances, and some wavelengths.

Angular Aperture

In optical microscopy, the angle between the most divergent rays that can pass through a lens to form the image of an object. See also *aperture (optical)*.

Angular-Contact Bearing

A ball bearing of the grooved type, designed in such a way that when under no load, a line through the outer and inner raceway contacts with the balls to form an angle with the plane perpendicular to the bearing axis.

Angularity

The conformity to, or deviation from, specified angular dimensions in the cross section of a shape or bar.

Anhydride

A compound from which water has been extracted. An oxide of a metal (basic anhydride) or of a nonmetal (acid anhydride) that forms a base or an acid, respectively, when united with water.

Aniline

An important organic base ($C_6H_5NH_2$) made by reacting chlorobenzene with aqueous ammonia in the presence of a catalyst. It is used in the production of aniline formaldehyde resins and in the manufacture of certain rubber accelerators and antioxidants.

Aniline Formaldehyde Resins

Members of the aminoplastics family made by the condensation of formaldehyde and aniline in an acid solution. The resins are thermoplastic and are used to a limited extent in the production of molded and laminated insulating materials. Products made from these resins have high dielectric strength and good chemical resistance.

Aniline Point

As applied to a petroleum product, the lowest temperature at which the product is completely miscible with an equal volume of freshly distilled aniline. The aniline point is a guide to the oil composition.

Anion

A negatively charged ion that migrates through the electrolyte toward the *anode* under the influence of a potential gradient. See also *cation* and *ion*.

Anisotropic

Exhibiting different properties when tested along axes in different directions. In magnetics, capable of being magnetized more readily in one direction than in a transverse direction.

Anisotropic Conductive Adhesive

Adhesive that can be made conductive in the vertical, or z, axis while remaining an insulator in the horizontal, or x and y, axes.

Anisotropic Laminate

One in which the properties are different in different directions along the laminate plane.

Anisotropy

The characteristic of exhibiting different values of a property in different directions with respect to a fixed reference system in the material.

Anisotropy of Laminates

The difference of the properties along the directions parallel to the length or width of the lamination planes and perpendicular to the lamination.

Anneal (Glass)

To prevent or remove objectionable stresses in glassware by controlled cooling from a suitable temperature.

Anneal to Temper (Metals)

A final partial anneal that softens a cold-worked nonferrous alloy to a specified level of hardness or tensile strength.

Annealed Powder

A metallic powder that is heat treated to render it soft and compactible.

Annealing (Abrasives)

A heating and cooling operation, implying usually a slow cooling to remove stress, to induce softness, to refine structure, or to alter used to secure one or more of the following effects physical properties.

Annealing (Glass)

A controlled cooling process for glass designed to reduce thermal residual stress to a commercially acceptable level and, in some cases, modify structure.

Annealing (Metals)

A generic term denoting a treatment consisting of heating to and holding at a suitable temperature followed by cooling at a suitable rate, used primarily to soften metallic materials, but also to simultaneously produce desired changes in other properties or in the microstructure. The purpose of such changes may be, but is not confined to improvement of machinability, facilitation of cold work, improvement of mechanical or electrical properties, and/or increase in stability of dimensions. When the term is used unqualifiedly, full annealing is implied. When applied only for the relief of stress, the process is properly called *stress relieving* or stress-relief annealing.

In ferrous alloys, annealing usually is done above the upper critical temperature, but the time-temperature cycles vary widely both in maximum temperature attained and in cooling rate employed, depending on composition, material condition, and results desired. When applicable, the following commercial process names should be used: *black annealing, blue annealing, box annealing, bright annealing, cycle annealing, flame annealing, full annealing, graphitizing*, in-process annealing, *isothermal annealing, malleabilizing*, orientation annealing, *process annealing, quench annealing, spheroidizing*, and *some critical annealing*.

And nonferrous alloys, annealing cycles are designed to (1) remove part or all of the effects of cold working (recrystallization may or may not be involved), (2) cause substantially complete coalescence of precipitates from solid solution in relatively coarse form, or (3) both, depending on composition of a material condition.

Specific process name of a commercial use are *final annealing*, *full annealing*, *intermediate annealing*, *partial annealing*, *recrystallization annealing*, *stress-relief annealing*, and *anneal to temper*.

Annealing (Plastics)

Heating to a temperature at which the molecules have significant mobility, permitting them to reorient to a configuration having less residual stress. In semicrystalline polymers, heating to a temperature at which retarded crystallization or recrystallization can occur.

Annealing Carbon

See *temper carbon*.

Annealing Point (Glass)

The temperature at which internal stresses in a glass are substantially relieved in a matter of minutes.

Annealing Range (Glass)

The range of glass temperature in which stress in glass can be relieved at a commercially practical rate. For purposes of comparing glasses, the annealing range is assumed to correspond with the temperature between the annealing point and the strain point.

Annealing Twin

A *twin* formed in a crystal during recrystallization.

Annealing Twin Bands

See *twin bands*.

Annular Bearing

(1) Usually a rolling bearing of short cylindrical form supporting a shaft carrying a radial load. (2) A flat disk-shaped bearing.

Anode

(1) The electrode of an electrolyte cell at which oxidation occurs. Electrons flow away from the anode in the external circuit. It is usually at the electrode that corrosion occurs and metal ions enter solution. (2) The positive (electron-deficient) electrode in an electrochemical circuit. In contrast with *cathode*.

Anode Aperture

In electron microscopy, the opening in the accelerating voltage anode shield of an electron gun through which the electrons must pass to illuminate or irradiate the specimen.

Anode Compartment

In an electrolytic cell, the enclosure formed by a diaphragm around the anodes.

Anode Copper

Special-shaped copper slabs, resulting from the refinement of *blister* copper in a reverberatory furnace, used as anodes in electrolytic refinement.

Anode Corrosion

The dissolution of a metal acting as an *anode*. Corrosion at the anode of an electrolytic cell, the normal site and wet environments.

Anode Corrosion Efficiency

The ratio of the actual corrosion (weight loss) of an *anode* to the theoretical corrosion (weight loss) calculated by *Faraday's law* from the quantity of electricity that is passed.

Anode Effect

The effect produced by polarization of the *anode* and electrolysis. It is characterized by a sudden increase in voltage and a corresponding decrease in amperage due to the anode becoming virtually separated from the electrolyte by a gas film.

Anode Efficiency

Current efficiency at the *anode*. The observed rate of anode dissolution as a percentage of the theoretical rate predicted by *Faraday's laws of electrolysis*.

Anode Film

(1) The portion of solution in immediate contact with the *anode*, especially if the concentration gradient is steep. (2) The outer layer of the anode itself.

Anode Metals

Metals used for the positive terminals in electroplating. They provide in whole or in part the source of the metal to be plated, and they are as pure as commercially possible, uniform in texture and composition, and have the skin removed by machining. They may be either cast or rolled, with their manufacture controlled to obtain a uniform grade and to exclude impurities, so that the anode will corrode uniformly in the plating bath and not polarize to form slimes or crusts. In some plating, such as for white bronze, the anode efficiency is much higher than the cathode efficiency, and a percentage of steel anodes is inserted to obtain a solution balance. In other cases, as in chromium plating, the metal is taken entirely from the solution, and insoluble anodes are employed. Chromium-plating anodes may be lead–antimony, with 6% antimony, or tin–lead, with 7% tin. In addition to pure single metals, various alloys are marketed in anode form. The usual brass is 80% copper and 20% zinc, but other compositions are used, some containing 1–2 tin. *Brass anodes* are called *platers' brass*. *Copper anodes* for metal plating are usually hot-rolled oval bars, 99.9% pure, while those for electro-type deposits may be hot-rolled plates, electrodeposited plates, or cast plates. *Copper ball anodes* are forged instead of cast to give a finer and more even grain. *Zinc anodes* are 99.99% pure. *Nickel anodes* are more than 99% pure, rolled or cast in iron molds, or 97% sand cast. *Bright nickel* anodes have 1% or more of cobalt. *Lead anodes* have low current-carrying capacity and may be made with a sawtooth or multiple-angled surface and ribs, to provide more area and give greater throwing power. Anodes of other metals are also made with sections gear-shaped, fluted, or barrel-shaped to give greater surface area and higher efficiency. *Rhodium anodes* are made in expanded-mesh form. *Platinum anodes*, also made in mesh form, have the platinum clad on tantalum wire. Special anode metals are marketed under trade names, usually accenting the color, hardness, and corrosion resistance of the deposit plate.

Anode Mud

Deposit of insoluble residue formed from the dissolution of the anode and commercial electrolysis. Sometimes called anode slime.

Anode Polarization

See *polarization*.

Anodic Cleaning

Pickling cleaning of a metal in an electrolytic cell in which the component forms the anode receiving an electrical current from an external source. Electrolytic cleaning in which the work is the anode. Also called reverse-current cleaning.

Anodic Coatings

Anodic oxidation or anodizing is the common commercial term used to designate the electrolytic treatment of metals in which stable oxide films or coatings are formed on surfaces. Aluminum and magnesium are anodized to the greatest extent on a commercial basis. Some other metals such as zinc, beryllium, titanium, zirconium, and thorium can also be anodized to form films of varying thicknesses, but they are not used to any large extent commercially.

Aluminum

It is a well-known fact that a thin oxide film forms on aluminum when it is exposed to the atmosphere. This thin, tenacious film provides excellent resistance to corrosion. The ability of aluminum to form an adherent oxide film led to the development of electrochemical processes to produce thicker and more effective protective and decorative coatings.

Anodic coatings can be formed on aluminum in a wide variety of electrolytes utilizing either AC or DC current or combinations of both. Electrolytes of sulfuric, oxalic, and chromic acid's are considered to be the most important commercially. Other electrolytes such as borates, citrates, chromates, bisulfates, phosphates, and carbonates can also be used for specialized applications.

Anodic coatings produced in a sulfuric acid type of electrolyte are generally translucent, can be produced with a wide range of properties by varying operating techniques, and have a greater number of pores that are smaller in diameter than other anodic coatings.

Coatings obtained by the chromic acid anodic process are much thinner than those produced by the sulfuric acid process based upon the same time of anodic oxidation. Although the coatings are thin, they provide high resistance to corrosion because of the presence of chromium compounds in combination with their relatively thick barrier layer. Because the conventional anodic coatings produced in a chromic acid electrolyte are thin, they have low resistance to abrasion but a high degree of flexibility.

The chromic acid anodic process is critical with respect to alloy composition. In general, the process is not recommended for aluminum alloys containing more than 7% Si or 4.5% Cu. Alloys containing over 7% total alloying elements are also not recommended. It is difficult to form films on these alloys, particularly on casting alloys. Anodic solution (surface attack) will generally occur unless the processing conditions are controlled very carefully.

On most aluminum alloys, anodic coatings of this type do not require sealing. However, sealing in a boiling die chromate or a dilute chromic acid solution produces coatings with the best resistance to salt spray corrosion on AA 2014-T6 and AA 7075-T6 alloys.

Anodic oxidation and dilute oxalic acid solutions produce coatings that are essentially transparent. Their color varies from a light yellow to bronze. Such coatings are dense and have little absorptive capacity but possess high resistance to abrasion.

Hard Coatings

Procedures have been developed that produce finishes of greater thickness and density than the conventional anodic coatings. They have high resistance to abrasion, erosion, and corrosion. The coatings have thicknesses in the range of 0.1–5 mils, depending upon the application.

Hard coatings are popular for applications requiring lightweight in combination with high resistance to wear, erosion, and corrosion. These applications include helicopter rotor blade surfaces, pistons, pinions, gears, cams, cylinders, impellers, turbines, and many others. Also, due to their attractive gray color, hard anodic coatings are now being used for architectural applications.

The processing conditions for obtaining hard coatings are such that thick coatings with maximum density can be obtained on most aluminum alloys. The selection of alloy is of utmost importance.

Variations of the anode oxidation process that produce conventional hard anodic coatings also form thick, dense, colored anodic coatings for architectural applications. Attractive bronze, gray, or black coatings are obtained by utilizing certain organic acids as electrolytes.

Other Coatings

Utilizing oxalic acid in combination with titanium, zirconium, and thorium salts for the electrolyte produces a dense oxide coating that has an opaque, light-gray appearance.

Anodic oxidation of aluminum in sulfamic acid produces coatings that are denser than those produced by the sulfuric acid process. This anodic process is expensive owing to the high cost of sulfamic acid.

Alternating current may be used to form anodic coatings on aluminum and alloys with all of the electrolytes previously mentioned, but the aluminum surface is anodic only half of the cycle so that an oxide coating is formed only at half the rate of coatings formed with direct current.

Superimposed AC–DC has also been used for anodizing and has produced hard, thick, anodic coatings.

Alloy Selection

The previous discussion on the anodic oxidation characteristics of various electrolytes was based upon the use of relatively pure aluminum. Alloy selection is important. Even aluminum of 99.3% minimum purity, such as the AA1100 alloy, is in a sense an aluminum alloy from the standpoint of anodic oxidation, because the other elements present have an effect on the characteristics of the coating.

The response of the different constituents of aluminum alloys varies considerably and is an important factor. Some constituents will be dissolved by the anodic reaction, whereas others may be unaffected. For alloys where the constituents are dissolved during the anodic treatment, the coatings will have voids that decrease the density of the coating and also lower its resistance to corrosive action and abrasion. Al–Cu alloys are an example of this type.

Silicon in Al–Si alloys is an example of a constituent that is unaffected by conventional anodic oxidation. The silicon particles remain unchanged in the coating in their original position.

Some constituents of aluminum alloys will themselves oxidize under anodic oxidation and the oxidation products will color the coating. For example, constituents such as manganese will produce a brownish opaque appearance due to the presence MnO_3. Also, chromium as a constituent will give a yellowish tint to the coating from oxidation products of chromium.

The lower the concentration of constituents present or the purer the aluminum, the more continuous and transparent will be the oxide coating. The so-called superpurity (99.99%) aluminum produces the most transparent oxide coating.

Properties and Applications

The oxide coatings produced by anodizing have many properties that make them commercially important. Anodic coatings are essentially Al_2O_3, which is a very hard substance. Because the Al_2O_3 is integral with the surface, it will not chip or peel from the surface; this outstanding characteristic is useful for architectural applications. The combination of high wear resistance and attractive satin sheen of the finish makes it a logical choice for aluminum hardware, handrails, moldings, and numerous other architectural components. Because the anodic finish reproduces the texture of the surface from which it is formed, a wide variety of attractive effects are possible by variation in the surface preparatory procedures. Many commercial architectural applications can use a natural aluminum finish, but for applications where it is desirable to preserve initial appearance, the anodic finish will require less maintenance.

Because the oxide coating is brittle compared with the aluminum underneath, it will crack if the coated article is bent. It is possible, however, to produce oxide coatings that are relatively flexible. In general, the thicker coatings formed in sulfuric and oxalic acid electrolytes will crack or craze to a much greater extent than oxide coatings formed in a chromic acid electrolyte. For many applications, this cracking may not be objectionable, because it is usually difficult to detect by visual observation. These fine cracks have an adverse effect on the bending properties of the metal, however, and may sometimes cause fracture if the bends are severe. For this reason, it is generally recommended that the finish be applied after forming.

If fatigue is a critical factor, then proper allowance must be made for the reduction in endurance limits produced by relatively thick oxide coatings. Fatigue tests indicate that a coating 0.1 mils thick on smooth surfaces will have little effect on fatigue strength. Thicker coatings in the range of 0.3–0.5 mils on smooth surfaces have a slight detrimental effect at high stresses.

Anodic coatings also provide substantial protection against corrosion. There are many factors that must be considered in this connection such as the continuity of the coating and the choice of alloy. Since continuity is dependent upon constituents present in the alloy, anodic coatings on high-purity aluminum are the most resistant to corrosion. On the other hand, anodic coatings formed from Al–Cu alloys have much lower resistance to corrosion.

Sealing of anodic coatings in dichromate solutions results in an appreciable improvement in resistance to corrosion, particularly by chlorides. The results of atmospheric exposure tests indicate that anodic coatings with a thickness of 0.4 mils or greater will provide greatly increased resistance to weathering.

The ability of anodic coatings to absorb coloring substances such as dyestuffs and pigments makes it possible to obtain finishes in a complete range of colors including black. The colors are unique because the luster of the underlying metals gives them a metallic sheen that is particularly attractive for applications that simulate metals such as gold, copper, bronze, and brass. Colorants are available that, when used to color anodic coatings on the proper alloys, will reproduce the natural colors of the metals listed earlier. Colored finishes can be used in a wide range of applications for nameplates, panels, and appliance trim, optical goods, cameras, fishing reels, instruments, giftware, and jewelry.

Anodic coatings have good electrical insulating properties. Anodic films produced in boric acid electrolytes are used commercially on aluminum foil for electrical capacitors. The voltage necessary to break down the anodic coatings is generally proportional to the thickness.

Anodic oxidation in oxalic acid and phosphoric acid electrolytes produces coatings that have been successfully used as preparatory treatments for electroplating. Copper, nickel, cadmium, silver, and iron have been successfully deposited over oxide coatings. Plating solutions that are highly alkaline should be avoided as they attack the coating and destroy the porous structure necessary for the best adhesion.

A "Krome-Alume" process utilizes an oxalic acid electrolyte to form the oxide coating, and subsequent modification of the coating with hydrofluoric acid produces a structure satisfactory for electroplating. Furthermore, the anodic oxidation process utilizing a phosphoric acid electrolyte produces a structure that requires no further modification to condition it for electroplating. The alloy has an important effect on coating structure, and in general, the phosphoric anodic process is not recommended for preparing high-purity aluminum, Al–Mg wrought alloys, and most die-casting alloys for electroplating.

Production Methods

Anodic coatings are applied to aluminum and its alloys by a variety of methods, including batch, bulk, continuous conveyor, and continuous strip.

The batch method of applying anodic coatings is similar to that used in electroplating except that the parts are anodic instead of cathodic. The continuous conveyor method is also similar to the conventional plating method.

The bulk method for applying anodic coatings to small parts such as rivets, washers, and screws is radically different from bulk electroplating methods. The barrel-finishing method is not suitable for applying anodic coatings because the initial flow of current forms an anodic coating on the parts and even if they contact each other during the rotation of the barrel, no current will flow because the coating is an insulator. The bulk methods employed for anodic coatings utilize special perforated nonmetallic cylindrical containers. Pressure, applied to the parts in the container through a threaded center contact post, maintains the initial contact between the surface of the parts.

The continuous strip process has been used commercially to apply anodic coatings to aluminum sheet that is subsequently roll formed into weather strip. In Europe, the same process is used to apply anodic coatings to aluminum sheet that is formed into food containers such as sardine cans.

Magnesium

Although all of the magnesium alloys in commercial use today have good resistance to corrosion, many parts are provided with maximum resistance to corrosion and abrasion through electrolytic treatments based on anodic treatments. The anodic treatments produce relatively thick and dense coatings with excellent adhesion and high resistance to corrosion and abrasion. As in the case of aluminum, anodic coatings on magnesium alloys are an excellent base for lacquers and enamels.

It is well known that magnesium alloys are attacked by most inorganic and organic acids. However, because magnesium alloys are resistant to alkalies, fluorides, borates, and chromates, the electrolytes for oxidizing are generally based upon these chemicals.

The simplest electrolyte for anodizing magnesium alloys is a 5% caustic soda solution. This electrolyte is used in a temperature range of 60°C–70°C. A voltage of 5–6 is satisfactory; anodic oxidation time is generally 30 min. All magnesium alloys will respond to this treatment. The coatings produced are approximately 0.3 mils thick and are essentially crystalline magnesium hydroxide. They have relatively high resistance to abrasion and are gray or tan in color in the as-formed condition. The coatings may be colored for decorative applications by immersion in water-soluble dyestuffs in much the same way as the coloring procedures used for anodic coatings on aluminum alloys. If maximum resistance to corrosion is required, immersion (sealing) in a 5% sodium chromate solution at 77°C–82°C is recommended.

Another anodizing treatment for magnesium alloys capable of producing a coating with many desirable properties is the "H.A.E." process. This process is also based upon an alkaline electrolyte. All magnesium alloys, both wrought and cast, respond to this process to give coatings with excellent resistance to corrosion, high dielectric strength, and high hardness.

Other Metals

Thin oxide films may be formed on beryllium by anodic oxidation in an electrolyte composed of 10% nitric acid containing 200 g/L of chromic acid. The anodic oxidation is carried out at approximately 25 A/ft². Such films retard high temperature oxidation and corrosion.

Thin oxide films can also be formed on zirconium, titanium and thorium by anodizing in an electrolyte composed of 70% glacial acetic and 30% nitric acid. The corrosion resistance of titanium is substantially improved by anodizing for 10–15 min in a 15%–22% (by weight) sulfuric acid solution (room temperature) at approximately 18 V DC. Anodic coatings on titanium are also used as a base for lubricants.

Zinc

Much of the work on anodic coatings for zinc has been conducted in alkaline electrolytes consisting of a two-stage process. The electrolyte for the first oxidation treatment is conducted in an alkali–carbonate solution followed by anodizing in a silicate solution.

Anodic Etching

Development of microstructure by selective dissolution of the polished surface under application of a direct current. Variation with layer formation; anodizing.

Anodic Inhibitor

A chemical substance or mixture that prevents or reduces the rate of the anodic or oxidation reaction. See also *inhibitor*.

Anodic Oxidation

See *anodizing*.

Anodic Pickling

Electrolytic pickling in which the work is the anode.

Anodic Polarization

The change of the electrode potential in the noble (positive) direction due to current flow. See also *polarization*.

Anodic Protection

(1) A technique to reduce the corrosion rate of a metal by polarizing it into its passive region, where dissolution rates are low. (2) Imposing an external electrical potential to protect a metal from corrosive attack. Normally, anodic areas would be expected to corrode. (Applicable only to metals that show active–passive behavior.) In contrast with *cathodic protection*.

Anodic Reaction

Electrode reaction equivalent to a transfer of positive charge from the electronic to the ionic conductor. An anodic reaction is an oxidation process. An example common in corrosion is $Me \sim Me^{n+} + ne^-$.

Anodizing

The process of forming a conversion coating on a metal surface by anodic oxidation; this oxide film of greater thickness than normal makes the component, usually aluminum, the anode, and electrolytic cell carrying an electric current from an external source. See *electrochemistry*. A freshly exposed surface of aluminum will normally develop a thin, hard, tenacious oxide film, which resists further corrosion. The thicker film produced by anodizing, typically 0.01 mm, has improved resistance to abrasion and corrosion and can be dyed various colors.

Anolyte

The electrolyte adjacent to the *anode* and *electrolytic cell*.

Antechamber

The entrance vestibule of a continuously operating sintering furnace.

Antielastic Deformation

The deformation experienced in the transverse plane when a beam is bent in the longitudinal plane. Bending induces compressive stresses at the exterior of the band and compression at the interior. In the case of a simple rectangular beam, the exterior face of the longitudinal curve tends to become concave and the interior face to become convex while the sides taper out toward the inner face.

Anticorrosive Additive

A lubricant additive used to reduce corrosion.

Antiextrusion Ring

A ring that is installed on the low-pressure side of a seal or packing, in order to prevent extrusion of the sealing material.

Antiferromagnetic Material

A material wherein interatomic force holds the elementary atomic magnets (electron spins) of a solid in alignment, a state similar to that of a *ferromagnetic material* but with the difference that equals numbers of elementary magnets (spins) face in opposite directions and are antiparallel, causing the solid to be weakly magnetic, which is paramagnetic, instead of ferromagnetic.

Antifoam Additive

An additive used to reduce or prevent foaming. Also known as *foam inhibitor.*

Antifouling

Intended to prevent fouling of underwater structures, such as the bottom of ships.

Antifouling Coatings

Applied to the immersed surfaces of vessels and other immersed structures to inhibit adhesion by marine organisms. The term usually implies materials that have a toxic component that is continuously released to poison or at least deter organisms.

Antifreeze Compounds

Materials employed in the cooling systems and radiators of internal combustion engines to ensure a liquid circulating medium at low temperatures to prevent damage from the formation of ice. The requirements are that the compound give a freezing point below that is likely to be encountered without lowering the boiling point much below that of water, that it does not corrode the metals or deteriorate rubber connections, that it be stable up to the boiling point, and that it be readily obtainable commercially. Calcium chloride was earlier used for automobile radiators but corroded the metals. It is still used in fire tanks, sodium chromate being added to retard corrosion. Oils were also used, but the high boiling points permitted overheating of the engine, and the oils softened the rubber. Denatured *ethyl alcohol* may be used, but *methanol* is less corrosive and less expensive. A 30% solution of ethyl alcohol in water has a freezing point of about 5°F (−15°C), and a 50% solution freezes at −24°F (−31°C). Alcohol, however, must be renewed frequently because of loss by evaporation.

Glycerol is also used as an antifreeze, a 40% solution in water lowering the freezing point to about 0°F (−18°C) and a 50% solution to −15°F (−25°C). It has the disadvantage of high viscosity, requiring forced circulation at low temperatures, but it does not evaporate easily. *Ethylene glycol* lowers the freezing point to a greater extent than alcohol but has a high boiling point so that it is not lost by evaporation, but it has a higher first cost and will soften ordinary natural rubber connections. *Acetamide* in water solution may also be used as an antifreeze. Antifreezes are sold under various trade names. *Zerone*, of DuPont, as a methanol base, while *Zerex* has a base of ethylene glycol. *Prestone*, marketed by Union Carbide Corp., is ethylene glycol antifreeze. *Pyro* is an antifreeze of U.S. Industrial Chemicals, Inc., with a low freezing point. *Ramp* is ethylene glycol with anticorrosion and antifungal agents added. *Antifreeze PFA55MB*, of Phillips Petroleum Co., used in jet engine fuels, is ethylene glycol monoethyl ether with 10% glycerin. *Dowtherm 209*, of Dow Chemical Co., is an antifreeze material of inhibited methoxypropanol, which boils off without forming gum. *Sierra*, a propylene glycol–based antifreeze of Safe Brands Corp., performs as well as ethylene glycol compounds but is less toxic. A 50–50 blend with water freezes at −26°F (−32°C).

Antifriction Alloys

Tin-, lead-, aluminum- and zinc-base alloys having a low coefficient of friction in contact with, particularly, steel.

Antifriction Bearing

A bearing containing a solid lubricant. An imprecise term often indicating only that the bearing option under discussion is better than the previous practice. In very general terms, friction reduces in the sequence metal to metal, plain bearing such as white metal, rolling element bearing, air bearing and magnetic levitation, ditto in vacuum, the first three being vastly improved by lubrication. See also *roller bearing*, *rolling-element bearing*, and *self-lubricating bearing*.

Antifriction Material

A material that exhibits low-friction or self-lubricating properties.

Antimonial Alloys

Alloys containing large quantities of antimony. The two classes of alloy best meriting the description are the very hard and relatively brittle antimony-based bearing alloys with about 25% copper or 15% tin and 5% copper and the lead-based alloy with up to about 30% antimony for chemical plant applications. The tin-based *Babbitt metals* also contain antimony but in lesser amounts.

Antimony

Antimony is a bluish-white metal, symbol Sb, with a crystalline scalelike structure that exhibits poor electrical and heat conductivity. It is brittle and easily reduced to powder. It is neither malleable nor ductile and is used only in alloys or in its chemical compounds. Like arsenic and bismuth, it is sometimes referred to as a metalloid, but in mineralogy, it is called a semimetal. The element is available commercially and 99.999+ percent purity and is finding increasing use in semiconductor technology.

Antimony is produced either by roasting the sulfide with iron or by roasting the sulfide and reducing the sublimate of Sb_4O_6 thus produced with carbon: high-purity antimony is produced by electrolytic refining. Antimony is one of the few elements that exhibit the unique property of expanding on solidification. Antimony is ordinarily stable and not readily attacked by air or moisture. Under controlled conditions, it will react with oxygen to form oxides. The chief uses of antimony are in alloys, particularly for hardening lead-base alloys.

Antimony imparts hardness and a smooth surface to soft-metal alloys, and alloys containing antimony expand on cooling, thus reproducing the fine details of the mold. This property makes it valuable for type metals. When alloyed with lead, tin, and copper, it forms the *Babbitt metals* used for machinery bearings. It is also much used in white alloys for pewter utensils. Its compounds are used widely for pigments.

Antioxidant

Any additive for the purpose of reducing the rate of oxidation and subsequent deterioration of a material. Antioxidants embrace a wide variety of materials, but in general, for antioxidant activity, the hydroxy groups must be substituted directly in an aromatic nucleus.

In the phenol group of antioxidants, the hydrogen atoms must be free. In the naphthol group, the alpha compound is a powerful antioxidant. Usually, only minute quantities of antioxidants are used to obtain the effect.

Since odor is a major component of flavor, and the development of unpleasant odors in edible fats arises from oxidation, the use of antioxidants is generally necessary, and such for use they are called *food stabilizers*. But degradation of some organic materials may not be a simple oxidation process. In polyvinyl chloride plastics, the initial stage of heat degradation is a dehydrochlorination with hydrogen chloride split out of the molecular chain to give a conjugated system subject to oxidation. Materials called *stabilizers* are thus used to prevent the initial release. Traces of iron and copper in vegetable oils promote rancidity, and citric acid is used as a stabilizer in food oils to suppress this action.

The term *corrosion inhibitors* usually refer to materials used to prevent or retard the oxidation of metals. They may be elements alloyed with the metal, such as columbium or titanium incorporated in stainless steels to stabilize the carbon and retard intergranular corrosion, or they may be material applied to the metal to retard oxygen attack from the air or from moisture. Many paint undercoats, especially the phosphate and chromate coatings applied to steel, are corrosion inhibitors. They may contain a ferrocyanide synergist.

Imidazole and *benzatriozole* systems are *copper antioxidants* used as alternatives to lead–tin solderable surfaces on printed wiring boards.

Antipitting Agents/Compound

An addition agent added to solutions used for pickling, electroplating, etc., to prevent pitting of the component being treated or large pores in the electrodeposit.

Antiscuffing Lubricant

A lubricant that is formulated to avoid scuffing. See also *extreme-pressure lubricant*.

Antiseizure Property

The ability of a bearing material to resist seizure during momentary lubrication failure.

Antislip Metals

Metals with abrasive grains cast or rolled into them, used for floor plates, stair treads, and car steps. They may be of any metal but are usually iron, steel, bronze, or aluminum. The abrasive may be sand, but it is more usually a hard and high-melting-point material such as aluminum oxide. In standard cast forms, antislip metals are marketed under trade names. *Alumalun* is the name of an aluminum alloy cast with abrasive grains. *Brozalum* is a similar product made of bronze. *Algrip steel* is steel plate with abrasive grains rolled into one face. It is used for loading platforms and ramps.

Antistatic Agents

Agents that, when added to a plastic molding material or applied to the surface of a molded object, make it less conductive, thus hindering the fixation of dust or the buildup of electrical charge.

Anti-Stokes Raman Line

A Raman line that has a frequency higher than that of the incident monochromatic radiation.

Antiwear Additive

A lubricant additive used to reduce wear.

Antiweld Characteristic

See *antiseizure property*.

Anvil

A large, heavy metal block that supports the frame structure and holds the stationary die of a forging hammer. Also, the metal block on which blacksmiths forgings are made. Alternatively, a pair of surfaces that both move to retain or manipulate a component.

Anvil Cap

Same as *sow block*.

Aperture (Electron)

See *anode aperture, condenser aperture,* and *physical objective aperture*.

Aperture (Optical)

In optical microscopy, the working diameter of a lens or a mirror. See also *angular aperture*.

Aperture Size

The opening of a mesh as in a sieve. See also *sieve analysis*.

Apiezon Oil

An oil of low vapor pressure used in vacuum technology.

API Gravity (API Degree)

A measure of density used in the U.S. petroleum industry where degrees API = (141.5/specific gravity at 60°F) − 131.5.

Aplanatic

Corrected for spherical aberration and coma.

Apochromatic Lens

A lens whose secondary chromatic aberrations have been substantially reduced. See also *achromatic*.

Apochromatic Objective

Objectives corrected chromatically for three colors and spherically for two colors are called apochromats. These corrections are superior to those of the achromatic series of lenses. Because apochromats are not well corrected for lateral color, special eyepieces are used to compensate. See also *achromatic*.

Apparent Area of Contact

In tribology, the area of contact between two solid surfaces defined by the boundaries of their macroscopic interface. In contrast with *real area of contact*. See also *Hertzian contact area* and *nominal contact area*.

Apparent Density

(1) The weight per unit volume of a powder, in contrast to the weight per unit volume of the individual particles. (2) The weight per unit volume of a porous solid, where the unit volume is determined from external dimensions of the mass. Apparent density is always less than the true density of the material itself.

Apparent Hardness

The value obtained by testing a sintered object with standard indentation hardness equipment. Because the reading is a composite of pores and solid material, it is usually lower than that of a wrought or cast material of the same composition and condition. Not to be confused with *particle hardness*.

Apparent Pore Volume

The total pore volume of a loose powder mass or a green compact. It may be calculated by subtracting the apparent density from the theoretical density of the substance.

Apparent Porosity

The relation of the open pore space to the bulk volume, expressed in percent.

Approach Distance

The linear distance, in the direction of feed, between the point of initial cutter contact and the point of full cutter contact.

Aqua Regia

A mixture of 25% nitric acid and 75% HCl by volume.

Aqueous

Relating to water, for example, a water environment or water-based solution.

Aramid (Aromatic Polyamide)

Aramid fibers are characterized by excellent environmental and thermal stability, static and dynamic fatigue resistance, and impact resistance. These fibers have a higher specific tensile strength (strength/density ratio) of any commercially available continuous-filament yarn. Aramid-reinforced thermoplastic composites have excellent wear resistance and near-isotropic properties—characteristics not available with glass or carbon-reinforced composites.

They are synthetic fibers mainly produced from long-chain polyamides (nylons) in which 85% of the amide linkages are attached directly to two aromatic rings. These are the *poly(isophthaloylchloride/m-phenylenediamine)* meta-aramid *Nomex* and the *poly(terephthaloylchloride/m-phenylenediamine)* para-aramid *Kevlar* from DuPont and the para-aramid *Twaron* from AkzoNobel. The fibers are not biodegradable, not toxic to aquatic life, and pose no unusual hazard in a spill or fire. *Nomex*, creamy white naturally but producible to customer color specifications, and gold-colored *Kevlar* are composed mainly of carbon, oxygen, nitrogen, and hydrogen. Both are insoluble in water and neither melts.

Aramid fiber, tradenamed *Kevlar*, is available in several grades and property levels for specific applications. The grade designated simply as *Kevlar* is made specifically to reinforce tires, hoses, and belting, such as V-belts and conveyor belts.

Kevlar 29 is similar to the basic *Kevlar* in properties but is designated specifically for use in ropes and cables, protective apparel, and as the substrate for coated fabrics. In short fiber or pulp form, *Kevlar 29* can substitute for asbestos in friction products or gaskets. Fabrics of *Kevlar 29* can be made into bullet-resistant vests. Clothing made from Kevlar 29 can be as heat resistant as that made from asbestos and also be extremely cut resistant.

Kevlar 49 has half the elongation (2.5%) and twice the modulus (117, 885 MPa) of *Kevlar 29*. Applications are principally in reinforcing plastic compounds used in lightweight aircraft, boat hulls, and sports equipment. Composites containing *Kevlar* also used as interior panels and secondary structural parts, such as fairings and doors on commercial aircraft.

Kevlar 149 is a highly crystalline aramid that has a modulus of elasticity 40% greater than that of *Kevlar 49* and a specific modulus nearly equal to that of high-tenacity graphite fibers. It is used to reinforce composites for aircraft components. See Table A.14.

Nomex aramid fiber is characterized by excellent high-temperature durability with low shrinkage. It will self-extinguish and

TABLE A.14
Properties of Aramid Fibers

Material	Density (g/cm³)	Filament Diameter		Tensile Modulus[a]		Tensile Strength[a]		Tensile Elongation (%)	Available Yarn Count, No. Filaments
		µm	µin.	GPa	10⁶ psi	GPa	10⁶ psi		
Kevlar 29 (high toughness)	1.44	12	470	83	12	3.6 2.8[b]	0.525–0.400[b]	4.0	134–10,000
Kevlar 49 (high modulus)	1.44	12	470	131	19	3.6–4.1	0.525–0.600	2.8	134–5,000
Kevlar 149 (ultrahigh modulus)	1.47	12	470	186	27	3.4	0.500	2.0	134–1,000

[a] ASTM D 2343, impregnated strand.
[b] ASTM D 885, unimpregnated strand.

does not melt, retaining a high percentage of its initial strength at elevated temperatures. It is available as continuous-filament yarn, staple, and tow. *Nomex* is used in military and civilian applications such as protective apparel, dry gas filtration, rubber reinforcement, and industrial fabrics. *Nomex* aramid fibers are also available as a paper for use in high-temperature electrical insulation and resilient, corrosion-proof honeycomb core for aerospace, and other transportation applications.

Nomex fiber products include staple and filament forms, the staple being the material totally, the material with as much as 60% *Kevlar*, or blends of those materials with antistatic fibers. *Nomex Omega* refers to firefighters' protective clothing of thermal inner liner, moisture barrier, air, and outer shell. Other forms, or products, are yarn, fabric, tow, and floc.

Arbitration Bar

A test bar, cast with a heat of material, used to determine chemical composition, hardness, tensile strength, and the deflection and strength under transverse loading in order to establish the state of acceptability of the casting. It is intended for destructive examination as a check on the component quality. In the case of castings, the bar will be from the same melt as the component. In other cases, a bar sourced from the same stock accompanies the component through treatment cycles such as hardening or coating.

Arbor

(1) In machine grinding, the spindle on which the wheel is mounted. (2) In machine cutting, a shaft or bar for holding and driving the cutter. (3) In foundry, a metal shape embedded in green sand or dry sand cores to support the sand or the applied load during casting.

Arbor Press

A machine used for forcing arbors or mandrels into drilled or bored parts preparatory to turning or grinding. Also used for forcing bushings, shafts, or pins into or out of holes.

Arborescent Powder

See preferred term *dendritic powder*.

Arbor-Type Cutter

A cutter having a hole for mounting on an arbor and usually having a keyway for a driving key.

Arc

The passage of an electrical current across the gap between two electrodes in vacuum, air, or liquid. The electrical transfer is accompanied by radiation of heat and light.

Arc Air Cutting/Gouging

Processes in which an electric arc melts the material, which is then ejected by at high velocity air stream. The arc may be struck between the component and electrode or between two electrodes. In welding, the undesirable deflection of the electric arc by magnetic fields usually arising from the welding currents.

Arc Blow

The deflection of an electric arc from its normal path because of magnetic forces.

Arc Brazing

A brazing process in which the heat required is obtained from an electric arc struck between two nonconsumable electrodes. See *twin carbon arc brazing*.

Arc Chamber

The confined space enclosing the anode and cathode in which the arc is struck.

Arc Cutting

A group of cutting processes that melts the metals to be cut with the heat of an arc between an electrode and the base metal. See *carbon arc cutting*, *metal arc cutting*, *gas metal arc cutting*, *gas tungsten arc cutting*, *plasma arc cutting*, and *air carbon arc cutting*. Compare with *oxygen arc cutting* and *arc air cutting*.

Arc Discharge

Arc discharge is a type of electrical conduction in gases characterized by high current density and low potential drop. A typical arc runs at a voltage drop of 100 V with a current drain of 10 A. The arc has negative resistance—the voltage drop decreases as the current increases—so a stabilizing resistor or inductor in series is required to maintain it. The high-temperature gas rises like a hot-air balloon, while it remains anchored to the current-feeding electrodes at its ends. It thereby acquires an upward-curving shape, which accounts for it being called an arc.

There are many applications of such an intensely hot object. The brilliant arc and the incandescent carbon adjacent to it form the standard light source for movie theater projectors. The electronic flash gun in a camera uses an intense pulsed arc in xenon gas, simulating sunlight. Since no solid-state material can withstand this temperature for long, the arc is used industrially for welding steel and other metals. Alternatively, it can be used for cutting metal very rapidly. Electric arcs form automatically when the contacts in electrical switches and power networks are opened, and much effort goes into controlling and extinguishing them. Lightning is an example of a naturally occurring electric arc.

Arc Force

The axial force developed by a plasma.

Arc Furnace

A furnace in which metal is melted either directly by an electric arc between the electrode and the work or indirectly by an arc between two electrodes adjacent to the metal.

Arc Gas

The gas introduced into the arc chamber and ionized by the arc to form a plasma.

Arc Gouging

An arc cutting process variation used to form a bevel or groove.

Arc Melting

Melting of a material by an electrical arc. The term normally refers to melting in a furnace rather than in processes such as arc welding.

Arc of Contact

The portion of the circumference of a grinding wheel or cutter touching the work being processed.

Arc Oxygen Cutting

See preferred term *oxygen arc cutting*.

Arc Plasma

See *plasma*.

Arc Plasma Welding

Same as *plasma arc welding*.

Arc Resistance

Ability to withstand exposure to an electric voltage. The total time in seconds that an intermittent arc may play across a plastic surface without rendering the surface conductive.

Arc Seam Weld

A seam weld made by an arc welding process.

Arc Spot Weld

A spot weld made by an arc welding process.

Arc Strike

A discontinuity consisting of any localized remelted metal, heat-affected metal, or change in the surface profile of any part of a weld or base metal resulting from an arc. In welding, the unintentional striking of an arc on parent material or previously deposited weld and the damage caused thereby. Such damage may range from superficial blemishes to serious cracking.

Arc Time

The time an arc is maintained in making an arc weld. Also known as *weld time*.

Arc Voltage

The voltage across the welding arc.

Arc Welding

A group of welding processes that produces coalescence of metals by heating them with arc, with or without the application of pressure and with or without the use of filler metal. See *electric arc welding*.

Arc Welding Electrode

A component of the welding circuit through which current is conducted between the electrode holder and the arc. See *arc welding*.

Arc Welding Gun

A device used in semiautomatic, machine, and automatic arc welding to transfer current, guide the consumable electrode, and direct shielding gas when used.

Ar_{cm}, Ar_1, Ar_3, Ar_4, Ar', Ar''

Defined under *transformation temperature*.

Area of Contact

A general term that, without other modifying terminology, is insufficiently specific to be defined precisely. Its use should therefore be avoided. Related terms are *actual area of contact*, *apparent area of contact*, *Hertzian contact area*, *nominal area of contact*, and *real area of contact*.

Areal Weight

The weight of a fiber reinforcement per unit area (width × length) of tape or fabric.

Argon

A chemical element (symbol Ar); Argon is the third member of the gaseous elements called a noble, inert, or rare gas, although argon is not actually rare. The earth's atmosphere is the only commercial argon source; however, traces of the gas are found in minerals and meteorites.

Argon is a colorless, odorless, and tasteless gas. The element is a gas under ordinary conditions, but it can be liquefied, solidified readily, and is a major industrial gas.

Argon does not form any chemical compounds in the ordinary sense of the word, although it does form some weekly bonded compounds with water, hydroquinone, and phenol. There is one atom in each molecule of gaseous argon.

Most argon is produced in air-separation plants. Air is liquefied and subjected to fractional distillation. Because the boiling point of argon is between that of N_2 and O_2, an argon-rich mixture can be taken from a tray near the center of the upper distillation column. The argon-rich mixture is further distilled and then warmed and catalytically burned with H_2 to remove O_2. A final distillation removes H_2 and N_2, yielding a very high purity.

The oldest largest-scale use for argon is in filling electric light bulbs. Welding and cutting metal consumes the largest amount of argon. Metallurgical processing constitutes an important application.

Argon and Ar–Kr mixtures are used, along with a little mercury vapor, to fill fluorescent lamps. The inert gases make the lamps easier to start, help to regulate the voltage, and supplement the radiation produced by the excited mercury vapor.

Argon mixed with a little neon is used to fill luminous electric-discharge tubes employed in advertising signs (similar to neon signs) when a blue or green color is desired instead of the red color of neon.

Argon is used to fill the space between the panes of higher-quality double-pane windows, reducing heat transfer by gaseous conduction by about 30% compared to air filling.

Argon is also used in gas-filled thyratrons. Geiger–Müller radiation counters, and electron tubes of various kinds. Argon atmospheres are used in dry boxes during manipulation of very reactive chemicals in the laboratory and in sealed-package shipments of such materials. In high-energy physics research, a tank of liquid argon can form a calorimeter to detect certain subatomic particles.

Argon Arc Welding

Electric arc welding in which the weld zone is protected by a shroud of argon gas.

Argon Oxygen Decarbonization

A secondary refining process for the controlled oxidation of carbon in a steel melt. In the argon oxygen decarbonization process, oxygen, argon, and nitrogen are injected into a molten metal bath through submerged, side-mounted tuyeres.

Argonaut Welding

A proprietary form of metal inert gas welding.

Arm (Resistance Welding)

A projecting beam extending from the frame of a resistance welding machine that transmits the electrode force and may conduct the welding current.

Armco Iron

A proprietary grade of commercially pure iron with low hardness, high ductility, and good corrosion resistance compared with carbon steel.

Armor Plate

An imprecise term usually indicating steel with a strength of about 1000 MPa. It may be through hardened, face hardened, or carburized on one face or composite, that is, in multilayer form. The term is also applied to nonferrous materials such as higher strength aluminum alloys when these are used as the skin of armored vehicles.

Aromatic Hydrocarbon

A hydrocarbon with a chemistry similar to that of benzene. Aromatic hydrocarbons are either benzenoid or nonbenzenoid. Benzenoid aromatic hydrocarbons contain one or more benzene rings and are by far the more common and the more important commercially. Nonbenzenoid aromatic hydrocarbons have carbon rings that are either smaller or larger than the six-membered benzene ring. Their importance arises mainly from a theoretical interest in understanding those structural features that imparts the property of aromaticity. Benzenoid aromatic hydrocarbons are also called arenes. Benzene itself is the prototypical arene.

Aromatic Polyester

A polyester derived from monomers in which all the hydroxyl and carboxyl groups are directly linked to aromatic nuclei.

Arrest Line (Ceramics)

Rib mark defining the crack front shape of an arrested crack prior to resumption of crack spread under an altered stress configuration. The duration of rest may be long to infinitesimal. See also *rib mark*.

Arrest Marks

Concentric marks on the surface of a fatigue fracture. They are visible to the unaided eye and mark the crack front at irregularities in the load cycling such as interruptions in cycling or an abnormal load. Also termed chonchoidal marks, shell marks, or beach marks. Compare with striations.

Arrest Points

The inflections observed on the curves plotted in the thermal analysis of metals and alloys. They reflect changes in the rate of heat release or absorption and hence identify the location of phase changes.

Arrhenius Equation

A semiempirical formula applicable to many rate control processes such as diffusion, some stages of creep, and some corrosion processes. It is usually presented as

$$\text{Rate} = A\mathrm{e}^{-E_a/RT}$$

where
    A is the Arrhenius constant
    E_a is the activation energy
    R is Boltzmann's constant
    T is temperature absolute

Arsenic

Arsenic (symbol As) is a soft, brittle, poisonous element of steel-gray color and metallic luster. In atomic structure, it is a semimetal, lacking plasticity, and is used only in alloys and compounds. The bulk of the arsenic used is employed in insecticides, rat poisons, and weed killers, but it has many industrial uses, especially in pigments. It is also used in poison gases for chemical warfare.

Metallic arsenic is stable in dry air. When exposed to human or moistened air, the surface first becomes coated with a superficial golden bronze tarnish, which on further exposure turns black. On heating in air, arsenic will vaporize and burn to As_2O_3.

Dermatitis can result from handling arsenical compounds; hence, it is desirable to use impervious gloves or protective creams. The best preventative for dermatitis is strict personal hygiene. In areas where arsenical dust and fumes are present, effective exhaust ventilation is necessary, or when impractical, a respirator should be used.

In the field of electronics, high-purity arsenic is finding an important use as a constituent of the III–V semiconductor compounds. The compounds of most interest are indium arsenide, which may be used for Hall effect devices, and gallium arsenide used for making diodes, transistors, and solar cells.

Arsenic has little commercial application in the pure form, but as an alloying element, it improves oxidation resistance and high-temperature strength in copper, inhibits dezincification in some brass, and hardens lead.

Arsenical Copper

Copper with up to 0.5% arsenic to improve high-temperature properties including creep strength and enhanced resistance to atmospheric oxidation.

Artifact

Generally, a manmade as opposed to a naturally occurring structure. And its metallurgical sense of feature not arising naturally in the material or microstructure in question, for example, an inadvertent scratch or a piece of dust on a metallographic specimen that could be mistaken for a structural feature or erroneously interpreted as a real feature. See also *abrasion artifact*, *mounting artifact*, and *polishing artifact*.

Artificial Aging (Heat Treatment)

Aging above room temperature. See *aging (heat treatment)*. Compare with *natural aging*.

Artificial Aging (Plastics)

The exposure of a plastic to conditions that accelerate the effects of time. Such conditions include heating, exposure to cold, flexing, application of electric field, exposure to chemicals, ultraviolet light radiation, and so forth. Typically, the conditions chosen for such testing reflect the conditions under which the plastic article will be used. Usually, the length of time the article is exposed to these test conditions is relatively short. Properties such as dimensional stability, mechanical fatigue, chemical resistance, stress cracking resistance, dielectric strength, and so forth, are evaluated in such testing. See also *aging*.

Artificial Intelligence

Artificial intelligence is the subfield of computer science concerned with understanding the nature of intelligence and constructing computer systems capable of intelligent action, abbreviated AI. It embodies the dual means of furthering basic scientific understanding and of making computers more sophisticated in the service of humanity.

AI is primarily concerned with symbolic representations of knowledge and heuristic methods of reasoning, that is, using common assumptions and rules of thumb. Two examples of problems studied in AI are planning how a robot, or person, might assemble a complicated device or move from one place to another and diagnosing the nature of a person's disease, or of a machine's malfunction, from the observable manifestations of the problem. In both cases, reasoning with symbolic descriptions predominates over calculating.

Artificial Weathering

The exposure of plastics to cyclic laboratory conditions, consisting of low temperatures, high and low relative humidities, and ultraviolet radiation energy, with or without direct water spray and moving air (wind), in an attempt to produce changes in the properties of the plastics similar to those observed after long-term continuous exposure outdoors. The laboratory exposure conditions are usually more intensified than those encountered in actual outdoor exposure in an attempt to achieve an accelerated effect. Also called accelerated aging.

Artificially Layered Structures

Manufactured, reproducibly layered structures with layer thicknesses approaching interatomic distances. Modern thin-film techniques are at a stage at which it is possible to fabricate these structures, also known as artificial crystals or superlattices, opening up the possibility of engineering new desirable properties into materials.

These structures serve as model systems and as a testing ground for theoretical models and for other naturally occurring materials that have similar structures. For example, ceramic superconductors consist of a variable number of conducting CuO_2 layers intercalated by various other oxide layers, and therefore, artificially layered structures may be used to study predictions of the behavior of suitably manufactured materials of this class.

Asbestos

Any of six naturally occurring minerals characterized by being extremely fibrous (asbestiform), strong, and incombustible. They are utilized in commerce for fire protection; as reinforcing material for tiles, plastics, and cements; for friction materials; and for thousands of other uses. Because of a great concern over the health effects of asbestos, many countries have promulgated strict regulations for its use. The six minerals designated as asbestos also occur in a nonfibrous form. In addition, there are many other minerals that morphologically mimic asbestos because of their fibrous nature.

The six naturally occurring minerals exploited commercially for their desirable physical properties, which are in part derived from their asbestiform habit, are chrysotile asbestos, a member of the serpentine mineral group, and grunerite asbestos, riebeckite asbestos, anthophyllite asbestos, tremolite asbestos, and actinolite asbestos, all members of the amphibole mineral group. Individual mineral particles, however, processed and regardless of their mineral name, are not demonstrated to be asbestos is the average length-to-width ratio which is less than 20:1.

Chrysotile is a hydrated magnesium silicate from the serpentine mineral group. Amosite and erocidolite are iron silicates from the amphibole mineral group. Anthophyllite is a magnesium silicate of the amphibole group with the magnesium isomorphically replaced by varying amounts of iron and aluminum.

The important characteristics of the asbestos minerals that make them unique are their fibrous form; high-strength and surface area; resistance to heat, acids, moisture, and weathering; and good bonding characteristics with most binders such as resins and cement.

Asbestos is used for many types of products because of its chemical and thermal stability, high tensile strength, flexibility, low electrical conductivity, and large surface area. Past uses of asbestos, such as sprayed-on insulation, where the fibers may become easily airborne, have been generally abandoned. Asbestos is used predominantly for the construction industry in the form of cement sheets, coatings, pipes, and roofing products. Additional important uses are for reinforcing plastics and tiles, for friction materials, and packings and gaskets.

- *Asbestos fabrics* are often a woven mix with some cotton. For brake linings and clutch facings, the asbestos is woven with fine metallic wire.
- *Asbestos shingles*, for fireproof roofing and siding of houses, are normally made of asbestos fibers and Portland cement forms on hydraulic pressure.

- *Asbestos board* is a construction or insulating material in sheets made of asbestos fibers and Portland cement molded under hydraulic pressure. Ordinary board for the siding and partitions of warehouses and utility buildings is in the natural mottled-gray color, but pigmented boards are also marketed in various colors.
- *Asbestos lumber* is asbestos–cement board molded in the form of boards for flooring and partitions, usually with imitation wood grain molded into the surface. Asbestos siding, for house construction, is grained to imitate cypress or other wood and is pigmented with titanium oxide to give a clear white color. Asbestos roofing materials may also be made with asphalt or other binder instead of cement.
- Asbestos paper is a thin asbestos sheeting made of asbestos fibers bonded usually with a solution of sodium silicate. It is strong, flexible, and white in color and is fireproof and a good heat insulator. For covering steam pipes and for insulating walls, it is made in sheets of two or three plies. For wall insulation, it is also made double with one-corrugated sheet to form air pockets when in place.

Other applications include insulating cements, fireproof garments, curtains, shields, brake linings, pipe coverings, and molded products.

Asbestos also appears granular, in crystals, compact such as nephrite, which is the jade of the Orient. Jade appears as a solid rock and is highly valued for making ornamental objects.

As-Brazed

The condition of brazements after brazing, prior to any subsequent thermal, mechanical, or chemical treatments.

As-Cast Condition

Castings as removed from the mold without subsequent heat treatment.

As-Cast/As-Welded

The condition reached upon completion of the indicated process but prior to heat treatment or other processing.

Ash Content

Proportion of the solid residue remaining after a reinforcing substance has been incinerated (charred or intensely heated).

Aspect Ratio

Depending upon context, various ratios such as length to diameter (fibers), lanes to width (crack), depth to diameter (corrosion pit), etc.

Asphalt

Asphalt refers to varieties of naturally occurring bitumen. Acetyl was also produced as a petroleum byproduct. Those substances are black and largely soluble in carbon disulfide. Asphalt are of variable consistency, ranging from a highly viscous fluid to a solid.

Asphalt is derived from petroleum in commercial quantities by removal of volatile components. It is an inexpensive construction material used primarily as a cementing and waterproofing agent.

Asphalt is composed of hydrocarbons and heterocyclic compounds containing N_2, sulfur, O_2; its components vary in molecular weight from about 400 to 5000. It is thermoplastic and viscoelastic; at high temperatures or over long loading times, it behaves as a viscous fluid; at low temperatures or short loading times as an elastic body.

The three distinct types of asphalt made from petroleum residues and their uses are described in Table A.15.

A-Spot

One of many small contact areas through which electrical current can pass when two rough, conductive solid surfaces are touching.

Assay

Determination of how much of a sample is the material indicated by the name. For example, foreign assay of $FeSO_4$ the analysis would determine both iron and SO_4^{2-} in the sample.

Assay Value

The measure of the metal content of an ore of precious metal. It may be expressed as a weight percentage or, traditionally, as troy ounces of metal per avoirdupois ton of ore.

TABLE A.15
Asphalts and Their Uses

Asphalt Type and % of Production	Manufacturing Process	Properties	Uses
Straight-run, 70%–75%	Distillation or solvent precipitation	Nearly viscous flow	Roads, airport runways, hydraulic works
Air-blown, 25%–30%	Reacting with air at 204°C–316°C	Resilient: viscosity less susceptible to temperature change than straight-run	Roofing, pipe coating, paints, underbody coatings, paper laminates
Cracked, less than 5%	Heating to 427°C–538°C	Nearly viscous flow; viscosity more susceptible to temperature change than that of straight-run asphalt	Insulation board saturant, dust laying

Source: *McGraw-Hill Encyclopedia of Science and Technology*, 8th edn., Vol. 2, McGraw-Hill, New York, p. 176. With permission.

Assembly

A number of parts or subassemblies or any combination thereof joined together.

Assembly Adhesive

An adhesive that can be used for bonding parts together, such as in the manufacture of a boat, an airplane, furniture, and the like. The term assembly adhesive is commonly used in the wood industry to distinguish such adhesives (formally called joint glues) from those used in making plywood (sometimes called veneer glues). It describes adhesive used in fabricating finished structures or goods, or subassemblies thereof, as differentiated from adhesives used in the production of sheet materials for sale as such, for example, plywood or laminates.

Assembly Line (Adhesives)

The time interval between the spreading of the adhesive on the adherend and the application of pressure or heat, or both, to the assembly. For some ways involving multiple layers or parts, the assembly time begins with the spreading of the adhesive on the first adherend. See also *closed assembly time* and *open assembly time*.

A-Stage

An early stage in the reaction of certain thermosetting resins in which the material is fusible and still soluble in certain liquids. Synonym for resole. Compare with *B-stage* and *C-stage*.

Astigmatism

A defect in a lens or optical system that causes rays in one plane parallel to the optical axis to focus at a distance different from those in the plane at right angles to it.

As-Welded

The condition of weld metal, welded joints, and weldments after welding but prior to any subsequent thermal, mechanical, or chemical treatments.

Atactic Stereoisomerism

A chain of molecules in which the position of the side chains or side atoms is more or less random. See also isotactic stereoisomerism and syndiotactic stereoisomerism.

Athermal

Not isothermal. Changing rather than constant temperature conditions.

Athermal Transformation

A reaction that proceeds without benefit of thermal fluctuations—that is, thermal activation is not required. Such reactions are diffusionless and can take place with great speed when the driving force is sufficiently high. For example, many martensitic transformations occur athermally on cooling, even at relatively low temperatures, because of the progressively increasing drive force. In contrast, a reaction that occurs at constant temperature is an *isothermal transformation*; thermal activation is necessary in this case and the reaction proceeds as a function of time.

Atmospheric Corrosion

The gradual degradation or alteration of a material by contact with substances present in the atmosphere, such as oxygen, carbon dioxide, water vapor, and sulfur and chlorine compounds.

Atmospheric Riser

A riser that uses atmospheric pressure to aid feeding. Essentially, a *blind riser* into which a small core or rod protrudes; the functions of the core or rod is to provide an open passage so that the molten interior of the riser will not be under a partial vacuum when metal is withdrawn to feed the casting but will always be under atmospheric pressure.

Atom

The smallest particle of an element that retains the characteristic properties and behavior of the element. Further division of the atom, termed fission or, colloquially, splitting the atom, reduces it to its constituent, subatomic, particles: protons, electrons, neutrons, etc. See also *atomic structure*, *isotope*, and *nuclear structure*.

Atom Probe

An instrument for measuring the mass of a single atom or molecule on a metal surface; it consists of a field ion microscope with a hole in its screen opening into a mass spectrometer; atoms are removed from the specimen by pulsed field evaporation, travel through the hole, and are detected in the mass spectrometer. See also *field-ion microscopy*.

Atomic Absorption Spectrometry

The measurement of light absorbed at the wavelength of resonance lines by the unexcited atoms of an element.

Atomic Arrangement

The way atoms are arranged in a crystal structure.

Atomic Fission

The breakup of the nucleus of an atom in which the combined weight of the fragments is less than that of the original nucleus, the difference being converted to a very large energy release.

Atomic Hydrogen

Hydrogen in a state in which the atoms are not bonded to each other. Normally hydrogen atoms bond strongly together in pairs to form molecules. If the hydrogen molecule, H_2, is raised to a high energy level by heating, it disassociates into two individual H atoms. When the atoms recombine, a large amount of heat is released.

Atomic Hydrogen (Welding) Process

An arc welding process that fuses metals together by heating them with an electric arc maintained between two metal electrodes enveloped in a stream of hydrogen. Shielding is provided by the hydrogen, which also carries heat by molecular disassociation and subsequent recombination. Pressure may or may not be used and filler metal may or may not be used. (This process is now of limited industrial significance.)

Atomic Mass

The mass of one atom of an element relative to the mass of a reference element. Originally, the reference element was hydrogen, atomic mass of 1.0, then oxygen of atomic mass 16.0, or carbon of atomic mass 12. See also *atomic weight*.

Atomic Number (*Z*)

The number of elementary positive charges (protons) contained within the nucleus of an atom. For an electrically neutral atom, the number of planetary electrons is also given by the atomic number. Atoms with the same *Z* (isotopes) may contain different numbers of neutrons. Also known as nuclear charge. See also *isotope* and *proton*.

Atomic Number Contrast

See *atomic number imaging*.

Atomic Number Imaging

In scanning electron microscopy, a technique in which contrast is controlled by atomic number (high atomic number areas appear light, while low atomic number areas appear dark). Usually obtained by imaging based on backscattered electron signal. See also *backscattered electron*.

Atomic Percentage

The composition of an alloy or mixture in terms of the number of atoms of each constituent for hundreds of mixture.

Atomic Replica

A thin replica devoid of structure on the molecular level. It is prepared by the vacuum or hydrolytic deposition of metals or simple compounds of low molecular weight. See also *replica*.

Atomic Scattering Factor, *f*

The ratio of the amplitude of the wave scattered by an atom to that scattered by a single electron.

Atomic Structure

The arrangement of the parts of an atom, which consists of a positively charged nucleus surrounded by a cloud of electrons arranged in orbits that can be described in terms of quantum mechanics.

Atomic Weight

(1) A number assigned to each chemical element that specifies the average mass of its atoms. Because an element may consist of two or more isotopes, each having atoms with well-defined but differing masses, the atomic weight of each element is the average of the masses of its naturally occurring isotopes weighted by the relative proportions of those isotopes. (2) The mean weight of the atom of an element in relation to $^{12}C = 12.000$.

Atomization (Materials Characterization)

The subdivision of a compound into individual atoms using heat or chemical reactions. This is a necessary step in atomic spectroscopy. See also *atomizer* and *nebulizer*.

Atomization (Powder Metallurgy)

The disintegration of a molten metal into particles by a rapidly moving gas or liquid stream or by other means.

Atomization (Thermal Spraying)

The division of molten material at the end of the wire or rod into fine particles.

Atomizer

A device that atomizes a sample, for example, a burner, plasma, or hydride reaction chamber. See also *atomization* and *nebulizer*.

Attack Polishing

Simultaneous etching and mechanical polishing.

Attenuation

(1) The fractional decrease of the intensity of energy flux, including the reduction of intensity resulting from geometrical spreading, absorption, and scattering. (2) The diminution of vibrations or energy over time or distance. The process of making thin and slender, as applied to the formation of fiber from molten glass. (3) The exponential decrease with distance in the amplitude of an electrical signal traveling along a very long, uniform transmission line, due to conductor and dielectric losses.

Attitude (Attitude Angle)

In a bearing, the angular position of the line joining the center of the journal to that of the bearing bore, relative to the direction of loading.

Attrition

Removal of small fragments of surface material during sliding contact.

Attritious Wear

Wear of abrasive grains in grinding such that the sharp edges gradually become grounded. A grinding wheel that has undergone such wear usually has a glazed appearance.

Attritor

A high-intensity ball mill whose drum is stationary and whose balls are agitated by rotating baffles, paddles, or rods at right angle to the drum axis.

Attritor Grinding

The intensive grinding or alloying in an attritor. Examples are milling of carbides and binder metal powders and mechanical alloying of hard dispersoid particles with softer metal or alloy powders. See also *mechanical alloying*.

Auger Chemical Shift

The displacement and energy of an Auger electron peak for an element due to a change in chemical bonding relative to a specified element or compound.

Auger Electron

An electron emitted from an atom with a vacancy and an inner shell. Auger electrons have a characteristic energy detected as peaks in the energy spectra of the secondary electrons generated.

Auger Electron Spectroscopy

A technique for chemical analysis of surface layers that identifies the atoms present in a layer by measuring the characteristic energies of their Auger electrons.

Auger Electron Yield

The probability that an atom with a vacancy in a particular inner shell will relax by an Auger process.

Auger Map

A 2D image of the specimen surface showing the location of emission of Auger electrons from a particular element. A map is normally produced by rastering the incident electron beam over the specimen surface and simultaneously recording the Auger signal strength for a particular transition as a function of position.

Auger Matrix Effects

Effects that cause changes in the shape of an Auger electron energy distribution or in the Auger signal strength for an element due to the physical environment of the emitting atom and not due to bonding with other elements or changes in concentration.

Auger Process

The radiationless relaxation of an atom, involving a vacancy and then enters the electron shell. An electron (known as an Auger electron) is emitted.

Auger Transition Designations

Transitions are designated by the electron shells involved. The first letter designates the shell containing the initial vacancy; the last two letters designate the shelves containing electron vacancies created by Auger emission (e.g., KLL and LMN).

Ausforming

Mechanical working of a suitable steel that has been rapidly cooled to retain the austenite phase below the lower critical temperature, 723°C in carbon steels. It is implicit that the steel composition is such that the austenite is unstable below this temperature and it will eventually transform to martensite or bainite. Such treatment, which effectively combines hardening by working with hardening by transformation can produce very high-strength components. See *steel* and *isothermal transformation*. Note that the term is not appropriate for steels that will remain austenitic.

Austempered Ductile Iron

A moderately alloyed *ductile iron* that is austempered for high strength with appreciable ductility. See also *austempering*.

Austempering

A heat treatment for ferrous alloys in which a part is quenched from the austenitizing temperature at a rate fast enough to avoid formation of ferrite or pearlite and then held at a temperature just above M_s until transformation to bainite is complete. Also designated as bainite in both austempered steel and austempered ductile iron, austempered steel consists of two phase mixtures containing ferrite and carbide, while austempered ductile iron consists of two phase mixtures containing ferrite and austenite.

Austenite

A solid solution of one or more elements interface-centered cubic iron (gamma iron). Unless otherwise designated (such as nickel austenite), the solute is generally assumed to be carbon.

Austenitic Corrosion-Resistant Stainless Steel

A highly alloyed corrosion-resistant stainless steel containing 16% or more chromium, a ferrite-stabilizing element, and sufficient austenite-stabilizing elements such as nickel (up to about 35%), manganese, and nitrogen. See also *stainless steels*.

Austenitic Grain Size

Generally, the grain size of steel when in the austenitic state. The term is also used in the sense of an inherent austenitic grain size, established by the original steelmaking practice, which influences the steel properties even after a number of subsequent working and heat treatment operations. This may be revealed by appropriate etching of cross sections after cooling to room temperature. The term "prior austenitic grain size" usually refers to the grain size when the steel was last in the austenitic condition.

Austenitic Manganese Steel

A cast, wear-resistant material containing about 1.2% C and 12% Mn. This usually implies a high content of nickel or manganese. The steel will be nonmagnetic but will be "stainless" only if it contains more than 12% chromium. Used primarily in the fields of earth moving, mining, quarrying, railroading, ore processing, lumbering, and in the manufacture of cement and clay products. Also known as Hadfield steel.

Austenitic Steel

An alloy steel whose structure is normally austenitic at room temperature.

Austenitizing

Forming austenite by heating a ferrous alloy into the transformation range (partial austenitizing) or above the transformation range (complete austenitizing). When used without qualification, the term implies complete austenitizing.

Autoclave

A close vessel for conducting and completing either a chemical reaction under pressure and heat or other operation, such as cooling. Widely used for bonding and curing reinforced plastic laminates.

Autoclave Molding

A process in which, after lay-up, winding, or wrapping, an entire assembly is placed in a heated autoclave, usually at 340–1380 kPa (50–200 psi). Additional pressure permits higher density and improved removal of volatiles from the resin. Lay-up is usually vacuum bagged with a bleeder and release cloth.

Autofrettage

Prestressing a hollow metal cylinder by the use of momentary internal pressure exceeding the yield strength.

Autogenous Weld

A weld made by the fusion of the parent components without the addition of filler metal.

Automated Image Analysis

See *image analysis.*

Automatic Brazing

Brazing with equipment that performs the brazing operation without constant observation and adjustment by a brazing operator. The equipment may or may not perform the loading and unloading of the work. See also *machine brazing.*

Automatic Gas Cutting

See preferred term *automatic oxygen cutting.*

Automatic Mold

A mold for injection or compression molding of plastics that repeatedly goes through the entire cycle, including ejection, without human assistance.

Automatic Oxygen Cutting

Oxygen cutting with equipment that performs the cutting operation without constant observation and adjustment of the controls by an operator. The equipment may or may not perform loading and unloading of the work. See also *machine oxygen cutting.*

Automatic Press (Metals)

A press in which the work is fed mechanically through the press in synchronism with the press action. An automation press is an automatic press that, in addition, is provided with built-in electrical and pneumatic control equipment.

Automatic Press (Plastics)

A hydraulic press for compression molding or an injection machine that operates continuously, being controlled mechanically (toggle) or hydraulically, or by a combination of these methods.

Automatic Press Stop

A machine-generated signal for stopping the action of a press, usually after a complete cycle, by disengaging the clutch mechanism and engaging the brake mechanism.

Automatic Welding

Welding with equipment that performs the welding operation without adjustment of the controls by a welding operator. The equipment may or may not perform the loading and unloading of the work. See also *machine welding.*

Automation Press

See *automatic press.*

Autoradiography

An inspection technique in which radiation spontaneously emitted by a material is recorded photographically. The radiation is emitted by radioisotopes that are (1) produced in a metal by bombarding it with neutrons, (2) added to a metal such as by alloying, or (3) contained within a cavity in a metal part. The technique serves to locate the position of the radioactive element or compound.

Auxiliary Anode

In electroplating, a supplementary *anode* positioned so as to raise the current density on a certain area of the *cathode* and thus obtain better distribution of plating.

Auxiliary Electrode

An *electrode* commonly used in polarization studies to pass current to or from a test electrode. It is usually made from a noncorroding material.

Auxiliary Magnifier or Enlarger (Eye Protection)

An additional lens or plate, associated with eye protection equipment, used to magnify or enlarge the field of vision.

Average Density

The density measured on an entire body or on a major number of its parts whose measurements are then averaged.

Average Erosion Rate

The cumulative erosion divided by the corresponding cumulative exposure duration, that is, the slope of a line from the origin to a specified point on the cumulative erosion-time curve.

Average Grain Diameter

The mean diameter of an equiaxed grain section whose size represents all the grain sections in the aggregate being measured. See also *grain size*.

Average Linear Strain

See *engineering strain*.

Average Molecular Weight

The molecular weight of the most typical chain in a given plastic; it is characteristic of neither the longest nor the shortest chain.

Average Particle Size

A single value representing the entire particle size distribution.

Avogadro's Number

The number of molecules (6.02×10^{23}) in a gram molecular weight of any substance. See also *gram molecular weight* and *mole*.

Axial

Longitudinal or parallel to the axis or center line of a part. Usually refers to axial compression or axial tension.

Axial Load Bearing

See *thrust bearing*.

Axial Relief

The relief for clearance behind the end cutting edge of a milling cutter. See also *face mill*.

Axial Rolls

And *ring rolling*, vertically displaceable, tape rolls mounted in a horizontally displaceable frame opposite to, but on the same centerline as, the main roll and rolling mandrel. The actual rolls control ring height during rolling.

Axial Winding

In filament-wound reinforced plastics, a winding with the filament parallel to, or at a small angle to, the axis (0° helix angle). See also *polar winding*.

Axis (Crystal)

The edge of the unit cell of a space lattice. Any one axis of any one lattice is defined in length and direction relative to other axes of that lattice.

B

B

The symbol for a repeating unit in a copolymer chain.

Babbitt Metal

The original name for tin–antimony–copper (Sn–Sb–Cu) white alloys used for machinery bearings; the term now applies to almost any white bearing alloy with either tin or lead base. The alloy consists of 88.9% tin, 7.4% antimony, and 3.7% copper. This alloy melts at 239°C. It has a Brinell hardness of 35 at 21°C and 15 at 100°C. As a general-utility-bearing metal, the original alloy has never been improved greatly, and makers frequently designate the tin-base alloys close to this composition as genuine babbitt.

Commercial white bearing metals now known as babbitt are of three general classes: tin-base, with more than 50% tin hardened with antimony and copper and used for heavy-duty service; intermediate, with 20%–50% tin, with lower compressive strength and more sluggish behavior as a bearing; and lead-base, made usually with antimonial lead with smaller amounts of tin together with other elements to hold the lead in solution.

Copper hardens and toughens the alloy and raises the melting point. Lead increases fluidity and raises antifriction qualities but softens the alloy and decreases its compressive strength. Antimony hardens the metal and forms hard crystals in a soft matrix, which improves the alloy as a bearing metal.

Alloys containing up to 1% arsenic (As) are harder at high temperatures and are fine-grained, but arsenic is used chiefly for holding lead in suspension. Zinc (Zn) increases hardness but decreases frictional qualities, and with much zinc the bearings are inclined to stick. Even minute quantities of iron (Fe) harden the alloys, and iron is not used except when zinc is present. Bismuth (Bi) reduces shrinkage and refines the grain but lowers the melting point and the strength at elevated temperatures. Cadmium (Cd) increases the strength and fatigue resistance, but any considerable amount lowers the frictional qualities, lowers the strength at higher temperatures, and causes corrosion. Nickel (Ni) is used to increase strength and raises the melting point. The normal amount of copper in babbitts is 3%–4%, at which point the maximum fatigue-resisting properties are obtained with about 7% antimony. More than 4% copper tends to weaken the alloy and raises the melting point. When the copper is very high, tin–copper crystals are formed and the alloy is more a bronze than a babbitt.

Because of the increased speeds and pressures in bearings and the trend to lighter weights, heavy-cast babbitt bearings are now little used despite their low cost and the ease of casting the alloys. The alloys are used mostly as antifriction metals in thin facings on steel backings, with the facing usually less than 0.03 cm thick, in order to increase their ability to sustain higher loads and dissipate heat.

Back Bead

See preferred term *back weld*.

Back Draft

A reverse taper on a casting pattern or a forging die that prevents the pattern or forged stock from being removed from the cavity.

Back Extrusion

See *backward extrusion*.

Back Gouging

The removal of weld metal and base metal from the other side of a partially welded joint to ensure complete penetration upon subsequent welding from that side.

Back Pressure

Resistance of a plastic, because of its viscosity, to continued flow when the mold is closing.

Back Rake

The angle on a single-point turning tool corresponding to axial rake in milling. It is the angle measured between the plane of the tool face and the reference plane and lies in a plane perpendicular to the axis of the work material and a base of the tool. See *single-point tool*.

Back Reflection

The diffraction of x-rays at a *Bragg angle* approaching 90°.

Back Reflection X-Ray Technique

A technique for studying the crystal structure in which a narrow, collimated beam of monochromatic x-rays is directed at the metal through a hole in a photographic film. *Crystallographic planes in the crystal structure reflect x-rays back onto the film producing dot patterns that can be interpreted to indicate grain size, cold work, etc. Sometimes termed von Laue analysis after the pioneer in such work.*

Back Side

The side of a plastic or composite part that is cured against the vacuum bag.

Back Step Sequence

A longitudinal sequence in which the weld bead increments are deposited in the direction opposite to the progress of welding the joint. See also *block sequence*, *cascade sequence*, *continuous sequence*, *joint building sequence*, and *longitudinal sequence*.

Back Taper

Reverse draft used in a mold to prevent the molded plastic article from drawing freely. See also *undercut*.

Back Weld

A weld deposited at the back of a single groove weld.

Backfill

Material placed in a drilled hole to fill space around anodes, vent pipe, and buried components of a cathodic protection system.

Backfire

The momentary recession of the flame into the welding tip or cutting tip followed by immediate reappearance or complete extinction of the flame. See also *flashback*.

Background

Any noise in the signal due to instabilities in the system or to environmental interferences. See also *signal-to-noise ratio*.

Backhand Welding

The welding technique in which the welding torch or gun is directed opposite to the progress of welding. Sometimes referred to as the "pull gun technique" in gas metal arc welding and flux-cored arc welding. Compare with *forehand welding*. See also *travel angle*, *work angle*, and *drag angle*.

Backing

(1) In grinding, the material (paper, cloth, or fiber) that serves as the base for coated abrasives. (2) In welding, a material placed under or behind a joint to enhance the quality of the weld at the root. It may be a metal backing ring or strip, a pass of weld metal, or a nonmetal such as carbon, granular flux, or a protective gas. (3) In plain bearings, that part of the bearing to which the bearing alloy is attached, normally by a metallurgical bond.

Backing Bead

See preferred term *backing weld*.

Backing Filler Metal

See *consumable insert*.

Backing Film (Metallography)

A film used as auxiliary support for the thin replica or specimen-supporting film.

Backing Pass (of Weld)

An initial pass of weld to form a surface on which the main weld is deposited. It may remain untouched or be machined away and, possibly, a *back weld deposited*.

Backing Plate

In plastic injection molding equipment, a heavy steel plate that is used as a support for the cavity blocks, guide pins, and bushings. In blow molding equipment, it is a steel plate on which the cavities (i.e., the bottle molds) are mounted. In casting, a second *bottom board* on which molds are open.

Backing Plate/Ring, Strip, etc. (of Weld)

Material placed behind the root of the weld to provide a surface onto which the first weld run is deposited. A temporary backing plate, etc., is one that undergoes little or no fusion and is subsequently removed. A permanent backing plate, etc., is one that is partly fused by the weld and may remain in position or be removed partly after welding. See also *fusible insert*.

Backing Strap

See preferred term *backing strip*.

Backing Strip

Backing in the form of a strip.

Backing Weld

Backing in the form of a weld.

Backing-Split Pipe

See *split pipe backing*.

Backlash

Loss of motion, play, or movement in moving parts such that the driving element (as a gear) can be reversed for some angle or distance before working contact is again made with a driven element.

Backoff

A rapid withdrawal of a grinding wheel or cutting tool from contact with a workpiece.

Back-Pressure Relief Port

An opening in an extrusion die for plastics that allows for the escape of excess material.

Backscatter Techniques/Gauge

The measurement of radiation, usually gamma or beta, reflected from some source by the component under test. Within limits, the amount of reflected radiation increases as a direct function of the thickness of the sample, so the technique is used for applications such as the continuous measurement of sheet during production.

Backscattered Electron

An information signal arising from elastic (electron–nucleus) collisions, wherein the incident electron rebounds from the interaction with a small energy loss. The backscattered electron yield is strongly dependent upon the atomic number, qualitatively describes the origin of characteristic rays, and reveals compositional and topographic information about the specimen. See also *atomic number imaging*.

Back-to-Back Ring Seal

An adaptation of the simple ring seal that employs two identical elements loaded actually by a spring placed between the rings. The spring forces the elements against mating rings on either side.

Backup (Flash and Upset Welding)

A locator used to transmit all or a portion of the upsetting force to the workpieces or to aid in preventing the workpieces from slipping during the upsetting.

Backup Coat

The ceramic slurry of dip coat that is applied to multiple layers to provide a ceramic shell of the desired thickness and strength for use as a casting mold.

Backward Extrusion

Same as *indirect extrusion*. See *extrusion*.

Baffle

A device used to restrict or divert the passage of fluid through a pipeline or channel.

Bag Molding

A method of molding for bonding plastics or composites involving the application of fluid pressure, usually by means of air, steam, water, or vacuum, to a flexible cover that is sometimes in conjunction with the rigid die and completely encloses the material to be bonded. Also called blanket molding. See also *vacuum bag molding*.

Bagging

Applying an impermeable layer of film over an uncured part and sealing the edges so that a vacuum can be drawn.

Bail

Hoop or arched connection between the core hook and ladle or between the crane hook and mold trunnions.

Bainite

A metastable aggregate of *ferrite and cementite* resulting from the transformation of *austenite* at temperatures below the pearlite range but above Ms, the martensite start temperature. Upper bainite is an aggregate that contains parallel lath-shape units of ferrite, produces the so-called "feathery" appearance in optical microscopy, and is formed above approximately 350°C (660°F). Lower bainite, which has an acicular appearance similar to tempered martensite, is formed below approximately 350°C (660°F). Bainite structures normally result from cooling, from the austenitic temperature range, at a rate too rapid to form pearlite but not fast enough to form martensite.

The precise structure is a function of the temperature at which the carbide precipitates. See *steel*.

Bainitic Hardening

Quench-hardening treatment resulting principally in the formation of *bainite*.

Bake

Heating in an oven to a low control temperature to remove gases or to harden a binder. The term is applied to various heating processes involving modest temperatures and no change to the primary properties of the component. For example, components being vitreous enamels require baking to fuse the coating, welding electrodes are baked to drive off moisture from the coating, and components containing high levels of hydrogen, introduced by treatments such as electroplating or welding, are baked to allow the potentially damaging gas to diffuse out.

Baked Core

A casting core that has been heated through sufficient time and temperature to produce the desired physical properties attainable from its oxidizing or thermal-setting binders.

Bakelite

A proprietary name for a phenolic thermosetting resin used as a plastic mounting material for metallographic samples. See *phenol formaldehyde resin*.

Baking

(1) Heating to a low temperature in order to remove gases. (2) Curing or hardening surface coatings such as paints by exposure to heat. (3) Heating to drive off moisture, as in baking of sand cores after molding.

Balance

(1) (Dynamic) condition existing where the principal inertial axis of a body coincides with its rotational axis. (2) (Static) condition existing where the center of gravity of a body lies on its rotational axis.

Balance Construction

In woven reinforcements, it is the equal parts of warp and fill fibers. Construction in which reactions to tension and compression loads result in extension or compression deformations only and in which flexural loads produce fewer bending of equal magnitude and axial and lateral directions.

Balanced Design

In filament-wound reinforced plastics, a winding pattern so designed that the stresses in all filaments are equal.

Balanced Laminate

A laminate in which all laminae at angles other than 0° and 90° occur only in ± pairs (not necessarily adjacent) and are symmetrical around the centerline. See also *symmetrical laminate*.

Balanced Twist

An arrangement of twists in a combination of two or more reinforcing strands that does not kink or twist when the yarn produced is held in the form of an open loop.

Balanced-in-Plane Contour

In a filament-wound part, a head contour in which the filaments are oriented within a plane and the radii of curvature are adjusted to balance the stresses along the filaments with the pressure loading.

Ball (Peen Hammer Ball)

The hammer typically carried by an engineering fitter. It has one smooth hemispherical face and usually a flat square face. The peen end is the likely origin of the term *peening*.

Ball Bearing

A rolling-element bearing in which the rolling elements are spherical. See also *rolling-element bearing*.

Ball Burnishing

(1) Same as *ball sizing*. (2) Removing burrs and polishing small stampings and small machine parts by *tumbling* in the presence of metal balls.

Ball Clay

A secondary clay, commonly characterized by the presence of organic matter, high plasticity, high dry strength, long vitrification range, and a light color when fired. Used extensively in traditional ceramics, such as whiteware, wall tile, and china. See also *traditional ceramics*.

Ball Complement

The number of balls contained in a ball bearing.

Ball Indented Bearing

A bearing with surface indentations serving as a lubricant reservoir.

Ball Mill

A machine consisting of a rotating hollow cylinder partly filled with metal balls (usually hardened steel or white cast iron) or sometimes pebbles; used to pulverize crushed ores or other substances such as pigments or ceramics. The balls may tumble freely in the cylinder or barrel as in the case of mills used for peening or some polarizing processes. Alternatively, in the case of some crushing mills, the balls may run in grooved tracks in a pair of rings. Typically the free-tumbling pulverizing ball would be small, say about 50 mm diameter when charged, while a track ball would be much larger, say about 400 mm.

Ball Milling

A method of grinding and mixing material, with or without liquid, in a rotating cylinder or conical mill partially filled with grinding media such as balls or pebbles.

Ball Sizing

Sizing and finishing a hole by forcing a ball of suitable size, finished, and hardness through the hole or by using a burnishing bar or broach consisting of a series of spherical lands of gradually increasing size coaxially arranged. Depending upon the application, various benefits ensue—burnishing the bore, developing an accurate size, and introducing a favorable pattern of residual stresses. No metal is removed, unlike broaching. Also called *ball burnishing* and sometimes ball broaching.

Balling

(1) Spheroidization. (2) The aggregation of a pasty mass of impure iron and slag in the *puddling process*.

Balling Up

The formation of globules of molten brazing filler metal or flux due to lack of wetting of the base metal.

Banbury

An apparatus for compounding polymeric materials. It is composed of a pair of contrarotating rotors that masticate the materials to form a homogeneous blend. This internal-type mixer produces excellent mixing.

Band Density

In filament winding of composites, the quantity of fiberglass reinforcements per inch of bandwidth, expressed as strands (or filaments) per inch.

Band Mark

An indentation in carbon steel or strip caused by external pressure on the packaging band around cut lengths or coils; it may occur in handling, transit, or storage.

Band Theory

The valence electrons of a metallic crystal are not tied to individual atoms but are shared as a general cloud. The atoms collectively impose a periodic electropotential function that causes the valence electrons to be confined to specific ranges, or bands, of energy level. Individual bands may be partially or completely filled or empty, and they are separated by forbidden bands in which the electrons cannot normally exist other than to jump across. The electrons in partly filled bands can move allowing conduction, hence the terms *conduction electrons* and *conduction band*. See also *atomic structure* and *semiconductor*.

Band Thickness

In filament winding of composites, the thickness of the reinforcement as it is applied to the mandrel.

Banded Structure

A segregated structure consisting of alternating, nearly parallel bands of different compositions, typically aligned in the direction of primary hot working.

Banding

Inhomogeneous distribution of alloying elements or phases aligned in filaments or plates parallel to the direction of working. See also *banded structure*, *ferrite–pearlite banding*, and *segregation banding*.

Bands

(1) Hot-rolled steel strip, usually produced for rerolling into thinner sheet or strip. Also known as hot bands or band steel. (2) See *electron bands*.

Bandwidth

In filament winding of composites, the width of the reinforcement as it is applied to the mandrel.

Bank Sand

Sedimentary deposits, usually containing less than 5% clay, occurring in banks or pits, used in core-making and in synthetic molding sands. See also *sand*.

Bar

(1) A section hot rolled from a *billet* to a form, such as round, hexagonal, octagonal, square, rectangular, with sharp or rounded corners or edges with a cross-sectional area of less than 105 cm^3 (16 in.^{2}). (2) A solid section that is long in relationship to its cross-sectional dimensions, having a completely symmetrical cross section and a width or greatest distance between terminal phases of 9.5 mm (3/8 in.) or more. (3) An obsolete unit of pressure equal to 100 kPa. (4) Material produced by rolling or drawing that is usually a simple section but that produced by extrusion may be extremely complex.

Bar Drawing

Tube drawing using a mandrel that travels with the tube.

Bar Folder

A machine in which a folding bar or wing is used to bend a metal sheet whose edge is clamped between the upper folding leaf and the lower stationary jaw into a narrow, sharp, close, and accurate fold along the edge. It is also capable of making rounded folds such as those used in wiring. A universal folder is more versatile in that it is limited to width only by the dimensions of the sheet.

Bar Mill

(1) A rolling mill for producing bar. (2) A barrel or similar container in which components are contained with short lengths of bar and possibly an abrasive. The container is rotated and the bar burnishes or abrades the components.

Bar Stock

Same as *bar*.

Barcol Hardness

A hardness value obtained by measuring the resistance to penetration of a sharp steel point under a spring load. The instrument, called the Barcol impressor, gives a direct reading on a 0–100 scale. The hardness value is often used as a measure of the degree of cure of a plastic.

Bare Electrode

A filler metal electrode consisting of a single metal or alloy that has been introduced into a wire, strip, or bar form and that has no coating or covering applied to it other than that which was incidental to its manufacturer or preservation.

Bare Glass

Glass in the form of yarns, rovings, and fabrics from which the sizing or finish has been removed. Also, such glass before the application of sizing or finish.

Bare Metal Arc Welding (BMAW)

An arc welding process that produces coalescence of metals by heating them with an electric arc between a bare or lightly coated metal electrode and the work. Neither shielding nor pressure is used as a filler metal is obtained from the electrode.

Barium

A metallic element of the alkaline earth group, symbol Ba. It occurs in combination with the minerals witherite and barite, which are widely distributed. The metal is silvery white and can be obtained by electrolysis from the chloride, but it oxidizes so easily that it is difficult to obtain in the metallic state. Powdered or granular barium is explosive when in contact with carbon tetrachloride, fluorochloromethanes, and other halogenated hydrocarbons. Its melting point is 1562°F (850°C) and its specific gravity 3.78. The most extensive use of barium is in the form of its compounds. The salts, which are soluble, such as sulfide and chloride, are toxic. And insoluble, nontoxic barium sulfate salt is used in radiography. Barium compounds are used as pigments, in chemical manufacturing, and in deoxidizing alloys of tin, copper, lead, and zinc. Barium is introduced into lead-bearing metals by electrolysis to harden the lead. When barium is heated to about 392°F (200°C) in hydrogen gas, it forms barium hydride, BaH_2, a gray powder that decomposes on contact with water and can be used as a source of nascent hydrogen for life rafts. Barium is also a key ingredient in ceramic superconductors.

Barium Chloride

A colorless crystalline material of composition $BaCl_2 \cdot 2H_2O$, or in anhydrous form without the water of crystallization. In the mechanical industries, it is used for heat-treating baths for steel, either alone or mixed with potassium chloride. The molten material is free from fuming and can be held at practically any temperature within the range needed for tempering steels. It is also used for making boiler compounds, for softening water, as a mordant in dyeing and printing inks, in tanning leather, in photographic chemicals, and in insecticides. Barium chlorate is a colorless crystalline powder, soluble in water. It is used in explosives as an oxygen carrier and in pyrotechnics for green-colored light. Barium fluoride, BaF_2, is used in crystal form for lasers. Doping with other elements gives diffused wavelengths for different communication beams. Barium cyanide is a poisonous, colorless, crystalline material melting at 1112°F (600°C). A 30% water solution is added to cyanide plating baths that removes carbonates and increases the current efficiency.

Barium Nitrate

Also called nitrobarite. A white crystalline powder of composition $Ba(NO_3)_2$ with specific gravity of 3.24, melting at 1098°F (592°C), and decomposing at higher temperatures. It is a barium salt of nitric acid obtained by roasting barite with coke, leaching out the precipitated barium sulfide, precipitating as a carbonate by the addition of soda ash, and then dissolving in dilute nitric acid. It has a bitter metallic taste and is poisonous. Barium nitrate is used in ceramic glazes, but its chief use is in pyrotechnics. It gives a pale-green flame in burning and is used for green signals and flares and for white flares in which the delicate green is blended with the light of other extremely luminous materials. It is also used as an oxygen carrier in flare powders and to control the time of burning of the aluminum or magnesium. Sparklers are composed of aluminum powder and steel filings with barium nitrate as the oxygen carrier. The steel filings produced the star-like sparks.

Barium Titanate (BaTiO₃)

The basic raw material used to make high dielectric constant ceramic capacitors. Used also in high thermal conductivity, thick-film ceramic pastes.

Bark

The decarburized layer just beneath the scale that results from heating steel in an oxidizing atmosphere.

Barrel Cleaning

Mechanical or electrolytic cleaning of metal and rotating equipment.

Barrel Distortion

See *negative distortion*.

Barrel Finishing

Improving the surface finish of workpieces by processing them in rotating equipment along with abrasive particles that may be suspended in a liquid. The barrel is normally loaded about 60% full with a mixture of parts, media, compound, and water.

Barrel Plating

Plating articles in a rotating container, usually a perforated cylinder that operates at least partially submerged in a solution.

Barreling

Convexity of the surfaces of cylindrical or conical bodies, often produced unintentionally during upsetting or as a natural consequence during compression testing. See also *compression test*.

Barrier Coat

An exterior coating applied to a composite filament-wound structure to provide protection. In fuel tanks, a coating applied to the inside of the tank to prevent fuel from permeating the sidewall.

Barrier Plastics

A general term applied to a group of lightweight, transparent, impact-resistant plastics, usually rigid copolymers with high acrylonitrile content. Barrier plastics are generally characterized by gas, aroma, and flavor with barrier characteristics approaching those of metal and glass.

Basal Plane

(1) That plane of a hexagonal or tetragonal crystal perpendicular to the axis of highest symmetry. Its Miller indices are (001). (2) A plane perpendicular to the principal axis (c axis) in a tetragonal or hexagonal structure.

Basalt

A dense, hard, dark-brown to black igneous rock, consisting of feldspar and augite and often containing crystals of green olivine. It occurs as trapped or as volcanic rock. Mrs is a basalt which is frequently found in columns or prisms, as in the celebrated basalt cliffs of Northern Ireland. It differs from granite in being a fine-grained extrusive rock and in having a high content of iron and magnesium. Basalt is used in the form of crushed stone for paving, as a building stone, and for making rock wool. A Russian cast basalt used for electrical insulators is called angarite. In Germany, cast basalt has been used as a building stone, for linings, and for industrial floors. It is made by melting the crushed and graded basalt and then tempering by slow cooling. The structure of the cast material is dense with needlelike crystals, and it has a Mohs hardness of 8–9. Basalt glass is not basalt, but pumice.

Basalt fiber is alkali resistant; thus it is compatible with concrete and perhaps suitable for infrastructure applications.

Base

(1) A chemical substance that yields hydroxyl ions (OH⁻) when dissolved in water. Compare with *acid*. (2) The surface on which a single-point tool rests when held in a tool post. Also known as heel. See also the term *single-point tool*. (3) In forging, see *anvil*.

Base Material

(1) The material to be welded, brazed, soldered, or cut. See also *base metal and substrate*. (2) For a printed circuit board, the insulating material upon which a conductor pattern may be formed. The base material may be rigid or flexible, and it may be a dielectric sheet or insulated metal sheet. See also *dielectrics*.

Base Metal

(1) The metal present in the largest proportion in an alloy; brass, for example, is a copper-base alloy. (2) The metal to be brazed, cut, soldered, or welded. (3) After welding, that part of the metal that was not melted. (4) A metal that readily oxidizes, or that dissolves to form ions. Contrast with *noble metal* (2).

Basic Bottom and Lining

The inner bottom and lining of a melting furnace, consisting of materials such as crushed burned dolomite, magnesite, magnesite bricks, or basic slag that gives a basic reaction at the operating temperature.

B

Basic Dynamic Load Capacity (of a Bearing)

The radial load that a rolling-element bearing can support for a rating life of one million revolutions (500 h at 331/3 rpm). See also *basic load rating*.

Basic Load Rating (C)

The radial load that a ball bearing can withstand for one million revolutions of the inner ring. The value of the basic load rating depends on bearing type, bearing geometry, accuracy of fabrication, and bearing material. See also *basic dynamic load capacity* and *dynamic load*.

Basic Metal

The original metal to which one or more coatings are applied.

Basic NMR Frequency

The frequency, measured in hertz, of the oscillating magnetic field applied to induce transitions between nuclear magnetic energy levels. See also *magnetic resonance*.

Basic Open Hearth Process

The largely obsolete open-hearth process utilizing a basic lining with a lime flux to deal with pig irons and higher phosphorus contents than could be dealt with in the acid open hearth.

Basic Oxygen Furnace

A large tiltable vessel lined with basic refractory material, which is the principal type of furnace for modern steelmaking. After the furnace is charged with molten pig iron (which usually comprises 65%–75% of the charge), scrap steel, and fluxes, a lance is brought down near the surface of the molten metal and a jet of high-velocity oxygen impinges on the metal. The oxygen reacts with carbon and other impurities in the steel to form liquid compounds that dissolve in the slag and gases that escape from the top of the vessel.

Basic Refractories

Refractories whose major constituent is lime, magnesia, or both, and which may react chemically with acid refractories, acid slags, or acid fluxes at high temperatures. Basic refractories are used for furnace linings. Compare with *acid refractory*.

Basic Static Load Rating (of the Bearing)

A load that, if exceeded on a nonrotating rolling-element bearing, produces a total permanent deformation of rolling elements and race at the most heavily stressed contact point of 0.0001 times the ball or roller diameter or greater.

Basic Steel

Steel melted in a furnace with a *basic bottom and lining* and under a slag containing an excess of a basic substance such as magnesia or lime. A steel produced by one of the basic steelmaking processes.

Basket Weave (Composites)

In this type of woven reinforcement, two or more wrap threads go over and under two or more filling threads in a repeat pattern. The basket weave is less stable than the *plain weave* but produces a flatter and stronger fabric. It is also a more pliable fabric than the plain weave and maintains a certain degree of porosity without too much sleaziness, although not as much as the plain weave.

Basket Weave (Titanium)

Alpha platelets with or without interleaved ß platelets that occur in colonies in a *Widmanstätten structure*.

Bastard

Usually, anything abnormal, unintended, or otherwise undesirable but sometimes merely a mixture or hybrid. Its use is usually vulgar but not always, for example, a bastard file is a recognized name for a file having a specific cut of medium coarseness.

Batch

A quantity of materials formed during the same process or in one continuous process and having identical characteristics throughout. See also *lot*.

Batch Furnace

A furnace used to heat-treat a single load at a time. Batch-type furnaces are necessary for large parts such as heavy forgings and are preferred for complex alloy grades requiring long cycles. See also *car furnace* and *horizontal batch furnace*.

Batch Sintering

Presintering or sintering in such a manner that compacts are sintered and removed from the furnace before additional unsintered compacts are placed in the furnace.

Bath

Molten metal on the hearth of a furnace, in a crucible, or in a ladle.

Bath Lubrication

See *flood lubrication*.

Batt

Felted fabrics. Structures built by the interlocking action of compressing fibers, without spinning, weaving, or knitting. Used for reinforcing plastic composite materials.

Battery

(1) An accumulator for storing electricity. (2) A factory containing a number of forges. (3) Multiple devices operating in concert.

Bauschinger Effect

The phenomenon by which plastic deformation increases yield strength in the direction of plastic flow and decreases it in another direction.

B

Bauxite

A whitish to reddish mineral composed largely of hydrates of *alumina* having a composition of $Al_2O_3 \cdot 2H_2O$, theoretically containing 74% alumina. It is the most important ore (source) of aluminum but is also used for making aluminum oxide abrasives, for alumina-based refractories, for white cement, and for decolorizing and filtering.

Bayer Process

A process of extracting alumina from bauxite ore before the electrolytic reduction. The bauxite is digested in a solution of sodium hydroxide, which adversely converts alumina to soluble aluminate. After the "red mud" residue has been filtered out, aluminum hydroxide is precipitated, filtered out and calcined to alumina.

B-Basis

The "B" mechanical property value is the value above which at least 90% of the population of values is expected to fall, with a confidence of 95%. See also *A-basis*, *S-basis*, and *typical basis*.

Beach Marks

Macroscopic progression marks on a fatigue fracture or stress-corrosion cracking surface that indicate successive positions of the advancing crack front. The classic appearances of irregular elliptical or semielliptical rings, radiating outward from one or more origins. Beach marks (also known as clamshell marks or arrest marks) are typically found on service fractures where the part is loaded randomly, intermittently, or with periodic variations in mean stress or alternating stress. Beach marks are visible to the unaided eye and mark the crack front at irregularities in a load cycling such as interruptions in cycling or an abnormal load. Also termed *conchoidal marks*, *shell marks*, or *arrest marks*. Compare with *striations* (metals).

Bead

(1) Half-round cavity in the mold, or half-round projection or molding on a casting. (2) A single deposit of weld metal produced by fusion.

Bead Weld

See preferred term *surfacing weld*.

Beaded Flange

A flange reinforced by a low ridge, used mostly around the hole.

Beading

Raising a ridge or projection on sheet metal.

Beam

(1) A length of material of constant cross section, usually wrought but occasionally cast. (2) A length of material carrying a bending load. (3) A projected stream of radiation.

Beam Hardening

The increase in effective energy of a polyenergetic (e.g., x-ray) beam with increasing attenuation of the beam. Beam hardening is due to the preferential attenuation of the lower-energy, or soft, radiation.

Bearing

A support carrying some component or structure. The term usually implies a small support area and a capacity to allow relative movement of the component being carried, for example, a bearing carrying a rotating shaft.

Bearing Area

(1) The projected bearing or load-carrying area when viewed in the direction of the load. Sometimes used as a synonym for *real area of contact* (this usage is not recommended). (2) The sum of the horizontal intercepts of a surface profile at a given level. (3) The product of the pin (or hole) diameter and the specimen thickness. See also *bearing test*.

Bearing Bronzes

Bronzes used for bearing applications. Two common types of bearing bronzes are copper-base alloys containing 5–20 wt.% tin and a small amount of phosphorus (*phosphor bronzes*) and copper-base alloys containing up to 10 wt.% tin and up to 30 wt.% lead (*leaded bronzes*).

Bearing Characteristic Number

A dimensionless number that is used to evaluate the operating conditions of plain bearings. See also *capacity number* and *Sommerfeld number*.

Bearing Materials

A large variety of metals and nonmetallic materials in monolithic and composite (laminate) form are used for bearings. Monolithic ferrous bearings are made of gray cast iron, pressed and sintered iron and steel powder, and many wrought steels, including low-carbon and high-carbon and plain-carbon steels, low-alloy steels, alloy steels, stainless steels, and tool steels. Most cast-iron bearings are made of gray iron because it combines strength with the lubricity of graphitic carbon. Pressed and sintered bearings can be made to control porosity and impregnated with oil for lubricity. Because of its wide use in ball and roller bearings, one of the best-known bearing steels is American Iron and Steel Institute (AISI) 52100 steel, a through-hardening 1% carbon –1.3% to 1.6% chromium alloy steel. Many steels, however, are simply surface-hardened for bearing applications. In recent years, the performance of bearing steels has been markedly improved by special melting practices that reduce the presence of nonmetallic inclusions.

Monolithic nonferrous bearings include copper (90%)–zinc (10%) bronze, leaded bronzes, and unleaded bronzes and an aluminum–tin alloy, containing about 6% tin as a principal alloying element. The bronze and aluminum alloy provide similar load-bearing capacity and fatigue resistance, but the bronze is somewhat better in resistance to corrosion by fatty acids that can form with petroleum-base oils. It is also less prone to seizure and abrasion from mating shafts, more able to embed foreign matter and, thus, prevent shaft wear, and more tolerant of shaft misalignment.

Applications include auto engine starter-motor bearings, or bushings, for the copper–zinc bronze; auto engine connecting-rod bearings for the aluminum alloy; and various bearings in motors, machine tools, and earth-moving equipment for the tin bronzes.

Monolithic bearings are also made of cemented tungsten and chromium carbides, plastics, carbon graphite, wood, and rubber. Plastics provide good combinations of inherent lubricity, corrosion resistance, and adequate strength at room to moderately elevated temperatures. Thermal conductivity and other performance features that may be required can be provided by metal and other fillers. Plastic bearings can be made of acetal, nylon, polyester, polytetrafluoroethylene, polysulfone, polyphenylene sulfide, polyimide, and polyamide-imide. Carbon–graphite bearings are more heat resistant but rather brittle, thus limited to nonimpact applications. Wood bearings are made of maple and the hard lignum vitae. Rubber bearings, usually steel backed, are used for bearings requiring resilience.

Nonferrous metals are widely used in dual- or trimetal systems. Dual-metal bearings comprise a soft, thin, inner liner metallurgically bonded to stronger backing metal. Steel lined with bronze containing 4%–10% lead provides the highest load-bearing capacity, 55 MPa, or about twice that of the bronze alone, and fatigue strength. However, the aluminum alloy with a steel backing provides the best corrosion resistance and only moderately less load-bearing capacity. Tin and lead babbitt linings excel in surface qualities conducive to free-sliding conditions and are used with either steel, bronze, or aluminum alloy backings.

Dual-metal systems cover a gamut of bearings for motors, pumps, piston pins, camshafts, and connecting rods.

Trimetal bearings, always steel backings, have an inner liner of tin or lead babbitt and an intermediate layer of a more-fatigue-resistant metal, such as leaded bronze, copper–lead, aluminum–tin, tin-free aluminum alloys, silver, or silver–lead. Load-bearing capacity ranges from 10 to 83 MPa. The silver bearing systems provide the best combination of load-bearing capacity, fatigue and corrosion resistance, and compatibility to mating materials, but a lead babbitt, medium-lead bronze, and steel system are a close second, sacrificing only a moderate reduction in corrosion resistance but at a reduction in cost. Applications include connecting rods, camshafts, and main bearings in auto engines and reciprocating aircraft engines.

Bearing Steels

Alloy steels have been used to produce *rolling-element bearings*. Typically, bearings have been manufactured from both high-carbon (1.00%) and low-carbon (0.20%) steels. The high-carbon steels are used in either a through-hardened or a surface induction-hardened condition. Low-carbon-bearing steels are carburized to provide the necessary surface hardness while maintaining desirable core properties.

Bearing Strain

The ratio of the deformation of the bearing hole, in the direction of the applied force, to the pin diameter. Also, the stretch or deformation strain for a sample under bearing load.

Bearing Strength

The maximum bearing stress that can be sustained. Also, the bearing stress at that point on the stress–strain curve at which the tangent is equal to the bearing stress divided by n% of the bearing hole diameter.

Bearing Stress (Metals)

The shear load on a mechanical joint (such as a pinned or riveted joint) divided by the effective bearing area. The effective bearing area of a riveted joint, for example, is the sum of the diameters of all rivets times the thickness of the loaded member.

Bearing Stress (Plastics)

The applied load in pounds divided by the bearing area. Maximum bearing stress is the maximum load in pounds sustained by the specimen during the test, divided by the original bearing area.

Bearing Test

A method of determining the response to stress (load) of sheet products that are subjected to riveting, bolting, or a similar fastening procedure. The purpose of the test is to determine the *bearing strength* of the material and to measure the *bearing stress* versus the deformation of the hole created by a pin or rod of circular cross section that pierces the sheet perpendicular to the surface.

Bearing Yield Strength

The *bearing stress* at which a material exhibits a specified limiting deviation from the proportionality of *bearing stress to bearing strain.*

Bed

(1) The stationary portion of a press structure that usually rests on the floor or foundation, forming the support for the remaining parts of the press and the pressing load. The *bolster* and sometimes the lower die are mounted on the top surface of the bed. (2) For machine tools, the portion of the mainframe that supports the tool, the work, or both. (3) Stationary part of the shear frame that supports the material being sheared and the fixed blade.

Bedding

Sinking a casting pattern down into the sand to the desired position and ramming the sand around it.

Bedding a Core

Placing an irregularly shaped casting core on a bed of sand for drying.

Beer's Law

A relationship in which the optical absorbance of a homogeneous sample containing an absorbing substance is directly proportional to the concentration of the absorbing substance. See also *absorptivity.*

Beilby Layer

A layer of metal disturbed by mechanical working, wear, or mechanical polishing presumed to be without regular crystalline structure (amorphous); originally applied to grain boundaries.

Bell

A jar-like enclosure for containing a vacuum or a controlled atmosphere in sintering equipment.

Bell Furnace

A furnace in the shape of an inverted bowl, that is, a bell. The bell is placed over a charge on a fixed hearth and the required heating carried out. The bell can then be transferred to a second and subsequent hearth while the charge on the first hearth cools and is removed. Occasionally, a simple unheated inner bell is left in place to minimize atmospheric contamination and drafts. A top hat furnace is similar.

Bellows

A thin-walled tube with circumferential corrugations that allow repeated longitudinal flexing.

Bellows Seal

A type of mechanical seal that utilizes bellows for providing secondary sealing.

Bell-Type Furnace

A furnace for the sintering of large batches of small pieces under a controlled atmosphere.

Belt Furnace

A continuous-type furnace that uses a mesh-type or cast-link belt to carry parts through the furnace.

Belt Grinding

Grinding with an *abrasive belt.*

Bench Allowance

The length of the ark of the neutral axis between the tangent points of a bend.

Bench Molding

Casting sand molds by hand tamping loose or production patterns at a bench without the assistance of air or hydraulic action.

Bench Press

Any small press that can be mounted on a bench or table.

Bend Angle

The angle through which a bending operation is performed, that is, the supplementary angle to that formed by the two bent tangent lines or planes.

Bend or Twist (Defect)

Distortion similar to warpage generally caused during forging or trimming operations. When the distortion is along the length of the part, it is termed bend; when across the width, it is termed twist. When bend or twist exceeds tolerance, it is considered a defect.

Corrective action consists of hand straightening, machine straightening, or cold restriking.

Bend Radius

(1) The inside radius of a bend section. (2) The radius of a tool around which metal is bent during fabrication.

Bend Tangent

A tangent point at which a bending arc ceases or changes.

Bend Test

A test for determining the relative ductility of metal that is to be formed (usually sheet, strip, plate, or wire) and for determining the soundness and toughness of metal (e.g., after welding). The specimen is usually bent over a specified diameter through a specified angle for a specified number of cycles.

In simpler tests, for example, of general quality or ductility, the test piece is bent in a vice or over an anvil of specified radius. Often the loading is not measured and the test piece is barely bent until it cracks, and to a specified angle or, in a reversed bend, bent to an angle and then bent back. Acceptance criteria depends on the particular test but may range from simply surviving intact to complete freedom from cracking. More complex tests where loading is measured include three-point bend testing in which the test piece is supported at its extremities and the load applied at a single central point. This induces a peak load at the line on the surface opposite the central loading point. For four-point bend testing, the test piece is supported at its extremities and loaded via two points symmetrically located between the support points. This induces an even stress over the length between the two loading points. In tests of fracture toughness, the test piece is precracked and the properties of interest include the stress required to extend the crack, as indicated by potential drop techniques, and the activity at the crack tip as reflected by crack opening displacement.

There are four general types of bend tests, named according to the manner in which the forces are applied to the specimen to make the bend: *free bend*, *guided bend*, *semiguided bend*, and *wraparound bend*.

Bender

Term denoting a die impression, tool, or mechanical device designed to bend forging stock to conform to the general configuration of die impressions to be subsequently used.

Bending

The straining of material, usually flat sheet or strip metal, by moving it around a straight axis lying in the neutral plane. Metal flow takes place within the plastic range of the metal, so that the part retains a *permanent set* after removal of the applied stress. The cross section of the bend inward from the neutral plane is in compression; the rest of the bend is in tension. See also *bending stress*.

Bending Brake

A form of open-frame single-action press that is comparatively wide between the housings, with a bed designed for holding long, narrow forming edges or dies. Used for bending and forming strip, plate, flake, and sheet (into boxes, panels, roof decks, and so on). Also known as *press brake*.

Bending Dies

Dies used in presses for bending sheet-metal or wire parts into various shapes. The work is done by the punch pushing the stock into cavities or depressions of similar shape in the die or by ancillary attachments operated by the descending punch.

Bending Moment

The algebraic sum of the couples or the moments of the external forces, or both, to the left or right of any section on a member subjected to bending by couples or transverse forces, or both.

Bending Rolls

Various types of machinery equipped with two or more rolls to form curve sheet and sections. They have their axes parallel but the center one is offset from the line joining the other two. Bar or sheet passing through the appropriately adjusted rolls is bent to a curve or cylinder.

Bending Stress

A stress involving tensile and compressive forces, which are not uniformly distributed. Its maximum value depends on the amount of flexure that a given application can accommodate. Resistance to bending can be turned to stiffness.

Bending Stress (Glass)

A stress system that simultaneously imposes a compressive component at one surface, graduating to an imposed tensile component at the opposite surface of a glass section.

Bending–Twisting Coupling

A property of certain classes of laminates that exhibit twisting curvatures when subjected to bending moments.

Beneficiation

Concentration or other preparation of ore for smelting.

Bentonite

A colloidal claylike substance derived from the decomposition of volcanic ash composed chiefly of the minerals of the montmorillonite family that are mainly hydrated calcium and magnesium silicates, capable of absorbing five times its weight of water. It is used for bonding molding sand.

Benzene

A colorless, liquid, inflammable, aromatic hydrocarbon that boils at 80.1°C and freezes at 5.4°C–5.5°C. Benzene is used as a solvent and particularly in Europe as a constituent of motor fuel. In the United States, the largest uses of benzene are for the manufacture of styrene and phenol. Other important outlets are in the production of chlorinated benzenes (used in DDT and moth flakes), and benzene hexachloride, and insecticides.

DDT (dichlorodiphenyltrichloroethane) is a colorless, crystalline, tasteless and almost odorless organo-chloride known for its insecticidal properties.

Benzene Ring

The six-carbon ring structure found in benzene, C_6H_6, and in organic compounds formed from benzene by replacement of one or more hydrogen atoms by other chemical atoms or radicals. The terms aromatic chemicals and aromatics refer to all the chemicals made from the benzene ring.

Nitrobenzene, $C_6H_5NO_2$, is a highly poisonous and inflammable liquid made by the action of nitric and sulfuric acids on benzene, used in soaps and cosmetics. It is called myrbane oil as a perfuming agent. The nitrated derivative called benzedrine, or amphetamine, originally used by wartime pilots to combat fatigue, is phenylamino-benzine. It is used in medicine to control obesity, but it is a stimulant to the central nervous system and is habituating. The isomer dextro-amphetamine is d-phenylamino-propane sulfate, commonly called Dexedrine. It causes a rise in blood pressure and stimulates cerebral activity that lasts several hours, but it has a depressant effect on the intestinal muscles, causing loss of appetite and delayed activity of the stomach with other side effects.

Beryl

At least 50 different beryllium-bearing minerals are known, but in only about 30 is it a regular constituent. The only beryllium-bearing mineral of industrial importance is beryl, a hexagonal beryllium-aluminum silicate with the ideal composition $Be_3Al_2Si_6O_{18}$, equivalent to 5% beryllium to 14% beryllium oxide, 19% Al_2O_3, and 67% SiO_2. The precious forms of beryl, emerald and aquamarine, approach the ideal composition. Emerald is a transparent, intensely green variety of beryl.

Beryllia

A colorless to white powder of the composition BeO used in the manufacture of hot-pressed ceramic parts—most notably basic refractories and substrates (heat sinks) in electronics. Also known as beryllium oxide.

Beryllium oxide is used in the manufacture of high-temperature refractory material and high-quality electrical porcelains, such as aircraft spark plugs and ultrahigh-frequency radar insulators. The high thermal conductivity of beryllium oxide and its good high-frequency electrical insulating properties find application in electrical and electronic fields.

Another use of beryllium oxide is as a slurry for coating graphite crucibles to insulate the graphite and to avoid contamination of melted alloys with carbon. Beryllium oxide crucibles are used where exceptionally high-purity or reactive metals are being melted. In the field of beryllium oxide ceramics, a type of beryllia has been developed that can be formed into custom shapes for electronic and microelectronic circuits. Beryllium oxide has a high thermal conductivity, equal to that of aluminum, and excellent insulating properties, which permits closer packing of semiconductor functions in silicon integrated circuits.

Berylliosis

A severe illness related to pneumonia resulting from the inhalation of beryllium.

Beryllium

Among structural metals, beryllium (Be) has a unique combination of properties. It has low density (two-thirds that of aluminum), high modulus per weight (five times that of ultrastrength steels), high specific heat, high strength per density, excellent dimensional stability, and transparency to x-rays. Beryllium is expensive, however, and its impact strength is low compared to values for most other metals.

Beryllium is a steel-gray lightweight metal, used mainly for its excellent physical properties rather than its mechanical properties. Except for magnesium, it is the lightest in weight of common metals, with a density of 1855 kg/m^3. It also has the highest specific heat (1833 J/kg K) and a melting point of 1290°C. It is nonmagnetic and has about 40% the electrical conductivity of copper, a thermal conductivity of 190 W/m K, high permeability to x-rays, and the lowest neutron cross section of any metal having a melting point above 500°C. Also its tensile modulus (28.9 × 10^4 MPa) is far greater than that of almost all metals.

Its ultimate tensile strength range is from 228 to 690 MPa and tensile elongation from 1% to 40%, depending on bill form. Thus, because of its low density, beryllium excels in specific strength and especially in specific stiffness. However, tensile properties, especially elongation, are extremely dependent on grain size and orientation and are highly anisotropic so that results based on uniaxial tensile tests have little significance in terms of useful ductility in fabrication or fracture toughness in structural applications. From this standpoint, the metal was considered to be quite brittle. Ductility, as measured by elongation and tensile tests, increases with increasing temperature to about 400°C and then decreases above about 500°C. Although resistant to atmospheric corrosion under normal conditions, beryllium is attacked by O_2 and N_2 at elevated temperatures and certain assets, depending on concentration, at room temperature.

Available forms include block, rod, sheet, plate, foil, extrusions, and wire. Machining blanks, which are machined from large vacuum hot pressings, make up the majority of beryllium purchases. However, shapes can also be produced directly from powder by processes such as cold press/sinter/coin, CIP/HIP, CIP/ sinter, CIP/hot press, and plasma spray/sinter. (CIP is cold isostatic press, and HIP is hot isostatic press.) Mechanical properties depend on powder characteristics, chemistry, consolidation process, and thermal treatment. Wrought forms, produced by hot working, have high strength in the working direction, but properties are usually anisotropic.

Beryllium parts can be hot-formed from cross-rolled sheet and plate as well as plate-machined from hot-pressed block. Forming rates are slower than for titanium, for example, but tooling and forming costs for production items are comparable.

Structural assemblies of beryllium components can be joined by most techniques such as mechanical fasteners, rivets, adhesive bonding, brazing, and diffusion bonding. Fusion-welding processes are generally avoided because they cause excessive grain growth and reduce mechanical properties.

Beryllium behaves like other light metals when exposed to air by forming a tenacious protective oxide film that provides corrosion protection. However, the bare metal corrodes readily when exposed for prolonged periods to tap water or seawater or to a corrosive environment that includes high humidity. The corrosion resistance of beryllium in both aqueous and gaseous environments can be improved by applying chemical conversion, metallic, or nonmetallic coatings. Beryllium can be electroless nickel-plated and flame or plasma sprayed.

All conventional machining operations are possible with beryllium, including EDM (Electric Discharge Machining) and ECM (Electric Chemical Machining). However, beryllium powder is toxic if inhaled. Because airborne beryllium particles and beryllium salts present a health hazard, the metal must be machined in specially equipped facilities for safety.

Machining damages the surface of beryllium parts. Strength is reduced by the formation of microcracks and "twinning." The depth of the damage can be limited during finish machining by taking several light machining cuts and sharpening cutting tools frequently or by using nonconventional metal-removal processes. For highly stressed structural parts, 0.05–0.10 mm should be removed from each surface by chemical etching or milling after machining. This process removes cracks and other surface damage caused by machining, thereby preventing premature failure. Precision parts should be machined with a sequence of light cuts and intermediate thermal stress reliefs to provide the greatest dimensional stability.

Beryllium is toxic if inhaled or ingested, necessitating special precaution in handling. Most applications are quite specialized and stem largely from the good thermal and electrical properties of the metal. Uses include precision mirrors and instruments, radiation detectors, x-ray windows, neutron sources, nuclear reactor reflectors, aircraft brakes, and rocket nozzles.

Beryllium typically appears in military aircraft and space shuttle brake systems and missile reentry body structures, missile guidance systems, and satellite structures. The modulus-to-density ratio is higher than that of unidirectionally reinforced, *high-modulus* boron, carbon, and graphite fiber composites. Beryllium has an additional advantage because its modulus of elasticity is isotropic.

The largest-volume uses of beryllium metal are in the manufacture of beryllium–copper alloys and in the development of beryllium-containing moderator and reflector materials for nuclear reactors. Addition of 2% beryllium to copper forms a nonmagnetic alloy that is six times stronger than copper. These beryllium–copper alloys find numerous applications in industry as nonsparking tools, as critical moving parts in aircraft engines, and in the key components of precision instruments, mechanical computers, electrical relays, and camera shutters. Beryllium–copper hammers, wrenches, and other tools are employed in petroleum refineries and other plants in which a spark from steel against steel might lead to an explosion or fire.

One of the largest uses of beryllium metal is in nuclear reactors as a moderator to lessen the speed of fission neutrons and as a reflector to reduce leakage of neutrons from the reactor core. Beryllium is useful in nuclear applications because of its relatively high neutron-scattering cross section, low neutron-absorption cross section, and low atomic weight.

Another large-scale use of beryllium is in the manufacture of beryllium bronze, which has high tensile strength and a capacity for being hardened by heat treatment. Beryllium–copper molds are used in manufacturing plastic furniture with the appearance of wood-grain surfaces.

A small but important use of beryllium is in foil form as window material in x-ray tubes. Beryllium transmits x-rays 17 times better than any equivalent thickness of aluminum and 6–10 times better than Lindemann glass. This, together with its high melting point, makes possible the use of x-ray beams of greater intensity.

The lightweight, very high elastic modulus, and heat stability of beryllium make it an attractive construction material for use in aircraft and missiles. However, its lack of ductility is a drawback. Were it not for its toxicity and scarcity, beryllium would find use as a rocket fuel because it produces more heat energy per unit volume than any other element. In multistage missiles, a small weight reduction in the final stage, such as might be achieved by using beryllium in place of steel, permits a much larger weight reduction in the earlier stages in terms of fuel and structure. Research in the utilization of beryllium metal and beryllium-containing materials for aircraft and missiles

B

has been carried out very actively. These and other still developing applications together with the continuing uses of beryllium in nuclear technologies sustain the ever-mounting production levels of beryllium.

The metal is also very expensive. It is also used as an alloying element to produce beryllium–aluminum, beryllium–copper, and beryllium–nickel alloys.

Beryllium Alloys

Dilute alloys of base metals that contain a few percent of beryllium in a precipitation-hardening system are the principal useful beryllium alloys manufactured today. Although beryllium has some solid solubility in copper, silver, gold, nickel, cobalt, platinum, palladium, and iron and form precipitation-hardening alloys with these metals, the copper–beryllium system and, to a considerably lesser degree, the nickel–beryllium alloys are the only ones used commercially.

Other than the precipitation-hardening systems, small amounts of beryllium are used in alloys of the dispersion type wherein there is little solid solubility (aluminum and magnesium). Various amounts of beryllium combine with most elements to form intermetallic compounds. Development of beryllium-rich alloys has been chiefly confined to the ductile matrix beryllium–aluminum, beryllium–copper solid solution alloy with up to 4% copper, and dispersed-phase-type alloys having relatively small amounts of compounds (0.25%–6%), chiefly as beryllium oxide or intermetallics, for dimensional stability, elevated temperature strength, and elastic limit control.

Beryllium–Copper Alloys

Commercial alloys of the beryllium–copper group are divided into cast or wrought types usually and ternary combination such as copper–beryllium–cobalt. Alloys with higher beryllium content have high strength while those with low beryllium have high electrical and thermal conductivity.

Age-hardenable copper–beryllium–cobalt alloys offer a wide range of properties because they are extremely ductile and workable in the annealed condition and are strong and hard after precipitation or aging treatment. Cobalt is added to inhibit grain growth and provide uniform heat treatment response. These alloys also have inherent characteristics of substantial electrical and thermal conductivity and resistance to corrosion and wear, being protected by beryllium oxide films that impart this property to all materials containing beryllium. These age-hardenable alloys resist dimensional change and fatigue.

Primary applications are found in the electronics, automotive, appliance, instrument, and temperature-control industries for electric-current-carrying springs, diaphragms, electrical switch blades, contacts, connectors, terminals, fuse clips, and bellows (foil, strip, and wire), as well as resistance-welding dies, electrodes, clutch rings, brake drums, and switch gear (rod, plate, and forgings). With 1.5% beryllium or more, the melting point of copper is severely depressed and a high degree of fluidity is encountered, allowing casting of intricate shapes with very fine detail. The characteristic is important for plastic injection molds.

For special applications, specific alloys have been developed. Free machining and nonmagnetic alloys have been made, as well as high-purity materials. A precipitation-hardening beryllium–Monel for oceanographic application, containing about 30% nickel, 0.5% beryllium, 0.5% silicon, and the remainder copper, illustrates one of a series of alloys having strength, corrosion resistance to seawater, and antifouling characteristics.

New applications for structural, aerospace, and nuclear fields are submarine repeater cable housings for transoceanic cable systems, wind tunnel throats, liners for magnetohydrodynamic generators for gas ionization, and scavenger tanks for propane–Freon bubble chambers in high-energy physics research. Important developing applications for beryllium–copper are trunnions and pivot bearing sleeves for the landing gear of heavy, cargo-carrying aircraft, because these alloys allow the highest stress of any sleeve bearing material.

Beryllium–copper master alloys are produced by direct reduction of beryllium oxide with carbon in the presence of copper in an arc furnace. Because Be_2C forms readily, the master alloy is usually limited to about 4.0%–4.25% beryllium. The master alloy is remelted with additional copper and other elements to form commercial alloys. Rolling billets up to 680 kg have been made by continuous-casting techniques.

Beryllium–copper alloys can be fabricated by all the industrial metalworking techniques to produce principally strip, foil, bar, wire, and forgings. They can be readily machined and can be joined by brazing, soldering, and welding. Annealing, to provide high plasticity at room temperature, is accomplished by heat treating (from 790°C to 802°C for the high-beryllium alloys or 900°C–930°C for the low-beryllium alloys) and water quenching. Precipitation hardening is accomplished by heating to 400°C–480°C (low beryllium) and 290°C–340°C (high beryllium).

Alloys with Nickel and Iron

Nickel containing 2% beryllium can be heat-treated to develop their tensile strength of 1700 MPa with 3%–4% elongation. Little commercial use is made of the hard nickel alloys although they have been employed, principally as precision castings, for aircraft fuel pumps, surgical instruments, and matrices and diamond drill bits. Another nickel alloy with 2.0%–2.3% beryllium, 0.5%–0.75% carbon, and the balance of nickel and refractory metals has been used for mold components and forming tools for glass pressware of optical and container quality. Thermal conductivity, wear resistance, and strength, coupled with unusual machinability for a nickel–beryllium alloy, make this alloy particularly advantageous for glassware tooling.

Wrought beryllium–nickel contains about 2% beryllium, 0.5% titanium, and the balance nickel. Casting alloys contain a bit more beryllium (2%–3%) and, in one alloy, 0.4% carbon. Arsenic additions in the case of beryllium-copper alloys has mechanical properties-that vary widely, depending on temper condition—from 310 to 1586 MPa in tensile yield strength and Rb 72 to Rc 55 in hardness at room temperature. The alloys retain considerable yield strength at high temperature: 896–1172 MPa at 538°C. They also have good corrosion resistance in general atmospheres and reducing media. Because beryllium is toxic, special precautions are required in many fabricating operations. The wrought alloy is used for springs, bellows, electrical contacts, and feather valves and the casting alloys for molding plastics and glass, pump parts, seal plates, and metal-forming tools.

Attempts have been made to add beryllium to a number of ferrous alloys. Small amounts are used to refine grains and deoxidize ferritic steels in Japan, and promising properties have been developed for austenitic and martensitic steels. Stainless steels (iron–nickel–chromium) may be made maraging by adding 0.15%–0.9% beryllium, developing strengths as high as 1800 MPa as cast or 2300 MPa as rolled while retaining their oxidation- and corrosion-resistant characteristics. Amounts of 0.04%–1.5% beryllium have been added to various iron alloys containing nickel, chromium, cobalt, molybdenum, and tungsten for special applications such as watch springs.

Beryllium-Base Alloys

Three types of beryllium-base alloys are of interest. These consist of dispersed-phase types containing up to 4% beryllium oxide; ductile-phase or duplex alloys of beryllium and aluminum, particularly 38% aluminum in beryllium; and solid solution alloys of up to 4% copper in beryllium.

Dispersed-phase alloys containing oxides, carbides, nitrites, borides, and intermetallic compounds in a beryllium matrix are chiefly characterized by increased strength and resistance to creep at elevated temperatures. Major commercial alloys in the series are of the fine-grain, high-beryllium-oxide (4.25%–6%), hot-pressed types such as materials used for inertial guidance instruments characterized by high dimensional stability, high-precision elastic limit (55–103 MPa), and good machinability.

The 62% beryllium–38% aluminum alloy previously discussed under aluminum alloys was developed as a structural aerospace alloy to combine high modulus and low density with the machining characteristics of the more common magnesium-base alloys. This alloy in sheet form has at room temperature about 344 MPa ultimate strength, 324 MPa yield strength, and about 8% elongation. It is also produced as extrusions. It has a duplex-type microstructure characterized by a semicontinuous aluminum phase. Other alloys of the beryllium–aluminum type have been reported, but the 62% beryllium–38% aluminum alloy is the most used.

Intermetallic Compounds

Beryllium, combined with most other elements, forms intermetallic compounds with high strength at high temperature (up to 552 MPa modulus of rupture at 1260°C), good thermal conductivity, high specific heat, and good oxidation resistance. The beryllium oxide film formed by surface oxidation is protective to volatile oxides (molybdenum) and two elements of high reactivity (zirconium and titanium) at temperatures 1100°C–1500°C and for short times up to 1650°C.

The beryllides are of interest to the nuclear field, to power generation, and to aerospace applications. Evaluation of the intermetallics as refractory coatings, reactor hardware, fuel elements, turbine buckets, and high-temperature bearings has been carried out.

Beryllium Aluminum

The 62% beryllium–38% aluminum alloy developed by Lockheed Aircraft in the 1960s and called Lockalloy. Noted primarily for its light weight and high tensile modulus, thus high specific rigidity, it was used in the form of extrusions for missile skin stiffeners and inrolled sheet on the YF-12 reconnaissance aircraft. The alloy has since become available as a powder-metallurgy product from Brush Wellman Inc. as AlBeMet AM 162 rolled sheet and extruded bar and AlBeMet AM 162H hot isostatic press billet, which contains by weight 60%–64% beryllium, 36%–40% aluminum, and as much as 1 oxygen, 0.1% carbon, and 0.2% each of other metallics. Typical minimum tensile properties of rolled sheet are 379, 276 MPa yield strength, and 5% elongation. The density etc. to 2.123 kg/m³. For the hot isostatically pressed billet, typical minimums are 262, 193 MPa, and 2%, respectively.

Beralcast alloys from Starmet, formerly Nuclear Metals, include beryllium–aluminum 363, 191, 310, and MGA alloys. Having a density of 2160 kg/m³ and a tensile modulus of 202 GPa, 363, 191, and 310 alloys have three to four times the specific rigidity of a 356 aluminum alloy, AZ 91 magnesium, and cast Ti–6Al–4V titanium alloy. They also have one order of magnitude better vibration-damping capacity than 6061 aluminum alloy. Alloys 363 and 191 are for investment casting and contain 61.1%–68.6% beryllium. The 363 also has 2.65%–3.35% silver, 0.65%–1.35% cobalt, and 0.55%–0.95% germanium, with balance aluminum. The 191 has 27.5%–34.5% aluminum, 1.65%–2.5% silicon, and 1.65%–2.35% silver. Ultimate tensile strength is 290 and 197 MPa, tensile yield strength is 214 and 140 MPa, and elongation in 1 in. (2.54 cm) is 3% and 1.7%, respectively.

Beryllium Bronze/Copper

Alloys of copper with up to 3% beryllium. They can be precipitation hardened to a high strength. See preferred term *beryllium–copper*.

Beryllium Oxide

A colorless to white crystalline powder of the composition beryllium oxide, also called beryllia. It has a specific gravity of 3.025, a high melting point, about 2585°C, and a Knoop hardness of 2000. It is used for polishing hard metals and for making hot-pressed ceramic parts. Its high heat resistance and thermal conductivity make it useful for crucibles, and its high dielectric strength makes it suitable for high-frequency insulators. Single-crystal beryllia fibers, or whiskers, have a tensile strength above 6800 MPa.

Ceramic parts with beryllia as the major constituent are noted for their high thermal conductivity, which is about three times that of steel and second only to that of the high-conductivity metals (silver, gold, and copper). They also have high strengths and good dielectric properties. Properties of typical grades of beryllia ceramics are tensile strength, 96 MPa; compressive strength, 2068 MPa; hardness (micro), 1300 Knoop; maximum service temperature, 2400°C; dielectric strength, 5.8. Beryllia ceramics are costly and difficult to work with. Above 1650°C, they react with water to form a volatile hydroxide. Also, because beryllia dust and particles are toxic, special handling precautions are required. Beryllia parts are used in electronic, aircraft, and missile equipment. A more recent application has been the use of beryllia as thermocouple insulators and vacuum furnace equipment operating below 1650°C.

Beryllium oxide powder is available in three particle size ranges: (1) submicron 1–2 mm, used for fabricating both ceramic components and $BeO–UO_2$ nuclear fuel elements; (2) 2–8 mm, used primarily for fabricating beryllia bodies of 96%–99.5% purity; and (3) ultrahigh density grains of specific size distribution for admixing with resins and other organics to provide very high thermal conductivity coatings and potting compounds.

Beryllia ceramics have these characteristics: outstanding resistance to wetting and corrosion by many metals and nonmetals; mechanical properties only slightly less than those of 96% Al_2O_3 ceramics; valuable nuclear properties, including an exceptionally low, thermal neutron absorption cross section; and ready availability and a wide variety of shapes and sizes. Like Al_2O_3 and some other ceramics, beryllia is readily metallized by a variety of thick- and thin-film techniques.

Major markets for beryllium oxide ceramics are microwave tube parts such as cathode supports, envelopes, spacers, helix supports, collector isolators, heat sinks, and windows; substrates, mounting pads, and packages for solid-state electronic devices; and bores or plasma envelopes for gas lasers.

Other uses include klystron and ceramic electron tube parts, radiation and antenna windows, and radar antennae. The exceptional resistance of beryllia to wetting (and thus corrosion) by many molten metals and slags makes it suitable for crucibles for melting uranium, thorium, and beryllium.

The high general corrosion resistance of beryllia has helped it capture new applications in the chemical and mechanical fields. And other uses in aircraft, rockets, and missiles are predicted.

Beryllium oxide is tapped for nuclear reactor service because of its refractoriness, high thermal conductivity, and ability to moderate (slow down) fast neutrons. The *thermal* neutrons that result are more efficient in causing fusion of uranium-235. Nuclear industry uses for beryllia include reflectors and a matrix material for fuel elements. When mixed with suitable nuclear poisons, beryllium oxide may be a new candidate for shielding and control rod assembly applications.

The market for electrically insulating, heat-conductive encapsulants based on beryllia grain–polymer mixtures is both small and restricted. Although these composites have thermal conductivities 10–20 times higher than those of other filled plastics, the handling restrictions necessitated by the presence of beryllia limit their use.

Beryllia is used in ceramics to produce gastight glazes. Thin films of the oxide are used on silver and other metals to protect the metal from discoloration. Very thin films are not visible, but heavier films give a faint iridescence. Two other beryllium compounds used especially in chemical manufacturing are beryllium chloride, $BeCl_2$, a water-soluble white powder melting at 824°F (440°C), and beryllium fluoride, BeF_2, melting at 1472°F (800°C). Another beryllium compound, useful for high-temperature, wear-resistant ceramics is beryllium carbide, BeC_2. Berlox, of the National Beryllia Corporation, is a beryllium oxide powder in particle sizes from 80 to 325 mesh for flame-sprayed, heat- and wear-resistant coatings.

Bessemer Process

A process for making steel by blowing air through molten pig iron contained in a refractory lined vessel so as to remove by oxidation most of the carbon, silicon, and manganese. This process is essentially obsolete in the United States.

Beta (β)

The high-temperature allotrope of titanium with a body-centered cubic crystal structure that occurs above the β transus.

Beta Annealing

Producing a beta phase by heating certain titanium alloys in the temperature range of which this phase forms followed by cooling at an appropriate rate to prevent its decomposition.

Beta Brass

Alloys of copper with zinc in the range of 46%–49% that at ambient temperature comprise beta intermetallic compound.

Beta Compounds

See *electron compounds*.

Beta Eutectoid Stabilizer

An alloying element in titanium that dissolves preferentially in the β phase, lowers the α–β to β transformation temperature, and results in β decomposition to α plus a compound. This eutectoid reaction can be sluggish for some alloys.

Beta Fleck

Alpha-lean region in the α–β titanium microstructure significantly larger than the primary α width. This β-rich area has a β transus measurably below that of the matrix. Beta flecks have reduced amounts of primary α that may exhibit a morphology different from the primary α and the surrounding α–β matrix.

Beta (β) Gage

A gage consisting of two facing elements, a β-ray-emitting source and a β-ray detector. Also called beta-ray gage.

Beta Isomorphous Stabilizer

An alloying element in titanium that dissolves preferentially in the β phase, lowers the α–β to β transformation temperature without a eutectoid reaction, and forms a continuous series of solid solutions with β-titanium.

Beta (β) Loss Peak

In dynamic mechanical or dielectric measurement, the second peak in the damping curve below the melt, in order of decreasing temperature or increasing frequency.

Beta Particles

Particles with a mass similar to that of an electron. They may be positively or negatively charged with an energy up to 4 MeV, a velocity approaching that of light but limited penetrating capacity. They are emitted by a radioactive source undergoing beta decay.

Beta Ray

A ray of electrons emitted during the spontaneous disintegration of certain atomic nuclei.

Beta Structure

A Hume-Rothery designation for structurally analogous body-centered cubic phases (similar to beta brass) or electron compounds that have ratios of three valence electrons to two atoms. Not to be confused with a beta phase on a constitution diagram.

Beta Transus

The minimum temperature above which equilibrium α does not exist in titanium alloys. For β eutectoid additions, the beta transus ordinarily is applied to hypoeutectoid compositions or those that lie to the left of the eutectoid composition.

Betts Process

A process for the electrolytic refining of lead in which the electrolyte contains lead fluosilicate and fluosilicic acid.

Bevel

See preferred term, *corner angle*, and also *face mill*.

Bevel Angle

The angle formed between the prepared edge of a member and a plane perpendicular to the surface of a member.

Bevel Flanging

Same as *flaring*.

Bevel Groove

See *groove weld*.

BHN

Brinell hardness number.

Bias

A systematic error inherent in a method (such as temperature effects and extraction inefficiencies) or caused by some artifact or idiosyncrasy of the measurement system (such as blanks, contamination, mechanical losses, and calibration errors). Bias may be both positive and negative, and several types can exist concurrently, so that the net bias is all that can be evaluated except under certain conditions.

Bias Fabric

Fabric consisting of warp and fill fibers at an angle to the length of the fabric.

Biaxial

Applying to or acting on two axes, for example, stresses.

Biaxial Load

A loading condition in which a specimen is stressed in two directions and its plane.

Biaxial Stress

A state of stress in which only one of the *principal stresses* is zero, the other two usually being in tension.

Biaxial Winding

In filament winding, a type of winding in which the helical band is laid in sequence, side by side, with crossover of the fibers eliminated.

Bidirectional Laminate

A reinforced plastic laminate with the fibers oriented in two directions in its plane. A cross laminate. See also *unidirectional laminate*.

Bidirectional Seal

A seal that is designed to seal equally well when the pressure is applied from either direction.

Bifilar Eyepiece

A filar eyepiece with motion in two mutually perpendicular directions.

Bifurcated Forked

A bifurcated rivet has a solid head and a forked shank that is inserted through the sheet materials being joined in the prongs of the fork bent apart.

Bifurcation

The separation of a material into two sections.

Big-End Bearing

A bearing at the larger (crankshaft) end of a connecting rod in an engine. Also known as *bottom-end bearing*, *crankpin bearing*, and *large-end bearing*. See also *little-end bearing*.

Billet

(1) A semifinished section that is hot rolled from a metal *ingot*, with a rectangular cross section usually ranging from 105 to 230 cm^2 (16–36 in.^{2}), the width being less than twice the thickness. Where the cross section exceeds 230 cm^2 (36 in.^{2}), the term *bloom* is properly but not universally used. Sizes smaller than 105 cm^2 (16 in.^{2}) are usually termed bars. (2) A solid semifinished round or square product that has been hot worked by forging, rolling, or extrusion. See also *bar*.

Billet Mill

A primary rolling mill used for making steel billets.

Bimetal Bearing

A bearing consisting of two layers. Bimetal bearings are usually made with a layer of bearing alloy on a bronze or steel backing.

Bimetal Casting

A casting made of two different metals, usually produced by *centrifugal casting*.

Bimetallic

Involving two metals.

Binary

Having two components. The term is often used casually in a metallurgical context in reference to binary phase diagrams.

Binary Alloy

An alloy containing only two component elements.

Binary System

The complete series of compositions produced by mixing a pair of components in all proportions.

Binder (Metals and Ceramics)

(1) In founding, a material other than water, added to foundry sand to bind the particles together, sometimes with the use of heat. (2) In powder technology, a cementing medium: either a material added to the powder to increase the green strength of the compact, which is expelled during sintering, or a material (usually of relatively low melting point) added to a powder mixture for the specific purpose of cementing together powder particles that alone would not sinter into a strong body.

Binder (Plastics)

The resin or cementing constituent (of a plastic compound) that holds the other components together. The agent applied to fiber mat or preforms to bond the fibers before laminating or molding. (2) A component of an adhesive composition that is primarily responsible for the adhesive forces that hold two bodies together. See also *extender and filler.*

Binder Metal

A metal used as a binder. An example would be cobalt in cemented carbides.

Binder Phase

The soft metallic phase that cements the carbide particles in cemented carbides. More generally, a phase in a heterogeneous sintered material that gives solid coherence to the other phase(s) present.

Bingham Solid

An idealized form of solid that begins to flow appreciably only when a certain stress, called the yield stress or yield point, has been exceeded. The solid subsequently flows at a rate proportional to the difference between the applied stress and this yield stress. Many greases can be regarded as Bingham solids.

Binodal Curve

In a 2D phase diagram, a continuous line consisting of both of the pair of conjugate boundaries of a two-phase equilibrium that join without inflection at a critical point. See also *miscibility gap.*

Biological Corrosion

Deterioration of metals as a result of the metabolic activity of microorganisms. Also known as biofouling.

Biomaterials

Materials used to repair, restore, or replace damaged or diseased tissue, or those that comprise parts of artificial organs, artificial tissues, or prostheses. The use of biomaterials dates back to antiquity. Hair, cotton, animal sinew, tree bark, and leather have been used as natural suture materials for almost 4000 years. Synthetic biomaterials are composed of metals, ceramics, polymers, and their composites, and they are often called biomedical materials to differentiate them from natural ones. Gold plates for skull repair were in use in 1000 BC and gold-wire sutures as early as 1550. Besides the mechanical properties of strength, elasticity, and durability, biomaterials need to be nontoxic, sterilizable, and biocompatible. Biocompatibility implies that the material will be inert when in contact with the body and not activate the immune system or lead to blood coagulation. Biomaterials are also categorized as bioabsorbable or nonbioabsorbable. Among the former are the peptides (collagen, fibrin, albumin, and gelatin), hemiacetals (starch, hyaluronic acid, chitin), esters (poly-β-hydroxybutyrate and polymalic acid), and phosphates (DNA and RNA). Synthetic polymers that are bioabsorbable are polylactides, polylactones, polycarbonates, poly-α-cyanoacrylates, polyphosphazenes, and polyanhydrides.

No natural suture material has been so prevalent as catgut, derived from the small intestines of animals, usually the outer serosal layer in cattle or the submucosal layer in sheep. For surgical catgut, or gut, the intestinal tracts of animals are slit lengthwise; the resultant ribbons are twisted into bonded strands and then sterilized by electron-beam irradiation. Catgut becomes too stiff to handle when dry, so it is packed in aqueous alcohol. When the material is treated with a chromic salt solution to prolong in vivo strength, it is called chromic catgut. For a more uniform surface, chromic catgut is treated with glycerin to form glycerin catgut. Catgut varies widely in quality and elicits relatively severe tissue reactions. It is absorbed by the body in 90 days.

A more uniform suture material has been prepared by wind-spinning dispersions of purified collagen into strands. Three kinds of synthetic, absorbable polyester sutures are now marketed. The first is a glycolide homopolymer, also known as poly(glycolic acid) (PGA), obtained by ring-opening polymerization of cyclic diester monomers at high vacuum using tin catalysts. The suture is formed by melt extrusion spinning, followed by hot drawing to give a high orientation and crystallinity, and annealed to improve dimensional stability. It is braided into a multifilament and can be dyed or coated. It is absorbed by the body in 90 days.

Polymer matrices are receiving attention for cell transplant devices to regenerate human tissue in wound healing and reconstruction surgery. Foreskin tissue, type I collagen, a protein that strengthens tissue and is available from animals, has been combined with chondroitin sulfate, a carbohydrate polymer, to form largely porous insoluble matrixes with controlled rate of degradation. Concern over adverse biological reactions of the collagen, however, has prompted the development of matrixes based on synthetic polymers, such as biodegradable polyester, which is widely used for absorbable sutures. Also being studied are hybrids of a degradable polymer such as hyaluronic acid, an anionic polysaccharide of the skin, covalently modified by a small peptide ligand to control biological response. For cartilage tissue, polyesters in the family of polylactides, polyglycolides, and their copolymers are of interest. They degrade mainly by hydrolysis to yield natural metabolic intermediates, most of which then convert to carbon dioxide and water. These polymers are sufficiently strong to make plates and screws for setting bone fractures.

B

Biofix screws, from a Bioscience of Finland, are made of either poly(L-lactide) (PLLA) semicrystalline polyesters or PGA thermoplastic resins and are used to reassemble broken bones. The screws, which are strengthened with sutures of the same materials, hold bone fragments in place during healing, then break down, and are absorbed by the body. An injectable paste, from monocalcium phosphate monohydrate, α-tricalcium phosphate, calcium carbonate, and sodium phosphate solution, speeds treating broken bones with less hardware and less-invasive procedures. After injection, the paste hardens within 10 min and, within 12 h, forms the middle phase of bone—a dahllite, or carbonated apatite—that is at least as strong as natural porous bone.

Joint replacement is one of the largest users of biomaterials. The first implants, artificial hips, were made from 316L low-carbon stainless steel, followed by *chromium–cobalt*, and in the 1970s by titanium. Ti-6Al-4V alloy is the most commonly used, but Kobe Steel's Ti-15Zr-4Ta-4Cb-0.2Pd-0.2O-0.05N is also biocompatible, stronger, and about as ductile. Initially, metal femurs with ball-shaped ends were inserted into the acetabulum, the hip socket, as a direct replacement for the natural femur. Now, the acetabulum itself may be made of metal cups coated with a self-lubricating plastic with good wear properties, such as high-molecular-weight (HMW) polyethylene. A bioactive silicate glass coating and an enameling process for applying ultrathin layers of it to cobalt and titanium alloy implants have been developed at Lawrence Berkeley National Laboratory to enable the implants to bond with bone. Europeans commonly used alumina for femoral balls and acetabular cups, and in the United States, Richards Medical is marketing an alumina-capped femur. Astro Met is developing a partially stabilized zirconia for the same application. Osteonics, of Stryker Corp., uses wrought cobalt alloy compositionally modulated multilayer (CMM) for the femoral cap of hip joints. The cap fits over the top of a hip stem made of either the cast or wrought alloy or a titanium alloy. The cap is within an acetabular cup made of ultrahigh-molecular-weight polyethylene that is mounted in a hemispherical titanium shell. The combination serves as the articulating couple of the hip joint. To reduce friction and wear of the cup, nitrogen ions are implanted in the cap surface. A thin coating of pliable titanium inhibits cracking of yttria-doped zirconia hip joints. It also lessens friction and its related wear debris. Lifecore Biomedical is testing graphite–polysulfone composites, and Zimmer is sheathing titanium in polysulfone or polyether ether ketone (PEEK). PEEK-Optima LT, implant version of Victrex's polyaryletherketone (PAEK) that can match bones' stiffness and toughness, is intended for hip, knee, spine, dental, heart valve, and other body parts. Hyaluronic acid is a viscoelastic liquid used to lubricate the traumatized joints in arthritic patients.

Thermoplastic aromatic polyetherurethanes are used for implants because of their resistance to flexural fatigue, self-bonding characteristics, and good tensile strength. Some grades, however, are degraded by enzymes. Potential uses include insulators for pacemaker leads, long-term in dwelling catheters, and pump diaphragms for artificial hearts. A composite of the copolymer and a silane is seen as a potential replacement for silicone breast implants. Being stronger, the composite is expected to have less tendency to rupture. Also, unlike silicone, it would permit x-ray examination.

A zirconium–palladium–ruthenium alloy developed at National Institute of Standards and Technology (NIST), to extend the life of medical and dental implants, holds promise for hip and knee implants. Extremely wear-resistant, it may reduce the amount of wear debris generated by other implant materials, which, even if inert, can degrade surrounding tissue. A dentin-regenerative protection based on OP-1, a protein that appears to stimulate uncommitted cells to lay down dentin, may preclude a root-canal operation.

Developed by Creative BioMolecules, Inc., it may also be useful in bone reconstruction and in treating osteoporosis. Bioglass, of U.S. Biomaterials, consists of silicon, sodium, and natural bone ingredients—calcium and phosphorus. It can be used to replace middle-ear bone and tooth roots.

To anchor artificial bones in place, they are layered with a porous mating that encourages the growth of natural bone tissue and blood vessels. Because these materials interact with the body, they are called bioactive materials. A porous coating of cobalt–chromium or titanium beads on metal femurs is bioactive. Calcium phosphate ceramics and coatings are osteophilic (bone-loving); their porosity provides the templates on which new bone growth can form for natural healing. Calcium phosphate cement, a dental material initially, can now be used for neurosurgical applications, such as rebuilding part of a person's face lost to cancer, NIST reports. The cement, which the body readily accepts, can replace a section of bone and serve as scaffolding around which new bone forms in the same shape. At present, use is limited to motionless and stress-free body areas.

Hard tissue replacement (HTR) polymer from HDR Sciences, a division of U.S. Surgical, is a porous plastic bead made from polyhydroxyethylmethacrylate and polymethylmethacrylate. It has a negative charge that attracts cell formation in bones and is hydrophilic, so it stays where surgeons place it. It is an open-cell lattice material reinforced with tantalum to replace damaged vertebral bodies and facilitate spinal fusion. It can be implanted between two vertebrae, where bone will infiltrate the porous structure and form a bony fusion at the joint. Hydroxyapatite, a ceramic similar to bone and which bonds to bone naturally, is used to coat bioactive implants. It is applied by plasma spraying, although high-velocity oxyfuel deposition also may be suitable and advantageous due to its lower processing temperature and ease of control. In coating titanium, a common implant material, good results were achieved with high velocity oxygen fuel (HVOF) at the Interdisciplinary Research Center in Biomaterials at Queen Mary College, London. Artificial blood is bioactive. One of the earliest and best known is Fluosol, a perfluorocarbon blood substitute made of perfluorodecalin and perfluorotripropylamine.

Pyrolytic carbon, a pure, glassy carbon developed originally to encapsulate fuel for nuclear reactors, is used for making heart valves because it is inert and extremely wear-resistant; it is also compatible with blood. The valves in the Jarvik-7 artificial heart are made from this material. Medtronics produces highly polished titanium for heart valves and for insulin delivery pumps.

Poly(methylmethacrylate) (PMMA) lenses are implants made of polymethylmethacrylate and are commonly used for patients suffering from severe cataracts. Recent improvements include lenses made of silicone rubber or of hydrogel, a viscous, jellylike polymer. These require a much smaller incision in the eye during insertion, because they are *foldable*. Silicone has been widely used for breast implants, a controversial use that has led to lawsuits by women, alleging leaks from the implants have injured their health. Sulfoxide hydrogel may extend the wear of soft contact lenses and reduce the chance of eye infection by enhancing water in the eye without increasing protein buildup. Excess water in the lens makes it softer, thus more comfortable, while providing oxygen to the eye.

Elastomeric polypeptides that will do work in response to changes in the body's chemical potential have been developed at the School of Medicine, University of Alabama. Polypentapeptide is the parent of this class of biomaterials. They are prepared by solution- and solid-phase peptide synthesis. Elastomeric polypeptides can match the compliance of natural biological tissues and can be modified to obtain desirable tissue reactions to the cellular and enzymatic levels. They can be made to contract and relax in response to changes in chemical potential analogous to the production of motion in living organisms. Some of their possible biomedical applications are synthetic arteries and ligaments, burn cover materials, and targeted drug delivery systems. In many other product areas, the materials can function as sensors, based on the thermomechanical and chemomechanical transduction properties.

Bipolar Electrode

An *electrode* in an *electrolytic cell* that is *not* mechanically connected to the power supply but is so placed in the electrolyte, between the *anode* and *cathode*, that the part nearer the anode becomes cathodic and the part nearer the cathode becomes anodic. Also called intermediate electrode.

Bipolar Field

A longitudinal magnetic field that creates two magnetic poles within a piece of material. Compare with *circular field*.

Birefringence

A double-refraction phenomenon in anisotropic materials in which an unpolarized beam of light is divided into two beams with different directions and relative velocities of propagation. The amount of energy transmitted along an optical path through a crystal that exhibits birefringence becomes a function of crystalline orientation.

Birefringent Crystal

A crystalline substance that is anisotropic with respect to the velocity of light.

Birotational Seal

A seal that is designed for applications in which a shaft rotation is in either direction.

Biscuit (Metals)

(1) An upset blank for drop forging. (2) A small cake of primary metal (such as uranium made from uranium tetrafluoride and magnesium by bomb reduction). Compare with *derby* and *dingot*.

Biscuit (Plastics)

See *cull* and *preform*.

Bismaleimide (BMI)

A type of polyamide that cures by an addition rather than a condensation reaction, thus avoiding problems with volatile formation, and which is produced by a vinyl-type polymerization of a prepolymer terminated with two maleimide groups. Intermediate in temperature capability between epoxy and polyimide.

Bismuth

Bismuth (Bi) is a brittle, crystalline metal with a high metallic luster with a distinctive pinkish tinge. The metal is easily cast but not readily formed by working. Within a narrow range of temperature, around 225°C, it can be extruded. Its crystal structure is rhombohedral.

It is one of the few metals that expand on solidification; expansion is 3.3%. The thermal conductivity of bismuth is lower than that of any metal, with the exception of mercury.

Bismuth is the most diamagnetic of all metals (mass susceptibility of -1.35×10^6). It shows the greatest Hall effect (increase in resistance under the influence of a magnetic field). It also has a low capture cross section for thermal neutrons (0.034 barn).

Bismuth is inert in dry air at room temperature, although it oxidizes slightly in moist air. It rapidly forms an oxide film at temperatures above its melting point, and it burns at red heat, forming the yellow oxide, Bi_2O_3. The metal combines directly with halogens and with sulfur, selenium, and tellurium; however, it does not combine directly with nitrogen or phosphorus.

Bismuth is not attacked at ordinary temperatures by air-free water, but it is slowly oxidized at red heat by water vapor. Bismuth does not dissolve in nonoxidizing acids but does dissolve in HNO_3 and in hot concentrated H_2SO_4. The formation of intermetallic compounds involves mainly the strongly electropositive metals.

Bismuth combined with a number of metallic elements forms a group of interesting and useful low-melting alloys. Some of the lowest melting of these are as follows:

	Melting Point (°C)
49.5 bismuth, 10.1 cadmium, 13.1 tin, 27.3 lead	70
49.0 bismuth, 12.0 tin, 18 lead, 20 indium	57
44.7 bismuth, 5.3 cadmium, 8.3 tin, 22.6 lead, 19.1 indium	47

The fusible alloys are used in many ingenious ways, for example, sprinkler-system triggering devices, bending pipes, anchoring tools during machining, and accurate die patterns.

Bismuth metal (0.1%) is also added to cast iron and steel to improve machineability and mechanical properties. An alloy of 50% bismuth and 50% lead is added to aluminum for screw machine stock, to increase machineability.

A permanent magnet (bismanol) with excellent resistance to demagnetization is produced from manganese and bismuth.

The development of refrigerating systems depending on the Peltier effect for cooling uses a bismuth–tellurium or selenium alloy for thermocouples. Bismuth telluride is used extensively for thermoelectric cooling and for low-temperature thermal electric power production.

Bismuth is playing an important role in nuclear research. Its high density gives it excellent shielding properties for gamma rays while its low thermal neutron capture cross section allows the neutrons to pass through. For investigations in which it is desired to irradiate objects, that is, animals, with neutrons that protect them from gamma rays, castings of bismuth are used as neutron windows in nuclear reactors.

Bismuth has been proposed as a solvent-coolant system for nuclear power reactors. The bismuth dissolves sufficient uranium so that, when the solvent and solute are pumped through a moderator (graphite), criticality is reached and fission takes place. The heat generated from the fission reaction raises the temperature of the bismuth. The heated bismuth is then pumped to conventional heat exchangers producing the steam power required for eventual conversion to electricity.

The advantages of such a reactor are that (1) it has potential for producing low-cost power, (2) it has an integrated fuel processing system, and (3) it converts thorium to fissionable uranium.

Another important use of bismuth is in the manufacture of pharmaceutical compounds. Various bismuth preparations have been employed in the treatment of skin injuries; alimentary diseases, such as diarrhea and ulcers; and syphilis. The oxide and basic nitrate are perhaps the most widely used compounds of bismuth. The trioxide is used in the manufacture of glass and ceramic products, and the basic nitrate is used in the porcelain painting to fire on gilt decoration.

Bit

A rotating tool with cutting edges at its tip for drilling holes.

Bit (Soldering)

That part of the soldering iron, usually made of copper, which actually transfers heat (and sometimes solder) to the joint.

Bit Soldering

See preferred term *iron soldering.*

Bitter Pattern

A technique for revealing the magnetic domains at an iron or steel surface. The surface is covered with a suspension of fine magnetic powder in a suitable carrier such as white spirit and the component is then magnetized. The magnetic powder then delineates the domain boundaries. The technique is essentially the same as that used for nondestructive crack detection.

Bitumen

Asphalt-like polymer.

Bituminous Coatings

Bitumens have been defined as mixtures of hydrocarbons of natural or pyrogenous origin or combinations of both (frequently accompanied by their nonmetallic derivatives), which can be gaseous, liquid, semisolid, or solid in which are completely soluble in carbon disulfide.

Bitumens used in the manufacture of coatings are of the semisolid and solid variety and are derived from three sources:

1. Asphalt produced by the distillation of petroleum
2. Naturally occurring asphalts and asphaltites
3. Coal tar produced by the destructive distillation of coal

It is customary to classify bituminous coatings by their application characteristics as well as by their generic composition, although the coatings can be divided into two classes depending on whether or not they require heating prior to application:

1. *Hot-applied coatings:* These are either 100% bitumen or bitumen coatings blended with selected fillers. A common loading for coatings employing fillers is 10%–20% filler. Hot-applied coatings are brought to the desired application viscosity by heating. The majority of buried pipelines are coated with this type of bituminous coating.
2. *Cold-applied coatings:* These employ both solvents and water to attain the desired application viscosity. A wide range of solvents is used and the choice depends mainly

on the drying characteristics desired and the solvent power required to dissolve the particular bitumen being used. Various fillers are also used in cold-applied coatings to obtain specific applications and end-use properties.

Bituminous coatings can be formulated from many combinations of bitumens, solvents, or dispersing agents and fillers. This makes possible a great variety of end products to meet application and service requirements. The coatings that can be produced range from thin-film (3 mils) coatings to protect machined parts in storage up to thick (100 mils), tough coatings to protect buried pipelines.

As with all other coatings, the conditions of the surface to which a bituminous coating is applied are an important, life-determining factor. A good sandblast is preferred, especially if the surface is badly corroded and the exposure is severe. In any case the surface should be free of moisture, grease, dust, salts, loose rust, and poorly adherent scale. A thin, penetrating bituminous primer can be beneficial on rusty surfaces that can only be cleaned by wire brushing and scraping.

End-Use Requirements

The application will dictate which performance properties of a coating should be given greatest consideration.

Service Temperature

Many applications of bituminous coating require a moderate surface temperature range, often no greater than that caused by weather changes. However, other applications, such as coatings for chemical processing vessels, may require much wider service temperature ranges. In any event, it is possible to obtain good performance with bituminous coatings over a range of −73°C to 163°C.

Thermal and Electrical Insulation

Bitumens themselves are relatively poor thermal insulators. However, by using low-density fillers, coatings with good insulation properties can be formulated. These coatings both protect from corrosion and provide thermal insulation. An added advantage of the coatings is that the insulating material (low-density filler) is completely surrounded by bitumen and is permanently protected from moisture. Thus, they are not subject to a loss in efficiency (as our conventional insulating systems) from damage or failure of the protecting vapor barrier.

Bitumens are naturally good electrical insulators. This is an important consideration in systems using cathodic protection as a complementary corrosion-prevention device.

Abrasion Resistance

Many bituminous coatings need to have high abrasion resistance. Automotive undercoatings, for example, need high abrasion resistance because they are continually buffeted by gravel and debris thrown up by the wheels of the vehicle.

Abrasion-resistant coatings are also used to protect interior surfaces of railroad cars or other vessels handling chemical solids or abrasive slurries.

Weathering Resistance

Asphalt-base bituminous coatings generally weather better than coal-tar-based coatings. Also, there is quite a wide difference in the performance of asphalt coatings derived from different petroleum crudes. By critically selecting the crude and the processing method, asphalt coatings can be formulated that will weather for many years.

In industrial areas, corrosive solids, solutions, and vapors will affect weathering performance. In general, the resistance of bituminous coatings to corrosive media is equal to that of the best organic coatings. Bituminous coatings have good resistance to dilute hydrochloric, sulfuric, and phosphoric acids as well as to sodium hydroxide. They also have good resistance to solutions of ammonium nitrate and ammonium sulfate. However, they have poor resistance to dilute nitric acid, and most coatings are not resistant to oils, greases, and petroleum solvents.

Mechanical Impact and Thermal Shock

Bituminous coatings generally have good adhesion when subjected to mechanical impact. However, where severe mechanical impact is expected, the coatings should first be field-tested or tested in the laboratory under simulated surface conditions.

Resistance to thermal shock is an important consideration in coatings used on some types of processing equipment. Laboratory tests in which a coated panel is transferred back and forth between hot and cold chambers can be used to predict field behavior. Bituminous coatings are available that will withstand thermal shock over a temperature range of −51°C to 60°C.

Types of Coatings

Because bitumens are very dark in color, heavy loadings of pigments are required to produce colors other than black. However, if color is an important consideration, certain colored paints, including whites, can be used as topcoats. Granules can also be blown into the coating before it has completely cured to produce a wide range of decorative effects.

Thin-Film Coatings

Coatings less than 6 mils thick are arbitrarily included in this group. For the most part, these are solvent cutback coatings. They are black except in the case of aluminum paints employing bituminous bases. The coatings are inexpensive and can be used to give good protection from corrosion when color is not important.

Asphalt coatings of this type are used extensively to protect machined parts in storage. Because the coatings retain their solubility of low-cost petroleum solvents even after long weathering, kerosene or similar solvents can be used to remove a coating from the protected part just prior to placing it in service. Coal-tar-based coatings are much more difficult to dissolve and cannot be used for this purpose. However, this property makes coal tar useful for protecting crude-oil tank bottoms and in other applications requiring resistance to petroleum fractions.

Industrial Coatings

Heavy-bodied industrial coatings incorporate low-density and fibrous mineral fillers; coatings can be formulated that will not slump or flow on vertical surfaces when applied as thick as 250 mils. However, they are usually used in thicknesses from 6 to 120 mils.

The coatings are used extensively in industrial plants to protect tanks and structural steel from such corrosive environments as acids, alkalis, salt solutions, ammonia, sulfur dioxide, and hydrogen sulfide gases.

Industrial coatings are also used in large volume by the railroad industry. Complete exteriors of tank cars carrying corrosive liquids are often coated and provide good protection of the saddle area where spillage is likely to occur. Coatings on the exteriors and interiors of hopper cars in dry chemical service also provide protection from both corrosive action and abrasive wear.

Railroad car bituminous cements are used to seal sills and joints in boxcars. An application of this material followed by an overcoating of granules makes an excellent roofing system for railroad cars.

Coatings for Use Over Insulation

Practically all industrial insulating materials must be protected from the weather and moisture; otherwise they would lose their efficiency. Bituminous coatings formulated to have low rates of moisture vapor transmission give best results on installations operating at low (−73°C) to moderate (82°C) temperatures. Coatings that allow a higher rate of moisture transmission (breathing type) are needed to protect the insulation on systems operating at 82°C and above. This is necessary so that moisture trapped beneath the coating can escape when the unit is brought to operating temperatures.

Thermal Insulating Coatings

Low-density fillers can be employed in bituminous mastics to produce coatings with relatively good insulating values; a k value of 0.6 Btu/ft^2/h/°F is typical. Insulating coatings are usually applied somewhat more quickly than conventional mastics to obtain the insulating value desired. They are commonly used in thicknesses of 252–275 mils, and because of their thickness and resiliency, they have excellent resistance to mechanical damage.

Automotive Underbody Coatings

These are mastic-type coatings containing fibrous and other fillers. They are used to coat the undersides of floor panels, fenders, gasoline tanks, and frames to protect against corrosion and provide sound deadening and joint sealing.

The coatings have high resistance to deicing salts, moisture, and water. They also have *sound-deadening properties that noticeably* reduce the noise level inside an automobile. This provides for a more pleasant and less fatiguing ride. The sealing and bridging action of the coatings is also especially effective in reducing drafts and dust infiltration.

Sound-Deadening Coatings

High-efficiency, sound-deadening coatings can be formulated from selected resinous bases and high-density fillers. They have better sound-deadening properties than automotive underbody coatings and are used on the wall, roof, and door panels of automotive equipment where sound deadening is the primary need, rather than abrasion resistance or protection from corrosion. They are also used on railroad passenger cars, house trailers, stamped bathtubs, kitchen sinks, air-conditioning cabinets, and ventilation ducts.

Pipe Coatings

Industrial coatings are excellent for protection of pipe aboveground. However, the environment of underground exposure and the complementary use of cathodic protection make it necessary to use specially designed coatings. The stresses created by shrinking and expanding soil require that the coating be very tough. Rocks and other sharp objects can be expected to cause high localized pressures on the coating surface. A coating must have good cold flow properties to resist penetration by objects, which can cause localized pressures as high as 690 MPa.

Cathodic protection (and impressed negative electrical potential) is widely used to prevent the corrosion processes from occurring at flaws

B

in the pipe coatings. Both asphalts and coal tars are good electrical insulators and make excellent coatings for cathodic protection applications.

Coatings for pipe are usually of the hot application type. Application can be made at the mill and a special pipe-coating yard, or over the ditch, depending on the terrain, size of pipe, and other factors. The coating may be given added strength by embedding it with a glass fabric while it is still hot. Outer wrappings of rag, asbestos, and glass felts are sometimes used to give added resistance to damage by soil stresses.

Application

Bituminous coatings that are cut back with solvents or emulsified with water can be produced to consistencies suitable for application by dipping, brushing, spraying, or troweling at ambient temperatures.

Dipping is usually used to coat small parts. As a rule coating viscosity is adjusted to produce a thickness of 1–6 mils.

Brushing is used on areas that cannot be reached by spraying and on jobs that do not warrant setting up spraying equipment. Coating thicknesses can range from 1 to 65 mils.

Spraying is the most popular method for applying cold coatings. The thickness required in one application determines the consistency of the formulation, and thicknesses of 1–250 mils are obtainable by spraying. Conventional paint-spray equipment can be used for coatings up to 6 mils thick. Heavier coatings require the use of mastic spray guns fed from pressure pots or heavy-duty pumps. Heated vessels and feed lines can also be used to decrease viscosity and permit faster application and buildup of thicker films in one application.

Troweling is usually used in inaccessible areas or where it is necessary to produce a very heavy coating in one application. Trowel coats are usually applied in thicknesses above 250 mils.

Bituminous coatings can also be applied hot without the need for any diluents. Such coatings are widely used on piping in thicknesses of about 95 mils. They are heated to 177°C–288°C and then pumped into a special apparatus that surrounds and travels along the pipe.

Bivariate Equilibrium

A stable state among several phases equal to the number of components in a system and in which any two of the external variables of temperature, pressure, or concentration may be varied without necessarily changing the number of phases. Sometimes termed divariant equilibrium.

Black Annealing

Box annealing or pot annealing ferrous alloy sheet, strip, or wire to impart a black color to the oxidized surface. See also *box annealing*.

Black Light

Electromagnetic radiation not visible to the human eye. The portion of the spectrum generally used in fluorescent inspection falls in the ultraviolet region between 330 and 400 nm, with the peak at 365 nm.

Black Liquor

The liquid material remaining from pulpwood cooking in the soda or sulfate papermaking process.

Black Marking

Black smudges on the surface of a pultruded plastic product that results from excessive pressures in the die when the pultrusion is

rubbing against it or unchromed die surfaces and that cannot be removed by cleaning or scrubbing or by wiping with solvent.

Black Oxide

A black finish on a metal produced by immersing it in hot oxidizing salts or salt solutions.

Black Spot(Ting)

Casting porosity filled with carbon-rich material. The term is usually used when the defects are exposed by machining.

Blackbody

A hypothetical "body" that completely absorbs all incident radiant energy, independent of wavelength and direction, that is, neither reflects nor transmits any of the incident radiant energy.

Blackbody Radiation

The radiation emitted by a theoretical perfect *blackbody*.

Blackheart Malleable

See *malleable cast iron*.

Blacking

Carbonaceous materials, such as graphite or powdered carbon, usually mixed with a binder and frequently carried in suspension in water or other liquid used as a thin facing applied to surfaces of molds or cores to improve casting finish.

Blacksmith Welding

See preferred term *forge welding*.

Bladder

An elastomeric lining for the containment of hydroproof or hydroburst pressurization medium and filament-wound structures.

Blade-Setting Angle

See preferred term *cone angle*.

Blank

(1) In forming, a piece of sheet metal, produced in cutting dies, that is usually subjected to further press operations. (2) A pressed, presintered, or fully sintered powder metallurgy compact, usually in the unfinished condition and requiring cutting, machining, or some other operations to produce the final shape. (3) A piece of stock from which a forging is made, often called a *slug* or *multiple*. (4) Any article of glass on which subsequent forming or finishing is required. (5) In various test procedures especially chemical analysis, a sample that does not contain the material being investigated or analyzed but which is subject to the full test procedure to identify any experimental error or contamination.

Blank Carburizing

Simulating the carburizing operation without introducing carbon. This is usually accomplished by using an inert material in place of the carburizing agent or by applying a suitable protective coating to the ferrous alloy.

Blank Nitriding

Simulating the nitriding operation without introducing nitrogen. This is usually accomplished by using an inert material in place of the nitriding agent or by applying a suitable protective coating to the ferrous alloy.

Blanket

Fiber or fabric plies that have been laid up in a complete assembly and placed on or in the mold all at one time (flexible bag process). Also, the type of bag in which the edges are sealed against the mold.

Blankholder

(1) The part of a drawing or forming die that holds the workpiece against the drawer ring to control metal flow. (2) The part of a drawing or forming die that restrains the movement of the workpiece to avoid wrinkling or tearing of the metal.

Blanking

The operation of punching, cutting, or shearing a piece out of stock to a predetermined shape.

Blanking Press

A machine for stamping out *blanks*.

Blast Furnace

A vertical shaft smelting furnace for producing pure pig iron from iron ore. The charge, loaded at the top, comprises iron ore, limestone as flux, coke as fuel, and reducing agent. The preheated air blast injected around the periphery at about a quarter height increases the temperature derived from burning the coke to about 1800°C. Carbon monoxide formed by the coke reduces the iron oxide to molten iron. The limestone combines with the contaminants to produce a fluid slag. The slag and impure pig iron are tapped off separately close to the bottom. Similar processes are used for some other metals. For example, where the temperature can be lower, as in smelting of copper, lead, and tin ores, a smaller furnace is economical, and preheating of the blast is not required.

Blasting or Blast Cleaning

A process for cleaning or finishing metal objects with air blast or centrifugal wheel that throws abrasive particles against the surface of the workpiece. Small, irregular particles of metal are used as the abrasive in grit blasting; sand, in sandblasting; and steel, in shot blasting.

Blebs

Blebbing growths of material on the exterior of ingots or castings as a result of inverse segregation. As they tend to contain a disproportionate amount of alloying and impurity, they may be relatively hard and brittle causing problems in subsequent operations.

Bleed

(1) To give up color when in contact with water or a solvent. Undesirable movement of certain materials in a plastic, such as plasticizers in vinyl, to the surface of the finished article or into an adjacent material; also called migration. (2) Refers to molten metal oozing out of a casting. It is stripped or removed from the mold before complete solidification.

Bleeder Cloth

A woven or nonwoven layer of material used in the manufacture of composite parts to allow the escape of excess gas and resin during cure. The bleeder cloth is removed after the curing process and is not part of the final composite.

Bleeding

(1) The removal of excess resin from a laminate during cure. The diffusion of color from a plastic part into the surrounding surface or part. (2) Separation of oil (or other fluid) from a grease. (3) Exudation of metal, usually high in impurities, from the newly solidified surface of a cooling ingot. See solidification. (4) The trail of red iron oxide seeping from an interface suffering from *fretting*.

Bleedout

The excess liquid resin that migrates to the surface of a winding. Primarily pertinent to filament winding.

Bleed-Out

The spread of adhesive away from the bond area.

Blemish

A nonspecific quality control term designating an imperfection that mars the appearance of a part but does not detract from its ability to perform its intended function.

Blend (Noun)

Thoroughly intermingled powders of the same nominal composition.

Blended Sand

A mixture of sands of different grain size and clay content that provides suitable characteristics for foundry use.

Blending

(1) In powder metallurgy, the thorough intermingling of powders of the same nominal composition (not to be confused with *mixing*). (2) The process of mixing mineral oils to obtain the desired consistency. Blending should be contrasted with compounding, which utilizes additives.

Blind Hole

A hole that is not drilled entirely through.

B

Blind Joint

A joint, no portion of which is visible.

Blind Riser

A *riser* that does not extend through the top of the mold.

Blind Rivet

A rivet that requires access to one side only for insertion and securing. For example, see *pop rivet*.

Blind Sample

A sample submitted for analysis whose composition is known to the submitter but unknown to the analyst, used to test the efficiency of a measurement process.

Blister (Adhesives)

An elevation of the surface of an adherend, the shape of which somewhat resembles a blister on the human skin. Its boundaries may be definitely outlined, and it may have burst and become flattened. A blister may be caused by insufficient adhesive; inadequate curing time, temperature, or pressure; or trapped air, water, or solvent vapor.

Blister (Ceramics)

A defect consisting of a bubble that forms during fusion and remains as porcelain enamel solidifies.

Blister (Metals)

(1) A casting defect, on or near the surface of the metal, resulting from the expansion of gas, almost invariably hydrogen, in a subsurface zone. It is characterized by a smooth bump on the surface of the casting and a hole inside the casting directly below the bump. (2) A raised area, often dome shaped, resulting from loss of adhesion between a coating or deposit and the basis metal.

Blister Copper

An impure intermediate product in the refining of copper, produced by blowing copper *matte* in a converter, the name being derived from the large blisters on the cast surface that result from the liberation of SO_2 and other gases.

Blister Steel

An early form of steel produced by recarburizing a good-quality wrought iron by heating in charcoal.

Blistering

The development during firing of enclosed or broken macroscopic vesicles or bubbles in a body, or in a glaze or other coating.

Block

A preliminary forging operation that roughly distributes metal preparatory for *finish*.

Block and Finish

The forging operation in which a part to be forged is blocked and finished in one heat through the use of tooling having both a block impression and a finish impression in the same die block.

Block Brazing

A brazing process in which the heat required is obtained from heated blocks applied to the parts to be joined.

Block Copolymer

An essentially linear copolymer consisting of a small number of repeated sequences of polymeric segments of different chemical structure.

Block Grease

A grease that is sufficiently hard to retain its shape and block or stick form.

Block Sequence

A combined longitudinal and buildup sequence for a continuous multiple-passed weld in which separated lengths are completely or partially built up in cross section before intervening lengths are deposited. See also *backstep sequence* and *longitudinal sequence*.

Blocked Curing Agent

A curing agent or hardener rendered unreacted, which can be reactivated as desired by physical or chemical means. Compare with *hardener*.

Blocker

The impression in the dies (often one of a series of impressions in a single die set) that imparts to the forging an intermediate shape, preparatory to forging of the final shape. Also called blocking impression.

Blocker Dies

Forging dies having generous contours, large radii, drift angles of 7° or more, and liberal finish allowances. See also *finish allowance*.

Blocker-Type Forging

A forging that approximates the general shape of the final part with relatively generous *finish allowance* and radii. Such forgings are sometimes specified to reduce die costs where only a small number of forgings are described and the cost of machining each part to its final shape is not excessive.

Block, first, second, and finish: The forging operation in which a part to be forged is passed in progressive order through three tools mounted in one forging machine; only one is involved for all three operations.

Blocking

In forging, a preliminary operation performed inclosed eyes, usually hot, to position metal properly so that in the finish operation, the dies will be filled correctly. Blocking can ensure proper working of the material and can increase die life.

Blocking (Adhesives)

An undesired adhesion between touching layers of a material, such as occurs under moderate pressure during storage or use.

Blocking (Glass)

(1) The process of shaping a gather of glass in a cavity of wood or metal. (2) The process of stirring and fining glass by immersion of a wooden block or other source of bubbles. (3) The process of reprocessing to remove surface imperfections. (4) The mounting of optical glass blanks in a shell for grinding and polishing operations. (5) The process wherein a furnace is idled at reduced temperatures. (6) The process of setting refractory blocks in a furnace.

Blocking Impression

Same as *blocker*.

Blocky Alpha

Alpha phase in titanium alloys that is considerably larger and more polygonal in appearance than the primary α in the sample. It may arise from extended exposure high in the α–β phase field or by slow cooling through the β transus during forging or heat treating. It may be removed by β recrystallization, or all-β working, followed by further α–β work, and may accompany grain boundary α.

Bloom

(1) A semifinished hot-rolled product, a rectangular cross section, produced on a blooming mill. See also *billet*. For steel, the width of a bloom is not more than twice the thickness, and the cross-sectional area is usually not less than 230 cm^3 (36 in.^{2}). Steel blooms are sometimes made by forging. (2) A visible exudation or efflorescence on the surface of an electroplating bath. (3) A bluish fluorescent cast to a painted surface caused by deposition of a thin film of smoke, dust, or oil. (4) A loose, flowerlike corrosion product that forms when certain metals are exposed to a moist environment. See *algae growth*.

Bloom (Plastics)

A noncontinuous surface coating on plastic products that comes from ingredients such as plasticizers, lubricants, and antistatic agents that are incorporated into the plastic resin or that occurs by atmospheric contamination. Bloom is the result of ingredients in the plastic coming out of solution and migrating to the surface.

Bloomer

The mill or other equipment used in reducing steel ingots to blooms.

Blooming Mill

A primary rolling mill for stage rolling of ingots used to make blooms.

Blotter

In grinding, a disk of compressible material, usually blotting-paper stock, used between the grinding wheel and its flanges to avoid concentrated stresses.

Blow

A term that describes the trapping of gas in castings, causing voids in the metal. Also (as in "*the blow*" and "*after blow*") the periods in which air or other gas is being injected during the Bessemer and other injections in steelmaking processes.

Blow Hole

(1) An internal void in cast metal, including weld metal, resulting from the entrapment of gas during solidification. Such holes may occur throughout the casting, within the grains, and then the boundaries. The term usually implies voids of substantial size, and in the case of weld metal in particular, voids smaller than about 1.5 mm diameter are usually termed *gas pores*. See also *porosity*. (2) A void in a solder connection caused by outgassing or a void in a fired dielectric.

Blow Holes

In casting technology, holes in the head plate or blow plate of a core blowing machine through which sand is blown from the reservoir into the *core box*.

Blow Molding

Essentially, blow molding involves trapping a hollow tube of thermoplastic material in a mold. Air pressure applied to the inside of the heated tube blows the tube out to take the shape of the mold. There are many variations on the basic technique.

In short, the process is an economical high-speed, high-production-rate method of forming thermoplastic parts of hollow shape, or parts that can be simply made from a hollow shape.

Uses include the container and toy field, where bottles and toys in many different shapes are formed in large quantities at low cost. The most commonly used material is polyethylene (PE).

Although any thermoplastic resin can be considered a candidate for blow molding, PE was the first used when blow molding started with low-density PE for blow-molded squeeze bottles. Now, low-, intermediate-, and high-density PE resins are used, as well as special ethylene copolymers designed to provide greatly improved stress cracking resistance compared with PE homopolymers, needed for detergent containers.

One of the main criteria of selection of a PE resin for blow molding is the proper balance of physical properties required for the specific use.

With the extension of blow molding into broader use in industrial products, the need for engineering properties other than those of PE has stimulated interest in other thermoplastics. The main plastics available for blow molding, other than PE, include cellulosics, polyamides (nylons), polyacetals, polycarbonates, polypropylene, and vinyls.

The cellulosic family of plastics includes acetate, butyrate, propionate, and ethyl cellulose. For blow molding, the cellulosics offer strength, stiffness, transparency, and high surface gloss. They have unlimited color possibilities. Chemical resistance and availability of nontoxic resins make them potentially suitable for medicine and food packaging. Their strength, stiffness, and transparency make them suitable for industrial parts, toys, and numerous decorative and novelty items.

Polyamides or nylons, although relatively high-cost materials, offer potential benefits in industrial parts in special containers, such as aerosols. Special developments have resulted in formulations tailored to special viscosity requirements for blow molding.

Polyacetal resins for blow molding offer toughness, rigidity, abrasion resistance, high heat distortion temperatures, and excellent resistance to organic solvents. Also, they are resistant to aliphatic and aromatic hydrocarbons, alcohols, ketones, strong detergents, weak organic acids, and some weak inorganic bases. Aerosol containers are another application for blow molding.

Polycarbonates have found their place in blow-molded industrial parts. Primarily they offer high toughness, strength, and heat resistance. They are transparent with almost unlimited colorability and are self-extinguishing.

Polypropylenes, somewhat similar to higher-density PEs but with lower specific gravity; higher rigidity, strength, and heat resistance; and lower permeability, offer interesting properties at a low cost. Because of their lower permeability, they are used in containers where PE is unsuited. They also have excellent stress-crack resistance.

Polyvinyl chloride (PVC) for blow-molded parts offers benefits in terms of variability of engineering properties. PVCs are available with properties ranging all the way from high rigidity in the unplasticized grades to highly flexible plasticized PVCs. The variability in performance resulting from the many possible formulations means that engineers must consult with the material supplier and attempt to obtain a formulation with the proper performance and processing characteristics to meet their needs.

The scope of blow-molded design has already broadened beyond that of round hollow objects. Production of such parts as housings by blowing a unit and sawing the item along the parting line to produce two housings is already a reality. As many as 10 cavities have been incorporated into a mold, using a wide tube and allowing air to pass through a hollow sprue or runner system between parts.

The future design possibilities of blow molding appear bright. The constant improvements in equipment design continue to add flexibility to the blow-molding process.

Secondary Operations

The equipment used will determine to a great extent the amount of finishing required on the parts. Parts must usually be trimmed and often decorated. Trimming may be made by hand, or may involve highly automated trimming, reaming, and cutting.

Decorating of parts depends on the shape of the part and the type of material used. But generally, a variety of techniques is available, including labeling, hot stamping, silk screening, and offset printing.

By far the largest market for blow-molding applications is in the container field, ranging from containers for food, drugs, and cosmetics to household and industrial chemicals. Toys and housewares represent sizable markets.

The industrial product area represents one of the biggest potential uses for the process. Present products include controls for television sets, rollers for lawnmowers, oil dispensers, toilet floats, molds for epoxy potting, and auto ducting for air-conditioning.

Blow Pin

Part of the tooling used to form hollow plastic objects or containers by the blow molding process. It is a tubular tool through which air pressure is introduced into the parison to create the air pressure necessary to form the parison into the shape of the mold. In some blow molding systems, it is a part of, or an extension of, the core pin.

Blow Pressure

The air pressure required to form the parison into the shape of the mold cavity, in a plastic blow molding operation.

Blow Rate

The rate of speed at which air enters, or the time required for air to enter, the parison during the *blow molding cycle*.

Blowdown

(1) In corrosion prevention, injection of air or water under high pressure through a tube to the anode area for the purpose of purging the annular space and possibly correcting high resistance caused by gas blocking. (2) In connection with boilers or cooling towers, the process of discharging a significant portion of the aqueous solution in order to remove accumulated salts, deposits, and other impurities.

Blowing Agent

A compounding ingredient used to produce gas by chemical or thermal action, or both, in the manufacture of hollow or cellular plastic articles.

Blown Oil

Fatty oil that is artificially thickened by blowing air through it.

Blown Tubing

A thermoplastic film that is produced by extruding a tube, applying a slight internal pressure to the tube to expand it while still molten, and subsequently cooling to set the tube. The tube is then flattened through guides and wound up flat on rolls. The size of blown tubing is determined by the flat width in inches as wound rather than by the diameter, as in the case of rigid types of tubing.

Blown-Film Extrusion

Technique for making film by extruding the plastic through a circular die, followed by expansion (by the pressure of internal air admitted through the center of the mandrel), cooling, and collapsing of the bubble.

Blowpipe (Brazing and Soldering)

A device used to obtain a small, accurately directed flame for fine work, such as in the dental and jewelry trades. Any flame may be used, a portion of it being blown to the desired location for the required time by the blowpipe that is usually mouth operated.

Blowpipe (Welding and Cutting)

See preferred terms *welding torch* and *cutting torch*.

Blow-Up Ratio

In *blow molding* of plastics, the ratio of the diameter of the product (usually its greatest diameter) to the diameter of the parison from which the product is formed. In blown film extrusion, the ratio between the diameter of the final film tube and the diameter of the die orifice.

Blue Anneal

The annealing of steel without any protection from the environment at a temperature of about 750°C–800°C. This avoids the severe oxidation associated with full annealing or normalizing and the resultant adherent dark blue oxide film may be a desirable feature.

Blue Brittleness

Brittleness exhibited by some steels after being heated to some temperature within the range of about 205°C–370°C (400°F–700°F), particularly if the steel is worked at the elevated temperature. Killed steels are virtually free of this kind of brittleness.

Blue Dip

A solution containing a mercury compound, once widely used to deposit mercury on a metal by immersion, usually prior to silver plating.

Blue Enamel

(1) In dry-process porcelain enameling, an area of enamel coating so thin that it appears blue in color. (2) In wet-process enameling, a cover coat applied too thinly to hide the substrate.

Blueing (Plastics)

A mold blemish in the form of a blue oxide film on the polished surface of a mold due to abnormally high mold temperatures.

Bluing (Metals)

Subjecting the scale-free surface of a ferrous alloy to the action of air, steam, or other agents at a suitable temperature, thus forming a thin blue film of oxide and improving the appearance and resistance to corrosion. This term is ordinarily applied to sheet, strip, or finished parts. It is used also to denote the heating of springs after fabrication to improve their properties.

Blushing

(1) Whitening and loss of gloss of a usually organic coating caused by moisture. Also called blooming. (2) The condensation of atmospheric moisture at the adhesive bond line interface.

B_4C Whiskers

These are single crystals of B_4C or $B_{13}C_2$ that range in diameter from 1 to 10 mm and average length of 30–300 mm. B_4C whiskers have been utilized in matrices of Al_2O_3, SiC, and Si_3N_4. Toughness values exceeding 9.0 MPa $m^{1/2}$ have been reported in a matrix of Al_2O_3.

B_4C whiskers have been tested in metal–matrix composites with high values reported in ductility.

Another boron compound used for the production of high-temperature ceramic parts by pressing and sintering is boron silicide (B_4Si). It is a black, free-flowing crystalline powder. The powder is microcrystalline, with particles about 75 μm in diameter, and the free silicon is less than 0.15%. This compound normally reacts at 1200°C to form B_6Si and silicon, but when compacted and centered, the ceramic forms a boron silicate oxygen protective coating, and the parts have a serviceable life in air at temperatures to 1400°C. Molded parts have high thermal shock resistance and can be water-quenched from 1093°C without shattering.

In combination with plastics or aluminum, boron provides an effective lightweight neutron-shielding material. Boron-containing shields are valuable because of their satisfactory mechanical properties and because boron absorbs neutrons without producing high-energy gamma rays. Rods and strips of boron steel have been used extensively as control rods in atomic reactors.

The physical properties that make boron attractive as a construction material in missile and rocket technology are its low density (15% lighter than aluminum), extreme hardness, high melting point, and remarkable tensile strength in filament form. Production of boron fibers by vapor deposition methods has been developed on a commercial scale. These fibers are being used in an epoxy (or other plastic) carrier material or matrix. The resulting composite is stronger and stiffer than steel at 25% lighter than aluminum. The balance of strength and stiffness of the composite makes it ideal for aircraft applications, where high performance is of primary importance. Another development in this area is the incorporation of boron filaments in metal matrices.

The use of boron has found its way in such diversified applications as in motor-starting devices, phonograph needles, lightning arresters, thermal cutouts for transformers, igniters in rectifier and control tubes, alloys resistant to high-temperature abrasion and scaling, constant-potential controllers, thermoelectric couples, and resistance thermometers.

Certain compounds of boron, such as borax and boric acid, have been known and used for a long time in glass, enamel, ceramic, and mining industries. Refined borax, $Na_2B_4O_7 \cdot H_2O$, is an important ingredient of a variety of detergents, soaps, water-softening compounds, laundry starches, adhesives, toiletry preparations, cosmetics, talcum powder, and glazed paper. It is also used in fireproofing, disinfecting of fruit and lumber, weed control, and insecticides, as well in the manufacture of leather, paper, and plastics.

BMC

See *bulk molding compound*.

BMI

See *bismaleimide*.

Board Hammer

A type of forging hammer in which the upper die and ram are attached to "boards" that are raised to the striking position by power-driven rollers and let fall by gravity. See also *gravity hammer*.

Boat

A box or container used to hold the green powder metallurgy compacts during passage through a continuous sintering furnace.

B

Body

(1) A loosely used term designating viscosity or consistency. (2) The consistency of an adhesive, which is a function of viscosity, plasticity, and rheological factors.

Body Putty

A pastelike mixture of plastic resin (polyester or epoxy) and talc used in repair of metal surfaces, such as auto bodies.

Body-Centered

Having an atom or group of atoms separated by a translation of 1/2, 1/2, 1/2 from a similar atom or group of atoms. The number of atoms in a body-centered cell must be a multiple of 2.

Body-Centered Cubic

A crystal lattice structure with an atom at the center and corners of the unit cube.

Bohr Atom

This regards an atom as a central nucleus with electrons and fixed orbits around it.

Boiler

A vessel or system of vessels and pipework in which water is boiled to produce steam. In many cases, the system will be under pressure thereby raising the boiling point considerably; 350°C is not uncommon for steam for power plant.

Boiler Plate

An imprecise term usually indicating a good-quality mild steel, typically 0.12–0.2 carbon, less than 1% manganese, and probably in the normalized condition.

Boiler Scale

An imprecise term usually referring to the deposits of calcium sulfate, possibly intermixed with magnetite, found on the water side of boiler circuits evaporating poor-quality water. The term has occasionally been used regarding the scale and deposits on the fireside of tubes.

Boiler Tube

The tubes in which water is heated to form steam in a boiler, as opposed to superheater tubes in which the steam is further heated. In a fire tube boiler, the flame and combustion products pass through the tubes that are immersed in the water. In a water tube boiler, the water is contained in the tubes that pass through the combustion chamber or, usually, form the wall of the combustion chamber, hence the term waterwall tubes.

Boiling

The vigorous evolution of bubbles as a fluid changes to a gas.

Bolster

(1) A plate to which dies may be fastened, the assembly being secured to the top surface of a press bed. In press forging, such a plate may also be attached to the ram.

Bolt

A threaded fastener having a head at one end, a threaded length at the other, and a plain shank between. Bright bolts are normally machined all over. The term black mold originally referred to bolts that were at least partially covered with a black scale; nowadays, the term usually indicates not color but a bolt with wider machining tolerances than a bright bolt.

Boltzmann Constant

The gas constant divided by Avogadro's number, 1.380×10^{-23} J/K.

Boltzmann Distribution

In materials characterization, a function giving the probability that a molecule of a gas in thermal equilibrium will have generalized position and momentum coordinates within a given infinitesimal range of values.

Bond

(1) In grinding wheels and other relatively rigid abrasive products, the material that holds the abrasive grains together. (2) In welding, brazing, or soldering, the junction of joined parts. Where filler metal is used, it is the junction of the fused metal and heat-affected base metal. (3) In an adhesive bonded or diffusion bonded joint, the line along which the faying surfaces are joined together. (4) In thermal spraying, the junction between the material deposited and the substrate, or strength. See also *adhesive bond*, *mechanical bond*, and *metallic bond*.

Bond Angle

The angle formed by the bonds of one atom to other atoms, for example, 109.5° for C–C bonds.

Bond Clay

Any clay suitable for use as a *bonding agent* in molding sand.

Bond Coat (Thermal Spraying)

A preliminary (or prime) coat of material that improves adherence of the subsequent thermal spray deposit.

Bond Face

The part or surface of an adherend that serves as a substrate for an adhesive.

Bond Length

The average distance between the centers of two atoms, for example, 0.154 nm (1.54 Å) for C–C bonds.

Bond Line

The cross section of the interface between thermal spray deposits and substrate, or the interface between an adhesive and an adherend in an adhesive bonded joint.

Bond Strength

(1) The unit load applied to tension, compression, flexure, peel, impact, cleavage, or shear required to break an adhesive assembly with failure occurring in or near the plane of the bond. The term *adherence* is frequently used in place of bond strength. (2) The force required to pull a coating free of a substrate. (3) The degree of cohesiveness that the *bonding agent* exhibits in holding sand grains together.

Bonded Film Lubricant

See *bonded solid lubricant*.

Bonded Solid Lubricant

A solid lubricant dispersed in a continuous matrix of a binder, or attached to a surface by an adhesive material.

Bonded-Phase Chromatography (BPC)

Liquid chromatography with a surface-reacted, that is, chemically bonded, organic stationary phase. See also *normal phase chromatography* and *reversed-phase chromatography*.

Bonderizing

A process for producing phosphate conversion coatings on steel components by immersion in a hot acid solution of zinc and manganese phosphates. The coatings improve corrosion resistance and paint adhesion and provide lubrication doing drawing.

Bonding

A method of holding the parts of an object together with or without the aid of an adhesive such as an epoxy or glue. Composite materials such as fiber-reinforced plastics require strong interfacial bonding between the reinforcement and the matrix. In the case of atomic or optical contact bonding, interatomic forces hold the parts together. An optical contact bonding, surface flatness, and cleanliness between the mating parts determine the bonding strength, and the atoms at the surface provide the necessary forces. The number of valence electrons in the atoms of a material determines the bonding strength between the group of atoms that constitutes a molecule. In these cases, the term *chemical bonding* is used.

Wire bonding as an interconnect technique widely used in microchip manufacturing to provide electrical continuity between the metal pads of an integrated circuit chip and the electrical leads of the package housing the chip. The two common methods of wire bonding are thermocompression and ultrasonic bonding. In these, a fine aluminum or gold wire is bonded at one end to the metal pad of the integrated circuit chip and at the other to the electrical lead of the package. There are three types of thermocompression bonds: wedge, stitch, and ball. In thermocompression bonding, a molecular metallurgical bond is formed at the two metal junctions—bond wire and IC metal pad and bond wire and package lead metal—by applying heat and pressure without melting. In ultrasonic bonding, the molecular metallurgical bond is achieved through a combination of ultrasonic energy and pressure. The bonding operation is done under pressure to break the few surface layers of the material and form the bond between the contamination-free surfaces. Thermocompression bonding has higher throughput and speed than ultrasonic bonding. The bonding wire is usually aluminum, which does not introduce any intermetallic problems.

Bonding Agent

Any material other than water that, when added to foundry sands, imparts strength in either the green, dry, or fired state.

Bonding Force

The force that holds two atoms together; it results from a decrease in energy as to atoms are brought closer to one another.

Bonding Materials

These are matrix materials in glass-fiber-reinforced plastics whereby the fiber glass reinforcement has been previously sized and supplied with these polymeric materials, which are catalysts, accelerators, fillers, inhibitors, and plasticizers.

- *Acrylic polymers*: Limited use as a laminating resin in the form of thermosetting acrylic syrup to make glazing panels. The monomer is used in conjunction with styrene as a cross-linking agent in polyesters to improve weathering resistance.
- *Alkyds*: Reaction products of polymerization of polyhydroxy alcohols and polycarboxylic acids, such as glycerol and phthallic anhydride. Modified grades are reinforced by glass fibers in chopped strand form.
- *Epoxy resins*: Reaction products of bisphenol A and epichlorohydrin. Epoxies are cross-linked by the addition of acids or amines to form laminating resins of exceptional bonding properties. These boast excellent electrical properties and high resistance to acids and alkalies and are used with fiberglass cloth and strands.
- *Melamine formaldehyde*: Melamine reacts with 1–6 mol of formaldehyde to form various compounds with good electrical characteristics. Used in flat electrical laminates, reinforced by glass cloth, these resins also are used for bonding glass fibers and thermal insulation.
- *Phenol formaldehyde resins*: Condensation polymerization products of phenol and formaldehyde are used primarily for bonding fiberglass insulation products. Also used in flat fiberglass cloth laminates for electrical applications.
- *Polyesters*: Unsaturated polyester resins are made by condensation polymerization of polyfunctional acids such as maleic acid and phthalic anhydride of cross-linking agents such as styrene. These represent the largest family of resins used in laminating and molding fiberglass cloth and strands.

Bone Oil

A fatty oil obtained by dry distillation of bones.

Book Mold

A split permanent mold hinged along one edge to open like a book.

Borate Glass

A glass in which the essential glass former is boron oxide instead of silica. See also *glass*.

Borax

A white or colorless crystalline mineral used in glass and ceramic and enamel mixes, as a scouring and cleansing agent, as a flux in melting metals and in soldering, brazing, and welding (it forms a hard glassy deposit on cooling), as a corrosion inhibitor in antifreeze liquids, as a constituent in fertilizers, in the production of many chemicals and pharmaceuticals, and as a source of boron. Borax is a hydrous sodium borate, or tetraborate.

Bore

(1) The hollow space within a tube or a similar vessel. (2) The internal diameter as a dimension, or the internal surface of a tube, gun, etc. (3) A hole or cylindrical cavity produced by a single-point or multipoint tool other than a drill.

Bore Seal

A device in which the outside diameter mates with a bore surface to provide sealing between the two surfaces.

Boriding

Thermochemical treatment involving the enrichment of the surface layer of an object with borides. This surface-hardening process is performed below the Ac_1 temperature. Also referred to as boronizing.

Boring

Enlarging a hole by removing metal with a single- or occasionally a multiple-point cutting tool moving parallel to the axis of rotation of the work or tool.

Boron

Boron (B) is a metallic element closely resembling silicon. Boron has a specific gravity of 2.31, with a melting point of about 2200°C, and a Knoop hardness of 2700–3200, equal to a Mohs hardness of about 9.3. At 600°C, boron ignites and burns with a brilliant green flame. Minute quantities of boron are used in steels for case hardening by the nitriding process to form boron nitride and in other steels to increase hardenability, or depth of hardness. In these boron steels, as little as 0.003% is beneficial, forming an iron boride, but with larger amounts, the steel becomes brittle and susceptible to being hot-short unless it contains titanium or some other element to stabilize the carbon. In cast iron, boron inhibits graph-atomization and also serves as a deoxidizer. It is added to iron and steel in the form of ferroboron.

Boriding, diffusing boron into metal surfaces, forms intermetallics that increase hardness, thus wear resistance. It can also improve corrosion resistance and the bonding and brazing of certain alloys. Boron diffusion can increase the surface hardness of ferrous alloys to 1500–1700 Knoop and their resistance to hydrochloric, hydrofluoric, and sulfuric acids. Resistance to attack by these acids also pertains to nickel alloys. For nickel- or cobalt-bonded tungsten carbide and titanium carbide, boron diffusion can markedly increase erosion resistance.

The free element is prepared in crystalline or amorphous form. The crystalline form is an extremely hard, brittle solid. It is jet-black to silvery-gray color with a metallic luster. One form of crystalline boron is bright red. The amorphous form is less dense than the crystalline and is a dark brown to black powder.

Crystalline boron, although relatively brittle compared to diamond, is second only to diamond and hardness.

The chemical properties of boron depend largely on the physical form as well as all of the purity of samples. Amorphous boron oxidizes slowly in the air even at room temperature and is spontaneously flammable at about 860°C.

Boron is not affected by either hydrochloric or hydrofluoric acids, even on prolonged boiling. Crystalline boron is quite stable to heat and oxidation even at relatively high temperatures. It is slowly attacked and oxidized by hot concentrated nitric acid and by mixtures of sodium dichromate and sulfuric acid. Hydrogen peroxide and ammonium persulfate also slowly oxidize crystalline boron.

All varieties of boron are completely oxidized by molten mixtures of alkali carbonates and hydroxides.

Boron reacts vigorously with sulfur at about 600°C to form a mixture of boron sulfides. Boron nitride is formed when boron is heated in a N_2 or NH_3 atmosphere above 1000°C. At high temperatures, boron combines with phosphorus and with arsenic to form a phosphide, BP, and an arsenide, BAs.

It reacts with silicon to form silicon borides at temperatures above 2000°C.

Boron reacts with the majority of metals and metal oxides at high temperatures to form metallic borides. Boron does not conform to the usual rules of valence in forming these compounds. The borides, along with the carbides and the nitrides, are called interstitial compounds. These compounds have crystal structures and properties very similar to those of the original metal.

Metallic borides are usually prepared by hot sintering of the elements. They may also be prepared by carbothermic and aluminothermic reduction of metal oxide–boron oxide mixtures. A number of metallic borides of high purity have been prepared on a semicommercial scale by vapor deposition methods, in which the volatile halides of the elements are deposited on a hot substrate, such as tungsten or tantalum wire, in an atmosphere of H_2. Borides have much in common with true metals. They have a high electrical conductivity, high melting points, and extreme hardness. Many metal borides are used as components of cermet compositions.

Boron fibers for reinforced structural composites are continuous fine filaments that are, themselves, composites. They are produced by vapor deposition of boron on a tungsten or carbon substrate. Their specific gravity is about 2.6, and they range from 0.10 to 0.15 mm in diameter. They have tensile strengths around 3450 MPa and a modulus of elasticity of nearly 0.5 million MPa. The fibers are used chiefly in aluminum or epoxy matrixes. Unidirectional boron–aluminum composites have tensile strengths ranging from 758 to over 1378 MPa. Their strength-to-weight ratio is about three times greater than that of high-strength aluminum alloys.

Boron compounds are employed for fluxes and deoxidizing agents and melting metals and for making special glasses. Boron, like silicon and carbon, has an immense capacity for forming compounds, although it has a different valence. The boron atom appears to have a lenticular shape, and two boron atoms can make a strong electromagnetic bond, with the boron acting like carbon but with a double ring.

Boron Carbide

Boron carbide (B_4C) is produced by the high-temperature (about 1371°C–2482°C) interaction of boric oxide, B_2O_3, and carbon in an electrical resistance-type furnace. It is a black, lustrous solid. It is used extensively as an abrasive because its hardness approaches that of the diamond. It is also used as an alloying agent, particularly in molybdenum steels.

Additionally, it is used in drawing dies and gauges, or into heat-resistant parts such as nozzles. The composition is either B_6C or B_4C; the former is the harder but usually contains an excess of graphite difficult to separate in the powder. It can be used thus as a deoxidizing agent for casting copper and also for lapping, since the graphite acts as a lubricant. Boroflux is B_4C with flake graphite, used as a casting flux.

B_4C parts are fabricated by hot pressing, sintering, and sinter-HIPing (hot isostatic press [HIP]). Industrially, densification is carried out by hot pressing (2100°C–2200°C, 20–40 MPa) in argon. The best properties are obtained when pure fine powder is densified without additives. Pressureless sintering to high density is possible using ultrafine powder, with additives (notably carbon). Less expensive than hot pressing, sintering also can be used for more complex shapes.

Special part formulations include bonding B_4C with fused sodium silicate, borate frits, glasses, plastics, or rubbers to lend strength, hardness, or abrasion resistance. B_4C-based cermets and MMC (especially Al/B_4C, Mg/B_4C, Ti/B_4C) and CMCs (e.g., TiB_2/B_4C) have unique properties, including superior ballistic performance, that make these materials suitable for highly specialized applications. High-temperature strength, light weight, corrosion resistance, and hardness make these composites especially attractive. B_4C shapes can be reaction-bonded using SiC as the bonding phase. B_4C–C mixtures are formed and then reacted with silicon to create the SiC bond. SiC also can be used as a sintering aid for B_4C and vice versa.

As an abrasive, B_4C is used for fine polishing and ultrasonic grinding and drilling, either as a loose powder or as a slurry. Its tendency to oxidize at workpiece temperatures precludes its use in bonded abrasive wheels. Abrasion-resistant parts made from B_4C include spray and blasting nozzles, bearing liners, and furnace parts. The refractory properties of B_4C, in addition to its abrasion resistance, are of value in the latter application.

B_4C is chemically inert, although it reacts with O_2 at elevated temperatures and with white hot or molten metals of the iron group and certain transition metals. B_4C reacts with halogens to form boron halides—precursors for the manufacture of most nonoxide boron chemicals. B_4C also is used in some reaction schemes to produce transition metal borides. Boronizing packings containing B_4C are used to form hard boride surface layers on metal parts.

B_4C and elemental boron are used for nuclear reactor control elements, radiation shields, and moderators.

B_4C platelets are single crystals of B_4C or $B_{13}C_2$ that are of 1–5 mm average diameter and about 0.2 mm thick. The tiny B_4C platelets have proved to be highly effective in reinforcing ceramic materials. Toughness values of 8.1 MPa $m^{1/2}$ have been reported in B_4C platelet-reinforced Al_2O_3 with corresponding four-point bed strengths of more than 650 MPa.

Boron Nitride

Boron nitride (BN) has many potential commercial applications. It is a white, fluffy powder with a greasy feel. It has an x-ray diffraction pattern almost identical with that of graphite, indicating a close similarity in structure. It is called white graphite. BN has a very low coefficient of friction, but unlike carbon, it is a nonconductor of electricity, and it is attacked by nitric acid. It sublimes at 3000°C. It reacts with carbon at about 2000°C to form boron carbide (B_4C). It is used for heat-resistant parts by molding and pressing the powder without a binder to a specific gravity of 2.1–2.25.

BN may be prepared in a variety of ways, for example, by the reaction of boron oxide with ammonia, alkali cyanides, and ammonium chloride, or of boron halides and ammonia. The usually high chemical and thermal stability, combined with high electrical resistance of BN, suggests numerous uses for this compound in the field of high-temperature technology. BN can be hot-pressed into molds and worked into desired shapes.

BN uses include as a lubricant for high-pressure bearings, for compacting into mechanical and electrical parts, and as an additive to plastics, potting compounds, oils, and grease to increase thermal conductivity and dielectric strength. Not wetted by and highly resistant to many molten metals and slags, it is used as a release agent and container material for molten salts, various metals, and nonlead glasses and for braze stop-off and weld-spatter release coatings.

BN powders can be used as mold-release agents, high-temperature lubricants, and additives in oils, rubbers, and epoxies to improve thermal conductance of dielectric compounds. Powders also are used in metal–matrix and ceramic–matrix composites (MMC and CMC) to improve thermal shock and to modify wetting characteristics.

Hexagonal BN may be hot-pressed into soft (Mohs 2) and easily machinable white or ivory billets having densities 90%–95% of theoretical (2.25 g/cm^3). Thermal conductivities of 17–58 W/m K and coefficients of thermal expansion of 0.4–5 × 10^{-6}/°C are obtained, depending on density, orientation with respect to pressing direction, and amount of boric oxide binder phase. Because of its porosity and relatively low elastic modulus (50–75 GPa), hot-pressed BN has outstanding thermal shock resistance and fair toughness. Pyrolytic BN, produced by chemical vapor deposition (CVD) on heated substrates, also is hexagonal; the process is used to produce coatings and shapes with thin cross sections.

Uses for hexagonal BN shapes include crucibles, parts for chemical and vacuum equipment, metal casting fixtures, boron sources for semiconductor processing, and transistor mounts.

A cubic form of BN, called borazon, has been prepared at pressures near 6500 MPa and temperatures near 1500°C. It is comparable to diamond in hardness and apparently has properties superior to those of diamond with regard to oxidation, electrical resistance, and thermal stability. This material will probably take its place with diamond for industrial grinding. Borazon is stable up to about 1925°C. This molded material is resistant to molten aluminum, is not wetted by molten silicon or glass, and is used for crucibles. Cubic BN tooling typically outlasts Al_2O_3 and carbide tooling and is preferred in applications where diamond is not appropriate, such as grinding of ferrous metals.

BN fibers are produced in diameters as small as 5–7 μm and in lengths to 0.38 m. The fibers have a tensile strength of 1378 MPa. They are used for filters for hot chemicals and as reinforcement to plastic lamination. BN HCJ is a fine powder, 99% pure, and is used as a filler in encapsulating and potting compounds to add thermal and electric conductivity. BN is basically the same as B_4C and can only be densified by hot pressing. Hot-pressed hexagonal BN is easy to machine and has a low density but must be kept free of boron oxide (or the boron oxide stabilized with CaO-containing additives), so BN is not destroyed by hydration during heating.

A tough, coherent, hard abrasive compact, consisting of a cubic form of the end bonded with B_4C, has been developed.

Boronizing

See *boriding*.

Borosilicate Glass

Any silicate glass having at least 5% of boron oxide (B_2O_3). See also *glass*.

Bort

(1) Natural diamond of a quality not suitable for gem use. (2) Industrial diamond.

Bosh

(1) The section of a blast furnace extending upward from the tuyeres to the plane of maximum diameter. (2) A lining of quartz that builds up during the smelting of copper ores and that decreases the diameter of the furnace at the tuyeres. (3) A tank, often with sloping sides, used for washing metal parts or for holding cleaned parts.

Boss

(1) A relatively short protrusion or projection from the surface of a forging or casting, often cylindrical in shape. Usually intended for drilling and tapping for attaching parts. (2) Projection on a plastic part designed to add strength, to facilitate alignment during assembly, or to provide for a fastening. (3) A raised area or reinforcement. It might, for example, act as a location, support point, or be drilled to act as a bearing.

Bottle

A partly pierced billet. See preferred term *cylinder*.

Bottle Bored

A bored forging with the chamber at some position along the bore. Large forgings for high-speed rotors for turbine generators, and similar applications are bored axially along their centerline to remove the poor-quality material remaining from the original ingot. In some cases, defects extend deeper than the, typically 100 mm diameter, bore. In such cases, a short length is machined deeper to cut out the defects forming a chamber or bottle, provided that the design can accommodate the variation.

Bottom Blow

A specific type of *blow molding* technique for plastics that forms hollow arteries by injecting the blowing air into the parison from the bottom of the mold (as opposed to introducing the blowing air at a container opening).

Bottom Board

In casting, a flat base for holding the *flask* in making sand molds.

Bottom Draft

Slope or taper in the bottom of a forge depression that tends to assist metal flow toward the sides of depressed areas.

Bottom Drill

A flat-ended twist drill used to convert a cone at the bottom of a drilled hole into a cylinder.

Bottom Pipe

An oxide-lined fold or cavity at the butt end of a slab, bloom, or billet; formed by folding the end of an ingot over on itself doing primary rolling. Bottom pipe is not *pipe*, in that it is not a shrinkage cavity, and in that sense, the term is a misnomer. Bottom pipe is similar to *extrusion pipe*. It is normally discarded when the slab, bloom, or billet is cropped following primary reduction.

Bottom Plate

In making of plastic parts, the part of the mold that contains the heel radius and the push-up.

Bottom Punch

In powder metallurgy, the part of the tool assembly that closes the die cavity at the bottom and transfers the pressure to the powder during compaction.

Bottom Running or Pouring

Filling of the casting mold cavity from the bottom by means of gates from the runner.

Bottom-End Bearing

See *big-end bearing*.

Bottoming Bending

Press-brake bending process in which the upper die (punch) enters the lower die and coins or sets the material to eliminate *spring back*.

Bottoming Tap

A tap with a chamfer of 1–1½ threads in length.

Bottom-Pour Ladle

A *ladle* from which metal, usually steel, flows through a *nozzle* located at the bottom.

Boundary Lubricant

A lubricant suitable for use in *boundary lubrication* conditions. Fatty acids and soaps are commonly used.

Boundary Lubrication

A condition of lubrication in which the friction and wear between two surfaces in relative motion are determined by the properties of the surfaces and by the properties of the lubricant other than bulk viscosity. The thin film of lubricant may be possibly only a molecule thick. See also the term *lubrication regimes*.

Bow

(1) A condition of longitudinal curvature in protruded plastic parts. (2) The tendency of material to curl downward during shearing, particularly when shearing long narrow strips.

Bowing

Deviation from flatness.

Box Annealing

Annealing a metal or alloy in a sealed container under conditions that minimize oxidation. In box annealing a ferrous alloy, the charges are usually heated slowly to a temperature below the transformation range, but sometimes above or within it, and are then cooled slowly; this process is also called close annealing or pot annealing. See also *black annealing*.

Box Furnace

A furnace used for batch sintering of powder metallurgy parts, normally utilizing a controlled atmosphere-containing sealed retort.

Boxing

The continuation of a fillet weld around a corner of a member as an extension of the principal weld.

Brackish Water

(1) Water having salinity values ranging from approximately 0.5 to 17 parts per thousand. (2) Water having less salt than seawater, but undrinkable.

Bragg Angle

The angle between the incident beam and the lattice planes considered.

Bragg Equation

See *Bragg's law*.

Bragg Method

A method of x-ray diffraction in which a single crystal is mounted on a spectrometer with a crystal face parallel to the axis of the instrument.

Bragg's Law

A statement of the conditions under which a crystal will diffract electromagnetic radiation. Bragg's law reads $n\lambda = 2d \sin \theta$, where n is the order of reflection, λ is the wavelength of x-rays, d is a distance between lattice planes, and θ is the Bragg angle, or the angular distance between the incident beam and the lattice planes considered.

Braiding

Intertwining two or more systems of yarns in the bias direction to form an integrated structure.

Brake

A device for bending sheet metal to a desired angle.

Brale

A diamond 120° indentor, of conical form with a hemispherical tip, used on the Rockwell hardness testing machine for testing hard materials. A 0.2 mm tip radius is typical.

Branched Polymer

In the molecular structure of polymers, a main chain with attached side chains, in contrast to a linear polymer. Two general types are recognized, short-chain and long-chain branching.

Branching

The presence of molecular branches in a polymer. The generation of branch crystals during the crystallization of a polymer.

Brass

Brass is an alloy of copper and zinc. In the manufacture, lump zinc is added to molten copper, and the mixture is poured either into castings ready for use or into billets for further working by rolling, extruding, forging, or a similar process. Brasses containing 75%–85% copper are red gold and malleable; those containing 60%–70% are yellow and also malleable; and those containing 50% or less copper are white, brittle, and not malleable.

Copper–zinc alloys whose zinc content ranges up to 40% with the copper crystal structure face-centered cubic (fcc) are considered brass. This solid solution alpha brass has good mechanical properties, combining strengths with ductility. Corrosion resistance is very good, but electrical conductivity is considerably lower than for copper.

Beta brass contains nearly equal proportions of copper and zinc. Specific brasses are designated as follows: gilding (95% copper; 5% zinc), red (85:15), low (80:20), and Admiralty (70:29, with balance of tin). Naval brass is 59%–62% copper, with about 1% tin, less than that of lead and iron, and the remainder zinc. With small amounts of other metals, other names are used: commercial bronze (10% zinc), jewelry bronze (12% zinc), red brass (15% zinc), yellow brass (35% zinc), and Muntz metal (40% zinc). As the zinc content increases in these alloys, the melting point, density, electric and thermal conductivity, and modulus of elasticity decrease while the coefficient of expansion, strength, and hardness increase. Work hardening also increases with zinc content. These brasses have a pleasing color, ranging from the rate of copper for the low-zinc alloys through bronze and gold colors to the yellow of high-zinc brasses. The color of jewelry bronze closely matches that of 14-karat gold, and this alloy and other low brasses are used in inexpensive jewelry.

The low-zinc brasses have good corrosion resistance along with moderate strength and good forming properties. Red brass, with its exceptionally high corrosion resistance, is widely used for condenser tubing. The high brasses (cartridge brass and yellow brass) have excellent ductility and high strength and are widely used for engineering and decorative parts fabricated by drawing, stamping, cold heading, spinning, and etching. Muntz metal, primarily a hot-working alloy, is used where cold working is not required.

Leaded brass is used for castings. These alloys have essentially the same range of zinc content as the straight brasses. Lead is present, ranging from less than 1% to 3.25%, to improve machinability and related operations. Lead also improves antifriction and bearing properties.

Another group, the tin brasses, is copper–zinc alloys with small amounts of tin. The tin improves corrosion resistance. Pleasing colors are also obtained when tin is added to the low brasses. Tin brasses and sheet and strip form, with 80% or more copper, are used widely as low-cost spring materials. Admiralty brass is a standard alloy for heat exchanger and condenser tubing. Naval brass and manganese bronze are widely used for products requiring good corrosion resistance and high strength, particularly in marine equipment.

Casting brasses are usually made from brass ingot metal and are medium plain copper–zinc alloys. In melting brass for casting, any superheating causes loss of zinc by vaporization, thus lowering the zinc amount. Small amounts of antimony or some arsenic are used to overcome this dezincification. The casting brass is roughly divided into two classes, *red casting brass* and *yellow casting brass*, which are various compositions of copper, tin, zinc, and lead to obtain the required balance of color, ease of casting, hardness, and machining qualities. *Copper–bismuth–selenium alloys*, C89510, C89520, and C89550, are lead-free or low-lead alternatives to lead-bearing red and yellow brasses, thereby meeting low-lead leach requirements of regulatory agencies for potable-water plumbing fixtures. The alloys are also called *EnviroBrass* I, II, and III, respectively, and because of their selenium and bismuth contents, *SeBiloys*.

Many brasses are now designated by alloy number and grouped into several families of standard alloys. There are three families of wrought brass: *copper–zinc brasses* (C20500–C28580); *copper–zinc–lead brasses*, or *leaded brasses* (C31200–C38590); and *copper–zinc–tin brasses*, or tin brasses (C40400–C49080). Casting alloys comprise four families: *copper–tin–zinc red brasses* and *copper–tin–zinc–lead brasses*, or *leaded red brasses* (C83300–C83810); *copper–zinc–tin–lead brasses*, *semired brasses*, and *leaded semired brasses* (C84200–C84800); *copper–zinc–lead–tin yellow brasses*, and *manganese-bearing high-strength yellow brasses* (C85200–C85800); and *copper–zinc–silicon brasses*, or *silicon brasses* (C87300–C87900). Wrought alloys are available in most mill forms, and casting alloys are amenable to most casting processes, including sand, permanent mold, investment, plaster, and centrifugal. A few alloys can be die-cast.

The mechanical properties of brasses vary widely. Strength and hardness depend on alloying and/or cold work. Tensile strengths of annealed grades are as low as 206 MPa, although some hard tempers approach 620 MPa.

Brass is annealed for drawing and bending by quenching in water from a temperature of about 538°C. Simple copper–zinc brasses are made in standard degrees of temper, or hardness. This hardness is obtained by cold rolling after the first anneal, and the degree of hardness depends on the percentage of cold reduction. When the thickness is reduced about 10.9%, the resulting sheet is 1/4 hard. The other grades are 1/2 hard, hard, extra hard, spring, and finally extra spring, which corresponds to a reduction of about 68.7% without intermediate annealing. Degrees of softness and annealed brass are measured by the grain size, and annealed brass is furnished in grain sizes from 0.010 to 0.150 mm.

Even slight additions of other elements to brass alter characteristics drastically. Slight additions of tin change the structure, increasing the hardness but reducing the ductility. Iron hardens the alloy and reduces the grain size, making it more suitable for forging but making it difficult to machine. Manganese increases strength, increases the solubility of iron in the alloy, and promotes stabilization of aluminum, but makes the brass extremely hard. Slight additions of silicon increase strength, but large amounts promote brittleness, loss of strength, and danger of oxide inclusion. Nickel increases strength and toughness, but when any silicon is present, the brass becomes extremely hard and more a bronze than a brass.

Brasses widely used in cartridge cases, plumbing fixtures, valves and pipes, screws, clocks, and musical instruments.

Brass Ingot Metal

Commercial ingots made in standard composition grades and employed for casting various articles designated as brass and bronze. They are seldom true brasses, but are composition metals intermediate between the brasses and the bronzes and their selection for any given purpose are based on a balance of the requirements in color, strength, hardness, ease of casting, and machinability. Brass ingot metal is usually made from secondary metals, but, in general, the grading is now so good that they will produce high-grade uniform castings. In producing the ingot metal, there is careful sorting of the scrap metals, and the impurities are removed by remeltings. An advantage of ingot metal over virgin metals is the ease of controlling mixtures in the foundry. The ATSM designates eight grades for brass ingot metal. Number 1 grade, the highest in copper, contains 88% copper, 6.5% zinc, 1.5% lead, and 4% tin, with only slight percentages of impurities. The No. 8 grade contains 63.5% copper, 34% zinc, 2.5% lead, and no tin. *Yellow ingot*, for plumbing fixtures, contains 65% copper, 1% tin, 2% lead, and the balance zinc. The most widely used ingot metal is *ASTM alloy No. 2*, which is the 85:5:5:5 alloy known as *composition metal*. Yellow brass, or *yellow casting brass*, is frequently cast from *ASTM alloy No. 6*, which contained 72% copper, 22% zinc, 4% lead, and 2% tin. It has a tensile strength of 138–172 MPa, elongation 15%–20%, and Brinell hardness 40–50. It is yellow and makes clean, dense castings suitable for various machine parts except bearings.

Braze

A joining process in which a metal is melted and applied to the interface between two components that are not melted in the process. It is implicit that the braze material has a composition different from, and a melting point lower than, those of the components being joined. The braze metal, sometimes termed the filler (metal), melts and bonds to the two components producing a strong joint. Use of the term is normally confined to joints where the interface to be joined is narrow but of large area relative to the cross section and where the molten filler is either drawn into the interface by capillary action or introduced as a preplaced strip or sleeve prior to heating. If this restriction is recognized, then joints made with bulky deposits of lower melting point filler metal are termed braze welds. Originally brazing referred to the use of brass to join copper or steel and it is still widely used in this narrow sense. However, the term is now applied to any joint meeting the aforementioned general description where the filler material melting range exceeds about 450°C. For example, aluminum may be brazed with a lower melting point of aluminum–silicon alloy. Where the filler melting range is below about 450°C, the process is normally termed soldering or soft soldering. In most brazing processes, a flux or other means of cleaning the component surfaces is required. Many processes are used including gas or flame brazing, furnace brazing, bath or dip brazing, and vacuum and induction brazing, the various terms indicating the heat source or the environment. The term hard soldering means brazing or silver soldering.

Braze Interface

See *weld interface.*

Braze Welding

A method of welding by using a filler metal having a liquidus above 450°C (840°F) and below the solidus of the base metals. Unlike *brazing*, in braze welding, the filler metal is not distributed in the joint by capillary action.

Brazeability

The capacity of a metal to be brazed under the fabrication conditions imposed into a specific suitably designed structure and to perform satisfactorily in the intended service.

Brazement

An assembly whose component parts are joined by brazing.

Brazing

A group of welding processes that join solid materials together by heating them to a suitable temperature and using a filler metal having a liquidus above 450°C (840°F) and below the solidus of the base materials. The filler metal is distributed between the closely fitted surfaces of the joint by capillary action.

Brazing provides advantages over other welding processes, especially because it permits joining of dissimilar metals that, because of metallurgical incompatibilities, cannot be joined by traditional fusion processes. Since base metals do not have to be melted to be joined, it does not matter that they have widely different melting points. Therefore, steel can be brazed to copper as easily as steel to steel.

Brazing also generally produces less thermally induced distortion (warping) than fusion welding does, because an entire part can be brought up to the same brazing temperature, thereby preventing the kind of localized heating that can cause distortion in welding.

Finally, and perhaps most important to the manufacturing specialist, brazing readily lends itself to mass-production techniques.

The important elements of the brazing process are filler-metal flow, base-metal and filler-metal characteristics, surface preparation, joint design and clearance, and temperature, time, rate, and source of heating.

A myriad of applications of brazing occur in many fields, including automotive and aircraft designs; engines and engine components; electron tubes; vacuum equipment and nuclear components, such as production of reliable ceramic-to-metal joints; and miscellaneous applications such as fabrication of food-service dispensers (scoops) used for ice cream and corrosion-resistant and leak-proof joints in stainless steel blood-cell washers.

Brazing Alloy

See preferred term *brazing filler metal.*

Brazing Filler Metal

(1) The metal that fills the capillary gap and introduced into the brazed joint (see *braze*) to form the bond between the two components. It has a liquidus above 450°C (840°F) but below the solidus of the base materials. (2) A nonferrous filler metal used in *brazing and braze welding.* (3) The composition of the filler metal depends on the materials being joined, but for steels and copper alloys, typical filler metals include phosphor–copper, various plain or complex brasses, and silver solders.

A common name for high-copper brass used for the casting of such articles as flanges that are to be brazed on copper pipe. Federal specifications for brazing metal call for 84%–86% copper and the balance zinc. *Brazing brass*, of the American Brass Co., has 75% copper and 25% zinc. Some alloys also contain up to 3% lead for ease of machining. It also makes the metal easier to cast. *Nickel welding rod* is often used to braze cast iron. So is *nickel silver* containing 46.3% copper, 43.4% zinc, 10% nickel, 0.1% silicon, and 0.02% phosphorous, which matches the color of cast iron.

The term "brazing metal" is also applied to *brazing rods* of brass or bronze, used for joining metals. A common brazing rod is the 50–50 brass alloy with a melting point of 1616°F (880°C). The SAE designates this alloy as *spelter solder.* The joints made with it are inclined to be brittle. *Brazing wire*, of Chase Brass and Copper Co., contains 59% copper and 41% zinc, while the *brazing solder*, for brazing high-zinc brasses, has 51% copper and 49% zinc. *Phos-copper* is a phosphor copper that gives joints 98% the electric conductivity of copper. It flows at 1382°F (750°C). *Trimetal* is a brazing metal consisting of sheet brass with a layer of silver rolled on each side. It is used for brazing carbide tips to steel tools. The silver ensures a tightly brazed joint, while the brass center acts as a shock absorber for the cutting tool.

Brazing rods for brazing brasses and bronzes are usually of a composition similar to that of the base metal. For brazing cast iron and steel, various bronzes, naval brass, manganese bronze, or silicon bronze may be used. Brass rods may contain some silicon. Small amounts of silver added to the high-copper brazing metals give greater fluidity and better penetration into small openings. Because of the toxicity of *cadmium*, cadmium-free brazing filler materials such as the silver–copper–zinc *BAg-5, BAg-6, BAg-20,* and *BAg-35 brazing filler metals* and silver–copper–zinc–tin *BAg-7, BAg-28, BAg-34, BAg-36,* and *BAg-37 brazing filler metals* have been developed as alternatives. For joining *tungsten carbide* tips to steel, silver–copper–zinc–nickel *BAg-4* and *BAg-24 brazing filler metals* are alternatives. For joining copper and copper alloys, there are the copper–phosphorus *BCuP-2, BCuP-3, BCuP-4, BCuP-5, BCuP-6,* and *BCuP-7 brazing filler metals* and silver–copper–phosphorus brazing filler metals, both of which are self-fluxing on copper-to-copper metals. With a brazing flux they also can braze brasses and bronzes. However, they should not be used to braze iron or nickel alloys, or copper alloys if the nickel content exceeds 10%.

Nicrobraz alloys, from Wall Colmonoy, are nickel-, cobalt-, and copper-alloy brazing filler metals. *Superbraze 250*, from Materials Resources International, is a *tin–silver–titanium brazing filler metal* that works at 480°F–540°F (250°C–280°C). Because of that low-temperature range, it can be considered more a solder than a braze. However, other ingredients activate conditions for surface wetting and joining, making it more a braze. In an aluminum–silicon–lithium brazing filler (aluminum alloy 4045 except for the lithium), introduced by Kaiser Aluminum and Chemical Corp., the lithium (0.015%–0.1%, up to 0.3% if vacuum brazing) allows a fluoride flux to wet the surfaces to be joined before magnesium from the core alloy can "poison" the flux. An *82% gold-18 indium brazing filler metal* having a solidus to liquidus temperatures of 843.8°F–905°F (451°C–485°C) was found by Sandia National Laboratory for brazing *Ferralium 255 duplex stainless steel* and silver–nickel–silver laminates to copper. Gold alloy brazing foils containing 33%–35% nickel, 3%–4% chromium, 1%–2% iron, and 1%–2% molybdenum (*SK-1 gold brazing filler metal*) and 34%–36% nickel, 4%–4.5% chromium, and 2%–3% iron (*SK-2 gold brazing filler metal*), developed by GTE Laboratories, have a brazing temperature range of 1922°F–2102°F (1040°C–1150°C). They have been effective in brazing titanium-coated PY6 ($6Y_2 O_3$-modified *silicon nitride*) to *Incoloy 909* superalloy with a nickel interlayer at the *ceramic–metal* interface.

Brazing Procedure

The detailed methods and practices including all joint brazing procedures involved in the production of a brazement. See also *joint brazing procedure*.

Brazing Sheet

Brazing filler metal in sheet form.

Brazing Technique

The details of a brazing operation that, within the limitations of the prescribed brazing procedure, are controlled by the brazer or the brazing operator.

Brazing Temperature

The temperature to which the base metal is heated to enable the filler metal to wet the base metal and form a brazed joint.

Brazing Temperature Range

The temperature range within which brazing can be conducted.

Break

The process of parting into two or more pieces usually by some form of loading but excluding processes such as cutting. There are ill-defined subtleties in the choice between the terms "break," "crack," "cracking," "fracture," and "rupture." This may reflect the fact that the first terms, with their Anglo-Saxon roots, or possibly seen as common usage, whereas the latter two with their Latin origins, may sound appropriately scientific to some ears or merely confusing jargon to others.

Break-In (Noun)

See *running in*.

Break On (Verb)

To operate a newly installed bearing, seal, or other tribocomponent in such a manner as to condition its surface(s) for improved functional operation. See also *run in* (verb).

Breakaway Oxidation

Oxidation in which the oxide repeatedly falls off. During the time that the oxide remains on the surface, the rate of oxidation falls progressively, but after a layer has spalled off, the rate immediately increases only to fall again as the oxide thickness builds again. As a result, over a long period, a continuing and fairly constant rate of attack is observed.

Breakaway Torque

See *starting torque*.

Breakdown

(1) An initial rolling or drawing operation applied to an ingot or a series of such operations, for the purpose of reducing a casting or extruded shape prior to the finish reduction to desired size. (2) A preliminary press forging operation.

Breakdown Potential

The least noble potential where *pitting* or *crevice corrosion*, or both, will initiate and propagate.

Breakdown Voltage

The voltage required, under specific conditions, to cause the failure of an insulating material. See also *dielectric strength* and *arc resistance*.

Breaker Plate

In plastic forming, a perforated plate located at the rear end of an extruder or at the nozzle end of an injector cylinder. It often supports the screens that prevent foreign particles from entering the die and is used to keep unplasticized material out of the nozzle and to improve distribution of color particles.

Breaking Extension

The elongation necessary to cause rupture of an adhesive and bonded test specimen. The tensile strain at the moment of rupture.

Breaking Factor

The breaking load divided by the original width of an adhesive and bonded test specimen, expressed in lb/in.

Breaking Load

The maximum load (or force) applied to a test specimen or structural member loaded to rupture.

Breaking Out

(1) Removal of a casting from its mold. (2) Fiber separation or break on surface plies at drilled or machined composite material edges.

Breaking Stress

Same as *fracture stress*.

Breaks

Creases or ridges usually in *untempered* or in aged material where the yield point has been exceeded. Depending on the origin of the breaks, they may be termed *cross breaks*, *coil breaks*, *edge breaks*, or *sticker breaks*.

Breather

A loosely woven material that serves as a continuous vacuum path over a part but is not in contact with the resin.

Breathing

The opening and closing of a mold to allow gas to escape early in the plastic molding cycle. Also called degassing; sometimes called bumping, in phenolic molding. When referring to plastic sheeting, the term breathing indicates permeability to air.

Breeze

Furnace ash of substantial form as opposed to the powder or fly ash. Such material can be bonded by cement to form building blocks.

Bremsstrahlung

See *continuum*.

Brick

Brick is a construction material usually made of clay and extruded or molded as a rectangular block. Three types of clay are used in the manufacture of bricks: surface clay, fire clay, and shale. Adobe brick is a sun-dried molded mix of clay, straw, and water, manufactured mainly in Mexico and some southern regions of the United States.

The first step in manufacture is crushing the clay. The clay is then ground, mixed with water, and shaped. Then the bricks are fired in a kiln at approximately 1093°C.

Substances in the clay such as Ferrous, magnesium, and calcium oxides impart color to the bricks during the firing process. The color may be uniform throughout the bricks, or the bricks may be manufactured with a coated face. The latter are classified as glazed, claycoat, or engobe. Engobes are ductile erosion behavior coatings, also called slurries, that are applied to plastic or dry body brick units to develop the desired color and texture. Claycoat is a type of engobe that is sprayed on as a coating of liquid clay and pigments.

Clay bricks are manufactured for various applications. The most commonly used brick product is known as facing brick. In addition to standard bricks, decorative bricks molded in special shapes are available in both standard and custom sizes. They are used to form certain architectural details such as water tables, arches, copings, and corners. Bricks are also used to create sculptures and murals.

Sand–lime bricks for fancy walls are of sand and lime pressed in an atmosphere of steam. They are not to be confused with the sand–lime bricks used for firebrick, which are of refractory silica sand with a lime band. Ceramic glazes and semiglazes are used on some building bricks especially on the yellow ones.

Bridge Die

In two-section extrusion die capable of producing tubing or intricate hollow shapes without the use of a separate mandrel.

Metal separates into two streams as it is extruded past a bridge section, which is attached to the main die section and holds a stub mandrel in the die opening; the metal then is rewelded by extrusion pressure before it enters the die opening. Compare with *porthole die*. See *extrusion*.

Bridging

(1) Premature solidification of metal across a mold section or ingot before the metal below or beyond solidifies. (2) Solidification of slag within a cupola at or just above the tuyeres. (3) Welding or mechanical locking of the charge in a down feed melting or smelting furnace. (4) In powder metallurgy, the formation of arched cavities in a powder mass. (5) In soldering, an unintended solder connection between two or more conductors, either securely or by mere contact. Also called a crossed joint or solder short.

Bright Annealing

Annealing in a protective medium to prevent discoloration of the bright surface.

Bright Bolt

See *bolt*.

Bright Dip

A solution that produces, through chemical action, a bright surface on an immersive metal.

Bright Finish

A high-quality finish produced on ground and polished rolls. Suitable for electroplating.

Bright Nitriding

Nitriding in a protective medium to prevent discoloration of the bright surface. Compare with *blank nitriding*.

Bright Plate

An electrodeposit that is lustrous in the as-plated condition.

Bright Range

The range of current densities, other conditions being constant, within which a given electroplating bath produces a bright plate.

Bright Stock

High-viscosity mineral oils that remain after vacuum distillation of crude oil.

Brightener

A material added to electroplating baths, etc., with the intention of producing a bright lustrous surface.

B

Bright-Field Illumination

For reflected light, the form of illumination that causes specularly reflected surfaces normal to the axis of the microscope to appear bright. For transmission electron microscopy, the illumination of an object so that it appears on a bright background. Compare with *dark-field illumination*.

Brillouin Zones

See *electron bands*.

Brine

Seawater containing a higher concentration of dissolved salt than that of the ordinary ocean.

Brine Quenching

A quench in which brine (saltwater chlorides, carbonates, and cyanides) is the quenching medium. The salt addition improves the efficiency of water at the vapor phase or hot stage of the quenching process.

Brinell Hardness Number (HB)

A number related to the applied load and to the surface area of the permanent impression made by a ball indenter computed from

$$HB = \frac{2P}{\pi D \sqrt{D - D^2 - d^2}}$$

where
    P is the applied load, kgf
    D is the diameter of the ball, mm
    d is the mean diameter of the impression, mm

Brinell Hardness Test

A test for determining the hardness of a material by forcing a hard steel or tungsten carbide ball of specified diameter (typically, 10 mm) into it under a specified load. The resultant indentation diameter is then measured and, by comparison with established tables, a hardness is determined. The result is expressed as the *Brinell hardness number*. See *hardness*.

Brinelling

(1) Indentation of the surface of a solid body by repeated local impact or impacts, or static overload. Brinelling may occur especially in a rolling-element bearing. (2) Damage to a solid bearing surface *characterized* by one or more plastically formed indentations brought about by overload.
    See also false Brinelling.

Briquette

A self-sustaining mass of powder of defined shape. See preferred term *compact* (noun).

Britannia Metal

An alloy of tin with, typically, about 7% antimony and 2% copper although there are variations. It is one of the range of pewters with mainly decorative applications.

British Association (Threads)

A series of thread sizes of standard profile used mainly in electrical and electronic applications rather than mechanical engineering.

British Thermal Unit

The quantity of heat required to raise the temperature of 1 lb of water by 1°F.

Brittle

(1) Permitting little or no plastic (permanent) deformation prior to fracture. (2) Failing without significant ductility. Sometimes, the term is used to imply susceptibility to failure under impact loading but materials can fail in a brittle manner even under static loads.

Brittle Crack Propagation

A very sudden propagation of a crack with the absorption of no energy except that stored elastically in the body. Microscopic examination may reveal some deformation even though it is not noticeable to the unaided eye. Contrast with *ductile crack propagation*.

Brittle Edge Technique

A technique for impact and bend test specimens to prevent the initiation of shear lips.

Brittle Erosion Behavior

Erosion behavior having characteristics (e.g., little or no plastic flow, the formation of cracks) that can be associated with *brittle fracture* of the exposed surface. The maximum removal occurs at an angle near 90°, in contrast approximately 25° for *ductile erosion behavior*.

Brittle Fracture

(1) Generally, cracking without associated ductility and, usually, with only a small energy input. In a narrower sense, the term is associated with the change from ductile to brittle behavior that some metals, particularly iron and other metals with a body-centered cubic structure, undergo as the temperature falls. In the case of many steels, this occurs at about ambient temperature. Above the brittle to ductile transition, this steel will accept deformation, that is, absorb energy and act in a ductile manner, before failing by overloading. Such behavior makes the metal tolerant of minor cracks and other stress-raising features, reduces sensitivity to impact and may give warning of imminent failure. Below the transition, the steel will be sensitive to cracks and impact loads and may fail without warning even under apparently static loads. Above the transition the fracture mode is shear, that is, planes of atoms sliding over each other, while at lower temperatures, the mode changes to cleavage in which the planes peel apart. Although the fractures are both transgranular, the surface produced by the two modes is quite distinct, one being a dull grey and the cleavage bright and crystalline, so the change from one to the other is readily observed on a series of test pieces broken over a range of temperatures. The % crystallinity of the test pieces broken at the various temperatures is noted to determine the fracture appearance transition temperatures (FATT), that is, the temperature at which the fracture is 50% ductile and 50% cleavage. The test pieces are often broken in an impact test and the energy required

to break the specimens is then plotted against the temperature. The large step change in toughness, as measured by the energy absorbed, takes place over a limited range, typically of about 50°C, but above and below this range, any change is gradual. Consequently, the terms upper shelf and lower shelf are used with reference to the respective parts of the curve and for behavior that is, respectively, fully ductile or fully brittle. Brittle fracture has been responsible for many major and dramatic failures of ships, bridges, pressure vessels and pipelines, etc. (2) Separation of a solid accompanied by little or no macroscopic plastic deformation. Typically, brittle fracture occurs by rapid crack propagation with less expenditure of energy than that for *ductile fracture*. Brittle tensile fractures have a bright, granular appearance and exhibit little or no necking. A *chevron pattern* may be present on the fracture surface, pointing toward the origin of the crack, especially in brittle fractures in flat platelike components. Examples of brittle fracture include *transgranular cracking* (*cleavage and quasi-cleavage fracture*) and *intergranular cracking* (*decohesive rupture*).

Brittleness

(1) A lack of ductility, sensitivity to cracks, and low absorption of energy during crack propagation. See *brittle fracture*. (2) The tendency of a material to fracture without first undergoing significant *plastic deformation*. Contrast with *ductility*.

Broaching

Cutting with a tool that consists of a bar having a single edge or a series of cutting edges (i.e., teeth) on its surface. The cutting edges of a multiple-tooth, or successive single-tooth, bridges increase in size and/or change in shape. The broach courage of a straight line or axial direction when relative motion is produced in relation to the workpiece, which may also be rotating. The entire cut is made in single or multiple passes over the workpiece to shape the required surface contour.

Broad Goods

Fiber woven to form fabric up to 1270 mm (50 in.) wide for reinforcement of plastics. It may or may not be impregnated with resin and is usually furnished in coils of 25–140 kg (50–300 lb).

Bromine

An elementary material, symbol Br. It is a reddish-brown liquid having a boiling point at 138°F (59°C). It gives off very irritating fumes and is highly corrosive. It is one of four elements called *halogens*, and its name is derived from Greek words meaning *salt producer*. They are fluorine, chlorine, iodine, and bromine. They are all chemically active, combining with hydrogen and most metals to form *halides*. Bromine is less active than chlorine but more so than iodine. It is moderately soluble in water. It never occurs free in nature, and it is obtained from natural bromide brines by oxidation and steaming or by electrolysis. It occurs in seawater to the extent of 65–70 parts per million and is extracted. It is marketed 99.7% minimum purity with specific gravity not less than 3.1, but dry elemental bromine, Br_2, is marketed 99.8% pure for use as a brominating and oxidizing agent. For these uses, also, bromine is available as a crystalline powder as *dibromodimethylhydantoin*, containing 55% bromine. *Brom 55* is this material. Bromine is also used as a flame retardant in plastics, although its use as such has been questioned due to environmental concerns.

It is used in the manufacture of agricultural chemicals, dyes, photographic chemicals, poison gases for chemical warfare, pharmaceuticals, disinfectants, and many chemicals. It is also employed in the extraction of gold. Bromine's major end use, as ethylene dibromide for scavenging lead antiknock compounds in gasoline, is decreasing as these environmentally hazardous additives are phased out.

Bronze

The term "bronze" is generally applied to any copper alloy that has as the principal alloying element a metal other than zinc or nickel. Originally the term was used to identify copper–tin alloys that had tin as the only, or principal, alloying element. Some brasses are called bronzes because of their color, or because they contain some tin. Most commercial copper–tin bronzes are now modified with zinc, lead, or other elements.

The *copper–tin bronzes* are a rather complicated alloy system. The alloy with up to about 10% tin has a single-phase structure. Above this percentage, a second phase, which is extremely brittle, can occur, making plastic deformation impossible. Thus high-tin bronzes are used only in cast form. Tin oxide also forms in the grain boundaries, causing decreased ductility, hot workability, and castability. Additions of small amounts of phosphorus, in the production of phosphor bronzes, eliminate the oxide and add strength. Because tin additions increase strength to a greater extent than zinc, the bronzes as a group have higher strength than brasses—from around 414 to 724 MPa in the cold-worked high-tin alloys. In addition, fatigue strength is high.

Bronzes containing more than 90% copper are reddish; below 90% the color changes to orange-yellow, which is a typical bronze color. Ductility rapidly decreases with increasing tin content. Above 20% tin, the alloy rapidly becomes white and loses the characteristics of bronze. A 90% copper and 10% tin bronze has a density of 0.317 lb/in.^{3} (8775 kg/m^3); an 80%–20% bronze has a density of 0.315 lb/in.^{3} (8719 g/m^3). The 80%–20% bronze melts at 1868°F (1020°C), and 95%–5% bronze melts at 2480°F (1360°C).

The family of *aluminum bronzes* is made up of alpha-aluminum bronzes (less than about 8% aluminum) and alpha-beta bronzes (8%–20% aluminum) plus other elements such as iron, silica, nickel, and manganese. Because of the considerable strengthening effect of aluminum, and the hard condition these bronzes are among the highest-strength copper alloys. Tensile strength approaches 690 MPa. Such strengths plus outstanding corrosion resistance make them excellent structural materials. They are also used in wear-resistance applications and for nonsparking tools. Phosphor bronzes have a tin content of 1.25%–10%. They have excellent mechanical and cold-working properties and a low coefficient of friction, making them suitable for springs, diaphragms, bearing plates, and fasteners. Their corrosion resistance is also excellent. In some environments, such as salt water, they are superior to copper. *Leaded phosphor bronzes* provide improved machinability. *Silicon bronzes* are similar to aluminum bronzes. Silicate content is usually between 1% and 4%. In some, zinc and manganese are also present. Besides raising strength, the presence of silicon sharply increases electrical resistivity. Aluminum–silicon bronze has exceptional strength and corrosion resistance and is particularly suited to hot working.

Gear bronze may be any bronze used for casting gears and worm wheels, but usually means a tin bronze of good strength deoxidized with phosphorus and containing some lead, to make it easy to machine and to lower the coefficient of friction. A typical gear bronze contains 88.5% copper, 11% tin, 0.25% lead, and 0.25% phosphorus. It has a tensile strength of up to 276 MPa, elongation 10%,

B

and Brinell hardness of 70%–80%, or up to 90% when chill-cast. The density is 0.306 lb/in.^{3} (8470 kg/m^3). This is *SAE bronze No. 65*. A *hard gear bronze*, or *hard bearing bronze*, contains 84%–86% copper, 13%–15% tin, up to 1.5% zinc, up to 0.75% nickel, and up to 0.5% phosphorus. Hard and strong bronzes for gears are often silicon bronze or manganese bronze.

In a modified 90–10 type of bronze, the zinc is usually from 2% to 4% and the lead up to 1%. A cast bronze of this type will have a tensile strength of about 276 MPa, elongation of 15%–25%, and a Brinell hardness of 60%–80%, those high in zinc being the stronger and more ductile and those high in lead being the weaker. Bronzes of this type are much used for general castings and are classified as *composition metal* in the United States. In England, they are called *engineer's bronze*.

Architectural bronze, or *art bronze*, is formulated for color and is very high in copper. One foundry formula for art bronze of a dull red color calls for 97% copper, 2% tin, and 1% zinc. For ease of casting, however, they are more likely to contain lead, and a *gold bronze* for architectural castings contains 89.5% copper, 2% tin, 5.5% zinc, and 3% lead. In leaded bronze, the hard copper–tin crystals aid in holding the lead in solution. These bronzes are resistant to acids and are grouped as a valve bronze, or as bearing bronze because of the hard crystals in a soft matrix. Federal specifications for bronze give 10 grades in wide variations of tin, zinc, and lead. The ASTM designates five grades of *bronze casting alloys*. *Alloy No. 1* contains 85% copper, 10% tin, and 5% lead; *alloy No. 5* contains 70% copper, 5% tin, and 25% lead. The British *coinage copper* is a bronze containing 95.5% copper, 3% tin, and 1.5% zinc.

Many of these bronzes are designated by alloy number and grouped into several families of standard alloys. There are four principal families of wrought bronze: *copper-tin-phosphorus alloys*, or *phosphor bronzes* (C 50100 to C 52400); *copper-tin-phosphorus-lead alloys*, or *leaded phosphor bronzes* (C53200 to C54800); *copper-aluminum alloys*, or *aluminum bronzes* (C60600 to C64400); and *copper-silicon alloys*, or *silicon bronzes* (C64700 to C66100). A few, such as two *manganese bronzes*, are included in the copper-zinc family of copper alloys. The aluminum bronze family is the largest, containing nearly three dozen standard compositions.

Casting alloys comprise five main families: *copper-zinc-manganese alloys*, or *manganese bronzes* (C86100 to C 86800); *copper-tin alloys*, or *tin bronzes* (C90200 to C91700); *copper-tin-lead alloys*, or *leaded* and *highly leaded tin bronzes* (C92200 to C94500); *copper-tin-nickel alloys*, or *nickel-tin bronzes* (C94700 to C 94900); and *copper-aluminum alloys*, or *aluminum bronzes* (C95200 to C 95900). *Copper-silicon bronzes* are included in the C87300 to C87900 family of copper-silicon alloys.

Flat products—0.12 in (3 mm) thick *aluminum bronze C61400*— have tensile yield strengths ranging from 310 MPa in the annealed (O60) temper to 414 MPa after cold working to the H04, or hard, temper, with ductility decreasing from 40% to 32% elongation, respectively. All casting alloys can be sand-cast, many can be centrifugally cast, and some can be permanent mold, plaster-, and investment-cast but not die-cast. Among the strongest is *manganese bronze C86300,* which, as cast in sand molds, provides a minimum tensile yield strength of 414 MPa and at least 12% elongation. Some bronzes, such as *nickel-bearing silicon bronzes* and aluminum bronzes, especially those containing more than 9% aluminum, can be strengthened by age-hardening.

Unfortunately, the term "bronze" has occasionally been coupled with a trade or brand name for some alloys having properties inferior to what might be inferred from the name. See also *gunmetal, phosphor bronze,* and *cold cast bronze.*

Bronze Powder

Pulverized or powdered bronze made in flake form by stamping from sheet metal. It is used chiefly as a paint pigment and as a dusting powder for printing. In making the powder, the sheets are worked into a thin foil, which becomes harder under the working and breaks into small flakes. Lubricant keeps flakes from sticking to one another. Usually stearic acid is used, but in the dusting powder, hot water or nonsticky lacquers are used. The powder is graded in standard screens and is then polished in revolving drums with a lubricant. This gives it the property of leafing or forming a metallic film in the paint vehicle. The leaf is also called *composition leaf*, or *Dutch metal leaf*, when used as a substitute for gold leaf. Flitters are made by reducing thin sheets to flakes, and they are not as fine as bronze powder. *Alpha bronze*, a prealloyed 91% copper, 10% tin powder from Makin Metal Powders of England for powder-metal parts, provides greater green strength and less wear and noise in bearing applications then a premixed 90–10 composition.

The compositions of bronze powder vary, and seven alloys form the chief commercial color grades from the reddest, called *pale gold*, which is 95% copper and 5% zinc, to the rich gold which has 70% copper and 30% zinc. Colors are also produced by heating to give oxides of deep red, crimson, or green-blue. The powder may also be dyed in colors, using tannic acid as a mordant, or treated with acetic acid or copper acetate to produce an antique finish. The color or tone of bronze powders may also be adjusted in paints by adding a proportion of mica powder. A *white bronze powder* is made from aluminum bronze, and the silvery colors are obtained with aluminum powder. The bronze powder of 400 mesh used for inks is designated as extra fine. The fine grade, for stencil work, is 325 mesh. Medium fine, for coated paper, has 85% of the particles passing through a 325-mesh screen and 15% retained on the screen. A 400-mesh powder has 500 million particles per gram. The old name for bronze powder is *gilding powder*. It is also called *gold powder* when used in cheap gold-colored paints, but bronze powders cannot replace gold for use in atmospheres containing sulfur, or for printing on leather where tannic acid would corrode the metal. *Gold pigments* used in plastics are bronze powders with oxygen stabilizers.

Bronze Welding

A process for Makin fusion joints with brazing filler metals, particularly using arc welding techniques, but forming large bulky deposits rather than the thin capillary film characteristic of true brazing. See *brazing*.

Bronzing

(1) Applying a chemical finish to copper or copper-alloy surfaces to alternate color. (2) Plating a copper-tin alloy on various materials.

Brush Anodizing

An *anodizing* process similar to *brush plating*.

Brush Plating

Plating with a concentrated solution or gel held in or fed to an absorbing medium, pad, or brush carrying the anode (usually insoluble). The brush is moved back and forth over the area of the cathode to be plated.

Brush Polishing (Electrolytic)

A method of electropolishing in which the electrolyte is applied with a pad or brush in contact with the part to be polished.

B-Stage

An intermediate stage in the reaction of certain thermosetting resins in which the material softens when heated and swells when in contact with certain liquids, but may not entirely fuse or dissolve. The resin is an uncured thermosetting adhesive, usually in the stage. Synonym for resitol. Compare with *A-stage* and *C-stage*.

B-Stage Resin

In a thermosetting reaction, a resin that is in an intermediate state of cure when it is sticky, or tacky, and capable of further flow. The cure is normally complete during the laminating cycle. See *C-stage resin* and *prepreg*.

Bubble

A spherical, internal void, or globule of air or other gas trapped within a plastic. See *void*.

Bubbler

In forming of plastics, a device inserted into a mold force, cavity, or core, which allows water to flow deep inside the hole into which it is inserted and to discharge through the open end of the hole. Uniform cooling of the mold and of isolated mold sections can be achieved in this matter.

Bubbler Mold Cooling

In injection molding of plastics, a method of uniformly cooling a mold; a stream of cooling liquid flows continuously into a cooling cavity equipped with a coolant outlet normally positioned at the opposite end.

Bubbly Oil

Oil containing bubbles of gas.

Buckle

(1) Bulging of a large, flat face of a casting; in investment casting, cause my *dip coat* peeling from the pattern. (2) An indentation in a casting, resulting from expansion of the sand, can be termed the start of an expansion defect. (3) A local waviness in metal bar or sheet, usually transverse to the direction of rolling.

Buckling

(1) A mode of failure generally characterized by an unstable lateral material deflection due to compressive action on the structural element involved. (2) In metal forming, a bulge, bend, kink, or other wavy condition of the workpiece caused by compressive stresses. See also *compressive stress*.

Buckyballs

Buckyballs, the soccer-shaped molecules discovered over a decade ago, are made entirely of carbon atoms linked with unusual chemical bonds.

Buckyballs have a spherical structure with their surfaces made up of from 28 to 450 carbon atoms and hexagonal and pentagonal arrays, with 60 carbon atoms being the most stable. This soccerball shaped configuration is reminiscent of R. Buckminster Fullers geodesic domes, hence the name Buckminster Fullerene or "buckyball."

Buckybowls, large fragments of buckyballs, have been synthesized. The molecules have the ability to take up electrons and give them back later, under the right conditions, and higher concentration than buckyballs.

The only practical application thus far has been the computer industry's use of another relative, the bulkytube, as in atomic-scale probe. Some researchers believe that the buckybowls may facilitate the development of plastic batteries that would be lighter, smaller, and more environmentally friendly than the rechargeable batteries now used in our cellular phones and laptop computers.

The shape of the molecule may also allow it to bond with other molecules. It fits over the buckyball much like a contact lens and may be able to serve as a medium that links other substances to the balls.

Bucky Diaphragm

An x-ray scanner reducing device originally intended for medical radiography but also applicable to industrial radiography in some circumstances. Thin strips of lead, with their widths held parallel to the primary radiation, are used to absorb scattered radiation preferentially; the array of strips is in motion during exposure to prevent formation of a pattern on the film.

Buffer

(1) A substance which by its addition or presence tends to minimize the physical and chemical effects of one or more of the substances in a mixture. Properties often buffered include pH, oxidation potential and flame or Pfizer temperatures. (2) A substance whose purpose is to maintain a constant hydrogen-ion concentration in water solutions, even where acids or alkalis are added. Each buffer has a characteristic limited range of pH over which it is affected.

Buffer Gas

A protective gas curtain at the charge or discharge end of a continuously operating sintering furnace.

Buffing

Polishing of a surface usually by a high-speed rotating soft fabric wheel, the Bob, carrying some polishing compound. The term may imply a process of surface mirroring rather than a clean cutting action.

Buffing Wheel

Buff sections assembled to the required face with for use on a rotating shaft between flanges. Sometimes called a buff.

B

B

Buff Section

A number of fabric, paper, or leather disks with concentric center holes held together by various types of sewing to provide degrees of flexibility or hardness. These sections are assembled to make wheels for polishing.

Builder

A material, such as an alkali, a buffer, or a water softener, added to a soap or synthetic surface-active agent to produce a mixture having enhanced detergency. Examples: (1) alkalis—caustic soda, soda ash, and trisodium phosphate. (2) *buffers*—sodium metasilicate and borax; and (3) water softeners—sodium tripolyphosphate, sodium tetraphosphate, sodium hexametaphosphate, and ethylene diamine tetraacetic acid.

Building Stone

Any stone used for building construction may be classified as building stone. *Granite and limestone* are among the most ancient of building materials and are extremely durable. Two million limestone and granite blocks, totaling nearly 8,000,000 long tons (8128 million kg), were used in the peer amid of Giza, built about 2980 B.C., the granite being used for casing. Availability, or a near supply, may determine the stone used in ordinary building, but for public buildings stone is transported long distances. Some sandstones, such as the *red sandstone* of the Connecticut Valley, weather badly and are likely to scale off with penetration of moisture and frost. Granite will take heavy pressures and is used for foundation tiers and columns. Limestones and well-segmented sandstones are employed extensively above the foundations. Nearly half of all the lifestyle used in United States and block form is Indiana limestone. Marble has a low crushing strength and is usually an architectural or facing stone.

Crushed stone is used for making concrete, for railway ballast, and for road making. The commercial stone is quarried, crushed, and graded. Much of the crushed stone used is granite, limestone, and *trap rock*. The last is a term used to designate basalt, gabbro, diorite, and other dark-colored, fine-grained igneous rocks. Graded crushed usually consists of only one kind of rock and is broken with sharp edges. *Granite granules* for making hard terrazzo floors are marketed in several sizes, and in pink, green, and other selected colors. *Roofing granules* are grated particles of crushed rock, slate, slag, porcelain, or tile used as surfacing on asphalt roofing and shingles. Granules have practically superseded gravel for this purpose. *Ceramic granules* are produced from clay or shale, fired and glazed with metallic salts. They are preferred because the color is uniform.

Building Up

To electroplate, metal spray, etc., to increase or restore dimensions. Compare with *surfacing*. Also the same as *buildup*.

Buildup

(1) Excessive electrodeposition that occurs on high current-density areas, such as corners or edges. (2) In welding, the sequence of deposition of multiple layers or beads to form a bulk weld.

Buildup Sequence

(1) The order in which the weld beads of a multi-pass weld are deposited with respect to the cross section of the joint. See also *block sequence* and *longitudinal sequence*.

Built-Up Edge

(1) Chip material adhering to the tool face adjacent to the cutting edge during cutting. (2) Material from the workpiece, especially a machining, which is stationary with respect to the tool. See also *wedge formation*.

Built Up Laminated Wood

An assembly made by joining layers of lumber with mechanical fastenings so that the grain of all laminations is essentially parallel.

Bulb Section

Rolled or extruded bar having one edge much thicker than the other, the transition being fairly abrupt rather than a continuous taper.

Bulging

(1) Expanding the walls of a cup, shell, or tube within an internally expanded segmented punch or a punch composed of air, liquids, or semiliquids such as waxes, rubber, and other elastomers. (2) The process of increasing the diameter of a cylindrical shell (usually to a spherical shape) or of expanding the outer walls of any shell or box shape whose walls were previously straight.

Bulk Adherend

With respect to interphase, the adherend, unaltered by the adhesive. Compare with *bulk adhesive*.

Bulk Adhesive

With respect to interphase, the adhesive, unaltered by the adherend. Compare with *bulk adherend*.

Bulk Density

(1) The weight of an object or material divided by its material volume less the volume of its open pores. (2) The density of a plastic molding material in loose form (granular, nodular, and so forth), expressed as a ratio of weight to volume. (3) Metal powder in a container or bin expressed in mass per volume.

Bulk Factor

The ratio of the volume of a raw plastic molding compound or powdered plastic to the volume of the finished, solid piece produced from it. The ratio of the density of the solid plastic object to the apparent, or bulk, density of the loose molding powder.

Bulk Forming

Forming processes, such as extrusion, forging, rolling, and drawing, in which the input material is in billet, ride, or slab form and a considerable increase in surface-to-volume ratio in the formed part occurs under the action of largely compressive loading. Compare with *sheet forming.*

Bulk Modulus of Elasticity (*K*)

The measure of resistance to change in volume; the ratio of hydrostatic stress to the corresponding unit change in volume. This elastic constant can be expressed by:

$$K = \frac{\sigma_m}{\Delta} = \frac{-p}{\Delta} = \frac{1}{\beta} = -$$

where *K* is the bulk modulus of elasticity, σ_m is hydrostatic or mean stress tensor, *p* is hydrostatic pressure, and ß is compressibility. Also known as bulk modulus, compression modulus, hydrostatic modulus, and volumetric modulus of elasticity.

Bulk Molding Compound

BMC is a puttylike mixture of thermosetting polyester, vinyl ester or phenolic resins, additives, fillers, pigments, and/or reinforcements generally extruded into shapes for compression, transfer, or injection molding. Bulk Molding Compounds Inc. makes a dozen product series compounds, including general-purpose, electrical, medium- and high-strength, food-contact, and corrosion-resistant types. Applications include electrical coil bobbins, brush holders and connectors, dishwares, pants, trays, tubs, and housings for headlamp reflectors, small appliances, auto parts, and hand-held power tools. **Nu-Stone**, of Industrial Dielectrics, is a BMC that looks like granite.

Bulk Sample

See *gross sample.*

Bulk Sampling

Obtaining a portion of a material that is representative of the entire lot.

Bulk Volume

The volume of the metal powder fill in the die cavity.

Bull Block

A machine with a power-driven revolving drum for cold drawing wire through a drawing die as the wire winds around the drum.

Bulldozer

Slow-acting horizontal *mechanical press* with a large bed used for bending and straightening. The work is done between dies and can be performed hot or cold. The machine is closely allied to a forging machine.

Bullion

(1) A semirefined alloy containing sufficient precious metal to make recovery profitable. (2) Refined gold or silver, uncoined. (3) In bulk, that is, as ingots or bar rather than a finished product.

Bulls-Eye Structure

The microstructure of malleable or ductile cast iron when graphite nodules are surrounded by a ferrite layer in a pearlitic matrix.

Bump Check

See *percussion cone.*

Bumper

A machine used for packing molding sand in a flask by repeated jarring or jolting. See also *jolt ramming.*

Bumping

(1) Forming a dish in metal by means of many repeated blows. (2) Forming a head. (3) Setting the seams on sheet metal parts. (4) Ramming sand in a flask by repeated jarring and jolting.

Bundle

A general term for a collection of essentially parallel filaments or fibers using composite materials.

Burden

(1) The total material charge into a furnace or other reaction vessel. (2) In submerged arc welding, the total amount of fused and unfused flux over the weld metal.

Buret

An instrument used to deliver variable and accurately known volumes of a liquid during titration or volumetric analysis. Burets are usually made from uniform-bore glass tubing in capacities of 5 to 100 mL, the most common being 50 mL. See also *titration* and *volumetric analysis.*

Burgers Vector

The total lattice translation resulting from the passage of a dislocation. See *dislocation.*

Burlap

A coarse, heavy cloth made of plane-woven jute, or jutelike fibers, and used for wrapping and bagging bulky articles, for upholstery linings, and as a backing fabric for linoleum. Finer grades are used for wall coverings. The standard burlap from India is largely from jute fibers, but some hibiscus fibers are used. For bags and wrappings, the weave is coarse and irregular, and the color is the natural tan. The coarse grades such as those used for wrapping cotton bales are sometimes called *gunny* in the United States, but *gunny* is a general name for all burlap in Great Britain. Dundee, Scotland, is the important center of burlap manufacture outside of India, but considerable quantities are made from native fibers in Brazil and other countries. Burlap is woven in widths up to 144 in (3.6 m), but 36, 40, and 50 in (0.91, 1.02, and 1.27 m) are the usual widths. Hessian is the name of a 9.5 oz (269 g), plane-woven finer burlap made to replace an older fabric of the same name woven from coarse and heavy flax fibers. When dyed in colors, it is used for linings, wall coverings, and upholstery. *Brattice cloth* is a very coarse, heavy, and tightly woven jute cloth, usually 20 oz (567 g) used for gas breaks in coal mines; but a heavy cotton duck substituted for the same purpose is called by the same name. Most burlap for commercial bags is 8, 9, 10, and 12 oz (226, 255, 283, at 340 g), feed bags being 8 oz (226 g), and grain bags 10 oz (283 g).

Burn

(1) The heat treatment to which refractory materials are subjected in the firing process. (2) To heat particulate material, grain, aggregate, refractory bricks, or ceramic ware in a furnace or kiln, sufficient to cause desired chemical, structural, or crystalline changes in the material though below the melting point of the major component. To fire, to calcine, or sinter; usually more intense than to dry or to activate.

Burn Off Rate

The rate at which a welding electrode or other consumable item is melted or otherwise consumed.

Burn Through (of Weld)

A hole completely through a welded joint due to local collapse of the weld pool.

Burnback Time

See preferred term *meltback time*.

Burned

Showing evidence of thermal decomposition or charring through some discoloration, distortion, destruction, or conversion of the surface of the plastic, sometimes to a carbonaceous char.

Burned Deposit

A dull, nodular electrodeposit resulting from excessive plating current density.

Burned-In Sand

A defect consisting of a mixture of sand and metal coherent to the surface of a casting.

Burned-On Sand

A mixture of sand and cast metal adhering to the surface of a casting. In some instances, may resemble *metal penetration*.

Burned Plating

See *burned deposit*.

Burned Sand

Foundry sand in which the binder or bond has been removed or impaired by contact with molten metal.

Burner

See preferred term *oxygen cutter*.

Burning

(1) Permanently damaging a metal or alloy by heating to cause either incipient melting or intergranular oxidation. See also *overheating* and *grain boundary liquidation*. The various forms of damage include severe grain growth, surface cracking, and grain boundary liquation. Decarburization, gross oxidation, and penetration of oxide intrusions down grain boundaries. (2) During subcritical annealing, particularly in continuous annealing, production of a severely decarburized and grain-coarsened surface layer that results from excessively prolonged heating to an excessively high temperature. (3) In grinding, getting the work hot enough to cause discoloration or to change the microstructure by tempering or hardening. (4) In sliding contacts, the oxidation of a surface due to local eating in an oxidizing environment. See also *metallurgical burn*. (5) An alternative term for flow welding techniques where bulk molten metal is poured into a substantial gap between the two phases to be joined. Sufficient heat input to melt the joint faces is insured by the molten metal being superheated or being allowed to flow to waste across the joint. Lead burning usually refers to the joining of led components, by autogenous welding, that is, without the addition of solder or any other filler. (6) The term is also colloquially used for flow weld techniques for localized repair in which new material is cast into place in a temporary mold formed around the damaged area of the component. In practice molten metal is usually allowed to flow through the mold to waste until the parent metal edges have use. It is essentially the same as burning (5) but "burning on" is particularly associated with marine repairs such as two propellers.

Burning (Firing) of Refractories

The final heat treatment in a kiln to which refractory brick and shapes are subjected in the process of manufacture for the purpose of developing bond and other necessary physical and chemical properties.

Burning In

See preferred term *flow welding*.

Burning Rate

The tendency of plastic articles to burn at given temperatures.

Burning (Welding)

See preferred term *oxygen cutting*.

Burnish

(1) To alter the original manufactured surface of a sliding or rolling surface to a more polished condition. (2) To apply a substance to a surface by rubbing.

Burnishing

Finished sizing and smooth finishing of surfaces (previously machined or ground) by displacement, rather than removal, of minute surface irregularities with smooth-point or line-contact fixed or rotating tools.

Burnoff

(1) Unintentional removal of an autocatalytic deposit from a non-conducting substrate, during subsequent electroplating operations, owing to the application of excessive current or a poor contact area. (2) Removal of volatile lubricants such as metallic stearates from metal powder compacts by heating immediately prior to sintering.

Burnoff Rate

See preferred term *melting rate*.

Burnout

In casting, firing a mold and a high temperature to remove pattern material residue.

Burn-Through

A term erroneously used to denote excessive melt-through or a hole. See also *melt-through*.

Burn-Through Weld

A term erroneously used to denote AC weld or spot weld.

Burnt

See *burning*.

Burnt Deposit

In electroplating, the nodules and scabs formed by faulty conditions such as excess current density. See also *burning*.

Burnt Weld

A weld that has suffered significant oxidation particularly penetration and grain boundaries. It usually results from heating for excessive periods or an excessive temperatures, particularly in the presence of higher than normal levels of oxygen.

Burr

(1) A thin ridge or roughness (turned over edge) left on a workpiece (e.g., forgings or sheet metal blanks) resulting from cutting, punching, or grinding. (2) A rotary tool having teeth similar to those on hand files.

Burring

Same as *deburring*.

Burst Strength

A measure of the ability of the material to withstand internal hydrostatic or gas dynamic pressure without rupture. Hydraulic pressure required to burst a vessel of given thickness.

Bus (Bar)

A metal conductor for electricity. Usually a high current carrying capacity and no insulation are applied.

Bush Bearing

A plain bearing in which the lining is closely fitted into the housing in the form of a bush, usually surfaced with a bearing alloy.

Bushing

A bearing or guide, for a shaft etc. usually carried in a housing.

Butadiene

Also called *divinyl*, *vital ethylene*, *erythrene*, and *pyrrolylene*.

A colorless gas used in the production of neoprene, nylon, latex paints, and residents. Butadiene has a boiling point of 26.6 °F (–3 °C) and a specific gravity of 0.6272. Commercial butadiene is at least 98% pure.

Butadiene is also obtained as an ethylene coproduct during the steam cracking of naphtha or gas oil. It is also made by oxidation the hydrogenation of *n*-butenes, the dehydrogenation of butanes, and conversion of ethyl alcohol. The largest use for butadiene is the production of elastomers, such as *polybutadiene, styrene-butadiene, poly chloroprene*, and *acrylonitrile-butadiene, or nitrile rubbers*.

Butadiene is insoluble in water but soluble in alcohol and ether, obtained from the cracking of petroleum, from coal tar been seeing, or from acetylene produced from coke and lime. It is widely used in the formation of copolymers with styrene, acrylonitrile, vinyl chloride, and other monomeric substances and imparts flexibility to the result moldings.

Butadiene-Styrene Thermosetting Resins

Butadiene-styrene resins are low-molecular-weight, all-hydrocarbon, thermosetting copolymers designed primarily for surface coatings. They form tough, inert films by numerous and different mechanisms: oxidation, polymerization, solvent evaporation, and through the use of cross-linking agents.

Proper selection of curing conditions and/or modifiers provides conversions ranging from seconds to hours at temperatures from 21.1 °C to 593 °C.

Butadiene-styrene resins are prepared in a variety of forms. The basic resin is a viscous, clear material with a high degree of on saturation. It is supplied in a solvent-free state for use in special applications such as camel linings and thin-film metal-strip coatings. It is readily soluble in hydrocarbons thinners and chlorinated hydrocarbons. Another resin is based on butadiene alone. It is similar in molecular weight of physical properties to the basic resin; however, it possesses certain metal wedding and fabrication characteristics that make it superior for such uses as template coatings. Other resins are produced by modifying the base resin to introduce such polar groups as hydroxyls, carbonyls, and carboxyls. As a result, they have a much more active chemical nature and are more compatible with other coding resins then either the previous two.

A wide range of film properties can be obtained by varying cure conditions or by modification with other resins and cross-linking agents. Basically, films have excellent hardness and flexibility as well as abrasion resistance equal or superior to other high-quality industrial coating resins.

B

Adhesion: Films provide excellent adhesion to steel, galvanized tin-plate, aluminum, brass, and die cast zinc. A wide variety of plastics, wood, glass, untreated concrete, and most other surfaces can also be coated successfully.

Chemical resistance: Butadiene-styrene films have excellent resistance to corrosive atmospheres, particularly salt-spray and detergents solutions. And chemical resistance, the films are generally equivalent or superior to other chemically resistant coating systems. Typical properties of baked butadiene-styrene films include extended resistance to water, salt water, acids, and ketones, alcohols, aromatics, and other solvents.

Some butadiene-styrene resins are frequently combined with other resins to obtain specific coating properties. They are used as modifiers for such leading conventional coding resins as alkyd, vinyl, acrylic, and epoxy. They can also be modified with other resins, primarily ureas, melamine, and nitrocellulose.

Butadiene-styrene resins can be cured by several systems—air-dry, low-temperature bake, high-temperature bake, and instant curing (2 s at 427 °C).

For means of instant cure have been found suitable: flame curing, high-density infrared heating, flame spraying, and induction heating. The principle is simply to supply a great deal of heat to the coating as rapidly as possible.

With the addition of metallic driers, butadiene-styrene resins will air-dry in 8 to 24 h. Curing time is rapid, however, when the resin is modified with nitrocellulose. This rapid lacquer-type drying composition is particularly suitable for wood furniture and hardboard finishes.

The resin is best suited for use in special thin-film coatings are up to 1.2 mils in thickness and are used in applications such as camel linings, flat steel primers, beverage-can base coats, as well as sanitary enamels. Additionally, automobile appliances, metal wall partitions, and preprimed steel sheeting are among the other uses for the resins in the primer field.

In the lacquer field, a resin modified with nitrocellulose produces non-penetrating, high-build, high-solids, chemically resistant finishes for such uses as wood furniture, metal cabinets, and plastics. This resin system is also suitable as the primer-surfacer for hardboard and wall paneling.

Other uses of butadiene-styrene resins include metal strip coatings and pipe coatings produced by instant cure methods and coatings for drum and tank liners.

Butler Finish

A semilustrous metal finish composed of fine, uniformly distributed parallel lines, usually produced with a soft abrasive buffing wheel; similar in appearance to the traditional hand-rubbed finish on silver.

Butter Coat

A commonly used term to describe a higher than usual amount of surface resin. See also *resin-rich area*.

Butterfly Extrusion

A technique for produce and deeply recessed sections by extruding a section with an open, flatter profile and then folding the product to produce the deep profile.

Buttering

A form of surfacing in which one or more layers of weld metal are deposited on the growth phase of one member (for example, a high-alloy will deposit on steel base metal that is to be welded to a dissimilar base metal). The buttering provides a suitable transition will deposit for subsequent completion of the butt weld (joint). The composition of the blood or usually differs from those of the main weld and the parent metal. The technique is used, for example, where there is a need for a progressive transition between differing parent materials or where some characteristic of the main weld metal is incompatible with the parent.

Butt Fusion

A method of joining pipe, sheet, or other similar forms of a thermoplastic resin in which the ends of the two pieces to be joined are needed to the molten state and then rapidly pressed together.

Butt Joint (Adhesives)

A type of edge joint in which the edge faces of the two adherends is at right angles to the other faces of the adherends.

Button

(1) A globule of metal remaining in an assaying crucible or cupel after fusion has been completed. (2) That part of a weld that tears out in destructive testing of a spot, seam, or projection welded specimen.

Butt Seam Welding

See *seam welding*.

Butt Weld

An erroneous term for a weld in a butt joint. See *butt joint*.

Butt Weld/Butt Joint (Welding)

These terms are used loosely as synonyms. More strictly, the "joint" term refers to the parent component geometry so "butt joint" indicates that the two components are welded edge to edge, usually defined as being at an angle of 135° to 180°. The "weld" term refers to the weld profile so "butt weld" indicates that the weld metal expands across the end face of at least one of the components. Usually there is no confusion for examples of butt welds of butt joints, fillet welds of butt joints and butt welds of T joints. See also *strap joint*.

Butt Welded Tube

Tube formed by rolling or otherwise curving strip to form a near circle and then welding the butting edges together by various welding processes.

Butt Welding

Welding a *butt joint*.

Butting a Tube

Manipulation process in which the two and is formed to a thicker wall and constant outside diameter. The thicker section can accommodate the local increased stresses due to joining processes such as brazing, welding, screwing, or tapping.

Butylene Plastics

Plastics based on resins made by the polymerization of butene or the copolymerization of butene with one or more unsaturated compounds, the butene being the greatest amount by weight.

Butyl Rubber

Butyl is a general-purpose synthetic rubber. There are two types of rubber in this category and both are based on crude oil. The first is polyisobutylene with an occasional isoprene unit inserted in the polymer chain to enhance vulcanization characteristics. The second is the same, except that chlorine is added (approximately 1.2% by weight), resulting in greater vulcanization flexibility and fewer compatibility with general-purpose rubbers.

Butyl rubbers have outstanding impermeability to gases and excellent oxidation and ozone resistance. The chemical inertness is further reflected in lack of molecular-weight breakdown during processing, thus permitting the use of hot-mixing techniques for better polymer/filler interaction.

Flex, tear, and abrasion resistance approach those of natural rubber, and moderate-strength (14.3 MPa) unreinforced compounds can be made at a competitive cost. Butyls lack the toughness and durability, however, of some of the general-purpose rubbers.

Butyl is superior tear natural rubber and styrene-butadiene resin (SBR) in resistance to aging and weathering, sunlight, heat, air, ozone and O_2, chemicals, flexing and cut-growth, and in impermeability to gases and moisture. Because of their resistance to deterioration, butyl vulcanizates retain their original properties in actual use longer than either natural rubber or SBR vulcanizates. On heating in an oxidative atmosphere, natural rubber and SBR tend to become hard and brittle, whereas butyl remains unchanged or become slightly softer. Butyl can be compounded to have abrasion resistance equal to that of natural rubber and SBR. Butyl also has excellent dielectric properties. In many of its performance properties butyl corresponds to the more expensive synthetics such as the polychloroprenes (neoprene), the polysulfides, and the various nitrile copolymers, although like natural rubber and SBR, butyl alone is neither oil and grease nor flame resistant. In common with other synthetics, butyl has advantages of uniformity and freedom from foreign matter, as compared with natural rubber.

Butyl has relatively high hysteresis loss, and thus unique dynamic properties that made it the preferred rubber in applications requiring shock, vibration, and sound absorption.

Butyl has found a great many applications in the automotive transport, mechanical goods, electrical, chemical, proffed-fabric, building, and consumer-goods fields.

The attribute responsible for the high-volume use of butyl rubber and automotive inner tubes and tubeless tire inner liners is its excellent permeability to air.

Butyl passenger car tires have been successful because the unique dynamic properties of butyl give a remarkably soft, cushioned right, no squeal on breaking or rounding corners, and higher coefficients of friction between tires and road, which result in shorter stopping distances. The abrasion resistance of butyl tires is compatible to that of other commercial tires.

The resistance of butyl to weathering and to cutting and chipping has been demonstrated in off-the-road machinery and farm tractor tires. It's resistance weathering, sunlight, and ozone have opened many applications in automotive weatherstrips, windshield gaskets, curtain-wall gaskets and scalers, caulking compounds, and the like. Butyl-coated fabrics are extensively used as convertible tops, tarpaulins, outdoor furniture covers, etc. Such fabrics have been used as liners for irrigation ditches, with the adjacent widths of fabric cemented together with butyl cement at the site to make linings wide enough for large ditches.

Resistance to heat is essential in many butyl applications, such as high-pressure steam hose. Butyl has become the preferred material for curing bags and ladders used in the vulcanization of tires.

Butyl is also used as the inner liner in many tubeless tires, particularly truck tires, because butyl liners remain in permeable even after exposure to the heat of several successive recapping operations.

The outstanding electrical properties of butyl, coupled with its age, ozone, and moisture resistance, have made it useful in many electrical applications. It's resistance to corona and tracking makes it a preferred insulation material for power cable, and because of its heat resistance, butyl-insulated cable can be used to carry more current than cable of equal diameter insulated with SBR or other rubbers. Butyl is also used in electrical encapsulation compounds, as bus-way, factory wire, and communications wire insulation, and in other miscellaneous electrical applications.

Many automotive and mechanical applications take advantage of the ability of butyl to absorb shock, vibration, and noise. These include axle and body bumpers, truck load cushions, boat dock bumpers, bowling alley bumpers, etc. Motor and machinery mounts, bridge pads, drive-shaft insulators, and gasketing applications also benefit from the good resistance of butyl to permanent set from prolonged compression and vibration. Although not usually considered oil resistant, butyl has given years of satisfactory under-the-hood service and spark plug nipples, motor mounts, grommets, radiator hose, and other parts occasionally exposed to oil and grease.

Chemical and heat resistance requirements are satisfied by butyl in applications such as chemical tank linings, gaskets, hose, diaphragms, etc. Butyl is the preferred gasketing material for use in contact with ester-type hydraulic fluids of low flammability. His acid resistance is used to advantage in dairy hose, and large rubber containers for the bulk transportation of foodstuffs. Chemical, heat, and abrasion resistance are essential in butyl conveyor belts used to transport hot granular solids and chemical plants. Resistance to chemicals and compression set in impermeability to gases are important in butyl food-jar seals and medicine bottle stoppers.

C

"C" Process

See *Croning process.*

CAD

Computer Aided (or assisted) Design.

CAM

Computer Aided (or Assisted) Machining (or Manufacture).

Cable

(1) A flexible electrical insulated conductor or group of conductors twisted (usually multistranded) together for the transmission of electrical energy. (2) Rope or chain, particularly in marine contexts.

Cadmium

Cadmium (symbol Cd) is a silvery-white crystalline metal that has a specific gravity of 8.6, is very ductile, and can be rolled or beaten into thin sheets. It resembles tin and gives the same characteristic cry when bent, but is harder than tin. A small addition of zinc makes it very brittle. It melts at 320°C and boils at 765°C. Cadmium is employed as an alloying element in soft solders and in fusible alloys, for hardening copper, as a white corrosion-resistant plating metal, and in its compounds for pigments and chemicals. It is also used for Ni–Cd batteries and to shield against neutrons in atomic equipment, but gamma rays are emitted when the neutrons are absorbed, and these rays require an additional shielding of lead.

Most of the consumption of cadmium is for electroplating. For a corrosion-resistant coating for iron or steel, a cadmium plate of 0.008 mm is equal in effect to a zinc coat of 0.025 mm. The plated metal has a silvery-white color with a bluish tinge, is denser than zinc, and harder than tin, but electroplated coatings are subject to H_2 embrittlement, and aircraft parts are usually coated by the vacuum process. Cadmium plating is not normally used on copper or brass since copper is electronegative to it, but when these metals are employed next to cadmium-plated steel a plate of cadmium may be used on the copper to lessen deterioration.

Cadmium oxide is used extensively in plastics. In conjunction with barium it forms a compound used to stabilize the color of finished plastics. It is also a major constituent of phosphors used in television tubes.

Cadmium is mutually soluble in a number of other metallic elements. It is combined with several of these elements to form a number of commercial alloys with special properties.

The alloys of cadmium with lead, tin, bismuth, and indium are unique because of their low melting points. For example:

Alloy	Melting Point, °C
33% Cd–67% Sn	176
18% Cd–51.2% Sn–30.6% Pb	145
40% Cd–60% Bi	144
20.2% Cd–25.9% Sn–53.9% Bi	103
10.1% Cd–13.1% Sn–49.5% Bi–27.3% Pb	70
5.3% Cd–12.6% Sn–47.5% Bi–25.4% Pb–19.1% In	47

These fusible alloys are used in applications ranging from fire-detection apparatus to accurate proof casting.

Cadmium has limited use in soft solders combined with tin, lead, and zinc but its major application in the field of joining is for joints requiring higher-temperature strength than can be obtained with the soft solders. Cadmium combined with silver, copper, and zinc forms several brazing filler metals. A typical filler metal contains 35% silver, 26% copper, 21% zinc, 18% cadmium and has a melting point of 608°C. Cadmium (0.7%–1%) is added to copper to make a strong ductile metal that has a high annealing temperature but no serious loss of electrical conductivity. (Trolley wire is an example of its use.) The copper above is called *cadmium copper* or *cadmium bronze*. *Hitenso* is a cadmium bronze of the American Brass Company. It has 35% greater strength than hard-drawn copper and 85% the conductivity of copper. The cadmium bronze known in England as *conductivity bronze*, used for electric wires, contains 0.8% cadmium and 0.6% tin. Tensile strength, hard-drawn, is 586 MPa, and conductivity is 50% that of copper. *Cadmium nitrate*, $Cd(NO_3)_2$, is a white powder used for making cadmium yellow and fluorescent pigments, and as a catalyst. *Cadmium sulfide*, CdS, is used as a yellow pigment, and when mixed with *cadmium selenide*, CdSe, a red powder, it gives a bright-orange pigment. The sulfide is used for growing *cadmium sulfide crystals* in plates and rods for semiconductor uses. Crystals grown at 1922°F (1050°C) are nearly transparent, but those grown at higher temperatures are dark amber. Cadmium, a carcinogen, can be extremely toxic, and caution is required not to create dust or fumes. Because of its toxicity, use in certain applications—pigments, for example—has declined considerably.

The importance of cadmium in the nuclear field depends on its high thermal neutron capture cross section. By absorbing neutrons, cadmium is employed to control the rate of fission in a nuclear reaction. Cadmium sulfide exhibits both photosensitivity and electroluminescence, that is, it can convert light to electricity and electricity to light.

Finally, the fumes of cadmium, its compounds, and solutions of its compounds are very toxic, and cadmium-plated articles should not be used in food, nor should cadmium-coated articles be welded or used in ovens.

Cage

In a bearing, a device that partly surrounds the rolling elements and travels with them, the main purpose of which is to space the rolling elements in proper relation to each other. See also *separator*.

Cake

(1) A copper or copper alloy casting, rectangular in cross section, used for rolling into sheet or strip. (2) A coalesced mass of unpressed metal powder.

Calcareous Coating or Deposit

A layer consisting of a mixture of calcium carbonate and magnesium hydroxide deposited on surfaces being cathodically protected against corrosion because of the increased pH adjacent to the protected surface.

Calcination

Heating ores, concentrates, precipitates, or residues to decompose carbonates, hydrates, or other compounds.

Calcine

(1) A ceramic material or mixture fired to less than fusion for use as a constituent in a ceramic composition. (2) Refractory material, often fireclay, that has been heated to eliminate volatile constituents and to produce desired physical changes.

Calcite

One of the most common and widely diffused materials, occurring in the form of limestones, marbles, chalks, calcareous marls, and calcareous sandstones. It is a *calcium carbonate*, $CaCO_3$, and the natural color is white or colorless, but it may be tinted to almost any shade with impurities. The specific gravity is about 2.72 and Mohs hardness 3. The common *black calcite*, containing manganese oxide, often also contains silver in proportions high enough to warrant chemical extraction of the metal.

It is a natural mineral, and in its pure form it is transparent and capable of polarizing light.

Calcium

Calcium (symbol Ca) is a metallic element belonging to the group of alkaline earths. It is one of the most abundant materials, occurring in combination in limestones and calcareous clays. The metal is obtained 98.6% pure by electrolysis of the fused anhydrous chloride. By further subliming, it is obtained 99.5% pure. Calcium metal is yellowish white in color. It oxidizes easily and, when heated in air, burns with a brilliant white light. It has a density of 1.5 g/cm³, a melting point of 838°C, and a boiling point of 1440°C. Its strong affinity for O_2 and sulfur is utilized as a cleanser for nonferrous alloys. As a deoxidizer and desulfurizer it is employed in the form of lumps or sticks of calcium metal or in ferroalloys and Ca–Cu.

Many compounds of calcium are employed industrially in fertilizers, foodstuffs, and medicine. It is an essential element in the formation of bones, teeth, shells, and plants. Oyster shells form an important commercial source of calcium for animal feeds. They are crushed, and the fine flour is marketed for stock feeds and the coarse for poultry feeds. The shell is calcium carbonate.

Edible calcium, for adding calcium to food products, is *calcium lactate*, a white powder of the composition $Ca(C_3H_5O_3)_2 \cdot 5H_2O$, derived from milk. *Calcium lactobionate* is a white powder that readily forms chlorides and other double salts, and is used as a suspending agent in pharmaceuticals. It contains 4.94% available calcium. *Calcium phosphate*, used in the foodstuffs industry and in medicine, is marketed in several forms. *Calcium diphosphate*, known as *phosphate of lime*, is $CaHPO_4 \cdot 2H_2O$, or in anhydrous form. It is soluble in dilute citric acid solutions and is used to add calcium and phosphorus to foods, and as a polishing agent in toothpastes.

Calcium oxide is made by the thermal decomposition of carbonate minerals in tall kilns using a continuous-feed process. Care must be taken during the heating to decompose the limestone at a low enough temperature so that the oxide will slake freely with water. If too high a temperature is used, so-called dead-burnt lime is formed. The oxide is used in high-intensity arc lights (limelights) because of its unusual spectral features and as an industrial dehydrating agent. At high temperatures, lime combines with sand and other siliceous material to form fusible slags; hence, the metallurgical industry makes wide use of it during the reduction of ferrous alloys.

Because of the low cost of *calcium hydroxide*, it is used in many applications where hydroxide ion is needed. Slaked lime is an excellent absorbent for *carbon dioxide* (CO_2) to produce the carbonate. Because of the great in solubility of the carbonate, gases are easily tested qualitatively for CO_2 by passing them through a saturated lime-water solution and watching for a carbonate cloudiness. The hydroxide is also used in the formation of mortar, which is composed of slaked lime (1 volume), sand (3–4 volumes), and enough water to make a thick paste. The mortar gradually hardens because of the evaporation of the water and the cementing action of the deposition of calcium hydroxide, and because of the absorption of CO_2.

Calcium silicide, CaSi, is an electric-furnace product made from lime, silica, and a carbonaceous reducing agent. This material is useful as a steel deoxidizer because of its ability to form calcium silicate, which has a low melting point.

Calcium carbide is a hard, crystalline substance of grayish-black color, used chiefly for the production of **acetylene** gas for welding and cutting torches and for lighting. It is made by reducing lime with coke in the electric furnace, at 2000°C–2200°C. It can also be made by heating crushed limestone to a temperature of about 1000°C, flowing a high-methane natural gas through it, and then heating to 1700°C. The composition is CaC_2, and the specific gravity is 2.26. It contains theoretically 37.5% carbon. When water is added to calcium carbide, acetylene gas is formed, leaving a residue of slaked lime. Although calcium carbide is principally used for making acetylene, this market is shrinking as acetylene is recovered increasingly as a by-product in petrochemical plants. A growing application for calcium carbide is desulfurization deoxidation of iron and steel. It is also a raw material for production of *calcium cyanamide*.

Calcium chloride is a white, crystalline, lumpy or flaky material of composition $CaCl_2$. The specific gravity is 2.15, the melting point is 1422°F (772°C), and it is highly hygroscopic and diliquescent with rapid solubility in water. The commercial product contains 75%–80% $CaCl_2$, with the balance chiefly water of crystallization. Some is marketed in anhydrous form for dehydrating gases. It is also sold in water solution containing 40% calcium chloride. Calcium chloride has been used on roads to aid in surfacing, absorb dust, and prevent cracking from freezing. It is also used for accelerating the setting of mortars, but more than 4% in concrete decreases the strength of the concrete. It is also employed as an antifreeze in fire tanks, for brine refrigeration, for storing solar energy, as an anti-ice agent on street pavements, as a food preservative, and in textile and paper sizes as a gelling agent. In petroleum production, it is used in drilling mud cementing operations, and workover or completion fluids. Calcium chloride is obtained from natural brines and dry lake beds, after *sodium chloride, bromide, and other products are extracted*. The *magnesium–calcium brine* remaining is marketed for dust control or purified into calcium chloride. It is a by-product of sodium bicarbonate production via the Solvay process and is made in small quantities by neutralizing waste of hydrochloric acid with line or limestone.

Calcium–silicon is an alloy of calcium and silica used as a deoxidizing agent for the elimination of sulfur in the production of steels and cast irons. Steels deoxidized or treated with calcium or calcium and silicon can have better machinability than those deoxidized with aluminum and silicon. It is marketed as low iron, containing 22%–28% calcium, 65%–70% silicon, and 5% maximum iron, and as high iron, containing 18%–22% calcium, 58%–60% silicon, and 15%–20% iron. It comes in crushed form and is added to the molten steel. At the temperature of molten steel, all the calcium passes off and leaves no residue in the steel. *Calcium–manganese–silicon* is another master alloy containing 17%–19% calcium, 8%–10% manganese, 55%–60% silicon, and 10% iron.

Calcium sulfate dihydrate is called *gypsum*. It constitutes the major portion of portland cement, and has been used to help reduce soil alkalinity. A hemihydrate of calcium sulfate, produced by heating gypsum at elevated temperatures, is known under the commercial name *plaster of paris*. When mixed with water, the hemihydrate reforms the dihydrate, evolving considerable heat and expanding in the process, so that a very sharp imprint of the mold is formed. Thus, plaster of paris finds use in the casting of small art objects and mold testing.

Calcium metal is employed as an alloying agent for aluminum-bearing metal, as an aid in removing bismuth from lead, and as a controller for graphitic carbon in cast iron. It is also used as a deoxidizer in the manufacture of many steels, as a reducing agent in preparation of such metals as chromium, thorium, zirconium, and uranium, and as a separating material for gaseous mixtures of N_2 and argon. When added to magnesium alloys (0.25%), it refines the grain structure, reduces their tendency to take fire, and modifies the strengthening heat treatment. It finds use also in the precipitation-hardening Pb–Ca alloys.

Calcium in the biosphere is an invariable constituent of all plants because it is essential for their growth. It is contained both as a structural constituent and as a physiological ion. The calcium ion is able to counteract the toxic effects of potassium, sodium, and magnesium ions. Calcium may also affect the growth of plants because its presence in soil affects the alkalinity of the latter.

Calcium is found in all animals in the soft tissues, in tissue fluid, and in the skeletal structures. The bones of vertebrates contain calcium as calcium fluoride, calcium carbonate, and calcium phosphate. In some lower animals, magnesium replaces either totally or partially the skeletal calcium. The importance of calcium in animals as a structural constituent is based on its abundance and on the low solubility of the three calcium salts just listed. Calcium is also essential in many biological functions of the vertebrates.

Calendered Sheet

Calendering is the process of forming a continuous sheet of controlled size by squeezing a softened thermoplastic material between two or more horizontal rolls. Along with extrusion and casting, calendering is one of the major techniques used to process thermoplastics into film and sheeting. It also is used in the manufacture of

flooring and to apply a plastic coating on paper, textiles, and other supporting materials.

The calendering process is used in the plastics, rubber, linoleum, paper, and metals industries for various roll-forming operations in the manufacture of sheeted materials. This coverage will be concerned only with calendering of thermoplastic materials (plastics and elastomers, the latter of which are calendered in a thermoplastic state prior to vulcanization).

The process consists of five steps: preblending, fluxing, calendering, cooling, and wind-up. Blending of the resin powder with plasticizers, stabilizers, lubricants, colorants, and fillers is usually done in large ribbon blenders. The compound is fluxed, that is heated and worked until it reaches a molten or dough-like consistency.

When a Banbury mill is used, the molten material from it is discharged to a two-roll horizontal mill and thence to the calender either in a continuous strip or in batches. When an extruder is used for fluxing, the extrudate is fed directly to the calender. Alternatively, the preblend can in some cases be fed directly to the calender.

After passage through the calender, the continuous sheet of hot plastic is stripped off the last calender roll with a small, higher-speed stripping roll. The hot sheet is cooled as it travels over a series of cooling drums. The film or sheeting is finally automatically cut into individual sheets or wound up in a continuous roll.

The desired surface finish on the calendered film or sheeting is imparted by the last pair of calender rolls and may range from a high gloss to a heavy matte finish. Extra-high-gloss or special engraved patterns can be made either by having a polished or engraved roll impinge on the last calender roll (contact embossing) or by passing the hot sheeting directly from the calender between an engraved metal roll and a rubber backup roll (in-line embossing). Many attractive patterns are made by these techniques.

The major groups of thermoplastics and elastomers that are calendered into film or sheeting are polyvinyl chloride (PVC) (plasticized and unplasticized) and natural and synthetic rubbers. ABS (acrylonitrile, butadiene, and styrene) polymers, polyolefins, and silicones are also calendered, but in much smaller quantities.

Calendered vinyl is used extensively for floor tile and continuous flooring, rainwear, shower curtains, table covers, pressure-sensitive tape, automotive and furniture upholstery, wall coverings, luminous ceilings, signs and displays, credit cards, etc.

In contrast, calendered rubber—except for some fabric coating—is mainly an intermediate product used in the manufacture of a multitude of articles such as automobile tires, footwear, molded mechanical goods, etc.

In applications that involve the use of calenders for paper and cloth coating, the paper or fabric is fed into the last calender nip so that the plastic or rubber coating is formed on top of the material (calender coating). Frictioning is a variation of this technique by which a thin layer of rubber is squeezed or rubbed into the fabric itself, a technique widely used in the rubber industry for the manufacture of friction tapes and cord fabric for tires.

Production speeds vary greatly with materials, from 10 to 300 fpm (feet per minute). Calenders are heavy, high-precision machines that are made in a variety of designs and with elaborate control equipment that automatically monitors film variations and readjusts the machine to produce the desired thickness.

Calibrate

To determine, by measurement or comparison with a standard, the correct value of each scale reading on a measuring (test) instrument.

Calibration

Determination of the values of the significant parameters by comparison with values indicated by a reference instrument or by a set of reference standards.

Calomel Electrode

(1) An *electrode* widely used as a reference electrode of known potential in electrometric measurement of acidity and alkalinity, corrosion studies, voltammetry, and measurement of the potentials of other electrodes. (2) A secondary reference electrode of the composition: Pt/Hg–Hg$_2$Cl$_2$/KCl solution. For 1.0 N KCl solution, its potential vs. a hydrogen electrode at 25°C (77°F) at 1 atm is +0.281 V.

Calorie

The quantity of heat required to raise the temperature of 1 g of water by 1°C. The exact value of this depends on the water temperature. The *calorie* (International Table) is 4.186 J. The dietitians calorie, sometimes termed the large calorie or the kilocalorie is 1000 cal.

Calorimeter

An instrument capable of making absolute measurements of energy deposition (or absorbed dose) in a material by measuring its change in temperature and imparting a knowledge of the characteristics of its material construction.

Calorimetry

The measurement in a calorimeter of heat energy relationships.

Calorizing

A process in which steel is coated with aluminum powder and heated to about 1000°C to form an aluminum oxide coating over an aluminum iron intermetallic layer. The coating offers good oxidation resistance up to about 950°C.

Cam Press

A mechanical forming press in which one or more of the slides are operated by cams; usually, a double-action press in which the blankholder slide is operated by cams through which the dwell is obtained.

Camber

(1) Deviation from edge straightness, usually referring to the greatest deviation of side edge from a straight line. (2) The tendency of material being sheared from sheet to bend away from the sheet in the same plane. (3) Sometimes used to denote crown in rolls where the center diameter has been increased to compensate for deflection caused by the rolling pressure. (4) The planar deflection of a flat cable or flexible laminate from a straight line of specified length. A flat cable or flexible laminate with camber is similar to the curve of an unbanked racetrack.

Campaign

The period in which a furnace or other equipment is operated between major overhaul such as the replacement of the refractory lining.

Can

A sheeting of soft metal that encloses a sintered metal billet for the purpose of hot working (hot isostatic pressing, hot extrusion) without undue oxidation.

Canada Balsam

The thin resin produced by Canadian firs. It has a refractive index similar to glass and hence is used as an adhesive for multielement lenses.

Cannel Coal

A variety of coal having some of the characteristics of petroleum, valued chiefly for its quick-firing qualities. It consists of coal-like matter intimately mixed with clay and shale, often containing fossil fishes, and probably derived from vegetable matter in lakes. It is compact in texture, dull black, and breaks along joints, often having an appearance similar to black shale. It burns with a long, luminous, smoky flame, from which it derives its old English name meaning candle.

Canning

(1) A dished distortion in a flat or nearly flat sheet metal surface, sometimes referred to as oil canning. (2) Enclosing a highly reactive metal within a relatively inert material for the purpose of hot working without undue oxidation of the active metal. (3) The materials used for canning nuclear fuel elements are selected for low neutron capture cross section, short half-life, good thermal conductivity, compatibility with the working environment, adequate mechanical properties, and forming characteristics. (4) The containment of foodstuffs and other materials in sealed cans made from tin plated steel, aluminum, or other materials.

Cantilever

A loaded beam supported at one end only.

Capacitance (C)

That property of a system of conductors and dielectrics that permits the storage of electorally separated charges when potential differences exist between the conductors. It is the ratio of a quantity, Q, of electricity to a potential difference, V. A capacitance value is always positive. The units are farads when the charge is expressed in coulombs and the potential in volts: $C = Q/V$.

Capacitor

A capacitor is an electrical device capable of storing electrical energy. In general, a capacitor consists of two metal plates insulated from each other by a dielectric. The capacitance of a capacitor depends primarily upon its shape and size and upon the relative permittivity ε_r of the medium between the plates. In vacuum, in air, and in most gases, ε_r ranges from one to several hundred.

One classification of capacitors comes from the physical state of their dielectrics, which may be gas (or vacuum), liquid, solid, or a combination of these. Each of these classifications may be subdivided according to the specific dielectric used. Capacitors may be further classified by their ability to be used in AC or DC circuits with various current levels.

TABLE C.1
Major Types of Capacitors

Type	Capacitance	Voltage (Working Voltage, DC), V[a]	Applications
Monolithic ceramics	1 pF–2.2 μF	50–200	Ultrahigh frequency, RF coupling, computers
Disk and tube ceramics	1pF–1 μF	50–1000	General, very high frequency
Paper	0.001–1 μF	200–1600	Motors, power supplies
Film			
Polypropylene	0.0001–0.47 μF	400–1600	Television vertical circuits, RF circuits
Polyester	0.001–4.7 μF	50–600	Entertainment electronics
Polystyrene	0.001–1 μF	100–200	General, high stability
Polycarbonate	0.01–18 μF	50–200	General
Metallized polypropylene	4–60 μF	400[a]	Alternating-current motors
Metallized polyester	0.001–22 μF	100–1000	Coupling, RF filtering
Electrolytic			
Aluminum	100,00 μF	5–500	Power supplies, filters
Tantalum	0.1–2200 μF	3–150	Small space requirement, low leakage
Gold	0.022–10 μF	2.5–5.5	Memory backup
Nonpolarized (either aluminum or tantalum)	0.47–1000 μF	10–200	Loudspeaker crossovers
Mica	330 pF–0.05 μF	50–1000	High frequency
Silver-mica	5–820 pF	50–500	High frequency
Variable			
Ceramic	1–5 to 16–100 pF	200	Radio, television, communications
Film	0.8–5 to 1.2–30 pF	50	Oscillators, antenna, RF circuits
Air	10–365 pF	50	Broadcast receiver
Poly(tetrafluoroethylene)	0.25–1.5 pF	2000	Very high frequency, ultrahigh frequency

Source: M. Schwartz, *McGraw-Hill Encyclopedia of Science and Technology*, 8th edn., Vol. 3, McGraw-Hill, New York, p. 215. With permission.

[a] Alternating-current voltage at 60 Hz.

Capacitors are also classified as fixed, adjustable, or variable. Capacitors made with air, gas, or vacuum as the dielectric between plates are constructed with flat parallel metallic plates (or cylindrical concentric metallic plates).

Solid-dielectric types of capacitors use one of several dielectrics such as a ceramic, mica, glass, or plastic film. Alternate plates of metal, or metallic foil, are stacked with the dielectric, or the dielectric may be metal-plated on both sides.

Plastic-film types are capacitors that use dielectrics such as polypropylene, polyester, polycarbonate, or polysulfone with a relative permittivity ranging from 2.2 to 3.2. This plastic film may be used alone or in combination with Kraft paper. The most common electrodes are aluminum or zinc vacuum-deposited on the film, although aluminum foil is also used. Other major types of capacitors can be seen in Table C.1.

The impedance of a capacitor is inversely proportional to the frequency of the voltage impressed; that is, it offers little resistance or impedance to high frequencies, but much to low frequencies.

Capacity

In tensile testing machines, the maximum load and/or displacement for which a machine is designed. Some testing machines have more than one load capacity. These are equipped with accessories that allow the capacity to be modified as desired.

Capacity Number (*Cn*)

The product of the *Sommerfeld number* and the square of the length-to-diameter ratio of a journal bearing.

Capillarity

The ability of the material to conduct liquids through its pore structure by the force of surface tension.

Capillary Action

(1) The phenomenon of intrusion of a liquid into interconnected small voids, pores, and channels in a solid, resulting from surface tension. (2) The force by which a liquid, in contact with the solid, is distributed between closely fitted faying surfaces of the joint to be brazed or soldered. (3) It is a consequence of surface tension and depends critically on the liquid wetting the solid surface.

Capillary Attraction

(1) The combined force of adhesion and cohesion that causes liquids, including molten metals, to flow between very closely spaced and solid surfaces, even against gravity. (2) In powder metallurgy, the driving force for the infiltration of the pores of a sintered compact by a liquid.

Capped Steel

A type of steel similar to rimmed steel, usually cast in a bottle-top ingot mold, in which the application of a mechanical or a chemical cap renders the rimming action incomplete by causing the top metal to solidify. The surface condition of capped steel is much like that of rimmed steel, but certain other characteristics are intermediate between those of *rimmed steel* and those of *semi-killed steel.*

Capping (of Abrasive Particles)

A mechanism of deterioration of abrasive points in which the points become covered by caps of adherent abrasion debris.

Capping (of Powder Metallurgy Compacts)

Partial or complete separation of a powder metallurgy compact into two or more portions by cracks that originate near the edges of the punch faces and that proceed diagonally into the compact.

Caprolactam

A cyclic amide type of compound containing six carbon atoms. When the range is open, caprolactam is polymerizable into a nylon resin known as type 6 nylon or poly-caprolactam.

Capture Cross Section

The measure of the capability of a material to absorb neutrons without undergoing fission. See *atomic structure.*

Capture Efficiency

See *collection efficiency.*

Car Furnace

A batch-type heat-treating furnace using a car on rails to enter and leave the furnace area. Core furnaces are used for lower stress relieving ranges.

Carat

(1) Units of purity of gold, 24 carat being pure. (2) Metric carat, for gemstones, etc. = 200 mg.

Carbanion Ion

Negatively charged organic compound (ion).

Carbide Tip/Tool

Cutting or forming tools of intermetallic compounds usually made from tungsten, titanium, tantalum, or niobium carbides, or a combination of them, in a matrix of cobalt, nickel, or other metals. Carbide tools are characterized by high hardnesses and compressive strengths and may be coated to improve wear resistance. See also *cemented carbides.*

Carbides

Of the several classes of carbides, only two types need to be considered for engineering applications: the covalent carbides of silicon and boron, and the interstitial carbides of the transition metals—titanium, zirconium, vanadium, niobium (columbium), tantalum, chromium, molybdenum, and tungsten. All of these may be characterized as hard, refractory materials of extreme chemical inertness. Other carbides such as those of aluminum, iron, and manganese are too reactive to be considered for engineering applications. Chemically, all the inert carbides, with the exception of SiC, are unusual in that they exist in their typical form over a range of composition. In this respect, they are more similar to alloys than true chemical compounds.

Structurally, *interstitial carbides* may be described as metal lattices into which the small carbon atoms have been inserted. In the ideal form, this leads to an fcc (face-centered cubic) structure for all of the metal carbide (MC) types except tungsten carbide (WC), which like the M_2C types has a simple hexagonal structure. B_4C is rhombohedral, and consists of a distorted B lattice in which part of the boron atoms have been replaced by carbon atoms. SiC exists in a number of crystalline polytypes, both cubic (diamond or zinc blend structures) and hexagonal (wurtzite type). All these structures may be described on the basis that every silicon atom is surrounded tetrahedrally by four carbon atoms and that every carbon atom is surrounded tetrahedrally by four silicon atoms.

Properties of carbides that make them unique are their extreme hardness, exceptional corrosion resistance, extreme refractoriness, high Young's modulus of elasticity, and high temperature strength.

The simplest method to produce carbides is a direct combination of the metal with carbon, used exclusively in the preparation of carbides where purity is more important than cost considerations. Most carbides are, however, prepared by reduction of metal oxides with carbon.

With the exception of SiC, all carbide bodies are fabricated using powder metallurgical techniques, that is, sintering and hot pressing. SiC bodies are formed by infiltrating a preformed carbonaceous body with elemental silicon, forming SiC in the shape desired.

Complete pump assemblies have been built to transfer molten metals (corrosion resistance); pipes for heat exchangers (high thermal conductivity); cyclone separators and sandblast nozzles (abrasion resistance); rocket nozzles (refractoriness and erosion resistance); electric-light sources (high melting temperature); and suction box covers for papermaking machines (ability to retain smooth finish without wear).

Several metal carbides mentioned above qualify as engineering ceramics. Most commonly used are B_4C and SiC.

B_4C is noted for its very high hardness and low density—unusual qualities for a brittle ceramic—which qualify this ceramic for lightweight, bulletproof armor plate. The material has the best abrasion resistance of any ceramic, so it is also specified for pressure-blasting nozzles and similar high-wear applications. A limitation of B_4C is its low strength at high temperatures.

Despite its higher cost, SiC is challenging Al_2O_3, particularly for the more critical applications. SiC is one of the high-temperature, high-strength "superstars" of the engineering ceramics. It is one of the strongest structural ceramics for high-temperature oxidation-resistant service. However, SiC does not easily self-bond. Consequently, many processing variations have been devised to fabricate parts from this material, creating a number of trade-offs in cost, fabric-ability, and properties. The ceramic can be consolidated by hot pressing. Under the combination of high temperature and pressure—with, in some cases, additives that act as bond-forming catalysts—fully dense material can be formed.

The hot-pressed ceramic is extremely strong and tough at high temperatures, but the manufacturing process is limited to simple shapes, bars, or billets. Complex parts made by hot pressing must be machined to shape—a slow and costly process of ultrasonic machining, EDM (if possible), or diamond grinding.

On the other hand, SiC particles can be bonded without pressure by a number of processes, variously called reaction bonding, recrystallization (for SiC), or reaction sintering. With these processes, "green" parts can be dry or isostatically pressed, extruded, slip-cast, or, in some cases, formed by conventional plastic molding techniques such as injection molding, then sintered. Complex shapes, close to finished size can be produced by these techniques, but the ceramic is only about 80% as dense as the hot-pressed counterpart and has lower strength and poorer thermal shock resistance. SiC—either hot pressed or reaction bonded—is not as strong as Si_3N_4 up to about 1427°C; Si_3N_4 grain boundaries soften, or creep, and strength drops. Above 1427°C, SiC is the stronger ceramic. At 1316°C, however, strength of hot-pressed ceramics nearly equals that of reaction-bonded ceramics.

Hot-pressed SiC is harder and more difficult to EDM than Si_3N_4, which has lower thermal expansion and better thermal shock resistance than SiC. Electrical resistivity of SiC is low at low frequencies and high at high frequencies—an unusual characteristic that qualifies this material as a semiconductor.

Refractory Hard Metals

Refractory hard metals (RHMs) are a ceramic-like class of materials made from metal-carbide particles bonded together by a metal matrix. Often classified as ceramics and sometimes called cemented or sintered carbides, these metals were developed for extreme hardness and wear resistance.

The RHMs are more ductile and have better thermal shock resistance and impact resistance than ceramics, but they have lower compressive strength at high temperatures and lower operating temperatures than most ceramics. Generally, properties of RHMs are between those of conventional metals and ceramics. Parts are made by conventional powder metallurgy compacting and sintering methods. Many metal carbides such as SiC and B_4C are not RHMs but are true ceramics. The fine distinction is in particle binding: RHMs are always bonded together by a metal matrix, whereas ceramic particles are self-bonded. Some ceramics have a second metal phase, but the metal is not used primarily for bonding.

Four RHM systems are used for structural applications and, in most cases, several grades are available within each system. WC with a 3%–20% matrix of cobalt is the most common structural RHM. The low-cobalt grades are used for applications requiring wear resistance; the high-cobalt grades serve where impact resistance is required.

TaC and WC combined in a matrix of nickel, cobalt, and/or chromium provide an RHM formulation especially suited for a combination of corrosion and wear resistance. Some grades are almost as corrosion resistant as platinum. Nozzles, orifice plates, and valve components are typical uses.

TiC in a molybdenum and nickel matrix is formulated for high-temperature service. Tensile and compressive strengths, hardness, and oxidation resistance are as high as 1093°C. Critical parts for welding and thermal metalworking tools, valves, seals, and high-temperature gauging equipment are made from grades of this RHM.

Tungsten–titanium carbide ($WTiC_2$) in cobalt is used primarily for metal-forming applications such as draw dies, tube-sizing mandrels, burnishing rolls, and flaring tools. The $WTiC_2$ is a gall resistant phase in the RHM containing WC as well as cobalt.

Carbon

Carbon is a nonmetallic element, symbol C, existing naturally in several allotropic forms, and in combination as one of the most widely distributed of all the elements. It is quadrivalent, and has the property of forming chain and ring compounds, and there are more varied and useful compounds of carbon than of all the other elements. The black amorphous carbon has a specific gravity of 1.88: the black crystalline carbon known as graphite has a specific gravity of 2.25; the transparent crystalline carbon, as in the diamond, has a specific gravity of 3.51. Amorphous carbon is not soluble in any known solvent. It is infusible, but sublimes at 3500°C, and is stable and chemically inactive at ordinary temperatures. At high temperatures it burns and absorbs O_2, forming the simple oxides CO and CO_2; the latter is the stable oxide present in the atmosphere and a natural plant food. Carbon dissolves easily in some molten metals, notably iron, exerting great influence on them. Steel, with small amounts of chemically combined carbon, and cast iron, with both combined carbon and graphitic carbon, are examples of this.

Carbon occurs as hydrocarbons in petroleum, and as carbohydrates in coal and plant life, and from these natural basic groupings an infinite number of carbon compounds can be made synthetically. Carbon for chemical, metallurgical, or industrial use is marketed in the form of compounds in a large number of different grades, sizes, and shapes; or in master alloys containing high percentages of carbon; or as activated carbons, charcoal, graphite, carbon black, coal-tar carbon, petroleum coke; or as pressed and molded bricks or formed parts with or without binders or metallic inclusions. Natural deposits of graphite, coal tar, and petroleum coke are important sources of elemental carbon.

Combined with carbonaceous binders, such as tars, pitches, and resins, the carbon is compacted by molding or extrusion, and baked at between 816°C and 1649°C to produce what is known as industrial carbon or baked carbon. Conventional industrial graphite (Gr) is made by mixing mined, natural graphite with carbon to produce in effect a C–Gr composite, or baked carbon can be heat treated at about 2985°C, at which temperature the carbon graphitizes.

Manufactured or artificial carbon has a two-phase structure consisting of carbon particles (or grains) in a matrix of binder carbon. Both phases consist essentially of disordered, or uncrystallized, carbon surrounding embryonic carbon crystallites.

Graphites, except for the pyrolytic types, have a two-phase structure similar to that of carbon, but as the result of high-temperature processing contain well-developed graphite crystallites in both phases. These multicrystalline graphites exhibit many of the properties of single-crystal graphite, such as high electrical conductivity, lubricity, and anisotropy. Compared to carbon, graphite has higher electrical and heat conductivity, better lubricity, and is easier to machine. Because of their more favorable properties, graphites have broader application as engineering materials than do carbons. See *graphite*.

Types and Forms of Carbon

Elemental carbon exists in two well-defined crystalline allotropic forms, diamond and graphite. Other forms, which are poorly developed in crystallinity, are charcoal, coke, and carbon black.

Charcoal is prepared by the ignition of wood, sugar, blood, and other carbon-containing compounds in the absence of air. X-ray diffraction studies reveal that it has a graphite structure but is not very well developed in crystallinity. The lack of crystallinity is the result of defects in the crystal structure and the high surface area. In the activated state, charcoal adsorbs gases, liquids, and solids.

Coke, another form of amorphous carbon, is prepared by heating coal in the absence of air. It is used primarily for the reduction of metal oxides to the free metals.

Chemically pure carbon is prepared by the thermal decomposition of sugar (sucrose) in the absence of air. Impurities in the carbon are removed by treatment with chlorine (Cl) gas at red heat. The substance is then washed with water and the low residual chlorine is removed by heating in an atmosphere of H_2 gas.

Carbon-13 is one of the isotopes of carbon, used as a tracer in biologic research where its heavy weight makes it distinguished from other carbon. Carbon-14, or radioactive carbon, has a longer life. It exists in the air, formed by the bombardment of N_2 by cosmic rays at high altitudes, and enters into the growth of plants. The half-life is about 6000 years. It is made from N_2 in a cyclotron.

Carbon fullerenes, such as C_{60}, are a new form of carbon, discovered in the mid-1980s, with considerable potential and diverse applications.

Carbon Fibers/Yarn/Fabric/Wool

These forms are made by pyrolysis of organic precursor fibers in an inert atmosphere. Pyrolysis temperatures can range from 1000°C to 3000°C; higher process temperatures generally lead to higher-modulus fibers. Only three precursor materials, rayon, polyacrylonitrile (PAN), and pitch have achieved significance in commercial production of carbon fibers (see Table C.2). The first high-strength and high-modulus carbon fibers were based on a rayon precursor. These fibers were obtained by being stretched to several times their original length at temperatures above 2800°C. The second generation of carbon fibers is based on a PAN precursor and has achieved market dominance. In their most common form, these carbon fibers have a tensile strength ranging from 2413 to 3102 MPa, a modulus of 0.2 to 0.5 GPa, and a shear strength of 90 to 117 MPa. This last property controls the traverse strength of composite materials. The high-modulus fibers are highly graphitic in crystalline structure after being processed from PAN at temperatures in excess of 1982°C.

TABLE C.2
Carbon Fibers

| | Nominal Fiber Modulus (10^6 psi) | | | | | | | |
| | Pan Fibers | | | | Pitch Fibers | | | |
	30[a]	40[b]	40[c]	50[d]	70[e]	50[f]	75[g]	100+[h]
Fiber strength (10^3 psi)	500	820	650	350	300	275	300	360
Tensile modulus (10^6 psi)	33	41	42	57	75	55	75	110
Composite strength (10^3 psi)								
Longitudinal	250	470	360	205	110	135	140	180
Transverse	10	10	10	5	4	5	5	3
Composite strength (10^6 psi)								
Longitudinal	20	25	25	35	44	35	49	72
Transverse	2	2	1.5	1	4	1	1	1

Source: Mach. Design Basics Eng. Design, 730, June 1993. With permission.

[a] Thornel 300 (Amoco).
[b] Thornel T-40.
[c] Thornel T650/42.
[d] Thornel 50 PAN.
[e] Celanese GY-70.
[f] Thornel P-55S.
[g] Thornel P-75S.
[h] Thornel P-100S.

Higher-strength fibers obtained at lower temperatures from rayon feature a higher carbon crystalline content. There are also carbon and graphite fibers of intermediate strength and modulus. The third generation of carbon fibers is based on pitch as a precursor. Ordinary pitch is an isotropic mixture of largely aromatic compounds. Fibers spun from this pitch have little or no preferred orientation and hence low strength and modulus. Pitch is a very inexpensive precursor compared with rayon and PAN. High-strength and high-modulus carbon fibers are obtained from a pitch that has first been converted into a mesophase (liquid crystal). These fibers have a tensile strength of more than 2069 MPa, and a Young's modulus ranging from 0.38 to 0.52 GPa.

The average filament diameter of continuous yarn is 0.008 mm. Pitch-based carbon and graphite fibers are expected to see essentially the same applications as the more costly PAN and rayon-derived fibers, for example, ablative, insulation, and friction materials and in metals and resin matrices.

Carbon fibers added to thermoplastic resins provide the highest strength, modulus, heat-deflection temperature, creep, and fatigue—endurance values commercially available in composites. These property improvements, coupled with greatly increased thermal conductivity and low friction coefficients, make carbon fibers ideal for wear and frictional applications where the higher cost can be tolerated. In applications where the abrasive nature of glass fibers wears the mating surface, the softer carbon fibers can be substituted to reduce the wear rate. Carbon fibers can also be used in conjunction with internal lubricants to further improve surface characteristics of most thermoplastic resin systems.

Another useful property of carbon-fiber-reinforced thermoplastics is their low volume and surface resistivities. Most resin systems reinforced with 15% or more carbon fibers can effectively dissipate static charge, which is a problem common to gears, slides, and bearings used in business machine, textile, electrical, and conveying equipment.

Thornel is a yarn made from these filaments for high-temperature fabrics. It retains its strength to temperatures above 1538°C. Carbon yarn is 99.5% pure carbon. It comes in plies from 2 to 30, with each ply composed of 720 continuous filaments of 0.0076 mm diameter. Each ply has a breaking strength of 0.91 kg. The fiber has the flexibility of wool and maintains dimensional stability to 3150°C. *Ucar* is a conductive carbon fabric made from carbon yarns woven with insulating glass yarns with resistivities from 0.2 to 30 Ω for operating temperatures to 288°C. Carbon wool, for filtering and insulation, is composed of pure carbon fibers made by carbonizing rayon. The fibers, 5–50 μm in diameter, are hard and strong, and can be made into rope and yarn, or the mat can be activated for filter use. *Thornel radiotranslucent carbon fibers*, from Amoco Polymers, allow electrical conduction while remaining invisible to x-rays, permitting babies monitoring equipment to stay intact during x-rays and MRIs. *K1lOOX* fiber, from Amoco Performance Products, Inc., is a pitch-carbon fiber for *prepreg* used to produce composites for thermal management systems in space satellites. *Carbon wool*, for filtering and insulation, is composed of pure-carbon fibers made by carbonizing rayon. *Avceram CS* of FMC Corp., is a composite rayon-silica fiber made with 40% dissolved sodium silicate. A highly heat-resistant fiber, *Avceram CS* is woven into fabric and then pyrolyzed to give a porous interlocked mesh of *carbon silica* fiber, with a tensile strength of 1138 MPa. *Dexsan*, of C.H. Dexter and Sons Co., for filtering hot gases and liquids, is a *carbon filter paper* made from carbon fibers pressed into a paper-like mat, 0.007–0.050 in. (0.18–0.127 mm) thick, and impregnated with activated carbon.

In a process developed by Mitsubishi Gas Chemical Co. (Japan), naphthalene is used as the feedstock for mesophase pitch, called

AR-Resin, to produce carbon fiber. Conoco, Inc. uses a mesophase pitch to make carbon-fiber mat. This pitch has an anisotropic molecular structure rather than the more amorphous one of the PAN precursor.

Carbon brushes for electric motors and generators and *carbon electrodes* are made of carbon in the form of graphite, petroleum coke, lampblack, or other nearly pure carbon, sometimes mixed with copper powder to increase the electric conductivity, and then pressed into blocks or shapes and sintered. Carbon–graphite brushes contain no metals but are made from *carbon–graphite powder* and, after pressing, are subjected to a temperature of 5000°F (2760°C), which produces a harder and denser structure, permitting current densities up to 125 A/in.^{2} (1538 A/m^2). *Carbon brick*, used as a lining in the chemical processing industries, is carbon compressed with a bituminous binder and then carbonized by sintering. If the binder is capable of being completely carbonized, the bricks are impervious and dense. *Graphite brick*, made in the same manner from graphite, is more resistant to oxidation than carbon bricks and has a higher thermal conductivity but it is softer. The binder may also be a furfural resin polymerized in the pores. *Karbate No. 1* is a carbon-base brick, and *Karbate No. 2* is a graphite brick. *Impervious carbon* is used for lining pumps, for valves, and for acid-resistant parts. It is carbon- or graphite-impregnated with a chemically resistant resin and molded to any shape. It can be machined. *Porous carbon* is used for the filtration of corrosive liquids and gases. It consists of uniform particles of carbon pressed into plates, tubes, or disks without a binder, leaving interconnecting pores of about 0.001–0.0075 in. (0.025–0.190 mm) in diameter. The porosity of the material is 48%, tensile strength of 1 MPa, and compressive strength about 3.5 MPa. *Porous graphite* has graphitic instead of carbon particles and is more resistant to oxidation but is lower in strength.

Diamond

Diamond is the cubic crystalline form of carbon. When pure, diamond is water clear, but impurities add shades of opaqueness including black. It is the hardest natural material with a hardness on the Knoop scale ranging from 5500 to 7000. It will scratch and be scratched by the hardest anthropogenic material Borozon. It has a specific gravity of 3.5. Diamond has a melting point of around 3871°C, at which point it will graphitize and then vaporize. Diamonds are generally electrical insulators and nonmagnetic. Synthetic diamonds are produced from graphite at extremely high pressures (5,444–12,359.9 MPa) and temperatures from 1204°C to 2427°C. They are up to 0.01 carat in size and are comparable to the quality of industrial diamonds. In powder form, they are used in cutting wheels. Of all diamonds mined, about 80% by weight are used in industry. Roughly 45% of the total industrial use is in grinding wheels. Tests have shown that under many conditions synthetic diamonds are better than mined diamonds in this application.

Composites

Carbon/carbon composites, which comprise carbon fibers in a carbon matrix, are noted for their heat resistance, high-temperature strength, high thermal conductivity, light weight, low thermal expansivity, and resistance to air/fuel mixtures. However, they are costly to produce. Also, they react with oxygen at temperatures above 800°F (427°C), necessitating oxygen-barrier coatings. *Silicon carbide*, 0.005–0.007 in. (0.127–0.178 mm) thick, serves as such a coating for applications in the nose cone and wing leading edges of the Space Shuttle. Other uses include the brakes of large commercial aircraft, clutches and brakes of Formula One race cars, and rocket nozzles.

Carbon films, usually made by chemical vapor deposition (CVD) at 2012°F (1100°C), can strengthen and toughen *ceramic–matrix composites* but are not readily adaptable to coating fibers, platelets,

or powder. The Japanese have developed what is said to be a more economical method using silicon carbide and other ceramics. Nanometer- to micrometer-thick films are formed on these forms including silicon carbide single crystals, by treating them with water under pressure at 572°F–1472°F (300°C–800°C). This treatment transforms the surface layer to carbon.

The so-called carbons used for electric-light **arc electrodes** are pressed from coal-tar carbon, but are usually mixed with other elements to bring the balance of light rays within the visible spectrum. Solid carbons have limited current-carrying capacity, but when the carbon has a center of metal compounds such as the fluorides of the rare earths, the current capacity is greatly increased. It then forms a deep positive crater in front of which is a flame five times the brilliance of that with a low-current arc. *The sunshine carbon*, used in electric-light carbons to give approximately the same spectrum as some light, is molded coal-tar carbon with a core of cerium metals to introduce more blue into the light. *Arc carbons* are also made to give other types of light, and to produce special rays for medicinal and other purposes. *B carbon*, of National Carbon Co., Inc., contains iron in the core and gives a strong emission of rays which are the antirachitic radiations. The light seen by the eye is only one-fourth the total radiation since the strong rays are invisible. *C carbon* contains iron, nickel, and aluminum in the core and gives off powerful lower-zone ultraviolet rays. It is used in light therapy and for industrial applications. *E carbon*, to produce penetrating infrared radiation, contains strontium. *Electrode carbon*, used for arc furnaces, is molded in various shapes from carbon paste.

Carbides

Carbon forms binary compounds known as carbides with elements less electronegative than carbon (C–H$_2$ compounds are excluded). Effectively then, carbides are composed of metal–carbon compounds if boron and silicon are included among the normal metals. Essentially no volatile compounds (except AlC) are known because decomposition sets in at higher temperatures before volatization of the carbide as such can occur.

Most carbides can be prepared by heating a mixture of the powdered metal and carbon, usually to high temperatures, but not necessarily as high as the melting point. Generally, the same result is possible by heating a mixture of the oxide of the metal with carbon.

It is useful to classify carbides as ionic (salt-like), metallic (interstitial), and covalent. The more electropositive elements (groups I, II, and III, and to some extent, members of the lanthanide and actinide series of the periodic system) form ionic carbides with transparent (or salt-like) crystals.

Highly refractory covalent carbides are formed with silicon and boron. SiC (diamond structure) is formed from a mixture of SiO$_2$ and coke. The very hard B$_4$C can be made similarly from its oxide; it is unusual both structurally and in having a fairly high electrical conductivity.

The carbides of chromium, manganese, iron, cobalt, and nickel are intermediate between the interstitial and covalent types, but are much nearer the former in properties. The presence of Fe$_3$C in iron is an important factor in the properties of steel.

Mechanical/Thermal/Electrical/Other Properties

Compared to metals and most polymers, room-temperature tensile strengths of conventional carbon and graphite are low, ranging from 6.8 to 9.16 MPa. Compressive strengths range from 20.6 to 54.9 MPa. These are generally with-grain values (i.e., specimen length is parallel to the grain).

Carbon and graphite do not have a true melting point. They sublime at 4399°C. Conventional graphite have exceptionally high thermal conductivity at room temperature, whereas carbon has only fair conductivity. Conductivity with the grain in graphite is comparable to that of aluminum; across the grain it is about the same as brass. Conductivity increases with temperature up to about 0°C; it then remains relatively high, but decreases slowly over a broad temperature range, before it drops sharply. In pyrolytic graphite, thermal conductivity with the grain approaches that of copper; across grain, it serves as a thermal insulator and is comparable to that of ceramics.

Thermal expansion of carbon and graphite is quite low (1 to 1.5 × 10^{-6}/°F)—less than one third that of many metals. Expansion of graphite across grain increases with increasing density, whereas along the grain it decreases with increasing density. But expansion increases in both directions with increasing temperatures.

The electrical characteristics associated with carbon and graphite use as electrodes or anodes are relatively well known. Carbon and graphite are actually semiconductors, with their electrical resistivity, or conductivity, falling between those of common metals and common semiconductors. At temperatures approaching absolute zero, carbon and graphite have few conducting electrons, the number increasing with increasing temperature. Thus, electrical resistivity decreases with increasing temperature. On the other hand, although increasing electron density tends to reduce resistivity as temperature rises, scattering effects may become dominant at certain temperatures in the range of 982°C and thus modify or even reverse this trend. Pyrolytic graphite with its higher density has improved electrical conductivity (along the grain). Further, its high degree of anisotropy results in a high degree of electrical resistivity across the grain.

Other properties of graphite are excellent lubricity and relatively low surface hardness; carbon has fair lubricity and relatively high surface hardness. Further, certain types of carbon graphitize relatively easily; others do not. Consequently, a wide variety of carbon, C–Gr, and graphite materials are available, each designed to provide specific types of surface characteristics for such uses as bearings and seals. Grades are also available impregnated with a wide variety of substances, from synthetic resin or oil to a bearing metal.

The nuclear grades of carbon and graphite are of exceptionally high purity. As a moderator and reflector in nuclear reactors, they have no equal because of their low thermal neutron absorption cross section and high scattering cross section coupled with high strength at elevated temperatures and thermal stability in nonoxidizing environments. In general, the properties of carbon and graphite are improved by exposure to nuclear radiation. Hardness and strength increase while thermal and electrical conductivity decrease.

Applications

The free element of carbon has many uses, ranging from ornamental applications of the diamond in jewelry to the black-colored pigment of carbon black in automobile tires and printing inks. Another form of carbon, graphite, is used for high-temperature crucibles, arc-light and dry-cell electrodes, lead pencils, and as a lubricant. Charcoal, an amorphous form of carbon, is used as an absorbent for gases and as a decolorizing agent.

The compounds of carbon find many uses. CO_2 is used for the carbonation of beverages, for fire extinguishers, and in the solid state as a refrigerant. CO finds use as a reducing agent for many metallurgical processes. Carbon tetrachloride and carbon disulfide are important solvents for industrial uses. Gaseous dichlorodifluoromethane, commonly known as *Freon*, is used in refrigeration devices. Calcium carbide is used to prepare acetylene, which is used

for the welding and cutting of metals as well as for the preparation of other organic compounds. Other metal carbides find important uses as refractories and metal cutters.

Carbon nanotubes may one day be used to construct extremely small-scale circuits. Scientists have observed ballistic conductance—electrons passing through a conductor without heating it—at room temperature in 5 μm carbon nanotubes. Because the nanotubes remain cool, extremely large current densities can flow through them.

The most common use of manufactured carbon is as sliding elements in mechanical devices. It is used as the primary rubbing face in most mechanical seals. Used as a brush, it transfers electrical current to the rotating commutator on small electric motors. Carbon vanes, piston rings, or cylinder liners are used in most small air pumps and drink-dispenser pumps. Carbon is also used for pistons in chemical-metering pumps and for metering valves in gasoline pumps. All these applications require the carbon to slide on metal with a coefficient of friction below 0.2 and a wear rate below 0.001 in. per million inches rubbed.

Bearings are another significant application area for manufactured carbon. In many cases, characteristics of the carbon bearing are tailored to satisfy a wide range of requirements. This is done by impregnating the porous, as-baked or as-graphitized carbon with various materials—for example, resin, babbitt, copper, or glass—or by combining impregnation with chemical conversion of the carbon surface (to hard SiC).

These self-lubricating materials are particularly suited for environments containing dust or lint, repeated steam cleaning, solvents or corrosive fluids, low or high temperatures, high static loads, or hard vacuums. Some of these conditions require impregnants in the carbon material. Other applications using manufactured carbon bearings are those inaccessible for lubrication or where product contamination (from lubricants) cannot be tolerated.

Carbon Arc Cutting (CAC)

An arc cutting process in which metals are severed by melting them with the heat of an arc between a carbon electrode and the base metal. To melt the metal along the line of cut.

Carbon Arc Welding (CAW)

An arc welding process that produces coalescence of metals by heating them with an arc between a carbon electrode and the work or struck between two carbon electrodes. No shielding is used although gases produced by the burning of the carbon may offer protection.

Carbon Black

An amorphous powdered carbon resulting from the incomplete combustion of a gas, usually deposited by contact of the flame on a metallic surface, but also made by the incomplete combustion of the gas and a chamber. The carbon black made by the first process is called *channel black*, taking the name from the channel iron used as the depositing surface. The modern method, called the *impingement* process, uses many small flames with the fineness of particles size controlled by flame size. The air-to-gas ratio is high, giving oxidized surfaces and acid properties. No water is used for cooling, keeping the ash content low. The supergrade of channel black has a particle size as low as 512 μin. (13 μm) and a pH of 3–4.2. Carbon black made by other processes is called *soft black* and is weaker in color strength, not so useful as a pigment. *Furnace black* is made with a larger flame in a confined chamber with the particles settling out in cyclone chambers. The particle size is oily, and the pH is high. Black Pearl 3700, 4350, and 4750 are high-purity furnace blacks from Cabot Corp. The 3700, with cleanliness and cable smoothness and cleanliness similar to acetylene, is intended as an alternative to the latter for semiconductor cable shields. The 4350 and 4750 could become the first furnace blacks used for single-service food packaging because of their low polyaromatic-hydrocarbon content and better dispersion and impact resistance than selective channel blacks approved for this application.

Because it possesses useful ultraviolet protective properties, it is also much used in molding compounds intended for outside weathering applications.

Carbon black from clean artificial gas is a glossy product with an intense color, but all the commercial carbon black is from natural gas. The finer grades of channel black are mostly used for color pigment, in paints, polishes, carbon paper, and printing and drawing inks. The larger use of carbon black is in automotive tires to increase the wear resistance of the rubber. The blacker blacks have a finer particle size than the grayer blacks, hence have more surface and absorptive capacity and compounding with rubber. Channel black is valued for rubber compounding because of its low acidity and low grit content.

In rubber compounding, the carbon black is evenly dispersed to become intimately attached to the rubber molecule. The fineness of the black determines the tensile strength of the rubber, the structure of the carbon particle determines the modulus, and the pH determines the cure behavior. Furnace blacks have a basic pH which activates the accelerator, and delaying-action chemicals are thus needed, but fine furnace blacks impart abrasion resistance to the rubber. Furnace black made with a confined flame with limited air has a neutral surface and a low volatility. Fineness is varied by temperature, size of flame, and time. Carbonate salts raise the pH. Most of the channel black for rubber compounding is made into dust-free pellets. *Color-grade black* for inks and paints is produced by the channel process or the impingement process. In general, carbon black for reinforcement has small particle size, and the electrically conductive grades, *CF carbon black* and *CC carbon black*, conductive furnace and conductive channel, have large particle sizes.

Acetylene black is a carbon black made by heat decomposition of acetylene. It is more graphite than ordinary carbon black with colloidal particles linked together in an irregular lattice structure and has high electrical conductivity and high liquid-absorption capacity. Particle size is intermediate between that of channel black and furnace black, with low ash content, nonoiliness, and a pH of 6.5. It is valued for use in dry cells and lubricants. *Ucet*, of Union Carbide Corp., is in the form of agglomerates of irregular fine crystals. The greater surface area gives higher thermal and electrical conductivity and high liquid absorption.

For electrically *conductive rubber*, the mixing of the black with the rubber is regulated so that carbon chain connections are not broken. Such conductive rubber is used for tabletops, conveyor belts, and coated filter fabrics to prevent static buildup. Carbon blacks are also made from liquid hydrocarbons, and from anthracite coal by treatment of the coal to liberate hydrogen and carbon monoxide and then high-temperature treatment with chlorine to remove impurities. The black made from anthracite has and open pore structure useful for holding gases and liquids.

NASA Propulsion Laboratories has determined that the addition of *Shawanigen carbon black* markedly increases the life of amorphous-carbon or graphite anodes in rechargeable lithium-ion electrochemical cells.

Carbon Dioxide

Also called *carbonic anhydride*, and in its solid state, *dry ice*. A colorless, odorless gas of composition CO_2, which liquefies at $-85°F$ ($-65°C$) and solidifies at $-108.8°F$ ($-78.2°C$). Release of CO_2 into the atmosphere by the burning of fossil fuels is said to be causing global warming by the process known as the *greenhouse effect*. It is recovered primarily as a by-product of the steam reforming of natural gas to make hydrogen or synthesis gas in petroleum and fertilizer plants. Smaller quantities are obtained by purifying flue gases generated from burning hydrocarbons or lime and from distilleries. Its biggest uses are captive, as a chemical raw material for making urea and in enhanced oil recovery operations in petroleum production. Merchant CO_2 is more than 99.5% pure, with less than 500 ppm (parts per million) of nonvolatile residues. In liquid form it is marketed in cylinders and is used in fire extinguishers, in spray painting, in refrigeration, for inert atmospheres, for the manufacture of *carbonated beverages*, and in many industrial processes. It is also marketed as dry ice, a white, snow-like solid used for refrigeration and transporting food products. *Cardox* is a trade name of Cardox Corp. for liquid carbon dioxide in storage units at 30 lb/in.^{2} (0.21 MPa) pressure for fire-fighting equipment. Other uses include hardening of foundry cores, neutralization of industrial wastes, and production of salicylic acid for aspirin. Carbon dioxide is a key lasing gas and *carbon dioxide lasers* and is also used as a shielding gas in some MIG welding applications for steel, and as a foaming agent in producing plastic foam products. It can behave as a *supercritical fluid*, in which state it can be used to foam plastics and extract hazardous substances and waste treatment processes and in soil remediation. CO_2 is used to wash brownstock in the pulp and paper industry, thereby sending cleaner pulp onto bleaching. In cooling systems, it is an alternative to halogenated-carbon refrigerants, **CO_2 "snow,"** pellets that is, is used to cool freshly laid eggs, cuts of meat and poultry, and flour in baking. Dry ice pellets are blasted on molds to clean them of plastic residuals. **Liquid carbon dioxide** is used in SuperFuge, an immersion system by Deflex Corp. to rid products of surface contaminants.

Carbon Dioxide Process (Sodium Silicate/CO₂)

In foundry practice, a process for hardening molds or cores in which carbon dioxide gas is blown through dry clay-free silica sand to precipitate silica in the form of a gel from the sodium silicate binder.

Carbon Dioxide Welding

A *gas metal-arc welding* (GMAW) process for steel in which carbon dioxide is fit into the arc vicinity to protect the weld zone from atmospheric contamination. Carbon dioxide is effectively inert in these circumstances and its choice, rather than argon, arises from its favorable effect on arc characteristics as well as the economic factor.

Carbon Edges

Carbonaceous deposits in a wavy pattern along the edges of a steel sheet or strip; also known as snaky edges.

Carbon Electrode

A nonfiller material electrode used in arc welding or cutting, consisting of a carbon or graphite rod, which may be coated with copper or other coatings.

Carbon Equivalent

(1) The measure of the effect of one or more alloying elements in cast iron or steel expressed in terms of the amount of carbon that would have the same effect. For example, the alloying elements in steel effect characteristics such as hardenability, and hence weldability, or microstructure in a way which reinforces or detracts from the effect of carbon. The element can therefore be ascribed as a factor reflecting its potency relative to carbon, taken as unity. This factor is termed the carbon equivalent of the element. In any iron or steel, individual alloy contents multiplied by their respective carbon equivalents can be added together to provide a total carbon equivalent for the material. (2) One of a number of formulae for predicting weldability of low alloy steels is:

$$\text{Carbon Equivalent} = C + \left(\frac{Mn}{20}\right) + \left(\frac{Ni}{15}\right) + \left(\frac{Cr, Mo, V}{10}\right)$$

Also see *Schaeffler Diagram* for an example of carbon equivalents related to microstructure. (3) For cast iron, an empirical relationship of the total carbon, silicon, and phosphorus contents expressed by the formula:

$$CE = \%C + 0.3(\%Si) + 0.33(\%P) - 0.027(\%Mn) + 0.4(\%S)$$

(4) For rating of weldability:

$$CE = C + \frac{Mn}{6} + \frac{Ni}{15} + \frac{Cu}{15} + \frac{Cr}{5} + \frac{Mo}{5} + \frac{V}{5}$$

Carbon Extraction Replica

See *extraction replica*.

Carbon Fiber

(1) Fiber produced by the pyrolysis of organic precursor fibers, such as rayon, polyacrylonitrile (PAN), and pitch, in an inert environment. The term is often used interchangeably with the term graphite; however, carbon fibers and graphite fibers differ. The basic differences lie in the temperature at which the fibers are made and heat-treated, and in the amount of elemental carbon produced. Carbon fibers typically are carbonized in the region of 1315°C (2400°F) and assay at 93%–95% carbon, while graphite fibers are graphitized at 1900°C–2480°C (3450°F–4500°F) and assay at more than 99% elemental carbon. (2) A filament comprised of nearly pure carbon typically up to 10 μm diameter used for reinforcement epoxy or other plastics to form carbon fiber composite materials. See also *pyrolysis*.

Carbon Flotation

In casting, segregation in which free graphite has separated from the molten iron. This defect tends to occur at the upper surfaces of the cope of the castings.

Carbon Monoxide

Carbon monoxide (CO) is a product of incomplete combustion, and is very reactive. It is one of the desirable products in synthesis gas for making chemicals; the synthesis gas made from coal contains at least 37% CO. It is also recovered from top-blown O_2 furnaces in steel mills. It reacts with H_2 to form methanol, which is then catalyzed by zeolites into gasoline. Acetic acid is made by methanol carbonylation, and acrylic acid results from the reaction of CO, acetylene, and methanol. CO forms a host of neutral, anionic, and cationic carbonyls, with such metals as iron, cobalt, nickel, molybdenum, chromium, rhodium, and ruthenium. These metals are from groups I, II, VI, VII, and VIII of the periodic table. The metal carbonyls can be prepared by the direct combination of the metal with CO, although several of the compounds require fairly high pressures. The metal carbonyls react with the halogens to produce metal carbonyl halides. With H_2, a similar reaction takes place to form metal carbonyl hydrides.

Nickel carbonyl finds application in purification and separation of nickel from other metals. Iron carbonyl has been used in antiknock gasoline preparations and to prepare high-purity iron metal.

CO is an intense poison when inhaled and is extremely toxic even in the small amounts from the exhausts of internal-combustion engines.

With chlorine, in the presence of sunlight, CO forms highly poisonous phosgene, $COCl_2$; with sulfur, carbonyl sulfide, COS, is obtained.

Carbon Nitride

Traditional materials synthesis techniques, which require high-temperature conditions to facilitate diffusion of atoms in the solid state, generally preclude the rational assembly of atoms since the products are restricted to those phases stable at high temperature. For example, heating graphite in molecular N_2 does not provide carbon nitride (C–N) solids. However, binary C–N solids are desirable materials to prepare and study because they may possess extreme hardness and high thermal conductivity. Extreme hardness is important in thin-film coatings of cutting tools used for machining and in bearing surfaces used in a variety of high-performance mechanical devices, whereas high thermal conductivity is important to the fabrication of advanced microelectronics devices. To overcome the limitations imposed by classical methodologies of solid-state chemistry and provide access to these potentially exciting materials requires better control of the reactants and reactions that might lead to C–N solids. A new experimental approach that meets these requirements and has provided access to C–N materials utilizes pulsed-laser evaporation of graphite to generate reactive carbon fragments and an atomic N_2 beam as a source of N_2 that can readily react with the carbon fragments.

The atomic N_2 beam is generated by using a radio-frequency discharge within an Al_2O_3 nozzle through which N_2 seeded in helium flows. This process produces a very high flux of atomic N_2. Furthermore, by varying the N_2:He ratio and the radio-frequency power, it is possible to control both the flux and energy of this critical reactant. Hence, this synthetic approach provides a ready means of producing reactants that are free from impurities and that have controllable energies.

Structure

The structural properties of the C–N materials produced by the laser ablation approach have been investigated and show that carbon and N_2 are bound covalently within this solid but cannot provide information about the three-dimensional structure. Diffraction ring patterns suggest that a single crystalline C–N phase was obtained by using the new synthetic strategy.

Other Key Properties

The new C–N materials also exhibit interesting physical properties that may be attractive for high-performance engineering applications. Qualitative scratch tests indicate that C–N materials produced by the new laser ablation technique are hard. For example, rubbing C–N and hard amorphous carbon surfaces against one another produces damage in the carbon but not the C–N material.

Studies have shown that C–N is an excellent electrical insulator, and that these electrical properties are stable to thermal cycling. Because C–N is also expected to exhibit good thermal conductivity, it could be an attractive candidate for the dielectric in advanced microelectronic devices where thermally conducting electrical insulators are needed to enable further miniaturization of devices.

Carbon Potential

A measure of the ability of an environment containing active carbon to alter or maintain, under prescribed conditions, the carbon level of the steel. In any particular environment, the carbon level attained will depend on such factors as temperature, time, and steel composition.

Carbon Refractory

A manufactured refractory comprised substantially or entirely of carbon (including graphite).

Carbon Restoration

Replacing the carbon lost in the surface layer from previous processing of the steel by carburizing this layer to substantially the original carbon level. Sometimes called re-carburizing.

Carbon Steel

Carbon (C) steel, also called plain carbon steel, is a malleable, iron-based metal containing carbon, small amounts of manganese, and other elements that are inherently present. The old shop names, machine steel and machinery steel, are still used to mean any easily worked low-carbon steel. By definition, plain carbon steels are those that contain up to about 1% carbon, not more than 1.65% manganese, 0.60% silicon, and 0.60% copper, and only residual amounts of other elements, such as sulfur (0.05% max) and phosphorus (0.04% max). They are identified by means of a four-digit numerical system established by the American Iron and Steel Institute (AISI). The digits are preceded by either "AISI" or "SAE." The first digit is the number 1 for all carbon steels. A 0 after the 1 indicates nonresulfurized grades, a 1 for the second digit indicates resulfurized grades, and the number 2 for the second digit indicates resulfurized and rephosphorized grades. The last two digits give the nominal (middle of the range) carbon content, in hundredths of a percent. For example, for grade 1040, the 40 represents a carbon range of 0.37%–0.44%. If no prefix letter is included in the designation, this steel was made by the basic open-hearth, basic O_2, or electric furnace process. The prefix B stands for the acid Bessemer process, which is obsolete, and the prefix M designates merchant quality. The letter L between the second and third digits identifies leaded steels, and the suffix H indicates that this steel was produced to hardenability limits. See Table C.3.

For all plain carbon steels, carbon is the principal determinant of many performance properties. Carbon has a strengthening and hardening effect. At the same time, it lowers ductility, as evidenced by a decrease in elongation and reduction of area. In addition, increasing carbon content decreases machinability and weldability, but improves wear resistance. The amount of carbon present also affects physical properties and corrosion resistance. With an increase in carbon content, thermal and electric conductivity decline, magnetic permeability decreases drastically, and corrosion resistance is less.

Carbon steels may be specified by chemical composition, mechanical properties, method of oxidation, or thermal treatment (and the resulting microstructure). Carbon steels are available in most wrought mill forms, including bar, sheet, plate, and tubing. Sheet is primarily a low-carbon-steel product, but virtually all grades are available in bar and plate. Plate, usually a low-carbon or medium-carbon product, is used mainly in the hot-finished condition, although it also can be supplied heat-treated. Bar products, such as rounds, squares, hexagonals, and flats (rectangular cross sections), are also mainly low-carbon and medium-carbon products and are supplied hot-rolled or hot-rolled and cold-finished. Cold finishing may be by drawing (cold-drawn bars are the most widely used); turning (machining) and polishing; drawing, grinding, and polishing; or turning, grinding, and polishing. Bar products are also available in various quality designations, such as merchant quality (M), cold-forging quality, cold-heading quality, and several others. Sheet products also have quality designations as noted in low-carbon steels, which follow. Plain carbon steels are commonly divided into three groups, according to carbon content: low carbon, up to 0.30%; medium carbon, 0.31%–0.55%; and high carbon, 0.56%–1%.

Low-carbon steels are the grades AISI 1005-1030. Sometimes referred to as mild steels, they are characterized by low strength and high ductility, and are nonhardenable by heat treatment except by surface-hardening processes. Because of their good ductility, low-carbon steels are readily formed into intricate shapes. These steels are also readily welded without danger of hardening and embrittlement in the weld zone. Although low-carbon steels cannot be through-hardened, they are frequently surface-hardened by various methods (carburizing, carbonitriding, and cyaniding, for example) that diffuse carbon into the surface. Upon quenching, a hard, wear-resistant surface is obtained.

Low-carbon sheet and strip steels (1008-1012) are widely used in cars, trucks, appliances, and many other applications. Hot-rolled products are usually produced on continuous hot strip mills. Cold-rolled products are then made from the hot-rolled products, reducing thickness and enhancing surface quality. Unless the fully work-hardened product is desired, it is then annealed to improve formability and temper-rolled to further enhance surface quality. Hot-rolled sheet and strip and cold-rolled sheet are designated commercial quality (CQ), drawing quality (DQ), drawing quality special killed (DQSK), and structural quality (SQ). The first three designations refer, respectively, to steels of increasing formability and mechanical property uniformity. SQ, which refers to steels produced to specified ranges of mechanical properties and/or hardenability values, does not pertain to cold-rolled strip, which is produced to several tempers related to hardness and bendability. Typically, the hardness of CQ hot-rolled sheet ranges from Rb 40 to 75 and tensile properties range from ultimate strengths of 276 to 469 MPa, yield strengths of 193 to 331 MPa and elongations of 14% to 43%. For DQ hot-rolled sheet: Rb 40 to 72, 276 to 414 MPa, 186 to 310 MPa, and 28% to 48%, respectively. For CQ cold-rolled sheet: Rb 35 to 60, 290 to 393 MPa, 159 to 262 MPa, and 30% to 45%. And for DQ cold-rolled sheet: Rb 32 to 52, 262 to 345 MPa, 138 to 234 MPa, and 34% to 46%.

Low-carbon steels 1018-1025 in cold-drawn bar (16–22 mm thick) have minimum tensile properties of about 483 MPa ultimate

C

TABLE C.3

Composition Ranges and Limits for AISI-SAE Standard Carbon Steels—Structural Shapes, Plate, Strip, Sheet, and Welded Tubing

AISI-SAE Designation	UNS Designation	Heat Composition Ranges and Limits,[a] %	
		C	Mn
1006	G10060	.....0.08 max	0.25–0.45
1008	G10080	.....0.10 max	0.25–0.50
1009	G10090	.....0.15 max	0.60 max
1010	G10100	.....0.08–0.13	0.30–0.60
1012	G10120	.....0.10–0.15	0.30–0.60
1015	G10150	.....0.12–0.18	0.30–0.60
1016	G10160	.....0.12–0.18	0.60–0.90
1017	G10170	.....0.14–0.20	0.30–0.60
1018	G10180	.....0.14–0.20	0.60–0.90
1019	G10190	.....0.14–0.20	0.70–1.00
1020	G10200	.....0.17–0.23	0.30–0.60
1021	G10210	.....0.17–0.23	0.60–0.90
1022	G10220	.....0.17–0.23	0.70–1.00
1023	G10230	.....0.19–0.25	0.30–0.60
1025	G10250	.....0.22–0.28	0.30–0.60
1026	G10260	.....0.22–0.28	0.60–0.90
1030	G10300	.....0.27–0.34	0.60–0.90
1033	G10330	.....0.29–0.36	0.70–1.00
1035	G10350	.....0.31–0.38	0.60–0.90
1037	G10370	.....0.31–0.38	0.70–1.00
1038	G10380	.....0.34–0.42	0.60–0.90
1039	G10390	.....0.36–0.44	0.70–1.00
1040	G10400	.....0.36–0.44	0.60–0.90
1042	G10420	.....0.39–0.47	0.60–0.90
1043	G10430	.....0.39–0.47	0.70–1.00
1045	G10450	.....0.42–0.50	0.60–0.90
1046	G10460	.....0.42–0.50	0.70–1.00
1049	G10490	.....0.45–0.53	0.60–0.90
1050	G10500	.....0.47–0.55	0.60–0.90
1055	G10550	.....0.52–0.60	0.60–0.90
1060	G10600	.....0.55–0.66	0.60–0.90
1064	G10640	.....0.59–0.70	0.50–0.80
1065	G10650	.....0.59–0.70	0.60–0.90
1070	G10700	.....0.65–0.76	0.60–0.90
1074	G10740	.....0.69–0.80	0.50–0.80
1078	G10780	.....0.72–0.86	0.30–0.60
1080	G10800	.....0.74–0.88	0.60–0.90
1084	G10840	.....0.80–0.94	0.60–0.90
1085	G10850	.....0.80–0.94	0.70–1.00
1086	G10860	.....0.80–0.94	0.30–0.50
1090	G10900	.....0.84–0.98	0.60–0.90
1095	G10950	.....0.90–1.04	0.30–0.50
1524[b]	G15240	.....0.18–0.25	1.30–1.65
1527[b]	G15270	.....0.22–0.29	1.20–1.55
1536[b]	G15360	.....0.30–0.38	1.20–1.55
1541[b]	G15410	.....0.36–0.45	1.30–1.65
1548[b]	G15480	.....0.43–0.52	1.05–1.40
1552[b]	G15520	.....0.46–0.55	1.20–1.55

[a] Limits on phosphorus and sulfur contents are typically 0.040% maximum phosphorus and 0.050% maximum sulfur. Silicon contents range from 0.080% to approximately 2%. Steels listed in this table can be produced with additions of lead or boron. Leaded steels typically contain 0.15%–0.35% lead and are identified by inserting the letter "L" in the designation—11L17; boron steels can be expected to contain 0.0005%–0.003% boron and are identified by inserting the letter "B" in the designation—15B41.

[b] Formerly designated 10xx grade.

strength, 413 MPa yield strength, and 18% elongation. Properties decrease somewhat with increasing section size, to, for example, 379 MPa, 310 MPa and 15%, respectively, for 50–76 mm cross sections.

Medium-carbon steels are the grades AISI 1030-1055. They usually are produced as killed, semikilled, or capped steels, and are hardenable by heat treatment. However, hardenability is limited to thin sections or to the thin outer layer on thick parts. Medium-carbon steels in the quenched and tempered condition provide a good balance of strength and ductility. Strength can be further increased by cold work. The highest hardness practical for medium-carbon steels is about 550 Bhn (Rockwell C55). Because of the good combination of properties, they are the most widely used steels for structural applications, where moderate mechanical properties are required. Quenched and tempered, their tensile strengths range from about 517 to over 1034 MPa.

Medium-carbon steel 1035 in cold-drawn bar 16–22 mm thick has minimum tensile properties of about 586 MPa ultimate strength, 517 MPa yield strength, and 13% elongation. Strength increases and ductility decreases with increasing carbon content, to, for example, 689 MPa, 621 MPa, and 11%, respectively, for medium-carbon steel 1050. Properties decrease somewhat with increasing section size, to, for example, 483 MPa, 414 MPa, and 10%, respectively, for 1035 steel 50–76 mm cross sections.

High-carbon steels are the grades AISI 1060-1095. They are, of course, hardenable with a maximum surface hardness of about Rockwell C64 achieved in the 1095 grade. These steels are thus suitable for wear-resistant parts. So-called spring steels are high-carbon steels available in annealed and pretempered strip and wire. In addition to their spring applications, these steels are used for such items as piano wire and saw blades. Quenched and tempered, high-carbon steels approach tensile strengths of 1378 MPa.

Free-machining carbon steels are low-carbon and medium-carbon grades with additions usually of sulfur (0.08%–0.13%), S–P combinations, and/or lead to improve machinability. They are AISI 1108-1151 for sulfur grades, and AISI 1211-1215 for phosphorus and sulfur grades. The latter may also contain bismuth and be lead free. The presence of relatively large amounts of sulfur and phosphorus can reduce ductility, cold formability, forgeability, weldability, as well as toughness and fatigue strength. Calcium deoxidized steels (carbon and alloy) have good machinability, and are used for carburized or through-hardened gears, worms, and pinions.

Low-temperature carbon steels have been developed chiefly for use in low-temperature equipment and especially for welded pressure vessels. They are low-carbon to medium-carbon (0.20%–0.30%), high-manganese (0.70%–1.60%), silicon (0.15%–0.60%) steels, which have a fine-grain structure with uniform carbide dispersion. They feature moderate strength with toughness down to −46°C.

For grain refinement and to improve formability and weldability, carbon steels may contain 0.01%–0.04% columbium. Called columbium steels, they are used for shafts, forgings, gears, machine parts, and dies and gauges. Up to 0.15% sulfur, or 0.045% phosphorus, makes them free-machining, but reduces strength.

Rail steel, for railway rails, is characterized by an increase of carbon with the weight of the rail. Railway engineering standards call for 0.50%–0.63% carbon and 0.60% manganese in a 27 kg rail, and 0.69%–0.82% carbon and 0.70%–1.0% manganese in a 64 kg rail. Rail steels are produced under rigid control conditions from deoxidized steels, with phosphorus kept below 0.04%, and silicon 0.10%–0.23%. Guaranteed minimum tensile strength of 551 MPa is specified, but it is usually much higher.

Sometimes, a machinery steel may be required with a small amount of alloying element to give a particular characteristic and

still not be marketed as an alloy steel, although trade names are usually applied to such steels. Superplastic steels, developed at Stanford University, with 1.3%–1.9% carbon, fall between high-carbon steels and cast irons. They have elongations approaching 500% at warm working temperatures of 538°C–650°C, and 4%–15% elongation at room temperature. Tensile strengths range from 1034 to over 1378 MPa. The extra-high ductility is a result of a fine, equiaxed grain structure obtained by special thermomechanical processing.

Damascus steels are 1%–2% carbon steels used for ancient swords made by blacksmiths using hot and warm forging, which developed layered patterns. The swords were eminent for their strength and sharp cutting edge. With carbon in the form of iron carbide, the forged products were free of surface markings. With carbon in the form of spherical carbide, the products could exhibit surface markings. So-called *welded Damascus steels*, also referred to as *pattern welded steels*, also exhibit surface markings. Superplasticity may be inherent in all of these steels.

Production Types

Steelmaking processes and methods used to produce mill products, such as plate, sheet, and bars, have an important effect on the properties and characteristics of the steel.

Deoxidation Practice

Steels are often identified in terms of the degree of deoxidation resulting during steel production. Killed steels, because they are strongly deoxidized, are characterized by high composition and property uniformity. They are used for forging, carburizing, and heat-treating applications. Semikilled steels have variable degrees of uniformity, intermediate between those of killed and rimmed steels. They are used for plate, structural sections, and galvanized sheets and strip. Rimmed steels are deoxidized only slightly during solidification. Carbon is highest at the center of the ingot. Because the outer layer of the ingot is relatively ductile, these steels are ideal for rolling. Sheet and strip made from rimmed steels have excellent surface quality and cold-forming characteristics. Capped steels have a thin low-carbon rim, which gives them surface qualities similar to rimmed steels. Their cross section uniformity approaches that of semikilled steels.

Melting Practice

Steels are also classified as air melted, vacuum melted, or vacuum degassed. Air-melted steels are produced by conventional melting methods, such as open hearth, basic O_2, and electric furnace. Vacuum-melted steels encompass those produced by induction vacuum melting and consumable electrode vacuum melting. Vacuum-degassed steels are air-melted steels that are vacuum processed before solidification. Compared with air-melted steels, those produced by vacuum-melting processes have lower gas content, fewer nonmetallic inclusions, and less center porosity and segregation. They are more costly, but have better mechanical properties, such as ductility and impact and fatigue strengths.

Rolling Practice

Steel mill products are produced from various primary forms such as heated blooms, billets, and slabs. These primary forms are first reduced to finished or semi-finished shape by hot-working operations. If the final shape is produced by hot-working processes, the steel is known as hot rolled. If it is finally shaped cold, the steel is known as cold finished, or more specifically as cold rolled or cold drawn. Hot-rolled mill products are usually limited to low and medium, non-heat-treated, carbon steel grades. They are the most economical steels, have good formability and weldability, and are used widely for large structural shapes. Cold-finish shapes, compared with hot-rolled products, have higher strength and hardness and better surface finish, but are lower in ductility.

Carbon Tetrachloride

A heavy, colorless liquid of composition CCl_4, tetrachloromethane, which is one of a group of chlorinated hydrocarbons. It is an important solvent for fats, asphalt, rubber, bitumens, and gums. It is more expensive than the aromatic solvents, but it is notable as a nonflammable solvent for many materials sold in solution and is widely used as a degreasing and cleaning agent in the dry-cleaning and textile industries. Since the *fumes* are highly toxic, it is no longer permitted in compounds for home use. It is used as a chemical in fire extinguishers such as Pyrene; but when it falls on hot metal, it forms a poisonous gas phosgene. It is also used as a disinfectant, and because of its high dielectric strength has been employed in transformers. Carbon tetrachloride is obtained by the chlorination of carbon disulfide. The specific gravity is 1.595, boiling point 169°F (76°C), and the freezing point 73°F (23°C). Chlorobromomethane, $Br \cdot CH_2 \cdot Cl$, is also used in fire extinguishers, as it is less corrosive and more than twice as efficient as an extinguisher. It is a colorless, heavy liquid with a sweet odor, a specific gravity 1.925, boiling point 153°F (67°C), and a freezing point −85°F (−65°C). It is also used as the high-gravity flotation agent.

Carbon Tube Furnace

An electric furnace that has a carbon retort for a resistor element and is especially suitable for the batch or continuous sintering of carbon-insensitive materials such as cement and carbides.

Carbonaceous

A material that contains carbon in any or all of its several allotropic forms.

Carbon–Carbon Composites

Carbon–carbon composites (C–C) provide high-strength, light weight, and resistance to high temperature and corrosion for pistons in both stationary and mobile engines.

Advantages in using carbon–carbon for both gasoline and diesel engines include increased performance (higher power output and faster engine acceleration); higher temperature operation for improved fuel economy; reduced levels of air pollutant emissions; and reduced noise, vibration, and harshness (NVH) levels. These benefits also could provide lower operational costs and reduced environmental impact.

Another carbon–carbon originally developed for use as rocket nozzles is now a replacement for graphite, quartz, and ceramic furnace components used in Czochralski crystal pulling, a processing step in semiconductor manufacturing. The material has a high strength-to-weight ratio, maintains its strength and rigidity despite high temperatures, resists wear, and is relatively chemically inert. When used in Czochralski furnaces, the thermal properties of the material lead to a 10% reduction in energy use and a 50% cut in cooling times. Because of the purity of the material, chipmakers achieve a 7% increase in wafer yields compared to furnaces with graphite components. See *composite materials*.

Carbonitriding

A *case hardening* process in which a suitable ferrous material is heated above the lower transformation temperature in a gaseous atmosphere of such composition as to cause simultaneous absorption of carbon and nitrogen by the surface and, by diffusion, create a concentration gradient. The heat-treating process is completed by cooling at a rate that produces the desired properties and the workpiece. See *case hardening*.

Carbonium Ion

Positively charged organic compound (ion).

Carbonization

The conversion of an organic substance into elemental carbon in an inert atmosphere at temperatures ranging from 800°C to 1600°C (1470°F to 2910°F) and higher, but usually at about 1315°C (2400°F). Range is influenced by precursor, processing of the individual manufacturer, and properties desired. Should not be confused with carburization.

Carbonizing Flame (in Gas Welding)

Same as carburizing flame. See preferred term *reducing flame*.

Carbonyl Powder

Metal powders prepared by the thermal decomposition of a metal carbonyl compound such as nickel tetracarbonyl $Ni(CO)_4$ or iron pentacarbonyl $Fe(CO)_5$. See also *thermal decomposition*.

Carbonyl Process

A process in which a metal carbonyl gas is heated to form a purified metal powder.

Carborundum

Silicon carbide, particularly as an abrasive.

Carburizing

(1) Absorption and diffusion of carbon into solid ferrous alloys by heating to a temperature usually above Ac_3, in contact with a suitable carbonaceous material. A form of *case hardening* that produces a carbon gradient extending inward from the surface, enabling the surface layer to be hardened either by quenching directly from the carburizing temperature or by cooling to room temperature, then reaustenitizing and quenching. (2) The process by which carbon diffuses into the steel. The process can be deliberately induced by appropriate treatments. It can occur accidentally during some manufacturing processes with potentially adverse consequences for the component.

Carburizing Flame

A gas flame that will introduce carbon into some heated metals, as during a gas welding operation. A carburizing flame is a *reducing flame*, but a reducing flame is not necessarily a carburizing flame. Carburizing flames have a long feathery inner cone and a relatively low temperature.

Carburizing Secondary-Hardening Steels

Case-carburized steels subsequently hardened and strengthened by precipitation of M_2C carbide. Three steels developed by QuesTek Innovations LLC, include *Ferrium C862 stainless steel, GearMet C61%, and GearMet C69%* for gears and bearings. Ferrium C862 nominally contains 15% cobalt, 9.0% chromium, 1.5% nickel, 0.2% vanadium, 0.08% core carbon, balance iron. It is targeted at matching the surface properties of standard nonstainless gear steels, maintaining sufficient core strength and toughness, and having better corrosion resistance than 440C stainless steel. Core hardness is 50 Rockwell C, core toughness 1740 MPa $mm^{1/2}$, and surface hardness 62 Rockwell C. GearMet C61% has 18 cobalt, 9.5 nickel, 3.5 chromium, 1.1 molybdenum, 0.16 core carbon, balance iron. It is designed to provide surface properties similar to conventional gear steels and an ultrahigh strength core with superior fracture toughness. Core hardness is 54 Rockwell C, court toughness of more than 2610 MPa $mm^{1/2}$, and surface hardness 61 Rockwell C. GearMet C69% has 27.8–28.2 cobalt, 5–5.2 chromium, 2.9–3.1 nickel, 2.4–2.6 molybdenum, 0.09–0.11 core carbon, 0.015–0.025 vanadium, balance iron. It combines a tough ductile core with an ultrahard case. Core hardness is 50 Rockwell C and surface hardness 69 Rockwell C.

Carrier

In emission spectra chemical analysis, a material added to a sample to facilitate its controlled vaporization into the analytical gap. See also *analytical gap.*

Carrier Gas

In thermal spraying, the gas is used to carry the powdered materials from the powder feeder or hopper to the gun.

Cartridge Brass

Basically, a 70% copper, 30% zinc wrought alloy, designated *brass alloy C26000*, which may also contain as much as 0.07% lead and 0.05% iron. Besides cartridge brass, a name resulting from its use in munitions, notably cartridge cases, it has been known as *brass alloy 70–30* brass, *spinning brass, spring brass, and extraquality brass.* Besides flat products, the alloy is available in bar, rod, wire, tubing, and, for cartridge cases, cups. It has excellent cold-forming characteristics and a machinability about 30% that of free-cutting brass. It is also relatively brazed and soldered and can be welded by oxyfuel and resistance methods. Its weldability by gas-metal-arc methods, however, is limited, and other welding methods are not advisable. Although corrosion-resistant in various waters and chemical solutions, the alloy may be susceptible to dezincification in stagnant or slow-moving, brackish waters and salt or slightly acidic solutions. Also, it is prone to stress-corrosion cracking, particularly in ammonia environments. Besides the additional applications, it is used for various stamped, spun, or drawn shapes, including lamp fixtures, shells and reflectors, autoradiator cores, locks, springs, fasteners, cylinder components, plumbing fixtures, and architectural grille work.

Cascade Separator

A special device to separate metal powder fractions of different particle size or specific gravity.

Cascade Sequence

A welding sequence in which a continuous multiple-pass weld is built up by depositing weld beads in overlapping layers, usually laid in a *back step sequence*. Compare with *block sequence*.

Case

In heat treating, that portion of the ferrous alloy, extending inward from the surface, whose composition has been altered during *case hardening*. Typically considered to be a portion of an alloy (a) whose composition has been measurably altered from the original composition, (b) that appears light when etched, or (c) that has a higher hardness value than the core. Contrast with *core*.

Case Crushing

A term used to denote longitudinal gouges arising from fracture in case-hardened gears.

Case-Hardening Materials

Case-hardening materials are those for adding carbon or other elements to the surface of low-carbon or medium-carbon steels or to iron so that upon quenching a hardened case is obtained, the center of the steel remaining soft and ductile. The material may be plain charcoal, raw bone, or mixtures marketed as carburizing compounds. A common mixture is about 60% charcoal and 40% barium carbonate. The latter decomposes, yielding CO_2, which is reduced to CO in contact with the hot charcoal. If charcoal is used alone, the action is slow and spotty. Coal or coke can be used, but the action is slow, and the sulfur in these materials is detrimental. Salt is sometimes added to aid the carburizing action. By proper selection of the carburizing material, the carbon content may be varied in the steel from 0.80% to 1.20%. The carburizing temperature for carbon steels typically ranges from 850°C to 950°C but maybe as low as 790°C or as high as 1095°C. The articles to be carburized for case hardening are packed in metallic boxes for heating in a furnace, and the process is called pack hardening, as distinct from the older method of burying the red-hot metal in charcoal.

Steels are also case-hardened by the diffusion of carbon and N_2, called carbonitriding, or N_2 alone, called nitriding. Carbonitriding, also known as dry cyaniding, gas cyaniding, liquid cyaniding, nicarbing, and nitrocarburizing, involves the diffusion of carbon and N_2 into the case. Nitriding also may be done by gas or liquid methods. In carbonitriding, the steel may be exposed to a carrier gas containing carbon and as much as 10% ammonia (NH_3), the N_2 source, or a molten cyanide salt, which provides both elements. NH_3, from gaseous or liquid salts, is also the N_2 source for nitriding. Although low- and medium-carbon steels are commonly used for carburizing and carbonitriding, nitriding is usually applied only to alloy steels containing nitride-forming elements, such as aluminum, chromium, molybdenum, or vanadium. In ion nitriding, or glow-discharge nitriding, electric current is used to ionize low-pressure N_2 gas. The ions are accelerated to the workpiece by the electric potential, and the workpiece is heated by the impinging ions, obviating an additional heat source. Of the three principal case-hardening methods, all provide a hard wear-resistant case. Carburizing, however, which gives the greater case depth, provides the best contact-load capacity. Nitriding provides the best dimensional control and carbonitriding is intermediate in this respect.

C

The principal liquid-carburizing material is sodium cyanide, which is melted in a pot that the articles are dipped in, or the cyanide is rubbed on the hot steel. Cyanide hardening gives an extremely hard but superficial case. N_2 as well as carbon is added to the steel by this process. Gases rich in carbon, such as methane, may also be used for carburizing, bypassing the gas through the box in the furnace. When NH_3 gas is used to impart N_2 to the steel, the process is not called carburizing but is referred to as nitriding.

Chromized steel is steel surface-alloyed with chromium by diffusion from a chromium salt at high temperature. The reaction of the salt produces an alloyed surface containing about 40% chromium.

Metalliding is a diffusion coating process involving an electrolytic technique similar to electroplating, but done at higher temperatures ($816°C–1093°C$). The process uses a molten fluoride salt bath to diffuse metals and metalloids into the surface of other metals and alloys. As many as 25 different metals have been used as diffusing metals, and more than 40 as substrates. For example, boride coatings are applied to steels, nickel-base alloys, and refractory metals. Beryllide coatings can also be applied to many different metals by this process. The coatings are pore-free and can be controlled to a tolerance of 0.025 mm.

Casein

A whitish to yellowish, granular or lumpy protein precipitated from skim milk by the action of a dilute acid, or coagulated by rennet, or precipitated with whey from a previous batch. The precipitated material is then filtered and dried. Cow's milk contains about 3% casein. It is insoluble in water and alcohol, but soluble in alkalies. Although the casein is usually removed from commercial milk, it is a valuable food accessory because it contains *methionine*, a complex mercaptobutyric acid which counteracts a tendency toward calcium hardening of the arteries. This acid is also found in the ovalbumin of egg white. Methionine, $CH_3 \cdot S \cdot CH_2CH_2CHNH \cdot COOH$, is one of the most useful of the amino acids, and it is used in medicine to cure protein deficiency and in dermatology to cure acne and falling hair. It converts dietary protein to tissue, maintains nitrogen balance, and speeds wound healing. It is now made synthetically for use in poultry feeds. Some casein is produced as a by-product in the production of lactic acid from whole milk, the casein precipitating at a pH of 4.5. It is treated with sodium hydroxide to yield *sodium caseinate*.

Most of the production of casein is by acid precipitation, and this casein has a moisture content of not more than 10% with no more than 2.25% fat and not over 4% ash. The casein made with rennet is the type used for making plastics. *Rennet* used for curdling cheese is an extract of an enzyme derived from the stomachs of calves and lambs and is closely related to pepsin. Rennet substitutes produced from pepsin and other vegetable sources are only partial replacements and often have undesirable off-flavors. *Whey* is a thin, sweet, watery part separated out when milk is coagulated with rennet. Whey solids are used in prepared meats and other foods to enhance flavor and in pastries to eliminate sogginess. *Tekniken* is a dry whey for use in margarine, chocolate, and cheese. *Orotic acid*, produced synthetically, is identical with the biotic *Lactobacillus bulgaricus* of *yogurt*, the fermented milk whey used as food. It is a vitamin-like material.

Argentina and the United States are the most important producers of casein. France, Norway, and the Netherlands are also large producers. Casein is employed for making plastics, adhesives, sizing for paper and textiles, washable interior paints, leather dressings, and as a diabetic food. *Casein glue* is a cold-worked, water-resistant paste made from casein by dispersion with a mild base such as ammonia. With a lime base it is more resistant but has a tendency to stain. It is marketed wet or dry, the dry powder being simply mixed with water for application. It is used largely for low-cost plywoods and in water paints, but it is not waterproof. Many gypsum wallboard cements are fortified with casein. Concentrated *milk protein*, available as *calcium caseinate* or *sodium caseinate*, is for adding proteins and for stabilizing prepared meats and bakery products. It contains eight amino acids and is high in lysine. *Sheftene* is this material.

Casein Adhesive

An aqueous colloidal dispersion of *casein* that may be prepared with or without heat that may contain modifiers, inhibitors, and secondary binders to provide specific adhesive properties; and that includes a subclass, usually identified as casein glue, that is based on a dry blend of casein, lime, and sodium salts, mixed with water prepared without heat.

Casein Plastics

A group of thermoplastic molding materials made usually by the action of formaldehyde on rennet casein. The process was invented in 1885, and the first commercial casein plastic was called *Galalith*, meaning *milk stone*. Casein plastics are easily molded, machine easily, are nonflammable, withstand temperatures up to 300°F (150°C) and are easily dyed to light shades. But they are soft, have a high water absorption (7%–14%), and so often when exposed to alkalies. They are thus not suitable for many mechanical or electrical parts. They are used for ornamental parts, buttons, and such articles as fountain-pen holders. *Casein fiber* is made by treating casein with chemicals to extract the albumen and salts, forcing it through spinnerets, and again treating it to make it soft and silk-like. The fiber is superior to wool in silkiness and resistant to moth attack, but is inferior in general properties. It is blended with wool and fabrics and in hat felts.

CASS Test

Abbreviation for *copper-accelerated salt-spray test*.

Cassette

A holder used to contain radiographic film storing exposure to x-rays or gamma rays, that may or may not contain intensifying or filter screens, or both. A distinction is often made between a cassette, which is a positive means for ensuring contact between screens and film and is usually rigid, and an exposure holder, which is rather flexible.

Cast

(1) The practice of pouring molten metal into a mold and allowing it to solidify. If the resultant item is of approximately the final shape, requiring only final machining and other treatments that do not involve significant deformation, it will be described as a casting. If the cast item is to be wrought to shape by rolling, forging, etc., it will usually be described as an ingot. See also *continuous casting*. (2) The natural curved form that a wire takes as it is drawn from a reel and allowed to lie unrestrained. Its measure is the diameter of the circle formed when a sufficiently is allowed to lie freely on a horizontal surface.

Cast (Plastics)

To form a "plastic" object by pouring a fluid monomer–polymer solution into an open mold where it finishes polymerizing. Forming plastic film and sheet by pouring the liquid resin onto a moving belt or by precipitation in a chemical bath.

Cast Corrosion-Resistant Stainless Steels

High chromium-containing (11%–30% Cr) cast steels have been specified for liquid corrosion service at temperatures below 650°C (1200°F). Corrosion-resistant, or C-type steel castings are classified on the basis of composition using the designation system of the High Alloy Product Group of the Steel Founder's Society of America. The first letter of the designation (C) indicates that the alloy is intended for liquid corrosion service. The second letter indicates the nickel content; as the nickel content increases, the second letter of the designation changes from A to Z.

Cast Film

A film made by depositing a layer of liquid plastic onto a surface and stabilizing this form by the evaporation of solvent, by fusing after deposition, or by allowing a melt to cool. Cast films are usually made from solutions or dispersions.

Cast Heat-Resistant Stainless Steels

Iron–chromium, iron–chromium–nickel, and iron–nickel–chromium steel castings have been specified for service at temperatures 650°C (1200°F) to approximately 980°C (1800°F). Designated similarly to *cast corrosion-resistant stainless steels*, H-type steels (the "H" denoting high-temperature service) have nickel contents ranging from 0% to 68%.

Cast Iron

Alloys of iron containing, at the time they are cast, carbon in the range from about 2% to 4%. This contrasts with steel which is also an iron carbon alloy but which contains considerably less carbon. It is common for the term "cast" to be omitted in description such as "Grey (cast) Iron," "Malleable (cast) Iron" or even "(cast) Irons." Iron containing about 3% carbon solidifies over a temperature range from about 1350°C to 1130°C, i.e., significantly lower than the 1535°C or so melting point of pure iron. This wide freezing range coupled with a high fluidity ensures that complex shapes are readily cast. Other favorable characteristics of the simple irons include good corrosion resistance (relative to carbon steel) and erosion resistance, a high damping capacity and good compressive strength. Cast irons can develop, under appropriate cooling conditions and by subsequent heat treatment, like the structures for steel. However, the very high carbon content has a significant effect and, in the as-cast condition, the structure will contain either soft, weak, free carbon, that is, carbon not combined with the iron, or large quantities of hard, brittle cementite (iron carbide). Factors such as precise composition and cooling rate during casting influence structures that initially develop. In many irons, the free carbon will exist as graphite flakes. This usually forms as a eutectic with the austenite at 1130°C but in adverse circumstances a coarse form, termed *Kish graphite*, can be precipitated at an earlier stage of solidification. However, even normal graphite flakes have a pronounced weakening effect although some improvement can be obtained by inoculation which involves adding a finely particulate refractory powder, such as calcium silicide, to the molten iron. The particles provide a larger number of nucleation sites for precipitation thereby refining the graphite. When irons containing graphite flakes are broken the fracture follows the weak flakes producing a characteristic grey fracture surface. Historically, irons have been categorized on the appearance of their fracture surface and hence such irons are termed *Grey Irons*. Apart from the graphite the microstructure will comprise a matrix of other phases such as pearlite or ferrite hence terms such as *Pearlitic Grey Iron* and *Ferritic Grey Iron*. Where the composition and cooling rate are such that no graphite is formed and the structure comprises cementite and pearlite the material will be very hard and brittle with a bright fracture surface, hence the term *White Iron*. The mechanical properties, particularly ductility and malleability, of white irons can be improved by various heat treatments to form *Malleable Irons*. Heating a white iron at about 850°C–950°C in a nonreactive environment for a long period (up to a week) causes the cementite to decompose producing a microstructure of ferrite with temper carbon, i.e., graphite in a nodular or rosette form which is not so damaging as flake. These carbon nodules are more prolific toward the center of this section so the appearance of the fracture surface leads to the term *Blackheart (Malleable) Iron*. A variation on this process requires a second heating to 900°C allowing some of the carbon to redissolve in the austenite so that pearlite is produced on cooling, hence the term *Pearlitic Malleable Iron*. Such irons can be heat-treated just like steel. An alternative malleablizing treatment involves heating white iron at about 950°C in an oxidizing environment for up to 100 h during which most of the carbon diffuses to the surface to react with oxygen-forming carbon dioxide. The resulting structure is largely ferritic or pearlitic, with a little temper carbon at the center (more in large castings), hence the name *Whitehear (Malleable) Iron*. An alternative approach to producing more ductile irons is to modify the form of the graphite in the as-cast state, usually by small additions of magnesium or cerium to the molten iron. These promote the formation of graphite in the nodular or spheroidal form which, compared with the flake form, has only a slightly adverse effect on the mechanical properties. These materials are referred to as *Ductile Irons, Nodular Irons, Spheroidal Graphite Irons (SG Irons)*. See Tables C.4 and C.5.

TABLE C.4
Range of Compositions for Typical Unalloyed Common Cast Irons

Type of Iron	Composition, %				
	C	Si	Mn	P	S
Gray (FG)	2.5–4.0	1.0–3.0	0.2–1.0	0.002–1.0	0.02–0.25
Compacted graphite (CG)	2.5–4.0	1.0–3.0	0.2–1.0	0.01–0.1	0.01–0.03
Ductile (SG)	3.0–4.0	1.8–2.8	0.1–1.0	0.01–0.1	0.01–0.03
White	1.8–3.6	0.5–1.9	0.25–0.8	0.06–0.2	0.06–0.2
Malleable (TG)	2.2–2.9	0.9–1.9	0.15–1.2	0.02–0.2	0.02–0.2

TABLE C.5

Classification of Cast Iron by Commercial Designation, Microstructure, and Fracture

Commercial Designation	Carbon-Rich Phase	Matrix[a]	Fracture	Final Structure After
Gray iron	Lamellar graphite	P	Gray	Solidification
Ductile iron	Spheroidal graphite	F, P, A	Silver-gray	Solidification or heat treatment
Compacted graphite iron	Compacted vermicular graphite	F, P	Gray	Solidification
White iron	Fe_3C	P, M	White	Solidification and heat treatment[b]
Mottled iron	Lamellar Gr + Fe_3C	P	Mottled	Solidification
Malleable iron	Temper graphite	F, P	Silver-gray	Heat treatment
Austempered ductile iron	Spheroidal graphite	At	Silver-gray	Heat treatment

[a] F, ferrite; P, pearlite; A, austenite; M, martensite; At, austempered (bainite).

[b] White irons are not usually heat treated, except for stress relief and to continue austenite transformation.

In some cases, iron microstructure may be all ferrite—the same constituent that makes low-carbon steels soft and easily machined. But the ferrite of iron is different because it contains sufficient dissolved silicon to eliminate the characteristic gummy nature of low-carbon steel. Thus, cast irons containing ferrite do not require sulfur or lead additions in order to be free machining.

Because the size and shape of a casting control its solidification rate and strength, design of the casting and the casting process involved must be considered in selecting the type of iron to be specified. Although most other metals are specified by a standard chemical analysis, a single analysis of cast iron can produce several entirely different types of iron, depending upon foundry practice and shape and size of the casting, all of which influence cooling rate. Thus, iron is usually specified by mechanical properties. For applications involving high temperatures or requiring specific corrosion resistance, however, some analysis requirements may also be specified.

Pattern making is no longer a necessary step in manufacturing cast iron parts. Many gray, ductile, and alloy-iron components can be machined directly from bar that is continuously cast to near-net shape. Not only does this "parts-without-patterns" method save the time and expense of pattern making, continuous cast iron also provides a uniformly dense, finer-grained structure, essentially free from porosity, sand, or other inclusions. Keys to the uniform microstructure of the metal are the ferrostatic pressure and the temperature-controlled solidification that are unique to the process.

For each basic type of cast iron, there are a number of grades with widely differing chemical properties. These variations are caused by differences in the microstructure of the metal that surrounds the graphite (or iron carbides). Two different structures can exist in the same casting. The microstructure of cast iron can be controlled by heat treatment, but once graphite is formed, it remains.

Pearlitic cast iron grades consist of alternating layers of soft ferrite and hard iron carbide. This laminated structure—called pearlite—is strong and wear resistant, but still quite machinable. As laminations become finer, hardness and strength of the iron increase. Laminations size can be controlled by heat treatment or cooling rate.

Cast irons that are flame-hardened, induction-hardened, or furnace-heated and subsequently oil-quenched contain a martensite structure. When tempered, this structure provides machinability with maximum strengths and good wear resistance.

Gray Iron

This is a supersaturated solution of carbon in an iron matrix. The excess carbon precipitates out in the form of graphite flakes.

Gray iron is specified by a two-digit designation; Class 20, for example, specifies a minimum tensile strength of 138 MPa. In addition, gray iron is specified by the cross section and minimum strength of a special test bar. Usually, the test bar cross section matches or is related to a particularly critical section of the casting. This second specification is necessary because the strength of gray iron is highly sensitive to cross section (the smaller the cross section, the faster the cooling rate and the higher the strength).

Impact strength of gray iron is lower than that of most other cast ferrous metals. In addition, gray iron does not have a distinct yield point (as defined by classical formulas) and should not be used when permanent, plastic deformation is preferred to fracture. Another important characteristic of gray iron—particularly for precision machinery—is its ability to damp vibration. Damping capacity is determined principally by the amount and type of graphite flakes. As graphite decreases, damping capacity also decreases.

The high compressive strength of gray iron—three to five times tensile strength—can be used to advantage in certain situations. For example, placing ribs on the compression side of a plate instead of the tension side produces a stronger, lighter component.

Gray irons have excellent wear resistance. Even the softer grades perform well under certain borderline lubrication conditions (as in the upper cylinder walls of internal-combustion engines, for example).

To increase the hardness of gray iron for abrasive-wear applications, alloying elements can be added, special foundry techniques can be used, or the iron can be heat-treated. Gray iron can be hardened by flame or induction methods, or the foundry can use a chill in the mold to produce hardened, "white-iron" surfaces.

Typical applications of gray iron include automotive engine blocks, gears, flywheels, brake disks and drums, and machine bases. Gray iron serves well in machinery applications because of its good fatigue resistance.

Ductile Iron

Ductile, or nodular iron, contains trace amounts of magnesium which, by reacting with the sulfur and oxygen in the molten iron, precipitates out carbon in the form of small spheres. These spheres improve the stiffness, strength, and shock resistance of ductile iron over gray iron. Different grades are produced by controlling the matrix structure around the graphite, either as-cast or by subsequent heat treatment.

A three-part designation system is used to specify ductile iron. The designation of a typical alloy, 60-40-18, for example, specifies a minimum tensile strength of 414 MPa, a minimum yield strength of 276 MPa, and 18% elongation in 5.08 mm.

Ductile iron is used in applications such as crankshafts because of its good machinability, fatigue strength, and high modulus of elasticity; in heavy-duty gears because of its high yield strength and wear resistance; and in automobile door hinges because of its ductility. Because it contains magnesium as an additional alloying element, ductile iron is stronger and more shock resistant than gray iron. But although ductile iron also has a higher modulus of elasticity, its damping capacity and thermal conductivity are lower than those of gray iron.

By weight, ductile iron castings are more expensive than gray iron. Because they offer higher strength and provide better impact resistance, however, overall part costs may be about the same.

Although it is not a new treatment for ductile iron, austempering has become increasingly known to the engineering community in the past 5–10 years. Austempering does not produce the same type of structure as it does in steel because of the high carbon and silicon content of iron. The matrix structure of austempered ductile iron (ADI) sets it apart from other cast irons, making it truly a separate class of engineering materials.

In terms of properties, the ADI matrix almost doubles the strength of conventional ductile iron while retaining its excellent toughness. Like ductile iron, ADI is not a single material; rather, it is a family of materials having various combinations of strength, toughness, and wear resistance. Unfortunately, the absence of a standard specification for the materials has restricted its widespread acceptance and use. To help eliminate this problem, the Ductile Iron Society has proposed property specifications for four grades of austempered ductile iron.

Most current applications for ADI are in transportation equipment—automobiles, trucks, and railroad and military vehicles. The same improved performance and cost savings are expected to make these materials attractive in equipment for other industries such as mining, earthmoving, agriculture, construction, and machine tools.

White Iron

White iron is produced by "chilling" selected areas of a casting in the mold, which prevents graphitic carbon from precipitating out. Both gray and ductile iron can be chilled to produce a surface of white iron, consisting of iron carbide, or cementite, which is hard and brittle. In castings that are white iron throughout, however, the composition of iron is selected according to part size to ensure that the volume of metal involved can solidify rapidly enough to produce the white-iron structure.

The principal disadvantage of white iron is its brittleness. This can be reduced somewhat by reducing the carbon content or by thoroughly stress-relieving the casting to spheroidize the carbides in the matrix. However, these measures increase cost and reduce hardness.

Chilling should not be confused with heat-treat hardening, which involves an entirely different metallurgical mechanism. White iron, so-called because of its very white structure, can be formed only by solidification. It will not soften except by extended annealing and it retains its hardness even above 538°C.

White irons are used primarily for applications requiring wear and abrasion resistance such as mill liners and shot-blasting nozzles. Other uses include railroad brake shoes, rolling-mill rolls, clay-mixing and brickmaking equipment, and crushers and pulverizers. Generally, plain (unalloyed) white iron costs less than other cast irons.

Compacted Graphite Iron

Until recently, compacted graphite iron (CGI), also known as vermicular iron, has been primarily a laboratory curiosity. Long known as an intermediate between gray and ductile iron, it possesses many of the favorable properties of each. However, because of process-control difficulties and the necessity of keeping alloy additions within very tight limits, CGI has been extremely difficult to produce successfully on a commercial scale. For example, if the magnesium addition varied by as little as 0.005%, results would be unsatisfactory.

Processing problems have been solved by an alloy-addition package that provides the essential alloying ingredients—magnesium, titanium, and rare earths—in exactly the right proportions.

Strength of CGI parts approaches that of ductile cast iron. CGI also offers high thermal conductivity, and its damping capacity is almost as good as that of gray iron; fatigue resistance and ductility are similar to those properties in ductile iron. Machinability is superior to that of ductile iron, and casting yields are high because shrinkage and feeding characteristics are more like gray iron.

The combination of high strength and high thermal conductivity suggests the use of CGI in engine blocks, brake drums, and exhaust manifolds of vehicles. CGI gear plates have replaced aluminum in high-pressure gear pumps because of the ability of the iron to maintain dimensional stability at pressures above 10.3 MPa.

Malleable Iron

Malleable iron castings are often used for heavy-duty bearing surfaces in automobiles, trucks, railroad rolling stock, and farm and construction machinery. Pearlitic grades are highly wear resistant, with hardnesses ranging from 152 to over 300 Bhn. Applications are limited, however, to relatively thin-sectioned castings because of the high shrinkage rate and the need for rapid cooling to produce white iron.

High-Alloy Irons

High alloy-irons are ductile, gray, or white irons that contain 3% to more than 30% alloy content. Properties are significantly different from those of unalloyed irons. These irons are usually specified by chemical composition as well as by various mechanical properties.

White high-alloy irons containing nickel and chromium develop a microstructure with a martensite matrix around primary chromium carbides. This structure provides a high hardness with extreme wear and abrasion resistance. High-chromium irons (typically, about 16%) combine wear and oxidation resistance with toughness. Irons containing from 14% to 24% nickel are austenitic; they provide excellent corrosion resistance for nonmagnetic applications. The 35% nickel irons have an extremely low coefficient of thermal expansion and are also nonmagnetic and corrosion resistant.

Cast Nonferrous Alloys

Nonferrous metals and alloys can be categorized as follows:

1. Aluminum-base
2. Copper-base (brasses and bronzes)
3. Lead-base
4. Magnesium-base
5. Nickel-base
6. Tin-base
7. Zinc-base
8. Titanium-base

Another common way of grouping nonferrous alloys is to divide them into heavy metals (copper, zinc, lead, and nickel) and light metals (aluminum, magnesium, and titanium base).

C

Aluminum and Its Alloys

Aluminum and its alloys continue to grow in acceptance, particularly in the automotive industry. The electronics industry is another major user of cast aluminum for chassis, enclosures, terminals, etc.

Most metallic elements can be alloyed readily with aluminum, but only a few are commercially important—among them copper, lithium, magnesium, manganese, silicon, and zinc. Many other elements serve as supplementary alloying additions to improve properties and metallurgical characteristics.

Aluminum can be cast by virtually all of the common casting processes—particularly diecasting, sand mold, permanent mold, expendable pattern (lost foam) casting, etc.

Aluminum alloy castings are produced by virtually all commercial processes in a range of compositions possessing a wide variety of useful engineering properties.

- Al–Mg alloys offer excellent corrosion resistance, good machinability, and an attractive appearance when anodized. Careful gating and risering are required.
- Binary Al–Si alloys exhibit good weldability, high corrosion resistance, and low specific gravity.
- Al–Zn alloys have good machinability characteristics and age at room temperature to moderately high strengths in a relatively short period of time without solution heat treatment.
- Al–Sn alloys were developed for bearings and bushings with high load-carrying capacity and fatigue strength. Corrosion resistance is superior, but there is a susceptibility to hot cracking.
- Al–Li alloys are of recent commercial significance because of their good mechanical properties. They are believed to have many potential aerospace applications, and alloy development is widespread.

Copper Casting Alloys

These alloys are grouped according to composition by these general categories: pure copper, high-copper alloys, brasses, leaded brasses, bronzes, aluminum bronzes, silicon bronzes, Cu–Ni alloys, and Cu–Ni–Zn alloys known as "nickel silvers." The UNS designations for copper-base casting alloys range from C80000 through C99900. In brasses, zinc is the principal alloying element. For cast brass, there are Cu–Zn–Sn alloys (red, semi-red, and yellow brasses), leaded and unleaded manganese bronze alloys (high-strength yellow brasses), and Cu–Zn–Si alloys.

Tin is the principal alloying element in cast bronze alloys, which consist of four families: tin bronzes, leaded and highly leaded tin bronzes, Ni–Sn bronzes, and aluminum bronzes.

Zinc Alloys

Zinc has a low melting temperature and is readily and economically diecast. In recent years, zinc "foundry alloys" known as the ZA-8, ZA-12, and ZA-27 (containing 8%, 12%, and 27% aluminum, respectively) have been developed. They are finding wide application as gravity-cast sand and permanent mold castings as well as diecastings.

Zinc alloys can be cast in thin sections and with tight dimensional control. The principal alloys used for diecastings contain low percentages of magnesium, 3.9%–4.3% aluminum, and small controlled quantities of impurities such as tin, lead, and cadmium. Copper, and nickel are significant alloying additions.

Magnesium Alloys

The usefulness of magnesium, lightest of the commercial metals, is enhanced considerably by alloying. Magnesium can be used for sand and permanent mold castings and diecastings. Hot-chamber diecasting is a more recent development that is finding wider application each year. With the help of different heat treatments, alloy tensile strengths range from 136 to 262 MPa, yield strengths from 69 to 262 MPa, and elongations from 1% to 15%.

Magnesium foundry technology has advanced significantly in recent years. Complex components are successfully made as sand castings, with wall thicknesses down to 3.5 mm and tolerances to ±0.6 mm for automotive, aircraft, and electronics applications.

Titanium Castings

Applications are broadening. Titanium is particularly suitable for withstanding corrosive environments or applications that take advantage of its light weight, high strength-to-weight ratio, and nonmagnetic properties. Long applied in military aircraft, titanium alloys are now solving problems in nonmilitary equipment and jewelry.

Metal Matrix Composites (MMC)

MMCs are an important new technology. These compounds consist of an inorganic reinforcement-particle filament, or whisker, in a metal matrix. Nearly all metals can be used as a matrix, but current applications center around aluminum, magnesium, and titanium. The lighter metals offer the best weight-to-strength ratio. MMCs also show improved wear resistance and allow adjustment of the coefficient of thermal expansion to meet varying requirements.

Superalloys

Superalloys are nickel, Fe–Ni, and cobalt-base alloys generally used at temperatures above about 538°C. The Fe–Ni-base superalloys are an extension of stainless steel technology and generally are wrought, whereas cobalt-base and nickel-base superalloys can be wrought or cast.

The more highly alloyed compositions normally are processed as castings. Fabricated cast structures can be built up by welding or brazing, but many highly alloyed compositions containing a high amount of hardening phase are difficult to weld. Properties can be controlled by adjustments in composition and by processing including heat treatment.

Cast Plastics

Plastics casting materials can be generally classified in two groups: (1) those resins, usually thermosetting, that are cast as liquids and cured by chemical cross-linking either at room temperature or elevated temperatures, and (2) thermoplastics that are supplied essentially in suspension or monomeric form and fused or polymerized at elevated temperatures.

Materials

Each casting resin has a unique combination of properties, such as heat resistance, strength, electrical properties, chemical resistance, cost, and shrinkage, which dictate their use for specific applications.

The most commonly used casting resins are phenolics, polyesters, and epoxies. Others are phenolics, acrylics, and urethane elastomers. Following is a brief description of each major type.

Phenolic Resins

Phenolic resins are used for low-cost parts requiring good electrical insulating properties, heat resistance, or chemical resistance. The average shelf life of this resin is about 1 month at 21.1°C. This can be extended by storing it in a refrigerator at 1.6°C–10°C. Varying the catalyst (according to the thickness of the cast) and raising the cure temperature to 93°C will alter the cure time from as long as 8 h to as short as 15 min.

Some shrinkage occurs in the finished casting (0.012–0.6 mm/mm), depending on the quantity of filler, amount of catalyst, and the rate of cure. Faster cure cycles produce a higher rate of shrinkage. Since the cure cycle can be accelerated, phenolics are used in short-run casting operations.

Cast phenolic parts are easily removed from the mold if the parting agents recommended by the supplier are used. Postcuring improves the basic properties of the finished casting.

Polyester Resins

Polyester resins are primarily used in large castings such as those required in the motion picture industry or for large sculptures for museums, parks, and display purposes. Since polyester shrinkage is about 0.024–0.032 mm/mm, castings are usually reinforced with glass cloth or mat and are generally cast in a flexible mold. Catalysts used to initiate the cure are peroxides or hydroperoxides and activators are cobalt naphthanate, alkyl mereaptans, or dialkyl aromatic amines. Recently, isophthalic polyesters have been introduced containing isophthalic acid, which provides improved heat, chemical, and impact resistance.

Clear polyester castings can be made using diallyl or triallyl cyanurate type polyesters. Triallyl cyanurate polyesters are used in casting clear sheets because they have excellent scratch and heat resistance.

Acrylic Resins

The process involved in casting acrylic resins is complex and forms a specialized field. The methyl-methacrylate monomer contains inhibitors that must be removed before adding the catalyst. The resin must be cured under very accurately controlled conditions. The primary use of cast acrylics is in optically clear sheet, rod, or tube stock and in the embedment of specimens for museums and display of industrial parts, as well as for the embedment of decorative motifs in the jewelry industry.

Vinyls

Plasticized vinyls (polyvinyl chloride, or PVC) are used in industry in a variety of plastisol processing techniques, for example, slush casting and rotational casting. Electroformed molds are commonly used for this purpose. Since the conversion of the semiliquid vinyl plastisol to a solid consists of fusing the suspended vinyl particles to each other, it is only necessary to raise the temperature of the mass to 82°C–177°C according to the formulation. Cast-vinyl prototype parts can be produced that are comparable with molded parts.

Epoxy Resins

Most cast epoxy resins, other than those used in plastic tooling, are used in encapsulating electrical components and in casting prototype parts. The variations in properties possible make them very versatile materials. They can be made to have almost infinite shelf life, can be varied from a liquid to a thixotropic gel, and can be highly flexibilized by the addition of polysulfides and polyamides. They can be formulated to provide heat resistance up to 260°C.

Cure cycles may range from 1 to 16 h, at which time the casting is removed from the mold. Epoxy resins for casting are available in transparent water-white, semitransparent, and opaque formulations. Room-temperature cures are effected when aliphatic amines are added to the resin in exact amounts. Heat resistance of such systems is about 82°C. Cure is relatively rapid; therefore, exothermic reaction produces relatively high temperatures. Thus, casting must generally be limited in thickness.

Proprietary amine hardeners and epoxy resin systems are available whose heat resistance is about 121°C–177°C. The use of liquid anhydrides yields castings with heat resistance above 204°C. Such systems permit casting in large masses because the exothermic reaction and the curing temperatures are low (about 121°C). The pot life is several days, because elevated temperatures are required to initiate the reaction.

Several facts are basic in the use of these systems. Because it is necessary to use acidic catalysts in order to achieve heat resistance, castings tend to be more brittle; also, more difficulty is encountered in releasing the cast from the mold. The use of the proper release agent with a given resin system is necessary to overcome this tendency.

Cast Replica

In *metallography*, a reproduction of a surface in plastic made by the evaporation of the solvent from a solution of the plastic or by polymerization of a monomer on the surface. See also *replica*.

Cast Steels

The general nature and characteristics of cast steels are, in most respects, closely comparable to wrought steels. Cast and wrought steels of equivalent composition respond similarly to heat treatment and have fairly similar properties. A major difference between them is that cast steel is more isotropic in structure. Therefore, properties tend to be more uniform in all directions in contrast to wrought steel, whose properties generally vary, depending on the direction of hot or cold working.

Five basic steel groups are available:

1. Carbon steels
2. Low-alloy steels
3. High-alloy steels
4. Tool steels
5. Stainless steels

The most common types of steels used in castings are the carbon steels, which contain only carbon as their principal alloying element. Other elements are present in small quantities, including those added for deoxidation. Silicon and manganese in cast carbon steels typically range from 0.25% to about 0.8% silicon and 0.50% to 1% manganese, respectively.

Cast plain carbon steels can be divided into three groups—low, medium, and high carbon. However, cast steel is usually specified by mechanical properties, primarily tensile strength, rather than composition. Standard classes are 414, 483, 586, and 690 MPa. Low-carbon grades, used mainly in the annealed and normalized conditions, have tensile strength ranging from 380 to 448 MPa. Medium-carbon grades, annealed and normalized, range from 483 to 690 MPa. When quenched and tempered, strength exceeds 690 MPa.

Ductility and impact properties of cast steels are comparable, on average, to those of wrought carbon steel. However, the longitudinal properties of rolled and forged steels are higher than those of cast steel. Endurance limit strength ranges between 40% and 50% of ultimate tensile strength.

By definition, low-alloy steels contain alloying elements, in addition to carbon, up to a total content of 8%. A casting containing more than 8% alloy content is classified as a high-alloy steel. Technically, tool steels and stainless steels are high-alloy steels but are normally classified separately.

Small quantities of titanium and aluminum are also used for grain refinement in cast low-alloy steels.

Stainless steels are grouped in three classes: martensitic, ferritic, and austenitic. They are all more resistant to corrosion than plain carbon steels or lower alloy steels, and they contain either significant amounts of chromium or chromium and nickel. Cast stainless steel grades are designated in general as either heat resistant or corrosion resistant. Stainless steel castings are specified by ACI designations.

The C series of ACI stainless grades designates corrosion-resistant steels; the H series designates heat-resistant steels that are suitable for service temperatures in the 649°C–1204°C range. Typical casting applications for C-series grades are valves, pumps, and fittings. H-series grades are used for furnace parts, turbine engine components, and other high-temperature requirements.

Interest has renewed in cast duplex stainless steels because the 50% ferritic and 50% austenitic metallographic structure of these grades makes them extremely resistant to stress corrosion cracking. In addition, they have twice the yield strength of austenitic grades and may cost less.

Although the basic properties of cast duplex alloys are determined primarily by the 50/50 mixture of ferrite and austenite, the metallurgy can be complex because of the numerous carbides, nitrides, and intermetallic phases or compounds that can form. Ferrite formation is a function of cooling rate and chemistry, and austenizing temperature affects strength.

Hardenability of cast steels does not vary significantly from that of wrought steels of similar composition. The one principal difference between wrought and cast steels is the effect of the casting surface. It may contain scale and oxides and may not be chemically or structurally equivalent to the base metal; see Table C.6.

A few of the industries and some of the specific products that are being made from cast steel are: automotive (frames, wheels, gears); electrical manufacturing (rotors, bases, housings, frames, shafts); transportation (couplings, drawbars, brake shoes, wheel truck frames); marine (rotors, sterns, anchor chains, ornamental fittings, capstans); off-the-road equipment (crawler side frames, levers, shafts, tread links, turntables, buckets, dipper teeth); municipal (fire hydrants, catch basins, manhole frames and covers); miscellaneous (ingot and pig molds, rolling mill rolls, blast-furnace ingot buggies, engine housings, cylinder blocks and heads, crankshafts, flanges and valves).

Cast Structure

The metallographic structure of a *casting* evidenced by shape and orientation of grains and by segregation of impurities. Other features include grains that are very coarse, possibly cored or dendritic, extensive columnar grains at the exterior and large quantities of internal voids.

Castability

(1) A complex combination of liquid-metal properties and solidification characteristics that promotes accurate and sound final castings.

(2) The relative ease with which a molten metal flows through a mold or casting die.

Castable

In casting, a combination of refractory grain and suitable bonding agent that, after the addition of a proper liquid, is generally poured into place to form a refractory shape or structure which becomes rigid because of chemical action.

Cast-Alloy Tool

A cutting tool made by casting a cobalt-based alloy and used at machining speeds between those for high-speed steels and cemented carbide. Nominal compositions for two commercially available grades are as follows in Table C.7.

Casting Copper

Fire-refined tough pitch copper usually cast from melted secondary metal into ingot bars only, and used for making foundry castings but not wrought products.

Casting Defect

Any imperfection in a casting that does not satisfy one or more of the required design or quality specifications. This term is often used in a limited sense for those floors formed by improper casting solidification.

Casting Section Thickness

The wall thickness of the casting. Because the casting may not have a uniform thickness, the section thickness may be specified at a specific place on the casting. Also, it is sometimes useful to use the average, minimum, or typical wall thickness to describe the casting.

Casting Shrinkage

The amount of dimensional change per unit length of the casting as it solidifies in the mold or die and cools to room temperature after removal from the mold or die. There are three distinct types of casting shrinkage. Liquid shrinkage refers to the reduction in volume or liquid metal as it cools to the liquidus. Solidification shrinkage is the reduction in volume of metal from the beginning to the end of solidification. Solid shrinkage involves a reduction in volume of metal from the solidus to room temperature.

Casting Strains

Strains in a casting caused by *casting stresses* that develop as a casting cools.

Casting Stresses

Residual stresses set up when the shape of a casting impedes contraction of the solidified casting during cooling.

Casting Thickness

See *casting section thickness.*

TABLE C.6
Cast Stainless Steels

	Corrosion-Resistant Grades (ACI Designation)									
	CA-15[a]	CB-30[b]	CC-50[c]	CE-30[d]	CF-8[e]	CH-20[d]			CG-6 MMN[h]	CF-10 MnN[h]
	Heat-Resistant Grades (ACI Designation)[f]									
			HC	HE	HF	HH	HT	HX		
	Equivalent AISI Grades[g]								UNS	
	410	442	446	312	304	309	330		S20910	S21800
Yield strength (10³ psi) grades	100	60	65	63	35–40	50	—	—	50	47
HR grades	—	—	75	45	45	50	40	25	40	—
Tensile strength (10³ psi)	115	95	70–110	87–92	75–92	80–88	69–70	56–75	94	96
Impact strength, Charpy, 70°F (ft lb)	35	2	45	10	70	15	4	—	71	134
Creep strength, 0.001% h⁻¹ (10³ psi) HR grades only, 1400°F	—	—	1.3	3.5	6	7.0	8.0	6.4	—	—
1800°F	—	—	0.336	1.0	3.2	2.1	2.0	1.6	—	—
Elongation (%) R grades	29	—	29	35	28	28	27	25	28	26
Hardness, Bhn CR grades	225	195	210	190	140	190	—	—	210	195
HR grades	—	—	223	200	165	185	180	176	—	—
Melting temperature (°F)	2700–2790	2700–2750	2650–2750	2600–2700	2550–2600	2500–2600	2400–2450	2350	—	—
Coefficient of thermal expansion (10⁻⁶ in./in. °F) 70°F–212°F	5.5	5.7	5.9	—	9.0	8.3	—	—	9.0	8.8
70°F–1000°F	6.4	6.5	6.4	9.6	10.9	9.6	9.8	9.5	10.2	10.0
Thermal conductivity (Btu ft/h ft² °F)	14.5	12.8	12.6	10.0	9.0	8.2	7.7	11.1	9.0	—
Density (lb/in.³)	0.275	0.272–0.274	0.272–0.277	0.276	0.280	0.279	0.286	0.294	0.285	0.276
Electrical resistivity, CR grades (μΩ cm)	78.0	76.0	77.0	85.0	76.2	84.0	—	—	82.0	98.2
Magnetic	Yes	Yes	Yes	Partially	CF, partially, HF, no	Partially	Partially	—	Slightly	Slightly

Source: *Mach. Design Basics Eng. Design*, 790, June 1993. With permission.

Note: CR, corrosion resistant; HR, heat resistant.

[a] 1800°F air cooled, 1200°F tempered.
[b] 1450°F air cooled.
[c] 1900°F air cooled.
[d] 2000°F air water quenched.
[e] >1900°F water quenched.
[f] As cast.
[g] Equivalent wrought grades are given for comparison only; the ACI designations, generally included in ASTM A743 and A297, are used to specify the cast stainless steel grades.
[h] Annealed.

Casting Volume

The total cubic units (mm³ or in.³) of cast metal in the casting.

Casting Wheel

A wheel carrying at its edge a number of molds which are filled as they pass beneath the molten metal stream. As the wheel rotates the molds are replaced or emptied if sufficiently cool.

Casting Yield

The weight of a casting(s) divided by the total weight of metal poured into the mold, expressed as a percentage.

Castings

Strength and performance of a cast part do not depend solely on part geometry. Proper alloy selection is crucial to a cost-effective casting design and trouble-free engineering and manufacturing. Materials

---

TABLE C.7
Two Cast-Alloy Tool Compositions

Element	Alloy	
	Tantung G, %	Tantung 144, %
Cobalt	42–47	40–45
Chromium	27–32	25–30
Tungsten	14–19	16–21
Carbon	2–4	2–4
Tantalum or niobium	2–7	3–8
Manganese	1–3	1–3
Iron	2–5	2–5
Nickel	7[a]	7[a]

[a] Maximum.

---

selection software has been developed to help meet the need for information about alloys. Because the packages are in a database format, users can search through the data to choose those materials that meet their requirements. Once the basic material selection has been made, three-dimensional (3-D) modeling, casting simulation, and finite-element analysis (FEA) software can be used to confirm the behavior of that material/part combination during manufacture and in service.

Rapid Prototyping

Computer technologies known as rapid prototyping allow manufacturers to fabricate 3-D models, prototypes, patterns, tooling, and production parts directly from computer-aided design (CAD) data in a fraction of the total time and cost of conventional methods. Several practical rapid prototyping systems are commercially available, and their use by OEM product parts designers as well as producers of cast metal parts and tooling shops are becoming widespread.

Each technology shares the same basic approach: A computer analyzes a 3-D CAD file that defines the object to be fabricated and "slices" the object into thin cross sections. The cross sections are then systematically recreated and combined to form a 3-D object.

Here's a thumbnail sketch of how they work: Stereolithography recreates the object by sequentially solidifying layers of photoactive liquid polymer by exposing the liquid to ultraviolet light.

Ultimately, scientists hope to develop numerical simulations that provide enough information to optimize all the variables in a casting operation. To improve understanding of the process and to improve gating design, researchers use an x-ray system that makes images of the molten metal as it fills the mold. Computer codes that predict mechanical properties are then compared with experimental results and modified to match the behavior of specific alloys.

Metal-Casting Processes

Expendable molds are for use only once (sand castings); other molds (or dies) are made of metal (permanent molding and diecasting) and can be used repeatedly. The pattern must be removable from the mold without damage, and the casting must be removable from the mold or die without damage to either the die or the casting.

Sand Casting Processes

More than 80% of all castings made in the United States are produced by green sand molding. (The term *green sand* does not refer to color, but to the fact that a raw sand and binder mixture has been tempered with water.) Sand molding is a versatile metal-forming process that provides freedom of design with respect to size, shape, and product quality.

Permanent Molding

In permanent mold casting, which also is referred to as gravity diecasting, a metal mold (or die) consisting of at least two parts is used repeatedly, usually for components that require relatively high production. Molds usually are made of cast iron, although steel, graphite, copper, and aluminum have been used as mold materials with varying degrees of success.

When molten metal is poured into a permanent mold, it cools more rapidly than in a sand mold and produces a finer-grained structure, a sounder and denser casting, and enhanced mechanical properties.

Diecasting Process

The diecasting process is used widely for high production of zinc, lead, tin, aluminum, copper, and magnesium cast components of intricate design. Molten alloy is poured manually or automatically into a shot well and injected into the die under pressure. An important factor in diecasting machine operation is the locking force (in tons), in which case the die is firmly closed against the injection pressure exerted by the plunger as it injects the molten metal.

There are two basic types of diecasting machines—hot chamber and cold chamber. The hot-chamber machine makes shots automatically and is used for low-melting-point materials, such as zinc alloys. The cold-chamber method, for higher-melting-point materials, such as aluminum and magnesium, holds molten metal at a constant temperature in a holding furnace of the bailout type. Metal is poured into the shot well either by hand or by automatic devices.

Vacuum diecasting sometimes is used to evacuate the die cavity. Its objectives are reduction of porosity, assisting metal flow in thin sections, and improving surface finish while at the same time permitting the use of injection pressures lower than those normally applied.

It should be noted that the meanings of the term *diecasting* in the United States and in European usage are different. Diecasting in Europe is any casting made in a metal mold. Pressure diecasting in Europe is a casting made in a metal mold in which the metal is injected under high pressure. In the United States, this is simply "diecasting." Gravity diecasting in Europe is a casting poured in a metal mold by gravity, with no application of pressure. In the United States, this is "permanent molding."

Investment Casting

In investment casting, a ceramic slurry is poured around a disposable pattern (normally of modified paraffin waxes, but also of plastics) and allowed to harden to form a disposable mold. The pattern is destroyed when it melts out during the firing of the ceramic mold. Later, molten metal is poured into the ceramic mold, and after the metal solidifies, the mold is broken up to remove the casting.

Two processes are used to produce investment casting molds—the solid mold and the ceramic shell methods. The ceramic shell method has become the predominant production technique today. The solid investment process is used primarily to produce dental and jewelry castings. Ceramic shell molds are used primarily for the investment casting of carbon and alloy steels, stainless steels, heat-resistant alloys, and other alloys with melting points above 1093°C. The process can be mechanized.

Almost any degree of external and internal complexity can be accommodated, the only limitation is the state of the art in ceramic core manufacturing. Many problems inherent in producing a component by forging, machining, or multiple-piece fabricated assembly can be solved by utilizing the investment casting process. Sheet metal components, assembled by riveting, brazing, soldering, or welding, have been investment-cast as a single unit. Advantages realized include weight savings and better soundness.

Shell Molding Process

The essential feature of the shell, or Croning, process is the use of thin-walled molds and cores. Thermosetting resin-bonded silica sand is placed on a heated pattern for a predetermined length of time. Heating cures the resin, causing the sand grains to adhere to each other to form a sturdy shell that constitutes half the mold. Because of pattern costs, this method is best suited to volume production of cast metal components. The complete assembly is preheated to 177°C–204°C and the surface treated with a parting compound (silicone emulsion). The cured resin binders are nonhygroscopic, permitting prolonged storage of shell molds and allowing flexibility for production scheduling.

Castings made by the shell molding process may be more accurate dimensionally than conventional sand castings. A high degree of reproducibility as well as dimensional accuracy can be achieved with a minimum of dependence on the craftsmanship that sometimes is required with other molding processes. Only metal patterns and metal core boxes can be used in the shell process.

Lost Foam Casting

This process is also referred to as expanded polystyrene (EPS) molding, expandable pattern casting, evaporative foam casting, the full mold process, the cavityless casting process, and the cavityless EPS casting process. The process is an economical method of producing complex, close-tolerance castings, and uses unbonded sand; the pattern material is EPS.

The process involves attaching patterns to allow heating systems also made of EPS, then applying a refractory coating to the total assembly. Molten metal poured into the down-sprue vaporizes the polystyrene instantly and reproduces the pattern exactly. Gases formed from the vaporized pattern escape through the pattern coating, the sand, and the flask vents. A separate pattern is required for each casting.

To the designer, a major advantage of the process is that no cores are required. Cast-in features and reduced finishing stock usually are benefits of using the lost foam process. Inserts can be cast into the metal, and bimetallic castings can be made commercially.

Vacuum Molding

The vacuum molding process, popularly known as the V-process, is a sand molding process in which no binders are used to retain the shape of the mold cavity. Instead, unbonded sand is positioned between two sheets of thin plastic that are held in place by the application of a vacuum.

Replicast Process

The replicast process is said to overcome the shortcomings of another process that is prone to cause the formation of lustrous carbon defects in steel castings, as well as undesirable carbon pickup. Outstanding features of the replicast process include surface finish comparable to that obtainable on investment castings, elimination of cores through the use of core inserts in pattern-making tooling, improved casting yields because of absence of spruce and runners, and high quality levels with regard to casting integrity and dimensional accuracy.

Other Casting Processes

Other molding systems and casting processes are used to make metal castings. For example, certain types of castings are produced in centrifugal casting machines.

Plaster mold casting is another specialized casting process used to produce nonferrous castings that is said to offer certain advantages over other processes.

Certain techniques are used to make castings in ceramic molds that are different from ceramic shell investment molding. The main difference is that the ceramic molds consist of a cope and a drag or, if the casting shape permits, a drag only.

Squeeze casting, also known as liquid-metal forging, is a process by which molten metal (ferrous or nonferrous) solidifies under pressure within closed dies positioned between the plates of a hydraulic press. The applied pressure and the instant contact of the molten metal with the die surface produce a rapid heat transfer condition that reportedly yields a pore-free, finer-grained casting with good mechanical properties.

CAT Scanning

See *computed tomography.*

Cat's Tongue (Surface)

A rough surface texture comprised of strong spikes and ridges separated by deep tapered pits. The texture often has a "lay," that is, it lies at an angle to the original surface and hence feels very rough when stroked in one direction but relatively smooth in the opposite direction, as does a cat's tongue. As the damage progresses the texture becomes increasingly coarse although, in terms of weight loss, the rate of damage may fall. The effect is characteristic of damage caused by repeated impact particles including water droplets. The damage mechanism is not a form of cutting abrasion but results from a fatigue action by the repeated impacts producing surface mechanical damage.

Catalyst

A substance capable of changing the rate of a reaction without itself undergoing any net change. A substance that markedly speeds up the *cure* of a plastic compound when added in minor quantity, compared to the amounts of primary reactants.

Small amounts of cocatalysts or promoters increase activity measurably. In the cracking of petroleum, activated carbon breaks the complex hydrocarbons into the entire range of fragments; activated alumina is more selective, producing a large yield of C_3 and C_4; and silica–alumina–zirconia is intermediate. *Contact catalysts* are the ones chiefly used in the chemical industry, and they may be in various forms. For bed reactors, the materials are pelleted. *Powdered catalysts* are used for liquid reactions such as the hydrogenation of oils. *Chemical catalysts* are usually liquid compounds, especially such acids as sulfuric and hydrochloric.

Various metals, especially platinum and nickel, are used to catalyze or promote chemical action in the manufacture of synthetics. Nitrogen in the presence of oxygen can be "fixed" or combined in chemicals at ordinary temperatures by the use of ruthenium as a catalyst. Acids may be used to aid in the polymerization of synthetic resins. Mineral soaps are used to speed up the oxidation of vegetable oils. Cobalt oxide is used for the oxidation of ammonia. Cobalt and thorium are used for synthesizing gasoline from coal. All of these are classified as *inorganic catalysts*. Sometimes, more complex chemicals are employed, silicate of soda being used as a catalyst for high-octane gasoline. In the use of *potassium persulfate*, $K_2S_2O_8$, as a catalyst in the manufacture of some synthetic rubbers, the material releases 5.8% active oxygen, and it is the nascent oxygen that is the catalyst. *Sodium methylate*, also called *sodium methoxide*, $CH_3 \cdot O \cdot Na$, used as a catalyst for ester-exchange reactions in the rearrangement of edible oils, is a white powder soluble in fats but violently decomposed in water. Transition-metal complexes, dispersed uniformly in solution, are called *homogeneous catalysts*. The most common ones are organometallic complexes, such as the carboxyls. They are more resistant to poisoning than solid *heterogeneous catalysts*, and they are highly active, specific, and selective. Magnetite, a magnetic iron ore, is used as a catalyst in the synthesis of ammonia. In a system from M.W. Kellogg Co., ruthenium, supported on a proprietary graphite structure, is more active, increasing ammonia production by 12%–16% over magnetite.

Metallocenes, organometallic coordination compounds obtained as cyclopentadienyl derivatives of a transition metal or metal hylide, are recent catalysts in the production of various plastics. Also referred to as single-site *catalysts*, they allow closer control of molecular weight and comonomer distribution, permitting monomers and comonomers previously considered incompatible to be combined. They also allow production of plastics in iso-tactic and syndiotactic forms and have been applied to polyethylene, ethylene copolymers, ethylene terpolymers (including ethylene–propylene–diene elastomers), polypropylene, and polystyrene.

Aluminum chloride, $AlCl_3$, in gray granular crystals, is used as a catalyst for high-octane gasoline and synthetic rubber and in the synthesis of dyes and pharmaceuticals. *Antimony trichloride*, $SbCl_3$, is a yellowish solid, used as a catalyst in petroleum processing to convert normal butane to isobutane. *Bead catalysts* of activated alumina have the alumina contained in beads of silica gel.

Molecular sieve zeolites are crystalline *aluminosilicates* of alkali and alkali-earth metals. The aluminum and silicon atoms form regular tetrahedral structures that have large voids interconnected by open three-dimensional channels. The microporous may amount to 50% of the volume, resulting in crystals with some of the highest internal surface areas. The alkaline cations are mobile and maybe ion-exchanged with metals with catalytic properties. Only reactants of the right molecular size may enter the channels and be catalyzed by the metal cations in the voids. As molecular sieves, zeolite catalysts are used as desiccants and absorbers and drying and purifying gases. Natural zeolites may be more effective than synthetic ones. *Catalyst carriers* are porous inert materials used to support the catalysts, usually in a bed through which a liquid or gas may flow. Materials used are generally alumina, silica carbide, or mullite, and they are usually in the form of graded porous granules or irregular polysurface pellets. High surface area, low bulk density, and good adherence of the catalyst are important qualities. Pellets are bonded with a ceramic that fuses around the granules with minute necks that hold the mass together as complex silicates and aluminates with no trace elements exposed to the action of a catalyst or chemicals. Catalyst carriers are usually bonded to make them about 40% porous. The pellets may be 15 mesh finer, or they may

be in sizes as large as 1 in. (2.5 cm). Platinum, palladium, and rhodium supported on activated alumina carriers are used in the catalytic converters of automobiles to clean up exhaust gases. A catalyst of precious metals supported on zeolite removes hydrocarbons, carbon monoxide, and nitrogen oxides from auto exhaust gases even in the presence of excess oxygen, as is the case for lean-burn engines. Developed by Mazda Motor of Japan, it could improve fuel efficiency of such engines by 5%–8%.

Catastrophic Failure

Sudden failure of a component or assembly that frequently results in extensive secondary damage to adjacent components or assemblies.

Catastrophic Period

In cavitation or liquid impingement erosion, a stage during which the erosion rate increases so dramatically that continued exposure threatens or causes gross disintegration of the exposed surface.

Catastrophic Wear

Sudden surface damage, deterioration, or change of shape caused by wear to such an extent that the life of the part is appreciably shortened or action is impaired.

Catchment Efficiency

See *collection efficiency*.

Catenary

A measure of the difference in length of reinforcing strands in a specified length of *roving* caused by unequal tension. The tendency of some strands in a taut horizontal roving to sag more than the others.

Cathode

The negative *electrode* of an *electrolytic cell* at which reduction is the principal reaction. (Electrons flow toward the cathode and the external circuit.) Typical cathodic processes are cations taking up electrons and being discharged, oxygen being reduced, and the reduction of an element or group of elements from a higher to a lower valence state. Contrast with *anode*.

Cathode Compartment

In an electrolytic cell, the enclosure formed by a diaphragm around the cathode.

Cathode Copper

Copper in slab form produced by electrolytic refining. It is subsequently re-melted to produce high conductivity grades such as electrolytic tough pitch, high conductivity copper, and oxygen-free high conductivity copper. It is also the basis for high conductivity copper castings and high conductivity alloys.

Cathode Efficiency

Current efficiency at the *cathode*.

Cathode Film

The portion of solution in immediate contact with the *cathode* during *electrolysis*.

Cathode-Ray Tube (CRT)

An electronic device in which a stream of electrons, that is, the negatively charged cathode ray, is formed and directed at a fluorescent screen. The path of the ray is deflected by electromagnetic coils to produce the screen display.

Cathodic Cleaning

Electrolytic cleaning in which the work is the *cathode*.

Cathodic Corrosion

Corrosion resulting from a cathodic condition of a structure usually caused by the reaction of an amphoteric metal and the alkaline products of *electrolysis*.

Cathodic Disbondment

The destruction of adhesion between a coating and its substrate by products of a *cathodic reaction*.

Cathodic Etching

See *ion etching*.

Cathodic Inhibitor

A corrosion protection, a chemical substance or mixture that prevents or reduces the rate of the cathodic or reduction reaction.

Cathodic Pickling

Electrolytic pickling in which the work is the *cathode*.

Cathodic Polarization

The change of the *electrode potential* in the active (negative) direction due to current flow. See also *polarization*.

Cathodic Protection

(1) Reduction of corrosion rate by shifting the *corrosion potential* of the electrode toward a less oxidizing potential by applying an external *electromotive force*. (2) Partial or complete protection of a metal from corrosion by making it a *cathode*, using either a galvanic or an impressed current. (3) Electrical currents can develop the metal components exposed to wet environments and corrosion will occur at the end anodic areas (see *electrochemistry*). However, attack may be prevented by inducing a reverse current that causes the component to be cathodic with respect to some external anode. One technique is to impose a DC electrical current from an external power source—termed Impressed Current cathodic protection. Alternatively, the component may be protected by attaching to it another metal that is more anodic. For example, steel was protected by zinc either in the form of plating, (galvanizing) or as large blocks of zinc bolted to, or buried close to, and in electrical circuit with, the component. These anodes will be progressively consumed by corrosion, hence the term *Sacrificial Cathodic Protection* but the rate of anode loss, in practical cases, is acceptably economic and the main component will be preserved. Contrast with *anodic protection*.

Cathodic Reaction

Electrode reaction equivalent to a transfer of negative charge from the electronic to the ionic conductor. A cathode reaction is a reduction process.

Cathodoluminescence

A radioactive transition wherein low-energy light photons are really storing electron irradiation.

Catholyte

The *electrolyte* adjacent to the cathode of an electrolytic cell.

Cation

A positively charged ion that migrates through the electrolyte toward the *cathode* under the influence of a potential gradient. See also *anion* and *ion*.

Cationic Detergent

A detergent in which the *cation* is the active part.

Caul

In adhesive bonding, a sheet of material employed singly or in pairs in the hot or cold pressing of assemblies being bonded. A caul is used to protect either the faces of the assembly or the press platens, or both, against marring and staining in order to prevent sticking, facilitate press loading, impact a desired surface texture or finish, and provide uniform pressure distribution. A caul may be made of any suitable material such as aluminum, stainless steel, hardboard, fiberboard, or plastic, the length and width dimensions generally being the same as those of the plates of the press where it is used.

Caul Plates

In fabrication of composites, smooth metal plates, free of surface defects, that are the same size and shape as a composite lay-up, and that contact the lay-up during the curing process in order to transmit normal pressure and temperature, and to provide a smooth surface on the finish laminate.

Cauliflower

A growth of material on the exterior, particularly the top, of an ingot as a result of inverse segregation.

Caulk

See *caulking*.

Caulk Weld

See preferred term *seal weld*.

Caustic

(1) Burning or corrosive. (2) A hydroxide of a light metal, such as sodium hydroxide or potassium hydroxide.

Caustic Attack

Generally, any corrosion caused by caustic alkali. The term often refers specifically to the attack on the water side of steam generating tubes and power plants. The cause is excessive sodium hydroxide resulting either from the introduction of quantities well in excess of that required for normal feed water treatment or from local concentration affects. Factors promoting concentration are very high levels of heat input and bore irregularities such as deposits or the crevice produced when the internal projection of a flash butt weld is smeared along the bore rather than being cleanly cut away. When such factors are present, the tube bore facing toward the fire develops a persistent steam blanket rather than the normal bubbles of steam which can be swept away. The caustic concentrates in this area and the normally protective magnetite film is disrupted leading to severe corrosion and in some cases to hydrogen embrittlement. In some cases, the corrosive attack produces deep grooves along the tube leading to terms such as *caustic grooving* or *gouging*.

Caustic Cracking

A form of *stress-corrosion cracking* most frequently encountered in carbon steels or iron–chromium–nickel alloys that are exposed to concentrated hydroxide solutions at temperatures of (200°C–250°C) (400°F–480°F). Also known as caustic embrittlement.

Caustic Dip

A strongly alkaline solution into which metal is immersed for etching, for neutralizing acid, or for removing organic materials such as greases or paints.

Caustic Embrittlement

(1) An obsolete historical term denoting a form of *stress-corrosion cracking* most frequently encountered in carbon steels or iron–chromium–nickel alloys that are exposed to concentrated hydroxide solutions at temperatures of (200°C–250°C) (400°F–480°F). (2) Cracking, intergranular stress corrosion cracking of steel in caustic, alkaline, solutions. It usually only occurs above about 70°C and in fairly high alkaline conditions. However, it can be a serious problem in cases where the solution is now only 2 weeks to have any significant effect but where an inadvertent concentration mechanism is active. For example, to allow a concentration to develop in the leak path interface as water evaporates at the exterior face.

Caustic Quenching

Quenching with aqueous solutions of 5%–10% sodium hydroxide.

Caustic Soda

Sodium hydroxide.

Cavitating Disk Apparatus

A flow cavitation test device in which cavitating wakes are produced by holes in, or protuberances on, a disk rotating within a liquid-filled chamber. Erosion test specimens are attached flush with the surface of the disc at the location where the bubbles are presumed to collapse.

Cavitation

The formation and collapse, within a liquid of cavities or bubbles that contain vapor or gas or both. In general, cavitation originates from a decrease in the static pressure in the liquid. It is distinguished in this way from boiling, which originates from an increase in the liquid temperature. There are certain situations where it may be difficult to make a clear distinction between cavitation in boiling, and the more general definition that is given here is therefore to be preferred. In order to erode a solid surface by cavitation, it is necessary for the cavitation bubbles to collapse on or close to that surface.

Cavitation Cloud

A collection of a large number of cavitation bubbles. The bubbles in a cloud are small, typically less than 1 mm in cross section.

Cavitation Corrosion

A process involving conjoint *corrosion* and *cavitation*.

Cavitation Damage

The degradation of a solid body resulting from its exposure to *cavitation*. This may include loss of material, surface deformation, or changes in properties or appearance.

Cavitation Erosion/Damage

The removal of metal from an immersed surface as a result of local severe pressure fluctuations associated with turbulent flow. Water subjected to a sudden pressure drop will form cavities, i.e., bubbles. These rapidly collapse producing shockwaves which damage adjacent surfaces by various mechanisms including a direct mechanical action, a fatigue action or by disrupting protective films. The development of local pits exacerbates cavity quality and size so metal loss is accelerated. Pumps and propellers are commonly affected.

Cavitation Tunnel

A flow cavitation test facility in which liquid is pumped through a pipe or tunnel, and cavitation is induced in a test section by conducting the flow through a constriction, or around an obstacle, or a combination of these.

Cavity (Metals)

The mold or die impression that gives a casting its external shape.

Cavity (Plastics)

The space inside a mold into which a resin or molding compound is poured or injected. The female portion of a mold. The portion of the mold that encloses the molded article (often referred to as the die). Depending on a number of such depressions, molds are designated as single cavity or multiple cavity.

Cavity Retainer Plates

In forming of plastics, plates in a mold that hold the cavities and forces. These plates are at the mold parting line and usually contain the guide pins and bushings. Also called force retainer plates.

CCP

Cubic close-packed crystallographic structure. Same as face-centered cubic.

CCT Diagram (Continuous Cooling Transformation Diagram)

See Isothermal Transformation Diagram.

Ceiling

The maximum that should not be exceeded in the process in question.

Cell

In honeycomb core, a cell is a single honeycomb unit, usually in a hexagonal shape.

Cell (Electrochemistry)

Electrochemical system consisting of an *anode* and a *cathode* immersed in an *electrolyte*. The anode and cathode may be separate metals or dissimilar areas on the same metal. The cell includes the external circuit, which permits the flow of electrons from the anode toward the cathode. See also *electrochemical cell*.

Cell (Plastics)

A single cavity formed by gaseous displacement in a plastic material. See also *cellular plastic*.

Cell Feed

The material supplied to the cell in the electrolytic production of metals.

Cell Size

The diameter of an inscribed circle within a cell of *honeycomb* core.

Cellular

A network comprising a large number of associated cells of broadly similar characteristics.

Cellular Adhesive

Synonym for foamed adhesive.

Cellular Plastic

A plastic with greatly decreased density because of the presence of numerous cells or bubbles dispersed throughout its mass. See also *cell (plastics)*, *foamed plastics*, and *syntactic cellular plastics*.

Cellulose

Cellulose is the main constituent of the structure of plants (natural polymer) that, when extracted, is employed for making paper, plastics, and in many combinations. Cellulose is made up of long-chain molecules in which the complex unit $C_5H_{10}O_6$ is repeated as many as 2000 times. It consists of glucose molecules with three hydroxyl groups for each glucose unit.

Cellulose is the most abundant of the nonprotein natural organic products. It is highly resistant to attack by the common microorganisms. However, the cellulose digests it easily, and this substance is used for making paper pulp, for clarifying beer and citrus juices, and for the production of citric acid and other chemicals from cellulose. Cellulose is a white powder insoluble in water, sodium hydroxide, or alcohol, but it is dissolved by sulfuric acid.

One of the simplest forms of cellulose used industrially is regenerated cellulose, in which the chemical composition of the finished product is similar to that of the original cellulose. It is made from wood or cotton pulp digested in a caustic solution. Cellophane is a regenerated cellulose in thin sheets for wrapping and other special uses include windings on wire and cable.

Cellulose Plastics

For plastics, pure cellulose from wood pulp or cotton linters (pieces too short for textile use) is reacted with acids or alkalis and alkyl halides to produce a basic flake. Depending upon the reactants, any one of four esters of cellulose (acetate, propionate, acetate butyrate, or nitrate) or a cellulose ether (ethyl cellulose) may result. The basic flake is used for producing both solvent cast films and molding powders.

Ethyl cellulose plastics are thermoplastic and are noted for their ease of molding, lightweight, and good dielectric strength, $15–20.5 \times 10^6$ V/m, and retention of flexibility over a wide range of temperature from $-57°C$ to $66°C$, the softening point. They are the toughest, the lightest, and have the lowest water absorption of the cellulosic plastics. But they are softer and lower in strength than cellulose-acetate plastics. Typical ethyl cellulose applications include football helmets, equipment housings, refrigerator parts, and luggage.

For molding powders, the flake is then compounded with plasticizers, pigments, and sometimes other additives. At this stage of manufacture, the plastics producer is able to adjust hardness, toughness, flow, and other processing characteristics and properties. In general, these qualities are spoken of together as flow grades. The flow of a cellulose plastic is determined by the temperature at which a specific amount of the material will flow through a standard orifice under a specified pressure. Manufacturers offer cellulosic molding materials in a large number of standard flow grades, and, for an application requiring a nonstandard combination of properties, are often able to tailor a compound to fit. Cellulose can be made into a film (cellophane) or into a fiber (rayon), but it must be chemically modified to produce a thermoplastic material.

Cellulosics are synthetic plastics, but they are not synthetic polymers; see Table C.8.

Because the cellulosics can be compounded with many different plasticizers in widely varying concentrations, property ranges are broad. These materials are normally specified by flow, defined in American Society for Testing and Materials (ASTM) D569, which is controlled by plasticizer content. Hard flows (low plasticizer content) are relatively hard, rigid, and strong. Soft flows (higher plasticizer content) are tough, but less hard, less rigid, and less strong. They also process at lower temperatures. Thus, within available

TABLE C.8
Properties of Cellulosics

ASTM or UL Test	Property	Cellulose Acetate	Cellulose Propionate	Cellulose Acetate Butyrate	Ethyl Cellulose
Physical					
D792	Specific gravity	1.22–1.34	1.16–1.24	1.15–1.22	1.09–1.17
D792	Specific volume (in.³/lb)	22.7–20.6	23.4–22.4	24.1–22.7	25.5–23.6
D570	Water absorption, 24 h, 1/8 in. thick (%)	1.7–4.5	1.2–2.8	0.9–2.2	0.8–1.8
Mechanical					
D638	Tensile strength (psi)	2,200–6,900	1,400–7,200	1,400–6,200	3,000–4,800
D638	Tensile modulus (10^5 psi)	0.65–4.0	0.6–2.15	0.5–2.0	2.2–2.5
D790	Flexural strength (psi)	2,500–10,400	1,700–10,600	1,800–9250	4,700–6800
D790	Flexural modulus (10^5 psi)	1.2–3.6	1.15–3.7	0.9–3.0	—
D256	Impact strength, Izod (ft lb/in. of notch)	1.0–7.3	1.0–10.3	1.1–9.1	3.0–8.0
D785	Hardness, Rockwell R	To 122	To 115	To 112	79–106
Thermal					
C177	Thermal conductivity (10^{-4} cal cm/s cm² °C)	4–8	4–8	4–8	3.8–7.0
D696	Coefficient of thermal expansion (10^{-5} in./in. °C)	8–16	11–17	11–17	10–20
D648	Deflection temperature (°F)				
	At 264 psi	111–195	111–228	113–202	115–190
	At 66 psi	120–209	147–250	130–227	170–180
UL94	Flammability rating	V-2, HB	HB	HB	—
Electrical					
D149	Dielectric strength[a] (V/mil) Short time, 1/8 in. thick	250–600	300–500	250–400	350–500
D150	Dielectric constant at 1 kHz	3.2–7.0	3.3–4.0	3.4–6.4	3.0–4.1
D150	Dissipation factor at 1 kHz	0.01–0.10	0.01–0.05	0.01–0.04	0.002–0.020
D257	Volume resistivity (Ω cm) at 73°F, 50% RH	10^{10}–10^{14}	10^{12}–10^{16}	10^{11}–10^{15}	10^{12}–10^{14}
D495	Arc resistance(s)	50–310	175–190	—	60–80
Optical					
D542	Refractive Index	1.46–1.50	1.46–1.49	1.46–1.49	—
D1003	Transmittance[b] (%)	80–92	80–92	80–92	—

Source: Mach. Design Basics Eng. Design, 684, June 1993. With permission.

[a] At 500 V/s rate of rise.

[b] For 1/8 in. thick specimen.

property ranges listed, no one formulation can provide all properties to the maximum degree. Most commonly used formulations are in the middle flow ranges.

Molded cellulosic parts can be used in service over broad temperature ranges and are particularly tough at very low temperatures. Ethyl cellulose is outstanding in this respect. These materials have low specific heat and low thermal conductivity—characteristics that give them a pleasant feel.

Dimensional stability of butyrate, propionate, and ethyl cellulose is excellent. Plasticizers used in these materials do not evaporate significantly and are virtually immune to extraction by water. Water absorption (which causes dimensional change) is also low, with that of ethyl cellulose the lowest. The plasticizers in acetate are not as permanent as those in other plastics, however, and water absorption of this material is slightly higher.

Butyrate and propionate are highly resistant to water and most aqueous solutions except strong acids and strong bases. They resist nonpolar materials such as aliphatic hydrocarbons and ethers, but they swell or dissolve in low-molecular-weight polar compounds such as alcohols, esters, and ketones, as well as in aromatic and chlorinated hydrocarbons. Acetate is slightly less resistant than butyrate and propionate to water and aqueous solutions, and slightly more resistant to organic materials. Ethyl cellulose dissolves in all the common solvents for this polymer, as well as in such solvents as cyclohexane and diethyl ether. Like the cellulose esters, ethyl cellulose is highly resistant to water.

Although unprotected cellulosics are generally not suitable for continuous outdoor use, special formulations of butyrate and propionate are available for such service. Acetate and ethyl cellulose are not recommended for outdoor use.

Applications

Acetate applications include extruded and cast film and sheet for packaging and thermoforming.

Cellulose Acetate

Cellulose acetate is an amber-colored, transparent material made by the reaction of cellulose and acetic acid or acetic anhydride in the presence of sulfuric acid.

It is thermoplastic and easily molded. The molded parts or sheets are tough, easily machined, and resistant to oils and many chemicals. In coatings and lacquers, the material is adhesive, tough, and resilient, and does not discolor easily. Cellulose acetate fiber for rayons can be made in fine filaments that are strong and flexible, nonflammable, mildew proof, and easily dyed. Standard cellulose acetate for molding is marketed in flake form.

In practical use, cellulose acetate moldings exhibit toughness superior to most other general-purpose plastics. Flame-resistant formulations are currently specified for small appliance housings and for other uses requiring this property. Uses for cellulose acetate molding materials include toys, buttons, knobs, and other parts where the combination of toughness and clear transparency is a requirement.

Extruded film and sheet of cellulose acetate packaging materials maintain their properties over long periods. Here also the toughness of the material is advantageously used in blister packages, skin packs, window boxes, and over wraps. It is a breathing wrap and is solvent and heat sealable.

Large end uses for cellulose acetate films and sheets include photographic film base, protective cover sheets for notebook pages and documents, index tabs, sound recording tape, as well as the laminating of book covers. The grease resistance of cellulose acetate sheet allows its use in packaging industrial parts with enclosed oil for protection.

For eyeglass frames, cellulose acetate is the material in widest current use. Because fashion requires varied and sometimes novel effects, sheets of clear, pearlescent, and colored cellulose acetate are laminated to make special sheets from which optical frames are fabricated.

The electrical properties of cellulosic films combined with their easy bonding, good aging, and available flame resistance bring about their specification for a broad range of electrical applications. Among these are as insulations for capacitors; communications cable; oil windings; in miniaturized components (where circuits may be vacuum metallized); and as fuse windows.

Cellulose triacetate is widely used as a solvent cast film of excellent physical properties and good dimensional stability. Used as photographic film base and for other critical dimensional work such as graphic arts, cellulose triacetate is not moldable.

Cellulose Propionate

Cellulose propionate, commonly called "CP" or propionate, is made by the same general method as cellulose acetate, but propionic acid is used in the reaction. Propionate offers several advantages over cellulose acetate for many applications. Because it is "internally" plasticized by the longer-chain propionate radical, it requires less plasticizer than is required for cellulose acetate of equivalent toughness.

Cellulose propionate absorbs much less moisture from the air and is thus more dimensionally stable than cellulose acetate. Because of better dimensional stability, cellulose propionate is often selected where metal inserts and close tolerances are specified.

Largest-volume uses for cellulose propionate are as industrial parts (automotive steering wheels, armrests, and knobs, etc.), telephones, toys, findings, ladies' shoe heels, pen and pencil barrels, and toothbrushes.

Cellulose Acetate Butyrate

Commonly called butyrate or CAB, it is somewhat tougher and has lower moisture absorption and a higher softening point than acetate. CAB is made by the esterification of cellulose with acetic acid and butyric acid in the presence of a catalyst. It is particularly valued for coatings, insulating types, varnishes, and lacquers.

Special formulations with good weathering characteristics plus transparency are used for outdoor applications such as signs, light globes, and lawn sprinklers. Clear sheets of butyrate are available for vacuum-forming applications. Other typical uses include transparent dial covers, television screen shields, tool handles, and typewriter keys. Extruded pipe is used for electric conduits, pneumatic tubing, and low-pressure waste lines. Cellulose acetate butyrate also is used for cable coverings and coatings. It is more soluble than cellulose acetate and more miscible with gums. It forms durable and flexible films. A liquid cellulose acetate butyrate is used for glossy lacquers, chemical-resistant fabric coatings, and wire-screen windows. It transmits ultraviolet light without yellowing or hazing and is weather-resistant.

Cellulose Acetate Propionate

This substance is similar to butyrate in both cost and properties. Some grades have slightly higher strength and modulus of elasticity. Propionate has better molding characteristics, but lower weatherability than butyrate. Molded parts include steering wheels, fuel filter bowls, and appliance housings. Transparent sheeting is used for blister packaging and food containers.

Cellulose Nitrate

Cellulose nitrates are materials made by treating cellulose with a mixture of nitric and sulfuric acids, washing free of acid, bleaching, stabilizing, and dehydrating. For sheets, rods, and tubes it is mixed with plasticizers and pigments and rolled or drawn to the shape desired. The lower nitrates are very inflammable, but they do not explode like the high nitrates, and they are the ones used for plastics, rayons, and lacquers, although their use for clothing fabrics is restricted by law. The names *cellulose nitrate* and *pyroxylin* are used for the compounds of lower nitration, and the term *nitrocellulose* is used for the explosives.

Cellulose nitrate is the toughest of the thermoplastics. It has a specific gravity of 1.35–1.45, tensile strength of 41–52 MPa, elongation 30%–50%, compressive strength 137–206 MPa, Brinell hardness 8–11, and dielectric strength 9.9–21.7×10^6 V/m. The softening point is 71°C, and it is easy to mold and easy to machine. It also is readily dyed to any color. It is not light stable, and is therefore no longer used for laminated glass. It is resistant to many chemicals, but has the disadvantage that it is inflammable. The molding is limited to pressing from flat shapes.

Among thermoplastics, it is remarkable for toughness. For many applications today, however, cellulose nitrate is not practical because of serious property shortcomings: heat sensitivity, poor outdoor aging, and very rapid burning.

Cellulose nitrate cannot be injection-molded or extruded by the nonsolvent process because it is unable to withstand the temperatures these processes require. It is sold as films, sheets, rods, or tubes, from which end products may then be fabricated.

Cellulose nitrate yellows with age; if continuously exposed to direct sunlight, it yellows faster and the surface cracks. Its rapid burning must be considered for each potential application to avoid unnecessary hazard.

The outstanding toughness properties of cellulose nitrate lead to its continuing use in such applications as optical frames, shoe eyelets, ping-pong balls, and pen barrels.

Cellulosic Electrode (for Welding)

See electrode (welding).

Cellulosic Plastics

Plastics based on cellulose compounds, such as esters (cellulose acetate) and ethers (ethyl cellulose). See Table C.8.

Celsius

The SI recommended scale of temperature based on reference points including 0°C as a freezing point of water and 100°C as the boiling point of water. It is essentially identical to centigrade.

Cement

Cement is a synthetic mineral mixture that, when ground to a powder and mixed with water, forms a stone-like mass. This mass results from a series of chemical reactions whereby the crystalline constituents hydrate, forming a material of high hardness that is extremely resistant to compressive loading. The main uses of cement are in civil engineering, for which, since the late nineteenth century, it has become indispensable.

The history of cement dates back to the Romans, who found that mixtures of volcanic ash, lime, and clay would harden when wet, and who used it extensively to build structures. In 1757, it was found that burned and ground high calcific clays would harden when placed in water. In 1824, a patent was granted to a British bricklayer who formulated a new type of cement with improved hardness. Because the color of the material reminded him of the limestone on the Isle of Portland, he named the product portland cement. This cement was made by lightly calcining small batches of lime and clay and grinding the product of fine powder.

The modern manufacturing process is very basic and has not been radically changed since its inception except for the use of computer-controlled equipment, which has greatly improved the consistency of the final product. The four basic cement processing operations are: (1) quarrying and crushing of raw materials, (2) grinding to high fineness and carefully proportioning the mineral constituents, (3) pyroprocessing the raw materials in a rotary calciner, and (4) cooling and grinding the calcined product, or clinker, to obtain a fine powder.

There are four main compounds that compose portland cement: tricalcium silicate, $3CaO \cdot SiO_2$ (C_3S); dicalcium silicate, $2CaO \cdot SiO_2$ (C_2S); tricalcium aluminate, $3CaO \cdot Al_2O_3$ (C_4A); and tetracalcium aluminoferrite, $4CaO \cdot Al_2O_3 \cdot Fe_2O_3$ (C_3AF). In the United States, portland cements are manufactured to comply with the ASTM Standard Specification for Portland Cement, ASTM C150. This specification defines five main types of portland cement. Type I, which is made in the greatest quantity, is intended for general-purpose use when the special properties of the other types are not required. The special properties of the other types when used in concrete are: Type II, moderate sulfate resistance or moderate heat of hydration; Type III, high early strength; Type IV, low heat of hydration; Type V, high sulfate resistance. The chemical and physical differences between the types that produce their special properties lie in the proportions of the cement compounds and in the fineness to which the cement is ground.

Oxychloride cement, or *Sorel cement*, is composed of magnesium chloride, $MgCl_2$, and calcined magnesia. It is strong and hard and, with various fillers, is used for floors and stucco. Magnesia cement is magnesium oxide, prepared by heating the chloride or carbonate to redness. When mixed with water, it sets to a friable mass but of sufficient strength for covering steam pipes or furnaces. It is usually mixed with asbestos fibers to give strength and added heat resistance. The term *85% magnesia* means 85% magnesia cement and 15% asbestos fibers. The cement will withstand temperatures up to 600°F (316°C).

Keene's cement, also known as *flooring cement* and *tiling plaster*, is a *gypsum cement*. It is made by burning gypsum at about 1100°F (593°C), to drive off the chemically combined water, grinding to a fine powder, and adding alum to accelerate the set. It will keep better than ordinary gypsum cement, has high strength, is white, and takes a good polish. *Parian cement* is similar, except that borax is used instead of alum. *Martin's cement* is made with potassium carbonate instead of alum. These cements are also called *hard-finish plaster*, and they will set very hard and white. They are used for flooring and to imitate tiling. An ancient natural cement is *pozzuolana cement*. It is a volcanic material found near Pozzuoli, Italy, and in several other places in Europe. It is a volcanic lava modified by steam or gases so that it is powdery and has acquired hydraulic properties. The chief components are silica and alumina, and the color varies greatly, being white, yellow, brown, or black. It has been employed as a construction cement since ancient times. *Slag cement* is made by grinding blast-furnace slag with portland cement. Pozzolans are siliceous materials which will combine with lime in the presence of water to form compounds having cementing properties. Fly ash is an artificial pozzolan composed principally of amorphous silica with varying amounts of the oxides of aluminum and iron and traces of other oxides. It is a fine, dark powder of spheroid particles produced as the by-product of combustion of pulverized coal, and collected at the base of the stack. As an admix, it improves the workability of concrete, and in large amounts its pozzolanic action adds to the compressive strength. A *fire-resistant cement*, developed by Arthur D. Little, Inc., is made of magnesium oxychlorides and magnesium oxysulfates. This inorganic resin foam cement contains 40%–50% bond water that is released when the material is exposed to high temperatures and absorbs heat. It is said not to burn, smoke, or produce poisonous fumes when subjected to a direct flame.

Cement Copper

In pure copper recovered by chemical deposition when iron (most often shredded steel scrap) is brought into prolonged contact with a dilute copper sulfate solution.

Cementation

The introduction of one or more elements into the outer portion of a metal object by means of the fusion at high temperature. See *carburizing*.

Cementation Process

An obsolete process in which wrought iron-activated charcoal is heated at about 900°C for a few days allowing carbon to diffuse into the low carbon iron to produce steel.

Cemented Carbides

Referred to as hard metals, belong to a class of hard, wear-resistant, refractory materials in which the hard carbide particles are bound together, or cemented, by a soft and ductile metal binder. The first cemented carbide produced was tungsten carbide (WC) with a cobalt binder. Over the years, the basic WC–Co material has been modified to produce a variety of cemented carbides, which are used in a wide range of applications, including metal cutting, mining, construction, rock drilling, metal-forming, structural components, and wear parts. Tungsten carbide-based materials with nickel or steel binders have also been produced for specialized applications.

Tungsten carbides are manufactured by powder metallurgy process consisting of (1) processing of the ore in the preparation of the WC powder, (2) preparation (ball milling) of WC powders and grade (alloying) powders, (3) the addition of suitable binder material, (4) powder consolidation, and (5) sintering of the compacted part at temperatures between 1300°C and 1600°C (2370°F and 2910°F), most often in vacuum. The sintered product can be directly used or can be ground, polished, and coated to suit a given application.

Approximately 50% of all carbide production is used for machining applications, and a wide variety of compositions are available. "Straight" grades, which consist of WC particles bonded with cobalt, generally contain 3%–12% Co and carbide grain sizes range from 0.5 to >5 μm. Alloy grades, or steel-cutting grades, contain titanium carbide (TiC), titanium carbonitride (TiCN), titanium nitride (TiN), and/or niobium carbide (NbC). Improved wear resistance of cemented carbide tools is achieved by multilayer hard coatings of TiC, TiCN, TiN, alumina (Al_2O_3), and occasionally hafnium carbide (HfC). These coatings are commonly applied by chemical vapor deposition.

Cemented carbides are also being used increasingly for nonmachining applications, such as metal and nonmetallic mining, oil and gas drilling, transportation and construction, metal-forming, structural and fluid-handling components, and forestry tools. Straight WC–Co grades are used for the majority of these applications. In general, cobalt contents range from 5% to 30% and WC grain sizes range from <1 to >8 μm.

Cementite

A hard (~800 HV), brittle compound of iron and carbon, known chemically as iron carbide and having the approximate chemical formula Fe_3C is characterized by an orthorhombic crystal structure. When it occurs as a phase in steel, the chemical composition will be altered by the presence of manganese and other carbide-forming elements. The highest cementite contents are observed in white cast irons, which are used in applications where high wear resistance is required.

Centane Number

A measure of the ignition quality of a fuel or petroleum product with reference to normal centane high-ignition quality fuel with an arbitrary number of 100.

Center Drilling

Drilling a short, conical hole in the end of a workpiece—a hole to be used to center the workpiece for turning on a lathe.

Center-Gated Mold

An injection or transfer mold in which the cavity is filled with plastic molding material, through a sprue or gate, directly into the center of the part.

Centering Plug

A plug fitting both spindle and cutter to ensure concentricity of the cutter mounting.

Centerless Grinding

Grinding the outside or inside diameter of a cylindrical piece which is supported on a work support blade instead of being held between centers and which is rotated by a so-called regulating or feed wheel.

Centerline Shrinkage

Shrinkage or porosity occurring along the central *plane or axis of a cast part.*

Centigrade

The original metric measure of temperature. Identical to Celsius in the SI system.

Centrifugal Casting (Plastics)

A method of forming thermoplastic resins in which the granular resin is placed in a rotatable container, heated to a molten condition by the transferor heat through the walls of the container, and rotated such that the centrifugal force induced will force the molten resin to conform to the configuration of the interior surface of the container. Used to fabricate large-diameter pipes and similar cylindrical items.

Centrifugal Castings

Centrifugal castings can be produced economically and with excellent soundness. They are used in the automotive, aviation, chemical, and process industries for a variety of parts having a hollow, cylindrical form or for sections or segments obtainable from such a form.

There are three modifications of centrifugal casting: (1) true centrifugal casting, (2) semicentrifugal casting, and (3) centrifuging.

1. True centrifugal casting is used for the production of cylindrical parts. The mold is rotated, usually in a horizontal plane, and the molten metal is held against the wall by centrifugal force until it solidifies.
2. Semicentrifugal casting is used for disk- and wheel-shaped parts. The mold is spun on a vertical axis, the metal is introduced at the center of the mold, and centrifugal force throws the metal to the periphery.
3. Centrifuging is used to produce irregular-shaped pieces. The method *differs from* static casting only in that the mold is rotated. Mold cavities are fastened at the periphery of a revolving turntable, the metal is introduced at the center, and thrown into the molds through radial ingates.

The nature of the centrifugal casting process ensures a dense, homogeneous cast structure free from porosity. Because the metal solidifies in a spinning mold under centrifugal force, it tends to be forced against the mold wall while impurities, such as sand, slag, and gases, are forced toward the inside of the tube. Another advantage of centrifugal casting is that recovery can run as high as 90% of the metal poured.

Certain types of castings are produced in centrifugal casting machines. There are essentially two types of those machines—the

C

horizontal type that rotates about a horizontal axis and the vertical type that rotates about a vertical axis. In general, horizontal machines are used to make pipe, tubes, bushings, cylinder sleeves, and other cylindrical or tubular castings that are simple in shape. Castings that are not cylindrical, or even symmetrical, can be made using vertical centrifugal casting machines.

Ferrous Castings

Centrifugal castings can be made of many of the ferrous metals—cast irons, carbon and low-alloy steels, and duplex metals.

Mechanical Properties

Regardless of alloy content, the tensile properties of irons cast centrifugally are reported to be higher than those of static castings produced from the same heat. Hydrostatic tests of cylinder liners produced by both methods show that centrifugally cast liners withstand about 20% more pressure than statically cast liners.

Freedom from directionality is one of the advantages that centrifugal castings have over forgings. Properties of longitudinal and tangential specimens of several stainless grades are substantially equal.

Shapes, Sizes, Tolerances

The external contours of centrifugal castings are not limited to circular forms. The contours can be elliptical, hexagonal, or fluted, for example. However, the nature of the true centrifugal casting process limits the bore to a circular cross section.

Iron and steel centrifugally cast tubes and cylinders are produced commercially with diameters ranging from 28.6 to 1500 mm, wall thickness of 0.25–102 mm, and in lengths up to 14.30 m. Generally it is impractical to produce castings with a ratio of the outside diameter to the inside diameter greater than about 4–1. The upper limit in size is governed by the cost of the massive equipment required to produce heavy castings.

As-cast tolerances for centrifugal castings are about the same as those for static castings. For example, tolerances on the outside diameter of centrifugally cast gray iron pipe range from 0.3 mm for 76 mm diameter to ±0.6 mm for 1.2 m diameter. Inside-diameter tolerances are greater, because they depend not only on the mold diameter, but also on the quantity of metal cast; the latter varies from one casting to another. These tolerances are generally about 50% greater than those on outside diameters. Casting tolerances depend to some extent also on the shrinkage allowance for the metal being cast.

The figures given earlier apply to castings to be used in the unmachined state. For castings requiring machining, it is customary to allow 2.35–3.2 mm on small castings and up to 6.4 mm on larger castings. If the end-use requires a sliding fit, broader tolerances are generally specified to permit additional machining on the inside surface.

Cast Irons

Large tonnages of gray iron are cast centrifugally. The relatively low pouring temperatures and good fluidity of the common grades make them readily adaptable to the process. Various alloy grades that yield pearlitic, acicular, and chill irons are also used. In addition, specialty iron alloys such as "Ni-Hard" and "Ni-Resist," have been cast successfully.

Carbon and Low-Alloy Steels

Centrifugal castings are produced from carbon steels having carbon contents ranging from 0.05% to 0.90%. Practically all of the AISI standard low-alloy grades have also been cast.

Small-diameter centrifugally cast tubing in the usual carbon steel grades is not competitive in price with mechanical tubing having normal wall thicknesses. However, centrifugally cast tubing is less expensive than statically cast material.

High-Alloy Steels

Most of the AISI stainless and heat-resisting grades can be cast centrifugally. A particular advantage of the process is its use in producing tubes and cylinders from alloy compositions that are difficult to pierce and to forge or roll.

The excellent ductility resulting in the stainless alloys from centrifugal casting makes it possible to reduce the rough cast tubes to smaller-diameter tubing by hot- or cold-working methods. For example, billets of 18-8 stainless steel, 114.5 mm outside diameter by 16 mm wall, have been reduced to 27-gauge capillary tubing without difficulty.

Duplex Metals

Centrifugal castings with one metal on the outside and another on the inside are also in commercial production. Combinations of hard and soft cast iron, carbon steel, and stainless steel have been produced successfully.

Duplex metal parts have been centrifugally cast by two methods. In one, the internal member of the pair is cast within a shell of the other. This method has been used to produce aircraft brake drums by centrifugally casting an iron liner into a steel shell.

In the second method, both sections of the casting are produced centrifugally; the metal that is to form the outer portion of the combination is poured into the mold and solidified and the second metal is introduced before the first has cooled. The major limitation of this method is that the solidification temperature of the second metal poured must be the same or lower than that of the first. This method is said to form a strongly bonded duplex casting.

The possibilities of this duplex method for producing tubing for corrosion-resistant applications and chemical pressure service have been developed.

Nonferrous Castings

Nonferrous centrifugal castings are produced from copper alloys, nickel alloys, and tin- and lead-base bearing metals. Only limited application of the process is made to light metals because it is questionable whether any property improvement is achieved; for example, differences in density between aluminum and its normal impurities are smaller than in the heavy metals and consequently separation of the oxides, a major advantage of the process, is not so successful.

Shapes, Sizes, Tolerances

As with ferrous alloys, the external shapes of nonferrous centrifugal castings can be elliptical, hexagonal, or fluted, as well as round. However, the greatest overall tonnage of nonferrous castings is produced in plain or semiplain cylinders. The inside diameter of the casting is limited to a straight bore or one that can be machined to the required contour with minimum machining cost.

Nonferrous castings are produced commercially in outside diameters ranging from about 25.4 mm to 1.8 m and in lengths up to 8.1 m. Weights of individual castings range from 0.2268 to 27,300 kg.

Although tolerances on as-cast parts are about the same as those for sand castings, most centrifugal castings are finished by machining. An advantage of centrifugal casting is that normally only a small machining allowance is required; this allowance varies from as little as 1.53 mm on small castings to 6.4 mm on the outside diameter of large-diameter castings. A slightly larger machining allowance is required on the bore to permit removal of dross and other impurities that segregate in this area.

Copper Alloys

A wide range of copper casting alloys is used in the production of centrifugal castings. The alloys include the plain brasses, leaded brasses and bronzes, tin bronzes, aluminum bronzes, silicon bronzes, manganese bronzes, nickel silvers, and beryllium copper. The ASTM lists 32 copper alloys for centrifugal casting; in addition, there are a number of proprietary compositions that are regularly produced by centrifugal casting.

Most of these alloys can be cast without difficulty. Some trouble with segregation has been reported in casting the high leaded (over 10% lead) alloys. However, alloys containing up to 20% lead are being cast by some foundries; the requirements are (1) rapid chilling to prevent excessive lead segregation and (2) close control of speed.

The mechanical properties of centrifugally cast copper alloys vary with the composition and are affected by the mold material used. Centrifugal castings produced in chill molds have higher mechanical properties than those obtained by casting in sand molds. However, centrifugal castings made in sand molds have properties about 10% higher than those obtained on equivalent sections of castings produced in static sand molds. (Castings produced in centrifugal chill molds have properties 20%–40% higher than those produced in static sand molds.)

Nickel Alloys

Centrifugal castings of nickel 210, 213, and 305; "Monel" alloys 410, 505, and 506; and "Inconel" alloys 610 and 705 are commercially available in cylindrical tubes. Centrifugal castings are also produced from the heat-resisting alloys 60% nickel–12% chromium and 66% nickel–17% chromium. These alloys should behave like other materials and show improved density with accompanying improvement in mechanical properties. The nickel alloys are employed for service under severe corrosion, abrasion, and galling conditions.

Bearing Metals

Centrifugal casting is a standard method of producing lined bearings. Steel cylinders, after being cleaned, pickled, and tinned, are rotated while tin- or lead-base bearing alloys are cast into them. The composite cylinder is then cut lengthwise, machined, and finished into split bearings.

Centrifuge

A mechanism in which a material is spun either by having a swirl induced in it or by being contained in a vessel that is spun at high speed. In either case, the intention is to apply high gravitational forces that separate materials of different densities.

Centrifuge Casting

A casting technique in which mold cavities are spaced symmetrically about a vertical axial common downgate. The entire assembly is rotated about that axis during pouring and solidification.

Ceramic (Adjective)

(1) Of or pertaining to ceramics, that is, inorganic nonmetallic as opposed to organic or metallic. (2) Pertaining to products manufactured from inorganic nonmetallic substances which are subjected to a high temperature during manufacture or use. (3) Pertaining to the manufacture or use of such articles or materials, such as ceramic process or ceramic science.

Ceramic Color Glaze

An opaque colored glass of satin or gloss finish obtained by spraying the clay body with a compound of metallic oxides, chemicals, and clays. It is fired at high temperatures, fusing the glaze to the body, making them inseparable.

Ceramic Fibers

Alumina–silica (Al_2O_3–SiO_2) fibers, frequently referred to as ceramic fibers, are formed by subjecting a molten stream to a fiberizing force. Such force may be developed by high-velocity gas jets or rotors or intricate combinations of these. The molten stream is produced by melting high-purity Al_2O_3 and SiO_2, plus suitable fluxing agents, and then pouring this melt through an orifice. The jet or rotor atomizes the molten stream and attenuates the small particles into fine fibers as supercooling occurs.

The resulting fibrous material is a versatile high-temperature insulation for continuous service in the 538°C–1260°C range. It thus bridges the gap between conventional inorganic fiber insulating materials (e.g., asbestos, mineral wool, and glass) and insulating refractories.

Al_2O_3–SiO_2 fibers have a maximum continuous use temperature of 1093°C–1260°C, and a melting point of over 1760°C for extended periods of time, a phenomenon called devitrification occurs. This is a change in the orientation of the molecular structure of the material from the amorphous state (random orientation) to the crystalline state (definitely arranged pattern). Insulating properties are not affected by this phase change but the material becomes more brittle.

Most ceramic fibers have an Al_2O_3 content from 40% to 60%, and an SiO_2 content from 40% to 60%. Also contained in the fibers are from 1.5% to 7% oxides of sodium, boron, magnesium, calcium, titanium, zirconium, and iron.

Fibers as formed resemble a cotton-like mass with individual fiber length varying from short to 254 mm, and diameters from less than 1 to 10 μm. Larger-diameter fibers are produced for specific applications. In all processes, some unfiberized particles are formed that have diameters up to 40 μm.

Low density, excellent thermal shock resistance, and very low thermal conductivity are the properties of Al_2O_3–SiO_2 fibers that make them an excellent high-temperature insulating material. Available in a variety of forms, ceramic fiber is in ever-increasing demand due to higher and higher temperatures now found in industrial and research processes.

Applications

Ceramic fibers were originally developed for application in insulating jet engines. Now, this is only one of numerous uses for this material. It can be found in aircraft and missile applications where a high-temperature insulating medium is necessary to withstand the searing heat developed by rockets and supersonic aircraft. Employed as a thermal-balance and pressure-distribution material, ceramic fiber in the form of paper has made possible the efficient brazing of metallic honeycomb-sandwich structures.

Successful trials have been conducted in aluminum processing where this versatile product in paper or molded form has been used to transport molten metal with very little heat loss. Such fibrous bodies are particularly useful in these applications because they are not readily wet by molten aluminum.

Industrial furnace manufacturers utilize lightweight ceramic fiber insulation between firebrick and the furnace shell. It is also used for "hot topping," heating element cushions, and as expansion joint packing to reduce heat loss and maintain uniform furnace temperatures.

Use of this new fiber as combustion chamber liners in oil-fired home heating units has materially improved heat-transfer efficiencies. The low heat capacity and light weight, compared to previously used firebrick, improve furnace performance and offer both consumer and manufacturer many benefits.

SiC Fibers

These fibers, capable of withstanding temperatures to about 1200°C, are manufactured from a polymer precursor. The polymer is spun into a fine thread, then pyrolized to form a 15 μm ceramic fiber consisting of fine SiC crystallites and an amorphous phase. An advantage of the process is that it uses technology developed for commercial fiber products such as nylon and polyester. Two commercial SiC fiber products are the Ube Industries Tyranno fiber and the Nippon Carbon Nicalon fiber.

Ceramic Glass Decorations

Ceramic glass enamels fused to glassware at temperatures above 245°C (800°F) to produce a decoration.

Ceramic Glass Enamels

Predominantly colored, silicate glass fluxes used to decorate glassware. Also referred to as ceramic enamels or glass enamels.

Ceramic Molding

A precision casting process that employs permanent patterns and fine-grain slurry for making molds. Unlike monolithic investment molds, which are similar in composition, ceramic molds consist of a *cope* and a *drag* or, if the casting shape permits a drag only.

Ceramic Printed Board

A printed board made from ceramic dielectric and cermet materials.

Ceramic Process

The production of articles or coatings from essentially inorganic, nonmetallic materials, the article or coating being made permanent and suitable for utilitarian and decorative purposes by the action of heat at temperatures sufficient to cause sintering, solid-state reactions, bonding, or conversion partially or wholly to the glassy state.

Ceramic Rod Flame Spraying

A thermal spraying process variation in which the material to be sprayed is a ceramic rod form. See also *flame spraying*.

Ceramic Tools

Cutting tools made from sintered, hot-pressed, or hot isostatically pressed alumina-based or silicon nitride-based ceramic materials. See also *alumina* and *silicon nitride*.

Ceramic Whiteware

A fired ware consisting of glazed or unglazed ceramic body which is commonly white and of fine texture. This term designates such product classifications as tile, china, porcelain, semi-vitreous ware, and earthenware. See also *traditional ceramics*.

Ceramic(s) (Noun)

Any of the class of inorganic nonmetallic products which are subjected to a high temperature during manufacture or use (high temperature usually means a temperature above a barely visible red, approximately 540°C, or 1000°F). Typically, but not exclusively, a ceramic is a metallic oxide, boride, carbide, or nitride, or a mixture or compound of such materials; that is, it includes anions that play important roles in atomic structures and properties.

See also *advanced ceramics*, *electronic ceramics*, *refractories*, *structural ceramics*, and *traditional ceramics*.

Ceramic–Matrix Composite

The class of materials known as ceramic–matrix composites, or CMCs, shows considerable promise for providing fracture-toughness values similar to those for metals such as cast iron. Two kinds of damage-tolerant ceramic–ceramic composites have been developed. One incorporates a continuous reinforcing phase, such as a fiber; the other, a discontinuous reinforcement, such as whiskers. The major difference between the two is in their failure behavior. Continuous-fiber-reinforced materials do not fail catastrophically. After matrix failure, the fiber can still support a load. A fibrous failure is similar to that which occurs in wood.

CMCs are candidate materials for high-performance engines and wear-resistant parts. Interest in ceramic composites has been stimulated by the realization that carbon–carbon composites (CCC) are difficult to protect from oxidation, and that metal–matrix composites (MMCs) have end-use temperature limitations that are below the level needed for engine components.

A wide variety of reinforcing materials, matrices, and corresponding processing methods have been studied. The most successful fiber-reinforced composites have been produced by hot pressing, chemical vapor infiltration, or directed metal oxidation, which is a process that uses accelerated oxidation reactions of molten metals to grow ceramic matrices around preplaced fiber or reinforcement material preforms. Much of the work has been on glass and glass-ceramic matrices reinforced with carbon fibers. Because of the low axial coefficient of thermal expansion (CTE) of carbon fibers and the requirements for CTE matching, the more successful composites have been produced with low CTE matrices, such as borosilicate glass (CTE = 3.5×10^{-6} K^{-1}) and lithium aluminosilicate glass-ceramics (CTE = 1.5×10^{-6} K^{-1}). Other fiber-reinforced ceramic composites include silicon carbide fibers in SiC produced by chemical vapor infiltration and deposition, and SiC fiber reinforced alumina (Al_2O_3) and zirconium diboride (ZrB_2) reinforced zirconium carbide (ZrC) composites produced by directed metal oxidation. Multidirectionally reinforced ceramics, such as fused quartz reinforced silica and Al_2O_3 reinforced silica, have also been produced, the latter material being used for larger radome structures on ballistic missiles.

Incorporating whiskers into a ceramic matrix improves resistance to crack growth, making the composite less sensitive to flaws. These materials are commonly described as being flaw tolerant. However, once a crack begins to propagate, failure is catastrophic.

Composed of fine equiaxed Al_2O_3 grains and needle-like SiC whiskers, SiC_w–Al_2O_3 composites exhibit promising fracture toughness (6.5 MPa$\sqrt{m}$, or 5.9 ksi$\sqrt{in}$.) and strength (600 MPa or 57 ksi) properties. SiC_w–Al_2O_3 composites have been used in cutting-tool applications. Composites with whisker loadings higher than 8 vol% must be hot pressed.

Of particular importance to the technology of toughened ceramics has been the development of high-temperature SiC reinforcements.

SiC Filaments

SiC filaments are prepared by chemical vapor deposition. A thick layer of SiC is deposited on a thin fiber substrate of tungsten or carbon. Diameter of the final product is about 140 μm.

Although developed initially to reinforce aluminum and titanium matrices SiC filaments have since been used as reinforcement in Si_3N_4.

SiC Whiskers

SiC whiskers consist of a fine (0.5–5 μm-diameter) single-crystal structure in lengths to 100 μm. The material is strong (to 15.9 GPa) and is stable at temperatures to 1800°C. Whiskers can be produced by heating SiO_2 and carbon sources with a metal catalyst in the proper environments.

Although these materials are relatively new, at least one successful commercial product is already being marketed. An SiC-whisker-reinforced Al_2O_3 cutting-tool material is being used to machine nickel-based superalloys. In addition, considerable interest has been generated in reinforcing other matrices such as mullite, SiC, and Si_3N_4 for possible applications in automotive and aerospace industries.

The excellent high-temperature strength, oxidation resistance, thermal shock resistance, and fracture toughness of silicon nitride has led to the development of SiC_w reinforced Si_3N_4. The major phase, Si_3N_4 offers many favorable properties, and the SiC whiskers provide significant improvement in the fracture toughness of the composite. Whisker-reinforced Si_3N_4 is a leading candidate material for hot-sections of ceramic-engine components.

DIMOX Process

CMCs are steadily moving from the laboratory to initial commercial applications. For example, engineers are currently evaluating these materials for use in the hot gas zones of gas turbine engines, because ceramics are known for their strength and favorable creep behavior at high temperatures. Advanced ceramics, for example, can potentially be used at temperatures 204°C–482°C above the maximum operating temperature for superalloys.

Until recently, however, there has been more evaluation than implementation of advanced ceramics for various reasons. Monolithic or single-component ceramics, for example, lack the required damage tolerance and toughness. Engine designers are put off by the potential of ceramic material for catastrophic, brittle failures. Although many CMCs have greater toughness, they are also difficult to process by traditional methods, and may not have the needed long-term high-temperature resistance.

A relatively new method for producing CMCs developed by Lanxide Corp. promises to overcome the limitations of other ceramic technologies. Called the DIMOX directed metal oxidation process, it is based on the reaction of a molten metal with an oxidant, usually a gas, to form the ceramic matrix. Unlike the sintering process, in which ceramic powders and fillers are consolidated under heat, directed metal oxidation grows the ceramic matrix material around the reinforcements.

Examples of ceramic matrices that can be produced by the DIMOX directed metal oxidation process include Al_2O_3, Al_2Ti_5, AlN, TiN, ZrN, TiC, and ZrC. Filler materials can be anything chemically compatible with the ceramic, parent metal, and growth atmosphere.

The first step in the process involves making a shaped preform of the filler material. Preforms consisting of particles are fabricated with traditional ceramic-forming techniques, while fiber preforms are made by weaving, braiding, or laying up woven cloth. Next, the preform is put in contact with the parent metal alloy. A gas-permeable growth barrier is applied to the surfaces of this assembly to limit its shape and size.

The assembly, supported in a suitable refractory container, is then heated in a furnace. For aluminum systems, temperatures typically range from 899°C to 1149°C. The parent metal reacts with the surrounding gas atmosphere to grow the ceramic reaction product through and around the filler to form a CMC.

Capillary action within the growing ceramic matrix continues to supply molten alloy to the growth front. There, the reaction continues until the growing matrix reaches the barrier. At this point, growth stops, and the part is cooled to ambient temperature. To recover the part, the growth barrier and any residual parent metal are removed. However, some of the parent metal (5%–15% by volume) remains within the final composite in micron-sized interconnected channels.

Traditional ceramic processes use sintering or hot pressing to make a solid CMC out of ceramic powders and filler. Part size and shapes are limited by equipment size and the shrinkage that occurs during densification of the powders can make sintering unfeasible. Larger parts pose the biggest shrinkage problem. Advantages of the directed metal oxidation process include low shrinkage because matrix formation occurs by a growth process. As a result, tolerance control and large part fabrication can be easier with directed metal oxidation.

In addition, the growth process forms a matrix whose grain boundaries are free of impurities or sintering aids. Traditional methods often incorporate these additives, which reduce high-temperature properties. And cost comparisons show the newer process is a promising replacement for traditional methods.

Ceramic-Metal Coating

A mixture of one or more ceramic materials in combination with a metallic phase applied to a metallic substrate which may or may not require heat treatment prior to service. This term may also be used for coatings applied to nonmetallic substrates, for example, graphite.

Ceramics

Ceramics are inorganic, nonmetallic materials processed or consolidated at high temperature. Ceramics, one of the three major material families, are crystalline compounds of metallic and nonmetallic elements. The ceramic family is large and varied, including such materials as refractories, glass, brick, cement and plaster, abrasives, sanitaryware, dinnerware, artware, porcelain enamel, ferroelectrics, ferrites, and dielectric insulators. There are other materials that, strictly speaking, are not ceramics, but that nevertheless are often included in this family. These are carbon and graphite, mica, and asbestos. Also, intermetallic compounds, such as aluminides and beryllides, which are classified as metals, and cermets, are usually thought of as ceramic materials because of similar physical characteristics to certain ceramics.

Ceramic materials can be subdivided into traditional and advanced ceramics. Traditional ceramics include clay-base materials such as brick, tile, sanitaryware, dinnerware, clay pipe, and electrical porcelain. Common-usage glass, cement, abrasives, and refractories are also important classes of traditional ceramics.

Advanced materials technology is often cited as an "enabling" technology, enabling engineers to design and build advanced systems for applications in fields such as aerospace, automotive, and electronics. Advanced ceramics are tailored to have premium properties through application of advanced materials science and technology to control composition and internal structure. Examples of advanced ceramic materials are Si_3N_4, SiC, toughened ZrO_2, ZrO_2-toughened Al_2O_3, AlN_3, PbMg niobate, PbLa titanate, SiC-whisker-reinforced Al_2O_3, carbon-fiber-reinforced glass ceramic, SiC-fiber-reinforced SiC, and high-temperature superconductors. Advanced ceramics can be viewed as a class of the broader field of advanced materials, which can be divided into ceramics, metals, polymers, composites, and electronic materials. There is considerable overlap among these classes of materials.

Advanced ceramics can be subdivided into structural and electronic ceramics based on primary function or application. Optical and magnetic materials are usually included in the electronic classification. Structural applications include engine components, cutting tools, bearings, valves, wear- and corrosion-resistant parts, heat exchangers, fibers and whiskers, and biological implants. The electronic-magnetic-optic functions include electronic substrates, electronic packages, capacitors, transducers, magnets, waveguides, lamp envelopes, displays, sensors, and ceramic superconductors. Thermal insulation, membranes, and filters are important advanced ceramic product areas that do not fit well into either the structural or the electronic class of advanced ceramics.

Advanced ceramics are differentiated from traditional ceramics such as brick and porcelain by their higher strength, higher operating temperatures, improved toughness, and tailorable properties. Also known as engineered ceramics, these materials are replacing metals in applications where reduced density and higher melting points can increase efficiency and speed of operation. The nature of the bond between ceramic particles helps differentiate engineering ceramics from conventional ceramics. Most particles within an engineering ceramic are self-bonded, that is, joined at grain boundaries by the same energy-equilibrium mechanism that bonds metal grains together. In contrast, most nonengineering ceramic particles are joined by a so-called ceramic bond, which is a weaker, mechanical linking or interlocking of particles. Generally, impurities and nonengineering ceramics prevent the particles from self-bonding.

A broad range of metallic and nonmetallic elements are the primary ingredients in ceramic materials. Some of the common metals are aluminum, silicon, magnesium, beryllium, titanium, and boron. Nonmetallic elements with which they are commonly combined are O_2, carbon, or N_2. Ceramics can be either simple, one-phase materials composed of one compound, or multiphase, consisting of a combination of two or more compounds. Two of the most common are single oxide ceramics, such as alumina (Al_2O_3) and magnesia (MgO), and mixed oxide ceramics, such as cordierite (magnesia alumina silica) and forsterite (magnesia silica). Other newer ceramic compounds include borides, nitrides, carbides, and silicides. Macrostructurally, there are essentially three types of ceramics: crystalline bodies with a glassy matrix; crystalline bodies, sometimes referred to as holocrystalline; and glasses.

The specific gravities of ceramics range roughly from 2 to 3. As a class, ceramics are low-tensile-strength, relatively brittle materials. A few have strengths above 172 MPa, but most have less than that. Ceramics are notable for the wide difference between their tensile and compressive strengths. They are normally much stronger under compressive loading than in tension. It is not unusual for a compressive strength to be 5–10 times that of the tensile strength. Tensile strength varies considerably depending on composition and porosity.

One of the major distinguishing characteristics of ceramics, as compared to metals, is their almost total absence of ductility. They fail in a brittle fashion. Lack of ductility is also reflected in low impact strength, although impact strength depends to a large extent on the shape of the part. Parts with thin or sharp edges or curves and with notches have considerably lower impact resistance than those with thick edges and gently curving contours.

Ceramics are the most rigid of all materials. A majority of them are stiffer than most metals, and the modulus of elasticity in tension of a number of types runs as high as 0.3–0.4 million MPa compared with 0.2 million MPa for steel. In general, they are considerably harder than most other materials, making them especially useful as wear-resistant parts and for abrasives and cutting tools.

Ceramics have the highest known melting points of materials. Hafnium and TaC, for example, have melting points slightly above 3870°C, compared to 3424°C for tungsten. The more conventional ceramic types, such as Al_2O_3, melt at temperatures above 1927°C, which is still considerably higher than the melting point of all commonly used metals. Thermal conductivities of ceramic materials fall between those of metals and polymers. However, thermal conductivity varies widely among ceramics. A two-order magnitude of variation is possible between different types, or even between different grades of the same ceramic. Compared to metals and plastics, the thermal expansion of ceramics is relatively low, although like thermal conductivity it varies widely between different types and grades. Because the compressive strengths of ceramic materials are 5–10 times greater than tensile strength, and because of relatively low heat conductivity, ceramics have fairly low thermal-shock resistance. However, in a number of ceramics, the low thermal expansion coefficient succeeds in counteracting to a considerable degree the effects of thermal conductivity and tensile–compressive-strength differences.

Practically all ceramic materials have excellent chemical resistance, and are relatively inert to all chemicals except hydrofluoric acid and, to some extent, hot caustic solutions. Organic solvents do not affect them. Their high surface hardness tends to prevent breakdown by abrasion, thereby retarding chemical attack. All technical ceramics will withstand prolonged heating at a minimum of 999°C. Therefore, atmospheres, gases, and chemicals cannot penetrate the material surface and produce internal reactions that are normally accelerated by heat.

Unlike metals, ceramics have relatively few free electrons and therefore are essentially nonconductive and considered to be dielectric. In general, dielectrical strengths, which range between 7.8×10^6 and 13.8×10^6 V/m, are lower than those of plastics. Electrical resistivity of many ceramics decreases rather than increases with an increase in impurities, and is markedly affected by temperature.

Fabrication Process

A wide variety of processes are used to fabricate ceramics. The process chosen for a particular product is based on the material, shape, complexity, property requirements, and cost. Ceramic fabrication processes can be divided into four generic categories: powder, vapor, chemical, and melt processes.

Powder Processes

Transitional clay-base ceramics and most refractories are fabricated by powder processes as are the majority of advanced ceramics. Powder processing involves a number of sequential steps. These are preparation of the starting powders, forming the desired shape (green forming), removal of water and organics, heating with or without application of pressure to densify the powder, and finishing.

Vapor Processes

The primary vapor processes used to fabricate ceramics are chemical vapor deposition (CVD) and sputtering. Vapor processes have been finding an increasing number of applications. CVD involves bringing gases containing the atoms to contact the ceramic and then into contact with a heated surface, where the gases react to form a coating. This process is used to apply ceramic coatings to metal and tungsten carbide cutting tools as well as to apply a wide variety of other coatings for wear, electronic, and corrosion applications. CVD can also be used to form monolithic ceramics by building up thick coatings. A form of CVD known as chemical vapor infiltration (CVI) has been developed to infiltrate and coat the surfaces of fibers in woven preforms.

Several variations of sputtering and other vacuum-coating processes can be used to form coatings of ceramic materials. The most common process is reactive sputtering, used to form coatings such as TiN on tool steel.

Chemical Processes

A number of different chemical processes are used to fabricate advanced ceramics. The CVD process described earlier as a vapor process is also a chemical process. Two other chemical processes finding increasing application in advanced ceramics are polymer pyrolysis and sol–gel technology.

Melt-Processes

These are used to manufacture glass, to fuse-cast refractories for use in furnace linings, and to grow single crystals. Thermal spraying can also be classified as a melt process. In this process a plasma-spray gun is used to apply ceramic coatings by melting and spraying powders onto a substrate.

Metal Oxide Ceramics

Although most metals form at least one chemical compound with O_2, only a few oxides are useful as the principal constituent of a ceramic. And of these, only three are used in their fairly pure form as engineering ceramics: Al_2O_3, BeO, and ZrO_2.

The natural alloying element in the Al_2O_3 system is SiO_2. However, Al_2O_3s can be alloyed with chromium (which forms a second phase with the Al_2O_3 and strengthens the ceramic) or with various oxides of silicon, magnesium, or calcium.

Al_2O_3s serve well at temperatures as high as 1925°C provided they are not exposed to thermal shock, impact, or highly corrosive atmospheres. Above 2038°C, strength of Al_2O_3 drops. Consequently, many applications are in steady-state, high-temperature environments, but not where abrupt temperature changes would cause failure from thermal shock. Al_2O_3s have good creep resistance up to about 816°C above which other ceramics perform better. In addition, Al_2O_3s are susceptible to corrosion from strong acids, steam, and sodium. See *Aluminum.*

BeO ceramics are efficient heat dissipaters and excellent electrical insulators. They are used in electrical and electronics applications, such as microelectric substrates, transistor bases, and resistor cores. BeO has excellent thermal shock resistance (some grades can withstand 816°C/s changes), a very low coefficient of thermal expansion (CTE), and a high thermal conductivity. It is expensive, however, and is an allergen to which some persons are sensitive. See *Beryllium.*

ZrO_2 is used primarily for its extreme inertness to most metals. ZrO_2 ceramics retain strength nearly up to their melting point—well over 2205°C, the highest of all ceramics. Applications for fused or sintered ZrO_2 include crucibles and furnace bricks. See *Zirconium.*

Transformation-toughened ZrO_2 ceramics are among the strongest and toughest ceramics made. These materials are of three main types: Mg-PSZ (ZrO_2 partially stabilized with MgO), Y-TZP (Y_2O_3 stabilized tetragonal ZrO_2 poly crystals), and ZTA (ZrO_2-toughened Al_2O_3).

Applications of Mg-PSZ ceramics are principally in low- and moderate-temperature abrasive and corrosive environments—pump and valve parts, seals, bushings, impellers, and knife blades. Y-TZP ceramics (stronger than Mg-PSZ but less flaw tolerant) are used for pump and valve components requiring wear and corrosion resistance in room-temperature service. ZTA ceramics, which have lower density, better thermal shock resistance, and lower cost than the other two, are used in transportation equipment where they need to withstand corrosion, erosion, abrasion, and thermal shock.

Many engineering ceramics have multioxide crystalline phases. An especially useful one is cordierite (MgO–Al_2O_3–SiO_3), which is used in cellular ceramic form as a support for a wash-coat and catalyst in catalytic converters in automobile emission systems. Its low CTE is a necessary property for resistance to thermal fracture.

Glass Ceramics

Glass ceramics are formed from molten glass and subsequently crystallized by heat treatment. They are composed of several oxides that form complex, multiphase microstructures. Glass ceramics do not have the strength-limiting porosity of conventional sintered ceramics. Properties can be tailored by control of the crystalline structure in the host glass matrix. Major applications are cooking vessels, tableware, smooth cooktops, and various technical products such as radomes.

The three common glass ceramics, Li–Al–SiO_3 (LAS, or beta spodumene), Mg–Al–SiO_3 (MAS, or cordierite), and Al–SiO_3 (AS, or aluminous keatite), are stable at high temperatures, have near-zero CTEs, and resist various forms of high-temperature corrosion, especially oxidation. LAS and AS have essentially no measurable thermal expansion up to 427°C. The high SiO_2 content of LAS is responsible for the low thermal expansion, but the SiO_2 also decreases strength. LAS is attacked by sulfur and sodium.

MAS is stronger and more corrosion resistant than LAS. A multiphased version of this material, MAS with $AlTiO_3$, has good corrosion resistance up to 1093°C.

AS, produced by leaching lithium out of LAS particles prior to forming, is both strong and corrosion resistant. It has been used, for example, in an experimental rotating regenerator for a turbine engine.

A proprietary ceramic (**Macor**) called machinable glass ceramic (MGC), is about as strong as Al_2O_3. It also has many of the high-temperature and electrical properties of the glass ceramics. The main virtue of this material is that it can be machined with conventional tools. It is available in bars, or it can be rough-formed, then finish-machined. Machined parts do not require firing.

A similar glass ceramic is based on chemically machinable glass that, in its initial state, is photosensitive. After the glass is sensitized by light to create a pattern, it is chemically machined (etched) to form the desired article. The part can then be used in its glassy state, or it can be fired to convert it to a glass ceramic. This material/process combination is used where precision tolerances are required and where a close match to thermal expansion characteristics of metals is needed. Typical applications are sliders for disk-memory read/write heads, wire guides for dot-matrix printers, cell sheets for gas-discharge displays, and substrates for thick-film and thin-film metallization.

Another ceramic-like material, glass-bonded mica, the moldable/machinable ceramic, is also called a "ceramoplastic" because its properties are similar to those of ceramics, but it can be machined and molded like a plastic material. A glass/mica mixture is pressed into a preform, heated to make the glass flow, then transfer- or compression-molded to the desired shape. The material is also formed into sheets and rods that can be machined with conventional carbide tooling. No firing is required after machining.

The thermal expansion coefficient of glass-bonded mica is close to that of many metals. This property, along with its extremely low shrinkage during molding, allows metal inserts to be molded into the material and also ensures close dimensional tolerances. Molding tolerances as close as ±0.01 mm/mm can be held. Continuous service temperatures for glass-bonded mica range from cryogenic to 371°C or 704°C depending upon material grade.

Aluminum-Ceramic Coatings

These coatings are used to protect aircraft-turbine and other turbomachinery parts from corrosion and heat at temperatures to 2000°F (1093°C) and greater. For compressor applications in ground-based turbines, aluminum-filled, chromate–phosphate coatings sealed with a ceramic topcoat have more than doubled service life. Aluminum-ceramic coatings are also alternatives to cadmium plating of fasteners and other products and used for galvanic protection of dissimilar materials. *Nickel-ceramic coatings*, with silicon carbide or silicon carbide and phosphorus added to the nickel matrix for hardness and hexagonal boron nitride or silicon nitride for lubricity are used in Japan on cylinder bores and pistons of outboard-marine, motorcycle, and snowmobile engines to increase wear resistance. Paintable ceramic coatings, especially of Zyp Coatings, Inc., combined corrosion resistance with heat resistance to 2000°F (1093°C).

Piezoelectric Ceramics

These ceramics produce voltage proportional to applied mechanical force and, conversely, mechanical force when electric voltage is applied. The ceramics are classified as materials into hard, soft, and custom groups. *Lead zirconate titanate* ceramics encompass both "hard" and "soft" groups. The hard, such as PZT-4, 4D, and 8, can withstand high levels of electrical excitation and stress. They are suited for high-voltage or high-powered generators and transducers. The soft, such as PZT-5A, 4B, 5H, 5J, and 5R as well as 7A and 7D, feature greater sensitivity and permittivity. Under high drive conditions, however, they are susceptible to self-heating beyond their operating temperature range. They are used in sensors, low-power motor-type transducers, receivers, low-power generators,

hydrophones, accelerometers, vibration pickups, inkjet printers, and towed array lines. Modified **lead metaniobate**, PN-1 and 2, features higher operating temperatures and is used in accelerometers, flow detectors, and thickness gages. All are available as rods, tubes, disks, plates, rings, and blocks as well as in custom shapes.

Because of their extreme hardness, hot hardness, wear resistance, and chemical inertness, ceramics are used for cutting tools, mainly in the form of inserts fixed to a tool-holder, to increase machining speeds or metal-removal rates, and to enhance machining of certain metals and alloys relative to traditional cutting-tool materials. On the other hand, the materials are more costly and brittle. The most commonly used ceramics for cutting tools are based on alumina or silicon nitride. Various other ceramics are added to the powder mix to enhance sintering mechanical properties, toughness primarily. Principal alumina-based materials, for example, contain titanium carbide, zirconia, or silicon carbide. Other additives include titanium nitride, titanium boride, titanium carbonitride, and zirconium carbonitride. Silicon nitride is generally stronger and tougher than the alumina but alumina, aluminum nitride, or silica is required as a sintering additive to achieve dense material. *SiALONs* consist of various amounts of alumina and silicon nitride, sometimes with zirconia or yttria additives.

Larsenite, of Blasch Precision Ceramics, Inc., is a ceramic composite of alumina and silicon carbide. It is more resistant to thermal shock than alumina and resists oxidation at higher temperatures over 3000°F (1649°C) than the carbide. It is made by firing alumina and a particular grain size of silicon carbide, which then forms a lattice and improves the thermal shock resistance of the alumina. The composite has been used instead of fused silica for nozzles used in atomizing metals into powder. *Sulfide ceramics*, developed at Argonne National Laboratory, hold promise for effective bonding of difficult-to-join materials, such as ceramics to metals. Because they form at lower temperatures than traditional welds, joints are stronger and less brittle. Materials having coefficients of thermal expansion differing by as much as 200% have been joined. The ceramics are candidates for use in lithium-iron sulfide batteries being developed for battery-powered cars.

Ecoceramics is the term given to *silicon carbide* ceramics developed from renewable resources and environmental waste (natural wood and sawdust) at the National Aeronautics and Space Administration Glenn Research Center. Parts are to net shape, pyrolyzed at 1800°F (982°C), and infiltrated with molten silicon or silicon alloys.

Cereal

An organic *binder*, usually cornflower.

Cerium

A chemical element, cerium (Ce) is the most abundant metallic element of the rare earth group in the periodic table. Cerium occurs mixed with other rare earths in many minerals, particularly monazite and blastnasite, and is found among the products of the fission of uranium, thorium, and plutonium.

It is a strong reducing agent used as a getter and thermionic valves and in a number of refining processes usually in the form of *Misch Metal*, and imprecise mixture of about 50% cerium and other rare earths. With 50% iron, it forms a pyrophoric alloy for products such as cigarette lighter "flints." Small additions are made to aluminum and magnesium to improve their strength and ductility and to cast irons to promote the formation of nodular graphite. See *cast iron*.

Ceric oxide, CeO_2, is the oxide usually obtained when cerium salts of volatile acids are heated. CeO_2 is an almost white powder that is insoluble in most acids, although it can be dissolved in sulfuric acid or other acids when a reducing agent is present. The metal is an iron-gray color and it oxidizes rapidly in air, forming a gray crust of oxide. Cerium has the interesting property that, at very low temperatures or when subjected to high pressures, it exhibits a face-centered cubic form, which is diamagnetic and 18% denser than the common form.

Cermets

A composite material made up of ceramic particles (or grains) dispersed in a metal matrix. Particle size is greater than 1 μm, and the volume fraction is over 25% and can go as high as 90%. Bonding between the constituents results from a small amount of mutual or partial solubility. Some systems, however, such as the metal oxides, exhibit poor bonding between phases and require additions to serve as bonding agents. Cermet parts are produced by powder–metallurgy (P/M) techniques. They have a wide range of properties, depending on the composition and relative volumes of the metal and ceramic constituents. Some cermets are also produced by impregnating a porous ceramic structure with a metallic matrix binder. Cermets can also be used in powder form as coatings. The powdered mixture is sprayed through an acetylene flame, and it fuses to the base material.

Although a great variety of cermets have been produced on a small scale, only a few types have significant commercial use. These fall into two main groups: oxide-base and carbide-base cermets. Other important types include the TiC-base cermets, Al_2O_3-base cermets, and UO_2 cermets specially developed for nuclear reactors.

The most common type of oxide-base cermets contains Al_2O_3 ceramic particles (ranging from 30% to 70% volume fraction) and a chromium or chromium-alloy matrix. In general, oxide-base cermets have specific gravities between 4.5 and 9.0, and tensile strengths ranging from 144 to 268 MPa. Their modulus of elasticity ranges between 0.25 and 0.34 million MPa, and their hardness range is A 70 to 90 on the Rockwell scale. The outstanding characteristic of oxide-base cermets is that the metal or ceramic can be either the particle or the matrix constituent. The 6MgO–94Cr cermets reverse the roles of the oxide and chromium, that is, the magnesium is added to improve the fabrication and performance of the chromium. Chromium is not ductile at room temperature. Adding MgO not only permits press-forging at room temperature but also increases oxidation resistance to five times that of pure chromium. Of the cermets, the oxide-base alloys are probably the simplest to fabricate. Normal P/M or ceramic techniques can be used to form shapes, but these materials can also be machined or forged. The oxide-base cermets are used as a tool material for high-speed cutting of difficult-to-machine materials. Other uses include thermocouple-protection tubes. Molten-metal-processing equipment parts, mechanical seals, gas-turbine flame holders (resistance to flame erosion), and flow control pins (because of Cr–Al_2O_3's resistance to wetting and erosion by many molten metals and to thermal shock).

There are three major groups of carbide-base cermets: tungsten, chromium, and titanium. Each of these groups is made up of a variety of compositional types or grades. WC cermets contain up to about 30% cobalt as the matrix binder. They are the heaviest type of cermet (specific gravity is 11–15). Their outstanding properties include high rigidity, compressive strength, hardness, and abrasion resistance. Their modulus of elasticity ranges between 0.45 and 0.65 million MPa, and they have a Rockwell hardness of about A90. Structural uses of WC–Co cermets include wire-drawing dies, precision rolls, gauges, and valve parts. Higher-impact grades can be applied where the steels were formally needed to withstand impact loading. Combined with superior abrasion resistance, the higher

impact strength results in die-life improvements as high, in some cases, as 5000%–7000%. Most TiC cermets have nickel or nickel alloys as the metallic matrix, which results in high-temperature resistance. They have relatively low density combined with high stiffness and strength at temperatures above 1204°C. Typical properties are specific gravity, 5.5–7.3; tensile strength, 517–1068 MPa; modulus of elasticity, 0.25–0.38 million MPa; and Rockwell hardness, A 70 to 90. Typical uses are integral turbine wheels, hot-upsetting anvils, hot-spinning tools, thermocouple protection tubes, gas-turbine nozzle vanes and buckets, torch tips, hot-mill-roll guides, valves, and valve seats. CrC cermets contain from 80% to 90% CrC, with the balance being either nickel or nickel alloys. Their tensile strength is about 241 MPa, and they have a tensile modulus of about 0.34–0.39 million MPa. Their Rockwell hardness is about A88. They have superior resistance to oxidation, excellent corrosion resistance, and relatively low density (specific gravity is 7.0). Their high rigidity and abrasion resistance make them suitable for gauges, oil-well check valves, valve liners, spray nozzles, bearing seal rings, and pump rotors.

Other cermets are barium-carbonate–nickel and tungsten–thoria, which are used in higher-power pulse magnetrons. Some proprietary compositions are used as friction materials. In brake applications, they combine the thermal conductivity and toughness of metals with the hardness and refractory properties of ceramics. UO_2 cermets have been developed for use in nuclear reactors. Cermets play an important role in sandwich-plate fuel elements, and the finished element is a siliconized SiC with a core containing UO_2. Control rods have been fabricated from B_4C-stainless steel and rare earth oxides-stainless steel. Other cermets developed for use in nuclear equipment include Cr–Al_2O_3 cermets, Ni–MgO cermets, and Fe–ZrC cermets. Nonmagnetic compositions can be formulated for use where magnetic materials cannot be tolerated.

Interactions

The reactions taking place between the metallic and ceramic components during fabrication of cermets may be briefly classified and described as follows:

1. Heterogeneous mixtures with no chemical reaction between the components, characterized by a mechanical interlocking of the components without formation of a new phase, no penetration of the metallic component into the ceramic component, and vice versa, and no alteration of either component (e.g., MgO–Ni).
2. Surface reaction resulting in the formation of a new phase as an interfacial layer that is not soluble in the component materials. The thickness of this layer depends on the diffusion rate, temperature, and time of the reaction (e.g., Al_2O_3–Be).
3. Complete reaction between the components, resulting in the formation of a solid solution characterized by a polyatomic structure of the ceramic and the metallic component (e.g., TiC–Ni).
4. Penetration along grain boundaries without the formation of interfacial layers (e.g., Al_2O_3–Mo).

Bonding Behavior

One important factor in the selection of metallic and ceramic components in cermets is their bonding behavior. Bonding maybe by surface interaction or by bulk interaction. In cermets of the oxide-metal type, for example, investigators differentiate among three forms of surface interaction: macrowetting, solid wetting, and wetting assisted by direct lattice fit.

Combinations

One distinguishes basically between four different combinations of metal and ceramic components: (1) the formation of continuous interlocking phases of the metallic and ceramic components, (2) the dispersion of the metallic component in the ceramic matrix, (3) the dispersion of the ceramic component in the metallic matrix, and (4) the interaction between the metallic and ceramic components.

Applications

Aside from the high-temperature applications in turbine buckets, nozzle vanes, and impellers for auxiliary power turbines, there is a wide variety of applications for cermets based on various other properties. One of the most successful applications for the TiC-base cermets is in elements of temperature sensing and controlling thermostats where their oxidation resistance together with their low coefficient of thermal expansion as compared with nickel-base alloys are the important properties. Their ability to be welded directly to the alloys is also important.

The TiC-base cermets are also used for bearings and thrust runners in liquid metal pumps, hot flash trimming and hot spinning tools, hot rod mill guides, antifriction and sleeve-type bearings, hot glass pinch jaws, rotary seals for hot gases, oil well valve balls, etc.

The size of the ceramic component varies, depending on the system and application. It can be as coarse as 50–100 μm, as in some type of cermets based on uranium dioxide (UO_2) that are used for nuclear reactor fuel elements, or as fine as 1–2 μm, as in the micrograin type of cemented carbides. It should be noted, technically, all metal-bonded tungsten carbide materials should fall into the category of cermets. However, it has been customary in the cutting tool industry to designate all cobalt-bonded tungsten carbide compositions as cemented carbides. If the ceramic component is even finer and is present in small amounts, the material is considered a dispersion-strengthened material and therefore falls outside the accepted definition of cermets.

Like cemented carbides, cermets contain a metal binder and are produced by powder metallurgy techniques. The metallic binder phase can consist of a variety of elements, alone or in combination, such as nickel, cobalt, iron, chromium, molybdenum, and tungsten; it can also contain other metals, such as stainless steel, superalloys, titanium, zirconium, or some of the lower-melting-point copper or aluminum alloys. The volume fraction of the binder phase depends entirely on the intended properties and end use of the material. It can range anywhere from 15% to 85%, but it is generally kept at the lower half of the scale (e.g., 10%–15%).

Cermets have proven their value in a variety of applications. The most important use of cermets is in cutting tools based on titanium carbide (TiC) or titanium carbonitride (TiC,N). Steel-bonded carbides consisting of 45 vol% TiC are used in wear-resistant parts and in dies and other forming equipment components. Cermets based on UO_2, as well as those based on uranium carbide (UC), offer potential for advanced fuel elements. Cermets based on zirconium boride (ZrB_2) or silicon carbide (SiC), and others containing alumina (Al_2O_3), silicon dioxide (SiO_2), boron carbide (B_4C), or refractory compounds combined with diamonds, possess unique properties. Several are used commercially in a wide range of applications, including hot-machining tools, shaft seals, valve components and wear parts, ultrahigh-temperature exposed ducts, nozzles, and other rocket engine components, furnace fixtures and hearth elements, grinding wheels, and diamond-containing drill heads and saw teeth.

Cesium

A chemical element, cesium (symbol Cs) is the heaviest of the alkali metals in group 1. It is a soft, light, very low melting temperature metal. It is the most reactive of the alkali metals and indeed is the most electropositive and the most reactive of all the elements.

Cesium oxidizes easily in the air, ignites at ordinary temperatures, and decomposes water with explosive violence. It can be contained in vacuum, inert gas, or anhydrous liquid hydrocarbons protected from O_2 and air. The specific gravity is 1.903, melting point 28.5°C, and boiling point 670°C. It is used in low-voltage tubes to scavenge the last traces of air. It is usually marketed in the form of its compounds such as cesium nitrate, $CsNO_3$, cesium fluoride, CsF, or cesium carbonate, Cs_2CO_3. In the form of cesium chloride, CsCl, it is used on the filaments of radio tubes to increase sensitivity. It interacts with the thorium of the filament to produce positive ions. In photoelectric cells, CsCl is used for a photosensitive deposit on the cathode, since cesium releases its outer electron under the action of ordinary light, and its color sensitivity is higher than that of other alkali metals. The high-voltage rectifying tube for changing AC to DC has cesium metal coated on the nickel cathode, and has cesium vapor for current carrying. The cesium metal gives off a copious flow of electrons and is continuously renewed from the vapor. Cesium vapor is also used in the infrared signaling lamp as it produces infrared waves without visible light. Cesium salts have been used medicinally as antishock agents after administration of arsenic drugs.

Cesium metal is generally made by thermochemical processes. The carbonate can be reduced by metallic magnesium, or the chloride can be reduced by CaC. Metallic cesium volatile lodges from the reaction mixture and is collected by cooling the vapor.

C-Frame Press

Same as a *gap-frame press*.

CFRP

Carbon Fiber Reinforced Plastics.

C-Glass

A glass with a soda-lime-borosilicate composition that is used for its chemical stability and corrosive environments.

Chafing

(1) Repeated rubbing between two solid bodies that can result in surface damage and/or wear. (2) Local abrasion or sometimes an alternative term for fretting.

Chafing Fatigue

Fatigue initiated in a surface damage by rubbing against another body. See also *fretting*.

Chain

(1) A length of interlinked loops intended to carry tensile loads and, sometimes transmit drive between toothed chain wheels. (2) A string of linked atoms forming, usually, an organic molecule.

Chain Intermittent Weld

A weld made intermittently along the two sides of a joint, for example, a T fillet weld, with the welds on the two sides lying opposite each other. Where the welds on one side lie opposite the gaps on the other the joint is termed a Staggered Intermittent Weld. Contrast with *staggered-intermittent fillet welding*.

Chain Length

In plastics, the length of the stretched linear macromolecule, most often expressed by the number of identical links.

Chain Reaction

Any continuing process occurring in repeating steps where each step initiates the next. The term often refers to nuclear reactions where fission of one atom releases neutrons that cause fission in further atoms and so on.

Chain Transfer Agent

In plastics, a molecule from which an atom, such as hydrogen, may be readily abstracted by a free radical.

Chalcogenide

A binary or ternary compound containing a chalcogen (sulfur, selenium, or tellurium) and a more electoral positive element. Ternary molybdenum chalcogenides, $M_x–Mo_6X_8$ where M is a cation and X is a chalcogen, are superconducting materials.

Chalk

A fine-grained limestone, or a soft, earthy form of *calcium carbonate*, $CaCO_3$, composed of finely pulverized marine shells. The natural chalk comes largely from the southern coast of England and the north of France, but high-calcium marbles and limestones are the sources of most U.S. chalk and precipitated calcium carbonate. Chalk is employed in putty, crayons, paints, rubber goods, linoleum, calcimine, and as a mild abrasive in polishes. *Whiting* and *Paris white* are names given to grades of chalk that have been ground and washed for use in paints, inks, and putty. French chalk is a high grade of massive talc cut to shape and used for marking. Chalk should be white, but it may be colored gray or yellowish by impurities. The commercial grades depend on the purity, color, and fineness of the grains. The specific gravity may be as low as 1.8.

Precipitated calcium carbonate is the whitest of the pigment extenders. *Kalite*, of Diamond Alkali Co., is a precipitated calcium carbonate of 39 μin. (1 μm) particle size, and *Suspenso, Surfex*, and *Nonferal* are grades with particle sizes from 197 to 394 μin. (5 to 10 μm). *Whitecarb RC*, of Witco Corp., for rubber compounding, is a fine-grained grade, 2.56 μin. (0.065 μm), coated to prevent dusting and for easy dispersion in the rubber. Purecal SC is a similar material. *Limeolith*, *Calcene*, of PPG Industries, and *Kalvan*,

of R.T. Vanderbilt Co., Inc., are precipitated calcium carbonates. A highly purified calcium carbonate for use in medicine as an antacid is *Amitone*.

Chalking

(1) Dry, chalk-like appearance of deposit on the surface of a plastic. (2) The development of loose removable powder at the surface of an organic coating usually caused by weathering.

Chamber Furnace

Powder metallurgy, a batch sintering furnace usually equipped with a retort that can be sealed gas tight.

Chamfer

(1) A beveled surface to eliminate an otherwise sharp corner. (2) A relieved angular cutting edge at a tooth corner.

Chamfer Angle

(1) The angle between a reference surface and the bevel. (2) On a milling contour, the angle between a beveled surface and the axis of the cutter.

Chamfering

Making a sloping surface on the edge of a member. Also called beveling. See also *bevel angle*.

Channeling

(1) The tendency of a grease or viscous oil to form air channels in a bearing or gear system, resulting in an incomplete lubricant film. (2) The tendency of a grease to form a channel by working down in a bearing or distribution system, leaving shoulders to act as a reservoir and seal.

Channeling Pattern

A pattern of lines observed in a scanning electron image of a single-crystal surface caused by preferential penetration, or channeling, of the incident beam between rows of atoms at certain orientations. The pattern provides information on the structure and orientation of the crystal.

Chaplet

Metal support that holds a core in place within a casting mold; molten metal solidifies around a chaplet and fuses it into the finished casting. They are made of similar metal to the casting.

Characteristic

A property of items in a sample or population that when measured, counted, or otherwise observed helps to distinguish between the items.

Characteristic Electron Energy Loss Phenomena

The inelastic scattering of electrons in solids that produces a discrete energy loss determined by the characteristics of the material. The most probable form is due to excitation of valence electrons.

Characteristic Radiation

Electromagnetic radiation of a particular set of wavelengths, produced by and characteristic of a particular element whenever its excitation potential is exceeded. Electromagnetic radiation is emitted as a result of electron transitions between the various energy levels (electron shells) of atoms; the spectrum consists of lines whose wavelengths depend only on the element concerned and the energy levels involved.

Charcoal

An amorphous form of carbon, made by enclosing billets in a retort and exposing them to a red heat for 4 or 5 h. It is also made by covering large heaps of wood with earth and permitting them to burn slowly for about a month. Much charcoal is now produced as a by-product in the distillation of wood, a retort charge of 10 cords of wood yielding an average of 2,650 gal (10,030 L) of pyroligneous liquor, 11,000 lb (4,950 kg) of gas, and 6 tons (5.4 metric tons) of charcoal. *Wood charcoal* is used as a fuel, for making black gunpowder, for carbonizing steel, and for making activated charcoal for filtering and absorbent purposes. *Gunpowder charcoal* is made from alder, willow, or hazelwood. Commercial wood charcoal is usually about 25% of the original weight of the wood and is not pure carbon. The average composition is 95% carbon and 3% ash. It is an excellent fuel, burning with a glow at low temperatures and with a pale-blue flame at high temperatures. Until about 1850, it was used in blast furnaces for melting iron, and it produces a superior iron with less sulfur and phosphorus than when coke is used. *Red charcoal* is an impure charcoal made at a low temperature that retains much oxygen and hydrogen.

Charge

(1) All the solid materials fed into a furnace. (2) Weights of various liquid and solid materials put into a furnace during one feeding cycle. (3) The weight of plastic material used to load a mold at one time or during one cycle. (4) In the case of smelting processes this includes, ore, flux and (solid) fuel but not liquid fuel or air. In the case of secondary production processes, it includes the components being treated and usually any individual support systems but not usually items such as bogies or removable hearths that are a basic part of the furnace. (5) Any process of adding material to a body or system, of imposing a static electric field or of introducing a quantity of electricity to a battery or similar equipment. See also *static charge*.

Charging

(1) For a lap, impregnating the surface with fine abrasive. (2) Placing materials into a furnace.

Charpy Test

An impact test in which a V-notched, keyhole-notched, or U-notched specimen, supported at both ends, is struck behind the notch by a striker mounted at the lower end of a bar that can swing as a pendulum. The energy that is absorbed in fracture is calculated from the height to which the striker would have risen had there been no specimen and the height to which it actually rises after fracture of the specimen. Contrast with *Izod test* and *impact test*.

Charring

The heating of a reinforced plastic or composite in air to reduce the polymer matrix to ash, allowing the fiber content to be determined by weight.

Chase (Machining)

To make a series of cuts each, except for the first, following in the path of the cut preceding it, as in chasing a thread.

Chase (Plastic)

In plastic part making, an enclosure of any shape, used to shrink-fit parts of a mold cavity in place, prevent spreading or distortion in hobbing, or enclose an assembly of two or more parts in a split cavity block.

Chatter

In machining or grinding, (1) A vibration of the tool, wheel, or workpiece producing a wavy surface on the work and causes noise or local surface damage. More specifically vibration and judder of a cutting tool at the point of metal removal. This causes poor surface quality termed chatter marks. (2) The finish produced by such vibration. (3) In tribology, elastic vibrations resulting from frictional or other instability.

Chatter Marks

Surface imperfections on the work being ground, usually caused by vibrations transferred from the wheel–work interface during grinding.

Check

The intermediate section of a flask that is used between the *cope* and the *drag* when molding a shape that requires more than one parting plane.

Checked Edges

Sawtooth edges seen after hot rolling and/or cold rolling.

Checkers

In a chamber associated with a metallurgical furnace, bricks stacked openly so that heat may be absorbed from the combustion products and later transferred to incoming air when the direction of flow is reversed.

Checking

The development of slight breaks in a coating that do not penetrate to the underlying surface. See also *checks* (1) and *craze cracking*.

Checks

(1) Numerous, very fine cracks in a coating or at the surface of a metal part. Checks may appear during processing or during service and are most often associated with thermal treatment or thermal cycling. (2) The shallow surface cracking, often extensive is the result of thermal shock resulting from the stresses associated with temperature transients. Also called check marks, *checking*, or *heat checks*. (3) Minute cracks in the surface of a casting caused by unequal expansion or contraction during cooling. (4) Cracks in a die impression corner, generally due to forging strains or pressure, localized at some relatively sharp corner. Die blocks too hard for the depth of the die impression have a tendency to check or develop cracks in impression corners. (5) A series of small cracks resulting from thermal fatigue of hot forging dies.

Cheesecloth

A thin, coarse-woven cotton fabric of plain weave, 40–32 count, and of coarse yarns. It was originally used for wrapping cheese, but is now employed for wrapping, lining, interlining, filtering, as a polishing cloth, and as a backing for lining and wrapping papers. The cloth is not sized and may be either bleached or unbleached. It comes usually 36 in. (0.91 m) wide. The grade known as *beef cloth*, originally used for wrapping meats, is also the preferred grade for polishing enameled parts. It is made of No. 22 yarn or finer. For covering meats the packing plants now use a heavily napped knitted fabric known as *stockinett*. It is made either as a flat fabric or in seamless tube form, and it is also used for covering inking and oiling rolls in machinery. Later grades of cheesecloth, with very open weave, known as *gauze*, are used for surgical dressings and for backings for paper and maps. *Baling paper* is made by coating cheesecloth with asphalt and pasting to one side of heavy craft or Manila paper. *Cable paper*, for wrapping cables, is sometimes made in the same way but with insulating varnish instead of asphalt. *Buckram* is a coarse, plain-woven open fabric similar to cheesecloth but heavier and highly sized with water-resistant resins. It is usually made of cotton, but may be of linen, and is white or in plain colors. It is used as a stiffening material, for bookbindings, inner soles, and interlinings. *Cotton bunting* is a thin, soft, flimsy fabric of finer yarn and tighter weave than cheesecloth, used for flags, industrial linings and declarations. It is dyed in solid colors or printed. But usually, the word *bunting* alone refers to a more durable, and nonfading, lightweight, worsted fabric in plain weave.

Chelant/Chelate Corrosion Attack

Corrosion on the water side of steam generating tubes resulting from excessive quantities, generally or locally, of chelating agent in the same manner as caustic attack. Corrosion may also be promoted by excess oxygen, turbulence, and high water velocities.

Chelate

(1) Five- or six-membered ring formation based on intramolecular attraction of H, O, or N atoms. (2) A molecular structure in which a heterocyclic ring can be formed by the unshared electrons of neighboring atoms. (3) A *coordination compound* in which a heterocyclic ring is formed by a metal down to two atoms of the associated *ligand*. See also *complexation*.

Chelating Agent

(1) An organic compound in which atoms form more than one coordinate bond with metals and solution. (2) A substance used in metal finishing to control or eliminate certain metallic ions present in undesirable quantities. Same as *chelant*.

Chelating Agents

Also called chelants and used to capture undesirable metal ions in water solutions, affect their chemical reactivity, dissolve metal compounds, increase color intensity in organic dyes, treat waters and organic acids, and preserve quality of food products and pharmaceuticals. Three major classes of organic chelants are *aminopolycarboxylic acids (APCAs), phosphonic acids*, and *polycarboxylic acids*. The polycarboxylic acids include *citrates*, *gluconates*, *polycrylates*, and *polyaspartates*. APCAs are stable at high temperatures and pH values, have a strong attraction for metals, and are not too costly. Their chelate stability surpasses that of the other two classes; they are useful in most industrial applications including metal cleaning, gas treatment by sulfur removal, and pulp and wood processing. The phosphonic acids are more costly but are stable over a wide range of temperature and pH values. They are used to treat waters to inhibit corrosion of storage vessels and for metals and plastics processing. The polycarboxylic acids are weak and less stable, but inexpensive and useful for alkaline-earth and hardness-ion control. In the United States, the major chelant produces are Dow Chemical, Akzo-Nobel, and BASF. Phosphates have been severely restricted for environmental reasons, especially in household detergents. *EDTA* has been implicated for raising metal concentrations in rivers by remobilizing metals in sludge. *Citrates*, which are biodegradable, are being used increasingly as substitutes for phosphates in liquid laundry detergents. *NTA*, a biodegradable member of EDTA, has largely replaced phosphates and detergents in Canada but is listed as a suspected carcinogen in the United States. *Zeolites*, though not chelants, serve as phosphate substitutes in detergents but are not as effective in removing magnesium. *Polyelectrolytes*, lightweight polymers of *acrylic acid* and *maleic anhydride*, reduce scale formation by dispersing calcium as fine particles.

Chelation

A chemical process involving formation of a heterocyclic ring compound that contains at least one metal cation or hydrogen ion in the ring.

Chemical Adsorption

See *chemisorption*.

Chemical Attack

An imprecise term implying any adverse effect resulting from a chemical reaction between a material and its environment. Such effects include chemical dissolution or combination with a component of the environment resulting in oxidation, scaling, rusting, pitting, etc., the term usually does not include damage mechanisms with a nonchemical component such as stress corrosion cracking or corrosion fatigue cracking.

Chemical Blowing Agent

In processing of plastics, an agent that readily decomposes to produce a gas.

Chemical Bonding

The joining together of atoms to form molecules. See also *molecule* and *interatomic bonding*.

Chemical Cleaning

The removal of surface films, rust and other contamination by the immersion in, or application of, appropriate chemicals.

Chemical Compound

A substance that comprises two or more elements joined in an interatomic bond in fixed weight ratios.

Chemical Conversion Coating

A protective or decorative nonmetallic coating produced *in situ* by chemical reaction of a metal with a chosen environment. It is often used to prepare the surface prior to the application of an organic coating.

Chemical Decomposition

The separating of a compound into its constituents.

Chemical Deposition

(1) The precipitation or plating-out of a metal from solutions of its salts through the introduction of another metal or reagent to the solution. (2) The deposition of a coating on an immersed surface by a chemical reaction between constituents of the solution or between the constituents in the surface.

Chemical Equivalent (of an Element)

The atomic weight divided by the valence.

Chemical Etching

The dissolution of the material of a surface by subjecting it to the corrosive action of an acid or an alkali.

Chemical Flux Cutting

An oxygen cutting process in which metals are severed using a chemical flux to facilitate cutting.

Chemical Indicators

Dyestuffs that have one color in acid solutions and a different color in basic or alkaline solutions. They are used to indicate the relative acidity of chemical solutions, as the different materials have different ranges of action on the acidity scale. The materials are mostly weak acids, but some are weak bases. The best known is *litmus*, which is red below a pH of 4.5 and blue above a pH of 8.3 and is used to test strong acids or alkalies. It is a natural dye prepared from several varieties of lichen, *Variolaria*, chiefly *Rocella tinctoria*, by allowing them to ferment in the presence of ammonia and potassium carbonate. When fermented, the mass has a blue-color and is mixed with chalk and made into tablets of papers. It is used also as a textile dye, wood stain, and food colorant.

Some coal-tar indicators are *malachite green*, which is yellow below a pH of 0.5 and green above 1.5; *phenolphthalein*, which is colorless below 8.3 and magenta above 10.0; and *methyl red*, which is red below 4.4 and yellow above 6.0. A universal indicator is a mixture of a number of indicators that gives a whole range of color changes, thereby indicating the entire pH range. But such indicators must be compared with a standard to determine the pH value.

The change in color is caused by a slight rearrangement of the atoms of the molecule. Some of the indicators, such as *thymol blue*, exhibit two color changes at different acidity ranges because of the presence of more than one chromophore arrangement of atoms. These can thus be used to indicate two separate ranges on the pH scale. Test papers are strips of absorbent paper that have been saturated with an indicator and dried. They are used for testing for acidic or basic solutions, and not for accurate determination of acidity range or hydrogen-ion concentration, such as is possible with direct use of the indicators. *Litmus paper* is used for acidity testing. Starch-iodide paper is paper dipped in starch paste containing potassium iodide. It is used to test for halogens and oxidizing agents such as hydrogen peroxide.

Chemical Lead

Lead, usually as sheet, high purity, 99.9% or better, for chemical reaction vessels or linings.

Chemical Machining

Removing metal stock by controlled selective chemical dissolution.

Chemical Metallurgy

See *process metallurgy*.

Chemical Milled Parts

Chemical milling is the process of producing metal parts to predetermined dimensions by chemical removal of metal from the surface. It is a machining process in which metal is formed into intricate shapes by masking certain portions and then etching away the unwanted material.

Acid or alkaline, pickling, or etching baths have been formulated to remove metal uniformly from surfaces without excessive etching, roughening, or grain boundary attack. Simple immersion of a metal part will result in uniform removal from all surfaces exposed to the chemical solution. Selective milling is accomplished by use of a mask to protect the areas where no metal is to be removed. By such means optimum strength per unit of construction weight is achieved. Nonuniform milling can be done by the protective masking procedure or by programmed withdrawal of the part from the milling bath. Complex milling is done by multiple masking and milling or withdrawal steps.

Versatility Offered

The aircraft industry, as an example, utilizes production chemical milling for weight reduction of large parts by means of precise etching. The process is the most economical means of removal of metal from large areas, nonplanar surfaces, or complex shapes. A further advantage is that metal is just as easily removed from fully hardened as from annealed parts. The advantages of chemical milling result from the fact that metal removal takes place on all surfaces contacted by the etching solution. The solution will easily mill inside and reentrant surfaces as well as thin metal parts or parts that are multiple racked. The method does not require elaborate fixturing or precision setups, and parts are just as easily milled after forming as in the flat. Job lots and salvage are treated, as well as production runs.

Maximum weight reduction is possible through a process of masking, milling, measuring, and remilling with steps repeated as necessary. Planned processing is the key to production of integrally stiffened structures milled so that optimum support of stresses is attained without the use of stiffening by attachment, welding, or riveting.

A level of ability comparable to that required for electroplating is necessary to produce chemically milled parts. Planned processing, solution control, and developed skill in masking and handling of the work are requisite to success. Periods to train personnel, however, are relatively short as compared to training for other precision metal-removal processes.

Tooling requirements are simple. Chemicals, tanks, racks, templates, a hoist, hangers, and a few special hand and measuring tools are required.

Although chemical milling skill can be acquired without extensive training, it is not feasible to expect to produce the extremes of complexity and precision without an accumulation of considerable experience. However, a number of organizations are available that will produce engineering quality parts on a job shop basis. The processes are well established, commercially, either in or out of the plant.

Specific Etchants Needed

It is anticipated that any metal or alloy can be chemically milled. On the other hand, it does take time to develop a specific process and only those metals can be milled for which an etchant has been developed, tested, and made available. Aluminum alloys have been milled for many years. Steel, stainless steel, nickel alloys, titanium alloys, and superalloys have been milled commercially and a great number of other metals and alloys have been milled experimentally or on a small commercial scale.

It is advantageous to be able to mill a metal without changing the heat-treated condition or temper, as can be done chemically. Defective or nonhomogeneous metal, however, can respond unfavorably. Porous castings will develop holes during milling and mechanically or thermally stressed parts will change in shape as stressed metal is removed. Good-quality metal and controlled heat treating, tempering, and stress relieving are essential to uniformity and reproducibility.

Process Characteristics

Almost any metal size or shape can be milled; limitations are imposed only by extreme conditions such as complex shapes with inverted pockets that will trap gases released during milling, or very thin metal foil that is too flimsy to handle. Shapes can be milled that are completely impractical to machine. For example, the inside of a bent pipe could easily be reduced in section by chemical removal of metal. This possibility is used to advantage to reduce weight on many difficult-to-machine areas such as webs of forgings or walls of tubing. Thin sections are produced by milling when alternate machining methods are excessively costly and the optimum in design demands thin metal shapes that are beyond commercial casting, drawing, or extruding capabilities.

Surface roughness is often reduced during milling from a rough-machined, cast, or forged surface to a semimatte finish. The milled finish may vary from about 30 to 250 μin., depending on the original finish, the alloy, and the etchant. In some instances, the production of an attractive finish reduces finishing steps and is a cost advantage. So-called etching that takes place during milling often causes a brightened finish and etchants have been developed that do not result in a loss of mechanical properties.

Complement Machining

Chemical milling has flourished in the aircraft industry where paring away of every ounce of weight is important. It has spread to instrument industries where weight or balance of working parts is important to the forces required to initiate and sustain motion. It has also become a factor in the design of modern weight-limited portable equipment.

A realistic appraisal of the limitations and advantages of the process is essential to optimum designs. The best designs result from complementing mechanical, thermal, and chemical processes. Chemical milling is not a substitute for mechanical methods but, rather, is more likely as an alternative where machining is difficult or economically unfeasible. It does not compete with low-cost, mass production mechanical methods but, rather, is successful where other methods are limited due to the configuration of the part.

Tolerances

It is good design practice to allow a complex shape to be manufactured by the most economical combinations of mechanical and chemical means. To allow this, print tolerances must reflect allowances that are necessary to apply chemical milling. Chemical milling will produce less well-defined cuts, radii, and surface finishes. The tolerance of a milled cut will vary with the depth of the cut. For 2.5 mm cut, a tolerance in depth of cut of ±0.10 m is commercial. This must be allowed in addition to the original sheet tolerance. Line definition (deviation from a straight line) is usually ±0.8 mm. Unmilled lands between two milled areas should be 0.004 mm minimum. Greater precision can be had at a premium price.

In general, milling rates are about 0.0004 mm/s and depth of cut is controlled by the immersion time. Cuts up to 12.7 mm are not unrealistic although costs should be investigated before designs are made that are dependent on deep cuts.

Limitations

There are limitations to the process. Deep cuts on opposite sides of a part should not be taken simultaneously. One side can be milled at a time but it is less costly to design for one cut rather than two. Complex parts can be made by step milling or by programmed removal of parts to produce tapers. In general, step milling is less expensive and more reliable. Chemical milling engineers should be consulted relative to the feasibility and cost of complex design. Very close tolerance parts can be produced by milling, checking, masking, and re-milling but such a multiple-step process could be more costly than machining.

Chemical Polishing

A process that produces a polished surface by the action of a chemical etching solution. The etching solution is compounded so that peaks in the topography of the surface are dissolved preferentially.

The chemical dissolution is performed without the assistance of abrasives or an external electric current.

Chemical Potential

In a thermodynamic system of several constituents, the rate of change of the Gibbs function of the system with respect to the change in the number of moles of a particular constituent.

Chemical Vapor Deposited (CVD) Carbon

Carbon deposited on a substrate by pyrolysis of a hydrocarbon, such as methane.

Chemical Vapor Deposition (CVD)

(1) A coating process, similar to gas carburizing and carbonnitriding, whereby a reactant atmosphere gas is fed into a processing chamber where it decomposes at the surface of the workpiece, liberating one material for either absorption by, or accumulation on, the workpiece. A second material is liberated in gas form and is removed from the processing chamber, along with excess atmosphere gas. (2) Process used in manufacture of several composite reinforcements, especially boron and silicon carbide, in which desired reinforcement material is deposited from vapor phase onto a continuous core, for example, boron on tungsten wire (core).

Chemical Wear

See *corrosive wear.*

Chemically Precipitated Powder

A metal powder that is produced as a fine precipitate by chemical displacement.

Chemically Strengthened

Glass that has been ion-exchanged to produce a compressive stress layer at the treated surface.

Chemisorption

(1) The taking up of a liquid or gas or of a dissolved substance, only one molecular layer in thickness, wherein a new chemical compound or bond is formed between the sorbent surface atoms and those of the sorbate. (2) The binding of an adsorbate to the surface of a solid by forces whose energy levels approximate those of a chemical bond. Contrast with *physisorption.*

Chevron Pattern

A fractographic pattern of radial marks (shear ledges) that look like nested letters "V"; sometimes called a herringbone pattern. Chevron patterns are typically found on brittle fracture surfaces in parts whose widths are considerably greater than their thicknesses. The points of the chevrons can be traced back to the fracture origin.

Chill

(1) A metal or graphite insert embedded in the surface of a casting sand mold or core or placed in a mold cavity to increase the cooling rate at that point. (2) White iron occurring on a gray or ductile iron casting, such as the chill in the wedge test. See also *chilled iron.* Compare with *inverse chill.*

Chill Casting

Any casting technique which promotes rapid cooling when particularly casting into metal molds. See also *splat casting.*

Chill Coating

In casting, applying a coating to a *chill* that forms part of the mold cavity so that the metal does not adhere to it, or applying a special coating to the sand surface of the mold that causes the iron to undercool.

Chill Crystals

The first fine crystals formed on the faces of castings as a result of solidification initiating at the large number of nuclei that result from the local undercooling, i.e., chilling, by the cold mold surface.

Chill Mark

A wrinkled surface condition on glassware resulting from uneven cooling in the forming process.

Chill Plate (in Welding)

A substantial piece of material held in good thermal contact with a more flimsy component to protect that component from overheating during welding.

Chill Ring

See preferred term *backing ring.*

Chill Roll

A cored roll, usually temperature controlled by circulating water, that cools the web before winding. For chill roll plastic (cast) film, the surface of the roll is highly polished. In extrusion coating, either a polished or matte surface may be used, depending on the surface desired on the finish coating.

Chill Roll Extrusion

The extruded plastic film is cooled while being drawn around two or more highly polished chill rolls cored for water cooling for exact temperature control. Also called cast film extrusion.

Chill Time

See preferred term *quenched time.*

Chilled Iron

Cast iron that is poured into a metal mold or against a mold insert so as to cause the rapid solidification that often tends to produce a white iron structure in the casting.

Chin (Ceramics and Glasses)

(1) Area along an edge or corner where the material has broken off. (2) An imperfection due to breakage of a small fragment out of an otherwise regular surface.

China

A glazed or unglazed vitreous ceramic whiteware use for nontechnical purposes. This term designates such products as dinnerware, sanitary ware, and artware when they are vitreous.

Chinese-Script Eutectic

A configuration of eutectic constituents, found particularly in some cast alloys of aluminum containing iron and silicon and in magnesium alloys containing silicon, that resembles the characters in Chinese script.

Chip

(1) Small metal particles cut away during machining. (2) A integrated circuit comprising a series of interconnected electronic devices such as transistors, resistors, etc. A large number of chips are formed together on a wafer, that is, a sheet of, usually, silicon.

Chip Breaker

(1) Notch or groove in the face of a tool parallel to the cutting edge, designed to break the continuity of the chip. (2) A step formed by an adjustable component clamped to the face of the cutting tool. (3) Features in the microstructure, particularly inclusions, which provide a plane of weakness causing breakup of the material removed during machining operations. See *free machining*. (4) Details on the profile of a machining tool that promote material removal as chips rather than continuous strands.

Chipping

(1) Removing seams and other surface imperfections in metals manually with a chisel or gouge, or by a continuous machine, before further processing. (2) Similarly, removing excessive metal.

Chips (Composites)

Minor damage to a protruded surface of a composite material that removes material but does not cause a crack or craze.

Chips (Metals)

Pieces of material removed from a workpiece by cutting tools or by an abrasive medium.

Chisel Steel

Any steel used for chisels and similar tools. Handheld carpenter chisels are typically high carbon steels, larger chisels for more onerous duty may be of various low alloy steels. They will be hardened and tempered as appropriate to the work. See *steel*.

Chitin

A cellulose-like polysaccharide, it holds together the shells of such crustaceans as shrimp, crab, and lobster; and it is also found in insects, mollusks, and even some mushrooms. It ranks after cellulose as natures most abundant polymer. Deacylation of chitin, a poly-*N*-acetyl glucose amine, yields *chitosan*, a cationic electrolyte that finds occasional use as a replacement for some cellulosic materials. Chitosan may serve as a flocculant and wastewater treatment, thickener or extender in foods, coagulant for healing wounds in medicine, and coating for moisture proof films. Chitin is insoluble in most solvents, whereas chitosan, although insoluble in water, organic solvents, and solutions above pH 6.5, it is soluble in most organic acids and dilute mineral acids. For removing heavy metals from wastewater, Manville Corp, immobilizes bacteria on diatomaceous earth and then coats the complex with chitosan; the bacteria degrade organic material, and the chitosan absorbs heavy metals, such as nickel, zinc, chromium, and arsenic.

Chloride of Lime

A white powder, a *calcium chloride hypochlorite*, having a strong chlorite order. It decomposes easily in water and is used as a source of chlorine for cleaning and bleaching. It is produced by passing chlorine gas through slaked lime. Chloride of lime, or *chlorinated lime*, is also known as *bleaching powder*, although commercial bleaching powder may also be a mixture of calcium chloride and calcium hypochlorite, and the term *bleach* is used for many chlorinated compounds.

Chlorinated Hydrocarbons

A large group of materials that have been used as solvents for oils and fats, for metal degreasing, dry cleaning of textiles, as refrigerants, in insecticides and fire extinguishers, and as foam-blowing agents. They range from the gaseous methyl chloride to the solid *hexachloroethane*, with most of them liquid. The increase in the number of chlorine atoms increases the specific gravity, boiling point, and some other properties. They may be divided into four groups: the methane group, including methyl chloride, chloroform, and carbon tetrachloride; the ethylene group, including dichloroethylene; the ethane group, including ethyl chloride and dichloroethane; and the propane group. All of these are toxic, and the fumes are injurious when breathed or absorbed through the skin. Some decompose in light and heat to form more toxic compounds. Some are very inflammable, while others do not support combustion. In general, they are corrosive to metals. Some have been implicated in the depletion of ozone in the stratosphere.

Chloroform, or trichloromethane or methenyl trichloride, is a liquid of composition $CHCl_3$. Used industrially as a solvent for greases and resins and in medicine as an anesthetic. It decomposes easily in the presence of light to form phosgene, and a small amount of ethyl alcohol is added to prevent decomposition. Ethyl chloride is a gas used in making ethyl fuel for gasoline, as a local anesthetic in dentistry, as a catalyst in rubber and plastics processing, and as a refrigerant and household refrigerators. It is marketed compressed into cylinders as a colorless liquid. Its disadvantage as a refrigerant is that it is highly inflammable, and there is no simple test for leaks. *Methyl chloride* is a gas which is compressed into cylinders as a colorless liquid. Methyl chloride is one of the simplest and cheapest chemicals for methylation. In water solution, it is a good solvent. It is also used as a catalyst in rubber processing, as a restraining gas in high-heat thermometers, and as a refrigerant. *Monochlorobenzene* is a colorless liquid, not soluble in water. It is used as a solvent for lacquers and resins, as a heat-transfer medium, and for making other chemicals. *Trichlor cumene*, or *isopropyl trichlorobenzene*, is valued as a hydraulic fluid and dielectric fluid because of its high dielectric strength, low solubility in water, and resistance to oxidation. It is a colorless liquid. *Halane*, used in processing textiles and paper, is a white powder containing 66% available chlorine.

Chlorinated Lubricant

A lubricant containing a chlorine compound that reacts with a rubbing surface at elevated temperatures to protect it from sliding damage. See also *extreme-pressure lubricant*, *sulfochlorinated lubricant*, and *sulfurized lubricant*.

Chlorinated Polyether

Chlorinated polyether is a thermoplastic resin used in the manufacture of process equipment. Chemically, it is a chlorinated polyether of high molecular weight, crystalline in character, and is extremely resistant to thermal degradation at molding and extrusion temperatures. It possesses a unique combination of mechanical, electrical, and chemical-resistant properties, and can be molded in conventional injection and extrusion equipment.

Properties

Chlorinated polyether provides a balance of properties to meet severe operating requirements. It is second only to the fluorocarbons in chemical and heat resistance and is suitable for high-temperature corrosion service.

Mechanical Properties

A major difference between chlorinated polyether and other thermoplastics is its ability to maintain its mechanical strength properties at elevated temperatures. Heat distortion temperatures are above those usually found in thermoplastics and dimensional stability is exceptional even under the adverse conditions found in chemical plant operations. Resistance to creep is significantly high and in sharp contrast to the lower values of other corrosion-resistant thermoplastics. Water absorption is negligible, assuring no change in molded shapes between wet and dry environments.

Chemical Properties

Chlorinated polyether offers resistance to more than 300 chemicals and chemical reagents, at temperatures up to 121°C and higher, depending on environmental conditions. It has a spectrum of corrosion resistance second only to certain of the fluorocarbons.

In steel construction of chemical processing equipment, chlorinated polyether liners or coatings on steel substrates provide the combination of protection against corrosion plus structural strength of metal.

Electrical Properties

Along with the mechanical capabilities and chemical resistance, chlorinated polyether has good dielectric properties. Loss factors are somewhat higher than those of polystyrenes, fluorocarbons, and polyethylenes, but are lower than many other thermoplastics. Dielectric strength is high and electrical values show a high degree of consistency over a range of frequencies and temperatures.

Fabrication

The material is available as a molding powder for injection-molding and extrusion applications. It can also be obtained in stock shapes such as sheet, rods, tubes, or pipe, and blocks for use in lining tanks and other equipment, and for machining gears, plugs, etc. In the form of a finely divided powder, it is used in a variety of different coating processes.

The material can be injection-molded by conventional procedures and equipment. Molding cycles are comparable to those of other thermoplastics. Rods, sheet, tubes, pipe, blocks, and wire coatings can be readily extruded on conventional equipment and by normal production techniques. Parts can be machined from blocks, rods, and tubes on conventional metalworking equipment.

Sheet can be used to convert carbon steel tanks into vessels capable of handling highly corrosive liquids at elevated temperatures. Using a conventional adhesive system and hot gas welding, sheet can be adhered to sandblasted metal surfaces.

Coatings of a finely divided powder can be applied by several coating processes and offer chemical processors an effective and economical means for corrosion control. Using the fluidized bed process, pretreated, preheated metal parts are dipped in an air-suspended bed of finely divided powder to produce coatings, which after baking are tough, pinhole free and highly resistant to abrasion and chemical attack. Parts clad by this process are protected against corrosion both internally and externally.

Uses

Complete anticorrosive systems are available with chlorinated polyether, and lined or coated components, including pipe and fittings, tanks and processing vessels, valves, pumps, and meters.

Rigid uniform pipe extruded from solid material is available in sizes ranging from 12.7 to 50.8 mm in either Schedule 40 or 80, and in lengths up to 6 m. This pipe can be used with injection-molded fittings with socket or threaded connections.

Lined tanks and vessels are useful in obtaining maximum corrosion and abrasion resistance in a broad range of chemical exposure conditions. Storage tanks, as well as processing vessels protected with this impervious barrier, offer a reasonably priced solution to many processing requirements.

A number of valve constructions can be readily obtained from leading valve manufacturers. Solid injection-molded ball valves, coated diaphragm, and plug valves are among the variety available. Also available are diaphragm valves with solid chlorinated polyether bodies.

Chlorinated Polyethylene

This family of elastomers is produced by the random chlorination of high-density polyethylene. Because of the high degree of chemical saturation of the polymer chain, the most desirable properties are obtained by cross-linking with the use of peroxides or by radiation. Sulfur donor cure systems are available that produce vulcanizates with only minor performance losses compared to that of peroxide cures. However, the free radical cross-linking by means of peroxides is most commonly used and permits easy and safe processing, with outstanding shelf stability and optimum cured properties.

Chlorinated polyethylene elastomers are used in automotive hose applications, premium hydraulic hose, chemical hose, tubing, belting, sheet packing, foams, wire and cable, and in a variety of molded products. Properties include excellent ozone and weather resistance, heat resistance to 149°C (to 177°C in many types of oil), dynamic flexing resistance, and good abrasion resistance.

Chlorinated Rubber

An ivory-colored or white powder produced by the reaction of chlorine and rubber. It contains about 67% by weight of rubber, although it is a mixture of two products, one having a CH_2 linkage instead of a CHCl. Chlorinated rubber is used in acid-resistant and corrosion-resistant paints, in adhesives, and in plastics.

The uncompounded film is brittle, and for paints chlorinated rubber is plasticized to produce a hard, tough, adhesive coating, resistant to oils, acids, and alkalies. It is soluble in hydrocarbons, carbon tetrachloride, and esters, but insoluble in water. The unplasticized material has a high dielectric strength, up to 2300 V/mil (90.6 × 10^6 V/m). *Pliofilm*, of Goodyear Tire and Rubber Company, is a rubber hydrochloride made by saturating the rubber molecule with hydrochloric acid. It is made into transparent sheet wrapping material which heat-seals at 221°F–266°F (105°C–130°C), or is used as a coating material for fabrics and paper. It gives a tough, flexible, water-resistant film. *Betacote 95* is a maintenance paint for chemical processing plants which is based on chlorinated rubber. It adheres to metals, cements, and wood and is rapid-drying; the coating is resistant to acids, alkalies, and solvents.

Chlorination

(1) Roasting ore in contact with chlorine or a chloride salt to produce chlorides. (2) Removing dissolved gases and entrapped oxides by passing chlorine gas through molten metal such as aluminum and magnesium. (3) The most common is chlorination of water to kill bacteria. (4) In a metallurgical context, chlorine may be injected into molten magnesium as a deoxidizing and degassing agent.

Chlorine

An elementary material, symbol Cl, which at ordinary temperatures is a gas. It occurs in nature in great abundance and combinations, and in such compounds as common salt. A yellowish-green gas, it has a powerful suffocating odor and is strongly corrosive to organic tissues and to metals. During World War I, it was used as a poison gas under the name *Bertholite*. An important use for liquid chlorine is for bleaching textiles and paper pulp, but is also used for the manufacture of many chemicals. It is a primary raw material for chlorinated hydrocarbons and for such inorganic chemicals as titanium tetrachloride. Chlorine is used extensively for treating potable, process, and wastewaters. Its use as a biocide has declined due to toxicological and safety issues. A key issue is the chlorinated organics, such as *trihalomethanes* (THMs), that form when chlorine reacts with organics and water. One alternative to chlorine biocides for process waters is FMC Corp.'s *tetra alkyl phosphonium chloride*, a strong biocide containing a surface-active agent that cleans surfaces fouled by biofilm. Use of chlorine in fluorocarbons has decreased as chlorofluorocarbons have been replaced with non-ozone-depleting compounds. Its use in *chlorofluorocarbons*, such as CFC-11 and CFC-12, is decreasing, as these are replaced with more environmentally acceptable refrigerants. Chlorine's used in bleach also has declined. For bleaching, it has been widely employed in the form of compounds easily broken up. The other two oxides of chlorine

are also unstable. *Chlorine monoxide*, or *hypochlorous anhydride*, Cl_2O, is a highly explosive gas. *Chlorine heptoxide*, or *perchloric anhydride*, Cl_2O_7, is an explosive liquid. The *chlorinating agents*, therefore, are largely limited to the more stable compounds. Dry chlorines are used in cleansing powders and for detinning steel, where the by-product is tin tetrachloride.

Chloride may be made by the electrolysis of common salt. The gas is an irritant and not a cumulative poison, but breathing large amounts destroys the tissues. Commercial chlorine is produced in making caustic soda, by treatment of salt with nitric acid, and as a by-product in the production of magnesium metal from seawater or brines. The chlorine yield is from 1.8 to 2.7 times the weight of the magnesium produced.

Chlorine Extraction

Removal of phases by formation of a volatile chloride. See also *extraction*.

Chlorofluorocarbon Plastics

Plastics based on polymers made with monomers composed of chlorine, fluorine, and carbon only.

Chlorofluorohydrocarbon Plastics

Plastics based on polymers made with monomers composed of chlorine, fluorine, hydrogen, and carbon only.

Chlorosulfonated Polyethylene

This material, more commonly known as *Hypalon*, can be compounded to have an excellent combination of properties including virtually total resistance to ozone and excellent resistance to abrasion, weather, heat, flame, oxidizing chemicals, and crack growth. In addition, the material has low moisture absorption, good dielectric properties, and can be made in a wide range of colors because it does not require carbon black for reinforcement. Resistance to oil is similar to that of neoprene. Low-temperature flexibility is fair at −40°C.

The material is made by reacting polyethylene with chlorine and SO_2 to yield chlorosulfonated polyethylene. The reaction changes the thermoplastic polyethylene into a synthetic elastomer that can be compounded and vulcanized. The basic polyethylene contributes chemical inertness, resistance to damage by moisture, and good dielectric strength. Inclusion of chlorine in the polymer increases its resistance to flame (makes it self-extinguishing) and contributes to its oil and weather resistance.

Selection

Hypalon is a special-purpose rubber, not particularly recommended for dynamic applications. The elastomer is produced in various types, with generally similar properties. The design engineer can best rely on the rubber formulator to select the appropriate type for a given application, based on the nature of the part, the properties required, the exposure, and the performance necessary for successful use.

In combination with properly selected compounding ingredients, the polymer can be extruded, molded, or calendered. In addition, it can be dissolved to form solutions suitable for protective or decorative coatings.

Initially used in pump and tank linings, tubing, and comparable applications where chemical resistance was of prime importance, this synthetic rubber is now finding many uses where its weatherability, its colorability, its heat, ozone, and abrasion resistance, and its electrical properties are of importance. Included are jacketing and insulation for utility distribution cable, control cable for atomic reactors, automotive primary and ignition wire, and linemen's blankets. Among heavy-duty applications are conveyor belts for high-temperature use and industrial rolls exposed to heat, chemicals, or abrasion.

Interior, exterior, and underhood parts for cars and commercial vehicles are an increasingly important area of use. Representative automotive applications are headliners, window seals, spark plug boots, and tractor seat coverings.

Chlorosulfonated polyethylene is used in a variety of mechanical goods, such as V-belts, motor mounts, O-rings, seals, and gaskets, as well as in consumer products like shoe soles and garden hose. It is also used in white sidewalls on automobile tires. In solution, it is used for fluid-applied roofing systems and pool liners, for masonry coatings, and various protective-coating applications. It can also be extruded as a protective and decorative veneer for such products as sealing and glazing strips.

Chord Modulus

The slope of the chord drawn between any two specific points on a *stress–strain curve*. See also *modulus of elasticity*.

Chromadizing

Improved paint adhesion on aluminum or aluminum alloys or magnesium alloys, mainly aircraft skins, by treatment with a solution of chromic acid. Also called chromidizing or chromatizing. Not to be confused with *chromating* or *chromizing*.

Chromate Treatment

A treatment of metal in a solution of a hexavalent chromium compound to produce a *conversion coating* consisting of trivalent and hexavalent chromium compounds.

Chromatic Aberration

A defect in a lens or optical lens system resulting in different focal lengths for radiation of different wavelengths. The dispersive power of a single positive lens focuses light from the blue end of the spectrum at a shorter distance than from the red end. An image produced by such a lens shows color fringes around the border of the image.

Chromating

The development of a metal/chromium conversion coating by immersion in a solution containing chromium compounds. The coating enhances corrosion resistance and improves paint adhesion. Performing a *chromate treatment*.

Chromatogram

In materials characterization, the visual display of the progress of a separation achieved by *chromatography*. A chromatogram shows the response of a chromatographic detector as a function of time.

Chromatography

The separation, especially of closely related compounds, caused by allowing a solution or mixture to seep through an absorbent (such as clay, gel, or paper) such that each **compound** becomes absorbed in a separate, often colored, layer.

Chrome Plating

(1) Producing a chromate conversion coating on magnesium for temporary protection or for a paint base. (2) The solution heat produces the conversion coating. (3) Is widely used where extreme hardness or resistance to corrosion is required. When plated on a highly polished metal, it gives a smooth surface that has no capillary attraction to water or oil, and chromium-plated bearing surfaces can be run without oil. For decorative purposes, chromium plates as thin as 0.0002 in. (0.0006 cm) may be used.

Chromel

(1) A 90Ni–10Cr alloy used in thermocouples. (2) A series of nickel–chromium alloys, some with iron, used for heat-resistant applications. (3) The most well-known that, with Alumel, is widely used for thermocouples.

Chromia

Formula Cr_2O_3, a compound having many properties and derivatives similar to those of *alumina*. Useful either pure or impure (e.g., as chrome ore) in both basic and high-alumina refractories.

Chromic Acid

A name given to the red, crystalline, strongly acid material of composition CrO_3 known also as *chromium trioxide* or as *chromic anhydride*. It is, in reality, not the acid until dissolved in water, forming a true chromic acid of composition H_2CrO_4. It is marketed in the form of porous lumps. It is produced by treating sodium or potassium dichromate with sulfuric acid. The dust is irritating and the fumes of the solutions are injurious to the nose and throat because the acid is a powerful oxidizing agent. Chromic acid is used in chromium-plating baths, for etching copper, and electric batteries, and in tanning leather.

Chrome oxide green is a *chromic oxide* in the form of dry powder or ground in oil, used in paints and lacquers and for coloring rubber. It is a bright-green crystalline powder of composition Cr_2O_3, and insoluble in water. The dry powder has a Cr_2O_3 content of 97% minimum and is 325 mesh. The paste contains 85% pigment and 15% linseed oil. Chrome oxide green is not as bright in color as chrome green but is more permanent.

Chromium

An elementary metal, chromium (symbol Cr) is used in stainless steels, heat-resistant alloys, high-strength alloy steels, electrical-resistance alloys, wear-resistant and decorative electroplating, and, in its compounds, for pigments, chemicals, and refractories. The specific gravity is 6.92, melting point 1510°C, and boiling point 2200°C. The color is silvery white with a bluish tinge. It is an extremely hard metal; the electrodeposited plates have a hardness of 9 Mohs. It is resistant to oxidation, is inert to HNO_3, but dissolves in HCl and slowly in H_2SO_4. At temperatures above 816°C, it is subject to intergranular corrosion.

Chromium occurs in nature only in combination. Its chief ore is chromite, from which it is obtained by reduction and electrolysis. It is marketed for use principally in the form of master alloys with iron or copper.

Most pure chromium is used for alloying purposes such as the production of Ni–Cr or other nonferrous alloys where the use of the cheaper ferrochrome grades of metal is not possible. In metallurgical operations such as the production of low-alloy and stainless steels, the chromium is added in the form of ferrochrome, an electric-arc furnace product that is the form in which most chromium is consumed.

Uses

Its bright color and resistance to corrosion make chromium highly desirable for plating plumbing fixtures, automobile radiators and bumpers, and other decorative pieces. Unfortunately, chrome plating is difficult and expensive. It must be done by electrolytic reduction of dichromate in H_2SO_4 solution. It is customary, therefore, to first plate the object with copper, then with nickel, and finally, with chromium.

Alloys

In alloys with iron, nickel, and other metals, chromium has many desirable properties. Chrome steel is hard and strong and resists corrosion to a marked degree. Stainless steel contains roughly 18% chromium and 8% nickel. Some chrome steels can be hardened by heat treatment and find use in cutlery; still others are used in jet engines. Nichrome and chromel consist largely of nickel and chromium; they have low electrical conductivity and resist corrosion, even at red heat, so they are used for heating coils in space heaters, toasters, and similar devices. Other important alloys are Hastelloy C—chromium, molybdenum, tungsten, iron, nickel—used in chemical equipment that is in contact with HCl, oxidizing acids, and hypochlorite. Stellite—cobalt, chromium, nickel, carbon, tungsten (or molybdenum)—noted for its hardness and abrasion resistance at high temperatures, is used for lathes and engine valves, and Inconel—chromium, iron, nickel—is used in heat treating and corrosion-resistant equipment in the chemical industry.

Biological Aspects

Chromium is essential to life. A deficiency (in rats and monkeys) has been shown to impair glucose tolerance, decrease glycogen reserve, and inhibit the utilization of amino acids. It has also been found that inclusion of chromium in the diet of humans sometimes, but not always, improves glucose tolerance.

On the other hand, chromates and dichromates are severe irritants to the skin and mucous membranes, so workers who handle large amounts of these materials must be protected against dusts and mists. Continued breathing of the dusts finally leads to ulceration and perforation of the nasal septum. Contact of cuts or abrasions with chromate may lead to serious ulceration. Even on normal skin, dermatitis frequently results.

Chromium Alloys and Steels

Chromium Copper

A name applied to master alloys of copper with chromium used in the foundry for introducing chromium into nonferrous alloys or to copper–chromium alloys, or chromium copper alloys, which are high-copper alloys. A chromium–copper master alloy, Electromet chromium copper, contains 8%–11% chromium, 88%–90% copper, and a maximum of 1% iron and 0.50% silicon.

Wrought chromium copper alloys are designated C18200, C18400, and C18500, and contain 0.4%–1.0% chromium. Soft, thus ductile, in the solution-treated condition, these alloys are readily cold-worked and can be subsequently precipitation-hardened. Depending on such treatments, tensile properties range from 35,000 to 70,000 lb/in.^{2} (241 to 483 MPa) ultimate strength, 15,000 to 62,000 lb/in.^{2} (103 to 427 MPa) yield strength, and 15% to 42% in elongation. Electrical conductivity ranges from 40% to 85% that of copper. Chromium copper alloys are used for resistance-welding electrodes, cable connectors and electrical parts.

Cr–Mo Steel

This is any alloy steel containing chromium and molybdenum as key alloying elements. However, the term usually refers specifically to steels in the American Iron and Steel Institute (AISI) 41XX series, which contain only 0.030%–1.20% chromium and 0.08%–0.35% molybdenum. Chromium imparts oxidation and corrosion resistance, hardenability, and high-temperature strength. Molybdenum also increases strength, controls hardenability, and reduces the tendency to temper embrittlement. AISI 4130 steel, which contains 0.30% carbon, and 4140 (0.40%) are probably the most common and can provide tensile strengths well above 1379 MPa. Many other steels have greater chromium and/or molybdenum content, including high-pressure boiler steels, most tool steels, and stainless steels. Croloy 2, which is used for boiler tubes for high-pressure superheated steam, contains 2% chromium and 0.50% max molybdenum, and is for temperatures to 621°C, and Croloy 5, which has 5% chromium and 0.50% max molybdenum, is for temperatures to 649°C and higher pressures as well as Croloy 7, which has 7% chromium and 0.50% molybdenum.

Chromium Steels

Any steel containing chromium as the predominating alloying element may be termed chromium steel, but the name usually refers to the hard, wear-resisting steels that derive the property chiefly from the chromium content. Straight chromium steels refer to low-alloy steels in the AISI 50XX, 51XX, and 61XX series. Chromium combines with the carbon of steel to form a hard CrC, and it restricts graphitization. When other carbide-forming elements are present, double or complex carbides are formed. Chromium refines the structure, provides deep-hardening, increases the elastic limit, and gives a slight red-hardness so that the steels retain their hardness at more elevated temperatures. Chromium steels have great resistance to wear. They also withstand quenching in oil or water without much deformation. Up to about 2% chromium may be included in tool steels to add hardness, wear resistance, and nondeforming qualities. When the chromium is high, the carbon may be much higher than in ordinary steels without making the steel brittle. Steels with 12%–17% chromium and about 2.5% carbon have remarkable wear-resisting qualities and are used for cold-forming dies for hard metals, for broaches, and for rolls. However, chromium narrows the hardening range of steels unless balanced with nickel. Such steels also work-harden rapidly unless modified with other elements. The high-chromium steels are corrosion resistant and heat resistant but are not to be confused with the high-chromium stainless steels that are low in carbon, although the non-nickel 4XX stainless steels are very definitely chromium steels. Thus, the term is indefinite but may be restricted to the high-chromium steels used for dies, and to those with lower chromium used for wear-resistant parts such as ball bearings.

Chromium steels are not especially corrosion resistant unless the chromium content is at least 4%. Plain chromium steels with more than 10% chromium are corrosion resistant even at elevated temperatures and are in the class of stainless steels, but are difficult to weld because of the formation of hard brittle martensite along the weld.

Chromium steels with about 1% chromium are used for gears, shafts, and bearings. One of the most widely used bearing steels is AISI 52100, which contains 1.3%–1.6% chromium. Many other chromium steels have greater chromium content and, often, appreciable amounts of other alloying elements. They are used mainly for applications requiring corrosion, heat, and/or wear resistance.

Cr–V Steels

Alloy steel containing a small amount of chromium and vanadium, the latter having the effect of intensifying the action of the chromium and the manganese in the steel and controlling grain growth. It also aids in formation of carbides, hardening the alloy, and in increasing ductility by the deoxidizing effect.

The amount of vanadium is usually 0.15%–0.25%. These steels are valued where a combination of strength and ductility is desired. They resemble those with chromium alone, with the advantage of the homogenizing influence of the vanadium. A Cr–V steel with 0.92% chromium, 0.20% vanadium, and 0.25% carbon has a tensile strength up to 689 MPa, and when heat-treated has a strength up to 1034 MPa and elongation 16%. Cr–V steels are used for such parts as crankshafts, propeller shafts, and locomotive frames. High-carbon Cr–V steels are the mild-alloy tool steels of high strength, toughness, and fatigue resistance. The chromium content is usually about 0.80%, with 0.20% vanadium, and with carbon up to 1%.

Chromium–Molybdenum Heat-Resistant Steels

Alloy steels containing 0.5%–9% Cr and 0.5%–1.10% Mo with a carbon content usually below 0.20%. The chromium provides improved oxidation and corrosion resistance, and the molybdenum increases strength at elevated temperatures. Chromium–molybdenum steels are widely used in the oil and gas industries and fossil fuel and nuclear power plants.

Chromizing

(1) A surface treatment at elevated temperature, generally carried out in pack, vapor, or salt baths, in which an alloy is formed by the inward diffusion of chromium into the base metal. (2) Usually at elevated temperature, which causes chromium to diffuse into the surface to improve corrosion resistance.

Chuck

A device for holding work or tools on a machine so that the part can be held or rotated during machining or grinding.

Chucking Hog

A projection forged or cast onto a part to act as a positive means of driving or locating the part during machining.

Chute

In powder metallurgy, a feeding trough for powder to pass from a fill hopper to the die cavity and an automatic press.

CIL Flow Test

A method of determining the rheology or flow properties of thermoplastic resins. In this test, the amount of the molten resin that is forced through a specified size orifice per unit of time when a specified variable force is applied gives a relative indication of the flow properties of various resins.

CIP

The acronym for *cold isostatic pressing.*

Circle Grid

A regular pattern of circles, often 2.5 mm (0.1 in.) in diameter, marked on a sheet metal blank.

Circle Grinding

Either *cylindrical grinding* or *internal grinding*; the preferred terms.

Circle Shear

A shearing machine with two rotary disc cutters mounted on parallel shafts driven in unison and equipped with an attachment for cutting circles where the desired piece of material is inside the circle. It cannot be employed to cut circles where the desired material is outside the circle.

Circle-Grid Analysis

The analysis of deformed circles to determine the severity with which a sheet metal blank has been deformed.

Circuit

(1) In filament winding of composites, one complete traverse of a winding band from one arbitrary point along the winding path to another point on a plane through the starting point and perpendicular to the axis. (2) The interconnection of a number of components in one or more closed pairs to perform a desired electrical or electronic function.

Circuit Board

In electronics, a sheet of insulating material laminated to foil that is etched to produce a circuit pattern on one or both sides. Also called printed circuit board or printed wiring board.

Circuit Breaker

A device designed to open and close a circuit by nonautomatic means and to open the circuit automatically on a predetermined overload of current, without injury to itself, when properly applied within its rating.

Circular Electrode

See *resistance welding electrode.*

Circular Field

The magnetic field that (a) surrounds a nonmagnetic conductor of electricity, (b) is completely contained within a magnetic conductor of electricity, or (c) both exists within and surrounds a magnetic conductor. Generally applied to the magnetic field within any magnetic conductor resulting from a current being passed through the part or through a section of the part. Compare with *bipolar field.*

Circular Mill

A measurement used to determine the area of wire. The area of a circle that is one one-thousandth inch in diameter.

Circular Resistance Seam Welding

See preferred term *transverse resistance seam welding.*

Circular-Step Bearing

A flat circular hydrostatic bearing with a central circular recess. See also *step bearing.*

Circumferential ("Circ") Winding

In filament-wound reinforced plastics, a winding with the filaments essentially perpendicular to the axis (90° or level winding).

Circumferential Resistance Seam Welding

See preferred term *transverse resistance seam welding.*

CIS Stereoisomer

In engineering plastics, a stereoisomer in which side chains or side atoms are arranged on the same side of a double bond present in a chain of atoms.

Citric Acid

$C_6H_8O_7$, produced from lemons, limes, and pineapples, is a colorless, odorless, crystalline powder of specific gravity 1.66 and melting point 307°F (153°C). It is also produced by the fermentation of blackstrap molasses. It is used as an acidulent in effervescent salts in medicine, and jams, jellies, and carbonated beverages in the food industry. *Acetyl tributyl citrate* is a vinyl resin plasticizer. It is also used in inks, etching, and as a resist in textile dyeing and printing. It is a good antioxidant and stabilizer for tableau and other fats and greases, but is poorly soluble in fats. Citric acid is also used as a preservative in frozen fruits to prevent discoloration and storage. Its salt, *sodium citrate*, is a water-soluble crystalline powder used in soft drinks to give a nippy saline taste, and it is also used in plating baths. Citric acid is a strong chelant and finds use in regenerating ion-exchange resins, recovering metals and spent baths, decontaminating radioactive materials, and controlling metal-ion catalysis. For example, it can be used to extract metal contaminants from incinerator ash and to treat uranium-contaminated soils.

Civil Transformation

A transformation from one solid phase to another in which the atoms at the advancing interface realign themselves to the new crystal lattice in a uncoordinated manner and without regard for the original grain boundaries.

Clad

The attachment of sheet to a structural framework.

Clad Brazing Sheet

A metal sheet on which one or both sides are clad with brazing filler metal.

Clad Metals

Cladding means the strong, permanent bonding of two or more metals or alloys metallurgically bonded together to combine the characteristic properties of each in composite form. Copper-clad steel, for example, is used to combine the electrical and thermal characteristics of copper with the strength of steel. A great variety of metals and alloys can be combined in two or more layers, and they are available in many forms, including sheet, strip, plate, tubing, wire, and rivets for application in electrical and electronic products, chemical-processing equipment, and decorative trim, including auto trim.

Cladding Processes

In the process a clad metal sheet is made by bonding or welding a thick facing to a slab of base metal; the composite plate is then rolled to the desired thickness. The relative thickness of the layers does not change during rolling. Cladding thickness is usually specified as a percentage of the total thickness, commonly 10%.

Other cladding techniques, including a vacuum brazing process, have been developed. The pack rolling process is still the most widely used, however.

Alloys

Generally speaking, the choice of alloys used in cladding is dictated by end-use requirements such as corrosion, abrasion, or strength.

Cladding supplies a combination of desired properties not found in any one metal. A base metal can be selected for cost or structural properties, and another metal added for surface protection or some special properties such as electrical conductivity. Thickness of the cladding can be made much heavier and more durable than obtainable by electroplating.

Combinations

The following clad materials are in common use:

- *Stainless steel on steel*: Provides corrosion resistance and attractive surface at low cost for food display cases, chemical-processing equipment, sterilizers, and decorative trim.
- *Stainless steel on copper*: Combines surface protection and high thermal conductivity for pots and pans, and for heat exchangers for chemical processes.

- *Copper on aluminum*: Reduces cost of electrical conductors and saves copper on appliance wiring.
- *Copper on steel*: Adds electrical conductivity and corrosion resistance needed in immersion heaters and electrical switch parts; facilitates soldering.
- *Nickel or Monel on steel*: Provides resistance to corrosion and erosion for furnace parts, blowers, chemical equipment, toys, brush ferrules, and many mechanical parts in industrial and business machines; more durable than electroplating.
- *Titanium on steel*: Supplies high-temperature corrosion resistance. Bonding requires a thin sheet of vanadium between titanium and steel.
- *Bronze on copper*: Usually clad on both sides, for current-carrying springs and switch blades; combines good electrical conductivity and good spring properties.
- *Silver on copper*: Provides oxidation resistance to surface of conductors, for high-frequency electrical coils, conductors, and braiding.
- *Silver on bronze or nickel*: Adds current-carrying capacity to low-conductivity spring material; cladding sometimes is in form of stripes or inlays with silver areas serving as built-in electrical contacts.
- *Gold on copper*: Supplies chemical resistance to a low-cost base metal for chemical processing equipment.
- *Gold on nickel or brass*: Adds chemical resistance to a stronger base metal than copper; also used for jewelry, wristbands, and watchcases.

Applications

Gold-filled jewelry has long been made by the cladding process: the surface is gold, the base metal bronze or brass with the cladding thickness usually 5%. The process is used to add corrosion resistance to steel and to add electrical or thermal conductivity, or good bearing properties, to strong metals. One of the first industrial applications was the use of a nickel-clad steel plate for a railroad tank car to transport caustic soda; stainless clad steels are used for food and pharmaceutical equipment. Corrosion-resistant pure aluminum is clad to a strong duralumin base, and many other combinations of metals are widely used in cladding; there is also a technique for cladding titanium to steel for jet engine parts.

Today's coinage uses clad metals as a replacement for rare silver. Dimes and quarters have been minted from composite sheet consisting of a copper core with Cu–Ni facing. The proportion of core and facing used duplicates the weight and electrical conductivity of silver so the composite coins are acceptable in vending machines.

A three-metal composite sole plate for domestic steam irons provides a thin layer of stainless steel on the outside to resist wear and corrosion. A thick core of aluminum contributes thermal conductivity and reduces weight, and a thin zinc layer on the inside aids in bonding the sole plate to the body of the iron during casting.

Clad metals have been applied in nuclear power reactor pressure vessels in submarines as well as in civilian power plants.

Other applications where use of clad is increasing are in such fields as fertilizer, chemicals, mining, food processing, and even seagoing wine tankers.

Producers see a market for clad metal curtain wall building panels, and even stainless clad bus and automobile bumpers.

Aluminum-clad wire for electric coils is copper wire coated with aluminum to prevent deterioration of the enamel insulation caused by copper oxide. Solder-clad aluminum strip, developed by Heraeus

Holding, GmbH, has soft solder adhesive-bonded to both sides and is intended for heat exchangers and other products.

DuPont cladding has been applied by explosive bonding and Bethlehem Lukens just as well by roll bonding. Clad plate includes carbon, alloy, and stainless-steel plate clad with stainless steel, copper or nickel alloys, titanium, tantalum, or zirconium. Clad-plate transition joints are made by DuPont. Other products include *clad wire, clad rivets*, and *clad welding tapes*.

Composite tool steel, used for shear blades and die parts, is not a laminated metal. The term refers to bar steel machined along the entire length and having an insert of tool steel welded to the backing of mild steel. Clad steels are available regularly in large sheets and plates. They are clad with nickel, stainless steel, Monel metal, aluminum, or special alloys on one or both sides of the sheet. Where heat and pressure are used in the processing, there is chemical bonding between the metals. For some uses the cladding metal on one side will be 10%–20% of the weight of the sheet. A composite plate having an 18-8 stainless steel cladding to a thickness of 20% on one side saves 100 and 144 lb (65 kg) of chromium and 64 lb (29 kg) of nickel per 1000 lb (454 kg) of total plate. The clads may also be extremely thin.

Stainless-clad copper is copper sheet with stainless steel on both sides, used for making cooking utensils and food processing equipment. With stainless steel alone, heat remains localized and causes sticking and burning of foodstuffs. Copper has high heat conductivity, is corroded by some foods, and has an injurious catalytic action on milk products. Thus, the stainless-clad copper gives a conductivity of copper with the protection of stainless steel. The internal layer of copper also makes the metal easier to draw and form.

Cladding

(1) A layer of material, usually metallic, that is mechanically or metallurgically bonded to a substrate. Cladding may be bonded to the substrate by any of several processes, such as roll-cladding and explosive forming. (2) A relatively thick layer (>1 mm, or 0.04 in.) of material applied by surfacing for the purpose of improved corrosion resistance or other properties. (3) The application of a substantial coating of one type of metal to a substrate of another. For example, a strong, corrosion prone aluminum alloy plate can be clad with a corrosion resistant but weak "pure" aluminum. Such cladding is accomplished by rolling either a pair of the materials or a sandwich comprising a slab of the alloy between two slabs of "pure" material. Rolled gold is another example. See also *coating, surfacing*, and *hard facing*.

Clamping Pressure

In injection molding and transfer molding of plastics, the pressure that is applied to the mold to keep it closed in opposition to the fluid pressure of the compressed molding material, within the mold cavity (cavities) and the runner system. In blow molding, the pressure exerted on the two mold halves (by the locking mechanism of the blowing table) to keep the mold closed during formation of the container. Normally, this pressure or force is expressed in tons.

Clamshell Marks

Same as *beach marks*.

Classification

(1) The separation of ores into fractions according to size and specific gravity, generally in accordance with Stokes' law of sedimentation. (2) Separation of a metal powder into fractions according to particle size.

Clay

Clay is composed of naturally occurring sediments that are produced by chemical actions resulting during the weathering of rocks. Often clay is the general term used to identify all earths that form a paste with water and harden when heated. The primary clays are those located in their place of formation. Secondary clays are those that have been moved after formation to other locations by natural forces, such as water, wind, and ice. Most clays are composed chiefly of SiO_2 and Al_2O_3. Clays are used for making pottery, tiles, brick, and pipes, but more particularly the better grade of clays are used for pottery and molded articles not including the fireclays and fine porcelain clays. Kaolins are the purest forms of clay. Clay is a natural mineral aggregate, consisting essentially of hydrous aluminum silicates. It is plastic when sufficiently wetted, rigid when dried en masse, and vitreous when fired to a sufficiently high temperature.

The fineness of the grain of a clay influences not only its plasticity but also such properties as drying performance, drying shrinkage, warping, and tensile, transverse, and bonding strength. For example, the greater the proportion of fine material, the slower the drying rate, the greater the shrinkage, and the greater the tendency to warping and cracking during this stage. Clays with a high fines content usually are mixed with coarser materials to avoid these problems.

For two clays with different degrees of plasticity, the more plastic one will require more water to make it workable, and water loss during drying will be more gradual because of its more extensive capillary system. The high-plasticity clay also will shrink more and will be more likely to crack.

The most important clays in the pottery industry are the ball clays and china clays (kaolin).

Commercial Clay

Commercial clays, or clays utilized as raw material in manufacturing, are among the most important nonmetallic mineral resources. The value of clays is related to their mineralogical and chemical composition, particularly the clay mineral constituents kaolinite, montmorillonite, illite, chlorite, and attapulgite. The presence of minor amounts of mineral or soluble salt impurities in clays can restrict their use. The more common mineral impurities are quartz, mica, carbonates, iron oxides and sulfides, and feldspar. In addition, many clays contain some organic material.

Mining and Processing

Almost all the commercial clays are mined by open-pit methods, with overburden-to-clay ratios ranging as high as 10^{-1}. The overburden is removed by motorized scrapers, bulldozers, shovels, or draglines. The clay is removed with draglines, shovels, or bucket loaders, and transported to the processing plants by truck, rail, aerial trainways, or belt conveyors, or as slurry in pipelines.

The clay is processed dry or, in some cases, wet. The dry process usually consists of crushing, drying, and pulverizing. The clay is crushed to egg or fist size or smaller and dried usually in rotary driers. After drying, it is pulverized to a specified mesh size such as 90% retained on a 200-mesh screen with the largest particle passing a 30-mesh screen. In other cases, the material may have to be pulverized to 99.9% finer than 325 mesh. The material is shipped in bulk or in bags. All clays are produced by this method.

Properties

Most clays become plastic when mixed with varying proportions of water. Plasticity of a material can be defined as the ability of the material to undergo permanent deformation in any direction without rupture under a stress beyond that of elastic yielding. Clays range from those that are very plastic, called fat clay, to those that are barely plastic, called lean clay. The type of clay mineral, particle size and shape, organic matter, soluble salts, adsorbed ions, and the amount and type of nonclay minerals all affect the plastic properties of a clay.

Strength

Green strength and dry strength properties are very important because most structural clay products are handled at least once and must be strong enough to maintain shape. Green strength is the strength of the clay material in the wet, plastic state. Dry strength is the strength of the clay after it has been dried.

Shrinkage

Both drying and firing shrinkages are important properties of clay used for structural clay products. Shrinkage is the loss in volume of a clay when it dries or when it is fired. Drying shrinkage is dependent on the water content, the character of the clay minerals, and the particle size of the constituents. Drying shrinkage is high in most very plastic clays and tends to produce cracking and warping. It is low in sandy clays or clays of low plasticity and tends to produce a weak, porous body.

Color

Color is important in most structural clay products, particularly the maintenance of uniform color. The color of a product is influenced by the state of oxidation of iron, the state of division of the iron minerals, the firing temperature and degree of vitrification, the proportion of Al_2O_3, lime, and MgO in the clay material, and the composition of the fire gases during the burning operation.

Uses

All types of clay and shale are used in the structural products industry but, in general, the clays that are used are considered to be relatively low grade. Clays that are used for conduit tile, glazed tile, and sewer pipe are underclays and shales that contain large proportions of kaolinite and illite.

Clays used for brick and drain tile must be plastic enough to be shaped. In addition, color and vitrification range are very important. For common brick, drain tile, and terra-cotta, shales and surface clays are usually suitable, but for high-quality face bricks, shales and underclays are used.

Clean Surface

A service that is free of foreign material, both visible and invisible.

Cleanup Allowance

See *finish allowance*.

Clearance

(1) The gap or space between two mating parts. (2) Space provided between the relief of a cutting tool and the surface that has been cut.

Clearance Angle

The angle between a plane containing the flank of the tool and a plane passing through the cutting edge in the direction of relative motion between the cutting edge and the work. See also the terms *face mill* and *single-point tool*.

Clearance Fit

Any of various classes of fit between mating parts where there is a positive allowance (gap) between the parts, even when they are made to the respective extremes of individual tolerances that enable the tightest fit between the parts. Contrast with *interference fit*.

Clearance Ratio

In a bearing, the ratio of radial clearance to shaft radius.

Cleavage

(1) Fracture of a crystal by crack propagation across a crystallographic plane of low index or low ductility fracture. (2) The tendency to cleave or split along definite crystallographic planes. (3) Breakage of covalent bonds. (4) The effect can be visualized as layers of atoms peeling apart in contrast to ductile shear failure where the layers slide across one another. The relevant atomic planes are referred to as cleavage planes and the resultant cracks as brittle, flat, or cleavage fractures.

Cleavage Crack (Crystalline)

A crack that proceeds across the grain, that is, a transgranular crack in a single crystal or in a single grain of a polycrystalline material.

Cleavage Crack (Glass)

Damage produced by the translation of a hard, sharp object across a glass surface. This fracture system typically includes a plastically deformed groove on the damage surface, together with median and lateral cracks emanating from this groove.

Cleavage Fracture

A fracture, usually of a polycrystalline metal, in which most of the grains have failed by cleavage, resulting in bright reflecting facets. It is one type of *crystalline fracture* and is associated with low-energy *brittle fracture*. Contrast was *sheer fracture*.

Cleavage Plane

A characteristic crystallographic plane or set of planes in a crystal on which *cleavage fracture* occurs easily.

Cleavage Strength

In testing of adhesive bonded assemblies, the tensile load and terms of kgf/mm (lbf/in.) of width required to cause the separation of a test specimen 25 mm (1 in.) in length.

Clenching/Clinching

The final process of tightening a mechanical joint by tightening a bolt, bending over the projecting point of a nail closing over the shank of a rivet.

Climb Cutting

Analogous to *climb milling*.

Climb Milling

Milling in which the cutter moves in the direction of feed at the point of contact.

Clink, Clinking

The noise of a crack occurring in metals, usually steel, during heating or cooling. The cracks themselves may be termed clinks. They result from restraint of thermal expansion or contraction or, during cooling from hydrogen damage.

Clinker

Generally a fused or partly fused by-product of the combustion of coal as opposed to fine ash but also including lava and portland cement clinker, and partly vitrified slag and brick.

Clip and Shave

In forging, a dual operation in which one cutting surface in the clipping die removes the *flash* and then another shapes and sizes the piece.

Close Packed Hexagonal

A crystal structure in which atoms are close packed, i.e., in contact with six others in the same layer and with three others on the layers above and below, and in which the atoms in alternate layers are aligned. Face-centered cubic, also termed cubic close packed, is similar except that every third layer is aligned.

Close(d) Annealing

Annealing in a sealed box to minimize reaction with air. In the case of steel the term may also imply subcritical annealing as opposed to full annealing. Same as *box annealing*.

Close(d) Joint

A joint in which the component faces to be welded are in contact prior to welding.

Closed Assembly Time

The time interval between completion of assembly of the parts for adhesive bonding and the application of pressure or heat, or both, to the assembly.

Closed Dies

Forging or forming impression dies designed to restrict the flow of metal to the cavity within the die set, as opposed to open dies, in which there is little or no restriction to lateral flow.

Closed Pass

A pass of metal through rolls where the bottom roll has a groove deeper than the bar being rolled and the top roll has a collar fitting into the groove, thus producing the desired shape free from *flash* or fin.

Closed Porosity

The volume fraction of all pores within a solid mass that are closed off by surrounding dense solid, and hence are inaccessible to each other and to the external surface; they thus are not detectable by gas or liquid penetration. In contrast, open pore material allows lubricant to permeate the material.

Closed-Cell Cellular Plastics

Cellular plastics in which almost all the cells are non-interconnecting.

Closed-Die Forging

The shaping of hot metal completely within the walls or cavities of two dies that come together to enclose the workpiece on all sides. The impression for the forging can be entirely either die or divided between the top and bottom dies. Impression-die forgings, often used interchangeably with the term closed-die forging, refers to a closed-die operation in which the dies contain a provision for controlling the flow of excess material, or *flash*, that is generated. By contrast, in flashless forging, the material is deformed in a cavity that allows little or no escape of excess material. See *forge*.

Close-Packed

A geometric arrangement in which a collection of equally sized spheres (atoms) may be packed together in a minimum total volume.

Close-Tolerance Forging

A forging held to unusually close dimensional tolerances so that little or no machining is required after forging. See also *precision forging*.

Closure

In fabricating of reinforced plastics, the complete coverage of a mandrel with one layer (two plies) of fiber. When the last tape circuit that completes mandrel coverage lays down adjacent to the first without gaps or overlaps, the wind pattern is said to have closed.

Cloth (Composites)

See *woven fabric* and *nonwoven fabric*.

Cloth (Powder Metallurgy)

Metallic or nonmetallic screen or fabric used for screening or classifying powders.

Cloud Point

The temperature at which a wax cloud first appears on cooling a mineral oil under specified conditions.

Cloudburst Treatment

A form of *shot peening*.

Cluster Mill

A rolling mill in which each of the two working rolls of small diameter is supported by two or more backup rolls.

CMOD

See crack opening displacement and fracture toughness.

CO_2

Carbon dioxide.

CO_2 Flux Welding Process

See CO_2 welding.

CO_2 Process

See *carbon dioxide process*.

CO_2 Process

A process for producing strong sand molds for casting. The sand is mixed first with a quantity of 3%–5% sodium silicate solution just sufficient to lightly coat the grains but allow normal molding. After the mold is formed CO_2 is blown through to react with the silicate and strongly bind the sand. Also termed the *sodium silicate* of the *silicate process*.

CO_2 Welding

See preferred term *gas metal arc welding*.

CO_2 Welding

A form of metal inert gas (MIG) welding or gas metal arc welding (GMAW) in which the electric arc is struck between the component to be welded and a continuously fed bare wire filler electrode. Carbon dioxide is delivered, usually via the wire feed system, to provide a protective gas shroud for the weld zone. The process is commonly used for steel with which CO_2 is effectively inert. CO_2 Flux Welding is similar except that the consumable electrode is a flux coated wire or a flux filled tube.

Coagulation

Precipitation of a polymer dispersed in a latex.

Coal

A general name for a black mineral formed of ancient vegetable matter, and employed as a fuel and for destructive distillation to obtain gas, coke, oils, and coal-tar chemicals. Coal is composed largely of carbon with smaller amounts of hydrogen, nitrogen, oxygen, and sulfur. It was formed in various geological ages and under varying conditions, and it occurs in several distinct forms. *Peat* is the first stage, followed by lignite, bituminous coal, and anthracite, with various intermediate grades. The mineral is widely distributed in many parts of the world. The value of coal for combustion purposes is judged by its fixed carbon content, volatile matter, and lack of ash. It is also graded by the size and percentage of lumps. The percentage of volatile matter declines from peat to anthracite, and the fixed carbon increases. A good grade of coal for industrial powerplant use should contain 55%–60% fixed carbon and not exceed 8% ash. Finely ground coal, or *powdered coal*, is used for burning in an air blast-like oil, or it may be mixed with oil. Coal in its natural state absorbs large amounts of water and also, because of the impurities in irregular sizes, is not so efficient a fuel as the *reconstructed coal* made by crushing and briquetting lignite or coal and waterproofing with a coating of pitch. *Anthracite powder* is used as a filler in plastics.

Low-sulfur coal burns cleaner than regular coal, but its heating value is much less so that it is uneconomical as a fuel. Increasing amounts of coal are being used for production of gas and chemicals. By the hydrogeneration of coal much greater quantities of phenols, cresols, aniline, and nitrogen-bearing amines can be obtained than by means of by-product coking, and low grades of coal can be used. The finely crushed coal is slurred to a paste with oil, mixed with a catalyst, and reacted at high temperature and pressure. *Synthesis gas*, used for producing gasoline and chemicals, is essentially a mixture of carbon monoxide and hydrogen. It is made from low-grade coals.

Coalesced Copper

Massive oxygen-free copper made by briquetting ground, brittle cathode copper, then sintering the briquettes in a pressurized reducing atmosphere, followed by hot working.

Coalescence

(1) The union of particles of a dispersed phase into larger units, usually affected at temperatures below the fusion point. (2) Growth of grains at the expense of the remainder by absorption or the growth of a phase or particle at the expense of the remainder by absorption or reprecipitation. (3) Examples include the coalescence of voids during creep or the coalescence of carbides in steel where a large number of fine particles can, by a process of diffusion, coalesced to a small number of coarse particles. (4) The bonding that results when powders are sintered. (5) In the context of welded or brazed joints, the term indicates a satisfactory bond between components.

Coarse Fraction

The large particles in a metal powder spectrum.

Coarse Grains

Grains larger than normal for the particular wrought metal or alloy or of a size that produces a surface roughening known as *orange peel* or alligator skin.

Coarsening

An increase in grain size, usually, but not necessarily, by *grain growth.*

Coarsening

The increase in grain size by the process of grain growth or, in steel, the increase in size, with associated reduction in number, of carbide particles or pearlite plates. See *steel*.

Coated Abrasive

An abrasive product (sandpaper, for example) in which a layer of abrasive particles is firmly attached to a paper, cloth, or fiber backing by means of glue or synthetic-resin adhesive.

Coated Electrode

See preferred term *covered electrode* and *lightly coated electrode*.

Coated Fabrics

The first coated fabric was a rubberized fabric produced in Scotland by Charles Macintosh in 1823 and known as *Macintosh cloth* for rainwear use. The cloth was made by coating two layers of fabric with rubber dissolved in naphtha and pressing them together, making a double fabric impervious to water. *Rubberized fabrics* are made by coating fabrics, usually cotton, with compounded rubber and passing between rollers under pressure. The vulcanized coating may be no more than 0.003 in. (0.008 cm) thick, and the resultant fabric is flexible and waterproof. But most coated fabrics are now made with synthetic rubbers or plastics, and the base fabric may be of synthetic fibers, or a thin plastic film may be laminated to the fabric.

Coated fabrics now have many uses in industrial applications, and the number of variations with different resins and backing materials is infinite. They are usually sold under trade names and are used for upholstery, linings, rainwear, bag covers, book covers, tarpaulins, outerwear, wall coverings, window shades, gaskets, and diaphragms. Vinyl-type resins are most commonly used, but for special purposes other resins are selected to give resistance to wear, oils, or chemicals.

Vinyl-coated fabrics are usually tough and elastic and are low cost, but unless specially compounded are not durable. Many plastics in the form of latex or emulsion are marketed especially for coating textiles. Water dispersions of acrylic resins are specifically made for this purpose. Coatings cure at room temperature, and have high heat and light stability, give softness flexibility to the fabric, and withstand repeated dry-cleaning. A water emulsion of a copolymer of vinyl pyrrolidone with ethyl acrylate forms an adherent, tough, and chemical-resistant coating. *Geon latex* is a water dispersion of polyvinyl chloride resin. Polyvinyl chloride of high molecular weight is resistant to staining, abrasion, and tearing and is used for upholstery fabrics. The base cloth may be of various weights from light sheetings to heavy ducks. They may be embossed with designs to imitate leather.

One of the first upholstery fabrics to replace leather was *Fabrikoid*, of DuPont. It was coated with a cellulose plastic and came in various weights, colors, and designs, especially for automobile seating and book covers. *Armalon* is twill or sateen fabric coated with ethylene plastic for upholstery. For some uses, such as for draperies or industrial fabrics, the fabric is not actually coated, but is impregnated, either in the fiber or in the finished cloth, to make it water-repellent, immune to insect attack, and easily cleaned.

Impregnated fabrics may have only a thin, almost undetectable surface coating on the fibers to make them water-repellent and immune to bacterial attack, or they may be treated with fungicides or with flame-resistant chemicals or waterproofing resins. *Stabilized fabrics*, however are not waterproofed or coated, but are fabrics of cotton, linen, or wool that have been treated with a water solution of urea formaldehyde or other thermosetting resin to give them greater resiliency with resistance to creasing and resistance to shrinking and washing. *Shrinkproof fabrics* are likewise not coated fabrics, but have a light impregnation of resins that usually remains only in the core of the fibers. The fabric retains its softness, texture, and appearance, but the fibers have increased stability. Various resin materials are marketed under trade names for creaseproofing and shrinkproofing fabrics.

Under the general name of protective fabrics, coated fabrics are now marketed by use characteristics rather than by coating designation since resin formulations vary greatly in quality. For example, the low-cost grades of vinyl resins may be hard and brittle at low temperatures and soft and rubbery in hot weather, and thus unsuitable for all-weather tarpaulins. Special weaves of fabrics are used to give a high tear strength with light weight, and the plastic may be impregnated, coated on one side or both, bonded with an adhesive or electronically bonded, or some combination of all these. Flame resistance and static-free qualities may also be needed. Many companies have complete lines to meet definite needs.

Coating

(1) A relatively thin layer (<1 mm, or 0.04 in.) of material applied by surfacing for the purpose of corrosion prevention, resistance to high-temperature scaling, wear resistance, lubrication, or other purposes. (2) Any process of applying a surface layer to a component, or the layer itself. The term is sometimes used in contrast to cladding to imply a relatively thin surface layer, less than about 1 mm. (3) The layer of flux and other materials on some welding electrodes.

Coating Density

The ratio of the determined density of a coating to the theoretical density of the material used in the coating process. Usually expressed as percent of theoretical density.

Coating Strength

A measure of the cohesive bond within a coating, as opposed to coating-to-coating substrate bond; the tensile strength of a coating.

Coating Stress

The stresses in a coating resulting from rapid cooling of molten or semi-molten particles as they impact the substrate.

Coatings

Plastic, metal, or ceramic coatings can be applied to the surface of a material in a variety of ways to achieve desired properties.

Coatings improve adherence, corrosion resistance, abrasion resistance, and electrical or optical properties. They can be applied by wet or dry techniques, with simple or complex equipment.

The choices are almost limitless because almost any coating material offers some degree of protection as long as it retains its integrity. If it provides a continuous barrier between the substrate and the environment, even a thin, decorative coating can do the job in a relatively dry and mild environment.

Metal Coatings

Many new materials have been developed, but steel remains the principal construction material for automobiles, appliances, and industrial machinery. Because of the vulnerability of steel to attack by aggressive chemical environments or even from simple atmospheric oxidation, coatings are necessary to provide various degrees of protection. They range from hot-dipped and electroplated metals to tough polymers and flame-sprayed ceramics.

In general, corrosive environments contain more than one active material, and the coating must resist penetration by a combination of oxidizers, solvents, or both. Thus, the best barrier is one that resists "broadband" corrosion.

Physical integrity of the coating is as important as its chemical barrier properties in many applications. For instance, coatings on impellers that mix abrasive slurries can be abraded quickly; coatings on pipe joints will cold-flow away from a loaded area if the creep rate is not low; and coatings on flanges and support brackets can be chipped or penetrated during assembly if impact strength is inadequate. Selecting the best coating for an application requires evaluating all effects of the specific environment, including thermal and mechanical conditions.

Zinc

One of the most common and inexpensive protection methods for steel is provided by zinc. Zinc-coated, or galvanized, steel is produced by various hot-dipping techniques, but more steel companies have moved into electrogalvanizing so they can provide both.

Oxidation protection of steel by zinc operates in two ways—first as a barrier coating, then as a sacrificial coating. If the zinc coating is scratched or penetrated, it continues to provide protection by galvanic action until the zinc layer is depleted. This sacrificial action also prevents corrosion around punched holes and at cut edges.

The grades of zinc-coated steel commercialized in recent years have been designed to overcome the drawbacks of traditional galvanized steel, which has been difficult to weld and to paint to a smooth finish. The newer materials are intended specifically for stamped automotive components, which are usually joined by spot welding and which require a smooth, Class A painted finish.

Aluminum

Two types of aluminum-coated steel are produced, each a different kind of corrosion protection. Type I has a hot-dipped Al–Si coating to provide resistance to both heat and corrosion. Type 2 has a hot-dipped coating of commercially pure aluminum, which provides excellent durability and protection from atmospheric corrosion. Both grades are usually used unpainted.

Type I aluminum-coated steel resist heat scaling to 677°C and has excellent heat reflectivity to 482°C. Nominal aluminum-alloy coating is about 1 mil on each side. The sheet is supplied with a soft, satiny finish. Typical applications include reflectors and housings for industrial heater panels, interior panels and heat exchangers for residential furnaces, microwave ovens, automobile and truck muffler systems, heat shields for catalytic converters, and pollution-control equipment.

C

Type 2 aluminized steel, with an aluminum coating of about 1.5 mil on each side, resists atmospheric corrosion and is claimed to outlast zinc-coated sheet in industrial environments by as much as 5–1. Typical applications are industrial and commercial roofing and siding, drying ovens, silo roofs, and housings for outdoor lighting fixtures and air-conditioners.

Electroplating

Use of protective electroplated metals has changed in recent years, mainly because of rulings by the Environmental Protection Agency (EPA). Cyanide plating solutions and cadmium and lead-bearing finishes are severely restricted or banned entirely. Chromium and nickel platings are much in use, however, applied both by conventional electroplating techniques and by new, more efficient methods such as fast rate electrodeposition (FRED), which has also been used successfully to deposit stainless steel on ferrous substrates.

Functional chromium, or "hard chrome," plating is used for anti-galling and low-friction characteristics as well as for corrosion protection. These platings are usually applied without copper or nickel underplates in thicknesses from about 0.3 to 2 mil. Hard-chrome plating is recommended for use in saline environments to protect ferrous components.

Nickel platings, in thickness from 0.12 to 3 mil, are used in food-handling equipment, on wear surfaces in packaging machinery, and for cladding in reaction vessels.

Electroless nickel plating, in contrast to conventional electroplating, operates chemically instead of using an electric current to deposit metal. The electroless process deposits a uniform coating regardless of substrate shape, overcoming a major drawback of electroplating—the difficulty of uniformly plating irregularly shaped components. Conforming anodes and complex fixturing are unnecessary in the electroless process. Deposit thickness is controlled simply by controlling immersion time. The deposition process is autocatalytic, producing thicknesses from 0.1 to 5 mil.

Proprietary electroless-plating systems contain, in addition to nickel, elements such as phosphorous, boron, and/or thallium. A relatively new composition, called the polyalloy, features three or four elements in the bath. These products are claimed to provide superior wear resistance, hardness, and other properties, compared with those of generic electroless-plating methods.

One polyalloy contains nickel, thallium, and boron. Originally developed for aircraft gas turbine engines, it offers excellent wear resistance. Comparative tests show that relative wear for a polyalloy-coated part is significantly less than that for hard chromium and Ni–P coatings.

In general, Ni–B coatings are nodular. As coating thickness increases, nodule size also increases. Because the columnar structure of the coating flexes as the substrate moves, Ni–B resists chipping and wear.

Adhesion quality depends on factors such as substrate material, part preparation, and contamination. Although it is excellent for tool steels, stainless steel, high-performance nickel- and cobalt-base alloys, and titanium, a few metal substrates are not compatible. These include metals with high zinc or molybdenum content, aluminum, magnesium, and tungsten carbide. Modifications can, however, eliminate this incompatibility.

Another trend in composite electroless plating appears to be toward codeposition of particulate matter within a metal matrix. These coatings are commercially available with just a few types of particulates—diamond, SiC, Al_2O_3, and polytetrafluoroethylene (PTFE)—with diamond heading the list in popularity.

These coatings can be applied to most metals, including iron, carbon steel, cast iron, aluminum alloys, copper, brass, bronze, stainless steel, and high-alloy steels.

Conversion Coatings

Electroless platings are more accurately described as conversion coatings, because they produce a protective layer or film on the metal surface by means of a chemical reaction. Another conversion process, the black oxide finish, has been making progress in applications ranging from fasteners to aerospace. Black oxide is gaining in popularity because it provides corrosion resistance and aesthetic appeal without changing part dimensions.

On a chemical level, black oxiding occurs when the Fe within the surface of the steel reacts to form magnetite (Fe_3O_4). Processors use inorganic blackening solutions to produce the reaction. Oxidizing salts are first dissolved in water, then boiled and held at 138°C–140°C. The product surface is cleaned in an alkaline soak and then rinsed before immersion in the blackening solution. After a second rinse, the finish is sealed with rust preventatives, which can produce finishes that vary from slightly oily to hard and dry.

Black oxiding produces a microporous surface that readily bonds with a topcoat. For example, a supplemental oil topcoat can be added to boost salt-spray resistance to the same level as that of zinc plate with a clear chrome coating (100–200 h).

Black oxide can be used with mild steel, stainless steel, brass, bronze, and copper. As long as parts are scale free and do not require pickling, the finish will not produce hydrogen embrittlement or change part dimensions. Operating temperatures range from cryogenic to 538°C.

Chromate conversion coatings are formed by the chemical reaction that takes place when certain metals are brought in contact with acidified aqueous solutions containing basically water-soluble chromium compounds in addition to other active radicals. Although the majority of the coatings are formed by simple immersion, a similar type of coating can be formed by an electrolytic method.

Protective chromate conversion coatings are available for zinc and zinc alloys, cadmium, aluminum and aluminum alloys, copper and copper alloys, silver, magnesium and magnesium alloys. The appearance and protective value of the coatings depends on the base metal and on the treatment used.

Chromate conversion coatings both protect metals against corrosion and provide decorative appeal. They also have the characteristics of low electrical resistance, excellent bonding characteristics with organic finishes, and can be applied easily and economically. For these reasons the coatings have developed rapidly, and they are now one of the most commonly used finishing systems. They are particularly applicable where metal is subjected to storage environments such as high humidity, salt, and marine conditions.

The greatest majority of chromate conversion coatings are supplied as proprietary materials and processes. These are available usually as liquid concentrates or powdered compounds that are mixed with water. In the case of the powdered compounds, they are often adjusted with additions of acid for normal operation.

Chromate conversion coatings are formed immersing the metal in an aqueous acidified chromate solution consisting substantially of chromic acid or water-soluble salts of H_2CrO_3 together with various catalysts or activators. The chromate solutions, which contain either organic or inorganic active radicals or both, must be acid and must be operated within a prescribed pH range.

Maximum corrosion protection is obtained by using drab or dark bronze coatings on zinc and cadmium surfaces, and yellow to brown-colored coatings on the other metals. Lighter iridescent yellow type coatings generally provide medium protection, and the clear-bright type coatings, produced either in one dip or by leaching, provide the least protection.

Chromate conversion coatings provide maximum corrosion protection in salt spray or marine types of environment, and in high humidity such as encountered in storage, particularly where stale air with entrapped water may be present. They also provide excellent protection against tarnishing, staining, and finger marking, or other conditions that normally produce surface oxidation.

Olive drab type coatings are widely used on military equipment because of their high degree of corrosion protection coupled with a nonreflective surface. Iridescent yellow coatings are widely used for corrosion protection where appearance is not a deciding factor. The clear-bright chemically polishing type coatings for zinc and cadmium have been widely used to simulate nickel and chromium electroplate and are primarily used for decorative appeal rather than corrosion protection. Where additional corrosion protection or abrasion resistance is desired, these clear coatings act as an excellent base for a subsequent clear organic finish.

Heavy olive drab and yellow coatings for zinc, cadmium, and aluminum can be dyed various colors. Generally speaking, the dyed colors are used for identification purposes only since they are not lightfast and will fade upon exposure to direct sunlight or other sources of ultraviolet.

Because of their low electrical resistance, chromate conversion coatings are widely used for electronics equipment. Surface resistance depends on the type and thickness of the film deposited, the pressure exerted at the contact, and the nature of the contact. Low-resistance coatings are particularly important on aluminum, silver, magnesium, and copper surfaces.

Chromate conversion coatings can also be soldered and welded. A chromate coating on aluminum, for example, facilitates heliac welding. Because of the slight increase in electrical resistance, an adjustment in current (depending upon the thickness of the coating) must be made to satisfactorily spot-weld. Soldering, using rosin fluxes, can be performed on cadmium-plated surfaces that have been treated with clear bright chromate conversion coatings. Clear, bright coatings on zinc-plate surfaces and colored coatings on both zinc and cadmium necessitate the use of an acid flux or removal of the film by an increase in soldering iron temperature, which burns through the coating, or by mechanical abrasion, which removes the film and provides a clean metal surface for the soldered joint.

Most chromate conversion treatments are applied by simple immersion in an acidified chromate solution. Because no electrical contacts need be made during immersion, the coatings can be applied by rack, bulk, or strip line operation. Under special situations, swabbing or brush coating can be used where small areas must be coated, as in a touch-up operation.

Chromate conversion coatings can also be applied by an electrolytic method in which the electrolyte is composed essentially of water-soluble chromium compounds and other radicals operated at neutral or slightly alkaline pH. This type of application is limited primarily to rack-type operation.

In general, processing can be placed in two categories: (1) over freshly electroplated surfaces; and (2) over electroplated surfaces that have been aged or oxidized, or other metal surfaces such as zinc diecastings, wrought metals, or hot-dipped surfaces.

Sputtering

Formally used primarily to produce integrated-circuit components, sputtering has moved on to large, production-line jobs such as "plating" of automotive trim parts. The process deposits thin, adherent films, usually of metal, in a plasma environment on virtually any substrate.

Sputtering offers several advantages to automotive manufacturers for an economical replacement for conventional chrome plating. Sputtering lines are less expensive to set up and operate than plating systems. And because sputtered coatings are uniform as well as thin, less coating material is required to produce an acceptable finish. Also, pollution controls are unnecessary because the process does not produce any effluents. Finally, sputtering requires less energy than conventional plating systems.

Chrome coating of plastics and metals is only one application for sputtering. The technique is not limited to depositing metal films. PTFE has successfully been sputtered on metal, glass, paper, and wood surfaces. In another application, cattle bone was sputtered on metallic prosthetic devices for use as hip-bone replacements. The sputtered bone film promotes bone growth and attachment to living bone.

Sputtering is the only deposition method that does not depend on melting points and vapor pressures of refractory compounds such as carbides, nitrides, silicides, and borides. As a result, films of these materials can be sputtered directly onto surfaces without altering substrate properties.

Much sputtering has been aimed at producing solid-film lubricants and hard, wear-resistant refractory compounds. NASA is interested in these tribological applications because coatings can be sputter-deposited without a binder, with strong adherence, and with controlled thickness on curved and complex-shaped surfaces such as gears and bearing retainers, races, and balls. Also, because sputtering is not limited by thermodynamic criteria (unlike most conventional processes that involve heat input), film properties can be tailored in ways not available with other deposition methods.

Most research on sputtered solid-lubricant films has been done with MoS_2. Other films that have been sputtered are WC, TiN, PbO_2, gold, silver, tin, lead, indium, cadmium, PTFE, and polyimide (PI). Of these coatings, the gold-colored TiN coatings are most prominent.

TiN coatings are changing both the appearance and performance of high-speed steel metal-cutting tools. Life of TiN-coated tools, according to producers' claims, increases by as much as tenfold, metal-removal rates can be doubled, and more regrinds are possible before a tool is discarded or rebuilt.

Sputter Coating Process

The SCX™ sputter coating process, a proprietary, computer-aided process developed by Engelhard-CLAL, Carteret, NJ, a producer of high-purity materials, enables the coating of base or refractory metals with precious metals. The source of the coating material can be almost any metallic composition. A major benefit of sputtering is the ability to deposit alloys or compounds that cannot be mechanically worked or alloyed as is required in the cladding process. By fabricating a segmented target comprising of two or more individual elements, a deposition can be made that is a uniformly dispersed "alloy" of the constituents

SCX sputtering is conducted at low temperatures (<300°C), permitting deposition on plastics and other temperature-sensitive materials in addition to metals. Conducted at reduced pressures of inert gases, entrapped gases are kept to a minimum. Finally, by replacing the inert gas with a reactive gas such as H_2, N_2, or O_2, a compound formed by the gas can be deposited. This permits the reaction of very interesting coatings, such as nitrides, hydrides, and oxides. The unique sputter coating process makes it possible to attain very thin as well as relatively thick coatings equally as well in the range of ½ to over 6 μm.

Typical substrate dimensions are wire: 0.08–1 mm diameter, continuous lengths up to 3000 m at 0.08 mm diameter, and 300 m at 0.89 mm diameter; ribbon: 0.017–0.50 mm thick and widths from 0.25 to 3.2 mm, continuous lengths from 120 to 500 m; rods: 3.2 mm diameter by 508 mm long; metallic foil: 0.05 mm thick and up, to 102 × 508 mm window dimension; polymeric: 0.25 mm thick and up, to 102 × 508 mm window; rigid metallic and nonmetallic: up to 127 mm thick and up to 102 × 508 mm window dimension. Flexible and discrete parts can be coated selectively on one or both sides.

Application to Power Tube Grids

Power tube grids control the flow of tube current by providing a bias between the cathode and anode. Semiconductor devices have replaced the bulk of electron tube usage, especially in receiving applications. However, in extremely high-power applications, the Triode style thermal emissions tube still finds global use. The electron tube is expected to provide long-lasting, high-quality performance throughout the typical frequency range of 20 kHz to over 20 GHz, with grid temperatures from 600°C to 1300°C, depending on the application. Secondary electron emission is of major concern to tube designers. Without controls or limits, a tube could easily become unstable and quickly self-destruct.

Platinum-coated molybdenum and tungsten are traditional materials for grid construction. Traditional platinum-clad molybdenum grid wire is produced with very thick precious metal coatings because it is difficult to produce claddings without base material breakthrough, and the molybdenum tends to diffuse through the platinum, embrittling the grid as well as causing an increase in emission. The SCX-PC sputter coating process accomplishes the same function as cladding, but with a precious metal savings of 15%–30%. This is achieved by introducing a diffusion barrier into the coating during the manufacturing process, which effectively prohibits the interdiffusion of the core and the coating.

Other unique coatings include SCXPZC, which includes zirconium in the deposition process to permit higher-temperature grid usage with closer cathode spacing, and SCX-TH, which produces titanium-hydride (TiH_2) coatings to control primary and secondary emissions and enables the grid to act as a getter for nascent gas molecules.

Ion Plating

The basic difference between sputtering and ion plating is that sputtered material is generated by impact evaporation and transferred by a momentum transfer process. In ion plating, the evaporant is generated by thermal evaporation. Ion plating combines the high throwing power of electroplating, the high deposition rates of thermal evaporation, and the high energy impingement of ions and energetic atoms of sputtering and ion-implantation processes.

The excellent film adherence of ion-plated films is attributed to the formation of a graded interface between the film and substrate, even where the two materials are incompatible. The graded interface also strengthens the surface and subsurface zones and increases fatigue life.

The high throwing power and excellent adherence makes possible the plating of complex three-dimensional configurations such as internal and external tubing, gear teeth, ball bearings, and fasteners. Gears for space applications, for example, have been ion plated with 0.12–0.2 μm of gold for lubrication and to prevent cold welding of the gear pitch line. Ion plating has also been used, on a production basis to plate aluminum on aircraft landing-gear components for corrosion protection.

Ion plating is also one of the two methods used to deposit diamond-like coatings (DLCs). A relative newcomer to the coatings field, DLCs are commonly made from hydrocarbon (often methane) and H_2 gases heated to 2000°C. The carbon coatings are prized for their wear resistance, as well as electrical and optical properties. Although they represent a huge potential, present DLCs are at the earliest stages of commercialization. However, their wide range of properties, along with their relatively low-cost, leads many to predict huge growth in DLCs.

Researchers have proposed that the coatings be used to improve wear resistance in tool bits, as electronic heat sinks, and to boost wear and corrosion resistance in optical materials.

Chemical vapor deposition (CVD) is the method most often used to deposit DLCs. Adjusting deposition conditions allows the processor to change the coating from graphite to diamond-like. One process used at Battelle deposits the DLC in a gas atmosphere at reduced pressure without a fixed target. This plasma-assisted CVD allows large workpieces to be coated on all sides without turning. However, substrates must be heated to roughly 800°C when using CVD.

Reduced substrate temperatures are offered by dual ion-beam-enhanced deposition. Substrate temperature reaches only 66°C, and the dual ion-beam process does not rely on epitaxial growth for its formation as CVD does. Epitaxial growth requires a crystalline substrate; because dual ion-beam processing is free of this need, it enables amorphous materials to be coated as well.

Materials that are compatible with the Diond process include ferrous and nonferrous metals, glasses, ceramics, plastics, and composites. In addition to the Diond coating, dual ion-beam enhanced deposition can apply metallic coatings to fiber-reinforced carbon-carbon materials.

The basic ion-implantation process sends beams of elemental atoms (produced in a particle accelerator) into the surface of the target component. With dual ion-beam enhanced deposition, two simultaneous beams are used. One beam continuously sputters carbon onto the surface, providing the carbon material necessary to grow a diamond film. A second beam, consisting of inert gas at higher energy, drives some of the diamond layer into the interface zone. Then, the energy of the second beam is reduced to allow diamond growth. Implanting diamond material within the interface zone optimizes adhesion.

Technologies developed for electronic and optical thin films are often transferred into the engineering coatings sector, leading to a wider use of ion- and plasma-based techniques.

A novel nonequilibrium plasma treatment method with unusual characteristics is now being developed by EA Technology Ltd. of Capenhurst, near Chester, Great Britain. The process acts like a low-pressure, nonthermal glow discharge. Although it appears to provide many of the conditions needed for plasma surface engineering, the process runs at atmospheric pressure. Atmospheric deposition is generally much simpler than traditional vacuum plasma deposition, and its higher reactant concentrations make it much faster and, therefore, more cost-effective.

The equipment needed for this new approach is little more than a modified commercial microwave oven in which the plasma is sustained within a flask by microwave energy. Processing can be done either within the plasma or in a downstream gas that flows through the flask. In-plasma treatment is more energetic. Those materials that can withstand high temperatures can be coated within the flask. Downstream processing is easier and particularly effective at coating epoxies and polymers such as polymethyl methacrylate with materials such as TiO_2. This technique improves the weather resistance of the surface. The plasma can even be used to break down noxious gases such as volatile organics with more than 97% efficiency.

Vacuum Plasma Processing

This is the heart of advanced physical vapor deposition (PVD) coatings used on tools such as molds, dies, drills, and cutters. Although TiN is the most widely used of these coatings, many demanding applications now use TiAlN. High-quality PVD coatings are produced when electron beams evaporate the coating material while N_2-rich plasma bombards the substrates. Electron beams cannot be used to evaporate alloys such as TiAl, since the vapor pressure of aluminum is 100 times higher than that of titanium. However, to overcome this problem, a new control technology has been developed for electron-beam deposition systems. This technology measures optical emissions from elements in the plasma, picking up characteristic titanium and aluminum emission lines. These are used to control individual electron-beam sources for alloy elements.

As a result, users can control the composition of the alloy coating—they can even modify the gradation of the coating chemistry, a technique not possible with alloy are and sputter sources.

C-coated components, when compared to nitrided, nickel, and chromium coatings, exhibit improved wear-resisting performance. In particular, Balinit® C WC/C coatings are claimed to offer the proper combination of low coefficient of friction and high hardness needed by highly loaded automotive and machine parts.

Balinit C WC/C coatings are made of hard WC particles in a soft amorphous carbon matrix. Ion bombardment of a WC target removes coating material for deposition onto component surfaces under controlled conditions. Several applications demonstrate the ability of the material to solve wear problems:

- A coating of WC/C, specifically developed for highly stressed machine components operating under less-than-optimum lubricating conditions, has given design engineers a way out of such predicaments. Produced by PVD, WC/C is said to improve seizure resistance and reduce failure due to particle-contaminated hydraulic oils.
- Pump pistons coated with Balinit C operate longer than nitrided pistons or nickel- and chromium-coated pistons. Replacement of sliding shoes made from bronze with Balinit C coated steel also cuts down on wear.
- Application results in a low coefficient of friction and "smoothing" of the surface of the part. Hardness measures 1000 VHN (25 g); thickness is approximately 3 μm.
- Balinit C WC/C coatings, with a maximum working temperature of 300°C, are at present used in racing, as well as industrial gear systems.

In tribological tests conducted on CrN, TiAlN, TiAlCN, TiCN, TiCN + C, TiN + C, TiB_2, WC, and molybdenum coatings for load capacitance, adhesion power, abrasion force, hardness, and fatigue strength showed TiAlN, the very hard, metallic coating used on cutting tool inserts, gave the best results. Unlike cutting tools, however, bearing components cannot be coated in a standard CVD process at temperatures over 500°C. Even with a specialty developed PVD process, which deposits hard coatings with high adhesion power at a temperature level of only 160°C, bearing rings must undergo special annealing after final grinding at a temperature of approximately 240°C to achieve the roundness deviation after coating within the normal manufacturing tolerance range.

Thermal Spraying

Arc spraying, a form of thermal spraying of metals, is done on a prepared (usually *grit-blasted*) metal surface, using a wire-arc gun.

The coating metal is in the form of two wires that are fed at rates that maintain a constant distance between their tips. An electric arc liquefies the metal, and an air spray propels it onto the substrate. Because particle velocity can be varied considerably, the process can produce a range of coating finishes from a fine to a coarse texture.

Arc-sprayed coatings are somewhat porous, as they are composed of many overlapping platelets. Used in applications where appearance is important, thermally sprayed coatings can be sealed with pigmented vinyl copolymers or paints, which usually increase the life of the metal coating. Arc-sprayed coatings are thicker than those applied by hot dipping, ranging from 3 to 5 mil for light-duty, low-temperature applications to 7 to 12 mil for severe service.

Because zinc and aluminum are, under most conditions, more corrosion resistant than steel, they are the most widely used spray-coating metals. In addition, since both metals are anodic to steel, they act galvanically to protect ferrous substrates.

In general, aluminum is more durable in acidic environments, and zinc performs better in alkaline conditions. For protecting steel in gas or chemical plants, where temperatures might reach 204°C, aluminum is recommended. Zinc is preferred for protecting steel in fresh, cold waters; in aqueous solutions above 66°C, aluminum is the usual choice.

For service to 538°C, a thermally sprayed aluminum coating should be sealed with a silicone–aluminum paint. Between 538°C and 899°C, the aluminum coating fuses and reacts with the steel base metal, forming a coating that, without being sealed, protects the structure from an oxidizing environment. And, for continuous service to 982°C, a nickel-chrome alloy is used, sometimes followed by aluminum.

In Europe, where thermally sprayed metal coatings for corrosion protection have been far more widely used than in the United States, many structures such as bridges are still in good condition after as long as 40 years, with minimum maintenance. Other applications include exhaust-gas stacks, boat hulls, masts, and many outdoor structures.

Thermal spraying has become much more than a process for rebuilding worn metal surfaces. Thanks to sophisticated equipment and precision control, it is now factored into the design process, producing uniform coatings of metals and ceramics. With some of the processes, even gradated coatings can be applied. This is done by coating the substrate with a material that provides a good bond and that has compatible expansion characteristics, then switching gradually to a second material to produce the required surface quality such as wear resistance, solderability, or thermal-barrier characteristics.

Plasma Spraying

Plasma-spray coating relies on a hot, high-speed plasma flame (N_2, H_2, or argon) to melt a powdered material and spray it onto the substrate. A DC arc is maintained to excite gases into the plasma state.

The high-heat plasma (in excess of 7075°C) enables this process to handle a variety of coating materials—most metals, ceramics, carbides, and plastics. Although most coating materials are heated to well beyond their melting points, substrate temperatures commonly remain below 121°C.

This process has found wide acceptance in the aircraft industry. Plasma-sprayed metallic coatings protect turbine blades from corrosion, and sprayed ceramics provide thermal-barrier protection for other engine parts.

Proprietary refinements in plasma-spray technology include a wear-resistance coating material that lends itself to forming amorphous/microcrystalline phases when plasma sprayed. The resultant coating provides excellent corrosion resistance with minimal oxidation at higher temperatures. This promises to eliminate problems of work-hardened crystalline coatings that chip or delaminate in response to stress, which have previously been taken care of by expensive alloying elements.

Another amorphous alloy development involves a crystalline material that, upon abrasive wear, transforms to an amorphous hard-phase alloy. The top layer, 3–5 µm thick, results in hardness levels over 1300 Vickers. Wear tests have indicated this material is superior to more expensive WC coatings.

Detonation gun coatings considered by many to be an industry standard, use a detonation wave to heat and accelerate powdered material to 732 mm/s. In the line-of-sight process, each individual detonation deposits a circle of coating with a 2.54 cm diameter and 2 µm thickness. Coatings, thus, consist of multiple layers of densely packed lenticular particles tightly bonded to the surface.

The Super D-Gun has been developed to increase particle velocities. New coatings (the UCAR 2000 Series) applied with the gun offer improved wear resistance without affecting fatigue performance. The system has been targeted for fatigue-sensitive aircraft components.

Other low-pressure plasma-spray (LPPS) coatings protect turbine vanes by improving the sulfidation and oxidation resistance of complex components. Inert-atmosphere LPPS systems are an effective means for applying complex corrosion-resistant coatings such as NiCoCrAlY to high-temperature engine components.

PS300 is a self-lubricating solid coating material for use in sliding contacts at temperatures up to 800°C. PS300 is a composite of metal-bonded CrO_2 with $BaFl_2/CaFl_2$ eutectic and silver as solid lubricant additives. The "PS" in the name of this and other self-lubricating, high-temperature composite materials signifies that the material is applied to a substrate by plasma spraying of a powder blend of its constituents.

Spray Coatings

Stabilized Zirconia (ZrO_2)

Yttria-stabilized zirconia (YZP) represents the bulk of all sprayed ceramics. This material is used primarily for thermal barrier coatings (TBCs) in aircraft, rocket, and reciprocating engines. TBCs are applied to engine components to lower substrate temperatures so that combustion gas temperatures can be higher, thereby increasing engine power and efficiency and lowering emissions. Stabilized ZrO_2 is unique for its high CTE and low thermal conductivity. The high CTE correlates well with the base metals to which ZrO_2 is commonly applied, reducing stresses that are induced by differential expansion. Plasma-sprayed YZP is also reasonably resistant to thermal fatigue and chemical attack.

Other elements used to stabilize ZrO_2 include the oxides of magnesium, calcium, and cerium. Phase stabilization is used to mitigate the large volume change that ZrO_2 undergoes during heating to and cooling from service temperatures. The phase transformation from low-temperature monoclinic to high-temperature tetragonal can be arrested by the inclusion of stabilizing components such as YO_2. Fully stabilized ZrO_2 maintains a cubic structure throughout heating. Partially stabilized ZrO_2, which has both cubic and tetragonal phases, is reported to be tougher and to have a better match of CTE with engine materials.

MCRALYS

Metallic coatings are used between the ceramic coating and substrate both to enhance bonding and to provide a barrier that prevents substrate oxidation and corrosion. As a class, these materials are denoted by the term MCRALY, which is derived from the components: a base metal (M), chromium (CR), aluminum (AL), and yttria (Y). The M component is iron (Fe), cobalt (Co), or nickel (Ni), singly or in combination. The coatings are then called FECRALY, COCRALY, NICRALY, and so forth.

Recent advances in thermal spraying have focused on controls. Process control includes barfeedstock, materials, and processing parameters. Historically, thermal spray coatings have been applied at a confidence level of around 70.

Applications

Common use of thermally sprayed YZP involves net-shape manufacturing. O_2 sensors for automobile emission control systems are manufactured by applying the coatings of YZP to remove mandrels. When the mandrel is removed, a free-standing shape is left.

Net- and near-net-shape (NNS) techniques facilitate the fabrication of parts that is not practical by other means. Freestanding net-shape ion engines have been manufactured from tungsten by using load chamber, plasma-arc spray. Plasma spraying in an argon atmosphere eliminates oxidation of reaction materials, such as tungsten. Precision Al_2O_3 tubes (0.75 mm) wall thickness, 75 mm diameter, and 1.2 m length would be difficult, if not impossible, to fabricate by casting and grinding; however, they have been successfully fabricated by thermal-spray net-shape techniques. On the other extreme, multilayer ceramic tubes with 1-mm inside diameter have been made to join blood vessels.

Ceramic coatings are applied to medical instruments used for endoscopic and other forms of minimal invasive surgery.

Polymer Coatings

Polymeric coatings designed for corrosion protection are usually tougher and are applied in heavier films than are appearance coatings. Requirements of such coatings are much more stringent: They must adhere well to the substrate and must not chip easily or degrade from heat, moisture, salt, or chemicals.

Environmental factors also drive the technology behind polymer coatings replacing chromium and cadmium coatings. This is partly due to increasing concern about heavy metals. Also, automakers must now contend with acid rain in addition to salt spray, and polymers surpass chromium and cadmium in acid rain resistance.

Acrylics and alkyds are widely used for farm equipment and industrial products requiring good corrosion protection at a moderate cost. Alkyd resins, particularly, play a major role in maintenance painting because of their good weathering characteristics and ease of application with low-cost, low-toxicity solvents. Alkyd paints are also relatively high in solids, permitting good buildup of a paint film with a minimum number of coats.

Silicone modification of organic resins improves overall weatherability and durability. Compared to organic coatings in general, silicones have greater heat stability, longer life, better resistance to deterioration from sunlight and moisture, and greater biological and chemical inertness.

For optimum weatherability, silicone content should be 25%–30%. Performance of the waterborne formulations is proving to be almost identical to that of the solvent-based coatings.

For coatings requiring higher heat resistance, silicone resins can be used alone for the paint vehicles, or they can be blended with various organic resins. These finishes are used on space heaters, clothes dryers, and barbecue grills. Similar formulations are used on smoke-stacks, incinerators, boilers, and jet engines. Performance of formulations containing ceramic frits approaches that of ceramic materials.

Polyurethane enamels are characterized by excellent toughness, durability, and corrosion resistance. These thermosetting materials, available in both one- and two-part formulations, cost more than the alkyds and acrylics.

Urethane chemistry is versatile enough to provide a hard, durable, environmentally resistant film, a tough, elastomeric coating, or a surface somewhere between. Urethanes have traditionally been available as solvent-based coatings containing 25%–45% solids, but environmental concerns have prompted manufacturers to also supply them in high solids, 100% solids, and waterborne formulations.

Coating thickness of polyurethanes ranges from about 2 mil for average requirements to as much as 30 mil for applications requiring impact and/or abrasion resistance as well as corrosion resistance. Typical uses are on conveyor equipment, aircraft radomes, tugboats, road-building machinery, and motorcycle parts. Abrasion-resistant coatings of urethanes are applied on railroad hopper cars, and linings are used in sandblasting cabinets and slurry pipes.

Epoxy finishes have better adhesion to metal substrates than do most other organic materials. Epoxies are attractive economically because they are effective against corrosion in thinner films than are most other finishing materials. They are often used as primers under other materials that have good barrier properties but marginal adhesive characteristics.

Coating thicknesses can vary from 1 mil for light-duty protection to as much as 20 mil for service involving the handling of corrosive chemicals or abrasive materials. Performance of epoxies is limited in the heavier thickness, however, because they are more brittle than other organic materials.

Nylon 11 coatings provide attractive appearance as well as protection from chemicals, abrasion, and impact. Applied by electrostatic spray in thicknesses from 2.5 to 8 mil, nylon coatings are used on office and outdoor furniture, hospital beds, vending-machine parts, and building railings. Heavier coatings—to 50 mil—are applied by the fluidized-bed method and are used to protect dishwasher baskets, food-processing machinery, farm and material-handling equipment, and industrial equipment such as pipe, fittings, and valves.

Fluorocarbons are more nearly inert to chemicals and solvents than all other polymers. The most effective barriers among the fluorocarbons for a variety of corrosive conditions are PFA, PTFE, ECTFE, FEP, and PVDF.

Ethylene-chlorotrifluoroethylene (ECTFE) is a member of the fluoropolymer family of resins. This high-temperature coating provides corrosion resistance and mechanical and electrical qualities up to 149°C. It is easy to apply and has excellent release and low-friction properties. Multiple layers of ECTFE can be applied up to 100 mil. Although more expensive than other powders, its performance often justifies a higher cost. ECTFE has a smooth surface as applied and is therefore used in water handling systems to minimize bacterial buildup.

Fluorinated ethylene propylene (FEP) is a soft fluoropolymer coating similar to PTFE and PFA. It has the best release properties among powder coatings. The corrosion-resistant qualities of FEP are better than PTFE, but FEP does not stand up to high temperatures as well.

Perfluoroalkoxy (PFA) is another member of the fluoropolymer family that resists corrosion and has better release and nonwetting qualities than PTFE. With wide temperature limits, PFA can be used in applications ranging from cryogenic levels to 260°C. It is typically used for coating molding cavities, and this food-grade-quality powder can also be used for baking surfaces.

Polytetrafluoroethylene (PTFE) is a soft and waxy material that has release properties similar to PFA. The high-temperature coating protects against corrosion in environments reaching 260°C.

However, its softness limits it to nonabrasive and moderately abrasive applications.

Polyvinylidenefluoride (PVDF) is the hardest powder coating with qualities similar to PPS, making it ideal for high-load and higher-modulus applications. The coating works well in corrosive environments for components in pulp mills, waste treatment plants, and petrochemical facilities. Once applied, PVDF coatings can be removed only with heat. The coating has a surface temperature limit of 130°C and is not commonly used for release and low-friction applications.

For impact service, PVDF and ECTFE coatings are recommended, in that order. PTFE, FEP, and PFA are also suitable, but they have a greater tendency to creep under load. For abrasive conditions, PVDF is outstanding among the fluorocarbons. Recommended for high surface temperatures—drying ovens and steam-handling equipment, for example—are PFA and PVDF. These materials are also used on engine components and welders. PVDF also has the highest compressive strength of the fluorocarbons. PTFE has the highest allowable service temperature (316°C) of the fluorocarbons.

Coatings based on PTFE are being used to reduce wear in the US automotive industry. Fluoropolymer coatings prevent binding and galling in disk brake systems at temperatures over 100°C. PTFE is also used as a dry lubricant. In addition, PTFE can be used as a coating on automotive fasteners, and a new process uses PTFE to prevent seizure in valve springs. This process is FluoroPlate impingement, a process whereby a mixture of inorganic and organic particles bombard the spring surface, thus relieving internal stresses and reducing surface flaws. The coating also helps the springs to repel oil.

A new class of coating—an alloy of fluoropolymer and other resins—has a different viscosity behavior than that of the earlier organics. Viscosity of "fastener-class" coating resins decreases sharply as film shear increases (as in application by the dip/spin process). Then, when the spinning basket stops, viscosity returns almost instantaneously to its original value. Thus, when applied to the dip/spin process, the coating clings to sharp edges, threads, and points.

Film thickness typically ranges from 0.5 to 0.7 mil, but formulations can be adjusted to provide films of 0.3–0.4 mil for parts with fine threads or other intricate features. Not only do these extremely tough coatings provide a more uniform barrier to corrosives, but they are also based on polymers that are inherently stable in the presence of a wide spectrum of acids, bases, and aqueous solutions.

Combination coatings blend the advantages of anodizing or hard-coat platings with the controlled infusion of low-friction polymers and/or dry lubricants. The coatings become an integral part of the top layers of metal substrates, providing increased hardness and other surface properties.

These coatings are different for each class of metals. For example, a Tufram coating for aluminum combines the hardness of Al_2O_3 and the protection of a fluorocarbon topcoat to impart increased hardness, wear and corrosion resistance, and permanent lubricity.

In the multistep process, the surface is first converted to Al_2O_3. Submicron particles of PTFE are then fused into the porous anodized surface, forming a continuous plastic/ceramic surface that does not chip, peel, or delaminate. The coating is claimed to have greater abrasion resistance than case-hardened steel or hard chromium plate.

Another proprietary coating that penetrates PTFE into precision hardcoat anodizing is Nituff. The coating achieves a self-curing, self-lubricating surface with low friction, high corrosion resistance, and dielectric properties superior to ordinary hardcoat anodizing. It is used extensively in aerospace, textile, food processing, packaging, and other industries, where it allows manufacturers to benefit from the lightweight and easy machinability of aluminum enhanced by the durability, cleanliness, and dry lubrication of the Nituff surface.

Other proprietary combination coatings have been developed for steel, stainless steel, copper, magnesium, and titanium that provide similar surface improvement. Coatings are also available that enhance specific properties such as lubricity, corrosion resistance, or wear resistance.

Powder Coatings

Powder coatings are generally much thicker than fluids, typically greater than 5 mil compared to 1 or 2 mil for fluids. Powders commonly protect substrates from corrosion and erosion wear, or provide release or aesthetic qualities. Fluid coatings, in contrast, are preferred for friction, abrasion, and spalling wear—corrosion applications that require low friction and release.

In addition to performance attributes, powder coatings also provide processing advantages over fluids. Powders are environmentally friendly because they do not contain volatile organic compounds (VOCs) that attack the ozone layer.

The powder coating combines properties of both plastics and paints. The coatings are manufactured using typical plastics-industry equipment. They are first sent through a melt-mix extruder and then ground. When applied as a coating, however, the powder becomes a coating film that is exactly like paint.

These coatings have been developed in response to pressures to reduce VOC emissions, which have increased over the past few years. Overspray from liquid paints contains solvents that are released into the atmosphere. Even with recovery systems, some volatile components escape. Powder coatings, on the other hand, are completely recyclable. Overspray can be collected easily and reused. If a small amount becomes too contaminated for recycling, safe disposal techniques are available.

Powder coatings also show promise as a substitute for clear coats in the automotive industry. Present solvent-based paints could be replaced by a clear powder coating that cures at roughly the same temperature as conventional paints. Powder coatings may also replace the baked-on porcelain enamel used for appliance parts. Washer and dryer lids are now powder-coated in industry.

Applications are not all that is new, however. Materials have changed. The majority of powder coatings have relied on either an epoxy or polyester resin base. Acrylics, however, are becoming more important, and other possible bases include nylon, vinyl, and various fluoropolymers.

Two processes for applying coatings have undergone refinements. With electrostatic spraying, the most popular method, powder is given a charge and sprayed onto electrically grounded parts. Baking them completes the cure. Nonconductive parts must be primed or heated to provide them with more electrostatic attraction.

In the fluid-bed process, air passes through a porous membrane at the bottom of a tank and aerates the powder so that it swirls around in the tank. A part is then heated and dipped into the tank, so that the powder melts on the surface. This process is used for thick-film protection coatings, and is suitable only for metal parts that can retain heat long enough to be coated.

Liquid Layers

Despite strict governmental restrictions that continue to thwart the amount of VOCs released into the atmosphere, the use of high-performance fluid coatings remains widespread. Unlike powder coatings, fluids can be applied in films as thin as 0.2 mil without affecting their integrity. Powder coatings, in contrast, require more clearance between components to accommodate thicknesses greater than 5 mil.

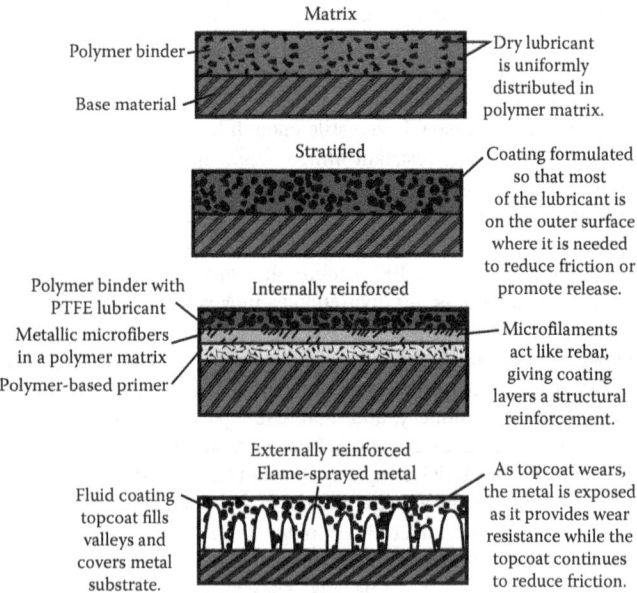

FIGURE C.1 A closer look at coatings. (From *Machine Designs*, September 10, 1998, p. 103. With permission.)

Another advantage of fluid coatings is that manufacturers can apply them by conventional and electrostatic spray processes, as well as by brush and dip methods. This is particularly important for components with deep recesses that are difficult to coat with powders using electrostatic means because the charged particles adhere to the outer edge of the recess. Probe-spray systems, in turn, are much more efficient for covering recesses with fluid coatings.

Following is a summary of the fluid coatings that are commonly used to protect engineering materials; see Figure C.1.

Matrix coatings have the highest mechanical properties, such as tensile strength and wear resistance, among fluid coatings. They consist of one or more polymer binders (PPS, polyimide) combined with a dry lubricant such as PTFE, PFA, MoS_2, or graphite. When applied, the dispersion of this composite mixture consists of lubricant evenly distributed within the binder. Typical applications include large-thread fasteners, actuators, pistons, bearings, impellers for mixers and superchargers, and rotating and sliding powder-metal components.

Stratified coatings also consist of binders and lubricants; however, their formulation keeps most of the low-friction, high-release agent on the bearing surface of the coating. With lubricant segregated to the outer surface, the top layer of stratified coatings is softer and has lower mechanical properties than matrix coatings. However, this concentration of lubricant improves release qualities, and with certain fluoropolymers, the coatings protect substrates from corrosion. Typical applications include components used in photocopiers, valves, and fuel-handling systems.

Hybrids are a new class of coatings that have greater strength than both powders and matrix-fluid coatings. To boost strength, the films are formulated with either internal or external reinforcements.

Internally reinforced coatings use microfilaments to provide a mechanical structure inside the film, acting in much the same way that rebar strengthens concrete structures. The overlapping filaments produce a film that has a wear limit that exceeds conventional measurement methods. Engineers typically use these coatings in applications with high loads such as cutting blades.

Externally reinforced coatings combine a rigid substrate such as stainless steel with a conventional high-release fluid coating containing a low-friction release agent such as high percentages of PTFE. The films consist of two layers. The first is a continuous layer of flame-sprayed stainless steel. Manufacturers then cover the stainless steel with a thick topcoat of low-friction fluid to create a total film thickness of 1–1.4 mil. The topcoat fills the depressions and smoothes over rough spots on the flame-sprayed metal coat.

Under high loads or extreme wear, the topcoat may wear away, exposing the substrate material. However, because the fluid coat remains in the asperities of the base metal, the composite coating maintains a low-friction surface while the steel helps resist wear. Typical applications are large locking and fastening mechanisms, tumblers, agitators, and parts subjected to high abrasive wear.

Conformal Coatings

Printed circuit board (PCB) assemblies used in avionics, marine, automotive, and military applications generally perform in environments affected by heat, moisture, industrial pollutants, manual handling, and process residues, which typically magnifies such effects.

The PCB designer must determine if and how the assembly must be protected with a worst-case environment in mind. Consequences ranging from malfunction to product failure will devolve from this decision. Other factors include feasibility and cost of implementation of such protection, but the designer will have several options: reducing circuit sensitivity through designed signal characteristics, increasing conductor spacing, and using buried vias. Other protective options include sealing the circuit in a box pressurized with inert gas or in a polymer via potting or molding, or employing an electrically insulating barrier between conductors and the ambient environment.

Using a polymeric-film barrier for board protection is the most common and practical approach. There are, however, limitations and peculiarities accompanying the two categories of coatings:

1. Bare-board coatings, or "permanent soldermasks," are applied during the board fabrication process as liquids or dry films over conductive finishes or bare copper, and polymerized (cured) either by heat or ultraviolet light.
2. Assembly coatings offer protective coverage against water in the form of atmospheric humidity and condensation. They are commonly referred to as "conformal" coatings, implying compliance of the film to the contour line of the assembly. However, this term may be a misnomer, since newer application techniques permit only specific areas be coated to eliminate the cost of masking.

The use of conformal coating materials is provided in four basic material types (acrylic, epoxy, silicone, and urethane resins) for liquid-applied coatings, and one (paraxylylene) for vapor deposition. As a result, a coating must be selected on the basis of electrical, thermal, mechanical, and other pertinent properties as dictated by requirements for circuit performance and characteristics, type and degree of environmental exposure, and consequences of failure.

Coaxing

Improvement of the fatigue strength of a specimen by the application of a gradually increasing stress amplitude, usually starting below the fatigue limit.

Cobalt

Cobalt (symbol Co) is a lustrous, silvery-blue metallic chemical element, resembling nickel but with a bluish tinge instead of the yellow of nickel. It is rarer and costlier than nickel and its price has varied widely in recent years. Although allied to nickel, it has distinctive differences. It is more active chemically than nickel. It is dissolved by dilute H_2SO_4, HNO_3, or HCl acids, and is attacked slowly by alkalis. The oxidation rate of pure cobalt is 25 times that of nickel. Its power of whitening copper alloys is inferior to that of nickel, but small amounts in Ni–Cu alloys will neutralize the yellowish tinge of the nickel and make them whiter. The metal is diamagnetic like nickel, but has three times the maximum permeability. Like tungsten, it imparts red-hardness to tool steels. It also hardens alloys to a greater extent than nickel, especially in the presence of carbon, and can form more chemical compounds in alloys than nickel.

Its chemical properties resemble, in part, those of both nickel and iron. Cobalt has the metal with the highest Curie temperature (1121°C) and the lowest allotropic transformation temperature (399°C). Below 421°C, cobalt is close-packed hexagonal; above, it is face-centered cubic.

Properties

Cobalt has a specific gravity of 8.756, a melting point of 1495°C, 85 Brinell hardness, and an electrical conductivity about 16% that of copper. The ultimate tensile strength of pure cast cobalt is 234 MPa, but with 0.25% carbon it is increased to 427 MPa. Strength can be increased slightly by annealing and appreciably by swaging or zone refining. The metal is used in tool-steel cutters, in magnet alloys, and high-permeability alloys, and as a catalyst, and its compounds are used as pigments and for producing many chemicals.

The natural cobalt is cobalt-59, which is stable and nonradioactive, but the other isotopes from 54 to 64 are all radioactive, emitting beta and gamma rays. Most have very short life except cobalt-57, which has a half-life of 270 days, cobalt-56 with a half-life of 80 days, and cobalt-58 with a half-life of 72 days. Cobalt-60, with a half-life of 5.3 years, is used for radiographic inspection. It is also used for irradiating plastics, and as a catalyst for the sulfonation of paraffin oils because the gamma rays cause the reaction of SO_2 and liquid paraffin. Co 60 emits gamma rays of 1.1–1.3 MeV energy, which gives high penetration for irradiation. The decay loss in a year is about 12%, the cobalt changing to nickel.

The best-known cobalt alloys are the cobalt-base superalloys used for aircraft turbine parts. The desirable high-temperature properties of low creep, high stress-rupture strength, and high thermal-shock resistance are attributed to the allotropic change of cobalt to a face-centered cubic structure at high temperatures. Besides containing 36%–65% cobalt, usually more than 50%, most of these alloys also contain about 20% chromium for oxidation resistance and substantial amounts of nickel, tungsten, tantalum, molybdenum, iron, and/or aluminum, and small amounts of still other ingredients. Carbon content is in the 0.05%–1% range. These alloys include L-605; S-816; V-36; WI-52; X-40; J-1650; Haynes 21 and 151; AiResist 13, 213, and 215; and MAR-M 302, 322, and 918. Their 1000 h stress-rupture strengths range from about 276 to 483 MPa at 649°C and from about 28 to 103 MPa at 982°C. Cobalt is also an important alloying element in some nickel-base superalloys, other high-temperature alloys, and alloy steels. Besides tool steels, the maraging steels are a good example. Although cobalt-free grades have been developed, due to the scarcity of this metal at times, most maraging steels contain cobalt, as much as 12%. Cobalt is also a key element in magnet steels, increasing residual magnetism and coercive force, and in nonferrous-base magnetic alloys.

An important group of cobalt alloys is the stellites. They contain chromium and various other elements such as tungsten, molybdenum, and silicon. The extremely hard alloy carbides in a fairly hard matrix give excellent abrasion and wear resistance and are used as hard-facing alloys and for aircraft jet engine parts.

The interesting properties of cobalt-containing permanent, soft, and constant-permeability magnets are a result of the electronic configuration of cobalt and its high curie temperature. In addition, cobalt in well-known Alnico magnet alloys decreases grain size and increases coercive force and residual magnetism.

Cobalt is a significant element in many glass-to-metal sealing alloys and low-expansion alloys. One iron-base alloy containing 31% nickel and 5% cobalt provides a lower CTE than the iron–36% nickel alloy called Invar, and is less sensitive to variations in heat treatment. Co–Cr alloys are used in dental and surgical applications because they are not attacked by body fluids. Alloys named Vitallium are used as bone replacements and are ductile enough to permit anchoring of dentures on neighboring teeth. They contain about 65% cobalt.

Cobalt is a necessary material in human and animal metabolism, and is used in fertilizers.

Applications

The major uses of cobalt are in cobalt-base and cobalt-containing materials for high-temperature alloys, permanent magnets, and steels. In addition, cemented carbides, which are considered cutting tools, are used in balls for ballpoint pens and high-temperature ball bearings.

The hard-facing alloys are useful because of their resistance to corrosion, abrasion, and oxidation at high temperatures for plowshares, oil bits, crushing equipment, tractor treads, rolling mill guides, knives, punches, shears, billet scrapers, valves for high-pressure steam, oil refineries, and diesel and auto engines.

The superalloys have found use as searchlight reflectors, and are also useful under the severe operating conditions of high-temperature nuclear reactors. Their superior elevated-temperature properties compensate, to some extent, for the high thermal-neutron absorption for cobalt. They are used in reactors in certain wear-resistant components as guides for control rods. In nuclear submarines, these alloys are used where severe wear in contact with seawater is encountered.

Although the main application of Alnico permanent magnets is in motors, generators, regulating devices, instruments, radar, and loudspeakers, some are used in games, novelties, and door latches.

Cobalt-containing tool and high-speed steels are also used for dies.

Cobalt Oxide

A steel-gray to blue-black powder employed as a base pigment for ceramic lasers on metal, as a colorant for glass, as a chemical catalyst. It gives excellent adhesion to metals and is valued as an undercoat for vitreous enamels. It is the most stable blue, as it is not changed by ordinary oxidizing or reducing conditions. It is also one of the most powerful colorants for glass, 1 part in 20,000 parts of a batch giving a distinct blue color. Cobalt oxide is produced from the cobalt-nickel and pyrite ores, and the commercial oxide may be a mixture of the three oxides. *Cobaltous oxide*, CoO, is called *gray cobalt oxide* but varies from greenish to reddish. It is the easiest to reduce to the metal, and it reacts easily with silica and alumina in ceramics. *Cobaltic oxide*, Co_2O_3, occurs in the mixture only as the unstable hydrate, and it changes to the stable *black cobalt oxide*, or *cobalto-cobaltic oxide*, Co_3O_4 on heating. Above about 1652°F (900°C) this oxide loses oxygen to form cobaltous oxide.

Cockling/Cockles

The wavy edge of rolled sheet and plate.

Cocoa

In *fretting wear*, a powdery form of debris, usually consisting of red iron oxides, that is expelled from a ferrous metal joint near the location where fretting wear is occurring. Also known as red mud. Fe_2O_3 is formed by fretting of iron and steel.

Co-Curing

The act of curing a composite laminate and simultaneously bonding it to some other prepared surface, or curing together an inner and outer tube of similar or dissimilar fiber–resin combinations after each has been wound or wrapped separately. See also *secondary bonding*.

COD Crack Opening Displacement

See fracture toughness.

Coefficient

A factor defining the relationship between two components. For example, the coefficient of thermal expansion is a measure of the change in dimensions resulting from a unit change in temperature.

Coefficient of Adhesion

(1) The ratio of the normal force required to separate two bodies to the normal load with which they were previously placed together. (2) In railway engineering, sometimes used to signify the coefficient of (limiting) static friction.

Coefficient of Compressibility

See *bulk modulus of elasticity*.

Coefficient of Elasticity

The reciprocal of Young's modulus in a tension test. See also *compliance*.

Coefficient of Expansion

A measure of the change in length or volume of an object, specifically, a change measured by the increase in length or volume of an object per unit length or volume.

Coefficient of Friction

The dimensionless ratio of the friction force (F) between two bodies to the normal force (N) pressing these bodies together: μ (or f) = (F/N).

Coefficient of Thermal Expansion

(1) Change in unit of length (or volume) accompanying a unit change of temperature, at a specified temperature. (2) The linear or volume expansion of a given material per degree rise of temperature, expressed at an arbitrary base temperature or as a more complicated equation applicable to a wide range of temperatures.

Coefficient of Wear

See *wear coefficient*.

Coercive Force

The magnetizing force that must be applied in the direction opposite to that of the previous magnetizing force in order to reduce magnetic flux density to zero; thus, a measure of the magnetic retentivity of magnetic materials.

Coextruded

An extrusion process in which two materials are extruded simultaneously so that one forms a surface layer around the other. The process is, for example, applied to tubes which require a combination of strength and corrosion resistance not obtainable in a single material.

Coextrusion Welding

A solid-state welding process which produces coalescence of the faying surfaces by heating and forcing materials through an extrusion die.

Cogging

The first reducing operation in working an ingot into a billet with a forging hammer or a forging press. See *cogging mill*.

Cogging Mill

A *blooming mill*.

Coherency

(1) The state in which the lattices of a precipitate and its surrounding parent matrix remains substantially in alignment but are distorted in the vicinity of the interface. (2) The parent phase (solvent) maintained by mutual strain and not separated by a phase boundary.

Coherent (Precipitate)

(1) An intermediate stage of precipitation, preceding formation of a distinct and separate phase, in which the lattices of the solution and the emerging precipitate are still in alignment but distorted. (2) A crystalline precipitate that forms from solid solution with an orientation that maintains continuity between the crystal lattice of the precipitate and the lattice of the matrix, usually accompanied by some strain in both lattices. Because the lattices fit at the interface between precipitate and matrix, there is no discernible phase boundary.

Coherent Radiation

Radiation in which the phase difference between any two points in the radiation field is constant throughout the duration of the radiation.

Coherent Scattering

In materials characterization, a type of x-ray or electron scattering in which the phase of the scattered beam has a definite (not random) relationship to the phase of the incident beam. Also termed unmodified scattering. See also *incoherent scattering*.

Cohesion

(1) The state in which the particles of a single substance are held together by primary or secondary valence forces. As used in the adhesive field, the state in which the particles of the adhesive (or adherend) are held together. (2) Force of attraction between the molecules (or atoms) within a single phase. Contrast with *adhesion*.

Cohesive Blocking

The blocking of two similar, potentially adhesive faces.

Cohesive Failure

Failure of an adhesive joint occurring primarily in an adhesive layer.

Cohesive Strength

(1) The hypothetical stress causing tensile fracture without plastic deformation. (2) The stress corresponding to the forces between atoms. (3) Intrinsic strength of an adhesive. (4) Same as *technical cohesive strength*. (5) Same as *disruptive strength*.

Coil

(1) An assembly consisting of one or more magnet wire windings. (2) Rolled metal sheet or strip.

Coil Breaks

Greases or ridges in sheet or strip that appear as parallel lines across the directions of rolling and that generally extend the full width of the sheet or strip.

Coil Winding

An electrically continuous length of insulated wire wound on a bobbin/spool, or form.

Coil with Support

A filler metal package type consisting of a continuous length of electrode in coil form wound on an internal support which is a simple cylindrical section without flanges.

Coil without Support

A filler metal package type consisting of a continuous length of electrode in coil form without an internal support. It is appropriately bound to maintain its shape.

Coin Silver

An alloy containing 90% silver, with copper being the usual alloying element.

Coin Straightening

A combination coining and straightening operation performed in special cavity dies assigned to impart a specific amount of working in specified areas of a forging to relieve the stresses developed during heat treatment.

Coinage Metals

Metals used for currency coinage because of their availability, durability and perceived value, originally gold, silver, copper and bronze, now presumably including aluminum, nickel, etc.

Coining

(1) A closed-die squeezing operation, usually perform cold, in which all surfaces of the work are confined or restrained, resulting in a well-defined imprint of the die upon the work. (2) A *restriking* operation used to sharpen or change an existing radius or profile. (3) The final pressing of a sintered powder metallurgy compact to obtain a definite surface configuration (not to be confused with *re-pressing* or *sizing*).

Coining Dies

Dies in which the coining or sizing operation is performed.

Coke

The poorest, gray, infusible residue left after the volatile matter is driven out of bituminous coal. The coal is heated to a temperature of 2192°F–2552°F (1200°C–1400°C), without allowing air to burn it, and the volatile matter expelled. The residue, which is mainly fixed carbon and ash, is a cellular mass of greater strength than the original coal. Its nature and structure make it a valuable fuel for blast furnaces, burning rapidly and supporting a heavy charge of metal without packing. Soft, or bituminous, coals are designated as coping or noncoking, according to their capacity for being converted to coke. Coal low in carbon and high in ash will produce a coke that is friable and not strong enough for furnace use, or the ash may have low-melting-point constituents that leave glassy slag in the coke. Coke is produced in the beehive and by-product ovens, or is a by-product of gas plants. Coke is also used as a fuel in cupola melting.

The fixed carbon of good coke should be at least 86%, and sulfur not more than 1%. The porosity may vary from 40% to 60%, and the apparent specific gravity should not be less than 0.8. *Foundry coke* should have an ignition point of about 1000°F (538°C), with sulfur below 0.7%, and the pieces should be strong enough to carry the burden of ore and limestone. Coke suitable for foundry use is also made from low-grade coals by reducing them to a semi-coke, or char, and briquetting, but *semi-coke* and *smokeless fuel* are generally coals carbonized at low temperatures and briquetted for house-hold use. These fuels are sold under trade names such as *Coalite* and *Carbolux*, and they are really by-products of the chemical industry since much greater quantities of liquids and more lighter fractions in the tar are obtained in the process.

Pitch coke, made by distilling coal tar, has a high carbon content, above 99%, with low sulfur and ash, and is used for making carbon electrodes. Petroleum coke is the final residue in the distillation of petroleum and forms about 5% of the weight of the crude oil. With the sand and impurities removed, it is about 99% pure carbon and is used for molded carbon products. *Calcined coke* is petroleum coke that has been calcined at 2400°F (1316°C) to remove volatile matter. It is used for electrodes. *Carbonite* is a natural coke found in England and in Virginia. It is a coke-like mineral formed by the baking action of igneous rocks on seams of bituminous coal.

Coke Breeze

In foundry practice, fines from coke screenings, used in blacking mixes after grinding; also briquetted for cupola use.

Coke Furnace

Type of pot or crucible furnace that uses coke as a fuel.

Coke Oven

The furnace in which coal is tightly packed and heated to form coke.

Coke Oven Gas

The gas produced in the manufacture of coke. It is a mixture of carbon monoxide, methane and hydrogen plus numerous other volatile constituents.

Coke Test

In foundry practice, the first layer of coke placed in the cupola. Also the coke used as a foundation and constructing a large mold in a *flask* or pit.

Cold Anodizing

A process for increasing the thickness of the normal oxide film on aluminum by immersing it in cold 50% nitric acid for a few minutes. Also see *anodizing*.

Cold Bend Test

A test in which a bar of metal is bent to some specified degree as a test of ductility or, especially in the case of tests that include a weld zone, to confirm freedom from cracking. The specimen may be plain or notched. See also *bend test*.

Cold Box Process

In foundry practice, a two-part organic resin binder system mixed in conventional mixers and blown into shell or solid core shapes at room temperature. A vapor mixed with air is blown into the core, permitting instant setting and immediate pouring of metal around it.

Cold Cast

A term used, dubiously, for a technique where a metal powder such as bronze is mixed with liquid resin and poured into a mold to solidify. The resultant item is then described as Cold Cast Bronze.

Cold Chamber Machine

A diecasting machine with an injection system that is charged with liquid metal from a separate furnace. Compare with *hot chamber machine*.

Cold Chisel

A chisel or relatively short length, substantial cross section and relatively obtuse angled cutting edge for cutting materials such as metal or brick as opposed to timber. It has no handle and the flat, beveled edge, head is struck with a hammer.

Cold Coined Forging

A forging that has been restruck cold in order to hold closer face distance tolerances, sharpen corners or outlines, reduce section thickness, flatten some particular surface, or in nonheat-treatable alloys, increase hardness.

Cold Compacting

See preferred term *cold pressing*.

Cold Cracking

(1) Cracks in cold or nearly cold cast metal due to excessive internal stress caused by contraction. Often brought about when the mold is too hard or the casting is of unsuitable design. (2) A type of weld cracking that usually occurs below 400°F (205°C). Cracking may occur during or after cooling to room temperature, sometimes with a considerable time delay. Three factors combine to produce cold cracks; stress (for example, from thermal expansion and contraction), hydrogen (from hydrogen-containing welding consumables), and a susceptible microstructure (plate martensite is most susceptible to cracking, ferritic and bainitic structures least susceptible). See also *hot cracking, lamellar tearing,* and *stress-relief cracking.*

Cold Die Quenching

A quench utilizing cold, flat, or shaped dies to extract heat from a part. Cold die quenching is slow, expensive, and is limited to smaller parts with large surface areas.

Cold Drawing

Technique for using standard metalworking equipment and systems for forming thermoplastic sheet at room temperature.

Cold Dry Die Quenching

Same as *cold die quenching.*

Cold Etching

Development of microstructure at room temperature and below.

Cold Extrusion

See *extrusion.*

Cold Extrusion/Forming/Drawing, etc.

See Cold Work, etc. Extrusion and Drawing, etc. In some cases of the terms there is an implication of close dimensional control of the product.

Cold Finger

In materials characterization, a liquid-nitrogen-cooled cold trap used to reduce contamination levels in vacuum chambers.

Cold Flow

The distortion that takes place in polymeric materials under continuous load at temperatures within the working range of the material without a phase or chemical change.

Cold Form Tapping

Producing internal threads by displacing material rather than removing it as either the tap or the workpiece is rotated. The thread form is produced by a tool, which has neither flutes nor cutting edges, that resembles a simple screw when viewed from the side but the end view shows that both the major and minor diameters have irregular contours for displacing the work material.

Cold Forming

See *cold working.*

Cold Galvanizing

Painting with zinc rich paint or a similar immersion process in an attempt to obtain the protection offered by hot dip or electrogalvanizing. The coating must have good adhesion and provide electrical continuity between zinc particles and the underlying steel if any cathodic protection is to be obtained. The term has also been used of electrogalvanizing although this latter term is normally to be preferred, if only because of the likely superior protection compared with a paint or similar film.

Cold Hardening—Natural Aging

See *precipitation hardening.*

Cold Heading

(1) Working metal at room temperature such that the cross-sectional area of a portion or all of the stock is increased. (2) Cold forging the heads of bolts by an upsetting process. See also *heading* and *upsetting.*

Cold Hearth Melting

The use of a furnace with a water cooled hearth, usually copper, on which to melt materials to avoid contamination by refractories.

Cold Inspection

A visual (usually final) inspection of forgings for visible imperfections, dimensions, weight, and surface condition at room temperature. The term may also be used to describe certain nondestructive tests such as magnetic-particle, dye-penetrant, and sonic inspection.

Cold Isostatic Pressing

Forming technique in which high fluid pressure is applied to a powder (metal or ceramic) part at ambient temperature. Water or oil is used as the pressure medium.

Cold Joint (Soldered or Brazed)

A joint with inadequate bonding resulting from insufficient heating of the parent components. See also *dry joint.*

Cold Junction

The cold end datum point of the pair of wires forming a thermocouple.

Cold Lap

(1) Wrinkled markings on the surface of an ingot or casting from incipient freezing of the surface and too low a casting temperature.

(2) A flaw that results when a workpiece fails to fill the die cavity during the first forging. A seam is formed as subsequent dies force metal over this gap to leave a seam on the workpiece surface. See also *cold shut*.

Cold Mill

A mill for cold rolling of sheet or strip.

Cold Parison Blow Molding

A plastic forming technique in which parisons are extruded or injection molded separately and then stored for subsequent transportation to the blow molding machine for blowing. See also *blow molding*.

Cold Pressing (Plastics)

A bonding operation in which a plastic assembly is subjected to pressure without the application of heat.

Cold Pressing (Powder Metallurgy)

Forming a powder metallurgy compact at a temperature low enough to avoid *sintering*, usually room temperature. Contrast with *hot pressing*.

Cold Rolled Sheets

A metal mill product produced from a hot-rolled pickled coil that has been given substantial cold reduction at room temperature. The resulting product usually requires further processing to make it suitable for most common applications. The usual end product is characterized by improved surface, greater uniformity in thickness, and improved mechanical properties compared with hot rolled sheet.

Cold Rolling

(1) Cold working, by rolling, of sheet, bar, etc., to affect a size change. (2) Local rolling, of surface features such as crankshaft radii, with hardened steel wheels or rollers to induce surface hardening with beneficial residual stresses.

Cold Short

Having poor ductility and hence a susceptibility to cracking during cold working operations.

Cold Shortness

Brittleness that exists in some metals at temperatures below the recrystallization temperature.

Cold Shot

(1) A portion of the surface of an ingot or casting showing premature solidification; caused by splashing of molten metal onto a cold mold wall during pouring. (2) Small globule of metal embedded in, but not entirely fused with, the casting.

Cold Shut

(1) A discontinuity that appears on the surface of cast metal as a result of two streams of liquid meeting and failing to unite. (2) A lap on the surface of a forging or billet that was closed without fusion

during deformation. (3) Freezing of the top surface of an ingot before the mold is full. (4) Casting defects characterized by a thin oxide film at the interface between neighboring areas solidifying in isolation. This occurs when the flow of metal is interrupted allowing surface solidification that fails to diffuse with subsequently poured metal or when molten metal flows in from two directions and the interfaces to oxidize to allow the flows to merge.

Cold Slug

The first plastic material to enter an injection mold; so-called because in passing through a sprue orifice it is cooled below the effective molding temperature.

Cold Soldered Joint

A joint with incomplete coalescence caused by insufficient application of heat to the base metal during soldering.

Cold Stretch

A pulling operation with little or no heat, usually on extruded filaments, to increase tensile properties of composite materials.

Cold Test

A test in which the pour point of an oil is determined.

Cold Treatment

Exposing steel to suitable subzero temperatures ($-85°C$ or $-120°F$) for the purposes of obtaining desired conditions or properties such as dimensional or microstructural stability. When the treatment involves the transformation of retained austenite, it is usually followed by tempering.

Cold Trimming

The removal of flash or excess metal from a forging at room temperature in a trimming press.

Cold Welding

(1) Welding, without any heating, between two surfaces in intimate contact. In practice high pressures and very clean, oxide free surfaces are required. Also termed *adhesion*, *galling* and *seizure*. (2) The application of special procedures allowing fusion welding, in particular arc welding, without preheat and, or postheat, in circumstances where they would normally be expected. For example, alloy steels in restrained thick sections. (3) A solid-state welding process in which pressure is used at room temperature to produce coalescence of metals with substantial deformation at the weld. Compare with *hot pressure welding*, *diffusion welding*, and *forge welding*.

Cold Work

(1) Any process of plastic deformation in which the component does not recrystallize but becomes progressively harder and stronger but less ductile up to some limit. The temperature will normally be at about ambient. If it is deliberately and significantly higher, but still below the recrystallization temperature, the process may be termed warm working. (2) Permanent strain in a metal accompanied by strain hardening.

Cold Working

Deforming metal plastically under conditions of temperature and strain rate that induce *strain hardening*. Usually, but not necessarily, conducted at room temperature. Contrast with *hot working*.

Cold-Molded Plastics

Cold-molded plastics are one of the oldest of the so-called plastic materials; they were introduced in the United States in 1908. For the first time they provided the electrical engineer with materials that could be molded into more complicated shapes than could porcelain or hard rubber, providing better heat resistance than hard rubber, and better impact strength than porcelain. They could also incorporate metal inserts.

A procedure in which a plastic is shaped at room temperature and subsequently cured by baking.

General Nature and Properties

So-called cold-molded plastics are formulated and mixed by the molder (usually in a proprietary formulation). The materials fall into two general categories: inorganic or refractory materials, and organic or nonrefractory materials.

Inorganic cold-molded plastics consist of asbestos fiber filler and either an SiO_2-lime cement or portland cement binder. Clay is sometimes added to improve plasticity. The SiO_2-lime materials are easier to mold although they are lower in strength than the portland cement types.

In general, advantages of these materials include high arc resistance, heat resistance, good dielectric properties, comparatively low-cost, rapid molding cycles, high production with single-cavity molds (thus low tool cost), and no need for heating of molds. On the other hand, they are relatively heavy, cannot be produced to highly accurate dimensions, are limited in color, and can be produced only with a relatively dull finish. They have been used generally for arc chutes, arc barriers, supports for heating coils, underground fuse shells, and similar applications.

Organic cold-molded plastics consist of asbestos fiber filler materials bound with bituminous (asphalt, pitches, and oils), phenolic, or melamine binders. The binder materials are mixed with solvents to obtain proper viscosities, then thoroughly mixed with the asbestos, ground and screened to form molding compounds. The bituminous-bound compounds are lowest in cost and can be molded more rapidly than the inorganic compounds; the phenolic and melamine-bound compounds have better mechanical and electrical properties than the bituminous compounds and have better surfaces as well as being lighter in color. Like the inorganic compounds, organic compounds are cold-molded, followed by oven curing.

Compounds with melamine binders are similar to the phenolics, except that melamines have greater arc resistance, lower water absorption, are nontracking, and have higher dielectric strength.

Major disadvantages of these materials, again, are relatively high specific gravity, limited colors, and inability to be molded to accurate dimensions. Also, they can be produced only with a relatively dull finish.

Compounds with bituminous binders are used for switch bases, wiring devices, connector plugs, handles, knobs, and fuse cores. Phenolic and melamine compounds are used for similar applications where better strength and electrical properties are required.

An important benefit of cold-molded plastics is the relatively low tooling cost usually involved for short-run production. Most molding is done in single-cavity molds, in conventional compression-molding presses equipped for manual, semiautomatic, or fully automatic operation.

The water-fillable plastics used to replace wood or plaster of Paris for ornamental articles, such as plaques, statuary, and lamp stands, and for model making, are thermoplastic resins that cure to closed-cell lattices that entrap water. The resin powders are mixed with water and a catalyst and poured into a mold without pressure. They give finer detail than plasters, do not crack or chip, are light in weight, and the cured material can be nailed and finished like wood. Water content can be varied from 50% to 80%.

Design Considerations

Cross sections are generally heavier than hot molded materials to provide durability in handling. Taper is not usually necessary on the part, except on projecting barriers or bosses, as well as on sides of recesses or depressions. Generous fillets should always be provided. Undercuts and reentrant angles should be avoided as they will increase mold costs and reduce production rate.

In molding, a variation of ±0.038 mm must be allowed in thickness of part. Also, because parts are cured out of the mold, dimensional tolerances cannot be held very closely.

Lettering, figures, and simple designs can be molded on surfaces; marking is usually of the raised type and is placed on recessed surfaces to prevent rubbing off.

Cold-Press Molding

A plastic molding process in which inexpensive plastic male and female molds are used with room temperature curing resins to produce accurate parts. Limited runs are possible.

Cold-Rolled Steel

Cold-rolled steels are flat steel products produced by cold-rolling hot-rolled products. The hot-rolled product is cleaned of oxide scale by pickling and passed through a cold-reduction mill to reduce and more uniformly control thickness and to enhance surface finish. Cold rolling also increases hardness, reducing ductility. Although the steel is sometimes used as rolled, it is often subsequently annealed to improve formability and then temper-rolled or roller-leveled for flatness. Cold-rolled steels are available in carbon and alloy grades as well as high-alloy grades, such as stainless steels. For plain carbon steels, carbon content is usually 0.25% maximum, often less. Quality designations include commercial quality (CQ) steel, which is produced from rimmed, capped, or semikilled steel; drawing quality (DQ), which is made from specially processed steel and is more ductile and uniform in forming characteristics; and drawing-quality special-killed (DQSK) steel, which is still more ductile and more uniform in forming characteristics. Cold-rolled structural-quality (SQ) steel refers to cold-rolled steel produced to specific mechanical properties. Bar and rod products are often cold-drawn through dies and called cold-drawn bar steel, or cold-finished in other ways and called cold-finished bar steel.

A series of SQ ultrahigh-strength steels featuring minimum tensile strength levels of between 1000 and 1400 MPa combine superior performance with low weight. Named Docol UHS, the new steels are particularly applicable to the automotive industry. Their high-energy-absorbing properties, for example, make them useful as structural members and for components used in a car's crumple zone.

Since the steel is hardened prior to leaving the factory, industries using these steels no longer require their own warm-up plants and hardening furnaces. Cutting, shaping, and welding are achieved with traditional methods. The Docol UHS series consists of three standard steels: Docol 1000 DP, Docol 1200 DP, and Docol 1400 DP. Numbers relate to maximum loads measured in megapascals, MPa.

Cold-Runner Molding

In plastic part making, a mold in which the sprue-and-runner system (the manifold section) is insulated from the rest of the mold and temperature-controlled to keep the plastic in the manifold fluid. This mold design eliminates scrap loss from spruces and runners.

Cold-Setting Adhesive

An adhesive that sets at temperatures below 20°C (68°F). See also *hot-setting adhesive*, *intermediate-temperature-setting adhesive*, and *room-temperature-setting adhesive*.

Cold-Setting Process

In foundry practice, any of several systems for bonding mold or core aggregates by means of organic binders, relying on the use of catalysts rather than heat for polymerization (setting).

Cold-Slug Well

In plastic part making, the space provided directly opposite the sprue opening in an injection mold to trap the cold slug.

Coldstream Process

In powder metallurgy, a method of producing cleavage fractures in hard particles through particle impingements in a high-velocity cold gas stream. Also referred to as *impact crushing*.

Cold-Worked Structure

A microstructure resulting from plastic deformation of the metal or alloy below its recrystallization temperature.

Collapse

Inadvertent densification of cellular plastic material during manufacture resulting from the breakdown of cell structure.

Collapsibility

The tendency of a sand mixture to break down under the pressures and temperatures developed during casting.

Collapsible Tool

A press tool that can be easily disabled.

Collar

The reinforcing metal of a nonpressure thermit weld.

Collar Oiler

A collar on a shaft that extends into the oil reservoir and carries oil into a bearing as the shaft rotates. Wipers are usually provided to direct the oil into the bearing.

Collaring (Thermal Spraying)

Adding a shoulder to a shaft or similar component as a protective confining wall for the thermal spray deposit.

Collection Efficiency

The cross-sectional area of undisturbed fluid containing particles that will ultimately impinge on a given solid surface, divided by the projected area of the solid surface. Also known as collision efficiency, capture efficiency, catchment efficiency, and impaction ratio.

Collet

A split sleeve used to hold work or tools during machining or grinding.

Colligative Properties

Properties of plastics based on the number of molecules present. Most important are certain solution properties extensively used in molecular weight characterization.

Collimate

To make parallel to a certain line or direction.

Collimated

Rendered parallel.

Collimated Roving

Roving for reinforced plastics that has been made using a special process (usually parallel wound), such that the strands are more parallel than in standard roving.

Collimation

The degree of parallelism of light rays from a given source. A light source with good collimation produces parallel light rays, whereas a poor light source produces divergent, nonparallel light rays.

Collimator

The x-ray system component that confines the x-ray beam to the required shape. An additional collimator can be located in front of the x-ray detector to further define the portion of the x-ray beam to be measured.

Collision Efficiency

See *collection efficiency*.

Collodian Replica

In metallography, a *replica* of a surface cast in nitrocellulose.

Colloid

A stable (nonsettling) finely particulate suspension of some material within a fluid host, the dimensions of the former usually being about ±1 μm. Fogs, smokes, foams, emulsions, sols, and gels are examples.

Colloidal

A state of suspension in a liquid medium in which extremely small particles are suspended and dispersed but not dissolved.

Colonies (Titanium)

Regions within prior-β grains with α Platelets having nearly identical orientations. In commercially pure titanium, colonies often have serrated boundaries. Colonies arise as transformation products during cooling from the β field at cooling rates that induce platelet nucleation and growth.

Colophony

See *rosin*.

Color Buffing

Producing a final high luster by buffing. Sometimes called *coloring*.

Color Center

In materials characterization, a point lattice defect that produces optical absorption bands in an otherwise transparent crystal.

Color Concentrate

A measured amount of dye or pigment incorporated into a predetermined amount of plastic. The pigmented or colored plastic is then mixed into larger quantities of plastic material to be used for molding. This mixture is added to the bulk of plastic and measured quantity in order to produce a precise, predetermined color of finished articles to be molded.

Color Filter

In metallography, a device that transmits principally a predetermined range of wavelengths. See also *contrast filter* and *filter*.

Color Temperature

The temperature in degrees Kelvin at which a blackbody must be operated to provide a color equivalent to that of the source in question. See also *blackbody*.

Colorimeter

An instrument for measuring the hue, purity, and brightness of a color.

Coloring

Producing desired colors on metal by a chemical or electrochemical reaction. See also *color buffing*.

Columbium and Alloys

One of the basic elements, columbium (Cb) is also known as niobium and occurs in the minerals columbite and tantalite. A refractory metal, it closely resembles tantalum, is yellowish-white in color, has a specific gravity of 8.57, a melting point of 2468°C, and an electrical conductivity 13.2% relative to copper. It is quite ductile when pure or essentially free of interstitials and impurities, notably N_2, O_2, and H_2, which are limited to very small amounts. Tensile properties depend largely on purity, and columbium, with a total interstitial content of 100–200 ppm, provides about 276 MPa ultimate strength, 207 MPa yield strength, 30% elongation, and 105,000 MPa elastic modulus. Drawn wire having an ultimate tensile strength of 896 MPa has been produced. The metal is corrosion resistant to many aqueous media, including dilute mineral and organic acids, and to some liquid metals, notably lithium, sodium, and NaK. It is strongly attacked, however, by strong dilute alkalis, hot concentrated mineral acids, and HFl acid. At elevated temperatures, gaseous atmospheres attack the metal primarily by oxidation even if O_2 content is low, with the attack especially severe at 399°C and higher temperatures, necessitating the use of protective coatings. Columbium tends to gall and seize easily in fabrication. Sulfonated tallow and various waxes are the preferred lubricants in forming, and carbon tetrachloride in machining. Ferro-columbium is used to add the metal to steel. Columbium is also an important alloying element in non-ferrous alloys.

Secondary Fabrication

Pure columbium is considered one of the most workable of the refractory metals, and commercially fabricated columbium can be forged, rolled, swaged, drawn, and stamped by existing commercial techniques. In the primary or mill fabrication, and ingot is hot-worked by forging or extruding, following which the surface is conditioned to remove the contaminated layer, annealed in vacuum or an inert atmosphere to obtain a recrystallized structure, and then cold-worked (with intermediate anneals, if required) by any desired technique to final shape and size. Columbium metal containing less than a total of 0.12% combined O_2, N_2, and carbon can be given a cold reduction of over 90% in cross-sectional area. Secondary fabrication is done cold to avoid O_2 contamination, and lubrication is used to minimize galling or sizing on the working tools.

Vapor degreasing is an effective way to remove oils or grease from columbium parts, while immersion in various hot acids can be used for surface cleaning.

Columbium Alloys

These alloys are noted mainly for their heat resistance at temperatures far greater than those that can be sustained by most metals, but protective coatings are required for oxidation resistance. Thus, they find use for aircraft turbine components and in rocket engines, aerospace reentry vehicles, and for thermal and radiation shields. Cb–Sn and Cb–Ti alloys have found use as superconductors, and Cb–1Zr, a columbium–1% zirconium alloy, as has been used for high-temperature components, liquid-metal containers, sodium or magnesium vapor-lamp parts, and nuclear applications. It has a tensile yield strength of about 255 MPa at 21°C and 165 MPa at 1093°C. Thin cold-rolled sheet of columbium alloy C-103, which contains 10% hafnium and 1% titanium, has a tensile yield strength of 648 MPa at 21°C and 172 MPa at 1093°C. After recrystallization at 1315°C, however, yield strength drops to 345 MPa at 21°C and 124 MPa at 1093°C. The room-temperature tensile properties of the 10% tungsten, 10% hafnium, 0.1% yttrium columbium alloy, known as columbium alloy C-129, are 620 MPa ultimate strength, 517 MPa yield strength, 25% elongation, and 110,000 MPa elastic modulus. Its strength falls rapidly with increasing temperatures; tensile yield strength declining to about 234 MPa at 1000°C. Other columbium alloys and their principal alloying elements are Cb-752 (10% tungsten, 2.5% zirconium), B-66 (5% molybdenum, 5% vanadium, 1% zirconium), Cb-132M (20% tantalum, 15% tungsten, 5% molybdenum, 1.5% zirconium, 0.12% carbon), FS-85 (28% tantalum, 10% tungsten, 1% zirconium), and SCb-291 (10% tantalum, 10% tungsten). Typical tensile properties of columbium alloy B-66 at room temperature and 1093°C, respectively, are 882 and 448 MPa ultimate strength, 745 and 400 MPa yield strength, 12% and 28% elongation, and 105,000 and 82,700 MPa elastic modulus. Additionally, $CbSe_2$, is more electrically conductive than graphite and forms adhesive lubricating film. It is used in powder form with silver, copper, or other metal powders for self lubricating bearings and gears. Columbium carbide, CbC, is an extremely hard crystalline powder, which can be molded with a metal binder and sintered for use in cutting tools. The melting point is about 3800°C (6872°F). It is made by sintering columbium powder and carbon in a hydrogen furnace.

The secondary fabrication of columbium-base alloys creates a problem because of their greater strength at high temperatures and as a result are more difficult to work than the unalloyed metal. Procedures have been developed for the primary and secondary fabrication. Breakdown of the initial ingot requires higher temperatures and finish cold rolling involves more frequent annealing or, in some cases, hot rolling.

Columnar Grains/Structure

A coarse structure of parallel elongated grains formed by unidirectional growth, most often observed in castings, but sometimes seen in structures resulting from diffusional growth accompanied by a solid-state transformation.

Coma

In materials characterization, a lens aberration occurring in that part of the image field that is some distance from the principal axis of the system. It results from different magnification in the various lens zones. Extra-axial object points appear as short cone-like images with the brighter small head toward the center of the field (positive coma) or away from the center (negative coma).

Combination Die

(1) A die-casting die having two or more different cavities for different castings. (2) For forming, see *compound die*.

Combination Mill

(1) An arrangement of a continuous mill for roughing and a *guide mill* or *looping mill* for shaping. (2) A rolling mill in which the initial cogging mill is followed immediately by a further series of rolls to complete the hot rolling process.

Combination Mold

See *family mold*.

Combine Stresses

Any state of stress that cannot be represented by a single component of stress; that is, one that is more complicated than simple tension, compression, or shear.

Combined Carbon

Carbon in iron or steel that is combined chemically with other elements; not in the free state as graphite or temper carbon. The difference between the total carbon and the graphite carbon analyses. Contrast with *free carbon*.

Combined Cyanide

The cyanide of a metal-cyanide complex ion.

Combing

Lining up of reinforcing fibers.

Comet Tails (on a Polished Surface)

A group of comparatively deep unidirectional scratches that form adjacent to a microstructural discontinuity doing mechanical polishing. They have the general shape of a comet tail. Comet tails form only when a unidirectional motion is maintained between the surface being polished and the polishing cloth.

Comminution

(1) The production of fine powder by various processes including pulverizing, attrition, chemical or electrochemical techniques. (2) Breaking up or grinding an ore into small fragments. (3) Reducing metal to powder by mechanical means. (4) The act or process of reduction of powder particle size, usually but not necessarily by grinding or milling. See also *pulverization*.

Common Brass

Alloy of 63% copper and 37% zinc. The highest zinc level and hence the cheapest alloy that remains single, (alpha) phase allowing it to be cold worked.

Commutator-Controlled Welding

Spot or projection welding in which several electrodes, in simultaneous contact with the work, function progressively under the control of an electrical commutating device.

Compact (Noun)

The object produced by the compression of metal powder, generally while confined in a die.

Compact (Verb)

The operation or process of producing a compact; sometimes called pressing.

Compact Tension Specimen (CTS)

See *fracture toughness*.

Compacted Graphite Cast Iron

Cast iron having a graphite shape intermediate between the flake form typical of gray cast iron and the spherical form of fully spherulitic ductile cast iron. An acceptable compacted graphite iron structure is one that contains no flake graphite, <20% spheroidal graphite, and 80% compacted graphite (ASTM A247, type IV). Also known as CG iron or vermicular iron, compacted graphite cast iron is produced in a manner similar to that for ductile cast iron, but using a technique that inhibits the formation of fully spherulitic graphite nodules. Typical nominal compositions of CG irons containing 3.1%–4.0% C, 1.7%–3.0% Si, and 0.1%–0.6% Mn.

Compactibility

See *compressibility*.

Compacting Crack

A crack in a powder metallurgy compact that is generated during the major phases of the pressing cycle, such as load application, load release, and ejection.

Compacting Force

The force that acts on a powder to be densified expressed in newtons or tons.

Compacting Pressure

In powder metallurgy, the specific compacting force related to the area of contact with the press punch expressed in megapascals, meganewtons per square meter, or tons per square inch.

Compacting Tool Set

See *die*.

Compaction

(1) The act of forcing particulate or granular material together (consolidation) under pressure or impact to yield a relatively dense mass or formed object. Usually followed by drying, curing, or firing in refractory or other ceramic or powder metallurgy processing. (2) In ceramics or powder metallurgy, the preparation of a compact or object produced by the compression of a powder, generally while confined in a die, with or without the inclusion of lubricants, binders, etc., and with or without the concurrent applications of heat. (3) In reinforced plastics and composites, the application of a temporary vacuum bag and vacuum to remove trapped air and compact the lay-up.

Comparison Standard

In metallography, a standard micrograph or a series of micrographs, usually taken at 75×–100×, used to determine grain size by direct comparison with the image.

Compatibility (Frictional)

In tribology, materials that exhibit good sliding behavior, including resistance to adhesive wear, are termed frictionally compatible. Under some conditions materials that are not normally considered compatible in a metallurgical sense (for example, silver and iron) may be very compatible in the frictional sense.

Compatibility (Lubricant)

In tribology, a measure of the degree to which lubricants or lubricant components can be mixed without harmful effects such as formation of deposits.

Compatibility (Metallurgical)

A measure of the extent to which materials are mutually soluble in the solid state.

Compatibility (Plastics)

The ability of two or more substances combined with one another to form a homogeneous composition having useful plastic properties; for example, the suitability of a sizing or finish for use with certain general resin types. Nonreactivity or negligible reactivity between materials and contact.

Compensating Eyepiece

In metallography, an eyepiece design for use with apochromatic objectives. They are also used to advantage with high-power (oil-immersion) achromatic objectives. Because apochromatic objectives are undercorrected chromatically, these eyepieces are overcorrected. See also *apochromatic objective*.

Compensating Leads

The electrical conductor wires between a thermocouple and the instrument measuring the e.m.f. The wire materials are selected to have the same thermoelectric characteristics as their respective thermocouple wires over the temperature range that the lead experiences. The benefits offered are that the lead can be stranded to improve flexibility and it can be of cheaper material than the thermocouple.

Complete Fusion

Fusion which has occurred over the entire base material surfaces intended for welding and between all layers and weld beads.

Complete Joint Fusion

Joint penetration in which the weld metal completely fills the groove and is fused to the base metal throughout its total thickness.

Complete Penetration

See preferred term *complete joint penetration*.

Complex Ion

An ion that may be formed by the additional reaction of two or more other ions.

Complex Modulus

The ratio of stress to strain in which each is a vector that may be represented by a complex number. May be measured in tension or flexure, compression, or shear.

Complex Shear Modulus

The vectorial sum of the shear modulus and the loss modulus.

Complex Silicate Inclusions

A general term describing silicate inclusions containing visible constituents in addition to the silicate matrix. An example is corundum or spinel crystals occurring in a silicate matrix in steel.

Complex Young's Modulus

The vectorial sum of Young's modulus and the loss modulus.

Complexation

The formation of complex chemical species by the coordination of groups of atoms termed ligands to a central ion, commonly a metal ion. Generally, the ligand coordinates by providing a pair of electrons that forms an ionic or covalent bond to the central ion. See also *chelate*, *coordination compound*, and *ligand*.

Complexing Agent

A substance that is an electron donor and that will combine with a metal ion to form a soluble complex ion.

Compliance

(1) Tensile compliance is the reciprocal of Young's modulus. Shear compliance is the reciprocal of shear modulus. The term is also used in the evaluation of stiffness and deflection. (2) The measure of the ability of a shaft or similar body to flex elastically.

Component

(1) One of the elements or compounds used to define a chemical (or alloy) system, including all phases, in terms of the fewest substances possible. (2) One of the individual parts of a vector as referred to a system of coordinates. (3) An individual functional element in a physically independent body that cannot be further reduced or divided without destroying its stated function, for example, a resistor, capacitor, diode, or transistor.

Component of Variance

A part of the total variance identified with a specified source of variability.

Composite

See *composite material*.

Composite Bearing Material

A solid material composed of a continuous or particulate solid lubricant dispersed throughout a load-bearing matrix to provide continuous replenishment of solid lubricant films as wear occurs, and effective heat transfer from the friction surface.

Composite Coating

A coating on a metal or nonmetal that consists of two or more components, one of which is often particulate in form. Example: a cermet composite coating on a cemented carbide cutting tool. Also known as multilayer coating.

Composite Compact

A metal powder compact consisting of two or more adhering layers, rings, or other shapes of different metals or alloys with each material retaining its original identity.

Composite Electrode

A welding electrode made from two or more distinct components, at least one of which is filler metal. A composite electrode may exist in any of various physical forms, such as stranded wires, filled tubes, or covered wire.

Composite Joint

A joint in which welding is used in conjunction with mechanical joining.

Composite Material

Composite materials are based on the controlled distribution of one or more reinforcement materials in a continuous matrix. Plastics are the most common matrix materials, although metals, ceramics, and intermetallics are also used. Reinforcements include ceramics, glass, polymers, carbon, and metals. They can be in the shape of filaments, spheres, irregularly shaped particles, short fibers known as whiskers, or flat particles known as flakes.

Composites are also found in nature. Wood is a composite of cellulose fibers bonded by a matrix of natural polymers, mainly lignin. Egyptians reinforced mud with straw to make bricks. Concrete can be classified as a ceramic composite in which stones are dispersed among cement. And in the 1940s, short glass fibers impregnated with thermosetting resins, known as fiberglass, became the first composite with a plastic matrix.

In a properly designed composite, the reinforcement compensates for low properties of the matrix. Furthermore, in many cases synergism enables the reinforcing material to improve properties in the matrix. Composites also offer the capability of placing specific properties where they are needed on the part.

All these developments mean a larger and more-complicated choice of materials. This diversity has made plastics applicable to a broad range of consumer, industrial, automotive, and aerospace products. It has also made the job of selecting the best materials from such a huge array of candidates quite challenging.

Definition

Composite materials are macroscopic combinations of two or more distinct materials with a discrete and recognizable interface separating them.

Constituents and Construction

In principle, composites can be constructed of any combination of two or more materials-metallic, organic, or inorganic; but the constituent forms are more restricted. The matrix is the body constituent serving to enclose the composite and give it bulk form. Major structural constituents are fibers, particles, laminae or layers, flakes, fillers, and matrices. They determine the internal structure of the composite. Usually, they are the additive phase.

Because the different constituents are intermixed or combined, there is always a contiguous region. It may simply be an interface, that is, the surface forming the common boundary of the constituents. An interface is in some ways analogous to the grain boundaries in monolithic materials. In some cases, however, the contiguous region is a distinct added phase, called an interphase. Examples are the coating on the glass fibers in reinforced plastics and the adhesive that bonds the layers of a laminate together. When such an interphase is present, there are two interfaces, one between the matrix and the interphase and one between the fiber and the interface.

Interfaces are among the most important yet least understood components of a composite material. In particular, there is a lack of understanding of processes occurring at the atomic level of interfaces, and how these processes influence the global material behavior. There is a close relationship between processes that occur on the atomic, microscopic, and macroscopic levels. In fact, knowledge of the sequence of events occurring on these different levels is important in understanding the nature of interfacial phenomena. Interfaces in composites, often considered as surfaces, are in fact zones of compositional, structural, and property gradients, typically varying in width from a single atom layer to micrometers. Characterization of the mechanical properties of interfacial zones is necessary for understanding mechanical behavior.

Nature and Performance

Several classification systems for composites have been developed, including classification by (1) basic material combinations, for example, metal-organic or metal-inorganic; (2) bulk-form characteristics, such as matrix systems or laminates; (3) distribution of the constituents, that is, continuous or discontinuous; and (4) function, for example, electrical or structural.

There are five classes under the classification of basic material combinations: (1) fiber composites, composed of fibers with or without a matrix; (2) flake composites, composed of flat flakes with or without a matrix; (3) particulate composites, composed of particles with or without a matrix; (4) filled (or skeletal) composites, composed of a continuous skeletal matrix filled by a second material; and (5) laminar composites, composed of layer or laminar constituents.

There is also a classification based on dimensions. The dimensions of some of the components of composite materials vary widely and overlap the dimensions of the microstructural features of common conventional materials. They range from extremely small particles or fine whiskers to the large aggregate particles or rods in reinforced concrete.

The behavior and properties of composites are determined by the composition, form and arrangements, and interaction between the constituents. The intrinsic properties of the materials of which the constituents are composed largely determine the general order or range of properties of the composite. Structural and geometrical characteristics—that is, the shape and size of the individual constituents, their structural arrangement and distribution, and the relative amount of each—contribute to overall performance. Of far-reaching importance are the effects produced by the combination and interaction of the constituents. The basic principle is that by using different constituents it is possible to obtain combinations of properties and property values that are different from those of the individual constituents.

A performance index is a property or group of properties that measures the effectiveness of a material in performing a given function. The values of performance indices for a composite differ from those of the constituents.

Fiber–Matrix Composites

Fiber–matrix composites have two constituents and usually a bonding phase as well.

Fibers

The performance of a fiber–matrix composite depends on orientation, length, shape, and composition of the fibers; mechanical properties of the matrix; and integrity of the bond between fibers and matrix. Of these, orientation of the fibers is perhaps most important.

Fiber orientation determines the mechanical strength of the composite and the direction of greatest strength. Fiber orientation can be one-dimensional, planar (two-dimensional), or three-dimensional. The one-dimensional type has maximum composite strength and modulus in the direction of the fiber axis. The planar type exhibits different strengths in each direction of fiber orientation; and the three-dimensional type is isotropic but has greatly decreased reinforcing values. The mechanical properties in any one direction are proportional to the amount of fiber by volume oriented in that direction. As fiber orientation becomes more random, the mechanical properties in any one direction become lower.

Fiber length also impacts mechanical properties. Fibers in the matrix can be either continuous or short. Composites made from short fibers, if they could be properly oriented, could have substantially greater strengths than those made from continuous fibers. This is particularly true of whiskers, which have uniform high tensile strengths. Both short and long fibers are also called chopped fibers. Fiber length also has a bearing on the processibility of the composite. In general, continuous fibers are easier to handle but have more design limitations than short fibers.

Bonding

Fiber composites are able to withstand higher stresses than their individual constituents because the fibers and matrix interact, resulting in redistribution of the stresses. The ability of constituents to exchange stresses depends on the effectiveness of the coupling or bonding between them. Bonding can sometimes be achieved by direct contact of the two phases, but usually a specially treated fiber must be used to ensure a receptive adherent surface. This requirement has led to the development of fiber finishes, known as coupling agents. Both chemical and mechanical bonding interactions occur for coupling agents.

Voids (air pockets) in the matrix are one cause of failure. A fiber passing through the void is not supported by resin. Under load, the fiber may buckle and transfer stress to the resin, which readily cracks. Another cause of early failure is weak or incomplete bonding. The fiber–matrix bond is often in a state of shear when the material is under load. When this bond is broken, the fiber separates from the matrix and leaves discontinuities that may cause failure. Coupling agents can be used to strengthen these bonds against shear forces.

Reinforced Plastics

Probably the greatest potential for lightweight high-strength composites is represented by the inorganic fiber-organic matrix composites, and no composite of this type has proved as successful as glass-fiber-reinforced composites. As a group, glass-fiber-plastic composites have the advantages of good physical properties, including strength, elasticity, impact resistance, and dimensional stability; high strength-to-weight ratio; good electrical properties; resistance to chemical attack and outdoor weathering; and resistance to moderately high temperatures (about 260°C).

A critical factor in reinforced plastics is the strength of the bond between the fiber and the polymer matrix; weak bonding causes fiber pullout and delamination of the structure, particularly under adverse environmental conditions. Bonding can be improved by coatings and the use of coupling agents. Glass fibers, for example, are treated with silane (SiH_4) for improved wetting and bonding between the fiber and the matrix.

Generally, the greatest stiffness and strength in reinforced plastics are obtained when the fibers are aligned in the direction of the tension force. Other properties of the composite, such as creep resistance, thermal and electrical conductivity, and thermal expansion, are anisotropic. The transverse properties of such a unidirectionally reinforced structure are much lower than the longitudinal. Seven mechanical and thermal properties are of direct interest in assessing the potential of a new composite: density, modulus, strength, toughness, thermal conductivity, expansion coefficient, and heat capacity; others, such as fracture toughness and thermal diffusivity, are calculated from them.

Tailoring Properties

The ideal way to develop a product made of composites is to model and analyze it extensively by computer before a prototype is built. But this is difficult because most computer programs were developed for metals and do not work well with composites.

Many new applications for composites are structural. Since the objective of structural parts generally is to maximize strength-to-weight ratios, a key design objective is to optimize configurations as well as materials.

After a design is defined, manufacturing is the next challenge. Building a single part normally is not technically taxing. The trick comes in fabricating composite parts reliably in mass production. Manufacturing operations tend to be expensive because fabrication is labor-intensive, and the labor must be skilled.

The processes for fabricating composites also may produce built-in defects. For this reason, provisions for nondestructive testing should go hand-in-hand with fabrication. Unfortunately, available methods for nondestructive testing often leave a lot to be desired.

All these problems are being combated. Better guidelines are being developed to help designers select a composite and define its shape. Software is being developed to cope with the analytical complexities posed by composites and to help with the optimization process. Finally, major efforts are being exerted to automate fabrication processes and refine nondestructive testing operations.

Thermoplastic Composites

No longer is product design a constraint to the property limits and performance characteristics of unmodified grades of resins. Thermoplastics that are reinforced with high-strength, high-modulus fibers provide dramatic increases in strength and stiffness, toughness, and dimensional stability. The performance gain of these composites usually more than compensates for their higher cost. Processing usually involves the same methods used for unreinforced resins.

Glass and Mineral Fibers

Glass fibers used in reinforced compounds are high-strength, textile-type fibers, coated with a binder and coupling agent to improve compatibility with the resin and a lubricant to minimize abrasion between filaments. Glass-reinforced thermoplastics are usually supplied as ready-to-mold compounds. Molded products may contain as little as 5% and as much as 60% glass by weight. Pultruded shapes (usually using a polyester matrix) sometimes have higher glass contents. Most molding compounds, for best cost/performance ratios, contain 20%–40% glass.

Practically all thermoplastic resins are available in glass-reinforced compounds. Those used in largest volumes are nylon, polypropylene, polystyrene, ABS, and SAN, probably because most experience with reinforced thermoplastics has been based on these resins. The higher-performance resins—PES, PEI, PPS, PEEK, and PEK, for example—are also available in glass-fiber-reinforced composites, and some with carbon or aramid fibers as well.

Glass-fiber reinforcement improves most mechanical properties of plastics by a factor of two or more. Tensile strength of nylon, for example, can be increased from about 70 MPa to over 210 MPa, and deflection temperature to almost 260°C, from 77°C. A 40% glass-fortified acetyl has a flexural modulus of 1.89 MPa, a tensile strength of 150.5 MPa, and a deflection temperature of 168°C. Reinforced polyester has double the tensile and impact strength and four times the flexural modulus of the unreinforced resin.

Also improved in reinforced compounds are tensile modulus, dimensional stability, hydrolytic stability, and fatigue endurance.

Fiber reinforcement of a resin always changes its impact behavior and notch sensitivity. The change may be in either direction, depending on the specific resin involved. But even when the change is an improvement, these properties may not be high enough for certain demanding applications. This need has led to the development of impact-modified compounds—specifically, nylon 6 and 6/6 alloys, a nylon 6/6 copolymer, and a polypropylene copolymer—with up to 50% improvement over reinforced unmodified compounds. Although the impact properties of a glass-reinforced compound are not always superior to those of the unreinforced compound, the reinforced modified compounds are always superior to the reinforced unmodified grades.

Applications

Molded glass-reinforced and mineral-reinforced plastics are used in a broad range of structural and mechanical parts. For example, glass-reinforced nylon, because of its strength and stiffness, is used in gears and automotive under-the-hood components, whereas mineral-reinforced nylon is used in housings and body parts because it is tougher and has low warpage characteristics. Polypropylene applications include automotive air-cleaner housings and dishwasher tubs and inner doors. Polycarbonate is used in housings for water meters and power tools. Polyester applications include motor components—brush holders and fans—high-voltage enclosures, television tuner gears, electrical connectors, and automobile exterior panels.

Advanced Composites

Advanced composites comprise structural materials that have been developed for high-technology applications, such as airframe structures, for which other materials are not sufficiently stiff. In these materials, extremely stiff and strong continuous or discontinuous fibers, whiskers, or small particles are dispersed in the matrix. A number of matrix materials are available including carbon, glass, ceramics, metals, and polymers. Advanced composites possess enhanced stiffness and lower density compared to fiberglass and conventional monolithic materials. Although composite strength is primarily a function of the reinforcement, the ability of the matrix to support the fibers or particles and to transfer load to the reinforcement, is equally important. Also, the matrix frequently dictates service conditions, for example, the upper temperature limit of the composite.

Reinforcements

Continuous filamentary materials that are used as reinforcing constituents in advanced composites are carbonaceous fibers, organic fibers, inorganic fibers, ceramic fibers, and metal wires. Reinforcing inorganic materials are used in the form of discontinuous fibers and whiskers.

Carbon and graphite fibers offer high modulus and the highest strength of all reinforcing fibers. These fibers are produced in a pyrolysis chamber from three different precursor materials—rayon, polyacrylonitrile (PAN), and pitch. High-modulus carbon fibers are available in an array of yarns and bundles of continuous filaments (tows) with differing moduli, strengths, cross-sectional areas, twists, and plies.

Almost any polymer fiber can be used in a composite structure, but the first one with high-enough tensile modulus and strength to be used as a reinforcement in advanced composites was an aramid, or aromatic polyamide, fiber. Aramid fibers have been the predominant organic reinforcing fiber; graphite is a close second.

The most important inorganic continuous fibers for reinforcement of advanced composites are boron and SiC, both of which exhibit high stiffness, high strength, and low density. Continuous fibers are made by chemical vapor deposition (CVD) processes. Other inorganic compounds that provide stiff, strong discontinuous fibers that predominate as reinforcements for metal–matrix composites (MMC) are SiC, Al_2O_3, graphite, Si_3N_4, TiC, and boron carbide.

Polycrystalline Al_2O_3 is a commercial continuous fiber that exhibits high stiffness, high strength, high melting point, and exceptional resistance to corrosive environments. One method to produce the fibers is dry spinning followed by heat treatment.

Whiskers are single crystals that exhibit fibrous characteristics. Compared to continuous or discontinuous polycrystalline fibers, they exhibit exceptionally high strength and stiffness. SiC whiskers are prepared by chemical processes or by pyrolysis of rice hulls. Whiskers made of Al_2O_3 and Si_3N_4 are also available. Particulates vary widely in size, characteristics, and function, and since particulate composites are usually isotropic, their distribution is usually random rather than controlled.

Organic–Matrix Composites

In many advanced composites the matrix is organic, but metal matrices are also used. Organic matrix materials are lighter than metals, adhere better to the fibers, and offer more flexibility in shaping and forming. Ceramic–matrix composites (CMC), carbon–carbon composites (C–C), and intermetallic–matrix composites (IMC) have applications where organic or metal matrix systems are unsuitable.

Materials

Epoxy resins have been used extensively as the matrix material. However, bismaleimide (BMI) resins and polyimide (PI) resins have been developed to enhance in-service temperatures. Thermoplastic resins, PEK, and polyphenylene sulfide (PPS) are in limited use.

The continuous reinforcing fibers for organic matrices are available in the forms of monofilaments, multifilament fiber bundles, unidirectional ribbons, roving (slightly twisted fiber), and single-layer and multilayer fabric mats. Frequently, the continuous reinforcing fibers and matrix resins are combined into a nonfinal form known as a prepreg.

The matrix material generally governs the service temperature. For PMCs, thermosets are the common matrix material. *Epoxy*, the most widely used, allows service temperatures up to about 300°F (149°C). *Bismaleimide* (BMI), which has replaced epoxy to some

extent in military aircraft applications, permits use to about 350°F (177°C). *Cycom 5250-4, 5260,* and *5270-1* are BMIs from Cytec Fiberite. The 5250-4 and toughened 5260 have service temperatures to about 350°F (177°C), the 5270-1 to as high as 450°F (232°C). *Cycom 5250-4 RTM* is for resin-transfer-molding applications.

Polyimide, with a maximum service temperature of at least 500°F (260°C), is used to a much more limited extent. The principal load-bearing elements, however, are the fibers, typically continuous, contained by the matrix. These include *aramid, Kevlar* mainly, *boron, glass,* and *graphite*. PMCs are lightweight, strong, and rigid, thus providing high strength-to-weight ratios (specific strength) and high rigidity-to-weight ratios (specific stiffness). Other thermosets include *cyanate esters*, which feature good moisture and heat resistance and better electrical properties; *polyetheramide* (PEA) from PEAR Industries for toughness and heat resistance; and, for aircraft interior parts, *phenolics*, which feature heat resistance and flame retardance. Thermoplastic matrices are not as commonly used but have potential advantages in moisture, heat, and impact resistance. These include polyamideimide (PAI), polyetheretherketone (PEEK), polyetherimide (PEI), and polyphenylene sulfide (PPS). Another advantage is that fiber direction can be oriented to suit applied load direction. Such composites are made by manual or automatic layout of thin (0.010 in. [0.254 mm]) prepreg plies or by filament winding, followed by curing in autoclaves or presses. Prepreg is a partially cured and somewhat tacky fiber-reinforced resin, which must be kept in refrigerated storage to keep from spoiling. Filament winding involves winding a tow of fibers or a series of tows (band) around a mandrel of the shape of the part to be produced. In "dry winding," tows of prepreg are used. In "wet winding," the tows or bands are first drawn through a resin bath.

C-Bar, or *composite rebar*, is a *PMC bar* developed by Marshall Industries Composites for reinforcing concrete. Intended to compete with epoxy-coated steel rebar, it consists of a protruded rod core of fiber-reinforced urethane-modified vinyl ester with a helically ribbed exterior of compression-molded, urethane-modified sheet molding compound to bond to concrete. The fibers, originally of E-glass, can also be aramid or graphite. The rebar is not conductive or corrodible, has a coefficient of thermal expansion closer to that of concrete than steel, and weighs about one-fourth as much as a comparable steel rod. Protruded fiber-reinforced epoxy plates are adhesive-bonded to form *glulams*—glued laminated beams—and used to locally reinforce wood glulams typically made of hemlock or Douglas fir plates.

Fabrication

Many processes are available for the fabrication of organic matrix composites. The first process is contact molding in order to orient the unidirectional layers at discrete angles to one another. Contact molding is a wet method, in which the reinforcement is impregnated with the resin at the time of molding. The simplest method is hand layup, whereby the materials are placed and formed in the mold by hand and the squeezing action expels any trapped air and compacts the part.

Molding may also be done by spraying, but these processes are relatively slow and labor costs are high, even though they can be automated. Many types of boats, as well as buckets for power-line servicing equipment, are made by this process.

Another process is vacuum-bag molding, where prepregs are laid in mold to form the desired shape. In this case, the pressure required to form the shape and achieve good bonding is obtained by covering the layup with a plastic bag and creating a vacuum. If additional heat and pressure are desired, the entire assembly is put into an autoclave. To prevent the resin from sticking to the vacuum bag and to facilitate removal of excess resin, various materials are placed on top of the prepreg sheets. The molds can be made of metal, usually aluminum, but more often are made from the same resin (with reinforcement) as the material to be cured. This eliminates any problem with differential thermal expansion between the mold and the part.

In filament winding, the resin and fibers are combined at the time of curing. Axisymmetric parts, such as pipes and storage tanks, are produced on a rotating mandrel. The reinforcing filament, tape, or roving is wrapped continuously around the form. The reinforcements are impregnated by passing them through a polymer bath. However, the process can be modified by wrapping the mandrel with prepreg material. The products made by filament winding are very strong because of their highly reinforced structure. For example, filament winding can be used directly over solid-rocket-propellant forms.

Pultrusion is a process used to produce long shapes with constant profiles, such as rods or tubing, similar to extruded metal products. Individual fibers are often combined into a tow, yarns, or roving, which consists of a number of tows or yarns collected into a parallel bundle without twisting (or only slightly so). Filaments can also be arranged in a parallel array called a tape and held together by a binder. Yarns or tows are often processed further by weaving, braiding, and knitting or by forming them into a sheet-like mat consisting of randomly oriented chopped fibers or swirled continuous fibers held together by a binder.

Weaving to produce a fabric is a very effective means of introducing fibers into a composite. There are five commonly used patterns, which include box or plain, basket, crowfoot, long-shaft, and leno weave. Although weaving is usually thought of as a two-dimensional process, three-dimensional weaving is often employed.

Knitting is a process of interlooping chains of tow or yarn. Advantages of this process are that the tow or yarn is not crimped as happens in weaving, and higher mechanical properties are often observed in the reinforced product. Also, knitted fabrics are easy to handle and can be cut without falling apart.

In braiding, layers of helically wound yarn or tow are interlaced in a cylindrical shape, and interlocks can be produced at every intersection of fibers. During the process, a mandrel is fed through the center of a braiding machine at a uniform rate, and the yarn or tow from carriers is braided around the mandrel at a controlled angle. The machine operates like a maypole, the carriers working in pairs to accomplish the over-and-under sequencing. The braiding process is most effective for cylindrical geometries. It is used for missile heat shields, lightweight ducts, fluid-sealing components such as packings and sleevings, and tubes for insulation.

Metal–Matrix Composites (MMCs)

MMCs are usually made with alloys of aluminum, magnesium, or titanium, and the reinforcement is typically a ceramic in the form of particulates, platelets, whiskers, or fibers, although other systems may be used. MMCs are often classified as discontinuous or continuous depending on the geometry of the reinforcement. Particulates, platelets, and whiskers are in the discontinuous category, whereas the continuous category is reserved for fibers and wires. The type of reinforcement is important in the selection of an MMC because it determines virtually every aspect of the product, including mechanical properties, cost, and processing method. The primary methods for processing of discontinuous MMCs are powder metallurgy, liquid metal infiltration, squeeze or pressure casting, and conventional casting; however, most of these methods do not result in finished parts. Therefore, most discontinuously reinforced MMCs require secondary processing, which includes conventional wrought metallurgy operations such as extrusion, forging, and rolling; standard and nonstandard machining operations; and joining techniques such as brazing and welding.

MMCs, like PMCs, were in use long before this term was coined. Examples include *cermets*, or ceramic-reinforced metals, such as *tungsten-carbide* particles in a cobalt matrix for cutting tools and *titanium-carbide* particles in steel for heat- and wear-resistant parts. MMCs may contain continuous or discontinuous fibers, particulates, whiskers or preforms as a reinforcing constituent. As a class, they are far more heat-resistant than PMCs. Among the MMCs that have been made are aluminum, copper, cobalt, lead, and magnesium reinforced with graphite. Boron has served as a reinforcement for aluminum, magnesium, and titanium; silicon carbide for aluminum, titanium, and tungsten; and alumina for aluminum. Compared with PMCs, applications so far have been limited, and these are largely limited to aluminum. Aluminum reinforced with continuous boron fibers is used for struts in the Space Shuttle, and aluminum reinforced with continuous graphite fibers is used for the Hubble telescope masts. *Fiber preforms* have been used to selectively reinforce cast aluminum products. Brake rotors made of 30% alumina in a 1%-magnesium aluminum alloy can operate at temperatures up to 1000°F (540°C) and 360 aluminum alloy with 30% silicon carbide has withstood 840°F (450°C). For semiconductor packaging, die-cast aluminum alloy with 70% silicon carbide provides low thermal expansion and high heat-dissipating thermal conductivity for superior reliability. *Titanium–matrix composites* are candidates for aircraft gas-turbine-engine parts. Pressure infiltration, mainly with either aluminum or magnesium alloys in porous ceramic, carbide, nitride, carbon, or graphite preforms, is used by Metal Matrix Cast Composites, Inc., to make MMCs. Pressureless infiltration is also used. For example, with the Primax Cast process, infiltrating a 30% by volume silicon carbide preform with *Lanxide 92-X-2050*, an aluminum, 10% silicon, 1% magnesium, 1% iron alloy results in an MMC with the density of 0.101 lb/in.^{3} (2796 kg/m^3). *Aluminum alloys* reinforced with *alumina, boron carbide*, or *silicon carbide* particulates are commercially available as wrought and foundry products.

Ceramic–Matrix Composites (CMCs)

One type of CMC incorporates a continuous fiber, and another type a discontinuous reinforcement such as whiskers. Both approaches enhance fracture resistance, but the mechanism is substantially different. Continuous-fiber-reinforced ceramics resist catastrophic failure because, after the matrix fails, the fiber supports the load. When whiskers are used as reinforcements, the resistance to crack propagation is enhanced and hence the composite is less sensitive to flaws. However, once a crack begins to propagate, the failure will be catastrophic.

CMCs and IMCs are largely developmental. Both are promising for still greater heat resistance, although the inherent brittleness of the CMCs may limit their use in structural applications. Allied Signal makes CMCs using directed metal oxidation or chemical vapor infiltration techniques. Components include silicon carbide-particulate-reinforced *alumina* tubes and connecting sleeves for high-temperature air heaters and *silicon carbide–reinforced silicon carbide* panels for the vortex finder of a cyclone high-performance particle separator. The SiC/SiC panels were made by fabricating fiber preforms woven, braided, or wound to shape and infiltrating them with chemical vapors reacting at high temperature to form the silicon carbide matrix on and between the fibers. Matrix materials for discontinuously reinforced CMCs made by Triton Systems include *silicon carbide, hafnium carbide, tantalum carbide, boron nitride,* or *silicon nitride*, and *refractory borides*. Continuous fiber CMCs include *carbon-reinforced silicon carbide, alumina–reinforced silicon carbide*, and *SiC/SiC. Silcomp*, from General Electric, comprises SiC fibers in an SiC and silicon matrix. It features low porosity, for oxidation and heat resistance, strength, and rigidity and may be suitable for gas-turbine-engine combustor liners and shrouds. A glass-fiber-reinforced CMC serves as armor in the U.S. Army's Crusader ground combat vehicle. Silicon nitride-coated fibers in a barium–strontium–aluminum-silicate glass that converts to a strong and tough glass ceramic on processing features low permittivity and electromagnetic absorption.

IMCs are seen as potential candidates for aircraft, aircraft-engine, and spacecraft components exposed to temperatures above 2000°F (1093°C). Promising matrix materials include *molybdenum disilicide* ($MoSi_2$), *nickel aluminides*, and *titanium aluminides*. Reinforcements include particles, whiskers, and continuous or discontinuous fibers of alumina or silicon carbide. $MoSi_2$, which excels in corrosion and oxidation resistance, has a brittle-to-ductile transition temperature of about 1832°F (1000°C), but alloying with tungsten disilicide (WSi_2) improves toughness at lower temperatures. The nickel aluminide, Ni_3Al, with 0.5% boron and reinforced with alumina fibers, has a potential service temperature of 1500°F (816°C) or greater. For titanium aluminide, TiAl, reinforced with alumina, this temperature may approach 1900°F (1038°C), and for Ti_3Al with columbium, reinforced with silicon-carbide fibers, it is within the range of 1472°F–1562°F (800°C–850°C). An *SiC/SiC composite* from Allied Chemical is highly resistant to high concentrations of potassium and sodium both in chlorides and sulfides as well as to more complex compounds such as coal ash at temperatures up to 2100°F (1150°C).

Carbon–Carbon Composites (C–C)

Carbon–carbon composite is a specialized material made by reinforcing a carbon matrix with continuous carbon fiber. This type of composite has outstanding properties over a wide range of temperatures in both vacuum and inert atmospheres. It will even perform well at elevated temperatures in an oxidizing environment for short times. It has high strength, modulus, and toughness up to 2000°C; high thermal conductivity; and a low coefficient of thermal expansion. A material with such properties is excellent for rocket motor nozzles and exit cones, which require high-temperature strength as well as resistance to thermal shock. Carbon–carbon composites are also used for aircraft and other high-performance brake applications that take advantage of the fact that C–C composites have the highest energy capability of any known material. If a carbon–carbon composite is exposed to an O_2-containing atmosphere above 600°C for an appreciable time, it oxidizes, and therefore it must be protected by coatings.

C–Cs are noted for their lightweight and good strength and low thermal expansion at temperatures greater than 3600°F (2000°C). A carbon-fiber-reinforced carbon piston developed at NASA Langley Research Center maintains high strength and stiffness at operating temperatures to over 2500°F (1371°C). C–Cs also have high thermal stability and nonoxidizing environments, are nonmelting and nonflammable, and possess low ablation and erosion rates. They are also tough and resistant to abrasion and corrosion, have a high thermal and electrical conductivity and high temperatures, and have excellent resistance to thermal shock. With *silicon carbide* as the oxygen-barrier coating, C–Cs service thermal-protection systems in the nose cone and wing leading edges of the Space Shuttle. Aircraft brake disks, 8–20 in. (200–500 mm) in diameter and 1–2 in. (25–50 mm) thick, are by far the largest-production use. Other applications include race-car brake and clutch components, heat sinks

for electronic circuit boards, solid- and liquid-propellant rocket-motor sections, aerospace-vehicle components, thermal insulation for spacecraft and vacuum or inert-gas furnaces, furnace trays and baskets, glass-production equipment, and high-temperature bolts, nuts, and rods.

Applications

The use of fiber-reinforced materials in engineering applications has grown rapidly. Selection of composites rather than monolithic materials is dictated by the choice of properties. The high values of specific stiffness and specific strength may be the determining factor, but in some applications wear resistance or strength retention at elevated temperatures is more important. A composite must be selected by more than one criterion, although one may dominate.

Components fabricated from advanced organic–matrix-fiber-reinforced composites are used extensively on commercial aircraft as well as for military transports, fighters, and bombers. The propulsion system, which includes engines and fuel, makes up a significant fraction of aircraft weight (frequently 50%) and must provide a good thrust-to-weight ratio and efficient fuel consumption. The primary means of improving engine efficiency are to take advantage of the high specific stiffness and strength of composites for weight reduction, especially in rotating components, where material density directly affects both stress levels and critical dynamic characteristics, such as natural frequency and flutter speed.

Composites consisting of resin matrices reinforced with discontinuous glass fibers and continuous glass-fiber mats are widely used in truck and automobile components bearing light loads, such as interior and exterior panels, pistons for diesel engines, drive shafts, rotors, brakes, leaf springs, wheels, and clutch plates.

The excellent electrical insulation, formability, and low cost of glass-fiber-reinforced plastics have led to their widespread use in electrical and electronic applications ranging from motors and generators to antennas and printed circuit boards.

Composites are also used for leisure and sporting products such as the frames of rackets, fishing rods, skis, golf club shafts, archery bows and arrows, sailboats, racing cars, and bicycles.

Advanced composites are used in a variety of other applications, including cutting tools for machining of superalloys and cast iron and laser mirrors for outer-space applications. They have made it possible to mimic the properties of human bone, leading to development of biocompatible prostheses for bone replacements and joint implants. In engineering, composites are used as replacements for fiber-reinforced cements and cables for suspension bridges. See also *carbon–carbon composites, ceramics–matrix composites, metal–matrix composites* and *resin–matrix composites*.

Composite Plate

An electrodeposit consisting of layers of at least two different compositions.

Composite Powder

A powder in which each particle consists of two or more different materials.

Composite Structure

A structural member (such as a panel, plate, pipe, or other shape) that is built up by bonding together two or more distinct components, each of which may be made of a metal, alloy, nonmetal, or *composite material*. Examples of composite structures include: honeycomb panels, clad plate, electrical contacts, sleeve bearings, carbide-tipped drills or lathe tools, and weldments constructed of two or more different alloys.

Composition Metal

Also called *composition brass*, although it does not have the characteristics of a true brass. A general name for casting alloys, such as *copper alloy C83600*, that are in a midposition between the brasses and bronzes. The most widely used standard composition metal is *ounce metal*, containing 85% copper, 5% zinc, 5% tin, and 5% lead. It makes a good average bearing metal, and because it gives a dense casting that will withstand liquid pressures, it is also used for valves, pumps, and carburetor parts. It cast well, machines easily, and takes a good polish, so that is widely employed for mechanical castings. It has about the same coefficient of expansion as copper and can thus be used for pipe fitting. *ASTM alloy No. 2* is this metal, and it may also contain up to 1% nickel and small amounts of iron, either as intentional additions to increase strength or as impurities. This alloy has also been called *red casting brass, hydraulic bronze*, and *steam brass*, and it has also been used for forgings, producing parts with a tensile strength of 227 MPa and 25% elongation.

Compositional Depth Profile

In materials characterization, the atomic concentration measured as a function of the perpendicular distance from the surface.

Compound

(1) In chemistry, a substance of relatively fixed composition and properties, whose ultimate structural unit (molecule or repeat unit) is comprised of atoms of two or more elements. The number of atoms of each kind in this ultimate unit is determined by natural laws and is part of the identification of the compound. (2) In reinforced plastics and composites, the intimate admixture of a polymer with other ingredients, such as fillers, softeners, plasticizers, reinforcements, catalysts, pigments, or dyes. A thermose compound usually contains all the ingredients necessary for the finished product, while a thermoplastic compound may require subsequent addition of pigments, blowing agents, and so forth.

Compound Compact

A powder metallurgy *compact* consisting of mixed metals, the particles of which are joined by pressing or sintering, or both, with each metal particle retaining substantially its original composition.

Compound Die

Any die designed to perform more than one operation on a part with one stroke of the press, such as blanking and piercing, in which all functions are performed simultaneously within the confines of the blank size being worked.

Compressibility

(1) The ability of a powder to be formed into a compact having well-defined contours and structural stability at a given temperature and pressure; and measure of the plasticity of powder particles.

(2) A density ratio determined under definite testing conditions. Also referred to as compactibility.

Compressibility Curve

A plot of the green density of a powder compact with increasing pressure.

Compressibility Test

In powder metallurgy, a test to determine the behavior of a powder under applied pressure. It indicates the degree of densification and cohesiveness of a compact as a function of the magnitude of the pressure.

Compression and Transfer Molded Plastics

Compression and transfer molding techniques are the most commonly used methods of molding thermosetting molding compounds as well as rubber parts. They may also be used for forming thermoplastic materials (e.g., compression molding of vinyl phonograph records), but usually other methods are more economical for molding thermoplastics.

The two processes are somewhat similar in terms of sizes and shapes produced. The major difference lies in the greater control of material flow permitted in transfer molding, allowing use of more delicate inserts and production of somewhat more complicated shapes.

Compression Molding

A technique of thermoset molding in which the plastic molding compound (generally preheated) is placed in the heated open mold cavity, the mold is closed under pressure (usually in a hydraulic press), causing the material to flow and completely fill the cavity, and then pressure is held until the material has cured.

The materials to be molded are generally softened by preheating in conventional ovens or, more frequently, in a dielectric preheating unit prior to placement in the mold.

Compression molding techniques are used most extensively for the manufacture of products made from thermosetting plastic materials and rubbers. These materials require the relatively high pressures and temperatures afforded by the compression molding process. Such materials include the phenolics, the melamines, the ureas, and the polyester resins. Under special circumstances, thermoplastics are compression-molded, but the injection molding process is usually more economical for the production of thermoplastic parts.

The process is ideal for the production of such items as thermosetting radio cabinets, television cabinets, trays, and other products that require resistance to heat. The size of the compression molded articles is generally limited only by the platen size and tonnage capacity of the presses used. Through the use of multicavity molds, small electrical components such as switch plates and terminal blocks may be produced economically. Compression molding is also used to produce extremely large parts such as fiberglass boat hulls and the complete fuselage for radio-controlled target aircraft.

Compression molding is not practical for many intricate products where complicated molds are required. Thermosetting materials are extremely stiff or viscous plastic masses during the mold-closing period. The internal pressures developed by these materials tend to distort or break delicate core pins and other small mold components. The process also may be unsatisfactory for the production of extremely close tolerance articles, particularly where critical dimensions are influenced by the mold parting line. Flash thickness produced at the mold parting line tends to vary from cycle to cycle, thus changing the dimensions in the direction of the stroke of the press.

Molding pressures range from as low as 0.35 MPa for certain polyester compounds to as high as 2.3 kg/mm² for stiff high-impact phenolic materials. Process temperatures range from 121°C to 177°C depending upon the material used.

Thickness of the molded part influences production rates. A general rule of thumb allows a minute of mold close-time for each 3.2 mm of wall thickness. Standard dimensional tolerances are usually figured at ±0.05 mm⁻¹, although closer tolerances can be held under special circumstances. Where phenolic and urea materials are to be molded, hardened steel is used for the mold construction. Such molds are usually carburized and hardened to about 50 Rockwell C. For use with polyester materials, where pressure requirements are not high, prehardened steel at about 38 Rockwell C is usually satisfactory.

Transfer Molding

Transfer molding is best described as a closed-mold technique wherein the material is injected or transferred into a closed mold through a gate or runner system. Essentially, transfer molding is a one-shot injection molding technique. The process is particularly adaptable again to thermosetting materials, since these materials retain their plastic condition after preheating for only a short length of time.

Generally, the comments made regarding compression molding apply to this process as well. The distinct advantage in the transfer molding process lies in the fact that the mold is completely closed and under clamping pressure before the material is injected into the mold cavity. This results in little or no flash and accurate control of dimensions.

A preweighted material charge is plasticized generally in a dielectric preheating unit. The charge is then placed in a pot that is usually positioned above the closed mold. A ram enters the pot and forces the material through an orifice into the closed mold. The transfer plunger and mold are kept under pressure for a predetermined time to allow the chemical/heat hardening process to proceed. When the mold is opened, the small amount of material remaining in the transfer pot and that which has filled the orifice is removed as a cull and discarded.

Transfer molding generally operates at faster cycles than compression molding. Because of the highly plastic condition of the material, complex part designs involving cores, undercuts, and moving die parts are best adapted to this process. Molded-in inserts can be held in position more easily in the transfer molding operation. Since the mold is closed during the entire molding operation, control of dimensions is more satisfactory.

With a properly designed mold, transfer molded articles require fewer finishing operations with a resultant lower net cost.

The process results in higher material costs due to the loss of material in the transfer pot and runner system. High-impact phenolic materials have generally lower physical properties when transfer molded compared with those obtained by compression molding.

Transfer molding techniques are used in the manufacture of a wide range of product shapes and sizes. Small, complex electrical components with molded-in terminals are made by this process. Radio and television cabinets weighing up to 1.7 kg have been transfer molded from phenolic materials.

Compression Crack

See *compacting crack.*

Compression Modulus

See *bulk modulus of elasticity.*

Compression Ratio (Plastics)

In an extruder screw, the ratio of the volume available in the first flight at the hopper to the volume at the last flight, at the end of the screw.

Compression Ratio (Powder Metallurgy)

The ratio of the volume of the loose powder to the volume of the compact made from it.

Compression Set

In seals, the difference between the thickness of a gasket of static seal before the seal is compressed and after it is released from compression. Compression set is normally expressed as a percentage of the total compression.

Compression Test

A method for assessing the ability of a material to withstand compressive loads. Analyses of structural behavior or metal-forming require knowledge of compression stress–strain properties. Failure modes include cracking, buckling, and general collapse.

Compressive

Pertaining to forces on a body or part of a body that tend to crush, or compress, the body.

Compressive Modulus

The ratio of compressive stress to compressive strain below the proportional limit. Theoretically equal to Young's modulus determined from tensile experiments.

Compressive Strength

The maximum compressive stress that a material is capable of developing, based on original area of cross section. If a material fails in compression by a shattering fracture, the compressive strength has a very definite value. If a material does not fail in compression by a shattering fracture, the value obtained for compressive strength is an arbitrary value depending upon the degree of distortion that is regarded as indicating complete failure of the material.

Compressive Stress

A stress that causes an elastic body to deform (shorten) in the direction of the applied load. Contrast with *tensile stress.*

Compressometer

Instrument for measuring change in length over a given gauge length caused by application or removal of a force. Commonly used in compression testing of metal specimens.

Compton Scattering

In materials characterization, the elastic scattering of photons by electrons. Contrast with *Rayleigh scattering*.

Computed Tomography (CT)

The collection of transmission data through an object and the subsequent reconstruction of an image corresponding to a cross section through this object. Also known as computerized axial tomography, computer-assisted tomography, CAT scanning, or industrial computed tomography.

Computer Aided (Assisted) Design (CAD)

Design techniques employing computer-based mathematical and graphics programs to assist in matters such as stress determination, computer layout, and product appearance.

Computer Aided (Assisted) Manufacture (Machining) (CAM)

Processes employing computers to assist in matters such as monitoring of product dimensions, rectification of deviations, etc.

ConCast

Continuously cast material or the continuous casting process.

Concave Grating

In materials characterization, a diffraction grating on a concave mirror surface. See also diffraction grating and plane grating.

Concave Root

An indentation along the line of the weld root. Same as root concavity.

Concave Root Surface

A root surface which is concave.

Concave Weld

Any weld including butt and fillet in which the exposed weld surface falls below a line joining the points where the weld metal meets the parent metal.

Concavity

The maximum distance from the face of a concave fillet weld perpendicular to a line joining the toes.

Concentration

(1) The mass of a substance contained in a unit volume of sample, for example, grams per liter. (2) A process for enrichment of an ore in valuable mineral content by separation and removal of waste material, or *gangue*. (3) The quantity of one substance contained in another, usually as a percentage.

Concentration Cell

An *electrolytic cell*, the *electromotive force* of which is caused by a difference in concentration of some component in the electrolyte. This difference leads to the formation of discrete *cathode* and *anode* regions.

Concentration Gradient

A variation, from point-to-point, and the proportion of two or more elements. The gradient may be a sharp step, as at the boundary between two phases, or it may be a shallow slope as in coring.

Concentration Polarization

That portion of the *polarization* of a cell produced by concentration changes resulting from passage of current through the electrolyte.

Conchoidal Markings

Concentric marks on the surface of a fatigue fracture. They are visible to the unaided eye and marked the crack front at irregularities in the load cycling such as interruptions in cycling or an abnormal load. Also termed *beach marks*, *shell marks* or *arrest marks*. Compare with *striations*.

Concrete

(1) A composite material that consists essentially of a binding medium within which are embedded particles or fragments of aggregates (maximum aggregate size >5 mm, or 0.2 in.). In hydraulic-cement concrete, the binder is formed from a mixture of hydraulic cement and water. (2) A homogenous mixture of portland cement, aggregates, and water and which may contain admixtures. See also *cement*.

Concrete is a construction material composed of portland cement and water combined with sand, gravel, crushed stone, or other inert material such as expanded slag or vermiculite. The cement and water form a paste that hardens by chemical reaction into a strong stone-like mass. The inert materials are called aggregates, and for economy no more cement paste is used than is necessary to coat all the aggregate surfaces and fill all the voids. The concrete paste is plastic and easily molded into any form or troweled to produce a smooth surface. Hardening begins immediately, but precautions are taken, usually by covering, to avoid rapid loss of moisture because the presence of water is necessary to continue the chemical reaction and increase the strength. Too much water, however, produces a concrete that is more porous and weaker. The quality of the paste formed by the cement and water largely determines the character of the concrete.

Proportioning of the ingredients of concrete is referred to as designing the mixture, and for most structural work the concrete is designed to give compressive strengths of 16–34 MPa. A rich mixture for columns may be in the proportion of 1 volume of cement to

1 of sand and 3 of stone, whereas a lean mixture for foundations may be in the proportion of 1:3:6. Concrete may be produced as a dense mass that is practically artificial rock, and chemicals may be added to make it waterproof, or it can be made porous and highly permeable for such use as filter beds. An air-entraining chemical may be added to produce minute bubbles for porosity or light weight. Normally, the full hardening period of concrete is at least 7 days. The gradual increase in strength is due to the hydration of the tricalcium aluminates and silicates. Sand used in concrete was originally specified as roughly angular, but rounded grains are now preferred. The stone is usually sharply broken. The weight of concrete varies with the type and amount of rock and sand. A concrete with trap rock may weigh 2483 kg/m^3. Concrete is stronger in compression than in tension, and steel bars or mesh are embedded in structural members to increase the tensile and flexural strengths. In addition to the structural uses, concrete is widely used in precast units such as block, tile, sewer and water pipe, and ornamental products.

Concretes are similar in composition to mortars, which are used to bond unit masonry. Mortars, however, are normally made with sand as the sole aggregate, whereas concretes contain much larger aggregates and thus usually have greater strength. As a result, concretes have a much wider range of structural applications, including pavements, footings, pipes, unit masonry, floor slabs, beams, columns, walls, dams, and tanks.

Classification

Concretes may be classified as flexible or rigid. These characteristics are determined mainly by the cementitious materials used to bond the aggregates. Flexible concretes tend to deform plastically under heavy loads or when heated. Rigid concretes are considerably stronger in compression than in tension and tend to be brittle. To overcome this deficiency, strong reinforcement may be incorporated in the concrete, or prestress may be applied to keep the concrete under compression.

Flexible Concretes

Usually, bituminous or asphaltic concretes are used when a flexible concrete is desired. The main use of such concretes is for pavements.

The aggregates generally used are sand, gravel or crushed stone, and mineral dust; the binder is asphalt cement, an asphalt specially refined for the purpose. A semisolid at normal temperatures, the asphalt cement may be heated until liquefied for bonding the aggregates. Ingredients usually are mixed mechanically in a pug mill, which has pairs of blades revolving in opposite directions. While the mix is still hot and plastic, it can be spread to a specified thickness and shaped with a paving machine and compacted with a roller or by tamping to a desired density. When the mix cools, it hardens sufficiently to withstand heavy loads.

Sulfur, rubber, or hydrated lime may be added to an asphalt-concrete mix to improve the performance of the product.

Rigid Concretes

Ordinary rigid concretes are made with portland cement, sand, and stone, or crushed gravel. The mixes incorporate water to hydrate the cement to bond the aggregates into a solid mass. Admixtures may be added to the mix to impart specific properties to it or to the hardened concrete.

Other types of rigid concretes include nailable concretes; insulating concretes; heavyweight concretes; lightweight concretes; fiber-reinforced concretes, in which short steel or glass fibers are embedded for resistance to tensile stresses; polymer and pozzolan

concretes, which exhibit improvement in several properties; and silica-fume concretes, which possess high strength. Air-entrained concrete formulations, in which tiny air bubbles have been incorporated, may be considered variations of ordinary concrete.

Stress and Reinforcement

Because ordinary concrete is much weaker in tension than in compression, it is usually reinforced or prestressed with a much stronger material, such as steel, to resist tension. Use of plain, or unreinforced, concrete is restricted to structures in which tensile stresses will be small, such as massive dams, heavy foundations, and unit-masonry walls. For reinforcement of other types of structures, steel bars or structural steel shapes may be incorporated in the concrete.

Cementitious Materials

These are for general-purpose cements, cements modified to achieve low heat of hydration and to produce a concrete resistant to sulfate attack, high-early-strength cement, and air-entraining cement.

Several other types of cement are sometimes used instead of portland cement for specific applications—for example, hydraulic cements, which can set and harden underwater. Included in this category, in addition to, are aluminous, natural, and white portland cements, and blends, such as portland blast-furnace slag, portland-pozzolan, and slag cements.

Aggregates for Ordinary Concrete

Aggregates should be inert, dimensionally stable particles, preferably hard, tough, and round or cubical. They should be free of clay, silt, organic matter, and salts.

Polymer Concretes

Polymers are used in several ways to improve concrete properties; they are impregnated in hardened concrete, are incorporated in a mix, or replace portland cement.

Impregnation is sometimes used for concrete road surfaces. It can more than double compressive strength and elastic modulus, decrease creep, and improve resistance to freezing and thawing.

Monomers and polymers added as admixtures are used for restoring and resurfacing deteriorated roads. The concrete hardens more rapidly than ordinary concrete, enabling faster return of roads to service.

Polymer concretes in which a polymer replaces portland cement possess strength and other properties similar to those of impregnated concrete. After curing for a relatively short time, for example, overnight at room temperatures, polymer concretes are ready for use; in comparison, ordinary concrete may have to cure for a week or more before exposure to service loads.

Casting

There are various methods employed for casting ordinary concrete. For very small projects, sacks of prepared mixes may be purchased and mixed on the site with water, usually in a drum-type, portable, mechanical mixer. For larger projects, mix ingredients are weighed separately and deposited in a stationary batch mixer, a truck mixer, or a continuous mixer. Concrete mixture agitated in a truck is called ready-mixed concrete. Paving mixers are self-propelled, track-mounted machines that move at about 0.45 m/s while mixing.

At the site, concrete may be conveyed to the forms in wheelbarrows, carts, trucks, chutes, conveyor belts, buckets, or pipelines. Where chutes or belts are used, the flow of material is maintained as continuous as possible. On large projects, concrete may be delivered to the formwork in dump buckets by cableways or by cranes. In many instances, direct pumping through pipelines is used to convey concrete from the mixer or hopper to the point of placement.

Concurrent Heating

The application of supplemental heat to a structure during a welding or cutting operation.

Condensate

Same as condensation except that the term usually implies bulk liquid such as would be produced in a condenser.

Condensation

(1) A chemical reaction in which two or more molecules combine, with the resulting separation of water or some other simple substance; the process is called polycondensation if a polymer is formed. (2) Liquid precipitated from the gaseous phase. The term may imply films were small volumes of liquid. See also *polymerization*.

Condensation Polymerization

In plastics technology, a stepwise chemical reaction in which two or more molecules combine, often but not necessarily accompanied by the separation of water or some other simple substance. If a polymer is formed, the process is called polycondensation. See also *polymerization*.

Condensation Ratio

A resin formed by polycondensation, for example, the alkyd, phenylaldehyde, and urea–formaldehyde resins.

Condenser

(1) A heat exchanger, in which a gas, often steam, is cooled causing it to revert to the liquid state. In a tube condenser cooling water is circulated through tubes to condense the steam exhausting from a steam turbine or other cycle. In a jet condenser the cooling water is injected into the exhaust steam; this requires high-quality cooling water. (2) An electrical capacitor. (3) A term applied to lenses or mirrors designed to collect, control, and concentrate radiation in an illumination system, such as an optical or scanning electron microscope.

Condenser Aperture

In electron microscopy, and opening in the condenser lens controlling the number of electrons entering the lens and the angular aperture of the illuminating beam.

Condenser Foil

Metal foil, usually aluminum or tin base, used for electrical capacitors, i.e., condensers. The foil may be varnished or, in the case of aluminum anodized, to provide an insulating surface to avoid the need for a nonconducting interleaf when rolled to form the capacitor.

Condenser Lens

See *condenser*.

Conditioning

Subjecting a material to a prescribed environmental and/or stress history before testing.

Conditioning Heat Treatment

A preliminary heat treatment used to prepare a material for a desired reaction to a subsequent heat treatment. For the term to be meaningful, the exact heat treatment must be specified.

Conditioning Time

See *joint-conditioning time*, *curing time*, and *setting time*.

Conductance (Electrical)

A measure of the ability of any material to conduct an electrical charge. Conductance is the ratio of the current flow to the potential difference. The reciprocal of electrical resistance.

Conductance (Thermal)

The time rate of heat flow through a unit area of a body, induced by a unit temperature difference between the body surfaces.

Conduction Band/Electrons

See *Band Theory*.

Conductive (Electrical) Polymers and Elastomers

These are polymers made electrically conductive by the addition of carbon black, nickel, silver, or other metals. Volume resistivities of plastics and rubbers, which normally are in excess of 10^8 Ω/cm, can be lowered to between 10^{-1} and 10^6 Ω/cm by addition of conductive materials. Carbon black is the most widely used filler. The relationship of carbon black loading and volume resistivity is not proportioned. Up to a 25% loading, conductivity significantly increases, but it falls off sharply thereafter. Generally, the addition of carbon black lowers the mechanical properties of the polymer. However, the use of carbon fibers to enhance conductivity improves mechanical properties.

Polyethylene and polyvinyl chloride resins loaded with carbon black are perhaps the most widely used conductive plastics. Plastics often made conductive by adding up to 30% carbon fiber are polysulfone, polyester, polyphenylene sulfide, nylon 6/6, ethylene tetrachloroethylene, and vinylidene fluoride–polytetrachloroethylene.

Although silicone is the most widely used base polymer for conductive rubber, other rubbers frequently used in compounding conductive elastomers include SBR, EPDM, TPR, and neoprene.

Another type of electrically conductive polymer includes materials that are doped with either electron acceptors, such as alkali metal ions and iodine, or electron donors, such as arsenic trifluoride. Also referred to as organic conductors, their conductivity is about one-hundredth that of copper. The most widely used are polyacetylene, polyparaphenylene, and polyparaphenylene sulfide.

Conductivity (Electrical)

The reciprocal of volume resistivity. The electrical or thermal conductance of a unit cube of any material (conductivity per unit volume).

Conductivity (Thermal)

The time rate of heat flow through unit thickness of an infinite slab of a homogeneous material in a direction perpendicular to the surface, induced by unit temperature difference. Recommended SI units: W/m K. See also IACS.

Conductors

This term is usually applied to materials (wire, cable, or other body), generally metals, used to conduct electric current, although heat conductors and sound conductors have important uses. Good conductors of electricity tend to be good conductors of heat, as well. Silver is the best conductor of electricity, but copper is the most commonly used. The conductivity of pure copper is 97.6% that of silver. The electric conductivity of metals is often expressed as a percentage of the electric conductivity of copper, which is arbitrarily set at 100%. Tough-pitch copper is the standard conductivity metal, and it is designated as the International Annealed Copper Standard (IACS).

Because of the low conductivity of zinc, the brasses have low current-carrying capacity, but are widely used for electric connections and parts because of their workability and strength. The electric conductivity of aluminum is only 63% that of copper, but it is higher than that of most brasses. Copper wire for electric conductors in high-temperature environments has a plating of heat-resistant metal. Aluminum wire with a steel core is used for power transmission because of the long spans possible. Steel has a conductivity only about 12% that of copper, but the current in a wire tends to travel near the surface, and the small steel core does not reduce greatly the current-carrying capacity. Aluminum is now widely used to replace brass in switches and other parts. Aluminum wire for electric equipment is usually commercially pure aluminum with small amounts of alloying elements, such as magnesium, which give strength without appreciably reducing the conductivity. Plastics, glass, and other nonconductors are given conductive capacity with coatings of transparent lacquer containing metal powder, but conductive glass usually is made by spraying on at high temperature an extremely thin invisible coating of SnO. Coated glass panels are available with various degrees of resistivity.

Cone

The conical part of an oxyfuel gas flame next to the orifice of the tip.

Cone Angle

The angle that the cutter axis makes with the direction along which the blades are moved for adjustment, as in adjustable-blade reamers where the base of the blade slides on a conical surface.

Cone Resistance Value (CVR)

A measure of the yield stress of a grease, obtained by static indentation with a cone. The equilibrium depth of penetration is measured, not the penetration in a given time, which is the penetration value. See also *penetration* (of a grease).

Confidence Interval

That range of values, calculated from estimates of the mean and standard deviation, which is expected to include the population mean with a stated level of confidence. Confidence intervals in the same context also can be calculated for standard deviations, lines, slopes, and points.

Configurations

In plastics technology, related chemical structures produced by the cleavage and reforming of covalent bonds.

Confirmations

Different shapes of polymer molecules resulting from rotation about single covalent bonds in the polymer chain.

Conformability

In tribology, that quality of a plain bearing material that allows it to adjust itself to shaft deflections and minor misalignments by deformation or by wearing away of bearing material without producing operating difficulties.

Conformal Surfaces

Surfaces whose centers of curvature are on the same side of the interface. In wear testing, it refers to the case where the curvature of both specimens matches such that the nominal contact area during the testing remains approximately constant. Contrast with *counterformal surfaces*.

Congruent Melting

An isothermal or isobaric melting in which both the solid and liquid phases have the same composition throughout the transformation.

Congruent Transformation

An isothermal or isobaric phase change in metals in which both of the phases concerned have the same composition throughout the process.

Conjugate Phases

In microstructural analysis, those states of matter of unique composition that coexist at equilibrium at a single point in temperature and pressure. For example, the two coexisting phases of a two-phase equilibrium.

Conjugate Planes

Two planes of an optical system such that one is the image of the other.

Connected Porosity

The volume fraction of all pores, voids, and channels within a solid mass that are interconnected with each other and communicate with the external surface, and thus are measurable by gas or liquid penetration. Contrast with *closed porosity*.

Consistency

(1) In adhesives, that property of a liquid adhesive by virtue of which it tends to resist deformation. Consistency is not a fundamental property but is composed of viscosity, plasticity, and other phenomena. See also *viscosity* and *viscosity coefficient*. (2) An imprecise measure of the degree to which a grease resists deformation under the application of a force.

Consolidation

In metal–matrix or thermoplastic composites, a processing step in which fiber and matrix are compressed by one of several methods to reduce voids and achieve desired density.

Constant Life Fatigue Diagram

In failure analysis, a plot (usually on rectangular coordinates) of a family of curves, each of which is for a single fatigue life (number of cycles), relating alternating stress, maximum stress, minimum stress, and mean stress. The constant life fatigue diagram is generally derived from a family of *S–N* curves, each of which represents a different stress ratio for a 50% probability of survival. See also *nominal stress*, *maximum stress*, *minimum stress*, *S–N curve*, *fatigue life*, and *stress ratio*.

Constantan

An alloy of 60% copper with 40% nickel having high electrical resistivity, about 48 µΩ cm, a low temperature coefficient of resistivity up to about 450°C and good corrosion resistance. It is used for various electrical purposes, such as resistors and thermocouples. See *resistance alloys*.

Constituent

(1) One of the ingredients that make up the chemical system. (2) A phase or a combination of phases that occurs in a characteristic configuration in an alloy microstructure. (3) In composites, the principal constituents are the fibers in the matrix.

Constraint

(1) Any restriction that limits the transverse contraction normally associated with a longitudinal tension, and that hence causes a secondary tension in the transverse direction; usually used in connection with welding. Contrast with *restraint*. (2) (of deformation) Localized restriction of deformation resulting from the Poisson effect, i.e., as the material is extended along one axis it has to contract across the transverse axes. At notches and other stress concentrations the material at the highly stressed notch tip is constrained from contraction in the transverse axes by the surrounding large volume of lower stress material. This gives rise to complex triaxial stresses at the notch tip. The effect is usually confined to small volumes of material while stresses are in the elastic range but large volumes may be involved if plastic deformation has occurred.

Constricted Arc (Plasma Arc Welding and Cutting)

A plasma arc column that is shaped by a constricting nozzle orifice. See also *plasma arc welding*.

Constricting Nozzle (Plasma Arc Welding and Cutting)

A water-cooled copper nozzle surrounding the electrode and containing the constricting orifice. See also *cutting torch (plasma arc)*.

Constricting Orifice (Plasma Arc Welding and Cutting)

The hole in the constricting nozzle through which the arc passes. See also *multi-port nozzle*.

Constriction Resistance

In electrical contact theory, the resistance that arises from the constriction of current flow lines in order to pass through small areas of contact (a-spots) at the interface of two connecting bodies. See also *a-spot*, *contact resistance*, and *film resistance*.

Consumable Electrode

A general term for any arc welding electrode made chiefly of filler metal. Use of specific names such as *covered electrode*, *bare electrode*, *flux-cored electrode*, and *lightly coated electrode* is preferred.

Consumable Guide Electroslag Welding

In electroslag welding process variation in which filler metal is supplied by an electrode and its guiding member. See also *electroslag welding*.

Consumable Insert (of Weld)

A preplaced filler having specific dimensions that is located in the root of a joint to be made from one side. It is intended to be fully fused to become an integral part of the joint and is not normally machined following welding. Same as *fusible insert*.

Consumable-Electrode Remelting

A process for refining metals in which an electric current passes between electrode made of the metal to be refined in an ingot of the refined metal, which is contained in a water-cooled mold. As a result of the passage of electric current, droplets of molten metal form on the electrode and fall to the ingot. The refining action occurs from contact with the atmosphere, vacuum, or slag through which the drop falls. See also *electro slag remelting* and *vacuum arc remelting*.

Contact Adhesive

An adhesive that is apparently dry to the touch and that will adhere to itself simultaneously upon contact. An adhesive that, when applied to both adherends and allowed to dry, develops a bond when the adherends are brought together without sustained pressure.

Contact Angle

(1) See *roll angle*. (2) The angle formed at the junction between a solid and a liquid. If the liquid "wets" the solid as with water on clean glass the angle is acute, if it does not wet as with water droplets on a wax car body the angle is obtuse. (3) (in a bearing). See *angle of contact*.

Contact Area

The common area between a conductor and a connector across which the flow of electricity takes place.

Contact Bond Adhesive

Synonym for *contact adhesive*.

Contact Corrosion

(1) Corrosion in the vicinity of the point of contact between dissimilar metals in electrolyte. (2) A term primarily used in Europe to describe galvanic corrosion between dissimilar metals.

Contact Electrode (for Welding)

An electrode, the coating of which during the welding operation maintains a cup profile projecting slightly beyond the tip of the metal electrode. The edge of the cup is rested against the component being welded to maintain a fixed arc length.

Contact Fatigue

Cracking and subsequent pitting of a surface subjected to alternating Hertzian stresses such as those produced under rolling contact or combined rolling and sliding. The phenomenon of contact fatigue is encountered most often in rolling-element bearings or in gears, where the surface stresses are high due to the concentrated loads and are repeated many times during normal operation.

Contact Infiltration

In powder metallurgy, the process of infiltration whereby the initially solid infiltrant is placed in direct contact with the compact and the pores are filled with the liquid phase by capillary force after the infiltrant has become molten. See also *infiltrant* and *infiltration*.

Contact Material

A metal, composite, or alloy made by the melt-cast method or manufactured by powder metallurgy that is used in devices that make and break electrical circuits or welding electrodes. A majority of contact applications in the electrical industry utilize silver-type contacts, which include the pure metal, alloys, and powder metal combinations. Silver, which has the highest electrical and thermal conductivity of all metals, is also used as a plated, brazed, or mechanically bonded overlay on other contact materials—notably, copper and copper-base materials. Other types of contacts used include the platinum group metals, tungsten, molybdenum, copper, copper alloys, and mercury. See also *electrode* (*welding*).

Contact Molding

A process for molding reinforced plastics in which reinforcement and resin are placed in a mold. Cure is either at room temperature using a catalyst-promoter system or by heating in an oven, without additional pressure. Also referred to as hand lay-up.

Contact Plating

(1) The plated-on material applied to the basic metal of an electrical contact to provide for required contact resistance and/or specified wear resistance characteristics. (2) A metal plating process wherein the plating current is provided by galvanic action between the work metal and a second metal, without the use of an external source of current.

Contact Potential

In corrosion technology, the potential difference at the junction of two dissimilar substances.

Contact Pressure Resins

Liquid resins that thicken or polymerize upon heating, and, when used for bonding laminates, require little or no pressure.

Contact Resistance

(1) It reflects aspects such as contact area, roughness, interfacial pressure and contamination. (2) The electrical resistance of metallic surfaces at their interface in the *contact area* under specified conditions. (3) The electrical resistance between two contacting bodies, which is the sum of the *constriction resistance* and the *film resistance*.

Contact Scanning

In ultrasonic inspection, a planned systematic movement of the beam relative to the object being inspected, the search unit being in contact with and coupled to this object by a thin film of coupling material.

Contact Stress

Stress that results near the surfaces of two contacting solid bodies when they are placed against one another under a nonzero normal force. This term is not sufficiently precise because it does not indicate the type of stress, although common usage usually implies elastic or Hertz stress.

Contact Stress (Glass)

The tensile stress component imposed at a glass surface immediately surrounding the contact area between the glass surface and an object generating a locally applied force.

Contact Tube

A device which transfers current to a continuous welding electrode.

C

Contact Weld

(1) The point of attachment of a contact to its support when accomplished by resistance welding. (2) A contacting failure due to the fusing of contacting surfaces under load conditions to the point that the contacts failed to separate when expected.

Contacting Ring Seal

A type of circumferential seal that utilizes a ring-spring loaded radially against the shaft. The ring is either gapped or segmented, in order to have radial flexibility. The seal has overlapping joints for blocking leakage at the gaps. An axial spring load seats the ring against the wall of its containing cartridge.

Container

The chamber into which an ingot or billet is inserted prior to extrusion. The container for backward extrusion of cups or cans is sometimes called a die.

Contaminant

An impurity or foreign substance present in a material or environment that affects one or more properties of the material.

Context Root Surface

A root surface which is convex.

Continuous Furnace (Glass)

A glass furnace in which the level of glass remains substantially constant because the feeding of batch continuously replaces the glass withdrawn.

Continuity Bond

In corrosion protection, a metallic connection that provides electrical continuity between metal structures.

Continuous Casting

Any system using a mold open at both ends such that solid metal is steadily drawn out of the bottom while molten metal continues to enter the top. The mold can be in various forms including a simple die or rolls and is usually water cooled. Often, equipment beyond the mold cuts the ingot, bar, tube, etc., into lengths allowing the process to continue virtually indefinitely. If the process is occasionally interrupted to allow removal of the ingot the process may be termed Semi-continuous casting. Used chiefly to produce semi-finished mill products such as billets, blooms, ingots, slabs, strip, and tubes. See also *strand casting*.

Continuous Compaction

In powder metallurgy, the production of relatively long compacts having a uniform cross-section, such as sheet, rod, tube, etc. By direct extrusion or rolling of loose powder.

Continuous Cooling Transformation (CCT) Diagram

Set of curves drawn using logarithmic time and linear temperature as coordinates, which define, for each cooling curve of an alloy, the beginning and end of the transformation of the initial phase. See *Isothermal Transformation Diagram*.

Continuous Covered Electrode

Welding electrode filler wire, supplied as coils, comprising a central main filler wire carrying an external coating and helical wire(s) which reinforced the coating and act as current pickup.

Continuous Filament Yarn

Yarn formed by twisting two or more continuous filaments into a single, continuous strand.

Continuous Furnace

A furnace which operates continually with materials being charged at one end and discharged at the other. A variation is the Rotary Hearth where the charge and discharge points are a short distance apart around the radius.

Continuous Furnace (Powder Metallurgy)

A furnace used for the uninterrupted sintering of compacts.

Continuous Mill

A rolling mill consisting of a number of strands of synchronized rolls (in tandem) in which metal undergoes successive reductions as it passes through the various strands.

Continuous Phase

In an alloy or portion of an alloy containing more than one phase, the phase that forms the matrix in which the other phase or phases are dispersed.

Continuous Precipitation

Precipitation from a supersaturated solid solution in which the precipitate particles grow by long-range diffusion without recrystallization of the matrix. Continuous precipitates grow from nuclei distributed more or less uniformly throughout the matrix. They usually are randomly oriented, but may form a *Widmanstätten structure*. Also called general precipitation. Should be compared with *discontinuous precipitation* and *localized precipitation*.

Continuous Sequence

A longitudinal welding sequence in which each pass is made continuously from one end of the joint to the other. See also *backstep sequence* and *longitudinal sequence*.

Continuous Sintering

In powder metallurgy, presintering or sintering in such a manner that the objects are advanced through a furnace at a fixed rate by manual or mechanical means; sometimes called stoking.

Continuous Spectrum (X-Rays)

In materials characterization, the polychromatic radiation emitted by the target of an x-ray tube. It contains all wavelengths above a certain minimum value, known as the short wavelength limit.

Continuous Weld

A weld extending continuously from one end of a joint to the other or, where the joint is essentially circular, completely around the joint. Contrast with *intermittent weld*.

Continuous-Type Furnace

A furnace used for heat treating materials that progress continuously through the furnace, entering one door and being discharged from another. See also *belt furnace*, *direct-fired tunnel-type furnace*, *rotary-retort furnace*, and *shaker-hearth furnace*.

Continuum

In materials characterization, the noncharacteristic rays emitted upon irradiation of a specimen and caused by deceleration of the incident electrons by interaction with the electrons and nuclei of the specimen. Also termed bremsstrahlung and white radiation.

Contour Forming

See *roll forming*, *stretch forming*, *tangent bending*, and *wiper forming*.

Contour Machining

Machining of irregular surfaces, such as those generated in tracer turning, tracer boring, and tracer milling.

Contour Milling

Milling of irregular surfaces. See also *tracer milling*.

Contraction

The volume change that occurs in metals and alloys upon solidification and cooling to room temperature. See *shrinkage*.

Contrast Enhancement (Electron Optics)

In electron microscopy, an improvement of electron image contrast by the use of an objective aperture diaphragm, shadow casting, or other means. See also *shadowing*.

Contrast Filter

In metallography, a color filter, usually with strong absorption, that uses the special absorption bands of the object to control the contrast of the image by exaggerating or diminishing the brightness difference between differently colored areas.

Contrast Perception

In metallography, the ability to differentiate various components of the object structure by various intensity levels in the image.

Controlled Atmosphere

(1) A specified inert gas or mixture of gases at a predetermined temperature in which selected processes take place. (2) As applied to sintering, to prevent oxidation and destruction of the powder compacts.

Controlled Atmosphere Chamber

An inert gas-filled enclosure or cabinet in which plasma spraying or welding can be performed to minimize (or prevent) oxidation of the coating or substrate. The enclosure is usually fitted with viewing ports, glove ports to permit manipulations, and a small separate airlock for introducing or removing components without loss of atmosphere.

Controlled Cooling

Any cooling at a prescribed rate but it is usually implicit that the rate is slower than would occur naturally, as opposed to quenching. It is imposed after heat treatment, welding or hot working for various reasons including avoiding hardening, minimizing residual stresses or allowing dissolved hydrogen to escape.

Controlled Etching

Electrolytic etching with selection of suitable etchant and voltage resulting in a balance between current and dissolved metal ions.

Controlled Rolling

A hot-rolling process in which the temperature of the steel is closely controlled, particularly during the final rolling passes, to produce a fine-grain microstructure.

Controlled Thermal Severity Test

A test of the weldability of steels, in particular the relationship between cooling rate and susceptibility to cracking. For the test, a 3 in.^{2} steel plate is bolted to a much larger and thicker steel plate and substantial fillet welds deposited along two opposite sides. Two test welds are then deposited, in sequence, on the remaining sides. The concept is that when the first test weld is deposited the cooling rate is controlled by the conduction of heat through the two assembly welds while the cooling rate for the second test weld is controlled by conduction through the previous three welds. The severity of the test is defined in terms of Thermal Severity Number, TSN, which is the number of conduction paths multiplied by the plate thickness in units of 0.25 in. For example, with quarter inch plate the TSNs of the test welds would be 2 and 3, and with half inch plate they would be 4 and 6.

Controlled-Potential Coulometry

Measurement of the number of coulombs required for an electrochemical reaction occurring under conditions where the working electrode potential is precisely controlled.

Controlled-Pressure Cycle

A forming cycle during which the hydraulic pressure in the forming cavity is controlled by an adjustable cam that is coordinated with the punch travel.

Convection

The motion resulting in a fluid from the differences in density and the action of gravity. In heat transmission, this meaning has been extended to include both forced and natural motion or circulation.

Conventional Forging

A forging characteristic by design complexity and tolerances that fall within the broad range of general forging practice.

Conventional Milling

Milling in which the cutter moves in the direction opposite to the feed at the point of contact. Contrast with *climb milling*.

Conventional Strain

See *engineering strain* and *strain*.

Conventional Stress

The stress calculated on the original cross-sectional area, as opposed to the Real Stress. See *tensile test* and *engineering stress* and *stress*.

Convergent-Beam Electron Diffraction (CBED)

In *analytical electron microscopy*, a technique of impinging a highly convergent electron beam on a crystal to produce a diffraction pattern composed of disks of intensity. In addition to *d*-spacing and crystal orientation information, the technique can provide information on crystallographic point or space group symmetry.

Conversion Coating

(1) A coating consisting of a compound of the surface metal, produced by chemical or electrochemical treatments of the metal. Examples include chromate coatings on zinc, cadmium, magnesium, and aluminum, and oxide and phosphate coatings on steel. It forms a compound between the applied material and the base metal rather than merely depositing a layer. Anodizing, chromizing, and phosphating are examples. See also *chromate treatment* and *phosphating*.

Conversion Coatings

Surface transformations formed naturally by chemical or electrochemical methods on ferrous and nonferrous metals and alloys. *Natural conversion coatings*, usually oxides, include *rust*, the ferric oxide and hydroxide that forms on iron and plain carbon steels in air or moisture, and the adherent and protective oxides that forms on aluminum, copper, and other metals. *Chemical conversion coatings* are mainly phosphates, chromates, and oxides induced by the reaction of specific chemicals with metal surfaces. *Electrochemical conversion coatings* are formed by anodic oxidation, or anodizing, in an electrolytic cell in which the metal being treated is the anode.

Phosphate conversion coatings are formed on iron, steel, galvanized steel, and aluminum by chemicals containing phosphoric acid and its salts. *Zinc phosphates*, one of three major types, are applied by immersion or spray in a wide variety of coating weights and crystal sizes. The microstructure type enhances paint adhesion, minimizes paint consumption, and improves bonding to plastics and rubber.

The heavy kind are quite absorbent, thus capable of retaining forming lubricants and rust-preventive oils. They also reduce friction and enhance wear resistance and corrosion resistance. *Iron phosphates*, similarly applied, are produced from alkali-metal phosphate solutions. When amorphous or of fine crystal size, they are used mainly to improve paint adhesion and to increase resistance to paint flaking from impact or flexing. *Manganese phosphates*, applied only by immersion, are used to retain oil so as to facilitate part break-in and prevent galling of mating surfaces. Phosphate coatings have specific colors, depending on the chemicals used and the metal to which they are applied. Special colors can be developed by pretreatments and posttreatments.

Chromate conversion coatings are formed on aluminum, cadmium, copper, magnesium, silver, zinc, and their alloys by immersion or spray using aqueous solutions of chromic acid, chromium salts such as sodium or potassium chromate or dichromate, and hydrofluoric, phosphoric, or other mineral acids. Generally, the basic ingredients are hexavalent chromium and sufficient acid for the desired pH. Solutions of trivalent chromium are used for clear coatings on electroplated cadmium and zinc. Chromate–phosphate mixtures are used to produce combination coatings on aluminum. Chromates are typically amorphous, pore-free, and gel-like initially but, on drying, harden and become hydrophobic, less soluble, and more abrasion-resistant. They are used primarily to increase corrosion resistance, especially in marine, humid, and tropical atmospheres; but they also serve as a good base for paint. Various colors can be provided, depending on the particular solution and posttreatments. Regarding hexavalent chromium, however, the aim is to eliminate the use because of its carcinogenicity.

Oxide chemical conversion coatings are the bluish, black, or brown oxides formed on iron or steel with hot caustic or alkaline solutions, and the black or brown oxides formed on cadmium, copper, iron, steel, or zinc alloys with acidic solutions. Although they are used mainly for abrasion resistance and anesthetics, some add a modest degree of corrosion protection. Alkaline chromate solutions as fused-salt solutions, for example, impart corrosion resistance and abrasion resistance to iron and steel.

Electrochemical conversion coatings, or *anodic coatings*, pertain mainly to aluminum alloys or magnesium alloys, although several other metals are also anodized. Anodized aluminum is produced primarily in aqueous solutions of sulfuric, chromic, or oxalic acid for a variety of reasons; to improve corrosion resistance, abrasion resistance, paint adhesion, adhesive bonding, or electroplating; to provide decorative finishes, including color; and to impart an electrically insulative surface or a base for photographic or lithographic emulsions. *Hard anodic coatings* on aluminum alloys, for abrasion resistance, are typically thicker than those for corrosion protection. Anodized magnesium is produced in aqueous solutions of ammonium bifluoride, sodium dichromate, and phosphoric acid or in aqueous solutions of potassium and aluminum hydroxide, trisodium phosphate, potassium fluoride, and potassium manganate or permanganate. Thin coatings serve mainly as a base for paint, thick ones for corrosion resistance and abrasion resistance. Other nonferrous metals are anodized primarily to increase corrosion resistance. *Anodizing zinc* is produced by immersion in an alkali-carbonate solution and then in a silicate solution, or in a single alkaline solution of sodium silicate, borate, and tungstate. *Anodized beryllium* is made in an aqueous solution of nitric and chromic acids, and *anodized titanium* and *anodized thorium* are made in mixtures of glacial acetic and nitric acids. *Anodized zirconium* also has been made in such mixtures although, for nuclear applications it and *anodized hafnium* are made in aqueous solutions of ethanol, glycerine, and lactic, phosphoric, and citric acids, followed by autoclaving.

Anodized columbium and *anodized tantalum* are produced in solutions of ammonium citrate or borate, with ammonium hydroxide for basicity.

Conversion coatings are also known by various terms and trade names. Among the latter by Allied-Kelite, of Witco Corp., are *Keycote phosphate coatings, Iridite chromate coatings, Iridize zinc and anodic coatings*, and *Irilac clear coatings*. The clear coatings can be applied to the anodic for additional corrosion protection. Other terms and key ingredients applicable to hard-anodizing aluminum are Martin Hard Coat (sulfuric acid). *Magnaplate HCR*, of General Magnaplate, is a surface treatment to improve the hardness, corrosion resistance, and abrasion resistance of aluminum and aluminum alloys. *Nituff*, a Teflon-impregnated hard-anodized aluminum, combines hardness and lubricity which markedly increases release properties and wear resistance. One application is chemical polishing tubs used to make sapphire optics.

Converter

A furnace in which air is blown through a bath of molten metal or matte, oxidizing the impurities and maintaining the temperature through the heat produced by the oxidation reaction. A typical converter is the *argon oxygen decarbonization* vessel.

Convex Fillet Weld

A fillet weld having a convex face.

Convexity

The maximum distance from the face of a convex fillet weld perpendicular to a line joining the toes.

Cool Time (Resistance Welding)

The time interval between successive heat times in multiple-impulse welding or in the making of seam welds.

Coolants

Liquids used to quench metals in heat treating, used to cool and lubricate cutting tools and workpieces in machining, or applied to forming tools and workpieces to assist in forming operations. In the case of machining, they are also called *cutting fluids*, and in the case of forming, *forming lubricants*. When water is used for the normal water-hardening steels, it may be modified with soda or other material to give a less drastic and more uniform cooling. A water bath containing 5% sodium hydroxide gives uniform, rapid cooling. Oils are used in cooling or quenching baths to provide a more moderate cooling effect. Quenching oils are usually compounded, although fish oils alone are sometimes employed. Fish oils, however, have offensive odors when heated. Vegetable oils alone are likely to oxidize and become gummy. Animal oils become rancid. Lard and palm oils give low cooling rates, while cottonseed, neatsfoot, and fish oils give more rapid cooling. Mineral oils compounded with fish, vegetable, or animal oils are sold under trade names and vary considerably in their content. *Oil-quenching baths* are usually kept at a temperature of not over 150°F (66°C) by providing cooling pipes. *Tempering oils* differ from quenching oils only in that they are compounded to withstand temperatures up to about 525°F (274°C).

Coolants for machining are classified into five groups: straight oils, soluble oils, chemical coolants, synthetics, and semisynthetics. *Straight oils*, which contain no water, are petroleum or mineral oils with or without additional compounding. Without further compounding, they are suitable for light-to moderate-duty cutting on readily machined metals. For more severe machining, they are typically compounded with up to 20% fatty oils, sulfur, chlorine, phosphorus, or combinations thereof. Sulfur, chlorine, and phosphorus are commonly called *extreme-pressure additives* (EP additives). For the most severe applications, compounding, mainly with chlorine and sulfurized fatty oils, may exceed 20%. *Soluble oils*, such as emulsified sulfonated mineral oils, are also suitable for light- and moderate-duty applications. Although they do not match the straight oils in lubricity, they, like water-dilutable fluids in general, are better heat dissipators. Because of their water content, they are usually formulated with additives to prevent corrosion of the workpiece and to resist microbial degradation and souring, necessitating maintenance and service to retain these characteristics. Heavy-duty soluble oils are suitable for most applications for which straight oils are used. Chemical coolants were originally *amine nitrites*, but *amine borates* are commonly used now because nitrites in contact with amine form *nitrosamine*, a suspected carcinogen. They are noted for excellent cooling capacity, inhibiting corrosion, and resisting microbial degradation and souring. They have limited lubricity, however, and are confined to light-duty operations, mainly light grinding. *Synthetic coolants*, which have been likened to soluble oils without oil, are water-dilutable systems designed for high-cooling capacity, lubricity, corrosion prevention, and easy maintenance. These synthesized materials are chemically similar to mineral-oil derivatives, but can be dispersed in water and are suitable for more severe operations than chemical coolants. They tend to defat human skin, however, causing dermatitis, thus necessitating that workers adhere to prescribed methods of personal hygiene. *Semisynthetic coolants* contain small dispersions of oil in an otherwise water-dilutable system, are almost transparent, are more broadly applicable than soluble oils, and are easier to maintain.

Straight oils, soluble oils, and synthetics are also used as forming lubricants. They also may contain a *wetting agent*, or *polarity agent*, such as animal fats, fatty acids, long-chain polymers, emulsifiers, and EP additives. The straight oils are the most varied in formulation and the most widely used. The soluble oils, however, can match their performance in many applications and, because of their superior cooling effect, sometimes provide better performance. The synthetics, which have been improved in recent years, also feature excellent cooling capacity as well as cleanliness and lubricity, and they have replaced both the straight and soluble oils and many applications. Because of their cleanliness, they are especially useful in forming precoated metals.

Cooling Channels

In plastic part making, chemicals or passageways located within the body of a mold through which a cooling medium can be circulated to control temperature on the mold surface. May also be used for heating a mold by circulating steam, hot oil, or other heated fluid through channels, as in the molding of thermosetting and some thermoplastic materials.

Cooling Curve

A graph showing the relationship between time and temperature during the cooling of a material. It is used to find the temperatures at which phase changes occur. A property or function other than time may occasionally be used—for example, thermal expansion.

Cooling Fixture

In plastic part making, block of metal or wood shape to hold a molded part to maintain the proper shape or dimensional accuracy of the part after it is removed from the mold until it is cooled enough to retain its shape without further appreciable distortion. Also called a shrink fixture.

Cooling Rate

The average slope of the time–temperature curve taken over a specified time and temperature interval.

Cooling Stresses

Residual stresses in castings resulting from nonuniform distribution of temperature during cooling.

Cooling Table

Same as a *hot bed*.

Coordinate Bonding

See *Interatomic Bonding*.

Coordination Catalysis

Ziegler-type of catalysis for processing plastics. See also *Ziegler–Natta catalysts*.

Coordination Compound

A compound with a central atom or ion bound to a group of ions or molecules surrounding it. Also called coordination complex. See also *chelate, complexation,* and *ligand*.

Coordination Number

(1) Number of items or radicals coordinated with the central atom in a complex covalent compound. (2) Number of nearest neighboring atoms to a selected atom and crystal structure. (3) It is 12 for the close packed hexagonal and face-centered cubic structures, and 8 for body centered cubic structures.

Cope

In casting, the upper or topmost section of a *flask, mold,* or *pattern*. The lower part is the drag.

Copolyesters

These thermoplastic elastomers are generally tougher over a broader temperature range than the urethanes. Also, they are easier and more forgiving in processing. Several grades include Hytrel, Riteflex, and Ecdel, ranging in hardness from 35 to 72 Shore D. These materials can be processed by injection molding, extrusion, rotational molding, flow molding, thermoforming, and melt casting. Powders are also available.

Copolyesters, which along with the urethanes are high-priced elastoplastics, have excellent dynamic properties, high modulus, good elongation and tear strength, and good resistance to flex fatigue at both low and high temperatures. Brittle temperature is below −68°C, and modulus at −40°C is only slightly higher than at room temperature. Heat resistance of 449°C is good.

Resistance of the copolyesters to nonoxidizing acids, some aliphatic hydrocarbons, aromatic fuels, sour gases, alkaline solutions, hydraulic fluids, and hot oils is good to excellent. Thus, they compete with rubbers such as nitriles, epichlorohydrins, and polyacrylates. However, hot polar materials, strong mineral acids and bases, chlorinated solvents, phenols, and cresols degrade the polyesters. Weathering resistance is low but can be improved considerably by compounding ultraviolet stabilizers or carbon blacks with the resin.

Copolyester elastomers are not direct substitutes for rubber in existing designs. Rather, such parts must be redesigned to use the higher strength and modulus, and to operate within the elastic limit. Thinner sections can usually be used—typically one-half to one-sixth that of a rubber part.

Applications of copolyester elastomers include hydraulic hose, fire hose, power-transmission belts, flexible couplings, diaphragms, gears, protective boots, seals, oil-field parts, sports-shoe soles, wire and cable insulation, fiber-optic jacketing, electrical connectors, fasteners, knobs, and bushings.

Copolymer

A long-chain molecule formed by the reaction of two or more dissimilar monomers. See also *polymer*.

Copolymerization

See *polymerization*.

Copper Acetate

Also known as crystals of Venus. A dark-green, crystalline, poisonous powder. It is soluble in water and alcohol. It is used as a pigment in paints, lacquers, linoleum, and inks and for making artificial verdigris or patina on copper articles. It is used as a catalyst in making phthalic anhydride plastics. When used for mildew-proofing cotton cloth, the copper precipitates out to form the *waxate*, or copper soap coating. *Verdigris* is an old name for basic copper acetate as a blue-grain pigment, but the name is now usually applied to the bluish-green corrosion crust on copper. The greenish-brown crust known as patina, forms on bronze, is esteemed as a characteristic of antiquity. It is a basic sulfate of copper, usually with oxides of tin, copper, and lead.

Copper and Copper Alloys

Copper (symbol Cu) is one of the most useful of the metals, and probably the one first used by humans. It is found native and in a large number of ores. Its apparent plentifulness is only because it is easy to separate from its ores and is often a by-product from silver and other mining. Copper has a face-centered cubic crystal structure. It is yellowish red in color, tough, ductile, and malleable; gives a brilliant luster when polished; and has a disagreeable taste and a peculiar odor. It is the best conductor of electricity next to silver, with a conductivity 97% that of silver. Copper refers to the metal at least 99.3% pure. Standard wrought grades number more than 50, many of which are more than 99.7% pure. They are represented by the C10XXX–C15XXX series of copper and copper alloy numbers of the Copper Development Association. These include O_2-free coppers, O_2-free-with-silver coppers, and O_2-bearing coppers (C11100–C10940); electrolytic-tough-pitch coppers and tough-pitch-with-silver coppers (C11100–C11907); phosphorus-deoxidizer coppers, fire-refined tough-pitch coppers (C12000–C13000); and certain coppers distinguished by very small amounts of specific ingredients such as cadmium copper (not to be confused with the high-copper alloys having a greater cadmium content), tellurium-bearing copper, sulfur-bearing copper, zirconium copper, and aluminum-bearing coppers (C14XXX–C15XXX). The highest-purity grade, O_2-free-electronic copper, is at least 99.99% pure. There are seven standard cast coppers (C80XXX–C81100); and their minimum purity ranges from 99.95% (C80100) to 99.70% (C81100).

O_2-free coppers (C10100 and C10200) have a melting point of 1082°C, a density of 8.94 mg/m^3, an electrical conductivity of 101%—or slightly greater than the 100% for electrolytic-tough-pitch copper (C11100) used as the International Annealed Copper Standard (IACS) for electrical conductivity—a thermal conductivity of 391 W/m K, and a specific heat of 385 J/kg K. Typical tensile properties of thin flat products and small-diameter rod and wire having an average grain size of 0.050 mm are 220 MPa ultimate strength, 69 MPa yield strength, 45%–50% elongation, and 117,000 MPa elastic modulus. Hardness is about RF 40. These properties are fairly typical of other wrought coppers as well. Strength increases appreciably with cold work, yield strengths reaching 345 MPa in the spring and hard-drawn conditions. Zirconium copper, which may be heat-treated after cold working, can provide yield strengths of 345–483 MPa in rod and wire forms and retains considerable strength at temperatures to 426°C. Cast coppers are suitable for sand, plaster, permanent-mold, investment, and centrifugal castings as well as for continuous casting. Regardless of grade, typical tensile properties are 172 MPa ultimate strength, 48 MPa yield strength, and 40% elongation. Hardness is typically Brinell 44.

Copper and copper-base alloys are unique in their desirable combination of physical and mechanical properties. The following properties are found or may be developed in these materials:

1. Moderate to high strength and hardness
2. Excellent corrosion resistance
3. Ease of workability both in primary and secondary fabrication operations
4. Pleasing color and wide color range
5. High electrical and thermal conductivity
6. Nonmagnetic properties
7. Superior properties at subnormal temperatures
8. Ease of finishing by polishing, plating
9. Good-to-excellent machinability
10. Excellent resistance to fatigue, abrasion, and wear
11. Relative ease of joining by soldering, brazing, and welding
12. Moderate cost
13. Availability in a wide variety of forms and tempers

The importance of copper and copper-based alloys is in no small part because many other metals, whether singly or in combination, may be alloyed with copper. There are many binary alloys, and ternary and quaternary alloy systems which are quite common. The copper content of alloys in commercial use ranges from about 58% to about 100%. Alloys of 63% to about 100% copper are of the alpha (single-phase) solid solution type, have excellent ductility, and are ideally suited for cold-working processes such as rolling, drawing, and press working.

Copper alloys in the lower range, 62% and less, are of the alpha plus beta (two-phase) type and are particularly suited to hot-working processes such as hot rolling, hot extrusion, and hot forging.

The chief elements alloyed with copper are zinc, tin, lead, nickel, silicon, and aluminum, and to a lesser extent, beryllium, phosphorus, cobalt, boron, arsenic, zirconium, antimony, cadmium, iron, manganese, chromium, and very recently, mercury.

Cast Products

Foundry alloys are used in the form in which they are originally molded. In general, cast alloys are not easily workable and subsequent operations are limited to such treatment as machining, electroplating, soldering, and brazing. The use of alloys in cast form permits the production of intricate and irregular shapes impractical or impossible to make by other means. The use of cores permits the making of hollow shapes. The alloying metals are generally greater in amount than for wrought alloys. Types of castings in commercial use include sand castings, diecastings, permanent mold, plaster mold, shell mold, and centrifugal and investment castings. The alloy involved, the properties desired, and the quantity to be made govern the choice of casting process. Where regularity of shape permits, a number of foundry alloys are available in the continuous-cast condition and may be either solid or hollow shapes. The classification of copper-base foundry alloys generally follows that used for wrought alloys. As in the wrought-alloy field, the number of cast alloys in commercial use runs well into the hundreds; many are modifications of standard alloys.

Wrought Products

Wrought products are those originally cast into starting shapes such as billets, cakes (slabs), and wire bar, which are further fabricated into useful forms such as sheet, strip, plate, rod, bar, seamless tubes, extruded shapes, and forgings by hot rolling, cold rolling, cold drawing, hot piercing, hot extrusion, or combinations of these primary fabricating processes. Secondary fabricating processes include such operations as machining, bending, spinning, cold heading, hot press forging, soldering, brazing, electroplating, cold extrusion, and operations normally associated with press working (blanking, deep drawing, coining, staking, embossing, drifting).

The various types of copper and the copper-base alloys are not fabricated with equal facility into usable forms or by secondary fabricating processes. This fact, together with the wide range and combination of physical and mechanical properties desirable for end use, accounts for the large number of commercial wrought copper-base alloys. Design considerations usually dictate a compromise between service properties desired, secondary fabricating properties, forms and sizes available, and cost.

The Alloys

In the copper industry, and in the brass mill section of that industry in particular, the alloys of copper have been designated by a sort of "word-terminology" that is often confusing and misleading. Names assigned to various alloys are not always descriptive or indicative of the composition. Several names may be in common use for the same composition. Numerous proprietary names are also used and, in some cases, several names in this category are in general use for essentially the same composition or type of alloy.

Since the more important alloying elements are used effectively in varying amounts, a convenient and logical classification is made possible by separating the alloys into well-recognized groups according to composition. Literally hundreds of wrought copper alloys are used. Those listed are standard alloys that account for the large majority of tonnage used. Alloys not listed are, in general, modifications of standard alloys often used for very special purposes.

Certain terms require clarification. Brass is an alloy of copper with zinc as the principal alloying element, used with or without additional elements. High brasses are Cu–Zn alloys containing more than 20% zinc, more properly called yellow brasses. Low brasses are alloys containing 20% zinc or less and low brass is a specific alloy, 80% copper and 20% zinc. Rich low brass is a term used for brass containing nominally 85% copper and 15% zinc, more properly called red brass. Straight brasses are those alloys containing copper and zinc only.

The term *bronze* in modern usage properly requires a modifying adjective. Early bronzes were alloys of copper with tin as the only, or principal, alloying metal, and thus the term *bronze* has been traditionally associated with alloys of copper and tin. The widespread use of numerous other alloy systems in recent times requires that a bronze be defined as an alloy of copper with some element, or elements, other than zinc as a chief alloying metal. Thus, "tin-bronze" is the proper name for alloys of copper and tin, silicon-bronze for alloys of copper and silicon, aluminum–silicon-bronze for an alloy of Cu–Al–Si. Tin-bronzes are also commonly called phosphor-bronzes because of the long-established practice of deoxidizing these alloys with phosphorus.

The "coppers" include pure copper and its modified forms, and also certain materials that might more properly be classed as alloys but that are commonly called "coppers" where the alloying element is less than about 1%, e.g., "TeCu"—0.5% tellurium, the balance, copper.

Refractory copper-base alloys are those that, because of their hardness or abrasiveness, require greater dimensional tolerances than those established for nonrefractory alloys. Examples of refractory alloys are the phosphor bronzes, silicon bronzes, aluminum bronzes, and nickel silvers.

Coppers

Many types and variations of commercial copper are available; the differences are slight or important depending upon the method of production or the intentional minor additions of other metals. Two types refer to the method of refining. Fire-refined copper is that finished by furnace refining. Electrolytically refined copper is that finished by electrolytic deposition where the original ore requires such refining because of certain undesirable impurities that can be satisfactorily removed only by this method, or where the original ore contains sufficient quantities of silver and gold, recoverable in electrolytic refining, to make the electrolytic process profitable. Over 85% of all copper is electrolytically refined. Three major types of commercial copper with respect to composition and method of casting are tough pitch, O_2-free, and phosphorus deoxidized. Tough-pitch copper contains a small but controlled amount of O_2 (about 0.04%) necessary to obtain a level set, i.e., the correct pitch, on the refinery casting (wire bar, cake, billet); the level set is necessary to provide adequate ductility for hot or cold primary fabricating operations, i.e., sufficient "toughness." Electrolytic tough-pitch copper is the standard copper of industry for electrical and many other purposes. Its low electrical resistance (0.15328 Ω/g m^2 at 20°C) equivalent to 100% IACS conductivity is the standard with which other metals are compared. Copper has the highest conductivity of any base metal and is exceeded only by that of silver.

Tough-pitch copper has one distinct disadvantage. When heated above about 399°C in atmospheres containing reducing gases, particularly H$_2$, the CuO$_2$ is reduced to copper and water vapor, the water changing to steam at the temperatures involved, thus "gassing" or embrittling the copper and destroying its ductility. Hence, the term *H$_2$ embrittlement*. Under circumstances where H$_2$ embrittlement is likely to occur (brazing, welding, annealing), O$_2$-free copper may be used without sacrifice of conductivity. In cases where some conductivity loss may be tolerated, various types of deoxidized copper may be used. Tough-pitch copper, while ductile enough for most fabricating work, is somewhat lacking in the ductility required for difficult forming operations (severe drawing, edgewise heading) and O$_2$-free copper is often used in these instances. Also, tough-pitch copper is not as free machining as the varieties made for that purpose, i.e., leaded copper, TeCu, sulfur copper. Silver is added to copper (without sacrifice of conductivity) for the purpose of increasing the recrystallization temperature, thus permitting soldering operations without reduction of strength and hardness. The amount of silver added varies with the application. Lake copper is a general term for silver-bearing copper, having varying but controlled amounts of silver up to about 30 oz/ton.

Phosphorized copper, with excellent hot and cold workability, is the second largest type in use with respect to tonnage and is widely used for pipe and tube for domestic water service and refrigeration equipment.

O$_2$-free copper is made either by melting and casting copper in the presence of carbon or carbonaceous gases or by extruding compacted, specially prepared cathode copper under a protective atmosphere, so that no O$_2$ is absorbed. No deoxidizing agent is required; therefore, no residual deoxidants are present, and optimum electrical conductivity is maintained. O$_2$-free copper is used extensively in the electrical and electronics industries, e.g., metal-to-glass seals.

Wrought Cu Alloys

Copper and zinc melted together in various proportions produce one of the most useful groups of copper alloys, known as the brasses. Six different phases are formed in the complete range of possible compositions. The relationship between composition and phases alpha, beta, gamma, delta, epsilon, and eta are graphically shown in the well-established constitution diagram for the Cu–Zn system. Brasses containing 5%–40% zinc constitute the largest volume of copper alloys. One important alloy, cartridge brass (70% copper, 30% zinc), has innumerable uses, including cartridge cases, automotive radiator cores and tanks, lighting fixtures, eyelets, rivets, screws, springs, and plumbing products. Tensile strength ranges from 310 MPa as annealed to 900 MPa for spring temper wire.

Lead is added to both copper and the brasses, forming an insoluble phase that improves machinability of the material. Free cutting brass (61% copper, 3% lead, 36% zinc) is the most important alloy in the group. In rod form it has a strength of 340–480 MPa, depending on temper and size. It is machined into parts on high-speed (10,000 rpm) automatic screw machines for a multiplicity of uses.

Increased strength and corrosion resistance are obtained by adding up to 2% tin or aluminum to various brasses. Admiralty brass (70% copper, 2% aluminum, 0.023% arsenic, 21.97% zinc) are two useful condenser-tube alloys. The presence of phosphorus, antimony, or arsenic effectively inhibits these alloys from dezincification corrosion.

Alloys of copper, nickel, and zinc are called nickel silvers. Typical alloys contain 65% copper, 10%–18% nickel, and the remainder zinc. Nickel is added to the Cu–Zn alloys primarily because of its influence on the color of the resulting alloys; color ranges from yellowish-white, to white with a yellowish tinge, to white. Because of their tarnish resistance, these alloys are used for table flatware, zippers, camera parts, costume jewelry, nameplates, and some electrical switch gear.

Copper forms a continuous series of alloys with nickel in all concentrations. The constitution diagram is a simple all alpha-phase system.

Nickel slightly hardens copper, increasing its strength without reducing its ductility. Copper with 10% nickel makes an alloy with a pink cast. More nickel makes the alloy appear white.

Three copper-base alloys, containing 10%, 20%, and 30% nickel, with small amounts of manganese and iron added to enhance casting qualities and corrosion resistance, are important commercially. These alloys are known as cupronickels and are well-suited for application in industrial and marine installations as condenser and heat-exchanger tubing because of their high corrosion resistance and particular resistance to impingement attack. Heat-exchanger tubes in desalinization plants use the cupronickel 10% alloy.

Cu–Sn alloys (3%–10% tin), deoxidized with phosphorus, form an important group known as phosphorus bronzes. Tin increases strength, hardness, and corrosion resistance, but at the expense of some workability. These alloys are widely used for springs and screens in papermaking machines.

Silicon (1.5%–3.0%), plus smaller amounts of other elements, such as tin, iron, or zinc, increases the strength of copper, making alloys useful for hardware, screws, bolts, and welding rods.

Sulfur (0.35%) and tellurium (0.50%) form insoluble compounds when alloyed with copper resulting in increased ease of machining.

Alloys of copper that can be precipitation-hardened have the common characteristic of a decreasing solid solubility of some phase or constituent with decreasing temperature. Precipitation is a decomposition of a solid solution leading to a new phase of different composition to be found in the matrix. In such alloys systems, cooling at the appropriate rapid rate (quenching) from an equilibrium temperature well within the all-alpha field will preserve the alloy as a single solid solution possessing relatively low hardness, strength, and electrical conductivity.

A second heat treatment (aging) at a lower temperature will cause precipitation of the unstable phase. The process is usually accompanied by an increase in hardness, strength, and electrical conductivity. Some 19 elements form copper-base binary alloys that can be age- or precipitation-hardened.

Two commercially important precipitation-hardenable alloys are Be–Cu and Cr–Cu. Be–Cu (2.0%–2.5% beryllium plus cobalt or nickel) can have a strength of 1400 MPa and an electrical conductivity of less than 50% of IACS. Cobalt adds high-temperature stability and nickel acts as a grain refiner. These alloys have found use as springs, diaphragms, and bearing plates and in other applications requiring high strength and resistance to shock and fatigue.

CuCr (1% chromium) can have a strength of 550 MPa and an electrical conductivity of 80%. Cu–Cr alloys are used to make resistance-welding electrodes, structural members for electrical switch gear, current-carrying members, and springs.

Cu–Ni, with silicon or phosphorus added, forms another series of precipitation-hardenable alloys. Typical composition is 2% nickel, 0.6% silicon. Strength of a 30 MPa can be obtained with high ductility and electrical conductivity of 32% IACS.

Zr–Cu is included in this group because it responds to heat treatment, although its strength is primarily developed through application of cold deformation or work. Heat treatment restores high electrical conductivity and ductility and increases surface hardness. Tensile strength of 500 MPa coupled with an electrical conductivity of 88% can be developed. Uses are resistance welding wheels and tips, stud bases for transistors and rectifiers, commutators, and electrical switch gear.

As little as 2% beryllium content, combined with the thermal conductivity of copper, has improved wear resistance to extend mold life and accelerated mold cycle times. Advancements in electrode materials are directly responsible for use of alloys like Be-Cu in mold making.

Conventional EDM performance in Be–Cu alloys with graphite electrodes is similar to that of other copper work metals. However, differences in EDM performance for various grades of Be–Cu alloys are evident at all power levels.

Be–Cu alloys with beryllium contents of 0.2%–0.6% and 1.6%–2.0% have shown that the higher-content group exhibited improved conventional EDM performance over the lower-content group. Testing of EDM with Be–Cu must still be performed relative to the possible health hazards involving fumes or gases, which must be recognized. Adequate EDM tank ventilation and dust collection for machining particulates could eliminate the hazards of inhaling any free beryllium or copper.

Cast Copper Alloys

Copper alloy castings of irregular and complex external and internal shapes can be produced by various casting methods, making possible the use of shapes for superior corrosion resistance, electrical conductivity, good bearing quality, and other attractive properties. High-copper alloys with varying amounts of tin, lead, and zinc account for a large percentage of all copper alloys used in the cast form. Tensile strength ranges from 250 to 330 MPa depending on composition and size. Leaded tin-bronzes with 6%–10% tin, about 1% lead and 4% zinc are used for high-grade pressure castings, valve bodies, gears, and ornamental work. Bronzes high in lead and tin (5%–25% lead and 5%–10% zinc) are mostly used for bearings. High tin content is preferred for heavy pressures or shock loading, but lower tin and higher lead is preferred for lighter loads, higher speeds, or where lubrication is less certain. A leaded red brass containing 85% copper and 5% each of tin, lead, and zinc is a popular alloy for general use.

High-strength brasses containing 57%–63% copper, small percentages of aluminum and iron, and the balance zinc have tensile strengths from 490 to 830 MPa, high hardness, good ductility, and resistance to corrosion. They are used for valve stems and machinery parts requiring high strength and toughness.

Copper alloys containing less than 2% alloying elements are used when relatively high electrical conductivity is needed. Strength of these alloys is usually notably less than that of other cast alloys.

Aluminum bronzes containing 5%–10.5% aluminum, small amounts of iron, and the balance copper have high strengths even at elevated temperature, high ductility, and excellent corrosion resistance. The higher-aluminum-content castings can be heat-treated, increasing their strength and hardness. These alloys are used for acid-resisting pump parts, pickling baskets, valve seats and guides, and marine propellers.

Additives impart special characteristics. Manganese is added as an alloying element for high-strength brasses where it forms intermetallic compounds with other elements, such as iron and aluminum. Nickel additions refine cast structures and add toughness, strength, and corrosion resistance. Silicon added to copper forms Cu–Si alloys of high strength and high corrosion resistance in acid media. Beryllium or chromium added to copper forms a series of age- or precipitation-hardenable alloys. Cu–Be alloys are among the strongest of the copper-base cast materials.

Copper Alloy Designations

Copper serves as the base metal for a great variety of wrought and cast alloys and the major wrought alloys and their designations, or alloy numbers, are high-copper alloys (C16200–C19750), which include CdCu, BeCu, and CuCr; CuZn brasses (C20500–C28580); Cu–Zn–Pb leaded brasses (C31200–C38590); Cu–Zn–Sn alloy or tin brasses (C40400–C49080); Cu–Sn phosphor bronzes (C50100–C52400); Cu–Sn–Pb bronze alloys or leaded phosphor bronzes (C53200–C54800); Cu–P alloys and Cu–Ag–P alloys (C55180–C55284); Cu–Al alloys or aluminum bronzes (C60600–C64400); Cu–Si alloys or silicon bronzes (C64700–C66100); miscellaneous Cu–Zn alloys (C66400–C69950); and Cu–Ni–Zn alloys or nickel silvers (C73150–C79900). All told, there are about 300 standard wrought alloys and many have cast counterparts (C81300–C99750). There are about 140 standard cast alloys. Many copper alloys are also available in powder form for powder-metal parts.

Cu–Ni wrought alloys are designated C70100–C72950; cast alloys, C96200–C96800. The alloys also have been referred to as cupronickels, Cu–Ni 20% (or whatever the percent nickel), also 80%–20% (or whatever the percent copper and nickel), nickel content may be as low as 2%–3% (C70200) or as high as 43%–46% (C72150), but intermediate amounts, nominally 10% (C70600 and C96200), 20% (C71000 and C96300), and 30% (C71500 and C96400), are the most common. Most of the alloys also contain small to moderate amounts of iron, zinc, manganese, and other alloying ingredients, and some contain substantial amounts of tin (1.8%–2.8% in C72500; 7.5%–8.5% in C72800).

The Brasses/Bronzes/Silvers/Cupronickels

The straight brasses are without question the most useful and widely used of all copper-base alloys. Zinc is used in substantial portions in five of the alloy groups listed. Most of the brasses are within the alpha field of the phase diagrams, thus maintaining good cold-working properties. With a higher zinc content of about 40%, the beta phase facilitates hot working. The addition of zinc decreases the melting point, density, electrical and thermal conductivities, and modulus of elasticity. It increases the coefficient of expansion, strength, and hardness. Work hardening increases with zinc content but 70–30 brass has the best combination of strength and ductility.

The low zinc brasses (20% zinc or less) have excellent corrosion resistance, are highly resistant to stress-corrosion cracking (SCC), and cannot dezincify. All of the low brasses are extensively used where these properties are required, along with good forming characteristics, and moderate strength. Commercial bronze is the standard alloy for domestic screen cloth, rotating bands on shells, and weatherstripping. Red brass is highly corrosion resistant, more so than copper in many cases, and is used for condenser tubes and process piping. The brasses have a pleasing color range varying from the red of copper through bronze and gold colors to the yellow of the high zinc brasses. The alloy 87.5% copper, 12.5% zinc very closely matches the color of 14-karat gold, and the low brasses in particular are used in inexpensive jewelry, for closures, and in other decorative items.

The high brasses, cartridge brass and yellow brass, have excellent ductility and high strength and have innumerable applications for structural and decorative parts that are to be fabricated by drawing, stamping, cold heading, spinning, and etching. Cartridge brass is able to withstand severe cold working in practically all fabricating operations, and derives its name from the deep-drawing operations necessary on small-arms munition cases, for which it is ideally suited. Muntz metal is primarily a hot-working alloy and

is cold-formed with difficulty. It is used in applications where cold spinning, drawing, or upsetting are not necessary. A typical use of Muntz metal is for condenser tube plates.

Silicon, Aluminum, and Manganese Brass

Silicon red brass has the corrosion resistance of the low brasses, with higher electrical resistance than that normally inherent in those brasses, and is especially suited to applications involving resistance welding. Aluminum brass is a moderate-cost alloy primarily made in tube form for use in condensers and heat exchangers, where its improved resistance to the corrosive and erosive action of high-velocity seawater is desired. Manganese brass serves the same purpose as silicon brass.

Leaded Brasses

The primary purpose in adding lead to any copper alloy is to improve machinability and related operations such as blanking or shearing. Lead also improves antifriction properties and is sometimes useful in that respect. Lead has an adverse effect in that both hot- and cold-working properties are hindered by reduced ductility. Optimum machinability is reached in free-cutting brass rod with 3.25% lead, ideally suited for automatic screw-machine work. Other leaded brasses in rod and other forms, with lead content ranging from 0.5% to over 2%, and with varying copper content, are available to permit varying combinations of corrosion resistance, strength, and hardness, and equally important, secondary fabricating processes (flaring, bending, drawing, stamping, thread rolling). Applications of all these alloys are found throughout industry.

Tin Brasses

Tin improves the corrosion resistance and strength of Cu–Zn alloys. Pleasing colors are obtainable when tin is alloyed with the low brasses. The tin brasses in sheet and strip form and with copper content of 80% or greater comprise a group of materials used as low-cost spring materials, for hardware, or for color effect alone, as in inexpensive jewelry, closures, and decorative items. Admiralty brass is the standard alloy for heat exchanger and condenser tubes, with good corrosion resistance to both seawater and domestic-water supplies. The addition of a small amount of antimony, phosphorus, or arsenic inhibits dezincification, a type of corrosion to which the high-zinc alloys are susceptible. Naval brass and manganese bronze are widely used for applications requiring good corrosion resistance and high strength, particularly in marine equipment for shafting and fastenings.

Phosphor Bronzes

As a group, these range in tin content from 1.25% to 10% nominal tin content. They have excellent cold-working characteristics, high strength, hardness, and endurance properties, low coefficient of friction, and excellent general corrosion resistance, and hence find wide use as springs, diaphragms, bearing plates, Fourdrinier wire, bellows, and fastenings.

Aluminum Bronzes

The alpha aluminum bronzes (containing less than about 8% aluminum) have good hot- and cold-working properties, while the alpha–beta alloys (8%–12% aluminum, with nickel, iron, silicon, and manganese) are readily hot-worked. All have excellent corrosion resistance, particularly to acids, and high strength and wear resistance. Certain alloys in this category may be hardened by a heat-treatment process similar to that used for steels. Certain alloys in this group are used for spark-resistant tools.

Cupronickels

These are straight Cu–Ni alloys with nickel content ranging from 2.5% to 30% nickel and have increased corrosion resistance with increasing nickel content. Cupronickels are markedly superior in their resistance to the corrosive and erosive effects of high-velocity seawater. They are moderately hard but quite tough and ductile, so that they are particularly suited to the manufacture of tubes for condenser and heat-exchanger use, and have flaring, rolling, and bending characteristics necessary for installation.

Cadmium Bronzes

Alloys of copper and cadmium, with or without small additions of tin, are primarily used for electrical conductors where high electrical conductivity and improved properties are found.

Nickel Silvers

Actually nickel brasses, these alloys contain varying amounts of copper, zinc, and nickel. They have a color ranging from ivory white in the lower nickel alloys to silver white in the higher nickel alloys. Their pleasing silver color, coupled with excellent corrosion resistance and cold formability, makes them ideally suited as a base for silver-plated flatware and hollow-ware, and structural and decorative parts in optical goods, cameras, and costume jewelry. The low-copper 18% nickel alloy is used for springs and other applications where high fatigue strength is desired.

Silicon Bronzes

The silicon bronzes are an extremely versatile series of alloys, having high strength, exceptional corrosion resistance, and excellent weldability, coupled with excellent hot and cold workability. The low silicon bronzes (1.5% silicon) are widely used for electrical hardware. The high silicon bronzes are used for structural parts throughout industry. Al–Si bronze is an important modification with exceptional strength and corrosion resistance, particularly suited to hot working and with better machinability than the regular silicon bronzes.

Copper Alloy Powders

Besides copper, there are a rather large range of compositions of copper alloy powders available, including the brasses, bronzes, and Cu–Ni. Brass powders are the most widely used for powder–metallurgy structural parts. Conventional grades are available with zinc contents from around 10% to 30%. Sintered brass parts have tensile strengths up to around 245 and 280 MPa and elongations of from 15% to around 40% depending on composition, design, and processing. In machinability they are comparable in cost with wrought brass stock of the same composition. Brass P/M parts are well-suited for applications requiring good corrosion resistance, and where free machining properties are desirable.

Cu–Ni, or Ni–Ag, powders contain 10% or 18% nickel. Their mechanical properties are rather similar to the brasses, with slightly higher hardness and corrosion resistance. Because they are easily polished, they find considerable use in decorative applications.

Copper and bronze powders are used for filters, bearings, and electrical and friction products. Bronze powders are relatively hard to press to densities to give satisfactory strength for structural parts. Probably the most commonly used bronze contains 10% tin. The strength properties are considerably lower than iron-base and brass powders, usually below 140 MPa.

Miscellaneous Copper Alloys

Narloy Z is a *copper–silver–zirconium alloy* for high temperature applications, such as vacuum-plasma-sprayed combustion chambers of rocket engines. It is also a candidate for engine inlets and wing leading edges of the U.S. national aerospace airplane. Powder-metal *copper–chromium–columbium alloys* Cu–8Cr–4Cb and Cu–6.5Cr–5.8Cb are also candidates for rocket-engine applications as well as heat exchangers, electrical contacts, and resistance welding electrodes. The latter alloy possesses high strength, creep resistance, and thermal conductivity at high temperatures. Besides their use for a great variety of parts, copper alloys are also used for surfacing ferrous and nonferrous parts for bearing applications, for corrosion and erosion resistance, and to rebuild worn parts. There are also *memory alloys*, or *shape-memory alloys*, which can be deformed and then revert to their original shape when heated to their transformation temperature. The reusable locknuts, from Memry Corp., are one application. With copper–aluminum–zinc or copper–aluminum–nickel alloy used for an answer, the insert can be deformed to lock the nut in place. When the nut is removed and heated to the alloys transformation temperature, the insert returns to its original shape so that the nut can be reused. An *89Cu–5Al–5Zn–1Sn* alloy, referred to as *Nordic Gold*, is used for several Euro coins. A series of *Thermitech copper–tungsten alloys*, from Mi-Tech Metals, Inc., are used for thermal heat sinks and electric circuits.

Copper Brazing

A term improperly used to denote brazing with a copper filler metal. See preferred terms *furnace brazing* and *braze welding*.

Copper Loss

The power loss resulting from the resistance in electrical circuit.

Copper Matte

An intermediate stage in the production of copper from sulfide ores; it contains about two-thirds copper.

Copper Oxide

There are several oxides of copper but usually the term refers to *red copper oxide*, or *cuprous oxide*, Cu_2O, a reddish crystalline powder formed by the oxidation of copper at high temperatures. It also occurs naturally in cuprite ore. It is insoluble in water but soluble in acids and alkalies. It is used in coloring glass and ceramics red, in electroplating, and in alternating-current rectifiers. *Black copper oxide*, or *cupric oxide*, CuO, is a brownish-black amorphous powder. It is used for coloring ceramics green or blue. In its natural ore form, it is called tenorite. Together with the red oxide, it is used as a copper paint for ships' bottoms.

Cupric oxide, when used in glazes, has a wide range of color. It may be used either as the raw oxide in a raw glaze, as the raw oxide in a fritted glaze, or as part of the frit itself. Cupric oxide is generally preferred in glazes and cuprous oxide in glasses.

Copper hydroxide, formed by the action of an alkali on the oxides, is also a poisonous blue powder and is used as a pigment.

Copper Shortening

See *shortening*.

Copper Steel

Steel containing up to 0.25% copper and very low in carbon, employed for construction work where mild resistance to corrosion is needed and where the cost of the more resistant chromium steels is not warranted. It is employed in sheet form for culverts, ducts, pipes, and such manufacturing purposes as washing-machine boilers. The alloy steels containing considerable copper for special purposes are not classified as copper steels. The copper neutralizes the corroding influence of the sulfur in the steel and aids in the formation of a fine-grained oxide that retards further corrosion. Copper is not added to unalloyed high-carbon steels because it causes brittleness and hot-shortness. Since the carbon content of copper steel is usually very low, the material is more a copper iron. Unless balancing elements, especially nickel, are present, more than 0.2% copper in steel may cause rolling defects. Molybdenum in small quantities may also be added to give additional corrosion resistance, and the percentage of carbon may be raised to 0.40% when about 0.05% molybdenum is added.

Copper Sulfate

Also called *bluestone*, *blue copperas*, and *blue vitriol*. An azure-blue, crystalline, lumpy material. It is soluble in water and insoluble in alcohol. It is produced as a by-product in copper refineries, or by the action of sulfuric acid on copper or copper oxide. A major market for copper sulfate is agriculture, where it is used in fungicides, micronutrients for fertilizers and animal feeds, and seed treatment. In chemical processes, it is used as an algicide in water treatment, for separating sulfide ores, in electroplating, in froth flotation, in leather tanning and hide preservation, and as a raw material for other salts and dies.

Copper-Accelerated Salt-Spray (CASS) Test

An *accelerated corrosion test* for some electrodeposits and for anodic or coatings on aluminum.

Copper-Constantan

A pair of metals used for thermocouples. See *Constantan*.

Copperhead

A reddish spot in a porcelain enamel coating caused by iron pickup during enameling, iron oxide left on poorly cleaned basic metal, or burrs on iron or steel basic metal that protrude through the coating and are oxidized during firing.

Cordierite

A talc-like mineral of composition $Mg_2Al_4Si_5O_{18}$ used in ceramics to make refractories, filters and diesel engines, and spark plug insulators. Cordierite has an unusually low coefficient of expansion (1.4–$2.6 \times 10^{-6}/°C$ from $25°C$ to $1000°C$) and high thermal shock resistance.

Core

(1) A specially formed material inserted in a mold to shape the interior other part of the casting that cannot be shaped as easily by the pattern. (2) To establish the internal shape of a hollow casting. (3) In a ferrous alloy prepared for *case hardening*, that portion of the alloy that is not part of the *case*. Typically considered to be the portion that (a) appears dark (with certain etchants) on an etched cross section (b) has an essentially unaltered chemical composition, or (c) has a hardness, after hardening, less than a specified value. (4) The central member of a *sandwich construction* (honeycomb material, foamed plastic, or solid sheet to which the faces of the sandwich are attached). (5) The inner material beneath a case hardened, or other surface layer. (6) The central mass of a transformer around which the windings are arranged.

Core (Plastics)

(1) In plastic part making, a channel in a mold for circulation of heat transfer media. (2) The part of a complex mold that molds undercut parts. Cores are usually withdrawn before the main sections of the mold are opened. Also called core pin.

Core Assembly

In casting, a complex score consisting of a number of sections.

Core Binder

In casting, any material used to hold the grains of core sand together.

Core Blow

A gas pocket in a casting adjacent to a cored cavity and caused by entrapped gases from the core.

Core Blower

A machine for making foundry cores using compressed air to blow and pack the sand into the core box.

Core Box

In casting, a wood, metal, or plastic structure containing a shaped cavity into which sand is packed to make a core.

Core Crush

A collapse, distortion, or compression of the core of a *sandwich construction*.

Core Depression

A localized indentation or gouge in the core of a *sandwich construction*.

Core Dryers

In casting, supports used to hold cores in shape during baking; constructed from metal or sand for conventional baking or from plastic material for use with the dielectric core-making equipment.

Core Filler

In casting, a material, such as coke, cinder, and sawdust, used in place of sand in the interiors of large cores; usually added to aid collapsibility.

Core Forging

(1) Displacing metal with a punch to fill a die cavity. (2) The product of such an operation.

Core Knockout Machine

In casting, a mechanical device for removing cores from castings.

Core Oil

In casting, a binder for core sand that sets when baked and is destroyed by the heat from the cooling casting.

Core Pin

In plastic part making, a pin used to mold a hole.

Core Pin Plate

In plastic part making, a plate holding core pins.

Core Plates

In casting, heat-resistant plates used to support cores during baking; may be metallic or nonmetallic, the latter being a requisite for dielectric core baking.

Core Plug

A plug inserted into the wall of a hollow casting to close a hole left by the core support or through which a sand core was extracted.

Core Print

In casting, projections attached to a pattern in order to form recesses in the mold at points where cores are to be supported.

Core Refining

The initial stage of hardening a carburized steel component. The carburizing treatment produces an undesirably large grain size in the component. Good quality work requires a fine grain size in both the low carbon core and the high carbon case. Initially, the component is heated to and quenched from about 880°C which is just within the austenitic region for the typical lower carbon core. This refines, hardens, and toughens the core. Subsequently, the component is heated to and quenched from about 750°C which is just within the austenitic region for the high carbon case. This hardens the case. See *Steel* and *Case Hardening*.

Core Rod

In powder metallurgy, a member of a die assembly used in molding a hole in a compact.

Core Sand

In casting, sand for making cores to which a binding material has been added to obtain good cohesion and permeability after drying; usually low in clays.

Core Separation

In a *sandwich construction*, a partial or complete breaking of the core node bond.

Core Shift

In casting, a variation from the specified dimensions of a cored casting section due to a change in position of the core or misalignment of cores in assembly.

Core Splicing

The joining of segments of a core of a *sandwich construction* by bonding, or by overlapping each segment and then driving them together.

Core Vents

(1) In casting, a wax product, round or oval in form, used to form the vent passage in a core. Also, a metal screen or slotted piece used to form the vent passage in the core box used in a core blowing machine. (2) Holes made in the core for the escape of gas.

Core Wash

In casting, a suspension of a fine refractory applied to cores by brushing, dipping, or spraying to improve the surface of the cored portion of the casting.

Core Wires or Rods

(1) In casting, reinforcing wires or rods for fragile cores, often preformed into special shapes. (2) (for welding) The central main wire of a continuous covered electrode.

Cored Bars

In powder metallurgy, a compact of bar shape heated by its own electrical resistance to a temperature high enough to melt its interior.

Cored Mold

In plastic part making, a mold incorporating passages for electrical heating elements, steam, or water.

Cored Solder

A solder wire or bar containing flux as a core.

Coreless Induction Furnace

An electric induction furnace for melting or holding molten metals that does not utilize a steel core to direct the magnetic field which stirs the melt.

Coring

(1) A condition of variable composition between the center and surface of a unit of microstructure (such as a dendrite, grain, carbide particle); results from nonequilibrium solidification, which occurs over a range of temperature. (2) A central cavity at the butt end of a rod extrusion, sometimes called *extrusion pipe*. (3) A variation in composition from center to edge of individual grains of an alloy solidifying as a solid solution. It is most pronounced where there is a large difference between the liquidus and the solidus and arises during casting because the first material to solidify is relatively rich in one element and the remaining molten material becomes progressively richer in the second element. The change in composition may be so extreme that a nonequilibrium phase precipitates in the latter stages of solidification. See Phase, Lever Rule and Solidification.

Coring (Plastics)

In plastic part making, the removal of excess material from the cross section of a molded part to attain a more uniform wall thickness.

Corner Angle

On face milling cutters, the angle between an angular cutting edge of a cutter tooth and the axis of the cutter, measured by rotation into an axial plane. See *face mill*.

Corner Joint

A joint between two members located approximately at right angles to each other in the form of an "L." Usually between 30° and 135°.

Corner-Flange Weld

A flange weld with only one member flanged at the location of welding.

Corona (Resistance Welding)

The area sometimes surrounding the nugget of a spot weld at the faying surfaces which provides a degree of solid-state welding.

Correction

In the case of a testing machine, the difference obtained by subtracting the indicated load from the correct value of the applied load.

Corrodkote Test

An *accelerated corrosion test* for electrodeposits.

Corrosion

Any process by which a metal reacts chemically or electrochemically with another element in the surrounding environment. It is usually implicit that the consequences will be damaging as opposed to processes such as anodizing. Damage includes loss of bulk section, pitting, rusting, and scaling. See also *Electrochemistry* and *Corrosion Resistance*.

Corrosion Effect

A change in any part of the *corrosion system* caused by *corrosion*.

Corrosion Embrittlement

Deeply penetrating corrosion damage, essentially cracking, often along grain boundaries. The damage is usually not obvious to a visual inspection but causes a major reduction in load carrying capacity not merely a reduction in toughness. See also *hydrogen damage*.

Corrosion Fatigue

A cracking mechanism in which corrosion and fatigue act concurrently to reduce significantly the total life that would be expected if fatigue loading and a corrosive environment had acted alone or consecutively. Purists sometimes argue that any fatigue, other than in vacuum, is corrosion fatigue. A more practical approach is to differentiate on the basis of the major difference in life seen in a clean and dry environment compared with that in a moist and contaminated one.

Corrosion Fatigue Strength

The maximum repeated stress that can be endured by a metal without failure under definite conditions of corrosion and fatigue and for a specific number of stress cycles and a specified period of time.

Corrosion Inhibitor

See *inhibitor*.

Corrosion Potential (E_{corr})

The electrical *potential* of a corroding surface in an electrolyte, relative to a *reference electrode*. Also called rest potential, open-circuit potential, or freely corroding potential.

Corrosion Product

Substance formed as a result of *corrosion*.

Corrosion Protection

Modification of a *corrosion system* so that corrosion damage is mitigated.

Corrosion Rate

Corrosion effect on a metal per unit of time. The type of corrosion rate used depends on the technical system and on the type of corrosion effect. Thus, corrosion rate may be expressed as an increase and corrosion depth per unit of time (penetration rate, for example, mil/year) or the mass of metal turned into corrosion products per unit area of surface per unit of time (weight loss, for example, g/m²/year). The corrosion effect may vary with time and may not be the same at all points of the corroding surface. Therefore, reports of corrosion rates should be accompanied by information on the type, time dependency, and location of the corrosion effect.

Corrosion Resistance

The ability of a metal to avoid unacceptable reactions with its environment. For metals, this could be pitting or rusting; for organic materials, it could be crazing. Generalizing, metals can be divided into three categories. Some metals such as gold and platinum have inherent resistance, that is, they do not react to any significant extent with the atmosphere and many other environments. Other metals such as aluminum or titanium react very rapidly when exposed to the atmosphere but the oxide film is thin, tightly bonded to the metal and forms an impervious barrier to further attack. This is often termed Acquired Resistance. Finally, metals such as iron and steel in moist environments form a bulky, loosely adherent and porous oxide layer. This does not significantly slow the rate of attack and it may even detach, exposing bare metal for further attack. See also *stainless steel*.

Corrosion System

System consisting of one or more metals and all parts of the environment that influence *corrosion*.

Corrosion–Erosion

A combination of corrosion and erosion. There are considerable similarities with impingement attack and some authorities recognize only one mechanism. Where the two are differentiated, and this may not be easy in some cases, impingement is regarded as having a major electrochemical component arising from the potential difference between uncorroded areas and bare metal exposed by erosion. The term corrosion–erosion would then be confined to cases where the erosion merely serves to remove the products of corrosion which, if undisturbed, could impede further attacks. See also *Electrochemistry* and *Cavitation*. See also *erosion–corrosion*.

Corrosion-Resistant Alloys

There are now several corrosion-resistant alloys available that are capable of meeting most difficult design requirements.

Cast Alloys

In general, these are the cast counterparts to 3XX and 4XX wrought stainless steels and, thus, are also referred to as cast stainless steels. Designations of the Alloy Casting Institute of the Steel Founders Society of America and the wrought designations to which they roughly correspond (compositions are not identical) include CA-15 (410), CA-40 (420), CB-30 (431), CC-50 (446), CE-30 (312), CF-3 (304L), CF-3M (316L), CF-8 (304), CF-8C (347), CF-8M (316), CF-12M (316), CF-16F (303), CF-20 (302), CG-8M (317), CH-20 (309), and CK-20 (310). There are also other alloys that do not correspond to wrought grades. The cast

alloys corresponding to 3XX wrought grades have chromium contents in the range of 17%–30% and nickel contents in the range of 8%–22%. Silicon content is usually 2.00% maximum (1.50 for CE-8M), manganese 1.50% maximum, and carbon 0.08%–0.30% maximum, depending on the alloy. Other common alloying elements include copper and molybdenum. Those corresponding to 4XX grades may contain as much chromium but much less nickel; 1%–5.5%, depending on the alloy. Manganese and silicon contents are also generally less and carbon may be 0.15%–0.50%, depending on the alloy. All of the alloys are Fe–Cr–Ni alloys and the widest used are CF-8 and CF-8M, which limit carbon content to 0.08%. CN-7M and CN-7MS contain more nickel than chromium and, thus, are referred to as Fe–Cr–Ni alloys.

The alloys are noted primarily for their outstanding corrosion resistance in aqueous solutions and hot gaseous and oxidizing environments. Oxidation resistance stems largely from the chromium. Nickel improves toughness and corrosion resistance in neutral chloride solutions and weak oxidizing acids. Molybdenum enhances resistance to pitting in chloride solutions. Copper increases strength, and permits precipitation hardening to still greater strength. After a 482°C age, for example, the room-temperature tensile properties of CB-7Cu are 1290 MPa ultimate strength, 1100 MPa yield strength, 10% elongation, and 196, 500 MPa elastic modulus. Higher aging temperatures, to 621°C, decrease strength somewhat but markedly increase impact strength. The alloys are widely used for pumps, impellers, housings, and valve bodies in the power-transmission, marine, and petroleum industries, and for chemical, food, pulp and paper, beverage, brewing, and mining equipment.

Corrosive Flux

A flux with a residue that chemically attacks the base metal. It may be composed of inorganic salts and acids, organic salts and acids, or activated rosins or resins.

Corrosive Wear

Wear in which chemical or electrochemical reaction with the environment is significant. See also *oxidative wear*.

Corrosivity

Tendency of an environment to cause *corrosion* in a given *corrosion system*.

Corrugating

The forming of sheet metal into a series of straight, parallel alternate ridges and grooves with a rolling mill equipped with matched roller dies or a *press brake* equipped with a specially shaped punch and die.

Corundum

Corundum is a very hard crystalline mineral used chiefly as an abrasive, especially for grinding and polishing optical glass. It is in the alpha, or hexagonal, crystal form, usually containing some lime and other impurities. The physical properties are theoretically the same as for synthetic alpha Al_2O_3, but they are not uniform. The melting point and the hardness are generally lower because of impurities, and the crystal structure also varies.

Pure corundum is transparent and colorless, but most specimens contain some transition elements substituting for aluminum, resulting in the presence of color. Substitution of chromium results in a deep red color; such red corundum is known as ruby. The term *sapphire* is used in both a restricted sense for the "cornflower blue" variety containing iron and titanium, and in a general sense for gem-quality corundums of any color other than red.

Corundum is synthesized by a variety of techniques for use as synthetic gems of a variety of colors and as a laser source (ruby). The most important technique is the Verneuil flame-fusion method, but others include the flux-fusion and Czochralski "crystal-pulling" techniques. Native alumina occurring as rhombohedral crystals and also in masses and variously colored grains. Corundum and its artificial counterparts are abrasives especially suited to the grinding of metals.

Cosmetic Pass

A weld pass made primarily for the purpose of enhancing appearance.

Cosworth Process

A process for producing high-quality precision castings. Quality is assured by extreme efforts to avoid entrainment of oxide or other contaminants, in particular, by the use of extended periods of holding the molten metal followed by pumped, up-hill feeding. Precision is achieved by the use of molds of zircon sand having a low expansion coefficient.

Cottoning

The formation of web-like filaments of adhesive between the applicator and substrate surface.

Cottrell Process

Removal of solid particulates from gases with electrostatic precipitation.

Coulomb

The SI unit of quantity of electricity, the electrical charge transported by 1 A flowing for 1 s.

Coulomb Friction

A term used to indicate that the frictional force is proportional to the normal load.

Coulometer

An electrolytic cell arranged to measure the quantity of electricity by the chemical action produced in accordance with Faraday's law.

Coulometry

In materials characterization, an electrochemical technique in which the total number of coulombs consumed in electrolysis is used to determine the amount of substance electrolyzed.

Coumarone

A colorless, oily liquid of composition C_8H_6O, used chiefly in making synthetic resins. It has a specific gravity of 1.096, is insoluble in water, and is easily oxidized. It can be reacted with butadiene to form an *indene–butadiene rubber* of superior properties. All of the *cumenes* are variants of benzene.

The indene resins are classified with the coumarone resins, but they are lighter in color and are used in varnishes. The coumarone resins are made by the action of sulfuric or phosphoric acid on coumarone, are very soluble in organic solvents and are used in lacquers, waterproofing compounds, molding, and adhesives.

Count

For fabric used in composites fabrication, the number of warp and filling yarns per inch and woven cloth. For yarn, size based on relation of length and weight.

Counterblock

A jog in the mating surfaces of dies to prevent lateral die shifting from side thrusts developed in forging irregularly shaped pieces.

Counterblow Equipment

Forging equipment with two opposed rams that are activated simultaneously to strike repeated blows on the workpiece placed midway between them.

Counterblow Forging Equipment

A category of forging equipment in which two opposed rams are activated simultaneously, striking repeated blows on the workpiece at a midway point. Action is vertical or horizontal.

Counterblow Hammer

A forging hammer in which both the *ram* and the *anvil* are driven simultaneously toward each other by air or steam pistons.

Counterboring

Removal of material to enlarge a hole for part of its depth with a rotary, pilot guided, end cutting tool having two or more cutting lips and usually having straight or helical flutes for the passage of chips and the admission of a cutting fluid.

Counterelectrode

In emission spectroscopy, the electrode that is used opposite to the self-electrode or supporting electrode and that is not composed of the sample to be analyzed. In voltammetry, the current between the working electrode and counterelectrodes is measured. See also *self-electrode*, *supporting electrode*, and *auxiliary electrode*.

Counterformal Surfaces

Surfaces whose centers of curvature are on the opposite sides of the interface, as in rolling-element bearings or gear teeth. In wear testing, this term is sometimes used to indicate that the test specimen surfaces are not conformal (for example, with a sphere-on-flat or flat block-on-rotating ring configuration). In such cases, the nominal contact area of at least one of the test piece surfaces increases as the amount of wear increases. Contrast with conformal surfaces.

Countersinking

Beveling or tapering the work material around the periphery of a hole creating a concentric surface at an angle less than 90° with the centerline of the hole for the purpose of chamfering holes or recessing screw and rivet heads.

Couple

The ability of dissimilar metals in electrical contact to produce electrical currents. Similar effects can occur in a single component because of variations, from point-to-point, and the composition of the material or of the environment. See Thermoelectric Effects and *galvanic corrosion.*

Coupling

(1) The degree of mutual interaction between two or more elements resulting from mechanical, acoustical, or electrical linkage. (2) A device or mechanism joining two components or structures, usually for some specific function such as transmitting a fluid, traction or power or such purpose.

Coupon

A piece of material from which a test specimen is to be prepared—often an extra piece (as on a casting or forging) or a separate piece made from test purposes (such as a test weldment).

Covalent Bond

A bond in which two atoms share a pair of electrons. Contrast with *ionic bond.* See Interatomic bonding.

Cover Core

(1) In casting, a core set in place during the ramming of a mold to cover and complete a cavity partly formed by the withdrawal of a loose part of the pattern. Also used to form part or all of the cope surface of the mold cavity. (2) A core placed over another core to create a flat parting line.

Cover Half

The stationary half of a die-casting die.

Cover Pass

The final layer of a multilayer well. The term often implies a deposit having a good appearance.

Cover Plate (Welding Eye Protection)

A removable pane of colorless glass, plastic-coated glass, or plastic that covers the filter plate and protects it from weld splatter, pitting, or scratching when used in a helmet, hood, or goggles.

Covered Electrode

A composite filler metal electrode consisting of a core of a bare electrode or metal cored electrode to which a covering sufficient to provide a slag layer on the weld metal has been applied. The covering may contain materials providing such functions as shielding from the atmosphere, deoxidation, and arc stabilization and can serve as a source of metallic additions to the weld. See *electrode welding.*

Covering Power

(1) The ability of a solution to give satisfactory plating at very low current densities, a condition that exists in recesses and pits. This term suggests an ability to cover, but not necessarily to build up, a uniform coating, whereas *throwing power* suggests the ability to obtain a coating of uniform thickness on an irregularly shaped object. (2) The degree to which a porcelain enamel coating obscures the underlying surface. (3) The ability of a glaze to uniformly and completely cover the surface of the fired ceramic ware.

Crack

(1) A fracture type discontinuity characterized by a sharp tip and high ratio of length and width to opening displacement. (2) A line of fracture without complete separation. (3) A planar defect that has developed in an area that was substantially defect-free. There is often an implication of limited ductility and "Crack" as opposed to "break" tends to be favored where the component has not parted completely. See also *brake, rupture, and fracture.*

Crack Arrest Temperature

The temperature at which a running brittle fracture in steel will cease to propagate. The term may be used in connection with some particular laboratory test such as the Robertson Test. More generally the term refers to the lowest service temperature of a steel component at which a brittle fracture will not propagate further.

Crack Arrestor

Any feature, intended or otherwise, that stops a crack from propagating further. Such features may be on a microscopic scale, for example, nonmetallic inclusions, or on a macro scale, for example, reinforcement plates, ductile materials are even holes placed across the path of an advancing crack.

Crack Branching

The separation of a material into two or more segments.

Crack Extension (Δa)

An increase in crack size. See also *crack length, effective crack size, original crack size,* and *physical crack size.*

Crack Growth

Rate of propagation of a crack through a material due to a static or dynamic applied load.

Crack Initiator

Any feature, such as a crack, section change or inclusion, which has a local stress raising effect and hence assist the initiation of cracking by any mechanism.

Crack Length (Depth) (a)

In *fatigue* and *stress-corrosion cracking*, the *physical crack size* used to determine the crack growth rate and the *stress-intensity factor*. For a compact-type specimen, crack length is measured from the line connecting the bearing points of load application. For a center-crack tension specimen, crack length is measured from the perpendicular bisector of the central crack. See also *crack size*.

Crack Mouth Opening Displacement (CMOD)

See *crack opening displacement*.

Crack Opening Displacement (COD)

On a K_{Ic} specimen, the opening displacement of the notch surfaces at the notch and in the direction perpendicular to the plane of the notch and the crack. The displacement at the tip is called the crack tip opening displacement (CTOD), at the mouth, it is called the crack mouth opening displacement (CMOD). See also *stress-intensity factor* for definition of K_{Ic}. See *fracture toughness* and *fracture mechanics*.

Crack Plane Orientation

An identification of the plane and direction of a fracture in relation to product geometry. This identification is designated by a hyphenated code, the first letter(s) representing the direction normal to the crack plane and the second letter(s) designated the expected direction of crack propagation.

Crack Size (a)

A lineal measure of a principal planar dimension of a crack. This measure is commonly used in the calculation of quantities descriptive of the stress and displacement fields. In practice, the value of crack size is obtained from procedures for measurement of *physical crack size*, *original crack size*, or *effective crack size*, as appropriate to the situation under consideration. See also *crack length (depth)*.

Crack Tip Opening Displacement (CTOD)

See *crack opening displacement*.

Cracked Gas

A generic term for a gas mixture obtained by thermal decomposition, with or without catalysis, of a gaseous compound. Examples: cracked ammonia (NH_3) is a mixture of nitrogen and hydrogen, and cracked natural gas hydrocarbons such as methane (CH_4) are a mixture of carbon and hydrogen. Cracked ammonia is also known as *disassociated ammonia*.

Crack-Extension Force (G)

The elastic energy per unit of new separation area that would be made available at the front of an ideal crack in an elastic solid during a virtual increment of forward crack extension. This definition is useful for either static cracks or running cracks. From past usage, crack extension force is commonly associated with linear-elastic methods of analysis. See also *J-integral*.

Crack-Extension Resistance (K_R)

A measure of the resistance of a material to *crack extension* expressed in terms of the *stress-intensity factor*, the *crack-extension force*, or values of *J* derived from the *J-integral* concept.

Cracking

In lubrication technology, that process of converting unwanted long-chain hydrocarbons to shorter molecules by thermal or catalytic action. See *fracture toughness* and *fracture mechanics*.

Crackled

Either a craze cracked surface or a severely wrinkled paint coating.

Crack-Tip Plane Strain

A stress–strain field near a crack tip that approaches plane strain to the degree required by empirical criterion.

Crank

Forging shape generally in the form of a "U" with projections at more or less right angles to the upper terminals. Crank shapes are designated by the number of *throws* (for example, *two-throw* crank).

Crank Press

A mechanical press who slides are actuated by a crankshaft.

Crankpin Bearing

See *big-end bearing*.

Crater

In arc welding, a depression or hole at the termination of a weld bead or in the molten weld pool resulting from local shrinkage during solidification.

Crater Crack

A crack in the crater of a weld bead.

Crater Fill Time

The time interval following weld time but prior to meltback time during which arc voltage or current reaches a preset value greater or less than welding values. Weld travel may or may not stop at this point.

Crater Fill Voltage

The arc voltage value during crater fill time.

Crater Wear

The wear that occurs on the rake face of a cutting tool due to contact with the material in the chip that is sliding along that face.

Cratering

Depressions on coated plastic surfaces caused by excess lubricant. Cratering results when paint is too thin and later ruptures, leaving pinholes and other voids. Use of less thinner in the coating can reduce or eliminate cratering, as can less lubricant on the part.

Craze Cracking

Irregular surface cracking of a metal associated with thermal cycling. This term is used more in the United Kingdom than in the United States, where the term checking is used instead.

Crazing

Multiple intersecting surface cracks, usually shallow, arising from various causes but particularly thermal shock.

Crazing (Ceramics)

The cracking which occurs in fired glazes or other ceramic coatings due to critical tensile stresses.

Crazing (Plastics)

Region of ultrafine cracks, which may extend in a network on or under the surface of a resin or plastic material. May appear as a white band. Often found in a filament-wound pressure vessel or bottle. In many plastics, craze growth precedes crack growth because crazes are load bearing.

Creel

In composites fabrication, a spool, along with its supporting structure, that holds the required number of roving balls or supply packages in a desired position for unwinding for the next processing step, that is, weaving, braiding, or filament winding.

Creep

Time-dependent strain occurring under stress. The creep strain occurring at a diminishing rate is called primary creep; that occurring at a minimum and almost constant rate, secondary creep; and that occurring at an accelerating rate, tertiary creep.

Creep Ductility

The elongation at failure of a component failing by creep. See Creep.

Creep Limit

(1) The maximum stress that will cause less than a specified quantity of creep in a given time. (2) The maximum nominal stress under which the creep strain rate decreases continuously with time under constant load and at constant temperature. (3) This term has a range of interpretations and some potential for confusion. The most basic definition is the maximum stress that can be allowed without causing unacceptable damage. Damage, in the context of Creep Limit, may

be defined as failure or as some level of Strain, in which cases it is necessary also to specify the service temperature and life required. Damage may also be specified as a rate of deformation in which case only the temperature need be specified. See Creep. Sometimes used synonymously with *creep strength*.

Creep Permanent

Progressive deformation, occurring over a period of time, at loads below the yield stress. Most metals at ambient temperature will carry a stress which is less than the yield stress for, effectively, an indefinite period; when the stress is removed the component will revert to its original dimensions. However, if the temperature of the component is raised above a critical level any significant stress will, over a period of time, cause progressive deformation that remains after the stress has been removed. If the load is maintained for a sufficient period internal voids and cracks will develop and ultimately the component will fail. A typical, perhaps idealized, creep test has the characteristics three stage, strain to time relationship, termed a Creep Curve. In the usually brief primary, or transient stage, deformation is rapid as it is dominated by mechanisms such as dislocation movement and grain boundary slide, i.e., the two neighboring grain faces rotating across each other. This high initial rate then slows as work hardening plays an increasing role. In the longer secondary, or steady-state stage, deformation continues at a lower uniform rate arising from the balance between work hardening and thermal softening. Toward the end of this stage, damage, in the form of grain boundary voids, starts to become visible under the microscope. In the final tertiary or runaway stage, deformation accelerates as voids join up, gross grain boundary cracks develop and the specimen necks prior to failure. This conventional view of creep being accompanied by major deformation is often confirmed by service experience but it is not universal; some metals fail by creep with negligible reduction in area and only a fraction of one percent elongation. Such behavior may be termed Creep Brittle. Creep failure is often termed Creep Rupture but this does not imply any special characteristic of failure; it could equally be termed "creep fracture" or even "creep breakage" although these phrases tend to sound inappropriate to most metallurgists. The temperature at which creep commences is broadly related to the melting point of the metal, it often being suggested that creep of pure metals is significant above half their absolute temperature melting point. For example, lead will creep, at least over decades or centuries, at normal ambient temperature and mild steel will creep at a significant rate above about 400°C. The rate of damage, above the critical temperature, rises with increasing stress and temperature. Creep deformation characteristics, often termed the Stress Rupture Properties, are determined by testing specimens of the material over ranges of stresses and temperatures to establish the time to reach a specified percentage creep strain or to fail. Unfortunately, the results obtained from testing identical specimens under identical conditions are rarely the same. Hence, the results of numerous tests plotted on a graph do not form a neat smooth curve but tend to be scattered over a fairly wide band. This gives rise to such terms as Creep Scatter Band while designs may be made on the basis of Lower Scatter Band (i.e., the lower line enclosing all test results) or 90% Scatter Band (the line enclosing 90% of all results) or Average Stress Rupture properties. See also Stress Rupture and Larsen Miller Parameter for examples of creep data presentation.

Creep Rate

The slope of the creep–time curve at a given time. Deflection with time under a given static load.

Creep Recovery

The time-dependent decrease in strain in a solid, following the removal of force.

Creep Relaxation

The reduction of imposed load as a result of deformation by creep, i.e., by the conversion of elastic strain to plastic strain by a creep mechanism. The phenomenon occurs in assemblies where the loading mechanism is rigid and does not move sufficiently to maintain the stress. One example is high temperature bolts which are tightened to induce a large amount of tensile elastic extension while the flange through which they pass, being of much larger cross section, experiences negligible elastic compression. If the bold creep, assuming it does not fail, the joint closure force will progressively fall. This contrasts with the case of, for example, a boiler tube which experiences a constant internal pressure. If the tube distends the internal pressure remains the same but is carried by a reducing wall thickness so the stress rises. See also preload and displacement control.

Creep Rupture

Failure by Creep. See also Larsen Miller Parameter.

Creep Strain

The time-dependent total strain (extension plus initial gauge length) produced by applied stress during a creep test.

Creep Strength

(1) The stress that will cause a given *creep strain* in a creep test at a given time in a specified constant environment. (2) Generally, the strength of the material in its creep range. More specifically, the same as creep limit.

Creep Stress

The constant load divided by the original cross-sectional area of the specimen.

Creep Test

A method of determining the extension of metals under a given load at a given temperature. The determination usually involves the plotting of time–elongation curves under constant load; a single test may extend over many months. The results are often expressed as the elongation (in millimeters or inches) per hour on a given gauge length (e.g., 25 mm or 1 in.).

Creep-Feed Grinding

See *grinding*.

Creep-Rupture Embrittlement

Embrittlement under creep conditions of, for example, aluminum alloys and steels that results in abnormally low rupture ductility. In aluminum alloys, iron and amounts above the solubility limit is known to cause such embrittlement; in steels, the phenomenon is related to the amount of impurities (for example, phosphorus, sulfur, copper, arsenic, antimony, and tin) present. In either case, failure occurs by *intergranular cracking* of the embrittled material.

Creep-Rupture Strength

The stress that causes fracture in a creep test at a given time, in a specified constant environment. This is sometimes referred to as the stress-rupture strength. In glass technology, this is termed the static fatigue strength.

Creep-Rupture Test

A test in which progressive specimen deformation and the time for rupture are both measured. In general, deformation is much greater than that developed during a creep test. Also known as stress-rupture test.

Creosote

Also called *dead oil* and *pitch oil*. A yellowish, poisonous oily liquid obtained from the distillation of coal tar. It has the odor of carbolic acid. The crude creosote oil is used as a wood preservative and as a harsh disinfectant, but its use in these applications is expected to decrease because it has been recently classified as a possible carcinogen. Other applications include use as a fluxing oil for coal-tar pitch and bitumen, production of carbon black, and use in sprays for dormant fruit plants. Creosote is also obtained in the distillation of pinewood tar and is then a yellowish liquid with a smoky odor, a mixture of phenols and derivatives.

Crescent Crack

Damage having the appearance of a crescent, produced in a glass surface by the frictive translation of a hard, blunt object across the glass surface. The crescent shape is concave toward the direction of translation on the damage surface.

Crevice Corrosion

(1) *Localized corrosion* of a metal surface at, or immediately adjacent to, an area that is shielded from full exposure to the environment because of close proximity between the metal and the surface of another material. (2) Corrosion, occurring within a narrow gap which is severe compared with that on a plain surface exposed to the same environment. The mechanism usually involves the development of variations in environment composition from point to point leading to electrochemical corrosion. The variation may be a concentration effect, for example, of dissolved salts, or it may be a depletion effect, for example, a reduction in oxygen within the crevice.

Crimping

(1) The formation of circumferential corrugations around one side of a tube to induce a bend, more usually termed wrinkle bending. (2) The deliberate kinking of a folded lap joint to lock the joint. (3) The introduction of angular ridges in a sheet surface to improve rigidity. (4) The introduction of multiple transverse indentations in bar to improve grip or to increase its length. See also *corrugating*.

Critical Anodic Current Density

The maximum anodic current density observed in the active region for a metal or alloy electrode that exhibits active–passive behavior in an environment.

Critical Cooling Rate

The minimum rate of continuous cooling for preventing undesirable transformations. For steel, unless otherwise specified, it is the slowest rate at which austenite can be cooled from above critical temperature to prevent its transformation above the martensite start temperature. See *steels*.

Critical Current Density

In an electrolytic process, a current density at which an abrupt change occurs in an operating variable or in the nature of an electrodeposit or electrode film.

Critical Curve

In a binary or higher order *phase diagram*, a line along which the phases of a heterogeneous equilibrium become identical.

Critical Damping

In dynamic mechanical measurement of plastics, that damping required for the borderline condition between oscillatory and non-oscillatory behavior.

Critical Diameter

Diameter of a steel bar that can be fully hardened with 50% martensite at its center.

Critical Dimension

A dimension on a part that must be held within the specified tolerance for the part to function in its application. A noncritical tolerance may be for cost or weight savings or for manufacturing convenience, but is not essential for the products.

Critical Flaw Size

The size of a flaw (defect) in a structure that will cause failure at a particular stress level.

Critical Frequency/Speed

The vibration frequency or rotational speed which coincides with a natural resonance frequency of the component. At this critical condition, perturbations which would normally be innocuous can rapidly increase in severity leading to damage or failure. Examples include gross instability of shafts, collapse of bridges, and rapid fatigue failure of turbine blades. Design and operating practices therefore aim to avoid completely, or to pass rapidly through, the critical stage.

Critical Humidity

The *relative humidity* above which the atmospheric corrosion rate of some metals increases sharply.

Critical Illumination

In metallography, the formation of an image of the light source in the object field.

Critical Length

In composites fabrication, the minimum fiber length required for shear loading to its ultimate strength by the matrix.

Critical Longitudinal Stress

The longitudinal stress necessary to cause internal slippage and separation of a spun yarn and a fiber-reinforced plastic. The stress necessary to overcome the interfiber friction developed as a result of twist.

Critical Mass

The minimum amount of fissile material required to sustain nuclear fission.

Critical Micelle Concentration

The concentration of a micelle at which the rate of increase of electrical conductance with increase in concentration levels off or proceeds at a much slower rate. See also *micelle*.

Critical Pitting Potential (E_{cp}, E_p, E_{pp})

The lowest value of oxidizing potential at which pits nucleate and grow. It is dependent on the test method used.

Critical Point

(1) The temperature or pressure at which a change in crystal structure, phase, or physical properties occurs. Also termed *transformation temperature*. (2) In an equilibrium diagram, that combination of composition, temperature, and pressure at which the phases of an inhomogeneous system are in equilibrium.

Critical Pressure

That pressure above which the liquid and vapor states are no longer distinguishable.

Critical Rake Angle

The rake angle in which the action of a V-point tool changes from cutting to plowing. See also *rake*.

Critical Range

In steels it is temperature range between the upper critical and lower critical temperatures. See *steel*.

Critical Shear Stress

The shear stress required to cause slip in a designated slip direction on a given slip plane. It is called the critical resolved shear stress if the shear stress is induced by tensile or compressive forces acting on the crystal.

Critical Strain

The minimum amount of plastic strain, i.e., deformation, required to induce recrystallization. Some metals, such as iron, can develop a completely new grain structure by heating them through a phase transformation. However, in the absence of such a transformation, a metal cannot form a new structure by heating below the melting point, unless it has been plastically deformed by more than a critical amount, the critical strain, typically a few percent. This process of developing a new grain structure with associated reduction in the hardness is termed annealing and, generalizing, the more severe the deformation the finer the subsequent grain size. Consequently, components deformed only slightly more than the critical amount can form structures with exceptionally large grain sizes when annealed. This effect is termed *critical strain grain growth*. (1) In mechanical testing, the strain at the yield point. (2) The strain just sufficient to cause recrystallization. Recrystallization takes place from only a few nuclei, which produces a recrystallized structure consisting of very large grains.

Critical Strain Energy Release Rate

The toughness as measured by the energy input required for a crack to continue propagating.

Critical Stress Intensity Factor

See *stress-intensity factor* and *fracture toughness*.

Critical Surface

In a ternary or higher order *phase diagram*, the area upon which the phases in equilibrium become identical.

Critical Temperature

That temperature above which the vapor phase cannot be condensed to liquid by an increase in pressure. Synonymous with *critical point* if pressure is constant.

Critical Temperature Range

Synonymous with transformation ranges, which is the preferred term.

Criticality of a Nuclear Reactor

The state at which the fission reaction is exactly in balance with the number of neutrons released being just sufficient to maintain the chain reaction.

Croning Process

In casting, a shell molding process that uses a phenolic resin binder. Sometimes referred to as *C* process or *Chronizing*.

Crop

(1) An end portion of an ingot that is cut off as scrap. (2) To shear a bar or billet.

Cropping

The practice of cutting off and discarding the ends of bars, ingots, etc., as these are the site of defects.

Cross Breaks

Same as *coil breaks*.

Cross Direction

See *transverse direction*.

Cross Forging

Preliminary working of forging stock in flat dies to develop mechanical properties, particularly in the center portions of heavy sections.

Cross Laminate

In composites fabrication, a laminate in which some of the layers of material are oriented approximately at right angles to the remaining layers with respect to the grain, or strongest direction in tension. A bidirectional laminate. See also *parallel laminate*.

Cross Link

Intermolecular bonds produced between long-chain molecules in a material to increase molecular size and weight by chemical or electron bombardment, resulting in a change in physical properties in the material, usually improved properties.

Cross Linking

With thermosetting and certain thermoplastic polymers, the setting up of chemical links between the molecular chains. When extensive, as in most thermosetting resins, cross-linking makes an infusible supermolecule of all the chains. In rubbers, the cross-linking is just enough to join all molecules into a network.

Cross Rolling

Rolling of metal or sheet or plate so that the direction of rolling is about 90° from the direction of a previous rolling.

Cross Slip

The extension of a dislocation from one slip plane to an adjacent plane.

Cross-Country Mill

A rolling mill in which the mill stands are so arranged that their tables are parallel with a transfer (or crossover) table connecting them. Such a mill is used for rolling structural shapes, rails, and any special form of bar stock not rolled in the ordinary bar mill.

Crossed Joint

See *bridging*.

Cross-Linking, Degree of

The fraction of cross-linked polymeric units in the entire system.

Cross-Ply Laminate

A laminate with plies usually oriented at 0° and 90° only.

Cross-Roll Straightener

A machine having paired rolls of special design for straightening round bars or tubes, the pass being made with the work parallel to the axes of the rolls.

Cross Section

(1) Of a bar, extrusion, etc., it is the shape and dimensions of the face formed by a full transverse cut. (2) In the context of nuclear interactions it is a function of the probability that a particle entering the material in question will meet a particle with which to react. The SI unit is the barn.

Cross-Wire Weld

A weld made at the junction between crossed wires or bars.

Crosswise Direction

In testing of plastics, crosswise refers to the cutting of specimens and to the application of load. For rods and tubes, crosswise is any direction perpendicular to the long axis. For other shapes or materials that are stronger in one direction than in another, crosswise is the direction that is weaker. For materials that are equally strong in both directions, crosswise is an arbitrarily designated direction at right angles to the lengthwise direction.

Crowfoot Satin

In this type of composite fabric weave, there is a three-by-one interlacing; that is, a filling thread floats over three warp threads and then under one. This type of fabric looks different on one side than the other. Fabrics with this weave are more pliable than either the *plain* or *basket weave* and, consequently, are easier to form around curves.

Crown

(1) The upper part (head) of a forming press frame. On hydraulic presses, the crown usually contains the cylinder; on mechanical presses, the crown contains the drive mechanism. See also *hydraulic press* and *mechanical press*. (2) A shape (crown) ground into a flat roll to ensure flatness of cold (and hot) rolled sheet and strip. (3) A contour on a sheet or roll where the thickness or diameter increases from edge to center.

Crucible

A vessel in which metal is melted.

Crucible Furnace

A melting or holding furnace in which the molten metal is contained in a pot-shaped (hemispherical) shell. Electric heaters or fuel-fired burners outside the shell generate the heat that passes through the shell (crucible) to the molten metal.

Crucible Steel

The term is sometimes still used to refer to steels made in small quantities and, possibly, claimed to have superior characteristics.

Cruciform

Having the geometry of a cross. A cruciform weld joint is formed when two bars or plates are welded in the same plane on either side of, and perpendicular to, another plate. A cruciform test is one in which such a joint is made, inspected and, usually, tested to destruction.

Crud

Originally an American term for solid particulate contamination in the boiler water of steam power plants. The term is now used for any solid suspension or deposition in any liquid.

Crush

(1) Buckling or breaking of a section of a casting mold due to incorrect register when the mold is closed. (2) An indentation in the surface of a casting due to displacement of sand when the mold was closed. (3) In a split-journal bearing, the amount by which a bearing half extends above the horizontal split of the bore before it is assembled. Also known as *nip* in the United Kingdom.

Crush Forming

Shaping a grinding wheel by forcing a rotating metal roll into its face so as to reproduce the desired contour.

Crushed Strip or Bead

In casting, an indentation in the *parting line* of a pattern plate that ensures that *cope* and *drag* will have good contact by producing a ridge of sand that crushes against the other surface of the mold or core.

Crushing

A process of comminuting large pieces of metal or ore into rough size fractions prior to grinding into powder. A typical machine for this operation is a jaw crusher.

Crushing Test

(1) A radial compressive test applied to tubing, sintered-metal bearings, or other similar products for determining radial crushing strength (maximum load in compression). (2) An axial compressive test for determining quality of tubing, such as soundness of weld and welded tubing.

Cryogenic Treatment

See *cold treatment*.

Cryogenics

Today's limited acceptance and use of deep-cryogenic treatment at liquid N_2 temperature ($-196°C$) is usually attributed to a lack of understanding of the technology, as well as to the absence of generally acceptable practices for deploying it.

When the above two factors are resolved the "science" of deep cryogenics and its potential contributions will be shown. For example, a deep-cryogenic treatment at $-196°C$ between quenching and tempering optimizes the mechanical properties of AISI T15 high-speed steel and dramatically improves its wear resistance, compared with standard heat treatments that do not incorporate a cryogenic step.

When the above is used, deep cryogenics does indeed become a science—results are predictable and the mechanical properties of the heat-treated parts are optimized. This is why it is believed that the use of the process during heat treating should now be standard rather than optional.

In future heat-treating guidelines and practices and procedures it is recommended that when using deep cryogenics as part of the heat treatment, the temperature of the subsequent temper be noted for the specific steel and the hardness vs. tempering temperature.

Cryolite

A mineral of composition Na_3AlF_6, found in commercial quantities in Greenland and used as a flux in the electrical production of aluminum, and the making of special glasses and porcelain, as a binder for abrasive wheels, and in insecticides. It acts as a powerful flux because of its solvent power on silicon, aluminum, and calcium oxides. In opal and milky classes, it forms a complex AlF_6 anion, retaining the alumina and preventing loss of the fluorine.

Cryopump

A type of vacuum pump that relies on the condensation of gas molecules and atoms on internal surfaces of the pump, which are maintained at extremely low temperatures.

Crystal

(1) A solid composed of atoms, ions, or molecules arranged in a pattern that is repetitive in three dimensions. (2) That form, or particle, or piece of a substance in which its atoms are distributed in one specific orderly geometrical array, called "lattice," essentially throughout. Crystals exhibit characteristic optical and other properties and growth or cleavage surfaces, in characteristic directions. (3) "Crystal" is often interchangeably with grain. Unfortunately, there are no widely accepted rules to guide usage but, generalizing, "crystal" tends to be favored in the context of features on an atomic scale, for example crystal lattice, whereas "grain" tends to be favored for larger scale matters, for example grain size. Note however, "Crystalline" is used, when describing a fracture surface, to indicate a bright angular faceted appearance resulting from cleavage along specific crystallographic planes, characteristic of brittle fracture. See *crystal structure*.

Crystal Analysis

A method for determining crystal structure, for example, the size and shape of the unit cell and the location of all atoms within the unit cell.

Crystal Lattice

See *crystal structure*.

Crystal Orientation

See *orientation*.

Crystal Structure

The arrangement of atoms in crystals including virtually all solid metals. Atoms in crystals are arranged in simple, repeating, three-dimensional patterns rather than in the random jumble of amorphous, i.e., noncrystalline, materials. The basic unit of the pattern can be visualized as comprising atoms located at specific points on a simple geometric body, for example the points of the cube. This concept leads to terms such as "Body Centered Cubic" or "Face Centered Cubic" in which atoms lie on the points of the cube and also, respectively, at the center of the body or the center of each face. The basic block is repeated for very large distances, measured on an atomic scale, and the atoms can then be regarded as lying in layers or on atomic planes. The three-dimensional arrangement is termed a lattice. An individual crystal or grain has all of its atoms and atomic planes in continuous alignment. Neighboring crystals will have the same patterns but their planes will not be in alignment. The interfaces between grains are referred to as the crystal boundaries or grain boundaries. Also see *ordering* and *dislocations*.

Crystal System

One of seven groups into which all crystals may be divided; triclinic, monoclinic, orthorhombic, hexagonal, rhombohedral, tetragonal, and cubic.

Crystalline

(1) That form of a substance which is comprised predominantly of (one or more) crystals, as opposed to glassy or amorphous. (2) A bright, angular faceted fracture surface. See *crystal*.

Crystalline Fracture

A pattern of highly reflecting crystal facets on the fracture surface of a polycrystalline metal, resulting from *cleavage fracture* of many individual crystals. Contrast with *fibrous fracture*, and *silky fracture*; see also *granular fracture*.

Crystalline Plastic

A material having an internal structure in which the atoms are arranged in an orderly three-dimensional configuration. More accurately referred to as a semicrystalline plastic because only a portion of the molecules are in crystalline form.

Crystallinity (Plastics)

A regular arrangement of the atoms of a solid in space. In most polymers, including cellulose, this state is usually imperfectly achieved. The crystalline regions (ordered regions) are submicroscopic volumes in which there is some degree of regularity in the arrangement of the component molecules. In these regions there is sufficient geometric order to obtain definite x-ray diffraction patterns.

Crystallite

An imprecise term which may mean no more than a very small crystal. Alternatively, the term may refer to sub crystals within a grain or small crystals or parts of crystals which do not exhibit all of the characteristics of a crystal.

Crystallization

(1) The separation, usually from a liquid phase on cooling, of a solid crystalline phase. (2) The progressive process in which crystals are first nucleated (started) and then grown in size within a host medium which supplies their atoms. The host may be gas, liquid, or of another crystalline form.

Crystallographic Cleavage

The separation of a crystal along a plane of fixed orientation relative to the three-dimensional crystal structure within which the separation process occurs, with the separation process causing the newly formed surfaces to move away from one another in directions containing major components of motion perpendicular to the fixed plane.

Crystallography

The study of crystal structure.

Crystobalite

An allotropy of quartz. See *quartz*.

C-Stage

In processing of plastics, the final stage in the reaction of certain thermosetting resins in which the material is practically insoluble and infusible. Also called *resite*. The resin in a fully cured thermoset molding is in this stage. See also *A-stage* and *B-stage*.

CTE Mismatch

The difference in the coefficients of thermal expansion (CTEs) of two materials or components joined together, producing strains and stresses at the joining interfaces or in the attachment structures (solder joints, leads, and so on).

CTOD

See *Cracking Opening Displacement*.

CTS

(1) Compact tension specimen. See *fracture toughness*. (2) Controlled thermal severity test.

Cube Texture

A texture found in wrought metals in the cubic system in which nearly all the crystal grains have a plane of the type (100) parallel or nearly parallel to the plane of working and a direction of the type (100) parallel or nearly parallel to the direction of elongation.

Cubic

Having three mutually perpendicular axes of equal length.

Cubic Boron Nitride (CBN)

An extremely hard ceramic material synthesized by high-pressure sintering of hexagonal *boron nitride*. CBN is used in the machining and grinding of ferrous material such as tool steels, cast irons, hardfacing alloys, and surface-hardened steels. See also *superabrasives*.

Cubic Close Packed

See *close packed hexagonal*.

Cubic Plane

A plane perpendicular to any one of the three crystallographic axes of the cubic (isometric) system; the *Miller Indices* are (100).

Cull

Plastic material in a transfer chamber after the mold has been filled. Unless there is a slight excess in the charge, the operator cannot be sure the cavity is filled. Charge is generally regulated to control thickness of the cull.

Cumulative Damage

The total amount of damage resulting from a number of damage mechanisms or from a number of exposures to a mechanism. The term is commonly used in the case of fatigue reflecting the variation and stress range or other variables that a component might experience. The basic concept in all cases is that each exposure consumes a fraction of the fatigue life and failure will occur when these fractions total unity.

Cumulative Erosion–Time Curve

A plot of cumulative erosion vs. cumulative exposure duration, usually obtained by periodic interruption of the erosion test and weighing of the specimen. This is the primary record of an erosion test used to derive other characteristics such as incubation period, maximum erosion rate, terminal erosion rate, and the erosion rate–time curve.

Cup

(1) A sheet metal part; the product of the first drawing operation. (2) Any cylindrical part or shell closed at one end.

Cupellation

Oxidation of molten lead containing gold and silver to produce lead oxide, thereby separating the precious metals from the base metal. The rate of oxidation is usually increased by an air jet.

Cupola

A vertical shaft, air blast furnace, lined with refractory, for remelting, in particular, cast iron. The charge of iron with metallurgical coke or charcoal as fuel and a flux are introduced at the top; the molten iron and slags are tapped at the bottom.

Cupping

(1) The first step in deep drawing. (2) Fracture of severely worked rods or wire where one end has the appearance of a cup and the other that of a cone.

Cupping Test

A mechanical test used to determine the ductility and stretching properties of sheet metal. It consists of measuring the maximum part depth that can be formed before fracture. The test is typically carried out by stretching the test piece clamped at its edges into a circular die using a punch with a hemispherical end. See also *Erichsen test*, *Olson ductility test*, and *Swift cup test*.

Cupronickel

Copper-based alloys containing significant amounts of nickel.

Cure

(1) To change the physical properties of a material (usually from a liquid to a solid) by chemical reaction or by the action of heat and catalysts, alone or in combination, with or without pressure. (2) To irreversibly change, usually at elevated temperatures, the properties of a thermosetting resin by chemical reaction, that is, by condensation, ring closure, or addition. Cure may be accomplished by the addition of curing (cross linking) agents, with or without heat and pressure.

Cure Cycle

The time/temperature/pressure cycle used to cure a thermosetting resin system or prepreg.

Cure Monitoring, Electrical

Use of electrical techniques to detect changes in the electrical properties and/or mobility of the resin molecules during cure. A measuring of resin cure.

Cure Stress

A residual internal stress produced during the curing cycle of a composite structure. Normally, these stresses originate when different components of a wet lay-up have different thermal coefficients of expansion. One Curie is the quantity of radionuclide in which the number of disintegrations in 1 s is 3.7×10^{10}.

Curie Point/Temperature

For a metal, it is the temperature below which is magnetic (more strictly ferromagnetic) and above which it is nonmagnetic (paramagnetic). For iron it is 770°C.

It is also the temperature at which piezoelectric materials lose their electromechanical characteristics. See also *ferromagnetism* and *paramagnetism*.

Curing Agent

A catalytic or reactive agent that causes cross linking of a plastic. Also called a hardener.

Curing Temperature

The temperature to which an adhesive or an assembly is subjected to cure the adhesive. The temperature attained by the adhesive in the process of curing (adhesive curing temperature) may differ from the temperature of the atmosphere surrounding the assembly (assembly curing temperature). See also *drying temperature* and *setting temperature*.

Curing Time

In plastic part making, the time between the instant of cessation of relative movement between the moving parts of a mold and the instant that pressure is released. Also called molding time.

Curing Time (No Bake)

In foundry practice, the period of time needed before a sand mass reaches maximum hardness.

Curling

Rounding the edge of sheet metal into a closed or partly closed loop.

Current

The net transfer of electric charge per unit time. Also called electric current. See also *current density*.

Current Decay

In spot, seam, or projection welding, the controlled reduction of the welding current from its peak amplitude to a lower value to prevent excessively rapid cooling of the weld nugget.

Current Density

The current in amps, flowing to or from a unit area of the electrode surface.

Current Efficiency

(1) The ratio of the electrochemical equivalent current density for a specific reaction to the total applied current density. (2) The proportion of current used in a given process to accomplish a desired result; in electroplating, the proportion used in depositing or dissolving metal. See *Faraday's Law*.

Current-Carrying Capacity

The maximum current that can be carried continuously, under specified conditions, by a conductor without causing objectionable degradation of electrical or mechanical properties.

Curtain Coating

A method of coating that may be employed with low-viscosity resins or solutions, suspensions, or emulsions of resins in which the substrate to be coated is passed through and perpendicular to a freely falling liquid curtain (or waterfall). The flow rate of the falling liquid and the linear speed of the substrate passing through the curtain are coordinated in accordance with the thickness of coating desired.

Curvature of Field

In metallography, a property of a lens that causes the image of a plane to be focused into a curved surface instead of a plane.

Cushion

Same as *die cushion*.

Cut

In lubricant technology, a product or fraction obtained by distillation within a specified temperature range.

Cut (Foundry Practice)

(1) To recondition molding sand by mixing on the floor with a shovel or blade-type machine. (2) To form the sprue cavity in a mold. (3) Defect in a casting resulting from erosion of the sand by metal flowing over the mold or cored surface. (4) Apart from obvious meanings it often implies the depth of cut at a single pass of the cutting tool.

Cut Edge

A mechanically sheared edge obtained by slitting, shearing, or blanking.

Cut Layers

With laminated plastics, a condition of the surface of machines or ground rods and tubes and of sanded sheets in which cut edges of the surface layer or lower laminations are revealed.

Cut-and-Carry Method

Stamping method wherein the part remains attached to the strip or is forced back into the strip to be fed through the succeeding stations of a progressive die.

Cutlery Steel

Originally approximately 1% carbon steel used for domestic cutlery and other blades but now usually 12% chromium, 0.3% carbon (martensitic stainless steel).

Cut-Off (Casting)

Removing a casting from the sprue by refractory wheel or saw, arc-air torch, or gas torch.

Cut-Off (Metal Forming)

A pair of blades positioned in dies or equipment (or a section of the die milled to produce the same effect as inserted blades) used to separate the forging from the bar after forging operations are completed. Used only when forgings are produced from relatively long bars instead of from individual, precut multiples or blanks. See also *blank* and *multiple*.

Cut-Off (Plastics)

The line where the two halves of a mold come together. Also called groove and pitch-off.

Cut-Off Wheel

A thin disk, coated or impregnated with abrasive, rotating at high speed and often water cooled, for cutting purposes. It can produce a narrow, clean-cut with minimal surface damage even in hard materials. Also variously termed cutting wheel, slitting wheel, elastic wheel, etc.

Cutting Alloys

Usually of complex Co–Cr–W–Fe–Si–C composition, used for lathes and planer tools for cutting hard metals. They form a class distinct from the cemented carbides, which are not true alloys; from the refractory hard metals, which are chemical compounds; and from the cobalt high-speed steels, which are high in iron and usually have less carbon. The hardness is inherent in the alloy and is not obtained by heat treatment, as with the tool steels. Cutting alloys are cast to shape and are usually marketed in the form of tool bits and shear blades. Complex alloys, however, may have heat-transition points at which the metal complexes change structure, limiting the range of use.

Since the development of balanced high-speed steels and cermet-type cutting tools, these alloys with a high proportion of the scarcer cobalt have lost their importance as cutting alloys and, because of their high corrosion, heat, and wear resistance, are used chiefly for weld-facing rods and heat-corrosion applications. The first of the commercial cobalt cutting alloys was *Stellite*, of Haynes Stellite Co., in various composition grades and with trade names, such as *J-metal* and *Star J-metal*. Other similar alloys contain boron for added wear resistance. This type of alloy is now used in *surgical alloys* for surgical tools and dental plates since they are not attacked by body acids and set up no electromotive currents. To make them more workable for this purpose, they usually contain a higher content of cobalt, 60% or more, with a smaller amount of molybdenum instead of tungsten, and with less carbon and silicon.

Cutting Attachment

A device for converting an oxyfuel gas welding torch into an oxygen cutting torch.

Cutting Compound/Fluid/Paste

A substance applied at the cutting edge during machining. Where it is a stream of fluid its primary function is cooling but all of these substances act as a lubricant and limit adhesion to the tool edge. They also contribute to surface finish, tool life and precision.

Cutting Down

Removing roughness or irregularities of a metal surface by abrasive action.

Cutting Edge

The leading edge of a cutting tool (such as a lathe tool, drill or milling cutter) where a line of contact is made with the work during machining. See *single-point tool.*

Cutting Fluid

A fluid used in metal cutting to improve finish, tool life, or dimensional accuracy. On being flowed over the tool and work, the fluid reduces friction, the heat generated, and tool wear, and prevents galling. It conducts the heat away from the point of generation and also serves to wash the *chips* away.

Cutting Head

The part of a cutting machine or automatic cutting equipment in which a cutting torch or tip is incorporated.

Cutting Nozzle

See preferred term *cutting tip.*

Cutting Process

A process which brings about the severing or removal of metals. See also *arc cutting* and *oxygen cutting.*

Cutting Speed

The linear or peripheral speed of relative motion between the tool and workpiece in the principal direction of cutting.

Cutting Tip

That part of an oxygen cutting torch from which the gases issue.

Cutting Tools

Polycrystalline diamond (PCD) and polycrystalline cubic boron nitride (PCBN) inserts could be the best means of increasing productivity despite their high cost. In the past, PCD and PCBN cutting tools were difficult to cost-justify, unless they were essential for the machining job.

Today, improvements in quality and reliability make these tools, although still costly, competitive in many machining applications in the automotive, aerospace, and medical equipment industries. More rigid machines and tooling setups enable manufacturers to take full advantage of the potential for improved productivity offered by PCD and PCBN inserts. Also, having more cutting-tool options gives manufacturing engineers an opportunity for cost-effective productivity improvement in various machining applications.

PCBN

New, thicker PCBN solid inserts with larger grain size possess improved wear and impact resistance—the key to effective machining of materials like cast irons containing less than 10% ferrite content. Such performance improvement is especially important when machining alloy cast irons for automotive applications. Previously, manufacturers had to grind these castings.

In roughing operations on alloy cast irons, tools must withstand interrupted cuts due to surface cracks, sand inclusions, and other surface discontinuities inherent to the casting process. Good wear resistance comes into play in finishing operations, where parts containing 28%–30% chromium have a hardness between R_c 68 and R_c 72.

Solid PCBN inserts provide multiple cutting edges on two sides, reducing insert cost per part produced. PCBN inserts also come in *full-face and tipped* types. The full-face type has a complete PCBN top face sintered onto a carbide substrate, and provides multiple cutting edges on one side only. These inserts are less expensive than solid PCBN inserts. The tipped style contains a small PCBN segment brazed onto one corner of a carbide insert, providing either a single cutting edge or double cutting edges. Most PCBN inserts used today are tipped. Both full-face and tipped PCBN inserts come in industry-standard sizes and, like the solid insert, can be used in the insert pockets of standard tool-holders and milling cutters.

Inserts made of PCBN work best in hard-part machining applications. In practice, the low end for part hardness falls at about R_c 45. Machining softer parts using PCBN produces insert cratering.

In roughing operations, maximum depths of cut using solid-style PCBN inserts range from about 4.76 mm for white iron and other hard, high-chromium irons to about 6.4 mm for unalloyed, "clean" cast irons. Finishing speeds range from 107 to 122 m/min on high-chromium irons to as high as 2134 m/min on gray cast irons. All PCBN operations require the use of very rigid tooling, work fixturing, machine spindles, and machine tools.

Currently, the largest growth area for PCBN inserts is in hard turning/finish turning of alloy steel automotive engine components such as gears, shafts, and bearings with hardnesses between R_c 60 and R_c 65. Traditionally, manufacturers ground these parts to obtain very tight dimensional tolerances and a fine surface finish. Hard turning gives the same results using a CNC lathe.

Insert edge preparation also strongly influences the success of PCBN machining. To promote good tool life, a reinforced cutting edge with the proper edge preparation is a must.

PCD

PCD inserts are cost-effective because they dramatically outperform carbide in most nonferrous applications, not because of any developments in the material itself.

Most PCD inserts used today are tipped. They employ a small segment of PCD brazed onto one corner of a carbide insert. PCD inserts come in industry-standard sizes for use in the insert pockets of standard tool-holders and milling cutters. Unlike carbide inserts, however, PCD-tipped inserts are not indexable, and provide only a single cutting-edge.

Standard-size, full-face-style PCD inserts provide a complete PCD top face sintered onto a carbide substrate.

More secure insert retention—especially in rotating tools—has played a key role in increasing the use of PCD inserts. Today's combinations of wedges and screws, as well as tapered inserts and wedges, enhance insert retention. Conventional screw or clamp designs, and cutters using direct-mount cartridges where the PCD segment is brazed directly onto the cartridge body, also provide the necessary rigidity. Rigid tooling setups provide more reliable results with PCD.

More secure part fixturing and more rigid machine tools and spindles also improve performance. Today's CNC lathes and machining centers provide all the rigidity required by PCD or PCBN cutting tools. And as companies install new equipment, the machine shop environment becomes more suited to high-performance PCD and PCBN cutting tools.

Cutters used for finish milling of aluminum automotive manifolds often combine PCD-tipped and plain carbide inserts to cut through the casting flash encountered at the insert depth-of-cut line.

Another major use of PCD-tipped inserts is roughing automotive parts in a process called "cubing." Foundries economize on shipping costs and recover aluminum chips for recycling by rough machining castings such as aluminum cylinder heads. They then semifinish the castings (top, bottom, side, and edge) at very high metal-removal rates using a PCD cutting tool. Machined parts are easier to handle and pack because of their rectangular shape, and shipping costs are lower for the smaller parts because they take up less space.

PCD-tipped inserts are preferred over carbide in applications where manufacturers want higher production rates, consistent part finish, and longer tool life (eliminating frequent tool changes). These inserts excel in machining abrasive high-silicon aluminum automotive parts like cylinder heads, engine blocks, manifolds, transmission cases, and wheels.

When using PCD inserts, good chip evacuation is essential. Because the inserts produce chips so rapidly, the system must remove them from the work zone quickly and continuously. Coolant, air, mist, refrigerated air, or combinations of all these methods will work. PCD cutting speeds as high as 3048 m/min are possible with effective chip evacuation and a rigid machine and operating set up. Today's CNC machining and turning centers easily achieve these speeds.

Cutting Torch (Oxyfuel Gas)

A device used for directing the preheating flame produced by the controlled combustion of fuel gases and to direct and control the cutting oxygen.

Cutting Torch (Plasma Arc)

A device used for plasma arc cutting to control the position of the electrode, to transfer current to the arc, and to direct the flow of plasma and shielding gas.

Cutting Wear

See *abrasive wear*.

CVD Carbon

See *chemical vapor deposited (CVD) carbon*.

Cyanate Resins

Thermosetting resins that are derived from bisphenols or polyphenols, and are available as monomers, oligomers, blends, and solutions. Also known as cyanate esters, cyanic esters, and triazine esters.

Cyanic Copper

Copper electrodeposited from an alkali-cyanide solution containing a complex ion made up of univalent copper and the cyanide radical; also the solution itself.

Cyanide Slimes

Finely divided metallic precipitates that are formed when precious metals are extracted from their ores using cyanide solutions.

Cyaniding

A case-hardening process in which a ferrous material is heated above the lower transformation temperature range at about 900°C in a molten salt containing cyanide to cause simultaneous absorption of carbon and nitrogen at the surface and, by diffusion, create a concentration gradient. *Quench hardening* completes the process. See *case hardening*.

Cyanoacrylate

A thermoplastic monomer adhesive characterized by excellent polymerizing and bonding strength.

Cycle

One complete operation of a plastic molding press from closing time, for example, in one cycle to closing time in the next cycle.

Cycle (N)

In fatigue, one complete sequence of values of applied load that is repeated periodically. See also *S–N curve*.

Cycle Annealing

An annealing process employing a predetermined and closely controlled time–temperature cycle to produce specific properties or microstructures.

Cycle Hydrocarbons

Cyclic or ring compounds; benzene (C_6H_6) is a classic example.

Cyclic Load

(1) Repetitive loading, as with regularly recurring stresses on a part, that sometimes leads to fatigue fracture. (2) Loads that change value by following a regular repeating sequence of change.

Cyclic Stressing

See *cyclic load*.

Cyclone

In powder metallurgy, a collector of fractions of a powder from air, water, or other gases or liquids; the device operates with the aid of centrifugal force that acts on the powder suspension in the fluid.

Cylinder Manifold

See preferred term *manifold*.

Cylindrical Grinding

Grinding the outer cylindrical surface of a rotating part.

Cylindrical Land

Land having zero relief.

D

Dalic Process

A portable electroplating system in which the plating solution is delivered through suitable pads or brushes.

Dalton's Atomic Theory

In essence, this stated that matter is comprised of atoms that are discrete, indivisible particles. All atoms of an element are the same. Chemical reactions are the result of atoms combining in simple proportions. Although it has subsequently been demonstrated that atoms can be divided into smaller particles and isotopes have been discovered, the general principles are the basis of chemistry.

Dam

In composites fabrication, a boundary support or ridge used to prevent excessive edge bleeding or resin runout of a laminate and to prevent crowning of the bag during cure.

Damage Tolerance

(1) A design measure of crack growth rate. Cracks in damage-tolerant designed structures are not permitted to grow to critical size during expected service life. (2) The ability of a part component, such as an aerospace engine, to resist failure due to the presence of flaws, cracks, or other damage for a specified period of usage. The damage tolerance approach is used extensively in the aerospace industry.

Damascening

The repeated folding and forge welding of layers of steel. Individual layers have varying carbon contents, deliberately or otherwise, which show as light and dark areas after polishing and etching to provide a decorative finish. The word derives from Damascus where the technique was used for high-quality swords and other blades. Generalizing, the greater the number of repetitions of folding and forging, the finer becomes the decorative effect and the stronger and tougher becomes the steel.

Damping

The loss in energy, as dissipated heat, that results when a material or material system is subjected to an oscillatory load or displacement.

Damping Capacity

(1) The ability of a material to absorb vibration (cyclical stresses) by internal friction, converting the mechanical energy into heat. (2) An induced strain where energy is dissipated internally. Materials having a low-damping-capacity *ring* when struck and those with a high-capacity sound dead and dull.

Dangler

The flexible electrode used in *barrel plating* to conduct current to the work.

Danner Process

A mechanical process for continuously drawing glass cane or tubing from a rotating mandrel.

Dark Reaction

In adhesives technology, the gelling of material in storage without light initiation.

Dark-Field Illumination

The illumination of an object such that it appears bright and the surrounding field dark. This results from illuminating the object with rays of sufficient obliquity so that none can enter the objective directly. In electron microscopy, the image is formed using only electrons scattered by the object. This is in contrast with *bright field illumination*.

Dashpot

A device used in hydraulic systems for damping down vibration. It consists of a piston attached to the part to be damped and fitted into a vessel containing fluid or air. It absorbs shocks by reducing the rate of change in the momentum of moving parts of machinery.

Daubing

In foundry practice, filling of cracks in molds or cores by specially prepared pastes or coatings to prevent penetration of metal into these cracks during pouring.

Daylight

The distance in the open position, between the moving and the fixed tables or the platens of a hydraulic press. In the case of a multiplaten press, daylight is the distance between adjacent platens. Daylight provides space for removal of the molded/formed part from the mold/die.

DBTT

Ductile/brittle transition temperature. See *impact test*.

DC Casting

This is the same as *direct chill casting*.

DC Direct

Direct (electric) current.

DC Intermittent Noncapacitative Arc

A low-voltage electric discharge used in spectrochemical analysis to vaporize the sample material. Each current pulse has the same polarity as the previous one and last for less than 0.1 s. See also *AC noncapacitative arc.*

DC Plasma Excitation

See *plasma jet excitation.*

DCPD

Direct current potential drop. A technique for measuring crack propagation. See *potential drop.*

Deactivation

The process of prior removal of the active corrosive constituents, usually oxygen, from a corrosive liquid by controlled corrosion of expendable metal or by other chemical means, thereby making the liquid less corrosive.

Dead Center

(1) A stationary center to hold rotating work. (2) Either of the two points in the path of a moving crank or connecting rod that lies at the ends of its stroke.

Dead Roast

A roasting process for complete elimination of sulfur. Also known as *sweet roast.*

Dead Soft

A *temper* of nonferrous alloys and some ferrous alloys corresponding to the condition of minimum hardness and tensile strength produced by *full annealing.*

Dead Time

The total time during which a spectrometer is processing information and is unavailable to accept input data.

Dead-Burned

The term applied to ceramic materials that have been fired to a temperature sufficiently high to render them relatively resistant to moisture and contraction.

Dealloying

The selective corrosion of one or more components of a solid solution alloy. This is also called parting or *selective leaching.* See also *decarburization, decobaltification, denickelification, dezincification,* and *graphitic corrosion.*

Dealuminification

A form of corrosion affecting aluminum bronzes, in which the aluminum is selectively removed leaving a weak, porous, copper-rich material. Similar to *dezincification of brass.*

Debond

In composites, a deliberate separation of a bonded joint or interface, usually for repair or rework purposes. Also, an unbonded or nonadherent region; a separation at the fiber–matrix interface due to strain and compatibility. In the United Kingdom, the term often refers to accidental damage. See also *disbond* and *delamination.*

Debossed

An indented or depressed design or lettering that is molded into a plastic article so that it is below the main outside surface of that article.

Debris

See *wear debris.*

Debulking

Compacting of a thick composite laminate under moderate heat and pressure and/or vacuum to remove most of the air, to ensure seating on the tool, and to prevent wrinkles.

Deburring

Removing bars, or sharp edges, or fins from metal parts by filing, grinding, or rolling the work in a barrel containing abrasives suspended in a suitable liquid medium. Sometimes called *barring.*

Debye Ring

A continuous circle, concentric about the undeviated being, produced by monochromatic x-ray diffraction from a randomly oriented crystalline powder. An analogous effect is obtained using electron diffraction.

Debye–Scherrer Method

A method of x-ray diffraction using monochromatic radiation and a polycrystalline specimen mounted on the axis of a cylindrical strip of film. See also *powder method.*

Decalescence

A phenomenon, associated with the transformation of alpha iron to gamma iron on the heating (super heating) of iron or steel, revealed by the darkening of the metal surface owing to the sudden decrease in temperature caused by the fast absorption of the latent heat of transformation. This is in contrast with *recalescence.*

Decarburization

Loss of carbon from the surface layer of a carbon-containing alloy due to reaction with one or more chemical substances in a medium that contacts a surface. It occurs by diffusion into, or reaction with, the environment. It is usually undesirable except in the case of certain treatments of cast iron. See also *dealloying*.

Decay Constant (λ)

The constant and the radioactive decay law $dN = -\lambda N dt$, where N is the number of radioactive nuclei present at time t. The decay constant is related to the half-life $t_{1/2}$ by the expression $t_{1/2} = \ln 2/\lambda$. See also *half-life*.

Deceleration Period

In cavitation or impingement erosion, the stage following the *acceleration period* or the *maximum rate period* (if any), during which the erosion rate has an overall decreasing trend although fluctuations may be superimposed on it.

Decibel (dB)

Unit that expresses differences of power level. Example: The decibel is 10 times the common logarithm of the power ratio. It is used to express power gain in amplifiers or power loss in passive circuits or cables.

Deckle Rod

In forming plastics, a small ride, or similar device, inserted at each end of the extrusion coating die that is used to adjust the length of the die opening.

Decobaltification

Corrosion in which cobalt is selectively leached from cobalt-based alloys, such as Stellite, or from cemented carbides. See also *dealloying* and *selective leaching*.

Decohesive Rupture

A *brittle fracture* that exhibits little or no bulk plastic deformation and does not occur by dimple rupture, cleavage, or fatigue. This type of fracture is generally the result of a reactive environment or unique microstructure and is associated almost exclusively with rupture along grain boundaries.

Decomposition

Separation of a compound into its chemical elements or components.

Decomposition Potential (or Voltage)

The *potential* of a metal surface necessary to decompose the electrolyte of a cell or a compound thereof.

Decoration (of Dislocations)

Segregation of solute atoms to the line of a dislocation in a crystal. In ferrite, the dislocations may be decorated with carbon or nitrogen atoms.

Deep Etching

In metallography, *macroetching*, especially for steels, to determine the overall character of the material, that is, the presence of imperfections, such as seams, forging bursts, shrinkage-void remnants, cracks, and coring.

Deep Groundbed

One or more *anodes* installed vertically at a nominal death of 1.5 m (50 ft) or more below the earth's surface in a drilled hole for the purpose of supplying *cathodic protection* for an underground or submerged metallic structure. See also *groundbed*.

Deep-Draw Mold (Plastics)

A mold having a core that is appreciably longer than the wall thickness.

Deep-Drawing

Forming deeply recessed parts by forcing sheet metal to undergo plastic flow between dies, usually without substantial thinning of the sheet. Pressing operations producing cups or similar shapes having a large cup depth relative to the diameter.

Defect

(1) A discontinuity whose size, shape, orientation, or location makes it detrimental to the useful service of the part in which it occurs. (2) A discontinuity or discontinuities that by nature or accumulated effect (e.g., total crack length) render a part or product unable to meet minimum applicable acceptance standards or specifications. (3) Any deviation from perfection. This term designates rejectability. See also *discontinuity* and *flaw*.

Defect Lattice

A crystal lattice in which vacant sites act in effect as foreign atoms and hence significantly influence the material properties and characteristics.

Defective

A quality control term, describing a unit of product or service containing at least one *defect*, or having several lesser imperfections that, in combination, cause the unit not to fulfill its anticipated function. The term defective is not synonymous with nonconforming (or rejectable) and should be applied only to those units incapable of performing their anticipated functions.

Defective Weld

A weld containing one or more defects.

Define (X-Rays)

To limit a beam of x-rays by passage through apertures to obtain a parallel, divergent, or convergent beam.

Deflagration

Rapid combustion with noise and flames.

Deflashing

A finishing technique used to remove the flash (excess, unwanted material) on a plastic molding.

Deflection

In metal forming and forging, the amount of deviation from a straight line or plane when a force is applied to a press member. Generally used to specify the allowable bending of the bed, slide, or frame at rated capacity with a load of predetermined distribution.

Deflection Temperature under Load (DTUL)

In testing of plastics, the temperature at which a simple cantilever beam deflects a given amount under load. Formerly called heat distortion temperature.

Deflocculant

An electrolyte adsorbed on colloidal particles in suspension that charges the particles to create repulsion forces that maintain the particles in a dispersed state, thus reducing the viscosity of the suspension.

Deformability

See *conformability*.

Deformation

A change in the form of a body due to stress, thermal change, change in moisture, or other causes. Measured in units of length.

Deformation Bands

Parts of a crystal that have rotated differently during deformation to produce bands of varied orientation without individual grains. This is the same as the *Lüders lines*.

Deformation Curve

See *stress–strain curve*.

Deformation Limit

In *drawing*, the limit of deformation is reached when the load required to deform the flange becomes greater than the load-carrying capacity of the cup wall. The deformation limit (limiting drawing ratio) is defined as the ratio of the maximum blank diameter, which can be drawn into a cup without failure, to the diameter of the punch.

Deformation Point

The temperature observed during the measurement of expansivity of glass by the interferometer method at which viscous flow exactly counteracts thermal expansion. The deformation point generally corresponds to a viscosity in the range of 1010–1011 Pa s.

Deformation Twins

This is the same as *mechanical twins*.

Deformation under Load

The dimensional change of a material under load for a specified time following the instantaneous elastic deformation caused by the initial application of the load. See also *cold flow* and *creep*.

Deformation Wear

Sliding wear involving plastic deformation of the wearing surface. Many forms of wear involve plastic deformation, so this term is imprecise and should not be used.

Degasification

See *degassing*.

Degasifier

A substance that can be added to molten metal to remove soluble gases that might otherwise be occluded or entrapped in the metal drawing solidification.

Degassing

Any process for removing gas from a solid or liquid. Heating allows hydrogen to diffuse from solid metals. Liquid metals may be degassed by, for example, adding powder to act as deoxidants or by bubbling through another gas that chemically reacts with or physically scavenges the unwanted gas.

Degassing (Metals)

(1) A chemical reaction resulting from a compound added to molten metal to remove gases from the metal. Inert gases are often used in this operation. (2) A fluxing procedure used for aluminum alloys in which nitrogen, chlorine, chlorine and nitrogen, and chlorine and argon are bubbled up through the metal to remove dissolved hydrogen gases and oxides from the alloy. See also *flux*.

Degassing (Plastics)

Opening and closing of a plastic mold to allow gases to escape during molding.

Degaussing

The elimination of all magnetic fields either by demagnetizing the material or by superimposing another field.

Degradable Plastics

Plastics that are decomposed by any of the three mechanisms— biodegradation, solubility, and photodegradation. *Biodegradable plastics* are those that are susceptible to being assimilated by microorganisms, such as fungi and bacteria, through enzyme action. The assimilating action requires heat, oxygen, and moisture. For all practical purposes, almost all synthetic polymers are immune to enzyme attack. Only *aliphatic polyesters* and *urethanes* derived from *aliphatic ester diols*, and low-molecular-weight (under 500) unbranched *polyethylene* derivatives can be assimilated.

Biodegradable packaging resins include cellulose acetate, caprolactones, polyesters, and polylactic acids (PLAs). Bionolle aliphatic polyester is considered superior to the other resins in biodegradability.

Aliphatic polyester works in polyethylene, polypropylene, polystyrene extruded, and blown film and foam for uses such as trash bags, beverage and cosmetic bottles, and diapers. *Green Block*, from JSP Corp. of Japan, is for foam applications. PLA products include shrink film, agricultural film, compost bags, and aluminum-laminated pharmaceutical packaging, which last because PLA does not readily absorb aromatic compounds contained in pharmaceuticals, thus precluding delamination. Polyester amide, from Germany's Bayer AG, is a candidate for garbage bags, disposable flowerpots, and mulch sheet. BASF (Badische Anilin – und Soda-Fabrik) of Germany offers a starch-based thermoplastic for household and packaging film. PHB, from Germany's PCD Polymere, retains flexibility at subzero temperatures.

The solubility of *water-soluble plastics* varies with formulations, molecular weight, and temperature. *Hydroxypropyl cellulose* is soluble in water above 115°F (46°C). Below this temperature, when immersed in water, it quickly forms a slippery gel on the outer surface. The gel layer must dissolve and wash away before further dissolving takes place. *Polyethylene oxides* are soluble in water above 150°F (66°C). They are nontoxic, eatable but nonnutritive, and nonchloric, and they wash through plumbing without damage or clogging. They are resistant to grease, oil, and petroleum hydrocarbons.

Photodegradable plastics are sensitive to ultraviolet light. Energy in the form of photons breaks down the bonds between the carbon and hydrogen atoms, and oxygen-reactive free radicals are formed. The free radicals react with oxygen in the environment to produce peroxide and hydroperoxides that decompose further to produce carbonyl groups, hydroxyl groups, water, and carbon dioxide. The best photodegradable materials are the linear, nonaromatic, molecular structured plastics. Unvulcanized syndiotactic polybutadiene is typical.

Degradation

A deleterious change in the chemical structure, physical properties, or appearance of a material.

Degrease

To remove oil and grease from *adherend* surfaces.

Degreasing

Removing oil or grease from a surface. See also *vapor degreasing.*

Degree of Polymerization

Number of structural units, or mers, in the average polymer molecule in a sample measure of molecular weight.

Degree of Saturation

The ratio of the weight of water vapor associated with a pound of dry air to the weight of water vapor associated with a pound of dry air saturated at the same temperature.

Degrees of Freedom

In metallography, the number of independent variables, such as temperature, pressure, or concentration, within the phases present that may be adjusted independently without causing a phase change in an alloy system at equilibrium.

Dehydration

The loss of water by processes such as natural evaporation or heating.

Deionized Water

Water that has been passed through columns of anion and cation ion-exchange resins to remove ionic contaminants. Such water will have low conductivity but may contain nonionic material such as air, sand, or bacteria.

Delamination

The separation of layers and a laminate because of failure of the adhesive, either in the adhesive itself or at the interface between the adhesive and the adherend.

Delamination Wear

A wear process in which thin layers of material are formed and removed from the wear surface.

Delayed Elasticity

This is the same as *elastic after effect.*

Delayed Field

(1) A phenomenon involving a delay in time between the application of a stress and the occurrence of the corresponding yield-point strain. (2) The period of a few milliseconds between the rapid imposition of a load exceeding yield and the onset of yielding. It is a result of the time taken for dislocations to disengage from lattice defects such as interstitials.

Delayed Fracture

The fracture, after a significant period of time, of a component subject to a bulk stress less than its tensile strength. The stress may be residual or externally imposed. The usual cause is hydrogen embrittlement.

Deliquescence

The absorption of atmospheric water vapor by a crystalline solid until the crystal eventually dissolves into a saturated solution.

Delta Ferrite

See *ferrite.*

Delta Iron

Solid phase of pure iron that is stable from 1400°C to 1539°C (2550°F–2800°F) and possesses the body-centered cubic lattice.

Demarest Process

A fluid forming process in which cylindrical and conical sheet metal parts are formed by a modified rubber bulging punch. The punch, equipped with a hydraulic cell, is placed inside the workpiece, which in turn is placed inside the die. Hydraulic pressure expands the punch.

Demixing

(1) The undesirable separation of one or more constituents of a powder mixture. (2) Segregation due to over mixing.

Dendrimers

Dendritic polymers, or dendrimers, consist of highly branched globular molecules grown from a core molecule and formed in stages, allowing the molecules to be built with specific diameters, weights, and surface characteristics for improved processability. *Paman*, for example, begins with ammonia molecule, which is reacted with methylacrylate and ethylenediamine. This results in a molecule with three branches, each ending in an amino group. As the process repeats, the dendrimer grows in layers, with each amino group reacting with two ethylenediamine molecules so that the new molecule has six branches ending in an amino group. Each successive reaction doubles the number of branches. The molecular weight of Paman varies only by as little as 0.005% and contrasts with as much as 5% for straight-chain polymers.

Dendrite

(1) A crystal that has a treelike branching pattern, being most evident in cast metals slowly cooled through the solidification range. (2) The pattern of growth adopted by a crystal during solidification; it involves the progressive outward growth, in three dimensions, of solidifying branches and multiple subbranches until the solid crystal is complete.

Dendritic Powder

Particles usually of electrolytic origin typically having the appearance of a pine tree.

Dendritic Segregation

Inhomogeneous distribution of alloying elements through the arms of dendrites.

Denickelification

Corrosion in which nickel is selectively leached from nickel-containing alloys. It is most commonly observed in copper–nickel alloys after extended service and freshwater. See also *dealloying* and *selective leaching*.

Denier

A yarn and filament numbering system in which the yarn number is numerically equal to the weight in grams of 9000 m (1 denier = 1.111×10^{-7} kg/m). Used for continuous filaments. The lower the denier, the finer the yarn.

Densification

Various techniques for increasing the density of porous materials including compression and sintering.

Density Ratio

The ratio of the determined density of a powder compact to the absolute density of metal of the same composition, usually expressed as a percentage. Also referred to as percent theoretical density.

Density, Absolute

The mass per unit volume of a solid material, expressed in g/cm^3, kg/m^3, or lb/ft^3.

Density, Dry

In powder metallurgy, the mass per unit volume of an unimpregnated sintered part.

Density, Wet

In powder metallurgy, the mass per unit volume of a sintered part impregnated with oil or other nonmetallic material.

Dental Materials

The field of dental materials is an interdisciplinary area that applies biology, chemistry, and physics to the development, understanding, and evaluation of materials used in the practice of dentistry. It is principally involved in restorative dentistry, prosthodontics, pedodontics, and orthodontics and, to a smaller extent, in the other areas of practice. The field of dental materials has advanced rapidly with the application of technologies such as genetic engineering of filling materials, computer designing and machining of restorations, adhesive bonding to teeth, dental implants, and new types of restorative materials.

Restorative Materials

The restoration of missing tooth structure is one of the most challenging areas for the use of dental materials. The enamel that covers the crown of the tooth is the hardest substance in the human body and is very resistant to physical and chemical attack. Direct restorations are put in place and hardened by a setting reaction, but require a replica of the tooth, which has to be fabricated.

Gold

Pure gold or gold foil has long been used as a restorative material. Pellets of pure gold are formed from very thin gold sheets or foils. The goal is compacted with compressed air or manually, and contoured and polished to the desired shape. These restorations are long-lasting and biologically inert. They are placed in low-stress areas because gold is soft.

Amalgam

Dental amalgam is the most frequently used direct restorative material for posterior (molar and premolar) teeth. It is formed by mixing mercury with an alloy powder of silver, tin, and copper. After mixing, there is time to insert and shape the material before it hardens.

Ag–Cu compounds and elements such as indium improve the alloy. Its advantages are ease of application, durability, and low cost; the disadvantages are chiefly aesthetic.

Resins

Composite resins are tooth-colored filling materials that were first developed for restoring anterior teeth (incisors and canines) but have found increased use as an attractive alternative to dental amalgam in posterior teeth. A composite resin consists of an organic polymer matrix reinforced with up to 80% of an inorganic filler such as SiO_2 or glass. Composite resin from a single paste is placed on the tooth and then polymerized with the help of a fiber-optic visible light source. This allows more time for contouring and gives a restoration with improved color stability. Composite resin restorations are strengthened by etching the enamel with 37% orthophosphoric acid prior to the placement of the resin to create a strong attachment to the enamel.

Glass ionomer

The only dental restorative material that forms a durable chemical bond to dentin is formed by the reaction of aluminosilicate glass with polyacrylic acid. Glass ionomer is particularly useful for restoring eroded or carious areas on exposed root surfaces with little or no enamel for bonding with composite resin. This material slowly releases fluoride to provide some additional protection from new decay. It is brittle and not used in areas of heavy stress.

Metal Castings

Crowns and inlays are made of metal castings and used when there is insufficient tooth structure to support filling. A negative of the tooth is prepared with synthetic rubber or agar, into which high-strength gypsum (plaster) is poured. This forms the replica of the tooth, to which wax is applied. An indirect lost wax process is used in fabricating dental castings. Alloy ingots used for the castings cover a range of compositions from 80% of noble metals—gold and the platinum group—combined with copper and other minor elements to base metal alloys of chromium, cobalt, and nickel. Traditionally, only noble alloys were used, but as the cost of gold increased, other alloys were developed.

Porcelain

Porcelain jacket crowns and metal crowns to which porcelain is bonded are used in visible areas. In one type, an inner core of high-strength aluminous (40%–50% Al_2O_3) porcelain is covered by feldspathic porcelain. The latter is applied to a refractory gypsum model covered with platinum foil and sintered under a vacuum at a high temperature. The porcelain powder contains metal oxides to color the crown. The other type is made from a glass ceramic of SiO_2 that is cast by the lost-wax process. After casting and contouring, the crown is heat treated to form an opaque crystal material, mica. Finally, color is applied by firing oxide stains on the surface. Porcelain and cast ceramics are used for a cosmetic veneer of the front surface of anterior teeth. The thin veneers are bonded to etched enamel with a composite resin cement. The third type of crown has a cast metal substructure for strength, onto which feldspathic porcelain is bonded by the same process described earlier.

Provisional Restoration

For provisional restorations, different materials are frequently used. While a permanent crown is being fabricated, a temporary or provisional crown protects the tooth. For best appearance, a custom-formed poly(methyl methacrylate) or polycarbonate crown is lined with acrylic resin; aluminum crowns lined with acrylic resin can be used with posterior teeth along with the temporary crowns. These crowns are easily removed, because the cement is used that sets when paste contains the active ingredients Zn_2O and eugenol that are combined. A Zn_2O–eugenol cement reinforced with poly(methyl methacrylate) beads is also used as a temporary direct filling material for patients with deep decay.

Prosthodontic Dental Materials

The materials used with fixed prosthodontics are similar to those described earlier under *metal castings* and *porcelain*.

Orthodontic Materials

The movement and stabilization of teeth can be controlled by several unique dental materials. Stainless steel alloys are commonly used for brackets and bands on teeth, but plastic brackets are also being used to improve appearance. Brackets are bonded to etched enamel with the help of composite resin without inorganic filler. Bands encircling posterior teeth are made of stainless steel and are generally secured with a cement formed by combining Zn_2O and HPO_4. Wires in different configurations are adapted to the dental arch to create the desired static forces for controlled tooth movement. The wires are commonly formed from wrought stainless steel, but alloys of Ni–Ti, Ti–Mo, and Co–Cr–Ni are also used.

Denudation

Local depletion of some element particularly when precipitates are formed.

Deoxidation

Removal of excess oxygen from the molten metal, usually accomplished by adding materials with a high affinity for oxygen and the removal of undesirable oxygen during melting and casting.

Deoxidation Products

Those nonmetallic inclusions that form as a result of adding the oxidizing agents to molten metal.

Deoxidized Copper

Copper from which cuprous oxide has been removed by adding a *deoxidizer*, such as phosphorous, to the molten bath.

Deoxidizer

A material that when added to molten metal combines referentially with dissolved or combined oxygen and, usually, forms a compound that becomes part of the slag.

Deoxidizing

(1) The removal of oxygen from molten metals through the use of a suitable *deoxidizer*. (2) Sometimes refers to the removal of undesirable elements other than oxygen through the introduction of elements or compounds that readily react with them. (3) In metal

finishing, the removal of oxide films from metal surfaces by chemical or electrochemical reaction.

Dephosphorization

The elimination of phosphorus from molten steel.

Depletion

(1) Selective removal of one component of an alloy, usually from the surface or preferentially from grain-boundary regions. (2) Reduction, particularly locally, of the quantity of alloying element. The term is used in various contexts and may indicate the loss of an element by diffusion through the matrix over significant distances (on a microscopical scale), or it may refer to localized diffusion, leading to precipitation of alloying rich precipitates. The surrounding material is then left depleted of the alloy element. See also *dealloying*.

Depolarization

A decrease in the *polarization* of an electrode.

Depolarizer

A substance that produces *depolarization*.

Deposit (Thermal Spraying)

See preferred term *spray deposit*.

Deposit Attack

Corrosion occurring beneath localized deposits, debris, etc. The deposit itself is inert but the area beneath it is low in oxygen relative to the surrounding surface and hence is relatively anodic. The metal surface beneath the deposit suffers electrolytic corrosion, and since the anode is small and the cathode large, penetration can be rapid. It is a form of *differential aeration corrosion*. See *deposit corrosion*.

Deposit Corrosion

Corrosion occurring under or around a discontinuous deposit on a metallic surface. It is also called *poultice corrosion*.

Deposit Sequence

See the preferred term *deposition sequence*.

Deposited Metal

Filler metal that has been added during a welding operation. Filler metal as opposed to fused parent metal.

Deposition

The process of applying a material to a base by means of vacuum, electrical, chemical, screening, or vapor methods, often with the assistance of a temperature and pressure container.

Deposition Efficiency (Arc Welding)

The ratio of the weight of deposited metal to the net weight of filler metal consumed, exclusive of stubs.

Deposition Efficiency (Thermal Spraying)

The ratio, usually expressed in percent, of the weight of spray deposit to the weight of material sprayed.

Deposition Rate (Thermal Spraying)

The weight of material deposited in a unit of time. It is usually expressed as kilograms per hour (kg/h) or pounds per hour (lb/h).

Deposition Sequence

The order in which the increments of weld metal are deposited. See also *longitudinal sequence*, *buildup sequence*, and *pass sequence*.

Depth Dose

The variation of absorbed dose with distance from the incident surface of a material exposed to radiation. Depth dose profiles give information about the distribution of absorbed energy in a specific material.

Depth of Cut

The thickness of material removed from a workpiece in a single machining part.

Depth of Field

The depth in the subject over which features can be seen to be acceptably in focus in the final image produced by a microscope.

Depth of Fusion/Fusion Penetration (of Weld)

The depth to which the parent material has fused, measured perpendicularly from the original surface. In the case of resistance welds, it is the maximum perpendicular distance from the original interface to the boundary of the nugget.

Depth of Penetration (Materials Characterization)

In various analytical techniques, the distance the probing radiation penetrates beneath the surface of a sample. See also *excitation volume*.

Depth of Penetration (Welding)

See *joint penetration* and *root penetration*.

Derby

A massive piece (intermediate in size, extending to more than 45 kg, or 100 lb, and usually cylindrical) of primary metal made by bomb reduction (such as uranium from uranium tetrafluoride reduced with magnesium). Compare with *biscuit* and *dingot*.

Descaling

(1) Removing the thick layer of oxides formed on some metals at elevated temperatures. (2) A chemical or mechanical process for removing scale or investment material from castings.

Deseaming

Analogous to *chipping*, the surface imperfections being removed by gas cutting.

Desiccant

A substance that can be used to dry materials because of its affinity for water and extract water or moisture from its environment.

Design Allowables

Statistically defined (by a test program) material property allowable strengths, usually referring to stress or strain. See also *A-basis*, *B-basis*, *S-basis*, and *typical basis*.

Desizing

The process of eliminating sizing, which is generally starch, from *gray* (also *greige*) *goods* before applying special finishes or bleaches (for yarn such as glass or cotton). Also, removing lubricant *size* following the weaving of a cloth.

Desorption

A process in which an absorbed material is released from another material. See also *absorption*, *adsorption*, and *chemisorption*.

Destannification

Corrosion in which tin is selectively removed from tin bronzes leaving a weak porous material. See *dezincification*, a similar but more extensively researched mechanism.

Destaticization

A treatment of plastic materials that minimizes the effects of static electricity on the surface either by treating the surface with specific materials or by incorporating materials in the molding compound. Minimization of surface static electricity prevents dust and dirt from being attracted to and/or clinging to the surface of the article.

Desulfurizing

The removal of sulfur from molten metal by reaction with a suitable slag or by the addition of suitable compounds.

Detection Limit

In an analytical method, the lowest massive concentration of an analyte that can be measured.

Detector, X-Ray

Sensor array used to measure the x-ray intensity. Typical detectors are high-pressure gas ionization, scintillator–photodiode detector arrays, and scintillator–photomultiplier tubes.

Detergent Additive

In lubrication technology, a surface-active additive that helps to keep solid particles suspended in an oil.

Detergent Oil

A heavy-duty oil containing a detergent additive. Detergent oils are used mainly in combustion engines.

Detergents

Materials that have a cleansing action like soap. Although soap itself is a detergent, as are the sodium silicates and the phosphates, the term usually applies to these synthetic chemicals, often referred to as *detergent soaps* or *soapless soaps*, which give this action. The targets may be the simple sulfonated fatty acids such as turkey-red oil, the *monopole soaps*, or highly sulfonated fatty acids or the *gardinols*, which are sulfonated fatty alcohols.

All the synthetic detergents are *surface-active agents*, or *surfactants*, with unsymmetric molecules that concentrate and orient at the interface of the solution to lower interfacial tension. They may be *anion-active agents*, with a positive-active ion; *cation-active agents*, with a negative-active ion; or *nonionic agents*. The anions and cations are sometimes called *gengenions*. Most of the household detergents are anion active and are powders. Most of the nonionics are liquids and are useful in textile processing since they minimize the difference in die affinity of various fibers. The cationics have lower detergency power and are usually skin irritants, but they have disinfectant properties and are used in washing machines and dairy cleansers. They are called *invert soaps* by the Germans. The synthetic detergents do not break down in the presence of acids or alkalies, and they do not form sludge and scum or precipitate salts in hard waters as soap does. They do not form quantities of suds as some soaps do, but suds contribute little to cleansing and are not desirable in automatic washing equipment. *Textile softeners* are different from surface-active agents. They are chemicals that attach themselves molecularly to the fibers, the polar, or charged and, of the cation orienting toward the fiber, with the fatty tails exposed to give the softness to the fabric.

Synthetic detergents have now largely replaced soaps for industrial uses. They are employed in textile washing, metal degreasing, paper-pulp processing, and industrial cleansing. They are also used in household cleansers, soapless shampoos, and toothpastes. *Biodegradable detergents* are those that can be chemically disintegrated by bacteria so that the discharge wastes do not contaminate the groundwaters. *Millox* is a group of biodegradable detergents made by the reaction of sucrose and fatty acids with a linking of ethylene oxide. This type of detergent is more powerful than petroleum-based detergents.

The detergents are more efficient than toilet soaps, but tend to leave the skin with an alkaline hardness. Lecithin may be used in detergent bars to reduce tackiness, and starch may be used for hardening.

To reduce package size, laundry detergents have become increasingly concentrated. *Phosphates*, once widely used, have lost favor because the water discharges can be environmentally damaging. *Aluminum silicate zeolites* are preferred but require additives for alkalinity, water softening, and equivalent cleaning. Containing water, however, they limit detergent concentration, and although they are effective and removing calcium, they only remove some of the magnesium, another hard water constituent.

Determination

The ascertainment of the quantity or concentration of a specific substance and a sample. See also *analysis*.

Detonation Flame Spraying

A thermal spraying process variation in which the controlled explosion of a mixture of fuel gas, oxygen, and powdered coating material is utilized to melt and propel the material to the workpiece.

Detritus

See *wear debris*.

Deuterium

An isotope of hydrogen. See *atomic structure*

Deuteron

The nucleus of the atom of heavy hydrogen, deuterium. The deuteron is composed of a proton and a neutron; it is the simplest multinucleon nucleus. Deuterons are used as projectiles in many nuclear bombardment experiments. See also *neutron* and *proton*.

Developed Blank

A sheet metal blank that yields a finished part without trimming or with the least amount of trimming.

Developer

See *dye penetrant examination*.

Deviation, X-Ray

The angle between the diffracted beam and the transmitted incident beam. It is equal to twice the Bragg angle θ.

Devitrification

(1) Crystallization in a glass. (2) The formation of crystals (seeds) in a glass melt, usually occurring when the melt is too cold. These crystals can appear as defects and glass fibers.

Devitrify

(1) To convert, partially or completely, from a glassy to a crystalline state, usually by controlled heating. (2) To deprive of glassy luster and transparency or to change from a vitreous to a crystalline condition.

Dewar flask

A vessel having double walls, the space between being evacuated to prevent the transfer of heat and the surfaces facing the vacuum being heat reflective, used to hold liquid gases and to study low-temperature phenomena.

Dewaxing

In casting, the process of removing the expendable wax pattern from an investment mold or shell mold, usually accomplished by melting out the application of heat or dissolving the wax with an appropriate solvent.

Dewetting

(1) The development and formation of a nonwetting condition after wetting has already commenced. (2) A condition that results when molten solder has coated a surface and then receded leaving irregularly shaped mounds of solder separated by areas covered with a thin solder film. The basis metal is not exposed, however.

Dewpoint Analyzer

An atmosphere monitoring device that measures the partial pressure of water vapor in an atmosphere.

Dewpoint Temperature

The temperature at which condensation of water vapor in a space begins for a given state of humidity and pressure as the vapor temperature is reduced; the temperature corresponding to saturation (100% relative humidity) for a given absolute humidity at constant pressure.

Dextrin

Also called *amylin*. A group of compounds with the same empirical formula as starch $(C_6H_{10}O_5)_x$, but with a smaller value of *x*. The compounds of stronger adhesive properties and used as pastes, particularly for envelopes, gummed paper, and postage stamps; for blending with gum arabic; in pyrotechnic compositions; and in textile finishing. Dextrin is a white, amorphous, odorless powder with a sweetish taste. It dissolves in water to form a syrupy liquid and is distinguished from starch by giving violet and red colors with iodine. Dextrin is made by moistening starch with a mixture of dilute nitric and hydrochloric acids and then exposing to a temperature of 212°F–257°F (100°C–125°C). Dextrin varies in grade chiefly owing to differences in the type of starch from which it is made. *British gum* is a name given to dextrins that give a high tack for paste use and are products containing partially converted starch.

Dezincification

A form of corrosion in warm moist environments affecting brass (copper–zinc alloys), which selectively removes the zinc, leaving weak, porous, redeposited copper of similar dimensions to the original. Localized, deeply penetrating attack is termed plug dezincification; more general surface attack is termed layer dezincification. Brass with less than about 15% zinc is not significantly affected and arsenic editions (0.02%–0.05%) inhibit attack in the higher-zinc-content, single-phase, alpha brasses, (up to 37% zinc). The alpha–beta duplex brasses can be made more resistant, but not immune, by the addition of tin, typically 1%, and arsenic up to 0.2%. The resistance of hot-worked, duplex alloys is improved by annealing at 525°C followed by water quenching. Cathodic protection can be effective. See also *dealloying* and *selective leaching*.

D-glass

A high-boron-content glass made especially for laminates requiring a precisely controlled dielectric constant.

Dilube

The removal of a lubricant from a powder compact, usually by burnout or alternatively by treatment with a chemical solvent.

Diadic Polyamide

Polyamide produced by condensation of diamine and a dicarboxylic acid.

Diamagnetic Material

A material whose specific permeability is less than unity and is therefore repelled weakly by a magnet. Compare with *ferromagnetic material* and *paramagnetic material*.

Diametrical Strength

A property that is calculated from the load required to crush a cylindrical sintered test specimen in the direction perpendicular to the axis. See also *radial crushing strength*.

Diamond

Diamond is a highly transparent and exceedingly hard crystalline stone of almost pure carbon. When pure, it is colorless, but it often shows tints of white, gray, blue, yellow, or green. It is the hardest known substance and is placed as 10 on the Mohs hardness scale. But the Mohs scale is only an approximation, and the hardness of the diamond ranges from 5500 to 7000 on the Knoop scale, compared with 2670 to 2940 for B_4C, which is designated as 9 on the Mohs scale. The diamond always occurs in crystals in the cubical system and has a specific gravity of 3.521 and a refractive index of 2.417.

In addition, diamonds are generally electrical insulators and nonmagnetic. They are quite transparent to x-rays and thus make useful pressure cell windows for x-ray measurements. Nearly, all diamonds show residual stresses in polarized light.

The melting point of diamond is given in terms of temperature and pressure because the phase diagram of carbon is such that unless pressure is maintained on diamond when at a temperature greater than 4000°C, the diamond will graphitize and then vaporize.

The diamond has been valued since ancient times as a gemstone, but it is used extensively as an abrasive, for cutting tools, and dies, for drawing wire. These industrial diamonds are diamonds that are too hard or too radial grained for good jewel cutting.

Synthetic Diamonds

These diamonds are produced from graphite at pressures from 5,512 to 12,402 MPa and temperatures from 1204°C to 2427°C. A molten metal catalyst of chromium, cobalt, nickel, or other metal is used, which forms a thin film between the graphite and the growing diamond crystal. Without the catalyst, much higher pressures and temperatures are needed. The shape of the crystal is controllable by the temperature. At the lower temperatures, cubes predominate, and at the upper limits, octahedra predominate; at the lower temperatures, the diamonds tend to be black, whereas at higher temperatures, they are yellow to white. The synthetic diamonds produced by the General Electric Co. are up to 0.01 carat in size and are of industrial quality comparable with natural diamond powders.

CVD Diamonds

A newer production method uses chemical vapor deposition (CVD) to produce a sheet of diamonds consisting of countless diamonds, each only a few micrometers across. The CVD diamonds have similar properties to those made by the older method. However, they conduct heat better than any known material—five times better than copper—making them useful as heatsinks to conduct heat away from electrical components.

The CVD process has been used to coat the cutting surfaces of tungsten carbide rotating tools such as routers, end mills, and drills. The technique should bring substantial productivity improvement to the machining of composites, carbon fiber, and glass/fiberglass-reinforced composites, epoxy resins, and graphite.

Tests show diamond-coated tools last up to 50 times longer than tungsten carbide ones and more than twice as long as polycrystalline diamond tools in aerospace applications. In addition to greater longevity, longer unattended runs, and fewer tool changes, diamond-coated tools also can run at higher speeds and feeds than conventional tools.

The CVD process allows round tools of virtually any shape or size to be diamond coated with no subsequent brazing or grinding.

Applications

Several types of grinding wheels (usually distinguished by the bond material) are made that carry diamonds as the abrasive material. Wheel bonds may be resinoid, vitrified, metal, etc. Diamond grit sizes in these wheels may vary from 0.0254 to 0.762 mm. In general, the diamond-carrying material is placed in a rather thin layer on the working surface of the wheel.

Wheel surface speeds of 22.5–33 m/s are recommended. Most grinding is done wet. By far, the greatest use for diamond wheels is in shaping and sharpening tungsten carbide tools. In addition, diamond wheels are used for edging and beveling glass and for lens grinding. Also, artificial sapphire is sliced and formed with diamond grinding wheels.

Diamond dust is a powder obtained by crushing the fragments of bort or from refuse from the cutting of gem diamonds. It is used as an abrasive for hard steels, for cutting other stones, and for making diamond wheels for grinding. Grit sizes for grinding wheels are 80–400. The coarsest, No. 80, for fast cutting, has about 1400 particles per square inch in the face of the bonded wheel.

Diamond drills range from small, diamond-tipped dental drills up to oil-well core bits containing 1000 carats. A 10.16 cm core bit may be set with 150–300 diamonds varying in size from 1/10 to 1/30 carat. As much as 3048 cm of limestone can be drilled before the diamonds require resetting in the hard metal matrix.

Diamond wire drawing dies are required for tungsten filament wire and preferred for many other wire materials. The low friction coefficient may contribute to the success of diamond in this application. Stones for wire drawing are large (up to 8 carats for 2.54 mm copper wire) and free from flaws and inclusions.

Single-point diamond lathe tools are made by setting a stone in the end of a steel shank suitable for mounting in a machine tool such as a lathe. The stone is mounted with consideration for its shape and crystallographic axis. After mounting, the tool is ground to the required shape on diamond grit wheels and polished on a scaife.

Single-point diamond tools are designed with angles between phases of 80° or greater to avoid the breakage that would occur if slender points were used. Diamond tools are not suitable for heavy cuts particularly on ferrous material. Accurate, fine finish cuts may be made on nonferrous metals and plastics. Tool wear is slow, so accuracy is maintained for many workpieces. A diamond tool maintains its cutting edge about 25 times as long as a high-speed steel tool and 3–4 times as long as a carbide tool.

Diamond Boring

Precision boring with a shaped diamond (but not with other tool materials).

Diamond Film

A carbon-composed film, usually deposited by chemical vapor deposition or related process, that has the following three characteristics: (1) a crystalline morphology that can be visually discerned by scanning electron or optical microscopy, (2) a single-phase crystalline structure identifiable by x-ray and/or electron diffraction, and (3) a Raman spectrum typical for crystalline diamond. See also *diamondlike film*.

Diamond Films

A new method has been developed for creating amorphous diamond films that are harder than any other coating, except for crystalline diamond. The new coatings are also smooth and can be created at room temperature, whereas crystalline diamond coatings need high temperatures and have rough surfaces.

This newly developed method uses a pulsed laser hitting a graphite target to deposit a carbon film containing a high percentage of diamondlike bonds. Heating the film illuminates internal stresses while allowing the film to retain diamondlike properties, including 90% of the stiffness and hardness of crystalline diamond. An added benefit is that researchers can control the level of stress. In fact, scientists have made stress-free films more than 7 μm thick and applied the coatings to plastic substrates as well as freestanding membranes more than an inch in diameter and less than 600 Å and tested to 800°C. Besides being incredibly hard, the films resist wear with low coefficient of friction and are almost inert to all chemicals. The films could find use as coatings for metal tools, auto parts, and biomedical devices.

Other methods have been developed that offer a lower-cost alternative by coating and chemically bonding an inexpensive substrate with a thin film of diamondlike carbon (DLC).

These diamond films have great potential as chemically inert protective coatings that make machine tools and parts last 10 times longer. Other applications include optical instruments, medical equipment, watch crystals, and eyeglasses.

This method is called direct ion-beam deposition for applying DLCs to a substrate. An ion generator creates a stream of ions from a hydrocarbon gas source; the C ions impinge directly on the target substrate and *grow* into a thin DLC film. This low-pressure, low-temperature process allows use of plastics and other substrates that cannot withstand the high pressures and temperatures normally used to synthesize diamonds.

Other commercial DLC applications include coatings for magnetic data-storage disks, surgical needles, and a diamond-coated ball for an artificial hip joint.

Diamond Pyramid Hardness Test

Hardness testing using a pyramid-shaped diamond indenter. See the *Vickers hardness test*.

Diamond Tool

(1) A diamond shape or form to the contour of a single-point cutting tool, for use in precision machining of nonferrous or nonmetallic materials. (2) An insert made from polycrystalline diamond compacts.

Diamond Wheels

A grinding wheel in which crushed and sized industrial diamonds are held in a resinoid, metal, or vitrified bond.

Diamondlike Film

A hard, noncrystalline carbon film, usually grown by chemical vapor deposition or related techniques, which contain predominantly sp^2 carbon–carbon bonds. See also *diamond film*.

Diaphragm

(1) A fixed or adjustable aperture in an optical system. Diaphragms are used to intercept scattered light, to limit field angles, or to limit image-forming bundles or rays. (2) A porous or permeable membrane separating anode and cathode compartments of an electrolytic cell from each other or from an intermediate compartment. (3) Universal die member made of rubber or similar material used to contain hydraulic fluid within the forming cavity and to transmit pressure to the part being formed.

Diaphragm Gate

A gate used in molding annular or tubular plastic articles. The gate forms a solid web across the opening of the part. This is also called *disk gate*.

Diatomaceous Earth

A class of compact, granular, or amorphous minerals composed of hydrated or *opaline silica*, used as an abrasive, for filtering, in metal polishes and soaps, as a filler in paints and molding plastics, for compacting into insulating bricks and boards, and in portland cement for fine detail work and for waterproofing. It is formed of fossil diatoms in great beds and is not earthy. In mineralogy, it is called *diatomite*.

After mining, the material is crushed and calcined. When pure, it is white; with impurities, it may be gray, brown, or greenish. The powder is marketed by fineness and chemical purity. Its high resistance to heat, chemical inertness, and dielectric strength and the good surface finish it imparts make it a desirable filler for plastics. For insulating purposes, bricks or blocks may be sawed from the solid or molded from the crushed materials, or it may be used in powder form. *Diatomite block* has a porosity of 90% of its volume and makes an excellent filter. *Celite*, of Celite Corp., is a 325-mesh, uncalcined, amorphous diatomaceous earth for portland cement mixtures and is used as a flattening agent in paints. *Sil-O-Cel* is diatomaceous earth in powder or in insulating block to withstand temperatures to 1600°F (871°C). *Compressible insulation*, to absorb the expansion stresses known as drag stress, between the firebrick and the steel shell of metallurgical furnaces, may be of diatomaceous silica. *Superex SG*, for blast furnaces, is a composite block with Superex on the hot side and a blanket of fine spun glass fibers on the cold side.

Dichromate Treatment

A chromate *conversion coating* produced on magnesium alloys and a boiling solution of sodium dichromate.

Didymium

A natural mixture of the rare earth elements praseodymium and neodymium, often given the quasi-chemical symbol Di.

Die

A tool, usually containing a cavity, that imparts shape to solid, molten, or powdered metal primarily because of the shape of the tool itself. It is used in many press operations (including blanking, drawing, forging, and forming), in die casting and in forming green powder metallurgy compacts. Die-casting and powder metallurgy dies are sometimes referred to as *molds*. See also *forging dies*.

Die

A master component that imposes its shape on another material. Manufacturing processes include drawing in which bar, tube, or wire is pulled through the die orifice. Extrusion, a process in which material is pushed through a die. Die stamping, forming, or forging in which material is pressed between shaped die faces and die casting in which molten metal is poured or injected into reusable shaped metal dies rather than into expendable sand molds.

Die Adapter

Informing plastics, the part of an extrusion die that holds the die block.

Die Assembly

The parts of a die stamp or press that hold the die and locate it for the punches.

Die Barrel

In powder metallurgy presses, a tubular liner for a die cavity.

Die Block

A block, often made of heat-treated steel, into which desired impressions are machined or sunk and from which closed-die forgings or sheet-metal stampings are produced using hammers or presses. In forging, die blocks are usually used in pairs, with part of the impression in one of the blocks and the rest of the impression in the other. In sheet-metal forming, the female die is used in conjunction with a male punch. See also *closed-die forging*.

Die Body

The stationary or fixed part of a powder pressing die.

Die Bolster

In powder metallurgy presses, the external steel ring that is shrunk fit around the hard parts comprising the die barrel.

Die Castings

Die castings are made by forcing molten metal under high pressure into a steel die containing an impression of the part to be made, which is called the die cavity. The process is most useful for the high-production manufacture of small- to medium-sized castings. Economies of the process are developed through the following:

1. *High speed of production*: The output from a die is many times more than four other casting processes.

2. *Strategic use of metal*: Through casting thinner walls, by coring holes and passages, and by contouring the die halves carefully, material can be saved and the casting made lighter.
3. *Dimensional accuracy*: The reproducibility is very good and dimensional tolerances can be held close enough to eliminate many machining operations.

Production

Die castings are made in a special machine containing a clamping mechanism to open and close the die and to hold the die halves together and an injection mechanism to force the metal into the die. Injection mechanisms can be of two types. The first system, the immersed plunger type, is suitable for alloys such as tin, lead, zinc, and sometimes magnesium where attack of the molten metal on the parts of the mechanism is insignificant. The die containing the cavity is closed and held tightly in the clamping mechanism, while the metal is injected by the downward motion of the plunger, which is operated by a hydraulic cylinder. The filling of the die is accomplished in less than 1/10 s. Dies can be opened in a few seconds, the casting removed by ejector pins, and the cycle repeated. Casting rates vary from 50/h for heavy parts to 15,000/h for small parts made on automatic machines.

Metals such as aluminum, brass, and magnesium, which because of either their solvent action or high melting temperatures are not adaptable to the immersed plunger system, can be cast by the second type, the cold chamber system, where the metal is melted in a separate unit and is hand-ladled or metered into a shot cylinder securely fastened to the die. The injection force is applied by a plunger that in turn is operated by a hydraulic cylinder. Higher pressures (200–1000 atm) are used in a cold chamber machine permitting improved density and mechanical properties of the die casting.

Characteristics

Die castings have smooth, dense surfaces, which require a minimum amount of finishing. A major advantage of die casting is the close tolerances that can be held. These substantially reduce machining operations and are largely responsible for the economy of the process.

Intricate casting shapes can be readily made through the use of cores and slides, ribs, and bosses. Special properties can be obtained by casting in inserts such as bronze sleeves, laminated pole shoes, magnets, and hardened steel plates.

One limitation of the process is the expense of the dies, which must be amortized over the expected production. However, most automotive, appliance, or business machine production is large enough that the tool and die costs become a small portion of the total casting cost and they are completely offset by manufacturing economies due to elimination of machining of parts after processing.

Average wall thickness for medium-sized parts and aluminum and magnesium is 2.2 mm and for large parts is 2.5–3.9 mm. Wall thickness for zinc will run approximately 75% of aluminum and magnesium because of the ease of casting of zinc in comparison to the light metals. Minimum wall thicknesses vary with the size of the part—1 mm is generally accepted for zinc and 1.3 mm for aluminum and magnesium.

Application

The largest single consumer of die castings is the automotive industry. Zinc is used for ornamentation and hardware such as radio grilles, instrument clusters, and housings, as well as for functional parts such as carburetors, fuel pumps, and speedometer housings. Zinc is used in appliances such as washing machines, dryers, electric ranges, and refrigerators for decorative and functional parts. It is also used for camera cases and projector housings and for toys. Military uses include precision fuse parts.

Because aluminum and magnesium can be used almost interchangeably, relative cost is often the governing selection factor. Automotive uses of aluminum die castings include transmission cases, torque converter housings and extensions, valve bodies, and carburetors. In the business machine field, typewriters are the largest application for aluminum die castings. The optical industry also uses aluminum die castings for binoculars, camera cases, projector frames, and other parts. Military uses are varied and include ordinance parts such as fuses, windshields, aircraft engine accessory parts, and airframe parts, as well as some small missile airframe parts. The electrical industry uses aluminum and magnesium for instrument housings and frames and for brush holders.

Materials

Dies are generally made of alloy steels and are relatively expensive. Dies for zinc can be made from prehardened steels of the SAE 4140 type. Dies for aluminum and magnesium are made from the Cr–Mo H-11 or H-13 steels heat-treated to C46–C48 Rockwell.

Zinc castings are made from high-purity Zn–Al alloys containing 4% aluminum with or without copper and containing approximately 0.03% magnesium. Impurities such as tin, lead, and cadmium must be kept at very low values, on the order of 0.005% or less, to avoid deterioration of the alloy in humid atmospheres. Zinc alloys have high tensile strength (292.4 MPa) and high impact strength (54.2 J). These properties change with decreasing temperature so that the useful temperature range for zinc is only recommended between 0°C and 93°C.

Aluminum die-casting alloys are made from the Al–Si, Al–Si–Cu, or Al–Mg series. These alloys have the best castability of physical properties for the die-casting process. Mechanical properties will vary between 276 and 325 MPa tensile strength, 135 and 172 MPa yield strength, and 3% and 5% elongation. Die castings of aluminum have high fatigue strength (135–142 MPa), which makes them particularly adaptable to structural uses where cyclic stresses are involved.

Magnesium alloys are of the Al–Zn type having tensile strength of 234 MPa, yield strengths of 142 MPa, and an elongation of 3%.

Copper alloys used are low-melting silicon brasses with tensile strength of 345–584 MPa, yield strength of 241–345 MPa, and an elongation of 15%–25%.

Die Cavity

The machine recess that gives a forging or stamping its shape.

Die Check

A crack in a die impression due to forging and thermal strains at relatively sharp corners. Upon forging, these cracks become filled with metal, producing sharp ragged edges on the part. Usual die wear is the gradual enlarging of the die impression due to erosion of the die material, generally occurring in areas subject to repeated high pressures during forging.

Die Clearance

Clearance between a mated punch and die, commonly expressed as clearance per side. This is also called clearance or punch-to-die clearance.

Die Cushion

A press accessory placed beneath or within a bolster plate or die block to provide an additional motion or pressure for stamping or forging operations, actuated by air, oil, rubber, springs, or a combination of these.

Die Cutting

Cutting shapes from metal sheet stock by sharply striking it with a shaped knife edge known as a steel rule die. Clicking and dinking are other names for die cutting of this kind.

Die Fill

A die cavity filled with powder.

Die Forging

A forging that is formed to the required shape and size through working in machined impressions in specially prepared dies.

Die Forming

The shaping of solid or powdered metal by forcing it into or through the *die cavity*.

Die Height

In forming, the distance between the fixed platinum and the moving one when dies are closed.

Die Holder

A plate or block, on which the die block is mounted, having holes or slots for fastening to the *bolster plate* or the *bed* of the press.

Die Impression

The portion of the die surface that shapes a forging or sheet metal part.

Die Insert

A relatively small die that contains part or all of the impression of a forging or sheet metal part and is fastened to the master *die block*.

Die Layout

The transfer of drawing or sketch dimensions to templates or die surfaces for use in sinking dies.

Die Life

The productive life of a *die impression*, usually expressed as the number of units produced before the impression has warned beyond permitted tolerances.

Die Lines

Lines or markings on form, drawing, or extruded metal parts caused by imperfections in the surface of the die.

Die Lubricant

(1) A lubricant applied to the working surfaces of dies and punches to facilitate drawing, pressing, stamping, and/or ejection. In powder metallurgy, the die lubricant is sometimes mixed into the powder before pressing into a compact. (2) A compound that is sprayed, swabbed, or otherwise applied on die surfaces or the workpiece during the forging or forming process to reduce friction. Lubricants also facilitate release of the part from the dies and provide thermal insulation. See also *lubricant.*

Die Match

The condition where dies, after having been set up in a press or other equipment, are in proper alignment relative to each other.

Die Opening

(1) In flash or upset welding, the distance between the electrodes, usually measured with the parts in contact before welding has commenced or immediately upon completion of the cycle but before upsetting. (2) In powder metallurgy, the entrance to the die cavity.

Die Pad

In forming, a movable plate or pad in a female die; usually used for part ejection by mechanical means, springs, or fluid cushions.

Die Plate

The base plate of a press into which the die is sunk.

Die Proof

A casting of a *die impression* made to confirm the accuracy of the impression.

Die Pull

The direction in which the solidified casting must move when it is removed from the die. The die pull direction must be selected such that all points on the surface of the casting move away from the die cavity surfaces.

Die Scalping

Removing surface layers from bar, rod, wire, or tube by drawing through a sharp-edged die to eliminate minor surface defects.

Die Separation

The space between the two halves of a die-casting die at the parting surface when the dies are close. The separation may be the result of the internal cavity pressure exceeding the locking force of the machine or warpage of the die due to thermal gradients in the die steel.

Die Set

(1) The assembly of the upper and lower die shoes (punch and die holders), usually including the *guide pins*, *guide pin bushings*, and *heel blocks*. This assembly takes many forms, shapes, and sizes and is frequently purchased as a commercially available unit. (2) Two machined dies used together during the production of a *die forging.*

Die Shaft

The condition that occurs after the dies have been set up in a forging unit in which a portion of the impression of one die is not in perfect alignment with the corresponding portion of the other die. This results in a mismatch in the forging, a condition that must be held within the specified tolerance.

Die Shoes

The upper and lower plates or castings that constitute a die set (punch and die holder). Also, a plate or block upon which a *die holder* is mounted, functioning primarily as a base for the complete *die assembly*. This plate or block is bolted or clamped to the *bolster plate* or the face of the press *slide.*

Die Sinking

The machining of the die impression to produce forgings of required shapes and dimensions.

Die Space

The maximum space (volume), or any part of the maximum space, within a forming or forging press for mounting a die.

Die Stamping

The general term for a sheet metal part that is formed, shaped, or cut by a die in a press in one or more operations.

Die Steels

Any of the various types of *tool steels* used for cold- and hot-forming dies, including forging, casting, and extrusion dies, stamping and trim dies, piercing tools and punches, molds, and mandrels. In general, all the major families of tool steels except the high-speed types are used for dies, including the hot-work, cold-work, shock-resisting, mold, special-purpose, and water-hardening types. The high-speed types, however, which are typically used for cutters, are also used for punches.

Die steel is any steel capable of achieving the necessary high strength without cracking or unacceptable distortion. Hence, they are usually, high-carbon low-alloy steels with corresponding good hardenability. For hot-working application, higher levels of chromium (5% or more) or tungsten (10% or more) are common.

Die Swell Ratio

In forming plastics, the ratio of the outer parison diameter (or parison thickness) to the outer diameter of the die (or die gap). Die swell ratio is influenced by head construction, land length, extrusion speed, and temperature. See also *parison* and *parison swell.*

Die Volume

See the preferred term *fill volume.*

Die Welding

See the preferred terms of *forge welding* and *cold welding*.

Dielectric

A nonconductor of electricity. It is the ability of a material to resist the flow of an electric current.

Dielectric Curing

The curing of a synthetic thermosetting resin by the passage of an electric charge (produced from a high-frequency generator) through the resin.

Dielectric Heating

The heating of plastic materials by dielectric loss in a high-frequency electrostatic field.

Dielectric Loss

A loss of energy evidenced by the rise in heat of a dielectric placed in an alternating electric field. It is usually observed as a frequency-dependent conductivity.

Dielectric Materials

These are materials that are electrical insulators or in which an electric field can be sustained with a minimum dissipation of power. A solid is a dielectric if its valence band is full and is separated from the conduction band by at least 3 eV.

Dielectrics are employed as insulation for wire cables and electrical equipment, as polarizable media for capacitors, in apparatus used for the propagation or reflection of electromagnetic waves, and for a variety of artifacts, such as rectifiers and semiconductor devices, piezoelectric transducers, dielectric amplifiers, and memory elements.

The ideal dielectric material does not exhibit electrical conductivity when an electric field is applied. The term *dielectric*, although it may be used for all phases of matter, is usually applied to solids and liquids. In practice, all dielectrics have some conductivity, which generally increases with increasing temperature and applied field. For a good dielectric, such as polytetrafluoroethylene, the low-field DC conductivity at room temperature may be lower than 10^{-16} Ω m, whereas the corresponding figure for some specimens of plasticized polyvinyl chloride (PVC) may be as high as 10^{-4} Ω m, similar to that of the low-conductivity semiconductors.

Breakdown

If the applied field is increased to some critical magnitude, the material abruptly becomes conducting, a large current flows (often accompanied by a visible spark), and local destruction occurs to an extent dependent upon the amount of energy that the source supplies to the low-conductivity path. This critical field depends on the geometry of the specimen, the shape and material of the electrodes, the time variation of the applied field, and other factors. Temperature instability can occur because of the heat generated through conductivity or dielectric losses, causing thermal breakdown.

Industrial Dielectric Materials

Many of the traditional materials are still in common use, and they compete well in some applications with newer materials regarding their electrical and mechanical properties, reliability, and cost. For example, oil-impregnated paper is still used for high-voltage cables; usually paper and various types of pressboard and mica, often as components of composite materials, are also in use. Elastomers and press-molded resins are also of considerable industrial significance. However, synthetic polymers such as polyethylene, polypropylene, polystyrene, polytetrafluoroethylene, PVC, polymethyl methacrylate, polyamide, and polyimide have become important, as has polycarbonate because it can be fabricated into very thin films.

Properties of Polymers

Generally, these have crystalline and amorphous regions, increasing crystallinity causing increased density, hardness, and resistance to chemical attack, but often producing brittleness. Many commercial plastics are amorphous copolymers, and often additives are incorporated in polymers to achieve certain characteristics or to improve their workability. Additives include inorganic fillers, antioxidants, stabilizers, and pigments.

Polyethylene

Polyethylene (C_nH_{2n+n}) is the most common and most investigated polymer. It is relatively inexpensive and is easily worked. Polyethylene readily oxidizes, and so it's working temperature is normally not above 70°C; although it may be worked at higher temperatures (e.g., up to 90°C) in suitable circumstances, it normally contains antioxidants.

Polyethylene Terephthalate

Polyethylene terephthalate is a useful material that has good resistance to acids and alkalies provided that they are diluted. It is insoluble in many common solvents, although it does dissolve in aromatic and chlorinated hydrocarbons, especially when warm. It softens above 250°C.

Polypropylene

Polypropylene is similar in structure to polyethylene, but every other carbon atom has one of the H_2 atoms replaced by a CH_2 group. Although electrically similar to polyethylene, polypropylene can be made in thinner films, say 5 μm as against about 25 μm for polyethylene. These films replace paper for impregnated capacitors, with reduced loss.

Polystyrene

Polystyrene is brittle at room temperature, becomes soft at 80°C, and is often modified by copolymerization. Traditionally, it is used in film form for capacitors, and it remains competitive for this application. Polystyrene is also used for coaxial cable insulation, but in round strip or bead form, because the solid is not very flexible.

Polytetrafluoroethylene

Polytetrafluoroethylene is similar to polyethylene, with all the H_2 atoms replaced by fluorine atoms. Highly crystalline (about 95), it resists twisting and bending and is mechanically tough, with a low coefficient of friction, 0.06. It is highly resistant to chemical attack and is useful for hostile environmental conditions.

Polyvinyl Chloride

PVC has a structure similar to that of polyethylene, but every other carbon atom has one of its H_2 atoms replaced by a chlorine atom. It is only about 10% crystalline and is compatible with a large number of other polymers. It is easily decomposed by heat and is not suitable for continuous use at 70°C. PVC is not well characterized and, depending on admixtures, may be molded; its main electrical use is coating wires.

Polymethyl Methacrylate

Polymethyl methacrylate is a clear solid, hard but easily scratched. Its dielectric properties are only moderate, and its use is restricted to undemanding conditions at normal ambient temperatures, usually in a molded form, where appearance is important.

Polyamide

Polyamide may be in linear form, known as nylon, or in aromatic form. Neither of these is of much value as a dielectric, but nylons, because of their toughness and resistance to solvents, are used to form coatings to protect insulation. The aromatic form can also make yarns, but it is most often used in impregnated board from which it gives low-voltage insulation capable of being used continuously at temperatures of 150°C, or even higher.

Polyimide

Polyimide is related to aromatic polyamide, but with additional aromatic groups. Expensive to produce, it can be made into films down to 25 μm thick and in varnish form for wire coating. It is very tough, does not burn but chars above 800°C, and remains flexible at liquid helium temperatures; it can be used continuously up to 250°C.

Resins

The resins are important members of the family of cross-linked polymers. Epoxy resins have a high mechanical strength, absorb very little water, and bond easily to most materials but not to polyethylene. They are used for bonding and encapsulation, and their properties can be modified by the curing process used and by the use of fillers and hardeners.

Dielectric Monitoring (of Plastics)

Monitoring the cure of thermosets by tracking the changes in their electrical properties during material processing.

Dielectric Shield

In a *cathodic protection* system, an electrically nonconductive material, such as a coating, or plastic sheet, or pipe, that is placed between an anode and an adjacent cathode to avoid current wastage and to improve current distribution, usually on the cathode.

Dielectric Strength

(1) The maximum voltage that a dielectric can withstand, under specified conditions, without resulting in a voltage breakdown (usually expressed as volts per unit dimension) and (2) A measure of the ability of a dielectric (insulator) to withstand a potential difference across it without electric discharge.

Dielectrometry

Use of electrical techniques to measure the changes in loss factor (dissipation) and in capacitance during cure of the resin and a laminate. This is also called dielectric spectroscopy.

Die-Parting Line

A lengthwise flash or depression on the surface of a protruded plastic part. The line occurs where separate pieces of the die join together to form the cavity.

Differential Aeration Cell

An *electrolytic cell*, the *electromagnetic force* of which is due to a difference in air (oxygen) concentration at one electrode as compared with that at another electrode of the same material. See also *concentration cell*.

Differential Aeration Corrosion

Corrosion resulting from variations in air, more specifically oxygen content at different points on an immersed component. The electrolytic corrosion occurs at the unnoticed, low-oxygen, areas. Such attack in crevices may be termed crevice corrosion or, if beneath debris, deposit attack.

Differential Coating

A coated product having a specified coating on one surface and a significantly lighter coating on the other surface (such as a hot-dipped galvanized product or electrolytic tin plate).

Differential Flotation

Separating a complex ore into two or more valuable minerals and *gangue* by *flotation*. This is also called selective flotation.

Differential Heating

Heating that intentionally produces a temperature gradient within an object such that, after cooling, a desired stress distribution or variation in properties is present within the object.

Differential Interference Contrast Illumination

A microscopic technique using a beam-splitting double-quartz prism, that is, a modified Wollaston prism placed ahead of the objective together with a polarizer and analyzer in the 90° crossed positions. The two light beams are made to coincide at the focal plane of the objective, revealing bright differences as variations in color. The prism can be moved, shifting the interference image through the range of Newtonian colors.

Differential Scanning Calorimetry (DSC)

Measurement of energy absorbed (endotherm) for production (exotherm). This may be applied to melting, crystallization, resin curing, loss of solvents, and other processes involving an energy change. It may also be applied to processes involving a change in heat capacity, such as the glass transition.

Differential Thermal Analysis (DTA)

An experiment analysis technique in which a specimen and a control are heated simultaneously and the difference in their temperatures is monitored. The difference in temperatures provides information on relative heat capacities, presence of solvents, changes in structure (i.e., phase changes, such as melting of one component in a resin system), and chemical reactions. See also *differential scanning calorimetry.*

Diffraction

(1) A modification that radiation undergoes, for example, in passing by the edge of opaque bodies or through narrow slits, in which the rays appear to be deflected. (2) Coherent scattering of x-rays by the atoms of a crystal that necessarily results in beams in characteristic directions. This is sometimes termed as reflection. (3) The scattering of electrons by any crystalline material through discrete angles depending only on the lattice spacings of the material in the velocity of the electrons.

Diffraction Contrast

In electron microscopy, contrast produced by intensity differences in Bragg diffracted beams from a crystalline material. These differences are caused by regions of varying crystal orientation.

Diffraction Grating

A series of a large number of narrow, close, equally spaced, diffracting slits or grooves capable of dispersing light into its spectrum. See also *concave grating* and *reflection grating*. Compare with *transmission grating.*

Diffraction Pattern

The spatial arrangement and relative intensities of diffracted beams.

Diffraction Ring

The diffraction pattern produced by a given set of planes from randomly oriented crystalline material. See also the *Debye ring.*

Diffuse Transmittance

In materials characterization, the transmittance value obtained when the measured radiant energy has experienced appreciable scattering in passing from the source to the receiver. See also *transmittance.*

Diffusion

(1) Spreading of a constituent in a gas, liquid, or solid, tending to make the composition of all parts uniform. (2) The spontaneous movement of atoms or molecules to new sites within a material. (3) The movement of a material, such as a gas or liquid, in the body of a plastic. If the gas or liquid is absorbed on one side of a piece of plastic and given off on the other side, the phenomenon is called permeability. Diffusion and permeability are not due to holes or pores in the plastic but are caused and control by chemical mechanisms. (4) The movement of individual atoms, ions, and molecules relative to the remainder of the material. The movement of individual atoms is random but where there is a concentration gradient in a large number of movements, there will be a net effect for atoms to move down the gradient. Ultimately, over a sufficient period of time, a uniform composition will develop. The rate of diffusion is fastest in gases and slower in fluids, but it is still significant in solid materials, and diffusion is a critical factor in many solid-state processes and phenomena. The rate of diffusion is accelerated by increasing the temperature.

Diffusion Aid

A solid filler metal sometimes used in *diffusion welding.*

Diffusion Bonding

Diffusion across the interface between solid contacting surfaces producing a bond of useful strength. Clean, close-fitting interfaces are vital. Heating is normally required as is a small amount of pressure. If the pressure is high, for example, sufficient to disrupt surface films, the joint is more correctly termed a *pressure weld*. See also *diffusion welding*, which is very similar to *diffusion brazing.*

Diffusion Brazing

A brazing process that produces coalescence of metals by heating them to suitable temperatures and by using a filler metal or an in situ liquid phase. The filler metal may be distributed by capillary action or may be placed or formed at the faying surfaces. The filler metal is diffused with the base metal to the extent that the joint properties have been changed to approach that of the base metal. Pressure may or may not be applied.

Diffusion Coating

Any process whereby a base metal or alloy is either (1) coated with another metal or alloy and heated to a sufficient temperature in a suitable environment or (2) exposed to a gaseous or liquid medium containing the other metal or alloy, thus causing *diffusion* of the coating or of the other metal or alloy into the base metal with resultant changes in the composition and properties of its surface.

Diffusion Coatings

A large number of elements can be diffused into the surface of metals to improve their hardness and resistance to wear, corrosion, and oxidation. Diffusion coatings (sometimes called cementation coatings) are applied by heating the base metal in an atmosphere of the coating material, which diffuses into the metal. The term may include sprayed or electroplated deposits that are subsequently heated to allow diffusion.

Calorized

Aluminum (calorized) coatings are applied by diffusion to carbon and alloy steels to improve their resistance to high-temperature oxidation. They can be applied by treating the metal in powdered aluminum compounds or in $AlCl_3$ vapor or by spraying the aluminum on and subsequently heat-treating it. The alloy coating formed (about 25% aluminum) protects the metal by stealing it from the surrounding air. The coatings range in depth from 5 to 40 mils and permit parts to remain serviceable for many years at temperatures up to 760°C. They have also been used for intermittent exposure as high as 927°C. Typical high-temperature uses are chemical and metal processing pots, bolts, air heater tubes, and parts for furnaces, steam superheaters, and oil and gas polymerizers.

Carburized

Carburizing allows steels to retain high internal strength and toughness and at the same time have high surface hardness. The hardened surface is produced by introducing carbon into a steel surface by heating the metal above the transformation temperature while it is in contact with carbonaceous material, which may be a solid, gas, or liquid.

In general, carburizing is limited to steels low enough in carbon (below about 0.45%) to take up that element readily. Plain carbon steels are generally used if surface hardness is the principal requirement and core properties are not too critical. Alloy steels must usually be used if high hardness and toughness are needed in the core. Typical applications are gears, cams, pawls, racks, and shafts.

Chromized

Chromizing is the process of diffusing chromium into ferrous metals to improve their resistance to corrosion, heat, and wear. Typical of the chromizing methods that have been developed is one in which the parts to be treated are packed in a proprietary powdered chromium compound and heated to 816°C–1038°C. This method produces high Cr–Fe alloy on ferrous metals with a low carbon content

The case (3 mils thick) exhibits good resistance to sealing and corrosion at high temperatures. A CrC case is produced on high-carbon materials such as cast iron, iron powder, tool steel, and plain carbon steels containing over 0.40% carbon. This case (1/2–2 mils thick) has a hardness of 1600–1800 vpn.

Cyanided, Carbonitrided

Both cyaniding and carbonitriding produce a hard and wear-resistant surface on low-carbon steels. Both methods cause carbon and N_2 to diffuse into the surface of the base metal. The case developed has high hardness after quenching. The methods differ in that a liquid bath is used in cyaniding, whereas a gas atmosphere is used in carbonitriding.

In general, cyaniding and carbonitriding are used with the same base metals and for the same applications as carburizing. Warpage is usually less serious than in carburizing. Quenching is usually required for full hardness, but file hardness can be obtained without quenching.

Ni–P Coated

With some exceptions, Ni–P coatings can be roughly classified as diffusion coatings. The coatings are prepared from NiO_2 to basic NH_3PO_4 and water and are applied to ferrous surfaces just like a paint. Subsequent heat treatment in a controlled atmosphere produces coatings with the degree of corrosion resistance approaching that of stainless steel and the high-nickel alloys. The coatings have little porosity and high resistance to heat and abrasion.

Nitrided

Nitriding is a means of improving wear resistance. In the most widely used process, steel is exposed to gaseous NH_4 at a temperature (about 538°C) suitable for the formation of metallic nitrites. The hardest cases are obtained with aluminum-bearing steels such as the Nitralloys. Where lower hardness is acceptable, steels containing aluminum, such as medium-carbon steels containing chromium and molybdenum, can be used.

Stainless steels can also be case hardened by nitriding. Straight chromium steels are more readily nitrided than Ni–Cr steels, although both are used. Tool steels can also be given a thin hard case.

Nitriding produces minimal distortion. Some growth occurs, but this can be allowed for. In general, nitriding is used for the same applications as carburizing.

Siliconized

Substantial improvements in the wear resistance and hardness of steel and iron parts can be obtained by impregnating with silicon (about 14%). The most wear-resistant cases are formed on low-carbon, low-sulfur steels. High-carbon, low-sulfur steels can also be impregnated, although treatment time is longer. White and malleable iron can also be siliconized; siliconizing of gray irons is not recommended.

The case of a siliconized surface (about 5–10 mils) is rather brittle, with hardness varying from Rockwell B80–B85. Siliconized surfaces are virtually nongalling and are especially effective in resisting combined wear and corrosion.

Diffusion Coefficient

A factor of proportionality representing the amount of substance diffusing across a unit area through a unit concentration gradient in unit time.

Diffusion Porosity

The porosity that is caused by the diffusion of one metal into another during sintering of a powder metallurgy compact. Also known as the Kirkendall porosity.

Diffusion Welding

(1) A solid-state welding process that produces coalescence of the faying surfaces by the application of pressure at elevated temperature. The process does not involve macroscopic deformation, melting, or relative motion of parts. A solid filler metal (diffusion aid) may or may not be inserted between the faying surfaces. (2) Welding resulting primarily from solid-state diffusion across the interface of the intended joint. Some interfacial force may be applied but no significant deformation, apart from crushing of minor asperities, occurs. Heating to a temperature below the melting point of the parent materials may be applied. If diffusion produces a low-melting-point phase that becomes molten, the process may be more precisely termed liquid-phase diffusion welding. If no molten phase is produced, the process may be termed solid-phase diffusion welding.

See also *forge welding, hot pressure welding*, and *cold welding*.

Diffusion Zone

The zone of variable composition at the junction between two different materials, such as in welds or between the surface layer in the core of clad materials or sleeve bearings, in which inner diffusion between the various components has taken place.

Diffusion-Limited Current Density

The *current density*, often referred to as *limiting current density*, that corresponds to the maximum transfer rate that a particular species can sustain because of the limitation of diffusion.

Digging

A sudden erratic increase in cutting depth, or in the load on a cutting tool, caused by unstable conditions in the machine setup. Usually, the machine is stalled or either the tool or the workpiece is destroyed.

Digital Radiography (DR)

Radiographic imaging technology in which a 2D set of x-ray transmission measurements is acquired and converted to an array of numbers for display with a computer. *Computed tomography* systems normally have DR imaging capabilities. Also known as digital radioscopy.

Dilatant

A reversible increase in viscosity with increasing shear stress. Compare with *pseudo-plastic behavior, rheopectic material,* and *thixotropy.*

Dilatometer

An instrument for measuring the linear expansion or contraction in a metal resulting from changes in such factors as temperature and allotropy.

Diluent

In an *organosol*, a liquid component that has little or no solvating action on the resin. Its purpose is to modify the action of the dispersant. The term diluent is commonly used in place of the term plasticizer.

Dilution

The change in chemical composition of a welding filler metal caused by the admixture of the base metal or previously deposited weld metal in the deposited weld bead. It is normally measured by the percentage of base metal or previously deposited weld metal in the weld bead.

Dilution Factor

When diluting a sample, the ratio of the final volume or mass after dilution to the volume or mass of the sample before dilution.

Dimensional Change

Object shrinkage or growth resulting from sintering a powder metallurgy compact.

Dimensional Stability

(1) A measure of dimensional change caused by such factors as temperature, humidity, chemical treatment, age, or stress (usually expressed as Δ units per unit). (2) The ability of a plastic part to retain the precise shape in which it was molded, cast, or otherwise fabricated.

Dimer

A substance (comprising molecules) form from two molecules of a *monomer.*

Dimerization

The formation of a *dimer.*

Dimethylglyoxime Test

A spot test to detect the presence of nickel in steel. A spot applied to the surface in question turns red in response to nickel.

Dimpled Rupture

A fractographic term describing *ductile fracture* that occurs through the formation and coalescence of microvoids along the fracture path. The fracture surface of such a ductile fracture appears dimpled when observed at high magnification and usually is most clearly resolved when viewed in a scanning electron microscope.

Dimples

Shallow indentations in a surface. They may be a manufacturing defect or they may be deliberate, for example, to locate the heads of fasteners. See also *dimpled fracture.*

Dimpling

(1) The stretching of a relatively small, shallow indentation into sheet metal. (2) In aircraft, the stretching of metal into a conical flange for a countersunk head rivet.

Dingot

An oversized *derby* (possibly a ton or more) of a metal produced in a bomb reaction (such as uranium from uranium tetrafluoride reduced with magnesium). For these metals, the term "ingot" is reserved for massive units produced in vacuum melting and casting. See also *biscuit* and *derby.*

Dinking

Cutting of nonmetallic materials or light-gauge soft metals by using a hollow punch with a knifelike edge acting against a wooden fiber or resiliently mounted metal plate.

Dip Brazing

A brazing process in which the heat required is furnished by a molten chemical or metal bath. When a molten chemical bath is used, the bath may act as a flux. When a molten metal bath is used, the bath provides the filler metal.

Dip Casting

In forming plastics, the process of submerging a hot mold into a resin. After cooling, the product is removed from the mold.

Dip Coat

(1) In the solid mold technique of investment casting, an extremely fine ceramic precoat applied as a slurry directly to the surface of the pattern to reproduce maximum surface smoothness. This coating is surrounded by coarser, less expensive, and more permeable investment to form the mold. (2) In the shell mold technique of investment casting, an extremely fine ceramic coating called the first coat, applied as a slurry directly to the surface of the pattern to reproduce maximum surface smoothness. The first coat is followed by other dip coats of different viscosity and usually containing different

grading of ceramic particles. After each dip, coarser stucco material is applied to the still wet coating. A buildup of several coats forms an investment shell mold. See also *investment casting*.

Dip Coating

Applying a plastic coating by dipping the article to be coated into a tank of melted resin or plastisol and then chilling the adhering metal.

Dip Plating

This is the same as *immersion plating*.

Dip Soldering

A soldering process in which the heat required is furnished by a molten metal bath that provides the solder filler metal.

Dip Transfer (in Welding)

See *metal transfer*.

Diphase Cleaning

Removing *soil* by an emulsion that produces two phases in the cleaning tank: a solvent phase and an aqueous phase. Cleaning is affected by both solvent action and emulsification.

Diphenyl Oxide Resins

Thermosetting resins based on diphenyl oxide and possessing excellent handling properties and heat resistance.

Dipole

A pair of equal and opposite separated charges, for example, a bar magnet.

Direct (Forward) Extrusion

See *extrusion*.

Direct Chill Casting

A continuous method of making ingots for rolling or extrusion by pouring the metal into a short mold. The base of the mold is a platform that is gradually lowered while the metal solidifies the frozen shell of metal acting as a retainer for the liquid metal below the wall of the mold. The ingot is usually cooled by the impingement of water directly on the mold or on the walls of the solid metal as it is lowered. The length of the ingot is limited by the depth in which the platform can be lowered; therefore, it is often called semicontinuous casting.

Direct Current Arc Furnace

An electric-arc furnace in which a single electrode positioned at the center of the furnace roof is the *cathode* of the system. Current passes from the electrode through the *charge* or bath to a cathode located at the bottom of the furnace. Current from the bottom of the furnace then passes through the furnace refractories to a copper base plate to outside cables. It is used in the production of ferroalloys, carbon and alloy steels, and stainless steels. See also *arc furnace*.

Direct Current Casting

This is the same as *cathodic cleaning*.

Direct Current Electrode Negative (DCEN)

The arrangement of direct current arc welding leads in which the work is the positive pole and the electrode is the negative pole of the welding arc. See also *straight polarity*.

Direct Current Electrode Positive (DCEP)

The arrangement of direct current arc welding leads in which the work is the negative pole and the electrode is the positive pole of the welding arc. See also *reverse polarity*.

Direct Current Reverse Polarity (DCRP)

See *reverse polarity* and *direct current electrode positive*.

Direct Current Straight Polarity (DCSP)

See *straight polarity* and *direct current electrode negative*.

Direct Fired

Furnaces in which the stream of fuel is injected into, and burnt in, the chamber and combustion products surround the charge.

Direct Injection Burner

A burner used in flame emission and atomic absorption spectroscopy in which the fuel and oxidizing gases emerge from separate ports and are mixed in the flame itself. One of the gases, usually the oxidant, is used for nebulizing the sample at the tip of the burner.

Direct Quenching

(1) Quenching carburized parts directly from the carburizing operation. (2) Also used for quenching pearlitic malleable parts directly from the malleabilizing operation. See *case hardening*.

Direct Sintering

In powder metallurgy, a method whereby the heat needed for sintering is generated in the body itself, such as by induction or resistance heating. This is in contrast with *indirect sintering*.

Direct Stress

Simple tensile or compressive stresses developed by axial loads, as opposed to the more complex stresses developed by torsion or bending.

Directed Metal Oxidation Process (DIMOX)

This process is based on the reaction of a molten metal with an oxidant, usually a gas, to form a ceramic matrix. Unlike the sintering process, in which the ceramic powders and fillers are consolidated under heat, directed metal oxidation grows the ceramic-matrix material around the reinforcements.

Examples of ceramic matrices that can be produced by the DIMOX-directed metal oxidation process include Al_2O_3, Al_2Ti_5, AlN, TiN, ZrN, TiC, and ZrC. Filler materials can be anything chemically compatible with the ceramic, parent metal, and growth atmosphere.

The first step in the process involves making a shaped preform of the filler material. Preforms consisting of particles are fabricated with traditional ceramic-forming techniques, while fiber preforms are made by weaving, braiding, or laying up woven cloth. Next, the preform is put in contact with the parent metal alloy. A gas-permeable growth barrier is applied to the surfaces of this assembly to limit its shape and size.

The assembly, supported in a suitable refractory container, is then heated in a furnace. For aluminum systems, temperatures typically range from 899°C to 1149°C. The parent metal reacts with the surrounding gas atmosphere to grow the ceramic reaction product through and around the filler to form a ceramic-matrix composite (CMC).

Capillary action within the growing ceramic matrix continues to supply molten alloy to the growth front. There, the reaction continues until the growing matrix reaches the barrier. At this point, growth stops, and the part is cooled to ambient temperature. To recover the part, the growth barrier and any residual parent metal are removed. However, some of the parent metal (5%–15% by volume) remains within the final composite in micron-sized interconnected channels.

Traditional ceramic processes use sintering or hot pressing to make a solid CMC out of ceramic powders and filler. Part size and shapes are limited by equipment size and the shrinkage that occurs during densification of the powders can make sintering unfeasible. Larger parts pose the biggest shrinkage problem. Advantages of the DIMOX process include no shrinkage because matrix formation occurs by a growth process. As a result, tolerance control and large part fabrication can be easier with directed metal oxidation.

In addition, the growth process forms a matrix whose grain boundaries are free of impurities or sintering aids. Traditional methods often incorporate these additives, which reduce high-temperature properties. And cost comparison shows the newer process is a promising replacement for traditional methods.

Direct-Fired Tunnel-Type Furnace

A continuous-type furnace is where the work is conveyed through a tunnel-type heating zone, and the parts are hung on hooks or fixtures to minimize distortion.

Directional Property

Property whose magnitude varies depending on the relation of the test axis to a specific direction within the metal. The variation results from preferred orientation or from fibering of constituents or inclusions.

Directional Solidification

Controlled solidification of molten metal in a casting so as to provide feed metal to the solidifying front of the casting.

Directionality

The variation in properties observed when some materials are tested in differing directions relative to the axis of the material. The material may be cast in which case any variation will merit this term. However, virtually all plate and sheet materials have inferior properties in the thickness direction, so in such cases the term usually refers to differences between properties along and across the directions. See also *preferred orientation*.

Directionally Solidified Castings

Investment castings produced by *directional solidification* that exhibit an oriented columnar structure.

Dirt Content

A measure of the size and concentration of foreign particles present in a lubricant. Dirt content is usually reported as the number of particles per cubic centimeter, for specified particle sizes.

Disappearing Filament Pyrometer

See *pyrometry*.

Disassociate Separate

Often a reversible process or brief duration of separation is implied.

Disassociated Ammonia

A frequently used sintering atmosphere. See also *cracked gas*.

Disassociation

As applied to heterogeneous equilibria, the transformation of one phase into two or more phases of different composition. Compare with *order–disorder transformation*.

Disassociation Pressure

At a designated temperature, the pressure at which a metallic phase will transform into two or more phases of different compositions.

Disbond

In adhesive bonded structures, an area within the bonded interface between two adherends in which in adhesion failure or separation has occurred. Also, colloquially, an area of separation between two laminae in the finish laminate (in this case, the term delamination is normally preferred). See also *debond*.

Disbondment

The destruction of adhesion between a coating and the surface coated.

Discard

Material scrapped from the tops of ingots, or the ends of extrusions and bars, etc., because of the inevitable deficiencies in quality at such positions.

Discontinuity

(1) Any interruption in the normal physical structure or configuration of a part, such as cracks, laps, seams, inclusions, or porosity. A discontinuity may or may not affect the utility of the part. (2) An interruption of the typical structure of a weldment, such as a lack of homogeneity in the mechanical, metallurgical, or physical characteristics of the material, or weldment. A discontinuity is not necessarily a defect. See also *defect* and *flaw*.

Discontinuous Precipitation

Precipitation from a supersaturated solid solution in which the precipitate particles grow by short-range diffusion, accompanied by recrystallization of the matrix in the region of precipitation. Discontinuous precipitates grow into the matrix from nuclei near grain boundaries, forming cells of alternate lamellae of precipitate and depleted (and recrystallized) matrix. It is often referred to as cellular or nodular precipitation. Compare with *continuous precipitation* and *localized precipitation*.

Discontinuous Sintering

Presintering or sintering of the subjects in a furnace according to a specified cycle that is tailored to the charge, for example, batch sintering, bell furnace sintering, box furnace sintering, and induction furnace sintering.

Discontinuous Yielding

(1) The nonuniform plastic flow of a metal exhibiting a yield point in which plastic deformation is inhomogeneously distributed along the gauge length. Under some circumstances, it may occur in metals not exhibiting a distinct yield point, either at the onset of or during plastic flow. (2) It occurs because grains are randomly oriented so some will deform before the others. The effect is most obvious with a coarse grain size and in material with a pronounced yield point.

Discrete

In its metallurgical context, this usually refers to features, such as particles in a microstructure, that are individually distinct and separate from their fellows.

Dished

Showing a symmetrical distortion of a flat or curved section of a plastic object so that, as normally viewed, it appears concave, or more concave than intended.

Dishing

Forming a shallow concave surface, the area being large compared to the depth.

Disinfectants

Materials used for killing germs, bacteria, or spore, thus eliminating causes of diseases or bad odors in factories and warehouses or in oils and compounds. The term *antiseptic* is employed in a similar sense in medicine, and the term *germicide* is often used for industrial disinfectants. Some disinfectants are also used as preservatives for leather and other materials, especially chlorine and chlorine compounds. *Phenol* is one of the best-known disinfectants, and the germ-killing power of other chemicals is usually based on a comparison with it. Practically, all bacteria are killed in a few minutes by a 3% solution of phenol in water, but phenol has the disadvantage of being irritating to skin. Industrial disinfectants are usually sold as concentrates to be diluted to the equivalent of a 3%–5% solution of phenol.

Too large a proportion of disinfectants in oils, solutions, or the air may be injurious to workers, so the advice of health officials is ordinarily obtained prior to general use. Creosote oil and cresylic acid are employed in emulsions and disinfecting sprays and dips, but continuous contact with creosote may be injurious. *Formaldehyde* has high germicidal power and is used for hides and leather, and some air sprays may contain chemicals such as chlorophyll that unite with moisture in the air to produce formaldehyde. But formaldehyde is not generally recommended for odor control, as it is an *anesthetizer*. It densities the olfactory receptors so that the individual is no longer able to detect the odor. *Masking agents*, which introduce a stronger, more pleasant odor, are likewise not a recommended method of disinfecting. They do not destroy the undesirable odor and may permit raising the total odor level to unhealthy proportions. Elimination of odors requires chemicals that neutralize or destroy the cause of the odors without causing undesirable effects.

The silver ion is an effective cleanser of water that contains bacteria that reduce sulfur-bearing enzymes, and silver sterilization is done with silver oxides on activated carbon or with organic silver compounds. The safe limit of silver and water for human consumption is specified by the U.S. Health Service as 10 ppb, and as with many other disinfectants, the use requires competent supervision. Among other metals, mercury is effective as an antibacterial agent in the form of *mercuric chloride*. Several organic mercurials are used as antiseptics, such as the *Mercurochrome* and *Metaphen* of Abbott Laboratories and *Merthiolate* of Eli Lilly and Co. Antiseptic atmospheres may be produced by spraying *chloramine T*, iodine, or argyrols. *Iodine* is a strong bactericidal antiseptic and is commonly used as a 2% tincture.

A 3%–5% solution of *hydrogen peroxide* is used as an antiseptic, although the chemical can cause corrosive burns at concentrations exceeding 25%. Compounds of various metals, such as iron manganese and cobalt, enhance peroxide's bactericidal action. Hydrogen peroxide is often trapped in *urea*, forming a solid containing up to 35% peroxide. *Ozone* generators are becoming popular for disinfecting swimming pools. Zinc peroxide and benzoyl peroxide are especially effective as antiseptic dressings.

The antiseptic throat lozenge known as *Sucrets* has a base of sugar and glucose, with hexylresorcinol and a flavor.

Disk Grinding

Grinding with the flat side of an abrasive disk or segmented wheel. It is also called *vertical-spindle surface grinding*.

Disk Machine

In tribology, a testing machine for rolling or rolling/sliding contact in which two disk-shaped rollers, with parallel axes of rotation, make tangential contact on their circumferences as they move relative to one another.

Disk Piercer

See *tube making*.

D

Dislocation

A linear imperfection in a crystalline array of atoms. Two basic types are recognized: (1) An edge dislocation corresponds to the row of mismatched atoms along the edge formed by an extra, partial plane of atoms within the body of a crystal. (2) A screw dislocation corresponds to the axis of a spiral structure in a crystal, characterized by a distortion that joins normally parallel planes together to form a continuous helical ramp (with a pitch of one interplanar distance) winding about the dislocation. The most prevalent is the so-called mixed dislocation, which is any combination of an edge dislocation and a screw dislocation. (3) Dislocation is a major, long-range, irregularity in the alignment of planes of atoms on the crystal lattice. Two forms of dislocations are the edge dislocation and the screw dislocation. A perfect crystal lattice would be extremely strong and resistant to plastic deformation. However, the presence of dislocations reduces the strain to about a thousandth of the theoretical level and facilitates plastic deformation. The principle can be visualized simplistically by considering two planes of atoms that are going to slide across each other. If the structures of the two planes match perfectly, then movement will not occur until the load is sufficient to overcome the bonds between every atom and its neighbor on the next plane. However, if a line of atoms is missing from one plane, then the load need only be sufficient to overcome the bonds on a single line of atoms. The dislocation will move one atomic spacing at a time, and when it has completed its movement across a crystal, it will leave a step with a height of one atomic space. An analogy often quoted to illustrate the low stress needed is the requirement to move a long strip of carpet for a short distance. If one end is pulled, a very large load is necessary to achieve any movement. However, a rod, representing a dislocation, can be put under the width of the carpet at one end and then pulled underneath the carpet to the other end. The load required is small and the carpet will be moved forward by a small increment. The movement of an individual dislocation is measured in terms of its Burgers vector, which identifies the total lattice translation resulting from the passage of the dislocation. Clearly, large numbers of dislocation movement are necessary to achieve any discernible deformation, but in practice metals contain, or can develop, immense numbers of dislocations. The movement of a series of dislocations along an individual plane is termed glide. However, as the dislocations move, they become mutually entangled or locked by grain boundaries and other microstructural features. This necessitates higher stresses to cause further deformation, that is, strain hardening occurs. Ultimately, further deformation is impossible and an increasing stress will initiate failure at the most severe *log jam* of dislocations. As mentioned earlier, although some dislocations are present in cast or annealed material, many others can be created by the imposed stress and associated deformation. The creation mechanism, termed a Frank–Rhead source, comprises an initial dislocation anchored at its ends. A shear load acting on the dislocation causes it to bulge outwards and then pivot round the anchored ends until it forms a complete dislocation loop. In the process, the original dislocation reestablishes itself to initiate further loops.

Dislocation Climb

A thermally activated, time-dependent mechanism in which the creation and diffusion of vacancies allows a dislocation to move or *climb* onto a neighboring plane.

Dislocation Etching

Etching of exit points of dislocations on a surface. It depends on the strain field ranging over a distance of several atoms. Crystal figures (edge pits) are formed at exit points. For example, edge pits for cubic materials are cube faces.

Dislocation Pending/Locking

The restraint on movement of a dislocation resulting from its interaction with some microstructural feature such as a precipitate, interstitial atom, or another dislocation.

Disordered Structure

The crystal structure of a solid solution in which the atoms of different elements are randomly distributed relative to the available lattice sites. This is in contrast with *ordered structure*.

Disordering

Forming a lattice arrangement in which the solute and solvent atoms of a solid solution occupy lattice sites at random. See also *ordering* and *superlattice*.

Dispersant Additive

In lubrication technology, an additive capable of dispersing cold oil sludge.

Dispersant Oil

A heavy-duty oil containing a dispersant additive.

Dispersing Agent

A substance that increases the stability of a suspension of powder particles in a liquid medium by the deflocculation of the primary particles.

Dispersion

Finely divided particles of a material in suspension and another substance.

Dispersion Hardening/Strengthening

Hardening induced by numerous fine inert particles introduced into the material and evenly distributed throughout the structure including within the grains. The particles are not soluble in the matrix even if the material is melted and hence are not an alloying addition in the usual sense of the term. However, they serve the same purpose as precipitates in that they can lock dislocations, impede slip, and increase strength, including creep strength. In addition, as they remain insoluble, they can impede grain growth and even inhibit recrystallization and so further enhance high-temperature performance. The particles may be introduced either by addition to the molten metal in the form of an inert powder or by a powder metal technique so that after processing, the material comprises the metal matrix plus oxide particles derived from the oxide film that originally covers the powder.

Dispersion-Strengthened Material

A metallic material that contains a fine dispersion of nonmetallic phase(s), such as Al_2O_3, MgO, SiO_2, CdO, ThO_2, Y_2O_3, or ZrO_2

singly or in combination, to increase the hot strength of the metallic matrix. Examples include dispersion-strengthened copper (Al_2O_3) used for welding electrodes, silver (CdO) used for electrical contacts, and nickel–chromium (Y_2O_3) superalloys used for gas turbine components. See also *mechanical alloying.*

Dispersion-Strengthened Metals

Particulate composites in which a stable material, usually an oxide, is disbursed throughout a metal matrix. The particles are less than 1 µm in size, and the particle volume fraction ranges from only 2% to 15%. The matrix is the primary load bearer, while the particles serve to block dislocation movement and cracking in the matrix. Therefore, for a given matrix material, the principal factors that affect mechanical properties are the particle size, the interparticle spacing, and the volume fraction of the particle phase. In general, strength, especially at high temperatures, improves as interparticle spacing decreases. Depending on the materials involved, dispersion-hardened alloys are produced by either powder metallurgy, liquefied-metal, or colloidal techniques. They differ from precipitation-hardened alloys in that the particle is usually added to the matrix by nonchemical means. Precipitation-hardened alloys derive their properties from compounds that are precipitated from the matrix through heat treatment.

There is a rather wide range of dispersion-hardened alloy systems. Those of aluminum, nickel, and tungsten, in particular, are commercially significant. Tungsten–thorium, a lamp-filament material, has been in use for more than 30 years. Dispersion-hardened aluminum alloys, known as SAP alloys, are composed of aluminum and Al_2O_3 and have good oxidation and corrosion resistance plus high-temperature stability and strength considerably greater than that of conventional high-strength aluminum alloys. Another dispersion-hardened metal, TD nickel, has dispersion of thoria in a nickel matrix. It is three to four times stronger than pure nickel at 871°C–1316°C. TD-Ni–Cr has also been produced for increased resistance to high-temperature oxidation. Other metals that have been dispersion strengthened include copper, lead, zinc, titanium, iron, and tungsten alloys. The copper is used for resistance-welding (spot-welding) tips.

Dispersoid

Finely divided particles of relatively insoluble constituents visible in the microstructure of certain metallic alloys.

Displacement

The distance that a chosen measurement point on the test specimen displaces normal to the crack plane. See also *crack opening displacement* and *crack plane orientation.*

Displacement Angle

In filament winding, the advancement distance of the winding ribbon on the equator after one complete circuit.

Displacement Controlled Loading

A term applied to forms of loading during testing (or service) where the load is applied by a rigid structure of, effectively, fixed dimensions so that if the test piece extends after initial loading, the load and stress fall. An example is a high-temperature flange joint with bolts passing through it. Because of the difference in cross sections,

the type bolts experience considerable tensile strain but the flanges are hardly compressed. As the bolts creep, converting elastic strain to permanent creep strain, the tensile stresses in the bolts fall so the closure forces on the flanges reduce, leading to leakage. The alternative is load controlled loading in which the load does not vary as strain occurs. An example is a weight hanging freely on a rod. If the rod extends, the load due to the weight obviously does not change, but if the rod cross section reduces by creep, the stress increases.

Display Resolution

Number of picture elements (pixels) per unit distance in the object.

Disproportionation

In the processing of plastics, termination by chain transfer between macroradicals to produce a saturated and an unsaturated polymer molecule.

Disrupted Strength

The stress at which a metal fractures under hydrostatic tension.

Dissimilar Metal Joint

In its general sense, any joint between different metals. Generalizing, such joints can be difficult to make by processes such as welding and often present problems in service because of differences in the chemical, electrochemical, and physical properties. The term is often used in a more specific sense regarding welded joints between austenitic and ferritic steel pipe work in power plants. The primary motivation to use two materials is cost; the low alloy ferritic steels are relatively cheap compared with the austenitic steels, but the latter have the superior creep properties required for the higher-temperature zones. Sound welds are fairly easy to achieve but high-temperature operation in the 500°C–600°C region causes considerable problems. Carbon tends to diffuse from the ferritic to the austenitic so the ferritic material is weakened; the expansion coefficients differ considerably so differential expansion and contraction causes large stresses that are superimposed on the stresses arising from internal pressure; and finally, the two materials have distinctly different oxidation characteristics. Collectively, these effects lead to premature creep damage, usually in the ferritic steel close to the weld and to a wedge-shaped oxide band penetrating from the exterior, again in the ferritic material close to the weld.

Dissipation Factor, Electrical

See *quality factor* (2).

Dissolution

The process of one material dissolving another.

Dissolution Etching

Development of microstructure by surface removal.

Distortion

Any deviation from an original size, shape, or contour that occurs because of the application of stress or the release of residual stress.

Distortion (Composites)

In fabric, the displacement of fill fiber from the 90° angle (right angle) relative to the warp fiber, while in a laminate, the displacement of the fibers (especially at radii), relative to their idealized location, due to motion during lay-up and cure.

Distributed Impact Test

In impingement erosion testing, an apparatus or method that produces a spatial distribution of impacts by liquid or solid bodies over an exposed surface of a specimen. Examples of such tests are those employing liquid sprays or simulated rainfields. If the impacts are distributed uniformly over the surface, the term uniformly distributed impact test may be used.

Distribution Contour

The shape of the particle size distribution curve.

Distribution Ratio

See *partition law.*

Disturbed Metal

The cold-work metal layer formed at a polished surface during the process of mechanical grinding and polishing.

Divariant Equilibrium

See *bivariate equilibrium.*

Divided Cell

A cell containing a diaphragm or other means for physically separating the *anolyte* from the *catholyte.*

Dividing Cone

A conical powder heap that is divided into quarters for the purpose of sample taking.

Divorced Eutectic

A demographic appearance in which the two constituents of a eutectic structure appear as massive phases rather than the finely divided mixture characteristic of normal eutectics. Often, one of the constituents of the eutectics is continuous and indistinguishable from an accompanying proeutectic constituent.

Divorced Pearlite

Pearlite whose original lamellar structure has broken down by diffusion, at an elevated temperature but below the lower critical, to form spheroidized carbides in a ferrite matrix. See *steel.*

DN Value

The product of bearing bore diameter in millimeters and speed in revolutions per minute.

Doctor Blade or Bar

In forming plastics, a straight piece of material used to spread resin, as in application of a thin film of resin for use in hot-melt prepregging or for use as an adhesive film. It is also called *paste metering blade.*

Doctor Roll

In applying adhesives, a roller mechanism that is revolving at a different surface speed, or in an opposite direction, resulting in a wiping action for regulating the adhesive supplied to the spreader roll.

Doily

In filament winding of composites, the planar reinforcement applied to a local area between windings to provide extra strength in an area where a cutout is to be made, for example, port openings. It is usually placed at the knuckle joints of cylinder to dome.

Dolly

A massive support held behind material being struck. Its inertia results in the material deforming rather than just bouncing away. The dolly is not normally shaped to the intended profile.

Dolomite

Dolomite is the carbonate material $CaMg(CO_3)_2$, often with small amounts of iron, manganese, or excess calcium, which replace some of the magnesium; cobalt, zinc, lead, and barium are more rarely found.

Dolomite is a very common mineral, occurring in a variety of geologic settings. It is often found in ultrabasic igneous rocks, notably in carbonates and serpentinites; in metamorphosed carbonate sediments, where it may crystallize to form dolomite marbles; and in hydrothermal veins. The primary occurrence of dolomite is in sedimentary deposits, where it constitutes the major component of dolomite rock and is often present in limestones.

Dolomite is normally white or colorless with a specific gravity of 2.9 and a hardness of 3.5–4 on Mohs scale.

Dolomite Brick

A calcium magnesium carbonate used as a refractory brick that is manufactured substantially or entirely of *dead-burned* dolomite.

Domain (Magnetic)

A substructure in a ferromagnetic material within which all the elementary magnets (electron spins) are held aligned in one direction by interatomic forces; if isolated, a domain would be a saturated permanent magnet.

Domain (Plastics)

A morphological term used in noncrystalline systems, such as block copolymers, in which the chemically different sections of the chain separate, generating two or more amorphous phases.

Dome

In filament winding, the portion of a cylindrical container that forms the spherical or elliptical shell ends of the container.

Domed

Showing a symmetrical distortion of a flat or curved section of a plastic object so that, as normally viewed, it appears convex, or more convex than intended.

Donnan Exclusion

In *ion chromatography*, the mechanism by which an ion-exchange resin can be made to act like a semipermeable membrane between an interstitial liquid and liquid occluded inside the resin particles. Highly ionized molecules are excluded from the resin particles by electrostatic forces; weakly ionized or nonionized molecules may pass through the membrane.

Donor

An element in a semiconductor. See *semiconductor*.

Dopant

(1) An impurity introduced under highly controlled conditions in very small but accurately known quantities into a semiconductor material, such as silicon. Dopants modify the electrical characteristics of the silicon by creating *p* or *n* regions and hence *p–n* junctions. (2) A material added to a polymer to change a physical property.

Doped Solder

A solder containing a small amount of an element intentionally added to ensure retention of one or more characteristics of the materials on which it is used.

Doping

In powder metallurgy, the addition of a small amount of an activator to promote sintering.

Doppler Effect

The change in the observed frequency of an acoustic or electromagnetic wave due to the relative motion of source and observer. See also the *Doppler shift*.

Doppler Shift

The amount of change in the observed frequency of a wave due to the Doppler effect, usually expressed in hertz. See also the *Doppler effect*.

Doré Silver

Crude silver containing a small amount of gold, obtained after removing lead in a cupelling furnace. This is the same as doré bullion and doré metal.

Dosimeter

A device for measuring radiation-induced signals that can be related to absorb the dose (or energy deposited) by radiation in materials and is calibrated in terms of the appropriate quantities and units. It is also called the dose meter.

Dot Map

See *x-ray map*.

Double Aging

Employment of two different aging treatments to control the type of precipitate formed from a supersaturated matrix in order to obtain the desired properties. The first aging treatment, sometimes referred to as intermediate or stabilizing, is usually carried out at higher temperature than the second.

Double Anneal

A double heat treatment cycle in which steel is first subject to a form of high temperature normalizing and then to a conventional anneal. The initial heating to well above the upper critical promotes homogenization and coalesces or blunts manganese sulfide inclusions. Rapid cooling to just below the lower critical then inhibits the formation of a coarse grain size. The steel is then immediately reheated to just above the upper critical and finally slow cooled. This produces a fully softened fine-grained material.

Double Arcing (Plasma Arc Welding and Cutting)

A condition in which the main arc does not pass through the constricting orifice but transfers to the inside surface of the nozzle. A secondary arc is simultaneously established between the outside surface of the nozzle and the workpiece. Double arcing usually damages the nozzle.

Double Cone Mixer

A vessel in the shape of two cones abutting at their base that rotates on an axis through the base and that provides thorough mixing or blending of a powder by cascading.

Double Etching

In metallography, use of two etching solutions in sequence. The second etchant emphasizes a particular microstructural feature.

Double Layer

The interface between an *electrode* or a suspended particle and an *electrolyte* created by charge–charge interaction leading to an alignment of oppositely charged ions at the surface of the electrode or particle. The simplest model is represented by a parallel plate condenser.

Double Pressing

A method whereby compaction of metal powders is carried out in two steps. It may involve removal of the compact from the die after the first pressing for the purpose of storage, drying, baking, presintering, sintering, or other treatment, before reinserting into a die for the second pressing.

Double Salt

A compound of two salts that crystallize together in a definite proportion.

Double Sintering

In powder metallurgy, a method consisting of two separate sintering operations with a shape change by machining or coining performed in between.

Double Skin

A layer of metal on the surface of a casting separated from the underlying material by an oxide film. It is produced during casting when, after solidification commences at the mold surface, the molten metal recedes temporarily exposing the newly solidified material to the atmosphere. Although the molten metal returns to fill the mold, the oxidized layer remains substantially intact as a major weakness.

Double Spread

The application of adhesives to both adherends of a joint.

Double Tempering

A treatment in which a quench-hardened ferrous metal is subjected to two complete tempering cycles, usually at substantially the same temperature, for the purpose of ensuring completion of the tempering reaction in promoting stability of the resulting microstructure.

Double-Acting Hammer

A forging hammer in which the ram is raised by admitting steam or air into a cylinder below the piston and the blow intensified by admitting steam or air above the piston on the downward stroke.

Double-Action Die

A die designed to perform more than one operation in a single stroke of the press.

Double-Action Forming

Forming or drawing in which more than one action is achieved in a single stroke of the press.

Double-Action Mechanical Press

A press having two independent parallel movements by means of two slides, one moving within the other. The inner slide or plunger is usually operated by a crankshaft; the outer or blankholder slide, which dwells during the drawing operation, is usually operated by a toggle mechanism or by cams. See also *slide*.

Double-Action Press

A press that provides pressure from two sides, usually opposite from each other, such as from top to bottom.

Double-Bevel Groove Weld

A groove weld in which the joint edge of one member is beveled from both sides.

Double-J Groove Weld

A groove weld in which the joint edge of one member is in the form of two Js, one from either side.

Doubler

In filament winding of composites, a local area with extra reinforcement, wound integrally with the part or wound separately and fastened to the part. See also *tabs*.

Double-Shot Molding

In forming plastics, a means of producing two color parts and/or two different thermoplastic materials by successive molding operations.

Double-U Groove Weld

A groove weld in which each joint edge is in the form of two Js or two half Us, and one from either side of the member.

Double-V Groove Weld

A groove weld in which each joint edge is beveled from both sides.

Dovetailing (Thermal Spraying)

A method of surface roughening involving angular undercutting to interlock the spray deposit.

Dow Process

A process for the production of magnesium by electrolysis of molten magnesium chloride.

Dowel

(1) In casting, a wooden or metal pin of various types used in the parting surface of parted patterns and core boxes. (2) In die-casting dies, metal pins to ensure correct registry of cover and ejector halves.

Down Cutting

See the preferred term *climb cutting*.

Down Milling

See the preferred term *climb milling*.

Downgate

This is the same as *sprue*.

Downhand

See the preferred term *flat position*.

Downhand Welding

See *flat position welding*, the usually preferred term.

Downslope Time (Automatic Arc Welding)

The time during which the current is changed continuously from final taper current or welding current to final current.

Downslope Time (Resistance Welding)

The time during which the welding current is continuously decreased.

Downsprue

This is the same as *sprue*.

Downtime

The period or total time that equipment or production facilities are not available. Plan maintenance activities may be included or excluded depending on site practice.

Draft

(1) An angle or taper on the surface of a pattern, core box, punch, or die (or of the parts made with them) that facilitates removal of the parts from a mold or die cavity or a core from a casting. (2) The change in cross section that occurs during rolling or cold drawing.

Draft Angle

(1) The angle of a taper or a mandrel or mold that facilitates removal of the finished plastic part. (2) The angle of taper, usually 5"–7", given to the sides of a forging in the sidewalls of the die impression. See the term *backhand welding*.

Drag

The bottom section of a *flask*, *mold*, or *pattern*. See *cope*.

Drag (Thermal Cutting)

The offset distance between the actual and the theoretical exit points of the cutting oxygen stream measured on the exit surface of the material.

Drag Angle

In welding, the angle between the axis of the electrode or torch and a line normal to the plane of the weld when welding is being done with the torch positioned ahead of the weld puddle. See the *backhand welding*.

Drag Lines/Markings

The characteristic linear steps or serrations on a thermally cut face.

Drag Soldering

A process in which supported, moving printed circuit assemblies or printed wiring assemblies are brought in contact with the surface of a static pool of molten solder.

Drag Technique

A method used in manual arc welding wherein the electrode is in contact with the assembly being welded without being in short circuit. The electrode is usually used without oscillation.

Drag-In

Water or solution carried into another solution by the work and its associated handling equipment.

Dragout

Material inadvertently dragged from a treatment solution or molten bath by the component or handling equipment.

Drainage

Conduction of electric current from an underground metallic structure by means of a metallic conductor. Forced drainage is that applied to underground metallic structures by means of an applied electromotive force or sacrificial anode. Natural drainage is that from an underground structure to a more negative (anodic) structure, such as the negative bus of a trolley substation.

Drape

In fabricating composites, the ability of a fabric or prepreg to conform to a contoured surface.

Drape Forming

A method of forming thermoplastic sheet in which the sheet is clamped into a movable frame, heated, and draped over high points of the male mold. Vacuum is then pulled to complete the forming operation. It is also known as basic male mold forming.

Draw

A term used to denote the shrinkage or void that appears on the surface of a casting or within a casting.

Draw Bead

(1) An insert or riblike projection on the draw ring or hold-down surfaces that aids in controlling the rate of metal flow during deep-draw operations. Draw beads are especially useful in controlling the rate of metal flow in irregularly shaped stampings.

Draw Forging

See *radial forging*.

Draw Forming

A method of curving bars, tubes, or rolled or extruded sections in which the stock is bent around a *rotating form block*. Stock is bent by clamping it to the form block and then rotating the form block while the stock is pressed between the form block and a pressure die held against the periphery of the form block.

Draw Head

Set of rolls or dies mounted on a drawbench for forming a section from strip, tubing, or solid stock. See also *Turk's head rolls* and *drawbench*.

Draw Marks

See *scoring*, *galling*, *pickup*, and *die line*.

Draw Plate

(1) In metal forming, a circular plate with a hole in the center contoured to fit a forming punch; used to support the *blank* during the forming cycle. (2) In casting, a plate attached to a pattern to facilitate drawing of a pattern from the mold.

Draw Radius

The radius at the edge of a die or punch over which sheet metal is drawn.

Draw Ring

A ring-shaped die part (either the die ring itself or a separate ring) over which the inner edge of sheet metal is drawn by the punch.

Draw Stock

The forging operation in which the length of a metal mass (stock) is increased at the expense of its cross section; no *upset* is involved. The operation includes converting ingot to press bar using "V," round, or flat dies.

Drawability

A measure of the formability of a sheet metal subject to a drawing process. The term is usually used to indicate the ability of a metal to be deep drawn. See also *drawing* and *deep-drawing*.

Drawbench

The machinery and associated equipment for drawing tube, wire, and rod, etc. It will typically include the table, that is, the bench, which carries the die block and replaceable die close rollers or slides, the gripping device (sometimes termed the head or the dog), the drawer chain, the drive motor, and associated equipment.

Draw-Down Ratio

In forming plastics, the ratio of the thickness of the die opening to the final thickness of the product.

Drawing

(1) A term used for a variety of forming operations, such as *deep-drawing* a sheet metal blank, redrawing a tubular part, and drawing rod, wire, and tube. The usual drawing process with regard to sheet metal working in a press is a method for producing a cuplike form from a sheet metal disc by holding it firmly between blankholding surfaces to prevent the formation of wrinkles while the punch travel produces the required shape. (2) The process of stretching a thermoplastic to reduce its cross-sectional area, thus creating a more orderly arrangement of polymer chains with respect to each other. (3) A misnomer for tempering (see *temper*).

Drawing (Pattern)

In foundry practice, removing a pattern from a mold or a mold from a pattern in production work.

Drawing Compound

(1) A substance applied to prevent *pickup* and *scoring* during drawing or pressing operations by preventing metal-to-metal contact of the work and die. It is also known as *die lubricant*. (2) In metalworking, a lubricant having extreme-pressure properties. See also *extreme-pressure lubricant*.

Drawing Out

A stretching operation resulting from a forging a series of upsets along the length of the workpiece.

Drawn Fiber

Fiber for reinforced plastics with a certain amount of orientation imparted by the joint process by which it is formed.

Drawn Shell

An article formed by drawing sheet metal into a hollow structure having a predetermined geometric configuration.

Dresser

(1) Various tools for hammering and forging operations. In most cases the final shaping operation is involved. (2) A tool with a head formed from a number of freely rotating ridged disks. It is held against a rotating grindstone to dress, that is, restore, the profile. (3) A tool used for *truing* and *dressing* a grinding wheel.

Dressing

(1) Cutting, breaking down, or crushing the surface of a grinding wheel to improve its cutting ability and accuracy. (2) Removing dulled grains from the cutting face of a grinding wheel to restore cutting quality. (3) (of weld) Machining or grinding of the surface of a weld to modify its profile. It may be undertaken for cosmetic or technical purposes, for example, to reduce the reinforcement height, to flush it with the parent surfaces, or to remove defects at the toe. See also *flushing*.

Driers

Materials used for increasing the rapidity of the drying of paints and varnishes. The chief function of driers is to absorb oxygen from the air and transfer it to the oil, thus accelerating its drying to a flexible film. They are in reality catalyzers. Excessive use of driers will destroy the toughness of the film and cause the pain to crack. Solutions of driers are called *liquid driers*; it is in this form that *paint driers* are most used. Certain oils, such as tung oil, have inherent drying properties and are classified as drying oils but not as driers. Driers may be oxides of metals, but the most driers are metallic salts of organic acids. *Manganese acetate* is a common paint drier. It is pinkish, crystalline powder soluble in water and in alcohol and is used in strengths of 6%, 9%, and 12% metal. *Sugar of lead*, used as a drier, is *lead acetate*, a white, crystalline powder with a faint acetic acid odor, also used as a mordant in textile printing. It is known as *plumbous acetate* and *Goulard's powder*. Lead oleate is a drier made by the action of a lead salt on oleic acid. It is used for thickening lubricants. Cobalt octoate, which has about 12% cobalt in combination with hexoic acid, is used as a drier. *Cobalt driers* are twice as rapid in drying power as manganese driers, but two rapid drying often makes a wrinkled film that is desirable for some finishes but not for others.

Naphthenate driers are metallic salts made with naphthenic acids instead of fatty-oil acids. They are usually more soluble in paint solvents, and since the naphthenic acids can be separated into a wide range of molecular weights by distillation, a wider variety of characteristics can be obtained. *Sodium naphthenate*, with 8.6% metal content, and *potassium naphthenate*, with 13.1%, are powders that are good bodying agents and emulsifiers as well as driers. *Tin naphthenate*, with 20% tin, may be added to lubricating oils as an antioxidant. *Mercuric naphthenate*, with 29% mercury, retards the growth of bacteria and molds when added to finishes. *Barium naphthenate*, with 22.6% barium, has binding and hardening properties and is used in adhesives and in linoleum.

Drift

(1) A flat piece of steel of tapering width used to remove taper shank drills and other tools from their holders. (2) A tapered rod used to force mismated holes into line for riveting or bolting. This is sometimes called a drift pin. (3) A small plain faced punch.

Drift test

A test in which a hard tapered mandrel is forced into a tube or hole to measure the materials capacity to distend without unacceptable damage.

Drill

(1) A rotary end-cutting tool used for making holes; it has one or more cutting lips and an equal number of helical or straight flutes for passage of chips and admission of cutting fluid. (2) A drill bit, that is, a rotating tool with cutting edges at its tip and flutes along the shank to lead away the swarf. (3) It is plunged axially into the workpiece to form, or drill, a cylindrical wall. (4) A drilling machine, that is, one in which a drill bit is mounted to drill holes in a workpiece.

Drill Rod

Tool-steel round rod made to a close degree of accuracy, generally not over or under 0.0005 in. (0.0127 mm) diameter size, and usually polished. It is employed for making drills, taps, reamers, and punches or for dowel pins, shafts, and rollers. Some mills also furnish square rods to the same accuracy under the name of drill rod. Common drill rod is of high-carbon steel hardened by quenching in water or in oil. The sizes are by the standard of drill gauges, with about 200 different diameters. The carbon content is usually from 0.9% to 1.05%, with 0.25 to 0.50 manganese, 0.10 to 0.50 silicon, and a maximum of 0.04 phosphorus or sulfur. It also comes in a high carbon with 1.50%–1.65% carbon and 0.15%–0.35% manganese. Drill rod can be obtained regularly in high-speed steels and in special alloy steels for dowel pins. *Needle wire* is a round tool-steel wire used for making needles, awls, and latch pins. *Needle tubing* for surgical instruments and radon implanters is stainless steel tubing. *Hypodermic tubing* is hard-drawn stainless steel tubing with a fine finish. Capillary tubing is also stainless steel but comes in lengths to 200 ft (61 m). *Drill steel*, for mine and quarry drills, comes in standard rounds, octagons, squares, and cruciform bars, solid or hollow, usually in carbon steel.

Drilling

Hole making with a rotary end-cutting tool having one or more cutting lips and one or more helical or straight flutes or tubes for the ejection of chips and the passage of a cutting fluid.

1. *Center drilling*: Drilling a conical hole in the end of a workpiece
2. *Core drilling*: Enlarging a hole with a chamfer-edged, multiple-flute drill
3. *Spade drilling*: Drilling with a flat blade drill tip
4. *Step drilling*: Using a multiple-diameter drill
5. *Gun drilling*: Using special straight flute drills with a single lip and cutting fluid at high pressures for deep hole drilling
6. *Oil hole or pressurized coolant drilling*: Using a drill with one or more continuous holes through its body and shank to permit the passage of a high pressure cutting fluid that emerges at the drill point and ejects chips

Drip Feed Lubrication

A system of lubrication in which the lubricant is supplied to the bearing surface in the form of drops at regular intervals. This is also known as *drop feed lubrication*.

Drip Test

Any test in which the material under test is subjected to a liquid drip of prescribed composition, size, and frequency to determine the response of the surface and any coating. The specimen is usually at a defined angle to the horizontal to ensure that the droplets run down the same track. The water drip test uses a drip of 0.05% sodium chloride in water.

Drive Fit/Driving Fit

This is the same as force fit except that more force, possibly in a press, may be required.

Drop

A casting imperfection due to a portion of the sand dropping from the cope or other overhanging sections of the mold.

Drop Etching

In metallography, placing of a drop of etchant on the polished surface.

Drop Forging/Hammer/Stamping

(1) Originally, a process in which the tup or hammerhead was allowed to fall under its own weight; now the term may be applied to fast acting vertical forges. The tup and anvil faces may be plain but usually carry dies to shape the component. (2) The forging obtained by hammering metal in a pair of closed dies to produce the form in the finishing impression under a *drop hammer*; forging method requiring special dies for each shape.

Drop Hammer

A term generally applied to forging hammers in which energy for forging is provided by gravity, steam, or compressed air. See also *airlift hammer*, *board hammer*, and *steam hammer*.

Drop Hammer Forming

A process for producing shapes by the progressive deformation of sheet metal in matched dies under the repetitive blows of a gravity-drop or power-drop hammer. The processes restricted to relatively shallow parts and thin sheet from approximately 0.6 to 1.6 mm (0.024 to 0.064 in.).

Droplet Erosion

Erosive wear caused by the impingement of liquid droplets on a solid surface. See also *erosion (erosive) wear*.

Drop Point

The temperature at which a drop falls from a grease through a specified orifice. The drop point does not necessarily represent the maximum operating temperature of a grease.

Drop-through

An undesirable sagging or surface irregularity, usually encountered when brazing or welding near the solidus of the base metal, caused by overheating with rapid diffusion or alloying between the filler metal and the base metal.

Dross

(1) The scum that forms on the surface of molten metal largely because of oxidation but sometimes because of the rising of impurities to the surface. (2) Oxide and other contaminants that form on the surface of molten solder.

Drum

A filler metal package type consisting of a continuous length of electrode wound or coiled within an enclosed cylindrical container.

Drum test

A test of the green strength of powder metallurgy compacts by tumbling them in a drum and examining the sharpness of the edges and corners.

Dry

To change the physical state of an adhesive on an adherend by the loss of solvent constituents by evaporation or absorption, or both. See also *cure* and *set*.

Dry and Baked Compression Test

An American Foundrymen's Society test for determining the maximum compressive stress that a baked sand mixture is capable of developing.

Dry Blend

Refers to a plastic molding compound containing all necessary ingredients mixed in a way that produces a dry, free-flowing, particulate material. This term is commonly used in connection with polyvinyl chloride molding compounds.

Dry Bond Adhesive

See *contact adhesive*.

Dry Coloring

Method commonly used by fabricators for coloring plastics by tumble blending uncolored particles of the plastic material with selected dyes and pigments.

Dry Copper

Tough pitch copper containing excessive quantities of copper oxide particles, hence of low ductility.

Dry Corrosion

See *gaseous corrosion*.

Dry Etching

In metallography, development of microstructure under the influence of gases.

Dry Fiber

In composites fabrication, a condition in which fibers are not fully encapsulated by resin during *pultrusion*.

Dry Friction

Friction that occurs between two bodies in the absence of lubrication. This term is inaccurate because it historically implies that there is no intentionally applied lubrication, when in fact solid lubrication conditions can be considered *dry*. Therefore, this term should not be used.

Dry Ice

Dry ice is a solid form of carbon dioxide, CO_2, which finds its largest application as a cooling agent in the transportation of perishables. It is nontoxic and noncorrosive and sublimes directly from a solid to a gas, leaving no residue. At atmospheric pressure it sublimes at $-78.7°C$, absorbing its latent heat of 573.1 kJ/kg. Including sensible heat absorption, the cooling effect per pound (kilogram) of dry ice is approximately 628.0 kJ at storage temperatures above $-9°C$ and 581.5 kJ at lower temperatures. Slabs of dry ice can easily be cut and used in shipping containers for frozen foods, in refrigerated trucks, and as a supplemental cooling agent in refrigerator cars.

The manufacture of CO_2 gases is a chemical process. The gas is liquefied by compressing it to 6.2–6.9 MPa gauge in three stages of reciprocating compressors and then condensing it in water-cooled condensers. The liquid is expanded to atmospheric pressure where the temperature is below the triple point $-56.6°C$.

Dry Joint (Soldered or Brazed)

A joint with inadequate bonding due to the failure of the solder or braze filler metal to wet the parent metal surface. It results from contamination or lacquer flux. See also *cold joint*.

Dry Laminate

A laminate containing insufficient resin for complete bonding of the reinforcement. See also *resin-starved area*.

Dry Lay-Up

Construction of a laminate by the layering of preimpregnated reinforcement (partly cured resin) in a female mold or on a male mold, usually followed by bag molding or autoclave molding. See also *vacuum bag molding*.

Dry Lubricant

See *solid lubricant*.

Dry Objective

Any microscope objective designed for use without liquid between the cover glass and the objective or, in the case of metallurgical objectives, in the space between objective and specimen.

Dry Permeability

In casting, the property of a molded mass of sand, bonded or unbonded, dried at −100°C to 110°C (−220°F to 230°F), and cooled to room temperature, which allows the transfer of gases resulting during the pouring of molten metal into a mold.

Dry Sand Casting

The process in which the sand molds are dried at above 100°C (212°F) before use.

Dry Sand Mold

A casting mold made of sand and then dried at 100°C (212°F) or above before being used. It is in contrast with *green sand mold*.

Dry Sliding Wear

Sliding wear in which there is no intentional lubricant or moisture introduced into the contact area. See also *unlubricated sliding*.

Dry Strength (Adhesives)

The strength of an adhesive joint determined immediately after drying under specified conditions or after a period of conditioning in the standard laboratory atmosphere. See also *wet strength*.

Dry Strength (Casting)

The maximum strength of a molded sand specimen that has been thoroughly dried at −100°C to 110°C (−220°F to 230°F) and cooled to room temperature. It is also known as dry bond strength.

Dry Tack

The property of certain adhesives, particularly nonvulcanizing rubber adhesives, to adhere on contact to themselves at a stage in the evaporation of volatile constituents, even though they seem dry to the touch. It is the synonym for aggressive tack.

Dry Winding

In composites fabrication, filament winding using preimpregnated roving, as differentiated from wet winding, in which unimpregnated roving is pulled through a resin bath just before being wound onto a mandrel. See also *wet winding*.

Dry-Bulb Temperature

The temperature of the air as indicated by an accurate thermometer, corrected for radiation if significant.

Dry-Film lubrication

Lubrication that involves the application of a thin film of solid lubricant to the surface or surfaces to be lubricated.

Drying

Removal, by evaporation, of uncombined water or other volatile substances from a ceramic raw material or product, usually expedited by low-temperature heating.

Drying Oils

Vegetable oils that are easily oxidized by exposure to air and thus suitable for producing a film in paints and varnishes, known as *paint oils*. The use of drying oils as the sole or main binder in alkyd coatings is steadily decreasing with the advent of water-based latex paints. Currently, it is limited to solvent-thinned exterior house paints and some metal paints. The oils are also used in oleoresinous varnishes and in the manufacture of synthetic resins for coating binders, epoxy ester resins, and oil-modified urethane resins. Small amounts are used in printing inks, linoleum, putty and caulking compounds, core oils, and hardboard. The best drying oils are those that contain the higher proportions of unsaturated acids, in which oxidation causes polymerization of the molecules. The drying of an oleoresinous varnish takes place in two stages. First, the reducer or solvent evaporates, leaving a continuous film composed of gums and drying oil. The drying oil is then oxidized by exposure, leaving a tough, hard skin. This oxidation is hastened by driers, but the drying oil itself is responsible for the film. The drying power of oils is measured by their *iodine value*, as their power of absorbing oxygen from the air is directly proportional to their power of absorbing iodine. Drying oils have typical iodine values about 140, semidry oils above 120, and nondrying oils below 120. Linseed oil is the most common of the drying oils, though tung oil and oiticica oil are faster in drying action. Linseed oil alone will take about 7 days to dry, but can be quickened to a few hours by the addition of driers. Linseed oil and other oils may be altered chemically to increase the drying power.

Conjugated oils are oils that have been altered catalytically by nickel, platinum, palladium, or carbon to give conjugated double bonds in place of isolated double bonds in the molecules of the fatty acids. The iodine value is 180, and the drying time is greatly reduced.

Castor oil, which is for drying properties, is dehydrated to form a good drying oil. Other methods are used to alter oils to increase the drying power, notably polymerization of the linoleic and some other acids in the oils; or oils may be fractionated and reblended to increase the percentage of acids that produce drying qualities.

Drying Temperature

The temperature to which an adhesive or an adherend, an adhesive in an assembly, or the assembly itself is subjected to dry the adhesive. The temperature attained by the adhesive in the process of drying (adhesive drying temperature) may differ from the temperature of the atmosphere surrounding the assembly (assembly drying temperature). See also *curing temperature* and *setting temperature*.

Drying Time

The period of time during which an adhesive or an adherend or an assembly is allowed to dry with or without the application of heat or pressure, or both. See also *curing time*, *joint conditioning time*, and *setting time*.

Dry-Running

In seals, running without liquid present at the seal surface.

Dry-Sand Rubber Wheel Test

In wear testing, the term used to describe a standard abrasive wear testing method in which a stream of dry quartz sand is passed between a rotating rubber wheel and a stationary test coupon that is held against it under specified normal force.

DSC

See *differential scanning calorimetry*.

DTA

See *differential thermal analysis*.

DTUL

See *deflection temperature under load*.

Dual-Metal Centrifugal Casting

Centrifugal castings produced by pouring a different metal into the rotating mold after the first metal poured has solidified. This is also referred to as *bimetal casting*.

Dual-Phase Steels

A new class of high-strength low-alloy steels characterized by a tensile strength value of approximately 550 MPa (80 ksi) and by a microstructure consisting of about 20% hard martensite particles dispersed in a soft ductile ferrite matrix. The term dual phase refers to the predominance in the microstructure of two phases, ferrite and martensite. However, small amounts of other phases, such as bainite, pearlite, or retained austenite, may also be present.

Duck

A strong, heavy cotton fabric employed for sails, awnings, tents, heavy bags, shoe uppers, machine coverings, and where a heavy and durable fabric is needed. It is woven plain, but with two threads together in the warp. It is made in various weights and is designated by the weight in ounces per running yard 22 in. (0.6 m) wide. It is marketed unbleached, bleached, or dyed in colors, and there are about 30 specific types with name designations usually for particular uses such as *sailcloth*. When woven with a colored stripe, it is called *awning dock*. *Russian duck* is a fine variety of *linen duck*. Large quantities of cotton duck are used for making laminated plastics and for plastics-coated fabrics and are then simply designated by the weight. *Belt duck*, for impregnated conveyor and transmission belts, is made in loosely woven, soft ducks and in hard-woven, fine-yarn hard fabric. *Conveyor belting* for foodstuff plants is usually a plastic fabric for cleanliness.

Hose duck, for rubber hose, is a soft-woven fabric of plied yarns. Canvas is duck of more open weave. The term is used loosely in the United States to designate heavy duck used for tarpaulins, bags, sails, and tents. But more properly, it is a heavy duck of square mesh weave more permeable than ordinary duck, such as the canvas used for paintings and for embroidery work. The word *duck* is from the Flemish *doeck*, meaning cloth, originally a heavy linen fabric. The word canvas is from the Latin cannabis, originally a coarse, heavy hempen cloth for tents.

Drill is a stout, twill cotton fabric used for linings and where a strong fabric lighter than duck is required. It differs from duck also in that it has a warp-flush weave that brings more warp than filling to the face of the cloth. It comes unbleached, bleached, or piece dyed, or it may be yarn dyed. It is made in various weights and is designated in ounces per yard, which is the same as duck. Tan-colored drill is called *khaki*. *Denim* is a heavy, twill-woven, warp-flush fabric usually lighter in weight than drill. The warp is yarn dyed. The filling is made with one black and one white yarn. It is much used for workers' clothing, and the lightweights for sportswear are called *jeans*. Denim is also used industrially where a tough fabric is needed. *Art denim*, in plain colors or woven with small figures, is used for upholstery.

Ductile Crack Propagation

Slow crack propagation that is accompanied by noticeable *plastic deformation* and requires energy to be supplied from outside the body. This is in contrast with *brittle crack propagation*.

Ductile Erosion Behavior

Erosion behavior having characteristic properties (i.e., considerable *plastic deformation*) that can be associated with *ductile fracture* of the exposed solid surface. A characteristic ripple pattern forms on the exposed surface at low values of angle of attack. This is in contrast with *brittle erosion behavior*.

Ductile Fracture

Fracture characterized by tearing of metal accompanied by appreciable gross plastic deformation and expenditure of considerable energy. This is in contrast with *brittle fracture*.

Ductile Iron

A *cast iron* that has been treated while molten with an element such as magnesium or cerium to induce the formation of free graphite as nodules or spherulites, which imparts a measurable degree of ductility to the cast metal. Ductile irons typically contain 3.0%–4.0% C, 1.8%–2.8% Si, 0.1%–1.0% Mn, 0.01%–0.1% P, and 0.01%–0.03% S. It is also known as nodular cast iron, spherulitic graphite cast iron, and spheroidal graphite iron.

Ductile iron is a graphite containing ferrous metal in which the graphite appears as rounded particles or spheroids rather than the usual plates, flakes, or clumps typical of other graphite ferrous metals.

In spherical form, the graphite exerts a minor influence on the properties of the steellike matrix because in that form the graphite-to-metal contact area is at a minimum. It is relatively economical to produce and has found applications in many industries.

Ductile iron is similar to other high-carbon, high-silicon alloys with respect to corrosion resistance, sliding wear, and machinability except insofar as the shape of the graphite alters these characteristics. The spheroidal form of the graphite, with lower surface area, alters corrosion resistance where the graphite is cathodic to the surrounding material. In comparison, with flake graphite iron, greater power will be required to remove metal in machining, while in metal-to-metal contact, the greater toughness resists removal of metal particles and therefore increases resistance to certain types of wear.

The spheroidal graphite irons have very good machining characteristics by virtue of their contained graphite and compare favorably with other materials at equivalent hardnesses and strengths. The soft annealed grade provides a strong, tough material that can be turned at very high speeds and feeds. Maximum turning speeds of well over 400 surface ft/min are attainable in this great. The cutting speed is, however, determined by the microstructure and maximum turning speed of the pearlitic grade reduced to about 150 surface ft/min and of the hardest oil-quenched grade to 100 surface ft/min. The chip produced in machining operations tends to be long and continuous, which complicates such operations as deep-hold drilling and internal broaching and reaming. For fast metal removal in these operations, special tool configurations are required to provide access for chip removal. Cutting fluids are recommended for all machining operations and, of course, production machining requires the use of carbide-tipped tools.

Welding

It is possible and practical to weld the spheroidal graphite irons by several welding processes. Most commonly used are the metallic arc and oxyacetylene processes. Due to the high carbon content and consequently high hardenability, the material must be handled very carefully to prevent excessive FeC from forming at the welding temperature. If maximum ductility and machinability are desired in the heat-affected zone, the welding operation should be followed by annealing. Filler rods, which undergo no phase changes, such as a 60% Ni–40% Fe rods, are the easiest and most practical to use. The operation requires the lowest practical input of amperage in the use of welding techniques that prevent overheating of the base metal in any location. The annealed grade, containing little or no combined carbon (to harden on cooling), is the grade on which welding is most easily performed.

Brazing

The joining of ductile iron castings to similar or dissimilar metals can be accomplished with silver or copper brazing filler metals, usually without resorting to special techniques in preparing the surfaces to be brazed. Overlay with hard surfacing alloys also presents no special problems.

Available Forms

Because its normal composition makes it eminently suited, ductile iron is essentially a casting alloy. In this condition, a wide variety of control is available on its properties through composition in heat treatment. Ferritic castings may be processed further by hot bending to shapes too difficult to cast conveniently. Tubular shapes may be formed by centrifugal casting and by the hot extruding of billets. The forging, rolling, extruding, and stamping processes may be used to produce bars, shafts, pipe, plate, angles, beams, gear blanks, etc., in order to utilize the good machinability and produced by explosive corrosion resistance of this high-silicon, high-carbon material. Certain shapes have been produced by explosive forming.

Applications

The incorporation into one material of high strength in all section sizes, excellent castability, wear resistance, machinability, and relatively simple control of properties suggests that such material would find application in many different industries and applications. It is lighter than nongraphic ferrous metals and, with its high strengths and good castability, permits the design of thin-section, lighter-weight machine components. The largest outages are used in automobile and diesel engines for crankshafts, rocker arms, and other engine and clutch components; in farm equipment and tractors for gears, sprockets, brackets, transmission casings, housings, and many structural components; in earthmoving machinery for the wheels, gears, and rope drums; and in heavy machinery for load-carrying components requiring good wear, easy machining, high strength, and antideflection properties. Other applications include fluid-handling devices such as valves and fittings; paper-mill machinery, roles, role heads, and gears; and steel mill rolls and mill equipment; as well as many others in a variety of industries.

Ductility

The capability of a metal to undergo plastic deformation during tensile loading. It is measured in the tensile test as the percentage reduction in cross-sectional area at the point of failure so material that can be drawn to a point before failing in tension has 100% ductility.

Ductility Transition, Ductile-to-Brittle Transition

(1) The temperature at which steel changes between ductile and brittle fractures about modes. (2) Some arbitrary level of energy absorption measured by impact testing. The actual energy level to be exceeded and, usually, temperature limitations are specified for this with most impact testing; the actual results do not provide a quantitative guide to surface performance. See *brittle fracture*.

Dullness

A lack of pultruded surface gloss or shine in plastic parts. This can be caused by insufficient cure locally or in large areas, resulting in the dull band created on a pultruded part within the die when the pultrusion process is interrupted briefly.

Dummy Block

In *extrusion*, a thick unattached disk placed between the ram and the billet to prevent overheating of the ram.

Dummy Cathode

(1) A *cathode*, usually corrugated to give variable current densities, that is plated at low current densities to preferentially remove impurities from a plating solution. (2) A substitute cathode that is used during adjustment of operating conditions.

Dummying

(1) An initial rough shaping in open dies prior to die forging. (2) Plating with *dummy cathodes*.

Dump Test

A test of malleability in which a columnar specimen is compressed longitudinally. It is typically applied to bar for rivets or bolts.

Duoplasmatron

A type of ion source in which a plasma created by an arc discharge is confined and compressed by a nonuniform magnetic field.

Duplex Alloys

Bearing alloys consisting of two phases, one much softer than the other.

Duplex Coating

See *composite plate*.

Duplex Grain Size

The simultaneous presence of two grain sizes in substantial amounts, with one grain size appreciably larger than the others. This is also termed as mixed grain size.

Duplex Microstructure

A two-phase structure.

Duplex Stainless Steels

(1) Stainless steels having a fine-grained mixed microstructure of ferrite and austenite with a composition centered around 26 Cr–6.5 Ni. The corrosion resistance of duplex stainless steels is like that of austenitic stainless steels. However, duplex stainless steels possess higher tensile and yield strengths and improve resistance to stress-corrosion cracking than their austenitic counterparts. (2) Grades of stainless steel containing high levels of chromium with nickel and, usually, small editions of molybdenum, for example, 22% chromium, 6% nickel, and 3% molybdenum. With further additions, for example, copper and tungsten, the steel may be termed superduplex: Tight control of composition and manufacturing variables produces a microstructure of austenite and ferrite in a specific ratio and distribution. These steels offer excellent corrosion resistance in certain circumstances. See also *steel* and *stainless steel*.

Duplexing

Any two-furnace melting or refining process. This is also called duplex melting or duplex processing.

Duplicate Measurement

A second measurement made on the same (or identical) sample material to assist in the evaluation of measurement variance.

Duplicate Sample

A second sample randomly selected from a population to assist in the evaluation of sample variance.

Duplicating

In machining and grinding, reproducing a form from a master with an appropriate type of machine tool, utilizing a suitable tracer or program-controlled mechanism.

Duralumin (Obsolete)

A term frequently applied to the class of age-hardenable aluminum–copper alloys containing manganese, magnesium, or silicon. This is also called Dural. It is in the proprietary range of aluminum alloys and sometimes used as a generic term for any aluminum alloy that can be straightened by heat treatment. See *precipitation hardening*.

Durometer Reading

An index that is used for ranking the relative hardness of elastomers. The durometer hardness test involves forcing a 30° tapered indenter into the surface of the specimen using calibrated loading springs. A dial gauge indicates the depth of penetration in durometer numbers, which are directly proportional to the load on the spring.

Durville Process

A casting process that allows rigid attachment of the mold in an inverted position above the crucible. The melt is poured by tilting the entire assembly, causing the metal to flow along a connecting *launder* and down the side of the mold. This minimizes contact with air and reduces the risk of dross and entrapment.

Dust

Specifically, a superfine metal powder having predominantly submicron-sized particles.

Dusting

(1) A phenomenon usually affecting carbon-based electrical motor brushes or other current-carrying contacts, wherein at low relative humidity or high applied current density, a powdery *dust* is produced during operation. (2) Applying a powder, such as sulfur to molten magnesium or graphite to a mold surface.

Duty

This specification giving the load, ambient temperature, and speed under which surfaces are required to move.

Duty Cycle

For electric welding equipment, the percentage of time that current flows during a specified period. In arc welding, the specified period is 10 min.

Duty Parameter

See *capacity number*.

Dwarf Width

A condition in which the crosswise (of the direction of pultrusion) dimension of a flat surface of a part is less than that which the die would normally yield for a particular plastic or composite. The condition is usually caused by a partial blockage of the pultrusion die cavity caused by buildup or particles of the composite adhering to the cavity surface. This condition is commonly called a lost edge, when the flat surface has a free edge that is altered by the buildup.

Dwell

(1) In forming plastics and composites, a pause in the application of pressure or temperature to a mold, made just before it is completely closed, to allow the escape of gas from the molding material. In filament winding, it is the time that the traverse mechanism is stationary while the mandrel continues to rotate to the appropriate point for the traverse to begin a new pass. In a standard autoclave cure cycle, it is an intermediate step in which the resin matrix is held at a temperature below the cure temperature for a specified period of time sufficient to produce a desired degree of staging. Used primarily to control resin flow. (2) The term is commonly used in the context of thermal fatigue testing where a load is applied, sustained for a period of hours or days, the dwell period, and then removed; this cycle is being repeated a number of times. The specimen is maintained at a high temperature during the dwell period so it sees not only a fatigue cycle but also suffers considerable creep.

Dwell Mark

A fracture surface marking that resembles a pronounced ripple mark, the presence of which indicates that the fracture paused at the location of the dwell mark for some indeterminable length of time. This is also known as arrest mark.

Dwell Time

In powder metallurgy, the time period during which maximum pressure is applied to a compact in cold pressing or hot pressing.

Dye Penetrant (Examination)

A liquid containing a dye capable of entering narrow cracks. Components suspected of containing cracks are immersed in, or swabbed or sprayed with, the penetrant. After the component has been wiped clean, any penetrant retained in a crack will seep out indicating the defect. Some dyes are fluorescent in the component when it is viewed, after application of the dye, under ultraviolet light. Sensitivity is usually improved by a developer, a powder, or quick-drying coating, applied immediately after wiping off excess penetrant, which absorbs and acts as a contrasting background for the dye.

Dyestuffs

Materials, also called *colorants*, used to color textiles, paper, leather, wood, or other products. They may be either natural or artificial. Many chemicals will stain and color other materials, but a product is not considered a dye unless it will impart a distinct color of some permanence to textiles. The natural dyestuffs may be mineral, animal, or vegetable, but the artificial dyes are derived mainly from coal-tar bases. Almost all naturally extracted dyes have been replaced by synthetic counterparts for commercial use. *Mineral dyes* now include ocher, chrome yellow, and Prussian blue. *Vegetable dyes* may be water solutions of woods, barks, leaves, fruits, or flowers. The buff and brown textile colors of early New England were made by boiling fresh green butternuts in water, while a dark-red dye was made by boiling the common red beet in water.

Synthetic dyes are mostly coal-tar or aniline colors. They are more intense, brighter, faster, and generally cheaper than natural dyestuffs. The dyes are complex chemicals, but they usually contain characteristic groups of atoms so that the color or change in color can be predicted.

Some of the synthetic dyes will color animal fibers well and not vegetable fibers, or vice versa, while some will color all fibers. As a result, it is possible to divide the bulk of the dyes into six classes, including *azoic dyes*. The *direct dyes* can be dyed directly, while others require a mordant. Some are permanent, or fast, while others are water soluble and will fade when the fabric is washed, where some may not be lightfast and will fade when exposed to light. Direct dyes usually have a week OH bond between the nitrogen in the dye and the fiber, usually cotton. In *reactive dyes*, the dye reacts with the fiber to produce both in OH and in oxygen linkage, which is the chlorine combining with the hydroxyl to form a strongly ether linkage. Such dyes are fast and very brilliant and are used for cotton, rayon, and nylon. *Acid dyes* contain a carboxylic or sulfuric acid group and operate best in an acid bath. They are used for drying protein fibers, such as wool, silk, and nylon, and sometimes for leather and paper. They are usually azo, *triarylmethane*, or anthraquinone complexes. Basic dyes are commonly amino and substituted amino compounds, such as triarylmethane or *xanthenes*. They are used for dyeing cotton with a mordant. *Vat dyes* are insoluble and are applied in the soluble colorless form and then reduced or oxidized to color. They usually have an anthraquinone or *indanthrene* structure and are solubilized by the reducing agent, a hydroxyl group, OH, diffusing into the fiber where it is fixed; the best-known example is *indigo*, the dye synonymous with the color of blue jeans, which has become one of the most important colorants because of the popularity of *denim* garments. Originally derived from plants, synthetic indigo dye now dominates the market. Synthesis, first commercialized by BASF of Germany more than 100 years ago, uses aniline formaldehyde and sodium cyanide as the raw materials. The process generates toxic wastewater. A biosynthesis process of Genecor International, Inc., however, is said to be environmentally benign. Pioneered by Amgen, Inc., and the University of Texas, it uses no petrochemical feedstocks, and biomass is the only waste product.

Color carriers, used to aid adherence of dyes to synthetic fibers, are usually chemicals that act as swelling agents to open the fiber structure, such as phenylphenol, benzoic acid, or dichlorobenzoic acid.

Dynamic

Moving or having high velocity. This is frequently used with high-strain-rate (<0.1 s^{-1}) testing of metal specimens. This is in contrast with *static*.

Dynamic Creep

Creep that occurs under conditions of fluctuating load or fluctuating temperature.

Dynamic Electrode Force

The force given in newtons (pounds force) between the electrodes during the actual welding cycle in making spot, seam, or projection welds by resistance welding.

Dynamic Friction

See *kinetic friction*.

Dynamic Hot Pressing

In powder metallurgy, a method of applying a vibrational load to the punches or die during hot pressing. See also *hot pressing* and *static hot pressing*.

Dynamic Load

An imposed force that is in motion, that is, one that may vary in magnitude, sense, and direction.

Dynamic Mechanical Measurement

A technique in which either the modulus and/or damping of a substance under oscillatory load or displacement is measured as a function of temperature, frequency, or time, or a combination thereof.

Dynamic Modulus

The ratio of stress to strain under cyclic conditions (calculated from data obtained from either free or forced vibration tests, in shear, compression, or tension).

Dynamic Seal

A seal that has rotating, oscillating, or reciprocating motion between its components, as opposed to stationery-type seal such as a gasket.

Dynamic Viscosity

See *viscosity*.

Dysprosium

A metallic element, in the rare earth group.

E

E

The elastic modulus, the ratio of extension to original length. See *tensile test*.

Ears/Earing

(1) The undesirable deeply scalloped edge of deep-drawn components due to preferred orientation. A typical cup-shaped component suffering from the defect will have four equally spaced deep scallops separating the ears that are aligned along and transverse to the rolling direction of the original sheet. (2) The formation of ears or scalloped edges around the top of a drawn shell, resulting from directional differences in the plastic-working properties of rolled metal, with, across, or at angles to the direction of rolling.

Earth

The nonpowered return path of an electrical circuit. It may not be via the soil.

Earthenware

A glazed or unglazed nonvitreous clay-based ceramic *whiteware*.

Eccentric

(1) Not concentric. Most to be produced by piercing and similar processes is significantly eccentric. (2) A disc set on a shaft with its center offset from the center of the shaft. (3) The offset portion of the driveshaft that governs the stroke or distance the crosshead moves on a mechanical or manual shear.

Eccentric Gear

A main press-drive gear with an eccentric (s) as an integral part. The unit rotates about a common shaft, with the eccentric transmitting the rotary motion of the gear into the vertical motion of the slide through a connection.

Eccentric Press

A *mechanical press* in which an eccentric, instead of a crankshaft, is used to move the *slide*.

Eccentricity

In *journal bearings*, the radial displacement of the journal center from the center of the bearing liner.

Eccentricity Ratio

In a bearing, the ratio of the eccentricity to the radial clearance.

ECM

An abbreviation for *electrochemical machining*.

Eddy Current Testing

A form of nondestructive testing in which eddy currents are induced in the component under test. The basic principle is that the probe that induces the eddy current moves relative to the component, or vice versa. Any changes in the current induced in the component are detected by the probe and transmitted to the monitoring equipment to indicate the location and size of the effects.

Edge Cracks

Shallow cracks in hardened steel containing high residual surface stresses, produced by etching in an embrittling acid.

Edge Dislocation

A line defect with its Burger's Vector perpendicular to the line of dislocation. See *dislocation*.

Edge Distance

The distance from the edge of a bearing test specimen to the center of the hole in the direction of applied force. See also *bearing test*.

Edge Distance Ratio

In a *bearing* test, the distance from the center of the bearing hole to the edge of a specimen in the direction of the principal stress, divided by the diameter of the hole.

Edge Effect (Thermal Spraying)

Loosening of the bond between the sprayed material and the base material at the edges, due to stresses set up in cooling.

Edge Joint (Adhesive Bonding)

A joint made by bonding the edge faces of two adherends.

Edge Joint (Welding)

A joint made between the edges of sheet or plate components particularly when they meet at an acute angle, usually less than 30°.

Edge Preparation

Surface preparation on the edge of a member for welding.

Edge Stability

An indicator of strength in a green powder metallurgy compact, as may be determined by tumbling in a drum. See also *drum test*.

Edge Strain

Transverse strain lines or Lüders lines ranging from 25 to 300 mm (1–12 in.) in the edges of cold rolled steel sheet or strip. See also *Lüders lines*.

Edge Strength

The resistance of the sharp edges of a powder metallurgy compact against abrasion, as may be determined by tumbling in a drum. See also *drum test*.

Edge Tool

A tool with one or more sharpened cutting edges.

Edge Weld

A weld in an *edge joint*.

Edge/Edging Rolls

Rolls that control the quality, dimensions, and shape of plate being rolled.

Edge-Flange Weld

A flange weld with two members flanged to the location of welding.

Edger (Edging Impression)

The portion of a die impression that distributes metal during forging into areas where it is most needed in order to facilitate filling the cavities of subsequent impressions to be used in the forging sequence. See also *fuller (fullering impression)*.

Edge-Trailing Technique

In metallography, a unidirectional motion perpendicular to and toward one edge of the specimen during abrasion or polishing used to improve edge retention.

Edging

(1) In sheet metal forming, reducing the flange radius by retracting the forming punch by a small amount after the stroke but before the release of the pressure. (2) In rolling, the working of metal in which the axis of the roll is parallel to the thickness dimension. It is also called edge rolling. (3) The forging operation of working a bar between contoured and dies while turning it 90° between blows to produce a varying rectangular cross-section. (4) In a forging, removing flash that is directed upward between dies, usually accomplished using a lathe.

EDM

Abbreviation for *electrical discharge machining*.

Effective Atomic Number

For an element, the number of protons in the nucleus of an atom. For mixtures, an effective atomic number can be calculated to represent a single element that would have attenuation properties identical to those of the mixture.

Effective Crack Size (α_e)

The *physical crack size* augmented for the effects of crack-tip plastic deformation. Sometimes, the effective crack size is calculated from a measured value of a physical crack size plus a calculated value of a plastic-zone adjustment. A preferred method for calculation of effective crack size compares compliance from the secant of a load-deflection trace with the elastic compliance from a calibration for the type of specimen.

Effective Draw

The maximum limits of forming depth that can be achieved with a multiple-action press, sometimes called maximum draw or maximum depth of draw.

Effective Leakage Area

The orifice flow area that will result in the same calculated flow for a given pressure drop as is measured for the seal in question. This concept is useful when comparing the leakage performance of seals of different sizes and designs, and of CEOs operating under different conditions.

Effective Length of Weld

The length of weld throughout which the correctly proportional cross-section exists. In a curve weld, it is measured along the axis of the weld.

Effective Modulus

The elastic modulus in circumstances where it is time-dependent or nonlinear. See also *secant modulus* and *tangent modulus*.

Effective Rake

The angle between a plane containing a tooth face and the axial plane through the tooth point as measured in the direction of chip flow through the tooth point. Thus, it is the rake resulting from both the cutter configuration and direction of chip flow.

Effective Throat

The minimum distance from the root of a weld to its face. See also *joint penetration*.

Effective Yield Strength

An assumed value of uniaxial yield strength that represents the influence of plastic yielding on fracture test parameters.

E-Glass

A family of glasses with a calcium aluminoborosilicate composition and a maximum alkali content of 2.0%. A general purpose fiber that is most often used in reinforced plastics, and is suitable for electrical laminates because of its high resistivity. It is also called electric glass. See also *glass fiber*.

Eight-Harness Satin

A type of fabric weave. The fabric has a seven-by-one weave pattern in which a filling thread floats over seven warp threads and then under one. Like the crowfoot weave, it looks different on one side than on the other. This weave is more pliable than any of the others and is especially adaptable to forming around compound curves, such as on radomes.

Ejection

Removal of a powder metallurgy compact after completion of pressing, whereby the compact is pushed through the die cavity by one of the punches. Also called *knockout*.

Ejection Mark

A surface mark on a plastic part caused by the ejector pin when it pushes the part out of the molded cavity.

Ejector

A device mounted in such a way that it removes or assists in removing a focused part from a die.

Ejector Half

The movable half of a die-casting die containing the ejector pins.

Ejector Punch

See *knockout punch*.

Ejector Rod

A rod used to push out a formed piece.

Elastic after Effect

The elastic strain that develops with time in a component subject to a steady stress below its elastic limit. Compared with the immediately observed strain, it is very small. After removal of the stress, a similar small amount of strain remains but disappears with time.

Elastic Calibration Device

A device for use in verifying the load readings of a testing machine consisting of an elastic member(s) to which loads may be applied, combined with a mechanism or device for indicating the magnitude for a quantity proportional to the magnitude of deformation under load.

Elastic Compliance

A condition under which two bodies in contact, which are subjected to a force, undergo small elastic displacement without slip.

Elastic Constants

The factors of proportionality that relate elastic displacement of a material to applied forces. See also *bulk modulus of elasticity*, *modulus of elasticity*, *Poisson's ratio*, and *shear modulus*.

Elastic constants are the Tensile Elastic Modulus (Young's modulus) (symbol E), Compression Modulus (K), Shear Modulus (G), and Poisson's Ratio (ð).

Elastic Constraint

See *constraint*.

Elastic Deformation

A change in dimensions directly proportional to and in phase with an increase or decrease in applied force.

Elastic Electron Scatter

The scatter of electrons by an object without loss of energy, usually in interaction between electrons and atoms.

Elastic Energy

The amount of energy required to deform a material within the elastic range of behavior, neglecting small heat losses due to internal friction. The energy absorbed by a specimen per unit volume of material contained within the gauge length being tested. It is determined by measuring the area under the stress–strain curve up to a specified elastic strain. See also *modulus of resilience* and *strain energy*.

Elastic Hysteresis

The phenomenon whereby the stress–strain curves on loading and unloading do not coincide other than at maximum and minimum stress. It reflects the energy consumed in the elastic strain process and is a consequence of the elastic after effect. See *tensile test*. More appropriately termed *mechanical hysteresis*.

Elastic Limit

The maximum stress that a material is capable of sustaining without any permanent strain (deformation) remaining upon complete release of the stress. A material is said to have passed its elastic limit when the load is sufficient to initiate plastic, or nonrecoverable, deformation. See also *proportional limit*.

Elastic Modulus

The ratio of stress to strain in the elastic range. See tensile test. Same as *modulus of elasticity*.

Elastic Ratio

Yield point divided by *tensile strength*.

Elastic Recovery

(1) The fraction of a given deformation that behaves elastically. A perfectly elastic material has an elastic recovery of 1; a perfectly

plastic material has an elastic recovery of zero. (2) In hardness testing, the shortening of the original dimensions of the indentation upon release of the applied load.

Elastic Resilience

The amount of energy absorbed in stressing a material up to the elastic limit; or the amount of energy that can be recovered when stress is released from the elastic limit.

Elastic Scattering

Collisions between particles that are totally described by conservation of energy and momentum. Contrast with *inelastic scattering*.

Elastic Strain

See *elastic deformation*.

Elastic Strain Energy

The energy expended by the action of external forces in deforming a body elastically. Essentially, all the work performed during elastic deformation is stored as elastic energy, and this energy is recovered upon release of the applied force.

Elastic True Strain (ε_e)

Elastic component of the *true strain*.

Elastic Waves

Mechanical vibrations in an elastic medium.

Elasticity

The property of a material by virtue of which deformation caused by stress disappears upon removal of the stress. A perfectly elastic body completely recovers its original shape and dimensions after release of stress.

Elastohydrodynamic Lubrication

A condition of lubrication in which the friction and film thickness between two bodies in relative motion are determined by the elastic properties of the bodies, in combination with the viscous properties of the lubricant at the prevailing pressure, temperature, and rate of shear. See also *boundary lubrication*, *plastohydrodynamic lubrication*, and *thin-film lubrication*.

Elastomers

Elastomers and rubber are differentiated from polymers by the mechanical property of returning to their original shape after being stretched to several times their length. The rubber industry differentiates between the terms *elastomer* and *rubber* on the basis of how long a deformed material sample requires to return to its approximate original size after a deforming force is removed, and on its extent of recovery. The American Society for Testing and Materials (ASTM) defines an elastomer as "a polymeric material which at room temperature can be stretched to at least twice its original length and upon immediate release of the stress will return quickly to approximately its original length." ASTM D 1566 is more specific and quantitative in defining rubber as "a material that is capable of recovering from large deformations quickly and forcibly ... (and which), and its modified state, free of diluents, retracts within one minute to less than 1.5 times its original length after being stretched at room temperature to twice its length and held for one minute before release."

The major distinguishing characteristic of elastomers is their great extensibility and high-energy storing capacity. Unlike many metals, for example, which cannot be strained more than a fraction of 1% without exceeding their elastic limit, elastomers of usable elongation is up to several hundred percent. Also, because of their capacity for storing energy, even after they are strained several hundred percent, virtually total recovery is achieved.

Before World War II, almost all rubber was natural. During the war, synthetic rubbers began to replace the scarce natural rubber, and since that time production of synthetics has increased and until now their use far surpasses that of natural rubber. There are thousands of different elastomer compounds. Not only are there many different classes of elastomers, but individual types can be modified with a variety of additives, fillers, and reinforcements. In addition, curing temperatures, pressures, and processing methods can be varied to produce elastomers tailored to the needs of specific applications.

Today, the term *rubber* means any material capable of extreme deformability, with more or less complete recovery upon removal of the deforming force. Synthetic materials such as neoprene, nitrile, styrene-butadiene (SBR), sometimes also called Buna S, and GR-S and butadiene rubber are now grouped with natural rubber. These materials serve engineering needs and fields dealing with shock absorption, noise and vibration control, sealing, corrosion protection, abrasion protection, friction production, electrical and thermal insulation, waterproofing, confining other materials, and load bearing.

Of all the available choices, SBR dominates the field, accounting for approximately one-half of all rubber—natural and synthetic—use in the United States. The demand for SBR has been responsible for the building of a massive production capability for this material. More than half of SBR production goes into passenger-car tires in the United States. Natural rubber is used almost exclusively in more demanding areas such as truck, bus, aircraft, and off-highway tires.

In the raw-material or crude stage, elastomers are thermoplastic. There are roughly 20 major classes of elastomers. For example, hard rubbers, which have the highest cross-linking of the elastomers, in many respects are similar to phenolics. In the unstretched state, elastomers are essentially amorphous because the polymers are randomly entangled and there is no special preferred geometrical pattern present. However, when stretched, the polymer chains tend to straighten and become aligned, thus increasing in crystallinity. This tendency to crystallize when stretched is related to the strength of an elastomer. Thus, as crystallinity increases strength also tends to increase (Table E.1). Neoprene, also known as chloroprene, has the distinction of being the first commercial synthetic rubber. It is chemically and structurally similar to natural rubber, and its mechanical properties are also similar. Its resistance to oils, chemicals, sunlight, weathering, aging, and ozone is outstanding. Also, it retains its properties at temperatures up to 121°C and is one of the few elastomers that does not support combustion, although it is consumed by fire. In addition, it has excellent resistance to permeability by gases, having about 1/4–1/10 the permeability of natural rubber, depending on the gas. Although it is slightly inferior to rubber in most mechanical properties, neoprene has superior resistance to compression set, particularly at elevated temperatures. It can be used for low-voltage insulation, but is relatively low in dielectric strength. Typical products made of chloroprene elastomers are heavy-duty conveyor belts,

TABLE E.1
Properties of Thermoplastic Elastomers

ASTM Test	Property	Polyurethane	Co-Polyester	Styrene Block Copolymer	Olefin
Physical					
D792	Specific gravity	1.10–1.24	1.15–1.25	0.93–1.10	0.88–1.0
D792	Specific volume (in.^{3}/lb)	26.5–22.0	—	37.4–27.5	31.5–27.7
D570	Water absorption. 24 h ½ in. ink (%)	0.1–0.3	0.3–1.6	0.19–0.39	0.01–0.1
Mechanical					
D638	Tensile strength (psi)	4.000–9.000	4.490–7.600	300–5.000	650–4.000
D638	Elongation (%)	225–570	250–800	250–1.350	180–600
D638	Tensile modulus (10^3 psi)	0.7–35	7–75	0.8–80	0.8–34
D790	Flexural strength (psi)	600–1.000	—	—	—
D790	Flexural modulus (10^3 psi)	—	5–75	4–150	1.5–20
D256	Impact strength Izod (ft. lbin. of notch)	No break	No break	No break	No break
D785	Hardness. Shore	65A–80D	35–72D	28–95A	60A–60D
Thermal					
C177	Thermal conductivity (10^{-4} cal cm/s cm^2 °C)	5	3.6–4.5	3.6	4.5–5
D696	Coefficient of thermal expansion (10^{-3} m/in. °F)	5.6–11	2.0–2.8	7.2–7.6	7.2–9.4
D648	Deflection temperature (°F)				
	At 264 psi	90	115–122	<75	85–90
	At 66 psi	145	129–284	95	120–180
Electrical					
D149	Dielectric strength (V/mil)	450–500	400–460	420–520	400–800
D150	Dielectric constant				
	At 1 kHz	6.7–7.5	3.9–5.1	2.5–3.4	2.2–3.1
D150	Dissipation factor				
	At 1 kHz	0.050–0.060	0.008–0.02	0.001–0.003	0.0006–0.003
D257	Volume resistivity (Ω cm)				
	At 73°F. 50% RH	2×10^{11}	1.1×10^{12}	2.5×10^{16}	$>10^{14}$
D495	Arc resistance (s)	122	—	95	120

Source: Machine Design Basics Eng. Design, p. 741, June 1993. With permission. Also Schwartz, M., *Encyclopedia of Materials, Parts, and Finishes*, 2nd edn., CRC Press LLC, P 198, 2002-ISBN 1-56676-661-3.

V belts, hose covers, footwear, break diaphragms, motor mounts, rolls, and gaskets. Butyl rubbers, also referred to as isobutylene–isoprene elastomers, are copolymers of isobutylene and 1%–3% isoprene. They are similar in many ways to natural rubber and are one of the lowest priced synthetics. They have excellent resistance to abrasion, tearing, and flexing. They are noted for low gas and air permeability (about 10 times better than natural rubber), and for this reason make a good material for tire inner tubes, hose, tubing, and diaphragms. Although butyls are non-oil-resistant, they have excellent resistance to sunlight and weathering, and generally have good chemical resistance. They also have good low-temperature flexibility and heat resistance up to around 149°C; however, they are not flame-resistant. They generally have lower mechanical properties such as tensile strength, resilience, abrasion resistance, and compression set, than the other elastomers. Because of their excellent dielectric strength, they are widely used for cable insulation, encapsulating compounds, and a variety of electrical applications. Other typical uses include weather stripping, coated fabrics, curtain wall gaskets, high-pressure steam hoses, machinery mounts, and seals for food jars and medicine bottles.

Isoprene is a synthetic natural rubber. It is processed like natural rubber and its properties are quite similar, although isoprene has somewhat higher extensibility. Like natural rubber, its notable characteristics are low hysteresis, low heat buildup, and high tear resistance. It also has excellent flow characteristics, and is easily injection-molded. Its uses complement those of natural rubber, and its good electrical properties plus low moisture absorption make it suitable for electrical insulation. Polyacrylate elastomers are based on polymers of butyl or ethyl acrylate. They are low-volume use, specialty elastomers, chiefly used in parts involving oils (especially sulfur-bearing) at elevated temperatures of 149°C and even as high as 204°C. A major use is for automobile transmission seals. Other oil-resistant uses are gaskets and O rings. Mechanical properties such as tensile strength and resilience are low. Further, except for recent new formulations, they lose much of their flexibility below −23°C. The new grades extend low-temperature service to −40°C. Polyacrylates have only fair dielectric strength, which improves, however, at elevated temperatures.

Nitrile elastomers, or NBR rubbers, known originally as Buna N, are copolymers of acrylonitrile and butadiene. They are principally known for their outstanding resistance to oil and fuels at both normal and elevated temperatures. Their properties can be altered by varying the ratio of the two monomers.

Most commercial grades range from 20% to 50% acrylonitrile. Those at the high end of the range are used where maximum resistance to fuels and oils is required, such as in oil well parts and fuel hose. Low-acrylonitrile grades are used where good flexibility at low temperatures is of primary importance. Medium-range types, which are the most widely used, find applications between these extremes. Typical products are flexible couplings, printing blankets, rubber rollers, and washing machine parts. Nitriles as a group are low in most mechanical properties. Because they do not crystallize appreciably when stretched, their tensile strength is low, and resilience is roughly one-third to one-half that of natural rubber. Depending on acrylonitrile content, low-temperature brittleness occurs at from −26°C to −60°C. Their electrical insulation quality varies from fair to poor. Polybutadiene elastomers are notable for their low-temperature performance. With the exception of silicone, they have the lowest brittle or glass transition temperature, −73°C, of all the elastomers. They are also one of the most resilient, and have excellent abrasion resistance. However, resistance to chemicals, sunlight, weathering, and permeability by gases is poor. Some uses are shoe heels, soles, gaskets, and belting. They are also often used in blends with other rubbers to provide improvements in resilience, abrasion resistance, and low-temperature flexibility.

Polysulfide elastomer is rated highest in resistance to oil and gasoline. It also has excellent solvent resistance, extremely low gas permeability, and good aging characteristics. Thus, it is used for such products as oil and gasoline hoses, gaskets, washers, and diaphragms. Its major use is for equipment and parts in the coating production and application field. It is also widely applied in liquid form in sealants for the aircraft and marine industries. Its mechanical properties, including strength, compression set, and resilience, are poor. Although it is poor in flame resistance, it can be used in temperatures up to 121°C. Ethylene–propylene elastomers, or EPR rubber, are available as copolymers and terpolymers. They offer good resilience, flexing characteristics, compression-set resistance, and hysteresis resistance, along with excellent resistance to weathering, oxidation, and sunlight. Although fair to poor in oil resistance, their resistance to chemicals is good. Their maximum continuous surface temperature is around 177°C. Typical applications are electrical insulation, footwear, auto hose, and belts. Urethane elastomers are copolymers of diisocyanate with their polyester or polyether. Both are produced in solid gum form and viscous liquid. With tensile strengths above 34 MPa and some grades approaching 49 MPa, urethanes are the strongest available elastomers. They are also the hardest, and have extremely good abrasion resistance. Other notable properties are low compression set, and good aging characteristics and oil and fuel resistance. The maximum temperature for continuous use is under 93°C, and their brittle point ranges from −51°C to −68°C. Their largest field of application is for parts requiring high wear resistance or strength. Typical products are forklift truck wheels, airplane tail wheels, shoe heels, bumpers on earth-moving machinery, and typewriter damping pads. Chlorosulfonated polyethylene elastomer, commonly known as *Hypalon*, contains about one-third chlorine and 1%–2% sulfur. It can be used by itself or blended with other elastomers. *Hypalon* is noted for its excellent resistance to oxidation, sunlight, weathering, ozone, and many chemicals. Some grades are satisfactory for continuous service at temperatures up to 177°C. It has moderate oil resistance. It also has unlimited colorability. Its mechanical properties are good but not outstanding, although abrasion resistance is excellent. *Hypalon* is frequently used in blends to improve oxidation and ozone resistance. Typical uses are tank linings, high-temperature conveyor belts, shoe soles and heels, seals, gaskets, and spark plug boots. Epichlorohydrin elastomers are noted for their good resistance to oils, and excellent resistance to ozone, weathering, and intermediate heat. The homopolymer has extremely low permeability to gases. The copolymer has excellent resistance at low temperatures.

Both have low heat buildup, making them attractive for parts subjected to repeated shocks and vibrations. Fluorocarbon elastomers, fluorine-containing elastomers, like their plastic counterparts, are highest of all the elastomers in resistance to oxidation, chemicals, oils, solvents, and heat, in they are also the highest in price. They can be used continuously at a temperature of 127°C and do not support combustion. Their brittle temperature, however, is only −23°C. Their mechanical and electrical properties are only moderate. Unreinforced types have tensile strengths of less than 13 MPa and only fair resilience. Typical applications are brake seals, O rings, diaphragms, and hose. The phosphonitrile plastics and elastomers have high elasticity and high-temperature resistance. They are derived from chlorophosphonitrile, or phosphonitrilic chloride, are highly resistant to oils and solvents, and remain flexible and serviceable at temperatures from −57°C to 177°C.

Viton is a vinylidene fluoride hexafluoropropylene tetrafluoroethylene copolymer as well as *Fluorel* with extra high resistance to solvents, hydrocarbons, steam, and water.

Silicone elastomers are polymers composed basically of silicon and oxygen atoms. There are four major elastomer composition groups. In terms of application, silicone elastomers can be divided roughly into the following types: general-purpose, low-temperature, high-temperature, low-compression set, high-tensile, high-tear, fluid-resistant, and room-temperature vulcanizing. All silicone elastomers are high-performance, high-price materials. The general-purpose grades, however are competitive with some of the other specialty rubbers, and are less costly than the fluorocarbon elastomers. Silicone elastomers are the most stable group of all the elastomers. They are outstanding in resistance to high and low temperatures, oils, and chemicals. High-temperature grades have maximum continuous service temperatures up to 316°C; low-temperature grades have glass transition temperatures of −118°C. Electrical properties, which are comparable to the best of the other elastomers, are maintained over a temperature range from −73°C to over 260°C. However, most grades have relatively poor mechanical properties. Tensile strength runs only around 8 MPa. However, grades have been developed with much improved strength, tear resistance, and compression set. Fluorosilicone elastomers have been developed that combine the outstanding characteristics of the fluorocarbons and silicones. However, they are expensive and require special precautions during processing. A unique characteristic of one of these elastomers is its relatively uniform modulus of elasticity over a wide temperature range and under a variety of conditions. Silicone elastomers are used extensively in products and components where high-performance is required. Typical uses are seals, gaskets, O rings, insulation for wire and cable, and encapsulation of electronic components.

Elastomeric Linings

Elastomeric lining materials are available in natural rubber, GR-S, butyl, neoprene, nitrile, and Hypalon synthetic rubbers as well as in polyvinyl chloride (PVC) and polyvinylidene chloride (**saran**) plastics and fluorocarbon rubbers. These basic materials, when formulated and processed into linings, are normally sold unapplied as heavy sheets at 4.7–6.4 mm thick.

Elastomeric linings are manufactured by plying up thin calendered films of 0.038–0.076 mm thickness to the full thickness of the lining.

They are applied in the soft and plastic uncured state to the base metal and are cured in place.

Natural Rubber

About 70% of the applied linings are estimated to be based on natural rubber. This is due to the fact that natural rubber can be formulated in an extremely wide range of physical properties. Linings from natural rubber are classified as soft (sulfur content at 2%–3%), semi-hard (sulfur content at 15%–30%), and hard (sulfur content at 30%–45%). As the sulfur content is increased, general chemical resistance increases from good to excellent, but at the same time the lining also becomes harder and less flexible.

Linings from natural rubber are among the lowest in cost, and all natural rubber linings can be applied without difficulty. Even hard rubber can be used satisfactorily on large vessels if the vessel is designed in sections that are then bolted together and the joints covered with lap straps. With proper usage and suitable repair when damaged, they can be expected to give service for 12–15 years or more, which makes for a very low overall cost.

Natural rubber linings give excellent service for many water solutions of chemicals and can be used for practically all plating solutions (except chromium). Hard rubber linings will also withstand paraffenic oils, greases, and fats at moderate temperatures. The linings will not withstand the solvating action of aniline, benzene, carbon disulfide, carbon tetrachloride, and other chlorinated or halogenated hydrocarbons. They are not resistant to the chemical attack of the oxidizing acids, HNO_3 above 10% or H_2SO_4 acid above 50% at 71°C.

GR-S

With the exception of gum (nonfilled) rubber linings, GR-S can be used to duplicate practically all of the linings that are now made from natural rubber and for the same type of service.

Butyl Rubber

Butyl rubber linings are available in physical characteristics approaching those of the soft natural rubbers.

Butyl linings have a higher degree of general chemical resistance than natural rubber, and are suitable for weak oxidizing acids and for organic chemicals that deteriorate soft rubber linings over a long period of time (e.g., fats, greases, and soaps).

Butyl linings are suitable for use above 71°C and for continued use at 149°C as a lining material. These characteristics allow butyl use in tank trucks and railroad tank cars where hot solutions are to be transported, and where the empty car can be returned without fear of rupture of the lining by shock in cold climates. Also, butyl is used where the continual vibration and shock encountered in railroad hauling illuminates the semihard and hard rubbers that would normally be used for such solutions. HF_3 and ethylene chlorohydrin can be handled with butyl linings.

Neoprene

Neoprene linings have gained acclaim for their resistance to caustic solutions and are recommended for use with fluorides and phosphates. Flexible neoprene linings also have good resistance to a large group of chemicals such as all fatty acids, oils, greases, and aliphatic hydrocarbons; they are superior to butyl rubber in this respect. Such resistance enables the neoprene linings to give good service and chemical processes where mixtures of acids with kerosene, oils, or other organic materials are involved.

Because neoprene is more expensive than natural rubber, it is normally used only where it gives service superior to natural rubber. Neoprene linings are specifically used where flexibility is needed above 71°C.

Nitrile Rubber

Nitrile rubber linings, because of their relatively high cost, are used only with specific organic solvents, where organic solvents are used in conjunction with acids and other water-based corrosive systems and where superior oil resistance is needed. The general chemical resistance of nitrile rubbers is only fair, but their resistance to oils is good.

Hypalon

Hypalon linings are flexible and are useful because of their resistance to heat as well as to oxidizing chemicals. Hypalon linings are especially suitable for use with hypochlorites and H_2SO_4; cold solutions of H_2SO_4 up to 90% exhibit very little attack upon Hypalon. It is probably the best lining available for hot solutions of H_2SO_4 below 50% concentration.

Polyvinyl Chloride (PVC)

PVC linings are available both in the flexible and rigid state. Flexible PVC is used almost exclusively where H_2CrO_3 solutions are involved in chromium plating, and where HNO_3 or mixtures of HNO_3 and HF acids are involved in the electropickling of stainless steels. No other linings are suitable for these operations. PVC linings are also suitable where mixtures of water solutions of corrosive chemicals with oils or specific organic solvents are involved.

Similar to PVC are the saran (polyvinylidene chloride) linings, which look like and are applied in much the same manner as PVC. They are good for use with HCl.

Application

Elastomeric linings are applied in much the same manner as wallpaper. In the application of rubber linings, it is important that the lining be applied when the rubber is uncured and is plastic and workable.

PVC and polyvinylidene chloride linings are applied similarly to rubber linings but they are not cured. There are applied with butt joints covered with a narrow sealing strip.

Uses

Elastomeric linings can be compounded to provide specific corrosion resistance for many applications and can be applied to almost any shape of equipment or surface under almost any set of conditions and at any location.

Those industries using the highest volume of linings include:

1. Chemical process industry, where practically every type of lining is utilized for a wide variety of chemicals
2. The steel and aluminum industries, where the linings are used in pickling and anodizing operations
3. In the automotive and appliance industries, for plating
4. In ordnance, where missile fuels are involved
5. The food processing industry, where specifically compounded linings are used in smaller volume.

Elastomeric Tooling

A tooling system that uses the thermal expansion of rubber materials to form reinforced plastic or composite parts during cure.

Electrical Ceramics

Electrical ceramics, like all other ceramics, consist of randomly oriented small crystallites, bonded together by either a glassy matrix or by close interlocking of one or several crystalline phases. The atomic structures of the crystalline and glassy phases determine the physical properties of the finished ceramic.

E

Raw Materials

Conventional raw materials for ceramic production are earthy, natural minerals, used in finely-ground form, such as clays, talc, feldspar, and flint. These are used for the production of electrical insulators, often carefully selected and refined from gross impurities by flotation or filtering processes. In many instances, these naturally occurring minerals are replaced by synthesized inorganic compounds, which are prepared from various oxides. This may be done by melting, recrystallizing, and grinding, or more frequently by high-temperature solid-state reactions.

These compounds may be composed of oxides of practically any metal, singly or in combination, or they may be refractory compounds such as carbides and silicides. The use of these inorganic synthetic raw materials has opened the field of likely ceramic compositions to practically infinite numbers, and the possibility of variation in physical and electrical properties is therefore equally staggering.

Fabrication Techniques

The compositions, which may consist of either finely ground minerals or synthetic compounds or combinations of both, are blended into so-called bodies. These are then shaped into desired forms and fired. The resultant products are hard, dense, and brittle. Further shaping is possible only by grinding. Other compositions can be formed by glass-shaping techniques, such as casting in the molten state or hot pressing.

The demand for complex shapes, close dimensional tolerances, and controlled physical and chemical properties has resulted in new developments in forming techniques. The older methods of slip casting and jiggering are still used for forming electrical porcelain, which contains a considerable amount of plastic clays.

Other methods of forming, particularly adaptable to nonplastic compositions and to line production techniques, are automatic dry pressing, extruding, and injection or compression molding. Filmed casting and stamping of ceramic sheets is another method used for the production of accurately formed thin ceramic shapes, such as capacitor dielectrics or vacuum tube spacers.

Materials

Electrical Porcelain

Porcelain is the outstanding ceramic insulator because it combines mechanical stability and strength with heat and arc resistance and the ability to resist the passage of electric current.

Conventional electrical porcelain is made of the natural minerals clay, flint, and feldspar and is very similar in composition to porcelain used for high-quality vitrified dinnerware. After forming, the ceramic is fired to complete vitrification to develop maximum mechanical and dielectric strength. For outdoor use and at locations exposed to humidity, it is desirable to use insulators with glazed surface. The glaze consists essentially of the same components as the porcelain itself, but in different proportions.

Electrical porcelain is a satisfactory insulator for low-tension electrical wiring systems, for outlets and switches, lamp sockets, and electrical appliances. It is unsurpassed for insulation of outdoor high-voltage power transmission and distribution systems, and is used extensively for suspension and pin-type insulators. Porcelain high-tension insulators also find numerous applications in transformers and as lead-in insulators.

High-Frequency Insulation

A dielectric material for use at high frequencies must have the additional characteristic of low dielectric loss, and its properties must not be affected by changes over the required temperature range.

Porcelain has a rather high dielectric loss under high-frequency conditions, but a number of low-loss ceramic materials have been developed especially for high-frequency insulation. It is generally agreed that a low-loss ceramic body should consist chiefly of uniformly small crystallites bonded with only a very small amount of glassy matrix. Typical special ceramics are frequently named according to the predominant crystalline phase in the ceramic structure, which is the chief contributor to the specific properties of the ceramic product.

Among the best-known high-frequency ceramics are *Steatite* ceramics ($MgO \cdot SiO_2$). They are based on the mineral talc or steatite, a hydrous magnesium silicate. Other low-loss ceramic compositions are *Forsterite* ($2MgO \cdot SiO_2$), *Wollastonite* ($CaO \cdot SiO_2$), *Mullite* ($3Al_2O_3 \cdot 2SiO_2$), and *Spinel* ($MgO \cdot Al_2O_3$).

Sintered Al_2O_3

Highest physical and dielectric strength and low dielectric loss over a wide temperature range is found in a sintered aluminum oxide or Al_2O_3 ceramic. The main crystalline phase is corundum Al_2O_3.

Sintered Al_2O_3 ceramics are impervious to gases and meet the very exacting requirements of spark plug insulation, of envelopes for ultra-high-frequency receiving and power vacuum tubes. Paper-thin wafers of sintered Al_2O_3 are used as wire supports in vacuum tubes, as a substitute for stamped natural mica spacers.

Thermal Shock-Resistant Electrical Ceramics

Numerous applications in the low- and high-frequency field demand ceramics that will withstand sudden heat shock. For example, insulation for electrical appliances such as toasters, ovens, also electric arc chambers and switch gears, thermocouple insulation, and supports for electrical resistors, are subjected to sudden thermal changes. Low thermal expansion, combined with high heat conductivity, is desirable in ceramics for such applications. The outstanding material in this group is Cordierite ($2MgO \cdot 2Al_2O_3 \cdot 5SiO_2$). Other ceramics of low thermal expansion are based on the mineral β-Spodume, a complex lithium aluminosilicate. Both groups of ceramics have lower mechanical strength than steatite and sintered Al_2O_3, but their excellent thermal shock resistance makes these materials very useful for the applications just mentioned.

Glass

A unique combination of properties has made glass and indispensable material of construction in the electrical industry. Among its most important properties are transmission of light, imperviousness to gases, ability to seal readily to metals, good dielectric properties, and ease of fabrication into many shapes. The chemical composition of glasses vary widely, but in general it can be stated that silica (SiO_2) is the foundation of most commercial glasses, fused with such metallic oxides as soda, potash, Ca, Pb, Ba, Mg, and B.

The properties of glasses are primarily determined by their composition. The most widely used glasses for electrical insulation are of the soda-lime-silicate type, the lead glasses, and the boron aluminosilicate (*Pyrex*) glasses. The major characteristics of the latter type are good mechanical strength, low thermal expansion, good weathering stability, and good dielectric properties.

Glass insulators are used extensively as line insulators for open wire lines, radio antennas, and for envelopes of incandescent lamps, vacuum tubes, and mercury switches, and other devices. In the electronic field, glass serves as an insulation basis for electronic components, such as resistors and inductors and as a dielectric for capacitors.

Certain glasses can be converted into finely crystalline bodies bonded together with a vitreous matrix or by fusion of the crystallites at their grain boundaries (**Pyroceram**). Articles to be converted from glass to a crystalline ceramic are first fabricated in the glassy state with special nucleating agents added to the glass composition to form crystallites in the body. After forming, crystallization is effected by either exposure to ultraviolet radiation and temperature treatment or by heat treatment alone. The outstanding advantage of this new group of material is that they can be fabricated into a variety of shapes by conventional glass-forming methods and have a higher strength and stability of crystalline ceramics.

Fibrous glass is an important insulating material for electrical equipment. It may be used in the form of yarn, tapes, sleeving, or for reinforcement of organic plastics. On the other hand, plastics improve abrasion resistance and dielectric strength of fibrous glass constructions.

Ceramic Papers

There are a number of ceramic papers on the market that serve as high-temperature electrical insulation in a similar way as fibrous glass. These papers are made from reconstituted mica flakes (*Samica, Mica Mat*), clays, or glass and mineral (mullite) fibers.

Glass-Bonded Mica

As the name suggests, this group of materials consists of either natural or synthetic mica flakes that are bonded under heat and pressure by a glassy matrix. Vitrified ceramics can only be lapped or ground with the hardest abrasives, such as SiC, corundum, or diamond powder. In contrast, glass-bonded mica can be subjected to all normal machining operations. The material is impervious to moisture, has high dielectric strength, and can be molded to accurate dimensions. Metal inserts can be molded directly into mica-bonded articles. This is a distinct advantage over other ceramics, which require assembly or attachment of metal parts after processing.

Electronic Ceramics

Great progress has been made during the past decades in the area of solid-state research. Solid-state components are finding many applications in electronics. Some of these, such as diodes, rectifiers, or transistors, are based on single crystals, germanium and silicon, grown from a melt and cut into required shape. These are not considered ceramics, but others based on polycrystalline systems and fabricated by ceramic processes are considered among electric ceramics.

Ferroelectric Ceramics

Ferroelectricity, the spontaneous alignment of electric dipoles under the influence of an electric field, is an outstanding property of certain ceramic materials, especially those based on barium titanate. Besides, barium titanate, additional ferroelectric materials were discovered in the field of niobates, tantalates, and zirconates. Ferroelectric ceramics exhibit a high dielectric constant, which makes them attractive for dielectrics in capacitors. The dielectric constant of these ceramics can be varied through compositional changes; values as high as 10,000 at room temperature can be obtained. Very high dielectric constant ceramics have rather steep negative temperature coefficients of capacity. The capacitance of special barium titanate capacitors is sensitive to applied voltage, which suggests their use and dielectric amplifiers. Nonlinear capacitors of this type have also been employed for frequency modulation and remote tuning devices. Ferroelectric ceramics can be made to show piezoelectric effects, that is, the ability to convert mechanical strain into electric charges and, conversely, the ability to transform a voltage into mechanical force. This is produced by exposing the material to an orienting electric field during the cooling after firing. Piezoelectric ceramics of the barium titanate type have found applications for photograph pickups, ultrasonic thickness gauges, accelerometers, ultrasonic cutting tools, and even in electromedical instruments for measuring heart conditions.

Ferromagnetic Ceramics

Ferromagnetic ceramics, also known as ferrites, are compounds of various metal oxides and have the general formula $MO \cdot Fe_2O_3$, where M stands for a bivalent metal ion, such as zinc, nickel, magnesium, and others. They are ceramic materials with a crystalline structure of the spinel ($MgO \cdot Al_2O_3$) type. The mineral magnetite ($FeO \cdot Fe_2O_3$) is the only naturally occurring mineral of this type and has been well-known and used for ages as lodestone for its magnetic properties. In contrast to metallic magnetic materials, ceramic ferrites have high-volume resistivity and high permeability. Their specific gravity is between 4 and 5, considerably less than iron (8). They can be made both into "soft" and "hard" permanent magnetic materials. The properties of soft ferrites can be varied over a wide range to meet specific application requirements. For cores of radio and television loop antennas they are made with emphasis on a high-quality (Q) factor, to attain optimum in reception quality and selectivity. For memory cores in electron computers, they are so compounded that they exhibit a square hysteresis loop of magnetization and have an extremely low switching time between magnetic saturation and demagnetization. Ceramic magnets are about 35% lower in density than metallic magnets, an important factor for military airborne applications.

Resistor Ceramics

Certain nonmetallics pass limited amounts of electricity and therefore are useful as electrical resistors. Some of these have advantages over metals in that they are more resistant to oxidation and are therefore useful at higher temperatures.

SiC, either self-bonded or bonded with silicon in the form of rods or tubes, is used for resistor heating elements in electric furnaces. These can be heated to 1400°C in air for indefinite periods of time and may be heated as high as 1600°C for shorter periods.

SiC as a resistor element shows nonlinear characteristics and is used where a nonohmic variation of current is desired (Thyrite resistors). Nonlinear resistors of this type are chiefly used for voltage regulation and current suppression, such as in lightning arrestors.

Many oxides become electrically conductive at elevated temperatures. The electrical conductivity of oxides is governed to a large degree by the amount of impurities present. For example, the electrical resistance of thoria is greatly reduced by the presence of such oxides as ceria, yttria, erbia, etc. Thoria and zirconia resistors have been used as electric furnace heating elements up to 2000°C. These oxides are nonlinear with temperature change and therefore have found applications as thermistors. These components are used as temperature sensors and pyrometers, temperature bridges, and microwave power meters.

Other Semiconductor Ceramics

Semiconductor ceramics of silicide, telluride, and oxide compositions have been used for thermoelectric devices, to convert heat directly into electrical energy (Seebeck effect) and for refrigeration (Peltier effect).

Electric Arc Furnace

See *arc furnace*.

Electric Arc Spraying

A thermal spraying process using as a heat source an electric arc between two consumable electrodes of a coating material and a compressed gas, which is used to atomize and propel the material to the substrate.

E

Electric Arc Welding

Any welding operation in which heat is developed by an electric arc. It is implicit that the heat generated is sufficient to melt the components being joined in contrast to arc brazing. The arc may be struck between two consumable electrodes, or between the component and one electrode which can be either consumable or nonconsumable. Where the electrode is consumable it can be a short length of metal rod or a continuous wire or tube. This metal forms part of the weld and is termed filler. The rod or two filler metals may have a coating or tube filling that, if present, contributes to the welding operation in various ways depending on the specific process. It forms a protective gaseous envelope around the arc and molten metal and provides flux to form a slag that removes contaminants from the molten weld metal; it may also contribute metal or alloying elements to the weld metal and modify the arc conditions or weld metal deposition characteristics. Nonconsumable electrodes are of refractory materials particularly tungsten although carbon is occasionally still used in some processes. Metal nonconsumable electrodes are normally inert gas such as argon, helium, or carbon dioxide (effectively inert with steel). The gas is injected into the arc area primarily to form a protective envelope around the arc and molten weld metal but also in some cases to modify arc and metal transfer characteristics. Although the electrode is not consumed, metal may be added to the joint by a separate filler rod. See also *manual metallic arc*, *tungsten inert gas*, *metal inert gas*, and *submerged arc*.

Electric Bonding (Thermal Spraying)

See preferred term *surfacing*.

Electric Brazing

See preferred term *resistance brazing* and *arc brazing*.

Electric Dipole

The result of a distribution of bound charges, that is, separated charges that are bound to their centers of equilibrium by an elastic force; equal numbers of positive and negative charges must be present in an uncharged medium.

Electric Dipole Moment

A quantity characteristic of a distribution of bound charges equal to the vector sum over the charges of the product of the charge and the position vector of the charge.

Electric Dipole Transition

A transition of an atom, molecule, or nucleus from one energy state to another, which results from the interaction of electromagnetic radiation with the dipole moment of the molecule, atom, or nucleus.

Electric Field Effect

See *Stark effect*.

Electric Furnace

A metal melting or holding furnace that produces heat from electricity. They operate on the resistance or induction principle. See also *induction furnace*.

Electric Resistance Welding

See *resistance welding*.

Electric Steel

(1) Steel made by one of the electric melting processes. There is often an implication of good quality, particularly freedom from inclusions. (2) Steel having low magnetic hysteresis, and hence used as sheet for the laminations of transformers and generator stators. Such steels are typically very low carbon with up to 5% silicon and are carefully cold rolled to maximize their magnetic characteristics.

Electric Strength

The maximum voltage that an insulating material can sustain without arcing or break down allowing the passage of a current.

Electrical Conductivity

See *conductivity, electrical*.

Electrical Discharge Grinding

Grinding by spark discharges between a negative electrode grinding wheel and a positive workpiece separated by a small gap containing a dielectric fluid such as petroleum oil.

Electrical Discharge Machining (EDM)

Metal removed by a rapid spark discharge between different polarity electrodes, one on the workpiece and the other on the tool separated by a gap distance of 0.013–0.9 mm (0.0005–0.035 in.). The gap is filled with dielectric fluid or paraffin and metal particles which are melted, in part vaporized and expelled from the gap. Complex shapes, deep narrow holes, and high precision can be achieved but surface damage due to the high local temperatures may occur.

Electrical Discharge Wire Cutting

A special form of electrical discharge machining wherein the electrode is a continuous moving conductive wire. Also referred to as traveling wire electrical discharge machining.

Electrical Disintegration

Metal removal by an electrical spark acting in air. It is not subject to precise control, the most common application being the removal of broken tools such as taps and drills.

Electrical Insulators

Any materials that retard the flow of electricity and are used to prevent the passage or escape of electric current from conductors. No materials are absolute nonconductors; those rating lowest on the scale of conductivity are therefore the best insulators. An important requirement of a good insulator is that it does not absorb moisture, which would lower its resistivity. Glass and porcelain are the most common line insulators because of low cost.

Synthetic rubbers and plastics have now replaced natural rubber for wire insulation, but some aluminum conductors are insulated only with an anodized coating of aluminum oxide. Wires to be coated with an organic insulator may first be treated with hydrogen fluoride, giving a coating of copper fluoride on copper wire and aluminum fluoride on aluminum wire. The thin film of fluoride has high dielectric strength and heat resistance.

Insulating oils are mineral oils of high dielectric strength and high flash point employed in circuit breakers, switches, transformers, and other electrical apparatus. An oil with a flashpoint of 285°F (140°C) and fire point of 310°F (154°C) is considered safe. A clean, well-refined oil will have high dielectric strength, but the presence of as low as 0.01% water will reduce the dielectric strength drastically. The insulating oils, therefore, cannot be stored for long periods because of the danger of absorbing moisture. Impurities such as acids or alkalies also detract from the strength of the oil. Since insulating oils are used for cooling as well for insulating, the viscosity should be low enough for free circulation, and they should not gum. *Askarel* is an ASTM designation for *insulating fluids* which give out only nonflammable gases if decomposed by an electric arc. They are usually chlorinated aromatic hydrocarbons such as trichlorobenzene (TCB), but fluorinated hydrocarbons are also used. Insulating gases are used to replace air in enclosed areas to insulate high-voltage equipment. The insulating oil, fluids, and gases are generally referred to as *dielectrics*, although this term embraces any insulator.

Insulation porcelain, or *electrical porcelain*, is not usually an ordinary porcelain except for common line insulators. For such uses as spark plugs, they may be molded silica, and for electronic insulation they may be molded steatite or specially compounded ceramics, more properly called *ceramic insulators*. Insulation porcelains compounded with varying percentages of zirconia and beryllia have a crystalline structure and good dielectric and mechanical strengths at temperatures as high as 2000°F (1093°C). These porcelains may have some magnesia, but are free of silica. However, zircon porcelain is made from zirconium silicate, and the molded and fired ceramic is equal to high-grade steatite for high-frequency insulation. The material called Nolex by the Naval Surface Weapons Center (NSWC) is made by hot-molding finely powdered synthetic fluorine mica. The molded parts are practically pure mica. They can be machined, have high-dimensional precision because they need no further heat treatment, and have high dielectric strength. Beryllia is a valued insulator for encapsulation coatings on heat-generating electronic devices as it has both high electrical resistivity and high heat conductivity.

Most ceramics are electrical insulators and are used widely for installation of electric power lines. The applications range from structural power insulators to electronic packaging and substrates. The ceramics serve as a structural and insulating base on which electronic components are deposited or attached. The requirement for good surface finish has led to the development of fine-grained alumina material that can be prepared with a very good finish. The requirement for high thermal conductivity in some applications has led to the use of *beryllia*, high-purity *alumina*, and more recently, *aluminum nitride* as substrate materials. Other materials are *silicon carbide* doped with beryllia to give electrical insulation and *glass ceramics* that can easily be produced in the complex shapes often needed. They can also be produced with a tailored thermal expansion coefficient *multilayer ceramic* (MLC) substrate for high-speed computer processing modules.

Electrical Isolation

The condition of being electrically separated from other metallic structures or the environment.

Electrical Pitting

The formation of surface cavities by removal of metal as a result of an electrical discharge across an interface.

Electrical Porcelain

Vitrified *whiteware* having an electrical insulating function.

Electrical Resistance Pyrometer

See *pyrometry*.

Electrical Resistance Strain Gauge

See *strain gauge*.

Electrical Resistance Welding (ERW)

See *resistance welding*.

Electrical Resistivity

The electrical resistance offered by a material to the flow of current, times the cross-sectional area of current flow and per unit length of current path; the reciprocal of the conductivity. It is measured in ohms but also see International Annealed Copper Standard. Also called *resistivity* or *specific resistance*.

Electrical-Contact Materials

These are materials used to make or break electrical contact, thus make-and-break electrical circuits, or to provide sliding or constant contact. Both require high electrical conductivity to ease current flow, high thermal conductivity to dissipate heat, high melting point or range to inhibit arc erosion and prevent sticking, corrosion and oxidation resistance to prevent formation of films that impede current flow, high hardness for wear resistance, and amenability to welding, brazing, or other means of joining. The sliding contact types also require low friction, and a lubricant is always required between the sliding material to prevent seizing and galling.

The materials used range from pure metals and alloys to composites, including those made by powder metallurgy methods. Copper is widely used but requires protection from oxidation and corrosion, such as by immersion in oil, coating, or vacuum sealing. Cu–W alloys or mixtures of Cu–Gr increase resistance to arcing and sticking and some copper alloys provide greater hardness, thus greater wear resistance and better spring characteristics. Silver is more oxidation-resistant in air and, pure or alloyed, is the most widely used metal for make-and-break contacts for application and currents to 600 A. Ag–Cu alloys provide greater hardness but less conductivity and oxidation resistance; Ag–Cd alloys increase resistance to arc erosion and sticking; and Ag–Pt, Ag–Pd, and Ag–Au alloys increase hardness, wear resistance, and oxidation resistance. All alloying elements, however, decrease conductivity. Gold has outstanding oxidation and sulfidation resistance but, being soft and prone to wear and arc erosion, is limited to low-current (0.5 A maximum) applications. To enhance these properties, gold alloys, such as Au–Ag, Au–Cu, Au–Ag–Pt, Au–Ag–Ni, and Au–Cu–Pt–Ag, are more commonly used. Platinum and palladium are also used for contacts, but, again, in alloy form more than as pure metals. Among the most common ones are Pt–Ir, Pt–Ru, Pt–Pd–Ru, Pd–Ru, Pd–Cu, and Pd–Ag. A Pd–Ag–Pt–Au alloy for brushes and sliding contacts is noted for its exceptional modulus of elasticity and high proportional limit. Aluminum, tungsten, and molybdenum are also used for electrical contacts but mainly in composite form. Al used for contacts provides an electrical conductivity of about 60% that of copper, but is prone to oxidation and, thus, is clad or plated with silver, tin, or copper. The refractory metals, although providing excellent resistance to wear and arc erosion, are poor conductors and oxidize readily.

The principal metals made in composite form by P/M methods are the refractory metals, including those in carbide form, and copper-base and silver-base metals. The refractory metals, notably tungsten and molybdenum or their carbides, usually serve as a base for infiltrating with copper or silver, thus combining electrical conductivity and resistance to wear and arc erosion in a single material. Many such composites are common, including W–Cu, W–Ag, WC–Ag, WC–Cu, W–Gr–Ag, and Mo–Ag. The amount of conductive metal may exceed or be less than that of the refractory metal or refractory-metal carbide. A common silver composite is Ag–CdO, which, for a given amount of silver, provides greater conductivity than Ag–Cd alloys as well as greater hardness and resistance to sticking. Others include Ag–Gr, Ag–Ni, and Ag–Fe. The Ag–Gr composites are used mainly for sliding or brush contacts.

Electrical-Resistance Metals and Alloys

This major family of metals, including alloys as well as pure metals, includes alloys used in controls and instruments to measure or regulate electrical performance, heating alloys used to generate heat, and thermostat metals used to convert heat to mechanical energy. There are seven types of electrical-resistance alloys: (1) radio alloys, which contains 78%–98% Cu with the balance Ni; (2) manganins, 87% Cu and 13% Mn or 83%–85% Cu, 10%–13% Mn, and 4% Ni; (3) constantans, 55%–57% Cu and 43%–45% Ni; (4) Ni–Cr–Al alloys, 72%–75% Ni, 20% Cr, 3% Al, and either 5% Mn or 2% Cu, Fe, or Mn; (5) Fe–Cr–Al alloys, 73%–81% Fe, 15%–22% Cr, 4%–5.5% Al; (6) various other alloys, mostly nickel-base alloys, some of which contain substantial amounts of chromium or iron, chromium and iron, chromium and silicon, and, in some cases, magnesium and aluminum, and (7) pure metals, notably aluminum, copper, iron, nickel, precious metals, and refractory metals.

Key characteristics of resistance alloys are uniform resistivity, stable resistance, reproducible temperature coefficients of resistance, and low thermal electric potential vs. copper. Less critical, but also important, are coefficient of thermal expansion (CTE); strength and ductility; corrosion resistance; and joinability to dissimilar metals by welding, brazing, or soldering. Heating alloys require high heat resistance, including resistance to oxidation and creep in particular environments, such as furnaces, in which they are widely used; high electrical resistivity; and reproducible temperature coefficients of resistance. Also desirable are high emissivity, low CTEs, and low modulus to minimize thermal fatigue, strength, and resistance to thermal shock and ductility for fabricability. Thermostat metals, two or more bonded materials of which one may be nonmetallic, are chosen based on different electrical resistivities and thermal expansivities so that applied heat can be converted to mechanical energy.

Electrical resistivity in ohm-circular nohm·m for the alloys encompassed by the seven groups range from 16 for silver to 1450 for an alloy made up of 72.5% Fe, and 5.5% Al. Iron has the highest resistivity, 970, of the pure metals, followed by Ta, 135; Pt, 106; Ni, 80; and W, 55. The radio alloys are in the 50–300 range, the manganins 380–480, the constantans 490–500, and most of the Ni–Cr–Al, Fe–Cr–Al, and various other alloys are in the 1015–1450 range. Temperature coefficients of resistance in parts per million per

degree Celsius range from ±10 to ±15 at 15–45°C for the manganins to +6000 at 20–35°C for pure nickel. Tensile strength and ductility also range widely depending on the alloy metal. Maximum operating temperature in air for the commonly used resistance-heating alloys range from 927°C for 43.5% Fe–35% Ni–20% Cr–1.5% Si alloy to 1374°C for 72.5% Fe–22% Cr–5.5% Al. For platinum, this temperature is 1510°C. The refractory metals are suitable for still higher temperatures and vacuum and, in the case of molybdenum and tungsten, in select environments.

Electrical-resistance alloys are mainly wire products, and the alloys have been known by a multitude of trade names. The standard alloy for electrical-resistance wire for heaters and electrical appliances in Ni–Cr, but Ni–Mn and other alloys are used. For high-temperature furnaces, tungsten, molybdenum, and alloys of the more expensive high-melting metals are employed. The much-used alloy with 80% Ni and 20% Cr resists scaling and oxidation to 1177°C, but it is subject to an inner granular corrosion, known as green rot, which may occur in chromium above 816°C unless modified with other elements. The 60%–80% Ni–Cr family of alloys are used in heater elements, resistors, rheostats, resistance thermometers, and in potentiometers.

The Cr–Al–Fe alloys have high resistivity and high oxidation resistance, but have a tendency to become brittle. The Kanthal alloys have 20%–25% chromium with some cobalt and aluminum, and the balance iron. Kanthal A, with 5% aluminum, will withstand temperatures to 1299°C and is resistant to H_2SO_4. The tensile strength is a 13 MPa with elongation of 12%–16%. Kanthal A-1, for furnaces, has an operating temperature to 1373°C.

Cu–Mn alloys have high resistivity and alloys with 96%–98% manganese, although they may be brittle and difficult to make into wire. Addition of nickel makes them ductile, but lowers resistivity. A typical alloy contains 35% Mn, 35% Ni, and 30% Cu.

Other alloys are used for rheostats, electrical heaters for 1093°C temperature operations, and thermocouples up to 1250°C.

Electro Force

The force between electrodes in a spot, seam, and projection weld. See also *dynamic electrode force*, *static electrode force*, and *theoretical electrode force*.

Electro Lead

The electrical conductor between the source of arc welding current and the electrode holder.

Electro-Beam Machining

Removing material by melting and vaporizing the workpiece at the point of impingement of a focused high-velocity beam of electrons. The machining is done at high vacuum to eliminate scattering of the electrons due to interaction with gas molecules. The most important use of electron beam machining is for hole drilling.

Electrochemical Admittance

The inverse of *electrochemical impedance*.

Electrochemical Cell

An electrochemical system consisting of an *anode* and a *cathode* in metallic contact and immersed in our *electrolyte*. The anode and cathode may be different metals or dissimilar areas on the same metal surface. See also *cathodic protection*.

Electrochemical Corrosion

Corrosion that is accompanied by a flow of electrons between cathodic and an anodic areas on metallic surfaces.

Electrochemical Discharge Machining

Metal removal by a combination of the processes of *electrochemical machining* and *electrical discharge machining*. Most of the metal removal occurs via an anodic dissolution (i.e., ECM action). Oxide films that form as a result of electrolytic action through an electrolytic fluid are removed by intermittent spark discharges (i.e., EDM action). Hence, the combination of the two actions.

Electrochemical (Chemical) Etching

General expression for all developments of microstructure through reduction and oxidation (redox reactions).

Electrochemical Equivalent

(1) The weight of an element or group of elements oxidized or reduced at 100% efficiency by the passage of the unit quantity of electricity. Usually expressed as grams per coulomb. (2) The mass liberated by 1 A of current flowing for 1 s. The electrochemical equivalent of any element is a function of its chemical equivalent which is its atomic weight divided by the valence. Silver, with an electrochemical equivalent of 0.0011183 g, is a reference from which the electrochemical equivalents of other elements are calculated from the formula:

$$\text{Electrochemical equivalent of element Y}$$

$$= \frac{0.0011183 \times \text{atomic weight of Y}}{\text{Valence of Y} \times \text{atomic weight of silver}}$$

Electrochemical Grinding

A process whereby metal is removed by deplating. The workpiece is the anode; the cathode is a conductive aluminum oxide-copper or metal-bonded diamond grinding wheel with abrasive particles. Most of the metal is removed by deplating; 0.05%–10% is removed by abrasive cutting.

Electrochemical Impedance

The frequency-dependent complex-valued proportionality factor ($\Delta E / \Delta i$) between the applied potential or current and the response signal. This factor is the total opposition (Ω or $\Omega \cdot cm^2$) of an electrochemical system to the passage of charge. The value is related to the *corrosion rate* under certain circumstances.

Electrochemical Machining

(1) Controlled metal removal by anodic dissolution. Direct current passes through flowing film of conductive solution which separates the workpiece from the electrode-tool. The workpiece is the *anode*, and the tool is the *cathode*. (2) Processes utilizing electrochemistry to selectively remove material to shape a component. The process may utilize the cathode of simple shape which traverses the component being formed in a preplanned pattern to produce the required shape. Alternatively, a cathode with a reverse image of the required form may be progressively moved toward the component. The concentration of current at the points of closest approach then produces the required image on the anode.

Electrochemical Process

The principles of electrochemistry may be adapted for use in the preparation of commercially important quantities of certain substances, both inorganic and organic in nature.

Inorganic Processes

Inorganic chemical processes can be classified as electrolytic, electrothermic, and miscellaneous processes including electric discharge through gases and separation by electrical means. In electrolytic processes, chemical and electrical energy are interchanged. Current passed through an electrolytic cell causes chemical reactions at the electrodes. Voltaic cells convert chemicals into electricity. Electrothermic processes use electricity to attain the necessary temperature for reaction.

Voltaic cells are used for the intermittent production of small amounts of electricity. When the chemicals involved are exhausted and must be replaced, the unit is called a primary cell. A special case of the primary cell is a fuel cell in which the fuel and anoxidizer are fed continuously to the cell, converted to electricity, and the products removed. If exhausted components can be revived by passing electricity backward through the unit, it is called a secondary cell, storage battery, or accumulator. Cells may be connected in parallel or in series to form a battery.

Organic Processes

Organic electrochemistry was once regarded as a tantalizing area with many important laboratory achievements but few successes in commercial practice. The situation has changed, however, in that electroorganic processes are commercially advantageous if they could fulfill either of two conditions: (1) performance under conditions of voltage corresponding thermodynamically to the conversion of an organic group to a reduced or oxidized group, with the cell products relatively easy to isolate and purify; (2) performance of a highly selective, specific technique to make an addition at a double bond, or to split a particular bond (e.g., between carbon atoms 17 and 18 of a complex molecule having 25 carbon atoms).

Selectivity and specificity are highly important in electroorganic processes for the manufacture of complicated molecules of vitamins and hormones—as well as for the medicinal products whose action on pathogenic organisms is a function of their spiral arrangement, steric forms, and resonance.

The electrolytic oxidation and reduction of organic compounds differ from the corresponding and more familiar inorganic reactions only in that organic reactions tend to be complex and have low yields. The electrochemical principles are precisely those of inorganic reactions, while the procedures for handling the chemicals are precisely those of organic chemistry.

Oxidations

Commercial success in inorganic electrochemistry has come about by well-engineered combinations of organic and inorganic techniques in areas where strictly chemical methods are either impossible or inefficient, for example, in catalytic hydrogenation or oxidation.

Reductions

Substances that are easy to reduce may be acted on, at the interface of cathodes, with low H_2 overvoltage. Hard-to-reduce materials may require much higher overvoltages, which are reached through either the cathode composition or the current density.

Electrochemical Potential

The partial derivative of the total electrochemical free energy of a constituent with respect to the number of moles of this constituent where all factors are kept constant. It is analogous to the *chemical potential* of a constituent except that it includes the electric as well as chemical contributions to the free energy. The *potential* of an electrode in an electrolyte relative to a *reference electrode* measured under open circuit conditions.

Electrochemical Reaction

A reaction caused by passage of an electric current through a medium that contains mobile ions (as in electrolysis); or, a spontaneous reaction made to cause current to flow in a conductor external to this medium (as in a galvanic cell). In either event, electrical connection is made to the external portion of the circuit via a pair of electrodes. See also *electrolyte*.

Electrochemical Series

Same as *electromotive force series*.

Electrochemical Shaping

This term and electrochemical machining are often used interchangeably but where they are differentiated "Machining" is usually recognized as a general term and "Shaping" refers to the process utilizing a reverse image cathode.

Electrochemistry

The study of the reactions occurring in a liquid electrolyte, that is, an ionic conductor, through which an electric current passes. This current may be externally imposed, it may arise between different metals that are in electrical contact directly as well as through the electrolyte, or it may develop between areas on a component that differ in some respect such as composition, hardness, or environment. This electrically induced activity is termed a voltaic cell, galvanic cell, or electrolytic cell. In such cells, positively charged cations (i.e., ions deficient in an electron) travel through the electrolyte from the anode to the cathode while a balancing flow of negatively charged anions (i.e., ions carrying an extra electron) travel from the cathode to the anode. The formation of metal cations at an anode involves loss of atoms from the metal surface, termed *electrolytic corrosion*, and their arrival at the cathode can involve deposition, that is, *electroplating*. The *electrode potential* or *electrochemical potential* is the voltage developed between the pairs of metals comprising the anode and cathode of the cell and its value depends on the individual metals involved. If a standard electrode is selected, the relative potential of all other metals, in terms of positive/negative and size, can be measured and tabulated. Eventually, the table is usually presented with the corrosion-resistant noble metals at the bottom and the most reactive base metals at the top. Such a table is termed an Electrochemical Series or Galvanic Series although the two are not identical. A metal may have two positions in the series,

and Active state high in the series and a Passive state where it forms an impervious surface corrosion film that resists further attack. The following list indicates the electropotential relative to the saturated calomel electrode, for various metals and seawater. Typically, differences of more than about 0.2 V may give rise to corrosion.

Material	Voltage (V)
Magnesium alloys	−1.6
Zinc alloys	−1.1
Aluminum	−0.9
Aluminum alloys	−0.6 to −0.8
Mild and carbon steels	−0.7
12% Chromium for ferritic steel	−0.45
18% Cr 8% Ni austenitic stainless	−0.2
Lead	−0.55
Tin and solders	−0.45
Copper and high copper alloys	−0.25
Nickel and high nickel alloys	−0.15
Silver	0
Titanium and high titanium alloys	0
Gold	+0.15
Platinum	+0.15

Electrocorrosive Wear

Wear of a solid surface that is accelerated by the presence of a corrosion-inducing electrical potential across the contact interface. This process is usually associated with wear in the presence of a liquid electrolyte in the interface. However, moisture from the air can also facilitate this type of wear when a galvanic wear couple exists and the contacting materials are sufficiently reactive.

Electroplated Nickel Silver (EPNS)

Silver electroplated onto nickel silver for cutlery and similar applications.

Electrode

Compressed graphite or carbon cylinder or rod used to conduct electric current in electric arc furnaces, arc lamps, and so forth.

Electrode (Electrochemistry)

One of a pair of conductors introduced into an electrochemical cell, between which the ions in the intervening medium flow in opposite directions and on whose surfaces reactions occur (when appropriate external connection is made). Indirect current operation, one electrode or "pole" is positively charged, the other negatively. See also *anode, cathode, electrochemical reaction*, and *electrolyte*.

Electrode (Welding)

(1) In arc welding, a current-carrying rod that supports the arc between the rod and work, or between two rods as in twin carbon-arc welding. It may or may not furnish filler metal. See also *bare electrode, covered electrode*, and *lightly coated electrode*. (2) In resistance welding, a part of a resistance welding machine through which current and, in most instances, pressure is applied directly to the work. The electrode may be in the form of a rotating

wheel, rotating roll, bar, cylinder, plate, clamp, chuck, or modification thereof. See also *resistance welding electrode*. (3) In arc and plasma spraying, the current-carrying components which supports the arc. (4) An electrical conductor from which the current leaves the supply circuit to enter the weld zone. In the case of pressure welding, the electrode is usually of a copper base alloy combining high conductivity with hot strength. An electrode for electric arc welding is defined as consumable or nonconsumable depending upon whether it is progressively consumed to provide filler metal to be fused and incorporated in the joint. A covered or coated (consumable) electrode carries an external coating that contributes greatly to the welding process by providing flux, protective gases, filler or alloying elements, or materials that modify the arc characteristics and control the slag characteristics. A cored or flux cored electrode contain similar constituents within the tubular electrode. A sheathed electrode has an external sheath surrounding the coating. There are various electrode classification systems which define the coating, welding characteristics and sometimes the expected properties or other features. Often, such classifications refer to the principal constituent of the coating. A rutile electrode has a coating or core that a substantial proportion of titanium oxide, i.e., rutile. These are general-purpose electrodes, relatively easy to handle in manual applications. A basic electrode has a coating or core that is substantially calcium carbonate and fluoride, that is, chemically basic. Such coatings, suitably handled to keep them dry, release only small quantities of hydrogen compared with rutile, so basic electrodes are sometimes termed Low Hydrogen Electrodes or, where the quantity of hydrogen is specified, Hydrogen Controlled Electrodes. Basic electrodes are used in more demanding applications such as highly restrained joints, alloy steels, etc. They tend to require more operative skill then rutile electrodes. Cellulosic electrodes have coatings or cores with high cellulose contents providing deep penetration and hence are sometimes termed Deep Penetration Electrodes. Iron powder and alloy powder electrodes have coatings or cores contributing iron or alloy elements to the weld metal. Nonconsumable electrodes are necessarily refractory conductors such as tungsten, tungsten alloy, or carbon.

Electrode Cable

Same as *electrode lead*.

Electrode Deposition

The weight of weld metal deposit obtained from a unit length of electrode.

Electrode Extension

The length of unmelted electrode extending beyond the end of the contact tube during welding.

Electrode Holder

A device used for mechanically holding the electrode while conducting current to it.

Electrode Negative

Direct current arc welding with the electrode connected to the negative pole of the supply. See also *reverse polarity* for a caution.

Electrode Polarization

Change of *electrode potential* with respect to a reference value. Often the *free corrosion potential* is used as the reference value. The change may be caused, for example, by the application of an external electrical current or by the addition of an oxidant or reductant.

Electrode Positive

Direct current arc welding with the electrode connected to the positive pole of the supply. See also *reverse polarity* for a caution.

Electrode Potential

The *potential* of an *electrode* in an *electrolysis* as measured against a *reference electrode*. The electrode potential does not include any resistance losses in potential and either the solution or external circuit. It represents the reversible work to move a unit charge from the electrode surface through the solution to the reference electrode.

Electrode Reaction

Interfacial reaction equivalent to a transfer of charge between electronic and ionic conductors. See also anodic reaction and cathodic reaction.

Electrode Setback

In plasma arc welding and cutting, the distance the electrode is recessed behind the constricting orifice measured from the outer face of the nozzle.

Electrode Skid

In spot, seam, or projection welding, the sliding of an electrode along the surface of the work.

Electrodeposition

(1) The deposition of a conductive material from a plating solution by the application of electrical current. (2) The deposition of a substance on an electrode by passing electric current through an electrolyte. *Electrochemical (plating), electroforming, electrorefining,* and *electrotwinning* result from electrodeposition.

Electroformed Molds

A mold for forming plastics made by electroplating metal on the reverse pattern of the cavity. Molten steel may then be sprayed on the back of the mold to increase its strength.

Electroformed Parts

Electroforming is essentially an electrolytic plating process for manufacturing metal parts. In general, it is best to consider electroforming for applications where a part is impossible or difficult to make by any other standard method or if the tooling required by another method such as forging or die-casting is extremely expensive. Electroforming is not generally used for large-quantity production because it is a batch operation. It is, however, valuable for short-run, simple parts or long-run, complicated parts because of its low tooling costs.

Electroforming is the technology of creating exact (mirror image) copies of uniquely shaped objects by electrodepositing a layer of heavy metal onto an original and subsequently separating the two.

Self-supporting parts can be made from such metals as Ni, Cu, Fe, Ag, Cr, and Au. In all instances, the pure metal has been found to be the only practical deposit obtainable. There are, however, a number of bimetallic deposits such as cobalt and/or nickel and tungsten.

Nickel is neither the toughest nor the most corrosion-resistant metal, and yet it is becoming more popular with plastic molders. Uniquely suited for one application in particular—mold insert electroforming—nickel makes possible the cost-effective mass production of many of today's advanced products. Today, the process is used to manufacture so-called Fresnel lenses, used in overhead projectors, as well as many other plastic optical products. A single precision-machined mandrel generates numerous inserts, assuring that each molded optical component exhibits the same geometric characteristics and surface finish essential for consistent optical performance. Nickel molds electroformed have high dimensional stability (meaning they display minimal warping), have an optical-quality surface finish, and exhibit the durability necessary to produce thousands of precision lenses using only a few electroformed mold inserts.

Among the easiest metals to electroplate out of environmentally benign aqueous solutions, nickel is the logical choice for electroformed tooling. Nickel is also sufficiently strong, hard, and tarnish-resistant to withstand molding conditions encountered in the processing of many popular plastics. Moreover, it is also easily machinable, brazeable, and weldable.

Advantages

Following are the chief advantages of electroforming:

1. Initial tooling costs are extremely low. This has several advantages. Short-run parts can be produced to determine market acceptability and, as is the case in missile and aircraft applications, small numbers of parts may be produced economically. The low initial tooling cost makes it economically possible to try various shapes and configurations without undue tooling charges.
2. Because this is an electrochemical process, the surface finish of the mandrel is duplicated exactly, thereby providing a means of producing parts to any surface finish—from a high polish to a surface resembling gravel. Many examples of these are found in industry. Among them are molds to make records, plastic tile resembling different fabric textures, and surface finish standards.
3. Complex shapes such as ducting and tubing may be made in one piece, thereby eliminating welding and soldering.
4. Parts requiring extremely close tolerances are easily made by the electroforming process, inasmuch as the mandrel or shape upon which the metal is deposited may be machined or produced to extremely close tolerances, inspected easily, and thereby duplicated exactly. Examples of extremely close tolerance parts that can be made are radar plumbing, hot-air ducting and tubing, wind-tunnel test nozzles and liners, reflectors, collectors, mirrors, nose cones, etc.

Sizes

Parts produced by electroforming range in size from miniature to extremely large. Typical miniature parts such as small electronic devices are 0.50–0.76 mm in diameter and 3.2 mm long with a wall thickness of 0.03 ± 0.01 mm. Large sizes may be as big as 4.8 m in diameter. Thicknesses also may vary from 0.03 to 25.4 mm or even thicker.

Parts such as wind-tunnel nozzles 4.8 m long, 762 mm in diameter, and varying in thickness, have been fabricated from 1.5 to 25.4 mm. Tolerances on a part like this are extremely close and are usually ±0.03 mm in the throat area and ±0.05 mm in the downstream sections. Tolerances on waveguide plumbing may be as close as ±0.01 mm.

The range of mechanical properties of the materials that may be deposited by electroforming are shown next:

Material	Brinell Hardness, Bhn	Ultimate Tensile Strength, MPa	Yield Strength, MPa	Elongation, % in 2 in.
Nickel	140–500	379–1552	276–862	2–20
Copper	51–170	248–552	83–276	21–39

Electroforming

The deposition, by electroplating, of metal onto a shaped former or die. The die is coated with some conducting release agent so that when a suitable thickness of metal has been deposited it can be lifted off providing an image of the die. Highly complex and detailed shapes can be produced.

Electrogalvanizing

Zinc coating of iron or steel by electroplating. See also *cathodic protection.*

Electrogas Welding (EGW)

(1) A *vertical position* arc welding process which produces coalescence of metals by heating them with an arc between a continuous filler metal (consumable) electrode and the work. Copper dams (molding shoes) are used to confine the molten weld metal. The electrodes may be either flux cored or solid wire. Shielding may or may not be obtained from an externally supplied gas or mixture. See also *gas metal arc welding* and *flux cored arc welding*. (2) Similar to electroslag welding except that the arc and weld pool are shielded by a gas rather than a slag and the arc is maintained throughout the process.

Electrohydraulic Forming

Processes in which thin components submerged in a liquid are formed by the shock wave produced by electrical discharge between a pair of submerged electrodes.

Electrokinetic Potential

This *potential*, sometimes called zeta potential, is a potential difference in the solution caused by residual, unbalanced charge distribution in the adjoining solution, producing a double layer. The electrokinetic potential is different from the *electrode potential* in that it occurs exclusively in the solution phase; that is, it represents the reversible work necessary to bring a unit charge from infinity in the solution up to the interface in question but not through the interface.

Electroless Plating

This is a chemical reduction process that, once initiated, is autocatalytic. The process is similar to electroplating except that no outside current is needed. The metal ions are reduced by chemical agents in the plating solutions, and deposited on the substrate. Electroless plating is used for coating nonmetallic parts. Decorative electroless plates are usually further coated with electrodeposited nickel and chromium. There are also applications for electroless deposits on metallic substrates, especially when irregularly shaped objects require uniform coating. Electroless copper is used extensively for printed circuits, which are produced either by coating the nonmetallic substrate with a very thin layer of electroless copper and electroplating to the desired thickness or by using the electroless process only. Electroless iron and cobalt have limited uses. Electroless gold is used for microcircuits and connections to solid-state components. Deeply recessed areas that are difficult to plate can be coated by the electroless process.

Nonmetallic surfaces and some metallic surfaces must be activated before electroless deposition can be initiated. Activation on nonmetals consists of the application of stannous and palladium chloride solutions. Once electroless plating is begun, it will continue to a desired thickness; that is, it is autocatalytic. The process does differ from a displacement reaction, in which a more noble metal is deposited while a less noble one goes into solution; this ceases when the more noble deposit, if pore-free, covers a less noble substrate.

Electroluminescence

The direct conversion of electrical energy into light.

Electrolysis

(1) Chemical change resulting from the passage of electric current through an *electrolyte*. (2) The separation of chemical components by the passage of current through an electrolyte.

Electrolyte

This is a material that conducts an electric current when it is fused or dissolved in a solvent, usually H_2O. Electrolytes are composed of positively charged species, called cations, and negatively charged species, called anions. For example, NaCl is an electrolyte composed of sodium cations (Na^+) and chlorine anions (Cl^-). The ratio of cations to anions is always such that the substance is electrically neutral. If two wires connected to a lightbulb and to a power source are placed in a beaker of H_2O, the lightbulb will not glow. If an electrolyte, such as NaCl, is dissolved in the H_2O, the lightbulb will glow because the solution can now conduct electricity. The amount of electric current that can be carried by an electrolyte solution is proportional to the number of ions dissolved. Thus, the bulb will glow more brightly if the amount of NaCl in the solution is increased.

Hydration

H_2O is a special solvent because its structure has two different sides. On one side is the O_2 atom, and on the other are two H_2 atoms. Covalent molecules, such as H_2O, are held together by covalent bonds, which are formed when two atoms share a single pair of electrons. However, when two different atoms form a covalent bond, the sharing of electrons is not always equal. An electron in a covalent bond between two different atoms might spend more time near one atom or the other. In H_2O, the electrons from the O–H bonds spend more time near the O_2 atom than near the H_2 atoms. As a result, the O_2 has somewhat more negative charge than the H_2 atoms. This phenomenon is not the same as ionization, where the electron is completely transferred from one atom to the other, so to indicate the

difference; the O_2 is said to have a partial negative charge (δ^-) and the H_2 to have a partial positive charge (δ^+).

When H_2O molecules surround dissolved ions, the ions are said to be hydrated. The electrostatic attraction associated with hydration provides the energetic driving force for dissolving ions.

Electrolytic Brightening

Same as *electropolishing*.

Electrolytic Cell

An assembly, consisting of a vessel, electrodes, and an electrolyte, in which *electrolysis* can be carried out. See *electrochemistry*.

Electrolytic Cleaning

A process of removing soil, scale, or corrosion products from a metal surface by subjecting it as an *electrode* to electric current in an electrolytic bath. See *electrochemistry*.

Electrolytic Copper

(1) Copper that has been refined by the electrolytic deposition, including cathodes that are the direct product of the refining operation, refinery shapes cast from melted cathodes, and, by extension, fabricators' products made therefrom. Usually, when this term is used alone, it refers to a electrolytic tough pitch copper without elements other than oxygen being present in significant amounts. See also *tough pitch copper*. (2) One of a number of grades of high purity copper, typically 99.95% or better, and hence having a high conductivity. They have in common an origin as *cathode copper*.

Electrolytic Corrosion

Corrosion by means of electrochemical or mechanical action. See *electrochemistry*.

Electrolytic Deposition

Same as *electrodeposition*.

Electrolytic Dissociation

The process in which a substance splits into positively and negatively charged ions. See *electrochemistry*.

Electrolytic Etching

See *anodic etching*.

Electrolytic Extraction

Removal of phases by using an electrolytic cell containing an electrolyte that preferentially dissolves the metal matrix. See also *extraction*.

Electrolytic Grinding

A combination of grinding and machining wherein a metal-bonded abrasive wheel, usually diamond, is the *cathode* in physical contact with the anodic workpiece, the contact being made beneath the surface of a suitable electrolyte. The abrasive particles produce grinding action as nonconducting spacers permitting simultaneous machining through electrolysis.

Electrolytic Lead

A high-purity grade of lead, 99.995% or better, produced by electrolytic refining.

Electrolytic Machining

The removal of material in a selective, controlled manner to achieve a predetermined shape and dimensions desired. See *electrochemistry* and *electrochemical machining*.

Electrolytic Pickling

Pickling in which electric current is used, the work being one of the electrodes.

Electrolytic Polishing

An electrochemical polishing process in which the metal to be polished is made the *anode* in an electrolytic cell where preferential dissolution at high points in the surface topography produces a specularly reflective surface. Also referred to as *electro-polishing*.

Electrolytic Powder

Powder produced by electrolytic deposition or by pulverizing of an electrodeposit.

Electrolytic Protection

See preferred term *cathodic protection*.

Electrolytic Tough Pitch

A term describing the method of raw copper preparation to ensure a good physical- and electrical-grade copper-finished product.

Electromagnetic Focusing Device

See *focusing device*.

Electromagnetic Forming

A process for forming metal by the direct application of an intense, transient magnetic field. The workpiece is formed without mechanical contact by the passage of a pulse of electric current through a forming coil. Also known as magnetic pulse forming.

Electromagnetic Interference

Interference related to accumulated electrostatic charge in a nonconductor.

Electromagnetic Lens

An electromagnet designed to produce a suitably shaped magnetic field for the focusing and deflection of electrons or other charged particles in electron-optical instrumentation.

Electromagnetic Radiation

Energy propagated at the speed of light by an electromagnetic field. The electromagnetic spectrum includes the following approximate wavelength regions: they range from gamma-rays with a wavelength (Å) of 0.005 to 1.40 (0.005–0.14 nm) to microwaves (3×10^6–1×10^{10}) (0.3 mm–1 m).

Electromagnetism

Magnetism caused by the flow of electric current.

Electromechanical Polishing

An attack-polishing method in which the chemical action of the polishing fluid is enhanced or controlled by the application of an electric current between the specimen and the polishing wheel.

Electrometallizing/Metallization

The deposition, by electrochemical techniques, of a metallic coating onto a nonmetallic and, usually, nonconducting substrate.

Electrometallurgy

An imprecise term referring to metallurgical processes that use electricity as a primary power source. In some contexts, the term is limited to processes based on electrochemistry, for example, electroplating or electrolytic refining. Other contexts include processes where electricity is the critical power source, for example, melting in an electric arc furnace. However, secondary uses of electricity, for example, for lighting, motor power or instrumentation or even electric arc welding would not be included.

Electrometric Titration

A family of techniques in which the location of the endpoint of a *titration* involves the measurement of, or observation of changes in, some electrical quantity. Examples of such quantities include potential, current, conductance, frequency, and phase.

Electromotive Force

(1) The force that determines the flow of electricity; a difference of electric potential. (2) Electrical potential; voltage.

Electromotive Force Series (emf Series)

A series of elements arranged according to their *standard electrode potentials*, with "noble" metals such as gold being positive and "active" metals such as zinc being negative. In corrosion studies, the analogous the more practical *galvanic series* of metals is generally used. The relative positions of a given metal are not necessarily the same in the two series.

Electron

An elementary particle that is the negatively charged constituent of ordinary matter. The electron is the lightest known particle processing an electric charge. Its rest mass is $m_e \equiv 9.1 \times 10^{-28}$ g, approximately 1/1836 of the mass of the proton or neutron, which are, respectively, the positively charged and neutral constituents of ordinary matter. The negatively charged subatomic particle that (simplistically) circles the nucleus of an atom. It has a charge of -1.6×10^{-19} coulomb. See *atom* and *inter atomic bonding*.

Electron Bands

Energy states for the free electrons in a metal, as described by the use of the band theory (zone theory) of electron structure. Also called Brillouin zones.

Electron Beam

A stream of electrons in electron-optical systems.

Electron Beam Analysis

Same as *electron probe analysis*.

Electron Beam Cutting

A cutting process that uses the heat obtained from a concentrated beam composed primarily of high-velocity electrons, which impinge upon the workpieces to be cut; it may or may not use an externally supplied gas.

Electron Beam Gun

A device for producing and accelerating electrons. Typical components include the emitter (also called the filament or cathode) that is heated to produce electrons and via thermionic emission, a cup (also called the grid or grid cup), and the anode.

Electron Beam Gun Column

The electron beam gun plus auxiliary mechanical and electrical components which may include beam alignment, focus, and deflection coils.

Electron Beam Welding

A welding process that produces coalescence of metals with the heat obtained from a concentrated beam composed primarily of high-velocity electrons impinging upon the surfaces to be joined. Welding can be carried out at atmospheric pressure (nonvacuum), medium vacuum (approximately 10^{-3} to 25 torr), or high vacuum (approximately 10^{-6} to 10^{-3} torr).

Electron Cloud

The valence electrons of a metallic crystal are not tied to individual atoms that are shared as a general cloud. See *interatomic bonding, band theory,* and *semiconductor*.

Electron Compound

An intermediate phase on a phase diagram, usually a binary phase, that has the same crystal structure in the same ratio of valence electrons to atoms as those of intermediate phases in several other systems. An electron compound is often a solid solution of variable composition and good metallic properties. Occasionally, an ordered arrangement of atoms is characteristic of the compound, in which

case the range of composition is usually small. Phase stability depends essentially on electron concentration and crystal structure and has been observed at valence-electron-to-atom ratios of 3/2, 21/13, and 7/4. Beta compounds with a ratio of 3/2 (i.e., three electrons to two atoms), Epsilon with 7/4, and Gamma with 21/13 are the most common examples.

Electron Density

The number of electrons in unit volume of material. Depending upon the context, the electrons counted may be either the total in the substance or merely the conduction electrons.

Electron Emission

The ejection of electrons from a free surface.

Electron Energy Loss Spectroscopy (EELS)

A spectrographic technique in the electron microscope that analyzes the energy distribution of the electrons transmitted through the specimen. The energy loss spectrum is characteristic of the chemical composition of the region being sampled.

Electron Flow

A movement of electrons in an external circuit connecting an *anode* and *cathode* in a corrosion cell; the current flow is arbitrarily considered to be in an opposite direction to the electron flow.

Electron Gun

A device for producing and accelerating a beam of electrons.

Electron Image

A representation of an object formed by a beam of electrons focused by an electron-optical system. See also *image*.

Electron Lens

A device for focusing an electron beam to produce an image of an object.

Electron Micrograph

A reproduction of an image formed by the action of an electron beam on a photographic emulsion.

Electron Microscope

An electron-optical device that produces a magnified image of an object. Detail may be revealed by selective transmission, reflection, or emission of electrons by the object. Instruments using an electron beam, in a vacuum, to examine specimens. The image is displayed on a screen similar to a television set but is also readily photographed. In a transmission electron microscope (TEM), the fixed beam travels through the very thin specimen. In a scanning electron microscope (SEM), the beam scans rapidly back and forth across, as reflected from, the specimen surface. Both instruments offer very high magnifications with resolution of up to about 1 Å compared with the 0.2 μ of the light microscope. The SEM has the major advantage that, unlike the conventional light microscope, it can examine surfaces that are rough or inclined to the beam, for example, fracture faces. Hence, much metallurgical SEM work is performed that magnification is below 1000x, that is, within the range of the light microscope. See also *Electron Probe Analysis*. See also *scanning electron microscope* and *transmission electron microscope*.

Electron Microscopy

The study of materials by means of an electron microscope.

Electron Microscopy Impression

See *impression*.

Electron Multiplier Phototube

See *photomultiplier tube*.

Electron Octet

The eight electrons forming the filled outer valence shell of an atom. This configuration is very stable; the elements having it are inert and the molecules having it are stable.

Electron Optical Axis

The path of an electron through an electron-optical system, along which it suffers no deflection due to lens fields. This axis does not necessarily coincide with the mechanical axis of the system.

Electron Optical System

A combination of parts capable of producing and controlling a beam of electrons to yield an image of an object.

Electron Pair

Two electrons having parallel but opposite spin axes to give a net magnetic moment of zero.

Electron Probe (Analysis)

(1) A facility on an electron microscope for analysis of small surface areas. (2) A narrow beam of electrons used to scan or illuminate an object or screen.

Electron Probe X-Ray Microanalysis (EPMA)

A technique in analytical chemistry in which a finely focused beam of electrons is used to excite an x-ray spectrum characteristic of the elements in a small region of the sample. Conventional mapping determines elemental location and concentration.

Electron Scattering

Any change in the direction of propagation or kinetic energy of an electron as a result of a collision.

Electron Spin

The angular rotational motion of an electron which results in the magnetic moment.

Electron Spin Residence (ESR) Spectroscopy

A form of spectroscopy similar to nuclear magnetic resonance, except that the species studied is an unpaired electron, not a magnetic nucleus.

Electron Trajectory

The path of an electron.

Electron Velocity

The rate of motion of an electron.

Electron Wavelength

The wavelength necessary to account for the deviation of electron rays in crystals by wave-diffraction theory. It is numerically equal to the quotient of *Planck's constant* divided by the electron momentum.

Electronegativity

The ability of the atom of an element to retain or gain electrons in interatomic bonds with other elements.

Electronic Beam Heat Treating

A selective surface hardening process that rapidly heats a surface by direct bombardment with an accelerated stream of electrons.

Electronic Beam Microprobe Analyzer

An instrument for selective analysis of a microscopic component or feature in which an electron beam bombards the point of interest in a vacuum at a given energy level. Scanning of a larger area permits determination of the distribution of selected elements. The analysis is made by measuring the wavelengths and intensities of secondary electromagnetic radiation resulting from the bombardment.

Electronic Ceramics

See *ceramics*.

Electronic Diffraction

The phenomenon or the technique of producing diffraction patterns through the incidence of electrons upon matter. See also *diffraction pattern*.

Electronic Heat Control

A device for adjusting the heating value (rms value) of the current in making a resistance weld by controlling the ignition or firing of the electronic devices in an electronic contactor. The current is initiated each half-cycle at an adjustable time with respect to the zero point of the voltage wave.

Electronic Microscope Column

The assembly of gun, lenses, specimen, and viewing and plate chambers. See also *scanning electron microscope.*

Electronic Packaging

The technical discipline of designing a protective enclosure for an electronic circuit so that it will both survive and perform under a plurality of environmental conditions.

Electronic Phenomena (Mechanisms)

These involve the activities of electrons on a subtle and individual scale and usually low current and are termed electronic; examples include thermionic valves, transistors and equipment where such components are the primary features such as television sets, computers, and modern cash machines. Phenomena and mechanisms that involve electron flow in bulk, often at high current, are better termed electrical; examples include motors and heaters although these may have control systems that are electronic.

Electrophoresis

The movement of colloidal particles, or other finely divided particulate matter suspended in a fluid, which is electrically charged by, and then responds to, an external electrical field.

Electrophoretic Plating

The deposition on an immersed surface of finely divided particulate or colloidal material by application of an electrical field.

Electroplate

The application of a metallic coating on a surface by means of an electrolytic action.

Electroplating

The electrodeposition of an adherent metallic coating on an object serving as a *cathode* for the purpose of securing a surface with properties or dimensions different from those of the basis metal.

Electroplated Coatings

Electroplating may be defined as the electrodeposition of an adherent metallic coating upon an electrode for the purpose of securing a surface with properties or dimensions different from those of the base metal. It must not be confused with electroforming or "electroless plating." "Brush plating," on the other hand, is a special method of electroplating.

In brush plating, the anode may be soluble or insoluble. It is covered with a cloth or similar spongelike material and moistened with the plating solution while it is gently moved back and forth across the surface to be plated. It is usually used for specialized applications.

Electroplating is extensively used to produce printed circuit boards. Its main advantage is that the circuit can be produced directly rather than having to be etched out of a piece of copper sheet. Electroplating is also widely used to impart corrosion resistance. Most parts of automobile bodies are zinc plated for corrosion resistance. Because zinc is more readily attacked by most corrosive agents that automobiles encounter than steel, it provides galvanic or sacrificial protection. An electrolytic cell is formed in which zinc, the less noble metal, is the anode and the steel, the more noble one, is the cathode. The anode corrodes and the cathode is protected. Zinc also provides a good base for paint. If a metal is more noble than the one on which it is electroplated, it provides protection against corrosion only if it is completely continuous. If a small area of the substrate is exposed such as under a pinhole, corrosion occurs there, rapidly forming a pit.

An example of an electroplated coating applied primarily for wear resistance is hard chromium on a rotating shaft. Electroplating is also used to build up worn or undersized parts. Gold-plated jewelry is an example of a decorative application. Gold and also palladium are electrodeposited on electrical contacts. Here, the absence of an oxide film avoids the rise in the electrical resistance of the contact. Nickel and aluminum are plated for some decorative applications; however, their wide use in the automotive industry has diminished considerably, primarily because of the associated environmental problems. Magnetic components made of such alloys as **Permalloy** can be manufactured by electroplating.

Process

The electroplating process consists essentially of connecting the parts to be plated to the negative terminal of a DC source and another piece of metal to the positive pole, and immersing both in a solution containing ions of the metal to be deposited. The part connected to the negative terminal becomes the cathode, and the other piece is the anode. In general, the anode is a piece of the same metal that is to be plated. Metal dissolves at the anode and is plated at the cathode. If the current is used only to dissolve and deposit the metal to be plated, the process is 100% efficient. Often, fractions of the applied current are diverted to other reactions such as the evolution of H_2 at the cathode; this usage results in lower efficiencies as well as changes in the acidity (pH) of the plating solution. In some processes, such as chromium plating, a piece of metal that is essentially insoluble in the plating solution is the anode. When such insoluble anodes are used, metal ions in the form of soluble compounds must also be added periodically to the plating solution. The anode area is generally about the same as that of the cathode; in some applications it is larger.

Most plating solutions are of the aqueous type. There is a limited use of fused salts or organic liquids as solvents. Nonaqueous solutions are employed for the deposition of metals such as aluminum that have overvoltages lower than H_2. Such metals cannot be plated in the presence of H_2O, as H_2 would be preferentially reduced.

In addition to metal ions, plating solutions contain relatively large quantities of various substances used to increase the electrical conductivity, to buffer, and in some instances to form complexes with the metal ions. Relatively small amounts of other substances, which are called addition agents, are also present in plating solutions to level and brighten the deposit, to reduce internal stress, to improve the mechanical properties, and to reduce the size of the metal crystals or grains or to change their orientation.

The quantity of metal deposited, that is, the thickness, depends on the current density (A/m), the plating time, and the cathode efficiency. The current is determined by the applied voltage, the electrical conductivity of the plating solution, the distance between anode and cathode, and polarization. Polarization potentials develop because of the various reactions and processes that occur at the anode and cathode, and depend on the rates of these reactions, that is, the current density. If the distance between anode and cathode varies because the part to be plated is irregular in shape, the thickness of the deposit may vary. A quantity

E

called the throwing power represents the degree to which a uniform deposit thickness is attained on areas of the cathode at varying distances from the anode. Good throwing power results if the plating efficiency is low because of polarization where the current density is high.

Plating of Specific Metals

Most metals can be electroplated from either aqueous or fused-salt solutions. The more important metals plated from aqueous baths are Cr, Cu, Au, Ni, Ag, Sn, and Zn. Alloys can also be electroplated. Electrodeposited Cu–Zn and Pb–Sn alloys are used extensively.

Chromium

Electroplated chromium is used primarily to produce wear and corrosion-resistant coatings. Chromium is not deposited for decorating purposes as extensively as in the past because of the associated pollution problem from the discharge of hexavalent chromium. Chromium plating solutions consist primarily of Cr_2O_3, H_2SO_4, and H_2O.

Copper

Electroplated copper is used extensively in the manufacture of printed circuit boards. Because copper cannot be directly plated on the insulating material substrates, they must be rendered electrically conductive first. The main advantage of using electrolytically deposited copper to produce printed circuits is that the areas of the board that are made conductive can be controlled. The actual circuit can be produced by selective etching or selective plating involving techniques such as photosensitizing, photoresist, and etch resist. Areas exposed through a mask can become conductive by coating them with electroless copper after activation with a solution of stannous and palladium chloride. Suitable organic materials are also used to render the board conductive. The areas made conductive serve as a substrate for copper electrodeposition. Some circuits are produced only with electroless plating of copper. Only the through holes are plated, generally with electroplated copper over electroless plated copper when the boards are made by laminating copper sheet to the plastic substrate. The copper segments of the printed circuit may be coated with an electroplated Sn–Pb alloy to facilitate subsequent soldering and also to protect them from oxidation. Gold deposits are also used for this purpose. There are other uses of electroplated copper in the electronic industry, such as in the production of microchips. Electroplated copper is also the undercoat for decorative Ni–Cr deposits.

Gold

Gold plating is used for electrical contacts that must remain free of oxides, connections for microcircuits, certain information-storage devices, solid-state components, and jewelry. The use of gold for electrical and electronic application exceeds that for decorative purposes. Electroplated palladium is being substituted for gold in some applications.

Nickel

Nickel coatings covered with chromium provide corrosion-resistant and decorative finishes for steel, brass, and zinc-base die castings. The most widely used plating solution is the Watts bath, which contains nickel sulfate, nickel chloride, and boric acid. All-chloride, sulfamate, and fluoborate plating solutions are also used. The sulfamate solution is used for low-stress applications. There are a number of compounds, mostly organic, which can be added to nickel-plating baths.

Silver

The principal use of silver electrodeposits is for tableware because of their corrosion resistance (except to sulfur-containing foods) and pleasing appearance. Other important uses are for bearings and electrical circuits, waveguides, and hot-gas seals. The plating solutions are of the cyanide type and generally contain additives that produce bright deposits.

Tin

Tin is used in electrodeposition as a component of solders. Solder is electroplated over copper to protect it from oxidation and to facilitate subsequent joining operations. The advantage of electroplating the solder is that it can be applied only where it is needed. Because of the desirability of eliminating lead in solders, those with higher tin contents are used.

Tin-plated steel for cans in which food is preserved has limited use, because lacquers are preferred to prevent contact between steel and food. Tin plating is employed for refrigerator coils and bearing surfaces. Bright pore-free surfaces can be produced by melting the electroplated tin and allowing it to "reflow."

Stannous chloride and stannous sulfate are the main components of acid tin-plating solutions.

Zinc

The sacrificial protection of steel against corrosion is the main reason for plating zinc. It is used extensively in the automobile industry for this purpose. Screws, bolts, and washers are barrel-plated with zinc also for corrosion protection. Continuous zinc electroplating of wire and strip is another application. The advantage of electro-depositing zinc over hot dipping is the ability to apply thinner coatings and higher purity.

Properties of Electrodeposits

In the various applications of electrodeposits, certain properties must be controlled and, therefore, must be measured. The properties of electroplated metal that should be considered, depending on the use of the deposit, are thickness, adhesion to substrate, brightness, corrosion resistance, wear resistance, the mechanical properties of yield strength, tensile strength, ductility, and hardness, internal stress, solderability, density, electrical conductivity, and the magnetic characteristics.

Electropolishing

A technique commonly used to prepare metallographic specimens, in which a high polish is produced in making the specimen the *anode* in an *electrolytic cell*, where preferential dissolution at high points smooths the surface. Also referred to as *electrolytic polishing*.

Electrorefining

Using electric or electrolytic methods to convert impure metal to purer metal, or to produce an alloy from impure or partly purified raw materials.

Electrorheological (ER) Fluids

Suspensions of fine particles, usually polymers, in nonconducting oils or other liquids. When electric current is passed through them, they turn from liquid to Jell-O-like solids or vice versa in 0.001–0.0001 s. With the amount of applied voltage governing the degree of solidity, the fluid itself can perform various control functions, such as damping shock and vibration. Particles include *aluminosilicate zeolites* and *polyacene quinones*. The most common fluids are *silicone oils*, *gasoline*, and *kerosene*.

Electroslag Remelting (ESR)

A *consumable-electrode remelting* process in which heat is generated by the passage of electric current through a conductive slag. The droplets of metal are refined by contact with the slag.

Electroslag Welding

A process in which the two faces to be welded, usually the edges of thick plates, are set vertically and spaced with a wide gap between them. The bottom of the gap is closed by a starter plate and the two sides are closed by movable water-cooled copper shoes. A granulated flux is placed in the enclosure and an arc struck between a filler wire or wires and the starter plate. The arc melts the flux, which then conducts the current, remains molten and melts the continuously fed filler. The arc is extinguished after the flux commences to carry the current. The molten filler beneath the flux layer fuses with the parent plates, the copper shoes commence moving upwards and a weld is formed between the two plates. The resistance heated slag not only melts filler-metal electrodes as they are fed into the slag layer, but also provides shielding for the massive weld puddle characteristic of the process.

Electroslag Welding Electrode

A filler metal component of the welding circuit through which current is conducted between the electrode guiding member and the molten slag (flux).

Electrostatic Lens

A lens producing a potential field capable of deflecting electron rays to form an image of an object.

Electrostatic Precipitator

Equipment in which a gas is passed across a high-voltage field to extract dust. The particles collect an electrical charge and are attracted to catcher troughs that direct them into hoppers.

Electrostrictive Effect

The reversible interaction, exhibited by some crystalline materials, between an elastic strain and an electric field. The direction of the strain is independent of the polarity of the field. Compare with *piezoelectric effect*.

Electrotinning

Electroplating tin on an object.

Electrotwinning

Recovery of a metal from an ore by means of electrochemical processes.

Electrotyping

The production of printing plates by *electroforming*.

Electrovalent Bonding

See *interatomic bonding*.

Element

A substance that can exist as a single atom. See *atomic structure*. There are 92 naturally occurring elements such as hydrogen, iron, chlorine, uranium, etc., plus a dozen or so man-made.

Elevation

In the context of engineering drawings, it is a joining of the appearance as viewed from the end (end elevation), or the side (side elevation), as opposed to alternative views such as plan (viewed from above) or section.

Elliptical Bearing

See *lemon bearing (elliptical bearing)*.

Elongated Alpha

A fibrous structure in titanium alloys brought about by unidirectional metalworking. It may be enhanced by the prior presence of blocky and/or grain boundary α.

Elongated Grain

A grain with one principal axis slightly longer than either of the other two.

Elongation

(1) A term used in mechanical testing to describe the amount of extension of a test piece when stressed. (2) In tensile testing, the increase in the gauge length, measured after fracture of the specimen within the gauge length, usually expressed as a percentage of the original gauge length. See also *elongation, percent*.

Elongation at Break

Elongation recorded at the moment of rupture of a specimen, often expressed as a percentage of the original length.

Elongation, Percent

The extension of a uniform section of a specimen expressed as a percentage of the original gauge length:

$$\text{Elongation}, \% = \frac{\left(L_x - L_o\right)}{L_o} - 100$$

where
   L_o is the original gauge length
   L_x is the final gauge length

Elutriation

A test for particle size in which the speed of a liquid or gas is used to suspend particles of a desired size, with larger sizes settling for removal and weighing, while smaller sizes are removed, collected, and weighed at certain time intervals.

Embeddability

The ability of a bearing material to embed harmful foreign particles and reduce their tendency to cause scoring or abrasion.

Embedded Abrasive

Fragments of abrasive particles forced into the surface of a workpiece during grinding, abrasion, or polishing.

Embossing

(1) Technique used to create depressions of a specific pattern in plastic film and sheeting. Such embossing in the form of surface patterns can be achieved on molded parts by the treatment of the mold surface with photoengraving or another process. (2) Raising a design in relief against a surface.

Embossing Die

A die used for producing embossed designs.

Embrittlement

The severe loss of *ductility* or *toughness* or both, of a material, usually a metal or alloy. Many forms of embrittlement can lead to *brittle fracture*. Many forms can occur during or thermal treatment or elevated temperature service (thermally induced embrittlement). Some of these forms of embrittlement, which affects steels, include *blue brittleness, 885°F (475°C) embrittlement, quench-age embrittlement, sigma-phase embrittlement, strain-age embrittlement, temper embrittlement, tempered martensite embrittlement,* and *thermal embrittlement*. In addition, steels and other metals and alloys can be embrittled by environmental conditions (environmentally assisted embrittlement). The forms of environmental embrittlement include *acid embrittlement, caustic embrittlement, corrosion embrittlement, creep-rupture embrittlement, hydrogen embrittlement, liquid metal embrittlement, neutron embrittlement, solder embrittlement, solid metal embrittlement,* and *stress-corrosion cracking*.

Embrittlement [885°F (475°C)]

Embrittlement of stainless steels upon extended exposure to temperatures between 400°C and 510°C (750°F and 950°F). This type of embrittlement is caused by fine, chromium-rich precipitates that segregate at grain boundaries; time and temperature directly influences the amount of segregation. Grain-boundary segregation of the chromium-rich precipitates increases strength and hardness, decreases ductility and toughness, and changes corrosion resistance. This type of embrittlement can be reversed by heating above the precipitation range.

Emery

A fine-grained, impure variety of the mineral corundum, with fine crystals of aluminum oxide embedded in a matrix of iron oxide. It usually contains only 55%–75% Al_2O_3. The specific gravity is 3.7–4.3 and Mohs hardness about 8. It occurs as a dark-brown, granular massive mineral. It is used as an abrasive either ground into powder or in blocks and wheels. In the material, the grains are irregular, giving a varying grinding performance. The grains are graded in sizes from 220 mesh, the finest to 20 mesh, the coarsest. *Emery paper* and cloth are usually graded from 24 to 120 mesh, and the grains are glued to one side of 9 by 11 in. (23 by 28 cm) sheets. *Flour of emery* is the finest powder, usually dust from the crushing. Emery cake is made for buffing and polishing is not likely to be made of emery but a graded combination of aluminum oxide and iron oxide, with a higher percentage of the hard aluminum oxide for buffing, and higher iron oxide for polishing.

EMF

An abbreviation for *electromotive force*.

Emission (of Electromagnetic Radiation)

The creation of radiant energy in matter, resulting in a corresponding decrease in the energy of the emitting system.

Emission Lines

Spectral lines resulting from emission of electromagnetic radiation by atoms, ions, or molecules during changes from excited states to states of lower energy.

Emission Spectrometer

An instrument that measures percent concentration of elements in samples of metals and other materials: when the sample is vaporized by an electric spark or arc, the characteristic wavelengths of light emitted by each element are measured with a diffraction grating and an array of photodetectors or photographic plates.

Emission Spectroscopy

The branch of spectroscopy treating the theory, interpretation, and application of spectra originating in the emission of electromagnetic radiation by atoms, ions, radicals, and molecules.

Emission Spectrum

An electromagnetic spectrum produced when radiation from any emitting source, excited by any of various forms of energy, is dispersed.

Emissive Electrode

A filler metal electrode consisting of a core of a bare electrode or a composite electrode to which a very light coating has been applied to produce a stable arc.

Emissivity

Ratio of the amount of energy or of energetic particles radiated from a unit area of a surface to the amount radiated from a unit area of an ideal emitter under the same conditions. It is normally necessary to specify circumstances such as wavelength, surface condition, temperature, etc.

Emulsifying Agents

Materials used to aid in the mixing of liquids that are not soluble in one another, or to stabilize the suspension of nonliquid materials in a liquid in which the nonliquid is not soluble. The suspension of droplets of one liquid in another liquid in which the first liquid is not soluble is called an *emulsion*. The emulsion of oil and water, used in machine shops as a cutting lubricant and work coolant, may be made with soap as the emulsifying agent. The emulsifying agent protects droplets of the dispersed medium from uniting and thus separating out. The oil itself may be treated so that it is self-emulsifying. Sulfonated oils contain strong negatively charged ester sulfate groups in the molecule and do not react and conglomerate with the molecules of a weakly charged liquid. They will thus form emulsions with water without any other agent.

In emulsions of a powder in a liquid, an emulsifying agent called a *protective colloid* may be used. This is usually a material of high molecular weight such as gelatin, and such materials form a protective film around each particle of the contained powder. A photographic emulsion is a suspension of finely divided silver halide grains in gelatin. The gelatin serves as a binder, protective colloid, and sensitizer for the silver halide. The emulsion consists of 35%–40% silver halide, 60–65 gelatin, with small amounts of stabilizers, antifoggers, and hardeners. For the suspension of drug materials and pharmaceutical mixtures, *gum arabic* or *tragacanth* may be used. Starches, egg albumin, and proteins are common emulsifying agents for food preparation. *Alginates* are among the best suspending agents for a wide range of emulsions because of the numerous repelling charges in the high molecular weight and the irregular configuration of the chain, but when added to protein-containing liquids such as many foodstuffs, the similar conditions of the algin and the protein molecules cause a neutralization reaction and a precipitation of the agglomerated particles.

Sucrose esters, used as emulsifiers for foods, cosmetics, and drugs, are made from sugar and palmitic, lauric, or other fatty acids. The monoesters are soluble in water and alcohol, and the diesters are oil-soluble.

Emulsion

(1) A teo-phase liquid system in which small droplets of one liquid (the internal phase) are immiscible in, and are dispersed uniformly throughout, a second continuous liquid phase (the external phase). The internal phase is sometimes described as the dispense phase. (2) A stable dispersion of one liquid in another, generally by means of an emulsifying agent that has affinity for both the continuous and discontinuous phases. The emulsifying agent, discontinuous phase, and continuous phase can together produce another phrase that serves as an enveloping (encapsulating) protective phase around the discontinuous phase.

Emulsion Calibration Curve

The plot of a function of the relative transmittance of the photographic emulsion versus a function of the exposure. The calibration curve is used in spectrographic analysis to calculate the relative intensity of a radiant source from the density of a photographically recorded image.

Emulsion Cleaner

A cleaner consisting of organic solvents dispersed in an aqueous medium with the aid of an emulsifying agent.

Emulsion Inversion

An *emulsion* is said to invert when, for example, a water-in-oil emulsion changes to an oil-in-water emulsion.

Emulsion Polymerization

Polymerization of monomers dispersed in an aqueous emulsion.

Enamel

A coating that upon hardening has an enameled or glossy face. *Pottery enamels*, *ceramic enamels*, or *ceramic coatings*, and *vitreous enamels* are composed chiefly of quartz, feldspar, clay, soda, and borax, with saltpeter or borax as fluxes. The quartz supplies the silica, and such enamels are fusible glasses. In acid-resisting enamels, alkali earths may be used instead of borates. To ensure enamels opaque, opacifiers are used. They may be tin oxide for white enamel, cobalt oxide for blue, or platinum oxide for gray. Enamel-making materials are prepared in the form of a powder which is called frit. The frit-making temperature is about 2400°F (1316°C), but the enamel application temperatures are from 1400°F to 1600°F (760°C–871°C). Each succeeding coat has a lower melting point than the one before it, so as not to destroy the preceding coat. It must also have about the same coefficient of expansion as the metal to prevent cracking. Enamels for aluminum usually have a high proportion of lead oxide to lower the melting point, and enamels for magnesium may be based on lithium oxide. Some enamels for low-melting-point metals have the ceramic frit bonded to the metal with monoaluminum phosphate at temperatures as low as 400°F (204°C).

The mineral oxide coatings Fuster metals are often called *porcelain enamels*, but they are not porcelain, and the term *vitreous enamel* is preferred in the industry, although ceramic-lined tanks and pipe are very often referred to as *glass-lined steel*. Vitreous enameled metals are used for cooking, utensils, signs, chemical tanks and piping, clock and instrument dials, and siding and roofing. Thick coatings on thin metals are fragile, but thin coatings on heavy metals are flexible enough to be bent. Standard porcelain-type enamel has a smooth, glossy surface with a light reflectance of at least 65% in the white color, but pebbly surfaces that breakup the reflected image may be used for architectural applications.

High-temperature coatings may contain a very high percentage of zirconium and will withstand temperatures to 1650°F (899°C). *Refractory enamels*, for coating superalloys to protect against the erosion of hot gases to 2500°F (1371°C), may be made with standard ceramic frits to which is added boron nitride with a lithium chromate or fluoride flux. Blue undercoats containing cobalt are generally used to obtain high adhesion on iron and steel, but some of the *enameling steels* do not require an undercoat, especially when a specially compounded frit or special fluxes used. When *sodium aluminum silicate* is used instead of borax, a white finish is produced without a ground coat.

Many trade names are applied to vitreous enamels and to enameled metals. *Vitric steel* is an enameled corrugated sheet steel for construction. Cloisonné enamel is an ancient decorative enamel produced by soldering thin strips of gold on the base metal to form cells into which the colored enamel is pressed and fused into place. It requires costly hand methods and is now imitated in synthetic plastics under names such as *Enameloid*.

The word *enamel* in the paint industry refers to glossy varnishes with pigments or to paints of oxide or sulfate pigments mixed with varnish to give a glossy face. They vary widely in composition, in color and appearance, in the properties. As a class, enamels are hard and tough and offer good mar and abrasion resistance. They can be formulated to resist attack by the most commonly encountered chemical agents and corrosive atmospheres. Because of their wide range of useful properties, enamels are one of the most widely used organic finishes in industry and are especially used as household appliance finishes. The modified phenolmelamine and alkyd-melamine synthetic resins produce tough and resistant enamel coatings. Quick-drying enamels are the cellulose lacquers with pigments. Fibrous enamel, used for painting roofs, is an asphalt solution in which asbestos fibers have been incorporated. When of heavy consistency and used for caulking metal roofs, it is called *roof putty*.

Enameling Iron

A low-carbon, cold-rolled sheet steel, produced specifically for use as a base metal for porcelain enamel.

Enantiotropy

The relation of crystal forms of the same substance in which one form is stable above a certain temperature and the other form is stable below that temperature. For example, ferrite and austenite are enantiotropic in ferrous alloys.

Encapsulated Adhesive

An adhesive in which the particles or droplets of one of the reactive components are enclosed in a protective film (microcapsules) to prevent cure until the film is destroyed by suitable means.

Encapsulation

The enclosure of an item in plastic. Sometimes used specifically in reference to the enclosure of capacitors or circuit board modules.

End

In composites fabrication, a strand of *roving* consisting of a given number of filaments gathered together. The group of filaments is considered an end, or strand, before twisting, and a yarn after twist has been applied. An individual warp, yarn, thread, fiber, or roving.

End Clearance Angle

See *clearance angle* and also the term *face mill* and *single-point tool*.

End Count

An exact number of ends supplied on a ball of roving. See also *end*.

End Cutting-Edge Angle

The angle of concavity between the face cutting edge and the face plane of the cutter. It serves as relief to prevent the face cutting edges from rubbing in the cut. See *face mill* and *single-point tool*.

End Mark

A roll mark caused by the end of a sheet marking the roll during hot or cold rolling.

End Mill

A rotating cylindrical cutting tool with multiple cutting edges on its flat (or nearly so) end face.

End Milling

A method of machining with a rotating cutting tool with cutting edges on both the face end and the periphery. See also *face milling* and *milling*.

End Relief

Defined by the term *single-point tool*.

End Return

See preferred term *boxing*.

End-Centered

Having an atom or group of atoms separated by a translation of the type ½, ½, 0 from a similar atom or group of atoms. The number of atoms in an end-centered cell must be a multiple of 2.

Endothermic Atmosphere

A gas mixture produced by the partial combustion of a hydrocarbon gas with air in an endothermic reaction. Also known as endogas.

Endothermic Reaction

Designating or pertaining to a reaction that involves the absorption of heat. See also *exothermic reaction*.

End-Quench Hardenability Test

A laboratory procedure for determining the hardenability of a steel or other ferrous alloy; widely referred to as the Jominy test. Hardenability is determined by heating a standard specimen above the upper critical temperature, placing the hot specimen and a fixture so that a stream of cold water impinges on one end, and, after cooling to room temperature is completed, measuring the hardness near the surface of the specimen at regularly spaced intervals along its length. The data are normally plotted as hardness versus distance from the quenched end. See the term *Jominy test*.

Endurance

The capacity of a material to withstand repeated application of stress.

Endurance Limit

(1) The maximum stress that a material can withstand for an infinitely large number of fatigue cycles. See also *fatigue limit* and *fatigue strength*. (2) An endurance limit may also be specified in the case of steels subjected to stresses which exceed their fatigue limit. See *fatigue*.

Endurance Ratio

(1) The ratio of the *endurance limit* for completely reversed flexural stress to the tensile strength of a given material. (2) The ratio of fatigue limit to ultimate tensile strength. For ferritic steels of U.T.S. up to about 1100 MPa (75 T/in.^{2}) it is approximately 0.5. See *fatigue*.

Energy Barrier

See *potential barrier*.

Energy Trough

See *potential barrier*.

Energy-Dispersive Spectroscopy (EDS)

A method of x-ray analysis that discriminates by energy levels the characteristic x-rays emitted from the sample. Compare with *wavelengths-dispersive spectroscopy*.

Engine Oil

An oil used to lubricate an internal combustion engine.

Engineering Adhesive

A bonding agent intended to join metal, plastics, wood, glass, ceramics, or rubber. The term differentiates such bonding agents from glues used to join paper and other nondurables.

Engineering Ceramics

Same as *advanced ceramics*.

Engineering Films

Specific properties that separate engineering films from their commodity counterparts include greater tensile and impact strength; improve moisture and gas barrier characteristics; good heat resistance and weatherability; better bonding and lamination; and improved electrical ratings. One or more of these properties can be obtained by choosing from a number of different polymer films.

Several melt-processable engineering thermoplastic films such as oriented polyester, oriented nylon, and unoriented nylon, exhibit high strength, especially at high temperatures. In addition, they provide toughness at low temperatures, stiffness and abrasion resistance, and good chemical resistance. Polyester film is made from the PET polymer. The monomer is polymerized, extruded, cast into a web, and biaxially oriented, forming a drawn polyester film.

Oriented polypropylene, a thermoplastic with low specific gravity, has excellent resistance, relatively high melting point, and good strength. Polycarbonate film is specified for its toughness, clarity, and high heat-deflection temperature.

Polyimides, both thermoplastics and thermosets, retain their principal properties over a wide temperature range. They have useful mechanical properties, even at cryogenic temperatures. At −162°C, the film can be bent around a 6.4-mm mandrel without breaking and, at 500°C, its tensile strength is 31 MPa. Room-temperature mechanical properties are comparable to those of polyester film.

To minimize transmission of moisture vapors, fluoropolymers are the best choice. This family of material has a general paraffinic structure with some or all of the hydrogen atoms replaced by fluorine.

Polyetheretherketone (PEEK), a high-performance thermoplastic, offers outstanding thermal properties as well as resistance to many solvents and proprietary fluids. This film can be used for interbonding or cladding in PEEK structural components. Thermoplastic and thermoset acrylics are noted for exceptional clarity and weatherability, and also offer favorable stiffness, density, chemical resistance, and toughness.

Plastic film can be manufactured from almost any resin, but not every resin produces an engineering film. Generally, the properties of a resin-based film are related to the chemistry of the basic polymer; however, properties may be further affected by subsequent processing techniques. Manufacturers can choose from, and end users can specify, a range of process treatments that significantly enhance heat stability, mechanical properties, electrical characteristics, barrier properties, and bondability.

Coatings are typically applied by emulsion, solvent, and dry methods. Results vary according to the formula used: PVDC coatings improve barrier properties; polyurethane improves abrasion resistance; and aluminum coatings alter electrical characteristics.

Some processors have developed proprietary antistatic coatings that are cured by electron-beam radiation. Metal coatings produce conductive capabilities and also enhance barrier properties. Often, metallization is used to improve moisture-barrier properties for biaxially oriented nylon and polypropylene.

Surface treatment, which removes low-molecular-weight residue, improves adhesion and appearance. Several methods may be used. Corona discharge techniques position the film between an electrically grounded roller and a high-voltage electrode. A continuous-arc discharge (corona) is generated to clean and activate the film surface.

In gas-plasma surface treatments, film is placed in a reaction chamber. After evacuation, the chamber is charged with O_2, A, helium, or N_2, while a radiofrequency field ionizes the gas. A resultant glow discharge creates free radicals on the surface, in adhesion.

Engineering Plastics

A general term covering all plastics, with or without fillers or reinforcements, that have mechanical, chemical, and thermal properties suitable for use as construction materials, machine components, and chemical processing equipment components. Included are acrylonitrile-butadiene-styrene, acetal, acrylic, fluorocarbon, nylon, phenoxy, polybutylene, polyaryl ether, polycarbonate, polyether (chlorinated), polyether sulfone, polyphenylene oxide, polysulfone, polyimide, rigid polyvinyl chloride, polyphenylene sulfide, thermoplastic urethane elastomers, and many other reinforced plastics.

Engineering Steels

In the United Kingdom, steels have a six-character designation comprising three initial digits, a letter and two final digits, for example, 080M40. These designations indicate, with some anomalies and to a limited extent, the type of steel, its condition of supply and its composition. The first three digits indicate alloy type. The fourth character, a letter, indicates the condition of supply. The final digits indicate for most grades except the austenitics the mean of the carbon content range.

Engineering Strain (e)

A term sometimes used for average linear strain in order to differentiate it from *true strain*. In tension testing, it is calculated by dividing the change in the gauge length by the original gauge length.

Engineering Stress (s)

A term sometimes used for conventional stress in order to differentiate it from *true stress*. In tension testing, it is calculated by dividing the breaking load applied to the specimen by the original cross-sectional area of the specimen.

Engler Viscosity

A commercial measure of viscosity expressed as the ratio between the time in seconds required for 200 cm^3 of a fluid to flow through the orifice of an Engler viscometer at a given temperature under specified conditions and the time required for 200 cm^3 of distilled water at 20°C (70°F) to flow through the orifice under the same conditions.

Engobe

A slip coating applied to a ceramic body for imparting color, opacity, or other characteristics, and subsequently covered with a glaze.

Enthalpy

The total internal and external energy content of a system.

Entraining Velocity

The velocity of a liquid at which bubbles of gas are carried along in the strain.

Entropy

The thermodynamic measure of the state of disorder in a system.

Entry Mark (Exit Mark)

A slight corrugation caused by the entry or exit rolls of a *roller leveling unit*.

Environment

The aggregate of all conditions (such as contamination, temperature, humidity, radiation, magnetic and electric fields, shock, and vibration) that externally influence the performance of a material or component.

Environmental Cracking

Brittle fracture of a normally ductile material in which the corrosive effect of the environment is a causative factor. Environmental cracking is a general term that includes *corrosion fatigue, high-temperature hydrogen attack, hydrogen blistering, hydrogen embrittlement, liquid metal embrittlement, solid metal embrittlement, stress-corrosion cracking,* and *sulfide stress cracking.* The following terms have been used in the past in connection with environmental cracking, but are becoming obsolete: caustic embrittlement, delayed fracture, season cracking, static fatigue, stepwise cracking, sulfide corrosion cracking, and sulfide stress-corrosion cracking. See also *embrittlement.*

Environmental Stress Cracking (alternatively, Stress Cracking)

Cracking occurring as a result of the effect of a corrosive environment and stress acting concurrently. This term is common with reference to polymeric plastic materials but is occasionally used in the case of metals as an alternative to stress corrosion. The stress may be externally imposed or it may be residual from manufacturing operations.

Environmentally Assisted Embrittlement

See *embrittlement.*

Epitaxis/Epitaxy

(1) The growth of a new crystal at the surface of a pre-existing crystal where the lattice orientation of the new crystal is dictated, across the intervening boundary, by the lattice of the preexisting crystal. (2) Growth of electrodeposit or vapor deposit in which the orientation of the crystals in the deposit are directly related to crystal orientations in the underlying crystalline substrate.

Epichlorohydrin

The basic epoxidizing resin intermediate in the production of epoxy resins. It contains an epoxy group and is highly reactive with polyhydric phenols such as bisphenol A.

Epoxide

Compound containing the oxirane structure, a three-member ring containing two carbon atoms and one oxygen atom. The most important members are ethylene oxide and propylene oxide.

Epoxy Plastic

A thermoset polymer containing one or more epoxide groups and curable by reaction with amines, alkyl malls, phenols, carboxylic acids, acid anhydrides, and mercaptans. An important matrix resin in reinforced composites and in structural adhesives.

Epoxy Resins

Epoxy (Ep) resins comprise an extremely broad and diverse family of materials. They are used in the form of protective coatings, adhesives, reinforced plastics, molding compounds, casting and potting compounds, and foams.

Epoxies, perhaps best known as adhesives, are premium thermosetting plastics, and are generally employed in high-performance uses where their high cost is justified. They are available in a wide variety of forms, both liquid and solid, and are cured into the finished plastic by a catalyst or with partners containing active H_2. Depending on the type, they are cured at either room temperature or elevated temperatures.

Liquid epoxies are used for casting, for potting or encapsulation, and for laminating. They are used unfilled or with any of a number of different mineral or metallic powders. Molding compounds are available as liquids, and also as powders with various types of fillers and reinforcements.

Properties

The epoxy resins have very high adhesion to metals and nonmetals, heat resistance from 177°C to 260°C, dielectric strength to 22 V/m, and hardness to Rockwell M110. The tensile strength may be up to 82 MPa, with elongation to 2%–5%, but some resilient encapsulating resins are made with elongation to 150% with lower tensile strengths. The resins have high resistance to common solvents, oils, and chemicals.

An unlimited variety of epoxy resins is possible by varying the basic reactions with different chemicals or different catalysts, or both, by combination with other resins, or by cross-linking with organic acids, amines, and other agents. To reduce costs when used as laminating adhesives, they may be blended with furfural resins, giving adhesives of high strength and high chemical resistance. Blends with polyamides have high dielectric strength, mold well, and are used for encapsulating electrical components. By using a polyamide curing agent, an epoxy can be made water emulsifiable for use in water-based paints. An epoxy resin with 19% bromine in the molecule is flame-resistant. Another grade, with 49% bromine, is a semisolid, used for heat-resistant adhesives and coatings.

The epoxies have excellent electrical, thermal, and chemical resistance. Their strength can be further increased with fibrous reinforcement or mineral fillers. The variety of combinations of epoxy resins and reinforcements provides a wide latitude in properties obtainable in molded parts.

Molded epoxy parts are hard, rigid, relatively brittle, and have excellent dimensional stability over a broad temperature range. Some fiber-reinforced formulations can withstand service temperatures above 206°C for brief periods. Their excellent electrical properties, in combination with high mechanical strength, qualify them for electrostructural applications. Resins based on bisphenol-A are adequate for most services. However, cycloaliphatics are recommended for parts subjected to arcing conditions or those requiring outdoor weatherability, see Table E.2.

Excellent adhesion in structural applications is another outstanding property of epoxy systems. Epoxy adhesives for bonding many dissimilar materials can be supplied either as one- or two-part systems. One-part systems require heat for curing; two-part systems usually cure at room temperature, but properties are improved when the materials are heat-cured. Some epoxy adhesive systems can withstand temperatures to 232°C, although properties at such temperatures are considerably lower than at room temperature.

Applications

- Casting resins are primarily used for potting or encapsulating electrical or electronic equipment. The excellent adhesion and extremely low shrinkage of the epoxies, coupled with their high dielectric properties, provide a well-sealed, voidless, well-insulated component.
- Foaming resins have also been used for electronic potting.
- Liquid resins are also formulated with a variety of fillers, such as metal powder, to provide effective patching and repair putties or pastes. Such compounds can be used to patch both metal and plastic surfaces.
- Liquid resin systems are used to produce low-pressure reinforced laminates and moldings, high-pressure industrial thermosetting laminates (NEMA Grade G-10 glass cloth base, as well as paper-based laminates for electrical uses), and filament wound shapes.
- Epoxy laminates are also widely used in plastic tools. They are used for drilling, checking, and locating fixtures, where dimensional accuracy is critical. They are also used to provide durable surfaces for metal-forming tools, such as draw dies for short-run production.
- Filament wound structures are used for such applications as rocket-motor booster cases, pressure vessels, and chemical tanks and pipe.
- Molding compounds provide the performance characteristics of epoxy resins with the automated speed and economy of compression and transfer molding. They are being used primarily for electrical components.

Handling and Resin Selection

Because the cure of an epoxy is brought about by a chemical reaction, and not by a simple process of solvent evaporation, it is quite

TABLE E.2
Properties of Epoxies

| | | Molding Compounds | | Casting Resins | | |
| | | | | Bisphenol A | | |
ASTM Test	Property	Glass Fiber Filler	Mineral Filler	No Filler	Silica Filler	Cycloaliphatic
Physical						
D792	Specific gravity	1.6–2.0	1.6–3.0	1.11–1.40	1.6–2.0	1.16–1.21
D792	Specific volume (in.^{3}/lb)	15.4–14.0	14.2–13.4	26.9–20.0	17.3–13.9	—
D570	Water absorption. 24 h ½ in. ink (%)	0.05–0.20	0.04	0.08–0.15	0.04–0.10	0.08–0.15
Mechanical						
D638	Tensile strength (psi)	10,000–20,000	5,000–10,000	4,000–13,000	7,000–13,000	10,000–20,000
D638	Elongation (%)	4	—	3–6	1.3	2–10
D638	Tensile modulus (10^3 psi)	30.4	—	3.5	—	5–9
D790	Flexural strength (psi)	10,000–60,000	8,000–15,000	13,300–21,000	8,000–14,000	15,000–32,000
D256	Impact strength Izod (ft. Ib/in. of notch)	2–30	0.3–0.4	0.2–1.0	0.3–0.45	0.2–1.0
D785	Hardness. Shore	100–110	100–110	80–110	85–120	85–120
Thermal						
C177	Thermal conductivity (10^{-4} cal cm/s cm^2 °C)	4–10	4–30	4.0–5.0	10–20	—
D696	Coefficient of thermal expansion (10^{-3} in./in. °F)	1.1–3.5	2.0–5.0	4.5–6.0	2.0–4.0	—
D648	Deflection temperature (°F) at 264 psi	250–500	250–500	115–550	160–550	200–450
Electrical						
D149	Dielectric strength, V/mil; short time, 1/2 in. thick	300–400	300–400	300–500	400–550	420–440
D150	Dielectric constant at kHz	3.5–5.0	3.5–5.0	3.5–4.5	3.2–4.0	3.9–4.5
D150	Dissipation factor at 1 kHz	0.01	0.01	0.002–0.02	0.008–0.03	—
D257	Volume resistivity (Ω cm) at 73°F. 50% RH	>10^{14}	>10^{14}	10^{12}–10^{11}	10^{12}–10^{10}	10^{15}
D495	Are resistance (s)	120–180	150–190	45–120	150–300	150–180

Source: Mach Design Basics Eng. Design, 688, June 1993. With permission.

important that users recognize at the outset that handling (particularly for the two-component, low-temperature curing compounds) requires more care than that of older materials. However, because this is, and probably will be, a problem for some time to come, procedures have been developed that keep difficulties in handling to a minimum. One simple and inexpensive way of eliminating frequent weighing is to use calibrated mixing containers where it is necessary only to fill to the first calibration with the base, to the second with a hardener, mix, and apply.

Formulators meet this problem by supplying the compounds packaged in preweighted containers, while equipment manufacturers use automatic mixers and dispensers. It is possible through the use of these machines to eject preweighted and premixed shots of compound as the operator requires.

One-component systems, as the name indicates, eliminate problems of weighing, mixing, and pot life, and are subdivided into solids, most often "B" staged epoxy powders, and liquids containing latent hardeners such as boron trifluoride, or anhydride/solvent solutions.

Powders have been used to some extent as adhesives and in potting, but are employed most widely for the fluidized-bed coating method or in compression and transfer molding. The epoxy fluidized-bed approach provides a relatively easy way of applying encapsulant coatings in uniform thickness over contours but is not suitable for impregnation, or for coating units that cannot be heated.

In resin selection, epoxy compounds can be categorized (although there are exceptions) as (1) two-component liquids that will cure at temperatures from 21.1°C to 60°C, but that have short pot lives; (2) two-component liquid systems requiring cure at temperatures up to 177°C, but that offer longer pot lives and improved operational characteristics; and (3) one-component liquids and powders that reduce handling problems, offer optimum operating characteristics, but that require higher curing temperatures (in the range of 177°C–204°C).

With respect to possible methods of application, with the exception of the powders, epoxies are supplied for use by spray, brush, spatula, roller coat, knife coat, dipping, filament winding, laminating, and casting. All these except casting are self-descriptive and this is simply the method of producing a defined shape by pouring a compound into a mold where it cures and from whence it can later be removed. On the other hand, if an object has been placed in the mold and epoxy cast around it, the process is called potting.

Epoxy Smear

See *resin smear*.

Epsilon (Ɛ)

Designation generally assigned to intermetallic, metal-metalloid, and metal-nonmetallic compounds found in ferrous alloy systems, for example, Fe_3Mo_2, $FeSi$, and Fe_3P.

Epsilon Carbide

Carbide with hexagonal close-packed lattice that precipitates during the first stage of tempering of primary martensite. Its composition corresponds to the empirical formula $Fe_{2.4}C$.

Epsilon Structure

Structurally analogous close-packed phases or electron compounds that have ratios of seven valence electrons to four atoms.

Equator

In *filament winding* of composites, the line in a pressure vessel described by the junction of a cylindrical portion and the end dome. Also called tangent line or point.

Equiaxed Grain Structure

A structure in which the grains have approximately the same dimensions in all directions.

Equicohesive Temperature

The temperature at which damage development within grains is matched by that at grain boundaries for a given strain rate. Above this temperature, a material fails in an intergranular mode, below in a transgranular mode.

Equilibrium

The dynamic condition of physical, chemical, mechanical, or atomic balance that appears to be a condition of rest rather than one of change.

Equilibrium Centrifugation

In resinography, a method for determining the distribution of a molecular weight by spinning a solution of the specimen at a speed such that the molecules of the specimen are not removed from the solvent but are held at a point where the (centrifugal) force tending to remove them is balanced by the dispersive forces caused by thermal agitation.

Equilibrium Diagram

A graph of the temperature, pressure, and composition limits of phase fields in an alloy system as they exist under conditions of thermal dynamical equilibrium. In metal systems, pressure is usually considered constant. Compare with *phase diagram*.

Equilibrium (Reversible) Potential

The *potential* of an electrode in an electrolytic solution when the forward rate of a given reaction is exactly equal to the reverse rate. The equilibrium potential can only be defined with respect to a specific electrochemical reaction.

Equilibrium Segregation

The tendency for solute atoms in a solid solution to diffuse to particular sites such as grain boundaries. This may result in depletion of the solute in the near grain boundary regions. See *solution*.

Equivalent/Equivalence Factor

Various factors applied to alloying elements to allow comparison of their relative effect on some aspect such as microstructure or hardenability. See carbon equivalent for example.

Equivalent Radial Load

The level of constant radial load on a rolling-element bearing that, when the bearing is stationary with respect to the outer race, will produce the same rating life as a given combination of radial and thrust loads under the same conditions of operation.

Equivalent Section/Round

In the context of hardenability of steel, the circular section equivalent to the noncircular section under consideration.

Equivalent Weight

The mass of a substance that will react, directly or indirectly, with unit mass of hydrogen.

Erg

The unit of energy in the non-SI metric (centimeter-gram-second) system; it is the work done by a force of one dyne moving 1 cm. 1 erg = 10^{-7} J.

Erichson (*Cupping*) Test

A test for determining the ductility of sheet metal particularly for cupping or deep drawing applications. The sheet under test is held by an outer ram against a die face having a central hole. A ball-ended cone plunger on an inner ram is then forced into the sheet until a crack appears. The high, in millimeters, of the cup at crack initiation is a measure of ductility.

Erosion

(1) Loss of material from a solid surface due to relative motion in contact with a fluid that contains solid particles. Erosion in which the relative motion of particles is nearly parallel to the solid surface is called *abrasive erosion*. Erosion in which the relative motion of the solid particles is nearly normal to the solid surface is called *impingement erosion* or impact erosion. (2) Progressive loss of original material from a solid surface due to mechanical interaction between that surface and a fluid, a multicomponent fluid, and impinging liquid, or solid particles. (3) Loss of material from the surface of an electrical contact due to an electrical discharge (arcing). See also *cavitation erosion*, *electrical pitting*, and *erosion-corrosion* and *cavitation*.

Erosion (Brazing)

A condition caused by dissolution of the base metal by molten filler metal resulting in a postbraze reduction in the thickness of the base metal.

Erosion Rate

Any determination of the rate of loss of material (erosion) with exposure duration. And certain contexts (e.g., ASTM erosion tests), it is given by the slope of the cumulative erosion–time curve.

Erosion Rate-Time Curve

A plot of instantaneous erosion rate versus exposure duration, usually obtained by numerical or graphical differentiation of the cumulative erosion–time curve.

Erosion-Corrosion

A conjoint action involving *corrosion* and *erosion* in the presence of a moving corrosive fluid, leading to the accelerator loss of material.

Erosive Wear

See *erosion*.

Erosivity

The characteristic of a collection of particles, liquid stream, or a slurry that expresses its tendency to cause erosive wear when forced against a solid surface under relative motion.

Error

Deviation from the correct value. In the case of a testing machine, the difference obtained by subtracting the load indicated by the calibration device from the load indicated by the testing machine.

Escape Peak

An artifact observed in x-ray analysis; manifested as a peak at energy 1.74 keV (the silicon Kα Peak) less than the major line detected. Escape peaks can be avoided by increasing the accelerating voltage.

Esters

Combination of alcohols with organic acids, which form several important groups of commercial materials. The esters occur naturally in vegetable and animal oils and fats as combinations of acids with the alcohol glycerin. The natural fats are usually mixtures of esters of many acids, coconut oil having no less than 14 acids. Stearic, oleic, palmitic, and linoleic acid esters are the common bases for most vegetable and animal fats, and the esters of the other acids such as linolenic, capric, and arachidic give the peculiar characteristics of the particular fat, although the physical characteristics and melting points may be governed by the basic esters. Esters occur also in waxes, the vegetable waxes being usually found on the outside of leaves and fruits to protect them from loss of water. The waxes differ from the fats in that they are combinations of monacids with monohydric, or simple, alcohols rather than with glycerin. They are harder than fats and have higher melting points. Esters of still lower molecular weights are also widely distributed in the essential oils of plants where they give the characteristic odors and tastes. All the esters have the characteristic formula ArCOOR or RCOOR, where R represents an *alkyl group*, and Ar an *aryl group*, that is, where R is a univalent straight-chain hydrocarbon and Ar is a univalent benzene ring. In the esters of low molecular weight that make the odors and flavors, the combination of different alcohols with the same acid yields oils of different flavor. Thus the esther methyl acetate is *peppermint oil*; amyl acetate is banana oil; and isoamyl acetate is *pear oil*. Esters are used as solvents, flavors, perfumes, waxes, oils, fats, fatty acids, pharmaceuticals, and in the manufacture of soaps and many chemicals. Esther liquid lubricants have good heat and oxidation resistance at high temperatures and good fluidity at low temperatures. They are widely used in jet aircraft.

The natural esters are recovered by pressing or extraction, and steam distillation. Synthetic esters are prepared by reacting an alcohol with an organic acid in the presence of a catalyst, such as sulfuric acid or *paratoluenesulfonic acid*. The product is purified with an azeotrope, such as benzene or toluene. A range of *cellulose acetate esters* are made by esterification of cellulose with acetate anhydride. *Cellulose nitrate ester* is obtained by reacting cellulose with nitric acid, *cellulose sulfate* from *chlorosulfonic acid* in pyridine solvent, and *cellulose phosphate* from phosphoric acid in the molten urea. *Alkoxysilanes* are silicon esters in which the silicon is connected to an organic group by oxygen.

Esther alcohols are intermediates that require less acid for esterification.

Estimate

This particular value, or values, of a parameter computed by an estimation procedure for a given sample.

Estimation

A procedure for making a statistical inference about the numerical values of one or more unknown population parameters from the observed values in a sample.

Etch Cleaning

Removing soil by dissolving away some of the underlying metal.

Etch Figuring

In metallography, deliberate etching to an extent that would normally be excessive to produce a pattern of pitting which reveals some feature of interest. Such features may be on a microscale, for example, the facets of the pits can reflect crystallographic planes, or on a macroscale as the distribution of pitting can indicate bulk segregation or flow.

Etch Pits

Pitting resulting from excessive etching unless etch figuring is intended.

Etch Rinsing

Pouring etchant over a tilted surface until the desired degree of attack is achieved. Use for excellence with severe gas formation.

Etchant

(1) A chemical solution used to etch a metal to reveal structural details. (2) A solution used to remove, by chemical reaction, the unwanted portion of material from a printed circuit board. (3) Hydrofluoric acid or other agents used to attack the surface of glass for marking or decoration.

Etched Metal Mask

A mask formed by etching apertures through a solid metal protected by a photo resist.

Etching

(1) Subjecting the surface of a metal to preferential chemical or electrolytic attack in order to reveal structural details for metallographic examination. (2) Chemically or electrochemically removing tenacious films from a metal surface to condition the surface for a subsequent treatment, such as painting or electroplating. (3) A process by which a printed pattern is formed on a printed circuit board by either chemical or chemical and electrolytic removal of the unwanted portion of conductive material bonded to a base. (4) Unless otherwise specified, the term usually implies chemical attack in a liquid. Other techniques include *electrochemical dissolution*, *selective oxidation in air at elevated temperature,* and *ion bombardment.*

Etching Materials

These are chemicals, usually acids, employed for cutting into, or etching, the surface of metals, glass, or other material. In the metal industries, they are called etchants. The usual method of etching is to coat the surface with a wax, asphalt, or other substance not acted upon by the acid; cut the design through with a sharp instrument; and then allow the acid to corrode or dissolve the exposed parts. For etching steel, a 25% solution of H_2SO_4 in water or an $FeCl_3$ solution may be used. For etching stainless steel a solution of $FeCl_3$ and HCl in water is used. For high-speed steels, brass, or nickel, a mixture of HNO_3 and HCl acids in water solution is used, or nickel may be etched with a 45% solution of H_2SO_4. Copper may be etched with a solution of $HCrO_3$ acid. Brass and nickel may be etched with an acid solution of $FeCl_3$ and $KClO_3$. For red brasses, deep etching is done with concentrated HNO_3 acid mixed with 10% HCl acid, with the latter added to keep the SnO_2 in solution and thus retain a surface exposed to the action of the acid. For etching aluminum, a 9% solution of $CuCl_3$ and 1% acetic acid, or a 20% solution of $FeCl_3$ may be used, followed by a wash with strong HNO_3 acid. NaOH, NH_4OH, or any alkaline solutions are also used for etching aluminum. Zinc is preferably etched with weak HNO_3 acid, but requires a frequent renewal of the acid. Strong acid is not used because of the heat generated, which destroys the wax coating. A 5% solution of HNO_3 acid will remove 0.005 cm of zinc/min compared with the removal of over 0.013 cm/min in most metal-etching processes. Glasses etch with HF acid or with white acid. White acid is a mixture of HF acid and ammonium bifluoride, a white crystalline material of the composition (NH_4) FHF.

The process in which the metal is removed chemically to give the desired finish as a substitute for mechanical machining is called chemical machining.

Ether

Ether is the common name for ethyl ether, or diethyl ether, a highly volatile, colorless liquid made from ethyl alcohol. It is used as a solvent for fats, greases, resins, and nitrocellulose, and in medicine as an anesthetic. The specific gravity is 0.720, and its vapor is heavier than air and is explosive.

Recently, the promotion for the production and use of cleaner-burning fuels was announced. As a result, methyl tertiary butyl ether (MTBF) became a very important petrochemical. MTBF is the most widely produced ether for oxygenates. It is commonly produced by the dehydrogenation of isobutane and the subsequent reaction of isobutylene with methanol.

Ethyl Silicate

A strong bonding agent for sand and refractories used in preparing molds in the investment casting process.

Ethylene Plastics

Plastics based on polymers of ethylene or copolymers of ethylene with other monomers, the ethylene being in greatest amount by mass.

Ethylene-Propylene Elastomer

Ethylene-propylene elastomer is a completely saturated copolymer made by solution polymerization. The remarkable properties of the material include exceptional ozone resistance, excellent electrical properties, good high- (149°C–163°C) and low-temperature properties, good stress–strain characteristics, and resistance to chemicals, light, and other types of aging.

Ethylene-propylene rubber has required a peroxide or peroxide-sulfur modified curing system. Sulfur improves the peroxide curing efficiency and assists in chemical cross-linking of the polymer chains, thereby importing better physical properties to the vulcanizate.

Some plasticizers used in other rubbers are not suitable for ethylene-propylene rubber. Most acceptable plasticizers are saturated materials of relatively low polarity such as paraffinic hydrocarbon oils and waxes.

Euler Angles

In crystallographic texture analysis, three angular parameters that specify the orientation of a body with respect to reference axes.

Eutectic

(1) An isothermal reversible reaction in which a liquid solution is converted into two or more intimately mixed solids on cooling, the number of solids formed being the same as a number of components in the system. (2) An alloy having the composition indicated by the eutectic point on a phase diagram. (3) An alloy structure of intermixed solid constituents formed by a eutectic reaction often in the form of regular arrays of lamellae or rods. (4) A eutectic is an intimate mixture of two solid solutions formed, on solidifying from a single liquid solution, at the eutectic temperature. Metals mixed together in different ways in both the liquid and the solid states. One simple case where two metals, A and B, are completely soluble in each other.

Eutectic Arrest

In a cooling or heating curve, an approximately isothermal segment corresponding to the time interval during which the heat transformation from the liquid phase to two or more solid phases is evolving.

Eutectic Bonding

The forming of a bond by bringing two solids to their lowest constant melting point at which the molten solids mix and become hard upon cooling.

Eutectic Carbides

Carbide formed during freezing as one of the mutually insoluble phases participating in the eutectic reaction of a hypereutectic tool steel. See also *hypereutectic alloy.*

Eutectic Melting

Melting of localized microscopic areas whose composition corresponds to that of the eutectic in the system.

Eutectic Point

The composition of a liquid phase in univariant equilibrium with two or more solid phases; the lowest melting alloy of a composition series.

Eutectic-Cell Etching

Development of eutectic cells (grains).

Eutectoid

(1) An isothermal reversible reaction in which a solid solution is converted into two or more intimately mixed solids on cooling, the number of solids formed being the same as the number of components in the system. (2) An alloy having composition indicated by the eutectoid point on a phase diagram. (3) An alloy structure of intermixed solid constituents formed by a eutectoid reaction.

Eutectoid Point

The composition of a solid phase that undergoes univariant transformation into two or more other solid phases upon cooling.

Evaporation

The vaporization of a material by heating, usually in a vacuum. In electron microscopy, this process is used for shadowing or to produce thin support films by condensation of the vapors of metals or salts.

Evaporative Deposition

The techniques of condensing a thin film of material on a substrate. The entire process takes place in a high vacuum. The source material may be radioactively heated by bombardment with electrons (electron beam) or may be heated by thermal-conduction techniques.

Even Tension

In composite reinforcing fibers, the process whereby each end of roving is kept in the same degree of tension as the other ends making up that ball of roving. See also *catenary.*

Ewald Sphere

In electron diffraction theory, a geometric construction, of radius equal to the reciprocal of the wavelength of the incident radiation, with its surface at the origin of the reciprocal lattice. Any crystal plane will reflect if the corresponding reciprocal lattice point lies on the surface of this sphere.

Exchange Current

When an electrode reaches dynamic equilibrium in a solution, the rate of anodic dissolution balances the rate of cathodic plating. The rate at which either positive or negative charges are entering or leaving the surface at this point is known as the exchange current.

Exchange Current Density

The rate of charge transfer per unit area when an electrode reaches dynamic equilibrium (at its reversible potential) in a solution; that is, the rate of anodic charge transfer (oxidation) balances the rate of cathodic charge transfer (reduction).

E

Excitation Index

In materials characterization, the ratio of the intensities of two selected spectral lines of an element having widely different excitation energies. This ratio serves to indicate the level of excitation energy in the source.

Excitation Potential (X-Ray)

The applied potential on an x-ray tube required to produce characteristic radiation from the target.

Excitation Volume

The volume within the sample in which data signals originate.

Exfoliation

Corrosion that proceeds laterally from the sites of initiation along planes parallel to the surface, generally at grain boundaries, forming corrosion products that force metal away from the body of the material, giving rise to a layered appearance.

Exogenous Inclusion

An *inclusion* that is derived from external causes. Slag, draws, entrapped mold materials, and refractories are examples of inclusions that would be classified as exogenous. In most cases, these inclusions are macroscopic or visible to the naked eye. Compare with *indigenous inclusion*.

Exotherm

The temperature/time curve of a chemical reaction or a phase change giving off heat, particularly the polymerization of casting resins. The amount of heat given off. The term has not been standardized with respect to sample size, ambient temperature, degree of mixing, and so forth.

Exothermic

Characterized by the liberation of heat.

Exothermic Atmosphere

A gas mixture produced by the partial combustion of a hydrocarbon gas with air in an exothermic reaction. Also known as *exogas*.

Exothermic Reaction

A reaction that liberates heat, such as the burning of fuel or when certain plastic resins are cured chemically.

Expandable Plastic

A plastic that can be made cellular by thermal, chemical, or mechanical means. Foam plastics such as expandable polystyrene and foamed polyurethane are examples.

Expanded Metal

Sheet metal that has been slit and expanded to form a mesh, which is used for reinforced-concrete work or plaster wall construction, and for making grills, vents, and such articles as trays, where stiffness is needed with light weight. The expanded metal has greater rigidity than the original metal sheet and permits welding of the concrete or plaster through the holes. It is made either with a plain diamond-shape mesh or with rectangular meshes. One type is made by slitting the sheet and stretching the slits into the diamond shape. The other variety is made by pushing out and expanding the metal in the meshes so that the flat surface of the cut strand is nearly at right angles to the surface of the sheet. Expanded metal is made from low-carbon steel, iron, or special metals. It is also marketed as *metal lath*, usually 2.5 m long and 0.4–0.5 m wide. *Rigidized metal*, or *textured metal*, is thin sheet that is not perforated, but has the designs rolled into the sheet so that the rigidity of the sheet is increased two to four times. Thus, extremely thin sheets of stainless steel can be made for novelties, small mechanical products, and paneling. *Perforated metals* are sheet metals with the perforations actually blanked out of the metal. They are marketed and sheets of carbon steel, stainless steel, or Monel metal, with a great variety of standard designs those with round, square, diamond, and rectangular designs are used for screens and for construction. *Agaloy* is perforated metal made into tube form.

Expanding

A process used to increase the diameter of a cup, shell, or tube. See also *bulging*.

Expansion

(1) An increase in size of a powder metallurgy compact, usually related to an increase in temperature. A decrease in temperature produces an opposite effect. (2) Sometimes used in mean growth.

Expansion Film

An *interference* or *force fit* made by placing a cold (subzero) inside member into a warmer outside member and allowing an equalization of temperature.

Expansive Metal

An alloy that expands on cooling from the liquid state. The expansive property of certain metals is an important characteristic in the production of accurate castings having full details of the mold such as type castings. The alloys are also used for proof-casting of forging dies, for sealing joints, for making duplicates of master patterns, for holding die parts and punches in place, and for filling defects in metal parts in castings. Antimony and bismuth are the metals most used to give expansion to the alloys.

Expendable Pattern

A pattern that is destroyed in making a casting. It is usually made of wax (*investment casting*) or expanded polystyrene (hot foam casting).

Explosion Welding

A solid-state welding process affected by a controlled detonation, which causes the parts to move together at high velocity. The resulting bond zone has a characteristic wavy appearance.

Explosive

A material that, upon application of a blow or by rise in temperature, is converted in a small space of time to other compounds more stable and occupying much more space. Commercial explosives are solids or liquids that can be instantaneously converted by friction, heat, shock, or spark to a large volume of gas, thereby developing a sudden rise in pressure which is utilized for blasting or propelling purposes. *Gunpowder* is the oldest form of commercial or military explosive, but this has been replaced for military purposes by more powerfully acting chemicals. *Smokeless powder* was a term used to designate nitrocellulose powders as distinguished from the smoky black gunpowder. *Blasting powders* are required to be relatively slow-acting to have a heaving or rending effect. Military explosive used as propellants must not give instantaneous detonation, which would burst the gun, but are arranged to burn slowly at first and not reach a maximum explosion until the projectile reaches the muzzle. This characteristic is also required in explosives used for the explosive forming of hard metals. The more rapid acting high explosives are generally used for bombs, torpedoes, boosters, and detonators. The *detonators* are extremely sensitive explosives, such as the fulminates, set off by a slight blow but too sensitive to be used in quantity as a charge. The *booster explosives* are extremely rapid but not as sensitive as the detonators. They are exploded by the detonators used and in turn set off the main charge of explosive. Some explosives such as nitroglycerin can be exploded by themselves, while others require oxygen carriers or carbon carriers mixed with them. In combination with *nitrocellulose*, it is the principal component of powders and solid rocket propellants. Together with *nitroglycol*, it is the major constituent of gelatinous industrial explosives. Other requirements of explosives are that they not react with the metal container, be stable at ordinary temperatures, and not decompose easily in storage or on exposure to air.

Shaped charges of high explosive give a penetrating effect, known as the *Monroe affect*, used in armor-piercing charges. A solid mass of explosives spends itself as a flat blast; but with a conical hole in the charge, and having the open end facing the target, a terrific piercing effect is generated by the converging detonation waves coming from the sides of the column. This effect drives a jet of hot gases through the steel armor.

Trinitrotoluene, or *trinitrotoluol*, also commonly known as TNT is the principal constituent of many explosives. It is safe in handling because it does not detonate easily but is exploded readily with mercury fulminate and is used for shrapnel, hand grenades, mines, and depth bombs. TNT is made by the titration of toluol with nitric and sulfuric acids.

Explosive Compacting

See *high-energy-rate compacting*.

Explosive Forming

The shaping of metal parts in which the forming pressure is generated by an explosive charge that takes the place of the punch in conventional forming. A single-element die is used with the blank held over it, and the explosive charge is suspended over the blank at a predetermined distance (standoff distance). The complete assembly is often immersed in a tank of water. See also *high-energy-rate forming*.

Exposed Underlayer

In composites fabrication, the underlying layer of mat or roving not covered by surface mat in a protrusion. This condition can be caused by reinforcement shifting, too narrow a surface mat, too wide an underlying mat, uneven slitting of the surface mat, necking down of the surface mat, or excessive tension in pulling the surface mat off the spindle.

Exposure

The product of the intensity of a radiant source in the time of irradiation.

Extend

The addition of fillers or low-cost materials to plastic resins as an economy measure. To add inert materials to improve void-filling characteristics and reduce crazing.

Extended X-Ray Absorption Fine Structure (EXAFS)

The weak oscillatory structure extending for several hundred electron volts away from an absorption edge. The oscillations occur because the electromagnetic wave produced by the ionization of the absorbing atom for some energy E has a wavelength $\lambda = 1.225/(E - E_k)^{1/2}$ nm, where E_k is the energy of the absorption edge. For example, a loss of 100 eV above an edge corresponds to a wavelength of 0.12 nm, which is of the order of atomic spacing. Consequently, the wave can be diffracted from neighboring atoms and return to interfere with the outgoing wave. An analysis of EXAFS data reveals important information about atomic arrangements and bonding. Either synchrotron x-radiation or the electron beam in the analytical transmission electron microscope can be used as the excitation source. See also *analytical electron microscopy* and *synchrotron radiation*.

Extenders

(1) Low-cost materials used to dilute or extend high-cost resins without significant lessening of properties. (2) Substances, generally having some adhesive action, added to an adhesive to reduce the amount of primary binder required per unit area. See also *binder*, *diluent*, *filler*, and *thinner*.

Extensibility

The ability of a material to extend or elongate upon application of sufficient force, expressed as a percent of the original length.

Extensional-Bending Composite

A property of certain classes of laminates that exhibit bending curvatures when subjected to extensional loading.

Extensometer

An instrument for measuring changes in length over a given gauge length caused by application or removal of a force. Commonly used in tension testing.

External Circuit

The wires, connectors, measuring devices, current sources, etc., that are used to bring about or measure the desired electrical conditions within the test cell. It is this portion of the cell through which electrons travel.

Externally Pressurized Seal

A seal that operates on a thin film at the interface with the mating surface. The film is formed by high-pressure fluid that is brought to the interface at some mid-dam location and that is at a pressure equal to, or higher than, the upstream seal pressure. See also *hydrostatic seal*.

Extinction

A decrease in the intensity of the diffracted x-ray beam caused by perfection or near perfection of crystal structure. See also *primary extinction* and *secondary extinction*.

Extra Hard

A *temper* of nonferrous alloys and some ferrous alloys characterized by values of tensile strength and hardness about one-third of the way from those of *full hard* to those of *extra spring temper*.

Extra Spring

A *temper* of nonferrous alloys and some ferrous alloys corresponding approximately to a cold worked state above *full hard* beyond which further cold work will not measurably increase strength or hardness.

Extraction

A general term denoting chemical methods of isolating phases from a metal matrix.

Extractive Metallurgy

The branch of *process metallurgy* dealing with the winning of metals from their ores. Compare with *refining*.

Extreme-Pressure Lubricant

A lubricant that imparts increased load-carrying capacity to rubbing surfaces under severe operating conditions. Extreme-pressure lubricants usually contain sulfur, halogens, or phosphorus. The term *antiscuffing lubricant* has been suggested as a replacement for extreme-pressure lubricant.

Extreme-Pressure Lubrication

A condition of lubrication in which the friction and wear between two surfaces in relative motion depend upon the reaction of the lubricant with a rubbing surface at elevated temperature.

Extruded Hole

A hole formed by a punch that first cleanly cuts a hole and then is pushed farther through to form a flange with an enlargement of the original coal.

Extruded Metals

Extrusion is the forcing of solid metal through a suitably shaped orifice under compressive forces. Extrusion is somewhat analogous to squeezing toothpaste through a tube, although some cold extrusion processes more nearly resemble forging, which also deforms metals by application of compressive forces. Most metals can be extruded, although the process may not be economically feasible for high-strength alloys.

Hot Extrusion

The most widely used method for producing extruded shapes is a direct, hot extrusion process. In this process, a heated billet of metal is placed in a cylindrical chamber and then compressed by a hydraulically operated ram. The opposite end of the cylinder contains a die having an orifice of the desired shape; as this die opening is the path of least resistance for the billet under pressure, the metal "squirts" out of the opening as a continuous bar with the same cross-sectional shape as the opening. By using two sets of dies, stepped extrusions can be made. The small section is extruded to the desired length, the small split die is replaced by the large die, and the large section is then extruded.

The most outstanding feature of the extrusion process is its ability to produce a wide variety of section configurations. Structural shapes can be extruded that have complex nonuniform and nonsymmetrical sections that would be difficult or impossible to roll.

In many instances, extrusions can replace bulky assemblies made up by joining, welding, or riveting rolled structural shapes, or sections previously machined from bar, plate, or pipe.

An extrusion die is relatively simple to make and inexpensive when compared to a pair of rolls or a set of forging dies. The low cost of dies and the short lead time for die changes make it possible to extrude small quantities more economically than by most other methods.

Lubricants are used to minimize friction and protect the die surfaces. Graphite is a common lubricant for nonferrous alloys, whereas for hot extrusion of steel, glass is an excellent lubricant.

Indirect, or inverted, extrusion was developed to overcome such difficulties as surface friction and entrainment of surface oxide of direct extrusion. In the indirect process the ram is hollow, the die opening is in the dummy block, and the opposite end of the cylinder is closed. As the ram advances, the billet does not move as in the case of direct extrusion, and the metal is extruded backward through the die and the hollow ram. However, the process is not very popular because the hollow ram is weaker, resulting in lower machine capacity; trouble-free operation requires that the extruded product be straight and not hit the inside of the ram.

Cold Extrusion

The extrusion of cold metal is variously termed *cold pressing, cold forging, cold extrusion forging, extrusion pressing,* and *impact extrusion*. The term *cold extrusion* has become popular in the steel fabrication industry, while *impact extrusion* is more widely used in the nonferrous field.

The original process (identified as *impact extrusion*) consists of the punch (generally moving at high velocity) striking a blank (or slug) of the metal to be extruded, which has been placed in a cavity of a die. Clearance is left between the punch in the die walls; as the punch comes in contact with the blank, the metal has nowhere to go except through the annular opening between punch and die. The punch moves a distance that is controlled by a press setting. This distance determines the base thickness of the finished part. The process is particularly adaptable to the production of thin-walled, tubular-shaped parts with thick bottoms, such as toothpaste tubes.

Advantages of cold extrusion are high strength because of severe strain-hardening, good finish and dimensional accuracy, and economy due to fewer operations and minimum machining required.

Metals Extruded

Extrusion can be used to fabricate practically all structural metals and alloys. Among the more common metals extruded on a commercial or semi-commercial basis are alloys in the following metal systems:

Magnesium	Carbon and alloy steels
Aluminum	Stainless steel
Brass	Iron superalloys
Copper	Nickel superalloys
Titanium	Columbium or Niobium
Zirconium	Molybdenum
Beryllium	Tantalum
Nickel	Tungsten

These materials span a range of working temperatures from about 316°C to 2205°C, in approximately the order shown above.

The wide range of extrusion temperatures gives rise to the major differences in processing that center around such variables as extrusion lubricants, die materials, die design, billet preparation, and extrusion speed. Present tool materials are capable of maintaining adequate strength and wear resistance for extrusion at temperatures only slightly higher than 538°C. At higher temperatures, lubricants are necessary not only to reduce friction but to insulate and protect the tooling surface from overheating. Also, the speed of extrusion must be more rapid to avoid prolonged contact between the tools and the hot billet. Thus, the extrusion method for magnesium and aluminum is quite different from that for the other metal systems.

The Light Alloys

Magnesium and aluminum are extruded at temperatures below 538°C with no lubrication and flat, sharp-cornered dies. Deformation of the billet occurs by shear flow, which is from within the billet so that the surface skin of the billet is retained in the container as discard. This type of turbulent flow is possible because of the ability of these materials to form sound welds when severely deformed, but requires comparatively slow pressing speeds, often less than 1.66 m/min. With clean tools and no lubricants there are no contaminants present to cause internal defects or laminations, and several sections can be extruded at one time by using multihole dies. Precise dimensional control is attained with ordinary hot-work tool-steel dies, which last for hundreds of extrusions.

Other Metals

With higher-temperature materials, for example, titanium, steels, refractory metals, it is necessary to use lubricants and die designs so that deformation occurs by uniform flow. In this case, the surface of the billet becomes the surface of the extrusion; otherwise, laminations and inclusions could occur. Graphitic lubricants are suitable for producing relatively short lengths at temperatures up to about 1093°C if the operation is performed at high speeds. The Ugine-Sejournet process in which molten glass serves as a lubricant is most widely used for high-temperature extrusion. Because of the insulating as well as lubricating properties of glass, overheating of tools does not occur and die life is increased. For titanium and steels, dies are usually made of tungsten hot-work tool steels. Ceramic coatings (Al_2O_3 or ZrO_2) on the dies are necessary at the temperatures required for refractory metals. Pressing speeds are usually in the range of 8.33–33.32 m/min.

Shape, Surface, and Tolerance Limitations

Extruded shapes are generally classified by configuration and include rod, bar, tube, and hollow, semihollow, and solid shape.

Although many asymmetrical shapes can be produced, probably the most important factor in the extrudability is symmetry. Hollow and semihollow shapes cost more than solid shapes and usually cannot be extruded with as thin sections. Semihollow shapes with long thin voids should be avoided. For best extrudability the length-to-width ratio of partially enclosed voids, channels, or grooves should not exceed 3:1 for aluminum and magnesium, 2:1 for brass, or 1:1 for copper, titanium, and steels. Wall thickness surrounding the voids should be as uniform as possible.

The size and weight of extruded shapes are limited both by the section configuration and by the material properties. The maximum size that can be extruded on a press of given capacity is determined by the circumscribing circle, which is the smallest circle that will enclose the shape. The circumscribing circle size controls the die size, which in turn is limited by the press size.

Thickness limitations are related to the size of the cross section as well as the type of material. As a rule, thicker sections are required with increased section size.

Sharp corners and edges are usually possible with aluminum and magnesium alloys, but 0.38 mm corner and fillet radii are preferred. Minimum fillet radii of 3.2 mm for steel and 4.5 mm for titanium are suggested by most extruders. Typical minimum corner radii are 0.8 mm for steel and 1.5 mm for titanium.

Smooth surfaces with finishes better than 30 μin. rms are readily attainable in aluminum and magnesium alloys. High-temperature alloys are characteristically rougher; and extruded finish of 125 μin. rms is generally considered acceptable for most steels and titanium alloys. Improved surface finishes can be produced by a cold-draw finishing operation.

Although extruded shapes minimize and often eliminate the need for machining, they do not possess the dimensional accuracy of machined parts. The tolerances of any given dimension vary somewhat depending on the size and type of shape, and the relative location of the dimension. Detailed standard tolerances covering straightness, flatness, twist, and cross-sectional dimensions such as section thickness, angles, contours, and corner and fillet radii have been established for magnesium, aluminum, copper, and brass by most extruders and are published in handbooks. Standard tolerances also have been established for steels and titanium alloys in simple sections, but in many instances these are subject to mill inquiry.

Extrudable Materials

Extrusion is a process for making articles of constant cross-section, called "continuous shapes," by forcing softened material through a hole approximating the desired shape. With plastics, the process is carried out by one of two methods: ram extrusion or screw extrusion.

In ram extrusion, the softened mass fills a cylinder to which the die—the shaped hole—is attached at one end. A closely fitting piston, the ram, enters the cylinder and pushes the mass through the die at pressures ranging up to 68 MPa. The product, or extrudate, is cooled or otherwise hardened shortly after leaving the die. Subsequent handling depends on the material and the shape. Ram extrusion is used chiefly for extruding TFE fluorocarbon (*tetrafluoroethylene*) resin (*Teflon*), which is damaged by the shearing action of screw extrusion, and for *cellulose nitrate*, whose extreme heat sensitivity and inflammability make screw extrusion dangerous.

Screw extrusion, by far the more economical and commercially important process, centers around the screw extruder. This consists of a heavy cylindrical barrel inside which turns a motor-driven screw, or worm. The screw is essentially a thick shaft with a helical blade, or flight, wrapped around it. At the rear end of the barrel, a feed hopper admits cold plastic particles that normally fall into the screw channel by gravity. As the screw rotates, the particles are dragged forward by frictional action between screw, plastic, and barrel. Electric band heaters on the outside of the barrel heat the plastic, which is further heated by the frictional action of the screw. Soon, the particles coalesce into a voidless mass that softens further to become a melt. This plastic melt is very viscous (a million times as viscous as water), so considerable pressure, on the order of 3.4–68 MPa, must be developed to force it through the die at the front of the extruder at economical rates.

Excludable Materials

All thermoplastics can be extruded by either ram or screw extrusion. Today, high-viscosity grades of type 66 nylon and other nylons are available and their extrusion presents no special difficulties. In extruding rigid polyvinyl chloride (PVC), extreme care must be taken not to overheat the resin since thermal decomposition, once started, snowballs. This simply means being careful to avoid extreme temperatures everywhere and to streamline meticulously all passages through which the melt must pass. To a lesser degree, CFE fluorocarbon (*trifluorochloroethylene*) resin (Kel-F et al.), cellulosics, nylons, and acetal (*polyoxymethylene*) are similarly heat-sensitive.

Kel-F® is a registered trade name of 3M Company. In 1996, 3M discontinued manufacturing of Kel-F and today, PCTFE resin is manufactured by Daikin under the trade name of Neoflon® or by Allied Signal under the trade name of Acion®. PolyChloroTriFluoroEthylene is a fluorocarbon-based polymer and is commonly abbreviated PCTFE. PCTFE (polychlorotrifluoroethylene), sometimes has been referred to as Kel-F®. PCTFE (polychlorotrifluoroethylene) is offered as sheet, rod, and tube.

Some thermosets can be extruded provided they are formulated to flow at temperature safely below the curing temperatures. The process has been used to make pipe and structural shapes, to coat wire with thermosetting compositions, and to prepare "rope" and pellets for compression molding. The extrusion of rubbers closely resembles plastics extrusion.

Extruder (Plastics)

A machine that accepts solid particles, such as pellets or powder, or liquid (molten) feed, conveys it through a surrounding barrel by means of a rotating screw, and pumps it, under pressure, through an orifice.

Extrusion (Ceramics)

(1) The process of forcing a mixture of plastic binder and ceramic powder through the opening (s) of a die at relatively high pressure. The material may thus be compacted and emerges in elongated cylindrical or ribbon (or wire, etc.) form having the cross-section of the die opening. Ordinarily followed by drying, curing, activating, or firing. (2) The process of forming clay products by forcing the plastic material through a die.

Extrusion Billet

A metal slug used as *extrusion stock*.

Extrusion Blow Molding

The most common blow molding process, in which a parison is extruded from a plastic melt and is then entrapped between the halves of a mold. The parison is expanded, under air pressure, against the mold cavity to form the part, and is then cooled, removed, and trimmed. See the term *blow molding*.

Extrusion Coating

Using a resin to coat a substrate by extruding a thin film of molten resin and pressing it onto or into the substrate, or both, without the use of an adhesive.

Extrusion Defect

Defects appearing in the last portion of an extrusion to emerge from the die. It is a characteristic of direct extrusion caused when the contamination on the exterior of the billet is effectively held back by friction against the cylinder wall and is then swept across the face of the ram to emerge initially at the center of the extrusion section. As extrusion continues, the defect forms an annulus progressively moving toward the exterior surface. Normally, the last 10%–20% is discarded because of this phenomenon. It is also termed *back end defect*. See preferred term *extrusion pipe*.

Extrusion Forging

(1) Forcing metal into or through a die opening by restricting flow in other directions. (2) A part made by the operation.

Extrusion Ingot

A cast metal slug uses *extrusion stock*.

Extrusion Pipe

A central oxide-lined discontinuity that occasionally occurs in the last 10%–20% of an extruded metal bar. It is caused by the oxidized outer surface of the billet flowing around the end of the billet and into the center of the bar during the final stages of extrusion. Also called *coring*.

Extrusion Stock

A rod, bar, or other section used to make extrusions.

Exudation

The action by which all or a portion of the low melting constituent of a powder metallurgy compact is forced to the surface during sintering; sometimes referred to as bleed out or sweating.

Eyeleted Hole

In an engineering context, this usually refers to a hole with a raised integral rim. Apart from any decorative aspect the rim, correctly formed, can reduce chaffing on material passing through or by the hole, improve the stiffness of sheet and reduce fatigue susceptibility. More generally, the term may refer to a hole which is protected by an added eyelet as in the eyelet applied to the lace holes of shoes.

Eyeleting

The displacing of material about an opening in sheet or plate so that a lip protruding above the surface is formed.

Eyepiece

A lens or system of lenses for increasing magnification in a microscope by magnifying the image formed by the objective.

F

Fabric Fill Face

That side of a woven fabric used in reinforced plastics on which the greatest number of yarns are perpendicular to the *selvage*.

Fabric Prepreg Batch

In composites processing, a prepreg containing fabric from one fabric batch and impregnated with one batch of resin and one continuous operation.

Fabric Warp Face

That side of a woven fabric used in reinforced plastics on which the greatest number of yarns are parallel to the *selvage*.

Fabrication Traces

Anomalous markings that may appear on crack surfaces of ceramics where the developing crack encounters regions of unusual weakness, density, or elastic modulus, introduced, usually inadvertently, during fabrication.

Fabrics, Nonwoven Bonded

Although there are several types of fabrics that are not woven, the term *nonwoven fabrics* is recognized in the textile trade as applying to those materials composed of a fibrous web held together with a bonding agent to obtain fabric-like qualities. These fabrics may be of a uniform, close-bonded fibrous structure or of a foraminous unitary construction.

They may be formed by processing on modifications either of textile type machines or papermaking equipment. In either case, the fibers as laid up in the basic web prior to bonding or to postforming may be oriented in one or more prescribed directions, or maybe distributed in a completely random fashion. They are secured in place by suitable adhesives incorporated in the web. The application of these adhesives may be controlled to coat and bond the fibers completely, or to bond them only in selected areas, or at points of individual fiber contact.

Nonwovens may be thick or thin and of either low or high density. The conditions under which nonwoven fabrics are manufactured and the possible combinations of fibers and adhesives permit the production of structures offering a wide range of physical and chemical properties.

Production Methods

Production of a nonwoven fabric may be divided into two basic steps: (1) formation of the web and (2) bonding the web. The most widely used means for forming the web is a series of cotton cards feeding to a common conveyor belt to build up a unidirectional composite web of the desired weight. The number of cards per line will depend on the maximum weight of the product to be produced. Each card in the line may be geared to produce webs ranging from some 35 grains up to 100 grains/yard at speeds of 66.66 m/min down to 15 m/min for the heavier material. Material from these lines is usually limited to 101.6 cm widths, with strength favoring the machine direction over the cross machine direction in ratios from 3:1 to as much as 20:1. Where wider material of heavier weight, or when the strength balanced in the machine and cross machine directions is required, a web production line consisting of a single breaker and a finisher garnet equipped with a cross lapper may be used to advantage.

The most versatile type of web for nonwoven fabrics is that produced by the air disposition of precarded fibers that are collected with a minimum of orientation as a uniform mat.

In contrast to nonwoven fabrics made from webs that have been dry-processed on modified textile equipment are those produced from wet-layup webs using papermaking machines. Such webs usually depend on adhesive additives or postbonding to impart the necessary physical properties.

Once the web has been formed by either the dry or wet-layup process, it may be further modified by techniques such as needle punching, aeration, or impingement with gaseous, liquid, or other means, to produce a patterned configuration of desired characteristics.

Nonwovens of such postformed webs may be characterized by added resistance to delamination, superior drape, flexibility, porosity, abrasion, and flame resistance or other desirable properties.

Types of Fibers

In the production of nonwoven fabrics, every type of natural and anthropogenic fiber can be used. Price, equipment, and quality, as well as chemical and physical requirements of the product, govern the particular fiber used as well as the bonding agent. The use of more virgin first-quality fiber, especially the anthropogenic cellulosics, is preferred, in everything from diapers to casket liners.

Rayon is the predominant fiber used for both utility as well as esthetic appeal. It is made in a wide variety of descriptions to fit the different manufacturing methods and end-use requirements. The finer deniers give the best tensile, tear, and bursting strength values. For special applications calling for particular chemical or electrical resistance, more use of the expensive synthetic fibers such as **Acrilan**, nylon, or **Dacron** may be warranted.

Bonding Agents

Properties of nonwoven fabrics are as dependent on the bonding agent as they are on the fiber that forms the foundation of the material. Both are selected with the end use in mind, and each must be compatible with the other.

Bonding agents may be grouped into three broad classifications: (1) liquid dispersions, (2) powdered adhesives, and (3) thermoplastic fibers.

Liquid Dispersions

Liquid dispersions are the most extensive type used. Among these are polyvinyl alcohol, generally used as a preliminary binder or where high strength and permanence are not essential; polyvinyl acetate for good strength and flexibility where freedom from odor and taste are important; polyvinyl chloride (PVC) for good wet and dry strength, and toughness; synthetic lattices of butadiene–acrylonitrile or butadiene–styrene for good adhesive and elastic properties where strength and a high degree of permanence are more important than color, stability, and odor; the acrylics for good strength, soft "hand," color stability, and permanence. These dispersions are applied (1) by spraying, generally used for low-density materials, (2) by saturator, for denser, more durable material, and (3) by printing, usually for selective bonding of localized areas in soft absorbent products.

Powdered Adhesives

These are usually of thermosetting or thermoplastic resin types and are sifted into the fiber web as formed. They are used especially in the low-density, high-bulk nonwovens where wetting by the binder or the application of pressure might cause excessive matting and compression of the material. Bonding is effected by heating either with or without the use of pressure.

Thermoplastic Fibers

The thermoplastic fiber binders have the advantage of constituting an integrated structural part of the fiber web that forms the fabric. To bond the web, they may be activated by solvents or by heat and pressure. By regulation of the amount of heat and pressure as well as the amount of thermoplastic fiber present, a wide variety of characteristics may be built into these nonwoven fabrics.

Applications

Construction and performance of nonwoven fabrics have not been standardized. They are usually constructed to fit a particular end-use requirement or are built around particular specifications.

Industrial products for which nonwoven fabrics have been used include acoustical curtains; artificial leather and chamois; automotive plumpers; backing for adhesive tapes; base for vinyl and rubber coatings; bagging; buffing wheels; cable and wire wrappings; electrical tapes; filters for air, gases, and liquids; insulation; laminate reinforcements; polishing and wiping cloths; and wall coverings.

Fabrics, Woven

By far the greatest volume of textile materials is used in consumer textiles, such as apparel. But textiles are extremely versatile materials, which have been applied to a large number of engineered uses, for example, thermal, acoustical, and electrical insulation; padding and packaging; barrier applications; filtration, both dry and wet; upholstery and seating; reinforcing for plastics or rubber; and various mechanical uses such as fire hose jackets, tenting, tarpaulins, parachutes, and marine lines.

Textiles are highly complex materials. Their properties depend not only on the fiber but also on the form in which it is used—whether the form be a felt, a bonded fabric, a woven or knit fabric, or cordage. Properties such as heat, chemical, and weather resistance depend primarily on the type of fiber used; properties such as mechanical strength, thermal transmission, and air or liquid permeability depend both on the fiber and the textile form.

The versatility of textiles stems from (1) the wide range of fibers that can be used, and (2) the range of complicated textile structures that can be formed from the fibers.

There are two important factors that should be considered in discussing textile needs with textile suppliers:

1. The types of finishes that can be applied to the finished textile product can substantially alter or modify the stability, "hand," and/or durability of the textile.
2. Combining of textiles with other materials, such as resins or rubber, either by impregnation or by coating, will substantially alter performance characteristics of the final composite.

Textile Constructions

Textile engineering materials can be classified generally as (1) nonwoven fabrics, including both felts and bonded fabrics; (2) woven or knit fabrics; and (3) cordage.

Woven and Knit Fabrics

Woven fabrics and knit fabrics are composed of webs of fiber yarns. The yarns may be of either filament (continuous) or staple (short) fibers. In knit fabrics, the yarns are fastened to each other by interlocking loops to form the web. In woven fabrics, the yarns are interlaced at right angles to each other to produce the web. The lengthwise yarns are called the warp, and the crosswise ones are the filling (or woof) yarns.

The many variations of woven fabrics can be grouped into four basic weaves. In the plain weave fabric, each filling yarn alternates up and under successive warp yarns. With a plain weave, the most yarn interlacings per square inch can be obtained for maximum density, "cover," and impermeability. The tightness or openness of the weave, of course, can be varied to any desired degree. In twill weave fabrics, a sharp diagonal line is produced by the warp yarn crossing over two or more filling yarns. Satin weave fabrics are characterized by regularly spaced interlacings at wide intervals. This weave produces a porous fabric with a smooth surface. Satins woven of cotton are called sateen. In the leno weave fabrics, the warp yarns are twisted and the filling yarns are threaded through the twist openings. This weave is used for meshed fabrics and nets.

Cordage

The term *cordage* includes all types of threads, twine, rope, and hawser. Essentially all cordage consists of fibers twisted together, plied, and in many cases cabled to produce essentially continuous strands of desired cross section and strength.

In addition to the type of fiber used, the most important determinants of the end properties of cordage are the type and degree of twist employed. The two major types of twist are (1) cable twist, in which the direction of twisting is alternated in each successive operation, that is, singles may be "S" twisted, plies "Z" twisted, and cables "S" twisted (a yarn or cord has an "S" twist if, when held in a vertical position, the spirals conform in direction of slope to the central portion of the letter "S," and a "Z" twist if the spirals conform in direction of slope to the central portion of the letter "Z"), and (2) hawser twist, in which the singles, plies, and cables are twisted "SSZ" or "ZZS." Hawser twist generally provides higher strength and resilience.

Applications

Astroquartz II and *III* fabrics are made of 95% fused-silica fiber filament yarns, featuring light weight, high strength, low dielectrics, and thermal and chemical stability.

While the largest single use of woven fabrics is, of course, for wearing apparel, they are used in many areas; in mechanical applications such as machine and conveyor belting, for filtration, for packaging, and as reinforcement for plastics and rubber.

Specifications

Textile specifications contain two important types of information: (1) descriptive information and (2) service property requirements.

Specifications that physically describe the textile fabric usually include (1) width, in inches, (2) weight, usually in ounces per square yard, (3) type of weave, such as twill, broken twill, leno, or satin, (4) thread count, both in warp and filling (e.g., 68×44 denotes 68 warp yarns/in. and 44 filling yarns/in.), (5) type of fiber and whether the yarn is to be filament or staple, (6) crimp, in percent, (7) twist per inch, and (8) yarn number both for warp and fill.

Yarn number designations are somewhat complex as they have been developed in a relatively unorganized fashion over the years, and different systems are used in different types of fibers. (Filament yarns are usually stated simply in denier, which is the weight in grams of 9000 m of yarn.)

Essentially, yarn numbers provide a measure of weight per unit length or length per unit weight. A typical yarn designation on a specification may appear as "210 (denier)/1 $\times$ 20/2 (cotton system)." This means that (1) the warp yarn is 210-denier single yarn, and (2) the filling yarn contains 2 plies, each of which is a 20 singles yarn (determined by the cotton numbering system).

A number of fabric-designation systems have been formalized by tradition. For example, sheetings, drills, twills, jeans, broken twills, and sateens are designated only by width in inches, number of linear yards per pound, and number of warp and filling threads per inch. "Specs" for equivalent synthetic fabrics also include fiber type, whether staple or filament.

Face (Crystal)

An idiomorphic plane surface on a crystal.

Face (Machine Tools)

In a lathe tool, the surface against which the chips bear as they are formed. See the term *single-point tool*.

Face Feed

The application of filler metal to the joint, usually by hand, during brazing and soldering.

Face Mill

See *axial rake angle*, *axial relief angle*, *radial rake angle*, *radial relief angle* and *corner* and *end cutting edge angle*, and *single relief angle*.

Face Milling

Milling a surface that is perpendicular to the cutter axis. See also *milling*.

Face of Weld

The exposed surface of an arc or gas weld on the side from which the welding was done. See also the term *fillet weld*.

Face Pressure

In seals, the face load divided by the contacting area of the sealing lip. The face load is the sum of the pneumatic or hydraulic force and the spring force. For lip seals and packings, the face load also includes the interference load. See also *face seal*.

Face Reinforcement

Reinforcement of a weld at the side of the joint from which welding was carried out. See also *root reinforcement*.

Face Seal

A device that prevents the leakage of fluids along rotating shafts, sealing is accomplished by a stationary primary-seal ring bearing against the face of a mating ring mounted on a shaft. Axial pressure maintains the contact between seal ring and mating ring.

Face Shield (Welding)

A device positioned in front of the eyes and a portion of, or all of, the face, whose predominant function is protection of the eyes and face during welding, brazing, or soldering.

Face-Centered

Having atoms or groups of atoms separated by translations of ½, ½, 0; ½, 0, ½; and 0, ½, ½ from a similar atom or group of atoms. The number of atoms in a face-centered cell must be a multiple of 4. See also the term *unit cell*.

Face-Centered Cubic

A crystal lattice in which atoms are close packed, that is, in contact with six others in the same layer and with three others on the layers above and below, and in which the layer alignment repeats every third layer. Close packed hexagonal is similar except that the layer alignment repeats on alternate layers.

Facet

A flat, usually brightly reflective, portion of surface on a crystal or fracture face.

Face-Type Cutters

Cutters that can be mounted directly on and driven from the machine spindle nose.

Facing

(1) In machining, generating a surface on a rotating workpiece by the traverse of a tool perpendicular to the axis of rotation. (2) In foundry practice, any material applied in a wet or dry condition to the face of a mold or core to improve the surface of the casting. See also *mold wash*. (3) For abrasion resistance, see preferred term *hardfacing*.

Factor of Safety

See *safety factor*.

Fadeometer

An apparatus for determining the resistance of resins and other materials to fading.

Fagot

In forging work, a bundle of iron bars or rods that will be heated and then hammered and welded to form a single bar. The term may imply some careful, tidy arrangement. Fagot (ed) iron was the result of rolling or forging carefully packed bundles of wrought iron.

Fail Safe

The concept whereby it is recognized that if a failure occurs it will do so in a manner that does not involve hazard to, for example, humans or the environment, although there could be acceptable economic cost. In some contexts, the term implies that the mode of failure is acceptable, as in the case of the braking system on some vehicles in which the brakes are applied if the brake system fails; in other cases, it implies that in the event of a failure some secondary systems will intervene or provide support before a hazard develops. See also *leak-before-break*.

Failure

A general term used to imply that a part in service (a) has become completely inoperable, (b) is still operable but incapable of satisfactorily performing its intended function, or (c) has deteriorated seriously, to the point that that it has become unreliable or unsafe for continued use.

Failure Criteria

The limiting conditions relating to the admissibility of the deviation from the characteristic value due to changes after the beginning of stress.

Failure Mechanism

A structural or chemical process, such as corrosion or fatigue, that causes failure.

Fairing

A secondary structure in airframes and ship hulls, the major function of which is to streamline the airflow or flow of fluid by producing a smooth outline and reducing drag.

Falling Weight Test

A crude form of impact testing in which a weight is allowed to fall onto the center of a bar or rail, etc., supported at its ends.

False Bottom

An *insert* put in either member of a die set to increase the strength and improve the life of the die.

False Brinelling

(1) Damage to a solid bearing surface characterized by indentations not caused by plastic deformation resulting from overload, but thought to be due to other causes such as *fretting corrosion*. (2) Local spots appearing when the protective film on a metal is broken continually by repeated impacts, usually in the presence of corrosive agents. The appearance is generally similar to that produced by *Brinelling* but corrosion products are usually visible. It may result from fretting corrosion. This term should be avoided when a more precise description is possible. False Brinelling (race fretting) can be distinguished from true *Brinelling* because in false Brinelling, surface material is removed so that original finishing marks are removed. The borders of a false Brinell mark are sharply defined, whereas a dent caused by a rolling element does not have sharp edges and the finishing marks are visible in the bottom of the dent. Contrast with Brinelling.

False Indication

In nondestructive inspection, an *indication* that may be interpreted erroneously as an *imperfection*. See also *artifact*.

False Wiring

Same as *curling*.

Family Mold

A multicavity mold used for forming of plastic in which each of the cavities form one of the component parts of the assembled finished object. The term is often applied to a mold in which parts from different customers are grouped together for economy. Sometimes called combination mold.

Farad (F)

A unit of electric capacity.

Faraday's Law

(1) The amount of any substance dissolved or deposited in electrolysis is proportional to the total electric charge passed. (2) The amounts of different substances dissolved or deposited by the passage of the same electric charge are proportional to their equivalent weights.

Far-Infrared Radiation

Infrared radiation in the wavelength range of 30–300 μm (3×10^5 to 3×10^6 Å).

Fast Fracture

Fracture occurring virtually instantaneously as opposed to mechanisms such as fatigue and creep which develop over a period of time. The term is also often used loosely as a synonym for brittle, as opposed to fast neutrons—neutrons with energies greater than 0.1 MeV.

Fasteners

Mechanical fasteners are among the most common components in fabricated products. Thus, fasteners are extremely important from both a manufacturing and a product standpoint. The proper selection of fasteners is necessary to provide the most value for the manufacturer as well as for the consumer.

The most common types of fasteners can be grouped into the following categories: bolts/screws, studs, pins, nuts, rivets, and holes with or without threads tapped into them.

Fastener Types and Materials Selection

Bolts and screws, two names for externally threaded fasteners, are certainly well-recognized fastener types. Like all fasteners, they are also a highly engineered method of joining. The significant features of a bolt are the tensile force on the bolt and the compressive loading of the joint members when applied.

The ability of a bolt to translate rotational motion into a clamping force has made it the standard fastener. The common bolt is also an ideal fastener from a reusability standpoint. Its helical thread form enables it to attach, detach, then reattach joint members an almost infinite number of times. The simple tools required are another reason that bolts have become so common. Bolts provide an excellent ratio of performance to cost, but during the design phase and installation, they require more consideration than other fastener types to achieve an optimized joint.

Many different materials types are available for fasteners. Standard metric bolts and nuts are made of steels with varying chemical compositions.

A rivet is a nonreusable fastener that is deformed to provide a mechanical clenching of joint members. Both aluminum and steel may be fabricated into rivets. In fact, the stem and body of a rivet can be easily made with two different materials. Low-carbon steel alloys in the AISI/SAE 1XXX series may be selected when cost is the most important factor, or when a higher-strength rivet is needed. Typical tensile strength values for joints made with steel rivets are 5–10 kN, which is about one-third to half the value of equivalently sized bolts. An aluminum rivet usually has about half to two-thirds the strength of steel rivets.

However, the shear resistance of both types of rivets can equal their tensile resistance, and can even exceed the value for a similar sized bolt. The typical aluminum alloys for rivets in automotive applications are AA 5052, 5056, 7075, 7178, and 2014. These alloys provide good strength, high corrosion resistance (without the need for additional corrosion protection), and light weight.

Some rivets are made of thermoplastics, namely, polyamide 6 (PA6), polyamide 66 (PA66), and polyoxymethylene (POM). These are semicrystalline polymers with high strength, low mass, and resistance to automotive fluids and salt environments. As a result, they are also ideal for the myriad pins and clips that connect plastic components.

Polyamides are also coated onto bolts and nuts and the plastic is added to reduce the allowance between the mating metallic threads, making them more resistant to vibrational loosening.

Fluoropolymers are commonly added to the threaded region of weld nuts prior to installation because of their low coefficient of friction.

Finally, polymer adhesives are often added to bolts. These bolts resist vibrational loosening through mechanical means, such as thermoplastic additions or locally deformed features. In fact, the application of adhesives is the only method that provides long-term resistance to loosening.

A final material being considered for fasteners is titanium. The mechanical properties of titanium are rather high. If low density is also required, as it is for automotive structures, then titanium appears to be the only current material that can challenge traditional steel alloys. The workhorse alloy titanium, 6% aluminum, 4% vanadium (Ti-6Al-4V) could definitely be chosen for fastener applications.

The advantages of titanium are its extremely high strength-to-weight ratio and its almost complete imperviousness to corrosive attack by chloride ions. This means that a lower-mass fastener can be made, one that does not require the expensive and time-consuming application of corrosion-resistant coatings. Many mechanical joints in aerospace structures currently rely on titanium alloys, including Ti-6Al-4V.

If the raw material costs of titanium can be reduced, it will become a viable alternative material for automotive applications, with fasteners at the top of the list.

Fatigue

The development of one or more cracks by the repeated application and removal of a stress significantly less than the tensile strength of the material. Components are normally expected to fail at a stress equivalent to their ultimate tensile strength, see tensile *test*. However, if a component is subjected to repeated cycles of application and removal of a lower stress it may ultimately fail by fatigue. Damage occurs by the initiation, usually at the surface, of a fatigue crack which, over time, propagates until the remaining intact section is insufficient to carry the load. The damage is often highly localized at major features such as manufacturing or welding defects or section changes. "Fatigue" or colloquially "metal fatigue" does not mean that the bulk material away from the crack has deteriorated in any general sense. Many millions of cycles of loading, perhaps over years, may be required to cause failure and, generalizing, in the absence of large initiating defects a considerable proportion of the total component life is occupied in developing a crack to a detectable size. Once initiated, cracks usually propagate at an increasing rate. The number of cycles to failure, in specified circumstances of loading, is termed the fatigue life and, as might be expected, the higher the stress the less the number of cycles required to cause failure. Consequently, failure occurring within a few thousand cycles is conventionally termed either low cycle fatigue or high strain fatigue. If more than about 10,000 cycles are required the mechanism can be termed either high cycle fatigue or low strain fatigue. The fatigue properties of the material are commonly determined by testing a series of test pieces over ranges of stress (more accurately stress amplitude as defined below). The results, in terms of the number of cycles to failure at each stress amplitude, are then plotted on an S/N curve. For most steels there is a level of stress amplitude below which a component will survive an infinite number of cycles. This stress is usually termed the fatigue limit. The ratio of fatigue limit to the tensile strength is termed the fatigue ratio and is typically about 50% for steels up to about 1100 MPa (75 tons/in.^{2}) U.T.S. For many other metals, a fatigue limit cannot be defined as failure will eventually occur if sufficient cycles are experienced. In these cases, it is normal to determine an endurance (or endurance limit). This is the maximum stress amplitude that can be tolerated to survive some specified number of cycles, often 10 million and hence endurance Ratio is the ratio of endurance to the U.T.S. The term endurance may also be used of steels, possibly as an alternative to fatigue limit but particularly where the stresses exceed the fatigue limit. Up to this point, it has been assumed that the fatigue cycle is very simple, merely the application and removal of a load. However, the cycle is often more complex than this. There is always a cyclic stress but it may range from tensile to compressive and there may also be an additional constant stress. In these complex circumstances the various stresses are normally resolved into the mean stress and the stress amplitude or stress range (nearly double the stress amplitude). Here are some examples. The hanging and other crane cable experiences a repeated stress or pulsating stress ranging from the point when unloaded (ignoring the hook) to high tensile when loaded—an example of high amplitude, moderate mean stress loading. A lift (elevator) cable experiences a fluctuating stress comprising a constant stress due to the cabin plus an additional stress due to passengers—an example of high mean stress and medium amplitude loading. A clock hair spring is deflected backwards and forwards either side of its unstressed rest position so a point on its periphery experience and Alternating stress or reversed stress ranging from highly tensile to highly compressive—an example of high amplitude, zero mean stress loading. The combined effect of the mean stress and the cyclic stress can be determined from various empirical relationships established, with varying degrees of conservatism. Reference may also be made to the stress ratio "R," which is the ratio of minimum stress to maximum stress or to stress, ratio "A," which is the ratio of alternating stress amplitude to mean stress.

Fatigue Crack Growth Rate (*da/dN*)

The rate of crack extension caused by constant-amplitude fatigue loading, expressed in terms of crack extension per cycle of load application, and plotted logarithmically against the stress intensity factor range, ΔK.

Fatigue Ductility (*D_f*)

The ability of a material to deform plastically before fracturing, determined from a constant-strain amplitude, low-cycle fatigue test. Usually expressed in percent in direct analogy with elongation and reduction in area ductility measures.

Fatigue Ductility Exponent (*c*)

The slope of a log–log plot of the plastic strain range and the fatigue life.

Fatigue Failure

Failure that occurs when a specimen undergoing *fatigue* completely fractures into two parts or has softened or been otherwise significantly reduced in stiffness by thermal heating or cracking.

Fatigue Life (*N*)

(1) The number of cycles of stress or strain of a specified character that a given specimen sustains before failure of a specified nature occurs. (2) The number of cycles of deformation required to bring about failure of a test specimen under a given set of oscillating conditions (stresses or strains). See also *S–N curve*.

Fatigue Life for *p*% Survival

An estimate of the fatigue life that *p*% of the population would attain or exceed at a given stress level. The observed value of the median fatigue life estimates the fatigue life for 50% survival. Fatigue life for *p*% survival values, where *p* is any number, such as 95, 90, etc., may also be estimated from the individual fatigue life values.

Fatigue Limit

The maximum stress that presumably leads to fatigue fracture in a specified number of stress cycles. The value of the *maximum stress* and the *stress ratio* also should be stated. See also *endurance limit*.

Fatigue Limit for *p*% Survival

The limiting value of fatigue strength for *p*% survival as *N* becomes very large; *p* may be any number, such as 95, 90, etc.

Fatigue Notch Factor (*K_f*)

The ratio of the *fatigue strength* of a unnotched specimen to a fatigue strength of a notched specimen of the same material and condition; both strengths are determined at the same number of *stress cycles*.

Fatigue Notch Sensitivity (*q*)

An estimate of the effect of a notch or hole of a given size and shape on the fatigue properties of a material, measured by $q = (K_f - 1)/(K_t - 1)$, where K_f is the *fatigue notch factor* and K_t is the *stress-concentration factor*. A material is said to be fully notch sensitive if *q* approaches a value of 1.0; it is not notch sensitive if the ratio approaches 0.

Fatigue Ratio

The ratio of fatigue strength to tensile strength. Mean stress and alternating stress must be stated.

Fatigue Strength

The maximum cyclical stress a material can withstand for a given number of cycles before failure occurs.

Fatigue Strength at *N* Cycles (*S_N*)

A hypothetical value of stress for failure at exactly *N* cycles as determined from an *S–N* diagram. The value of S_N that is commonly found in the literature is the hypothetical value of maximum stress, S_{max}, minimum stress S_{min} or stress amplitude, S_a at which 50% of the specimens of a given sample could survive *N* stress cycles in which the mean stress $S_m = 0$. This is also known as the *median fatigue strength at N cycles*. See also *S–N curve*.

Fatigue Strength for *p*% Survival at *N* Cycles

An estimate of the stress level at which *p*% of the population would survive *N* cycles; *p* may be any number, such as 95, 90, etc. The estimates of the fatigue strength for *p*% survival values are derived from particular points of the fatigue life distribution since there is no test procedure by which a frequency distribution of fatigue strength at *N* cycles can be directly observed.

Fatigue Striation (Metals)

Parallel lines frequently observed in electron microscope fractographs or fatigue fracture surfaces. The lines are transverse to the direction of local crack propagation; the distance between successive lines represents the advance of the crack front during the one cycle of stress variation.

Fatigue Test

A method for determining the range of alternating (fluctuating) stresses a material can withstand without failing.

Fatigue Wear

(1) Removal of particles detached by fatigue arising from cyclic stress variations. (2) Wear of a solid surface caused by fracture arising from material fatigue. See also *spalling*.

Fatigue-Strength Reduction Factor

The ratio of the fatigue strength of a member or specimen with no stress concentration to the fatigue strength with stress concentration. This factor has no meaning unless the stress range and the shape, size, and material of the member or specimen are stated.

Fats

Natural combinations of glycerin with fatty acids, so-called *triglycerides*, some fats having as many as 10 or more different fatty acids in the combination. At ambient temperature fats are solids; if liquid, they are normally called *fat oils*. Fats are also known as *lipids*. Waxes differ slightly in composition from fats and are mixed esters of polyhydric alcohols, other than glycerin, and fatty acids. *Animal fats* are *butter*, *lard*, and edible and inedible *tallows*. Fats contain less than 5% of phospholipids, pigments, vitamins, antioxidants, and sterols. They are derived from animal or vegetable sources, the latter source being chiefly the seeds or nuts of plants. Fats in a pure state would be odorless, tasteless, and colorless, but the natural fats always contain other substances that give characteristic odors and tastes. Fats are used directly in foods and in the making of various foodstuffs. They are used in making soaps, candles, and lubricants, and in the compounding of resins and coatings. They are also distilled or chemically split to obtain the fatty acids. Crude fats are refined to remove nonglyceride impurities, including free fatty acids, phosphatides, and proteinaceous and mucilaginous matter, by treatment with strong caustic soda. The fatty acids are converted to oil-insoluble soaps known as foots or soap stock.

Fats are most important for food, containing more than twice the fuel value of other foods. They are also important carriers of glycerin necessary to the human system. Metabolism, or absorption of fats into the system, is not a simple process and is varied with the presence of other food materials. The fats with melting points above 45°C are not readily absorbed into the system. The heavy fats are called tallow. Lack of certain fats or fatty acids causes skin diseases, scaly skin, and other conditions. Some fatty acids are poisonous alone, but in the glyceride form in the fats they may not be poisonous but beneficial. Fats can be made synthetically from petroleum or coal. The world resources of natural fats are potentially unlimited, especially from tropical knots, forming a cheap source of fatty acids in readily available form.

Margarine, *shortening*, *confectionery fat*, and other edible fats are made by hydrogenating a variety of semisolid or liquid fats. The hardening process converts unsaturated fatty glycerides to more saturated forms. About one cubic meter of hydrogen is needed per metric ton of oil to reduce the oils iodine number one unit. The catalyst is nickel.

Engineered, invitation, or *artificial fats* represent a new market for products that have low-calorie saturated fat, or cholesterol contents. They are predominately mixtures of such hydrogenated vegetable oils as soybean, corn, peanut, palm, and cottonseed.

Fatty Acids

A series of *organic acids* deriving the name from the fact that the higher members of the series, the most common ones, occur naturally in animal fats, but fatty acids are readily synthesized, and the possible variety is almost infinite. All these acids contain the *carboxyl group*·COOH. The acids are used for making soaps, candles, and coating compounds; as plasticizers; and for the production of plastics and many chemicals. The hydrogen atom of the group can be replaced by metals or alkyl radicals with the formation of salts or esters, and other derivatives such as the halides, anhydrides, peroxides, and amides can also be made. Some of the fatty acids can be polymerized to form plastics. Various derivatives of the acids are used as flavors, perfumes, driers, pharmaceuticals, and antiseptics. Certain fatty acids, such as oleic and stearic, are common to most fats and oils regardless of their source, while others, such as arachidic and ericic, are characteristic only of specific fats and oils.

Saturated acids are acids that contain all the hydrogen with which they can combine. They have high melting points. *Unsaturated acids*, such as oleic, linoleic, are liquid at room temperature and are less stable than saturated acids. Fatty acid glycerides in the form of animal and vegetable fats form an essential group of human foods. Fats of the highly unsaturated acids are necessary in the metabolism of the human body, the glycerides of the saturated acids such as palmitic being insufficient alone for food.

Polyunsaturated acids of the linoleic type with more than one double bond lower blood cholesterol, but saturated acids with no double bond do not. *Arachidonic acid* with four double bonds lowers blood cholesterol greatly. It is manufactured in the body from linoleic acid if vitamin B_6 is present. *Linoleic acid*, the characteristic unsaturated food acid, has two double bonds. *Linoleic acid*, found in linseed oil, has three double bonds.

The names of the fatty acids often suggest their natural sources, though commercially they may be derived from other sources or made synthetically. *Butyric acid* is the characteristic acid of butter. Also called *butanoic acid* and *ethylacetic acid*, it is made synthetically as a colorless liquid with a strong odor and completely soluble in water. With alcohols its forms butyrates of pleasant fruity odors used as flavors.

Faying Surface

The surfaces of materials in contact with each other and joined or about to be joined.

F.C.C.

See *face-centered cubic*.

Feasibility

The ability of a grease to flow to the suction of a pump.

Feather (Ceramics)

A striation having the appearance of a feather. See also *striation*.

Feathering

The tapering of an *adherend* on one side to form a wedge section, as used in an adhesively bonded scarf joint.

Feed

(1) The material entering a process or its rate of introduction. (2) The rate at which material is cut. Depending on context, it may refer to the depth of cut or the rate at which a tool traverses a surface or a combination of the two.

Feed Hopper

A container used for holding metal powder prior to compacting in a press.

Feed Lines

Lynn earmarks on a machine or ground surface that are spaced at intervals equal to the *feed* per revolution or per stroke.

Feed Rate (Thermal Spraying)

The rate at which material passes through the gun in a unit of time. A synonym for *spray rate*.

Feeder (Feeder Head, Feed Head)

A channel through which molten metal is fed to a casting. A reservoir of material that remains molten to feed shrinkage in the solidifying casting below. In foundry practice, a *riser*.

Feeding

(1) In casting, providing molten metal to a region undergoing solidification, usually at a rate sufficient to fill the mold cavity ahead of the solidification front and to compensate for any shrinkage accompanied solidification. (2) Conveying metal stock or workpieces to a location for use or processing, such as wire to a consumable electrode, strip to a die, or workpieces to an assembler.

Feeler Gauge

A strip of metal of high precision thickness to measure or set gaps.

Feldspar

A group of alumina silicate minerals consisting chiefly of microcline ($K_2O \cdot Al_2O_3 \cdot 6SiO_2$), albite ($Na_2O \cdot Al_2O_3 \cdot 6SiO_2$), and/or anorthite ($CaO \cdot Al_2O_3 \cdot 2SiO_2$) used in the production of glass and ceramic whiteware. Feldspar constitutes 60% of the outer 13–17 km of the earth's crust and is the most common mineral in crystalline rocks. Feldspars are aluminum silicates of barium, calcium, sodium, and potassium.

The importance of the many feldspars that occur so widely in igneous, metamorphic, and some sedimentary rocks cannot be underestimated, especially from the viewpoint of a petrologist attempting to unravel Earth history.

With weathering, feldspars form commercially important clay materials. Economically, feldspars are valued as raw material for the ceramic and glass industries, as fluxes in iron smelting, and as constituents of scouring powders. Occasionally, their luster or colors qualify them as semiprecious gemstones. Some decorative building and monument stones are predominantly composed of weather-resistant feldspars.

Knowledge of the composition of a feldspar and its crystal structure is indispensable to an understanding of its properties. However, it is the distribution of the aluminum and silicon atoms among the available tetrahedral sites in each chemical species that is essential to a complete classification scheme, and is of great importance in unraveling clues to the crystallization and thermal history of many igneous and metamorphic rocks.

Felt Metal

A substantial felt fabric reinforced with fine metal strands. The strands may be loosely woven or bonded for improved strength. The term may be extended to similar materials where the reinforcement is plastic.

Felts

Classes of felts in use as engineering materials are wool and synthetic fiber felts.

Wool felt is a fabric obtained as a result of the interlocking of wool fibers under suitable combinations of mechanical work, chemical action, moisture, and heat, alone or in combination with other fibers.

Synthetic fiber felt is a fabric obtained as a result of interlocking of synthetic fibers by mechanical action. Other fibers include asbestos, cotton, glass, and so forth. See also *batt*.

Properties

Neither wool nor synthetic fiber felts require binders and exist as 100% fibrous materials.

Felts generally exhibit the same chemical properties as do the fibers of which the felt is composed. Wool felts are characterized by excellent resistance to acids but are damaged by exposure to strong alkalies. They exhibit remarkable resistance to atmospheric aging and are as a class probably the most inert of all nonmetallic engineering materials with respect to nonaqueous liquids, oils, or solvents. Wool felts are not generally recommended for stressed dry uses at temperatures in excess of 82°C, because of changes in physical properties, but are used as dry spacers and gaskets in many applications up to 149°C, and the use of the materials as oil wicks and lubricating system components at ambient temperatures up to 149°C is common.

Synthetic fiber felts are available in virtually all classes of fiber composition including regenerated cellulose, cellulose acetate, cellulose triacetate, polyamide, polyester, acrylic, modacrylic, olefin, and CFC fluorocarbon (tetrafluoroethylene). The variety of types available provides a virtually infinite range of physical and chemical properties for application beyond the natural versatility of wool fiber felts. Outstanding among the properties of this class of engineering material are chemical, solvent, thermal, and biological stability as well as low moisture absorption, quick drying after aqueous wetting, abrasion resistance, and frictional and dielectrical features.

Forms

Wool felts are produced as "sheet" stock in a standard 91.4 × 91.4 cm size in thickness ranging from 1.6 to 76 mm. There are a number of density classifications based on the weight for a 914 × 914 mm sheet in 25.4 mm thickness from 5.4 to 14.4 kg with the weight for any given thickness in the density class proportioned to the weight per square meter at the 25.4 mm thickness. "Roll" felts are produced in either 1522 or 1830 mm widths in lengths up to 160 m and standard thicknesses range from 0.8 to 25.4 mm.

Synthetic fiber felt widths are available from 600 to 1830 mm in thicknesses from 0.8 to 19 mm.

Both wool and synthetic fiber felts are nonforming and nonraveling and thus provide considerable ease of cutting and fabricating. In addition, wool felt lends itself to most grinding, cutting, shaping, extruding, and other machining operations so that special shaped parts such as polishing laps, round wicking, ink rollers, and others are produced.

Applications

The structural elasticity of felts as a class makes these materials suitable for molding and forming and parts of these types are produced to provide special shaped gaskets, seals, fillers, instrument covers, and the like. In many cases, these shaped parts are stabilized through the use of resinous or rubber impregnants and this modified class of felt is finding ever-increasing use as flat stock for special gasketing, sealing, and other applications. Combinations of felt and plastic and elastomer sheet materials are also produced for application where resilience plus nonpermeability is desired as in sealing and frictional uses.

Use of felt as an engineering material covers a broad spectrum of application. Major contributing properties include resilience, mechanical, thermal, and acoustic energy absorption; high porosity-to-weight ratio; resistance to aging; thermal and chemical stability; high effective surface area per unit volume; and solvent resistance.

Applied uses include wet and dry filtration, thermal and acoustical insulation, vibration isolation, impact absorption, cushioning and packaging, polishing, frictional surfacing, liquid absorption and reservoirs, wicking, gasketing, sealing, and percussion mechanical dampening.

FEP

See *fluorinated ethylene propylene*.

Ferric Oxide

The red *iron oxide*, Fe_2O_3, also called gamma ferric oxide, found in abundance as the ore *hematite*, or made by calcining the sulfate. It has a dark-red color and comes in powder or lumps. It is used as a paint pigment under such names as *Indian red*, *Persian red*, and *Persian gulf oxide*. In cosmetics and in polishing compounds it is called *rouge*.

The names *metallic red* and *metallic brown* are applied to pigments from Pennsylvania ores containing a high percentage of *red iron oxide*. *Venetian red* is a name for red iron oxide pigments mixed with various fillers, most commonly an equal proportion of the pigment extender calcium sulfate. Commercially, it is made by heating ferrous sulfate with quicklime in a furnace. Venetian red is a permanent and inert pigment that is generally used on wood. It cannot be used on many metals, including iron, because the calcium sulfate can cause corrosion. The *Tuscan red* treatments are red iron oxide blended with up to 75% of lakes, but may also be barium sulfate with lakes. *Ferric oxide pigments* make low-priced paints and are much used as base coats for structural steel work.

Yellow iron oxide, known as *ferrite yellow* and *Mars yellow*, is used as a paint pigment. It is made by precipitating ferrous hydroxide from iron sulfate and lime and then oxidizing to the yellow oxide. *Black ferric oxide* is a reddish-black amorphous powder and is used as a paint pigment for polishing compounds and for decarburizing steel.

Ferrimagnetic Material

(1) A material that macroscopically has properties similar to those of a *ferromagnetic material* but that microscopically also resembles an antiferromagnetic material in that some of the elementary magnetic moments are aligned antiparallel. If the moments are of different magnitudes, the material may still have a large resultant magnetization. (2) A material in which unequal magnetic moments are lined up antiparallel to each other. Permeabilities are of the same order of magnitude as those of ferromagnetic materials, but are lower than they would be if all atomic moments were parallel and in the same direction. Under ordinary conditions, the magnetic characteristics of ferrimagnetic materials are quite similar to those of ferromagnetic material.

Ferrite

(1) A solid solution of one or more elements in body-centered cubic iron. Unless otherwise designated (for instance, as chromium ferrite), the solute is generally assumed to be carbon. On some equilibrium diagrams, there are two ferrite regions separated by an austenite area. The lower area is alpha ferrite; the upper, delta ferrite. If there is no designation, alpha ferrite is assumed. (2) An essentially carbon-free solid solution in which alpha iron is the solvent, and which is characterized by a body-centered cubic crystal structure. Fully ferritic steels are only obtained when the carbon content is quite low. The most obvious microstructural features of such metals are the ferrite grain boundaries.

Ferrite

A ferrite is any of the class of magnetic oxides. Typically, the ferrites have a crystal structure that has more than one type of site for the cations. Usually, the magnetic moments of the metal ions on sites of one type are parallel to each other, and antiparallel to the moments on at least one site of another type. Thus, ferrites exhibit ferromagnetism.

A term referring to magnetic oxides in general, and especially to material having the formula $MOFe_2O_3$, where M is a divalent metal ion or a combination of such ions. Certain ferrites, magnetically "soft," are useful for core applications at radio and higher frequencies because of their advantageous magnetic properties and high volume resistivity. Other ferrites, magnetically "hard" in character, have desirable permanent magnet properties.

Commercial Types

There are three important classes of commercial ferrites. One class has the spinel structure. The second class of commercially important ferrites has the garnet structure. Yttrium-based garnets are used in microwave devices. Thin monocrystalline films of complex garnets have been developed for bubble domain memory devices. The third class of ferrites has a hexagonal structure of the magneto-plumbite type. Because of their large magneto crystalline anisotropy, the hexagonal ferrites develop high coercivity and are an important member of the permanent magnet family.

Properties

The important intrinsic parameters of a ferrite are the saturation magnetization, Curie temperature, and magnetocrystalline (K_1) and magnetostrictive (Ω_x) anisotropies. These properties are determined by the choice of the cations and their distribution in the various sites.

In addition to the intrinsic magnetic parameters, microstructure plays an equally important role in determining device properties. Thus, grain size, porosity, chemical homogeneity, and foreign inclusions dictate in part such technical properties as permeability, line width, remanence, and coercivity in polycrystalline ceramics. In garnet films for bubble domain device applications, the film must essentially be free of all defects such as inclusions, growth pits, and dislocations.

Preparation

Polycrystalline ferrites are most economically prepared by ceramic techniques. Component oxides or carbonates are mixed, calcined at elevated temperatures for partial compound formation, and then granulated by ball milling. Dispersants, plasticizers, and lubricants are added, and the resultant slurry is spray-dried, followed by pressing to desired shape and sintering. The last step completes the chemical reaction to the desired magnetic structure and effects homogenization, densification, and grain growth of the compact. It is perhaps the most critical step in optimizing the magnetic properties of commercial ferrites.

Ferrite Banding

Parallel bands of free ferrite aligned in the direction of working. Sometimes referred to as *ferrite streaks*.

Ferrite Devices

These are electrical devices whose principle of operation is based on the use and properties of ferrites, which are magnetic oxides. Ferrite devices are divided into two categories, depending on whether the ferrite is magnetically soft (low coercivity) or hard (high coercivity). Soft ferrites are used primarily as transformers, inductors, and recording heads, and in microwave devices. Since the electrical resistivity of soft ferrites is typically 10^6–10^{11} times that of metals, ferrite components have much lower eddy current losses and hence are used at frequencies generally above about 10 kHz. Hard ferrites are used in permanent-magnet motors, loudspeakers, and holding devices, and as storage media in magnetic recording devices.

Chemistry and Crystal Structure

Soft ferrite devices are spinels with the general formula of MFe_2O_4, in which M is a divalent metal ion. The commercially practical ferrites are those in which the divalent ion represents one or more magnesium, manganese, iron, cobalt, nickel, copper, zinc, and cadmium ions. The trivalent iron ion may also be substituted by other trivalent ions such as aluminum. The compositions are carefully adjusted to optimize the device requirements, such as permeability, loss, ferromagnetic resonance line width, and so forth.

The ferromagnetic garnets have the general formula of $M_3Fe_5O_{12}$, in which M is a rare earth or yttrium ion. Single-crystal garnet films form the basis of bubble domain device technology. Bulk garnets have applications in microwave devices.

Hard ferrites for permanent-magnetic device applications have the hexagonal magneto-plumbite structure, with the general formula $MFe_{12}O_{19}$, where M is usually barium or strontium.

The material of choice for magnetic recording is δ-Fe_2O_3, which has a spinel structure.

With the exception of some single crystals used in recording heads and special microwave applications, and particulates used as storage media in magnetic recording, all ferrites are prepared in polycrystalline form by ceramic techniques.

Applications

Applications of ferrites may be divided into nonmicrowave, microwave, and magnetic recording applications. Further, the nonmicrowave applications may be divided into categories determined by the magnetic properties based on the B–H behavior, that is, the variation of the magnetic induction or flux density B with magnetic field strength H. The categories are linear B–H, with low flux density, and nonlinear B–H, with medium to high flux density. The highly nonlinear B–H, with a square or rectangular hysteresis loop, was once exploited in computer memory cores.

Linear B–H Devices

In the linear region, the most important devices are high-quality inductors, particularly those used in filters in frequency-division multiplex telecommunications systems and low-power wideband and pulse transformers. Virtually all such devices are made of either MnZn ferrite or NiZn ferrite, although predominantly the former.

Nonlinear B–H Devices

The largest usage of ferrite measured in terms of material weight is in the nonlinear B–H range and is found in the form of deflecting yokes and flyback transformers for television receivers. Again, MnZn and NiZn ferrites dominate the use in these devices.

A rapidly growing use of ferrites is in the power area, where ferrite transformers are extensively used in switched mode (AC–DC) and converter mode (DC–DC) power supplies. Such power supplies are widely used in various computer peripheral equipment and private exchange telephone systems.

Microwave Devices

Microwave devices make use of the reciprocal propagation characterisics of ferrites close to or at a gyromagnetic resonance frequency in the range of 1–100 GHz. The most important of such devices are isolators and circulators. The garnets have highly desirable, small, ferromagnetic-resonance line widths, particularly in single-crystal form.

Magnetic Recording Devices

Vast amounts of audio and video information and digital data from computers are stored in magnetic tapes and disks. Here magnetic recording materials function as hard magnetic materials. The most widely used particles in magnetic recording are δ-Fe_2O_3 and co-modified δ-Fe_2O_3. The basic δ-Fe_2O_3 is generally used in audio recording, while the higher-coercivity Co-δ-Fe_2O_3 dominates video recording.

Ferrite Number

An arbitrary, standardized value designating the ferrite content of an austenitic stainless steel weld metal. This value directly replaces percent ferrite or volume percent ferrite and is determined by the magnetic test described in AWS A4.2.

Ferrite Streaks

Same as *ferrite banding*.

Ferrite–Pearlite Banding

Inhomogeneous distribution of ferrite and pearlite aligned in filaments or plates parallel to the direction of working. See the term *banding*.

Ferritic Grain Size

The grain size of the ferritic matrix of a steel.

Ferritic Malleable

See *malleable cast iron.*

Ferritic Stainless Steel

See *stainless steels.*

Ferritic Steels

Steels in which the predominant phase is ferrite. Such steels range from virtually pure iron to complex high alloys. They are magnetic and, with sufficient carbon, can be hardened by rapid cooling. See *steel.*

Ferritizing Anneal

A treatment given as-cast gray or ductile (nodular) iron to produce an essentially ferritic matrix. For the term to be meaningful, the final microstructure desired or the time–temperature cycle used must be specified.

Ferroalloys

Ferroalloys compose an important group of metallic raw materials required for the steel industry. Ferroalloys are the principal source of such additions as silicon and manganese, which are required for even the simplest plain-carbon steels, and chromium, vanadium, tungsten, titanium, and molybdenum, which are used in both low- and high-alloy steels. Also included are many other more complex alloys. Ferroalloys are unique in that they are brittle and otherwise unsuited for any service application, but they are important as the most economical source of these elements for use in the manufacture of the engineering alloys. These same elements can also be obtained, at much greater cost in most cases, as essentially pure metals. The ferroalloys contain significant amounts of iron and usually have a lower melting range than the pure metals and are therefore dissolved by the molten steel more readily than the pure metal. In other cases, the other elements in though ferroalloy serve to protect the critical element against oxidation during solution and thereby give higher recoveries. Ferroalloys are used both as deoxidizers and as a specified addition to give particular properties to the steel.

Many ferroalloys contain combinations of two or more desirable alloy additions, and well over 100 commercial grades and combinations are available. Although of less general importance, other sources of these elements for steel making are metallic nickel, nickel, SiC, Mo_2O_5, and even misch metal (a mixture of rare earths).

The three ferroalloys that account for the major tonnage in this class are the various grades of silicon, manganese, and chromium. For example, 5.9 kg of manganese is used on the average in the United States for every ton of open-hearth steel produced.

Elements supplied as ferroalloys are among the most difficult metals to reduce from ore.

Ferrochromium

A high-chromium iron master alloy used for adding chromium to irons and steel, ferrochromium is also called ferrochrome. It is made from chromite ore by smelting with lime, silica, or fluorspar in an electric furnace. High-carbon-ferrochrome is used for making tool steels, ball-bearing steels, and other alloy steels. It melts at about 1250°C. Low-carbon-ferrochrome is used for making stainless steels and acid-resistant steels. Simplex ferrochrome comes in pellet form and is used for making low-carbon stainless steels. Low-carbon ferrochrome is also preferred for alloy steel mixtures where much scrap is used because it keeps down the carbon and inhibits the formation of hard chromium carbides. The various grades of ferrochromium are also marketed as high-N_2 ferrochrome, and for use in making high-chromium cast steels where the N_2 refines the grain and increases the strength. Foundry-grade ferrochrome is used for making cast irons, as well as for ladle additions to cast iron to give uniform structure and increase the strength and hardness.

Ferroelectric

A crystalline material that exhibits spontaneous electrical polarization, hysteresis, and piezoelectric properties.

Ferroelectric Effect

The phenomenon whereby certain crystals may exhibit a spontaneous dipole moment (which is called ferroelectric by analogy with ferromagnetism exhibiting a permanent magnetic moment). Ferroelectric crystals often show several Curie points, domain structures, and hysteresis, much as do ferromagnetic crystals.

Ferrograph

An instrument used to determine the size distribution of wear particles in lubricating oils of mechanical systems. The technique relies on the debris being capable of being attracted to a magnet. In addition, it displays on a screen, the magnetic characteristics of a material, particularly hysteresis curves.

Ferromagnetic Material

A material that in general exhibits the phenomena of hysteresis and saturation, and whose permeability is dependent on the magnetizing force. Microscopically, the elementary magnets are aligned parallel in volumes called *domains*. The unmagnetized condition of a ferromagnetic material results from the overall neutralization of the magnetization of the domains to produce zero external magnetization.

Ferromagnetic Resonance

Magnetic resonance of a ferromagnetic material. See also *ferromagnetism* and *magnetic resonance.*

Ferromagnetism

A property exhibited by certain metals, alloys, and compounds of the transition (iron group), rare earth, and actinide elements in which, below a certain temperature termed the Curie temperature, the atomic magnetic moments tend to line up in a common direction. Ferromagnetism is characterized by the strong attraction of one magnetized body for another. See also *Curie temperature*. Compare with *paramagnetism*.

Ferromanganese

This is a master alloy of manganese and iron used for deoxidizing steels, and for adding manganese to iron and steel alloys and bronzes. Manganese is the common deoxidizer and cleanser of steel, forming oxides and sulfides that are carried off in the slag. Ferromanganese is made from the ores in either the blast furnace or the electric furnace. Spiegeleisen is a form of low-manganese ferromanganese and the German name, meaning mirror iron, is derived from the fact that the crystals of the fractured face shine like mirrors. Spiegeleisen has the advantage that it can be made from low-grade manganese ores, but the quantity needed to obtain the required proportion of manganese in the steel is so great that it must be premelted before adding to the steel. It was used for making irons and steels by the Bessemer process.

Ferrophosphorus

This substance is an iron containing a high percentage of phosphorus, used for adding phosphorus to steels. Small amounts of phosphorus are used in open-hearth steels to make them free-cutting, and phosphorus is also employed in tinplate steels to prevent the sheets from sticking together in annealing. Ferrophosphorus is made by melting phosphate rock together with the ore in making pig iron. There is also a master alloy, ferroselenium, for adding selenium to steels, especially stainless steels, to give free-machining qualities.

Ferrosilicon

This is a high-silicon master alloy used for making silicon steels, and for adding silicon to transformer irons and steels. It is made in the electric furnace by fusing quartz or silica with iron turnings and carbon. Silicon is often added to steels in combination alloys with deoxidizers or other alloying elements. Ferrosilicon aluminum is a more effective deoxidizer for steel than aluminum alone. It is also used for adding silicon to aluminum casting alloys. The alloy serves as a deoxidizer, fluxes the slag inclusions, and also controls the grain size of the steel.

Ferrotitanium

A master alloy of titanium with iron, ferrotitanium is used as a purifying agent for irons and steel owing to the great affinity of titanium for O_2 and N_2 at temperatures above 800°C. The value of the alloy is as a cleanser, and little or no titanium remains in the steel unless the percentage is gauged to leave a residue. The ferrocarbon titanium is made from ilmenite in the electric furnace, and the carbon-free alloy is made by reduction of the ore with aluminum. Ferrotitanium comes in lumps, crushed, or screened, and it is used for ladle additions for cleansing steel. Low-carbon ferrotitanium is used as a deoxidizer and as a carbide stabilizer in high-chromium steels. Graphidox improves the fluidity of steel, increases machinability, and adds a small amount of titanium to increase the yield strength, and the Grainal alloys that are used to control alloy steels and have various compositions.

MnTi is used as a deoxidizer for high-grade steels and for nonferrous alloys.

NiTi is used for hard nonferrous alloys and columbium is used for adding columbium to steel. It has an exothermic reaction that prevents chilling of the molten metal.

Ferrous

Metallic materials in which the principal component is iron.

Ferrous Oxide

Black iron oxide, FeO.

Ferrous P/M

Specifying powder-metallurgy (P/M) parts and their consolidation process used to be a simple process: Design the part, select the metal powders and lubricants that provide the required properties, compact the powders into a briquette, and sinter the briquette into its finished form. Through this procedure, millions of parts have been produced for applications ranging from automobiles to appliances and from business to farm and garden machines.

However, the needs of industries have changed significantly. Removing weight from all products has risen to primary importance. Energy, tooling, and materials costs now figure prominently in parts design, and productivity has emerged as the new watchword.

With these changes have come changes in P/M technology. Through the many manufacturing processes, improvements have been made in the powders themselves—improvements such as lower levels of inclusions and higher compressibility. In addition to conventional iron and steel metals, the list of available powders has been expanded to include new classes of tool steel, as well as materials such as cermets and alloys of titanium, nickel, and aluminum.

Accompanying these developments has been the growth of new consolidation technologies. As a result, design engineers need current information on which P/M technologies are viable, cost-effective, and production-effective, and which have potentially wide application.

Although P/M is used to fabricate parts from just about any metal, the most commonly used metals are the iron-based alloys. Low-density iron P/M parts (5.6–6.0 g/cm^3), with a typical tensile strength of 108.8 MPa, are usually used in bearing applications. Copper is commonly added to improve both strength and bearing properties. Alloy-steel powders are sometimes hot-forged to high or nearly theoretical density to form parts with improved mechanical properties that, when heat-treated, may have tensile strengths to 1156 MPa. Powder forging (P/F) is now established as a serious contender for parts formerly made as wrought forgings or machined from mill forms.

Iron P/M or sintered Fe–Cu alloy strength can be varied by adjusting density, carbon content (up to 15%), or all three to satisfy specific design requirements. The mechanical properties of ferrous powder parts can be considerably improved by impregnating or infiltrating them with any one of a number of different materials, both metallic and nonmetallic, such as oil, wax, resins, copper, lead, and babbit.

Low-density P/M parts are used in bearing applications because they provide porosity for oil storage. Impregnating sintered-metal bearings with oil usually eliminates the need for relubrication.

For higher-strength needs, alloyed (frequently prealloyed nickel–molybdenum–iron) iron, compacted to a higher density, is used.

When carbon or other alloying elements are mixed with the iron powders and densities exceed 6.2 g/cm^3, the parts are considered to be steel rather than iron. As carbon content is increased up to 1%, the strength of steel P/M parts increases, just as the strength of wrought steel increases with higher carbon content.

Ferrous-base P/M parts can range in size from about 2.5 mm thick and 3.2 mm in diameter to 50.8 mm thick and over 612 mm in diameter. Because they can be mass-produced at relatively low-cost, iron-base P/M parts find a wide variety of uses in such high-volume products as appliances, business machines, power tools, and automobiles. Typical parts are gears, bearings, rotors, valves, valve plates, cams, levers, ratchets, and sprockets.

Additional applications can be accommodated by sealing the pores in iron P/M parts. The sealing materials used are copper, polyesters, and anaerobics; each requires a different processing system to impregnate the parts. Impregnation of sintered P/M parts is done for any of several reasons:

- To serve in pressure-tight applications
- To improve surface finish (impregnated parts are platable)
- To improve machinability
- To improve corrosion resistance

Although high precision has been achieved in P/M parts for many years, their application was once restricted because of mechanical property limitations. Now, however, the mechanical properties can be 100% increased in steel P/M parts by hot forging in closed dies. Properties of P/M parts forged to 100% theoretical density in production conditions are claimed to be equal, and sometimes superior, to those of wrought steels of similar composition.

Ferrule

A cylindrical reinforcement or sleeve.

Fettling

Various processes for improving or restoring a component or equipment. For example, trimming flash or other excess material from castings or locally repairing components, furnace linings, etc.

Fiber

(1) The variation in mechanical properties, structure, etc., observed when wrought material is tested or examined in different directions relative to the primary direction of working. For example, plate has a greater strength and ductility in the longitudinal direction than in the thickness direction. (2) Filaments. Thus fiber metal comprises metal fibers packed and sintered together possibly with another material filling the interstices.

Fiber (Composites)

A general term used to refer to filamentary materials. Often, fiber is used synonymously with filament. It is a general term for a filament with a finite length that is at least 100 times its diameter, which is typically 0.10–0.13 mm (0.004–0.005 in.). In most cases, it is prepared by drawing from a molten bath, spinning, or depositing on a substrate. Fibers can be continuous or specific short lengths (discontinuous), normally no less than 3.2 mm (1/8 in.).

Fiber (Metals)

(1) The characteristic of wrought metal that indicates *directional properties* and is revealed by etching of a longitudinal section or is manifested by the fibrous or woody appearance of the fracture. It is caused chiefly by extension of the constituents of the metal, both metallic and nonmetallic, in the direction of working. (2) The pattern of preferred orientation of metal crystals after a given deformation process, usually wiredrawing. See also *fibering* and *preferred orientation*.

Fiber Bridging (Composites)

Reinforcing fiber material that bridges an inside radius of a protruded product. This condition is caused by shrinkage stresses around such a radius during cure.

Fiber Content

The amount of fiber present in reinforced plastics and composites, usually expressed as a percentage volume fraction or weight fraction.

Fiber Count

The number of fibers per unit width of ply present in a specified section of a reinforced plastic or composite.

Fiber Diameter

The measurement (expressed in micrometers or microinches) of the diameter of individual filaments used in reinforced plastics or composites.

Fiber Direction (Composites)

The orientation or alignment of the longitudinal axis of a fiber with respect to a stated reference axis.

Fiber Exposure (Composites)

A condition in which reinforcing fibers within the base material are exposed in machined, abraded, or chemically attacked areas. See also *weave exposure*.

Fiber Grease

A type of grease having a pronounced fibrous structure.

Fiber Metallurgy

The technology of producing solid bodies from fibers or chopped filaments, with or without a metal matrix. The fibers may consist of such nonmetals as graphite or aluminum oxide, or of such metals as tungsten or boron. See also *metal-matrix composites*.

Fiber Pattern

Visible fibers on the surface of laminates or molding. The thread size and weave of glass cloth.

Fiber Show

Strands or bundles of fibers that are not covered by plastic because they are at or above the surface of a reinforced plastic or composite.

Fiber Stress

The stress acting along the main axis of a fiber. The fiber may be an individual filament or it may be a local unit section of a component that is not uniformly stressed.

Fiber Texture (Metals)

A texture characterized by having only one preferred crystallographic direction.

Fiber Wash (Composites)

Splaying out of woven or nonwoven fibers from the general reinforcement direction. Fibers are carried along with bleeding resin during cure.

Fiberboard

Heavy sheet material of fibers matted and pressed or rolled to form a strong board, used for making containers and partitions and for construction purposes. Almost any organic fiber may be used, with or without a binder. The soft boards are made by felting wood pulp, wood chips, or bagasse, usually without a binder. *Masonite* is produced from by-product woodchips reduced to the cellulose fibers by high steam pressure. The long fibers and the lignin adhesive of the wood are retained, and no chemicals are used in pressing the pulp into boards. Masonite quarter board, for paneling, is made in boards 0.25 in. (0.64 cm) thick. *Presdwood* is a grainless grade made by compressing under hydraulic pressure and is dense and strong.

These types belong to the class known as *hardboard*, in the processing of which the carbohydrates and soluble constituents of the original wood are dissolved out and the relative proportion of lignin is increased, resulting in a grainless, hard, stiff, and water-resistant board free from shrinkage. The specific gravity of most hard boards is greater than 1.0, and the modulus of rupture is from 5,000 to 15,000 lb/in.^{2} (34–103 MPa). The lignin acts as a binder for the fibers, but some hard boards are made harder and more resistant by adding a percentage of an insoluble resin. Densified hardboard, made with high pressure, is 85 lb/ft^3 (1362 kg/m^3) or greater. Hard boards have uniform strength in all directions and have smooth surfaces.

Irradiated wood is natural wood impregnated with residents of low molecular weight and irradiated with gamma rays from cobalt 60 which cross-links the resin molecules and binds them to the fibers of the wood. The resins add strength and hardness without changing the grain structure and color of the wood. Maple, impregnated with 0.5% by weight of methyl methacrylate and irradiated, is three times as hard as the natural wood but can be worked with ordinary tools. *Particleboards*, made with wood particles, have lower density and have greater flexibility but lower strength than hardboard. The process is not limited to the making of boards. Wood particles are also used for low-cost molded parts, with up to 90% wood particles and the balance urea, phenolic, or melamine resin. Birch and maple particles are preferred.

Wood molding powder is made by the same method of treating with fibers with steam pressure and hydrolyzing the hemicellulose, leaving the lignin free as a binder. Hardboard is used for countertops, flooring, furniture, and jigs and templates.

Hardwood is a hardboard made from hardwood waste compressed into sheets under heat and hydraulic pressure. The surface is hard with a high polish.

Fiberglass

An individual filament made by drawing molten glass. A continuous filament is a glass fiber of great or indefinite length. A staple fiber is a glass fiber of relatively short length, generally less than 430 mm (17 in.), the length depending on the forming or spinning process used. The four main glasses used for Fiberglas are high-alkali glass (A-glass), electrical grade glass (E-glass), a modified E-glass that is chemically resistant (ECR-glass), and high-strength glass (S-glass).

Fiberglass Reinforcement

Major material used to reinforce a plastic. Available as mat, roving, fabric, and so forth, it is incorporated into both thermosets and thermoplastics.

Fibering

Elongation and alignment of internal boundaries, second phases, and inclusions in particular directions corresponding to the direction of metal flow during deformation processing.

Fiber-Reinforced Composite

A material consisting of two or more discrete physical phases, in which a fibrous phase is dispersed in a continuous matrix phase. The fibrous phase may be macro-, micro-, or submicroscopic, but it must retain its physical identity so that it could conceivably be removed from the matrix intact.

Fiber-Reinforced Glass

Fiber-reinforced glass (FRG) composites are glass or glass ceramic matrices reinforced with long fibers of carbon or SiC. These composites are lighter than steel but just as strong as many steel grades, and can resist higher temperatures. They also have outstanding resistance to impact, thermal shock, and wear, and can be formulated to control thermal and electrical conductivity. With proper tooling, operations such as drilling, grinding, and turning can be completed in half the time required for non-reinforced glass.

Currently, FRG components are primarily used for handling hot glass or molten aluminum during manufacturing operations. FRG is also under test as an engineering material in a variety of markets, including the aerospace, automotive, and semiconductor industries.

Properties

Glass and glass ceramics are versatile materials because of such characteristics as high chemical resistance and special electrical behavior, but fragility under tensile stress has limited their structural applications. However, the addition of continuous fibers of carbon and SiC into glass produces a material that withstands very high mechanical stresses and loads. As a result, it is now possible to manufacture glass matrix composites that are capable of structural roles.

The reinforcement process provides an increase in both admissible maximum stress and ultimate elongation: brittle fracture is replaced by an almost ductile behavior. These properties, together with others, make FRG composites advantageous as a material in the machine-building industry. For example, low density and high modulus of elasticity allow extremely strong and stiff structures.

Another important characteristic of long-fiber-reinforced glasses is that their mechanical properties are largely independent of the condition at the surface. This means that they can be drilled, even near edges, and joined with other parts by means of screws and bolts.

In addition, FRG composites have very high resistance to temperature changes, low coefficient of thermal expansion, high specific heat capacity, and good chemical resistance.

They also exhibit good tribological and wear properties. However, because of the great variety of tribological and tribomechanical processes and applications, each service environment must be evaluated before selecting an FRG composite.

Fiber Effects

Performance of FRG composites under various operating conditions depends primarily on the type of fiber, the amount of fiber, and its orientation within the matrix. It also depends on the operating environment and the duration or cycle of the specific application.

Fiber Types

Carbon fibers are stable up to >2000°C under inert atmospheres, but few applications require such performance. Moreover, at those temperature levels, the matrix itself becomes soft. In air, carbon fibers remain stable to about 450°C. This temperature limit also applies to composites containing carbon fibers, unless the fibers have been completely sealed.

By contrast, SiC fibers are stable in air up to about 1200°C. In this case, the heat resistance of the matrix is the limiting factor. Therefore, SiC FRG products can be considered stable to 500°C if the matrix consists of the alkali–borosilicate–glass, *Duran*.

On the other hand, if an alkali-free aluminosilicate glass is the matrix, the SiC-fiber-containing composite is stable up to 750°C.

When quenched from 350°C to 20°C, most nonreinforced glasses break. However, SiC FRG is capable of withstanding a 60-fold quench (thermal shock) from 550°C to ambient temperatures. Furthermore, the product has exhibited good fracture toughness at temperatures as low as −200°C.

Amount of Fiber

Performance at high temperatures and resistance to thermal shock can be improved by increasing the percentage of fiber within the composite. However, fiber is the chief cost factor in producing FRG. Therefore, performance advantages must continue to outweigh costs as fiber count is increased.

Fiber Orientation

Depending on the arrangement of the fibers, the properties of a FRG composite can be either isotropic or anisotropic. For example, fibers may be arranged so that heat conductivity is low in one direction, but high in the perpendicular direction. Absolute values of heat conductivity range from 1.7 to ~25 W/m K or half the value of steel. As in high-temperature performance, heat conductivity depends on the direction, type, and amount of fiber in the composite.

Relatively high conductivity values result when the composite contains carbon fibers aligned longitudinally in the direction of heat travel; otherwise, the material may work as a heat barrier.

Similarly, the electrical resistance of composites has been measured at room temperature in a composite 40% fiber by volume. In the fiber direction, specific electrical resistance for the SiC product was 10 Ω cm, while for the carbon product, it was 0.01 Ω cm. Normal to the fibers, the specific electrical resistance may be several orders of magnitude higher.

Potential Applications

FRG components are used as replacements for asbestos and other materials for handling hot glass during glass manufacturing operations. These include pushers, grips, transporters, and takeout pads as well as other parts in contact with hot glowing glass. It is an application characterized by extreme conditions in terms of heat and thermal shock, in which FRG composites have a great advantage over traditional materials.

Within the metals industry, FRG pads act as thermal buffers in the handling of molten aluminum and its alloys. In this application, they form insulation pads between the melt vessel and supporting structural work, thereby minimizing heat conductivity away from the melt.

Other applications being exploited are the replacement for fragile glass substrates that hold silicon wafers during chemical vapor deposition (CVD) coating processes. Required properties include high fracture toughness for higher yields, a low coefficient of thermal expansion, and high stiffness.

Automotive applications include piston inserts, valve-control components, and other parts subjected to thermal and mechanical shock. FRG may also be applied in products for the protection of areas that must be insulated from heat, or where fracture toughness is needed.

The tribological and wear properties of FRG suggest applications as bearings and seals in pump manufacturing, and for off-road industrial vehicle and equipment brakes. Other potential applications include aerospace components with glass ceramic matrices for temperatures above 1000°C; scanning-mirror substrates for space and missile systems; and protective inserts for parts requiring high impact resistance. As FRG manufacturing technology advances and costs are reduced, the composites would be suitable for the construction of safes and strong rooms, and as armor in vehicles.

FRG components are being produced in several configurations, which include plates, disks, or rings that have a maximum dimension (length, width, or diameter) of 400 mm, and thickness of 50 mm. The minimum thickness of unprocessed forms is 0.5 mm. Semifinished material can be reprocessed into end products with complex geometric forms.

Fiber-Reinforced Plastics

Fiber-reinforced plastics (FRPs) comprise a broad group of composite materials composed of fibers embedded in a plastic resin matrix. In general, they have relatively high strength-to-weight ratios and excellent corrosion resistance compared to metals. They can be formed economically into virtually any shape and size. In size, FRP products range from tiny electronic components to large boat hulls. Between these extremes, there are a wide variety of FRP gears, bearings, housings, and parts used in all product industries.

FRPs are composed of three major components—matrix, fiber, and bonding agent. The plastic resin serves as the matrix in which are embedded the fibers. Adherence between matrix and fibers is achieved by a bonding agent or binder, sometimes called a coupling agent. Most plastics, both thermosets and thermoplastics, can be the matrix material. In addition to these three major components, a wide variety of additives—fillers, catalysts, inhibitors, stabilizers, pigments, and fire retardants—can be used to fit specific application needs.

Fibers

Glass is by far the most-used fiber in FRPs. Glass-fiber-reinforced plastics are often referred to as GFRP or GRP. Asbestos fiber has some use, but is largely limited in applications where maximum thermal insulation or fire resistance is required. Other fibrous materials used as reinforcements are paper, sisal, cotton, nylon, and Kevlar. For high-performance parts and components, more costly fibers, such as boron, carbon, and graphite, can be specified.

The standard glass fiber used in GRP is a borosilicate type, known as E-glass (E-Gl). The fibers are spun as single glass filaments with diameters ranging from 0.01 to 0.03 mm. These filaments, collected into strands, usually around 200 per strand, are manufactured into many forms of reinforcement. The E-Gl fibers have a tensile strength of 3447 MPa. Another glass fiber, known as S-Gl, is higher in strength, but because of its higher cost its use is limited to advanced, high-performance applications. In general, in reinforced thermoplastics, glass content runs between 20% and 40%; with thermosets it runs as high as 80% in the case of filament-wound structures.

There are a number of standard forms in which glass fiber is produced and applied in GRP.

1. Continuous strands of glass supplied either as twisted, single-end strands (yarn) or as untwisted multistrands (continuous) roving
2. Fabrics woven from yarns in a variety of type, weights, and widths
3. Woven rovings—continuous rovings worked into a coarse, heavy, drapable fabric
4. Chopped strands made from either continuous or spun roving cut into 3.2–12.7 mm lengths
5. Reinforcing mats made of either chopped at random or continuous strand laid down in a random pattern
6. Surfacing mats composed of continuous glass filaments in random patterns

Resins

Although a number of different plastic resins are used as the matrix for reinforced plastics, thermosetting polyester resins are the most common. The combination of polyester and glass provides a good balance of mechanical properties as well as corrosion resistance, low cost, and good dimensional stability. In addition, curing can be done at room temperature without pressure, thus making for low processing equipment costs.

Polyesters are also available as casting resins, both in water-extended formulations for low-cost castings, and in compounds filled with ground wood or pecan-shell flour for furniture components.

Low-profile molding resins are mixtures of polyester resins, thermoplastic polymers, and glass-fiber reinforcements. These are used to mold parts with smooth surfaces that can be painted without the need for prior sanding.

For high-volume production, special sheet-molding compounds (SMC) are available in continuous-sheet form. Resin mixtures of thermoplastics with polyesters have been developed to produce high-quality surfaces in the finished molding.

These polyester resins are available for use as coatings for curing by ultraviolet radiation. These 100%-solids materials cure in a matter of seconds and release no solvents. Although some styrene may be lost upon exposure to ultraviolet radiation, the amount is small.

Prepregs are partially cured thermoset resin-coated reinforcing fabrics in roll or sheet. The prepregs (short for preimpregnated) can be laid or wrapped in place and then fully cured by heat.

Other glass-reinforced thermosets include phenolics and epoxies. GR (glass-reinforced) phenolics are noted for their low cost and good overall performance in low-strength applications. Because of their good electrical resistivity and low water absorption, they are widely used for electrical housings, circuit boards, and gears. Since epoxies are more expensive than polyesters and phenolics, GR epoxies are limited to high-performance parts where their excellent strength, thermal stability, chemical resistance, and dielectric strength are required.

Initially, GRP materials were largely limited to thermosetting plastics. Today, however more than 1000 different types and grades of reinforced thermoplastics or GR+ P are commercially available. Leaders in volume are nylon and the styrenes. Unlike thermosetting resins, GR+ P parts can be made in standard injection-molding machines. The resin can be supplied as pellets containing chopped glass fibers. As a general rule, a GR+ P with chopped fibers at least doubles the tensile strength and stiffness of the plastic. Glass-reinforced thermoplastics are also produced as the materials for forming on metal-stamping equipment and compression-molding machines.

Processing Methods

Matched Metal Die Molding

This is the most efficient and economical method for mass-producing high-strength parts. Parts are press-molded in matched male and female molds at pressures of 1380–2068 MPa and at heats of 113°C–127°C.

Four main forms of thermosetting resin reinforcement are used:

1. Chopped fiber preforms, shaped like the part, are saturated with resin at the mold. They are best for deep draw, compound curvature parts.
2. Flat mat, saturated with resin at the mold, is used for shallow parts with simple curvature.
3. SMC, a preimpregnated material, has advantages for parts with varying thickness. SMCs consist of polyester resin, long glass fibers (to 5.08 cm), a catalyst, and other additives. They are supplied in rolls, sandwiched between polyethylene carrier films. Uniformity of SMC materials is closely controlled, making these materials especially suited for automated production. Structural SMCs contain up to 65% glass in continuous, as well as random, fiber orientation. This compares with 20%–35% glass, in random orientation only, of conventional SMC.
4. Bulk molding compound (BMC), a premix of polyester resin, short glass fibers (3.2–6.4 mm long), filler, catalyst, and other additives for specific properties. BMC is supplied in bulk form or as extruded rope for ease of handling and it is used for parts similar to castings.

Injection Molding

In this high-volume process, a mix of short fibers and resin is forced by a screw or plunger through an orifice into the heated cavity of a closed matched metal mold. It is the major method for forming reinforced thermoplastics and is used for thermoplastic-modified thermosetting BMCs.

Hand Layup

This is the simplest of all methods of forming thermosetting composites. It is best employed for quantities under 1000, for prototypes and sample runs, for extremely large parts, and for larger volume where model changes are frequent, as in boats. In hand layup, only one mold is used, usually female, which can be made of low-cost wood or plaster. Duplicate molds are inexpensive. The reinforcing mat or fabric is cut to fit, laid in the mold, and saturated with resin by hand, using a brush, roller, or spray gun. Layers are built up to the required thickness; then the laminate is cured to permanent hardness, generally at room temperature.

Spray-Up

Like hand layup, the spray-up method uses a single mold, but it can introduce a degree of automation. This method is good for complex thermoset moldings, and its portable equipment eases on-site fabrication and repair. Short lengths of reinforcement and resin are projected by a specially designed spray gun so they are deposited simultaneously on the surface of the mold. Cure is usually accomplished by a catalyst in the resin at room temperature.

Filament Winding

This method produces moldings with the highest strength-to-weight ratio of any reinforced thermoset because of its high glass-to-resin ratio. It is generally limited to surfaces of revolution—round, oval, tapered, or rectangular—but it can achieve a high degree of automation. Continuous fiber strands are wound on a suitably shaped mandrel or core and precisely positioned in predetermined patterns. The mandrel may be left in place permanently or removed after cure. The strands may be preimpregnated or the resin may be applied during or after winding. Heat is used to effect final cure.

Centrifugal Casting

This is another method of producing round, oval, tapered, or rectangular parts. It offers low labor and tooling costs, uniform wall thicknesses, and good inner and outer surfaces. Chopped fibers and resin are placed inside a mandrel and uniformly distributed as the mandrel is rotated inside an oven.

Continuous Laminating

Continuous laminating is the most economical method of producing flat and corrugated panels, glazing, and similar products in large volume. Reinforcing mat or fabric is impregnated with resin, run-through laminating rolls between cellophane sheets to control thickness and resin content, and then cured in a heating zone.

Pultrusion

Pultrusion produces shapes with high unidirectional strength such as "I" beams, flat stock for building siding, fishing rods, and shafts for golf clubs. Continuous fiber strands, combined with mat or woven fibers for cross strength, are impregnated with resin and pulled through a long heated steel die. The die shapes the product and controls resin content.

Properties

The mechanical properties of fiber composites are dependent on a number of complex factors. Two of the dominating ones in GRPs are the length of fibers and the glass content by weight. In general, strength increases with fiber length. For example, reinforcing a thermoplastic with chopped glass fibers at least doubles the strength of the plastic, whereas long-glass fiber reinforced the plastics

thermoplastics exhibit increases of 300% and 400%. Also heat distortion temperatures usually increase by about 37.8°C, and impact strengths are appreciably increased. Similarly, as a general rule, an increase in glass content results in the following property changes:

- Tensile and impact strength increase.
- Modulus of elasticity increases.
- Heat deflection temperature increases, sometimes as much as 149°C.
- Creep decreases and dimensional stability increases.
- Thermal expansion decreases.

Properties of thermoset polyesters are so dependent on type, compounding, and processing method that a complete listing covering all combinations would be almost impossible. However, typical strength ranges obtainable in parts fabricated from various forms of polyester/glass compounds and processed by several methods can be found in *Encyclopedia of Materials, Parts, and Finishes*, Second Edition, p. 240, CRC Press, 2002.

Fibers

By definition, a fiber has a length at least 100 times its diameter or width, and its length must be at least 0.5 cm. Length also determines whether a fiber is classified as staple or filament. Filaments are long and/or continuous fibers. Staple fibers are relatively short, and, in practical applications, range from under 2.5 to 15.2 cm long (except for rope, where the fibers can run to several centimeters). Of the natural fibers, only silk exists in filament form; synthetics are produced as both staple and filaments.

The internal, microscopic structure of fibers is basically no different from that of other polymeric materials. Each fiber is composed of an aggregate of thousands of polymer molecules. However, in contrast to bulk plastic forms, the polymers in fibers are generally longer and aligned linearly, more or less parallel to the fiber axis. Thus, fibers are generally more crystalline than are bulk forms.

Also in contrast to bulk forms, fibers are not used alone, but either in assemblies or aggregates such as yarn or textiles or as a constituent with other materials, such as in composites. Also, compared with other materials, the properties and behavior of both fibers and textile forms are more critically dependent on their geometry. Hence, fibers are sometimes characterized as tiny microscopic beams, and, as such, their structural properties are dependent on such factors as cross-sectional area and shape, and length. The cross-sectional shape and diameter of fibers vary widely. Glass, nylon, *Dynel*, and *Dacron*, for example, are essentially circular. Some other synthetics are oval, and others are irregular and serrated round. Cotton fibers are round tubes, and silk is triangular.

Fiber diameters range from about 0.01 to 0.04 mm. Because of the irregular cross-section of many fibers, it is common practice to specify diameter or cross-sectional area in terms of fineness, which is defined as a weight-to-length or linear density relationship. One exception is wool, which is graded in micrometers. The common measure of linear density is the denier, which is the weight in grams of a 9000 m length of fiber. Another measure is the tex, which is defined as grams per 1 km. A millitex is the number of grams per 1000 km.

Of course, the linear density, or denier, is also directly related to fiber density. This is expressed as the denier/density value, commonly referred to as denier per unit density, which represents the equivalent denier for a fiber with the same cross-sectional area and a density of 1.

The cross-sectional diameter or area generally has a major influence on fiber and textile properties. It affects, for example, yarn packing, weave tightness, fabric stiffness, fabric thickness and weight, and cost relationships. Similarly, the cross-sectional shape affects yarn packing, stiffness, and twisting characteristics. It also affects the surface area, which in turn determines the fiber contact area, air permeability, and other properties.

For efficient production, a number of filaments are pulled simultaneously from several orifices in the bushing. These filaments (usually numbering about 204) are collected into a bundle, called a "strand," at a gathering device, where a "size" is applied to the filament surfaces. The strand is then wound into a forming package called a "cake." From this cake, shippable forms of fibrous glass are produced.

Fibers Fracture

A gray and amorphous *fracture* that results when a metal is sufficiently ductile for the crystals to elongate before fracture occurs. When a fibrous fracture is obtained in an impact test, it may be regarded as definite evidence of toughness of the metal. Silky, dull gray and ductile as opposed to bright and crystalline. See also *crystalline fracture* and *silky fracture*.

Fibrillation

Production of fiber from film.

Fibrous Glass

The primary engineering benefits of glass fibers are their (1) inorganic nature, which makes them highly inert; (2) high strength-to-weight ratio; (3) nonflammability; and (4) resistance to heat, fungi, and rotting.

Glass fibers are produced in both filament and staple form. Their major engineering uses are (1) thermal and/or acoustical insulation and (2) as reinforcements, primarily for plastics.

Types

The largest volume of glass fibers used for engineering applications are so-called "E," made from a lime-Al_2O_3 borosilicate glass that is relatively soda-free. Although its initial strength at the bushing may be about 2758–3447 MPa, surface damage to fibers (both mechanical damage in handling and effects of moisture) reduces usable strengths to 1034–1380 MPa. But at least 1,380 MPa tensile strength, the relatively low density of glass (0.092 lb/ft³) produces a strength-to-weight ratio of about 5, 511, 800 cm, superior to that of a 3,060 MPa tensile strength steel. Modulus of E-glass fibers is about 68,666 MPa. Although essentially unaffected by low temperatures, E-glass is limited to a maximum continuous operating temperature of about 316°C.

Other specialized types of glass (primarily used in specialty reinforced plastics applications) include:

1. *High silica, leached glass fiber*: Fibers with silica content of 96%–99% are produced by leaching glass fibers. Such fibers provide excellent heat resistance, but relatively low strengths. They are usually used in short-fiber form for molding compounds.
2. *Silica or quartz fiber*: Fibers of pure silica provide optimum heat resistance (to about 93°C), although strength is somewhat lower than that of conventional E-glass.

3. *High modulus fibers*: Fibers of a beryllia-containing glass have been developed (primarily for filament winding use) with modulus of about 109,866–123,599 MPa.

Production Methods

Most fibrous glass is produced either by air, steam, or flame blowing or by mechanical pulling or drawing. Blowing produces relatively short staple fibers; mechanical drawing produces continuous monofilaments.

In blowing, steam or air jets impinge upon and breakup molten streams of glass, forming fibers. The type of fiber produced depends on the pressure of the steam or air and the temperature and viscosity of the molten glass.

In the mechanical drawing process, the molten glass is fed into a "bushing," which contains a number of orifices through which the glass flows. Continuous filaments are then drawn from the molten stream. During the early stages of cooling, the stream is attenuated into filaments by being pulled at very high speeds—usually ranging from 25 to 50 m/s.

Insulation

In general, fibrous glass insulation is available in densities ranging from 0.5 to 12 lb/ft³. Maximum operating temperature is about 316°C–1093°C, depending on type of glass. It provides high sound absorption, relatively high tensile strength, and resistance to moisture, fire, rotting, and fungi and bacteria growth. It is available in either flexible or rigid form.

The excellent insulating properties of fibrous glass are due to the large pockets of air between the fibers. These air pockets take up considerable volume.

Fibrous glass is not affected by low temperatures and has been used satisfactorily at temperatures as low as −212°C. Heat resistance depends on type of glass: borosilicate glass is generally limited to operating temperatures of 316°C–538°C; high silica glasses are capable of operating at 999°C; silica (quartz) fibers are usable up to 1093°C. The heat resistance of bonded insulations is normally limited by the heat resistance of the binder (maximum, about 232°C–316°C).

Forms

Following are the various forms in which fibrous glass is used in reinforced plastics:

1. Rovings consist of a number of strands (usually 60) gathered together from cake packages and wound on a tube to form a cylindrical package. Rovings have very little or no twist. They are used either to provide completely unidirectional strength characteristics, such as in filament winding, or are chopped into predetermined lengths for preform-matched metal or spray molding.
2. Chopped strand consists of strands that have been cut into short lengths (usually 12.7–50.8 mm) in a manner similar to chopped roving, for use in preform-matched metal or spray molding, or to make molding compounds. It is the least expensive form of fibrous-glass reinforcement.
3. Milled fibers are produced from continuous strands that are hammer-milled into small modules of filamented

glass (nominal lengths of 0.8–3.2 mm). Largely used for filler reinforcement in casting resins and in resin adhesives, they provide greater body and dimensional stability.

4. Yarns are twisted from either filaments or staple fibers on standard textile equipment. Although primarily an intermediate form from which woven fabrics are made, yarns are used for making rod stock, and for some very high strength, unidirectionally reinforced shapes. A common form in which yarn is available is the "warp beam," where many parallel yarns are wrapped on a mandrel.

5. Nonwoven mats are available both as reinforcing mats and as surfacing or overlay mats. Reinforcing mats are made of either chopped strands or swirled continuous strands laid down in a random pattern. Strands are held together by resinous binders. In laminates, mats provide relatively low strength levels, but strengths are isotropic. Surfacing or overlay mats are both thin mats of staple monofilaments. They provide practically no reinforcing, but serve to stabilize the surface resin coat, providing better appearance.

6. Woven fabrics and rovings provide the highest strength characteristics to reinforced plastic laminates (except for filament-wound structures), although strengths are orthotropic. A wide variety of fabrics and weaves are available both in woven yarns and woven rovings. Probably the most common types used are plain, basket, crowfoot satin, long shaft satin, unidirectional, and leno weaves.

Fibers Structure (Metals)

(1) In forgings, a structure revealed as laminations, not necessarily detrimental, on an etched section or as a ropy appearance on a fracture. It is not to be confused with silky or ductile fracture of a clean metal. (2) In wrought iron, a structure consisting of slag fibers embedded in ferrite. (3) In rolled steel plate stock, a uniform, fine-grained structure on a fractured surface, free of laminations or shale-type discontinuities.

Fibers Surface

A rough surface with linear features similar to rope.

Fick's Law

The quantity of atoms diffusing across a unit plane in unit time is inversely proportional to the concentration gradient.

Field Ion Microscopy

An analytical technique in which atoms are ionized by an electric field near a sharp specimen tip; the field then forces the ions to a fluorescent screen which shows an enlarged image of the tip, and individual atoms are made visible. See also *atom probe*.

Field Ionization

The ionization of gaseous atoms and molecules by an intense electric field, often at the surface of a solid.

Field-Emission Microscopy

An image-forming analytical technique in which a strong electrostatic field causes emission of electrons from a sharply rounded

point or from a specimen that has been placed on that point. The electrons are accelerated to a phosphorescent screen, or photographic film, producing a visible picture of the variation of emission over the specimen surface.

Field-of-View

The maximum diameter of an object that can be imaged by a microscope or other analytic technique.

Filament

The smallest unit of fibrous material. The basic units formed during drawing and spinning, which are gathered into strands of fiber for use as reinforcements. Filaments usually are of extreme length and very small diameter, usually less than 25 μm (1 mils). Normally, filaments are not used individually. Some textile filaments can function as a reinforcing yarn when they are of sufficient strength and flexibility.

Filament Winding

A process for fabricating a reinforced plastic or composite structure in which continuous reinforcements (filament, wire, yarn, tape, and the like), either previously impregnated with a matrix material or impregnated during the winding, are placed over a rotating and removable form or mandrel in a prescribed way to meet certain stress conditions. Generally, the shape is a surface of revolution and may or may not include end closures. When the required number of layers is applied, the wound form is cured and the mandrel is removed. See also *helical winding* and *polar winding*.

Filament-Wound Reinforced Plastics

The true fiberglass filament-wound structure may be more appropriately termed a resin-bonded filament-wound structure because it comprises approximately 80% glass fibers by weight and 20% bonding resin. Fibers are generally oriented to resist the principal stresses and the resin protects while secondarily supporting the fiber system. Filament winding is well adapted to the fabrication of internal pressure vessels; it has also performed well under external pressure and can be designed to function efficiently as a column or beam. This structure has the tensile strength of moderately heat-treated alloy steel and one-quarter of the weight.

The Winding Process

Bands of parallel glass filaments (usually in the form of roving) are wound over a mandrel following a precise pattern in such a way that subsequent bands lie adjacent, progressively covering the mandrel in successive layers, thus generating a shell structure. Liquid resin is simultaneously applied, generally by passing the filament band through a bath of catalyzed resin.

Tension generates a running load between the curved work surface and filament band which forces out air and excess resin and allows each successive layer ultimately to rest on solid material while the remaining interstices are filled with resin. Precision of filament placement plus tension and viscosity control are primary controlling factors in the attainment of high fiber content, which is generally desired for high strength.

Preimpregnation of the fiber strands is also used as a means of applying the resin binder. Such prepregs must fuse on contact with

F

the work to accomplish a bond so the fundamental relationships remain the same. The fiber bands are parallel strands only, because a cross weave or other structural filler would not bear primary loads and would preclude the true maintenance of equal tension on fibers in the band or roving when winding over crowned surfaces.

Tension serves only the purpose of accomplishing high fiber content and cannot be considered as accomplishing any prestressing. This is primarily true because the dry strength of glass fibers, as wound, is only about one-quarter of the ultimate resin consolidated strength. Also, unless some structural component is to remain within the wound shell (rather than a removable mandrel), there is no member against which a prestress can be maintained.

Range and Accuracy

Large or small structures are easily fabricated by adhering to the basic principles. Winding precision is important as is the relation of filament tension, resin viscosity, and radius of curvature of the filament path on the work surface. Small tubes have been made down to 0.65 cm in diameter, and there appear to be no fixed limitations in either direction. Wall thicknesses may be several centimeters or more because the normal bonding resins contain no volatile components and the glass content is so high that there is little danger of the exothermic heat becoming excessive.

Dimensional control in winding depends upon mandrel accuracy as well as both material and process control. The bands of filaments are generally 0.13 mm thick, and a full layer requires coverage by both right- and left-hand helices making the layer thickness 0.25 mm. Glass fiber thickness is subject to some variation, and resin content variation will also affect thickness. Wall thickness can generally be held to ±5%. Length and diameter are easily held to 1/10% of 1% and less because there is little resin shrinkage or wound-in strain due to winding tension.

Machining may be accomplished by carbide tools or grinding techniques. Tolerances can be held as closely as in metals. The inner surface as-wound against a good steel mandrel can have a finish of approximately 30 μin., and normal machining or grinding will produce a 40–60 μin. finish. Cutting of surface fibers in machining does not weaken the structure.

Component Materials

There are three primary materials in this composite: the glass fiber, fiber finish, and the bonding resin. Glass fibers are continuous and each "end" contains 204 monofilaments approximately 0.01 mm in diameter. These ends are plied together without twist to form a strand of "roving"; 12-end, equally tensioned, roving is generally used for the winding of high-performance structures.

The fiber finish generally includes compounds with a chemical affinity for the glass surface and the bonding resin. These are called coupling agents. Other functional components are lubricants and "film formers" for the generation of strand integrity. Both improve handling properties but do not contribute to the performance of filament structures.

The resin binder materials are generally liquid at room temperatures or are wound hot for liquid integration of the system. Best results are obtained with strong tough resins such as the epoxies. Polyesters have also been extensively used (primarily for radomes where electrical properties are critical) and any bonding resin should be free of polymerization products of a volatile nature, which would have to escape through the cured structure.

Chemical resistance and electrical properties of the constituents are usually critical in selecting the proper resin, as applications of the filament-wound structure are found in both the electrical and chemical industries.

Filamentary Shrinkage

A fine network of shrinkage cavities, occasionally found in steel castings, that produces a radiographic image resembling lace.

Filar Eyepiece

In an optical microscope, an eyepiece having in its focal plane a fiducial line that can be moved using a calibrated micrometer screw. Useful for accurate determination of linear dimensions. Also termed filar micrometer.

File Hardness

Hardness is determined by the use of a steel file of standardized hardness on the assumption that a material that cannot be cut with the file is as hard as, or harder than, the file. Files covering a range of hardnesses may be employed; the most common are files heat-treated to approximately 67–70 HRC.

Filiform Corrosion

Corrosion beneath surface coats, paint, etc., progressing as lines or fingers from initiation points which commonly are sites of coating breakdown or of precoating contamination.

Filigree

Delicate ornamental wire work.

Fill (Composites)

Reinforcing yarn oriented at right angles to the warp in a woven fabric.

Fill Density

See preferred term *apparent density*.

Fill Depth

Synonymous with *fill height*.

Fill Factor

In powder metallurgy, the quotient of the fill volume of a powder over the volume of the green compact after ejection from the die. It is the same as the quotient of the powder fill height over the height of the compact. Inverse parameter of *compression ratio*.

Fill Height

In powder metallurgy, the distance between the lower punch face and the top plane of the die body in the fill position of the press tool.

Fill Position

In powder metallurgy, the position of the press tool that enables the filling of the desired amount of powder into the die cavity.

Fill Ratio

See *compression ratio* (powder metallurgy).

Fill Shoe

See preferred term *feed shoe*.

Fill Volume

The volume that a metal powder fills after flowing loosely into a space that is open at the top, such as a die cavity or a measuring receptacle.

Fill-and-Wipe

Technique used with plastic parts that are molded with depressed designs; after application of paint, the surplus is wiped off, leaving paint only in the depressed areas. Sometimes called *wipe-ins*.

Filled Shell

See *atomic structure*.

Filler

(1) A relatively inert substance added to a plastic to alter its physical, mechanical, thermal, electrical, or other properties, or to lower cost or density. Sometimes, the term is used specifically to mean particulate additives. See also *inert filler* and *reinforced plastics*. (2) A relatively non-adhesive substance added to an adhesive to improve its working properties, permanence, strength, or other qualities. See also *binder* and *extruder*. (3) In lubrication, a substance such as lime, talc, mica, and other powders, added to a grease to increase its consistency or to an oil to increase its viscosity. (4) Additional material added in the molten state to a welded, brazed or soldered joint by a filler rod or by a consumable electrode. (5) Bulking material such as powder or fiber added to plastics.

Filler Metal

Metal added in making a brazed, soldered, or welded joint.

Filler Rod

Rod or wire melted in welding, brazing, or soldering processes to become incorporated in the joint. Filler rods are used in processes such as oxyacetylene acetylene welding or tungsten inert gas welding (T.I.G.) and are often held in one hand while the other holds the torch.

Filler Sheet

A sheet of deformable or resilient material that, when placed between the assembly to be adhesively bonded and pressure applicator, or when distributed within a stack of assemblies, aids in providing uniform application of pressure over the area to be bonded.

Fillet (Adhesive Bonding)

A rounded filling or adhesive that fills the corner or angle where two adherends are joined.

Fillet (Metals)

(1) Concave corner piece usually used at the intersection of casting sections. Also the radius of metal at such junctions as opposed to an abrupt angular junction. (2) A radius (curvature) imparted to inside needing surfaces.

Fillet Radius

Bend radius between two abutting walls.

Fillet Weld

A weld, approximately triangular in cross-section, joining two surfaces, essentially at right angles to each other in a lap, tee, or corner joint.

Fillet Weld Size

See preferred term *size of weld*.

Filling Yarn

The transverse threads or fibers in a woven fabric used in reinforced plastics or composites. Those fibers running perpendicular to the warp. Also called *weft*.

Film Adhesive

A synthetic resin adhesive, usually of thermosetting type, in the form of a thin, dry film of resin with or without a paper or glass carrier.

Film Resistance

The electrical resistance that results from films at contacting surfaces, such as oxides and contaminants, that prevent pure metallic contact.

Film Strength

An imprecise term denoting ability of a surface film to resist rupture by the penetration of asperities during sliding or rolling. A high film strength is primarily inferred from a high load-carrying capacity and is seldom directly measured. It is recommended that this term should not be used.

Film Stress

The compressive or tensile forces appearing in a film, such as internal film stress, which is the intrinsic stress of a film related to its mechanical structure and deposition parameters, or induced film stress, which is the component of film stress related to an external force such as mismatched mechanical properties of the substrate.

Film Thickness

In a dynamic seal, the distance separating the two surfaces that form the primary seal.

Filter

(1) A porous article or material for separating suspended particulate matter from liquids by passing the liquid through the pores in the filter and sieving out the solids. (2) Any transmission network used in electrical systems for the selective enhancement of a given class of input signals. Also known as electric filter; electric-wave filter. (3) A device employed to reject sound in a particular range of frequencies while passing sound in another range of frequencies. Also known as acoustic filter. (4) A semitransparent optical element capable of absorbing unwanted electromagnetic radiation and transmitting the remainder. A neutral density filter attenuates relatively uniformly from the ultraviolet to the infrared, but in many applications highly wavelength-selective filters are used. See also *neutral filter*.

Filter Fabrics

Any fabric used for filtering liquids, gases, or vapors, but, because of the heat and chemical resistance usually required, generally synthetic or metal fibers. Weave is an important consideration. Plain weave permits maximum interlacings, and a tight weave gives high impermeability to particles. Twill weave has lower interlacings in sharp diagonal lines and gives a more selective porosity for some materials. Satin weave has fewer interlacings, is spaced widely but regularly, and is used for dust collection and gaseous filtration.

Fibers are chosen for their particular chemical resistance, heat resistance, and strength. Dacron has good acid resistance except for concentrated sulfuric or nitric acid. It can be used to 325°F (162°C). High-density polyethylene has good strength and abrasion resistance, and its smooth surface minimizes clogging of the filter, but it has an operating temperature only to 230°F (110°C). Polypropylene can be used to 275°F (134°C). Nylon gives high strength and abrasion resistance. It has high solvent resistance, but low acid resistance. Its operating limit is about 250°F (121°C). Teflon is exceptionally resistant to a wide variety of chemicals. It can be operated above 400°F (204°C), and its waxy, non-sticking surface prevents clogging and makes it easy to clean, but the fiber is available only in single-filament form.

Filter Glass

See preferred term *filter plate*.

Filter Plate (Eye Protection)

An optical material that protects the eyes against excessive ultraviolet, infrared, and visible radiation.

Filter Sand

A natural sand employed for filtration, especially of water. Much of the specially prepared filter sand comes from New Jersey, Illinois, and Minnesota and is from ocean beaches, lake deposits, and sandbanks. The specifications for filter sand require that it be of fairly uniform size, free from clay and organic matter, and chemically pure, containing not more than 2% combined carbonates. The most common grain sizes are 0.014–0.026 in. (0.35–0.65 mm). Very fine sand clogs the filter. *Greensand*, produced from extensive beds in New Jersey, is used as a water softener. It is a type of marl classed as *zeolite* and consists largely of *glauconite*, which is a greenish granular mineral containing up to 25% iron, with a large percentage of silica and some potash and alumina. Synthetic zeolite is a *sodium alumina silicate* made by reacting caustic soda with bauxite

to form sodium aluminate and then reacting with sodium silicate. In addition to filtering, the greensand softener extracts the calcium and magnesium from the water.

Molecular sieves are synthetic crystalline zeolites whose molecules are arranged in a crystal lattice so that there are a large number of small cavities interconnected by smaller pores of uniform size, the network of cavities and pores being up to 50% of the volume of the crystal. This consists of three-dimensional frameworks of SiO_5 and AlO_4 tetrahedra. Electrovalence of each tetrahedron is balanced by the inclusion in the crystal of a metal cation of Na, Ca, or Mg.

Filtration

The separation of a liquid plus solid mixture by passing it through a filter. The liquid produced is the filtrate.

Fin

(1) Excess material left on a molded plastic object at those places where the molds or dies mated. Also, the web of material remaining in holes or opening in a molded part, which must be removed and finished. (2) Metal on a casting caused by an imperfect joint in the mold or die. (3) A thin projecting piece of material. It may be deliberate as in a cooling fin or it may be inadvertent for example, the fin or flash formed on a die forging at the interface of badly fitting dies.

Final Annealing

An imprecise term used to denote the last anneal given to a nonferrous alloy prior to shipment.

Final Density

The density of a sintered product.

Final Polishing

A polishing process in which the primary objective is to produce a final surface suitable for microscopic examination.

Fine Ceramics

Same as *advanced ceramics*.

Fine Gilt

A thin gold surface layer applied by coating items with a mercury gold amalgam and then heating them at about 360°C causing the mercury to evaporate leaving an adherent gold film.

Fine Gold

24 carat gold.

Fine Grain Steel

Steels having an intrinsic tendency to retain a fine austenitic grain size during normalizing and annealing at the temperature appropriate to the composition. The characteristic is common in steels deoxidized with aluminum which leaves a large quantity of very fine aluminum nitride particles which impede grain growth.

Fine Hackle

See *hackle.*

Fine Silver

Silver with a fineness of 999; equivalent to a minimum content of 99.9% Ag with the remaining content unrestricted.

Fineness

A measure of the purity of gold or silver expressed in parts per thousand.

Fines (Ceramics, Metals, Ores)

(1) The product that passes through the finest screen in sorting crushed or ground material. (2) Sand grains that are substantially smaller than the predominating size in a batch or lot of foundry sand. (3) The portion of a powder composed of particles smaller than a specified size, usually 44 μm (−325 mesh).

Fines (Plastics)

Very small particles (usually under 200 mesh) accompanying larger grains, usually of molding powder.

Finish (Composites)

A mixture of materials for treating glass or other fibers. It contains a coupling agent to improve the bond of resin to the fiber and usually includes a lubricant to prevent abrasion, as well as a binder to promote strand integrity. With graphite or other filaments, it may perform any or all of the above functions.

Finish (Metals)

(1) Surface condition, quality, or appearance of a metal. (2) Stock on a forging or casting to be removed and finish machining. (3) The forging operation in which the part is forged into its final shape in the finish die. If only one finish operation is scheduled to be performed in the finish die, this operation will be identified simply as finish; first, second, or third finish designations are so termed when one or more finish operations are to be performed in the same finish die.

Finish (Plastics)

To complete the secondary work on a molded plastic part so that it is ready for use. Operations such as filling, deflashing, buffing, drilling, tapping, and degating are commonly called finishing operations.

Finish Allowance

(1) The amount of excess metal surrounding the intended final configuration of a formed part; sometimes called forging envelope, machining allowance, or cleanup allowance. (2) Amount of stock left on the surface of a casting for machining.

Finish Annealing

A *subcritical annealing* treatment applied to cold-worked low- or medium-carbon steel. Finish annealing, which is a compromise treatment, lowers residual stresses, thereby minimizing the risk of distortion in machining while retaining most of the benefits of machinability contributed by cold working. Compare with *final annealing.*

Finish Grinding

The final grinding action on a workpiece, of which the objectives are surface finish and dimensional accuracy.

Finish Machining

A machining process analogous to *finish grinding.*

Finish Trim

Flash removal from a forging; usually performed by trimming, but sometimes by hand sawing or similar techniques.

Finished Steel

Steel that is ready for the market and has been processed beyond the stages of billets, blooms, sheet bars, slabs, and wire rods.

Finisher (Finishing Impression)

The die impression that imparts the final shape to a forged part.

Finishing Die

The die set used in the last forging step.

Finishing Temperature

The temperature at which *hot working* is completed.

Finite Element Analysis

An important technique for determining the levels and distribution of stress in components. Its approach is to develop a mathematical model, in the form of a grid or mesh, which divides the structure into many small, simple segments, usually termed "elements." These are analyzed individually and collectively, usually by powerful computer programs, to determine the stress induced in each element when a force is applied. The technique can deal with complex and cracked three-dimensional structures and components. See also *fracture mechanics.*

Fire Bars

The grading supporting a solid fuel fire.

Fire Box

The combination chamber of a fire tube boiler.

Fire Cracking

Cracking during heating.

Fire Extinguishers

Materials used for extinguishing fires, usually referring to chemicals in special containers rather than the materials, like water, used in quantity for cooling and soaking the fuel with a noncombustible liquid. There are three general types of fire extinguishers: those for smothering, such as carbon dioxide; those for insulating the fuel from the oxygen supply, such as licorice and protein foams, which also include mineral powders that melt and insulate metallic fires; and chemicals which react with the combustion products to terminate the chain reaction of combustion, such as *bromotri-fluoromethane*. $CBrF_3$, a nontoxic colorless gas liquefied in cylinders. *Freon FE 1301*, of DuPont, is this chemical, while *Freon 13B1* of the same company is *monobromotrifluoromethane* gas pressurized with nitrogen. The relative effectiveness of extinguishers varies with the type of fuel in the fire, but on average, with bromotrifluoromethane taken as 100%, dibromodifluoromethane would be almost 67, the dry chemical sodium hydrogen carbonate 66, carbon tetrachloride 34, and carbon dioxide about 33. Others in the family include the *Halons*, halogenated hydrocarbons. *Halon 1211* is *bromochlorodifluoromethane*, *Halon 1301* is *bromotrifluoromethane*, *Halon CTFE* is *polychlorotrifluoroethylene*, and *Halon TFE* is *polytetrafluoroethylene*. The brominated compounds destroy stratospheric ozone even more drastically than do the *chlorofluorocarbons*, but since smaller quantities are released, their use is not being curtailed as severely as the CFCs.

Fire Point

The temperature at which a material will continue to burn for at least 5 s without the benefit of an outside flame.

Fire Scale

Intergranular copper oxide remaining below the surface of silver-copper alloys that have been annealed and pickled.

Fire Side Corrosion

Severe corrosion occurring on the fire side of, usually, a tube in a steam raising boiler or similar plant as a result of a reaction between the metal and the gas stream or more often, aggressive ash deposits on the surface.

Fire Tube Boiler

See *boiler tube*.

Fire Welding

See *forge welding*.

Fireclay

Clays that will withstand high temperatures without melting or cracking have been used for lining furnaces, flues, and for making firebricks and lining tiles. Common fireclays are usually *silicate of alumina*. Theoretically these clays contain 45.87% alumina and 54.13% silica, but in general they contain considerable iron oxide, lime, and other impurities. The clays are grouped as low-duty, intermediate-duty, high-duty, and super-duty. The low-duty has low alumina and silica with high impurities and is limited to a temperature of 1600°F (871°C). Standard types are good for temperatures

of 2400°F–2700°F (1316°C–1482°C), and the super-duty to temperatures of 2700°F–3000°F (1482°C–1649°C). Keown-Bern Clay should have a balanced proportion of coarse, intermediate, and fine grain sizes. Clays with an excess of silica are also used.

Firecracker Welding

A variation of the *shielded metal arc welding* process in which a length of covered electrode is placed along the joint in contact with the parts to be welded; during the welding operation, the stationary electrode is consumed as the arc travels the length of the electrode.

Fired Mold

A shell mold or solid mold that has been heated to a high temperature and is ready for casting.

Fire-Refined Copper

Copper that has been refined by the use of a furnace process only, including refinery shapes and, by extension, fabricators' products made therefrom. Usually, when this term is used alone it refers to fire-refined tough pitch copper without elements other than oxygen being present in significant amounts. Fire refining involves melting the metal first under oxidizing conditions to oxidize impurities and then under reducing conditions to reduce the excess oxygen.

Firing

The controlled heat treatment of ceramic ware in a kiln or furnace, during the process of manufacture, to develop the desired properties.

Firing Range

(1) The range of fired temperature within which a ceramic composition develops properties that render it commercially useful. (2) The time–temperature interval in which a porcelain enamel or ceramic coating is satisfactorily matured.

First Block, Second Block, and Finish

The forging operation in which the part to be forged is passed in progressive order through three tools mounted in one forging machine; only one heat is involved for all three operations.

First-Degree Blocking

An adherence between adhesively bonded surfaces under test of such degree that when the upper specimen is lifted, the lower specimen will cling thereto, but may be parted with no evidence of damage to either surface.

First-Order Transition

A change of state associated with crystallization, melting, or a change in crystal structure of a polymer.

Fir-Tree Crystal

A type of *dendrite*.

Fisheye (Metals)

An area on a steel fracture surface having a characteristic white crystalline appearance.

Fisheye (Weld Defect)

A discontinuity found on the fracture surface of a weld in steel that consists of a small pore or inclusion surrounded by an approximately round, bright area.

Fisheye Fracture

Subsurface fatigue cracks that have initiated failure by some other mechanism, perhaps simple overload but often brittle fracture. Against a contrasting surface texture, the dull concentric fatigue markings have a fisheye appearance. Note the potential confusion with *fisheyes*.

Fisheyes

Bright marks, typically a millimeter or two in diameter, observes also fracture surfaces and evidence of hydrogen embrittlement. The hydrogen embrittlement cracks may initiate other damage mechanisms including fatigue increasing the potential confusion with fisheye fracture.

Fishmouthing

See *alligatoring*.

Fishscale

A scaly appearance in a porcelain enamel coating in which the evolution of hydrogen from the basis metal (iron or steel) causes loss of adhesion between the enamel and the basis metal. Individual scales are usually small, but have been observed in sizes up to 25 mm (1 in.) or more in diameter. The scales are somewhat like blisters that have cracked partway around the perimeter but still remain attached to the coating around the rest of the perimeter.

Fishtail

(1) In roll forging, the excess trailing end of a forging. It is often used, before being trimmed off, as a tong hold for a subsequent forging operation. (2) In hot rolling or extrusion, the imperfectly shaped trailing end of a bar or special section that must be cut off and discarded as mill scrap.

Fission

Splitting into y number of parts. See *nuclear*.

Fission Fragments

Subatomic particles resulting from nuclear fission. The term often implies larger particles such as nuclei comprising multiple protons and neutrons.

Fission Poisons

Materials and nuclear fuel, particularly those produced in the nuclear fission process, which have a high neutron capture cross section and hence wastefully absorbs neutrons which would otherwise contribute to the chain reaction.

Fission Products

Elements produced by the nuclear fission of the heavy elements such as uranium, usually elements of atomic number 34–58. The term often implies radioactive isotopes of these elements.

Fissure

A small crack-like weld discontinuity with only slight separation (opening displacement) of the fracture surfaces. The prefixes macro or micro indicate relative size.

Fit

The amount of clearance or interference between mating parts is called actual fit. Fit is the preferable term for the range of clearance or interference that may result from the specified limits on dimensions (limits of size): Referred to ANSI standards.

Fixed Oil

An imprecise term denoting an oil that is difficult to distill without decomposition.

Fixed Position Welding

Welding in which the work is held in a stationary position.

Fixed-Feed Grinding

Grinding in which the wheel was fed into the work, or vice versa, by given increments or at a given rate.

Fixed-Land Bearing

See *fixed-pad bearing*.

Fixed-Load or Fix-Displacement Crack Extension Force Curves

Curves obtained from a fracture mechanics analysis for the test configuration, assuming a fixed applied load or displacement and generating a curve of *crack extension force* vs. the *effective crack size* as the independent variable.

Fixed-Pad Bearing

An axial- or radial-load bearing equipped with fixed pads, the surfaces of which are contoured to promote hydrodynamic lubrication.

Fixture

A device designed to hold parts to be joined in proper relation to each other.

Fixture Time

The shortest time required by an adhesive to develop handling strength such that test specimens can be removed from fixtures, unclamped, or handled without stressing the bond and thereby affecting bond strength. Also referred to as set time.

Fixturing

The placing of parts to be heat-treated in a constraining or semi-constraining apparatus to avoid heat-related distortions. See also *racking*.

Flake (Metals)

A short, discontinuous internal crack in ferrous metals attributed to stresses produced by localized transformation and hydrogen-solubility effects during cooling after hot working. In fracture surfaces, flakes appear as bright, silvery areas with a coarse texture. In deep acid-etched transverse sections, they appear as discontinuities that are usually in the midway to center location of the section. Also termed hairline cracks and shatter cracks. Also termed *fisheyes* when they are exposed on a fracture surface.

Flake (Plastics)

A term used to denote the dry, unplasticized base of cellulosic plastics.

Flake Graphite

Graphitic carbon, in the form of platelets, occurring in the microstructure of *gray iron*.

Flake Powder

A flat or scale-like particles whose thickness is small compared to the other dimensions.

Flaking

(1) The removal of material from a surface in the form of flakes or scale-like particles. (2) A form of pitting resulting from fatigue. See also *spalling*.

Flame Annealing

Annealing in which the heat is applied directly by a flame.

Flame Cleaning

The use of a soft, broad, oxy-fuel gas flame to remove deposits of paint, grease, rust, etc., from components of large structures prior to painting.

Flame Cutting

Processes utilizing a torch similar to that used for gas welding but in which the primary cutting action is the chemical reaction of the material with oxygen. A fuel gas is also involved but its main contribution is to raise the component to reaction temperature at the commencement of the cut. See also preferred term *oxy (gen) cutting*.

Flame Gouging

See *gouging*.

Flame Hardening

A process for hardening the surfaces of hardenable ferrous alloys in which an intense flame is used to heat the surface layers above the upper transformation temperature, whereupon the workpiece is immediately quenched.

Flame Plating

A process in which an oxy-acetylene mixture carrying a suspension of powdered refractory material is detonated to project the powder at high temperature and high velocity onto the surface to be coated.

Flame Resistance

Ability of a material to extinguish flame once the source of heat is removed. See also *self-extinguishing resin*.

Flame Retardants

Certain chemicals that are used to reduce or eliminate the tendency of a resin to burn.

Flame Scaling

(1) Descaling by a process similar to flame cleaning. (2) Flame heating freshly galvanized steel wire to promote intermetallic bonding and improve surface finish.

Flame Spraying

Thermal spraying in which a coating material is fed into an oxyfuel gas flame, where it is melted. Compressed gas may or may not be used to atomize the coating material and propel it onto the substrate. The sprayed material is originally in the form of wire or powder. See the terms *powder flame spraying* and *wire flame spraying*. The term flame spraying is usually used when referring to a combination-spraying process, as differentiated from *plasma spraying*.

Flame Spraying (Plastics)

Method of applying a plastic coating in which finely powdered fragments of the plastic, together with suitable fluxes, are projected through a cone of flame onto a surface.

Flame Straightening

Correcting distortion in metal structures by localized heating with a gas flame.

Flame Temperature

The maximum temperature within a flame.

Flame Treating (Plastics)

A method of rendering inert thermoplastic objects receptive to inks, and lacquers, paints, adhesives, and so forth, in which the object is bathed in an open flame to promote oxidation of the surface of the article.

Flame-Sprayed Coatings

Processes

Flame-spraying methods are used to produce coatings of a wide variety of heat-fusible materials, including metals, metallic compounds, and ceramics. There are three basic flame-spray processes.

Wire-Type Guns

These guns are used to produce coatings of metals, alloys, and, in some cases, ceramics. With this type of equipment, wire or rod from 0.76 to 4.7 mm in diameter is fed axially through the center of a fuel gas/O_2 flame at a controlled rate. The flame is surrounded by an annular blast of air, which imparts high velocity to the burning gases and provids the kinetic energy needed to atomize the metal as it melts. Acetylene is the most commonly used fuel gas, although propane, natural gas, and H_2 are also used.

Wire-type guns are the most widely used, because of their versatility and ease of operation.

Handheld guns are used to coat large areas such as bridges, ship hulls, and tanks. Machine-mounted guns are used principally for machine element applications, as in surfacing rolls or salvaging journals by building up worn areas. In many production applications, one or more electronically controlled guns are operated and cycled automatically by a central console.

Powder-Type Guns

A variety of materials that cannot be readily produced in the form of wire or rod utilize this type of gun. The guns are also used for spraying low-melting metals that are readily available in powder form.

With the powder gun, metal or ceramic powders are fed axially through a fuel gas/O_2 flame at a controlled rate. The powders are entrained in a carrier gas, which may be air, acetylene, or O_2. Because the powders are finely divided and dispersed as they enter the flame, further atomization is not required and annular air blast is not needed.

The powder supply for guns of this type may be a reservoir connected to the gun by a powder tube or hose, or it may be a canister attached directly to the gun.

Powder guns are used for flame-spraying a wide variety of ceramics, and for coating with "self-fluxing" alloys, which are subsequently fused to the base material. They have also been used for many years to apply zinc and aluminum coatings for corrosion prevention.

Plasma Guns

The plasma flame is produced by passing suitable gases through a confined arc, where dissociation and ionization occur. The ionized gases form a conductive path within a water-cooled nozzle, so that an arc of considerable length is maintained. The gases most commonly used are N_2, H_2, and argon.

Temperature of the plasma flame depends on the type and volume of gas used, the size of the nozzle, and the amount of current used. For flame-spraying purposes temperature ranges of 5482.4°C–8232.4°C are generally employed, although much higher temperatures may be attained if desired. Plasma flame-spray guns usually operate at 20–40 kW, using 47–141 L/min of gas.

In addition to their high-temperature capabilities, plasma guns have other advantages. Extremely high velocities are possible, and favorable environmental conditions can be obtained by proper selection of gases. Spraying within a controlled atmosphere chamber permits the production of oxide-free coatings.

With the plasma gun, metal or ceramic powders are fed into the flame at a point downstream from the actual arc path. Current, gas flow, and powder flow must be adjusted for different coating materials, many of which would otherwise be completely vaporized.

Surface Preparation

Regardless of the flame-spray method used or the type of coating material being applied, some sort of surface preparation is usually required. Bond to the base is often largely mechanical, and, in general, the greater the degree of surface roughness, the better the bond. Thin coatings require less elaborate preparation than thick coatings, and some coating materials require much more thorough preparation than other materials.

Bonding methods used include abrasive blasting, rough threading, molybdenum bonding coats, heating, or combinations of these steps. Abrasive blasting is probably the most widely used.

Finishing Methods

Flame-sprayed deposits may be finished by machining or by grinding, depending on the hardness of the particular coating material. Sintered carbide tools are generally used for machining because even the softer materials contain some amount of abrasive oxides that may cause rapid wear of tool steel. For those materials that must be finished by grinding, specific wheel recommendations are available from flame-spray equipment manufacturers.

Flame-sprayed coatings may require sealing, depending on the type of coating and service requirements. A wide variety of impregnants are used to reduce porosity, enhance physical properties, or to improve friction characteristics.

Flammability

Measure of the extent to which a material will support combustion.

Flange

A projecting rim or edge of a part; usually narrow and of approximately constant width for stiffening or fastening. Usually perpendicular to the principal axis and often applied to tubes.

Flange (d) Joint

A joint formed between two components one or both of which is flanged. The joint may be made by bolting, welding, or brazing.

Flange Weld

A weld made on the edges of two or more members to be joined, usually light gage metal, at least one of the members being flanged. Flange weld, flare-bevel, and flare-V-groove welds may be confused because they have similar geometry before welding. A flange is welded on the edge and a flare is welded in the groove. See the term edge-flange weld which compares *flange*, *flare-bevel*, and *flare-V-groove welds*.

Flank

The end surface of a tool that is adjacent to the cutting edge and below it when the tool is in a horizontal position, as for turning. See single-*point tool.*

Flank Wear

The loss of relief on the flank of the tool behind the cutting edge due to rubbing contact between the work and the tool during cutting; measured in terms of linear dimension behind the original cutting edge. See the term *crater wear.*

Flapping

In copper refining, hastening oxidation of molten copper by striking through the slag-covered surface of the melt with a *rabble* just before the bath is poled.

Flare Test

A test applied to tubing, involving tapered expansion over a cone. Similar to *pin expansion test.*

Flare-Bevel Groove Weld

A weld in a groove formed by a member with a curved surface in contact with a planar member. Compare with *flange weld.* See also the term *edge-flange weld.*

Flare-V-Groove Weld

A weld in a groove formed by two members with curved surfaces. Compare with *flange weld.* See the term *edge-flange weld.*

Flaring

(1) Forming an outward acute-angle flange on a tubular part. (2) Forming a flange by using the head of a hydraulic press.

Flash (Metals)

(1) In forging, metal in excess of that required to fill the blocking or finishing forging impression of a set of dies completely. Flash extends out from the body of the forging as a thin plate at the line where the dies meet and is subsequently removed by trimming. Because it cools faster than the body of the component during forging, flash can serve to restrict metal flow at the line where dies meet, thus ensuring complete filling of the impression. See also *closed-die forging.* (2) In casting, a fin of metal that results from leakage between mating mold surfaces. (3) In welding, the material which is expelled or squeezed out of a weld joint and which forms around the weld.

Flash (Plastics)

The portion of the charge that flows from or is extruded from the mold cavity during the molding. Extra plastic attached to a molding along the parting line, which must be removed before the part is considered finished.

Flash Anneal

Annealing of short duration applied to wire or thin sheet.

Flash Coat

A thin metallic coating usually less than 0.05 mm (0.002 in.) in thickness.

Flash Extension

That portion of *flash* remaining on a forged part after trimming; usually included in the normal forging tolerances.

Flash Land

Configuration and the blocking or finishing impression of forging dies designed to restrict or to encourage the growth of *flash* at the parting line, whichever may be required in a particular case to ensure complete filling of the impression.

Flash Line

The line left on a forging after the flash has been trimmed off.

Flash Mold (Plastics)

A mold in which the mold faces are perpendicular to the clamping action of the press, so that the greater the clamping force, the tighter the mold seam.

Flash Plate

A very thin final electrode deposited film of metal.

Flash Point

(1) The temperature to which a material must be heated to give off sufficient vapor to form a flammable mixture. (2) The lowest temperature at which the vapor of a lubricant can be ignited under specified conditions.

Flash Temperature

The maximum local temperature generated at some point in a sliding contact. The flash temperature occurs at areas of real contact due to the frictional heat dissipated at these areas. The duration of the flash temperature is often of the order of a microsecond. The term flash temperature may also mean the average temperature over a restricted contact area (for example, between gear teeth).

Flash Weld

A weld made by *flash welding.*

Flash Welding

A resistance welding process that joins metals by first heating abutting surfaces by passage of electric current across the joint, then forcing the surfaces together by the application of pressure.

Flashing and upsetting are accompanied by expulsion of metal from the joint.

Flash (Butt) Welding

An electric welding process in which the joint faces are first brought together to strike an arc and then held in light contact or even drawn slightly apart to allow heating before they are finally forced together again. It is essentially a pressure weld as any molten metal is injected in the final closure. See preferred term *flash welding*.

Flashback

A recession of the welding or cutting torch flame into or back of the mixing chamber of the torch.

Flashback Arrestor

A device incorporated into an oxygen or oxyfuel welding or cutting torch to limit damage from a *flashback* by preventing propagation of the flame front beyond the point at which the arrestor is installed.

Flashing

In *flash welding*, the heating portion of the cycle, consisting of a series of rapidly recurring localized short-circuits followed by molten metal expulsions, during which time the surfaces to be welded are moved one toward the other at a predetermined speed.

Flashing Time

The time during which the flashing action is taking place in *flash welding*.

Flash-Off Time

See preferred term *flashing time*.

Flashover

The passage of electric current across the surface of an insulator.

Flask

(1) Any vessel for containing fluid. (2) In the context of nuclear waste transportation, a metal vessel of massive proportions, high integrity and tightly sealed. (3) A metal or wood frame used for making and holding a sand mold. The upper part is called the *cope*; the lower, the *drag*. See the term *blind riser*.

Flat Drill

A rotary end-cutting tool constructed from a flat piece of material, provided with suitable cutting lips at the cutting end. See also the term *drill*.

Flat Edge Trimmer

A machine for trimming notched edges on shells. The slide is cam driven so as to obtain a brief dwell at the bottom of the stroke, at which time the die, sometimes called a shimmy die, oscillates to trim the part.

Flat Fracture

A fracture surface approximately perpendicular to the principal stress and substantially level. The fracture mode may be crystalline, as in a brittle fracture, or it may comprise multiple, microscopic ductile shear dimpled fractures. The term may be used in contrast to ductile or sheer fracture where the fracture mode is ductile shear but the plane of fracture is at approximately 45° to the principal stress.

Flat Glass

A general term covering sheet glass, plate glass, and various forms of rolled glass.

Flat Position Welding

The position in which the components being welded and, or the weld surface are approximately horizontal and the weld is made from above. This is usually taken to mean a weld slope not greater than 5° and a weld rotation not greater than 10°. Down hand position is the same. See term *welding position*.

Flat Wire

A roughly rectangular or square mill product, narrower than *strip*, in which all surfaces are rolled or drawn without any previous slitting, shearing, or sawing.

Flat-Die Forging

Forging metal between flat or simple-contour dies by repeated strokes and manipulation of the workpiece. Also known as *open-die forging*, *hand forging*, and *smith forging*.

Flats

(1) A longitudinal, flat area on a normally convex surface of a protruded plastic, caused by shifting of the reinforcement, lack of sufficient reinforcement, or local fouling of the die surface. (2) An imprecise term usually indicating rolled steel products 3–6 mm thick and up to about 600 mm wide.

Flattening

(1) A preliminary operation performed on forging stock to position the metal for a subsequent forging operation. (2) The removal of irregularities or distortion in sheets or plates by a method such as *roller leveling* or *stretcher leveling*.

Flattening Dies

Dies used to flatten sheet metal hems; that is, dies that can flatten a bend by closing it. These dies consist of a top and bottom die with a flat surface that can close one section (flange) to another (hem, seam): See the term *press-brake forming*.

Flattening Mill

A rolling mill producing sheet metal.

Flattening Test

A test in which a hollow component, weld, tube, etc., is flattened to a specified degree that may or may not be in contact with internal surfaces. Acceptance criteria depends on the application but usually specifies limitations on cracking.

Flaw

A nonspecific term often used to imply a crack-like discontinuity. See also preferred terms *discontinuity*, *imperfection*, and *defect*.

Flaw Detection

Any technique for detecting defects such as cracks and voids. See *nondestructive testing*.

Flax

A fiber obtained from the flax, or linseed, plant, *Linum usitatissimum*, used for making the fabrics known as linens and for thread, twine, and cordage. It is valued because of its strength and durability. It is finer than cotton and very soft, and the fibers are usually about 20 in. (50.8 cm) long. Flax consists of the **bast fibers**, or those in the layer underneath the outer bark, which are of fine texture. The plants are pulled up by the roots, retted, or partly decayed, scraped, and the fibers combed out and bleached in the sun. The plants that are grown for the oil seed yield a poor fiber and are not employed to produce flax.

Flex

(1) Flexible insulated cable for conducting electricity particularly for domestic equipment. (2) Bend usually with the implication that the degree of bending is within the elastic range so that a component returns to shape following removal of load.

Flex Roll

A movable jump roll designed to push up against a metal sheet as it passes through a roller leveler. The flex roll can be adjusted to deflect the sheet any amount up to the roll diameter.

Flex Rolling

Passing metal sheets through a flex roll unit to minimize yield-point elongation in order to reduce the tendency for *stretcher strains* to appear during forming.

Flexibility

The quality or state of a material that allows it to be flexed or bent repeatedly without undergoing rupture. See also *flexure*.

Flexibilizer

An additive that makes a finished plastic more flexible or tough. See also *plasticizer*.

Flexible Cam

An adjustable pressure-control cam of spring steel strips used to obtain varying pressure during a forming cycle.

Flexible Hinge

See *flexure pivot bearing*.

Flexible Manufacturing System

A number of machines and other plant items, often computer-controlled, that can be readily arranged and utilized to produce a variety of components.

Flexible Molds

Molds made of rubber or elastomeric plastics, used for casting plastics. They can be stretched to remove cured pieces having undercuts.

Flexural Failure

A material failure caused by repeated flexing.

Flexural Modulus

The ratio, within the elastic limit, of the applied stress on a reinforced plastic test specimen in flexure to the corresponding strain in the outermost fibers of the specimen.

Flexural Strength

(1) A property of solid material that indicates its ability to withstand a flexural or transverse load. (2) The tensile strength and related properties deduced from bending tests. Brittle materials likely to break in the grips of a tensile machine are often tested in this matter.

Flexural Strength (Composites)

The maximum stress that can be borne by the surface fibers in a beam in bending. The flexural strength is the unit resistance to the maximum load before failure by bending, usually expressed in force per unit area.

Flexure

A term used in the study of strength of materials to indicate the property of a body, usually a rod or beam, to bend without fracture. See also *flexibility*.

Flexure Pivot Bearing

A type of bearing guiding the moving parts by flexure of an elastic member or members rather than by rolling or sliding. Only limited movement is possible with a flexure pivot.

Flexure Stress (Glass)

The tensile component of the bending stress produced on the surface of a glass section opposite to that experiencing a locally impinging force.

Flint

Flint is SiO_2 and is a black, gray, or brown cryptocrystalline variety of quartz. In the United States, ceramists often employ the term *flint* to include other siliceous minerals in addition to true flint.

Calcined and ground flint is used in pottery to reduce shrinkage in drying and firing and to give the body a certain rigidity. Flint is employed in the manufacture of whiteware, such as fine earthenware, bone china, and porcelain.

An opaque variety of *chalcedony* or nearly pure amorphous *quartz* which shows no visible structure. It is deposited from colloidal solution and is an intimate mixture of quartz and opal. It contains 96%–99% silica and may be colored to dull colors by impurities. Thin plates are translucent. When heated, it becomes white. Flint is finally crystalline. It breaks or chips with a convex, undulating surface. The hardness is Mohs 7, and the specific gravity is 2.6. It was the prehistoric utility material for tools, and was later used with steel to give sparks on percussion. *Gun flints* are still made from a type of flint mined at Brandon, England, for special uses. *Lydian stone*, or *touchstone*, was a cherry flint used for testing gold. Flint is now chiefly used as an abrasive and in pottery and glass manufacture. *Flint paper* for abrasive use contains crushed flint grades from 20 to 240 mesh. Flint is also used in the form of grinding pebbles. *Potters' flint*, used for mixing in ceramics to reduce the firing and drying shrinkage and to prevent deformation, is ground flint of about 140 mesh made from white *French pebbles*.

Flotation

A technique for separating material suspended in a liquid, in particular to concentrate the mineral content of ores. The fine particulate ore is mixed into the fluid with oil or similar agent in the mixture and agitated. The oil attaches to the mineral and air bubbles assist in floating the particle to the surface where it is retained by surface tension before being skimmed off.

The concentration of valuable minerals from ores by agitation of the ground material with water, oil, and floatation chemicals. The valuable minerals are generally wetted by the oil, lifted to the surface by clinging air bubbles, and then floated off.

Floating Bearing

A bearing designed or mounted to permit axial displacement between shaft and housing.

Floating Chase

In forming of plastics, a mold member, free to move vertically, that fits over a lower plug or cavity, and into which an upper plug telescopes.

Floating Die

(1) In metal forming, a die mounted in a die holder or punch mounted in its holder such that a slight amount of motion compensates for tolerance of the die parts, the work, or the press. (2) A die mounted on heavy springs to allow vertical motion in some trimming, shearing, and forming operations.

Floating Die Pressing

The compaction of a metal powder in a floating die, resulting in densification at opposite ends of the compact. Analogous to double action pressing.

Floating Plug

In tube drawing, an unsupported mandrel that locates itself at the die inside the tube, causing a reduction in wall thickness while the die is reducing the outside diameter of the tube.

Floating-Ring Bearing

A type of journal bearing that includes a thin ring between the journal and the bearing. The ring floats and rotates at a fraction of the journal rotational speed.

Flocculant

An electrolyte added to a colloidal suspension to cause the particles to aggregate and settle out as a result of reduction in repulsion between the particles.

Flocculate

A grouping of primary particles, aggregates, or agglomerates having weaker bonding than either the aggregate or agglomerate structures. Flocculates are usually formed in a gas or liquid suspension and those formed in a liquid can generally be broken up by gentle shaking and stirring.

Flocculation

Agglomeration of particles in a suspension causing them to settle out.

Flock

A material obtained by reducing textile fibers to fragments as by cutting, tearing, or grinding, to give various degrees of comminution. Flock can either be fibers in entangled, small masses or beads, usually of irregular broken fibers, or comminuted (powdered) fibers.

Flock Point

A measure of the tendency of a lubricant to precipitate wax or other solids from solution. Depending on the test used, the flock point is the temperature required for precipitation, or the time required at a given temperature for precipitation.

Flocking

A method of coating by spraying finely dispersed textile powders or fibers.

Flong Papier-Mâché (Pulped Paper)

Used as a molding material forecasting low melting-point metals, in particular tin-based printer-type metals.

Flood Lubrication

A system of lubrication in which the lubricant is supplied in a continuous stream at low pressure and subsequently drains away. Also known as *bath lubrication*.

Floor Molding

In foundry practice, making sand molds from loose or production patterns of such size that they cannot be satisfactorily handled on a bench or molding machine, the equipment being located on the floor during the entire operation of making the mold.

Flop Forging

A forging in which the top and bottom die impressions are identical, permitting the forging to be turned upside down during the forging operation.

Floppers

On metals, lines or ridges that are transverse to the direction of rolling and generally confined to the section midway between the edges of a coil as rolled.

Flospinning

Forming cylindrical, conical, and curvilinear shaped parts by power spinning over a rotating mandrel. See also *spinning*.

Flow

Movement (slipping or sliding) of essentially parallel planes within an element of a material in parallel directions; occurs under the action of *shear stress*. Continuous action in this matter, and at constant volume and without disintegration of the material, is termed *yield*, *creep*, or *plastic deformation*.

Flow (Adhesives)

Movement of an adhesive during the bonding process, before the adhesive is set.

Flow (Plastics)

The movement of resin under pressure, allowing it to fill all parts of a mold.

Flow Brazing

Brazing by pouring hot molten nonferrous filler metal over a joint until the brazing temperature is attained. The filler metal is distributed in the joint by capillary action.

Flow Brightener

(1) Melting of an electrodeposit, followed by solidification, especially of tin plate. (2) Fusion (melting) of a chemically or mechanically deposited metallic coating on a substrate, particularly as it pertains to soldering. (3) It is practiced on electrodeposited tin coatings on steel or copper base items.

Flow Cavitation

Cavitation caused by a decrease in static pressure induced by changes in the velocity of a flowing liquid. Typically this may be caused by flow around an obstacle or through a constriction, or relative to a blade or foil.

Flow Factor

See preferred term *flow rate*.

Flow Lines

(1) Texture showing the direction of metal flow during hot or cold working. Flow lines can often be revealed by etching the surface or a section of a metal part. (2) In mechanical metallurgy, paths followed by minute volumes of metal during deformation. (3) The lines on a polished and etched section through a raw material that reveals the deformation pattern during previous working operations. They are caused by the varying etching characteristics of areas with minor differences in composition or by lines of inclusions. See also *sulfur print* in *nature print*.

Flow Marks

Wavy surface appearance of an object molded from thermoplastic resins, caused by improper flow of the resin into the mold.

Flow Meter

(1) A device for indicating the rate of gas flow in a system. (2) In powder metallurgy, a metal cylinder whose interior is funnel shaped and whose bottom has a calibrated orifice of standard dimensions to permit passage of a powder and the determination of the *flow rate*.

Flow Molding

The technique of producing leather-like materials by placing a die-cut plastic blank (solid or expanded vinyl or vinyl-coated substrate) in a mold cavity (usually silicone rubber molds) and applying power via a high-frequency radio frequency generator to melt the plastic such that it flows into the mold to the desired shape and with the desired texture.

Flow Rate

The time required for a metal powder sample of standard weight to flow through an orifice in a standard instrument according to a specified procedure.

Flow Soldering

See *wave soldering*.

Flow Spinning

See *spinning*.

Flow Stress

The true stress at which plastic strain commences. It is increased by prior plastic strain. Flow stress and fracture stress both fall as temperature rises but the rates of fall may differ. Some materials, for example steel, have a flow stress higher than the fracture stress at low temperature but the reverse at higher temperatures. Thus at low temperatures they fracture before plastic strain can commence, that is, they fail in a brittle manner but at the higher temperatures they plastically deform before failing, i.e., they behave in a ductile matter. See *tensile test*.

Flow Test

A standardized test to measure how readily a metal powder flows. See also *flow rate*.

Flow Through

A forging defect caused by metal flow past the base of a rib with resulting rupture of the grain structure.

Flow Welding

(1) A welding process that produces coalescence of metals by heating them with molten filler metal poured over the surfaces to be welded until the welding temperature is attained and until the required filler metal has been added. The filler metal is not distributed in the joint by capillary action. (2) A process in which bulk molten metal is poured into the joint gap which is usually of substantial width. The weld metal needs sufficient superheat to melt the joint faces to form a fusion weld. Alternatively, the weld metal may be allowed to overflow to waste until the joint faces have fused sufficiently. See also *burning*, *burning on*, and *thermic welding*.

Flowability

(1) In casting, a characteristic of a foundry sand mixture that enables it to move under pressure or vibration so that it makes intimate contact with all surfaces of the pattern or core box. (2) In welding, brazing, or soldering, the ability of molten filler metal to flow or spread over a metal surface.

Flower (s)

The bright spangle on hot dip galvanized components.

Fluid

Any liquid or gas. Its shape is defined by the containment vessel.

Fluid Bearing

See *hydrostatic bearing*.

Fluid Erosion

See *liquid impingement erosion*.

Fluid Forming

A modification of the *Guerin process*, fluid forming differs from the fluid-cell process in that the die cavity, called a pressure dome, is not completely filled with rubber, but with hydraulic fluid retained by cup-shaped rubber diaphragm. See also *rubber-pad forming*.

Fluid Friction

Frictional resistance due to the viscous or rheological flow of fluids.

Fluid Lubrication

The condition where closely approaching surfaces in relative motion are maintained apart by a film of fluid lubricant. See *oil wedge*.

Fluid-Cell Process

A modification of the *Guerin process* for forming sheet metal, the fluid-cell process uses higher pressure and is primarily designed for forming slightly deeper parts, using a rubber pad as either the die or punch. A flexible hydraulic fluid cell forces an auxiliary rubber pad to follow the contour of the form block and exert a nearly uniform pressure at all points on the workpiece. See also *fluid forming* and *rubber-pad forming*.

Fluidity

(1) The ability of liquid metal to run into and fill a mold cavity. (2) The reciprocal of *viscosity*.

Fluidized Bed

A bath of granular material through which gas is blown, usually via a porous bottom plate, to make the material flow like a liquid. For heat treatment applications the component is quenched into the bed to achieve fairly high rates of heat transfer, similar to an oil bath but cleaner. The gas may be air or an inert gas. Fluidized beds are also utilized as a combustion device in which coal particles are fluidized and burned by the injected air.

Fluidized-Bed Coatings

The fluidized-bed process is used to apply organic coatings to parts by first preheating the parts and then immersing them in a tank of finely divided plastic powders which are held in a suspended state by a rising current of air. In this suspended state, the powders behave and feel like a fluid. The method produces an extremely uniform, high-quality, fusion bond coating that offers many technical and economic advantages.

Fusion bond coatings are generally applied to metal parts, although other substrates have been successfully coated. The major fields of application are electrical, chemical, and mechanical equipment as well as household appliances. The process is now being used in every major industrial country in the world for applying many different plastics to a variety of parts.

Coating Application

Objects to be coated are preheated in an oven to a temperature above the melting point of the plastic coating material. The preheat temperature depends on the type of plastic used, the thickness of the coating to be applied, and the mass of the article to be coated.

After preheating, the parts are immersed with suitable motion in the fluidized bed. Air used to fluidize the plastic powders enters the tank through a specially designed porous plate located at the bottom of the unit.

When the powder particles contact the heated part, they fuse and adhere to the surface, forming a continuous and extremely uniform coating. In many cases, the part is postheated to coalesce the coating completely and improve its appearance. Thickness of the coating depends on the temperature of the part surface and how long it is immersed in the fluidized bed.

Coating Types

Cellulosic

Cellulosic fluidized-bed coatings are noted for their all-around combination of properties and are especially popular for decorative/protective applications. They combine good impact and abrasion resistance with outstanding electrical insulation. The coatings have excellent weathering properties, salt-spray resistance, high gloss, and can be made in an almost unlimited range of colors. They can be solvent-etched to provide a satin finish and heat-embossed for additional effects.

The economy of cellulosic coatings combined with their excellent appearance and durability are particularly useful for such applications as indoor and outdoor furniture, kitchen fixtures, home and marine hardware, metal stampings, fan guards, and sporting goods. Major uses of cellulosic fusion bond finishes are coated transformer tanks and covers, reclosure tanks and covers, outdoor electrical equipment housings, and many pole line hardware parts.

Vinyl

Vinyl fluidized-bed coatings have a good combination of chemical resistance, decorative appeal, flexibility, toughness, and low-frequency insulating properties. They have excellent salt-spray resistance, outstanding outdoor weathering characteristics, and can be used for general-purpose electrical insulation. The vinyl fusion bond coatings are claimed to have better uniformity and edge coverage than plasticol coatings.

Fusion bond vinyl coatings have been especially successful on wire goods for applications such as dishwasher racks, washing-machine parts, and refrigerator shelves. They are also being used on wire furniture and hardware, and in industrial applications such as bus bars, pump impellers, transformer tanks and covers, auto battery brackets, conveyor rollers, and other material-handling equipment. Cast iron, die castings, and expanded metal parts can be readily coated.

Epoxy

Both thermoplastic and thermosetting materials can be applied. Epoxy coatings have a smooth, hard surface and exceptionally good electrical-insulation properties over a wide temperature range. They are available in rigid and semiflexible variations with different combinations of electrical, physical, and chemical properties. Epoxy coatings on electrical motor laminations provide good dielectric strength and uniform coverage over sharp edges. When properly applied, epoxy coatings have good impact resistance and do not sacrifice toughness for surface hardness. Other electrical applications include torroidal cores, wound coils, encapsulated printed circuit boards, bus bars, watt hour meter coils, resistors, and capacitors.

Nylon

The combination of properties that have made the polyamide plastics unique for molded and extruded parts are obtainable in coatings, and nylon used as a solution coating technique applied by the fluidized-bed process offers many advantages to the design engineer.

Nylon fusion bond finishes combine a decorative, smooth, and glossy appearance with low surface friction and excellent bearing and wear properties. They minimize scratching and cut down undesirable noise. The frictional heat developed on nylon is dissipated more rapidly when used as a coating over a conductive metal surface than when used as a solid plastics member. Coated metal parts offer increased dimensional stability. Because of the unique properties of nylon-coated metal parts, many users of the process are able to reduce the number of metal component parts at substantial savings.

By an additional immersion of the heated and coated part into a fluid bed of whirling molybdenum sulfide or graphite, an impregnated nylon surface can be produced with unusual bearing, frictional, and wear characteristics.

Nylon fusion bonds are effectively used in machine shop fixtures, modern indoor furniture to simulate a wrought iron finish, aircraft instrument panels, ball-joint suspensions, collars, guards, and slide valves used in textile and farm equipment, knitting machine parts, switch box cover panels, tractor control handles, and radar and calculator component parts.

Polyethylene

Polyethylene coatings combine low water absorption and excellent chemical resistance with good electrical insulation properties. They can be applied successfully in thicknesses of 10–60 mils by the fluidized-bed process. The primary uses for polyethylene coatings are for protecting chemical processing equipment and on food-handling equipment. Typical chemical applications include pipe and fittings, pump and motor housings, valves, battery hold-downs, fans, and electroplating jigs.

Chlorinated Polyether

This fluidized-bed coating has an excellent combination of mechanical, chemical, thermal, and electrical properties. It has good resistance to wear and abrasion. It provides good electrical insulation even under high humidity and high temperature conditions, and has very low moisture absorption. It can be used continuously at 121°C and even up to 149°C.

Chlorinated polyether coatings have excellent chemical resistance and are widely used to coat equipment for the chemical industry such as valves, pipe and pipe fittings, pump housings and impellers, electroplating jigs and fixtures, cams, and bushings. However, chlorinated polyether coatings should not be used in contact with some chlorinated organic solvents, or with fuming sulfuric and nitric acids.

Fluidized-Bed Heating

Heating carried out in a medium of solid particles suspended in a flow of gas.

Fluidized-Bed Reduction

The finely divided solid is a powdered ore or reducible oxide, and the moving gasis reducing; the operation is carried out at elevated temperature in a furnace.

Fluorescence

(1) Emission of electromagnetic radiation that is caused by the flow of some form of energy into the emitting body and which ceases abruptly when the excitation ceases. (2) Emission of electromagnetic radiation that is caused by the flow of some form of energy into the emitting body and whose decay, when the excitation ceases, is temperature-independent. (3) A type of photoluminescence in which the time interval between the absorption and re-emission of light is very short. (4) If emission continues after radiation has ceased the phenomenon is termed phosphorescence. Contrast with *phosphorescence.*

Fluorescent Magnetic-Particle Inspection

Inspection with either dry magnetic particles or those in a liquid suspension, the particles being coated with a fluorescent substance to increase the visibility of the indications.

Fluorescent Penetrant Inspection

Inspection using a fluorescent liquid that will penetrate any surface opening; after the surface has been wiped clean, the location of any surface flaws may be detected by the fluorescence, under ultraviolet light, of back-seepage of the fluid.

Fluorimetry

See *fluorometric analysis.*

Fluorinated Ethylene Propylene (FEP)

A member of the fluorocarbon family of plastics that is a copolymer of tetrafluoroethylene and hexafluoroethylene, possessing most of the properties of poly-tetrafluoroethylene, and having a melt viscosity low enough to permit conventional thermoplastic processing. Available in pellet form for molding and extrusion, and as dispersions for spray or dip coating processes.

Fluorine

An elementary material, symbol F, which at ordinary temperatures is an irritating pale-yellow gas, F_2. Fluorine gas is obtained by the reduction and electrolysis of fluorspar and cryolite. It is used in the manufacture of fluorine compounds. It combines violently with water to form hydrofluoric acid, and it also reacts strongly with silicon and most metals. Liquid fluorine is used as an oxidizer for liquid rocket fuels. In combustion, a pound of fluorine produces a pound of hydrogen fluoride which is highly corrosive. Fluorine is one of the most useful of the halogens.

The gas *sulfur hexafluoride*, SF_6, resembles nitrogen in its inactivity. It is odorless, colorless, nonflammable, nontoxic, and five times as heavy as air. It is used as a refrigerant, as a dielectric medium in high-voltage equipment, as an insecticide propellant, and as a gaseous diluent. *Aluminum fluoride*, AlF_3, is a white crystalline solid used in ceramic glazes and for fluxing nonferrous metals. Chlorine- and bromine-free *perfluorocarbon fluids*, of 3 M, are intended for spot-free drying of metal parts, replacing those ondepleting chlorofluorocarbons.

Fluorocarbons

The family of plastics including polytetrafluoroethylene, polychlorotrifluoroethylene, polyvinylidene, and fluorinated ethylene propylene. They are characterized by good thermal and chemical resistance, nonadhesiveness, low dissipation factor, and low dielectric constant. They are available in a variety of forms, such as molding materials, extrusion materials, dispersions, film, or tape, depending on the particular fluorocarbon.

These are any of the organic compounds in which all of the hydrogen atoms attached to a carbon atom have been replaced by fluorine. Fluorocarbons are usually gases or liquids at room temperature, depending on the number of carbon atoms in the molecule. A major use of gaseous fluorocarbons is in radiation-induced etching processes for the microelectronics industry; the most common one is tetrafluoromethane. Liquid fluorocarbons possess a unique combination of properties that has led to their use as inert fluids for cooling of electronic devices and soldering. Solubility of gases in fluorocarbons has also been used in biological cultures requiring O_2, and as liquid barrier filters for purifying air.

Fluorocarbons may be made part hydrocarbon and part fluorocarbon, or may contain chlorine. The fluorocarbons used as plastic resins may contain as much as 65% fluorine and also chlorine, but

are very stable. Liquid fluorocarbons are used as heat-transfer agents, hydraulic fluids, and fire extinguishers. Benzene-base fluorocarbons are used for solvents, dielectric fluids, lubricants, and for making dyes, germicides, and drugs. Synthetic lubricants of the fluorine type consist of solid particles of a fluorine polymer in a high-molecular weight fluorocarbon liquid. Chlorine reacts with fluorocarbons to form chlorofluorocarbons, commonly referred to as CFCs. CFC 11 is used as a foam-blowing agent, and CFC 12 is employed as a refrigerant. CFC 113 is a degreasant in semiconductor manufacturing. Because they are strong depletants of stratospheric ozone, the use of CFCs as aerosol propellants has been banned in the United States since 1978, and is being phased out in Europe. Alternatives to CFCs are being sought for other applications by partially substituting the chlorine with other elements. CFC 22, which has 95% less ozone-depleting capacity than CFC 12, is a potential candidate to replace CFC 12.

Fluorometric Analysis

A method of chemical analysis that measures the fluorescence intensity of the analyte or a reaction product of the analyte and a chemical reagent.

Fluoroplastics

Plastics based on polymers with monomers containing one or more atoms of fluorine, or copolymers of such monomers with other monomers, with the fluorine-containing monomer (s) being in greatest amount by mass.

Also termed fluoropolymers, fluorocarbon resins, and fluorine plastics, fluoroplastics are a group of high-performance, high-price engineering plastics. They are composed basically of linear polymers in which some or all of the hydrogen atoms are replaced with fluorine, and are characterized by relatively high crystallinity and molecular weight. All fluoroplastics are natural white and have a waxy feel. They range from semirigid to flexible. As a class, they rank among the best of the plastics in chemical resistance and elevated temperature performance. Their maximum service temperature ranges up to about 260°C. They also have excellent frictional properties and cannot be wet by many liquids. Their dielectric strength is high and is relatively insensitive to temperature and power frequency. Mechanical properties, including tensile creep and fatigue strength, are only fair, although impact strength is relatively high.

PTFE, FEP, and PFA

There are three major classes of fluoroplastics. In order of decreasing fluorine replacement of hydrogen, they are fluorocarbons, chlorotrifluoroethylene, and fluorohydrocarbons. There are two fluorocarbon types: tetrafluoroethylene (PTFE or TFE) and fluorinated ethylene propylene (FEP). PTFE is the most widely used fluoroplastic. It has the highest useful surface temperature, 260°C, and chemical resistance.

Their high melt viscosity prevents PTFE resins from being processed by conventional extrusion and molding techniques. Instead, molding resins are processed by press-and-sinter methods similar to those of powder metallurgy or by lubricated extrusion and sintering. All other fluoroplastics are melt processable by techniques commonly used with other thermoplastics.

PTFE resins are opaque, crystalline, and malleable. When heated above 341°C, however, they are transparent, amorphous, relatively intractable, and they fracture if severely deformed. They return to their original state when cooled.

F

The chief advantage of FEP is its low melt viscosity, which permits it to be conventionally molded. FEP resins offer nearly all of the desirable properties of PTFE, except thermal stability. Maximum recommended surface temperature for these resins is lower by about 37.8°C. Perfluoro-alkoxy (PFA) fluorocarbon resins are easier to process than FEP and have higher mechanical properties at elevated temperatures. Service temperature capabilities are the same as those of PTFE.

PTFE resins are supplied as granular molding powders for compression molding or ram extrusion, as powders for lubricated extrusion, and as aqueous dispersions for dip coating and impregnating. FEP and PFA resins are supplied in pellet form for melt extrusion and molding. FEP resin is also available as an aqueous dispersion.

Teflon is a tetrafluoroethylene of specific gravity up to 2.3. The tensile strength is up to 23.5 MPa, elongation 250%–350%, dielectric strength 39.4×10^6 V/m, and melting point 312°C. It is water resistant and highly chemical resistant. Teflon S is a liquid resin of 22% solids, sprayed by conventional methods and curable at low temperatures. It gives a hard, abrasion-resistant coating for such uses as conveyors and chutes. Its temperature service range is up to 204°C. Teflon fiber is the plastic in extruded monofilament, down to 0.03 cm in diameter, oriented to give high strength. It is used for heat- and chemical-resistant filters. Teflon tubing is also made in fine sizes down to 0.25 cm in diameter with wall thickness of 0.03 cm. Teflon 41-X is a colloidal water dispersion of negatively charged particles of Teflon, used for coating metal parts by electrodeposition. Teflon FEP is fluorinated ethylenepropylene in thin-film, down to 0.001 cm thick, for capacitors and coil insulation. The 0.003 cm film has a dielectric strength of 126×10^6 V/m, tensile strength of 20 MPa, and elongation of 250%.

Properties

Outstanding characteristics of the fluoroplastics are chemical inertness, high- and low-temperature stability, excellent electrical properties, and low friction. However, the resins are fairly soft and resistance to wear and creep is low. These characteristics are improved by compounding the resins with inorganic fibers or particulate materials. For example, the poor wear resistance of PTFE as a bearing material is overcome by adding glass fiber, carbon, bronze, or metallic oxide. Wear resistance is improved by as much as 1000 times, and the friction coefficient increases only slightly. As a result, the wear resistance of filled PTFE is superior, in its operating range, to that of any other plastic bearing material and is equaled only by some forms of carbon.

The static coefficient of friction for PTFE resins decreases with increasing load. Thus, PTFE bearing surfaces do not seize, even under extremely high loads. Sliding speed has a marked effect on friction characteristics of unreinforced PTFE resins; temperature has very little effect.

PTFE resins have an unusual thermal expansion characteristic. A transition at 18°C produces a volume increase of over 1%. Thus, a machined part, produced within tolerances at a temperature on either side of this transition zone, will change dimensionally if heated or cooled through the zone.

Electrical properties of PTFE, FEP, and FPA are excellent, and they remain stable over a wide range of frequency and environmental conditions. Dielectric constant, for example, is 2.1 from 60 to 10^9 Hz. Heat-aging tests at 300°C for 6 months show no change in this value. Dissipation factor of PTFE remains below 0.0003 up to 10^8 Hz. The factor for FEP and PFA resins is below 0.001 over the same range. Dielectric strength and surface arc resistance of fluorocarbon resins are high and do not vary with temperature or thermal aging.

CTFE or CFE

Chlorotrifluoroethylene (CTFE or CFE) is stronger and stiffer than the fluorocarbons and has better creep resistance. Like FEP and unlike PTFE, it can be molded by conventional methods.

Sensitivity to processing conditions is greater in CTFE resins than in most polymers. Molding and extruding operations require accurate temperature control, flow channel streamlining, and high pressure because of the high melt viscosity of these materials. With too little heat, the plastic is unworkable; too much heat degrades the polymer. Degradation begins at about 274°C. Because of the lower temperatures involved in compression molding, this process produces CTFE parts with the best properties.

Thin parts such as films and coil forms must be made from partially degraded resin. The degree of degradation is directly related to the reduction in viscosity necessary to process a part. Although normal, partial degradation does not greatly affect properties, seriously degraded CTFE becomes highly crystalline, and physical properties are reduced. Extended usage above 121°C also increases crystallinity.

CTFE plastic is often compounded with various fillers. When plasticized with low-molecular-weight CTFE oils, it becomes a soft, extensible, easily shaped material. Filled with glass fiber, CTFE is harder, more brittle, and has better high-temperature properties.

Properties

CTFE plastics are characterized by chemical inertness, thermal stability, and good electrical properties, and are usable from 400 to −400°C. Nothing adheres to these materials, and they absorb practically no moisture. CTFE components do not carbonize or support combustion. Up to thicknesses of about 3.2 mm, CTFE plastics can be made optically clear. Ultraviolet absorption is very low, which contributes to its good weatherability.

Compared with PTFE, FEP, and PFA fluorocarbon resins, CTFE materials are harder, more resistant to creep, and less permeable; they have lower melting points, higher coefficients of friction, and are less resistant to swelling by solvents than the other fluorocarbons.

Tensile strength of CTFE moldings is moderate, compressive strength is high and the material has good resistance to abrasion and cold flow. CTFE plastic has the lowest permeability to moisture vapor of any plastic. It is also impermeable to many liquids and gases, particularly in thin sections.

PVF_2 and PVF

The fluorohydrocarbons are of two kinds: polyvinylidene fluoride (PVF_2) and polyvinyl fluoride (PVF). Although similar to the other fluoroplastics, they have somewhat lower heat resistance and considerably higher tensile and compressive strength.

Except for PTFE, the fluoroplastics can be formed by molding, extruding, and other conventional methods. However, processing must be carefully controlled. Because PTFE cannot exist in a true molten state, it cannot be conventionally molded. The common method of fabrication is by compacting the resin in powder form and then sintering.

PVF_2, the toughest of the fluoroplastic resins, is available as pellets for extrusion and molding and as powders and dispersions for corrosion-resistant coatings. This high-molecular-weight homopolymer has excellent resistance to stress fatigue, abrasion, and to cold flow. Although insulating properties and chemical inertness of PVDF are not as good as those of the fully fluorinated polymers, PTFE and FEP, the balance of properties available in PVDF qualifies this resin for many engineering applications. It can be used over the temperature range from –73°C to 149°C and has excellent resistance to abrasion.

PVDF can be used with halogens, acids, bases, and strong oxidizing agents, but it is not recommended for use in contact with ketones, esters, amines, and some organic acids.

Although electrical properties of PVDF are not as good as those of other fluoroplastics, it is widely used to insulate wire and cable in computer and other electrical and electronic equipment. Heat-shrinkable tubing of PVDF is used as a protective cover on resistors and diodes, as an encapsulant over soldered joints.

Valves, piping, and other solid and lined components are typical applications of PVDF in chemical-processing equipment. It is the only fluoroplastic available in rigid pipe form. Woven cloth made from PVDF monofilament is used for chemical filtration applications.

A significant application area for PVDF materials is as a protective coating for metal panels used in outdoor service. Blended with pigments, the resin is applied, usually by coil-coating equipment, to aluminum or galvanized steel. The coil is subsequently formed into panels for industrial and commercial buildings.

A recently developed capability of PVDF film is based on the unique piezoelectric characteristics of the film in its so-called beta phase. Beta-phase PVDF is produced from ultrapure film by stretching it as it emerges from the extruder. Both surfaces are then metallized, and the material is subjected to a high voltage to polarize the atomic structure.

When compressed or stretched, polarized PVDF generates a voltage from one metallized surface to the other, proportional to the induced strain. Infrared light on one of the surfaces has the same effect. Conversely, a voltage applied between metallized surfaces expands or contracts the material, depending on the polarity of the voltage.

PFA, ECTFE, and ETFE

The following three fluoroplastics are melt processable. Perfluoroalkoxy (PFA) can be injection-molded, extruded, and rotationally molded. Compared to FEP, PFA has slightly greater mechanical properties at temperatures over 150°C and can be used up to 260°C.

Ethylene-chlorotrifluoroethylene (ECTFE) copolymer resins also are melt processable with a melting point of 240°C. Their mechanical properties—strength, wear resistance, and creep resistance, in particular—are much greater than those of PTFE, FEP, and PFA, but their upper temperature limit is about 165°C. ECTFE also has excellent property retention at cryogenic temperatures.

Ethylene-tetrafluoroethylene (ETFE) copolymer resin is another melt-processable fluoroplastic with a melting point of 270°C. It is an impact-resistant, tough material that can be used at temperatures ranging from cryogenic up to about 179°C.

One of the advantages of the copolymers of ethylene and TFE—called modified ETFE—and of ethylene and CTFE—called ECTFE—compared with PTFE and CTFE is their ease of processing. Unlike their predecessors, they can be processed by conventional thermoplastic techniques. Various grades can be made into film or sheet, into a monofilament, or used as a powder coating; all grades can be heat-sealed or welded.

Although these resins have lower heat resistance than PTFE or CTFE, they offer a combination of properties and processability that is unattainable in the predecessor resins. Maximum service temperature for no-load applications is in the range of 149°C–199°C for ETFE and ECTFE, compared with 199°C for CTFE and 288°C for PTFE. Glass reinforcement increases these values by 10°C.

Both tensile strength and toughness of these resins are higher than those of the other fluoropolymers; they are rated "no break"

in notched Izod tests. The modulus of ECTFE is higher than that of ETFE up to about 100°C; above 150°C, ETFE has a higher modulus.

As with other fluoroplastics, these resins are compatible with most chemicals, even at high temperatures. ETFE is not attacked by most solvents to temperatures as high as 199°C. ECTFE is similar to 121°C, but is attacked by chlorinated solvents at higher temperatures. ETFE has better chemical stress-crack resistance.

Applications for these resins include wire and cable insulation, chemical-resistant linings and molded parts, laboratory ware, and molded electrostructural parts.

Fluoroscopy

An inspection procedure in which the radiographic image of the subject is viewed on a fluorescent screen, normally limited to low-density materials or thin sections of metals because of the low light output of the fluorescent screen at safe levels of radiation.

Fluorspar

Also called fluorite, fluorspar is a crystalline or massive granular mineral of the composition CaF_2, used as a flux in the making of steel, for making hydrofluoric acid, in opalescent glass, in ceramic enamels, for snaking artificial cryolite, as a binder for vitreous abrasive wheels, and in the production of white cement. It is a better flux for steel than limestone, making a fluid slag, and freeing the iron of sulfur and phosphorus. About 2.5 kg of fluorspar is used per ton of basic open-hearth steel.

Acid spar is a grade used in making hydrofluoric acid. It is also used for making refrigerants, plastics, and chemicals, and for aluminum reduction. Optical fluorspar is the highest grade but is not common. Fluoride crystals for optical lenses are grown artificially from acid-grade fluorspar. Pure calcium fluoride, Ca_2F_6, is a colorless crystalline powder used for etching glass, in enamels, and for reducing friction in machine bearings. It is also used for ceramic parts resistant to hydrofluoric acid and most other acids. Calcium fluorite has silicon in the molecule and is a crystalline powder used for enamels. The clear rhombic fluoride crystals used for transforming electric energy into light are lead fluoride, PbF_2.

Flush Weld

A weld with surfaces following the profile of the parent materials. The profile may be formed directly by welding or, more often, by subsequent grinding, machining, etc.

Flushing (of Weld)

Machining or grinding the surface of a weld to bring it flush with the parent surfaces. The term underflushing is ambiguous, being used to indicate either insufficient or excessive flushing.

Flute

(1) As applied to drills, reamers, and taps, the channels or grooves formed in the body of the tool to provide cutting edges and to permit passage of cutting fluid and chips. (2) As applied to milling cutters and hobs, the chip space between the back of one tooth and the face of the following tooth.

Fluted Bearing

A sleeve bearing with oil grooves generally in an axial direction.

Fluted Core

An integrally woven reinforcement material consisting of ribs between two skins in a unitized *sandwich construction*.

Flutes

Elongation grooves or voids that connect widely spaced cleavage planes.

Fluting

(1) Forming longitudinal recesses in a cylindrical part, or radial recesses in a conical part. (2) A series of sharp parallel kinks or creases occurring in the arc when sheet metal is roll formed into a cylindrical shape. (3) Grinding the grooves of a twist drill or tap. (4) In bearings, a form of pitting in which the pits occur in a regular pattern so as to form grooves. Ridges may occur with or without burnt craters. The general cause involves vibration together with excessive wear or excessive load. (5) Electric discharge pitting in a rolling-contact bearing subject to vibration. (6) A fracture process whereby *flutes* are produced.

Flux

(1) In metal refining, a material added to a melt to remove undesirable substances, like sand, ash, or dirt. Fluxing of the melt facilitates the agglomeration and separation of such undesirable constituents from the melt. It is also used as a protective covering for certain molten metal baths. Lime or limestone is generally used to remove sand, as in iron smelting; sand, to remove iron oxide in copper refining. (2) In brazing, cutting, soldering, or welding, material used to prevent the formation of, or to dissolve and facilitate removal of, oxides and other undesirable substances. (3) A material or compound applied or introduced in various processes, usually involving melting or heating, to react with undesirable materials such as impurities in the bulk material or contaminants on the surface. The material formed by the combination of the flux and the contaminants forms a slag which is readily separated from the product. (4) The density of magnetic lines of force. (5) A measure of nuclear intensity, the number of particles per unit volume times the mean particle velocity. (6) Rate of heat input. Units include J/s and Btu/h.

Flux is a substance added to a refractory material to aid in its fusion. A secondary action of a flux, which may also be a primary reason for its use, is as a reducing agent to deoxidize impurities.

Any material that lowers the melting temperature of another material or mixture of materials is a flux. Fluxing substances may occur as natural impurities in a raw material. Thus, the alkali content of a clay will flux the clay. In other cases, fluxes are separate raw materials. Example: use of feldspar to flux a mixture of clays and flint.

An auxiliary flux is a third component that may make the primary flux more effective. Thus, addition of 2% dolomite, talc, or fluorspar to a whiteware mixture that contains 25% feldspar will produce a substantial decrease in vitrification temperature. The auxiliary constituent may be incapable of producing the same result (or too expensive to use) as the sole flux.

Compounds of alkali metals (sodium, potassium, and lithium) are popular fluxes for clay bodies. Compounds of alkaline earth metals (calcium, magnesium, and, to a lesser extent, barium and strontium) are common auxiliary fluxes. However, they also may be primary fluxes for such products as low-loss dielectrics. Lead and boron compounds are important fluxes for glasses, glazes, and enamels. And pre-melted glasses or frits may be used to flux clay or other bodies.

The term *flux* also may be used to specify a low-melting glass used in decorating glass products or an overglaze for clayware. Pigments are mixed with the powdered glass flux and then applied to the object to produce a vitrifiable coating at temperatures <650°C.

Fluxing alloys for brasses and bronzes are phosphor tin, phosphor copper, or silicon copper. They deoxidize the metals at the same time that alloying elements are added. For tinning steel, palm oil is used as a flux. For ordinary soldering, zinc chloride is a common flux. Tallow, rosin, or olive oil may also be used for soldering. Acetamide is used for soldering painted metals. For silver braze filler metals, borax is a common flux. For soldering stainless steel, borax is mixed with boric acid, or pastes are made with zinc chloride and borax. Borax may also be used as a welding flux. *White flux* is a mixture of sodium nitrate and nitrite and is a strong oxidizer used for welding. Other fluxes used in brazing contain potassium chloride, lithium fluoride, boric acid, borates, and fluoroborates.

Welding fluxes for high-temperature welding are usually coated on the rod and contain a deoxidizer and a slag former. *Lithium fluoride*, LiF, is a powerful flux with the fluxing action of both lithium and fluorine, and it gives a low-melting-point liquid slag. Deoxidizers may be ferromanganese or silicon-manganese. *Slag formers* are titanium dioxide, magnesium carbonate, feldspar, asbestos, or silica. Soluble silicate is a binder, while cellulose may be used for shielding the arc.

Flux Cored Arc Welding (FCAW)

An arc welding process that joins metal by heating them with an arc between a continuous tubular filler-metal electrode and the work. Shielding is provided by a flux contained within the consumable tubular electrode. Additional shielding may or may not be obtained from an externally supplied gas or gas mixture. See also *electrogas welding*.

Flux Cored Electrode

A composite filler metal electrode consisting of a metal tube or other hollow configuration containing ingredients to provide such functions as shielding atmosphere, deoxidation, arc stabilization, and slag formation. Alloying materials may be included in the core. External shielding may or may not be used.

Flux Cover

In metal bath dip brazing and dip soldering, a cover of flux over the molten filler metal bath.

Flux Density

In magnetism, the number of *flux lines* per unit area passing through a cross-section at right angles. It is given by $B = \mu H$, where μ and H are permeability and magnetic-field intensity, respectively.

Flux Lines

Imaginary lines used as a means of explaining the behavior of magnetic and other fields. Their concept is based on the pattern of lines produced when magnetic particles are sprinkled over a permanent magnet. Sometimes called magnetic lines of force.

Flux Oxygen Cutting

Oxygen cutting with the aid of a flux.

Flux Solder Connection

A solder joint characterized by entrapped flux that often causes high electrical resistance in electronic component.

Fly Ash

A finely divided siliceous material formed during the combustion of coal, coke, or other solid fuels.

Fly Cutting

Cutting with a single-tooth milling cutter.

Flying Shear

A machine for cutting continuous rolled products to length that does not require a halt in rolling, but rather moves along the runout table at the same speed as the product while performing the cutting, and then returns to the starting point in time to cut the next piece.

Foam Inhibitor

A surface-active chemical compound used in minute quantities to prevent or reduce *foaming*. Silicone fluids are frequently used as foam inhibitors.

Foam Materials

These are materials with a sponge-like, cellular structure. They include the well-known sponge rubber, plastic foams, glass foams, refractory foams, and a few metal foams. Ordinary chemically blown sponge rubber is made up of interconnecting cells in a labyrinth-like formation. When made by heating latex, it may show spherical cells with the porous walls perforated by the evaporation of moisture. It is also called foam rubber. Special processes are used to produce cell-tight and gas-tight cellular rubber, which is nonabsorbent.

Cellular rubber comes in sheets of any thickness for gaskets, seals, weather stripping, vibration insulation, and refrigerator insulation. Most of the so-called sponge rubbers are not made of natural rubber but are produced from synthetic rubbers or plastics, and may be called by a type classification, such as urethane foam.

Others include phenolic foam, which is made by incorporating sodium bicarbonate and an acid catalyst into liquid phenol resin. The reaction liberates CO_2 gas, expanding the plastic.

Cellular cellulose acetate is expanded with air-filled cells for use as insulation and as a buoyancy material for floats. It is tough and resilient. *Strux* is cellulose acetate foam, made by extruding the plastic mixed with barium sulfate in an alcohol-acetone solvent. When the pressure is removed, it expands into a light, cellular structure. Polystyrene foam is widely used for packaging and for building insulation. It is available as prefoamed board or sheet, or as beads that expand when heated.

Styrofoam is polystyrene expanded into a multicellular mass 42 times the original size. It has only one-sixth the weight of cork, but will withstand hot water or temperatures above 77°C, as it is thermoplastic. It is used for cold-storage insulation, and is resistant to mold. Polyethylene foam and polypropylene foam are also available. Compared with expanded polystyrene, expanded polyolefins have greater toughness and can be molded more easily. Current uses include automotive bumper cores, sun visor cores, and electronic packaging.

Other foams include foamed vinyl plastisols and expanded polyvinyl chloride. Polyester foam is odorless, flame-resistant, and resistant to oils and solvents. It has only half the weight of foamed rubber with greater strength and high resistance to oxidation. It is used for upholstery and insulation.

Silicone foam, used for insulation, is silicone rubber foamed into a uniform unicellular structure. Vinyl foams are widely used and they are made from various types of vinyl resins with the general physical properties of the resin used. Epoxy foams come as a powder consisting of an epoxy resin mixed with diaminodiphenyl sulfone. Glass foam is used as thermal insulation for buildings, industrial equipment, and piping. Ceramic foams of alumina, silica, and mullite are used principally for high-temperature insulation. Aluminum foam is a metal foam that has found appreciable industrial use as a core material in sandwich composites. Foamed zinc is a lightweight structural metal with equal strength in all directions, made by foaming with an inert gas into a closed cell structure. It is used particularly for shock and vibration insulation. Foaming agents for metals are essentially the same as those used for plastics. They are chemical additives that release a gas to expand the material by forming closed bubbles. Or they may be used to cause froth as in detergents or firefighting foams.

Foam Polymers

Most polymers, thermoplastics, or thermosets can be expanded into a foam by the addition of a foaming or blowing agent. They can be foamed to desired density, ranging from near the weight of solid resin, at over 64.08×10^{-2} g/cm^3, to extremely light, at 1.602×10^{-2} g/cm^3. The foamed product will maintain the general characteristics of the base resin as its mechanical properties decrease along with its density.

Thermoplastic Foams

Thermoplastic foams can be extruded by heating the resin and blowing agent to a threshold temperature, causing the blowing agent to react and expand into a gas. The gas combines with the resin, forming a frothy mass that is extruded through a shaping die. The resulting buns, logs, or sheets can then be cut into shapes for inclusion in composite panels and other applications.

Bead foaming is accomplished by heating resin beads that include the blowing agent in the formulation. A mold cavity is partially filled with beads and heated to the temperature at which the blowing agent gives off gas. This causes the softened beads to expand, filling the mold and welding them together. The process commonly used to make polystyrene foam creates slab stock or shapes molded to specific dimensions. Foam stock forms can be mechanically knife-cut or band-sawed into sheets, blocks, or other desired shapes.

Thermoset Foams

Some type of liquid foaming is necessary to manufacture thermoset foams, used primarily for the rigid structural foam-type applications mentioned earlier. The liquid formulation for these foams

F

contains not only the base resin and blowing agent, but also the catalyst required for curing in thermoset chemistry. Liquid foaming techniques, which include spraying, can be employed for open- or closed-mold processes and often include a reinforcement or filler material.

Polyurethane

Polyurethane, long considered a workhorse of low-cost foams, can be formed by first reacting two liquid components together. Water is the principle blowing agent (fluorocarbon is also used). This reaction produces CO_2, which is trapped in closed cells, giving good thermal insulation to the foam. Depending on the reaction and the chemicals or additives used, the foams can either be soft and flexible or tough, hard, and rigid.

Cell Structure and Density

Polymer foams have either open- or closed-cell structures, depending primarily on the type of resin and foaming process employed. Closed-cell foams, which dominate the marine industry, are impermeable, preventing water ingress. Depending on the resin chemistry used, closed-cell foams may also be impervious to corrosive liquid chemicals and solvents, making them useful as an insulating material for industrial tanks.

Most low-density (1.05–3.20×10^{-2} g/cm^3) rigid urethane foams have a closed-cell content of 90% or greater in slab form. Under controlled conditions, the closed-cell content can be increased to 99%.

Medium- to high-density polyurethane foams (4.81–64.08×10^{-2} g/cm^3) can be formulated with various characteristics and diverse high-performance capabilities that range from insulating and isolating toxic industrial wastes and radioactive nuclear materials to sandwich panels requiring higher modulus and high-temperature resistance for use in aircraft.

Higher-density, heavier foams, sometimes formulated with phenolic resins, are more dimensionally stable at elevated temperatures. Low-density foams may swell when exposed to elevated temperatures because of gas expansion and increased pressure from inner cell air and moisture. If this happens, foam cells will then erupt and collapse, since internal pressure is necessary to maintain cell structure. Contraction can continue until the foam becomes so dense that it resembles a solid more than a foamed plastic.

Foam Additives and Reinforcements

A variety of additives can be included in foam formulations that greatly impact the density, cost, manufacturability, and performance characteristics of the finished product. Randomly distributed short-fiber reinforcements can be blown into foam during spray-up, which become an integral part of the cell walls. Fiber in mat, fabric, or long-fiber form can also be introduced into the mold. Mechanical properties of the finished product will increase dramatically with the introduction of fiber reinforcements. From low-cost mineral fillers, to microspheres and oriented fiber webs, new reinforced foam products are being developed that offer elegant, low-cost solutions to esoteric application requirements.

Structural Foam

Structural foam, SRIM-type constructions, are similar in configuration to laminated sandwich panels. They have full-density outer skins and lightweight cellular cores, and the relative benefits inherent in this type of configuration, such as high stiffness-to-weight and strength-to-weight ratios, are similar to sandwich panels as well. These rigid foam products, usually made with polyurethane-based chemistry, are often reinforced, enhancing their mechanical properties.

Structural foams are made using both high- and low-pressure processes. When products are made using high pressure, the mold is solidly filled under pressure with unfoamed resin; then it is expanded (or a core removed) and the foaming agent is activated or injected to form the core layer. High-pressure processes are able to provide dense, smooth skins and achieve high levels of surface detail. Tooling for these processes is usually relatively expensive.

In low-pressure methods, the mold is partially filled with molten resin that expands to fill the mold and forms a skin upon contact with the walls. In the past, there has been some concern about surface quality with self-skinning open-molded structural foams but recent advances in resin chemistry and molding techniques are producing class A finishes for demanding automotive applications under low pressures.

Applications

Sandwich Panel Construction

One of the oldest and best-understood uses for foam core materials is in the construction of structural sandwich panels. The typical sandwich panel consists of strong outer laminate skins made with oriented glass, carbon, or aramid fiber and a lightweight inner core.

Tailoring foam core selection to the specific application is especially important. When failure of a sandwich panel occurs, it is usually in the core, because plastic foams have low shear rigidity compared to the skins. The core material must be strong enough to stabilize the sandwich structure, providing a shear load path between the laminate skins to prevent buckling.

An inexpensive foam product, a lightweight, non-chlorofluorocarbon-blown polyisocyanurate, *ELFOAM* is a product for insulation applications, where nonwoven fiberglass incorporates webs within the foam to carry the structural load in the design for structural sandwich panel construction. The foam is co-cured by resin infusion or pressurized processes and the dry fiber provides channels through which resin is drawn into the core. This offers high strength-to-weight and stiffness-to-weight ratios, creating tough, thick, lightweight structures ideal for industrial or construction applications.

Boatbuilding

Commercial boatbuilding has used composite materials over the years. Tough, durable thermoplastic PVC foam formulations are the most commonly used marine panel applications. In fact, the educated consumer has grown to appreciate and demand the lightweight, self-insulating, low-condensation improvements provided by two fiberglass skins over a foam core.

Another resin option designed for boat hull applications that must withstand repeated exposure to dynamic loads is linear structural foam. The new chemistry addresses the discrepancy between anti-modulus core with no flexure and a flexural core with low modulus. *Core-Cell* is a linear formulation that offers the stiffness of cross-linked PVCs, while maintaining its plastic range for high impact and damage tolerance. Core-Cell has heat distortion temperature comparable to cross-linked PVC foams, making it suitable for deck and superstructure laminates.

There is also a spray-applied foam for marine applications that is based on polyester/polyurethane hybrid chemistry. For use in open-mold processes only, the unsaturated polyester provides stiffness and hardness, and the polyurethane enables a fast cure and internal toughness.

The spray-applied foam provides high flexural modulus and flexural strength. It can be sprayed at a 0.38–12.7 mm thickness to provide structural strength at less weight than fiber-reinforced plastics/polyester. Ultracore foams are bondable with epoxy, vinyl ester, and polyurethane resins.

High-Performance Foams

High-performance foam core materials are typically striving for exceptionally high-temperature resistances or strength-to-weight ratios, or for other esoteric properties. These are generally derived from state-of-the-art resin chemistry or additives. Cores made from resins with exotic chemistries like polymethacrylimide (PMI) or bismaleimide (BMI) lend themselves to use in aerospace applications that demand dimensional stability at temperatures up to 149°C. For example, *Rohacell PMI* rigid foams have been selected for the interstage and payload fairings of the Delta II–IV rockets. These components are co-cured in thermoformed-to-shape sandwich cores. Rohacell XT is the newest, closed-cell isotropic foam with an extended temperature grade available that is suitable for use with BMI prepreg skins, co-curing at temperatures up to 190°C with a 232°C postcure.

Syntactic foams are an entirely different concept in foam technology. Cell structure is created by incorporating hollow microspheres into a compound (usually a liquid resin), eliminating the need for a blowing agent. The microspheres can be made from a variety of materials such as glass, diatomaceous earth, or plastic, with foam density controlled by how much resin is used to hold the microspheres together. Deriving properties primarily from the microsphere material, syntactic foams are highly customizable for esoteric applications.

Foam-Filled Honeycomb

Honeycomb cores, filled with foam and sandwiched between laminates, offer both high static-load and dynamic-load properties, as well as high fatigue endurance, strength and stiffness, and resistance to localized compression of random loads. Combining honeycomb and foam technologies in a single product raises the mechanical properties significantly. Additionally, the foam technology provides flexibility, memory, good elongation properties, and shear strength of 1.7 MPa and above.

Foaming

In tribology, the production and coalescence of gas bubbles on a liquid lubricant surface.

Foaming Agent

Chemicals added to plastics and rubbers that generate inert gases, such as nitrogen, upon heating, causing the resin to form a cellular structure.

Focal Length

In an optical microscope, the distance from the second principal point to the point on the axis at which parallel rays entering the lens will converge or focus.

Focal Point

See preferred term *focal spot*.

Focal Spot

(1) That area on the target of an x-ray tube that is bombarded by electrons. (2) The area of the x-ray tube from which the x-rays originate. The effective focal-spot size is the apparent size of this area when viewed from the detector. (3) In electron beam and laser beam welding, a spot at which an energy beam has the most concentrated energy level and the smallest cross-sectional area.

Focus

In microscopy, a point at which rays originating from a point in the object converge or from which they diverge or appear to diverge under the influence of a lens or distracting system.

Focusing Device, Electrons

A device that effectively increases the angular aperture of the electron beam illuminating the object, rendering the focusing more critical.

Focusing, X-Rays

The operation of producing a convergent beam in which all rays meet in a point or line.

Fog Quenching

Quenching in fine mist or spray. It gives a cooling rate intermediate between air cooling and quenching into oil.

Foil

Foil is very thin sheet metal used chiefly for wrapping, laminating, packaging, insulation, and electrical applications. Tinfoil is higher in cost than some other foils, but is valued for wrapping food products because it is not poisonous; it has now been replaced by other foils such as aluminum.

Lead foil, used for wrapping tobacco and other nonedible products, is rolled to the same thickness as tinfoil, but because of its higher specific gravity gives less coverage. Stainless steel foil is produced in thicknesses from 0.005 to 0.038 cm for laminating and for pressure-sensing bellows and diaphragms.

Aluminum foil has high luster, but is not as silvery as tinfoil. The tinfoil usually has a bright side and a matte side because two sheets are rolled at one time. Aluminum foil also comes with a satin finish, or in colors or embossed designs. For electrical use the foil is 99.999% pure aluminum, but foil for rigid containers is usually aluminum alloy 3003, with 1%–1.5% manganese, and most other foil is of aluminum alloy 1145, with 99.45% aluminum. The tear resistance of thin aluminum foil is low, and it is often laminated with paper for food packaging. Trifoil is aluminum foil coated on one side with Teflon and on the other with an adhesive. It is used as coatings for tables and conveyors or liners in chemical and food plants. Since polished aluminum reflects 96% of radiant heat waves, this foil is applied to building boards or used in crumpled form in walls for insulation.

Aluminum yarn, for weaving ribbons, draperies, and dress goods, is made from aluminum foil, 0.003–0.008 cm thick, by gang-slitting to widths from 0.0317 to 0.3175 cm and winding the thread on spools. Gold foil is called gold leaf and is not normally classed as foil. It is used for architectural coverings and for hot-embossed printing on leather. It is made by hammering in books, and can be made as thin as 0.0000083 cm, a gram of gold covering 3.4 m^2. Usually, goldleaf contains 2% silver, and copper for hardening.

Metal film, or metal foil, for overlays for plastics and for special surfacing on metals or composites, comes in thicknesses from flexible foils as thin as 0.005 cm to more rigid sheets for blanking and forming casings for intricately shaped parts. There are many types of hot-stamping foils, including metallic pigment foils, printed foils, and vacuum metallized foils. Composite metal films come in almost any metal or alloy such as film of tungsten carbide in a matrix of nickel alloy for wear-resistant overlays.

Foil Bearing

A bearing in which the housing is replaced by a flexible foil held under tension against a partition of the journal periphery, lubricant being retained between the journal and the foil.

Foil Decorating

Molding paper, textile, or plastic foils printed with compatible inks directly into a plastic part so that the foil is visible below the surface of the part as integral declaration.

Fold

(1) A defect in metal, usually on or near the surface, caused by continued fabrication of overlapping surfaces. (2) A forging defect caused by folding metal back onto its own surface during its flow in the die cavity. See also *lap*.

Folded Chain

The confirmation of a flexible polymer when present in a crystal. The molecule exits and reenters the same crystal, frequently generating folds.

Folic Acid

Sodium folate, a yellow to yellow-orange liquid used in medicine for folic acid deficiency. It is known as *folic acid sodium salt*, *folacin*, and, in its biologically active form, *folate*.

It is a B vitamin known to prevent spina bifida and anencephaly, devastating birth defects and may be effective in preventing cardiovascular diseases and common cancers. Foods rich in the nutrient include chicken liver in various grains, cereals, beans, vegetables, nuts, and fruit juices.

Follow Board

In foundry practice, a board contoured to a pattern to facilitate the making of a sand mold.

Follow Die

A *progressive die* consisting of two or more parts in a single holder; used with a separate lower die to perform more than one operation (such as piercing and blanking) on a part in two or more stations.

Follower Plate

A plate fitted to the top surface of a grease dispenser.

Fools Gold

Iron pyrites, FeS_2, which can resemble gold.

Foot Press

A small press with a low capacity actuated by foot pressure on a treadle.

Force Fit

Any of various interference fits between parts assembled under various amounts of force. See *interference fit*.

Force Plug

The male half of the mold for making plastics that enters the cavity, exerting pressure on the resin and causing it to flow. Also called punch. Sometimes called a core, plunger, or ram.

Forced-Air Quench

A quench utilizing blasts of compressed air against relatively small parts such as a gear.

Force-Feed Lubrication

See *pressure lubrication*.

Forehand Welding

(1) The technique of manual gas welding in which the flame both points and moves forward to the unwelded joint. Any filler rod points back toward the completed weld. Also termed forward welding and leftward welding although the latter may give rise to some confusion in the case of left-handed operators. (2) Welding in which the palm of the principal hand (torch or electrode hand) of the welder faces the direction of travel. It has special significance in oxyfuel gas welding in that the flame is directed at the head of the weld bead, which provides *preheating*. Contrast with *backhand welding*. See also *push angle*, *travel angle*, and *work angle*.

Foreign Atom

Atoms of an element other than that forming the bulk of the crystal structure.

Foreign Structure

Any metallic structure that is not intended as part of a *cathodic protection* system of interest.

Forensic Science

The application of scientific techniques to assist in matters of civil and criminal law.

Forge Delay Time

In spot, seam, or projection welding, the time between the start of the welding, current or weld interval and the application of forging pressure.

Forge Roll Scleroscope Hardness Number (HFRSc or HFRSd)

A number related to the height of rebound a diamond-tipped hammer dropped on a forged steel roll. It is measured on a scale determined by dividing into 100 units the average rebound of a hammer from a forged steel roll of accepted maximum hardness. See also *Scleroscope hardness number* and *Scleroscope hardness test.*

Forge Structure

The microstructure through a suitable section of a forging that reveals direction of working.

Forge Welding

Solid-state welding in which metals are heated in a forge (in air) and then welded together by applying pressure or blows sufficient to cause permanent deformation at the interface. The process is most commonly applied to the butt welding of steel.

Forge Welding Die

A device used in *forge welding* primarily to form the work while hot and apply the necessary pressure.

Forgeability

Term used to describe the relative ability of material to deform without fracture. Also describes the resistance to flow from deformation. See also *formability.*

Forging

The process of working metal to a desired shape by impact or pressure in hammers, forging machines (upsetters), presses, rolls, and related forming equipment. Forging hammers, counter-blow equipment, and high-energy-rate forging machines apply impact to the workpiece, while most other types of forging equipment apply squeeze pressure in shaping the stock. Some metals can be forged at room temperature, but most are made more plastic for forging by heating. Specific forging processes include *closed-die forging, high-energy-rate forging, hot upset forging, isothermal forging, open-die forging, powder forging, precision forging, radial forging, ring rolling, roll forging, rotary forging,* and *rotary swaging.*

Forging Billet

A wrought metal slug used as *forging stock.*

Forging Dies

Forms for making forgings; they generally consist of a top and bottom die. The simplest will form a completed forging in a single impression; the most complex, consisting of several die inserts, may have a number of impressions for the progressive working of complicated shapes. Forging dies are usually in pairs, with part of the impression in one of the blocks and the rest of the impression in the other block.

Forging Envelope

See *finish allowance.*

Forging Ingot

A cast metal slug used as *forging stock.*

Forging Machine (Upsetter or Header)

A type of forging equipment, related to the mechanical press, in which the principal forming energy is applied horizontally to the workpiece, which is gripped and held by prior action of the dies. See also *heading, hot upset forging,* and *upsetting.*

Forging Plane

In forging, the plane that includes the principal die face and that is perpendicular to the direction of ram travel. When parting surfaces of the dies are flat, the forging plane coincides with the parting line. Contrast with *parting plane.*

Forging Quality

Term used to describe stock of sufficient quality to make it suitable for commercially satisfactory forgings.

Forging Range

Temperature range in which a metal can be forged successfully. See the term *forgeability.*

Forging Rolls

Power-driven rolls used in preforming bar or billet stock that have shaped contours and notches for introduction of the work. See also *roll forging.*

Forging Stock

A wrought rod, bar, or other section suitable for subsequent change in cross-section by forging.

Forgings

Forging is a process of plastic deformation, usually at elevated temperatures, of ingot, bar, or billet, mill product, or metal powder to produce a desired shape and mechanical properties. Forging develops a metallurgical sound, uniform, and stable material that will have optima properties in the operating component after being completely processed and assembled.

During the forging process, metal is distributed within the required shape as it is needed according to function and stress requirements for purposes of improved design, material utilization, producibility, and cost. Forged integral components permit reduction in the number of mechanical and welded joints and elimination of welds in critical areas, with resultant reduction in stress concentrations, increased reliability and weight savings.

Forging processes are usually classified either by the type of equipment used or by the geometry of the end product. The simplest forging operation is upsetting, which is carried out by compressing the metal between two flat parallel platens. From this simple operation, the process can be developed into more complicated geometries with the use of dies. A number of variables are involved in forging; among major ones are properties of the workpiece and die materials, temperature, friction, speed of deformation, die geometry, and dimensions of the workpiece. One basic principle in forging is the fact that the material flows in the direction of least resistance.

Forgeability

In practice, forgeability is related to the strength, ductility, and friction of the material. Because of the great number of factors involved, no standard forgeability test has been devised. For steels, which constitute the majority of forgings, torsion tests at elevated temperatures are the most predictable; the greater the number of twists of a round rod before failure, the greater is its ability to be forged. A number of other tests, such as simple upsetting, tension, bending, and impact tests, have also been used.

In terms of factors such as ductility, strength, temperature, friction, and quality of forging, various engineering materials can be listed as follows in order of decreasing forgeability; aluminum alloys, magnesium alloys, copper alloys, carbon and low-alloy steels, stainless steels, titanium alloys, iron-base superalloys, cobalt-base superalloys, columbium alloys, tantalum alloys, molybdenum alloys, nickel-base alloys, tungsten alloys, and beryllium.

Types of Forging Methods

All methods of forging are basically related to hammering or pressing; the main difference between the two is the speed of pressure application. At times, the two methods may be interchangeable, depending on the availability of equipment and forging characteristics of the alloy. Certain metals and alloys resist rapid deformation and require the normally slower pressing operation.

Hammers are energized by gravity, air, or steam, and repeated blows of a vertically guided ram on metal (usually heated) resting on the anvil cause the metal to change shape. The steam hammer is generally used in the modern forge shop because of its flexibility of operation.

Pressing causes metal to move as the result of a slowly applied force. Presses operate by hydraulic, air, or steam action or by mechanical means such as crank or screw. Hydraulic presses have slower action than hammers, and their pressure application may be more closely controlled, allowing the maintenance of sustained pressures. Mechanical action presses, particularly of the crank type, may closely approach the speed of hammer blows and cannot maintain sustained pressures.

Although presses are usually assumed to operate in a vertical plane, upsetters, which fall into the category of mechanical presses, operate in a horizontal plane with side or gripper dies moving in coordination and at 90° to the basic press movement. Ring rolling machines come in various designs, but fundamentally operate as presses, with moving rolls forcing the ring to the desired shape and size. Extrusion presses may operate either vertically or horizontally.

Swaging or, as it is also called, rotary swaging consists of two or four dies which are activated radially by blocks in contact with a series of rollers. The rotation of the block die assembly causes the curved ends of the blocks (either circular or modified sine curve) to be in contact with the rollers; relative motion between the blocks and the roller housing then gives a reciprocating motion to the dies. The hammering action on the outer surface of the stock reduces its diameter; the stock is generally prevented from rotating.

Die Types

In forging, force is transmitted to the workpiece through dies generally made of chromium–molybdenum–vanadium steels, sometimes modified by addition of nickel or tungsten. In open dies, the primary force (compression) is applied locally, and different parts of the forging are progressively worked. In closed dies, the primary force is applied on the entire surface and the metal is forced into a cavity for forging to shape.

Open die (or hand) forgings are produced either in a hammer or a press, using a minimum of tooling. This method offers low tool cost and fast initial delivery, but relatively poor utilization of material and slow production rates. Closed die forgings involve higher tool cost for small quantities of finished parts, but offer relatively good utilization of material, generally better properties, close tolerances, good production rates, and good reproducibility.

Flat die forgings are produced on the anvil and offer simple configurations. Blocker-type forgings have fairly generous design tolerances and are usually made in one set of closed dies. Normal (or commercial) dies impart a more exact configuration to the workpiece that may be machined to required tolerances. No-draft close-tolerance dies impart a finished shape and size to the forging, which requires little or no machining. Generally, precision is inversely proportional to size, melting point of the metal, forging temperature, and the reactive tendencies of the metal surface with the atmosphere and lubricants at the forging temperature. Precision is directly proportional to the number of dies used.

With most configurations, therefore, economics dictate that it is cheaper to machine a commercial forging than to continue to approach the finish tolerances with die forgings alone.

Advantages of Forgings

Orientation of crystal structure of the base metal and the flow pattern distribution of secondary phases (nonmetallics and alloy segregation) aligned in the direction of working is called grain flow. Metals in the solid state are a crystalline aggregate and are, therefore, anisotropic to a degree in properties such as strength, ductility, and impact and fatigue resistance. This anisotropy or directionality can be employed to the desired extent by orienting the metal during forging (through die design) in the direction requiring maximum strength.

It is possible to develop the maximum strength potential of a particular alloy in the forging process. Only quality-controlled rolled bar or cast ingot stock is used. Quality is determined by chemical analysis, microstructure, macrostructure, ultrasonics, and mechanical testing. Further improvement comes during the forging process, because work on the material itself achieves recrystallization and grain refinement to produce material for optimum heat-treatment response.

Forgings are better than cast material for many applications because of their greater strength and ductility in a particular alloy as well as greater soundness, uniformity in chemistry, and finer grain size. There is no drastic change of state or volume in forging as there is in castings during solidification.

Forgings are highly reproducible because of this, and because of the use of carefully controlled material, controlled working temperatures, and controlled metal flow in specially designed permanent die cavities. Forged components are also stronger than welded fabrications because weld efficiencies rarely equal 100%. Ordinary sintered products do not develop the full strength potential of an alloy because of porosity in the sintered part. As a result, a larger section is required for equivalent strength, but even here, ductility will be appreciably lower than in a part forged from material of the same chemistry.

Forgings, designed to approach the desired finished part configuration, make better utilization of material than parts machined from plate or bar stock. The closeness of this approach will be determined by the amount of money desired to be spent on forge tools rather than machining capacity. Also, bar or plate stock has only one direction of grain flow. As a result, changes in section size cut across flow lines and render the material more notch, fatigue, and stress-corrosion sensitive.

Other Comparisons

Forging vs. Stamping

Stampings are suitable at most levels of production, but are most economical where annual production is high. Press productivity and die costs logically depend on the number of dies and presses required. The list price of sheet stock is competitive with forging stock, but stamping is not usually as material efficient. Energy consumption, however, is low since the stock is not heated.

Six factors give forging advantages over sheet metal stamping.

1. The engineered scrap rate for some types of stampings that are alternatives to forging maybe as high as 50%.
2. Most stampings are made in stages requiring separate dies. Processing costs climb as the number of dies increases.
3. Many applications require several stampings to be separately formed and joined in an assembly. Production costs climb with the amount of tooling, fixturing, and joining required. In some cases forgings have been chosen over one-piece stampings to achieve weight advantages or to gain secondary benefits from shapes that cannot be stamped.
4. Stampings are usually made from stock of uniform thickness. The stock thickness of a stamping is driven by the mechanical requirements of one critical feature. Alternatives, such as added reinforcements or tailored blanks, require separate parts to be processed or joined. Forging allows designers to tailor feature thickness to functional requirements, reducing overall component weight.
5. Stamping assemblies require features, such as flanges, to facilitate joining. These features usually increase the amount of purchase stock and the weight of the end product.
6. Stamping processes work-harden metal to some degree, increasing strength and hardness and decreasing ductility in some areas of the product. These increases are driven by the process, and usually cannot be optimized to the application as can be done in forging. In some cases, work hardening requires intermediate annealing.

Forgings vs. Weldments

Weldments are generally made from bar, tubing, and plate. Part shapes are made by burning, laser cutting, shearing, or sawing, depending on complexity and thickness. Tooling cost is very low but cutting and welding can be labor-intensive.

Weldments may offer an advantage over forgings in low production quantities, but this economic advantage decreases as quantities increase. Applications where production requirements start low can often be introduced as weldments and converted to forgings as production grows.

Forgings can also become part of a weldment. For example, when special features are added to a forging it is sometimes more economical to weld two forgings together than to forge the entire part. A good example is friction welding a bar of steel to a flange to form a long axial shaft.

Forging vs. Foundry Casting

Forgings offer significant advantages over castings in applications where reliability, high tensile strength, or fatigue strength are required. Forgings are free from porosity, which is difficult to eliminate in castings. This is particularly true in areas where geometric transitions occur, which are typically areas of stress concentration.

The superior fracture toughness of forgings often must be considered when designing equivalent castings by applying a "casting factor." This casting factor imposes a weight penalty that is often enough to make forgings the more economical choice.

Forging vs. Investment Casting

Investment casting, sometimes known as the "lost-wax" process, is used with a wide range of alloys, including carbon and alloy steels, stainless steels, titanium, nickel, cobalt, and aluminum alloys. Tools consist of aluminum molds for injection molding the wax patterns. They are relatively inexpensive and require very little maintenance.

The investment casting process is more labor-intensive than forging, and is more suited to lower production quantities. Investment casting is most advantageous for small- and medium-size castings of highly complex shapes that require extremely tight dimensional precision and good surface quality.

Forging vs. CNC Machining

There is virtually no limit to the shapes that can be produced by CNC machines. In most cases, standard cutting tools are used, eliminating the need to purchase special tooling. Processing and material costs are generally high because much of the purchased stock is removed. Hogouts are useful for complex or precision components in very low quantities. They can also be used on a limited basis for prototype forgings. Forgings generally exhibit superior directional properties and fatigue performance due to grain flow.

Forging vs. Powder Metallurgy

Conventional powder metallurgy (P/M), metal injection molding (MIM), and powder forging (P/F) are the three most commonly specified powder metallurgy processes. MIM is currently limited to very small components, up to approximately 100 g of complex configuration. It is rarely an alternative to forging. Conventional P/M and P/F may be alternatives in some applications.

Conventional P/M produces very close dimensional precision, but the process is characterized by porosity, which reduces mechanical properties. Tensile and yield strengths decline approximately in proportion to the level of porosity. Ductility and dynamic properties, such as impact toughness, fracture toughness, and fatigue strength, are usually much lower than for forgings. Additional processes, such as infiltration and special sintering procedures, can improve these properties. However, the metallurgical properties of such products are not equivalent to forgings made from similar alloys.

Material properties of P/M parts approach those of forgings only when the porosity is reduced to 0.5% or less. Generally, this can only be achieved by P/F. P/F is an alternative to impression die forging for small- and medium-sized components with a high degree of symmetry and high production volumes, such as automotive connecting rods.

Forgings vs. Reinforced Plastics and Composites

Reinforced plastics and composites generally utilize thermoset plastics, and occasionally thermoplastics, as a matrix. Reinforcing fibers of glass, mineral, carbon, and aramid are added to increase strength and stiffness. These materials are well established in applications where low weight is essential and increased cost can be tolerated, such as aerospace applications. Regardless of the reinforcing fibers used, the operating temperature range of reinforced plastics and composites is limited by the polymer matrix materials.

Forgings offer advantages of lower cost and higher production rates. Forging materials outperform composites in almost all physical and mechanical properties, especially impact toughness, fracture toughness, and compression strength.

Forking

A phenomenon in which a propagating fracture in a ceramic system branches into two or more new fractures, each separated from its immediate neighbor by an acute angle.

Form Block

Tooling, usually the male part, used for forming sheet-metal contours; generally used in *rubber-pad forming.*

Form Cutter

Any cutter, profile sharpened or cam relieved, shaped to produce a specified form on the work.

Form Die

A die used to change the shape of a sheet metal blank with minimal plastic flow.

Form Grinding

Grinding with a wheel having a contour on its cutting face that is a mating fit to the desired form.

Form Machining/Tool Machining

Processes which utilize a cutting tool shaped to the final intended profile.

Form Rolling

Hot rolling to produce bars having contoured cross-sections; not to be confused with *roll forming* of sheet metal or with *roll forging.*

Form Tool

A single-edge, nonrotating cutting tool, circular or flat, that produces its inverse or reverse form counterpart upon a workpiece.

Formability

The ease with which a metal can be shaped through plastic deformation. Evaluation of the formability of a metal involves measurement of strength, ductility, and the amount of deformation required to cause fracture. The term workability is used interchangeably with formability; however, formability refers to the shaping of sheet metal, while workability refers to shaping materials by *bulk forming.* See also *forgeability.*

Formaldehyde

Also called *methylene oxide.* A colorless, poisonous gas of composition HCHO. It is very soluble in water and is marketed as a 40% solution by volume, 37 by weight, under the name of *formalin.* The commercial formalin is a clear, colorless liquid with a specific gravity of 1.075–1.081. The material is obtained by oxidation from methyl alcohol. It is used in making plastics, as a reducing agent, as a disinfectant, and in the production of other chemicals, such as ethylene glycol, hexamethylenetetramine, pentaerythritol, and butadiene. *Trioxane* is polymerized formaldehyde, or a ring compound of anhydrous formaldehyde. It is marketed as colorless crystals of a pleasant ether-alcohol odor. It is used as a source of dry formaldehyde gas, as a tanning agent, and as a solvent. It burns with a hot, odorless flame, and is used in tablet form to replace solidified alcohol for heating.

Form-and-Spray

Technique for thermoforming plastic sheet into an end product and then backing up the sheet with *spray-up* reinforced plastics.

Forming (Ceramics and Glasses)

(1) The shaping or molding of ceramic ware. (2) The shaping of hot glass.

Forming (Metals)

(1) Making a change, with the exception of shearing or blanking, in the shape or contour of a metal part without intentionally altering its thickness. (2) The plastic deformation of a billet or a blank sheet between tools (dies) to obtain the final configuration. Metal-forming processes are typically classified as *bulk forming* and *sheet forming.* Also referred to as metalworking.

Forming (Plastics)

A process in which the shape of plastic pieces such as sheets, rods, or tubes is changed to a desired configuration. See also *thermoforming.* The use of the term forming in plastics technology does not include such operations as molding, casting, or extrusion, in which shapes or pieces are made from molding materials or liquids.

Forming Limit Diagram (FLD)

A diagram in which the major strains at the onset of necking in sheet-metal are plotted vertically and the corresponding minor strains are plotted horizontally. The onset-of-failure line divides all possible strain combinations into two zones; the safe zone (in which failure during forming is not expected) and the failure zone (in which failure during forming is expected).

Forming Processes

Those processes that change the shape of material along its major axis without substantially changing its cross-section, for example, bending, coiling, or twisting.

Form-Relieved Cutter

A cutter so relieved that by grinding only the tooth face the original form is maintained throughout its life.

Formvar

A plastic material used for the preparation of replicas or for specimen-supporting membranes.

Formvar Replica

A reproduction of a surface in a plastic Formvar film. See also *replica.*

Forsterite

A magnesium silicate mineral of composition $2MgO \cdot SiO_2$, which is usually produced synthetically as a ceramic raw material, but it may also be a reaction-produced phase and fired ceramics used in refractory brick.

Forward Extrusion

Same as direct extrusion. See *extrusion.*

Forward Welding

See *forehand welding.*

Fouling

(1) Unwanted deposits particularly barnacles and weeds on the submerged surface of ships or other marine equipment. (2) Unwanted mineral or other deposits on, for example, the interior of pipework or boiler gas passages. (3) An accumulation of deposits. This term includes accumulation and growth of marine organisms on a submerged metal surface and also includes the accumulation of deposits (usually inorganic) on heat exchanger tubing. See also *biological corrosion.*

Fouling Organisms

Any aquatic organism with a sessile adult stage that attaches to and fouls underwater structures of ships.

Foundry

A commercial establishment or building where metal castings are produced.

Foundry Returns

Metal in the form of gates, sprues, runners, risers, and scrapped castings of known composition returned to the furnace for remelting.

Four Point Bending

A test, or other loading situation, in which the component, usually a bar, is supported at two points toward its extremities and the load is applied through two points symmetrically located between the supports. A feature of such loading is that the stress is uniform over the length between the two inner points, unlike three-point bending.

Four-Harness Satin

A fabric weave, also called *crowfoot satin* because the weaving pattern, when laid out on cloth design paper, resembles the imprint of crow's foot. In this type of weave there is a three-by-one interlacing. That is, a filling thread floats over the three warp threads and then under one. The two sides of the fabric have different appearances. Fabrics with this weave are more pliable than either the plain or basket weaves and, consequently, are easier to form around curves. See the term *crowfoot satin.*

Four-High Mill

A type of rolling mill, commonly used for flat-rolled mill products, in which too large-diameter backup rolls are employed to reinforce two smaller work rolls, which are in contact with the product. Either the work rolls or the backup rolls may be driven. Compare with *two-high mill* and *cluster mill.*

Fourier Transform

(1) A mathematical process for changing the description of a function by giving its value in terms of its sinusoidal spatial (or temporal) frequency components instead of its spatial coordinates (or vice versa). (2) An analytical method used automatically in advanced forms of spectroscopic analysis such as infrared and nuclear magnetic resonance spectroscopy.

Fourier Transform Infrared (FT-IR) Spectrometry

A form of infrared spectrometry in which data are obtained as an interferogram, which is then Fourier transformed to obtain an amplitude vs. wavenumber (or wavelength) spectrum.

Four-Point Press

A press whose slide is actuated by four connections and four cranks, eccentrics, or cylinders, the chief merit being to equalize the pressure at the corners of the slides.

FP Fiber

A polycrystalline all-alumina fiber (>99% Al_2O_3). A ceramic fiber useful for high temperature (1370°C–1650°C, or 2500°F) composites.

Fraction

In powder metallurgy, the portion of a powder sample that lies between two stated particle sizes.

Fractography

(1) Descriptive treatment of fracture of materials, with specific reference to photographs of the fracture surface. Macrofractography

involves photographs at low magnification (<25×); microfractography, photographs at high magnification (>25×). (2) The examination of fracture surfaces under a conventional light microscope or in a scanning electron microscope. Normally no preparation of the surface, apart from cleaning, is involved although replication is common.

Fracture

A crack or break or the process of cracking, see *break*. Often "fracture" without qualification has implications of low ductility whereas "rupture" may imply significant ductility but there are no fully reliable rules and "ductile fracture" is quite acceptable.

Fracture (Composites)

The separation of a body. Defined both as rupture of the surface without complete separation of the laminate and as complete separation of a body because of external or internal forces. Fractures in continuous fiber reinforced composites can be divided into three basic fracture types: intralaminar, interlaminar, and translaminar. Translaminar fractures are those oriented transverse to the laminated plane in which conditions of fiber fracture are generated. Interlaminar fracture, on the other hand, describes failures oriented between plies, whereas intralaminar fractures are those located internally within a *ply*.

Fracture (Metals)

The irregular surface produced when a piece of metal is broken. See also *brittle fracture, cleavage fracture, crystalline fracture, decohesive rupture, dimple rupture, ductile fracture, fibrous fracture, granular fracture, intergranular fracture, silky fracture,* and *transgranular fracture.*

Fracture Analysis

Any technique, with particularly numerical ones, based on the data derived from examining and measuring features on a fracture surface. For example, detailed measurements of the number and spacing of beach marks on a fatigue fracture can indicate key factors such as the date of initiation, the number of cycles and the levels of stress.

Fracture Appearance Transition Temperature (FATT)

The temperature at which a fracture faces 50% cleavage and 50% ductile. See *impact test.*

Fracture Ductility

The true plastic strain of fracture.

Fracture Energy

The energy input required to produce unit area increase of crack face. Since any crack has two faces the fracture energy is half the fracture toughness.

Fracture Grain Size

Grain size determined by comparing a fracture of a specimen with a set of standard fractures. For steel, a fully martensitic specimen is generally used, and the depth of hardening and the prior austenitic grain size are determined.

Fracture Mechanics—Also Termed Linear Elastic Fracture Mechanics (LEFM)

The study of the mechanics of crack growth, in particular stress-related matters in the zone of material at the crack tip. Note that although this definition excludes crack initiation in a defect-free material, many apparently sound components contain crack-like features. Examples include, on a macro scale, the surface irregularities on welds (see *weld defects*) or, on a microscopical scale, the inclusions inevitable in many metals. The principles of fracture mechanics are most readily applicable to low ductility mechanical damage mechanisms such as brittle fracture and fatigue but they are also applied, perhaps with more difficulty to other mechanisms such as stress corrosion and creep. Clearly, the severity of loading in the presence of a crack is influenced by both the size of the externally imposed load and the size of the crack, and hence, a basic requirement is to develop a parameter reflecting these two variables that is measurable and calculable in a manner analogous to stress in an uncracked component. This parameter is termed stress intensity factor, "K" and can be defined as the measure of the elastic stress field in the vicinity of a crack tip. Stress intensity should not, therefore, be confused with stress concentration factor or even stress intensification factor since the latter two terms are not readily measurable and, at best, are quantified only as a multiplication factor. Cracks can extend in three modes depending on the form of loading: mode I—the crack opening mode (simple tensile loading), mode II—the edge sliding mode (in-plane shear loading), and mode III—the shear mode (out-of-plane shear loading). Of these, the opening mode is by far the most common but, where necessary, the relevant mode is indicated by the appropriate subscript, for example, K_I, K_{II}, or K_{III}. Basic fracture mechanics deals with idealized circumstances where all stresses in the crack tip zone are elastic and K increases to a critical stress intensity, termed K_C (or, if appropriate, K_{IC}, etc.), at which the crack will extend without any plastic deformation. Such behavior is, by definition brittle and for brittle materials K_C is a measure of fracture toughness. In practice, some plastic deformation is inevitable but, where it is slight, the basic concepts can be applied with adequate accuracy. However, where the amount of plasticity is large, relative to the crack size, then more complex treatments are required. These are usually based on the J integral which can be considered analogous to K with corrections for plasticity. Effectively elastic conditions are encountered in components which are thick relative to the crack size. In these circumstances, the potential local plasticity at the highly stressed crack tip is constrained to a very small volume by the surrounding bulk of lower stress material. This causes stresses in the transverse directions and is described as a condition of plane strain, hence the use of the term plane strain fracture toughness when referring to K_{IC}. The term plane stress refers to circumstances where the tensile stress is uniaxial, that is, there is no stress in the transverse directions and all material in the section carries the net section stress, that is, the total applied load divided by the full cross section. Such circumstances are encountered in crack-free material or at a crack that is large relative to the plate thickness. In the latter case of bulk yielding ahead of the crack allows stress redistribution to eliminate stresses in the transverse directions. Generalizing, if the net section stress does not exceed about 0.8 of the yield strength, the plastics will be sufficiently small for plane strain to apply and hence for calculations based on K to be reasonably accurate. See also *fracture toughness; linear elastic fracture mechanics.*

Fracture Strength

The normal stress at the beginning of fracture. Calculated from the load at the beginning of fracture during a tension test and the original cross-sectional area of the specimen.

Fracture Stress

The true, normal stress on the minimum cross-sectional area at the beginning of fracture. The term usually applies to tension tests of unnotched specimens.

Fracture Surface Markings

Fracture surface features that may be used to determine the fracture origin location and the nature of the stress that produced the fracture.

Fracture System

That family of related fracture surfaces lying within an object, having a common cause and origin.

Fracture Test

Test in which a specimen is broken and its fracture surface is examined with the unaided eye or with a low-power microscope to determine such factors as composition, grain size, case depth, or discontinuities.

Fracture Toughness

A generic term for measures of resistance to extension of a crack. The term is sometimes restricted to results of *fracture mechanics* tests, which are directly applicable in fracture control. However, the term commonly includes results from simple tests of notched or pre-cracked specimens not based on fracture mechanics analysis. Results from tests of the latter type are often useful for fracture control, based on either service experience or empirical correlations with fracture mechanics tests. See also *stress-intensity factor*.

Fracture Transition

Same as *fracture appearance transition*.

Fragmentation

The subdivision of a grain into small, discrete crystallite outlined by a heavenly deformed network of intersecting slip bands as a result of cold working. These small crystals or fragments differ in orientation and tend to rotate to a stable orientation determined by the slip systems.

Fragmented Powder

A powder obtained by fragmentation and mechanical comminution into fine particles.

Frame

The main structure of a forming or forging press.

Francium

A metallic element, one of the alkali group.

Frank-Condon Principle

The principle that states that the transition from one energy state to another is so rapid that the nuclei of the atoms involved can be considered stationary during the transition.

Frank-Rhead Source

A mechanism capable of generating dislocations. The mechanism can be visualized as an edge dislocation anchored at its two ends and subjected to a stress that causes it to bow outwards forming loops around the anchor points. As the loops grow, the structure becomes unstable and jumps to form a new continuous loop dislocation round the original source, which remains to trigger further loops.

Freckling

A type of segregation revealed as dark spots on a macroetched specimen of a consumable-electrode vacuum-arc-remelted alloy.

Free Bend

The bend obtained by applying forces to the ends of the specimen without the application of force at the point of maximum bending.

Free Carbon

The part of the total carbon in steel or cast iron that is present in elemental form as graphite or *temper carbon*. Contrast with *combined carbon*.

Free Corrosion Potential

Corrosion potential in the absence of net electrical current flowing to or from the metal surface.

Free Cutting/Machining

A metal that has been treated usually by the deliberate introduction of impurities, to improve its machining characteristics, in particular to ensure that material being cut away is released as small chips rather than continuous strands. The impurities form an even dispersion of particles of a size large enough to serve as chip breakers but not so large that they have an acceptable effect on mechanical properties. The effect is achieved, for example, in steels by appropriate quantities of manganese sulfide inclusions and in brass by additions of lead. Machining characteristics can also, in some cases, be improved by heat treatments that influence the distribution and size of precipitates or that induce some beneficial grain structure. For example, some steels can develop a blocky structure with a favorable ferrite/pearlite distribution. The effect of such treatments may not be so pronounced as that induced by inclusions but it is useful where any inclusions are unacceptable, as in the case of steels for ball and roller bearings.

Free Electron Theory

This suggests that the valence electrons in a metallic bond form an electron cloud or electron gas in which there is no constraint on their movement. Subsequently, the energy band theory was developed to better explain phenomena such as semiconduction.

Free Ferrite

(1) Ferrite that is formed directly from the decomposition of hypoeutectoid austenite during cooling, without the simultaneous formation of cementite. (2) Ferrite formed into separate grains and not intimately associated with carbides as is pearlite. Also called *proeutectoid ferrite*.

Free Fit

Any of various clearance fits for assembly by hand and free rotation of parts. See also *running fit*.

Free Machining

Pertains to the machining characteristics of an alloy to which one or more ingredients have been introduced to produce small broken chips, lower power consumption, better surface finish, and longer tool life; among such additions are sulfur or lead to steel, lead to brass, lead and bismuth to aluminum, and sulfur or selenium to stainless steel.

Free Radical

Any molecule or atom that possesses one unpaired electron. In chemical notation, a free radical is symbolized by a single dot (to denote the odd electron) to the right of the chemical symbol.

Free Rolling

Rolling in which no traction is deliberately applied between a rolling element and another surface.

Free Rotation

The rotation of atoms, particularly carbon atoms, about a single bond. Because the energy requirement is only a few kcal, the rotation is said to be free if sufficient thermal energy is available.

Free Vibration

A technique for performing dynamic mechanical measurements in which the sample is deformed, released, and allowed to oscillate freely at the natural resonant frequency of the system. Elastic modulus is calculated from the measured resonant frequency, and damping is calculated from the rate at which the amplitude of the oscillation decays.

Free Wall

The portion of a *honeycomb* cell wall that is not connected to another cell.

Free-Energy Diagram

A graph of the variation with concentration of the Gibbs free energy at constant pressure and temperature.

Free-Energy Surface

In a ternary or higher order free-energy diagram, the locus of points representing the Gibbs free energy as a function of concentration, with pressure and temperature constant.

Free-Radical Polymerization

A type of polymerization in which the propagating species is a long-chain free radical initiated by the introduction of free radicals from thermal or photochemical decomposition of an initiator molecule.

Freezing Point

See preferred term *liquidus* and *solidus*. See also *melting point*.

Freezing Range

That temperature range between *liquidus* and *solidus* temperatures in which molten and solid constituents coexist.

Frequency

The number of cycles per unit time. The recommended unit is the hertz, Hz, which is equal to 1 cycle/s.

Frequency Distribution

The way in which the frequencies of occurrence of members of a population, or a sample, are distributed according to the values of the variable under consideration.

Fresnel Fringes

A class of diffraction fringes formed when the source of illumination and the viewing screen are at a finite distance from a diffracting edge. In the electron microscope, these fringes are best seen when the object is slightly out of focus.

Fretting

A type of wear that occurs between tight-fitting surfaces subjected to cyclic relative motion of extremely small amplitude. Usually, fretting is accompanied by corrosion, especially of the very fine wear debris. Also referred to as *fretting corrosion* and *false Brinelling* (in rolling-element bearings).

Fretting Corrosion

(1) The accelerated deterioration at the interface between contacting surfaces as the result of corrosion and slight oscillatory movement between the two surfaces. (2) A form of *fretting* in which chemical reaction predominates. Fretting corrosion is often characterized by the removal of particles and subsequent formation of oxides, which are often abrasive and so increase the wear. Fretting corrosion can involve other chemical reaction products, which may not be abrasive.

Fretting Fatigue

Fatigue fracture that initiates at a surface area where fretting has occurred. The progressive damage to a solid surface that arises from fretting. *Note:* If particles of wear debris are produced, then the term *fretting wear* may be applied.

Fretting Wear

Wear arising as a result of *fretting*.

Friction

The resulting force tangential to the common boundary between two bodies when, under the action of an external force, one body moves or tends to move relative to the surface of the other. The term friction is also used, incorrectly, to denote *coefficient of friction*. It is vague and imprecise unless accompanied by the appropriate modifiers, such as *dry friction* or *kinetic friction*. See also *static coefficient of friction*. The force acting between two contacting bodies in motion relative to each other. The Laws of Friction are (1) The static frictional force is proportional to the load acting normal to the interface. (2) The frictional force is independent of the interfacial area. (3) Sliding friction is less than the limiting static friction.

Friction Coating/Surfacing

The deposition of a coating by a rod of material which rotates as it bears heavily against the component to be coated. The rotation initially causes friction heating to a temperature near the melting point of the rod material and secondly scours the receiving surface allowing a good bond with the deposited metal. A wide range of materials including hard facing alloys can be deposited.

Friction Coefficient

See *coefficient of friction*.

Friction Cutting/Sawing

Operations in which frictional heating plays a significant role in melting or softening the material being cut. In the case of friction sawing the blade is usually toothed to assist metal removal but in friction cutting the blade may be plain edged.

Friction Force

The resisting force tangential to the interface between two bodies when, under the action of an external force, one body moves or tends to move relative to the other.

Friction Grip Bolting

A bolted joint which relies on having sufficient preload in the bolts to develop a high level of friction between the mating faces of the joint. Loads across the joint are then carried by friction in the parent materials rather than by the bolts and shear.

Friction Material

A sintered material exhibiting a high coefficient of friction design for use where rubbing or frictional wear is encountered—for example, aircraft brake linings and clutch facings on tractors, heavy trucks, and earth-moving equipment. Friction materials consist of a dispersion of friction-producing ingredients in a metallic matrix.

Friction Polymer

An amorphous organic deposit that is produced when certain metals are rubbed together in the presence of organic liquids or gases. Friction polymer often forms on moving electrical contacts exposed to industrial environments. The varnish-like film will attenuate or modify transmitted signals.

Friction Soldering

See preferred term *abrasion soldering*.

Friction Welding

Welding in which heat is produced by the friction between two contacting components moving relatively to each other. The term usually refers to a production process in which two components are forced together while one is rapidly rotated or otherwise moves relative to the other. The relative movement causes heating, disrupts surface films allowing contact between clean surfaces and produces local deformation. When these effects reach an optimum stage, the movement is sharply halted and the interfacial force is considerably increased leading, and suitable cases, to strong joints even between materials not readily welded by other techniques. The friction welding effect may occur unintentionally as a result of adhesive wear in which case it is usually termed *seizure*.

Friction Welding (Metals)

A solid-state process in which welds are made by holding a nonrotating workpiece in contact with a rotating workpiece under constant or gradually increasing pressure until the interface reaches the welding temperature and rotation can be stopped.

Friction Welding (Plastics)

A method of welding thermoplastic materials in which the heat necessary to soften the components is provided by friction.

Frictive Track

In glass, a series of crescent cracks lying along a common axis, paralleling the direction of frictive contact. Also known as a chatter sleek.

Fringes

Bands of light of varying intensity or color produced by interference effects.

Frit

A glass produced by *fritting*, which contains fluxing material and is employed as a constituent in a glaze, body, or other ceramic composition.

Fritting

The rapid chilling of the molten glassy material to produce frit. See *sintering*.

Frost Line

In the extrusion of polyethylene lay-flat film, a ring-shape zone located at the point where the film reaches its final diameter. This zone is characterized by a frosty appearance on the film caused when the film temperature falls below the softening range of the resin.

Frosted Area

See *hackle*.

Frosting

A form of ball bearing groove damage, and tearing as a frosted area, suggestive that surface distress has occurred.

Frothing

A technique for applying urethane foam in which blowing agents or tiny air bubbles are introduced under pressure into the liquid mixture of foam ingredients.

Frozen Equilibrium

A condition in which a material is unable to progress to the theoretical equilibrium state because there is insufficient thermal energy to allow diffusion.

Frozen Stress

A photo-elastic stress analysis technique.

FRP

See *fiber-reinforced plastic*.

Fuel Briquettes

Also termed *coal briquettes*. Various-shaped briquettes made by compressing powdered coal, usually with an asphalt or starch binder, but sometimes as *smokeless fuel* without a binder. They are sometimes also made waterproof by coating with pitch or coal tar. They have a great advantage over raw coal that they do not take up large amounts of water, as coal does, and thus have uniformity of firing. Fuel briquettes are made from anthracite screenings usually mixed with bituminous screenings, as the bituminous coals require no binders. The usual forms of the briquettes are pillow-shaped, cubic, cylindrical, ovoid, and rectangular. The term *packaged fuel* is used for cube-shaped briquettes wrapped in paper packages, used for hand firing in domestic furnaces. *Charcoal briquettes* for home fuels are charcoal powders pressed with a starch binder. *Fuelettes*, solid fuel made from the low-grade nonrecyclable portion of the mix-paper waste, are intended for cofiring with coal in industrial boilers. They are best suited for solid-fuel and fluidized-bed boilers. Because they are sulfur-free, their use reduces emissions of sulfur dioxide.

Fuel Cells

Solid Oxide Fuel Cells

Solid oxide fuel cells (SOFC) can be classified in terms of their structures into tubular, planar, and honeycomb types. Among them, a planar SOFC will be the most suitable for large power plant utilization because the higher power density can be expected. Furthermore, well-established ceramic processing methods such as the tape-casting method can be applied to fabricate planar SOFC components cheaply and easily.

SOFC are high-temperature (900°C–1000°C), ceramic, electrochemical reactors that directly convert chemical into electrical energy. The basic unit of a cell consists of two porous gas-diffusion electrodes separated by a gastight (i.e., dense), oxygen-ion-conducting electrolyte. Yttria-stabilized zirconia (YSZ) has been most popularly used as electrolyte material, because of its relatively high ionic conductivity and high chemical stability both in reducing and oxidizing atmospheres. The perovskite $La_{1-x}Sr_xMnO_{3-y}$ (SLM, with $0 < x < 0.5$) has been used mostly as cathode material and the cermet Ni/8 mol% yttria-stabilized zirconia (8YSZ) has been used as anode material.

Two well-known methods for attaching the LSM to the YSZ electrolyte are plasma spraying and sputtering. Because these are gas-phase processing methods, significantly good contact conditions between the electrode and the electrolyte can be obtained; a thin and homogenous electrode layer can be produced on the electrolyte.

For optimum SOFC performance, the electrodes require good lateral conductivity, electrochemical activity, and chemical stability toward the electrolyte and gas environment. These factors depend on the composition and microstructure of the electrode, which is determined by the nature of the starting powder in the applied manufacturing technique.

Application

The high-temperature SOFC using zirconia electrolyte are being considered for power generation because they are expected to provide high energy efficiency, yet produce a low level of pollutant gases. The SOFC plants will be operated at high temperatures near 1000°C, and different cell materials will be exposed to either oxidizing or reducing atmosphere over a long time. Hence, the degradation of the cell performance due to the reaction between electrode materials and the zirconia electrolyte may become an important factor in determining the service life.

Other Cells

An environmentally cleaner way to generate electricity, known as molten carbonate fuel cell technology, could significantly increase the demand for nickel over the next decade.

Stationary plants that will use H_2 and O_2 to generate electrical power, with water and heat as the sole byproducts, could be built. These quiet, nonpolluting plants could become an important part of the power industry in the twenty-first century.

If the present design concepts hold up, the anodes and cathodes of these massive power plants will be made of porous nickel alloys. Typically, the anode is a Ni–Cr alloy, and the cathode is composed of a lithiated NiO. Nickel catalysts (supported either on MgO or Li–Al–dioxide) are also required to reform hydrocarbon fuel such as natural gas, producing H_2 gas, which is needed in the fuel cell.

Fuel cells may be a preferred technology of the future for perfluorinated polymers, which are critical components in membrane fuel cells. Membranes made with perfluorinated polymers act as a separator and electrolyte to allow the fuel cell to run at high current densities and voltages.

Last are the proton exchange membrane (PEM) fuel cells for major office buildings and for the automotive industry.

The fuel cells in a power plant, for example, if installed in a 52-story office building, will convert natural gas to electrical energy through a series of chemical processes. Combined, two cells will generate 400 kW of building power, including external lighting and signage for the building façade and hot water for thermal heat.

Similar to the common battery, the fuel cell uses an electrochemical process to convert chemical energy found in H_2 and O_2 into electricity and water. Fuel cells are more efficient than internal-combustion engines and are virtually pollution- and noise-free. Stationary fuel cells can be powered by gasoline, natural gas, or methanol. Cells designed for cars are powered by gasoline.

The transportation market is the next frontier—to advance PEM technology for so-called electric cars. The resulting fuel cell can produce more than 50 kW of power, enough to power a midsize car. The extraordinary progress being made in using technology to reduce greenhouse gases, improve the air that we breathe, and use our energy more efficiently is being advanced in automotive technology.

Fuel Element

A component for a nuclear reactor comprising a fissionable material in a metal sheath that provides protection and support and retains any fission products.

Fuel Gases

Gases usually used with oxygen for heating such as acetylene, natural gas, hydrogen, propane, methyl-acetylene propadiene stabilized, butane, methane, coal gas and other synthetic fuels are hydrocarbons.

Fuel Oil

Distillates of petroleum or shale oil used in diesel engines and in oil-burning furnaces. True fuel oils are the heavier hydrocarbons in kerosene, but the light or distillate oils are used largely for heating and the heavy or residual oils for industrial fuels. In some cases, only the light oils, naphtha and gasoline, are distilled from petroleum, and the residue is used for fuel oil, but this is wasteful of the lighter oils. Commercial grades of *furnace oil* for household use and *diesel oil* for trucks may be low grades of kerosene. *Gas oil*, which receives its name from its use to enrich fuel gas and increase the luminance of the flame, is also used as a fuel oil in engines. *Bunker C oil*, for diesel engines, is a viscous black oil.

Fuels

The term normally covers a wide range, since innumerable organic materials can be used as fuel. Coal, oil, or natural gas or products derived from them are the basic industrial fuels, but other materials are basic in special situations, such as sawdust in lumbering areas and bagasse in sugarcane areas.

But modern technical reference to fuels generally applies to *high-energy fuels* for jet engines, rockets, and special-use propellants; and the comparisons of these fuels are in terms of *specific impulse*, which is the thrust in pounds per pound of propellant per second. The molecular weight of the products produced by the reaction of a fuel must be extremely low to give high specific impulse, that is, above 400. Hydrogen gives a high specific-impulse rating, but it has very low density in the liquid state and other unfavorable properties, so that it is usually employed in compounds. The initial impulse of a rocket is proportional to the square root of the combustion temperature of the fuel. *Hydrogen fuels* reacted with pure oxygen produce temperatures above 5000°F (2760°C), and some fuels may react as high as 9000°F (4982°C). Aluminum powder or lithium added to hydrogen increases the efficiency. *Boron fuels* in general release 50% more thermal energy than petroleum hydrocarbons. The first Saturn space rocket had kerosene–liquid oxygen in the first stage and liquid hydrogen–oxygen in the following three stages. *Solid rocket fuels*, designed for easier handling, have a binder of polyurethane or other plastic. *Fuel oxidizers*, for supplying oxygen for combustion, may be ammonium, lithium, or potassium perchlorates. In solid fuels, oxidizers make up as much as 80% of the total.

A *monopropellant*, high-energy fuel is a chemical compound which, when ignited under pressure, undergoes an exothermic reaction to yield high-temperature gases. Examples are nitromethane, methyl acetylene, ethylene oxide, and hydrogen peroxide. Gasoline oxidized by hydrogen peroxide gives a specific impulse of 248, while pentaboranes under pressure and oxidized with hydrogen peroxide give a specific impulse of 363. *ASTM fuel A*, for jet engines, is isooctane, and *ASTM fuel B* is isooctane and toluene. *Turbine jet fuel*, or *JP fuel*, has been the naphtha-based *JP-4* for military aircraft and the kerosene-based *JP-8* for commercial aircraft, with JP-4 having a lower flashpoint, or ignition temperature. However, military aircraft are switching to JP-8 in the interest of greater safety on impact. Besides its higher flashpoint, flame spread of a pool of JP-8 is much slower. The *naphthalenes*, such as *decahydronaphthalene*, have high thermal stability, and they have a high density which gives high thermal energy per unit volume. *Bio Oil*, made from sugarcane bagasse, is a clean fuel for gas turbines, diesels, and boilers.

Sodium boron hydride, a white crystalline solid of composition $NaBH_4$, made by reacting sodium hydride with methyl borate, is also used to produce the bore range for fuels. Any element or chemical which causes spontaneous ignition of a rocket fuel is called a *hypergolic material*.

Chemical radicals are potential high-energy fuels, as the recombining of them produces high specific impulses. But chemical radicals normally exit only momentarily and are thus not stable materials and, in general, are not commercial fuels. *Ion propellants* operate on the principle that like charges repel each other, and the fuel is in ion-plasma jet actually formed outside the engine. The original fuel is a metal such as cesium from which electrons can be stripped by passing the vapor through a hot screen, leaving positive cesium ions, which are formed into a beam and exhausted from the jet thrust to be electronically neutralized in the ionized plasma.

Fugitive Binder

An organic substance added to a metal powder to enhance the bond between the particles during compaction and thereby increase the green strength of the compact, and which decomposes during the early stages of the sintering cycle.

Full Annealing

(1) An imprecise term that denotes an annealing cycle to produce minimum strength and hardness. For the term to be meaningful, the composition and starting condition of the material at the time–temperature cycle used must be stated. (2) Annealing steel by heating into the austenitic region and cooling slowly, usually in the furnace, as opposed to normalizing or subcritical annealing. See *steel*.

Full Center

Mild waviness down the center of a metal sheet or strip.

Full Fillet Weld

A fillet weld whose size is equal to the thickness of the thinner member joined.

Full Hard

A *temper* of nonferrous alloys and some ferrous alloys corresponding approximately to a cold-worked state beyond which the material can no longer be formed by bending. In specifications, a full hard temper is commonly defined in terms of minimum hardness or minimum tensile strength (or, alternatively, a range of hardness or strength) corresponding to a specific percentage of cold reduction following a full anneal. For aluminum, a full hard temper is equivalent to a reduction of 75% from *dead soft*; for austenitic stainless steels, a reduction of about 50%–55%.

Full Journal Bearing

A journal bearing that surrounds the journal by a full 360°.

Full Mold

A trade name for an expendable pattern casting process in which the polystyrene pattern is vaporized by the molten metal as the mold is poured. The metal rises to fill the mold which can be of highly complex form. See the term *lost foam casting*.

Full Width at Half Maximum (FWHM)

A measure of resolution of a spectrum or chromatogram determined by measuring the peak width of a spectral or chromatographic peak at half its maximum height.

Full-Automatic Plating

Electroplating in which the work is automatically conveyed through the complete cycle.

Full-Contour Length

The length of a fully extended polymer chain.

Fuller (Fullering Impression)

Portion of the die used in hammer forging primarily to reduce the cross section and to lengthen a portion of the forging stock. The fullering impression is often used in conjunction with an *edger* (*edging impression*).

Fullerenes

Fullerenes are a family of molecules that contain an even number of carbon atoms in a closed cage. The molecule is a hollow, pure carbon molecule in which the atoms lie at the vertices of a polyhedron with 12 pentagonal faces and any number (other than one) of hexagonal faces. The fullerenes were discovered as a consequence of astrophysically motivated chemical physics experiments that were interpreted by using geodesic architectural concepts. Fullerene chemistry, a field that appears to hold much promise for materials development and other applied areas, was born from pure fundamental science.

Buckminster fullerene (C_{60} or fullerene-60) is the archetypal member of the fullerenes. Other stable members of the fullerene family have similar structures. The fullerenes can be considered, after graphite and diamond, to be the third well-defined allotrope of carbon.

The fullerenes promise to have synthetic, pharmaceutical, and industrial applications. Derivatives have been found to exhibit fascinating electrical and magnetic behavior, in particular superconductivity and ferromagnetism.

Structures

All 60 atoms in fullerene-60 are equivalent and lie on the surface of a sphere distributed with the symmetry of a truncated icosahedron. The molecule was named after R. Buckminster Fuller, the inventor of geodesic domes, which conform to the same underlying structural formula.

Chemistry

Fullerene-60 behaves as a soft electrophile, a molecule that readily accepts electrons during a primary reaction step. It can accept three electrons readily and perhaps even more. The molecule can be multiply hydrogenated, methylated, ammoniated, and fluorinated.

Superconductivity

On exposure of C_{60} to certain alkali and alkaline earth metals, exohedrally doped crystalline materials are produced that exhibit superconductivity at relatively high temperatures (10–33 K).

Previously only metallic and ceramic materials exhibited superconductivity at temperatures much greater than a few kelvins. This discovery has opened the field of superconductivity to a different range of substances—in this case, molecular superconductors.

Nanoparticles and Nanotubes

The discovery that graphite networks (single sheets of hexagonally interconnected carbon atoms) can close readily has revolutionized general understanding of certain types of graphitic materials.

Carbon microparticles can spontaneously rearrange at high temperatures to form onion-like structures in which the concentric shells are fullerene or giant fullerenes. This phenomenon reveals the dynamics of carbon "melting."

Most importantly, carbon nanofibers consisting of concentric graphene tubes can form. These structures are essentially elongated giant fullerenes and apparently form quite readily. They will be significant in the production of carbon-fiber composite materials.

Applications

The properties of fullerene materials that have been determined suggest that there is likely to be a wide range of areas in which the fullerenes or their derivatives will have uses. The facility for acceptance and release of electrons suggests a possible role as a charge carrier in batteries.

Fullerene nanotubes, tiny, tubular carbon fibers, were recently cut into open-ended pipes for the first time. This allows them to be chemically manipulated for use in nanotechnologies and materials. The attachment of molecules to the ends of the pipes lets them serve as means of binding to other chemical groups or surfaces.

The properties of graphite suggest that lubricative as well as tensile and other mechanical properties of the fullerenes are worthy of investigation. Liquid solutions exhibit excellent properties of optical harmonic generation. The high temperature at which superconducting behavior is observed suggests possible applications in microelectronics devices, as does the detection of ferromagnetism in other fullerene derivatives.

Full-Film Lubrication

A type of lubrication wherein the solid surfaces are separated completely by an elastohydrodynamic fluid film. See also *elastohydrodynamic lubrication* and *lubrication regimes.*

Fully Killed Steel

Steel that is fully deoxidized, mainly by silicon but also manganese, aluminum or titanium as appropriate.

Fully Penetrating Weld

A weld in which there is complete fusion through the thickness including the root.

Fulminates

Explosives used in percussion caps and detonators because of their sensitivity. They may be called *cap powder* in cartridge caps and detonators when used for detonating or exploding artillery shells. *Mercury fulminate*, a gray or brown, sandy powder, is the basis for many detonating compositions. It is made by the action of nitric acid on mercury and alcohol and is 10 times more sensitive than picric acid. It may be mixed with potassium chlorite and antimony sulfide for percussion caps. The fulminates may be neutralized with a sodium thiosulfate. The azides are a group of explosives containing no oxygen. They are compounds of hydrogen or a metal and a monovalent N_3 radical. *Hydrogen azide*, HN_3, or *axoic acid*, and its sodium salt are soluble in water. *Lead azide* is used as a substitute for fulminate detonators. It is much more sensitive than mercury fulminate and in large crystals is subject to spontaneous explosion, but it is precipitated as a 93% pure product to suppress crystal formation and to form a free-flowing powder less sensitive to handling. Lead azide detonators for use in coal mining have *copper detonators*; for all other blastings *aluminum caps* are used.

Functional Group

A chemical radical or structure that has characteristic properties; examples are hydroxyl and carboxyl groups.

Functionality

The average number of reaction sites on an individual polymer chain.

Fungus Resistance

The resistance of a material to attack by fungi in conditions promoting their growth.

Furan Resins

Dark-colored thermosetting resins available primarily as liquids ranging from low-viscosity polymers to thick, heavy syrups, which cure to highly cross-linked, brittle substances. Made primarily by polycondensation of furfural alcohol in the presence of strong acids, sometimes in combination with formaldehyde or furfuraldehyde.

Furfural

Also known as furfuraldehyde, furol, and pyromucleadehyde, furfural is a yellowish liquid with an aromatic odor, soluble in water and in alcohol, but not in petroleum hydrocarbons. On exposure, it darkens and gradually decomposes. Furfural occurs in different forms in various plant life and is obtained from complex carbohydrates known as pentosans, which occur in such agricultural wastes as cornstalks, corncobs, straw, oat husks, peanut shells, bagasse, and rice. Furfural is used for making synthetic plastics, as a plasticizer in other synthetic resins, as a preservative in weed killers, and as a selective solvent especially for removing aromatic and sulfur compounds from lubricating oils. It is also used for the making of butadiene, adiponitrile, and other chemicals.

Various derivatives of furfural are not used, and these, known collectively as furans, are now made synthetically from formaldehyde and acetylene, which react to form butyl nedole.

The *Tygon resins* are furfural resins used for brush application as protective coatings for such purposes as chemical tank linings. They cure by self-polymerization, will withstand temperatures to 350°F (177°C), and are resistant to acids, alkalies, alcohols, and hydrocarbons.

Furane plastics have high adhesion and chemical resistance, but they do not have high dielectric strength, and are black or dark in color. They are used for pipe, fittings, and chemical equipment parts and for adhesives and coatings.

Furfural Resins

A dark-colored synthetic resin of the thermosetting variety obtained by the condensation of furfural with phenol or its homologs. It is used in the manufacture of molding materials, adhesives, and impregnating varnishes. Properties include high resistance to acids and alkalies.

Furnace

A vessel or chamber in which a reaction or heat treatment occurs at a considerably elevated temperature, for example, blast furnace and annealing furnace. For lower temperatures "oven" tends to be favored although there are obvious exceptions such as coke oven. There is a large variety of furnaces. Heating may be any fuel including gas, oil, and electricity. The atmosphere may be air, combustion products, controlled (with respect to, for example, its oxidation or carbonization characteristics), or vacuum. The charger may be inserted from above, the end or the side, by hand, by a charging machine or on a rail car hearth. The charger may remain stationary or move forward progressively. See *rotary hearth* and *walking beam.*

Furnace Brazing

A mass-production *brazing* process in which the filler metal is preplaced on the joint, then the entire assembly is heated to brazing temperature in a furnace. Usually, a protective furnace atmosphere is required, and wetting of the joint surfaces is accomplished without using a brazing flux.

Furnace Soldering

A soldering process in which the parts to be joined are placed in a furnace heated to a suitable temperature.

Fuse

(1) A length of wire, or similar, in a suitable insulating and heat-resistant carrier that is installed in electrical circuits to protect the circuit if an excessive current is drawn. The size of the fuse wire is selected to match the maximum continuous current limit. A surge fuse has a similar function but is intended to protect against short-term excursions to high current. (2) Generally, to melt.

Fused Coating

A metallic coating (usually tin or solder alloy) that has been melted and solidified, forming a metallurgical bond to the basis metal.

Fused Silica

A glass made either by flame hydrolysis of silicon tetrachloride or by melting silica, usually in the form of granular quartz.

Fused Spray Deposit

A self-fluxing spray deposit which is deposited by conventional *thermal spraying* and subsequently fused using either a heating torch or a furnace. The coatings are usually made of nickel and cobalt alloys to which hard particles, such as tungsten carbide may be added for increased wear resistance.

Fused Zone

See preferred terms *fusion zone*, *nugget*, and w*eld interface*.

Fusible Alloys

A group of binary, ternary, quaternary, and quinary alloys containing bismuth, lead, tin, cadmium, and indium. The term "fusible alloy" refers to any of more than 100 alloys that melt at relatively low temperatures, that is, below the melting point of tin-lead solder (183°C or 360°F). The melting points of these alloys range as low as 47°C (116°F).

Fusible alloys are those with melting points below the boiling point of water (100°C). They are used as binding plugs in automatic sprinkler systems, for low-temperature boiler plugs, for soldering pewter and other soft metals, for tube bending, and for casting pattern and many ornamental articles and toys. They are also used for holding optical lenses and other parts for grinding and polishing. They consist generally of mixtures of lead, tin, cadmium, and bismuth. The general rule is that an alloy of two metals has a melting point lower than that of either metal alone. By adding still other low-fusing metals to the alloy a metal can be obtained with almost any desired low melting point.

Newton's metal, used as a solder for pewter, contains 50% bismuth, 25% cadmium, and 25% tin. It melts at 95°C, and will dissolve in boiling water. Lipowitz alloy, another early metal, contains 3 parts cadmium, 4 tin, 15 bismuth, and 8 lead. It melts at 70°C, is very ductile, and takes a fine polish. It was employed for casting fine ornaments, but now has many industrial uses. A small amount of indium increases the brilliance and lowers the melting point 1.45°C for each 1% of indium up to a maximum of 18%. Wood's alloy, or Wood's fusible metal, was the first metal used for automatic sprinkler plugs. It contained 7–8 parts bismuth, 4 lead, 2 tin, and 1–2 cadmium. It melts at 71°C, and this point was adopted as the operating temperature of sprinkler plugs in the United States; in England it is 68°C. The alloy designated as Wood's metal contains 50% bismuth, 25% lead, 2.5% tin, and 12.5% cadmium. It melts at 70°C.

Cerrobend, or Bendalloy, is a fusible alloy for tube bending that melts at 71°C. Cerrocast is a bismuth–tin alloy with pouring range of 138°C–170°C, and shrinkage of only 0.0025 cm/cm, used for making pattern molds. Cerrosafe, or Safalloy, is a fusible metal used for toy casting sets, as the molten metal will not burn wood or cause fires. Alloys with very low melting points are sometimes used for this reason for pattern and toy casting. A fusible alloy with a melting point at 60°C contains 26.5% lead, 13.5% tin, 50% bismuth, and 10% cadmium. These alloys expand on cooling and make accurate impressions of the molds.

The Tempil pellets are alloy pellets made with melting points in steps of –10.8°C, 10°C, and 37.7°C for measuring temperatures from 45°C to 1371°C. The Semalloy metals cover a wide range of fusible alloys with various melting points. Semalloy 1010 with a melting point at 47°C can be used where the melting point must be below that of thermoplastics. It contains 45% bismuth, 23% lead, 19% indium, 8% tin, and 5% cadmium. Semalloy 1280, for uses where the desired melting is near the boiling point of water, melts at 96°C. It contains 52% bismuth, 32% lead, and 16% tin.

Fusible Core

A core for injection molding manufactured from a low melting temperature alloy.

Fusible Insert (of Weld)

A preplaced filler having specific dimensions to locate snugly in the root of a joint to be made from one side. It is intended to be fully fused to become an integral part of the joint and is not normally machined following welding. Same as *consumable insert*.

Fusible Plug

A plug in the wall of a pressure vessel that is intended to melt and release the pressure if an excessive temperature is reached.

Fusing

The melting of a metallic coating (usually electrodeposited) by means of a heat-transfer medium, followed by solidification.

Fusion

In a nuclear reaction, the joining of light elements to form heavy elements with associated energy release.

Fusion (Plastics)

In vinyl dispersions, the heating of a dispersion to produce a homogeneous mixture.

Fusion (Welding)

The melting together of filler metal and base metal (substrate), or of base metal only, which results in coalescence. See also *depth of fusion*.

Fusion Cutting/Sawing

Same as *friction cutting/sawing*.

Fusion Face

A surface of the base metal which will be melted during welding.

Fusion Penetration (of Weld)

Same as *depth of fusion*.

Fusion Spray (Thermal Spraying)

The process in which the coating is completely fused to the base metal, resulting in a metallurgically bonded, essentially void free coating.

Fusion Welding

Any welding process in which the filler metal and base metal (substrate), or base metal only, are melted together to complete the weld.

Fusion Zone

The area of base metal melted as determined on the cross section of a weld.

Fuzz

Accumulation of short, broken filaments after passing glass strands, yarns, or rovings over a contact point. Often, weighted and used as an inverse measure of abrasion resistance.

G

Gadolinium

A metallic element, one of the rare earth group.

Gag

A metal spaces inserted so as to render a floating tool or punch inoperative.

Gage

(1) The thickness of sheet or the diameter of wire. The various standards are arbitrary and differ with regard to ferrous and nonferrous products as well as sheet and wire. (2) An aid for visual inspection that enables an inspector to determine more reliably whether the size or contour of a formed part meets dimensional requirements. (3) An instrument used to measure thickness or length. (4) *Gauge* (United States).

Gagger

In foundry practice, an irregularly shaped piece of metal used for reinforcement and support in a sand mold.

Galena

Lead sulfide (PnS) particularly as the naturally occurring ore.

Gall

To damage the surface of a powder-metallurgy compact or die part, caused by adhesion of powder to the die cavity wall or a punch surface.

Galling

(1) A condition whereby excessive friction between high spots results in localized welding with subsequent *spalling* and a further roughening of the rubbing surfaces of one or both of two mating parts. (2) A severe form of scuffing associated with gross damage to the surfaces or failure. Galling has been used in many ways in tribology; therefore, each time it is encountered, its meaning must be ascertained from the specific context of the usage. See also *scoring* and *scuffing*.

Gallium

An elementary metal, symbol Ga, gallium is silvery white, resembling mercury in appearance but having chemical properties more nearly like aluminum. It melts at 30°C and boils at 2403°C, and this wide liquid range makes it useful for high-temperature thermometers. Like bismuth, the metal expands on freezing, and the expansion amounting to about 3.8%. Pure gallium is resistant to mineral acids and dissolves with difficulty in caustic alkali. Commercial gallium

has a purity of 99.9%. In the molten state, it attacks other metals, and small amounts have been used in Pb–Sn solders to aid wetting and decrease oxidation, but it is expensive for this purpose.

Gallium alloys readily with most metals at elevated temperatures. It alloys with tin, zinc, cadmium, aluminum, silver, magnesium, copper, and others. Tantalum resists attack up to 450°C and tungsten to 800°C. Gallium does not attack graphite at any temperature and silicon-based refractories are satisfactory up to about 1000°C.

Applications

Ga–Sn alloy has been used when a low-melting metal is needed. It is also used in rectifiers to operate to 316°C. The material has high electron mobility. This material in single-crystal bars is produced for lasers and modulators.

GaAs is an interesting new material because both gallium and arsenic are available in the state of extreme purity required for semiconductor applications and because the finished gallium arsenide (GaAs) in the proper state of purity can be used in transistors at high frequencies and high temperatures.

Another device using GaAs is a tunnel diode. Basically, a tunnel diode is a heavily doped junction diode that displays a quantum-mechanical tunneling effect under forward bias. This effect leads to an interesting negative resistance effect. Several applications for tunnel diodes are replacement for phase-locked oscillators, switching circuits, frequency-modulated transmitter circuits, and amplifiers.

GaAs is also used in increasing quantities for solar cells. A *paddle wheel* satellite in orbit demonstrates a dramatic commercial application of solar batteries. The NASA Deep Space 1 probe has a unique lens system that will enable its solar panels to generate power from a solar-cell area 1/16th the size of conventional silicon devices. A pair of solar-array wings uses refractive Fresnel lenses to concentrate light onto the cells; thus, less material is required. The panels are critical to the mission. They supply power not only for electronics but for the electric propulsion system of the vehicle as well. Each of the 5232 × 1600 mm GaAs-based array wings will produce 1.3 kW. The light-concentrating Fresnel lenses are silicon with a thin glass coating. Although more expensive per unit area than conventional silicon arrays, the more efficient (at 23%) multijunction GaAs cells further reduce the require solar-cell area, which, in turn, cut spacecraft size and mass. Net result: cost is half that of conventional planar panels.

In addition to GaAs, several gallium compounds have found application in the semiconductor field. GaO has been used for vapor-phase doping of other semiconductor materials, and the oxide and halides have application in epitaxial growth of GaAs and gallium phosphide (GaP). Gallium itself has been used as a dopant for semiconductors. Gallium ammonium chloride has been used in plating baths for the electrodeposition of gallium onto whisker wires used as leads for transistors.

GaSe, GaSc, GaI$_3$, and other compounds are also used in electronic applications.

A new solar cell being developed boasts 50% more power than traditional designs. The new cells, based on the two-junction,

G

Ga–In–P on GaAs technology, also have a longer life and are more resistant to radiation than silicon-based cells. This makes them well suited for communication satellites. (Silicon solar cells lose half their efficiency after 5 years in space.)

Other gallium alloys are suitable as dental alloys, and gallium is used in gold–platinum–indium alloys for dental restoration. Because of its low vapor pressure, gallium is being used as a sealant for glass joints in laboratory equipment, particularly mass spectrometers. Certain alloys (principally with cadmium and zinc) are used as cathodes in specialized vapor-arc lamps. Hard gallium alloys are used as low-resistance contact electrodes for bonding thermocouples and other wires to ferrites and semiconductors.

A new generation of transistors based on gallium nitride promises to deliver up to 100 times more power at microwave frequencies than current semiconductors. Tests have shown gallium nitride transistors with an output of up to 2.2 W/mm at a frequency of 4 GHz. This compares with about 1 W/mm at 10 GHz for GaAs transistors. By combining four devices, it may be possible to make transistors with an output power of 12.5 W/mm, with each device 2 mm long, on a monolithic integrated circuit to make a chip with an output power of 100 W at a frequency of 10 GHz.

The crystal from which the chips are made is grown on a heat sink made of either silicon carbide or sapphire (Al_2O_3). Instead of doping, in which small amounts of another material are added to the crystal, the researchers chose another method. In this technology, a thin layer of gallium aluminum nitride is placed on top of a base of gallium nitride. The bond between the two layers places a strain on the upper layers, which enables free electrons to flow into the gallium nitride layer, producing the holes that make it a semiconductor.

Applications could include military radar, portable satellite phones, and satellite transmitters. The high-power transistors may also save money. Satellites equipped with the devices could operate in higher orbits. This could usher the way for fewer satellites flying higher above the earth to give the same ground coverage.

The principal use of gallium is in the manufacture of semiconducting compounds. More than 90% of the gallium consumed in the United States is used for optoelectronic devices and integrated circuits. Optoelectronic devices—light-emitting diodes (LEDs), laser diodes, photodiodes, and solar (photovoltaic) cells—take advantage of the ability of GaAs to convert electrical energy into optical energy, and vice versa. An LED, which is a semiconductor that emits light when an electric current is passed through it, consists of layers of epitaxially grown material on a substrate. These epitaxial layers are normally gallium aluminum arsenide (GaAlAs), GaAs phosphide (GaAsP), or indium GaAsP (InGaAsP); the substrate material is either GaAs or GaP. Laser diodes operate on the same principle as LEDs, but they convert electrical energy to a coherent light output. Laser diodes principally consist of an epitaxial layer of GaAs, GaAlAs, or InGaAsP on a GaAs substrate. Photodiodes are used to detect a light impulse generated by a source, such as an LED or laser diode, and convert it to an electrical impulse. Photodiodes are fabricated from the same materials as LEDs. GaAs solar cells have been demonstrated to convert 22% of the available sunlight to electricity, compared with about 16% for silicon solar cells.

Although ICs currently represent a smaller share of the GaAs market than optoelectronic devices, they are important for military and defense applications. Two types of ICs are produced commercially: analog and digital. Analog ICs are designed to process signals generated by military radar systems, as well as those generated by satellite communications systems. Digital ICs essentially function as memory and logic elements in computers.

Nonsemiconducting applications include the use of gallium oxide for making single-crystal garnets—such as gallium gadolinium garnet (GGG), which is used as the substrate for magnetic domain (bubble) memory devices. Small quantities of metallic gallium are used for low-melting-point alloys, for dental alloys, and as an alloying element in some magnesium, cadmium, and titanium alloys. Gallium is also used in high-temperature thermometers and as a substitute for mercury in switches. Gallium-based superconducting compounds, such as GaV_3, have also been developed.

Galvanic Anode

A metal that, because of its relative position in the galvanic series, provides *sacrificial protection* to metals that are more noble in the series, when coupled in an electrolyte. See also *cathodic protection*.

Galvanic Cell

(1) A cell in which chemical change is the source of electrical energy. It usually consists of two dissimilar conductors in contact with each other and with an electrolyte or of two similar conductors in contact with each other and with dissimilar electrolytes. (2) A cell or system in which a spontaneous oxidation–reduction reaction occurs, the resulting flow of electrons being conducted in an external part of the circuit. See the term *cathodic protection*.

Galvanic Corrosion

Corrosion associated with the current of a galvanic cell consisting of two dissimilar conductors in electrolyte or two similar conductors in dissimilar electrolytes. Where the two dissimilar metals are in contact, the resulting reaction is referred to as couple action.

Galvanic Couple

A pair of dissimilar conductors, commonly metals, in electrical contact. See also *galvanic corrosion*.

Galvanic Couple Potential

See *mixed potential*.

Galvanic Current

The electric current that flows between metals or conductive nonmetals in a *galvanic couple*.

Galvanic Series

A list of metals and alloys arranged according to their relative corrosion potentials in a given environment. Compare with *electromotive force series*.

Galvanized Steel and Iron

Galvanizing is the process of coating irons and steels with zinc for corrosion protection. The zinc may be applied by immersing the substrate in a bath of the molten metal (hot-dip galvanizing), by electroplating the metal on the substrate (*electrogalvanizing*), or by spraying atomized particles of the metal onto otherwise finished parts. The zinc protects the substrate in two ways: (1) as a barrier to atmospheric attack and (2) galvanically, that is, if the coating is broken, exposing the substrate, the coating will corrode sacrificially or in preference to the substrate.

Both hot-dip galvanizing and electrogalvanizing are continuous processes applied in the production of galvanized steel, and the coating may be applied on one or both sides of the steel. In the case of hot-dip galvanized steel, the zinc at the steel face alloys with about 25% iron from the steel. Iron alloying decreases progressively to a region that is 100% zinc. Electrogalvanized steel typically has a more homogeneous but thinner coating of pure zinc and is somewhat more formidable than the hot-dip variety.

A spangled surface has long been characteristic of traditional hot-dip galvanized steel. Although that effect can be minimized, concern by automakers that the spangles might show through on painted external body panels gave rise to the development of *Zincrometal* in the early 1970s and increased use of electrogalvanized steel.

The American Iron and Steel Institute lists eight types of galvanized steel for auto applications in addition to Zincrometal. Five are of the hot-dip variety: (1) regular and minimum spangle, (2) fully coated but one side having a substantially lower weight, or thickness, of zinc than the other; (4) differentially zinc-iron-coated (same as differentially zinc coated except that the side with a thin coating is heat treated or wiped to produce a fully alloyed zinc–iron coating; and (5) one-side-coated (one side is zinc-free). The three types of electrogalvanized sheet steels are (1) electrolytic flash-coated [0.10–0.20 oz/ft^2 (30–60 g/m^2) on both sides for minimal corrosion protection]; (2) electroplated zinc coated (coated on one side or both sides, the latter with equal differential coating weights, with as much as 0.65 oz/ft^2 (200 g/m^2) total coating) and (3) electroplated iron-zinc coated (coated on one or both sides, the latter with equal or differential coating weights, by simultaneous electroplating of zinc and iron to form an alloy coating).

Galfan

Another two-side-coated hot-dip galvanized steel. The coating with 95% zinc and 5% aluminum and mischmetal is said to provide superior corrosion resistance compared to conventional galvanized steels in rural, marine, and severe-marine atmospheres.

Galvanizing

Galvanizing is the process of coating irons and steels with zinc for corrosion protection. From the standpoint of barrier protection alone, a coating weight of 400 g/m^2 on sheet steel will provide a service life of about 30 years in rural atmosphere and about 5 years in severe industrial atmosphere.

Hot dipping is widely practiced with mild steel sheet for garbage cans and corrugated sheets for roofing, sheeting, culverts, and iron pipe and with fencing wire. The electroplating method is also used for wire, as well as for applications requiring deep drawing. An alloy layer does not form; hence, the smooth electroplated coating does not flake in the drawing die.

It can be achieved by a hot-dip process, that is, immersion in molten zinc at about 450°C beneath a molten zinc chloride flux. The relatively thick coating comprises a zinc outer layer and a zinc iron diffusion layer at the interface. It is strongly bonded to the steel and is particularly suitable for substantial sections for structural applications. Electroplating or electrogalvanizing provides a thin or more accurately controlled coating suitable for smaller precision items. Metal spraying can offer a range of thicknesses and it can be applied on site if necessary, but the adhesion to the steel is critically dependent on the quality of preparation and application. Sherardizing involves baking components in zinc powder at just below the 419°C melting point of zinc to promote the formation of a thin diffusion coating. The term cold galvanizing is sometimes used for the application of paints containing a high proportion of metallic zinc particles. It is claimed that the quantity of zinc and zinc particles and the substrate are in sufficiently good electrical contact to provide some cathodic protection.

Galvanizing Embrittlement

The embrittlement of steel resulting from the heating involved during hot-dip galvanizing at about 450°C. A form of *strain age hardening*.

Galvanneal

To produce a zinc–iron alloy coating on iron or steel by keeping the coating molten after hot-dip galvanizing until the zinc alloys completely with the basis metal.

Galvanometer

An instrument for indicating or measuring a small electric current by means of a mechanical motion derived from electromagnetic or electrodynamic forces produced by the current.

Galvanostatic

An experimental technique whereby an *electrode* is maintained at a constant current in an *electrolyte*.

Gamma

The Greek alphabetical reference to various phases in alloy systems, particularly gamma iron–austenite. See *steel*.

Gamma Iron

The face-centered cubic form of pure iron, stable from 910°C to 1400°C (1670°F to 2550°F).

Gamma Loop

The closed loop of the austenitic (i.e., gamma) region of iron chromium alloys.

Gamma Radiation

The electromagnetic radiation emitted by some radioactive materials as a result of their nuclear activity.

Gamma Ray

A high-energy photon, especially as emitted by a nucleus in a transition between two energy levels. It is similar to x-rays but of a nuclear origin; camera rays have a range of wavelengths from about 0.0005 to 0.14 nm. See the term *electromagnetic radiation*.

Gamma-Ray Spectrometry

See *gamma-ray spectroscopy*.

Gamma-Ray Spectroscopy

Determination of the energy distribution of γ-rays emitted by a nucleus. This is also known as *gamma-ray spectrometry*.

Gamma Structure

Structurally analogous phases or electron compounds having ratios of 21 valence electrons to 13 atoms. This is generally a large, complex cubic structure.

Gamma Transition

See *glass transition*.

Gang Milling

Milling with several cutters mounted on the same arbor or with workpieces similarly positioned for cutting either simultaneously or consecutively during a single setup.

Gang Slitter

A machine with a number of pairs of rotary cutters spaced on two parallel shafts, used for *slitting* metal into strips or for trimming the edges of sheets.

Gangue

The worthless portion of an ore that is separated from the desired part before smelting is commenced.

Ganister

A siliceous sandstone used as an acidic furnace lining.

Gap (Composites)

(1) In filament winding, the space between successive windings in which windings are usually intended to lay next to each other. It is the separation between fibers within a filament winding band. (2) The distance between adjacent plies in a lay-up of unidirectional tape materials.

Gap (in Welding)

(1) The minimum distance between surfaces to be joined as in root gap and the distance between the surfaces at the root. (2) The root opening in a weld joint. (3) The distance between the electrode and the workpiece over which an electric arc is struck, as an air gap.

Gap-Filling Adhesive

An adhesive subject to reduce shrinkage upon setting, used as a sealant.

Gap-Frame Press

A general classification of press in which the uprights or housings are made in the form of a letter C, thus making three sides of the die space accessible.

Garnet

A generic name for a related group of mineral silicates that have the general chemical formula $A_3B_2(SiO_4)_3$, where A is Fe^{2+}, Mn^{2+}, Mg, or Ca and B is Al, Fe^{3+}, Cr^{3+}, or Ti^{3+}. Garnet is used for coating abrasive paper or cloth, bearing pivots in watches, electronics, and the finer specimens for gemstones. The hardness of garnet varies from 6 to 8 Mohs (1360 Knoop), the latter being used for abrasive applications.

Garnets are trisilicates of alumina, magnesia, calcia, ferrous oxide, manganese oxide, or chromic oxide. Garnet-coated paper and cloth are preferred to quartz for the woodworking industries, because garnet is harder and gives sharper cutting edges, but Al_2O_3 is often substituted for garnet.

Synthetic garnets for electronic application are usually rare earth garnets with a rare earth metal substituted for the calcium and iron substituted for the aluminum and the silicate. Yttrium garnet is thus $Yt_3Fe_3(FeO_4)_3$. Yt–Al garnets of 3 mm diameter are used for lasers. Gadolinium garnet has been chosen for microwave use. Gd–Ga garnet made from GdO and GaO is used for computer bubble memories.

Gas

The state of matter in which a material is in a low density, highly fluid, and in a plastically compressible form.

Gas Atomization

An *atomization* process whereby molten metal is broken up into particles by a rapidly moving inert gas stream. The resulting particles are nearly spherical with attached satellites.

Gas Bearing

A journal or thrust bearing lubricated with gas.

Gas Brazing

See the preferred term *torch brazing*.

Gas Carbon Arc Welding

A carbon arc welding process variation that produces coalescence of metals by heating them with an electric arc between a single-carbon electrode and the work. Shielding is obtained from a gas or gas mixture.

Gas Carburizing

See *case hardening*.

Gas Chromatography

A separation method involving passage of a gaseous mobile phase through a column containing a stationary adsorbent phase; used principally as a quantitative analytical technique for volatile compounds. See also *chromatography*, *ion chromatography*, and *liquid chromatography*.

Gas Classification

The separation of a powder into its particle size fractions by means of a gas stream of controlled velocity flowing counterstream to the gravity-induced fall of the particles. The method is used to classify submesh-size particles.

Gas Classifier

A device for gas classification; it may be of laboratory size for quality control testing or of industrial capacity to accommodate powder production requirement.

Gas Constant

The constant of proportionality appearing in the equation of state of an ideal gas, equal to the pressure of the gas multiplied by its molar volume divided by its temperature. This is also known as *universal gas constant*. The constant, *R*, for 1 mol of gas is given by

$$pV = RT$$

where
  p is the pressure
  V is the volume
  T is the temperature, absolute

In SI units, $R = 8.31$ J/K/mol.

Gas Cutting

The use of oxygen/fuel gas torches, frequently oxyacetylene, for cutting metal, primarily steel. The general principle is that initially an intense gas/fuel flame heats the steel locally to its ignition temperature, 750°C–825°C, and then an additional large quantity of high-pressure oxygen is introduced to cause combustion of the steel. The temperature is sufficient to produce a molten oxide slag that is vigorously ejected by the gas stream giving, under suitable conditions, a clean square edged cut. The equipment for manual use is somewhat similar to that used for gas welding except that the burner head has a pair of concentric nozzles, the outer providing the initial mixed fuel and the inner delivering the main oxygen blast. This main oxygen blast is usually controlled by a lever valve beneath the handgrip. The burner head, or multiple heads, may also be machine mounted and programmed to cut components of complex form with, if necessary, beveled edges. These machines are often termed profile cutters. See the preferred term *oxygen cutting*.

Gas Cyaniding

A misnomer for *carbonitriding*.

Gaseous Corrosion

Corrosion with gas as the only corrosive agent and without any aqueous phase on the surface of the metal. Also called *dry corrosion*. See also *hot corrosion* and *sulfidation*.

Gaseous Reduction

(1) The reaction of a metal compound with a reducing gas to produce the metal. (2) The conversion of metal compounds to metallic particles by the use of a reducing gas.

Gas Gouging

See the preferred term *oxygen gouging*.

Gas Holes

Holes in castings or welds that are formed by gas escaping from molten metal as it solidifies. Gas holes may occur individually, in clusters, or throughout the solidified metal.

Gasket Materials

These are any sheet material used for sealing joints between metal parts to prevent leakage, but gaskets may also be in the form of cordage or molded shapes. The simplest gaskets are waxed paper or thin copper. A usual requirement is the material will not deteriorate by the action of water, oils, or chemicals. Gasket materials are usually marketed under trade names. There are sheets of paper or fiber, 0.025–0.318 cm thick, coated to withstand oils and gasoline.

To resist high heat and pressure, there are sheets of metal coated with graphited asbestos, with a sheet metal punched with small tongues to hold the asbestos. There is felt impregnated with zinc chromate to prevent corrosion and electrolysis between dissimilar metal surfaces.

Foamed synthetic rubbers in sheet form, and also plastic impregnants, are widely used for gaskets. Some of the specialty plastics, selected for heat resistance or chemical resistance, are used alone or with fillers or as binders for fibrous materials.

A gasketing sheet to withstand hot oils and super octane gasolines is based on *Viton*, a copolymer of vinylidene fluoride and hexafluoropropylene. It contains about 65% fluorine, has a tensile strength of 13 MPa with elongation of 400%, and will withstand operating temperatures to 204°C with intermittent temperatures to 316°C.

Since the decline in the use of asbestos as a gasket material, several high-temperature alternatives have been developed. Aramid fibers, such as Akzo Fibers' *Twaron* and DuPont's *Kevlar*, contained within a nitrile–butadiene binder provide gaskets that will withstand temperatures up to 450°F (232°C). These products are also available coated with tetrafluoroethylene for use as braided packings. *Carbon-fiber gaskets* with the same binder will resist temperatures to 800°F–900°F (427°C–482°C). For temperatures 1100°F–1200°F (593°C–649°C), gaskets of graphite flakes compressed without a binder are used. *Grafoil* gaskets are made of *flexible graphite*. Intended for high-temperature uses, they also resist fire, acids, alkalis, salt solutions, halogens, and various organics. The *G-9900* gaskets rely on graphite fibers for heat resistance to 1004°F (540°C). They also resist saturated steam, hot oils, gasoline, aliphatic gases, and hydrocarbons. *Flexi-Braid 5000* gaskets are based on graphite ribbon braided yarn. They are used for packings in pumps and valves.

Gas Lubrication

A system of lubrication at which the shape and relative motion of the sliding surfaces cause the formation of a gas film having sufficient pressure to separate the surfaces. See also *pressurized gas lubrication*.

Gas Mass Spectrometry

An analytical technique that provides quantitative analysis of gas mixtures through the complete range of elemental and molecular gases.

Gas Metal Arc Cutting

An arc-cutting process used to sever metals by melting them with the heat of an arc between a continuous metal (consumable) electrode and the work. Shielding is obtained internally from an externally supplied gas or gas mixture. This is also known as *metal inert gas welding*.

Gas Metal Arc Welding (GMAW)

An arc welding process that produces coalescence of metals by heating them with an arc between a continuous filler metal (consumable) electrode and the work. Shielding is obtained entirely from an externally supplied gas or gas mixture. Variations of the process include short-circuit arc GMAW, in which the consumable electrode is deposited during repeated short circuits, and pulsed arc GMAW, in which the current is pulsed. See also *globular transfer*, *short-circuiting transfer*, and *spray transfer*.

Gasoline

Gasoline is a colorless liquid hydrocarbon obtained in the fractional distillation of petroleum. It is used chiefly as motor fuel, but also as a solvent. Ordinary gasoline consists of the hydrocarbons between C_6H_{14} and $C_{10}H_{22}$, which distill off between the temperatures 69°C and 174°C, usually having the light limit at heptane, C_7H_{16}, or octane, C_8H_{18}. The octane number is the standard of measure of detonation in the engine. Motor fuel, or the general name gasoline, before the wide use of high-octane gasolines obtained by catalytic cracking meant any hydrocarbon mixture that could be used as a fuel in an internal combustion engine by spark ignition without being sucked in as a liquid and without being so volatile as to cause imperfect combustion and carbon deposition. These included also mixtures of gasoline without call or benzol.

Gas Plating

Same as *vapor plating*.

Gas Pocket/Hole/Porosity

A cavity caused by entrapped gas during solidification.

Gas Pore (of Weld)

Relatively small internal voids formed by the entrapment of gases during solidification of the weld metal. Voids larger than about 1.5 mm are usually termed *weld zone* from contamination. Note that, strictly, this excludes processes in which shielding gases are produced by materials on, or in, the electrodes. Also see, *arc welding* and *gas metal arc welding*.

Gas Porosity

Five holes or pores within a metal that are caused by entrapped gas or by the evolution of dissolved gas during solidification.

Gas Regulator

See the preferred term *regulator*.

Gas Shielded Arc Welding

A general term used to describe gas metal arc welding, gas tungsten arc welding, and flux cored arc welding when gas shielding is employed. Typical gases employed include argon, helium, argon–hydrogen mixture, or carbon dioxide.

Gas Shielded Stud Welding

See *stud arc welding*.

Gassing

(1) Absorption of gas by a metal. (2) Evolution of gas from a metal during melting operations or upon solidification. (3) Evolution of gas from an electrode during electrolysis.

Gas Torch

See the preferred terms *welding torch* and *cutting torch*.

Gas Tungsten Arc Cutting

An arc-cutting process in which metals are severed by melting them with an arc between a single tungsten (nonconsumable) electrode and the work. Shielding is obtained from a gas or gas mixture.

Gas Tungsten Arc Welding (GTAW)

An arc welding process that produces coalescence of metals by heating them with an arc between a tungsten (nonconsumable) electrode and the work. Shielding is obtained from a gas or gas mixture. Pressure may or may not be used and filler metal may or may not be used.

Gas Welding

Any fusion welding process in which the heat source is provided by the combustion of a fuel gas with oxygen including air. Acetylene is the most common fuel gas but alternatives include propane, butane, coal gas, and hydrogen. In a typical handheld gas torch or blowpipe, the two gases are mixed in the burner head and exit through a nozzle. A pair of screw valves to control gas supply is located close to the grip that is separated from the burner by metal tubing of length sufficient to avoid overheating the grip area and the operator. The torch is connected by substantial flexible rubber tubes to the gas supplies that may be cylinders or some more permanent installation. Additional filler may be introduced to the joint by a plain metal rod manipulated by the second hand of the operator.

See the preferred term *oxyfuel gas welding*.

Gate (Casting)

The portion of the runner in a mold through which molten metal enters the mold cavity. The generic term is sometimes applied to the entire network of connecting channels that conduct metal into the mold cavity. See the term *gating system*.

Gate (Plastics)

In injection and transfer molding of plastics, the orifice through which the melt enters the mold cavity. The gate can have a variety of configurations, depending on product design.

Gated Pattern

In foundry practice, a *pattern* that includes not only the contours of the part to be cast but also the *gates*.

Gate Mark

A surface discontinuity on a molded plastic part caused by the gate through which material enters the cavity.

Gathering

A forging operation that increases the cross section of part of the stock, usually a preliminary operation.

Gathering Stock

Any operation whereby the cross section of a portion of the forging stock is increased beyond its original size.

Gating System

The completed assembly of sprues, runners, and gates in a mold through which metal flows to enter casting cavity. The term is also applied to equivalent portions of the *pattern*.

Gauge (Gage in the United States)

(1) A measuring device. It may be an instrument or other device with some system for reading off dimensions, pressure, etc., or it may be a simple precision shape, for example, a feeler gauge. (2) A dimension such as thickness of sheet, wall thickness of tube, diameter of wire, and bar or spacing between tracks. There are numerous national and industry systems relating gauge numbers or letters to dimensions.

Gauge Length

The parallel length of a tensile or creep specimen on which elongation is measured.

Gauge Length

The original length of that portion of the specimen over which strength, change of length, and other characteristics are measured.

Gauss

The unit in the non–SI metric system of magnetic flux density defined as unit magnetic pole subjected to a force of $1°$. 1 G (gauss) = 10^{-4} T (tesla).

Gear Cutting

Producing tooth profiles of equal spacing on the periphery, internal surface, or face of a workpiece by means of an alternate shear gear-form cutter or a gear generator.

Geared Press

A press whose main crank or eccentric shaft is connected by gears to the driving source.

Gear Hobbing

Gear cutting by use of a tool resembling a worm gear in appearance, having helically cutting teeth. In a single-thread hob, the rows of teeth advance exactly one pitch as the hob makes one revolution. With only one hob, it is possible to cut interchangeable gears of a given pitch of any number of teeth within the range of the hobbing machine.

Gear Milling

Gear cutting with a milling cutter that has been formed to the shape of the tooth space to be cut. The tooth spaces are machined one at a time.

Gear Shaping

Gear cutting with a reciprocating gear-shaped cutter rotating in mesh with the work blank.

Gear Shaving

A finishing operation performed with a serrated rack or gear-like cutter in mesh with the gear, but with their axis skewed.

Geiger Counter

A device for detecting ionizing radiation. Various systems are employed but the common feature is that each ionizing event that is detected produces an electrical signal that may be recorded or emitted as noise.

Gel

A gel is a continuous solid network and enveloped in a continuous liquid phase; the solid phase typically occupies less than 10 vol% of the gel. Gels can be classified in terms of the network structure. The network may consist of agglomerated particles formed, for example, by destabilization of a colloidal suspension, a *house of cards* consisting of plates (as in a clay), fiber polymers joined by small crystalline regions, and polymers linked by covalent bonds.

In a gel, the liquid phase does not consist of isolated pockets, which is continuous. Consequently, salts can diffuse into the gel almost as fast as they disperse in a dish of free liquid. Thus, the gel seems to resemble a saturated household sponge, but it is distinguished by its colloidal size scale: the dimensions of the open spaces and of the solid objects constituting the network are smaller (usually much smaller) than a micrometer. This means that the interface joining the solid and liquid phases has an area on the order of 1000 m/g of solid. As a result, the properties of a gel are controlled by interfacial and short-range forces, such as van der Waals, electrostatic, and hydrogen-bonding forces. Factors that influence these forces, such as introduction of salts or another solvent, application of an electric field, or changes in pH or temperature, affect the interaction between the solid and liquid phases. Variations in these parameters can induce huge changes in volume as the gel imbibes or expels liquid, and this phenomenon is exploited to make mechanical actuators or hosts for controlled release of drugs from gels. For example, a polyacrylamide gel (a polymer linked by covalent bonds) shrinks dramatically when it is transferred from a dish of water (a good solvent) to a dish of acetone (a poor solvent), because the polymer chains tend to favor contact with one another rather than with acetone, so the network collapses onto itself. Conversely, the reason that water cannot be gently squeezed out of such a gel (as from a sponge) is that the network has a strong affinity for the liquid, and virtually all of the molecules of the liquid are close enough to the solid–liquid interface to be influenced by those attractive forces.

Gelation

The most striking feature of a gel, which results from the presence of a continuous solid network, is elasticity: if the surface of a gel is displaced slightly, it springs back to its original position. If the displacement is too large, gels (except those with polymers linked by covalent bonds) may suffer some permanent plastic deformation, because the network is weak. The process of gelation, which transforms a liquid into an elastic gel, occurs with no change in color or transparency and no evolution of heat. It may begin with a change in pH that removes repulsive forces between the particles in a colloidal suspension or a decrease in temperature that favors crystallization of a solution of polymers or the initiation of a chemical reaction that creates or links polymers.

In many cases, both the liquid and solid phases of a gel are of practical importance (as in timed release of drugs or a gelatin dessert). More often (as in preparation of catalyst supports, chromatographic columns, or desiccants), it is only the solid network that is of use. If the liquid is allowed to evaporate from a gel, large capillary tension develops in the liquid, and the suction causes shrinkage of the network. The poorest network remaining after evaporation of the liquid is called a xerogel.

To maximize the ferocity of the dried product, the gel can be heated to a temperature and pressure greater than the critical point of the liquid phase, where capillary pressures do not exist. The fluid can then be removed with little or no shrinkage occurring. The resulting solid is called an aerogel, and the process is called supercritical drying.

Sol–Gel Processing

Sol–gel processing comprises a variety of techniques for preparing inorganic materials by starting with a sol, then gelling, drying, and (usually) firing. Many inorganic gels can be made from solutions of salts or metalorganic compounds and thus offers several advantages in ceramics processing: the reactants are readily purified; the compounds can be intimately mixed in the solution or sol stage; the sols can be applied as coatings, drawn into fibers, emulsified or spray-dried to make particles, or molded and gelled into shapes; xerogels can be sintered into dense solids at relatively low temperatures, because of their small pore size.

Gel (Polymers)

The initial jellylike solid phase that develops during the formation of a resin from a liquid. With respect to vinyl plastisols, a state between liquid and solid that occurs in the initial states heating, or upon prolonged storage. In general, gels have very low strengths and do not flow like a liquid. They are soft and flexible and may rupture under their own weight unless supported externally. In a cross-link thermoplastic, gel is the fraction of polymeric material present in the network.

Gelatin

A colorless to yellowish, water-soluble, tasteless colloidal hemicellulose obtained from bones or skins and used as a dispersing agent, sizing medium, coating for photographic films, and stabilizer for foodstuffs and pharmaceutical preparations. It is also flavored for use as a food jelly, and it is a high-protein, low-calorie foodstuff. While albumin has a weak, continuous molecular structure that is cross-linked and rigidized by heating, gelatin has an ionic or hydrogen bonding in which the molecules are brought together into large aggregates, and it sets to a firmer solid. Gelatin differs from glue only in the purity. *Photographic gelatin* is made from skins. *Vegetable gelatin* is not true gelatin, but is algin from seaweed.

Collagen is the gelatin-bearing protein in bone and skins. The bone is dissolved in hydrochloric acid to separate out the calcium phosphate and is washed to remove the acid. The organic residue is called *osseine* and is the product used to produce gelatin and glue. About 25% of the weight of the bone is osseine, and the gelatin yield is about 65% of the osseine. One short ton (0.9 metric ton) of green bones, after being degreased and dried, yields about 300 lb (136 kg) of gelatin. When skins are used, they are steeped in a weak acid solution to swell the tissues so that the collagen may be washed out. The gelatin is extracted with hot water, filtered, evaporated, dried, and ground or flaked.

Gelatin Replica

A reproduction of a surface prepared in a film composed of gelatin. See also *replica*.

Gelation Time

(1) That interval of time, in connection with the use of synthetic thermosetting resins, extending from the introduction of a catalyst into a liquid adhesive system until the start of gel formation. (2) The time under application of load for a resin to reach a solid state.

Gel Coat

A quick-setting resin applied to the surface of a mold and gelled before *lay-up*. The gel coat becomes an integral part of the finished laminate and is usually used to improve surface and bonding.

Gelling Agent

See *thickener*.

Gel-Permeation Chromatography

See *size-exclusion chromatography*.

Gel Point

(1) The point at which a thermosetting system attains an infinite value of its average molecular weight. (2) The viscosity at which a liquid begins to exhibit pseudoelastic properties. This stage may be conveniently observed from the inflection point on a viscosity–time plot.

Gel Time

The period of time from the initial mixing of the reactants of a liquid material composition to the point in time when *gelation* occurs, as defined by a specific test method.

General Corrosion

(1) A form of deterioration that is distributed more or less uniformly over a surface. (2) *Corrosion* dominated by uniform thinning that proceeds without appreciable localized attack. See also *uniform corrosion*. (3) Corrosion acting over most or all of a specified area of

a component and producing a fairly even attack with no significant pitting and no associated cracking.

General Precipitate

A precipitate that is dispersed throughout the metallic matrix.

Geodesic

The shortest distance between two points on a surface.

Geodesic Isotensoid

Constant stress level in any given *filament* at all points in its path.

Geodesic-Isotensoid Contour

In filament-wound reinforced plastic pressure vessels, a dome contour in which the filaments are placed on geodesic paths so that the filaments exhibit uniform tensions throughout their length under pressure loading.

Gerber Diagram

See *fatigue*.

German Silver

Various alpha, single-phase alloys typified by 52% copper, 26% zinc, and 22% nickel with no silver content.

Germanium

A rare elemental metal, germanium (Ge) has a grayish white crystalline appearance and has great hardness of 6.25 Mohs. Its specific gravity is 5.35, and its melting point is 937°C. It is resistant to acids and alkalies. It has metallic-appearing crystals with diamond structure, gives greater hardness and strength to aluminum and magnesium alloys, and as little as 0.35% in tin will double the hardness. It is not used, commonly in alloys, however, because of its rarity and great cost. It is used chiefly as metal in rectifiers and transistors. An Au–Ge alloy, with about 12% germanium, has a melting point of 359°C and has been used for soldering jewelry.

Germanium is obtained as a by-product from flue dust of the zinc industry, or it can be obtained by reduction of its oxide from the ores and is marketed in small irregular lumps. Germanium crystals are grown in rods up to 3.49 cm in diameter for use in making transistor wafers. High-purity crystals are used for both *P* and *N* semiconductors. They are easier to purify and have a lower melting point than other semiconductors, specifically silicon.

Alloys

Two alloys of importance to the semiconductor industry are Ge–Al and Ge–Au. At 55 wt.% germanium, the Ge–Al system forms a eutectic that melts at 423°C. At 12 wt.% germanium, the Ge–Au system forms a eutectic that melts at 356°C. The two alloys are very useful in forming electrical contact systems for germanium transistors, diodes, and rectifiers. The Ge–Au system has been evaluated regarding its use for dental alloys because of its good dimensional stability upon cooling, thus allowing precision castings to be made.

The Ge–Si alloy system forms a continuous series of solid solutions. It has been explored quite extensively from the standpoint of its semiconductor characteristics.

Germanium has been investigated as a possible alloying agent for zirconium in the development of corrosion-resistant, high-strength zirconium alloys. Small additions of germanium are known to give increased hardness and strength to copper, aluminum, and magnesium. In addition, small quantities of germanium improve the rolling properties of some alloys and do not create any appreciable increase in production costs. Fundamental studies on the magnetic properties of Fe–Ge and Mn–Ge alloys have been made. These materials are ferromagnetic, as are the alloys UGe_2, PuGe, and CrGe.

Uses

The properties of germanium are such that there are several important applications for this element.

Semiconductor

During World War II, germanium was intensively investigated for its use in the rectification of microwaves for radar applications.

A major development in electronics occurred in 1948 with the invention of the transistor. This solid-state device, which was first made of germanium, had a profound influence on all electronic applications. The transistor captured the hearing aid and radio markets and then moved into industrial applications, such as computers and guidance and control systems for missile and antimissile systems. Transistors cover signal processing from dc to gigahertz frequencies, with power-handling capabilities from microwatts to hundreds of watts.

The development of germanium semiconductor technology quite closely parallels the development of the germanium transistor. In fabricating a transistor, a need arose for high-purity polycrystalline germanium to be used in the growth of single-crystal germanium since it was soon discovered that not only purity but also good crystal structure was required for good device performance. The growth processes for single crystals of germanium had to be developed. Several techniques have been extensively studied. Two of the most common are the Czochralski or Teal–Little method and zone leveling. The general method proceeds by allowing germanium to slowly freeze or crystallize onto a single crystal seed, which may be rotated and withdrawn from the melt. The growth (freeze) rate is controlled by a combination of the temperature of the melt and the amount of heat lost from the crystal by conduction up the seed and by radiation from its surfaces. Crystals from 0.16 to 15 cm in diameter have been grown by this process. Growth rates vary with diameter and desired crystal properties but are usually in the inches-per-hour range.

In the zone-leveling technique, or horizontal method, the seed and polycrystalline charge or ingot are usually loaded into a quartz boat. The boat and contents are then placed into the quartz tube of the zone leveler. With this technique, melting occurs on the leading edge of the molten zone, and freezing occurs on the trailing edge as the furnace is moved. Growth rates vary with crystal properties desired but are usually in the inches-per-hour range. The advantage of this process over the Teal–Little for germanium is the extremely uniform resistivity profile of the crystal due to the uniform mass of molten germanium and the growth process. The constant mass of molten germanium enables the ratio of dopant to germanium to remain the same, thus producing uniform resistivity.

In the fabrication of germanium devices, the single-crystal material must be sawed, lapped, or ground and then polished into thin-slice forms with flat, damage-free surfaces. Various techniques have been developed to produce slices with these characteristics. Some of these are slicing with diamond wheels, grinding with diamond, or lapping with an abrasive slurry. These techniques induce damage in the form of microcracks and fissures, which must be removed by chemical, electrochemical, or chemical–mechanical polishing methods.

Two other growth techniques (modifications of melt growth) have been explored. One is dendritic growth, which depends on the growth characteristics of twinned crystals, and the other in shaped crystal growth, which is dependent on a shaped heat zone. Both techniques can produce thin ribbons of germanium with desired characteristics for some type of devices.

Vapor-phase crystal growth at temperatures well below the melting point is used for growing thin films or layers onto slices and is usually referred to as epitaxial growth. The primary advantage of gas-phase growth is that it allows doping impurities to be changed quite rapidly so that thin (micrometer range) layers of quite different resistivities can be grown sequentially.

Optoelectronics

Another use that surfaced as a result of the cost of material such as gallium arsenide and the inability to control its growth is the use of Czochralski- or Teal–Little-grown single-crystal germanium as a substrate for vapor-phase growth of GaAs and gallium arsenide phosphide (GaAsP) thin films used in some light-emitting diodes (LEDs). These devices have been used in digital displays for calculators, watches, and so on.

Infrared Optical Materials

Germanium lenses and filters have been used in instruments that operate in the infrared region of the spectrum. Windows and lenses of germanium are vital components of some laser and infrared guidance or detection systems. Glasses prepared with germanium dioxide have a higher refractivity and dispersion than do comparable silicate glasses and may be used in wide-angle camera lenses and microscopes.

Getter

(1) A special metal or alloy that is placed in a vacuum tube during manufacture and vaporized after the tube has been evacuated; when the vaporized metal condenses, it absorbs residual gases. (2) In powder metallurgy, a substance that is used in a sintering furnace for the purpose of absorbing or chemically binding elements or compounds from the sintering atmosphere that are damaging to the final product. (3) A reactive material enclosed in sealed vessels such as thermionic valves to scavenge the last trace of oxygen or other undesirable gases. The getter material forms, or is located, on a filament that is heated to incandescence after the vessel has been sealed. Various metals are used including magnesium, cerium, zirconium, and titanium.

Gettering Box

In powder metallurgy, a container for the getter substance that is readily accessible to the atmosphere and prevents contamination of the sintered product by direct contact.

Ghost Bands

Bands of ferrite within the ferrite/pearlite matrix of some wrought pearlitic steels. They are residual effect of casting segregation and usually contain high phosphorus.

Ghost Lines

Lines running parallel to the rolling direction that appear in a sheet-metal panel when it is stretched. These lines may not be evident unless the panel has been sanded or painted. Not to be confused with *leveler lines*.

Gibbs Free Energy

The thermodynamic function $\Delta G = \Delta H - T\Delta S$, where H is the enthalpy, T is the absolute temperature, and S is the entropy. This is also called *free energy*, free enthalpy, or the Gibbs function.

Gibbs Triangle

An equilateral triangle used for plotting composition in a ternary system.

Gibs

Guides or shoes that ensure the proper parallelism, squareness, and sliding fit between metal-forming press components such as the slide and the frame. They are usually adjustable to compensate for wear and to establish operating clearance.

Gilding

The application of gold leaf or sheet to a surface for decorative effect.

Gilding Metal

Brass with 90% copper and 10% zinc and no gold. It is similar in color to gold and is usually used in wrought bulk form rather than as a surface coating. It can also be treated to produce a brownish tint similar in appearance to bronze.

Gilsonite

A natural asphalt used for roofing, paving, floor tiles, storage battery cases, and coatings and for adding to heavy fuel oils. It is a lustrous, black, almost odorless, brittle solid, having a specific gravity of 1.10. Gilsonite is one of the purest asphalts and has high molecular weight. It is soluble in alcohol, turpentine, and mineral spirits.

Gilt

Silver or other metal that has been gilded, that is, surface with a thin layer of gold.

Girder

A beam for structural purposes, typically of I section to give maximum resistance to bending for a given weight, usually steel.

Glancing Angle

In materials characterization, the angle (usually small) between an incident x-ray beam and the surface of the specimen.

Glass

Glass, one of the oldest and most extensively used materials, is made from the most abundant of earth's natural resources—silica sand. For centuries considered as a decorative, fragile material suitable for only glazing and art objects, today glasses produced in thousands of compositions and grades for a wide range of consumer and industrial applications.

Composition and Structure

As just stated, the basic ingredient of glasses is silica (silicon dioxide), which is present in various amounts, ranging from about 50% to almost 100%. Other common ingredients are oxides of metals, such as lead, boron, aluminum, sodium, and potassium.

Unlike most other ceramic materials, glass is noncrystalline. In its manufacture, a mixture of silica and other oxides is melted and then cooled to a rigid condition. The glass does not change from a liquid to a solid at a fixed temperature but remains in a vitreous, noncrystalline state and is considered as a supercooled liquid. Thus, because the relative positions of the atoms are similar to those of liquids, the structure of glass has short-range order. However, glass has some distinct differences compared to a supercooled liquid. Glass has a 3D framework and the atoms occupy definite positions. There are covalent bonds present that are the same as those found in many solids. Therefore, there is a tendency toward an ordered structure in that there is present in glass a continuous network of strongly bonded atoms.

Properties and Processing

Glass is an amorphous solid made by fusing silica (silicon dioxide) with a basic oxide. Its characteristic properties are its transparency, its hardness and rigidity at ordinary temperatures, its capacity for plastic working at elevated temperatures, and its resistance to weathering and to most chemicals except hydrofluoric acid. The major steps in producing glass products are (1) melting and refining, (2) forming and shaping, (3) heat treating, and (4) finishing. The mixed batch of raw materials, along with broken or reclining glass, called cullet, is fed into one end of a continuous-type furnace where it melts and it remains molten at around 1499°C. Molten glass is drawn continuously from the furnace and runs in troughs to the working area, where it is drawn off for fabrication at a temperature of about 999°C. When small amounts are involved, glass is melted in pots.

Types of Glass

There are a number of general families of glasses, some of which have many of hundreds of variations in composition.

Soda–Lime Glasses

The soda–lime family is the oldest, lowest in cost, easiest to work, and most widely used. It accounts for about 90% of the glass used in this country. Soda–lime glasses have only fair to moderate corrosion resistance and are useful up to about 460°C, annealed, and up to 249°C, tempered. Thermal expansion is high and thermal shock resistance is low compared with other glasses. They are the glass of ordinary windows, bottles, and tumblers.

Lead Glasses

Lead or lead alkali glasses are produced with lead contents ranging from low to high. They are relatively inexpensive and are noted for

high electrical resistivity and high refractory index. Corrosion resistance varies with lead content, but they are all poor in acid resistance compared with other glasses. Thermal properties also vary with lead content. The coefficient of expansion, for example, increases with lead content. High lead grades are heaviest of the commercial glasses. As a group, lead classes are the lowest in rigidity. They are used in many optical components, for neon sign tubing, and for electric light bulb stems.

Borosilicate Glasses

Borosilicate glasses are most versatile and are noted for their excellent chemical durability, for resistance to heat and thermal shock, and for low coefficients of thermal expansion. There are six basic kinds. The low-expansion type is best known as the Pyrex brand ovenware. The low-electrical-loss types have a dielectric loss factor only second to fused silica and some grades of 96% silica glass. Sealing types, including the well-known Kovar, are used in glass-to-metal sealing applications. Optical grades, which are referred to as crowns, are characterized by high light transmission and good corrosion resistance. Ultraviolet-transmitting and laboratory apparatus grades are two other borosilicate-type glasses. Because of this wide range of types and compositions, borosilicate glasses find use in such products as sites and gauges, piping, seals to low-expansion metals, telescope mirrors, electronic tubes, laboratory glassware, ovenware, and pump impellers.

Aluminosilicate Glasses

These glasses are roughly three times more costly than borosilicate types, but are useful at higher temperatures and have greater thermal shock resistance. Maximum service temperature for annealed condition is about 649°C. Corrosion resistance to weathering, water, and chemicals is excellent, although acid resistance is only fair compared with other glasses. Compared to 96% silica glass, which it resembles in some respects, it is more easily worked and is lower in cost. It is used for high-performance power tubes, traveling wave tubes, high-temperature thermometers, combustion tubes, and stovetop cookware.

Fused Silica

Fused silica is 100% silicon dioxide. If naturally occurring, the glass is known as fused quartz. There are many types and grades of both glasses, depending on impurities present and manufacturing method. Because of its high purity level, fused silica is one of the most transparent glasses. It is also the most heat resistant of all glasses and it can be used up to 899°C in continuous service and to 1260°C for short-term exposure. In addition, it has outstanding resistance to thermal shock, maximum transmittance to ultraviolet, and excellent resistance to chemicals. Unlike most glasses, its modulus of elasticity increases with temperature. Because fused silica is high in cost and difficult to shape, its applications are restricted to such specialty applications as laboratory optical systems and instruments and crucibles for crystal growing. Because of the unique ability to transmit ultrasonic elastic waves with little distortion or absorption, fused silica is used in the delay lines in radar installations.

96% Silica Glass

These glasses are similar in many ways to fused silica. Although less expensive than fused silica, they are still more costly than most other glasses. Compared to fused silica, they are easier to fabricate and have a slightly higher coefficient of expansion, about 30% lower thermal stress resistance, and a lower softening point. They can be used continuously up to 799°C. Uses include chemical glassware and windows and heat shields for space vehicles.

Phosphate Glass

Phosphate glass will resist the action of hydrofluoric acid and fluorine chemicals. It contains no silica, but is composed of P_2O_5 with some alumina and magnesia. It is transparent and can be worked like ordinary glass, but it is not resistant to water.

Sodium-Aluminosilicate Glass

These glasses are chemically strengthened and are used in premium applications, such as aircraft windshields. Molten salt baths are used as the strengthening process.

Industrial Glass

This is a general name usually meaning any glass molded into shape for product parts. The lime glasses are the most frequently used because of low cost, ease of molding, and adaptability to fired colors. For such uses as light lenses and condenser cases, the borosilicate heat-resistant glasses may be used.

Boric Oxide Glasses

These glasses are transparent to ultraviolet rays. The so-called invisible glass is a borax glass surface treated with a thin film of sodium fluoride. It transmits 99.6% of all visible light rays, thus casting back only slight reflection and giving the impression of invisibility. Ordinary soda and potash glasses will not transmit ultraviolet light. Glass containing 2%–4% ceric oxide absorbs ultraviolet rays and is also used for x-ray shields. Glass capable of absorbing high-energy x-rays or gamma rays may contain tungsten phosphate, while the glass used to absorb slow neutrons in atomic-energy work contains cadmium borosilicate with fluorides. The shields for rocket capsule radio antennas are made of 96% silica glass.

Optical Glass

Optical glass is a highly refined glass that is usually a flint glass of special composition, or made from rock crystal, used for lenses and prisms. It is cast, rolled, or pressed. In addition to the regular glassmaking elements, silica and soda, optical glass contains barium, boron, and lead. The high-refracting glasses contain abundant silica or boron oxide. A requirement of optical glasses is transparency and freedom from color.

Plate Glass

This is any glass that has been cast or rolled into a sheet and then ground and polished. However, the good grades of plate glass are, next to optical glass, the most carefully prepared and the closest to perfect of all the commercial glasses. It generally contains slightly less calcium oxide and slightly more sodium oxide than window glass, and small amounts of agents to give special properties may be added, such as agents to absorb ultraviolet or infrared rays, but inclusions that are considered impurities are kept to a minimum. The largest use of plate glass is for storefronts and office partitions.

Plate glass is now made on a large scale on continuous machines by pouring on a casting table at a temperature of about 1000°C, smoothing with a roller, annealing, setting rigidly on a grinding table, and grinding to a polished surface.

Conductive Glass

This glass, which is employed for windshields to prevent icing and for uses where the conductive coating dissipates static charges, is plate glass with a thin coating of stannic oxide produced by spraying glass, at 482°C–704°C, with a solution of stannic chloride. Coating thicknesses are 50–550 nm, and the coating will carry current densities of 9.300 W/m^2 indefinitely. The coatings are hard and resistant to solvents. The light transmission is 70%–88% that of the original glass, and the index of refraction is 2.0, compared with 1.53 for glass.

Transparent Mirrors

These are made by coating plate glass on one side with a thin film of chromium. The glass is a reflecting mirror when the light behind the glass is less than in front and is transparent when the light intensity is higher behind the glass. Photosensitive glass is made by mixing submicroscopic metallic particles in the glass. When ultraviolet light is passed through the negative on the glass, it precipitates these particles out of solution, and since the shadowed areas of the negative permit deeper penetration into the glass than the high light areas, the picture is in three dimensions and in color. The photograph is developed by heating the glass to 538°C.

Other Glasses

Colored glasses: Made by adding small amounts of colorants to glass batches and are used in lamp bulbs, sunglasses, light filters, and signalware.

Opal glasses: Contain small particles dispersed in transparent glass. The particles disperse the light passing through the glass, producing an opalescent appearance.

Polarized glass for polarizing lenses: Made by adding minute crystals of tourmaline or peridot to the molten glass and stretching the glass while still plastic to bring the axes of the crystals into parallel alignment.

Porous glass: Made of silica sand, boric acid, oxides of alkali metals (sodium, potassium, etc.), and a small amount of zirconia. This glass has 500 times better resistance to alkali solution that has otherwise been achieved to date. The glass is made by first melting, with the heat treatment of 650°C–800°C, then separating into two phases, one composed mainly of silica and the other containing boric acid and alkali metal oxides; this is known as phase splitting. The boric acid–alkali oxides phase elutes in heat treatment, producing small holes that give porous glass its name. The glass is heat resistant, transmits gas, and permits adhesion of many substances to its surface. It is, therefore, used at high temperatures as a gas-separating membrane and as a carrier of various substances.

Oxycarbide Glass

This glass has been developed in which substituting carbon for oxygen or even nitrogen can create a whole new category of high-strength glasses. In a 1.0% carbon–glass system of Mg-Si-Al-O, Vickers hardness was increased significantly as well as glass transition temperature. The oxycarbide glass was prepared firing SiO_2, Al_2O_3, MgO, and SiC in a molybdenum crucible at 1800°C for 2 h.

Oxycarbide glass systems based on Si–metal–O–C and Al–metal–O–C are likely to exist and could potentially be used to produce refractory glasses. Controlled recrystallization of oxycarbides could lead to stable glass–ceramic matrices for ceramic-reinforced composites.

Products

The most common glass products are containers and flat glass. The latter is principally used in transportation and architecture. Flat glass is produced by the float process, developed in the 1950s, and considered one of the important technological achievements of the twentieth century.

Containers are largely made on automatic machines; some special pieces may be made by hand blowing into molds, and some very special ware may be made by the free or offhand blowing of a skilled glassblower.

Electric light bulb envelopes are molded on a special machine that converts a fast-moving ribbon of glass into over 10,000 bulbs/min.

Glass fibers are either continuous for reinforcement or discontinuous for insulation; see *glass fiber*.

Glass wool is most often made by spinning, or forcing molten glass centrifugally out of small holes (around 20,000) in the periphery of a rapidly rotating steel spinner, and attenuating the fibers by entrainment with gas burners. As the fibers fall to a conveyor, they are sprayed with a binder that preserves their open structure. The resulting glass wool mat is mainly composed of still air, which accounts for its excellent insulation properties.

Glass–Ceramics

A family of finer-grained crystalline materials made by a process of controlled crystallization from special glass compositions containing nucleating agents; glass–ceramics are sometimes referred to as nucleated glass, devitrified ceramic, or vitro ceramics. Because they are mixed oxides, different degrees of crystallinity can be produced by varying compositions and heat treatments. Some of the types produced are cellular foams, coatings, adhesives, and photosensitive compositions.

Glass–ceramics are nonporous and generally either opaque white or transparent. Although not ductile, they have much greater impact strength than commercial glasses and ceramics. However, softening temperatures are lower than those for ceramics, and they are generally not useful above 1093°C. Thermal expansion varies from negative to positive values depending on composition. Excellent thermal shock resistance and good dimensional stability can be obtained if desired. These characteristics are used to advantage in *heatproof* skillets and range tops. Like chemical glasses, these materials have excellent corrosion and oxidation resistance. They are electrical insulators and are suitable for high-temperature, high-frequency applications in the electronics field.

Pyroceram is a hard, strong, opaque-white nucleated glass with a flexural strength to above 206 MPa, a density of 2.4–2.62, a softening point at 1349°C, and high thermal shock resistance. It is used for molded mechanical and electrical parts, heat-exchanger tubes, and coatings. *Macor* is a machinable glass–ceramic. Axles for mechanisms that provide power in cardiac pacemakers have been chosen from Macor due to its chemical inertness and light weight. It is also used as welding fixtures and as welding nozzles. *Nucerite*, which is used for lining tanks, pipes, and valves, is nucleated glass. It has about four times the abrasion resistance of a hard glass, will withstand sudden temperature differences of 649°C, and has high impact resistance. It also has high heat-transfer efficiency.

Applications

The first use of glass–ceramics was in radomes for supersonic missiles, where a radar-transmitting material with a combination of strength, hardness, temperature and thermal shock resistance, uniform quality, and precision finishing is required.

The second large-scale application is in "heatproof" skillets and saucepans, again taking advantage of the thermal shock resistance.

One of the products of potential interest to mechanical engineers is bearings for operation at high temperatures without lubrication or in corrosive liquids. Another is a lightweight, dimensionally stable honeycomb structure, which has promise for use in heat regenerators for gas turbines operating at high temperatures. Still a third possibility is precision gauges and machine tool parts whose dimensions do not change with temperature.

Bearings

Although metal bearings are perfectly satisfactory for most applications, there are conditions such as high temperature, dry (unlubricated) operation, and presence of corrosive media in which even the best metal bearings may perform poorly. Also, the metals that do perform best are expensive, difficult to fabricate, and heavy.

Since glass–ceramic bearings might be expected to be stable at high temperatures, resistant to oxidation and to corrosive conditions, as well as light in weight, a thorough evaluation of their characteristics is being made by a number of laboratories and bearing manufacturers. One of the surprising findings has been that glass–ceramic bearings can be finished almost as easily as steel bearings and in the same general type of equipment.

High-Temperature Heat Exchangers

The thermal shock resistance and dimensional stability of the low-expansion glass–ceramics make them useful in various kinds of heat exchangers. One interesting type has been developed for use in high-temperature turbine engines.

Precision Uses

As requirements for precision instruments and machines become more stringent, the gauges and machine tools required to make them must become still more precise and dimensions. If these are made of relatively high expansion metals such as steel, the dimensions vary with temperature.

Glass Cloth

Woven glass-fiber material. See also *scrim*.

Glass Electrode

A glass membrane electrode used to measure pH or hydrogen-ion activity.

Glass Fiber

Fine flexible fibers made from glass are used for heat and sound insulation, fireproof textiles, acid-resistant fabrics, retainer mats for storage batteries, panel board, filters, and electrical insulating tape, cloth, and rope. Molten glass strings out easily into thread-like strands, and this spun glass was early used for ornamental purposes, but the first long fibers of fairly uniform diameter were made in England by spinning ordinary molten glass on revolving drums. The original fiber, about 0.003 cm in diameter, was called glass silk and glass wool, and the loose blankets for insulating purposes were called *navy wool*. The term navy wool is still used for the insulating blankets faced on both sides with flameproof fabric, employed for duct and pipe insulation and for soundproofing.

Glass fibers are now made by letting the molten glass drop through tiny orifices and blowing with air or steam to attenuate the fibers. The usual composition is that of soda–lime glass, but it may be varied for different purposes. The glasses low in alkali have high electrical resistance, whereas those of higher alkali are more acid resistant. They have very high tensile strengths, up to about 2757 MPa.

The standard glass fiber used in glass-reinforced plastics is a borosilicate type known as *E-glass*. The fibers spun as single glass filaments, with diameters ranging from 0.0005 to 0.003 cm, are collected into strands that are manufactured into many forms of reinforcement. E-glass fibers have a tensile strength of 3445 MPa. Another type, *S-glass*, is higher in strength, about one-third stronger than E-glass, but because of higher cost, 18 times more costly per pound vs. E-glass, its use is limited to advanced, higher-performance products.

Staple glass fiber is usually from 0.0007 to 0.0009 cm, is very flexible and silky, and can be spun and woven on regular textile machines. Glass-fiber yarns are marketed in various sizes and twists, in continuous or staple fiber, and with glass compositions varied to suit chemical or electrical requirements.

Halide glass fibers are composed of compounds containing fluorine and various metals such as barium, zirconium, thorium, and lanthanum. They appear to be promising for fiber-optic communication systems. Their light-transmitting capability is many times better than that of the best silica glasses now being used.

Glass-fiber cloth and glass-fiber tape are made in satin, broken twill, and plain weaves; satin-wear cloth is 0.02 cm thick weighing 0.24 kg/m². Glass cloth of plain weave of either continuous fiber or staple fiber is much used for laminated plastics. The usual thicknesses are from 0.005 to 0.058 cm in weights from 0.05 to 0.50 kg/m². Cloth woven of monofilament fiber in loose rovings to give better penetration of the impregnating resin is also used. Glass mat, composed of fine fibers felted or intertwined in random orientation, is used to make sheets and boards by impregnation and pressure. Fluffed glass fibers are tough, twisted glass fibers. For filters and insulation the cell withstands temperatures to 538°C. Chopped glass, consisting of glass fiber cut to very short lengths, is used as a filler for molded plastics. Translucent corrugated building sheet is usually made of glass-fiber mat with a resin binder. All these products, including chopped fiber, mat, and fabric preimpregnated with resin, and the finished sheet and board, are sold under a wide variety of trade names. Glass fiber bonded with a thermosetting resin can be preformed for pipe and other insulation coverings. Glass-fiber block is also available to withstand temperatures to 316°C. Glass filter cloth is made in twill and satin weaves in various thicknesses and porosities for chemical filtering. Glass belting, for conveyor belts that handle hot and corrosive materials, is made with various resin impregnations. Many synthetic resins do not adhere well to glass, and the fiber is sized with vinyl chlorosilane or other chemicals.

Four major principles should be recognized in using glass fibers as composite reinforcements. Mechanical properties depend on the combined effect of the amount of glass-fiber reinforcement used and its arrangement in the finished composite. The strength of the finished object is directly related to the amount of glass in it. Generally speaking, strength increases directly in relation to the amount of glass. A part containing 80 wt% glass and 20 wt% resin is almost four times stronger than a part containing the opposite amounts of these two materials. Chemical, electrical, and thermal performance are influenced by the resin system used as a matrix. Materials selection, design, and production requirements determine the proper fabrication process to be used. Finally, the cost–performance value achieved in the finished composite depends on good design and judicious selection of raw materials and processes. See also *fiberglass*.

Glass Filament

A form of glass that has been drawn to a small diameter and extreme length. It is standard practice in the fiberglass industry to refer to a specific filament diameter by a specific alphabet designation. Fine fibers, which are used in textile applications, range from D (–6 µm) through G (–10 µm). Conventional plastics reinforcement, however, uses filament diameters that range from G to T (–24 µm).

Glass Filament Bushing

The units through which molten glass is drawn in making glass filaments.

Glass Finish

A material applied to the surface of a glass reinforcement to improve the bond between the glass and the plastic resin matrix.

Glass Flake

Thin, irregularly shaped flakes of glass used as a reinforcement in composites.

Glass Former

An oxide that forms a glass easily and also one that contributes to the network of silica glass when added to it.

Glass Paper

An abrasive material produced by bonding a layer of graded powdered glass to a strong paper backing. More useful for wood and similarly soft materials rather than metal.

Glass, Percent by Volume

The product of the specific gravity of a laminate and the percent glass by weight, divided by the specific gravity of the glass.

Glass Sand

Sand employed in glassmaking. Glass sands are all screened and usually washed, to remove fine grains and organic matter. Sand for first-quality optical glass should contain 99.8% SiO_2 and a maximum of 0.1 Al_2O_3 and 0.02 Fe_2O_3. *Potters' sand* is usually a good grade of glass sand of uniform grain employed for packing to keep the wares apart.

Glass Stress

In a filament-wound part, usually a pressure vessel, the stress calculated using the load and the cross-sectional area of the reinforcement only.

Glass Transition Temperature (T_g)

The temperature at which an amorphous polymer (or the amorphous regions in a partially crystalline polymer) changes from a hard and relatively brittle condition to a viscous or rubbery condition. In this temperature region, many physical properties, such as hardness, brittleness, thermal expansion, and specific heat, undergo significant rapid changes.

Glassy

A state or matter that is amorphous or disordered like a liquid in structure, hence capable of continuous composition variation and lacking a true melting point, but softening gradually with increasing temperature. Glasses of commerce are mainly complex silicates in chemical combination with numerous other oxidic substances, made by melting the source materials together, forming in various ways while fluid, and allowing to cool.

Glaze

(1) A ceramic coating matured to the glassy state on a formed ceramic article or the material or mixture from which the coating is made. (2) In tribology, a ceramic or other hard, smooth surface film produced by sliding. (3) The production of a hard smooth surface or the surface so formed. The term may be used of a surface that has itself become glazed as a result of some process or of material applied as a liquid or powder that is then heated to form a glaze over the substrate. (4) The installation of glass sheet into a framework of glazing bars.

Glazing

Glazing involves the application of finely ground glass, or glass-forming materials, or a mixture of both, to a ceramic body and then heating (firing) to a temperature where the material or materials melt, forming a coating of glass on the surface of the ware. Glazes are used to decorate the ware, to protect against moisture absorption, to give an easily cleaned, sanitary surface, and to hide a poor body color.

Glazes are classified and described by the following characteristics: surface, glossy or matte; optical properties, transparent or opaque; method of preparation, fritted or raw; composition, such as lead, tin, or boron; maturing temperatures; and color. Opaque glazes contain small crystals embedded in the glass, but special glazes in which a few crystals grow to recognizable size are called crystalline glazes. A glaze may be applied during the firing; such a glaze is called salt glaze. Common salt, NaCl, or borax, $Na_2B_4O_7 \cdot 10H_2O$, or a mixture of both is introduced into the kiln at the finishing temperature. The salt evaporates and reacts with the hot ware to form the glaze. This type of glaze has been applied to sewer pipe and some fine stoneware.

The most important factor in compounding a glaze, after a suitable maturing temperature has been obtained, is the matching of the coefficient of thermal expansion of the glaze and the body on which it is applied. A slightly lower coefficient for the glaze will place it in compression (the desired condition) when the ware cools.

The reverse state (with the glaze in tension because it has a higher coefficient) leads to the formation of fine hairline cracks, a condition known as crazing.

Glaze Materials

As reaction times, melting points, temperatures, and substrates have changed in the whiteware industry, the materials used in glaze formulation have also changed over time.

Silica is a major glaze component and is added in many forms, such as quartz, feldspar, or wollastonite. Silica acts as a glass former and is used to control thermal expansion and help impart acid resistance to the glaze.

Clay, such as kaolin, ball clay, china clay, or bentonite, continues to be the primary suspending agent used in ceramic glazes. The rheology characteristics required by the application method, as well as physical properties such as glaze drying time or shrinkage characteristics, need to be taken into account when selecting the clay to be used in a ceramic glaze. For example, glazing wet column brick requires glazes with up to 25% clay, while only 5%–10% clay is needed for glaze suspension.

Feldspathic minerals, such as soda and potash feldspar and nepheline syenite, remain some of the most commonly used raw materials. These materials are a major source of silica and alkali fluxes in a glaze. Feldspar can be used as either a flux or a refractory material in a glaze, depending on the firing temperature.

Alumina is normally added as calcined alumina or alumina trihydrate, although both clay and feldspars are also sources of alumina in the glaze. The alumina is used to improve the scratch resistance or abrasion resistance of the glaze and also influences the gloss level.

Alkaline earth oxide materials, such as calcium carbonate, wollastonite, and zinc oxide, are also generally added as raw materials. Other alkaline earth oxides, such as lead oxide, strontium oxide, barium oxide, and magnesium oxide, are more typically added in a fritted form. The alkaline earth oxides are beneficial because they provide fluxing action without having a major effect on glaze thermal expansion.

Zirconium silicate is the major opacifier used in ceramic glazes. However, tin oxide is used by some manufacturers, particularly if chrome–tin pigments are being used in the glaze. Using zircon-opacified frits to provide some or all of the zirconium silicate needed in a glaze is also becoming more common. This is especially true in fast-firing cycles, where the use of a high percentage of a refractory material (such as zirconium silicate) is not desirable.

Ceramic frits play a major part in glaze formulation. Frits continue to be a source of highly soluble oxides, such as soda, potassium, or boron. As firing cycles have grown shorter, however, material such as zircon, calcium, alumina, or barium are commonly added in the fritted form. In addition, frit producers are able to tailor frits formulations for particular uses and processes.

Glide

(1) Same as *slip*. (2) A noncrystallographic shearing movement, such as of one grain over another. (3) Referring to deformation within a grain, the term usually means plastic deformation by slip along the main planes, usually of closest packing. See *dislocation*.

Globular Cementite

Cementation in discrete globules rather than combined and pearlite. See *steel*.

Globular Transfer

In consumable electrode arc welding, a type of metal transfer in which molten filler metal passes across the arc as large droplets. Compare with *spray transfer* and short-*circuiting transfer*. See the term *short-circuiting transfer*.

Glow Discharge Spectroscopy

A surface analysis technique in which a glow discharge lamp provides a low-pressure argon environment for the specimen that forms the cathode of the high-voltage (800–1200 V) circuit. The plasma, or glow discharge, of Ar^+ ions bombards the specimen releasing material in a high energy state. The wavelength and intensity of the resultant radiation are measured to provide a quantitative analysis of the surface material.

Glucose

A syrupy liquid that is a monosaccharide, or simple sugar, occurring naturally in fruits and in animal blood or made by the hydrolysis of starch. It is also produced as a dry, white solid by evaporation of the syrup. Glucose is made readily from cornstarch by heating the starch with dilute hydrochloric acid, which is essentially the same process as occurs in the human body. Commercial glucose is made from cornstarch, potato starch, and other starches, but in Japan it is also produced from wood. Glucose is only 70% as sweet as sugarcane and has a slightly different flavor. It is used in confectionery and other foodstuffs for blending with sugarcane and syrups to prevent crystallization on cooling and it is usually cheaper than sugar. It is used in tobacco and inks to prevent drying and in tanning as a reducing agent. The name *glucose* is usually avoided by the manufacturers of edible products, because of prejudices against its substitution for sugar; but in reality it is a simple form of sugar easily digested. It is used in medicine as a blood nutrient and to strengthen heart action, and it may be harmful only in great excess. When free from starch, it is called *dextrose*. It is also marketed as corn syrup, but corn syrup is not usually pure glucose, as it contains some dextrine and maltose. The *maltose*, or *malt sugar*, or the combination has the empirical formula $C_{12}H_{22}O_{11}$. When hydrolyzed in digestion, it breaks down easily to glucose. It is produced from starch by enzyme action. When purified, it is transparent and free of malt flavor. It is not as sweet as the sucrose of sugarcane, but is used in confectionery and as an extender of sugarcane. *Dry corn syrup* is in colorless glasslike flakes. It is made by instantaneous drying and quick cooling of the syrup. Sweetose is a crystal-clear enzyme-converted corn syrup used in confectionery to enhance flavor and increase brightness.

Glucose derived from grapes is called *grape sugar*. The glucose in fruits is called *fruit sugar, levulose,* or *fructose*. This is dextroglucose, and when separated out, it is in colorless needles that melt at 219°F (104°C). It is used for intravenous feeding and is absorbed faster than glucose. It is also used in low-calorie foods and in honey to prevent crystallization. It is normally expensive but is made synthetically. It can be made from corn and is superior to corn syrup as a sweetening agent. *Maple syrup* is prepared by concentrating sap from the maple tree. *Molasses* is a by-product of sugarcane manufacture. *Corn sugar* is also a solid white powder, consisting of glucose with one molecule of water crystallization. When the refined liquor is cooled, the corn sugar crystallizes in a mother liquor known as *hydrol* or *corn-sugar molasses*. Ethyl glucoside is marketed as a colorless syrup

in water solution with 80% solids and a specific gravity of 1.272. It is noncrystallizing and is used as a humectant and plasticizer in adhesives and sizes.

Glue

Originally, a hard gelatin obtained from hides, tendons, cartilage, bones, skins, and heads of fish and of other animals and also an adhesive prepared from this substance by heating with water. Through general use, the term is now synonymous with the terms bond and adhesive. See also *adhesive, gum, mucilage, paste, resin,* and *sizing*.

The term *animal glue* is limited to *hide glue, extracted bone glue,* and *green bone glue*. Fish glue is not usually classified with animal glue nor is casein glue. The vegetable glues are also misnamed, being classified with the mucilages. Synthetic resin glues are more properly classified with adhesive cements. Animal glues are *hot-work glues* that are applied hot and bind on cooling. Good grades of glue are semitransparent, free from spots and cloudiness, and not brittle at ordinary temperature. *Bone glue* is usually light amber; the strong hide and sinew glues are light brown. The stiffening quality of glue depends upon the evaporation of water, and it will not bind in cold weather. Glues made from blood, known as *albumin glues*, and from casein are used for some plywood; however, they do not have the strength of the best grades of animal glue and are not resistant to mold or fungi. *Marine glue* is a glue insoluble in water, made from solutions of rubber or resins, or both. The strong and water-resistant plywoods are now made with synthetic resin adhesives.

Animal glue has been made since ancient times and is now employed for cementing wood, paper, and paperboard. It will not withstand dampness, but white lead or other material may sometimes be added to make it partly waterproof. Casein glues and other protein glues are more water resistant. *Soybean glue* is made from soybean cake and is used for plywood. It is marketed as dry. It has greater adhesive power than other vegetable glues, or pastes, and is more water resistant than other vegetable pastes. Hide glue is used in the manufacture of furniture, abrasive papers and cloth, gummed paper and tape, matches, and print rollers. The bone glues are used either alone or blended in the manufacture of cartons and paper boxes. Green bond glue is used chiefly for gummed paper and tape for cartons. In making bone glue, the bones are crushed, the greases are extracted by solvents, and the mineral salts are removed by dilute hydrochloric acid. The bones are then cooked to extract the glue. Glues are graded according to the quality of the raw material, method of extraction, and blend.

There are 16 grades of hide glue and 15 grades of bone glue. Those with high viscosity are usually the best. Most glue is sold in ground form but also as flake or pearl. Glues for such uses as holding abrasive grains to paper must have flexibility as well as strength, obtained by adding glycerin. The *animal protein colloid* of Swift & Co. is a highly purified bone glue is specially adapted for use as an emulsifier and for sizing, water paints, stiffening, and adhesives. Hoof and horn pith glue is the same as bone glue and is inferior to hide glue. *Fish glue* is made from the jelly separated from fish oil or from solutions of the skins. The best fish glue is made from Russian isinglass. Fish glues do not form gelatin well and are usually made into liquid glues for photographic mounting, gummed paper, household use, and use in paints and sizes. Liquid glues are also made by treating other glues with a weak acid. Pungent odors indicate defective glue. Glues made from decomposed materials are weak. Preservative such as sulfur dioxide or chlorinated phenol may be used. The melting point is usually about 140°F (60°C).

Glut

Metal bar or section providing filler material in forge welding. The glut becomes fully incorporated as a load-bearing member rather than merely occupying space.

Glue-Laminated Wood

An assembly made by bonding layers of veneer or lumber with an adhesive so that the grain of all laminations is essentially parallel.

Glue Line

Synonym for *bond line*.

Glue Line Thickness

Thickness of layer of cured adhesive.

Glycerin

A colorless, syrupy liquid with a sweet, burning taste, soluble in water and in ethyl alcohol. It is the simplest *trihydroxy alcohol*, with composition $C_3H_5(OH)_3$. It has a specific gravity 1.26, a boiling point of 554°F (290°C), and a freezing point of 68°F (20°C). It is also called *glycerol* and was used as a lotion under the name of *sweet oil* for more than a century after its discovery in 1783.

Glycerin occurs as *glycerides*, or combinations of glycerin with fatty acids, and vegetables and animal oils and fats and is a by-product in the manufacture of soaps and in the fractionation of fats, and is also made synthetically from propylene. Coconut oil yields about 14% glycerin, 11% palm oil, 10% tallow, 10% soybean oil, and 9% fish oils. It does not evaporate easily and has a strong affinity for water, and it is used as a moistening agent in products that must be kept from drying, such as tobacco, cosmetics, foodstuffs, and inks. As it is nontoxic, it is used as a solvent in pharmaceuticals, as an antiseptic in surgical dressings, as an emollient in throat medicines, and in cosmetics. Since a different type of group can replace anyone of or all three *hydroxyl groups* (OH), a large number of derivatives can be formed, and it is thus a valuable intermediate chemical, especially in the making of plastics. Commercially, the most important are the *alkyd resins*. It is also used as a plasticizer in resins and to control flexibility in adhesives and coatings. An important use is in nitroglycerin and dynamite.

Gob

(1) A portion of hot glass delivered by a feeder. (2) A portion of hot glass gathered on a punty or pipe.

Gob Process

A process whereby glasses delivered to a forming unit in *gob* form.

Gold and Gold Alloys

Gold (Au) is a soft, ductile, yellow metal, known since ancient times as a precious metal on which all material trade values are based. Commercially pure gold is 99.97% pure, and higher purity material is available.

The outstanding useful property of gold is its oxidation resistance. It is not attacked by the common acids when used singly. It does, however, dissolve an aqua regia (nitric acid plus hydrochloric acid) and cyanide solutions and is attacked by chlorine above 80°C. It is resistant to dry fluorine up to about 300°C, hydrogen fluoride, dry hydrogen chloride, and dry iodine. It is also resistant to sulfuric acid, sulfur, and sulfur dioxide.

Gold is found widely distributed in all parts of the world. It is used chiefly for coinage, ornaments, jewelry, and gilding.

Gold is the most malleable of metals and can be beaten into extremely thin sheets. In most cases, gold is alloyed to increase its hardness without appreciable loss of oxidation resistance. Copper is a common alloying element along with silver and small amounts of the platinum metals. Some of these alloys can be heat treated to relatively high strengths. Gold and its alloys are worked into all the usual forms of sheet, wire, ribbon, and tubing.

A gram of gold can be worked into leaf covering 0.6 m², and only 0.0000084 cm thick, or into a wire 2.5 km in length. Precision casting is also used to form gold alloys, particularly for jewelry. The expense of gold and its low hardness are often offset by using it as a laminate or plating on base metals. It can also be applied to metals, ceramics, and some plastics by the thermal decomposition of certain gold compounds.

Cast gold has a tensile strength of 137 MPa. The specific gravity is 19.32 and the melting point 1063°C. It is not attacked by nitric, hydrochloric, or sulfuric acid, but is dissolved by aqua regia, or by a solution of azoimide, and is attacked by sodium and potassium cyanide plus oxygen. The metal does not corrode in air, only a transparent oxide film forming on the surface.

Gold alloys (Au–Ag–Cu and Au–Ni–Cu–Zn) can be made in a range of colors from white to many shades of yellow. For this reason, gold is widely used for jewelry and other decorative applications. Similar alloys (Au–Ag–Cu–Pt–Pd) are used in dentistry, the nobility of the alloys and their response to heat treatment hardening being of concern here. Gold–silver alloys have been used in low-current electrical contacts (under 0.5 A). Gold is often used in electrical and other equipment, which is used for standards and where stability is of prime concern.

Gold–gallium and gold–antimony alloys for electronic uses come in wire as fine as 0.013 cm in diameter and in sheet as thin as 0.003 cm. The maximum content of antimony in workable gold alloys is 0.7%. A gold–gallium alloy with 2.5% gallium has a resistivity of 15×10^{-8} Ω m and has a tensile strength of 379 MPa and 22% elongation. Gold powder and gold sheet, for soldering semiconductors, are 99.999% pure. The gold wets silicon easily at a temperature of 371°C.

The corrosion resistance and melting point of gold also make it useful as a brazing material. It is used as well in chemical equipment, where its susceptibility to chlorine attack is not a problem. In particular, it is used to line reaction vessels and as a gasketing material.

Gold may be readily applied by electroplating from cyanide and other solutions. It may also be applied to some metals by simple immersion in special plating solutions. Plating has, of course, many decorative applications.

Gold plating is also used to make reflectors, particularly for the infrared wavelengths. Electrical components are often gold plated, especially for high frequencies, because of the low electrical resistance of gold. Vacuum tube grids may be gold plated to reduce electron emission and some electrical contacts are gold plated.

Gold may also be applied by using *liquid bright golds*. These are varnishlike solutions of gold compounds that may be applied at any suitable manner—brushing, spraying, printing, etc.—to metals, ceramics, and some plastics. After being applied, the material is heated to decompose the compound, depositing a tightly adhering

G

gold layer. Some gold alloys may also be applied in this matter. This method is used for the decoration of china and glassware, as well as for printed circuits and electrical resistance elements.

Gold is one of the noble metals, and it is one of the only two colored metals (copper being the other). It is soft and, being the most malleable of all metals, can be beaten into ultrathin gold leaf. The purity of gold is measured as a percentage or, for jewelry, coinage, and similar applications, in carats (parts per 24) or fineness (parts per 1000). Apart from decorative and coinage uses and its external *hoarding* value, it has considerable industrial application where its corrosion resistance justifies its cost.

Gold Bronze

An alloy of copper with about 4% aluminum (no gold) having a gold color.

Gold Filled

Covered on one or more surfaces with a layer of gold alloy to form a clad or composite material. Gold-filled dental restorations are an example of such materials.

Gold Leaf

Thin sheet gold of high purity typically 0.075–0.125 μm, 3–5 μin. thick.

Goldschmidt Process

The thermite reaction when used for extraction of metal from its ore.

Goniometer

(1) In *x-ray spectrometry*, an instrument devised for measuring the angle through which a specimen is rotated or for orienting a sample (e.g., a single crystal) in a specific way. (2) An instrument for measuring the orientation of the surfaces and planes of crystalline materials.

Goodman Diagram

See *fatigue*.

Gooseneck

In die casting, a spout connecting a molten metal holding pot, or chamber, with a nozzle or sprue hole in the die and containing a passage through which molten metal is forced on its way to the die. It is the metal injection mechanism in a *hot chamber machine*.

Gouging

In welding practice, the forming of a bevel or groove by material removal. See also *back gouging*, *arc gouging*, and *oxygen gouging*. A severe form of localized wear in which material is removed in a single pass or cut leaving a deep groove. Flame gouging is the use of some oxyfuel gas flame to melt defects or other unwanted features on a component surface. The molten material is vigorously ejected by the main gas stream or an additional stream leaving a deep groove. Thermal gouging is similar but includes processes in which heating is by an electric arc.

Gouging Abrasion

A form of high-stress abrasion in which easily observable grooves or gouges are created on the surface. See also *abrasion* and *low-stress abrasion*.

G-P Zone

A *Guinier–Preston zone*.

Gradated Coating

A thermal sprayed deposit composed of mixed materials in successive layers that progressively change in composition from the constituent material of the substrate to the surface of the sprayed deposit.

Grade

In powder metallurgy, a specific, nominal chemical analysis powder identified by a code number, for example, cemented carbide manufacturers grade 74 M 60 FWC (74 is usage; M is equipment manufacturer; 60 is nominal HRC; FWC is fine cut tungsten carbide).

Graded Abrasive

An abrasive powder in which the sizes of the individual particles are confined to certain specified limits. See also *grit size*.

Graded Coating

A thermal spray coating consisting of several successive layers of different materials, for example, starting with 100% metal, followed by one or more layers of metal–ceramic mixtures, and finishing with 100% ceramic.

Gradient Coating

See *gradated coating*.

Gradient Elution

A technique for improving the efficiency of separations achieved by *liquid chromatography*. It refers to a stepwise or continuous change with time in the mobile phase composition.

Graft Copolymers

A chain of one type of polymer to which side chains of a different type are attached or grafted.

Grain

(1) A portion of a metal or alloy having all of its atoms in alignment forming a crystal structure. Grain is largely synonymous with crystal. Metals are comprised of grains, much as a lump of sugar is, except that metals have no gaps at the interfaces and bonding between grains is strong. The interface between grains is termed the grain boundary that may be visualized as a band of material in which the atoms have a virtually random arrangement as they blend from the lattice of one grain to that of the neighboring grain. In most cases, the grain boundaries are not a weak zone. The exceptions

include high temperatures, see *creep*, in certain cases of *embrittlement*. (2) A unit of weight, the smallest on the apothecaries' scale, 1 grain = 0.00208 oz troy = 0.00229 oz av. = 64.7989 mg. (3) The elongated fibrous cell structure of wood. (4) An individual crystal in a polycrystalline material; it may or may not contain twinned regions and subgrains.

Grain Boundary

A narrow zone in a metal or ceramic corresponding to the transition from one crystallographic orientation to another, the separating one *grain* from another; the atoms in each grain are arranged in an orderly pattern.

Grain-Boundary Corrosion

Same as *intergranular corrosion*. See also *interdendritic corrosion*.

Grain-Boundary Diffusion

One of the diffusion mechanisms in *sintering*. It is characterized by a very high diffusion rate because of an abundance of imperfections in the grain boundaries. See also *surface diffusion* and *volume diffusion*.

Grain-Boundary Etching

In metallography, the development of intersections of grain faces with a polished surface. Because of severe, localized crystal deformation, grain boundaries have higher dissolution potential than grains themselves. Accumulation of impurities in grain boundaries increases this effect.

Grain-Boundary Liquation

An advanced stage of overheating of metals in which material in the region of austenitic grain boundaries melts. Also termed *burning*.

Grain-Boundary Sulfide Precipitation

An intermediate state of overheating of metals in which sulfide inclusions are redistributed to the austenitic grain boundaries by partial solution at the overheating temperature and reprecipitation during subsequent cooling.

Grain Coarsening

A heat treatment that produces excessively large austenitic grains in metals.

Grain Contrast

Variations in appearance of the grains on a prepared surface. The variation may be produced by various techniques including etching or heat tinting, and examination may be by normal light or polarized light.

Grain-Contrast Etching

In metallography, the development of grain surfaces lying in the polished surface of the microsection. These become visible through differences in reflectivity caused by reaction products on the surface or by differences in roughness.

Grain Fineness Number

A system developed by the American Foundry Society for rapidly expressing the average grain size of a given sand. It approximates the number of meshes per inch of that sieve that would just pass the sample.

Grain Flow

Fiberlike lines on polished and etched sections of forgings caused by orientation of the constituents of the metal in the direction of working during forging. Grain flow produced by proper die design can improve required *mechanical properties* of forgings. See also *flow lines* and *forged structure*.

Grain Growth

(1) An increase in the average size of the grains in polycrystalline material, usually as a result of heating at elevated temperature. (2) In polycrystalline materials, a phenomenon occurring fairly close below the melting point in which the larger grains grow still larger, while the smallest ones gradually diminish and disappear. See also *recrystallization*.

Graining

The process of vigorously stirring or agitating a partially solidified material to develop large grains having a thin oxide coating.

Grain Refinement (Metals)

The manipulation of the solidification process to cause more (and therefore smaller) grains to be formed and/or to cause the grains to form specific shapes. The term refinement is usually used to denote a chemical addition to the metal but can refer to control of the cooling rate.

Grain Refiner

A material added to a molten metal to induce a finer-than-normal grain size in the final structure.

Grain Size

(1) For metals, a measure of the areas or volumes of grains in a polycrystalline material, usually expressed as an average when the individual sizes are fairly uniform. In metals containing two or more phases, grain size refers to that of the matrix unless otherwise specified. Grain size is reported in terms of number of grains per unit area or volume, in terms of average diameter, or as a grain size number derived from area measurements. (2) For grinding wheels, see the preferred term *grit size*.

Grain Size Analysis

The measurement of grain size in metals or the measurement of the particle size of powders.

Grain Size Distribution

Measures of the characteristic grain or crystalline dimensions (usually, diameters) in a polycrystalline solid or of their populations by size increments from minimum to maximum. This is usually determined by microscopy.

Gram-Equivalent Weight

The mass in grams of a reactant that contains or reacts with Avogadro's number of hydrogen atoms. See also *Avogadro's number.*

Gram-Molecular Weight

The mass of a compound in grams equal to its molecular weight.

Granite

A coarse-grained, igneous rock having an even texture and consisting largely of quartz and feldspar with often small amounts of mica and other minerals. There are many varieties. Granite is very hard and compact, and it takes a fine polish, showing the beauty of the crystals. It is the most important building stone and is also used as an ornamental stone. An important use is for large rolls in pulp and paper mills. Granite surface plates, for machine-shop layout work, are made in sizes up to 30 × 72 in. (76 × 183 cm) and 10 in. (25 cm) thick, ground and highly polished to close accuracy. It is extremely durable, and since it does not absorb moisture, as limestone and sandstone do, it does not weather or crack as the stones do. The colors are usually reddish, greenish, or gray. The hard composite igneous rock *diabase*, called "black granite," is used for making precision parallels for machine-shop work.

Granular

(1) With reference to volume comprising separate, approximately equiaxed grains. (2) With reference to surfaces having a rough irregular texture, rather like coarse-grained lump sugar.

Granular Fracture

A type of irregular surface produced when metal is broken that is characterized by a rough, grainlike appearance, rather than a smooth or fibrous one. It can be subclassified as *transgranular fracture* or *intergranular fracture*. This type of fracture is frequently called *crystalline fracture*; however, the interference that the metal broke because it *crystallized* is not justified, because all metals are crystalline in the solid state. See also *fibrous fracture* and *silky fracture.*

Granular Powder

A powder having equidimensional but nonspherical particles.

Granular Structure

Nonuniform appearance of finished plastic material due to retention of, or incomplete fusion of, particles of composition, either within the mass or on the surface.

Granulated Metal

Small pellets produced by pouring liquid metal through a screen or by dropping it onto a revolving disk and, in both instances, by chilling with water.

Granulation

Any process for producing metals as granules including pouring through a mesh or onto a spinning disk.

Graphite

Graphite is a form of carbon. It was formerly known as black lead, and when first used for pencils, it was called Flanders' stone. It is a natural variety of elemental carbon with a grayish, black color in a metallic tinge.

Carbon and graphite have been used in industry for many years, primarily as electrodes, arc carbons, brush carbons, and bearings. In the last decade or so, the development of new types and emergence of graphite fibers as a promising reinforcement for high-performance composites have significantly increased the versatility of this family of materials.

Types of Graphite

Recrystallized graphite is produced by a proprietary hot-working process that yields recrystallized or *densified* graphite with specific gravities in the 1.85–2.15 range, as compared with 1.4–1.7 for conventional graphites. The major attributes of the material are a high degree of quality reproducibility, improved resistance to creep, and grain orientation that can be controlled from highly anisotropic to relatively isotropic, lower permeability than usual, absence of structural macroflaw, and ability to take a fine surface finish.

Graphite fibers are produced from organic fibers. One line of development used rayon as the precursor, and the other used polyacrylonitrile (PAN). Although the detailed processing conditions for converting cellulose or PAN into carbon and graphite fibers differ in detail, they both consisted fundamentally of a sequence of thermal treatments to convert the precursor into carbon by breaking the organic compound to leave a *carbon polymer.* The fibrous carbon formed by the controlled pyrolysis of organic precursor fiber was viscous rayon or acrylonitrile. Carbon fibers produced by the rayon-precursor method have a fine-grained, relatively disordered microstructure, which remains even after treatment at temperatures up to 3000°C. Graphite crystallites with a long-range 3D order do not develop. In both the rayon and PAN processes, a high degree of preferred crystal orientation was responsible for the high elastic modulus and tensile strength.

Although the names *carbon* and *graphite* are used interchangeably when related to fibers, there is a difference. Typically, PAN-based carbon fibers are 93%–95% carbon by elemental analysis, whereas graphite fibers are usually 99+%. The basic difference is the temperature at which the fibers are made or heat treated. PAN-based carbon is produced at about 1316°C, whereas higher-modulus graphite fibers are graphitized at 1899°C–3010°C. This also applies to carbon and graphite cloths. Unfortunately, with only rare exceptions, none of the carbon fibers is ever converted into classic graphite regardless of the heat treatment.

When used in composites, the fibers are generally made into yarn containing some 10,000 fibers. Depending on the precursor fiber, their tensile strength ranges from 1378 to nearly 3445 MPa, and their modulus of elasticity is from 0.2 million to 0.5 million MPa.

Graphite fiber–reinforced graphite composites can be used at temperatures in excess of 3500°C. No compatibility problems exist because the graphite fiber or filament is in a graphite matrix. This composite system is good in reducing environments; in air or oxidizing atmospheres, special protective coatings are sometimes needed.

The graphite matrix is produced by the pyrolytic decomposition of polymeric systems in which the graphite fiber or filaments are originally embedded. Although many of the matrix starting materials are considered proprietary, usable polymer systems include phenolic, furfuryl ester, and epoxy resins. Graphite–graphite composites, even those fabricated with low-modulus materials, are up to 20 times stronger than conventional carbon and graphite materials. At a density of approximately 1384 kg/m^3, they also are about 30% lighter than conventional carbons. They provide a very high strength-to-weight ratio at temperatures to 3300°C and exhibit superior thermal stability. Most Gr–Gr composites are more than 99% carbon (carbon content about 99.5%–99.9%). This high purity provides a good chemical inertness and corrosion resistance. Gr–Gr composites are not wetted by molten metals, which makes them ideally suited to metallurgical applications where high strength, lightweight, erosion resistance, and good thermal conductivity are important. Typical properties are tensile strength, 56.5 MPa; flexural strength, 76 MPa; compressive strength, 276 MPa; and modulus, 17.2 GPa.

Graphite fiber–epoxy composites provide exceptionally high strength and stiffness, and because of their lightweight are finding an increasing use of these composites for golf club shafts, tennis racquet frames, and a multitude of sports equipment, as well as extensive use in the aerospace industry (wings, engine casings, fittings etc.), is found.

PT graphites are graphite fibers impregnated or bonded with an organic resin (such as furfural) and then carbonized. The result is a graphite-reinforced carbonaceous material with a high degree of thermal stability. The composite has a low density (0.93–1.2 specific gravity) and what is reported to be the highest strength-to-weight ratio of the material at temperatures in the 2204°C–2706°C range.

Colloidal graphite consists of natural or artificial graphite in very fine particle form, coated with a protective colloid and dispersed in a liquid. The selection of the liquid—water, oils, or synthetics—is made on the basis of intended use of the product. Significant characteristics of colloidal graphite dispersions are that the graphite particles remain in suspension indefinitely and the particles *wick*— that is, they are carded by the liquid to most places penetrated by the liquid.

The supergraphite used for rocket casings and other heat-resistant parts is recrystallized molded graphite. It will withstand temperatures to 3038°C. Pyrolytic graphite is an oriented graphite. It has high density, with a specific gravity of 2.22; has exceptionally high heat conductivity along the surface, making it very flame resistant; is impermeable to gases; and will withstand temperatures to 3704°C. It is made by deposition of carbon from a stream of methane on heated graphite and the growing crystals formed with thin planes parallel to the existing surface. The structure consists of close-packed columns of graphite crystals joined to each other by strong bonds along the flat planes, but with weak bonds between layers. This weak and strong electron bonding provides a laminal structure. The material conducts heat and electricity many times faster along the surface than through the material. The flexural stresses 172 MPa compared with less than 55 MPa for the best conventional graphite. At 2760°C the tensile strength is 275 MPa. Sheets as thin as 0.003 cm are impervious to liquids or gases. It is used for nozzle inserts and reentry parts for spacecraft, as well as for atomic shielding with an addition of boron.

Mechanical Properties

The degree of anisotropy in graphites varies, but cross grain strengths are usually substantially lower. Tensile strengths of the newer engineering graphites are substantially higher, with that of pyrolytic graphite reaching around 95.2 MPa. See *carbon* and *carbon composites.*

Graphitic Carbon

Carbon- and iron-base materials that are in the free form as graphite rather than combined with iron or another element.

Graphitic Corrosion

A form of corrosion affecting graphitic cast irons in aggressive aqueous environments, particularly soils containing sulfate-reducing bacteria. An electrolytic cell is formed in which the iron matrix is progressively corroded, leaving a weak mass of graphite and iron oxide. The corrosion product is the same volume as the original material and does not readily detach so even extensive deep graphitic corrosion may not be obvious to a visual examination. It occurs in relatively mild aqueous solutions and on buried pipe and fittings.

Graphitic Steel

Alloy steel made so that part of the carbon is present as graphite.

Graphitization

The formation of free carbon, that is, graphite in iron and steel. When it occurs during solidification, it is termed primary graphitization. When it occurs in the solid state, it is termed secondary graphitization. The process may be deliberately induced in the production of cast irons. In steels, graphitization may occur as a result of very prolonged exposure to high temperatures, for example, about 100,000 h in the temperature range 850°C–900°C for mild steel, perhaps 50°C higher for 0.5% molybdenum steel. Aluminum deoxidized steel is relatively more susceptible and imposed stress accelerates the rate of graphitization. The term has also been used, perhaps not very accurately, with reference to graphitic corrosion.

Graphitization (Organic Materials)

The process of pyrolyzation in an inert atmosphere at temperatures in excess of 1925°C (3500°F), usually as high as 2480°C (4500°F), and sometimes as high as 5400°C (9750°F), converting carbon to its crystalline allotropic form. Temperature depends on precursor and properties desired.

Graphitizing

Annealing a ferrous alloy such that some or all of the carbon precipitates as graphite.

Graticule

A grating, network, or scale on a lens or disk inserted in the microscope. Its image is superimposed on the material being examined allowing features of interest to be measured or their positions plotted.

Gravel

A natural material composed of small, usually smooth, rounded stones or *pebbles*. It is distinguished from sand by the size of the grain, which is usually above 0.25 in. (0.64 cm); but gravel may contain large stones up to 3 in. (7.62 cm) in diameter and some sand. It will also contain pieces of shale, sandstone, and other rock materials. Gravel is used in making concrete for construction and as a loose paving material. Commercial gravel is washed to remove the clay and organic material and is screened. *Pea gravel* is screened gravel used for surfacing with asphalt or for roofing. Gravel is sold in cubic yards (cubic meters) or in tons (metric tons) and shipped by weight.

Gravity Die Casting

See *permanent mold*.

Gravity Hammer

A class of forging hammer in which energy for forging is obtained by the mass and velocity of a freely falling ram and the attached upper die. Examples are the *board hammer* and *air-list hammer*.

Gravity Segregation

The variable composition of a casting or ingot caused by settling out of heavy constituents, or rising of light constituents, before or during solidification.

Gravity Welding

Processes in which a metal arc electrode is carried by a device that locates the electrode tip at the start of the joint and, once the arc has been struck, directs the falling electrode along the line of the joint without further assistance from the operative. The device is usually very simple, for example, no more than a bipod set across the joint.

Gray Body

A body having the same spectral emittance at all wavelengths.

Gray Cast Iron

See *gray iron*.

Gray Iron

A broad class of ferrous casting alloys (*cast irons*) normally characterized by a microstructure of flake graphite in a ferrous matrix. Gray irons usually contain 2.5%–4% C, 1%–3% Si, and additions of manganese, depending on the desired microstructure (as low as 0.1% Mn in ferritic gray irons and as high as 1.2% in pearlitics). Sulfur and phosphorus are also present in small amounts as residual impurities. See the term *flake graphite*.

Gray Irons

Gray iron is characterized by the presence of flakes of graphite supported in a matrix of ferrite, pearlite, austenite, or any other matrix attainable in steel. The major dimensions of the flakes may vary from about 0.05 to 1.0 mm. Because of their low density, the graphite flakes occupy about 10% of the metal volume. The flakes interrupt the continuity of the matrix and have a large effect on the properties of gray iron. In addition, the flakes give a fractured surface that is great. This is responsible for the name "gray iron."

High carbon content and the flakes of graphite give gray iron unique properties as follows:

1. Lowest melting point of the ferrous alloys, so that low-cost refractories can be used for molds
2. High fluidity in the molten state, so that complex and thin designs can be cast
3. Excellent machinability, better than steel
4. High damping capacity and ability to absorb vibrations
5. High resistance to wear involving sliding
6. Low ductility and low impact strength when compared with steel

Gray iron is by far the most common and widely used cast iron.

Gray iron is encountered almost exclusively as shaped castings used either with or without machining. Typical applications include

1. Pipe for underground service for water or gas
2. Ingot molds into which steel and other metals are cast
3. Cylinder blocks and heads for internal combustion engines
4. Frames and end bells for electric motors
5. Bases, frames, and supports for machine tools
6. Sanitaryware such as sinks and bathtubs (usually coated with porcelain enamel)
7. Pumps, car wheels, and transmission cases

The major industries that consume gray iron castings are as follows: automotive, building and construction, utilities, machine tools, architectural, rolling mills (steel plants), general machinery, household appliances, and heating equipment.

For engineering applications where tensile strength is important, gray iron usually is classified on the basis of minimum tensile strength in a specimen machine from a separately cast test bar.

Mechanical and Physical Properties

Compressive strength: Unusually high; at least three times the tensile strength.

Modulus of elasticity: Increases with tensile strength; about 0.87×10^5 MPa for a tensile strength of 136 MPa and up to about 1.45×10^5 MPa for a tensile strength of 480 MPa.

Endurance limit: About 35%–50% of tensile strength. Gray iron is relatively insensitive to the effect of notches.

Damping capacity: Very high, especially in irons of high carbon content. Specific damping capacity is about 10 times that of steel.

Specific gravity: Varies from about 6.8 for high-carbon, low-strength irons to about 7.6 for low-carbon, high-strength irons.

Coefficient of thermal expansion: Slightly lower than that of steel.

Coefficient of thermal conductivity: About the same as many other ferrous alloys; about 0.11–0.14 in CGS units. It can be lowered appreciably by adding alloying elements.

Grease

A lubricant composed of an oil thickened with a soap or other thickener to a semisolid or solid consistency. A lime-base grease is prepared from lubricating oil and calcium soap. Sodium-, barium-, lithium-, and aluminum-base greases are also used. Greases may contain various additives. The liquid phase may also be a synthetic fluid.

Green

This term is often used to indicate a product and a preliminary state of preparation, for example, green compact—the pressed powder compact prior to sintering—or green strength, the strength of such a compact. See unsintered (not sintered).

Green Ceramic

An unsintered ceramic.

Green Compact

An unsintered powder-metallurgy or ceramic compact.

Green Liquor

The liquor resulting from dissolving molten smelt from the kraft recovery furnace in water. See also *kraft process* and *smelt*.

Green Rot

A form of high-temperature attack on stainless steels, nickel–chromium alloys, and nickel–chromium–iron alloys subjected to simultaneous oxidation and carburization. Basically, attack occurs first by precipitation of chromium as chromium carbide and then by oxidation of the carbide particles.

Green Sand

A naturally bonded sand, or a compounded molding sand mixture, that has been *tempered* with water and that is used while still moist.

Green Sand Core

(1) A *core* made of *green sand* and used as rammed. (2) A sand core that is used in the unbaked condition.

Green Sand Mold

A casting mold composed of moist prepared molding sand. This is in contrast with *dry sand mold*.

Green Strength (Foundry Sands)

The strength of a tempered sand mixture at room temperature.

Green Strength (Plastics)

The mechanical strength of material that, while cure is not complete, allows removal from the mold and handling without tearing or permanent distortion.

Green Strength (Powder Compacts)

(1) The ability of a *green compact* to maintain its size and shape during handling and storage prior to *sintering*. (2) The tensile or compressive strength of a green compact.

Greenware

A term for formed ceramic articles in the unfired condition.

Greige, Gray Goods

Any fabric before finishing, as well as any yarn or fiber before bleaching or dyeing; therefore, fabric with no finish or size.

Grey (Cast) Iron

See *cast iron*.

Grey Tin

The allotrope of tin stable below −13°C that is hard and friable. In practice, it forms when tin is cooled below about −20°C and is inhibited by alloying additions of lead above 5% or antimony above 0.1%.

Griffith Critical Crack

The sharp tipped crack, assumed to be present in brittle materials and having, at its tip, a stress concentration sufficient to explain the difference between the high theoretical strength of the material and a much lower strength observed in practice.

Grignards

Reagents formed by the reaction of metallic magnesium and an organic halide—chloride, bromide, or iodide—in the presence of an ether solvent and the absence of water. Grignards are used to make products, including organometallics, pharmaceuticals, fungicides, and aromatic phosphenes. Potential applications include polymerization catalysts for methyl methacrylate, polyethylene, and polypropylene as well as biotechnology products.

Grindability

Relative ease of grinding, which is analogous to *machinability*.

Grindability Index

A measure of the grindability of a material under specified grinding conditions, expressed in terms of volume of material removed per unit volume of wheel wear.

Grinding

Removing material from a workpiece with a grinding wheel or abrasive belt.

1. *Surface grinding*: Producing a flat surface with a rotating grinding wheel as the workpiece passes under.
2. *Creep-feed grinding*: A subset of surface grinding, creep-feed grinding produces deeper (full) depths of cut at slow traverse rates.

G

G

3. *Cylindrical grinding*: Grinding the outside diameters of cylindrical pieces held between centers.
4. *Internal grinding*: Grinding the inside of a rotating workpiece by use of a wheel spindle that rotates and reciprocates through the length of depth of the hole being ground.
5. *Centerless grinding*: Grinding cylindrical surfaces without use of fixed centers to rotate the work. The work is supported and rotates between three fundamental machine components: a grinding wheel, the regulating wheel, and the work guide blade.
6. *Gear (form) grinding*: Removal of material to obtain correct gear tooth form by grinding. This is one of the more exact methods of finishing gears.
7. *Thread grinding*: Thread cutting by use of suitably formed grinding wheel.

Grinding Burn

Surface damage caused resulting from frictional heating during grinding and similar operations. Damage may include cracking, grain-boundary oxidation, local softening, local rehardening with brittle martensite (in the case of steel), and the introduction of residual stresses.

Grinding Cracks

Surface cracking resulting from severe grinding. The frictional heating causes local surface deformation that on cooling induces high levels of tensile residual stress. In addition, steel components may be locally hardened. These various consequences render the material liable to low ductility cracking on cooling. See also *grinding sensitivity*.

Grinding Fluid

An oil- or water-based fluid introduced into grinding operations to (1) reduce and transfer heat during grinding, (2) lubricate during chip formation, (3) wash loose chips or swarf from the grinding belt or wheel, and (4) chemically aid the grinding action or machine maintenance.

Grinding Oil

An oil-type grinding fluid; it may contain additives, but not water.

Grinding Pebbles

Hard and tough, rounded small stones, usually of flint, employed in cylindrical mills for grinding ores, minerals, and cement. Quantities of flint pebbles come from Denmark and Greenland for use in tube mills. They are smooth, round pebbles formed by the washing of the sea on the chalk cliffs, and they come from the islands off the Danish coast. Small pebbles, 0.5 in. (1.27 cm) in diameter, are used for polishing iron castings by tumbling.

The *tumbling abrasives*, for use in tumbling barrels, come in aluminum oxide or silicon carbide preformed balls, cubes, triangles, or cylinders of various sizes to conform to the parts being tumble polished.

Grinding Relief

A groove or recess located at the boundary of a surface to permit the corner of the wheel to overhang during grinding.

Grinding Sensitivity

Susceptibility of a material to surface damage such as *grinding cracks*; it can be affected by such factors as hardness, microstructure, hydrogen content, and residual stress.

Grinding Stress

Residual stress, generated by grinding, in the surface layer of work. It may be tensile or compressive, or both.

Grinding Wheel

A cutting tool of circular shape made of abrasive grains bonded together. See the term *diamond wheel*.

Grindstones

Sandstones employed for grinding purposes. Grindstones are generally used for the sharpening of edged tools and do not compete with the hard emery, aluminum oxide, and silicon carbide abrasive wheels that are run at high speeds for rapid cutting. Grindstones are quarried from the sandstone deposits and made into wheels usually ranging from 1 to about 6 ft (0.3 to about 1.8 m) in diameter and up to 16 in. (41 cm) in thickness. They are always operated at low speeds because of their inability to withstand high centrifugal stresses. The grades vary from coarse to fine. Good grindstones have sharp grains, without an excess of cementing material that will cause the stone to glaze in grinding. The texture must also be uniform so that the wheel will wear evenly. The hard silica grains are naturally cemented together by limonite, clay, calcite, quartz, or mixtures. Too much clay causes crumbling, while too much calcite results in disintegration in the atmosphere. An excess of silica results in a stone that is too hard.

Gripper Dies

The lateral or clamping dies used in a forging machine or mechanical upsetter.

Grit

Crushed ferrous or synthetic abrasive material in various mesh sizes that is used in abrasive blasting equipment to clean castings. For materials used for grinding belts or grinding wheels, the term *abrasive* is preferred. See also *blasting* or *blast cleaning*.

Grit Blasting

(1) *Abrasive blasting* with small irregular pieces of steel, malleable cast iron, or hard nonmetallic materials. (2) Processes in which grit is projected at a surface to remove scale, paint, etc. The metal or mineral grit may be entrained in a high velocity air stream or flung from a high-speed wheel (a Wheelabrator). If the grid is not trained in water, various terms such as Hydra blasting are used.

Grit Size

Nominal size of abrasive particles in a grinding wheel, corresponding to the number of openings per linear inch in a screen through which the particles can pass.

Grizzly

A set of parallel bars (or grating) use for coarse separation or screening of ores, rock, or other material.

Grog

A hard granular material added to a refractory to reduce shrinkage and improve resistance to thermal shock.

Grommet/Grummett

A link of material, for example, twisted rope rowlocks or a pad of material to prevent chafing such as the circumferentially slit range of rubber, inserted to sit astride the edge of a hole in sheet metal.

Groove (Thermal Spraying)

A method of surface roughening in which grooves are made and the original surface roughened and spread. This is also called rotary roughening.

Groove (Welding)

An opening or channel in the surface of a part or between two components that provides space to contain a weld.

Groove Angle

The total included angle of the groove between parts to be joined. Thus, this is the sum of two bevel angles, either or both of which may be 0°.

Groove Face

The portion of a surface or surfaces of a member included in a groove. See also the term *root of joint*.

Groove Radius

The radius used to form the shape of a J- or U-groove weld joint.

Groove Type

The geometric configuration of a groove.

Groove Weld

A weld made in the groove between two members. The standard types are square, single bevel, single flare bevel, single flare-V, single-J, single-U, single-V, double bevel, double flare bevel, double flare-V, double-J, double-U, and double-V. See the terms *double-bevel groove weld* and *single-bevel groove weld*.

Gross Energy Requirement

The total energy expended in producing a material or component from the ore mining stage to its installation in service.

Grossmann Number (*H*)

A ratio describing the ability of a quenching medium to extract heat from a hot steel workpiece in comparison to still water defined by the following equation:

$$H = \frac{h}{2k}$$

where
   h is the heat-transfer coefficient
   k is the conductivity of the metal

Gross Porosity

In weld metal or in a casting, pores, gas holes, or globular voids that are larger and in much greater numbers than those obtained and good practice.

Gross Sample

One or more increments of material taken from a larger quantity (lot) of material for assay or record purposes. This is also termed bulk sample or lot sample. See also *increment* and *lot*.

Groundbed

A buried item, such as junk steel or graphite rods, that serves as the *anode* for the *cathodic protection* of pipelines or other buried structures. See also *deep groundbed*.

Ground Connection

In arc welding, a device used for attaching the work lead (ground cable) to the work.

Ground-Support Cable

A cable construction, usually rugged and heavy, for use in control or power systems.

Group

The specimens tested at one time, or consecutively, at one stress level. A group may comprise one or more specimens.

Group (of Elements)

See *atomic structure*.

Growth (Cast Iron)

A permanent increase in the dimensions of cast iron resulting from repeated or prolonged heating at temperatures above 480°C (900°F) due to either graphitizing of carbides or oxidation. At such temperatures, the cementite breaks down forming ferrite plus graphite and oxygen permeates along the graphite causing internal oxidation. Both effects cause a volume increase that may lead to cracking.

Guard

(1) A device, often made of sheet metal or wire screening, that prevents accidental contact with moving parts of machinery. (2) In electroplating, it is the same as *robber*.

Guerin Process

A *rubber-pad forming* process for forming sheet metal. The principal tools are the rubber pad and form block, or punch.

Guide

The parts of a drop hammer or press that guide the up-and-down motion of the ram in a true vertical direction.

Guide Bearing

A bearing used for positioning a slide or for axial alignment of a long rotating shaft.

Guided Bend

The bend obtained by use of a plunger to force the specimen into a die in order to produce the desired contour of the outside and inside surfaces of the specimen.

Guided Bend Test

A test in which the specimen is bent to a definite shape by means of a punch (mandrel) and a bottom block.

Guide Mill

A small handbill with several stands in a train and with guides for the work at the entrance to the rolls.

Guide Pin Bushings

Bushings, pressed into a die shoe, that allow the *guide pins* to enter in order to maintain punch-to-die alignment.

Guide Pins

Hardened, ground round pins or posts that maintain alignment between punch and die doing the fabrication, setup, operation, and storage. If the press slide is out of alignment, the guide pins cannot make the necessary correction unless heel plates are engaged before the pins enter the bushings. See also *heel block*.

Guillet Diagram

A diagram predicting a steel microstructure on the basis of the alloy contents that are defined in terms of their equivalent effects relative to nickel and carbon.

Guillet Equivalent

Generally, the factor by which the percentage quantity of an element has to be multiplied to indicate its effect on some characteristic such as microstructure. This term is used particularly regarding the structure of brass evaluated in terms of the equivalent percent of zinc. The Guillet equivalents for brass are 5× aluminum, 2× tin, 2× magnesium, 10× silicon, 1× lead, 0.9× iron, 0.5× manganese, and −1.2× nickel. The products of each element percentage times its factor are added to the actual zinc content to give the zinc equivalent.

Guinier–Preston (G–P) Zone

A small precipitation domain in a supersaturated metallic solid solution. A G-P zone has no well-defined crystalline structure of its own and contains an abnormally high concentration of solute atoms. The formation of G-P zones constitutes the first stage of precipitation and is usually accompanied by a change in properties of the solid solution in which they occur.

Gum (Adhesives)

Any of the class of colloidal substances exuded by or prepared from plants, sticky when moist, composed of complex carbohydrates and organic acids, which are soluble or swell in water. The term gum is sometimes used loosely to denote various materials that exhibit gummy characteristics under certain conditions, for example, gum balata, gum benzoin, and gum asphaltum. Gums are included by some of the category of natural resins. See also *adhesive, glue,* and *resin.*

Gum (Lubrication)

In lubrication, a rubberlike, sticky deposit, black or dark brown in color, that results from the oxidation and/or polymerization of fuels and lubricating oils. Harder deposits are described as *lacquers* or *varnishes.*

Gun

See the preferred terms *arc welding gun, electron beam gun, resistance welding gun, soldering gun,* and *thermal spraying gun.*

Gun Drill

A drill, usually with one or more flutes and with coolant passages through the drill body, used for deep hole drilling. See the term *drill.*

Gun Extension (Thermal Spraying)

An extension tube attached in front of a thermal spraying device to permit spraying within confined areas or deep recesses.

Gunmetal

The name for a *casting bronze,* C90500, containing 88% copper, 10% tin, and 2% zinc. It was originally used for small cannons but is now used where the golden color and strong, crystalline structure are desired. It casts and machines well and is suitable for making steam and hydraulic castings, valves, and gears. It has a tensile strength of 221–310 MPa, with elongation 15%–30%. This alloy is similar to *G bronze* (C90300), which contains 88% copper, 8% tin, and 4% zinc. *Gunmetal ingot* may have the zinc replaced by 2% lead. Such an alloy is easier to machine but has less strength. Modified gunmetal contains lead in addition to the zinc. It is used for gears and bearings. A typical modified gunmetal contains 86% copper, 9.5% tin, 2.5% lead, and 2% zinc.

Gunpowder

Also known as *black powder*. An explosive extensively used for blasting purposes and for fireworks. It was introduced into Europe prior to 1250 and was the only propellant used in guns until 1870. It is now superseded for military uses by high explosives. Black powder deteriorates easily in air from the absorption of moisture. It is a mechanical mixture of potassium nitrate, charcoal, and sulfur, in the usual proportions of 75%, 15%, and 10%. More saltpeter increases the rate of burning; additional charcoal decreases the rate. A slow-burning powder for fireworks rockets may have only 54% saltpeter and 32% charcoal. Commercial black powder comes in grains of graded sizes and is glazed with graphite. The grain sizes are known as *pebble powder*, large grain or fine grain, *sporting powder*, *mining powder*, *Spanish spherical powder*, and *cocoa powder*. A temperature of about 3712°F (2100°C) is produced by the explosive. Gunpowder is the slowest acting of all the explosives, and it has a heaving, not a shattering, effect. Hence, it is effective for blasting and breaking up stone. Blasting powder is divided by DuPont into two grades: A and B. The *A powder* contains saltpeter; the *B powder* contains nitrate of soda. The saltpeter concentration varies from 64% to 74% in commercial formulations. The other ingredients are the usual sulfur and charcoal. The B powder is not so strong or water resistant as A powder but is cheaper and is extensively used. *Pellet powder* is blasting powder made up in cylindrical cartridges for easier use in mining. *White gunpowder* is a powder in which the saltpeter is replaced by potassium chlorate. It is very sensitive and explodes with violence. It is used only for percussion caps and fireworks.

Gusset

A reinforcement, usually a plate of triangular form set between two surfaces meeting at an angle.

Gutter

A depression around the periphery of a forging *die impression* outside the flash pan that allows space for the excess metal, surrounds the finishing impression, and provides room for the excess metal used to ensure a sound forging. It is a shallow impression outside the parting line.

Gypsum

Gypsum is the most common sulfate mineral, characterized by the chemical formula $CaSO_4 \cdot 2H_2O$; it shows little variation from this composition.

Gypsum is one of the several evaporite minerals. This mineral group includes chlorides, carbonates, borates, nitrates, and sulfates. These minerals precipitate in seas, lakes, caves, and salt flats due to concentration of ions by evaporation. When heated or subjected to solutions with very large salinities, chips and converts to bassanite ($CaSO_4 \cdot H_2O$) or anhydrite ($CaSO_4$). Under equilibrium conditions, this conversion to anhydrite is direct. The conversion occurs above 42°C in pure water.

Gypsum is used for making building plaster and wallboard tiles, as an absorbent for chemicals, as a paint pigment and extender, and for coating papers. Natural gypsum of California, containing 15%–20% sulfur, is used for producing ammonium sulfate for fertilizer. Gypsum is also used to make sulfuric acid by heating to 1093°C in an air-limited furnace. The resultant calcium sulfide is reacted to yield lime and sulfuric acid. Raw gypsum is also used to mix with portland cement to retard the set. Compact massive types of the mineral are used as building stones. The color is naturally white, but it may be colored by impurities to gray, brown, or red. The specific gravity is 2.28–2.33 and the hardness 1.5–2. It dehydrates when heated to about 190°C, forming the hemihydrate $2CaSO_4 \cdot H_2O$, which is the basis of most gypsum plasters. It is called calcined gypsum, or when used for making ornaments or casts, it is called plaster of Paris. When mixed with water, it again forms the hydrated sulfate that will solidify and set firmly owing to interlocking crystallization. Theoretically, 18% of water is needed for mixing, but actually more is necessary. Insufficient water causes cracking. Water solutions of synthetic resins are mixed with gypsum for casting strong, waterproof articles.

Much calcined gypsum, or plaster of Paris, is used as gypsum plaster for wall finish. For such use, it may be mixed in lime water or glue water and with sand. Because of its solubility in water, it cannot be used for outside work.

The presence of halite (NaCl) or other sulfates in the solution lowers this temperature, although metastable gypsum exists at higher temperatures.

Crystals of gypsum are commonly tabular, diamond shaped, or lenticular; swallow-tailed twins are also common. The mineral is monoclinic with symmetry 2/m. Gypsum is the index mineral chosen for hardness 2 on the Mohs scale. In addition to free crystals, the common forms of gypsum are satin spar (fibrous), alabaster (finely crystalline), and selenite (massive crystalline).

Gypsum is used for a variety of purposes, but chiefly in the manufacture of plaster of Paris, in the production of wallboard, in agriculture to loosen clay-rich soils, and in the manufacture of fertilizer. Plaster of Paris is made by heating gypsum to 200°C in air.

Gypsum deposits play an important role in the petroleum industry. The organic material commonly associated with its formation is considered a source of hydrocarbon (oil and gas) generation. In addition, these deposits act as a seal for many petroleum reservoirs, preventing the escape of gas and oil.

Habit Plane

The plane or system of planes of a crystalline phase along which some phenomenon, such as twinning or transformation, occurs. See *crystal structure*.

Hackle

(1) A line on a glass crack surface, running parallel to the local direction of cracking, separating parallel but noncoplanar portions of the crack surface. (2) A finely structured fracture surface marking that gives a matte or roughened appearance to the surface, having varying degrees of coarseness. Finely structured hackle is variously known as fine hackle, frosted area, gray area, matte, mist, and stippled area. Coarsely structured hackle is also known as *striation*. See also *mist hackle*, *shear hackle*, *twist tackle*, and *wake hackle*.

Hackle Marks

Fine ridges on the fracture surface of a glass, parallel to the direction of propagation of the fracture.

Hackle Surface

A rough textured surface characteristic of the fast tensile failure of brittle metals.

Hadfield's (Manganese) Steel

An austenitic steel with about 13% manganese and 1.2% carbon. It is tough and wear resistant particularly to gouging abrasion. It is usually quenched from about 1000°C to retain the austenitic phase, in which condition it is relatively soft, about 200 Hv. However, cold deformation such as abrasion induces a local formation of martensite with a hardness of about 600 Hv. See *austenitic manganese steel*.

Hematite

Red iron oxide, Fe_2O_3.

Hafnium and Alloys

Hafnium (Hf), the heaviest of the three metals comprising the Group IV transition metals, is now in production. Because of the startling similarity in their chemical properties, zirconium and hafnium always occur together in nature. In their respective ability to absorb neutrons, however, they differ greatly and this difference has led to their use and surprisingly different ways in nuclear reactors. Zirconium, with a low neutron absorption cross section (0.18 barn), is highly desirable as a structural material in water-cooled nuclear reactor cores. Hafnium, on the other hand, because of its high neutron absorption cross section (105 barns), can be used as a neutron-absorbing control material in the same nuclear reactor cores. Thus, the two elements, which occur together so intimately in nature that they are very difficult to separate, are used as individual and important but contrasting components in the course of nuclear reactors.

Properties

Pure hafnium is a lustrous, silvery metal that is not so ductile nor so easily worked as zirconium; nevertheless, hafnium can be hot- and cold-rolled on the same equipment and with similar techniques as those used for zirconium. All zirconium chemicals and alloys may contain some hafnium, and hafnium metal usually contains about 2% zirconium. The melting point, 2222°C, is higher than that of zirconium, and heat-resistant parts for special purposes have been made by compacting hafnium powder to a density of 98%. The metal has a close-packed hexagonal structure. The electric conductivity is about 6% that of copper. It has excellent resistance to a wide range of corrosive environments.

Hafnium Alloys and Compounds

Hafnium forms refractory compounds with carbon, nitrogen, boron, and oxygen. Hafnium oxide or hafnia, HfO_2, is a better refractory ceramic than zirconia but is costly.

Hafnium carbide, HfC, produced by reacting hafnium oxide and carbon at high temperature, is obtained as a loosely coherent mass of blue-black crystals. The crystals have a hardness of 2910 Vickers and a melting point of 4160°C. It is thus one of the most refractory materials known. Heat-resistant ceramics are made from hafnium titanate by pressing and sintering the powder. The material has the general composition $x(TiO_2) \cdot n(HfO_2)$, with varying values of x and n. Parts are made with 18% titania and 82% hafnia have a density of 7.197 kg/m³, a melting point of about 2204°C, a low coefficient of thermal expansion, good shock resistance, and a rupture strength above 68 MPs at 1093°C. Hafnium nitride, with a melting point of 3300°C, has the highest melting point of any nitride and hafnium boride and, with a melting point of 3260°C, has a melting point higher than any other boride. The alloy Ta_4HfC_5 has the highest melting point of any substance known, about 4215°C.

Hair

The fibrous covering of skins of various animals, used for making coarse fabrics and for stuffing purposes. It is distinguished from wool in having no epidermal scales. It cannot be spun readily, although certain hairs such as camel hair are noted for great softness and can be made into fine fabrics. *Horsehair* is from the manes and tails and is used as a brush fiber and for making haircloth. It is largely imported from China and Argentina, cleaned and sorted. The imported hair from live animals is more resilient than domestic hair from dead animals. *Cattle hair* is taken from slaughtered animals in packing plants. The body hair is used as a binder in plaster and cements, for hair felt, and for stuffing. The tail hair is used for upholstery, filter cloth, and stuffing. The *ear hair* is used for brushes. It has a strong body and fine, tapered point suitable for poster brushes. In the brush industry, it is known as *ox hair*.

H

Artificial horsehair, or *monofil*, is a single-filament cellulose acetate fiber, used for braids, laces, hairnets, rugs, and pile fabrics. *Haircloth* is a stiff, wiry fabric with a cotton or linen wrap and a filling of horsehair. It is elastic and firm and is used as a stiffening and interlining material. The colors are black, gray, and white. The fabric is difficult to weave and disintegrates easily, as the hairs cannot be made into a single strand and must be woven separately. *Press cloth*, used for filtering oils, was made from human hair, which has high tensile strength, resiliency, and resistance to heat. The hair came from China, but filter fabrics are now made from synthetic fibers. *Rabbit hair* from Europe and Australia is used for making felt hats and is referred to as *rabbit fur*, although it does not felt as wool does. The white rabbit hair known as *Angora wool* is from the Angora rabbit of France and Belgium, called *Belgian hare*. The hairs are clipped or plucked four times a year when they are up to 3 in. (8 cm) long. They are soft and lustrous, dye easily to delicate shades, and are used for soft wearing apparel. Because of its fluffiness and hairy characteristics, the wool is difficult to spin and is usually employed in mixtures.

Hair Grease

A grease containing horsehair or wool fiber.

Hairline Cracks

Fine cracking particularly those due to hydrogen damage. See *flake*.

Hairline Craze

Multiple fine surface separation cracks in composites that exceed ¼ in. (6 mm) in length and do not penetrate in depth the equivalent of a full ply of reinforcement. See also *crazing*.

Half Cell

An *electrode* immersed in a suitable *electrolyte*, designed for measurements of *electrode potential*.

Half Hard

A *temper* of nonferrous alloys and some ferrous alloys characterized by tensile strength about midway between those of *dead soft* and *full hard tempers*.

A grade of copper or other nonferrous material. Many such materials are obtainable in various strength levels achieved by varying the severity of the final cold work operation. As the amount of work increases, so does the hardness and strength, leading to terms such as quarter hard, half hard, and full hard.

Half Journal Bearing

A journal bearing extending 180° around a journal.

Half-Life ($t_{1/2}$)

The time required for one-half of an initial (large) number of atoms of a radioactive isotope to decay. Half-life is related to the decay constant λ. By the expression $t_{1/2} = \ln 2/\lambda$. See also *decay constant*.

Half Thickness

The thickness of material that reduces the level of radioactivity by one half. It may be expressed as thickness, that is, mm, or weight per unit area, that is, g/cm^2.

Halides

Fluorides, chlorides, bromides, iodides, and astatides.

Hall Effect

The phenomenon whereby a current-carrying conductor in a magnetic field develops a charge perpendicular to the plane containing the conductor and the field.

Hall–Petch Relationship

The relationship between the grain size and yield strength.

$$\text{Yield strength} = \sigma 0 = kd$$

where
 d is the average grain size
 k and $\sigma 0$ are constants for the material

Typically, the yield strength is inversely proportional to the square root of the grain size.

Hall Process

A commercial process for winning aluminum from alumina by electrolytic reduction of a fused bath of alumina dissolved in cryolite.

Halocarbon Plastics

Plastics based on resins made by the polymerization of monomers composed only of carbon and a halogen or halogens.

Halogen

Any of the elements of the halogen family, consisting of fluorine, chlorine, bromine, iodine, and astatine.

Hammer

A machine that applies a sharp blow to the work area through the fall of a ram onto an anvil. The ram can be driven by gravity or power. See also *gravity hammer* and *power-driven hammer*.

Hammer Forging

Forging in which the work is deformed by repeated blows. Compare with *press forging*.

Hammering

The working of metal sheet into a desired shape over a form or on a high-speed hammer and a similar anvil to produce the required dishing or thinning.

Hammer Welding

Forge welding by hammering.

Hand Brake

A small manual folding machine designed to bend sheet metal, similar in design and purpose to a press brake.

Hand Forge (Smith Forge)

A forging operation in which forming is accomplished on dies that are generally flat. The piece is shaped roughly to the required contour with little or no lateral confinement; operations evolving mandrels are included. The term hand forge refers to the operation performed, while hand forging applies to the part produced.

Hand Lay-Up

In composite processing, it is the process of manually placing (and working) successive plies of reinforcing material or resin-impregnated reinforcement in position on a mold.

Handling Brakes

Irregular *brakes* caused by improper handling of metal sheets during processing. These brakes result from bending or sagging of the sheets during handling.

Handling Strength

A low level of strength initially obtained by an adhesive that allows specimens to be handled, moved, or unclamped without causing disruption of the curing process or affecting bond strength.

Hand Straightening

A straightening operation performed on a surface plate to bring a forging within straightness tolerance. A bottom die from a set of finish dies is often used instead of a surface plate. Hand tools used include mallets, sledges, blocks, jacks, and oil gear presses in addition to regular inspection tools.

Hansgirg Process

A process for producing magnesium by reduction of magnesium oxide with carbon.

Hard Chromium

Chromium electrodeposited for engineering purposes (such as to increase the wear resistance of sliding metal surfaces) rather than as a decorative coating. It is usually applied directly to basis metal and is customarily thicker (>1.2 μm or 0.05 mils) under a decorative deposit, but not necessarily harder.

Hard Drawn

An imprecise term applied to drawn products, such as wire and tubing, that indicates substantial cold reduction without subsequent annealing. Compare with *light drawn*.

Hard-Drawn Copper Wire

Copper wire that has been drawn to size and not annealed.

Hardenability

The relative ability of a ferrous alloy to form martensite when quenched from a temperature above the upper critical temperature. Hardenability is commonly measured as the distance below a quenched surface at which the metal exhibits a specific hardness (e.g., 50 HRC) or a specific percentage of martensite in the microstructure. HRC represents the Rockwell hardness number and 50 HRC is 50 on the Rockwell hardness C scale. See *Rockwell*.

Hardener (Metals)

An alloy rich in one or more alloying elements that is added to a melt to permit closer control of composition than is possible by the addition of pure metals, or to introduce refractory elements not readily alloyed with the base metal. Sometimes called *master alloy* or rich alloy.

Hardener (Plastics)

A substance or mixture added to a plastic composition to promote or control the curing action by taking part in it.

Hardening

Increasing hardness of metals by suitable treatment, usually involving heating and cooling. When applicable, the following more specific terms should be used: *age hardening, case hardening, flame hardening, induction hardening, precipitation hardening,* and *quench hardening.*

Hard Face

A seal facing of high hardness that is applied to a softer material, such as by flame spraying, plasma spraying, electroplating, nitriding, carburizing, or welding.

Hard Facings

Hard facing is a technique by which a wear-resistant overlay is welded on a softer and usually rougher base metal. The method is versatile and has a number of advantages:

1. Wear resistance can be added exactly where it is needed on the surface.
2. Hard compounds and special alloys are easy to apply.
3. Hard facings can be applied in the field as well as in the plant.
4. Expensive alloying elements can be economically used.
5. Protection can be provided in depth.
6. A unique and useful structure is provided by the hard-surfaced, tough-core composite.

Many of the merits of hard facing stem from the hardness of the special materials used. For example, ordinary weld deposits range in hardness up to about 200 Brinell, hardened steels have a hardness up to 700–800 Vickers, and special carbides have a hardness of to about 3000 Vickers. However, it is important to note that the hardness of the materials does not always correlate with wear resistance. Thus, special tests should be performed to determine the resistance of the material to impact, gouging abrasion, grinding (high-stress abrasion), erosion (low-stress scratching abrasion), seizing or galling, and hot wear.

H

Another important point is that durable overlays are not necessarily hard. Most surfacing is used to protect base metals against abrasion, friction, and impact. However, many *hard facings* such as the stainless steels, related nickel-base alloys, and copper alloys are used for corrosion-resistant applications where hardness may not be a factor. Also, the relatively soft leaded bronzes may be used for bearing surfaces. Other facings are also used for heat- and oxidation-resistant applications.

Methods of Application

Hard facings can be applied by

1. Manual, semiautomatic, and automatic method using bare or flux-coated electrodes
2. Submerged arc welding
3. Inert-gas shielded arc welding (both consumable and tungsten electrode types)
4. Oxyacetylene and oxyhydrogen gas welding
5. Metal spraying
6. Welded or brazed on inserts

Gas welding and spraying usually provide higher quality and precise placement of surfaces; arc welding is less expensive. Automatic or semiautomatic methods are preferred where large areas are to be covered, or where repetitive operations favor automation.

Surfacing filler metals are available in the form of drawn wire, cast rods, powders, and steel tubes filled with ferroalloys or hard compounds (e.g., tungsten carbide). The electrodes may take the form of filled tubes or alloyed wires, stick types, or coils specially designed for automated operations. The stick-type electrodes may have a simple steel core and a thick coating containing the special alloys. In submerged arc welding, the alloys may be introduced through a special flux blanket. In spray coating, the materials are used in the form of powders or bonded wire.

Sprayed facings are advantageous in producing thin layers and in following surface contours. With this method, it is usually necessary to fuse the sprayed layer in place after deposition to obtain good abrasion resistance. However, under boundary lubrication conditions, the as-sprayed porosity of the facings may aid against frictional wear.

Hard facings are used in thicknesses from 0.031 to 25.4 cm or more. The thinnest layers are usually deposited by gas welding, usually with low-melting alloys that solidify with many free carbide or other hard compound crystals. The thick deposits are usually made from air-hardening or austenitic steels.

Hard overlays are usually strong in compression but weak in tension. Thus, they perform better in pockets, grooves, or low ridges. Edges and corners must be treated cautiously unless the deposit is tough. Brittle overlay should be deposited over a base of sufficient strength to prevent subsurface flow under excessive compression.

Gas welding is a useful method for depositing small, precisely located surfacings in applications where the base metal can withstand the welding temperatures (e.g., steam valve trim and exhaust valve facings). On the other hand, heavy layers and large areas may be impossible to surface without cracks with the harder, more wear-resistant alloys because of the severe thermal stresses that are encountered (e.g., usually in arc welding). Thus, the opposing factors of wear resistance and freedom from cracking frequently require a compromise in process and material selection.

Materials

Basically, hard-facing materials are alloys that lend themselves to weld fusion and provide hardness or other properties without special heat treatment. Thus, for hard surfacing, the steels and the matrices of high-carbon irons contain enough alloys to cause the hardening transformation during weld cooling, rather than after a quenching treatment.

The properties of the iron-, nickel-, and cobalt-base alloys are strongly affected by carbon content and somewhat by the welding technique used. For example, gas welding usually provides superior abrasion resistance, although carbon pickup may lower corrosion resistance. Arc welding tends to burn out carbon and alloys, thereby lowering abrasion resistance but increasing toughness; high thermal stresses from arc welding may also accentuate cracking tendencies.

The martensitic irons, martensitic steels, and austenitic manganese steels are suited for light-, medium-, and heavy-impact applications, respectively. Gouging abrasion applications usually require an austenitic manganese steel because of the associated heavy impact. Grinding abrasion is well resisted by the martensitic irons and steels. Erosion is most effectively resisted by a good volume of the very hard compounds (e.g., high-chromium irons). Tungsten carbide composites have outstanding resistance to abrasion where heavy impact is not present, but deposits may develop a rough surface.

Selection of materials for hot-wear applications is complicated by oxidation, tempering, softening, and creep factors. Oxidation resistance is provided by using a minimum of 25% chromium. Tempering resistance (up to 593°C) is provided by chromium, molybdenum, tungsten, etc. Creep resistance is provided by the austenitic structure and nickel- or cobalt-bearing alloys. The chromium–cobalt–tungsten grade of materials usually provides a good combination of properties above 649°C.

Hard Head

A hard, brittle, white residue obtained in refining of tin by liquation, containing, among other things, tin, iron, arsenic, and copper. Also a refractory lump of ore, only partly smelted.

Hard Metal

A collective term that designates a sintered material with high hardness, strength, and wear resistance and is characterized by a tough metallic binder phase and particles of carbides, borides, or nitrides of the refractory metals. The term is in general use in Europe, while for the carbides, the term "cemented carbides" is preferred in the United States, and the boride and nitride materials are usually categorized as cermets.

Hardness

A measure of the resistance of a material to surface indentation or abrasion; may be thought of as a function of the stress required to produce some specified type of surface deformation. There is no absolute scale for hardness; therefore, to express hardness quantitatively, each type of test has its own scale of arbitrarily defined hardness. Indentation hardness can be measured by *Brinell*, *Rockwell*, *Vickers*, *Knoop*, and *scleroscope hardness tests*.

Hardness Profile

Hardness as a function of distance from a fixed reference point (usually from the surface).

Hardness Satin

A fabric weaving pattern producing a satin appearance. See also *eight-hardness satin* and *four-hardness satin*.

Hard Rubber

Hard rubber is a plastic. It is a resinous material mixed with a polymerizing or curing agent and fillers and can be formed under heat and pressure to practically any desired shape.

The bulk of today's hard rubber is made with SBR (styrene-butadiene rubber) synthetic rubber. Other types of synthetic rubbers, such as butyl or nitrite or, in rare cases, silicone or polyacrylic, can also be used.

Once it has gone through the process of heat and pressure, hard rubber cannot be returned to its original state and therefore falls into the class of thermosetting plastics, that is, those that undergo chemical change under heat and pressure. It differs, however, from other commercial thermosetting plastics such as the phenolics and ureas and that after it has gone through the thermosetting process it will still soften somewhat under heat. In this characteristic, it mostly resembles the thermoplastic acetates, polystyrenes, and vinyls. It differs from all others in that it is available in pliable sheet form before vulcanization and is therefore adaptable to many shapes for which molds and presses are not necessary. Because of this feature and because it can be softened again after vulcanization, it falls into a class by itself in the field of plastics.

The term "hard rubber" is self-descriptive. The hardness is measured on the Shore D scale, which is several orders of magnitude higher than the Shore A scale used for conventional rubbers and elastomers. Similar in composition to soft rubber, it contains a much higher percentage of sulfur, up to a saturation point of 47% of the weight of the rubber in the compound. If sulfur is present rubber compounds in amounts over 18% of the weight of rubber in the compound when the material is completely vulcanized, the product will be generally known as hard rubber.

Properties and Fabrication

The most important properties of hard rubber are the combination of relatively high tensile strength, low elongation, and extremely low water absorption.

Hard rubber may be compression-, transfer-, or injection-molded. In sheet form it can be hand-fabricated into many shapes. Its machining qualities are comparable to brass, and it may be drilled and tapped. The material lends itself readily to permanent or temporary sealing with hot or cold cements and sealing compounds.

The size and shape of a hard rubber part are dependent only upon the size of press equipment and vulcanizers available.

Uses

Perhaps the largest application for hard rubber is in the manufacture of battery boxes. The water-meter industry is also a large user. Hard rubber linings in coatings either molded or hand laid-up account for large amounts of material. In the electrical industry, hard rubber is used for terminal blocks, insulating materials, and connector protectors. The chemical, electroplating, and photographic industries use large quantities of hard rubber for acid-handling devices.

Hard Solder

A term used erroneously to denote silver-base brazing filler metals.

Hard-Surfaced Polymers

Polymers, developed at the U.S. Department of Energy's Oak Ridge National Laboratories, which combine the flexibility and corrosion resistance of plastics with the strength and durability of metals. They are made by high-energy ion-beam irradiation, which displaces some atoms while ionizing others, forming new bonds in creating a highly cross-linked microstructure of much greater hardness, wear resistance, and abrasion resistance. Chemical resistance and oxidation resistance are also improved.

Hard Surfacing

See preferred terms *surfacing* or *hard facing*.

Hard Temper

Same as *full hard temper*.

Hard Water

Water that contains certain salts, such as those of calcium or magnesium, which form insoluble deposits in boilers and form precipitates with soap.

Haring Cell

A four-electrode cell for measurement of electrolyte resistance and electrode polarization during electrolysis.

Hartmann Lines

See *Lüders lines*.

Haze

Cloudy appearance under or on the surface of a plastic, not describable by the terms *chalking* or *bloom*.

H-Band Steel

Carbon, carbon–boron, or alloy steel produced to specified limits of hardenability; the chemical composition range may be slightly different from that of the corresponding grade of ordinary carbon or alloy steel.

HDPE

See *high-density polyethylene*.

Header

See *upsetter*.

Heading

The upsetting of wire, rod, or bar stock in dies to form parts that usually contain portions that are greater in cross-sectional area than the original wire, rod, or bar.

Head-to-Head

On a polymer chain, a type of configuration in which the functional groups are on adjacent carbon atoms.

Head-to-Tail

On a polymer chain, a type of configuration in which the functional groups or adjacent polymers are as far apart as possible.

Healed-over Scratch

A scratch in a metallic object that occurred in an earlier mill operation and was partially masked in subsequent rolling. It may open up during forming.

Hearth

The bottom portions of certain furnaces, such as blast furnaces, air furnaces, and other reverberatory furnaces, that support the charge and sometimes collect and hold molten metal.

Heat

A stated tonnage of metal obtained from a period of continuous melting in a cupola or furnace, or the melting period required to handle this tonnage.

Heat-Activated Adhesive

A dry adhesive that is rendered tacky or fluid by application of heat, or heat and pressure, to the assembly.

Heat-Affected Zone (HAZ)

That portion of the base metal that was not melted during brazing, cutting, or welding but whose microstructure and mechanical properties were altered by the heat.

Heat Buildup

In processing of plastics, the rise in temperature in a part resulting from the dissipation of applied strain energy as heat or from applied mold cure heat. See also *hysteresis*.

Heat Check

A pattern of parallel surface cracks that are formed by alternate rapid heating and cooling of the extreme surface metal, sometimes found on forging dies and piercing punches. There may be two sets of parallel cracks, one set perpendicular to the other.

Heat Checking

A process in which fine cracks are formed on the surface of a body in sliding contact due to the buildup of excessive frictional heat.

Heat Cleaned

A condition in which glass or other fibers are exposed to elevated temperatures to remove preliminary sizings or binders not compatible with the resin system to be applied.

Heat-Deflection Temperature

The temperature at which a standard plastic test bar deflects a specified amount under a stated load. Now called deflection temperature under load (DTUL).

Heat-Disposable Pattern

In foundry practice, a *pattern* formed from a wax- or plastic-base material that is melted from the mold cavity by the application of heat.

Heat Distortion

Distortion or flow of a material or configuration due to the application of heat.

Heat Distortion Point

The temperature at which a standard plastic test bar deflects a specified amount under a stated load. Now called deflection temperature.

Heat-Fail Temperature

The temperature at which delamination of an adhesively bonded structure occurs under static loading and shear.

Heat Forming

See *thermoforming*.

Heat Insulators

Materials having high resistance to heat rays, or low heat conductivity, used as protective insulation against either hot or cold influences. The materials are also called "thermal insulators." Insulators for extremely high external temperatures, as on aerospace vehicles, are of ablative materials. Efficiency of heat insulators is measured relatively in Btu/(h·ft·F) {W/(m·K)}, known as the "K factor." The thermal conductivity of air and gases is low, and the efficiency of some insulators, especially fibrous ones, is partly due to the airspaces. On the other hand, the thermal conductivity of a porous insulator may be increased if water is absorbed into the spaces.

A wide variety of materials are used as thermal insulators in the form of powder or granules for loose fill, blanket batts of fibrous materials for wall insulation, and sheets or blocks. Although metals are generally high-heat conductors, the polished white metals may reflect as much as 95% of the heat waves and make good *reflective insulators*. But for this purpose the bright surface must be exposed to airspace. Aluminum has a high K factor, up to 130, but crumbled alumina foil is an efficient thermal insulator as a fill in walls. Wool and hair, either loose or as felt, with a K factor of 0.021 are among the best of the insulators, but organic materials are usable only for low temperatures, and they are now largely replaced by mineral wool or ceramic fibers. Mineral wool has a low K factor, 0.0225. *Tipersul*, of DuPont, is a *potassium titanate fiber* used loose or in batts, blocks, or sheets. Its melting point is 2500°F (1371°C), and it withstands continuous temperatures to 2200°F (1204°C). Another ceramic fiber, called *Fibrox*, for the same purpose, is a *silicon oxycarbide*, SiCO, in light, fluffy fibers.

Magnesia or asbestos, or combinations of the two, is much used for insulation of hot pipelines, while organic fibrous materials are used for cold lines. *High-heat insulators*, for furnaces and boilers, are usually made of refractory ceramics such as chromite. For intermediate temperatures, expanded glass, such as *Foamglass* of Pittsburgh-Corning Corp., may be used. Some rigid materials of good structural strength serve as structural parts as well as insulators. *Roofinsul*, used for roof decks, is a lightweight board compressed from wood fibers. *Ludlite board*, of Allegheny

Ludlum Corp., for paneling, is a thin stainless steel backed with a magnesia–asbestos composition. Insulators in sheets, shapes, and other forms are sold under a great variety of trade names. *Dry Zero*, for refrigeration insulation, consists of kapok batts enclosed in fiberboard. *Balsam wool* is wood fibers chemically treated to prevent moisture absorption.

Heat-Resistant Alloys (Cast)

Cast alloys suitable for use at service temperatures to at least 538°C and, for some alloys, to 1093°C are classed as *heat-resistant* or *high-temperature* alloys. They have the characteristic of corroding at very slow rates compared with unalloyed or low-alloy cast iron or steel in the atmospheres to which they are exposed, and they offer sufficient strength at the operating temperature to be useful as load-carrying engineering structures. Iron-base and nickel-base alloys comprise the bulk of production, but cobalt-base, chromium-base, molybdenum-base, and columbium-base alloys are also made.

Although some cast heat-resistant alloys are available in composition similar to wrought alloys, it is necessary to differentiate between them. Cast alloys are made to somewhat different chemical specifications than wrought alloys; physical and mechanical properties for each group are also somewhat different.

For these reasons, it is advisable to follow the alloys designated as the H series by the Alloy Casting Institute of the American Steel Founders Society as well as nickel-base alloys and cobalt-base alloys. Most of the nickel-base and cobalt-base alloys are also known as superalloys because of their exceptional high-temperature stress-rupture strength and creep resistance as well as corrosion and oxidation resistance.

There are, moreover, a number of heat-resistant cast alloys that are not available in wrought form; this is frequently of advantage in meeting special conditions of high-temperature service. In addition to the grades HA to HX discussed later, the industry produces special heat-resistant compositions. Many of these are modifications of the standard types, but some are wholly different and are designed to meet unique service conditions.

Selection of a particular alloy, of course, is dependent upon the application, and in this article the composition, structure, and properties of the various cast heat-resistant alloys are discussed from this point of view.

Proper selection of an alloy for a specific high-temperature service involves consideration of some or all of the following factors: (1) required life of the part, (2) range and speed of temperature cycling, (3) the atmosphere and its contaminants, (4) complexity of casting design, and (5) further fabrication of the casting. The criteria that should be used to compare alloys depend on the factors enumerated, and the designer will be aided in the choice by providing the foundry with as much pertinent information as possible on intended operating conditions before reaching a definite decision to use a particular alloy type.

Physical and Mechanical Properties

For high-temperature design purposes, a frequently used design stress is 50% of the stress that will produce a creep rate of 0.0001%/h maximum operating temperature. Such a value should be applied only under conditions of direct axial static loading and essentially uniform temperature or slow temperature variation. When impact loading or rapid temperature cycles are involved, a considerably lower percentage of the limiting creep stress should be used. In the selection of design stresses, safety factor should be higher if the parts are inaccessible, nonuniformly loaded, or of complex design;

they may be lower if the parts are accessible for replacement, fully supported or rotating, and of simple design with little or no thermal gradient.

H-Series Cast Alloys

The H-series cast alloys include iron–chromium, iron–chromium–nickel, and iron–nickel–chromium alloys also containing 0.20%–0.75% carbon, 1%–2.5% silicon, and 0.35%–2% manganese. A letter (A–X) following the H is used to distinguish alloy compositions more closely. The iron–chromium cast alloys (HA, HC, and HD) contain as much as 30% chromium and under 7% nickel. The iron–chromium–nickel cast alloys (HE, HF, HH, HI, HK, and HL) contain as much as 32% chromium and 22% nickel. And the iron–nickel–chromium cast alloys (HN, HP, HP-50 WZ, HT, HU, HW, and HX) contain as much as 68% nickel (HX) and 32% chromium (HN) so that some of these alloys are actually nickel-base instead of iron-base alloys.

In selecting alloys from this group, the following factors are considered:

1. Increasing nickel content increases resistance to carburization, decreases hot strength somewhat, and increases resistance to thermal shock.
2. Increasing chromium content increases resistance to corrosion and oxidation.
3. Increasing carbon content increases hot strength.
4. Increasing silicon content increases resistance to carburization but decreases hot strength.

All are noted primarily for their oxidation resistance and ability to withstand moderate to severe temperature changes. Most are heat treatable by aging room-temperature tensile properties in the aging condition ranging from 503 to 793 MPa in terms of ultimate strength, 297–552 MPa in yield strength, and 4%–25% in elongation. Hardness of the aged alloys ranges from Brinell 185 to 270. Applications include heat-treating fixtures, furnace parts, oil-refinery and chemical processing equipment, gas turbine components, and equipment used in manufacturing steel, glass, and rubber.

Both the nickel-base and cobalt-base alloys are probably best known for their use in aircraft turbine engines for disks, blades, vanes, and other components. The nickel alloys contain 50%–75% nickel and usually 10%–20% chromium and substantial amounts of cobalt, molybdenum, aluminum, and titanium and small amounts of zirconium, boron, and, in some cases, hafnium. Carbon content ranges from less than 0.1% to 0.20%. Because of their complex compositions, they are best known by trade names, such as B-1900; Hastelloy X; IN-100, IN-738X, and IN-792; Rene 77, Rene 80, and Rene 100; Inconel 713C, Inconel 713 L C, Inconel 718, and Inconel X-750; MAR-M 200, MAR-M 246, and MAR-M 247; Udimet 500, Udimet 700, and Udimet 710; and Waspaloy. The high-temperature strength of most of these alloys is attributed to the presence of refractory metals, which provide solid solution strengthening; the presence of grain-boundary-strengthening elements, such as carbon, boron, hafnium, and zirconium; and, because of the presence of aluminum and titanium, strengthening by precipitation of an $Ni_3(Al, Ti)$ compound known as "gamma prime" during age hardening. Many of these alloys provide 1000 h stress-rupture strengths in the range of 690–759 MPa at 649°C and 55–124 MPa at 982°C.

The cobalt alloys contain 36%–65% cobalt, usually more than 50%, and usually about 20% chromium and substantial amounts of nickel, tungsten, tantalum, molybdenum, iron and/or aluminum and small amounts of still other ingredients. Carbon content is

H

0.05%–1% Erie, although not generally as strong as the nickel alloys, some may provide better corrosion and oxidation resistance at high temperatures. These alloys include L-605; S-816; V-36; WI-52; X-40; J-1650; Haynes 21 and Haynes 151; AiResist 13, AiResist 213, and AiResist 215; and MAR-M 302, MAR-M 322, and MAR-M 918. Their 1000 h stress-rupture strengths range from about 276 to 483 MPa at 649°C and from about 28 to 103 MPa at 982°C.

Heat-Resistant Plastics (Superpolymers)

Several different plastics developed in recent years that maintain mechanical and chemical integrity above 204°C for extended periods are frequently referred to as superpolymers. They are polyimide, polysulfone, polyphenylene sulfide, polyarylsulfone, and aromatic polyester.

In addition to their high-temperature resistance, all these materials have in common high strength and modulus and excellent resistance to solvents, oils, and corrosive environments. They are also among the highest-priced plastics, and a major disadvantage is processing difficulty. Molding temperatures and pressures are extremely high compared with conventional plastics. Some of them, including polyimides and aromatic polyester, are not molded conventionally. Because they do not melt, the molding process is more of a sintering operation. Because of their high price, suprapolymers are largely used in specialized applications in the aerospace and nuclear energy field.

Indicative of their high-temperature resistance, the superpolymers have a glass transition temperature well over 260°C as compared to less than 177°C for most conventional plastics. In the case of polyimides, the glass temperature is greater than 427°C, and the material decomposes rather than softens when heated excessively.

Polysulfone has the highest service temperature of any melt-processable thermoplastic. Its flexural modulus stays above 2040 MPa at up to 160°C. At such temperatures, it does not discolor or degrade.

Aromatic polyester does not melt, but at 427°C can be made to flow in a nonviscous manner similar to metals. Thus, filled and unfilled forms and parts can be made by hot sintering, high-velocity forging, and plasma spraying. Notable properties are high thermal stability, good strength at 316°C, high thermal conductivity, good wear resistance, and extrahigh compressive strength.

Heat Sealing

A method of joining plastic films by simultaneous application of heat and pressure to areas in contact.

Heat-Sealing Adhesive

A thermoplastic film adhesive that is melted between the adherend surfaces by heat application to one or both of the surfaces.

Heat Shock

A test to determine the stability of a material by sudden exposure to a significantly higher or lower temperature for a short period of time.

Heat Sink

A material that absorbs or transfers heat away from a critical element or part.

Heat Time

In resistance welding, the time that the current flows during any one impulse.

Heat Tinting

Coloration of a metal surface through oxidation by heating to reveal details of the microstructure.

Heat Transfer

Flow of heat by conduction, convection, or radiation.

Heat-Transfer Agents

Liquids or gases used as intermediate agents for the transport of heat or cold between the heat source in the process, or for dissipating heat by radiation. Water, steam, and air are the most common heat-transfer agents, but the term is usually applied only to special materials. Air can be used over the entire range of industrially important temperatures, but it is a poor heat-transfer medium. Water can be used only between its freezing and boiling points, unless high pressures are employed to keep the water liquid. A liquid agent should have a wide liquid range, be noncorrosive and nontoxic, and have low vapor pressure to minimize operational loss.

Gallium, with a freezing point of 85.6°F (29°C) and boiling point of 3600°F (1982°C), offers an exceptionally wide liquid range; but it is too costly for ordinary use, and in the liquid metal it also attacks other metals. Mercury is used for heat transfer but is costly and toxic, and at 1200°F (649°C), it exerts a vapor pressure of 500 lb/in.^{2} (3.4 MPa). Among commercial products, the heat-transfer fluid spans the range in chemical structure from *alkylated benzenes, alkylated biphenyls, alkylated naphthalenes, unhydrogenated polyphenols, benzylated aromatics, diphenyl-diphenyl oxide eutectics, polyalkylene glycols, dicarboxylic acid esters, polymethyl siloxanes, mineral oils*, and inorganic nitrate salts.

Brine solutions of sodium or calcium chlorides are used for heat transfer for temperatures down to −6°F (−21°C), but are corrosive to metals. Molten sodium and potassium salts are used for temperatures from 1112°F to 2552°F (600°C–1400°C), but are corrosive to metals.

Heat-Treatable Alloy

An alloy that can be hardened by heat treatment.

Heat-Treating Film

A thin coating or film, usually an oxide, formed on the surface of a metal during heat treatment.

Heat Treatment

Heating and cooling a solid metal or alloy in such a way as to obtain desired conditions or properties. Heating for the sole purpose of hot working is excluded from the meaning of this definition.

Heavy Alloy

This is a name applied to tungsten–nickel alloy produced by pressing and sintering the metallic powders. It is used for screens for x-ray tubes and radioactivity units, for contact surfaces for circuit breakers, and for balances for high-speed machinery. The original composition was 90% tungsten and 10% nickel, but a proportion of copper is used to lower sintering temperature and give better binding as the copper wets the tungsten. Too large a proportion of copper makes the product porous. In general, the alloys weigh nearly 50% more than the last, permitting space-saving in counterweights and balances, and they are more efficient as gamma-ray absorbers than lead. They are highly heat resistant, retain a tensile strength of about 137 MPa at 1093°C, have an electric conductivity about 15% that of copper, and can be machined and brazed with silver solder.

An alloy of 90% tungsten, 7.5% nickel, and 2.5% copper has a tensile strength of 930 MPa, elongation 15%, Rockwall hardness C30, and weight of 16.885 kg/m³. *Kenertium* has this composition. *Fansteel 77 metal* contains 89% tungsten, 7% nickel, and 4% copper. Specific gravity is 16.7, tensile strength 586 MPa, elongation 17%, and Brinell hardness 280. The coefficient of expansion is low, $6.5 \times 10^{-6} \text{ K}^{-1}$. *Heavy metal powder*, for making parts by powder metallurgy, is prealloyed with the tungsten in a matrix of copper–nickel to prevent settling out of the heavy tungsten.

Heavy-Duty Oil

An oil that is stable against oxidation, protects bearings from corrosion, and has detergent and dispersant properties. Heavy-duty oils are suitable for use in gasoline and diesel engines.

Heel

Synonymous with *base*.

Heel Block

A block or plate usually mounted on or attached to a lower die in a forming or forging press that serves to prevent or minimize the deflection of punches or cams.

Helical Winding

In filament-wound items, a winding in which a filament band advances along a helical path, not necessarily at a constant angle, except in the case of a cylinder.

Helium

Helium is a colorless, odorless, elementary gas, He, with a specific gravity of 0.1368, liquefying at −268.9°C, freezing at −272.2°C. It has a valency of zero and forms no electron-bonded compounds. It has the highest ionization potential of any element. The lifting power of helium is only 92% that of hydrogen, but it is preferred for balloons because it is inert and nonflammable and is used in weather balloons. It is also used instead of air to inflate large tires for aircraft to save weight. Because of its low density, it is also used for diluting oxygen in the treatment of respiratory diseases. Its heat conductivity is about six times that of air, and it is used as a shielding gas in welding and in the vacuum tube of electric lamps. Because of its inertness, helium can also be used to hold free chemical radicals, which, when released, give high energy and thrust for missile propulsion. When electric current is passed through helium, it gives a pinkish-violet light and is thus used in advertising signs. Helium can be obtained from atmospheric nitrogen but comes chiefly from natural gas.

Properties

Helium has the lowest solubility in water than any known gas. It is the least reactive element. The density and the viscosity of helium vapor are low. Thermal conductivity and heat content are exceptionally high. Helium can be liquefied, but its condensation temperature is the lowest of any known substance. At pressures below 2.5 MPa, helium remains liquid even at absolute zero.

Applications

Gases

Helium was first used as a lifting gas in balloons and dirigibles. This use continues for high-altitude research and for weather balloons.

Welding

The principal use of helium is in inert gas–shielded arc welding. Using helium instead of argon permits a greater heat release, which is useful in welding very heavy sections or in high-speed machine welding of long seams. By mixing helium and argon, the optimum heat release can be obtained for different welding jobs.

Superconductive Devices

The greatest potential for helium use continues to emerge from extremely-low-temperature applications. Helium is the only refrigerant capable of reaching temperatures below 14 K. In the laboratory many fundamental properties of matter are studied at temperatures near absolute zero with helium refrigeration. Infrared detectors and masers operate with exceptionally low noise distortion at these low temperatures. The chief value of ultralow temperature is a development of the state of superconductivity, in which there is virtually zero resistance to the flow of electricity. Very large currents are carried by even small conductors with little loss of voltage. Electromagnets producing immensely powerful magnetic fields can be made small and light and are energized with modest amounts of electric power through the use of superconducting windings. These magnets are already used in particle accelerators, bubble chambers, and plasma confinement for nuclear physics research. Thermonuclear and magnetohydrodynamic (MHD) power plants are expected to use superconducting magnets. Additional applications are electric motors and generators. Superconductive devices make highly sensitive detectors of electric voltage and frequency, magnetic field strength, and temperature, especially at low-temperature levels.

Lasers

Helium is also used in gas-discharge lasers. Energy is transferred by helium to the lasing gas, carbon dioxide, or neon, for example.

Rockets

Consumption of helium as a pressurizing gas in liquid-fueled rockets declined with the completion of the Apollo space program. Because it is light, inert, and relatively insoluble in the fuel and oxidizer fluids, helium is an ideal material to fill the tankage as the liquids are consumed.

Breathing Mixtures

Use of helium–oxygen breathing mixtures for divers at great depths is required to eliminate the narcotic effects of nitrogen. The low density and low viscosity of helium also reduce the work of breathing. Similarly, helium–oxygen breathing mixtures promote both intake of oxygen and removal of carbon dioxide for persons whose breathing passages are constricted.

Nuclear Reactors

Inertness and heat-transfer capability make helium an excellent working fluid for gas-cooled nuclear power reactors. Because the reactor core is composed of graphite and ceramic materials, very high temperatures can be attained without damage. Helium-cooled reactors operate with the highest efficiency of all reactor types. In addition to electric power generation, with helium, working fluid nuclear reactors can provide the process heat for coal gasification, steelmaking, and various chemical processes.

Chemical Analysis

Helium is the most frequently used carrier gas for chemical analysis by gas chromatography. It is the most sensitive leak detection fluid and can be used at extremes of high and low temperature.

Hematite

(1) An iron mineral crystallizing in the rhombohedral system; the most important ore of iron. (2) An iron oxide, Fe_2O_3, corresponding to an iron content of approximately 70%. Also known as red hematite, red iron ore, and rhombohedral iron ore.

Hemming

A bend of 180° made in two steps. First, a sharp-angle bend is made; next, the bend is closed using a flat punch and a die.

Henry

An electrical unit denoting the inductance of a circuit in which a current varying at the rate of 1 A/s produces an electromotive force of 1 V.

HERF

A common abbreviation for *high-energy-rate forging* or *high-energy-rate forming*.

Hermetic

Sealed so that the object is gastight. The test for hermeticity is to fill the object with a test gas, often helium, and observe leak rates when the object is placed in a vacuum. Plastic encapsulation is not hermetic because it allows permeation by gases.

Herringbone Bearing

Any plain, sleeve, or thrust bearing with herringbone-shaped oil grooves.

Herringbone Pattern

Same as *chevron pattern*.

Hertz

A designation of electrical frequency that denotes cycles per second. Abbreviated Hz.

Hertzian Cone Crack

See *percussion cone*.

Hertzian Contact Area

(1) The contact area (also, diameter or radius of contact) between two bodies calculated according to Hertz's equations of elastic deformation. (2) The apparent area of contact between two nonconforming solid bodies pressed against each other, as calculated from Hertz's equations of elastic deformation.

Hertzian Contact Pressure

(1) The pressure at a contact between two solid bodies calculated according to Hertz's equations of elastic deformation. (2) The magnitude of the pressure at any specified location in a Hertzian contact area, as calculated from Hertz's equations of elastic deformation.

Hertzian Stress

See *contact stress*.

Heterogeneity

The degree of nonuniformity of composition or properties. Compare with *homogeneity*.

Heterogeneous

Of a body of material or matter, comprised of more than one phase (solid, liquid, and gas) separated by boundaries; similarly of a solid, comprised of more than one chemical, crystalline, and/or glassy species, separated by boundaries.

Heterogeneous Equilibrium

In a chemical system, a state of dynamic balance among two or more homogeneous phases capable of stable coexistence in mutual or sequential contact.

Heterogeneous Nucleation

In the crystallization of polymers, the growth of crystals on vessel surfaces, dust, or added nucleating agents.

Heyn Stresses

Same as *microscopic stresses*.

Hexa

An abbreviated form of hexamethylenetetramine, a source of reactive methylene for curing *novolacs*.

Hexagonal (Lattices for Crystals)

Having two equal coplanar axes, a_1 and a_2, at 120° to each other and the third axis, c, at right angles to the other two; c may or may not equal a_1 and a_2.

Hexagonal Close-Packed

(1) A structure containing two atoms per unit cell located at (0, 0, 0) and (1/3, 2/3, 1/2) or (2/3, 1/3, 1/2). (2) One of the two ways in which

spherical objects can be most closely packed together so that the close-packed planes are alternately staggered in the order A-B-A-B-A-B. See the term *unit cell*.

High Aluminum Defect

An α-stabilized region in titanium containing an abnormally large amount of aluminum that may span a large number of β grains. It contains an inordinate fraction of primary α, but has a microhardness only slightly higher than the adjacent matrix. Also termed type II defects.

High-Conductivity Copper

Copper that, in the annealed condition, has a minimum electrical conductivity of 100% *IACS* as determined by ASTM test methods.

High-Cycle Fatigue

Fatigue that occurs at relatively large numbers of cycles. The arbitrary, but commonly accepted, dividing line between high-cycle fatigue and *low-cycle fatigue* is considered to be about 10^4–10^5 cycles. In practice, this distinction is made by determining whether the dominant component of the *strain* imposed during cyclic loading is elastic (high cycle) or plastic (low cycle), which in turn depends on the properties of the metal and on the magnitude of the nominal *stress*.

High-Density Polyethylene (HDPE)

High-density polyethylenes (HDPEs) are thermoplastic materials that are solid in their natural state. Under extrusion conditions of heat, pressure, and mechanical shear, they soften into a highly viscous, molten mass and take the shape of the desired end product. The polymer is characterized by its opacity, chemical inertness, toughness at both low and high temperatures, and moisture barrier and electrical-insulating properties.

The term HDPE generally includes polyethylene ranging in density from about 0.94 to 0.965 g/cm³. While molecules in low-density polyethylene are branched and linked randomly, those in the higher-density polyethylene are linked in longer chains with fewer side branches, resulting in a more rigid material with greater strength, hardness, and chemical resistance and a higher softening temperature.

The physical properties of HDPE are also affected by the weight-average molecular weight (MW) of the polymer. As the MW increases, the mechanical properties also increase significantly, but the polymer becomes more difficult to process. Polymer grades with MW in the 200,000–500,000 range are considered high-performance, high-molecular-weight HDPEs (HMW HDPEs). The combination of HMW and high density provides even higher stiffness and abrasion resistance and extended product surface life in critical environmental applications. HMW resins also provide excellent environmental stress-corrosion cracking resistance. Polyethylenes with a MW 10 times that of HMW HDPE are also available. These materials, referred to as ultrahigh-molecular-weight polyethylenes (UHMWPEs), have the highest abrasion resistance and highest impact strength of any plastic.

The four major end-use markets of HMW HDPE are pipe, large-part blow molding, film, and sheet. HMW pipe is used in applications that require resistance to environmental cracking an excellent impact resistance at temperatures as low as −50°C (−60°F).

The major application areas for HMW HDPE blow-molded parts are shipping containers in sizes up to 210 L (55 gal), refuse containers, and bulk storage containers (1040 L or 275 gal). The largest of the film applications are grocery bags and garbage can liners. Typical applications for HMW sheet include truck bed liners, shipping pallets, and pond liners.

High-Energy-Rate Compacting

Compacting of a powder at a very rapid rate by the use of explosives in a closed die.

High-Energy-Rate Forging (HERF)

A closed-die hot- or cold-forging process in which the stored energy of the high-pressure gas is used to accelerate a ram to unusually high velocities in order to effect deformation of the workpiece. Ideally, the final configuration of the forging is developed in one blow or, at most, a few blows. In high-energy-rate forging, the velocity of the ram, rather than its mass, generates the major forging force. Also known as HERF processing, high-velocity forging, and high-speed forging.

High-Energy-Rate Forming

A group of forming processes that applies a high rate of strain to the material being formed through the application of high rates of energy transfer. See also *explosive forming*, *high-energy-rate forging*, and *electromagnetic forming*.

High-Energy-Rate Forming (Explosive Forming)

In a high-energy-rate forming (HERF), parts are shaped by the extremely rapid application of high pressures. Pressures as high as 13,600 MPa and speeds as high as 914 m/s may be used.

The principal advantages of HERF are as follows:

1. Parts can be formed that cannot be formed by conventional methods.
2. Exotic metals, which do not readily lend themselves to conventional forming processes, may be formed over a wide range of sizes and configurations.
3. The method is excellent for restrike operations.
4. Springback after forming is reduced to a minimum.
5. Dimensional tolerances are generally excellent.
6. Variations from part to part are held to a minimum.
7. Scrap rate is low.
8. Less equipment and fewer dies cut down on production lead time.

Explosive Forming

There are three different explosive forming techniques that have been used: free forming, bulk-head forming, and cylinder forming. Both free forming and bulkhead forming allow the workpiece to be heated before forming. Although air can be used as a coupling medium between the explosive and the workpiece, in most cases water is used. Efficiency in air is approximately 4%; in water, 33%.

Changes in mechanical properties caused by explosive forming correlate closely with those obtained in material cold-worked to the same degree.

In HERF of nickel alloys and the 300 series of stainless steel, strength and hardness are increased and, as expected, ductility is decreased.

Studies indicate that explosive impact hardening is useful with materials hardened by cold work, for example, austenitic stainless steel, Hadfield steel, nickel, and molybdenum. An interesting application is the possibility of restoring mechanical properties of parts that have been welded or heat-treated and thus softened.

Simple forgings can be made by explosive forming techniques. One study showed that aluminum alloys could be explosive-forged if the design had no extreme contours.

Copper has been welded to copper by the application of explosive force. The joint was metallurgical, not mechanical.

Expanding Gas

Gases generated by the burning of propellant powders in a closed container produce the pressures required to form metals. Most of the work that uses propellant powder gases as the energy source is classified as bulge-type forming.

Dynapak

Dynapak is another forming method and machine that uses a gas-powered die to form a part and the operation is under close control.

Dynapak can form low-alloy steels such as AISI 4340, austenitic steels of the 200 and 300 series, titanium, and the refractory metals, to name a few examples. It can also be used to compact powders to a density higher than normally obtained with conventional powder-metallurgy processes.

Parts have been extruded with excellent surface finish and close dimensional tolerances. Web thicknesses of 0.25 mm can be obtained and wire 0.50 mm in diameter has been extruded directly from a 25.4 mm billet.

Hot and cold forgings of various materials can be produced with zero draft angles and minimum radii. The smooth, close tolerance surfaces that are produced often minimize finish machining requirements.

Capacitor Discharge Techniques

Explosives and compressed gases are not the only means of achieving high deformation rates. In one type of device, a spark is discharged in a nonconducting liquid medium and generates a shock wave that travels at the speed of sound from the spark source to the workpiece.

This forming technique has several advantages:

1. Explosives, with their potential safety hazard, are eliminated.
2. Parts can be sized into a die by several applications of energy impulses. Since the devices is electrical, components of the system do not have to be repositioned after each shot.
3. A standard machine tool, based on this principle, can be constructed for about 1/10 the cost of a conventional hydraulic press and occupy only a fraction of the floor space.

One problem encountered with capacitor discharge techniques is the containment of high voltages, because stored electrical energy increases with the square of the voltage ($E = CV^2$, where C is the capacitance and V the voltage). There are two other problems: corona and arcing. Normal safety procedures for handling high voltage must be followed.

Magnetic Forming

Use of the pressure generated by a magnetic field permits parts to be formed in 6 s. Of this time, only 10–20 µs may be needed for the forming operation; the balance is taken up by setup and removal of the part from the apparatus.

The many possible magnetic coil configurations permit a wide variety of forming operations from expanding to compressing to forming flat stock. Coils usually are massively supported since they must be able to withstand the high pressures generated by the magnetic field. For example, at a flux density of 300,000 Gauss, the pressure is approximately 340 MPa. At higher fields (up to 1 million Gauss), magnetic forming devices may generate pressures exceeding 3400 MPa.

The value of magnetic forming methods lies in the ability to perform quickly and economically many conventional operations such as swaging, bulging, expanding, and assembly.

High-Frequency Heating

The heating of materials by dielectric loss in a high-frequency electrostatic field. The material is exposed between electrodes and is heated quickly and uniformly by absorption of energy from the electrical field.

High-Frequency Resistance Welding

A resistance welding process that produces coalescence of metals with the heat generated from the resistance of the workpieces to a high-frequency alternating current in the 10–500 kHz range and the rapid application of an upsetting force after heating is substantially completed. The path of the current in the workpiece is controlled by use of the proximity effect (the feed current follows closely the return current conductor).

High-Impact Polystyrenes (HIPSs)

High-impact polystyrenes (HIPSs) are thermoplastic resins produced by dissolving polybutadiene rubber in a styrene monomer before polymerizing. Polystyrene (PS) forms the continuous phase, with the rubber phase existing as discrete particles having inclusions of PS. Different production techniques allow the rubber phase to be tailored to a wide range of properties. With advances in rubber morphology, control of molecular weight distribution, and additives, technology has enabled producers to offer a wide range of standard and specialty HIPS products, which offer good dimensional stability, low-temperature impact properties, and high rigidity. Relative disadvantages of HIPS are their poor high-temperature properties and lower chemical resistance compared with most crystalline polymers.

HIPSs are used in myriad applications and industries because of their ease of processing, performance, and low cost. Major industries and markets include packaging and disposables, appliances and consumer electronics, toys and recreation, buildings, and furnishings. In recent years, product development has focused on specialty products. Grades are now available that provide improved resistance to stress cracking from fats, oils, and chlorofluorocarbon blowing agents. Also available are high-gloss/high-toughness, ignition-resistant, glass-filled (up to 40 wt.% type E glass), and very-high-impact products.

All of the conventional processing technologies for thermoplastics can be used on HIPS. These include injection molding, structural-foam molding, extrusion, thermoforming, and injection blow molding.

High Interstitial Defect

Interstitially stabilized α-phase region in titanium of substantially higher hardness than surrounding material. It arises from very high local nitrogen or oxygen concentrations that increase the β transus and produce the high-hardness, often brittle α phase. Such a defect is often accompanied by a void resulting from thermomechanical working. Also termed type I or low-density interstitial defects, although they are not necessarily low density.

Highlighting

Buffing or polishing selected areas of a complex shape to increase the luster or change the color of those areas.

Highly Deformed Layer

In tribology, a layer of severely plastically deformed material that results from the shear stresses imposed on that region during sliding contact. See also *Beilby layer* and *white layer*.

High Polymer

A macromolecular substance that, as indicated by the polymer by which it is identified, consists of molecules that are multiples of the low molecular unit and have a molecular weight of at least 20,000.

High-Pressure Laminates

Laminates molded and cured at pressures not lower than 6.9 MPa (1.0 ksi) and more commonly in the range of 8.3–13.8 MPa (1.2–2.0 ksi).

High-Pressure Molding

A plastic molding or laminating process in which the pressure used is greater than 1400 kPa (200 psi), but commonly 7000 kPa (1000 psi).

High-Pressure Spot

See *resin-starved area.*

High Pulse Current

In welding, current levels during the high pulse time that produces the high-heat level.

High Pulse Time

In welding, the duration of the high current pulse time.

High-Residual-Phosphorus Copper

Deoxidized copper with residual phosphorus present in amounts (usually 0.013%–0.04%) generally sufficient to decrease appreciably the conductivity of the copper.

High-Speed Machining

High-productivity machining processes that achieve cutting speeds in excess of 600 m/min (2,000 sfm) and up to 18,000 m/min (60,000 sfm). Such speeds result in segmented shear-localized chips rather than the continuous chip formation associated with lower-speed machining processes.

High-Speed Steels

High-speed steels are those alloy steels developed and used primarily for metal-cutting tools. They are characterized by being heat-treatable to very high hardness (usually Rockwell C64 and over) and by retaining their hardness and cutting ability at temperatures as high as 538°C, thus permitting truly high-speed machining. Above 538°C, they rapidly soften and lose cutting ability.

All high-speed steels are based on either tungsten or molybdenum (or both) as the primary heat-resisting additive, with carbon for high hardness; chromium for ease of heat treating; vanadium for grain refining and, in amounts over 1%, for abrasion resistance; and sometimes cobalt for additional hardness or resistance to heat softening.

Popular compositions of high-speed steels readily available in the United States, the most common characteristics, and recommend temperature ranges are shown in Table H.1. By far the most

TABLE H.1
High-Speed Steel

AISI Type	Austenitizing Temperature, °C	Tempering Temperature, °C	Resistance to Decarburization[a]	Characteristics
T1	1260–1288	551–580	Excellent	General purpose
T5	1274–1301	551–580	Fair	Extra heat resistance
T15	1218–1245	538–566	Good	Heaviest duty
M1	1177–1218	551–566	Poor	General purpose
M2	1190–1232	551–566	Fair	General purpose
M3	1204–1232	538–551	Fair	Heavy duty, abrasive materials
M7	1190–1227	538–566	Poor	Special applications
M10	1177–1218	538–566	Poor	General purpose
M36	1218–1245	551–580	Fair	Extra heat resistance

[a] During heat treatment.

important of these are M1, M2, and M10. Nearly all twist drills, taps, chasers, reamers, saw blades, and high-speed steel hand tools are made from M1 or M10, whereas the more complex tools such as milling cutters, gear hobs, broaches, and form tools utilize the M2 type.

Special applications, such as machining of hard heat-treated materials, call for a more abrasion-resistant type (such as M3 or T15), and extra heavy-duty cutting involving maximum heat generation leads to the use of the high-cobalt types M36 or T5.

Properties

High-speed steel possesses the highest hardness after heat treating of any well-known ferrous alloy. The value of high-speed steel lies in its ability to retain this hardness under considerable exposure to heat and to retain a sharp cutting edge when exposed to abrasive wear.

The hardness of high-speed steel when heated-treated is usually Rockwell C64–C66, equivalent to Brinell 725–760. It is brittle at this hardness, particularly in the cobalt-bearing grades, and must be sharpened and handled carefully. This high hardness is obtained by somewhat special heat-treating techniques as compared with lower-alloyed steels. Temperatures much in excess of normal steel heat-treating temperatures are employed.

Heat Treatment

In general, for maximum hardness and heat resistance, it is necessary to heat-treat at as high a temperature as possible short of the point of initial fusion or grain growth. This would normally result in severe surface damage when done in conventional furnace atmospheres, so furnace protection by special gaseous atmospheres or by molten salt (usually $BaCl_2$) is required for production of quality tools.

After quenching in oil, or a salt eutectic (KCl–NaCl–BaCl₂) held at about 566°C–593°C, and air cooling to 52°C, high-speed steel is tempered in the *secondary* hardening range (523°C–566°C) to develop maximum hardness and cutting life. Such tampering is usually done two or three times successively for best results.

Occasionally, a shallow surface treatment (under 0.03 mm) is imparted by nitriding in salt, or by one of several proprietary methods, to elevate surface hardness and reduce friction. These treatments are often very useful in improving cutting life.

Use and Selection

High-speed steels are used for all types of cutting tools, particularly those powered by machines, such as drills, taps, reamers, milling cutters of all types, form cutting tools, shavers, broaches, and lathe, planar, and shaper bits. They have preference over cemented carbide when the tool is difficult to form (high-speed steel is machinable before heat treating), when subject to shock loading or vibration (high-speed steel was tough and resistant to fatigue, considering its hardness level), or when the machining problem is not particularly difficult. High-speed steel is considerably less expensive than carbides and much simpler to form into complex tools, but it does not have a high hardness, abrasion resistance, or tool life and severe high-speed cutting applications associated with cemented carbide. On the other hand, high-speed steel, with good heat resistance, consistently cuts far better than carbon steel, or one of the *fast-finishing* types.

Other uses for high-speed steel are in forming dies, drawing dies, inserted heading dies, knives, chisels, high-temperature bearings, and pump parts. In these applications, use is made of the combination of high hardness, heat resistance, and abrasion resistance rather than cutting ability.

Among the types of high-speed steel listed in Table H.1, common mass-produced tools are made from M1 or M10. These grades have the lowest cost and are easiest to machine, heat-treat, and sharpen. They also are the toughest when hard and thus withstand the abuse often given common tooling drills, taps, threading dies, etc.

More complex, expensive tools are usually made from M2 high-speed steel. It has better abrasion resistance and is easier to heat-treat in complex shapes. Most milling cutters, gear hobs, broaches, and similar multiple-point tools are in this category. M7 is also becoming popular for specialized applications.

Occasionally, extremely difficult machining operations are encountered, such as cutting plastic, synthetic wood, and paperboard products, or hardened alloy wheels. Better tool life can then be obtained by use of M3 or T15, the high-vanadium-speed steels. They are more expensive and considerably more difficult to sharpen and maintain because of resistance to grinding, but these factors are often outweighed by the superior tool life developed.

The high-cobalt high-speed steels T5 and M36 have the best heat resistance and therefore are particularly suited for tools cutting heavy castings or forgings, where cutting speeds are relatively slow, but cuts that are deep in the cutting edge get very hot. T5 and M36 are more expensive than other grades and thus have limited use, but are the most economical for some operations.

Fabrication

High-speed steels in the annealed condition are machinable by all common techniques. Their machinability rating is about 30% of Bessemer screw stock, and they must be cut slowly and carefully. The recent development of machining high-speed steels has eased this situation, but considerable care is still required.

Ordinarily, tools are machined from bar stock or forgings, either singly for complex tools or in automatic screw machines for mass production items (taps, twist drills, etc.). After finishing almost to final size, the tools are heat-treated to final hardness and then finish-ground. The manufacture of unground tools machined to final size before heat treating is growing because of improvements in heat-treating facilities and better cutting ability of a properly hardened unground surface.

After high-speed steel tools become dull, they can easily be resharpened, with some care given to selection of the proper grinding wheel and technique. They are rarely softened by annealing and re-heat-treated, since this may produce a brittle grain structure in the steel unless great care is employed. High-speed steels are never welded after hardening, and tools are seldom repaired by welding because of extreme brittleness in the weld. Often high-speed steel inserts are brazed to alloy steel bodies, or flash-welded to alloy steel shanks (heavy drills, taps, and reamers).

High-Strength Hydrogen-Resistant Alloy

NASA-23 is a hydrogen-resistant alloy that has been developed for applications in which there are requirements for high strength and high resistance to corrosion.

Adequate resistance to corrosion is necessary for the survival of alloys in hydrogen environments. Alloys for use in such environments typically contain a minimum of 10% chromium. The unique feature of NASA-23 is that it combines high strength with resistance to corrosion and can readily be made into parts that are suitable for use in hydrogen environments.

The use of high-strength alloys that have low resistance to hydrogen results in frequent replacement of components because hydrogen severely degrades mechanical properties of these alloys. At present, coatings that serve as barriers to hydrogen/metal interactions are often applied to such alloys used in hydrogen. However, the use of coatings thus results in higher production costs. Additionally, alloys that have lower resistance to hydrogen can be used by increasing section sizes; however, this method results in an increase in component weight.

NASA-23 can be used for components that encounter temperatures up to 649°C in hydrogen environments.

Innovators of NASA-23 have carefully chosen alloying elements of iron, nickel, cobalt, chromium, niobium, titanium, and aluminum to ensure adequate precipitation hardening for strength and minimum precipitation of detrimental grain-boundary precipitates.

High-Strength Low-Alloy (HSLA) Steels

Steels designed to provide better mechanical properties and/or greater resistance to atmospheric corrosion than the conventional carbon steels. They are not considered to be alloy steels in the normal sense because they are designed to meet specific mechanical properties rather than a chemical composition (HSLA steels have yield strengths greater than 275 MPa, or 40 ksi). The chemical composition of a specific HSLA steel may vary for different product thicknesses to meet mechanical property requirements. The HSLA steels have a low carbon content (0.05% to −0.25% C) in order to produce adequate formability and weldability, and they have manganese contents of up to 2.0%. Small quantities of chromium, nickel, molybdenum, copper, nitrogen, vanadium, niobium, titanium, and zirconium are used in various combinations. The types of HSLA steels commonly used include the following:

1. Weathering steels, designed to exhibit superior atmospheric corrosion resistance.
2. Control-rolled steels, hot rolled according to a predetermined rolling schedule designed to develop a highly deformed austenite structure that will transform to a very fine equiaxed ferrite structure on cooling.
3. Pearlite-reduced steels, strengthened by very-fine-grain ferrite and precipitation hardening but with low carbon content and therefore little or no pearlite in the microstructure.
4. Microalloyed steels, with very small additions (generally <0.10% each) of such elements as niobium, vanadium, and/or titanium for refinement of grain size and/or precipitation hardening.
5. Acicular ferrite steel, very-low-carbon steels with sufficient hardenability to transform one cooling to a very fine high-strength acicular ferrite (low-carbon bainite) structure rather than the usual polygonal ferrite structure.
6. Dual-phase steels, processed to a microstructure of ferrite containing small, uniformly distributed regions of high-carbon martensite, resulting in a product with low yield strength and a high rate of work hardening, thus providing a high-strength steel with superior formability.

High-Stress Abrasion

A form of abrasion in which relatively large cutting forces are imposed on the particles or protuberances causing the abrasion and that produces significant cutting and deformation of the wearing surface. In metals, high-stress abrasion can result in significant surface strain hardening. This form of abrasion is common in mining and agricultural equipment and in highly loaded bearings where hard particles are trapped between mating surfaces. See also *low-stress abrasion*.

High-Temperature Combination

An analytical technique for determining the concentrations of carbon and sulfur in samples. The sample is burned in a graphite crucible in the presence of oxygen, which causes carbon and sulfur to leave the sample as carbon dioxide and sulfur dioxide. These gases are then detected by infrared or thermal conductive means.

High-Temperature Hydrogen Attack

A loss of strength and ductility of steel by high-temperature reaction of absorbed hydrogen with carbides in the steel resulting in decarburization and internal fissuring.

High-Temperature Materials

These are materials that serve above about 540°C. In the broad sense, high-temperature materials can be identified by the following classes of construction solids: stainless steel (limited), austenitic superalloys, refractory metals, ceramics and ceramic composites, metal-matrix composites, and graphite composites. The first three classes are well proven in industrial use, although stainless steel is served slightly above 540°C and refractory metals are usually limited to nonoxidizing atmospheric conditions. The other classes are under extensive worldwide research to establish whether they can be utilized to replace and extend the capabilities of austenitic superalloys, which are the mainstay of high-temperature service.

The most demanding applications for high-temperature materials are found in the aircraft jet engines, industrial gas turbines, and nuclear reactors. However, many furnaces, ductings, and electronic and lighting devices operate at such high temperatures. To perform successfully and economically at high temperatures, a material must have two essential characteristics: It must be strong, because increasing temperature tends to reduce strength, and it must have resistance to its environment, because oxidation and corrosion also increase with temperature.

High-temperature materials have acquired their importance because of the pressing need to provide society with energy and transportation. Machinery that produces electricity or some other form of power from a heat source operates according to a series of thermodynamic cycles, including the basic Carnot cycle and the Brayton cycle, where the efficiency of the device depends on the difference between its highest operating temperature and its lowest temperature. Thus, the greater this difference, the more efficient is the device—a result providing great impetus to the creation of materials that operate at very high temperatures.

Metallic Materials Designs

Alloys used at high temperatures in heat engines are composed of several elements. High-temperature metallurgists, using both theoretical knowledge and application of empirical experimental techniques, depend upon three principal methods for developing and maintaining strength. The alloys are composed of grains of regular crystalline arrays of atoms and show usable ductility and toughness when one plane of atoms slips controllably over the next (dislocation movement); excessive dislocation movement leads to weak alloys. Thus, in terms of achieving mechanical strength, alloy design attempt to inhibit but not completely block dislocation movement.

The most common technique is solid solution strengthening. Foreign atoms of a size different from that of the parent group cause the crystal lattice to strain. This distortion impedes the tendency of the lattice to slip and thus increases strength.

A second significant mechanism is dispersion strengthening. This is the introduction into the alloy lattice of extremely hard and fine foreign phases such as carbide particles or oxide particles. Carbide dispersions can usually be created by a solid-state chemical reaction within the alloy to precipitate the particles, whereas oxide particles are best added mechanically. These hard particles impede slippage of the metal lattice simply by intercepting and locking the dislocations in place.

Still another strengthening technique is coherent-phase precipitation; a foreign phase component of similar but modestly differing crystalline structure is introduced into the alloy—always by a chemical solid-state precipitation mechanism. The phase develops significant binding strength with the mother alloy in which it resides (it is *coherent*) and, like the other mechanisms, then impedes and controls dislocation flow through the metal lattice. Coherent phase precipitation is a special case within precipitation hardening.

However, certain advances in processing have had significant effects on these classical strengthening approaches in recent years.

Metallic Materials Systems

These include superalloys, stainless steel, eutectic alloys, and intermetallic compounds.

Superalloys

High-temperature alloys also must be resistant to chemical attack by the atmosphere in which they exist, more often than not a high-speed, high-pressure gas of highly oxidizing nature. Protection from the atmosphere is achieved by causing the material to develop a tough diffusion-resistant oxide film from various elements in the alloy, or by applying a protective coating to the material.

From all this, a high degree of success has been achieved with the superalloy class of materials.

Superalloys serve under tough conditions of mechanical stressing and atmospheric attack to over 1100°C, which often is within a few hundred degrees of their melting point. These alloys are capable of supporting modern aircraft jet engines.

Undoubtedly, the most complex and sophisticated group of high-temperature alloys are the superalloys. A superalloy is defined as an alloy developed for elevated temperature service, where relatively severe mechanical stressing is encountered and high surface stability is frequently required. Superalloys, classically, are those utilized in the hottest parts of aircraft jet engines and industrial turbines; in fact, the demand of these technically sophisticated applications created the need for superalloys.

The superalloys are strengthened by all of the methods described earlier, as well as by some even more subtle ones, such as control of the boundaries between the grains of the crystalline metal by minor additives and other little understood solid-state chemical reactions.

Stainless Alloys

Stainless steels are strengthened principally by carbide precipitations and solid solution strengthening; because they cannot form

the coherent precipitate, their use is limited to about 650°C, except where strength may not be needed.

Refractory Alloys

Refractory metal alloys are based on elements that have extremely high melting points—graded in 1650°C. These elements—tungsten, tantalum, molybdenum, and niobium (columbium)—are strengthened principally by a combination of solid solution strengthening, carbide precipitation, and unique metalworking. However, they cannot be used in modern heat engines because no commercially successful method of preventing their extensive reaction with oxidizing environments for broad application has been found.

Eutectic Alloys

These metal-based systems have been under study since the late 1960s, with service applications that have not been developed. They are usually similar to or based on superalloys, but they are processed by directional solidification so that a particularly strong stable phase forms as a needlelike structure, getting unusual strength. However, they are expensive to process, and industrial use is questionable.

Intermetallic Compounds

These are metal-based systems centered on the fixed atomic compositions occurring in metallic systems of aluminum with nickel, titanium, and niobium, such as Ni_3Al, Ti_3Al, $TiAl$, and Nb_3Al. Intermetallic compounds are of interest because they often possess a lattice arrangement of atoms that leads to higher melting points and less ease of deformation. Some have shown ductility and toughness potential, and alloying to optimize properties is under significant study. For instance, Ti_3Al is fabricable and ductile, particularly when alloyed with niobium. None is in gas turbine service, but there is a possibility that Ti_3Al will become acceptable if the danger of titanium combustion and hydrogen embrittlement can be avoided.

Oxidation and Corrosion

In addition to possessing strength, high-temperature alloys must resist chemical attack from the environment in which they serve. Most commonly this attack is characterized by a simple oxidation of the surface. However, in machines that utilize crude or residual oils or coal or its products for fuel, natural or acquired contaminants can cause severe and complex chemical attack. Involved reaction with sulfur, sodium, potassium, vanadium, and other elements that appear in these fuels can destroy high-temperature metals rapidly—sometimes in a matter of a few hours. This is known as hot corrosion, and it is a major problem facing otherwise well-suited alloys for service in turbines operating in the combustion products of coal.

Some nuclear reactors operate above 540°C, and the working fluid is often not an oxidizing gas. For example, high-temperature gas-cooled reactors in the development stages utilize high-temperature, high-pressure helium as a working fluid. However, the helium inevitably contains very low levels of impurities—and it does not contain enough oxygen to form a protective oxide film on the metals that contain it. Impurities enter these alloys and reduce strength by precipitating excessive amounts of oxide and carbide phases and by other deleterious effects. Refractory metals, nominally of high interest here, have shown long-term degradation by carbide-related mechanisms, and so metallurgists have been working to develop resistance superalloys.

In other reactor applications such as the liquid metal fast-breeder reactor, construction metals must resist the high-temperature liquid metal to transfer heat from hot uranium–plutonium fuel.

The liquid metal is usually sodium or potassium. All classes of high-temperature alloys—stainless steels, superalloys, and refractory metals—had been under evaluation for service. As in helium gas, small amounts of impurities are the critical item. They can react with the metal at one temperature and transfer it to a component operating at a lower temperature. Stainless steels remain the prime construction materials for these breeder reactors.

The most significant problem, however, remains that of resistance to oxidation and high-temperature corrosion in present heat engines. Superalloys contain small-to-moderate amounts of highly reactive elements such as chromium and aluminum, which react easily with oxygen to form a thick tenacious semiplastic oxide surface film. This prevents further reaction of the aggressive environment with the underlying metal. This is the reason stainless steels are *stainless*—all contain a minimum of 10% chromium. Superalloys follow approximately the same rule but often also contain about 5% aluminum, which further enhances oxidation resistance. It has been found that small (<1.0%) additions of rare earth elements (such as yttrium) further improve oxidation resistance by reducing oxide spalling during the inevitable thermal cycles that gas turbines experience.

The natural protective system works well when oxidation is the only or primary type of attack. However, when the contaminants—vanadium and others—are present, they react in myriad ways in the 540°C–1100°C temperature range to destroy the protective oxide and eventually the alloy. Coating is a method used to combat this problem.

Superalloy Processes

Metallurgists have been seeking ways to increase the capability of superalloys through the development of new processes.

Oxide Dispersion Strengthening

One technique to generate improved strength is known as oxide dispersion strengthening. The objective is to distribute very fine, uniformly dispersed nonreactive oxide particles throughout the alloy. The processing usually involves starting with very fine particles of the metal itself, to which the oxide is added by a chemical or mechanical process step. The alloy is then consolidated by a mechanical pressing operation and forged into a final useful shape. Oxide dispersion strengthening materials are characterized by unusual creep resistance at very high temperatures; however, intermediate-temperature creep and all tensile properties are mediocre.

Success has been obtained in combining some of the classic strengthening factors described previously with oxide dispersion strengthening. This gives a balance of property enhancement that can be particularly useful to turbine metallurgists—good high-temperature creep strength from oxide dispersion strengthening and good intermediate-temperature strength from the other mechanisms.

Directional Solidification

A technique adaptable to investment casting is that of directional solidification, which is particularly suitable to the complex shapes of airfoil parts used in gas turbines. It has been found that by commencing the freezing of molten superalloys (held in a ceramic mold of the shape desired) at the bottom and then allowing the freezing process slowly to proceed up through the shape to be cast, the grains of the structure require a long slender shape in the direction of freezing. This significantly increases the ability of the structure to withstand mechanical load in the freezing direction and particularly increases resistance of the superalloy to a complex phenomenon involving stress and temperature cycling or temperature gradients, known as thermal fatigue. Thermal fatigue failures commence at transverse grain boundaries by this process.

Another approach to advanced airfoil materials is to eliminate grain boundaries entirely. In directional solidification, a number of grains nucleate in the first solid to freeze, and the few with the most preferred orientation for growth choke off solidification of the others and grow along the length of the airfoil. By reducing the cross section of the first solid to form, fewer grains are nucleated. If the ceramic mold also has a short length of spiral cavity before the airfoil shape, solidification of metal through this spiral "pigtail" will allow only a single grain to grow into the airfoil section. If a particular crystal orientation with respect to the airfoil is required, a seed crystal is used instead of a pigtail. The seed placed at the bottom of the mold is long enough that the very bottom portion is not melted in the directional solidification furnace. As the mold-containing molten alloy is withdrawn, solidification occurs on the seed, with essentially the same crystal orientation as the seed. The seed can be cut from the solidified airfoil and used to seed additional airfoils. These single crystals, also known as monocrystals, allow elimination of grain-boundary strengtheners such as zirconium, carbon, boron, or hafnium. This may allow more flexibility in the remainder of the composition and higher heat treatment temperatures to optimize alloy mechanical properties and microstructure. Further freezing occurs in the 001 crystallographic direction preferred because it is in the low-elastic-modulus direction for the face-centered cubic alloys.

Powder Metallurgy

An old process, powder metallurgy (P/M), has found increased use for high-temperature superalloys and offers the possibility of added flexibility in alloy chemistry. Alloys originally developed for cast and wrought processing have been modified slightly for use as P/M materials. For large structures such as turbine disks, this has resulted in much more homogeneous chemistry and microstructure and achievement of outstanding thermal fatigue resistance. Metal flow resistance in isothermal forging is reduced because of nearly superplastic grain-boundary sliding.

Coatings

Superalloys and stainless steels must possess a balance of properties for high-temperature service. The most oxidation- and corrosion-resistant alloys do not have acceptable strength for most structural applications. Therefore, a viable solution is to utilize strong alloys for airfoil structures but then coat them to create environmental stability. This is done by adding elements such as chromium and aluminum into the surface of the alloy; these elements react with the aggressive environment to form highly protective oxides. The coating is applied by chemically reacting the turbine part with an atmosphere containing chromium or aluminum halides at very high temperature so that the active elements diffuse into the surface of the alloy to form the protective layer. Thus, ultimately, the coating layer is composed mainly of nickel from the alloy, and aluminum or chromium or both added in the coating process. Other methods developed overlayer coatings, which form not by reaction with the nickel substrate but by deposition from sources of the desired overlayer chemistry that are made by processes such as physical vapor deposition from an electron-beam-heated source, or by low-pressure plasma spraying of powders of desired coating compositions. Coatings formed by any of these methods are bonded to the substrate and the high-temperature heat treatment to cause the necessary interdiffusion. Such aluminum- and chromium-rich coatings can triple the life of industrial gas turbine parts at temperatures such as 870°C in oxidizing atmospheres.

Eventually, however, the coatings fail by oxidizing away or by further interdiffusion with the superalloy underneath. If a very thin layer of platinum is used, the concentration of aluminum in the surface appears to be enhanced, creating an even greater measure of protection.

Nonmetallic Materials

Considerable activity involving attempts to adapt ceramics to high-temperature applications has occurred since the mid-1960s. These are ceramics of the covalent-bonded type, such as silicon carbide and silicon nitride. Oxide ceramics, such as aluminum oxide and zirconium oxide, possess ionically bonded structures and so tend not to possess usable high-temperature creep resistance. Turbine designers and materials engineers are struggling with the problems of utilizing the covalent ceramics, because they possess great strength. It has also become apparent that these ceramics possess great oxidation and corrosion resistance—features that would be useful in turbine equipment that must handle high-temperature products of combustion from coal. Data have shown that silicon carbide and silicon nitride are attacked to only 0.05–0.08 mm in depth after 6000–8000 h exposure in corrosive atmospheres; most high-temperature metals or alloys cannot meet this performance level. However, their complete lack of ductility means that new design techniques are required to prevent early and catastrophic failure, and only very small parts have been shown to be useful so far. It is not certain that such ceramics will be usable in heat engines.

To bypass the brittleness problem, the concept of ceramic composites has emerged. In these materials, a ceramic matrix is filled with ceramic fibers, which strengthen the whole body. When subjected to a high mechanical load, the fibers pull against the matrix and slip slightly, giving the material a certain level of tolerance. This is known as fiber pullout. However, the effect is not elastic, so that it does not approach the deformation tolerance normally seen by metallic alloys that is essential for safe structures in tension.

The use of graphite and graphite/graphite compositions has also been explored and developed. Graphite has a strikingly unique combination of very low density and high elastic modulus, and it demonstrates increasing mechanical strength with increasing temperature. It is capable of mechanical service at temperatures of 2200°C or higher. However, since graphite is a form of carbon, it has virtually no oxidation resistance. Therefore, attempts to utilize graphite must involve truly extraordinary success in protecting it from oxidation. So far, protective coatings have had but limited success, and it appears that use may well be limited to rather low temperatures.

Hindered Contraction

Contraction where the shape will not permit a metal casting to contract in certain regions in keeping with the coefficient of expansion.

Hinge Stress

The tensile component of the bending stress generated on the same surface of a glass section, but not displaced from the site of a locally impinging force.

HIP

See *hot isostatic pressing*.

Histogram

A plot of frequency of occurrence versus the measured parameter.

Hob (Machine Tool)

A rotary cutting tool with its teeth arranged along a helical thread, used for generating gear teeth or other evenly spaced forms on the periphery of a cylindrical workpiece. The hob and the workpiece are rotated in timed relationship to each other while the hob is fed axially or tangentially across or radially into the workpiece. Hobs should not be confused with multiple-thread milling cutters, rack cutters, and similar tools, where the teeth are not arranged along a helical thread. See the term *gear hobbing*.

Hob (Plastic Molding)

A master model used to sink the shape of a mold into a soft steel block.

Hogging

Machining a part from bar stock, plate, or a simple forging in which much of the original stock is removed.

Hohman A-6 Wear Machine

A widely used type of wear and friction testing machine in which a rotating ring specimen is squeezed between two diametrically opposed rub blocks. This design is said to eliminate shaft flexure such as that found in other machines whose load application from the rub block to the ring is from one side only. Block geometry can be changed from flat to conforming or V-block. This type of machine is designed for use with either lubricated or unlubricated specimens.

Hold-Down Plate (Pressure Pad)

A pressurized plate designed to hold the workpiece down during a press operation. In practice, this plate often serves as a *stripper* and is also called a stripper plate.

Holding

In heat treating of metals, that portion of the thermal cycle during which the temperature of the object is maintained constant.

Holding Furnace

A furnace into which molten metal can be transferred to be held at the proper temperature until it can be used to make castings.

Holding Temperature

In heat treating of metals, the constant temperature at which the object is maintained.

Holding Time (Heat Treating)

Time for which the temperature of the heat-treated metal object is maintained constant.

Holding Time (Joining)

In brazing and soldering, the amount of time a joint is held within a specified temperature range.

Hold Time (Welding)

In resistance welding, the time during which pressure is applied to the work after the current ceases.

Hole Expansion Test

A simulative test in which a flat metal sheet specimen with a circular hole in its center is clamped between angular die plates and be formed by a punch, which expands and ultimately cracks the edge of the hole.

Hole Flanging

The forming of an integral collar around the periphery of a previously formed hole in a sheet metal part.

Hole Sawing

The use of a cylindrical saw having end teeth that cut a circular slot through the workpiece leaving a core.

Holidays

Discontinuities in a coating (such as porosity, cracks, gaps, and similar flaws) that allow areas of basis metal to be exposed to many corrosive environments that contacts the coated surface.

Hollow Milling

Using a special end-cutting mill so designed to leave a core after feeding into or through the workpiece.

Holography

A technique for recording, and later reconstructing, the amplitude and phase distributions of a wave disturbance; widely used as a method of 3D optical image formation and also with acoustical and radio waves. In optical image formation, the technique is accomplished by recording on a photographic plate the pattern of interference between coherent light reflected from the object of interest and light that comes directly from the same source or is reflected from a mirror. In acoustical holography, acoustic beams form an interference pattern of an object and a beam of light interacts with this pattern and is focused to form an optical image.

Homogeneity

The degree of uniformity of composition or properties. Contrast with *heterogeneity*.

Homogeneous

A body of material or matter, alike throughout; hence, comprised of only one chemical composition and phase, without internal boundaries.

Homogeneous Carburizing

Use of a carburizing process to convert a low-carbon ferrous alloy to one of uniform and higher carbon content throughout the section.

Homogeneous Nucleation

In the crystallization of polymers, the primary nucleated species generated by the polymer molecules.

Homogenizing

A heat-treating practice whereby a metal object is held at high temperature to eliminate or decrease chemical segregation by diffusion.

Homologous

Belonging to or consisting of a series of organic compounds differentiated by the number of methylene groups (CH_2).

Homopolymer

A polymer resulting from polymerization of a single monomer.

Honeycomb

Manufactured product of resin-impregnated sheet material (paper, fiberglass, and so on) or metal (aluminum, titanium, and corrosion-resistant alloys) foil, formed into hexagonal-shaped cells. Used as a core material in composite sandwich constructions. See also *sandwich construction*.

Honing

A low-speed finishing process used chiefly to produce uniform high-dimensional accuracy and fine finish, most often on inside cylindrical surfaces. In honing, very thin layers of stock are removed by simultaneously rotating and reciprocating a bonded abrasive stone or stick that is pressed against the surface being honed with lighter force than is typical of grinding.

Hooker Process

Extrusion of a hollow billet or cup through an annulus formed by the die aperture and the mandrel or pilot to form a tube or long cup.

Hooke's Law

A generalization applicable to all solid material, which states that stress is directly proportional to strain and is expressed as

$$\frac{\text{Stress}}{\text{Strain}} = \frac{\sigma}{\varepsilon} = \text{constant} = E$$

where E is the modulus of elasticity or Young's modulus. The constant relationship between stress and strain applies only below the proportional limit. See also *modulus of elasticity*.

Hoopes Process

An electrolytic refining process for aluminum, using three liquid layers in the reduction cell.

Hopper Dryer

A combination feeding and drying device for extrusion and injection molding of thermoplastics. Hot air flows upward through the hopper containing the feed pellets. See also the term *extruder*.

Hopper Loader

A curved pipe through which molding plastic powders are pneumatically conveyed from shipping drums to machine hoppers.

Hoop Stress

The circumferential stress in a material of cylindrical form subjected to internal or external pressure.

Horizontal Batch Furnace

A versatile batch-type heat-treating furnace that can give light or deep case depths, and because the parts are not exposed to air, horizontal batch furnaces can give surfaces almost entirely free of oxides.

Horizontal-Position Welding

(1) Making a fillet weld on the upper side of the intersection of a vertical surface and a horizontal surface. (2) Making a horizontal groove weld on a vertical surface.

Horizontal Fixed Position (Pipe Welding)

The position of a pipe joint in which the axis of the pipe is essentially horizontal and the pipe is not rotated during welding. See also *vertical position* (pipe welding).

Horizontal Rolled Position (Pipe Welding)

The position of a pipe joint in which the axis of the pipe is essentially horizontal, and welding is carried out as the pipe is rotated about its axis. See also *vertical position* (pipe welding).

Horn

(1) In a resistance welding machine, a cylindrical arm or beam that transmits the electrode pressure and usually conducts the welding current. (2) A cone-shaped member that transmits ultrasonic energy from a transducer to a welding or machining tool. See also *ultrasonic impact grinding* and *ultrasonic welding.*

Horn Press

A mechanical metal-forming press equipped with or arranged for a cantilever block or horn that acts as the die or support for the die, used in forming, piercing, setting down, or riveting hollow cylinders and odd-shaped work.

Horn Spacing

The distance between adjacent surfaces of the horns of a resistance welding machine.

Horseshoe Thrust Bearing

A tilting-pad thrust bearing in which the top pads are omitted, making an incomplete annulus.

Hot Bed

An area adjacent to the *runout table* where hot-rolled metal is placed to cool. Sometimes called a cooling table.

Hot Box Process

In foundry practice, resin-based (furan or phenolic) binder process for molding sands similar to shell core making; cores produced with it are solid unless mandrelled out.

Hot Brake Forming

A forming process where the blank is heated up and formed in a cold tool. This is usually done very quickly, and the spring back of the material is similar to cold forming.

Hot Cathode Gun

See thermi*onic cathode gun.*

Hot Chamber Machine

A *die-casting* machine in which the metal chamber under pressure is immersed in the molten metal in a furnace. The chamber is sometimes called a gooseneck, and the machine is sometimes called a gooseneck machine.

Hot-Cold Working

(1) A high-temperature thermomechanical treatment consisting of deforming a metal above its transformation temperature and cooling fast enough to preserve some or all of the deformed structure. (2) A general term synonymous with the *warm working.*

Hot Corrosion

An accelerated corrosion of metal surfaces that results from the combined effect of oxidation and reactions with sulfur compounds and other contaminants, such as chlorides, to form a molten salt on a metal surface that fluxes, destroys, or disrupts the normal protective oxide. See also *gaseous corrosion.*

Hot Cracking

(1) A crack formed in a weldment caused by the segregation at grain boundaries of low-melting constituents in the weld metal. This can result in grain-boundary tearing under thermal contraction stresses. Hot cracking can be minimized by the use of low-impurity welding materials and proper joint design. (2) A crack formed in a cast metal because of internal stress developed upon cooling following solidification. A hot crack is less open than a *hot tear* and usually exhibits less oxidation and decarburization along the fracture surface. See also *cold cracking, lamellar tearing*, and *stress-relief cracking.*

Hot Densification

Rapid deformation of a heated metal powder preform in a die assembly for the purpose of reducing porosity. Metal is usually deformed in the direction of the punch travel. See also *hot pressing.*

H

Hot-Die Forging

A hot forging process in which both the dies and the forging stock are heated; typical die temperatures are 110°C–225°C (200°F–400°F) lower than the temperature of the stock. Compare with *isothermal forging*.

Hot Dip

Covering a surface by dipping the surface to be coated into a molten bath of a coating material. See also *hot dip coating*.

Hot-Dip Coatings

A hot-dip coating is produced by immersing a base metal in a bath of the molten coating metal. Adhesion results from the tendency of the coating metal to diffuse into the base metal and form an alloy layer. Most hot-dip coatings consist of at least two distinct layers: an alloy layer and a layer of relatively pure coating metal. The alloy layer is usually a brittle intermetallic compound. Hot-dip coatings in which the alloy layer is relatively thick are not readily deformable, but modern techniques make it possible to keep the alloy layer quite thin.

Fairly thick coatings of inexpensive metals can be obtained more cheaply by hot dipping than by electroplating. Except on simple shapes, however, hot-dip coatings are nonuniform and wasteful of material. The nature of the process is such that coating metals are restricted to relatively low-melting metals, and base metals are limited to high-melting metals such as cast iron, steel, and copper.

Zinc

Properties and Uses

Hot-dipped or galvanized zinc coatings have been popular for many years for protecting ferrous products because of their ideal combination of high corrosion protection at low cost. Their corrosion protection stems from three important factors; (1) zinc has a slower rate of corrosion than iron, (2) zinc corrosion products are white and nonstaining, and (3) zinc affords electrolytic protection to iron.

The amount of protection against corrosion depends largely upon coating weight—the heavier the coating, the longer the life of the base metal. For example, a coating 0.04 mm thick is expected to have a life of 25 years in rural atmospheres, whereas a 0.88 mm coating will last 50 years. The life of zinc coatings may be 5–10 times greater in rural atmospheres than in industrial atmospheres containing sulfur and acid gases. Nevertheless, the coatings are still popular for industrial use because of their low cost.

Hot dipping is particularly valuable for zinc coating parts that cannot conveniently be made of galvanized sheet. Thus, it is quite popular for structural parts, castings, bolts, nuts, nails, poleline hardware, heater and condenser coils, windlasses, and many other products.

Application Procedures

Hot-dip zinc coatings must be applied to absolutely clean metal. Consequently, surfaces are usually cleaned in a caustic or degreasing medium and then pickled. After rinsing, parts must be fluxed to promote bonding of the zinc coating. The coating itself is applied at 449°C–460°C. A zinc-iron alloy layer is formed between the base metal and coating. Immersion time depends on the thickness of coating desired, with most coatings applied in less than 1 min.

Lead

Properties and Uses

Hot-dip lead coatings provide many important advantages over ferrous metals. They are relatively inexpensive, provide very good protection against indoor and outdoor atmospheric corrosion, and can be used in contact with many chemicals such as sulfuric, hydrochloric, hydrofluoric, phosphoric, and chromic acids. Their atmospheric corrosion resistance stems from the formation of a superficial oxide film, which is relatively impervious to corrosion.

Because of their softness, lead coatings can withstand severe deformation. Poor adhesion may be a problem since the bond is mechanical, but this problem can be minimized by adding alloying elements. Pinholes formed during application may be potential sources of corrosion but they can be eliminated by slight working or burnishing. Typical successful applications for the coatings are wire, pole line hardware, nuts, bolts, washers, tanks, barrels, cans, and miscellaneous air ducts.

Application Procedure

Because pure lead will not alloy with ferrous metals it is usually combined with an alloying agent such as tin, which alloys with iron, forming an interface between the base metal and the coating. A typical sequence of operations for coating small parts involves solvent cleaning, electrocleaning, pickling, predipping, fluxing, coating, and quenching.

Tin

Properties and Uses

Hot-dip coatings can be applied to fabricated parts made of mild and alloy steels, cast iron, and copper and copper alloys to improve appearance and corrosion resistance. Like zinc, the coatings consist of two layers—a relatively pure outer layer and an intermediate alloy layer.

An invisible surface film of stannic oxide is formed during exposure, which helps to retard, but does not completely prevent, corrosion. The coatings have good resistance to tarnishing and staining indoors, and in most rural, marine, and industrial atmospheres. They also resist foods. Corrosion resistance in all cases can be markedly improved by increasing thickness and controlling porosity. Typical applications where they can be used are milk cans, condenser and transformer cans, food and beverage containers, and various items of sanitary equipment such as cast iron mincing machines and grinders.

Application Procedure

Steel products first must be thoroughly cleaned and fluxed. They are then immersed in a preliminary tinning pot, followed by immersion in a second pot at lower temperature. Finally, the parts are withdrawn through palm oil or are dipped in a separate oil pot. Small parts are handled in one pot, centrifuged to remove excess tin, and then quenched or air-cooled. Thickness of the coatings varies from 0.3 to 0.5 mm. The aforementioned treatments are typical and variations are used for cast iron and copper and copper alloy products.

Aluminum

Aluminum hot-dip coatings (usually about 1 mm) are more expensive but much more atmospheric-resistant than zinc. The coatings are also highly heat-reflective and the aluminum-iron alloy layer is

H

highly refractory. Although these coatings seem promising for more general use in outdoor (especially industrial) atmospheres, they are currently used primarily to protect steel from high-temperature oxidation; typical applications include aircraft fire walls, toasters, automobile mufflers, and water heater casings. Hot-dip aluminum-coated (aluminized) steel sheet, strip, and wire are commercially available.

Hot Etching

Development and stabilization of the microstructure at elevated temperature in etchants or gases.

Hot Extrusion

A process whereby a heated *billet* is forced to flow through a shaped die opening. The temperature at which extrusion is performed depends on the material being extruded. Hot extrusion is used to produce long, straight metal products of constant cross section, such as bars, solid and hollow sections, tubes, wires, and strips, from materials that cannot be formed by cold extrusion.

Hot Forging

(1) A forging process in which the die and/or forging stock are heated. See also *hot-die forging* and *isothermal forging*. (2) The plastic deformation of a pressed and/or sintered powder compact in at least two directions at temperatures above the recrystallization temperature.

Hot Forming

See *hot working*.

Hot-Gas Welding

A technique for joining thermoplastic materials (usually sheet) in which the materials are softened by a jet of hot air from a welding torch and joined together at the softened points. Generally, a thin rod of the same material is used to fill and consolidate the gap.

Hot/heated Manifold Mold

A thermoplastic injection mold in which the portion of the mold (the manifold) that contains the runner system has its own heating elements, which keep the molding material in a plastic state ready for injection into the cavities, from which the manifold is insulated. See also *thermoplastic injection molding*.

Hot Isostatic Pressing

(1) A process for simultaneously heating and forming a compact in which the powder is contained in a sealed flexible sheet metal or glass enclosure and the so-contained powder is subjected to equal pressure from all directions at a temperature high enough to permit plastic deformation and sintering to take place. (2) A process that subjects a component (casting, powder forgings, etc.) to both elevated temperature and isostatic gas pressure in an autoclave. The most widely used pressurizing gas is argon. When castings are hot isostatically pressed, the simultaneous application of heat and pressure virtually eliminates internal voids and microporosity through a combination of plastic deformation, creep, and diffusion.

Hot Isostatic Pressure Welding

A diffusion welding method that produces coalescence of materials by heating and applying hot inert gas under pressure.

Hot Machining

Machining in which the workpiece shear zone is heated by auxiliary means to reduce the shear strength and increase the machinability of the material.

Hot-Melt Adhesives

Hot-melt adhesives are 100% solid thermoplastics that are applied in a molten state and form a bond after cooling to a solid state. In contrast to other adhesives, which achieve the solid state through evaporation of solvents or chemical cure, hot-melt adhesives achieve a solid state and resultant strength by cooling. In general, hot-melt adhesives are solid at temperatures below 79°C (175°F). Ideally, as the temperature is increased beyond this point, the material rapidly melts to a low-viscosity fluid that can be easily applied. Upon cooling, the adhesive sets rapidly. Because these adhesives are thermoplastics, the melting–resolidification process is repeatable with the addition and removal of the required amount of heat. Typical application temperatures of hot-melt adhesives are 150°C–290°C (300°F–550°F).

Materials that are primarily used as hot-melt adhesives include ethylene and vinyl acetate (EVA) copolymers; polyvinyl acetates (PVAs); polyethylene (PE); amorphous polypropylene; thermoplastic elastomers such as polyurethane, polyether amide, and block copolymers (e.g., styrene–butadiene–styrene, styrene–isoprene–styrene, and styrene–olefin–styrene); polyamides; and polyesters. Hot-melt adhesives can also be divided into nonpressure-sensitive and pressure-sensitive types. Nonpressure-sensitive adhesives include those for direct bonding and heat sealing. Pressure-sensitive hot-melt adhesives are tacky to the touch and can be bonded by the application of pressure alone at room temperature.

Hot-melt adhesives are used to bond all types of substrates, including metals, glass, plastics, ceramics, rubbers, and wood. Primary areas of application include packaging, bookbinding, assembly bonding (such as air filters and footwear), and industrial bonding (such as carpet tape and backings).

Hot Mill

A production line or facility for hot rolling of metals.

Hot Mill Gouges

A series of short scratches caused by slippage of one surface of a coil relative to another during hot coiling.

Hot Press Forging

Plastically deforming metals between dies in presses at temperatures high enough to avoid strain hardening.

Hot Pressing

Simultaneous heating and forming of a powder compact. See also *pressure sintering*.

Hot Pressure Welding

A solid-state welding process that produces coalescence of materials with heat and application of pressure sufficient to produce macrodeformation of the base material. Vacuum or other shielding media may be used. See also *forge welding* and *diffusion welding*. Compare with *cold welding*.

Hot Quenching

An imprecise term for various quenching procedures in which a quenching medium is maintained at a prescribed temperature above 70°C (160°F).

Hot Rod

Same as *wire rod*.

Hot-Rolled Sheet

Steel sheet reduced to the required thickness at a temperature above the point of scaling and therefore carrying hot mill oxide. The sheet may be flattened by cold rolling without appreciable reduction in thickness or by roller leveling, or both. Depending on the requirements, hot-rolled sheet can be pickled to remove hot mill oxide and is so produced when specified.

Hot-Runner Mold

A thermoplastic injection mold in which the runners are insulated from the chilled cavities and remain hot so that the center of the runner never cools during the molding cycle. Contrary to usual practice, the runners are not ejected with the molded pieces. Also called insulated-runner mold. See also *thermoplastic injection molding*.

Hot-Setting Adhesive

An adhesive that requires a temperature at or above 100°C (212°F) to set.

Hot Shortness

A tendency for some alloys to separate along grain boundaries when stressed or deformed at temperatures near the melting point. Hot shortness is caused by a low-melting constituent, often present only in minute amounts, that is segregated at grain boundaries.

Hot Size

A process where a preformed part is placed into a hot die above the annealing temperature to set the shape and remove spring back tendencies.

Hot Stamping

Engraving operation for plastics in which a design is stamped with heated metal dyes onto the face of the plastics.

Hot Strip or Pickle Line Scratch

Scratches that are superficially similar to slivers or skin laminations but originate from mechanical scoring of the strip in the hot mill, pickle line, or slitter.

Hot Top

(1) A reservoir, thermally insulated or heated, that holds molten metal on top of a mold for feeding of the ingot or casting as it contracts on solidifying, thus preventing formation of *pipe* or *voids*. (2) A refractory-lined steel or iron casting that is inserted into the tip of the mold and is supported at various heights to feed the ingot as it solidifies.

Hot Trimming

The removal of *flash* or excess metal from a hot part (such as a forging) in a trimming press.

Hot Upset Forging

A *bulk forming* process for enlarging and reshaping some of the cross-sectional area of a bar, tube, or other product form of uniform (usually round) section. It is accomplished by holding the heated forging stock between grooved dies and applying pressure to the end of the stock, in the direction of its axis, by the use of a heading tool, which spreads (upsets) the end by metal displacement. Also called hot heading or hot upsetting. See also *heading* and *upsetting*.

Hot-Wire Analyzer

An electrical atmosphere analysis device that is based on the fact that the electrical resistivity of steel is a linear function of carbon content over a range from 0.05% C to saturation. The device measures the carbon potential of furnace atmospheres (typically). This term is not to be confused with the *hot-wire test* that measures heat extraction rates.

Hot-Wire Test

Method used to test heat extraction rates of various quenchants. Faster heat-extracting quenchants will permit more electric current to pass through a standard wire because it is cooled more quickly. Compare with *hot-wire analyzer*.

Hot-Wire Welding

A variation of arc welding processes in which a filler metal wire is resistance heated as it is fed into the molten weld pool.

Hot-Worked Structure

The structure of a material worked at a temperature higher than the recrystallization temperature.

Hot Working

(1) The plastic deformation of metal at such a temperature and strain rate that recrystallization takes place simultaneously with the deformation, thus avoiding any *strain hardening*. Also referred to as hot forging and hot forming. (2) Controlled mechanical operations for

shaping a product at temperatures above the recrystallization temperature. Contrast with *cold working*.

Hot Zone

The part of a continuous furnace or kiln that is held at maximum temperature. Other zones are the preheat zone and cooling zone.

HSLA Steels

An interesting class of alloys known as high-strength, low-alloy (HSLA) steels has emerged in response to requirements for weight reduction of vehicles. The compositions of many commercial HSLA steels are proprietary and they are specified by mechanical properties rather than composition. A typical example, however, might contain 0.2 wt.% carbon and about 1 wt.% or less of such elements as manganese, phosphorus, silicon, chromium, nickel, or molybdenum. The high strength of HSLA steels is the result of optimal alloy selection and carefully controlled processing such as hot rolling (deformation at temperatures sufficiently elevated to allow some stress relief).

Housing Press

A type of hydraulic press in which the crown and bed are separated by uprights or spaces extending from front to back through which strain rods pass.

Hub

A *boss* that is in the center of a forging and forms a part of the body of the forging.

Hubbing

The production of forging die cavities by pressing a male master plug, known as a *hub*, into a block of metal.

Hull Cell

A special electrodeposition cell giving a range of known current densities for test work.

Humidity, Absolute

See *absolute humidity*.

Humidity Ratio

In a mixture of water vapor and air, the mass of water vapor per unit mass of dry air.

Humidity, Relative

See *relative humidity*.

Humidity, Specific

See *specific humidity*.

Humidity Test

A corrosion test involving exposure of specimens at controlled levels of humidity and temperature. Contrast with *salt fog test*.

Hybrid

A composite laminate consisting of laminae of two or more composite material systems. A combination of two or more different fibers, such as carbon and glass or carbon and aramid, in a structure. Tapes, fabrics, and other forms may be combined; usually only the fibers differ. See also *interply hybrid* and *intraply hybrid*.

Hybrid Materials

A hybrid material is a combination of two or more different material systems that results from the attachment of one material to another. For example, a hybrid composite consists of two or more types of reinforcing fibers in one or more types of matrices. By bringing together the different properties of the reinforcements, a hybrid composite can achieve improved performance with a balance in cost. Hybrid composites have been applied successfully in transportation vehicles including race cars, helicopters, powerboats, and other load-bearing components since the late 1960s. Recently, hybrid composites have been considered in infrastructure applications because of enhanced performance compared to all glass-fiber composites. It has been shown that combining carbon-fiber and glass-fiber composites will result in increased tensile and compressive stiffness, flexural stiffness and strength, fatigue performance, impact properties, and environmental resistance. The tensile failure strain of a glass–carbon hybrid composite is increased compared to all carbon-fiber composites.

Several studies on modeling of hybrid composites have demonstrated that by using an analytical model, a carbon–glass hybrid sheet molding compound (SMC) composite can have a Young's modulus up to three times as high and a density up to 15% lower than the glass SMC. For equal Young's modulus, a carbon–glass hybrid SMC with unfilled matrix costs considerably less than all carbon-fiber SMC.

Because carbon fibers are resistant to water and dilute acids, carbon–glass hybrid composites are more resistant to stress corrosion than all-glass fiber composites. The corrosion resistance of a hybrid composite is also dependent on laminate construction. Additionally, fatigue performance of hybrid composite was enhanced compared to all-glass composites.

Polymets

Polymets are a new class of hybrid materials composed of metals and polymers. Recent research has demonstrated the feasibility of extruding metal–polymer composites (polymets) of aluminum with poly(etheretherketone) and of aluminum with a liquid-crystal polymer (Al-LCP). Extrusion through either a high-sheer 90° die or a converging conical die improves the properties of these aluminum polymer composites. The improvement in properties is believed to be the result of texture development, that is, the bulk and molecular orientation of the polymer in the extrusion direction. The yield strengths of these metal–polymer composites were found to be 13%–18% greater than that predicted by the rule of mixtures and their specific yield strengths 14%–21% greater than that of the aluminum control specimen.

Tensile Properties

The strength of the polymets decreases with increasing polymer content. Ductility also decreases; tensile elongation declines from 1.6% to 1.2%. Thus, the properties of the polymets are affected by the extrusion ratio; high extrusion ratios decrease the yield strength, increase ultimate tensile strength, and improve ductility of the polymets.

Continued research on Al-LCP polymets has yielded the following:

1. Formation of in-place polymer fibrils and films results from extrusion processing.
2. The ultimate tensile strength and ductility of Al-LCP aromatic polyester polymets is improved by increasing the extrusion ratio from 10.7/1 to 114/1.
3. The yield strengths of the polymets decline by increasing the extrusion ratio, but the normalized yield strengths of the polymets increase an average of 26.7%.
4. Increases in the ultimate tensile strengths and normalized yield strengths of the polymets with extrusion ratio are attributable to formation of polymer fibrils and films.
5. The energy required to produce tensile failure is significantly enhanced by increasing the extrusion ratio from 10.7/1 to 114/1.

Hydration

The chemical reaction between cement and water, forming new compounds, most of which have strength-producing properties.

Hydraulic Cement

A cement that sets and hardness by chemical interaction with water and that is capable of doing so under water.

Hydraulic Fluid

A fluid used for transmission of hydraulic pressure or action, not necessarily involving lubricant properties. Hydraulic fluids can be based on oil, water, or synthetic (fire-resistant) liquids.

Hydraulic Hammer

A gravity-drop forging hammer that uses hydraulic pressure to lift the hammer between strokes.

Hydraulic–Mechanical Press Brake

A mechanical *press brake* that uses hydraulic cylinders attached to mechanical linkages to power the ram through its working stroke.

Hydraulic Press (Metal Forming)

A press in which fluid pressure is used to actuate and control the ram. Hydraulic presses are used for both open- and closed-die forging.

Hydraulic Press (Plastic Molding)

A press in which the molding force is created by the pressure exerted by a fluid.

Hydraulic Press Brake

A *press brake* in which the ram is actuated directly by hydraulic cylinders.

Hydraulic Shear

A shear in which the crosshead is actuated by hydraulic cylinders.

Hydrazine

NH:NH is a colorless liquid boiling at 236°F (113.5°C) and freezing at 36°F (2°C). It is used as a propellant for rockets, yielding exhaust products of high temperature and low molecular weight. With a nickel catalyst it decomposes to nitrogen and hydrogen. It is a strong reducing agent and has been used in soldering fluxes, for corrosion control in boilers, in metal plating, in noble metal catalysts, and in organic syntheses. However, since it has been declared a suspected carcinogen, its use has declined. It is a starting material for antioxidants and herbicides. Reacted with citric acid, it produces the antituberculosis drug *cotinazin*, which is isonicotinic acid hydrazine. It is also used as a blowing agent for foamed rubber and for the production of plastics. For industrial applications it may be used in the form of dihydrazine sulfate, a white, crystalline, water-soluble flake. Hydrazine hydrate is a colorless, water-miscible liquid. Hydrazine is made by reacting chlorine and caustic soda and treating with ammonia.

Hydride Descaling

Descaling by the action of a hydride in a fused alkali.

Hydride Phase

The phase TiH_x formed in titanium when the hydrogen content exceeds the solubility limit, generally locally due to some special circumstance.

Hydride Powder

A powder produced by removal of the hydrogen from a metal hydride.

Hydride Process

In powder metallurgy, the hydrogenation of such reactive metals as titanium and zirconium, followed by comminution of the brittle compound and vacuum treatment to remove the hydrogen from the powder.

Hydrocarbon Plastics

Plastics based on resins made by the polymerization of monomers composed of carbon and hydrogen only.

Hydrocarbons

Organic compounds of hydrogen and oxygen. Most organic compounds are hydrocarbons. Aliphatic hydrocarbons are straight-chained structures. Aromatic hydrocarbons are ringed structures based on the benzene ring. Methyl alcohol and trichloroethylene are among the aliphatic; benzene, xylene, and toluene are among the aromatic.

Hydrochloric Acid

Hydrochloric acid (HCl) is soluble in water and is a strong mineral acid made by the action of sulfuric acid on common salt, or as a by-product of the chlorination of hydrocarbons such as benzene.

Uses

HCl is used to some extent in pickling of metal prior to porcelain enameling. Pickling solutions generally contain 5%–10% HCl and should be contained in steel tanks lined with acid-proof brick. Rubber-lined tanks may be used for small jobs. Pickling temperatures range from room temperature up to $-93°C$; pickling times range from 2 to 15 min, depending on metal condition, acid strength, and temperature.

HCl will pickle faster than sulfuric acid. Also, the metal surface will be cleaner because the iron salts formed during pickling rinse off more readily. The danger of overpickling is greatly reduced by the use of HCl, because chlorine radical does not promote defects as readily as does the sulfate radical of H_2SO_4. HCl requires no steam for heating. Metal sheet should not be immersed in HCl longer than 12–15 min.

Also called *muriatic acid* and originally called *spirits of salt*. An inorganic acid used for pickling and cleaning metal parts; producing glues and gelatin from bones; manufacturing chlorine, chlorine dioxide, pharmaceuticals, dies, and pyrotechnics; recovering zinc from galvanized iron scrap; making high-fructose corn syrup; tanning, etching, and reclaiming rubber; and treating oils and fats. It is a water solution of hydrogen chloride and is a colorless or yellowish fuming liquid with pungent, poisonous fumes. Fuming HCl has a specific gravity of 1.194 and contains about 37% hydrogen chloride gas. Reagent-grade HCl is usually of this high strength and is clear and colorless, unlike the impure fuming variety. HCl is shipped in class carboys. Anhydrous hydrogen chloride gas is also marketed in steel cylinders for use as a catalyst. The acid known as "aqua regia," used for dissolving or testing gold and platinum, is a mixture of 3 parts HCl and 1 nitric acid. It is a yellowish liquid with suffocating fumes.

Hydrocyanic Acid

Also called *prussic acid*, *formonitrile*, and *hydrogen cyanide*. A colorless, highly poisonous gas of composition HCN. The specific gravity is 0.697, the liquefying point is 79°F (26°C), and it is soluble in water and in alcohol. It is usually marketed in water solutions of 2%–10%. It is used for the production of acrylonitrile and adiponitrile and for making sodium cyanide. It is also employed as a disinfectant and fumigant, as a military poison gas, and in mining and metallurgy in the cyanide process. It is so poisonous that death may result within a few seconds after it is taken into the body. It was used as a poison by the Egyptians and Romans, who obtained it by crushing and moistening peach kernels. It is produced synthetically from natural gas. The French war gas known as *vincennite* was hydrocyanic acid mixed with stannic chloride. *Manganite* was a mixture with arsenic trichloride. HCN discoids are cellulose disks impregnated with 98% hydrocyanic acid, used for fumigating in closed warehouses.

Hydrodynamic Lubrication

A system of lubrication in which the shape and relative motion of the sliding surfaces cause the formation of a fluid film that has sufficient pressure to separate the surfaces. See also *elastohydrodynamic lubrication* and *gas lubrication*.

Hydrodynamic Seal

A seal that has special geometric features on one of the mating faces. These features are designed to produce interfacial lift, which arises solely from the relative motion between the stationary and rotating portions of the seal.

Hydrofluoric Acid

A water solution of *hydrogen fluoride*, HF. It is a colorless, fuming liquid, highly corrosive and caustic. It dissolves most metals, except gold and platinum, and glass, stoneware, and organic material. The choking fumes are highly injurious. It is widely used in the chemical industry, for etching glass, and for cleaning metals. In cleaning iron castings, it dissolves the sand from the castings. The specific gravity of the gas is 0.99, and the boiling point is 67°F (19.5°C). Hydrofluoric acid is made by treating calcium fluoride or fluorspar with sulfuric acid. It is marketed in solution strengths of 30%, 52%, 60%, and 80%. The anhydrous material, HF, is used as an alkylation catalyst. *Hydrobromic acid*, HBr, is a strong acid that reacts with organic bases to form bromides that are generally more reactive than chlorides. The technical 48% grade has a specific gravity of 1.488.

Hydrogen

A colorless, odorless, tasteless elementary gas, with an atomic weight of 1.008, hydrogen is the lightest known substance. The specific gravity is 0.0695, and its density ratio in relation to air is 1:14.38. It is liquefied by cooling under pressure, and its boiling point at atmospheric pressure is $-252.7°C$. Its light weight makes it useful for filling balloons, but because of its flammable nature, it is normally used only for signal balloons, for which use the hydrogen is produced easily and quickly from hydrides. Hydrogen (H_2) produces high heat and is used for welding and cutting torches. For this purpose it is used in atomic form rather than the usual H_2 molecular form. Its high thrust value makes it an important rocket fuel. It is also used for the hydrogenation of oil and coal, for the production of ammonia and many other chemicals, and for water gas, a fuel mixture of hydrogen and carbon monoxide made by passing steam through hot coke.

Hydrogen is so easily obtained in quantity by the disassociation of water and as a by-product in the production of alkalies by the electrolysis of brine solutions that it appears as a superabundant material, but its occurrence in nature is much less than that of many of the other elements. It occurs in the atmosphere to the extent of only about 0.01% and in the Earth's crust to the extent of about 0.2%, or about half that of the metal titanium. However, it constitutes about 1/9 of all water, from which it is easily obtained by high heat or by electrolysis.

Hydrogen has three isotopes. Hydrogen-2, called deuterium, occurs naturally in ordinary hydrogen-1, called protium, to the extent of one part in about 5000. Deuterium has one proton and one neutron in the nucleus, with one orbital electron revolving around. A gamma ray will split off the neutron, leaving a single electron revolving around a single proton. Deuterium is also called double-weight hydrogen. Deuterium oxide is known as heavy water. The formula is H_2O, but with the double-weight hydrogen, the molecular weight is 20 instead of the 18 for ordinary water.

Heavy water is used for shielding atomic reactors because it is more effective than graphite in slowing down fast neutrons. Hydrogen-3 is triple-weight hydrogen and is called tritium. It has two neutrons and one proton in the nucleus and is radioactive. It is a by-product of nuclear fission reactors, and most commercial production is from this source. It is a beta emitter with little harmful secondary ray emission, which makes it useful and self-luminous phosphorous. It is a solid at very low temperatures. Liquid hydrogen for rocket fuel use is made from ordinary hydrogen.

Hydrides are metals that contain hydrogen in a reduced state and as a solid solution in their lattice. Titanium hydride and zirconium hydride have catalytic activity. Lanthanum and cerium react with hydrogen at room temperature forming hydrides and are used for storing hydrogen. Sodium borohydride is used commercially as a reducing agent, for removing trace impurities from organic chemicals, in the synthesis of pharmaceuticals, for wood-pulp bleaching, for brightening clay, and for recovering trace metals in effluents. Silicate hydride, also known as silane, is a gas used for manufacturing ultrapure silicon for fabrication into semiconductors. *Gelled hydrogen* for rocket fuel is liquid hydrogen thickened with silica powder.

Hydrogen-Assisted Cracking (HAC)

See *hydrogen embrittlement.*

Hydrogen-Assisted Stress-Corrosion Cracking (HSCC)

See *hydrogen embrittlement.*

Hydrogen Blistering

The formation of blisters on or below a metal surface from excessive internal hydrogen pressure. Hydrogen may be formed during cleaning, plating, or corrosion.

Hydrogen Brazing

A term sometimes used to denote brazing in a hydrogen-containing atmosphere, usually in a furnace; use of the appropriate process name is preferred.

Hydrogen Damage

A general term for the embrittlement, cracking, blistering, and hydride formation that can occur when hydrogen is present in some metals.

Hydrogen Electrode

A reference electrode against which other materials are compared to determine their electrode potential. See *electrochemistry.*

Hydrogen Embrittlement

A process resulting in a decrease of the *toughness* or *ductility* of a metal due to the presence of atomic hydrogen. Hydrogen embrittlement has been recognized classically as being of two types.

The first, known as internal hydrogen embrittlement, occurs when the hydrogen enters molten metal that becomes supersaturated with hydrogen immediately after solidification. The second type, environmental hydrogen embrittlement, results from hydrogen being absorbed by solid metals. This can occur during elevated temperature thermal treatments and in service during electroplating, contact with maintenance chemicals, corrosion reactions, cathodic protection, and operating in high-pressure hydrogen. In the absence of residual stress of external loading, environmental hydrogen embrittlement is manifested in various forms, such as blistering, internal cracking, hydride formation, and reduced ductility. With a tensile stress or stress-intensity factor exceeding a specific threshold, the atomic hydrogen interacts with the metal to induce subcritical crack growth leading to fracture. In the absence of a corrosion reaction (polarized cathodically), the usual term used is hydrogen-assisted cracking (HAC) or hydrogen stress cracking (HSC). In the presence of active corrosion, usually as Spitz or crevices (polarized anodically), the cracking is generally called *stress-corrosion cracking* (SCC), which should more properly be called hydrogen-assisted stress-corrosion cracking (HSCC). Thus, HSC and electrochemically anodic SCC can operate separately or in combination (HSCC). In some metals, such as high-strength steels, the mechanism is believed to be all, or nearly all, HSC. The participating mechanism of HSC is not always recognized and may be evaluated under the generic heading of SCC.

Even when hydrogen produces no readily observable crack, it may reduce the ductility of certain steels—termed hydrogen embrittlement—and it can lead to delayed fracture, also termed sustained load failure, under an otherwise acceptable steady load. Steel boiler tubes charged with hydrogen as a result of severe waterside corrosion suffer a reaction between a carbon of the steel and the hydrogen to form methane, CH_4. This causes local decarbonization and internal cracking leading to significant weakening. The resultant failure is of low ductility leading again to the term hydrogen embrittlement, although this particular usage is not usually recognized outside the power industry. Tough pitch copper, that is, copper containing oxygen as finally distributed copper oxide particles, is damaged if exposed to hydrogen-rich atmospheres during heat treatment, welding, brazing, or service. The hydrogen reacts with the oxide to form water that causes internal fissuring with serious effects on the strength and toughness of the material. The term gassing is specific to this form of attack although the terms hydrogen embrittlement and hydrogen damage are occasionally used. Some titanium alloys suffer a considerable reduction of impact strength when charged with hydrogen due to the formation of hydride platelets that impede slip. See also *fisheyes.*

Hydrogen Fluoride

Hydrogen fluoride is the hydride of fluorine and the first member of the family of halogen acids. Anhydrous hydrogen fluoride is a mobile, colorless liquid that fumes strongly in air. It has the empirical formula HF, melts at −83°C, and boils at 19.8°C. The vapor is highly aggregated, and gaseous hydrogen fluoride deviates from perfect gas behavior to a greater extent than any other gaseous substance known.

Hydrogen fluoride is prepared on the large industrial scale by treating fluorspar (calcium fluoride, CaF_2) with concentrated sulfuric acid. The crude product is purified by fractional distillation to yield a product containing more than 99.5% hydrogen fluoride.

Properties

Anhydrous hydrogen fluoride is an extremely powerful acid, exceeded in this respect only by 100% sulfuric acid. Like water, hydrogen fluoride is a liquid of high dielectric constant that undergoes self-ionization and forms conducting solutions with many solutes. Because anhydrous hydrogen fluoride is a superacid, many organic solutes dissolve in it to form stable carbonium ions. Alkali metal fluorides and silver fluoride dissolve readily in hydrogen fluoride to form conducting solutions. The alkali metal fluorides are bases in the hydrogen fluoride systems and correspond to solutions of alkali metal hydroxides in water.

Conversely, antimony pentafluoride and boron trifluoride act as acids in hydrogen fluoride and accentuate the already strong acid properties of the solvent.

Uses

Hydrogen fluoride is a widely used industrial chemical. It was formally used in the petroleum refining industry for the isomerization of aliphatic hydrocarbons to form more desirable automotive fuels, but this application has been superseded by other methods. The largest industrial use a hydrogen fluoride is in making fluorine-containing refrigerants (Freons, Genetrons).

Another important use of hydrogen fluoride is in the preparation of organic fluorocarbon compounds by the Simons electrochemical process. In this procedure, an organic compound is dissolved in hydrogen fluoride, and an electric current is passed through the solution, whereupon the hydrogen atoms in the organic solute are replaced by fluorine. Hydrogen fluoride is employed in the electrochemical preparation of fluorine and for the preparation of inorganic fluorides. Thus, hydrogen fluoride is used for the conversion of uranium dioxide to uranium tetrafluoride and intermediate in the preparation of uranium metal and uranium hexafluoride. With the great increase in nuclear energy–produced electricity, this represents an important use of hydrogen fluoride.

Both hydrogen fluoride and hydrofluoric acid cause unusually severe burns; appropriate precautions must be taken to prevent any contact of the skin or eyes with either the liquid or the vapor.

Hydrogen-Induced Cracking (HIC)

Same as *hydrogen embrittlement.*

Hydrogen-Induced Delayed Cracking

A term sometimes used to identify a form of *hydrogen embrittlement* in which a metal appears to fracture spontaneously under a steady stress less than the *yield stress.* There is usually a delay between the application of stress (or exposure of the stressed metal to hydrogen) and the onset of cracking. Also referred to as static fatigue.

Hydrogen Loss

The loss in weight of metal powder of a compact caused by heating a representative sample according to a specified procedure in a purified hydrogen atmosphere. Broadly, a measure of the oxygen content of the sample when applied to materials containing only such oxides as are reducible with hydrogen and no hydride-forming element.

Hydrogen Overvoltage

In electroplating, *overvoltage* associated with the liberation of hydrogen gas.

Hydrogen Peroxide

A liquid that readily yields oxygen for bleaching and oxidizing purposes. The C. P. grade of hydrogen peroxide is a colorless liquid with 90% H_2O_2 and 10 water. The specific gravity is 1.39. It contains 42% active oxygen by weight, and 1 volume yields 410 volumes of oxygen gas. Grades for oxidation and bleaching contain 27.5% and 35% H_2O_2. It is also used as an oxidizer for liquid fuels. A variety of chemicals are used for providing oxygen for chemical reactions. These are known as *oxidizers* or *oxidants*, and they may be peroxides or superoxides, which are compounds with the oxygen atoms singly linked. They break down into pure oxygen and a more stable reduced oxide. Sodium peroxide is used in submarines to absorb carbon dioxide and water vapor and to give off oxygen to restore the air. To provide oxygen and rockets and missiles, *lithium nitrate*, $LiNO_3$, with 70% available oxygen, and *lithium perchlorate*, $LiClO_4 \cdot 3H_2O$, with 60% available oxygen, are used. Another rocket fuel oxidizer that is liquid under moderate pressure and is easily stored is *perchloryl fluoride*, ClO_3F, normally boiling at $-52°F$ ($-47°C$).

Liquid air was first used in the first V-2 rockets, with alcohol, potassium permanganate, and hydrogen peroxide. Liquid air is used in the chemical industry and for cold-treating. It is atmospheric air liquefied under pressure, and it contains more than 20% free oxygen.

Oxidizers and *reducers*, or *reductants*, are used in solutions for water disinfection, bleaching, cyanide destruction, chromium reduction, and metal etching. Common oxidizers include *chlorine, bromine, ozone, sodium hypochlorite*, and *hydrogen peroxide*. Well-known reductants include *sodium bisulfate*, or *sodium metabisulfate*, and *sulfur dioxide*. Ozone is about twice as effective as bromine. Because of environmental concerns in recent years, hydrogen peroxide has displaced chlorine in various water treatment and pollution-control applications.

Hydrogen-Reduced Powder

Metal powder produced by hydrogen reduction of a compound.

Hydrogen Stress Cracking (HSC)

See *hydrogen embrittlement.*

Hydrolysis

(1) Decomposition or alteration of a chemical substance by water. (2) In aqueous solutions of electrolytes, the reactions of cations with water to produce a weak base or of anions to produce a weak acid.

Hydrolytic Stability

The ability of an organic or polymeric material to withstand an irreversible change of state when exposed to elevated temperature and humidity.

Hydromatic Welding

See preferred term *pressure-controlled welding.*

Hydromechanical Press

A press in which forces are created partly by a mechanical system and partly by a hydraulic system.

Hydrometallurgy

Industrial *winning* or *refining* of metals using water or an aqueous solution.

Hydrometer

Instruments for measuring the density of liquid.

Hydrophilic

Having an affinity for water; easily wetted by water. Contrast with *hydrophobic*.

Hydrophobic

Lacking an affinity for, repelling, or failing to absorb water; poorly wetted by water. Contrast with *hydrophilic*.

Hydrostatic

Acting equally in all directions. In hydrostatic tension, the three principal stresses are equal.

Hydrostatic Bearing

A bearing in which the solid bodies are separated and supported by hydrostatic pressure, applied by an external source, to a compressible or incompressible fluid interposed between those bodies.

Hydrostatic Compaction

Pressing processes in which powders intended for sintering are enclosed in a flexible bag or similar container and subjected to a high-pressure environment, usually hydraulic. This allows the use of a higher but more even pressure than conventional pressing and reduces wear of the equipment. See *hydrostatic pressing*.

Hydrostatic Extrusion

A method of extruding a *billet* through a die by pressurized fluid instead of the ram used in conventional *extrusion*.

Hydrostatic Lubrication

A system of lubrication in which the lubricant is supplied under sufficient external pressure to separate the opposing surfaces by a fluid film. See also *pressurized gas lubrication*.

Hydrostatic Modulus

See *bulk modulus of elasticity*.

Hydrostatic Mold

In powder metallurgy, a sealed flexible mold made of rubber, a polymer, or pliable sheet made from a low-melting metal such as aluminum.

Hydrostatic Pressing

A special case of isostatic pressing that uses a liquid such as water or oil as a pressure transducing medium and is therefore limited to near room temperature operation.

Hydrostatic Seal

A seal incorporating features that maintain an interfacial film thickness by means of pressure. The pressure is provided either by an external source or by the pressure differential across the seal. The interfacial pressure profile of a seal face is normally speed-dependent; the interfacial pressure profile of the hydrostatic seal is not speed-dependent.

Hydrostatic Tension

Three equal and mutually perpendicular tensile stresses.

Hydroxyl Group

A chemical group consisting of one hydrogen atom and one oxygen atom. The OH portion of a compound occurring in solution as the OH^- ion.

Hygroscopic

(1) Capable of attracting, absorbing, and retaining atmospheric moisture. (2) Possessing a marked ability to accelerate the condensation of water vapor; applied to condensation nuclei composed of salts that yield aqueous solutions of a very-low-equilibrium vapor pressure compared with that of pure water at the same temperature. (3) Pertaining to a substance whose physical characteristics are appreciably altered by effects of water vapor. (4) Pertaining to water absorbed by dry soil minerals from the atmosphere; the amounts depend on the physicochemical character of the surfaces, and increase with rising relative humidity.

Hygrothermal Effect

Change in properties of the material (particularly plastics) due to moisture absorption and temperature change.

Hypereutectic Alloy

In an alloy system exhibiting a *eutectic*, any alloy whose composition has an excess of alloying element compared with the eutectic composition and whose equilibrium microstructure contains more eutectic structure.

Hypereutectoid Alloy

In an alloy system exhibiting a *eutectoid*, any alloy whose composition has an excess of alloying element compared with the eutectoid composition and whose equilibrium microstructure contains some eutectoid structure.

Hypoeutectic Alloy

In an alloy system exhibiting a *eutectic*, any alloy whose composition has an excess of base metal compared with the eutectic composition and whose equilibrium microstructure contains some eutectic structure.

Hypoeutectoid Alloy

In an alloy system exhibiting a *eutectoid*, any alloy whose composition has an excess of base metal compared with the eutectoid composition and whose equilibrium microstructure contains some eutectoid structure.

Hypoid Bevel Gear

A gear that transmits rotary motion between parallel shafts that are usually at 90° to each other. In a hypoid gear set, the axis of the pinion is offset somewhat from the axis of the gear.

Hypoid Gear Lubricant (Hypoid Oil)

A gear lubricant with extreme-pressure characteristics used in hypoid gears.

Hysteresis (Magnetic)

The lag or delay in the change of one variable as another related variable changes. For example, magnetic induction lags behind the change in the magnetizing force. The hysteresis loss is the loss arising from a single cycle of the hysteresis loop.

Hysteresis (Mechanical)

The phenomenon of permanently absorbed or lost energy that occurs during any cycle of loading or unloading when a material is subjected to repeated loading.

Hysteresis Loop (Magnetic)

A closed curve that characterizes the magnetization/demagnetization characteristics as a function of the applied magnetic field for magnetic materials.

Hysteresis Loop (Mechanical)

In dynamic mechanical measurement, the closed curve representing successive stress–strain status of a material during a deformation cycle.

Hz

Hertz.

I

IACS

International annealed copper standard; a standard reference used in reporting electrical conductivity. The conductivity of a material, in %IACS, is equal to 1724.1 divided by the electrical resistivity of the material in $n\Omega \cdot m$.

Iceland Spar

A transparent form of calcite.

I.D.

Internal diameter.

Ideal Crack

A simplified model of a crack used in elastic stress analysis. In a stress-free body, the crack has two smooth surfaces that are coincident and join within the body along a smooth curve called the crack front; in 2D representations, the crack front is called the crack tip.

Ideal-Crack-Tip Stress Field

The singular stress field, infinitesimally close to the crack front, that results from the dominant influence of an ideal crack in an elastic body that is the form. In a linear-elastic homogeneous body, the significant stress components vary inversely as the square root of the distance from the crack tip. In a linear-elastic body, the crack-tip stress field can be regarded as the superposition of three component stress fields called *modes*.

Ideal Critical Diameter (D_1)

Under an ideal quench condition, the bar diameter that has 50% martensite at the center of the bar when the surface is cooled at an infinitely rapid rate (i.e., when H = infinity, where H is the quench severity factor or *Grossman number*).

Identification Etching

Etching to expose particular microconstituents; all others remain unaffected.

Idiomorphic Crystal

An individual crystal that has grown without restraint so that the habit planes are clearly developed. Crystals having a shape corresponding to their crystal lattice. They occur when there is no external constraint to their growth. Compare with *allotriomorphic crystal*.

Ignition Loss

(1) The difference in weight before and after burning. (2) The burning off of *binder* or *size*.

Ignition Temperature

The temperature at which combustion can commence. See also *gas cutting*.

Ihrigizing

A process for forming a hard corrosion-resistant silicon coating on steel by heating at high temperature in a silicon tetrachloride vapor.

Illumination

The luminous flux density incident on a surface; the ratio of flux to area of illuminated surface.

Image

A representation of an object produced by radiation, usually with a lens or mirror system.

Image Analysis

Measurement of the size, shape, and distributional parameters of microstructural features by electronic scanning methods, usually automatic or semiautomatic. Image analysis data output can provide individual measurements on each separate feature (feature specific) or field totals for each measured parameter (field specific).

Image Contrast

A measure of the degree of detectable difference in intensity within an image.

Image Rotation

In electron optics, the angular shift of the electron image of an object about the optic axis induced by the tangential component of force exerted on the electrons perpendicular to the direction of motion in the field of a magnetic lens.

Immersed-Electrode Furnace

A furnace used for liquid carburizing of parts by heating molten salt baths with the use of electrodes immersed in the liquid. See also *submerged-electrode furnace*.

Immersion Cleaning

Cleaning in which the work is immersed in a liquid solution.

Immersion and Chemical Coatings

Immersion coatings are applied without electricity by immersing parts in a chemical solution or bath containing the metal to be deposited. Deposition can take place either by displacement, when the metal in solution displaces the base metal, or by reduction, where the base metal does not enter into the reaction.

Although many metals can be deposited in immersion baths, comparatively few have proved acceptable for decorative or functional applications. These are nickel, tin, copper, gold, and silver.

Electroless Nickel

Properties

Electroless nickel is generally more expensive than electroplated nickel. For this reason, it is used primarily for its functional properties; although it is very smooth, bright deposit can be obtained on buffed ferrous and nonferrous metals.

Because of its amorphous structure and phosphorus content (8%–10%), the coatings are said to have better corrosion resistance than electrolytic or wrought nickel. Hardness of the coatings is relatively high—about 50 R_c—and can be raised to 64 R_c by heat treatment. Thickness of the coatings ranges from 1 to 5 mils, depending on end use.

Applications

The most important uses for electroless nickel are to protect parts from corrosion and to prevent product contamination. The coatings are widely used on tank-car interiors to protect caustic soda, ethylene oxide, tetraethyl lead, tall oil, and many other liquids from contamination. Other similar applications include oil refinery air compressors, missile fuel for plates, gas storage bottles for liquid, and pumps for petroleum and related products.

The hardness of the coatings is particularly valuable in increasing the life of rotating and reciprocating surfaces in gas compressors, pumps, hydraulic cylinders, sheaves, and armatures. The coatings are also used on aluminum electronic devices to facilitate soldering, on stainless steel to facilitate brazing, on moving metal parts to prevent galling, and on stainless steel equipment to prevent stress corrosion cracking.

Tin

Properties

Tin immersion coatings are especially noted for their low cost, bright appearance, good frictional properties, and ease of application to many common metals such as copper, brass, bronze, aluminum, and steel. However, their corrosion resistance is only fair.

As with some other immersion coatings, plating usually stops when the base metal is completely covered. Thus, thickness for common decorative uses is limited to about 0.015 mils. However, thicknesses up to 2 mils for heavy-duty applications have been produced by placing the base metal in contact with a dissimilar metal, thereby generating current and promoting additional plating.

Applications

Tin immersion coatings are popular for decorative finishing of small parts such as safety pins, thimbles, and buckles. They are also applied to copper tubing to prevent discoloration from water and to aluminum engine pistons to provide lubrication during break-in periods.

Copper

Properties

The most important characteristics of copper immersion coatings are their high electrical conductivity, good lubrication properties, and unique appearance. In addition to steel, they can be applied to brass and aluminum and to printed circuit boards. Usual thickness range is 0.1–1 mils.

Applications

Because of their conductivity, copper immersion coatings have proved particularly useful for printed circuits. They are not especially noted for their decorative appeal, but can be used in applications where a particular appearance is required, for example, inexpensive, decorative hardware such as casket parts. Because of their good lubrication properties, they can also be used on steel wire in die-forming operations.

Gold

Properties

Gold immersion coatings are relatively inexpensive because of their extreme thinness—about 0.001 mil. The coatings have good electrical conductivity and emissivity characteristics and a bright, attractive appearance. As deposited, they are not especially resistant to discoloration and abrasion; however, they can be protected with a clear lacquer finish. They are used on a wide variety of ferrous and nonferrous metals and on copper printed circuit boards.

Applications

Because of their good appearance, gold immersion coatings are principally used on costume jewelry, trophies, auto trim, and inexpensive novelties. Their conductivity and solderability are used to advantage in electrical applications such as printed circuits, transistors, and connectors. Also, the unique emissivity properties of the coatings have proved useful in missile applications.

Silver

Properties

Like gold, silver immersion coatings are relatively inexpensive because of their extreme thinness. The coatings have a bright, attractive appearance when first deposited. Their resistance to tarnishing and abuse is poor; however, they can be protected somewhat with a clear lacquer coating.

Silver immersion coatings can be applied to most base metals except lead, zinc, aluminum, and very active metals. They perform best on copper, nickel, and steel. Usual thickness is about 0.001 mil, but 0.03 mil can be deposited in some cases.

Applications

Because of their poor durability, silver immersion coatings are not too popular; the only applications are cheap decorative products, minor electronic parts, and maintenance plating.

Immersion Etching

Method in which a microsection is dipped face up into etching solution and is moved around during etching. This is the most common etching method.

Immersion Lens

See *immersion objective*.

Immersion Objective

In optical microscopy, an objective in which a medium of high refractive index is used in the object space to increase the numerical aperture and therefore the resolving power of the lens.

Immersion Objective (Electron Optics)

A lens system in which the object space is at a potential or in a medium of index of refraction different from that of the image space.

Immersion Oil

An oil that is interposed between the specimen and the objective lens of a microscope. It has a high refractive index compared with air and hence improves the resolving power.

Immersion Plating

Depositing a metallic coating on a metal immersed in a liquid solution, without the aid of an external electric current. Also called *dip plating*.

Immiscible

(1) Of two phases, the inability to dissolve in one another to form a single solution; mutually insoluble. (2) With respect to two or more fluids, not mutually soluble; incapable of attaining homogeneity.

Immunity

A state of resistance to corrosion or anodic dissolution of a metal caused by thermodynamic stability of the metal.

Impact Bruise

See *percussion cone*.

Impact Energy

The amount of energy, usually given in joules or foot-pound force, required to fracture a material, usually measured by means of an *Izod test* or *Charpy test*. The type of specimen and test conditions affects the values and therefore should be specified.

Impact Extrusions

Impact extrusion consists of subjecting metallic materials to very high pressures at room temperature. Under these pressures, the metals become "plastic" and assume predetermined shapes. Whereas in coining this process takes place within a closed die, typical impact extrusions allow a portion of the metal to be "squirted" or "squeezed" out of the die cavity, thereby forming an integral part of the desired shape.

From a practical point of view, impact extrusion of suitable parts may result in substantially lower unit manufacturing cost because it permits

1. High production rates
2. Substantial material savings (up to 75%)
3. Little or no machining
4. Low initial tool costs and long tool life
5. Bright and smooth surface finish ready for decorating

There are two types. Impact extrusion of the slug takes place between punch and die. Under pressure, the metal may be forced to flow counter to the direction of punch travel (backward extrusion) or in the direction of punch travel (forward extrusion). Frequently, parts require a combination of both types.

Starting Material

The raw material for the process is usually referred to as a slug. Slugs may be blanked from sheet or plate, sawed from bar stock, or cast. The cross section of a slug—round, oval, square, or rectangular—fits into the die bottom; its height is determined by the volume of metal required to produce the part.

Equipment Selection

The pressures necessary for impact extrusion are available on mechanical or hydraulic presses. Mechanical presses are usually preferred when higher production rates are required. Both toggle and crank presses are used—depending on load and performance characteristics required. Hydraulic presses have primarily been used for forward extrusions requiring particularly high pressures and heavy cross sections. Factors bearing on the selection of equipment for impact extrusion are dimensional characteristics of the part to be produced, tonnage, and speed.

Dimensional characteristics of the part (diameter, height, etc.) determine die size of the press and length of stroke required.

Tonnage required for metal flow must be carefully determined in advance. It is predicated on the relative plastic deformation required for the part, that is, the ratio of the cross-sectional area of the extruded part to the area of the slug. Maximum limits of plastic deformation vary from 90%–95% for 99.5% aluminum, lead, and tin to 70%–80% for mild steel and brasses.

Design Criteria

Impact extrusion should be specified by the designer primarily for the following:

1. Parts that are essentially hollow shells consisting of a wall and a bottom or partial bottom section. While in drawing the ratio between height and bottom diameter is limited to approximately 2 to 3:1, it is possible to impact-extrude (backward) up to 8:1. Forward extrusions many times as long as the diameter have also been made.
2. Parts with straight, no-draft walls.
3. Parts requiring high-strength characteristics.

4. Parts with longitudinal ribs, flutes, splines, etc., or with bosses, cavities, etc., in the bottom section.
5. Parts made in large quantities calling for low unit costs.

Applications

It is significant for this process that the flow of metal takes place primarily in a direction parallel to punch travel. Parts produced by impact extrusion are essentially longitudinally oriented, for example, collapsible tubes and cans.

Originally used only on soft materials (lead, tin, etc.) to make collapsible tubes, the process has, in recent years, found rapidly increasing applications in the field of metallic containers, as well as in the production of a wide range of automotive, electrical, and hardware components. Aluminum and its alloys, copper, high brasses, and mild steel are impact-extruded commercially in large quantities.

Since it is a "chipless" metalworking process, impact extrusion competes in many applications with the automatic screw machine, deep drawing, die casting, hot forging, cold upsetting, and other operations. To achieve optimum results, it is important that likely parts be designed with an understanding of characteristics of metal flow, die design, and proper distribution of pressures in impact extrusion.

The impact extrusion of unheated slugs is often called *cold extrusion*.

Impact (er) Forging

A forging process in which the component is forged between a pair of moving die faces so both sides of the component are worked.

Impact Line

A blemish on a drawn sheet-metal part caused by a slight change in metal thickness. The mark is called an impact line when it results from the impact of the punch on the blank; it is called a recoil line when it results from transfer of the blank from the die to the punch during forming or from a reaction to the blank being pulled sharply through the *draw ring*.

Impact Load

An especially severe shock load such as that caused by instantaneous arrest of a falling mass, by shock meeting of two parts (e.g., in a mechanical hammer), or by explosive impact, in which there can be an exceptionally rapid buildup of stress.

Impact Sintering

An instantaneous sintering process during *high-energy rate compacting* that causes localized heating, welding, or fusion at the particle contacts.

Impact Strength

A measure of the resiliency or toughness of a solid. The maximum force or energy of a blow (given by a fixed procedure) that can be withstood without fracture, as opposed to fracture strength under a steady applied force.

Impact Test

A test for determining the energy absorbed in fracturing a test piece at high velocity, as distinct from static test. The test may be carried out in tension, bending, or torsion, and the test bar may be notched or unnotched. See also *Charpy test*, *impact energy*, and *Izod test*.

Generally, any test in which the load is dynamic rather than static. More specifically, a test in which a specially machined test piece is dynamically loaded in a purpose-made machine that registers the energy absorbed in causing failure as a measure of the material toughness. The test pieces usually have a single V notch to initiate the fracture, hence the term V notch test, but less common variations include plain bar tests and keyhole-shaped notch tests. The two most popular testing machines are the Charpy and the Izod. Both of these utilize a weighted pendulum carrying a striker that impacts the test piece located at the low point of the swing. The Izod striking energy is 167 J (120 ft lb) and that of the Charpy 300 J (220 ft lb). The test is commonly used for steels as many exhibit a dramatic reduction in toughness as the temperature falls. This reduction, reflecting a change in fracture mode from ductile to brittle, is measured by recording the energy absorbed by a series of test pieces broken over a range of temperature. In addition, the characteristics of the fracture faces may be examined to determine the relative proportions of the ductile and brittle (or crystalline) zones. The observations provide another measure of the change from brittle to ductile behavior by identifying the temperature at which a fracture is 50% ductile and 50% brittle. This temperature is termed either the fracture appearance transition temperature or the ductile/brittle transition temperature. Reference may also be made to the nil ductility transition temperature, which is the highest temperature at which no ductile fracture is observed. This latter term is sometimes also applied to the temperature at which some specified low level of impact strength is recorded, for example, 10 ft lb in the Charpy test. See *brittle fracture*.

Impact Tube

Same as *Pitot tube*.

Impact Value

The energy absorbed by a specimen of standard design when sheared by a single blow from a testing machine hammer. Expressed in J/m^2 or ft·lb in.^{2}

Impact Velocity

The relative velocity between the surface of a solid body and an impacting liquid or solid particle. To describe this velocity completely, it is necessary to specify the direction of motion of a particle relative to the solid surface in addition to the magnitude of the velocity. The following related terms are also in use: (1) absolute impact velocity, the magnitude of the impact velocity, and (2) normal impact velocity, the component of the impact velocity that is perpendicular to the surface of the test solid at the point of impact.

Impact Wear

Wear of a solid surface resulting from repeated collisions between that surface and another solid body. The term *erosion (erosive) wear* is preferred in the case of multiple impacts and when the impacting body or bodies are very small relative to the surface being impacted.

Impaction Ratio

See *collection efficiency*.

Imperfection

(1) When referring to the physical condition of a part or metal product, any departure of a quality characteristic from its intended level or state. The existence of an imperfection does not imply nonconformance (see *nonconforming*) nor does it have any implication as to the usability of a product or service. An imperfection must be rated on a scale of severity, in accordance with applicable specifications, to establish whether or not the part or metal product is of acceptable quality. (2) Generally, any departure from an ideal design, state, or condition. (3) (In crystal lattice) Any deviation from perfection in lattice alignment such as a dislocation or any atomic imperfection such as a substitutional or interstitial atom in a solid solution or a vacancy.

Impingement

A process resulting in a continuing succession of impacts between liquid or solid particles in a solid surface. In preferred usage, impingement also connotes that the impacting particles are smaller than the solid surface and that the impacts are distributed over the surface or a portion of the surface. If all impacts are superimposed on the same point or zone, then the term repeated impact is preferred.

Impingement Corrosion Attack

A form of surface damage occurring in flowing liquids resulting from entrained particles or turbulent flow and associated bubbles abrading or disrupting the normally protective surface oxide film. The exposed metal is anodic to the neighboring oxide and hence is preferentially corroded; see *electrochemistry*. Generalizing the likelihood and severity of attack is increased by increases in fluid velocity, fluid contamination, (suspended, dissolved, or deposited), and surface roughness and irregularities in the system such as bends and inlets to tubes. See also *corrosion–erosion* and *cavitation*.

Impingement Erosion

Loss of material from a solid surface due to *liquid impingement*. See also *erosion*.

Impingement Umbrella

The partial screening of the surface of a solid specimen subjected to solid impingement that sometimes occurs when some of the solid particles rebound from the surface and impede the motion of other impinging particles.

Implant

A manufactured device inserted into the human body to replace or reinforce some structure or undertake some function.

Impregnate

In reinforcing plastics, to saturate the reinforcement with a resin.

Impregnated Fabric

A fabric impregnated with a synthetic resin. See also *prepreg*.

Impregnated Wood

Also called *compressed wood* or *densified wood*. Many types are forms of laminated wood. *Compreg*, developed by U.S. Forest Products Laboratory, consists of many layers of 0.0625 in. (0.16 cm), rotary cut, yellow-birch plies bonded with about 30% resin under a pressure of 600–1500 lb/in.^{2} (4–10 MPa).

Wood impregnated with *polyethylene glycol* is known as "Peg." This treatment is used for walnut gunstocks for high-quality rifles and for tabletops. This impregnant can be used to reduce checking of green wood during drying. Wood can also be vacuum-impregnated with certain liquid vinyl monomers and then treated by radiation or catalyst heat systems, which transform the vinyl to a plastic. *Methyl methacrylate*, or *acrylic*, is a common resin used to produce this type of product, known as *wood–plastic combinations* (*WPCs*). A principal commercial use of this modified wood is as parquet flooring and for sporting goods such as archery bows. WPC material resists indentation from rolling, concentrated, and impact loads better than white oak. This is largely due to improved hardness, which is increased 40%. Abrasion resistance is no better than that of white oak.

Impregnating Materials for Castings

Castings are impregnated for several reasons, the most obvious of which is to prevent leakage. Other important reasons for impregnating are to prevent corrosion, improve surface finish, remove sites that may lodge food particles and cause bacterial growth, and prevent back seepage of occluded fluids. Leakage may be avoided by blocking the pores at any point along their length, whereas all of the other aims can be met only when the pores are blocked at the surface as well as in depth.

Impregnating Process

There are four stages in the impregnation process. First, a porous casting may have different types of porosity (continuous, noncontinuous or blind, and isolated). After evacuation and impregnation under pressure, an ideal condition that is rarely attained may be found; however, it is more likely that there is bleeding.

Bleeding will lead to incompletely filled pores near the surface; and contraction in curing will lead to incomplete filling along the length of the pore. The latter condition may arise from loss of solvent from impregnants introduced as a solution, loss of water from vehicles such as sodium silicate, and decomposition of organic impregnants by exposure to excessive temperatures (carbonization). Some compensation for the loss of solvent or water can be obtained if the residue can be made to expand on curing, for example, by oxidation of metallic particles carried in sodium silicate.

It has been demonstrated that the sealing action of an impregnant generally decreases as the amount of volatile in the formulation increases. One exception to this rule has been demonstrated, in which a self-fluxing brazing filler metal powder is fused in the pores after evaporation of the vehicle. The molten braze filler metal runs into the narrowest part of the pore under capillary action and gives optimum sealing in spite of incomplete filling of the pores.

Impregnating Materials

The oldest method of reducing leakage, particularly in cast iron, is to rust or oxidize the pores, sometimes after "impregnation" with mud. Natural drying oils such as tung and linseed were also among the first impregnant materials to be used. Another early type of impregnant was based on the readily available water glass (sodium silicate).

The contraction of sodium silicate impregnants during drying was reduced by the addition of inert particles such as asbestos, chalk, or oxides. Further development has led to the so-called metallic impregnants in which a large proportion of the solids are metal powders (e.g., copper or iron). Special agents are added to the sodium silicate to reduce sedimentation. During the curing of these "metallic" impregnants, the metal particles oxidize so that a certain amount of expansion occurs to offset the considerable contraction. However, these impregnants have several disadvantages; their resistance to high temperatures is poor (e.g., 149°C max); the impregnants are attacked by steam; and the abrasive oxides make them unsuitable for bearings or machining.

There are several differences between the sodium silicate and plastic impregnants. In general, because plastics or oils do not wet castings readily, a vapor degreasing is usually necessary, followed by pumping under a vacuum of at least 712 mmHg for periods of 30 min (one impregnator has achieved success in difficult cases by holding under a vacuum of 10 μm for 2 h). A second difference results from the higher viscosity of plastics. During the pressure stage of impregnation, a minimum of 30 min under pressure may be necessary to force the plastics to flow deeply into the pores. One solution to the problem of high viscosity is to thin the plastic by solvents, but this leads to bleeding and contraction.

The cost of impregnating with sodium silicate is usually less (about one quarter less) than with plastics because prior cleaning of castings needs to be less thorough and excess impregnant is washed away with water.

The most important types of plastic impregnants are based on the styrene monomer. A typical composition of this substitute is 80% styrene monomer with 20% linseed oil and a small quantity of organic oxidizer as a catalyst. The instability of the catalyst in the presence of lead, zinc, and copper restricts the use of this type of impregnant primarily to the light metals. The castings are heated for 2 h at 135°C to polymerize the impregnant.

The thermosetting of plastics formed by copolymerization of the styrene monomer with polyesters represents an advance on the styrene–linseed oil polymers because it is possible to introduce liquid bath curing. The surfaces are cured rapidly when they make contact with the liquid curing agent, so that subsequent bleeding (which occurs on curing in ovens) is prevented. Other improvements that have been introduced with the styrene–polyester plastics include detergent washing (does not wash the impregnant from pores), freedom from inhibition by copper, and low contraction on curing (e.g., 6%–7%). A typical curing cycle is 1 h at 135°C. These impregnants will withstand 260°C for short periods and 204°C in continuous service.

Phenolics, epoxy, and furfural plastics can also be used for impregnation. The phenolics are dissolved in a solvent such as alcohol and require curing into stages, first to remove the alcohol (e.g., 1 h at 80°C) and second to cure the plastic (e.g., 3 h at190°C). Phenolic resins do not have good adherence to metal, and they seal according to bleeding and contraction. Therefore, they do not provide the highest quality seals, although they may be suitable for many types of work. Epoxy resins have good adhesion to metals and resistance to chemicals, but many have to be thinned with solvents to achieve low viscosity so that the impregnation occurs readily. Furfural-type plastics have been developed for resistance to alkalies.

Polyester–styrene and sodium silicate are the most widely used impregnants. The plastics give the higher-quality seal, but sodium silicate continues to be used because of its low cost.

Impregnation

(1) Treatment of porous castings with a sealing medium to stop pressure leaks. (2) The process of filling the pores of a sintered compact, usually with a liquid such as a lubricant. (3) The process of mixing particles of a nonmetallic substance in a cemented carbide matrix, as in diamond-impregnated tools.

Impressed Current

Direct current supplied by a device employing a power source external to the electrode system of a *cathodic protection* insulation.

Impression

(1) In electron microscopy, the reproduction of the surface contours of a specimen formed in a plastic material after the application of pressure, heat, or both. (2) In hardness testing, the imprint or dent made in the specimen by the indenter of a hardness-measuring device. See also *indentation hardness* and *indenter*. (3) A cavity machined into a forging die to produce a desired configuration in the workpiece during forging.

Impression-Die Forging

A forging that is formed to the required shape and size by machined impressions in specially prepared dies that exert 3D control on the workpiece.

Impression Replica

A surface replica made by impression. See also *impression* and *replica*.

Impulse (Resistance Welding)

An impulse of welding current consisting of a single pulse or a series of pulses, separated only by an interpulse time.

Impurities

(1) Elements or compounds with presence in a material that is undesirable. (2) In a chemical war material, minor constituent (s) or component (s) not included deliberately; usually, to some degree or above some level, undesirable.

Inadequate Joint Penetration

Weld joint penetration that is less than that specified.

Incandescence

The omission of light radiation by a substance at elevated temperature, above 540°C in dim background light.

Inclinable Press

A press that can be inclined to facilitate handling of the formed parts. See also *open-back inclinable press*.

Inclined Position

The position of a pipe joint in which the axis of the pipe is at an angle that is approximately 45° to the horizontal and the pipe is not rotated during welding. A loose term for welding positions not lying within the limits of the four basic positions: flat, vertical, horizontal, and overhead.

Included Angle

See preferred term *groove angle*.

Inclusion

(1) A physical and mechanical discontinuity occurring within a material or part, usually consisting of solid, encapsulated foreign material. Inclusions are often capable of transmitting some structural stresses and energy fields, but to a noticeably different degree than from the parent metal. (2) Particles of foreign material in a metallic matrix. The particles are usually compounds, such as oxides, sulfides, or silicates, but may be of any substance that is foreign to (and essentially insoluble in) the matrix. See also *exogenous inclusion*, *indigenous inclusion*, and *stringer*.

Inclusion Count

Determination of the number, kind, size, and distribution of nonmetallic inclusions in metals.

Incoherent Scattering

In materials characterization, the deflection of electrons by electrons or atoms that results in a loss of kinetic energy by the incident electrons. See also *coherent scattering*.

Incoloy and Inconel

Ranges of proprietary alloys based on nickel with major additions of other elements. In particular, chromium and iron but also molybdenum, cobalt, titanium, aluminum, etc. Generalizing, they have good corrosion resistance including resistance to oxidation at high temperatures.

Incomplete Fill (of Weld)

An insufficient quantity of filler material in a welded joint, particularly a butt joint. Same as *underfill*.

Incomplete Fusion (of Weld)

Same as *lack of fusion*.

Increment

An individual portion of a material collected by a single operation of a sampling device from parts of a lot separated in time or space. Increments may be tested individually or combined (composited) and tested as a unit.

Incubation Period

(1) A period prior to the detection of corrosion while the metal is in contact with a corrodent. (2) In cavitation and impingement erosion, the initial stage of the erosion rate-time pattern during which the erosion rate is zero or negligible compared to later stages. (3) In cavitation and impingement erosion, the exposure duration associated with the initial stage of the erosion rate-time pattern during which the erosion rate is zero or negligible compared to later stages. See also *cavitation erosion* and *impingement erosion*.

Indentation

In a spot, seam, or projection welding, the depression on the exterior surface of the base metal.

Indentation Hardness

(1) The resistance of a material to indentation. This is the usual type of hardness test, in which a pointed or rounded indenter is pressed into a surface under a substantially static load. (2) Resistance of a solid surface to the penetration of a second, usually harder, body under prescribed conditions. Numerical values used to express indentation hardness are not absolute physical quantities, but depend on the hardness scale used to express hardness. See also *Brinell hardness test*, *Knoop hardness test*, *nanohardness test*, *Rockwell hardness test*, and *Vickers hardness test*.

Indenter

In hardness testing, a solid body of prescribed geometry, usually chosen for its high hardness, that is used to determine the resistance of a solid surface to penetration.

Index of Refraction

See *refractive index*.

Indication

In inspection, a response to a nondestructive stimulus that implies the presence of an *imperfection*. The indication must be interpreted to determine if (a) it is a true indication or a *false indication* and (b) whether or not a true indication represents an unacceptable deviation.

Indicator

A substance that, through some visible change such as color, indicates the condition of a solution or other material as to the presence of free acid, alkali, or other substance.

Indices

See *Miller indices*.

Indigenous Inclusion

An *inclusion* that is native, innate, or inherent in the molten metal treatment. Indigenous inclusions include sulfides, nitrides, and oxides derived from the chemical reaction of the molten metal with the local environment. Such inclusions are small and require microscopic magnification for identification. Compare with *exogenous inclusion*.

Indigo

One of the most important of all vegetable dyestuffs and valued for the beauty and permanence of its color. It is widely used to color *denim* for clothing. Commercial blue indigo is obtained from plants and several other species, of India and Java, and the plants of Europe, by steeping the freshly cut plants in water, and after decomposition of the glucoside *indican*, the liquid is run into beating vats where the indigo separates out in flakes that are pressed into cakes. *Indigo red* is a crimson dyestuff obtained in the proportion of 1%–5% in the manufacture of indigo by extraction in inorganic solvents. *Indigo brown* is an impurity that occurs during the manufacture of indigo, but has little influence in the dyeing process. Other constituents are *indigo yellow* and *indigo gluten*. *Indigo white* is obtained by reducing agents and an alkali.

Indirect-Arc Furnace

An electric-arc furnace in which the metallic charge is not one of the poles of the arc.

Indirect (Backward) Extrusion

See *extrusion*.

Indirect Sintering

A process whereby the heat needed for sintering is generated outside the body and transferred to the powder compact by conduction, convection, radiation, etc. Contrast with *direct sintering*.

Indium

Indium (symbol In) is a silvery white metal with a bluish hue, whiter than tin. It has a specific gravity of 7.31, tensile strength of 103 MPa, and elongation 22%. It is very ductile and does not work-harden, because its recrystallization point is below normal room temperature, and it softens during rolling. The metal is not easily oxidized, but above its melting point, 157°C, it oxidizes and burns with a violet flame.

Indium is now obtained as a by-product from a variety of ores. Because of its bright color, light reflectance, and corrosion resistance, it is valued as a plating metal, especially for reflectors. It is softer than lead, but a hard surface is obtained by heating the plated part to diffuse the indium into the base metal. It has high adhesion to other metals. When added to chromium plating baths, it reduces brittleness of the chromium.

In spite of its softness, small amounts of indium will hard copper, tin, or lead alloys and increase the strength. About 1% in lead will double the hardness of the lead. In solders and fusible alloys, it improves wetting and lowers the melting point. In lead-base alloys, a small amount of indium helps to retain oil film and increases the resistance to corrosion from the oil acids. Small amounts may be used in gold and silver dental alloys to increase the hardness, strength, and smoothness. Small amounts are also used in silver–lead and silver–copper aircraft-engine bearing alloys. Lead–indium alloys are highly resistant to corrosion and are used for chemical-processing equipment parts. Gold–indium alloys have a high fluidity, a smooth lustrous color, and good bonding strength. An alloy of 77.5% gold and 22.5 indium, with the working temperature of about 500°C, is used for brazing metal objects with glass inserts. Silver–indium alloys have a high hardness and a fine silvery color. A silver–indium alloy used for nuclear control rods contains 80% silver, 15% indium, and 5% cadmium. The melting point is 746°C, tensile strength 289 MPa, and elongation 67%, and it retains a strength of 120 MPa at 316°C. It is stable to irradiation and is corrosion resistant

to high-pressure water up to 360°C. The thermal expansion is about six times that of steel.

Mechanical Properties

Typical properties of annealed indium are as follows: tensile strength, 262 MPa; hardness, 0.9 Bhn; and compressive strength, 215 MPa.

Because indium does not work-harden and almost all of the deformation occurring in the tensile test is localized, deformation is very low for such a low-strength material.

Reactivity

Indium is stable in dry air at room temperature. The metal boils at 2000°C but sublimes when heated in hydrogen or in vacuum. Because a thin, tenacious oxide film forms on its surface, indium resists oxidation up to, and a little beyond, its melting point. The film, however, dissolves in dilute hydrochloric acid.

Indium can be slowly dissolved in dilute mineral acids and more readily in hot dilute acids. It unites with the halogens directly when warm. Concentrated mineral acids react vigorously with indium, but there is no attack by solutions of strong alkalies. The metal dissolves in oxalic acid, but not in acetic acid.

There is no evidence that the metal is toxic and it has no action as a skin irritant.

Applications

The three largest uses of indium are in semiconductor devices, bearings, and low-melting-point alloys.

Semiconductors

Indium is used to form p–n junctions in germanium. Two characteristics—the fact that indium readily wets the germanium and dissolves it at 500°C–550°C, as well as the fact that after alloying, the indium does not set up contraction stresses in the germanium—make it suitable for this application.

If a piece of n-type germanium is dissolved in indium, germanium containing excess indium recrystallizes after subsequent cooling. The excess indium changes the germanium from n-type to p-type.

Other compounds include InSb where interest is twofold. Its small energy gap (0.18 eV) makes it valuable as a photodetector in wavelengths from the infrared up to ~8 mm. InSb also has a very high electron mobility—up to 80,000 cm^2/Vs at room temperature and up to 800,000 cm^2/Vs at liquid nitrogen temperature. Consequently, the material can be used in devices based on magneto resistance or the Hall effect (gyrators, switching elements, magnetometers, analog computers, etc.).

InAs is of interest as a semiconductor: energy gap ~0.47 eV and electron mobility ~50,000 cm^2/Vs (room temperature). The energy gap is too small to make InAs practical as a transistor material. It is, however, a candidate for infrared photoconductor applications. The relatively high mobility and small dependence of electrical properties on temperature make InAs a particular interest for Hall effect and magnetoresistance devices.

InO_3 is an n-type semiconductor finding use as a resistive element in integrated circuits, and InP is a semiconductor that is useful for rectifiers and transistors and is a promising material for electronic devices that operate at intermediate temperatures.

Bearings

Impregnating the surface of steel-backed lead–silver bearings increases the strength and hardness, improves resistance to corrosion by acids in the lubricants, and permits better retention of the bearing oil film. Indium-coated bearings can be used for high-duty service such as found in aircraft engines and diesel engines.

Alloying

A common glass sealing alloy contains approximately 50% tin–50% indium. A solder alloy containing 37.5% lead, 37.5% tin, and 25% indium has greater resistance to alkalies than the 50% lead–50% tin solder.

Adding indium to gold for use in dental alloys increases the tensile strength and ductility of gold, improves resistance to discoloration, and improves bonding characteristics.

Individuals

Conceivable constituent parts of a population. See also *population*.

Induction Bonding

The use of high-frequency (5–7 MHz) electromagnetic fields to heat a bonding agent placed between the plastic parts to be joined. The bonding agent consists of microsized ferromagnetic particles dispersed in a thermoplastic matrix, preferably the parent material of the parts to be bonded. When this binder is exposed to the high-frequency source, the ferromagnetic particles respond and melt the surrounding plastic matrix, which in turn melts the interface surfaces of the parts to be joined.

Induction Brazing

A brazing process in which the surfaces of components to be joined are selectively heated to brazing temperature by electrical energy transmitted to the workpiece by induction, rather than by a direct electrical connection, using an inductor or work coil.

Induction Furnace

An alternating current electric furnace in which the primary conductor is coiled and generates, by electromagnetic induction, a secondary current that develops heat within the metal charge. There are two classifications of induction furnaces: coreless and channel. In a coreless furnace, the refractory-lined crucible is completely surrounded by a water-cooled copper coil, while in the channel furnace, the coil surrounds only a small appendage of the unit, called an inductor. The term "channel" refers to the channel that the molten metal forms as a loop within the inductor. It is this metal loop that forms the secondary of the electrical circuit, with the surrounding copper coil being the primary. In a coreless furnace, the entire metal content of the crucible is the secondary. See also *coreless induction furnace*.

Induction Hardening/Heating/Melting

Processes in which heating is by the electrical induction of eddy currents in the material being treated. The general principle is that the material is surrounded by a water-cooled conducting coil carrying ac current. Eddy currents are induced in the material causing

heating. Low- and medium-frequency currents up to about 100 kHz cause general heating through the thickness or volume of material and hence are used for melting and billet heating applications. Higher frequencies induce current at the surface, the so-called skin effect, and hence are useful for case hardening.

Induction Heating

Heating by combined electrical resistance and hysteresis losses induced by subjecting a metal to the varying magnetic field surrounding a coil carrying alternating current. See Table I.1.

Induction Melting

Melting in an *induction furnace*.

Induction Sintering

Sintering in which the required heat is generated by subjecting the compact to electromagnetic induction. See also *direct sintering*.

Induction Soldering

A soldering process in which the heat required is obtained from the resistance of the work to induce electric current.

Induction Tempering

Tempering of steel using low-frequency electrical *induction heating*.

Induction Welding

Processes in which the heat for welding is induced by an alternating electric current. A forging action is also normally applied to effect the joint. If the ac frequency is 10 Hz or higher, the process can be termed inelasticity. See *anelasticity*.

Induction Work Coil

The *inductor* used when induction heating and melting as well as induction welding, brazing, and soldering.

Inductively Coupled Plasma

An argon plasma excitation source for atomic emission spectroscopy or mass spectroscopy. It is operated at atmospheric pressure and sustained by inductive coupling to a radio-frequency electromagnetic field. See also *radiofrequency*.

Inductor

A device consisting of one or more associated windings, with or without a magnetic core, for introducing inductance into an electric circuit.

Industrial Atmosphere

An atmosphere in an area of heavy industry with soot, fly ash, and sulfur compounds as the principal constituents.

Industrial Jewels

Hard stones, usually *ruby* and *sapphire*, used for bearings and impulse pins in instruments and for recording needles. *Ring jewels* are divided into large and small. The large rings are about 0.050 in. (0.127 cm) in diameter and 0.012 in. (0.030 cm) thick with holes above 0.006 in. (0.015 cm) in diameter. Ring jewels are used as pivot bearings in instruments, timepieces, and dial indicators. From 2 to 14 are used in a watch. *Vee jewels* are used in compasses and electrical instruments. *Cup jewels* are used for electric meters and compasses. *End stones* are flat, undrilled stones with polish faces to serve as end bearings. *Pallet stones* are rectangular impulse stones for watch escapements. *Jewel pins* are cylindrical impulse stones for timepiece escapements. In making *bearing jewels*, the synthetic sapphire boules are split in half, secured to wooden blocks, and then sawed to square blanks. These are then rounded on centerless grinding machines and flat-ground to thickness by means of copper wheels and diamond powder. *Quartz bearings* are made from fused quartz rods. A notch is grounded in the end of the rod and then polished and cut off, repeating the process for each bearing. Quartz has a Mohs hardness of only 7, while the synthetic ruby and sapphire have a Mohs hardness of 8.8, but Quartz has the advantage of low thermal expansion.

The making of industrial jewels was formally a relatively small, specialized industry, and a national stockpile of cut jewels was maintained for wartime emergencies. But the process of slicing and shaping hard crystals for semiconductors and other electronic uses is essentially the same, and stones of any required composition and cut to any desired shape are now regularly manufactured.

Industrial Thermosetting Laminates

So-called industrial thermosetting laminates are those in which a reinforcing material has been impregnated with thermosetting resins, and the laminates have usually been formed at relatively high pressures.

Resins most commonly used are phenolic, polyester, melamine, epoxy, and silicone; reinforcements are usually paper, woven cotton, asbestos, glass cloth, or glass mat.

The National Electrical Manufacturers' Association has published standards covering many standard grades of laminates. Each manufacturer, in addition to these, normally provides a range of special grades with altered or modified properties or fabricating characteristics. Emphasis here is on the standard grades available.

Laminates are available in the form of sheet, rod, and rolled or molded tubing. Laminated shapes can also be specified; these are generally custom-molded by the laminate producer.

General Properties

Resin binders generally provide the following characteristics: (1) phenolics are low in cost, have good mechanical and electrical properties, and are somewhat resistant to flame; (2) polyesters used with glass mat provide a low-cost laminate for general-purpose uses; (3) epoxy resins are more expensive, but provide a high degree of resistance to acids, alkalies, and solvents, as well as extremely low moisture absorption, resulting in excellent mechanical and electrical property retention under humid conditions; (4) silicones, high in cost, provide optimum heat resistance; and (5) melamines provide resistance to flame, alkalies, arcing, and tracking, as well as good colorability.

Mechanical Properties

Tensile strengths of paper, asbestos, and cotton fabric-base laminates vary from about 136 to over 552 MPa; flexural strengths are somewhat higher and the moduli of elasticity are about 7,000–14,000 MPa.

Other Properties

In general, moisture absorption is about 0.3%–1.3% (24 h, 1.5 mm thickness), depending on resin and reinforcement.

Heat resistance depends on both resin and reinforcements. Most standard laminates are designed for a variety of insulation requirements. Generally, maximum continuous service temperatures are about 121°C for phenolics reinforced with organic reinforcements; 149°C for phenolics, melamines, and epoxies reinforced with inorganic materials; and 260°C for silicones reinforced with inorganic materials.

Flame retardance of most standard grades other than melamines is relatively poor; but special flame-resistant or self-extinguishing grades are available.

Glass cloth reinforcements provide tensile strength values in the 92–255 MPa range; modulus of elasticity can approach 20,600 MPa. Typical flatwise compression strength values fall in the 160–340 MPa range.

Electrical Properties

Dielectric strengths (perpendicular to laminates) generally range between 400 and 700 V/mil and up to 1000 V/mil for paper-based phenolics. Dissipation factor, as-received and at 10^6 cps, ranges from as high as 0.055 for cotton-based phenolics to as low as 0.0015 for a glass-reinforced silicone laminate.

Inelastic Electron Scatter

See *incoherent scattering.*

Inelastic Scattering

Any collision or interaction that changes the energy of an incident particle. Contrast with *elastic scattering.*

Inert

Unreactive, stable, or indifferent to the presence of other materials. A relative term, usually applying under some limited set of circumstances or application.

Inert Anode

An *anode* that is insoluble in the *electrolyte* under the conditions prevailing in the *electrolysis.*

Inert Filler

A Filler material added to a plastic to alter the end-item properties through physical rather than chemical means.

Inert Gas

(1) A gas, such as helium, argon, or nitrogen, that is stable, does not support combustion, and does not form reaction products with other materials. (2) In welding, a gas that does not normally combine chemically with the base metal or filler metal. (3) In processing of

plastics, a gas (usually nitrogen) that does not absorb or react with ultraviolet light in a curing chamber.

Inert Gas Fusion

An analytical technique for determining the concentrations of oxygen, hydrogen, and nitrogen in a sample. The sample is melted in a graphite crucible in an inert gas atmosphere; individual component concentrations are detected by infrared or thermal conductive methods.

Inert Gas Metal Arc Welding

See preferred term *gas metal arc welding*.

Inert Gas Shielded Cutting/Welding

Processes in which the heat source is an electric arc and the cutting/welding zone is protected from the environment by an inert gas such as argon.

Inert Gas Tungsten Arc Welding

See preferred term *gas tungsten arc welding*.

Infiltrant

Material used to infiltrate a porous powder compact. The infiltrant as positioned on the compact is called a slug. See also *contact infiltration*.

Infiltration

(1) The process of filling the pores of a sintered or unsintered compact with a metal or alloy of lower melting temperature. See also *contact infiltration*. (2) In particular, a metal introduced in the molten state and entering by capillary action.

Inflection Point

Position on a curved line, such as a phase boundary on a *phase diagram*, at which the direction of curvature is reversed.

Infrared

Pertaining to that part of the electromagnetic spectrum between the visible light range and the radar range. Radiant heat is in this range, and infrared heaters are frequently used in the thermoforming and curing of plastics and composites. Infrared analysis is used for identification of polymer constituents.

Infrared Analyzer

An atmosphere-monitoring device that measures a gas (usually carbon monoxide, carbon dioxide, and methane) presence based on specific wavelength absorption of infrared energy.

Infrared Brazing

A brazing process in which the heat required is furnished by infrared radiation.

Infrared Radiation

Electromagnetic radiation in the wavelength range of 0.78–300 μm (7800 to 3×10^6 Å). See also *electromagnetic radiation*, *far-infrared radiation*, *middle-infrared radiation*, and *near-infrared radiation*.

Infrared Soldering

A soldering process in which the heat required is furnished by infrared radiation.

Infrared Spectrometer

A device used to measure the amplitude of electromagnetic radiation of wavelengths between visible light and microwaves.

Infrared Spectroscopy

The study of the interaction of material systems with electromagnetic radiation in the infrared region of the spectrum. The technique is useful for determining the molecular structure of organic and inorganic compounds by identifying the rotational and vibrational energy levels associated with the various molecules.

Infrared Spectrum

(1) The range of wavelengths of infrared radiation. (2) A display or graph of the intensity of infrared radiation emitted or absorbed by a material as a function of wavelength or some related parameter.

Ingate

Same as *gate*.

Ingot

A casting of simple shape that will be subjected to further working operations like hot working or remelting. The as-cast ingot is normally cropped to remove the top section containing impurities and other defects.

Ingot Iron

Nearly chemically pure *iron* made by the basic open-hearth process and highly refined, remaining in the furnace 1–4 h longer than the ordinary time and maintained at a temperature of 2900°F–3100°F (1593°C–1704°C). In England, it is referred to as *mild steel*, but in the United States, the line between iron and steel is placed arbitrarily at about 0.15% carbon content. Ingot iron has as low as 0.02% carbon. It is obtainable regularly in grades 99.8%–99.9% pure iron. Ingot iron is cast into ingots and then rolled into plates or shapes and bars. It is used for construction work where a ductile, rust-resistant metal is required, especially for tanks, boilers, enameled ware, and galvanized culvert sheets. The tensile strength, hot rolled, is 48,000 lb/in.^{2} (331 MPa), elongation 30%, and Brinell hardness 82–100. Dead soft, the tensile strength is 38,500 lb/in.^{2} (265 MPa), elongation 45%, and Brinell hardness 67. *Armco ingot iron*, of Armco Steel Corp., is 99.94% pure, with 0.013% carbon and 0.017% manganese. It is used as a rust-resistant construction material, for electromagnetic cores, and as a raw material for making special steels. The specific

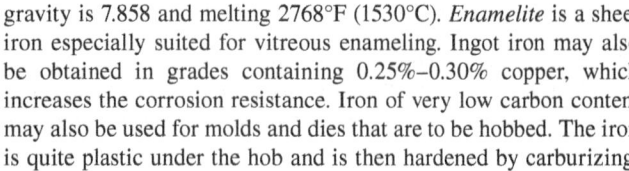

gravity is 7.858 and melting 2768°F (1530°C). *Enamelite* is a sheet iron especially suited for vitreous enameling. Ingot iron may also be obtained in grades containing 0.25%–0.30% copper, which increases the corrosion resistance. Iron of very low carbon content may also be used for molds and dies that are to be hobbed. The iron is quite plastic under the hob and is then hardened by carburizing. *Plastiron* is such an iron.

Ingot Mold

The substantial vessel into which molten metal is poured to solidify as an ingot.

Ingotism

Excessively large grain size in castings, usually the result of casting at too high a temperature.

Inherent (Austenitic) Grain Size

The austenitic grain size developed in steel when cast.

Inhibitor

A substance that retards some specific chemical reaction. Picking inhibitors retard the dissolution of metal without hindering the removal of scale from steel. Inhibitors are also used in certain types of monomers and resins to prolong storage life.

Initial Current

The current after starting, but before establishment of welding current.

Initial Modulus

The slope of the initial straight portion of a stress–strain or load-elongation curve. See also *Young's modulus*.

Initial Pitting

Surface fatigue occurring during the early stages of gear operation, associated with the removal of highly stressed local areas and *running-in*.

Initial Recovery

The decrease in strain in a solid during the removal of force before any creep recovery takes place, usually determined at constant temperature. Sometimes referred to as instantaneous recovery.

Initial Strain

The strain produced in a specimen by given loading conditions before creep occurs.

Initial Stress

The stress produced by strain in a specimen before stress relaxation occurs. Also called instantaneous stress.

Initial Tangent Modulus

The slope of the stress–strain curve at the beginning of loading. See also *modulus of elasticity*.

Initiation of Stable Crack Growth

The initiation of slow stable crack advance from the blunted crack tip.

Initiator

Sources of free radicals, often peroxides or azo compounds. They are used in free-radical polymerizations, for curing thermosetting resins, and as cross-linking agents for elastomers and cross-linked polyethylene.

Injection Blow Molding

A *blow molding* process in which an injection-molded preform is used instead of an extruded parison.

Injection Molding (Ceramics)

A process for forming ceramic articles in which a granular ceramic-binder mix is heated until softened and then forced into a mold cavity, where it cools and resolidifies to produce a part of the desired shape.

Injection Molding (Metals)

A process similar to plastic injection molding using a plastic-coated metal powder of fine particle size (~10 μm).

Injection Molding (Plastics)

Method of forming a plastic to the desired shape by forcing the heat-softened plastic into a relatively cool cavity under pressure. See also *thermoplastic injection molding* and *thermoset injection molding*.

Injection Molding

Injection molding of plastics is analogous to die casting of metals. The plastic is heated to a fluid state in one chamber and then forced at high speed into a relatively cold, closed mold where it cools and solidifies to the desired shape. This method of processing plastics became a commercial reality in the early 1930s. It is the fastest and most economical of all commercial processes for the molding of thermoplastic materials and is applicable to the production of articles of intricate as well as simple design.

A slightly modified version of injection molding known as "jet molding" is applicable to the molding of thermosetting materials. The principal difference between the two processes is the function of the temperature of the nozzle and mold on the material. Injection molding employs a relatively cold mold to solidify thermoplastics by chilling the mass below the melting point. Jet molding employs a relatively hot nozzle and mold to harden the thermosetting material by completing the cure.

Process

The sequence of operation known as the molding cycle is as follows:

1. Two mold halves that, when closed together, combine to form one or more negative forms of the article to be molded are tightly clamped between the platens of an injection-molding machine.
2. The closed mold is brought into contact with the nozzle orifice of a heating chamber. The heating chamber, known as the plastifying cylinder, is of sufficient size to carry an inventory of material equal to several volumes required to fill the mold. This permits gradual heating of the plastic to fluidity.
3. An automatically weighed or measured quantity of granulated thermoplastic, sufficient to fill the mold cavity, is fed into the rear of the plastifying cylinder.
4. A reciprocating plunger actuated by a hydraulically operated piston forces the material into the plastifying cylinder. An equal quantity of fluid plastic is thus forced out of the front of the cylinder through the nozzle orifice and into the mold.
5. A pressure of several thousand megapascals is maintained on the material within the mold until the plastic cools and solidifies.
6. After the molded item has hardened sufficiently to permit removal from the mold without distortion, the mold is opened and the part ejected.

Injection molding offers several advantages over other methods of molding. Some of the more important of these are as follows:

1. The process lends itself to complete automation for the molding of a great number of parts.
2. Mold parts require little if any postmolding operations.
3. High rates of production are made possible by the high thermal efficiency of the operation and by the short molding cycles possible.
4. There is a low ratio of mold-to-part cost in large volume production.
5. Long tool and machine life requires a minimum of maintenance and relatively low amortization costs.
6. Reuse of material is possible in most applications.

Molding machines vary considerably in design as well as capacity. For example, the clamp mechanism may be hydraulically operated, hydraulic-mechanically operated, or entirely mechanical. The injection of the material into the mold may be accomplished by a rotating screw as well as a plunger.

Machines are rated according to the weight of plastic that can be injected into the mold (the shot) with one stroke of the injection plunger. The capacity of commercially available machines covers a range from a fraction of a gram to 12.5 kg.

Limitations

The injection-molding process is subject to the following limitations:

Material: Any thermoplastic material may be a candidate for injection molding. Upon heating, it can be rendered sufficiently fluid to permit injection into a mold, and the resulting molded article retains all the desired properties.

Geometry of part: Any part, regardless of geometry, can be removed from a mold without damage to the mold or part is moldable.

Weight of part: The weight of the part must be within the "shot" rating of the particular machine being used.

Projected area and wall thickness of part: The limitation on these dimensions will be governed by several factors including the relative fluidity of the plastic being molded, the pressure necessary to fill the mold, the rigidity of the part to permit ejection from the mold without deformation, and sufficient clamp force available to hold the mold in a closed position when the necessary injection pressure is developed within the cavity.

Production economy: Economy can be realized only when relatively large quantities are produced. In addition to the material consumed and the cost of the molding operation, the number of pieces produced must also bear the cost of the mold in which they are cast. Depending on the size and complexity of the geometry of the part to be molded, mold costs will vary from a few hundred to many thousands of dollars. Again depending on size and complexity of the part, the rate of production will vary from a few minutes for a single part to several hundreds per minute.

In addition to the high production rates made possible by injection molding, articles of high quality and relatively precise dimensions can be produced. The production of industrial parts held 2D tolerances of ± 0.002 in./in. Is quite common.

The injection-molding process has made possible the development of an extremely large family of plastics. The basic types of thermoplastics and the human-modified compositions available are analogous to metals and alloys. New compositions as well as process variations are constantly being developed to meet the demands of particular situations.

Molds

The molds used in injection-molding machines for producing such large automotive components as body panels can now be made more compact and lighter than ever before. What makes this possible is nickel vapor deposition.

For car manufacturers, this means the capital invested in production molding machines can be significantly reduced, or, in more practical terms, it means smaller equipment is needed to handle the molds.

In the manufacture of shell molds, nickel carbonyl vapor is fed into a low-pressure chamber where, at a temperature of 180°C, the vapor decomposes and nickel is deposited, atom by atom, onto a heated metallic master, or mandrel. Carbon monoxide and any unused carbonyl gases are recycled.

The process creates a layer of nickel that is 99.9% pure, at a deposition rate of 0.25 mm/h, faithfully reproducing all surface features. Delivery time for the production of large molds is greatly reduced, compared with conventional mold-making techniques. Uniform wall thicknesses of 1.5–25 mm have been achieved using the vapor deposition method.

A surface hardness of up to 42 R_c to a depth of 0.25 mm gives the shell molds excellent wear resistance.

Also, the excellence thermal conductivity of nickel reduces cycle times in the injection-molding process because heat is quickly transferred from the shell to water-cooling channels in the mold assembly.

Nickel has other engineering advantages in shell molds, including its outstanding weldability, the uniform wall thicknesses possible, and the absence of residual stresses.

Recently, a 68,080 kg injection mold was created for the side panels of a prototype composite concept vehicle. Larger components, up to 1.2×2.5 m, can be easily accommodated.

Updated Processes

Injection–compression molding (ICM) has been around for years—in fact, it is how old vinyl records were made. However, with the recent availability of high-speed microprocessors and advanced software to precisely control the molding cycle, the process now works with long-fiber-reinforced thermoplastics. This offers the prospect of stronger and lighter parts, lower costs, and better part-to-part consistency. While standard injection-molding produces reinforced thermoplastic parts with good success, ICM can optimize the performance of long-fiber-reinforced thermoplastic materials.

The difference between ICM and conventional injection molding is that the shot is injected at low pressure into a partially open tool, as opposed to a closed one. The mold closes to compress and distribute the melt into the far reaches of the cavity, thus completing the filling and packing phase. This eliminates molded-in stresses resulting from high injection pressures.

There are two basic types of ICM: sequential and simultaneous. In the latter, compression can begin at any point during injection, and cycle times are similar to those for conventional injection molding. With sequential ICM, the injection stroke ends before compression begins, and cycle times are 1–2 s longer to accommodate secondary clamping motions.

ICM uses significantly lower injection pressures than standard injection molding so longer fibers remain in the finished part. That translates into better mechanical properties. For example, ICM maintains a 12.7 mm fiber length of commercial composites, resulting in finished parts with higher impact strength and more isotropic mechanical properties. In typical applications, impact strength improves 15%–20% in 3.2 mm wall thickness and over 50% and 1.5 mm wall thickness.

Thus, one major benefit of ICM is that it can produce thinner-walled, long-fiber-reinforced parts previously unattainable with injection molding. This is an important consideration in automotive applications where companies want to reduce weight and still maintain high stiffness and impact strength.

However, ICM start-up costs may be higher. The tooling, for example, depending on part and size could cost 10%–15% more. ICM also requires a press with a second-stage compression stroke, as well as precise clamping, accurate shot-size control, and speed control during secondary clamping.

Costs to convert standard injection-molding presses to ICM vary with the type of machine, age, sophistication of the controller, and whether it has precision linear-position encoders to determine clamp and screw locations. In most cases, the cost to upgrade a newer injection-molding machine is justified by the reduction in part cost.

Parts can also incorporate ribs, bosses, gussets, and through-holes. ICM is particularly suited to relatively flat parts, such as automobile load floors, sunroof liners, seat backs, and door panels.

Another new process is the plastic injection-molding method of forming in which heat-softened material is forced under pressure into a cavity, where it cools and takes the shape of the cavity. The molding operation cannot be completed in one step because details such as screw threads and ribs may be easily integrated into the

mold. High process repeatability is important to economical operation, because it means that few variables must be monitored and adjusted during the process. The key benefit of electric molding technology (EMT) is that it delivers this repeatability, while improving productivity and quality.

Repeatability improvement is the reason that will likely also pull EMT into the mainstream of injection molding. This success is based on the fact that electric machines can control variables much more closely than is possible at comparable costs with hydraulic machinery. This tighter process capability translates into a variety of benefits, including less scrap, lower labor costs, and higher quality.

The essential *repeatability potential* for EMT is inherently higher than that of hydraulic power. The reason for this is that hydraulic drives are typically distributed systems, involving a compressible fluid and a complex network of hoses, tubes, and valves that enable one or two pumps to drive all machine axes. Therefore, any conditions that affect fluid or flow properties also affect positioning of the machine.

By contrast, an electric machine has a motor for each axis. In this case, "axis" may be defined as any motion that is controlled through a feedback loop on the basis of variables such as time, position, pressure, or velocity. On an injection-molding machine, the main linear axes involve clamping, ejection, injection, and sled pull-in. Rotary axes may control extrusion and, in some cases, the height adjustment.

An all-electric power train on one of these axes may consist of a belt, two pulleys, and a ballscrew. With a separate motor for each axis, electric machines have the ability to drive and coordinate all axes of motion simultaneously, a significant advantage over hydraulic machines in some applications. Whatever its configuration, the electromechanical power train is rigid, "solid on solid."

Currently, EMT is more expensive than hydraulic machines, as is true with most newer technology. However, when comparing prices with hydraulic machines, specifications and capabilities should be balanced. By the time circuits and controls on a hydraulic machine are enhanced to approach to performance of a general-application electric, the cost difference narrows significantly—and hydraulic technology remains less precise.

Because it impacts molding costs in so many ways, hydraulic oil will be seen as a business liability as well as an environmental hazard. Any increase in electricity prices will also drive demand for EMT. In the mold-building industry, a corresponding increase is anticipated in the production of servo-electric systems to actuate core pulls and other functions. The environment of the molding plant will likely change considerably in a relatively few years, along with the standard of quality that is expected from the process.

Ink

Colored liquid or paste for writing, drawing, marking, and printing. Black writing inks usually contain gallotannate of iron that is obtained by adding an infusion of nutgalls to a solution of ferrous sulfate. Good *writing ink* is a clear, filterable solution, not a suspension. It flows easily from the pen without clogging; gives a smooth, varnishlike coating; and adheres to inner fibers of the paper without penetrating through the paper. It must have an intense color that does not bleach out. Ink is essentially a pigment in a liquefying and adhesive medium, but the *iron–gallotannate* writing inks develop their full color by chemical action and become insoluble in the outer fibers of the paper. For the proper development of the black color in gallotannate inks, a high percentage of iron is needed, and this requires a liberal use of acid, which will tend to injure the paper. It is thus usually the practice to reduce the amount of iron and bring up the color with dyes or pigments. Gums or adhesive materials may also be added.

Carbon inks are composed of lamp black or carbon black in solutions of gums or glutenous materials. *India ink* is a heavy-bodied drawing ink. The original India ink, or *Chinese ink*, was made with a jet-black carbon pigment produced by burning tung oil with insufficient air. The pigment was imported into Europe in compressed sticks known as *indicum*. India ink was originally only black ink, but the name is now used also for colored, heavy drawing inks made with various mineral pigments. *Marking inks* are usually solutions of dyes that are fast to laundering, but they may also be made with silver salts that develop full color and stability by the application of heat. *Fountain-pen ink* is not a special composition ink, but is a writing ink free of sediment and tendency to gum. It usually contains tannic, gallic, and hydrochloric acids with a pH above 2 to avoid corrosion. *Permanent inks* contain dissolved iron, not over 1%, to avoid sludge. *Ballpoint ink* is usually a paste and is a true solution with 40%–50% dye concentration. It must be stable to air, noncorrosive, and a good lubricant. An *encaustic ink* is a special writing ink that will penetrate the fibers of the paper and set chemically to make erasure difficult, but an *indelible ink* for textiles is a marking ink. *Invisible-writing inks*, or *sympathetic inks*, are inks that remain invisible until the writing is brought out by the application of heat or with another chemical that develops the color. They are made with sal ammoniac or salts of metals. *Magnetic ink* for use on bank checks to permit mechanical processing contains 50%–70% of a magnetic powder that is smaller than 197 μin. (5 μm) in particle size. The powder may be hydrogen-reduced iron, carbonyl iron, or electrolytic iron.

Printing inks are in general made with carbon black, lampblack, or other pigment suspended in an oil vehicle, with a resin, solvent, adhesive, and drier. But there are innumerable modifications of printing inks to meet different conditions or printing and varieties of surfaces. The oil or chemical vehicles are innumerable, and the pigments, resin straighteners and gloss formers, adhesives, tackifiers, and driers, vary greatly to suit the nature of surface coating of the base material. Oils may dry by oxidation, polymerization, absorption, or solvent evaporation; and resins may be used to add gloss, strength, hardness, and color fastness or to increase the speed of drying. It is estimated that there are about 8000 variables in an ink, and thus printing ink is a prescription product for any given job. They are not normally purchased on composition specification, but on ability to meet the requirements of the printing.

Inoculant

Materials that, when added to molten metal, modify the structure and thus change the physical and mechanical properties to a degree not explained on the basis of the change in composition resulting from their use. Ferrosilicon-base alloys are commonly used to inoculate gray irons and ductile irons.

Inoculated (Cast) Iron

Cast iron with a fine, evenly dispersed graphite flake structure giving improved mechanical properties compared with a normal gray cast iron. This structure is developed by inoculation, that is, adding a small quantity of finely powdered refractory material, such as calcium silicide, to the molten metal immediately prior to pouring. The powder particles provide many nucleation sites for the graphite and hence ensure the development of large numbers of fine graphite flakes.

Inoculation

The addition of a material to molten metal to form nuclei for recrystallization. See also *inoculant*.

Inorganic

Being or composed of matter other than hydrocarbons and their derivatives or matter that is not of plant or animal origin. Contrast with *organic*.

Inorganic Pigments

Natural or synthetic metallic oxides, sulfides, and other salts that impart heat and light stability, weathering resistance, color, and migration resistance to plastics.

Inorganic Polymer

An inorganic polymer is defined as a giant molecule linked by covalent bonds but with an absence or near absence of hydrocarbon units in the main molecular backbone; these may be included as pendant side chains. Carbon fibers, graphite, and so forth are considered inorganic polymers. Much of inorganic chemistry is the chemistry of high polymers.

For compounds that do not melt or dissolve without chemical change, both the absence of an equilibrium vapor pressure and the observation of a disassociation pressure resulting from depolymerization bring them into the framework of the definition.

Properties

Some special characteristics of many inorganic polymers are higher Young's modulus and a lower failure strain compared with organic polymers. Relatively few inorganic polymers dissolve in the true sense, or alternatively, if they swell; few can revert. Crystallinity and high glass transition temperatures are also much more common than in organic polymers. In highly cross-linked inorganic polymers, stress relaxation frequently involves bond interchange.

The properties of inorganic polymers require a different technology from that of their organic counterparts. Such technology is either completely new (such as reconstructive processing—the spinning of an inorganic compound on an inorganic support or binder subsequently removed by oxidation/volatilization), or it has been adapted from other fields, for example, glass technology. Thus, reconstructed vermiculite can give flexible sheets. Yarn, paper, woven cloth, and even textiles can be made from alumina and zirconia fibers by the spinning/volatilization process. A mica-forming glass ceramic is resistant to thermal and mechanical shock and can be worked with conventional metalworking tools.

Classification

Inorganic polymers can be classified in a number of ways. Some are based on the composition of the backbone, such as the silicones (Si–O), the phosphazenes (P–N), and polymeric sulfur (S–S). Others are based on their connectivity, that is, the number of network bonds linking the repeating unit into the network. Thus, the silicones based on R_2SiO, the phosphazenes based on NPX_2, and polymeric sulfur each have a connection of two, while boric oxide based on B_2O_3 has a connectivity of three, and amorphous silica based on SiO_2 has a connection of four.

Types

The number of inorganic polymers is very large. Sulfur, selenium, and tellurium all form high polymers. Polymers of sulfur are usually elastomeric, and those of selenium and thorium are generally crystalline. In the melt at 220°C, the molecular weight of the sulfur polymer is about 12,000,000 and that of selenium about 800,000.

Silicones

Perhaps the best known of all the synthetic polymers based on inorganic molecular structures are the silicones, which are derived from the basic units.

Chalcogenide Glasses

These are amorphous cross-linked polymers with a connectivity of three. Probably the best known is arsenic sulfide, $(As_2S_3)_n$, which can be used for infrared transparent windows. Threshold and memory switching are also interesting properties of these glasses. Ultraphosphate glasses resemble glassy organic plastics and can be processed by the same methods, such as extrusion and injection molding. They are used for antifouling surfaces for marine applications and in the manufacture of nonmisting spectacle lenses.

Graphite

This is a well-known 2D polymer with lubricating and electrical properties. Intercalation compounds of graphite can have supermetallic anisotropic properties.

Boron Polymers

Structurally related to graphite is hexagonal boron nitride. Like graphite, it has lubricating properties, reflecting the relationship between molecular structure and physical properties, but unlike graphite, it is an electrical insulator. Molybdenum disulfide, $(MoS_2)_n$, with a similar and related structure, is also a solid lubricant. Both graphite and hexagonal boron nitride can be readily machined. Outstanding properties of the latter include high thermal and chemical stability and good dielectric properties. Crucibles and such items as nuts and bolts can be made from this material.

Borate glasses with comparatively low softening points are used as solder and sealing glasses and can be prepared by fusing mixtures of metal oxides with boric oxide, $(B_2O_3)_n$.

Silicate Polymers

The silicates, both crystalline and amorphous, supply a very large number of inorganic polymers. Examples include the naturally occurring fiberlike asbestos and sheetlike mica. The industrially important water-soluble alkali metal silicates can give highly viscous polymeric solutions. Borosilicate glasses form another important group of silicate polymers. The Pyrex type is well known for its resistance to thermal shock; the leached Vycor type is porous and can be used for filtering bacteria and viruses. Asbestos occurs as a ladder polymer, of which crocidolite is the most important, and as layer polymer, exemplified by chrysotile. The zeolites, many of which have been found naturally or have been synthesized, are 3D network polymers. Their uses as molecular sieves are well known.

Other Polymers

Silicon nitride, $(Si_3N_4)_n$, is another macromolecule with interesting properties. Prepared by heating of silicon powder in an atmosphere of nitrogen (nitridation) at above 1200°C, the product is a material that can be machined readily and with good thermal shock resistance and creep resistance at high temperatures, which is further improved by admixture of another inorganic macromolecule silicon carbide, making it useful for applications in gas turbine, diesel engines, thermocouple sheaths, and a variety of components.

Allotropic forms of carbon boron nitride are diamond and cubic boron nitride, both preparable by high-temperature syntheses and characterized by extreme hardness, which make them useful industrially in cutting and grinding tools.

Importance

Inorganic polymeric materials are growing in importance as a result of a combination of two major factors: the depletion of the world's fossil fuel reserves (the basis of the petrochemical industry) and the ever-increasing demands of modern technology, coupled with environmental and health regulations, such as flame retardancy and nonflammability.

Inorganic Zinc-Rich Paint

Coating containing a zinc powder pigment in an *inorganic* vehicle.

Insecticides

Chemicals, either natural or synthetic, used to kill or control insects, particularly agricultural pests. They are also referred to as *pesticides*. Of about 800,000 known species of insects, one-half feed directly on plants, retarding growth of the plant and causing low yields and interior crops. The production of insecticides is now one of the important branches of the chemical industry, and increasing quantities are used; but the specification and use of insecticides require much skill because of the cumulative effect on the earth and animal or plant life. Indiscriminate use may destroy honeybees and other useful insects, produce sterility of soils by killing worms and anaerobic life, and poison the waters of lakes and streams. Dichlorodiphenyltrichloroethane (DDT), for example, is highly valuable for the control of malaria and other insect-borne diseases, but its uncontrolled use as an insecticide has been disastrous to wildlife.

Insecticides are generally classified as *stomach poisons* and *contact poisons*. Stomach poisons include *calcium arsenate*, a white powder, which constitutes about one-half of all insecticides used, and also paris green, lead arsenate, and white arsenic. *Cryolite*, or *sodium fluoroaluminate*, and sodium fluoride are used occasionally. An *antimetabolite* is not a direct poison, but acts on the insect to stop the desire for food so that the insect dies from starvation.

A *larvicide* is a chemical, such as chloropicrin, used to destroy fungi and nematodes in soils and insect eggs and organisms in warehouses. Chemicals used against fungi and bacteria are called *fungicides* and *bactericides*, and those used to control plant diseases caused by viruses are called *viricides*. None of these are properly classified as insecticides, but are often used with them. Herbicides are used to kill weeds, usually by overstimulating cell growth. Most of the pest-control chemicals are cumulative toxic poisons. *Benzene hexachloride* destroys bone marrow, and all the chlorinated hydrocarbons affect the liver. *Deodorants* may have an insect-kill action, but are usually chemicals such as chlorophyll that combine with impurities in the air to eliminate unpleasant odors.

Insecticides may be solids or liquids, and the solids may be applied as a fine powder, usually in dilution and a dusting powder, or the powder may be suspended in a liquid carrier. Usually, the proportion of poison mixed with a metal powder is no more than 5%. The mineral carrier, or *dusting powder*, for this purpose should be gritless and inert to the insecticide. Ordinary *dusting clay* is a light, fluffy, air-floated kaolin, or it may be a finely ground, soft limestone.

Some materials, such as citronella oil, used as mosquito repellents in households, have little or no value as insecticides for the eradication of mosquitoes in important applications such as at military sites or mining and lumbering camps. The *aerosol bomb* employed during the World War II contained 3% DDT, 2–20 pyrethrum concentrate, 5 cyclohexanone, 5 mineral oil, and the balance a carrier gas. *Dimethyl phthalate*, a liquid, is a mosquito repellent having an effect lasting 1.5 h in the open air.

The insecticide called *DDT* was used effectively during the World War II against flies, mosquitoes, body lice, and agricultural pests. It has no noxious odor, but it is cumulative and in concentration is toxic to humans and other warm-blooded animals. Oil paint containing 0.5%–5% DDT kills flies on walls painted with it. Because it is highly resistant to degradation, DDT is now limited to essential uses.

Insert

A general term referring to small devices or piece of material set into something larger by casting in, pressing, screwing, etc., to fulfill some special local function such as a bearing and screw location.

Insert (Metals)

(1) A part formed from a second material, usually a metal, which is placed in the molds and appears as an integral structural part of the final casting. (2) A removable portion of a die or mold.

Insert (Plastics)

An integral part of a plastic molding consisting of metal or other material that may be molded or precedent to position after the molding is completed.

Insert Die

A relatively small die that contains part or all of the impression of a forging and is fastened to a master die block.

Inserted-Blade Cutters

Cutters having replaceable blades that are either solid or tipped and are usually adjustable.

In Situ—In its Normal Place

In a metallurgical context, this Latin term indicates that the item of interest was examined in its normal location rather than being removed to a laboratory.

Instantaneous Erosion Rate

The slope of a tangent to the cumulative erosion-time curve at a specified point on that curve.

Instantaneous Recovery

The decrease in strain occurring immediately upon unloading a specimen. A more reproducible value is obtained if the decrease in strain is measured after a given small increment of time (such as 1 min) following unloading. The value is expressed in the same units as strain, that is, the decrease in weight divided by the gauge length, usually in inches per inch.

Instantaneous Strain

The strain occurring immediately upon loading a creep specimen. Value is measured after loading to obtain a more reproducible value. The value is expressed in the same units as strain, that is, the extension divided by the gauge length, usually in inches per inch.

Instrument Response Time

The time required for an indicating or detecting device to attain a defined percentage of its steady-state value following an abrupt change in the quantity being measured.

Instrumented Impact Test

An impact test in which the load on the specimen is continually recorded as a function of time and/or specimen deflection prior to fracture.

Insulating Pads and Sleeves

In foundry practice, insulating material, such as gypsum and diatomaceous earth, used to lower the rate of solidification. As sleeves on open risers, they are used to keep the metal liquid, thus increasing the feeding efficiency. Contrast with *chill*.

Insulators

These are any materials that retard the flow of electricity and are used to prevent the passage or escape of electric current from conductors. No materials are absolute nonconductors; those rating lowest on the scale of conductivity are therefore the best insulators. An important requirement of a good insulator is that it does not absorb moisture, which would lower its resistivity. Glass and porcelain are the most common line insulators because of low cost. Pure silica glass has an average dielectric strength of 20×10^6 V/m and glass-bonded mica about 17.7×10^6 V/m, while ordinary porcelain maybe as low as 8×10^6 V/m and steatite about 9.4×10^6 V/m. Slate, steatite, and stone slabs are still used for panel boards, but there is now a great variety of insulating boards made by compressing glass fibers, quartz, or minerals with binders, or standard laminated plastics of good dielectric strength may be used.

Synthetic rubbers and plastics have now replaced natural rubber for wire insulation, but some aluminum conductors are insulated only with an anodized coating of aluminum oxide. Wires to be coated with an organic insulator may first be treated with hydrogen fluoride, giving a coating of copper fluoride on copper wire and aluminum fluoride on aluminum wire.

Integral Composite Structure

Composite structure in which several structural elements, which would conventionally be assembled together by bonding or mechanical fasteners after separate fabrication, are instead laid up and cured as a single, complex, continuous structure. The term is sometimes applied more loosely to any composite structure not assembled by mechanical fasteners. Or some parts of the assembly may be cocured.

Integral Skin Foam

Urethane foam with a cellular core structure and a relatively smooth skin.

Integrated Circuit

An interconnected array of active and passive elements integrated with a single semiconductor substrate or deposited on the substrate by a continuous series of compatible processes and capable of performing at least one complete electronic circuit function.

Integrated circuits, based on gallium arsenide (GaAs), have come into increasing use since the late 1970s. The major advantage of these circuits is their fast switching speed.

Gallium Arsenide FET

The gallium arsenide field effect transistor (GaAs FET) is a majority carrier device in which the cross-sectional area of the conducting path of the carriers is varied by the potential applied to the gate. Unlike the MOSFET, the gate of the GaAs FET is a Schottky barrier composed of metal and gallium arsenide.

As noted earlier, the major advantage of gallium arsenide integrated circuits over silicon integrated circuits is the faster switching speed of the logic gate. The reason for the improvement of the switching speed of GaAs FETs with short gate lengths (less than 1 μm) over silicon FETs of comparable size has been the subject of controversy. In essence, the speed or gain-bandwidth product of a FET is determined by the velocity with which the electrons pass under the gate. The saturated drift velocity of electrons in gallium arsenide is twice that of electrons in silicon; therefore, the switching speed of gallium arsenide might be expected to be only twice as fast.

Intelligent Materials

See *smart materials*.

Intense Quenching

Quenching in which the quenching medium is cooling the part at a rate of at least 2½ times faster than still water. See also *Grossmann number*.

Intensiostatic

See *galvanostatic*.

Intensity

The energy per unit area per unit time incident on a surface.

Intensity Ratio

The ratio of two (relative) intensities.

Intensity of Scattering

The energy per unit time per unit area of the general radiation diffracted by matter. Its value depends on the scattering power of the individual atoms of the material, the scattering angle, and the wavelength of the radiation.

Intensity, x-Rays

The energy per unit of time of a beam per unit area perpendicular to the direction of propagation.

Interatomic Bonding

The manner in which atoms bond together. In electrovalent bonding, also termed ionic or heteropolar bonding, an element, typically a metal, having a small number of valence electrons combines with one having a large number, typically a nonmetal. For example, sodium as a single valence electron in its outer shell and eight electrons in its full second shell. Chlorine has seven valence electrons in its outer shell. When these two elements combine, a sodium atom donates the single valence electron in its outer shell to the outer shell of a chlorine atom. Both atoms then have full, and hence stable, outer shells. The atoms, respectively, depleted an increased buy-in electron, and their associated charge is termed ions, represented as Cl^{-1} and Na^{+1}. These opposite electrostatic charges powerfully attract forming a strong bond for the solid compound. In a solution, ionically bonded materials can ionize allowing the two ions to move independently within the solution. Such movement is the phenomenon underlying *electrochemistry*. In covalent bonding, also termed homopolar bonding, the atoms share electrons to fill their outer shells to form fairly stable molecules. A simple example is hydrogen with its single valence electron; molecules of hydrogen are formed by two atoms each contributing its electron to a shared outer shell filled with two electrons. Coordinate or dative covalent bonding is a variation in which the bond is formed by the sharing of a pair of electrons donated from one of the atoms. In metallic bonding, the valence electrons of all the atoms are pooled, effectively forming a cloud shared between the mass of atoms; see *band theory*. Apart from these relatively powerful bonds that hold atoms together, there are lesser secondary forces capable of holding molecules together. These are termed *van der Waals forces*.

Intercalation Compounds

These are crystalline or partially crystalline solids consisting of a host lattice containing voids into which guest atoms or molecules are inserted. Candidate hosts for intercalation reactions may be classified by the number of directions (0–3) along which the lattice is strongly bonded and thus unaffected by the intercalation reaction. Isotropic, 3D lattices (including many oxides and zeolites) contain large voids that can accept multiple guest atoms or molecules. Layer-type, 2D lattices (graphite and clays) swell up perpendicular to the layers when the guest atoms enter. The chains in 1D structures (polymers such as polyacetylene) rotate cooperatively about their axes during the intercalation reaction to form channels that are occupied by the guest atoms.

Properties

The physical properties of the host are often dramatically altered by intercalation. In carbon-based hosts such as graphite, polyacetylene, and solid C_{60}, the charge donated to the host by the guest is delocalized in at least one direction, leading to large enhancements and electrical conductivity. Some hosts can be intercalated by electrodonating or electron-accepting guests, leading, respectively, to n-type or p-type synthetic metals, analogous to doped semiconductors. The chemical bonding between guest and host can be exploited to fine-tune the chemical reactivity of the guest. In many cases, superconductors can be created out of nonsuperconducting constituents.

Applications

Many applications of intercalation compounds derive from the reversibility of the intercalation reaction. The best-known example is pottery. Water intercalated between the silicate sheets makes wet clay plastic, while driving the water out during firing results in a dense, hard, durable material. Many intercalation compounds are good ionic conductors and are thus useful as electrodes in batteries and fuel cells. A technology for lightweight rechargeable batteries employs lithium ions, which shuttle back and forth between two different intercalation electrodes as the battery is charged and discharged: vanadium oxide (3D) and graphite (2D). Zeolites containing metal atoms remain sufficiently porous to serve as catalysts for gas-phase reactions. Many compounds can be used as convenient storage media, releasing the gas molecules in a controlled manner by mild heating.

Intercept Method

Techniques for measuring microstructural features. A line is projected across the structure, and the intercept points at which it crosses the various phase boundaries are noted. The distance between intercepts is totaled for each phase to give a ratio of the volumes of the phases. Alternatively, the number of features, such as grain boundaries or inclusions, that intercept the line may be recorded to allow calculation of grain size or the quantity of inclusions.

Intercommunicating Porosity

See preferred term *interconnected porosity*.

Interconnected Pore Volume

The volume fraction of all pores that are interconnected within the entire pore system of a powder compact or sintered product.

Interconnected Porosity

A network of connecting pores in a sintered object that permits a fluid or gas to pass through the object. Also referred to as interlocking or open porosity.

Intercritical Annealing

Any annealing treatment that involves heating to, and holding at, a temperature between the upper and lower critical temperatures to obtain partial austenitization, followed by either slow cooling or holding at a temperature below the lower critical temperature.

Intercrystalline

Any feature at the interface between the crystals, or grains, of a polycrystalline material.

Intercrystalline Corrosion

See *intergranular corrosion*.

Intercrystalline Cracking

See *intergranular cracking*.

Interdendritic

Occurring or located at the boundaries between dendrites.

Interdendritic Corrosion

Corrosive attack that progresses preferentially along interdendritic paths. This type of attack results from local differences in composition, such as coring commonly encountered in alloy castings.

Interdendritic Porosity

Voids occurring between the dendrites in cast metal.

Interface

The boundary between any two phases. Among the three phases (gas, liquid, and solid), there are five types of interfaces: gas–liquid, gas–solid, liquid–liquid, liquid–solid, and solid–solid.

Interface (Composites)

The boundary or surface between two different, physically distinguishable media. With fibers, the contact area between the fibers and the sizing or finish. In a laminate, the contact area between the reinforcement and the laminating resin.

Interface Activity

A measure of the chemical potential between the contacting surfaces of two particles in a compact or two grains in a sintered body.

Interface of Phases

The abrupt transition from one phase to another at the boundaries of gas, liquid, and solid even though subject to the kinetic effects of molecular motion, is statistically a surface only one or two molecules thick.

A unique property of the surfaces of the phases that adjoin at an interface is the surface energy that is the result of unbalanced molecular fields existing at the surfaces of the two phases. Within the bulk of a given phase, the intermolecular forces are uniform because each molecule enjoys a statistically homogeneous field produced by neighboring molecules of the same substance. Molecules in the surface of a phase, however, are bounded on one side by an entirely different environment, with the result that there are intermolecular forces that then tend to pull the surface molecules toward the bulk of the phase. A drop of water, as a result, tends to assume a spherical shape to reduce the surface area of the droplet to a minimum.

Interfacial Tension

The contractile force of an interface between two phases.

Interference

The difference in lateral dimensions at room temperature between two mating components before assembly by expansion, shrinking, or press fitting. Can be expressed in absolute or in relative terms.

Interference

In light, and other energy forms transmitted as a waveform, the alternate light and dark bands that result when two transmissions meet with the waves out of phase.

Interference Filter

A combination of several thin optical films to form a layered coating for transmitting or reflecting a narrow band of wavelengths by interference effects.

Interference Fit

(1) A joint or mating of two parts in which the male part has an external dimension larger than the internal dimension of the mating female part. Distension of the female by the male creates a stress, which supplies the bonding force for the joint. (2) Any of the various classes of fit between mating parts where there is nominally a negative or zero allowance between the parts and where there is either part interference or no gap when the mating parts are made to the respective extremes of individual tolerances that ensure the tightest fit between the parts. Contrast with *clearance fit*. (3) A joint in which one component fits inside and is slightly larger than the receiving orifice and the other. They may be simply pushed together—a force fit—or the outer may be heated (and/or but less common, the inner cooled) to allow the inner to be inserted easily—a shrink fit. A typical shrink fit interference would be a difference in dimensions of one in one thousand. Less common, and less reliable, is the wrong fit in which the interfaces are slightly tapered and the components are rotated relative to each other as they are pushed together. They are then held together by a combination of interference and galling.

Interference Techniques

Techniques of microscopical examination that utilize a single beam of light that is split into two, one of which is reflected onto the surface being examined and the other onto a high-quality reference surface. The return beams from the two surfaces are then recombined so that any irregularities on the surface being examined are revealed by interference patterns.

Interference of Waves

The process whereby two or more waves of the same frequency or wavelength are combined to form a wave with amplitude that is the sum of the amplitudes of the interfering waves.

Interferometer

An instrument in which the light from a source is split into two or more beams, which are subsequently reunited and interfered after traveling over different paths.

Intergranular

Occurring or located at the interface between crystals or grains. Also called intercrystalline. Contrast with *transgranular*.

Intergranular Beta

In titanium alloys, β phase situated between α grains. It may be at grain corners, as in the case of equiaxed α-type microstructures in alloys having low β-stabilizer contents.

Intergranular Corrosion

Corrosion occurring preferentially at grain boundaries, usually with slight or negligible attack on the adjacent grains. See also *interdendritic corrosion*.

Intergranular Cracking

Cracking or fracturing that occurs between the grains or crystals in a polycrystalline aggregate. Also called intercrystalline cracking. Contrast with *transgranular cracking*.

Intergranular Fracture

Brittle fracture of a polycrystalline material in which the fracture is between the grains, or crystals, that form the material. Also called intercrystalline fracture. Contrast with *transgranular fracture*.

Intergranular Penetration

In welding, the penetration of a filler metal along the grain boundaries of a base metal.

Intergranular Stress-Corrosion Cracking

Stress-corrosion cracking in which the cracking occurs along grain boundaries.

Interlaminar

Between two adjacent laminae, for example, an object (such as a flaw), an event (such as a fracture), or a potential field (such as shear stress). See also *fracture* (composites).

Interlaminar Shear

Shearing force tending to produce a relative displacement between two laminae in a laminate along the plane of their interface.

Intermediate Annealing

Annealing wrought metals at one or more stages during manufacture and before final treatment.

Intermediate Electrode

Same as *bipolar electrode*.

Intermediate Flux

A soldering flux with a residue that generally does not attack the base metal. The original composition may be corrosive.

Intermediate Heat Treatment/Annealing

Heat treatment carried out between stages of cold working.

Intermediate Phase

In an alloy or a chemical system, a distinguishable homogeneous phase with composition range that does not extend to any of the pure components of the system.

Intermediate-Temperature Setting Adhesive

An adhesive that sets in the temperature range from 30°C to 100°C (87°F–211°F).

Intermetallic Compounds

Intermetallic compounds are materials composed of two or more types of metal atoms, which exist as homogeneous, composites substances and differ discontinuously in structure from that of the constituent metals. They are also called, preferably, intermetallic phases. Their properties cannot be transformed continuously into those of their constituents by changes of composition alone, and they form distinct crystalline species separated by phase boundaries from their metallic components and mixed crystals of these components. It is generally not possible to establish formulas for intermetallic compounds on the sole basis of analytical data, so formulas are determined in conjunction with crystallographic structural information.

The term *alloy* is generally applied to any homogeneous molten mixture of two or more metals, as well as to the solid material that crystallizes from such a homogeneous liquid phase. Alloys may also be formed from solid-state reactions. In the liquid phase, alloys are essentially solutions of metals in one another, although liquid compounds may also be present. Alloys containing mercury are usually referred to as amalgams. Solid alloys may vary greatly in range of composition, structure, properties, and behavior.

Phase Transformations

Much of the accumulated experimental information about the nature of the interaction and the phase transformation in systems composed of two or more metals is contained in phase or equilibrium diagrams. For example, in the phase relationships in the copper–zinc (brass) system, such phase diagrams, even for binary metal systems, may be of all degrees of complexity, ranging from systems showing the formation of simple solid solutions to dozens of intermetallic phases exhibiting structural (polymorphic), order–disorder, magnetic, or bond-type or deformation-type transformations as a function of composition or temperature, or both. Intermetallic compounds are composed of two or more metals. They may be stable over only a very narrow or over a relatively wide range of composition, which may be stoichiometric, as for the compounds GaAs, $PdCu_{13}$, $Zr_{57}Al_{43}$, and $Nb_{48}Ni_{39}Al_{13}$, or non-stoichiometric, as for the compounds $Co_{1-x}Te_x$ (where x extends continuously from 0.5 to 0.67) and $Al_{-0.08}Ge_{-0.2}Nb_3$ (an important superconducting alloy).

Crystal Structure

The crystal structures found for intermetallic compounds may likewise range from the simple rock-salt structure displayed by BeTe to the extremely complex arrangement found for $NaCd_2$, $Mg_{32}(Al,Zn)_{49}$, and Cu_4Cd_3.

Family Groups

The intermetallic compound families include the borides, hydrides, nitrides, silicides, and aluminides of the transition metal elements of groups V and VI in the periodic table combined with semimetallic elements of small diameter. More than 1000 possible compounds fit this definition, with approximately 200 of them having melting points above 1500°C.

Methods of preparation for silicides and aluminides are very similar to those used for carbides, nitrides, and borides: (1) synthesis by fusion or sintering, (2) reduction of the metal oxide by silicon or aluminum, (3) reaction of the metal oxide with SiO_2 and carbon, (4) reaction of the metal with silica and halide, or (5) fused salt electrolysis. The simplest preparation method consists of mixing the metal powders in proper ratio and then heating the mixture in vacuum or inert atmosphere to the temperature where reaction begins. At that point, the exothermic reaction furnishes the heat needed to drive it to completion.

The best forming methods are hot pressing and vacuum sintering. Some intermetallic compounds—ZrB_4 with $MoSi_2$, Mo_2NiB_2, and Mo_2CoB, for example—are as hard as tungsten carbide. Borides of zirconium or titanium added in small amounts to Al_2O_3, BeO, ZrO_2, or ThO_2 improve their resistance to heat checking. Nitrides of aluminum, titanium, silicon, and boron are used as refractory materials.

Fibers and whiskers of intermetallic materials also are being considered for strength-enhancing fillers in metal- and ceramic-matrix composites.

Intermetallic Materials

Ordered intermetallics bridge the gap between metals and ceramics. They exhibit physical properties characteristic of metals and can be processed by conventional casting techniques, but their mechanical properties resemble those of ceramics. Some intermetallics also have poor resistance to oxidation and sulfidation attacks at elevated temperatures.

The primary barrier to their use in load-carrying applications is their brittleness or low crack tolerance. The reasons for this behavior include complex crystal structure and deformation behavior, embrittlement due to defects (interstitials, vacancies) and to hydrogen, and notch sensitivity. The major effort in research and development on intermetallics is concerned with increasing the ductility and toughness while maintaining good high-temperature strength, stiffness, creep resistance, and resistance to oxidation and sulfidation.

Ordered Structures

An ordered intermetallic is an alloy of two elements, A and B, in specific ratios (e.g., AB, AB_2, AB_3, A_2B, A_3B). In the fully ordered condition, the two species are arranged on specific lattice sites. Depending on their constitution, ordered intermetallics are divided into groups such as aluminides, silicides, and beryllides.

Nickel Aluminides

The two major nickel aluminides are Ni_3Al and NiAl.

In the crystal structure of Ni_3Al, the nickel atoms occupy the face-centered sites, and the aluminum atoms occupy the corner sites of a face-centered cubic unit cell. Ni_3Al has excellent elevated-temperature strength and good oxidation resistance; there is an anomaly in the yield strength in that it increases, rather than decreases (as is the case with most other structural materials), with increasing temperature. Single crystals of Ni_3Al are highly ductile, whereas polycrystalline materials are brittle at ambient temperatures because of the brittleness of the grain boundaries. Polycrystalline Ni_3Al is also prone to environmental embrittlement at both ambient and elevated temperatures.

The mechanical properties of Ni_3Al can be improved by alloying additions of boron, chromium, hafnium, molybdenum, and zirconium. Binary Ni_3Al alloys are being examined for alternative applications, such as turbocharger rotors and diesel engine trucks.

The alloy NiAl has two-thirds the density of nickel-base superalloys and, in addition, exhibits significantly higher (four to eight times depending on composition and temperature) thermal conductivity and excellent oxidation resistance. Although binary NiAl is brittle even in single-crystal form, minor alloying with iron, molybdenum, and gallium has recently been shown to significantly improve tensile ductility.

Iron Aluminides

The two major iron aluminides are Fe_3Al and FeAl. Both have excellent oxidation resistance and corrosion resistance because they form protective scales at elevated temperatures in hostile environments. They also are low in cost and have low density and good strength of two intermediate temperatures. Their main drawbacks are poor ductility and brittle fracture at ambient temperatures and poor strength and creep resistance above 650°C. The poor ductility at ambient temperatures is caused by the environmental embrittlement due to hydrogen from the moisture in the air. These compounds are being investigated as potential replacements for stainless steel in selected applications in the chemical, petroleum, and coal industries.

Titanium Aluminides

The major titanium aluminides are Ti_3Al and TiAl. They are well suited for aerospace applications because of their low density compared to the standard engine materials, nickel-base superalloys. The density of titanium aluminides is about half that of superalloys. Titanium aluminides also have good high-temperature strength but poor ductility and toughness at ambient temperatures.

Alloys corresponding to Ti_3Al have been developed as two-phase alloys with compositions based on titanium with 23%–25% aluminum and 10%–30% niobium (a phase is a constituent of an alloy that is physically distinct and is homogeneous and chemical composition). The oxygen impurity level is a very important consideration, because it has a large effect on the ductility. Ti_3Al has limited ductility at ambient temperatures. The mechanical properties depend on the composition and the microstructure. Increasing the niobium content increases the ductility, but the toughness and creep resistance remain low. Increasing the aluminum content increases the strength. The microstructure can be altered significantly by thermal mechanical processing.

Promising materials corresponding to TiAl have the following compositions: titanium with 46%–52% aluminum and 1%–10% M, where M is at least one element from the group of metals chromium, molybdenum, niobium, tantalum, vanadium, and tungsten. The materials are produced as either single-phase or two-phase materials. The two-phase materials contained both the Ti_3Al and the TiAl phases. Additions of niobium or tantalum increase the strength and oxidation resistance of single-phase materials.

The mechanical properties of these alloys are sensitive to these microstructural modifications. Unlike the NiAl alloys, TiAl alloys are produced in the polycrystalline form, and investment casting appears to be the preferred route to producing net-shape components.

A number of gas-turbine engine components have been identified for TiAl-base alloy applications, rotational and stationary compressor components such as high-pressure compressor blades, turbine components such as low-pressure turbine blades, combustor components such as diffuser case and swirler, and nozzle components such as flaps and outer skins. A low-pressure turbine wheel with cast TiAl-base alloy blades has passed a rigorous simulated engine test.

Molybdenum Disilicide

High-temperature silicides are a new class of materials with potential applications in the temperature range 1200°C–1600°C. However, there are several issues to be solved before they can be used as structural materials. Such issues include (1) improving the intermediate-temperature oxidation behavior to avoid pesting (the disintegration of a silicide material in the oxygen-bearing environment, generally occurring between 300°C and 900°C), (2) increasing the ambient-temperature ductility and toughness, and (3) improving the high-temperature creep resistance. The silicide that seems most promising is molybdenum disilicide ($MoSi_2$) because of its high melting temperature (2200°C) and excellent oxidation resistance. It has a tetragonal crystal structure. A major problem is its absence of ductility at temperatures up to 1000°C. It also has poor high-temperature strength due to the presence of a grain-boundary silicon-rich phase that may become viscous at very high temperatures. Recent research has shown that this problem may be solved by the addition of carbon into the material.

Production

The primary processing methods for production of intermetallics are melting, casting, ingot processing, and powder metallurgy. The secondary processes include forming, machining, and chemical milling.

Primary Processing

The two major methods of producing intermetallics are by melting and casting or by the production of powders. In the melting of aluminides, the large difference between the melting temperature of aluminum and that of iron, nickel, and titanium must be considered. Other pertinent factors are (1) the large amount of aluminum in the intermetallic, (2) the melting temperature of the intermetallic (it may be higher than that of either constituent), and (3) the general reactivity of the elements to be melted. Similar issues exist for $MoSi_2$, except that the difference in the melting temperatures of the elements is smaller.

The melting methods used are air induction melting, vacuum induction melting (VIM), vacuum arc remelting (VAR), vacuum arc double electrode remelting (VADER), electroslag remelting (ESR), plasma melting, and electron-beam melting. Ni_3Al and the iron aluminides can be melted in air, although the high amount of aluminum promotes the rapid formation of a continuous aluminum oxide film on the top of the molten materials. For many applications, it is best to use processes that do not involve air melting, such as VIM, VAR,

VADER, and ESR. The extreme reactivity of molten titanium with moisture and with oxidizing and carburizing environments makes the melting of titanium aluminides in conventional melting crucibles impossible. Titanium aluminides are melted by a process known as the induction skull method, which combines features of consumable arc melting and conventional ceramic crucible induction melting. Also, the VAR process and plasma arc cold hearth melting process have been used to melt titanium aluminides as well as $MoSi_2$. Many applications of intermetallics require cast components. Casting methods include sand, investment, centrifugal, directional solidification, and near-net-shape methods.

Finer-grained wrought products are required in other applications. The commercial use of intermetallics at competitive cost requires fabrication by conventional hot-working operations. The primary processing of cast ingots is feasible, but the requirement for hot working is more stringent than for commercial metallic alloys. The secondary processing of intermetallics is very difficult and varies from intermetallic to intermetallic.

Powder metallurgy offers the most flexibility in producing intermetallics. The problem of using powder metallurgy for this purpose is that these production methods often result in surface contamination of the powders. Each of the powder consolidation methods for producing intermetallics from powders has processing difficulties. These methods are hot pressing, hot isostatic pressing, powder injection molding, extrusion, and explosive compaction. The reaction synthesis process has been uniquely applicable for intermetallics and has been used to produce many different materials.

Secondary Processing

Secondary steps such as machining and joining are critical in using these advanced materials for various applications, and extensive efforts have been used to develop these technologies. Innovative joining techniques such as friction welding, capacitor discharge welding, flash welding, laser welding, welding using the combustion synthesis concept, welding using microwaves and infrared waves, electron-beam welding, brazing, and diffusion bonding have been evaluated. A variety of machining techniques including electrodischarge machining, water-jet cutting, ultrasonic machining, and laser cutting are available to precision-machined complex geometries and contours. Conventional grinding, diamond drilling, and boring techniques have seen limited applications in machining TiAl alloys.

Intermetallic Phases

Compounds, or intermediate solid solutions, containing two or more metals, which usually have compositions, characteristic properties, and crystal structures different from those of the pure components of the system.

Intermetallic Matrix Composites (IMCs)

IMCs have recently received considerable attention, and a variety of matrices and reinforcements have been examined to date. Reinforcement type, volume fraction, size, shape, and distribution have been shown to affect microstructure and mechanical properties. Several innovative approaches ranging from conventional techniques, such as mechanical alloying, to more exotic techniques, such as reactive consolidation and magnetron sputtering, have been used to produce these composites.

Significant advances in characterization have been made and continuously reinforced—based alloys (SiC fibers in Ti_3Al + Nb alloys) and particulate-reinforced TiAl alloys (TiAl + TiB_2 particulates), in directionally solidified eutectic's of NiAl, and in discontinuously reinforced $MoSi_2$.

Intermittent Weld

A joint formed, deliberately or otherwise, by series of longitudinal welds with gaps between. Where such a weld is deliberate, a chain intermittent weld has the welds on one side aligned with the welds on the other and a staggered intermittent weld has the gaps on one side aligned with the welds on the other.

Internal Friction

The ability of the material to absorb vibration-induced strain energy and dissipate it internally as heat. Similar to *damping capacity*. The conversion of energy into heat via materials subjected to fluctuating stress.

Internal Grinding

Grinding an internal surface such as that inside a cylinder or hole. See also the term *grinding*.

Internal Oxidation

The oxidation of parent material or other phases within the body of the component due to diffusion of oxygen from the exterior. The effect is usually damaging but it may be deliberately induced to produce large quantities of fine oxide particles having a dispersion hardening effect. Also called *subscale formation*.

Internal Shrinkage

A void or network of voids within a casting caused by inadequate feeding of that section during solidification.

Internal Shrinkage Cracks

Longitudinal cracks in a composite pultrusion that are found within sections of roving reinforcement. This condition is caused by shrinkage strains during cure that show up in the roving portion of the pultrusion, where transverse strength is low.

Internal Standard

In spectroscopy, a material present in or added to samples that serves as an intensity reference for measurements, used to compensate for variations in sample excitation and photographic processing in emission spectroscopy.

Internal Standard Line

In spectroscopy, a spectral line of an *internal standard*, with which the radiant energy of an analytical line is compared.

Internal Stress

See preferred term *residual stress*.

International Annealed Copper Standard (IACS)

A measure of electrical resistance and its reciprocal, conductivity. The standard defines pure annealed copper as having, at 20°C, a specific resistance of 1.7241 μΩ-cm and thus a conductivity of 0.58001 reciprocal μΩ-cm. This conductivity is then defined as 100% IACS as the basis for all other metals. As examples, pure annealed silver would have an IACS of 105% and a cold worked cadmium copper an IACS of 85%.

International Screw Thread

An internationally agreed system of screw threads of standard profile with a radiused root and a truncated crest and dimensions in millimeters.

Interpass Temperature

In a multiple-pass weld, the temperature (minimum or maximum as specified) of the deposited weld metal before the next pass is started.

Interphase

The boundary region between a bulk resin or polymer and an adherend in which the polymer has a high degree of orientation to the adherend on a molecular basis. It plays a major role in the load transfer process between the bulk of the adhesive and the adherend or the fiber and the laminate matrix resin.

Interplanar Distance

The perpendicular distance between adjacent parallel lattice planes.

Interply Hybrid

A reinforced plastic laminate in which adjacent laminae are composed of different materials.

Interpulse Time (Resistance Welding)

The time between successive pulses of current within the same impulse.

Interrupted Aging

Aging at two or more temperatures, by steps, and cooling to room temperature after each step. See also *aging*, and compare with *progressive aging* and *step aging*.

Interrupted-Current Plating

Plating in which the flow of current is discontinued for periodic short intervals to decrease anode polarization and elevate the *critical current density*. It is most commonly used in cyanide copper plating.

Interrupted Quenching

A quenching procedure in which the workpiece is removed from the first quench at a temperature substantially higher than that of the quenchant and is then subjected to a second quenching system having a different cooling rate than the first.

Intersection Scarp

A line, of any shape, that is the locus of intersection of two portions of a crack with one another. This is exemplified by intersection of a portion of a slow crack running wet with a portion not wetted. See also *transition scarp*.

Interstice

Generally, any gap situated between two features. More specifically, in a metallurgical context, the space between the atoms forming the main crystal lattice.

Interstitial

The space, or an atom located in the space, between the atoms forming the main crystal lattice. Interstitial atoms are much smaller than the primary atoms, for example, carbon and nitrogen are interstitials in iron. See *solution*.

Interstitial Solid Solution

A type of solid solution that sometimes forms in alloy systems having two elements of widely different atomic sizes. Elements of small atomic size, such as carbon, hydrogen, and nitrogen, often dissolve in solid metals to form this solid solution. The space lattice is similar to that of the pure metal, and the atoms of carbon, hydrogen, and nitrogen occupy the spaces or interstices between the metal atoms.

Intersystem Crossing

A transition between electronic states that differ in total spin quantum number.

Interval Erosion Rate

The slope of a line connecting two specified points on the cumulative erosion-time curve.

Internal Estimate

The estimate of a parameter given by two statistics, defining the endpoints of an interval.

Interval Test

Method used to test heat extraction rates of various quenchants. This test measures the increase in temperature of a quenchant when a standard bar of metal is quenched for 5 s. Faster quenchants will exhibit greater temperature increases.

Intracrystalline

Within or across the crystals or grains of a metal; same as transcrystalline and transgranular.

Intracrystalline Cracking

See *transgranular cracking*.

Intragranular

Any effect such as precipitation occurring within individual grains. Not to be confused with intergranular (between grains) or transgranular (across grains).

Intralaminar

Within a single lamina, for example, an object (such as a void), an event (such as a fracture), or a potential field (such as a temperature gradient). See also the term *interlaminar*.

Intraply Hybrid

A reinforced plastic laminate in which more than one material is used within a specific layer.

Intrinsic Viscosity

For a polymer, the limiting value of infinite dilution of the ratio of the specific viscosity of the polymer solution to its concentration in moles per liter.

Introfaction

The change in fluidity and wetting properties of a polymeric impregnating material, produced by the addition of an *introfier*.

Introfier

A chemical that converts a colloidal solution into a molecular one. See also *introfaction*.

Intrusion

Surface notches and steps formed by local slip during fatigue.

Intumescence

The swelling or bubbling of the coating usually because of heating (term currently used in space and fire protection applications).

Invar

An alloy of iron with 36% nickel that has a very low thermal expansion over a range of temperature from about ambient to about 140°C.

Inverse Chill

The condition in a casting section in which the interior is mottled or white, while the other sections are gray iron. Also known as reverse chill, internal chill, and inverted chill.

Inverse Rate Cooling Curve

A graph plotting temperature on the vertical scale against time taken for incremental falls or rises in temperature (e.g., 1 degree intervals) on the horizontal scale. Such graphs are plotted for metals being heated or cooled, and inflections in the curve indicate the temperature of critical points such as phase changes.

Inverse Segregation

Segregation where, contrary to normal, an excess of the lower melting point metal is located at the outer surface of a casting.

Inverted Microscope

A microscope arranged so that the line of sight is directed upward through the objective to the object.

Inverted "V" Segregation

See *segregation*.

Investing

In *investment casting*, the process of pouring the investment slurry into a flask surrounding the pattern to form the mold.

Investment

A flowable mixture, or slurry, of a graded refractory filler, a binder, and a liquid vehicle that, when poured around the patterns, conforms to their shape and subsequently sets hard to form the investment mold.

Investment Casting

(1) Casting metal into a mold produced by surrounding, or *investing*, an expendable pattern with a refractory slurry coating that sets at room temperature, after which the wax or plastic pattern is removed through the use of heat prior to filling the mold with liquid metal. Also called *precision casting* or *lost wax process*. (2) A part made by the investment casting process.

Investment Castings

The investment-casting process derives its name from one operation of the overall process, that of investing, enclosing, or encasing a disposable pattern or pattern cluster within a refractory slurry that subsequently hardens to form a mold. This method of mold making and casting in a hot mold (871°C–1038°C) distinguishes investment casting from other casting procedures. Other casting processes, such as sand casting, die casting, and shell mold casting, employee a split mold-making technique.

The investment-casting process is also known as the lost-wax or precision casting process. (Another technique akin to it is the frozen mercury process.) Investment castings are now used in many industries.

General Description

Investment casting consists of three major operations—pattern production, mold production, and casting:

1. Castings are produced by using disposable patterns. Patterns are formed of wax or plastic by injecting these materials, which are fluid, into a die cavity or cavities. The shape and detail of the part intended for reproduction are cast or machined into the die.

   A pattern die is constructed in two parts so that the pattern, or in the case of a multiple-cavity die patterns, can be removed. Machined-pattern dies may be cut from a solid piece of steel or may be built in components for assembly. Cast pattern dies require that a metal master pattern first be introduced, which is in turn used to cast a soft metal die cavity or cavities. These dies are usually cast from a low-melting alloy such as tin–bismuth. Steel dies are normally used for injection of plastics. In many instances, a part of the foundry gating system (e.g., feeders and runners) is built into the die cavity to eliminate subsequent assembly work in making up a pattern cluster. A pattern die can be made to produce a single pattern, multiple patterns, or only a portion of a pattern of a given part. In the latter case, more than one die would be used and these components would be put together to make a complete pattern.

After the patterns are produced, the next step is making the pattern assembly or pattern cluster with subsequently forming the mold cavity. The pattern assembly consists of the part pattern or patterns and the gating system, which includes feeders, runners, and metal reservoir.

2. The first operation in mold making is to dip the pattern cluster into a liquid mix of silica flour to wet the patterns and then cover them with a dry mix of flour and sand. This forms the dip coat of the patterns and is the start of the mold buildup. There are now two major methods of completing the molds.

The first method is to seal the pattern cluster to a mold board and place a container on the board that encloses the pattern cluster. A refractory slurry is poured into the container, which invests (encloses) the pattern cluster except at that area at the bottom of the mold board where the pattern cluster is stuck to the board. The fluids in the slurry harden chemically and the result is a very dense, hard, heat-resistant mold with the pattern cluster embedded. The mold board is then removed, exposing the end of wax pattern clusters.

An alternative method of completing the mold after the initial dip coating is to continue with the procedure of alternately dipping in a heavier slurry and coating with dry refractory particles until a ceramic shell is formed around the pattern cluster. Drying is necessary between cycles and the number of cycles depends on the strength requirements to contain the molten metal when poured into the mold.

3. The casting operation includes three steps: (1) removing the pattern cluster from the mold, (2) heating the metal to a molten state and pouring it into the mold, and (3) cooling the mold and removing the cast parts from it. The pattern material (wax or plastic) is removed from the mold by a combination of melting and burning. Approximately an 8 h heating cycle at temperatures of about 871°C–1038°C is used to fire the ceramic mold and remove the pattern material.

This, then, leaves a cavity in the mold. The molds are removed from the furnace shortly before casting so they are near furnace temperature when poured.

Depending on the alloy from which the castings are to be made, various melting–casting techniques are used. Alloys are melted while exposed to normal atmosphere, while covered protectively by an inert gas or while the melting crucible and mold are in a vacuum chamber. Melting is usually accomplished by indirect arc or induction heating. Molds may be cast by "floor pouring" to fill the mold directly from the furnace or from furnace to ladle to mold. The floor-pouring method is used primarily for large molds. The more prevalent way is to secure the mold on top of the melting furnace and invert both furnace and mold so that the alloy enters directly into the mold from the crucible, normally under slight pressure.

After solidification and cooling have taken place, the mold material is mechanically or hydraulically broken away from the castings. The gates are cut from the casting and the casting is processed through other operations of finishing and inspection as required.

Materials

The majority of the alloys cast by this process are iron-, nickel-, or cobalt-base alloys. The iron-base alloys include the complete range from plain carbon steels to the stainless and high-alloy steels. Many nonferrous alloys can also be used, such as gold, brasses, and bronzes, as well as gold, aluminum, and copper alloys.

The alloy to be cast is generally not considered as a restriction, because the process is sufficiently versatile to accommodate almost all types. However, not all manufacturers cast all alloys. The use of inert gas and vacuum protection for melting and pouring has enabled the industry to cast the ultrahigh-temperature iron- and nickel-base precipitation-hardening alloys.

Sizes and Configurations

The size of investment castings is often thought of in terms of castings weighing less than 2.8 g to castings weighing about 1500 g; this size range includes a majority of the castings poured. However, the capability of the process permits the successful casting of parts over 50,000 g in over 500 mm in diameter with considerable accuracy and detail. Small castings have been produced with edges down to 0.38 mm, holes 3.2 mm in diameter by 152–203 mm long, and shorter holes to 0.076 mm. The type of alloy selected and the corresponding pouring practice may modify the maximum or minimum sizes and tolerances.

Applications

One of the major uses of investment-cast parts is in gas turbines. Investment-cast parts have proved very successful here because the shapes required (airfoils) are particularly difficult to machine and the high-strength alloy being used is difficult to finish, even for fairly simple operations. Many of the alloy compositions develop their outstanding properties in the cast form and may not be amenable to working in the wrought form.

No single industry, other than the gas-turbine industry, uses a large volume of investment castings. However, practically all industries are using some investment castings in their products or processes.

Advantages of the Process

There are several important advantages in the mass production of metal parts by the investment-casting process, not only by comparison with other casting processes but also with other production methods such as machining or forging.

Compared with other casting processes (such as sand or shell molding), investment casting provides the following:

1. An improved casting from the standpoint of cleanliness of the alloy and metallurgical soundness. This is especially true when an inert gas or vacuum is used to protect the alloy.
2. Considerably more versatility in producing intricate shapes and cast detail. This feature opens the possibility of combining two or more components of an assembly into a single casting.
3. Greater dimensional accuracy, which in many cases eliminates or minimizes the need for finishing. This certainly reduces the necessary stock allowance.
4. Better cast surfaces, which are beneficial from the standpoint of quality and appearance. This may further eliminate or minimize finishing operations.

The advantages of investment casting compared with other manufacturing processes are the following:

1. Many alloy compositions develop their outstanding properties in the cast form only. Also, many of the compositions cannot be produced other than in cast form.
2. There has always been controversy whether cast parts could compete with the consistent high quality and uniformity of forged parts. The investment-casting industry has proved this can be done, primarily, by consistently producing to the stringent quality requirements of the aircraft industry.
3. There have been many instances on both small and large parts where castings have been very beneficial in eliminating great amounts of machining, either starting with a wrought mill stock or from forged parts. Castings also have greater versatility and accommodating changes in section and shape.

Investment Compound

A mixture of a graded refractory filler, a binder, and a liquid vehicle, used to make molds for *investment casting*.

Investment Precoat

An extremely fine investment coating applied as a thin slurry directly to the surface of the pattern to reproduce maximum surface smoothness. The coating is surrounded by a coarser, cheaper, and more permeable investment to form the mold. See also *dip coat* and *investment casting*.

Investment Shell

Ceramic mold obtained by alternately dipping a pattern set up in *dip coat* slurry and stuccoing with coarse ceramic particles until the shell of desired thickness is obtained. See also *investment casting*.

Iodine

Iodine is a purplish-black, crystalline, poisonous elementary solid, with chemical symbol I, best known for its use as a strong antiseptic in medicine, but also used in many chemical compounds and war gases. In tablet form, it is used for sterilizing drinking water and has lesser odor and taste than chlorine for this purpose. It is also used in cattle feeds. Although poisonous in quantity, iodine is essential to proper cell growth in the human body and is found in every cell in a normal body, with larger concentration in the thyroid gland.

A wide range of compounds are made for electronic and chemical uses. Iodine is also a chemical reagent, used for reducing vanadium pentoxide and zirconium oxide into high-purity metals.

Ion

An atom, or group of atoms, that by loss or gain of one or more electrons has acquired an electric charge. If the ion is formed from an atom of hydrogen or an atom of a metal, it is usually positively charged; if the ion is formed from an atom of a nonmetal or from a group of atoms, it is usually negatively charged. The number of electronic charges carried by an ion is termed its electrovalence. The charges are denoted by superscripts that give their sign and number, for example, a sodium ion, which carries one positive charge, is denoted by Na^+ and a sulfate ion, which carries two negative charges, by $SO_4^{2=}$. See also *atomic structure* and *chemical bonding*.

Ion Beam–Assisted Deposition

See *ion implantation*.

Ion Beam Mixing

See *ion implantation*.

Ion Beam Sputtering

See *ion implantation*.

Ion (ic) Bombardment/Plating/Scrubbing

Process in which a component in a vacuum or suitable gas at low pressure forms the cathode of an electrostatic circuit and is impacted by positive ions as a means of revealing structure, plating, or cleaning.

Ion Carburizing

A method of surface hardening in which carbon ions are diffused into a workpiece in a vacuum through the use of high-voltage electrical energy. Synonymous with plasma carburizing or glow-discharge carburizing.

Ion Chromatography

An area of high-performance liquid chromatography that uses ion-exchange resins to separate various species of irons in solution and elute them to a suitable detector for analysis. See also *chromatography*, *gas chromatography*, and *liquid chromatography*.

Ion Etching

Surface removal by bombarding with accelerated ions in the vacuum (1–10 kV).

Ion Exchange

The reversible interchange of ions between a liquid and solid, with no substantial structural changes in the solid.

Ion Exchange (Ceramics)

An exchange of ions between two materials in intimate contact, usually accomplished by the application of heat.

Ion-Exchange Chromatography

Liquid chromatography with a stationary phase that possesses charged functional groups. This technique is applicable to the separation of ionic (charged) compounds. See also *ion chromatography*.

Ion-Exchange Resins

Cross-linked polymers that form salts with ions from aqueous solutions.

Ion Implantation

The process of modifying the physical or chemical properties of the near surface of a solid (target) by embedding appropriate atoms into it from a beam of ionized particles. The properties to be modified maybe electrical, optical, or mechanical, and they may relate to the semiconducting behavior of the material or its corrosion behavior. The solid may be crystalline, polycrystalline, or amorphous and need not be homogeneous. Related techniques are also used in conjunction with ion implantation to increase the ratio of material introduced into the substrate per unit area, to provide appropriate mixtures of materials, or to overcome other difficulties involved in surface modification by ion implantation alone. These techniques include the following:

1. *Ion beam sputtering*: An ion beam of argon or xenon directed at a target sputters material from the target to a substrate; the sputtered material arrives at the substrate with enough energy to promote good adhesion of the coating to substrate.
2. *Ion beam mixing*: Deposited layers (electroplating, sputtering) tens or hundreds of nanometers thick are mixed and bonded to the substrate by an argon or xenon ion beam.
3. *Plasma ion deposition*: Ion beams are used to create coatings having special phases, especially ion beam–formed carbon coatings in the diamond phase or ion beam–formed boron nitride coatings.
4. *Ion beam–assisted deposition*: Ion beams are combined with physical vapor deposition.

Ion Migration

The movement the free ions within a material or across the boundary between two materials under the influence of an applied electric field.

Ion Neutralization

The generic term for a class of charge-exchange processes in which an ion is neutralized by passage through a gas or by interaction with a material surface.

Ion Nitriding

A method of surface hardening in which nitrogen ions are diffused into a workpiece in a vacuum through the use of high-voltage electrical energy. Synonymous with plasma nitriding or glow-discharge nitriding.

Ion Plating

A generic term applied to atomistic film deposition processes in which the substrate surface and/or the depositing film is subjected to a flux of high-energy particles (usually gas ions) sufficient to cause changes in the interfacial region or film properties. Such changes may be in film adhesion to the substrate, film morphology, film density, film stress, or surface coverage by the depositing film material.

Ion Species

Type and charge of an ion. If an isotope is used, it should be specified.

Ionic Bond

(1) A type of chemical bonding in which one or more electrons are transferred completely from one atom to another, thus converting the neutral atoms into electrically charged ions. These ions are approximately spherical and attract each other because of their opposite charges. (2) A primary bond arising from the electrostatic attraction between two oppositely charged ions. Contrast with co*valent bond*. See *interatomic bonding*.

Ionic Charge

The positive or negative charge of an ion.

Ionic Conduction

Conduction arising from the movement of ions rather than electrons or holes.

Ionic Crystal

A crystal in which atomic bonds are *ionic bonds*. This type of atomic linkage, also known as (hetero) polar bonding, is characteristic of many compounds (sodium chloride, for instance).

Ionization

(1) The process in which neutral atoms become charged by gaining or losing an electron. (2) The act of splitting into, or producing, ions.

Ionization Chamber

An enclosure containing two or more electrodes surrounded by a gas capable of conducting an electric current when it is ionized by x-rays or other ionizing rays. It is commonly used for measuring the intensity of such radiation.

Ionomer Resins

A polymer that has ethylene as its major component, but that contains both covalent and ionic bonds. The polymer exhibits very strong interchain ionic forces. The anions hang from the hydrocarbon chain, and the cations are metallic, for example, sodium, potassium, or magnesium. These resins have many of the same features as polyethylene, plus high transparency, tenacity, resilience, and increased resistance to oils, greases, and solvents. Fabrication is accomplished as it is with polyethylene.

Ion-Pair Chromatography

Liquid chromatography with a mobile phase containing an ion that combines with sample ions, creating neutral ion pairs. The ion pairs are typically separated using bonded-phase chromatography. See also *bonded-phase chromatography*.

Ion-Scattering Spectrometry

A technique to elucidate composition and structure of the outermost atomic layers of a solid material, in which principally monoenergetic, singly charged, low-energy (less than 10 keV) probe ions are scattered from the surface and are subsequently detected and recorded as a function of the energy.

IRG Transition Diagram

Developed by the International Research Group (IRG) on Wear of Engineering Materials of the Organization of Economic Cooperation and Development, it is a plot of normal force in newtons (ordinate) versus sliding velocity in meters per second (abscissa) wherein boundaries identify three distinct regions of varying lubricant effectiveness.

Iridescence

Loss of brilliance in metallized plastics and development of multicolor reflectance. Iridescence is caused by the cold flow of plastic or of coating and by excess heat during vacuum metallizing.

Iridium

Iridium (symbol Ir) is a grayish-white metal of extreme hardness. It is insoluble in all acids and in aqua region. The melting point is 2447°C, and the specific gravity is 22.50. It occurs naturally with the metal osmium as an alloy, known as osmiridium, 30%–60% osmium, used chiefly for making fountain-pen points and instrument pivots.

Iridium is employed as a hardener for platinum, the jewelry alloys usually containing 10%. With 35% iridium, the tensile strength of platinum is increased to 965 MPa. Iridium wire is used in spark plugs because it resists attack of leaded aviation fuels. Iridium–tungsten alloys are used for springs operating at temperatures to 800°C. Iridium intermetallic compounds such as Cb_3Ir_2, Ti_3Ir, and $ZrIr_2$ are superconductors.

Upon atmospheric exposure, the surface of the metal is covered with a relatively thick layer of iridium dioxide (IrO_2). Only when heated to redness is the metal attacked by oxygen (to give IrO_2) and by the halogens. The vapor pressure of IrO_2 is atm (10^5 Pa) at 1120°C, and consequently pieces of iridium metal lose mass at elevated temperatures in an oxygen atmosphere.

Metallic iridium is available in powder, sponge, wire, and sheet forms. Iridium is usually difficult to forge and fabricate, but it is ductile when hot and consequently is worked hot. It remains ductile when cooled as long as it is not annealed. Annealing causes loss of ductility and brittleness. In the absence of oxygen, iridium is not attacked by many molten metals, including lithium, sodium, potassium, gold, lead, and tin.

Uses

Because of its scarcity and high cost, applications of iridium are severely limited. Although iridium metal and many of its complex compounds are good catalysts, no large-scale commercial application for these has been developed. In general, other platinum metals have superior catalytic properties. The high degree of thermal stability of elemental iridium and the stability it imparts to its alloys do give rise to those applications where it has found success. Particularly relevant are its high melting point (2443°C), its oxidation resistance, and the fact that it is the only metal with good mechanical properties that survives atmospheric exposure above 1600°C. Iridium is alloyed with platinum to increase tensile strength, hardness, and corrosion resistance. However, the workability of these alloys is decreased. These alloys find use as electrodes for anodic oxidation, for containing and manipulating corrosive chemicals, for electrical contacts that are exposed to corrosive chemicals, and as primary standards for weight and length. Platinum–iridium alloys are used for electrodes in spark plugs that are unusually resistant to fouling by antiknock lead additives. Iridium–rhodium thermocouples are used for high-temperature applications, where they have unique stability. Very pure iridium crucibles are used for growing single crystals of gadolinium gallium garnet for computer memory devices and of yttrium aluminum garnet for solid-state lasers.

Iridium–columbium intermetallic compounds have been identified as a potential material for high-temperature aircraft-turbine parts because of its high melting temperature and room-temperature ductility; it has a specific gravity of 15.2 and melts at 1900°C.

Iron

Iron (symbol Fe) is one of the most common of the commercial metals. It has been in use since the most remote times, but it does not occur native except in the form of meteorites—early tools of Egypt were apparently made from nickel irons from this source. The common iron ores are magnetic pyrites, magnetite, hematite, and carbonates of iron. To obtain the iron, the ores are fused to drive off the oxygen, sulfur, and impurities. The melting is done in a blast furnace directly in contact with the fuel and with limestone as a flux. The latter combines with the quartz and clay, forming a slag that is readily removed. The resulting product is crude pig iron, which requires subsequent remelting and refining to obtain commercially pure iron. Sintered iron and steel are also produced without blast-furnace reduction by compressing purified iron oxide in rollers, heating to 1204°C, and hot strip rolling. The final cold-rolled product is similar to conventional iron and steel.

Originally, all iron was made with charcoal, but because of the relative scarcity of wood and the greater expense, charcoal is now seldom used in the blast furnace.

Iron is a grayish metal, which until recently was never used pure. It melts at 1525°C and boils at 2450°C. Even very small additions of carbon reduce the melting point. It has a specific gravity of 7.85. Iron containing more than 0.15% chemically combined carbon is termed steel. When the carbon is increased to above about 0.40%, the metal will harden when cooled suddenly from a red heat.

Iron, when pure, is very ductile, but a small amount of sulfur, as little as 0.03%, will make it hot-short, or brittle at red heat. As little as 0.25% of phosphors will make iron cold-short, or brittle when cold. Iron forms carbonates, chlorides, oxides, sulfides, and other compounds. It oxidizes easily and is also attacked by many acids.

Because pure iron is allotropic, it can exist in a solid in two different crystal forms. From subzero temperatures up to 910°C, it has a body-centered cubic structure and has been identified as alpha (α) iron. Between 910°C and 1400°C, the crystal structure is face-centered cubic. This form is known as gamma (γ) iron. At 1400°C and up to its melting point of 1540°C, the structure again becomes body-centered cubic. This last form, called delta (δ) iron, has no practical use. The transformation from one allotropic form to another is reversible. Thus, when iron is heated to above 910°C, the alpha body-centered cubic crystal changes into face-centered cubic crystals of gamma iron. When cooled below this temperature, the metal again reverts back to a body-centered cubic structure. These allotropic phase changes inherent in iron make possible the wide variety of properties obtainable in ferrous alloys by various heat-treating processes.

Types

Electrolytic iron is a chemically pure iron produced by the deposition of iron in a manner similar to electroplating. Bars of cast iron are used as anodes and dissolved in electrolyte of ferrous chloride. The current precipitates almost pure iron on the cathodes, which are hollow steel cylinders. The deposited iron tube is removed by hydraulic pressure or by splitting and then annealed and rolled into plates. The iron is 99.9% pure and is used for magnetic cores and where ductility and purity are needed.

Iron powder is made by reducing iron ore by the action of carbon monoxide at a temperature below the melting point of the iron and below the reduction point of the other metallic oxides in the ore. In the United States, it is made by the reduction of iron oxide mill scale, by electrolysis of steel borings and turnings in an electrolyte of ferric chloride, or by atomization. Iron powders are widely used for pressed and sintered structural parts, commonly referred to as powder metallurgy (P/M) parts. Pure iron powders are seldom used alone for such parts. Small additions of carbon in the form of graphite and/or copper are used to improve performance properties.

Iron–copper powders contain 2%–11% copper. Small amounts of graphite are sometimes added. Copper increases strength and hardness and improves corrosion resistance, but lowers ductility somewhat. Tensile strength ranges from 206 to 680 MPa, depending on density and heat treatment.

Iron–carbon powders contain up to 1% graphite. When pressed and sintered, internal carburization results and produces a carbon–steel structure, although some free carbon remains. In general, the density of structural or mechanical iron–carbon steel P/M parts is 80%–95% that of wrought steels, and the greater the density, the greater the strength and other mechanical properties. These carbon–steel P/M parts have a higher strength and hardness than those of iron, but they are usually more brittle. As-sintered strengths range from about 241 to 482 MPa depending on density. By heat treatment, strengths up to 861 MPa are achieved. P/M parts can be made to control porosity for filters or for filling with oil or other materials for surface lubricity.

Wrought iron refers usually, in a metallurgical context, to low-carbon iron containing elongated inclusions and made, in the final stages, by repeated folding and forging. See *wrought iron*.

Cast irons are iron alloys with carbon contents of about 2%–4%. Note that the cast-iron industry commonly omits the term "cast" in descriptions such as "malleable (cast) iron," "gray (cast) iron," and even (cast) "irons." See *cast iron*. See also *blast furnace*.

Iron Alloys

These are solutions of metals, where one metal is iron. A great number of commercial alloys have iron as an intentional constituent. Iron is the major constituent of wrought and cast iron and wrought and cast steel. Alloyed with usually large amounts of silicon, manganese, chromium, vanadium, molybdenum, niobium, selenium, titanium, phosphorus, or other elements, singly or sometimes in combination, iron forms a large group of materials known as ferroalloys that are important as addition agents in steelmaking. Iron is also a major constituent of many special-purpose alloys developed to have exceptional characteristics with respect to magnetic properties, electrical resistance, heat resistance, corrosion resistance, and thermal expansion. See Table I.1.

Iron–Aluminum Alloys

Although pure iron has ideal magnetic properties in many ways, its low thermal resistivity makes it unsuitable for use in ac magnetic circuits. Addition of aluminum in fairly large amounts increases the electrical sensitivity of iron, making the resulting alloys useful in such circuits.

Three commercial iron–aluminum alloys with moderately high permeability at low field strengths and high electrical resistance nominally contain 12% aluminum, 16% aluminum, and 16% aluminum with 3.5% molybdenum, respectively. These three alloys are classified as magnetically soft materials, that is, they become magnetized in a magnetic field but are easily demagnetized when the field is removed.

The addition of more than 8% aluminum to iron results in alloys that are too brittle for many uses because of difficulties in fabrication. However, addition of aluminum to iron markedly increases its resistance to oxidation. One steel containing 6% aluminum possesses good oxidation resistance up to 1300°C.

Iron–Carbon Alloys

The principal iron–carbon alloys or wrought iron, cast iron, and steel.

Wrought iron of good quality is nearly pure iron; its carbon content seldom exceeds 0.035%. In addition, it contains 0.075%–0.15% silicon, 0.10 to less than 0.25% phosphorus, less than 0.02% sulfur, and 0.06%–0.10% manganese. Not all of these elements are alloyed with the iron; part of them maybe associated with the intermingled slag that is a characteristic of this product. Because of its low carbon content, the properties of wrought iron cannot be altered in any useful way by heat treatment.

Cast iron may contain 2%–4% carbon and varying amounts of silicon, manganese, phosphorus, and sulfur to obtain a wide range of physical mechanical properties. Alloying elements (silicon, nickel, chromium, molybdenum, copper, titanium, and so on) may be added in amounts varying from a few tenths to 30% or more. Many of the alloy cast irons have proprietary compositions.

Steel is a generic name for a large group of iron alloys that include the plain carbon and alloy steels. The plain carbon steels represent the most important group of engineering materials known. Although any iron–carbon alloy containing less than about 2% carbon can be considered a steel, the American Iron and Steel Institute (AISI) standard carbon steels embrace a range of carbon contents from 0.06% maximum to about 1%. In the early days of the American steel industry, hundreds of steels with different chemical compositions were produced to meet individual demands of purchasers. Many of these steels differed only slightly from each other in chemical composition.

TABLE I.1
Some Typical Composition Percent Ranges of Iron Alloys Classified by Important Uses[a]

Type	Fe	C	Mn	Si	Cr	Ni	Co	W	Mo	Al	Cu	Ti
Heat-resistant alloy castings	Bal.[b]	0.30–0.50	—	1–2	8–30	0–7						
	Bal	0.20–0.75	—	2–2.5	10–30	8–41						
Heat-resistant cast irons	Bal.	1.8–3.0	0.3–1.5	0.5–2.5	15–35	5 max						
	Bal.	1.8–3.0	0.4–1.5	1.0–2.75	1.75–5.5	14–30	—	—	1	—	7	
Corrosion-resistant alloy castings	Bal.	0.15–0.50	1 max	1	11.5–30	0–1	—	0.5max				
	Bal.	0.03–0.20	1.5 max	1.5–2.0	18–27	8–31						
Corrosion-resistant cast irons	Bal.	1.2–4.0	0.3–1.5	0.5–3.0	12–35	5 max	—	—	4 max		3 max	
	Bal.	1.8–3.0	0.4–1.5	1.0–2.75	1.75–5.5	14–32	—	—	1 max		7 max	
Magnetically soft materials	Bal.	—	—	0.5–4.5								
	Bal.	—	—	—	—	—	—	—	3.5	16		
	Bal.	—	—	—	—	—	—	—	—	16		
	Bal.	—	—	—	—	—	—	—	—	12		
Permanent magnet materials	Bal.	—	—	—	—	—	12	—	17			
	Bal.	—	—	—	—	—	12	—	20			
	Bal.	—	—	—	—	20	5	—	—	12		
	Bal.	—	—	—	—	17	12.5	—	—	10	6	
	Bal.	—	—	—	—	25	—	—	—	12		
	Bal.	—	—	—	—	28	5	—	—	12		
	Bal.	—	—	—	—	14	24	—	—	8	3	
	Bal.	—	—	—	—	15	24	—	—	8	3	1.25
Low-expansion alloys	Bal.	—	0.15	0.33	—	36						
	Bal.	—	0.24	0.03	—	42						
	61–53	0.5–2.0	0.5–2.0	0.5–2.0	4–5	33–35	—	1–3				

Source: Parker, S.P., *McGraw-Hill Encyclopedia of Science and Technology*, 8th ed., Vol. 9, McGraw Hill, New York, 1997, pp. 445–457. With permission.

[a] This table does not include any AISI standard carbon steels, alloy steels, or stainless and heat-resistant steels or plain or alloy cast iron for ordinary engineering uses; it includes only alloys containing at least 50% iron, with a few exceptions.

[b] Bal., balance percent or composition.

Studies were undertaken to provide a simplified list of fewer steels that would still serve the various needs of fabricators and users of steel products. The Society of Automotive Engineers (SAE) and the AISI both periodically publish lists of steels, called standard steels, classified by chemical composition. These lists are published in the SAE Handbook and the AISI Steel Products Manuals. The lists are altered periodically to accommodate new steels and to provide for changes in consumer requirements. There are minor differences between some of the steels listed by the AISI and SAE.

A numerical system is used to indicate grades of standard steels. Provision also is made to use certain letters of the alphabet to indicate the steelmaking process, certain special additions, and steels that are tentatively standard. The basic numerals for the AISI classification and corresponding types of steels in this system include the first digit of the series designation, which indicates the type to which a steel belongs; thus, 1 indicates a carbon steel, 2 indicates a nickel steel, and 3 indicates a nickel–chromium steel. In the case of simple alloy steels, the second usually indicates the percentage of the predominating alloying element. Usually, the last two (or three) digits indicate the average carbon content in points, or hundredths of a percent. Thus, 2340 indicates a nickel steel containing about 3% nickel and 0.40% carbon.

All carbon steels contain minor amounts of manganese, silicon, sulfur, phosphorus, and sometimes other elements. At all carbon levels, the mechanical properties of carbon steel can be varied to a useful degree by heat treatments that alter its microstructure. Above about 0.25%, carbon steel can be hardened by heat treatment. However, most of the carbon steel produced is used without a final heat treatment.

Alloy steels are steels with enhanced properties attributable to the presence of one or more special elements or of larger proportions of manganese or silicon than are present ordinarily in carbon steel. The major classifications of alloy steels are high strength, low alloy, AISI alloy, alloy tool, heat resisting, electrical, and austenitic manganese. Some of these iron alloys are discussed briefly in the following.

Iron–Chromium Alloys

An important class of iron–chromium alloys is exemplified by the wrought stainless and heat-resisting steels of the type 400 series of the AISI standard steels, all of which contain at least 12% chromium, which is about the minimum chromium content that will confer stainlessness. However, considerably less than 12% chromium will improve the oxidation resistance of steel for up to 650°C, as is true of AISI types 501 and 502 steels that nominally contain about 5% chromium and 0.5% molybdenum. A comparable group of heat- and corrosion-resistant alloys, generally similar to the 400 series of

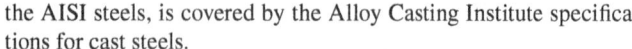

the AISI steels, is covered by the Alloy Casting Institute specifications for cast steels.

Corrosion-resistant cast irons alloyed with chromium contain 12%–35% of that element and up to 5% nickel. Cast irons classified as heat resistant contain 15%–35% chromium and up to 5% nickel.

During World War I, a high-carbon steel used for making permanent magnets contained 1%–6% chromium (usually around 3.5%); it was developed to replace the magnet steels containing tungsten that had been formerly but could not then be made because of a shortage of tungsten.

Iron–Chromium–Nickel Alloys

The wrought stainless and heat-resisting steels represented by the type 200 and the type 300 series of the AISI standard steels are important class of iron–chromium–nickel alloys. A comparable series of heat- and corrosion-resistant alloys is covered by specifications of the Alloy Casting Institute. Heat- and corrosion-resistant cast irons containing 15%–35% chromium and up to 5% nickel.

Iron–Chromium–Aluminum Alloys

Electrical-resistance heating elements are made of several iron alloys of this type. Nominal compositions are 72% iron, 23% chromium, 5% aluminum, and 55% iron, 37.5% chromium, 7.5% aluminum. The iron–chromium–aluminum alloys (with or without 0.5%–2% cobalt) have higher electrical resistivity and lower density than nickel–chromium alloys used for the same purpose. When used as heating elements in furnaces, the iron–chromium–aluminum alloys can be operated at temperatures of 2350°C maximum. These alloys are somewhat brittle after elevated-temperature use and have a tendency to grow or increase in length while at temperature, so that heating elements made from them should have additional mechanical support. Addition of niobium reduces the tendency to grow.

Because of its high electrical resistance, the 72% iron, 23% chromium, 5% aluminum alloy (with 0.5% cobalt) can be used for semiprecision resistors in, for example, potentiometers and rheostats.

Iron–Cobalt Alloys

Magnetically soft iron alloys containing up to 65% cobalt have higher saturation values than pure iron. The cost of cobalt limits the use of these alloys to some extent. The alloys also are characterized by low electrical resistivity and high hysteresis loss. Alloys containing more than 30% cobalt are brittle unless modified by additional alloying and special processing. Two commercial alloys with high permeability at high field strengths (in the annealed condition) contain 49% cobalt with 2% vanadium and 35% cobalt with 1% chromium. The latter alloy can be cold-rolled to a strip that is sufficiently ductile to permit punching and shearing. In the annealed state, these alloys can be used in either ac or dc applications. The alloy of 49% cobalt with 2% vanadium has been used in pole tips, magnet yokes, telephone diaphragms, special transformers, and ultrasonic equipment. The alloy of 35% cobalt with 1% chromium has been used in high-flux-density motors and transformers as well as in some of the applications listed for the higher-cobalt alloy.

Although seldom used now, two high-carbon alloys called cobalt steel were used formerly for making permanent magnets. These were both high-carbon steels. One contained 17% cobalt, 2.5% chromium, 8.25% tungsten; the other contains 36% cobalt, 5.75% chromium, 3.75% tungsten. These are considered magnetically hard materials as compared to the magnetically soft materials.

Iron–Manganese Alloys

The important commercial alloy in this class is an austenitic manganese steel (sometimes called Hadfield manganese steel after its inventor) that nominally contains 1.2% and 12%–13% manganese. This steel is highly resistant to abrasion, impact, and shock.

Iron–Nickel Alloys

The iron–nickel alloys exhibit a wide range of properties related to their nickel contents.

Nickel content of a group of magnetically soft materials ranges from 40% to 60%; however, the higher saturation value is obtained at about 50%. Alloys with nickel content of 45%–60% are characterized by high permeability and low magnetic losses. They are used in such applications as audio transformers, magnetic amplifiers, magnetic shields, coils, relays, contact rectifiers, and choke coils. The properties of the alloys can be altered to meet specific requirements by special processing techniques involving annealing in hydrogen to minimize the effects of impurities, grain-orientation treatments, and so on.

Another group of iron–nickel alloys, those containing about 30% nickel, are used for compensating changes that occur in magnetic circuits due to temperature changes. The permeability of the alloys decreases predictably with increasing temperature.

Low-expansion alloys are so-called because they have low thermal coefficients of linear expansion. Consequently, they are valuable for use as standards of length, surveyors' rods and tapes, compensating pendulums, balance wheels in timepieces, glass-to-metal seals, thermostats, jet-engine parts, electronic devices, and similar applications.

The first alloy of this type contains 36% nickel with small amounts of carbon, silicon, and manganese (totaling less than 1%). Subsequently, a 39% nickel alloy with a coefficient of expansion equal to that of low-expansion glasses and a 46% nickel alloy with a coefficient equal to that of platinum were developed. Another important alloy is one containing 42% nickel that can be used to replace platinum as lead-in wire in light bulbs and vacuum tubes by first coating the alloy with copper. An alloy containing 36% nickel and 12% chromium has a constant modulus of elasticity and low expansivity over a broad range of temperatures. Substitution of 5% cobalt for 5% nickel in the 36% nickel alloy decreases its expansivity. Small amounts of other elements affect the coefficient linear expansion, as the variations in heat treatment, cold-working, and other processing procedures.

A 9% nickel steel is useful in cryogenic and similar applications because of good mechanical properties at low temperatures. Two steels (one containing 10%–11% nickel, 3%–5% chromium, about 3% molybdenum, and lesser amounts of titanium and aluminum and another with 17%–19% nickel, 8%–9% cobalt, 3%–3.5% molybdenum, and small amounts of titanium and aluminum) have exceptional strength in the heat-treated (aged) condition. These are known as maraging steels.

Cast irons containing 14%–30% nickel and 1.75%–5.5% chromium possess good resistance to heat and corrosion.

Iron–Silicon Alloys

There are two types of iron–silicon alloys that are commercially important: the magnetically soft materials designated silicon or electrical steel and the corrosion-resistant, high-silicon cast irons.

Most silicon steels used in magnetic circuits contain 0.5%–5% silicon. Alloys with these amounts of silicon have high permeability, high electrical resistance, and low hysteresis loss compared with relatively pure iron. Most silicon steel is produced in flat-rolled (sheet) form and is used in transformer cores, stators, and rotors of motors that are built up in laminated-sheet form to reduce eddy current losses. Silicon-steel electrical sheets, as they are called commercially, are made in two general classifications: grain oriented and nonoriented.

The grain-oriented steels are rolled and heat-treated in special ways to cause the edges of most of the unit cubes of the metal lattice to align themselves in the preferred direction of optimum magnetic properties. Magnetic cores are designed with the main flux path in the preferred direction, thereby taking advantage of the directional properties. The grain-oriented steels contain about 3.25% silicon, and they are used in the highest-efficiency distribution and power transformers and in large turbine generators.

The nonoriented steels may be subdivided into low-, intermediate-, and high-silicon classes. Low-silicon steels contain about 0.5%–1.5% silicon and are used principally in rotors and stators of motors and generators; steels containing about 1% silicon are also used for reactors, relays, and small intermittent-duty transformers. Intermediate-silicon steels contain about 2.5%–3.5% silicon and are used in motors and generators of average to high efficiency and small- to medium-size intermittent-duty transformers, reactors, and motors. High-silicon steels contain about 3.75%–5% silicon and are used in highest-efficiency motors, generators, and transformers.

High-silicon cast irons containing 14%–17% silicon and sometimes up to 3.5% molybdenum possess corrosion resistance that makes them useful for acid-handling equipment and for laboratory drainpipes.

Iron–Tungsten Alloys

Although tungsten is used in several types of relatively complex alloys (including high-speed steels), the only commercial alloy made up principally of iron and tungsten was a tungsten steel containing 0.5% chromium in addition to 6% tungsten that was used up to the time of World War I for making permanent magnets.

Hard-Facing Alloys

Hard-facing consists of welding a layer of metal of special composition on a metal surface to impart some special property not possessed by the original surface. The deposited metal may be more resistant to abrasion, corrosion, heat, or erosion than the metal to which it is applied. A considerable number of hard-facing alloys are available commercially. Many of these would not be considered iron alloys by the 50% iron content criterion adopted for the iron alloys. Among the iron alloys are low-alloy facing materials containing chromium as a chief alloying element, with smaller amounts of manganese, silicon, molybdenum, vanadium, tungsten, and in some cases nickel to make a total alloy content of up to 12%, with the balance iron. High-alloy ferrous materials containing a total of 12%–25% alloying elements form another group of hard-facing alloys; a third group contains 26%–50% alloying elements. Chromium, molybdenum, and manganese are the principal alloying elements of the 12%–25% group; smaller amounts of molybdenum, vanadium, nickel, and in some cases titanium are present in various proportions. In the 26%–50% alloys, chromium (and in some cases tungsten) is the principal alloying element, with manganese, silicon, nickel, molybdenum, vanadium, niobium, and boron as the elements from which a selection is made to bring the total alloy content within the 26%–50% range.

Permanent-Magnet Alloys

These are magnetically hard ferrous alloys, many of which are too complex to fit the simple compositional classification used earlier for other iron alloys. As already mentioned in discussing iron–cobalt and iron–tungsten alloys, the high-carbon steels (with or without alloying elements) are now little used for permanent magnets. These have been supplanted by a group of sometimes complex alloys with much higher retentivities. The ones considered here are all proprietary compositions. Two of the alloys contain 12% cobalt–17% molybdenum and 12% cobalt–20% molybdenum.

Members of a group of six related alloys contain iron, nickel, aluminum, and with one exception cobalt; in addition, three of the cobalt-containing alloys contain copper and one has copper and titanium. Unlike magnet steels, these alloys resist demagnetization by shock, vibration, or temperature variations. They are used in magnets for speakers, watt-hour meters, magnetrons, torque motors, panel and switchboard instruments, and so on, where constancy and degree of magnet strength are important.

Iron Carbide

A hard, brittle intermetallic compound of iron and carbon, having the approximate composition Fe_3C. Is also called *cementite*, which dominates *white cast iron*, a type of cast iron formed by rapid solidification of molten iron. Iron carbide has a density that is slightly less than that of pure iron. It is extremely hard, thus exceptionally wear resistant, retains high hardness up to red heat, has high compressive strength, and exhibits a white crystalline fracture. With suitable compositions, it can be induced to form in select regions of otherwise softer irons by the use of chills in these regions of a casting mold to affect more rapid solidification.

Iron Loss

The loss of power in an electrical circuit arising from eddy currents in the magnetic iron core of transformer.

Iron Powder Electrode

A welding electrode with a covering containing up to about 50% iron powder, some of which becomes part of the deposit.

Iron Pyrite

A mineral sometimes mined for the zinc, gold, or copper associated with it, but chiefly used for producing sulfur, sulfuric acid, and copperas. It is an *iron disulfide*, FeS_2, containing 53.4% sulfur. It often occurs in crystals, also massive or granular. It is brittle, with a Mohs hardness of 6–6.5 and specific gravity 4.95–5.1. The color is brassy yellow and it is called *fool's gold* because of the common error made in detection. In the Italian method of roasting pyrites, oxygen combines with sulfur to form sulfur dioxide and with iron. *Pyrite* is found in rocks of all ages associated with different minerals. The pyrites mined in Missouri, known as *marcasite*, also used for gemstones, have the formula FeS, and the gem specimens have a yellow color with a greenish tinge. *Sulfur pyrites*, found in great quantity in Shasta County, California, are roasted to produce sulfuric acid, and the residue is used for making cement and ferric sulfate. In Japan, pyrite is chlorinated, producing sulfur, copper, gold, silver, and zinc, and the residue of sintered pellets of Fe_2O_3 is used in blast furnaces.

Iron Rot

Deterioration of wood in contact with iron-base alloys.

Iron Shot

An abrasive material made by running molten iron into water. It is employed in tumbling barrels and in the cutting and grinding of stones. The round sizes between No. 30 and No. 20 are used for the shot peening of mechanical parts to increase fatigue resistance. *Peening shot* is marketed by SAE numbers from 6 to 157, the size numbers being the diameter in thousandths of an inch. *Steel grit* is made by forcing molten iron through a steam jet so that the metal forms into small and irregular pieces, and large globules are crushed into steel grit in sizes from No. 8 to No. 80. It is preferred to sand for sandblasting some materials. *Steelblast* is this material for tumbling and sandblasting. All the materials for metal cleaning are termed *blasting shot.*

Iron Soldering

A soldering process in which the heat required is obtained from a soldering iron.

Ironing

An operation used to increase the length of a tube or cup through reduction of wall thickness and outside diameter, the inner diameter remaining unchanged.

Ironstone

A mineral, usually iron carbonate, containing a large proportion of iron.

Irradiance

The radiant flux density incident on a surface; the ratio of flux to area of irradiated surface.

Irradiation

The exposure of a material or object to x-rays, gamma-rays, ultraviolet rays, or other ionizing radiation.

Irradiation (Plastics)

The bombardment of plastics with a variety of subatomic particles, usually alpha-, beta-, or gamma-rays. Used to initiate the polymerization and copolymerization of plastics and in some cases to bring about changes in the physical properties of a plastic.

Irregular Powder

Particles lacking symmetry.

Irreversible (Plastics)

Not capable of redissolving or remelting. Chemical reactions that proceed in a single direction and are not capable of reversal (as applied to thermosetting resins).

Isinglass

A gelatin made from the dried swimming bladders, or *fish sounds*, of sturgeon and other fishes. *Russian isinglass* from the sturgeon is the most valued grade and is one of the best of the water-soluble adhesives. It is used in glues and cements and in printing inks. It is also used for clarifying wines and other liquids. It is prepared by softening the bladder in water and cutting it into long strips. These dry to a dull-gray, horny, or stringy material. It is used as an antiseptic astringent. The name *isinglass* was also used for mica when employed as a transparent material for stove doors.

ISO

International Standards Organization.

ISO Metric Fasteners

An ISO-approved system of sizes, tolerances, and materials for threaded fasteners. For steel fasteners, the material strength is identified by two numbers, the second being a decimal, for example, "12.9." The first number is 1/10 of the ultimate tensile strength (UTS) of the material in kgf/mm^2; the second number is the yield to ultimate ratio as a decimal. Thus, in the example quoted, 12.9, the material has a UTS of 120 kgf/mm^2 and the yield strength is the UTS times 0.9, that is, 108 kgf/mm^2.

Isobar

(1) In atomic physics, one of two or more atoms that have a common mass number A, but differ in atomic number Z. Thus, although isobars possess approximately equal masses, they have different chemical properties; they are atoms of different elements. See also *nuclear structure*. (2) A line joining points of constant pressure.

Isocorrosion Diagram

A graph or chart that shows constant corrosion behavior with changing solution (environment) composition and temperature.

Isocratic Elution

In liquid chromatography, the use of a mobile phase with composition that is unchanged throughout the course of the separation process.

Isocyanate Plastics

Plastics based on resins made by the condensation of organic isocyanates and other compounds. Generally reacted with polyols on a polyester or polyether backbone molecule, with the reactants being joined by the formation of the urethane linkage. See also *polyurethane* and *urethane plastics.*

Isomer

A compound, radical, ion, or nuclide that contains the same number of atoms of the same elements but differs in structural arrangement and properties. See also *stereoisomer.*

Isometric

A crystal form in which the unit dimension on all three axes is the same.

Isomorphous

Having the same crystal structure. This usually refers to intermediate phases that form a continuous series of solid solutions.

Isomorphous System

A complete series of mixtures in all proportions of two or more components in which unlimited mutual solubility exists in the liquid and solid states.

Isopolyester Composites

An isopolyester composite is used to retrofit concrete causeway support columns. The Snap-Tite composite jacket will use isopolyester made with purified isophthalic acid. Isopolyesters have a proven record of delivering corrosion resistance and mechanical strength at less cost. The resin encapsulates the reinforcing network, which for most applications is conventional e-glass woven roving and bidirectional fabric. As a result, the compressive strength of the columns will reportedly more than double, 283 MPa or more, and the ductility will improve from 3% to 10%.

Isostatic

At equal pressure. In some, perhaps casual, usages, the term is used merely to indicate that pressure does not vary with time, but in a stricter sense, in terms such as hot isostatic pressing, it indicates that the pressure acts equally in all directions.

Isostatic Mold

A sealed container of glass or sheet of carbon steel, stainless steel, or a nickel-based alloy. See also *isostatic pressing*.

Isostatic Pressing

The isostatic pressing process was pioneered in the mid-1950s and has steadily grown from a research curiosity to a viable production tool. Many industries apply this technique for consolidation of powders or defect healing of castings. The process is used for a range of materials, including ceramics, metals, composites, plastics, and carbon.

Isostatic pressing applies a uniform, equal force over the entire product, regardless of shape or size. It thus offers unique benefits for ceramic and refractory applications. The ability to form product shapes to precise tolerances (reducing costly machining) has been a major driving force for its commercial development.

There are three basic types of isostatic pressing. Cold isostatic pressing (CIP) is applied to consolidate ceramic or refractory powders loaded into elastomeric bags. Warm isostatic pressing differs from CIP only in that shapes are pressed at warm temperature to about 100°C. Hot isostatic pressing involves both temperature and pressure applied simultaneously to obtain fully dense parts (to 100% theoretical density) and is used mainly for engineered ceramics requiring optimum properties for high-performance applications.

Cold Isostatic Pressing (CIP)

CIP is mainly a powder-compacting process for obtaining 60%–80% theoretically dense parts ready for sintering. Because of the good green strength obtained with this forming method, premachining before sintering is feasible without causing breakage.

When comparing uniaxial pressing to isostatic pressing, one can say that uniaxial pressing is more suitable for small shapes at high production rates. Die wall friction may result in nonuniform densities, especially for large aspect ratios (greater than 3:1).

CIP is slower than uniaxial pressing but can be used for small or large, simple or complex shapes. The uniform green density offers more even shrinkage during sintering, which is very important for good shape control and uniform properties. In addition, CIP does not require a wax binder as does uniaxial pressing, thus eliminating the waxing operations.

Low-cost elastomer tooling is used for isostatic pressing, but close tolerances can only be obtained for surfaces that are pressed against a highly accurate steel mandrel. Surfaces in contact with the elastomer tooling may require post machining when tight tolerances and good surface finishes are specified.

Wet-Bag CIP

Two types of CIP methods have evolved over the years; wet-bag and dry-bag. The so-called wet-bag method is used for producing mixed shapes. It is estimated that there are more than 3000 wet-bag presses in use worldwide today, ranging in size from 50 to 2000 mm in diameter.

A typical cycle time for a production press ranges from 5 to 30 min, depending mainly on size, powder volumetric compaction ratio, and pump selected. This speed is rather slow but can be improved by higher-volume pumps, better vessel use, and improved loading mechanisms. A 5 min cycle calculates to 24,000 annual cycles based on a 1-shift, 8 h operation, or 240,000 cycles in 10 years. That should be the minimum design value of a vessel. Therefore, it is very important to select the proper vessel design to meet the end user's specified fatigue requirements as proven by the supplier's theoretical analysis and past performance.

Dry-Bag CIP

The dry-bag process is more applicable to producing same-shaped parts. Automated dry-bag isostatic pressing equipment was developed in the 1930s for compacting spark plug insulators, which are exclusively produced that way today for worldwide distribution.

The dry-bag method involves the same flex-bag technology for powder containment as the wet bag except that a stationary polyurethane "master bag," called the membrane, is inserted inside the pressure vessel. The pressure force (water) is transmitted through the membrane to the mold and then to the powder, thus keeping the mold dry. Generally one part is pressed at a time, and loading occurs from the bottom.

A minimum of three cassettes (each cassette typically consisting of the mandrel, mold, and powder) are in motion simultaneously; one is pressed, one is filled with powder, and one is debagged. The cycle time is 1 min or less, which calculates to 120,000 cycles/year on a 1-shift, 8 h basis. This rate of cycling places a much higher demand on the pressure vessel fatigue, and proper design is critical to withstand the higher pressures. The dry-bag method is preferred over the wet-bag for automated production of same-size or same-shaped parts and lots of about 50 parts per hour or above.

Warm Isostatic Pressing (WIP)

WIP follows the same path as CIP except the parts are compacted both at pressure and low temperature to 100°C. The pressing fluid water may be substituted with oil. To date, there are a few applications for manufacturers in the electronics industry as a cost-effective means of compacting different shaped parts.

Traditionally, a heated platen press has been applied in these applications. The problem associated with this method is the lack of uniform pressure all around the parts, resulting in dimensional variations from one side to the other. WIP, on the other hand, is a well-suited alternative for applying equal and uniform pressure on all surfaces.

Hot Isostatic Pressing (HIP)

HIP is a densification method for powders, compacts, or castings. It applies a gas pressure of 100–200 MPa and temperatures to 2200°C. An inert gas, most commonly argon, is used as the pressing fluid. The goal was to improve the performance of critical parts by eliminating defects and porosity resulting in fully dense compacts. Improvements in mechanical and physical properties, fatigue, surface finish, reliability, and/or rejection rate are possible using HIP.

Two HIP methods are used today for compacting parts: direct HIP, which applies to encapsulated powders, and post-HIP, which applies to presintered compacts without interconnected porosity. (The HIP and CIP processes can also be combined, sometimes called CHIP. In CHIP, loose powder is cold-compacted, then sintered, and then post-HIPed to achieve fully dense parts.)

Direct HIP involves a ceramic powder enclosed in a container usually made from an impermeable barrier, such as glass, ceramic, or refractory metal. Glass is the most common barrier today for direct compacting of ceramic powders and serves in two ways: (1) it acts as a barrier for consolidation and (2) it isolates and protects the powder from the processing gas.

Several precompacting methods are available for direct HIP, including injection molding, slip casting, CIP, or dry pressing. After preforming, the parts are glass encapsulated prior to HIP. A major goal is to form parts to near-net configurations to minimize final machining, which usually requires expensive diamond tooling.

Post-HIP is an alternative method and is a widely practiced way to HIP products such as oxide ceramics, tool bits, and ferrites. This method is only valid for materials that are able to sinter without pressure to close surface-connected porosity, equivalent to about 92%–95% of theoretical density. The powder is shaped-formed by either casting of CIP, is then pressureless sintered, and finally post-HIP is employed for full densification. Post-HIP is also chosen when material decomposition is to be avoided; the post-HIP process allows the pressure fluid to contact the partially dense parts for reactive processing. For example, a mixture of 95% argon and 5% oxygen is preferred for oxide ceramics, whereas nitrogen gas is preferred for nitride ceramics.

The HIP process is rather slow, and a cycle may take 10–15 h, depending on part size, material, and furnace design. One of the reasons for the long cycle is that most installations use conventional furnaces that cool not only slowly but also not uniformly. Advanced furnaces are available with uniform rapid cooling provisions that have been used for metals since early 1990. Shorter cycle times improve throughput, thus reducing processing costs.

The rate of cooling is programmable from 1°C/min to 50°C/min. A variable-speed fan is used to select the appropriate rate to avoid cracking when processing thermally sensitive ceramic parts. This feature can reduce HIP cycles to less than 8 h but adds only 10%–15% to the equipment price.

One of the largest HIP units for processing ceramics is 0.64 m in diameter and is rated at 1850°C.

Versatile Process

Isostatic processing has found wide use for many different materials used in a variety of applications. CIP is mainly a powder consolidation process using inexpensive molds as barriers for compacting simple to complex shapes to 60%–80% densities. The selection of a "wet-bag" versus "dry-bag" method depends on the type, mix, and production lots of parts produced.

Warm isostatic pressing has found a niche in certain industries where combined pressure and low temperatures to 100°C are specified.

The HIP process is gaining momentum in the engineered ceramics field for obtaining near-net-shape and fully dense ceramics for high-performance applications. Either direct HIP or post-HIP may be selected, depending on the material or process specified.

Isotactic Stereoisomerism

A type of polymeric molecular structure containing a sequence of regularly spaced asymmetric groups arranged in like configuration in a polymer chain. Isotactic (and syndiotactic) polymers are crystallizable.

Isotherm

A line joining points of equal temperature.

Isothermal

Occurring at a single temperature. Isothermal reactions occur over a period of time and involve an input or output of heat.

Isothermal Annealing

Austenitizing a ferrous alloy and then cooling to and holding at a temperature at which austenite transforms to a relatively soft ferrite–carbide aggregate. See also *austenitizing of steel*.

Isothermal Forging

To meet the increasing forging demands of the aerospace turbine engine industry, a highly integrated, state-of-the-art isothermal forging cell has been developed. The cell has the capacity to meet 100% of the current and projected industry needs for superalloys, isothermally forged turbine engine components.

Equipment within the isothermal forging cell includes an 8000 ton clearing press, a 1204°C heat treatment furnace, and customized, computer-controlled ultrasonic inspection equipment. The cell can completely produce and process (forge, machine, heat treat, and inspect) isothermally forged engine components within one facility.

Process

A hot forging process in which a constant and uniform temperature is maintained in the workpiece during forging by heating the die to the same temperature as the workpiece. The process permits the use of extremely slow strain rates, thus taking advantage of the strain rate sensitivity of flow stress for certain alloys (e.g., titanium- and nickel-base alloys). The process is capable of producing net-shape forgings that are ready to use without machining or near-net-shape forgings that require minimal secondary machining.

Advantages of ISO

The aircraft industry is increasing its use of nickel-based powder alloys. These alloys allow for the manufacture of higher-performing aircraft components that can withstand higher temperatures and greater stress than those manufactured from more traditional wrought alloys. Powder alloys must be extruded and consolidated into a billet, which is then forged and machined to its final net shape.

In addition to these qualitative advantages of utilizing powder alloys, isothermal forging offers a number of performance and production capacity advantages. Isothermal forging is a process in which the billet and the forging dies are heated to the same temperatures. This allows for a nearer net-shape forging, thus reducing the amount of material necessary to produce the part as always reducing machining requirements. A reduction in machining requirements reduces materials waste and improves throughput.

The use of an ISO cell has resulted in a significant increase in productivity where the average cycle time has been reduced by 30%–40%, with machining times reduced by as much as 50%. Forging press setup time has been reduced by up to 50% and downtime is less than 5%. All total, these enhancements to productivity have resulted in significant unit cost reductions.

Isothermal Transformation

A change in phase that takes place at a constant temperature. The time required for transformation to be completed and, in some instances, the time delay before transformation begins depends on the amount of supercooling below (or superheating above) the equilibrium temperature for the same transformation.

Isothermal Transformation Diagram

A diagram indicating the phase changes that occur in a steel quenched from the austenitic state to various temperatures. See *steel*. These diagrams are also termed time, temperature, transformation diagrams for obvious reasons or "S" curves because of the shape of the transformation boundaries in some of the diagrams. The concept of the diagram is that it indicates the changes undergone by a steel after it has been quenched to the temperature in question and held at that level until all transformation is complete or equilibrium is reached. Transformation to ferrite/carbide structures, including bainites, is diffusion controlled and hence time dependent, so the diagram indicates the progress of transformation to these phases as time elapses. See also *austenite*. Transformation to martensite is, effectively, an instantaneous shear process, not time controlled but temperature dependent. The diagram therefore indicates the percentage transformation to martensite at any temperature. As noted, the diagram assumes instantaneous quenching from the austenitic region to the temperature in question. Such idealized conditions are not met in commercial practice where bulk components cannot be instantaneously quenched and where cooling is continuous rather than isothermal. Consequently, the diagrams have to be used with some caution. However, if it is possible to superimpose on the diagram, curves correspond to the cooling of, for example, the exterior and center of a particular size bar. Such diagrams, termed continuous cooling transformation diagrams will give a good indication of the structure and properties likely to be achieved. The basic diagram is also of value in devising treatments such as martempering, austempering, and ausforming in which the steel is rapidly cooled to an intermediate temperature, held for a period insufficient for transformation to commence, and then cooled to ambient.

Isotone

One of two or more atoms that display a constant difference $A–Z$ between their mass number A and their atomic number Z. Thus, despite differences in the total number of nuclear constituents, the number of neutrons in the nuclei of isotones is the same. See also *nuclear structure*.

Isotope

One of two or more nuclidic species of an element having an identical number of protons (Z) in the nucleus, but a different number of neutrons (N). Isotopes differ in mass, but chemically are the same element. See also *nuclear structure*.

Isotropic

Having uniform properties in all directions. The measured properties of an isotropic material are independent of the axis of testing.

Isotropy

The condition of having the same values of properties in all directions.

Item

(1) An object or quantity of material on which a set of observations can be made. (2) Been observed value or test result obtained from an object or quantity of material.

Ivory

The material that composes the tusks and teeth of the elephant. It is employed mostly for ornamental parts, art objects, and piano keys, although the latter are now usually a plastic which does not yellow. The color is the characteristic ivory white, which yellows with age. The specific gravity is 1.87. The best grades are from the heavy tusks weighing more than 55 lb (25 kg), sometimes 6 ft (1.8 m) long. The ivory from animals long dead is a gray color and inferior. The softer ivory from elephants living in the highlands is more valuable than the hard and more brittle ivory of the low marshes. The west coast of Africa, India, and southern Asia are the chief sources of ivory. The tusks of the hippopotamus, walrus, and other animals, as well as the fossil mammoth of Siberia, also furnish ivory, although of inferior grades. *Odontolite* is ivory from fossil mammoths of Russia. Ivory can be sawed readily and is made into thin veneers for ornamental uses. It takes a fine polish. Artificial ivory is usually celluloid or synthetic resins.

Izod Test

A type of impact test in which a V-notched specimen, mounted vertically, is subjected to a sudden blow delivered by the weight at the end of a pendulum arm. The energy required to break off the free end is a measure of the impact strength or toughness of the material. Contrast with *Charpy test*.

J

Jacquet Polishing

Electrolytic polishing particularly for metallographic examination. See *electrochemistry*.

Jam Nut

See *lock nut*.

Jaw Crusher

A machine for the primary disintegration of metal pieces, ores, or agglomerates into coarse powder. See also *crushing*.

Jeffries' Method

A method for determining grain size based on counting grains in a prescribed area.

Jet Molding

A processing technique for plastics characterized by the fact that most of the heat is applied to the material as it passes through the nozzle or jet, rather than in a heating cylinder, as is done in conventional processes.

Jet Pulverizer

A machine that comminutes metal pieces, ores, or agglomerates by means of pressurized air or steam injected into a chamber.

Jet Test

Various laboratory tests in which a jet of liquid impinges on the surface of the material in question to determine its corrosion characteristics under flowing conditions. See *B.N.F. jet impingement test*.

Jetting (Plastics)

The turbulent flow of resin from an undersized gate or thin section into a thicker mold cavity, as opposed to the laminar flow of material progressing radially from a gate to the extremities of the cavity.

Jetting (Welding)

The flow process and expulsion of the metal surface that takes place during *explosion welding*.

Jewel Bearing

A bearing made of diamond, sapphire, or a hard substitute metal.

Jewelers Rouge

Red iron oxide, hematite, Fe_2O_3, in fine powder form used as an abrasive and polish. It is now little used for metallography as it smears rather than cutting cleanly like diamond paste.

Jewelry Alloys

An indefinite term that refers to the casting alloys used for novelties to be plated and to the copper and other nonferrous sheet and strip alloys for stamped and turned articles. The base-metal jewelry industry, as distinct from the precious-metal industry, produces costume jewelry, trinkets, souvenirs, premium and trade goods, clothing accessories known as notions, and low-cost religious goods such as medals and statuettes. The soft white alloys of this type are not now as important as they were before the advent of plastics, and articles now produced in metals are likely to be made from standard brasses, nickel brasses, and copper–nickel alloys, but laminated and composite sheet metals are much used to eliminate plating. With the base metals clad or plated with gold, they are called *gold-filled metal* if the gold alloy is 10-karat or above and the amount used is at least 5% of the total weight. When the coating is less than 5%, it is called *rolled-gold plate*. Formally, a great variety of trade names were used for jewelry alloys. *Platinoid* was a nickel–silver–tungsten alloy with a small amount of tungsten added to the melt in the form of phosphor tungsten. *Proplatinum*, another substitute for platinum, was a nickel–silver–bismuth alloy. *Nuremberg gold*, with a nontarnishing golden color, contained 90% copper, 7.5% aluminum, and 2.5% gold.

Jig

A mechanism for holding a part and guiding the tool during machining or assembly operation.

Jig Boring

Boring with a single-point tool where the work is positioned upon a table that can be located so as to bring any desired part of the work under the tool. Thus, holes can be accurately spaced. This type of boring can be done on milling machines or jig borers.

Jig Grinding

Analogous to *jig boring*, where the holes are ground rather than machined.

J-Integral

A mathematical expression; a line or surface integral that encloses the crack front from one crack surface to the other, used to characterize the *fracture toughness* of a material having appreciable plasticity before fracture. The *J-integral* eliminates the need to describe the

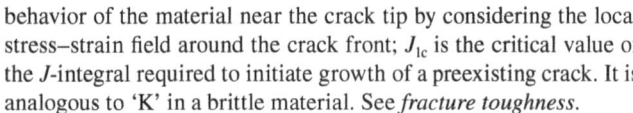

behavior of the material near the crack tip by considering the local stress–strain field around the crack front; J_{1c} is the critical value of the *J*-integral required to initiate growth of a preexisting crack. It is analogous to 'K' in a brittle material. See *fracture toughness*.

Jog

A step in a dislocation produced when the dislocation traversed another.

Joggle

An offset in a flat plain consisting of two parallel bends at the same angle but in opposite directions.

Johnson Noise

See *thermal noise*.

Joint

The location where two or more members are to be or have been fastened together mechanically or by welding, brazing, soldering, or adhesive bonding.

Joint Buildup Sequence

See *buildup sequence*.

Joint Clearance

The distance between the faying surfaces of a joint. In brazing, this distance is referred to as that which is present before brazing, at the brazing temperature, or after brazing is completed.

Joint-Conditioning Time

In adhesive bonding, the time interval between the removal of the joint from the conditions of heat or pressure, or both, used to accomplish bonding and the attainment of approximately maximum bond strength. See also *curing time*, *drying time*, and *setting time*.

Joint Design

The joint geometry together with the required dimensions of the welded joint.

Joint Efficiency

The strength of a welded joint expressed as a percentage of the strength of the unwelded base metal.

Joint Geometry

The shape and dimensions of a joint in cross section prior to welding.

Joint Penetration

The minimum depth to which a groove or flange weld extends from its face into the joint, exclusive of reinforcement. Joint penetration may include *root penetration*.

Joint Severity (of Weld)

See *weldability*.

Jolt Ramming

Packing sand in a mold by raising and dropping the sand, pattern, and flask on a table. Jolt-type, jolt-squeezer, jarring, and jolt-ramming machines are using this principle. Also called a jar ramming.

Jolt-Squeezer Machine

A combination machine that employs a jolt action followed by a squeezing action to compact the sand around the pattern.

Jominy (End Quench) Test

(1) A test for assessing the hardenability of steel in which a standard bar specimen is heated to the austenitic range and then one end is quenched by a jet of water. This results in a progressive variation in cooling rate along the bar. Following cooling the bar is tested for hardness along its length to determine the extent of hardening, and it may also be metallographically examined to establish the limit of the martensitic and bainitic zones. (2) A laboratory test for determining the hardenability of steel that involves heating the test specimen to the proper *austenitizing* temperature and then transferring it to a quenching fixture so designed that the specimen is held vertically 12.7 mm (0.5 in.) above an opening through which a column of water can be directed against the bottom face of the specimen. While the bottom end is being quenched by the column of water, the opposite end is cooling slowly in air, and intermediate positions along the specimen are cooling at intermediate rates. After the specimen has been quenched, parallel flats 180° apart are ground 0.38 mm (0.015 in.) deep on the cylindrical surface. Rockwell C hardness is measured at intervals of 1/16 in. (1.6 mm) for alloy steels and 1/32 in. (0.8 mm) for carbon steels, starting from the water-quenched end. A typical plot of these hardness values and their positions on the test bar indicates the relation between hardness and cooling rate, which in effect is the hardenability of the steel. Also referred to as *end-quench hardenability test*.

Joule

The SI unit of energy, symbol J, quantity of heat or work done. One joule is the work done when the point of application of a force of 1 N is displaced by 1 m. Also, in other terms, 1 J = 0.728 ft lb = 1 W s = 0.239 cal.

Joule Effect

(1) The heat resulting from the passage of an electric current. It is expressed as $J = I^2RT$, where I is the current in amps, R is the resistance of the conductor in ohms, and T is the time in seconds. (2) See *magnetostriction*.

Joule–Thompson Effect

A change in temperature in a gas undergoing Joule–Thompson expansion. See also *Joule–Thompson expansion*.

Joule–Thompson Expansion

The adiabatic, irreversible expansion of a gas flowing through a porous plug or partially open file. See also *Joule–Thompson effect.*

Journal

The part of a shaft or axle that rotates or oscillates relative to a radical bearing. A journal was part of a larger unit, for example, a crankshaft or lineshaft, and it is preferred that the term shaft be kept for the whole unit.

Journal Bearing

A sliding-type bearing in which a journal rotates or oscillates relative to its housing. A full journal bearing extends 360°, but partial bearings may extend, for example, over 180° or 120°. See also the term *solid lubrication.*

Jute

A fiber employed for making burlap, sacks, cordage, ropes, and upholstery fabrics. It is obtained from several plants in India where it is most widely cultivated, growing in a hot, steamy climate. Also fiber from Brazil is called *juta indiana,* or *Indian jute.* Most of the commercial jute comes from Bengal. The plant grows in tall, slender stalks like hemp, and the fiber is obtained by retting and cleaning. The fibers are long, soft, and lustrous, but are not as strong as hemp. It also loses its strength when damp, but is widely used because of its low cost and because of the ease with which it can be spun. Its chemical composition is intermediate between those of hemp and kapok. It contains 60% alpha cellulose, 15% pentosan, 13% lignin, and 4.5% *uronic anhydride,* which is also a constituent of kapok. The butts, or short ends of the stalks, and the rough fibers are used for paper stock. *Jute paper,* used for cement bags, is a strong paper made of these fibers usually mixed with old rope and old burlap in the pulping. It is tan in color. *Jute fiber* is also used for machine packings.

J

K

K

Temperature in Kelvin.

K, K_{1C}, etc.

K is the stress-intensity factor defining the relationship, at the crack tip, between stress and crack size. The numerical subscripts identify the mode of cracking, for example, "1" indicates the common case of the crack being pulled open. The subscript "C" indicates a critical level of K at which a crack will extend. See *fracture toughness*.

Kaldo Process

One of the first oxygen steelmaking processes. In the process, a stream of oxygen impinges on the surface of molten steel in a rotating converter vessel. The vessel has a basic lining with a lime flux, and the process removes both carbon and phosphorus from high-phosphorus pig iron. The use of oxygen is of choice and nitrogen pick up associated with air lancing.

Kaolin (China Clay)

The two terms are used interchangeably to describe a type of clay that fires to a white color. The name kaolin comes from the two Chinese words kao–ling, meaning high ridge, and was originally a local term used to describe the region from which the clay was obtained.

Kaolin ($Al_2O_3 \cdot 2SiO_2 \cdot 2H_2O$) usually contains less than 2% alkalies and smaller quantities of iron, lime, magnesium, and titanium. Because of its purity, kaolin has a high fusion point and is the most refractory of all clays. Lone kaolins are widely used in casting sanitary ware, ceramics, and refractories.

Georgia china clay is one of the most uniform kaolins to be found. Generally speaking, there are two types of Georgia-sourced kaolin, both of which are widely used for casting and other processes. One type imparts unusually high strength and plasticity and is used for both casting and jiggering where a high degree of workability is required. The other type typically is a fractionated controlled particle size clay that also behaves well in casting, dries uniformly, and reduces cracking of ware.

Calcined Kaolin

This commercial product is made from a specially prepared kaolin that is low in iron and alkalies. Analyses show the calcined product to be principally mullite ($3Al_2O_3 \cdot 2SiO_2$) in association with an amorphous siliceous material. In fact, of the alumina present, 96% is converted to mullite. Iron content is not only low but in such a state of oxidation as to facilitate solid solution with the alumina.

Outstanding properties of refractories containing this calcined kaolin are high refractoriness and retention of shape under load; a high resistance to corrosion by slags, glasses, and glaze or enamel frits; resistance to thermal shock; and high mechanical strength.

It is being used in thermal shock bodies, refractories subjected to reducing atmospheres, kiln furniture compositions, thermal insulation bodies, low-expansion bodies, permeable ceramic compositions, high-temperature castables, and investment molds for precision casting; as a placing medium, as a kiln wash, and as gripping sand for high-tension insulators; and in many other special refractory applications.

The *Cornwall kaolin* of England and the *Limoges kaolin* of France are the best known. English china clays contain little or no iron oxide, and the yellow clays contain only organic materials that can be bleached out. The best grades of English clay are used for coating and filling paper. Cornish clay, known as *china stone*, is used for the best grades of porcelain glazes. Cheaper grades of kaolin, called *mica clay*, are used for earthenware glazes and as an absorbent and oil purifying. *Kaolin fiber* of extreme fineness, with average diameter of 118 μin. (3 μm), is made from kaolin containing about 46% alumina, 51% silica, and 3% iron and titanium oxide. It withstands continuous temperatures to 2000°F (1093°C). Kaowool, of Babcock and Wilcox, is kaolin fiber in the form of insulating blankets, and *kaolin paper* is made from the fibers compressed to thicknesses up to 0.08 in. (0.20 cm) with or without a binder. It withstands temperatures of up to 2000°F (1093°C) and is used for filters, separators, and gaskets.

Kapok

A silky fiber obtained from the seed pods of the silk-cotton trees, now grown in most tropical countries. It is employed for insulation and fine padding work. It is extremely light and resilient. The chemical constituents are the same as those of jute, but the proportions are different. Kapok is low in alpha cellulose, 43%, and high in pentosans, 24%; lignin, 15%; and uronic anhydride, 6.6%. Most of the commercial kapok normally comes from Java.

Karat

A unit for designating the fineness of gold in an alloy. In this system, 24 karat (24k) is 1000 fine or pure gold. The most popular jewelry golds are as follows:

Karat Designation	Gold Content
24k	100% Au (99.95% min)
18k	18/24th, or 75% Au
14k	14/24th, or 58.33% Au
10k	10/24th, or 41.67% Au

Keel Block

A standard test casting, for steel and other shrinkage alloys, consisting of a rectangular bar that resembles the keel of a boat, attached to the bottom of a large riser, or shrinkhead. Keel blocks that have only one bar are often called Y blocks; keel blocks having two bars, double keel blocks. Test specimens are machined from the rectangular bar, and the shrinkhead is discarded.

Keeper

A piece of magnetically soft iron set across the poles of a permanent magnet to close the magnetic circuit and hence preserve the magnetism.

Keller's Etch

2.5 mL nitric acid, 1.5 mL hydrochloric acid, 1 mL hydrofluoric acid, and 100 mL water. Used for aluminum alloys.

Kelvin

A scale of absolute temperatures in which zero is approximately −273.16°C (−459.69°F). The color temperature of light is measured in kelvins. See *absolute temperature*.

Kelvin Effect

The electromotive force produced by a difference of temperature between two sites of a conductor and the release or absorption of heat associated with an electrical current flowing between two sites on the conductor that are at different temperatures. See also *thermoelectric effect*.

Kennedy Friction

Friction under conditions of macroscopic relative motion between two bodies. This term is sometimes used as a synonym for *kinetic coefficient of friction*; however, it can also be used merely to indicate that the type of friction being indicated is associated with macroscopic motion rather than static conditions.

Kerf

The width of a saw cut or other gap left when material has been cut away. Also *the cut face*.

Kerosene

Originally the name of illuminating oil distilled from coal, and also called *coal oil*. It is now a light, oily liquid obtained in the fractional distillation of petroleum. It is distilled off after gasoline between the limits of 345°F and 550°F (174°C and 288°C). It is a hydrocarbon with a composition $C_{10}H_{22}$ to $C_{16}H_{34}$, with a specific gravity between 0.747 and 0.775. Kerosene is employed for illuminating and heating purposes, as a fuel in internal combustion engines, and for turbine jet fuels. The heaviest distillate, known as *range oil*, is sufficiently volatile to burn freely in the wick of a heating range, but not as volatile as to be explosive. It is nearly free from odor and smoke. *Deodorized kerosene*, used in insect sprays, is a kerosene highly refined by treatment with activated earth or activated carbon. *Fialasol* is a *nitrated kerosene* used as a solvent for the scouring of wool. It is less flammable and has a slower rate of evaporation than kerosene, and it is odorless.

Kettle

A small vessel with a loose top in which small quantities of low-melting-temperature metals are melted.

Kevlar

An organic polymer composed of aromatic polyamides (*aramids*) having a paratype orientation (parallel chain with bonds extending from each aromatic nucleus). Often used as a reinforcing fiber. It has high tensile strength and elastic modulus, making it particularly suitable for reinforcement of composites. The term is also loosely used of the composite material containing the Kevlar fiber.

Key

(1) In a metallurgical context, this term refers to the ability of a surface to provide good adhesion for paint or other coatings. The key may be improved by mechanical means such as abrading or shot blasting or by chemical treatment. (2) A piece of material set longitudinally, half and half, and the interface between, typically, a shaft and a body mounted on it. It is inserted in the keyways in the two components to maintain location or provide drive.

Keyhole

A technique of welding in which a concentrated heat source, such as a plasma arc, penetrates completely through a workpiece forming a hole at the leading edge of the molten weld metal. As the heat source progresses, the molten metal fills in behind the hole to form the weld bead.

Keyhole Notch

A notch, in an impact test specimen, which has a narrow neck at the surface and terminates within the material and a round hole. The fracture is intended to initiate at the hole opposite the neck.

Keyhole Specimen

A type of specimen containing a hole-and-slot notch, shaped like a keyhole, usually used in impact bend tests. See also *Charpy test* and *Izod test*.

k-Factor

The ratio between the unknown and standard x-ray intensities used in quantitative analyses.

Kikuchi Lines

Light and dark lines superimposed on the background of a single-crystal electron diffraction pattern caused by diffraction of diffusely scattered electrons within the crystal: the pattern provides structural information on the crystal.

Kill/Killing

See *killed steel*.

Killed Spirits

Saturated solution of zinc chloride used as a soldering flux.

Killed Steel

(1) Usually, a steel that has been fully deoxidized with a strong deoxidizer such as silicon, manganese, or aluminum preventing any

reaction between carbon and oxygen during solidification. (2) Less commonly than (1), a steel that has been lightly worked to eliminate the pronounced yield point that gives rise to Lüders lines. See *capped steel.*

Kiln

(1) Any vessel or enclosed space in which some heating process is carried out. Choice between "kiln," "oven," and "furnace" is more a matter of preference in the industry concern rather than any consistent technical difference. It is used for baking, drying, or burning firebrick or refractories or for calcining ores or other substances.

Kiloelectron Volt (keV)

A unit of energy usually associated with individual particles. The energy gained by an electron when accelerated across 1000 V.

Kinematic Viscosity

See *viscosity.*

Kinematic Wear Marks

In ball bearings, a series of short curved marks on the surface of a bearing race due to the kinematic action of embedded particles or asperities rolling and spinning at the ball or roller contact points. The length and curvature of these marks depend on the degree of spinning and on the distance from the spinning axis of the rolling element.

Kinetic Coefficient of Friction

The *coefficient of friction* under conditions of macroscopic relative motion between two bodies.

Kinetic Energy

The energy that a body possesses because of its motion; in classical mechanics, equal to one-half of the body's mass times the square of its speed.

Kinetic Theory

At a temperature of absolute zero, atoms and molecules are at rest. As the temperature rises, the atoms and molecules absorb energy in which they retain as kinetic energy associated with their vibration. As the energy increases, the atoms and molecules vibrate increasingly until at some stage they are able to flow over each other, that is, the material becomes liquid. Further increase in energy and vibrations allow individual atoms or molecules to escape from the bulk, that is, the material becomes a gas. The impact of the vibrating atoms on the containment is manifested as pressure.

Kingsbury Bearing

See *tilting-pad bearing.*

Kink Band (Deformation)

In polycrystalline materials, a volume of crystal that has rotated physically to accommodate differential deformation between adjoining parts of a grain while the band itself has deformed homogeneously. This occurs by regular bending of the slip lamellae along the boundaries of the band. Deformation bands associated with slip on multiple parallel crystallographic planes.

Kirkendall Voids

The formation of voids by diffusion across the interface between two different materials, with the material having a greater diffusion rate diffusing into the other.

K_{ISCC}

Abbreviation for the critical value of the plane strain *stress-intensity factor* that will produce crack propagation for *stress-corrosion cracking* of a given material in a given environment.

Kish

Free graphite that forms in molten hypereutectic cast iron as it cools. In castings, the kish may segregate toward the cope surface, where it lodges at or immediately beneath the casting surface and it is very weakening.

Klystron

An evacuated electron-beam tube in which an initial velocity modulation imparted to electrons in the beam subsequently results in density modulation of the beam. This device is used as an amplifier or oscillator in the microwave region.

Knee (Resistance Welding)

The lower arm supporting structure in a resistance welding machine.

Knife Coating

A method of coating a substrate (usually paper or fabric) in which the substrate, in the form of a continuous moving web, is coated with a material, the thickness of which is controlled by an adjustable knife or bar set at a suitable angle to the substrate.

Knife-Line Attack

Intergranular corrosion of an alloy, usually stabilized stainless steel, along a line adjoining or in contact with a weld after heating into the sensitization temperature range. Local attack in the heat-affected zone of welds in some austenitic stainless steels. Similar to weld decay except that it is located in the temperature zone that closely approaches the solidus.

Knitted and Woven Metals

Although they are commonly referred to in some context and are used for some of the same applications, there is a considerable difference between knitted and woven metals. As their name implies, knitted metals are knitted into a mesh structure in much the same way as stockings or sweaters. The structure of woven wire, on the other hand, is usually simpler, consisting of interwoven strands of wire. Whereas weaving usually produces a symmetrical mesh, usually with square openings and parallel wires, knitting produces an asymmetrical mesh of interlocking loops.

Knitted Metals

An important advantage of knitting is that it produces a mesh of interlocking loops, each of which acts as a small spring and provides resiliency. Because of this resiliency, knitted metals are usually able to withstand greater loads and deflections without being permanently deformed.

Knitted wire also has both a large surface area and a high percentage of free space. Thus, knitting permits construction of a mesh using wire with a maximum surface area and with interstices of almost any desirable size, regardless of wire diameter. Fine wire, for example, has been knitted into a mesh with as few as 3–5 openings to the inch.

The free volume of knitted wire can be controlled between 50% and 98%, depending on the interstice size, and regardless of the wire size used. Even when the wires are widely spaced to produce a potential volume of 98%, the structure retains its shape; a similar spacing with woven wire would result in a shape that would be almost impossible to handle.

Another important property of knitted metals is their ability to maintain their dimensions during temperature cycling. When the meshes are slightly stretched in every direction and expansion occurs, the sides of the loops are merely forced closer together without changing the overall dimensions of the plane of the surface.

Knitted metals can be produced in wire dimensions of 0.01–0.6 mm in such materials as steel, copper, brass, aluminum, stainless steel, and various nickel alloys including Monel. In fact, almost any metal that can be drawn into wire can be knitted. Thus, knitted parts can be made in a wide range of strength, corrosion resistance, wear resistance, electrical shielding, and heat resistance properties.

The asymmetrical mesh of knitted parts is advantageous for electronic shielding because the continuous-loop structure apparently causes induced currents to cancel themselves. Apart from this application, however, the biggest use for knitted wire is to remove one material from another. Thus, it can be used to separate two phases of the same material, to separate two admissible liquids, or to separate a solid from a liquid or a gas. The degree of separation can be controlled by varying the compression used to form or shape the knitted part and by controlling the size and shape of the wire and the size of the loops.

A good example of this kind of application is a mist eliminator in which the knitted mesh separates the liquid phase from the gas phase. Gases bearing droplets from such processes as distillation, evaporation, scrubbing, cleaning, or absorption are passed through built-up layers of knitted fabrics of up to 152 mm thick. The droplets collect on the loops and slowly run to the bottom where they accumulate. Liquid particles as small as 5–20 μm can be separated with efficiencies as high as 98%–100%. Although small eliminators are usually made in one piece, eliminators can be made up to 8.4 m in diameter by building up layers of crimped mesh.

Knitted metals are also used in many other applications where their special structure and properties are useful. Fuel line filters, for example, of knitted wire are resilient and do not require precise machining for sidewall fit. Gaskets of knitted wire have excellent conductivity and sufficient resiliency to produce tight joints on uneven surfaces, thereby preventing radio frequency leakage. Heat dissipation sleeves of knitted metal for subminiature glass tube envelopes provide high cooling efficiency. Knitted metals also provide good shock and vibration control when used as mountings for airborne and industrial equipment.

Woven Metals

Woven wire cloth is used in a wide range of applications for grading materials, filtering, straining, washing, guarding, reinforcing, and decorating. A variety of meshes in different materials and sizes is available to meet these applications.

Like knitted wire, woven wire cloth can be produced in almost any metal that can be drawn into wire, including carbon and stainless steel, copper, brass, bronze, nickel, Monel, Inconel, and aluminum. Plain steel wire cloth is one of the most economical and generally used materials. However, it may require a protective coating to prevent rusting.

Corrosion can be prevented by using tinned or galvanized wire; tinned wire is preferred for handling food products, galvanized for all other applications. Where required, the cloth can also be provided with a protective electroplate of cadmium, chromium, or tin. Phosphate coatings and paints can also be used.

Where severe corrosion conditions are encountered, the weave can be made from such materials as stainless steel, Monel, phosphor bronze, and silicon bronze; some of these materials are available only in limited sizes. Abrasive-resistant steels are available for applications where abrasive materials have to be handled. Many grades of stainless can also be used to withstand high temperatures, for example, 347 stainless (up to 760°C), 309 and 310 (up to 871°C), and 310 stainless as well as Inconel and nichrome (up to 1003°C).

Woven wire cloth is widely used for filtering and straining, particularly in automotive and aviation applications for carburetors, air screens, and oil and fuel strainers. They are also used to grade materials and for wire baskets, insect screening, and safety guards.

Knitted Fabrics

Fabrics produced by inner looping chains of yarn.

Knock

In a spark ignition engine, uneven burning of the air/fuel charge that causes violent, explosive combustion and an audible metallic hammering noise. Knock results from premature ignition of the last part of the charge to burn.

Knockout

(1) Removal of sand cores from a casting. (2) Jarring of an investment casting mold to remove the casting and investment from the flask. (3) A mechanism for freeing formed parts from a die used for stamping, blanking, drawing, forging, or heading operations. (4) A partially pierced hole in a sheet metal part, where the slug remains in the hole and can be forced out by hand if a hole is needed.

Knockout Mark

A small protrusion, such as a button or ring of flash, resulting from depression of the *knockout pin* from the forging pressure or the entrance of metal between the knockout pin and the die.

Knockout Pin

A power-operated plunger installed in a die to aid removal of the finished forging.

Knockout Punch

A punch used for ejecting powder compacts.

Knoop Hardness Number (HK)

A number related to the applied load and to the projected area of the permanent impression made by a rhombic-based pyramidal diamond indenter having included edge angles of 172°30' and 130°0' computed from the equation

$$HK = \frac{P}{0.07028d^2}$$

where
   P is the applied load, kgf
   d is the long diagonal of the impression, mm

In reporting Knoop hardness numbers, the test load is stated. See *hardness test.*

Knoop Hardness Test

An indentation hardness test using calibrated machines to force a rhombic-based pyramidal diamond indenter having specified edge angles, under specified conditions, into the surface of the material under test and to measure the long diagonal after removal of the load.

Knuckle Area

In reinforced plastics, the area of transition between sections of different geometry in a filament-wound part, for example, where the skin joins the cylinder of the pressure vessel. Also called Y joint.

Knuckle-Lever Press

A heavy short-stroke press in which the slide is directly actuated by a single toggle joint that is opened and closed by a connection and crack. It is used for embossing, coining, sizing, heading, swaging, and extruding.

Knurling

Impressing a design into a metallic surface. A pattern of deep indentations, typically a diamond grid and similar in appearance to a file face, formed on a surface, usually cylindrical, to improve grip. The effect is achieved by forcing hard rollers, machined to the required pattern, against the rotating surface.

K Radiation

Characteristic x-rays produced by an atom or ion when a vacancy in the K shell is filled by an electron from another shell.

Kraft Process

A wood-pulping process in which sodium sulfate is used in the caustic soda pulp-digestion liquor. Also called kraft pulping or sulfate pulping.

Kroll Process

A process for the production of metallic titanium sponge by reduction of titanium tetrachloride with a more active metal, such as magnesium or sodium. The sponges are further processed to granules or powder.

K-Series

The set of characteristic x-ray wavelengths making up K-radiation for the various elements.

K Shell

The innermost shell of electrons surrounding the atomic nucleus, having electrons characterized by the principal quantum number 1.

Krypton

An element, one of the inert gases.

Kyanite

A natural aluminum silicate, $Al_2O_3 \cdot SiO_2$, used as a refractory especially for linings of glass furnaces and furnaces for nonferrous metals. Molded or cast ceramic parts having nearly 0 thermal expansion up to 2300°F (1260°C). It is a common material but occurs disseminated with other minerals and is found in commercial quantities and grades in only a few places. Most of the world production has been in eastern India, but high-grade kyanite is now obtained from Kenya. It is also mined in North Carolina, Georgia, Virginia, and California. Low-grade kyanite is used in glassmaking as a source of alumina to increase strength and chemical and heat resistance. Aluminosilicate materials are widely distributed in nature combined in complex forms, and the aluminosilicate extracted from them is called *synthetic.* Kyanite powder, produced from Florida beach sands, has round, single-crystal grains of kyanite and sillimanite in nearly equal proportions. The beach sands are screened and graded and marketed as granules of various sizes for compacting and sintering into refractory ceramic parts.

L

L_{10}-Life

See *rating life*.

Labile

Unstable.

Laboratory Sample

A sample, intended for testing or analysis, prepared from a gross sample or otherwise obtained; the laboratory sample must retain the composition of the gross sample. Reduction in particle size is often necessary in the course of reducing the quantity.

Lack of Fusion (of Weld)

The absence of the intended fusion bond in which fusion is less than complete.

Lack of Penetration (LOP)

A condition in a welded joint in which joint penetration is less than that specified.

Lack of Resin Fill-Out

In reinforced plastics, a condition in which an area contains reinforcement not wetted with a sufficient quantity of resin. This condition usually occurs at the edge of a pultrusion.

Lacquer

(1) A coating formulation based on thermoplastic film-forming material dissolved in organic solvent. The coating dries primarily by evaporation of the solvent. Typical lacquers include those based on lac, nitrocellulose, and other cellulose derivatives, vinyl resins, acrylic resins, and so forth. (2) In lubrication, a deposit resulting from the oxidation and/or polymerization of fuels and lubricants when exposed to high temperatures. Softer deposits are described as varnishes or gums.

Ladder Polymer

A polymer with two polymer chains cross-linked at intervals.

Ladle

A vessel for transporting molten metal from one stage to the next, particularly from the melting furnace to the mold. It may range from a handheld "spoon" to very large mechanized or crane-carried vessels holding many tons of metal. Types include hand, bull, crane, bottom-pour, holding, teapot, shank, and lip-pour.

Ladle Addition/Treatment/Process

Processes for the final refinement and composition control of metal before it is cast, particularly treatment in a ladle or holding bath with no additional heating.

Ladle Brick

Refractory brick suitable for lining ladles used to hold molten metal.

Ladle Coating

The material used to coat metal ladles to prevent iron pickup in aluminum alloys. The material can consist of sodium silicate, iron oxide, and water, applied to the ladle when it is heated.

Ladle Metallurgy

Degassing processes for steel carried out in a *ladle*.

Ladle Preheating

The process of heating a ladle prior to the addition of molten metal. This procedure reduces metal heat loss and eliminates moisture-steam safety hazards.

Lagging

Thermal insulation particularly when it is applied on the exterior of the vessel being protected rather than on the interior and in contact with the reaction environment.

Lamella

The basic morphological unit of a crystalline polymer, usually ribbonlike or platelike in shape. Generally, about 10 mm thick, 1 μm long, and 0.1 μm wide, if ribbonlike.

Lamellae

Multiple thin plates stacked as laminations.

Lamellar Corrosion/Attack

The loss of material, as layers, from a metal surface. Corrosion proceeds along a plane just below and parallel with the surface and the release layer is substantially uncorroded. The term *exfoliating corrosion* is usually applied to cases where the sheets of material released are entirely or predominantly corrosion product. See *exfoliation corrosion*.

L

Lamellar Cracking/Tearing Cracking

This cracking of rolled products such as plate or bar, can lie beneath and parallel with the surface. It usually follows planes of weakness, particularly areas of nonmetallic inclusions, introduced at the original casting stage and deformed to a planar form during rolling. The planes of weakness have little effect on the longitudinal or transverse strength but considerably reduce the strength in the through-thickness direction. Loads acting in this direction are imposed when a weld is applied to the surface of a plate or bar. The initial defect can develop as a result solely of the stresses introduced by the welding operation and in this case they will often be confined to the parent metal beneath the weld. However, when a service load is applied the cracks are likely to propagate leading to failure. See also *cold cracking*, *hot cracking*, and *stress-relief cracking*.

Lamellar Thickness

A characteristic morphological parameter in plastics, usually estimated from x-ray studies or electron microscopy, which is usually 10–50 nm. The average thickness of lamellae in a specimen. See also *lamella*.

Lamina

A single ply or layer in a laminate, which is made up of a series of layers.

Laminar Flow

The flow of plastic resin in a mold, characterized by solidification of the first layer to contact the mold surface, which then acts as an insulating tube through which material flows to fill the remainder of the cavity. This type of flow is essential to duplication of the mold surface. The movement of this fluid stream is such that no lateral intermixing occurs, effectively *streamline flow*.

Laminate

To unite laminate with a bonding material, usually with pressure and heat (normally refers to flat sheets, but also includes rods and tubes); a product made by such bonding. Two or more layers of material bonded together. The term can apply to preformed layers joined by adhesives or by heat and pressure. The term also applies to composites of plastic films with other films, or with foil or paper, even though they have been made by spread coating or by extrusion coating. A reinforced laminate usually refers to superimposed layers of resin-impregnated or resin-coated fabrics or fibrous reinforcements that have been bonded, especially by heat and pressure. Products produced with little or no pressure, such as hand layups, filament-wound structures, and spray-ups, are sometimes called contact-pressure laminate. A single resin-impregnated sheet of paper, fabric, or glass mat is not considered a laminate. Such a single-sheet construction may be called a *lamina*. See also *bidirectional laminate*, *cross laminate*, *parallel laminate*, and *unidirectional laminate*.

Laminate Coordinates

A reference coordinate system used to describe the properties of a laminate, generally in the direction of principal axes, when they exist.

Laminate Orientation

The geometric configuration of a cross-plied composite laminate with regard to the angles of cross-plying, the number of laminae at each angle, and the exact sequence of laminae layup.

Laminate Ply

One fabric-resin or fiber-resin layer (of a product) that is bonded to adjacent layers in the curing process.

Laminate Void

The absence of resin in an area that normally contains resin.

Laminated Glass

Two or more glasses that are fused together to produce a material with properties that would have been either difficult or impossible to obtain in a single glass. There are many multistep lamination processes. Usually, a glass is formed during one process and is subsequently laminated with another glass during a separate process that involves melting the two glasses in separate furnaces and then passing them through a cladding delivery system.

Laminated Plastics

Laminated plastics are a special form of polymer matrix composites (PMC) consisting of layers of reinforcing materials that have been impregnated with thermosetting resins, bonded together, and cured under heat and pressure. The cured laminates, called high-pressure laminates, are produced to more than 70 standard grades.

Laminated plastics are available in sheet, tube, and rod shapes that are cut or machined for various end uses. The same base materials are also used in molded-laminated and molded-macerated parts. The molded-laminated method is used to produce shapes that would be uneconomical to machine from flat laminates, where production quantities are sufficient to warrant mold costs.

Strength of a molded shape is higher than that of a machined shape because the reinforcing plies are not cut as they are in a machined part. The molded-macerated method is used for similar parts that require uniform strength properties in all directions.

Other common forms of laminated plastics are composite sheet laminate that incorporate a third material bonded to one or both surfaces of the laminate. Metals most often used in composites are copper, aluminum, nickel, and steel. Copper-clad sheets (one or both sides) for printed-circuit and multilayer boards comprise the largest volume of metal composite sheet laminate. Nonmetallics include elastomers, vulcanized fiber, and cork. Composite metal/plastic materials are also produced in rods and tubes.

Vulcanized fiber is another product often classified with the laminated plastics because end uses are similar. Vulcanized fiber is made from regenerated cotton cellulose and paper, processed to form a dense material (usually in sheet form) that retains the fibrous structure. The material is tough and has good resistance to abrasion, flame, and impact.

Phenolics are the most widely used resin in laminated plastics. These low-cost resins have good mechanical and electrical properties and resistance to heat, flame, moisture, mild acids, and alkalies. Most paper- and cloth-reinforced laminates are made with phenolics.

Lamination

(1) A type of discontinuity with separation or weakness generally aligned parallel to the work surface of the metal. May be the result of pipe, blisters, seams, inclusions, or segregation elongated and made directional by working. Laminations may also occur in powder metallurgy compacts. (2) Generally, any sheets intended for assembly as a pile. More particularly, varnished (for insulation) steel sheets piled and clamped together to form the core of transformers, generator stators, and some rotors. This construction, together with the use of special steels minimizes iron loss.

Lampblack

A soot formed by the smudge process of burning oil, coal tar, resin, or other carbonaceous substances in an insufficient supply of air, the soot being allowed to settle on the walls or floors of the collecting chambers. Lampblack is practically pure carbon, but inferior grades may contain unburned oil. It is chemically the same as carbon black made from gas, but since it may contain as high as 2.5% oil, it is not generally used in rubber. The particle size is large, 2559–3937 µin. (65–100 µm), and the pH is low, 3–3.5. However, some *amorphous carbons*, made from either tar oil or crude oil by spraying the oil and air into a close retort at 3000°F (6049°C) to obtain partial combustion, are equal to carbon black for many uses. Lampblack is used in making paints, lead pencils, metal polishes, electric brush carbons, crayons, and carbon papers. Carbon brushes for use in electric machinery are made by mixing lampblack with pitch. Petroleum coke or graphite is added to import special properties. For special arc lights, a mixture of lampblack, rare-earth oxides, fluorides, and coal as a binder is extruded. The combination of rare earths determines the type of radiation emitted. It is grayish black, flaky, and granular. The color is not as intensely black as carbon black. *Lampblack oil* is a coal-tar product marketed for making lampblack.

Fine soot used in the reduction and carburization of tungsten trioxide and titanium dioxide to produce tungsten carbide and titanium carbide powder, respectively.

Lampworking

Forming glass articles from tubing and cane by heating in a gas flame.

Lance

A tube directing some fluid onto a process vessel.

Lancing

(1) A press operation in which a single-line cut is made in strip stock without producing a detached slug. Chiefly used to free metal for forming, or to cut partial contours for blank parts, particularly in progressive dies. (2) A misnomer for *oxyfuel gas cutting.*

Land (Metals)

(1) For profile-sharpened milling cutters, the relieved portion immediately behind the cutting-edge. (2) For reamers, drills, and taps, the solid section between the flutes. (3) On punches, the portion adjacent to the nose that is parallel to the axis and of maximum diameter. (4) A flat surface on component, particularly one machined to provide a seat or similar function.

Land (Plastics)

(1) The horizontal bearing surface of a semi-positive or flash mold by which excess material escapes. (2) The bearing surface along the top of the flights of a screw in a screw extruder. (3) The surface of an extrusion die parallel to the direction of melt flow. (4) The land region of a nozzle used in injection molding.

Lang's Lay

One of the more common systems of laying, that is, assembling, the individual fibers or wires forming each strand of a multi-strand rope or cable. In essence all wires and strands follow the same twist direction.

Langelier Saturation Index

An index calculated from total dissolved solids, calcium concentration, total alkalinity, pH, and solution temperature that shows a tendency of a water solution to precipitate or dissolve calcium carbonate.

Lanthanides

The rare earth metallic elements with atomic numbers 57 (lanthanum) to 71.

Lanthanum

A chemical element, lanthanum, symbol La, the second most abundant element in the rare earth group, is a metal. The naturally occurring element is made up of the isotopes and is one of the radioactive products of the fission of uranium, thorium, or plutonium. Lanthanum is the most basic of the rare earths and can be separated rapidly from other members of the rare earth series by fractional crystallization. Considerable quantities of it are separated commercially, because it is an important ingredient in glass manufacture. Lanthanum imparts a high refractive index to the glass and is used in the manufacture of expensive lenses. The metal is readily attacked in air and is rapidly converted to a white powder.

Lanthanum becomes a superconductor below about −267°C in both the hexagonal and face-centered crystal forms.

Lanthanum Oxide

Lanthanum oxide (La_2O_3) has a melting point of 2250°C, and is soluble in acids, and very slightly soluble in water. This oxide of a rare earth element occurs in monazite and bastnasite. It is marketed as the oxide or as other salts such as the oxalate, nitrate, or hydrate. It quickly absorbs water and carbon dioxide from the atmosphere.

Its chief use is as an ingredient in nonsilica, rare earth optical glass with oxides of tungsten, tantalum, and thorium. Lanthanum increases refractive index, decreases dispersion, and is also used in x-ray image intensifying screens that speed up x-ray exposure as much as 2–10 times so that diagnostic dosages may be reduced by as much as 80% with fewer retakes. It is also used in barium titanate capacitors.

Lanxide Process

A composite material formation process that involves pressureless metal infiltration into a ceramic preform. The molten metal is exposed to an oxidizing atmosphere resulting in a matrix material composed of a mixture of the oxidation reaction product and unre-acted metal.

Lanxides

A composite formed in the reaction between a molten metal and oxygen in air to some other vapor-phase oxidant. Normally, such a situation produces an unwelcome scum on the surface of the metal. However, by controlling the temperature of the molten metal and by adding traces of suitable dopant metals, an inch/cm-thick layer of a metal oxide composite can be grown on the surface of the liquid. Composites up to 10 cm thick and weighing up to 18 kg have been grown by this method, with no falling off in growth rate. Under the right conditions, a lanxide composite is considerably stronger than sintered alumina. This makes such a composite potentially useful for armor plating, rocket or jet engines, and other applications.

Lap (Composites)

(1) In filament winding, the amount of overlay between successive windings, usually intended to minimize gapping. (2) In adhesive bonding, the distance one adherend covers another adherend.

Lap (Metals)

(1) A surface imperfection, with the appearance of a seam, caused by hot metal, fins, or sharp corners being folded over and then being rolled or forged into the surface but without being welded. (2) A planar manufacturing defect, either a small cold shut from casting or a fold, produced during some working operation that has been flattened but not welded or bonded. (3) A polishing pad carrying embedded abrasives.

Lap (Shear) Test

A tensile test in the plane of the parent plates of a lap joint so that the load is carried by the full joint, as opposed to peel test.

Lap Joint

The joint formed between two overlapping components such as plate or sheet. The joint may be affected by riveting, brazing, or various welding techniques, including spot welding through the interface or fillet welding the internal corner between edge and face.

Lapping

A finishing operation using fine abrasive grits loaded into a lapping material such as cast iron. Lapping provides major refinements in the workpiece, including extreme accuracy of dimension, correction of minor imperfections of shape, refinement of surface finish, and close fit between mating surfaces.

Large-End Bearing

See *big-end bearing.*

Larmor Frequency

The classical frequency at which a charged body processes in a uni-form magnetic field. $\omega_L = -eB/2mc$, where e is the electron charge,

B is the magnetic field intensity, m is mass, and c is the velocity of light. See also *Larmor period.*

Larmor Period

The inverse of the Larmor frequency. See also *Larmor frequency.*

Larson Miller Parameter

An empirical relationship for predicting the complete range of creep rupture properties of a material on the basis of a limited number of tests. Creep life is a function of stress and temperature and, for a particular material at a fixed stress, the parameter P is given:

$$P = T(c + \log t)$$

where

T is the temperature in degrees absolute
t is the time to failure
c is a constant for each material

Normal practice is to undertake a series of tests over a fairly broad range of stress levels and plot the P value against the logarithm of the stress. This produces a smooth curve, which can be correlated with rupture life and temperature. Thus when any two variables are specified, the third can be deduced. For example, a designer might need to know the maximum stress allowable to achieve a life of 10,000 h at 500°C for the material in question. To determine this, a horizontal line is drawn across at the 500°C level to intersect the 10,000 line and from this intersection a vertical line is drawn down to intersect the P curve. A Z line drawn horizontally from the P line intersects across the stress scale at the allowable stress.

Laser

Laser Alloying

This is a material processing method that utilizes the high power density available from focused laser sources to melt metal coatings and a portion of the underlying substrate. Since the melting occurs in a very short time and only at the surface, the bulk of the material remains cool, thus serving as an intimate heat sink. Large tempera-ture gradients exist across the boundary between the melted surface region and the underlying solid substrate. The result is rapid self-quenching and resolidification.

Transitions

The sequence in the transitioning that occurs during and follow-ing an individual laser exposure. Initially, the metal substrate (B) is coated with a thin metal foil (A) and is irradiated with a laser pulse. For metal surfaces and most laser wavelengths, a significant frac-tion of the incident light will be specularly or diffusely scattered away. The absorbed energy is "instantaneously" (10^{-12} s) transferred to the lattice. The near-surface region very rapidly reaches the melt-ing point, and a liquid–solid interface starts to move through the film. The liquid–solid interface has swept through the original thin-film–substrate interface. Interdiffusion of the film and substrate ele-ments starts. The laser pulse is nearly terminated, and the surface has remained below the vaporization temperature. At this point, the maximum melt depth has been reached, and interdiffusion con-tinues. The resolidification interface velocity is momentarily zero and then rapidly increases. The resolidification interface has moved approximately halfway back to the surface from the melt depth.

Interdiffusion into the liquid continues, but the re-solidified metal behind the liquid–solid interface cools so rapidly that solid-state diffusion may be neglected. Finally, the material is completely resolidified, and a "surface alloy" has been produced.

What makes laser surface alloying both attractive and interesting is the wide variety of chemical and microstructural states that can be retained because of the rapid quench from the liquid phase. These include chemical profiles where the "alloyed" element A is highly concentrated near the atomic surface and decreases in concentration over shallow depths (hundreds of nanometers), and uniform profiles where the concentration of A in B is the same throughout the entire melted region. The types of microstructures observed include extended solid solutions (the concentration of A in B greatly exceeds equilibrium values), metastable crystalline phases (high-temperature phases retained because of the rapid return to room temperature), and metallic glasses.

For a pulsed or Q-switched laser (high-power, short-pulse laser) source, there also is an optical and effective spot size, and the "pulse length" is the time of exposure. The laser pulses are emitted in a train of pulses characterized by a repetition rate. For these lasers, the effective spot size, degree of overlap, and repetition rate determine the area per unit time that can be produced.

For all the laser sources (continuous-wave pulsed and Q-switched), the exposure time (dwell time or pulse length) strongly influences the depth that will be melted. Because deeper melting means a longer total time in the molten state, that means more time available for diffusion of the one or more alloying elements into the molten portion of the substrate. Deeper melting and longer melt times result in more dilute surface alloys, whereas shallow melting and shorter melt times result in more concentrated surface alloys. It is also evident that in some instances convection, surface tension, and plasma effects can enhance the mixing within the liquid state and drive the melt toward homogenization.

In making laser alloys, many other processing variables need to be considered. In addition to the exposure time just discussed, these include the laser power, the thickness of the film put down prior to laser melting, and in some instances the nature of the gaseous ambient during the laser processing. The processing variables are interrelated, and one variable cannot be freely changed without affecting another. Another consideration is that laser alloying is a liquid state-rapid quenching phenomenon. The near-surface region must be melted and yet vaporization avoided. Different minimum and maximum energy densities are thus defined for each laser exposure time. In addition to these processing constraints, there are certain properties of matter that strongly influence whether or not certain element combinations may be laser-alloyed. For example, it is not possible to laser-alloy a low-melting-point/high-vapor-pressure element like zinc into a high-melting-point/low-vapor-pressure metal substrate such as tungsten. The zinc would vaporize before the underlying tungsten could be melted. Because liquid-state intermixing is required, suitable systems must exhibit miscibility in the molten state. Binary systems like silver–nickel, iron–lead, copper–molybdenum and aluminum–bismuth have miscibility gaps in the liquid state spanning nearly all compositions. Such systems cannot be laser-alloyed.

Advantages

Surface alloys, particularly laser surface alloys, have advantages that include tailoring surface properties, conservation of materials, and creation of new metal surfaces.

Surface alloying allows alteration of the metal surface to achieve characteristics best suited to the surface environment. For example,

in an ordinary saw blade the material requirements for the cutting teeth are very different from those of the length of the blade. In many applications requiring corrosion or oxidation resistance, that resistance is needed only on the external surface of the material. It is possible to design the bulk of the material with characteristics most suitable for structural needs or ease of fabrication and to tailor the metal surface for interface requirements. Material conservation is another reason often cited for considering surface alloying. In the case of stainless steels and superalloys, the elements added to impart the special properties are often strategic elements for which there are sparse domestic supplies. For example, chromium and nickel may not be necessary on the inside of knives, forks, and spoons. Further, precious metals are expensive and in relatively short supply. Thus surface alloying can be very cost-effective.

These first two advantages—tailoring for surface properties and material conservation—are generally important considerations in almost any coating technology. However, laser alloying differs from coating technologies in that the near-surface region is a continuous extension of the interior of the metal. There is no interface between bulk and "coating" the laser alloy is a mixture of bulk and surface elements. Also, problems in regard to porosity and adherence do not exist.

Laser alloying involves very large temperature gradients and quenching from the liquid state. In this way, it resembles other rapid-solidification technologies.

It is anticipated that laser alloying will produce many new alloys that could not have been produced by conventional methods.

Laser Cooling

Reducing the thermal motion of atoms with the force exerted by a laser beam is laser cooling. Typically, such cooling is used to reduce the temperature of a gas of atoms, or the spread of atoms in an atomic beam.

The keys to using such repeated kicks to reduce the random, thermal motion of a gas of atoms are the monochromatic nature of laser light, the selectivity of absorption of light by atoms, and the Doppler effect.

The viscosity of laser cooling does not stop atoms. Although this might seem to be a fundamental limit, techniques have been developed that go even farther, by arranging that when, by chance, an atom achieves a very low velocity it stops absorbing light and remains cold. In this way, atoms have been cooled below the recoil limit.

Applications

Improving atomic clocks, where the thermal motion of atoms reduces the precision and accuracy, was a major motivation to developing laser cooling. Clocks using laser-cooled trapped ions or free natural atoms rival the performance of the best conventional atomic clocks. Laser cooling is also used in atom optics, where well-collimated, monoenergetic atomic beams are more easily and effectively manipulated. In addition, laser cooling has been used to study collisions between very slow atoms.

Laser Cutting

Types of Lasers for Metal Cutting

The continuous-wave (CW) CO_2 laser leads the way for metal cutting. The CW laser beam builds up in small-diameter gas-containing tubes with electrodes at each end. High-voltage direct current flows between electrodes, exciting gas atoms to develop the lasing effect.

The workhorse for metal cutting is the fast-axial-flow (FAF) CO_2 laser. The majority of these units are DC design, an economical laser excitation technique for cutting thin-gauge materials. The FAF laser produces a stable beam for high-quality cuts. Fast-flow technology moves gas rapidly through the discharge area to operate between 500 W and 6 kW. The fast flow efficiently cools the laser gas, making more power available in more compact units.

In the transverse flow (TF) CO_2 laser, gas flows in a transverse direction to the laser cavity axis. The laser gas circulates at high speed and cools through a heat exchanger. This design results in an output power in the 5–45 kW range. The TF laser produces a beam with a mixed mode for high-speed quality cuts. The compact TF design can enclose the laser in a workstation.

A solid-state laser describes an optically pure material that will lase when doped with specific elements. Neodymium-doped yttrium aluminum garnet (Nd–YAG) rates as the most popular solid-state laser for industrial use in either rod- or rectangular-shaped design.

Types of Cutting Machines

Flying-optics machines are systems that move the beam along the x–y axes and the workpiece is stationary. In the gantry design, a stationary workpiece improves positioning tolerances, especially for heavier workpieces. The space-efficient design and the unobstructed beam motion allow easy integration into production lines. Beam-motion accuracy holds constant over the entire working volume at 0.010 mm with a repeatability of 0.05 mm. Traverse rates can double other machine configurations.

The stationary head, or fixed-beam design, a widely used metal-cutting machine, integrates x and y workpiece motion and under a fixed-position laser beam. The beam design requires a minimum number of mirrors to bend the beam and maintain beam characteristics and quality over the travel distance.

The hybrid machine moves the workpiece in the x axis and the optics move in the y axis. To maintain beam accuracy, hybrid units offer shorter beam-delivery paths than full moving-beam systems. Moving the workpiece in only one axis permits more-accurate movement of heavier workpieces than with flying optics and holds position accurately to ±0.010 mm and repeatability to ±0.05 mm.

The advantages of cutting using robot-beam manipulation, the classic articulated arm with cylindrical or spherical coordinate designs, are obvious. Robots with internal optics for CO_2 laser cutting can cut very complex parts, but require good mechanics because the many joints and mirrors make alignment difficult. Fiber optics used to convey an Nd:YAG beam align more easily, but the minimum bending radius of the fiber can limit mobility. Units can hold 0.05 mm point-to-point repeatability, with dynamic path repeatability of 1 mm at 0.86 mm/s.

Laser Inspection

Laser inspection involves an inspection system in which lasers check that the soldering on printed circuit boards has been developed. The three-dimensional solder shape inspection machine detects faulty soldering by the non-contact three-dimensional shape measuring system. The system is based on laser measuring technology, and is said to provide accuracy 10 times higher than conventional technology in judging faulty soldering. The system reportedly also reduces the time required for preoperational arrangement to 1/10 that of normal.

The system utilizes a thin laser light 5 µm thick, which scans the outline of solder paths on the principal board and compares the shape with a standard. Together with a high-speed camera, the system pictures the solder outline visualized through laser reflection from the solder surface, then processes the picture at a speed of 0.016 s/frame. The system then makes quantitative evaluation based on measured data about the size and the adhesion angle of the solder and the component height.

Laser Lithography

A line narrowed by 193 nm laser radiation has been demonstrated with 21 W output power. This is reportedly the highest ever demonstrated from a single-stage line narrowed argon-fluorine excimer laser oscillator.

The laser was operated with 1 kHz repetition rate and yielded more than 20 mJ/pulse with a bandwidth of less than 0.6 pm, FWHM. The achievement is a result of ongoing studies to improve the efficiency of the line narrowing resonator and laser tube design.

The argon-fluorine excimer lasers will likely be used for the next generation of deep ultraviolet lithography tools for semiconductor production of 1 Gbit memory chips and advanced microprocessors. Development efforts are underway to transfer the new technology into products suitable for 193 nm lithography production, particularly due to the reduced cost operation achievable.

Laser Peening

A laser peening method is being perfected that could significantly improve the durability of critical jet engine components. Although the laser peening concept is not new, a neodymium doped glass laser with 600 W of average power makes the process faster and economically feasible. The lasers capable of firing 10 pulses/s, compared with only 1 pulse every 2 s from commercial lasers. Laser peening, which uses concentrated light to generate a small, sudden force equivalent to about 30,000 times the pressure of the atmosphere, can leave a compressive residual stress in engine fan blades, disks, rotors, and shafts up to 1 mm deep. That is four times deeper than conventional shot peening. The compression significantly retards metal fatigue and corrosion, and improves damage tolerance.

Laser Alloying

See *laser surface processing*.

Laser Beam Cutting

(1) A cutting process that severs materials with the heat obtained from the application of a concentrated coherent light beam impinging upon the workpiece to be cut. The process can be used with (gas-assisted laser beam cutting) or without an externally supplied gas. (2) The cutting process utilizes a laser beam as a heat source. The high energy intensity of the focused beam facilitates deeply penetrating cuts and welds with minimum distortion. In the case of cutting, material is removed by some combination of melting with ejection of molten material and vaporization. Gas jet laser cutting utilizes an additional gas jet to assist ejection of molten material.

Laser Beam Machining

Use of a highly focused mono-frequency collimated beam of light to melt or sublime material at the point of impingement on a workpiece.

Laser Beam Welding

A welding process that joins metal parts using the heat obtained by directing a beam from a *laser* onto the weld joint.

Laser Hardening

A surface-hardening process that uses a laser to quickly heat a surface. Heat conduction into the interior of the part will quickly cool the surface, leaving a shallow martensitic layer.

Laser Surface Processing

The use of lasers with continuous outputs of 0.5–10 kW to modify the metallurgical structure of a surface and to tailor the surface properties without adversely affecting the bulk properties. The surface modification can take the following three forms. The first is transformation hardening in which a surface is heated so that thermal diffusion and solid-state transformations can take place. The second is surface melting, which results in a refinement of the structure due to the rapid quenching from the melt. The third is surface (laser) alloying, in which alloying elements are added to the melt pool to change the composition of the surface. The novel structures produced by laser surface melting and alloying can exhibit improved electrochemical and tribological behavior.

Latent Curing Agent

A curing agent for plastics that produces long-term stability at room temperature but rapid cure at elevated temperatures.

Latent Heat

Thermal energy absorbed or released when a substance undergoes a phase change.

Latent Heat of Fusion

Heat given off by a liquid freezing to a solid, or gained by a solid in melting to a liquid, without a change in temperature.

Latent Heat of Vaporization

Heat given off by a vapor condensing to a liquid, or gained by a liquid evaporating to a vapor, without a change in temperature.

Lateral Contraction

The contraction across the width and thickness of a material being loaded in tension. See *Poissons Ratio* and *anti-clastic*.

Lateral Crack

A crack produced beneath and generally paralleling a glass surface during the unloading phase of mechanical contact with a hard, sharp object. See also *cleavage crack*.

Lateral Extrusion

In operation in which the product is extruded sideways through an orifice in the container wall.

Lateral Runout

Same as *axial runout*.

Latex

The milklike juice of the rubber tree, now much used instead of the cured crude rubber for many rubber applications such as adhesives, rubber compounds, and rubber powder. The properties of latex vary with the type of tree, age of tree, method of tapping, and climate. Latex from young trees is less stable than that from old trees. Intensive tapping of the trees results in less rubber content, which may vary from 20% to 50%. For shipping, a preservative and anticoagulant are added to the latex, usually ammonia or sodium sulfate. Concentrated 60% latex is a stable liquid of cream like consistency. *Latex foam* is a cellular sponge rubber made by whipping air into latex, pouring into molds, and vulcanizing. *Artificial latex* is a water dispersion of reclaimed rubber. Water dispersions of crude or reclaim rubbers are produced by swelling and dissolving the rubber in an organic solvent, treating with an organic acid or with ammonia, and emulsifying. They resemble latex, but are softer and tackier and are used for adhesives. Latex foams are now usually made by incorporating a chemical, which releases a gas to form the cells.

The term *latex* now also refers to water dispersions of synthetic rubbers and rubberlike plastics. *Neoprene latex* is a water dispersion of neoprene rubber, and it has dispersed particles smaller than those of natural latex, giving better penetration including paper and textiles. Butadiene-styrene latex with 68% solids is used for producing foamed rubber. *Latex water paints* are now usually made with synthetic rubber or plastic dispersions.

Lath Martensite

Martensite formed partly in steels containing less than approximately 1.0% C and solely in steels containing less than approximately 0.5% C as parallel arrays of packets of lath-shape units 0.1–0.3 μm thick.

Lathe

A machine tool in which the workpiece is gripped in the rotating chuck and possibly supported at its far end by a tail stock. A lathe tool carried in the tool head of the saddle cuts the workpiece as the saddle travels along the bed or the tool head traverses the saddle.

Lattice

(1) A space lattice is a set of equal and adjoining parallelopipeds formed by dividing space by three sets of parallel planes, the planes in any one set being equally spaced. There are seven ways of so dividing space, corresponding to the seven crystal systems. The unit parallelopiped is usually chosen as the unit cell of the system. See also *crystal system* and *unit cell*. (2) A point lattice is a set of points in space located so that each point has identical surroundings. There are 14 ways of so arranging points in space, corresponding to the 14 Bravais lattices. See *7 crystal systems* and *corresponding 14 space Bravais lattices*. (3) The three-dimensional grid on which it is visualized, atoms are arranged in the crystal structure of solid metal.

Lattice Constants

See *lattice parameter*.

Lattice Construction/Girder, etc.

An assembly of slender girders, beams, etc., joined in a triangulated manner to provide a lightweight but high-strength construction.

Lattice Defects

Defects in the crystal structure such as dislocations, vacancies, and foreign atoms.

Lattice Diffusion

See *volume diffusion*.

Lattice Parameter

The length of any side of a unit cell of a given crystal structure. The term is also used for the fractional coordinates x, y, and z of lattice points when these are variables.

Lattice Pattern

A pattern of a filament winding with a fixed arrangement of voids.

Laue Method (for Crystal Analysis)

A method of x-ray diffraction using a beam of white radiation, a fixed single crystal specimen, and a flat photographic film usually normal to the incident beam. If the film is located on the same side of the specimen as the x-ray source, the method is known as the back reflection Laue method; if on the other side, as the transmission Laue method.

Launder

(1) A channel for transporting molten metal. (2) A box conduit conveying particles suspended in water.

Lava

Lava is molten rock material that reaches the Earth's surface through volcanic vents and fissures; also, the igneous rock formed by consolidation of such molten material. Relatively rapid cooling of the Earth's surface may transform fluid lava into a dense-textured volcanic rock composed of tiny crystals or glass or both.

The temperature of liquid lava ranges widely but generally does not exceed 1200°C. Basaltic lavas are usually hotter than rhyolitic ones. The viscosity of lava depends largely upon the temperature, composition, and gas content.

As lava cools, it becomes more viscous and the rate of flow decreases. Rapid cooling, as at the surface of a flow, promotes the formation of glass. Slower cooling as near the center of a flow, favors the growth of crystals.

During many volcanic eruptions, the lava is so rapidly ejected that it is blown to bits by the explosive force of expanding gases. The small masses rapidly congeal and settle to the Earth to form thick blankets of volcanic tuff and related pyroclastic rock. Lava flows and volcanic tuffs cover large areas of the Earth's surface and may form more or less alternating layers totaling many thousands of feet in thickness.

Lava is also a name given to ceramic material used for molding gas-burner tips, electrical insulating parts, nozzles, and handles. It may be calcined talc, steatite, or other material. It is molded from magnesium oxide, and it is hardened by heat treatment after shaping and cutting. It is baked at 1093°C. The compressive strength is from 138 to 207 MPa. It will resist moisture and has high dielectric strength. Rods as small as 0.05 cm in diameter can be made. Alsimag is a trade name of lava that is produced from ground talc and sodium silicate, but *Alsimag 602* is phosphate-bonded steatite talc, used for thyratron tubes and as a substitute for mica in receiving tube spacers.

Laves' Phases

Intermediate phases formed from two metals in the ratio of one atom of one to two of the other, such as $MgCu_2$, and where the two atoms differ in size by about 22.5% allowing them to form a close packed structure.

Law of Convervation of Matter

Matter is not created or destroyed during any chemical or physical process.

Lay

Direction of predominant surface pattern remaining after cutting, grinding, lapping, or other processing.

Lay (Composites)

The length of twist produced by stranding filaments, such as fibers, wires, or roving. The angle that such filaments make with the axis of the strand during a stranding operation. The length of twist of a filament is usually measured as the distance parallel to the axis of the strand between successive turns of the filament.

Layer

A stratum of weld metal or servicing material. The layer may consist of one or more weld beads laid side-by-side.

Layer Bearing

A bearing constructed in layers. See also *bimetal bearing* and *trimetal bearing*.

Layer Level Wound

See preferred term *level wound*.

Layer Wound

See preferred term *level wound*.

Layer-Lattice Material

Any material having a layerlike crystal structure, but particularly *solid lubricants* of this type.

LayUp

Reinforcing material that is placed in position in the mold. The process of placing the reinforcing material in position and the mold. The resin-impregnated reinforcement and a description of the component materials, geometry, and so forth, of a laminate.

LBBT

See *low blow brittle transition test.*

L/D Ratio

In bearing technology, the ratio of the axial length of a plain bearing to its diameter.

LD Process

See *Linz-Donawitz.*

L-Direction

The ribbon direction, that is, the direction of the continuous sheets of honeycomb.

Leaching

Extracting an element or compound from a solid alloy or mixture by preferential dissolution in a suitable liquid.

Lead Alkali Metals

Lead hardened with small amounts of alkali metals used chiefly as bearing metals. Alloys of this type are also called *tempered lead* and *alkali lead*. *Calcium–lead alloy* is made by electrolysis of the fused alkali salts, using a molten lead cathode and contains up to 1% calcium. The calcium forms a chemical compound, Pb_3Ca, with part of the lead, and the crystals of this compound are throughout the lead matrix. The Brinell hardness is 35, and the compressive strength is 172 MPa, being thus superior in strength to the high-tin bearing metals. But the metal is difficult to melt without oxidation and is easily corroded. The melting point is 370°C, and it retains its bearing strength at more elevated temperatures than the babbitts. *Calcium lead*, with about 0.04% calcium, is used as cable sheathing to replace antimonial lead, giving greater fatigue resistance. It is also used for grids in storage batteries and has a lower rate of self-discharge than antimonial lead. The small amount of lithium in Bahnmetall (original alkali lead) is intended to prevent the corrosion set up by the calcium. It also increases the compressive strength to about 207 MPa.

Lead and Alloys

A soft, heavy, bluish-gray metal (symbol Pb), lead is obtained chiefly from the mineral galena. It surface-oxidizes easily but is then very resistant to corrosion. It is soluble in nitric acid but not in sulfuric or hydrochloric, and is one of the most stable of the metals. Its crystal structure is face-centered cubic. It is very malleable, but it becomes hard and brittle on repeated melting because of the formation of oxides. The specific gravity of the cast metal was 11.34, and that of the rolled is 11.37. The melting point is 327°C, and boiling point is 1750°C. The tensile strength is low, that of the rolled metal being about 25 MPa, with the elongation of 30%. The coefficient of expansion is 0.0000183, and the thermal conductivity is 8.2% that of silver. The electric conductivity is only 7.8% that of copper. When used in storage batteries, the metal is largely returned as scrap after a period and is remelted and marketed as secondary lead, as is also that from pipes and cable coverings. Lead is highly toxic and, thus, poses a health hazard. Inhalation of dust and fumes should be avoided, and it should not be used in contact with food or drink products.

Not only is lead the most impervious of all common metals to x-rays and gamma radiation, it also resists attack by many corrosive chemicals, most types of soil, and marine and industrial environments. Although it is one of the heaviest metals, only a few applications are based primarily on its high density. The main reasons for using lead often include low melting temperature, ease of casting and forming, good sound and vibration absorption, and ease of salvaging from scrap.

With its high internal damping characteristics, lead is one of the most efficient sound attenuators for industrial, commercial, and residential applications. Sheet lead, lead-loaded vinyls, lead composites, and lead-containing laminates are used to reduce machinery noise. Lead sheet with asbestos or rubber sandwich pads are commonly used in vibration control.

The natural lubricity and wear resistance of lead make the metal suitable, and alloys, for heavy-duty bearing applications such as railroad-car journal bearings and piston-engine crank bearings. Lead is also widely used as a constituent in solders. Most common solders are the lead–tin alloys; melting temperature can be as low as 183°C.

Forms

Sheet and Foil

Because of its malleability, lead and its alloys are readily rolled to any desired thickness down to 0.01 mm. Sheets are usually fabricated by burning or soldering. Standard widths run to 2.4 m or more for sheet and sheets may be cut to any desired size. Blanks for impact extrusion, gaskets, washers, or other purposes may be stamped out. Tin-coated lead can be produced by rolling lead and tin together.

Extrusions

Lead is easily extruded in the form of pipe, rod, wire, or any desired cross section like window cames (H-shaped), rounds, hollow stars, rectangular duct. Commercially available extrusions range in size from 612 mm pipe down to solder wire 0.25 mm in diameter. Lead is extruded over paper, rubber, or plastic in making electrical cable and around steel bars. Common flux-cored solder is a lead extrusion; toothpaste tubes are impact extrusions.

Castings

One of the simplest metals to cast— lead is used in tiny die castings and massive cast counterweights. Type metal, renowned for its ability to reproduce minute detail, is a lead alloy. Lead grids for most batteries are die-cast. Casting temperature (usually about 316°C) is moderate. Arsenic, antimony, or tin are frequently alloyed to impart strength or special properties. Small die castings can have wall thicknesses as low as 1.3 mm and "as-cast" dimensions are reproducible to 0.03 mm.

Coatings

Protection of underlying iron and steel is the main objective in most lead coating. In the purely protective class one finds terne-plate for roofing, fireproof frames and doors, automotive parts, and containers for paint and oil. Lubricity imparted by the coating cases drawing and stamping operations and produces an excellent surface for soldering—hence, television chassis and automotive gas tanks are made of terne. Other hot-dip processes as well as electroplating and flame spraying are also used for outdoor hardware automotive mufflers, bearings, bushings, nuts, bolts, and for maintenance as well.

Laminations

Developed originally for x-ray protection, a large family of laminated lead materials now exists. In addition to their original niche, these are finding increasing use in sound isolation and noise control. Typical examples include lead–plywood, lead–gypsum board, lead–cinder block, leaded plastic–fabric laminates, leaded plastic, and glass–fiber combinations.

Cladding

Metallic lead in thicknesses from 3.2 to 305 mm or more may be bonded to other metals. Thus, for example, lead and steel may be combined for corrosion resistance and strength or lead and copper for gamma shielding and heat transfer. In many instances, a product such as a tank or chemical reactor is completely or partially fabricated in steel and then clad with bonded lead as a unit.

Powder

Spheres, irregular grains, and flakes of lead from 4 μm diameter and above find use in special greases, as a constituent of bearings, brake, and clutch facings and is filling plastics and rubber and in paints and pile-joint compounds. Wire rope is usually treated with such a powder to lubricate it and to fill any nicks in the filter, thus renewing it with self-lubricating surfaces.

Shot

This form of lead is produced in abundance—about $30,360 \times 10^3$ kg go into shotgun ammunition each year. Ammunition sizes range from 1 to 11.3 mm; small shot is made for other uses. Easily handled, it is a preferred form when mass shielding is required inside an irregular enclosure. It is also used in making free-machining steels.

Wool

By passing molten lead through a fine sieve and allowing it to solidify in the air, a loose rope of filters is produced. Under pressure, usually by being driven into a crevice with a caulking iron and hammer, the fibers weld into a homogeneous mass. This permits the forming of a solid metal seal where temperature or explosion hazards prohibit joining procedures requiring heat. Continuous lead fiber is also produced by being spun on textile machines.

Alloys

In its unalloyed form as 99.95% minimum, lead is soft and weak; it requires support for mechanical applications. This "chemical lead" is used primarily in corrosive chemical-handling applications such as tank linings.

"Hard lead"—lead alloyed with 1%–13% antimony—has sufficient tensile strength, fatigue resistance, and hardness for many mechanical applications. These alloys can be cast, rolled, or extruded and are especially suited for castings requiring good detail and moderate strength. Rolled antimonial alloys are harder and stronger than the cast alloys. Battery-plate lead contains 7%–12% antimony.

Calcium (0.03%–0.12%) forms another series of mechanically suitable alloys with lead. These alloys age-harden naturally at room temperature—usually for 30–60 days—after being cast or worked. Properties of wrought Pb–Ca alloys are somewhat directional, being greater in the longitudinal direction. Uses include cable sheathing and grids in storage batteries.

Tin, added to Pb–Ca alloys in amounts to about 1.5%, raises tensile strength and stress-rupture resistance but increases aging time to 180 days. Tin is often used to reduce the coefficient of friction for bearing applications. Higher-tin-bearing alloys are primarily used in solders, which normally contain from 40% to 60% tin.

Lead alloys may exhibit greatly improved mechanical or chemical properties as compared to pure lead. The major alloying additions to lead are antimony and tin. The solubilities of most other elements in lead are small, but even fractional weight percent additions of some of these elements, notably copper and arsenic, can alter properties appreciably.

Cable-Sheathing Alloys

Lead is used as a sheath over the electrical components to protect power and telephone cable from moisture. Alloys containing 1% antimony are used for telephone cable, and lead–arsenical alloys, containing 0.15% arsenic, 0.1% tin, and 0.1% bismuth, for example, are used for power cable. Aluminum and plastic cable sheathing have replaced lead alloy sheathing in many applications.

Battery-Grid Alloys

Lead alloy grids are used in the lead-acid storage battery (the type used in automobiles) to support the active material composing the plates. Lead grid alloys contain 6%–12% antimony for strength, small amounts of tin to improve castability, and one or more other minor additions to retard dimensional change in service. No lead alloys capable of replacing the lead–antimony alloys in automobile batteries have been developed. An alloy containing 0.03% calcium for use in large stationary batteries has had success.

Chemical-Resistant Alloys

Lead alloys are used extensively in many applications requiring resistance to water, atmosphere, or chemical corrosion. They are noted for their resistance to attack by sulfuric acid. Alloys most commonly used contain 0.06% copper, or 1%–12% antimony, where greater strength is needed. The presence of antimony lowers corrosion resistance to some degree.

Type Metals

Type metals contain 2.5%–12% tin and 2.5%–25% antimony. Antimony increases hardness and reduces shrinkage during solidification. Tin improves fluidity and reproduction of detail. Both elements lower the melting temperature of the alloy. Common type metals melt at 238°C–246°C.

Bearing Metals

Lead bearing metals (babbitt metals) contain 10%–15% antimony, 5%–10% tin, and for some applications small amounts of arsenic or copper. Tin and antimony combine to form a compound that provides wear resistance. These alloys find frequent application in cast sleeve bearings and are used extensively in freight-car journal bearings. In some cast bearing bronzes, the lead content may exceed 25%.

Solders

A large number of lead-base solder compositions have been developed. Most contain large amounts of tin with selected minor additions to provide specific benefits, such as improved wetting characteristics.

Free-Machining Brasses, Brasses, Steels

Lead is added in amounts from 1% to 25% for brasses and bronzes to improve machining characteristics. Lead remains as discrete particles in these alloys. It is also added to some construction steel products to increase machinability. Only about 0.1% is needed, but the tonnage involved is so large that this forms an important use for lead.

Uses

Lead wool is lead in a shredded form used for caulking. Sheet lead is produced by cold-rolling and is used as a sound barrier in building construction.

Lead has a high capacity for the capture of neutrons and gamma rays and is used for radiation shielding in the form of sheet lead or as metal powder in ceramic mortars and blocks, paints, and in plastic composite structures. DS Lead is a dispersion-strengthened lead containing up to 1.5% lead monoxide evenly distributed through the structure. The oxide combines chemically with the lead, doubling the strength and stiffness of the metal, but increasing its brittleness. It is used for chemical piping and fittings. A neoprene-lead fabric is a neoprene fabric impregnated with lead powder. It has a radiation shielding capacity one third that of solid lead sheet. It comes in thicknesses of 0.08–0.64 cm, and its flexibility makes it suitable for protective clothing and curtains. Shielding cements for x-ray and nuclear installation shielding are metallic mortars containing a high percentage of lead powder with ceramic oxides as binders and other elements for selective shielding. They are mixed with water to form plasters or for casting into sections and blocks. The formulation varies with the intended use for capture, attenuation, or dissipation of neutrons, gamma rays, and other radiation. Shielding paints are blended in the same manner.

Battery-plate lead for the grid plates of storage batteries and silver alloy for positive-plate grids are additional applications. Lead-coated copper used for roofing and for acid-resistant tanks as well as frangible bullets, which shatter on striking a target surface and are used for aerial gunner practice, are two other uses.

Antimonial lead is an alloy containing up to 25% antimony with the balance lead, used for storage-battery plates, type metal, bullets, tank linings, pipes, cable coverings, bearing metals, roofing, collapsible tubes, toys, and small cast articles. The alloy is also known as hard lead.

One hard lead application has 10% antimony and 90% lead and melts at 252°C. Called cable lead, or sheathing lead, it is used to cover telephone and power cables to protect against moisture and mechanical injury. Terne-plate, as the coated steel is called, is widely used for automobile gasoline tanks and also has been used for roofing on buildings.

Lead Angle

In cutting tools, the helix angle of the flutes.

Lead Bronze

An alloy of copper with up to 30% lead use for high load bearings. Contrast with *leaded bronze* and *bearing bronzes*.

Lead Glass

Glass containing a substantial proportion of lead oxide (PbO).

Lead Oxide

Two types of lead oxide are used in ceramics: litharge, or lead monoxide, and red lead.

Litharge: Lead monoxide (PbO) has a specific gravity of 9.3–9.7; a melting point of 888°C. It is insoluble in water but soluble in alkalies, certain acids, and some chloride solutions.

Red Lead: Lead oxide (Pb_3O_4) has a specific gravity of 9.0–9.2; it decomposes between 500°C and 530°C. It is insoluble in water and is decomposed in some acids, leaving insoluble lead peroxide, PbO_2.

Lead oxide is used quite extensively in optical glass, electrical glass, and tableware. It increases the density and refractive index of glass. In addition, it can be cut more easily than other glasses and has superior brilliance, both of which make it good for cut glass.

Lead glasses may be formulated with a wide variety of electrical and acid-resistant characteristics; desirable properties, such as weather resistance, electrical resistivity, etc., will depend upon the total composition of the glass.

Lead has many advantages as a glaze ingredient. The superiority of lead glazes lies in their brilliance, luster, and smoothness, which are due to their lower fusion point and viscosity. Lead glazes are, in general, highly resistant to water solubility and chipping, and have few faults in texture and bond. They have high mobility, refractivity, and elasticity and are softer than leadless glazes.

All investigators agree that the use of fritted glazes, in which all of the lead is fritted, has important health advantages. In this way the raw lead oxide is converted into relatively harmless lead silicates, which are much less soluble in dilute acids or gastric juices. Lead silicates are more easily absorbed after entering the respiratory system, and can be eliminated with less lead absorption.

Lead oxide also is used in enamels. With an increase in lead and a corresponding decrease in potash, with flint constant, enamels become more fusible, have less tendency to craze, and become more refractive, but are less durable in acid fumes.

Lead Pigments

Chemical compounds of lead used in paints to give color. They are to be distinguished from the lead compounds such as lead oleate, used as driers for paints. *Lead flake* is useful in exterior primers, where it exhibits excellent durability and rust inhibition. But because of restrictions on the use of lead, lead flake and many other lead compounds are being phased out as pigments. *White lead* is the common name for *basic lead carbonate*, the oldest and most important lead paint pigment, also used in putty and in ceramics. It is a white, amorphous, poisonous powder. It is insoluble in water and decomposes on heating. The specific gravity is 6.7. It is made from metallic lead and is marketed dry, or mixed with linseed oil and turpentine in paste form. *Lead carbonate* is used as a pigment in the same way as the basic compound, but it discolors more easily. Basic lead sulfate, called *sublimed white lead*, makes a fine white pigment. Commercial sublimed white lead contains 75% lead sulfate, 20% lead oxide, and 5% zinc oxide. Commercial white lead may be mixed with lithopone, magnesium oxide, antimony oxide, witherite, or other materials.

The basic *silicate white lead* is made by fusing silica sand with litharge and hydrating by ball-milling with water. It has corrosion-inhibiting properties and is used in metal-protective paints for underwater service. It is also used in ceramics and as a stabilizer in vinyl plastics. *Basic lead silicate* is made from silica and litharge with sulfuric acid as a catalyst. The material has a core of silica with a surface coating of basic lead silicate, giving a pigment of lower weight per unit of volume but retaining the activity of the silicate white lead. *Chrome yellow*, or *Leipzig yellow*, is *lead chromate*, and it comes in yellow, poisonous crystals. Specific gravity is 6.123. It is insoluble in water and decomposes at 600°C. *Basic lead chromate*, is red and is used for anti-corrosive base coats for steel. *Red lead* is a bright orange pigment used exclusively for corrosion inhibition; because of its poor opacity it is combined with iron oxide. *Basic lead silicochromate* is another rust inhibitor. *American vermillion*, also called *Chinese scarlet* and *chrome red*, is basic lead chromate made from white lead.

Lead thiosulfate is a white insoluble powder used chiefly in matches. All of the lead compounds are poisonous when absorbed through skin or taken internally. *Lead sulfide* is used as a feeler in missiles as it is very sensitive to heat rays. Molybdate oranges are lead chromate, lead molybdate, and lead sulfate. DuPont Co's *Kroler* is a heat-resistant variety suitable for employment in plastics.

Litharge is the yellow *lead monoxide*, PbO, also called *massicot*. It is a yellow powder used as a pigment and in the manufacture of glass, and for the fluxing of earthenware. An important use is as a filler in rubber. With clustering, litharge is used as a plumber's cement. The specific gravity is 9.375. Litharge is produced by heating lead in a reverberatory furnace and then grinding the lumps. For storage-battery use the black oxide, or suboxide of lead, is now largely substituted. Lead dioxide, or lead peroxide, comes as a brown, crystalline powder and is used in making matches, dyes, and pyrotechnics, as a mordant, and as an oxidizing agent.

Lead Proof

See *die proof*.

Lead Zirconate Titanate

Lead zirconate titanate, $Pb(Zr_{0.4}Ti_{0.3})O_3$ to $Pb(Zr_{0.9}Ti_{0.1})O_3$, is also known as PZT ceramics. It is used mainly in the manufacture of piezoelectric ceramic elements.

The wide range of possible compositions provides a wide range of dielectric constant values, piezo-electric activity, and primary transition temperatures. These are normally greater than is possible with barium titanate. Small additions of niobium, strontium, barium, and antimony serve as modifiers.

To techniques for producing PZT powders are used today. Calcining (CMO) is the more common method of powder production, and the calcine is then formed and fired in its final shape.

PZT is the most widely used polycrystalline piezoelectric material. Its electrical output can measure pressure. It is used in hydrophones, which permit listening to sound transmitted through water.

One variation of the PZT ceramics is PLZT ceramics (lead–lanthanum zirconate titanate). They are a range of ferroelectric, optically active, transparent ceramics based on the $PbZrO_2TiO_2$ system. When hot pressed, the material can reach 99.8% theoretical density. For device applications of PLZT ceramics, the ratio that is important is x:65:35, where x is lanthanum (8%–10%) and 65:35 is the $PbZrO_3$:$PbTiO_3$ ratio. They have basically the same characteristics as PZT ceramics and are frequently finding use as filters, oscillators, and vibrators in many areas. They are also used for optical shutters.

Leaded Brass and Bronze

Any of the copper–zinc, copper–zinc–tin, and copper–tin alloys containing generally 1%–6% lead, sometimes more. The lead improves machinability, providing "free-machining" quality, but sacrifices ductility somewhat. Specific alloys are many, and they are available as wrought and cast alloys. Among the most common are the *leaded wrought brasses* (C31400–C38600), *leaded cast brasses* (C83300–C85800), *leaded phosphor bronzes* (C53200–C54800), and *leaded tin bronzes* (C92200–C92900).

Leaded Copper/Brass/Bronze, etc.

Metals to which lead has been added, primarily to make it free machining.

Leaded Steel

A free-machining steel containing about 0.25% lead. Lead does not alloy with iron, but when a stream of finely divided lead is shot at the stream of molten steel to the ingot mold, the lead is distributed in the steel in tiny particles. The lead eases machining without imparting to the steel the unfavorable characteristics given by sulfur or phosphorus. There is little weakening of the physical properties of the steel. The lead forms a layer of liquid lubricant at the chip–tool interface, thus reducing the stress required to overcome friction. Also, by slightly embrittling the steel, lead reduces the deformation stress and serves to initiate microcracks to produce small chips. *Ledloy*, of Inland Steel Co., contains 0.15%–0.30% lead in the regular SAE grades of steel.

Leak Testing

A nondestructive test for determining the escape or entry of liquids or gases from pressurized or into evacuated components of systems intended to hold these liquids. Leak testing systems, which employ a variety of gas detectors, are used for locating (detecting and pinpointing) leaks, determining the rate of leakage from one leak or from a system, or monitoring for leakage.

Leakage Field

The magnetic field that leaves or enters a magnetized part at a magnetic pole.

Leak-Before-Break

This term recognizes that some forms of failure are acceptable, others are not. In particular, a small leak is often acceptable; a major burst never is. Hence, a pressurized component known to be developing a crack might be considered acceptable to continue operation if it could be established that the crack would announce its development to an unacceptable stage by a slowly developing leak allowing ample time to shut down the plant before there was a possibility of a major explosion.

Lean Mixture

A gas/air mixture in which too much air is present, so that burning is difficult. The flame is usually noisy, frail, very light blue, short, and inefficient.

Least Count

The smallest value that can be read from an instrument having a graduated scale. Except on instruments provided with a *vernier*, the least count is that fraction of the smallest division that can be conveniently and reliably estimated; this fraction is ordinarily one-fifth or one-tenth, except where the graduations are very closely spaced. Also known as least reading.

Least Reading

See *least count*.

Leather

Leather is made from the hides or skins of animals, birds, reptiles, and fish. Two main steps are involved in this process. First, the hides or skins are cured or dressed to prepare them for tanning. This curing process removes all the flesh, hair, and foreign matter. The hides are then tanned to produce durable, useful leather.

There are two general tanning methods—(1) bark or vegetable and (2) chrome tanning. In chrome tanning, salts of chromium are used to process the hides. Chrome-tanned leather cannot be tooled, but it can be dyed and embossed.

Vegetable tanning makes use of the tannic acid that is found in bark, leaves, and other vegetable products. Leather made by vegetable tanning is toolable and is, in general, the most suitable for leather craft.

Tanned leather, which is about the color of human skin, is smoother grained, but can be further finished by dyeing, glazing, buffing, graining, or embossing.

While the better leathers are usually left smooth, those having scratches, flaws, or other surface defects after tanning are generally embossed with a grain.

Leather can be purchased in many sizes and thicknesses. The thickness of leather is expressed either in ounces or fractions of an inch. When ounce is used it is really a measure of thickness, and not of weight. Leathers range from about 1 to 8 oz. The weight to use depends on the end-service requirements.

Types

Cowhide

This is a strong, tough, durable leather available in a very wide variety of grades, types, weights, finishes, and colors. Heavy vegetable-tanned cowhide back leather, ranging from 6 to 8 oz, is widely used for tooling, carving, and stamping. Heavy cowhide sides, sometimes called strap leather, are also suitable for tooling deep designs. Shoulder pieces are widely used for belts. Thinner cowhide, both vegetable and chrome tanned, is used in leather craft kits. Cowhide belly, lowest in cost and quality, can be used where appearance and durability are not important.

Calfskin

Calf has the finest grain of all tooling leathers and is therefore best for tooling. It ranges in weight from 1.5 to 3.5 oz; it is more expensive than cowhide. When moistened and tooled on the grain side, the designs will remain permanently. Calfskin is available in natural finish or in a variety of colors. Chrome-tan grades are less expensive than the tooling grades and are excellent for lining. (Vegetable-tanned calfskin is known as saddle leather.)

Steerhide

Steerhide has a crinkly surface, ranges in weight from about 2.5 to 4.5 oz, tools easily, and is not readily marred it is available in many colors and in three-toned combination finish.

Sheepskin

Sheepskin is available in both tooling and nontooling grades. The largest nontooling use is for linings. Several types are available; suede, the best of lining leathers; skivers, thin and lightweight; glazed, with a fine, smooth glossy finish, and firm texture. All these are available in many different colors and are generally inexpensive.

The tooling grade is economical and tools without difficulty. Relatively deep tooling is also possible, but the leather should be dampened only very lightly. Sheepskin is often embossed or dyed to resemble calfskin.

Goatskin or "Morocco"

Although goat is naturally a fine-grained, smooth leather, it is most commonly seen with the pebbly or crinkly "morocco" grain produced by boarding. It is stronger and more attractive than sheepskin of the same finish. In the heavier weights, it is suitable for tooling line design and initials. Morocco leather is used extensively for book covers. Glazed goatskin has a high gloss, firm finish, and is excellent for lacing things.

Kidskin

Exceptionally thin, smooth, and strong, kidskin is characterized by tiny holes on the grain side. Excellent for lining, and available in various colors, it is used for lining, handbags, book covers, and wallets.

Pigskin

This leather has a fine grain and smooth surface with visible holes in groups of three. The finest grades are usually imported. Tooling of pigskin is very difficult and not recommended. It is an expensive leather and therefore its use is confined largely to small parts such as billfolds, card holders, and pocket secretaries.

Suede

Suede can be made from many kinds of leather, but most commonly from sheepskin. It has a soft velvet finish, it is nontooling and is suitable as a lining as well as for garments. Velvet Persian is a high-quality suede from skins of Persian sheep.

Alligator

Alligator is very expensive and is therefore restricted to small products. Besides genuine alligator, simulated alligator-grain leather is also available. It is usually made from calfskin or cowhide. It is attractive, strong, nontooling, available in many colors, and relatively inexpensive.

Snake and Lizard

Reptile skin is specialty leathers noted for the beauty of their markings. The skins are durable as well as beautiful. Because they are quite expensive, they are used only for very small items.

Ledeburite

The eutectic of the iron–carbon system, the constituents of which are *austenite* and *cementite*. The austenite decomposes into *ferrite* and cementite on cooling becomes Ar_1, the temperature at which transformation of austenite to ferrite, plus cementite is completed during cooling. See *steel*.

LEFM (Linear Elastic Fracture Mechanics)

See *fracture mechanics*.

Left-Hand Cutting Tool

A cutter all of whose flutes twist away in a counterclockwise direction when viewed from either end.

Leftward Welding

See *forward welding*.

Leg (Length)

In welding, the size of a fillet weld measured from the intersection of the two components to the toe at which the fused metal meets the parent metal.

Legging

The drawing of filaments or strings when adhesive-bonded substances are separated. Compare with *teeth*. See also *stringiness* and *webbing*.

Leg of Fillet Weld

(1) *Actual*: The distance from the root of the joint to the toe of the fillet weld. (2) *Nominal*: The length of the side of the largest right triangle that can be inscribed in the cross section of the weld. See also the terms *concave fillet weld* and *convex fillet weld*.

Lehigh (Slow) Bend Test

A test of the brittle fracture characteristics of steel usually after welding. It is basically a three point bend test using a notched specimen.

Lehigh (Weld) Cracking Test

A test of the susceptibility of steel to cracking during welding. The standard test piece is a plate 12 × 8 in., having a series of slots at 1 in. intervals along its two long sides and a central longitudinal slot to accept the weld. The longitudinal slot as of standard profile, but its length depends on the plate thickness. The depth of the side slots is selected to produce the required degree of restraint.

Leidenfrost Phenomenon

Slow cooling rates associated with a hot vapor blanket that surrounds a part being quenched in a liquid medium such as water. The gaseous water envelope acts as an insulator, thus slowing the cooling rate.

Lemon Bearing (Elliptical Bearing)

A two-lobed bearing.

Leno Weave

A locking-type weave in which two or more warp threads cross over each other and interlace with one or more filling threads. It is used primarily to prevent the shifting of fibers and open-weave fabrics.

Lens

A transparent optical element, so constructed that it serves to change the degree of convergence or divergence of the transmitted rays.

Lenticular

Shaped like a double convex lens.

Let-Go

An area in laminated glass over which an initial adhesion between interlayer and glass has been lost.

Letting Down

Tempering of steel particularly when carried out in a relatively crude matter with temperature estimated by color.

Level Winding

See *circumferential winding*.

Level Wound

Spooled or coiled weld filler metal that has been wound in distinct layers such that adjacent turns touch.

Leveler Lines

Lines on sheet or strip running transverse to the direction of *roller leveling*. These lines may be seen upon stoning or light sanding after leveling (but before drawing) and can usually be removed by moderate stretching.

Leveling

(1) Flattening of rolled sheet, strip, or plate by reducing or eliminating distortions. See also *stretcher leveling* and *roller leaveling*. (2) Various processes for flattening sheet by light deformation either by uniform tensile loading or by passing the sheet through a series of staggered rollers.

Leveling Action

Action exhibited by a plating solution yielding a plated surface smoother than the basis metal.

Leveling Agents

Substances added to electrolytic plating solutions to minimize variations in plate thickness.

Lever Rule

A method that can be applied to any two-phase field of a binary *phase diagram* to determine the amounts of different phases present at a given temperature in a given alloy. A horizontal line, referred to as a tie line, represents the lever, and the alloy composition its fulcrum. The intersection of the tie line with the boundaries of the two-phase field fixes the compositions of the coexisting phases, and the amounts of the phases are proportional to the segments of the tie line between the alloy and the phase compositions.

Levigation

(1) Separation of fine powder from coarser material by forming a suspension of the fine material in a liquid. (2) A means of classifying a material as to particle size by the rate of settling from a suspension.

Levitation Melting

An *induction melting* process in which the metal being melted is suspended by the electromagnetic field and is not in contact with a container.

Life Assessment

Systems for estimating the remaining life of components such as power plant and other high-temperature, high duty plants. The term may be used of a simple calculation of creep life for comparison with the life experience. However, it more often refers to a comprehensive review of all information that can be obtained including original materials data and manufacturers certificates, site inspection and measurements, removal of samples for laboratory testing and examination, review of historical plant data, and reappraisal of design utilizing improved computer-based stress analysis techniques.

Life Fraction

The portion of life consumed by a particular set of circumstances. The concept is sometimes applied in damage mechanisms where the variables change. For example, in creep both the load and the temperature can vary, in fatigue to cyclic and steady loads may both vary. The life fraction approach involves calculating for each set of circumstances the life consumed as a fraction of the total life in those circumstances. Failure is assumed to occur when the fractions total unity.

Lift Beam Furnace

A continuously operating heat treating or sintering furnace. The term in general use by the U.S. furnace industry is *walking beam furnace*.

Lift Rod

Part of the press tooling used for the raising or lifting of one or more punches.

Liftout

The mechanism also known as *knockout*.

Ligament

In an engineering context, a load bearing part of a structure, for example, the metal remaining between a number of holes set across the line of loading.

Ligand

The molecule, ion, or group bound to the central atom in a *chelate* or a *coordination compound*.

Light

Radiant energy in a spectral range visible to the normal human eye (~380–780 nm, or 3800–7800 Å). See also *electromagnetic radiation*.

Light Drawn

An imprecise term, applied to drawn products such as wire and tubing, that indicates a lesser amount of cold reduction than for *hard drawn* products.

Light Filter

See *color filter*.

Light Fraction

The first liquid produced during the distillation of a crude oil.

Light Metal/Alloy

Any alloy based on the low density metals aluminum, magnesium, and, less common, lithium. Titanium and its alloys have a density intermediate between these and iron and copper and, depending on context, may be included in the category. Alloys of magnesium, beryllium, and lithium are sometimes termed ultra light. Examples of low-density metals, such as aluminum (~2.7 g/cm^3), magnesium (~1.7 g/cm^3), titanium (~4.4 g/cm^3), beryllium (~1.8 g/cm^3), or their alloys.

Light-Emitting Diodes (LEDs)

LEDs have been developed for numerous applications and systems for special lighting technology. For example, LEDs developed for experiments to activate light-sensitive, tumor-treating drugs. The light source, consisting of 144 of the tiny diodes, measures only 12.7 mm diameter—about the size of a small human finger. With its cooling system, the entire light source is the size of a medium suitcase. This type of therapy system involves injecting into the patient's bloodstream a drug, which attaches to the unwanted tissues and permeates into them, without affecting the surrounding tissues. A solid-state LED probe is placed near the affected tissue to illuminate the tumor and activate the drug. Once activated by the light, the drug destroys the tumor cells.

Light-Field Illumination

See *bright-field illumination*.

Lightly Coated Electrode

A filler-metal electrode used in arc welding, consisting of a metal wire with a light coating, usually of metal oxides and silicates, applied subsequent to the drawing operation primarily for stabilizing the arc. Contrast with *covered electrode*.

Lignin

A colorless to brown crystalline product recovered from paper-pulp sulfite liquor, and used in furfural plastics, as an extender of phenol in phenolic plastics, as a corrosion inhibitor, in adhesives and coatings, as a natural binder for compressed-wood products, and for the production of synthetic vanilla when it contains coniferin. It is also used as a fertilizer, providing humus and organic material and some sulfur to the soil. For this use, it may be mixed with phosphates. Lignin is coprecipitated with natural and synthetic rubbers to produce stronger and more lightweight products. When lignin is incorporated into nitrile rubber by coprecipitation, the rubber has higher tensile and tear strengths and greater elongation than with an equal loading of carbon black.

Lignite

Also called *brown coal*. A variety of organic mineral of more recent age than coal, occurring in rocks of tertiary age, and intermediate in composition between wood and coal. It is widely distributed over Europe and found in many parts of the world. Freshly cut lignite often contains a large quantity of water, up to 40%, and is sometimes also high in ash. When dried, it breaks up into fine lumps and powder. Dry lignite contains 75% carbon, 10%–30% oxygen, and 5%–7% hydrogen. It kindles easily but burns with a low caloric power and a smoky flame. In retort gas production, lignite loses its gas in half the time required for gas removal from bituminous coal, with a temperature of 688°C, compared with 901°C for bituminous. The *pitch coal* is brownish black, breaks with a pitchlike fracture, and shows no woody structure. Lignite is briquetted by crushing and pressing with a binder under heat. Belgian lignite briquettes have a binder of 8% asphalt, with 2% flour to assist binding.

Lime

Lime, which is calcium oxide (CaO), has a specific gravity of 3.4, a melting point of 2572°C, boils at 2850°C, and is soluble. It is introduced into ceramic mixtures in several different forms. In the pottery bodies and glazes, it is bought as whiting (calcium carbonate) or dolomite (calcium carbonate and magnesium carbonate). In glass batches, it is introduced by limestone, burned lime (calcined limestone), and dolomite. In the enameling industry, it is used in the form of whiting.

Lime as CaO is not found in nature. Calcium carbonate, which is the chief source of lime, is found in the form of the minerals calcite and araganite.

The chemical requirements of lime used in glass vary with the type of ware produced. The combined CaO and MgO should be at least 89% for bottle glass, 91% for sheet glass, 93% for blown glass, 96% for rolled glass, and 99% for optical glass. The iron oxide should be practically zero for optical glass, whereas in bottle glass as much as 0.5% is permissible, with nearly the same limits for blown or sheet glass. The silica or alumina may run as high as 15% for bottle glass, which should be vented much less for other grades.

Lime

A general term that includes the various chemical and physical forms of quicklime, hydrated lime, and hydraulic lime. It may be high-calcium, magnesium, or dolomitic. The chemical forms of CaO, calcium hydroxide (Ca(OH)$_2$), magnesium oxide (MgO), or magnesium hydroxide (Mg(OH)$_2$), alone or in combination may be produced either primarily or as a by-product of materials other than limestone, for example, Ca(OH)$_2$ formed by acetylene generation from calcium carbide (CaC$_2$) and water treatment sludges.

Lime Glass

A glass containing a substantial proportion of lime, usually associated with soda and silica. See also *glass*.

Limestone

A number of terms are in general use for the different varieties of limestone based on differences of origin, texture, composition, etc. *Marble* is a limestone that is more or less distinctly crystalline.

Chalk is a fine-grained aragonite limestone composed of finely divided calcium carbonate usually from marine shell sources.

It may be said, regardless of the impurities that are found in limestone, that lime is in all cases practically the only base found in a pure theoretical limestone. In glass, lime is one of the most important of the common batch ingredients.

Lime gives to glass, when added in proper quantities, stability or permanency, hardness, viscosity, and tenacity, and facilitate melting and refining. Line decreases the viscosity at high temperatures but increases the rate of setting in working range.

Magnesium lime or dolomitic lime is largely used because of its low iron content. Dolomitic lime seems to have a more powerful fluxing action and a glass using dolomitic lime is set to fine or plain up quicker than one using lime from another source.

Calcium carbonate is not used to the same extent in enamels as it is in glasses and glazes. This is probably because the average burning range of enamels is lower than that of either glasses or glazes, and calcium carbonate exerts strong fluxing action only at high temperatures.

Limestone is a sedimentary rock consisting chiefly of calcium carbonate or of the carbonates of calcium and magnesium. (1) Dolomitic limestone contains from 35% to 46% magnesium carbonate. (2) Magnesium limestone contains from 5% to 35% magnesium carbonate. (3) High-calcium limestone contains from 0% to 5% magnesium carbonate. Limestone is used extensively in glass formulations.

Limit of Proportionality

The stress below which a change in stress produces a linearly proportionate change in strain. See *tensile test*.

Limited Solid Solution

A crystalline miscibility series whose composition range does not extend all the way between the components of the system; that is, the system is not *isomorphous*.

Limited-Coordination Specification (or Standard)

A specification (or standard) that has not been fully coordinated and accepted by all interested parties. Limited-coordination specifications and standards are issued to cover the need for requirements unique to one particular department. This applies primarily to military agency documents.

Limiting Creep Stress

Originally, this referred to the stress below which no measurable creep occurs. Since this stress will depend on the sensitivity of the measuring equipment any remaining use of the term is usually specific to the application and the level of creep that is acceptable.

Limiting Current Density

The maximum current density that can be used to obtain a desired electrode reaction without undue interference such as from *polarization*.

Limiting Dome Height (LDH) Test

A mechanical test, usually performed unlubricated on sheet metal, that simulates the fracture conditions in a practical press-forming operation. The results are dependent on the sheet thickness.

Limiting Drawing Ratio (LDR)

See *deformation limit.*

Limiting Static Friction

The resistance to the force tangential to the interface that is just sufficient to initiate relative motion between two bodies under load. The term static friction, which properly describes a tangential resistance called into operation by a force less than this, should not be substituted for limiting static friction.

Limiting Stress Range

Same as *fatigue limit.*

Line Defect/Imperfection

Sometimes used as a synonym for *dislocation.*

Line Indices

The *Miller indices* of the set of planes producing a diffraction line.

Line Pair

In spectroscopy, an analytical line and the internal standard line with which it is compared. See also *internal standard line.*

Line Reaming

Simultaneous *reaming* of coaxial holes in various sections of a workpiece with a reamer having cutting faces or piloted surfaces with the desired alignment.

Lineage Structure

(1) Deviations from perfect alignment of parallel arms of a columnar dendrite as a result of interdendritic shrinkage during solidification from a liquid. This type of deviation may vary in orientation from a few minutes to as much as 2° of arc. (2) A type of substructure consisting of elongated subgrains. Individual areas have differences in lattice orientation that are detectable but not enough to constitute a grain boundary.

Linear (Tensile or Compressive) Strain

The change per unit length due to force in an original linear dimension. An increase in length is considered positive.

Linear Attenuation Coefficient

The fraction of an x-ray beam per unit thickness that a thin object will absorb or scatter (attenuate). A property proportional to the physical density and dependent on the atomic number of the material and the energy of the x-ray beam.

Linear Damage Law

A crude estimate of life fraction consumed is calculated. Failure occurs when the life fractions total unity.

Linear Dispersion

In spectroscopy, the derivative $dx/d\lambda$, where x is the distance along the spectrum and λ is the wavelength. Linear dispersion is usually expressed as mm/Å.

Linear Elastic Fracture Mechanics

A method of fracture analysis that can determine the stress (or load) required to induce fracture instability in a structure containing a cracklike flaw of known size and shape. See also *fracture mechanics* and *stress-intensity factor.*

Linear Expansion

The increase of a given dimension measured by the expansion of a specimen or component subject to a temperature gradient. See also *coefficient of thermal expansion.*

Linear Shrinkage

The shrinkage in one dimension of a powder compact during sintering. Contrast with *volume shrinkage.*

Linen

A general name for the yarns spun from the fiber of the variety of flax plant cultivated for its fiber, or for the cloth woven from the yarn. Linen yarns and fabrics have been made from the earliest times, and the ancient Egyptian linen fabrics were of exceeding fineness, containing 540 warp threads per inch, not equaled in Europe until the twentieth century. Ireland, Belgium, and France are the principal producers of linen. Linen yarns are used for the best grades of cordage, and linen fabrics are employed industrially wherever a fine, even, and strong cloth is required. *Linen fabrics* are sold under a wide variety of trade names. They are graded chiefly according to the fineness of the yarns and the class of weave and may contain some fine hemp fibers. *Lisle* was formally a fine, hard linen thread, made at Lille, France, but is now a fine, smooth yarn made of long-staple cotton spun tightly in a moist condition. *Tow yarns* are the coarsest linen yarns, use for making *crash*, a coarse, plain-woven fabric used for towels or covers. *Chintz* is a plain-woven linen fabric in brightly covered designs. It is also made as cotton chintz. The name is from the Hindu word meaning color. Damask is a jacquard reversible woven linen fabric for table linen. It is now also made in cotton, silk, or rayon.

Liner (Composite)

In a filament-wound pressure vessel, the continuous, usually flexible coating on the inside surface of the vessel, used to protect the laminate from chemical attack or to prevent leakage under stress.

Liner (Metals)

(1) The slab of coating metal that is placed on the core alloy and is subsequently rolled down to flat sheet as a composite. (2) In extrusion, a removable alloy steel cylindrical chamber, having an outside longitudinal taper firmly positioned in the container or main body of the press, into which the billet is placed for extrusion.

Liners

Thin strips of metal inserted between the dies and the units into which the dies are fastened.

Lining

Internal refractory layer of firebrick, clay, sand, or other material in a furnace or ladle.

Linishing

A method of finishing by grinding on a continuous abrasive belt.

Linoleum

A general name for floor-covering material consisting of a mixture of ground cork or wood flour, rosin or other gum, blown linseed oil, pigments, with sometimes a filler such as lithopone, on a fabric backing of burlap or canvas, rolled under pressure to give a hard, glossy surface. It is pigmented or dyed in plain colors, or printed with designs. The cork is usually in about 50-mesh particles. *Battleship linoleum*, a very heavy grade in plain colors or mosaics, is described in federal specifications as made with oxidized linseed oil, fossil or other resins or oxidized rosin, and an oleoresinous binder, mixed with ground cork, wood flour, and pigments, processed on a burlap back. Because of its resiliency, battleship linoleum is still valued for high-grade flooring and is coated with an adherent, wear-resistant lacquer based on cellulose ester and alkyd resins.

Linseed Oil

This oil is the most common of the drying oils, and it is widely used for paints, varnishes, linoleum, printing inks, and soaps. It is obtained by pressure from the seeds of the flax plant, which is cultivated for oil purposes. It is sometimes referred to as flaxseed, though this name properly belongs only to the seed of the flax plant for producing flax fiber. The varieties producing linen fiber do not yield much seed.

Linters

Short fibers adhere to the cottonseed if there ginning. Used in rayon manufacture as fillers for plastics and as a base for the manufacture of cellulosic plastics.

Linz-Donnawitz Process

The first (probably) commercial process to produce steel on a large scale by injecting oxygen into molten pig (impure) iron.

Lipophilic

Having an affinity for oil. See also *hydrophilic* and *hydrophobic*.

Lip-Pour Ladle

Ladle in which the molten metal is poured over a lip, much as water is poured out of a bucket. See the term *ladle*.

Liquation

(1) The separation of a low melting constituent of an alloy from the remaining constituents, usually apparent in alloys having a wide melting range. (2) Partial melting of an alloy, usually as a result of *coring* or other compositional heterogeneities.

Liquation (Cracking)

In its broadest sense this is a synonym for melting but usually implies transient localized melting of an alloy heated just to its solidus temperature. It normally occurs at the grain boundaries and may lead to other forms of damage such as oxide penetration down the boundaries or to cracking immediately or in subsequent service.

Liquation Temperature

The lowest temperature at which partial melting can occur in an alloy that exhibits the greatest possible degree of segregation.

Liquefied Petroleum (LP) Gas

Gases, such as propane and butane, which are usually stored as a liquid under pressure, but are released for use as a gas by a regulator.

Liquid Carburizing

Surface hardening of steel by immersion into a molten bath consisting of cyanides and other salts.

Liquid Chromatography

A separation method based on the distribution of sample compounds between a stationary phase and a liquid mobile phase. Used extensively in the characterization of organic, inorganic, pharmaceutical, and biochemical compounds. See also *chromatography*, *gas chromatography*, and *ion chromatography*.

Liquid Crystal Polymers

Liquid-crystal polymers (LCP) are a unique class of wholly aromatic polyester polymers that provide previously unavailable high-performance properties. Particularly outstanding is their heat-deflection temperature of 1.79 MPa at 238°C–318°C. Structure of the LCPs consists of densely packed fibrous polymer "chains" that provide self-reinforcement almost to the melting point.

Before the commercial introduction of LCP resins, LCPs could not be injection-molded. Today's resin can be melt-processed on conventional equipment into thin-wall as well as heavy-wall components at fast speeds with excellent replication of mold details and efficient use of regrind.

Commercial melt-processable resins now available are Xydar formulations (biphenyl based) and Vectra resins (naphthaline based). Like most thermoplastics, molding these high-temperature resins requires heated tools and equipment capable of producing melt temperatures of 230°C–338°C for Vectra resins and 371°C–454°C for Xydar materials.

Properties

LCP resins are characterized by outstanding strength at extreme temperatures; excellent mechanical property retention after exposure to weathering and radiation; good dielectric strength, arc resistance, and dimensional stability; low coefficient of thermal expansion; excellent flame retardance; and easy processibility. Underwriters' Laboratories continuous-use rating for electrical properties is as high as 240°C and, for mechanical properties, 220°C. The high heat-deflection value of biphenyl-based resins permits molded parts to be exposed to intermittent temperatures as high as 315°C without affecting properties. Resistance to high-temperature flexural creep is excellent, as are fracture-toughness characteristics.

LCPs are exceptionally inert. They resist stress cracking in the presence of most chemicals at elevated temperatures, including aromatic or halogenated hydrocarbons, strong acids, bases, ketones, and other aggressive industrial substances. Hydrolytic stability in boiling water is excellent. Environments that determine the polymers are high-temperature steam, concentrated sulfuric acid, and boiling caustic materials.

The oxygen index of LCP resins ranges from 35% to 50%. When exposed to open flame, the material forms an intumescent char that prevents dripping and results in extremely low generation of smoke containing no toxic by-products.

Easy processibility of the resins is attributed to its liquid-crystal molecular structure, which provides high melt flow and fast setup in molded parts. However, molded parts are highly anisotropic, and knit lines are much weaker than other areas. Properties are not affected by minor variations in processing conditions, and no post curing is required.

Liquid Crystals

These are nonisotropic materials—neither crystalline nor liquid—that are composed of long molecules parallel to each other in large clusters and that have properties intermediate between those of crystalline solids and liquids. It is estimated that 1 in every 200 organic compounds has the capability of being produced in the liquid crystal form.

There are three principal types of liquid crystals, based on the arrangement of the molecules. In the smectic type, the molecules are parallel with their ends in line, forming layers that are usually curved or distorted, but are still capable of movement over one another. In the nematic type, the molecules are essentially parallel, but there is no regular alignment of their ends. The cholesteric type is formed by optically active compounds that have the capability for molecular organizations of the nematic type.

Liquid crystals have some of the properties of liquids, such as fluidity, and some of the properties of crystals, such as optical anisotropy. A major use of liquid crystals is for digital displays, which consist of two sheets of glass separated by a sealed-in transparent liquid crystal material. The outer surface of the glass sheet is coated with a transparent conductive coating, with the viewing side coating etched into character-forming segments. A voltage applied between the two glass sheets disrupts the orderly arrangement of the molecules, thus darkening the liquid to form visible characters. Other typical applications of cholesteric liquid crystals are in skin thermography for tumor detection, in electronics for temperature mapping of circuits, and in nondestructive testing of laminates.

Liquid Disintegration

The process of producing powders by pouring molten metal on a rotating surface.

Liquid Honing

Producing a finely polished finish by directing an air-ejected chemical emulsion containing fine abrasives against the surface to be finished.

Liquid Impact Erosion

See *erosion* (*erosive wear*).

Liquid Impingement Erosion

See *erosion* (*erosive wear*).

Liquid Injection Molding (LIM)

A process that involves an integrated system for proportioning, mixing, and dispensing two-component liquid resin formulations and directly injecting the resultant mix into a mold, which is clamped under pressure. Generally used for the encapsulation of electrical and electronic devices. Also, variations on *reaction injection molding*, using mechanical mixing rather than a high-pressure impingement mixer. However, unlike mechanical mixing in other systems, the mixer here does not need to be flushed because a special feed system automatically dilutes the residue in the mixer with part of the polyol needed for the next shot, thereby keeping the ingredients from reacting.

Liquid Metal Attack/Embrittlement (ME)

Generally, any form of damage involving a liquid metal by corrosion, erosion, or dissolution. More specifically, especially when termed "embrittlement," the term implies a mechanism in which liquid metal penetrates along grain boundaries or crystallographic planes of a higher melting temperature solid metal. Such attack is most likely when the metal is under stress, either applied, or residual. The thin planes of penetrating material may initiate cracks immediately or during subsequent service. For example, steel is usually brazed without problems but if, during brazing, it is subject to high levels of stress, either applied or residual, it is susceptible to deep brazing penetration at the grain boundaries.

Liquid Metal Infiltration

Process for immersion of metal fibers in a molten metal bath to achieve a metal–matrix composite; for example, graphite fibers in molten aluminum.

Liquid Nitriding

A method of surface hardening in which molten nitrogen-bearing, fused-salt baths containing both cyanides and cyanates are exposed to parts at some critical temperatures. A typical commercial bath for liquid nitriding is composed of a mixture of sodium and potassium salts. The sodium salts, which comprise 60%–70% (by weight) of the total mixture, consist of 96.5% $NaCN$, 2.5% Na_2CO_3, and 0.5% $NaCNO$. The potassium salts, 30%–40% (by weight) of the mixture, consists of 96% KCN, 0.6% K_2CO_3, 0.75% $KCNO$, and 0.5% KCl. The operating temperature of the salt bath is 565°C.

Liquid Nitrocarburizing

A nitrocarburizing process (where both carbon and nitrogen are absorbed into the surface) utilizing molten liquid salt baths below the lower critical temperature. Liquid nitro carburizing processes are used to improve wear resistance and fatigue properties of steels and cast irons.

Liquid Penetrant Inspection

A type of nondestructive inspection that locates discontinuities that are open to the surface of a metal by first allowing a penetrating dye or fluorescent liquid to infiltrate the discontinuity, removing the excess penetrant, and then applying a developing agent that causes the penetrant to seep back out of the discontinuity and register as an indication. Liquid penetrant inspection is suitable for both ferrous and nonferrous materials, but is limited to the detection of open surface discontinuities in nonporous solids.

Liquid Phase Diffusion Welding

See *diffusion welding*.

Liquid Phase Sintering

Sintering of a compact or loose powder aggregate under conditions where a liquid phase is present during part of the sintering cycle.

Liquid Resin

An organic, polymeric liquid that becomes a solid when converted to its final state for use.

Liquid Shim

Material used to position components in an assembly where dimensional alignment is critical. For example, epoxy adhesive is introduced into gaps after the assembly is placed in the desired configuration.

Liquid Shrinkage

The reduction in volume of liquid metal as it cools to the liquidus.

Liquid Spray Quench

Same as *spray quenching*.

Liquid–Liquid Chromatography (LLC)

Liquid chromatography with a stationary phase composed of a liquid dispersed onto an inert supporting material. Liquid–liquid chromatography has been used in the separation of phenols, aromatic alcohols, organometallic compounds, steroids, drugs, and food products. Also termed liquid-partition chromatography.

Liquid-Partition Chromatography (LPC)

See *liquid–liquid chromatography*.

Liquid–Solid Chromatography (LSC)

Liquid chromatography with silica or alumina as the stationary phase. See also *adsorption chromatography*.

Liquidus (Glass)

The maximum temperature at which equilibrium exists between the molten glass and its primary crystalline phase.

Liquidus (Metals)

(1) The lowest temperature at which a metal or an alloy is completely liquid. (2) In a *phase diagram*, the locus of points representing the temperatures at which the various compositions in the system begin to freeze on cooling or finish melting on heating. See also *solidus*. (3) The line on a phase diagram above which only the liquid phase is stable and below which solidification commences.

Liquor Finish

A smooth, bright finish characteristic of wet-drawn wire. Formerly produced by using liquor from fermented grain mash as a drawing lubricant.

Lithia

Lithia is the oxide of lithium, (LiO_2), usually added to ceramic batches by means of chemically prepared lithium compounds.

Lithia is a very powerful flux, especially when used in conjunction with potash and soda feldspars. It is a valuable component in glasses having a low thermal expansion where its use permits the total alkali content to be kept to a minimum. The low thermal expansion properties also are exploited in flameproof ceramic bodies and glass ceramics where the formation of beta spodumene is the basis for oven-to-tableware production. It also enables the production of certain glasses having high electrical resistance and desirable working properties. A relatively high content of lithia allows the production of glasses that transmit ultraviolet light.

Glasses containing lithia are much more fluid in the molten state than those containing proportional amounts of sodium or potassium, and a successful use of lithia in glassmaking lies in the fact that much smaller amounts are required to produce a glass of the necessary fluidity for working without sacrificing the desired physical and chemical properties. In addition, lithia is being utilized to increase furnace capacity, decrease melting temperatures, and increase production capacities.

Uses

Although lithium batteries boast the highest energy density of any rechargeable, cobalt in the cathode keeps cost high—a lithium battery for an electric vehicle cost about $20,000. Computer modeling predicts a less expensive replacement material. Follow-on tests verify that a cathode made from a mixture of lithium aluminum oxide and lithium cobalt oxide could not only decrease battery cost by a significant margin, but also increase cell voltage.

Aluminum–lithium alloys are basically 2XXX and 7XXX aluminum alloys containing up to about 3% lithium. Because of the extremely light weight of lithium, they provide higher stiffness-to-density ratios that traditional structural aluminum alloys and, thus, have potential for aircraft applications. Because of the low weight, lithium compounds give the highest content of hydrogen, oxygen, or chlorine. Lithium hydride, LiH, a white or gray powder, is used for the production of hydrogen for signal balloons and floats. Lithium aluminum hydride, or lithium alanate, $LiAlH_4$, is used in the chemical industry for one-step reduction of esters without heat. Lithium metal is very sensitive to light, and is also used in light-sensitive cells.

Lithium is soluble in most commercial metals only to a slight extent; it is a powerful deoxidizer and desulfurizer of steel, but no lithium is left in the lithium-treated steel. In stainless steels it increases fluidity to produce dense castings. Cast iron treated with lithium has a fine grain structure and increased density with high impact value. Not more than 0.01% remains in the casting when treated with lithium–copper. In magnesium alloys the tensile strength is increased greatly by the addition of 0.05% lithium.

Lithium copper is a high conductivity, high density copper containing a minute quantity of residual lithium, 0.005%–0.008%, made by treating copper with a lithium–calcium–master alloy.

Lithia has been widely used in the production of pottery glazes of high quality. The addition of 1% lithium carbonate in the frit or the fluoride or silicate in the mill to dinnerware, electrical porcelain, and sanitaryware glazes has been found to increase the resulting gloss to a marked degree.

Lithium

This lightest of all metals, symbol Li, has a specific gravity of 0.534. It is found in more than 40 minerals, but is obtained chiefly from lepidolite, spodumene, and salt brines.

Lithiz melts at 186°C and boils at 1342°C. It is unstable chemically and burns in the air with a dazzling white flame when heated to just above its melting point. The melt is silvery white but tarnishes quickly in the air. The metal is kept submerged in kerosene. Lithium resemble sodium, barium, and potassium but has a wider reactive power than the other alkali metals. It combines easily with oxygen, nitrogen, and sulfur to form low melting-point compounds that pass off as gases, and is thus useful as a deoxidizer and degasifier of metals. In glass the small ionic radius of lithium permits a lithium ion coupled with an aluminum iron to displace two magnesium liens in the spinel structure. Lithium cobaltite, $LiCoO_2$, and lithium zirconate, Li_2ZrO_3, are also used in ceramics. Lithium carbonate, Li_2CO_3, is a powerful fluxing agent for ceramics, and is used in low-melting ceramic enamels for coating aluminum. It is used in medicine to treat mental depression.

Uses

Lithium metal, 99.4% pure, is produced by the reduction of lithium chloride, LiCl. The salts of lithium burn with a crimson flame, and lithium chloride is used in pyrotechnics. It is also used for dehumidifying air for industrial drying and for air-conditioning, as it absorbs water rapidly. It is also employed in welding fluxes for aluminum and its storage batteries. The anode is lithium, the cathode is a lithium–tellurium alloy, and the electrolyte is a molten bath of lithium salts at 427°C. Lithium ribbon, for high-energy battery use, is 99.96% pure metal in continuous strip form, 0.05 cm thick. It comes on spools packed dry under argon. An anhydrous form of lithium hexafluoroarsenate powder is used as the anode in dry batteries.

Little-End Bearing

A bearing at the smaller (piston) end of a connecting rod in an engine. See also **big-end bearing**.

Live Center

A lathe or grinder center that holds, yet rotates with, the work. It is used in either the headstock or tail stock of a machine to prevent wear and reduce the driving torque.

Live Load

The variable portion of a load system, for example, the passengers in an elevator are the live load; the cabin is the fixed base or static load.

L Shell

The second shell of electrons surrounding the nucleus of an atom, having electrons with principal quantum number 2.

Load

(1) In the case of testing machines, a force applied to a test piece that is measured in units such as pound-force, newton, or kilogram-force. (2) In tribology, the force applied normal to the surface of one body by another contacting body or bodies. The term normal force is more precise and therefore preferred; however, the term normal load is also in use. If applied vertically, the load can be expressed in mass units, but it is preferable to use force units such as newtons (N). (3) The weight or force applied. It is not synonymous with stress—see *tensile test*.

Load Cell

A mechanism for measuring applied loads and, usually, embedding a signal for remote display or recording.

Load Controlled Loading

See *displacement controlled loading*.

Load Range (*P*)

In fatigue, the algebraic difference between the maximum and minimum loads in a fatigue cycle.

Load Ratio (*R*)

In fatigue, the algebraic ratio of the minimum to maximum load in a fatigue cycle, that is, $R = P_{min}/P_{max}$. Also known as *stress ratio*.

Load-Carrying Capacity (of a Lubricant)

(1) The maximum load that a sliding or rolling system can support without failure. (2) The maximum load or pressure that can be sustained by a lubricant (when used in a given system under specific conditions) without failure of moving bearings or sliding contact surfaces as evidenced by seizure or welding.

Load–Deflection Curve

A curve in which the increasing tension, compression, or flexural loads are plotted on the ordinate axis and the deflections caused by those loads are plotted on the abscissa axis.

Load–Extension Curve

The stress strain curve, as produced in the tensile test.

Loading

(1) In cutting, building up of a cutting tool back of the cutting edge by undesired adherence of material removed from the work. (2) In grinding, filling the pores of a grinding wheel with material from the work, usually resulting in a decrease in production and quality of finish. (3) In powder metallurgy, filling of the die cavity with powder.

Loading Sheet

In powder metallurgy, the part of a die assembly used as a container for a specific amount of powder to be fed into the die cavity. Sometimes it is part of the feed shoe.

Loading Weight

See preferred term *apparent density*.

Loam

A molding material consisting of sand, silt, and clay, used over brick work or other structural backup material for making massive castings, usually of iron or steel.

Lobed Bearing

A journal bearing which two or more lobes, around its periphery produced by machining or by elastic distortion to increase stability or to provide adjustable clearance.

Local Action

Corrosion due to the action of "local cells," that is, galvanic cells resulting from inhomogeneities between adjacent areas on a metal surface exposed to *electrolyte*. The electrolytic corrosion occurring at small, localized variations in composition or hardness.

Local Cell

A *galvanic cell* resulting from inhomogeneities between areas on a metal surface in an *electrolyte*. The inhomogeneities may be of physical or chemical nature in either the metal or its environment.

Local Current Density

Current density at a point or on a small area.

Local Preheating

Preheating a specific portion of a structure.

Local Stress Relief Heat Treatment

Stress relief heat treatment of a specific portion of a structure.

Localized Corrosion

Corrosion at discrete sites, for example, *crevice corrosion*, *pitting*, and *stress-corrosion cracking*.

Localized Precipitation

Precipitation from a supersaturated solid solution similar to *continuous precipitation*, except that the precipitate particles form at preferred locations, such as along slip planes, grain boundaries, or incoherent twin boundaries.

Locating Boss

A *boss*-shaped feature on a casting to help locate the casting to an assembly or to locate the casting during secondary tooling operations.

Locating Ring

In injection molding machine, a ring that serves to align the nozzle of an injection cylinder with the entrance of the sprue bushing and the mold to the machine platen.

Locational Fit

A clearance or interference *fit* intended for locating mating parts.

Lock

In forging, a condition in which the flash line is not entirely in one plane. Where two or more plane changes occur it is called compound lock. Where a lock is placed in the die to compensate for die shift caused by a steep lock, it is called a counterlock.

Lock (ed) Fit

A severe interference fit not capable of being parted.

Lock Nut

A second solid nut tightened hard against the first nut on a threaded component to prevent loosening in service as a result of vibration or other inadvertent loads. Apart from the obvious risks associated with a nut falling off, even small reductions in pre-load increases the risk of fatigue failure. Where the nuts differ in thickness the thicker should be fitted last as it carries the full tightening torque and pre-load. It is incorrect but not uncommon for the thin nut to be fitted second. Alternative names are *jam nut* and *thin nut*. Another form of lock nut is of a thin hard sheet with edges raised to receive the wrench. These are not tightened to full pre-load torque and are fitted second. Also see *self locking nut*.

Locked Dies

Dies with mating faces that lie in more than one plane.

Logarithmic Decrement (Log Decrement)

The natural logarithm of the ratio of successive amplitudes of vibration of a member in free oscillation. It is equal to one/half the specific damping capacity.

Long Period

A morphological parameter for plastics obtained from small-angle x-ray scattering. It is usually equated to the sum of the *lamellar thickness* and the amorphous thickness.

Long Range Order

An ordered structure extending over a large proportion of a crystal.

Long Ton

The Imperial ton of 2240 lb as opposed to the U.S. "Short Ton" of 2000 lb.

Long Transverse

See *transverse*.

Long-Chain Branching

A form of molecular branching found in addition polymers as a result of an internal transfer reaction. It primarily influences the melt flow properties.

Longitudinal

In a metallurgical context, it is the line along which a material is principally extended during a working operation.

Longitudinal Direction

That direction parallel to the direction of maximum elongation in a work material. See also *normal direction* and *transverse direction*.

Longitudinal Field

A magnetic field that extends within a magnetized part from one or more poles to one or more other poles and that is completed through a path external to the part.

Longitudinal Resistance Seam Welding

The making of a resistance seam welding and a direction essentially parallel to the throat depth of a resistance welding machine.

Longitudinal Sequence

The order in which the increments of a continuous weld are deposited with respect to its length. See also *backstep sequence* and *block sequence*.

Long-Line Current

Current that flows through the earth from an anodic to a cathodic area of a continuous metallic structure. Usually used only where the areas are separated by considerable distance and where the current results from concentration-cell action.

Longos

Low-angle helical or longitudinal filament windings.

Long-Term Etching

Etching times of a few minutes to hours.

Loop Classifier

A cyclone-type classifier sometimes connected with a conical ball mill in an airtight system.

Loop Tenacity

The tenacity or strength value obtained by pulling two loops, such as two links in a chain, against each other to demonstrate the susceptibility of a fibrous material to cutting or crushing; loop strength.

Looping Mill

An arrangement of hot rolling stands such that a hot bar, while being discharged from one stand, is fed into a second stand in the opposite direction. See *rolling mill*.

Loose Metal

Refers to an area in a formed panel that is not stiff enough to hold its shape, may be confused with *oil canning*.

Loose Powder

Uncompacted powder.

Loose Powder Sintering

Sintering of uncompacted powder using no external pressure.

Loss Factor

The product of the dissipation factor and the dielectric constant of a dielectric material. See also *tan delta*.

Loss Modulus

A quantitative measure of energy dissipation in polymers, defined as the ratio of stress 90° out of phase with oscillating strain to the magnitude of strain. The loss modulus may be measured in tension or flexure, compression, or shear. See also *complex modulus*.

Loss On Ignition (LOI)

(1) The fractional or percentage weight loss of a material on heating in air from an initial defined state (usually, dried) to a specified temperature, such as 1000°C (1830°F), and holding therefore a specified period, such as 1 h. Fixed procedures are designed, usually, such that LOI represents the loss of combined H_2O, CO_2, certain other volatile inorganics, and combustible organic matter. (2) Weight loss, usually expressed as a percent of the total, after burning off an organic sizing from glass fibers, or an organic resin from a glass fiber laminate.

Lost Foam Casting

An *expendable pattern* process in which an expandable polystyrene pattern surrounded by the unbonded sand, is vaporized during pouring of the molten metal. Also referred to as evaporative pattern casting, evaporative foam casting, the lost pattern process, the cavity-less expanded polystyrene casting process, expanded polystyrene molding, or the full mold process.

Lost Wax Process

An *investment casting* process in which a wax pattern is used.

Lot

(1) A specific amount of material produced at one time using one process and constant conditions of manufacture, and offered for sale as a unit quantity. (2) A quantity of material that is thought to be uniform in one or more stated properties such as isotopic, chemical, or physical characteristics. (3) A quantity of bulk material of similar composition whose properties are under study. (4) A definite quantity of a product or material accumulated under conditions that are considered uniform for *sampling* purposes. Compare with *batch*.

Lot Sample

See *gross sample*.

Low Blow Brittle Transition Test

Similar to the Charpy Impact test except that the standard notched test pieces are pre-cracked by an initial impact at a temperature above the transition temperature. A series of test pieces is then tested, as normal, over a range of temperatures to produce the usual graph of energy absorbed against temperature. The technique is expected to produce a graph with a distinct inflection point at the temperature which, it is claimed, is the lowest safe service temperature for the batch of material. This temperature is termed the Low Blow Brittle Transition Temperature (*LBBT*).

Low Frequency Resistance Welding Cycle

One positive and one negative pulse of current within the same weld or heat time at a frequency lower than the power supply frequency from which it is obtained.

Low Hysteresis Steel

See *transformer steel*.

Low Melting Temperature Alloys

Various alloys with a low melting point, the criterion usually being 232°C, the melting point of tin. The constituents are usually two or more from tin, lead, bismuth, cadmium, gallium, thallium, indium and zinc. Compositions are often eutectics to ensure a sharp melting point. Some of the alloys are molten at ambient temperature. Applications include fusible plugs, sprinkler activators, tube bending fillers, and Jaeger die applications. Also termed *Fusible Alloys*. See *Wood's Metal* as an example.

Low-Alloy Carbon Steels

Also known by other terms, including alloy constructional steels, these are generally limited to a maximum alloy content of 5%. One or more of the following elements may be present: manganese, nickel, chromium, molybdenum, vanadium, and silicon. Of these, nickel, chromium, and molybdenum are the most common. The steels are designated by a numerical code prefixed by AISI (American Iron and Steel Institute) or SAE (formally Society of Automotive Engineers). The last two digits show the nominal carbon content. The first two digits identify the major alloying element(s) or group. For example, 2317 is a nickel–alloy steel with a nominal carbon content of 0.17%.

Whereas surface hardness attainable by quenching is largely a function of carbon content, the depth of hardness depends in addition on alloy content. Therefore, a principal feature of low-alloy steels is their enhanced hardenability compared to plain carbon steels. Like plain carbon steels, however, the mechanical properties of low-alloy steels are closely related to carbon content. In heat-treated, low-alloy steels, the alloying elements contribute to the mechanical properties through a secondary hardening process that involves the formation of finely divided alloy carbides. Therefore, for a given carbon content, tensile strengths of low-alloy steels can often be double those of comparable plain carbon steels.

Low-alloy steels may be surface-hardening (carburizing) or through-hardening grades. The former are comparable in carbon content to low-carbon steels. Grades such as 4023, 4118, and 5015 are used for parts requiring better core properties than are obtainable with the surface-hardening grades of plain carbon steel. The higher-alloy grades, such as 3120, 4320, 4620, 5120, and 8620, are used for still better strength and core toughness.

Most through-hardening grades are medium in carbon content and are quenched and tempered to specific strength and hardness levels. These steels also can be produced to meet specific hardenability limits as determined by end quenched tests. Identified as H steels, they afford steel producers more latitude in chemical composition limits. The boron steels, which contain very small amounts of boron, are also H steels. They are identified by the letter B after the first two digits.

A few low-alloy steels are available with high carbon content. These are mainly spring-steel grades 9260, 6150, 5160, 4160, and 8655, and bearing steels 52100 and 51100. The principal advantages of low-alloy spring steels are their high degree of hardenability and toughness. The bearing steels, because of their combination of high hardness, wear resistance, and strength, are used for a number of other parts, in addition to bearings.

Low-Alloy, High-Strength Steels

High-strength, low-alloy (HSLA) steels are low- to medium-carbon (0.10%–0.30%)/manganese (0.6%–1.70%) steels containing small amounts of alloying elements, such as aluminum, boron, chromium, columbium, copper, molybdenum, nickel, nitrogen, phosphorus, rare earth metals, titanium, vanadium, and zirconium. Because of the small amount of some of these elements, these steels have been referred to as microalloyed steels. The chemical compositions and minimum mechanical properties of the steels are commonly designated by minimum tensile yield strength, which ranges from about 241 MPa to more than 552 MPa. They are available in most mill forms with hot-rolled sheet and plate probably the most common, and they are typically used in the as-supplied condition. Thus, they provide high-strength without heat treatment by users, and that is the principal reason for their use, which includes structural applications in cars and cargo vessels, rail cars, and agricultural, earthmoving, and materials handling equipment as well as office buildings and highway bridges.

HSLA steels are tougher than plain carbon steels; they are not quite as formable, although sheet grades having yield strengths of 345 MPa can be formed at room temperature to 1T (one times thickness) to 2T bends, depending on thickness. The most formable are those produced with inclusion-shape control. That is, with the use of special alloying ingredients, such as rare earth metals, titanium, and zirconium, and controlled-cooling practice, resulting inclusions are small dispersed globules rather than stringer-like in shape. They are also relatively welded by all common methods, and can be brazed and soldered. Most of the steels are two to eight times more resistant to atmospheric corrosion than plain carbon steels, and those commonly called weathering steels naturally acquire a deep purple-brown corrosion-inhibiting surface that precludes painting for corrosion protection. The color is considered attractive, especially in rural areas, and, thus, the steels have found considerable use for exposed building members and highway applications. Dualphase HSLA steel has a deformable martensite phase in the ferrite matrix and exhibits a high rate of strain hardening during cold working. In the as-rolled condition in which it is supplied, it has a tensile yield strength of about 345 MPa and, thus, the ductility (about 30% tensile elongation) and formability of conventional HSLA steels of this strength level. But strains of 2%–3% during forming operations will increase yield strength in the strained regions to 552 MPa or greater. Thus, the steel provides a formability of medium-strength HSLA steels and the opportunity to achieve strength levels in selected regions equivalent to those of stronger, but less formable, as-supplied grades.

As contrasted to the HSLA steels, quenched-and-tempered steels are usually treated at the steel mill to develop optimum properties. Generally low in carbon, with an upper limit of 0.2%, they have minimum yield strengths from 551 to 861 MPa. Some two dozen types of proprietary steels of this type are produced. Many are available as three or four different strengths or hardness levels. In addition, there are several special abrasion-resistant grades. Mechanical properties are significantly influenced by section size. Hardenability is chiefly controlled by the alloying elements. Roughly, an increase in alloy content counteracts the decline of strength and toughness as section size increases. Thus, specifications for the steels take section size into account. In general, the higher-strength grades have endurance limits of about 60% of their tensile strength. Although their toughness is acceptable, they do not have the ductility of HSLA steels. Their atmospheric-corrosion resistance in general is comparable, and in some grades, it is better. Most quenched-and-tempered steels are readily welded by conventional methods.

Applications

High-strength steels can be used advantageously in any structural application where their greater strength can be utilized either to decrease the weight or increase the durability of the structure.

Although high-strength steels find application in all recognized market classifications, the largest single field of application has been in the manufacture of construction machinery and transportation equipment. One of the leading grades of high-strength steel has been used in the construction of railroad freight cars and railroad passenger cars.

In bridges, designers are increasingly recognizing the importance of reducing deadweight by using high-strength steels, particularly for bridges involving long spans in which a reduction of weight at the center permits additional savings in the weight of supporting members. High-strength steels also lend themselves to economical lower construction where the properties permit the use of section smaller than would be required in structural carbon steel. This advantage

is important to tall television towers where dynamic loading due to wind resistance is lessened by use of smaller sections, and in transmission towers where lighter weight is a substantial advantage in reducing freight and handling costs.

Another use of high-strength steels has been for columns in high-rise buildings. Judicious use of high-strength steels in place of, and in combination with, structural carbon steel can result in substantial cost savings and an increase in usable floor area. High-strength steels are also being used to advantage in framing members of industrial and farm buildings.

The weight of containers for liquefied petroleum gas has been reduced appreciably by the use of high-strength steel, making them easier and less costly to handle and ship. Almost all such containers are now made of high-strength steel.

Other applications of high-strength steels include the inner bottoms, floors, tanks, and hatch covers of ore boats; hulls and other structural members of small tankers, barges, tugs, launches, and riverboats; coal bunkers; street lighting poles; portable oil-drilling rigs; jet-blast fences; cable reels; automobile bumpers; pole-line hardware; air-conditioning equipment; stokers; agricultural-machinery parts; earthmoving equipment; military and domestic shipping containers; and air-preheater units.

Low-Carbon Ferritic Steels

Low-carbon ferritic steels, which were developed by Inco, Ltd., are low-alloy steels containing nickel, copper, and columbium. They are precipitation-hardened and have yield strengths from 482 to 689 MPa and sections up to 1.9 cm. They possess excellent welding and cold-forming characteristics. A major use of these steels has been for vehicle frame members. Atmospheric corrosion resistance is roughly three or four times that of carbon steels.

Low-Cycle Fatigue

Fatigue that occurs at relatively small numbers of cycles ($<10^4$ cycles). Low-cycle fatigue may be accompanied by some plastic, or permanent, deformation. Compare with *high-cycle fatigue*.

Low-Energy Electron Diffraction

A technique for studying the atomic structure of single-crystal surfaces, in which electrons of uniform energy in the approximate range of 5–500 eV are scattered from a surface. Those scattered electrons that have lost no energy are selected and accelerated to a fluorescent screen where the diffraction pattern from the surface is observed.

Lower Bainite

Bainite formed at relatively low temperature as a result of cooling at a rate not quite sufficient to form martensite. See *steel* and *isothermal transformation diagram*.

Lower Ram

The part of a pneumatic or hydraulic press that is moving in a lower cylinder and transmits pressure to the lower punch. See the term *hydraulic press*.

Lower Shelf

See *brittle fracture*.

Lower Yield

See *tensile test*.

Low-Expansion Alloys

These are alloys, mainly of iron and nickel, having low coefficients of thermal expansion, usually within a specific temperature range. Uses include precision-instrument parts requiring dimensional stability at various temperatures and glass-to-metal ceiling applications, in which the thermal expansivity of the metal must closely match that of the glass. The best-known alloy Invar, also known as Nilvar, an iron–36% nickel composition also containing (as impurities) minute amounts of carbon, manganese, and silicon. It has a lowest coefficient of thermal expansion of all metals in the –273°C to 177°C range. In the annealed condition, the alloy has the coefficient of thermal expansion ranging from about 1.44 m/m/K × 10⁻⁶ at –17.8°C to 25°C. At 149°C, the value is still only 1.8 m/m/K × 10⁻⁶. Expansivity is affected by heat treatment and cold work. Quenching from about 830°C, for example, reduces the coefficient of thermal expansion below that of annealed material, as does cold forming. A combination of quenching and cold work can even result in zero or negative coefficients. Invar has a thermal conductivity of 11 W/m K from room temperature to 100°C and is quite soft, with a hardness of about Brinell 160. Tensile properties are about 517 MPa ultimate strength, 345 MPa yield strength, and 35%–40% elongation. The alloy is ferromagnetic at room temperature but becomes paramagnetic with increasing temperature. Because the thermal expansivity of the alloy is rather constant within a specific temperature range, Invar is also known as a controlled-expansion alloy.

There are many other such alloys, each suited for specific coefficients of thermal expansion within certain temperature ranges. They include iron with 39% nickel, or Fe–39Ni, Fe–42Ni (Dumet and Alloy 42), Fe–48Ni (Platinite), Fe-48.5Ni, Fe-50.5Ni, Fe-42Ni-6Cr, Fe-45Ni-6Cr, Fe-36Ni-12Cr (Elinvar), Fe-22Ni-3Cr, and Fe-42Ni-5.5Cr-2.5Ti-0.40Al (NiSpan C and Elinvar Extra). Besides its low coefficient of thermal expansion, Elinvar is noted for its constant modulus of elasticity over a wide temperature range.

Cobalt in iron–nickel alloys increases the coefficient of thermal expansion at room temperature but enhances thermal stability over a wider temperature range. Kovar and Fernico, Fe-28Ni-18Co alloys, and Fernichrome (Fe-30Ni-25Co-8Cr) are used for applications requiring vacuum sealing to glass. A Co54-Fe37-Cr9 alloy is noted for its near-zero and sometimes negative coefficient of thermal expansion in the 0°C–100°C range. Elgiloy (40Co-20Cr-15.5Ni-15.3Fe-7Mo-2Mn-0.15C-0.04Be), originally a watch-spring alloy, has found many more spring applications. Besides dimensional stability, the alloy is noted for its good fatigue strength, corrosion and heat resistance, and nonmagnetic characteristics. Incoloy 903 (42Fe-38Ni-15Co-3Cb-1.4Ti-0.7Al), which is also heat-treatable, is noted for a new constant coefficient of thermal expansion, about 7.2 m/m/K × 10⁻⁶ from 100°C to 427°C and a near-constant modulus of elasticity from –196°C to 649°C.

Other low-expansion or controlled-expansion alloys that have been developed include Nivar, which contains 54% cobalt; the Swiss alloys Nivarox (Fe-37Ni-8Cr with small amounts of manganese, beryllium, silicon, and carbon) and Contracid (60Ni-15Cr-15Fe-7Mo-2Mn and small amounts of beryllium and silicon); the French iron–nickel alloys Dilvar and Adr; Super-Invar from Japan, a 5% cobalt; iron–nickel alloy; Sylvania 4 (Fe-42Ni-5.7Cr with small amounts of manganese, silicon, carbon, and aluminum) and the similar Sealmet HC-4; Niron 52 (52Ni-48Fe); Rodar (Fe-29Ni-17Co-0.3Mn); and Nicromet (54Fe-46Ni).

Low-Hydrogen Electrode

A covered arc welding electrode that provides an atmosphere around the arc and molten weld metal that is low in hydrogen.

Low-Pressure Laminates

In general, composite laminate's molded and cured in the range of pressures from 2760 kPa (400 psi) down to and including pressure obtained by the mere contact of the plies.

Low-Pressure Molding

The distribution of relatively uniform low-pressure (1400 kPa, or 200 psi, or less) over a resin-bearing fibrous assembly of cellulose, glass, asbestos, or other material, with or without application of heat from an external source, to form a structure possessing specific physical properties.

Low-Profile Resins

Special polyester resin systems for reinforced plastics that are combinations of thermoset and thermoplastic resins. Although the terms of low-profile and low-shrink are sometimes used interchangeably, there is a difference. Low shrink resins contain up to 30 wt.% thermoplastic polymer, while low-profile resins contain from 30 to 50 wt.%. Low shrink offers minimum surface waviness in the molded part (as low as 25 µm/25 mm, or 1 mil/in., mold shrinkage); low profile offers no surface waviness (from 12.7 to 0 µm/25 mm, or 0.5 to 0 mils/in., mold shrinkage).

Low-Residual-Phosphorus Copper

Deoxidized copper with residual phosphorus present in amounts (usually 0.004%–0.012%) generally too small to decrease appreciably the electrical conductivity of the copper.

Low-Shaft Furnace

A short shaft-type blast furnace used to produce pig iron and Ferro alloys from low-grade ores, using low-grade fuel. The air blast is often enriched with oxygen. Also used for making a variety of other products such as alumina, cementmaking slags, and ammonia synthesis gas.

Low-Shrink Resins

See *low-profile resins*.

Low-Stress Abrasion

A form of abrasion in which relatively low contact pressures on the abrading particles or protuberances cause only fine scratches and microscopic cutting chips to be produced. See also *high-stress abrasion*.

LPG

Liquid Petroleum Gas.

L-Radiation

Characteristic x-rays produced by an atom or ion when a vacancy in the L shell is filled by an electron from another shell.

L-Series

The set of characteristic x-ray wavelengths making up L-radiation for the various elements.

Lubricant

Materials imposed at the interface between surfaces in relative motion. Their primary function is to keep the surfaces apart and hence avoid surface damage and seizure. Clearly they should minimize friction and they may also act as coolants. (1) Any substance interposed between two surfaces in relative motion for the purpose of reducing the friction or wear between them. This definition implies intentional addition of a substance to an interface; however, species such as oxides and tarnishes on certain metals can also act as lubricants even though they were not added to the system intentionally. (2) A material applied to dies, molds, plungers, or workpieces that promotes the flow of metal, reduces friction and wear, and aids in the release of the finished part. (3) Lubricants may be solids including graphite and PTFE, semi solids including grease and animal fats, liquid including oil and water (in appropriate cases), glass (*see Ugine Process*) or gas including air.

Lubricants

Lubricants are substances that facilitate the flow of nonplastic, or poorly plastic, materials in the formation of dense compacts under pressure. Lubricants, which may be liquids or solids, either organic or inorganic, are particularly useful in dry pressing. Pieces formed under high pressure are apt to stick to the die, and even more so if the die is of intricate design.

It has been shown that proper lubrication of the die and of the powders to be pressed does much to equalize pressure in the piece. Lubricants reduce the friction between particles, and particles and die surfaces. This results in denser compacts, possible use of lower forming pressures, and easier ejection.

According to some investigators, the major cause of pressure variation is due to die surface friction. It has been shown by them that a stearic acid lubricant applied to the die walls completely eliminated pressure variations. Since it is impractical, in many cases, to lubricate the die after each operation, lubricants must be added to the powders.

Lubricants can be added directly to the powder batch, or, in other cases, special techniques must be used, such as hot mixing to disperse low-melting-point solids. The total amount of lubricants, with other special additives, ranges from ~0.1% to ~10% of batch weight. The lubricants themselves usually are used in less than 5% amounts. Actual concentrations depend on the basic nature of the body and the complexity of the shape.

Some lubricants also serve as binders and plasticizers. These are beneficial secondary functions and might serve as a basis for selecting one lubricant over another. Other factors to be considered are the compatibility of the lubricant with the body and the manufacturing processes, possible discoloration in the fired state, undesirable residues, and the effect on glaze application.

Types

Descriptions of typical lubricants that have been used in industry follow:

- *Alginates*: Colloidal carbohydrate compounds, water-soluble, that thicken ceramic bodies and facilitate pressing and extrusion operations.
- *Camphor*: A whitish, water insoluble material with a melting point of 174°C–197°C. It is soluble in several organic liquids.
- *Cetyl Alcohol*: A water insoluble, white crystalline powder with a melting point of 49.3°C. It also has a lubricating value as a result of its fatty nature.
- *Graphite, Talc, Clay, and Mica*: These are useful lubricants, particularly when finely pulverized, because of their platy nature. The plates tend to slide over one another and also deter sharp, hard particles from being embedded in the die surfaces.
- *Kerosene–Lard Oil*: These mixtures, sometimes known as die oil, can be added directly to dry powders, or else applied to die surfaces. It is the lard oil that provides the lubricating properties. This is a relatively inexpensive lubricant but it probably should not be used where a glaze is to be sprayed on the unfired piece.
- *Lignosulfonates*: An organic material derived from wood pulping that can provide improved plasticity and reduced forming friction. Some lignosulfonate products contain additives to enhance lubrication further and also to function as binders.
- *Methyl Cellulose*: Synthetic gum increases the viscosity of water phases and gives the body increased workability.
- *Mineral Oils:* Petroleum products that have viscosities similar to other orderly liquids.
- *No. 4 Fuel Oil*: A petroleum product of moderate viscosity, which might be used as are kerosene-lard oil mixtures.
- *Polyvinyl Acetate*: Available in powder or emulsion form and can be made to be stable with water.
- *Starches:* Specially prepared for the ceramic industry; reportedly serve as lubricants due to a retained superficial water film.

Lubricant Compatibility

See *compatibility (lubricant)*.

Lubricant Residue

The carbonaceous residue resulting from lubricant that is burned onto the surface of a hot forged part.

Lubricating Grease

Usually a compound of a mineral oil with a soap, lubricating grease is employed for lubricating machinery where the speed is slow or where it would be difficult to retain a free-flowing oil. The soap is one that is made from animal or vegetable oils high in stearic, oleic, and palmitic acids. The lime soaps give water resistance, or a mineral soap may be added for this purpose. Aluminum stearate gives high film strength to the grease. All of these greases are more properly designated as mineral lubricating grease. Originally, grease for lubricating purposes was hog fat or the inedible grades of lard, varying in color from white to brown. Some of these greases were stiffened with fillers of rosin, wax, or talc, which were not good lubricants. The stiffness of such a grease should be obtained with a mineral soap. ASTM specifications for heavy journal bearing grease require 45% soap content. About 2% calcium benzoate increases the melting point. Mineral lubricating grease may contain from 80% to 90% mineral oil and the remainder a lime soap.

Uses

Oronite GA-10, for example, is a sodium salt of terephthalic acid used as a gelling agent in high-temperature greases. It adds water resistance and stabilizes against the emulsion. Ortholeum 300 is a mixture of complex amines, and small amounts added to a grease will give high heat stability. Braycote 617 is a synthetic grease for rockets subject to both heat and cold.

The lubricating grease known as trough grease, used in food plants for greasing trays, tables, and conveyors, contains no mineral oil and is edible.

Lime greases do not emulsify as readily as those made with a soda base and are thus more suitable for use where water may be present.

Graphite grease contains 2%–10% amorphous graphite and is used for bearings, especially in damp places. For large ball and roller bearings a low-lime grease is used, sometimes mixed with a small percentage of graphite. Cylinder grease is made of about 85% mineral oil or mineral grease and 50% tallow. Compounded greases are also marketed containing animal and vegetable oils, or are made with blown oils and compounded with mineral oils. The fatty acids in vegetable and animal oils, however, are likely to corrode metals. Tannin holds graphite in solution, in gear grease *Metaline* is a compound of powdered antifriction metal oxide, and the gums, which is packed in holes in the bearings to form self-lubricating bearings. *Lead-Lube* grease has finally powdered lead metal suspended in the grease for heavy-duty lubrication.

Sett greases are mixtures of the calcium soaps of rosin acids with various grades of mineral oils. They are low-cost semisolid greases used for lubricating heavy gears or for greasing skidways. Clay fillers may be added to improve the film strength, or copper or lead powders may be incorporated for heavy load conditions. *Solidified oil* is also a name given to grease made from lubricating oil with a soda soap and tallow, used for heavy bearings.

Lubrication

(1) The reduction of frictional resistance and wear, or other forms of surface deterioration, between two load-bearing surfaces by the application of a *lubricant*. (2) Mixing or incorporating a lubricant with a powder to facilitate compacting and ejecting of the compact from the die cavity; also, applying a lubricant to die walls and/or punch surfaces.

Lubrication Regimes

Ranges of operating conditions for lubricated *tribosystems* that can be distinguished by their frictional characteristics and/or by the manner and amount of separation of the bearing surfaces. See also *boundary lubrication, elastiohydrodynamic lubrication, full-film lubrication, hydrodynamic lubrication, quasi-hydrodynamic lubrication,* and *thin-film lubrication.*

Lubricious (Lubricous)

Relating to a substance or surface condition that tends to produce relatively low friction.

Lubricity

The ability of a lubricant to reduce wear and friction, other than by its purely viscous properties.

Lüders Lines

Elongated surface markings or depressions in sheet metal, also visible with the unaided eye, caused by discontinuous (inhomogeneous) yielding. Also known as *Lüders bands, Hartmann lines, Piobert lines,* or *stretcher strains.* Lines formed on the surface of components due to uneven deformation when only a portion of the component exceeds the yield point. The lines form the boundary of the yielded zones and traverse the surface as the material continues to yield. Steels having a pronounced yield point are particularly susceptible. The lines are aligned at 45° to the tension axis and are observed on components such as lightly drawn tube and pressed sheet. Although not damaging, they may be unsightly but can usually be prevented by prior uniform light working.

Luggin Probe

A small tube or capillary filled with electrolyte, terminating close to the metal surface under study, and used to provide an ionically conducting path without diffusion between an *electrode* under study and a *reference electrode.*

Luminous

Luminescent emitting light or other radiation by phosphorescence and such phenomenon.

Luminous Flame

A flame that emits significant light due to incandescent carbon particles remaining from incomplete combustion.

Luminous Pigments

Pigments used in paints to make surfaces visible in the dark and in coatings for electronic purposes. They are used for signs, watch and instrument hands, airfield markings, and signals. They are of two general classes. The *permanent* ones are the *radioactive paints,* which give off light without activation, and the *phosphorescent paints,* or *fluorescent paints,* which require activation from an outside source of light. The radioactive paints contain a radioactive element that emits alpha and beta rays, which strike the phosphors and produce visible light. Radium, sometimes used for paints for watch hands, gives a greenish-blue light, but it emits dangerous gamma rays. Also, the intense alpha rays of radium destroy the phosphors quickly, reducing the light. Strontium 90 gives a yellow-green light and has a long half-life of 25 years, but it emits both beta and gamma rays and is dangerous. *Tritium paints,* with a tritium isotope and a phosphor in the resin-solvent paint base, have a half-life of 12.5 years and require no shielding. The self-luminous phosphors for clock and instrument dials contain tritium, which gives off beta rays with only low secondary emission so that the glass or plastic covering is sufficient shielding. Other materials used are *krypton 85,* with a half-life of 10.27 years, *promethium 147,* with a half-life of 2.36 years, and *thallium 204,* with a half-life of 2.7 years.

Fluorescent paints depend upon the ability of the chemical to absorb energy from light and to emit it again in the form of photons of light. This variety usually has a base of calcium, strontium, or barium sulfide and traces of other metal salts to improve luminosity, and the vehicle contains a moisture proof gum or oil. Temporarily luminous paints may be visible for long periods after the activating light is withdrawn. A paint activated by 5 min exposure to sunlight may absorb sufficient energy for 24 h of luminosity. *Luminous wall paints* used for operating rooms to eliminate shadows are made by mixing small amounts of zinc or cadmium sulfide into ordinary paints. After being activated with ultraviolet rays, they will give off light for 1.5 h.

Phosphorescent paints are lower in cost than radioactive paints and may be obtained in various colors. In general, the yellow and orange phosphorescent pigments are combinations of zinc and cadmium sulfide, the green is zinc sulfide, and the violet and blue pigments are combinations of calcium and strontium sulfides. They are marketed in powder form to be stirred into the paint or ink vehicle, since mixing by grinding lowers the phosphorescence. The natural minerals are not used, as the pigments must be of a high degree of purity, as little as a millionth part of iron, cobalt, or nickel killing the luminosity of zinc sulfide. These phosphorescent pigments are called *phosphors*, but technically they are incomplete phosphors; and copper, silver, or manganese is coprecipitated with the sulfide as an activator or to change the color of the emitted light. The metals that are used as activators are called *phosphorogens*, and their atoms diffuse into the lattice of the sulfide. For fluorescent screens the phosphors must have a rapid rate of extinguishment so that there will be no time lag in the appearance of the events. For television, electron microscope, and radar screens, the phosphors must cease to flow 0.02 s after withdrawal of excitation. They must also be of very minute particle size so as not to give a blurred image. For a white television screen, mixtures of blue zinc sulfide with silver and yellow zinc–beryllium silicates are used. For color television screen is completely covered with a mixture of various colored phosphors, especially rare-earth metal combinations. For scintillation counters for gamma-ray-detection phosphors, the pulses should be of longer duration, and for this purpose crystals of cadmium or cadmium tungstate are used.

Fluorescent fabrics for signal flags and luminescent clothing are impregnated with fluorescent chemicals, which can be activated by an ultraviolet light that is not seen with the eye. Some fluorescent paints contain a small amount of luminous pigment to increase the vividness of the color by absorbing the ultraviolet light and emitting it as visible color. *Fluorescein*, made from phallic anhydride and resorcinol, has the property of fluorescence in a solvent. Since cellulose acetate will keep it in a permanently solid-state, acetate rayon is used as a carrier fabric. Signal panels are distinguishable from a plane at great heights even through a haze, and at night they give a brilliant glow when activated with ultraviolet rays. Fluorescent paints for signs may have a white undercoat to reflect the light passing through the semitransparent pigment. In passing through the color pigments the shorter violet and blue wavelengths are changed to orange, red, and yellow hues, and the reflected visible light is greater than the original light. *Uranine*, the sodium salt of fluorescein, is used by flyers to mark spots in the ocean. One pound (0.45 kg) of uranine will cover 1 acre (4047 m^2) of water to a brilliant, yellowish green easily seen from the air. One part of uranine is detectable in 16×10^6 parts of water. *Luminous plastic* for aircraft markings is coated on the inside with radioactive material to give visibility in the dark.

The fluorescent pigments almost always consist of particles of a colorless resin containing a color-fluorescing dye. Two well-known dyes are *Potomac Yellow* and *Alberta Yellow* from Day-Glo Color Corp. and BASF AG makes a series under the trade name *Rhodamine*.

Whitening agents, optical whiteners, or brightening agents, used to increase the whiteness of paper and textiles, are fluorescent materials that convert some of the ultraviolet of sunlight to visible light. The materials are colorless, but the additional light supplied is blue, and it neutralizes yellow discolorations and enhances the whiteness.

Luminous materials also occur in nature as organic materials, with bioluminescence thought to be a form of chemiluminescence. Fireflies, bacteria, glow worms, and some luminous fish are capable of this feat. It occurs by the mixing of two substances present in the organism; one is *luciferin*, which oxidizes the second, an enzyme known as *luciferase*. The reaction produces an excited form of luciferase, which emits light when it returns to its normal state.

Luster Finish

A bright as-rolled finish, produced on ground metal rolls; it is suitable for decorative painting or plating, but usually must undergo additional surface preparation after forming.

Lute

(1) A mixture of fireclay used to seal cracks between a crucible and its cover or between container and cover when heat is to be applied. (2) To seal with clay or other plastic material.

Lutes

Adhesive substances, usually of earth composition, deriving the name from the Latin *lutum*, meaning "mud." A clay cement was used by the Romans for cementing iron posts into stone. Although lutes often contain a high percentage of silica sand or clays, the active ingredient is usually sulfur. They may also contain other reactive ingredients such as lead monoxide or magnesium compounds. *Plumber's lutes* are used for pipe joints and seams and for coating pipes to withstand high temperatures. Plaster of paris mixed with a weak glue will withstand a dull-red heat. *Sulfur cements*, or lutes, usually have fillers of silica or carbon to improve the strength. They form a class of acid-proof cements used for ceramic pipe connections. Modern lutes for very high heat resistance do not contain elemental sulfur. Industrial lutes are used for sealing in wires and connections in electrical apparatus, and are compounded to give good bonding to ceramics and metals. A lute cement for adhering knife blades to handles is composed of magnesium acid sulfates, calcined magnesia, with fine silica or powder. The term *sealant* generally refers to a wide range of mineral-filled plastics formulated with a high proportion of filler for application by troweling or air gun.

Lye

Concentrated solutions of either sodium hydroxide or potassium hydroxide.

Lyotropic Liquid Crystal

A type of liquid crystalline polymer that can be processed only from solution.

M

Macerate

To chop or shred, as fabric, for use as a filler for a molding resin.

Machinability

The relative ease of machining a metal.

Machinability Index

A relative measure of the machinability of an engineering material under specified standard conditions. Also known as *machinability rating*.

Machinability of Metals

The ease and economy with which a metal may be cut under average conditions is its machinability. Frequently, no truly quantitative assessment is made, but rather a rating or an index is established vis-à-vis a reference material. More quantitative comparisons are based on tool life. For example, maximum cutting speeds for a given tool life may be used as a rating of machinability. Alternatively, tool wear rate may be the basis for a machinability rating. Surface finishes sometimes used for assessing machinability.

Machining Process

The wide range of metal-cutting processes may be represented, with some oversimplification, by the orthogonal cutting process.

Cutting Speed

The machining response of ductile metals is very sensitive to cutting speed. Below about 0.02 m/s, chips form continuously, metal chunks are lifted out of the surface, and the surface is scalloped or pockmarked. When the speed is in the 0.1 m/s range, chips are formed continuously, the shear zone is narrow, and the chip slides on the tools face. Under these conditions, a cutting fluid can lubricate both the rake and flank faces of the tool.

Cutting Fluid

The interaction between the tool and the workpiece is considerably affected by the presence of cutting fluids. Cutting fluid has two primary functions. First, as long as the cutting speed is slow, the cutting fluid can act as a lubricant between the chip and the tool face. Even at higher speeds, some lubricating effect at the flank face may be present. Second, and perhaps more importantly, the cutting fluid serves as a coolant. In most instances, the cutting fluid will be an emulsion of a lubricating phase (oil, graphite, and so on) in water since water is the best heat-transfer medium readily available. Beyond lubrication and cooling, the cutting fluid can be used to flush out the cutting zone.

Surface Quality and Tool Wear

Cutting speed thus affects the all-important machinability considerations of surface quality and tool wear. Surface finish is best with a well-lubricated, moderately low-speed operation or high-speed cutting with no built-up edge. The high pressures and temperatures of operation, abetted by shock loading and vibrations, can lead to rapid tool wear. Tool wear is often sufficiently rapid to make tool replacement a major factor in machining economics. Tools must be replaced when they break or when they have worn to the point of producing an unacceptable surface finish or an unacceptable degree of surface heating.

Metal Properties

A machining operation can be optimized to affect metal removal with the least energy, or the best surface, or the longest tool life, or a reasonable compromise among these factors. Even so, it remains that the optimal ease and economy of machining some alloys is vastly different from that of others, and some basic attributes of easily machined alloys can be set forth.

Toughness

To ensure that chip separation occurs after minimum sliding, low ductility is required. To minimize cutting force, low strength and low ductility combine to mean low toughness. Toughness is generally defined as energy per unit volume consumed en route to fracture. Ironically, materials of maximized toughness are desirable for most engineering applications, and, thus, some of the most attractive alloys, such as austenitic stainless steels, are difficult to machine.

Adhesion

The degree to which the metal adheres to the tool material is important to its machinability.

Actually this attribute can work to advantage or disadvantage. If diffusion results, the tool can be weakened and rapid wear occurs. Otherwise, high adhesion will stabilize the secondary shear zone.

Workpiece Second Phases

Small particles or inclusions in the metal can have a marked effect on machinability. Hard sharp oxides, carbides, and certain intermetallic compounds abrade tooling and accelerate tool wear. On the other hand, soft second phases are beneficial because they promote localized shear and chip breakage.

Thermal Conductivity

In some cases, workpiece thermal conductivity can be important to machinability. A low thermal conductivity generally results in high shear-zone temperature. This can be advantageous in reducing the strength of the metal or in softening second-phase particles. Of course, if adhesion and diffusional depletion of tool alloy content result, the high temperature is a problem. The workpiece temperature can be managed by cutting-speed and cutting-fluid manipulation.

Alloy Systems

Commercial alloys can be grouped into two categories, namely, those designed for ease of machining (so-called free machining grades), and the vast majority, which are of widely varying but generally less than optimum machinability. Considering these latter, ordinary alloys, it can be shown that their machinability may be considerably improved by metallurgical operations that limit strength or ductility or both. Of course, it is not often possible to reduce both strength and ductility simultaneously. Even so, machinability often can be improved by grossly reducing one or the other property.

Machine Forging

Forging performed in upsetters or horizontal forging machines.

Machine Shot Capacity

The maximum weight of thermoplastic resin that can be displaced or injected by the injection (molding) ram in a single stroke.

Machine Welding

Welding with equipment that performs under the continual observation and control of a welding operator. The equipment may or may not load the work. Compare with *automatic welding*.

Machining

Any one of a group of operations that change the shape, surface finish, or mechanical properties of a material by the application of special tools and equipment. Machining almost always is a process in which a cutting tool removes material to affect the desired change in the workpiece. Typically, powered machinery is required to operate the cutting tools.

Although various machining operations may appear to be very different, most are very similar; they make chips. These chips vary in size from the long continuous ribbons produced on a lathe to the microfine sludge produced by lapping or grinding. These chips are formed by shearing away the workpiece material by the action of a cutting tool. Cylindrical holes can be produced in a workpiece by drilling, milling, reaming, turning, and electric-discharge machining (EDM). Rectangular (or nonround) holes and slots may be produced by broaching, EDM, milling, grinding, and nibbling; and cylinders may be produced on lathes and grinders. Special geometries, such as threads and gears, are produced with special tooling and equipment utilizing the same turning and grinding mentioned earlier. Polishing, lapping, and buffing are variants of grinding where a very small amount of stock is removed from the workpiece to produce a high-quality surface.

In almost every case, machining accuracy, economics, and production rates are controlled by the careful evaluation and selection of tooling and equipment. Speed of cut, depth of cut, cutting-tool material selection, and machine tool selection have a tremendous impact on machining. In general, the more rigid and vibration-free a machining tool is, the better it will perform. Jigs and fixtures are often used to support the workpiece. Since it relies on the plastic deformation and shearing of the workpiece by the cutting tool, machining generates heat that must be dissipated before it damages the workpiece or tooling. Coolants, which also act as lubricants, are often used.

Boring

This machining operation increases the size of an existing hole in a workpiece. The usual purpose of boring is to produce a hole with an accurate diameter and good surface finish. Boring can be performed on a special machine or a lathe, with either the workpiece or the boring tool being on a movable table. A rotating spindle, holding either a single-point cutting tool or the workpiece, is fed into the work. As the spindle rotates, the cutting tool engages the interior of the existing hole, and chips are formed as a tool cuts into the workpiece. The actual cutting action of a boring tool is very similar to a lathe turning tool.

Broaching

This is the removal of material to produce a slot (or other formed shape) in a workpiece by moving a multiple-tooth, barlike tool across the workpiece. The cutting action results from the configuration of each tooth being progressively higher than the preceding one. Tooth of the brooch removes a small, predetermined amount of stock, the chip. Broaching is a very economical machining operation, although tool costs can be high; accordingly, it is applied most often too high-volume production.

Drilling

One of the most common machining operations, drilling is a method of producing a cylindrical hole in a workpiece. A typical twist drill consists of a helically grooved steel rod with two cutting edges on the end. The helical flutes or grooves in the drill allow the chips to be removed from the cutting edge, conduct coolant to the cutting lips, and form part of the cutting-edge geometry. Although drills may appear to be simple, their geometries are carefully controlled. Drilling is a very fast and economical process, but it usually does not produce a very accurate hole diameter or a fine surface finish.

Turning

This type of machining is performed on a lathe. The process involves the removal of material from a workpiece by rotating the workpiece under power against a cutting tool. The cutting tool is held in a tool post that is supported on a cross slide and carriage. The tool may be moved radially or longitudinally in relation to the turning axis of the workpiece. Forms such as cones, spheres, and related workpieces of concentric shape as well as true cylinders can be turned on a lathe.

The most common lathe is an engine, where the workpiece may be rotated and held between tapered centers or by means of a collet or chuck. A turret lathe is an engine lathe that has a multisided indexing tool holder or turret instead of a rail stock center. This adds to the versatility of the machine by allowing a greater variety of cutting tools to be applied to the rotating workpiece.

Automatic Screw Machines

Automatic screw machines are sophisticated lathes that have been designed to perform several turning operations automatically in rapid succession without removing the workpiece from the machine. They are used when the volume and complexity of the required workpieces justify the expense of setting up and operating these very versatile but complex machines. A screw machine (named for the screw manufacturing role for which it was created) is cam-controlled or driven by computer numerical control (CNC).

Milling

This process removes material by feeding a workpiece through the periphery of a rotating circular cutter. Each tooth of the rotating multitoothed milling cutter removes a portion of material from the passing workpiece.

Milling cutters are designed for particular operations and are classified as either shell type or end type. Shell mils are disk shaped and usually produce continuous slots. When well supported in the machine tool to minimize vibration, they perform well and economically. The more versatile end mill is held by its shank only. End mills can be used for slotting just like a shell mill, but they also can cut pockets, contours, and even cylindrical holes. End mills with four or six flutes are stronger and more rigid than two-flute mills, and so are better able to machine tougher materials.

Reaming

This machining operation enlarges an existing hole by a few thousandths of an inch and produces a hole whose diameter is very accurately controlled. The cutting edges of a reamer may be ground on the apexes between longitudinal flutes or grooves, or cutting may take place on chamfered edges at the end of the reamer. Reaming is performed either manually or by machine, often as a finishing operation after drilling. Reamed holes have good surface finish.

Sawing

This is the parting of material by using blades, bands, or abrasive disks as the cutting tools. In the most common type of saw, a toothed blade is passed across the workpiece in either a reciprocating or continuous motion. The teeth can be mounted on continuous bands, short steel blades, or the periphery of a disk. Friction sawing is a rapid process used to cut steel as well as certain plastics. A very high-speed blade softens the workpiece material with frictional heat. The material is then wiped away from the workpiece by the cutting blade. Since many teeth engage the workpiece, no single tooth overheats. Abrasive sawing looks similar to friction sawing, except that a thin rubber or bakelite bonded abrasive disk grinds the material away instead of simply softening and wiping it.

Nibbling

Nibbling is the operation that cuts away small pieces of material by the action of a reciprocating punch. A nibbler takes repeated small bites from the workpiece (which is usually a thin sheet of material) utilizing a quickly reciprocating punch system. As the work is passed beneath the punch, a small nibble of material is removed by the punch during each punch cycle. After each quick nibble, the workpiece is advanced under the punch to allow another small bite to be taken away. In this manner, a great deal of material can be removed from the workpiece after several tens or hundreds of individual punching cycles.

Shaping

The process cuts flat or contoured surfaces by reciprocating a single-point tool across the workpiece. The tool is mounted on a hinged unit known as a clapper box, which lifts up to disengage the tool from the surface on the return stroke. The cutting action of a shaper is actually similar to turning, except that the single-point cutting tool moves straight across a workpiece, instead of having the workpiece rotate against the cutting tool.

Grinding

This process removes material by the cutting action of a solid rotating, grinding wheel. The abrasive grains of the wheel perform a multitude of minute machining cuts on the workpiece. Although grinding is sometimes used as the sole machining operation on a surface, it is generally considered a finishing process used to obtain a fine surface and extremely accurate dimensions.

Grinding is used to machine a wide range of metals, carbide materials, stone, and ceramics. The grinding process may be used on metals too hard to machine, otherwise because commercial grinding abrasives are many times harder than the metals to be machined.

Grinding wheels are composed of abrasive grains plus a bonding material. The wheels are often very porous, with homogeneous open areas between the grains. The abrasives most commonly used are silicon carbide and aluminum oxide. Coarse-grained wheels are used for rapid removal of stock; wheels with fine grains cut more slowly but give smoother finishes. Coolants are applied to the grinding point to dissipate the heat generated and to flush away the fine chips.

Honing

Honing is a grinding process that removes a small amount of material from a workpiece by means of abrasive stones. It is able to produce extremely close dimensional tolerances and very fine surface finishes. The abrading action of the fine-grit stones occurs on a wide surface area rather than on a line of contact as in grinding.

Lapping

This is a precision abrading process used to finish a surface to a desired state of refinement or dimensional accuracy by removing an extremely small amount of material. Lapping is accomplished by abrading a surface with a fine abrasive grit rubbed about it in a random manner. A loose unbonded grit is used. It is traversed about on a lap, made of a somewhat softer material than the workpiece. The unbonded grit is mixed with a vehicle such as oil, grease, or soap and water.

Polishing

This is a smoothing of a surface by the cutting action of an abrasive grit that is either glued to or impregnated in a flexible wheel or belt. Polishing is not a precision process; it removes stock until the desired surface condition is obtained.

Superfinishing

For a mirror-like surface, a precision abrading process known as superfinishing is used. Superfinishing removes minute flaws or inequalities because it is performed with an extremely fine-grit abrasive stone, shaped to match and cover a large portion of work surface.

Buffing

This is the smoothing and brightening of a surface by rubbing it with a fine abrasive compound carried in a soft wheel or belt. The abrasive is a fine powder or flour mixed with tallow or wax to form a smooth paste. Buffing is performed on the same type of machines as polishing; frequently, both buffing and polishing wheels are included on the same machine. Buffing differs from polishing in that a finer grit is used and less material is removed from the workpiece.

Ultrasonic Machining

In this process, material is removed by abrasive bombardment and crushing in which a vibrating tool drives an abrasive grit against the workpiece. In ultrasonic machining, the tool never directly contacts the workpiece; rather, the vibrations (typically 20,000 Hz) drive the abrasive, which is suspended in a liquid, against the workpiece. Impressions may be economically sunk in glass, ceramics, carbides, and hard brittle metals by this method. The shape of the tool basically determines the shape of the impression.

Electric-Discharge Machining

This is a machining process in which electrically conductive materials can be removed by repeated electric sparks. EDM is used to form holes of varied shape and materials of poor machinability. Unlike other machining operations, it does not rely on a cutting tool to shear away the workpiece. Instead, it uses electrical energy to melt or vaporize small areas on the workpiece. The sparks created by an EDM unit (at a rate greater than 20,000 per second) are discharged through the space between the tool (cathode) and the workpiece (anode). The small gap between the tool and workpiece is filled with a circulating dielectric hydrocarbon oil, which serves as a cooling medium and flushes away metal particles. Although not as fast as other machining processes, EDM has the unique ability to remove hard materials that otherwise would not be machinable.

Machining Allowance

See *finish allowance*.

Machining Damage (Ceramics and Glasses)

A typical or excessively large surface microcracks or damage resulting from the machining process; for example, striations, scratches, and impact cracks. Small surface and subsurface damage is intrinsic to the machining damage.

Machining Damage (Metals)

See *surface alterations* (metals) and *surface integrity* (metals).

Machining Stress

Residual stress caused by machining.

Machining, High Speed

Machining operations such as turning, milling, and drilling involve the modification of a workpiece through the removal of material. In machining of metals, this material removal occurs via a concentrated shear flow initiated at the point of contact between the workpiece and a wedge-shaped tool (e.g., turning insert and milling flute) to produce a chip and a machined surface. The cutting action is generated by the rotation of the machine spindle. Increasing the power and speed of the spindle and axes has several advantages, which include (1) shorter machining time, (2) improved surface finish, (3) reduced thermal and mechanical stresses on the workpiece and tool, and (4) increased dynamic stability. These potential advantages have driven a recent, rapid increase in the industrial adoption of high-speed machining processes and technology.

The definition of high-speed machining is not static but instead changes with time depending on available tool materials and machine technology. Reliable machining spindles capable of running continuously at speeds of up to 40,000 revolutions per minute (rpm) with a power output of 30 kW are now available on production equipment. In addition, axis speeds of production machining centers now exceeded 60 m/min with accelerations of greater than $1g$ (g = gravitational acceleration 9.8 m/s^2). The marked increase in material removal rates and the reduction in production time and costs afforded by this new technology have the potential of revolutionizing the manufacturing process. However, because of rapid tool wear, not all materials can be machined at high speeds. In general, machining speeds on difficult-to-machine materials such as titanium and hardened steel are limited by the availability of suitable tool materials, whereas machining speeds and more machinable material such as aluminum and plastic are limited by the available machine technology.

Tooling Materials

The increased use of high-speed machining has been enhanced by the development of suitable tool materials. In general, the higher the yield strength and melting temperature of the material being machined, the greater the stresses and temperatures that will be generated at the tool–chip interface. These increased stresses and temperatures cause increased rates of chemical and mechanical tool wears, but with the introduction of Al_2O_3-TiC, silica nitride, and SiC whisker-reinforced Al_2O_3 ceramic cutting tools, the high-speed machining of certain materials becomes practical. Speeds as high as 5000 sfm have been reported, and speeds 1000–3000 sfm are routine with essentially pure ceramics (i.e., low metallic impurities). However, the older cermets (ceramic-coated cemented carbides) have been limited to speeds below 1400 sfm. This is because plastic deformation of the metallic phase in these carbides occurs at temperatures as low as 600°C, which limits the cutting speed.

High toughness, strength, and thermal conductivity result in tools that have improved resistance to chipping, which makes it possible to do interrupted cutting when machining metal. High hardness generally correlates with improved wear resistance, although chemical compatibility is essential for good performance. Since temperatures can be extremely high at the tools/workpiece interface, it is important to retain high hardness and toughness as the temperature increases. The greater efficiency of many ceramics, compared with cermets, is that they maintain wear resistance (i.e., hardness) and strength at these elevated temperatures.

Controllers: Data Processing

The increased complexity of machined parts leads to a similar increase in the size and complexity of the programs required to generate necessary machine motion. The flow rate of the data required to control tool motions over complex surfaces has grown dramatically as spindle speeds have increased. This is especially true as high-speed machining becomes practical in the die and mold industry, where the tool motion over the workpiece is described in a series of point-to-point movements and the volume of data required to geometrically define a part is quite large. As a result, machine tool controllers must be able to store large amounts of data and process the data quickly. Most high-speed machines are now equipped with large memories, powerful computers, and network connections.

Macro

In reinforced plastics, the gross properties of a composite as a structural element without consideration of the individual properties are the identity of the constituents.

Macroetching

Etching a metal surface to accentuate gross structural details, such as grain flow, segregation, porosity, or cracks, for observation by the unaided eye or at magnification to 25×.

Macrograph

A graphic representation of the surface of a prepared specimen at a magnification not exceeding 25×. When photographed, the reproduction is known as a photomacrograph.

Macrohardness Test

A term applied to such hardness testing procedures as the Rockwell or Brinell hardness tests to distinguish them from microindentation hardness tests such as the Knoop or Vickers tests. See also *microindentation* and *microindentation hardness number*.

Macropore

Pores in pressed or sintered powder compacts that are visible with the naked eye.

Macroscopic

Visible at magnification at or below 25×.

Macroscopic Stress

Residual stress in a material in a distance comparable to the gauge length of strain measurement devices (as opposed to stresses within very small, specific regions, such as individual grains). Compare with *microscopic stress*.

Macroshrinkage

Isolated, clustered, or interconnected voids in a casting that are detectable macroscopically. Such voids are usually associated with abrupt changes in section size and are caused by feeding that is insufficient to compensate for solidification shrinkage.

Macroslip

A type of sliding in which all points on one side of the interface are moving relatively to those on the other side in a direction parallel to the interface. See also *microslip*.

Macrostrain

The main strain over any finite gage length of measurement large in comparison with interatomic distances. This can be measured by several methods, including electrical-resistant strain gages and mechanical or optical extensometers. Elastic macrostrain can be measured by x-ray diffraction.

Macrostress

Same as *macroscopic stress*.

Macrostructure

The structure of metals as revealed by macroscopic examination of the etched surface of a polished specimen.

Magnesia

A fine, white powder of **magnesium oxide**, MgO, obtained by calcining magnesite or dolomite and refining chemically and use principally in *basic refractories*. It is used in pharmaceuticals, in cosmetics, in rubbers as a scorch-resistant filler, in soaps, and in ceramics. The powder is converted from magnesium hydroxide. **Maglite**, is of Whittaker, Clark and Daniels, Inc., used for rubbers, is produced from seawater. **Magox magnesia**, of Basic Chemicals, is 98% pure MgO extracted from seawater. It comes in particle sizes to 325 mesh and high- and low-activity grades for rubber, textile, and chemical uses. A very pure magnesia is also produced by reducing magnesium nitrate.

Magnesia ceramic parts, such as crucibles and refractory parts, are generally made from magnesia that is usually electrically fused and crushed from the large cubic crystals. The crystals have ductility and can be bent. The particle size and shape are easily controlled in the crushing to fit the needs of the molded article. Pressed and sintered parts have a melting point of about 5070°F (2765°C) and can be employed to 4172°F (2300°C) in oxidizing atmospheres or to 3092°F (1700°C) in reducing atmospheres. The material is inert to molten steels and two basic slags. Magnafrax 0340, of Carborundum Co., is magnesia in the form of plates, tubes, bars, and disks. The material has a specific gravity of 3.3 and a thermal conductivity twice that of alumina. Its vitreous structure gives it about the same characteristics as a single crystal for electronic purposes. Magnorite, of Norton Co., is fused magnesia in granular crystals with a melting point of 5072°F (2800°C), used for making ceramic parts and for sheathing electric heating elements. **K-Grain magnesia**, of Kaiser Aluminum and Chemical Corp., is 98% magnesia, containing no more than 0.4 silica. The magnesium ceramic, of Corning Inc., is 99.8% pure.

Magnesia refractory brick consists mainly of the mineral periclase (MgO) and is available in chemically bonded, pitch-bonded, burned or fired, and burned and pitch-impregnated forms. Historically, natural magnesite ($MgCO_3$) that was calcined provided the raw material for this brick, but with the increased demands for higher temperatures and the introduction of fewer process impurities, more high-purity magnesia from seawater or underground brines has been used. In the seawater and brine processes, the magnesia is obtained by calcining precipitated $Mg(OH)_2$, which provides MgO of purity up to 98%. See also *refractories*.

Magnesite

A white to bluish-gray mineral used in the manufacture of bricks for basic refractory furnace linings and as an ore of magnesium.

The ground, burned magnesite is a light powder, shaped into bricks at high pressure and baked in kilns. Magnesite is a magnesium carbonate with some iron carbonate and ferric oxide. Magnesite releases carbon dioxide on heating and forms magnesia. When heated further, it forms a crystalline structure known as *periclase*. The fused magnesia made in the arc furnace is actually synthetic periclase. The synthetic material is in transparent crystals up to 2 in. (5 cm), which are crushed to powder for thermal insulation and for making refractory parts. Magnesite in compact, earthy form or granular masses has a vitreous luster, and the color may be white, gray, yellow, or brown. Mohs hardness is 3.5–4.5, and a specific gravity is about 3.1.

The product known as *dead-burned magnesite* is in the form of dense particles used for refractories. It is produced by calcining magnesite at 2642°F–2732°F (1450°C–1500°C). Caustic magnesite is a product resulting from calcination at 1292°F–2192°F (700°C–1200°C), which leaves from 2% to 7% carbon dioxide in the material and gives sufficient cementing properties for use as a refractory cement.

Magnesite for use in producing magnesium metal should have at least 40% MgO, with not over 4.5% CaO and 2% FeO. Magnesite is a valued refractory material for crucibles, furnace brick and linings, and high-temperature electrical insulation because of its basic character, chemical resistance, high softening point, and high electrical resistance. Its chief disadvantage is its low resistance to heat shock. Magnesite brick and refractory products are marketed under a variety of trade names, such as *Ritex*, of General Refractories Co., and *Ramix*. It is also used as a covering for hot piping. The German artificial stone called *Kunststein* is magnesite.

Magnesium and Alloys

A silvery-white metal, symbol Mg, magnesium is the lightest metal that is stable under ordinary conditions and produced in quantity. It is the sixth most abundant element, with a specific gravity of 1.74, melting point 650°C, boiling point about 1110°C, and electrical conductivity about 40% that of copper. Ultimate tensile strengths are about 90 MPa as cast, at least 158 MPa for annealed sheet, and 179 MPa for hard-rolled sheet, with corresponding elongations of about 4%, 10%, and 15%, respectively. The strength is somewhat higher in the forged metal. Magnesium has a close-packed hexagonal structure that makes it difficult to roll cold, and its narrow plastic range requires close control in forging. Repeated reheating causes grain growth. Sheet is usually formed at 150°C–200°C. It is the easiest to machine of the metals.

Characteristics

The excellent machinability of magnesium makes its use economical in parts where weight saving may not be of primary importance, but where much costly machining is required. Such parts, when made of magnesium, can be machined at higher speeds and with greater economy than would be possible with most other commonly used metals. Chemical milling can be used on magnesium.

Magnesium can be cast and fabricated by practically every method known to the metal worker. The metal is cast in sand or permanent molds to obtain lightweight castings with good strength, stiffness, and resistance to impact or shock loading. Magnesium sand and permanent-mold castings are heat treatable to further improve mechanical properties.

The die-casting process is similarly applicable to magnesium, and this method of casting should always be considered when the quantities desired are in the range that indicates its use. Both hot and cold chamber processes are usable.

The metal can also be cast by some of the less common methods, including plaster mold, centrifugal, shell molding, and investment processes.

Magnesium is rolled into sheet and plate and can be extruded into rods, bars, tubing, and an almost endless variety of structural and special shapes.

Sheet and extrusions are very easily formed using techniques that have been developed especially for magnesium. Stamping, deep and shallow drawing, blanking, coining, spinning, and impact extrusion are just a few of the production forming operations regularly used on magnesium, and these indicate the adaptability of the metal to a large variety of metal working procedures.

The forging of magnesium is accomplished by methods much the same as those used for forging other metals. Both press and hammer equipment are used, but the former is most commonly employed because the physical structure of magnesium makes the metal better adapted to the squeezing action of the forging press. Magnesium forgings are chosen when a high strength-to-weight ratio, rigidity, or pressure tightness is required. The selection of forging, however, is governed by the fact that, like permanent mold and die castings, a sufficient number of parts must be needed to justify the cost of die equipment.

Magnesium parts can be joined by any of the common methods. Arc and electric-resistance welding, adhesive bonding, and mechanical fastening are in daily production use. Brazing and gas welding, although not as frequently used as the other methods, are also suitable ways of joining magnesium.

Magnesium possesses relatively high thermal and electrical conductivities, very high damping capacities, and is nonferromagnetic. It has good stability to atmospheric exposure and good resistance to attack by alkalies, chromic and hydrofluoric acids, and many organic chemicals, including hydrocarbons, aldehydes, alcohols (except methyl), phenols, amines, esters, and most oils. Bare magnesium surfaces are nonsticking (snow, ice, sand, etc.) and nonmarking. Magnesium also has a low sparking tendency.

Magnesium develops a corrosion-inhibiting film upon exposure to clean atmospheres and fresh water, but that film breaks down in the presence of chlorides, sulfates, and other media, necessitating corrosion protection in many applications. Many protective treatments have been developed for this purpose. It is also rapidly attacked by mineral acids, except chromic and hydrofluoric acids, but is resistant to dilute alkalies; aliphatic and aromatic hydrocarbons; certain alcohols; and dry bromine, chlorine, and fluorine gases. Anodized magnesium is produced by immersing in a solution of ammonium fluoride and applying a current of 120 V. The fluoride film has a thickness of only 0.0003 cm, but it removes cathodic impurities from the surface of the magnesium, giving greater corrosion resistance and also better paint adhesion.

Magnesium is valued chiefly for parts where light weight is needed. It is a major constituent in many aluminum alloys, and very light alloys have been made by alloying magnesium with lithium.

The pure metal ignites easily, and even when alloyed with other metals, the fine chips must be guarded against fire. In alloying, it cannot be mixed directly into molten metals because of flashing, but is used in the form of master alloys. The metal is not very fluid just above its melting point, and casting is done at temperatures considerably above the melting point so that there is danger of burning and formation of oxides. A small amount of beryllium added to magnesium alloys reduces the tendency of the molten metal to oxidize and burn. The solubility of beryllium in magnesium is only about 0.05%. As little as 0.001% lithium also reduces fire risk in melting and working the metal. Molten magnesium decomposes water so that green-sand molds cannot be used because explosive hydrogen gas is liberated. For the same reason, water sprays cannot be used to extinguish magnesium fires. The affinity of magnesium for oxygen, however, makes the metal a good deoxidizer in the casting of other metals.

Production

Magnesium is produced commercially by the electrolysis of a fused chloride, or fluoride obtained either from brine or from a mineral ore, or it can be vaporized from some ores.

These two major methods of producing magnesium are used throughout the world. These can be seen in Figures M.1 and M.2 (electrolytic and the silicothermic processes). In the electrolytic process, the electrolysis of magnesium chloride to yield chlorine and metallic magnesium is the basis of this process. Although magnesite, dolomite, and natural brines have been used as raw materials, the principal source is seawater, which contains about 0.13% magnesium.

The ferrosilicon process (silicothermic) was developed commercially during the World War II in Canada.

Alloys and Alloy Designation

Although magnesium alloys are moderate in strength and rigidity, they have high specific strength and rigidity because of their low density, which, as it is in the range of 1771–1827 kg/m³, is the lowest of common metals. Modulus of elasticity in tension is typically 44,800 MPa, and ultimate tensile strength ranges from 152 to 379 MPa, depending on the alloy and form. Both wrought and cast alloys are available, the former in sheet, plate, rod, bar, extrusions, and forgings, and the latter for sand, permanent-mold, investment, and die castings. Alloys are designated by a series of letters and numbers followed by a temper designation. The first part of the

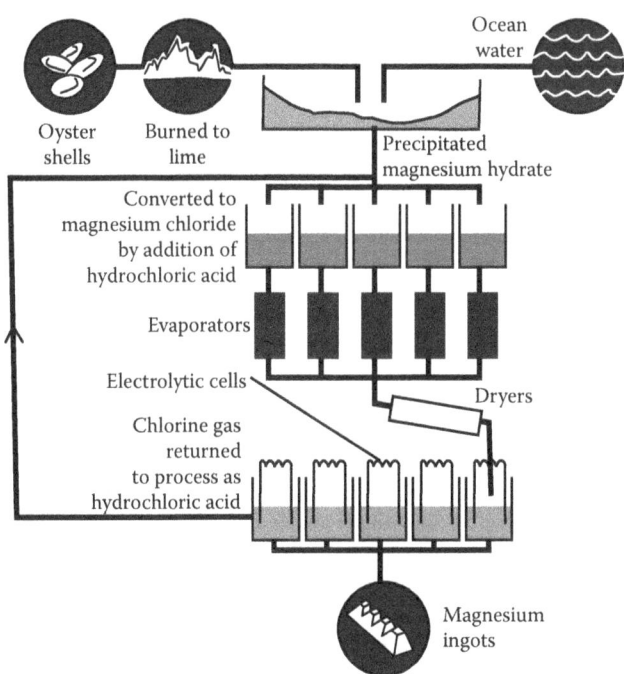

FIGURE M.1 The electrolytic extraction of magnesium from seawater. (From Schwartz, M., ed., *McGraw-Hill Encyclopedia of Science and Technology*, Vol. 10, McGraw-Hill, New York, 1992, p. 295. With permission.)

alloy designation indicates by letters the two principal alloying elements (or one if the alloy contains only one alloying element: A for aluminum, E, rare earth elements; H, thorium; K, zirconium; M, manganese; Q, silver; S, silicon; T, tin; and Z, zinc. The two (or one) numbers that follow indicate the amount (in percent rounded off to whole numbers) of these elements. These numbers are followed by a letter to distinguish among alloys having the same amount of these elements. The temper designations that follow are similar to those for aluminum alloys; F (as fabricated); O (annealed); H10 and H11 (slightly strain hardened); H23, H24, and H26 (strain hardened and partially annealed); T4 (solution heat treated); T5 (artificially aged); T6 (solution heat treated and artificially aged); and T8 (solution heat treated, cold worked, and artificially aged). Thus, AZ91C-T6 is the designation for an alloy containing 8.7% aluminum and 0.7% zinc as the major alloying elements. The letter C indicates that it is the third such alloy to be standardized, and, in this case, it is the solution heat-treated and artificially aged temper. These designations, however, do not distinguish between wrought and cast alloys.

In addition to AZ91C, other magnesium sheet and plate alloys include AZ31B, HK31A, and HM21A, of which AZ31B is the strongest at room temperature and the most commonly used. In the H24 temper, it has an ultimate tensile strength of 290 MPa, a tensile yield strength of 220 MPa, 15% elongation and be used at service temperatures to about 93°C. The others, however, especially HM21A, are more heat resistant and can sustain temperatures to about 315°C. For HM21A, the 100 h creep strength for 0.1% deformation is 86 MPa at 214°C and 52 MPa at 316°C. None of the alloys are especially formable, with a minimum bend radii for AZ31B-O, the most formable, ranging from about five times thickness ($5T$) at room temperature to $2T$ at 260°C. Thus, heat is often required in forming operations, especially deep-drawing.

AZ31B is also widely used in the form of bar and other extruded shapes. The alloys for bar in extruded shapes are generally of two kinds: those alloyed principally with aluminum and zinc, and those alloyed with zinc and a bit of zirconium. In the former, strength increases with increasing aluminum content, and in the latter with increasing zinc content. Some of each kind responds to artificial aging, providing in the T5 temper ultimate tensile strengths in the 345–379 MPa range. AZ31B and many of the alloys for bar and extrusions are also suitable for forging. The hot-working range may be as low as 232°C–371°C or 293°C–538°C.

There are several magnesium die-casting alloys but more than a dozen magnesium sand casting alloys and magnesium permanent-mold casting alloys. One composition, AZ91, is available in four grades; AZ91A, B, C, and D. AZ91C, for sand- and permanent-mold castings, contains 8.7% aluminum, as opposed to 9% in the die-casting alloys (A and B). Each also contains 0.7% zinc and 0.13% manganese. As die-cast, AZ91A and AZ91B provide an ultimate tensile strength of 228 MPa, a tensile yield strength of 152 MPa, and 24% elongation. For AZ91C-T6, these values are 276 MPa, 145 MPa, and 6%, respectively. AZ91D is a high-purity version, containing extremely low residual contents of iron, copper, and nickel, which markedly improves corrosion resistance, precluding the need for protective treatments in certain applications, such as auto underbody parts. The benefits of high purity are apparently applicable to other alloys as well, such as AS41, a more heat-resistant die-casting alloy.

The sand- and permanent-mold casting alloys generally use either aluminum, zinc, and manganese or zinc, thorium, zirconium, and, in some cases, silver and rare earth elements, as alloying elements. Almost all these alloys respond to artificial aging or solution heat treating and artificial aging. The strongest, ZE63A, in the T6

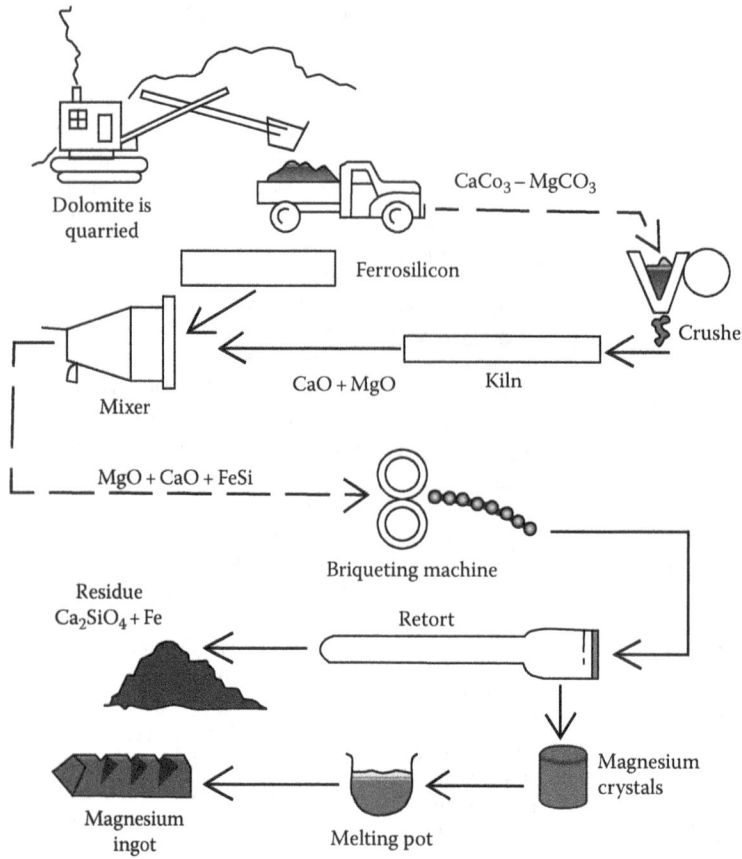

Dolomite is
quarried

$CaCo_3 - MgCO_3$

Crusher

Ferrosilicon

$CaO + MgO$ Kiln

Mixer

$MgO + CaO + FeSi$

Briqueting machine

Residue
$Ca_2SiO_4 + Fe$

Retort

Magnesium
crystals

Magnesium
ingot

Melting pot

FIGURE M.2 Extraction of magnesium metal by silicothermic process. (From Schwartz, M., ed., *McGraw-Hill Encyclopedia of Science and Technology*, Vol. 10, McGraw-Hill, New York, 1992, p. 296. With permission.)

temper, and ZK61A, in the T5 or T6 temper, have an ultimate tensile strength of about 310 MPa, a tensile yield strength of about 193 MPa, and about 10% elongation. The alloys containing zirconium and rare earth elements—ZE33A, ZE41A, and ZE63A—are more creep resistant at higher temperatures than the aluminum-, zinc-, and manganese-bearing alloys, but are more difficult to cast. The magnesium–silver-bearing alloys, QE22A and QH21A, also are superior in elevated-temperature performance, have good cast ability and weldability, but are quite costly.

Like their base metal, magnesium alloys have outstanding machinability, providing faster cutting speeds and greater depths of cut at less power than all commonly machined metals. However, dust, chips, and turnings can pose a fire hazard, necessitating special precautions.

Although lithium is no longer used in magnesium alloys, a series of magnesium–lithium alloys, developed in the past for aerospace applications and ammunition containers, were noted for their extremely light weight and moderate ultimate tensile strengths—100–248 MPa. One such alloy is LA141A, at a density of only 1245 kg/m³, or three quarters that of magnesium.

Magnesium–nickel is a master alloy of magnesium and nickel used for adding nickel to magnesium alloys and for deoxidizing nickel and nickel alloys. One such alloy, contains about 50% of each metal, is silvery white in color, and is furnished in round bar form. Magnesium–Monel contains 50% magnesium and 50% Monel metal. Alloys of magnesium with nickel, Monel, zinc, copper, or aluminum, used for deoxidizing nonferrous metals, are called stabilizer alloys.

Properties

Magnesium–aluminum–zinc alloys often have sufficient strength to give satisfactory service at temperatures as high as 177°C. For still higher-temperature services (and sometimes lower-temperature service), the alloys especially developed for such service are usually required. These are the magnesium–rare earth metal and the magnesium–thorium alloys. They have good short-time mechanical strength, good strength under long-time loading (creep strength), and good modulus of elasticity, at these higher temperatures. In addition, they retain these good properties throughout long exposures to elevated temperatures.

Magnesium alloys are similar to other nonferrous alloys in that they do not exhibit a definite endurance limit when subjected to fatigue loading. Instead, their fatigue strength continues to drop as the required life of the part increases. But the "signal–noise curves" for magnesium are much more horizontal at long lives than the curves of aluminum alloys, for example. Therefore, magnesium alloys are especially suited to applications requiring a high number of cycles of fatigue loading.

All magnesium alloys have excellent damping capacity compared to the same form in other metals. As with other metals, sand castings have the highest, die castings have lower, and the wrought forms have the lowest damping capacity.

K1A

A magnesium–zinc alloy, K1A, was developed to give the greatest damping capacity in an alloy that has improved castability and better strength.

K1A magnesium casting alloy gives the greatest damping capacity in both sand and die castings. Where greater strength is required, especially creep strength, the elevated-temperature casting alloys can be used, as they have about the next highest damping capacity.

EZ33A-T5 sand castings where rare earth metals and zirconium with or without zinc are added to magnesium, results in casting alloys for use at temperatures falling roughly between 177°C and 260°C. This alloy is often used for its high damping capacity combined with good strength.

It has damping capacity comparable to that of gray cast-iron. The magnesium–zinc–zirconium and the magnesium–aluminum–zinc casting alloys have lower, but still good, damping capacity. No special high-damping wrought alloys have been developed, as the common alloys, AZ31B and ZE10A sheet and AZ31B extrusions, already have as high damping capacity as appears possible for wrought magnesium alloys.

Applications

Reviewing the nonstructural uses of magnesium provides an accurate perspective of the entire industry. Nonstructural uses are those in which magnesium is used for its particular chemical, electrochemical, and metallurgical properties. The principal nonstructural uses are as an alloying constituent in other metals (principally aluminum); nodularizing agent for graphite in cast-iron; reducing agent in the production of other metals, such as uranium, titanium, zirconium, beryllium, and hafnium; desulfurizing agent and iron and steel production; sacrificial anode in the protection of other metals against corrosion; and anode material in reserve batteries and dry cells.

As structural materials, magnesium alloys are best known for their light weight and high strength-to-weight ratio. Accordingly, they are used generally in applications where weight is a critical factor and where high mechanical integrity is needed.

Some automobiles contain magnesium parts such as the distributor diaphragm, steering column brackets, and a lever cover plate. Magnesium alloy die castings are also used on chainsaws, portable power tools, cameras and projectors, office and business machines, tape reels, sporting goods, luggage frames, and many other products.

Use of sand and permanent-mold castings are confined largely to aircraft engine and airframe components. Engine parts include gear boxes, compressor housings, diffusers, fan thrust reversers, and miscellaneous brackets. On the airframe, magnesium sand castings are used for leading edge flaps, control pulleys and brackets, and entry door gates, and various cockpit components.

Magnesium is used extensively in wheels, auxiliary equipment, flooring, seating, electronic and instrument cases, etc. Other military uses are in ground handling equipment for aircraft and missiles, ordinance equipment such as vehicles and weapons, and portable shelters and hand-carried communications equipment.

In ground transportation vehicles, both military and commercial, magnesium is used in the engines, transmissions, differentials, pumps, and other parts of the power plant. It has also been used in the floors and body panels of trucks and trailers, while for many years magnesium wheels have been used on all the Indianapolis "500" racing cars.

Additionally, the use of magnesium wrought products has become rather limited. Magnesium sheet, extrusions, and forgings continue to be used in a few limited airframe applications. The temperature alloys are used in at least 20 different missiles (including ICBMs) and on various spacecraft.

In consumer goods, magnesium is used in a wide variety of equipment. Among the household goods in which magnesium is used are portable appliances, furniture, luggage, griddles, ladders, and lawn mowers. Among types of office equipment that use magnesium are typewriters, dictating machines, adding machines, calculators, and furniture. In sporting goods, magnesium is used in such items as sleds, high-jump and pole vault cross bars, and baseball masks. Magnesium is also used in instruments and for such items as binoculars and camera bodies.

Because of its rapid, yet controlled, etching characteristics as well as its lightness, strength, and wear characteristics, magnesium finds usage in photo-engraved printing plates.

In addition to the many structural uses, magnesium has several nonstructural uses. A large nonstructural use is in the cathodic protection of other metals from corrosion. It also functions well in dry-cell battery construction. Magnesium also has chemical and metallurgical uses. Among these are the Grignard reaction; pyrotechnics; high-energy fuels; alloying in aluminum, zinc, and lead alloys; and as an additive in the manufacture of nodular cast iron, lead, nickel alloys, and copper alloys. It is also used in the production of titanium, zirconium, beryllium, uranium, and hafnium.

Magnesium Carbonate

A white, insoluble powder of composition $MgCO_3$, containing also water recrystallization. The specific gravity is 3.10. It is made by calcining dolomite with coke, slaking with water, saturating with carbonic acid gas, and crystallizing out the magnesium carbonate. It is employed as an insulating covering for steam pipes and furnaces, for making oxychloride cement, in boiler compounds, and as a filler for rubber and paper. *Montax*, used as a filler, is a mixture of hydrated magnesium carbonate and silica powder. Magnesium carbonate is a good heat insulator because of the great number of microscopic dead-air cells in the material. The insulating material known as 85% magnesia has a density of 192 kg/m^3. Hydrated magnesium carbonate is a fine, white powder called **magnesia alba levis**, slightly soluble in water, and used in medicine.

Magnesium Oxide

Magnesium oxide (MgO) is a synthetic mineral produced in electric arc furnaces or by sintering of amorphous powder (periclase). Refractory applications consume a large quantity of MgO. Both brick and shapes are fabricated at least partially of sintered grain for use primarily in the metal-processing industries. Heating unit insulation is another major application for periclase. Principal advantages of periclase are its thermal conductivity and electrical resistivity at elevated temperatures.

Specialty crucibles and shapes also are fabricated from MgO. These are used in pyrometallurgical and other purifying processes for specialty metals. Both slip-casting and pressing techniques are employed to manufacture shapes.

Thermocouple insulation comprises still another outlet for periclase. Since most of these go into nuclear applications, a high-purity product is required. MgO is also an important glaze constituent.

Single crystals of MgO have received attention because of their use in ductile ceramic studies. Extreme purity is required in this area. Periclase windows are also of potential interest in infrared applications because of their transmission characteristics.

Magnesium Sulfate

A colorless to white, bitter-tasting material occurring in sparkling, needle-shaped crystals of composition, $MgSO_4 \cdot 7H_2O$. The natural mineral was called *epsomite*, from an Epsom Spa, Surrey, England. In medicine, it is called *epsom salt*. It is used in leather tanning, as a mordant in dyeing and printing textiles, as a filler for cotton cloth, for sizing paper, in water-resistant and fireproof magnesia cements, and as a laxative. It has been sold and canisters as a heat-storage medium for solar energy by Pfizer Co. It can also be obtained in the anhydrous form as a white powder. The specific gravity of the hydrous material is 1.678 and of the anhydrous 2.65. It occurs naturally as deposits from spring waters and is also made by treating magnesite with sulfuric acid.

Magnet Wire

Insulated wire for the winding of electromagnets and for coils for transformers and other electrical applications. Since compactness is usually a prime consideration, *high-conductivity copper* is used in the wire, but where weight saving is important, aluminum wire may be used and has the advantage that an extremely thin, anodized coating of oxide serves as the insulation either alone or with a thin varnish coating. Square or rectangular wires may be used, but ordinary magnet wire is round copper wire covered with cotton and an enamel. *Vitrotex* is a magnet wire coated with a resin enamel and covered with alkali-free glass fiber. It withstands temperatures up to 130°C, and the glass fibers dissipate heat rapidly. *Silotex* has a silicon resin and glass-fiber insulation and withstands temperatures up to 300°C. Various types of synthetic resins are used as insulation to give a high dielectric strength and heat and abrasion resistance. Heat resistance is usually designated by the AIEE class standards.

Magnetic

Having the capacity to be attracted by, or repelled by, other magnetized materials or fields. Commonly, metals such as iron are described as magnetic as they are strongly attracted by magnetic fields. More strictly, such metals are ferromagnetic since some other metals, termed **paramagnetic**, have the same characteristics but to a very weak extent. The remaining metals and all nonmetals are repelled by magnetic fields and are termed *diamagnetic*. "Magnetic" indicates the commonly understood form, that is, *ferromagnetic*. A material is termed magnetically hard if it is ferromagnetic and it retains most of its magnetism after the removal of the magnetizing field, termed *permanent magnetism*. If it loses its magnetism, it is termed *soft*. The magnetic properties of elements arise from their atomic structures. An atom can be considered as comprising a central nucleus around which the electrons orbit. Each electron is spinning about its own axis so it develops its individual magnetic field. However, no more than two electrons can exist at a particular energy level. Were they do exist, they are of opposite spin, so their fields cancel out, and the atoms having all their electrons paired do not independently align to produce powerful magnetic effects. These are the diamagnetic materials, and in a magnetic field, their atoms realign to oppose the external field. They carry a slightly smaller magnetic flux than would a vacuum and have a permeability of less than unity. In those elements having single electrons at particular energy levels, the unpaired electrons have magnetic moments that are not canceled, out and hence the individual atoms have magnetic moments. When exposed to an external magnetic field, these atoms realign to produce a magnetic flux greater than that in a vacuum, hence they have a permeability greater than unity, and they are attracted to the field. Elements having this characteristic to a small extent are termed paramagnetic; those having it to major extent—iron, cobalt, nickel, and gadolinium, are ferromagnetic. See also *ferrites*.

Magnetic Alignment

An alignment of the electron-optical axis of the electron microscope so that the image rotates about a point in the center of the viewing section when the current flowing through her lens is varied. See also *alignment*.

Magnetic Annealing

The annealing of steel followed by slow cooling in a powerful magnetic field to develop magnetic characteristics such as increased permeability.

Magnetic Bearing

A type of bearing in which the force that separates the relatively moving surfaces is produced by a magnetic field.

Magnetic Ceramics

Inorganic nonmetallic materials having properties associated with the phenomena of magnetism, that is, these materials can produce or conduct magnetic lines of force capable of interacting with electric fields or other magnetic fields. Magnetic ceramics form the basis for numerous devices that rely on soft or hard (permanent) magnets. The soft magnets include materials such as the *ferrites* and *garnets*, while the hard magnets include magnetoplumbites and γ-Fe_2O_3. The applications are diverse, from items such as microwave components to recording tape. They are particularly useful for high-frequency devices and can be found in numerous television and radio applications. In thin films applied to nonmagnetic substrates, these types of ceramics form the basis for magnetic bubble memories for computers.

Magnetic Circuit

The complete path followed by magnetic lines of force from pole to pole.

Magnetic Contrast

In electron microscopy, contrast that arises from the interaction of the electrons in the beam with the magnetic fields of individual magnetic domains in ferromagnetic materials. Special instrumentation is required for this type of work.

Magnetic Core

The central core of an electro magnet. It is formed from a solid piece or laminations of a magnetically soft iron or steel and is magnetic to a significant extent only while there is an electrical current in this coil surrounding it.

Magnetic Domain

The zone of material with its atoms all having the same magnetic characteristics. An individual grain will comprise multiple domains.

Magnetic Etching

The use of magnetized particles in a colloidal suspension (ferrofluid) to reveal specific features in the microstructure of fully or partially magnetic materials. Used primarily to observe domain of magnetic materials in order to relate metallographic and domain structures to properties. See also *domain, magnetic.*

Magnetic Ferroelectrics

These are materials that display both magnetic order and spontaneous electric polarization. Research of these materials has enabled considerable advances to be made in understanding the interplay between magnetism and ferroelectricity. The existence of both linear and higher-order coupling terms has been confirmed, and their consequences studied. They have given rise, in particular, to a number of magnetically induced polar anomalies and have even provided an example of a ferromagnetic whose magnetic moment per unit volume is totally induced by its coupling via linear terms to a spontaneous electric dipole moment.

Most known ferromagnetic materials are metals or alloys. Ferroelectric materials, on the other hand, are nonmetals by definition because they are materials that can maintain a spontaneous electric moment per unit volume (called the polarization), which can be reversed by the application of an external electric field. It, therefore, comes as no surprise to find that there are no known room-temperature ferromagnetic ferroelectrics. In fact, there are no well-characterized materials that are known to be both strongly ferromagnetic and ferroelectric at any temperature. This is unfortunate because not only would a study of the interactions between ferromagnetism and ferroelectricity be valuable as basic research but such interplay could well give rise to important device applications.

Most antiferromagnetic materials, however, are nonmetals, so there is no apparent reason ferroelectricity and antiferromagnetism should not coexist. A study of antiferromagnetic ferroelectrics would also provide much information concerning the interplay of magnetic and ferroelectric characteristics even if the device potential were very much reduced. Somewhat unaccountably, antiferromagnetic ferroelectrics are also comparative rarities in nature.

The antimagnetic ferroelectrics $BaMnF_4$, $BaFeF_4$, $BaNiF_4$, and $BaCoF_4$ and their nonmagnetic magnesium and zinc counterparts are orthorhombic and all spontaneously polar (i.e., pyroelectric) at room temperature. For all except the iron and manganese materials, which have a higher electrical conductivity than the others, the polarization has been reversed by the application of an electric field, so that they are correctly classified as ferroelectric, although their ferroelectric transition (or Curie) temperatures are in general higher than their melting points.

The importance of these magnetic ferroelectrics is the opportunity they provide to study and to separate the effects of a variety of magnetic and nonmagnetic excitations upon the ferroelectric properties and particularly upon the spontaneous polarization.

Magnetic Field

The region within which a body or current experiences magnetic force.

Magnetic Flux

The rate of flow of magnetic energy across or through a surface.

Magnetic Forming

The use of short-duration, high-intensity magnetic fields to induce deformation. The material to be shaped, which must be a conductor,

is subject to a rapidly changing magnetic field that induces, in the body of the material, large eddy currents, and associated magnetic fields. The interaction of the induced and external magnetic fields produces large forces capable of deforming the material.

Magnetic Heating

Magnetic heating is bidding for recognition as a mainstream thermal technology with the introduction of the Coreflux process. The equipment generates an intense low-frequency magnetic field throughout the cross section of magnetic and paramagnetic materials.

The process is sometimes called UMH for "uniform magnetic heating," because it heats the entire cross section of a workpiece simultaneously. Temperature distribution is uniform; there is little or no thermal gradient from the surface to the center.

However, UMH should not be confused with low-frequency induction heating. Although both processes require magnetic fields, that is where the similarity ends. Induction heating is a surface process. The induction coil generates magnetic eddy currents that are stronger on the surface of the workpiece than at the center. This produces a skin heating effect, and conductivity carries heat into the interior of the part.

Low-frequency induction heating systems (around 60 Hz) can penetrate components to a depth of no more than 25 mm for most materials. This type of system still uses eddy currents to generate the heat—so, uniform through-heating of thicker parts is not possible.

Process Principles

UMH is based on the principle of hysteresis loss and heats without relying on thermal conductivity to transmit heat from the surface to the center. With UMH, an alternating magnetic field causes domains or crystals in the metal to align and strain in reaction to the field (in the direction of the field for ferrous metals; in the opposite direction for nonferrous metals). As field polarity is reversed, these domains realign; the lag in the process, called hysteresis, is the mechanism that creates heat. This heat has spread rapidly and uniformly through the metal component.

Process Differences

Both induction heating and UMH use coils, but in very different ways. A part to be heated by induction is placed within the loops of a single coil. The magnetic field for UMH is generated by two coils that induce magnetic fields in two cores laminated from high-permeability, low-reluctance directional steel. The cores direct magnetic flux into the workpiece, and pneumatic cylinders clamp the workpiece between the laminated cores to prevent vibration when the field is energized.

For large masses that require uniform heating to produce optimum results, UMH is a logical alternative to induction.

Generating the field for uniform magnetic heating requires less energy and equipment than induction heating and can produce a field that can be varied from 40 to 400 Hz using a standard, variable-frequency motor drive inverter. It is less costly than the power supply for induction heating.

Another advantage is that UMH has no plumbing, so the installation is simplified, and one of the other biggest advantages of this power supply is that water cooling is eliminated; only 480 VAC three-phase electric service and 90 psi shop air are needed. Also, a huge amount of induction heat is lost in water-cooling the coil. With the air-cooled coils, most of the energy flows to the workpiece.

UMH also has a tooling advantage. The same laminated cores can be used for a variety of parts. Induction heating, however, requires a separate coil customized for each component.

Temperatures can be maintained to an accuracy of ±2°C, which is generally a tighter tolerance than most users can measure. A proportional integral derivative (PID) controller tapers the percentage output of the power supply as the workpiece reaches the required temperature, to avoid overshooting the set point. The programmable logic controller includes a touchscreen interface and multiple thermocouple inputs.

Processing

UMH is capable of heating up to 1100°C, well beyond the Curie point, at which magnetic materials lose their magnetic properties. But its benefits compared to induction heating are said to be more tangible at sub-Curie temperatures, at which induction tends to overheat part surfaces.

UMH can through-heat to 260°C–316°C very efficiently, while induction really struggles at these low temperatures because localized surface heating is very pronounced and there is little penetration. Above the Curie point, the eddy currents of induction heating penetrate deeper into the workpiece.

Applications

UMH machines can be applied to tampering, stress relieving, and hardening of parts as well as bonding and shrink fitting and are all sound applications. UMH can also preheat billets for extrusion and semisolid forming, and dies for extrusion and forging. Sintering of powder metals and brazing are among its other capabilities.

Magnetic Irons and Steels

All magnetic materials fall into two general classes: they are either permanently or nonpermanently magnetized. The permanently magnetized materials retain their magnetization after being placed in a magnetic field and can be used as a constant source of magnetic field. On the other hand, the nonpermanently magnetized, or soft magnetic materials, retain their magnetization only while a magnetic field is applied to the material.

The basic magnetization process and hysteresis cycle are best understood in terms of the static domain theory, which is applicable under DC excitation. However, most applications of magnetic materials are for AC use. The predominant frequency commercially used in the United States is 60 cycles. For many special applications, particularly in the aircraft and military field, 400 cycles is a common frequency. The AC losses are not simple multiples of the DC hysteresis loss, for in addition to the basic magnetic losses.

Applications

The magnetic alloys used for AC applications are used in sheet form and are most often alloyed so that maximum resistivity is obtained commensurate with the induction required, ability to hot- and cold-work the material, minimum hysteresis loss, and cost.

The number of applications of magnetic materials is so large that a complete listing is not feasible. Furthermore, even for the same general application, such as a motor, many acceptable designs could be made, each requiring different quality levels of magnetic material. Therefore, no attempt will be made to evaluate all the uses for

a given alloy. However, several typical uses will be given. In some instances, very special properties of a magnetic material may be required. The most common magnetic materials with their characteristic AC and DC properties can be found in Tables M.3 and M.4, pp. 378–379, Second Edition, *Encyclopedia of Materials, Parts, and Finishes*—CRC Press (Mel Schwartz).

Tables M.3 and M.4 are designed to give the reader a ready reference source to identify the kinds of materials available and a brief resume of their most important properties. After preliminary selection, reference should be made to the detailed curves available from reputable suppliers before any design is anticipated. Many of the materials listed require special handling and annealing techniques to achieve optimum properties.

Magnetic Lens

A device for focusing an electron beam using a magnetic field.

Magnetic Materials (Hard and Soft)

These are materials exhibiting ferromagnetism. The magnetic properties of all materials make them respond in some way to a magnetic field, but most materials are diamagnetic or paramagnetic and show almost no response.

The materials that are most important to magnetic technology are ferromagnetic and ferrimagnetic materials. Their response to a field H is to create an internal contribution to the magnetic induction B proportional to H, expressed as $B = \mu H$, where μ is the permeability, varies with H for ferromagnetic materials. Ferromagnetic materials are the elements iron, cobalt, nickel, and their alloys, some manganese compounds, and some rare earths. Ferrimagnetic materials are spinels of the general composition MFe_2O_4, and garnets, $M_3Fe_5O_{12}$, where M represents a metal.

Ferromagnetic materials are characterized by a Curie temperature above which thermal agitation destroys the magnetic coupling giving rise to the alignment of the elementary magnets (electron spins) of adjacent atoms in a crystal lattice. Below the Curie temperature, ferromagnetism appears spontaneously in small volumes called domains. In the absence of a magnetic field, the domain arrangement minimizes the external energy and the bulk material appears unmagnetized.

Magnetic materials are further classified as soft or hard according to the ease of magnetization. Soft materials are used in devices in which change in the magnetization during operation is desirable, sometimes rapidly, as in AC generators and transformers. Hard materials are used to supply a fixed field either to act alone, as in a magnetic separator, or to interact with others, as in loudspeakers and instruments. Both soft and hard materials are characterized by their magnetic hysteresis curve.

These two broad groups—soft magnetic materials, sometimes called electromagnets—do not retain their magnetism when removed from a magnetic field; hard magnetic materials, sometimes referred to as permanent magnets, retain their magnetism when removed from a magnetic field. Cobalt is the major element used for obtaining magnetic properties in hard magnetic alloys.

Soft Materials

These materials are characterized by their low loss and high permeability. There are a variety of alloys used with various combinations of magnetic properties, mechanical properties, and cost.

There are seven major groups of commercially important materials: iron and low-carbon steels, iron–silicon alloys, iron–aluminum–silicon alloys, nickel–iron alloys, iron–cobalt alloys, ferrites, and amorphous alloys.

The behavior of soft materials is controlled by the pinning of domain walls at heterogeneities such as grain boundaries and inclusions. Thus, the common goal in their production is to minimize such heterogeneities. In addition, eddy current loss is minimized through alloying additions that increase the electrical resistivity. Initial permeability, important in electronic transformers and inductors, is improved by minimizing all sources of magnetic anisotropy, for example, by using amorphous metallic alloys and by using zero magnetostrictive alloys. A high maximum permeability, which is necessary for motors and power transformers, is increased by the alignment of the anisotropy, for example, through development of crystal texture or magnetically induced anisotropy.

The class of alloys used in the largest volume is by far iron and 1%–3.5% silicon–iron for applications in motors and large transformers. In these applications, the cost of the material is often the dominant factor, with losses and excitation power secondary but still important. Thus, the improvement over the years in these alloys has been in developing lower losses without increases in cost.

Soft magnetic materials are iron, iron–silicon alloys, and nickel–iron alloys. Irons are widely used for their magnetic properties because of their relatively low cost. Common iron–silicon magnetic alloys contain 1%, 2%, 4%, and 5% silicon. There are about six types of nickel–irons, sometimes called permeability alloys, used in magnetic applications. For maximum magnetostriction, the two preferred nickel contents are 42% and 79%. Additions of molybdenum give higher resistivities, and additions of copper result in higher initial permeability and resistivity.

Soft magnetic ceramics, also referred to as ceramic magnets, ferromagnetic ceramics, and ferrites (soft), were originally made of an iron oxide, Fe_2O_3, with one or more divalent oxides such as NiO, MgO, or ZnO. The mixture is calcined, ground to a fine powder, pressed to shape, and sintered. Ceramic and intermetal types of magnets have a square hysteresis loop and high resistance to demagnetization and are valued for magnets for computing machines where a high remanence is desired. A ferrite with a square loop for switching in high-speed computers contains 40% Fe_2O_3, 40% MnO, and 20% CdO. Some intermetallic compounds, such as zirconium–zinc, $ZrZn_2$, which are not magnetic at ordinary temperatures, become ferromagnetic with properties similar to ferrites at very low temperatures and are useful in computers in connection with subzero superconductors. Some compounds, however, are the reverse of this, being magnetic at ordinary temperatures and nonmagnetic below their transition temperature point. This transition temperature, or Curie point, can be arranged by the compounding to vary from subzero temperatures to above 100°C. Chromium-manganese–antimonide, $Cr_xMn_{2x}Sb$, is such a material. Chromium manganese alone is ferromagnetic, but the antimonide has a transition point varying with the value of x.

Vectolite is a lightweight magnet made by molding and sintering ferric and ferrous oxides and cobalt oxide. The weight is 3.2 g/cm³. It has high coercive force and has such high electrical resistance that it may be considered as a nonconductor. *Magnadur* was made from barium carbonate and ferric oxide. *Indox* and *Ferroxdure* are similar. This type of magnet has a coercive force to 127,200 A/m, with initial force to 206,700 A/m, high electrical resistivity, high resistance to demagnetization, and light weight, with specific gravity from 4.5 to 4.9. *Ferrimag* and *Cromag* are ceramic magnets.

Strontium carbonate is superior to barium carbonate for magnets but is more costly. *Lodex* magnets are extremely fine particles of iron–cobalt in lead powder made into any desired shape by powder metallurgy.

Hard Materials

Permanent magnets, or hard magnetic materials, strongly resist demagnetization once magnetized. They are used, for example, in motors, loudspeakers, meters, and holding devices, and have coercivities H_x from several hundreds to many thousands of oersteds (10 to over 100 kA/m). The bulk of commercial permanent magnets are of the ceramic type, followed by the Alnicos and the cobalt–samarium, iron–neodymium, iron–chromium–cobalt, and elongated single-domain (ESD) types in decreasing sequence of usage. The overall quality of a permanent magnet is represented by the highest-energy product $(BH)_m$; but depending on the design considerations, high H_c, high residual induction B_r (the magnetic induction when H is reduced to zero), and reversibility of permeability may also be controlling factors.

To understand the relationship between the resistance to demagnetization, that is, the coercivity, and the metallurgical microstructure, it is necessary to understand the mechanisms of magnetization reversal. The two major mechanisms are reversal against a shape anisotropy and reversal through nucleation and growth of reverse magnetic domains against crystal anisotropy. The Alnicos, the iron–chromium–cobalt alloys, and the ESD Lodex alloys are examples of materials of the shape anisotropy structure, whereas barium ferrites, the cobalt–samarium alloys, and the iron–neodymium–boron alloys are examples of the crystal anisotropy-controlled materials.

Neodymium–iron–boron is a powerful magnetic material that works best at room temperature. It begins losing its properties at higher temperatures, and at 312°C it loses its magnetism completely.

The use of magnets in a Stirling engine to power deep space probes has been investigated. Stirling engines have a sealed cylinder in which hot gases move two pistons back and forth. By placing magnets on the ends of the pistons, and surrounding the cylinder with wire coils, the magnets would induce a current flow.

Research is being conducted whereby a magnetic material that would work at elevated temperature is being produced in particles that each have two different compositions—one at the outer edge to resist demagnetization, and the other at the core to retain magnetic power at higher temperatures. The "functionally graded" material would essentially have the outside material protecting the inside material.

The new magnets would be attractive for a variety of applications on Earth as well—cars, electronics, computers, and power tools.

Magnet steels, now largely obsolete, includ plain high-carbon (0.65% or 1%) steels or high-carbon (0.7%–1%) compositions containing 3.5% chromium–chromium magnet steels; 0.5% chromium and 6% tungsten–tungsten magnet steel; or chromium, tungsten, and substantial cobalt (17% or 36%)–cobalt magnet steels. They were largely replaced by ternary alloys of iron, cobalt, and molybdenum, or tungsten. *Comol* has 17% molybdenum, 20% cobalt, and 71% iron. *Indalloy* and *Remalloy* have similar compositions: about 20% molybdenum, 12% cobalt, and 68% iron. *Chromindur* has 28% chromium, 15% cobalt, and the remainder iron, with small amounts of other elements that give it improved strength and magnetic properties. In contrast to Indalloy and Remalloy, which must be processed at temperatures as high as 1250°C, Chromindur can be cold-formed.

Some cobalt magnet steels contain 1.5%–3% chromium, 3%–5% tungsten, and 0.50%–0.80% carbon, with high cobalt. *Alfer* magnet alloys, first developed in Japan to save cobalt, were iron–aluminum alloys. *MK* alloy had 25% nickel, 12% aluminum, and the balance iron, close to the formula Fe_2NiAl.

Cunife is a nickel–cobalt–copper alloy that can be cast, rolled, and machined. It is not magnetically directional like the tungsten magnets, and thus gives flexibility in design. The electric conductivity is 7.1% that of copper, and it has good coercive force. Cunife 1 contains 50% copper, 21% nickel, and 29% cobalt. Cunife 2, with 60% copper, 20% nickel, and 20% iron, is more malleable. This alloy, heat-treated at 593°C, is used in wire form for permanent magnets for miniature apparatus. It has a coercive force of 39,750 A/m. *Hipernom* is a high-permeability nickel–molybdenum magnet alloy containing 79% nickel, 4% molybdenum, and the balance iron. It has a Curie temperature of 460°C and is used for relays, amplifiers, and transformers.

In the Alnico alloys, a precipitation hardening occurs with AlNi crystals dissolved in the metal and aligned in the direction of magnetization to give greater coercive force. This type of magnet is usually magnetized after setting in place. Alnico 1 contains 21% nickel, 12% aluminum, 5% cobalt, 3% copper, and the balance iron. The alloy is cast to shape, is hard and brittle, and cannot be machined. The coercive force is 31,800 A/m. Alnico 2, a cast alloy with 19% nickel, 12.5% cobalt, 10% aluminum, 3% copper, and the balance iron, has a coercive force of 44,520 A/m. The cast alloys have higher magnetic properties, but the sintered alloys are fine-grained and stronger. Alnico 4 contains 12% aluminum, 27% nickel, 5% cobalt, and the balance iron; it has a coercive force of 55,650 A/m, or 10 times that of a plain tungsten magnet steel. Alnico 8 has 35% cobalt, 34% iron, 15% nickel, 7% aluminum, 5% titanium, and 4% copper. The coercive force is 115,275 A/m. It has a hardness of Rockwell C 59. The magnets are cast to shape and finished by grinding. *Hyflux Alnico 9*, of the same coercive force, has an energy product of 75,620 T A/m. The magnets of this material, made by Indiana General Corp., are cylinders, rectangles, and prisms, usually magnetized and oriented in place. The *Alnicus* magnets are Alnico-type alloys with the grain structure oriented by directional solidification in the casting, which increases the maximum energy output. *Ticonal*, *Alcomax*, and *Hycomax* are Alnico-type magnet alloys produced in Europe.

Cobalt–platinum, as an intermetallic rather than an alloy, has a coercive force above 341,850 A/m and a residual induction of 0.645 T. It contains 76.8% by weight of platinum and is expensive, but is used for tiny magnets for electric wristwatches and instruments. *Placovar* is a similar alloy that retains 90% of its magnetization flux up to 343°C. It is used for miniature relays and focusing magnets. *Ultramag* is a platinum–cobalt magnet material with a coercive force of 381,600 A/m. The Curie temperature is about 500°C, and it has only slight loss of magnetism at 350°C, whereas cobalt–chromium magnets lose their magnetism above 150°C. The material is easily machined. *Alloy 1751* is a cobalt–platinum intermetallic with a coercive force of 341,850 A/m, or of 540,600 A/m in single-crystal form. The metal is not brittle and can be worked easily. It is used for the motor and index magnets of electric watches.

Ceramic permanent magnets are compounds of iron oxide with oxides of other elements. The most used are barium ferrite, oriented barium ferrite, and strontium ferrite. Yttrium–iron garnets (YIG) and yttrium–aluminum garnets (YAG) are used for microwave applications.

Flexible magnets are made with magnetic powder bonded to tape or impregnated in plastic or rubber in sheets, strip, or forms. Magnetic tape for recorders may be made by coating a strong, durable plastic tape, such as a polyester, with a magnetic ferrite powder. For high-duty service such as for spacecraft, the tape may be of stainless steel. For recording heads, the ferrite crystals must be hard and wear resistant.

Ferrocube is manganese zinc. The tiny crystals or compacted with a ceramic bond for pole pieces for recorders. *Plastiform* is a barium ferrite bonded with rubber in sheets and strips. *Magnyl* is a vinyl resin tape with the fine magnetic powder only on one side. It is used for door seals and display devices.

Rare earth magnetic materials, used for permanent magnets and computers and signaling devices, have coercive forces up to 10 times that of ordinary magnets. They are of several types. Rare earth–cobalt magnets are made by compacting and extruding the powders with a binder of plastic or soft metal into small precision shapes. They have high permanency. Samarium–cobalt and cesium–cobalt magnets are cast from vacuum melts and are chemical compounds, $SmCo_5$ and $CeCo_5$. These magnets have intrinsic coercive forces up to 2.2 million A/m. The magnetooptic magnets for memory systems and components are made in thin wafers, often no more than a spot in size. These are ferromagnetic ceramics of europium-chalcogenides. Spot-size magnets of europium oxide only 4 μm in diameter perform reading and writing operations efficiently. Films of this ceramic less than a wavelength in thickness are used as memory storage mediums.

Magnetic fluids consist of solid magnetic particles in a carrier fluid. When a magnetic field is applied, the ultramicroscopic iron oxide particles become instantly oriented. When the field is removed, the particles demagnetize within microseconds. Typical carrier fluids are water, hydrocarbons, fluorocarbons, diesters, organometallics, and polyphenylene ethers. Magnetic fluids can be specially formulated for specific applications such as damping, sealing, and lubrication.

Magnetic Pole

The area on a magnetized part at which the magnetic field leaves or enters the part. It is a point of maximum attraction in a magnet.

Magnetic Quenchometer Test

Method used to test heat extraction rates of various quenchants. The test works by utilizing the change in magnetic properties of metals at the Curie point—the temperature above which metals lose their magnetism. See also *Curie temperature*.

Magnetic Resonance

A phenomenon in which the magnetic spin systems of certain atoms absorb electromagnetic energy at specific (resonant) natural frequencies of the system.

Magnetic Resonance Imaging (MRI)

A technique in which an object placed in a specially varying magnetic field is subjected to a pulse of radiofrequency radiation, and the resulting nuclear magnetic resonance spectra are combined to give cross-sectional images.

Magnetic Seal

A seal that uses magnetic material, instead of springs or bellows, to provide the closing force.

Magnetic Separator

A device used to separate magnetic from less magnetic or nonmagnetic materials. The crushed material is conveyed on a belt past a magnet.

Magnetic Shielding

In electron microscopy, shielding for the purpose of preventing extraneous magnetic fields from affecting the electron beam in the microscope.

Magnetic Writing

In magnetic-particle inspection, a *false indication* caused by contact between a magnetized part and another piece of magnetic material.

Magnetically Hard Alloy

A ferromagnetic alloy capable of being magnetized permanently because of its ability to retain induced magnetization and magnetic poles after removal of externally applied fields; an alloy with high coercive force. The name is based on the fact that the quality of the early permanent magnets was related to their hardness. See also *permanent magnet materials*.

Magnetically Soft Alloy

A ferromagnetic alloy that becomes magnetized readily upon application of the field and that returns to practically a nonmagnetic condition when the field is removed; an alloy with the properties of high magnetic permeability, low coercive force, and low magnetic hysteresis loss. See also *soft magnetic materials*.

Magnetic-Analysis Inspection

A nondestructive method of inspection to determine the existence of variations in magnetic flux in ferromagnetic materials of constant cross section, such as might be caused by discontinuities and variations in hardness. The variations are usually indicated by a change in pattern on an oscilloscope screen.

Magnetic-Particle Inspection

A nondestructive method of inspection for determining the existence and extent of surface cracks and similar imperfections in ferromagnetic materials.

The use of magnetic particles to detect surface defects in iron and steel. The fine particles, as a dry powder or suspended in fluid, are applied to the surface under test, and a magnetic field is imposed by magnets or electrical coils. Any defect aligned across the lines of magnetic force effectively forms a pair of magnets that attract the particles hence revealing the location of the defect. Prior to application of the particles, it is, to apply a quick drying white coating, referred to as a developer, to improve contrast.

Magnetic-Shielding Metals

The materials that fall under this category are used in airport environments, which deal with what is known in the industry as ELF (extremely low frequency) radiation interference to ticketing computer terminals caused by the close proximity of building utility electrical faults as well as to main switchgear rooms.

Another application is in the production of handheld electronic equipment with and without radio cards.

Material Types

MuShield works primarily with an 80% nickel alloy. The alloy with such trade names as *HyMu 80*, *Permalloy*, and *MuMetal* shields from low-frequency electromagnetic interference. For high-frequency shielding, refer to fabricators that work with copper, aluminum, and other conductive materials.

The term "B-40" is an arbitrary starting point for high-permeability alloys because it is a good average value of the flux density induced within a piece of high-permeability material.

It has also been found that time-varying (AC) magnetic fields need to be attenuated below 7 mG for most monitor applications. Also, larger monitors are more sensitive to EMI.

Magnetite Black

Magnetic iron oxide, Fe_3O_4, in contrast to rust, magnetite, can form a protective coating. For example, in high-temperature water of good quality, it forms on the surface of steels being hot worked, in which case it is commonly termed *mill scale*.

Magnetizing Force

A force field, resulting from the flow of electric currents or from magnetized bodies, that produces magnetic induction.

Magnetoelasticity

Less common term for magnetostriction.

Magnetohydrodynamic Lubrication

Hydrodynamic lubrication in which a significant force contribution arises from electromagnetic interaction. Magnetohydrodynamic bearings have been proposed for very high-temperature operation, for example, in liquid sodium.

Magnetometer

An instrument for measuring the magnitude and sometimes also the direction of a magnetic field, such as the Earth's magnetic field. See also *torque-coil magnetometer*.

Magneton

A unit of magnetic moment used for atomic, molecular, or nuclear magnets. The Bohr magneton (μ_B), which has the value of the classical magnetic moment of the electron, can theoretically be calculated as

$$\mu_B = \frac{eh}{2mc} = 9.2731 \times 10^{-20} \text{ erg/G}$$

where
e and m are the electronic charge and mass, respectively
h is Planck's constant divided by 2π
c is the velocity of light

See also *Planck's constant*.

Magnetostriction

The reversible dimensional change of a material due to magnetization. The change may be expansion or contraction, volumetric (Barret effect), or linear (Joule effect).

Magnetostriction occurs where in a position sensor, a pulse is induced in a magnetostrictive waveguide by the momentary interaction of two magnetic fields: one from a magnet passing along the outside of the sensor tube, and the other field from a current pulse launched along a waveguide within the tube. The interaction produces a strain pulse (twisting the waveguide) that travels at sonic speeds down the waveguide until detected at the sensor head. Measuring the elapsed time between the launching of the electronic pulse and the arrival of the strain pulse, or pulses, precisely determines the position of one, or more, magnets.

Such noncontact position sensing produces no wear in the sensing elements, cutting maintenance and extending sensor life. The encapsulated waveguide and electronics also provide durability in severe environments. And, modularity gives mounting flexibility and easy integration.

Magnetostrictive Cavitation Test Device

A vibratory cavitation test device driven by a magnetostrictive transducer.

Magnification

In metallography, the ratio of the image diameter to the object diameter. It approximates to the product of the magnifications of the eyepiece and the objective lenses of the microscope.

Magnolia Metal

A cheap white metal bearing metal suitable only for light duty but retaining its properties to a relatively high temperature. Typically 80% lead, 15% antimony, and 5% tin.

Magnox Alloys

Nuclear fuel canning alloys based primarily on magnesium with small amounts of elements such as aluminum and beryllium added to reduce flammability, limit grain size, and enhance resistance to high-temperature oxidation in carbon dioxide, the cooling medium in magnox reactors.

Magnox Reactors

Nuclear reactors of a design cooled by carbon dioxide and having the fuel canned in magnox alloy.

Main Bearing

A bearing supporting the main power-transmitting shaft.

Major Segregation

See *segregation*.

Male

A term used to identify the item in an assembly that fits inside another. It may be regarded as not "politically correct."

Malleability

The characteristic of metals that permits *plastic deformation* in compression without fracture. See also *ductility*. The capacity of a metal to be deformed by predominately compressive loads in processes such as rolling, forging, or beating.

Malleabilizing

Annealing *white iron* in such a way that some or all of the combined carbon is transformed into graphite or, in some cases, so that part of the carbon is removed completely.

Malleable Iron

A cast iron made by prolonged annealing of *white iron* in which decarburization, graphitization, or both take place to eliminate some or all of the cementite. The graphite is in the form of temper carbon. If decarburization is the predominant reaction, the product will exhibit a light fracture surface; hence, whiteheart malleable. Otherwise, the fracture surface will be dark; hence, blackheart malleable. Only the blackheart malleable is produced in the United States. Ferritic malleable has a predominately ferritic matrix; pearlitic malleable may contain pearlite, spheroidite, or tempered martensite, depending on heat treatment and desired hardness. Small amounts of chromium (0.01%–0.03%), boron (0.0020%), copper (~1.0%), nickel (0.5%–0.8%), and molybdenum (0.35%–0.5%) are also sometimes present.

Malleable Irons

The malleable irons are a family of cast alloys—consisting primarily of iron, carbon, and silicon—which are cast as hard, brittle white iron and then rendered tough and ductile through a controlled heat conversion process. Because of their unique metallurgical structure, they possess a wide range of desirable engineering properties including strength, toughness, ductility, resistance to corrosion, machinability, and castability.

Three principal types of malleable iron are in wide use in this country; ferritic, pearlitic, and alloy malleable iron. A fourth type, called cupola malleable because of the method of manufacture, is also produced but only in small tonnages. Most important from the standpoint of production volume and use is standard malleable iron, which has a ferrite matrix. Pearlitic malleable, which, as the name implies, has a pearlitic matrix, is being produced in ever-increasing quantities. Alloy malleable iron is basically a specialty type iron with higher strength and corrosion resistance, finding primary use in railroad parts.

By far the largest tonnage of malleable castings normally is consumed by the automotive industry. The railroad, agricultural implement, electrical line hardware, pipe fittings, and detachable chain industries, and most other basic industries use standard and pearlitic malleable castings.

Advantages

Important attributes of malleable iron can be summarized as follows:

1. Malleable iron can be produced with a high yield strength, which is a static mechanical property upon which most mechanical design is based.
2. Ferritic and pearlitic malleable irons have a high ratio of yield strength to tensile strength. This means that the engineer can design to high applied strength values in service

for materials of construction, concomitant with good machinability, and low production cost for the final part.

3. Pearlitic malleable irons can be produced to a wide range of mechanical properties through carefully controlled heat treatments.

4. Malleable irons have a high modulus of elasticity and a low coefficient of thermal expansion, compared with the nonferrous metals.

5. Malleable and pearlitic malleable irons exhibit a low nil ductility transition temperature for brittle fracture.

6. Compared with steel, malleable and pearlitic malleable irons have considerably better damping capacity, which makes operation of moving components less prone to noise because of resonant vibration.

7. Pearlitic malleable irons exhibit good wear resistance and can be selectively hardened by flame, induction, or the carbonitriding process.

8. Pearlitic malleable irons will take a high-quality finish. Honed surfaces with 2–3 μin. finish and hardness values of 197–207 Bhn have been reported.

9. The uniformity of properties from surface to center is excellent, particularly in oil-quenched and tempered pearlitic malleable iron.

10. All malleable irons are substantially free from residual stresses as a result of long heat treatments at high temperatures.

11. Pearlitic malleable iron provides the properties of medium to high carbon steel coupled with a machinability rating unequaled for a material of similar hardness.

With respect to mechanical properties, minimum specification values are generally exceeded by a comfortable margin in the better-controlled malleable foundries.

Brinell hardness of ferritic malleable irons varies from about 110 to 145. Pearlitic malleable and alloyed malleable iron grades have higher values, ranging usually between 160 and 280 Bhn. Both hardness and tensile strength increase with combined carbon content.

Since the final properties of malleable iron castings are the result of thermal treatments, section thickness has no appreciable effect on strength. Therefore, mechanical properties will be essentially the same throughout the entire cross section.

Manufacture

The manufacture of malleable iron castings is fundamentally a two-phase operation. Phase 1 consists of producing the white iron castings, and the second phase involves the controlled heat treatment of these castings to obtain the desired finished product.

Structurally, malleable iron castings consist essentially of carbon-free iron (ferrite) and uniformly dispersed nodules of temper carbon. This combination of soft, ductile ferrite and nodular temper carbon accounts for the desirable mechanical properties of malleable iron. In pearlitic malleable iron, the matrix is essentially pearlitic, resembling that of a medium-carbon steel.

Properties

Effect of Temperature

Studies of the behavior of ferritic malleable iron at both high and lower temperatures demonstrate, in general, that this material is well suited to applications in a temperature range from −51°C to 649°C.

Low-temperature investigations have been concerned primarily with impact resistance and notch sensitivity; high-temperature studies have focused principally on tensile strength, yield point, elongation, stress rupture, and creep behavior.

Results of research have indicated a high level of performance at elevated temperatures, equal or superior to other ferritic materials for which data are available, particularly at 427°C. Strength at 538°C is adequate for many applications, and strength is retained even at 649°C. No evidence was found in any of the investigations of changes in structure or performance during the test periods, which extended from 1 to over 2000 h.

Surface Hardness

Many structural parts require high surface hardness backed up by a strong, tough core. In steel components, this can be accomplished by carburizing or nitriding after machining, followed by a suitable heat treatment. In pearlitic malleable iron, the combined carbon content is adequate for production of high surface hardnesses through quenching after either induction or flame heating. Many parts are preferentially hardened on wearing surfaces.

Mandrel (Metals)

(1) A blunt-ended tool or rod used to retain the cavity in a hollow metal product during working. (2) A metal bar around which other metal may be cast, bent, formed, or shaped. (3) A shaft or bar for holding work to be machined. (4) A form, such as a mold or matrix, used as a cathode in electroforming.

Mandrel (Plastics and Reinforced Plastics)

(1) In blow molding of thermoplastics, part of the tooling that forms the inside of the container neck through which air is forced to form the hot parison to the shape of the mold. (2) In extrusion of thermoplastics, the solid, cylindrical part of the die that forms tubing or pipe. (3) In filament winding of reinforced plastic, the form (usually cylindrical) around which the filaments are wound.

Mandrel Forging

The process of rolling or forging a hollow blank over a mandrel to produce a weldless, seamless ring or tube. See also *radial forging*.

Manganese and Alloys

A metallic element, symbol Mn, manganese, is found in the minerals manganite and pyrolusite and with most iron ores and traces in most rocks. Manganese has a silvery-white color with purplish shades.

It is brittle but hard enough to scratch glass. The specific gravity is 7.42, melting point 1245°C, and weight is 7418 kg/m^3. It decomposes in water slowly. It is not used alone as a construction metal. The electrical resistivity is 100 times that of copper or 3 times that of 18-8 stainless steel. It also has a damping capacity 25 times that of steel and can be used to reduce the residence of other metals.

It is the 12th most abundant element in the Earth's crust (approximately 0.1%) and occurs naturally in several forms, primarily as a silicate ($MnSiO_3$) but also as the carbonate ($MnCO_3$) in a variety of oxides, including pyrolusite (MnO_2) and hausmannite (Mn_3O_4). Pyrolusite is the most common and has been used in glassmaking since the time of the pharaohs in Egypt. Weathering of land deposits has led to large amounts of the oxide being washed out to sea, where they have aggregated into the so-called manganese nodules containing 15%–30% manganese. Vast deposits, estimated at over

M

10^{12} metric tons, have been detected on the seabed, and a further 10^7 metric tons is deposited every year. The nodules also contain smaller amounts of the oxides of other metals such as iron, cobalt, nickel, and copper. The economic importance of the nodules as a source of these important metals is enormous.

Uses

All steels contain some manganese, the major advantage being an increase in hardness, although it also serves as a scavenger of oxygen and sulfur impurities that would induce defects and consequent brittleness in the steel. The manganese ore is first converted to a metallic alloy with iron known as ferromanganese.

Ferromanganese has a significantly lower melting point than pure manganese and is therefore more readily dissolved by molten iron for steel production. The demand for manganese by the steel industry is so great that about 95% of mined manganese ores are converted to ferromanganese.

In addition to its importance to the steel industry, manganese has a number of other industrial uses. Manganese dioxide (MnO_2) is employed in the manufacture of dry-cell batteries, most commonly of the carbon–zinc Leclanché type. It acts to suppress undesirable formation of hydrogen gas at the carbon (positive) electrode. Only very high-quality manganese dioxide ore can be used directly in these batteries; consequently, increasing use is being made of synthetic manganese dioxide obtained by the electrolysis of manganese sulfate ($MnSO_4$) solutions. Manganese dioxide is also used in the brick industry to provide a range of red-to-brown and gray tints. Its venerable use in glassmaking has been mentioned; its role is to neutralize the effects of iron impurities that would impart a greenish tinge to the glass. The good oxidizing properties of manganese dioxide also make it useful for the oxidation of aniline to quinone, important as a photographic developer and also in the production of paints and dyes. Manganese even has some use in the electronics industry, where manganese dioxide, either natural or synthetic, is employed to produce manganese compounds possessing high electrical resistivity; among other applications, these are utilized as components in every television set.

Manganese metal has very high sound-absorbing properties, and copper–manganese alloys with high percentages of manganese are used as sound-damping alloys for thrust collars for jackhammers and other power tools.

Alloys

A manganese alloy with 72% manganese, 18% copper, and 10% nickel is noted for its high coefficient of thermal expansion, electrical resistivity, strength, and vibration dampening. It is used for rheostat resistors and electrically heated expansion elements.

Manganese–aluminum is a hardener alloy employed for making additions of manganese to aluminum alloys. Manganese lowers the thermal conductivity of aluminum but increases strength. Manganese up to 1.2% is used in aluminum alloys when strength and stiffness are required. One manganese–aluminum contains 25% manganese and 75% aluminum. Manganese–boron is used for deoxidizing and hardening bronzes. It contains 20%–25% boron, with small amounts of iron, silicon, and aluminum. For deoxidizing and hardening brasses, nickel bronze, and copper–nickel alloys, manganese copper, or copper manganese, may be used. The alloys used contain 25%–30% manganese and the balance copper. The best grades of manganese copper are made from metallic manganese and are free from iron. For nickel bronzes and nickel alloys, the manganese copper must be free of both iron and carbon, but grades

containing up to 5% iron can be used for manganese bronze. Grades made from ferromanganese contain iron.

Manganese bronze is wrought as well as cast alloys of copper and zinc mainly, with lesser amounts of iron, aluminum, silicon, tin, and lead. The two standard alloys, C67000 (2.5%–5.0% manganese) and C67500 (0.05%–0.50% manganese), were formally designated manganese bronze B and A, respectively. Manganese bronze cast alloys constitute the C86100–C86800 series, some of which also may contain as much as 5% manganese. Some manganese bronzes were formally designated high-strength yellow brasses and leaded high-strength yellow brasses.

C67500, which contains 58.5% copper, 39% zinc, 1.4% iron, 1% tin, and 0.1% manganese, has an ultimate tensile strength of 448 MPa, a tensile yield strength of 207 MPa, and a tensile elongation of 33% in the annealed condition. The alloy is weldable; has good brazing and soldering characteristics; and has good resistance to corrosion in rural, industrial, and marine atmospheres. Available in rod and shapes, it is used in pumps, clutches, and valves. Most of the cast alloys are castable by various methods, with C86200 the most versatile in this respect. As sand cast, the alloys provide typical ultimate tensile strength ranging from 448 to 793 MPa, and they are rather ductile as indicated by tensile elongations of 15%–30%. The alloys are not hardenable by heat treatment; weldability, including brazing and soldering, is generally poor or fair; and their machinability is 8%–65% that of free-cutting brass, with C86400 and C86700 the best in this respect.

Manganese Brass

A term with various interpretations, it usually indicates a brass based on 60% copper and 40% zinc with additions of about 1.5% manganese and often a percent or so of iron and aluminum. A better term for this alloy is *High-Tensile Brass*.

Manganese Dioxide

Manganese dioxide (MnO_2) is soluble in water and nitric acid and also in hydrochloric acid. It occurs in nature as the blue-black mineral pyrolusite.

In glass, manganese dioxide is used as a colorant and decolorizer. As a coloring oxide in lead potash and glasses, manganese produces an amethyst color, while in soda glass a reddish-violet is produced. Manganese suitable for such purposes should contain at least 85% MnO_2 and not more than 1% iron oxide.

The major use of manganese oxides is an ore of manganese for the manufacturing of steel; manganese serves to increase the hardness and decrease the brittleness of steel. Another important use of manganese oxides is as the cathode material of common zinc/carbon and alkaline batteries (such as flashlight batteries).

Manganese Nodules

These are concentrations of manganese and iron oxides found on the floors of many oceans. The complex growth histories of manganese nodules are revealed by the texture of module interiors.

Nodules from certain regions are significantly enriched in nickel, copper, cobalt, zinc, molybdenum, and other elements so as to make them important reserves for the strategic metals. Modern oceanographic surveys have delineated areas of the world's seafloors where nodule abundances and metal concentrations are highest.

Although manganiferous modules and crusts have been sampled or observed on most seafloors, attention has been focused on the nickel–copper-rich nodules (2–3 wt.%) from the northern equatorial

Pacific belt stretching from Southeast Hawaii to Baja California, as well as the high-cobalt modules and crusts from seamounts in the Central Pacific. Manganese nodules from the Atlantic Ocean and from higher latitudes in the Pacific Ocean have significantly lower concentrations of the minor strategic metals. Surveys of the Indian Ocean have revealed metal-enrichment trends comparable to those found in the Pacific Ocean nodules; high nickel–cobalt–copper-bearing nodules are found near the Equator. The ferromanganese nodules and crusts associated with submarine hydrothermal deposits have extremely low concentrations of nickel–cobalt–copper.

Chemistry

Marine manganese nodules are usually classified by mode of formation into hydrogenetic, diagenetic, and hydrothermal types. A fourth, mixed type also exists. Hydrogenetic nodules form by direct precipitation of manganese–iron oxide phases onto an existing nucleus at the sediment–water interface, diagenetic nodules are believed to be biologically driven, and hydrothermal nodules form by submarine hydrothermal activity. Hydrothermal nodules are rarely reported; in submarine hydrothermal environments, manganese–iron oxides occur mainly as crusts. Cobalt is the only strategic metal reported in hydrogenetic nodules that have relatively low manganese/iron ratios. Nodules with high nickel, cobalt, and copper contents are diagenetic and have relatively high manganese/iron ratios.

Manganese Steels

All commercial steels contain some manganese that has been introduced in the process of the oxidizing and the sulfurizing with ferromanganese, but the name was originally applied to steels containing from 10% to 15% manganese. Steels with from 1.0% to 1.5% manganese are known as carbon–manganese steel, pearlitic manganese steel, or intermediate manganese steel. Medium manganese steels, with manganese from 2% to 9% are brittle and are not ordinarily used, but steels with 1% to 2% manganese and with or without small amounts of chromium or molybdenum are used for air-hardening and oil-hardening cold-work tool steels. The original Hadfield manganese steel made in 1883 contained 10%–14.5% manganese and 1% carbon.

Manganese increases the hardness and tensile strength of steel. In the absence of carbon, manganese up to 1.5% has only slight influence on iron; as the carbon content increases, the effect intensifies.

High-manganese steels are not commercially machinable with ordinary tools, but can be cut and drilled with tungsten carbide and high-speed steel tools. The austenitic steels, with about 12% manganese, are exceedingly abrasion resistant and harden under the action of tools. They are nonmagnetic.

High-manganese steels are brittle when cast and must be heat-treated. For castings of thin sections or irregular shapes where the drastic water quenching might cause distortion, nickel of 5% may be added. The manganese–nickel steels have approximately the same characteristics as the straight manganese steels.

The manganese steels used for dipper teeth, tractor shoes, and wear-resistant castings contained 10%–14% manganese, 1%–1.4% carbon, and 0.30%–1% silicon. The tensile strength is up to 861 MPa, elongation 45%–55%, weight is 7916 kg/m^3, and a Brinell hardness, when heat-treated, of 185–200.

A manganese–aluminum steel, has 30% manganese, 9% aluminum, 1% silicon, and 1% carbon. Its tensile strength is 840 MPa with the elongation of 18%, but it work-hardens rapidly, and when cold-rolled and heat-aged the tensile strength is 2068 MPa with a yield strength of 1999 MPa. This alloy forms a special type of stainless steel, with high resistance to oxidation and sulfur gases to 760°C.

Tank car steel M-128 is a manganese–vanadium steel with 0.25% carbon, up to 1.5% manganese, and 0.02% or more vanadium. It has a minimum tensile strength of 558 MPa with the elongation of 18%. This type of steel with up to 1.75% manganese is used for forgings.

Manganese Sulfide

A very common nonmetallic inclusion in steel.

Manifold

A pipe or other vessel with multiple inlet and/or outlet connections.

Manipulator

A mechanical device for handling an ingot or billet during forging.

Man-Made (Synthetic) Diamond

A manufactured diamond, darker, blockier, and considered to be more friable than most natural diamonds.

Mannesmann Process

A process for piercing tube billets in making seamless tubing. The billet is rotated between two heavy rolls mounted at an angle and is forced over a fixed mandrel. See *tube making*.

Manson–Haferd Relationship

An empirical method for extrapolating creep data based on a linear relationship between the reciprocal of the absolute temperature and the logarithm (base 10) of the time to rupture.

Manual Metal(lic) Arc Welding (MMA)

Metal arc welding in which the consumable electrode rod, installed in a suitable holder, is manipulated by the operator during the welding operation.

Manual Welding

Welding in which the operator holds key items of the equipment such as the welding torch or the electrode holder and, in addition, controls key parameters such as welding current or gas flame, and the entire welding operation is performed and controlled by hand.

Manufactured Unit

A quantity of finished adhesive or finished adhesively bonded component, processed at one time. The manufactured unit may be a bunch or a part thereof.

Maraging

A hardening process applicable to a precipitation-hardening treatment applied to a special group of iron-base alloys to precipitate one or more intermetallic compounds in a matrix of essentially carbon-free martensite. See also *maraging steels*.

The term is derived from "martensite" and "age hardening." Such steels, on cooling from the austenitic region, that is, about 800°C, developed a very low-carbon martensite that is softer and tougher than that formed in conventional steels. However, on heating to about 480°C, the highly alloyed martensite can undergo a precipitation hardening process as intermetallic particles precipitate. This produces a steel combining very high strength with good fracture toughness. The term was first used for a range of steels based on about 18% nickel, with typically, about 9% cobalt, 5% molybdenum, 0.6% titanium, and 0.1% aluminum but is now used at other alloys ranges offering the same characteristics.

Maraging Steels

The maraging steels develop unique combinations of properties that have not been obtained in conventional low-alloy steels. Some of these properties are (1) useful yield strengths to and above 2040 MPa, (2) high toughness and impact energy even at the 2040 MPa yield strength level, (3) low nil ductility temperature (NDT), (4) exceptional stress-corrosion resistance, (5) through hardening without quenching, (6) simple heat treatment, (7) good formability without prolonged softening treatments, (8) good machinability, (9) low distortion during maraging after forming or machining, (10) good weldability, and (11) freedom from decarburization problems.

The maraging steels are a family of low carbon, high-alloy steels typically containing 12%–18% nickel, 3%–5% molybdenum, 0%–12% cobalt, 0.2%–1.6% titanium, and 0.1%–0.3% aluminum (one cobalt-free grade also contains 5% chromium). They are noted for their high strength and toughness, simple heat treatment, dimensional stability during heat treating, good machinability, and excellent weldability. The term *maraging* refers to the martensitic structure that forms during heat treatment, which is a precipitation-hardening, or aging, treatment usually at 482°C. The 18% nickel steels, the most well known, are produced in four grades to provide tensile yield strength of 1379, 1724, 2069, or 2413 MPa. Although the 18% nickel steels were originally developed for aerospace applications primarily, they also are now used for die-casting dies, cold-forming dies, and molds for forming plastics.

The 9% nickel and 4% cobalt alloys were designed to provide high strength and toughness at room temperature as well as at moderately elevated temperatures—to about 427°C. Weldability and fracture toughness are good, but the alloys are susceptible to hydrogen embrittlement. The steels are used in airframes, gears, and large aircraft parts.

Marble

Marble is a term applied commercially to any limestone or dolomite taking polish. Marble is a compact crystalline limestone used for ornamental building, for large slabs for electric power panels, and for ornaments and statuary. In the broad sense, marble includes any limestone they can be polished, including breccia, onyx, and others. Pure limestone would naturally be white, but marble is usually streaked and variegated in many colors. Carrara marble, from Italy, is a famous white marble, being of delicate texture, very white, and hard. In the United States, the marbles of Vermont are noted and occur in white, gray, light green, dark green, red, black, and mottled.

Marforming Process

A *rubber-pad forming* process developed to form wrinkle-free shrink flanges and deep-drawn shells. It differs from the *Guerin process* in that the sheet metal blank is clamped between the rubber pad and the blankholder before forming begins.

Margin

The cylindrical portion of the *land* of a drill that is not cut away to provide clearance.

Marquenching

See *martempering*.

Martempering

(1) A hardening procedure in which an austenitized ferrous material is quenched into an appropriate medium at a temperature just above the martensite start temperature of the material, held in the medium until the temperature is uniform throughout, although not long enough for bainite to form, then cooled in air. The treatment is frequently followed by tempering. (2) When the process is applied to carburized material, the controlling martensite start temperature is that of the case. This variation of the process is frequently called *marquenching*. (3) In martempering, the part is held at the martensite start temperature for a limited time and then further cooling, which can be at a relatively slow rate, causes transformation to martensite. The initial fast cooling suppresses the normal pearlite transformation, and the hold at the intermediate temperature allows temperature gradients to even out, thereby reducing residual stresses. The virtual absence of residual stress and the relatively slow final cooling minimizes the risk of cracking. Austempering is similar except that it allows transformation to occur at the hold temperature, thereby forming a bainite structure. See *steel* and *isothermal transformation diagram*.

Martensite

A generic term for microstructures formed by diffusionless phase transformation in which the parent and product phases have a specific crystallographic relationship. Martensite is characterized by an acicular pattern in the microstructure in both ferrous and non-ferrous alloys. In alloys where the solute atoms occupy interstitial positions in the martensitic lattice (such as carbon in iron), the structure is hard and highly strained; but where the solute atoms occupy substitutional positions (such as nickel in iron), the martensite is soft and ductile. The amount of high-temperature phase that transforms to martensite on cooling depends to a large extent on the lowest temperature attained, there being a rather distinct beginning temperature (M_s) and a temperature at which the transformation is essentially complete (M_f). See also *lath martensite*, *plate martensite*, and *tempered martensite*.

The structure formed in steels by a shear mechanism, rather than a diffusion process, when the steel is cooled, from the austenitic range, faster than the critical rate. The fast cooling retains the carbon is a supersaturated solid solution so the face centered cubic ferrite structure cannot form. Instead, the austenite transforms by shear to a body-centered tetragonal structure that, under the microscope has an acicular, that is, needle-shaped, appearance. See *steel*. The term is used for similar crystallographic forms in other alloys.

Martensitic

A platelike constituent having an appearance and a mechanism of formation similar to that of martensite. See also *lath martensite* and *plate martensite*.

Martensitic Range

The interval between the martensite start M_s and the martensite finish M_f temperatures.

Martensitic Stainless Steel

See *stainless steels*.

Martensitic Transformation

A reaction that takes place in some metals on cooling, with the formation of an acicular structure called *martensite*.

Mash Resistance Seam Welding

Resistance seam welding in which the weld is made in a lap joint, the thickness at the lap being reduced plastically to approximately the thickness of one of the lapped parts. A resistance seam weld in which the similar thickness sheet components are set up to form an overlap that is mashed, that is, crushed, during welding so that the final weld thickness approximates the thickness of a single sheet.

Mask

A device for protecting a surface from the effects of blasting and/or coating. Masks are generally either reusable or disposable.

Masking Tape

A tape used as a *resist for stopping-off purposes*.

Masonry Cement

A hydraulic cement for use in mortars for masonry construction, containing one or more of the following materials: portland cement, portland blast furnace slag cement, portland-pozzolan cement, natural cement, slag cement, or hydraulic lime. In addition, masonry cement usually contains one or more materials such as hydrated lime, limestone, chalk, calcareous shell, or talc, slag, or clay as prepared for this purpose. See also *cement*.

Mass Absorption Coefficient

The linear absorption coefficient divided by the density of the medium.

Mass Concentration (in a Slurry)

The mass of solid particles per unit mass of mixture, expressed in percent.

Mass Effect (in Steels)

The effect of cooling rate resulting from an increase in mass. See *hardenability*.

Mass–Energy Relationship

The Einstein equation $E = mc^2$, where E is the energy in ergs, m is the mass in g, and c is the velocity of light in cm/s.

Mass Spectrograph

An analytical technique in which ions of the material under test are passed across magnetic and electrostatic fields. The path deflection is recorded as a measure of their composition.

Mass Spectrometer

An instrument that is capable of separating ionized molecules of different mass/charge ratios and measuring the respective ion currents.

Mass Spectrometry

An analytical technique for identification of chemical structures, and analysis of mixtures, and quantitative elemental analysis, based on application of the *mass spectrometer*.

Mass Spectrum

A record, graph, or table that shows the relative number of ions of various masses that are produced when a given substance is processed in a *mass spectrometer*.

Master Alloy

An alloy, rich in one or more desired addition elements, that is added to a metal melt to raise the percentage of a desired constituent.

Master Alloy Powder

A prealloyed metal powder of high concentration of alloy content, designed to be diluted when mixed with a base powder to produce the desired composition. See also *prealloyed powder*.

Master Block

A forging *die block* used primarily to hold insert dies. See also *die insert*.

Master Pattern

In foundry practice, a pattern embodying a double contraction allowance in its construction, used for making castings to be employed as patterns in production work.

Mastic

The gum exudation of the tree called **Chios mastic** and from **Bombay mastic**, both small evergreens native to the Mediterranean countries. It contains an ethereal oil (2%) that is mainly **pinene**. In ancient times, the resin was highly valued for artists' paints and coating lacquers, adhesives, and for incense, dental cements, and as a chewing gum from which use it derives its name. Because of high cost, its use is now largely limited to fine art paints and lacquers and as an astringent in medicine.

Mat

A fibrous glass material used as a plastic reinforcement and consisting of randomly oriented chopped filaments, short fibers (with or without a carrier fabric), or swirled filaments loosely held together with a binder. Available in blankets of various widths, weights, and lengths. Also, a sheet formed by filament winding a single-hoop ply of fiber on a mandrel, cutting across its width and laying out a flat sheet.

Match

A condition in which a point in one metal forming or forging die half is aligned properly with the corresponding point in the opposite die half within specified tolerance.

Match Plate

A plate of metal or other material on which patterns for metal casting are mounted (or formed as an integral part) to facilitate molding. The pattern is divided along its *parting plane* by the plate.

Matched Edges

Two edges of the die face that are machined exactly at 90° to each other, and from which all dimensions are taken in laying out the die impression and aligning the dies in the forging equipment. Also referred to as match lines.

Matched Metal Die Molding

A reinforced plastic manufacturing process in which matching male and female metal molds are used (e.g., in compression molding) to form the part, with time, pressure, and heat.

Matching Draft

The adjustment of draft angles (usually involving an increase) on parts with asymmetrical ribs and side walls to make the surfaces of a forging meet at the parting line.

Materials Characterization

The use of various analytical methods (spectroscopy, microscopy, chromatography, etc.) to describe those features of composition (both bulk and surface) and structure (including defects) of a material that are significant for a particular preparation, study of properties, or use. Test methods that yield information primarily related to materials properties, such as thermal, electrical, and mechanical properties, are excluded from this definition.

Materials Handling

Materials handling involves the loading, moving, and unloading of materials. The loading, moving, and unloading of ore from a mine to a mill and of garments within a factory are examples of materials handling. There are hundreds of different ways of handling materials. These are generally classified according to the type of equipment used. For example, the International Materials Management Society has classified equipment as (1) conveyor; (2) cranes, elevators, and hoists; (3) positioning, weighing, and control equipment; (4) industrial vehicles; (5) motor vehicles; (6) railroad cars; (7) marine carriers; (8) aircraft; and (9) containers and supports.

Every materials-handling problem starts with the material—its dimensions, its nature, and its characteristics. Engineers who fail to start here usually end up trying to justify equipment rather than achieving safe and economical movement of the material. The quantity to be moved—both in total and the rate of moving desired—is next in selecting the appropriate handling method. Then comes the sequence of operations or the routing. Basically, this what, when (how much and how often), and where is the minimum information needed to evaluate or determine any handling system or equipment.

Materials handling is both a planning and an operating activities. These two activities are generally separated in industry; an analytical group designs or selects the system or equipment, and the operating group puts it to use.

Equipment

This refers to devices used for handling materials in an industrial distribution activity. The equipment moves products as discrete articles, in suitable containers, or as solid bulk materials that are relatively free-flowing. Such equipment does not include the means employed to control the flow of fluids.

Many different types of machines result from combinations and permutations of the following factors:

1. The route over which the product is moved may be fixed or variable.
2. The path of travel may be horizontal, inclined, declined, or vertical.
3. Motion may be imparted to the product manually, by the force of gravity, by air pressure, by vacuum, by vibration, or by power-actuated components of the machine.
4. The motion may be continuous or intermittent (reciprocating).
5. The product may be supported or carried suspended during the handling operation.

Based on their most common characteristics, materials-handling equipment is classified into broad categories: bulk-handling machines, elevating machines, hoisting machines, industrial trucks, and monorail.

Improvements in handling techniques stem from the wide adoption by industry of palleting and of the forklift truck. These innovations have produced far-reaching effects. Among these are radical changes in plant layout, elevator design for multistory operations, and the increasing trend to single-story facilities.

Automation, in the sense of feedback control and advanced mechanization in the fabrication and transfer of products from one operation to the next, brings together two major industrial technologies.

Computer-controlled electronic data-processing devices facilitate the compiling of inventory records and the handling of orders. Use of photoelectric devices for counting and controlling the action of doors, conveyors, and other materials-handling equipment is another example of how electronic techniques are applied. Television and two-way radio improve the communication for materials handling and plants and yards.

Materials Processing

One of the most promising, fast, low-cost methods for processing materials is known as self-propagating high-temperature synthesis (SHS). Table M.1 presents a comparison of this combustion synthesis process and the other synthetic processes.

Combustion Synthesis

Nonoxide materials (such as carbides, nitrites, borides, chalcogenides, hydrides, intermetallic compounds, and sulfides) have been

TABLE M.1
Comparison of Various Synthesis Processes

Process	Advantages	Disadvantages	Compositions
Carbothermal reduction	Possible automation; some control of chemistry	Can be somewhat energy and capital cost-intensive; usually requires milling, which can produce impurities; large scale-up may be difficult; expensive raw materials	Titanium boride, silicon carbide, other nonoxides
Solid–solid, solid–gas, combustion synthesis	Usually self-propagating (requiring no external heat source) and fast (within seconds)	Exothermic, volatile reactions, sometimes low density or low yields, densification may require high pressures, addition of dopants may be required	Titanium boride, titanium carbide, silicon nitride, other nonoxides, composites
Vapor-phase synthesis	High purity, no aggregation, ease of preparation, narrow size distribution, versatility, homogeneity	Limited chemistry, ternary compounds difficult; low yields; reactant gases expensive	Oxides, nonoxides (nitrides, carbides), metals, binary compounds
Laser synthesis	Wide range of composition, short reaction times, uniform heating rates, improved process control, uniform size distribution, minimum agglomeration	Volatile reactants, expensive equipment, powder yields can be low, contamination a problem for certain reactions	Refractory materials (nitrides, borides, silicides, carbides), transition metal compounds, oxides with other emission lives
Plasma synthesis	Highly efficient, simple, continuous; homogeneous mixtures; high surface areas; oxides with wide range of available starting materials; very fast quench rates	Requires high power (10^3 kW), large capital and operating costs, higher surface areas cause greater pyrophoricity, health hazards due to inhaling, carbides and nitrides are sensitive, low powder yields, some agglomeration, nonreproducibility	Oxides, carbides, nitrides, mixtures (silicon oxide + aluminum nitride, silicon carbide + silicon carbide, aluminum + aluminum nitride)

Source: Adv. Mater. Proc., 131(4), 53, 1987. With permission.

produced by combustion synthesis. This SHS, originally developed in the late 1960s, usually involves exothermic reactions above 2500°C. The combustion process itself can be stable or unstable and does not require a furnace. The high temperatures remove any volatile contaminants by vaporization.

Combustion synthesis can involve several different types of reactions. An oxidation–reduction, or thermite, can produce multiphase composition such as cermets. For example, thermite occurs when a mixture of aluminum powder and iron oxide powder reacts, causing strong heating and yielding aluminum oxide plus a white-hot molten mass of metallic iron.

The thermite reactions have been used to weld metals, as in the welding of cracks in railroad rails.

Synthesis of Refractory Materials

An example of this process is the reaction between transition metals and carbon powders. Upon ignition of the cold-pressed compact, a combustion wave rapidly (several seconds) propagates through the mass, converting the reactants to metal carbides (Figure M.3). Besides carbides, β-sialon, and micro composites, silicon carbide + aluminum oxide and titanium carbide + aluminum oxide have been formed.

As a class of refractory materials, the nitrides, such as aluminum nitride, titanium nitride, zirconium nitride, hafnium nitride, boron nitride, and silicon nitride, have all been formed by SHS. The process for synthesizing nitrides proceeds in a fairly predictable manner. The metal powders are cold-pressed into porous compacts. Subsequently, these compacts are ignited in nitrogen gas or liquid nitrogen by a resistance-heated tungsten wire. The self-propagating combustion wave quickly transforms a metal powder into porous solid nitrides.

Control of Synthesis

Of four innovations to control the synthesis reactions, one is used to slow down (kinetic braking) and three to intensify the reactions (thermal explosion, chemical furnace, and chemical activators).

Other Self-Propagating High-Temperature Synthesis Processes

- Reaction hot pressing, which utilizes SHS reactions in place in a uniaxial hot press to form densified ceramic products. Reaction hot pressing takes advantage of the favorable thermodynamics of SHS reactions to rapidly form an densify product phases.
- Reactive sintering, which occurs when pressure is applied during sintering. The process is also known as reactive hot isostatic pressing.
- Shock compression, where chemical reactions can also be initiated by shock compression of powder mixtures. This type of dynamic processing technique, referred to as shock-induced reaction synthesis, utilizes the simultaneous application of very high pressure and temperature generated during the passage of shock waves through a powder mixture.

Materials Science and Engineering

This is a multidisciplinary field concerned with the generation and application of knowledge relative to the composition, structure, and processing of materials and their properties and uses. The field encompasses the complete knowledge spectrum for materials ranging from the basic end (materials science) to the applied end (materials engineering). It forms a bridge of knowledge from the basic sciences (and mathematics) to various engineering disciplines. New materials with special properties are constantly being discovered and developed, and thus materials science and engineering is a continually expanding field.

Metallic Materials

The study of metallic materials constitutes a major division of the material science and engineering field. In general, metallic materials are inorganic substances composed of one or more metallic elements, but they may also contain nonmetallic elements. Most metals have a crystalline structure of closely packed atoms arranged in an orderly manner. Metals in general are good electrical and thermal conductors. Many are relatively strong at room temperature and retain good strength at elevated temperatures. Metals are commonly alloyed together in the liquid state so that, upon solidification, new solid metallic structures with different properties can be produced. Metals and alloys are often cast into the nearly final shape in which they will be used, and these products are called castings. However, most metals and alloys are first cast into shapes such as sheet ingots or extrusion billets, which are subsequently worked by processes such as rolling and extrusion into wrought products, for example, sheet, plate, and extrusions.

Ceramic Materials

The study of ceramic materials forms a second major division of the field of materials science and engineering. Ceramics are inorganic materials consisting of metallic and nonmetallic elements chemically bonded together. They can be crystalline, noncrystalline, or mixtures of both. Most ceramic materials have high hardness, high-temperature strength, and good chemical resistance; however, they tend to be brittle.

Ceramics in general have low electrical and thermal conductivities, which makes them useful for electrical and thermal insulative applications. Most ceramic materials can be classified into three groups: traditional ceramics, technical ceramics, and glasses.

Traditional Ceramics

These consist of three basic components; clay, silica, and feldspar. The clay provides the workability of the ceramic before it is hardened by the firing process. Clay makes up the major body material; it consists mainly of hydrated aluminum silicates ($Al_2O_3 \cdot SiO_2 \cdot H_2O$) with smaller amounts of other oxide impurities. The silica (SiO_2) has a high melting temperature and provides the refractory component of traditional ceramics. The third component, feldspar ($K_2O \cdot Al_2O_3 \cdot 6H_2O$), has a low melting temperature and produces a glass when the ceramic mix is fired; it bonds the refractory components together. Traditional ceramic products fabricated from whiteware such as electrical porcelain and sanitaryware are made from components of clay, silica, and feldspar for which the composition is controlled.

Technical Ceramics

Technical ceramics are based on pure or nearly pure ceramic components alone or in combination. The raw materials for technical ceramics must be processed carefully so that a controlled product can be produced, and they are made by using various composition mixes and processing procedures. Examples of technical ceramics are aluminum oxide (Al_2O_3), zirconia (ZrO_2), silicon carbide (SiC), silicon nitride (Si_3N_4), and barium titanate ($BaTiO_3$). Applications for technical ceramics include aluminas for auto spark-plug insulators and substrates for electronic circuitry, dielectric materials for capacitors, ceramic tool bits for machining, and high-performance ball bearings.

Glasses

Glasses differ from the other ceramic materials in that their constituents are heated to fusion and then cooled to a rigid state without crystallization. A characteristic of a glass is that it has a noncrystalline structure with no long-range order. Many inorganic classes are based on the glass-forming silicon oxide, silica (SiO_2). About 90% of the glass produced is soda-lime glass, which is the basic composition of 1%–3% SiO_2, 12%–14% sodium oxide (Na_2O), and 10%–12% calcium oxide (CaO). The sodium oxide and calcium oxide are added to lower the viscosity of the glass so that it becomes easier to work. Soda-lime glass is used, for example, for flat glass, containers, and light products, where high chemical durability and heat resistance are not required. Many other types of glasses with different compositions are produced for specialized applications.

Polymeric Materials

The study of polymeric materials forms a third major division of materials science and engineering. Most of these materials consist of carbon-containing long molecular chains or networks. Structurally, most of them are noncrystalline, but some are partly crystalline. The strength and ductility of polymeric materials vary greatly. Most polymers have low densities and relatively low softening or decomposition temperatures. Many are good thermal and electrical insulators. Polymeric materials have replaced metals and glasses for many applications.

Most polymeric materials can be classified as thermoplastics, thermosets, or elastomers.

Thermoplastics

These are polymeric materials that have a structure usually consisting of very long chains of carbon atoms strongly (covalently) bonded together, sometimes with other atoms, such as nitrogen, oxygen, and sulfur, also covalently bonded in the molecular chains. Weaker secondary bonds bind the chains together into a solid mass. Because of their weak intermolecular bonds, thermoplastics can be heated to a soft, viscous condition for forming into a desired shape and then cooled to a rigid state to retain that shape. Typical thermoplastics are polyethylenes, polyvinyl chlorides, and polyamides (nylons). Some examples of applications for thermoplastics are containers, electrical insulation, automotive internal parts, and appliance housings.

Thermosets

These are polymeric materials that have a network of mainly carbon atoms covalently bonded together to form a rigid solid. Sometimes, nitrogen, oxygen, or other atoms are also covalently bonded into the network. Thermosets are formed into a permanent shape and are cured (set) by a chemical reaction that may require heat and pressure. Thermoset plastics cannot be remelted or reformed into another shape after curing. Thermosets such as phenolics (e.g., Bakelite) and epoxies are used for handles, knobs, electrical connectors, and matrix materials for fiber-reinforced plastics.

Elastomers

This class of materials, also known as rubbers, can be deformed elastically by a large amount when a force is applied to them, and then they can return to approximately the same shape when the force is removed. Most elastomers consist of long, carbon-containing molecular chains with periodic strong bond links between the chains. Elastomers include both natural and synthetic rubbers, which are used for auto tires, electrical insulation, and industrial hoses and belts.

Composite Materials

A fourth major division of materials science and engineering comprises the study of composite materials. A composite material is a mixture of two or more materials that differ in form and chemical composition and are essentially insoluble in each other, and most are produced synthetically by combining various types of fibers with different matrices to increase strength, toughness, and other properties. Three important types of composite materials have polymeric, metallic, and ceramic matrices.

Polymeric-matrix composites are the most common type and find most applications where the temperature does not exceed about 100°C–200°C. For example, glass-fiber materials produced with short glass fibers embedded in a polyester plastic matrix are used in appliances, boats, and car bodies because of their light weight, ease of fabrication into complex shapes, corrosion resistance, and moderate cost. Other more expensive advanced composites made with stronger carbon or aramid fibers, usually embedded in heat-resistant thermoset polymeric matrices, are used for aircraft surface material and structural members. Advanced polymeric-matrix composites are also finding use in sports equipment and other products, but their high cost has limited their use.

Metal–matrix composites (MMCs) have also been developed by embedding fibers such as silicon carbide and aluminum oxide into aluminum, magnesium, and other metal alloy matrices. The fibers strengthen the metal alloys and increase high-temperature stability. MMCs are used, for example, for automotive pistons and missile guidance systems. Ceramic–matrix composites have also been considered to develop new and tougher ceramic materials. For example, the reinforcement of alumina with silicon carbide whiskers significantly improves its fracture toughness and strength.

Other Materials

In addition to metallic, ceramic, polymeric, and composite materials, materials science and engineering is also concerned with the research and development of other special classes of materials that are based on applications. Some major types of these materials are electronic materials, optical materials, magnetic materials, superconducting materials, dielectric materials, nuclear materials, biomedical materials, and building materials.

Matrix (Adhesives)

The part of an adhesive that surrounds or engulfs embedded filler or reinforcing particles and filaments.

Matrix (Composites)

The essentially homogeneous plastic resin in which the fiber reinforcement is embedded. Both thermoplastic and thermoset resins may be used.

Matrix (Metals)

The continuous or principal phase in which another constituent is dispersed.

Matrix Isolation

A technique for maintaining molecules at low temperature for spectroscopic study; this method is particularly well suited for preserving reactive species in a solid, inert environment.

Matrix Metal

The continuous phase of a polyphase alloy, mechanical mixture, or *MMC*; the physically continuous metallic constituent in which separate particles of another constituent are embedded.

Matte

(1) An intermediate product of *smelting*; an impure metallic sulfide mixture made by melting a roasted sulfide ore, such as an ore of copper, lead, or nickel. (2) A matte surface is level but dull and diffuses rather than reflects light.

Matte Dip

An etching solution used to produce a dull finish on metal.

Matte Finish

(1) A dull texture produced by rolling sheet or strip between rolls that have been roughened by blasting. (2) A dull finish characteristic of some electrodeposits, such as cadmium or tin.

Mauritius Hemp

The fiber obtained from the fleshy leaves of the plant grown in Mauritius, Nigeria, and Ghana, used for rope and cordage. The product from West Africa is often erroneously termed *sisal*. The plant belongs to the lily family. The leaves yield up to 3.5% of their green weight in dry fiber, which is coarser than henequen but is used for coffee-bag fabric. The fibers are used extensively for burlap for bagging. **Fique fiber**, of Columbia used for rope and coffee bags, is from the leaves of *F. macrophylla*. The leaves are longer than those of henequen, and the fiber is finer and more lustrous.

Maximum Elongation

The elongation at the time of fracture, including both elastic and plastic deformation of the tensile specimen. Applicable to rubber, plastic, and some metallic materials. Maximum elongation is also called ultimate elongation or break elongation.

Maximum Erosion Rate

The maximum instantaneous erosion rate in a test that exhibits such a maximum followed by decreasing erosion rates. Occurrence of such a maximum is typical of many cavitation and liquid impingement tests. In some instances, it occurs as an instantaneous maximum; in others, it occurs as a steady-state maximum that persists for some time.

M

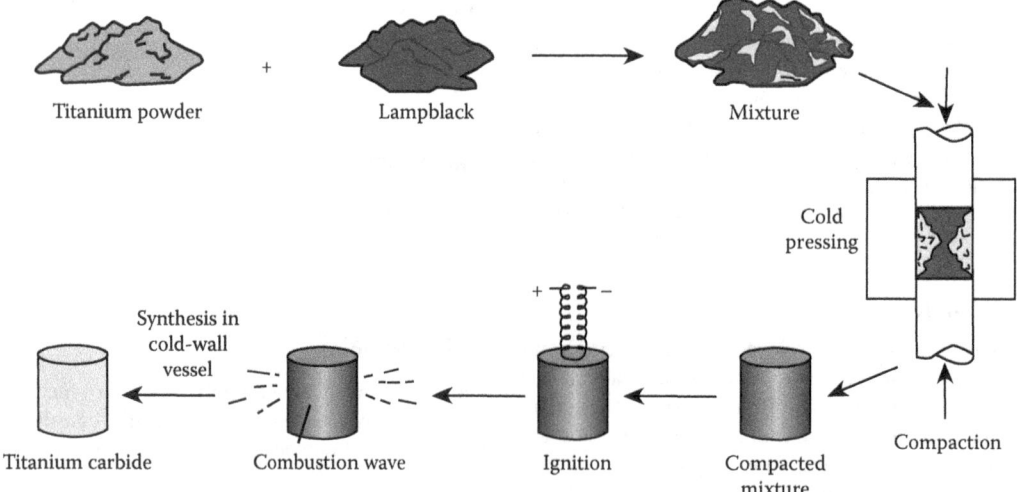

FIGURE M.3   Illustration of SHS. (From *McGraw-Hill Handbook of Structural Ceramics,* McGraw-Hill, New York, 1992, p. 423. With permission.)

Maximum Load Rate (P_{max})

(1) The load having the higher algebraic value in the load cycle. Tensile loads are considered positive and compressive loads negative. (2) Used to determine the strength of a structural member; the load that can be borne before failure is apparent.

Maximum Pore Size

See *absolute pore size.*

Maximum Rate Period

In cavitation and liquid impingement erosion, a stage following the acceleration period, during which the erosion rate remains constant (or nearly so) at its maximum value.

Maximum Strength

See *ultimate strength.*

Maximum Stress (S_{max})

The stress having the highest algebraic value in the stress cycle, tensile stress being considered positive and compressive stress negative. The *nominal stress* is used most commonly.

Maximum Stress Intensity Factor (K_{max})

The maximum value of the *stress-intensity factor* in a fatigue cycle.

Maximum Tensile Stress

Same as *ultimate tensile strength.* See *tensile test.*

Mazac Alloy

A proprietary range of alloys based on zinc with about 4% aluminum and minor additions of copper and magnesium. They have excellent die casting characteristics.

McQuaid-Ehn Grain Size

The austenitic grain size developed in steels by carburizing at 927°C (1700°F) followed by slow cooling. Eight standard McQuaid-Ehn grain sizes rate the structure, from No. 8, the finest, to No. 1, the coarsest. The use of standardized ASTM methods for determining grain size is recommended. The prior austenitic grain size of the Hyper Eutectoid carburizing case is readily established by metallographic examination.

Mean Free Path

The average distance between collisions for an electron or molecule.

Mean Stress (S_m)

The algebraic average of the maximum and minimum stresses in one cycle, that is, $S_m = (S_{max} + S_{min})/2$. Also referred to as steady component of stress. See *fatigue.*

Mechanical Activation

The acceleration or initiation of a chemical reaction by mechanical exposure of a nascent solid surface. Metal cutting (machining) is an effective method of exposing large areas of fresh surface.

Mechanical Adhesion

Adhesion between surfaces produced solely by the interlocking of protuberances on those surfaces. See also *adherence.*

Mechanical Alignment

A method of aligning the geometrical axis of the electron microscope by relative physical movement of the components, usually as a step preceding magnetic or voltage alignment. See also *alignment.*

Mechanical Alloying

The application of powder metallurgy (P/M) techniques has greatly increased in recent years. In many cases, P/M allows for metal parts to be produced at lower cost with improved properties. Another advantage that P/M offers is that it can produce alloys and microstructures that are not possible to produce by standard metallurgical practices such as casting and forging. An important area of P/M is the production of the metal powders used. The physical and chemical nature of the initial powders will determine many of the final properties of the P/M parts produced.

Mechanical alloying is a materials-processing method that involves the repeated welding, fracturing, and rewelding of a mixture of powder particles, generally in a high-energy ball mill, to produce a controlled, extremely fine microstructure. The mechanical alloying technique allows alloying of elements that are difficult or impossible to combine by conventional melting methods. In general, the process can be viewed as a means of assembling metal constituents with a controlled microstructure. If two metals will form a solid solution, mechanical alloying can be used to achieve this state without the need for a high-temperature excursion. Conversely, if the two metals are insoluble in the liquid or solid state, an extremely fine dispersion of one of the metals in the other can be accomplished. The process of mechanical alloying was originally developed as a means of overcoming the disadvantages associated with using P/M to alloy elements that are difficult to combine. By using P/M, homogeneity is dictated by the size of the particle, but contamination and fire hazards become a concern when particle size is very small.

Process

The process of mechanical alloying consists of repeated flattening, fracture, and rewelding of the powder particles in a high-energy ball charge. Every time two steel balls collide, they trap powder particles between them. This deforms the particles and creates minimal foreign species on the surface so that welding together of adjacent surfaces can occur easily. The alloying is affected by the ball-powder-ball collisions in several stages. In the first stage, intense cold welding predominates, and layered composite particulates of the starting constituents form. This is followed by a hardening of the particles and fracturing and cold welding leading to a finer composite particle size. Solid-solution formation begins at this stage. Next, in a moderate cold welding period, the microstructure of the particles gets finer, with a typical spacing between adjacent regions of 1 μm the compositions of the individual particles converge to the original blend composition, with the rate of refinement of the internal structure (reduction in scale of the microstructure) of the particles approximately logarithmic with processing time. The final stage is a steady-state period. By the time processing has been completed, the particles have an extremely deformed metastable structure, which can contain dispersoids. Generally, at this point, the microstructural scale is significantly below the micrometer level.

Applications

Some oxides are insoluble in molten metals. Mechanical alloying was originally developed to provide a means of dispersing these oxides in the metals. Examples are nickel-based superalloys that strengthened with dispersed thorium oxide or yttrium oxide (Y_2O_3), the latter being preferred. These superalloys have excellent strength and corrosion resistance at elevated temperatures, making them attractive candidate materials for use in applications such as jet-engine turbine blades, vanes, and combustors. This class of alloys is known as oxide-dispersion-strengthened (ODS) nickel-base superalloys. Iron- and aluminum-base dispersion-strengthened alloys have also been developed. Additionally, a number of other potential applications of mechanical alloying materials has been explored, including powders for coating applications, alloys of immiscible systems,

amorphous alloys, intermetallics, cermets, and organic–ceramic–metallic materials systems in general.

Alloys like MA 754 (Ni-20%, Cr-0.3%, Al-0.5%, Ti-0.6%, Y_2O_3) and MA 6000 (Ni-15%, Cr-4%, W-2%, Mo-2%, Ta-4.5%, Al-2.5%, Ti-1.1%, Y_2O_3) are commercially available as bar and plate for use in gas turbine vanes and turbine blades and for other applications in oxidizing or corrosive atmospheres. The vane alloy MA 754 is capable of withstanding a surface temperature of about 1100°C and is used in military engines. The iron-base alloy MA 956 (Fe-20%, Cr-4.5%, Al-0.5%, Ti-0.5%, Y_2O_3), available as plate sheet, bar, and wire, is capable of withstanding temperatures up to 1300°C or even higher in corrosive environments; it has applications in aircraft and industrial gas turbine combustors, in swirlers and heat exchangers of power generation equipment, and in heat-treatment equipment.

Conventional coatings in advanced gas turbine engines fail by loss of aluminum in the coating or by interdiffusion between the coating and substrate. Coatings possessing a large amount of Y_2O_3, produced by mechanical alloying, could solve such problems.

Liquid or solid immiscible systems are difficult to process by conventional pyrometallurgy, for example, the copper–lead or copper–iron systems. In these cases, mechanical alloying provides a route to obtain a homogeneous distribution in the solid phase. Mechanical alloying has also been used to fabricate the superconducting intermetallic niobium–tin compound Nb_3Sn, which is difficult to produce conventionally because of the large melting point difference between niobium and tin. Mechanical alloying can also be used to produce supercorroding magnesium-base alloys. These alloys are designed to corrode at a controlled rate for release of deep-sea equipment at specific depths. The requirement is to have the anode and cathode in proximity by using mechanical alloying. In another interesting application of mechanical alloying, titanium and magnesium have been combined by using mechanical alloying; usually, it is very difficult to produce such an alloy because the boiling point of magnesium is lower than the melting point of titanium. This achievement could result in a lower-density titanium alloy for aerospace applications.

Amorphous alloys can also be produced by using the mechanical alloying technique. These include Nb_3Sn starting from elemental powders, various rare earth–cobalt combinations, and titanium–copper and titanium–copper–palladium systems. See also *attritor grinding* and *dispersion-strengthened material*.

Mechanical Bond (Thermal Spraying)

The adherence of a thermal sprayed deposit to a roughened surface by the mechanism of particle interlocking.

Mechanical (Cold) Crack

A crack or fracture in a casting resulting from rough handling or from thermal shock, such as may occur at shakeout or during heat treatment.

Mechanical Disintegration

See preferred terms *comminution* and *pulverization*.

Mechanical Finishes

Finishing with Coated Abrasives

Coated abrasives or "sandpaper" can be used in many ways on many materials to produce a wide variety of finished surfaces. The method of application as well as the type of abrasive used greatly influences the results obtained. Following is a description of the three principal methods of using endless abrasive belts and the application for each method.

Contact Wheel Application

The contact wheel method produces the best abrasive efficiency, including the fastest rate of cut and the best abrasive life. The finish produced by this method is characterized by short individual scratches that are uniform across the width of contact and in line with the running direction of the belt. This method can be used on all types of parts having flat or lightly contoured surfaces such as hand tools, small household appliances, cutlery, and stainless steel sheet for decorative trim and kitchens, hospitals, etc.

Platen Method

This method is used for finishing work with a flat surface that is either continuous or interrupted. Because individual abrasive grain remains in contact with the work from top to bottom, the method produces a continuous scratch running the length of the part in contact with the belt. The advantage of the platen method is that the whole surface is contacted simultaneously and it easily produces flatness with a high degree of accuracy.

Slack Belt

This method is used on work that is highly contoured such as metal furniture, light fixtures, and plumbing fixtures. The flexibility of the belt is useful in polishing hard-to-reach surfaces where contours and shapes are involved. The length of scratch pattern produced is between the short scratch pattern of the contact wheel method and the long scratch pattern of the platen.

Types of Abrasives and Finishes

Two principal types of abrasives are used for coated-abrasive types of finishing operations—aluminum oxide and silicon carbide. The former has a blocky shape and is very tough. It is used for grinding and finishing materials having a high tensile strength. Silicon carbide is harder and sharper but is more friable or brittle; consequently, it is used for grinding and finishing materials having low tensile strength.

Coated abrasives are used extensively for a vast multitude of grinding and finishing operations because of their inherent versatility and economy. The finishes produced with coated abrasives are primarily decorative but can also be used for functional purposes in cases where geometry of the finished parts is important.

Finishing with Loose Abrasives

Finishes produced by loose abrasives, such as aluminum oxide and silicon carbide, are available in many variations of surface roughness and appearance, depending on the methods used. The methods used to produce or polish a surface may include polishing wheels (referred to as "setup" wheels), lapping, pressure blasting, and barrel finishing (often referred to as tumbling).

Setup wheels can be likened to bonded grinding wheels. The basic difference is that setup wheels are resilient because they are made of cloth laminations that are stitched so as to give a certain rigidity or resiliency. Abrasive grain is cemented on the periphery of setup wheels with cold cement or hot glue. Because of its "cushioning" effect, the resiliency of setup wheels simplifies the blending of contoured surfaces.

Buffing

Buffing with loose abrasives usually follows a polishing operation, with the latter done with either setup wheels or coated belts. Buffing generates smooth surfaces that are essentially free of scratches and have high reflectivity. It is accomplished by bringing the surface into contact with the periphery of a buffing wheel (usually cloth laminations) to which abrasive materials have been applied in composition form. Buffing generates smooth and reflective surfaces by removing or displacing relatively small amounts of metal. It is not designed to establish or retain dimensional accuracy.

Pressure Blasting

Pressure blasting is the process in which solid abrasive particles are propelled against a surface by means of expanding compressed air. The method always produces a matte finish. The finer the abrasive, the finer the finish, the coarser the abrasive, the faster the rate of stock removal and the rougher the finish.

Pressure blasting can be used for many finishing purposes. These include removal of heat-treat scale and oxides from metallic surfaces, preparation of surfaces for protective coatings, removal of contamination from nuclear devices, generation of decorative finishes, preparation of surfaces for visual inspection for surface defects, and removal of small burrs.

Lapping

Lapping can be used to produce very smooth finishes and geometrically true surfaces. Like grinding, the degree of finish obtained by lapping can be controlled by the particle size of the abrasive. Coarse grits plow deep furrows and therefore remove stock faster than fine particles. Fine particles, conversely, remove stock at a lower rate but produce finishes with a lower root mean square (rms). To reduce the surface roughness from, for example, 30–5 rms, usually requires using 2–3 progressively finer abrasives.

The finishes generated by lapping are functional rather than decorative. The surface smoothness dictates a low rms where wearing parts are involved. Flatness and scratch-free parts are essential for components such as high-pressure valves, and germanium and silicon transistor wafers.

Barrel Finishing

Barrel finishing is a surface-conditioning operation in which a mixture of metallic or nonmetallic parts, abrasive media, and various compounds is placed in a receptacle and rotated for the purpose of rounding corners, deburring, improving surface finish, cleaning, derusting, burnishing for high luster, and producing low microinch finishes. Many of these finishing operations are accomplished in simultaneous operations.

Unlike grinding wheel, setup wheel, or abrasive belt operations, tumbling produces a random scratch pattern. This is because the media and parts have a random motion while the barrel is rotating. But like grinding wheels and other types of abrasive finishing methods, the types and size of abrasive media (sometimes called nuggets, pellets, chips, etc.) influence the roughness and rate of stock removal.

Finishing with Bonded Abrasives

Finishes produced by grinding wheels vary over a wide range both in surface roughness and in surface appearance. These finishes can be divided into two major classifications: those produced by rough grinding and those produced by finish grinding.

Rough-grinding operation is primarily intended for stock removal without any particular regard for appearance. The finish produced can be described as having the appearance of a plowed field with the direction of the furrows in the same direction as the path of the grinding wheel. The depth of the furrows is in direct proportion to the roughness of the grind. The rougher the grind, the deeper the furrow; the less severe the grind, the shallower the furrow. The surface roughness produced by rough grinding varies over a wide range, beginning in the 30 μin. area and becoming progressively rougher.

Finish grinding, as its name implies, is used for the production of finish and, in addition, the generation of geometry. This finish may be one that has a low surface roughness or a high luster and, in some cases, both a low roughness and a high luster.

The appearance of a finely ground surface can either be a dull matte finish, or it can be a highly reflective surface. A distinction is made between the roughness of the dull matte ground finish and that of a highly reflective ground surface. A dull matte finish is generally smooth and many times, smoother than a highly reflective surface. The matte finish is a truly ground surface and can be described as a plowed field where the furrows are very shallow in depth and all the ridges have been removed.

A highly reflective surface, on the other hand, can be described as a plowed field where some of the ridges have fallen into the adjacent furrow and are pushed into that furrow by the grinding wheel as it passes over it. This produces a luster but not necessarily a low surface roughness.

Mechanical Hysteresis

Energy absorbed in a complete cycle of loading and unloading within the elastic limit and represented by the closed loop of the stress–strain curves for loading and unloading. Sometimes referred to as elastic, but more properly, mechanical. See also the term *hysteresis loop* (mechanical).

The phenomenon by which a material deformed in tension within its elastic range does not return immediately to zero strain on removal of the load. Further loading in elastic compression and then repeating the tensile load cause the stress–strain relationship to follow a hysteresis loop. The larger the area contained in the loop, the greater is the energy consumed and the greater the damping. The effect is related to the *elastic after effect*.

Mechanical Metallurgy

The science and technology dealing with the behavior of metals when subjected to applied forces; often considered to be restricted to plastic working or shaping of metals.

Mechanical Plating

Plating wherein fine metal powders are peened onto the work by *tumbling* or other means. The process is used primarily to provide ferrous parts with coatings of zinc, cadmium, tin, and alloys of these metals in various combinations. The process takes place in a rotating barrel with balls or powder of the plating metal. The metal plating is applied by mechanical means rather than electroplating or dip processes.

Mechanical Polishing

A process that yields a specularly reflecting surface entirely by the action of machining tools, which are usually the points of abrasive particles suspended in a liquid among the fibers of a polishing cloth.

Mechanical Press

A press to slide is operated by a crank, eccentric, cam, toggle links, or other mechanical device. See the term *eccentric gear, knuckle-lever press,* and *toggle press.*

Mechanical Press Brake

A *press brake* using a mechanical drive consisting of a motor, flywheel, crankshaft, clutch, and eccentric to generate vertical motion.

Mechanical Properties

The properties of a material that reveal its elastic and inelastic behaviors when force is applied, thereby indicating its suitability for mechanical applications; for example, modulus of elasticity, tensile strength, elongation, hardness, and fatigue limit, bend, impact, and torsion. Compare with *physical properties.*

Mechanical Seal

See *face seal.*

Mechanical Stability (of a Grease)

Grease shear stability tested in a standard rolling tester.

Mechanical Stage

In microscopy, a device provided for adjusting the position of a specimen, usually by translation into directions at right angles to each other.

Mechanical Testing

Any testing to determine the mechanical properties of a material, particularly tensile testing but including bend, impact, fatigue, and creep testing. The term usually includes hardness testing since this can give an accurate indication of some mechanical properties.

Mechanical Twin

A *twin* formed in a crystal by simple shear on the external loading. See *twin.*

Mechanical Upsetter

A three-element forging press, with two gripper dies and a forming tool, for flanging or forming relatively deep recesses.

Mechanical Wear

Removal of material due to mechanical processes under conditions of sliding, rolling, or repeated impact. The term mechanical wear includes *adhesive wear, abrasive wear,* and *fatigue wear.* Compare with *corrosive wear* and *thermal wear.*

Mechanical Working

The subjecting of metals to pressure exerted by rolls, hammers, or presses in order to change the shape or physical properties of the metal.

Mechanically Formed Plastic

A cellular plastic having a structure produced by physically incorporated gases.

Median Crack

Damage produced in glass by the static or translational contact of a hard, sharp object on the glass surface. The crack propagation into the glass perpendicular to the original surface.

Median Fatigue Life

The middle value when all of the observed fatigue life values of the individual specimens in a group tested under identical conditions are arranged in order of magnitude. When an even number of specimens are tested, the average of the two middlemost values is used. Use of the sample median rather than the arithmetic mean (i.e., the average) is usually preferred.

Median Fatigue Strength at N Cycles

An estimate of the stress level at which 50% of the population would survive N cycles. The estimate is derived from a particular point of the fatigue life distribution since there is no test procedure by which a frequency distribution of fatigue strengths at N cycles can be directly observed. Also known as *fatigue strength at N cycles*.

Meehanite

Proprietary name for a range of cast irons. The term is used casually referring to inoculated irons.

Meerschaum

A soft, white or gray, claylike mineral of composition $3SiO_2 \cdot 2MgO \cdot 2H_2O$, used for making pipes and cigar holders, but also employed for making various other articles, as it can be cut easily when wet and withstands heat. When fresh, the mineral absorbs grease and makes a lather, its German name means seafoam. It is used as a filler in soaps in Germany. The hardness is about Mohs 2 and the specific gravity 1.28. Most of the commercial meerschaum comes from Asia Minor, and the mines at Eskisehir have been worked for 20 centuries. **Artificial meerschaum** is made from meerschaum shavings, kieselguhr, and from silicates of aluminum, calcium, and magnesium.

Mega

As a prefix to units of the SI and other systems of units, it is 1 million. For example, Mega-Pascal.

Megaelectron Volt (MeV)

A unit of energy usually associated with a particle. The energy gained by an electron accelerated across 1,000,000 V.

Melamine Formaldehyde Plastics

The outstanding features of melamines, one of the amino resins, include excellent colorability, high hardness, and good electrical characteristics, including exceptional arc resistance.

Primary use is for melamine plastics as engineering materials are in decorative molded dinnerware, decorative thermosetting laminates, molded housings, and wiring devices, closures, and buttons.

The resins are produced by reaction of melamine and formaldehyde. Crystals of melamine, an amino chemical, are reacted with formaldehyde and mixed with highly purified filler, dried, and ground. Color pigments, plasticizers, lubricant, and accelerators are added to produce the molding compound. Melamine is similar in many respects to urea–formaldehyde. Melamine resists water absorption and chemical attack to a greater extent than urea, and its cost is higher.

Applications

Molding Compounds

For molding applications, the hardness and colorability of melamine are outstanding advantages. Melamine is the hardest of plastics (Rockwell M118–M124) and has unlimited colorability. Parts can be produced in any color in the spectrum (both translucent and opaque). Color quality ranges from pastels to bright, jewel-like tones, as well as two tones in the same object. Products such as dishes and tableware can be decorated by foil inlays.

Dinnerware (dishes, plates, cups, and saucers) make up 90% of the uses for melamine molding compounds. Hardness and colorability as well as resistance to attack by foods, detergents, and hot water account for its popularity in the home.

Melamine molding compounds are widely used as handles for flatware and kitchen utensils because of hardness, color, and resistance to heat and water. Handles can now be decorated with patterns applied integrally by foil.

Agitators for washing machines are being molded of melamines because of their resistance to hot water and detergents and excellent appearance. A colored agitator is more attractive than a black one.

Nondecorative applications for molding compounds include wiring fixtures and other electrical uses, where the arc resistance and nontracking characteristics of melamine, coupled with its nonflammability, make it useful. For such uses, strength and impact resistance are improved by incorporating fabrics and glass-fiber reinforcements.

High-Pressure Laminates

The majority of top-quality decorative thermosetting laminates, or high-pressure laminates (e.g., "Formica," "Micarta," and "Textolite"), use melamine resins impregnated in alpha-cellulose paper or other clear reinforcements to provide their decorative surfaces. With the proper resin selection, such laminates can be burn and stain resistant. Further, the high hardness of melamines makes them scuff and mar resistant.

Melamine Resin

This is a synthetic resin of the alkyd type made by reacting melamine with formaldehyde. The resin is thermosetting, colorless, odorless, and resistant to organic solvents. It is more resistant to alkalies and acids than urea resins, has better heat and color stability, and is harder. The melamine resins have the greater general uses of molding plastics and also are valued for dishes for hot foods or acid juices, and they will not soften or warp when washed in hot water.

The melamine resins have good adhesiveness but are too hard for use alone in coatings and varnishes. They are combined with alcohol-modified urea–formaldehyde resins to give coating materials of good color and gloss, flexibility, and chemical resistance. Urea-modified melamine–formaldehyde resins are used for coatings and varnishes. Melamine–formaldehyde molding resin, with cellulose filler, has a tensile strength of 51 MPa and dielectric strength per mil of 12.8×10^6 V/m.

Melamine–urea–formaldehyde resin with a lignin extender is used as an adhesive for water-resistant plywood. Phenyl-modified melamine–formaldehyde resin solution is used for laminating fibrous materials. Highly translucent melamine–formaldehyde resin is used for molding high-gloss buttons. Methylol–melamine, made by alkylating a melamine–formaldehyde resin with methyl alcohol, is used for shrink-proofing woolen fabrics.

Melt

(1) To change a solid to a liquid by the application of heat. (2) A charge of molten metal or plastic. (3) In submerged arc welding, the molten flux above the weld metal.

Melt Index

The amount, in grams, of a thermoplastic resin that can be forced through a 2.0955 mm (0.0825 in.) orifice when subjected to 20.7 N (2160 gf) in 10 min at 190°C (375°F).

Melt Lubrication

Lubrication provided by steady melting of a lubricating species. Also known as *phase-change lubrication*.

Melt Run (of Weld)

A line of parent metal that has been fused without the addition of filler.

Melt Strength

The strength of a plastic while in the molten state.

Meltback Time

In arc welding, the time interval at the end of crater fill time to arc outage during which the electrode feed is stopped. Arc voltage and arc length increase and current decreases to zero to prevent the electrode from freezing in the weld deposit.

Melting Point (Metals)

The temperature at which a pure metal, compound, or eutectic changes from solid to liquid; the temperature at which the liquid and the solid are at equilibrium.

Melting Point (Plastics)

The term that refers to the first-order transition in crystalline polymers. The fixed point between the solid and liquid phases of a material when approached from the solid phase under a pressure of 101.325 kPa (1 atm).

Melting Pressure (Metals)

At a stated temperature, the pressure at which the solid phases of an element or congruently melting compound may coexist at equilibrium with liquid of the same composition.

Melting Range (Metals)

The range of temperatures over which an alloy other than a compound or eutectic changes from solid to liquid; the range of temperatures from *solidus* to *liquidus* at any given composition on a *phase diagram*.

Melting Rate

In electric arc welding, the weight or length of electrode melted in a unit of time. Sometimes called melt-off rate or burn-off rate.

Melting Temperature (Glass)

The range of furnace temperatures within which melting takes place at a commercially desirable rate and within which the resulting glass generally has a viscosity of $10^{0.5}–10^{1.5}$ Pa s ($10^{1.5}–10^{2.5}$ P). To compare the melting temperatures of glasses, it is assumed that a glass at its melting temperature has a viscosity of 10 Pa s (10^2 P).

Melting Temperature (Metals)

See *melting point* (metals).

Melt-Off Rate

See *melting rate*.

Melt-Through

Complete joint penetration for a joint welded from one side. To prevent melt-through, the welding current and the width of the root opening should be reduced and travel speed increased.

Membrane

Any thin sheet or layer.

Memory Alloys

See *shape memory*.

Mendeléev's (Periodic) Table

See *Periodic Table*.

Menstruum Method

A method of producing multicarbide powder, such as WC + TiC solid solution, by introducing the individual elements into a molten bath of a noncarbide-forming metal such as cobalt or nickel. The multicarbide is formed above 2100°C (3800°F), slowly cooled in the dispersed condition in the Menstruum to room temperature, and finally processed by chemical separation.

Mer

The repeating structural unit of any polymer.

Merchant Iron

Multiple forged and hence high-grade wrought iron.

M

Mercury

Mercury (symbol Hg), also called quicksilver, is a metallic element. It is the only metal that is a liquid at room temperature. Mercury has a silvery-white color and a high luster. Its specific gravity is 13.596. The solidifying point is –40°C, and its boiling point at 1 atm is 357°C. It does not oxidize at ordinary temperatures, but when heated to near its boiling point it absorbs oxygen and is converted into mercuric oxide, HgO, used as a pigment in marine paints.

Metallic mercury is used as a liquid contact material for electrical switches, in vacuum technology as a working fluid of diffusion pumps, for the manufacture of mercury–vapor rectifiers, thermometers, barometers, tachometers, and thermostats, and for the manufacture of mercury–vapor lamps. Mercury–vapor lamps serve as sources for ultraviolet light.

For micro gas analyses, mercury is most often used as a sealing liquid for the evolved gases. Very large amounts of mercury are used as the electrode material for the electrolysis of aqueous solutions of alkali halides for the manufacture of chlorine and sodium hydroxide. Also, it finds application for the manufacture of silver amalgams for tooth fillings in dentistry.

Some mercury salts serve as catalysts for organic chemical reactions. Fulminate of mercury is used as a primer for explosives. Some complex salts find application as temperature indicators.

A large number of mercury compounds have been used for hundreds of years as medicines. The compounds of mercury with some organic substances are powerful diuretic substances and have applications as a disinfectant, and ammoniated mercury is used in the treatment of various diseases.

Of importance and electrochemistry are the standard calomel electrode, used as a reference electrode for the measurement of potentials and for potentiometric titrations, and the Weston standard cell, metallic mercury is used in contact with solutions that contain mercury salts as the solid phase.

Mercury is used for separating gold and silver from their ores, for coating mirrors, in tanning, in batteries, for the frozen-mercury molding process, mercury–vapor motors, as a circulating medium in atomic reactors, and in its compounds for paint pigments and explosives.

The metal is marketed in steel flasks holding 75 lb (34 kg). The black *mercurous oxide*, Hg_2O, is used in skin ointments. *Mercuric chloride*, or *corrosive sublimate*, $HgCl_2$, is an extremely poisonous, white, crystalline powder soluble in water and in alcohol, used as a wood preservative, as an insecticide and rat poison, in tanning, as a mordant, and as a caustic antiseptic in medicine. *Vermillion red*, one of the oldest paint pigments, is red *mercury sulfide*, HgS, made directly by heating mercury and sulfur. It is a brilliant, water-insoluble, red powder of specific gravity 8.1. Because of its expense, it is often mixed with other red pigments. *Mercurochrome* is a green crystalline powder that gives a deep-red solution in water and is used as an antiseptic. Mercury forms a vast number of compounds, all of which are poisonous and some of which are explosive. *Mercury 203* is radioactive. A mercury distillation process, developed by Lumalampan AB of Switzerland and called the *MRT System*, recovers mercury from fluorescent light bulbs, button batteries, amalgams, and electrical devices.

Mercury Seal

A gas-tight seal provided when a shaft enters equipment via a pool of mercury.

Mercury Switch

An electrical switch in which a rocking tubular vessel contains a limited amount of mercury that, at one rock position, covers the electrical contacts.

Mesh

(1) The number of screen openings per linear inch of screen; also called *mesh size*. (2) The screen number on the finest screen of a specified standard screen scale through which almost all of the particles of a powder sample will pass. See also *sieve analysis* and *sieve classification*. (3) Apart from its common meanings, in a metallurgical context, this term usually refers to the series of numbered sieves of decreasing mesh hole size used for grading powders.

Mesh Size

(1) The opening(s) or size of opening(s) in a designated sieve or screen, hence the approximate diameter of particles below which they will pass through and above which they will be retained on the screen. Mesh sizes are given as the number of wires per inch of standard screen construction, for example, Tyler or U.S.; these are translated by tables into equivalent particle diameters in inches (in.), millimeters (mm), or micrometers (μm). (2) The width of the aperture in a cloth or wire screen. See also *sieve analysis* and *sieve classification*.

Mesh-Belt Conveyor Furnace

A continuously operating furnace that uses a conveyor belt for the transport of the charge. Mesh-belt conveyor furnaces consist of a belt-driven table, a burn-off zone, a slow-cooling zone, a final-cooling zone, and a discharge table.

Mesophase

An intermediate phase in the formation of carbon from a pitch precursor. This is a liquid crystal phase in the form of microspheres, which upon prolonged heating above 400°C (750°F), coalesce, solidify, and form regions of extended order. Heating to above 2000°C (3630°F) leads to the formation of graphite structure.

Metal

(1) An imprecise term. Generally material showing most if not all of the following characteristics—crystalline with the atoms in a simple lattice with shared valence electrons, good strength, ductility and malleability, lustrous, and high electrical and thermal conductivities. (2) An opaque lustrous elemental chemical substance that is a good conductor of heat and electricity and, when polished, a good reflector of light. Most elemental metals are malleable and ductile and are, in general, denser than the other elemental substances. (3) As to structure, metals may be distinguished from nonmetals by their atomic binding and electron availability. Metallic atoms tend to lose electrons from the outer shells, the positive ions thus formed being held together by the electron gas produced by the separation. The ability of these "free electrons" to carry electric current, and the fact that this ability decreases as temperature increases, establishes a prime distinctions of a metallic solid. (4) From a chemical viewpoint, an elemental substance whose hydroxide is alkaline. (5) An alloy.

Metal

A metal is a solid or liquid (molten), opaque material with a lustrous surface and good electrical and thermal conductivities. Solid metal is usually crystalline and ductile and can be permanently deformed by shear on crystal planes; permanent deformation is accompanied by an increase in strength (work hardening). Metallic properties are related to the arrangements of positively charged ions bonded through a surrounding field (sea) of free electrons that draw the ions into a close-packed crystalline structure with planes appropriate for slip. Liquids are nearly close-packed noncrystalline structures, with a thermal energy great enough to activate random, free movements of atoms.

The rapid developments of the metals industry since the late nineteenth century required an in-depth understanding of the fundamentals of metallurgy to improve metal properties and expand their usefulness.

Characteristics of Solids

The unique characteristics of the different classes of solids are directly related to the electron pattern and the types of bonds between the atoms of the solid. Atomic bonding is classified into two major categories: strong attraction (covalent, ionic, or metallic) and weaker, secondary bonding (van der Waals' or hydrogen bonding).

Directional Bonds

A covalent bond between two or more atoms is a negative charge directed in space and equivalent in magnitude to one or more electrons. As a consequence, at the lowest energy state, the corresponding bond is directional and usually strong, and considerable energy must be supplied to the material to free an electron from this bond or shift an ion from one fixed position to another. For covalent materials, bonding valence electrons require large energies to transport a current composed of many electrons through the aggregate atoms. Thus, electrical conductivity and thermal conductivity by electrons are low.

Metallic Bonding

In contrast to directional bonds, metallic bonds result from linking ions each having centro-symmetrical attraction to a surrounding atmosphere of the distributed charges of nearly free electrons. These electrons, usually valence electrons, have the same likelihood of being associated with any ion in the aggregate.

Crystal Structure

To visualize groups of atoms in an aggregate, the atoms are considered as spheres of the same radius.

This representation of metal atoms is reasonable since bonds originating from metal atom cores with the superimposition of the electron charges in the intervening space are manifested by a nearly spherically symmetrical attraction between the ions.

Among metals, the face-centered cubic crystalline structure is found in aluminum, nickel, copper, rhodium, palladium, silver, iridium, platinum, gold, and lead.

Because some metals display characteristics of other bonding types, they occur in other crystal structures primarily because of the interference of electrons at energy levels that are not centrally symmetric. All close-packed hexagonal metals belong to two categories that can be illustrated by packing elongated or compressed spheres.

An ideal hexagonal close-packed structure illustrates the hexagonal stacking of spheres, whereas spheres compressed in a direction perpendicular to a close-packed plane have an axial ratio less than the ideal of 1.633. Examples are cobalt, rhenium, titanium, gadolinium, hafnium, and beryllium as well as lanthanum, cadmium, and zinc.

Other metals such as iron, titanium, tungsten, molybdenum, niobium, tantalum, potassium, sodium, vanadium, chromium, and zirconium (alkali and transition metals) are slightly less closely packed, with a body-centered cubic crystal structure that is 8.1% lower in packing density than face-centered cubic crystals. Furthermore, the equilibrium crystal structure for many metals is dependent on temperature and pressure. Consequently, a large number of elements are allotropic; that is, the element will appear as different crystal structures at various ranges of pressure–temperature conditions. Included in this group are calcium, titanium, manganese, iron, cobalt, yttrium, zirconium, tin, hafnium, thallium, polonium, thorium, and uranium.

Polycrystalline Metals

Common metallurgical production methods inevitably lead to solids made up of many small crystals separated by higher-energy boundaries containing nonlattice-positioned arrangement of atoms. The crystals in these polycrystalline metals are called grains, and the boundaries are grain boundaries. Since the grain boundaries interrupt periodicity of the lattice, they have a large effect on many of the physical and mechanical properties. They are origins (sources) and terminators (sinks) to all other defects, especially vacancies, solute atoms, and dislocations.

In an ideal polycrystalline arrangement composed of many randomly oriented grains, the anisotropic features of each grain are averaged, and the resulting metal solid would have isotropic properties. Actually, metal manufacturing processes such as casting and deformation forming (forging, rolling, and so forth) tend to develop grains with specific crystallographic directions distributed about certain directions that are distinctive of the forming process and the geometry of the bulk product. This nonrandom arrangement of grains, called preferred orientation, contributes to anisotropy of properties in the final polycrystalline product.

Metal Alloys

An alloy has metallic features and is composed of two or more chemical elements. Macroscopically, it appears homogeneous, but microscopically various thermodynamically stable arrangements of the chemical elements are possible, depending on the characteristics of the components.

Metal Casting

Metal casting involves the introduction of molten metal into a cavity or mold where, upon solidification, it becomes an object whose shape is determined by mold configuration. Casting offers several advantages over other methods of metal forming. It is adaptable to intricate shapes, to extremely large pieces, and to mass production; it can provide parts with uniform physical and mechanical properties throughout; and, depending on the particular material being cast, the design of the part, and the quantity being produced, its economic advantages can surpass other processes.

Categories

Two broad categories of metal-casting processes exist: ingot casting (which includes continuous casting) and casting to shape.

Ingot castings are produced by pouring molten metal into permanent or reusable molds. Following solidification, these ingots (or bars, slabs, or billets, as the case may be) are then further processed mechanically into many new shapes. Casting to shape involves pouring molten metal into molds in which the cavity provides the final useful shape, followed only by machining or welding for the specific application. Another mode of metal casting, zero gravity casting, has been developed for processing in a space station.

Ingot Casting

Ingot castings make up the majority of all metal castings and are separated into three categories: static cast ingots, semi-continuous or direct-chill cast ingots, and continuous cast ingots.

Static Cast Ingots: Static ingot casting simply involves pouring molten metal into a permanent mold. After solidification, the ingot is withdrawn from the mold and the mold can be reused. This method is used to produce millions of tons of steel annually.

Semicontinuous Cast Ingots: A semicontinuous casting process is employed in the aluminum industry to produce most of the cast alloys from which rod, sheet, strip, and plate configurations are made. In this process, molten aluminum is transferred to a water-cooled permanent mold that has a movable base mounted on a long piston. After solidification has progressed from the mold surface so that a solid "skin" is formed, the piston is moved down, and more metal continues to fill the reservoir. However, technological advances have allowed major aluminum alloy producers to replace the metal mold (at least in part) by an electromagnetic field so that molten metal touches the metal mold only briefly, thereby making a product with a much smoother finish than that produced conventionally.

Continuous Cast Ingots: Continuous casting provides a major source of cast material in the steel and copper industry and is growing rapidly in the aluminum industry. In this process, molten metal is delivered to a permanent mold, and the casting begins much in the same way as in semi-continuous casting. However, instead of the process ceasing after a certain length of time, the solidified ingot is continually sheared or cut into lengths and removed during casting. Thus, the process is continuous, with the solidified bar or strip being removed as rapidly as it is cast. The method has many economic advantages over the more conventional casting techniques; as a result, all modern steel mills produce continuous cast products.

Cast to Shape

Casting to shape is generally classified according to the molding process, molding material, or method of feeding the mold. There are four basic types of these casting processes: sand, permanent-mold, die, and centrifugal.

Sand Casting: This is the traditional method that still produces the largest volume of cast-to-shape pieces. It utilizes a mixture of sand grains, water, clay, and other materials to make high-quality molds for use with molten metal. Other casting processes that utilize sands as a basic component are the shell, carbon dioxide, investment casting, ceramic molding, and plaster molding processes. In addition, there are a large number of chemically bonded sands that are becoming increasingly important.

Permanent Mold Casting: Many high-quality castings are obtained by pouring molten metal into a mold made of cast iron, steel, or bronze. Semipermanent mold materials such as aluminum, silicon carbide, and graphite may also be used. The mold cavity and the gating system are machined to the desired dimensions after the mold is cast; the smooth surface from machining thus gives a good surface finish and dimensional accuracy to the casting.

Die Casting: A further development of the permanent molding process is die casting. Molten metal is forced into a die cavity under pressures of 0.7–700 MPa. Two basic types of die-casting machines are hot chamber and cold chamber. In the hot-chamber machine, a portion of the molten metal is forced into the cavity at pressures up to about 14 MPa. The process is used for casting low-melting point alloys such as lead, zinc, and tin. In the cold-chamber process, the molten metal is ladled into the injection cylinder and forced into the cavity under pressures that are about 10 times those in the hot-chamber process. High-melting-point alloys such as aluminum-, magnesium-, and copper-base alloys are used in this process. Die casting has the advantages of high production rates, high quality and strength, surface finish on the order of 1.0–2.5 μm rms, and close tolerances with thin sections. Rheocasting is the casting of a mixture of solid and liquid. In this process, the alloy to be cast is melted and then allowed to cool until it is about 50% solid and 50% liquid. Vigorous stirring promotes liquidlike properties of this mixture so that it can be injected in a die-casting operation. A major advantage of this type of casting is expected to be much reduced die erosion due to the lower casting temperatures.

Centrifugal Casting: Inertial forces of rotation distribute molten metal into the mold cavities during centrifugal casting, of which there are three categories: true centrifugal casting, semi-centrifugal casting, and centrifuging. The first two processes produce hollow cylindrical shapes and parts with rotational symmetry, respectively. In the third process, the mold cavities are spun at a certain radius from the axis of rotation; the centrifugal force thus increases the pressure in the mold cavity.

Evaporative Cooling

This is another casting process that utilizes sand as a basic component. However, it is unique in that the casting shape is not defined by molding the sand. Rather, a polystyrene foam pattern or replica is formed into the shape of the component to be cast and determines the size and shape of the final casting.

The evaporative casting process is compatible with bronze, steel, aluminum, and iron castings. In general, microstructural characteristics of castings made by the evaporative casting process tend to be the same as those of castings made by conventional sand casting processes.

The major advantages of the evaporative casting process are reduced costs and greater overall flexibility. Cost savings occur both in lower investment costs and lower cost to make a casting. The lower investment costs come about because the equipment to make polystyrene foam replicas is less expensive than the equipment to make conventional sand molds and cores. Also, the polystyrene molding equipment operates at higher speeds. Finally, the tooling to make polystyrene foam replicas is also less expensive and has a longer life than tooling for conventional sand molds and cores. The costs for making a single casting by the evaporative casting process are lower because this process does not require cores and allows for easy reuse of the casting sand (no resins are added to the sand). Also, the casting made by this process is easier to clean than a conventional sand casting because these are no fins and parting lines that need to be ground down.

Zero-Gravity Casting

Another mode of metal casting, zero-gravity casting, has come of age with the advent of space stations containing laboratory equipment with metal-processing capabilities. In the absence of gravity, the concept of melting and casting is drastically changed. The disadvantages of Earth-based processing that appear to be diminished or eliminated by processing under conditions of weightlessness are thermal and solutal convection and sedimentation, contamination of and contact with the container during processing, and hydrostatic pressure effects and deformation at high temperatures. Thus, there appears to be potential for processing liquid metals at zero gravity. However, commercial production is expected to be limited initially to small quantities of high-cost materials whose solidification characteristics are greatly improved by the lack of gravity. Semi-conductor materials are thought to be excellent candidates for initial development efforts.

Principles and Practice

Successful operation of any metal-casting process requires careful consideration of mold design and metallurgical factors. These include design consideration, heating systems, riser design, ease of pattern extraction, machining considerations, and, finally, metallurgical considerations, where the relationship between the processing history and the solidification event for a part that is cast to shape.

Applications

Ferrous alloys, steels, and cast irons constitute the largest volume of metals cast. Aluminum-, copper-, zinc-, titanium-, cobalt-, and nickel-base alloys are also cast into many forms, but in much smaller quantity than iron and steel.

Aluminum alloy castings have advantages such as resistance to corrosion, high electrical conductivity, ease of machining, and architectural and decorative uses. Magnesium alloy castings have the lowest density of all commercial casting alloys. Copper alloy castings, although costly, have advantages such as corrosion resistance, high thermal and electrical conductivities, and wear properties suitable for antifriction bearing materials. In steel, castings have more uniform (isotropic) properties than the same component obtained by mechanical working. On the other hand, because of high temperatures required, casting of steel is relatively expensive and requires considerable knowledge and experience. Cast irons, which constitute the largest quantity of all metals cast to shape, and properties such as hardness, wear resistance, machinability, and corrosion resistance. Gray iron castings are commonly produced for their low cost, machinability, good damping capacity, and uniformity. Nodular cast irons have significantly higher strengths than gray irons but are more costly and not as machinable. There is an intermediate grade of cast-iron called CG (compacted graphite) iron that shares the best properties of both gray and nodular irons.

Metal Coatings

These are thin films of material bonded to metals to add specific surface properties, such as corrosion or oxidation resistance, color, attractive appearance, wear resistance, optical properties, electrical resistance, or thermal protection. In all cases, proper surface preparation is essential to effective bonding between coating and basis metal so that coated metals can function as duplex materials. The various methods of applying either metallic coatings (hot dip, electroplating, spray, cementation, vapor deposition, cladding, and immersion) or nonmetallic coatings, such as vitreous enamel and ceramics, and the conversion of surfaces to suitable reaction-product coatings.

Hot-Dip

Low-melting metals provide inexpensive protection to the surfaces of a variety of steel articles. To form hot-dip coatings, thoroughly cleaned work is immersed in a molten bath of the coating metal. The coating consists of a thin alloy layer together with a relatively pure coating metal that adheres to the work as it is withdrawn from the bath. Sheet, strip, and wire are process on a continuous basis at speeds of several hundred feet per minute. On the other hand, hardware and holloware are handled individually or in batches.

Galvanized steel, that is, steel hot-dipped in zinc, is used for roofing, structural shapes, hardware, sheet, strip, and wire products. Hot-dipped tinplate is now largely supplanted by electroplated tinplate for tin cans. Terne plate, with coatings up to 20 μm, is used for roofing, chemical cabinets, and gasoline tanks. Hot-dip aluminum-coated (aluminized) steel with coatings up to 10 μm thick are used for oil refinery equipment and furnace and appliance parts where protection at temperatures up to 538°C is required.

Sprayed

A particular advantage of sprayed coatings is that they can be applied with portable equipment. The technique permits the coating of assembled steel structures to obtain corrosion resistance, the building up of worn machine parts for rejuvenation, and the application of highly refractory coatings with melting points in excess of 1650°C.

Nearly any metal or refractory compound can be applied by spraying. Coating material in the form of wire or powder is fed through a specially designed gun, where it is melted and subjected to a high-velocity gas blast that propels the atomized particles against the surface to be coated.

Coatings such as zinc and aluminum are applied with a gun that provides heat by burning acetylene, propane, or hydrogen in oxygen. Highly refractory coating materials, such as oxides, carbides, and nitrides, can be applied by plasma-arc spraying. In this process, temperatures of 12,000°C or more may be produced by partially ionizing a gas (nitrogen or argon) in an electric arc and passing the gas through a small orifice to produce a jet of hot gas moving at high velocity. Another variation for applying refractory coatings is detonation-flame plating. In this process, a mixture of oxygen and gas-suspended fine particles is fired four times per second by a timed spark.

Sprayed zinc or aluminum coatings up to 250 μm are used to protect towers, tanks, and bridges. Such coatings are normally sealed with an organic resin to enhance protection.

Sprayed refractory coatings have been developed for the high temperatures experienced in aerospace applications. They are also used for wear resistance, heat resistance, and electrical insulation.

Cementation

There are surface alloys formed by diffusion of the coating metal into the base metal, producing little dimensional change. Parts are heated in contact with powder coating material that diffuses into the surface to form an alloy coating, whose thickness depends on the time and the temperature of treatment. A zinc alloy coating of 25 μm is formed on steel in 2–3 h at 375°C. A chromium alloy (chromized) coating of 100 μm is formed in 1 h at 1000°C.

Chromized coatings on steel protect aircraft parts and combustion equipment. Sherardized (zinc–iron alloy) coatings are used in threaded parts and castings. Calorized (aluminum–iron alloy) coatings protect chemical equipment and furnace parts. Diffusion coatings are used to provide oxidation resistance to refractory metals, such as molybdenum and tungsten, in aerospace applications where reentry temperatures may exceed 1650°C. In addition to the pack process described earlier, such coatings may be applied in a fluidized bed. In forming disilicide coatings on molybdenum, the bed consists of silicon particles suspended in a stream of heated argon flowing at 0.15 m/s, to which a small amount of iodine is added. The hot gases react with the silicon to form SiI_4, which in turn reacts with the molybdenum to form $MoSi_2$.

Vapor Deposition

A thin specular coating is formed on metals, plastics, paper, glass, and even fabrics. Coatings form by condensation of metal vapor originating from molten metal, from high-voltage (500–2000 V) discharge between electrodes (cathode sputtering), or from chemical means such as hydrogen reduction or thermal decomposition (gas plating) of metal halides. Vacuums up to 10^{-4} Pa often are required.

Aluminum coatings of 0.125 μm are formed on zinc, steel, costume jewelry, plastics, and optical reflectors. Chemical methods can form relatively thick coatings, up to 250 μm.

Immersion

Either by direct chemical displacement or, for thicker coatings, by chemical reduction (electroless coating), metal ions plate out of solution onto the workpiece.

Tin coatings are displaced onto brass and steel notions and on aluminum-alloy pistons as an aid during the breaking-in period. Displacement nickel coatings of 1.25 μm are formed on steel articles. Electroless nickel, involving the reduction of a nickel salt to metallic nickel (actually a nickel–phosphorus alloy), permits the formation of relatively thick uniform coatings up to 250 μm on parts with recessed or hidden surfaces difficult to reach by electroplating.

Vitreous Enamel

Glassy but noncrystalline coatings for attractive durable service in chemical, atmospheric, or moderately high-temperature environments are provided by enamel or porcelain coating.

Dry enameling is used for castings, such as bathtubs. The casting is heated to a high temperature, and then dry enamel powder is sprinkled over the surface, where it fuses.

Firing temperatures for conventional enameling of iron or steel range up to 870°C. Low-temperature enamels have been developed, permitting the enameling of aluminum and magnesium.

Coatings of 75–500 μm are used for kitchenware, bathroom fixtures, highway signs, and water heaters. Vitreous coatings with crystalline refractory additives protect stainless steel equipment at temperatures up to 950°C.

Ceramic

Essentially crystalline, ceramic coatings are used for high-temperature protection above 1100°C. The coatings may be formed by spraying refractory material such as aluminum oxide or zirconium oxide, or by the cementation processes for coatings of intermetallic compounds such as molybdenum disilicide. Cermets are intimate mixtures of ceramic and metal, such as zirconium boride particles, dispersed throughout an electroplated coating of chromium.

Surface Conversion

An insulating barrier of low solubility is formed on steel, zinc, aluminum, or magnesium without electric current. The article to be coated is either immersed in or sprayed with an aqueous solution, which converts the surface into a phosphate, an oxide, or a chromate. Modern solutions react so rapidly that sheet and strip materials can be treated on continuous lines.

Phosphate coatings, equivalent to 1–4 g/m^2, are applied to bare or galvanized steel and to zinc-base die castings as preparation for painting. The coating enhances paint adhesion and prevents underfilm corrosion. Phosphate coatings, containing up to 40 g/m^2 (lubricated), serve as an aid in deep-drawing steel and in other friction-producing processes or applications. Iridescent chromate coatings on zinc-coated steel improve appearance and reduce zinc corrosion. Chromate, phosphate, and oxide coatings on aluminum or magnesium are used to prepare the surface before painting.

Anodic

Coatings of protective oxide may be formed on aluminum or magnesium by making them the anode in an electrolytic cell. Anodized coatings on aluminum up to 75 μm thick are formed in sulfuric acid to form a porous oxide that may be sealed in boiling water or steam to provide a clear, abrasion-resistant, protective coating.

Such coatings are used widely on aluminum furniture, automobile trim, and architectural shapes. Thin, nonporous, electrically resistant coatings are formed on aluminum in a boric acid bath in the production of electrolytic capacitors. Anodized coatings on magnesium are thicker (up to 75 μm) and harder than those formed by chemical conversion. Anodic coatings of 7.5 μm thick are often used as a paint base.

Powder

The term *powder coatings* refers to a process whereby organic polymers such as acrylic, polyester, and epoxies are applied to substrates for protection and beautification. It is essentially an industrial painting process that uses a powdered version (25–50 μm particle size) of the resin rather than the solvent solution represented by industrial baking enamels.

In the widest type of commercial usage, these powders are applied to electrically grounded substrates by means of an electrostatic spray gun. In these guns, high-voltage (60–100 kV), low-amperage charges are applied to the powders. The powder particles are attracted to and adhere to the substrate until it can be transported to an oven, where the powder particles melt, coalesce, flow, and form a smooth coating. Typical baking temperatures are 90°C for 20 min.

Coatings can be applied at thicknesses ranging from 25 to 250 μm, depending upon use requirements.

Although electrostatic spray is the principal method of commercial application, other techniques exist, including fluidized-bed coating, where a preheated part is immersed into an aerated bed of powder particles, and electrostatic fluidized bed, where a substrate is passed through an "electrified" cloud of the powder.

Powder coating uses no organic solvent, and the oversprayed material can be recaptured and reused. Thus, there are several notable advantages over conventional industrial solvent-based painting:

1. There are no hydrocarbon baking byproducts, and the process is environmentally acceptable.
2. Material conservation and economics are optimized because of the recycling of oversprayed material.
3. Powders are easy to apply and readily lend themselves to automation.

A wide variety of parts are finished with powder coatings. Outdoor lawn and patio furniture coated in this process display good weathering and abuse resistance. Powder-coated electrical transformers are insulated electrically and provided with corrosion protection. Powder coatings have also been developed for finishing major appliances and for automobile coatings. Also, laundry tops and lids have displayed excellent detergent resistance where organic powder coatings have replaced porcelain finishes.

Metal-Cored Electrode

A composite filler metal welding electrode consisting of a metal tube or other hollow configuration containing alloying ingredients. Minor amounts of ingredients facilitate arc stabilization and fluxing of oxides. External shielding gas may or may not be used.

Metal Dusting

Accelerated deterioration of metals in carbonaceous gases at elevated temperatures to form a dust-like corrosion product.

Metal Electrode

An electrode used in arc welding or cutting that consists of a metal wire or rod that is either bare or covered with a suitable covering or coating.

Metal Fatigue

The same as fatigue. The shorter term is usually used by metallurgists.

Metal Foams

Metal foams are metallic cellular materials that have a high porosity fraction, typically ranging from 40 to 90 vol%. Because of their high stiffness and low specific weight, cellular materials are applied in construction: packaging, insulation, noise and vibration damping, and filtering. They are considered by many to be a new class of engineering material.

Typical foaming processes include casting, powder pressing, metallic deposition, and sputter deposition. Metal foams can be fabricated in a variety of different ways, and many attempts have been made in the past to develop good foam structure. However, the choice frequently seems to be between high cost and poor quality.

A new powder method allows for direct net-shape fabrication of phone parts with a relatively homogeneous pore structure. Metallic foams fabricated by this approach exhibit a closed-cell microstructure with a higher mechanical strength than open-cell foams. This type of microstructure is particularly appropriate for applications requiring reduced weight and energy-absorption capabilities.

The powder-metallurgy production method makes it possible to build metallic foam parts that have complex geometry. Sandwich structures composed of a porous metallic foam core and metallic face sheets can also be produced, with options exploiting combined materials and shapes. These foams enlarge the application range of cellular materials because of their excellent physical and mechanical properties, as well as their relative recyclability.

Processing

The process consists of mixing metal powders (either prealloyed metal powders or powder blends) with a small amount of foaming agent. When the agent is a metal hydride, a content of less than 1% is generally sufficient. After the foaming agent is uniformly distributed within the matrix powders, the mixture is compacted into a dense, semifinished product with no residual open porosity. Typical compaction methods include uniaxial pressing, extrusion, and powder rolling. The foamable material may be further shaped through subsequent metal-working processes such as rolling, swaging, or extrusion.

Potential Applications

The following is a list of applications currently being investigated:

- *Automotive industry*: Light, stiff structures made of aluminum foam and foam sandwich panels could help to reduce weight and increase stiffness. Examples are hoods, trunk lids, and sliding roofs. With regard to energy absorption, it is possible to engineer controlled deformation into the crash zone of cars and trains with maximum impact-energy dissipation. Possible applications include elements for side and front impact protection. In general, foam filling leads to higher deformation forces when profiles are bent and to higher energy absorption when profiles are axially crushed. Potential applications include bumpers, underside protection of trucks, A- and B-pillars, and other elements subjected to large deformation.
- *Aerospace industry*: Metal foam sandwich panels offer good potential to replace expensive honeycomb structures in the aerospace industry. The attributes of metal foams include the isotropy of the foam material and fire retardation to maintain the integrity of the structure. Roll cladding for direct metallic bonding can eliminate the need for adhesive bonding in the sandwich panels.
- *Building industry*: Many construction applications require light, stiff, and fire-resistant elements, or supports for such elements. Foam sandwich panels could help to reduce the energy consumption for elevators by trimming the deck weight. Combining energy absorption and high specific stiffness, the foamed sandwich panels may be a good candidate for these applications. Another application of the metal foams is to fasten plugs into concrete walls. To fill the gap between the plugs and the wall, the foamable materials can be inserted into the gap and locally heated. The foamable material will expand and fill the space between the plugs and the concrete wall, provided that the resultant foam density is high enough.
- *Further applications*: Additional applications can capitalize on the properties of foams made from other metals. For example, lead and nickel foams may be suitable for batteries, and gold or silver foams may be applicable in art and jewelry. Furthermore, open-cell foams can be fabricated

by modifying the process scheme. This material would be excellent for several applications, such as heat exchangers, filters, and catalyst carriers. Foaming of high-temperature alloys such as nickel and titanium will also help expand the application range, especially for aerospace and bio-medical components.

Metal Forming

Metal forming is any manufacturing process by which parts or components are fabricated from metal stock. In the specific technical sense, metal forming involves changing the shape of a piece of metal. In general terms, however, it may be classified roughly into five categories: (1) mechanical working, such as foraging, extrusion, rolling, drawing, and various sheet-forming processes; (2) casting; (3) powder and fiber metal forming; (4) electroforming; and (5) joining processes. The selection of a process, or combination of processes, requires a knowledge of all possible methods of producing the part if a serviceable part is to be produced at the lowest overall cost.

Metal Hydrides

A metal hydride is a compound in which hydrogen is bonded chemically to a metal or metalloid element. The compounds are classified generally as ionic, transition metal, and covalent hydrides.

Ionic Hydrides

The most reactive metals (alkali and alkaline earth metals, with the exception of magnesium and beryllium) combine directly on heating with hydrogen gas at a pressure of 10^2 kPa. Magnesium reacts at elevated hydrogen pressure.

Transition Metal Hydrides

This group of compounds is less well understood than the ionic and covalent hydrides. When lanthanum is heated in hydrogen, the gas is readily taken up, forming first a material whose composition is approximately $LaH_{1.86}$ (the "dihydride" phase) and finally LaH_3.

The ionic hydrides, such as NaH, exhibit much more limited non-stoichiometry. When sodium is converted to pure NaH, all electrons in the conduction band are consumed and the product is white and insulating.

Copper hydride, CuH, is unique in that it is the only hydride that can be precipitated from aqueous solution.

Titanium, zirconium, vanadium, and niobium all react with hydrogen, forming dihydrides at the limiting compositions. The materials are conducting at all compositions.

Common metals such as steel and copper dissolve small quantities of hydrogen at elevated temperatures. On cooling, the gas comes out of solution and results in severe degradation of the mechanical properties of the metals. This is called hydrogen embrittlement and can be prevented by degassing the metal while it is still molten.

A number of alloys or intermediate compounds are known that react with hydrogen and form ternary hydrides. Examples are $LaNi_5H_6$, Mg_2NiH_4, $AlTh_2H_2$, and $CaAg_2H$.

Covalent Hydrides

Most evidence indicates that in ionic and metallic hydrides, an electronic pair is associated primarily with the hydrogen as H^-, while in covalent hydrides, the electron pair is shared between the hydrogen atom and an atom of another element. In these compounds, hydrogen is considerably smaller (radius 0.03 nm) than in the ionic hydrides. Covalent hydrides usually consist of small molecules, in which case they are gases (SiH_4 and SbH_3), but some form high polymers, in which case they are nonvolatile solids (AlH_3 and ZnH_2). All tend to decompose irreversibly rather easily on heating. Covalent hydrides are generally synthesized indirectly, not from direct combination of the elements. Gaseous hydrides such as SiH_4, PH_3, and AsH_3 can be generated by heating a solid mixture of the corresponding oxide and $LiAlH_4$.

Metal Inert-Gas Welding (MIG)

Gas metal arc welding using an inert gas such as argon as a shielding gas.

Metal Inject Ion Molding

Injection molding is one of the most productive techniques for shaping materials. Until recently, injection molding was restricted to thermoplastic polymers (polymers that melt on heating). However, metals have property advantages over polymers. They are stronger, stiffer, are electrically and thermally conductive, can be magnetic, and are more wear and heat resistant. The concept of metal injection molding (MIM) combines metal powders in a thermoplastic binder to allow shaping of complex objects in the high-productivity manufacturing setting associated with injection molding. After shaping, the metal powders are sintered to densities close to those listed in handbooks; that is, they have very few pores. Accordingly, the process delivers materials with metallic properties in an efficient manner.

MIM, and the related process of ceramic injection molding, is a derivative of powder metallurgy. Metal powders can be shaped in a semifluid state (powders that are poured into containers take on the shape of the container), but after heating to high temperatures, the particles bond into a strong, coherent mass. In ceramics, this is analogous to the manufacture of clay pottery, where shaping occurs with a water–clay system at the potter's wheel, but after kiln firing, the structure is strong and rigid.

MIM is a process in which metal in a powdered form is combined with a polymer binder to produce a homogeneous mixture known as feedstock. The feedstock is processed in a manner similar to plastic injection molding, except that parts are designed slightly larger to account for shrinkage during the final sintering step. The molded part is called the "green" part.

Once molded, three methods are available to remove the binder from the green parts: solvent, thermal, and catalytic debinding. Differences in these methods affect the outcome of the product. The debinding phase can be the most time-consuming aspect of the process. After debinding, parts (known as "brown" parts) are sintered—a process in which the components are exposed to heat over an extended period.

Applications

MIM is applied in the production of high-performance components that have complex shapes. The most successful applications involve components that have complex shapes and also must be inexpensive to produce while providing high relative performance. Generally, MIM can be used for all shapes that can be formed by plastic injection molding, especially for very small complex geometries.

M

Many applications have been developed for MIM. They include components for use in diverse fields ranging from surgical tools to microelectronic packaging. Recently, automotive components such as sensor parts in the air-bag actuator mechanism have moved into production. The technology has been extended to utilize a wide range of materials, including steels, stainless steels, nickel, copper, cobalt alloys, tungsten alloys, niobium alloys, nickel-base superalloys, and intermetallics. Additionally, most ceramics can be processed in the same manner; these include silica, alumina, zirconia, or silicon nitride, silicon carbide, aluminum nitride, cemented carbides, and various electronic ceramics.

Although MIM is generally viable for all shapes that can be formed by plastic injection molding, it is not cost competitive for relatively simple or asymmetric geometries. Large components require larger molding and sintering devices that are more difficult to control. Accordingly, MIM is largely applied to smaller shapes with masses below 100 g.

As the engineering community better appreciates its favorable attributes, MIM applications will grow. Although the current industry is relatively small, the anticipated growth is impressive. One area of keen interest is in the processing of titanium alloys for biomedical applications. Another is in the co-molding of different materials, that is, forming part of a component from one material and then another portion from a second material. This option has merit for forming corrosion barriers, wear surfaces, and electrical interconnections in ceramics. Other major growth areas include high-performance magnets, technical ceramics, materials associated with heat dissipation electronic circuits, hard materials, ultrahigh-strength metals, and biocompatible materials.

Metal Leaf

Thin metal sheet, usually thinner than foil, and traditionally produced by beating rather than by rolling.

Metal Penetration

A surface condition in metal castings in which metal or metal oxides have filled voids between sand grains without displacing them.

Metal Powder Cutting

A technique that supplements an oxyfuel torch with a stream of iron or blended iron–aluminum powder to facilitate flame cutting of difficult-to-cut materials. The powdered material propagates and accelerates the oxidation reaction, as well as the melting and spalling action of the materials to be cut.

Metal Powders

By definition, a metal powder is an aggregate of discrete metal particles that are usually in the size range of 1–1000 μm.

The metal powders in common use are lead–tin alloys for solders; iron-, nickel-, and cobalt-base alloys for hard facing; copper-, silver-, and nickel-base alloys for brazing; aluminum, bronze, and stainless steel for paint pigments; magnesium for pyrotechnics; iron for welding rods, torch cutting and scarfing, and metal powder parts; copper for metal-powder parts; and carbide for tools. Used in smaller quantities are iron powders for radio and television tuning cores; nickel–iron and silicon–iron alloy powders for other soft magnetic parts; aluminum, iron, nickel, and cobalt powders for small permanent magnets in the "Alnico" series; nickel and cobalt

powders as binders in the production of carbide tools and alloy and stainless steel powders for high-strength and special property parts.

Of these uses, the fabrication of structural parts or machinery components accounts for the largest single use of metal powders and competes with such other metal-forming methods as machining, casting, stamping, and forging. The powder-metallurgy technique may be selected by the designer as the best way to make a particular part for one of several reasons:

1. The process is ideal for mass-producing machine components at low unit cost.
2. Residual porosity can be controlled to provide long-wearing qualities through the self-lubricating feature of the "oil-less" sleeve bearing.
3. Metal combinations are possible through powder metallurgy that cannot be melted.
4. Powder-metallurgy techniques provide the only practical method of forming high-melting point metals.

The good surface finish and close dimensional tolerances possible are additional reasons why the designer may specify metal powder parts.

Although most metal powder parts are made from iron, copper, or mixtures of these primary powders with or without graphite additions, many other powders are used to develop special properties like high strength, magnetic or electrical properties, corrosion resistance, or oxidation resistance. The special powders include brass, bronze, alloy steels, stainless steel, and various nickel-base alloys. Many of the special powders are also employed in the production of metal filters.

Production Methods

Electrolytic

These methods are used to make most of the copper powder used for metal powder parts and a special, high-purity iron powder often used for making-dense iron powder parts.

Atomization

This is a molten metal production method applied to a wide range of powders including aluminum, magnesium, brass, bronze, lead, tin, nickel–silver, and stainless and low-alloy steels. Other more complex iron-, nickel-, and cobalt-base alloys are also made by atomization.

Metal Shadowing

The enhancement of contrast in a microscope by vacuum depositing a dense metal onto the specimen at an angle generally not perpendicular to the surface of the specimen. See also *shadowing*.

Metal Spraying

Any process for coating a surface by spraying it with a molten metal as small droplets. The flattened droplets only mechanically attached as there is insufficient time for diffusion to develop a stronger bond. See *thermal spraying* and *flame spraying*.

Metal Stitching

A mechanical technique for repairing cracks, for example, pairs of holes jig drilled on the two sides of a crack. The ligament between the pairs is then slit and "dumbbells" inserted. The dumbbells can be a more complex shape, and they may be of a material matching the parent or they may be of a lower expansion material to provide a crack closure effect if the component is heated. If a pressure tight repair is required, a further hole is drilled on the line of the crack. This hole is tapped (i.e., threaded) and filled with a threaded plug. This is repeated along the crack with each plug slightly overlapping its predecessor and the dumbbells. The technique can never restore full strength, and it is usually confined to low ductility fractures, but it can be performed on-site and it does not require heating.

Metal Transfer (in Welding)

The manner in which metal is transferred across an electric arc from a consumable electrode to the weld pool. The term is particularly aimed at metal inert gas welding (MIG). High current cause spray transfer in which individual droplets, not greater in diameter than the electrode diameter, are projected across the arc gap. Lower currents cause globular transfer in which the molten metal accumulates at the electrode tip until eventually a droplet appreciably larger in diameter than the electrode is projected across the gap. The lower currents result in dip transfer in which an arc is initiated in the steadily advancing electrode dips into the weld pool causing a short circuit. The electrical surge melts a portion of the tip that detaches to enter the weld pool. The ark re-strikes, and the process repeats.

Metal(lic) Active Gas Welding (MAG)

Welding in which the electric arc is struck between the component being welded and a continuously fed consumable filler wire and the weld zone is shielded by a gas that is active rather than inert during the welding operation.

Metal Arc Cutting

Any of a group of arc cutting processes that sever metals by melting them with the heat of an arc between a metal electrode and the base metal. See also *shielded metal arc cutting* and *gas metal arc cutting*.

Metal Arc Welding

Any of a group of arc welding processes in which metals are fused together using the heat of an arc between a metal electrode and the work. Use of the specific process name is preferred.

Metal–Matrix Composites

A MMC is a material in which a continuous metallic phase (the matrix) is combined with another phase (the reinforcement) that constitutes a few percent to around 50% of the total volume of the material. In the strictest sense, MMC materials are not produced by conventional alloying. This feature differentiates most MMCs from any other multiphase metallic materials, such as pearlitic steels or hypereutectic aluminum–silicon alloys.

The particular benefits exhibited by MMCs, such as lower density, increased specific strength and stiffness, increased high-temperature performance limits, and improved wear-abrasion resistance, are dependent on the properties of the matrix alloy and of the reinforcing phase. The selection of the matrix is empirically based, using readily available alloys, and the major consideration is the nature of the reinforcing phase.

Matrices and Reinforcements

A large variety of MMC materials exist. The reinforcing phase can be fibrous, platelike, or equiaxed (having equal dimensions in all directions), and its size can also vary widely, from about 0.1 to more than 100 µm. Matrices based on most engineering metals have been explored, including aluminum, magnesium, zinc, copper, titanium, nickel, cobalt, iron, and various aluminides. This wide variety of systems has led to an equally wide spectrum of properties for these materials and of processing methods used for their fabrication.

Reinforcements used in MMCs fall in five categories: continuous fibers, short fibers, whiskers, equiaxed particles, and interconnected networks.

Continuous Fibers

Several continuous fibers or filaments are used in MMCs. Their elastic moduli vary significantly, depending on the nature of the fiber and its fabrication process. For example, silica–alumina spinels and micro crystalline or amorphous polycarbosilane-derived fibers possess significantly lower elastic moduli than do pure alumina or crystalline β-silicon carbide produced by chemical vapor deposition. Carbon fiber strength and modulus also vary significantly with processing, depending on the level of graphitization of the microstructure.

Short Fibers

Short fibers are less expensive, especially when they are mass-produced for other applications such as high-temperature thermal insulation. Their physical properties can be similar to those of continuous fibers; however, their reinforcing efficiency in the matrix is also far lower. Short fibers used in engineering practice include chopped carbon fibers and alumina–silica fibers.

Whiskers

Whiskers are single-crystal short fibers, produced to feature highly desirable mechanical properties due to lack of microstructural defects. Whiskers have typically been made of silicon carbide, and they are often priced far higher than short fibers. The high price and toxicity of most whiskers have prevented their application in engineering practice.

Single-crystal whiskers, because of the absence of grain boundary defects, offer much higher tensile strength than other types of discontinuous reinforcements, and thus they are preferred for certain applications of discontinuously reinforced MMCs. The whiskers can be aligned to a preferred orientation by conventional metallurgical processes; higher directional strengths can be achieved in finished components where fabrication is by extrusion, rolling, forging, or superplastic forming. Whiskers tend to produce anisotropic properties due to their alignment during processing, whereas particular materials usually produce essentially isotropic properties.

Equiaxed Particles

Equiaxed particles of several ceramics, including those containing silicon carbide, aluminum oxide, boron carbide, and tungsten carbide, do not provide the possibility for preferential strengthening of the matrix along selected directions; however, their price is low and their combination with the metal is relatively easier. These reinforcements are, therefore, used in many MMC systems, including mass-produced aluminum–matrix composites.

Interconnected Cellular Networks

These can be produced by several methods, such as by chemical vapor deposition of ceramic onto a pyrolizable polymer foam or by conversion of a preceramic polymer foam prior to infiltration with the molten matrix. Alternatively, some processing techniques for in-place MMCs, including directional oxidation of aluminum melts, produce interconnected reinforcing networks.

Microstructures

The microstructure of a MMC comprises the structure of matrix and reinforcement, that is, the interface and the distribution of the reinforcement within the matrix.

Composite Properties

Composite properties depend first and foremost on the nature of the composite; however, certain detailed microstructural features of the composite can exert a significant influence on its behavior.

Physical properties of the metal, which can be significantly altered by addition of a reinforcement, are chiefly dependent on the reinforcement distribution. A good example is aluminum–silicon carbide composites, for which the presence of the ceramic increases substantially the elastic modulus of the metal without greatly affecting its density. Elastic moduli for 6061 aluminum–matrix composites reinforced with discrete silicon carbide particles or whiskers have been calculated by using the rule of mixtures for the same matrix reinforced with two types of commercial continuous silicon carbide fibers. As a result, several general facts become apparent. First, modulus improvements are significant, even with equiaxed silicon carbide particles, which are far less expensive than fibers or whiskers. However, the level of improvement depends on the shape and alignment of the silicon carbide. Also, it depends on the processing of the reinforcement; for the same reinforcement shape (continuous fibers), microcrystalline polycarbosilane-derived silicon carbide fibers yield much lower improvements than do crystalline β-silicon carbide fibers. These features, which influence reinforcement shape, orientation, and processing of modules, are quite general; they are also observed, for example, in MMCs reinforced with aluminum oxide or carbon.

Other properties, such as the strength of MMCs, depend in a much more complex manner on composite microstructure. The strength of a fiber-reinforced composite, for example, is determined by fracture processes, themselves governed by a combination of microstructural phenomena and features. These include plastic deformation of the matrix, the presence of brittle phases in the matrix, the strength of the interface, distribution of flaws in the reinforcement, and the distribution of the reinforcement within the composite. Consequently, predicting the strength of the composites from that of its constituent phases is generally difficult.

Production

A variety of techniques are available for the production of continuous or discontinuous MMCs. These may be broadly classified as diffusion processes, deposition processes, and liquid processes.

Fabrication

Composite processing methods combine the reinforcement with the matrix. This is accomplished while the matrix is either solid or liquid.

Typical liquid-state processes include the dispersion processes, which are casting techniques. A second set of processes involve liquid-metal impregnation; these include squeeze casting, where a preform or a bed of dispersoids is impregnated by molten alloy under hydraulic pressure. A third set comprises spray processes. In one of these, a molten metal stream is fragmented by means of a high-speed cold inert-gas jet passing through a spray gun, and dispersoid powders are simultaneously injected. A stream of molten droplets and dispersoid powders is directed toward a collector substrate where droplets recombine and solidify to form a high-density deposit.

Depending on the process, the desired microstructure, and the desired part, MMCs can be produced to net or near-net shape; or alternatively, they can be produced as billet or ingot material for secondary shaping and processing.

Applications

The combined attributes of MMCs, together with the costs of fabrication, vary widely with the nature of the material, the processing methods, and the quality of the product. In engineering, the type of composite used and its application vary significantly, as do the attributes that drive the choice of MMCs in design. For example, high specific modulus, low cost, and high weldability of extruded aluminum oxide particle-reinforced aluminum are the properties desirable for bicycle frames. High wear resistance, low weight, low cost, improved high-temperature properties, and the possibility for incorporation in a larger part of unreinforced aluminum are the considerations for design of diesel engine pistons.

Metallic Bond

The principal bond between metal atoms, which arises from the increased spatial extension of valence-electron wave functions when an aggregate of metal atoms is brought close together. An example is the bond formed between base metals and filler metals in all welding processes. See also *covalent bond* and *ionic bond*. See also *interatomic bonding*.

Metallic Fiber

A manufactured fiber composed of metal, plastic-coated metal, metal-coated plastic, or a core completely covered by metal.

Metallic Glass

A noncrystalline metal or alloy, commonly produced by drastic supercooling of a molten alloy, by molecular deposition, which involves growth from the vapor phase (e.g., thermal evaporation and sputtering) or from a liquid phase (e.g., electroless deposition and electrodeposition), or by external section techniques (e.g., ion implantation and ion beam mixing). Glassy alloys can be grouped into two major categories. The first group includes the transition metal–metal binary alloy systems, such as Cu–Zr, Ni–Ti, W–Si, and Ni–Nb. The second-class consists of transition metal-metalloid alloys. These alloys are usually iron-, nickel-, or cobalt-base systems, may contain film formers (such as chromium and titanium), and normally contain approximately 20 at.% P, B, Si, and/or C as the metalloid component. Also called amorphous alloy or metal. See also *amorphous solid*.

These are alloys with an amorphous and/or glassy structure. A glass is a solid material obtained from a liquid, which does not crystallize during cooling. It is, therefore, an amorphous solid, which means that the atoms are packed in a more or less random fashion similar to that in the liquid state. The word *glass* is generally associated with the familiar transparent silicate glasses containing mostly silica and other oxides of aluminum, magnesium, sodium, and so on. These classes are not metallic; they are electrical insulators and do not exhibit ferromagnetism. One obvious answer to obtaining a glass having metallic properties is to start from a metal containing metallic elements instead of oxides.

Electrical and Superconductivity Properties

The electrical resistivity of metallic glasses is high, for example, 100 $\mu\Omega$ cm and higher, which is in the same range as the familiar nichrome alloys widely used as resistance elements in electric circuits. Another interesting characteristic of the electrical resistivity of metallic glasses is that it does not vary much with temperature. The temperature coefficient is of the order of 10^{-4} K^{-1} and can even be zero or negative, in which case the resistivity decreases with increasing temperature. Because of their insensitivity temperature variations, metallic glasses are suitable for applications in electronic circuits for which the property is an essential requirement.

Superconducting metallic glasses are much more stable, and some of them do not crystallize at temperatures as high as 500°C. Some superconducting metallic glasses contain only two metals, such as $Zr_{75}Rh_{25}$, and some are more complex alloys in which there is approximately 20% of metalloid elements, mostly boron, silicon, or phosphorus.

One of the main reasons for continuing research on new superconducting glasses is their projected usefulness in high-field electromagnets, which will be required to contain the high-temperature plasma in fusion reactors. The present requirement for these magnets is a field of at least 100,000 G, which is not attainable with conventional copper-wound electromagnets. Observations and results have shown that the electrical resistivity of metallic glasses is not affected by radiation.

Magnetic Properties

The ferromagnetic properties of metallic glasses have received a great deal of attention, probably because of the possibility that these materials can be used as transformer cores. One class of ferromagnetic metallic glasses is based on transition metals with zirconia or hafnium, for example, $(Co,Ni,Fe)_{90}Zr_{10}$.

Because metallic glasses do not have crystalline anisotropy, the core maker is able to anneal in a desired domain pattern. For example, metallic glasses can provide extremely high magnetic efficiency in motors because their anisotropy can be built in during a final annealing to develop a low-reluctance path in the exact configuration required by the motor. The use of metallic glasses in motors can reduce core loss by as much as 90% as compared with conventional crystalline magnets. Alternately, for transducer applications, one can just as well anneal in a domain structure that is transverse to the ribbon length by applying a transverse magnetic field during the annealing cycle. This will maximize the rotation of domains and hence the magnetostrain during any subsequent longitudinal excitation.

Mechanical and Other Properties

Although a metallic glass under tension will fracture with very small permanent deformation, scanning electron micrography reveals that a large local shear deformation occurs before rupture.

Metallic glasses thus are intrinsically ductile. The interest in the mechanical properties of metallic glasses is motivated by their high rupture strength and toughness. The fracture strength of metallic glasses approaches the theoretical strength that is about 1/50 of Young's modulus.

Future and Applications

Considering the unusual physical and chemical properties of metallic glasses, there is no doubt that they will play an important role as an engineering material in the future. The melt-spinning process of producing metallic glasses should also result in substantial savings in labor and energy when compared with the present technology because it can produce a thin sheet of material by direct casting from the liquid state.

Possible applications of metallic glasses have already been demonstrated on audio and video magnetic tape recording heads, sensitive and quick-response magnetic sensors or transducers, security systems, motors, and power transformer cores. The combination of excellent strength, resistance to corrosion and wear, and magnetic properties may lead to interesting applications, for example, the use of such glasses as inductors in magnetic separation equipment.

Besides the straightforward applications, there remain many less obvious potential applications of metallic glasses: the freedom from constraints imposed by the equilibrium phase diagram may well allow metallurgists to devise chemically interesting glassy alloys that could not be made otherwise in single-phase form. Glasses based on palladium–phosphorus alloys show a considerably higher catalytic activity for oxidation of methanol, and they are more stable than the fine-grained platinum electrodes. Another potential application of metallic glasses is their use when intentionally crystallized. Since glasses are highly supercooled liquids, crystallization can be affected via copious nucleation, which results in the formation of a fine-grained crystalline product that is microstructurally very homogeneous. An example is the use of a glass sheet of the alloy $Ni_{69}Cr_6Fe_3B_{14}Si_8$ as a brazing filler metal; the glass sheet can be stamped to shape and inserted between the parts to be joined, yielding excellent control of the joint properties after fusion of the products.

Another development is an application of powder-metallurgy methods to the fabrication of amorphous alloys. In the process of "atomizing" liquid alloys into a very fine powder, the rate of cooling may be high enough to produce an amorphous powder.

By using explosive loading, it is indeed possible to subject the powder to very high pressure, and probably some rather high temperature, for a very short enough time to avoid recrystallization, and to achieve near-theoretical densities. Massive ingots of metallic glasses, susceptible of being forged and rolled into various shapes, may become available within the not-too-distant future. Finally, since metallic glasses may be regarded as liquids whose structure has been frozen, they constitute ideal materials for low-temperature transport and critical behavior studies, and they are most suited for the study of electrons in noncrystalline metals.

Metallic Materials

About three quarters of the elements available can be classified as metals, and about half of these are of at least some industrial or commercial importance. Although the word *metal*, by strict definition, is limited to the pure metal elements, common usage gives it wider scope to include metal alloys. Although pure metallic elements have a broad range of properties, they are quite limited in commercial use. Metal alloys, which are combinations of two or more elements, are far more versatile and for this reason are the form in which most metals are used by industry.

Metallic materials are crystalline solids. Individual crystals are composed of unit cells repeated in a regular pattern to form a three-dimensional crystal lattice structure. A piece of metal is an aggregate of many thousands of interlocking crystals (grains) immersed in a cloud of negative-valence electrons detached from the atoms of the crystals. These loose electrons serve to hold the crystal structures together because of their electrostatic attraction to the positively charged metal atoms (ions). The bonding forces, which are large because of the close-packed nature of metallic crystal structures, account for the generally good mechanical properties of metals. Also, the electron cloud makes most metals good conductors of heat and electricity.

Metals are often identified as to the method used to produce the forms in which they are used. When a metal has been formed or shaped in the solid, plastic state, it is referred to as a wrought metal. Metal shapes that have been produced by pouring liquid metal into a mold are referred to as cast metals.

There are two families of metallic materials—ferrous and nonferrous. The basic ingredient of all ferrous metals is the element iron. These metals range from cast irons and carbon steels, with over 90% iron, to specialty iron alloys, containing a variety of other elements that add up to nearly half the total composition.

Except for commercially pure iron, all ferrous materials, both irons and steels, are considered to be primarily iron–carbon alloy systems. Although the carbon content is small (less than 1% in steel and not more than 4% in cast irons) and often less than other alloying elements, it nevertheless is the predominant factor in the development and control of most mechanical properties.

By definition, metallic materials that do not have iron as their major ingredient are considered to be nonferrous metals. There are roughly a dozen nonferrous metals in relatively wide industrial use. At the top of the list is aluminum, which next to steel is the most widely used structural metal today. It and magnesium, titanium, and beryllium are often characterized as light metals because their density is considerably below that of steel.

Copper alloys are the second nonferrous material in terms of consumption. There are two major groups of copper alloys: brass, which is basically a binary alloy system of copper and zinc, and bronze, which was originally a copper–tin alloy system. Today, the bronzes include other copper-alloy systems.

Zinc, tin, and lead, with melting points below 427°C, are often classified as low-melting alloys. Zinc, whose major structural use is in die castings, ranks third to aluminum and copper in total consumption.

Lead and tin are rather limited to applications where their low melting points and other special properties are required. Other low-melting alloys are bismuth, antimony, cadmium, and indium.

Another broad group of nonferrous alloys is referred to as refractory metals. Such metals as tungsten, molybdenum, and chromium, with melting points above 1649°C, are used in products that must resist unusually high temperatures. Although nickel and cobalt have melting points below 1649°C, they serve as the base metal or as alloying elements of many heat-resistant alloys.

Finally, the precious metals, or noble metals, have a common characteristic of high cost. In addition, they generally have high corrosion resistance, many useful physical properties, and generally high density.

Metallic Soap

This is a term used to designate compounds of the fatty acids of vegetable and animal oils with metals other than sodium or potassium.

They are not definite chemical compounds like the alkali soaps, but may contain complex mixtures of free fatty acid, combined fatty acid, and free metallic oxides or hydroxides. The name distinguishes the water-insoluble soaps from the soluble soaps made with potash or soda. Metallic soaps are made by heating a fatty acid in the presence of a metallic oxide or carbonate, and are used in lacquers, leather and textiles, paints, inks, ceramics, and grease. They have the properties of being driers, thickening agents, and flattening agents. They are characterized by inability to gel in solvents and oils and by their catalytic action in speeding the oxidation of vegetable oils.

When made with fatty acids having high iodine values, the metallic soaps are liquid, such as the oleates and linoleates, but the resinates and tungates are unstable powders. The stearates are fine, very stable powders. They are found as barium, strontium, chromium, manganese, cerium, nickel, lead, and lithium stearates. The fatty acid determines the physical properties, but the metal determines the chemical properties. Aluminum stearate is the most widely used metallic soap for colloid products. Aluminum soaps are used in polishing compounds, in printing inks and paints, for water proofing textiles, and for thickening lubricating oils. The resinates, linoleates, and naphthenates are used as driers; the lead, cobalt, and manganese are the most common.

Metallic Wear

Typically, wear due to rubbing or sliding contact between metallic materials that exhibits the characteristics of severe wear, for example, significant plastic deformation, material transfer, and indications that cold welding of asperities possibly has taken place as part of the wear process. See also *adhesive wear* and *severe wear*.

Metallic Whisker

A fiber composed of a single crystal of metal. See also *whisker*.

Metallization

A deposited or plated thin metallic film used for its protective or electrical properties.

Metallizing

A process for applying a metal coating to the surface of a component that may be metal or other material. The term usually implies application by hot spraying or similar mechanical process rather than by a plating or diffusion processes. However, terms such as aluminizing or chromizing usually refer to diffusion processes.

Metallizing is the application of a metallic coating to a ceramic to permit subsequent brazing to a mating part. Various techniques are employed, but the basic steps follow the common practice of ceramic decorating except that materials must be carefully chosen.

Metallizing Systems

Metals

Reactivity being desirable, most metal powders are purchased as—325 mesh to ensure a high surface-to-volume ratio.

Copper

May be mixed, as flake, with coarse glass particles that melt and seal to the ceramic preserving the integrity of the flake. This continuous electrical path through the glass seal is useful in the manufacture of spark plugs. Copper, as the oxide, will bond to a ceramic under precise firing conditions and, when mixed with silver, may be used as a metallizing preparation.

Gold

Frequently combined with reactive glass for uses in the air-fired paste and solder category.

Iron

Reactive material originally combined with tungsten. Very common additive for sintered powder process.

Manganese

Reactive material combined with molybdenum and is a very common additive.

Molybdenum

Most used as basic metal for sintered powder metallizing. Oxidation potential allows control of oxidation state in controlled atmosphere furnace. Coefficient of expansion of the metal and its reaction products are favorable.

Palladium

Similar to platinum and sometimes added to platinum–frit mixtures for air firing.

Platinum

Used with reactive glasses in air-fired metallizing.

Silver

Basic metal for many air-fired pastes. Mixed as granular or flake material with a reactive glass; for example, borosilicate.

Tin

Basis for direct chemical bonding to numerous ceramics and high-temperature metals.

Titanium

Basis of the active metals process. May be used as powder or foil prior to braze alloying. Titanium-bearing brazes will wet and flow over ceramic; in vacuum, almost as well as solder over copper. Frequently applied as the hydride, which disassociates at <800°C, providing nascent hydrogen, which tends to scour the surface to be wetted. May be added to sintered powder compositions to promote reaction.

Tungsten

Similar in metallizing properties to molybdenum.

Zirconium

Performs similarly to titanium, but with less activity. Has lower coefficient of expansion.

Miscellaneous Reactive Powders

Many companies have evolved proprietary sintered powder metallizing systems that may be very complicated. Some of the likely additives to these compositions are aluminum, barium, boron, cadmium, magnesium, rare earths, and silicon. Most of these would be added as an oxide.

Glass Compositions

Most metallizing suppliers purchase glass-bearing metallizing compositions from commercial vendors. The glass compositions are regarded as proprietary and are not generally known.

Binders

Binders are fugitive materials that are required to increase viscosity and density of the vehicles developed for suspending metal powders during application of the ceramic. They must leave no deleterious residue following the firing operation and must hold the metal *in situ* until the particles become somewhat adherent. Typical binders are represented by acrylic polymers, commercial binders, nitrocellulose, and pyroxyline resins.

Solvents

Solvents are required to dissolve the binder material and usually have a high vapor pressure, so they are effectively lost before firing of the metallizing begins. These materials, with the binders, constitute the vehicle. The best vehicles allow a smooth application with controlled thickness and dry to a dense abrasion-resistant layer prior to firing. Solvents are selected typically from the following families: acetates, alcohols, commercial solvents, ethers, and ketones.

Application Methods

Pastes, when used, are prepared by selecting the metal powders and blending with additives in a ball mill. Acetone is frequently used as the carrier during the milling operation. Attention is given to the desired degree of particle size reduction. The milled material is removed, dried, pulverized, and combined with the chosen binder and solvent. This preparation may be milled further to assure complete blending. The paste is then ready to be applied to the ceramic.

Many methods of application are in use. They vary depending on size of the order, configuration of the part, the precision required, and operating economy.

Brushing

A paintable slurry is prepared and applied by brush to the desired area of the ceramic. Useful for small lots or unusual configurations.

Decalcomania

Commercial sources offer patterns prepared for transfer to ceramic.

Dipping

Applicable where complete coverage is permissible or where subsequent methods of recovering a pattern are feasible.

Evaporation

Vacuum evaporation of metals offers a readily controllable metal thickness for thin film and other applications not requiring high strength.

Printing

Banding or printing equipment may be successfully used with certain pastes. Limitations result from ink thickness requirements.

Screening

A slurry is forced through a fine-mesh screen by a squeegee. This technique yields excellent patterns on cylindrical or flat surfaces.

M

Solution Metallizing

A liquid solution of the desired metal is supplied to the ceramic, subsequently dissociated, and then reduced to provide a metallic coating.

Spraying

Paint-spray techniques are adapted to metallizing. Masking is usually required on the part. High-volume production may be maintained.

Tapes

Metallizing compositions manufactured as a tape of controlled thickness are being offered commercially. The tape may be applied to the ceramic, adhered by solvent action, and fired.

Vapor Deposition

Reduction of chemical vapors, for example, nickel carbonyl, produces a uniform metallic layer on the ceramic. Deposition can be conventional or electrostatic firing.

Air Furnaces

Useful for glass-fritted pastes containing noble metal powders. Firing temperatures vary up to 1000°C.

Controlled Atmosphere Furnaces

Hydrogen or cracked ammonia atmospheres, with dew point and temperature controls, are typically used to react sintered-powder metallized coatings with ceramic. These furnaces operate at 1300°C–1700°C.

Vacuum Furnaces

Vacuum of 10^{-4} torr, or less, is required for metallizing and sealing by the active metals method. Vacuum evaporators operating in the 10^{-3}–10^{-10} torr range are at present in use, experimentally, for studies of ultrapure metals deposited for two-dimensional circuitry requirements.

Others

While not yet of full commercial significance, work in plasma-jet technology and electron beam machining has significant potential.

Electroplating

Following firing, the active metal assembly hardware may be electroplated for ease of installation by user. Air-fired silver may be electroplated to improve solderability. Sintered-powder metallized ceramics must ordinarily be electroplated to allow wetting by solders or brazes. Many methods of electroplating are in use. Barrel plating is more economical than rack plating and is the preferred method. The thickness, purity, and type of electroplated metal have a significant effect on the effective strength of sintered-powder metallizing.

Etching

A chemical process used in conjunction with a masking system to create fine lines.

Metallograph

An optical instrument designed for visual observation and photomicrography of prepared surfaces of opaque materials at magnifications of 25× to approximately 2000×. The instrument consists of a high-intensity illuminating source, a microscope, and a camera bellows. On some instruments, provisions are made for examination of specimen surfaces using polarized light, phase contrast, oblique illumination, dark-field illumination, and bright-field illumination.

Metallography

The practice of examining metals by microscopic techniques to establish their structure, reveal defects, and provide information on damage mechanisms or other matters of interest. The term macroscopic examination is often used for preliminary examination at low magnifications, say up to about 30×. This term is sometimes used to imply that the item being examined is in the as-received state, that is, it has not been polished or etched. Microscopic examination usually implies the examination, and a light microscope, of mounted and prepared specimens cut from a component. Typical practice involves cutting a piece of metal, say 10 mm² × 5 mm thick (1/2 × 1/4 in.), from the component under investigation. This is then encased, with the face of interest exposed, in a plastic resin to facilitate handling and to protect the edges that are often of particular interest. The face of interest is then ground flat prior to polishing, either mechanically or electrolytically. Mechanical polishing involves careful abrasion on a series of progressively finer grinding papers and polishing pads to provide a flat, mirror finish. Microscopic examination in this state provides some information on microstructure and damage mechanisms. However, some detail may be obscured by the surface deformation and smearing, termed the Beilby layer, produced by mechanical polishing. Such detail is revealed by etching whereby the specimen is exposed to acid or other reagents that remove the surface film or otherwise react selectively with features of interest. In some cases, the specimen may be electrolytically etched. This is essentially similar to the alternative polishing technique. Electrolytic polishing that involves making the specimen the anode of an electrolytic cell; see *electrochemistry*. With careful selection of reagents and control of the voltage and current, a flat, high-quality surface can be achieved. If necessary, the specimen may finally be etched. A conventional light microscope is a useful magnification of up to about 1000×. Examination at higher magnification normally requires the use of an electron microscope that uses an electron beam instead of light to examine the specimen. The system for producing the electron beam and projecting the image onto a screen is essentially a cathode ray tube similar to a television set. In the case of a transmission electron microscope (TEM), the metal samples to be examined are very thin and the electron beam travels through the specimen for projection on the screen or photographic film. In the case of a scanning electron microscope (SEM), the beam scans across, and is reflected from, the specimen surface prior to projection on the screen. Apart from the higher magnification, which is often not utilized, the SEM offers the major advantage that the electron beam has a very large depth of focus, so it can deal with a rough, irregular surface lying at an angle to the beam. The light microscope has only a shallow depth of focus, so it is effectively confined to flat surfaces aligned normal to the beam. This characteristic makes the SEM a valuable tool for examining fracture surfaces. The disadvantage of electron microscopy is that the specimen has to be small enough to be inserted into the instruments' vacuum chamber. With suitable ancillary equipment, the SEM can undertake analysis either of large areas to give a bulk composition or of small features such as inclusions or precipitates.

Metalloid

A material having some but not all of the characteristics of a metal, also termed a semi-metal.

Metallurgical Bond

Adherence of a coating to the base material characterized by diffusion, alloying, or intermolecular or intergranular attractions at the interface between the coating and the base material.

Metallurgical Burn

Modification of the microstructure near the contact surface due to frictional temperature rise.

Metallurgical Coke

A coke, usually low in sulfur, having a very high compressive strength at elevated temperatures; used in metallurgical furnaces not only as fuel, but also to support the weight of the charge.

Metallurgy

The science and technology of metals and alloys. Process metallurgy is concerned with the extraction of metals from their ores and with refining of metals; physical metallurgy, with the physical and mechanical properties of metals as affected by composition, processing, and environmental conditions; and mechanical metallurgy, with the response of metals to applied forces.

Metalock

Similar terms including some proprietary names. See *metal stitching.*

Metalworking

See *forming.*

Metastable

(1) Of a material not truly stable with respect to some transition, conversion, or reaction but stabilized kinetically either by rapid cooling or by some molecular characteristics as, for example, by the extremely high viscosity of polymers. (2) Possessing a state of pseudo-equilibrium that has a free energy higher than that of the true equilibrium state. (3) A nonequilibrium state in which a metal appears to be stable because it does not transform spontaneously (even over centuries). Essentially similar to unstable.

Metastable Beta

A β-phase composition in titanium alloys that can be partially or completely transformed to martensite, α, or eutectoid decomposition products with thermal or strain-energy activation during subsequent processing or service exposure.

Meter

(1) Metre. (2) A general term for a measuring instrument.

Methane

Also known as marsh gas, in coal mines as firedamp, and chemically as methyl hydride, methane is a colorless, odorless gas, CH_4, employed for carbonizing steel, in the manufacture of formaldehyde, and as a starting point for many chemical compounds. The molecule has no free electrons and is the only stable carbon hydride, although it reacts easily on the no. 1 and 2 electrons of the carbon to form the hexagonal molecule called the benzene ring. It may thus be considered as the simplest of the vast group of hydrocarbons to rise from petroleum, coal, and natural gas. Methane occurs naturally from the decomposition plant and animal life and is also one of the chief constituents of illuminating gas. It is made synthetically by the direct union of carbon or carbon monoxide with hydrogen. It is also produced by the action of water on aluminum carbide.

Methane is also the chief constituent of natural gas from oil fields. Natural gas contains usually at least 75% methane. A new source of methane is a result of environmental cleanups. Methane is generated as a by-product during the biodegradation of polluted industrial waters and waste treatment plants. It is also obtained by tapping the gases produced as biomass matter degrades in municipal landfills.

Since 1990, to reduce air pollution, U.S. automakers have built cars, vans, and trucks fueled with natural gas for urban-area fleets. *Re-formed gas,* used for copper refining, contains 86% methane and is free of hydrogen sulfide and the higher homologs of methane. *Synthetic gas,* or oil gas, made from crude oil, can be had with an energy value equivalent to that of natural gas. *Sour gas* is natural gas with more than 1 g of hydrogen sulfide/100 ft³. This hydrogen sulfide is removed to eliminate the odor before it is piped. The propane, butane, and heavier hydrocarbons may also be removed for the production of chemicals.

Methyl Methacrylate

A colorless, volatile liquid derived from acetone, cyanohydrin, methanol, and dilute sulfuric acid and used in the production of *acrylic resins.*

Metre

The SI unit of length.

Metric System

The system of single standardized units for all properties, dimensions, etc., with all multiples or divisions to base 10. Now replaced by the SI system.

Metrology

The science and practice of measurement.

Meyer Index

A measure of the work hardening characteristics of a material determined by performing a series of hardness tests over a range of loads using a ball indentor. The index is then derived from a graph of the load against indentation diameter.

M_f Temperature

For any alloy system, the temperature at which martensite formation on cooling is essentially finished. See also *transformation temperature* for the definition applicable to ferrous alloys.

M-Glass

A high beryllia (BeO_2) content glass designed especially for high modulus of elasticity.

MHO

An electrical unit of conductivity that is the conductivity of a body with the resistance of one ohm.

MIC

Mineral insulated cable.

Mica

Mica has been known as an electrical material of excellent insulating and fabricating properties for many years. Almost everyone has seen or used mica in one form or another, yet it is surprising how little even the average technically trained person knows about its composition, origin, and properties.

It is, of course, a naturally occurring mineral having a wide range of possible chemical compositions and properties. All true mica, however, belongs to one mineral class of silicates having a sheet type of structure, and found in certain areas of the world growing from pegmatite deposits in "book" form. In recent years, the increasing demands for high-quality sheet mica have led to the development and commercial production of synthetic mica as an electric furnace product.

Synthetic mica has recently been put into commercial production and is available in various powdered forms and organically bonded paper sheets; and a three-dimensional, hot-pressed, machinable synthetic mica ceramic; and in sheet form of limited area. Synthetic mica has unusually high thermal stability.

Of considerable interest to the refractories field is vermiculite, a form of mica that, on rapid heating, expands 16 times accordion-fashion to form a light and inert material suitable for use as a grog in refractories or as an aggregate for the production of lightweight casting refractory concretes.

Types

Mica comprises a group of hydrous aluminum silicate minerals with platy morphology in perfect basal (micaceous) cleavage. This group of silicate minerals has monoclinic crystals that break off easily into thin, tough scales, varying from colorless to black. Muscovite is the common variety of mica and is called potash mica or potash silicate. It has superior dielectric properties and is valued for radio capacitors.

Properties

The most singular outstanding property of mica is its physical structure. As a sheet mineral, it can be split into strong, flexible films with good high-temperature resistance and electrical insulating properties.

Uses

Commercial mica is of two main types: sheet, and scrap or flake. Sheet muscovite is used as a dielectric in capacitors in vacuum tubes and electronic equipment. Lower-quality muscovite is used as an insulator in home electrical products such as hot plates, toasters, and irons. Scrap and flake mica is ground for use in coatings on roofing materials and waterproof fabrics, and in paint, wallpaper, joint cement, plastics, cosmetics, well-drilling products, and a variety of agricultural products.

For many years, glass-bonded mica has been used in every type of electrical and electronic system where the insulation requirements are preferably low-dissipation factor at high frequencies, a high-insulation resistance and dielectric-breakdown strength, along with extreme dimensional stability. Glass-bonded micas are made in both machinable grades and precision-moldable grades. Basically, the material consists of natural mica flake bonded with a low-loss electrical glass.

The availability of synthetic mica resulted in the development of so-called ceramoplastics, consisting of high-temperature electrical glass filled with synthetic mica. Ceramoplastics provide an increase in the electrical characteristics over those of natural mica, and, in addition, are more easily molded and have greater thermal stability.

Glass-bonded mica and ceramoplastics have found use in many advanced components such as telemetering commutation plates, molded printed circuitry, high-reliability relay spacers and bobbins, coil forms, transducer housing-miniature-switch cases, and innumerable other component applications.

Mica Powder

Mica is very small flakes used as a filler in plastics, in paints, in roofing and asphalt shingles, and for making glass-bonded mica. When produced by grinding the small scrap pieces from mica workings, it is known as **ground mica**.

Mixed with aluminum powder, mica produces a finished paint superior to aluminum alone. It has some activating properties and is a substitute for zinc oxide in rubber. It is also used as a filler in plastics and replaces graphite as a foundry core and mold wash. **Water-ground mica** is ground to a fineness of 90% through a 325-mesh screen. It is for paint and rubber use. **Micronized mica** is a powder of a fineness of 400–1000 mesh, used as a filler. **Mica flake**, used in the manufacture of shingles and roofing, is washed from pegmatite deposits, but the mica flake used for molding into mica ceramic electric insulators is ground phlogopite scrap, or from various compositions of synthetic mica.

Micelle

A submicroscopic unit of structure built up from ions or polymeric molecules.

Micro

In relation to reinforced plastics and composites, the properties of the constituents only, that is, matrix, reinforcements, and interface, and their affects on the properties of the composite.

Micro Alloyed Steel

Steels with alloying additions in small, but carefully selected and controlled, quantities. Under closely controlled manufacturing conditions, these steels are capable of giving a very fine grain size with high strength and with good ductility. Typical additions include aluminum, vanadium, titanium, and niobium of up to about 0.1% each. Some of these steels are also able to retain the high strength following welding, unlike more conventional steels that suffer softening of the zone adjacent to the weld. Also termed *High Strength Low Alloy Steel (HSLA)*.

Micro Analyzer

An analytical instrument essentially similar to, and often part of, an electron microscope. It is capable of focusing on and analyzing very small features on rolled or prepared specimens. Also see *metallography*.

Micro Cracking

Cracks formed in composites when thermal stresses locally exceed the strength of the matrix. Since most microcracks do not penetrate the reinforcing fibers, microcracks in a cross-plied laminate or in a laminate made from cloth prepreg are usually limited to the thickness of a single ply.

Micro Inch

One millionth of an inch symbol μin.

Micro Metre

One millionth of a metre, a micron, symbol μm.

Microanalysis

(1) The analysis of samples smaller than 1 mg. (2) Various techniques for analyzing small quantities of material. A microprobe analyzer and similar terms including microanalyzer is an instrument associated with an electron microscope for analyzing small areas of surface.

Microbands

Long, straight bands of highly concentrated *slip* lying on the *slip planes* of individual grains in metals. They are usually 0.1–0.2 μm thick, traverse an entire grain, and correspond to the *slip bands* seen on a polished surface.

Microbial Attack/Corrosion

Corrosion caused, or assisted, by the presence of microbes, that is, bacteria, etc., in the surrounding environment. Microbes do not directly attack the metal, but their presence may modify the environment to make it more aggressive. See *biological corrosion*.

Microcap Shrinkage/Porosity, etc.

Phenomena observable only under a microscope or occurring on a very small scale.

Microcrack/Microfissure

A small crack not visible to the naked eye examination nor to the simpler form of nondistractive crack detection techniques.

Microdynamometer

An instrument for measuring mechanical force and observing the change in microscopic appearance of a small specimen.

Microelectromechanical Systems

A new discipline in miniaturization has emerged and is referred to as microelectromechanical systems (**MEMS**).

Three regions of "smallness" can now be identified. The "micro" region spans the dimensional range from micrometers to millimeters and is the most traditional. Here, MEMS/MST systems are dominant and contain such elements as micropressure sensors, accelerometers, microactuators, biochips, microswitches, and traditional MEMS and optical MEMS (*MOEMS*) components.

Another region that is beginning to attract considerable interest is the "nano" regime. It spans the nominal range from 100 nm down to 0.5 nm, a region in which the properties of many materials transition from those of macroscopic systems to become dominated by the discrete electronic states of molecular and atomic domains.

The nano regime has rich potential for offering useful new materials characteristics including enhanced strength and hardness, reduced resistivity, superplasticity, self-assembly, enhanced catalytic intensity, and increased magnetic and dielectric performance, to name a few.

A very short list of potential applications for nanostructured devices include field-emitter tips, carbon-fullerene nanotubular structures with electronic and superconducting potential, nanophase components for high-strength composite materials, more effective giant magnetoresistive structures for high-capacity magnetic data storage applications, a variety of proximal probe methods for high-density data reading, plus many more in the fields of chemistry, biology, and medicine.

Still a third regime of smallness, the mesoscale region, is beginning to achieve recognition. With dimensions in the submillimeter to centimeter range, mesoscale structures encompass complex composite systems of microcomponents. Examples, including some now being developed, including a mesoscale heat exchanger and a mesoscale vacuum pump. Both of these consist of arrays of individual microcomponents acting in concert. In addition, many microinstrumentation packages such as those now under development for medical applications and space exploration may qualify for re-categorization as mesosystems.

Microelectronics

The area of electronic technology associated with or applied to the realization of electronic systems from extremely small electronic parts or elements.

Microengineering

Microengineering is the design and production of small, three-dimensional objects, usually for manufacturing high volumes at low cost. Such designs have a wide range of applications, including automobiles, medicine, aircraft, printing, security, and insurance. The techniques employed come from a wide range of disciplines, including biochemical, chemical, electrical, electronic, fluidic, mechanical, and optical engineering. Present designs occupy a volume of about 1 mm³ (40 mils³) and are made in tolerances of a few micrometers (about 50 μin.).

The applications of the described sensors cover air-bag sensors, which detects crashes and sets off an air-bag.

Microengineering is also used to provide automatic braking systems and intelligent suspension for smoother rides, and to optimize the flame profile in an engine to reduce fuel consumption.

The process uses synchrotron radiation to produce sharp cuts in thick-film photoresists, which may then be plated with plastics, metals, or other materials. The use of nickel produces strong metal parts in substantial quantities. In one design, these are assembled to produce a very small motor, which is suitable for use in microsurgery, watches, and camcorders.

Fiber clamps, gyroscopes, fuses, water-quality monitoring, patient monitoring, eye surgery microsurgery, printing, aviation, and food production are just some of the applications and processing methods being utilized and considered under microengineering. In the United States, MEMS covers the electronic and mechanical parts of microengineering. In Japan, the term *micromachines* is used to cover a small specific subset that includes sensors and robots. In Europe, the terms *microsystems* and *microsystems technology* cover systems and the technologies used to make them, but usually only the use of semiconductor-like processing; they also exclude the many small devices that operate independently.

The term *nanoengineering* is usually used to describe even smaller devices, nominally 1000 times smaller, which are usually much more expensive and are not normally mass produced. The term *mechatronics* covers the use of mechanical and electronic engineering irrespective of size.

Microetching

Development of microstructure for microscopic examination. The usual magnification exceeds 25× (50× in Europe).

Microfissure

A crack of microscopic proportions.

Micrograph

A graphic reproduction of the surface of a specimen and a magnification greater than 25×. If produced by photographic means, it is called a photomicrograph (not a microphotograph).

Microgrinding

This technology offers benefits for optical materials manufacturers where lapping and finishing alternatives can provide reduced cycle time and increased quality in the manufacture of optical components.

In deterministic microgrinding, the infeed rate, rather than the nominal pressure, is kept constant. This manufacturing operation uses rigid, computer-controlled machining centers and high-speed tool spindles. The system eliminates the specialized tooling, special skills, and long cycle times required with conventional equipment—cycle time reductions of 50% have been routinely achieved.

The equipment is capable of surfacing, edging, centering, and beveling in one setup sequence. The grinding geometry uses a bound abrasive cup wheel that is rotated at high speed and tilted at a very precise angle. Specular surfaces, resulting after less than 5 min of deterministic microgrinding, have a typical rms microroughness of less than 20 nm.

Because the equipment uses sequential cuts with coarse, medium, and fine grinding wheels, it takes only about 10 min to grind each surface. The fine wheel was run at 15,000 rpm and is fed into the lens at 6 µm/min. In less than 3 min, a total of 15 µm of material was removed, which is equal to the depth of the subsurface cracks produced by the medium tool. When used at slow infeed rates and for appropriate removal tasks, the 2–4 µm tools can cut hundreds of parts without needing to be redressed.

Finishing can be defined as the production of the surface to within 0.25 µm (peak to valley) of the specified shape and under 20 Å rms finish. Conventional finishing processes are time-consuming and labor-intensive and can require much rework.

A new finishing technology has been developed that addresses these problems.

Called magnetorheological (MR) finishing, the method is based on an MR fluid, a suspension of noncolloidal magnetic particles (usually carbonyl iron in water with smaller concentrations of stabilizers) and finishing abrasives (cerium oxide). This viscous liquid becomes a solid when a DC magnetic field is applied to it; stiffness is directly proportional to field strength. The MR fluid can be thought of as a compliant replacement for the conventional rigid lap in the loose abrasive grinding and finishing process. A magnetic field applied to the fluid creates a temporary finishing surface, which can be controlled in real time by varying the strength and direction of the field.

Surface smoothing, removal of subsurface damage, and figure correction are accomplished by rotating the part on a spindle at a constant speed while sweeping the lens about its radius of curvature through the stiffened finishing zone.

With the standard MR fluid, MRF reduces the surface microroughness of fused silica and other optical materials to less than 10 Å rms. The time required to do this varies from 5 to 60 min, depending on the material and its initial roughness. If the initial rms surface microroughness is less than 30 nm, smoothing occurs in 5–10 min.

Microhardness

The hardness of a material as determined by forcing an indenter such as a Vickers or Knoop indenter into the surface of a material under very light load; usually, the indentations are so small that they must be measured with a microscope. Capable of determining hardness is of different microconstituents within a structure, or of measuring steep hardness gradients such as those encountered in *case hardening*. See also *microhardness test*.

Microhardness Number

A commonly used term for the more technically correct term *microindentation hardness number*.

Microhardness Test

A microindentation hardness test using a calibrated machine to force a diamond indenter of specific geometry, under a test load of 1–1000 gf, into the surface of the test material and to measure the diagonal or diagonals optically. See also *Knoop hardness test* and *Vickers hardness test*.

Microindentation

(1) In hardness testing, the small residual impression left in a solid surface when an indenter, typically a pyramidal diamond stylus, is withdrawn after penetrating the surface. Typically, the dimensions of the microindentations are measured to determine microindentation hardness number, but newer methods measure the displacement of the indenter during the indentation process to use in the hardness calculation. The precise size required to qualify as a microindentation has not been clearly defined; however, typical measurements of the diagonals of such impressions range from approximately 10 to 200 µm, depending on normal force and material. (2) The process of indenting a solid surface, using a hard stylus of prescribed geometry and under a slowly applied normal force, usually for the purpose of determining its microindentation hardness number. See also *Knoop hardness number*, m*icroindentation hardness number*, and *Vickers hardness number*.

M

Microindentation Hardness Number

A numerical quantity, usually stated in units of pressure (kg/mm^2), that expresses the resistance to penetration of a solid surface by a hard indenter a prescribed geometry and under a specified, slowly applied normal force. The preface "micro" indicates that the indentations produced are typically between 10.0 and 200.0 μm across. See also *Knoop hardness number*, *nanohardness test*, and *Vickers hardness number*.

Microinfiltrated Macrolaminated Composites

Ceramics offer attractive properties, including good high-temperature strength and resistance to wear and oxidation. However, the major limitation to their use in structural applications is their inherent low fracture toughness, that is, the tendency to break (fracture) and produce low values. Ceramics have low fracture toughness, whereas steels and superalloys have better fracture toughness. Methods currently being used to improve the toughness of ceramics involve the incorporation of reinforcing whiskers and fibers; inclusion of a phase that undergoes transformation within the stress field associated with a crack; and cermet technology (cermet/metal composites), in which the tough metallic component absorbs energy. Improved toughness is attributed to various mechanisms, including crack branching, transformation-induced residual stresses, crack bridging, and energy absorption by plastic flow.

Another possible approach to improving the toughness of ceramics is via laminated construction. Sophisticated coating techniques have been used to produce laminated microstructures of ceramics and metals, but the cost of producing a bulk-laminated composite of useful size is prohibitive. Fabrication of microinfiltrated macrolaminated composites offers an economically feasible approach to produce a variety of cermet/metallic and ceramic/metallic bulk-laminated composites. The basic architecture of a microinfiltrated macrolaminated composite is a double-layer structure. One layer consists of a soft, ductile material having a low modulus of elasticity, low strength, and high toughness; the second layer consists of a hard, brittle material having a high modulus of elasticity and low toughness. This double layer is repeated as many times as necessary to form the bulk composite; in addition, the brittle material is infiltrated with the ductile constituent.

Advantages

The architecture of the microinfiltrated macrolaminated composite offers a compromise to the conventional composite microstructure by providing repeated alternating layers of bulk tough metallic and brittle cermet and ceramic materials. Any crack introduced into the brittle constituent will, upon entering the metallic interlayer, be subjected to a potential crack, stopping higher toughness.

Composition and Fabrication

Fabrication of microinfiltrated macrolaminated composite materials offers a processing route to economical manufacture of large, bulk laminated composites in which an interpenetrating microstructure is achievable. Use of both ceramic and metal powders in the process opens the door to a potentially wide range of material combinations. The same fundamental processing steps can be used to produce a wide variety of microinfiltrated macrolaminated composites, including ceramic/metal, metal/intermetallic, and ceramic/intermetallic systems.

Various combinations of composite properties, including hardness, strength, ductility, and fracture toughness, are possible by varying the laminate layer thicknesses. Microinfiltrated macrolaminated composites, when used in bulk form, are expected to have properties far superior to those of the individual monolithic constituents of the composite.

Alternative processing routes for microinfiltrated macrolaminated composites are available if the composite constituents have some solubility for one another, such as in the tungsten carbide–cobalt/cobalt (WC–Co/Co) and tungsten–nickel–iron heavy alloy/nickel (W–Ni–Fe/Ni) systems. In general, to obtain a large composite of this type, the best approach is to use the tape-casting process (well known in the ceramic industry) to produce large thin tapes (sheets) of the material.

Consolidation (compaction) helps the material achieve its full density by removing gas voids, porosity, and so forth. An attractive alternative for producing sheets consists of rolling the powder material followed by sintering to the required percent of theoretical density, retaining a certain level of interconnected porosity.

The fabrication steps used to make a W–Ni–Fe/Ni microinfiltrated macrolaminated composite illustrate the process. Tungsten powder is tape cast, and the sheets are sintered to 60%–70% of theoretical density, which produces totally interconnected porosity. Small amounts of nickel can be used together with tungsten to promote activated sintering and to produce a porous tungsten sheet that can be handled safely. Sheet thickness ranges from 1 to 10 mm.

An 80:20 ratio of nickel and iron powders also is tape-cast and sintered to full density. Sheets of porous tungsten and fully dense nickel–iron alloy material are laid up in alternate layers and heated to about 1475°C, which is about 40°C above the melting point of the nickel–iron alloy. The molten Ni–Fe alloy infiltrates the porous tungsten sheets and takes into solution a fraction of the tungsten. A thin layer of the liquid phase can be retained between the tungsten sheets after the infiltration process is complete. This retention is possible if the tungsten sheet contains porosity levels that are lower than the volume of liquid formed by melting the nickel–iron sheets.

Applications

The design of microinfiltrated macrolaminated composites could prove advantageous in applications involving impact or ballistic penetration. The metallic interlayer would function to hold together damaged portions of the brittle constituent. Thus, the development of ceramic armor having multihit capability becomes possible. Current metal/ceramic composite armor is layered on an extremely macroscopic scale. An approach involving microinfiltrated macrolaminated composites permits optimization of layer frequency and thickness to resist specific threats, as in antipersonal weapons or armored vehicles.

In addition, high-temperature composites, consisting of a ceramic and intermetallic compound, such as aluminum oxide/nickel aluminide (Ni_3Al) plus boron, could be fabricated in the form of microinfiltrated macrolaminated composites to yield combinations with new properties. Ultrahigh-temperature composites incorporating-temperature ceramics and ductile niobium is another potential application to obtain various combinations of high wear resistance and toughness. The hard ceramic outer layer would provide a wear-resistant surface, with toughness provided by the ductile material. A similar concept could lead to a new generation of cutting tools, in which the outer most cutting layer could be made of ultrahard materials suitably layered with a soft, ductile constituent for toughness.

Other potential applications are in heat- and oxidation-resistant, low-density structural components and aerospace parts having low density, high strength and modulus, and good fracture toughness.

Finally, the reinforcement of concrete with carbon fibers is a recently developed technique for the construction of structures such as buildings, bridges, and tunnels. Advantages of carbon-fiber reinforcement include outstanding electromagnetic shielding, high resistance to corrosive environments, light weight, and high mechanical property strength. Two methods are used to achieve the reinforcement: mixing carbon fibers, either chopped or mat, directly into cement (carbon fiber-reinforced concrete); and using carbon fiber/plastic composites in rod or tape form for strengthening the concrete.

Micromachining

Magnetron plasma etching has been found to be a promising technique for micromachining of single-crystal silicon carbide to fabricate microscopic structures comprising integrated mechanical, electronic, and optical devices. Silicon carbide offers several advantages over silicon for the development of future devices. In comparison with silicon, silicon carbide is harder and stiffer, is less chemically reactive, and has greater thermal conductivity; moreover, silicon carbide-based electronic devices can operate at temperatures higher than silicon-based electronic devices can withstand. Etching techniques for micromachining of silicon are well known, but the lesser chemical reactivity of silicon carbide makes it necessary to devise alternative etching techniques.

In the present magnetron-plasma-etching technique as in the ECR etching technique, a magnetic field is used to increase the density of the plasma. Unlike in ECR etching, the main body of the plasma is confined in proximity to the sample. Such confinement results in a high concentration of reactive chemical species together with a high flux of bombarding ions at the sample, making it possible to achieve rapid etching.

Magnetron plasma etching of SiC was demonstrated in preliminary experiments in a vacuum chamber using a magnetron sputter gun as a cathode. The duration of each edge was 7 min. At a radiofrequency power of 250 W, a maximum edge depth of 4.5 μm was achieved with a gas mixture of $0.6CHF_3 + 0.4O_2$ at a total pressure of about 2.7 Pa. This depth corresponds to an etch rate of about 640 nm/min, which is a relatively high rate, significantly higher than that which has been achieved using an ECR plasma. The etched surface was found to have a rms surface roughness of only 20 Å. The reaction of aluminum and fluorine yields a nonvolatile product, which makes aluminum suitable for use as a mask material to provide selective etching in fluorine-based plasmas. At the radiofrequency power of 250 W, aluminum films were etched at a high rate. At a radiofrequency power of 50 W, silicon carbide was etched at a rate of 170 nm/min, while aluminum was etched at 1/12 that rate.

Micromesh

A sieve with precisely square openings in the range of 10–120 μm produced by electroforming.

Micromesh Sizing

The process of sizing micromesh particles using an air or a liquid suspension process.

Micrometer

(1) Any instrument for measuring earlier dimensions at high precision. (2) A dimension of 0.001 mm, written with the abbreviation μm. Also referred to as micron (not recommended).

Micron

One millionth of a meter, symbol μm. This is not an SI-approved name.

Micropore

The pores in a sintered product that can only be detected under a microscope.

Microporosity

Extremely fine porosity in castings.

Microprobe Analyzer

Same as *micro analyzer*.

Micropulverizer

A machine that disintegrates powder agglomerates by strong impacts from small hammers fastened to a solid disk that rotates at very high velocity.

Microradiography

The technique of passing x-rays through a thin section of a material in contact with a fine-grained photographic film and then viewing the radiograph at 50× to 100× to observe the distribution of constituents and/or defects.

Microscope

Unless otherwise indicated, this term usually refers to the light microscope. This is a multilens instrument for examining a material at a high magnification. Biological microscopes project a beam of light through the specimen under examination. This light is then focused through a system of magnifying lenses to form an image at the observer's eye. Basically two lenses are involved, the objective close to the specimen which produces the initial image that is then magnified by the ocular or eyepiece close to the observer. In practice, age of the lenses is of compound form, that is, form from multiple individual lenses, some spaced apart, some bonded together. The total magnification is approximately the multiple of the ocular and objective. A metallurgical microscope is similar except that the light beam is injected into the body of the microscope between the objective and eyepiece. It is initially reflected by a clear glass plate reflector onto the specimen and then reflected back along its original path through the objective, straight through the clear reflector and ocular into the eye. In any microscope, the objective lens is responsible for developing the initial image so that its numerical aperture, that is, its efficiency in gathering light is critical and the wavelength of light is such that details finer than about 0.2 m cannot be resolved by the objective lens. This corresponds to an objective lens magnification of about 100× diameters, so useful magnification of the microscope ranges up to about 1000× diameters. At the highest magnifications, an immersion oil may be interposed in the very small gap between the specimen and the objective lens to approve the resolution. See also *electron microscope* and *metallography*.

Microscopic

Invisible at magnifications above 25×.

Microscopic Stress

Residual stress in a material within a distance comparable to the grain size. See also *macroscopic stress*.

Microscopical Examination

See *metallography* and *microscope*.

Microscopy

The science of the interpretive use and applications of microscopes and examining materials under a microscope.

Microsection

A sample of metal cut from a component, etc., for metallographic examination.

Microsegregation

Segregation within a grain, crystal, or small particle. See also *coring*.

Microshrinkage

A casting imperfection, not detectable microscopically, consisting of inner dendritic voids. Microshrinkage results from contraction during solidification where the opportunity to supply filler material is inadequate to compensate for shrinkage. Alloys with wide ranges in solidification temperature are particularly susceptible.

Microshrinkage Cavity

A fine void found microscopically in the low melting metal phase of infiltrated powder metallurgy compacts due to contraction during solidification.

Microslip

(1) Small relative tangential displacement in a contacting area at an interface, when the remainder of the interface in the contacting area is not relatively displaced tangentially. Microslip can occur in both rolling and stationary contacts. (2) The term microslip is sometimes used to denote the microslip velocity. This usage is not recommended. See also *macroslip* and *slip*.

Microspheres

Spherical particles used in plastics and other materials as fillers and reinforcing agents. They are also used to encapsulate various materials. Those used as fillers or reinforcements are made of glass or ceramics or resins. There are two different kinds of glass microspheres—solid and hollow. Solid spheres, made of soda-line glass, range in size from 157 to 197,000 μin. (4 to 5,000 μm) in diameter and have a specific gravity of about 2.5. Hollow glass microspheres have densities ranging from 5 to 50 lb/ft³ (80 to 801 kg/m³ and diameters from 790 to 7900 μin. [20 to 200 μm]). The spheres in plastics improve tensile, flexural, and compressive strengths and lower elongation in water absorption. Hollow, glass, **microballoon microspheres**, of Emerson and Cuming can be as small as 200 μin. (5 μm). **Plastic microspheres** are used mostly in the production of syntactic foams. *Polyvinylidene chloride microspheres* are excellent resin extenders and are used in sandwich construction of boat hulls. *Epoxy microspheres* are used as low-density bulk fillers for plastics and ceramics and were developed for use in submerged deep-water floats. *Phenolic microspheres* filled with nitrogen are used for production of polyester foams and syntactic epoxy foams. Polystyrene microspheres are also used to produce syntactic foams.

In the pharmaceutical industry, microspheres are used to encapsulate active ingredients. Multilayered *liposome microspheres* are typically made of fatty acids or lipid materials. Liposome formation takes advantage of the amphipathic nature of fat molecules, which allows fats to form liposomes spontaneously in turbulent water. For injectable drugs, nontoxic and biocompatible phospholipids, such as lecithin, cephalin, and phosphatidyl inositol, are used for less demanding applications, liposomes can be made from fatty acids, alkyl derivatives, and amines. Microsponges, made from cross-linked polymers, are used as time-release delivery systems of process ingredients.

Microstrain

The strain over a gauge length comparable to interatomic distances. These are the strains being averaged by the *macrostrain* measurement. Microstrain is not measurable by existing techniques. The variance of the microstrip distribution can, however, be measured by x-ray diffraction.

Microstress

Same as micro*scopic stress*.

Microstress/Microstrain

Stresses and strains that either are of very low magnitude or act over very short distances relative to the context.

Microstructure

(1) The structure of an object, organism, or material as revealed by a microscope at magnifications greater than 25×. (2) The arrangement and features of grains, phases, inclusions, and other features within a metal. It is normally examined by light or electron microscope.

Microtome

Equipment for slicing a very thin film or sheet of material from some larger item, usually for some form of examination or test.

Microwave Heating

A microwave-heating technique provides for batch processing of multiple, identically sized and shaped samples of the same material. The technique involves (1) excitation of asymmetrical electromagnetic mode or modes in a asymmetrical microwave cavity and (2) positioning the samples symmetrically in the cavity so that all samples are exposed to the same electromagnetic field conditions and thus the same heating conditions. Typically, the electromagnetic modes and the pattern for mounting the samples are chosen to maximize the heating effect and make it as nearly spatially uniform as possible.

The same principle can be applied to microwave heating of multiple spherical or disk-shaped samples.

Microwave Radiation

Electromagnetic radiation in the wavelength range of 0.3 mm to 1 m (3×10^6 to 10^{10} Å). See also *electromagnetic radiation*.

Middle-Infrared Radiation

Infrared radiation in the wavelength range of 3–30 μm (3×10^4 to 3×10^5 Å). See also *infrared radiation* and *electromagnetic radiation*.

Middling

A product intermediate between concentrate and tailing and containing enough of a valuable mineral to make retreatment profitable.

MIG Welding

See preferred term *gas metal arc welding*.

Migration

Movement of entities (such as electrons, ions, atoms, molecules, vacancies, and grain boundaries) from one place to another under the influence of a driving force (such as an electrical potential or a concentration gradient).

Mil

(1) An English measure of thickness or diameter equal to 0.0254 mm (0.001 in.). A common designation of wire size, coating thickness, or corrosion loss. (2) Either 1000th of an inch or 1 mm. Both uses are informal, and the potential for confusion is obvious.

Mild Steel

Carbon steel with a maximum of about 0.25% C and containing 0.4%–0.7% Mn, 0.1%–0.5% Si, and some residuals of sulfur, phosphorus, and/or other elements.

Mild Wear

A form of wear characterized by the removal of material in very small fragments. Mild wear is an imprecise term, frequently used in research and contrasted with severe wear. In fact, the phenomena studied usually involve the transition from mild to severe wear and the factors that influence this transition. Mild wear maybe appreciably greater than can be tolerated in practice. With metallic sliders, mild wear debris usually consists of oxide particles. See also *normal wear* and *severe wear*.

Military Transformation

A transformation from one phase to another in which the atoms at the advancing interface act in a relatively restricted manner and can be constrained to certain crystallographic orientations or confined by some of the original phase grain boundaries.

Mill

(1) A factory in which metals are hot worked, cold worked, or melted and cast into standard shapes suitable for secondary fabrication into commercial products. (2) A production line, usually of four or more *stands*, for hot or cold rolling metal into standard shapes such as bar, rod, plate, sheet, or strip. (3) A single machine for hot rolling, cold rolling, or extruding metal; examples include *blooming mill*, *cluster mill*, *four-high mill*, and *Sendzimer mill*. (4) A shop term for a milling cutter. (5) A machine or group of machines for grinding or crushing ores and other minerals. (6) A machine for grinding or mixing material, for example, a ball mill and a paint mill. (7) Grinding or mixing a material, for example, milling a powder metallurgy material.

Mill Edge

The normal edge produced in hot rolling of sheet metal. This edge is customarily removed when hot-rolled sheets are further processed into cold-rolled sheets.

Mill Finish

(1) A nonstandard (and typically nonuniform) surface finish on *mill products* that are delivered without being subjected to a special surface treatment (other than a corrosion preventative treatment) after the final working or a heat-treating step. (2) Material in the state in which it left the last major manufacturing operation, such as rolling drawing or extrusion. As examples, mill finished hot rolled plate will have wide tolerances on dimensions and carry a heavy oxide coating, termed mill scale, while mill finished cold-rolled material will have a smooth surface or even a deliberately textured or embossed surface but will not have been polished and perhaps not thoroughly cleaned. See also *milled finish*.

Mill Product

Any commercial product of a *mill*.

Mill Scale

The heavy oxide layer that forms during the hot fabrication or heat treatment of metals. See *magnetite* and *mill finish*.

Milled Fiber

Continuous glass strands hammer milled into very short glass fibers. Useful as inexpensive filler or anticrazing reinforcing fillers for adhesives and engineering plastics.

Milled Finish

Material that has been machined on a milling machine and hence will have a bright surface and, probably, precise dimensions. In some cases, the repetitive pattern of circular swirl marks characteristic of some milling operations is seen as aesthetically pleasing. Note, however, that a similar effect is sometimes achieved by abrading a sheet surface with a flat end polishing bob. See also *mill finish*.

Miller Bravais Indices

See *Miller Indices*.

Miller Indices

A system for identifying planes and directions in any crystal system by means of sets of integers. The indices of a plane are related to the intercepts of that plane with the axes of a unit cell; the indices of a direction, to the multiples of lattice parameter that represent the coordinates of a point on a line parallel to the direction and passing through the arbitrary chosen origin of a unit cell.

This is a system for identifying planes and orientations in a crystal lattice. It is best considered in three steps. In the simple case of a cubic system, three axes X, Y, and Z radiate from an origin at the corner of the cube unit cell. At step 1, the plane of interest is specified by measuring the distance, from the origin, at which it intercepts the three axes. Such distances are measured in terms of the unit cell size.

Step 2 is to take the reciprocal of the three intercept units from step 1, giving 1.0.0 and 1.1.1, which are the *Miller Indices*. Step 3 only necessary for planes less simple than those just discussed, requires these reciprocals to be factored to give the smallest possible whole numbers, that is, X, 1/2, 1, would be factored to 0, 1, 2, and 2, 4, 6, would be factored to 1, 2, 3. Miller Indices are conventionally enclosed in parentheses, the use of simple brackets (111) indicating all planes lying parallel to the reference plane, while in irregular brackets (111) indicate all planes having similar atomic structures but lying in all directions.

Directions, derived by identifying the intercepts of the line on the axes, are indicated by square brackets [111]. If any of the intercepts at step 1 had been negative, it would be identified in the Index by a bar above it. The more complex case of hexagonal structures is dealt with by Miller–Bravais Indices by reference to four axes, the W, X, and Y axes at 120° on the basal plane and the Z axis normal to the basal plane.

Miller Number

A measure of slurry abrasivity as related to the instantaneous mass-loss rate of a standard metal wear block at a specific time on the cumulative abrasion–corrosion time curve. See also *slurry abrasion response number.*

Milling (Machining)

Using a rotary tool with one or more teeth that engage the workpiece and remove material as the workpiece moves past the rotating cutter.

1. *Face milling*: Milling a surface perpendicular to the axis of the cutter. Peripheral cutting edges remove the bulk of the material, while the face cutting edges provide the finish of the surface being generated.
2. *End milling*: Milling accomplished with a tool having cutting edges on its cylindrical surfaces as well as on its end. In peripheral end milling, the peripheral cutting edges on the cylindrical surface are used; while in slotting, both end and peripheral cutting edges remove metal.
3. *Side and slot milling*: Milling of the side or slot of a workpiece using a peripheral cutter.
4. *Slab milling*: Milling of a surface parallel to the axis of a helical, multiple-toothed cutter mounted on an arbor.
5. *Straddle milling*: Peripheral milling a workpiece on both sides at once using two cutters as required.

Milling (Powder Technology)

The mechanical comminution of a material, usually in a ball mill, to alter the size or shape of the individual particles, to coat one component of a mixture with another, or to create uniform distributions of components.

Milling Cutter

A rotary cutting tool provided with one or more cutting elements, called teeth, which intermittently engage the workpiece and remove material by relative movement of the workpiece and cutter.

Milling Fluid

An organic liquid, such as hexane, in which *ball milling* is carried out. The liquid serves to reduce the heat of friction and resulting surface oxidation of the particles during grinding, and to provide protection from other surface contamination.

Millstone

Any stone employed for grinding paint, cement, grain, or minerals. Millstones are made from sandstone, basalt, granite, or quartz conglomerate. *Burrstone* is a millstone made from chalcedony silica of cellular texture, usually yellowish. The stone is used as a building stone.

Pulpstones are blocks of sandstone cut into wheels and used for grinding, chiefly for the grinding of wood pulp in paper manufacture. The sandstones for pulpstones must be uniform in texture, have sharp grains, have medium hardness, and be composed of even quartz grains of which 85% will be retained on a 150-mesh screen, and 90% on a 200-mesh screen. The cement milling material may be siliceous, calcareous, or argillaceous, but must be firm enough to hold the stone together when working under pressure, and soft enough to wear faster than the quartz grains and prevent glazing. Large pulpstones for paper mills are now made of silicon carbide or aluminum oxide in fitted sections.

Mineral Dressing

(1) Preliminary actions to prepare an ore for the main process of extracting the metal. The term includes physical processes and, usually, initial chemical concentrating processes. (2) Physical and chemical concentrations of raw ore into a product from which a metal can be recovered at a profit.

Mineral Insulated Cable (MIC)

An electrical cable comprising an outer tubular copper sheath and one or more bare copper conductors, the gap between them being filled with compacted magnesium oxide powder as insulant.

Mineral Oil

A refined hydrocarbon oil without animal or vegetable additives.

Mineral Wool

A fibrous material employed as a heat insulator in walls or as a sound insulator. It was first obtained as a natural product from volcanic craters in Hawaii and was known as *Pele's hair*. It is made by mixing stone with molten slag from blast furnaces and blowing steam through it. Slags from copper and lead furnaces are also used. *Slag wool* is made from slag without the rock. A lead slag containing 30%–50% calcium and magnesium oxides makes a mineral wool that withstands temperatures up to 1500°F (816°C) when made into blocks or boards. Mineral wool usually consists of a mass of fine, pliant, vitreous fibers, which are incombustible and non conductors of heat. *Rock wool* is made by blowing molten rock in the same manner and is more uniform than common mineral wool, with physical qualities depending, however, on the class of rock used. *Rock cork* is a name for a low-temperature insulating material made of rock wool molded in sheet form with a waterproof binder, used for walls in cold-storage rooms. *Mineral-wool board* is a moisture-resistant board for cold-storage insulation. *Mono-Block* is rock wool made into standard blocks and slabs by a felting process.

Granulated mineral wool is the fiber milled into pellets, and these pellets can be poured into a space for insulation, or they may be pressed into insulating blocks. *Rockwool quilt* consists of felted fibers stitched between layers of Kraft paper for wall insulation.

Selected minerals, to give various characteristics, may be used to make rock wools. Wollastonite is melted to produce a rock wool. It is also ground to a white vesicular powder of 30–350 mesh for use as a filler in molded plastics. The grains are minute fibers. *Ceramic fiber* differs from mineral wool in that it is made of special composition. *Aluminum silicate fiber* is made by melting alumina and silica and passing the molten material through a stream of high-pressure air that produces a fluffy mass of extremely fine fibers up to 3 in. (7.62 cm) long. The melting point of the fibers about 3000°F (1649°C), but the maximum usable temperature is about 2300°F (1260°C), as the fibers devitrify at higher temperatures. The fibers are used for thin insulating paper, panel board, filler for plastics, and chemical-resistant rope or are woven into fabrics. Aluminum silicate fabrics are used for filters, gaskets, belting, insulation, and protective clothing. Fiberfrax, of Carborundum Co., is an aluminum silicate fiber.

Miners Rule

A means for estimating fatigue life when a component is subjected to a series of different stresses. For each stress level, the number of cycles experienced is expressed as a fraction of the number of cycles to failure at that stress. Failure occurs when the sum of the fractions is unity.

Minimized Spangle

A hot dip galvanized coating of very small grain size, which makes the *spangle* less visible when the part is subsequently painted.

Minimum Bend Radius

The minimum radius over which a metal product can be bent to a given angle without fracture. See also the term *bend radius*.

Minimum Load (P_{min})

In fatigue, the least algebraic value of applied load in a cycle.

Minimum Stress (S_{min})

In fatigue, the stress having the lowest algebraic value in the cycle, tensile stress being considered positive and compressive stress negative.

Minimum Stress-Intensity Factor (K_{min})

In fatigue, the minimum value of the *stress-intensity factor* in a cycle. This value corresponds to the *minimum load* when the *load ratio* >0 and is taken to be zero when the load ratio is ≤0.

Minor Segregation

See *segregation*.

Minus Sieve

The portion of a powder sample that passes through a standard sieve of a specified number. See also *plus sieve* and *sieve analysis*.

Mirror Illumination

A thin, half-round opaque mirror interposed in a microscope for directing an intense oblique beam of light to the object. The light incident on the object passes through one half of the aperture of the objective, and the light reflected from the object passes through the other half aperture of the objective.

Mirror Region

The comparatively smooth region that symmetrically surrounds a fracture origin in ceramics and glasses. The mirror region ends in a microscopically irregular manner at the beginning of the mist region. See also *mist hackle*.

Mischmetal

A natural mixture of rare earth elements (atomic numbers 57–71) in metallic form. It contains about 50% cerium, the remainder being principally 40% lanthanum and neodymium. Mischmetal is used as an alloying additive in ferrous alloys to scavenge sulfur, oxygen, and other impurities and in magnesium alloys to improve high-temperature strength. Alloys with iron or magnesium, it is pyrophoric, and used for lighter "flints."

Miscibility

The ability of two or more liquids to mix forming a single homogeneous solution.

Miscibility Gap

A region of multiphase equilibrium. It is commonly applied to the specific case in which an otherwise continuous series of liquid or solid solutions is interrupted over a limited temperature range by a two-phase field terminating at a critical point. See also *binodal curve*.

Miscible

Of two phases, the ability of each to dissolve in the other. May occur in a limited range of ratios of the two, or in any ratio.

Mismatch

The misalignment or error in register of a pair of forging dies; also applied to the condition of the resulting forging. The acceptable amount of this displacement is governed by blueprint or specification tolerances. Within tolerances, mismatch is a condition; in excess of tolerance, it is a serious defect. Defective forgings can be salvaged by hot-reforging operations.

Misrun

Denotes an irregularity on a cast metal surface caused by incomplete filling of the mold due to low pouring temperatures, gas back pressure from inadequate venting of the mold, and inadequate gating and cold shuts.

Mist

An area of dull flat tensile fracture observed on some brittle fractures.

Mist Hackle

Markings on the surface of a crack in ceramics and glasses accelerating close to the effective terminal velocity, observable first as a mist on the surface and with increasing velocity revealing a fibrous texture elongated in the direction of cracking and coarsening up to the stage at which the crack bifurcates. Velocity bifurcation or velocity forking is the splitting of a single crack into two mature diverging cracks or at near the effective terminal velocity of about half the transverse speed of sound in the material. See also *bifurcation* and *mirror region*.

Mist Lubrication

Lubrication by an oil mist produced by injecting oil into a gas stream.

Mitchell Bearing

See *tilting-pad bearing*.

Mixed Dislocation

See *dislocation*.

Mixed Grain Size

See *duplex grain size*.

Mixed Lubrication

See *quasi-hydrodynamic lubrication*.

Mixed Potential

The potential of a specimen (or specimens in a *galvanic couple*) when two or more electrochemical reactions are occurring. Also called galvanic couple potential.

Mixing

In powder metallurgy, the thorough intermingling of powders of two or more different materials (not *blending*).

Mixing Chamber

The part of a torch or furnace burner in which gases are mixed.

MMA Welding

See *Manual Metal Arc Welding*.

Mobile Phase

In chromatography, the gas or liquid that flows through the chromatographic column. A sample compound in the mobile phase moves through the column and is separated from compounds residing in the stationary phase. See also *stationary phase*.

Mock

When used as prefix to gold, silver, etc., it indicates a wide range of alloys of base metals claimed to have a superficial resemblance to the metal implied.

Mock Leno Weave

An open fabric weave for composites that resembles a leno and is accomplished by a system of interlacings that draws a group of threads together and leaves a space between that group and the next. The warp threads do not actually cross each other as in a real leno and, therefore, no special attachments are required for the loom. This type of weave is generally used when a high thread count is required for strength, and the fabric must remain porous. See the term *leno weave*.

Mode

One of the three classes of crack (surface) displacements adjacent to the crack tip. These displacement modes are associated with stress–strain fields around the crack tip and are designated I, II, and III. See also *crack-tip plane strain*, *crack opening displacement*, and *fracture mechanics*.

Moderator

In a nuclear reactor context, a material having a low atomic weight and low capture cross section that slows down fast neutrons allowing them to be captured by the fuel causing fission.

Modification

Treatment of molten hypoeutectic (8%–13% Si) or hypereutectic (13%–19% Si) alumina–silicon alloys to improve mechanical properties of the solid alloy by refinement of the size and distribution of the silicon phase. Involves additions of small percentages of sodium, about 0.02%–0.05%, strontium, or calcium (hypoeutectic alloys) or of phosphorus (hypereutectic alloys), as well as sodium chloride or fluoride, is added to the molten metal to refine the normally coarse eutectic structure that develops on solidification.

Modified Acrylic

A thermoplastic polymer that has been altered to eliminate mixing, curing ovens, and odor and that cures rapidly at room temperature.

Modifier

Any chemically inert ingredient added to an adhesive formulation to change its properties. Compare with *filler*, *plasticizer*, and *extender*.

Modulus of Elasticity (*E*)

(1) The measure of rigidity or stiffness of a material; the ratio of stress, below the proportional limit, to the corresponding strain. In the tensile stress of 13.8 MPa (2.0 ksi) results in an elongation of 1.0%, the modulus of elasticity is 13.8 MPa (2.0 ksi) divided by 0.01, or 1380 MPa (200 ksi). (2) In terms of the *stress–strain curve*, the modulus of elasticity is the slope of the stress–strain curve in the range of linear proportionality of stress to strain. Also known as *Young's modulus*. For materials that do not conform to Hooke's law throughout the elastic range, the slope of either the tangent to the stress–strain curve at the origin or at low stress, the secant drawn from the origin to any specified point on the stress–strain curve, or the chord connecting any two specific points on the stress–strain curve is usually taken to be the modulus of elasticity. In these cases, the modulus is referred to as the *tangent modulus, secant modulus,* or *chord modulus,* respectively.

Modulus of Resilience

The amount of energy stored in a material when loaded to its elastic limit. It is determined by measuring the area under the *stress–strain curve* up to the *elastic limit.* See also *elastic energy, resilience,* and *strain energy.* See also *Limit of Proportionality.*

Modulus of Rigidity

See *shear modulus.*

Modulus of Rupture

(1) A measure of the strength of a material determined in a three-point bend test. The test is usually applied to brittle materials as it avoids the problem associated with gripping the specimen that arises in the tensile test. (2) Nominal stress at fracture in a bend test or torsion tests. In bending, modulus of rupture is the bending moment at fracture divided by the section modulus. In torsion, modulus of rupture is the torque at fracture divided by the polar section modulus. See also *modulus of rupture in bending* and *modulus of rupture in torsion.*

Modulus of Rupture in Bending (*S*ₛ)

The value of maximum tensile or compressive stress (whichever causes failure) in the extreme fiber of a beam loaded to failure in bending computed from the flexure equation:

$$S_b = \frac{M_c}{I}$$

where

    M is the maximum bending moment, computed from the maximum load and the original moment arm
    c is the initial distance from the neutral axis to the extreme fiber where failure occurs
    I is the initial moment of inertia of the cross section about the neutral axis

See also *modulus of rupture.*

Modulus of Rupture in Torsion (*S*ₛ)

The value of maximum shear stress in the extreme fiber of a member of circular cross section loaded to failure in torsion computed from the equation:

$$S_s = \frac{T_r}{J}$$

where
    T is the maximum twisting moment
    r is the original outer radius
    J is the polar moment of inertia of the original cross section

See also *modulus of rupture.*

Modulus of Strain Hardening

See preferred term *rate of strain hardening.*

Modulus of Toughness

The amount of work per unit volume of a material required to carry that material to failure under static loading. See also *toughness.*

Mohs Hardness Number

A scale of hardness according to a scale proposed by Mohs, based on 10 minerals, each of which would scratch the one below it. These metals, in decreasing order of hardness, are as follows:

Diamond	10	Apatite	5
Corundum	9	Fluorite	4
Topaz	8	Calcite	3
Quartz	7	Gypsum	2
Feldspar	6	Talc	1

Moiety

A portion of a molecule, generally complex, having a characteristic chemical property.

Moiré Pattern

A pattern developed from interference or light blocking when gratings, screens, or regularly spaced patterns are superimposed on one another.

Moisture Absorption

The pickup of water vapor from air by a polymeric material, in reference to vapor withdrawn from the air only, as distinguished from water absorption, which is the gain in weight due to the absorption of water by immersion.

Moisture Content

The amount of moisture in a polymeric material determined under prescribed conditions and expressed as a percent of the mass of the moist specimen, that is, the mass of the dry substance plus the moisture.

Moisture Equilibrium

The condition reached by a plastic sample when it no longer takes up moisture from, or gives up moisture to, the surrounding environment.

Moisture Regain

The moisture in a polymeric material determined under prescribed conditions and expressed as a percent of the weight of the moisture-free specimens. Moisture regain may result from either sorption or desorption, and it differs from moisture content only in the basis used for calculation.

Moisture Vapor Transmission (MVT)

A rate at which water vapor passes through a polymeric material at a specified temperature and relative humidity.

Molal Solution

Concentration of a solution expressed in moles of solute divided by 1000 g of solvent.

Molality

The number of gram-molecular weights of a compound dissolved in 1 L of solvent. See also *gram-molecular weight*. Compare with *molarity* and *normality*.

Molar Conductivity

The electrical conductivity of a solution containing the molecular weight in grams dissolved in 1 L of water.

Molar Heat

The molecular weight in grams × the specific heat in grams per °C.

Molar Solution

Aqueous solution that contains 1 mole (gram-molecular weight) of solute in 1 L of the solution.

Molarity

The number of gram-molecular weights of a compound dissolved in 1 L of solution. See also *gram-molecular weight*. Compare with *molality* and *normality*.

Mold (Metals)

(1) The form, made of sand, metal, or refractory material, that contains the cavity into which molten metal is poured to produce a casting of desired shape. (2) A die.

Mold (Plastics)

The cavity into which, or matrix on which, the plastic composition is placed and from which it takes form. To shape plastic parts or finished articles by heat and pressure. The assembly of all the parts that function collectively in the molding process.

Mold Cavity (Metals)

The space in a mold that is filled with liquid metal to form the casting upon solidification. The channels through which liquid metal enters the mold cavity (sprue, runner, gates) and reservoirs for liquid metal (risers) are not considered part of the mold cavity proper.

Mold Coating (Metals)

(1) Coating to prevent surface defects on permanent mold castings and die castings. (2) Coating on sand molds to prevent metal penetration and to improve metal finish. Also called mold facing or mold dressing.

Mold Jacket (Metals)

Wood or metal form that is slipped over a sand mold for support during pouring of a casting.

Mold Release Agent (Plastics)

A lubricant, liquid, or powder (often silicon oils and waxes) used to prevent sticking of molded plastic articles in the cavity.

Mold Shift (Metals)

A casting defect that results when the parts of the mold do not match at the parting line.

Mold Shrinkage (Plastics)

(1) The immediate shrinkage that a molded plastic part undergoes when it is removed from a mold and cooled to room temperature. (2) The difference in dimensions, expressed in inches per inch, between a molding and the mold cavity in which it was molded (at normal-temperature measurement). (3) The incremental difference between the dimensions of the molding and the mold from which it was made, expressed as a percentage of the mold dimensions.

Mold Steel

This is a term that generally refers to the steels used for making molds for molding plastics. Mold steel should have a uniform texture that will machine readily with die-sinking tools. It must have no microscopic porosity and must be capable of polishing to a mirrorlike surface. When annealed, it should be soft enough to take the deep imprint of a bob, and when hardened, it must be able to withstand high pressure without sinking and have sufficient tensile strength to prevent breakage of thin mold sections. And, it should be dimensionally stable during heat treatment and corrosion-resistant to the plastics being formed.

The principal mold steels are the P-type mold steels, specifically P6, which contains 0.10% carbon, 3.50% nickel, and 1.50% chromium; P20 (0.35% carbon, 1.70% chromium, and 0.40% molybdenum); and P21 (0.20% carbon, 4.00% nickel, and 1.20% aluminum). P6 is a carburizing steel having moderate hardenability, P20 also has only moderate hardenability, but P21 is a deep-hardening steel. P20 and P21 are usually supplied hardened to R_c 30–36, in which condition they can be machined to complex shapes. P21, a precipitation-hardening steel, can be supplied moderately hard, then machined and post hardened by a moderate-temperature age.

Mold Surface (Plastics)

The side of a laminate that faced the mold (tool) during cure in an autoclave or hydroclave.

Mold Wash (Metals)

An aqueous or alcoholic emulsion or suspension of various materials used to coat the surface of a casting mold cavity.

Various fluids, for example, a suspension of graphite in water or paraffin, which is applied to the faces of a mold prior to casting to improve surface finish, reduce the damage to reusable molds, and assist stripping. Release agents are similar, but this term tends to be applied to more sophisticated products for smaller precision products.

Molded Edge (Plastics)

An edge on a plastic part that is not physically altered after molding for use in final form, particularly one that does not have fiber ends along its length.

Molded Net (Plastics)

Description of a molded plastic part that requires no additional processing to meet dimensional requirements.

Molding (Plastics)

The forming of a polymer or composite into a solid mass of prescribed shape and size by the application of pressure and heat for a given time. The finished part.

Molding Compound (Plastics)

Plastic material in varying stages of pellet form or granulation (powder), consisting of resin, filler, pigments, reinforcements, plasticizers, and other ingredients, ready for use in the molding operation. Also called molding powder.

Molding Cycle (Plastics)

The period of time required for the complete sequence of operations on a molding press to produce one set of plastic moldings. The operations necessary to produce a set of moldings without reference to the total time.

Molding Machine (Metals)

A machine for making sand molds by mechanically compacting sand around a pattern. See the term *jolt ramming* and *jolt-squeezer machine*.

Molding Press (Metals)

A press used to form powder metallurgy *compacts*.

Molding Pressure (Plastics)

The pressure applied to the ram of an injection machine, compression press, or transfer press to force the softened plastic to fill the mold cavities completely.

Molding Sands

Foundry sands containing over 5% natural clay, usually between 8% and 20%. See also *naturally bonded molding sand*.

Mole

One mole is the mass numerically equal (in grams) to the relative molecular mass of a substance. It is the amount of substance of a system that contains as many elementary units (6.02×10^{23}) as there are atoms of carbon in 0.012 kg of the nuclide ^{12}C; the elementary unit must be specified and may be in an atom, molecule, ion, electronic, photon, or even a specified group of such units.

Mole Fraction

The number of atoms of one element, as a percentage of the total number of atoms in a material or phase.

Molecular Fluorescence Spectroscopy

An analytical technique that measures the fluorescence emission characteristic of a molecular, as opposed to an atomic species. The emission results from electronic transitions between molecular states and can be used to detect and/or measure trace amounts of molecular species.

Molecular Mass

The sum of the atomic mass of all atoms in a molecule. In *high polymers*, because the molecular masses of individual molecules vary widely, they must be expressed as averages. The average molecular mass of polymers may be expressed as number-average molecular mass or mass-average molecular map.

Molecular Materials

Matter of nanoscale dimensions often displays properties unlike those of isolated gas-phase molecules or the bulk liquids or solids that they make up. These unusual properties are often attributable to the unique bonding, structure, and morphology that small systems of finite dimensions adopt, whereas in other cases they result because the small sizes bring about new phenomenon arising from what is termed *quantum confinement*, the restriction of space available to electrons. Because of the special properties that these cluster systems (aggregates of atoms or molecules) display, they have become an active subject of basic investigation in recent years. Often, impetus for such research has arisen from the promise that new advanced materials might be produced by assembling nanoscale systems.

Investigations into the reactions of small hydrocarbons with transition metal ions, atoms, and clusters led to the discovery of an unexpected dehydrogenation reaction leading to the assembly of a new molecular cluster family termed metallocarbohedrenes (met-cars).

Met-Car Composition and Structure

The surprising finding that a heretofore unknown compound composed of transition metal and carbon atoms was very stable, and displayed unique bonding far different than that of the well-known cubic structure of bulk titanium carbide, prompted a search for other clusters with similar compositions.

The first successful alternative method involved the laser vaporization of well-mixed powders of selected metals and graphite, followed by the entrainment of the formed product species in pulses of helium gas. Met-cars were readily produced with appropriate conditions of laser power and metal-to-carbon ratio mixtures. Thereafter, it was found that met-cars could be produced from a mixture of metal carbides along with carbon and metal powders by vaporizing them in a small crucible suspended in vacuum, without any need for a gas to transport the plasma or assist in its cooling.

An important advance toward exploring the potential use of these molecular clusters as new materials came from successful attempts to synthesize bulk quantities of met-cars. Synthesizing was first accomplished with an arc-discharge technique similar to ones that have been used to produce so-called buckyballs (fullerenes).

Future Status

Met-cars have been established as a new class of molecular cluster materials that are likely to have unusual properties and many potential applications. It has been observed that these molecular clusters readily ionize even at very low intensities of light of widely varying wavelengths, suggesting that they have a substantially delocalized electronic character and that they may display unique optical and perhaps electrical properties.

All the evidence points to their structure as one in which the metal and carbon atoms are bound in a well-defined lattice with one type of metal site. The formation of these interesting materials appears to be dominated by kinetic effects involving MC_2 building units. The clusters are found to grow by a unique mechanism leading to the development of a multicaged extended network structure.

Work is actively being pursued to isolate large amounts of bulk material and study the exact geometry and properties of met-cars using conventional chemical and physical techniques. Of particular interest is elucidating their optical, electronic, and reactive properties, through the use of molecular beams, lasers, and flow reactors.

Molecular Seal

A seal that is basically of the windback type, but that is used for sealing vapors or gases.

Molecular Spectrum

The spectrum of electromagnetic radiation emitted or absorbed by a collection of molecules as a function of frequency, wave number, or some related quantity.

Molecular Structure

The manner in which electrons and nuclei interact to form a molecule, as elucidated by quantum mechanics and the study of molecular spectra.

Molecular Weight

(1) The sum of the atomic weights of all the atoms in a molecule. Atomic weights (and therefore molecular weights) are relative weights arbitrarily referred to an assigned atomic weight of exactly 12.0000 for the most abundant isotope of carbon, ^{12}C. See also *atomic weight*. (2) The weight of the smallest quantity of a substance that has all the physical characteristics of the substance, referred to the weight of a molecule of hydrogen defined as 2 or a molecule of oxygen defined as 32 or a molecule of carbon defined as 12.

Molecule

(1) A molecule may be thought of either as a structure built of atoms bound together by chemical forces or as a structure in which two or more positively charged nuclei are maintained in some definite geometrical configuration by attractive forces from the surrounding cloud of electrons. Besides chemically stable molecules, short-lived molecular fragments termed free radicals can be observed under special circumstances. See also *chemical bonding*, *free radical*, and *molecular structure*. (2) Two or more atoms bonded together by sharing electrons to form the smallest quantity of a substance that has all the physical characteristics of the substance. The two or more atoms may be the same element, for example, two atoms of hydrogen will normally join together to form a hydrogen molecule, H_2. Alternatively, the atoms may be different, for example, two atoms of hydrogen will join with one atom of oxygen to form water, H_2O. See also *inter-atomic bonding*.

Molten Metal Flame Spraying

A thermal spraying process variation in which the metallic material to be sprayed is in the molten condition. See also *flame spraying*.

Molten Metal Penetration

Penetration, usually at grain boundaries, of molten metal into a solid metal, usually under stress, either applied or residual. Such penetration is usually a plane of weakness, and cracks may form immediately or later in service. There are many examples including lead, solder, copper, brass (which is sufficiently common to have its own term brazing penetration) in sync penetration into steel (it has been suggested that austenitic steel is particularly sensitive to zinc), and of the lower melting point materials into copper and brass. It is a particular form of liquid-metal attack.

Molten Weld Pool

The liquid state of a weld prior to solidification as weld metal.

Molybdenum and Alloys

Molybdenum (Mo) is a silvery-white metal, occurring chiefly in the mineral molybdenite but also obtained as a by-product from copper ores. The metal has a specific gravity of 10.2 and a melting point of 2610°C. It is ductile, softer than tungsten, and is readily worked or drawn into very fine wire. It cannot be hardened by heat treatment, but only by working.

Its major use is in alloy steels, for example, as tool steels ($\leq$10% molybdenum), stainless steel, and armor plate. Up to 3% molybdenum is added to cast iron to increase strength. Up to 30% molybdenum may be added to iron-, cobalt-, and nickel-base alloys designed for severe heat- and corrosion-resistant applications. It may be used in filaments for light bulbs, and it has many applications in electronic circuitry.

Molybdenum forms mirrors and films on glass when it is produced by gas-phase reduction or decomposition of volatile molybdenum compounds in glass tubes. Also, because the molybdenum metal surface binds strongly to oxygen without weakening the metal structure, and because the metal has such a low coefficient of thermal expansion—virtually equal to hard glass—molybdenum metal is an extraordinarily good material for use in glass-metal seals. Molybdenum trioxide (MoO_3) dissolves in glass, allowing strong binding of molten glass with preoxidized metal surfaces. Annealing is very effective, with little or no difference in thermal expansion at the metal–glass interface. Molybdenum found early use in filaments for electric light bulbs and later in the construction of electronic devices (e.g., in vacuum tubes, contacts, electrodes, and transistors).

Protection from Oxidation

At temperatures over about 538°C, unprotected molybdenum oxidizes so rapidly in air or oxidizing atmospheres that its continued use under these conditions is impractical. Uncoated molybdenum is, however, used satisfactorily where very short lives are involved (as in some missile parts) or where the surrounding atmosphere is non-oxidizing (as in hydrogen and vacuum furnaces). Protective coatings seem to be the answer where oxidation is a problem. Various coatings differing in maximum time–temperature capabilities and in physical and mechanical characteristics are available. Selection of the proper coating for a specific application involves consideration of a number of factors, foremost among which is the service temperature. For temperatures up to 1204°C, nickel-base alloys applied as cladding or sprayed coatings, and chromium–nickel electroplates appear most generally suitable. For temperatures up to 1538°C, or short periods at higher temperatures, modified chromized coatings and sprayed aluminum–chromium–silicon are predominant. For longer periods at higher temperatures, the choice would probably rest between siliconizing and ceramic coatings. Component tests have been found most reliable for final selection of suitable coatings.

Applications

The applications of molybdenum are generally those based on its high melting point, high modulus of elasticity, high strength at elevated temperatures, high thermal conductivity, high resistance to corrosion, low specific heat, and low coefficient of expansion. Long-established uses are found in the electric and electronic industries for applications such as mandrels and supports in lamp fabrication, anodes and grids in electronic tubes, resistance elements and radiation shields in high-temperature furnaces, and electrical contacts and electrodes. In recent years, molybdenum has become increasingly important in the missile field for nozzles, nozzle inserts, leading edges of control surfaces, and support vanes. Uses in the metalworking industry include boring bars, grinding quills, resistance-welding electrodes, and thermocouples. The nuclear-energy industry has been active in developing molybdenum heat exchangers, piping, heat shields, and structural parts. The glass industry is a major user of large molybdenum parts, especially for melting electrodes and stirrers. The chemical and petrochemical industry is beginning to use molybdenum for corrosion resistance. The usefulness of molybdenum as a metallized coating, either to improve bonding between the base metal and a sprayed top coating, or alone for improved wear resistance, is no longer confined to maintenance but is extending to original equipment.

Alloys

Four main classes of commercial molybdenum-base alloys exist. One class relies on the formation of fine metal carbides that strengthen the material by dispersion hardening and extend the resistance of the microstructure to recrystallization above that of pure molybdenum.

The most common of the carbide-strengthened alloys is known as TZM, containing about 0.5% titanium, 0.08% zirconium, and 0.03% carbon. Other alloys in this class include TZC (1.2% titanium, 0.3% zirconium, 0.1% carbon), MHC (1.2% hafnium, 0.05% carbon), and ZHM (1.2% hafnium, 0.4% zirconium, 0.12% carbon). The high-temperature strength imparted by these alloys is their main reason for existence.

Both TZM and MHC have found application as metalworking tool materials. Their high-temperature strength and high thermal conductivity make them quite resistant to the collapse and thermal cracking that are common failure mechanisms for tooling materials. A particularly demanding application is the isothermal forging process used to manufacture nickel-base superalloy gas-turbine engine components. In this process, the dies and workpiece are both heated to the hot working temperature, and the forging is performed in a vacuum by using large hydraulic presses.

A second class relies on solid-solution hardening to strengthen molybdenum. These two classes of materials are typically produced in both vacuum-arc-casting and powder metallurgy grades. In the solid-solution class, tungsten and rhenium are the two primary alloying additions. The most common compositions are 30% tungsten (Mo-30% W), 5% rhenium (Mo-5% Re), 41% rhenium (Mo-41% Re), and 47.5% rhenium (Mo-50% Re). With the exception of the Mo-30% W alloy, which is available as a vacuum-arc-cast product, these alloys are normally produced by powder metallurgy. The tungsten-containing alloys find application as components in systems handling molten zinc because of their resistance to this medium. They were developed as a lower-cost, lighter-weight alternative to pure tungsten and have served these applications well over the years. The 5% rhenium alloy is used primarily as thermocouple wire, whereas the 41% and 47.5% alloys are used in structural aerospace applications.

The third class uses combinations of carbide formers and solution partners to provide improved high-temperature strength. This class of alloys is normally produced by powder metallurgy techniques, but some of the alloys are also amenable to vacuum-arc-casting processing.

The beneficial effects of solid-solution hardening and dispersion hardening found in the carbide-strengthened alloys have been combined in the HWM-25 alloy (25% tungsten, 1% hafnium, 0.07% carbon). This alloy offers high-temperature strength greater than that of carbide-strengthened molybdenum, but it has not found wide commercial application because of the added cost of tungsten and the expense of processing the material.

The final class of alloys, known as dispersion-strengthened alloys, relies on second-phase particles (usually an oxide of a ceramic material) introduced or produced during powder processing to provide resistance to recrystallization and to stabilize the recrystallized grain structure, and enhancing high-temperature strength and improving low-temperature ductility. These latter materials by their very nature must be produced by powder metallurgy techniques.

The dispersion-strengthened alloys rely exclusively on powder metallurgy manufacturing techniques. This allows the production of fine stable dispersions of second phases that stabilize the wrought structure against recrystallization, resulting in a material having improved high-temperature creep strength as compared to pure molybdenum. Once recrystallization occurs, the dispersoids also stabilize the interlocked recrystallized grain structure. This latter effect produces significant improvements in the ductility of the recrystallized material.

The potassium- and silicon-doped alloys such as MH (150 ppm potassium, 300 ppm silicon) and KW (200 ppm potassium, 300 ppm silicon, 100 ppm aluminum) are the oldest of this category; they are analogous to the doped tungsten alloys in common use for tungsten lamp filaments. They were first developed to satisfy the requirement of the lighting industry for creep-resistant molybdenum components. They do not possess particularly high strength at low temperatures, but they are quite resistant to recrystallization and they possess excellent creep resistance due to their stable, interlocked recrystallized grain structure. They are used in applications requiring high-temperature creep resistance such as nuclear fuel sintering boats.

Composites

A family of copper–molybdenum–copper (CMC) laminates has been developed to serve the needs of the high-performance electronics industry. Laminating molybdenum with copper raises the coefficient of thermal expansion and improves the heat-transfer characteristics of the composite relative to pure molybdenum. This allows better matching between the composite that serves as a packaging or mounting material and the solid-state devices being employed.

Molybdenum Disilicide

The commercial product contains approximately 60% molybdenum, 31% silicon, 8% iron, with small amounts of carbon, sulfur, vanadium, phosphorus, and copper. Its principal use is for alloying iron and steel, but it has ceramic uses as well. Its outstanding resistance to oxidation in air at temperatures up to 1700°C makes it potentially useful for high-temperature applications.

The excellent resistance to oxidation at high temperatures combined with fairly good elevated temperature strength makes $MoSi_2$ promising for elevated temperature structural applications where impact resistance is not a major consideration. Possible uses are gas turbine nozzle blades and furnace heating elements. Also, a metal–ceramic combination of $MoSi_2$ and Al_2O_3 has been considered for use as kiln furniture, saggers, sand blast nozzles, hot-draw or hot-pressed dies, and induction braising fixtures.

Molybdenum Steel

Next to carbon, molybdenum is the most effective hardening element for steel. It also has the property, like tungsten, of giving steel the quality of red hardness, requiring a smaller amount for the same effect. It is also used in hot-work steels, and to replace part of the tungsten in high-speed steels. It is added to heat-resistant irons and steels to make them resistant to deformation at high temperatures and to creep at moderate temperatures. It increases the corrosion resistance of stainless steels at high temperatures. Molybdenum in small amounts also increases the elastic limit of steel, reduces grain size, strengthens crystalline structure, and gives deep hardening. It goes into solid solution, but when other elements are present, it may form carbides and harden the steel, giving greater wear resistance. It also widens the heat-treating range of tool steels. As it decreases the temper brittleness of aluminum steels, small amounts are added to nitriding steels.

Plain carbon–molybdenum steels are easier to machine than other steels of equal hardness. Molybdenum structural steels usually have from 0.20% to 0.75% molybdenum. SAE (formally Society of Automotive Engineers) steels 4130 and 4140 contain about 1.0% chromium and 0.20% molybdenum and are high-strength forging steels for such uses as connecting rods. Hollow-head screws are made of SAE 4150 steel, and SAE steels 4615 and 4650 have no chromium but contain about 1.75% nickel and 0.25% molybdenum. SAE 4615 is used for molds and dies to be hobbed. It is easily worked and has a tensile strength of 1482 MPa. Finally, the old Damascus steel and Toledo steel were molybdenum steels; the molybdenum was in the original ore.

Moly–Manganese Process

A common method of joining oxide ceramics (most commonly of alumina) whereby a mixture of molybdenum and manganese powders (typically 10 at% Mn) is applied on the ceramic surface, and then sintered to bond to the ceramic. The sintering temperature is usually above 1400°C (2500°F) in a hydrogen atmosphere with a controlled dewpoint. Subsequent nickel plating in conjunction with this Mo–Mn metallized layer facilitates the application and adherence of brazing filler metals that would not bond to the original ceramic substrate.

Moment

The turning action of a force acting about a point. It is the product of the force and the perpendicular distance from the line of action of the force to the point.

Mond Process

A process for extracting and purifying nickel. The main features consists of forming nickel carbonyl by reaction of finely divided reduced metal with carbon monoxide, then decomposing the nickel carbonyl to deposit purified nickel on small nickel pellets.

Monel

Alloys of nickel containing about 30% copper with small quantities of iron, manganese, and aluminum. Various alloys generally offer good mechanical properties and good corrosion resistance.

Monobloc

Manufactured in one piece as opposed to comprising a number of sub-members joined by bolting, welding, etc.

Monochromatic

Consisting of electromagnetic radiation having a single wavelength or an extremely small range of wavelengths, or particles having a single energy or an extremely small range of energies.

Monochromatic Objective

An *objective* in a microscope, usually a fused quartz, that has been corrected for use with monochromatic light only.

Monochromator

A device for isolating radiation of one or nearly one wavelength from a beam of many wavelengths.

Monoclinic

A crystal structure having three axes of any length, with two included angles equal to 90° and one included angle not equal to 90°. See also *unit cell*.

Monocoque

A form of construction in which all of the principal components carry their share of the applied load. For example, the bodywork of a monocoque vehicle is loadbearing rather than merely being carried by a substantial chassis.

Monofilament

A single fiber or filament of indefinite length that is strong enough to function as a yarn in commercial textile operation.

M

Monolayer

(1) The basic laminate unit from which cross-plied or other laminate types are constructed. (2) A "single" layer of atoms or molecules adsorbed on or applied to a surface.

Monolithic

An object comprised entirely of one massive piece (although polycrystalline or even heterogeneous) as opposed to being built up of preformed units.

Monolithic Lining/Coating

A lining or coating in one piece formed by techniques such as pouring, spraying, or plastering rather than being built up from bricks or slabs. The terms are particularly applied to refractory applications.

Monolithic Refractory

A refractory that may be installed *in situ* without joints to form an integral structure.

Monomer

A single molecule that can react with like or unlike molecules to form a polymer. The smallest repeating structure of a polymer (mer). For addition polymers, this represents the original unpolymerized compound.

Monotectic

An isothermal reversible reaction in a binary system, in which a liquid on cooling decomposes into a second liquid of a different composition and a solid. It differs from a *eutectic* in that only one of the two products of the reaction is below its freezing range.

Monotropism

The ability of a solid to exist in two or more forms (crystal structures), but in which one form is the stable modification at all temperatures and pressures. *Ferrite* and *martensite* are a monotropic pair below the temperature at which *austenite* begins to form, for example, in steels. Alternate spelling is monotrophism.

Monte Carlo Technique

Calculation of the trajectory of incident electrons within a given matrix and the pathway of the x-rays generated during interaction.

Mordant

A substance used in dyeing for fixing the color (the word means *bitter*). A mordant must have an affinity for the material being dyed and be chemically reactive; that is, the molecule must have free electrons that combine with the dyestuff. The vegetable fibers, such as cotton and linen, frequently require mordants. Basic aniline dyes require a mordant for cotton or rayon, and the water-soluble acid dyes need a mordant for vegetable fibers. The mordant may be applied first, usually in a hot solution or simultaneously with the dye. *Mordant dyes* are dyes chemically able to accept and hold a metal atom, the metal being added in the mordant. The metal atom is held to the oxygen atoms and to one of the nitrogen atoms in each chromophoric group of the die, forming highly insoluble materials known as *lakes*.

Besides fixing the color, mordants sometimes increased the brilliance of the dye, particularly when chelation occurs. Common mordants are *alum* and *sodium bichromate*, but salts of aluminum and other metals are used. *Sodium stannate*, used both as a mordant and for fireproofing textiles, is a great-white, water-soluble powder made by treating tin oxide with caustic soda. *Chromium acetate*, used as a mordant for chrome colors and to produce khaki shades with iron solutions, is a grayish-green powder. It is soluble in water. In gilding, the term *mordant* is used to mean a viscous or sticky substance employed to make the gold leaf adhere, but such a material does not have the chemical action of a mordant.

Morphology

The characteristic shape, form, or surface texture or contours of the crystals, grains, or particles of (or in) a material, generally on a microscopic scale.

Mortar

A plastic mixture of cementitious materials, fine aggregates, and water in ratios of about 1:3:0.5 by mass.

Mosaic Crystal

An imperfect single crystal composed of regions that are slightly disoriented relative to each other.

Mosaic Structure

In crystals, a substructure in which adjoining regions have only slightly different orientations.

Mössbauer Effect

The process in which γ-radiation is emitted or absorbed by nuclei in solid matter without imparting recoil energy to the nucleus and without Doppler broadening of the γ-ray energy.

Mössbauer Spectroscopy

An analytical technique that measures recoilless absorption of γ-rays that have been emitted from a radioactive source as a function of the relative velocity between the absorber and the source.

Mössbauer Spectrum

A plot of the relative absorption of γ-rays versus the relative velocity between an absorber and a source of γ-rays.

Mother of Pearl

The hard, brilliant-colored internal layer of the *pearl oystershell* and of certain other marine shellfish. It is employed for knife handles, buttons, inlay work, and other articles. The iridescent appearance of the pearl oyster is due to the structure of the nacre coating, and the shells are also called *nacre*. Mother of pearl is brittle but can be worked with steel saws and drills using a weak acid lubricant. The shells are thick and heavy, and disks cut from them are split for buttons. A beautiful pink nacre occurs on the inner surface of the conch shell; a sea snail that grows in great abundance off the Caicos Island near Haiti. There are about 56 species, and they grow up to 12 in. (30 cm) long and 8 in. (20 cm) wide. The shells were formally much used for making cameos. *Mussel shell*, from freshwater mussels of the Mississippi River, is also called *pearl shell*, but it does not have the iridescence of mother of pearl. There is a large production in Iowa, used for buttons. The waste shell from the manufacturer is crushed and marketed for poultry feed. *Pearl essence*, used for making *imitation pearls* and in plastics and lacquers, is a motley-silver compound extracted from the scales of fish, notably the sardines, herring, and alewife. Only a few types of fish have iridescent scales. The chief constituent is guanine, a chemical related to caffeine. *Synthetic pearl essence* may be ground nacreous shells in a liquid vehicle, or it may be produced chemically. Compounds of basic lead carbonate and lead mono-hydrogen phosphate have multiple reflectivity and give an optical effect resembling that of mother of pearl. *H-scale*, of Hoechst Celanese, and *Ko-Pearl*, of all Ultra Ray Pearl Essence Corp., are synthetic pearl essences.

Mottled Cast Iron

Iron that consists of a mixture of variable proportions of gray cast-iron and white cast iron; such a material has a mottled fracture appearance. It also has graphite flakes in some parts, but not in others.

Mounting

A means by which a specimen for metallographic examination may be held during preparation of a section surface. The specimen can be embedded in plastic or secured mechanically in clamps.

Mounting Artifact

A false structure introduced during the mounting stages of a metallographic surface-preparation sequence.

Mounting Resin

Thermosetting (e.g., Bakelite or diallyl phthalate) or thermoplastic (e.g., methyl methacrylate or polyvinyl chloride) resins used to mount metallographic specimens. Standard plastic mounts usually measure 25 mm (1 in.), 32 mm (1.25 in.), or 38 mm (1.5 in.) in diameter; mount thickness is approximately 1/2 the mount diameter.

M Shell

The third layer of electrons surrounding the nucleus of an atom, having electrons characterized by the principal quantum number 3.

MPa

MegaPascal. See *Pascal*.

MPI

Magnetic particle inspection.

M_s Temperature

For any alloy system, the temperature at which martensite starts to form one cooling. See *transformation temperature* for the definition applicable to ferrous alloys.

Mucilage

A sticky paste obtained from linseed and other seeds by precipitation from a hot infusion and used as a light adhesive for paper and as a thickening agent. It contains arabinose, glucose, and galactose, and it is easily soluble in water. Mucilage as a general name also includes *water-soluble gums* from various parts of many plants and has the same uses. It is the stored reserve food of plants. There are two types: the cell-content mucilage, which acts as a disorganization product of some of the carbohydrates, and membrane mucilage, which acts as a thickening agent to the cell wall. Membrane mucilage occurs in the acacias, algae (seaweeds), linum (flax plants), ulmus (elms), and astragalus. When it is collected in the form of exudations from the trees, it is called a gum. *Cherry gum* is this type of water-soluble gum, as is *medlar gum*, from a small tree of the same family as the cherry, grown in Europe and the Near East. See also *adhesive, glue, paste,* and *sizing*.

Muck Rolled Iron

See Puddling process.

Muffle Furnace

A common type of *sintering* furnace for continuous or discontinuous operation. The muffle may be ceramic to support electric heating elements or resistor wire windings, or may be a gas-tight metallic retort to retain the furnace atmosphere and support the work trays; both kinds may be used in the same furnace.

Mulling

The mixing and kneading of foundry molding sand with moisture and clay to develop suitable properties for molding.

Mullite

Artificial mullite, or synthetic mullite, a ceramic material made by a prolonged fusing in the electric furnace of a mixture of silica sand and diasphoric clay and bauxite, has the composition $3Al_2O_3 \cdot 2SiO_2$, has a melting point of 1810°C, and softens at 1650°C.

Mullite occurs in nearly all ceramic products containing alumina and silica but, with the exception of refractories, is seldom introduced as such except as calcined kyanite.

The bricks are resistant to flame and to molten ash, and they have a low, uniform coefficient of thermal expansion and a heat conductivity only slightly above that of fireclays. Normally, mullite has very fine crystals that change form and become enlarged after prolonged heating, making the product porous and permeable. For stable high-temperature refractories, the mullite is prefused to produce larger crystals. At very high temperatures, mullite tends to decompose to form corundum and alkali-silicate minerals of lower heat resistance. Mullite is also used for making spark plugs, chemical crucibles, and extruding dies, and a foamed mullite is used as a uniformly latticed honeycomb structure for lightweight, heat-resistant structural parts.

Multiaxial Stresses

(1) Any stress state in which two or three principal stresses are not zero.
(2) Any state in which more than one principal stresses are applied.

Multifilament Mesh

Woven material with multiple-strand threads.

Multifilament Yarn

A large number (500–2000) of fine, continuous filaments (consisting of 5–100 individual filaments) usually with some twist in the yarn to facilitate handling.

Multigrade Oil

An oil having relatively little change in viscosity over a specified temperature range.

Multiple

A piece of stock for forging that is cut from bar or billet lengths to provide the exact amount of material for a single workpiece.

Multiple Die Pressing

The simultaneous compaction of powder into several identical parts with a press tool consisting of a number of components.

Multiple Etching

Sequential etching of a microsection, with specific reagents attacking distinct microconstituents.

Multiple Punch Press

A mechanical or hydraulic press that actuates several punches individually and independently of each other.

Multiple Spot Welding

Spot welding in which several spots are made during one complete cycle of the welding machine.

Multiple-Impulse Welding

Spot, projection, or upset welding with more than one impulse of current during a single machine cycle. Sometimes called *pulsation welding*.

Multiple-Layer Adhesive

A dry-film adhesive, usually supported, with a different adhesive composition on each side; designed to bond dissimilar materials such as the core two-faced bond of a sandwich composite. See also *honeycomb* and *sandwich construction*.

Multiple-Pass Weld

A weld made by depositing filler metal with two or more successive passes.

Multiple-Screw Extruder

An extruder machine for processing thermoplastics that has two or four screws (conical or constant depth), in contrast to conventional single-screw extruders. Types include machines with intermeshing counter-rotating screws, intermeshing corotating screws, and nonintermeshing counter-rotating screws.

Multiple-Slide Press

A press with individual slides, built into the main slide or connected to individual eccentrics on the main shaft, that can be adjusted to vary the length of stroke and the timing. See also *slide*.

Multiplier Phototube

See *photomultiplier tube.*

Multiport Nozzle (Plasma Arc Welding and Cutting)

A constricting nozzle containing two or more orifices located in a configuration to achieve a degree of control over the arc shape.

Muntz Metal

Montz metal is a yellow brass containing—60% copper and 40% zinc; invented in 1832, it is also called yellow metal and malleable brass. Muntz metal is also modified with small amounts of iron and manganese. Iron above 0.35% forms a separate iron-rich constituent that is stable and gives hardness and high strength to the alloys. The addition of manganese helps to absorb the iron and also hardens the alloy. These alloys were also called high-strength brass and were employed for such uses as hydraulic cylinders and marine forgings, but are now largely replaced by manganese bronze or aluminum bronze.

Current standard wrought alloys are C28000 (59%–63% copper, as much as 0.30% lead, 0.07% iron, and the balance zinc); four leaded alloys—C36500, C36600, C36700, and C36800—each containing 58%–61% copper, 0.25%–0.70% lead, and as much as 0.15% iron and 0.25% tin, with the balance zinc except for 0.02%–0.06% arsenic in C36600, 0.02%–0.10% antimony in C36700, and 0.02%–0.10% phosphorus in C36800.

The alloys are available in most wrought forms, are hardenable only by cold working, and are used for forging parts, heat-exchanger tubing and tube sheets, baffles, valve stems, and fasteners. The corrosion resistance of C28000 is said to be generally similar to that of copper and better than that of higher-copper alloys to sulfur-bearing compounds.

Mushets Steel

The first (probably) steel developed for cutting at high temperature, that is, a high-speed steel; it was manganese deoxidized and probably had 1%–2% manganese with about 5%–10% tungsten and 2% carbon.

Music Wire

A high-grade, uniform steel originally intended for strings in musical instruments, but now employed for the manufacture of spiral springs. It is the highest grade of *spring wire* and is free from slag or dirt, and low in sulfur and phosphorus. For springs, it is usually *steel 1085* reduced about 80% in 8 or 10 drawing passes, but piano wire may have as many as 40 draws from a No. 7 rod. The tensile strength of spring wire, when hard drawn, is 225,000–400,000 lb/in.^{2} (1551–2758 MPa), but it should be formable enough to bend 180° flat without cracking, or wind into a close helix with inside diameter 1–1.5 times the diameter of the wire. The smaller size of music wire is 0.003 in. (0.008 cm) in diameter. *Piano wire*, intended for piano strings but much used for springs, is a cold-drawing, high-quality wire formerly only drawn from Swedish billets or rods but now made from U.S. steels.

Other spring steels and alloys include *austenitic stainless steels*, such as 302 and A286, and *iron-base superalloy*, containing about 25% nickel, 15% chromium, 2% titanium, 1.5% manganese, 1.25% molybdenum, 0.75% silicon, 0.3% vanadium, 0.08% maximum carbon, and not over 0.01% boron, and are used for springs in aircraft turbine engines. The tensile strength of the wire alloy reduced 30% is 1365 MPa at 732°C. *Alloy 355*, for highly stressed springs and jet engines, has a nominal composition of 15.65% chromium, 4.38% nickel, 2.68% molybdenum, 1% manganese, 0.32% silicon, 0.12% copper, 0.124% nitrogen, and 0.142% carbon. See *piano wire*.

m-Value

See *strain-rate sensitivity*.

N

"N"-Type Semiconductor

See *semi* conductor.

N.A.M.A.S.

National Measurement Accreditation Service. The system for calibrating and certificating, measuring, and test equipment in the United Kingdom.

NaK

Sodium (Na) and potassium (K) alloy that is liquid at ambient temperature and so is used as a circulating coolant in some nuclear reactors.

Nanohardness Test

An indentation hardness testing procedure, usually relying on indentation force versus tip displacement data, to make assessments of the resistance of surfaces to penetrations of the order of 10–1000 nm deep. The prefix "nano-" normally would imply hardnesses 1000 times smaller than "microhardness"; however, the use of this prefix was primarily designed as a means to distinguish this technique from the more traditional microindentation hardness procedures. Most nanohardness testing procedures used three-sided pyramidal diamond indenters.

Nanophase Materials

Emerging materials, often multilayered in composite form, produced in building blocks so small as to be measured in nanos, or 10^{-9} units, and having superior physical or mechanical properties compared with your conventional, or "bulk," counterparts. They are made by controlling the arrangement of matter on an atomic or molecular scale so as to create materials comprising particles or grains 39–3900 nin. (1–100 nm) wide or of layers or filaments of this thickness. Near-perfect materials, with critical flaw sizes less than the layer thickness, can be made, thereby creating materials approaching theoretical strengths. An early use, an extension of thin-film deposition, was in mirrors for soft x-rays.

Nanophase materials include metals, plastics, elastomers, carbon, and ceramics. The powder can be used in several ways: as a dispersion in fluid media, on supporting substrates, for consolidation into parts and forms, as a sintering aid, and as a feedstock for films and coatings. Among the applications are polishing slurries with nanoparticles of *alumina, silica,* and *cerium oxide*; titanium or *zinc oxide* particles in sunscreens for greater protection from ultraviolet rays; Fe_2O_3 pigments in cosmetics and silver or copper particles in *conductive polymers* and *inks*; *titanium oxide* and *tin oxide* films for gas sensors; nanostructured *tungsten carbide cobalt* tool bits for cutting and oil exploration; doped oxides for low-voltage displays; *silicon carbide* and *silicon nitride* engine parts; magnetic devices; and alumina and aluminum nitride infrared window applications. Sintering temperatures with nanopowders are typically several hundred degrees less than with conventional powders. For electrically conductive polymers, the use of *silver* nanopowder rather than flake fillers reduces metal embrittlement and accommodates larger differences in thermal expansion.

As grain size gets smaller, metals tend to get harder, stronger, and tougher. *Danaloy DSC copper*, a *nanophase copper* from Ultram International, is more than twice as hard as pure, annealed, conventional copper and is about 90% as conductive. Nanophase *superplastic aluminum alloys* and *superplastic nickel–aluminum alloys* exhibit superplasticity at lower temperatures than the conventional superplastic alloys. *Nanograin tungsten carbide* is harder and more abrasion and crack resistant than micrograin material, can provide sharper cutting tools, and can be coated with cobalt or nickel. Nanodyne, Inc. uses a solution-spray conversion process to make tungsten carbide–cobalt powder, which can be used for hard facing by thermal spraying or for cutting tools. The process also can be used to produce *cermets,* such as *chromium carbide nickel* and *iron titanium carbide,* as well as *oxide-dispersion-strengthened (ODS) ceramics.*

Nanomers, from Nancor, Inc., are *montmorillonite* minerals surface treated with compatibilizing agents for dispersion in exfoliated form in nanoscale size in a resin matrix. In this form, the nanomer particles have a flexural sheet-like structure about 394 nin. (1 nm) thick and a length and width of 59 μin. (1.5 μm) and less. One grain has more than 1 million particles. *Nanomer 1.30 TC,* intended for use with *nylon 6* and designed for extrusion compounding, doubles heat-deflection temperature (HDT), especially of low-viscosity grades, and reduces gas permeability while increasing flame retardance. Nanoscale particles can also selectively hydrolyze polymers to form microcrystal polymers. Micro Poly Corp. has focused microcrystal *amylose, nylon,* and *polyester* on food and pharmaceutical products. The polyesters also aimed at replacing urea formaldehyde resins in making plywood and particle board. Other potential products are *dry lubricants, chelants, inks, dyes, conductive carbon* for electrodes, and *activated carbon,* the latter having 10 times the adsorptive surface of ordinary carbon. *Nanosilicate-coated high-density polyethylene* (HDPE) has greater barrier resistance to gasoline than HDPE alone. *Nanophase-clay* platelets in polymer-silicate-layered nanocomposites promise stronger polyester, polypropylene, and fluoropolymer packaging films of greater barrier resistance and flame retardance. *Silicon elastomer nanocomposites* containing 5% *fluorohectorite clay* are stronger and tougher than traditional silicone elastomers.

Carbon nanophase materials, also called *carbon nanotubes,* consist of a graphite sheet that has been rolled into a tube and having *fullerene* end caps. They contain millions of carbon atoms, each in an assigned place, and are only a few atoms in circumference. Their tensile strength is said to be at least 10 times that of traditional carbon fibers, making them the strongest of material in this respect. They are light in weight, quite rigid, and flexible and resilient normal to the tube surface or length, thus promising for reinforcements of *polymer–matrix composites* in aircraft and aerospace applications. *Carbon nanofibers* as small as 394 nin. (100 nm) diameter have been made from polyacrylonitrile as well as mesophase pitch by electrospinning.

Because of their fine size, they can provide a greater surface-area-to-mass ratio than the currently used 394 μin. (10 μm) fibers. *Pyrograf-III carbon*, from Pyrograf Products, is a vapor-grown *carbon nanofiber* made under license from General Motors. Because of their fineness, nanofiber-reinforced polymers can provide a Class A autobody panel finish, unlike the current carbon fibers that yield rougher finishes. Also, being electrically conductive, such composites can be electrostatically painted to such a finish.

Lengthwise, carbon nanotubes are about as electrically conductive as copper, though much less so transverse to length. At Samsung Advanced Institute of Technology (South Korea), they have been made into cathodes for field-emission color displays that require only half the power of *liquid crystal* displays. The nanotubes are as thermally conductive as diamond and behavior as metals or semiconductors. Changing their environment from oxygen to vacuum, changes molded from positive to negative, and the reaction is reversible.

Thus, the nanotubes can serve as diodes and transistors as well as oxygen sensors. Other potential uses are high-temperature catalyst supports, thermal-management devices, and electrostatic dissipation of various products. Hyperion Catalysis International offers resin masterbatches of *nylon, polybutylene terephthalate, polyethylene terephthalate, polycarbonate, polystyrene,* and *polyetheretherketone* containing 15%–20% "graphite fibrils," or multiwall carbon nanotubes. Target applications include nylon auto fuel lines, computer disk drive components and chip trays, and handheld industrial bar scanners. General Electric Plastics' nanotube-filled *Noryl GTX*, a *phenylene oxide–nylon alloy*, has been used for painted automirror housings and fuel filler caps and is a candidate for electrostatically painted fenders and other autopanels.

Many potential applications are also emerging for *nanoceramics*. Use of nanosize ceramic particles reduces sintering temperature and shrinkage during sintering and subsequent brittleness while increasing formability and machinability. Reaction-sintered formulations of equal amounts of very pure natural *dolomite* and synthesized *zirconia* doped with 0.5% by weight *lithium fluoride* are candidates for high-temperature filters. *Gallium oxide* nanoceramic sensors are intended for measuring air–fuel ratio in exhaust gas flow of advanced turbine engines. *Sialon* silicon nitride and alumina ceramic, modified with calcium, could be used in nanophase form for catalyst carriers and microfilters. Nanosize FeS_2 powder, formulated with eutectic salt and silicon oxide, is a candidate for thermal battery cathode disks. Nanosize *titanium dioxide* could be used in cosmetics, photochromic coatings, fibers, and, for better oxygen permeability, contact lenses.

At BIRL, the industrial research laboratory of Northwestern University, magnetron vacuum reactive sputtering has led to nanophase *composite coatings* twice as hard as either constituent material. Within alternating layers of columbium nitride titanium nitride and vanadium nitride titanium nitride, Vickers hardness values half that of diamond are achieved, portending potential applications for cutting tools and wear-resistant parts. Pratt and Whitney are exploring various metal/metal, metal/ceramic, and ceramic/ceramic coatings for greater heat, wear, or erosion resistance than conventional coatings can provide. In these combinations, ductile layers could mitigate cracking of hard and wear-resistant but brittle layers, or the various layers could be graded to better match the substrate's coefficient of thermal expansion. *Dylyn coatings*, from Advanced refractory coatings, are diamond-like carbon–silicon coatings of low friction coefficient, extreme hardness, and high thermal stability.

Nanostructure

This is a material structure assembled from a layer or cluster of atoms with size on the order of nanometers. Interest in the physics of condensed matter at size scales larger than that of atoms and smaller than that of bulk solids (mesoscopic physics) has grown rapidly since the 1970s, owing to the increasing realization that the properties of these mesoscopic atomic ensembles are different from those of conventional solids. As a consequence, interest in artificially assembling materials from nanometer-sized building blocks, whether layers or clusters of atoms, arose from discoveries that by controlling the sizes in the range of 1–100 nm and the assembly of such constituents, it was possible to begin to alter and prescribe the properties of the assembled nanostructures. (Many examples of naturally formed nanostructures can be found in biological systems, from seashells to the human body.)

Nanostructured materials are modulated over nanometer length scales and zero to three dimensions. They can be assembled with modulation dimensionalities of zero (atom clusters or filaments), one (multilayers), two (ultrafine-grained overlayers or coatings or buried layers), and three (nanophase materials), or with intermediate dimensionalities. Thus, nanocomposite materials containing multiple phases can range from the most conventional case in which a nanoscale phase is embedded in a phase of conventional sizes, to the case in which all the constituent phases are of nanoscale dimensions. All nanostructured materials share three features: atomic domains (grains, layers, or phases) spatially confined to less than 100 nm in at least one dimension, significant atom fractions associated with interfacial environments, and interaction between their constituent domains.

Multilayers and Clusters

Multilayer materials have had the longest history among the various artificially synthesized nanostructures, with applications to semiconductor devices, strained-layer superlattices, and magnetic multilayers. Recognizing the technological potential of multilayered quantum heterostructure semiconductor devices helped to drive the rapid advances in the electronics and computer industries. A variety of electronic and photonic devices could be engineered utilizing the low-dimensional quantum states in these multilayers for applications in high-speed field-effect transistors antiefficiency lasers, for example. Subsequently, a variety of nonlinear optoelectronic devices, such as lasers and light-emitting diodes, have been created by nanostructuring multilayers.

Examples

Examples of nanostructured materials that have been characterized include multilayers, individual and assembled atom clusters, and cluster-consolidated nanophase materials.

Magnetic Multilayers

Magnetic multilayers, such as those formed by alternating layers of ferromagnetic iron and chromium, can be nanostructured so that the electrical resistance is significantly decreased (by up to a factor of 2 depending on the chromium layer thickness) by the application of a magnetic field. Such an effect, called giant magnetoresistance, occurs when the magnetic moments of the neighboring alternating layers are arranged in an antiparallel fashion so that application of the magnetic field overcomes the antiferromagnetic coupling and aligns the layers into a condition of parallel ferromagnetic ordering, strongly reducing the electron scattering in the system. Magnetoresistive materials have been introduced in the magnetic recording industry as read heads because of their lower noise and improved signal-handling capabilities. Other nanostructured materials besides multilayers also exhibit giant magnetoresistance, such as magnetic cobalt clusters embedded in a nonmagnetic matrix of copper or silver, or magnetic plate-like nickel–iron deposits embedded in silver.

Optical Properties of Cluster Assemblages

Noninteracting assemblages of small semiconductor clusters have optical properties of both scientific and technological importance. The optical absorption behavior of cadmium sulfide clusters with diameters in the nanometer size regime made by any of a variety of methods, including chemical precipitation in solutions or in zeolite supports, is different from that for bulk cadmium sulfide.

Chemical Reactivity

Chemical reactivity of nanostructured materials, with their potentially high surface areas compared to conventional materials, can also be significantly altered and enhanced. Since clusters can be assembled by means of a variety of methods, there can be an excellent degree of control over the total available surface area and the resulting self-supported ensembles. Thus, it is possible to maximize porosity to obtain very high surface areas, remove most of it via consolidation but retain some to facilitate low-temperature doping or other processing, or fully densify the nanophase material.

Nanophase Materials

The assembly of larger atom clusters into bulk nanophase materials can also have dramatic effects upon properties. In this case, the clusters interact fully with one another, yet the effects of cluster size are still very important. Clusters of metals or ceramics in the size range of 5–25 nm have been consolidated to form ultrafine-grained polycrystals that have mechanical properties remarkably different and improved relative to their conventional course-grained counterparts.

Nanotechnology

This involves systems for transforming matter, energy, and information, based on nanometer-scale components with precisely defined molecular features. The term "nanotechnology" has also been used more broadly to refer to techniques that produce or measure features less than 100 nm in size; this meaning embraces advanced microfabrication and metrology. Although complex systems with precise molecular features cannot be made with existing techniques, they can be designed and analyzed. Studies of nanotechnology in this sense remain theoretical but are intended to guide the development of practical technological systems.

Basic Principles

Nanotechnology based on molecular manufacturing requires a combination of familiar chemical and mechanical principles in unfamiliar applications. In conventional chemistry, molecules move by diffusion and encounter each other in all possible positions and orientations. The resulting chemical reactions are accordingly hard to direct. Molecular manufacturing, in contrast, can exploit mechanosynthesis, that is, using mechanical devices to guide the motions of reactive molecules. By applying the conventional mechanical principle of grasping and positioning to conventional chemical reactions, mechanosynthesis can provide an unconventional ability to cause molecular changes to occur at precise locations in a precise sequence. Reliable positioning is required in order for mechanosynthetic processes to construct objects with millions to billions of precisely arranged atoms.

Naphtha

A light, colorless to straw-colored liquid, which distills off from petroleum between 158°F and 194°F (70°C and 90°C). The specific gravity is from 0.631 to 0.660 or slightly higher, ranging from C_6H_{14} to C_7H_{18}. The lightest of the distillates used as solvents for fats, rubber, and resins approaches petroleum ether; the heaviest distillates approach benzin and gasoline and are used for fuel.

The name *naphtha* is also applied to heavier distillates from petroleum and to various grades of light oils obtained in the distillation of coal tar. They are widely used because of their low volatility and safety in handling. These include *solvent naphtha,* having a specific gravity of 0.862–0.872, with a boiling point below 320°F (160°C), and *heavy naphtha,* a dark liquid of specific gravity between 0.925 and 0.950 and boiling point between 320°F and 428°F (160°C and 220°C). *High-flash naphtha* for solvent purposes is a petroleum fraction with a specific gravity within the range of the gasolines. *VM&P,* or varnish maker's and painter's naphtha, is a specialty product that is available from Ashland Chemical, Inc. Various trade names are given to the light petroleum distillates used as solvents and in paints and harnesses such as Naphtholite.

At high temperatures, and in the absence of air, naphtha is cracked to ethylene, the most widely used commodity organic. In this application, naphtha is replacing ethane and propane as the choice of feedstock.

Naphthalene

Also called *tar camphor.* A white solid that is one of the heavy distillates from coal tar; it may also be obtained from petroleum. Naphthalene burns with a smoky flame and is soluble in benzene and in hot alcohol but not in water. Refined naphthalene comes in balls, flakes, and pellets, largely for use as an insect repellent, and most of the material is sold in technical grades for use in making dyestuffs, synthetic resins, coatings, tanning agents, and celluloid. The largest use is for production of phthalic anhydride. It is also the source of carbaryl, a *methyl carbamate* used extensively as a replacement for DDT. Union Carbide Corp.'s *Sevin* is such a product. *Naphthalene crystals* are very transparent to fluorescent radiation, and light produced in thick layers can escape and reach a photo surface. Low pulses can be obtained from the absorption of beta and gamma rays in naphthalene. It is thus used in photomultiplier tubes as a gamma-ray detector.

Napped Cloth

A woven metallurgical polishing cloth in which some fibers are aligned approximately normal to one of its surfaces.

Nascent Surface

A completely uncontaminated surface produced, for example, by cleavage fracture under ideal vacuum conditions.

National Bureau of Standards

The body in the United States responsible for establishing standards for scientific and technical data.

Native Metal

(1) Any deposit in the earth's crust consisting of uncombined metal. (2) The metal in such a deposit. (3) Metals found naturally in bulk and near pure form, including gold, silver, and copper. Sometimes, the term is extended to include extraterrestrial material such as meteorites.

Natural Aging

(1) Spontaneous aging of a supersaturated solid solution at room temperature. See also *aging*. Compare with *artificial aging*. (2) Generally, any time-related deterioration but more specific. See *precipitation hardening*.

Natural Cement

A *hydraulic cement* produced by calcining a naturally occurring argillaceous limestone at a temperature below the sintering point and then grinding to a fine powder.

Natural Diamond

The densest form of crystallized carbon and the hardest substance known, natural diamond occurs most commonly as well-developed crystals in volcanic pipes or in alluvial deposits. Bort sometimes refers to all diamonds not suitable for gems, or it may refer to off-color flawed or impure diamonds not fit for use for gems or most other industrial applications but suitable for the preparation of diamond grain and powder for use in lapping or the manufacture of most diamond grinding wheels. This type of bort is also called crushing bort or fragmented bort. See also *diamond* and *man-made diamond*.

Natural Draft

Taper on the sides of a forging, due to its shape or position in the die, that makes added *draft* unnecessary.

Natural Fiber

A natural fiber is one obtained from a plant, animal, or mineral. The commercially important natural fibers are those cellulosic fibers obtained from the seed hairs, stems, and leaves of plants; protein fibers obtained from the hair, fur, or cocoons of animals; and the crystalline mineral asbestos.

The natural fibers may be classified by their origin as cellulosic (from plants), protein (from animals), and mineral. The plant fibers may be further ordered as seed hairs, such as cotton; bast (stem) fibers, such as linen from the flax plant; hard (leaf) fibers, such as sisal; and husk fibers, such as coconut. The animal fibers are grouped under the categories of hair, such as wood; fur, such as angora; or secretions, such as silk. The only important mineral fiber is asbestos, which because of its carcinogenic nature has been banned from consumer textiles.

Natural Gas

Fuel gas obtained from natural subterranean sources and comprising a wide range of hydrocarbon gases.

Natural Rubber

Rubber is characterized as being a highly elastic or resilient material, and the natural product is obtained mainly as a latex from cuts in the trunks of the *Hevea brasiliensis* tree. The latex consists of small particles (averaging about 2500 Å units in diameter) of rubber suspended in an aqueous medium (and about 35% solids content). The system also contains about 6%–8% nonrubber constituents, some of which are emulsifiers, naturally occurring antioxidants, and proteins.

The rubber part of the composition, which is obtained from the latex by coagulation, washing, and drying, consists of long linear polymeric molecules of high average molecular weight. One report states that natural rubber contains molecules of molecular weights from 50,000 to 3,000,000 with 60% of it over 1,300,000. The rubber is built up of C_6H_8 units, each containing a double bond, and the overall structure is called *cis*-polyisoprene. Although a product almost identical to natural rubber has been made by polymerizing isoprene, there is no indication that a tree makes rubber from isoprene.

Vulcanization

Raw rubber as obtained from the latex is not useful directly in many commercial applications because of stiffening at low temperatures and softening at high temperatures. It must be vulcanized to give a wider temperature range for application and good physical properties. The most usual vulcanization is by the use of sulfur and organic accelerators of vulcanization in the presence of zinc oxide at temperatures in the neighborhood of 100°C–204°C depending on the curing system.

Also, for many applications, the high rubber vulcanizates are not tough enough for good service. To improve the situation, reinforcing fillers are used for compounding the rubber, such as finely divided carbon black, silica, and silicates including clays. In addition, many other nonreinforcing fillers, such as whiting (calcium carbonate) and barytes, are used to lower cost or to lend special properties to finished articles.

Environmental Effects

Under service conditions, natural rubber articles are subject to some deteriorating effects, and the rubber must be compounded to minimize these effects. For example, it is subject to deterioration by the action of oxygen, ozone, heat, and light and to fatigue cracking under dynamic conditions. Antioxidants are used in compounds to retard air and heat deterioration, and special chemicals are added to protect against light deterioration, as well.

Processing

Natural rubber is, for most applications, masticated and compounded on mills or in Banbury mixers; the masticated material is then passed through a tubing machine to form various articles or parts of articles that are then vulcanized at an elevated temperature in a mold or under other suitable conditions.

Natural rubber is furnished to fabricators in the form of large bales or as latex. The bale rubber comes in a number of different grades, which are priced according to quality. The latex is sold at different solids contents; the higher solids contents are preferred partly because of shipping costs from the plantations.

Applications

Natural rubber is used for making many types of articles. Because of its abrasion-resistant quality and low hysteresis in reinforced compounds, it is used in truck-tire tread stocks and in conveyor belts that are employed in conveying abrasive material such as coal, crushed rock, ore, and cinders. In large tires, it has found application in carcass compounds because of the tack and building qualities of the raw polymer. It has also been used in carcass compounds because of the low heat buildup (low hysteresis) of the carcass compound vulcanizate during severe service conditions in tire usage.

Some of the many other applications for natural rubber include waterproof clothing and footwear, and wire insulation, as well as mechanical goods and sundries, products such as hot-water bottles, surgical goods, rubber bands and thread, engine mounts, tank tread blocks, and gaskets, sound and vibration damping equipment, and road underlay. In fact, it can be used almost anywhere a highly resilient, elastic, durable material is required. A few applications in the latex field are (1) in making latex thread by coagulating a fine stream of latex compound coming from a nozzle and then vulcanizing it, (2) in making foam sponge such as mattresses by gelling and vulcanizing a foamed latex, (3) in tire cord dipping, (4) in making such articles as surgeon's gloves by the latex dipping method, and (5) in carpets where it is used as a backing material to provide anchorage of the yarns and to give dimensional stability.

Natural Strain

See *true strain*.

Naturally Bonded Molding Sand

A foundry sand containing sufficient bonding material as mined to be suitable for molding purposes.

Nature Print

A technique for revealing flow lines in forgings or other wrought material. The forging is sectioned, polished, and heavily etched. It is then thinly coated with ink and a print obtained by pressing paper against a surface. See also *sulfur print*.

N.D.T.T.

Nil Ductility Transition Temperature. See *impact test*.

NDE

See *nondestructive evaluation*.

NDI

See *nondestructive inspection*.

NDT

(1) See *nondestructive testing*. (2) Variation on the next item.

Near-Infrared Radiation

Infrared radiation in the wavelength range of 0.78–3 μm (7,800–30,000 Å). See also *electromagnetic radiation* and *infrared radiation*.

Near-Net Shape

See *net shape*.

Neat Oil

Hydrocarbon oil with or without additives used undiluted. This term is used particularly in metal cutting to distinguish these fluids from soluble oils (emulsions).

Neat Resin

Resin to which nothing (additives, reinforcements, and so on) has been added.

Nebulizer

A device for converting a sample solution into a gas–liquid aerosol for atomic absorption, emission, and fluorescence analysis. This may be combined with a burner to form a nebulizer burner. See also *atomizer*.

Neck

The contact area between abutting particles in a powder compact undergoing sintering.

Neck Formation

The growth of interparticle contacts through diffusion processes during sintering.

Neck-in

In *extrusion coating* of plastics, the difference between the width of the extruded web as it leaves the die and the width of the coating on the substrate.

Necking

(1) The reduction in the cross-sectional area of a material in a localized area by uniaxial tension or by stretching. (2) The reduction in the diameter of a portion of the length of a cylindrical shell or tube. (3) Local reduction in diameter just prior to failure during tensile loading. It occurs when the rate of work hardening ceases to increase in proportion to the increase in strain. As a consequence, further deformation is confined to the neck, which thins until the available ductility is exhausted and the material fails. Also termed as *waisting*.

Necking Down

Localized reduction in area of a specimen during tensile deformation.

Necking Strain

Same as *uniform strain*.

Needle Bearing

A bearing in which the relatively moving parts are separated by long, thin rollers that have a length-to-diameter ratio exceeding 5.0.

Needle Blow

A specific *blow molding* technique in which the blowing air is injected into the hollow article through a sharpened, hollow needle that pierces the parison.

Needled Mat

A *mat* for reinforcing plastics formed of strands cut to a short length, then felted together in a needle loom with or without a carrier.

Needles

Elongated or rodlike powder particles with a high aspect ratio.

Neel Point

See *Neel temperature*.

Neel Temperature

The temperature below which spins in an antiferromagnetic material are ordered antiparallel so that there is zero net magnetic moment. Also known as *neel point*.

Negative Creep

A contraction in dimensions resulting from an alloy undergoing an order/disorder change. It has been reported in Nimonic components in service at about 550°C but is not a creep phenomenon, although it might be argued that the contraction increases the creep stress.

Negative Distortion

The distortion in a microscopic image that occurs when the magnification in the center of the field exceeds that in the edge of the field. Also termed as barrel distortion. In contrast with *positive distortion*.

Negative Eyepiece

An eyepiece in an optical microscope in which the real image of the object forms between the lens elements of the eyepiece.

Negative Ion

An ion that can be an atom, molecule, or radical that has gained one or more electrons and hence is negatively charged.

Negative Rake

Describes a tooth face and rotation whose cutting edge lags the surface of the tooth face. See the term *face mill*.

Negative Replica

A method of reproducing a surface obtained by the direct contact of the replicating material with the specimen. Using this technique, the contour of the replica surface is reversed with respect to that of the original. See also *replica*.

Neodymium

A metallic element, one of the rare earth group.

Neon

A gaseous element, one of the rare or noble group.

Neoprene Rubber

Except for polybutadiene and polyisoprene, neoprene is perhaps the most natural rubberlike of all, particularly with regard to dynamic response. Neoprenes are a large family of rubbers that have a property profile approaching that of natural rubber and with better resistance to oils, ozone, oxidation, and flame. They age better and do not soften on heat exposure, although high-temperature tensile strength may be lower than that of natural rubber.

These materials, like natural rubber, can be used to make soft, high-strength compounds. A significant difference is that, in addition to neoprene being more costly than natural rubber by the pound, its density is about 25% greater than that of natural rubber. Neoprenes do not have the low-temperature flexibility of natural rubber, which detracts from their use of low-temperature shock or impact applications.

General-purpose neoprenes are used in hose, belting, wire and cable, footwear, coated fabrics, tires, mountings, bearing pads, pump impellers, adhesives, seal windows and curtain wall panels, and flashing and roofing. Neoprene latex is used for adhesives, dip-coated goods, and cellular cushioning jackets.

Nernst Equation

An equation that expresses the exact *electromotive force* of a cell in terms of the activities of products and reactants of the cell. The Nernst equation is given by

$$E = E^0 - \frac{RT}{nF} \ln \frac{(\text{ox})}{(\text{red})}$$

where
  E is the electrode potential
  E^0 is the standard electrode potential
  R is the gas constant (1.987 cal/K mol)
  T is the absolute temperature (in K)
  n is the number of moles of electrons transferred into a half-cell reaction
  F is the Faraday constant ($F = 23,060$ cal/V equivalent)
  (ox) and (red) are the activities of the oxidized and reduced species, respectively

Nernst Layer, Nernst Thickness

The diffusion layer or the hypothetical thickness of this layer is given by the theory of Nernst. It is defined by

$$i_\text{d} = nFD \left\{ \frac{(C_\text{o} - C)}{\delta} \right\}$$

where
  i_d is the diffusion coefficient
  C_o is the concentration at the electrode surface
  δ is the Nernst thickness (0.5 mm in many cases of unstirred aqueous electrolytes)

Nesting

In reinforced plastics, the placing of plies of fabric such that the yarns of one ply lie in the valleys beneath the yarns of the adjacent ply. Also called *nested cloth*.

Net Positive Suction Head

The difference between total pressure and vapor pressure in a fluid flow expressed in terms of equivalent height of fluid, or "head," by the following equation:

$$NPSH = \left\{ \frac{P_0}{w} + \frac{V^2}{2g} - \frac{P_v}{w} \right\}$$

where

P_0 is the static pressure
P_v is the vapor pressure
V is the flow velocity
w is the specific weight of fluid
g is the gravitational acceleration

This quantity is used in pump design as a measure of the tendency for cavitation to occur at the pump inlet.

Net Section Stress

The nominal stress calculated from the total load divided by the total cross section. It makes no allowance for stress raisers or other sources of uneven distribution of stress.

Net Shape

The shape of a powder metallurgy part, casting, or forging that conforms closely to specified dimensions. Such a part requires no secondary machining or finishing. A near-net shape part can be either one in which some but not all of the surfaces are net or one in which the surfaces require only minimal machining or finishing.

Netting Analysis

The analysis of filament-wound structures that assumes that the stresses induced in the structure are carried entirely by the filaments and the strength of the resin is neglected and that assumes also that the filaments possess no bending or shearing stiffness and carry only the axial tensile loads.

Network Etching

Formation of networks, especially in *mild steels*, after etching in nitric acid. These networks relate to subgrain boundaries.

Network Structure

A metallic structure in which one constituent occurs primarily at the grain boundaries, partially or completely enveloping the grains of the other constituents.

Neutral Axis/Plane

The line or plane, in a component subject to a complex stress pattern, where stress passes through zero as it changes from positive to negative. For example, a symmetrical beam being bent from its natural form has a tensile longitudinal stress at the bend exterior and a compressive longitudinal stress at the inner surface. At approximately mid-thickness the neutral axis has zero longitudinal stress.

Neutral Filter

A filter that attenuates the radiant power reaching the detector by the same factor at all wavelengths within a prescribed wavelength region. See also *filter*.

Neutral Flame

(1) A gas flame in which there is no excess of either fuel or oxygen in the inner flame. Oxygen from ambient air is used to complete the combustion of CO_2 and H_2 produced in the inner flame. (2) An oxyfuel gas flame in which the portion used is neither oxidizing nor reducing. See also *carburizing flame*, *oxidizing flame*, and *reducing flame*.

Neutral Flux

A flux that is neither acidic nor basic but is added to modify some physical characteristic particularly fluidity.

Neutral Oil

A lubricating oil obtained by distillation, not treated with acid or with alkali.

Neutralization Number

An ASTM number given to quenching oils that reflects the oil's tendency toward oxidation and sludging. See also *saponification number*.

Neutron

An elementary particle that has approximately the same mass as the proton, but no electric charge. Rest mass is 1.67495×10^{27} kg. An unbound (extranuclear) neutron is unstable and β-decays with a half-life of 10.6 min. See *atom*.

Neutron Absorber

A material in which a significant number of neutrons entering combine with nuclei and are not reemitted.

Neutron Absorption

A process in which the collision of a neutron with a nucleus results in the absorption of the neutron into the nucleus with the emission of one or more prompt γ-rays, and in certain cases, emission of α-particles, protons, or other neutrons or fission of the nucleus results. Also known as *neutron capture*.

Neutron Activation Analysis

Activation analysis in which the specimen is bombarded with neutrons; identification is made by measuring the resulting radioisotopes. See also *activation analysis*.

Neutron Capture

See *neutron absorption*.

Neutron Cross Section

A measure of the probability that an interaction of a given kind will take place between a nucleus and an incident neutron; it is an area such that the number of interactions that occur in a sample exposed to a beam of neutrons is equal to the product of the cross section, the number of nuclei per unit volume in the sample, the thickness of the sample, and the number of neutrons in the beam that would enter the sample if their velocities were perpendicular to it. The usual unit is the barn (10^{24} cm^2). See also *barn*.

Neutron Detector

Any device that detects passing neutrons, for example, by observing the charged particles or γ-rays released in nuclear reactions induced by the neutrons or by observing the recoil of charged particles caused by collisions with neutrons.

Neutron Diffraction

The phenomenon associated with the interference processes that occur when neutrons are scattered by the atoms within solids, liquids, and gases.

Neutron Embrittlement

Embrittlement resulting from bombardment with neutrons, usually encountered in metals that have been exposed to a neutron flux in the core of the reactor. In steels, neutron bombardment is evidenced by a rise in the ductile-to-brittle *transition temperature*.

Neutron Flux

The number of neutrons passing through an area in a unit of time.

Neutron Radiography

Radiography that uses neutrons generated by a nuclear reactor, accelerator, or by certain radioactive isotopes to form a radiographic image of a test piece. The neutrons are detected by placing a conventional x-ray film next to a converter screen composed of potentially radioactive materials or next to a transfer screen composed of prompt emission materials, which convert the neutron radiation to other types of radiation more easily detected by the film.

Neutron Spectrometry

See *neutron spectroscopy*.

Neutron Spectroscopy

Determination of the energy distribution of neutrons. Scintillation detectors, proportional counters, activation foils, and proton recoil are used.

Neutron Spectrum

The distribution by energy of neutrons impinging on a surface, which can be measured by neutron spectroscopy techniques or sometimes from knowledge of the neutron source. See also *dosimeter*.

Newman Band/Line/Lamellae

A mechanical twin in ferrite, usually indicative of high strain rates. See *twins*.

Newton

The SI-derived unit of force, symbol N equivalent to kg m/s^2.

Newton's Laws of Motion

(1) A body will remain static or will maintain its motion in a straight line unless an external impressed force acts upon it. (2) The change of momentum occurs at a rate proportional to the impressed force and in the same direction as the force. (3) Action and reaction are equal and opposite.

Newtonian Fluid

A fluid exhibiting Newtonian viscosity wherein the shear stress is proportional to the rate of shear. Compare with *dilatant*, *rheopeptic material*, and *thixotropy*.

Nib

(1) A pressed, preheated, shaped, sintered, hot-pressed, Russ drilled, or finished compact. (2) A generic term for a piece of cemented carbide intended for use as a wire drawing die.

Nibbling

(1) Contour cutting of sheet metal by use of a rapidly reciprocating punch that makes numerous small cuts. (2) Cutting sheet by punching multiple small overlapping holes with a powered punch and die set reciprocating at high speeds.

Nichrome

A range of proprietary alloys, the most well known of which is 80% nickel and 20% chromium. The alloys have good high-temperature strength and oxidation resistance, while their electrical resistance is low and their temperature coefficient of electrical resistivity is high. They are used for electrical heating elements and furnace furniture for service up to about 100°C.

Nick Break Test

A tensile or bend test in which fracture is initiated in a notch. The term usually refers to tests of welds where fracture is directed through a particular weld zone.

Nickel Aluminide

Nickel aluminide (NiAl) is available commercially in powder form in various mesh sizes from granular material to an average particle size of a few micrometers. Nickel aluminide is fairly strong at room temperature with the modulus of transverse rupture values ranging from 204 to 952 MPa, depending on fabrication methods, while at 1000°C the corresponding values are approximately half the room temperature strength. Creep resistance at these temperatures is extremely poor. Thermal shock resistance is quite good.

Nickel aluminide can be formed by hot pressing or by cold pressing and sintering, but the former method produced by far the best results.

Excellent oxidation resistance and fairly good strength make this material of interest for turbine blading or other combustion chamber applications. It has impact resistance that is better than most ceramics, intermetallic compounds, and some cermets. NiAl is also resistant to attack by molten glass and red and white fuming nitric acids, which suggests possible uses in the glass-processing industry or in some high-temperature chemical processes.

Nickel and Alloys

A silvery white metal, nickel (Ni) has been used in alloy form with copper since ancient times. Nickel has a specific gravity of 8.902, a density of 8.902 g/cm³, and is magnetic up to 360°C. The metal is highly resistant to atmospheric corrosion and resists most acids, although it is attacked by oxidizing acids, such as nitric. Its principal use is as an alloying element to stainless steels and nonferrous metals. It is also used for electroplating.

There are several commercial high-purity nickels, containing at least 99% of the metal plus trace amounts of complying cobalt and small amounts of other elements, such as carbon, copper, iron, manganese, silicon, and sulfur. They find use in electronic and aerospace applications, chemical- and food-processing equipment, and for anodes and cathodes, caustic evaporators, and heat shields. Nickel 200, also known as commercially pure nickel, may contain as much as 0.15% carbon, 0.25% copper, 0.40% iron, 0.35% manganese, 0.35% silicon, and 0.01% sulfur.

It is especially resistant to caustics, high-temperature halogens and hydrogen halides, and salts but not to oxidizing halides or ammonium hydroxide. Because it is susceptible to temperature embrittlement by carbon precipitation, it is restricted to a maximum service temperature of 316°C. Nickel 201, is low-carbon (0.02% maximum) counterpart, can be used at higher temperatures. The cast grade, designated CZ-100, is recommended for use at temperatures above 316°C.

Nickel 270 is the highest purity grade, being at least 99.97% nickel and containing no more than 0.02% carbon, 0.005% iron, and 0.001% each of other ingredients. Although similar to the other nickels in general corrosion resistance, it may be more prone to sulfur embrittlement under some conditions.

At cryogenic temperatures, nickel alloys are strong and ductile. Several nickel-base superalloys are specified for high-strength applications at temperatures to 1093°C. High-carbon nickel-based casting alloys are commonly used at moderate stresses above 1204°C.

Dispersion-Strengthened Nickel

Dispersion-strengthened nickel, or DS nickel, contains about 2.2% thoria as the dispersion-strengthening phase and, thus, has also been called TD nickel. The thoria markedly increases high-temperature strength. Tensile strengths, about 483 MPa ultimate and 331 MPa yield at room temperature, are on the order of 117 MPa at 1093°C. The material also retains about 60% of its room temperature tensile modulus, about 138×10^3 MPa, at 1093°C. DS nickel has poor oxidation resistance and, thus, requires coating for sustained use at high temperatures, such as for aircraft-turbine components and furnace equipment. It has been alloyed with about 20% chromium to improve oxidation resistance but still may require coating for prolonged use at high temperatures.

Ferrous Alloys

The largest commercial application for nickel is as an alloying element in both ferrous and nonferrous alloys. In unhardened nonaustenitic steels, nickel has the effect of slightly strengthening ferrite and mildly improving corrosion resistance. In balanced combinations with small amounts of chromium and manganese; or chromium, molybdenum, and manganese; or molybdenum and manganese (AISI 3100, 3300, 4300, 4600, 4800, 8600, 8700, 9300, 9800 series), hardenabilities of steels are greatly increased. In carburizing steels (AISI 2300 and 2500 series), nickel is used to strengthen and toughen the core.

Nickel is a strong austenite stabilizer and with chromium is used to form the important AISI 300 series of nonmagnetic austenitic stainless steels. In these steels, chromium is the principal alloying element and is the element responsible for their excellent corrosion resistance. Nickel contributes to this corrosion resistance, but its primary function is to promote and retain an austenitic structure at all recommended temperatures of use and under most conditions of plastic deformation. The mechanical properties of an austenitic steel are generally much less adversely affected by moderately elevated temperatures and very low temperatures that are ferritic or martensitic steels. Wrought stainless steels (AISI listing) with chromium and nickel content up to 22% and 26%, respectively, and in some cases with the addition of molybdenum, have been designed for specific heat- and corrosion-resisting service over and above the standard 18-8 type. Some cast stainless steels with extremely high chromium and nickel contents (up to 28% chromium, 10% nickel and 17% chromium, 66% nickel) are used for severe corrosion or heat-resisting applications. Nickel is also widely used as a property-improving addition to cast iron and as one of the many alloying elements and some ferrous high-temperature, high-strength superalloys.

Magnetic Alloys

Nickel and iron form a series of alloys with thermal expansion and magnetic characteristics of commercial importance. The alloy of 36% nickel ("Invar") has a thermal coefficient of expansion of almost zero over a small temperature range. The addition of 12% chromium, in lieu of some of the iron, produces an alloy ("Elinvar") with an invariable modulus of elasticity over a considerable temperature range as well as a fairly low coefficient of expansion. Variations in the content of nickel and other metals have produced a large number of alloys with particular coefficients for specific applications such as geodetic tapes, balance wheels in watches, tuning forks, and glass-to-metal seals. The alloy with 78% nickel (Permalloy) and many modifications of this composition have very high magnetic permeabilities at low field strengths and find application in the electronics field. Nickel also plays an important role in permanent magnets of the Alnico type, which contain 14%–18% nickel plus aluminum, cobalt, and sometimes copper.

Nonferrous Alloys

The principal groups of nonferrous alloys are those of slightly alloyed nickel, nickel and copper, nickel and chromium, and the superalloys. The slightly alloyed nickels ("Duranickel," "Permanickel," E-nickel, electronic nickel, etc.) in most cases contain more than 94% nickel.

Duranickel 301, a precipitation-hardened, 94% nickel alloy, has excellent spring properties to 316°C. During thermal treatment, Ni_3 AlTi particles precipitate throughout the matrix. This action enhances alloy strength. Corrosion resistance is similar to that of commercially pure wrought nickel.

Permanickel 300 is a high-nickel alloy, containing at least 97% nickel and small amounts of titanium and magnesium and most of the other elements found in high-purity nickels. It is used for high-strength parts in electrical and electronic applications. It has a tensile modulus of 207,000 MPa and, depending on form and condition, tensile yield strengths of 241–1034 MPa.

The aforementioned alloys retain much of the corrosion resistance and physical properties of nickel, but they are alloyed to make them heat treatable by age hardening, more resistant to a specific form of corrosion (manganese to reduce susceptibility to sulfur embrittlement), or to produce desirable electronic characteristics.

Wrought beryllium nickel containing 97.5% nickel, 2% beryllium, and 0.5% titanium is used for springs, switches, bellows, diaphragms, and small valves. The age-hardenable alloy provides tensile yield strengths of 310–1586 MPa, 896 MPa at 538°C and has good corrosion resistance in general atmospheres and to reducing media. Casting alloys, containing 2%–3% beryllium and also age hardenable, are about equally strong and are used for molds to form glasses and plastics and for metal-forming tools, aircraft fuel-pump impellers, and seal plates and aircraft engines.

Binary Nickel Alloys

The primary wrought alloys in this category are the Ni–Cu grades known as Monel alloy 400 (Ni–31%Cu) and K-500 (Ni–29%Cu), which also contain small amounts of aluminum, iron, and titanium. The Ni–Cu alloys differ from nickel 200 and 201 because their strength and hardness can be increased by age hardening. Although the Ni–Cu alloys share many of the corrosion characteristics of commercially pure nickel, their resistance to sulfuric and hydrofluoric acids and brine is better. Handling of waters, including seawater and brackish water, is a major application. Monel alloys 400 and K-500 are immune to chloride-ion stress-corrosion cracking, which is often considered in their selection.

Other commercially important binary nickel compositions are Ni–Mo and Ni–Si. One binary type, *Hastelloy alloy B-2* (Ni–28%Mo), offers superior resistance to hydrochloric acid, aluminum chloride catalysts, and other strongly reducing chemicals. It also has excellent high-temperature strength in inert atmospheres and in vacuum.

Cast nickel–copper alloys comprise a low and high silicon grade, M-35-1 and QQ-N-288, and Grades A and E (1.5% silicon) are commonly used in conjunction with wrought nickel–copper in pumps, valves, and fittings. A higher silicon grade, QQ-N-288, Grade B (3.5% silicon), is used for rotating parts and wear rings because it combines corrosion resistance with high strength and wear resistance. Grade D (4.0% silicon) offers exceptional galling resistance.

Two other binary cast alloys are ACI N-12M-1 and N-12M-2. These Ni–Mo alloys are commonly used for handling hydrochloric acid in all concentrations at temperatures up to the boiling point. These alloys are produced commercially under the trade names *Hastelloy alloy B* and *Chlorimet 2*.

Many chemical-resistant nickel alloys, both wrought and cast, are noted mainly for their resistance to chemicals that are aggressive toward other metals, and many of these also possess substantial strength at elevated temperature. Although trade names abound, the *Hastelloys* and the *Incoloys* and *Inconels* are probably the most well known. Many of these alloys are resistant to normally aggressive acids, such as hydrofluoric and sulfuric, as well as to acetic and phosphoric acids, mixed acids, chlorides, solvents, and high-temperature oxidation and, thus, are widely used for chemical-processing equipment.

Among the more common alloys and their principal ingredients are *Hastelloy B,* which contains 26%–30% molybdenum, 46% iron, 2.5% cobalt, and 1% chromium; *Hastelloy B-2,* which contains the same amount of molybdenum and chromium but only 1% iron and 1% cobalt; *Hastelloy C,* having 15%–18% molybdenum and about as much chromium 4%–7% iron, 3%–5% tungsten, and 2.5% cobalt; *Hastelloy C-4,* with molybdenum and chromium content similar to those of *Hastelloy C* but no more than 3% iron and 2% cobalt; *Hastelloy D,* 8.5%–10% silicon, 2%–4% copper, and as much as 2% iron, 1.5% cobalt, and 1% chromium; *Hastelloy G,* 21%–23% chromium, 18%–21% iron, 5.5%–7.5% molybdenum, 1.75%–2.5% columbium plus tantalum, 1.5%–2.5% copper, and as much is 2.5% cobalt; *Hastelloy S,* 14.5%–17% chromium, 14%–16.5% molybdenum, and as much as 3% iron and 2% cobalt; and *Hastelloy C-276,* 15%–17% molybdenum, 14.5%–16.5% chromium, 4%–7% iron, 3%–4.5% tungsten, and as much is 2.5% cobalt. *Incoloy 800,* although iron base, is often grouped with these alloys. It contains 46% iron, 32.5% nickel, and 21% chromium. Nickel-base (42%) *Incoloy 825* contains 30% iron, 21.5% chromium, 3% molybdenum, and 2.3% copper.

Hastelloy X, which provides substantial strength and oxidation resistance at temperatures to about 1204°C, contains 20%–23% chromium, 17%–20% iron, 8%–10% molybdenum, 0.5%–2.5% cobalt, and 0.2%–1% tungsten. Solution-treated rapidly cooled sheet has room temperature tensile properties of 786 MPa ultimate strength, 359 MPa yield strength, 43% elongation, and 197×10^3 MPa modulus. At 982°C, these properties are 155 MPa, 110 MPa, 45%, and 126×10^3 MPa, respectively. The alloy is widely used for gas-turbine parts and other applications requiring heat and oxidation resistance. Although mainly a wrought alloy, it also can be investment cast.

Ternary Nickel Alloys

Two primary wrought and cast compositions are Ni–Cr–Fe and Ni–Cr–Mo. Ni–Cr–Fe is known commercially as *Haynes alloys 214 and 556* and *Inconel alloy 600* and *Incoloy alloy 800*. *Haynes'* new alloy *No. 214* (Ni–16%Cr–2.5%Fe–4.5%Al–Y) has excellent resistance to oxidation to 1204°C and resists carburizing and chlorine-contaminated atmospheres. *Haynes alloy 556* (Fe–20%Ni–22%Cr–18%Co) combines effective resistance to sulfidizing, carburizing, and chlorine-bearing environments with good oxidation resistance, fabricability, and high-temperature strength. Inconel alloys 600 (Ni–15.5%Cr–8%Fe) has good resistance to oxidizing and reducing environments. Intended for severely corrosive conditions at elevated temperatures, Incoloy 800 (Ni–46%Fe–21%Cr) has good resistance to oxidation and carburization at elevated temperatures, and it resists sulfur attack, internal oxidation, scaling, and corrosion in many atmospheres.

A cast Ni–Cr–Fe alloy CY-40, known as Inconel, has higher carbon, manganese, and silicon contents than the corresponding wrought grade. In the as-cast condition, the alloy is insensitive to the type of intergranular attack encountered in as-cast or sensitized stainless steels.

Significant additions of molybdenum make Ni–Cr–Mo alloys highly resistant to pitting. They retain high strength and oxidation resistance at elevated temperatures, but they are used in the chemical industry primarily for their resistance to a wide variety of aqueous corrosives. In many applications, these alloys are considered the only materials capable of withstanding the severe corrosion conditions encountered.

In this group, the primary commercial materials are *C-276, Hastelloy alloy C-22,* and *Inconel alloy 625*. Hastelloy alloy C-22

(Ni–22%Cr–13%Mo–3%W–3%Fe) has better overall corrosion resistance and versatility than any other Ni–Cr–Mo alloy. Alloy C-276 (57%Ni–15.5%Cr–60%Mo) has excellent resistance to strong oxidizing and reducing corrosives, acids, and chlorine-contaminated hydrocarbons. Alloy C-276 is also one of the few materials that withstand the corrosive effects of wet chlorine gas, hypochlorite, and chlorine dioxide. Hastelloy alloy C-22, the newest alloy in this group, has outstanding resistance to pitting, crevice corrosion, and stress corrosion cracking. Present applications include the pulp and paper industry, various pickling acid processes, and production of pesticides and various agrichemicals.

Superalloys

There is a great variety of high-temperature, high-strength nickel alloys; these are called superalloys because of their outstanding strength, creep resistance, stress-rupture strength, and oxidation resistance at high temperatures. They are widely used for gas turbines, especially aircraft engines. Most of these alloys contain substantial chromium for oxidation resistance; refractory metals for solid-solution strengthening; small amounts of grain-boundary-strengthening elements, such as carbon, boron, hafnium, and/or zirconium; and aluminum and titanium for strengthening by precipitation of an Ni(Al, Ti) compound known as "gamma prime" during age hardening. Among the well-known wrought alloys are *D-979*; *GMR-235-D*; *IN 102*; *Inconel 625, 700, 706, 718, 722, X750*, and *751*; *MAR-M 200* and *421*; *René 41, 95*, and *100*;

Udimet 500 and *700*; and *Waspaloy*. Cast alloys include *B-1900*; *GMR-235-D*; *IN 100, 162, 738*, and *792*; *M252*; *MAR-M-200, 246*, and *421*; *Nicrotung*; *René 41, 77, 80*, and *100*; and *Udimet 500* and *700*. Some wrought alloys are also suitable for casting, primarily investment casting.

A third class of superalloys includes ODS alloys such as *IN MA-754* (Ni–20%Cr–0.6% yttria) and *IN MA-6000* (Ni–15%Cr–2%Mo–4%W–2.5%Ti–4.5%Al), which are strengthened by dispersions such as yttria coupled (in some cases) with gamma prime precipitation (MA-6000).

An additional dimension of nickel-base superalloys has been the introduction of grain–aspect ratio and orientation as a means of controlling properties. In some instances, grain boundaries have been removed. Wrought powder-metallurgy alloys of the ODS class and cast alloys such as *MAR M-247* have demonstrated property improvements due to grain morphology control by directional crystallization or solidification. Virtually all uses of the cast and wrought nickel-based superalloys are for gas-turbine components.

Specialty Nickel Alloys

In addition to the aforementioned families, there are specialty nickel alloys for glass sealing and other applications. A series of paramagnetic alloys, called *nitinol*, are intermetallic compounds of nickel and titanium rather than nickel–titanium alloys. The compound NiTi contains theoretically 54.5% nickel, but the alloys may contain Ti_2Ni and $TiNi_3$ with about 50%–60% nickel. The NiTi and nickel-rich alloys are paramagnetic, with a permeability value of 1.002, compared with the unity value of a vacuum. A 54.5% nickel alloy has a tensile strength of 758 MPa with the elongation of about 15% and hardness of Rockwell C 35. The alloys close to the NiTi composition are ductile and can be cold-rolled.

The nitinols, with nickel content ranging from 53% to 57%, are known as memory alloys because of their ability to "remember" or return to a previous shape upon being heated. This unusual behavior stems from a diffusionless transformation of the alloy. These shape-memory alloys have excellent fatigue strength, and damping around room temperature is reported to be one of the highest ever measured in a metal.

Nickel Brass

A number of alloys of copper, nickel, and zinc are termed nickel brass. Nickel–silicon brass contains very small percentages of silicon, usually about 0.60%, which forms a nickel silicide, Ni_2Si, increasing the strength and giving heat-treating properties. Rolled nickel–silicon brass, containing 30% zinc, 2.5% nickel, and 0.65% silicon, has a tensile strength of 785 MPa. Imitation silver, for hardware and fittings, was a nickel brass containing 57% copper, 25% zinc, 15% nickel, and 3% cobalt. The bluish color of the cobalt neutralizes the yellow cast of the nickel and produces a silver-white alloy. *Silvel* was a nickel brass containing 67.5% copper, 26% zinc, and 6.5% nickel, with sometimes a little cobalt. Nickel brasses of alloy are used where white color and corrosion resistance are desired.

Nickel Bronze

This is a name given to bronzes containing nickel, which usually replaces part of the tin, producing a tough, fine-grained, and corrosion-resistant metal. A common nickel bronze containing 88% copper, 5% tin, 5% nickel, and 2% zinc has a tensile strength of 330 MPa, elongation 42%, and Brinell hardness 86 as cast. When heat treated or age hardened, the tensile strength is 599 MPa, elongation 10%, and Brinell hardness 196. Small amounts of lead take away the age-hardening quality of the alloy and also lower the ductility. But small amounts of nickel added to bearing bronzes increase the resistance to compression and shock without impairing the plasticity. A bearing bronze of this nature contains 73%–80% copper, 15%–20% lead, 5%–10% tin, and 1% nickel.

Nickel Oxide

Two of the nickel oxides are useful as colorants in ceramics: (1) nickelous oxide or green nickel oxide, NiO, and (2) black nickel oxide, Ni_2O_3. At 400°C, it oxidizes to Ni_2O_3, and at 600°C is reduced back to NiO.

Nickel produces a bluish-violet color in potash glasses and a violet tending toward brown in soda glasses. Nickel rates as one of the more powerful colorants, since 1 part in 50,000 produces a recognizable tin.

Nickel oxide is sometimes used to decolorize potash glass. Nickel oxide and nickel silicate have an advantage over manganese dioxide for decolorizing purposes in that they are not as sensitive in changing oxidizing and reducing environments.

Nickel Silver

Nickel silver is a name applied to an alloy of copper, nickel, and zinc, which is practically identical with alloys known in the silverware trade as *German silver*. *Packfong*, meaning white copper, is an old name for these alloys. The very early nickel silvers contained some silver and were used for silverware.

Some three dozen standard wrought alloys (C 73150–C 79900) and for standard cast alloys (C 97300–C 97800) are designated nickel silvers. Depending on the alloy, copper content of wrought alloys ranges from 48% to 80% and nickel content from about 7% to 25%, with zinc (1%–5%) except for smaller quantities of other elements, mainly manganese, iron, and lead. The cast alloys range from about 55% to 65% copper, 12% to 25% nickel, 2.5% to 21% zinc, 2% to 10% lead, with lesser amounts of other elements.

The most common alloy, nickel silver C 75200, nominally contains 65% copper and 18% nickel and, thus, is often referred to as nickel silver 65-18. The electrical conductivity of the alloy is about 6% that of copper, and its thermal conductivity is 33 W/(m K). Tensile properties for thin flat products in the annealed condition are about 414 MPa ultimate strength, 207 MPa yield strength, and 30% elongation. Cold working to the hard temper triples yield strength and markedly reduces ductility. Wire, which has similar tensile properties annealed, can be cold worked to still greater tensile strength. Modulus of elasticity in tension is 124,110 MPa. All the cast alloys are suitable for sand and investment casting and some also for centrifugal and permanent mold casting. The strongest of these alloys, nickel silver C 97800, has typical tensile properties of 379 MPa ultimate strength, 207 MPa yield strength, 15% elongation, and 131,000 MPa modulus. Applications for wrought alloys include holloware and tableware, watch and camera parts, hardware, diary equipment, costume jewelry, nameplates, keys, fasteners, and springs. The cast alloys are used for fittings, valves, ornaments, pump parts, and marine equipment.

Over the years, nickel silvers have been known by a variety of names. *Benedict metal* originally had 12.5% nickel, with two parts copper to one part zinc, but the alloy used for hardware and plumbing fixtures contains about 57% copper, 2% tin, 9% lead, 20% zinc, and 12% nickel. The cast metal has a strength of 241 MPa with the elongation of 15%.

Nickel Steels

This is steel containing nickel as the predominant alloying element. The first nickel–steel armor plate, with 3.5% nickel, was known as *Harveyized steel.*

Nickel added to carbon steel increases the strength, elastic limit, hardness, and toughness. It narrows the hardening range but lowers the critical range of steel, reducing danger of warpage and cracking, and balances the intensive deep-hardening effect of chromium. The nickel steels are also of finer structure than ordinary steels, and then nickel retards grain growth. When the percentage of nickel is high, the steel is very resistant to corrosion. At high nickel contents, the metals are referred to as iron–nickel alloys or nickel–iron alloys. The steel is nonmagnetic above 29% nickel, and the maximum permeability is at about 78% nickel. The lowest thermal expansion is at 36% nickel. The percentage of nickel in nickel steels usually varies from 1.5% to 5%, with up to 0.80% manganese. The bulk of nickel steels contains 2% and 3.5% nickel. They are used for armor plate, structural shapes, rails, heavy-duty machine parts, gears, automobile parts, and ordnance.

The standard ASTM structural nickel steel used for building construction contains 3.25% nickel, 0.45% carbon, and 0.70% manganese. This steel has tensile strength of 586–689 MPa and a minimum elongation of 18%. An automobile steel contains 0.10%–0.20% carbon, 3.25%–3.75% nickel, 0.30%–0.60% manganese, and 0.15%–0.30% silicon. When heat treated, it has a tensile strength up to 551 MPa and an elongation of 25%–35%. Forgings for locomotive crank pins, containing 2.5% nickel, 0.27% carbon, and 0.88% manganese, have a tensile strength of 572 MPa, elongation 30%, and reduction of area 62%. A nickel–vanadium steel, used for high-strength cast parts, contains 1.5% nickel, 1% manganese, 0.28% carbon, and 0.10% vanadium. The tensile strength is 620 MPa and elongation 25%. *Univan steel* for high-strength locomotive castings is a nickel–vanadium steel of this type. *Unionaloy steel* is an abrasion-resistant steel; the federal specifications for 3.5% nickel

carbon steel call for 3.25%–3.75% nickel and 0.25%–0.30% carbon. This steel has a tensile strength of 586 MPa and elongation 18%. When oil quenched, a hot rolled 3.5% nickel medium-carbon steel, SAE steel 2330, develops a tensile strength up to 1516 MPa, and a Brinell hardness of 223–424, depending on the drawing temperature. Standard 3.5% and 5% nickel steels are regular products of the steel mills, although they are often sold under trade names. Steels with more than 3.5% nickel are too expensive for ordinary structural use. Steels with more than 5% nickel are difficult to forge, but the very high nickel steels are used when corrosion-resistant properties are required. *Nicloy,* used for tubing to resist the corrosive action of paper-mill liquors and oil-well brines, contains 9% nickel, 0.10% chromium, 0.05% molybdenum, 0.35% copper, 0.45% manganese, 0.20% silicon, and 0.09% max carbon. The heat-treated steel has a tensile strength of 758 MPa, with elongation 35%. The cryogenic steels, or low-temperature steels, for such uses as liquid-oxygen vessels, are usually high-nickel steels. ASTM steel A-353, for liquid-oxygen tanks at temperatures up to −196°C, contains 9% nickel, 0.85% manganese, 0.25% silicon, and 0.13% carbon. It has a tensile strength of 654 MPa with an elongation of 20%. A 9% nickel steel, for temperatures down to −196°C, contains 9% nickel, 0.80% manganese, 0.30% silicon, and not over 0.13% carbon. It has a minimum tensile strength of 620 MPa and an elongation of 22%.

Nickel Sulfate

Nickel sulfate is the most widely used salt for nickel-plating baths and is known in the plating industry as single nickel salt. It is easily produced by the reaction of sulfuric acid on nickel and comes in pea green water-soluble crystalline pellets of the composition $NiSO_4 \cdot 7H_2O$, of a specific gravity of 1.98, melting at about 100°C. Double nickel salt is nickel ammonium sulfate, used especially for plating on zinc. To produce a harder and whiter finish in nickel plating, cobaltous sulfamate, a water-soluble powder, is used with the nickel sulfate. Nickel plate has a normal hardness of Brinell 90–140, but by controlled processes, file-hard plates can be obtained from sulfate baths. In the electroless plating, nickel sulfate, a reducing agent, a pH adjuster, and complexing and stabilizing agents are combined to deposit metallic nickel on an immersed object. The electroless nickel coating is comparable to electrolytic chrome.

Nickel-Bonded Titanium Carbide

This is a hard, strong material that retains its strength at temperatures up to 1100°C. It is most commonly used in the so-called cermet materials for high-temperature structural applications.

Nickel-bonded titanium carbide is available in powder form. Nickel-bonded titanium carbides have hardnesses ranging from HRA 80 to 89, depending on binder content. The melting point is that of the binder material, and the density is 5.0 g/cm³ min but can be higher depending on the amount and nature of the binder material.

Nickel-bonded titanium carbide is usually made by cold-pressing and sintering methods, but it can be hot pressed and, if the binder content is very high, it can even be cast.

The material may be used for turbine blading, and such high-temperature structural applications as tool bits, and as a high-temperature bearing and seal material, where high compressive strength, low coefficient of friction, and high wear resistance are needed at elevated temperatures. Oxidation resistance of these materials is fair and can be improved somewhat by adding niobium or tantalum carbide to the hard phase or by adding cobalt or chromium to the nickel binder.

Nickel–Chromium Steel

This is defined as steel containing both nickel and chromium, usually in a ratio of 2–3 parts nickel to 1 part chromium. The 2:1 ratio gives great toughness, and the nickel and chromium are intended to balance each other and physical effects. The steels are especially suited for large sections that require heat treatment because of their deep and uniform hardening. Hardness and toughness are the characteristic properties of these steels. Nickel–chromium steel containing 1%–1.5% nickel, 0.45%–0.75% chromium, and 0.38%–0.80% manganese is used throughout the carbon ranges for case-hardened parts and for forgings where high tensile strength and great hardness are required. Low nickel–chromium steels, having more carbon (0.60%–0.80%), are used for drop-forging dies and other tools.

Nickel–chromium steels may have temper brittleness or low impact resistance when improperly cooled after heat treatment. A small amount of molybdenum is sometimes added to prevent this brittleness. A nickel–chromium coin steel, used by the Italian government for coins, was a stainless steel type containing 22% chromium, 12% nickel, and some molybdenum.

Low-carbon nickel–chromium steels are water hardening, but those with appreciable amounts of alloying elements require oil quenching. Air-hardening steels contain up to 4.5% nickel and 1.6% chromium but are brittle unless tempered in oil to strengths below 1378 MPa. The alloy known as *Krupp analysis steel* contains 4% nickel and 1.5% chromium.

Nickel–Molybdenum Steel

This alloy steel is mostly used in compositions of 1.5% nickel and 0.15%–0.25% molybdenum, with varying percentages of carbon up to 0.50%. The steels are characterized by uniform properties and are readily forged and heat treated. Molybdenum toughens the steels and, in the case-hardening steels, gives a tough core. Roller bearings are made of this class of steel. Superalloy steel is 3160 steel. A 5% nickel steel with 0.30% carbon and 0.60% molybdenum has a tensile strength of 1206–1585 MPa with the elongation of 12%–22%, depending on the heat treatment. Molybdenum is more frequently added to the steels containing also chromium, the molybdenum giving air-hardening properties, reducing distortion, and making the steels more resistant to oxidation.

Nicol Prism

A prism used in optical microscopes made by cementing together with Canada balsam, two pieces of a diagonally cut calcite crystal. In such a prism, the ordinary ray is totally reflected at the calcite/cement interface, while the orthogonally polarized extraordinary ray is transmitted. The prism can thus be used to polarize light or analyze the polarization of light.

Nicrosilal

A proprietary, austenitic, growth-resistant cast iron with about 18% nickel, 5% chromium, 5% silicon, and only 2% carbon.

Nihard

A proprietary white cast iron with about 3% carbon, 4% nickel, 2% chromium, 0.5% silicon, and 0.5% manganese. It has a high hardness, about 600 Hv, but poor toughness and is used for grinding balls and rings, etc.

Ni-Hard

See *Nihard*.

Nil Ductility Transition Temperature

See *impact test*.

Nimocast

See *nimonic*.

Nimonic

A range of proprietary alloys, and cast or wrought form, based on nickel with major additions of other elements, in particular chromium, typically about 20%, with aluminum and titanium plus in some cases cobalt, molybdenum, etc. These alloys have good high-temperature characteristics and particular resistance to oxidation and good creep strength. The nimocast materials are similar materials in the cast form.

Niobium and Alloys

Note should be made of the fact that in the United States, this element was originally called columbium (symbol Cb). The Nomenclature Committee of the International Union of Pure and Applied Chemistry in 1951 adopted a recommendation to name this element niobium (symbol Nb). American chemists use this name, but the metallurgists and metals industries still use the name columbium. Most niobium is used in special stainless steels, high-temperature alloys, and superconducting alloys such as Nb_3Sn. The low cross-section capture of niobium for thermal neutrons of only 1.1 barn makes it suitable for use in nuclear piles.

Niobium is a tough, shiny, silver-gray, soft, ductile metal that somewhat resembles stainless steel in appearance. Niobium is relatively low density, yet can maintain its strength at high temperatures. It has excellent corrosion resistance to liquid metals and can be easily fabricated into wrought products.

Over 95% of all niobium is used as additions to steel and nickel alloys for increasing strength. Only 1%–2% of niobium is in the form of niobium-based alloys or pure niobium metal. Superconducting niobium–titanium alloys account for over half of that, and high-temperature and corrosion applications account for the remainder.

The density of niobium at 8.57 g/cm³ is moderate compared with most other high-melting-point metals. It is less than molybdenum at 10.2 g/cm³ and half that of tantalum at 16.6 g/cm³.

Alloys

Commercial niobium alloys are relatively low in strength and extremely ductile and can be cold worked over 70% before annealing becomes necessary. The resulting ease of fabrication into complex parts combined with relatively low density frequently favors the selection of niobium alloys over other refractory metals, such as molybdenum, tantalum, or tungsten.

High-temperature niobium alloys were developed in the 1960s for nuclear and aerospace applications and today serve in communications satellites, human body imaging equipment, and a variety of high-temperature components. Although niobium alloys have useful strength at temperatures hundreds of degrees above nickel-based superalloys, applications have been limited by their susceptibility to oxidation and too long-term creep.

Production

Typical production electron beam furnaces generate 500–5000 kW of power and are capable of purifying ingots with diameters of 305–508 mm and weights over 2 m. "Drip-melting" is the current standard electron beam method, but hearth melting may eventually become practical as higher-power furnaces are developed.

Alloys of niobium are made by vacuum-arc remelting with the appropriate elemental additions. The most common alloying additions are zirconium, titanium, and hafnium, which readily go into solution during arc melting.

Properties

The most common high-temperature niobium alloys are all hardened and primarily by solid-solution strengthening; however, small amounts of second-phase particles are present. The composition of these particles varies, but they are generally associated with interstitial impurities that form oxides, nitrites, and carbides. These particles are important because the size and distribution of second phases can often have a strong influence on mechanical properties and recrystallization behavior. For example, a variation of the Nb–1%Zr alloy, commonly known as PWC-11, contains an intentional addition of 0.1 wt% carbon, specifically to form carbide precipitates that significantly improve high-temperature creep properties.

Another alloy, WC-3009, normally contains ~0.10 wt% oxygen, which is approximately five times more oxygen than other niobium alloys. This high level of oxygen, which is introduced during powder processing, is not deleterious to mechanical properties because the oxygen combines with hafnium in the alloy to form stable hafnium oxide precipitates. The WC-3009 alloy is unique in that it exhibits an oxidation rate less than 1/10 that of most other niobium alloys. When WC-3009 was developed, it was speculated that such an alloy could survive a short supersonic mission even if its protective coating failed.

In general, niobium alloys are much less tolerant of impurity pickup than other reactive metals such as titanium and zirconium. Alloys containing second-phase particles that form a continuous boundary between grains can exhibit drastically reduced tensile elongation. This condition is usually caused by contamination or improper heat treatment. Copper, which can accidentally be introduced during welding, is particularly disastrous to mechanical properties. In fact, the total permissible interstitial oxygen, hydrogen, carbon, and nitrogen content of niobium alloys is typically 1/5–1/10 that of titanium or zirconium alloys.

All of the commercial alloys are quite ductile at room temperature. The highest tensile strength at room temperature and elevated temperature are exhibited by the WC-3009 alloy. Even though WC-3009 clearly exhibits the highest tensile strength, FS-85 has superior creep strength. Its high creep strength is due to its higher melting point, which is elevated by its high concentration of tantalum and tungsten.

Elastic modulus, thermal conductivity, coefficient of thermal expansion, and total hemispherical emissivity are properties given in Tables N.2 and N.3 in *Adv. Mater. Proc.*, 154(6), 27–31, 1998 as well as tensile properties of the common alloys at 20°C. The emissivity data are for smooth and nonoxidized surfaces, which exhibit much lower emissivity values than oxidized material. Also shown is an emissivity value of 0.7–0.824 silicide-coated C-103. This value is for a common Si–20%Fe–20%Cr coating applied by the "slurry coat and fusion" method.

Applications

The most common application for niobium alloys is in sodium vapor lamps. The Nb–1%Zr alloy demonstrates excellent formability, weldability, and long life in a sodium vapor environment.

These bulbs are used throughout the world for highway lighting, because of their high electrical efficiency and long life, which is typically in excess of 25,000 h. The high-carbon version of this alloy, PWC-11, has a creep rate approximately five times lower than that of Nb–1%Zr at 1100°C, because of the effects of carbide precipitates.

For aerospace applications at 1100°C–1500°C, alloy C-103 has been the workhorse of the niobium industry because of its higher strength. Excellent cold-forming and welding characteristics enable fabricators to construct very complex shapes, such as thrust cones and high-temperature valves. Closed die forgings are also easily produced. Most of these components and propulsion systems are exposed for relatively short times to temperatures between 1200°C and 1400°C.

The service environment for propulsion systems often is less oxidizing than the normal atmosphere. Because C-103 has virtually no oxidation resistance, components are extensively coated with silicides.

Another very successful application for coated C-103 is thrust augmentor flaps in a turbine engine. These flaps, placed at the tail end of the engine to form a high-temperature liner in the afterburner section, typically reach 1200°C–1300°C and last for ~100 h of afterburner time.

Niobium alloys have also been evaluated for various high-temperature components of the National Aerospace Plane. Hypersonic leading edges and nose cones were fabricated to function as heat-pipe thermal management systems. The heat-pipe concept was designed to transport extreme heat away from hot spots, such as hypersonic leading edges, to cooler areas where heat could be expelled by radiation. A typical 500 g niobium heat pipe can dissipate over 10 kW of heat and operate isothermally at 125°C–1350°C. These devices were successfully tested in combustion torches, high-velocity jet fuel burners, tungsten–quartz lamps, and even electric welding arcs and heat flux as well over 1000 W/cm^2.

Nip

In a bearing, the amount by which the outer circumference of a pair of bearing shells exceeds the inner circumference of the housing. Also known as *crush*.

Nip Angle

See *angle of bite*.

Niresist

A proprietary, austenitic, corrosion resistant, cast iron with about 15% nickel, 2% chromium, and a percentage of copper.

Ni-Resist

See *Niresist*.

Nital

Nitric acid (2% unless otherwise stated) in alcohol for etching steels.

Nitinol

A proprietary alloy of 55% nickel and 45% titanium with shape-memory characteristics.

Nitralloy

A proprietary range of steels designed for nitriding and containing aluminum, chromium, and/or molybdenum in the 1% or so range.

Nitric Acid

Also called aqua fortis and azotic acid, nitric acid is a colorless to reddish fuming liquid of the composition HNO_3, having a wide variety of uses for pickling metals, etching, and in the manufacture of nitrocellulose, plastics, dyestuffs, and explosives. It has a specific gravity of 1.502 (95% acid) and a boiling point of 86°C and is soluble in water. Its fumes have a suffocating action, and it is highly corrosive and caustic. Fuming nitric acid is any water solution containing more than 86% acid and having a specific gravity above 1.480. Nitric acid is made by the action of sulfuric acid on sodium nitrate and condensation of the fumes. It is also made from ammonia by catalytic oxidation or from the nitric oxide produced from air.

Nitride–Carbide Inclusion Types

A compound with the general formula $M_x(C,N)$, observed generally as colored idiomorphic cubic crystals, where M includes titanium, niobium, tantalum, and zirconium.

Nitrides

Nitrides are less stable than the oxides, carbides, and sulfides, and their use in air at elevated temperature is limited because of their tendency to oxidize. However, in several instances, the oxide film is protective and deterioration is slow. Despite their limitations, nitrides have interesting properties and are sure to find many specialized uses as technology becomes more complex.

Stable Nitrides

Aluminum Nitride

Aluminum nitride is conveniently prepared by an electric arc between aluminum electrodes in a nitrogen atmosphere. Crucibles of the pressed powder, and sintered at 1985°C, are resistant to liquid aluminum at 1985°C, to liquid gallium at 1316°C, and to liquid boron oxide at 1093°C. Aluminum nitride has good thermal shock resistance and is only slowly oxidized in air (1.3% converted to Al_2O_3 and 30 h at 1427°C). It is inert to hydrogen at 1705°C but is attacked by chlorine at 593°C.

Aluminum nitride is an excellent substrate for creating wide bandgap semiconductors for wireless communications and power industry applications. Since aluminum nitride is stable at very high temperatures, this substrate material can be used for microelectronic devices on jet engines. Such substrates could also improve the production of blue and ultraviolet lasers, which are used to squeeze a full-length movie onto a CD. Aluminum nitride crystals have also been grown in a tungsten crucible at 2300°C.

Boron Nitride

The crystal structure of boron nitride is similar to that of graphite, giving the powder the same greasy feel. The platy habit of the particles and the fact that born nitride is not wet by glass favor the use of the powder as a mold wash, for example, in the fabrication of high-tension insulators. It is also useful as thermal insulation in induction heating.

Boron nitride can be hot pressed to strong ivory-white bodies that are easily machined. Hot-pressed boron nitride is stable in air to about 704°C. From 704°C to 982°C, the rate of oxidation increases moderately. It is also stable in chlorine up to 704°C. However, at 982°C, it is attacked rapidly.

Like commercial graphite, hot-pressed boron nitride is anisotropic. Thermal expansion parallel to the direction of pressing is 10 times than in the perpendicular direction. The ratio for modulus of rupture is 2:1.

Although boron nitride resembles graphite in many respects, it differs uniquely and electrical characteristics, having high resistivity and high dielectric strength even at elevated temperatures. This feature, combined with easy machinability, has led to extensive use in high-temperature electronics.

Boron nitride is available as –325-mesh powder.

A cubic form of boron nitride (Borazon) similar to diamond in hardness and structure has been synthesized by the high-temperature, high-pressure process for making synthetic diamonds. Any uses it may find as a substitute for diamonds will depend on its greatly superior oxidation resistance.

Cutting tool materials like mixed ceramics or CBN cutting tools are already available for hard machining. To identify the proper cutting tool material, one must analyze the application.

In case of cutting interruptions, CBN cutting tools will be the appropriate choice. Continuous cuts allow the use of mixed ceramics or coated mixed ceramics for better efficiency. When producing a gear wheel, for example, turning with a CBN-tipped insert reduces the cost per wheel by more than 60%, compared to grinding. At the same time, disposal costs for grinding sludge vanish, because hard turning does not require coolant.

Silicon Nitride

Silicon nitride is most easily prepared by direct reaction of nitrogen at about 1316°C with finely divided elemental silicon (≤150 mesh), either as loose powder or as a slip-cast or otherwise preformed part. Conversion of silicon particles to the nitride Si_3N_4 is accompanied by the growth of a felt of interlocking needles in the void space between particles. Despite an overall porosity of 15%–25%, silicon nitride bodies are effectively impervious in many applications because of the microscopic size of the pores.

Although silicon nitride is not machinable in its final form except by grinding, the partially converted body can be machined by conventional methods after which conversion can be completed without dimensional change.

Silicon nitride is indefinitely resistant to air oxidation up to 1649°C but begins to sublime at about 1925°C. It is not attacked by chlorine at 899°C or hydrogen sulfide at 982°C nor by the common acids. Because of a low coefficient of thermal expansion, resistance to thermal shock is relatively good.

Uncoated and coated silicon nitride cutting tools dominate the high-performance end of gray cast iron machining. They typically offer metal removal rates at least three times higher than coated carbide grades.

The newly developed cutting tool combines a 6% cobalt substrate with a 10 μm thick, medium-temperature $TiCN/Al_2O_3/TiN$ coating. Medium-temperature chemical vapor deposition TiCN coatings show a reduced tendency for the forming of eta phase at the interface between coating and carbide substrate.

Silicon nitride cutting tools have a substantially improved fracture resistance. Because of their insufficient chemical wear resistance, they have a limited use in machining nodular cast irons, mainly in areas of severe cutting interruptions at higher speeds (>400 m/min).

Titanium and Zirconium Nitride

Titanium and zirconium nitrides for use in refractory bodies are most conveniently prepared by treating the corresponding metal hydrides with ammonia at 1000°C.

Sintered TiN can be heated to a bright red heat with only superficial oxidation and then plunged into water without cracking; ZrN is less resistant to oxidation.

Combination coatings involving both chemical vapor deposition and physical vapor deposition technologies provide the wear-resistance advantages of chemical vapor deposition TiCN coatings with the compressive residual stress advantages of physical vapor deposition TiN coatings. The net result is improved wear and chipping resistance. These coatings increase the speed capabilities of carbide cutting tools and titanium turning by a factor of 2.

Eliminating coolants can turn and easy-to-machine material into a difficult drilling problem, when using standard cutting tools. The introduction of TiAlN coatings represents a significant step toward dry drilling.

The goal in all dry machining is to develop cutting tools with higher resistance to thermal load and fatigue. Cermet tools may be one of the most suitable materials for these applications.

Nitriding

Introducing nitrogen into the surface layer of a solid ferrous alloy by holding at a suitable temperature (below Ac_1 for ferritic steels) in contact with a nitrogenous material, usually ammonia or molten cyanide of appropriate composition. Quenching is not required to produce a hard case. See also *aerated bath nitriding*, *bright nitriding*, and *liquid nitriding* and *case hardening*.

Nitriding Steels

Nitriding steels are alloy steels (low- and medium-carbon steels with combinations of chromium and aluminum or nickel, chromium, and aluminum) design particularly for optimum results when they are subjected to the nitriding operation. The composition is such that the required microstructure for optimum nitriding is produced after heat treatment. Nitrided parts made from nitriding steels have extremely high surface hardness of about 92–95 Rockwell N scale, wear resistance, and resistance to certain types of corrosion.

Processing and Applications

Nitriding consists of exposing steel parts to gaseous ammonia at about 538°C to form metallic nitrides at the surface. The hardest coatings are obtained with the aluminum-bearing steels. Nitriding of stainless steel was known as Malcomizing. The nitride layer also has considerable resistance to corrosion from alkalies, the atmosphere, crude oil, natural gas, combustion products, tap water, and still saltwater. Nitrided parts usually grow about 0.003–0.005 cm during nitriding. The growth can be removed by grinding or lapping, which also removes the brittle surface layer. Most uses of nitrided steels are based on resistance to wear. The steels can also be used at temperatures as high as 538°C for long periods without softening. The slick, hard, and tough nitrided surface also resists seizing, galling, and spalling. Typical applications are cylinder liners and barrels for aircraft engines, bushings, shafts, spindles, and thread guides,

cams, rolls, piston pins, rubber and paper-mill products rolls, special oil tool equipment, bearings, rollers, etc.

Fabrication

The fabrication characteristics of nitriding steels are basically the same as those of other steels of similar alloy content. They can be drilled, broached, tapped, milled, sawed, or ground.

Light feeds and depth of cuts are recommended. Welding is done with rod or wire of similar composition. Flash welding is permissible.

If very heavy cuts are involved, they usually are made prior to heat treatment. Normal machining is done on heat-treated material and is followed by a stress-relieving treatment of not less than 37.8°C above the nitriding temperature, before finish machining or grinding. It is essential that in machining, sufficient removal be allowed to remove all decarburization from the surface prior to nitriding. The surface also must be clean and free of any surface contamination.

Because nitriding is a low-temperature treatment, little or no warpage is encountered. If it is necessary to straighten because of residual stress, the part should be heated to 538°C–593°C to prevent surface cracking.

Nitrided parts normally grow about 0.03–0.05 mm during nitriding. This may be removed by grinding or lapping. This also has the advantage of removing a brittle layer on the surface and exposes a slightly harder layer immediately beneath it. This operation, however, will reduce corrosion resistance to a large degree.

Nitriding steels are available in all standard steel forms. They can be purchased heat treated or annealed to desired physical properties.

Most uses of nitriding steels are based on resistance to wear. An outstanding property is that these steels can be heated to as high as 538°C for long periods without softening.

The slick, hard surface produced also makes it ideal to prevent seizing, galling, and spalling, and it is not readily attacked by combustion products.

Nitrile Rubber

The nitriles are copolymers of butadiene and acrylonitrile, used primarily for applications requiring resistance to petroleum oils and gasoline. Resistance to aromatic hydrocarbons is better than that of neoprene but not as good as that of polysulfide. *Nitrile butyl rubber (NBR)* has excellent resistance to mineral and vegetable oils but relatively poor resistance to the swelling action of oxygenated solvents such as acetone, methyl ethyl ketone, and other ketones. It has good resistance to acids and bases except those having strong oxidizing effects. Resistance to heat aging is good, often a key advantage over natural rubber.

With higher acrylonitrile content, the solvent resistance of an NBR compound is increased but low-temperature flexibility is decreased. Low-temperature resistance is inferior to that of natural rubber, and although NBR can be compounded to give improved performance in this area, the gain is usually at the expense of oil and solvent resistance. As with SBR, this material does not crystallize on stretching, and reinforcing materials are required to obtain high strength. With compounding, nitrile rubbers can provide a good balance of low creep, good resilience, low permanent set, and good abrasion resistance.

Tear resistance is inferior to that of natural rubber, and electrical insulation is lower. NBR is used instead of natural rubber where increased resistance to petroleum oils, gasoline, or aromatic hydrocarbons is required. Uses of NVR, NBR include carburetor and fuel-pump diaphragms and aircraft hoses and gaskets. In many of these applications, the nitriles compete with polysulfides and neoprenes.

Essentially, the same techniques of emulsion polymerization employed in manufacture of general-purpose synthetic rubber may be used for production of nitrile polymers. Nitrile rubbers are supplied in various physical forms including sheet, crumb, powder, and liquid. The sheet is the most widely used type, with the other varieties offered for specialty applications.

Blends

An outstanding feature of nitrile rubber is its compatibility with many different types of resins permitting it to be easily blended with them. In combination with phenolic resins, it provides adhesives with especially high strengths. Other resins used include resorcinol formaldehyde, urea formaldehyde, alkyd, epoxy, and polyvinyl chloride (to produce Type 2 rigid PVC). Both slab- and crumb-type nitrile rubbers are used in this type of application, with the crumb type directly soluble and of special interest to adhesive manufacturers who do not have rubber-mixing equipment. Nitrile rubber-phenolic resin solvent solutions are used in shoe sole attaching adhesives, for structural bonding in aircraft, adhering automotive brake lining to brake shoes, and many other industrial applications.

The powder-type rubbers were also developed for blending with phenolic resins, primarily for the manufacture of improved impact phenolic molding powders.

The liquid nitrile polymer finds use as a tackifier and nonextractable plasticizer in molded rubber parts, cements, friction, and calendered stocks.

Both the liquid and powder are of interest as curing-type plasticizers and vinyl plastisols. Nitrile rubber–PVC blends of various types are used in many other fields including cable jacket, retractable cord, abrasion-resistant shoe soles, industrial face masks, boat bumpers, and fuel lines.

Composition

The ratio of butadiene to acrylonitrile in the commercially available rubbers ranges from a low of about 20% to as high as 50% acrylonitrile. The various grades are usually referred to as high, medium-high, medium-low, and low acrylonitrile content.

The high-acrylonitrile polymers are used in applications requiring maximum resistance to aromatic fuels, oils, and solvents. This would include oil well parts, fuel-cell liners, fuel hose, and other similar applications. The low acrylonitrile grade finds use in those areas requiring good flexibility at very low temperatures where oil resistance is of secondary importance. The medium types are most widely used and are satisfactory for all oil-resistant applications between these two extremes. Typical applications include conveyor belts, flexible couplings, soles, heels, floor mats, printing blankets, rubber rollers, sealing strips, aerosol bomb gaskets, milking inflations, seals, diaphragms, O-rings, packings, hose, washing machine parts, valves, and grinding wheels. These established uses give only a slight idea of products that are made of nitrile rubbers.

Physical properties of cured nitrile rubber parts are directly related to the ratio of butadiene and acrylonitrile in the polymer, as indicated in the following.

As acrylonitrile content increases

1. Oil and solvent resistance improves
2. Tensile strength increases
3. Hardness increases
4. Abrasion resistance improves
5. Gas impermeability improves
6. Heat resistance improves

As acrylonitrile content decreases

1. Low-temperature resistance improves
2. Resilience increases
3. Plasticizer compatibility increases

Properties

The polymer with the highest acrylonitrile content produces the highest tensile strength and hardness; it also exhibits the best resistance to fuels and oils. As the percentage of acrylonitrile decreases, there is a corresponding decrease in resistance to fuels and oils; at the same time, low-temperature flexibility characteristics are improved. Resiliency also increases. The lowest acrylonitrile polymer exhibits only moderate resistance to swelling in aromatic fluids but remains flexible at very low temperatures in the range of $-57°C$ to $-62°C$.

Thus, properly compounded nitrile polymers will provide high tensile strength, excellent resistance to abrasion, low compression set, very good aging under severe operating conditions, and excellent resistance to a wide range of fuels, oils, and solvents. They are practically unaffected by alkaline solutions, saturated salt solutions, and aliphatic hydrocarbons, both saturated and unsaturated. They are affected little by fatty acids found in vegetable fats and oils or by aliphatic alcohols, glycols, or glycerols.

Nitrile rubbers are not recommended, generally, for use in the presence of strong oxidizing agents, ketones, acetates, and a few other chemicals.

Nitrocarburizing

Salt bath nitrocarburizing is a thermochemical process for improving the properties of ferrous metals. However, some tools and other high-alloy steels are susceptible to reductions in core hardness after standard nitrocarburizing. To prevent such losses, a low-temperature salt bath nitrocarburizing process has been developed with treatment temperatures as low as 480°C; this process not only maintains core hardness but can also sometimes increase core hardness.

Processing

During salt bath nitrocarburizing, the part is immersed in a vessel of molten salt. Nitrogen and carbon in the salt react with the iron on the surface, forming a compound layer with an underlying diffusion zone. The compound layer consists of iron nitrides, chromium nitrides, or other such compounds, depending on the alloying elements in the steel, and small amounts of carbides.

Ranging in depth from 2.5 to 20 μm, the compound layer provides improvements in wear and corrosion resistance, as well as in service behavior and hot strength. Hardness of the compound layer, measured on a cross section, ranges from 700 HV on unalloyed steels up to 6000 HV on high chromium steels. Note that, this layer is formed from the base metal and is an integral part of it and is therefore not a coating. The diffusion zone can extend as deep as 1 mm, depending on the steel. This diffusion zone causes an increase in rotating–bending strength and rolling fatigue strength as well as pressure loadability.

Salt bath nitrocarburizing may be applied to a wide range of ferrous metals, from low-carbon to tool steels, cast iron to stainless steels. Specifically, the process

• Improves wear and corrosion resistance
• Reduces or eliminates galling and seizing
• Increases fatigue strength

- Raises surface hardness
- Provides highly predictable, repeatable results
- Performs consistently, even with varying contours and thicknesses within the same part or load
- Maintains dimensional integrity
- Shortens cycle times
- Offers flexibility and ease of operation

Conventional treatment temperatures are in the range of 580°C, but for highly alloyed steels as well as stainless steel and tool steels, this temperature can cause a reduction in core hardness. The aforementioned benefits, derived both from the nitrogen and carbon diffused into the metal surface, as well as the processing in a liquid bath, are often necessary for applications in which a reduction in core hardness is not acceptable. For this reason, a new low-temperature process was developed.

The low-temperature process normally takes place at 480°C, although it can operate at 480°C–520°C. This process has the following specific advantages:

- Core hardness and tensile strength are maintained in the tempered condition.
- Very thin compound layers can be formed.
- Distortion is extremely low.
- Formation of a compound layer on high-speed steels can be suppressed.
- Hardness of surface and diffusion layers can be customized.

This low-temperature process is beneficial for high-alloy steels such as stainless, tool, die, and high-speed steels. The steels include D2, D3, AISI 420, H11, H13, HNV3, 17-4PH, HSS36, and HSS M2.

Nitrocellulose

A compound made by treating cellulose with nitric acid, using sulfuric acid as a catalyst. Since cotton is almost pure cellulose, it was originally the raw material used, but alpha cellulose made from wood is now employed. The cellulose molecule will unite with from one to six molecules of nitric acid. *Trinitrocellulose* contains 9.13% nitrogen and is the product used for plastics, and lacquers, adhesives, and celluloid. It is classified as cellulose nitrate. The higher nitrates, or *pyrocellulose*, are employed for making explosives. *FNH powder*, or *flashless powder*, is nitrocellulose, which is nonhygroscopic and which contains a partially inert coolant, such as potassium sulfate, to reduce the muzzle flash of the gun. *Ballistite* is a rapid-burning, double-base powder used in shotgun shells and as a propellant in rockets. It is composed of 60% nitrocellulose and 40 nitroglycerin made into square flakes 0.005 in. (0.013 cm) thick or extruded in cruciform blocks.

Nitrogen

Nitrogen is an element (symbol N) that at ordinary temperatures is an odorless and colorless gas. The atmosphere contains 78% nitrogen in the free state. It is nonpoisonous and does not support combustion. Nitrogen is often called an inert gas and is used for some inert atmospheres for metal treating and in light bulbs to prevent arcing, but it is not chemically inert. It is a necessary element in animal and plant life, and as a constituent of many useful compounds. Lightning forms small amounts of nitric oxide from the air, which is converted into nitric acid and nitrates, and bacteria continuously convert atmospheric nitrogen into nitrates.

Nitrogen combines with many metals to form hard nitrides useful as wear-resistant metals. Small amounts of nitrogen in steels inhibit grain growth at high temperatures and also increase the strength of some steels. It is also used to produce a hard surface on steels.

Applications

Because of the importance of nitrogen compounds in agriculture and chemical industry, much of the industrial interest in elementary nitrogen has been in processes for converting elemental nitrogen into nitrogen compounds. The principal methods for doing this are the direct synthesis of ammonia from nitrogen and hydrogen, the electric arc process, which involves the direct combination of N_2 and O_2 to nitric oxide, and the cyanamide process.

Nitrogen Blanking

The use of nitrogen to produce an inert atmosphere.

Nitrogen Case Hardening

Nitriding. See *case hardening*.

Nitroglycerin

A heavy, oily liquid known chemically as *glyceryl trinitrate*. It is made by the action of mixed acid (90% nitric at 25–30 oleum) on very pure glycerol in the presence of sulfuric acid. It is highly explosive, detonating upon concussion. Liquid nitroglycerin when exploded forms carbonic acid, CO_2, water vapor, nitrogen, and oxygen; 1 lb (0.45 kg) is converted into 156.7 ft³ (4.4 m³) of gas. The temperature of explosion is about 628°F (330°C). For use as a commercial explosive, it is mixed with absorbents, usually kieselguhr or wood flour, under the name of *dynamite*. Cartridges of high density explode with greater shattering effect than those of low density. By varying the density and the mixture of the nitroglycerin with ammonium nitrate, which gives a heaving action, a great diversity in properties can be obtained. *Ethylene glycol dinitrate* (nitroglycol) and *diethylene glycol dinitrate* are also explosives. They are generally used to plasticize nitrocellulose.

Dynamites are rated on the percentage, by weight, of nitroglycerin that they contain. A 25% dynamite has 25% by weight of nitroglycerin. The regular grades range from 25% to 60%. *Ditching dynamite* has a 50% grade. *Extra dynamite* has half of the nitroglycerin replaced by ammonium nitrate. It is not so quick and shattering, and not as water-resistant, but is lower in cost. It is used for quarrying, stump and boulder blasting, and highway work.

Gelatin dynamite is made by dissolving a special grade of nitrocotton in nitroglycerin. It has less fumes, it is more water resistant, and its plasticity makes it more adaptable for loading solidly in holes for underground work. It is marketed as straight gelatin or as *ammonium gelatin*, called *gelatin extra*. The gelatin dynamites come in grades from 20% to 90%. *Blasting gelatin*, called *oil-well explosive*, is a 100% dense and waterproof gelatin with the appearance of crude rubber.

Dynamite is also sometimes used for explosive metal forming, as it releases energy at a constant rate regardless of confinement, and produces pressures up to 2×10^6 lb/in.² (1.379 MPa). For bonding *metal laminates*, a thin sheet, or film, of the explosive is placed on top of the composite, and the progressive burning of the explosive across the film produces an explosive force downward and in vectors that produces a microscopic wave, or ripple, in the alloy bond that strengthens the bond but is not visible on a laminated sheet.

N Shell

The fourth layer of electrons surrounding the nucleus of an atom, having electrons with the principal quantum number 4.

NLGI Number

Abbreviation for the National Lubricating Grease Institute number, which is the numerical classification of the consistency of greases, based on the ASTM D 217 test.

NMR

See *nuclear magnetic resonance*.

No-Bake Binder

In foundry practice, a synthetic liquid resin sand binder that hardens completely at room temperature, generally not requiring baking; used in a *cold-setting process*.

Noble

(1) The positive direction of *electrode potential*, thus resembling noble metals such as gold and platinum. (2) In electrochemistry, the more cathodic of a pair of metals in the electrochemical series of relative potentials.

Noble Gases

The six inert gases, helium, neon, argon, krypton, xenon, and radon. More usually termed the rare gases.

Noble Metal

(1) A metal whose *potential* is highly positive relative to the hydrogen electrode. (2) A metal with marked resistance to chemical reaction, particularly to oxidation and to solution by inorganic acids. The term as often used is synonymous with *precious metal*. (3) Generally, metals that do not readily react with common environments, for example, gold, platinum, and palladium. Silver is usually also included.

Noble Potential

A *potential* more cathodic (positive) than the standard hydrogen potential.

Node

The connected portion of adjacent ribbons of honeycomb.

No-Draft (Draftless) Forging

A forging with extremely close tolerances and little or no *draft* that requires minimal machining to produce the final part. Mechanical properties can be enhanced by closer control of grain flow and by retention of surface material in the final component.

Nodular Cast Iron

Nodular cast irons, such as GGG 40, GGG 50, and GGG 60, have become popular for parts such as housings, wheel parts, crankshafts, and camshafts. These metals offer higher strength and toughness than other cast irons, a result of spherical inclusions of carbon in the metal matrix. Generally easy to machine, GGG 40 irons with higher ferrite content tend to produce built-up edges on the cutting tool. For materials such as GGG 60 and higher, abrasiveness increases as the pearlite content increases, which can result in rapid insert wear. These nodular iron grades present unique machining characteristics. See preferred term *ductile iron*.

Nodular Graphite

(1) Graphite in the nodular form as opposed to flake form (see *flake graphite*). Nodular graphite is characteristic of *malleable iron*. The graphite of nodular or *ductile iron* is spherulitic in form but called nodular. (2) One of the seven graphite shapes used to classify cast irons.

Nodular Pearlite

Pearlite that has grown as a colony with an approximately spherical morphology.

Nodular Powder

Irregularly shaped metal powder having knotted, rounded, or similar shapes.

Noise

Any undesired signal that tends to interfere with the normal reception or processing of a desired signal.

NOL Ring

A parallel filament- or tape-wound hoop test specimen developed by the Naval Ordnance Laboratory (NOL) (now the Naval Surface Weapons Laboratory) for measuring various mechanical strength properties of a material, such as tension and compression, by testing the entire ring or segments of it. Also known as parallel fiber-reinforced ring.

Nomarski Interference Technique

A microscopical technique utilizing polarized light to reveal slight topographical variations.

Nominal

This term is used vaguely in technical contexts to indicate concepts such as a token amount or a broad overall view.

Nominal Area of Contact

The area bounded by the periphery of the region in which macroscopic contact between two solid bodies is occurring. This is often taken to mean the area enclosed by the boundaries of a wear scar, even though the real area of contact, in which the solids are touching instantaneously, is usually much smaller. See also *area of contact* and *apparent area of contact*.

Nominal Dimension

The size of the dimension to which the tolerance is applied. For example, if a dimension is 50 ± 0.5 mm (2.00 ± 0.02 in.), the 50 mm (2.00 in.) is the nominal dimension and the ± 0.5 mm (± 0.02 in.) is the tolerance.

Nominal Strength

See *ultimate strength*.

Nominal Stress

(1) The stress at a point calculated on the net cross section without taking into consideration the effect on stress of geometric discontinuities, such as holes, grooves, fillets, and so forth. The calculation is made using simple classic theory. (2) The stress is calculated on the basis of the external load acting on the minimum original dimensions of the component or test piece and ignoring matters such as stress raising effects at section changes or deformation under stress.

Nominal Value

A value assigned for the purpose of a convenient designation. A nominal value exists in name only. In dimensions, it is often an average number with a tolerance in order to fit together with adjacent parts.

Nonconformal Surfaces

(1) Surfaces whose centers of curvature are on the opposite sides of the interface, as in rolling element bearings or gear teeth. (2) In wear testing, a geometric configuration in which a "point" or "line" of contact is initially established between specimens before the test is started. Examples of nonconformal contacts are ball-on-ring and flat block-on-ring geometries.

Nonconforming

A quality control term describing a unit of product or service that does not meet normal acceptance criteria for the specific product or service. A nonconforming unit is not necessarily *defective*.

Noncontact Bearing

A bearing in which no solid contact occurs between relatively moving surfaces. Strictly speaking, a bearing in which *full-film lubrication* is occurring would be considered a noncontact bearing; however, this term is more typically applied to gas bearings and magnetic bearings. See also *gas lubrication* and *magnetic bearing*.

Noncorrosive Flux

A rosin-based soldering flux that, in neither its original nor residual form, chemically attacks the base metal.

Nondestructive Evaluation (NDE)

Broadly considered synonymous with nondestructive inspection (NDI). More specifically, the quantitative analysis of NDI findings to determine whether the material will be acceptable for its function, despite the presence of discontinuities. With NDE, a discontinuity can be classified by its size, shape, type, and location, allowing the investigator to determine whether or not the flaw(s) is acceptable. Damage-tolerant design approaches are based on the philosophy of ensuring safe operation in the presence of flaws.

Nondestructive Inspection

A process or procedure, such as ultrasonic or radiographic inspection, for determining the quality or characteristics of a material, part, or assembly, without permanently altering the subject or its properties. Used to find internal anomalies in a structure without degrading its properties or impairing its serviceability.

Nondestructive Testing/Examination

Any test or inspection system that provides information regarding the properties and characteristics of a component without causing unacceptable damage. These terms are most commonly used with respect to flaw detection systems, for example, dye penetrant and magnetic particle testing for surface defects and ultrasonic testing or radiography for internal defects. However, they may also include a wide range of techniques meeting the "no unacceptable damage" criterion including chemical spot tests or other analytical techniques, hardness measurement, and dimensional checks particularly when these are performed on-site.

Nonelectrolyte

A substance that does not ionize in solution. See *inter atomic bonding*.

Nonferrous

Metals and alloys not based on iron.

Nonfill

A forging condition that occurs when the die impression is not completely filled with metal. Also referred to as *underfill*.

Nonhygroscopic

Lacking the property of absorbing and retaining an appreciable quantity of moisture (water vapor) from the air.

Nonmagnetic

An imprecise term usually indicating that the material is not ferromagnetic. See *magnetic*.

Nonmagnetic Steels

These are steel and iron alloys used where magnetic effects cannot be tolerated. Manganese steel containing 14% manganese is nonmagnetic and casts readily but is not machinable. Nickel steels and iron–nickel alloys containing high nickel content are nonmagnetic. Many mills regularly produce nonmagnetic steels containing from 20% to 30% nickel. Manganese–nickel steels and manganese–nickel–chromium steels are nonmagnetic and may be formulated to combine desirable features of the nickel and manganese steels. One nonmagnetic steel with a composition of 10.5%–12.5% manganese, 7%–8% nickel, and 0.25%–0.40% carbon has low magnetic permeability at low eddy current loss, can be machined readily, and work-hardens only slightly. The tensile strength is 551–2758 MPa, elongation 25%–50%, and specific gravity 8.02. It is austenitic and cannot be hardened. The 18–8 austenitic chromium–nickel steels are also nonmagnetic. A nonmagnetic alloy steel for watch gears and escapement wheels is not a steel but is a copper–nickel–manganese alloy containing 60% copper, 20% nickel, and 20% manganese. It is very hard but can be machined with diamond tools.

Nonmetallic Inclusions

See *inclusions*.

Non-Newtonian Viscosity

The apparent viscosity of a material in which the shear stresses are not proportional to the rate of *shear*.

Nonpropagating Cracks

Cracks that have ceased to grow. The term is most common in fatigue where a number of cracks initiate but only one continues to extend. However, many forms of cracks can become dormant because of some change in the stress field local to the crack tip or because of a change in the environment.

Nonresonant Forced and Vibration Technique

A technique for performing dynamic mechanical measurements in which a plastic sample is oscillated mechanically at a fixed frequency. Storage modulus and damping are calculated from the applied strain and the result of stress and shift in phase angle.

Nonrigid Plastic

For purposes of general classification, a plastic that has a modulus of elasticity either in flexure or in tension of not over 70 MPa (10 ksi) at 23°C (73°F) and 50% relative humidity.

Nonshattering Glass

Also referred to as *shatterproof glass*, *laminated glass*, or *safety glass*, and when used in armored cars, it is known as *bulletproof glass*. A material composed of two sheets of plate glass with a sheet of transparent resinoid between; the whole molded together under heat and pressure. When subjected to a severe blow, it will crack without shattering. The first of these was a German product marketed under the name of *Kinonglas*, which consisted of two clear glass plates with a cellulose nitrate sheet between, and it was first used for protective shields against chips from machines. Nonshattering glass is now largely used for automobile and car windows. The original cellulose nitrate interlining sheets had the disadvantage that they were not stable to light and became cloudy. Cellulose acetate was later substituted. It is opaque to actinic rays and prevents sunstroke but

has the disadvantage of opening in cold weather, permitting moisture to enter between the layers. The acrylic resins are notable for their stability in this use; in some cases, they are used alone without the plate glass, especially for aircraft windows. Polyvinyl acetal resins, as interlining for safety glass, are weather resistant and will not discolor. Polyvinyl butyral is much used as an interlayer, but an airplane glass at about 150°F (66°C) tends to bubble and ripple. Silicone resins used for this purpose withstand heat up to 350°F (177°C), and they are not brittle at subzero temperatures. Standard bulletproof glass is from 1.5 in. (3.81 cm), 3 ply, to 6 in., 5 or more ply.

Nonshielded Welding

Metal arc welding processes in which no shielding medium is applied to the weld zone to protect from contamination, etc.

Nonsoap Grease

A grease made with a thickener other than soap, such as clay or asbestos.

Nontransferred Arc

In plasma arc welding, cutting, and thermal spraying, an arc established between the electrode and the constricting nozzle. The workpiece is not in the electrical circuit. Compare with *transferred arc*.

Nonwetting

(1) The lack of metallurgical *wetting* between molten solder and a metallic surface due to the presence of a physical barrier on the metallic surface. (2) A condition in which a surface has contacted molten solder, but the solder has not adhered to all of the surfaces; base metal remains exposed.

Nonwoven Fabric

A planar textile structure for reinforced plastics produced by loosely compressing to gather fibers, yarns, rovings, and so forth, with or without a scrim cloth carrier. Accompanied by mechanical, chemical, thermal, or solvent means, or combinations thereof.

In the most general sense, nonwoven fabrics are fibrous-sheet materials consisting of fibers mechanically bonded together by interlocking or entanglement, by fusion, or by an adhesive. They are characterized by the absence of any patterned interlooping or interlacing of the yarns. In the textile trade, the terms *nonwovens* and *bonded fabrics* are applied to fabrics composed of a fibrous web held together by a bonding agent, as distinguished from felts, in which the fibers are interlocked mechanically without the use of a bonding agent. There are three major kinds of nonwovens based on the method of manufacture. Dry-laid nonwovens are produced by textile machines. The web of fibers is formed by mechanical or air-laying techniques, and bonding is accomplished by fusion bonding the fibers or by the use of adhesives or needle punching. Either natural or synthetic fibers, usually 2.5–7.6 cm in length, are used. Wet-laid nonwovens are made on modified papermaking equipment. Either synthetic fibers or combinations of synthetic fibers and wood pulp can be used. The fibers are often much shorter than those used in dry-laid fabrics, ranging from 0.64 to 1.27 cm in length. Bonding is usually accomplished by a fibrous binder or an adhesive. Wet-laid nonwovens can also be produced as composites, for example, tissue paper laminates bonded to a reinforcing substrate of scrim. Spin-bonded nonwovens are produced by allowing the filaments emerging from the fiber-producing extruder to form into a random web, which is then usually thermally bonded. These nonwovens are limited commercially to thermoplastic synthetics such as nylons, polyesters, and polyolefins. They have exceptional strength because the filaments are continuous and bonded to each other without an auxiliary bonding agent. Fibers in nonwovens can be arranged in a great variety of configurations that are basically variations of three patterns: parallel or unidirectional, crossed, and random. The parallel pattern provides maximum strength in the direction of fiber alignment but relatively low strength in other directions. Cross-laid patterns (like wovens) have maximum strength in the direction of the fiber alignments and less strength in other directions. Random nonwovens have relatively uniform strength in all directions.

Normal

An imaginary line forming right angles with a surface or other lines sometimes called the perpendicular. It is used as a basis for determining angles of incidence reflection and refraction.

Normal Direction

That direction perpendicular to the plane of working in a worked material. See also *longitudinal direction* and *transverse direction*.

Normal Force

See *load*.

Normal Load

See *load*.

Normal Segregation

Segregation of the lower-melting-point constituents into the areas of a casting that solidify last. See *inverse segregation*.

Normal Solution

A solution having the equivalent weight in grams of the substance dissolved in 1 L of water.

Normal Stress

The stress component that is perpendicular to the plane on which the forces act. Normal stress may be either *tensile* or *compressive*.

Normal Temperature and Pressure

A reference environment conventionally, 0°C and 760 mm mercury.

Normal Wear

Loss of material within the design limits expected for the specific intended application. The concept of normal wear depends on economic factors, such as the expendability of a worn part.

Normality

A measure of the number of gram-equivalent weights of a compound per liter of solution. Compare with *molarity*.

Normalized Erosion Resistance

The volume loss rate of a specified reference material divided by the volume loss rate of a test material obtained under similar testing and analysis conditions. "Similar testing and analysis conditions" means that the volume loss rates of the two materials are determined at the corresponding portions of the erosion rate-time pattern, for example, the maximum erosion rate or the terminal erosion rate.

Normalizing

(1) Heating a ferrous alloy to a suitable temperature above the transformation range and then cooling in air to a temperature substantially below the transformation range. (2) The practice of heating to the austenitic range and cooling in air. This achieves recrystallization, eliminates residual stresses, and renders the steel relatively soft and ductile, although not quite as soft as the fully annealed condition. For steel capable of hardening upon air cooling, the term is inappropriate. See *steel*.

Normal-Phase Chromatography

This refers to liquid–solid chromatography or to bonded-phase chromatography with a polar stationary phase and a nonpolar mobile phase. See also *bonded-phase chromatography* and *liquid–solid chromatography*.

Nose Radius

The radius of the rounded portion of the cutting edge of a tool. See the term *single-point tool*.

Nosing

Closing in the end of a tubular shape to a desired curve contour.

Notch

See *stress concentration*.

Notch Acuity

Relates to the severity of the *stress concentration* produced by a given notch in a particular structure. If the depth of the notch is very small compared with the width (or diameter) of the narrowest cross section, acuity may be expressed as the ratio of the notch depth to the notch root radius. Otherwise, acuity is defined as the ratio of one-half the width (or diameter) of the narrowest cross section to the notch root radius. Same as *notch sharpness*.

Notch Brittle/Ductile, etc.

These terms refer to the notch sensitivity of a material, that is, its behavior and strength in the presence of a notch, which may be intentional, as in an impact test or inadvertent, as with a fatigue crack in service. The basic concept in determining notch sensitivity is the difference in strength obtained from testing two bars of the material in question, one of which is plain and the other is notched. The strength of the notched bar is calculated on the remaining intact cross section beyond the notch. If the notched bar gives a higher strength, the material is termed notch ductile or notch strengthening; if lower, it is termed notch brittle or notch weakening. Alternatively, the two strengths may be identified in terms of the notch strength ratio whereby a value less than 100% is, conventionally, notch weak behavior. It should be noted, however, that the strength of the notched bar may be sensitive to notch geometry, particularly notch tip radius. The term notch ductile steel is sometimes used specifically for steels that meet some impact test specification.

Notch Constriction Ratio

A measure of the severity of a notch in terms of notch depth relative to section thickness. See also *notch sharpness*.

Notch Depth

The distance from the surface of a test specimen to the bottom of the notch. In a cylindrical test specimen, the percentage of the original cross-sectional area removed by machining an annular groove.

Notch Ductile

See *notch brittleness* or *notch brittle*.

Notch Ductility

The percentage reduction in area after complete separation of the metal in a tensile test of a *notched specimen*. The percentage reduction in cross-section area is calculated from the original cross-section area at the tip of the notch.

Notch Factor

Ratio of the resilience determined on a plain specimen to the resilience determined on a notched specimen.

Notch Rupture Ductility

Same as *notch ductility*.

Notch Rupture Strength

The ratio of applied load to original area of the minimum cross section in a *stress-rupture test* of a *notched specimen*.

Notch Sensitivity

The extent to which the sensitivity of a material to fracture is increased by the presence of a *stress concentration*, such as a notch, a sudden change in cross section, a crack, or a scratch. Low notch sensitivity is usually associated with ductile materials, and high notch sensitivity is usually associated with brittle materials. See *notch brittle*.

Notch Sharpness/Acuity

The sharpness at the tip of a notch or crack. It may be defined as a notch tip radius, some ratio between the radius and the specimen geometry, often specimen diameter: 2× notch tip radius. See *notch acuity* and *notch constriction ratio*.

Notch Specimen

A test specimen that has been deliberately cut or notched, usually in a V-shape, to induce and locate point of failure.

Notch Strength

The maximum load on a notched test specimen divided by the minimum cross-sectional area (the area at the root of the notch). Also called *notch tensile strength*. See *notch brittle*.

Notch Tensile Strength

See *notch strength*.

Notch Toughness

An imprecise term variously referring to the results of notched bar impact tests or to notch ductility or notch sensitivity. See *notch brittle*.

Notch Weak

See *notch brittle*.

Notched Bar Test

Any test in which the specimen has a notch positioned across the line of principal stress. In some contexts, the term is taken to refer specifically to bend test, but more generally, it includes impact, tensile, creep, bend and drop weight tests.

Notching Press

A mechanical press used for notching internal and external circumferences and also for notching along a straight line. These presses are equipped with automatic feeds because only one notch is made per stroke.

Novolac

A linear, thermoplastic, two-stage phenolic resin, which, in the presence of methylene or other cross-linking groups, reacts to form a thermoset phenolic. See also *phenolics*.

Nozzle (Metals)

(1) Pouring spout of a bottom-pour ladle. (2) On a hot-chamber die-casting machine, the thick-wall tube that carries the pressurizer molten metal from the gooseneck to the die. (3) A device that directs shielding media in a welding or cutting torch. See also *bottom-pour ladle*, *hot chamber machine*, and *welding torch (arc)*.

Nozzle (Plastics)

The hollow-cored metal nose screwed into the injection end of either the heating cylinder of an injection machine or a transfer chamber when this is a separate structure. A nozzle is designed to form under pressure a seal between the heating cylinder or the transfer chamber and the mold. The shape of the front end of a nozzle may be either flat or spherical. See also *reciprocating screw injection molding*.

Nuclear

Referring to the nucleus of the atom.

Nuclear Cermet Fuel

A sintered fuel rod composed of a fissile carbide or oxide constituent and a metallic matrix.

Nuclear Charge

See *atomic number*.

Nuclear Cross Section (σ)

The probability that a nuclear reaction will occur between a nucleus and a particle, expressed in units of area (usually barns). See also *barn*.

Nuclear Magnetic Resonance

A phenomenon exhibited by a large number of atomic nuclei that is based on the existence of nuclear magnetic moments associated with quantized nuclear spins. These nuclear moments, when placed in a magnetic field, give rise to distinct nuclear Zeeman energy levels between which spectroscopic transitions can be induced by radio-frequency radiation. Plots of these transition frequencies, termed spectra, furnish important information about molecular structure and sample composition. See also *Zeeman effect*.

Nuclear Power

The controlled use of nuclear energy released by nuclear fission, that is, splitting of the atomic nucleus, or by chain reaction. In a nuclear reactor, the heat energy is used to produce steam to drive a turbine generator producing electrical power. Nuclear fusion occurs when elements with only a few neutrons per nucleus are treated to a very high temperature, which provides sufficient energy to initiate the fusion of the two simple nuclei to form an element with a larger nucleus. Again, the process is accompanied by the release of a large quantity of heat. In this fusion, hydrogen with only one neutron fuses to form helium with two. On earth, the process is the basis of the hydrogen bomb, but its development for civil use remains elusive.

Nuclear Structure

The atomic nucleus at the center of the atom, containing more than 99.975% of the total mass of the atom. Its average density is approximately 3×10^{11} kg/cm³, its diameter is approximately 10^{-12} cm, and thus is much smaller than the diameter of the atom, which is approximately 10^{-8} cm. The nucleus is composed of protons and neutrons. The number of protons is denoted by Z and the number of neutrons by N. The total number protons and neutrons in a nucleus is termed the mass number and is denoted by $A = N + Z$. See also *atomic structure*, *electrons*, *neutrons*, *protons*, *isobar*, *isotone*, and *isotope*.

Nucleate Boiling (of Water)

The production, at a heated surface, of steam as discrete bubbles rather than as a broad persistent film as occurs in steam blanketing.

Nucleating Agent (Plastics)

A foreign substance, often crystalline, usually added to a crystallizable polymer to increase its rate of solidification during processing.

Nucleation (Metals)

The initiation of a phase transformation at discrete sites, with the new phase growing on the nuclei. See also *nucleus* (2).

Nucleus

(1) The heavy central core of an atom, in which most of the mass and the total positive electric charge are concentrated. (2) The first structurally stable particle capable of initiating recrystallization of a phase or the growth of a new phase and possessing an interface with the parent metallic matrix. The term is also applied to a foreign particle that initiates such action.

Nuclide

A species of atom distinguished by the constitution of its nucleus. Nuclear constitution is specified by the number of protons, the number of neutrons, and energy content or by atomic number, mass number, and atomic mass. See also *isotope*.

Nugget

1. A small mass of metal, such as gold or silver, found free in nature.
2. The weld metal in a spot, seam, or projection weld.

Nugget Size

The diameter or width of the *nugget* obtained during resistance welding that is measured in the plane of the interface between the pieces joined.

Numerical Aperture

A measure of the ability of the objective lens of a microscope to gather light from a wide angle. It is a function of the angle of the cone of light that the lens will accept and the refractive index of the medium between the lens and the specimen, usually air but sometimes oil. For a given magnifying power and focal length, the larger the numerical aperture, the better the resolution, that is, the fineness of detail that can be discerned.

Numerical Sphere

The measure of the light-collecting ability of an objective lens in a light microscope. It is defined as

$$NA = n \sin \alpha$$

where
   n is the minimum refraction index of the medium (air or oil) between the specimen and the lens
   α is the half-angle of the most oblique light rays that enter the front lens of the objective

n-Value

See strain-hardening exponent.

Nylon

The generic name, by common usage, for all synthetic polyamides.

Nylon Plastics

Although nylon polymers are most familiar in fiber form, their combination of excellent chemical and mechanical properties, plus their ability to be molded and extruded into precise forms, has permitted their use in a wide variety of nontextile applications. These range from hammerheads, gears, and rifle stocks to miniature coil forms and delicately colored personal products.

Characteristics

Nylons are a group of polyamide resins that are long-chain polymeric amides in which the amide groups form an integral part of the main polymer chain and that have the characteristic that when formed into a filament the structural elements are oriented in the direction of the axis. Nylon was originally developed as a textile fiber, and high tensile strengths, above 344 MPa, are obtainable in the fibers and films. But this high strength is not obtained in the molded or extruded resins because of the lack of oriented stretching. When nylon powder that has not precipitated from solution is pressed and sintered, the parts have high crystallinity and very high compressive strength, but they are not as tough as molded nylon.

Production

Nylons are produced from the polymerization of a dibasic acid and a diamine. The most common one of the group is that obtained by the reaction of adipic acid with hexamethylenediamine.

More specifically, nylons may be made by the amidation of diamines with dibasic acids, for example, hexamethylenediamine plus adipic acid (Type 6/6 nylon) or hexamethylenediamine and sebacic acid (Type 6/10 nylon). They can also be made by the polymerization of amino acids or their derivatives, for example, polycaprolactam (Type 6 nylon) and polymerized 11-aminoun-decanoic acid (Type 11 nylon).

Types

Nylon 6/6 is the most widely used of the nylon plastics because of its overall balance of properties. The second most widely used of the nylon family is nylon 6. Type 6/6 nylon resins have higher heat resistance, abrasion resistance, stiffness, and hardness than type 6 nylons. The type 6 nylon resins are tougher and more flexible than type 6/6 nylons, and they have a wider processing window.

Nylon 6/12 absorbs less moisture and, therefore, maintains most mechanical and electrical properties better in high-humidity environments. But the reduced moisture sensitivity is accompanied by the lower strength, low or stiffness, lower use temperatures, and higher costs.

Nylons 11 and 12 have lower moisture absorption combined with superior resistance to fuels, hydraulic oils, and most automotive fluids. The melting point of nylons 11 and 12 (180°C–185°C) is the lowest of the commercial polyamides. These two polyamides are often combined with plasticizers to generate a flexible, tough material suitable for tubing extrusion. Recently, nylon 12/12 was introduced with a slightly higher use temperature while maintaining good fuel resistance. Nylon 6/6T resins have low moisture absorption, and they are much stronger, stiffer, tougher, fatigue resistant, and more heat resistant than type 6/6 nylons. The Ultramid type 6/6 T resins also have the better resistance to hot oils and fats than type 6/6 nylons. Reinforced grades of type 6/6T nylon resins also are available.

Nylon 4/6 is the latest version of the short-repeat-unit polyamides. Its melting point of 295°C is 12.2°C above that for nylon 6/6 and is the highest in the polyamide family. The inherent molecular symmetry of nylons 4/6 results in self-nucleation, rapid crystal growth, and, thus, a higher level of crystallinity. This higher level of crystallinity leads to faster setup and, hence, faster injection-molding cycles, up to 30% faster than for 6/6. Nylon 4/6 absorbs more water than nylon; however, its dimensional stability is similar to nylon 6/6 due to its high crystallinity.

Higher crystallinity has a major effect on nearly all properties leading to higher strength, higher stiffness, high HDT, high fatigue resistance, high wear resistance, and high creep resistance. Semicrystalline polymers maintain useful properties above the glass transition in contrast to amorphous polymers, which transformed into a viscous mass. Nylon 4/6, with its unusually high crystallinity, maintains a higher level of performance at elevated temperatures. The HDT for reinforced nylon 4/6 is 545°C.

Nylon resin is available in a wide range of reinforcement levels, filler types, toughening agents, stabilizers, and flame-retardant additives. Newer flame retardants can provide good flammability ratings (UL 94V-0) while maintaining acceptable electrical properties.

Toughening technology has reduced sensitivity providing notched Izod values over 15 ft lb/in.

Properties

Property comparisons among commercial grades of nylon vary widely because so many formulations are available. In general, however, nylons of excellence fatigue resistance, low coefficient of friction, good toughness (depending on degree of crystallinity), and they resist a wide spectrum of fuels, oils, and chemicals. They are inert to biological attack and have adequate electrical properties for most voltages and frequencies.

The crystalline structure of nylons, which can be controlled to some degree in processing, affects stiffness, strength, and heat resistance. Low crystallinity imparts greater toughness, elongation, and impact resistance but at the sacrifice of tensile strength and stiffness.

Nylon 6/6, 6/6T, and 4/6 have the lowest permeability of nylons by gasoline, mineral oil, and fluorocarbon refrigerants. Nylons 6/12 and 6/6T are used where lower moisture absorption (and better dimensional stability) is needed.

All nylons absorb moisture from the environment; however, type 6/6T nylon has much lower moisture absorption than any other type of nylon resin. Moisture absorption leads to dimensional and property changes dependent upon the equilibrium level absorbed. At elevated temperatures, the moisture equilibrium level decreases above 88°C, nylons begin to dry out by a combination of internal diffusion and surface volatile emission. Extended exposure to temperatures above 121°C will reduce moisture content to about 0.1% (similar to dry-as-molded content).

Nylons are sensitive to ultraviolet radiation. Weatherability will be reduced unless ultraviolet stabilizers are incorporated into the formulation. Carbon black is the most commonly used ultraviolet stabilizer. Carbon black lowers the ductility and toughness as tradeoff for ultraviolet stability.

Nylons have good resistance to creep and cold flow compared with many less rigid thermoplastics. Creep resistance is better at higher levels of crystallinity as demonstrated in nylon 4/6. Creep can be calculated from long-term apparent modulus under load data.

Processing

Nylons are generally fabricated by injection molding or by extrusion. Precise, intricate shapes of a variety of colors can be molded with little or no finishing required. These can often replace an assembly of several metal parts. Thus, even where nylons cost more on a per-volume basis than the common die-casting metals, economies in finishing and assembly often result in lower ultimate costs.

Tubing and rod stock manufacture, plus the coating of wire and cable, are the major forms of nylon extrusion. Film and relatively complex cross sections are also made but in less volume. In general, tubing, film, and other unsupported shapes require higher melt viscosity than is desirable for injection molding. Most manufacturers of nylon supply these high-viscosity grades.

Applications

Nylons are usually specified because of their combination of properties. Gears, bearings, cams, clutch facings, and similar mechanical parts require their strength, stiffness, low coefficient of friction, and resistance to fatigue and abrasion. In cases where oiling or greasing is apt to be neglected, as in home appliances, or is undesirable from a contamination standpoint, as in textile and food-handling machinery, nylon parts usually perform satisfactorily without any lubrication whatsoever.

Designers often utilize the mechanical properties of nylons, plus one or more characteristics of particular value. For example, the nylon housing for an electric drill must be tough, stiff, dimensionally stable, and resistant to commonly encountered lubricants and solvents. However, its electrical nonconductivity and safety are the critical advantages.

Similarly, one manufacturer makes an entire rifle stock, plus many of the moving parts of the rifle, out of nylon. It is far lighter and tougher than wood and provides moving surfaces that do not need lubrication. The ability of nylon to be molded into precise sections thus permits custom-quality guns to be mass-produced.

By utilizing different properties, marine electrical stuffing tubes of nylon capitalize on the durability, lightness, resistance to corrosion, and cost advantage of the resin over machined brass. Washing machine mixing valves and valves for the dispensing of hot beverages require the mechanical properties of nylon and its excellent resistance to the effects of hot water.

The coating of wire and cable construction is the most important extrusion application of nylon. Although it is most commonly used as a jacket over a primary insulator such as polyvinyl chloride or polyethylene to impart resistance to lubricants, abrasion, and to the effects of high temperature, its electrical properties are adequate for low-voltage uses.

Exercise bikes used in health clubs undergo about 10 h of use per day. Further, to be competitive in today's hot exercise-equipment marketplace, one must design bikes to last from 5 to 7 years. The jump on the competition is a newly designed thermoplastic composite pillow block, the structure that supports the driveshaft of the bike. The material selected was a long-glass-fiber-reinforced nylon 6/6 structural composite. When compared to conventional short-fiber-reinforced thermoplastics, then nylon 6/6 structural composites have enhanced mechanical properties, while remaining lightweight. As an added bonus, this combination of properties reduces the weight of the pillow block from 3.68 to 0.69 g.

A nylon foot brace for kayakers has become the standard of excellence. The brace consists of a rail screwed into the bottom of the boat, and a footpad that slides up and down the rail to fit the kayaker's leg. Today, most kayak makers feature the brace as standard equipment. It consists of glass-reinforced, nylon 6, injection-molded resins.

The nylon combines high-strength, stiffness, and heat-deflection characteristics while extending the retention of these properties at high temperatures and a lower cost than competitive materials.

A new sports car features a lightweight, high-performance nylon air-intake manifold with a molded-in fuel rail. The automaker claims the integrated air-fuel module with two fuel passages into the body of the intake manifold breaks the mold when it comes to under-the-hood components. It uses a nylon manifold/fuel rail using these fusible-core, injection-molding techniques. The manifold, with its molded-in fuel rail, weighs about 3.68 g. That is nearly 50% less than a similar design based on pressure cast aluminum would weigh. The smooth inner walls of the injection-molded manifold increase airflow. This, combined with the low thermal conductivity of the nylon manifold, helps improve engine performance up to 5% more than that of an aluminum counterpart. And the nylon manifold insulates the air inside from engine heat, allowing high-density air to flow into the engine. The specially formulated 6/6 nylon resists hot engine temperatures and attacks from oil, gasoline, and battery vapors. It also produces a manifold that better withstands engine vibration stresses while reducing engine noise.

As a wire insulation, nylon is valued for its toughness and solvent resistance. Nylon fibers are strong, tough, and elastic and have a high gloss. The finer fibers are easily spun into yarns for weaving or knitting either alone or in blends with other fibers, and they can be crimped and heat-set. For making carpets, nylon staple fiber, lofted or wrinkled, is used to give the carpet a bulky texture resembling wool. Tire cord, made from nylon 6 of high molecular weight, has the yarn drawn to four or five times its original length to orient the polymer and give one half twist per inch. Nylon film was made in thicknesses down to 0.005 cm for heat-sealed wrapping, especially for food products where tight impermeable enclosures are needed.

Nylon sheet, for gaskets and laminated facings, comes transparent or in colors in thicknesses from 0.013 to 0.152 cm. Nylon monofilament is used for brushes, surgical sutures, tennis strings, and fishing lines. Filament and fiber, when stretched, have a low specific gravity down to 1.068, and the tensile strength may be well above 344 MPa. Nylon fibers made by condensation with oxalic acid esters have high resistance to fatigue when wet.

N

O

Objective

The primary magnifying system of a microscope. A system, generally of lenses, less frequently of mirrors, forming a real, inverted, and magnified image of the object.

Objective Aperture

See aperture (*electron*); an aperture (*optical*).

Objective Lens

The lens in a microscope closest to the specimen being examined. It forms the initial magnified image that in turn is further magnified by the eyepiece or ocular lens. The objective lens, therefore, controls the resolution that can be achieved. See *microscope* and *numerical aperture*.

Oblique Evaporation Shadowing

The condensation of evaporated material onto a substrate that is inclined to the direct line of the vapor stream to produce shadows. See also *shadowing*.

Oblique Illumination

Illumination from light inclined at an oblique angle to the optical axis.

Observed Value

The particular value of a characteristic determined as a result of a test or measurement.

Obsidian

A jet-black volcanic natural glass formed by rapid cooling of viscous lava. Although its composition varies, it is highly siliceous.

Occluded

Closed or shut off or, in a chemical context, absorbed.

Ochre

A compact form of earth used for paint pigments and as a filler for linoleum, also spelled *ocher*. It is an argillaceous and siliceous material, often containing compounds of barium or calcium, and owing the yellow, brown, or red colors to hydrated iron oxide. The tints depend chiefly upon the proportions of silica, white clay, and iron oxide. Ochres are very stable as pigments. They are prepared by careful selection, washing, and grinding in oil. They are inert and are not affected by light, air, or ordinary gases. They are rarely adulterated, because of their cheapness, but are sometimes mixed with other minerals to alter the colors. *Chinese yellow* and many other names are applied to the ochres. *Golden ochre* is ochre mixed with chrome yellow. *White ochre* is an ordinary clay. *Sienna* is a brownish-yellow ochre found in Italy and Cyprus. The material in its natural state is called *raw sienna*. *Burnt sienna* is a material calcined to a chestnut color. Indian red and Venetian red are hematite ochres.

Octahedral

Eight sided. The octahedral planes of a cubic structure are those having Miller indices of (111).

Octahedral Plane

In cubic crystals, a plane with equal intercepts on all three axes.

Ocular

See *eyepiece*.

Ocular Lens

The lens in a microscope closest to the eye. It magnifies the image initially produced by the objective lens. See *microscope*.

Ocvirk Number

A dimensionless number used to evaluate the performance of journal bearings and defined by the following equation:

$$\text{Ocvirk number} = \frac{P}{\eta U}\left\{\frac{c}{r}\right\}^2\left\{\frac{d}{b}\right\}^2$$

where
    P is the load per unit width
    η is the dynamic viscosity
    U is the surface velocity
    c is the radial clearance
    r is the bearing surface
    b is the bearing length
    d is the bearing diameter

This number may be used in its inverted form and is related to the *Sommerfeld number*.

O.D.

Outside diameter.

Off Time

In resistance welding, the time that the electrodes are off the work. This term is generally applied where the welding cycle is repetitive.

Offal

The material trimmed from blanks or formed panels.

Offhand Grinding

Grinding where the operator manually forces the wheel against the work, or vice versa. It often implies casual manipulation of either grinder or work to achieve the desired result. Dimensions and tolerances frequently are not specified or are only loosely specified; the operator relies mainly on visual inspection to determine how much grinding should be done. Contrast with *precision grinding*.

Offset (Yield/Stress/Strength)

(1) Same as proof stress. (2) The distance along the strain coordinate between the initial portion of a stress–strain curve and a parallel line that intersects the stress–strain curve at a value of stress (commonly 0.2%) that is used as a measure of the *yield strength*. Used for materials that have no obvious *yield point*. See *tensile test* and *stress–strain curve*.

Offset Yield Strength

The stress at which the strain exceeds by a specific amount (the *offset*); an extension of the initial, approximately linear, proportional portion of the stress–strain curve. It is expressed in force per unit area.

Ohm (Ω)

The SI unit of electrical resistance equal to the resistance through which a current of 1 amp will flow when there is a potential difference of 1 V across it.

Ohm's Law

In a conductor, volts = amps × ohms, where the voltages are the potential drop across the conductor, amps is the current flowing in it, and ohms is its resistance. It is applicable to direct current circuits but not usually to alternating current circuits.

Oil

A liquid of vegetable, animal, mineral, or synthetic origin that feels slippery to the touch.

Oil Canning

See *canning*.

Oil Content

The amount of oil that an impregnated part, such as a self-lubricating bearing, retains.

Oil Cup

A device connected to a bearing that uses a wick, valve, or other means to provide a regulated flow of a lubricant.

Oil Flow Rate

The rate at which a specified oil will pass through a porous sintered powder compact under specified test conditions.

Oil Fog Lubrication

See *mist lubrication*.

Oil Groove

A channel or channels cut in a bearing to improve oil flow through the bearing. A similar groove may be used for grease-filled bearings.

Oil Hardening

The process of hardening steels by quenching into oil. See *steel*.

Oil Hardening Steel

Steel containing a sufficient quantity of alloying elements to cause it to fully harden when quenched into oil. Unless specified to the contrary, it is usually implicit that full hardening should be achieved in a round bar of 50 mm (2 in.) diameter. See *steel* and *hardenability*.

Oil Impregnation

The filling of a sintered skeleton body with oil by capillary attraction or under influence of an external pressure or a vacuum.

Oil Less Bearing

(1) A bearing manufactured by sintering metal powders with a proportion of graphite as a lubricant. (2) Sintered metal, produced so as to leave a considerable volume of voids, and subsequently impregnated prior to installation with a lubricant that may be a grease or even thick oil. This usage is a bit dubious as it indicates not an absence of oil but only that the bearing does not need further oil addition in service. See also *oilite bearing* and preferred term *self-lubricating bearing*.

Oil Permeability

A measure of the capacity of a sintered powder metallurgy bearing to allow the flow of an oil through its open pore system.

Oil Pocket

A depression designed to retain oil on a sliding surface.

Oil Quenching

Hardening of carbon steel in an oil bath. Oils are categorized as conventional, fast, and martempering.

O

Oil Ring Lubrication

A system of lubrication for horizontal shafts. A ring of larger diameter rotates with the shaft and collects oils from a container beneath.

Oil Starvation

A condition in which a bearing, or other *tribocomponent*, receives an inadequate supply of lubricant.

Oil Wedge

When at rest, the shaft journal carried in a plain bearing may penetrate the oil film and contact the bearing material. However, assuming a satisfactory design and adequate lubricant supply, a rotating shaft induces hydrodynamic forces in the lubricant that develop a standing wedge of oil slightly offset from the bottom centerline. The shaft and rise on this wedge. The bore of the bearing may contain a circumferential recess of complex form over much of its length to develop and stabilize the wedge.

Oil Whirl

Instability of a rotating shaft associated with instability in the lubricant film. Oil whirl should be distinguished from shaft whirl, which depends only on the stiffness of the shaft.

Oilcloth

A fabric of woven cotton, jute, or hemp, heavily coated with turpentine and resin compositions, usually ornamented with printed patterns, and varnished. It was employed chiefly as a floor covering but a light, flexible variety having a foundation of muslin is used as a covering material. This class comes in plain colors or in printed designs. It was formally the standard military material for coverings and ground protection but has been replaced by synthetic fabrics. *Oilskin* is a cotton or linen fabric impregnated with linseed oil to make it waterproof. It was used for coverings for cargo and for waterproof coats but has now been replaced by coated fabrics. *Oiled silk* is a thin silk fabric impregnated with blown linseed oil that is oxidized and polymerized by heat. It is waterproof, very pliable, and semitransparent. It was much used for linings but has now been replaced by fabrics coated with synthetics.

Oiliness

See *lubricity*.

Oilite Bearing

A sintered powder metal bearing containing a considerable volume of voids, typically 30%. It may be impregnated with oil prior to installation or may be fed in service, the void contents providing a reservoir during oil flow interruption.

Oils

A large group of fatty substances that are divided into three general classes: vegetable oils, animal oils, and mineral oils. The *vegetable oils* are either fixed or volatile oils. The fixed oils are present in the plant in combined form and are largely glycerides of stearic, oleic, palmitic, and other acids, and they vary in consistency from light fluidity to solid fats. They nearly all boil at 500°F–600°F (260°C–316°C), decomposing into other compounds. The volatile, or essential, oils are present in uncombined form and bear distillation without chemical change.

Seed oils, or *oilseeds*, obtained from various plant seeds, are fatty acids of varying chain lengths containing hydroxy, keto, epoxy, and other functional groups. The oils are chemically very pure. Among important uses of these oils are for polymers, surface coatings, plasticizers, surfactants, and lubricants. The seeds of the *Chinese tallow tree* are coated with a semisolid fat. An oil similar to linseed oil is inside the kernel. The oil can be used as a substitute for cocoa butter and for fatty acids in cosmetics.

Fish oils are thick, with a strong odor. Vegetable and animal oils are obtained by pressing, extraction, or distillation. Oils that absorb oxygen easily and become thick are known as *drying oils* and are valued for varnishes, because on drying they form a hard, elastic, waterproof film. Unsaturation is proportional to the number of double bonds, and in food oils, these govern the cholesterol depressant effect of the oil. Oils and fats are distinguished by consistency only, but waxes are not oils. *Mineral oils* are derived from petroleum or shale and are classified separately. The most prolific sources of vegetable oils are palm kernels and copra.

Under comparable aggressive plantation work, 10–20 times more palm and coconut oil can be produced per acre than peanut or soybean oil. Babassu oil is almost chemically identical with coconut oil, and vast quantities of babassu nuts grow wild in northeast Brazil.

Blown oils are fatty oils that have been oxidized by blowing air through them while hot, thereby thickening the oil. They are mixed with mineral oils to form special heavy lubricating oils, such as marine engine oil, or are employed in cutting oils. They are also used in paints and varnishes, as the drying power is increased by the oxidation, the flashpoint and the iodine value are both lowered by the blowing. The oils usually blown are rapeseed, cottonseed, linseed, fish, and whale oils.

Oilstone

A fine-grained, slaty silica rock for sharpening edged tools. The bluish-white and opaque white oilstones, fine grains from Arkansas, are called *novaculite*, and they received their name because they were originally used for razor sharpening. They are composed of 99.5% chalcedony silica and are very hard and fine grain. Novaculite is a deposit from hot springs. It is fine grained, and the ordinary grades are employed for the production of silica refractories. Arkansas oilstones are either hard or soft and have a waxy luster. They are shipped in large slabs or blocks or in chips for tumbling barrel finishing. *Artificial oilstones* are also produced of aluminum oxide. *India oilstone* was originally blocks of emery, but the name now may refer to aluminum oxide stones.

Olefin

A group of unsaturated hydrocarbons of the general formula C_nH_{2n} and named after the corresponding paraffins by the addition of -ene or sometimes -ylene to the root.

Olefin Copolymers

The principal olefin copolymers are the polyallomers, ionomers, and ethylene copolymers. The polyallomers, which are highly crystalline, can be formulated to provide high stiffness and medium impact strength, moderately high stiffness and high impact strength, high impact strength, or extrahigh impact strength. Polyallomers, with their unusually high resistance to flexing fatigue, have "hinge" properties better than those of polypropylenes. They have the characteristic milky color of polyolefins, are softer than polypropylene, but have greater abrasion resistance. Commonly injection molded, extruded, and thermoformed, polyallomers are used for such items as typewriter cases, snap clasps, threaded container closures, embossed luggage shells, and food containers.

Ionomers are nonrigid plastics characterized by low density, transparency, and toughness. Unlike polyethylenes, density and properties are not crystalline dependent. Their flexibility, resilience, and high molecular weight combine to provide high abrasion resistance. They have outstanding low-temperature flexural properties but upper temperature use is limited to 71°C. Resistance to attack from organic solvents and stress cracking chemicals is high. Ionomers have high melt strength for thermoforming and extrusion coating and a broad temperature range for blow molding and injection molding. Representative ionomer parts include injection-molded containers, housewares, tool handles, and closures; extruded film, sheet, electrical insulation, and tubing; and low-molded containers and packaging.

There are four commercial ethylene copolymers, of which ethylene vinyl acetate (EVA) and ethylene ethyl acrylate (EEA) are the most common; EVA copolymers approach elastomers and flexibility and softness, although they are processed like other thermoplastics. Many of their properties are density dependent, but in a different way from that of polyethylenes. Softening temperature and modulus of elasticity decrease as density increases, which is contrary to the behavior of polyethylene. Similarly, the transparency of EVA increases with density to a maximum that is higher than that of polyethylenes, which become opaque when density increases above around 0.935 g/cc. Although the electrical properties of EVA are not as good as those of low-density polyethylene, they are competitive with vinyl and elastomers normally used for electrical products. The major limitation of EVA plastics is the relatively low resistance to heat and solvents; the Vicat softening point is 64°C. EVA copolymers can be injection, blow, compression, transfer, and rotationally molded; they can also be extruded. Molded parts include appliance bumpers in a variety of seals, gaskets, and bushings. Extruded tubing is used in beverage vending machines and for hoses for air-operated tools and paint spray equipment.

EEA is similar to EVA in its density–property relationships. It is also generally similar to EVA in high-temperature resistance, and like EVA, it is not resistant to aliphatic and aromatic hydrocarbons or their chlorinated versions. However, EEA is superior to EVA in environmental stress cracking and resistance to ultraviolet radiation. Similar to EVA, most of the applications of EEA are related to the outstanding flexibility and toughness of the plastic. Typical uses are household products such as trash cans, dishwasher trays, flexible hose and water pipe, and film packaging.

Two other ethylene copolymers are ethylene hexane (EH) and ethylene butane (EB). Compared with the other two, these copolymers have greater high-temperature resistance; their useful service range is between 66°C and 88°C. They are also stronger and stiffer and, therefore, less flexible than EVA and EEA. In general, EH and EB are more resistant to chemicals and solvents than the other two, but their resistance to environmental stress cracking is not as good.

Oleic Acid

Also called *red oil*, *elaine oil*, *octadecenoic acid*, and *rapic acid*, although the latter is a misnomer based on a former belief that it was the same as the erucic acid of rapeseed. It occurs in most natural fats and oils in the form of glyceride, and it is obtained in the process of saponification or by distillation. Much of this acid is obtained from lard and other animal fats, but *Emery 3758-R* is produced from soybean or other vegetable sources by hydrolysis of *glycerol trioleate*. It is an oily liquid with a specific gravity of 0.890, and below 57°F (14°C), it forms colorless needles if heated to the boiling point of water, and it reacts with oxygen to form a complex mixture of assets, including a small percentage of acetic and formic acids. When reacted with potassium hydroxide, it is converted to an acetate and a palmitate. It is also readily converted to pelargonic and other acids for making plastics. Oleic acid is a basic foodstuff in the form of the glyceride, and the acid has a wide use for making soaps, as a chemical raw material, and for finishing textiles. In soluble oils and cutting compounds, it forms sodium oleate. The two commercial grades of oleic acid, yellow and red, are known as *distilled red oil* and *saponified*. Aluminum oleate is used for thickening lubricating oils.

Oligomer

A polymer consisting of only a few monomer units, for example, a dimer, trimer, and tetramer, or their mixtures.

Oliver

A blacksmith's forge with two or more dies, the lower halves of which were set on a single anvil, the upper halves being foot actuated by individual treadles.

Olivine

A translucent mineral, usually occurring in granular form, employed as a refractory. The formula is usually given as $(Mg \cdot Fe)_2 \cdot SiO_4$, but it is a solid solution of *forsterite*, $2MgO \cdot SiO_2$, and *fayalite*, $2FeO \cdot SiO_2$. The fayalite lowers the refractory quality, but forsterite is not found alone. The mineral is also called *chrysolite*, and the choice green stones used as gems are called *peridot*. The melting point of forsterite is 3470°F (1912°C). When it is mixed with chrome ore, the low-fusing elements form a black glass that presents a nonporous face. Some refractory material marketed as forsterite may be olivine blended with magnesite or may be serpentine treated with magnesite. *Forsterite firebrick* in the back walls of basic open-hearth steel furnaces gives longer life than silica brick but only two-thirds that of chrome–magnesite brick. Forsterite refractories are usually made from olivine rock to which MgO is added to adjust the composition to $2MgO \cdot SiO_2$. *Olivine sand* is substituted for silica sand as a foundry sand, where silica is expensive. When olivine is used as a foundry sand, it is noted that the heat-resisting qualities decrease with particle size. Olivine contains 27%–30% magnesium metal and is also used to produce magnesium by the electrolysis of the chloride. *Magnesium phosphate* fertilizer is made by fusing olivine with phosphate rock at 2912°F (1600°C), tapping off the iron, and spray cooling and crushing the residue. It contains 20% citric-acid-soluble phosphate and is useful for acid soils.

Olsen Ductility Test

A *cupping test* in which a piece of sheet metal, restrained except at the center, is deformed by a standard steel ball until fracture occurs. The height of the cup at the time of fracture is a measure of the ductility. Similar to the *Erichsen test*.

Omega Phase

A nonequilibrium, submicroscopic phase that forms in titanium alloys as a nucleation growth product; often thought to be a transition phase during the formation of α from β. It occurs in metastable β alloys and can lead to severe embrittlement. It typically occurs during aging at low temperature but can also be induced by high hydrostatic pressures.

One-Component Adhesive

Adhesive material incorporating a latent hardener or catalyst that is activated by heat.

One-Shot Molding

In the urethane foam field, a system in which the isocyanate, polyol, catalyst, and other additives are mixed together directly, and a foam is produced immediately. Compare with *prepolymer molding*.

Onyx

A variety of chalcedony silica mineral differing from agate only in the straightness of the layers. The alternate bands of color are usually white and black or white and red. Onyx is artificially colored in the same way as agate. It is used as an ornamental building stone, usually cut into slabs, and used as decorative articles. *Onyx marble* is limestone with impurities arranged and banded in layers. *Opalized wood* is an onyx-like petrified wood from Idaho. It is cut into ornaments.

Opacifiers

Materials used in ceramic glazes and vitreous enamels primarily to make them nontransparent, but opacifiers may also enhance the luster, control the texture, promote craze resistance, or stabilize the color of the glaze. An opacifier must have fire resistance so as not to vitrify or decrease the luster. Tin oxide is a widely used white opacifier, and up to 3% also increases the fusibility of the glaze or enamel. Titanium oxide adds scratch hardness and high acid resistance to the enamel. It also increases the flow, making possible thinner coats that minimize chipping. Opacifiers may also serve as the pigment colors. Thus, cobalt oxide gives a blue collar, and platinum oxide gives a gray collar. Lead chromate gives an attractive red color on glazes fired at 1652°F (900°C), but when fired at 1832°F (1000°C), the lead chromate decomposes and a green chromium oxide is formed. If the glaze is acid, the basic lead chromate is altered and the color tends toward green. The *zirconium opacifiers* have a wide range of use from ordinary dishes to high-heat electrical porcelain and sanitaryware enamels. The amount of zirconium oxide used is a minimum of 3%. Lead oxide is used to lower the melting point of a glaze. Matte effects are obtained by adding barium oxide, magnesium, or other materials to the opacifier.

Open Arc Welding

Welding operations during which the electric arc is visible.

Open Assembly Time

The time interval between the spreading of the adhesive on the adherend and the completion of the assembly of the parts for bonding.

Open Dies

Dies with flat surfaces that are used for preforming stock or producing and forgings.

Open (Gap) Joint

A joint to be welded in which the components have a significant gap deliberately set between them prior to welding. Filler metal is deposited in the gap.

Open Pore

A pore open to the surface of a powder compact. See also *intercommunicating porosity*.

Open Porosity

See *interconnected porosity*.

Open Rod Press

A *hydraulic press* in which the slide is guided by vertical, cylindrical rods (usually four) that also serve to hold the crown and bed in position.

Open-Back Inclinable Press

A vertical crank press that can be inclined so that the bed will have an inclination generally varying from 0° to 30°. The formed parts slide off through an opening in the back. It is often called an OBI press.

Open-Cell Cellular Plastic

Foamed or cellular plastic with cells that are generally interconnected.

Open-Cell Foam

Foamed or cellular polymeric material with cells that are generally interconnected. Closed cell refers to cells that are not interconnected.

Open-Circuit Potential

The *potential* of an electrode measured with respect to a reference electrode or another electrode when no current flows to or from it.

Open-Circuit Voltage

The voltage between the output terminals of a welding machine when no current is flowing in the welding current.

Open-Die Forging

The hot mechanical forming of metals between flat or shaped dies in which metal flow is not completely restricted. Also known as hand or smith forging. See also *hand forge (smith forge)*.

Open-Gap Upset Welding

A form of *forge welding* in which the weld interfaces are heated with a fuel gas flame and then forced into intimate contact by the application of force. Not to be confused with *upset welding*, which is a resistance welding process.

Open-Hearth Furnace

A reverberatory melting furnace with a shallow hearth and a low roof. The flame passes over the charge on the hearth, causing the charge to be heated both by direct flame and by radiation from the roof and side walls of the furnace. See also *reverberatory furnace*.

Open-Hearth Steel

Steel made by the process of melting pig iron and steel or iron scrap in a lined regenerative furnace and boiling the mixture with the addition of pure lump iron ore until the carbon is reduced. The boiling is continued for 3–4 h. The process was developed in 1861 by Siemens in England. The furnaces contain regenerative chambers for the circulation and reversal of the gas and air. The fuels used are natural gas, fuel oil, coke-oven gas, or powdered coal. Both the acid- and the basic-lined open-hearth furnaces are used, but most steel made in the United States is basic open hearth. Ganister is used as a lining in the acid furnaces and magnesite in the basic.

An advantage of the open-hearth furnace is the ability to handle raw materials that vary greatly and to employ scrap. Iron low in silicon requires less heating time. The duplex process consists in melting the steel in an acid Bessemer furnace until the silicon, manganese, and part of the carbon have been oxidized and then transferring to a basic open-hearth furnace where the phosphorus and the remainder of the carbon are removed. Open-hearth steel is of uniform quality and is produced in practically all types.

Opening Mode

See *fracture mechanics*.

Open-Sand Casting

Any casting made in a mold that has no *cope* or other covering.

Operating Stress

The stress to which a structural unit is subjected in service.

Optical Emission Spectroscopy

Pertaining to *emission spectroscopy* in the near-ultraviolet, visible, or near-infrared wavelength regions of the electromagnetic spectrum. See also *electromagnetic radiation*.

Optical Etching

Development of microstructure under application of special illumination techniques, such as *dark-field illumination*, *phase contrast illumination*, *differential interference contrast illumination*, and *polarized light illumination*.

Optical Fibers

These are flexible transparent fiber devices, sometimes called light guides, used for either image or information transmission, in which light is propagated by total internal reflection. In its simplest form, the optical fiber or light guide consists of a core of material with a refractive index higher than the surrounding cladding. The optical fiber properties and requirements for image transfer, in which information is continuously transmitted over relatively short distances, are quite different from those for information transmission, where typically digital encoding of information into on–off pulses of light (on = 1; off = 0) is used to transmit audio, video, or data over much longer distances at high bit rates. Another application for optical fibers is in sensors, where a change in light transmission properties is used to sense or detect a change in some property, such as temperature, pressure, or magnetic field.

Fiber Designs

There are three basic types of optical fibers. Propagation in these light guides is most easily understood by ray optics, although the wave or modal description must be used for an exact description. In a multimode, step-refractive-index-profile fiber, the number of rays or modes of light that are guided, and thus the amount of light power coupled into the light guide is determined by the core size and the core-cladding refractive index difference. Such fibers, used for conventional image transfer, are limited to short distances for information transmission due to pulse broadening. An initially sharp pulse made up of many modes broadens as it travels long distances in the fiber, because high-angle modes have a longer distance to travel relative to the low-angle modes. This limits the bit rate and distance because it determines how closely input pulses can be spaced without overlap at the output end. At the detector, the presence or absence of a pulse of light in a given time slot determines whether this bit of information is a zero or one.

A graded-index multimode fiber, where the core refractive index varies across the core diameter, is used to minimize pulse broadening due to intermodal dispersion. Since light travels more slowly in the high-index region of the fiber relative to the low-index region, significant equalization of the transit time for the various modes can be achieved to reduce pulse broadening. This type of fiber is suitable for intermediate-distance, intermediate-bit-rate transmission systems. For both fiber types, light from a laser or light-emitting diode can be effectively coupled into the fiber.

A single-mode fiber is designed with a core diameter and refractive index distribution such that only one fundamental mode is guided, thus eliminating intermodal pulse-broadening effects. Material and waveguide dispersion effects cause some pulse broadening, which increases with the spectral width of the light source. These fibers are best suited for use with a laser source to couple light efficiently into the small core of the light guide and to enable information transmission over long distances at very high bit rates. The specific fiber design and the ability to manufacture it with controlled refractive index and dimensions determine its ultimate band with information carrying capacity.

Attenuation

The attenuation or loss of light intensity is an important property of the light guide because it limits the achievable transmission distance and is caused by light absorption and scattering. Every material has some fundamental absorption due to the atoms or molecules composing it. In addition, the presence of other elements as impurities can cause strong absorption of light at specific wavelengths. Fluctuations in a material on a molecular scale cause intrinsic Rayleigh scattering of light. In actual fiber devices, fiber-core-diameter variations or the presence of defects such as bubbles can cause additional scattering light loss.

Optical fibers based on silica glass have an intrinsic transmission window at near-infrared wavelengths with extremely low losses. Such fibers are used with solid-state lasers and light-emitting diodes for information transmission, especially for long distance (greater than 1 km). Plastic fibers exhibit much higher intrinsic as well as total losses and are more commonly used for image transmission, illumination, or very short-distance data links.

Many other fiber properties are also important, and their specifications and control are dictated by the particular application. Good mechanical properties are essential for handling: plastic fibers are ductile, whereas glass fibers, intrinsically brittle, are coated with a protective plastic to preserve their strength. Glass fibers have much better chemical durability and can operate at higher temperatures than plastics.

Optical Glass

Glass of high quality and having closely specified optical properties, used in the manufacture of plane or curved windows and refractive and reflective elements for precision instruments and devices. Most products are silicates designed for maximal transmittance to the visible spectrum. Some glasses are prepared from extremely pure raw materials under special conditions to yield high ultraviolet transmission. Among infrared transmitting glasses, the chalcide glass As_2S_3, fused silica (waterfree, mostly prepared from natural quartz), and calcium–aluminate glasses are most common. See *pyrometry*.

Optical Microscope

An instrument used to obtain an enlarged image of a small object, utilizing visible light. In general, it consists of a light source, a condenser, an objective lens, an ocular or eyepiece, and a mechanical stage for focusing and moving the specimen. Magnification capability of the optical microscope ranges from 1× to 1500×.

Optical Pyrometer

An instrument for measuring the temperature of heated material by comparing the intensity of light emitted with a known intensity of an incandescent lamp filament.

Optoelectronic Device

A device that detects and/or is responsive to electromagnetic radiation (light) in the visible, infrared, and/or ultraviolet spectra regions, emits or modifies noncoherent or coherent electromagnetic radiation in these same regions, or utilizes such electromagnetic radiation for its internal operation.

Orange Peel

A coarse-textured surface with dimples that are rounded similar to the exterior of an orange rather than angular like the surface of a cube of sugar. It is observed on badly sprayed paint or on coarse-grained sheet metal that has been deformed. In the latter case,

it results from the slightly differing deformation characteristics of neighboring grains.

Orange Peel (Ceramics)

(1) A surface condition characterized by an irregular waviness of the porcelain enamel resembling and an orange skin texture; sometimes considered a defect. (2) A pitted texture of a fired glaze resembling the surface of rough orange peel.

Orange Peel (Metals)

A surface roughening in the form of a pebble-grained pattern that occurs when a metal of unusually coarse-grained size is stressed beyond its elastic limit. Also called pebbles and alligator skin.

Orange Peel (Plastics)

In injection molding of plastics, a part with an undesirable uneven surface somewhat resembling the skin of an orange.

Orbital Forging

See *rotary forging*.

Order (In x-Ray Reflection)

(1) The factor *n* in *Bragg's law*. In x-ray reflection from a crystal, the order is an integral number that is the path difference measured in wavelengths between reflections from adjacent planes. (2) In the context of crystal structure, it is the phenomenon in certain alloys in which the atoms of the elements in a solid solution arrange themselves in a repeated simple pattern on the atomic lattice often termed a superlattice. For example, two elements present in the appropriate proportion could position themselves at alternate sites on the 3D lattice.

Order Hardening

A low-temperature *annealing* treatment for metals that permits short-range ordering of solute atoms within a matrix, which greatly impedes dislocation motion.

Order–Disorder Transformation

A phase change among two *solid solutions* having the same crystal structure, but in which the atoms of one phase (disordered) are randomly distributed; in the other, the different kinds of atoms occur in a regular sequence upon the crystal lattice, that is, in an ordered arrangement.

Ordered Intermetallics

See *intermetallics*.

Ordered Structure

The crystal structure of a *solid solution* in which the atoms of different elements seek preferred lattice positions. Contrast with *disordered structure*.

Ordering

Forming a *superlattice*.

Ore

A metal-bearing mineral from which a metal or metallic compound can be extracted commercially. Earths and rocks containing metals that cannot be extracted at a profit are not rated as ores. Ores are named according to their leading useful metals. The ores may be oxides, sulfides, halides, or oxygen salts. A few metals also occur native in veins in the minerals. Ores are usually crushed and separated and concentrated from the *gangue* with which they are associated and then shipped as *concentrates* based on a definite metal or metal oxide content. The metal content to make an ore commercial varies widely with the current price of the metal and with the content of other metals present in the ore. Normally, a sulfide copper ore should have 1.5% copper in the unconcentrated ore, but if gold or silver is present, an ore with much less copper is workable, or if the deposit can be handled by high-production methods, a mineral of very low metal content can be utilized as ore. Low-grade lead minerals can be worked if silver is recoverable, and low-grade manganese minerals become commercial when prices are high. Thus, the term ore is only relative, and in other different economic conditions, minerals that are not considered ores in one country may be much used as ores in another.

Ore Dressing

A preliminary stage to the extraction of metal from an ore in which material with no metal content such as earth is removed. Same as *mineral dressing*.

Organic

Being or composed of hydrocarbons or their derivatives, or matter of plant or animal origin. Contrast with *inorganic*.

Organic Acid

A chemical compound with one or more carboxyl radicals (COOH) in its structure; examples are butyric acid, maleic acid, and benzoic acid.

Organic Coatings

Organic coatings are chiefly additive-type finishes that find use on almost all types of materials. They can be monolithic consisting simply of one layer, or coat, or they can be composed of two or more layers. The total thickness of coating systems varies widely. Some run less than 1 mil thick. Others go as high as 10 and 15 mils thick. Generally, by definition, coatings that are more than 10 mils thick are referred to as linings, films, or mastics.

To function as a protective barrier against corrosion and oxidation, organic coatings depend presently on their chemical inertness and impermeability. In addition, however, some coatings provide protection with the use of inhibiting pigments that have a passivating action, particularly on metal surfaces. Also, some coatings contain metallic pigments that give electrochemical protection to metals.

Coating Application and Drying

Organic coatings are commonly applied by the following mechanisms: (1) evaporation or loss of solvent, (2) oxidation, and (3) polymerization. After application of the coating, the volatile ingredients, which are almost always present in at least a small amount, evaporate. Some finishes, such as lacquers, dry completely by evaporation of solvents. After evaporation, coatings that do dry solely by evaporation are still in a semifluid state and depend on oxidation or polymerization, or a combination of both, to convert to their final form.

Drying by oxidation, which is usually done at room temperature, is the slowest of the three methods. Polymerization, which involves polymer chain-forming mechanism, can be done at normal or at elevated temperatures. Polymerization is speeded by use of heat. Radiation curing involving the use of an electron beam to polymerize the coating in a few seconds has found production usage.

Coating Types and Systems

Coating Composition

An organic coating is made up of two principal components: a vehicle and a pigment. The vehicle is always there. It contains the film-forming ingredients that enable the coating to convert from a mobile liquid to a solid film. It also acts as a carrier and suspending agent for the pigment. Pigments, which may or may not be present, are the coloring agents and, in addition, contribute a number of other important properties.

Organic coatings are commonly divided into about a half-dozen broad categories based on the types and combinations of vehicle and pigment used in their formulation. They are paints, enamels, varnishes, lacquers, dispersion coatings, emulsion coatings, and latex coatings. However, with the complexity in modern formulations, distinctions between these various types are often difficult to make.

As mentioned earlier, organic finish systems are frequently composed of more than one layer or coat. These various layers are commonly classified as primers, intermediate coats, and finish coats.

Primers

These are the first coatings placed on the surface (except for fillers, in some cases). Where chemical pretreatments are used, primer coats may often be unnecessary. Primers for industrial or production finishing are of two types: air-dry types and baking types. The air-dry types have drying oil vehicle bases and are usually referred to as paints. They may or may not be modified with resins. They are not used as extensively as the baking-type primers, which have resin or varnish vehicle bases and dry chiefly by polymerization. Some primers, known as flash primers, are applied by spraying and dry by solvent evaporation within 10 min. In practically all primers, the pigments impart most of the anticorrosion properties to the primer and, along with the vehicle, determine its compatibility and adherence with the base metal.

Intermediate Coats

These are fillers, surfacers, and sealers. They can be applied either before or after the primer, but more often after the primer and sometimes after the surfacer coat. Their function is to fill in large irregularities in the surface or local imperfections. They are usually puttylike substances, and a variety of materials are used. Their chief characteristics are as follows: (1) must harden with a minimum shrinkage, (2) must have good adhesion, (3) must have good sanding properties, and (4) must work smoothly and easily.

Surfacers are often similar to primers; they usually have the same composition as the priming coat, except that more pigment is present. Surfacers are applied over the priming coat to cover all minor irregularities in the surface. Sealers as a rule are used over either the fillers or surfacers. The chief function of sealers is to fill up the pores of the undercoat to avoid "striking in" of the finish coats. This filling in of the porous surfacer or sealer also tends to strengthen the entire coating system. The sealers when used over surfacers are usually formulated with the same type pigment and vehicle as used in the final coat.

Finish Coats

Finish or top coats are usually the decorative or functional part of a paint system. However, they often also have a protective function. The primer coats may require protection against the service conditions, because although the pigments used in primers are satisfactory for corrosion protection of the metal, they are frequently not satisfactory as top coats. Their color retention upon weathering or their physical durability may be poor. There are also one-coat applications where the finish coats are applied directly to the base material surface and, therefore, provide the sole protective medium.

Vehicles

Vehicles are composed of film-forming materials and various other ingredients, including thinners (volatile solvents), which control viscosity, flow, and film thickness, and driers, which facilitate applications and improve drying qualities.

The main concern is chiefly the film-forming part of the vehicle, because it is that part of the vehicle that to a large extent determines the quality and character of an organic finish. It determines the possible ways in which the finish can be applied and how the "wet" finish will dry to a hard film; it provides for adhesion to the metal surface; and it usually influences the durability of the finish.

Vehicles can be divided into three main types: (1) oil, (2) resin, and (3) varnish. The simplest and among the oldest vehicles are the straight drying oil types. Resins, as a class, can serve as vehicles in their own right or can be used with drying oils to make varnish-type vehicles. Varnish vehicles are composed of resins and either drying or nondrying oils, together with required amounts of thinners and driers. They are often used alone as a full-fledged organic finish.

Drying Oils

Vehicles consisting of oil only are used to a limited extent in industrial finishes. Linseed oil is probably the most widely used oil. There are a number of different kinds that differ in rate of drying and in such properties as water resistance, color, and hardness.

Tung oil or China wood oil, when properly treated, excels all other drying oils and speed of drying, hardening, and water resistance. Oiticica oil is similar to tung oil in many of its properties. Dehydrated castor oil dries better than linseed oil, but slower than tung oil. Some of its advantages are good color and color retention and flexibility. The oils from some fish are also used as drying oils. If processed properly, they dry reasonably well and have little odor. They are often used in combination with other oils. Perilla oil is quite similar in properties to fast-drying linseed oil. Its use is largely dependent upon its price and availability.

Soybean oil has the slowest drying rate in the drying oil classes and is usually used in combination with some faster-drying oils such as linseed oil.

Resins

Although both natural and synthetic resins can serve as organic coating vehicles, today the natural types, such as rosin, have been largely replaced by plastic resins. Nearly all the plastic resins—both thermosets and thermoplastics as well as many elastomers—can be used as film formers, and frequently two or more kinds are combined to give the set of properties desired. Typical thermoplastics used in vehicles are acrylics, acetates, butyrates, and vinyls. Commonly used thermosets for vehicles include phenolics, alkyds, melamines, ureas, and epoxies.

Pigments

Pigments are the second of the two principal components that make up most organic finishes. They contribute a number of important characteristics to a coating. They, first of all, serve a decorative function. The choice of color and shades of color by use of one or combinations of pigments are practically unlimited. Closely associated with color is the hiding power function or their ability to obscure the surface of the material being finished. In many primers, the principal function of pigments is to prevent corrosion of the base metal. In other cases, they may be added to counteract the destructive action of ultraviolet light rays. Pigments also help give body and good flow characteristics to the finish. And, finally, some pigments may give to organic coatings what is termed package stability, that is, they keep the coating material in usable condition in the container until ready for use.

Pigments can be conveniently divided into three classes as follows: (1) white hiding pigments, (2) colored pigments, and (3) extender or inert pigments. White pigments are used not only in making white paints and enamels but also in making white bases for the tinted and light shades. Colored pigments furnish the finish with both opacity and color. They may be used by the cells to form solid colors, or in combination with whites to produce tints, and often provide rust-inhibitive properties. For example, red lead, certain lead chromates, zinc chromates, and blue lead are used in iron and steel primers as rust inhibitors. There are two general classes of colored pigments—earth colors, which are very stable and are not readily affected by acids and alkalies, heat, light, and moisture, and chemical colors, which are produced under control conditions by chemical reaction. The metallic pigments can also be included within this class. Aluminum powder is perhaps the best known.

The chief functions of extender pigments are to help control consistency, gloss, smoothness, and filling qualities and leveling and check resistance. Thus, particle size and shape, oil absorption, and flatting power are important selection considerations. Extender pigments are for the most part chemically inactive. They usually have little or no hiding power.

Enamels

By definition, enamels are an intimate dispersion of pigments in a varnish or a resin vehicle or in a combination of both. Enamels may dry by oxidation at room temperatures and/or by polymerization at room or elevated temperatures. They vary widely in composition, in color and appearance, and in properties and are available in all colors and shades. Although they generally give a high-gloss finish, there are some that give a semigloss or eggshell finish and still others that give a flat finish. Enamels as a class are hard and tough and offer good mar and abrasion resistance. They can be formulated to resist attack of most commonly encountered chemical agents in corrosive atmospheres.

Because of their wide range of useful properties, enamels are probably the most widely used organic coating in industry. One of their largest fields of use is for coating household appliances—washing machines, stoves, kitchen cabinets, and the like. A large portion of refrigerators, for example, are finished with synthetic baking enamels. These appliance enamels are usually white, and, therefore, most have a high degree of color and gloss retention when subjected to light and heat. Other products finished with enamels include automotive products, railway equipment, office equipment, toys and sports supplies, industrial equipment, and novelties.

Lacquers

The word *lacquer* comes from lac resin, which is the base of common shellac. Lac resin dissolved in alcohol was one of the first lacquers and has been in use for many centuries. Nowadays, shellac is called spirit lacquer. It is only one of several different kinds of lacquers; these, except for spirit lacquer, are named after the chief film-forming ingredient. The most common ones are cellulose acetate, cellulose acetate butyrate, ethyl cellulose, vinyl, and nitrocellulose.

A distinguishing characteristic of lacquers is that they dry by evaporation of the solvents or thinners in which the vehicle is dissolved. This is in contrast to oils, varnishes, or resin base finishes, which are converted to a hard film chiefly through oxidation or polymerization.

Because many modern lacquers have high resin content, the gap between lacquer and synthetic-type varnishes diminishes until finally one has modified synthetic air-drying varnishes. They may dry chiefly by oxidation or polymerization.

Lacquers normally dry hard and dust-free in a very few minutes at room temperature. In production line work, forced drying is often used. It is possible, therefore, to do a multicoat job without having to lose time between coats. Because of the speed of drying and the fact that they are permanently soluble in the solvents used for application, lacquers usually are not applied by brush. Spray application or dipping is the usual procedure.

Lacquers can be either clear and transparent or pigmented, and the color range is practically unlimited. Lacquers in themselves have good color retention, but sometimes the added pigments, modifying resins, and plasticizers may adversely affect this property. They are hard and mar resistant. Inherently, they lack good adhesion to metal, but modern lacquer formulations have greatly improved their adhesion properties. Lacquers can be made to be resistant to a large variety of chemicals, including water and moisture, alcohol, gasoline, vegetable oil, animal and mineral oils, and mild acids in alkalies. Because of the volatile solvents, lacquers are inflammable in storage and during application, and this sometimes limits their application.

Because of their fast drying speeds, lacquers find wide application in the protection and decoration of products that can be dipped, spray, roller coated, or flow coated. They are especially advantageous for coating metal hardware and fixtures, toys, and other articles that, because of volume production, must dry hard enough to handle and pack in a short period of time. Lacquers

are widely used in automobile finishing and especially for refinishing autos and commercial vehicles where fast drying without baking equipment is a requirement. Lacquers also compete with enamels for coating metal stampings and castings, including die castings.

Varnishes

Varnishes consist of thermosetting resins and either drying or nondrying oils. They are clear and unpigmented and can be used alone as a coating. However, their major use in industrial finishing is as a vehicle to which pigments are added, thus forming other types of organic coatings.

The drying mechanisms of varnishes all follow the same general pattern. First, any volatile solvents that are present evaporate; then, drying by oxidation and/or polymerization takes place, depending on the nature of the resin and oil. At high temperatures, of course, there is more tendency to polymerize. So, varnishes can be formulated for either air or bake drying. Varnishes may be applied by brushing or by any of the production methods.

It is evident that with the large variety of raw materials to choose from and the unlimited number of combinations possible, varnishes have an extensive range of properties and characteristics. They range from almost clear white to a deep gold; they are transparent, lacking any appreciable amount of opacity. Japan, a hard-baked black-looking varnish, is an exception. It is opaque, due to carbon and carbonaceous material being present.

There are some distinctions in properties between oil-modified alkyd varnishes and the other types. In general, oil-modified alkyds have better gloss and color retention and better resistance to weathering. They form a harder, tougher, more durable film and dry faster. On the other hand, they have less alkali resistance than the other varnishes. In such things as adhesion and inhibitiveness, there is no distinctive difference.

The major use of varnishes, as coatings in their own right, is for food containers, closures such as bottle caps, and bandings of various kinds. Another large application is as a clear finish coat over lithographic coatings.

Paints

The word *paints* is sometimes used broadly to refer to all types of organic coatings. However, by definition, a paint is a dispersion of a pigment or pigments in drying oil vehicle. They find little use these days as industrial finishes. Their principal use is for primers. Paints dry by oxidation at room temperature. Compared to enamels and lacquers, their drying rate is slow; they are relatively soft and tend to chalk with age.

Other Organic Coatings

Dispersion and Emulsion Coatings

In recent years, these coatings have become known as water-based paints or coatings because many of them consist essentially of finely divided ingredients, including plastic resins, fillers, and pigments, suspended in water. An organic media may also be involved. There are three types of water-based coatings. Emulsions, or latexes, are aqueous dispersions of high-molecular-weight resins. Strictly speaking, latex coatings are dispersions of resins in water, whereas emulsion coatings are suspensions of an oil phase in water.

Emulsion and latex coatings are clear to milky in appearance and have low gloss, excellent resistance to weathering, and good impact resistance. Chemical and stain resistance varies with composition. Dispersion coatings consist of ultrafine fine, insoluble resin particles present as a colloidal dispersion in an aqueous medium. They are clear or nearly clear. Weathering properties, toughness, and gloss are roughly equal to those of conventional solvent paints.

Water-soluble types, which contained low-molecular-weight resins, are clear finishes, and they can be formulated to have a high gloss, fair to good chemical and weathering resistance, and high toughness. Of the three types, they handle and flow most like conventional solvent coatings.

Plastic Powder Coatings

Several different methods have been developed to apply plastic powder coatings. In the most popular process—fluidized bed—parts are preheated and then immersed in a tank of finely divided plastic powders, which are held in a suspended state by a rising current of air. When the powder particles contact the heated part, they fuse and adhere to the surface, forming a continuous, uniform coating.

Another process, electrostatic spraying, works on the principle that oppositely charged materials attract each other. Powder is fed through a gun, where an electrostatic charge is applied opposite that applied to the part to be coated. When the charged particles leave the gun, they are attracted to the part where they cling until fused together as a plastic coating. Other powder application methods include flock coating, flow coating, flame and plasma spraying, and a cloud chamber technique.

Although many different plastic powders can be applied by these techniques, vinyl, epoxy, and nylon are most often used. Vinyl and epoxy provide good corrosion and weather resistance as well as good electrical insulation. Nylon is used chiefly for its outstanding wear and abrasion resistance. Other plastics frequently used in powder coating include chlorinated polyether, polycarbonate, acetal, cellulosics, acrylic, and fluorocarbons.

Hot Melt Coatings

These consist of thermoplastic materials that solidify on the metal surface from the molten state. The plastic is either applied in solid form and then melted and flowed over the surface or is applied molten by spraying or flow coating. Since no solvent is involved, fixed single coats are possible. Bituminous coatings are also commonly applied by the hot melt process.

Lining and Sheeting

Sheet, film, and tapes of various plastics and elastomers cemented to material shapes and parts are used to provide corrosion and abrasion resistance. Thickness usually ranges from 3.2 to 12.7 mm. The most widely used materials are polyvinyl chloride, polyethylene butadiene–styrene rubber, and neoprene.

Specialty Finishes

An almost infinite number of specialty or novelty finishes are available. Most of them are really lacquers or enamels to which special ingredients have been added or which are processed in some unique way to give the effects desired.

One of the most common types is those giving a roughened or wrinkled appearance, which is obtained by the use of high percentages of driers causing wrinkling when the finish is baked. Another group of specialty finishes gives a crystalline effect. They are enamels in which impurities are purposely introduced during the baking process by retaining the products of combustion in the oven while the coating dries. The wrinkle and crystalline finishes are widely used on instrument panels, office equipment, and a variety of other industrial and consumer products.

Other unusual finishes are obtained by adding special ingredients to lacquers that give them a stringy or "veiled" appearance when applied by spraying. The application of the silk-screen process to organic finishing of metals has also resulted in unique finishes with multicolored effects.

Organic Fiber

A fiber derived or composed of matter originating in plant or animal life or composed of chemicals of hydrocarbon origin, either natural or synthetic.

Organic Zinc-Rich Plant

Coating containing zinc powder pigment and an *organic* resin.

Organometallic Compounds

These are members of a broad class of compounds with structures that contain both carbon (C) and a metal (M). Although not a required characteristic of organometallic compounds, the nature of the formal carbon–metal bonds can be of the covalent type.

The term *organometallic* chemistry is essentially synonymous with organotransition metal chemistry; it is associated with a specific portion of the periodic table ranging from groups 3 through 11 and also includes the lanthanides.

This particular area has experienced exceptional growth since the mid-1970s largely because of the continuous discovery of novel structures as elucidated mainly by x-ray crystallographic analyses, the importance of catalytic processes in the chemical industry, and the development of synthetic methods based on transition metals that focus on carbon–carbon bond constructions. From the prospective of inorganic chemistry, organometallics afford seemingly endless opportunities for structural variations due to changes in the metal coordination number, alterations in ligand–metal attachments, mixed-metal cluster formation, and so forth. From the viewpoint of organic chemistry, organometallics allow for manipulations in the functional groups that in unique ways often result in rapid and efficient elaborations of carbon frameworks for which no comparable direct pathway using nontransition organometallic compounds exists.

As one moves across the periodic table, the metals that have found application include

- Titanium
- Zirconium
- Chromium
- Molybdenum
- Tungsten
- Ruthenium
- Osmium

- Cobalt
- Rhodium
- Palladium
- Copper

Of interest to the ceramic industry are those organometallics, the hydroxide-free alkoxides, that can be used for the vapor-phase synthesis of hard ceramic oxide coatings, films, or free-standing bodies. Very fine particulate oxides also can be formed from these chemicals.

Compounds now available include aluminum isopropoxide, aluminum hexafluoroisoproproxide, lithium hexafluoroisoproproxide, sodium hexafluoroisoproproxide, zirconium hexafluoroisoproproxide, and zirconium tertiary amyloxide.

Organosol

A suspension of a finely divided resin in a volatile, organic liquid. The resin does not dissolve appreciably in the organic liquid at room temperature but does at elevated temperatures. The liquid evaporates at an elevated temperature, and the residue upon cooling is a homogenous plastic mass. Plasticizers may be dissolved in the volatile liquid.

Organosol Coatings

Organosol coatings are coatings in which the resin (usually polyvinyl chloride) is suspended rather than dissolved in an organic fluid. The dispersion technique permits the use of high-molecular-weight, relatively insoluble resins without the use of expensive solvents. In organosols, the fluid, or dispersant, consists of plasticizers together with a blend of inexpensive volatile diluents selected to give the desired fluidity, speed of fusion, and physical properties. The dispersant provides little or no solvating action on the resin particles until a critical temperature is reached at which point the resin is dissolved in the dispersant to form a single-phase solid solution. Since a portion of a liquid is made up of volatile diluents, diffusion process results in a proportional shrinkage. (Nonvolatile dispersions using only plasticizer dispersants are termed *plastisols*.)

Organosols are available with a wide range of flow characteristics and consequently may be formulated for application by any of the conventional techniques. Because they contain substantial quantities of volatile diluents, their thickness is limited to 10–12 mils per coat. Single-coat applications of greater thickness blister during baking are a result of trapped solvents.

Baking is generally accomplished in two stages. The volatiles are removed in the first stage at temperatures of 93°C–121°C, but fusion does not occur until a temperature of 149°C–191°C is reached. The fusion stage accomplishes the union of the discrete vinyl particles into a single-phase solid. In addition to polyvinyl chloride, the organosol technique can be used with acrylonitrile-vinyl and polychlorotrifluoroethylene resins. A balance must be achieved in the baking operation between the removal of the volatiles and the solvation of the resins. Rapid heating results in solvent blistering, whereas the reverse causes a mud-cracking effect.

Application Methods

Organosol coatings can be applied by several methods.

Spread Coating

The bulk of the organosols is applied by spread coating methods to fabrics and paper. Several plants are used as spread coaters for the application of organosols and plastisols to strip steel to provide materials competitiveness with the light metals and plastics. Two basic processes are available: the knife coater and the roll coater. Knife coating is simple and fast, but the product lacks uniformity afforded by roller coating. Fusion is accomplished in a tunnel oven, and embossing rolls may be employed at the oven exit to impart a texture or pattern to the hot-gelled coating.

Strand Coating

Wires or filaments may be coated with organosols by first passing the strand through a dip tank and then through a wiping die to set the thickness. Fusion is accomplished in a drying tower. As many as 9–10 passes may be required to build a thickness of 20 mils.

Dip Coating

One of the major problems encountered in dip coating with organosols is the tendency of dried organosol to fall back into the dip tank and cause coating rejects. For this reason, much of the dip coating is done with modified plastisols rather than organosols. A dipping formulation must have low viscosity with high yield value to prevent sags and drains. The rate at which the article is withdrawn from the dip tank is a determining factor in the thickness and quality of deposit. Withdrawal rates generally range between 1.7 and 5.5 mm/s. Special techniques such as inversion of the dipped article just prior to fusion may be employed to alleviate the drip problem when it is particularly troublesome.

Spray Coating

Organosols are readily handled in either suction or pressure spray equipment. For production-line spraying, pressure systems afford rapid delivery and are generally preferred. There is an increasing tendency to use electronic spray processes for handling organosols. A number of electrostatic processes are available including several "handguns."

Properties

Organosols have the characteristic vinyl properties of toughness and moisture resistance. However, although the coatings possess good electrical resistance and are frequently used as secondary insulation to reduce shock hazards, they seldom meet the needs of primary wire insulation. Adhesive primers are required to bond the materials to metals and other dense substrates. Prolonged exposure to temperatures greater than 93°C causes thermal degradation, which is generally evidenced by a gradual darkening.

Uses

Vinyl organosols have found their widest usage in the coating of paper and fabric stock where their ease of handling has permitted the use of simplified, low-cost application equipment. Where fabric coating strike-through is to be avoided, as in open weave, thixotropic plastisols are employed rather than organosols. Closer weaves require some penetration.

Textured finishes for metals are a growing use and offer competition to the vinyl laminates and other decorative finishes.

Typical applications and related properties of organosols are shown in the following.

Application	Related Properties
Automotive interiors: Station-wagon flooring, roof liners, dashboards, cowling, kick plates	Ease of application; uniformity of color and texture; resistance to gasoline, grease, and polishes; resistance to impact damage
Commercial vehicles: Seatbacks, trim, interior paneling, luggage racks, sill plates	*Durability*: Resistance to abrasion and scuffing; cleanability; resistance to moisture and staining
Appliance finishes: Television and radio cabinets, slide projectors, refrigerator panels	*Aesthetic qualities*: Novelty of appearance; durability; resistance to abrasion and scratching; resistance to moisture and staining
Business machines: Typewriters, calculators, electronic computers, laboratory instruments	*Durability*: Resistance to abrasion, scuffing, and chipping; resistance to chemicals and perspiration; sound-deadening qualities
Architectural applications: Paneling, partitions, shower stalls, elevator doors, bathroom wall sections	*Cleanability*: Resistance to moisture and detergents; aesthetic qualities; durability; resistance to abrasion and scuffing; sound-deadening qualities
Office furniture: Desks, file cabinets, showcases, counters, wastebaskets, chair finishes	*Durability*: Resistance to abrasion, impact, and scuffing; resistance to moisture and staining
Luggage	*Durability*: Toughness, resistance to abrasion and scuffing; aesthetic qualities
Paper and fabrics: Floor and wall coverings, place mats, bottlecap liners, containers for food packaging, bandage dressings, upholstery fabrics, safety clothing, glove coatings	*Ease of application*: Economy and simplicity of application equipment; resistance to abrasion and tearing; resistance to moisture and staining
Glass coatings: Perfume bottles, bleach and chemical reagent bottles, photo flash bulbs	Resiliency, feel, cohesion strength; resistance to alcohol, moisture, and chemicals; ease of application

Orientation

The alignment of some feature relative to its surroundings, for example, in a metallurgical context, the alignment of the crystal lattice relative to the axis of the material. In an as-cast material, the orientation of the lattice in individual grains will, in most cases, be random, but working processes such as rolling tend to bring the orientation of all grains into alignment. This is termed *preferred orientation*. It causes properties such as ductility to vary in different directions. For example, a rolled sheet may have more ductility in the direction of rolling than at an angle of 45° to the rolling direction. This can cause problems in processes such as deep drawing.

Orientation (Crystal)

Arrangements in space of the axes of the lattice of a crystal with respect to a chosen reference or coordinate system. See also *preferred orientation*.

Orientation (Plastics)

The alignment of the crystalline structure in polymeric materials to produce a highly aligned structure. Orientation can be accomplished by cold drawing or stretching and fabrication. Orientation can generally be divided into two classes: uniaxial and biaxial.

Oriented Materials (Plastics)

Polymeric materials with molecules and/or macroconstituents that are aligned in a specific way. Oriented materials are anisotropic.

Orifice Gas

In plasma arc welding and cutting, the gas that is directed into the torch to surround the electrode. It becomes ionized in the arc to form the plasma and issues from the orifice in the torch nozzle as the plasma jet. See the terms *nontransferred arc* and *plasma arc welding*.

Original Crack Size (α_0)

The physical crack size at the start of testing.

Orsat Analyzer

A furnace atmosphere analysis device in which gases are absorbed selectively (volumetric basis) by passing them through a series of preselected solvents.

Orthochromatic

Sensitive to all colors except orange and red.

Orthorhombic

Having three mutually perpendicular axes of unequal lengths.

Orthotropic

Having three mutually perpendicular planes of elastic symmetry.

Oscillating Die Press

A small high-speed metal forming press in which the die and punch move horizontally with the strip during the working stroke. Through a reciprocating motion, the die punch returns to their original positions to begin the next stroke.

Osmiridium

An alloy felt as a native metal is platinum ore. As the name indicates, it is predominantly osmium (about 30%–40%) and iridium (about 50%–60%) with, usually, some lead as well as other elements of the platinum group. It is very hard and wear resistant and is used for small high-value applications such as pen tips.

Osmium

A platinum-group metal, symbol Os, osmium is noted for its high hardness, about 400 Brinell. The heaviest known metal, it has a high specific gravity, 22.65, and a high melting point, 2698°C. The boiling point is about 5468°C. Osmium has a close-packed hexagonal crystal structure and forms solid solution alloys with platinum, having more than double the hardening power of iridium in platinum. However, it is seldom used to replace iridium as a hardener except for fountain-pen tips where the alloy is called *osmiridium*.

Osmium is not affected by the common acids and is not dissolved by aqua regia. It is practically unworkable, and its chief use is as a catalyst.

Uses

Osmium tetraoxide, a commercially available yellow solid (melting point 40°C), is used commercially as a stain for tissue in microscopy. It is poisonous and attacks the eyes. Osmium metal is catalytically active, but it is not commonly used for this purpose because of its high price. Osmium and its alloys are hard and resistant to corrosion and wear (particularly to rubbing wear). Alloyed with other platinum metals, osmium has been used in needles for record players, fountain-pen tips, and mechanical parts.

Osmosis

The percolation of solvent through a semipermeable membrane into a more concentrated solution so that the concentrations on the two sides of the membrane tend to equalize.

Osprey Spray Forming

Copper alloy semifinished products, made by the Osprey spray-forming process, feature the ability to combine properties such as high strength, good corrosion resistance, and excellent machinability.

In the Osprey process, powder particles are liquefied and sprayed by an inert gas onto a substrate, where they are shaped into a billet. Because the droplets solidify very rapidly, the microstructure is much finer than that of conventionally cast material.

For example, the high yield strength and low elastic modulus of Cu-14%–Sn alloy (with a small lead addition) make this alloy (BO5) ideal for spring applications. Its elastic modulus is 80–90 GPa, yield strength is 800–950 MPa, and ultimate tensile strength is 900–1000 MPa, with hardness of 240–280 HV.

For the fabrication of contact elements by stamping and bending, alloy Cu-15%–Ni-8%–Sn (CN8) would be a good choice. It has elastic modulus of 110–120 GPa, yield strength of 800–1300 MPa, ultimate tensile strength of 900–1400 MPa, and hardness of 240–400 HV.

Out of Position Welding

Any position other than flat.

Out Time

The time a *prepreg* is exposed to ambient temperature, namely, the total amount of time the prepreg is out of the freezer. The primary effects of out time or a decrease in the *drape* and *tack* of the prepreg and the absorption of moisture from the air.

Outgassing

(1) The release of adsorbed or occluded gases or water vapor, usually by heating, as from a vacuum tube or other vacuum system. (2) Devolatilization of plastics or applied coatings during exposure to vacuum in vacuum metallizing. Resulting parts show voids or thin spots in plating with reduced and spotty brilliance. Additional drying prior to metallizing is helpful, but outgassing is inherent to plastic materials and coating ingredients, including plasticizers and volatile components.

Outlier

In a set of data, a value so far removed from other values in the distribution that is probably not a bona fide measurement. There are statistical methods for classifying a data point as an outlier and removing it from a data set.

Ovaloid

A surface of revolution symmetrical about the polar axis that forms the end closure for a filament-wound cylinder.

Oven Dry

The condition of a plastic material that has been heated under prescribed conditions of temperature and humidity until there is no further significant changes in its mass.

Oven Soldering

See preferred term *furnace soldering*.

Ovenware

Glass and glass–ceramic materials able to withstand thermal downshock of 150°C (270°F) without breakage that are used for culinary oven use. See *glass* and *glass ceramics*.

Overaging

Aging under conditions of time and temperature greater than those required to obtain maximum change in a certain property, so that the property is altered in the direction of the initial value. See *precipitation hardening*.

Overbasing

A technique for increasing the basicity of lubricants. Overbased lubricants are used to assist in neutralizing acidic oxidation products.

Overbending

Bending metal through a greater arc than that required in the finished part to compensate for springback.

Overdraft

A condition wherein a metal curves upward on leaving the rolls because of the higher speed of the lower roll.

Overfill

The fill of a die cavity with an amount of powder in excess of specification.

Overglaze

A glass coating over another component or element, normally used to give physical or electrical protection.

Overhead Welding Position

The position in which the weld run is approximately horizontal but is made from beneath the components. This is usually taken to mean a weld slope not greater than 15° and a weld rotation between 168° and 180° for a butt weld and between 115° and 180° for a fillet.

Overhead-Drive Press

A mechanical press with the driving mechanism mounted in or on the crown or upper parts of the uprights. See the term *straight-side press*.

Overheating

Heating a metal or alloy to such a high temperature that its properties are impaired. When the original properties cannot be restored by further heat treating, by mechanical working, or by a combination of working and heat treating, the overheating is known as *burning*.

Overlap

(1) Pultrusion of weld metal beyond the toe, face, or root of a weld. (2) In resistance seam welding, the area in a given weld remelted by the succeeding weld. (3) A weld defect caused by molten metal running onto adjacent parent metal but not fusing to it. See also *face of weld*, *root of weld*, and *toe of weld*.

Overlay Sheet

A nonwoven fibrous mat (e.g., of glass synthetic fiber) used as the top layer in a cloth or mat *layup* to provide a smoother finish, minimize the appearance of the fibrous pattern, or permit machining or grinding to a precise dimension. Also called *surfacing mat*.

Overlaying

See preferred term *surfacing*.

Overmix

Mixing of a powder longer than necessary to produce adequate distribution of powder particles. Overmixing may cause particle size segregation.

Oversinter

The sintering of a powder compact at higher temperature or for longer time periods than necessary to obtain the desired microstructure or physical properties. It often leads to swelling due to excessive pore formation.

Oversize Powder

Powder particles larger than the maximum permitted by a particle size specification.

Overspray

The excess spray material that is not deposited on the part during *thermal spraying*.

Overstraining

The application of a strain beyond yield particularly where this is deliberate to develop beneficial residual stresses.

Overstressing

(1) In fatigue testing, cycling at a stress level higher than that used at the end of the test. (2) Any excessive stress but particularly the damaging application of one or more cycles of fatigue with a stress in excess of the fatigue limit.

Overvoltage

The difference between the actual electrode potential when appreciable electrolysis begins and the reversible electrode potential.

Oxalic Acid

Also known as *ethane diacid*. A strong organic acid that crystallizes as the ortho acid. It reduces iron compounds and is thus used in writing inks, stain removers, and metal polishes. When it absorbs oxygen, it is converted to the volatile carbon dioxide and to water, and it is used as a bleaching agent, as a mordant in dyeing, and in detergents. Oxalic acid occurs naturally in some vegetables, notably *Swiss chard*, and is useful in carrying off excess calcium in the blood. The acid is produced by heating sodium formate and treating their resulting oxides with sulfuric acid, or it can be obtained by the action of nitric acid on sugar or strong alkalies on sawdust. It comes in colorless crystals with a specific gravity of 1.653, containing about 71% of the anhydrous acid, and soluble in water and alcohol. It is used in metal cleaning, dying, photography, and pulp bleaching. *Oxamide* is a stable anhydrous derivative with a high melting point 786°F (419°C). It is a white crystalline powder used in flame proofing and in wood treatment. Potassium ferric oxalate, $K_3Fe(C_2O_4)_3$, is stable in the dark but is reduced by the action of light and is used in photography.

Oxidation

In its metallurgical sense, the reaction of a metal with oxygen to form its oxide. In its electrochemical sense, the loss of electrons from a metal during any chemical reaction. The term encompasses a wide range of reaction rates, extending from the slow surface oxidation of iron in air, through burning to explosions.

Oxidation (Carbon Fibers)

In carbon/graphite fiber processing, the step of reacting the precursor polymer (rayon, polyacrylonitrile, or pitch) with oxygen, resulting in stabilization of the structure for the hot stretching operation.

Oxidation (Metals)

(1) A reaction in which there is an increase in valence resulting from a loss of electrons. Contrast with *reduction*. (2) A corrosion reaction in which corroded metal forms an oxide; usually applied to reaction with a gas containing elemental oxygen, such as air. (3) A chemical reaction in which one substance is changed to another by oxygen combining with the substance. Much of the dross from holding and melting furnaces is the result of oxidation of the alloy held in the furnace.

Oxidation Grain Size

(1) Grain size determined by holding a metallic specimen at a suitably elevated temperature in a mildly oxidizing atmosphere. The specimen is polished before oxidation and etched afterward. (2) Refers to the method involving heating a polished steel specimen to a specified temperature, followed by quenching and repolishing. The grain boundaries are sharply defined by the presence of iron oxide.

Oxidation Losses

Reduction in the amount of metal or alloy through *oxidation*. Such losses are usually the largest factor in melting loss.

Oxidative Wear

(1) A *corrosive wear* process in which chemical reaction with oxygen or oxidizing environment predominates. (2) A type of *wear* resulting from the sliding action between two metallic components that generate oxide films on the metal surfaces. These oxide films prevent the formation of a metallic bond between the sliding surfaces, resulting in fine wear debris and low rates. See also *fretting*.

Oxide

A compound of any element with oxygen. The manner in which metals form oxides determines the corrosion characteristics of a metal. The noble metals such as gold do not readily form oxides and hence are considered corrosion resistant. Other metals such as aluminum are highly reactive with oxygen and hence form an oxide film as soon as they are exposed to air. However, the oxide film is thin, tenacious, and impervious to oxygen and so further attack is stifled. Such metals, which include aluminum, titanium, and chromium, are also regarded as corrosion resistant in the environments in which the protective oxide is formed and maintained. Finally, some metals such as iron in a moist environment react with oxygen to form an oxide that is bulky, cracks, and is readily detached so it does not significantly impede continued attack.

Oxide Ceramics

Oxide ceramics can be divided into two groups—single oxides that contain one metallic element and mixed or complex oxides that contain two or more elements. Examples include alumina, beryllia, magnesia, zirconia, and thoria. As a class, they are low in cost compared to other technical ceramics, except for thoria and beryllia. Each of them can be produced in a variety of compositions, porosity, and microstructure, to meet specific property requirements.

Oxide ceramic parts are produced by slip casting or pressing or extrusion and then fired at about 1800°C. They are more difficult to fabricate than other types of ceramics, because of the usual requirement to obtain a high-density body with minimum distortion and dimensional error, except in the case of porous bodies for use as thermal insulation. Powder pressing produces bodies with the lowest porosity and higher strength, because of the high pressures and the small amount of binder required.

Single Oxides

Aluminum Oxide (Alumina)

Alumina is the most widely used oxide, chiefly because it is plentiful, relatively low in cost, and equal to or better than most oxides in mechanical properties. Density can be varied over a wide range, as can purity—down to about 90% alumina—to meet specific application requirements. Alumina ceramics are the hardest, strongest, and stiffest of the oxides. They are also outstanding and with electrical resistivity and dielectric strength, are resistant to a wide variety of chemicals, and are unaffected by air, water vapor, and sulfurous atmospheres. However, with a melting point of only 2039°C, they are relatively low in refractoriness and at 1371°C retain only about 10% of room-temperature strength. In addition to its wide use as electrical insulators and its chemical and aerospace applications, the high hardness and close dimensional tolerance capability of alumina make this ceramic suitable for such abrasion-resistant parts as textile guides, pump plungers, chute linings, discharge orifices, dies, and bearings.

Beryllium Oxide (Beryllia)

Beryllia is noted for its high thermal conductivity, which is about 10 times that of a dense alumina (at 499°C), three times that of steel, and second only to that of the high-conductivity metals (silver, gold, and copper). It also has high strength and good dielectric properties. However, beryllia is costly and is difficult to work with. Above 1649°C, it reacts with water to form a volatile hydroxide. Also, because beryllia dust and particles are toxic, special handling precautions are required. The combination of strength, rigidity, and dimensional stability makes beryllia suitable for use in gyroscopes; and because of high thermal conductivity, it is widely used for transistors, resistors, and substrate cooling in electronic equipment.

Magnesium Oxide (Magnesia)

Magnesia is not as widely useful as alumina and beryllia. It is not as strong, and because of high thermal expansion, it is susceptible to thermal shock. Although it has better high-temperature oxidation resistance than alumina, it is less stable in contact with most metals at temperatures above 1705°C in reducing atmospheres or in a vacuum.

Zirconium Oxide (Zirconia)

There are several types of zirconia: a pure (monoclinic) oxide and a stabilized (cubic) form and a number of variations such as yttria- and magnesia-stabilized zirconia and nuclear grades. Stabilized zirconia has a high melting point, about 2760°C, low thermal conductivity, and is generally unaffected by oxidizing and reducing atmospheres and most chemicals. Yttria- and magnesia-stabilized zirconias are widely used for equipment and vessels in contact with liquid metals. Monoclinic nuclear zirconia is used for nuclear fuel elements, reactor hardware, and related applications where high purity (99.7%) is needed. Zirconia has the distinction of being an electrical insulator at low temperatures, gradually becoming a conductor as temperatures increase.

Thorium Oxide (Thoria)

Thoria, the most chemically stable oxide ceramic, is only attacked by some earth alkali metals under some conditions. It has the highest melting point (3315°C) of the oxide ceramics. Like beryllia, it is costly. Also, it has high thermal expansion and poor thermal shock resistance.

Mixed Oxides

Cordierite ($2MgO \cdot 2Al_2O_3 \cdot ASiO_2$)

Cordierite is most widely used in extruded form for insulators such as heating elements and thermocouples. It has low thermal expansion, excellent resistance to thermal shock, and good dielectric strength. There are three traditional groups of cordierite ceramics:

1. Porous bodies that have relatively little mechanical strength due to limited crystalline intergrowth and absence of ceramic bond. With long thermal endurance and low thermal expansion, they are used for radiant elements in furnaces, resistor tubes, and rheostat parts.
2. Low porosity bodies, developed principally for use as furnace refractory brick.
3. Vitrified bodies used for exposed electrical devices that are subjected to thermal variations.

Forsterite ($2MgO \cdot SiO_2$)

This mixed oxide has high thermal shock resistance but good electrical properties and good mechanical strength. It is somewhat difficult to form and requires grinding to meet close tolerances.

Steatite

Steatites are noted for their excellent electrical properties and low cost. They are easily formed and fired at relatively low temperatures. However, compositions containing little or no clay or plastic material present fabricating problems because of the narrow firing range. Steatite parts are vacuum tight, can be readily bonded to other materials, and can be glazed or ground to high-quality surfaces.

Zircon ($ZrO_2 \cdot SiO_2$)

This mixed oxide provides ceramics with strength, low thermal expansion, and relatively high thermal conductivity and thermal endurance. Its high thermal endurance is used to advantage in various porous-type ceramics.

Oxide Coatings

The black oxide finish on steel was one of the most widely used black or blue–black finishes. Some of the advantages of this type of finish are (1) attractive black color; (2) no dimensional changes; (3) corrosion resistance, depending on the final finish dip used; (4) nongalling surface; (5) no flaking, chipping, or peeling because the finish becomes an integral part of the metal surface; (6) lubricating qualities due to its ability to absorb and adsorb the final oil or wax dips; (7) ease and economy of application; and (8) nonelectrolytic solutions and a minimum of plain steel tank equipment required.

The black oxide finish produced on steel is composed essentially of the black oxide of iron (Fe_3O_4) and is considered by many to be a combination of FeO and Fe_2O_3. It can be produced by several methods: the browning process, carbonia process, heat treatment, and the aqueous alkali–nitrate process. Each of these processes produces a black oxide of iron finish, although the finish produced by each particular process differs in some characteristics. The chemical dip aqueous alkali–nitrate process is the most widely used in applying a black oxide finish on steel.

Aqueous Alkali–Nitrate Process

In this process, a blackening solution is used that is highly alkaline and that also contains strong oxidizing chemicals. Refinements such as penetrants and rectifiers are also used to promote ease of operation, faster blackening, and trouble-free processing. At specific concentrations and boiling temperatures, these solutions will react with the iron in the steel to form the black oxide of iron (Fe_3O_4).

Because the reaction is directly with the iron in the steel, the finish becomes an integral part of the metal itself and, therefore, cannot flake, chip, or peel. For all practical purposes, there are also no dimensional changes. In transforming iron to black iron oxide, it is correct to assume that there is a change in volume. However, because of the highly alkaline nature of the blackening solution and because of the operating temperature, the blackening solution will dissolve a small amount of iron. Therefore, the amount of iron lost in this manner is compensated for by the buildup in volume from the change of iron to iron oxide—resulting in, for all intents and purposes, no dimensional changes. Extremely close measurements have shown that the actual change amounts to a buildup of only about 5 millionths of an inch.

There are types and conditions of steel and many different products that may require special procedures or additional steps.

It is important that only steel or steel alloys be immersed in the solution because other metals such as copper, zinc, cadmium, and aluminum will contaminate it. Many improvements have been incorporated into some proprietary black oxide salts and the latest one will rectify approximately 50 times more contaminants than heretofore. This particular product causes the contaminants to boil to the top, from which they can be skimmed or dragged out and rinsed away during processing.

After a black oxide blackening solution has been mixed, there is a "breaking-in" period that can run from 24 to 48 h, depending on the volume of blackening solution and the amount of work being processed. During this period, if it occurs, erratic blackening may be encountered, resulting in some work being blackened and some remaining partially or totally unblackened.

Chemical black oxide finishes today are being used on a wide variety of consumer and military parts. Some the most important applications are guns, firearms and components, metal stampings, toys, screws, spark plugs, machine parts, screw machine products, typewriter and calculating machine parts, auto accessories and parts, tools, gauges, and textile machinery parts. A black oxide finish can normally be used for indoor or semioutdoor applications on metal parts or fabrications that require an economical attractive finish, nominal corrosion resistance, and, in many cases, where *dimensional changes* cannot be tolerated.

Heat Treatment Methods

These methods can be divided into three classes: (1) oven or furnace heating, (2) molten salt bath immersion, and (3) steam heat process.

Oven or Furnace Method

The parts are heated to a temperature of 316°C–371°C at which temperature the metal surface is oxidized to a bluish-black color. The shaded color depends on the temperature and the analyses of the steel.

Molten Salt Bath Method

The oxide finish can be obtained in several ways, depending on the manufacturing or processing requirements:

1. In a molten salt bath composed essentially of nitrate salts maintained at 316°C–371°C, the pieces are first cleaned of any oils, greases, or objectionable oxides and then immersed in the molten bath. They will take on a blue–black finish, after which they are quenched in clean water and given a final oil dip.
2. In a molten nitrate bath in an austempering operation, the pieces are heated to the hardening temperature in a neutral salt hardening bath, after which they are quenched in the molten nitrate bath at 316°C–371°C. They are then removed, cleaned, and given a final oil dip.

Steam Process

In this method, the steel is placed in a retort and heated to a minimum temperature of 316°C. The retort is then purged with steam. Under these conditions, a black oxide is formed on the metal surface.

Browning Process

The browning process is commonly known as a rusting process. The pieces are first thoroughly cleaned and then swabbed with an acidic solution. After drying in a dry atmosphere approximately 77°C, the pieces are then placed in an oven with an atmosphere of 100% humidity and temperature at around 77°C for about 1.5 h. They then become quite rusty and the surface is rubbed down to remove the loose rust.

This procedure is carried out three or four times, after which the surface will have taken on a bluish-black finish. The pieces are then given a final oil dip.

Carbonia Process

To apply a black oxide finish by this method, the pieces are placed in a rotary furnace heated to around 316°C–371°C. Charred bone or other carbonaceous material is placed in the furnace along with a thick oil known as carbonia oil. The floor or cover of the furnace is occasionally opened and closed to allow circulation of air. After the parts have been treated for approximately 4 h, they are removed and then immersed in oil. This finish is essentially a black oxide of iron, but because the metal surface is in contact with carbonaceous material and oil, some black carbon penetrates into the metal surface.

Processes for Nonferrous Metals

The following are typical methods for applying black oxide coatings to nonferrous metals:

1. *Stainless steel*: Aqueous alkali–nitrate and molten dichromate methods. In the aqueous alkali–nitrate method, a solution is made up by using approximately 2.07–2.3 kg of blackening salt mixture to make up a gallon of blackening solution, which is operated at a boiling point of 124°C–127°C. The parts, after cleaning and acid pickling, are immersed in the boiling solution and will take on a black color. They are then given a rust-preventative oil dip. In the molten dichromate method, a molten bath of sodium dichromate or a mixture of sodium and potassium dichromates is used at a molten temperature of 316°C–399°C. The parts are immersed in the molten bath until they take on a blue or blue–black color, after which they are removed and cooled in oil or water. They are then cleaned to remove the salt and oil (is cooled in oil), after which they are given a dip in a clean, rust-preventative oil.
2. *Zinc, zinc plate, zinc-base die castings, and cadmium plate*: Hot molybdate blackening method, chromate and black die method, and the black nickel plate method.
3. *Copper and copper alloys (brasses and bronzes)*: Alkali–chlorite aqueous solution method, cuprammonium carbonate method, anodic oxidation method, and aryl-sulfone monochloramide-sodium hydroxide method.
4. *Aluminum and aluminum alloys*: Anodize and die method.

Oxide Film Replica

A thin film of an oxide of the specimen to be examined. The replica is prepared by air, oxygen, chemical, or electrochemical oxidation of the parent metal and is subsequently freed mechanically or chemically for examination. See also *replica*.

Oxide Rooting

Oxidation penetrating along grain boundaries at and close to the surface of a component.

Oxide Wedging

(1) The buildup of a bulky hard scale in a crack or interface that forces the surfaces apart. The simplest case is where two components are clamped together, and the interface develops a bulky oxide or other solid corrosion product with sufficient volume and compressive strength to force the components apart and perhaps break the bolts or other fasteners. (2) This term is also used to describe the damage observed in welded joints between austenitic and ferritic steels used for some pipework in power plants. Such dissimilar metal joints can develop a wedge-shaped oxide band in the ferritic material immediately adjacent to the weld. However, the damage mechanism in this case is more complicated than a simple jacking action because of (a) carbon diffusion from ferritic to austenitic steels, (b) the complex stress system resulting from the pipe internal pressure and from the expansion differential between ferritic and austenitic materials, and (c) the difference in oxidation characteristics between the austenitic and ferritic steels.

Oxide-Type Inclusions

Oxide compounds occurring as nonmetallic inclusions in metals, usually as a result of deoxidizing additions. In wrought steel products, they may occur as a stringer formation composed of distinct granular or crystalline-appearing particles.

Oxidized Steel Surface

Surface having a thin, tightly adhering oxidize skin (from straw to blue in color), extending inward from the edge of a coil or sheet.

Oxidizing Agent

A compound that causes *oxidation*, thereby itself being reduced.

Oxidizing Atmosphere

A furnace atmosphere with an oversupply of oxygen that tends to oxidize materials placed in it.

Oxidizing Flame

A gas flame produced with excess oxygen in the inner flame that has an oxidizing effect. See also the term *neutral flame*.

Oxyacetylene Welding/Cutting, etc.

Any process of fusion welding, cutting, etc., in which the heat source, at least initially, is provided by combustion of acetylene with oxygen. See *gas cutting and powder cutting*.

Oxyfuel Gas Cutting

Any of a group of processes used to sever metals by means of chemical reaction between hot base metal and a fine stream of oxygen. The necessary metal temperature is maintained by gas flames resulting from combustion of a specific fuel gas such as acetylene, hydrogen, natural gas, propane, propylene, or Mapp gas (stabilize methylacetylene–propadiene). See also *oxygen cutting* and the term *cutting torch (oxyfuel gas)*.

Oxyfuel Gas Spraying

See preferred term *flame spraying*.

Oxyfuel Gas Welding

Any of a group of processes used to fuse metals together by heating them with gas flames resulting from combustion of a specific fuel gas such as acetylene, hydrogen, natural gas, or propane. The process may be used with or without the application of pressure to the joint and with or without adding any filler metal. See the term welding torch (*oxyfuel gas*).

Oxygas Cutting

See preferred term *oxygen cutting*.

Oxygen

An abundant element, constituting about 89% of all water, 33% of the earth's crust, and 21% of the atmosphere. It combines readily with most of the other elements, forming their oxides. It is a colorless and odorless gas and can be produced easily by the electrolysis of water, which produces both oxygen and hydrogen, or by chilling air below –300°F (–184°C), which produces both oxygen and nitrogen. The specific gravity of oxygen is 1.1056. It liquefies at –171°F (–113°C) at 59 atm. *Liquid oxygen* is a pale-blue, transparent, mobile liquid. As gas, oxygen occupies 862 times as much space as the liquid. Oxygen is one of the most useful elements and is marketed in steel cylinders under pressure, although most of the industrial uses are in the form of its compounds. An important direct use is in welding and metal cutting, for which it should be at least 99.5% pure. Oxygen-enriched air is used in a number of oxidation and combustion processes in the steel, cement, glass, petrochemical, refining, and paper-and-pulp industries, and it has potential economic and environmental benefits in waste combustion. Oxygen enrichment improves overall combustion by raising oxygen partial pressure, thus increasing the combustion temperature and waste destruction.

Oxygen is the least refractive of all gases. It is the only gas capable of supporting respiration but is harmful if inhaled pure for a long time. *Ozone* is an allotropic form of oxygen with three atoms of oxygen, O_3. It is formed in the air by lightning or during the evaporation of water, particularly of spray in the sea. In minute quantities in the air, it is an exhilarant, but pure ozone is an intense poison. It has a peculiar odor, which can be detected with 1 part in 20 million parts of air. Ozone is a powerful oxidizer, capable of breaking down most organic compounds and bleaching vegetable colors. Its use as an alternative to chlorine for treating water and wastewater has risen in recent years because it is a safer and stronger oxidizing agent. *Liquid ozone* explodes violently in contrast with almost any organic substance. It is bright blue and is not attracted by magnet, although liquid oxygen is attracted. Ozone absorbs ultraviolet rays, and a normal blanket in the upper ozonosphere at heights of 60,000–140,000 ft (18,288–42,672 m), with 1 part per 100,000 parts of air, shields the earth from excess shortwave radiation from the sun. The destruction of the stratospheric ozone layer is resulting in the cutting back in usage of chlorofluorocarbons and other inert chlorinated compounds. At ground levels, however, ozone concentrations are rising due to increased use of fossil fuels; such higher ozone levels are viewed as deleterious to health. As an oxidizer in the rubber industry, ozone is known as *activated oxygen*. It is used widely as a catalyst in chemical reactions. It is made commercially by bombardment of oxygen with high-speed electrons.

Oxygen Arc Cutting

An oxygen cutting process used to sever metals by means of the chemical reaction of oxygen with the base metal at elevated temperatures. The necessary temperature is maintained by an arc between a consumable tubular electrode and the base metal. See *gas cutting*, *powder cutting*, and also *thermic lance*.

Oxygen Concentration Cell

See *differential aeration cell*.

Oxygen Cutting

Metal cutting by directing a fine stream of oxygen against a hot metal. The chemical reaction between oxygen and the base metal furnishes heat for localized melting, hence cutting. In the case of oxidation-resistant metals, the reaction is facilitated by the use of a chemical flux or metal powder. See also *metal powder cutting.*

Oxygen Deficiency

A form of *crevice corrosion* in which galvanic corrosion proceeds because oxygen is prevented from diffusing into the crevice.

Oxygen Gouging

Oxygen cutting in which a bevel or groove is formed.

Oxygen Grooving

See preferred term *oxygen gouging.*

Oxygen Lance

A length of pipe used to convey oxygen either beneath or on top of the melt in a steelmaking furnace or to the point of cutting in *oxygen lance cutting.* See also the terms *argon oxygen decarburization* and *basic oxygen furnace.*

Oxygen Lance Cutting

An oxygen cutting process used to sever metals with oxygen supplied through a consumable lance; the preheat to start the cutting is obtained by other means.

Oxygen Lance/Blown Processes

Various processes for producing refined steel from pig iron and/or scrap, in which the molten metal is exposed to oxygen injected from beneath, as in the Bessemer process or is delivered from above by a water-cooled tube, that is, a lance.

Oxygen Probe

An atmosphere-monitoring device that electronically measures the difference between the partial pressure of oxygen in a furnace or furnace supply atmosphere and the external air.

Oxygen-Free Copper

Electrolytic copper free from cuprous oxide, produced without the use of residual metallic or metalloidal deoxidizers.

Oxyhydrogen Cutting

An *oxyfuel gas cutting* process in which the fuel gas is hydrogen.

Oxyhydrogen Welding

An *oxyfuel gas welding* process in which the fuel gas is hydrogen.

Oxynatural Gas Cutting

An *oxyfuel gas cutting* process in which the fuel gas is natural gas.

Oxypropane Welding

An *oxyfuel gas welding* process in which the fuel gas is propane.

Ozone

A powerfully oxidizing allotropic form of the element oxygen. Both liquid and solid ozones are opaque blue–black in color, similar to that of ink. See *oxygen.*

Ozone Test

Exposure of material to a high concentration of ozone to give an accelerated indication of degradation expected in normal environments.

P

"p"-Type Semiconductor

See *semiconductor*.

PA

See *polyamides*.

Pack Carburizing

A method of surface hardening of steel in which parts are packed in a steel box with a carburizing compound and heated to elevated temperatures. Common carburizing compounds contain 8%–10% alkali or alkaline earth metal carbonates (e.g., barium carbonate, $BaCO_3$) bound to hardwood charcoal or to coke by oil, tar, or molasses. This process has been largely supplanted by gas and liquid carburizing processes. See *case hardening*.

Pack Nitriding

A method of surface hardening of steel in which parts are packed in a steel box with a nitriding compound and heated to elevated temperatures.

Package (Composites)

Yarn, roving, and so forth for reinforcing plastics in the form of units capable of being unwound and suitable for handling, storing, shipping, and use.

Package (Electronics)

In the electronics/microelectronics industry, an enclosure for a single element, and integrated circuit, or a hybrid circuit.

Packed Density

See preferred term *tap density*.

Packed Rolling

Hot rolling a pack of two or more sheets of metal; scalar prevents their being welded together. Unlike cladding, there is no intention that the sheets will bond to each other, this being inhibited by oxide or oil films.

Packing Density

See preferred term *apparent density*.

Packing Material

Any material in which powder-metallurgy compacts are embedded during the presintering or sintering operation. The material may act as a getter to protect the compacts from contamination. See also *getter*.

Pad

The general term used for that part of a die that delivers holding pressure to the metal being worked.

Pad Lubrication

A system of lubrication in which the lubricant is delivered to a bearing surface by a pattern of felt or similar material.

Padding

In foundry practice, the process of adding metal to the cross section of a casting wall, usually extending from a riser, to ensure adequate feed metal to a localized area during solidification where a shrink would occur if the added metal were not present.

Paddle Mixer

A mixture that uses paddles mounted on a rotating shaft or disk to move and mix the metal powder.

PAEK

See *polyaryletherketone*.

PAI

See *polyamide-imide*.

Paint

Paint is a term used since the dawn of history to designate cosmetics, marking chalks and pastes, tempera plaster, and colored fluids applied to surfaces for artistic, decorative, or weatherproofing purposes. The term survives today as a general marketing designation for decorative and protective formulations used in architectural, commercial, and industrial applications, with such diverse materials as lacquer, varnish, baking finishes, and specialty coating systems covered in a single category.

There is a general definition that suits most products: paint is a fluid, with viscosity, drying time, and flowing properties dictated by formulation, normally consisting of a vehicle or binder, a pigment, a solvent or thinner, and a drier, which may be applied in relatively thin layers and which changes to a solid in time. The change to a solid may or may not be reversible and may occur by evaporation of the solvent, by chemical reaction, or by a combination of the two.

Paint is a general name sometimes used broadly to refer to all types of organic coatings. However, by definition, paint refers to a solution of a pigment in water, oil, or organic solvent, used to cover wood or metal articles either for protection or for appearance.

Solutions of gums or resins, known as varnishes, are not paints although their application is usually termed painting. Enamels and lacquers, in the general sense, are under the classification of paints, but specifically the true paints do not contain gums or resins. Stain is a varnish containing enough pigment or dye to alter the appearance or tone of wood in imitation of another wood or to equalize the color in wood. It is usually a dye rather than a paint.

In modern technology, paint is classified in three major categories because of differing performance requirements: architectural paints, commercial finishes, and industrial coatings. A fourth category, artistic media, now admits inks, cements, pastes, dyes, plastics, semisolids, and conventional pigmented oils as acceptable materials.

Architectural Paints

These are air-drying materials applied by brush or spray to architectural and structural surfaces and forms for decorative and protective purposes. Materials are classified by formulation type as solvent thinned and water thinned.

Solvent-Thinned Paints

The drying mechanism of solvent-thinned paints predominantly may be by solvent evaporation, oxidation, or a combination of the two, and paints in this classification are subdivided accordingly.

Solvent-thinned paints that dry essentially by solvent evaporation rely on a fairly hard resin as the vehicle. Resins include shellac, cellulose derivatives, acrylic resins, vinyl resins, and bitumens. Shellac is usually dissolved in alcohol and is commonly used as shellac varnish. Paints based on nitrocellulose or other cellulose derivatives are usually called lacquers. Paints derived from acrylic and vinyl resins usually require a solvent such as ketone, and their architectural applications are limited. However, in addition to the formulation of agents that result in emulsion, polymerization has produced products with extensive architectural use. Bitumens or asphalts of petroleum or coal-tar derivation are most often used in roofing, and waterproofing applications where heavy layers are required an opportunity for renewal may be limited.

In paints that dry by oxidation, the vehicle is usually an oil or an oil-based varnish. These usually contain driers to accelerate drying of the oil. Paints based essentially on linseed oil with suitable pigments such as titanium dioxide and zinc oxide extenders once were the conventional exterior house paints. However, the successful development of polyvinyl acetate and acrylic emulsion types of paint reached the point where these materials became dominant in the exterior house paint market.

Water-Thinned Paints

This group of paints may be subdivided into those in which the vehicle is dissolved in water and those in which it is dispersed in emulsion form.

Paints with water-soluble vehicles include the calcimines, in which the vehicle is glue, and casein paints, in which the vehicle is casein or soybean protein. These paints are water sensitive and have only limited use. Synthetic resins soluble to water and treatments rendering drying oils soluble in water are relatively recent developments. Water evaporates from paints having these materials in their formulation, and further chemical change, either oxidation or heat polymerization, converts the vehicle so that the film no longer is water sensitive. These paints have a limited market because of the widespread utilization of latex emulsions.

Nearly any solvent-thinned paint may be emulsified by the addition of a suitable emulsifier and adequate agitation. The use of fugitive emulsifiers and of vehicles especially processed for use in emulsion paints has produced materials with excellent water resistance, color retention, and durability when cured. Materials formed by emulsion polymerization are described as a latex, and products are called latex paints. The most common latexes are made from a copolymer of butadiene and styrene, from polyvinyl acetate, and from acrylic resin. Properties and performance differ slightly among these paints, but as a group, they dominate the architectural market.

Enamel Paints

Enamel paint is an intimate dispersion of pigments in either a varnish or a resin vehicle or in a combination of both. Enamels may dry by oxidation at room temperature and/or by polymerization at room or elevated temperatures. They vary widely in composition, in color and appearance, and in properties. Although they generally give a high-gloss finish, some give a semigloss or eggshell finish, and still others give a flat finish. Enamels as a class are hard and tough and offer good mar and abrasion resistance. They can be formulated to resist attack by the most commonly encountered chemical agents in corrosive atmospheres and have good weathering characteristics.

Because of their wide range of useful properties, enamels are probably the most widely used organic coating in industry. One of their largest areas of use is as coatings for household appliances—washing machines, stoves, kitchen cabinets, and the like. A large proportion of refrigerators, for example, are finished with synthetic baked enamels. These appliance enamels are usually white and therefore must have a high degree of color and gloss retention when subjected to light and heat. Other products finished with enamels include automotive products; railway, office, sports, and industrial equipment; toys; and novelties.

House Paints

House paint for outside work consists of high-grade pigment and linseed oil, with a small percentage of a thinner and drier. The volatile thinner in paints is for ease of application, the drying oil determines the character of the film, the drier is to speed the drying rate, and the pigment gives color and hiding power. Part or all of the oil may be replaced by a synthetic resin. Many of the newer house paints are water-based paints.

Paints are marketed in many grades, some containing pigments extended with silica, talc, barytes, gypsum, or other material, fish oil or inferior semidrying oils in place of linseed oil, and mineral oils in place of turpentine. Metal paints contain basic pigments such as red lead, ground in linseed oil, and should not contain sulfur compounds. Red lead is a rest inhibitor and is a good primer paint for iron and steel, although it is now largely replaced by chromate primers. White lead has a plasticizing affect that increases adhesion. It is stable and not subject to flaking. Between some pigments and the vehicle, there is a reaction that results in progressive hardening of the film with consequent flaking or chalking, or there may be a development of water-soluble compounds. Linseed oil reacts with some basic pigments, giving chalking and flaking. Fading of a paint is usually from chalking. The composition of paints is based on relative volumes because the weights of pigments vary greatly, although the custom is to specify pounds of dry pigment per gallon of oil.

Bituminous Paints

Bituminous paints are usually coal tar or asphalt in mineral spirits, used for the protection of piping and tanks and for waterproofing concrete. For line pipe, heavy pitch coatings are applied hot, but a bitumen primer is first applied cold. The bituminous paints also have poor solvent resistance.

Commercial Finishes

These include air-drying or baking-cured materials applied by brush, spray, or magnetic agglomeration to kitchen and laundry appliances, automobile, machinery, and furniture and are used in highway marking materials. Paints in this group are subdivided by their drying mechanism.

Air-Drying Finishes

These materials once were the conventional factory-applied finishes. In the finishing of furniture, lacquers, varnishes, and shellac are still extensively used, but epoxy and unsaturated polyester materials have been adopted in recent years. Automobile manufacturers used in lacquers in air-drying alkyd enamels until the adoption of baked acrylic and urea finishes. Epoxy, urethane, and polyester resins, converted at room temperature with a suitable catalyst, are beginning to replace conventional solvent-thinned paints as machinery finishes. The marking of center lines on highways and other painted areas for the control of traffic requires a finish that dries rapidly, adheres well to both asphalt and concrete, and resists abrasion and staining. Solvent-thinned materials especially formulated for this service from alkyds, modified rubbers, and other resins are now extensively used, but some latex formulations have been tried with success.

Baking Finishes

Urea and melamine resins polymerize by heat and are used in baking finishes where extreme hardness, chemical resistance, and color retention are required as on kitchen and laundry appliances. Baked acrylic resin formulations now dominate the automobile finish market. Certain phenolic resins are converted by heat to produce finishes with excellent water and chemical resistance.

Industrial Coatings

Industrial coatings are subdivided by their intended service: corrosion-resistant coatings, high-temperature coatings, and coatings for immersion service. Materials in each subdivision are applied in a system that usually requires a base coat or primer, and intermediate coat or coats, and a top or finish coat.

Corrosion-Resistant Coatings

These are materials generally inert when cured to deterioration by acidic, alkaline, or other corrosive substances and applied in a system as a protective layer over steel or other substrates susceptible to corrosion attack.

The base, or prime, coat in the corrosion-resistant system is applied to dry surfaces prepared by abrasive blasting or other methods to a specified degree of cleanliness and toughness. The prime coat provides adhesion to the substrate for the entire coating system: adherence is by mechanical anchorage, chemical reaction, or a combination of the two. Prime coat materials once were predominantly a red lead pigment dispersed in linseed oil and cured by oxidation. Today, the prime coat material, commonly specified for application to steel surfaces is a dispersion of zinc in a suitable inorganic or organic vehicle. Although these zinc-rich primers now dominate the market where protection of steel is concerned, prime coat materials generically similar to subsequent coats in the system are specified when protection of nonferrous or nonmetallic substrates from corrosion attack is required.

An intermediate coat in the corrosion-resistant system is not always required. When used, intermediate coat materials usually are high-build layers of the same generic type specified for the top or surface coat and are applied only to increase the dry-film thickness of the protective area in places where airborne corrosive fumes, particulates, or droplets are heavily concentrated or where splash and spillage of corrosive fluids are relatively frequent.

A top or surface coat in the corrosion-resistant system may be selected from a variety of formulations with vehicles that include phenolic resins, chlorinated rubber, coal tar and epoxy combination, epoxy resin cured from a solvent solution with polyfunctional amines, polyamide resins, vinyl resin in solvent solution, elastomers, polyesters, and polyurethanes.

Top coat materials are required to have good scaling properties; high resistance to corrosive deterioration, oxidation, erosion, and ultraviolet degradation; and relative freedom, when cured, from pinholes, blisters, and crevices. Color retention is a desirable but not a mandatory requirement.

Selection of materials able to withstand corrosive attack in a given environment and to maintain cohesion and integrity among the various coats in system has become a technological specialty.

High-Temperature Coatings

These are materials used to alleviate or prevent corrosion, thermal shock, fatigue, oxide sublimation, and embrittlement of metals at high temperature.

Formulation of high-temperature coatings is a relatively recent development in the art. Material problems encountered since World War II such as high-temperature chemical reactions, production of steel by the oxygen lance process, aerodynamic heating to complex high-speed aircraft, and aerospace vehicle launching and reentry have been alleviated to an extensive degree by the results of research toward finding high-strength resistant metals and cements and toward producing coatings serviceable at high temperatures. Research continues as higher and higher temperatures are required by technological developments and radiation effects in outer space are better understood.

Available high-temperature coatings include inorganic zinc dispersed in a suitable vehicle, serviceable to 400°C; a phosphate bonding system in which ceramic fillers are mixed with an aqueous solution of monoaluminum phosphate, applied by spraying, brushing, or dipping, and serviceable to 1538°C after curing at 204°C; and a "ceramic gold" coating used on jet engine shrouds and having a temperature limit near 538°C.

Ablative formulations that absorb heat through melting, sublimation decomposition, and vaporization, or that expand upon heating to form a foamlike insulation that replenishes itself until the coating is depleted, employ silicone rubber or silicone resins or polyamide and tetrafluoroethylene polymers to provide short-term heat protection from 147°C to 538°C.

Coatings for Immersion Service

Included here are materials used to coat or line interior surfaces of vessels of containers holding or storing corrosive fluids, pipelines to which corrosive fluids are the flowing medium, and hoppers or bins conveying or holding abrasive or corrosive pellets or particulates.

These coatings are usually applied in relatively high-build systems to carefully cleaned and prepared surfaces. Air-drying formulations are used when extensive coverage is required and where polymerization by automatic heating apparatus is impractical. Small vessels, pipe spools, and assembly components may be protected by coatings cured by oven baking.

Coatings for immersion service may be selected from a variety of formulations that include asphalt, chemically cured coal tar, thermoplastic coal tar, epoxy–furans, amine-cured epoxies, fluorocarbons, furfuryl alcohol resins, neoprene, baked unmodified phenolics, unsaturated polyesters, polyether resins, low-density polyethylene, chlorosulfonated polyethylene, polyvinyl chloride plastisols, resinous cements, rubber, and urethanes, among other vehicles.

Selection and specification of coatings for immersion service has become a technological specialty.

Paint Removers

Paint removers, for removing old paint from surfaces before refinishing, are either strong chemical solvents or strong caustic solutions. In general, the more effective they are in removing the paint quickly, the more damaging they are likely to be to the wood or other organic material base. The hiding power of a paint is measured by the quantity that must be applied to a given area of a black and white background to obtain nearly uniform complete hiding. The hiding power is largely in the pigment, but when some fillers of practically no hiding power alone, such as silica, are ground to microfine-particle size, they may increase the hiding power greatly. Paint making is a highly developed art, and the variables are so many, and the possibilities for altering the characteristics by slight changes in the combinations are so great that the procurement specifications for paints are usually by usage requirements rather than by composition.

Palladium

Palladium (symbol Pd), a rare metal, is found in the ores of platinum. It resembles platinum, but is slightly harder and lighter in weight and has a more beautiful silvery luster. It is only half as plentiful but is less costly. The specific gravity is 12.10 and a melting point is 1552°C. Annealed, the metal has a hardness of Brinell 40 and a tensile strength of 186 MPa.

It is highly resistant to corrosion and to attack by acids, but, like gold, it is dissolved in aqua regia. It alloys readily with gold and is used in some white golds. It alloys in all proportions with platinum and the alloys are harder than either constituent.

Physical and Chemical Properties

Palladium is soft and ductile and can be fabricated into wire and sheet. The metal forms ductile alloys with a broad range of elements. Palladium is not tarnished by dry or moist air at ordinary temperatures. At temperatures from 350°C to 790°C, a thin protective oxide forms in air, but at temperatures from 790°C, this film decomposes by oxygen loss, leaving the bright metal. In the presence of industrial sulfur-containing gases, a slight brownish tarnish develops; however, the alloying palladium with small amounts of iridium or rhodium prevents this action.

At room temperature, palladium is resistant to nonoxidizing acids such as sulfuric acid, hydrochloric acid, hydrofluoric acid, and acetic acid, but the metal is attacked by nitric acid. Palladium is also attacked by moist chlorine and bromine.

In its pure form, it is malleable and corrosion resistant although it oxidizes in air above about 400°C. It can absorb about 1000 times its own volume of hydrogen at atmospheric pressure, and it is readily permeated by the gas above about 300°C, so it is used as a filter, typically in fine tube form, to purify hydrogen. It is used for jewelry in its pure form or for larger items alloyed with about 3% of either ruthenium or molybdenum. With about 25% copper, 5% nickel, and 1% nickel, it forms palladium copper that has good machining characteristics, reasonable hardness, and excellent corrosion resistance. With 50% gold, it forms palladium gold used as a dental material.

Uses

The major applications of palladium are in the electronics industry, where it is used as an alloy with silver for electrical contacts or in pastes in miniature solid-state devices and in integrated circuits. Palladium is widely used in dentistry as a substitute for gold. Other consumer applications are in automobile exhaust catalysts and jewelry.

The palladium–silver–gold alloys offer a series of noble brazing materials covering a wide range of melting temperatures. A palladium–silver alloy can be used as a diffusion septum for the separation of hydrogen from gas mixtures.

Palladium supported on carbon or alumina is used as a catalyst for hydrogenation and dehydrogenation in both liquid- and gas-phase reactions. Palladium finds widespread use in catalysis because it is frequently very active under ambient conditions and it can yield very high selectivities. Palladium catalyzes the reaction of hydrogen with oxygen to give water. Palladium also catalyzes isomerization and fragmentation reactions.

Palladium alloys are also used for instrument parts and wires, dental plates, and Felton-10 nibs. Palladium is valued for electroplating because it has a fine white color that is resistant to tarnishing even in sulfur atmospheres.

Although palladium has low electric conductivity, 16% that of copper, it is valued for its resistance to oxidation and corrosion. Palladium-rich alloys are widely used for low-voltage electric contacts. Palladium–silver alloys, with 30%–50% silver, for relay contacts, have 3%–5% the conductivity of copper. A palladium–silver alloy with 25% silver is used as a catalyst in powder or wire-mesh form. A palladium–copper alloy for sliding contacts has 40% copper with a conductivity of 5% that of copper. Many of the palladium salts, such as sodium palladium chloride, are easily reduced to the metal by hydrogen or carbon monoxide and are used in coatings and electroplating.

Palladium leaf is palladium beaten into extremely thin foil and used for ornamental work like gold leaf. Hydrogen forms solid solutions with palladium, forming *palladium sponge*, which is used for gas lighters. Palladium powder is made by chemical reduction and has a purity of 99.9% with amorphous particles 12–138 μ in (0.3–3.5 μm) in diameter. Atomized powder has spherical particles of 50–200 mesh and is free flowing. The powders are used for coatings and parts for service temperatures to 2300°F (1260°C). *Palladium flake* has tiny laminar platelets with average diameter of 118 μin. (3 μm) and thickness of 3.9 μin. (0.1 μm). The particles form an overlapping film in coatings. For autocatalytic converters, a "palladium-only" catalyst developed by Allied Signal can reduce precious metal caused by excluding platinum and rhodium.

Palm Oil

An oil obtained from the fleshy covering of the seed nuts of several species of palm trees, native to tropical Africa, but also grown in Central America. The oil is used as a fluxing dip in the manufacture of tinplate; for soaps, candles, and margarine; and for the production of palmitic acid. About 10% weight of the palm oil is recovered as by-product glycerin in making soaps or in producing the acid.

Palm oil contains 50%–70% palmitic acid, which in the form of glyceride is an ingredient of many fats. When isolated, it is a white crystalline powder of specific gravity 0.866 and melting point 65°C, soluble in hot water. It is used for soaps, cosmetics, pharmaceuticals, and food emulsifiers and in making plastics. *Greco 55L*, which is a white crystalline solid, for cosmetics and soaps, is 50% palmitic acid with the balance stearic acid.

The oil from the kernel of the palm nut, known as *palm kernel oil*, has different characteristics from palm oil. It contains about 50% lauric acid, 15% myristic acid, 16% oleic acid, and 7% palmitic acid, together with capric and caprylic acids found in coconut oil, while palm oil is very high in palmitic and oleic acids. The specific gravity is 0.873, iodine number 16–23, saponification value 244–255, and melting point 75°F–86°F (24°C–30°C).

PAN

See *polyacrylonitrile*.

Pancake Forging

A rough forged shape, usually flat, which can be obtained quickly with minimal tooling. Considerable machining is usually required to attain the finish size.

Pancake Grain Structure

A metallic structure in which the lengths and widths of individual grains are large compared to their thicknesses.

Paper

Paper is a flexible web or mat of fibers isolated from wood or other plant materials by the operation of pulping. Nonwovens are webs or mats made from synthetic polymers, such as high-strength polyethylene fibers, that substitute for paper in large envelopes and tote bags.

Paper is made with additives to control the process and modify the properties of the final product. The fibers may be whitened by bleaching, and the fibers are prepared for papermaking by the process of refining. Stock preparation involves removal of dirt from the fiber slurry and mixing of various additives to the pulp prior to papermaking. Papermaking is accomplished by applying a dilute slurry of fibers in water to a continuous wire or screen; the rest of the machine removes water from the fiber mat. The steps can be demonstrated by laboratory handsheet making, which is used for process control.

Although there is no distinct line to be drawn between papers and paperboard, paper is usually considered to be less than 0.15 mm thick. Most all fiber sheets over 0.30 mm thick are considered to be board. In the borderline range of 0.15–0.30 mm, most are considered to be papers, although some are classified as board.

Although paper has numerous specialized uses in products as diverse as cigarettes, capacitors, and countertops (resin-impregnated laminates), it is principally used in packaging (~50%), printing (~40%), and sanitary (~7%) applications. Paper was manufactured entirely by hand before the development of the continuous paper machine; this development allowed the U.S. production of paper to increase by a factor of 10 during the nineteenth century and by another factor of 50 during the twentieth century. In 1960, the global production of paper was 70 million tons (50% by the United States); in 1990, it was 210 million tons (30% by the United States). The annual per capita paper use in the United States is 300 kg, and about 40% is recovered for reuse.

Material of basis weight greater than 200 g/m^2 is classified as paperboard; lighter material is called paper. Production by weight is about equal for these two classes. Paperboard is used in corrugated boxes; corrugated material consists of top and bottom layers of paperboard called linerboard, separated by fluted corrugating paper. Paperboard also includes chipboard (a salad material used in many cold-cereal boxes, shoeboxes, and the backs of paper tablets) and food containers.

Mechanical pulp is used in newsprint, catalog, and other short-lived papers; they are only moderately white and yellow quickly with age because the lignin is not removed. A mild bleaching treatment (called brightening) with hydrogen peroxide or sodium dithionite (or both) masks some of the color of the lignin without lignin removal. Paper made with mechanical pulp and coated with clay to improve brightness and gloss is used in 70% of magazines and catalogs, and in some enamel grades. Bleach chemical pulps are used in higher grades of printing paper used for xerography, typing paper, tablets, and envelopes; these papers are termed uncoated wood-free (meaning free of mechanical pulp). Coated wood-free papers are of high to very high grade and are used in applications such as high-quality magazines and annual reports; they are coated with calcium carbonate, clay, or titanium dioxide.

Like wood, paper is a hygroscopic material; that is, it absorbs water from, and also releases water into, the air. It has an equilibrium moisture content of about 7%–9% at room temperature and 50% relative humidity. In low humidities, paper is brittle; in high humidities, it has poor strength properties.

General Features and Uses

The major attribute of paper is its extreme versatility. A wide range of end properties can be obtained by control of the variables in (1) original selection of the type and size of fiber, (2) the various pulp processing methods, (3) the actual web-forming operation, and (4) the treatments that can be applied after the paper has been produced.

Papers have been specifically developed for a number of engineered applications. These include gasketing; electrical, thermal, acoustical, and vibration insulation; liquid and air filtration; composite structural assemblies; simulated leathers and backing materials; cord or twine; and as yarns for paper and textiles.

Wood

Wood, a diverse, variable material, is the source of about 90% of the plant fiber used globally to make paper. Straw, grasses, canes, bast, seed hairs, and reeds are used to make pulp, and in many regards, their pulp is similar to wood pulp. Fibers are tubular elements of plants and contain cellulose as the principal constituent. Softwoods (gymnosperms) have fibers that are about 3–5 mm long, while in hardwoods (angiosperms), they are about 0.8–1.6 mm long. In both cases, the length is typically about 100 times the width. Softwoods are used in papers such as linerboard where strength is the principal intent. Hardwoods are used in papers such as tissue and printing to contribute to smoothness. Many papers also include some softwood pulp for strength and some hardwood pulp for smoothness.

Wood consists of three major components: cellulose, hemicellulose, and lignin. The first two are white polysaccharides of high molecular weight that are desirable in paper. Cotton is over 98% cellulose, while wood is about 45%. Hemicellulose, although not water soluble, is similar to starch. The hydroxyl groups of these materials allow fibers to be held together in paper by hydrogen bonding. Adhesives are not required to form paper, but some starch is usually used and has a similar effect to the hemicelluloses in helping the fibers bond together. Lignin makes up about 25%–35% of softwoods and 18%–25% of hardwoods. It is concentrated between fibers.

Paper Production

Most papers are made from crude fibrous wood pulps.

Types of Pulp

The type of pulp to a large extent determines the type of paper produced. Pulps are generally classified as mechanical or chemical wood pulps.

Mechanical wood pulps produced by mechanical processes include

1. Ground wood, which is used in a number of papers where absorbency, bulk, opacity, and compressibility are primary requirements and permanence and strength are secondary
2. Defibrated pulps, which are used for insulating board, hardboard, or roofing felts where good felting properties are required
3. Exploded pulps, used for building and insulation hardboards, or so-called wood composition materials

Chemical wood pulps are produced by "cooking" the fibrous material and various chemicals to provide certain characteristics. They include the following:

1. Sulfite pulps, used in the bleached or unbleached state for papers ranging from very soft or weak to strong grades. There are about 12 grades of sulfite pulps.
2. Neutral sulfite or monosulfite pulps, used for strong papers for bags, wrappings, and envelopes.
3. Sulfate or kraft pulps, providing high strength, fair cleanliness, and, in some instances, high absorbency. Such pulps are used for strong grades of unbleached, semibleached, or bleached paper (called kraft paper) and board.
4. Soda pulps, used principally in combination with bleached sulfite or bleached sulfate pulp for book-printing papers.
5. Semichemical pulps, used for specialty boards, corrugating papers, glassine and greaseproof papers, test liners, and insulating boards and wallboards.
6. Screenings, used principally for coarse grades of paper and board, such as mill wrapper, and as a substitute for chipboard, corrugated papers, and insulation board.

Papermaking

The two basic types of papermaking machines are the Fourdrinier and the cylinder machines; the Fourdrinier machine is the most commonly used.

In Fourdrinier papermaking, the pulp, mixed to a consistency of 97.5%–99.5% water, is fed continuously to Fourdrinier, which consists of an endless belt of fine-mesh screen called the "wire."

As the pulp web travels along the wire, water drains from it into suction boxes. As it leaves the wire (at about 83% water), it passes through presses, and usually a variety of other types of equipment, such as dryers and calender rolls, depending on the type of paper produced.

Cylinder machine papermaking differs from Fourdrinier in that the web of pulp is formed on a cylindrical mold surface instead of a continuous wire covered with fine-wire cloth, which revolves in a vat of paper stock or pulp. The felt conveyor carries the resulting web to the press and dryers.

The cylinder machine is used to produce a greater variety of paper thicknesses, ranging from the thinnest tissue to the thickest building board.

Bleaching

Chemical pulp for printing paper has 3%–6% lignin, which gives the pulp a brown color. This lignin is removed with bleaching chemicals and in 4–8 stages. Each stage consists of a pump to mix the bleaching agent with the pulp, a retention tower to allow the chemical to react with the pulp for 30 min to several hours, and a washing unit to remove the solubilized lignin and residual chemicals from the pulp. Usually an oxidizing material is followed by a stage of alkali extraction, because lignin becomes more soluble at high pH. A common bleaching sequence involves elemental chlorine, alkali extraction, and, finally, chlorine dioxide. Sodium hypochlorite, also found in household liquid bleach, is sometimes used. There is some pressure for the bleaching process to be elemental chlorine-free. This is possible by using oxygen as the first bleaching chemical, chlorine dioxide in place of chlorine, and hydrogen peroxide as a bleaching agent.

Types of Paper

In the broadest classification, there are three basic types of papers: cellulose fiber, inorganic fiber, and synthetic organic fiber papers.

Cellulose Fiber Papers

These papers, made from wood pulp, constitute by far the largest number of papers produced. A great many of the engineering papers are produced from kraft or sulfate pulps. The term *kraft* is used broadly today for all types of sulfate papers, although it is primarily descriptive of the basic grades of unbleached sulfate papers, where strength is the chief factor and cleanliness and color are secondary. By various treatments, kraft can be altered to produce various grades of condenser, insulating, and sheathing papers.

Other types of vegetable fibers used to produce papers include

1. Rope, used for strong, pliable papers, such as those required in cable insulation, gasketing, bags, abrasive papers, and pattern papers
2. Jute, used for papers possessing excellent strength and durability
3. Bagasse, used for paper for wallboard and insulation, usually where strength is not a primary requirement
4. Esparto, used for high-grade book or printing papers (A number of other types of pulps are also used for these papers.)

Inorganic Fiber Papers

There are three major types of papers made from inorganic fibers:

1. Asbestos is the most widely used inorganic fiber for papers. Asbestos papers are nonflammable and resistant to elevated temperatures and have good thermal insulating characteristics. They are available with or without binders and can be used for electrical insulation or for high-temperature reinforced plastics.
2. Fibrous glass can be used to produce porous and non-hydrating papers. Such papers are used for filtration and thermal and electrical insulation and are available with or without binders. High-purity silica glass papers are also available for high-temperature applications.
3. Ceramic fiber (aluminum silicate) papers provide good resistance to high temperatures and low thermal conductivity, have good dielectric properties, and can be produced with good filtering characteristics.

Synthetic Organic Fiber Papers

A great deal of research has been carried out on the use of such synthetic textile fibers as nylon, polyester, and acrylic fibers in papers. Some of the earliest appear highly promising for electrical insulating uses. Others appear promising for chemical or mechanical applications. They are most commonly combined with other fibers in a paper, primarily to add strength.

P

Paper Treatments

Papers can be impregnated or saturated, coated, laminated, or mechanically treated. The major treatments used are covered and indicate the extent of treatments available.

Impregnation or Saturation

Impregnation or saturation can be carried out either at the beater stage in the processing of the pulp or after the paper web has been formed. Beater saturation permits saturation of nonporous or adsorbent papers, whereas paper saturated after manufacture must be of the absorbent type to permit complete impregnation by the saturant.

Papers can be saturated or impregnated with almost any known resin or binder. Probably the most commonly used are asphalt for moisture resistance; waxes for moisture vapor and water resistance; phenolic resins for strength and rigidity; melamine and certain ureas for wet strength (not to be confused with moisture resistance); rubber latexes, both natural and synthetic, for resilience, flexibility, strength, and moisture resistance; epoxy or silicone resins for dielectric characteristics at elevated temperatures; and ammonium salts, or other materials, for flame proofing.

A number of proprietary beater-saturated papers are currently available. They are used primarily for gasketing, filtration, simulated leathers, and backing materials. Most of these consist of cellulose or asbestos fibers blended with natural or synthetic rubbers. In some cases, cork is added to the blend for increase compressibility. Another type of proprietary beater-saturated paper consists of leather fibers blended with rubber latexes.

Coatings

Papers may be coated either by the paper manufacturer or by converters. Coating materials, which also impregnate the paper to a greater or lesser degree, include practically every known resin or binder and pigment used in the paint industry. Coatings can be applied in solvent or water solutions, water emulsions, hot melts, and extrusion coatings or in the form of plastisols or organosols.

The most important properties provided by coatings are (1) gas and water vapor resistance; (2) water, liquid, and grease resistance; (3) flexibility; (4) heat sealability; (5) chemical resistance; (6) scuff resistance; (7) dielectric properties; (8) structural strength; (9) mold resistance; (10) avoidance of fiber contamination; and (11) protection of printing.

Coating materials range from the older asphalts, waxes, starches, casein, shellac, and natural gums to the newer polyethylenes, vinyl copolymers, acrylics, polystyrenes, alkyds, polyamides, cellulosics, and natural or synthetic rubbers.

Laminations

Paper can be laminated to other papers or to other films to provide a variety of composite structures.

Probably, the most common types of paper laminates are those composed of layers of paper laminated with asphalt to provide moisture resistance and strength. Simple laminations of paper can be so oriented that overall characteristics of the composite are isotropic.

Laminating paper with plastics or other types of films or with metal foils will, in many cases, combine the desirable properties of the film or foil with those of the paper.

Scrim is a mat of fibers, usually laminated as a "core" material between two faces of paper. It is usually used to provide strength but can also provide bulk for cushioning, or a degree of "hand" to the composite material.

Mechanical Treatments

Several mechanical treatments can be applied to papers to provide particular special properties.

Crimping, which can be done either on the paper web or on the individual fibers of the paper, essentially adds stretch or extensibility. Crimping the paper web results in crepe paper with improved strength, stretch, bulk, and conformability and with texture similar to that of cloth. The creping process usually consists of "crowding" the paper into small pleats or folds with a "doctor." Typical range of elongation or stretch obtainable is 20%–300%. Cross-creping can provide controllable stretch in directions perpendicular to each other, further improving drapability.

A high degree of stretch, conformability, and flexibility is produced by a patented process that differs from creping in that the individual fibers in the paper web are crimped, rather than the web itself. The amount of stretch is variable, but about 10% stretch in the machine direction seems to be optimal for most industrial applications. The major advantages of this type of paper are reported to be a high degree of toughness, combined with a smooth surface, and a high resistance to tearing or punching.

Twisting

Twisting is used to convert paper to twine or yarns. Such yarns have substantially higher strength than the paper from which they are made.

Twisting papers are usually sulfate papers, either bleached or unbleached. High tensile strength is required in the machine direction, and the heavier-weight papers should usually be soft and pliable. Treatments to impart such characteristics as wear and moisture resistance can be applied during or after the spinning operation.

Embossing and Other Techniques

Decorative papers can be produced by embossing in a variety of patterns. Embossing does not usually improve strength significantly. Embossing or "dimpling" in certain patterns, followed by lamination, can produce composites with added strength as well as bulk and thermal insulation.

Other mechanical methods include (1) shredding for bulk or padding; (2) pleating, used as a forming aid and for strength in paper cups and plates; (3) die cutting and punching; and (4) molding, which consists of compressing the wet pulp web in a mold to form a finished shape, such as an egg crate.

Applications

As an engineering material, paper has several important applications.

As filtration material, paper can be used either as a labyrinth barrier material to guide the fluid or gas to be filtered or, more commonly, as the filtering medium itself. Paper is used for filtering automotive air and oil, and air in room air conditioners, as well as for industrial plant filtration, filtering liquids in teabags, in addition to machine-cooling oil and domestic hot water filters.

Papers are used for both light- and heavy-duty gaskets, for such applications as high- and low-pressure steam and water, high- and low-temperature oil, aromatic and nonaromatic fuel systems, and sealing both rough and machined surfaces.

Electrical insulation represents one of the largest engineering uses of papers. For electrical uses, special types include coil papers or layer insulation, cable paper or turn insulation, capacitor papers, condenser papers, and high-temperature, inorganic insulating papers.

A large and growing use for paper is in structural sandwich materials, where papers are impregnated with a resin such as phenolic and formed in the shape of a honeycomb. The honeycomb is used as a core between facing sheets of a variety of materials including paperboard, reinforced plastics, and aluminum.

Another large-volume application of papers is as backing material for decorative films or other surfacing materials. These provide bulk and depth to the product in simulating leather or fabrics.

Paper Plants

Cellulose for papermaking is obtained from a wide variety of plant life, made directly into *paper pulp*, or obtained from old rags that were originally made from vegetable fibers. Animal fibers incorporated into some papers are fillers for special purposes, not papermaking materials. *Rice paper* of China came from the *Tetrapanax papyriferum*, but the so-called rice paper used for cigarettes in the United States is made from flax fiber. *Cigarette paper* is also made from ramie and sunn hemp. The distinction between cigarette paper and the tissue paper used for wrapping is that it must be free of any substance that would impart a disagreeable flavor to the smoke, and it must be opaque and pure white, must burn at the same rate as tobacco, and must be tasteless.

Wood pulp is now the most important papermaking material. Spruce is the chief wood used for the sulfite process, but hemlock and balsam fir are also used. Aspen and other hardwoods are used in the soda process, and also southern pine. White fir is readily pulped by any process, but western red cedar is high in lignin content, about 30%, and reduced with difficulty by the sulfate process to a dark-colored pulp. It is pulped by the kraft process. Its fibers are fine and short, yielding a paper of high bursting strength. Normally, the *pulpwoods* of the West Coast are western hemlock, white fir, and Sitka spruce, leaving the Douglas fir to the lumber mills. Western hemlock, balsam, and spruce are the chief pulpwoods of Canada. Pines are used extensively in the United States, especially for kraft paper, paperboard, and book paper. More than 50% of all pulpwood used in United States is now from the southern states, and about 10% of this is salvage from lumber mills. But, in general, special methods are used for pulping pine since conventional sulfite liquor does not free the fibers as the phenolic compounds in the heartwood condense to form insoluble compounds.

Kraft paper is sulfate-pulped from a mixture of 50% western hemlock, 25% western red cedar, and 25% Douglas fir. The fir has a coarse fiber that gives high tear strength; cedar has a long, thin fiber that gives a smooth surface; and hemlock is abundant and used as a filler. Poplars are also used for pulpwood, and Scott Paper Co. uses fast-growing scrub alder. Newsprint made from hardwoods has a bursting strength 20% higher than that made from softwoods, and the brightness value is higher, but the pulping of hardwoods is usually a more involved chemical process.

In England, fine printing papers are made by the soda process from *esparto* grass. It gives a soft, opaque, light paper, although the cellulose content is less than 50%. Some grades of cardboard and some newsprint are made from straw. *Deluwang paper* of the East Indies is made from the scrapped and beaten bark of the *paper mulberry* tree. It is an ancient industry in Java, and the paper is used for lampshades and fancy articles. Under the name of *tapa cloth*, the sheets are dried and used as a muslinlike fabric by the Polynesians. The strips are welded together by overlapping and beating together the wet material. Bagasse is of increasing importance as a papermaking material in the sugar-growing areas of the world.

Parabolic Reflector

A reflector for ultraviolet (UV) curing of adhesives that projects parallel light beams perpendicular to the assembly, but is not focused. See also *UV electron beam–cured adhesives*.

Paraffin

A general name often applied to paraffin wax, but more correctly referring to a great group of hydrocarbons obtained from petroleum. Paraffin compounds begin with methane, CH_4, and are sometimes called the *methane group*. The compounds in the series have the general formula C_nH_{2n+n} and include gases methane and ethane; the products naphtha, benzene, gasoline, lubricating oils, and jellies; and the common paraffin. The name *paraffin* indicates little affinity for reaction with other substances. And, in practice, the name is limited to the waxes that follow petroleum jelly in the distillation of petroleum. These waxes melt at 104°F–140°F (40°C–60°C) and consist of the hydrocarbons between C_{22} and C_{27}; the refined waxes may range up to 194°F (90°C). They burn readily in air. Paraffin occurs to some extent in some plant products, but its only commercial source is from natural petroleum.

Paraffin oil is drip oil from the wax presses in the process of extracting paraffin wax from the wax-bearing distillate in refining of petroleum. The oil is treated, redistilled, and separated into various grades of lubricating oils from light to heavy. They may be treated and bleached with sulfuric acid and neutralized with alkali. When decolorized with acid and sold as filtered, they are brilliant liquids, but are not suitable in places where they may be in contact with water, since the sulfo compounds present cause emulsification. *Triton oil* is 100% pure paraffin oil.

Paraffin and Whitewash Test

An early form of penetrant test in which the suspect component was immersed in paraffin, usually warm, withdrawn and wiped dry, and then painted with whitewash. Any cracks were revealed by paraffin seeping from them to stain the whitewash.

Paraffin Wax

The first distillate taken from petroleum after the cracking process is referred to as *wax bearing* and is put through a filter press and separated from the oils. The wax collected on the plates is called *slack wax*, and it contains 50% wax and 50% oil. This is chilled to free it from oil. The yellow waxes are filtered to make a white, semitranslucent, refined wax, which is odorless and tasteless. For large-scale operations, solvent methods of wax extraction are used. Paraffin waxes are soluble in ether, benzene, and essential oils.

Petrolatum is a two-phase colloidal system or gel consisting of high-molecular-weight hydrocarbon oils dispersed in microcrystalline waxes. *Wax tailing* is a name for the distillate that comes from petroleum after the wax-bearing distillate is removed. It contains no wax, but at ordinary temperatures looks like beeswax. It is very adhesive and is employed in roofings and for waterproof coatings.

Synthetic paraffin wax, called *Ruhr wax*, is made in Germany from low-grade coals and other hydrocarbon sources. They are used as additives in paper coatings and printing inks and in mixing with refined paraffin.

Ozokerite, also known as *mineral wax*, and as *earth wax*, is a natural paraffin found in Utah and in central Europe and used as a substitute or extender of beeswax and in polishes, candles, printing inks, crayons, sealing waxes, phonograph records, and insulation.

Parallel Laminate

(1) A composite laminate of woven fabric in which the plies are aligned in the same position as they were on the fabric roll. (2) A series of flat or curved cloth–resin layers stacked uniformly one on top of the other.

Paramagnetic

See *magnetic*.

Paramagnetic Material

(1) A material whose specific permeability is greater than unity and is practically independent of the magnetizing force. (2) Material with a small positive susceptibility due to the interaction and independent alignment of permanent atomic and electronic magnetic moments with the applied field. Compare with *ferromagnetic material*.

Paramagnetism

A property exhibited by substances that, when placed in a magnetic field, are magnetized parallel to the field to an extent proportional to the field (except at very low temperatures or in extremely large magnetic fields). Compare with *ferromagnetism*.

Parameter

In statistics, a constant (usually unknown) defining some property of the frequency distribution of a population, such as a population median or a population standard deviation.

Parameter (in Crystals)

See *lattice parameter*.

Parchment

Originally, goatskin or sheepskin specially tanned and prepared with a smooth, hard finish for writing purposes. It was used for legal documents, maps, and fancy books, being more durable than the old papers. The extremely thin, high-quality parchment that was used for documents and handmade books was rubbed with pumice and flattened with lead. Parchment now is usually *vegetable parchment*. It is made from a base paper of cotton rags or alpha cellulose called *water leaf* that contains no sizing or filling materials. The water leaf is treated with sulfuric acid that converts a part in the cellulose to a gelatinlike amyloid. When the acid is washed off, the amyloid film hardens on the fibers and the interstices of the paper. The strength of the paper is increased, and it will not disintegrate even when fully wet. The paper now has a wide usage in food packaging as well as for documents as a competitor of the resin-treated wet-strong papers. The wet strength and grease resistance are varied by differences in acid treatment and subsequent sizing. *Vellum* is a thick grade of writing paper made from high-grade rag pulp pebbled to imitate the original calfskin parchment called vellum.

Parent Metal

In welding, the original unjoined metal and its microstructure. See preferred term *base metal*.

Parfocal Eyepiece

Microscope eyepieces, with common focal planes, which are interchangeable without refocusing.

Parfocal Lenses

Lenses that have the same focal length so that they can be interchanged without the need to refocus the microscope or other optical equipment.

Parison

The hollow plastic tube for which a plastic component is blow molded. See also *blow molding*.

Parison Swell

In blow molding of plastics, the tendency of the *parison* to enlarge as it emerges from the die. It is expressed as the ratio of the cross-sectional area of the parison to the cross-sectional area of the die opening. See also *blow molding*.

Parkerizing

A proprietary treatment for forming corrosion-resistant phosphate conversion coating on steel.

Parkes Process

A process used to recover precious metals from lead and based on the principle that if 1% to 2% zinc is stirred into the molten lead, a compound of zinc with gold and silver separates out and can be skimmed off.

Partial Annealing

An imprecise term used to denote a treatment given cold-worked metallic material to reduce its strength to a controlled level or to affect stress relief. To be meaningful, the type of material, the degree of cold work, and the time–temperature schedule must be stated.

Partial Hydrodynamic Lubrication

See *quasihydrodynamic lubrication*.

Partial Joint Penetration

Weld joint penetration which is less than complete. Compare with *complete joint penetration*.

Partial Journal Bearing

A bearing in which the bore extends not more than half the circumference of the journal.

Partial Pressure

The pressure exerted by individual gases in a mixture.

Partially Stabilized Zirconia

Zirconia (ZrO_2) that contains a mixture of cubic and tetragonal and/or monoclinic phases produced by the addition of small amounts of magnesium oxide (MgO), calcium oxide (CaO), or yttrium oxide (Y_2O_3). These materials are used in structural applications where their high strength (600–700 MPa, or 87–100 ksi) and moderate fracture toughness (11–14 MPa√m, or 10–13 ksi√in.) can be exploited. See also *zirconia*.

Particle

(1) Any small part of matter, such as a molecule, atom, or electron. (2) Any small subdivision of matter, such as a metal or ceramic powder particle.

Particle Accelerator

A device that raises the velocities of charged atomic or subatomic particles to high values.

Particle Hardness

The hardness of an individual ceramic or metal powder particle as measured by a Knoop- or Vickers-type microhardness indentation test.

Particle Morphology

The form and structure of an individual metal or ceramic powder particle.

Particle Shape

The appearance of a metal particle, such as spherical, rounded, angular, acicular, dendritic, irregular, porous, fragmented, blocky, rod, flake, nodular, or plate.

Particle Size

The controlling lineal dimension of an individual particle as determined by analysis with screens or other suitable instruments. See also *sieve analysis* and *sieve classification*.

Particle Size Analysis

See preferred term *sieve analysis*.

Particle Size Classification

See preferred term *sieve classification*.

Particle Size Distribution

The percentage, by weight or by number, of each fraction into which a powder or sand sample has been classified with respect to sieve number or *particle size*. See *sieve classification*.

Particle Size Range

(1) The limits between which variation in particle size is allowed. (2) Classification of spray powders defined by an upper and lower size limit; for example, −200 + 300 mesh: a quantity of powder, the largest particles of which will pass through a 200-mesh sieve and the smallest of which will not pass through a 325-mesh sieve. See also *sieve analysis* and *sieve classification*.

Particle Sizing

Segregation of granular material into specified particle size ranges.

Particle Spacing

The distance between the surfaces of two or more adjacent particles in a loose powder or a compact.

Particle-Induced x-Ray Emission

A method of trace elemental analysis in which a beam of ions (usually protons) is directed at a thin foil on which the sample to be analyzed has been deposited; the energy spectra of the resulting x-rays is measured. See also *particle accelerator* and *proton*.

Particulate Composite

Material consisting of one or more constituents suspended in a matrix of another material. These particles are either metallic or nonmetallic.

Particulates

Particulates are solids or liquids in a subdivided state. Because of this subdivision, particulates exhibit special characteristics that are negligible in the bulk material. Normally, particulates will exist only in the presence of another continuous phase, which may influence the properties of the particulates. A particulate may comprise several phases. They can be categorized into particulate systems that relate them to commonly recognized designations. Fine-particle technology deals with particulate systems in which the particulate phase is subject to change or motion. Particulate dispersions in solids have limited specialized properties and are conventionally treated in disciplines other than fine-particle technology.

The universe is made up of particles, ranging in size from the huge masses in outer space (such as galaxies, stars, and planets) to the known minute building blocks of matter (molecules, atoms, protons, neutrons, electrons, neutrinos, and so on). Fine-particle technology is concerned with those particles that are tangible to human senses, yet small compared to the human environment—particles that are larger than molecules but smaller than gravel. Fine particles are in abundance in nature (as in rain, soil, sand, minerals, dust, pollen, bacteria, and viruses) and in industry (as in paint pigments, insecticides, powdered milk, soap, powder, cosmetics, and inks). Particulates are involved in such undesirable forms as fumes, fly ash, dust, and smog and in military strategy in the form of signal flares, biological and chemical warfare, explosives, and rocket fuels.

Processing and Uses

There are many cases where a particulate is either a necessary or an inadvertent intermediary in an operation. Areas that may be involved in processing particulates may be classified as follows:

> *Size reduction*
>   Mechanical (starting with bulk material)
>   Grinding
>   Atomization
>   Emulsification
>   Physicochemical (conversion to molecular dispersion)
>   Phase change (spray drying, condensation)
>   Chemical reaction

P

Size enlargement (agglomeration, compaction)
    Pelletizing
    Briquetting
    Nodulizing
    Sintering
Separation or classification
    Ore beneficiation
    Protein shift
Deposition (collection, removal)
Coating or Encapsulation
Handling
    Powders
    Gas suspensions (pneumatic conveying)
    Liquid suspensions (non-Newtonian fluids)

End products in which particulate properties themselves are utilized include the following:

Mass or heat-transfer agents (used in fluidized beds)
Recording (memory) agents
    Electrostatic printing powders and toners (for optical images)
    Magnetic recording media (for electronic images)
Coating agents (paints)
Nucleating agents
Control agents (for servomechanisms)
    Electric fluids
    Magnetic fluids
Charge carriers (used in propulsion, magnetohydrodynamics)
Chemical reagents
    Pesticides, fertilizers
    Fuels (coal, oil)
    Soap powders
    Drugs
    Explosives
Food products

Characterization

The processing or use of particulates will usually involve one or more of the following characteristics:

Physical
    Size (and size distribution)
    Shape
    Density
    Packing or concentration
Chemical
    Composition
    Surface character
Physiochemical (including adhesive, cohesive)
Mechanical or dynamic
    Inertial
    Diffusional
    Fluid drag
    Dilute suspensions (Stokes' law, Cunningham factor, drag coefficient)
    Concentrated suspensions (hindered settling, rheology, fluidization)
Optical (scattering, transmission, absorption)
    Refractive index (including absorption)
    Reflectivity

Electrical
    Conductivity
    Charge
Magnetic
Thermal
    Insulation (conductivity, absorptivity)
    Thermophoresis

Many of the characteristics of particulates are influenced to a major extent by the particle size. For this reason, particle size has been accepted as a primary basis for characterizing particulates. However, with anything but homogeneous spherical particles, the measured "particle size" is not necessarily a unique property of the particulate but may be influenced by the technique used. Consequently, it is important that the techniques used for size analysis be closely allied to the utilization phenomenon for which the analysis is desired.

Particle Size

Size is generally expressed in terms of some representative, average, or effective dimension of the particle. The most widely used unit of the particle size is the micrometer (μm), equal to 0.001 mm. Another common method is to designate the screen mesh that has an aperture corresponding to the particle size. The screen mesh normally refers to the number of screen openings per unit length or area; several screen standards are in general use; the two most common in the United States are the U.S. standard and the Tyler standard screen scales.

Particulate systems are often complex. Primary particulates may exist as loosely adhering (as by van der Waals forces) particles called flocs or as strongly adhering (as by chemical bonds) particulates called agglomerates. Primary particles are those whose size can only be reduced by the forceful shearing of crystalline or molecular bonds. Some particulates can be dispersed and can be reversed with application of light shearing forces.

Parting

(1) In the recovery of precious metals, the separation of silver from gold. (2) The zone of separation between *cope* and *drag* portions of the mold or flask in sand casting. (3) A composition sometimes used in sand molding to facilitate the removal of the pattern. (4) Cutting simultaneously along two parallel lines or along two lines that balance each other in side thrust. (5) A shearing operation used to produce two or more parts from a stamping.

Parting Agent

See *mold release agent.*

Parting Compound

A material dusted or sprayed on foundry (casting) patterns to prevent adherence of sand and to promote easy separation of *cope* and *drag* parting surfaces when the cope is lifted from the drag.

Parting Line

(1) The intersection of the parting plane of a casting or plastic mold or the parting plane between forging dies with the mold or die cavity.

(2) A raised line or projection on the surface of a casting, plastic part, or forging that corresponds to the said intersection.

Parting Sand

In foundry practice, a fine sand for dusting on sand mold surfaces that are to be separated.

Partition Coefficient

See *partition law*.

Partition Law

This states that "the concentrations of any individual molecular species into immiscible phases maintain a constant ratio to each other at constant temperature." It is of course implicit that the two phases are sufficiently mixed or otherwise in contact for a time sufficient for equilibrium to be reached. The ratio of the two concentrations is the partition coefficient or distribution ratio.

Parts Former

A type of upsetter designed to work on short billets instead of bars and tubes, usually for cold forging.

Parts Per Billion

A measure of proportion by weight, equivalent to one unit weight of a material per billion (10^9) unit weights of compound.

Parts Per Million

A measure of proportion by weight, equivalent to one unit weight of a material per million (10^6) unit weights of compound.

PAS

See *polyaryl sulfone*.

Pascal

The SI unit of pressure, stress. 1 Pa = 1 N/m². 1 N (Newton, the unit of force) = 0.224809 lbf.

Pass

(1) One movement of the material through a pair of rolls or alternatively through the series of pairs of rolls comprising a rolling mill. (2) One line of one layer of weld deposit. Some processes such as the metal inert gas process can deposit a complete pass without interruption, but in the manual metallic arc process, a number of electrodes may be required to make one pass. (3) The open space between two grooved rolls through which metal is processed. See also *weld pass*.

Pass Sequence

See *deposition sequence*.

Passivation

(1) A reduction of the anodic reaction rate of an electrode involved in corrosion. (2) The process in metal corrosion by which metals become *passive*. (3) The changing of a chemically active surface of a metal to a much less reactive state. Contrast with *activation*. (4) The formation of an insulating layer directly over the semiconductor surface to protect the surface from contaminants, moisture, and so forth.

Passivator

A type of corrosion *inhibitor* that appreciably changes the potential of a metal to a more noble (positive) value.

Passive

(1) A metal corroding under the control of a surface reaction product. (2) The state of the metal surface characterized by low corrosion rates in a potential region that is strongly oxidizing for the metal. (3) Generally, a metal not reacting significantly with the environment, particularly in electrochemistry. More specifically, a metal that can be reactive with the environment but that has developed some surface characteristic such as a protective oxide film that prevents further attack.

Passive–Active Cell

A corrosion cell in which the *anode* is a metal in the *active* state and the *cathode* is the same metal in the *passive* state.

Passivity

A condition in which a piece of metal, because of an impervious covering of oxide or other compound, has a *potential* much more positive than that of the metal in the active state.

Paste

An adhesive compound having a characteristic plastic-type consistency, that is, a high order of yield value, such as that prepared by heating a mixture of starch and water and subsequently cooling the hydrolyzed product. See also *adhesive*, *glue*, *mucilage*, and *sizing*.

Paste Brazing Filler Metal

A mixture of finely divided brazing filler metal with an organic or inorganic flux or neutral vehicle or carrier. Brazing paste mixtures are usually prepared for *furnace brazing* in a *protective atmosphere*.

Paste Extrude

An extrusion method for plastics in which the executable material, a fine powder form mixed with a lubricant, is forced through a die of given size, without heat, as opposed to melt extrude. See also *extruder* (plastics).

Paste Soldering Filler Metal

A mixture of finely divided metallic solder with an organic or inorganic flux or neutral vehicle or carrier.

Pasteboard

A class of thick paper used chiefly for making boxes and cartons and for spacing and lining. It may be made by pasting together several single sheets, but more usually by macerating old paper and rolling into heavy sheets. It may also be made of straw, certain grasses, and other low-cellulose paper materials, and it is then known as *strawboard*. Colloquially, the term *pasteboard* applies to any paper stock board used for making boxes, including the hard and stiff boards made entirely from pulp, but the term *pasteboard* is not liked in the paper industry. The bulk of packaging boards are now pulp boards treated with resin and are called *carton boards*. *Cardboard* is usually a good-quality chemical pulp or rag pasteboard used for cards, signs, or printed material or for the best-quality boxes. Ivory board, for art printing and menu cards, is a highly finished cardboard clay coated on both sides. *Jute board*, used for folding boxes, is a regular product of the paper mills and is a strong, solid board made of kraft pulp. *Chipboard* is a cheap board made from mixed scrap paper, used for boxes and book covers. When made with a percentage of mechanical wood pulp, it is called *pulpboard*. A heavy, rope–pulp paper or board, usually reddish and used for large expansion filing envelopes, is called *paperoid*.

Pastes

Conductor, resistor, dielectric, seal glass, polymer, and soldering compositions are available in paste or ink form. They are used to produce hybrid circuits, networks, and ceramic capacitors. The materials are often called thick-film compositions.

Types

Conductor pastes consist of metallic elements and binders suspended in an organic vehicle. Primarily, precious metals such as gold, platinum, palladium, silver, copper, and nickel are used singularly or in combination as the conductive element. The adhesion mechanism to the substrate is provided by either a frit bond, reactive bond, or mixed bond.

Important properties of conductor pastes include wire bondability, conductivity, solderability, solder leach resistance, and line definition.

Thick-film resistor plates are composed of a combination of class frit, metal, and oxides. These pastes are used in microcircuits, voltage dividers, resistor networks, chip resistors, and potentiometers.

Dielectric compounds are used as insulators for the fabrication of multilayer circuits and crossovers or as protective coverings.

Solder pastes are one of the more common component attach products. They consists of finally divided solder powders of all common alloys of tin, lead, silver, gold, etc., suspended in a vehicle–flux system. The fluxes maybe nonactivated or completely activated. The most popular is RMA (rosin, mildly activated).

Patent Leveling

Same as *stretcher leveling*.

Patenting

An isothermal heat treatment process applied to medium- and high-carbon steel wire prior to its final drawing operation. The steel is heated to the austenitic region, typically 950°C, and then cooled to about 480°C–550°C in a salt or lead bath to produce structure of upper bainite, that is, ferrite with a fine distribution of carbide; see *steel*. This has a very high ductility and can be cold drawn with total reductions in diameter of 90%, nearly double that possible with annealed wire. This produces a wire with very high tensile strength, in excess of 1600 MPa, 230,000 psi, with adequate ductility.

Patina

(1) Generally, surface discoloration or a coating arising from atmospheric deposition or corrosion. The term usually implies that the effect is more decorative than deleterious; in some cases, it may even be induced by chemical treatment. More specifically, the term refers to the green corrosion product formed on copper- and sulfur-rich atmospheres. (2) Also used to describe the appearance of a weathered surface of any metal.

Pattern

(1) A form of wood, metal, or other material around which molding material is placed to make a mold for casting metals. (2) A form of wax- or plastic-base material around which refractory material is placed to make a mold for casting metals. (3) A full-scale reproduction of a part used as a guide in cutting.

Pattern Draft

In foundry practice, the paper allowed on the vertical faces of a pattern to permit easy withdrawal of the pattern from the mold or die.

Pattern Layout

A full-size drawing of a foundry (casting) pattern showing its arrangement and structural features.

Pattern Metals

Various low melting temperature alloys used for patterns. They are cast in a master mold or die, and, after the working mold has been formed around them, they are melted and poured out for reuse.

Patternmaker's Allowance

See *shrinkage allowance*.

Patternmaker's Rule (r)

A rule (r) with calibrations oversize by the shrinkage allowance appropriate to the intended cast metal.

Patternmaker's Shrinkage

Contraction allowance made on patterns to compensate for the decrease in dimensions as the solidified casting cools in the mold from the freezing temperature of the metal to room temperature.

The pattern is made larger by the amount of contraction that is characteristic of a particular metal to be used. See *shrinkage*.

Pattinson Process

A process for separating silver from lead, in which the molten lead is slowly cooled so that crystals poorer in silver solidify out and are removed, leaving the melt richer in silver.

Pauli Exclusion Principle

An electron state can contain no more than two electrons and they will have opposite spins.

PBI

See *polybenzimidazole*.

PBT

See *polybutylene terephthalate*.

PC

See *polycarbonates*.

Peak Overlap

In materials characterization techniques such as electron probe x-ray microanalysis or Auger electron spectroscopy, the formation of a single peak when two closely spaced x-ray peaks cannot be resolved; the energy (or wavelength) of the peak is some average of the characteristic energies (or wavelengths) of the original two peaks. See also *full width at half maximum*.

Pearlite

A metastable lamellar aggregate of *ferrite* and *cementite* resulting from the transformation of *austenite* at temperatures above the *bainite* range. The iron/iron carbide eutectoid. See steel.

Pearlitic Iron/Steel

See *cast iron and steel*.

Pearlitic Malleable

See *malleable cast iron*.

Pearlitic Structure

A microstructure resembling that of the pearlite constituent in steel. Therefore, it is a lamellar structure of varying degrees of coarseness.

Pebble Surface/Pebbling

A rounded surface texture similar to, but usually coarser than, orange peel effect.

Pebbles

See *orange peel (metals)*.

Pedestal Bearing

A bearing that is supported on a column or pedestal rather than on the main body of the machine.

PEEK

See *polyether ether ketone*.

Peel Ply

In composites fabrication, a layer of open-weave material, usually fiberglass or heat set nylon, applied directly to the surface of a prepreg lay-up. The peel ply is removed from the cured laminate immediately before bonding operations, leaving a clean, resin-rich surface that needs no further preparation for bonding, other than the application of a primer if one is required. See also lay-up and prepreg.

Peel Strength (Adhesives)

The average load per unit width of bond line required to separate progressively one member from the other over the adhered surfaces at a separation angle of approximately 180° in a separation rate of 152 mm (6 in.) per minute. It is expressed in force per unit width of specimen (N/mm, or lbf/in.).

Peel Test (Adhesives)

See peel strength (adhesives).

Peel Test (Weldments)

(1) A destructive method of inspection that mechanically separates a lap joint by peeling. (2) A test of spot, seam, and projection welds in which the two joined components are pried or peeled apart in such a manner that the parting load is carried by individual spot welds or at one point on the seam. Also termed a *slug test*.

Peeling

The detaching of one layer of a coating from another, or from the base metal, because of poor adherence.

Peen Plating

See *tumbling*.

Peening

Repeated local impacting to produce plastic deformation of the surface layers. It may be performed by handheld, round-end hammer, for example, on weld deposits and the adjacent areas. Alternatively, the component, or specific areas, may be bombarded with peening shot in automatic machines, termed *shot peening*. Performed correctly, peening induces beneficial compressive residual stresses in the surface although there are balancing tensile stresses below. Significant deformation, deliberate or otherwise, can occur in thin components. Peening aims to deform the surface to a carefully judged extent but without scratches or notches. Consequently, the shot is spherical and of controlled size, and broken particles are removed to avoid surface damage. In contrast, shot blasting is a relatively crude process intended to remove surface scale or to roughen the surface and accordingly may use broken, irregular shot or even angular, sharp edged grit.

Peening Wear

Removal of material from a solid surface caused by repeated impacts on very small areas.

PEI

See *polyetherimide*.

Peierls/(Nabarro) Force

The force necessary to move a dislocation from one equilibrium position to another.

Pellet

In powder metallurgy, a small rounded or spherical solid body that is similar to a shotted particle. See also *shotting*.

Pellini Drop Test

A comparative test of the brittle fracture resistance of plate or strip. The test piece comprises a full thickness strip having, on one face, a brittle weld deposit notched to a depth corresponding with the original plate face. It is supported, notched down, at its two ends and impacted on its centerline by a falling weight. A key feature is that the height of the supports is selected to O'Reilly deflection of the plate of 5° at the stage where the specimen face contacts the rig base. The weld is judged to initiate a crack at 3° deflection, and hence, the test determines whether the plate can deflect a further 2° without fracturing at the test temperature.

Pellini Explosive Test

Similar in concept to the *Pellini drop test* except that the load is applied by the shock wave from an explosion and the explosive charge is shaped and positioned to apply load over a large area of the specimen rather than just the centerline.

Peltier Effect

The release or absorption of heat as an electric current crosses the junction from one conductor to a dissimilar one. See *thermoelectric effect*.

Penetrameter

A device for assessing sensitivity in radiography. It comprises a stepped plate or a stepped series of holes in a plate, and it is placed on the component under test during irradiation. The extent to which the steps can be resolved indicating the size of defect that can be detected.

Penetrant

A liquid with low surface tension used in *liquid penetrant inspection* to flow into surface openings of parts being inspected. See *dye penetrant*.

Penetrant Inspection

See preferred term *liquid penetrant inspection*.

Penetration (Adhesives)

The entering of an adhesive into an adherend. This property of a system is measured by the depth of penetration of the adhesive into the adherend.

Penetration (Metals)

(1) In founding, an *imperfection* on a casting surface caused by metal running into voids between sand grains, usually referred to as *metal penetration*. (2) In welding, the distance from the original surface of the base metal to that point at which fusion ceased. (3) In spot, seam welds, etc., the depth of the nugget measured from the interface. See also *joint penetration* and *root penetration*.

Penetration (of a Grease)

The depth in 1/10 mm that a standard cone penetrates the sample in a standard cup under prescribed conditions of weight, time (5 s), and temperature (25°C or 77°F). The result depends on whether or not the grease has been subjected to shear. In unworked penetration, the grease is transferred with as little deformation as possible or is tested in its container. In worked penetration, the grease is subjected to 60 double strokes in a standard device. In prolonged worked penetration, the grease is worked for a specified period before the 60 strokes. The results may be quoted as penetration number or penetration value.

Penetration Bead/Run/Pass (of Weld)

The first run of a multiple run weld deposited from one side that penetrates and fully fuses the root.

Penetration Hardness

Same as *indentation hardness*.

Penetration Hardness Number

Any numerical value expressing the resistance of a body to the penetration of a second, usually harder, body.

Penetrometer

In grease technology, an instrument for measuring the consistency of a grease by allowing a cone to penetrate into the grease under control conditions. See also *penetration (of a grease)*.

Percent Conductivity

The conductivity of a material expressed as a percentage of that of copper. See also *IACS*.

P

Percent Error

In the case of a mechanical testing machine, the ratio, expressed as a percentage, of the error to the correct value of the applied load.

Percent Ferrite

See preferred term *ferrite number*.

Percent Theoretical Density

See *density ratio*.

Percentage

The proportion relative to one hundred.

Percentile

The boundary line cutting off the indicated percentage. For example, a large number of creep rupture tests performed under identical conditions will fail after a wide scatter of lives. The 5 percentile line is the line drawn to exclude the 5% that failed in the shortest time.

Percussion Cone

Damage produced by contact stresses generated by mechanical contact of a hard, blunt object with a glass surface. Typically, it has the appearance of a semicircular or circular crack on the damaged surface, propagating into the glass, flaring out with increasing depth into a cone-shaped crack. Also called impact bruise, butterfly bruise, bump check, and Hertzian cone crack.

Percussion Welding

A resistance welding process that produces coalescence of abutting surfaces using heat from an arc produced by a rapid discharge of electrical energy. Pressure is applied percussively during or immediately following the electrical discharge.

Perforating

The punching of many holes, usually identical and arranged in a regular pattern, in a sheet, workpiece blank, or previously formed part. The holes are usually round, but may be any shape. The operation is also called multiple punching. See also *piercing*.

Perfume Oils

Volatile oils obtained by distillation or by solvent extraction from the leaves, flowers, gums, or woods of plant life, although a few are of animal origin. Perfumes have been used since the earliest times, not only for aesthetic value but also for antiseptic value and for religious purposes. Simple perfumes usually take their names from the name of the plant, but the most esteemed perfumes are blends, and the blending is considered a high art. It is done by tones imparted by many ingredients. Some oils with repugnant odors have an attractive fragrance in extreme dilution and a persistence that is valued in blends and for stabilization. Some oils with heavy odors, such as *coumarin*, are used in dilution to give body. Since many of the odors come directly from esters, aldehydes, and ketones, they can be made synthetically from coal-tar hydrocarbons and alcohols. Synthetics are now most used in perfumes, although some natural

odors have not yet been duplicated synthetically, and about 30,000 aromatics have been developed. A perfume may contain 50 components, sometimes as high as 300, and the average perfume manufacturer employs about 3000 components. Some of the chemicals are not odors, but give lasting qualities or enhance odor. Some are used as fixatives or blending agents. *Hydroquinone dimethyl ether* has an odor of sweet clover but is used as a fixative in other perfumes.

In general, the aldehyde odors are fugitive, and some become acid in the presence of light or oxygen. Ketones are more stable. Esters are usually stable, but some are saponified in hot solution and cannot be used for soaps. Some esters, made from complex high alcohols, are used to give a fresh top note to floral perfumes. *Linalyl acetate*, produced from citral, is an example. Acid perfumes neutralize free alkali and cannot be used in soaps. Phenol odors alter the color of soaps, and the odor may also become disagreeable.

Some odors are never extracted from the flowers, but are compounded. *Crab apple*, for example, which is a peculiarly sweet odor, is compounded of 16 oils, including bois de rose, ylang-ylang, nutmeg oil, jasmine, musk, heliotropin, and coumarin. *Wisteria* is the honey-like odor of the mauve and white flowers of the climbing plant *Wisteria sinensis*. The oil is never extracted but is compounded from geranium, Peru balsam, benzoin, bois de rose, and synthetics. Some oils, such as *lavender*, have no value when used alone but require skillful blending to develop the pleasant odor. Apple and peach odors are *allyl cinnamate*. Synthetic *rose* is the ester of phenyl ethyl alcohol made from benzene and ethylene oxide. Although the natural rose odor is readily extracted, it is more expensive.

Fixatives are used for the finer perfumes. They are essential oils that are less volatile and thus delay evaporation. The animal oils, such as musk and civet, are of this class, and also the balsam oils. Musk is from the male musk deer of Tibet. It is one of the most expensive materials. Synthetic musk is as powerful as the natural. The synthetic musk of DuPont is called *astrotone*.

Versilide is a cyclic ketone synthetic musk that is very stable in soaps and cosmetics and does not discolor.

Attar of rose is one of the most ancient and popular perfume oils. The name is derived from the Persian *attar* and is sometimes incorrectly given as *ottar* but with the same French pronunciation. The finest attar of rose is from Bulgaria, where it is distilled from the flowers of the *damask rose*. The fresh oil is colorless but turns yellowish green. In France, the oil is obtained from the *Rosa centifolia*. *Rosewater* is the scented water left after distillation or is made by dissolving attar in water.

As with jewelry, the manufacture of perfumes is normally classified as a luxury industry. But the vital test of essentiality comes in wartime, and these anesthetic materials are always considered essential to the public morale, and hence basic. Even under desperate wartime conditions, France never stopped the manufacture of perfumes, and during the life struggle of England in World War II, the restrictive regulations placed upon perfume manufacture had to be abandoned quickly because of public pressure. Wartime restrictions placed on the imports of perfume oils into the United States during World War II were immediately abandoned.

Period/Periodicity (of Elements)

See *atomic structure* and *periodic table*.

Periodic Reverse

Pertains to periodic changes in the direction of flow of the current in electrolysis. It applies to the process and also the machine that controls the time for both directions.

Periodic Table

A table of the elements presented in periods, that is, with elements having similar chemical characteristics arranged in vertical rows. See *atomic structure*.

Peripheral Clearance Angle

See *clearance angle*.

Peripheral Milling

Milling a surface parallel to the axis of the cutter. See also *milling* and *milling cutter*.

Peripheral Speed

See preferred term *cutting speed*.

Peritectic

An isothermal reversible reaction in metals in which a liquid phase reacts with a solid phase to produce a single (and different) solid phase on cooling.

Peritectic Equilibrium

A reversible univariant transformation in which a solid phase stable only at lower temperature decomposes into a liquid and a solid phase that are conjugate at higher temperature.

Peritectoid

An isothermal reversible reaction in which a solid phase reacts with a second solid phase to produce a single (and different) solid phase on cooling.

Peritectoid Equilibrium

A reversible univariant transformation in which a solid phase stable only at low temperature decomposes with rising temperature into two or more conjugate solid phases.

Permanence

(1) The property of a plastic that describes its resistance to appreciable change in characteristics with time and environment. (2) The resistance of an adhesive bond to deteriorating influences.

Permanent Load

The loading due to self weight and other nonvarying sources as opposed to additional loads due to varying applied loads.

Permanent Magnet Materials

Permanent magnet is the term used to describe solid materials capable of being magnetized permanently because of their ability to retain induced magnetization and magnetic poles after removal of externally applied fields. Such materials, which are characterized by their high magnetic induction, high resistance to demagnetization,

and maximum energy content, include a variety of alloys, intermetallics, and ceramics. Commonly included are certain steels, alnico (a cast or sintered iron-base alloy containing 7%–10% Al, 15%–19% Ni, 13%–35% Co, 3%–4% Cu, with an optional 1%–5% Ti); Cunife (60Cu–20Ni–20Fe); Fe–Co alloys containing vanadium or chromium, platinum–cobalt, and hard ferrites ($SrO–Fe_2O_3$ or BaO-6 and Fe_2O_3); cobalt–rare earth alloys ($SmCo_5$ or Sm_2Co_{17}); and Nd–Fe–B alloys made by powder-metallurgy processing. Each type of magnet material possesses unique magnetic and mechanical properties, corrosion resistance, temperature sensitivity, fabrication limitations, and cost. These factors provide designers with a wide range of options in designing magnetic parts.

Permanent magnet materials are based on the cooperation of atomic and molecular moments within a magnet body to produce a high magnetic induction. This induced magnetization is retained because of a strong resistance to demagnetization. These materials are classified as ferromagnetic or ferrimagnetic and do not include diamagnetic or paramagnetic materials. The natural ferromagnetic elements are iron, nickel, and cobalt. Other elements, such as manganese or chromium, can be made ferromagnetic by alloying to induce proper atomic spacing. Ferromagnetic metals are combined with other metals or with oxides to form ferrimagnetic substances; ceramic magnets are of this type.

Uses

Permanent magnet materials are marketed under a variety of trade names and designations throughout the world. Over the years, the number and range of applications utilizing permanent magnets has increased dramatically. Some of the more predominant applications include aircraft magnetos, alternators, magnetos for lawnmowers, garden tractors, and outboard motors, small and large direct current (dc) motors (including automotive motors), acoustic transducers, magnetic couplings, magnetic resonance imaging, magnetic focusing systems, ammeters and voltmeters, and watt-hour meters.

Permanent Mold

A metal, graphite, or ceramic mold (other than an ingot mold) of two or more parts that is used repeatedly for the production of many *castings* of the same form. Liquid metal is usually poured in by gravity.

Permanent Mold Castings

Permanent mold casting is performed in a mold, generally made of metal, which is not destroyed by removing the casting. Several types of casting can be included in the description of pressure die casting, centrifugal casting, and gravity die casting.

Advantages and Disadvantages

Gravity die castings are dense and fine grained and can be made with better surfaces and closer tolerances than sand castings. Tolerances are wider than for pressure die castings and plaster mold castings, but narrower than for sand castings. Production rates are lower than those obtainable by die casting.

The process stands somewhere between sand and die casting with respect to possible complexity, dimensional accuracy, mold or die casts, etc. For many parts, it provides an attractive compromise when the ultimate—whether in complexity of one-piece construction, narrow tolerances, or ultrahigh production rates—need not be met.

Casting Alloys

Lead-, zinc-, aluminum-, magnesium-, and copper-base alloys, as well as gray cast iron, can be cast by permanent molding. Less than 1% of total gray iron production is permanent mold cast, 5%–7% of copper and magnesium alloy, and up to 40% of total aluminum casting production. Here's a breakdown of problems to expect:

Gray iron: Mechanical properties depend on thickness.

Aluminum: Gates must be made larger to take the low specific gravity of aluminum into account.

Magnesium: Extreme caution is needed when removing metal from the furnace. Ladles must be kept at red heat to exclude moisture and prevent an explosion.

Copper: Aluminum bronze is the most popular permanent-mold-cast copper-base alloy. Turbulence must be prevented in the die cavity to reduce oxidation of the molten metal during solidification.

Permanent Set

The deformation remaining after a specimen has been stressed a prescribed amount in tension, compression, or shear for a specified time period and released for a specified time period. For the test, the residual unrecoverable deformation after the load causing the creep has been removed for a substantial and specified period of time. Also, the increase in length, expressed as a percentage of the original length, by which an elastic material fails to return to its original length after being stressed for a standard period of time.

Permeability

(1) The passage or diffusion (or rate of passage) of a gas, vapor, liquid, or solid through a material (often porous) without physically or chemically affecting it; the measure of fluid flow (gas or liquid) through a material. (2) A general term used to express various relationships between magnetic induction and magnetizing force. These relationships are either "absolute permeability," which is a change in magnetic induction divided by the corresponding change in magnetizing force, or "specific (relative) permeability," the ratio of the absolute permeability to the permeability of free space. (3) In metal casting, the characteristic of molding materials that permit gas is to pass through them. "Permeability number" is determined by a standard test.

Permeability Alloys

This is a general name for a group of nickel–iron alloys with special magnetic properties. These soft magnetic materials possess a magnetic susceptibility much greater than iron. An early alloy was composed of 78.5% nickel and 21.5% iron. It also contained about 0.37% cobalt, 0.1% copper, 0.04% carbon, 0.03% silicon, and 0.22% manganese. It is produced sometimes with chromium or molybdenum, under the name of permalloy, and is used in magnetic cores for apparatus that operates on feeble electric currents and in the loading of submarine cables. It has very little magnetic hysteresis.

Types and Uses

Supermalloy for transformers contains 79% nickel, 15% iron, 5% molybdenum, and 0.5% manganese, with total carbon, silicon, and sulfur kept below 0.5%. It is melted in vacuum and poured in an inert atmosphere. It can be rolled as thin as 0.00064 cm. The alloy has an initial permeability 500 times that of iron.

Supermendur contains 49% iron, 49% cobalt, and 2% vanadium. It is highly malleable and has very high permeability with low hysteresis loss and high flux density. Duraperm is a high flux magnetic alloy containing 84.5% iron, 9.5% silicon, and 6% aluminum. Perminvar is an alloy containing 45% nickel and 25% cobalt, intended to give a constant magnetic permeability for variable magnetic fields. A-metal is a nickel–iron alloy containing 44% nickel and a small amount of copper. It is used in transformers and loudspeakers to give nondistortion characteristics when magnetized.

Alfenol contains no nickel, but has 16% aluminum and 84% iron. It is brittle and cannot be rolled cold, but can be rolled into thin sheets at a temperature of 575°C. It is lighter than other permeability alloys and has superior characteristics for transformer cores and tape-recorder heads. A modification of this alloy, called thermenol, contains 3.3% molybdenum without change in the single-phase solid solution of the binary alloy. The permeability and coercive force are varied by heat treatment. At 18% aluminum, the alloy is practically paramagnetic. The annealed alloy with 17.2% aluminum has constant permeability.

Aluminum–iron alloys with 13%–17% aluminum are produced in sheet form for transformers and relays. They have magnetic properties equal to the 50–50 nickel–iron alloys and to the silicon–iron alloys, and they maintain their magnetic characteristics under changes in ambient temperature.

Iron–nickel permeability alloys are used as loading in cable by wrapping layers around the full length of the cable. When nickel–copper alloys are used, they are employed as a core for the cable. Magnetostrictive alloys are iron–nickel alloys that will resonate when the frequency of the applied current corresponds to the natural frequency of the alloy. They are used in radios to control the frequency of the oscillating circuit. Magnetostriction is the stress that occurs in a magnetic material when the induction changes. In transducers, it transforms electromagnetic energy into mechanical energy. Temperature-compensator alloys are iron–nickel alloys with about 30% nickel. They are fully magnetic at −29°C but lose their magnetic permeability in proportion to rise in temperature, until, at about 54°C, they are nonmagnetic. Upon cooling, they regain permeability at the same rate. They are used in shunts in electrical instruments to compensate for errors due to temperature changes in the magnets.

Permissible Variation

In mechanical testing machines, the maximum allowable error in the value of the quantity indicated. It is convenient to express permissible variation in terms of the *percent error*. See also *tolerance*.

Permittivity

The ratio of the capacity of a condenser with and without the dielectric material in position between the plates. Also termed the *relative dielectric constant*.

Perovskites

A naturally occurring mineral with a structure type that includes no less than 150 compounds; the crystal structure is ideally cubic. Perovskites are noted for their ferroelectricity, piezoelectricity, pyroelectricity, electrostriction, high permittivity, and optical and electro-optic properties. The hardness of perovskites is approximately 5.5 on the Mohs scale.

Peroxy Compounds

Polymeric compounds containing O–O linkage.

Perpendicular Section

A section cut perpendicular to a surface of interest in a medley graphic specimen. Compare with *taper section.*

PESV

See *polyethersulfone.*

PET

See *polyethylene terephthalate.*

Petroff Equation

An equation describing the viscous power loss in a concentric bearing full of lubricant. The resisting torque on the shaft (T_0 = shear stress × shaft radius × bearing area) is given by

$$T_0 = \frac{\pi \eta NLD^2}{2c}$$

where
  π is equal to 3.1416
  η is the dynamic viscosity
  N is the shaft speed
  L is the bearing length
  D is the shaft diameter
  $2c$ is the diametrical clearance

Petrography

The study of nonmetallic matter under suitable microscopes to determine structural relationships and to identify the phases or minerals present. With transparent materials, the determination of the optical properties, such as the indices of refraction and the behavior in transmitted polarized light, is a means of identification. With opaque materials, the color, hardness, reflectivity, shape, and etching behavior in polished sections are means of identification.

Petrolatum

A jellylike substance obtained in the fractional distillation of petroleum. Its composition is between $C_{17}H_{36}$ and $C_{21}H_{44}$, and it distills off above 577°F (303°C). It is also called petroleum jelly. It is used for lubricating purposes and for compounding with rubber and resins. When highly refined for the pharmacy trade, it is used as an ointment. The specific gravity is from 0.820 to 0.865. It is insoluble in water but readily soluble in benzene and turpentine. For lubricating purposes, it should be refined by filtration only and not with acids, and it should not be adulterated with paraffin. *Sherolatum* and *Vaseline* are brands of petroleum jelly, but petroleum jellies for pharmaceutical uses may be compounded with other materials.

Petroleum

A heavy, liquid, flammable oil stored under the surface of the earth and originally formed as the by-product of the action of bacteria on marine plants and animals. It consists chiefly of carbon and hydrogen in the form of hydrocarbons, including most of the liquids of the *paraffin series*, C_5H_{12} to $C_{16}H_{34}$, together with some of the gases, CH_4 to C_4H_{10}, and most of the solids of the series from $C_{17}H_{36}$ to $C_{27}H_{56}$. It also contains hydrocarbons of other series. While petroleum is used primarily for the production of fuels and lubricating oils, it is one of the most valuable raw materials for a very wide range of chemicals. The name *petrochemicals* is used in a general sense to mean chemicals derived from petroleum, but it does not mean any particular class of chemicals. Sulfur and helium are by-products from petroleum working.

Petroleums from different localities differ in composition, but tests of oils from all parts of the world give the limits as 83%–87% carbon and 11%–14% hydrogen, with sulfur, nitrogen, and oxygen in amounts from traces to 3%. Liquefied petroleum gases, including *propane*, *butane*, *pentane*, or mixtures, are marketed under pressure in steel cylinders as bottled gas. Propane is used in cook stoves. Butane is used to enrich illuminating and heating gas. *Liquid gas* is also used for internal combustion engines, as a solvent and for making many chemicals.

Certain highly refined oils used in medicine as laxatives and used in other specified applications are referred to as *mineral oils*. *White mineral oil* is petroleum highly refined to color. Russian and Rumanian oils with high naphthene content were particularly suitable for this refining, and the oil was called *Russian oil*. It is used as a laxative, as a carrier of many drugs, in textile spinning, as a plasticizer for synthetic resins, and for sodium dispersions where the alkali metal would normally react with any impurities.

Petroleum Oil

See *mineral oil.*

Pewter

Pewter is a very old name for tin–lead alloys used for dishes and ornamental articles; the term now refers to the use rather than to the composition of the alloy. Tin was the original base metal of the alloy, the ancient Roman pewter having about 70% tin and 30% lead, although iron and other elements were present as impurities. Pewter, or latten ware, of the sixteenth century contained as much as 90% tin, and a strong and hard English pewter contain 91% tin and 9% antimony. This alloy is easily cold-rolled and spun and can be hardened by long annealing at 225°C and quenching in cold water and tempering at 110°C. Pewter is now likely to contain lead and antimony, and very much less tin; when the proportion of tin is less than about 65%, the alloys are unsuited for vessels to contain food products, because of the separation of the poisonous lead.

Early pewter, with high lead content, darkened with age. With less than 35% lead, pewter was used for decanters, mugs, tankards, bowls, dishes, candlesticks, and canisters. The lead remained in solid solution with the tin so that the alloy was resistant to the weak acids in foods.

The addition of copper increases ductility; additions of antimony increase hardness. Pewter high (e.g., in the 91% tin and 9% antimony or antimony and copper) has been used for ceremonial objects, such as religious communion plates and chalices, and for cruets, civic symbolic cups, and flagons.

Pewter is a tin-base *white metal* containing antimony and copper. Originally, pewter was defined as an alloy of tin and lead, but to avoid toxicity and dullness of finish, lead is excluded from modern pewter. These modern compositions contain 1%–8% antimony and 0.25%–3% copper. Typical pewter products include coffee and tea services, trays, steins, candy dishes, jewelry, plates, vases, compotes, and cordial cups.

pH

(1) The negative logarithm of the hydrogen-ion activity; it denotes the degree of acidity or basicity of a solution. At 25°C (77°F), 7.0 is the neutral value. Decreasing values below 7.0 indicate increasing acidity; increasing values above 7.0 indicate increasing basicity. The pH values range from 0 to 14. (2) The "p" derives from the German "potenz" for power and the H, obviously, from hydrogen.

Phase

A portion of a system that is chemically homogenous and physically distinct. For example, steam (liquid), water, and ice are phases of water. If a small amount of salt is dissolved in liquid water, a single phase solution will exist. However, at a certain level, the solution will be saturated and no further salt can normally be dissolved. Addition of further salt will result in two phases, the saturated solution and salt particles. In abnormal circumstances, it is possible for the amount of salt that is dissolved to exceed the saturation limit. Such solutions are termed supersaturated. Similar effects occur in solid metals where a quantity of one metal can dissolve in another to form a solid solution that is a single phase of completely uniform composition and that appears featureless under the microscope, apart from grain boundaries. The simplest case is where the two metals are completely soluble in each other at all compositions, an example being the copper–nickel system, mentioned again in later text. Most systems are, however, less simple, and, as the composition is changed, a series of different phases will be formed. In certain cases, two or more elements will combine with each other in specific ratios, for example, one atom of copper with two atoms of aluminum forms $CuAl_2$. These are termed intermetallic compounds rather than solid solutions but are still phases. With many solid solutions, the solubility of the secondary component decreases as the temperature falls. If such an alloy containing the maximum amount of the second element is allowed to cool slowly, the second element will come out of solution by a precipitation process. The resultant particles or precipitates form a second phase although if cooling is rapid, precipitation may be impeded resulting in a supersaturated solid solution. A phase diagram or equilibrium diagram is a graphical presentation of the phases that are stable, under equilibrium conditions, for a range of temperatures and compositions of an alloy. The simplest phase diagram is the binary in which composition ranging from 100% of one element to 100% of the second element is plotted against temperature. The copper–nickel is an example of this simple phase diagram. More complex systems are considered in ternary diagrams (three elements), quaternary diagrams (four elements), and so on. Other forms of phase diagrams depict the phases arising under nonequilibrium conditions. See *isothermal transformation diagram*.

Phase Change

The transition from one physical state to another, such as gas to liquid, liquid to solid, gas to solid, or vice versa.

Phase Contrast (Electronic Microscopy)

(1) Contrast in high-resolution transmission electron microscope images arising from interference effects between the transmitted beam and one or more different active beans. (2) A technique of microscopical examination that utilizes the small differences in phase between light reflected from irregular surfaces to provide an image of surface features.

Phase Contrast Illumination (Optical Microscopy)

A special method of controlled illumination ideally suited to observing thin, transparent objects where structural details vary only slightly in thickness or refractive index. This can also be applied to the examination of opaque materials to determine surface elevation changes.

Phase Diagram

A graphical representation of the temperature and composition limits of phase fields in an alloy or ceramic system as they actually exist under the specific conditions of heating or cooling. A phase diagram may be an equilibrium diagram, an approximation to an equilibrium diagram, or a representation of metastable conditions or phases. Synonymous with constitution diagram. Compare with *equilibrium diagram*.

Phase Rule

(1) The maximum number of phases (P) that may exist at equilibrium is two, plus the number of components (C) in the mixture, minus the number of degrees of freedom (F). (2) (Gibb's) rule—for the general case and equilibrium:

$$P + F = C + E$$

where

P is the number of phases present
F is the number of degrees of freedom or variants
C is the number of components, that is, stable chemical substances
E is the number of environmental factors, that is, two— temperature and pressure

In the case of metal reactions, the consequences of varying pressure are insignificant and can be ignored leaving temperature as the only environmental factor so the equation becomes

$$P + F = C + 1.$$

In this case, the components are the pure metals or other elements that comprise the alloy.

Phase Separation

The formation of a second liquid portion from a previously homogenous liquid over time.

Phase-Change Lubrication

See *melt lubrication*.

Phenol

Phenol is the simplest member of a class of organic compounds possessing a hydroxyl group attached to a benzene ring or to a more complex aromatic ring system.

Also known as carbolic acid or monohydroxybenzene, phenol is a colorless to white crystalline material of sweet odor, having the composition C_6H_5OH, obtained from the distillation of coal tar and as a by-product of coke ovens.

Phenol has broad biocidal properties, and dilute aqueous solutions have long been used as an antiseptic. At higher concentrations, it causes severe skin burns; it is a violent systemic poison. It is a valuable chemical raw material for the production of plastics, dyes, pharmaceuticals, syntans, and other products.

Properties

Phenol melts at about 43°C and boils at 183°C. The pure grades have melting point of 39°C, 39.5°C, and 40°C. The technical grades contain 82%–84% and 90%–92% phenol. The crystallization point is given as 40.41°C. The specific gravity is 1.066. It dissolves in most organic solvents. By melting the crystals and adding water, liquid phenol is produced, which remains liquid at ordinary temperatures. Phenol has the unusual property of penetrating living tissues and forming a valuable antiseptic. It is also used industrially in cutting oils and compounds and in tanneries. The value of other disinfectants and antiseptics is usually measured by comparison with phenol.

Uses and Derivatives

Phenol is one of the most versatile industrial organic chemicals. It is the starting point for many diverse products used in the home and industry. A partial list includes nylon, epoxy resins, surface active agents, synthetic detergents, plasticizers, antioxidants, lube oil additives, phenolic resins (with formaldehyde, furfural, and so on), cyclohexanol, adipic acid, polyurethanes, aspirin, dyes, wood preservatives, herbicides, drugs, fungicides, gasoline additives, inhibitors, explosives, and pesticides.

Phenol Formaldehyde Resin

This is a synthetic resin, commonly known as phenolic, made by the reaction of phenol and formaldehyde, and employed as a molding material for making of mechanical and electrical parts. It was the earliest type of hard, thermoset synthetic resins, and its favorable combination of strength, chemical resistance, electrical properties, glossy finish, and nonstrategic abundance of low-cost raw materials has maintained the resin, with its many modifications and variations, as one of the most widely employed groups of plastics for a variety of products. The resins are also used for laminating, coatings, and casting resins.

Phenolic resins are used most extensively as thermosetting plastic materials, as there are only a few uses as thermoplastics. The polymer is composed of carbon, hydrogen, oxygen, and sometimes nitrogen. Its molecular weight varies from a very low value during its early state of formation to almost infinity in its final state of cure. The chemical configuration, in the thermoset state, is usually represented by a 3D network in which the phenolic nuclei are linked by methylene groups. The completely cross-linked network requires three methylene groups to two phenolic groups. A lesser degree of cross-linking is attainable either by varying the proportions of the ingredients or by blocking some of the reactive positions of the phenolic nucleus by other groups, such as methyl and butyl. Reactivity can be enhanced by increasing the hydroxyl groups on the phenolic nuclei, for example, by the use of resorcinol.

Characteristics

The outstanding characteristics of phenolics are good electrical properties, very rigid set, good tensile strength, excellent heat resistance, good rigidity at elevated temperature, good aging properties, and also good resistance to water, organic solvents, weak bases, and weak acids. All these characteristics are coupled with relatively low-cost.

Phenolics are used in applications that differ widely in nature. For example, wood is impregnated to make "impreg" and "compreg"; paper is treated to make battery separators and oil and air filters; specific chemical radicals can be added to the molecule to make an ion-exchange material. Phenolics are also widely used in protective coating.

The hundreds of different phenolic molding compounds can be divided into six groups on the basis of major performance characteristics. General-purpose phenolics are low-cost compounds with fillers such as wood flour and flock and are formulated for noncritical functional requirements. They provide a balance of moderately good mechanical and electrical properties and are generally suitable in temperatures up to 149°C. Impact-resistant grades are higher in cost. They are designed for use in electrical and structural components subject to impact loads. The fillers are usually paper, chopped fabric, or glass fibers. Electrical grades, with mineral fillers, have high electrical resistivity plus good arc resistance, and they retain their resistivity under high-temperature and high-humidity conditions. Heat-resistant grades are usually mineral- or glass-filled compounds that retain their mechanical properties and the 190°C–260°C temperature range.

Special-purpose grades are formulated for service applications requiring exceptional resistance to chemicals or water, or combinations of conditions such as impact loading and a chemical environment. The chemical-resistant grades, for example, are inert to most common solvents and weak acids, and their alkali resistance is good. Nonbleeding grades are compounded specially for use in container closures and for cosmetic cases.

Fillers   Proper balance of fillers is important, because too large a quantity may produce brittleness. Organic fillers absorb the resin and tend to brittleness and reduce flexural strength, although organic fibers and fabrics generally give high impact strength. Wood flour is the most usual filler for general-service products, but prepared compounds may have mineral powders, mica, asbestos, organic fibers, macerated fabrics, or mixtures of organic and mineral materials. Bakelite was the original name for phenol plastics, but trade names now usually cover a range of different plastics, and the types and grades are designated by numbers.

The specific gravity of filled phenol plastics may be as high as 1.70. The natural color is amber, and, as the resin tends to discolor, it is usually pigmented with dark colors. Normal phenol resin cures to single-carbon methylene groups between the phenolic groups, and the molded part tends to be brittle. Thus, many of the innumerable variations of phenol are now used to produce the resins, and modern phenol resins may also be blended or cross-linked with other resins to give higher mechanical and electrical characteristics. Furfural is frequently blended with the formaldehyde to give better flow, lower specific gravity, and reduced cost. The alkylated phenols give higher physical properties.

Phenol resins may also be cast and then hardened by heating. The cast resins usually have a higher percentage of formaldehyde and do not have fillers. They are poured in syrupy state in lead molds and hardened in a slow oven.

Some of the uses for phenolic resins are for making precisely molded articles, such as telephone parts, for manufacturing strong and durable laminated boards, or for impregnating fabrics, wood, or paper. Phenolic resins are also widely used as adhesives, as the binder for grinding wheels, as thermal insulation panels, as ion-exchange resins, and in paints and varnishes.

Molding Compounds   The largest single use for phenolic resins is in molding compounds. To make these products, either one- or two-stage resins are compounded with fillers, lubricants, dyes, plasticizers, etc. Wood flour is used as an inexpensive reinforcing agent in the general-purpose type of compounds. Cotton flock, chopped fabric, and sisal and glass fibers are used to improve strength characteristics; mineral fillers such as asbestos and mica are used where improvements in dimensional stability, heat resistance, or electrical properties are desired. The compounds are usually produced in granular, macerated, or nodular forms, depending on the type of filler used. Since the color of the base resin is not stable to light, molding compounds are commonly produced only in dark colors such as black and brown.

Molding compounds are usually processed in hardened steel molds, and molds can be designed to operate using the compression, transfer, or plunger-molding techniques, depending on the design of the article to be fabricated. Molded parts are drilled, tapped, or machined.

About one-third of all phenolic resins produced is processed into parts by molding. Compression and transfer molding are the principal processes used, but they can also be extruded and injection molded.

Molded phenolic parts are used in bottle caps, automotive ignition and engine components, electrical wiring devices, washing machine agitators, pump impellers, electronic tubes and components, utensil handles, and a multitude of other products.

Adhesives   The thermosetting nature and good water- and fungus-resistant qualities of phenolic resins make them ideal for adhesive applications. Almost all exterior-grade plywood is phenolic resin bonded. This constitutes the second largest market for phenolics. The essential ingredient in many metal-to-metal and metal-to-plastic adhesives is a phenolic resin. One-stage phenol formaldehyde resins are used predominantly for hot-pressed plywood. Special resorcinol-formaldehyde resins curing at room temperature are employed for fabricating laminated timber.

Laminates   The third largest use for phenolic resins is in the manufacture of laminated materials. Many variations of paper, from cheap kraft to high-quality alpha cellulose, in addition to asbestos, cotton, linen, nylon, and glass fabrics are most commonly used for reinforcing filler sheets. The laminate is formed by combining, under a temperature of 177°C and a pressure of 3.4–14 MPa, multiple layers of the various reinforcing sheet, after saturation with phenolic resin, generally of the one-stage type dissolved in alcohol.

Paper laminates are used most extensively in the electrical and decorative fields. A large number of the laminate for the electrical industry are of the punching grade, making it possible to fabricate all kinds of small parts in a punch press. The laminate used for decorative purposes usually contains a surface sheet of melamine resin–treated paper for providing unlimited color or design configurations. Other fillers are used for special applications where superior dimensional stability, or water, fire, or chemical resistance, or extra strength is required.

Casting Resins

Plastic parts can also be manufactured by pouring resin into molds and heat-curing without pressure. Two basic grades of casting resins are manufactured commercially. Both are of the one-stage type and are cured under neutral to strongly acidic conditions depending on the application. The first grade is manufactured for its variegated color and artistic possibilities and is used primarily in the cutlery and decorative field. Since this type of cast material is noted for ease of machining, it is well adapted for small production runs or where machined prototypes are desired.

The second grade of casting resins includes all those modified by fillers and reinforcing agents. Designed primarily to have low-shrinkage characteristics during cure, they are usually set with a strong acid catalyst to obtain a low-temperature set. Uses include containers, jigs, fixtures, and metal-forming dies.

Bonding Agents   Phenolic resins are noted for their excellent bond strength characteristics under elevated temperature conditions and thus are used in such applications as thermal and acoustical insulation, grinding wheels, coated abrasives, brake linings, and clutch facings. Glass wool insulation is manufactured by spraying water-soluble one-stage resins on the glass fibers as they are formed. Heat given off by the fibers as they cool is sufficient to set the resin. Where organic fibers are used, finely pulverized, one-stage resins are distributed between the fibers, either by a mixing or a dusting operation, followed by an oven treatment.

Grinding wheels bonded with phenolic resin are commonly known as resinoid-bonded wheels. By combining a liquid one-stage resin and a powdered two-stage resin, the material can be evenly distributed with abrasive grit and fillers so that the mixture can be pressed into wheels and baked in ovens. Some wheels are also hot-pressed and cured directly in a press. Resinoid-bonded wheels are used primarily where the application requires a bond of exceptional strength such as in cutoff and snagging wheels.

Coated adhesives are manufactured by bonding abrasive grit to paper, fabric, or fiberboard by means of phenolic resins of the liquid one-stage type. Sander disks and belts are common applications.

Brake linings and clutch facings are made by bonding asbestos, fillers, metal shavings, and friction modifiers with phenolic resin, usually performed by a mixing, forming, and baking operation. Most resins for these applications are specially formulated, but simple two-stage resins are sometimes used. Formulation for each friction lining or clutch face must be determined carefully. Wood composition products of all descriptions are manufactured by hot-pressing sawdust, wood chips, or wood flour containing 8%–20% resin. Two-stage resins are most frequently used in composition boards. One-stage resins are employed for special applications where the yellow color or the slight ammonia odor of two-stage resins is undesirable.

Foundry Use   A relatively new application holding great promise for phenolic resins is in the shell mold process for the foundry industry. The basic principle involves binding sand grains with resin. The process is equally well adapted to making cores and is satisfactory for practically all metals, including magnesium and chrome alloys.

The process has many intriguing possibilities. Molds thus made can be reproduced in exact composition, detail, and size; they are rigid and, having no affinity for water, can be racked and stored indefinitely, with or without cores. Castings from these molds have excellent surface finish and detail and can be held too close to dimensional tolerances. The process can be completely mechanized, yields more castings per ton of melt than sand casting, simplifies cleaning of castings, minimizes problems with sand control and handling, and is a relatively clean operation. It is revolutionizing the foundry industry.

Consumption of resin in the foundry industry has grown rapidly, and conceivably this could become the largest single outlet for phenolics. However, markets for phenolic resins and plywood adhesive, insulation, and wood composition board applications have also expanded.

Phenoxy Resin

A high-molecular-weight thermoplastic polyester resin (polyhydroxy ether) based on bisphenol A and epichlorohydrin. The material is available in grades suitable for injection molding and extrusion and for applications such as coatings and adhesives.

Phenylsilane Resins

Thermosetting copolymers of silicone and phenolic resins. Furnished in solution form.

Phosgene

The common name for carbonyl chloride, $COCl_2$, a colorless, poisonous gas made by the action of chlorine on carbon monoxide. It was used as a poison war gas. But it is now used in the manufacture of metal chlorides and anhydrides, pharmaceuticals, perfumes, isocyanate resins, and for blending in synthetic rubbers. It liquefies at 7.6°C and solidifies at −118°C. It is decomposed by water. When chloroform is exposed to light and air, it decomposes into phosgene. One part in 10,000 parts of air is a toxic poison, causing pulmonary edema. For chemical warfare, it is compressed into a liquid in shells.

Because of its toxicity, most phosgene is produced and employed immediately in captive applications. The biggest use of the material is for toluene diisocyanate, which is then reacted into polyurethane resins for foams, elastomers, and coatings. About 0.9 metric ton of phosgene is consumed to make a metric ton of polymethylene polyphenylisocyanate, also used for making polyurethane resins for rigid foams. Polycarbonate manufacturers require 0.42 metric ton phosgene per ton of product resin. Polycarbonate is used for making break-resistant housings, signs, glazings, and electrical tools. Phosgene also is a reactant for isocyanates that are used in pesticides and di- and polyisocyanates in adhesives, coatings, and elastomers.

Phosphate Conversion Coatings

Phosphate coatings have been used commercially for approximately 50 years. They are used on iron, steel, zinc, and aluminum surfaces to increase corrosion protection, provide a base for paint, reduce wear on bearing parts, and aid in the cold forming and extrusion of metals.

Phosphate coatings are formed by chemically reacting a clean metal surface with an aqueous solution of a soluble metal phosphate of zinc, iron, or manganese, accelerating agents, and free phosphoric acid. For example, when steel is treated, the surface is converted into a crystalline coating consisting of secondary and tertiary phosphates, adherent to and integral with the base metal.

The performance of a phosphate coating depends largely on the unique properties of the coating, which is integrally bound to the base metal and acts as a nonmetallic, adsorptive layer to hold a subsequent finish of oil or wax, paint, or lubricant. Heavy phosphate coatings are normally used in conjunction with an oil or wax for corrosion resistance. The combination of the coating with the oil film gives a synergistic effect, which affords much greater protection than that obtained by the sum of the two taken separately. The stable, nonmetallic, nonreactive phosphate coating provides an excellent base for paint. It is chemically combined with the metal surface, which results in increased adsorption of paint and materially reduces electrochemical corrosion normally occurring between the paint film and the metal.

Oil-absorptive phosphate coatings are useful in holding and maintaining a continuous oil film between metal-to-metal moving parts. They also permit rapid break-in of new bearing surfaces. Even after the coatings have been worn away, the controlled etched condition of the metal surface continues to hold the oil film between the moving parts. The ability of a well-anchored coating to hold a soap or oil-type lubricant is used in the cold forming and drawing of metals.

Processes

Most phosphate coatings are formed from heated solutions following a hot cleaning cycle. Effective coatings, both zinc phosphate and the iron phosphate types, are now produced by the relatively cold system with up to 70% savings in heating costs.

All steel must be thoroughly cleaned prior to phosphate coating to remove grease, oil, rust, and undesirable soils on the steel surface, which prevent or alter the formation of a satisfactory phosphate coating and interfere with paint adhesion. The cleaner used to remove oily soil prior to the formation of the zinc phosphate coating is a light-duty, mildly alkaline material especially formulated to function effectively at low temperature. The cleaner that is best suited for use in connection with low-temperature iron phosphate coating process may be either a light-duty, low-foaming, mildly acidic, or mildly alkaline cleaner, depending upon the conditions of operation.

Proper formulation of the phosphating solutions allows the coatings to form rapidly at low temperature on the cleaned steel surface. Simple, on-the-job, chemical controls enable the operator to adjust the addition of coating chemicals to the requirements of the steel being treated.

Following the coating operation, a cold water rinse is used to remove excess coating chemicals. The flow of water through the rinse is regulated with the rate of production so that contamination of the main body of the rinse is minimized. An acidified rinse containing hexavalent chromium compounds follows the water rinse. This lapse has a specific effect of enhancing the corrosion resistance of the coating. An oven dry-off to remove surface moisture completes the process.

Typical products being treated by the cold phosphate system are automotive body and sheet-metal parts, refrigerator cabinets, office furniture, lighting fixtures, commercial air conditioners, home heating equipment, home laundries, kitchen cabinets, desk and filing cabinets, steel drums, and window sash.

There are advantages of the cold phosphate system over conventional methods that require higher-temperature operation:

1. Direct heat savings—up to 70%
2. Less heat-up time
3. Less maintenance due to decreased load on heating coils, steam traps, etc.
4. Reduce downtime (maintenance personnel can enter the units immediately after shutdown to make adjustments or repairs)
5. Increased worker comfort near the installation
6. Reduced use of water through decreased evaporation
7. Elimination of exhaust fans

Applications

Phosphate conversion coatings have a wide range of application. Several examples follow.

Base for Plastisol Coatings

The domestic appliance industry has pointed toward the use of plastisol (polyvinyl chloride) coatings as a replacement for more costly porcelain enamel. During this period, several major manufacturers of domestic dishwashers have standardized on plastisol films for coating tubs, lids, dish racks, etc.

The metal preparation method meeting this requirement was a zinc phosphate coating system. The fabricated parts are cleaned in a conditioned cleaner, phosphated, and rinsed thoroughly. The rinse procedure includes an acidified chromic–phosphoric acid rinse followed by a closely controlled deionized water rinse and thorough drying to remove surface moisture. This treatment produces a continuous, uniform, fine-grained coating of approximately 150–250 mg of zinc phosphate coating per square foot of surface area treated.

After the steel has been zinc phosphate coated and especially formulated, a primer is applied to a very thin and closely controlled film build and followed by the subsequent application of 12–15 mils of plastisol with an intermediate and final baking operation.

Bond for Vinyl Coatings

Vinyl films of 5–15 mils thickness are applied to both steel and aluminum sheets by lamination of calendered and decorated films or by spray or roller coating. For the vinyl laminate as with the plastisol application, the base metal must be chemically treated to provide the necessary bond for laminate to metal surface. A controlled, accelerated iron phosphate treatment followed by an acidified chromate rinse has proved to be best for the preparation of steel for vinyl lamination to sheet or coil on a continuous line. This method is now being used by several fabricators and rollers of steel.

Aluminum can also be finished with vinyl laminates and plastisols. Chromate conversion coatings of the "gold" oxide type provide excellent adhesion of the vinyl film as well as excellent corrosion resistance on the unprotected surface.

Metal preparation for both steel and aluminum is usually done by spray application in five-stage equipment in the following sequence:

1. Alkali claim—30 s (smutty steel may require brush scrubbing)
2. Water rinse—10 s
3. Phosphate coating—10 s
4. Water rinse—5 s
5. Acidified chromate rinse—6 s

The same sequence and time cycles are used for aluminum except that an appropriate chromate conversion coating solution is in stage 3. Consequently, installations have been engineered to prepare both steel and aluminum for subsequent application of vinyl laminates by providing interchangeable coating solution tanks at that stage, thereby providing efficient and versatile "in-line" operation for this new decorative treatment.

Bolt Making

Phosphate coatings are also finding wide acceptance in the fastener industry as applied to rod to facilitate the forming of bolts. The coating is produced from a dilute, especially accelerated, zinc acid phosphate solution that reacts chemically with the surface of the rod to form an insoluble nonmetallic phosphate coating integral with the surface of the rod. The coating forms a porous bond to carry the extruding and heading lubricant and prevent metal-to-metal contact and subsequent heading operations. The result is longer tool life, increased percentage of reduction, and improved surface appearance of the finished product.

The methods of application vary from the conventional pickle-house immersion method to the in-line strand processes. In the immersion method, the coils of scaled rods are dipped in the treating solutions in the following sequence of operations:

1. Acid pickle—10 min to 2.5 h
2. Water rinse–cold, overflowing, or spray

3. Hot water rinse
4. Phosphate coating—5–15 min, 71°C–93°C
5. Cold water rinse
6. Neutralizing rinse—line
7. Soap dip
8. Bake dry

It has been found that a dip in a high-strength soap solution improves die life and reduces knockout pressures. The soap solution follows the neutralizing rinse, or in some instances, it is combined with the neutralizing rinse.

The in-line strand process eliminates intermediate handling from scaled rod to drawn wire and the necessity of acid disposal. The coils are handled continuously. Line speeds up to ~300 million m/s are required with this process so that the phosphate solution must deposit the desired heavy coating in about 15 s.

Even during this extremely short processing time, a line speed up to ~300 million m/s requires a special coating tank. Compact prerinse and neutralizing rinse stages are incorporated in this unit from which it is possible to obtain better than 750 mg/ft^2 of coating and 15 s by use of specially formulated phosphate solutions adapted to strand processing.

Wire produced by the strand method of processing produces wire superior in quality to that produced by conventional cleaning and drawing methods and does so at lower cost. As a consequence of the better quality, phosphate-base-lubricated material will produce close tolerance bolts with full heads and sharp shoulders.

Wire Drawing

Phosphate coatings are also gaining acceptance as a lubricant carrier in the wire industry to permit increased drawing speeds and prolonged die life. This is particularly true in connection with both the dry and wet drawing of high carbon wire. Other advantages that the zinc phosphate coatings afford to the wire industry are increased corrosion resistance after drawing and closer dimensional tolerances. Improved dimensional tolerance is a major advantage in forming springs from spring wire.

The zinc phosphate coating may be applied by immersion in the conventional pickle house installation, or by the newly developed fast coating, continuous strand method. The continuous strand method fits in well with fast, in-line travel of the wire. This adaptability and the lower labor costs involved warrant the recommendation of strand phosphate lines in wire processing.

Extrusion

Another important development is the use of zinc phosphate coatings as an aid in the rapidly developing field of cold extrusion of steel and aluminum. In this application, the phosphate coating solution is formulated to deposit a considerably heavier zinc phosphate coating than heretofore mentioned. The coating is then chemically reacted with a soap-base lubricant to form a water-insoluble lubricating film.

The use of the zinc phosphate coating is considered a basic requirement in the cold extrusion field where the pressure exerted by the tools on the steel being formed may be in excess of 2040 MPa. The phosphate and lubricant coating withstands the high unit pressure and temperatures developed in this type of cold forming. At the same time, it maintains the required separating film between the tools and workpiece being extruded to prevent scoring, galling, and tool breakage. A specially formulated phosphate coating solution produces a zinc phosphate coating on aluminum and is gaining wide acceptance as an aid in the cold extrusion of the heat-treatable alloys of this metal.

Phosphate Glass

A glass in which the essential glass former in phosphorus pentoxide (P_2O_5) instead of silica. Also used in *glass–ceramics* and *glass tableware*.

Phosphating

(1) Forming an adherent phosphate coating on a metal by immersion in a suitable aqueous phosphate solution. Also called phosphatizing. (2) The production of a thin conversion coating of metal phosphate on the surface of, usually, steel. The coating offers useful corrosion resistance on finished components and improves lubrication/friction characteristics of material requiring further cold work such as drawing. See also *conversion coating*.

Phosphor

A substance capable of luminescence, that is, absorbing energy such as x-rays, ultraviolet radiation, electrons, or alpha particles, and emitting a portion of that energy in the visible or near-visible region. If emission ceases within 10^{-4} s after excitation, the phosphoric is fluorescent; if it persists, it is called phosphorescent. See also *fluorescence* and *phosphorescence*.

Phosphor Bronze

This, a copper-base alloy with low phosphorus content, originally called steel bronze, was 92/8 bronze deoxidized with phosphorus and cast in an iron mold. It is now any bronze deoxidized by the addition of phosphorus to the molten metal. It may or may not contain residual phosphorus in the final state. Ordinary bronze frequently contains cuprous oxide formed by the oxidation of the copper during fusion. By the addition of phosphorus, a partial reducing agent, a complete reduction of the oxide takes place. Phosphor bronzes have excellent mechanical and cold-working properties and low coefficient of friction, making them suitable for springs, diaphragms, bearing plates, and fasteners. In some environments, such as saltwater, they are superior to copper.

At present, there are 18 standard wrought phosphor bronzes, designated C50100–C54800. Tin, which ranges from as much as 0.8%–1% depending on the alloy, is the principal alloying element, although leaded alloys may contain as much lead (e.g., 4%–6%) as tin. Phosphate content is typically on the order of 0.1%–0.35%, zinc 0%–0.3% (1.5%–4.5% in C 54400), iron 0%–0.1%, and lead 0%–0.05% (0.8%–6% in leaded alloys). The principal alloys were formally known by letter designation representing nominal tin content: phosphor bronze A, 5% tin (C51000); phosphor bronze B, 4.75% tin (C53200); phosphor bronze C, 8% tin (C52 100); phosphor bronze D, 10% tin (C52400); and phosphor bronze E, 1.25% tin (C50500). Phosphor bronze E, almost 99% copper, is one of the leanest of these bronzes in the way of alloying ingredients and is used for electrical contacts, pole-line hardware, and flexible tubing. Its electrical conductivity is about half that of copper, and it is readily formed, soldered, brazed, and flash-welded. Thin flat products have tensile yield strengths ranging from about 83 MPa in the annealed condition to 517 MPa in the extra spring temper. More highly alloyed C54400 (4% tin, 4% lead, and 3% zinc, nominally) is about 1/5 as electrically conductive as copper and has good forming characteristics and 80% of the machinability of C36000, a free-machining brass. Its ultimate tensile strength ranges from about 331 MPa in the annealed condition to 690 MPa in the extra spring temper. Its uses include bearings, bushings, gear shafts, valve components, and screw-machined products. See *bearing bronzes*.

Phosphor Copper

An alloy of phosphorus and copper, phosphor copper was used instead of pure phosphorus for deoxidizing brass and bronze and for adding phosphorus in making phosphor bronze. It comes in 5%, 10%, and 15% grades and is added directly to the molten metal. It serves as a powerful deoxidizer, and the phosphorus also hardens the bronze. Even slight additions of phosphorus to copper or bronze increase fatigue strength. Phosphor copper is made by forcing cakes of phosphorus into molten copper and holding until the reaction ceases. Phosphorus is soluble in copper up to 8.27%, forming Cu_3P, which has a melting point of about 707°C. A 10% phosphor copper melts at 850°C and a 15% at about 1022°C. Alloys richer than 15% are unstable. Phosphor copper is marketed in notched slabs or in shot.

Phosphor tin is a master alloy of tin and phosphorus used for adding to molten bronze in the making of phosphor bronze. It usually contains up to 5% phosphorus and should not contain lead. It has an appearance like antimony, with large glittering crystals, and is marketed in slabs.

Phosphorescence

A type of photoluminescence in which the time period between absorption and reemission of light is relatively long (of the order of 10^{-4} to 10 s or longer). See also *photoluminescence*. Contrast with *fluorescence*.

Phosphoric Acid

Also known as orthophosphoric acid, phosphoric acid is colorless, superbly liquid of the composition H_3PO_4 used for pickling and rustproofing metals, for the manufacture of phosphates, pyrotechnics, and fertilizers, as a latex coagulant, as a textile mordant, as an acidulating agent in jellies and beverages, and as a clarifying agent in sugar syrup. The specific gravity is 1.65 and melting point 73.6°C, and it is soluble in water. The usual grades are 90%, 85%, and 75%, technical 50%, and dilute 10%. As a cleanser for metals, phosphoric acid produces a light edge on steel, aluminum, or zinc, which aids paint adhesion. *Deoxidine* is a phosphoric acid cleanser for metals. *Nielite D* is a phosphoric acid with a rust inhibitor, used as a nonfuming pickling acid for steel. *Albrite* is available in 75%, 80%, and 85% concentrations in food and electronic grades, both high-purity specifications. *DAB* and *Phosbrite* are called bright dip grades, for cleaning applications. Phosphoric anhydride, or phosphorus pentoxide, P_2O_5, is a white, water-soluble powder used as a dehydrating agent and also as an opalizer for glass. It is also used as a catalyst in asphalt coatings to prevent softening at elevated temperatures and brittleness at low temperatures.

Phosphorus

Phosphorus, a chemical element, symbol P, forms the basis of a very large number of compounds; the most important class is phosphates. For every form of life, phosphates play an essential role in all energy-transfer processes such as metabolism, photosynthesis, nerve function, and muscle action. The nucleic acids that, among other things, make up the hereditary material (the chromosomes) are phosphates, as are a number of coenzymes. Animal skeletons consist of a calcium phosphate.

Uses

About 90% of the total phosphorus (in all of its chemical forms) used in the United States goes into fertilizers.

P

Commercial phosphorus is obtained from phosphate rock by reduction in the electric furnace with carbon or from bones by burning and treating with sulfuric acid. Phosphate rock occurs in the form of land pebbles and as hard rock.

The superphosphate used for fertilizers is made by treating phosphate minerals with concentrated sulfuric acid. It is not a simple compound, but may be a mixture of calcium acid phosphate, $CaHPO_4$, and calcium sulfate. Nitrophosphate for fertilizer is made by acidulating phosphate rock with a mixture of nitric and phosphoric acids or with nitric acid and then ammoniation and addition of potassium or ammonium sulfate.

Other important uses are as binders for detergents, nutrient supplements for animal feeds, water softeners, additives for foods and pharmaceuticals, coating agents for metal-surface treatment, additives and metallurgy, plasticizers, insecticides, and additives for petroleum products. Except for the last four items, these uses involve phosphates.

Sodium tripolyphosphate is the major compound used in building synthetic detergents to achieve improved cleaning, primarily by dispersing inorganic soil and softening the water. The average phosphate-containing household detergent produced in the United States for washing clothes consists of about 40% by weight of sodium tripolyphosphate, $Na_5P_3O_{10}$. This compound is used extensively in water softening, as are other members of the homologous series of chain phosphates. The large-volume usage of phosphates in detergents has led to unwanted growth of algae in inland waters (lakes and rivers) into which the dirty dishwasher discharges are disposed. As a result of this fertilizing action, phosphates are considered water pollutants in those areas where such discharges occur, and in some areas, phosphates have been eliminated from detergents by law. For reasonably fast-flowing rivers that discharge directly into the ocean, phosphates are not a problem.

An interesting water-softening application is found in "threshold treatment" in which tiny traces of a chain phosphate (much less than would be used in sequestering) are used to prevent the formation of pipe scale from hard waters. The application is related to the dispersing action of the phosphates, because traces of phosphate absorb on the growing surface of the pipe scale as it begins to form, and this inhibits its further growth.

A major pharmaceutical use of phosphates is in toothpastes, in which dicalcium phosphate is the most popular polishing agent. Monocalcium phosphate and sodium acid pyrophosphate, $Na_2H_2P_2O$ (the pyrophosphate is the second member of the phosphate family) are employed as leavening agents in cake mixes, refrigerated biscuits, self-rising flour, and baking powder.

Automobile bodies, for example, are generally phosphatized before they are painted to prevent rusting in use. Orthophosphate esters find wide use as plasticizers that have flameproofing properties and as gasoline and oil additives.

The phosphorus compound of major biological importance is adenosine triphosphate, which is an ester of the salt sodium tripolyphosphate, widely employed in detergents and water-softening compounds. Practically, every reaction in metabolism and photosynthesis involves the hydrolysis of this tripolyphosphate to its pyrophosphate derivative, called adenosine diphosphate.

Phosphorus is an essential element in the human body; a normal person has more than a pound of it in the system, but it can be taken into the system only in certain compounds. Nerve gases used in chemical warfare contain phosphorus, which combines with and inactivates the cholinesterase enzyme of the brain. This enzyme controls the supply of the hormone that transmits nerve impulses, and when it is inactivated, the excess hormone causes paralysis of the nerves and cuts off breathing. Organic phosphates are widely used in food, textile, and chemical industries. Other phosphates are used as a plasticizer in plastics and as an antifoaming agent in paper coatings and in textile sizings. They are also employed for scale and corrosion control, ore flotation, pigment dispersion, and detergents.

Flour and other foodstuffs are fortified with ferric phosphate, $FePO_4 \cdot 2H_2O$. Iron phosphate is used as an extender in paints. Tricalcium phosphate, $Ca_2(PO_4)_2$, is used as an anticaking agent in salt, sugar, and other food products and to provide a source of phosphorus. The tricalcium phosphate, used in toothpastes as a polishing agent and to reduce the staining of chlorophyll, is a fine white powder. Dicalcium phosphate, used in animal feeds, is precipitated from the bones used for making gelatin, but is also made by treating lime with phosphoric acid made from phosphate rock. Diammonium phosphate, $(NH_4)_2HPO_4$, is a mildly alkaline, white crystalline powder used in ammoniated dentifrices, for pH control in bakery products, for making phosphors, for the prevention of afterglow in matches, and for flameproofing paper.

Forms

There are two common forms of phosphorus, yellow and red. The former, also called white phosphorus, P_4, is a light-yellow waxlike solid, phosphorescent in the dark, and exceedingly poisonous. Its specific gravity is 1.83 and it melts at 44°C. It is used for smokescreens in warfare and for rat poisons and matches. Yellow phosphorus is produced directly from phosphorus rock in the electric furnace. It is cast to cakes of 0.45–1.36 kg each. Red phosphorus is a reddish-brown amorphous powder, with a specific gravity of 2.20 and a melting point of 725°C. Red phosphorus is made by holding white phosphorus at its boiling point for several hours in a reaction vessel. Both forms ignite easily. Amorphous phosphorus, or crystalline black phosphorus, is made by heating white phosphorus for extended periods. It resembles graphite and is less reactive than the red or white forms, which can ignite spontaneously in air. Black phosphorus is made by this process. Phosphorus sulfide, P_4S_3, may be used instead of white phosphorus in making matches. Phosphorus pentasulfide, P_2S_3, is a canary-yellow powder with a specific gravity of 1.30 or solid with a specific gravity of 2.0, containing 27.8% phosphorus, used in making oil additives and insecticides. It is decomposed by water.

Phosphorized Copper

General term applied to copper deoxidized with phosphorus. The most commonly used deoxidized copper. The small amount of residual phosphorus, typically 0.02%, seriously reduces electric conductivity.

Phosphorus Banding

See *ghost banding*.

Photoelastic Modeling/Stress Analysis

A technique in which a model of some component is formed in a suitable plastic and deformed while being viewed under polarized light. The bright and dark areas that are produced in the plastic by interference effects indicate the state and direction of stresses. It is possible to stress the model, which can be, for example, a complex pressure vessel, while warm and then allow it to cool under stress to "freeze in" the strain. The model of sections cut from it may then be examined at leisure.

Photoelasticity

An optical method for evaluating the magnitude and distribution of stresses, using a transparent model of a part or a thick film of photoelastic material bonded to a real part.

Photoelectric Effect

The liberation of electrons by electromagnetic radiation incident on a substance.

Photoelectric Electron Multiplier Tube

See *photomultiplier tube.*

Photographic Materials

These are the light-sensitive recording materials of photography, that is, photographic films, plates, and papers. They consist primarily of a support of plastic sheeting, glass, or paper, respectively, and a thin, light-sensitive layer, commonly called the emulsion, in which the image will be formed and stored. The material will usually embody additional layers to enhance its photographic or physical properties.

Supports

Film support, for many years made mostly of flammable cellulose nitrate, is now exclusively made of slow-burning "safety" materials, usually cellulose triacetate or polyester terephthalate, which are manufactured to provide thin, flexible, transparent, colorless, optically uniform, tear-resistant sheeting. Polyester supports, which offer added advantages of toughness and dimensional stability, are widely used for films intended for technical applications. Film supports usually range in thickness from 0.06 to 0.23 mm and iron maiden rolls up to 1.5 in. wide and 1800 m long.

Glass is the predominant substrate for photographic plates, although methacrylate sheet, fused quartz, and other rigid material are sometimes used. Plate supports are selected for optical clarity and flatness. Thickness, ranging usually from 1 to 6 mm, is increased with plate size as needed to resist breakage and retain flatness. The edges of some plates are especially ground to facilitate precise registration.

Photographic paper is made from bleached wood pulp of high α-cellulose content, free from ground wood and chemical impurities. It is often coated with a suspension of barium sulfate in gelatin for improved reflectance and may be calendered for high smoothness. Fluorescent brighteners may be added to increase the appearance of whiteness. Many paper supports are coated on both sides with water-repellent synthetic polymers to preclude wetting of the paper fibers during processing. This treatment hastens drying after processing and provides improved dimensional stability and flatness.

Emulsions

Most emulsions are basically a suspension of silver halide crystals in gelatin. The crystals, ranging in size from 2.0 to less than 0.05 μm, are formed by precipitation by mixing a solution of silver nitrate with a solution containing one or more soluble halides in the presence of a protective colloid. The salts used in these emulsions are chlorides, bromides, and iodides. During manufacture, the emulsion is ripened to control crystal size and structure. Chemicals are added in small but significant amounts to control speed, image tone, contrast, spectral sensitivity, keeping the qualities, fog, and hardness; to facilitate uniform coating; and, in the case of color films and papers,

to participate in the eventual formation of dye instead of metallic silver images upon development. The gelatin, sometimes modified by the addition of synthetic polymers, is more than a single vehicle for the silver halide crystals. It interacts with the silver halide crystals during manufacture, exposure, and processing and contributes to the stability of the latent image.

After being coated on a support, the emulsion is chilled so that it will set and then dried to a specific moisture content. Many films receive more than one high-sensitive coating, with individual layers as thin as 1.0 μm. Overall thickness of the coatings may range from 5 to 25 μm, depending on the product. Most x-ray films are sensitized on both sides, and some black-and-white films are double coated on one side. Color films and papers are coated with at least three emulsion layers and sometimes six or more plus filter and barrier layers. A thin, nonsensitized gelatin layer is commonly placed over film emotions to protect against abrasion during handling. A thicker gelatin layer is coated on the back of most sheet films and some roll films to counteract the tendency to curl, which is caused by the effect of changes in relative humidity on the gelatin emulsion. Certain films are treated to reduce electrification by friction because static discharges can expose the emulsion. The emulsion coatings on photographic papers are generally thinner and more highly hard than those on film products.

Another class of silver-based emulsions relies on silver behenate compounds. These materials require roughly 10 times more exposure than silver halide emulsions having comparable image structure properties (resolving power, granularity); are less versatile in terms of contrast, maximum density, and spectral sensitivity; and are less stable both before exposure and after development. However, they have the distinct advantage of being processed through the application of heat (typically at 116°C–127°C) rather than a sequence of wet chemicals. Hence, products of this type are called dry silver films and papers.

Photoluminescence

Reemission of light absorbed by an atom or molecule. The light is emitted in random directions. There are two types of photoluminescence: *fluorescence* and *phosphorescence.*

Photomacrograph

A macrograph produced by photographic means.

Photometer

A device so designed that it measures the ratio of the radiant power of two electromagnetic beams.

Photomicrograph

A micrograph produced by photographic means.

Photomicrograph/Macrograph

Photographs taken with a microscope at high/low magnifications.

Photomultiplier Tube

(1) A device in which incident electromagnetic radiation creates electrons by the *photoelectric effect.* These electrons are accelerated by a series of electrodes called dynodes, with secondary emission

adding electrons to the stream at each dynode. (2) A light-sensitive vacuum tube with multiple electrodes providing a highly amplified electrical output.

Photon

(1) A particle representation of the electromagnetic field. The energy of the photon equals hv, where v is the frequency of the light in hertz and h is Planck's constant. See also *Planck's constant*. (2) A quantum, that is, a single indivisible portion, of light energy.

Photopolymer

A polymer that changes characteristics when exposed to light of a given frequency.

Photoresist

(1) A radiation-sensitive material which, when properly applied to a variety of substrates and then properly exposed and developed, masks portions of the substrate with a high degree of integrity. (2) A photosensitive coating that is applied to a laminate and subsequently exposed through a photo tool (film) and developed to create a pattern that can be either plated or etched.

Photosensitive

Sensitive to light.

Phthalate Esters

The most widely used group of *plasticizers*, produced by the direct action of alcohol on phthalic anhydride. The phthalates are generally characterized by moderate cost, good stability, and good all-around properties.

Phthalic Anhydride

A white, crystalline material with a melting point of 267°F (130.8°C) soluble in water and in alcohol. It is made by oxidizing naphthalene, or it is produced from orthoxylene derived from petroleum. BASF Corp. markets the product both as flake and in the molten state. It is used in the manufacture of alkyd resins and for the production of dibutyl phthalate and other plasticizers, dyes, and many chemicals. *Chlorinated phthalic anhydride* is also used as a compounding medium in plastics. It is a white, odorless, nonhygroscopic, stable powder containing 50% chlorine. It gives higher-temperature resistance and increased ability to plastics. *Terephthalic acid* may be obtained as a by-product in the production of phthalic anhydride from petroleum. It has a long-chain alkyl group having an amide linkage on one end and a methyl ester on the other. It is used for producing *polyethylene terephthalate* and other polyesters. The esters can also be made from *dimethyl terephthalate*, a molten material but burns readily when ignited. The dust can form explosive mixtures with air. The terephthalates are useful in textile and tire cord fibers, plastic tape, and food-packaging polymers. Their sodium salt is used as a gelling agent for lubricating grease used in high temperatures up to 600°F (316°C). It forms fine *crystallites* of soft, flexible fibers in the grease.

Maleic anhydride is also a building block for L-aspartame, use for making *NutraSweet*, the aspartame synthetic sweetener.

Maleic anhydride was traditionally made from benzene, but *n*-butane has become the feedstock of choice because of its lower cost and because benzene is a carcinogen. In the United States, the transition was completed in the late 1980s. Recent uses for maleic anhydride, and *maleic acid* recovered from the catalytic oxidation of butane to maleic anhydride, are the production of chemical intermediates.

Physical Adsorption

See *physisorption*.

Physical Blowing Agent

A gas, such as a fluorocarbon, which is pumped into a mold during the forming of plastics.

Physical Crack Size (α_p)

In fracture mechanics, the distance from a reference plane to the observed crack front. This distance may represent an average of several measurements along the crack front. The reference plane depends on the specimen form, and it is normally taken to be either the boundary or a plane containing either the load line or the centerline of a specimen or plate.

Physical Etching

Development of microstructure through removal of atoms from the surface or lowering the grain-surface potential.

Physical Metallurgy

The science and technology dealing with the properties of metals and alloys and of the effects of composition, processing, and environment on those properties at an atomic and microscopical level.

Physical Objective Aperture

In electron microscopy, a metal diaphragm centrally pierced with a small hole used to limit the cone of electrons accepted by the objective lens. This improves image contrast, because highly scattered electrons are prevented from arriving at the Gaussian image plane and therefore cannot contribute to background noise.

Physical Properties

Properties of a material that are relatively insensitive to structure and can be measured without the application of force; for example, density, electrical conductivity, coefficient of thermal expansion, magnetic permeability, and lattice parameter. Does not include chemical reactivity. Compare with *mechanical properties*.

Physical Testing

Methods used to determine the entire range of a material's *physical properties*. In addition to density and thermal, electrical, and magnetic properties, physical testing methods may be used to assess simple fundamental physical properties such as color, crystalline form, and melting point.

Physical Vapor Deposition (PVD)

A coating process whereby the deposition species are transferred and deposited in the form of individual atoms or molecules. The most common PVD methods are sputtering and evaporation. Sputtering, which is the principal PVD process, involves the transport of a material from a source (target) to a substrate by means of the bombardment of the target by gas ions that have been accelerated by a high voltage. Atoms from the target are ejected by momentum transfer between the incident ions in the target. These ejected particles move across the vacuum chamber to be deposited on the substrate. Evaporation, which was the first PVD process used, involves the transfer of material to form a coating by physical means alone, essentially the vaporization. The streaming vapor is generated by melting and evaporating a coating material source bar, by an electron beam in a vacuum chamber. Because both of these methods are line-of-sight processes, it is necessary to use specially shaped targets or multiple evaporation sources and to rotate or move the substrate uniformly to expose all areas. PVD coatings are used to improve the wear, friction, and hardness properties of cutting tools and as corrosion-resistant coatings.

Physisorption

The binding of an adsorbate to the surface of a solid by forces whose energy levels approximate those of condensation. Contrast with *chemisorption*.

PI

See *polyimide*.

pi Bonding

Covalent bonding of atoms in which the atomic orbitals overlap along a plane perpendicular to the sigma bond(s) joining the nuclei of two or more atoms. See also *sigma bonding*.

pi Electron

An electron that participates in *pi bonding*.

Piano Wire

High tensile (up to about 2000 MPa, 130 tons/in.²), high carbon (about 0.9%) steel wire produced by patenting.

PIC

See *pressure impregnation–carbonization*.

Pick Count

In reinforced plastics, the number of *tows*/mm (in.) of woven fabric in both the *warp* and *fill* directions.

Pickle

The chemical removal of surface oxides (scale) and other contaminants such as dirt from iron and steel by immersion in an aqueous acid solution. The most common pickling solutions are sulfuric and hydrochloric acids.

Pickle Embrittlement

See *pickling*.

Pickle Liquor

A spent acid-pickling bath.

Pickle Patch

A tightly adhering oxide or scale coating not properly removed during *pickling*.

Pickle Stain

Discoloration of metal due to chemical cleaning without adequate washing and drying.

Pickling

The removal of scale and other contaminants from the surface of a metal component by immersion in appropriate chemicals, commonly acids or alkalis or electrochemical reaction. It is implicit that the metal component is not significantly attacked and often inhibitors are added to achieve this. If carried out, steel, particularly of high hardness, can be charged with hydrogen from the surface reaction causing a form of hydrogen embrittlement called *pickle embrittlement* or similar terms. The hydrogen can be allowed to diffuse out by baking at about 200°C for some hours.

Pickling Acids

Acid used for pickling or cleaning castings or metal articles. The common pickling bath for iron and steel is composed of a solution of sulfuric acid and water, one part acid to 5–10 parts water being used. This acid attacks the metal and cleans it from oxides and sand by loosening them. For pickling scale from stainless steels, a 25% cold solution of hydrochloric or sulfuric acid is used, or hydrofluoric acid with the addition of anhydrous ferric sulfate is used. Hydrofluoric acid solutions are sometimes used for pickling iron castings. This acid attacks and dissolves away the sand itself. For bright-cleaning brass, a mixture of sulfuric acid and nitric acid is used. For a matte finish, the mixed acid is used with a small amount of zinc sulfate. Copper and copper alloys can be pickled with sulfuric acid to which anhydrous ferric sulfate is added to speed the action, or sodium bichromate is added to the sulfuric acid to remove red cuprous oxide stains. Brass forgings are pickled in nitric acid to bring out the color. Since all of these acids form salts rapidly by the chemical action with a metal, they must be renewed with frequent additions of fresh acid. The French pickling solution known as *framanol*, used for aluminum, is a mixture of chromium phosphate and triethanolamine. The latter emulsifies the grease and oil, and the aluminum oxide film is dissolved by the phosphoric acid, leaving the metal with a thin film of chromic oxide.

The temperature of most pickling is from 140°F to 180°F (60°C–82°C). An increase of 20°F (11°C) will double the rate of pickling. *Acid brittleness* after pickling is due to the absorption of hydrogen when the acid acts on iron and is reduced by shortening the pickling time. *Inhibitors* are chemicals added to reduce the time of pickling by permitting higher temperatures and stronger solutions without hydrogen absorption. In plating baths, fluoboric acid, HBF_4, has high throwing power and has a cleansing effect by dissolving sand and silicides from iron castings and steel surfaces. It is a colorless liquid with a specific gravity of 1.33.

Phosphoric acid is employed in a hot solution as a dip bath for steel parts to be finished to a rough or etched surface. It leaves a basic iron phosphate coating on a steel which is resistant to corrosion and gives a rough base for the finish. *Parkerized steel* is rustproof steel treated in a bath of iron and manganese phosphates. *Bonderized steel* is steel treated with phosphoric acid and a catalyst to give a rough, tough, rust-resistant base for paints. *Granodized steel* is produced with zinc phosphate.

Metals can be treated with alkaline solutions, too. Rust can be removed by caustic soda baths in which a sequestrant, such as *sodium gluconate* or *ethylenediaminetetraacetic acid*, is mixed to complex the dissolved iron and keep it from precipitating.

Pick-Off

An automatic device for removing a finished part from the press die after it has been stripped.

Pickup

(1) Transfer of metal from tools to part or from part to tools during a forming operation. (2) Small particles of oxidized metal adhering to the surface of a *mill product*. (3) Undesirable material collected or entrained during some process. Elements entering the weld metal during fusion of the parent metal.

Pickup Roll

A spreading device where the roll for picking up the adhesive runs in a reservoir of adhesive.

Picral

Picric acid in alcohol used to etch ferrous materials particularly pearlitic steels and mostly cast irons. Unless otherwise stated, a solution of 4% picric acid is usually implied.

Pidgeon Process

A process for production of magnesium by reduction of magnesium oxide with ferrosilicon.

Piercing

The general term for cutting (shearing or punching) openings, such as holes and slots, in sheet material, plate, or parts. This operation is similar to *blanking*; the difference is that the slug or pierce produced by piercing is scrap, while the blank produced by blanking is a useful part.

Piezoelectric

A material or crystal that becomes polarized and its surface becomes charged when a stress is applied to it. Conversely, if the material is subject to an electric field, it will expand in one direction and contract in another.

Piezoelectric Effect

The reversible interaction, exhibited by some crystalline materials, between an elastic strain and an electric field. The direction of the strain depends on the polarity of the field or vice versa. Compare with *electrostrictive effect*.

Piezoelectric Polymers

Polymers that spontaneously give an electric charge when mechanically stressed or that develop a mechanical response when an electric field is applied. Used as transducers or acoustic sensors.

Piezoelectricity

Piezoelectricity is electricity, or electric polarity, resulting from the application of mechanical pressure on a dielectric crystal. The application of mechanical stress produces in certain dielectric (electrically nonconducting) crystals an electric polarization (electric dipole moment per cubic meter) that is proportional to this stress. If the crystal is isolated, this polarization manifests itself as a voltage across the crystal, and if the crystal is short circuited, a flow of charge can be observed during loading. Conversely, application of a voltage between certain faces of the crystal produces a mechanical distortion of the material. This reciprocal relationship is referred to as the *piezoelectric effect*. The phenomenon of generation of a voltage under mechanical stress is referred to as the direct piezoelectric effect, and the mechanical strain produced in the crystal under electric stress is called the converse piezoelectric effect.

Piezoelectric materials are used extensively in transducers for converting a mechanical strain into an electrical signal. Such devices include microphones, phonograph pickups, and vibration-sensing elements. The converse effect, in which a mechanical output is derived from an electrical signal input, is also widely used in such devices as sonic and ultrasonic transducers, headphones, loudspeakers, and cutting heads for disk recording. Both the direct and converse effects are employed in devices in which the mechanical resonance frequency of the crystal is of importance. Such devices include electric wave filters and frequency-control elements in electronic oscillator circuits.

Pig

A metal casting used in remelting.

Pig Bed/Casting Machine

The bed of sand or the continuous belt of molds into which molten iron is poured to solidify as pig iron.

Pig Iron

Pig iron is the iron produced from the first melting of the ore. The melt of the blast furnace is run off into rectangular molds, forming, when cold, ingots called pigs. So-called because the iron was originally run off from the furnace into multiple branched channels formed in a bed of sand. The multiple channels were reminiscent of a family of piglets feeding from the sow. Pig iron contains small percentages of silicon, sulfur, manganese, and phosphorus, besides carbon. It is useful only for resmelting to make cast iron or wrought iron. Pig iron is either sand-cast or machine-cast. When it is sand-cast, it has sand adhering and fused into the surface, giving more slag in the melting. Machine-cast pig iron is cast in steel forms and has a fine-grained chilled structure, with lower melting point. Pig irons are classified as Bessemer or non-Bessemer, according to whether the phosphorus content is below or above 0.10%. There are six general grades of pig iron: low-phosphorus pig iron, with less than 0.03%, used for making steel for steel castings and for crucible steelmaking; Bessemer pig iron, with less than 0.10% phosphorus, used for Bessemer steel and for acid open-hearth steel; malleable pig iron, with less than 0.20%, used for making malleable iron; foundry pig iron, with from 0.5% to 1%, and low-silicon, less than 1%, for basic open-hearth steel; and basic Bessemer, with from 2% to 3%, used for making steel by the basic Bessemer process employed in England.

Because silicon is likely to dissolve the basic furnace lining, it is kept as low as possible, 0.70%–0.90%, with sulfur not usually over 0.095%. Pig irons are also specified on the basis of other elements, especially sulfur. The sulfur may be from 0.04% to 0.10%, but high-sulfur pig iron cannot be used for the best castings. The manganese content is usually from 0.60% to 1%. Most of the iron for steelmaking is now not cast but is carried directly to the steel mill in car ladles. It is called direct metal.

Pigment (Material)

A finely divided material, pigment contributes to optical and other properties of paint, finishes, and coatings. Pigments are insoluble in the coating material, whereas dyes dissolve in and color the coating. Pigments are mechanically mixed with a coating and are deposited when the coating dries. Their physical properties generally are not changed by incorporation in and deposition from the vehicle. Pigments may be classified according to composition (inorganic or organic) or by source (natural or synthetic). However, the most useful classification is by color (white, transparent, or colored) and by function.

White Pigments

These pigments are essentially transparent to visible light. Because of the difference in refractive index between the pigment particles and the vehicles, white pigments refract the light from a multitude of surfaces and return a substantial portion in the direction of illumination without significant change in the spectral composition of the light.

The common white pigments are titanium dioxide, derived from titanium ores; white lead, from corrosion of metallic lead; zinc oxide, from burning of zinc metal; and lithopone, a mixture of zinc sulfide and barium sulfate. Pure zinc sulfide and antimony oxide are less commonly used.

Titanium dioxide may be crystallized in the rutile or anatase form, depending on the method of production. It may be further modified by surface treatment to control the rate of chalking and other properties. Rutile titanium dioxide has a higher refractive index than anatase and therefore higher hiding power, but it has a somewhat yellow color. Anatase titanium dioxide provides a purer white.

White lead pigments are the oldest of white pigments and were used extensively to provide excellent hiding power, flexibility, and durability to interior and exterior paints and enamels. Consumer protection rulings have removed white lead paints from the market, because leaded paint particles were ingested by children, with toxic effects.

Zinc oxide and lithopone pigments were extensively used in paint formulation, but have been superseded by titanium dioxide. Pure zinc oxide pigment is rarely used. Antimony oxide pigment is used chiefly in certain fire-retardant paints.

Transparent Pigments

The refractive indices of these pigments are very close to the index of the paint vehicle (about 1.54). They are used to provide bulk, control setting, and contribute to the hardness, durability, and abrasion resistance of the paint film. Because they are commonly used to add bulk to other pigments, they are called extenders. Most transparent pigments are natural minerals reduced to pigment particle size. Among the most commonly used transparent pigments are calcium carbonate (ground limestone, whiting, or chalk), magnesium silicate, bentonite clay, silicate, or barites (barium sulfate). Transparent pigments often constitute a substantial portion of a protective coating.

Colored Pigments

These pigments are available in a wide variety of colors and properties, depending upon the end use. Several hundreds have been used; the following are the most common:

Red: Iron oxides, often classified by color, include Indian red, Spanish red, Persian Gulf red, and Venetian red, a mixture of iron oxide and calcium red sulfate. Other red pigments include cadmium red (cadmium selenide) and organic reds, which are usually coal-tar derivatives either precipitated in pigment form (toners) or deposited on a transparent pigment (lakes). Organic reds include toluidines and lithols.

Orange: Chrome orange (basic lead chromate), molybdate orange (lead chromate–molybdate), and various organic toners and lakes are the most common orange pigments.

Brown: Browns are nearly always iron oxides, although certain lakes and toners are used for special purposes.

Yellow: These pigments include natural iron oxides such as ocher or sienna, or synthetic iron oxides, which are stronger and brighter, such as chrome yellow (normal lead chromate) and cadmium yellow (cadmium sulfide), and organic toners and lakes such as Hansa yellow and benzidine yellow.

Green: The most important green pigments are chrome green, a mixture of chrome yellow and Prussian blue; chromium oxide, duller but more permanent; phthalocyanine green, an organic pigment containing copper; and various other organic toners or lakes, often precipitated with phosphotungstic or phosphomolybdic acid.

Blue: The blue pigments include Prussian blue (ferric ferrocyanide, sometimes called milori or Chinese blue, depending on the shade); ultramarine, an inorganic pigment made by fusing soda sulfur and other materials under controlled conditions; phthalocyanine blue, an organic pigment containing copper; and numerous organic toners and lakes.

Purple and violet: These are nearly all organic toners or lakes. Manganese phosphate is a very weak, inorganic purple pigment.

Black: The vast majority of black pigments consist of finely divided carbon—carbon black, lampblack, and bone black—usually obtained by allowing a smoky flame to impinge on a cold surface. Black iron oxide and certain organic pigments are used with special properties required.

Special Pigments

Anticorrosive pigments are used to prevent the formation or spread of rust on iron when the metal is exposed by a break in the coating. The most common are red lead, an oxide of lead, and zinc yellow or zinc chromate, a basic chromate of zinc. Other colored chromates are sometimes used. The color of red leads fades rapidly, and the anticorrosive chromates are usually very weak in tinting strength. Metallic lead is sometimes used for anticorrosive paint.

Metallic pigments are small, usually flat particles of metal, prepared for dispersion and coatings. Aluminum is most commonly used because it leafs and forms a smooth, metallic film. The flakes are sometimes colored. Bronze, copper, lead, nickel, stainless steel, and silver appear occasionally. Zinc dust, or powdered zinc, is used more often because of its excellent adhesion to galvanized iron than because of its appearance.

Luminous pigments radiate visible light when exposed to ultraviolet light. Phosphorescent pigments continue to glow for a period of time even after the exciting light has been removed; these are usually sulfides of zinc and other materials, with small amounts of additives that control the phosphorescent properties. Fluorescent pigments lose luminosity as soon as the exciting light is removed; these pigments may be sulfides, although many organic pigments have this property.

Other specialized pigments include pigments that change color at some predetermined temperature, used to indicate hot areas on motors; pigments that give a pearly appearance; and pigments that conduct electricity for printed circuits.

Coarse material such as pumice is often added when a nonslippery coating is required. Glass beads give a very high degree of refractivity in the direction of illumination and are often used in centerline paints or for signs where night visibility is required. Intumescent pigments puff up under heat, giving a fire-resistant coating.

Pilger Tube-Reducing Process

See *tube reducing*.

Pilot Arc

A low-current continuous arc between the electrode and the constricting nozzle of a plasma arc welding torch used to ionize the gas and facilitate the start of the main welding arc.

Pimple

An imperfection, such as a small protuberance of varied shape on the surface of a plastic product.

Pin (for Bend Testing)

The plunger or tool used in making semiguided, guided, or wraparound bend tests to apply the bending force to the inside surface of the bend. In free bends or semiguided bends to an angle of 180°, a shim or block of the proper thickness may be placed between the legs of the specimen as bending is completed. This shim or block is also referred to as a pin or mandrel. See also *mandrel* (metals).

Pin Expansion Test

A test for determining the ability of a tube to be expanded or for revealing the presence of cracks or other longitudinal weaknesses, made by forcing a tapered pin into the open end of the tube.

Pin Joint

A structural joint that is free to hinge so that it does not transmit a moment.

Pin Metals

Metals for pins, originally cold-drawn common brass wire, sometimes chromium plated, now more usually hard-drawn medium-carbon steel chromium plated or for higher-quality purposes hard-drawn austenitic stainless steel.

Pinch Pass

A rolling pass producing a very small thickness reduction. Usually it is the last pass in the production of plate or sheet and is intended to provide close dimensional control. For steel, it can avoid the Lüders lines during subsequent processing.

Pinch Trimming

The trimming of the edge of a tubular metal part by pushing or pinching the flange or lip over the cutting edge of a stationary punch or over the cutting edge of a draw punch.

Pinchbeck

Brass with 5%–15% zinc, remainder copper, for cheap decorative applications, hence a derogatory term for cheap jewelry.

Pinchers

Surface disturbances on metal sheet or strip that result from rolling processes and that ordinarily appear as fernlike ripples running diagonally to the direction of rolling.

Pinch-Off

In blow molding of plastics, a raised edge around the cavity in the mold that seals off the part and separates the excess material as the mold closes around the *parison*.

Pincushion Distribution

See *positive distortion*.

Pine Tree Crystal

A type of *dendrite*.

Pinhead Blister

See *blister*.

Pinhole Eyepiece

An *eyepiece* in an *optical microscope*, or a cap to place over an eyepiece, which has a small central aperture instead of eye lens. It is used in adjusting or aligning microscopes.

Pinhole Ocular

See *pinhole eyepiece*.

Pinhole Porosity

Porosity consisting of numerous small gas holes distributed throughout a metal, found in weld metal, castings, and electrodeposited metal.

Pinholes

(1) Very small holes that are sometimes found as a type of porosity in a casting because of the microshrinkage or gas evolution during solidification. In wrought products, due to removal of inclusions or microconstituents during macroetching of transverse sections. (2) Small cavities that penetrate the surface of a cured composite or plastic part. (3) In photography, a very small circular aperture.

Pinion

The smaller of two mating gears.

Pinning

The locking into position of a dislocation by some features such as an interstitial atom.

Pin-On-Disk Machine

A tribometer in which one or more relatively moving styli (i.e., the "pin" specimen) is loaded against a flat-disk specimen surface such that the direction of loading is parallel to the axis of rotation of either the disk or the pin-holding shaft, and a circular wear path is described by the pin motion. The typical pin-on-disk arrangement resembles that of a traditional phonograph. Either the disk rotates or the pin specimen holder rotates so as to produce a circular path on the disk surface. An arrangement wherein the pin specimen is loaded against the curved circumferential surface of a flat disk is not generally considered to be a pin-on-disk machine.

Piobert Lines

See the *Lüders lines*.

Pipe

(1) The central cavity formed by contraction in metal, especially ingots, during solidification. (2) An imperfection in wrought or cast products resulting from such a cavity. (3) A tubular metal product, cast or wrought. (4) The elongated vertical cavity at the top central zone of a casting or ingot. When a molten metal solidifies and cools, it contracts. The metal tends to solidify first at the sides and bottom of the casting leaving insufficient molten metal to fill the top central area and a cavity or pipe is formed. (5) A tube or hollow section. Various terms are largely interchangeable, but "pipe" usually implies a fairly large diameter and a simple cross section. See also *extrusion pipe*.

Pipe Tap

A *tap* for making internal *pipe threads* within pipe fittings or holes.

Pipe Threads

Internal or external machine threads, usually tapered, of a design intended for making pressure-tight mechanical joints in piping systems.

Pipet

A tube, usually made of glass or plastic, used almost exclusively to deliver accurately known volumes of liquids or solutions during titration or volumetric analysis.

Pirani Gauge

An instrument used to measure the pressure inside a vacuum chamber. The gauge measures electrical resistance in a wire filament that will change in temperature depending on atmospheric pressure.

Piston-Pin Bearing

See *little-end bearing*.

Pit

A small, regular or irregular, crater in the surface of a material created by exposure to the environment, for example, corrosion, wear, or thermal cycling. See also *pitting*.

Pit Casting/Molding

Casting or molding by pouring into a refractory lined pit.

Pit Molding

Molding method in which the *drag* is made in a pit or hole in the floor.

Pitch

A high-molecular-weight material that is a residue from the destructive distillation of coal and petroleum products. Pitches are used as base material for the manufacture of certain high-modulus carbon fibers.

Pitot Tube

An instrument that measures the stagnation pressure of a flowing fluid, consisting of an open tube pointing into the fluid and connected to a pressure-indicating device. Also known as *impact tube*.

Pitting

(1) Forming small sharp cavities in a surface by corrosion, wear, or other mechanically assisted degradation. (2) *Localized corrosion* of a metal surface, confined to a point of a small area, which takes the form of cavities. (3) Loss of metal, usually by corrosion process, which penetrates to a considerable depth relative to the surface extent. Also a form of surface fatigue causing damage to gear teeth and similar sliding surfaces.

Pitting Factor

Ratio of the depth of the deepest pit resulting from corrosion divided by the average penetration calculated from weight loss.

Pivot Bearing

An axial-load, radial-load type bearing that supports the end of a rotating shaft or pivot.

Pivoted-Pad Bearing

See *tilting-pad bearing*.

Pixel

Shortened term for picture element. A pixel is the smallest display-able element of a digitized image; a single value of the image matrix. A pixel corresponds to the measurement of a volume element (*voxel*) in the object.

Plain Bearing

Any simple sliding-type bearing, as distinct from pad- or rolling-type bearings. See also *oil wedge*.

Plain Journal Bearing

A plain bearing in which the relatively sliding surfaces are cylindri-cal and in which there is relative angular motion. One surface is usually stationary and the force acts perpendicularly to the axis of rotation.

Plain Thrust Bearing

A plain bearing of the axial-load type, with or without grooves.

Plain Weave

A weaving pattern in which the warp and fill fibers alternate; that is, the repeat pattern is warp/fill/warp/fill, and so on. The two sides, or faces, of a plain weave are identical. Properties are signifi-cantly reduced relative to a weaving pattern with fewer crossovers. See *basket weave* (composites), *crowfoot weave*, and *leno weave*.

Plain-Carbon Steel

Steel containing sufficient carbon to allow it to be hardened by rapid cooling but not containing a significant amount of any other alloying element. See *steel*.

Planar

Lying essentially in a single plane.

Planar Anisotropy

A variation in physical and/or mechanical properties with respect to direction within the plane of material in sheet form. See also *plastic strain ratio*.

Planar Helix Winding

In *filament winding* of composites, a winding in which the filament path on each dome lies on a plane that intersects the dome, while a helical path over the cylindrical section is connected to the dome paths. See *helical winding*.

Planar Winding

In *filament winding* of composites, a winding in which the filament path lies on a plane that intersects the winding surface. See also *polar winding*.

Planchet

A metal disk with milled edges, ready for coining.

Planck's Constant (*h*)

A fundamental physical constant, the elementary quantum of action; the ratio of the energy of a photon to its frequency, which is equal to $6.62620 \pm 0.00005 \times 10^{-34}$ J·s.

Plane (Crystal)

An idiomorphic face of a crystal.

Plane Glass Illuminator

A thin, transparent flat glass disk interposed in a microscope or a lens imaging system to direct light to the object without reducing the useful aperture of the lens system.

Plane Grating

In materials characterization, an optical component used to disperse light into its component wavelengths by diffraction off a series of finely spaced equidistant ridges. A plane grating has a flat substrate. See also *concave grating*, *diffraction grating*, *reflection grating*, and *transmission grating*.

Plane Strain

The stress condition in *linear elastic fracture mechanics* in which there is zero strain in a direction normal to both the axis of applied tensile stress and the direction of crack growth (i.e., parallel to the crack front), most nearly achieved in loading thick plates along a direction parallel to the plate surface. Under plane-strain conditions, the plane of fracture instability is normal to the axis of the principal tensile stress. See *fracture mechanics*.

Plane Stress

The stress condition in *linear elastic fracture mechanics* in which the stress in the thickness direction is zero, most nearly achieved in loading very thin sheet along a direction parallel to the surface of the sheet. Under plane-stress conditions, the plane of fracture insta-bility is inclined 45° to the axis of the principal tensile stress.

Plane-Strain Fracture Toughness (K_{Ic})

The crack-extension resistance under conditions of *crack-tip plane strain*. See also *stress-intensity factor*.

Plane-Stress Fracture Toughness (K_c)

In *linear elastic fracture mechanics*, the value of the crack-extension resistance at the instability condition determined from the tangency between the *R-curve* and the critical crack-extension force curve of the specimen. See also *stress-intensity factor*.

Planimetric Method

A method of measuring grain size in which the grains within a defi-nite area are counted. See also *Jeffries' method*.

Planing

Producing flat surfaces by linear reciprocal motion of work and the table to which it is attached, relative to a stationary single-point cutting tool.

Planishing

Producing a smooth finish on metal by a rapid succession of blows delivered by highly polished dies or by a hammer designed for the purpose or by rolling in a planishing mill.

Planishing Hammer

A hammer having a working face (or two) that is square and slightly convex.

Plasma

(1) A gas of sufficient energy so that a large fraction of the species present is ionized and thus conducts electricity. Plasmas may be generated by the passage of a current between electrodes, by induction, or by a combination of these methods. (2) The condition when, at very high temperature, the atoms in a gas become highly ionized, that is, partially disassociated releasing some electrons, sometimes termed a fourth state of matter (with solid, liquid, and gas). A band in which the plasma carries the current in an electric arc. See also *plasma arc welding*.

Plasma Arc Coatings

In this process, a flow of gas, such as argon, is directed through the nozzle of a device called the plasma arc torch. When a high-current electric arc is struck within the torch between a negative tungsten electrode and the positive water-cooled copper nozzle, electrical and aerodynamical effects force the arc through the nozzle, which concentrates and stabilizes it. A substantial portion of the gas flows through the arc and is heated to temperatures as high as 16,649°C and accelerated to supersonic speeds to form an ionized gas jet called plasma. A cool layer of gas next to the nozzle wall effectively insulates the torch from the tremendous heating effect of the arc column.

Particles of refractory coating material, introduced into the plasma in either powder or wire form, are melted and accelerated to high velocity. When these molten particles strike the workpiece, they impact to form a dense, high-purity coating. Spraying of cold carbon dioxide gas, played on the workpiece, keeps it from overheating during the process and protects the purity of the coating from air oxidation.

Characteristics

The primary advantage of the process is its ability to combine the bulk properties of a base material with the surface properties of a refractory material. Furthermore, the application of the thin, tenacious coatings can be limited to the specific areas of the base material where a coding is needed, and warpage or distortion of precision parts is eliminated because of the low temperature of the base material maintained during coating.

Whether as-coated or finished, the refractory coatings have extremely good resistance to wear, abrasion, and corrosion and erosion, even under the adverse conditions of high temperature, high load, and lack of lubrication and cooling. When ground and lapped, the coatings give superior performance under conditions of fretting corrosion. Finished coating, when mated with proper materials, have generally lower coefficients of friction than most metal-to-metal combinations. This ratio is also true at elevated temperatures.

The coatings have a porosity of less than 1% and an as-coated surface finish of approximately 150 µin. rms (root mean square), which can be finished down to better than 1 µin. rms.

Fabrication

Coatings can be applied in practically any desired thickness. But only the areas that allow the particles free access will be coated evenly. This limitation excludes narrow holes, blind cavities, and deep V-shaped grooves. All corners and edges should be rounded by a minimum 0.38 mm rad or have a minimum chamfer of 0.38 mm by 45° to prevent weak spots.

Several types of parts that can be coated are

1. Long external cylindrical parts
2. Short external cylindrical parts
3. Internal diameters
4. Rectangular flat surfaces
5. Circular flat surfaces

Since plasma arc coatings can be deposited in practically any desired thickness, it is also possible to fabricate parts by this method. The required thickness is built up on a mandrel formed to the desired internal shape of the finished part, and the mandrel is then removed chemically from the part with acid or caustic.

This method allows intricate shapes to be made of materials that are normally difficult to fabricate. But, as with flame spraying, only areas that allow the particles of coating material sufficient access will be plated evenly.

Materials

Almost any base material can be coated, even certain reinforced plastics, and any known inorganic solid that will melt without decomposition can be used as a coating material. Many basic coatings have already been established including tantalum, palladium, platinum, molybdenum, tungsten, alumina, zirconium diboride, and oxide and the three combinations of tungsten with additives to improve its properties. These additives are zirconia, chromium, and alumina.

Other coatings include the refractory metals such as columbium; some of the refractory metal compounds such as the borides of tungsten, columbium, tantalum, titanium, and chromium; the refractory carbides of columbium, hafnium, tantalum, zirconium, titanium, tungsten, and vanadium; the refractory oxides of thorium, hafnium, magnesium, cerium, and aluminum; and other pure metals such as aluminum, copper, nickel, chromium, and boron.

Properties

The properties of the coatings or parts made from all of these materials are equivalent to those of the pure materials themselves.

Plasma Arc Cutting

An arc cutting process that severs metals by melting a localized area with heat from a constricted arc and removing the molten metal with a high-velocity jet of hot, ionized gas issuing from the plasma torch. See *cutting torch* (plasma arc).

Plasma Arc Welding

(1) An arc welding process that produces coalescence of metals by heating them with a constricted arc between an electrode and the

workpiece (transferred arc) and the electrode and the constricting nozzle (nontransferred arc). Shielding is obtained from hot, ionized gas issuing from an orifice surrounding the electrode and may be supplemented by an auxiliary source of shielding gas, which may be an inert gas or a mixture of gases. Pressure may or may not be used, and filler metal may or may not be supplied. (2) Somewhat similar to tungsten inert gas (TIG) welding in that an arc is struck from a nonconsumed tungsten electrode carried in the torch that also supplies inert gas to protect the weld zone. However, in the plasma processes, the electrode is surrounded by a water-cooled nozzle that constricts the arc increasing the arc current density. This increases the temperature of the portion of the gas passing through the arc giving a "flame" temperature to some 15,000°C. This very high temperature compared with the normal TIG "flame" allows much deeper penetration, and hence, thicker plate can be welded in a single pass. If the constricted arc is struck between the electrode and the constricting nozzle, it is termed a nontransferred arc; if between the electrode and workpiece, it is termed a transferred arc. See also *nontransferred arc*, *transferred arc*, and *welding torch* (arc).

Plasma Carburizing

Same as *ion carburizing*.

Plasma Flame (Welding)

The zone of intense heat and light emanating from the orifice of the arc chamber resulting from energy liberated as the charged gas particles (ions) recombine with electrons. See *plasma arc welding*.

Plasma-Forming Gas

The gas, in the plasma torch, which is heated to the high-temperature plasma state by the electric arc. See also *welding torch* (arc).

Plasma Ion Deposition

See *ion implantation*.

Plasma Metallizing

See preferred term *plasma spraying*.

Plasma Nitriding

Same as *ion nitriding*.

Plasma Spraying

A *thermal spraying* process in which the coating material is melted with heat from a plasma torch that generates a nontransferred arc; molten powder coating material is propelled against the base metal by the hot, ionized gas issuing from the torch.

Plasma-Jet Excitation

In materials characterization, the use of a high-temperature plasma jet to excite an element in a sample, for example, for inductively coupled plasma atomic emission spectroscopy. Also known as dc plasma excitation.

Plasmon

A quantum of a collective longitudinal wave in the electron gas of a solid.

Plaster Mold Castings

Plaster mold casting is primarily used for producing parts in quantities that are too small to justify the use of permanent molds, yet large enough to outweigh the machining costs of sand castings. The process is noted for its ability to produce parts with high dimensional accuracy, smooth and intricate surfaces, and low porosity. On the other hand, it is limited to nonferrous metals (aluminum and copper alloys) and relatively small parts. Also, production times are relatively high because the molds take relatively long to make and are not reusable.

Process

The plaster used for molding generally consists of water mixtures of gypsum or plaster of Paris (calcium sulfate) and strengthening binders such as asbestos, magnesium silicate, and silicate flour. Impurities such as salts in section thickness are about 40–60 mils, and bosses and undercuts can be incorporated into the design.

Applications

Plaster mold castings are usually used for medium-production applications, and their cost falls between sand castings and permanent mold castings. Typical parts where the process has been used include gears, ratchet teeth, cams, handles, small housings, pistons, wing nuts, locks, valves, hand tools, and radar parts for aircraft, railroad, household, and electrical uses.

Plaster of Paris

The material (calcined gypsum), $CaSO_4 \cdot O \cdot 5H_2Os$, is a white, gray, or pinkish-colored powder prepared by heating gypsum ($CaSO_4 \cdot 2H_2O$) to remove 75% of its water of crystallization.

When mixed with water and allowed to rehydrate to the dihydrate ($CaSO_4 \cdot 2H_2O$), there is no apparent action at first, but soon a slight stiffening takes place, and shortly after that, it "sets" to a solid mass. As set progresses, the mass begins to heat and expand, and final set is not reached until the evolution of heat has ceased and expansion is complete.

Through changes in the manufacturing process, the time of set can be varied widely (from a few minutes to many hours), and linear setting expansion also is controllable from 0.05% to 2.0%. The normal linear setting expansion of pottery plasters is −0.20% in all directions if the cast is unconfined, but under conditions of confinement, all the setting expansion may take place in one direction only.

Applications

Plasters are used in a variety of ceramic industry applications:

1. In a limited way, as chemical additives to glazes, supplying neutral, slightly soluble calcium and sulfate sulfur.
2. As a glass batching material to replace part or all of the salt cake when combined with soda ash in proper proportions. Here, the use of plaster eliminates saltwater scumming retaining the desirable fluxing property of salt cake.

3. As a bedding and leveling agent in grinding and polishing plate glass, plasters cement the glass to the grinding bed during the operation while also being easy to remove from the glass surface.
4. Optical glass mounting. Used to retain optical glass, lenses, prisms, and oculars in position while surfaces are formed to the desired curves by grinding and polishing.
5. Model making. Used in the ceramic industry generally for preparing original models.
6. Metal mold making. When suitably compounded with refractory substances, molds for the casting of nonferrous alloys such as white metal, brass, and aluminum alloys are made with plaster.
7. Low-density insulation. Used to provide green strength to mixtures of clays, nonplastic refractories, and organics.
8. Potter mold and die making. This use constitutes the principal ceramic application of plasters.

Plastic

A material that contains as an essential ingredient an organic polymer of large molecular weight, is solid in its finished state, and, at some stage in its manufacture or its processing into finished articles, can be shaped by flow. Although materials such as rubber, textiles, adhesives, and paint may in some cases meet this definition, they are not considered plastics. The terms plastic, resin, and polymer are somewhat synonymous, but the terms resins and polymers most often to note the basic material as polymerized, while the term plastic encompasses compounds containing plasticizers, stabilizers, fillers, and other additives.

Plastic Alloys and Blends

Plastics, like metals, can be alloyed. And like metal alloys, the resulting materials have different, and often better, properties than those of the base materials making up the alloys.

These alloys consist of two thermoplastics compounded into a single resin. The two polymers must be melt compatible. Some polymers are naturally compatible; others require the use of compatibilizing agents. The purpose of alloying polymers is to achieve a combination of properties not available in any single resin. There are a great many alloys available, and the list continues to grow. At present, some of the more widely used alloys are acrylonitrile, butadiene, and styrene (ABS)/polycarbonate, ABS/polyurethane, polyvinyl chloride (PVC)/acrylic, PVC/chlorinated polyethylene (CPE), polyphenylene oxide (PPO)/polystyrene, nylon/ABS, PBO/polybutylene terephthalate (PBT) thermoplastic polyester, polycarbonate/PBT thermoplastic polyester, polycarbonate/acrylate–styrene–acrylonitrile, and polysulfone/ABS.

The plastics most widely used in alloys today are PVC, ABS, and polycarbonate. The street plastics can be combined with each other or with other types of polymers.

ABS

ABS, in addition to its use with polycarbonate, can also be alloyed with polyurethane. ABS–polycarbonate alloys extend the exceptionally high impact strength of carbonate plastics to section thicknesses over 0.16 cm. ABS–polyurethane alloys combine the excellent abrasion resistance and toughness of the urethanes with the lower cost and rigidity of ABS.

The materials can be injection molded into large parts but cannot be extruded. Typical applications for which they are suitable include such parts as wheel treads, pulleys, low load gears, gaskets, automotive grilles, and bumper assemblies.

ABS is also being successfully combined with PVC and is available commercially in several grades. One of the established grades provide self-extinguishing properties, thus eliminating the need for intumescent (nonburning) coatings in present ABS applications, such as power tool housings, where self-extinguishing materials are required. A second grade possesses an impact strength about 30% higher than general-purpose ABS. This improvement, plus its ability to be readily molded, has resulted in its use for automobile grilles.

PVC

ABS–PVC alloys are available commercially in several grades. Two of the established grades were described earlier. ABS–PVC alloys also can be produced in sheet form. The sheet materials have improved hot strength, which allows deeper draws than are possible with standard rubber-modified PVC base sheet. They also are non-fogging when exposed to the heat of sunlight. Some properties of ABS–PVC alloys are lower than those of the base resins. Rigidity, in general, is somewhat lower, and tensile strength is more or less dependent on the type and amount of ABS in the alloy.

Another sheet material, an alloy of about 80% PVC and the rest acrylic plastic, combines the nonburning properties, chemical resistance, and toughness of vinyl plastics with the rigidity and deep drawing merits of the acrylics. The PVC–acrylic alloy approaches some metals in its ability to withstand repeated blows. Because of its unusually high rigidity, sheets ranging in thickness from 1.5 to 0.5 cm can be formed into thin-walled, deeply drawn parts.

PVC is also alloyed with CPE to gain materials with improved outdoor weathering or to obtain better low-temperature flexibility. The PVC–CPE alloy applications include wire and cable jacketing, extruded and molded shapes, and film sheeting. Acrylic-base alloys with a polybutadiene additive have also been developed, chiefly for blow-molded products. The acrylic content can range from 50% to 95%, depending on the application. Besides blow-molded bottles, the alloys are suitable for thermoformed products such as tubs, trays, and blister pods. The material is rigid and tough and has good heat distortion resistance up to 82°C.

PPO

Another group of plastics, PPO, can be blended with polystyrene to produce a PPO–polystyrene alloyed with improved processing traits and lower cost than nonalloyed PPO. The addition of polystyrene reduces tensile strength and heat-deflection temperature and increases thermal expansion.

Plastic Bronze

A name once applied by makers of bearing bronzes to copper alloys that are sufficiently pliable to assume the shape of the shaft and make a good bearing by running in. These bronzes have a variety of compositions, but the plasticity is always obtained by the addition of lead, which in turn weakens the bearing. In some cases, the lead content is so high, and the tin content is so low, and the alloy is not a bronze. These copper–lead alloys have been referred to as *red metals*. The plastic bronze ingot marketed by one large foundry for journal bearings contain 65%–75% copper, 5%–7% tin, and the balance lead. *Semiplastic bronze* usually contains above 75% copper and not more than 15% lead. *ASTM alloy No. 7* has about 10% lead, 10% tin, 1% zinc, 1% antimony, and 78% copper. The compressive strength is 12,500 lb/in.[2] (85 MPa).

Plastic Constraint

See *constraint*.

Plastic Deformation

(1) The permanent (inelastic) distortion of materials under applied stresses that strain the material beyond its *elastic limit*. (2) The permanent change in dimensions remaining after all stresses have been removed. See *tensile test*.

Plastic Flow (Metals)

The phenomenon that takes place when metals are stretched or compressed permanently without rupture.

Plastic Flow (Plastics)

(1) Deformation of plastics under the action of a sustained hot or cold force. (2) Flow of semisolids in the molding of plastics.

Plastic Foam

See preferred term *cellular plastic*.

Plastic Instability

The stage of deformation in a tensile test where the plastic flow becomes nonuniform and *necking* begins. See *tensile test*.

Plastic Laminates

These are resin-impregnated paper or fabric, produced under heat in high pressure; they are also referred to as high-pressure plastic laminates. Two major categories are decorative thermosetting laminates and industrial thermosetting laminates. Most of the decorative thermosetting laminates are paper base and are known generically as papreg. Decorative laminates are usually composed of a combination of phenolic- and melamine-impregnated sheets of paper. The final properties of the laminate are related directly to the properties of the paper from which the laminate is made.

Early laminates were designed by trade names, such as Bakelite, Textolite, Micarta, Condensite, Dilecto, Phenolite, Haveg, Spauldite, Synthane, and Formica. These are designated as various types of laminates with a decorative facing layer for such uses as tabletops. Trade names now usually include a number or symbol to describe the type and grade. Textolite, for example, embraces more than 70 categories of laminates subdivided into use-specification grades, all produced in many sizes and thicknesses. Textolite 11711 is an electronic laminate for such uses as multilayer circuit boards. It is made with polyphenylene oxide resin and may have a copper or aluminum cladding.

Forms

Industrial thermosetting laminates are available in the form of sheet, rod, and rolled or molded tubing. Impregnating resins commonly used are phenolic, polyester, melamine, epoxy, and silicone. The base material or reinforcement is usually one of the following: paper, woven cotton or linen, asbestos, glass cloth, or glass mat.

Laminating resins may be marked under one trade name by the resin producer and other names by the molders of the laminate. Paraplex P resins, for example, comprise a series of polyester solutions in monomeric styrene that can be blended with other resins to give varied qualities. But panelyte refers to the laminates that are made with phenolic, melamine, silicone, or other resin, for a variety of applications.

Plastic Memory

The tendency of a thermoplastic material that has been stretched while hot to return to its unstretched shape upon being reheated.

Plastic Powder Coatings

Although many different plastic powders can be applied as coatings, vinyl, epoxy, and nylon are most often used. Vinyl and epoxy provide good corrosion and weather resistance as well as good electrical insulation. Nylon is used chiefly for its outstanding wear and abrasion resistance. Other plastics frequently used in powder coating include chlorinated polyethers, polycarbonates, acetals, cellulosics, acrylics, and fluorocarbons.

Several different methods have been developed to apply these coatings. In the most popular process, fluidized bed, parts are preheated and then immersed in a tank of finely divided plastic powders, which are held in a suspended state by a rising current of air. When the powder particles contact the heated part, they fuse and adhere to the surface, forming a continuous, uniform coating. Another process, electrostatic spraying, works on the principle that oppositely charged particles attract each other. Powder is fed through a gun, which applies an electrostatic charge opposite to that applied to the part to be coated. When the charged particles leave the gun, they are attracted to the part where they cling until fused together as a plastic coating. Other powder application methods include flock and flow coating, flame and plasma spraying, and a cloud chamber technique.

Plastic Replica

In fractography and metallography, a reproduction in plastic of the surface to be studied. It is prepared by evaporation of the solvent from a solution of plastic, polymerization of a monomer, or solidification of a plastic on the surface. See *replica*.

Plastics

Plastics are a major group of materials that are primarily noncrystalline hydrocarbon substances composed of large molecular chains whose major element is carbon. The three terms—*plastics*, *polymers*, and *resins*—are sometimes used interchangeably to identify these materials. However, the term *plastics* has now come to be the commonly used designation.

The first commercial plastic, celluloid, was developed in 1868 to replace ivory for billiard balls. Phenolic plastics, developed by Baekeland and named bakelite after him, were introduced around the turn of the century. A plastic material, as defined by the Society of the Plastics Industry, is "any one of a large group of materials consisting wholly or in part of combinations of carbon with oxygen, hydrogen, nitrogen, and other organic and inorganic elements which, while solid in the finished state, at some stage in its manufacture is made liquid, and thus capable of being formed into various shapes, most usually through the application, either singly or together, of heat and pressure."

There are two basic types of plastics based on intermolecular bonding: thermoplastics, because of little or no cross-bonding between molecules, soften when heated and harden when cooled, no matter how often the process is repeated, and thermosets, on the other hand, having strong, intermolecular bonding. Therefore, once the plastic is set into permanent shape under heat and pressure, reheating will not soften it.

Within these major classes, plastics are commonly classified on the basis of base monomers. There are over two dozen such monomer families or groups. Plastics are also sometimes classified roughly into three stiffness categories: rigid, flexible, and elastic. Another method of classification is by the "level" of performance or the general area of application, using such categories as engineering, general purpose, and specialty plastics, or the two broad categories of engineering and commodity plastics.

The following are some major characteristics of plastics that distinguish them from other materials, particularly metals:

1. They are essentially noncrystalline in structure.
2. They are nonconductors of electricity and are relatively low in heat conductance.
3. They are, with some important exceptions, resistant to chemical and corrosive environments.
4. They have relatively low softening temperatures.
5. They are readily formed into complex shapes.
6. They exhibit viscoelastic behavior—that is, after an applied load is removed, plastics tend to continue to exhibit strain or deformation with time.

Polymers can be built of one, two, or even three different monomers and are termed homopolymers, copolymers, and terpolymers, respectively. The geometrical form can be linear or branched. Linear or unbranched polymers are composed of monomers linked end-to-end to form a molecular chain that is like a simple string of beads or a piece of spaghetti. Branched polymers have side chains of molecules attached to the main linear polymer. These branches can be composed either of the basic linear monomer or of a different one. If the side molecules are arranged randomly, the polymer is atactic; if they branch out on one side of the linear chain in the same plane, the polymer is isotactic, and if they alternate from one side to the other, the polymer is syndiotactic.

Plastics are produced in a variety of different forms. Most common are plastic moldings, which range in size from 2 cm to several meters. Thermoplastics, such as polyvinyl chloride (PVC) and polyethylene, are widely used in the form of plastic film and plastic sheeting. The term *film* is used for thicknesses up to and including 0.25 cm, while sheeting refers to thicknesses over that.

Both thermosetting and thermoplastic materials are used as plastic coatings on metal, wood, paper, fabric, leather, glass, concrete, ceramics, or other plastics. There are many coating processes, including knife or spread coating, spraying, roller coating, dipping, brushing, calendering, and the fluidized-bed process. Thermosetting plastics are used in high-pressure laminates to hold together the reinforcing materials that comprise the body of the finished product. The reinforcing materials may be cloth, paper, wood, or glass fibers. The end product may be plain flat sheets or decorative sheets as in countertops, rods, tubes, or formed shapes.

Plastic Additives

Almost all plastics contain one or more additive materials to improve their physical properties and processing characteristics or to reduce costs. There is a wide range of additives for use with plastics, including antimicrobials, antistatic agents, clarifiers, colorants, fillers, flame retardants, foaming agents, heat stabilizers, impact modifiers, light stabilizers, lubricants, mold release agents, odorants, plasticizers, reinforcements, and smoke retardants.

Fillers

Fillers are probably the most common of the additives. They are usually used to either provide bulk or modify certain properties. Generally, they are inert and thus do not react chemically with the resin during processing. The fillers are often cheap and serve to reduce costs by increasing bulk. For example, wood flour, a common low-cost filler, sometimes makes up 50% of a plastic compound. Other typical fillers are chopped fabrics, asbestos, talc, gypsum, and milled glass. Besides lowering costs, fillers can improve properties. For example, asbestos increases heat resistance, and cotton fibers improve toughness.

Plasticizers

Plasticizers are added to plastic compounds either to improve flow during processing by reducing the glass transition temperature or to improve properties such as flexibility. Plasticizers are usually liquids that have high boiling points, such as certain phthalates. Substances that are themselves polymers of low molecular weight, such as polyesters, are also used as plasticizers.

Stabilizers

Stabilizers are added to plastics to help prevent breakdown or deterioration during molding or when the polymers are exposed to sunlight, heat, oxygen, ozone, or combinations of these. Thus there is a wide range of compounds, each designated for a specific function. Stabilizers can be metal compounds, based on tin, lead, cadmium, barium, and others. And phenols and amines are added antioxidants that protect the plastic by diverting the oxidation reactions to themselves.

Catalysts

Catalysts, by controlling the rate and extent of the polymerization process in the resin, allow the curing cycle to be tailored to the processing requirements of the application. Catalysts also affect the shelf life of the plastics. Both metallic and organic chemical compounds are used as catalysts.

Colorants

Colorants, added to plastics for decorative purposes, come in a wide variety of pigments and dyestuffs. The traditional colorants are metal-based pigments such as cadmium, lead, and selenium. More recently, liquid colorants, composed of dispersions of pigments in a liquid, have been developed.

Flame Retardants

Flame retardants are added to plastic products that must meet fire-retardant requirements, because polymer resins are generally flammable, except for such notable exceptions as PVC. In general, the function of the fire retardants is limited to the spread of fire. They do not normally increase heat resistance or prevent the plastic from charring or melting. Some fire-retardant additives include compounds containing chlorine or bromine, phosphate–ester compounds, antimony trioxide, alumina trihydrate, and zinc borate.

Reinforced Materials

Reinforcement materials in plastics are not normally considered additives. Usually in fiber or mat form, they are used primarily to improve mechanical properties, particularly strength. Although asbestos and some other materials are used, glass fibers are the predominant reinforcement for plastics.

Plastics Processing

Plastics processing includes those methods and techniques used to convert plastic materials in the form of pellets, granules, powders, sheets, fluids, or preforms into formed shapes or parts. The plastic materials may contain a variety of additives that influence the properties as well as the process ability of the plastics. After forming, the part may be subjected to a variety of ancillary operations such as welding, adhesive bonding, and surface decorating (painting, metallizing).

As with other materials of construction, processing of plastics is but one step in the normal design-to-finished-part sequence. The choice of process is influenced by economic considerations, number and size of finished parts, and complexity of postfinishing operations, as well as the adaptability of the plastics to the process.

Injection Molding

This process consists of heating and homogenizing plastic granules in a cylinder until they are in the fluid form sufficient to allow pressure injection into a relatively cold mold where they solidify and take the shape of the mold cavity. For thermoplastics, no chemical changes occur within the plastic, and consequently, the process is repeatable. Injection molding of thermosetting resins differs primarily in that the cylinder heating is designed to homogenize and preheat the reactive materials and the mold is heated to complete the chemical cross-linking reaction to form an intractable solid. Solid particles, in the form of pellets or granules, constitute the main feed for injection moldable plastics. The major advantages of the injection molding process are the speed of production, minimal requirements for postmolding operations, and simultaneous multipart molding.

The development of reaction injection molding (RIM) allowed the rapid molding of liquid materials. In these processes, cold or warm, two highly reactive, low-molecular-weight, low-viscosity resin systems are first injected into a mixing head and from there into a heated mold, where the reaction to a solid is completed.

Polymerization and cross-linking occur in the mold. This process has proved particularly effective for high-speed molding of such materials as polyurethanes, epoxies, polyesters, and nylons.

Extrusion

In this process, plastic pellets or granules are fluidized, homogenized, and continuously formed. Products made this way include tubing, pipe, sheet, wire and substrate coatings, and profile shapes. The process is used to form very long shapes or a large number of small shapes that can be cut from the long shapes. The homogenizing capability of extruders is used for plastic blending and compounding. Pellets used for other processing methods, such as injection molding, are made by chopping long filaments of extruded plastic.

Blow Molding

This process consists of forming a tube (called a parison) and introducing air or other gas to cause the tube to expand into a free-blown hollow object or against a mold for forming into a hollow object with a definite size and shape. The parison is traditionally made by extrusion, although injection-molded tubes have gained prominence because they do not require postfinishing, have better dimensional

tolerances and wall thicknesses, and can be made unsymmetrical and in a higher volume production.

Thermoforming

Thermoforming is the forming of plastic sheets into parts through the application of heat and pressure. Pressure can be obtained through the use of pneumatics (air) or compression (tooling) or vacuum. Tooling for this process is the most inexpensive compared to other plastic processes, accounting for the popularity of the method. It can also accommodate very large parts as well as small parts, which are useful in low-cost prototype fabrication.

Rotational Molding

In this process, finely ground powders are heated in a rotating mold until melting or fusion occurs. If liquid materials, such as vinyl plastisols, are used, the process is often called slush molding. The melted or fused resin uniformly coats the inner surface of the mold. When cooled, a hollow finished part is removed. The processes require relatively inexpensive tooling, are scrap-free, and are adaptable to large, double-walled, hollow parts that are strain-free and of uniform thickness. The processes can be performed by relatively unskilled labor. On the other hand, the finely ground plastic powders are more expensive than pellets, or sheet, thin-walled parts cannot be easily made, and the process is not suited for large production runs of small parts.

Compression and Transfer Molding

Compression molding is one of the oldest molding techniques and consists of charging a plastic powder or preformed plug into a mold cavity, closing a mating mold half, and applying pressure to compress, heat, and cause flow of the plastic to conform to the cavity shape. The process is primarily used for thermosets, and consequently, the mold is heated to accelerate the chemical cross-linking.

Transfer molding is an adaptation of compression molding in that the molding powder or preform is charged to a separate preheating chamber and, when appropriately fluidized, injected into a closed mold. The process predates, yet closely parallels, the early techniques of ram injection molding of thermoplastics. It is mostly used for thermosets and is somewhat faster than compression molding. In addition, parts are more uniform and more dimensionally accurate than those made by compression molding.

Foam Processes

Foamed plastic materials have achieved a high degree of importance in the plastic industry. Foams can be made in a range from soft and flexible to hard and rigid. There are three types of cellular plastics: blown (expanded matrix, such as a natural sponge), syntactic (the encapsulation of hollow organic or inorganic microspheres in the matrix), and structural (dense outer skin surrounding a foamed core).

There are seven basic processes used to generate plastic foams. They include the incorporation of a chemical blowing agent that generates gas (through thermal decomposition) in the polymer liquid or melt; gas injection into the melt, which expands during pressure relief; generation of gas as a by-product of a chemical condensation reaction during cross-linking; volatilization of a low-boiling liquid (e.g., Freon) through the exothermic heat of reaction; mechanical dispersion of air by mechanical means (whipped cream); incorporation of nonchemical gas-liberating agents (adsorbed gas or finely divided carbon) into the resin mix, which is released by heating; and expansion of small beads of thermoplastic resin containing a blowing agent through the external application of heat.

Structural foam differs from other foams in that the part is produced with a hard integral skin on the outer surfaces in a cellular core in the interior. They are made by injection-molding liquefied resins containing chemical blowing agents. The initial high injection pressure causes the skin to solidify against the mold surface without undergoing expansion. The subsequent reduction in pressure allows the remaining material to expand and fill the mold. Coinjection (sandwich) molding permits injection molding of parts containing a thermoplastic core within an integral skin of another thermoplastic material. When the core is foam, an advanced form of structural foam is produced.

Reinforced Plastics/Composites

These are plastics whose mechanical properties are significantly improved because of the inclusion of fibrous reinforcements. The wide variety of resins and reinforcements that constitute this group of materials led to the more generalized description "composites."

Composites consist of two main components, the fibrous material in various physical forms and the fluidized resin, which will convert to a solid. There are fiber-reinforced thermoplastic materials, and

these are typically processed in standard thermoplastic processing equipment.

The first step in any composite fabrication procedure is the impregnation of the reinforcement with the resin. The simplest method is to pass the reinforcement through a resin bath and use the wet impregnate directly. For easier handling and storage, the impregnated reinforcement can be subjected to heat to remove impregnating solvents or advance the resin cure to a slightly tacky or dry state. The composite in this form is called a prepreg. This B-stage condition allows the composite to be handled, yet the cross-linking reaction has not proceeded so far as to preclude final flow and conversion into a simultaneous part, when further heat or pressure is applied.

Premixes, often called bulk molding compounds, are mixtures of resin, inert fillers, reinforcements, and other formulation additives that form a puttylike rope, sheet, or preformed shape.

Converting these various forms of composite precursors to final-part shape is achieved in a number of ways. Hand lay-up techniques entail an open mold onto which the impregnated reinforcement or prepreg is applied layer by layer until the desired thicknesses and contours are achieved; see Figure P.1a depicting techniques for producing reinforced plastics and composites. The thermoset resin is then allowed to harden (cure). Often the entire configuration will be enclosed in a transparent sealed bag (vacuum bag) so that

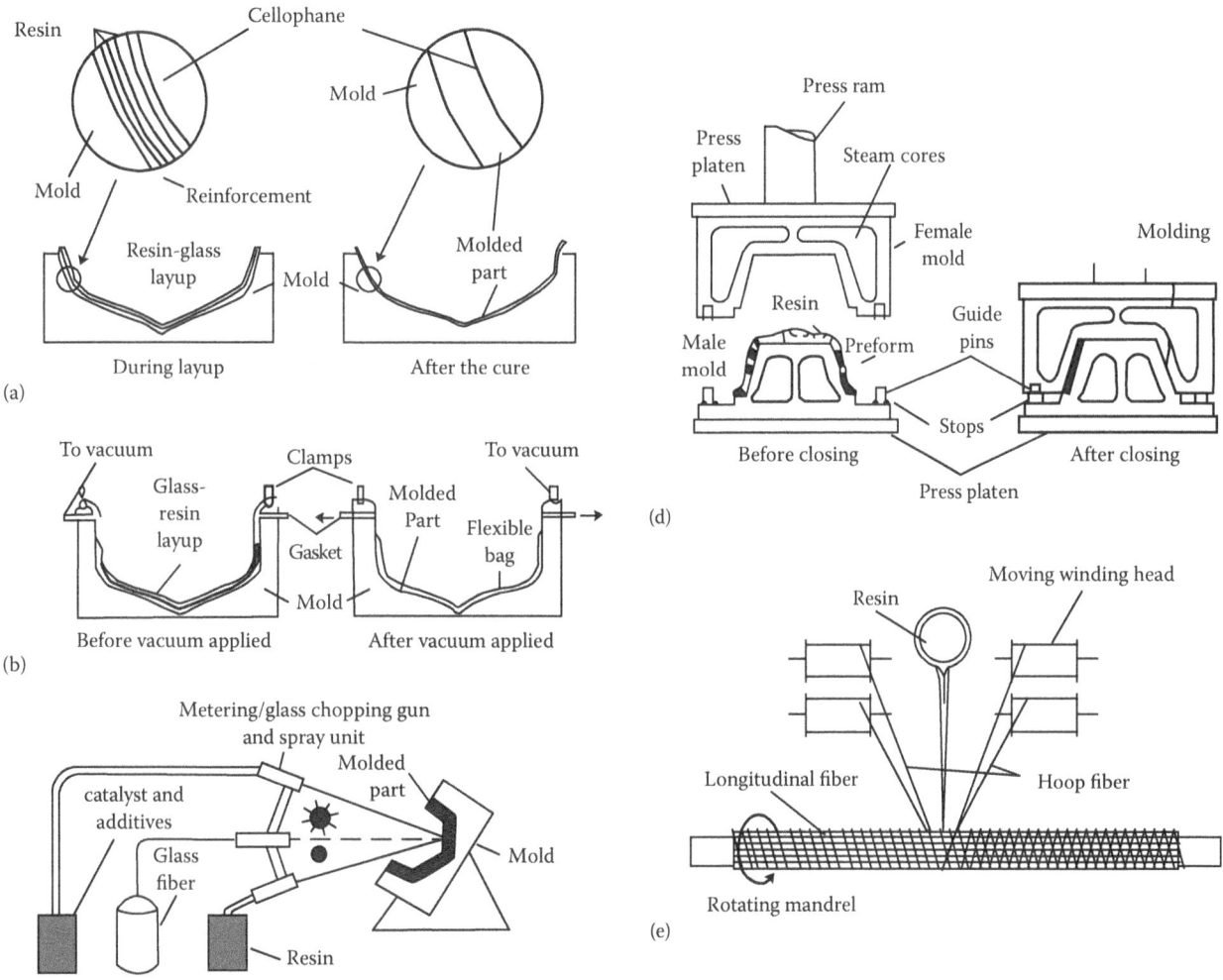

FIGURE P.1 Techniques for producing reinforced plastics and composites. (a) Hand lay-up technique for reinforced thermosets, (b) vacuum bag molding method, (c) spray-up method, (d) matched metal die molding, (e) filament winding. (From *McGraw-Hill Encyclopedia of Science and Technology*, 8th edn., Vol. 14, McGraw-Hill, New York, p. 45. With permission.)

a vacuum can be applied to remove unwarranted volatile ingredients and entrained air for improved densification of the composite (Figure P.1b). External heat may be applied to accelerate the process. Often a bagged laminate will be inserted into an autoclave so that the synergistic effects of heat, vacuum, and pressure can be obtained. At times, a specially designed spray apparatus is used that simultaneously mixes and applies a coating of resin and shop reinforcement to a mold surface (Figure P.1c). This technique is particularly useful for large structures such as boat hulls and truck cabs, covering complex shapes as readily as simple configurations.

Matched die compression molding resembles normal compression molding, although the pressures are considerably lower (Figure P.1d). Premix molding is essentially the same process, except that premix compounds are used. *Pultrusion* is a term coined to describe the process for the continuous extrusion of reinforced plastic profiles. Strands of reinforcement are drawn (pulled) through an impregnating tank, the forming die, and finally a curing area (radio-frequency exposure). Filament winding is a process in which the continuous strands of reinforcement are drawn through an impregnating bath and then wound around a mandrel to form the part (Figure P.1e). This technique is mostly used for the formation of hollow objects such as chemical storage tanks or chemically resistant pipe. Advanced automated processes, such as ply cutting, tape laying and contouring, and ply laminating, are providing improved parts and reduced costs particularly in the aerospace industry.

Casting and Encapsulation

Casting is a low-pressure process requiring nothing more than a container in the shape of the desired part. For thermoplastics, liquid monomer is poured into the mold and, with heat, allowed to polymerize in place to a solid mass. For vinyl plastisols, the liquid is fused with heat. Thermosets, usually composed of liquid resins with appropriate curatives and property-modifying additives, are poured into a heated mold in which the cross-linking reaction completes the conversion to a solid. Often a vacuum is applied to gasify the resultant part for improved homogeneity.

Encapsulation and potting are terms for casting processes in which a unit or assembly is encased or unimpregnated, respectively, with a liquid plastic, which is subsequently hardened by fusion or chemical reaction. There are several low-pressure plastic processes: casting, potting, encapsulating, and sealing. These processes are predominant in the electrical and electronic industries for the insulation and protection of components.

Calendering

In the calendering process, a plastic is masticated between two rolls that squeeze it out into a film that then passes around one or more additional rolls before being stripped off as a continuous film. Fabric or paper may be fed through the latter rolls, so that they become impregnated with the plastic.

Plastic Strain

Dimensional change that does not disappear when the initiating stress is removed. Usually accompanied by some *elastic deformation*.

Plasticity

The property of a material that allows it to be repeatedly deformed without rupture when acted upon by a force sufficient to cause deformation and which allows it to retain its shape after the applied force has been removed.

Plasticizer

(1) A material incorporated in a plastic to increase its workability, flexibility, or distensibility. Normally used in thermoplastics. (2) A material added to a plastic (or polymer) of lower molecular weight to reduce stiffness and brittleness, resulting in a lower glass transition temperature for the polymer. (3) A material added to an adhesive to cause a reduction in melt viscosity, lower the temperature of the second-order transition, or lower the elastic modulus of the solidified adhesive. See also *modifier*.

Plastic–Strain Ratio (*r*-Value)

In formability testing of metals, the ratio of the true width strain to the truth thickness strain in a sheet tensile test,

$$r = \varepsilon_w \varepsilon_t$$

r = plastic strain ratio
ε_w = plastic strain in-plane or in width

A formability parameter that relates to drawing and is also known as the anisotropy factor. A high *r*-value indicates a material with good drawing properties.

Plastic-Zone Adjustment (*r*$_Y$)

In *linear elastic fracture mechanics*, in addition to the *physical crack size* to account for plastic crack-tip deformation enclosed by a linear-elastic stress field. See also *crack-extension resistance* (K_R).

Plastigel

A plastisol exhibiting gel-like flow properties. One having an effective yield value.

Plastisol Coatings

Vinyl plastisols, or pastes, as they are described in Europe, are suspensions of vinyl resin in nonvolatile oily liquids known as plasticizers. They vary in viscosity from a motor oil consistency to a puttylike dough. In the more viscous state, the plastisol is termed *plastigel*, while the more fluid materials to which volatile diluents have been added are known as *modified plastisols*. Modified plastisols differ from organosols in the function of the volatile components. In organosols, the volatiles are used as resin dispersants, whereas in modified plastisols, they serve as diluents to adjust fluidity and are generally present in small quantities.

The polyvinyl chloride resin, resembling confectioner's sugar, is blended into a mixture of one or more plasticizers to form a suspension. This fluid remains essentially unchanged until heat is applied. During the heating process, the dispersion first sets or gels; this is followed by solution or fusion of the resin in the hot plasticizer to form a single-phase solid solution. Upon cooling, the coating assumes the properties of a tough, rubbery plastic.

These plastisols have no adhesion to metals or dense, nonporous substrates and consequently require the use of adhesive primers for bonding. To a large extent, the nature of these primers determines the suitability of plastisol coatings for specific applications.

Application Methods

The fluidity or absence of fluidity in the liquid plastisol is sometimes deceiving. These materials are supplied at very high solid content and consequently exhibit non-Newtonian flow (viscosity varies with applied shear). While the viscosity cup is satisfactory for many paints and lacquers and may even suffice for organosols, it can only serve to mislead the plastisol user. Viscosity of plastisols should be specified and measured using a viscometer capable of operating over a range of shear rates preferably within the area of use.

Spread Coating

Roller and knife coaters are the two major types of spread coating equipment used for handling plastisols. Fabric, paper, and even strip steel are all being coated with this type of process. Compound viscosity characteristics, speed of coating, clearance between the web and the knife or roll, and the type and angle of the knife are all factors in determining the quality of the coating. Heavy paper and fabric coatings, which will withstand folding and forming, may be applied to porous stock without danger of penetration or strike-through. The momentary application of heat to the coated side of the stock will fuse the plastisol with a minimum of thermal action on the paper.

Plastisol-coated strip steel is currently being produced by roller-coating processes for use by appliance and other manufacturers.

Dip Coating

Two dip-coating processes are available. In hot dipping, the object for coating is prebaked, prior to immersion in the plastisol. The heat content in the article serves to gel a deposit on the surface of the object. This gelled coating must then be fused by baking. Plastisol formulation and temperature, dipping rate, mass, shape, and heat content of the article to be coated all serve to determine the thickness and nature of deposit.

Cold-dip processes permit the application of from 1 to 60 mils per coat without the necessity of a prebake operation. Cold dips lend themselves to conveyor line coating of products. These coatings have a high yield value and permit controlled film thicknesses without the presence of sags or drips to mar the appearance.

Spray Coating

Plastisols may be spray applied through either pressure or suction guns, but generally the pressure equipment is preferred because it permits faster delivery with a minimal use of volatile diluents. Airless spray equipment is currently available that operates at fluid pressures in excess of 13.6 MPa and requires no atomizing air. This airless spray process is reported to give extreme smoothness to highly thixotropic formulations.

Properties

Although plastisols may be modified with slight additions of volatile diluents, their fluidity is mainly due to the presence of large quantities of plasticizer. Unlike the fluid phase of the organosol, which is largely volatile, the plasticizers remain behind after baking, as a portion of the fused film. Thus, the plastisol tends to be softer and more resilient than the organosol. Its low volatile content (or its absence altogether) permits wide baking latitude by eliminating the problem of mud cracking and reducing the solvent entrapment tendency found in organosols. Film thicknesses may range from 2 to 250 mils per coat.

Uses

Plastisols, because of their ability to be applied readily in heavy thicknesses, found early success in the electroplating field as rack coatings. The plastisol serves as an insulation, confining current to the work being plated, and is resistant to chemical attack by plating solutions. The use of plastisols as linings for tanks, chemical equipment, and steel drums.

In fabric coating, they have replaced solution coatings by eliminating the need for expensive solvents lost in the baking operation. Rubber has been replaced by vinyl plastisols and organosols as coatings for wire baskets because of their superior resistance to moisture and detergents.

One of the most dramatic applications of plastisols is as a lining for kitchen dishwashers. The use of a plastisol lining permitted a lightweight tub design not possible with porcelain enameling in which firing resulted in buckling and warping of light-gauge steel. Plastisols also served to reduce scrap units since defects may be readily patched and repaired. The resistance to impact damage, etching, and enamel erosion are other factors that prompted manufacturers to select plastisols for this application.

A list of applications on the properties related to the specific application follow:

Application	Related Property
Industrial: Tool handles, stair treads, conveyor hooks, conveyor rollers, railings	Resiliency, thermal and electrical insulating qualities, resistance to abrasion
Electrical: Bus bar conduit boxes, battery clamps and cases, toggle switches, electroplating racks, and plating barrels	Dielectric strength, electrical resistivity, resistance to moisture and chemicals
Linings: Tanks, ductwork, pumps, filter presses, centrifugal cleaners, dishwasher tubs, piping, drums, and shipping containers	Resistance to abrasion and impact, resistance to moisture and chemicals
Wire goods: Dish-drain baskets, egg baskets, deep-freeze baskets, refrigerator shelves, record racks, clothes hangers	Resistance to moisture, detergents, staining, and resiliency
Miscellaneous: Bottles and glassware, glove coating, bobby pin coatings	Resiliency and aesthetic qualities, abrasion resistance, softness

Plastisols

Mixtures of vinyl resins and plasticizers that can be molded, cast, or converted to continuous films by the application of heat. If the mixtures contain volatile thinners as well, they are known as *organosols*.

Plastisols are dispersions of high-molecular-weight vinyl chloride polymer or copolymer resins in nonaqueous liquid plasticizers, which do not dissolve the resin at room temperature. Plastisols are converted from liquids to solids by fusing under heat, which causes the resin to dissolve in the plasticizers.

There are many advantages in molding with plastisols each varying in importance according to the particular type of molding application. Vinyl plastisol is supplied as a liquid and consequently is easy to handle. The material requires no catalysts or curing agents to convert it to a solid, only moderate heat in the range of 149°C–204°C. Vinyl plastisol does not require a long baking cycle nor high pressures to fuse and shape it. Consequently, lightweight inexpensive molds are suitable for molding. It can be formulated to have a virtually indefinite shelf life. Plastisols are usually 100% solids and shrinkage from the mold is at an absolute minimum, thus assuring that the molded object is exact and consistent.

P

Properties

Chemical and physical properties of plastisols can be varied throughout a wide range. This versatility makes plastisols adaptable to a multitude of end uses.

The following general ranges indicate the properties that may be compounded into plastisols:

Specific gravity: 1.05–1.35

Tensile strength: As required to 27.2 MPa

Elongation: As required to 600%

Flexibility: Good to a temperature as low as −55°C

Hardness: From 10 to 100 on the Shore A durameter scale; up to 80 on the Shore D durameter scale

Chemical resistance: Outstanding to most acids, alkalies, detergents, oils, and solvents

Heat resistance: Can resist 107°C for as long as 2000 h and 232°C for over 2 h

Electrical properties: Dielectric strength at a minimum of 400 V/mil in thicknesses of 3 mils and over

Flammability: Slow burning to self-extinguishing

Colors: All colors available including phosphorescent and fluorescent shades

Molding Methods

There are several different methods by which plastisols may be molded.

Pour and Injection

Two of the simplest methods are pour molding and low-pressure injection molding. The first method entails merely pouring plastisol into a cavity until it is filled and subsequently fusing the compound. The system is used in manufacturing products such as plastic doilies, sink stoppers, and display plaques.

If the mold is closed, a low-pressure injection system such as a grease gun can be used to inject the liquid plastisol into the cavity. A low-pressure injection mold should be designed with bleeders at the extremities of the cavity to ensure complete filling of the mold as well as to relieve the minor pressure on the mold surface caused by expansion during the heating. Laboratory models, novelties, and electrical harnesses are products that are commonly low-pressure injection molded with vinyl plastisol.

Heating sources used for molding plastisols vary according to the particular product under consideration. For shallow, open molds, such as those for plastic doilies, radiant heat would be unsatisfactory. When this type of heat is used, the material thickness should not be so great that the open surface exposed to the heat overfuses in the time taken for the temperature to reach the mold surface of the part. Conductive heat is another source for fusing plastisol, in particular when closed molds are employed. Immersing the mold in a hot bath or using cartridge heaters are two conductive heating methods. A more commonly used method of fusion is convection heat. The advantage of a convection oven, in particular a forced air type, is that the entire inner area of the oven, and consequently the entire surface of the mold, is maintained at a constant temperature, ensuring a more even heat transfer through the mold.

Pour and low-pressure injection molds usually are made of aluminum, electroformed copper, brass, or steel. The thickness of the metal should be kept to a minimum for good heat transfer yet should be thick enough to withstand expansion pressure during fusing.

In-Place Molding

Another molding process, which is somewhat similar to the foregoing methods, is in-place molding. This method permanently attaches plastisol to another component during the fusing process. The combination serves an important functional purpose and usually eliminates the need for several assembly steps. In-place molding is most commonly used in forming seals and gaskets of various sorts.

Gaskets are applied to vitrified clay pipe by pouring plastisol into special molds on the bell and spigot ends. Such gaskets compensate for inherent out of roundness of the pipe. Flowed-in gaskets also are applied to bottle caps and jar lids.

Dip Molding

When a hollow object is to be molded and the internal dimensions are of importance, many times a dip molding process is employed. The metal molds are shaped according to the interior design of the molded object. They are usually solid and are made of cast or machined aluminum, machined brass, steel, or ceramic. These molds are preheated to a temperature in the range of 149°C–204°C, dipped into the plastisol, and allowed to dwell until the proper thickness has gelled on the mold. To eliminate drips or sags, the mold is withdrawn at a rate that does not exceed the rate at which the liquid residue drains from the gelled coating. The thickness of this coating can be varied by altering the preheating time and temperature as well as the dwell time. A dip molding system can be easily conveyorized. In such a system, the mandrels holding the molds would be conveyed through a preheat oven, dipping station, fusing oven, cooling station, and stripping station.

Slush Molding

Another method for molding hollow pieces is slush molding. In this process, an open-end metal mold is heated to a temperature in the range of 149°C–204°C and then filled with plastisol. The plastisol is allowed to dwell until the desired thickness has gelled on the inner surface of the mold and then the remaining liquid in the mold is poured back into a reservoir for use again. The mold with the gelled inner coating is placed in an oven where the plastisol is fused. Upon cooling, the plastisol part is stripped from the mold, retaining the design of its inner surface on the exterior of the piece. Molds of electroformed copper or fine sand-cast aluminum are usually used for slush molding.

This process is generally known as the single-pour system of slush molding. For a mold of intricate detail, a two-pour method often is used. In this process, the mold is filled when it is cold, vibrated to remove bubbles, and then emptied, leaving a thin film of plastisol on the inner surface. In this way, the plastisol does not have a chance to gel before flowing into the mold extremities.

Rotational Molding

Completely enclosed hollow parts can be produced by rotational molding. A measured amount of plastisol is poured into one-half of a two-piece mold. The mold is closed and rotated in two or more planes while being heated. During this rotation, the plastisol flows, gels, and fuses evenly over the interior walls of the mold. Molds for this operation are either electroformed copper or cast or machined aluminum, and the molds are arranged in clusters or "gangs" so that the maximum number of molds can be operated per spindle.

In rotational molding, it is possible to vary the thickness of the walls of the molded piece. One way this can be accomplished is by rotating the mold more in one plane than in another.

A few familiar products manufactured by this process are toys and novelties such as dolls and beach balls, swimming pool floats, and artificial fruit.

Combinations

In many cases, several of the aforementioned molding methods are combined to produce a product made from plastisol. For example, vinyl foam products such as armrests, toys, or electrical harnesses are manufactured by first forming a tough vinyl skin by spraying, slush molding, or rotational molding. The interior then is formed by casting, low-pressure injecting, or rotational molding a vinyl plastisol foam within the pregelled skin.

Plastohydrodynamic Lubrication

A condition of lubrication in which the friction and film thickness between two bodies in relative motion are determined by plastic deformation of the bodies in combination with the viscous properties of the lubricant at the prevailing pressure, temperature, and rate of shear. Compare with *elastohydrodynamic lubrication.*

Plastometer

An instrument for determining the flow properties of a thermoplastic resin by forcing the molten resin through a die or orifice of specific size at a specified temperature and pressure.

Plate

A flat-rolled metal product of some minimum thickness and width arbitrarily dependent on the type of metal. Plate thicknesses commonly range from 6 to 300 mm (0.25–12 in.) and widths from 200 to 2000 mm (8–80 in.). See also *plating.*

Plate Glass

Flat glass formed by a rolling process, ground and polished on both sides, with surfaces essentially plain and parallel.

Plate Martensite

Martensite formed partly in steel containing more than approximately 0.5% C and solely in steel containing more than approximately 1.0% C that appears as lenticular-shaped plates (crystals).

Platelet Alpha Structure

In titanium alloys, *acicular alpha* of a coarser variety, usually with low aspect ratios. This microstructure arises from cooling α or α-β alloys from temperatures at which a significant fraction of β phase exists.

Platelets

Flat particles of metal powder having considerable thickness. The thickness, however, is smaller when compared with the length and widths of the particles. See also *particle shape.*

Platen

(1) The sliding member, *slide,* or *ram* of a metal-forming press. (2) The mounting plates of a plastic-forming press, to which the entire mold assembly is bolted. (3) A part of a resistance welding, mechanical testing, or other machine with a flat surface to which dies, fixtures, backups, or electrode holders are attached, which transmits pressure or force. (4) A flat working surface of a press or other equipment. It may contact the workpiece or carry tooling and movable jigs, etc. (5) A series of tubes or bars joined by welding to form a plate.

Platen Force

The force available at the movable platen in a resistance welding machine that causes upsetting in flash or upset welding.

Plating

(1) Forming an adherent layer of metal on an object, often used as a shop term for *electroplating.* (2) Metal in plate form. (3) The application of a film or coating of some material to another material, for example, chromium is applied to brass for mainly decorative purposes and zinc is applied to steel to provide corrosion resistance. See also *electroplating, galvanizing, electrodeposition* and *electroless plating.*

Plating Rack

A fixture used to hold work and conduct current to it during *electroplating.*

Plating Range

The current-density range over which a satisfactory electroplate can be deposited.

Platinum

Platinum, a whitish-gray metal, symbol Pt, is more ductile than silver, gold, or copper and is heavier than gold. The melting point is 1769°C, and the specific gravity is 21.45. The hardness of the annealed metal is 45 Brinell, and its tensile strength is 117 MPa; when hard rolled, the Brinell hardness is 97 and tensile strength 34 MPa. Electrical conductivity is about 16% that of copper. The metal has a face-centered cubic lattice structure and it is very ductile and malleable. It is resistant to acids and alkalies, but dissolves in aqua regia. Platinum is widely used in jewelry, but because of its heat resistance and chemical resistance, it is also valued for electric contacts and resistance wire, thermocouples, standard weights, and laboratory dishes. Generally too soft for use alone, it is almost always alloyed with harder metals of the same group, such as osmium, rhodium, and iridium. An important use of the metal, in the form of gauze, is as a catalyst. Platinum gauze is of high purity in standard meshes of 18–34/cm, with wire from 0.020 to 0.008 cm in diameter. Dental foil is 99.99% pure and of maximum softness. Platinum foil for other uses is made in thicknesses as thin as 0.0005 cm. Platinum powder comes in fine submesh particle size. It is made by chemical reduction and is at least 99.9% pure, with amorphous particles 0.3–3.5 μm in diameter. Platinum flake has the powder particle in the form of tiny laminar platelets that overlap in the coating film.

Because of the high resistance of the metal to atmospheric corrosion even in sulfur environments, platinum coatings and electroplating are used on springs and other functioning parts of instruments and electronic devices where precise operation is essential. Coatings are also produced by vapor deposition on platinum compounds; thin coatings, 0.0005 cm or less, are made by painting the surface with a solution of platinum powder in an organic vehicle and then firing to drive off the organic material, leaving an adherent coating of platinum metal.

Platinum is sometimes used in glazes to obtain luster and metallic effects. Liquid bright platinum and liquid bright palladium (an element of the platinum group) are preparations used in metallic decorations. As platinum produces a better silver affect than silver itself and is less likely to tarnish, platinum is preferred to that metal. A luster produced from a strong solution of platinum chloride and spirits or oil of lavender upon firing gives a steely appearance that is nearly opaque. Another method consists of precipitating the metal from its solution in water by heating it with a solution of caustic soda and glucose. The metal is mixed with 5% bismuth subnitrate, applied to the ware by painting, and fired in a reducing atmosphere.

Platinum Alloys

Platinum is alloyed to obtain greater hardness, strength, and electrical resistivity. Because most applications require freedom from corrosion, the other platinum metals are usually employed as alloying agents.

Platinum–Iridium

Iridium is added to platinum most often used to provide improved mechanical properties. It increases resistance to corrosion while the alloy retains its workability. Up to 20% iridium, the alloys are quite ductile. With higher iridium content, fabrication becomes difficult.

Platinum–iridium alloys are employed for instruments, magneto contacts, and jewelry. The alloys are hard, tough, and noncorrosive. An alloy of 95% platinum and 5% iridium, when hardworked, has a Brinell hardness of 170; an alloy with 30% iridium has a hardness of 400. Alloys containing 5% of platinum and 10% of iridium are used for jewelry manufacture, while alloys containing 25% of platinum and 30% of iridium are employed for making surgical instruments. An alloy of 80% platinum and 20% iridium is used for magneto contact points, and an alloy of 90% platinum and 10% iridium is widely used for electric contacts in industrial control devices. The addition of iridium does not alter the color of the platinum. Alloy with 5% platinum dissolves readily in aqua regia, while an alloy with 30% iridium dissolves slowly.

Platinum–Rhodium

The addition of rhodium to platinum also provides improved mechanical properties to platinum and increases its resistance to corrosion. For applications at high temperatures, platinum–rhodium alloys are preferred because of retention of good mechanical properties including good hot strength and very little tendency toward volatilization or oxidation.

Platinum–rhodium alloys are used for thermocouples for temperatures above 1100°C. The standard thermocouple is platinum versus platinum with 10% rhodium. Other thermocouples for higher operating temperatures use platinum–rhodium alloys in both elements. The alloys of platinum–rhodium are widely used in the glass industry, particularly as glass-fiber extrusion bushings. Rhodium increases the high-temperature strength of platinum without reducing its resistance to oxidation. Platinum–rhodium gauze for use as a catalyst in producing nitric acid from ammonium contains 90% platinum and 10% rhodium.

Platinum–Ruthenium

Alloying platinum with ruthenium has the most marked effect upon both hardness and resistivity. However, the limit of workability is reached at 15% ruthenium. The lower cost and the lower specific gravity of ruthenium offer an appreciable economic benefit as an alternate to other platinum alloys.

Platinum–ruthenium alloy, with 10% ruthenium, has a melting point of 1800°C and electrical conductivity 4% that of copper.

Platinum–Gold

Platinum–gold alloys cover a wide range of compositions and provide distinct chemical and physical characteristics.

Platinum–Cobalt and Platinum–Nickel

Platinum–cobalt alloys, with about 23% cobalt, are used for permanent magnets. Platinum–nickel alloys, with as much as 20% nickel, are noted for high strength. With 5% nickel, for example, tensile strength of the annealed alloy is about 621 MPa, and with 15%, it increases to 896 MPa. Strength almost doubles with appreciable cold work.

Platinum–Rhenium

These alloys are efficient catalyst for reforming operations on aromatic compounds. The platinum alloys have lower electric conductivity than pure platinum, but are generally harder and more wear resistant, and have high melting points. A platinum–rhenium alloy with 10% rhenium has an electrical conductivity of only 5.5% that of copper compared with 16% for pure platinum. Its melting point is 1850°C, and the Rockwell T hardness of the cold-rolled metal is 91 compared with 78 for cold-rolled platinum.

Platinum–Tungsten

These alloys, with 2%–8% tungsten, have been used for aircraft engine spark plug electrodes, radar-tube grids, strain gauges, glow wires, switches, and heating elements. Tungsten markedly increases electrical resistivity while decreasing the temperature coefficient of resistivity. It also substantially increases tensile strength—to 896 MPa for platinum (8% tungsten alloy in the annealed condition)—and tensile strength more than doubled with appreciable cold work.

Platinum Black

(1) A finely divided form of platinum powder of a dull black color, usually, but not necessarily, produced by reduction of salts in an aqueous solution. (2) A very effective catalyst of some chemical reactions.

Platinum Group Metals

The platinum group metals—ruthenium, rhodium, palladium, osmium, iridium, and platinum—are found in the second and third long period in group VIII of the periodic table. Platinum and palladium are the most abundant of the group although all are generally found together.

The outstanding characteristics of platinum, the most important member of the group, are its remarkable resistance to corrosion and chemical attack, high melting point, retention of mechanical strength, and resistance to oxidation in air, even at very high temperatures. These qualities, together with the ability of the metal to greatly influence the rates of reaction in a large number of chemical processes, are the basis of nearly all its technical applications. The other five metals of the platinum group are also characterized by high melting points, good stability, and resistance to corrosion. Addition of these metals to platinum forms a series of alloys that provide a wide range of useful physical properties combined with the high resistance to corrosion that is characteristic of the parent metals.

Platinum (Pt)

When heated to redness, platinum softens and is easily worked. It is virtually nonoxidizable and is soluble only in liquids generating free chlorine, such as aqua regia. At red heat, platinum is attacked by cyanides, hydroxides, sulfides, and phosphides. When heated in an atmosphere of chlorine, platinum volatilizes and condenses as the crystalline chloride. Reduction of platinum chloride with zinc gives platinum black, which has a high adsorptive capacity for hydrogen. Platinum sponge is finely divided platinum.

Palladium (Pd)

Palladium is silvery white, very ductile, and slightly harder than platinum. It is readily soluble in aqua regia and is attacked by boiling nitric and sulfuric acids. Palladium has a remarkable ability to occlude large quantities of hydrogen. When properly alloyed, it can be used for the commercial separation and purification of hydrogen. Palladium and platinum can both be worked by normal metalworking processes.

Iridium (Ir)

This is the most corrosion-resistant element known. It is a very hard, brittle, tin-colored metal with a melting point higher than that of platinum. It is soluble in aqua regia only when alloyed with sufficient platinum. Iridium has its greatest value in platinum alloys where it acts as a hardening agent. By itself, it can be worked only with difficulty.

Rhodium (Rh)

Rhodium serves an important role in applications with high temperatures up to 1649°C. Platinum–rhodium thermocouple wire makes possible high-temperature measurement with great accuracy. Rhodium and rhodium alloys are used in furnace windings and in crucibles at temperatures too high for platinum. It is a very hard, white metal and is workable only under certain conditions and then with difficulty. Applied to a base metal by electroplating, it forms a hard, wear-resistant, permanently brilliant surface. Solubility is light even in aqua regia.

Ruthenium (Ru)

Ruthenium is hard and brittle with a silver-gray luster. Its tetroxide is very volatile and poisonous. When alloyed with platinum, its effect on hardness and resistivity is the greatest of all the metals in the group. It is unworkable in the pure state.

Osmium (Os)

This element has the highest specific gravity and melting point of the platinum metals. It oxidizes readily when heated in air to form a very volatile and poisonous tetroxide. Application has been predominantly in the field of catalysis. As a metal, it is also practically unworkable.

Plenum

The space between the inside wall of the constricting nozzle and the electrode in a plasma arc torch used for welding, cutting, or thermal spraying.

Plied Yarn

Yarn for reinforced plastics made by collecting two or more single yarns. Normally, the yarns are twisted together, though sometimes they are collected without a twist.

Plowing

(1) In tribology, the formation of grooves by *plastic deformation* of the softer of two surfaces in relative motion. (2) Deep grooving by severe gouging and usually involving significant local mechanical deformation.

Plug

(1) A rod or mandrel over which a pierced tube is forced. (2) A rod or mandrel that fills a tube as it is drawn through a die. (3) A punch or mandrel over which a cup is drawn. (4) A protruding portion of a die impression for forming a corresponding recess in the forging. (5) A false bottom in a die. (6) A short mandrel over which a tube is drawn in the plug drawing process, sizing the bore at the same time as the die sizes the exterior. The plug may be retained in the die by a small mandrel in the tube bore or it may be designed to self-locate in position termed a *floating plug*. See also *tube making*.

Plug Corrosion

See *dezincification*.

Plug Tap

A *tap* with *chamfer* extending from 3 to 5 threads.

Plug Weld

A circular weld made through a hole in one member of a lap or tee joint. Neither a fillet-welded hole nor a spot weld is to be construed as a plug weld. The hole may be partially or completely filled with weld metal.

Plug-Assist Forming

A *thermoforming* process in which a plug or male mold is used to partially preform the plastic part before forming is completed using vacuum or pressure.

Plumbage

A special quality of powdered graphite used to coat molds and, in a mixture of clay, to make crucibles and bulk items.

Plumber's Solder

An alloy of about one-third tin, two-thirds lead. It solidifies over a wide range, 250°C–183°C, so during cooling, it has a long pasty stage allowing it to be wiped to form a smoothly curved surface. See *soft solder*.

Plunge Grinding

Grinding wherein the only relative motion of the wheel is radially toward the work.

Plunge Milling

Machining using a milling cutter that cuts only on its end face rather than as it traverses.

P

Plunger

Ram or piston that forces molten metal into a die in a *die-casting* machine. Plunger machines are those having a plunger in continuous contact with molten metal. See *cold chamber machine* and *hot chamber machine.*

Plus Mesh

The powder sample retained on a screen of stated size, identified by the retaining mesh number. See also *sieve analysis* and *sieve classification.*

Plus Sieve

The portion of a sample of a granular substance (such as metal powder) retained on a standard sieve of specified number. Contrast with *minus sieve.* See also *sieve analysis* and *sieve classification.*

Plutonium

One of the man-made transuranic elements. It is radioactive with a half-life of 24,000 years and is toxic. Its applications are largely confined to nuclear power and weapons.

Ply

A single layer in a *laminate*. In general, fabrics or felts consisting of one or more layers. Yarn resulting from a twisting operation (e.g., three-ply yarn consists of three strands of yarn twisted together). In *filament winding*, a ply is a single pass (two plies forming one layer).

Plymetal

Sheet consisting of bonded layers of dissimilar metals.

Plywood

Plywood is a term generally used to designate glued wood panels made up of layers, or plies, with the grain of one or more layers at an angle, usually 90°, with the grain of the other. The outside plies are called faces or face and back, the center plies are called the core, and plies immediately below the face and back, laid at right angles to them, are called the crossbands.

The core may be veneer, lumber, or various combinations of veneer and lumber, the total thickness may be less than 1.5 mm or more than 76 mm; the different plies may vary as to number, thickness, and wood species. Also, the shape of the members may vary. The crossbands and their arrangement generally govern both the properties (particularly warping characteristics) and uses of all such constructions.

Plywood is an outgrowth of the laminated wood known as veneer, which consists of an outside sheet of hardwood glued to a base of lower-cost wood. The term *veneer* actually refers only to the facing layer of selected wood, used for artistic effect or for economy in the use of expensive woods. Veneers are generally marketed in strip form and thicknesses of less than 0.32 cm in mahogany, oak, cedar, and other woods. The usual purpose of plywood now is not aesthetic but to obtain high strength with low weight. The term *laminated wood* generally means heavier laminate for special purposes, and such laminates usually contain a heavy impregnation of bonding resin that gives them more of the characteristics of the resin than of the wood.

Composition

The composition of a plywood panel is generally dependent on the end use for which it is intended. The number of plywood constructions is almost endless when one considers the number of wood species available, the many thicknesses of wood veneers used in the outer plies or cores, the placement of the adjacent plies, the type of adhesives and their qualities, various manufacturing processes, and more technical variations.

Conventional plywood generally consists of an odd number of plies with the grains of the alternate layers perpendicular to each other. The use of an odd number permits an arrangement that gives a substantially balanced effect; that is, when three plies are glued together with the grain of the outer two plies at right angles to that of the center ply, the stresses are balanced and the panel tends to remain flat with changes in moisture content. These forces may be similarly balanced with five, seven, or some other uneven number of plies. If only two plies are glued together with the grain of one ply at right angles to the other, each ply tends to distort the other when changes in moisture content occur; cupping will result.

Low-cost plywoods may be bonded with starch pastes, animal glues, or casein and are not water resistant, but are useful for boxes and for interior work. Waterproof plywood for paneling and general construction is now bonded with synthetic resins, but when the plies are heavily impregnated with the resin and the whole cured into a solid sheet, the material is known as a hardboard or as a laminated plastic rather than a plywood.

Grades and Types

Broadly speaking, two classes of plywood are available—hardwood and softwood. Most softwood plywood is composed of Douglas fir, but western hemlock, white fir, ponderosa pine, redwood, and other wood species are also used. Hardwood plywood is made of many wood species.

Various grades and types of plywood are manufactured. "Grade" is determined by the quality of the veneer and "type" by the moisture resistance of the glue line. For example, there are two types of Douglas fir plywood—interior and exterior. The interior type is expected to retain its form and strength properties when occasionally subjected to wetting and drying periods. It is commonly bonded with urea–formaldehyde resin adhesives. On the other hand, the exterior type is expected to retain its form and strength properties when subjected to cyclic wetting and drying and to be suitable for permanent exterior use. It is commonly bonded with hot-pressed phenolic resin glues.

For construction purposes, where plywood is employed because of its unit strength and nonwarping characteristics, the plies may be of a single type of wood and without a hardwood face. The Douglas Fir Plywood Association sets up four classes of construction plywood under general trade names. Plywall is plywood in wallboard grade; Plypanel is plywood in three standard grades for general uses; Plyscord is unsanded plywood with defects plugged and patched on one side; and Plyform is plywood in a grade for use in concrete forms.

The bulk of commercial plywood comes within these classes; the variations are in the type of wood used, the type of bonding adhesive, or the finish. Each wood, for example, is a paneling plywood with the face wire brushed to remove the soft fibers and leave the hard grain for two-tone finish. Paneling plywoods with faces of mahogany, walnut, or other expensive wood have cores of lower-cost woods, but the woods of good physical qualities are usually chosen.

Engineering Properties

The mechanical and physical properties of plywood are dependent upon the particular construction employed. Plywood may be designed for beauty, durability, rigidity, strength, cost, or many other properties. With practically an unlimited variety of constructions to choose from, there is a wide range of differing characteristics in any given plywood panel. Most important among properties are the following:

High Strength–Weight Ratio

Perhaps the most notable feature of plywood is its high strength–weight ratio. Plywood is given special consideration whenever lightness and strength are desired. Plywood is widely used for concrete form work, floor underlayment, roof decks, siding, and many other applications because of its high strength–weight ratio. A comparison between birch plywood and other structural materials shows that its strength–weight ratio was 1.52 times that of 100,000 lb test heat–treated steel and 1.36 times that of 10,000 lb test aluminum.

Bending Properties

A most desirable characteristic of plywood is its flatness, but it can and will support substantial curvatures without appreciable loss of strength. Standard construction plywood can be bent or shaped to nominal radii and held in place with adhesives, nails, screws, or other fixing methods.

The radius of curvature to which a panel can be formed varies with panel thickness and the species of wood employed in the panel construction. The arc of curvature is limited by the tension force in the outer plies of the convex perimeter and by the compression forces in the outer plies of the concave perimeter.

Waterproof plywood, soaked or steamed before bending, exhibits approximately 50% greater flexibility than panels bent when dry.

Resistant to Splitting

Because plywood has no line of cleavage, it cannot split. This is an exceptional property when one considers its effect on fastening. The crisscross arrangement of wood plies in plywood construction develops extraordinary resistance to pull-through of nail or screw heads.

Resistance to Impact

The absence of a cleavage line has a pronounced effect on the impact resistance of plywood. Plywood will fracture only when the impact force is greater than the tensile strength of the wood fibers in the panel composition. Under an impact force, the side of the panel opposite the impact point will rupture along the long grain fiber followed by successive shattering of the various plies. Splintering usually does not take place because pressure is dispersed throughout the panel at the point of impact. Under similar conditions, solid lumber will show complete rupture.

Beauty

Plywood has certain intrinsic qualities that add much to any structure or construction in which it is used. Because of improved modern methods of manufacture, there is practically no limit to the decorative potential of plywood.

The entire range of fine woods is at the designer's disposal; they vary in shade from golden yellow to ebony, from pastels to reds and browns. The use of bleaches, toners, and stains in manufacturing procedures gives the designer even greater latitude with respect to design freedom.

The fine woods employed in the manufacture of plywood offer warmth and charm to any decorative scheme because the surface of the wood variously absorbs, reflexes, and refracts light rays, giving the wood pattern depth and making it restful to the eye. This phenomenon accounts for the play of color and pattern when a plywood panel is viewed from different angles.

Dimensional Stability

The absorption of water causes wood to swell, and this movement is much greater across the grain then along the grain. The alternating layers of veneers in standard plywood construction inhibit this cross-grain movement because the cross-grain weakness is reinforced by the long-grain stability. Therefore, the dimensional stability of a plywood panel can be controlled by controlling its moisture content. In the field, this control can be achieved by applying coatings such as paints, lacquers, and sealers of various types.

Thermal Insulating Qualities

The thermal insulating qualities of plywood are the same as those of the wood of which it is composed.

The use of plywood as an insulating material can be attributed to two factors: (1) the use of large sheets reduces the numbers of cracks and joints and thereby inhibits with leakage; and (2) the resistance of plywood to moisture vapor transmission stabilizes the moisture content of the trapped air and maintains its insulating qualities.

Fire Resistance

Fire-resistant plywood is manufactured by impregnating the core stock with a salt solution, which, upon evaporation, leaves a salt deposit in the wood. Plywood or wood treated in this manner will not support combustion but will char when heated beyond the normal charring point of the wood.

Nonimpregnated plywood can be made fire resistant by applying surface coatings such as intumescent paint and chemicals such as borax. An intumescent paint has a silicate of soda base that bubbles or intumesces in the presence of heat, thus forming a protective coating. At high temperatures, borax releases a gas such as carbon dioxide, which blankets the fire. It must be noted that fire-resistant coatings are effective only in direct ratio to their thickness. Highly resistant plywood can be manufactured by using an incombustible core such as asbestos.

Resistance to Borers

Plywood panels are subject to attack by borers to the same extent as the wood species of which they are composed, but phenol formaldehyde resin glue lines are fairly effective barriers against further penetration. Panels may also be treated with pentachlorophenol for increased resistance to these pests.

Fatigue Resistance

Plywood has the same resistance to fatigue as the wood of which it is composed.

Fabrication

The fact that almost anyone can use plywood has contributed greatly to its wide acceptance in many varied applications. The utilization of plywood does not require special tools, special skills, or safeguards, and practically anyone capable of handling a saw and hammer can make use of its inherent engineering properties.

Since plywood does not exhibit the typical cross-grain weakness of lumber, it can very often be used in place of lumber for various applications. For example, 0.312 mm thick plywood replaces conventional 0.750 mm sheathing; 0.250 mm plywood can be used for interior wall paneling without sheathing; and 0.375 mm plywood can be used for shipping containers, furniture, and case goods instead of 0.750 mm lumber. The use of the thinner plywood reduces weight and bulkiness and is less fatiguing for tradespeople to handle. The use of large plywood sheets instead of narrow boards also reduces the amount of cutting, fitting, and fastening involved in a particular job.

Plywood is especially adaptable to the portable power-driven saws, drills, and automatic hammers normally used on production or construction jobs. Multiple cutting with band saws may be done with assurance because even the thinnest plywood has strength in all directions, and the danger of splitting or chipping is reduced to a minimum. This quality is particularly important when fine fitting is required.

Whenever plywood is employed in a structure, fewer and smaller fastenings can be specified because consideration need be given only to the holding power of the fastener and the tensile strength of the fastening itself.

Available Forms, Sizes, and Shapes

Plywood is available in practically any size, but the 1.2×2.4 m panel has become the standard production unit of the industry. Larger panels usually demand a price premium and are available on special order. Panels with continuous cores and faces can be produced in one piece up to 4 m long.

Oversize panels up to 2.4 m in width and of unlimited length can be manufactured by scarf jointing. (A scarf is an angling joint, made either in veneers or plywood, where pieces are spliced or lapped together. The length of the scarf is usually 12–20 times the thickness. When properly made, scarf joints are as strong as the adjacent unspliced material.)

Decorative plywood is now commercially available with a plant-applied finish. Prefinished wall paneling is supplied with a finish varying from offset printing to polyester film.

General Fields of Application

There are many uses and applications for plywood in industry today. Owing to the wide diversity of plywood applications, only the more prominent ones are mentioned here:

Architectural	Marine construction
Aviation	Mock-ups, models
Boatbuilding	Paddles
Building construction	Panel boards
Cabinet work	Patterns
Concrete forms	Prefabrication
Containers, cases	Remodeling
Die boards	Sheathing
Display	Signs
Fixtures	Sporting goods
Floor underlayment	Tabletops
Furniture	Toys
Hampers	Trays
Luggage	Truck floors, bodies
Machine bases	Wall paneling

P/M

The acronym for *powder metallurgy*.

PMMA

See *polymethyl methacrylate*.

PMR Polyimides

A novel class of high-temperature-resistant polymers. PMR represents in situ polymerization of monomer reactants. See also *polyimide*.

Pneumatic Press

A press that uses air or a gas to deliver the pressure to the upper and lower rams.

Pocket

In a *rolling-element bearing*, the portion of the case that is shaped to receive the rolling element. Compare with *oil pocket*.

Pocket Thrust Bearing

An externally pressurized thrust bearing having three or more hydrostatic pads with central relieved chambers of pocket supplied with pressurized oil.

Point Angle

In general, the angle at the point of a cutting tool. Most commonly, the included angle at the point of a twist drill, the general-purpose angle being 118°.

Point Defects

Irregularities in the crystal lattice that affect only a single site or local area, for example, a vacancy or a foreign atom, either substitutional or interstitial.

Point Estimate

The estimate of a *parameter* given by a single statistic.

Pointing

(1) Reducing the diameter of wire, rod, or tubing over a short length at the end by swaging or hammer forging, turning, or squeezing to facilitate entry into a drawing die and gripping in the drawhead. (2) The operation in automatic machines of chamfering or rounding the threaded end or the head of a bolt.

Poise (P)

The centimeter–gram–second (cgs) unit of dynamic viscosity ($1\ P = 0.1\ Pa \cdot s$).

Poiseuille

The meter–kilogram–second (mks) or Système International d' Unités unit of dynamic viscosity ($1\ Pl = 10\ P$).

Poison Gas

Poison gases are classified according to their main effect on the human system, but one gas may have several effects. They are grouped as follows: *lethal gases*, intended to kill, such as phosgene; *lachrymators*, or *tear gases*, which have a powerful irritating effect on the eyes, causing temporary blindness and swelling of the eyes with a copious flow of tears; *vesicants*, or skin blisters, such as lewisite and mustard gas; *sternutatory gases*, which induce sneezing, and *camouflage gases*, which are harmless, but cause soldiers to suffer the inconvenience of wearing gas masks and thus reduce their morale. Some of the gases have a sour, irritating odor and are also classified as *harassing agents*. Gases are also sometimes designated as *casualty agents* and further subdivided into persistent and nonpersistent. A *systemic gas* is one that interferes with one phase of the system, such as carbon monoxide, which paralyzes the respiratory function of the blood. A *labyrinthic gas*, which contains dichloromethyl *ether*, is one that affects an organ of the body, which affects the ears.

Poison Gases

Substances employed in chemical warfare for disabling people and in some cases used industrially as fumigants. They are all popularly called gases, but many are liquids or solids. Normally, information on military gases is kept secret, but the tear gases used by police are also poisonous, often causing serious damage to the eyes, throat, and lungs. *Anesthetic gases* have not been used so far in chemical warfare, but are used in medicine. One of the simplest of these, *nitrous oxide*, N_2O, called *laughing gas*, produces a deep sleep. *Fluothane*, or *ethyl fluoride*, is a volatile liquid like ether, but is nonexplosive, and it is used to replace ether in surgery.

Poisson's Ratio

The ratio of transverse strain to longitudinal strain. When a metal is pulled in tension, it extends longitudinally and contracts in the transverse direction. The ratio of the two strains is between 0.25 and 0.5 for elastic deformation and 0.5 and 0.5 for plastic deformation.

Poke Welding

Same as *push welding*.

Polar Bonding

Same as *ionic bonding*; see *interatomic bonding*.

Polar Winding

In filament winding of composites, a winding in which the filament path passes tangent to the polar opening at one end of the chamber and tangent to the opposite side of the polar opening at the other end. A one-circuit pattern is inherent in the system. See also *helical winding*.

Polarimeter

An instrument used to determine the rotation of the plane of polarization of plane polarized light when it passes through a substance; the light is linearly polarized by a polarizer (such as a *Nicol prism*), passes through the material being analyzed, and then passes through an analyzer.

Polariscope

Any of several instruments used to determine the effects of substances on polarized light, in which linearly or elliptically polarized light passes through the substance being studied, and then through an analyzer.

Polarity (Welding)

See *direct current electrode negative*, *direct current electrode positive*, *straight polarity*, and *reverse polarity*.

Polarization

(1) The change from the open-circuit electrode potential as the result of the passage of current. (2) A change in the *potential* of an electrode during electrolysis, such that the potential of an *anode* becomes more noble, and that of a *cathode* more active, than their respective reversible potentials. Often accomplished by formation of a film on the electrode surface. (3) In the context of light and optics, polarized light is comprised of waves in a single plane as opposed to normal light that has waves in all planes.

Polarization Admittance

The reciprocal of *polarization resistance (di/dE)*.

Polarization Curve

A plot of *current density* versus *electrode potential* for a specific electrode–electrolyte combination.

Polarization Resistance

The slope (*dE/di*) at the *corrosion potential* of a potential (€)/current density (*i*) curve. Also used to describe the method of measuring corrosion rates using the slope.

Polarized Light Illumination

A method of illumination in which the incident light is plane polarized before it impinges on the specimen. See also *optical microscope*.

Polarizer

In an optical microscope, a *Nicol prism*, polarizing film, or similar device into which a normal light passes and from which polarized light emerges.

Polarizing Element

A general term for a device for producing or analyzing plane-polarized light. It may be a *Nicol prism*, some other form of calcite prism, a reflecting surface, or a polarizing filter. See also *polarizer*.

Polarography

An electroanalytical technique in which the current between a dropping mercury electrode (DME) and a counterelectrode (both of which are immersed in electrolyte) is measured as a function of the potential difference between the DME and a reference electrode.

Poldi Test

A simple handheld hardness testing device in which a 10 mm diameter, hard steel ball is located in a head piece so as to register on the component surface to be tested and, at the same time, on the surface of a square bar of known hardness. The head is struck causing simultaneous indentations on the bar and on the test surface. The two indentations are then measured and the results read against charts to provide a measure of hardness of the component.

Pole

(1) A means of designating the orientation of a crystal plane by stereographically plotting its normal. For example, the North Pole defines the equatorial plane. Either of the two regions of a permanent magnet or electromagnet where most of the lines of induction enter or leave.

Pole Figure

A stereoscopic projection of a polycrystalline aggregate showing the distribution of poles, or plane normals, of a specific crystalline plane, using specimen axes as reference axes. Pole figures are used to characterize preferred orientation in polycrystalline materials. See *orientation*.

Pole Piece

The magnetic material, forming the extension of a magnet, which is intended to concentrate and direct the magnetic flux.

Polepiece

In reinforced plastics, the supporting part of the mandrel used in filament winding, usually on one of the axes of rotation.

Poling

A step in the fire refining of copper to reduce the oxygen content to tolerable limits by covering the bath with coal or coke and thrusting green wood poles below the surface. There is a vigorous evolution of reducing gases, which combine with the oxygen contained in the metal.

Polished Surface

A surface prepared for metallographic inspection that reflects a large proportion of the incident light in a specular manner. See also *polishing* (4).

Polishing

(1) A surface-finishing process for ceramics and metals utilizing successive grades of abrasive. (2) Smoothing metal surfaces, often to a high luster by rubbing the surface with a fine abrasive, usually contained in a cloth or other soft lap. Results in microscopic flow of some surface metal together with actual removal of a small amount of surface metal. (3) Removal of material by the action of

abrasive grains carried to the work by a flexible support, generally either a wheel or a coated abrasive belt. (4) A mechanical, chemical, or electrolytic process or combination thereof used to prepare a smooth, reflective surface suitable for microstructural examination that is free of artifacts or damage introduced during prior sectioning or grinding. See also *electrolytic polishing* and *electropolishing*.

Polishing Artifact

A false structure introduced during a polishing stage of a surface-preparation sequence.

Polishing Lap, Stick, Pad, etc.

The device or structure on which a polishing medium is carried to polish some component.

Polishing Rate

The rate at which material is removed from a surface during polishing. It is usually expressed in terms of the thickness removed per unit of time or distance traversed.

Polishing Wear

An extremely mild form of wear for which the mechanism has not been clearly identified, but that may involve extremely fine-scale abrasion, plastic smearing of micro-asperities, and/or tribochemical material removal.

Polonium

Polonium (symbol Po) is a rare metallic element belonging to the group of radioactive metals, but emitting only alpha rays. The melting point of the metal is about 254°C. It is used in meteorological stations for measuring the electrical potential of the air. Polonium-plated metal in strip and rod forms has been employed as a static dissipator in textile-coating machines. The alpha rays ionize the air near the strip, making it a conductor and drawing off static electric charges. Polonium-210 is obtained by irradiating bismuth; 45 kg yields 1 g of polonium-210. It is used as a heat source for emergency auxiliary power such as in spacecraft. The metal is expensive, but can be produced in quantity from bismuth.

Poly-

As a prefix, it means many. In most cases, it refers to various types of polymer.

Polyacrylate Resin

Useful polymers can be obtained from a variety of acrylic monomers, such as acrylic in methacrylic acids, and their salts, esters, amides, and the corresponding nitriles. Polymethyl methacrylate (PMMA), polyethylene acrylate, and a few other derivatives are the most widely used.

PMMA is a hard, transparent polymer with high optical clarity, high refractive index, and good resistance to the effects of light

and aging. It and its copolymers are useful for lenses, signs, indirect lighting fixtures, transparent domes and skylights, dentures, and protective coatings.

Solutions of PMMA and its copolymers are useful as lacquers. Aqueous latexes formed by the emulsion polymerization of methyl methacrylate with other monomers are useful as water-based paints and in the treatment of textiles and leather.

Polyethyl acrylate is a tough, somewhat rubbery product. The monomer is used mainly as a plasticizing or softening component of copolymers. Ethyl acrylate is usually produced by the dehydration and ethanolysis of ethylene cyanohydrin.

Modified acrylic resins with high impact strength can be prepared. Blends or "alloys" with polyvinyl chloride are used for thermoforming impact-resistant sheets.

Methyl methacrylate is of interest as a polymerizable impregnant for concrete: usually a cross-linking acrylic monomer is also incorporated.

Polymers of methyl acrylate or acrylamide are water soluble and useful for sizes and finishes. Addition of polylauryl methacrylate to petroleum lubricating oil improves the flowing properties of the oil at low temperatures and the resistance to thinning at high temperatures.

Polyacrylic Rubber

The first types of polyacrylic rubbers were proposed as oxidation-resistant elastomeric materials. Chemically, they were polyethyl acrylate, and a copolymer of ethyl acrylate and 2-chloroethyl vinyl ether.

The development of polyacrylic rubbers was accelerated by the interest expressed throughout the automotive industry in the potential applications of this type of polymer in special types of seals. An effective seal for today's modern lubricants must be resistant not only to the action of the lubricant but to increasingly severe temperature conditions. It must also resist attack of highly active chemical additives that are incorporated in the lubricant to protect it from deterioration at extreme temperature.

Polyacrylic rubber compounds were developed to provide a rubber part that would function in applications where oils and/or temperatures as high as 204°C would be encountered. These were also very resistant to attack by sulfur-bearing chemical additives in the oil. These properties have resulted in general use of polyacrylic rubber compounds for automotive rubber parts as seals for automatic transmission fluids and extreme-pressure lubricants.

Polyacrylic rubber will prove most useful in fields where these special properties are used to the maximum. It is recommended for products such as automatic transmission seals, extreme-pressure lubricant seals, searchlight gaskets, belting, rolls, tank linings, hose, O-rings and seals, white- or pastel-colored rubber parts, solution coatings, and pigment binders on paper, textiles, and fibrous glass.

Curing

A typical polyacrylic rubber, such as the copolymer of ethyl acrylate and chloroethyl vinyl ether, is supplied as a crude rubber in the form of white sheets having a specific gravity of approximately 1.1. It may be mixed and processed according to conventional rubber practice.

However, polyacrylic rubber is chemically saturated and cannot be cured in the same manner as conventional rubbers. Sulfur and sulfur-bearing materials act as retarders of cure and function as a form of age resistor in most formulations. Polyacrylic rubber is cured with amines; "Trimene Base" and triethylene tetramine are

most widely used. Aging properties may be altered by balancing the effect of the amine and the sulfur.

Like other rubber polymers, reinforcing agents such as carbon black or certain white pigments are necessary to develop optimum physical properties in a polyacrylic rubber vulcanizate. Selection of the pigments is more critical in that acidic materials, which would react with the basic amine curing systems, must be avoided. The SAF (Super Abrasion Furnace), ASTM designation N110 or FEF (Fast Extruding Furnace), ASTM designation N550 carbon blacks are most widely used, while hydrated silica or precipitated calcium silicate are recommended for light-colored stocks.

Typical curing temperatures are from 143°C to 166°C and cure times of 10–45 min depending on the thickness of the part. Polished, chromium-plated molds are recommended. For maximum overall physical properties, the cured parts should be tempered in an air oven for 24 h at 149°C.

Forming

To obtain smooth extrusions, more loading and lubrication are necessary than for molded goods, because of the inherent nerve of the polymer. Temperatures of 43°C in the barrel and 77°C on the die are recommended.

Generally, those compounds that extrude well are also good calendering stocks. Suggested temperatures for calendering are in the range of 37.8°C–54°C. Higher temperatures will result in sticking of the stock to the rolls. Under optimum conditions, 15 mil films may be obtained.

Polyacrylic rubber may be coated on nylon either by calendering or from solvent solution. It also has excellent adhesion to cotton and is often used as a solvent solution applied to cotton duck to be used as belting. Solvents generally used include methylethyl ketone, toluene, xylene, or benzene.

Polyacrylic rubber is most widely used in many types of seals because of its excellent resistance to sulfur-bearing oils and lubricants.

In general, polyacrylic rubber vulcanizates are resistant to petroleum products and animal and vegetable fats and oils. They will swell in aromatic hydrocarbons, alcohols, and ketones. Polyacrylic rubber is not recommended for use in water, steam, ethylene glycol, or alkaline media.

Laboratory tests indicate that polyacrylic vulcanizates become stiff and brittle at a temperature of −23°C. But in actual service, these same polyacrylic rubbers have been found to provide satisfactory performance at engine start-up and operation in oil at temperatures as low as −40°C.

For those applications requiring improvement in low-temperature brittleness by as much as −4.0°C and that can tolerate considerable sacrifice in overall chemical oil and heat resistance, a copolymer of butyl acrylate and acrylonitrile may be used.

Polyacrylonitrile (PAN)

A base material or precursor used in the manufacture of certain carbon fibers. PAN-based carbon fibers have a ribbonlike structure and possess high strength (400 GPa, or 10^6 psi, tensile modulus).

Polyacrylonitrile (PAN) Resins

The PAN resins are hard, horny, relatively insoluble, and high-melting materials. PAN (*polyvinyl cyanide*) is used almost entirely in copolymers. The copolymers fall into three groups: fibers, plastics, and rubbers. The presence of acrylonitrile in a polymeric

P

composition tends to increase its resistance to temperature, chemicals, impact, and flexing.

Acrylonitrile is generally prepared by several methods, including the catalyzed addition of hydrogen cyanide to acetylene. The polymerization of acrylonitrile can be readily initiated by means of the conventional free-radical catalysts such as peroxides, by irradiation, or by the use of alkaline metal catalysts. Although polymerization in bulk proceeds rapidly to be commercially feasible, satisfactory control of a polymerization or copolymerization may be achieved in suspension and in emulsion, and in aqueous solutions from which the polymer precipitates. Copolymers containing acrylonitrile may be fabricated in the manner of thermoplastic resins.

The major use of acrylonitrile is in the form of fibers. By definition, an acrylic fiber must contain at least 85% acrylonitrile; a modacrylic fiber may contain less than 35%–85% acrylonitrile. The high strength; high softening temperature; resistance to aging, chemicals, water, and cleaning solvents; and the soft wood-like feel of fabrics have made the product popular for many uses such as sails, cordage, blankets, and various types of clothing. Commercial forms of the fiber probably are copolymers containing minor amounts of other vinyl derivatives, such as vinyl pyrrolidone, vinyl acetate, maleic anhydride, or acrylamide. The comonomers are included to produce specific effects, such as improvement of dyeing qualities.

Copolymers of vinylidene chloride with small proportions of acrylonitrile are useful as tough, impermeable, and heat-sealable packaging films.

Copolymers for extensive use are made of acrylonitrile with butadiene, often called NBR (formerly Buna N) rubbers, which contain 15% acrylonitrile. Minor amounts of other unsaturated esters, such as ethyl acrylate, which yield carboxyl groups on hydrolysis, may be incorporated to improve the curing properties. The NBR rubbers are resistant to hydrocarbon solvents such as gasoline and abrasion and in some cases show high flexibility at low temperatures.

In the 1960s, development of blends and interpolymers of acrylonitrile-containing resins and rubbers represented a significant advance in polymer technology. The products, usually called *acrylonitrile, butadiene, and styrene (ABS) resins*, typically are made by blending acrylonitrile–styrene copolymers with a butadiene–acrylonitrile rubber or by interpolymerizing polybutadiene with styrene and acrylonitrile. Specific properties depend on the proportions of the comonomer, on the degree of grattings, and on molecular weight. In general, the ABS resins combine the advantages of hardness and strength of the vinyl resin component with toughness and impact resistance of the rubbery component. Certain grades of the ABS resin are used for blending with brittle thermoplastic resins such as polyvinyl chloride to improve impact strength.

The combination of low cost, good mechanical properties, and ease of fabrication by a variety of methods, including typical metal working methods such as cold stamping, led to the rapid development of new uses for ABS resins. Applications include products requiring high impact strength, such as pipe, and sheets for structural uses, such as industrial duct work and components of automobile bodies. ABS resins are also used for housewares and appliances, because of their ability to be electroplated for decorative items in general.

Polyamide Plastic

See *nylon plastics* and *polyamides*.

Polyamide-Imide Resins

A family of polymers based on the combination of trimellitic anhydride with aromatic diamines. In the uncured form (ortho-amic acid), the polymers are soluble in polar organic solvents. The imide linkage is formed by heating, producing an infusible resin with thermal stability up to 260°C (500°F). These resins, which also include graphite- (powder and fiber) and glass-fiber-reinforced grades, are used in the automotive (friction and wear parts), aerospace (fasteners and housings), and electronic (connector industries) applications.

Polyamides (PA)

PA resins are thermoplastic polymers that are most commonly regarded as being synonymous with nylons, that is, synthetic polymers that contain an amide group, –CONH–, as a recurring part of the chain. Nylons are made from diamines and dibasic acids, ω-amino acids, or lactams. Nylons are commonly identified by numbers corresponding to the number of C atoms in the monomers. Thus, resins are known as nylon 6/6, nylon 11, and nylon 6, respectively. The molecular weights of nylons range from 11,000 to 34,000. They are usually semicrystalline polymers with melting points in the range of 175°C–300°C (350°F–570°F).

Nylon 6/6 and nylon 6 are the most important commercial products. Their melting points are 269°C and 228°C (516°F and 440°F), respectively. Other commercial nylons are 6/9, 6/10, 6/12, 11, 12, and 4/6. The more C atoms, that is, the lower the concentration of amide groups, the lower the melting point. Nylons are very readily modified by the use of monomer mixtures leading to copolymers. These are normally less crystalline, more flexible, and more soluble than the previously mentioned homopolymers.

Additives are used in nylons to improve thermal and photolytic stability, facilitate processing, increase flammability resistance, increase lubricity, and generally improve whatever specific property is required for a specific application. Susceptibility to modification is an important asset. Fiber and mineral reinforcements are widely used. Blending with elastomeric modifiers has yielded nylons with improved toughness (impact strength).

Molded and extruded nylons are used in virtually every industry and market. Transportation represents the largest single market for nylons (~30% of the market). Applications for unreinforced materials include electrical connectors, wire jackets, and light-duty gears for windshield wipers and speedometers. Toughened nylons are used as stone shields and trim clips. Glass-reinforced nylons are used for engine fans, radiator headers, brake, and power-steering fluid reservoirs, as well as other uses. Mineral-reinforced resins are used for mirror housings. Nylons containing both glass and minerals are used as fender extensions. Nylons for electrical and electronic applications constitute about 11% of the market. Applications include plugs, connectors, relays, and antenna-mounting devices. Because of their excellent resistance to fatigue and repeated impact, nylons are widely used in industrial products such as unlubricated gears, bearings, and antifriction parts. Consumer products include toughened nylons for ski boots, ice and roller skate supports, racket sports equipment, kitchen utensils, and toys. Extruded nylon film is widely used for packaging meals and cheeses and for cook-in bags and pouches. Other applications include wiring cable jacketing, tubing used to convey fluids, and nylon filaments for paintbrush bristles, fishing lines, and sewing thread.

Polyamide-Imides (PAI)

PAIs are engineering thermoplastics characterized by excellent dimensional stability, high strength at high temperature, and good impact resistance. Molded parts can maintain structural integrity and continuous use at temperatures to 260°C.

PAI, produced and called *Torlon*, is available in several grades including a general-purpose, injection-molding grade; three polytetrafluoroethylene (PTFE)/graphite wear-resistant compounds; a 30% graphite-fiber-reinforced grade; and a 30% glass-fiber-reinforced grade. Additional grades are developed to meet special requirements.

Torlon resins are moldable on screw-injection-molding machines. Molds must be heated to 218°C, and the barrel and nozzle should be capable of being heated to about 371°C. High injection speed and pressure (136 MPa or greater) are desirable. Developing optimum physical properties in injection-molded or extruded parts requires postcuring through an extended, closely controlled temperature program gradually reaching 260°C. The specific time/temperature program depends on part configuration and thickness.

Properties

Room-temperature tensile strength of unfilled PAI is about 190 MPa and compressive strength is 213 MPa. At 232°C, tensile strength is about 61.2 MPa—as strong as many engineering plastics at room temperature—and continued exposure at 260°C for up to 8000 h produces no significant decline in tensile properties.

Flexural modulus of 36.5 MPa of the unfilled grade is increased, with graphite-fiber reinforcement, to 19,913 MPa. Retention of modulus at a temperature of 260°C is on the order of 80% for the reinforced grade. Creep resistance, even at high temperature and under load, is among the best of the thermoplastics; dimensional stability is extremely good.

PAI is extremely resistant to flame and has very low smoke generation. Reinforced grades have surpassed Federal Aviation Administration requirements for flammability, smoke density, and toxic gas emission.

Radiation resistance of PAI is good; tensile strength drops only about 5% after exposure to 10^9 rad of gamma radiation. Chemical resistance is good, the resin being virtually unaffected by aliphatic and aromatic hydrocarbons, halogenated solvents, and most acid and base solutions. It is attacked, however, by some acids at high temperature, steam at high pressure and high temperature, and strong bases.

Torlon moldings absorb moisture in human environments or when immersed in water, but the rate is low, and the process is reversible. Parts can be restored to original dimensions by drying.

Polyaryl Sulfone

A thermoplastic resin consisting mainly of phenyl and biphenyl groups linked by thermally stable ether and sulfone groups. Its most outstanding property is resistance to high and low temperatures (from −240°C to 260°C, or −400°F to 500°F). It also has good impact resistance, resistance to chemical oils and most solvents, and good electrical insulating properties. It can be processed by injection molding, extrusion, compression molding, ultrasonic welding, and machining.

Polyarylates

These high-heat-resistant thermoplastics are derived from aromatic dicarboxylic acids and diphenols. When molded, they become amorphous, providing a combination of toughness, dimensional stability, high dielectric properties, and ultraviolet (UV) stability. Polyarylates have heat-deflection temperatures up to 175°C at 264 lb/in.² The resins can be injection molded, extruded, and blow molded, and sheet can be thermoformed. These resins also are blended with other engineering thermoplastics and reinforcements. They have excellent dimensional and UV stability, electrical properties, flame retardants, and warp resistance.

Polyarylene Ether Benzimidazoles (PAEBI)

PAEBI polymers, developed at NASA-Langley in 1991, common by high-gloss transition temperature and optical transmission with inherent resistance to high-energy particles, especially atomic oxygen, thus being suitable for thermal control in harsh space environments, such as low earth orbit. Though reacting similarly to other plastics on initial exposure to atomic oxygen, PAEBI incorporates a phosphorus–oxygen linkage that subsequently forms an in situ protective coating, making the plastic more than 15 times as resistant as *Kapton* polyimide to erosion by atomic oxygen. The polymer is produced in the form of *TOR* and *TOR-LM castable films* by Triton Systems Inc. and, for thermal-control applications, can be metallized with aluminum, silver, or Inconel.

Polyaryletherketones (PAEK)

A glass-fiber-reinforced (PAEK) semicrystalline polymer has been designed into the sensor housing for a new conductivity-measuring cell.

Conductivity-measuring cells are used to determine the electrolytic conductivity of media in the food and pharmaceutical industries. The primary requirements for the cell are resistance to corrosive media and biocompatible surface quality to remain in compliance with U.S. and European hygienic requirements. The PAEK polymer is insoluble in all common solvents and can be immersed for thousands of hours at temperatures more than 250°C in steam or high-pressure water environments without significant degradation. The addition of glass fibers increases the chemical resistance of the base resin as well as mechanical strength at elevated temperatures. This definition is also germane to *polyetheretherketone*, *polyetherketone*, and *polyetherketoneketone*.

Polybenzimidazole (PBI)

A thermoplastic resin that is strong and stable, has a high molecular weight, and contains recurring aromatic units. The resin is produced by the high-temperature, melt condensation reaction of aromatic bis-ortho-diamines and aromatic dicarboxylates (acids, esters, or amides). Although a wide variety of PBIs have been synthesized, the most common is (2.2′–(m-phenylene)-5.5′-bibenzimidazole). Parts made from PBI are used in petrochemical, geothermal, chemical process, power generation, aerospace, and transportation markets.

PBI has been used in a variety of applications for the U.S. space program, including flight suits and other protective clothing, webbings, straps, and tethers. Research of raw materials and process development has continued, along with applications development that began in the 1960s and 1970s.

In 1983, commercial production of PBI fiber commenced. Today, PBI fiber has been used successfully in firefighters' gear, industrial protective clothing, fire-blocking layers for aircraft seats, braided pump packings, and other high-performance products. Research has continued to develop other forms of PBI. In addition to *Celazole* molded parts, these forms include polymer additives, films, fibrids, papers, microporous resin, sizing, and coatings.

P

Scientists say the material has no known melting point and can withstand pressures of up to 394 MPa. It also has demonstrated resistance to steam at 343°C.

Polybutylene Terephthalate

A member of the polyalkyleneterephthalate family that is produced by the transesterification of dimethyl terephthalate with butanediol by means of a catalyzed melt polycondensation. Properties include high strength, dimensional stability, low moisture absorption, good electrical characteristics, and resistance to heat and chemicals when suitably modified.

Polycarbonates

Polycarbonate resins offer a combination of properties that extends the usefulness and fields of application for thermoplastic materials. This relatively new plastic material is characterized by very high impact strength, superior heat resistance, and good electrical properties.

In addition, the low water absorption, high heat distortion point, low and uniform mold shrinkage, and excellent creep resistance of the material result in especially good dimensional stability. Of value for many applications are its transparency, shear strength, stain resistance, color ability and loss, oil resistance, machinability, and maintenance of good properties over a broad temperature range from less than −73°C to 132°C to 138°C. The fact that polycarbonate resin is self-extinguishing is important in many applications.

Polycarbonates are amorphous engineering thermoplastics that offer exceptional toughness over a wide temperature range. The natural resins are water-clear and transparent.

Polycarbonate resins are available in general-purpose molding and extrusion grades and in special grades that provide specific properties or processing characteristics. These include flame-resistant formulations as well as grades that meet Food and Drug Administration regulations for parts used in food-contact and medical applications. Other special grades are used for blow molding, weather and UV-resistance, glass-reinforcement, EMI, RFI, ESD-shielding, and structural-foam applications. Polycarbonate is also available in extruded sheet and film.

Composition

The polycarbonate name is taken from the carbonate linkage that joins the organic units in the polymer. This is the first commercially useful thermoplastic material that incorporates the carbonate radical as an integral part of the main polymer chain.

There are several ways polycarbonates can be made. One method involves a bifunctional phenol, bisphenol A, which combines with carbonyl chloride by splitting out hydrochloric acid to give a linear polymer consisting of bisphenol groups joined together by carbonate linkages. Bisphenol A, which is the condensation product of phenol and acetone, is the basic building block used also in the preparation of epoxy resins.

Other members of the polycarbonate family may be made by using other phenols and other ketones to modify the isopropylidene group or to replace this bridge entirely. Substitutions on the benzene ring offer further possibilities for variations.

Properties

Polycarbonate is a linear, low-crystalline, transparent, high-molecular-weight plastic. It is generally considered to be the toughest of all plastics. In thin sections, up to about 0.478 cm, its impact strength is as high as 24 kg. In addition, polycarbonate is one of the hardest plastics. It also has good strength and rigidity and, because of its high modulus of elasticity, is resistant to creep. These properties, along with its excellent electrical resistivity, are maintained over a temperature range of about −170°C to 121°C. It has negligible moisture absorption, but it also has poor solvent resistance and, in a stressed condition, will craze or crack when exposed to some chemicals. It is generally unaffected by greases, oils, and acids. Polycarbonate plastics are easily processed by extrusion; by injection, blow, and rotational molding; and by vacuum forming. They have very low and uniform mold shrinkage. With a light transmission of almost 90% and high impact resistance, they are good glazing materials. They have more than 30 times the impact resistance of safety glass. Other typical applications are safety shields and lenses. Besides glazing, the high impact strength of polycarbonate makes it useful for air-conditioner housings, filter bowls, portable tool housings, marine propellers, and housings for small appliances and food-dispensing machines.

Humidity changes have little effect on dimensions or properties of molded parts. Even boiling water exposure does not change dimensions more than 0.30 mm/mm if the parts return to room temperature. Creep resistance is excellent throughout a broad temperature range and is improved by a factor of 2–3 in glass-reinforced compounds.

The insulating and other electrical characteristics of polycarbonate are excellent and almost unchanged by temperature and humidity conditions. One exception is arc resistance, which is lower than that of many other plastics.

Polycarbonates are generally unaffected by greases, oils, and acids. Nevertheless, compatibility with specific substances in a service environment should be checked with the resin supplier. Water at room temperature has no effect; the continuous exposure in hot (65°C) water causes gradual embrittlement. The resins are soluble in chlorinated hydrocarbons and are attacked by most aromatic solvents, esters, and ketones, which cause crazing and cracking in stressed parts. Grades with improved chemical resistance are available, and special coating systems can be applied to provide additional chemical protection.

Fabrication

Polycarbonate resin has been molded in standard injection equipment using existing molds designed for nylon, polystyrene, and acrylic, or other thermoplastic materials. Differences in mold shrinkage must be considered. And, in fabrication, the polycarbonate does have its own unique processing characteristics. Most important among these are the broad plastic range and high melt viscosity of the resin. Production runs in molds designed for nylon or acetal resin are not recommended.

Like other amorphous polymers, polycarbonate resin has no precise melting point. It softens and begins to melt over a range from 216°C to 227°C. Optimal molding temperatures lie above 271°C. The most desirable range of cylinder temperatures for molding the resin is in the area of 275°C to 316°C.

Mold design must take into consideration the high melt viscosity of the material. Large sprues, large full-round runners, and generous gates with short lands usually give best results.

Tab gating is good for filling large thin sections. The gate to the tab should be large.

In injection molding of polycarbonate, the following conditions are desirable:

1. *Heated molds*: Generally, hot water heat is adequate for heating molds, with typical molding temperatures ranging from 77°C to 93°C. Molds for large areas, thin sections,

or complex shapes or multiple cavity molds with long runners may require higher temperatures. Some molds have run best at 110°C–121°C.

2. *Cylinder temperatures*: The most usual cylinder temperature for molding polycarbonates is in the range of 275°C–316°C. Few parts require cylinder temperatures above 316°C. A heated nozzle and adequate molding temperature are helpful in keeping cylinder temperatures below 316°C. In most cases, rear cylinder temperatures higher than front cylinder temperatures give best results.

3. *Heated nozzle*: In general, nozzle temperature equal to front cylinder temperature gives good results.

4. *Adequate injection pressure*: Injection pressures used in molding polycarbonate resin range from 80 to 204 MPa. Most usual range is in the 103–136 MPa range. Typical pressure setting is 3/4 to full pressure capacity of the press.

5. *Fast fill time*: For most parts molded, a fast ram travel time has been found desirable for thick as well as thin sections. For very thick sections, it is better to utilize somewhat slower ram speed.

Polycarbonate must be well dried to obtain optimum properties in the molded part. For this reason, the resin is packaged in sealed containers.

Preheating pellets in the can to 121°C for 4–8 h and using copper heaters at 121°C are recommended for production operations to prevent moisture pickup.

Although polycarbonates are fabricated primarily by injection molding, other fabricating techniques may be used. Rod, tubing, shapes, film, and sheet may be extruded by conventional techniques. Films and coatings may be cast from solution. Parts can readily be machined from rod or standard shapes. Cementing, painting, metallizing, heat sealing, welding, machining operations, and other standard finishing operations may be employed. Film and sheet can be vacuum formed or cold formed.

Applications

The properties of polycarbonate resin make this new plastic suitable for a wide variety of applications. It is now being used in business machine parts, electrical and electronic parts, military components, and aircraft parts and is finding increasing use in automotive, instrument, pump, appliance, communication equipment, and many other varied industrial and consumer applications.

One of the applications is in molded coil forms, which take advantage of the electrical properties, heat and oxidation resistance, dimensional stability, and resistance to deformation under stress of the resin.

A transparent plastic with the heat resistance, the dimensional stability, and the impact resistance of polycarbonate resin has created considerable interest for optical parts, such as outdoor lenses, instrument covers, lenses, and lighting devices.

Housings make use of the impact resistance of the material and its attractive appearance and colorability. In many cases, also, heat resistance and dimensional stability are important.

An interesting application area for plastic materials is the use of polycarbonate resin for fabricating fasteners of various types. Such uses as grommets, rivets, nails, staples, and nuts are in production or under evaluation. The ability of polycarbonate parts to be coldheaded has developed considerable interest in rivet applications.

Terminal blocks, connectors, switch housings, and other electrical parts may advantageously be molded of polycarbonate resin to take advantage both of the electrical properties of the material and the unusual physical properties, which give strength and toughness

to the parts over a range of temperatures. Because of the heat resistance of polycarbonate, and the fact that it is self-extinguishing, molded parts can be used in current-carrying support applications.

Another application is in bushings, cams, and gears. Here, dimensional stability is important as is the high impact strength of the resin. Good physical properties over a broad temperature range, low water absorption, resistance to deformation under load, and resistance to creep suggest its use for many applications of this type. However, the resin has a higher coefficient of friction and a lower fatigue endurance limit than do some other plastics used in these types of applications. For this reason, it should not be considered a general-purpose gear and bearing material, but might be considered for applications subject to light loading or to heavier but intermittent loading.

In most heart bypass operations, the saphenous vein from a patient's leg replaces blocked blood vessels in the heart. The separate surgical procedure, performed during the heart bypass operation, involves removing the vein through a long incision. Following the surgery, patients frequently complain of ongoing leg pain, potentially leading to reduced mobility and delayed rehabilitation while the large incision heals.

A new generation of surgical instruments not only makes the procedure less invasive, it helps speed the patient's return to normal activities. The system uses endoscopic techniques to harvest the vein. This, in turn, requires smaller incisions. The potential benefits are less postoperative pain, fewer wound healing complications, minimal scarring, and quicker recovery.

The materials chosen for these applications have provided several significant benefits, and, from the medical side, the materials were biocompatible. The materials resist chemicals and withstand gamma sterilization. The balloon amount is made with polycarbonate. On the orbital dissection cannula, the end piece embodies a polycarbonate resin.

Polybutylenes

A group of polymers consisting of isotactic, stereoregular, highly crystalline polymers based on butylene 1. Their properties are similar to those of polypropylene and linear polyethylene, with superior toughness, creep resistance, and flexibility.

Polybutadiene Rubber

Polybutadiene may be prepared in several ways to yield different products. The method of polymerization can have a marked effect on polymer structure, which in turn controls the properties of the polymer and thus its ultimate end use.

Composition

When polybutadiene is made, the butadiene molecule may enter the polymer chain by either 1,4-addition or 1,2-addition. In 1,4-addition, the unsaturated bonds may be either of *cis* or of *trans* configuration. Polybutadienes containing a more or less random mixture of these polymer units can be prepared with alkali metal catalysts or with emulsion polymerization systems, but these have not achieved complete commercial significance as general-purpose rubbers in the United States.

In recent years, new catalysts have been developed that allow the structure of the polymer chain to be controlled. Polybutadienes containing an excess of 85% of either *cis*, *trans*, or vinyl unsaturation have been studied and, in the case of *cis* and *trans* polymers, produced commercially. It is also possible to prepare a number of other polymers with various combinations of these three component structures.

Of current importance are those types commercially available, which are (1) polymers of high *cis*-1,4 content (more than 85%) with low *trans* and vinyl content; (2) polymers of more than 80% *trans*, low *cis*, and low vinyl content; and (3) polymers of intermediate *cis* content (approximately 40% *cis*, 50% *trans*) with low vinyl content.

High-cis Polybutadiene

Outstanding properties of high-*cis* polybutadiene are high resilience and high resistance to abrasion. The high resilience is indicative of low hysteresis loss under dynamic conditions, that is, low heat rise in the polymer under rapid, repeated deformations. In this respect, *cis*-polybutadiene is similar to natural rubber. The combination of good hysteresis properties and resistance to abrasion makes this polymer attractive for use in tires, especially in heavy-duty tires where heat generation is a problem.

Another favorable factor, particularly for use in tire bodies, is that high-*cis* polybutadiene imparts good resistance to heat degradation under heavy loads in dynamic applications. Tensile strength in reinforced stock is lower than for natural rubber or styrene–butadiene rubber (SBR) but is adequate for many uses. Modulus depends on the degree of cross-linking, but *cis*-polybutadiene generally requires relatively low stress to reach a given elongation (at slow deformation). Hardness may be varied depending on the compound formulation and is similar to that of SBR and natural rubbers. Ozone resistance is typical of unsaturated polymers and is inferior to that of saturated rubbers. Oil resistance is comparable to that of SBR and natural rubber. Permeability to gases is higher than that of most other rubbers, which may be either an advantage or a disadvantage, depending on the application. Freezing point is quite low; *cis*-polybutadiene tread compounds do not become brittle until very low temperatures are reached, lower than −101°C.

Abrasion resistance of the high-*cis* polybutadiene varies with the type of service and in nearly all cases is superior to that of natural rubber or SBR. The advantage for these polybutadienes in relation to natural rubber and SBR increases as the severity of the service increases. From 30% to as high as 100% improvement in wear resistance has been reported through the substitution of these polybutadienes for natural rubber in tire treads.

Processing

Polybutadiene rubbers in general are more difficult to process in conventional equipment than natural rubber, particularly with regard to milling and extrusion operations. The processing problems have been overcome in many cases by treating the polymer, changing compounding formulations, or blending with other rubbers. Blends of natural rubber and a high-*cis* polybutadiene are particularly attractive. The presence of natural rubber alleviates the processing difficulties and improves tensile and tear properties of the polybutadiene, while the latter improves abrasion resistance and complements the already good hysteresis properties of natural rubber.

Processability can also be improved by the use of higher levels of reinforcing fillers and oils than normally employed. Consideration must also be given to changes in quality, but extension with 35–70 phr oil is possible with retention of properties suitable for tires and many rubber goods. Even in blends with natural rubber, where processing is not a problem, increases in carbon black and oil content have proved practical and often desirable.

Applications

For most uses, vulcanization is necessary to develop the desired strength and elastic qualities. The polymer chain is unsaturated and can be readily vulcanized with sulfur (in conjunction with the usual activators) or with other cross-linking agents such as peroxides. With some exceptions, admixture of the rubber with a reinforcing pigment is required to obtain high strength. Antioxidants should be used for most applications. Lower than normal sulfur levels provide a better balance of properties in many instances, particularly with blends of high-*cis* polybutadiene and natural rubber.

In large tires, the use of blends of high-*cis* polybutadiene and natural rubber in treads improves both abrasion resistance and resistance to tread groove cracking compared to natural rubber alone. Substitution of high-*cis* polybutadiene for a portion of the natural rubber in the tire body has improved resistance to blowout or other heat failures (in some cases to a remarkable degree).

The high-*cis polybutadienes* are also used in increasing quantity in blends with SBR for the production of tires for passenger cars and small trucks. In this use, advantages include improved resistance to abrasion and cracking as well as adaptability to extension with large amounts of oil and carbon black.

As indicated, properties of high-*cis* polybutadiene are well suited for tire use, and this is expected to be the major application. Its use should be considered in other areas where high resilience, resistance to abrasion, or low-temperature resistance is required. Sponge stocks, footwear, gaskets and seals, and conveyor belts are possible applications. In shoe heels, high-*cis* polybutadiene may be used to provide good resilience and better abrasion resistance than realized with other commonly used rubbers. In shoe soles designed to be soft and resilient, high-*cis* polybutadiene has been used as a partial or total replacement for natural rubber to maintain high resilience, improve abrasion resistance, and provide better resistance to crack growth.

An important use for high-*cis* polybutadiene is in blends with other polymers to improve low-temperature properties. Replacement of 35% of an acrylonitrile (nitrile) rubber in a compound with *cis* polybutadiene can reduce the brittleness failure temperature from −40°C to −55°C. Similarly, it is possible to reduce the brittle point of neoprene compounds by some −6.7°C with the substitution of *cis* polybutadiene for one-fourth of the neoprene rubber. Such substitutions usually reduce resistance to swelling in hydrocarbons, but this effect can be lessened by compounding with oil-resistant resins or by using *cis* polybutadiene as a replacement for plasticizer in the compound rather than as a replacement for the polymer. In many cases, of course, oil resistance is not required, and *cis* polybutadiene may be blended with various polymers to give low-temperature properties approaching those of special arctic rubbers such as butadiene–styrene copolymers with low styrene content.

Partial substitution of *cis* polybutadiene for other rubbers used in light-colored or black-reinforced mechanical goods can provide a product that displays better slap. Also, resilience, abrasion resistance, and low-temperature properties are usually improved; tensile strength and tear strength may be reduced but shows less decline after aging.

High-*cis* polybutadienes have been used successfully as base components of caulking compounds and sealants. Another use is as a base polymer for graft polymerization of styrene to produce high-impact polystyrene.

Commercially available *cis* polybutadiene rubbers are supplied in talc form. The rubber contains a small amount of antioxidant to provide stability during storage.

High-Trans Polybutadiene

In contrast to the soft, rubbery nature of *cis* polybutadiene, polybutadiene of high *trans* content is a hard, horny material at room temperature. It is thermoplastic and thus can be molded without

P

addition of vulcanization agents. The softening point, hardness, and tensile strength increase with increasing *trans* content. A polymer of approximately 90% *trans* content, for example, displays a softening point near 93°C, Shore A hardness of about 98, and tensile strength of more than 6.8 MPa without addition of other ingredients.

High-*trans* polybutadiene can be readily vulcanized with sulfur or peroxides. It remains a hard, partially thermoplastic material when lightly vulcanized but can be made rubbery by increasing the curative level. Tensile strength is increased by the addition of reinforcing pigments (carbon black, silica, or clay). Compounded in such a manner, high-*trans* polybutadienes are characterized by high modulus, tensile strength, elongation, and hardness, by moderate resilience, and by excellent abrasion resistance.

In vulcanized stocks, properties such as hardness, resilience, and heat buildup can be modified.

Uses

Applications include those where balata has been used such as in golf ball covers and wire and cable coverings. Properties have also been demonstrated to be suitable for shoe soles (high hardness, good abrasion resistance), floor tile (high hardness, low compression set), gasket stocks, blown sponge compounds, and other molded or extruded items. High-*trans* polybutadiene requires processing temperatures above its softening point but is relatively easy to mix, mill, or extrude at these temperatures. Care should be taken to avoid scorch or precure when handling stocks containing curatives at elevated temperatures.

Polybutadiene of Intermediate cis Content

This type of product has a structure of approximately 40% *cis*, 50% *trans*, and 10% (or less) vinyl content.

The raw polymer is often somewhat waxlike in character. This polybutadiene displays high resilience, high dynamic modulus, and a low brittle point.

Processing difficulties in milling and extrusion operations may be encountered, and it is usually recommended that polybutadienes of intermediate *cis* content be used in blends with other rubbers. In tire treads, substitution of 40% *cis*-polybutadiene for a portion of the natural rubber or SBR 1500 improves resilience and abrasion resistance and gives some reduction in operating temperature. Other applications include products where similar changes in the aforementioned properties are desired or products where improvements in low-temperature properties are desired.

Liquid Polymers

Liquid polybutadienes can be prepared in most of the systems used to make solid rubbers. Such polymers may be cross-linked with chemicals or solidified with heat. One type of liquid polybutadiene has been manufactured on a small-scale utilizing sodium catalyst. Uses for this type of polymer include coatings, binders, adhesives, potting agents, casting and laminating resin, or vulcanizable plasticizer for rubber.

Polychromator

A *spectrometer* that has many (typically 20–50) detectors for simultaneously measuring light from many spectral lines. Polychromators are commonly used in atomic emission spectroscopy. See also *monochromator*.

Polycondensation

See *condensation polymerization*.

Polycrystalline

Pertaining to a solid comprised of many crystals or crystallites, intimately bonded together. May be homogeneous (one substance) or heterogeneous (two or more crystal types or compositions).

Polyester Film

Polyester film is a transparent, flexible film, ranging from 0.15 to 14 mils and thickness, used as a product component, in industrial processes, and for packaging. The most widely used type is produced from polyethylene terephthalate (i.e., *Mylar*). Polyester films based on other polymers or copolymers or manufactured by other methods are not identical, although they are similar in nature.

It is the strongest of all plastic films and strength is probably the outstanding property. However, it is useful as an engineering material because of its combination of desirable physical, chemical, electrical, and thermal properties. For example, strength combined with heat resistance and electrical properties makes it a good material for motor slot liners.

Polyester film was made by the condensation of terephthalic acid and ethylene glycol. The extremely thin film, 0.00063–0.0013 cm, used for capacitors and for insulation of motors and transformers, has a high dielectric strength, up to 236×10^6 V/m. It has a tensile strength of 137 MPa with the elongation of 70%. It is highly resistant to chemicals and has low water absorption. The material is thermoplastic, with a melting point at about 254°C. Polyester fibers are widely used in clothing fabrics. The textile fiber produced from dimethyl terephthalate is known as *Dacron*.

For magnetic sound-recording tape, polyester tape has the molecules oriented by stretching to give high strength. The 0.013 cm tape has a breaking strength of 3.4 kg/0.64 cm in width. Electronic tape may also have a magnetic-powder coating on the polyester. But where high temperatures may be encountered, as in spacecraft, the magnetic coating is applied to metal tapes.

Properties

Polyester film has excellent resistance to attack and penetration by solvents, greases, and oils and many of the commonly used electrical varnishes. At room temperature, permeability to such silence as ethanol, ethyl acetate, carbon tetrachloride, hexane, benzene, acetone, and acetic acid is very low. It is degraded by some strong alkali compounds and embrittles under severe hydrolysis conditions.

Moisture absorption is less than 0.8% after immersion for a week at 25°C. Water vapor permeability is similar to that of polyethylene film and permeability to gases is very low. The film is not subject to fungus attack and copper corrosion is negligible.

An outstanding feature of the film is the fact that good physical and mechanical properties are retained over a wide temperature range. Service temperature range is –60°C to 150°C. The effect of temperature is relatively small between –20°C and 80°C. No embrittlement occurs at temperatures as low as –60°C, and useful properties are retained up to 150°C–175°C. Tensile modulus drops off sharply at 80°C–90°C.

Melting point is 250°C–255°C, thermal coefficient of expansion is 15×10^{-6} in./in./°F; and shrinkage at 150°C is 2%–3%.

Fabrication and Forms

Polyester film can be printed, laminated, metallized, coated, embossed, and dyed. It can be split into extremely narrow tapes (0.038 mm and narrower), and light gauges can be wound into spiral tubing. Heavy gauges can be formed by stamping or vacuum (thermo-) forming.

P

Matte finishes can be applied, and adhesives for bonding the film to itself and practically any other material are available.

Because of its desirable thermal characteristics, polyester film is not inherently heat sealable. However, some coated forms of the film can be heat-sealed, and satisfactory seals can be obtained on the standard film by the use of a benzyl alcohol, heat, and pressure.

Polyethylene terephthalate film is available in several different types:

A. General-purpose and electrical film for wide variety of uses
C. Special electrical applications requiring high insulation resistance
D. Highly transparent film, minimum surface defects
K. Coated with a polymer for heat sealability and outstanding gas and moisture impermeability
HS. Shrinks uniformly about 30% when heated to approximately 100°C (after shrinking, it has substantially the same characteristics as the standard film)
T. A film with high tensile strength (available in some thin gauges) with superior strength characteristics in the machine direction, designed for use in tapes requiring high-strength properties
W. For outdoor applications, resistant to degradation by ultraviolet light

Applications

The range of properties outlined previously has made polyester film functional in many totally different industrial applications and suggests its use in numerous other ways.

The largest current user of the film is the electrical and electronic market, which uses it as slot liners in motors and as the dielectric for capacitors, replacing other materials that are less effective, bulkier, and more expensive. It is found in hundreds of wire and cable types, sometimes used primarily as an insulating material, sometimes for its mechanical and physical contributions to wire and cable construction. Reduction of cost of materials and processing can improve cable performance result.

Magnetic recording tapes for both audio and instrumentation uses are based on polyester film. In audio application, they contribute toughness, durability, and long play, or for instrumentation tapes, the film ensures maximum reliability. The film has proved to be a highly successful new material for the textile industry since it can be used to produce metallic yarns that are nontarnishing and unusually strong. They can be run unsupported, knit, dyed at the boil, and either laundered or dry-cleaned. Yarns are made by laminating the film to both sides of aluminum foil or by laminating metallized and transparent film. The structure is then slit into the required yarn widths.

As a surfacing material, polyester film is used on both flexible and rigid substrates for both protective and decorative purposes. Metallized, laminated to vinyl, and embossed, the film becomes interior trim for automobiles, for example.

With a special coating, the film becomes a drafting material that is tougher and longer lasting than drafting cloth. It is used for mapmaking, templates, and other applications in which its dimensional stability becomes a significant factor.

Strength in thin sections makes the film advantageous for sheet protectors, cardholders, sheet reinforcers, and similar stationery products. Pressure-sensitive adhesives make the properties of the film available for uses ranging from decorative trim to movie splicing.

The weatherable form (type W) has a life of 4–7 years. It is principally used in greenhouses, where it cuts construction costs by as much as two-thirds because a simple, inexpensive structure suffices and maintenance costs are at a minimum.

In the packaging field, polyester film serves in areas where other materials fail or have functional disadvantages. In window cartons, it lasts longer and does not break as other materials do; its toughness permits transparent packaging of heavy items, and, coated with polyethylene, it has made possible the "heat-in-the-bag" method of frozen-food preparation.

An optically clear form of polyester film in thicknesses of 4–7 mils is used as a base for the coating of light-sensitive emulsions in the manufacture of photographic film. The outstanding qualities of toughness and dimensional stability make this film especially well suited as a base for graphic arts films, motion picture film, engineering reproduction films, and microfilm. Other advantages include excellent storage and aging characteristics.

Polyester Plastics

The materials are commonly called polyester resins, but this simple name does not distinguish between at least two major classes of commercial materials. Also, the same name is used within the unsaturated class to designate both the cured and uncured state. These plastics may be defined by identifying the materials as "unsaturated polyester resins that, when cured, yield thermoset products as opposed to thermoplastic products." The latter, as exemplified by *Dacron* and *Videne*, are saturated polyesters.

Composition

Unsaturated polyester resins of commerce are composed of two major components, a linear, unsaturated polyester and a polymerizable monomer. The former is a condensation type of polymer prepared by esterification of an unsaturated dibasic acid with a glycol. Actually, most polyesters are made from two or more dibasic acids.

The most commonly used unsaturated acids are maleic or fumaric with limited quantities of others used to provide special properties. The second acid does not contain reactive unsaturation. Phthalic anhydride is most commonly used, but adipic acid and, more recently, isophthalic acid are employed. The properties of the final product can be varied widely, from flexible to rigid, by changing the ratios and components in the polyester portion of the resin.

The polyesters vary from viscous liquids to hard, brittle solids but, with a few exceptions, are never sold in this form. Instead, the polyester is dissolved in the other major component, a polymerizable monomer. This is usually styrene; diallyl phthalate and vinyl toluene are used to a lesser extent. Other monomers are used for special applications.

Polyester resins may contain numerous minor components such as light stabilizers and accelerators, but must contain inhibitory components so that storage stability is achieved. Otherwise, polymerization will take place at room temperature.

The most widely used polyester resins are supplied with viscosities in the range of 300–5000 centipoise. They are clear liquids varying in color from nearly water-white to amber. They can be colored with certain common pigments, which are available ground in a vehicle for ease of dispersion. Inert, inorganic fillers such as clays, talcs, and calcium carbonates are often added, usually by the fabricator, to reduce shrinkage and lower costs. Resins with thixotropic properties are also available.

Curing

The liquid resins are cured by the use of peroxides with or without heat to form solid materials. During the cure, the monomer copolymerizes with the double bonds of the unsaturated polyester. The resulting copolymer is thermoset and does not flow easily again under heat and pressure. Heat is involved during the cure. This must be considered when thick sections are made. Also, a volume shrinkage occurs and the density increases to 1.20–1.30. The amount of shrinkage varies between 5% and 10% depending on monomer content and degree of unsaturation in the polyester. The most popular catalyst is benzoyl peroxide but methylethyl ketone peroxide, cumene hydroperoxide, and others find application.

Reinforcement

The development of this field of commercial resins owes a great deal to the commercial production of two other products: The first was the production of low-cost styrene for the synthetic rubber program. Other cheap monomers will work but not nearly as well. Styrene and its homologues have relatively high boiling points and fast polymerization rates. Both properties are important in this field.

The second development was the commercial production of fibrous glass. Polyester resins are not widely used as cast materials nor are the physical properties of such castings particularly outstanding.

The unique property of polyesters is their ability to change from a liquid to a hard solid in a very short time under the influence of a catalyst and heat. This property was not available in any of the earlier plastic materials. Polyesters flow easily in a mold with little or no pressure so that expensive, high-pressure molds are not required. Alternatively, very large parts can be made because the total pressure required to form the material is low.

Glass fibers of fine diameter have high tensile strengths, good electrical resistance properties, and low specific gravity when compared to metals. When such fibers are used as reinforcement for polyester resins (like steel in concrete), the resulting product possesses greatly enhanced properties. Specific physical properties of the polyester resins may be increased by a factor varying from 2 to 10. Naturally, the increase in physical strengths obtained will depend upon both the amount of glass fibers used and their form. Strengths approaching those of metals on an equal weight basis are obtained with some constructions.

Fabrication Techniques

Hand Lay-Up

This method involves the use of either male or female molds. Products requiring the highest physical properties are made with glass cloth. The cloth may be precoated with resin, but usually, one ply is laid on or in the mold, coated with resin by brushing or spraying, and then a second ply of cloth laid on top of the first. The process is continued until the desired thickness is built up. Aircraft parts, such as radomes, usually require close tolerances on dimensions and resin-to-glass weight ratio. After the lay-up has been completed, pressure is applied either by covering the assembly with flexible, extensible blanket or drawing it down by vacuum, or the mold is so made that a rubber bag can be contained above the part. This is then blown up to apply pressure on the laminate. Thereafter, the cure is accomplished by heating in an oven, by infrared lamps, or by heating means built into the mold.

Corrugated Sheet Molding

Glass-reinforced polyester sheets are sold in large volume and are made by a relatively simple intermittent or continuous method. The process consists of placing a resin-impregnated mat between cellophane sheets, rolling or squeegeeing out the air, placing the assembly between steel and aluminum molds of the desired corrugation, and curing the assembly in an oven.

Matched Die Molding

This method produces parts rapidly and generally of uniform quality. The molds used are somewhat similar to those employed in compression molding, usually of two-piece, mating construction. The process consists of two steps. A "preform" of glass fibers is prepared by collecting fibers on a screen, which has the shape of the finished article. Suction is used to hold the fibers on the screen; fibers are either blown at it or fall on it from a cutter. Commercial equipment is available for this operation. When the desired weight of glass fibers has been collected, a resinous binder is applied. The preform and screen are then baked to cure the binder, after which the preform is ready for use. The second step is the actual molding. The preform is placed in the mold, the catalyzed resin poured on the preform incorrect amount, the mold closed, and the cure affected by heat. Design of the mold is very important, and some features are different from those in other molding fields.

Molding cycles vary from 2 to 5 min at a temperature from 110°C to 149°C. Trimming, sanding, and buffing are usually required at the flash line.

Premix Molding

This branch of the field provides parts for automotive and similar end uses. The parts are strictly functional and are usually pigmented black. The strength properties required are relatively low except that impact strength must be good. As the name infers, the unsaturated resin is first mixed with fillers, fibers, and catalyst to pursue a nontacky compound. The mixers used are of the heavy-duty Day or Baker Perkins type. The "premix" is usually extruded to provide a "rope" or strip of material easily handled at the press. The fibers used most extensively are cut sisal, but glass and asbestos are also used, and frequently all three are present in a compound. The fillers are clays, carbonates, and similar cheap inorganic materials. A typical premix will contain 38% catalyzed unsaturated resin, 12% total fiber, and 50% filler. However, rather wide variations in composition are practiced to obtain specific end-use properties.

The premix is molded at pressures of 103–350 MPa and at temperatures ranging from 121°C to 154°C. Cure cycles are short, usually from 30 to 90 s. Again, the fact that the resin starts as a liquid makes possible the molding of intricate parts because of ease of flow in the mold. Heater housings for autos are the largest use, but housings of many types and electrical parts are produced in volume. This process provides a cheap but very serviceable molded material.

Properties

The strengths obtainable in the finished product are of prime interest to the engineer. However, the numerous forms of reinforcement materials and the variations possible in the polyester constituent present a whole spectrum of obtainable properties. As an example, products are available that will resist heat for long periods of time at temperatures varying from 66°C to 177°C, but the higher the temperature, the more expensive the resin.

In general, the commercial resins have good electrical properties and are resistant to dilute chemicals. Alkali resistance is poor, as is resistance to strong acids. The strength-to-weight ratio of polyester parts and their impact resistance are outstanding physical properties.

Available Forms

Probably, over 95% of the unsaturated polyester resins sold are liquids in the uncured state. However, certain types are available in solid or paste form for special uses. The cured resins are available in laminate form as corrugated sheeting, which is sold widely for partitions, windows, patio roofs, etc. Rod stock may be purchased for fishing rods and electrical applications. Paper and glass cloth laminates are sold to fabricators. The boat end use is making the material more familiar to the general public. However, a large part of the industrial production is concerned with custom-molded parts. These are perhaps best classified by industries rather than specific products. The aircraft industry uses substantial quantities but automotive end uses are larger. The chemical industry is using increasing amounts in fume ducts and corrosion-resistant containers. The electrical industry uses the material in laminate and molded forms and as an encapsulation medium. Furniture applications are growing. The machinery industry uses moldings as housings and guards; a recent volume use is motorboat shrouds. There are few fields that have not found the material useful in some applications. See also *thermoplastic polymers*, *thermosetting polyesters,* and *unsaturated polyesters.*

Polyester Thermoplastic Resins

There are several types of melt-processable thermoplastics, including polybutylene terephthalate, polyethylene terephthalate, and aromatic copolyesters.

Polybutylene Terephthalate (PBT)

This plastic material is made by the transesterification of dimethyl terephthalate with butanediol through a catalyzed melt polycondensation. These molding and extrusion resins have good resistance to chemicals, low moisture absorption, relatively high continuous-use temperature, and good electrical properties (track resistance and dielectric strength). PBT resin is sensitive to alkalies, oxidizing acids, aromatics, and strong bases. Various additives, fillers, and fiber reinforcements are used with PBT resins, in particular flame retardants, mineral fillers, and glass fibers. PBT resins and compounds are used extensively in automotive, electrical–electronic, appliance, military, communications, and consumer product applications.

Polyethylene Terephthalate (PET)

This is a widely used thermoplastic packaging material. Beverage bottles and food trays for microwave and convection oven use are the most prominent applications. PET resins are made from ethylene glycol and either terephthalate acid or the dimethyl ester of terephthalic acid. Most uses for PET require the molecular structure of the material to be oriented. Orientation of PET significantly increases tensile strength and reduces gas permeability and water vapor transmission. For packaging uses, PET is processed by blow molding and sheet extrusion. Typical trade names are *Cleartuf, Traytuf, Tenite,* and *Kodapak.* Injection-molding grades of PET, with fillers and/or reinforcements, also are available for making industrial products.

Aromatic polyesters (liquid crystal polymers [LCP]) have high mechanical properties and heat resistance. Commercial grades are aromatic polyesters, which have a highly ordered or liquid crystalline structure in solution and molten states. A high degree of molecular orientation develops during processing and, hence, anisotropy in properties. Typically, these melt-processable resins can be molded or extruded to form products capable of being used at temperatures over 260°C. Tensile strengths up to 240.3 MPa and flexural moduli up to 3586.3 MPa are reported for LCPs. Chemical resistance also was excellent. Trade names for LCPs include *Vectra* and *Xydur Granlar.* Applications are in chemical processing, electronic, medical, and automotive components.

Polyester Thermosetting Resins

These are a large group of synthetic resins produced by condensation of acids such as maleic, phthalic, or itaconic with an alcohol or glycol such as allyl alcohol or ethylene glycol to form an unsaturated polyester that, when polymerized, will give a cross-linked, 3D molecular structure, which in turn will copolymerize with an unsaturated hydrocarbon, such as styrene or cyclopentadiene, to form a copolymer of complex structure of several monomers linked and cross-linked. At least one of the acids or alcohols of the first reaction must be unsaturated. The polyesters made with saturated acids and saturated hydroxyl compounds are called alkyd resins, and these are largely limited to the production of protective coatings and are not copolymerized.

The resins undergo polymerization during cure without liberation of water and do not require high pressure for curing. Through the secondary stage of modification with hydrocarbons, a very wide range of characteristics can be obtained. The most important use of the polyesters is as laminating and molding materials, especially for glass-fiber-reinforced plastic products. The resins have high strength, good chemical resistance, high adhesion, and capacity to take bright colors. They are also used, without fillers, as casting resins, for filling and strengthening porous material such as ceramics and plaster of Paris articles and for sealing the pores in metal castings. Some of the resins have great toughness and are used to produce textile fibers and thin plastic sheet and film. Others resins are used with fillers to produce molding powders that cure at low-pressure of 3–6 MPa with fast operating cycles.

Polyester Laminates

These are usually made with a high proportion of glass-fiber mat or glass fabric, and high-strength reinforced moldings may also contain a high proportion of filler. A resin slurry may contain as high as 70% calcium carbonate or calcium sulfate, with only about 11% of glass fiber added, giving an impact strength of 165 MPa in the cured material. Bars and structural shapes of glass-fiber-reinforced polyester resins of high tensile and flexural strengths are made by having the glass fibers parallel in the direction of the extrusion. Rods and tubes are made by having the glass-fiber rovings carded under tension and then passing through an impregnating tank, an extruding die, and a heat-curing die. The rods contain 65% glass fiber and 35% resin. They have a flexural strength of 441 MPa and a Rockwell M hardness of 65.

Physical properties of polyester moldings vary with the type of raw materials used and the type of reinforcing agents. A standard glass-fiber-filled molding may have a specific gravity from 1.7 to 2.0, a tensile strength of 27–68 MPa with the elongation of 16%–20%, a flexural strength to 206 MPa, a dielectric strength to about 16×10^6 V/m, and a heat distortion temperature of 177°C–204°C. The moldings have good acid and alkali resistance. But, because an almost unlimited number of fatty-type acids are available from natural fatty oils or by synthesis from petroleum and the possibilities of variation by combination with alcohols, glycols, and other materials are also unlimited, the polyester forms an ever-expanding group of plastics.

Some of the polyester-type resins have rubberlike properties, with higher tensile strengths than the rubbers and superior resistance to oxidation. These resins have higher wear resistance and chemical resistance than GRS (Government Rubber Styrene), a copolymer of butadiene and styrene rubber. They are made by reacting adipic acid with ethylene glycol and propylene glycol and then adding diisocyanate to control the solidifying action. They can be processed like rubber, but solidified more rapidly.

Polyetheretherketones (PEEKs)

The latest developments in PEEK resin formulations let manufacturers take advantage of its wear and strength qualities for pump applications.

Uses

Pumps with mating components made of cast iron, stainless steel, and bronze have long been a source of problems for manufacturers and end users. The abrasive quality of metals leads to significant wear in pump parts, which frequently results in failures from galling and seizing. As anyone working in a manufacturing environment knows, one's worst nightmare can come true when a pump runs dry or a bearing fails or shutting down an entire assembly line or plant operation. And the costs are hardly trivial for repairing all of the pump components destroyed or carrying out periodic maintenance on the equipment to handle such a nightmare.

Engineers are starting to replace metals with advanced plastics for pump components such as bushings, line shaft bearings, and case and impeller wear rings. The resins they use are based on PEEK chemistry and were initially developed for aerospace and defense applications. Using these engineered resins, pump manufacturers improve performance, boost output, and cut costs by taking advantage of better wear and friction qualities, mechanical strength, and the ability to resist chemicals.

The chemical makeup of PEEK resins helps ease engineers' concerns about galling and seizing and are safer to work with. Taking advantage of these qualities, along with an understanding of some basic design guidelines, opens new doors in pump performance and reliability.

Types, Fabrication, and Properties

Engineers often choose PEEK resin because of its overall balance of qualities, including hydrolysis resistance, dimensional stability, wear resistance, temperature stability, strength, and chemical resistance. However, switching from metal to plastic in pump parts requires engineers to change their mind-sets.

PEEK can be injection molded, compression molded, or extruded into stock shapes, which are typically machined to final-part dimensions. Most pump components can be injection molded as long as they do not require tight tolerances. The strict tolerances required for bushings and wear rings require parts to be machined from molded tube stock.

Although manufacturers may be used to machining metals, speeds and feeds are much different for plastics and the process generally calls for high-end carbide and diamond tooling. Machining operations vary depending on which fillers and reinforcements are used in each resin grade.

Even though plastics can be machined more quickly than metals, it is not as easy to hold tolerances because plastics tend to spring when parted off. Tolerances in the range of ±0.05 mm can be held on parts with diameters up to 254 mm.

The most common blends of PEEK resin used in pump applications are carbon filled and carbon-fiber reinforced. Carbon-filled PEEK compounds can be molded in solids or tubes up to 101.6 cm in diameter. Carbon-fiber-reinforced resins, in contrast, are robotically

wound and formed around a mandrel. This limits fiber-reinforced resins to hollow shapes no smaller than 9.5 mm in diameter. Splitting hollow shapes is not recommended because the strength of the composite relies on the carbon fibers being molded and kept in tension.

Another difference between carbon-filled and carbon-fiber-reinforced PEEK resins is how they react to temperature changes. Carbon-filled compounds have a radial thermal expansion rate of 16×10^{-6} in./in./°F, while that of fiber-reinforced resins is less than 3×10^{-6} in./in./°F. These two values fall on either side of the thermal expansion rate of carbon steel, which is 6×10^{-6} in./in./°F, making the choice between the two resins critical when combining PEEK resins with metal pump housings and components.

Carbon-filled resins are recommended for press-in applications such as bushings and case wear rings. As temperatures rise, PEEK components will become tighter because they are expanding at a faster rate than the steels surrounding the bushings or rings. In contrast, carbon-fiber-reinforced resins work better for press-on applications such as impeller wear rings. The inside diameter of the wear ring fits on the impeller skirt, which has a slightly larger outside diameter than the ring inside diameter. As operating temperatures rise, the steel impeller skirt expands more quickly than the PEEK wear rings, creating a tighter fit.

In summary, when compared to carbon-fiber-reinforced resin, carbon-filled PEEK is less expensive and easier to machine and can be split without compromising physical properties. However, fiber-reinforced resins have the advantage when it comes to higher operating temperatures. Reinforced resins are also stronger than carbon-filled blends. Fiber-reinforced PEEK is a true metal replacement because it combines the physical properties of steel with the processing ease and flexibility associated with thermoplastics.

Applications

One area where metal-to-plastic conversions are reaping benefits is in the petroleum industry where multistage horizontal pumps often experience high vibration levels that frequently cause seals to fail. Vibrations wear down bronze bushings and case wear rings, which, in turn, reduce overall pump efficiency and increase maintenance costs.

Replacing bronze and steel wear rings with carbon-filled or carbon-reinforced PEEK components, allows manufacturers to increase pump efficiency by achieving tighter clearances. PEEK resins also reduce vibrations caused by fluid flow across wear rings and excess rotor runout.

The ductility of plastic pump components helps them outperform metal versions when it comes to reducing vibration, as well as resisting impact. For example, a petroleum refinery that once considered vibrations of 6.67 mm/s acceptable for two multistage pipeline diesel-fuel feed pumps was able to reduce vibration by more than 80% using carbon-filled PEEK case wear rings and bushings.

An added benefit when designing pump components from plastic is safety, particularly for industries working with flammable fluids. Plastics eliminate the risk of sparks when pumps run dry because of momentary suction losses during system upsets, which is a concern with metal-on-metal contact in conventional designs.

Another area is boiler-feed water pumps. Galling and even seizure are common concerns when using pumps with steel components and wear parts. This is especially true in water pumps where, because of the corrosion resistance of the material, stainless steels are used for pump parts as well as wear components. Although they protect pumps from corrosion, stainless steels gall more easily than other metals.

Engineers often design boiler-feed water pumps with tight diametrical running clearances to maximize discharge pressures. However, the tight clearances on stainless steel wear rings and bushing severely damage pumps storing brief periods of cavitation. Cavitation takes place when suction pressures drop below the net positive suction head required by a pump. Momentary cavitations are common in pumps running close to the vapor pressure of a fluid, which is often the case in boiler-feed water pumps.

Wear, or hard rubs, resulting from cavitation often show up initially in the first-stage wear rings and the center-stage bushings. As a result, rotors deflect more, which causes pumps to seize.

Impeller wear rings and sleeves made of fiber-reinforced PEEK eliminate galling because the material can run for short periods without lubrication. As a result, PEEK wear parts and mating steel components remain undamaged during momentary system upsets. See also *polyaryletherketone*.

Polyetherimides (PEIs)

PEI is an amorphous engineering thermoplastic characterized by high heat resistance, high strength and modulus, excellent electrical properties that remain stable over a wide range of temperatures and frequencies, and excellent processability. Unmodified PEI resin is transparent and has inherent flame resistance and low-smoke evolution. The resin is produced as *Ultem*.

Polyetherimide resin is available in an unreinforced grade for general-purpose injection molding, blow molding, foam molding, and extrusion, in four glass-fiber-reinforced grades (10%, 20%, 30%, and 40% glass), in bearing grades, and in several high-temperature grades. The unreinforced grade is available as a transparent resin and in standard and custom colors.

The resin can be processed on conventional thermoplastic molding equipment. Melting temperatures of 349°C–427°C are typical for injection-molding applications. Molding temperatures of 66°C–177°C are required.

Polyetherimide is extruded to produce profiles, coated wire, sheet, and film. Film thicknesses as low as 0.25 mil are obtained by solvent-casting techniques. Molded and extruded parts can be machined using either conventional or laser techniques and can be bonded together or to dissimilar materials using ultrasonic, adhesive, or solvent methods.

Properties

The Underwriters Laboratories' continuous-use listing of PEI is 170°C, glass transition temperature is 215°C, and heat-deflection temperature is 200°C at 1.78 MPa, contributing to its high strength and modulus retention for service under load at elevated temperatures.

A key feature of polyetherimide is maintenance of properties at elevated temperatures. For example, at 179°C, tensile strength and flexural modulus are 48 and 2040 MPa. Moduli and strengths of the glass-reinforced grades are still higher. For example, flexural modulus is 8840 MPa with 30% glass reinforcement, and more than 80% of this is retained at 179°C.

Polyetherimide has good creep resistance as indicated by its apparent modulus of 2403 MPa after 1000 h at 82°C under an initial applied load of 34.3 MPa.

The resin resists a broad range of chemicals under varied conditions of stress and temperatures. Compatibility has been demonstrated with aliphatic hydrocarbons and alcohols including gasoline and gasohol, mineral-salt solutions, dilute bases, and fully halogenated hydrocarbons. Resistance to mineral acids is outstanding. The polymer is attacked by partially halogenated solvents such as methylene chloride and trichloroethane and by strong bases.

Resistance to ultraviolet radiation is good; change in tensile strength after 1000 h of xenon arc exposure is negligible. Resistance to gamma radiation is also good; strength loss is less than 6% after 500 Mrad exposure to cobalt-60 at the rate of one Mrad/h.

Hydrolytic stability tests show that more than 85% of tensile strength is retained after 10,000 hr of boiling water immersion. The material is also suitable for applications requiring short-term or repeated steam exposure.

Polyether Resins

These are thermoplastic or thermosetting materials that contain ether–oxygen linkages, –C–O–C–, in the polymer chain. Depending upon the nature of the reactants and reaction conditions, a large number of polyethers with a wide range of properties may be prepared.

The main groups of polyethers in use are epoxy resins, prepared by the polymerization and cross-linking of aromatic diepoxy compounds; phenoxy resins, high-molecular-weight epoxy resins; polyethylene oxide and polypropylene oxide (PPO) resins; polyoxymethylene, a high polymer of formaldehyde; and polyphenylene oxides, polymers of xylenols.

Epoxy Resins

The epoxy resins forms an important and versatile class of cross-linked polyethers characterized by excellent chemical resistance, adhesion to glass and metals, electrical insulating properties, and ease and precision of fabrication.

The type of curing agent employed has a marked effect on the optimum temperature of curing and has some influence on the final physical properties of the product. By judicious selection of the curing system, the curing operation can be carried out at almost any temperature from 0°C to 200°C.

Various fillers such as calcium carbonate, metal fibers and powders, and glass fibers are commonly used in epoxy formulations to improve such properties as the strength and resistance to abrasion and high temperatures.

Because of their good adhesion to substrates and good physical properties, epoxies are commonly used in protective coatings. Because of the small density change on curing and because of their excellent electrical properties, the epoxy resins are used as potting or encapsulating compositions for the protection of delicate electronic assemblies from the thermal and mechanical shock of rocket flight. Because of their dimensional stability and toughness, the epoxies are used extensively as dies for stamping metal forms, such as automobile gasoline tanks, from metal sheeting. Foams are also made.

By combining epoxies, especially the higher-performance types based on polyfunctional monomers, with fibers such as glass or carbon, exceedingly high moduli and strengths may be obtained. Thus, a typical carbon-fiber-reinforced epoxy may have a tensile modulus and strength of 500 and 2 GPa, respectively. Such composites are of importance in aerospace applications where high ratios of a property to density are desired, as well as in such domestic applications of sports equipment.

The adhesive properties of the resins for metals and other substrates and their relatively high resistance to heat and to chemicals have made the epoxy resins useful for protective coatings and for metal-to-metal bonding.

Polyolefin Oxide Resins

Polyethylene oxide and PPO are thermoplastic products whose properties are greatly influenced by molecular weight. Oxides, such as tetrahydrofuran, can also be polymerized to give polyethers.

Low- to moderate-molecular-weight polyethylene oxides vary in form from oils to waxlike solids. They have relatively nonvolatile, are soluble in a variety of solvents, and have found many uses as thickening agents, plasticizers, lubricants for textile fibers, and components of various sizing, coating, and cosmetic preparations. The PPOs of similar molecular weight have somewhat similar properties, but tend to be more oil-soluble (hydrophobic) and less water soluble (hydrophilic).

Phenoxy Resins

Phenoxy resins differ from the structurally similar epoxy resins based on the reaction of epichlorohydrin with bisphenol A maybe by possessing a much higher molecular weight, about 25,000. The polymers are transparent, strong, ductile, and resistance to creep and, in general, resemble polycarbonates in their behavior. Cross-linking may be affected by the use of curing agents that can react with the OH groups. Molding and extrusion may be used for fabrication. The major application is as a component in protective coatings, especially in metal primers.

Polyphenylene Oxide (PPO) Resins

PPO is the basis for an engineering plastic characterized by chemical, thermal, and dimensional stability. The PPO resin is normally blended with another compatible but cheaper resin such as high-impact polystyrene. The blends are cheaper and more processable than PPO and still retain many of the advantages of PPO by itself.

PPO is outstanding in its resistance to water and at its maximum usable temperature range (about 170°C–300°C). In spite of the high softening point, the resins, including glass-reinforced compositions, can be molded and extruded in conventional equipment. Uses include medical instruments, pump parts, and insulation. Structurally modified resins are also of interest and the general family of resins may be expected to replace other materials in many applications.

Commercial products can be processed by injection or compression molding as well as by slurry or electrostatic coating; reinforcement with glass fibers is also commonly practiced. The combination of properties is useful in such applications as electronic pumps and automotive components and coatings, especially where environmental stability is required.

Polyoxymethylene (POM)

POM, or polyacetal, resins are polymers of formaldehyde. With high molecular weights and high degrees of crystallinity, they are strong and tough and are established in the general class of engineering thermoplastics.

Although somewhat similar to polyethylene in general molecular structure, polyacetal molecules pack more closely and attract each other to a much greater extent, so that the polymer is harder and has higher melting point than polyethylene. Polyacetals are typically strong and tough and resistant to fatigue, creep, organic chemicals (but not strong acids or bases) and have low coefficients of friction. Electrical properties are also good. Improved properties for particular application may be attained by reinforcement with fibers of glass or polytetrafluoroethylene.

The combination of properties has led to many uses such as plumbing fittings, pump and valve components, bearings and gears, computer hardware, automobile body parts, and appliance housings. Other aldehydes may be polymerized in a similar way.

Polyethersulfone

A high-temperature engineering thermoplastic consisting of repeating phenyl groups linked by thermally stable ether and sulfone groups. The material has good transparency and flame resistance and is one of the lowest smoke-emitting materials available.

Both polymer and reinforced grades are available in granular form for extrusion or injection molding.

Polyethylene

Polyethylene thermoplastic resins include low-density polyethylenes (LDPE); linear low-density polyethylenes (LLDPE); high-density polyethylenes (HDPE); and ethylene copolymers, such as ethylene–vinyl acetate (EVA) and ethylene–ethyl acrylate (EEA); and ultra-high-molecular-weight polyethylenes (UHMWPE). In general, the advantages gained with polyethylenes are lightweight, outstanding chemical resistance, good toughness, excellent dielectric properties, and relatively low cost compared to other plastics. The basic properties of polyethylenes can be modified with a broad range of fillers, reinforcements, and chemical modifiers, such as thermal stabilizers, colorants, flame retardants, and blowing agents. Further, polyethylenes are considered to be very easy to process by such means as injection molding, sheet extrusion, film extrusion, extrusion coating, wire and cable extrusion coating, blow molding, rotational molding, fiber extrusion, pipe and tubing extrusion, and powder coating. Major application areas for polyethylenes are packaging, industrial containers, automotive, materials handling, consumer products, medical products, wire and cable insulation, furniture, housewares, toys, and novelties.

Polyethylene is a water-repellent, white, tough, leathery, thermoplastic resin very similar in appearance to paraffin wax. Properties vary from a viscous liquid at low molecular weights to a hard wax-like substance and high molecular weights. It is used as a coating for glass bottles and fiberglass fabrics (special treatments for glass are required to obtain good adhesion between polyethylene and glass) and is also used as an injection-molding material for ceramics.

The basic building blocks for polyethylenes are hydrogen and carbon atoms. These atoms are combined to form the ethylene monomer, C_2H_4, that is, two carbon atoms and four hydrogen atoms. In the polymerization process, the double bond connecting the carbon atoms is broken. Under the right conditions, these bonds reform with other ethylene molecules to form long molecular chains. Ethylene copolymers, EVA, and EEA, are made by the polymerization of ethylene units with randomly distributed comonomer groups, such as vinyl acetate and ethyl acrylate.

Properties

Three basic molecular properties affect most polyethylene properties: crystallinity (density), average molecular weight, and molecular weight distribution.

Molecular chains in polyethylenes in crystalline areas are arranged somewhat parallel to each other. In amorphous areas, they are randomly arranged. High-density polyethylene resins have molecular chains with comparatively few side chain branches. Therefore, the chains are packed more closely together. The result is crystallinity up to 95%. LDPE resins have crystallinity from 60% to 75%. LLDPE resins have crystallinity from 60% to 85%. For polyethylenes, the higher the degree of crystallinity, the higher the resin density. Higher density, in turn, influences numerous properties. With increasing density, heat-softening point, resistance to gas and moisture vapor permeation, and stiffness increase. However, increased density results in a reduction of stress cracking resistance and low-temperature toughness. LLDPE resins range from 0.910 to 0.930 g/cm³; LDPE resins range from 0.9102 0.930 g/cm³ and HDPE from 0.941 to 0.965 g/cm³.

High-molecular-weight LDPE resins are used to extrude high clarity, tough film used in shrink packaging and for making heavy-duty bags. High-molecular-weight HDPE resins (ave. MW 200,000–500,000) have excellent environmental stress cracking resistance, toughness, high moisture barrier properties, and high strength and stiffness. The high-molecular-weight HDPE resins are used in film, pressure pipe, large blow moldings, and sheet for thermoforming. UHMWPEs generally are considered to be those resins with molecular weights greater than two million (materials with MW up to six million are available). These resins have excellent abrasion resistance, stress cracking resistance, and toughness. They are, however, much more difficult to process than standard polyethylenes and require special forming techniques. UHMWPE resins are used in applications requiring high wear resistance, chemical resistance, and low coefficient of friction.

Polyethylene rubbers are rubberlike materials made by cross-linking with chlorine and sulfur, or they are ethylene copolymers. Chlorosulfonated polyethylene is white spongy material, in which chlorine atoms and sulfonyl chloride groups are spaced along the molecule. It is used to blend with rubber to add stiffness, abrasion resistance, and resistance to ozone and also for wire covering. Ethylene-propylene rubber, produced by various companies, is a chemically resistant rubber of high tear strength. The ethylene butadiene resin can be vulcanized with sulfur to give high hardness and wide temperature range. For greater elongation, a terpolymer with butene can be made.

Polyethylene of low molecular weight is used for extending and modifying waxes and also in coating compounds especially to add toughness, gloss, and heat-sealing properties. Such materials are called *polyethylene wax*, but they are not chemical waxes. They can be made emulsifiable by oxidation, and they can be given additional properties by copolymerization with other plastics. The polymethylene waxes are microcrystalline and have sharper melting points than the ethylene waxes. They are more costly, but have a high luster and durability. Polybutylene plastics are rubberlike polyolefins with superior resistance to creep and stress cracking. Films of this resin have high tear resistance, toughness, and flexibility and are used widely for industrial refuse bags. Chemical and electrical properties are similar to those of polyethylene and polypropylene plastics. Polymethyl pentene is a moderately crystalline polyolefin plastic resin that is transparent even in thick sections. Almost optically clear, it has a light transmission value of 90%. Parts molded of this plastic are hard and shiny with good impact strength down to −29°C. Its specific gravity (0.83) is the lowest of any commercial solid plastic. A major use is for molded food containers for quick frozen foods that are later heated by the consumer.

LDPE

LDPE, the first of the polyethylenes to be developed, has good toughness, flexibility, low-temperature impact resistance, clarity in film form, and relatively low heat resistance. Like the higher-density grades, LDPE has good resistance to chemical attack. At room temperature, it is insoluble in most organic solvents but is attacked by strong oxidizing acids. At higher temperatures, it becomes increasingly more susceptible to attack by aromatic, chlorinated, and aliphatic hydrocarbons.

Polyethylene is susceptible to environmental and some chemical stress cracking. Wetting agents such as detergents accelerate stress cracking. Some copolymers of LDPE are available with improved stress-crack resistance.

About half of LDPE production goes into packaging applications such as industrial bags, shrink bundling, soft goods, and produce and garment bags. Other applications include blow-molded containers and toys, hot-melt adhesives, injection-molded housewares, paperboard coatings, and wire insulation. LDPE resins are rotationally molded into large agricultural tanks, chemical shipping containers, tote boxes, and battery jars.

One of the fastest-growing plastics is linear LLDPE, used mainly in film applications but also suitable for injection, rotational, and blow molding. Properties of LLDPE are different from those of conventional LDPE and HDPE in that impact, tear, and heat-seal strengths and environmental stress-crack resistance of LLDPE are significantly higher. Major uses at present are grocery bags, industrial trash bags, liners, and heavy-duty shipping bags for such products as plastic resin pellets.

HDPE

Rigidity and tensile strength of the HDPE resins are considerably higher than those properties in the low- and medium-density materials. Impact strength is slightly lower, as is to be expected in a stiffer material, but values are high, especially at low temperatures, compared with those of many other thermoplastics.

HDPE resins are available with broad, intermediate, and narrow molecular weight distribution, which provides a selection to meet specific performance requirements. As with the other polyethylene grades, very-high-molecular-weight copolymers of HDPE resins are available with improved resistance to stress cracking.

Applications of HDPE range from film products to large, blow-molded industrial containers. The largest market area is in blow-molded containers for packaging milk, fruit juices, water, detergents, and household and industrial liquid products. Other major uses include high-quality, injection-molded housewares, industrial pails, food containers, and tote boxes; extruded water and gas-distribution pipe, and wire insulation; and structural-foam housings.

HDPE resins are also used to rotationally mold large, complex-shaped products such as fuel tanks, trash containers, dump carts, pallets, agricultural tanks, highway barriers, and water and waste tanks for recreational vehicles.

A special category of HDPE known as high-molecular-weight HDPE (HMW-HDPE) offers outstanding toughness and durability, particularly at low temperatures. These characteristics result from a unique combination of high average molecular weight (250,000–500,000) and a bimodal molecular weight distribution.

In blow-molding applications, HMW-HDPE allows drum manufacturers to meet Department of Transportation and Occupational Health and Safety Administration specifications. In pipe production, HMW-HDPE meets PE-3408, currently the highest strength rating for polyethylene pipe. Extruded sheet applications include pond liners, truck bed liners, and outdoor leisure products. The primary market, however, for HMW-HDPE is in film applications, where its toughness allows down-gauging in merchandise bags and trash bags. The material is also suited for use in T-shirt-type grocery sacks requiring high handle strength.

UHMWPE

UHMWPE was originally defined as a polyethylene whose average molecular weight, as measured by the solution-viscosity method, is greater than 2,000,000. (Molecular weight of HDPE ranges from 100,000 to 500,000.) Over the years, producers and processors of UHMWPE materials have tried to reach agreement on just how high is "ultrahigh." The values proposed in the past have ranged from as low as 1,000,000 to over 3,500,000. Also in dispute was the question of the relationship between molecular weight and properties of UHMWPE in finished parts.

Several years ago, a value of 3,100,000 was agreed upon for the molecular weight as the dividing line, above which the UHMW description should apply. Resin properties increase with increasing molecular weight and start to level off at the 3,100,000 value.

Also, processability is more difficult above that dividing line, and material cost rises more rapidly.

As with most high-performance polymers, processing of UHMWPE is not easy. Because of its high melt viscosity (it does not register a melt-flow index), conventional molding and extrusion processes would break the long molecular chains that give the material its excellent properties. Methods used currently are compression molding, ram extrusion, and warm forging of extruded slugs. Developmental work has been done on injection molding of UHMWPE resins, but the process forces the polymer to behave in a manner that is not conducive to maintaining its molecular structure. Compression-molded sheets as large as 1.43 × 4 m are available.

UHMWPE has outstanding abrasion resistance and a low coefficient of friction. Impact strength is high, and chemical resistance is excellent. The material does not break in impact strength tests using standard notched specimens; double-notched specimens break at 20 ft lb/in. Crystalline melting point of the material is 130°C. Recommended maximum service temperature is about 93°C, however, because of a high coefficient of thermal expansion.

Applications of UHMWPE include conveyor wear strips and guide rails, paper-machine suction-box covers, chute linings, and snowmobile-track sprockets, bearings, parts for textile looms, pipe for distribution of slurry materials, and other components requiring abrasion resistance, impact strength, and a low-friction coefficient. These plastics have outperformed other materials, including metals, in these applications.

Polyethylene Glycol

Polyethylene glycols or water-soluble, nonvolatile liquids and solids. Polyethylene glycols dissolve in water to form transparent solutions and are also soluble in many organic solvents. They do not hydrolyze or deteriorate.

Characteristics

The intermediate members of the series with average molecular weights of 200–20,000 are produced as residue products by the sodium or potassium hydroxide-catalyzed batch polymerization of ethylene oxide onto water or monomer diethylene glycol. These polymers are formed by stepwise anionic addition polymerization and, therefore, possess a distribution of molecular weights. Examples of commercial uses for products in this range are in ceramic, metal forming, and rubber-processing operations; as drug suppository bases and in cosmetic creams, lotions, and deodorants; as lubricants; as disbursements for casein, gelatins, and inks; and as antistatic agents. These polyethylene glycols generally have low human toxicity.

The highest members of the series have molecular weights from 100,000 to 10,000,000. They are produced by special anionic polymerization catalysts that incorporate metal such as aluminum, calcium, zinc, and iron and coordinated ligands such as amides, nitriles, or ethers. These members of the polyethylene glycol series are of interest because of their ability at very low concentrations to reduce friction of flowing water.

Applications

Polyethylene glycol has a range of properties making it suitable for medical and biotechnical applications. These polymers of ethylene oxide are frequently described as amphiphilic, meaning that they are soluble both in water and in most organic solvents. The medical and biotechnical applications derived from this amphiphilicity, from a lack of toxicity and immunogenicity, and from a tendency to avoid other polymers and particles also present in aqueous solution.

Attaching polyethylene glycol to another molecule provides the latter with enhanced solubility in solvents in which polyethylene glycol is soluble, and thus polyethylene glycol is attached to drugs to enhance water and blood solubility. Similarly, polyethylene glycol is attached to insoluble enzymes to impart solubility in organic solvents. These polyethylene glycol enzymes are used as catalysts for industrial reactions in organic solvents.

The tendency to avoid other polymers results in formation of two immiscible aqueous layers when a solution of a polyethylene glycol is mixed with certain other polymer solutions. These aqueous two-phase systems are used to purify biological materials such as proteins, and nucleic acids, and cells by partitioning of desired and undesired materials between the two phases.

The tendency of polyethylene glycols to avoid interaction with cellular and molecular components of the immune system results in the material being nonimmunogenic.

Ceramic applications are varied and interesting. For example, they may be used to bind glass tubes into bundles prior to cutting into lengths. They also have been used as components of ceramic slips and glazes for the manufacture of porcelain signs and other vitreous coatings.

As binders for colors, excellent adhesion is obtained when the coatings are sprayed on the ceramic surface. A crayon for applying identification marks or decorations to ceramic particles, prior to firing, may be made from pigmented polyethylene glycol. As an extrusion binder, it aids in slurrying the clay, promotes die life, and gives good binding properties. The automatic extrusion of tile, pipe, and electrical insulators has been improved through its use. Polyethylene glycol is an effective additive for the electrophoretic coating of certain components for vacuum tubes.

Polyethylene Plastics

Thermoplastic materials composed of ethylene. They are normally translucent, tough, waxy solids that are unaffected by water and by a large range of chemicals. In common usage, these plastics have no less than 85% ethylene and no less than 95% total olefins. See also *high-density polyethylenes* and *ultrahigh-molecular-weight polyethylene*.

Polyethylene Terephthalate

A saturated thermoplastic polyester resins made by condensing ethylene glycol and terephthalic acid and used for fibers, films, and injection-molded parts. It is extremely hard, wear resistant, eventually stable, and resistant to chemicals, and it has good dielectric properties. Also known as polyethylene glycol terephthalate.

Polyfluorolefin Resins

These resins are distinguished by their resistance to heat and chemicals and by the ability to crystallize to a high degree. Several name products are based on tetrafluoroethylene (TFE), hexafluoropropylene (HFP), and monochlorotrifluoroethylene.

Copolymers of TFE and HFP with each other and of TFE and HEP with ethylene are available commercially.

Polytetrafluoroethylene (PTFE)

The polymer is insoluble and resistant to heat (up to 275°C) and chemical attack and, in addition, has a lowest coefficient of friction of any solid. Because of its resistance to heat, the fabrication of PTFE requires modification of conventional methods. After molding the powdered polymer using a cold-press, the moldings are sintered at 316°C–400°C by procedures similar to those used in powder metallurgy. The sintered product can be machined or punched. Extrusion is possible if the powder is compounded with a lubricating material. Aqueous suspensions of the polymer can also be used for coating various articles. However, special surface treatments are required to ensure adhesion because PTFE does not adhere well to anything.

PTFE resin is useful for applications under extreme conditions of heat and chemical activity. PTFE bearings, valves seats, packings, gaskets, coatings, and tubing can withstand relatively severe conditions. Fillers such as carbon, inorganic fibers, and metal powder may be incorporated to modify the mechanical and thermal properties.

Because of its excellent electrical properties, PTFE is useful when a dielectric material is required for service at a high temperature. The nonadhesive quality is often turned to advantage in the use of PTFE to coat articles such as rolls and cookware to which materials might otherwise adhere.

Polymonochlorotrifluoroethylene

The properties of polymonochlorotrifluoroethylene (CTFE resin) are generally similar to those for polymonochlorotrifluoroethylene; however, the presence of the chlorine atoms in the former causes the polymer to be a little less resistant to heat and to chemicals. The polymonochlorotrifluoroethylene can be shaped by use of conventional molding and extrusion equipment, and it is obtained in a transparent, noncrystalline condition by quenching. Dispersions of the polymer in organic media may be used for coating.

The applications of polymonochlorotrifluoroethylene are in general similar to those for polytetrafluoroethylene. Because of its stability and inertness, the polymer is useful in the manufacture of gaskets, linings, and valve seats that must withstand hot and corrosive conditions. It is also used as a dielectric material, as a vapor and liquid barrier, and for microporous filters.

Copolymers

Copolymers of TFE and HFP propylene (fluorinated ethylene propylene [FEP] resins) and copolymers of TFE with ethylene are often used in cases where ease of fabrication is desirable. The copolymers can be processed by conventional thermoplastic techniques, and except for some diminution in the level of some properties, properties generally resemble those of the TFE homopolymer.

Copolymers of ethylene with CTFE can also be processed by conventional methods and have better mechanical properties than TFE, FEP, and PFA resins.

Polyvinylidene Fluoride

The properties are generally similar to those of the other fluorinated resins: relative inertness, low dielectric constant, and thermal stability (up to about 150°C). The resins (PVF$_2$ resins) are, however, stronger and less susceptible to creep and abrasion than TFE and CTFE resins.

Applications of polyvinylidene fluoride are mainly as electrical insulation, in piping and process equipment, and as a protective coating in the form of a liquid dispersion.

Fluorinated Elastomers

Several types of fluorinated, noncrystallizing elastomers were developed to meet needs (usually military) for rubbers that possess good low-temperature behavior with a high degree of resistance to oils and to heat, radiation, and weathering.

Copolymers of hexafluoropropylene with vinylidene fluoride make up an important class with such applications as gaskets and seals. Copolymers of nitrosomethane with tetrafluoroethylene ethylene has shown considerable promise for similar applications.

Polyimides (PIs)

Available both as thermoplastic and thermoset resins, PIs are a family of some of the most heat- and fire-resistant polymers known. Moldings and laminates are generally based on thermoset resins, although some are made from thermoplastic grades. Unlike most plastics, PIs are available as laminates and shapes, molded parts, and stock shapes from some material producers. Thin-film products—enamel, adhesives, and coatings—are usually derived from thermoplastic PI resins.

Laminates are based on continuous reinforcements including woven glass and quartz fabrics or fibers of graphite, boron, quartz, or organic materials. Molding compounds, on the other hand, contain discrete fibers such as chopped glass or asbestos or particulate fillers such as graphite powders, MoS$_2$, or polytetrafluoroethylene (PTFE).

PI films and wire enamels are generally unfilled. Coatings may be pigmented or filled with particles such as PTFE for lubricity. Most adhesives contain aluminum powder to provide a closer match to the thermal expansion characteristics of metal substrates and to improve heat dissipation.

PI parts are fabricated by techniques that ranged from powder-metallurgy methods to conventional injection, transfer, and compression molding and extrusion methods. Porous PI parts are also available. Generally, those compounds that are the most difficult to fabricate have the highest heat resistance.

Properties

PI parts and laminates can serve continuously in air at 260°C; service temperature for intermittent exposure can range from cryogenic to as high as 482°C. Glass-fiber-reinforced versions retain over 70% of their flexural strength and modulus at 249°C. Creep is almost nonexistent, even at high temperatures, and deformation under load (27.2 MPa) is less than 0.05% at room temperature for 24 h.

These materials have good wear resistance and low coefficients of friction, both of which are further improved by PTFE fillers. Self-lubricating parts containing graphite powders have flexural strengths above 68 MPa, which is considerably higher than those of typical thermoplastic bearing compounds.

Electrical properties of PI moldings are outstanding over a wide range of temperature and humidity conditions.

PI parts are unaffected by exposure to dilute acids, aromatic and aliphatic hydrocarbons, esters, ethers, alcohols, Freons, hydraulic fluids, JP-4 fuel, and kerosene. They are attacked, however, by dilute alkalies and concentrated inorganic acids.

PI adhesives maintain useful properties for over 12,000 h at 260°C, 9,000 h at 302°C, 500 h at 343°C, and 100 h at 371°C. Resistance of these adhesives to combined heat (to 302°C) and salt-water exposure is excellent.

PI is similar to a polyamide, differing only in the number of hydrogen molecules contained in the groupings. This polymer is suitable for use as a binder or adhesive, and its exceptional thermo-mechanical properties make it suitable for many high-temperature applications. See also *thermoplastic polyimides*.

P

Polyketones

Polyketones are partially crystalline engineering thermoplastics that can be used at high temperatures. They also have excellent chemical resistance, high strength, and excellent resistance to burning. Although they require high melting temperatures, polyketones can be extruded and injection-molded with standard processing equipment.

Several polyketones are commercially available:

- Polyaryletherketones, repeating ether and ketone groups combined by phenyl rings (PAEK or PEK).
- Polyetheretherketones, repeating monomers of two ether groups and a ketone group (PEEK).
- Polyetherketoneketones, repeating monomers of one ether group and two ketone groups (PEKK).

In addition, other materials with various combinations of the ether and ketone groups are designed to provide a balance of high heat resistance and good processability, for example, poly[ether][ketone][ether][ketone][ketone].

Polyketone resins are available as natural resins and glass- or carbon-fiber-reinforced and mineral-filled grades.

Properties

Glass transition temperatures (T_g) and melting temperatures (T_m) for polyketones depend on the ratio of ketone to ether groups. With increasing ether groups, both temperatures decrease. With increasing ketone groups, the temperatures go up. The polyether ether ketone (PEEK) T_g is 148°C and its T_m is 335°C, the polyaryletherketone T_g is 190°C and its T_m 380°C, and PEKEEK has a T_g of 177°C and a T_m of 375°C.

Polyketones are stronger and more rigid than most other engineering plastics. They are tough and impact resistant over a wide range of temperatures. Polyketones have very high fatigue strength. Both coefficients of friction and wear rates for polyketones are very low.

The thermal oxidative stability of polyketones is excellent. Typically, continuous-use temperatures are above 249°C (i.e., 50% of strength and rigidity are retained up to this temperature). Moduli of polyketones remain almost constant until the temperature is close to the T_g. Polyketones have comparatively low thermal coefficients of linear expansion. Polyketones have excellent resistance to burning and very low flame spread. These materials have good dielectric properties, with high volume and surface resistivities, and high dielectric strength.

Polyketones are extremely resistant to numerous inorganic and organic chemicals. They are dissolved or decomposed by concentrated, anhydrous, or strong oxidizing acids. Common solvents do not attack polyketones even at elevated temperatures. They have very good resistance to hydrolysis, even in hot water. Like most plastics composed of aromatic building blocks, polyketones are affected by ultraviolet radiation. Polyketones of extremely high resistance to beta rays, gamma rays, and x-rays over a wide range of temperatures.

Processing

Applicable processing methods are injection molding, extrusion, rotational molding, and powder coating. Conventional injection-molding machines can be used with polyketones. Melting temperatures for polyketones vary depending on the resin type. PEK melting temperatures range from 390°C to 430°C; for PEEK, the range is 380°C–399°C. Molding temperatures from 186°C to 216°C are recommended for molding PEK, and 180°C–198°C for PEEK. Parts for high-temperature applications should be molded at the high end of the temperature range.

Polyketones can be extruded to form sheet, cast film, stock for machining, and wire coatings. Typical melting temperatures for extrusion are 399°C–430°C.

Polyketone parts can be assembled using various adhesives and welding techniques. The adhesives can be epoxies, cyanoacrylates, polyurethanes, or silicones. Welding techniques include heated-tool welding (399°C–538°C for 10–90 s), spin welding, hot-air welding (449°C–499°C), and ultrasonic welding.

Polymer

The terms *polymer, high polymer, macromolecule,* and *giant molecule* are used to designate high-molecular-weight materials of either synthetic or natural origin. Plastics are relatively stiff at room temperature, rubbers or elastomers are flexible and retract quickly after stretching, and fibers are especially strong filamentary materials. Coatings are generally either plastics or rubbers that have been applied as a thin layer on a substrate. In practice, plastics, rubbers, fibers, and coatings are used as formulations of the polymers with other ingredients such as fillers, pigments, plasticizers, flow improvers, and stabilizers against aging and degradation.

Historical Development

The first modified natural polymers, cellulose nitrate and casein–formaldehyde, were commercially produced about 1860, and the first fully synthetic polymer, phenol formaldehyde was made about 1910. The major development of present polymer science and technology has taken place since about 1920.

Interest in the synthesis of products similar to natural products but possessing more useful properties has been continually stimulated by the successful synthesis of polyamide fibers and rubbers equivalent to natural rubber and by increasing understanding of the nature of proteins, nucleoproteins, carbohydrates, and enzymes in living tissues.

Several striking advances have been the development of cheap raw materials for plastics and other new polymerization processes, and remarkable advances in understanding the relationships among molecular structure, morphology, and physical and chemical behavior. As properties have been improved, plastics have been developed that can be readily and economically fabricated and that can be used for hitherto inappropriate engineering purposes such as gears, bearings, and structural members. Such engineering plastics may frequently be used advantageously to replace metals or other materials.

There has been considerable interest in polymers that can withstand even more extreme environments or possess other specialized properties, for use under the sea, in space, and in biological systems. These interests have emphasized the continuing need for still deeper understanding of bonding (in organic, inorganic, an organic–inorganic systems) and of the implications of bonding and structure to morphology and, in turn, to the overall response to applied stresses and given environments.

A structure that can be represented by a repeated small unit the mer which is a high molecular weight compound, natural or synthetic. Examples include polyethylene, rubber, and cellulose. Synthetic polymers are formed by addition or condensation polymerization of monomers. Some polymers are *elastomers*, some are *plastics*, and some are *fibers*. When two or more dissimilar monomers are involved, the product is called a copolymer. The chain lengths of commercial thermoplastics vary from –1,000 to > 100,000 repeating units. Thermosetting polymers approach infinity after curing, but their resin precursors, often called prepolymers, may be relatively short—6 to 100 repeating units—before curing. The lengths of polymer chains, usually measured by molecular weight, have very significant effects on the performance properties of plastics and profound effects on processability.

Properties

The properties of polymeric materials are determined by the molecular properties of the macromolecules, the morphology, and the type of formulation involving plasticizers or fillers. The morphology in turn depends on the conditions of fabrication in which molecular orientation or crystallization may be induced. Properties also depend on the temperature and elapsed time of the measurement.

Environmental Stability

The stability of a polymer depends on the chemical nature, on morphology, and on molecular properties. Thus, polyamides are susceptible to hydrolysis on long exposures to dilute acids at high temperatures. Resistance to high temperatures under oxidizing conditions is enhanced by minimizing the hydrogen content, by increasing the molecular softness, and by increasing the content of aromatic structures or heterocyclic rings.

Compounding

Plastic masses, rubber formulations, coatings, and other polymeric compositions may contain age inhibitors and strengthening and coloring pigments, flow improvers, and plasticizers or softening agents. Roll mill, sigma blade, and dough mixers are generally employed to mix the resin with plasticizers and pigments at temperatures usually between 50°C and 250°C.

Additives

The addition of plasticizers to a polymeric material causes it to be softer and more rubbery in character. Plasticizers are held in association with the polymer chains by secondary valence forces. They separate the molecules, thus reducing the effective intermolecular attractive forces.

Age inhibitors are almost always incorporated in polymeric compositions. Oxygen, ozone, light, and electric discharge produce free radicals that cause degradation of the polymer chains. Free-radical inhibitors and light-masking agents are therefore commonly used.

Polyblends

These are produced by the addition of small amounts of a rubbery polymer to a polymeric glass. The impact strength of the glass thus is substantially increased. The rubbery polymer is not truly compatible and exists as a finely dispersed separate phase. Interpenetrating polymer networks, in which two different networks are intertwined, constitute an interesting special case of polyblends. The phenomenon has somewhat similar counterparts in inorganic glass technology and in physical metallurgy. The study of polyblends combines the rigor of physics, for example, fracture mechanics, with organic in physical chemistry and the engineering behavior of the systems of interest.

Fabrication

Polymer formulations can be fabricated into useful forces or articles by a variety of methods.

In the use of molded thermosetting compositions in which heat and pressure are required for the production of sound articles, some form of compression molding is usually employed. The physically compacted composition containing the resin in the fusible stage is forced into a mold cavity of the desired shape and is held under heat and pressure until the curing or vulcanization is complete. When high-pressure is not required, as in the preparation of epoxy compositions and polyester-styrene-glass fiber compounds, it is desirable to use moderate pressure to obtain a uniform molding.

Various forms of injection molding are used for the shaping of thermoplastic (permanently fusible) compositions. The composition is first softened temporarily by forcing the compounded resin granules through a heated chamber, after which it is driven into a relatively cold mold. Under proper conditions, the resin remains soft long enough to fill the mold completely and then rapidly hardens as it cools. The injection molding of engineering plastics often makes it possible to replace an assembly of metal parts by one plastic piece. Variations of the injection molding process are used in the extrusion of films, rods, and pipe and in the spinning of fibers.

Blow molding, a process in which air is blown into a hot tube held in a mold, is used for the manufacture of bottles and other shapes. Certain articles are molded conveniently by rotating a hot mold filled with resin.

In extrusion, a hot melt is forced through a die with an opening shape to produce the cross section desired. Extrusion and molding techniques may be combined to form foams into sheets and other shapes.

Films are also produced by extrusion of a tube into which air is forced. The process is called bubble extrusion. The pipe expands because of the air pressure, to a wall thickness equivalent to the film thickness desired. The walls of the expanded bubble are pressed together in nip rolls, and later the large, thin-walled, collapsed pipe is slipped to yield flat film. Films and sheets are also produced by calendaring; in this process, the hot resin is forced between tightly fitted rolls.

A technique useful for shaping sheets into various shapes is thermoforming, in which a heated sheet is forced against the contours of a mold by a positive pressure or vacuum.

In the casting process, fluid compositions are poured into molds of the desired shape and then allowed to cool or cure. This process is used for the production of foams and in encapsulation, such as in the protection of electronic components or the mounting of biological specimens. In broad terms, coating, slush molding, and painting may be considered to be casting operations.

Some polymers can be formed by methods similar to those used in metallurgical forming. Forging and cold forming are adaptable to the mass production of an increasing number of plastics. Sintering techniques and variations are also used for coating metals and forming polymers that cannot be formed by conventional processes.

In the shaping of thermoplastics particularly, the conditions of molding (temperature, time, and pressure) have a marked effect on the properties of the final product. Uneven cooling produces strains and the flow of the plasticized resin can cause some orientation of the molecules. In bubble molding of film, biaxial orientation is produced. In the drawing of films and fibers, either uniaxial or biaxial orientation may be obtained with equipment similar to the tentors used in the textile industry. Orientation results in a substantial increase in the strength of the product in the direction of stretching; in many polymers, crystallization is induced during cold-drawing or stretching. On occasion, discontinuities or areas of strain are produced by partially orienting the molecules, and the product becomes more subject to stress cracking in the presence of solvents or other agents. Thus, the optimum condition of a product results from a judicious combination of mechanical treatment and thermal annealing.

Polymeric Composite

A polymeric composite is any of the combinations or compositions that comprise two or more materials as separate phases, at least one of which is a polymer. By combining a polymer with another material, such as glass, carbon, or another polymer, it is often possible to obtain unique combinations or levels of properties. Typical examples of synthetic polymeric composites include glass-, carbon-, or polymer-fiber-reinforced thermoplastic or thermosetting resins, carbon-reinforced rubber, polymer blends, silica- or mica-reinforced resins, and polymer-bonded or polymer-impregnated concrete or wood. It is also often useful to consider as composites such materials as coatings (pigment-binder combinations) and crystalline polymers (crystallites in a polymer matrix). Typical naturally occurring composites include wood (cellulosic fibers bonded with lignin) and bone (minerals bonded with collagen). On the other hand, polymeric compositions compounded with a plasticizer or very low proportions of pigments or processing aids are not ordinarily considered composites.

Typically, the goal was to improve strength, stiffness, or toughness, or dimensional stability by embedding particles or fibers in a matrix or binding phase. A second goal is to use inexpensive, readily available fillers to extend a more expensive or scarce resin; this goal is increasingly important as petroleum supplies become costlier and less reliable. Still other applications include use of some fillers such as glass spheres to improve processability, the incorporation of dry-lubricant particles such as molybdenum sulfide to make a self-lubricating bearing, and the use of fillers to reduce permeability.

Emphasis on the development of the polymeric composites has been stimulated by the need for greatly improved mechanical and environmental behavior, especially on a strength- or stiffness-to-weight basis. Such composites are also often more efficient in their energy requirements for production than traditional materials. The high absolute and specific (per unit of weight) values of properties such as strength and stiffness have made composites ideal candidates for new applications in aircraft and boats, in passenger vehicles and farm equipment, and in machinery tools and appliances. Composites based on chemically resistant matrixes are used in chemical process equipment.

Mechanical Properties

The behavior of composites depends on the volume fractions of the phases, their shape, and the nature of the constituents and their interfaces. With anisotropic phases, the orientation with respect to the direction of stressing or exposure to permeants is also important. In general, given an appropriate preferred direction, the greater the anisotropy, the greater the effect on a given property, at least up to some point. Thus, all high-modulus reinforcements will stiffen a lower-modulus matrix, but fibers and platelets are more effective than spheres; a similar role of shape holds for the ability to reduce permeability at right angles to the anisotropic particles. Anisotropic high-modulus inclusions invariably increase strength if adhesion is good, but the effect is more complex with particulate fillers such as spheres. Rubbery inclusions lower the stiffness of a high-modulus matrix, but may enhance toughness by stimulating a combination of localized crazing and shear deformation.

Many fibers, for example, glass and carbon, are very stiff and strong. However, the maximum strength of these brittle materials cannot be realized in practical objects because of a high sensitivity to the inevitable small cracks and flaws that are ordinarily present. Most polymers are much less sensitive to such flaws, even though they are inherently less strong. More energy is needed to fracture a polymer than a ceramic or glass, and a crack tends to grow much less readily in a polymer.

If, for example, the mass of a strong but flaw-sensitive ceramic or glass is divided into many parts, typically into a fiber, and embedded in a polymer matrix, a growing crack may break one fiber, but its progress may be hindered by the matrix or diverted along the interfaces. Thus, even though the matrix contributes only insignificantly to the total strength, it permits a closer approach to the theoretical maximum strength of the glass. Similar considerations apply to short fibers and, with some qualifications, to other forms of reinforcement.

The matrix has several other functions besides the dissipation of energy, which would otherwise cause a catastrophic failure. It protects the fiber against damage by mechanical action, such as rubbing, or by environmental agents, such as water. It must also transfer an applied stress or force to the filament so that they, being much stronger, can bear most of the load. To transfer the stress, the matrix must adhere well to the fiber, although the optimum strength of the interfacial bond desired may vary depending on the application.

The strength of such a composite depends on the orientation of the fibers with respect to the applied force and on the nature of the stress (tensile or compressive). In the selection of materials and design for a composite, the ultimate application must be known. For example, strength of a composite based on longitudinal aligned fibers will be greatest in the direction of the fibers. Rupture under tension will require the pulling out of many fibers; this implies a fiber matrix bond that can yield fairly readily and in yielding increase the energy required for rupture in compression. On the other hand, a stronger interfacial bond is needed to prevent buckling. Compromises in design are often necessary. Thus, at the expense of some strength, fibers are often crisscrossed to minimize the directionality of strength.

Reinforced Thermosetting Resins

The most common fiber-reinforced polymer composites are based on glass fibers, cloth, mat, or roving embedded in a matrix of an epoxy or polyester resin. Usually, the glass surface is treated with a coupling agent to promote adhesion to the matrix. Fabrication may be affected in several ways. Frequently resin-dipped continuous filaments are wound on a mandrel in one of several patterns. Tapes or bundles of fiber coated with resin can also be positioned by hand or machine in layers or pulled through a die. After the desired shape is obtained, curing is affected under heat and pressure.

Boron, polyaramids, and especially carbon fibers confer especially high levels of strength and stiffness. Hybrid fiber combinations, for example, glass with carbon, are often used to achieve a desired balance between cost and performance.

Some examples of properties for several experimental composites (unidirectional fibers) in comparison with metals are striking. In fact, the relative properties of the polymer composites can exceed those of other materials (see *McGraw-Hill Encyclopedia of Science and Technology*, 8th edn., Vol. 14, McGraw-Hill, New York, p. 182; edited by M.M. Schwartz.). Although stiffness is less favorable for glass-fiber composites, carbon-fiber composites have a relative stiffness five times that of steel. Because of these excellent properties, many applications are uniquely suited for epoxy and polyester composites, such as components in new jet aircraft, parts for automobiles, boat hulls, rocket motor cases, and chemical reaction vessels.

Reinforced Thermoplastic Resins

Although the most dramatic properties are found with reinforced thermoplastic resins such as epoxy and polyester resins, significant

P

improvements can be obtained with many thermoplastics. Usually short fibers of glass are used—about 3–50 mm in length.

Nylon resins are particularly acceptable. Reinforcements of nylon-6,-10 with 20% fiber glass raises the tensile strength and heat-deflection temperature from 59 to 138 MPa and from 150°C to 210°C.

Polycarbonates, acetals, polyethylene, and polyesters are among the resins available as glass-reinforced composition.

The combination of inexpensive, one-step fabrication, by injection molding, with improved properties has made it possible for reinforced thermoplastics to replace metals in many applications in appliances, instruments, automobiles, and tools.

Other Composites

In the development of other composite systems, various matrices are possible; for example, polyimide resins are excellent matrices for glass fibers and give a high-performance composite. Different fibers are of potential interest, including polymers (such as polyvinyl alcohol), single-crystal ceramic whiskers (such as sapphire), and various metallic fibers.

Polymer Matrix

The resin portion of a reinforced or filled plastic. See also *resin-matrix composites*.

Polymeric Transflective Materials

These materials consist of hundreds or more of alternating co-extruded layers of two dissimilar, optically clear thermoplastics, such as acrylic and polycarbonate, for light management and decorative effects. Developed by Dow Plastics, they are referred to as *transflective materials* because they simultaneously transmit and reflect light. The composite structure provides a reflective surface ranging from diffuse to spectral or mirrorlike. Changes to the front and back layers create various effects, such as pearlescence.

Polymerization

(1) Polymerization is the linking of small molecules (monomers) to make larger molecules. Polymerization requires that each small molecule have at least two reaction points or functional groups. There are two distinct major types of polymerization processes: condensation polymerization, in which the chain growth is accompanied by elimination of small molecules such as water or methanol, and addition polymerization, in which the polymer is formed without the loss of other materials. (2) A chemical reaction in which the molecules of a *monomer* are linked together to form large molecules with a molecular weight that is a multiple of the molecular weight of the original substance. When two or more monomers are involved, the process is called copolymerization. (3) The forming creates extended chains and cross-links.

Polymethyl Methacrylate

A thermoplastic polymer synthesized from methyl methacrylate. It is a transparent solid with exceptional optical properties. Available in the form of sheets, granules, solutions, and emulsions, it is used as facing material in certain composite constructions. See also *acrylic plastic*.

Polymorph

In crystallography, one crystal form of a polymorphic material. See also *polymorphism*.

Polymorphism

A general term for the ability of a solid to exist in more than one form. In metals, alloys, and similar substances, this usually means the ability to exist in two or more crystal structures or in an amorphous state and at least one crystal structure. See also *allotropy*, *enantiotropy*, and *monotropism*.

Polyol

An alcohol having many hydroxyl groups. Also known as polyhydric alcohol or polyalcohol. In cellular plastic usage, the term includes compounds containing alcoholic hydroxyl groups such as polyethers, glycols, polyesters, and castor oil used in urethane foams, and other polyurethanes. See also *alcohols*.

Polyolefin Resins

These resins are polymers derived from hydrocarbon molecules that possess one or more alkenyl (or olefinic) groups. The term *polyolefin* typically is applied to polymers derived from ethylene, propylene, and other alpha-olefins, isobutylene, cyclic olefins, and butadiene and other diolefins. Polymers produced from other olefinic monomers, such as styrene, vinyl chloride, and tetrafluoroethylene, generally are considered separately.

Polyolefin homopolymers are made from ethylene, propylene, butylene, and methyl pentene. Other olefinic monomers such as pentene and hexane are used to make copolymers.

Because the chemical and electrical properties of all olefins are similar, they often compete for the same applications. They differ from each other primarily in their crystalline structure. However, since strength properties vary with the type and degree of crystallinity, the tensile, flexural, and impact strength of each polyolefin may be quite different. Stress-crack resistance and useful temperature range also vary with crystalline structure.

In addition to the solid polyolefin resins, these materials are also available as beads from which very low density (1.25–5.0 lb/ft³) foam shapes and blocks are produced. Resilience and energy-absorption properties of these products are exceptional compared with those of conventional polystyrene foams. Polymers available as moldable beads include polyethylene, polypropylene (PP), and a polyethylene/polystyrene copolymer alloy.

The bead forms can be processed by the same methods used for expandable polystyrene. After the beads are expanded (20–40 times that of the solid resin) and conditioned, they are poured into a mold and heated; usually by direct injection of steam. This softens, expands further, and fuses the particles together, forming a uniform, void-free, closed-cell shape. After molding, the shapes are usually annealed by being stored at 49°C–71°C to stabilize shape and dimensions.

Because they contain 80%–95% air by volume, the foamed shapes are not nearly as strong as solid moldings. They are used primarily to cushion impact, insulate thermally, and provide high stiffness-to-weight core materials and composite components. An important application for PP foam is in bumper cores for automobiles. A 76–102 mm thick section of foam at 2–4 lb/ft³ can absorb the energy of a 5 mph impact. Package cushioning for fragile and valuable products such as electronic or audio components is another application for either polyethylene or PP foam. The toughness of polyethylene/polystyrene alloy foams qualifies them for material-handling applications where repeated use is required.

The principal resins of the polyolefin family are polyethylene and PP. Other polyolefin polymers and copolymers are ethylene–vinyl acetate, ionomer, polybutylene, and polymethyl pentene.

Polyoxymethylene

Acetal plastics based on polymers in which oxymethylene is essentially the sole repeating structural unit in the chains. See also *acetal resins*.

Polyphenylene Ether (PPE)

Alloys, or blends of PPE and polystyrene in various proportions, are marketed under the trade names of *Noryl* and *Prevex*. The resins are processed by conventional injection-molding, or extruding, and thermoforming methods. Structural-foam parts are processed in standard foam-molding systems, using either direct induction of nitrogen gas or conventional chemical blowing agents.

PPE is produced by a process based on oxidative coupling of phenolic monomers. The result, a resin that has good mechanical properties and thermal stability, is then blended with a polystyrene to improve processability. Available grades of molding resins include glass-reinforced and platable compounds and heat-resistant grades (containing nylon), in addition to extrusion and foamable grades. All resins can be furnished in a wide range of colors.

Properties

PPE blends are characterized by outstanding dimensional stability, the lowest water absorption of the engineering thermoplastics, broad temperature ranges, excellent mechanical and thermal properties, and excellent dielectric properties over a wide range of frequencies and temperatures.

Because of their excellent hydrolytic stability, both at room and elevated temperatures, PPE blend parts can be repeatedly steam-sterilized with no significant change in properties. In exposure to aqueous environments, dimensional changes are low and predictable. Resistance to acids, bases, and detergents is excellent. The material is attacked, however, by many halogenated or aromatic hydrocarbons. Prototype testing of components requiring exposure to such environments is recommended.

Polyphenylene Oxide

Polyphenylene oxide is a plastic that is notable for its high strength and broad temperature resistance. There are two major types: phenylene oxide (PPO) and modified phenylene oxide (Noryl). These materials have a deflection temperature ranging from 100°C to 174°C at 2 MPa. Their coefficients of linear thermal expansion are among the lowest for engineering thermoplastics. Room-temperature strength and modulus of elasticity are high and creep is low. In addition, they have good electrical resistivity. Their ability to withstand steam sterilization and their hydrolytic stability make them suitable for medical instruments, electric dishwashers, and food dispensers. They are also used in the electrical or electronic fields and for business machine housings.

Tensile strength and modulus of phenylene oxides rank high among engineering thermoplastics. They are processed by injection-molding, extrusion, and thermoforming techniques. The foam grades, with their high rigidity, are suitable for large structural parts. Because of good dimensional stability at high temperatures and under moisture conditions, these plastics are readily plated without blistering. Also known as *polyphenylene ether*.

Polyphenylene Sulfide (PPS)

PPS is a crystalline, high-performance engineering thermoplastic characterized by outstanding high-temperature stability, inherent flame resistance, and broad chemical resistance. PPS resins and compounds manufactured in the United States are trade named *Ryton*.

A wide range of injection-molding grades of PPS is available. The series with various glass-fiber levels (designated R-3, R-4, and R-5) is recommended for mechanical electronic applications requiring high mechanical strength, impact resistance, and insulating characteristics. All other compounds contain various mineral fillers in addition to glass-fiber reinforcement. Ryton R-7 and R-8 are suitable for electrical applications requiring high arc resistance and low arc tracking. The R-10 series of pigmented compounds include several grades suitable for support of current-carrying parts in electrical components. PPS is essentially transparent to microwave radiation, so the R-11 series is suited specifically for microwave ovenware and appliance components.

Unreinforced PPS resins are also available as powders for slurry coating and electrostatic spraying. The resin coatings are suitable for food-contact applications as well as for chemical-processing equipment.

The injection-moldable PPS compounds require processing temperature of 316°C–343°C. Molding temperatures can range from 37.8°C to 135°C to control the crystallinity. Cold-molded parts deliver optimum mechanical strength, and hot-molded highly crystalline parts provide optimum dimensional stability at high temperatures.

PPS is also available in long-fiber-reinforced forms. One type, called stampable sheet, contains fiber-mat reinforcement for processing by compression molding. The other form contains reinforcement and is designed for laminating and thermoforming. Reinforcement in both forms can be glass or carbon fiber.

Properties

Most PPS compounds are used for their combination of high-temperature stability, chemical resistance, dimensional reliability, and flame retardancy. The compounds all have excellent stability at very high temperatures. Underwriters Laboratories (UL) temperature index is 200°C–240°C, depending on compound, thickness, and end use. In short-term excursions, the compounds have heat-deflection temperatures of 260°C or higher, depending on the crystallinity of the molding. Mechanical strength of the compounds remains high at high temperatures. For example, the flexural modulus of Ryton R-4 is as high as 260°C as that of ABS (acrylonitrile, butadiene, styrene) at room temperature.

PPS has excellent resistance to a broad variety of chemicals, even at high temperatures. In fact, the resin has no known solvent below 204°C. The resins are flame retardant without additives (UL 94V-0/5). The oxygen index of the resin is 44, and indices of the compounds range from 47 to 53. Because flame retardancy is inherent, regrind is as flame resistant as virgin material.

Mechanical properties of the various PPS compounds are tailored for target applications. The balance of properties can be controlled by the degree of crystallinity of a molded part. Amorphous moldings have optimum mechanical strength at room temperature, and crystalline moldings deliver optimum dimensional stability at high temperatures.

Applications

A metal plate spins at 7200 rpm inside a computer's hard disk drive. Less than three virus widths above that surface resides a rock-hard ceramic block on the end of an aluminum arm. The temperature inside is 260°C–316°C. Under these conditions, a new grade of PPS, Ryton, R-4-230 NA, provided connectors with the proper electrical insulation, dimensional stability, flame retardancy, rigidity, and creep resistance for the drives. It is also a material that would not outgas inside the drives.

The Ryton PPS stiffness and creep resistance also make the material ideal for load-bearing applications, such as the actuator latch and stop.

Polyphthalamide (PPA)

PPA is a family of semicrystalline thermoplastics of Amoco Performance Products, with the trade name *Amodel*, having advantages over conventional nylons in resistance to humidity, auto fuels, water/glycol mixtures, and heat. The melting temperature is 590°F (310°C), and unreinforced, the resin has a specific gravity of 1.17%, 24 h water absorption of 0.81%, a tensile yield strength of 104 MPa, a flexural modulus of 3275 MPa, a heat-deflection temperature of 248°F (120°C) at 1.8 MPa, and Izod notched impact strength of 53 J/m. An extra-tough grade is slightly lighter in weight, less water absorbent (0.65%), and considerably less strong and rigid, but has an impact strength of 1068 J/m. *Thermocomp UF-1006*, of LNP Engineering Plastics, is a 30% glass-reinforced grade based on Amodel. RTP Co. makes a line of lubricated compounds based on PPA for gears and bearings. Polytetrafluoroethylene and/or silicone serve as lubricants and glass, carbon, and/or aramid fibers as reinforcements. And *RTP 4000* grade with 15% aramid fiber, 14% PTFE, and 10% silicone is the most wear resistant.

Polypropylene (PP) Plastics

PP is made by polymerization of propylene using catalysts and similar to the low-pressure polymerization of ethylene. It is a linear polymer, more than 95% of which has a spatially ordered structure. The commercial polymer contains over 99% C_3H_6, with the remainder stabilizing additives. Molded pieces or molding pellets of unpigmented material may be identified by density (0.905) and by maximum crystalline melting point (168°C–170°C).

Produced from propylene gas, PP resins are semitranslucent and milky white in color and have excellent colorability. Most PP parts are produced by injection molding, blow molding, or extrusion of either unmodified or reinforced compounds. Other applicable processes are structural-foam molding and solid-phase and hot-flow stamping of glass-reinforced sheet stock (a product of *Azdel*).

PP plastics are an important group of synthetic plastics employed for molding resins, film, and texture fibers. Propylene is a methyl ethylene, $CH_3CH:CH_2$, and is produced in the cracking of petroleum. It belongs to the class of unsaturated hydrocarbons known as olefins, which are designated by the word ending -*ene*. Thus, propylene is known as propene as distinct from propane, the corresponding saturated compound of the group of alkanes from petroleum and natural gas. These unsaturated hydrocarbons tend to polymerize and form gums and are thus not used in fuels although they have antiknock properties.

Properties

PP is a low-density resin that offers a good balance of thermal, chemical, and electrical properties, along with moderate strength and moderate cost. Strength properties are increased significantly with glass-fiber reinforcement. Increased toughness is provided in special, high-molecular-weight, rubber-modified grades.

Electrical properties of PP moldings are affected to varying degrees by service temperature. Dielectric constant is essentially unchanged, but dielectric strength increases and volume resistivity decreases with increased temperature.

PP has limited heat resistance, but heat-stabilized grades are available for applications requiring prolonged use at elevated temperatures. Useful life of parts molded from such grades may be as long as 5 years at 121°C, 10 years at 110°C, and 20 years at 99°C. Specially stabilized grades are rated by Underwriters Laboratories at 120°C for continuous service.

PP resins are unstable in the presence of oxidative conditions and ultraviolet (UV) radiation. Although all grades are stabilized to some extent, specific stabilization systems are often used to suit a formulation for a particular environment. PPs resist chemical attack and staining and are unaffected by aqueous solutions of inorganic salts or mineral acids and bases, even at high temperatures. They are not attacked by most organic chemicals, and there is no solvent for the resin at room temperature. The resins are attacked, however, by halogens, fuming nitric acid, and other active oxidizing agents and by aromatic and chlorinated hydrocarbons at high temperatures.

Polypropylenes is low in weight. The molded plastic has a density of 0.910; a tensile strength of 34 MPa, with an elongation of 150%; and hardness of Rockwell R95. The dielectric strength is 59×10^6 V/m, dielectric constant 2.3, and softening point 150°C. Blow-molded bottles of PP have good clarity and are nontoxic. The melt flow is superior to that of ethylene. A unique property is their ability in thin sections to withstand prolonged flexing. This characteristic has made PPs popular for "living hinge" applications. In tests, they have been flexed over 70 million times without failure.

Molecular Types

In PP plastics, each carbon atom linked in the molecular chains between CH_2 units has a CH_3 and an H attached as side links, with the bulky side group spiraled regularly around the closely packed chain. The resulting plastic has a crystalline structure with increased hardness and toughness and a higher melting point. This type of stereosymmetric plastic has been called isostatic plastic. It can also be produced with butylene or styrene, and the general term for the plastics is *polyolefins*.

Grades

The many different grades of PPs fall into three basic groups: homopolymers, copolymers, and reinforced and polymer blends. Properties of the homopolymers vary with molecular weight distribution and the degree of crystallinity. Commonly, copolymers are produced by adding other types of olefin monomers to the propylene monomers to improve properties such as low-temperature toughness. PPs are frequently reinforced with glass fibers and fillers to improve mechanical properties and increase resistance to deformation at elevated temperatures. Biaxially oriented polypropylene film (film stretched in two ways) has greatly improved moisture resistance, clarity, and stiffness. It is used for packaging tobacco products, snack foods, baked goods, and pharmaceuticals. Metallized grays also are available for packaging design for extended shelf life.

Foamed PPs include expandable polypropylene (EPP) bead and injection-molded structural foam. EPP provides greater energy absorption and flexibility than expandable polystyrene. Structural PP foam moldings consist of a solid outer skin and foamed core. They are used to achieve greater stiffness in larger, lightweight parts (strength-to-weight ratios are three to four times greater than solid parts).

PP fiber unless modified is more brittle at low temperatures and has less light stability than polyethylene, but it has about twice the strength of high-density linear polyethylene. Monofilament fibers are used for filter fabrics and have high abrasion resistance and a melting point at 154°C. Multifilament yarns are used for textiles and rope. PP rope is used for marine hawsers, will float on water, and does not absorb water like Manila rope. It has a permanent elongation, or set, of 20%, compared with 19% for nylon and 11% for Manila rope, but the working elasticity is 16%, compared with 25% for nylon and 8% for Manila. The tensile strength of the rope is 406 MPa. Fine-denier multifilament PP yarn for weaving and knitting dyes easily and comes in many colors. Chlorinated PP is used in coatings, paper sizing, and adhesives. It has good heat and light stability, high abrasion resistance, and high chemical resistance.

Fabrication

Compression Molding

This is seldom used with PP except for making heavy slab and multiple daylight presses.

Injection Molding

Standard techniques for molding apply to PP. Cylinder temperatures of 288°C or less, and fast rams operating in the neighborhood of half the available machine pressure, generally give good molding at faster cycles. No special metals are needed for cylinders or molds. Since PP shows a definite change in melt viscosity between 232°C and 274°C, the indicated cylinder temperature should be balanced with the machine heater capacity, the size of each shot, and the cycle time to maintain such melting temperatures.

Molds for PPs should embody the best techniques used with thermoplastics; uniform wall thicknesses; avoidance of heavy ribs, bosses, and fillets; use of channels or curved walls instead of ribs to increase rigidity.

Extrusion

Heavy sheets, shapes, thin film, and monofilament are all being produced commercially by extrusion. Equipment required can be made of the usual steel alloys with no danger of corrosive degradation products. For even melting, screws of length-to-diameter ratio of 20:1 are best.

Uses

PP homopolymer, random copolymer, and impact copolymer resins are tailored for specific polymer applications and fabrication methods and also to achieve desired end-product performance. Low-molecular-weight resins, used for melt-spun and melt-blown fibers and for injection-molding applications, are produced by oxidative degradation of a higher-molecular-weight polymers at elevated temperatures. These materials, often called controlled rheology resins, have narrower molecular weight distributions and lower viscoelasticity. The brittleness of PPs homopolymer, particularly at temperatures below 0°C, is greatly reduced by blending it with ethylene-propylene rubber. Compounding with mineral fillers and glass fibers improves product stiffness and other properties. Higher-stiffness resins also are produced by increasing the stereoregularity of the polymer or by the addition of nucleating agents. PP resins are used in extrusion and blow-molding processes and to make cast, slip, and oriented films. Stabilizers are added to PPs to protect it from attack by oxygen, UV light, and thermal degradation; other additives improve resin clarity, flame retardancy, or radiation resistance.

New applications for PP include in-line skates. The low-cost gate consists of a boot and frame injection molded as a single unit. Using *ACC-TUF* PP impact copolymer to make the one-piece skate has produced a significant cost savings, while retaining the performance attributes of the more expensive material. ACCTUF PP impact copolymer was a product line of Amoco Polymers. BP Amoco Acctuf® 3950 impact copolymer PP and a series of PP copolymers however, have been discontinued. It can be ordered under special arrangements.

The copolymer solved the problem of balance between impact resistance and stiffness in thermoplastic materials. Typically, the greater the impact resistance of a material, the less its stiffness. The balance properties of ACC-TUF result in a high degree of toughness, while resisting low-temperature impacts and thermal deformation.

In another application, a PP composite was used to keep boats ice-free; it is used as a submersible circulation unit that is able to keep water around boats and docks from freezing. With a powerful flow of water around a boat or dock, the formation of ice is presented. The submersible, motor-driven unit features a one-piece plastic shroud made with a long-glass-fiber-reinforced PP composite material.

Because the shroud protects the propeller blade from damage by wood and other debris, a material was needed with excellent strength, stiffness, and impact resistance.

Suspended by two lines over the side of the boat, the D-Icer unit works by continuously propelling warmer subsurface water up to the surface. Because it lies under the water, a material that was available in bright colors for easy visibility was needed. It also had to be UV stabilized so that any UV light that penetrates the water would not destroy it.

In addition, the structural composites have good dimensional stability to help maintain mechanical performance in wet environments. This is important because the shroud forms the upper half of the motor housing, which contains a gasket. The stable material helps prevent leaks. And, most important, it resists subzero temperatures.

Finally, recently introduced TSL snowshoes, made of lightweight, injection-molded PP, combine strength and reliability with a convex shape that prevents snow buildup on either side. Each model offers several other features that ensure trouble-free hiking, even in challenging snow conditions and over difficult terrain. They include steel crampons attached to the shoe base that resists temperatures as low as −20°C for traction on icy or hard-packed snow, radial snow fins that aid walking in deep powder, and a binding system that exhibits excellent grip in transverse or when scrambling up steep slopes.

An assortment of models makes the shoes an easy fit for various users and conditions. For example, the TSL 510 fits children, the TSL 710 adapts to lighter-weight adults or firmer packed snow, the TSL 810 accommodates heavier weights or powder-snow use, and the TSL 225 multipurpose shoe is designed for steep, irregular terrain.

Different bindings also adapt to differing needs. The Trappeur model, a high-strength, expandable rubber shoe, conforms to any kind of boot. The Rando binding, designed for use with hiking, pack, and after-ski boots, features a climbing arc for steep ascents and a heel block for descent and jumping. And the Aventure binding, built for extreme terrains, adapts to mountaineering boots notched in the heel and toe.

Polystyrenes

Polystyrenes comprise one of the largest and most widely used families of plastics. Often called the "workhorse" of the thermoplastics, polystyrenes consist of the basic general-purpose materials, plus a wide variety of modified grades of polymers, copolymers, and blends.

All polystyrenes generally have several features in common: (1) low cost, (2) unexcelled electrical insulating properties, (3) virtually unlimited colorability, (4) ability to be made crystal clear (in general-purpose grades), and (5) high hardness and gloss.

Polystyrenes have excellent molding and extrusion characteristics and are formed readily and inexpensively by any of the thermoplastic forming methods. Types available include molding and extrusion grades, foams, sheet and film, and fiber.

Characteristics and Properties

A thermoplastic resin used for molding, in lacquers, and for coatings, polystyrene is formed by the polymerization of the monomeric styrene, which is a colorless liquid of the composition $C_6H_5CH:CH_2$, with a specific gravity of 0.906 and boiling point of 145°C. It is made from ethylene, in which ethylene has one of the hydrogen atoms replaced by a phenyl group. It is also called phenyl ethylene and vinyl benzene. As it can be made by heating cinnamic acid, an acid found in natural balsams and resins, it is also called cinnamene. In the form of vinyl toluene, which consists of mixed isomers of methyl styrene, the material is reacted with drying oils to form alkyd resins for paints and coatings.

The polymerized resin is a transparent solid very light in weight with a specific gravity of 1.054–1.070. The tensile strength is 27–68 MPa, compressive strength 82–117 MPa, and dielectric strength $18–24 \times 10^6$ V/m. Polystyrene is notable for its water resistance and high dimensional stability. It is also tougher and stronger at low temperatures than most other plastics. It is valued as an electrical insulating material, and the films are used for cable wrapping.

Types

General-Purpose Types

So-called general-purpose polystyrenes are characterized by clarity, luster, colorability, rigidity, unexcelled dielectric properties, and moldability. They are used where rigidity and appearance are important, but where toughness is not required. Typical uses include wall tile, container lids, and brush backs.

Grades are available with a wide range of processing characteristics. Highly heat-resistant types, which also have improved toughness, are available in crystal and a full range of transparent, translucent, and opaque colors.

Impact Grades

To overcome the relatively low impact strength of general-purpose polystyrene, combinations of polystyrene and rubber provide grades whose impact strength depends on the proportion of rubber added. Grades are generally characterized as medium-, high-, and extra-high-impact types. As impact strength increases, rigidity or modulus decreases. Such materials are available in virtually unlimited colors, but cannot reproduce crystal clear.

Medium-impact polystyrenes are used where moderate toughness plus good translucency is required. They are used for such products as containers, closures, and table–model radio cabinets. Types of medium-impact polystyrenes are available with improved heat resistance, surface gloss, and moldability.

High-impact polystyrenes include special grades with improved heat resistance, moldability, and surface gloss. They are used for refrigerator inner-door liners, crisper trays, containers, appliance housings, and toys. Highly heat resistant, high-impact grades (generally suitable for sections greater than 2.2 mm) are used for television masks and housings, portable radio cabinets, auto heater ducts, and automatic washer soap dispensers.

Extra-high-impact strength grades have relatively low moduli about 1360–1700 MPa and are used primarily where resistance to high-speed loading is required.

Chemical-Resistant Grades

Copolymers of styrene and acrylonitrile provide resistance to chemicals such as carbon tetrachloride, aliphatic hydrocarbons, and food stains and provide much better stress-crack resistance than polystyrene. The copolymers are transparent and haze-free, but are slightly yellow; they are available also in a wide variety of colors.

Primary use for the copolymers are in drinking tumblers and cups that have high resistance to crazing by butter fat and staining by coffee oils. Industrial uses include water filter parts, oil filter bowls, storage battery containers, and washing machine parts.

Other Special Grades

Light-stabilized types of styrene–methyl methacrylate copolymers and glass-reinforced molding materials are other special grades of polystyrene.

Light-stabilized grades were developed specifically for applications involving exposure to intense fluorescent light radiation, for example, "egg crate" light diffusers. Such formulations prevent the yellowing that occurs in unstabilized polystyrene when it is exposed to fluorescent light.

Styrene–methyl methacrylate copolymers were developed to provide a material with weatherability approaching that of acrylics, but at a lower cost. They have been used primarily for escutcheons, instrument panels, decorative medallions, brush blocks, auto tail light lenses, and advertising signs.

Glass-reinforced polystyrenes (incorporating chopped glass fibers) are available for injection molding or extrusion and provide higher strength, greater durability, and higher strength at elevated temperatures than do unreinforced polystyrenes.

Styrene can be polymerized with butadiene, acrylonitrile, and other resins. The terpolymer, acrylonitrile–butadiene–styrene (abbreviated ABS), is one of the common combinations. Styrene–acrylonitrile has excellent resistance to acids, bases, salts, and some solvents. It also is among the stiffest of the thermoplastics with a tensile modulus of 2757–3791 MPa. Acrylate–styrene–acrylonitrile has very good weathering resistance (nonyellowing), good toughness, and stress cracking resistance. Styrene–butylene resins are copolymers that mold easily and produce thermoplastic products of low water absorption and good electrical properties. They have strength equal to the vinyls with greater elongation.

Foamed Polystyrene

It is used in many forms, including extruded sheet (which is then thermoformed, e.g., egg cartons, trays); expandable polystyrene beads, which contain a blow agent (usually pentane) and which are processed into low-density foamed products, such as hot- and cold-drink cups and protective packaging; and block and heavy sheet, used for thermal insulation. See also *high-impact polystyrenes*.

Polysulfide Resins

These resins vary in properties from the viscous liquids to rubber-like solids. Organic polysulfide resins are prepared by the condensation of organic dihalides with a polysulfide.

The linear, high-molecular-weight polymers can be cross-linked or cured by reaction with zinc oxide. Compounding and fabrication of the rubbery polymers can be handled on conventional rubber machinery. The polysulfide rubbers are distinguished by their resistance to solvents such as gasoline and to oxygen and ozone. The polymers are relatively impermeable to gases. The products are used to form coatings that are chemically resistant and special rubber articles, such as gasoline bags.

Applications

The polysulfide rubbers were among the very first commercial synthetic rubbers. Although the products are not as strong as other rubbers, the chemical resistance makes them useful in various applications. The polysulfide rubbers were also among the first polymers to be used in solid-fuel compositions for rockets.

Polysulfide Rubber

Basically, polysulfide rubbers are chemically saturated polymers with reactive terminals through which conversion to a thermoset, elastomeric state can be affected by means of suitable catalysts or curing agents. More specifically, these rubbers are the products of a condensation polymerization in which one or more organic dihalides are reacted with an aqueous solution of sodium polysulfide.

Properties

As with a number of other synthetic elastomers, polysulfide rubbers require reinforcing fillers to achieve optimum physical properties. However, the high tensile values possible with unsaturated rubbers are not obtainable with polysulfides. For practical purposes, values of 10.3–12.4 MPa are the upper maximum limit.

The primary assets of these rubbers are outstanding oil, gasoline, and solvent resistance, as well as very low permeability to gases and solvent vapors. Also because of their saturated structure, polysulfides possess excellent resistance to oxidation, ozone, and weathering. Performance as regards temperature is somewhat dependent on polymer structure, but with few exceptions, the serviceability range is from −51°C to 121°C and intermittently up to 149°C. The principal limitation of polysulfide rubber compounds is relatively low resistance to compression set.

Available Forms

Polysulfide rubbers are novel in that they are available not only as solids but also as liquids. The liquids are 100% polymer, unadulterated with solvents or diluents, which also can be converted to highly elastic rubbers with properties closely approaching those of the cured, solid polysulfides. Viscosities range from a very fluid 5 poise to a heavy molasses-like 700 poise.

Fabrication and Applications

Design with both polysulfide crudes and liquids is quite similar, but because of the difference in physical form, the products made from these materials are processed and fabricated somewhat differently.

Crudes or Solid Polymers

Polysulfide crude rubbers are processed in the same way as other synthetic rubbers, on conventional mixing and fabricating equipment. The incorporation of reinforcing fillers, curing agents, and other additives is accomplished on two roll mills or Banbury-type internal mixers. Subsequent operations such as extruding, calendering, molding, or steam vulcanization can be carried out in the normal manner, except that somewhat closer factory control must be exercised than might be necessary with a larger volume, general-purpose rubbers.

Because polysulfide crudes may be classified as specialty rubber, their use is usually limited to those applications that demand exceptional solvent resistance. Such products include gaskets, washers, diaphragms, various types of oil and gasoline hose, and other mechanical rubber goods items. However, the solvent combinations encountered in various paints, coatings, and inks are responsible for the major consumption of these rubbers. A great number of the rollers employed for can lacquering, wood and metal coating, and the application of quick-drying inks are fabricated with a polysulfide rubber covering. Much of the hose use with hot lacquer and paint spraying equipment is made with a polysulfide tube or inner liner. One other unique commercial use of crudes is in the form of nonhardening putties, which make effective solvent-resistant seals for static type joints.

Liquid Polymers

Liquid polymer compounds are mixed by a three-roll paint mill, colloid mill, ball mill, or internal mixer. The resulting products may be applied by brushing, spraying, casting, or caulking gun depending on the characteristics of the specific compound involved and the type of application for which it was designed. Curing or conversion of these materials to highly elastic rubbers can be accomplished over a relatively wide temperature range, but their ability to cure at room temperature has been a primary factor in the employment of these polymers for a diversity of industrial applications. Despite their similarity in performance properties, there has been little overlapping use of the polysulfide crudes and liquid polymers in the same application areas; however, because of versatility in end-use product fabrication, liquids are threatening to intrude into solid-rubber fields of application.

One of the major uses of liquid polysulfide polymers is in the manufacture of stimulants in the aircraft, building, and marine industries. Such products may be compounded to bond to most building materials and provide flexible, elastic seals for joints that are subject to a high degree of movement and vibration. Curing or conversion to the rubbery state can be regulated to occur in minutes or hours at normal atmospheric temperatures depending on the demands of the application techniques involved.

Other applications include cold-setting casting, potting, and molding compounds, which exhibit flexibility, very low shrinkage, and excellent dimensional stability. The impregnation of leather with polysulfide liquid polymers imparts water and solvent resistance without loss of pliability. The liquid polysulfides may also be employed as coatings and adhesives, but more commonly, they are used as modifiers of epoxy resins for these and many other industrial applications. Modification is chemical rather than physical and results from an addition reaction between the polysulfide thiol terminals and epoxide groups. Greatly improved flexibility and impact resistance, lower shrinkage, less internal strain, better wetting properties, and lower moisture vapor transmission are the advantages gained.

Polysulfone (PSU) Resins

These are plastics whose molecules contain sulfone groups ($-SO_2-$) to the main chain, as well as a variety of aromatic or aliphatic constituents such as ether or isopropylidene groups. They have been made by reacting bisphenol A and 4,4′-dichlorodiphensulfone with potassium hydroxide in dimethyl sulfoxide at 265°F–285°F (130°C–140°C). The structure of the polymer is benzene rings or phenylene units linked by three different chemical groups: a sulfone group, an ether linkage, and an isopropylidene group. They may be processed by extrusion, injection molding, or blow molding.

PSUs based on aromatic backbones constitute a useful class of engineering plastics, owing to their high strength, stiffness, and toughness together with high thermal and oxidative stability, low creep, transparency, and the ability to be processed by standard techniques for thermoplastics. Aliphatic PSUs are less stable, for example, to hydrolysis, but are of interest for some biomedical applications. Four major uses of PSUs have been of commercial interest: (1) PSU, (2) polyarylsulfone, (3) polyethersulfone, and (4) polyphenylsulfone.

The aromatic structural elements and the presence of sulfone groups are responsible for the resistance to heat and oxidation; ether and isopropylidene groups contribute some chain flexibility. Aromatic PSUs can be used over wide temperature ranges, to ~150°C for PSU and to 200°C for polyethersulfone. In fact, the high-temperature performance of polyethersulfones is surpassed by few other polymers. Resistance to hydrolysis at high temperatures and to most acids, alkalies, and nonpolar organic solvents is excellent; however, the resins may be attacked or dissolved by bipolar solvents, especially if the component is under stress. Although resistance to ionizing radiation is high, protection against ultraviolet light is recommended for outdoor applications. Dimensional stability and electrical properties such as dielectric loss and strength are retained well during service, and flammability as well.

Because of the combination of properties discussed, polyethersulfone resins find many applications in electronic and automotive parts, medical instrumentation subject to sterilization, chemical and food processing equipment, and various plumbing and home appliance items. Coating formulations are also available, as well as grades reinforced with glass beads or fibers.

Polyterephthalate

A thermoplastic polyester in which the terephthalate group is a repeating structural unit in the chain, the terephthalate being greater in amount than other dicarboxylates that may be present.

Polytetrafluoroethylene (PTFE)

PTFE lubricants dispersed into a thermoplastic base resin greatly improve surface-wear characteristics. Molecular weight and particle size of the PTFE lubricant are designed to provide optimum improvements and wear, friction, and *PV* values for selected resin systems. PTFE has the lowest coefficient of friction (0.02) of any known internal lubricant. Its static coefficient of friction is lower than its dynamic coefficient, which accounts for the slip/stick properties associated with PTFE/metal sliding action. During the initial break-in period, the PTFE particles embedded in a thermoplastic matrix shear to form a high-lubricity film over the mating surface. The PTFE cushions asperities from shock and minimizes fatigue failure.

Forms

Glass fibers are frequently used in combination with silicone and PTFE lubricants, which offset the negative wear effects that the glass fibers have on surface characteristics. The use of silicone only, in conjunction with glass fibers, is not recommended; however, PTFE provides far more protection to the mating surface and should be used (with or without silicone) if the wear rate of the mating surface is important.

Uses

The first thermoplastics that were recognized for their inherent lubricity were nylon, acetal, and PTFE. These materials perform well, but for the more critical uses, their coefficients of friction may be too high, or wear may be too rapid.

The next generation of self-lubricated thermoplastics was formulated of various space resins that contained molybdenum disulfide,

graphite, or PTFE particles to improve both lubrication and wear characteristics. Although wear resistance was indeed upgraded considerably, mechanical strength and dimensional stability of these compounds are often insufficient.

To minimize these deficiencies, reinforcing fibers of glass or carbon are added. The resulting composites are several times stronger than the unreinforced materials, and they are extremely stable in a wide range of service environments. But these materials too have a shortcoming: In service, a period of time is required for the internal lubricant to become exposed and to be burnished over the wear surfaces. During this run-in period, as a bearing or wear member is put into service, the unlubricated surfaces are in contact, and damage may occur.

Two approaches to eliminating these problems use silicone fluids to provide the lubrication function.

Thermoplastic composites can also be internally lubricated with a variety of systems to improve wear resistance. PTFE and silicone, separately or in combination, provide the best improvements and wear characteristics. Graphite powder and molybdenum disulfide are also used, primarily in nylons. The PTFE lubricants are specially modified to enhance their lubricious nature in the compound. The optimum level of lubricating filler varies depending on filler type and resin, but typical ranges follow:

PTFE	15%–20%
Silicone	1%–5%
PTFE/silicone	15%–20%
Graphite	10%
MoS_2	2%–5%

Addition of these lubricants further improves wear characteristics of good bearing materials such as nylon and acetal. The lubricants also allow the use of poor wearing, but close-tolerance materials, such as polycarbonate, in gear or bearing applications. Lubricants can be used by themselves or in conjunction with glass or carbon-fiber reinforcements.

PTFE and silicone fluids; glass, aramid, and carbon fibers; and graphite powder are the primary reinforcements and lubricants used in internally lubricated composites. The composites are based on engineering resins for injection-molded wear and structural parts.

Polyurethanes

Extremely wide variations in forms and in physical and mechanical properties are available in polyurethanes. Grades can range in density from 13.88 g/cm³ in cellular form, to over 1937.6 g/cm³ in solid form, and in hardness from rigid solids at 85 Shore D to soft, elastomeric compounds.

Polyurethane polymers, produced by the reaction of polyisocyanates with polyester or polyether-based resins can be either thermoplastic or thermosetting. They have outstanding flex life, cut resistance, and abrasion resistance. Some formulations are as much as 20 times more resistant to abrasion than metals.

The noncellular grades—millable gums and viscous, castable, liquid urethanes—are elastomeric thermoset types, processed by conventional rubber methods.

Processing

There are three major types of polyurethane elastomers. One type is based on ether- or ester-type prepolymers that are chain-extended and cross-linked using polyhydroxyl compounds or amines;

alternatively, the unsaturated groups may be introduced to permit vulcanization with common curing agents such as peroxides. All of these can be processed by methods, commonly used for rubber. A second type is obtained by first casting a mixture of prepolymer with chain-extending and cross-linking agents and then cross-linking further by heating. The third type is prepared by reacting a dihydroxy ester– or ether-type prepolymer, or a diacid, with a diisocyanate such as diphenyl-methane diisocyanate and a diol; these thermoplastic elastomers can be processed on conventional plastics equipment. In general, urethane elastomers are characterized by outstanding mechanical properties and resistance to ozone, although they may be degraded by acids, alkalies, and steam.

A wide variety of tough and abrasion-resistant urethane coatings are available. Many are based on the reaction of castor oil, a triol, with an excess of diisocyanate; the resulting triisocyanate undergoes cross-linking by reaction with atmospheric moisture. Urethane alkyds can also be made by reacting an unsaturated drying oil with glycerol and then reacting the product with a diisocyanate; curing is affected by atmospheric oxidation of the double bonds. Still other coatings are based on the use of prepolymers. Polyurethanes are also used as adhesives, for example, in the bonding of rubber and of nylon.

Foams

Polyurethane foams are thermoset materials that can be made soft and flexible or firm and rigid at equivalent densities. These foams, made from either polyester- or polyether-type compounds, are strong, even at low density, and have good chemical resistance. Polyether-based foams have greater hydrolysis resistance, are easier to process, and cost less. Polyester-based foams have higher mechanical properties, better oil resistance, and more uniform cell structure. Both types can be sprayed, molded, foamed in place, or furnished as sheets cut from slab stock.

Flexible Foams

Glass transition temperatures (the temperature at which an elastomeric material become stiff and brittle) of flexible foams are well below room temperature. The foams can be pigmented to any color, but, regardless of pigmentation, they yellow when exposed to air and light. Some types of flexible foams are excellent liquid-absorbing media and can hold up to 40 times their weight of water.

Polyether-type foams are not affected by high-temperature aging, either wet or dry, but ultraviolet exposure produces brittleness and reduces properties. In use, these foams are always covered with a fabric or other material.

Most solvents and corrosive solutions decrease tear resistance and tensile strength and cause swelling of flexible foams. Swelling is not permanent, however, if the solvent is removed and the foam dried. However, the foams can be destroyed by strong oxidizing agents and hydrolyzed in strong acids or bases. Generally, the polyether foams are more resistant to hydrolytic degradation; the polyester foams are more resistant to oxidative attack.

Applications for polyester flexible urethane foam include gasketing, air filters, sound-absorbing elements, and clothing interliners (laminated to a textile material). The polyether types are used in automobile and recreational-vehicle seats, carpet underlay, furniture upholstering, bedding, and packaging.

Rigid Foams

Bases for rigid foams are polymers having glass transition temperatures higher than room temperature. The cells of rigid foam are about the same size and uniformity as those of flexible foam, but rigid foams usually consist of 90% closed cells. For this reason,

water absorption is low. Compressing the foam beyond its elastic limit damages the cellular structure.

Rigid foams are blown with either carbon dioxide or fluorocarbons. Gas generated by vaporization of fluorocarbons, entrapped in the closed cells, gives the foam a very low thermal conductivity of 0.11–0.14 Btu in./h ft^2 °F. Conductivity increases with age, however, to a constant value of about 0.16.

Rigid urethane foams are used for thermal insulation of refrigerators, refrigerated trucks and railroad cars, cold-storage warehouses, and process tanks because of their low conductivity and high-strength-to-weight ratio. Other applications include flotation devices, encapsulation, structural and decorative furniture components, and sheathing and roof insulation for buildings.

Integral Skin Foam

Urethane foams that are formed with integral skins ranging from soft and flexible types to impact-absorbing grades and rigid foams used in structural parts. Color can be added, but since the foams yellow on aging, black is most practical for the surface color. If other colors are required, coatings are recommended. The tough, high-density, integral skin is formed against the mold surface, and the low-density core is produced by a blowing agent—usually a fluorocarbon.

Elastomeric foams of this type are used in automotive bumper and fascia systems and, most recently (reinforced with milled glass fibers), in fenders and other exterior body panels. The semirigid types are used in athletic protective gear, in automotive crash-protection areas, horn buttons, sun visors, and armrests. Applications for the rigid structural foams include housings for computer systems, chair shells, furniture drawers, and sports equipment.

Part Fabrication

A low-pressure molding process—reaction injection molding (RIM)—is used almost exclusively to produce urethane parts, some weighing as much as 45.4 kg. In the process, two were more highly reactive liquid systems injected with high-pressure impingement mixed into a closed mold at low pressure, where they react to form a finished polymer. Depending on formulation, the polymer can be a rigid, integral skin, microcellular urethane foam with a flexural modulus of over 680 MPa or a soft, flexible elastomer with a flexural modulus as low as 49.06 MPa or a rigid structural foam having a density of 830.4 g/cm^3. Cycle time is short; parts can be demolded in less than a minute.

Reinforcement in the form of milled glass fiber, glass flake, or mineral filler increases the stiffness, the thermal properties, and dimensional stability of RIM parts. Maximum glass content in reinforced reaction injection molding (RRIM) is about 25%—a limit determined by the increased viscosity with increasing glass. Natural color of unpigmented RIM urethane parts is tan.

Applications

Reticulated polyester foams are used for explosion suppression. The material protects aircraft fuel tanks from small arms gunfire. Over the years, several technical advances were made to the foam, lowering weight and increasing service life and conductivity (to prevent static buildup).

The material is an open-pore, reticulated polyurethane foam that contains a network of skeletal strands with 98% void space at any pore size. The material functions essentially as a 3D fire screen similar to a safety fire screen over a lighted Bunsen burner.

In a fuel tank, the empty space above the ullage may readily contain an explosive mixture of fuel vapor and air. Since the liquid fuel

P

itself does not explode, a completely filled tank is far less likely to explode than one that is not. The lower the fuel level in the tank, the greater amount of explosive vapor present. When an ignition source is present, the vapor adjacent to the spark ignites rapidly. This ignition, in turn, ignites the vapor around it, creating a chain reaction as the ignition, or flame front, gets larger and moves faster as it propagates through the vapor. The rapid ignition and propagation of the flame results in an ever-growing compression wave in front of it, compressing the unignited vapor, thus adding even greater force to an explosion. This sequence occurs in milliseconds.

This chain reaction is prevented from occurring; instead, vapor ignition is confined to the area immediately around the ignition source. Flame and wave propagation are mitigated by the foam, thus preventing an explosion.

The foam may be easily fabricated into any configuration by conforming to the inside of complex, rigid or flexible, bladder-type fuel tanks. Once installed, the foam can remain in the fuel tank for many years without degradation or loss of physical properties and performance characteristics. They can be readily removed for internal fuel tank maintenance and reinstalled repeatedly.

In the Volkswagen Beetle, polyurethane raw materials protect the finishes of plastic and metal components, both exterior and interior. Polyurethane foam is used inside the doors and instrument panel, and polyurethane raw materials help protect the instrument panel.

Polyvinyl Acetals

Members of the family of vinyl plastics. Polyvinyl acetal is a general name for resins produced from a condensation of polyvinyl alcohol with an aldehyde. There are three main groups: polyvinyl acetal itself, poly*vinyl butyral*, and *polyvinyl formal*. Polyvinyl acetal resins are thermoplastics that can be processed by casting, extruding, molding, and coating, but their main uses are in adhesives, lacquers, coatings, and films.

Polyvinyl Acetate

A thermoplastic material composed of polymers of vinyl acetate in the form of a colorless solid. It is obtainable in the form of granules, solutions, lattices, and pastes and is used extensively in adhesives, for paper and fabric coatings, and in bases for inks and lacquers.

Polyvinyl Acetate Emulsion Adhesive

A latex adhesive in which the polymeric portion comprises polyvinyl acetate, copolymers based mainly on polyvinyl acetate, or a mixture of these and which may contain modifiers and secondary binders to provide specific properties.

Polyvinyl Alcohol

A thermoplastic material composed of polymers of the hypothetical vinyl alcohol. Usually a colorless solid, insoluble in most organic solvents and oils, but soluble in water when the content of hydroxyl groups in the polymer is sufficiently high; the product is normally granular. It is obtained by the partial hydrolysis or by the complete hydrolysis of polyvinyl esters, usually by the complete hydrolysis of polyvinyl acetate. It is mainly used for adhesives and coatings.

Polyvinyl Butyral

A thermoplastic material derived from a polyvinyl ester in which some or all of the acid groups have been replaced by hydroxyl groups and some or all of these hydroxyl groups have been replaced by butyral groups by reaction with butyraldehyde. It is a colorless, flexible, tough solid used primarily in interlayers for laminated safety glass. See also *polyvinyl acetals*.

Polyvinyl Carbazole

A thermoplastic resin, brown in color, obtained by reacting an acetylene with carbazole. The resin has excellent electrical properties and good heat and chemical resistance. It is used as an impregnant for paper capacitors.

Polyvinyl Chloride Acetate

A thermoplastic material composed of copolymers of vinyl chloride and vinyl acetate. It is a colorless solid with good resistance to water as well as concentrated acids and alkalies. It is obtainable in the form of granules, solutions, and emulsions. Compounded with plasticizers, it yields a flexible material superior to rubber in aging properties. It is widely used for cable and wire coverings, in chemical plants, and in protective garments.

Polyvinyl Chloride (PVC)

Among the vinyl polymers and copolymers, the PVC thermoplastics are the most commercially significant. With various plasticizers, fillers, stabilizers, lubricants, and impact modifiers, PVC is compounded to be flexible or rigid, opaque or transparent, to have high or low modulus, or to have any of a wide spectrum of properties or processing characteristics.

PVC resin can also be chlorinated and it can be alloyed with other polymers such as ABS (acrylonitrile–butadiene–styrene), acrylic, polyurethane, and a nitrile rubber to improve impact resistance, tear strength, resilience, heat-deflection temperature, or processability.

Forms

PVC is a hard, flame-resistant, and chemical-resistant thermoplastic resin. The resin is available in the powder form, as a latex, or in the form of plastisol, PVC resin, pigments, and stabilizers are milled into plasticizers to form a viscous coating material (plastisol) that polymerizes into a tough elastic film when heated. Plastisols are used extensively for coating glass bottles and glass fabrics. The dispersion types of resins are used in flexible molding compounds. Such formulations consist of a vinyl paste resin, a suitable plasticizer such as dioctyl phthalate, and a stabilizer (usually a compound of lead). Flexible molds are widely applied to plaster casting and encapsulation of electronic circuits with epoxy resins.

Processing

PVC compounds are processed by extrusion, injection molding, calendering, compression molding, and blow molding. PVC coatings are applied by fluidized-bed and electrostatic powder-coating methods. The resins are also used for dip molding and coating, in the form of plastisols and organosol dispersions or water dispersions (latexes). Cellular PVC products are made by introducing gas into the resin during molding or extrusion. Foams can be open or closed cell and can be elastomeric or rigid, depending on plasticizer content.

PVC compounds can be made water white in flexible compounds and very clear in rigid compounds, and they can be pigmented to almost any color.

Properties

With so many property variations attainable by compounding methods, no single compound can be considered typical of PVC. For example, creep rate of rigid compounds is so low and predictable that they can be used to make pressure pipe for water distribution; flexible compounds can be soft enough, yet impermeable, so that they are used for baby pants and for an excellent imitation suede, or they can be transparent, nontoxic, and tough enough to be used for mineral water bottles.

Rigid PVC, sometimes called the "poor man's engineering plastic," is a hard, tough material that can be compounded to a wide range of properties. Noteworthy among its properties is low combustibility; it has high resistance to ignition and is self-extinguishing. It also provides good corrosion and stain resistance, thermal and electrical insulation, and weatherability. However, PVC is attacked by aromatic solvents, and ketones, aldehydes, naphthalenes, and some chloride, acetate, and acrylic esters. Some impact modifiers used in rigid PVC reduce chemical resistance. In general, normal-impact grades have better chemical resistance than the high-impact grades.

Most PVC compounds are not recommended for continuous use above 60°C. Chlorination increases heat-deflection temperature, flame retardancy, and density and extends the continuous-use temperature to 80°C–100°C, depending on the amount of chlorination.

Polyvinyl Formal

One of the groups of polyvinyl acetal resins made by the condensation of formaldehyde in the presence of polyvinyl alcohol. It is used mainly in combination with cresylic phenolics, for wire coatings, and for impregnations, but can also be molded, extruded, or cast. It is resistant to greases and oils.

Polyvinylidene Chloride

A thermoplastic material composed of polymers of vinylidene chloride (1,1-dichloroethylene). It is a white powder with a softening temperature at 185°C–200°C (365°F–390°F). The material is also supplied as a copolymer with acrylonitrile or vinyl chloride, giving products that range from the soft, flexible type to the rigid type. Also known as *saran*.

Polyvinylidene Fluoride

This recent member of the fluorocarbon family of plastics is a homopolymer of polyvinylidene fluoride. It is supplied as powders and pellets for molding and extrusion and in solution form for casting. The resin has good tensile and compressive strength and high impact strength.

Polyvinyl Resins

These resins are polymeric materials generally considered to include polymers derived from monomers.

Many of the monomers can be prepared by addition of the appropriate compound to acetylene. For example, vinyl chloride, vinyl fluoride, vinyl acetate, and vinyl methyl ether may be formed by the reactions of acetylene with HCl, HF, CH_3OOH, and CH_3OH, respectively. Processes based on ethylene as a raw material have also become common for the preparation of vinyl chloride and vinyl acetate.

The polyvinyl resins may be characterized as a group of thermoplastics, which, in many cases, are inexpensive and capable of being handled by solution, dispersion, injection-molding, and extrusion techniques. The properties vary with chemical structure, crystallinity, and molecular weight.

Polyvinyl Acetals

These are relatively soft, water-insoluble thermoplastic products obtained by the reaction of polyvinyl alcohol (PVAL) with aldehydes. Properties depend on the extent to which alcohol groups are reacted. Polyvinyl butyral (PVB) is rubber and tough and is used primarily in plasticized form as the inner layer and binder for safety glass. Polyvinyl formal is the hardest of the group. It is used mainly in adhesive, primer, and wire-coating formulations, especially when blended with a phenolic resin.

PVB is usually obtained by the reaction of butyraldehyde with PVAL. The former can be produced by the same process, but is more conveniently obtained by the reaction of formaldehyde with polyvinyl acetate in acetic acid solution.

Polyvinyl Acetate

Polyvinyl acetate is a leathery, colorless thermoplastic material that softens at relatively low temperatures and that is relatively stable to light and oxygen. The polymers are clear and noncrystalline. The chief applications of polyvinyl acetate are in adhesives and binders for water-based or emulsion paints.

Vinyl acetate is conveniently prepared by the reaction of acetylene with acetic acid.

Polyvinyl Alcohol (PVAL)

PVAL is a tough, whitish polymer that can be formed into strong films, tubes, and fibers that are highly resistant to hydrocarbon solvents. Although PVAL is one of the few water-soluble polymers, it can be rendered insoluble in water by drawing or by the use of cross-linking agents.

Polyvinyl Chloride (PVC)

PVC is a tough, strong thermoplastic material that has an excellent combination of physical and electrical properties. The products are usually characterized as plasticized or rigid types. PVC (and copolymers) is the second most commonly used polyvinyl resin and one of the most versatile plastics.

Polyvinylidene Chloride (PVDC)

PVDC is a tough, horny thermoplastic with properties generally similar to those of PVC. In comparison with the latter, PVDC is softer and less soluble; it softens and decomposes at lower temperatures, crystallizes more readily, and is more resistant to burning.

Because of its relatively low solubility and decomposition temperature, the material is most widely used in the form of copolymers with other vinyl monomers, such as vinyl chloride. The copolymers are employed as packaging film, rigid pipe, and as filaments for upholstery and window screens.

Films of PVDC, and especially the copolymer containing about 15% of vinyl chloride, are resistant to moisture and gases. Also, they can be heat-sealed and, when oriented, have the properties of shrinking on heating. By warming a food product wrapped loosely with a film of the polymer, a skin tight, tough, resistant coating is produced.

By cold-drawing, the degree of crystallinity, strength, and chemical resistance of sheets, filaments, and even piping can be greatly increased.

Polyvinyl Ethers

Polyvinyl ethers exist in several forms varying from soft, balsam-like semisolids to tough, rubbery masses, all of which are readily soluble in organic solvents. Polymers of the alkyl vinyl ethers are used in adhesive formulations and a softening or flexibilizing agents for other polymers.

Polyvinyl Fluoride (PVF)

PVF is a tough, partially crystalline thermoplastic material that has a higher softening temperature than PVC. Films and sheets are characterized by high resistance to weathering.

Films are used in industrial and architectural applications. Coatings, for example, on pipe, are resistant to highly corrosive media.

POM

See *polyoxymethylene.*

Pop Riveting

A technique for making riveted lap joints in sheet or other thin material with access from one side only. The tee-shaped hollow rivet carries a pin-headed mandrel through its bore, the mandrel shaft emerging at the head end of the rivet. A hole, of a diameter just sufficient to allow passage of the rivet shank and mandrel head, is drilled through the pair of sheets. The shank of the rivet, with the mandrel head projecting beyond it, is inserted through the hole and the mandrel is then drawn back so that its head expands the projecting shaft bore. The rivet firmly nips the sheets before the mandrel breaks leaving its head on the foreface of the sheets.

Pop-Off

Loss of small portions of a porcelain enamel coating. The usual cause is outgassing of hydrogen or other gases from the basis metal during firing, but pop-off may also occur because of oxide particles or other debris on the surface of the basis metal. Usually, the pits are minute and cone shaped, but when pop-off is the result of severe *fish scale*, the pits may be much larger and irregular.

Population

In statistics, a generic term denoting any finite or infinite collection of individual samples or data points in the broadest concept; an aggregate determined by some property that distinguishes samples that do and do not belong.

Porcelain

Porcelains and stoneware are highly vitrified ceramics that are widely used in chemical and electrical products. Electrical porcelains, which are basically classified clay-type ceramics, are conventionally divided into low- and high-voltage types. The high-voltage grades are suitable for voltages of 500 and higher and are capable of withstanding extremes of climatic conditions. Chemical porcelains and stoneware are produced from blends of clay, quartz, feldspar, kaolin, and certain other materials. Porcelain is more vitrified than stoneware and is white in color. A hard glaze is generally applied. Stoneware can be classified into two types: a dense, vitrified body

for use with corrosive liquids and a less dense body for use in contact with corrosive fumes. Chemical stoneware may range from 30% to 70% clay, 5% to 25% feldspar, and 30% to 60% silica. The vitrified and glazed product will have a tensile strength up to 16 MPa and a compressive strength of 551 MPa. Industrial stoneware is made from specially selected or blended clays to give desired properties.

Both chemical porcelains and stoneware resist all acids except hydrofluoric. Strong, hot, caustic alkalies mildly attack the surface. These ceramics generally show low thermal shock resistance and tensile strength. Their universal chemical resistance explains their wide use in the chemical and processing industries for tanks, reactor chambers, condensers, pipes, cooling coils, fittings, pumps, ducks, blenders, filters, and so on.

Ceratherm is an acid-resistant and heat-shock-resistant ceramic with a base of high-alumina clay. It is strong and nonporous and is used for pump and chemical-equipment linings.

Properties

Porcelain is distinguished from other fine ceramic ware, such as china, by the fact that the firing of the unglazed ware (the bisque firing) is done at a lower temperature (1000°C–1200°C) than the final or gloss firing, which may be as high as 1500°C. In other words, the ware reaches its final state of maturity at the maturing temperature of the glaze.

The white color is obtained by using very pure white-firing kaolin or china clay and other pure materials, the low absorption results from the high firing temperature, and the translucency results from the glass phase.

This term is frequently used as a synonym for *china.*

Porcelain Enamels and Ceramic Coatings

Earlier definitions of ceramic materials usually stressed their mineral origin and the need for heat to convert them into useful form. As a consequence, only porcelain enamels and glazes were recognized as ceramic coatings until recently when the principles of phase relations, bonding mechanisms, and crystal structure were applied to ceramic materials and to coatings made from them. In consequence, ceramic materials can now be most safely defined as solid substances that are neither metallic nor organic in nature, a definition that is somewhat more inclusive than older ones, but more accurately reflects modern scientific usage.

Most ceramics are metal oxides or mixtures and solutions of such oxides. Certain ceramic materials, however, contain little or no oxygen. As a whole, ceramic materials are harder, more inert, and more brittle than organic or metallic substances. Most ceramic coatings are employed to exploit the first two properties while minimizing the third.

Low-Temperature Coatings

The outstanding resistance to corrosion of certain metals, notably aluminum and chromium, is attributable to the remarkable adherence of their oxide films. Aluminum does not corrode because its oxidation product, unlike that of iron, is a highly protective coating. It was once believed that some mysterious kinship between a metal and its own oxide was needed for this protection, but recent knowledge relating to the structure of metals and metal oxides has enabled metallurgists to develop alloys that form even more stable and adherent films. Methods for thickening or stabilizing these oxide coatings by heat treatment, electrolysis, or chemical reaction are widely accepted.

The presence of such coatings on metals and alloys strongly influences such properties as their emission and absorption of radiation, frictional and wetting characteristics, and electrical and electrochemical properties.

Usually, ceramic coatings contain compounds of metals other than the substrate. Most of these are oxides, which are amorphous or cryptocrystalline in nature. They can be divided into two groups: chemical reactants (in which a new compound or complex is formed) and inorganic colloids.

Chemical Reactants

These coatings usually involve a chemical modification of the natural metal oxide into a coating that is stabler or denser. This classification includes the treatment of aluminum, tin, and zinc, with chromates, electrolytic "anodizing," the treatment of iron with soluble phosphates, etc. The films that are formed are more inert to most chemicals than normal oxide films, and the process is termed "passivation."

Inorganic Colloids

Colloidal particles are sufficiently small so that their surface energy is sufficient for bonding. These particles can be made into colloidal suspensions called "slurries" or "slips." They contain natural colloids such as hydrophilic clays, or finely ground materials of a fibrous or platelike nature such as asbestos, mica, graphite, potassium titanate, alumina monohydrate, zirconia hydrate, or molybdenum sulfide.

Some inorganic colloids are so finely dispersed as to be true sols. "Water glass" is a familiar example of the colloidal sol; depending on its sodium content, alkalinity, and dilution, it may be a thin fluid, a sticky, viscous liquid, or a translucent semisolid. Newer aqueous sols include other alkali silicates, aluminum acid hydrates and phosphates, alkyl titanates and silicates, and lime hydrate.

These coatings are usually processed by drying or flocculation. The dry film and substrate are usually heated sufficiently to drive off all traces of moisture and irreversibly "set" the coating for its final use, as for lubricants, thermionic emitters, and fluorescent lamp coatings. The relatively weak bonding power of these dried films can be used to hold them in place for further heating; this is the basis of "wet process" porcelain enameling.

Moderate Temperature Processes

In the temperature range between 538°C and 1093°C, most silicate glasses melt. In this range, the fusion of a vitreous powder (usually called a frit) or the formation of a glass from its component materials is then used as the basis for porcelain enameling and glazing processes.

Glazes differ from enamels principally in the substrates on which they are applied. When the substrate is metallic, the vitreous coating is called an enamel; when applied to ceramic bodies such as porcelain, china, terra cotta, pottery, and electrical ceramics, the coating is termed a glaze. Since these coatings are vitreous, they may be transparent, but most enamels and many glazes contain finely divided crystalline materials that color them and make them more or less opaque.

Enamels are used primarily to provide resistance to corrosion, heat, and/or abrasion and wear; they are frequently employed for their attractive appearance as well.

Vitreous enamels cannot easily be classified; a typical enameling slip may contain 10 or more ingredients, including those that are glassy or form a vitreous network (feldspars, frits, borax, and other mineral sources of silicates, phosphates, and borates). The modifiers usually consist of alkali and alkaline earth compounds or lead. They also contain fluxing agents, opacifiers, suspending agents, and clays,

as well as refractory oxides intended to dissolve in the melting glass and increase its viscosity. The oxides or other compounds of cobalt, nickel, manganese, arsenic, or antimony may be included to promote adhesion between glass and metal.

Most porcelain enamels consist of two or more layers of glass, separately applied and fused. The first layer is called the ground coat or base coat; its purpose is to attach firmly to the metal substrate and prevent undesirable interactions between substrate and enamel or the evolution of gas from the metal. It is in the ground coat that the adherence-promoting additives are used, and it is these additives that produce the blue, brown, or black color of most ground coats. Where light colors are not required (range parts, heat-resisting coatings, and certain abrasion-resistant applications), a ground coat may suffice over the base metal.

Although "cover coat" may be required to resist chemical attack, abrasion, impact, heat, or weathering, enamels are most commonly used to provide a durable and attractive finish for steelhead ferrous alloys. Most of these are white, hence the name "porcelain" enamels. Earlier enamels contain antimony or zircon to provide opacity and whiteness, but zirconium oxide and titanium oxide are now most frequently used. The latter not only provides superior opacity but usually improves acid resistant as well. As a consequence, titanium-bearing enamels may be applied in a single coating, not only over a suitable base coat but also directly upon special steels. Such enamels can now be fused at less than 760°C to titanium-bearing steels, to steels precoated with a thin nickel film, or to steels pretreated with iron phosphate.

On cast iron, enamels are used for range parts and a high-grade sanitary ware. Certain chemical ware (tanks, pumps, etc.) are also made with cast iron. The rigidity and good acoustical damping of cast iron, together with its resistance to distortion by heat, permits heavier layers of the protective glass to be used than are possible on steel or enameling iron.

Special enamels have been developed for the chemical industry and hot water tanks. Such equipment is frequently called "glass lined," and can contain all acids except hydrofluoric and hot phosphoric, moderately alkaline solutions up to their boiling point, and water under pressures up to 3500 MPa.

Enamels are not limited to use on ferrous substrates; the earliest decorative enamels were used on precious metals and copper hardware and jewelry. Enamels have recently been developed suitable for aluminum. Such enamels can be fired at temperatures well below 538°C. Although not so hard or corrosion resistant as sheet-steel enamels, these aluminum enamels provide attractive and durable finishes for sheet, extrusions, and castings.

Although most enamels are applied to the substrate by a wet-process technique, some coatings may be applied by dusting or sieving the powdered composition directly upon the heated surface. This "dry process" is principally used with chemical ware and cast iron sanitary ware.

The mechanical properties of enamels are strongly influenced by the composition, thickness, and geometry of the substrate as well as by the kind, thickness, method of application, and firing conditions of the enamel layer or layers. In general, thin enamel coatings are best able to resist thermal and mechanical shock and stress. Very thin (1–4 mils thick) coatings on steel or aluminum may even be bent, punched, sheared, or drilled without damage. Thicker coatings are usually required for ordinary applications (in appliances, curtain walls or structural panels, and signs); the thickness of the glassy layers is usually between 3 and 20 mils. For cast iron sanitary ware, the vitreous coating may be 40 mils thick or more, and some chemical tanks employ 6.4 mm of protective glass.

For maximum resistance to chipping, the enamel must be supported by a hard, relatively thick substrate. Where the enamel's article must resist bending or twisting, thinner and more ductile substrates are indicated.

The selection of metal for porcelain enamel and its preparation strongly influences the properties of the composite. "Enameling irons" are essentially very low carbon, basic open-hearth, rimmed steels. Regular SAE (Society of Auto Engineers) and AISI (American Iron and Steel Institute) low-carbon steels can seldom be perfectly enameled especially when hot-rolled. Premium enameling stock may contain titanium sufficient to further lower the available carbon content and improve resistance to warping or sagging.

For thicker substrates, a basic open-hearth plate steel of flange quality may be used. Higher impurity levels, however, may require that the enamel be fired in an inert atmosphere to eliminate "boiling" defects. In some cases, aluminized steel can be used in moderately oxidizing furnace atmospheres.

Cast steel and cast iron can be enameled acceptably; carbon is oxidized from the surface during the relatively long firing period needed to fuse and consolidate the ground coat. Purity of the metal is not so stringent a requirement for cast iron as for steel.

Surface treatment of the metal usually requires the removal of all scale and dirt (this may require sandblasting for heavy-gauge stock) followed by a light pickling for ultimate control of the oxide layer thickness. For some metals and cast irons, sandblasting or grit blasting alone may be sufficient pretreatment. Special enamels may require phosphate bath treatment or the deposition of a nickel-, copper-, or aluminum-base coating over the metal.

Metals to be enameled should be reasonably strain-free. Burrs, sharp edges, or small external radii and large variations in substrate thickness should be avoided. Welds must be sound and metallurgically similar to the parent area.

Refractory Enamels and Glazes

Because the fused porcelain enamel cools with its substrate, it is important that the total contraction that occurs during cooling be approximately equal in both metal and glass. If the metal contracts more than the enamel, the latter will be forced into compression and may shatter or chip easily; if the enamel contracts more than the metal, it will crack or craze.

Among the silicate glasses, the most refractory generally have a low thermal expansion coefficient, and they must therefore be used on metals that have low expansion. Most of the refractory enamels contain finely powdered silica and chromium oxide, which dissolve in the glass on fusion, further lowering expansion. In consequence, such enamels are largely restricted to refractory substrate metals such as certain stainless steels and nickel and chromium alloys. The service temperature of enamel-coated stainless steel and nickel alloys is about 954°C in aircraft engine exhaust manifolds, turbosupercharger linings, jet engine combustors, and commercial burners.

Thin, electrically conductive ceramic coatings can be used as resistance heaters on aircraft windshields and the like. Most of these consist of a mixture of the oxides and suboxides of tin with traces of bismuth, antimony, cadmium, or arsenic. They are applied by heating the glass to a dull red heat and spraying it with a fusible tin halide under slightly reducing conditions. As it cools in air, the tin halide decomposes into the conductive complex. When sufficiently thin, these coatings are quite transparent.

High-Temperature Coatings

To attach a ceramic coating to any substrate by enameling, both substrate and glass must be heated to the fusion temperature of the glass. However, fusion methods have not been successful for the more refractory materials. Because most refractory ceramic coatings are amorphous or crystalline in nature, they have to be applied by relatively novel techniques.

Although most ceramic materials are refractory, some of them can be vaporized in an electric arc or hot vacuum. Thin coatings of amorphous silica can be applied readily to relatively cool substrates by vaporizing metallic silicon or silicon halides in the presence of small quantities of oxygen. Apparently, the transferor is accomplished largely as silicon monoxide, which recombines with oxygen on cooling. The process is used to obtain thin, protective, optically transparent films on lenses, certain electrical components, and metal reflectors.

Other ceramic coatings may be produced by vaporizing one or more components of the coating. In this way, coatings of their respective carbides of silicon, boron, aluminum, and chromium can be deposited on graphite, silicon nitride can be formed on metallic silicon, and silicide coatings can be deposited on metals such as tungsten and molybdenum. These processes are necessarily expensive and are poorly adaptable to large specimens or complex shapes.

Flame or Arc Spraying

Many metallic oxides and interstitial compounds can be heated to or above their melting temperatures with a chemical flame or electrical arc. The coatings obtained by directing a spray of nearly molten ceramic particles toward an otherwise unheated substrate are interesting and useful for a variety of purposes. Most of the processes are proprietary and differ chiefly in the form in which material is fed into the heat source.

The coatings obtained in this way may be quite porous, or they may approach the theoretical density of the material being sprayed. Since the substrate need not be heated, the need for a close match of thermal expansion is less important than with vitreous coatings. The porous coatings are surprisingly immune to thermal and mechanical shock, but confer little chemical protection.

Substrates must usually be roughened before application of these coatings, but heat-resistant glasses, glazes, and some porcelain enamels make excellent substrates. Adherence seems to be largely mechanical, and the adherence tests used for porcelain enamels are not applicable to these coatings. No standards for testing or performance have yet been established.

Coatings obtained by flame spraying may consist of pure ceramic materials, metals, and some organic polymers; a modification of flame spraying can be used to produce pyrolytic graphitic coatings of high density and resistance to hot gas erosion. Mixtures of ceramic and metal powders may be used to produce cermet coatings or even graded coatings, and unusual electrical, magnetic, and dielectric properties can be obtained with such mixtures or with multilayer application.

Because each particle is suddenly and individually chilled, the structure of flame-sprayed ceramics may be unlike those of the bulk materials. Nonstoichiometry is common, and the stresses in the coating and between the coating and substrate are complex. Nevertheless, certain flame-sprayed and arc plasma–sprayed coatings have already found acceptance in missile technology and metalworking in foundry applications and for heat- or wear-resistant coatings on metallic, ceramic, and polymer substrates.

Pore

(1) Voids in a material. Usually the term implies voids that individually are of small size relative to the section and particularly those resulting from gas entrapment during solidification. In the case of a highly porous materials, the porosity is often reported as a percentage

by volume. (2) A small opening, void, interstice, or channel within a consolidated solid mass or agglomerate, usually larger than atomic or molecular dimensions. (3) A minute cavity in a powder-metallurgy compact, sometimes added intentionally. See also *porous P/M parts*. (4) A minute perforation in an electroplated coating.

Pore Area

The effective surface porosity of a sintered powder-metallurgy compact to determine the permeability to a test fluid.

Pore Channels

The connections between pores in a sintered body.

Pore Formation

The natural formation of pores during compaction and/or sintering. See also *pore-forming material*.

Pore Size

Width of a pore in a compacted and/or sintered metal powder or within a particle.

Pore Size Distribution

Indicates the volume fractions of different pore size categories in a sintered body, which are determined metallographically.

Pore Size Range

The limits between which a variation in pore size and a sintered body is allowed.

Pore Structure

Pattern of pores in a solid body indicating such characteristics as pore shape, *pore size*, and *pore size distribution*.

Pore-Forming Material

A substance included in a metal powder mixture that volatilizes during sintering and thereby produces a desired kind of porosity in a finished compact.

Porosimeter

A test apparatus to measure the interconnected porosity in a sintered compact by means of determining its permeability through use of a test fluid such as mercury, which either partially or completely fills the open pores.

Porosity

(1) Fine holes or pores within a solid; the amount of these pores is expressed as a percentage of the total volume of the solid. (2) Cavity-type discontinuities in weldments formed by gas entrapment during solidification. (3) Trapped air or shrinkage in a casting. A characteristic of being porous, with voids or pores resulting from trapped air or shrinkage in a casting. See also *gas porosity* and *pinhole porosity*.

Porous Bearing

A bearing made from porous material such as compressed and sintered metal powder. The pores may act as reservoirs or passages for supplying fluid lubricant, or the bearing may be impregnated with solid lubricant. See also *self-lubricating bearing*.

Porous Metals

These are metals with uniformly controlled pore sizes, in the form of sheets, tubes, and shapes, used for filtering liquids and gases. They are commonly made by powder metallurgy, and the pore size and density are controlled by the particle size and the pressure used. Stainless steel, nickel, bronze, silver, and other metal powders are used, depending on the corrosion resistance required of the filter. Pore sizes can be as small as 0.2 µm, but the most generally used filters have pores of 4, 8, 12, and 25 µm. Pore sizes have a uniformity within 10%. The density range is from 40% to 50% of the theoretical density of the metal. Standard filter sheet is 0.76–1.52 cm, but thinner sheets are available. Sheet as thin as 0.010 cm, and with void fractions as high as 90%, have been made for fuel cells in catalytic reactors. Porous steel is made from 18-8 stainless steel, with pore openings from 20 to 65 µm. The fine-pore sheet has a minimum tensile strength of 69 MPa, and the coarse sheet has a strength of 48 MPa. Felted metal is porous sheet made by felting metal fibers, pressing and sintering. It gives a high-strength-to-porosity ratio, and the porosity can be controlled over a wide range. In this type of porous metal, the pores may be from 0.003 to 0.038 cm in diameter and of any metal to suit the filtering conditions. A felted fiber filter of 430 stainless steel with 25% porosity has a tensile strength of 172 MPa.

Porous Molds

Molds for forming plastics that are made up of bonded or fused aggregate (powered metal, coarse pellets, and so forth) such that the resulting mass contains numerous open interstices of regular or irregular size allowing either air or liquids to pass through the mass of the mold.

Porous Powder-Metallurgy (P/M) Parts

P/M components that are characterized by *interconnected porosity*. Primarily, application areas for porous P/M parts are filters, damping devices, storage reservoirs for liquids (including self-lubricating bearing), and battery elements. Bronzes, stainless steels, nickel-base alloys, titanium, and aluminum are used in P/M porous metal applications.

Porous Region (Ceramics)

A 3D zone of porosity or microporosity of higher concentration than is normally found in the ceramic matrix.

Porous Seam (Ceramics)

A 2D area of porosity or microporosity of higher concentration than is normally found in the ceramic matrix.

Port

The opening through which molten metal enters the injection cylinder of a *die-casting* plunger machine or is ladled into the injection cylinder of a cold chamber machine. See also *cold chamber machine* and *plunger*.

Portable Hardness Testing

Techniques and equipment for hardness testing on site rather than using large unwieldy laboratory equipment. With due care, the quality of the results is adequate for most purposes.

Porthole Die

A multiple-section extrusion die capable of producing tubing or intricate hollow shapes without the use of a separate mandrel. Metal is extruded in separate streams through holes in each section and is rewelded by extrusion pressure before it leaves the die. Compare with *bridge die*. See also *tube making*.

Positioned Weld

A weld made in a joint that has been oriented to facilitate making the weld.

Positive Distortion

In an optical microscope, the distortion in the image that results when the magnification in the center of the field is less than that at the edge of the field. Also termed pincushion distortion. See also *negative distortion*.

Positive eyepiece

In an optical microscope, an eyepiece in which the real image of the object is formed below the lower lens elements of the eyepiece. See also *eyepiece*.

Positive Ion

An ion, which may be an atom, molecule, or radical, which has lost one or more electrons and hence carries a positive charge.

Positive mold

A mold for forming plastics designed to trap all the molding material when it closes.

Positive Rake

Describes a tooth face in rotation whose cutting edge leads the surface of the tooth face. See also *face mill*.

Positive replica

A replica whose contours correspond directly to the surface being replicated. Contrast with *negative replica*.

Positive-Contact Bushing

A bushing, the inside diameter of which has direct contact with the outside diameter of a shaft or sleeve. Radial or axial clearances are provided in the hole.

Positive-Contact Seal

A seal, the primary function of which is achieved by one surface mating with another. Examples include lip, circumferential, and face-type seals. See also *face seal*.

Positron

Subatomic particles having the same mass as an electron but a positive charge. See *atom*.

Post (Weld)-Heat Treatment

Any process of heating following welding or some other process. Such heating induces stress relief, softening of hardened zones, diffusion of hydrogen, etc. This term usually refers to circumstances where the weldment is allowed to cool fully before being reheated perhaps in some special facility away from the welding site; compare with *postheating*.

Postcure (Adhesives)

A treatment (normally involving heat) applied to an adhesive assembly following the initial cure, to modify specific properties. To expose an adhesive assembly to an additional cure, following the initial cure, for the purpose of modifying specific properties.

Postcure (Plastics)

Additional elevated-temperature cure of a plastic, usually without pressure, to improve final properties and/or complete the cure or decrease the percentage of volatiles in the compound. In certain resins, complete cure and ultimate mechanical properties are attached only by exposure of the cured resin to higher temperatures than those of curing. See also *cure*.

Postforming

The forming, bending, or shaping of fully cured, C-staged thermoset laminates that have been heated to make them flexible. Upon cooling, the formed laminate retains the contours and shape of the mold over which it has been formed. See also *C-stage*.

Postheat Current

In resistance welding, the current through the welding circuit during *postheat time*.

Postheat Time

In resistance welding, the time from the end of weld heat time to the end of weld time. See also *postheat current*.

Postheating

Heating weldments immediately after welding, for tempering, for stress relieving, or for providing a controlled rate of cooling to prevent formation of a hard or brittle structure, hydrogen diffusion, etc. If the welded component is allowed to cool before being reheated, the term post (weld)-heat treatment is more appropriate.

Pot (Metals)

(1) A vessel for holding molten metal. (2) The electrolytic reduction cell used to make such metals as aluminum from a fused electrolyte.

Pot (Plastics)

To embed a component or assembly in liquid resin, using a shell, can, or case that remains an integral part of the product after the resin is cured.

Pot Annealing

Same as *box annealing*.

Pot Die Forming

Forming products from sheet or plate through the use of a hollow die and internal pressure that causes the preformed workpiece to assume the contour of the die.

Pot Life

The length of time that a catalyzed thermosetting resin system retains a viscosity low enough to be used in processing. Also called working life.

Potash

Also called *pearl ash*. A white, alkaline, granular powder, which is a potassium carbonate, K_2CO_3, used in soft soaps, for wool washing, and in glass manufacture. It is produced from natural deposits and also produced from wood and plant ashes. The U.S. production is largely from potash salts of New Mexico, from the brines of Searles Lake, California, and from solar evaporation in Utah, and usually contains iron, clay, and salt as impurities. *Kalium* is a high-purity grade that results from an evaporation–crystallization step; it has about 62.4% K_2O and less than 1% salt. The material is a free-flowing, white powder of 91%–94% K_2CO_3 or is the hydrate at 84% or calcined at 99% purity. The specific gravity of potash is 2.33 and melting point 1668°F (909°C). The sylvite ore mined at Carlsbad, New Mexico, contains KCl and NaCl. It is electrically refined to 99.95% KCl and is used to produce caustic potash. Electrolysis of the chloride solution yields caustic potash.

Potassium Alloys and Compounds

An elementary metal, symbol K, and atomic weight 39.1, potassium is also known as kalium. It is silvery white in color, but it oxidizes rapidly in the air and must be kept submerged in ether or kerosene.

It stands in the middle of the alkali metal family, below sodium and above rubidium, in group II of the periodic table of the elements. It is a lightweight, soft, low-melting reactive metal. It is very similar to sodium in its behavior and its metallic form, and its uses are limited by the availability of low-cost sodium in large volume.

Physical Properties and Alloys

Potassium has a low melting point of 63°C and a boiling point of 756°C. The specific gravity is 0.855 at 20°C. It is soluble in alcohol and in acids. It decomposes water with great violence. Potassium is obtained by the electrolysis of potassium chloride. Potassium metal is used in combination with sodium as a heat-exchange fluid in atomic reactors and high-temperature processing equipment.

A potassium–sodium alloy contains 78% potassium and 22% sodium. It has a melting point of –11°C and a boiling point of 756°C and is a silvery mobile liquid. Cesium–potassium–sodium alloys are called BZ alloys. Potassium hydride is used for the photosensitive deposit on the cathode of some photoelectric cells. It is extremely sensitive and will emit electrons under a flash so weak and so rapid as to be imperceptible to the eye. Potassium diphosphate, KH_2PO_4, a colorless, crystalline, or white powder soluble in water, is used as a lubricant for wool fibers to replace olive oil in spinning wool.

It has the advantages that it does not become rancid like oil and can be removed without scouring. Potassium, like sodium, has a broad range of use in its compounds, giving strong bonds. Metallurgically, it is listed as having a body-centered cubic structure, but the atoms arrange themselves in pairs in the metal as K_2, and the structure is cryptocrystalline.

Chemical Properties

Potassium is even more reactive than sodium. It reacts vigorously with the oxygen in the air to form the monoxide, K_2O, and the peroxide, K_2O_2. In the presence of excess oxygen, it readily forms the superoxide, KO_2 (formerly believed to be K_2O_4).

Potassium does not react with nitrogen to form a nitride, even at elevated temperatures. With hydrogen, potassium reacts slowly at 200°C and rapidly at 350°C–400°C. It forms the least stable hydride of all the alkali metals.

Principal Compounds

Potassium chloride, KCl, is the most important potassium compound. It is not the only form in which potassium is often found in nature, but it is the form in which potash is used as a fertilizer.

Potassium hydroxide, KOH, is also known as caustic potash. It is usually made by the electrolysis of aqueous solutions of potassium chloride.

Potassium carbonate, K_2CO_3, is made from potassium hydroxide and carbon dioxide. It cannot be made by the Solvay process used for sodium carbonate, because potassium bicarbonate is too soluble in ammonium chloride solution.

Potassium nitrate, KNO_3, is made from fractional crystallization of an aqueous solution containing sodium nitrate and potassium chloride.

Handling

Handling of potassium metal is much the same as that of sodium metal, with two major exceptions. First, the formation of the superoxide, KO_2, causes difficulties because it can react vigorously with hydrocarbons and other organic matter. Second, potassium is generally more reactive than sodium. Potassium forms an explosive carbonyl with carbon monoxide, and the metal deteriorates in contact with bromine. Usually sodium, potassium, and sodium–potassium (NaK) alloys are considered to be in the same general class of reactivity, allowing for the chemical differences outlined previously and for the liquid (and hence more reactive) nature of the NaK alloys over a wide range of composition.

Uses

Potassium chloride finds its main use in fertilizer mixtures. It also serves as the raw material for the manufacture of other potassium compounds.

Potassium hydroxide is used in the manufacture of liquid soaps, and potassium carbonate in making soft soaps.

Potassium carbonate is also an important raw material for the glass industry.

Potassium nitrate is used in matches, and pyrotechnics, and in similar items that require an oxidizing agent.

Biological Activity

The recognition of potassium ions by biological molecules is optimal when there is a good match between the cavity size provided by the host molecule and the ionic diameter of the potassium ion; other factors, such as the water structure surrounding the potassium ion and the number and nature of donor atoms in the host molecule, may, however, be important. Novel physical methods, such as nuclear magnetic resonance spectroscopy based on the ^{39}K isotope, are being used to help achieve an improved understanding of potassium ion transport in biological systems.

Potassium deficiency may occur in several conditions including malnutrition and excessive vomiting or diarrhea and in patients undergoing dialysis; supplementation with potassium salts is sometimes required. Although toxicity caused by therapeutic doses of potassium is rare, it may lead to cardiac arrest if left untreated; potassium supplementation should be administered with caution to patients suffering from cardiovascular diseases or those with impaired renal function.

Potassium Compounds

Potash

See *potash* earlier.

The compound K_2O is very soluble in water and other solvents. The most important original source of commercial potash is natural potassium salts. These are prepared as potassium nitrate and potassium carbonate for use in ceramics, but most of the potash is automatically introduced into batches in feld-soda glass; this is especially true with the use of manganese, nickel oxide, and selenium. In the potash glasses, much less cobalt oxide is required in connection with manganese to secure a good neutral tint for crystal glass. Similarly, nickel is a suitable decolorizer for glasses high in potash, whereas its effect in the soda glasses is decidedly ugly.

The alkali content of commercial glasses runs about 15% in window glass, 15%–17% and container glass, and 20% in thin blown glass. Most of the alkali is soda, and while a higher potash content is often desirable, its greater cost limits its wider application. The growth of the American potash industry may allow a price reduction that will make this material more available to glass manufacturers, who now limit its general use to the more expensive glass products.

In optical glass, a ratio of 7 parts potash to 3 parts soda gives good durability and color to a number of commercial compositions, in which the total potash content of the glass may vary from 7% to 16% for some crowned types. It probably is not possible to derive a potash–soda ratio suitable for all optical glasses. Some high-lead glasses, for example, contain no soda at all, yet show high durability. The discoloring effect of ferrous iron is much less noticeable in a potash–soda optical glass than in a high-soda glass.

It has been found that glasses containing both Na_2O and K_2O give lower thermal conduction than either alone; the minimum conductivity is obtained with a potash–soda ratio of 4:1. This factor is becoming increasingly important in view of developments in fiberglass.

The behavior of colorants and colored glass is often superior in potash glass to that in spar.

In enamels, the alkali content averages 10% in sheet ground coats, 20% in cover coats, 15% in cast iron enamels, and as much as 36% in jewelry enamels. In the last type, all the alkali is potash, which is believed to increase brilliance and luster, but in other enamels, all or most of the potash is merely accessory to alumina in the addition of feldspar. The same may be said for the potash content of glazes. As a flux and glazes, potash is only about 85% as active as soda. If present in excess, K_2O may cause peeling and crazing if the other constituents are not in suitable proportions. Potash is reported not as conductive as soda to the formation of crystals and crystalline glazes.

Potash in the hydroxide or carbonate form is an important deflocculating agent. It is used at ordinary temperatures to prepare casting slips, glazed slips, and engobes, to purify clay, to reduce the plasticity of excessively plastic clays, and to neutralize any acid present.

Potassium Carbonate

In glass manufacture, potassium carbonate is supplied in both calcined and hydrated form. The product sold to the glass industry is easy to handle as it is of granular particle form and has entirely eliminated the dusty material, formally supplied from abroad, with its irritating handling problems.

The present domestic materials are supplied with very low chloride and sulfate content and are entirely suitable for all types of glass production. Although the viscosity of the potash glasses is high, thus making them somewhat difficult to work, the viscosity is easily remedied by introducing lead oxide. Hence, the combination of potash and lead oxide leads to the production of a glass that lends itself well to handworking.

This combination possesses a long working range. All of the potash glasses, which from their nature must be melted in closed pots, exert a different sort of corrosive action on the clay wall from that exhibited by soda glasses.

The corrosion by soda glass proceeds quite smoothly, but the potash glasses produce a honeycombing or pitting effect, and the thin partitions between these pits finally reduced to small pinnacles, float out into the glass, forming stones. It seems to be an almost unavoidable characteristic of the potash-lead glasses to produce a great deal of stony ware.

In its influence on the physical properties of glass, potash does not differ greatly from soda. Compared with soda in equal weight percentages, potash seems to confer a little more density, less hardness, and less tenacity. The two are by far the most expansible oxides in glass.

In glazes, potassium carbonate appears as an ingredient when it is desirable to modify the effect of a colorant such as copper oxide, which may be brought through tints of green toward yellow.

When potassium carbonate is used in glazes in combination with sodium oxide, lead oxide, or calcium oxide, the potassium oxide derivative cannot exceed 0.15 equivalent without affecting the color.

When potassium carbonate is used in colored glazes, it is advisable to frit about 90% of the clay, but none of the color.

In enamels, potassium carbonate tends to produce high luster, but it decreases strength and elasticity, making the enamel soft.

In general, enamels containing potassium are more readily fusible than those with sodium. Potassium carbonate has largely been replaced in enamels, however, by sodium carbonate, due to the difference in price, except in occasional cases where it is used to alter colors.

Potassium Chlorate

Also known as chlorate of potash and potassium oxymuriate, this is a white crystalline powder, of lustrous crystalline substance, of the composition $KClO_3$, employed in explosives, chiefly as a source of oxygen. It is also used as an oxidizing agent in the chemical industry, as a cardiac stimulant in medicine, and in toothpaste. It melts at 357°C and decomposes at 400°C with the rapid evolution of oxygen. It is odorless but has a slightly bitter saline taste. The specific gravity is 2.337. It is not hygroscopic, but is soluble in water. It imparts a violet color to the flame in pyrotechnic compositions.

Potassium Chloride

Potassium chloride is a colorless or white crystalline compound of the composition KCl, used for molten salt baths for the heat treatment of steels. The specific gravity is 1.987. A bath composed of three parts potassium chloride and two parts barium chloride is used for hardening carbon steel drills and other tools. Steel tools heated in this bath and quenched in a 3% sulfuric acid solution have a very bright surface. A common bath is made up of potassium chloride and common salt and can be used in high temperatures up to 900°C.

Potassium chloride is used in the porcelain enamel industry as a setting-up agent in titanium cover coats. In general, the quantities of potassium chloride, when used as an electrolyte, will be approximately the same as sodium nitrite, which it replaces. However, KCl does not aid tearing resistance as does nitrite. The main advantage in using potassium chloride is the freedom from yellowing or creaming when used in a blue-white enamel. Potassium chloride may exert an adverse affect on the gloss and may cause a slight decrease in the acid-resisting properties of the enamel, although the latter effect is somewhat debatable.

Potassium Cyanide

A white amorphous or crystalline solid of the composition KCN, potassium cyanide is employed for carburizing steel for case hardening and for electroplating. The specific gravity is 1.52, and it melts at about 843°C. It is soluble in water and is extremely poisonous, giving off the deadly hydrocyanic acid gas. For cyaniding steel, the latter is immersed in a bath of molten cyanide and then quenched in water, or the cyanide is rubbed on the red-hot steel.

Commercial potassium cyanide is likely to contain a proportion of sodium cyanide. Potassium ferrocyanide or yellow prussiate of potash can also be used for case-hardening steel. Nitrogen as well as carbon enters the steel to form the hard case. Potassium ferricyanide, or red prussiate of potash, is a bright-red granular powder used in photographic reducing solutions, in etching solutions, in blueprint paper, and in silvering mirrors. *Redsol* crystal is the name of this chemical uses as a reducer and mild oxidizing agent or toner for photography. Potassium cyanate, KCNO, is a white crystalline solid used for the production of organic chemicals and drugs. It melts at 310°C. The potassium silver cyanide used for silver plating comes in white, water-soluble crystals, known as *Sel-Rex*. Potassium gold cyanide has a similar function in gold plating. *Platina* comes as colorless tablets that are soluble in both water and alcohol.

Potassium Dichromate

This material, $K_2Cr_2O_7$, decomposes at 500°C. Bright yellowish-red crystals that are soluble and poisonous. Sometimes, potassium chromate and dichromate are utilized in ceramics as coloring agents.

Potassium dichromate is used in glass for aventurine effects. It is said that 20–21 parts to 100 parts sand will give a chrome aventurine. This glass is characterized by glittering metallic scales of chromium oxide. Potassium dichromate is also used in glass to give a green color. However, it has been shown that it may cause considerable trouble by formation of black, chrome corundum crystals in the glass. Air-floated chromite is suggested to avoid this problem.

Potassium dichromate is used in glazes to produce chrome-tin pinks, low-fire reds, greens, and purplish-red colors.

Potassium Nitrate

Potassium nitrate is also called niter and saltpeter, although these usually referred to the native mineral. A substance of the composition KNO_3, it is used in explosives, for bluing steel, and in fertilizers. A mixture of potassium nitrate and sodium nitrate is used for steel-tempering baths. The mixture melts at 250°C. Potassium nitrate is made by the action of potassium chloride on sodium nitrate. It occurs in colorless prismatic crystals or as a crystalline white powder. It has a sharp saline taste and is soluble in water. The specific gravity is 2.1 and the melting point is 337°C.

Potassium nitrate contains a large percentage of oxygen, which is readily given up and is well adapted for pyrotechnic compounds. It gives a beautiful violet flame in burning. It is used in flares and in signal rockets.

Most enamels contain some oxidizing agent in the form of potassium or sodium nitrate. Only a small amount of nitrate is necessary; 2%–4% is sufficient to maintain oxidizing conditions in most smelting operations.

In glazes, it is sometimes used as a flux in place of potassium oxide, but, owing to its cost and solubility, very little of it is contained in glaze. Where conditions prevent the use of sufficient potash feldspar, potassium oxide is introduced into the mix, usually in the form of the nitrate in a frit.

Potassium nitrite is a solid of the composition KNO_2 used as a rust inhibitor, for the regeneration of heat-transfer salts, and for the manufacture of dyes.

Potential

Any of various functions from which intensity or velocity at any point in a field may be calculated. The driving influence of an electrochemical reaction. See also *active potential, chemical potential, corrosion potential, critical pitting potential, decomposition potential, electrochemical potential, electrode potential, electrokinetic potential, equilibrium (reversible) potential, free corrosion potential, global potential, open-circuit potential, protective potential, redox potential,* and *standard electrode potential.*

Potential Barrier

The atoms in a crystal lattice interact so that each is normally located at the minimum energy level it can reach; it is said to lie in an energy trough or potential trough. If an atom has to be moved, it needs an injection of energy to initiate its movement out of the trough and over the potential barrier (or energy barrier) into the next trough.

Potential Difference

The difference in potential, measured in volts, between two points in the electrical circuit when subject to an electromotive force or carrying a current.

Potential Drop Techniques

A series of techniques for measuring the propagation of a crack by measuring the voltage drop in a constant current imposed across the crack zone. Four contacts are made to the crack zone. One pair placed either side of the crack applies the current; the second pair, similarly placed, measures the potential drop. Depending upon the power supply, the techniques are termed alternating current potential drop and direct current potential drop, usually abbreviated to ACPD and DCPD, respectively.

Potential Energy

The energy contained in some body by virtue of its position relative to another body in circumstances such as a gravitational field.

Potential Trough

See *potential barrier*.

Potential-pH Diagram

See *Pourbaix (potential-pH) diagram*.

Potentiodynamic (Potentiokinetic)

The technique for varying the *potential* of an electrode in a continuous manner at a preset rate.

Potentiometer

An instrument that measures electromotive force by balancing against it an equal and opposite electromotive force across a calibrated resistance carrying a definite current.

Potentiometric Membrane Electrodes

Electrochemical sensing device that can be used to quantify cationic and anionic substances and gaseous species in aqueous solutions. These devices are also used for analytical titrations. See also *titration*.

Potentiostat

(1) An instrument for automatically maintaining an electrode in an electrolyte at a constant potential or controlled potentials with respect to a suitable reference electrode. (2) An instrument for supplying electric current of constant voltage but variable amperage.

Potentiostatic

The technique for maintaining a constant *electrode potential*.

Potentiostatic Etching

The selective corrosion of one or more morphological features of a microstructure that results from holding the metal to be etched in a suitable etching electrolyte at a controlled potential relative to a reference electrode. Adjusting the potential makes possible a defined etching of singular phases.

Pottery

A generic term for all fired ceramic wares that contain clay, except for technical, structural, and refractory products. Specifically, however, pottery describes the low-temperature fired porous ware that is usually colored. The term is properly applied to the clay products of primitive peoples or to decorative art products made of unrefined clays by using unsophisticated methods. See also *traditional ceramics*.

Poultice Corrosion

A term used in the automotive industry to describe the corrosion of vehicle body parts due to the collection of road salts and debris on ledges and in pockets that are kept moist by weather and washing. Also called deposit corrosion or attack. Differential aeration corrosion is a major contributory factor to the attack. It is a common cause of corrosion on the under surfaces of road vehicles.

Pour Point

The lowest temperature at which a lubricant can be observed to flow under specified conditions.

Pour Welding

A flow welding process. See *burning*.

Pourbaix (Potential-pH) Diagram

A plot of the *redox potential* of a corroding system versus the pH of the system, compiled using thermodynamic data and the *Nernst equation*. The diagram shows regions within which the metal itself or some of its compounds are stable. The diagram identifies corrosion characteristics of a metal in aqueous solutions. The diagram plots electrode potential on the vertical axis and pH on the horizontal. Zones on the diagram indicate circumstances in which the metal acts in a passive, immune, or a corrosive manner.

Pouring

The transfer of molten metal from furnace to ladle, ladle to ladle, or ladle to molds.

Pouring Basin

In metal casting, a basin on top of a mold that receives the molten metal before it enters the sprue or downgate. See *gating system*.

Pour-Point Depressant

An additive that lowers the pour point of a lubricant.

Powder

An aggregate of discrete particles that are usually in the size range of 1–1000 μm.

Powder Cutting/Injection

Oxygen cutting in which a powder, usually iron, is injected into the cutting stream. The iron assists the exothermic reaction as well as providing fluxing and scouring actions. The process can cut high chromium and stainless steel that because of their resistance to oxidation are not readily cut by the normal equipment. See *chemical flux cutting* and *metal powder cutting*.

Powder Feeder

A mechanical device designed to introduce a controlled flow of powder into the plasma-spray torch. See also *plasma spraying*.

Powder Fill

In powder metallurgy, the filling of the die cavity with powder prior to compaction.

Powder Flame Spraying

A thermal spraying process variation in which the material to be sprayed is in powder form. See also *flame spraying*.

Powder Forging

The plastic deformation of a powder-metallurgy *compact* or *preform* into a fully dense finished shape by using compressive force; usually done hot and within closed dies.

Powder Lubricant

In powder metallurgy, an agent or component incorporated into a mixture to facilitate compacting and ejecting of the compact from its mold.

Powder Metallizing

See preferred term *powder flame spraying*.

Powder Metallurgy

(1) The technology and art of producing metal powders and utilizing metal powders for production of massive materials and shaped objects. (2) The manufacture of metal powders and, from them, bulk components by processes such as sintering rather than melting.

Powder Method

Any method of *x-ray diffraction* involving a polycrystalline and preferably randomly oriented powder specimen and a narrow beam of monochromatic radiation. The powder method is best known as a phase characterization tool because it can routinely differentiate between phases having some chemical composition but different crystal structures (polymorphs). See also *polymorphism*.

Powder Molding

General term used to denote several techniques, such as injection molding, for producing objects of varying size and shape by melting plastic powder, usually against the inside of a mold. The molds are either stationary (e.g., in variations of slush molding techniques) or rotating (e.g., in variations of rotational molding).

Powder Plastics Moldings

The term *powder molding* broadly describes a technique of sintering or fusing finely divided thermoplastic materials to conform to the surface of a mold.

Processing

There are a number of processes and techniques loosely termed powder molding. The Engel or Thermofusion process was developed in Europe and sublicensed in the United States in 1960. This patented technique employs inexpensive sheet-metal molds and a hot-air oven heated by either gas or electricity. In essence, the process consists of filling the mold with powdered thermoplastic material, fusing the layer of material next to the mold walls, removing the excess powder, and then smoothing the inside surface with heat.

Another older technique, known as the Heisler process, involves rotating a heated mold. Both of these patented techniques employ an *excess* of powdered material over that required for the object being fabricated; the extra material is removed as one of the processing steps.

More recently, it was discovered that powdered polyethylene behaves enough like a liquid to permit use of rotational casting equipment designed for use with vinyl plastisols. The technique involves multiaxial rotation of a closed mold filled with a *measured* charge of material, all of which is fused during the heating step. This permits fabrication of totally enclosed articles. A hobby horse for children is the most striking example of the versatility of the rotational casting process. There is at present a variety of rotational casting apparatus available. Now engineered specifically for use with powdered thermoplastic materials, it is anticipated that this technique will find widespread acceptance.

The selection of a particular powder-molding technique depends on the application under consideration. The selection depends on factors such as size, wall thickness, geometric configuration, and quantity of parts desired. Rotational casting with a measured charge is obviously indicated where automation for a large volume of production is desired, due to the reduction of material-handling requirements.

Advantages

Because of low tooling costs, all of these various powder-molding processes offer significant economies over conventional fabricating techniques such as injection or blow molding in small or moderately large quantities (up to 250,000 pieces, depending on the application concerned).

Typically, inexpensive sheet-metal or cast aluminum molds may be used, as compared with expensive matched metal molds made of tool steels. The heat source is usually an electric or gas-fired air oven, a small capital cost in comparison to an injection-molding machine or extruder. Moreover, the large size of some objects that are fabricated by powder molding defy production by most other techniques. The size of an object to be fabricated by this particular process is limited only by the size of the oven. Large tanks, refuse containers, and even boats are included in the wide variety of products being fabricated by this method. These parts range in wall thickness from 1.5 to 6.4 mm with a tolerance of ±10% considered reasonable.

Materials Used

Although powder-molding processes are theoretically applicable to any thermoplastic material, current practice has been restricted primarily to polyethylenes of low and intermediate density. As soon as effective stabilizing systems are found for other materials, particularly the broader range of polyolefin materials, it is anticipated that such materials will also find use where their particular properties offered advantages in specific applications. The particles of powder range in size from 30 to 100 mesh (U.S. standard), with the peak of the bell-shaped distribution at approximately 60 mesh. The powder is made by grinding the material in pellet form as an additional manufacturing step.

Production cycles can be cited only generally, as they depend on the application concerned. Cycles range from roughly 3 min for small, thin-walled parts made by rotational casting, to as much as 30 min for very large, heavy-walled parts made by the Engel process.

Powder Production

The process by which a metal powder is produced, such as machining, milling, atomization, condensation, reduction, oxide decomposition, carbonyl decomposition, electrolytic deposition, or precipitation from a solution.

Powder Rolling

See preferred term *roll compacting*.

Powder Technology

A broad term encompassing the production and utilization of both metal and nonmetal powders.

Powder-Feed Rate

In thermal spraying, the quantity of powder introduced into the arc per unit time; expressed in pounds/hour or grams/minute.

Powder-Metallurgy Forging

See *powder forging*.

Powder-Metallurgy (P/M) Parts

Powder-metallurgy parts, commonly referred to as P/M parts, are produced by the P/M process, which involves blending of powders, pressing the mixture in a die, and then sintering or heating the compact in a controlled atmosphere to bond the contacting surfaces of the particles. Where desirable, parts can be sized, coined, or repressed to closer tolerances; they can be impregnated with oil or plastic or infiltrated with a lower-melting metal; and they can be heat-treated, plated, and machined. Production rates range from several hundred to several thousand per hour.

Shapes that can be fabricated in conventional P/M equipment range up to about 16.1 kg. Parts of over 460 kg can be produced with special techniques such as isostatic compacting and extrusion. However, most P/M parts weigh less than 2.3 kg. While most of the early P/M parts were simple shapes, such as bearings and washers, developments over the years in equipment and materials now make economical the production of more intricate and stronger parts. And shapes with flanges, hubs cores, counterbores, and combinations of these are fairly commonplace.

P/M parts are made from a wide range of materials, including combinations not available in wrought or cast form. And these materials can be processed by P/M techniques to provide tailored densities and parts ranging from porous components to high-density structural and mechanical parts. In addition, almost any conceivable alloy system under equilibrium or nonequilibrium conditions can be achieved, and segregation effects (nonhomogeneities) are avoided or minimized.

Most metal powders are produced by atomization, reduction of oxides, electrolysis, or chemical reduction. Metals available include iron, nickel, copper, and aluminum, as well as refractory and reactive metals. These metals can be blended together to form different alloy compositions during sintering. Also, prealloys such as low-alloy steels, bronze, brass, nickel–silver, and stainless steel are produced in which each particle is itself an alloy, thus ensuring a homogeneous metallurgical structure in the part. And it is possible to combine metal and nonmetal powders to provide composite materials with the desirable properties of both in the finished part.

The method selected is dictated primarily by composition, intended application, and cost. Typically, metal powders for commercial usage range from 1 to 1200 μm. Depending on the method of production, metal powders exhibited diversity of shapes from spherical to acicular. Particle shape is an important property, because it influences the surface area of the powder, its permeability and flow, and its density after compaction.

Chemical composition and purity also affect the compaction behavior of powders. For most applications, powder purity is higher than 99.5%.

The P/M process is being used to produce many thousands of different parts in most product and equipment manufacturing industries, including automotive, business machines, aircraft, consumer products, electrical and electronic, agricultural equipment, machinery, ordnance, and atomic energy.

The application of P/M parts in these industries falls into two main groups. The first are those applications in which the part is impossible to make by any other method. For example, parts made of refractory metals like tungsten and molybdenum or of materials such as tungsten carbide cannot be made efficiently by any other means. Porous bearings and many types of magnetic cores are exclusively products of the P/M process. The second group of uses consists of mechanical and structural parts that compete with other types of metal forms such as machine parts, castings, and forgings.

Processes

P/M processes include pressing and sintering, powder injection molding, and full-density processing.

Pressing and Sintering

The basic P/M process requires two steps involving powder compaction followed by sintering. First, the rigid steel die (mold) is filled with powder. Pressure is then applied uniaxially at room temperature via steel punches located above and below the powder, after which the compact is ejected from the mold. Commercial compaction pressures are normally in the range from 140 to 900 MPa. The powder compact is porous. Its density depends on compaction pressure and the resistance of the powder particles to deformation; compact densities about 90% of the theoretical level are common.

In the sintering process, the powder compact is heated below its melting point to promote bonding of the solid powder particles. The major purpose of sintering is to develop strength in the compact. Normally an increase in the density of the compact also occurs, but the compact still contains pores. The internal architecture (microstructure) of the material is developed during sintering. As a rule of thumb, the sintering temperature must be higher than one half the melting temperature (Kelvin scale) of the powder.

Normally, parts made by pressing and sintering require no further treatment. However, properties, tolerances, and surface finish can be enhanced by secondary operations. Examples of secondary or finishing operations are repressing, resintering, machining, heat treatment, and various surface treatments such as deburring, plating, and sealing.

Powder Injection Molding

This is a process that builds on established injection-molding technology used to fabricate plastics into complex shapes at low cost. The metal powder is first mixed with a binder consisting of waxes, polymers, oils, lubricants, and surfactants. After granulation, the powder binder mix is injection molded into the shape desired. Then, the binder is removed and the remaining metal powder skeleton sintered. Secondary operations may be performed on the sintered part, similar to those for conventional press-plus-sintered parts. The viscosity of the powder–binder mix should be below 100 Pa·s (1000 poise) with about 40% by volume of binder, and spherical powders less than 20 μm in diameter are preferred. Powder injection molding produces parts that have the shape and precision of injection-molded plastics but that exhibit superior mechanical properties such as strength, toughness, and ductility.

Full-density Powder Processing

Parts fabricated by pressing and sintering are used in many applications. However, their performance is limited because of the presence of porosity. To increase properties and performance and to better compete with products manufactured by other metal working methods (such as casting and forging), several P/M techniques have been developed that result in fully dense materials; that is, all porosity is eliminated. Three examples of full-density processing are hot isostatic pressing, powder forging, and spray forming.

In the hot isostatic pressing process, the metal (or ceramic) powder is sealed in a metallic or glass container, which is then subjected to isostatic pressure at elevated temperature. Temperatures up to 2200°C and pressures up to 200 MPa are possible in modern presses. After complete densification of the powder, the compact is removed and the container stripped from the densified compact. Hot isostatic pressing is applied to the consolidation of nickel and titanium alloys, tool steels, and composites.

Powder forging is an adaptation of conventional forging in which a compact is prepared that is then sintered to retain about 20% by volume of porosity. This preform is transferred to a forging press and formed in a closed die cavity to its final shape with one stroke of the press. All porosity is removed in the forging operation. Powder forging is used primarily to fabricate components from low-alloy steels.

In the spray-forming process, a stream of liquid metal is gas-atomized into a spray of droplets that impinge on a substrate. The spraying conditions are such that the droplets arrive at the substrate in the semisolid state. By control of the geometry and motion of the substrate, it is possible to fabricate sheet or plate, tubes, and circular billets. It is a rapid process with a deposition rate up to about 2 kg/s. Applications include aluminum, copper, iron, and nickel alloys.

Characteristics and Applications

P/M competes with several more conventional metalworking methods in the fabrication of parts, including casting, machining, and stamping. Characteristic advantages of P/M are close tolerances, low cost, net shaping, high production rates, and controlled properties. Other attractive features include compositional flexibility, low tooling costs, available shape complexity, and a relatively small number of steps in most P/M production operations. Control of the level of porosity is the intrinsic feature of P/M that enables the parts produced to predict and specify physical mechanical properties during fabrication. Broad areas of usage for P/M parts include structural (load bearing) components and controlled-porosity, electrical, magnetic, thermal, friction, corrosion-resistant, and wear-resistant applications. Industries that make extensive use of P/M parts are aerospace, agricultural, automotive, biomedical, chemical processing, and electrical. In addition, the fabricators of domestic appliances and office equipment are dependent on the availability of a wide range of sizes and geometries of P/M parts exhibiting unique combinations of physical and mechanical properties.

Ferrous parts fabricated by pressing and sintering make up the largest segment of the P/M industry, and a majority of the applications are for the automotive market. Representative P/M parts include gears, bearings, rod guides, pistons, fuel filters, and valve plates. In the nonautomotive sector, examples of parts include gears, levers, and cams, lawn tractors and garden appliances, and gears, bearings, and sprockets in office equipment.

Full-density processing has been used in the powder forging of steel automobile connecting rods. The iron–copper–carbon rod is forged to a minimum density of 7.84 g/cm³.

Important characteristics of powder forging include straightness, illumination of surface defects inherent in conventional forging, uniform microstructure, dimensional control, and superior machinability. Secondary operations include deflashing, shot peening, and machining.

An example of a part manufactured by powder injection molding is a miniature read/write latch arm made from stainless steel powder and used in a small-capacity hard disk drive. The geometric form of the entire outside profile of the part is very critical relative to the pivot hole. Coining is the only secondary operation.

Safety and Health Considerations

Because metal powders possess a high surface area per unit mass, they can be thermally unstable in the presence of oxygen. Very fine metal powders can burn in air (pyrophoricity) and are potentially explosive. Therefore, the clean handling of powder is essential; methods include venting, controlled oxidation to passivate particle surfaces, surface coating, and minimization of sparks or heat sources. Some respirable fine powders pose a health concern and can cause disease or lung dysfunction; the smaller the particle size, the greater the potential health hazard. Control is exercised by the use of protective equipment and safe handling systems such as glove boxes. There is no recognized hazard associated with the normal handling of the common grades of metal and alloy powders such as copper and iron.

Power Factor (Electricity)

(1) The ratio of the power to the effective values of the electromotive force multiplied by the effective value of current, in volts and amperes, respectively. The cosine of the angle between voltage applied and the resulting current. (2) The multiplication factor applicable to voltage (volts) × current (amperes) to give power (watts) in an alternating current circuit. In an alternating current circuit with any significant amount of inductance or capacitance, the wave forms for voltage and current will be out of phase with each other. The vector of the degree of out of phase is represented by f and the power factor is then $\cos f$. Hence, power = volts × amps × $\cos f$.

Power Reel

A reel that is driven by an electric motor or some other source of power, used to wind or coil strip or wire as it is drawn through a continuous normalizing furnace, through a die, or through cool rolls.

Power-Driven Hammer

A forging hammer with a steam or air cylinder for raising the ram and augmenting its downward blow. Also known as power drop hammer. See also *drop hammer* and *gravity hammer*.

PP

See *polypropylenes.*

PPE

Polyphenylene ether. See also *polyphenylene oxides.*

PPO

See *polyphenylene oxides.*

PPS

See *polyphenylene sulfides.*

Praseodymium

A metallic element, one of the rare earths group.

Prealloyed Powder

A metallic powder composed of two or more elements that are alloyed in the powder manufacturing process and in which the particles are of the same nominal composition throughout.

Prearc (or Prespark) Period

In emission spectroscopy, the time interval after the initiation of an arc (or spark) discharge during which the emitted radiation is not recorded for analytical purposes.

Prebond Treatment

Synonym for surface preparation prior to *adhesive bonding.*

Precharge

In metal forming, the pressure introduced into the cavity prior to forming of the part.

Precious Metals

These are the metals gold, silver, and platinum, which are used for coinage, jewelry, and ornaments and also for industrial applications. Expense or rarity alone is not the determining factor; rather, a value is set by law, with the coinage having an intrinsic metal value as distinct from a copper coin, which is merely a token with little metal value. The term noble metal is not synonymous, although a metal may be both precious and noble, as platinum. Although platinum was once used in Russia for coinage, only gold and silver fulfill the three requisites for coinage metals. Platinum does not have the necessary wide distribution of source. The noble metals are gold, platinum, iridium, rhodium, osmium, and ruthenium. Unalloyed, they are highly resistant to acids and corrosion. Radium and certain other metals are more expensive than platinum but are not classed as precious metals. Because of the expense of the noble metals, they may be alloyed with gold for use in chemical crucibles.

The eight precious metals, listed in order of their atomic number found in periods 5 and 6 (groups VIII and Ib) of the periodic table, are ruthenium, rhodium, palladium, silver, osmium, iridium, platinum, and gold. Precious metals, also referred to as noble metals, are of inestimable value to modern civilization. Their functions in coins, jewelry, and bullion, and as catalysts in devices to control auto exhaust emissions, are widely understood. But in certain applications, their functions are not as spectacular and, although vital to the application, are largely unknown except to the users. For example, precious metals are used in dental restorations and dental fillings; thin precious metal films are used to form electronic circuits, and certain organometallic compounds containing platinum are significant drugs for cancer chemotherapy.

Silver is a bright, white metal that, next to gold, is the most easily fabricated metal in the periodic table. It is very soft and ductile in the annealed condition. Silver does not oxidize at room temperature, but it is attacked by sulfur. Nitric, hydrochloric, and sulfuric acids attack silver, but the metal is resistant to many organic acids and to sodium and potassium hydroxide. The primary application for silver (about 50% of the silver demand) is its use for photographic emulsions. The use of silver in photography is based on the ability of the exposed silver halide salts to undergo a secondary image amplification process called development. The second largest use is in the electrical and electronic industries for electrical contacts, conductors, and primary batteries. Other applications include brazing filler metals, dental alloys, electroplated ware, sterling wear (see *sterling silver*), and jewelry and coins.

Gold is a bright, yellow, soft, and very ductile metal. Its special properties include corrosion resistance, good reflectance, resistance to sulfidation and oxidation, and higher electrical and thermal conductivity. Because gold is easy to fashion, has a bright pleasing color, is nonallergenic, and remains tarnish-free, definitely, it is used extensively in jewelry (about 55% of the gold market). For much the same reasons, it is used in dental alloys and appliances. Gold is also used to a considerable extent in electronic devices, particularly in printed circuit boards, connectors, keyboard contactors, and miniaturized circuitry. Other applications include gold films used as a reflector of infrared radiation in thermal barrier windows for large buildings and space vehicles, fired-on gold organometallic compounds used for decorating glass in china, sliding electrical contacts, and brazing filler metals.

The six remaining precious metals are referred to as the platinum group metals because they are closely related and commonly occur together in nature. Ruthenium, rhodium, and palladium each have a density of approximately 12 g/cm^3; osmium, iridium, and platinum each have a density of about 22 g/cm^3. The most distinctive trait of the platinum group metals is their exceptional resistance to corrosion. Of the six metals, platinum has the most outstanding properties and is mostly used, the primary application being its use as an automobile exhaust emission catalyst. Second in industrial importance is palladium, which is used primarily in electrical applications. Rhodium and ruthenium are used as alloying elements in platinum and palladium, while osmium and iridium are used for wear-resistant and heat-resistant applications, respectively.

Precipitate

Particles that emerge from a supersaturated solution. See *phase* and *solution.*

Precipitation

In metals, the separation of a new phase from solid or liquid solution, usually with changing conditions of temperature, pressure, or both.

Precipitation (Deposit) Etching

Development of microstructure in a metallographic specimen through formation of reaction products at the surface of the micro-section. See also *staining*.

Precipitation Hardening

Hardening of an alloy as a result of a phase precipitating, or partially precipitating, from a supersaturated solid solution. Some alloys have compositions selected so that all or most of the minor constituents are in solid solution at high temperature but form a second phase at ambient temperature under equilibrium conditions. These characteristics can be manipulated to induce hardening of the alloy by a heat treatment cycle. The effects can be illustrated by reference to aluminum alloys containing, for example, small amounts of copper, silicon, and magnesium. The initial stage of the cycle is solution treatment in which the metal is heated to about 480°C for about 30 min and then cooled rapidly. This retains the alloying elements in supersaturated solid solution. In this condition, the material is relatively soft and ductile allowing it to be formed to final shape. The material subsequently undergoes a precipitation process in which it hardens progressively, with an associated loss of ductility, as the alloying elements commence to come out of solution. This can occur, in certain alloys, over a few days at normal temperature, a process referred to as natural aging. Alternatively, it may be induced by heating for a few hours at about 170°C referred to as artificial aging, precipitation treatment, warm hardening, warm aging, or similar terms. Maximum hardening is achieved when the growing precipitates remain partially coherent with the parent phase lattice. This approximates to the stage just before the emerging precipitates become visible under an optical microscope. See also *Guinier–Preston zones*. If the temperature or time is excessive, the precipitation process continues to completion, the material progressively softens, and it is termed overage. See *age hardening and aging*.

Precipitation Heat Treatment

Artificial aging of metals in which a constituent precipitates from a supersaturated solid solution.

Precision

The reproducibility of measurements within a set, that is, the scatter or dispersion of a set of data points about its central axis. Generally expressed as standard deviation or relative standard deviation.

Precision Casting

(1) A metal casting of reproducible, accurate dimensions, regardless of how it is made. Often used interchangeably with *investment casting*. (2) The lost wax process.

Precision Forging

A forging produced to closer tolerances than normally considered standard by the industry. With precision forging, a net shape, or at least a near-net shape, can be produced in the as-forge condition. See also *net shape*.

Precision Grinding

Machine grinding to specified dimensions and low *tolerances*.

Precoat

(1) In investment casting, a special refractory slurry applied to a wax or plastic expendable pattern to form a thin coating that serves as a desirable base for application of the main slurry. See *investment casting*. (2) To make the thin coating. (3) The same coating itself.

Precoated Metal Products

Mill products that have a metallic, organic, or conversion coating applied to their surfaces before they are fabricated into parts.

Precoating (Joining)

Coating the base metal in the joint by dipping, electroplating, or other applicable means prior to brazing or soldering.

Preconditioning

Any preliminary exposure of a plastic to specified atmospheric conditions for the purpose of favorably approaching equilibrium with a prescribed atmosphere.

Precracked Specimen

A mechanical test specimen that is notched and subjected to alternating stresses until a crack has developed at the root of the notch.

Precure

The full or partial setting of a synthetic resin or adhesive in a joint before the clamping operation is complete or before pressure is applied. See also *cure*.

Precursor

With respect to carbon or graphite fiber, the rayon, polyacrylonitrile, or pitch fibers from which carbon and graphite fibers are derived. See also *carbon fiber*, *pitch*, and *polyacrylonitrile (PAN)*.

Preferred Orientation

A condition of a polycrystalline aggregate in which the crystal orientations are not random, but rather exhibit a tendency for alignment with a specific direction in the bulk material, commonly related to the direction of working. See also *texture* and *orientation*.

Prefinished Metals

Prefinished metals are sheet metals that are precoated or treated at the mill so as to eliminate or minimize final finishing by the user. The metals are made in a ready-to-use form with a decorative and/or functional finish already applied. Prefinished metals provide many advantages including: (1) better product appearance, (2) lower product cost, (3) greater product uniformity, and (4) improve product function. (See Table P.1, which reflects types of prefinished metals.) It is possible to obtain base metal preplated with almost any decorative or functional metal, from bright, shiny chromium to dull, rich-looking brass. Similarly, it is possible to obtain sheet with prepainted surfaces in almost every color and in a wide variety of special-purpose plastic resins. Also, where extra durability or a special decorative effect is needed, some of these resins, notably polyvinyl

TABLE P.1
Types of Prefinished Metals

Material	Surface Composition, Appearance	Base Metals
Preplated metals	**Nickel, chromium:** Excellent appearance; available in dull, satin, and bright (sometimes provided with a clear lacquer coating for added protection) finishes, many of which can be embossed with wide range of patterns).	Steel, zinc, brass, copper, aluminum.
	Brass, copper: Excellent appearance; available in dull, satin, and highly polished finishes, many of which can be embossed with wide range of patterns.	Steel, zinc.
	Zinc: Natural grayish finish that can be used to improve product appearance by providing with semilustrous finish and coating with clear lacquer.	Steel.
Prepainted metals	Almost every organic coating is available or can be ordered in prepainted form. Selection of coating resin depends on end-use requirements. Five most popular prepainted metals now in use are alkyds, acrylics, vinyls, epoxies, and epoxy–phenolics. Many coatings can be pigmented to provide metallike appearance.	Most common ferrous and nonferrous metals. For added protection, the reverse side of ferrous metals is usually given a rust-preventive treatment or provided with an organic or metallic coating. Selection of base metal depends on cost, appearance, and product life requirements. Most popular base metals are cold-rolled steel, tinplate, tin mill backplate, hot-dipped and electrogalvanized steel, and standard aluminum alloys.
Plastic–metal laminates	**Vinyl (PVC)–metal:** PVC sheet laminated to base metal with thermosetting adhesives under heat and pressure (25–60 psi).	Can be applied to most metals; popular are steel, provides strength at low cost and the reverse side can be painted or provided with corrosion-resistant coating; aluminum, lightweight and/or corrosion resistant; magnesium, lightweight.
	Vinyl (PVF)–metal: PVF film laminated to base metal with thermosetting adhesive under heat and pressure.	Cold-rolled steel, provides strength at low cost; galvanized, aluminized, and tin-plated steel, corrosion resistant; aluminum, lightweight and corrosion resistant.
	Polyester–metal: Polyester film laminated to base metal.	Usually steel.
Textured and embossed metals	Surfaces available in hundreds of different textures and patterns. Available with texture on one surface only or with pattern that extends completely through cross section of metal. Also available in perforated form. Surfaces can be provided with dull satin or highly polished finish or combinations thereof (e.g., dull background with polished highlights). Can also be painted, porcelain enameled, or oxidized; these finishes can be buffed off high spots to provide two-tone effect.	All common sheet metals, including carbon steel, strength at low cost; stainless steel, corrosion resistance plus strength; aluminum, lightweight plus corrosion resistant; copper, pleasing appearance plus strength and corrosion resistance.
Hot-dipped other plated metals	**Galvanized (zinc coated):** Zinc surface with intermediate zinc–iron alloy layer.	Steel, ingot iron.
	Aluminized: Aluminum surface; intermediate aluminum–iron layer forms above 900°F.	Steel.
	Tin coated: Hot dipped or electroplated tin.	Mild carbon steel.
	Terne or lead coated: Lead–tin alloys, pure lead.	Steel.
Specially finished aluminum	**Color anodized:** Anodized aluminum available in clear, yellow gold (70:30 brass), red gold, rich low brass (85:15 brass), copper, blue, green, red, and black. Colors are obtainable over standard mill, satin, and bright finishes, as well as over embossed and perforated textures.	All commercial aluminum alloys and tempers.
	Spangled: Uncoated surface containing large grains that stand out in relief and facets that break up and reflect light. Available in wide variety of colors and mill finishes.	Wrought aluminum alloys.

P

chloride (PVC), are available in sheet or film, laminated to several different metals.

Furthermore, almost every metal is available in a limitless number of textured, patterned, and embossed finishes right from the mill. These textured metals can be used as is, but even they can be supplied preplated, prepainted, or even with a colored preanodized finish, as in the case of aluminum.

Many sheet metals are also available with galvanized, aluminized, tin, and terne coatings. These materials, especially galvanized, were among the first prefinished metals to be developed and remain today as basic prefinished metals.

Functional Advantages

Many prefinished metals are used for functional applications. Practically, all zinc-plated sheet, for example, is used in functional applications where good corrosion resistance, rather than a bright, decorative finish is wanted (e.g., condenser cans and hidden parts and door locks). The zinc coating also provides a good paint base, provided the surface is first given a chemical conversion treatment.

Copper-plated steel is another good example of a functional finish. It provides a good lubricating surface for deep-drawing operations and also makes a good base for further electroplating. The material is also used for its electrical conductivity, for its usefulness in low-temperature tinning and high-temperature brazing operations, and as a stop-off coating in carburizing operations.

Prepainted metals are also popular for functional applications; this is borne out by the wide number of functional resins that are now available. In addition to providing decorative appearance, these coatings prevent the base metal from corroding and can provide good resistance to chemicals and foods (epoxies), toughness and resistance to forming damage (vinyls), and good resistance to outdoor exposure (acrylics).

Types and Applications

Prepainted metals are produced using various organic coatings on many common ferrous and nonferrous metals. Extra durability or special decorative effects are provided by plastic–metal laminates. PVC, polyvinyl fluoride, and polyester are the plastics commonly used.

Black-coated steel is used to give a high thermal emittance in electronic equipment. The base metal is aluminum-deoxidized steel containing 0.13% carbon, 0.45% manganese, 0.04% max phosphorus, and 0.05% max sulfur. The steel is coated with a 5% by weight layer of nickel oxide, which is reduced in a hydrogen furnace to form a spongy layer of nickel. This sponge is impregnated with a carbon slurry to form a black carbonized surface.

Preplated metals consist of a thin electrodeposited plate of one metal or alloy on a base, or substrate, usually sheet, of another metal. Steel is the most common base metal, and it is commonly plated with brass, chromium, copper, nickel, nickel–zinc, or zinc. Other common preplated metals are chromium-plated brass, copper, nickel, or zinc, and nickel-plated brass, copper, or zinc. The surface of the plate may be mirror bright, satin, bright satin, embossed, antique, or black or have some other finish. The plate, usually 0.025–0.127 mm thick, is sufficiently ductile to withstand shearing, bending, drawing, and stamping operations. Joining methods include lock seaming, stud welding, adhesive bonding, and spot welding. One of the earliest groups of preplated metals included Brassoid, Nickeloid, and Chromaloid, which were brass-, nickel-, and chromium-plated zinc sheet.

Prefinished metals are now available with almost any metal plated or bonded to almost any other metal, or single metals may be prefinished in colors and patterns. They come in bright or matte finishes and usually have a thin paper coating on the polished side, which is usually stripped off before or after forming. The metals are sold under a variety of trade names and are used for decorative articles, appliances, advertising displays, panels, and mechanical parts.

Prefit

A process for checking the fit of mating detail parts in an assembly prior to *adhesive bonding* to ensure proper bond lines. Mechanically fastened structures are sometimes prefitted to establish shimming requirements.

Preform (Ceramics)

A porous ceramic mass in the shape of the desired final part that is infiltrated with metal to form ceramic–metal composite. See also *Lanxide process*.

Preform (Composites and Plastics)

A preshaped fibrous reinforcement formed by the distribution of chopped fibers or cloth by air, water floatation, or vacuum over the surface of a perforated screen to the approximate contour and thickness desired in the finished part. Also, a preshaped fibrous reinforcement of mat or cloth formed to the desired shape on a mandrel or mock-up before being placed in a mold press. (2) A compressed tablet or biscuit of plastic composition used for efficiency in handling and accuracy in weighing materials. (3) To make plastic molding powder into pellets or tablets.

Preform (Joining)

Brazing or soldering filler metal fabricated in a shape or form for a specific application.

Preform Binder

A resin applied to the chopped strands of a *preform*, usually during its formation, and cured so that the preform will retain its shape and can be handled.

Preformed Ceramic Core

In foundry practice, a preformed refractory aggregate inserted in a wax or plastic pattern to shape the interior of that part of a casting that cannot be shaped by the pattern. The wax is sometimes injected around the preformed core. See also *investment casting*.

Preforming

(1) The initial pressing of a metal powder to form a compact that is to be subjected to a subsequent pressing operation other than coining or sizing. (2) Preliminary forming operations, especially for impression die forging.

Pregel

An unintentional, extra layer of cured resin on part of the surface of a reinforced plastic. Not related to gel coat.

Precipitation Hardening

Precipitation hardening is the method now employed to impart high strength at high temperatures to most of the superalloys used for critical components of aircraft gas turbines. This includes alloys ranging from 25% nickel A-286 wheel alloys to such recent and complex high-temperature nickel-base wrought and cast turbine blade alloys as "Udimet" 700, "Nimonic" alloy 115, and IN-100.

The presence of aluminum and titanium, usually jointly, in a nickel–chromium base, with or without iron, imparts unique age-hardening characteristics through the precipitation of gamma-prime phase {$Ni_3(Al,Ti)$}.

The outstanding difference between the previously used age-hardening systems, typified by duralumin or beryllium copper, and those utilizing gamma-prime hardening lies in the fact that an alloy of the latter type can be heated in service appreciably above the optimum aging temperature without permanent loss of strength. In the conventional critical dispersion type age-hardening alloys, strength can only be restored by a complete cycle of heat treatment involving a high-temperature solution treatment and reaging.

The difference between the two types of age-hardening systems is that the aluminum–titanium hardened alloy retains most of its initial hardness (and high-temperature properties) when the overheat is removed. However, an alloy typical of the other type of system (such as beryllium copper or beryllium nickel) overages and does not regain its hardness when the overheat is removed. It is this "reversible" aging behavior, along with high resistance to overaging (agglomeration of gamma-prime hardening phase), that has led to the widespread use of aluminum–titanium age-hardened nickel-base alloys for first-stage turbine blades in commercial and advanced design jet engines. The somewhat less complex iron–nickel–chromium alloys, such as A-286 and "Incoloy" alloy 901, used for turbine disks, are similarly age-hardened, with Ni_3Ti comprising most of the age-hardening component in these two alloys. Sufficient aluminum is also present in the gamma-prime precipitate to improve structural stability.

Carbide Hardening

Carbides provide a major source of dispersed-phase strengthening and cobalt-base alloys. These alloys do not respond to age hardening with aluminum and titanium since the gamma-prime phase does not form unless substantial amounts of nickel are present. Although gamma prime {$Ni_3(Al,Ti)$} is the principal strengthener in the age-hardenable nickel-base alloys, important auxiliary strengthening can be obtained by precipitation of quite complex carbides. The nature of the carbides formed and their mode of distribution can usually be controlled by alloy formation in heat treatment.

Deoxidizers and Malleabilizers

In addition to the major alloying elements, small but effective amounts of malleabilizers such as boron and zirconium must be present to neutralize the effects of impurities that adversely affect hot ductility. As little as 0.005% boron is highly effective in nickel-base alloys, and zirconium, in a concentration about 10 times that of boron, is also useful. In air melting of nickel-base alloys, magnesium is also added to "fix" any sulfur picked up during melting.

The high-temperature properties of the age-hardenable alloys are governed to a large degree by the hardener (aluminum plus titanium) content. However, as the hardener increases, the temperature of incipient fusion decreases and a lower-temperature limit of forgeability rises, because of increased high-temperature strength, until forging is no longer practical by conventional procedures. The use of cast turbine blade alloys to meet very high-temperature requirements is a natural consequence.

While other co-present elements and the amount of hot and cold workability required are also factors, it may be considered that alloys with up to 4% total hardener (aluminum plus titanium) are generally available in all common mill forms and/or fabricable by conventional methods. As the hardener increases to about 8% or 9%, wrought alloys are still available, but in progressively fewer forms, ultimately being limited to small forgings such as turbine blades, at increasingly greater cost. The commercial cast turbine blade alloys contain from about 7% to 11% total hardener and are often chosen over forgings for reason of cost or necessity or both.

Phases

Superalloys consist of a face-centered cubic (fcc) austenitic gamma-phase matrix plus a variety of secondary phases. The principal secondary phases are carbides (MC, $M_{23}C_6$, M_6C, and then M_7C_3) in all superalloy types and gamma-prime (γ) fcc ordered $Ni_3(Al,Ti)$ intermetallic compound in Ni- and Fe–Ni-based alloys. In alloys containing niobium and tantalum, the primary strengthening phase is gamma double prime (γ), a body-centered tetragonal phase. Superalloys derive their strength from solid-solution hardeners and precipitating phases. Carbides may provide limited strengthening directly (e.g., through dispersion hardening) or, more commonly, indirectly (e.g., by stabilizing grain boundaries against excessive shear). In addition to those elements that produce solid-solution hardening and promote carbide and γ formation, other elements (e.g., boron, zirconium, hafnium, and cerium) are added to enhance mechanical and/or chemical properties.

The three types of superalloys (Fe–Ni-, Ni-, and Co-base) are further subdivided into wrought, cast, and powder metallurgy alloys. Cast alloys can be further broken down into polycrystalline, directionally solidified, and single-crystal superalloys. The most important class of Fe–Ni-base superalloys includes those alloys that are strengthening by intermetallic-compound precipitation in an fcc matrix. The most common precipitate is γ, typified by alloys A-286 and Incoloy 901, but some alloys precipitate γ, typified by Inconel 718. Another class of cast Fe–Ni-base superalloys is hardened by carbides, nitrides, and carbonitrides; some tungsten and molybdenum may be added to produce solid-solution hardening. Other Fe–Ni-base alloys are modified in stainless steels primarily strengthened by solid-solution hardening.

The most important class of Ni-base superalloys is strengthened by intermetallic-compound precipitation in an fcc matrix. The strengthening precipitate is γ, typified by Waspaloy and Udimet 700. Another class is represented by Hastelloy X, which is essentially solid-solution strengthened, but which also derives some strengthening from carbide precipitation produced through a working-plus-aging schedule. A third class includes oxide dispersion strengthened (ODS) alloys, which are strengthened by dispersions of inert particles such as yttria.

Cobalt-base superalloys are strengthened by solid-solution alloying and carbide precipitation. Unlike the Fe–Ni- and Ni-base alloys, no intermetallic phase has been found that will strengthen Co-base alloys to the same degree that γ or γ' strengthens the other superalloys.

Compositions and Properties

Although factors such as tensile properties, corrosion resistance, fatigue strength, expansion characteristics, etc., are important, the creep-rupture characteristics are usually the prime requisite in the selection of a superalloy.

Vacuum Melting

Vacuum induction-melted superalloys are generally more ductile than air-melted material, permitting the use of higher aluminum plus titanium levels while still retaining adequate ductility. Vacuum induction melting accomplishes refining and also permits closer control of composition than does air melting. The vacuum arc (consumable electrode) method is widely used for superalloys, especially those employed for turbine disks. Often the electrode for the vacuum arc-melting charge is obtained from a vacuum induction melt, the end product thus combining the benefits of refining occurring in vacuum induction melting with the controlled solidification and sound the ingot structure associated with the vacuum arc process. Some refining is also accomplished in the vacuum arc process, but to a lesser degree than in vacuum induction melting.

Iron-Base Superalloys

The iron-base superalloys include solid-solution alloys and precipitation hardening (PH), or precipitation-strengthened, alloys. Solid-solution types are alloyed primarily with nickel (20%–36%) and chromium (16%–21%), although other elements are also present in lesser amounts. Superalloy 16-25-6, for example, the alloy designation indicating its chromium, nickel, and molybdenum contents, respectively, also contain small amounts of manganese (1.35%), silicon (0.7%), nitrogen (0.15%), and carbon (0.06%). Superalloy 20Cb-3 contains 34% nickel, 20% chromium, 3.5% copper, 2.5% molybdenum, as much as 1% columbium, and 0.07% carbon. Incoloy 800, 801, and 802 contain slightly less nickel and slightly more chromium with small amounts of titanium, aluminum, and carbon. N-155 or Multimet, and early sheet alloy, contains about equal amounts of chromium, nickel, and cobalt (20% each), +3% molybdenum, 2.5% tungsten, 1% columbium, and small amounts of carbon, nitrogen, lanthanum, and zirconium. At 732°C, this alloy has a 1000 h stress-rupture strength of about 165 MPa.

PH iron-base super alloys provide greater strengthening by precipitation of a nickel–aluminum–titanium phase. One such alloy, which may be the most well-known of all iron-base superalloys, is A-286. It contains 26% nickel, 15% chromium, 2% titanium, 1.25% molybdenum, 0.3% vanadium, 0.2% aluminum, 0.04% carbon, and 0.005% boron. At room temperature, it has a tensile yield strength of about 690 MPa and a tensile modulus of 145×10^3 MPa. At 649°C, tensile yield strength declines only slightly, to 607 MPa, and its modulus is about the same or slightly greater. It has a 1000 h stress-rupture strength of about 145 MPa at 732°C. Other PH iron-base superalloys are Discoloy, Haynes 556 (whose chromium, nickel, cobalt, molybdenum, and tungsten content is similar to that N-155); Incoloy 903 and Pyromet CTX-1, which are virtually chromium-free but high in nickel (37%–38%) and cobalt (15%–16%); and V-57 and W-545, which contain about 14% chromium, 26%–27% nickel, about 3% titanium, 1%–1.5% molybdenum, plus aluminum, carbon, and boron. V-57 has a 1000 h stress-rupture strength of about 172 MPa at 732°C and greater tensile strength but similar ductility, than A-286 at room and elevated temperatures.

Nickel-Base Superalloys

Nickel-base superalloys are solid-solution, precipitation, or oxide-dispersion strengthened. All contain substantial amounts of chromium, 9%–25%, which, combined with the nickel, accounts for their excellent high-temperature oxidation resistance. Other common alloying elements include molybdenum, tungsten, cobalt, iron, columbium, aluminum, and titanium. Typical solid-solution alloys

include Hastelloy X (22%–23% chromium, 17%–20% iron, 8%–10% molybdenum, 0.5%–2.5% cobalt, 2% aluminum, 0.2%–1% tungsten, and 0.15% carbon); Inconel 600 (15.5% chromium, 8% iron, 0.25% copper maximum, 0.08% carbon); and Inconel 601, 604, 617, and 625, the latter containing 21.5% chromium, 9% molybdenum, 3.6% columbium, 2.5% iron, 0.2% titanium, 0.2% aluminum, and 0.05% carbon. At 732°C, wrought Hastelloy X (it is also available for castings) has a 1000 h stress-rupture strength of about 124 MPa, and has high oxidation resistance at temperatures to 1204°C.

The precipitation-strengthened alloys, which are the most numerous, contain aluminum and titanium for the precipitation of a second strengthening phase, the intermetallic $Ni_3(Al,Ti)$ known as gamma prime (γ') or the intermetallic Ni_3Cb known as gamma double prime (γ''), during heat treatment. One such alloy, Inconel X-750 (15.5% chromium, 7% iron, 2.5% titanium, 1% columbium, 0.7% aluminum, 0.25% copper maximum, and 0.04% carbon), has more than twice a tensile yield strength Inconel 600 at room temperature and nearly three times as much at 760°C. Its 1000 h stress-rupture strength at 760°C is in the range of 138–207 MPa. Still great tensile yield strength at room at elevated temperatures and a 172-MPa stress-rupture strength at 760°C are provided by Inconel 718 (19% chromium, 18.5% iron, 5.1% columbium, 3% molybdenum, 0.9% titanium, 0.5% aluminum, 0.15% copper maximum, 0.08% carbon maximum), a wrought alloy originally that also has been used for castings. Among the strongest alloys in terms of stress-rupture strength is the wrought or cast IN-100 (10% chromium, 15% cobalt, 5.5% aluminum, 4.7% titanium, 3% molybdenum, 1% vanadium, less than 0.6% iron, 0.15% carbon, 0.06% zirconium, 0.015% boron). Investment cast, it provides a 1000 h stress-rupture strength of 517 MPa at 760°C, 255 MPa at 871°C, and 103 MPa at 982°C. Other precipitation-strengthened wrought alloys include Astroloy; D-979; IN 102; Inconel 706 and 751; M 252; Nimonic 80A, 90, 95, 100, 105, 115, and 263; René 41, 95, and 100; Udimet 500, 520, 630, 700, and 710; Unitemp AF2-1DA; and Waspaloy. Other cast alloys, mainly investment-cast, include B-1900; IN-738X; In-792; Inconel 713C; M 252; MAR-M 200, 246, 247, and 421; NX-188; René 77, 80, and 100; Udimet 500, 700, and 710; Waspaloy; and WAZ-20.

A few of the cast alloys, such as MAR-M 200, are used to produce directionally solidified castings, that is, investment castings in which the grain runs only unidirectionally, as long the length of turbine blades. Eliminating transverse grains improves stress-rupture properties and fatigue resistance. Grain-free alloys, or single-crystal alloys, also have been cast, further improving high-temperature creep resistance. Regarding powder-metallurgy techniques, emphasis has been the use of pre-alloy powder made by rapid solidification techniques (RST) and mechanical alloying (MA), a high-energy milling process using attrition mills or special ball mills. Dispersion-strengthened nickel alloys are alloys strengthened by a dispersed oxide phase, such as thoria, which markedly increases strength at very high temperatures but only moderately so at intermediate elevated temperatures, thus limiting applications. TD-nickel, or thoria-dispersed nickel, was the first of such superalloys, and it was subsequently modified with about 20% chromium, TD-NiCr, for greater oxidation resistance. MA 754 and MA 6000E alloys combine dispersion strengthening with yttria and gamma-prime strengthening.

Cobalt-Base Superalloys

Cobalt-base superalloys are for the most part solid-solution alloys, which, when aged, are strengthened by precipitation of carbide or intermetallic phases. Most contain 20%–25% chromium, substantial nickel and tungsten and/or molybdenum, and other elements, such as iron, columbium, aluminum, or titanium. One of the most well-known, L-605, or Haynes 25, is mainly a wrought alloy, although it is also used for castings. In wrought form, it contains 20% chromium, 15% tungsten, 10% nickel, 3% iron, 1.5% manganese, and 0.1% carbon. At room temperature, it has a tensile yield strength of about 462 MPa, and at 871°C about 241 MPa. Its 1000 h stress-rupture strength at 815°C is 124 MPa. The more recent Haynes 188 (22% chromium, 22% nickel, 14.5% tungsten, 3% iron, 1.5% manganese, 0.9% lanthanum, 0.35% silicon, and 0.1% carbon), which was developed for aircraft-turbine sheet components, provides roughly similar strength and high oxidation resistance to about 1093°C. MP35N (35% nickel, 20 chromium, 10 molybdenum) is a work-hardening alloy used mainly for temperature corrosion-resistant fasteners. Another alloy, S-816, contains equal amounts of chromium and nickel (20% each), equal amounts of molybdenum, tungsten, columbium, and iron (4% each), and 0.38% carbon. Primarily a wrought alloy, although also used for castings, it has a 1000 h stress-rupture strength of 145 MPa at 815°C. Other casting alloys include AiResist 13, 213, and 215; Haynes 21 and 31, the latter also known as X-40; Haynes 151; J-1650; MAR-M 302, 322, 509, and 918; V-36; and WI-52. Their chromium content ranges from 19% (AiResist 215) to 27% (Haynes 21) and some are nickel-free or low in nickel. Most contain substantial amounts of tungsten or tantalum, and various other alloying elements. Among the strongest in terms of 1000 h stress-rupture strength at 815°C are Haynes 21 and 31–290 MPa and 352 MPa, respectively.

Fabrication

Forging and Hot Working

Many of the commercial nickel-base age-hardenable superalloys can be forged or hot-worked with varying degrees of ease. As has been indicated, the top side of the forging temperature range is limited by such considerations as incipient fusion temperature, grain size requirements, tendency for "bursts," etc., and the lower side by the stiffness and ductility of the alloy. The recommendations of the metal producer should be sought for optimum forging practice for a given alloy.

The hot extrusion process increases the gamut of superalloys compositions that can be hot-worked. By the use of a suitable sheath on the extrusion billet, otherwise unworkable alloys can be reduced to bar by hot extrusion. In some instances, mild steel has been used for a sheath, while in others nickel–chromium alloys such as Nimonic alloy 75 and Inconel alloy 600 have been employed. Some of the very high-speed hot extrusion processes may be of value in a hot-working the more refractory superalloys.

Heat Treatment

The conventional heat-treating equipment and fixtures are generally suitable for nickel alloys, and austenitic stainless steels are also applicable to the nickel-base high-temperature alloys. Nickel-base alloys are more susceptible to sulfur and lead embrittlement than iron-base alloys. It is therefore essential that all foreign material, such as grease, oil, cutting lubricants, marking paints, etc., be removed by suitable solvents, vapor greasing, or other methods, before heat treatment.

When fabricated parts made from thin sheet or strip of age-hardening alloys such as Inconel alloy X-750 must be annealed during and after fabrication, it is desirable, especially in light gauges, to provide a protective atmosphere such as argon or dry hydrogen to lessen the possibility of surface depletion of the age-hardening elements.

This precaution may not be as necessary in heavier sections, since the surface oxidation involves a much smaller proportion of the effective cross section.

It is usually necessary after severe forming, or after welding, to apply a stress relief anneal (above 899°C) to assemblies fabricated from aluminum–titanium age-hardenable nickel-base alloys prior to aging. It is vitally important to heat the structure rapidly through the age-hardening temperature range of 649°C–760°C (which is also the low ductility range) so that stress relief can be achieved before any appreciable aging takes place. This is conveniently done by charging into a furnace at or above the desired annealing temperature. It has been found at times that the efficacy of this procedure has been vitiated in large welded structures by charging onto a cold car, resulting in a slower and nonuniform heating of the fabricated part when run into the hot furnace. Contrary to expectations, little difficulty has been encountered with distortion under the above rapid heating conditions. In fact, distortion of weldments of substantial size has been reported to be less than by conventional slow-heating methods.

Forming

All of the wrought nickel-base alloys available as sheet can be formed successfully into quite complex shapes and loving much plastic flow. The lower-strength Inconel alloy 600 and Nimonic alloy 75 offer few problems. The high-strength age-hardening varieties, processed in the annealed condition, can be subjected to a surprising amount of cold work and deformation, provided sufficient power is available. Explosive forming has also been successfully employed on a number of nickel-base alloys.

Machining

All of the alloys discussed can be machined, the strongest and highest hardener content materials causing the most difficulty. The recommendations of the metal producer should be followed with respect to Altamont condition of heat treatment, type of tool, speed and feed, cutting lubricant, etc. Wrought alloys of quite high hardener content, such as Inconel alloy 700 and Udimet 500, although difficult to handle, can be machined with reasonable facility using high-speed-steel tools of the tungsten–cobalt type, and cemented carbide tools of the tungsten–cobalt and tungsten–tantalum–cobalt type.

Various electroerosion processes have been successfully used on a number of the age-hardened superalloys and, at hardener levels, may be necessary for some operations such as drilling.

Welding

Inconel alloy 600, Nimonic alloy 75, and other nickel-base alloys of the predominantly solid-solution strengthened type offer no serious problems in welding. All of the common resistance and fusion welding processes (except submerged arc) are regularly and successfully employee.

In handling the wrought superalloys age-hardened with gamma prime, it is necessary to observe certain precautions. Material should be welded in the annealed condition to minimize the hazard of cracking in weld or parent metal. If the components to be joined to have been severely worked or deformed they should be stress-relief-annealed before welding by charging into a hot furnace to ensure rapid heating to this stress-relieving temperature. Similarly, weldments should be stress-relieved before attempting to apply the 704°C age-hardening treatment.

Where subassemblies must be joined in the age-hardened condition, the practice of "safe ending" with a compatible nonaging material prior to age hardening can be usefully employed. The final weldments joining the fully age-hardened components are then made on the "safe ends."

Such new welding processes as "short arc," electron, and laser beam have come into increasing use and have been helpful in joining some of the very high hardener content alloys.

Brazing

The solid-solution-type chromium-containing alloys, such as Inconel alloy 600, are quite readily brazed, using techniques and brazing filler metals applicable to the austenitic stainless steels. Generally speaking, it is desirable to braze annealed (stress-free) material to avoid embrittlement by the molten braze metal. Where brazing filler metals are employed that melt above the stress-relieving temperature, a prior anneal is usually not needed. As with the stainless steels, dry hydrogen, argon, and helium atmospheres are used successfully; and vacuum brazing is also very successfully employed.

The age-hardened nickel-base alloys containing titanium and aluminum are rather difficult to braze, unless some method of fluxing, solid or gaseous, is used. Alternatively, the common practice is to preplate the areas to be furnace braised with 0.01–0.03 mm of nickel, which prevents the formation of aluminum or titanium oxide films and permits ready wetting by the brazing filler metal.

Silver brazing filler metals can be used for lower-temperature applications. However, since the nickel-base superalloys are usually employed for high-temperature applications, the higher melting point and stronger and more oxidation-resistant brazing filler metals of the Ni–Cr–Si–B type are generally used. The silver–palladium–manganese and palladium–nickel filler metals also provide useful brazing materials for intermediate service temperatures.

Preheat

See *preheating* and *preheat temperature.*

Preheat Current

In resistance welding, an impulse or series of current impulses that occurs prior to and is separated from welding current.

Preheat Temperature

A specified temperature that the base metal must attain in the welding, brazing, soldering, thermal spraying, or cutting area immediately before these operations are performed.

Preheat Time

In resistance welding, a portion of the preweld interval during which preheat current occurs.

Preimpregnated Decorative Foil

Special papers preimpregnated with melamine resin make possible a broad range of molded-in decoration for thermoset plastic products. By far, the leading application to date is the use of melamine-impregnated rayon paper. It is used to decorate about 70% of all melamine dinnerware. Development of new materials

and techniques to shorten the production cycle is extending use of prepreg paper in closures, cutlery handles, and other markets.

Fabrication

The rayon paper is impregnated with resin under closely controlled conditions at the impregnator's plant to provide a foil with precise resin content, even weight and resin distribution, specified penetration, and absorption. The melamine resin is partially cured, then held at a stable.

Preheating (metals)

(1) Any deliberate heating before some further thermal or mechanical treatment. Intended to confer some benefit. It is normally implicit that the preheat temperature will be exceeded during the subsequent process. See also *weldability.* For tool steel, heating to an intermediate temperature immediately before final austenitizing. For some nonferrous alloys, heating to a high temperature for a long time, in order to homogenize the structure before working. (2) In welding and related processes, heating to an intermediate temperature for a short time immediately before welding, brazing, soldering, cutting, or thermal spraying. (3) In powder metallurgy, an early stage in the sintering procedure when, in a continuous furnace, lubricant or binder burn-off occurs without atmosphere protection prior to actual sintering in the protective atmosphere of the high heat chamber.

Preheating (Plastics)

The heating of a polymeric compound before molding or casting to facilitate the operation or reduce the molding cycle.

Preheat Temperature

A specified temperature that the base metal must attain in the welding, brazing, soldering, thermal spraying, or cutting area immediately before these operations are performed.

Preheat Time

In resistance welding, a portion of the preweld interval during which preheat current occurs.

Preimpregnated Decorative Foil

Special papers preimpregnated with melamine resin make possible a broad range of molded-in decoration for thermoset plastic products. By far, the leading application to date is the use of melamine-impregnated rayon paper. It is used to decorate about 70% of all melamine dinnerware. The development of new materials and techniques to shorten the production cycle is extending the use of prepreg paper in closures, cutlery handles, and other markets.

Fabrication

Rayon paper is impregnated with resin under closely controlled conditions at the impregnator's plant to provide a foil with precise resin content, even weight and resin distribution, specified penetration, and absorption. Melamine resin is partially cured, then held at a stable, flexible B-stage, cut into sheets convenient for printing, and shipped to printers for decoration.

Designs are printed by either lithography or silkscreen with special inks and techniques developed by specialty printers. Virtually any

P

design can be reproduced, including photographs in black and white or full color. Even moldable silver and gold inks have been perfected.

The printed foil is cut to final size and shape and delivered to the molder ready for the mold. Normally, the product to be decorated is formed and partially cured; then the mold is opened, the foil inserted, and the mold closed to fuse the foil to the product and complete the cure. Except when metallic inks are used, paper is inserted so that the designs face the product. The foil becomes transparent but forms a protective melamine surface.

Decorating Closures

A relatively new development is the use of prepreg foils to decorate closures such as cosmetic jar covers. Previously, decoration was limited to the one or two colors possible with hot stamping and direct screening or to lithography on metal, which is subject to rust. Molding the foil into a thermosetting plastic closure gives the package designer full color range and combination and later permits the change of design without altering the molds.

The demand of the closure producer for high-volume, high-speed molding encouraged a research program that has yielded a new one-shot decorative foil. Instead of preforming the product and opening the mold for the insertion of the foil, the new one-shot foil is put into the mold first, the molding powder poured on top of it, and the mold closed for a complete, uninterrupted cycle. As yet, the new technique has been used only for closures, but the same principles are believed to be technically feasible for larger products such as dinnerware. More immediate markets for the one-shot foil are control knobs and drawer pulls.

Materials

Prepreg foils can be used to apply decoration to a wide variety of thermosetting resins, including all melamines, all ureas, and most phenolics.

Although the self-effacing rayon foil is most popular for dinnerware and most melamine and urea products, opaque prepreg paper is favored for such products as cutlery handles, clock faces, switch plates, and trays. The opaque paper shields the darker resins and even permits use of the varicolored odds and ends in a molder's inventory for such applications as a clock face. Often, of course, the opaque paper is used to provide a decorative color contrast.

The opaque print-base paper is made with a high-purity alpha pulp, fillers, pigments, and other additives. Like the rayon foils, this paper is preimpregnated with the resin and advanced to a stable B-stage for shipment and printing. Among other products for which the opaque prepreg papers are suitable are handles for pots and pans, organ keys, and wall tile.

Preimpregnated Materials for Reinforced Plastics

The so-called *prepregs* are ready-to-mold reinforced plastics with resin and reinforcement pre-combined into one easy-to-handle material. They are fabricated by what is called dry layup techniques; the alternative is the wet layup method of combining a liquid resin and reinforcement at the mold.

Fabrication

Prepregs are produced by impregnating continuous webs of fabric or fiber with synthetic resins under close control. The resins are then partially cured to the B-stage, or partly polymerized. At this stage, the preimpregnated material remains stable, and may be shipped or stored, ready for forming into the final shape by heat and pressure.

Most of the controllable variables in a reinforced plastics structure are taken into account in the production of a prepreg, for example, resin type and content, reinforcement, and finish, which are largely determined by the requirements of the final laminate. Prepregs are preengineered to meet performance and processing requirements.

Major Advantages

The major advantages of using preimpregnated materials for molding reinforced plastics products are (1) high and more uniform strength, (2) uniform quality, (3) simplified production, and (4) design freedom.

Key to the majority of the advantages of prepreg is the close control maintained through all stages of the preimpregnated process. For example, the amount of resin solution picked up by the reinforcing web as it passes through one or more resin baths is influenced not only by resin viscosity and resin solids, but also by web tension and speed (which regulate the time the reinforcement is in the bath) and by the temperature of the resin. Because of the high degree of control afforded by preimpregnating equipment over tension, speed, and temperature, resin content can be controlled ±2%.

Such control also permits the production of prepregs with a reinforcement content from as high as 85% to as low as 20%. It is the high reinforcement content possible with prepregs that accounts for the high strength of these materials. Further, since the reinforcement is thoroughly saturated during preimpregnation, little air is entrapped in the reinforced plastic material, assuring better quality in the molded part.

Handling characteristics of prepregs can also be closely controlled by controlling the temperature and speed of the material as it passes through the second stage of the process, where resin is partially cured. Temperature of the drying oven, speed of travel, and resin type and content determine the degree of polymerization or B-stage, which, in turn, determines such handling and molding characteristics as resin flow, gel time, tack, and drape. In prepregs, these factors can be tailored to suit a variety of production molding requirements.

Prepregs also facilitate the use of such resins as phenolics, melamines, and silicones, which, when available in liquid form, are usually solvent solutions (except for newer solventless silicones). Wet layup of solvent solutions of resins is usually unsatisfactory, at best, because solvents are released during cure. As B-staged prepregs, such resin systems are virtually solvent-free.

Prepregs can also be made using resins whose viscosities are so low that they would be difficult, or impossible, to handle using wet layup techniques.

Prepregs can be partially advantageous to the smaller reinforced plastics fabricator, eliminating the need for resin formulation; scrap loss can be minimized by chopping up or macerating leftover prepreg to form molding compound.

Because the resin is already properly distributed throughout the reinforcement, prepregs offer distinct advantages in molding odd-shaped parts, with varying thicknesses, undercuts, flanges, etc. They preclude the problem of excessive flow of resin, eliminating resin-rich or resin-starved areas of such parts. Cut to the proper pattern, the prepreg can be accurately prepositioned in the mold to provide optimum finished parts.

Possibly the greatest advantage of prepregs is that they lend themselves to automated, high-production molding. The continuous prepreg web can be prepared for molding by cutting, slitting, or blanking operations. Often parts can be molded directly from such blanks. Die-cut parts can also be used to preassemble a complete lay up in advance of molding, thus minimizing press time.

Materials Used

A wide range of reinforcing materials and resins are available in prepreg form. Principal resins are polyesters, epoxies, phenolics, melamines, silicones, and several elastomers. The most commonly used reinforcements are glass cloth, asbestos, paper, and cotton. Specialty fibrous reinforcements include high-silica glass, nylon, rayon, and graphite.

Reinforcing Materials

Reinforcing materials available in prepregs today include the following:

1. Glass fiber is by far the largest volume reinforcing material in use today. Prepregs incorporating glass fiber are available in the form of preimpregnated roving (for filament winding), cloth, and mat. It provides a good balance of properties: outstanding strength-to-weight characteristics, high tensile strength, high modulus of elasticity compared to other fibers, resilience, and excellent dimensional stability. It is used for such products as airplane and missile parts, ducts, trays, electrical components, truck body panels, and construction materials.
2. Asbestos is used in prepregs in the form of paper, felt, and fabrics. All provide good thermal insulation. Papers provide cost advantage where structural characteristics are not critical. Felts provide maximum tensile and flexural strength and good ablation resistance. Asbestos fabrics are used principally in flat laminates. Prepregs employing asbestos webs have won an important place in rocket and missile components in good part because of their remarkable short-term resistance to extremes of temperature and flame exposure.
3. Paper, a leading reinforcement for high-pressure prepreg laminates, is inexpensive, prints well, and has adequate strength for its principal uses as counter surfacing, furniture and desktops, and wall boards.
4. Specialty reinforcements, like high-silica glass, quartz, graphite, nylon, have their main use in missile and rocket parts. Some exhibit unusual high-temperature resistance; some are good ablative materials.

Resins

The major plastic resins used in prepregs are polyesters, epoxies, phenolics, melamines, and silicones, although any thermosetting resin or elastomer may be used.

1. Polyester resin prepregs are comparatively low in cost and easy to mold at low temperatures. They have good mechanical, chemical, and electrical properties, and some types are flame resistant.
2. Epoxy prepregs provide high mechanical strength, excellent dimensional stability, corrosion resistance and interlaminar bond strength, good electrical properties, and very low water absorption.
3. Phenolic prepregs have high mechanical strength, excellent resistance to high temperature, good thermal insulation and electrical properties, and high chemical resistance.
4. Melamine prepregs have excellent color range and color retention, high abrasion resistance, good electrical properties, and are resistant to alkalies, and flame.
5. Silicone prepregs provide the highest electrical properties available in reinforced plastic and are the most heat stable. They retain strength and electrical characteristics under long-term exposure to 260°C–316°C.

Actually any thermosetting resin or elastomer may be used and any continuous-length reinforcement. Even sheet material not capable of supporting its own impregnated weight in a drying oven can be used. The web may be a woven fabric, nonwoven sheet, or continuous strand or roving. After impregnation, the prepreg may be slit into a tape, chopped, macerated, or cut to a specified pattern, depending on the requirements of the part and the desire of the molders.

Molding

The one requirement for using prepregs is that some heat and pressure must be used to mold them to a final shape. They cannot be room-temperature cured. Prepregs can be formulated to fit all other methods of processing; the four major methods are vacuum-bag molding, pressure-bag molding, matched metal-die molding, and filament and tape winding.

The choice of method of forming prepregs is most often governed by production volume. The exception is the critical part, like a missile component, where the higher performance qualities achieved by matched metal-die molding justifies the use of this process even for a few hundred parts.

The vacuum-bag molding method uses the simplest and most economical molds and equipment and is used where the number of parts needed is low or where frequent design changes occur.

The pressure-bag method of molding uses pressures up to 0.70 MPa to achieve greater strength than is possible with the vacuum-bag method. Mold and equipment costs remain comparatively low. In both vacuum-bag and pressure-bag molding, additional pressure with resulting higher strengths can be achieved through the use of an autoclave.

Matched metal-die molding is used for long runs where it gives the lowest cost and the highest production rates. This method permits an accurate control of dimensions, density, rate of cure, and surface smoothness.

Tape and filament winding are used for cylindrical, spherical, and conical shapes and can produce high physical properties and high strength-to-weight characteristics. Techniques range from simple winding to highly automated operations. In fact, it comes closest to automation of current fabrication methods and achieves homogeneous, void-free products. It does not involve the expense of matched metal dies.

Preimpregnation (Reinforced Plastics)

The practice of mixing resin and reinforcement and effecting partial cure before use or shipment to the user. See also *prepreg*.

P

Preload/Prestress

The tightening load and resultant stress applied to threaded fasteners, such as bolts, screws and studs, which clamp components together. A high level of preload maintains pressure tightness, largely eliminates the risk of fastener fatigue, reduces the risk of the fastener becoming unscrewed because of vibration, and, in structural steel work, ensures that the load between members is carried by interfacial friction rather than by shear loading of the fastener. Usually, maximum mechanical and economic efficiency are achieved when the specified prestress matches the materials yield stress. However, in practice, tightening equipment and techniques are of limited precision, and hence, prestress is commonly specified as about 80% of yield stress to avoid the risk of gross yielding leading to necking and failure. An alternative approach, in special cases, is to develop a technique that deliberately extends the fastener beyond yield by a controlled amount without causing necking. Fasteners operating in their creep range are usually tightened to lower levels of prestress selected for the particular application. At first consideration, it might be thought that the preload is directly additive to service loads applied to the assembly, for example, by the internal pressure acting on a pipe flange. However, this is not usually the case, and, generalizing, only a small proportion of the service load is additive to the preload. This is a consequence of the small cross-sectional area of the fastener relative to the much larger load-bearing area of the flange and the associated differences in elastic strain. As a result, until such time as the service stress nearly matches the prestress, the joint will remain pressure tight. It follows that large fluctuations in service stresses will induce only small fluctuations in the fastener stress, so fatigue will be avoided provided the service stress does not closely approach the prestress. Some control of preload is provided automatically by proportioning wrench size to fastener diameter, but many other techniques are available for measuring preload. These include the torque wrench, mechanical or ultrasonic measurement of fastener extension, measurement of turn of nut angle after an initial "nip" has been achieved, and the observation of deformable washers or collapsible projections on nut faces. All have their respective advantages and problems.

Premimn

Material or treatment better than the normal standard, or the additional price for such material or treatment.

Premix (Metals)

(1) A uniform mixture of components prepared by a metal powder producer for direct use in compacting. (2) A term sometimes applied to the preparation of a premix; see preferred term *mixing*.

Premix (Plastics)

A plastic molding compound prepared prior to and apart from the molding operations and containing all components required for molding: resin, reinforcement, fillers, catalysts, release agents, and other ingredients.

Premix Burner

A burner used in flame emission and atomic absorption spectroscopy in which the fuel gas is mixed with the oxidizing gas before reaching the combustion zone. See the term *nebulizer*.

Premix Moldings

Premix molding materials are physical mixtures of a reactive thermosetting resin (usually polyester), chopped fibrous reinforcement (usually fibrous glass, asbestos, or sisal), and powdered fillers (usually carbonates or clays). Such mixtures, when properly formulated, can exhibit a wide range of performance properties at variable costs. In general, the resin type will determine the corrosion-resistance properties of the premix; increasing amounts of glass reinforcement will increase strength and increasing amounts of filler will reduce costs and corrosion resistance.

Comparative Benefits

In comparison with such corrosion-resistant metals as brass, bronze, and stainless steels, premixes offer, primarily, cost reductions due to the ease of manufacture of complex shapes, and reductions in actual material costs. In some cases, design latitude, colorability, resistance to abrasion, and reduction in weight are also important factors in favor of the premix parts.

In comparison with aluminum, premixes can offer superior corrosion resistance, especially to alkaline detergents, in addition to the advantages noted earlier.

In comparison with thermoplastic resins, premix materials offer considerably greater hardness, rigidity, and heat resistance. Premix materials are also stronger. Thermoplastics usually have superior colorability, surface smoothness, gloss, and a broader range of corrosion resistance. Thermoplastics are also somewhat lower in cost for the molded item when the number of parts is large. In general, tooling costs for injection-molded thermoplastics are higher, but molding, material-handling, and finishing costs are lower for thermoplastics.

Phenolics, ureas, and melamines are considerably less strong than glass fiber–reinforced premixes, and, in general, they do not have as good a range of resistances to aqueous solutions as do properly formulated premixes. Phenolics, of course, also have limited color possibilities.

In comparison with most other plastics, premixes are usually more difficult to handle because they are more difficult to meter or preweigh automatically. Generally, they must be preweighed by hand to an exact mold charge. Recently, however, there has been a trend toward extrusion and chopping of premix to logs of predetermined charge weight.

Molding

Because premixes are essentially heterogeneous materials, certain well-known molding phenomena such as weld lines and orientation during flow take on added significance. Fiber orientation tends to give somewhat greater effects of anisotropy, and more care must be taken with premix molded parts to prevent excessive orientation. Similarly, weld lines in premix parts tend to be proportionately weaker than weld lines in thermoplastics. Much of the "art" of premix molding is concerned with reducing or eliminating orientation and weld line effects by proper mold design, molding conditions (closing speed, pressure, temperature), and by selection of charge shape and location in the mold. Again, recent advances in glass fiber manufacturing have resulted in superior glass fibers, which reduce orientation and weld-line effects.

Premixes are generally molded in conventional compression molds and presses using pressures on the order of 3.5–13.6 MPa and temperatures in the range of 121–177°C.

Premixes are also occasionally molded in transfer presses, usually at somewhat higher pressures. Transfer molding tends to reduce strength properties by degrading the glass fibers in the compounds.

Premixes are suitable for molding very complex shapes over a wide range of sizes. Filter plates and frames weighing over 45.4 kg have been molded successfully.

A large volume of premix is consumed in automotive air-conditioner and heater housings and ducts. These parts are low-cost, low-strength items requiring high rigidity and good heat resistance. Other large-volume applications include electrical insulators and housings of various types. Most premixes are manufactured captively; that is, molders prepare their own premix material. The advantages of captive premix are lower initial raw material costs, reduced packaging costs, ability to formulate faster curing compounds, and ability to tailor-make special compounds for each part. The major disadvantage of the premix very often is lack of good quality control. Another disadvantage of the captive operation is that not enough compound development effort is spent in working out the "bugs" in a new formulation before such a formulation becomes commercial.

Compounding

Premixes are prepared by mixing glass, filler, and resin mix in a 180° spiral double-arm dough mixer. Blade clearance should be about 6.4–12.7 mm. The resin mix is usually prepared separately and includes resident, lubricant, catalyst, inhibitor, and other minor additives. The resin mix is charged to the mixer, followed by the filler. After the filler is thoroughly blended (usually 10 min is required) the fibrous material can be added. If the fibrous material is glass, care must be taken to disperse the fiber to prevent clumping; close control of mixing time must be maintained to prevent fiber breakdown. Recently, a high strand integrity fiber has been introduced, which reduces fiber-breakdown tendency.

After mixing, the premix must be stored in airtight containers to prevent styrene monomer loss. Cellophane bags may be used to store small amounts of premix (6.7–9 kg), and larger bins or hoppers may be used for larger quantities.

Premix of high filler loadings may be extruded through screw-type or ram-type extruders and automatically chopped to logs of predetermined weight.

The storage life of premix that is sealed against styrene loss may vary from 1 or 2 days to 1 year depending upon formulation. The premix with short storage life can be molded at lower temperatures and shorter cycles than the more stable premix.

Premolding

In composites fabrication, the *layup* and partial *cure* at an immediate cure temperature of a laminated or chopped fiber detail part to stabilize its configuration for handling and assembly with other parts for final cure.

Prep.

Preparation (of weld).

Preparation as in Joint Preparation or Weld Preparation

(1) The various activities involved in preparing a joint for welding or the geometry and dimensions of the surfaces to be welded.

Preplasticization

Technique of premelting injection molding plastic powders in a separate chamber, then transferring the melt to the injection chamber. See also *injection molding* (plastics).

Preply

A composite material *lamina* in the raw material stage, ready to be fabricated into a finished *laminate*. The lamina is usually combined with other raw laminae before fabrication. A preply encloses a fiber system that is placed in position relative to all or part of the required matrix material to constitute the finished lamina. An organic matrix preply is called a *prepreg*.

Prepolymer

A chemical intermediate with a molecular weight between that of the *monomer* or monomers and the final *polymer* or resin.

Prepolymer Molding

In the urethane foam field, a system in which a portion of the *polyol* is pre-reacted with the isocyanate to form a liquid *prepolymer* with a viscosity range suitable for pumping or metering. This component is supplied to end users with a second premixed blend of additional polyol, catalyst, blowing agent, and so forth. When the two components are subsequently mixed, foaming occurs. See also *one-shot molding*.

Prepreg

In composites fabrication, either ready-to-mold material in sheet form or ready-to-wind material in *roving* form, which may be cloth, mat, unidirectional fiber, or paper, impregnated with resin and stored for use. The resin is partially cured to a *B-stage* and supplied to the fabricator, who lays up the finished shape and completes the cure with heat and pressure. The two distinct types of prepreg available are commercial prepregs, in which the roving is coated with a hot melt or solvent system to produce a specific product to meet specific customer requirements; and wet prepreg, in which the basic resin is installed without solvents or preservatives but has limited room temperature shelf life.

Preproduction Test

A test or series of tests conducted by an adhesive manufacturer to determine conformity of an adhesive batch to established production standards, or by a fabricator to determine the quality of an adhesive before parts are produced, or by an adhesive specification custodian to determine conformance of an adhesive to the requirements of a specification not requiring qualification test. Compare with *acceptance test* and *qualification test*.

Preservatives

Chemicals used to prevent oxidation, fermentation, or other deterioration of foodstuffs. The antioxidants, inhibitors, and stabilizers used to retard deterioration of industrial chemicals are not usually called preservatives. The most usual function of a preservative is to kill bacteria, and this may be accomplished by an acid, an alcohol, an aldehyde, or a salt. A legal requirement under the Food and

Drug Act is that a preservative be nontoxic in the quantities permitted. **Sugar** is the most commonly used preservative for fruit products. **Sodium chloride** is used for protein foods. **Sodium nitrate** is reduced as sodium nitrite in curing meats, and the nitrite as an inhibitory action on bacterial growth, the effect being greatest in acid flesh. **Potassium sorbate,** a white, water-soluble powder, inhibits the growth of many molds, yeasts, and bacteria which cause food deterioration, and it is used in cheese, syrups, pickles, and other prepared foods.

The inhibitory effect of organic acids is due chiefly to the undissociated molecule. **Acetic acid** is normally more toxic to bacteria than lactic acid, but when sugar is present, the reverse is true, and citric acid then has little toxicity. The inhibitory action of inorganic acids is due mainly to the pH change that they produce.

Only small quantities of chemicals are usually needed for preservation. Preservatives are also marketed for external application to foodstuffs in storage, though these are more properly classified as fumigants. Beverages containing fruit juices or little carbonation are preserved with less than 0.05% **sodium benzoate.** Methyl, propyl, butyl, and ethyl **parabens; potassium sorbate; sodium dihydroacetate;** and **imidazolidinyl urea** are all industrial **microbials** used in the food, cosmetic, and pharmaceutical industries.

Preshadowed Replica

A replica for fractographic or metallographic inspection that is formed by the application of shadowing material to the surface to be replicated. It is formed before the thin replica film is cast or otherwise deposited on the surface. See also *shadowing.*

Presintered Blank

A metal powder compact sintered at a low temperature but at a long enough time to make it sufficiently strong for metal working. See also *presintering.*

Presintered Density

The relative density of a presintered compact. See also *presintering.*

Press

A machine tool having a stationary bed and a slide or ram that has reciprocating motion at right angles to the bed surface, the slide being guided in the frame of the machine. See also *hydraulic press, mechanical press, slide,* and *straight-side press.*

Press Brake

An open-frame single-action press used to bend, blank, corrugate, curl, notch, perforate, pierce, or punch sheet metal or plate. See also the term *mechanical press brake.*

Press-Brake Forming

A metal forming process in which the workpiece is placed over an open die and pressed down into the die by a punch that is actuated by the ram portion of a *press brake.* The process is most widely used for the forming of relatively long, narrow parts that are not adaptable to *press forming* and for applications in which production quantities are too small to warrant the tooling cost for contour *roll forming.*

Pressed Capacity

The rated force a *press* is designed to exert at a predetermined distance above the bottom of the stroke of the *slide.*

Press Clave

In composites fabrication, a simulated autoclave made by using the platens of a press to seal the ends of an open chamber, providing both the force required to prevent loss of the pressurizing medium and the heat required to cure the laminate inside.

Pressed Bar

(1) A powder metallurgy compact in the form of a bar. (2) A green (unsintered) rectangular compact.

Pressed Density

The weight per unit volume of an unsintered compact. Same as green density.

Press Fit

An interference or *force fit* made through the use of a press. See *interference fit.*

Press Forging

(1) The forging of metal between dies by mechanical or hydraulic pressure; usually accompanied with a single work stroke of the press for each die station. (2) This achieves a more uniform deformation throughout the section compared with drop forging.

Press Forming

Any sheet metal forming operation performed with tooling by means of a mechanical or hydraulic press.

Pressing

(1) In metal working, the product or process of shallow drawing of sheet or plate. (2) Forming a powder metallurgy part with compressive force. See also *compacting.*

Pressing Area

The clear distance (left to right) between housings, stops, gibs, gibways, or shoulders of strain rods, multiplied by the total distance from front to back on the bed of a metal forming *press.* Sometimes called working area.

Pressing Crack

A rupture in a green powder metallurgy compact that develops during ejection of the compact from the die. Sometimes referred to as a slip crack.

Pressing Skin

The surface of a powder metallurgy compact that is superficially more deformed than the interior due to a preferential alignment of the particles caused by contact with the die wall and punch faces.

Press Load

The amount of force exerted in a given forging or forming operation.

Press Quenching

A quench in which hot dies are pressed and aligned with a part before the quenching process begins. Then the part is placed in contact with a quenching medium in a controlled manner. This process avoids part distortion.

Press Slide

See *slide*.

Press Tool

The complete tool assembly used for forming powder metallurgy compacts that consists of the die, a die adapter, the punches, and, when required, a core rod.

Pressure Bag Molding

A process for molding reinforced plastics in which a tailored, flexible bag is placed over the contact lay-up on the mold, sealed and clamped in place. Fluid pressure, usually provided by compressed air or water, is placed against the bag, and the part is cured. See also *vacuum bag molding*.

Pressure Break

In laminated plastics, a break in one or more outer sheets of the paper, fabric, or other base, which is visible through the surface layer of resin that covers it.

Pressure Bubble Plug-Assist Forming

A *thermoforming* process in which a heated plastic sheet is positioned over the mold and air is blown up through the base plate channel causing the sheet to billow upward. The plug assist pushes the sheet into the mold cavity and vacuum is applied to transfer the sheet from the plug to the mold surface.

Pressure Casting

(1) Making castings with pressure on the molten or plastic metal, as in *injection molding, die casting, centrifugal casting,* cold chamber pressure casting, and *squeeze casting.* (2) A casting made with pressure applied to the molten or plastic metal. (3) Casting processes in which the molten metal is injected under pressure into reusable metal molds. The process offers dimensional accuracy, good surface detail, and finish with a fine grain size and minimum porosity. It is suitable for low melting point alloys, particularly those of zinc and aluminum, having relatively low shrinkage during solidification. Pressures can be as high as 10,000 psi (670 bars).

Pressure-Controlled Welding

A resistance welding process variation in which a number of spot or projection welds are made with several electrodes functioning progressively under the control of a pressure-sequencing device.

Pressure Gas Welding

An oxyfuel gas welding process that produces coalescence simultaneously over the entire area of abutting surfaces by heating them with gas flames obtained from the combustion of a fuel gas with oxygen and by application of pressure, without the use of filler metal.

Pressure-Impregnation-Carbonization (PIC)

A densification process for carbon–carbon composites involving pitch impregnation and carbonization under high temperature and isostatic pressure conditions. This process is carried out in hot isostatic pressing equipment. See also *carbon–carbon composites*.

Pressure Intensifier

In composites fabrication, a layer of flexible material (usually a high-temperature rubber) used to ensure the application of sufficient pressure to a location, such as a radius, in a lay-up being cured.

Pressureless Sintering

Sintering of loose powder.

Pressure Lubrication

A system of lubrication in which the lubricant is supplied to a bearing under pressure.

Pressure Plate

A plate located beneath the *bolster plate* and a metal forming *press* that acts against the resistance of a group of cylinders mounted to the pressure plate to provide uniform pressure throughout the press stroke when the press is symmetrically loaded.

Pressure-Sensitive Adhesives

Advanced pressure-sensitive adhesives (PSAs) are finding their way into every aspect of modern automobile manufacturing. Automakers and their suppliers are turning to new assembly methods to speed manufacturing, improve quality, lower costs, and meet more stringent environmental regulations. Automotive applications for today's adhesives include mirrors, brakes, headliners, flexible circuit bonding, sound deadening, carpets, door panels, instrument panels, seats, and windows. In a remarkable example of the potential of automotive applications, advanced structural adhesives have even replaced certain chassis-to-body panel welds in new automobiles.

Automotive air bags are yet another component where PSAs play a critical role. In fact, today's air bags are operating more reliably and safely, thanks, in part, to the use of the latest adhesive formulations.

The millions of air bags in service today must resist years of shock, vibration, temperature fluctuations, and humidity extremes and then work flawlessly when needed.

Various adhesives are used to assemble and seal the components of air bags, initiators, igniters, gaskets, O-rings, high pressure seals, and vibration dampers.

In general, the adhesives serve to

- Position and set the air bag in place
- Hermetically seal the propellant so that it remains protected from moisture
- Attach the propellant cover, which limits the chance of the ignition during a vehicle fire or other non-impact event

Assembly Benefits

Although acrylic and heat-activated adhesives offer excellent benefits in other applications, the advantages of silicone PSAs increasingly make them the adhesive system of choice for air bag inflators. Manufacturers' testing and field operation indicate that for air bag inflators, silicone PSAs offer numerous advantages over acrylic and heat-activated adhesives. Silicone PSAs simply provide the best seal for air bag inflators. In this application, silicone PSA benefits also include

- Thermal stability not found with acrylic or heat-activated adhesives
- Uniform wet-out and adhesion
- Noncombustibility
- Lower outgassing
- Fewer seal leaks

PSA technology lends itself to manual, semiautomatic, and fully automated manufacturing. Environmental releases of volatile organic compounds are also lower with the new adhesives, good for both worker safety and air quality.

Many manufacturers find that utilizing PSAs in place of heat-activated adhesives saves them money. That is because when using silicone, the reject rate due to leaks, adhesive voids, high outgassing, and other manufacturing problems are also dramatically lower. Postassembly testing shows a 15-fold decline in rejects after postassembly testing.

Added together, the cost factor of utilizing silicone PSAs (not to mention the performance benefits) makes them an attractive option for this application.

Other PSAs

Polymers used in the preparation of PSAs can be divided into two general classes of raw materials. In the first class are polymers that undergo chemical cross-linking following coating to achieve high cohesive strength. Examples include solvent-based acrylic PSAs and solvent-based natural rubber PSAs. Polymers in the second general class do not undergo chemical cross-linking, but instead have high molecular weight (e.g., water-based acrylic, SBR) or physical cross-links (e.g., hot melt styrenic block copolymers) to achieve adequate cohesive strengths.

Pressure Sintering

A hot pressing technique that usually employs low loads, high sintering temperatures, continuous or discontinuous sintering, and simple molds

to contain the powder. Although the terms *pressure sintering* and *hot pressing* are used interchangeably, distinct differences exist between the two processes. In pressure sintering, the emphasis is on thermal processing; in hot pressing, applied pressure is the main process variable.

Pressure-Viscosity Coefficient

The slope of a graph showing variation in the logarithm of viscosity with pressure. The use of the term pressure-viscosity coefficient assumes a linear relationship.

Pressure Welding

Any welding process in which the joint is affected primarily by the interfacial force between components. The effects of the force include disruption of surface films, exposure of clean material, plastic deformation, and mechanical mixing. Heating may be involved but little or no fused metal is incorporated in the joint. Examples include *electrical resistance spot welding* and *flash butt welding*. See preferred terms *cold welding, diffusion welding, forge welding, hot pressure welding, pressure gas welding,* and *solid-state welding*.

Pressurized Gas Lubrication

A system of lubrication in which a gaseous lubricant is supplied under sufficient external pressure to separate the opposing surfaces by a gas film.

Prestress

See *preload*.

Prestressed/Pretensioned Concrete

Concrete containing steel bars that are held under tension while the concrete is setting. The tension bars develop a compressive stress in the concrete improving its load carrying capacity.

Pretinning

See preferred term *precoating*.

Prevailing Torque Fasteners

Threaded fasteners that do not rely solely on the preload to prevent loosening. Systems include local thread deformation, non-matching of mating threads forms, nonmetallic inserts, or high friction coatings. See also *lock nuts*.

Preweld Interval

In resistance welding, the time elapsing between the end of squeeze time and the beginning of welding current in making spot welds and in projection or upset welding. In flash welding, it is the time during which the material is preheated.

Primary Alloy

An alloy whose major constituent has been refined directly from ore, not recycled scrap metal. Compare with *secondary alloy*.

Primary Alpha

Alpha phase is a titanium alloy crystallographic structure that is retained from the last high-temperature α-β working or heat treatment. The morphology of α is influenced by the prior thermomechanical history.

Primary Creep

The first, or initial, stage of *creep,* or time-dependent deformation. See also the term *creep.*

Primary Crystals

The first type of crystals that separate from a melt during solidification.

Primary Current Distribution

The current distribution in an *electrolytic cell* that is free of *polarization.*

Primary Extinction

In x-ray diffraction, a decrease in intensity of a diffracted x-ray beam caused by perfection of crystal structure extending over such a distance (approximately 1 μm or greater) that interference between multiple reflected beams inside the crystal decreases the intensity of the externally diffracted beam.

Primary Gas

In thermal spraying, the gas constituting the major constituent of the arc gas fed to the gun to produce the plasma.

Primary Leakage

In seals, the leakage of a mechanical seal, with the fluid escaping from the region between the end faces of the primary sealing elements.

Primary Metal

Metal extracted from minerals or ore, or other natural source and free of reclaimed metal scrap. Compare with *secondary metal* and *native metal.*

Primary Mill

A mill for rolling ingots or the rolled products of ingots to blooms, billets, or slabs. This type of mill is often called a *blooming mill* and sometimes called a cogging mill.

Primary Nucleation

The mechanism by which crystallization is initiated in plastics, often by an added nucleation agent.

Primary Passive Potential (Passivation Potential)

The potential corresponding to the maximum active current density (critical anodic current density) of an electrode that exhibits active-passive corrosion behavior.

Primary Recrystallization

A process by which nucleation and growth of a new generation of strain-free grains occur in a matrix that has been plastically deformed. See also *secondary recrystallization.*

Primary Solid Solution

See *solid solution.*

Primary Stress

A directly applied tension or compression stress as opposed to secondary tensile stresses resulting from bending or torsion.

Primary X-Ray

The emergent beam from the x-ray source.

Primer

A coating applied to a surface, prior to the application of an adhesive to improve the performance of the bond. The coating can be a low-viscosity fluid that is typically a 10% solution of the adhesive in an organic solvent, which can wet out the adherend surface to leave a coating over which the adhesive can readily flow. See also the term *adhesive bonding.*

A surfacing material employed in painting or finishing to provide an anchorage or adhesion of the finishing material. A primer maybe colorless, or it may have color. In the latter case, it is sometimes called an **undercoat.** A primer is distinct from the filler coat used on woods to fill the pores and thus economize on the more expensive finish. Primers for industrial or production finishing are of two types: air-dry and baking. The air-dry types have drying-oil vehicle bases and are usually called **paints.** They may be modified with resins. They are not used as extensively as the baking-type primers, which have resin or varnish vehicle basis. These dry chiefly by polymerization. Some primers, known as **flash primers,** are applied by spraying and dry by solvent evaporation within 10 min. In practically all primers, the pigments in part most of the anti-corrosion properties to the primer and, along with the vehicle, determine its compatibility with and adherence to the base metal.

Primer coats of **red lead paint** were formally much used on construction steel to give corrosion resistance, but chromate or phosphate primers are now more common. **Barium potassium chromate,** which gives a pale yellow coating, has good anticorrosion properties for steel, aluminum, and magnesium. Zinc yellow paints may also be used as primer coats on metal. **Zinc chromate** is used as a primer on steel. It dissolves when moisture penetrates the paint, and the dissolved chromate retards the corrosion of the steel. **Zinc phosphate primers** applied to iron and steel give corrosion resistance and improve paint adhesion. **Manganese phosphate** forms a dense, crystalline coating on steel, which acts as a corrosion-resistant base for paint. A mixture of 95% zinc powder and 5% of epoxy resin binder in a solvent gives a gray, metallic finish, and the zinc blocks corrosion by galvanic action. In addition to the pigment, various corrosion inhibitors may be used in primer paints. **Ammonium ferrous phosphate** has a platelike structure, which gives impermeability to the film as well as adds corrosion resistance. It is greenish. A primer is especially required in the finishing of sheet-metal objects that are likely to receive dents or severe service, but it is not usually necessary for castings or roughened surfaces.

For sheet-metal work, baked enamels were formally much used for the primers for lacquer finishes, but synthetic resin primers give good adhesion and are less expensive.

High-performance solvent- or waterborne epoxy primers with chromate corrosion inhibitors have traditionally protected U.S. Navy aircraft aluminum components from the corrosive effects of seawater and high humidity. With chromates implicated as carcinogenic, nontoxic inhibitors have been sought. Thus far, inorganic nonchromated solutions based on permanganate, cobalamine, and cericion as the active corrosion-inhibiting agent have proved only marginally comparable at best in corrosion resistance and paint adhesion. The Navy-developed **Chromium III pretreatment** (SJS6-7) has demonstrated comparable corrosion resistance and paint adhesion in laboratory tests. And **sol-gel formulations** of organic or inorganic polymers, based on the hydrolysis and condensation of metal alkoxides, show promise. Other alternatives are special aliphatic polyurethane topcoats; SJS12, a lead-and chromium-free, high-solids **polyurethane self-priming topcoat** (SPT); and a waterborne polyurethane topcoat.

Waterborne epoxy-based **BR6747 primer,** from Cytec Engineered Materials, matches the long-term corrosion-inhibiting performance of its solvent-based counterpart, **BR127 primer.**

Primes

Metal products, principally sheet and plate, over the highest quality and free from blemishes or other visible imperfections.

Principal Stress (Normal)

The maximum or minimum value of the *normal stress* at a point in a plane considered with respect to all possible orientations of the considered plane. On such principal planes, the shear stress is zero. There are three principal stresses on three mutually perpendicular planes. The state of stress at a point may be (1) uniaxial, a state of stress in which two of the three principal stresses are zero, (2) biaxial, a state of stress in which only one of the three principal stresses is zero, and (3) triaxial, a state of stress in which none of the principal stresses is zero. Multiaxial stress refers to either biaxial or triaxial stress.

Printed Circuit

An electronic circuit produced by printing an electrically conductive pattern, wiring, or components on a supporting dielectric substrate, which may be either rigid or flexible.

Printing

A metallographic method in which a carrier material is saturated with an etchant and pressed against the surface of the specimen. The etchant reacts with one of the phases, and substances form that react with the carrier material, leaving behind a life-size image. Used for exposing particular elements—for example, sulfur (sulfur prints).

Prior Austenite Grain Boundary

The locations, in the phase now existing, of the grain boundaries of the austenite that existed previously.

Prism

A transparent optical element whose entrance and exit apertures are polished plane faces. Using refraction and/or internal reflection, prisms are used to change the direction of propagation of monochromatic light and to disperse polychromatic light into its component wavelengths.

Prismatic Plane

In noncubic crystals, any plane that is parallel to the principal axis (c axis).

Probe Analyzer

An analytical facility on an electron microscope.

Probe Ion

In materials characterization, an ionic species intentionally produced by an ion source and directed onto the specimen surface at a known incident angle and a known energy.

Process Annealing

(1) A heat treatment used to soften metal for further cold working. In ferrous sheet and wire industries, heating to a temperature close to but below the lower limit of the transformation range and subsequently cooling for working. In the nonferrous industries, heating above the recrystallization temperatures at a time and temperature sufficient to permit the desired subsequent cold working. (2) Same as sub-critical annealing of steel. It is so called because it is often applied part way through processing when work hardened steel requires softening with minimum scaling.

Processing Window

In forming of plastics, the range of processing conditions, such as stock (melt) temperature, pressure, shear rate, and so on, within which a particular grade of plastic can be fabricated with optimum or acceptable properties by a particular fabrication process, such as extrusion, injection molding, sheet molding, and so forth. The processing window for a particular plastic can vary significantly with design of the part and the mold, with the fabricating machinery used, and with the severity of the end-use stresses.

Process Metallurgy

The science and technology of winning metals from their ores and purifying metals; sometimes referred to as chemical metallurgy. Its two chief branches are *extractive metallurgy* and *refining* and may be extended to include secondary manufacturing processes such as working and heat treatment.

Process Scrap

See *scrap.*

Process Tolerance

The dimensional variations of a part characteristic of a specific process, once the setup is made.

Proeutectoid Carbide

Primary crystals of cementite formed directly in ferrous alloys from the decomposition of *austenite* exclusive of that cementite resulting from the eutectoid reaction. See also *eutectoid*.

Proeutectoid Ferrite

Primary crystals of *ferrite* formed directly in ferrous alloys from the decomposition of *austenite* exclusive of that ferrite resulting from the eutectoid reaction. See also *eutectoid*.

Proeutectoid Phase

Particles of a phase in ferrous alloys that precipitate during cooling after austenitizing but before the eutectoid transformation takes place. See also *eutectoid*.

Profile (Contour) Rolling

In ring rolling, a process used to produce seamless rolled rings with a predesigned shape on the outside or the inside diameter, requiring less volume of material and less machining to produce finished parts.

Profile Examination

(1) The examination of a component in profile, that is, in silhouette (2) The examination of a surface illuminated by a beam of light impinging at a shallow angle so that the shadows magnify surface irregularities.

Profile Gauge

A device comprising a large number of needles arranged in-plane like the teeth of a comb but held in a spine through which they can slide when pressed firmly. The edge formed by the needle tips is pressed against the surface of interest so that the needles slide to record the profile.

Profiling

Any operation that produces an irregular contour on a workpiece, for which a tracer or template-controlled duplicating equipment usually is employed.

Progression

In metal-forming equipment, the constant dimension between adjacent stations in a progressive die.

Progressive Aging

Aging by increasing the temperature in steps or continuously during the aging cycle. See also *aging* and compare with *interrupted aging* and *step aging*.

Progressive Block Sequence

A welding sequence during which successive blocks are completed progressively along the joint, either from one end to the other or from the center of the joint toward either end. See also the term *block sequence*.

Progressive Die

A *die* with two or more stations arranged in line for performing two or more operations on a part; one operation is usually performed at each station.

Progressive Forming

See cultural forming at consecutive stations with a single die or separate dies.

Progressive Fracture

Cracking that extends slowly such as fatigue as opposed to fast brittle fracture.

Projected Area

The area of a cavity, or portion of a cavity, in a mold or die casting and the die measured from the projection on a plane that is normal to the direction of the mold or die opening.

Projection Lens

The final lens in the electron microscope corresponding to an ocular or projector in a compound optical microscope. This lens forms a real image on the viewing screen or photographic film.

Projection Welding

A resistance welding process that produces coalescence of metals with the heat obtained from resistance to electric current through the work parts held together under pressure by electrodes. The resulting welds are localized at predetermined points by projections, embossments, or intersections. The projections are usually crushed flat during welding. See also the term *nugget*.

Promoter

A chemical, itself a feeble catalyst, that greatly increases the activity of a given catalyst. Used in formulating plastics. See also *accelerator*.

Proof

(1) To test a component or system at its peak operating load or pressure. (2) Any reproduction of a *die impression* in any material; often a lead or plaster cast. (3) The test product of a die to confirm accuracy and quality. See also *die proof*.

Proof Load

A load, in excess of the maximum designed service load, applied to a component or structure to prove fitness for service. The amount, ratio, or percentage by which the proof load exceeds the maximum service load may be referred to as the *safety factor*.

Proof Pressure

The test pressure that pressurized components must sustain without detrimental deformation or damage. The proof pressure test is used to give evidence of satisfactory workmanship and material quality.

Proof Stress

(1) A specified stress to be applied to a member or structure to indicate its ability to withstand service loads. (2) The stress that will cause a specified small *permanent set* in a material. (3) The stress required to induce a specified percentage permanent strain. See *tensile test* and compare with *proof load.*

Proof Test

The test carried out on a component or assembly to confirm that it is capable of performing to the design intent. The term usually implies mechanical testing, see *proof load,* but can include electrical and other tests.

Propagation

In a metallurgical context, often the growth of a crack.

Proportional Limit

The greatest stress a material is capable of developing without a deviation from straight-line proportionality between stress and strain. See also *elastic limit* and *Hooke's law.* See also *tensile test.*

Propylene Plastics

Plastics based on polymers of propylene or copolymers of propylene with other monomers, the propylene being the greatest amount by mass. See also *polypropylenes (PP).*

Prosthesis

A manufactured component for insertion into the human body to replace or support a defective item.

Protective Atmosphere

(1) A gas envelope surrounding the part to be brazed, welded, or thermal sprayed, with the gas composition controlled with respect to chemical composition, dewpoint, pressure, flow rate, and so forth. Examples are inert gases, combusted fuel gases, hydrogen, and vacuum. (2) The atmosphere and a heat treating or sintering furnace designed to protect the parts or compacts from oxidation, nitridation, or other contamination from the environment.

Protective Potential

The threshold value of the *corrosion potential* that has to be reached to enter a *protective potential range.*

Protective Potential Range

A range of *corrosion potential* values in which an acceptable corrosion resistance is achieved for a particular purpose.

Protein

A nitrogen organic compound of high molecular weight, from three thousand to many millions. Proteins are made up of complex combinations of simple amino acids, and they occur in all animal and vegetable matter, but are also made synthetically. They form a necessary constituent of foods and feeds, but are also used for many commercial products, but some proteins are highly poisonous. The poison of the cobra and that of the jellyfish are proteins.

Different types of plant and animal life have different types of proteins. At least 10 different proteins are known to be essential to human body growth and maintenance, but many others may have subsidiary functions since the amino acids are selective chelating agents, separating copper, iron, and other elements from the common sodium, calcium, and potassium compounds entering the system. The simple proteins are made up entirely of the acids, but the complex or conjugated proteins also contain carbohydrates and special groups, while the cystine of hair and wool also contain sulfur. The constituent amino acids of the protein molecule are linked together with a peptide bond, and the linkage forms the backbone of the molecule, but the arrangement is not similar to the high polymers usually associated with plastics with one type of polymer, or group, repeating itself. The linkage is formed by the loss of carbon dioxide rather than by the loss of water, as in plastics.

The simplest proteins are the **protamines** with molecular weights down to about 3000. They are strongly basic, are water-soluble, and contain no sulfur. **Clupeine** in herring and **salmine** in salmon and trout are examples. The **histones** that occur in white blood corpuscles contain sulfur and are more complex. The albumens of eggs and milk are soluble in water and coagulate with heat. They also occur in plants, as in the **leucosin** of wheat. **Prolamines** are **vegetable proteins,** as the zein of corn and the gliadin of wheat. They are not an adequate human protein food without **animal proteins.**

Proteins plastics are produced by the isolation of precipitation of proteins from animal or vegetable products and hardening or condensing into stable compounds that can be molded into sheet or fiber. The oldest of the protein plastics is **casein plastic,** which was used to replace bone for toothbrush handles and buttons. It is still an important molding plastic, as it is tough and can be made in pastel shades, but it is more costly than many other molding plastics. The proteins from soybean meal or other vegetable products are condensed with aldehydes or with various mineral salts or acids to form plastics. These plastics are distinct from those made from the fatty acids of soybean or other oils.

Proton

See *atom.*

Prototype

A model suitable for use in the complete evaluation of form, design, performance, and material processing.

Protuberances

See *asperities.*

Prow Formation

See *wedge formation.*

PS

See *polystyrenes.*

P.S. Proof Stress

See *tensile test*.

Pseudobinary System

(1) A three-component or ternary alloy system in which an intermediate phase acts as a component. (2) A vertical section through a ternary diagram.

Pseudocarburizing

See *blank carburizing*.

Pseudonitriding

See *blank nitriding*.

Pseudoplastic Behavior

A decrease in viscosity with increasing shear stress. Compare with *dilatant, rheopectic material,* and *thixotropy*.

PSU

See *polysulfones*.

P-T Diagram

A two-dimensional graph of phase relationships in a system of any order by means of the pressure and temperature variables.

P.T.F.E.-Polytetrafluoroethylene

A thermoplastic having a relatively high melting point, good mechanical properties, good resistance to chemical attack and very low coefficient of friction. Above about 300°C it can produce fluorine and if melted may produce hydrofluoric acid.

P-T-X Diagram

A three-dimensional graph of phase relationships in a binary system by means of the pressure, temperature, and concentration variables.

Puckering

Wrinkling or buckling in a drawn shell in an area originally inside the drawer ring.

Puckers

Local areas on prepreg material where the material has blistered and pulled away from the separator film or release paper. See also *prepreg*.

Puddle

See preferred term *molten weld pool*.

Pull Cracks

In a casting, cracks that are caused by residual stresses produced during cooling, and that result from the shape of the object.

Pulled Compact

A powder metallurgy compact expanded by internal gas pressure.

Pulled Surface

In laminated plastics, imperfections in the surface, ranging from a slight breaking or lifting in localized areas to pronounced separation of the surface from the body.

Pullover Mill

A non-reversing rolling mill through which the material is passed a number of times. Between passes the material is pulled over the top of the mill.

Pulp Molding

The process by which a resin-impregnated pulp material is preformed by application of a vacuum and subsequent molding or curing.

Pulsation Welding

Sometimes used as a synonym for *multiple-impulse welding*.

Pulse

A current of controlled duration through a resistance welding circuit.

Pulsed Arc Welding

Various electric arc welding techniques in which repeated pulses of high current are superimposed on a steady, lesser background current. Any arc welding process in which the power is cyclically varied to give short-duration pulses of either voltage or current that are significantly different from the average value. In M.I.G. (metal inert gas) welding, the background current provides the main heat source and the pulses control the mode of metal transfer from feed wire to weld pool. In T.I.G. (tungsten inert gas) welding, the background current maintains arc continuity and the pulses control the heat input into the weld zone.

Pulse Time

The duration of a pulse during resistance welding.

Pultrusion

A continuous process for manufacturing composites that have a constant cross-sectional shape. The process consists of pulling a fiber-reinforcing material through a resin impregnation bath and then through a shaping die, where the resin is subsequently cured.

Pulverization

The process of reducing metal powder particle sizes by mechanical means; also called comminution or mechanical disintegration.

Pumice

A porous, frothlike volcanic glass, which did not crystallize due to rapid cooling, and frothed with the sudden release of dissolved gases. Powder or ground pumice is used as an abrasive for fine polishing, in metal polishes, in scouring compounds and soaps, and in plaster and light weight concrete and pozzolanic cement. In very fine powder it is called **pounce,** when used for preparing parchment and tracing cloth. **Pouncing paper** is paper coated with pumice used for pouncing, or polishing felt hats. Pumice is grayish white, and the fine powder will float on the surface of water. The natural lump pumice contains 65–75% silica, 12–15 alumina, and 4–5 each of soda and potash.

Pumping Efficiency

In a bearing, the ratio of actual oil flow to the maximum theoretical flow for a bearing with a 180° oil film operating at an eccentricity ratio of unity.

Punch

(1) The male part of a die – as distinguished from the female part, which is called the die. The punch is usually the upper member of the complete die assembly and is mounted on the *slide* or in a *die set* for alignment (except in the inverted die). (2) In double-action draw dies, the punch is the inner portion of the upper die, which is mounted on the plunger (inner slide) and does the drawing. (3) The act of piercing or punching a hole. Also referred to as *punching.* (4) The movable tool that forces material into the die and powder molding and most metal-forming operations. (5) The movable die in a trimming press or a forging machine. (6) The tool that forces the stock through the die in rod and tube extrusion and forms the internal surface in can or cup extrusion.

Punching

(1) The die shearing of a closed contour in which the sheared out sheet metal part is scrap. (2) Producing a hole by die shearing, in which the shape of the hole is controlled by the shape of the punch and its mating die. Multiple punching of small holes is called *perforating.* See also *piercing.*

Punch Press

(1) In general, any mechanical press. (2) In particular, and end wheel gap-frame press with a fixed bed, used in piercing.

Punch Radius

The radius on the end of the punch that first contacts the work, sometimes called *nose radius.* See also the term *press.*

Punch-To-Die Clearance

See *die clearance.*

PUR

See *polyurethanes.*

Purge

The removal of air from a sintering furnace chamber by replacing it with a vacuum or an inert gas prior to the introduction of the sintering atmosphere.

Purnell Process

Similar to *austempering.*

Purple Plague

A failure mechanism in electronic components that involves the formation of brittle intermetallic compounds at aluminum wire/gold bonding pad intersections. Both moisture and temperature enhance the formation of such compounds. The term "purple plague" is used because one of the five compounds that can form appears purple to the eye when viewed through a microscope.

Push Angle

The angle between a welding electrode and a line normal to the face of the weld when the electrode is pointing forward along the weld joint. See the term *forehand welding.*

Push Bench

Equipment used for drawing moderately heavy-gage tubes by cupping sheet metal and forcing it through a die by pressure exerted against the inside bottom of the cup. See *tube manufacture.*

Pusher Furnace

A type of continuous furnace in which parts to be heated are periodically charged into the furnace in containers that are pushed along the hearth against a line of previously charged containers, thus advancing the containers toward the discharge end of the furnace, where they are removed. Pusher furnaces are widely used for heat treating and sintering of metals.

Pusher-Type Seal

A mechanical seal in which the secondary seal is pushed along the shaft or sleeve to compensate for face wear.

Push Fit

A tight fit without slack but which can be made by hand. It lies between interference fit and sliding fit. Similar to *snug fit.*

Push-Pull

Any system of loading where the load alternates between tension and compression and acts axially rather than in bending or torsion.

Push Welding

Spot or projection welding in which the force is applied manually to one electrode, and the work or backing plates take the place of the other electrode.

Putty

A mixture of calcium carbonate with linseed oil, with sometimes white lead added. It is used for cementing window glass in place and as a filler for patterns. Litharge is often added to putty for steel sash. Another putty for steel contains red lead, calcium carbonate, and linseed oil. The dry pigment for putty, **whiting putty,** according to ASTM specifications, contains 95% calcium carbonate and 5 tinting, pigment. **White lead putty** contains 10% or more white lead mixed with the calcium carbonate. Putty powder is a mixture of lead and tin oxides, or a mixture of tin oxide and oxalic acid, or it may be merely an impure form of tin oxide. It is used in the enameling, for polishing stone and glass, and as a mild abrasive for dental polishes. **Caulking putty,** used for setting window and door frames, is made of asbestos fibers, pigments, and drying oils, or with rubber or resins. **Caulking compounds** are now composed of synthetics with usually a polysulfide rubber and a lead peroxide curing agent. They are heavy pastes of 75% to 95% solids. **Koplac 1251-5** for automobile body patching and general utility putties is a low-viscosity polyester resin with 32% styrene. In putty formulations it is used with 50% talc and a catalyst. It hardens rapidly at room temperature.

PVAC. See *polyvinyl acetate.*
PVAL. See *polyvinyl alcohol.*
PVB. See *polyvinyl butyral.*
PVC. See *polyvinyl chloride.*
PVDC. See *polyvinylidene chloride.*
PVDF. See *polyvinylidene fluoride.*
PVF. See *polyvinyl formal.*

PV Factor

The product of bearing pressure and surface velocity traditionally expressed in terms of $(lb \sqrt{in^2}) \times ft/min$; (the ISO equivalent is Pa·m/s).

Pyramidal Plane

In noncubic crystals, any plane that intersects all three axes.

Pyrites

Iron pyrites, iron sulfide, FeS, "Fools Gold."

Pyrolysis

With respect to fibers, the thermal process by which organic precursor fiber materials, such as rayon, polyacrylonitrile (PAN), and pitch, are chemically changed into carbon fiber by the action of heat in an inert atmosphere. Pyrolysis temperatures can range from 800 to 2800°C (1470–5070°F), depending on the precursor. Higher processing graphitization temperatures of 1900–3000°C (3450–5430°F) generally lead to higher-modulus carbon fibers, usually referred to as graphite fibers. During the pyrolytic process, molecules containing oxygen, hydrogen, and nitrogen are driven from the precursor fiber, leaving continuous chains of carbon. See also *carbon fiber* and *carbon–carbon composites.*

Pyrolytic Graphite

Pyrolytic graphite is a high-purity form of carbon produced by thermal decomposition of carbonaceous gases.

The commercial manufacture of pyrolytic graphite products is a relatively new division of the graphite industry. Although this material has been known for some 50 years (glance coal, deposited carbon in gas retorts, etc.), it is only within the last 30 years that the necessary production techniques have been developed.

The manufacturing process essentially consists of bringing relatively cold, carbonaceous gases into contact with a heated surface (mandrel) and thus extracting the carbon, in the form of graphite, directly from the gas. This process is not to be confused with the pyrolysis of resins and pitches, which is involved in the manufacture of the more common bulk or polycrystalline graphites.

Pyrolytic Materials

Essentially, pyrolytic deposition (literally, deposition by thermal decomposition) is a form of the so-called gas or vapor plating. Gas or vapor plating can be accomplished by (1) hydrogen reduction, (2) displacement, or (3) thermal decomposition. Pyrolytic deposition is accomplished by the last mechanism.

The process involves passing the vapors of a compound over a surface maintained at a temperature above the decomposition temperature of the compound in a vacuum furnace. The surface provides a source for nucleation of the desired material, which is built up to the desired section thickness.

Elemental materials are deposited from single-compound vapors, for example, carbon, from a hydrocarbon; metal, from its halide; compound materials, from mixtures of compounds, for example, boron nitride from boron halide and ammonia.

In producing coatings, the substrate serves as the surface on which the coating is deposited. Introducing self-supporting parts, the substrate serves as a mandrel or mold from which the pyrolytic material is removed after deposition.

The pyrolytic deposition process is used to produce pyrolytic graphite, as well as coatings or self-supporting structures of an extremely broad range of materials. Theoretically, the only limitation on the type of material that can be produced is that (1) it must be available in the form of a compound whose vaporization temperature is below its decomposition temperature, and (2) the desired material must separate cleanly from the vapor of the compound.

General Properties

The major benefits of pyrolytic materials are as follows:

1. Highly directional properties are obtained in some materials by the substantial degree of orientation of crystals or grains, (*Note:* Highly directional properties are only obtained in materials such as pyrolytic graphite and boron nitride, which possess the unique anisotropic graphite crystal structure).
2. High densities, equivalent to theoretical densities, are obtainable.
3. High purity of material and close control of ingredients in "alloys" are obtainable by control of the reactant gases.

All initial work has been aimed at producing high-temperature materials, primarily for aerospace use. The unique properties of these materials make them attractive for a number of commercial applications.

Materials and Forms

To date, pyrolytic graphite has been the largest-volume material produced, but newer materials include the following:

1. Graphite-boron compounds: pyrolytic graphite to which less than 2% boron has been added for increased strength and oxidation resistance as well as lower electrical resistivity.
2. Other graphite compounds: pyrolytic graphite to which varying percentages of columbium, molybdenum, or tungsten have been added.
3. Boron nitride: pyrolytic boron nitride containing 50 atm% of boron and nitrogen.

4. Carbides: pyrolytic carbides of tantalum, columbium, hafnium, and zirconium.
5. Tungsten: pyrolytic tungsten has been produced in the form of coatings and parts such as crucibles and tubes.

Coatings can be deposited on complex surfaces, providing gas impermeability in extremely thin sections. On the other hand, only those surfaces of the shape that can be exposed to the flow of gases will be coated. Also, differences in coefficients of thermal expansion between substrate and coating materials must be carefully considered.

Self-supporting shapes are limited because the material must be produced by deposition on a mandrel that must be removed after the part is formed.

Directionality Depends on Crystal Structure

The directional properties of the produced part or coating depend on the inherent crystal structure of the material. In general, materials are either highly anisotropic or nearly isotropic.

Pyrolytic deposition of material with hexagonal graphite crystal structures (i.e., pyrolytic graphite and boron nitride) results in preferred crystal orientation producing a high degree of directionality of properties. The hexagonal or layer plane alignment of the grains is essentially parallel to the substrate surface. Directionality results from strong atomic bonds within layer planes and weak bonds between layer planes, and also from the mode of heat transfer through the material, which is predominantly by lattice vibration.

Pyrolytic deposition of materials with face-centered or body-centered cubic structures (i.e., carbides and tungsten) results in relatively isotropic properties. Although such materials can have a high degree of crystal orientation, orientation does not necessarily produce directional properties.

Properties of Anisotropic Materials

Pyrolytic graphite and its compounds and boron nitride have substantial directionality of properties. The ratio of the number of crystallites with layer planes parallel to the deposition surface (i.e., α axis) to the number normal to the surface (i.e., c axis) can be varied by process control. For example, in graphite, ratios may range from 100 to 1000 to 1, compared with ratios of 3 or 4 for some commercial graphites. Orientation obtained in boron nitride has been as high as 1900 to 1.

Conductivity

One of the most useful properties of pyrolytic graphite is its insulating ability. In the direction normal to the deposition surface, pyrolytic graphite is a better insulator than the most refractory ceramic materials. In addition, the thermal conductivity parallel to the deposition surface is comparable to the more *conductive* metals, tungsten and copper. This high conductivity evens out hotspots over the total surface.

The high destruction temperature of pyrolytic graphite combined with low conductivity normal to the surface allows a surface temperature to become very high. This cuts down heat absorbed by the component by reradiating heat back to the atmosphere, thus acting as a "hyperinsulator."

Tensile Strength Improved

Tensile strengths in the α axis are orders of magnitude higher than in the c axis. For example, at room temperature, pyrolytic graphite has an α axis average tensile strength of about 95.2 MPa, compared with about 3.5 MPa in the c direction. The graphite-boron compound has room temperature tensile strengths of 112.2 and 4.9 MPa in the α and c axes; boron nitride has α – and c-axis tensile strengths of 84 and 4.5 MPa, respectively.

Pyrolytic graphite (like conventional graphite) offers a singular advantage of increasing in strength with increasing temperature. Preliminary data indicate that the c-axis strength decreases with temperature.

Addition of alloying elements has been found to improve c-axis strength. Additions of low concentration of less than 1% tungsten and molybdenum have increased c-axis tensile strength by 50% to 90%.

Oxidation Resistance Improved

Pyrolytic graphite has somewhat greater oxidation resistance than normal graphite due largely to its imperviousness. Addition of boron improves oxidation resistance of pyrolytic graphite by a factor of 1. Oxidation resistance of boron nitride is superior to other pyrolytic materials produced to date specifically at temperatures below 2000°C.

Other Properties

Owing to the atom-by-atom deposition process, pyrolytic materials are all near theoretical density, are impervious, and have extremely high purity levels. The materials do exhibit substantial directionality of thermal expansion, which must be carefully considered in designing components.

Properties of Isotropic Materials

Performance data on the isotropic pyrolytic materials, tungsten, and the carbides of tantalum, hafnium, and columbium are much more limited than those for the anisotropic pyrolytic materials.

As mentioned before, although such materials are considered to be isotropic (in comparison with the anisotropic materials), crystal growth does tend to provide a preferred orientation in a plane normal to the deposition surface. Thus, generally speaking, the strength of such materials is greater in the plane normal to the surface than in a plane parallel to the surface.

Carbides

The pyrolytic deposition process can be controlled to produce carbides of varying metal-to-carbon ratio, resulting in carbides of differing microhardness. The following are Knoop hardness number (K_{100}) for three carbides, hardness increasing with increasing carbon-to-metal ratio:

TaC 1400–3500
HfC 2000–3600
NbC 1700–4000

The very hard grades of carbide are extremely brittle and difficult to handle. Their strength is low but they promise to be useful as thin, well-bonded coatings.

As hardness decreases, the ductility and strength of the carbides increase. Bend strengths, except for the high-hardness grade, were found to fall in the range of 68–272 MPa; a value as high as 1030 MPa was observed for tantalum carbide low in carbon.

Tungsten

Pyrolytic tungsten, which can now be produced in thickness over 6.4 mm as coatings on components to 457 mm in diameter, has been evaluated for missile application. The increased strength and lower impurity level of pyrolytic tungsten is believed to be a major factor permitting the production of sound coatings on large rocket nozzles.

Pyrometallurgy

High-temperature *winning* or *refining* of metals.

Pyrometer

A device for measuring temperatures above the range of liquid thermometers.

Pyrometry

The measurement of temperatures, for example, by measuring the electrical resistance of wire, the thermoelectric force of a couple, the expansion of solids, liquids, or gases, the specific heat of solids, or the intensity of radiant energy per unit area.

The Measurement of High Temperature

The term usually implies temperatures above about 300°C which is about the limit of a simple mercury-in-glass thermometer although more sophisticated mercury/special glass devices can operate at higher levels. A wide variety of devices are available. The simplest technique is merely to judge the color by eye but refinements of the technique utilize standard color charts or a series of comparison filters. Such techniques are useful from about 550°C (in dull light) to about 1200°C although the results are operator sensitive. Optical pyrometers superimpose a heated filament on a background of the light from the source being measured; the calibrated filament is then heated until it just disappears, hence the description *Disappearing Filament Pyrometer*. They are useful from about 700°C to about 2000°C. Radiation pyrometer focuses the radiation on a calibrated thermocouple and has no theoretical upper limit. Electrical resistance pyrometers measure the change in resistance of a wire element, commonly platinum, that is exposed to the environment in question. They have a maximum operating temperature of about 1000°C. Thermocouples utilize the thermoelectric effects associated with pairs of metals in contact at their high and low temperature functions and have an operating range beyond 2500°C. Pyrometric cones ("Seger" cones) are tall combs of selected mineral compounds that commence to collapse at the tip at specific temperatures up to about 200°C. A series with a range of collapse temperatures is loaded into a furnace with the charge and their progressive collapse observed.

Pyrophoric Alloys

Metals that produce sparks when struck by steel, used chiefly for gas and cigarette lighters. The original pyrophoric alloy, or **sparking metal**, was known as **Auer metal**. A very durable alloy for cigarette lighters is **zirconium-lead** alloy containing 50% of each metal. Titanium can replace part of the zirconium, and tin can replace part of the lead, but alloys with less than 25% zirconium are not pyrophoric. The 50-50 alloy has a crystalline structure. Some metals are pyrophoric. Trimethyl aluminum, $Al(CH_3)_3$, a colorless alloy made by sodium reduction of methyl aluminum chloride, is used as pyrophoric fuel.

Pyrophoricity

The property of a substance with a large surface area to self-ignite and burn when exposed to oxygen or air.

Pyrophoric Powder

A powder whose particle self-ignite and burn upon exposure to oxygen or air. Example: fine zirconium powder.

P-X Diagram

A two-dimensional graph of the isothermal phase relationships in a binary system; the coordinates of the graph are pressure and concentration.

P-X Projection

A two-dimensional graph of the phase relationships in a binary system produced by making an orthographic projection of the phase boundaries of a *P-T-X diagram* upon a pressure-concentration plane.

Q

Quadrivariant Equilibrium

In metals, a stable state among several conjugate phases equal to two less than the number of components, that is, having four degrees of freedom.

Qualification Test

A series of tests conducted by the procuring activity, or an agent thereof, to determine the conformance of materials, or materials system, to the requirements of a specification, normally resulting in a qualified products list under the specification. Generally, qualification under a specification requires a conformance to all tests in the specification, or it may be limited to conformance to a specific type of class, or both, under the specification. Compare with *acceptance test* and *preproduction test.*

Qualified Products List (QPL)

A list of commercial plastic products that have been pretested and found to meet the requirements of a specification, especially a government classification.

Qualitative Analysis

An analysis in which some or all of the components of a sample are identified. Compare with *quantitative analysis.*

Quality

(1) The totality of features and characteristics of a product or service that bear on its ability to satisfy a given need (fitness-for-use concept of quality). (2) Degree of excellence of a product or service (comparative concept). Often determined subjectively by comparison against an ideal standard or against similar products or services available from other sources. (3) A quantitative evaluation of the features and characteristics of a product or service (quantitative concept).

Quality Assurance

Systems and procedures in the widest sense for measuring that appropriate quality levels are identified, working practices defined, records maintained, and product quality monitored. They can be considered as the system established in advance to ensure that problems will not arise.

Quality Characteristics

Any dimension, mechanical property, physical property, functional characteristic, or appearance characteristic that can be used as a basis for measuring the quality of a unit of product or service.

Quality Control Systems

Include testing procedures and physical measurements, for ensuring that established quality standards are met. They can be considered as the system of checking that confirms that a product is of intended quality.

Quality Factor

(1) In plastics, the ratio of elastic modulus to loss modulus, measured in tension, compression, flexure, or shear. This is a nondimensional term and is the reciprocal of tan delta. (2) The reciprocal of the electrical dissipation factor, which is the ratio of the power loss in a dielectric material to the total power transmitted through it, thus the imperfection of the dielectric. Equal to the tangent of the loss angle or to the ratio of the dielectric loss to the dielectric constant.

Quality Heat Treatment

The final heat treatment intended to induce the properties required by the component in service. Previous heat treatments could have intended, for example, to produce homogenization, grain refinement, or softening between stages of cold working.

Quantitative Analysis

A measurement in which the amount of one or more components of a sample is determined. Contrast with *qualitative analysis.*

Quantitative Metallography

Determination of specific characteristics of a microstructure by quantitative measurements on micrographs or metallographic images. Quantities so measured include volume concentration of phases, grain size, particle size, mean free path between like particles or secondary phases, and surface-area-to-volume ratio of microconstituents, particles, or grains. See also *image analysis.*

Quantometer/Quantograph Analytical Instruments

They are spectrographs in which an arc is struck between an electrode and a sample of the material to be identified, sometimes in a vacuum. Either the light emitted is presented as a spectrum that is then displayed on a photographic plate to be measured on associated equipment or, alternatively, selected lines of the spectrum are measured within the instrument and the results printed. In the latter case, analysis of an item is a matter of seconds provided that it falls within the standardized range of the instrument.

Quantum

The minimum quantity or increment of physical energy.

Quantum Mechanics

(1) The theoretical concepts necessary to describe and explain activities and reactions on an atomic scale in the same way as Newton's laws and other physical and mechanical laws deal with such matters on a bulk scale. (2) The modern theory of matter, of electromagnetic radiation, and of the interaction between matter and radiation; also, the mechanics of phenomena to which this theory may be applied. Quantum mechanics, also termed wave mechanics, generalizes and supersedes the older classical mechanics and Maxwell's electromagnetic theory.

Quantum Number

One of the quantities, usually discrete with integer or half-integer values, needed to characterize a quantum state of a physical system.

Quantum Theory

See *quantum mechanics*.

Quarter Hard

A *temper* of nonferrous alloys and some ferrous alloys characterized by tensile strength about midway between that of *dead soft* and *half hard* tempers.

Quartering

A method of sampling a metal powder by dividing a cone-shaped heap into four parts, selecting one of them randomly, dividing this again into four parts, and repeating the procedure until the sample is small enough for particle size analysis.

Quartz

One of the several crystalline forms of silica (SiO_2); others include cristobalite and tridymite. All occur as minerals, also synthetic. Quartz is harder than most minerals, being 7 on the Mohs scale, and the crushed material is used as an abrasive. Fused quartz or quartz glass is used for light bulbs, optical glass, crucibles, and tubes and rods in furnaces. Typical values for the maximum operating temperatures for quartz glass and competing glasses are as follows:

Glass	Temperature	
	°C	°F
Borosilicate	500	930
Aluminosilicate glass	650	1200
Doped quartz glass	800	1470
Vycor	900	1650
Quartz glass	1100	2010

The most common variety of silica. It occurs mostly in grains or in masses of a white or gray color, but often colored by impurities. Pure crystalline quartz is colorless and is called *rock crystal*. Quartz usually crystallizes in hexagonal prisms or pyramids. Many crystals are obtained from modules, called geodes, which are rounded, hollow rocks with the crystals grown on the inside surface of the cavity. These rocks range in size from very small to hundreds of pounds. The crystals are not always quartz but may be grown from minor constituents of the rocks. A geode in limestone usually has a shell of silica, and the interior crystals are of quartz or calcite, but some geodes contain crystals of gem quality containing metal-coloring constituents. The hard, rigid *beta quartz crystals* have a latticelike molecular structure in which each silicon atom is linked to four separate oxygen atoms, each oxygen atoms being linked to two different silicon atoms. The grains in sand are often less than 0.04 in. (0.10 cm), but crystals up to 20 in. (51 cm) have been found. The specific gravity is 2.65.

Quartz crystals have the property of generating an electric force when placed under pressure and, conversely, of changing dimensions when electric field is applied. This property is termed *piezoelectric*. A *piezoelectric crystal* is made up of molecules that lack both centers and planes of symmetry. Many materials other than quartz have this property, such as Rochelle salt and ammonium dihydrogen phosphate, but most are water soluble or lack hardness. The best quartz crystals are hexagonal prisms with three large and three small cap faces. For electric use, the crystals must have no bubbles, cracks, or floors, and they should be free from twinning or change in the atomic plane. Piezoelectric quartz is used in microsensors that monitor emissions of spraying solvents to clean printed circuit boards.

Owing to which particular refractive powers, quartz crystal is also employed for the plates in polarization instruments and in lenses. Quartz crystals for radio-frequency control are marketed in three forms: rough-sawed blanks, cut to specified angles; semifinished blanks, machine-lapped to approximate size; and electrically finished blanks, finished by hand and electrically tested. *Synthetic quartz crystals* of large size and high purity and uniformity are grown from seed crystals suspended in an alkaline silica solution at high temperature and pressure. The synthetic crystals are purified by imposing an electric current across the crystal at 932°F (500°C), which sweeps out the sodium lamp and lithium impurities by electrolysis. The addition of lithium nitride to the sodium hydroxide solution used in the hydrothermal growing process increases the *Q* value of the crystal to the range of natural quartz crystals. The synthetic crystals are used in precision oscillators and highly selective wave filters.

Barium titanate crystals are used to replace quartz for electronic use.

Quartz is harder than most minerals, and the crushed material is much used for abrasive purposes. Finely ground quartz is also used as a filler, and powdered quartz is employed as a flux in melting metals. When quartz is fused, it loses its crystalline structure and becomes a *silica glass* with a specific gravity of 2.2. The chemical formula of this material is sometimes given as SiO_3 but is really SiO_2 repeated in a lattice structure but different from that of quartz crystal. *Fused quartz*, or *quartz glass*, withstands rapid changes of temperature without breaking. Fused quartz made from rock crystal is transparent to visible light, while fused silica is normally transparent or opaque. *Vitreosil* is fused quartz, containing 99.8% silica. It is opaque, translucent, and transparent. It transmits ultraviolet and short wavelengths, has high electrical resistance, and has a coefficient of expansion about 1/17 that of ordinary glass. *Quartz tubing* for electronic use comes in round, square, hexagonal, and other shapes. Tubing as small as 0.003 in. (0.008 cm) is flexible and as strong as steel.

Quartz fiber originally was made by extruding the molten quartz through a stream of high-pressure hot air that reduces a fluffy mass of fine fibers of random lengths. Quartz fibers are now made with many differing compositions and methods of manufacture. Fibers used for wall or mat have a diameter of 39–591 µin. (1–15 µm). *Astroquartz* and Astroquartz II fiber are 99.95% fused silica.

The specific gravity is 2.2, tensile strength 6,000 MPa, elastic modulus 69,000 MPa, dielectric strength 3.78, and thermal expansion over zero. It is an excellent electrical insulator. It is insoluble in water, nonhygroscopic, and resistant to halogens and most common acids but not hydrofluoric or hot phosphoric. It should not be used in strong alkali concentrations. It is used in high-temperature composites, radome and antenna applications, high-speed printed circuit boards, and insulation blankets.

Quartz yarn made from these filaments is used for weaving into tape and fabric. *Quartz paper*, or *ceramic paper*, developed by the Naval Surface Weapons Center and used to replace mica for electrical insulation, is made from quartz fiber by mixing with bentonite and sheeting on a papermaking machine. *Microquartz* is felted, fine quartz fibers for insulation.

Since quartz crystallizes more slowly than many other minerals, the natural crystals may include other minerals that were crystallized previously.

Quartzite is a rock composed of quartz grains cemented together by silica. It is firm and compact and brakes with uneven, splintery fractures. Most of the quartzites used are made up of angular grains of quartz and are white or light in color with a glistening appearance. It often resembles marble, but is harder and does not effervesce in acid. Quartzite is employed for making silica brick, abrasives, and siliceous linings for tube mills. It is also rather widely used as a structural stone and as a broken stone for roads. It is a rock and widely distributed.

Quasi-Binary System

In a ternary or higher-order system, a linear composition series between two substances, each of which exhibits congruent melting, wherein all equilibria, at all temperatures or pressures, involve only phases having compositions occurring in the linear series, so that the series may be represented as a binary on a *phase diagram*.

Quasi-Cleavage Fracture

A fracture mode that combines the characteristics of *cleavage fracture* and *dimple fracture*. An intermediate type of fracture found in certain high-strength metals.

Quasi-Hydrodynamic Lubrication

A loosely defined regime of lubrication, especially in metal working, where thin-film lubrication predominates.

Quasi-Isotropic

See *isotropic*.

Quasi-Isotropic Laminate

A laminate approximating isotropy by orientation of plies in several or more directions.

Quaternary System

The complex series of compositions produced by mixing four components in all proportions.

Quench (Plastics)

A process of shock-cooling thermoplastic materials from the molten state.

Quench

Rapidly cool in various media. Common media include, in increasing order of severity, still air, air blast, oil, water, and brine. The process is perhaps most commonly associated with a hardening of steel, but it is also encountered in other circumstances such as the preliminary steps of precipitation hardening, or it may be utilized to avoid undue oxidation at the high-temperature or to induce scale to spall off.

Quench-Age Embrittlement

Embrittlement of low-carbon steels resulting from precipitation of solute carbon at existing dislocations and from precipitation hardening of the steel caused by differences in the solid solubility of carbon in ferrite at different temperatures. Quench-age embrittlement usually is caused by rapid cooling of the steel from temperatures slightly below Ac_1 (the temperature at which austenite begins to form) and can be minimized by quenching from lower temperatures.

Quench Aging

Aging induced by rapid cooling after *solution heat treatment*. See *precipitation hardening*.

Quench Annealing

Annealing an austenitic ferrous alloy by *solution heat treatment* followed by rapid quenching.

Quench Cracking

(1) Fracture of a metal during quenching from elevated temperature. Most frequently observed in a hard-carbon steel, alloy steel, or tool steel parts of high hardness and low toughness. Cracks often emanate from fillets, holes, corners, or other stress raisers and result from high stresses due to the volume changes accompanying transformation to martensite. (2) Cracks occurring during or immediately following quenching as a result of stresses induced by uneven thermal contraction and/or dimensional changes associated with phase changes.

Quench Hardening

(1) Hardening suitable for alpha–beta alloys (most often certain copper to titanium alloys) by solution treating and quenching to develop a martensitic-like structure. (2) In ferrous alloys, hardening by austenitizing and then cooling at a rate such that a substantial amount of austenite transforms to martensite.

Quenching

Rapid cooling of metals (often steels) from a suitable elevated temperature. This generally is accomplished by immersion in water, oil, polymer solution, or salt, although forced air is sometimes used. See also *brine quenching, caustic quenching, direct quenching, fog quenching, forced-air quenching, hot quenching, intense quenching, interrupted quenching, oil quenching, press quenching, selective quenching, spray quenching, time quenching*, and *water quenching*.

Quenching Crack

A crack formed in a metal as a result of thermal stresses produced by rapid cooling from a high temperature.

Quenching Oil

Oil used for quenching metals during a heat-treating operation. See the term *oil quenching*.

Quench Time

In resistance welding, the time from the finish of the welding operation to the beginning of tempering. Also called chill time.

Quicking

The deposition of a loosely adherent mercury film on copper alloy components by dipping in a bath of, or swabbing with, various mercury-containing solutions. Normally, no electrical current is applied. The mercury inhibits the initial nonelectrolytic, nonadherent deposition of silver when the component is introduced to the silver plating bath. The liquid mercury film can cause liquid metal attack at grain boundaries if high levels of residual stress are present, and hence, variations in the process are sometimes used as a test to confirm freedom from such stresses.

Quicksilver

Mercury (historical).

Quill

(1) A hollow tubular shaft, designed to slide or revolve, carrying a rotating member within itself. (2) Removable spindle projection for supporting a cutting tool or grinding wheel.

Q Value

A quality factor of a magnetic core material, also called energy factor or coiled magnification factor. It represents the ratio of the reactance of a coil to its series resistance. Its specific change with the frequency is a function of the type and composition of the magnetic powder used for the core.

R

R

The stress ratio. See *fatigue*.

RA

See *reduction in area*.

Rabbit

A hoelike bladed tool or similar device used for stirring molten metal.

Rabbit Ear

Recess in the corner of a metal forming die to allow for wrinkling or folding of the blank.

Race (or Raceway)

The groove or path in which the rolling elements in a rolling-contact bearing operate. See the term *rolling-element bearing*.

Racking

A term used to describe the placing of metal parts to be heat treated on a rack or tray. This is done to keep parts in a proper position to avoid heat-related distortions and to keep the parts separated. See also *fixturing*.

Rad

The measure of absorbed radiation. 1 rad = 100 ergs/g = 0.01 J/kg.

Radial Crack

Damage produced in brittle materials by a hard, sharp, object pressed onto the surface. The resulting crack shape is semi-elliptical and generally perpendicular to the surface.

Radial Crushing Strength

The relative capacity of a powder metallurgy bearing specimen to resist fracture induced by a force applied between flat parallel planes in a direction perpendicular to the axis of the specimen.

Radial Distribution Function Analysis

An *x-ray diffraction* method that gives the distribution of interatomic distances present in a sample along with information concerning the frequency with which the particular distances occur.

Radial Draw Forming

The forming of sheet metals by the simultaneous application of tangential stretch and radial compression forces. The operation is done gradually by tangential contact with the die member. That type of forming is characterized by very close dimensional control.

Radial Forging

A process using two or more moving anvils or dies for producing shafts with constant or varying diameters along their length or tubes with internal or external variations. Often incorrectly referred to as *rotary forging*. See also the term *mandrel forging*.

Radial Lip Seal

A radial type of seal that features a flexible sealing member, referred to as a lip. The lip is usually of an elastomeric material. It exerts radial sealing pressure on a mating shaft in order to retain fluids and/or exclude foreign matter.

Radial Marks

Lines on a fracture surface that radiate from the fracture origin and are visible to the unaided eye or at low magnification. Radial marks result from the intersection and connection of brittle fractures propagating at different levels. Also known as shear ledges. See also *chevron pattern*.

Radial Rake

The angle between the tooth face and a radial line passing through the cutting edge in a plate perpendicular to the cutter axis.

Radial Roll

The primary driven roll of the rolling mill for rolling rings in the radial pass. The roll is supported at both ends. Also referred to as main roll. See also the terms *axial rolls* and *ring rolling*.

Radial Rolling Force

In *ring rolling*, the action produced by the horizontal pressing force of the rolling mandrel acting against the ring and the main roll.

Radial Runout

For any rotating element, the total variation from true radial position, taken in a plane perpendicular to the axis of rotation. Compare with *axial runout*.

Radial-Load Bearing

A bearing in which the load acts in a radial direction with respect to the axis of rotation.

Radiant Energy

Energy transmitted as *electromagnetic radiation.*

Radiant Heating

Alloy Characteristics, Other Alloys

Traditionally, radiant heating tubes have been fabricated from Ni–Cr–Fe alloys and are welded, cast, or extruded. These materials have the benefit of possessing good heat/strength properties and are very easy to weld and handle.

The protective oxide layer in these alloys consists mainly of Cr_2O_3. This layer is easily displaced revealing a fresh surface of metal which, in turn, oxidizes. However, this results in a thinning of the tube walls and contamination of the tube interior by loose oxides.

The oxide sealing from the tube affects the heat transfer in a negative way, which could affect the function of the recuperator and burner. It might also be exhausted into the furnace atmosphere. The useful life of the tube is also limited through the carburization of the wall or by stress cracking due to reduced strength in the remaining material.

Another common material is mullite, which is used for ceramic tubes. These are especially popular in the straight-through design as the material has a higher temperature capability than Ni–Cr–Fe. However, the ceramic tubes suffer from thermal shock sensitivity. Also, fabrication is more complicated than metallic materials since mullite cannot be welded to fit flanges and end closures.

Silicon carbide (SiC) tubes are increasing in popularity, both in straight-through applications and in single-ended recuperative (SER) systems. The advantage of this material is the possibility of increasing load. However, the limitations are fragility and difficulties of fabrication and handling.

Combinations of metallic and SiC materials in SER systems exploit the benefits of both materials and offer a range of interesting possibilities.

Alloy Characterization

A metallic Fe–Cr–Al-based alloy (APM) with a higher maximum temperature, better loading capabilities, and longer life than Ni–Cr alloys has to be utilized for the production of radiant tubes for gas-fired heating systems. The material offers better ductility and thermal shock resistance than ceramic materials.

APM is a powder metal, dispersion-strengthened ferritic alloy (Fe-21% Cr-5.7% Al), and able to withstand a maximum temperature of 1250°C in tube form. The radiant tubes are seamless and extruded.

This alloy forms a thin adherent Al_2O_3 surface oxide, which provides a good protective layer in most corrosive atmospheres, especially those that have a high carbon potential or that contain sulfur, which is common in heat-treatment processes.

A particular advantage of the Al_2O_3 layer is that it does not scale, and, therefore, there is no contamination of the burner or the product.

The high-temperature creep strength is also improved compared with conventional Fe–Cr–Al alloys. Therefore, the tubes can be used both vertically and horizontally, in straight-through as well as SER applications.

The majority of the tubes in use today are of the SER design, where the burner and exhaust system are placed on the same side of the tube. With this solution, the exhaust gases are used to preheat the gas and air, which is required for combustion resulting in major improvements in system efficiency.

Other Alloys

There is a wide array of resistance material that easily covers the whole range of industrial requirements.

Nikrothal® and other Nickel-based Alloys

Ni–Cr–Fe formulations in wire, ribbon, strip, and foil form for temperatures up to 1200°C.

Cuprothal® Alloys

Copper-based alloys, in wire and ribbon form for temperatures up to 600°C.

Kanthal SUPER®

Molybdenum disulfide elements covering a range of temperatures up to 1900°C. SUPER is ideal for the high-heating power applications and increased furnace capacities, or atmospheres that preclude many metallic elements.

Superthal® Modules

SUPER element mounted in vacuum-formed, ceramic fiber modules provides compact, high density heating power to 1550°C.

Radiant Power (Flux)

The energy emitted by a source or transported in a beam per unit time, expressed, for example, in ergs per second or watts.

Radiant Tube Furnace

A furnace in which combustion of the fuel (oil or gas) occurs in tubes, which passed through the furnace chamber. The charge is heated by radiation but not contaminated by the products of combustion.

Radiant Tubes

Tubes in which fuel is burned for supplying radiant heat to a furnace. Radiant tubes are made from heat-resistant cast or wrought iron–nickel–chromium alloys or wrought nickel–chromium alloys.

Radiation Damage

(1) A general term for the alteration of properties of a material arising from exposure to ionizing radiation (penetrating radiation), such as x-rays, gamma rays, neutrons, heavy-particle radiation, or fission fragments in nuclear fuel material. See also *neutron embrittlement.* (2) Damage usually on a microscopic level but possibly affecting bulk properties, resulting from radiation of any sort. However, unless otherwise qualified the term usually implies radiation by subatomic particles in particular neutrons released by radioactive decay in nuclear reactions. Effects may include embrittlement, increase in hardness, tensile strength, and yield strength and reduction of creep strength.

Radiation Dose

A cumulative exposure to ionizing radiation during a specified period of time.

Radiation Energy

The energy of a given photon or particle in a beam of radiation, often expressed in electron volts.

Radiation Gage

An instrument for measuring the intensity and quantity of ionizing radiation.

Radiation Hardening

A form of radiation damage resulting from the formation of defects in the crystal structure, such as vacant sites or interstitial atoms, which hinders the movement of dislocations.

Radiation Intensity

In general, the quantity of radiant energy at a specified location passing perpendicularly through unit area in unit time. It may be given as number of particles or photons per square centimeter per second, or in energy units as J/m^3 s.

Radiation Monitoring

The continuous or periodic measurement of the intensity of radiation received by personnel or present in any particular area.

Radiation Pyrometer

See *pyrometry*.

Radiation Quality

A term describing roughly the spectrum of radiation produced by a radiation source, with respect to its penetrating power or its suitability for a given application.

Radical

A very reactive chemical intermediate.

Radio Frequency

A frequency at which coherent *electromagnetic radiation* of energy is useful for communication purposes; roughly the range is from 10 kHz to 100 GHz.

Radio Frequency (RF) Preheating

A method of preheating plastic molding materials to facilitate the molding operation and/or reduce molding cycle time. The frequencies most commonly used are those between 10 and 100 MHz.

Radio Frequency Spectrometer

An instrument that measures the intensity of radiation emitted or absorbed by atoms or molecules as a function of frequency at frequencies from 10 kHz to 100 GHz.

Radio Frequency Spectroscopy

The branch of *spectroscopy* concerned with the measurement of the intervals between atomic and molecular energy levels that are separated by frequencies from about 10^5 to 10^9 Hz as compared to the frequencies that separate optical energy levels of about 10×10^{14} Hz.

Radio Frequency Welding

A method of welding thermoplastics using a radio frequency field to apply the necessary heat. Also known as high-frequency welding.

Radio Interference

Undesired conducted or radiated electrical disturbances, including transients, that may interfere with the operation of electrical or electronic communications equipment or other electronic equipment.

Radio Tracing

Same as *radioactive tracing*.

Radioactive

Having an isotope that spontaneously disintegrates embedding particles or radiation.

Radioactive Element

An element that has at least one isotope that undergoes spontaneous nuclear disintegration to emit positive alpha particles, negative beta particles, or gamma rays.

Radioactive Metals

Metallic elements that admit radiations that are capable of penetrating matter opaque to ordinary light. They give out light and appear luminous, also having an effect on photographic plates. The metal radium is the most radioactive of all the natural elements and was much used for luminous paints for the hands of watches and instrument pointers. Because of the mission of dangerous gamma rays, however, it has been replaced but for this purpose by radioactive isotopes of other metals. These isotopes, such as *cobalt 60*, used as a source of gamma rays, and *krypton 85*, for beta rays, are marketed selectively. Radioactive metals are used in medicine for luminous paints, for ionization, for breaking particle bonds in powdering minerals, for polymerization and other chemical reactions, and for various electronic applications.

The metals that are naturally radioactive, such as uranium and thorium, all have high atomic weights. The radiating power is atomic and is unaffected in combinations. *Radium* and other radioactive metals are changing substances. Radium gives out three types of rays; some of the other elements give out only one or two. The measure of the rate of radioactivity is the curie, which is the equivalent of the radioactivity of 0.0022 lb (1 g).

Each radioactive metal has a definite breakdown measured in half-life. *Actinium*, which is *element 89*, has a half-life of 21.7 years. It emits alpha particles to decay to *actinium K*, which is the radioactive isotope of *Francium*, and then it emits beta particles. Radioactive metals breakdown successively into other elements. By comparison of changing atomic weights, it has been deduced that the metal lead is the ultimate product, and uranium the parent metal under existing stability conditions. But heavier metals, now no longer stable under present conditions, have been produced synthetically, notably plutonium. The heavy *element 103* was first produced in 1961 and named *lawrencium* in honor of the inventor of the cyclotron. Not all radiation produces radioactive materials, and by controlled radiation useful elements may be introduced into alloys in a manner not possible by metallurgy. The crystal lattice of an alloy can be expanded, or atoms displaced in the lattice, thus altering the properties of the alloy. In like manner, the molecules of plastics may be cross-linked or otherwise modified by the application of radiation. For example, ethylene bottles may be irradiated after blowing to give higher strength and stiffness. Radioactive isotopes are also used widely in chemistry and in medicine and as sources of electric power.

Radioactive Tracing

The introduction of a radioactive material to another so that the radiation can be monitored to reveal the progress of reactions or movements in the system.

Radioactivity

(1) The property of the nuclei of some isotopes to spontaneously decay (lose energy). Usual mechanisms are emission of α, β, or other particles and splitting (fissioning). Gamma rays are frequently, but not always, given off the process. (2) A particular component from a radioactive source, such as β radioactivity. See also *isotope* and *radioisotope*.

Radioanalysis

An analytical chemistry technique that uses the radiation properties of a radioactive isotope of an element for its detection and quantitative determination. It can be applied to the detection and measurement of natural and artificial radioactive isotopes. See also *isotope* and *radioisotope*.

Radiograph

A photographic shadow image resulting from uneven absorption of penetrating radiation in a test object. See also *radiography*.

Radiography

A method of nondestructive inspection in which a test object is exposed to a beam of x-rays or gamma rays and the resulting shadow image of the object is recorded on photographic film placed behind the object, or displayed on a viewing screen or television monitor (real-time radiography). Internal discontinuities are detected by observing and interpreting variations in the image caused by differences in thickness, density, or absorption within the test object. Variations of radiography include *computed tomography, fluoroscopy*, and *neutron radiography*. See also *real-time radiography* and *nondestructive testing*.

Radioisotope

An isotope that emits ionizing radiation during its spontaneous decay; a radioactive isotope. Also known as *radionuclide*. See also *isotope* and *radio activity*.

Radiology

The general term given to material inspection methods that are based on the differential absorption of penetrating radiation—either electromagnetic radiation of very short wavelength or particulate radiation—by the part or test piece (object) being inspected. Because of differences in density and variations in thickness of the part or differences in absorption characteristics caused by variations in composition, different portions of a testpiece absorb different amounts of penetrating radiation. These variations in the absorption of the penetrating radiation can be monitored by detecting the unabsorbed radiation that passes through the test piece. See also *radiography*.

Radionuclide

See *radioisotope*.

Radionuclide Tracer Element

A radioactive isotope or an element used to study the movement and behavior of atoms by observing the distribution and intensity of radioactivity.

Radium

The best-known radioactive metal, symbol Ra, scattered in minute quantities throughout almost all classes of rocks, but commercially obtainable only from the uranium ores monazite, carnotite, and uraninite. It is a breakdown product, and it disintegrates with a half-life of 1590 years. The metal is white, but it tarnishes rapidly in air. The melting point is about 1292°F (700°C). It was discovered in 1898 by Curie, and the original source was from the pitchblende of the Sudetenland area of Austria after extraction of thorium oxide, but most of the present supply comes from the carnotite of Zaire and from the pitchblende of western Canada. One gram of radium and 7800 lb (3538 kg) of uranium are obtained from 370 tons (336 metric tons) of pitchblende. The ratio of radium to uranium in any uranium ore is about 1:3,000,000. Radium is marketed in the form of bromides or sulfate in tubes and is extremely radioactive in these forms.

In a given interval of time, a definite proportion of the atoms break up with the expulsion of α, β, and γ rays. When an alpha particle is emitted from radium, the atoms from which it is emitted become a new substance, the inert gas *radon*, or *element 86*, with a half-life of 3.82 days. During its short life, it is a definite elemental gas, but it deposits as three isotopes in solid particles, decaying through polonium to lead. Radium is most widely known for its use in therapeutic medicine. It is also used for inspecting metal castings for flaws. *Radium–beryllium powder* is marketed for use as a neutron source.

In the metallic form, its commercial applications usually utilize its radioactive characteristics, for example, radiography and radiotherapy.

Radius

(1) A line segment joining the center of a circle and light point on the circumference or surface of the circle. (2) The length of such a line. (3) To remove the sharp edge or corner of forging stock by means of a radius or form tool.

Radius of Bend

The radius of the cylindrical surface of the pin or mandrel that comes in contact with the inside surface of the bend during bending.

In the case of free or semiguided bends to 180° in which a shim or block is used, the radius of bend is one-half the thickness of the shim or block. See also the term *bend radius*.

Radon

An inert gaseous element produced naturally by the decay of radium. It is radioactive with a half-life of about 4 days.

Rag Burr

The rough edge left on a cut surface.

Rain Erosion

A form of liquid impingement erosion in which the impinging liquid particles are raindrops. This form of the erosion is of particular concern to designers and materials selectors for external surfaces of rotary-wing and fixed-wing aircraft. See also *erosion* (*erosive wear*).

Rake

The angular relationship between the tooth face, or a tangent to the tooth face at a given point, and a given reference plane or line. See also the terms *face mill*, *radial rake*, and *single-point tool*.

Rake (of Tool)

The angle between the approach (usually the top) surface of a tool and a line perpendicular to the component surface at the cutting edge. An angle greater than 90°, termed Positive Rake, usually gives a good surface. An angle less than 90°, termed Negative Rake, allows more rapid metal removal at the expense of surface finish.

Ram

The moving or falling part of a drop hammer or press to which one of the dies is attached; sometimes applied to the upper flat die of a steam hammer. Also referred to as the *slide*. See the terms *drop hammer* and *drop hammer forging*.

Ram Travel

In forming of plastics, the distance the injection ram moves to fill the mold in injection or transfer molding.

Raman Line (Band)

A line (band) that is part of a *Raman spectrum* and corresponds to a characteristic vibrational frequency of the molecule being probed.

Raman Shift

The displacement in wave number of a *Raman line* (*band*) from the wave number of the incident monochromatic beam. Raman shifts are usually expressed in units of cm^{-1}. They correspond to differences between molecular vibrational, rotational, or electronic energy levels.

Raman Spectroscopy

An analysis of the intensity of Raman scattering a monochromatic light as a function of frequency of the scattered light.

Raman Spectrum

The spectrum of the modified frequencies resulting from inelastic scattering when matter is irradiated by a monochromatic beam of radiant energy. Raman spectra normally consist of lines (bands) at frequencies higher and lower than that of the incident monochromatic beam.

Ramie

A fiber use for cordage and for various kinds of coarse fabrics, and found in temperate climates as well as tropical climates. The plants grow in tall, slender stalks like hemp and belong to the nettle family. The bast fibers underneath the bark are used but are more difficult to separate than hemp fiber owing to the insolubility of the adhesive gums. The fibers are eight times stronger than cotton, four times stronger than flax, and nearly three times stronger than hemp. They are fine and white and are as silky as jute. They are not very flexible and are not, in general, suitable for weaving, but their high wet strength, absorbent qualities, and resistance to mildew make the fibers suitable for warp yarns in wool and rayon fabrics. The yarn is used also for strong, wear-resistant canvas for such products as fire hose. The fiber is valued for marine gland packings and for twine. The composition is almost pure cellulose, and the tow and waste are used for making cigarette paper. *China grass* is the hand-cleaned but not degummed fiber. It is stiff and greenish yellow. *Grass cloth* is woven fabric made in China from ramie. *Swatow grass cloth*, imported into the United States, is made of ramie fibers in parallel strands, not twisted into yarns.

Ramming

(1) Packing foundry sand, refractory, or other material into a compact mass. (2) The compacting of molding (foundry) sand in forming a mold.

Random Intermittent Welds

Intermittent welds on one or both sides of a joint in which the weld increments are deposited without regard to spacing.

Random Orientation

A condition of a polycrystalline aggregate in which the orientations of the constituent crystals are completely random relative to each other. Contrast with *preferred orientation*.

Random Pattern

(1) In *filament winding* of composites, a winding with no fixed pattern. If a large number of circuits are required for the pattern to repeat, a random pattern is approached. (2) A winding in which the filaments do not lie in an even pattern.

Random Sequence

A longitudinal welding sequence wherein the weld-bead increments are deposited at random to minimize distortion.

Range

In inspection, the difference between the highest and lowest values of a given *quality characteristic* within a single *sample*.

Range of Stress (S_r)

The algebraic difference between the maximum and minimum stress in one cycle—that is, the range of stress. Stress range, see *fatigue*.

$$S_r = S_{max} - S_{min}$$

Rankine

See *absolute temperature*.

Rapid Prototyping

The terms *agility* and an *agile manufacturing* have been applied to the operation of forward-looking companies. These terms imply a number of characteristics about the organization and operation of a company that provide for quick response to changes in customer needs, and especially the ability to minimize the time between conception and marketing of new models or new products.

One of the most important tools of an agile manufacturing organization is rapid prototyping (RP). Rapid prototyping refers to the production of a part directly from a computer file generated with a computer-aided design program. While rapid prototyping is sometimes considered to include computer numerical control machining; in most cases, it refers to building up a part from a stack of thin layers, each of which is patterned from the computer-generated three-dimensional model of the part—a process that is also called "solid free-form fabrication."

Rapid Tooling

Rapid prototyping is evolving from just a means of making prototypes to a technique for making production tooling, chiefly dies for plastic parts.

Initially, the prototypes were quite fragile, and the various techniques produced products that could only show form. As materials improved the prototypes became stronger, the products could be tried for fit and measured. Today, some systems produce parts strong enough to run briefly in a machine and do low-volume production of parts. The good news is that there are a wide variety of rapid tooling systems; the bad news is no black-and-white answers exist on how to match a project with a system.

There are three levels of rapid tooling. Some users need only a few prototypes. Called soft tooling, these parts can usually be made from room-temperature-vulcanizing materials. The next, "bridge tooling," includes those production situations that require up to several hundred items to cover the time between early prototype and full production. The third type, hard tooling, employees tools for actual production.

Future

Recent developments aim at reducing the amount of postprocessing required for rapidly produced molds. Researchers are also experimenting with new materials such as thermoplastics, composites, and perhaps even cement. Bolstering this work are improvements in underlying rapid prototyping hardware, particularly lasers.

Other developments in rapid tooling focus on processing new materials in existing machines. For example, one experimental rapid-production method showing good results so far marries powder injection-metal feedstocks, from injection-metal molding methods, with a rapid prototyping machine. Powder-injection metals contain a polymer that lubricates and suspends the powder as it is deposited in the rapid prototyping machine.

Other efforts in rapid tooling concern the making of production molds out of ordinary SLA resins. Although SLA epoxies would melt a typical injection-molding temperatures, judicious cooling can let molds last long enough to make a few dozen pads.

Rapid Prototyping Materials

Materials used primarily to rapidly create models of prototype parts using computer-based systems and computer-aided-design data. They are also used to make tools and limited quantities of parts.

Several systems are laser based. Stereolithography, of 3D Systems, Inc., involves sequential curing of a liquid *photopolymer* on a descending platform by an ultraviolet laser beam. Two such polymers are Ciba Geigy's **Cibatool 5170** epoxy for use with helium–cadmium lasers and **Cibatool 5180** for use with argon lasers. **Exactomer** resins, of Allied Signal, use *vinyl ethers* and cationic photoinitiators to start polymerization. DTM Corp. uses a low-power CO_2 laser beam to trace part outlines and a thin layer of powder and to sinter the powder particles. Again, layer upon layer of material was built up on a descending platform until the model is completed. Materials, called **Laserite**, include *investment-casting wax*, *nylon*, and *polycarbonate*. This system can also be used to form *iron–matrix composite* tools. In this case, thermoplastic-coated carbon-steel powder particles are tacked in place and transferred to a furnace where the coating is burned off and the powder sintered to a porous shape, and the shape infiltrated with copper. A similar laser is used by Helisys, Inc., in what is called *laminated object manufacturing*. Here, the laser cuts thin paper or film, dispensed on a roll, to the part outline, and the cut pieces are deposited on a descending table until the model is completed. The materials are heat-activated, adhesive-coated, bleached *kraft paper*, or polyester film in various colors. The paper results in a model resembling wood. The film creates water-resistant models.

In fused deposition modeling, by Stratasys, Inc., a thermoplastic or wax filament from a spool is heated, extruded, and deposited in thin layers onto a base by a robot-held dispensing head. The model is built from the base up. The materials, which include *machinable wax, investment-casting wax*, **Plastic P200** polyolefin, and nylonlike Plastic **P300** and **Plastic P301**, are heated just sufficiently to flow and solidify instantly upon deposition. *Acrylonitrile–butadiene–styrene* parts, having a tensile strength of 21 MPa, and *polyester* parts also have been made. A robotic extrusion system developed by IBM accepts thermoplastic pellets, including an elastomer and a machinable nylonlike material. In solid ground modeling, by Cubital Ltd. (Israel) and Cubital America, successive layers of a liquid photopolymer are exposed and cured by an ultraviolet lamp through a glass mask generated ionographically using toner and representing part cross sections. The exposed resin is then removed, and the cavity is filled with *water-soluble wax*. The layer is cooled and milled to accurate thickness, and the next layer is formed. When all the layers have been built up, the wax is dissolved, leaving the model shape. Direct shell production casting, developed at MIT and marketed by Soligen Technology, Inc., is based on three-dimensional printing. This system bypasses the modeling stage and is used directly to make metal-infiltrated, *ceramic–matrix composite* cores and molds for casting parts. A jet, similar to an ink jet but using a *colloidal silica* binder, sprays successive patterns of the part shape onto *alumina* powder, which is compacted between powder charges. When the final layer is formed, the partially consolidated shape is removed, the binder burned off, and the shape centered and infiltrated with metal. Aluminum and stainless steel parts have been cast using molds and cores made in this way.

Rapid Solidification .

The cooling or quenching of liquid (molten) metals at rates that range from 104°C s⁻¹ to 108°C s⁻¹.

Rapping

Repeated light tapping, for example, to release a pattern from a sand mold or to encourage the flow of dust deposited on the collector plates of an electrostatic precipitator.

Rare Earth Elements/Metals

The rare earths are a closely related group of highly reactive metals comprising about one-sixth of the known elements. The rare earths form a transition series including the elements of atomic number 57–71, all having three outer electrons and differing only in the inner electronic structure. Because chemical properties are determined by the outer electronic structure, it is evident why these metals are chemically alike. Although not truly members of the series, scandium and yttrium (atomic numbers 21 and 39, respectively) are frequently included with this grouping that shows the rare earth elements.

Element	Atomic No.	Element	Atomic No.
Scandium	21	Gadolinium	64
Yttrium	39	Terbium	65
Lanthanum	57	Dysprosium	66
Cerium	58	Holmium	67
Praseodymium	59	Erbium	68
Neodymium	60	Thulium	69
Promethium	61	Ytterbium	70
Samarium	62	Lutetium	71
Europium	63		

Scandium and yttrium occur together in nature with the rare earths and are similar in properties.

The rare earths are neither rare (even the scarcest are more abundant than cadmium or silver) nor earths. The term *earth* stems from the oxide mineral in which these elements were first discovered. More suitable names such as lanthanons (after lanthanum, the first member of the group) have been proposed, but the term *rare earths* persists.

Two Groups

This series of metals is frequently divided into two groups based on atomic weight and chemical properties. The "light" rare earths consist of elements with atomic number 57–63 and may be called the cerium group. The "heavy" rare earths consist of elements 64–71 as well as scandium and yttrium because of similar chemical behavior. The metals of the heavy or yttrium group are more difficult to separate from each other and, as a result, have found commercial interest only recently. Other historical terms will be filed in commercial usage and can lead to confusion. For example, *didymium* is not actually another metal in this series, but it refers to the neodymium-rich rare earth mixture left after lanthanum and cerium are removed.

Properties

The rare earth elements are metals possessing distinct individual properties that make them potentially valuable as alloying agents. They are usually reduced thermally by treating the anhydrous hydride with calcium, lithium, or other alkali metals and then remelting under vacuum to volatilize the last traces of the reductant. They can also be reduced electrolytically from fused-salt baths.

Mechanical

For the various metals of 99.5% purity, the room-temperature ultimate tensile strength ranges from 103 to 272 MPa with the elongations of 5%–25%. Strength at 427°C is about one-half that at room temperature. Therefore, the rare earths do not appear to offer any outstanding mechanical properties that would indicate their use as a base for structural alloys. Scandium has a density similar to aluminum, but the potential for high-strength-to-density materials is unknown. Yttrium provides an interesting combination of properties: density similar to titanium, a melting point of 1549°C, transparency two neutrons, and formation of one of the most stable hydrides.

Fabricability

The rare earths may be hot worked, and some of them can be fabricated cold. Small arc-cast ingots of yttrium have been reduced 95% at room temperature. The metals are poor conductors of electricity. All the metals are paramagnetic, and some are strongly ferromagnetic below room temperature.

Oxidation Resistance

Yttrium and the higher atomic number rare earths maintain a typical metallic appearance at room temperature. Lanthanum, cerium, and europium oxidize rapidly under ordinary atmospheric conditions; the other metals of the light group form a thin oxide film. Some of the rare earths, notably samarium, form a stable protective oxide in air at temperatures up to at least 593°C.

Applications

Applications of the rare earths may be divided into two general categories: the long-established uses and the newer developments that frequently require the higher-purity separated elements. Three of the older applications that still account for three-fourths of total output are rare earth-cored carbons for arc lighting; lighter flints, which are misch metal-iron alloys; and cerium oxide for polishing of glass and also salts for coloring or decolorizing of glass.

Yttrium oxide and cerium sulfide are high-melting refractories. Yttrium-stabilized zirconia ceramic materials with excellent properties are commercially available.

Misch metal is used in making aluminum alloys, and in some steels and irons. In cast iron, it opposes graphitization and produces a malleablized iron. It removes the sulfur and the oxides and completely degasifies steel. In stainless steel, it is used as a precipitation-hardening agent. An important use of misch metal is in magnesium alloys for castings. From 3% to 4% of misch metal is used with 0.2%–0.6% zirconium, both of which refine the grain and give sound castings of complex shapes. The cerium metals also added heat resistance to magnesium castings.

Ceria, cerium oxide, or ceric oxide, CeO_2, is used in coloring ceramics and glass fiber, producing distortion-free optical glass. It is used also for decolorizing crystal glass, but, when the glass contains titania, it produces a canary-yellow color.

Neodymium is used in magnesium alloys to increase strength at elevated temperatures and is used in some glasses to reduce glare. Neodymium glass, containing small amounts of the neodymium oxide, is used for color television filter plates because it transmits 90% of the blue, green, and red light rays and no more than 10% of the yellow. It thus produces truer colors and sharper contrasts in the pictures and decreases the tendency toward gray tones. Neodymium is also a dopant for yttrium–aluminum-garnet, or YAG, lasers as well as for glass lasers.

Lanthanum oxide, La_2O_3, is a white powder used for absorbing gases in vacuum tubes. Lanthanum boride, LaB_6, is a crystalline powder used as an electron emitter for maintaining a constant, active cathode surface. It has high electrical conductivity.

Dysprosium has a corrosion resistance that is higher than that of other cerium metals. It also has good neutron-absorption ability, with a neutron cross section of 1100 barn. The metal is paramagnetic. It is used in nuclear-reactor control rods, in magnetic alloys, and in ferrites for microwave use. It is also used in mercury vapor lamps. With argon gas in the arc area, it balances the color spectrum and gives a higher light output. Samarium has a higher neutron cross section, 5500 barn, and is used for neutron absorption in reactors.

Ytterbium metal is produced in lumps and ingots. Yttrium is more abundant in nature than lead but is difficult to extract. It is found associated with the elements 57–71, although its atomic number is 39. It is the lightest of the cerium metals except scandium. The metal is corrosion resistant to 400°C. It has a hexagonal close-packed crystal structure. Ytterbium oxide, Yb_2O_3, and yttrium oxide, Y_2O_3, are the usual commercial forms of these metals.

Rare earths are also used in the petroleum industry as catalysts.

Another very important use of individual rare earths is in the manufacture of solid-state microwave devices widely used in radar and communications systems. Yttrium–iron garnets are especially good, because they transmit short wave energy with low energy losses. The devices, however, are very small, so the total use of rare earths is not large.

Still another important use of individual rare earths is in the construction of lasers. A high percentage of the patent applications for new lasers involves the use of a rare earth as the active constituent of the laser.

The rare earths show interesting magnetic properties. Alloys of cobalt with the rare earths, such as cobalt–samarium, produce permanent magnets that are far superior to most of the varieties on the market, and many uses for these magnets are developing.

Many phosphors contained rare earths, and barium–phosphate–europium phosphor finds applications in x-ray films that form satisfactory images with only half the exposure time of conventional x-rays.

Yttrium–aluminum-garnets (YAG) are used in the jewelry trade as artificial diamonds. Single crystals of YAG have a very high refractive index, similar to diamond, and when these crystals are cut in the form of diamonds, they sparkle in the same manner as a diamond. Also, they are very hard so that they scratch glass, and only an expert can tell the difference between a YAG diamond and a real one.

Rare Gases

The six inert gases are helium, neon, argon, krypton, xenon, and radon. Sometimes termed the *noble gases.*

Also known as inert gases in the metallurgical industry, and as noble gases, rare gases is a general name applied to the five elements helium, neon, argon, krypton, and xenon. They are rare in that they are highly rarefied gases at ordinary temperatures and are found dissipated in minute quantities in the atmosphere and in some substances. All have zero valence and normally form no chemical combinations. The rare gases are colorless, odorless, and tasteless at ambient temperatures. However, they exhibit very different properties when cooled to extreme low temperatures. When saturated helium, or helium-4, is cooled to 2.17 K, it becomes a superfluid. One unique property of superfluid is the ability to pass undetected through very small openings. Helium-3, the unsaturated counterpart, differs in that it is magnetic.

Types

Neon

Neon is also used in voltage-regulating tubes for radio apparatus and will respond to low voltages. In television, the neon lamp will give fluctuations from full brilliancy to total darkness as many as 100,000 times a second. Colored electric advertising signs are often referred to as neon signs, but the colors other than orange are produced by different gases.

Argon

Argon is obtained by passing atmospheric nitrogen over red-hot magnesium, forming magnesium nitride and free argon. It is also obtained by separation from industrial gases. Argon is employed in incandescent lamps to give increased light and to prevent vaporization of the filament and is used instead of helium for shielding electrodes in arc welding and as an inert blanket for nuclear fuels.

Krypton

Krypton, which occurs in the air to the extent of 1 part in 1 million, is a heavy gas used as a filler for fluorescent lamps to decrease filament evaporation and heat loss and to permit higher temperatures in the lamp. Krypton-85, obtained from atomic reactions, is a beta-ray emitter used in luminous paints for activating phosphors and also as a source of radiation.

Xenon

Xenon, another gas occurring in the air to the extent of 1 part in 11 million, is the heaviest of the rare gases. When atomic reactors are operated at high power, xenon tends to build up as a reaction product, poisoning the fuel and reducing the reactivity. Xenon lamps for military use give a clear white light known as sunlight plus north-sky light. This color does not change with the voltage, and thus the lamps require no voltage regulators. Xenon is a mild anesthetic; the accumulation from air helps to induce natural sleep, but it cannot be used in surgery since the quantity needed produces asphyxiation.

Rare Metals

These metals that are rare in the sense that they are difficult to extract and are rare and expensive commercially. They include the elements astatine, technetium, and francium. The silvery metal technetium, element 43, has been produced by bombardment of molybdenum with neutrons. Although radium is a widely distributed metal, it is classed as a rare metal. All of the ultra-heavy metallic elements, such as plutonium, which are produced synthetically, are classed as rare metals. They are called transuranic metals because they are above the heavy metal uranium in weight. They are all radioactive.

Element 99, called einsteinium, was originally named ekaholmium because it appears to have chemical properties similar to holmium. It is produced by bombarding uranium-238 with stripped nitrogen atoms. It decays rapidly to form the lighter berkelium, or element 97. Neptunium (element 93), californium (element 98), and illinium (element 61) are also made atomically. The latter also has the names florentium and promethium.

Plutonium is made from uranium-238 by absorption of neutrons from recycled fuel. The metal, 99.8% pure, is obtained by reduction of plutonium fluoride, PuF_4, or plutonium chloride, $PuCl_3$. Plutonium-238 has a low radiation level and is used as a heat source for small water-circulating heat exchangers for naval undersea diving suits.

Plutonium-241 emits beta and gamma rays. Because all the allotropic forms are radioactive, it is a pure nuclear fuel in contrast to uranium, which is only 0.7% directly useful for fission. It is thus necessary to dilute plutonium for control. For fuel elements, it may be dispersed in stainless steel and pressed into pellets at about 871°C, or pellets may be made of plutonium carbide. Plutonium–iron alloy, with 9.5% iron, melts at 410°C. It is encased in a tantalum tube for use as a reactor fuel. Plutonium–aluminum alloy is also used.

Element 102, called nobelium, has a half-life of only 12 min. Other transuranic metals produced synthetically are americium element (95) and curium (element 96). Curium is used as a heat source in remote applications. Curium 244 is obtained as curium nitrate in the reprocessing of spent reactor fuel. It is converted to curium oxide. The by-product americium is used as a component in neutron sources. Other transuranic metals that have been produced by nuclear reactions and synthesis include fermium (element 100), mendelevium (element 101), lawrencium (element 103), rutherfordium or kurchatovium element (104), and hahnium or nielsbohrium (element 105).

Ratchet Marks

Lines or markings on a fatigue fracture surface that results from the intersection and connection of fatigue fractures propagating from multiple origins. Ratchet marks are parallel to the overall direction of crack propagation and are visible to the unaided eye or at low magnification. The term is most commonly used in the case of fatigue cracks having multiple origins such as are associated with high stresses and stress concentrations. The term is also used to describe some multi-origin ductile failures such as occur on threaded fasteners torqued until they fail.

Ratcheting

Progressive cycle inelastic deformation (growth, for example) that occurs when a component or structure is subjected to a cyclic secondary stress superimposed on a sustained primary stress. The process is called thermal ratcheting when cyclic strain is induced by cyclic changes in temperature, and isothermal ratcheting when cyclic strain is mechanical in origin (even though accompanied by cyclic changes in temperature). See *shortening* and *thermal fatigue*.

Rate of Flame Propagation

The speed at which a flame travels through a mixture of gases.

Rate of Oil Flow

The rate at which specified oil will pass through a porous sintered powder metallurgy compact under specific test conditions.

Rate of Strain Hardening

Rate of change of *true stress* with respect to *true strain* in the plastic range.

Rating Life

Currently, the fatigue life in millions of revolutions or hours at a given operating speed that 90% of a group of substantially identical rolling-element bearings will survive under a given load. The 90% rating life is frequently referred to as L_{10}-life or B_{10}-life. The rating life in revolutions can be obtained from the following equation:

$$L_{10} = \left[\frac{C}{P}\right]^k \times 10^6$$

where

C is the basic load rating (in lb)
P is the equivalent radial load (in lb)
k is a constant (3 for ball bearings and 10/3 for roller bearings)

For a rating life in hours use

$$L_{10} = \frac{16,700}{N}\left[\frac{C}{P}\right]^k$$

where N is the rotational speed in rev/min.

Rattail

A surface imperfection on a casting, occurring as one or more irregular lines, caused by expansion of sand in the mold. Compare with *buckle*.

Rayleigh Scattering

Scattering *electromagnetic radiation* by independent particles that are smaller than the wavelength of radiation. Contrast with *Compton scattering*.

Rayleigh Step Bearing

A stepped-pad bearing having one step only in each pad.

Rayon

Rayon is a general name for artificial-silk textile fibers or yarns made from cellulose nitrate, cellulose acetate, or cellulose derivatives. In general, the name *rayon* is limited to the viscose, cuprammonium, and acetate fibers, or to fibers having a cellulose base. Other synthetic-fiber groups have their own group names, such as azlon for the protein fibers and nylon for the polymeric amine fibers, in addition to individual trade names.

Viscose rayon is made by treating the cellulose with caustic soda and then with carbon disulfide to form cellulose xanthate, which is dissolved in a weak caustic solution to form the viscose.

Rayons manufactured by the different processes vary both chemically and physically. They are resistant to caustic solutions that would destroy natural silk. They are also mildewproof, durable, and easily cleaned. But they do not have the permeability and soft feel of silk. The acetate rayons are more resistant than the viscose or cuprammonium. Malacca permeability of the fibers is partly overcome by having superfine fibers so that the yarns are permeable.

High-tenacity rayon is produced by stretching the fibers so that the molecular chains run parallel to the filament axis, and a number of small crystalline regions act as anchors for the cellulose chains. Tire cord stretched in this way has greater tensile strength.

Razor Steel

Any steel used for razors or razor blades. Open razors and earlier "safety" blades were usually plain, high-carbon steel (typically 1.5% C), more recently "stainless" grades with about 12%–14% chromium and about 1% carbon have become the norm. All types of steel are used in the hard and likely tempered condition.

R-Curve

In fracture mechanics, a plot of crack-extension resistance as a function of stable crack extension, which is the difference between either the *physical crack size* or the *effective crack size* and the *original crack size*. *R*-curves normally depend on specimen thickness and, for some materials, on temperature and strain rate.

Reactance

That part of the impedance of an alternating current circuit that is due to capacitance or inductance.

Reaction Sintering

The sintering of a metal powder mixture consisting of at least two components that chemically react during the treatment.

Reaction Stress (Welding)

The *residual stress* that could not otherwise exist if the members or parts being welded were isolated as free bodies without connection to other parts of the structure.

Reaction-Injection Molding

Reaction-injection molding (RIM) is a process in which two or more liquid–chemical components from separate tanks are metered through-pressure supply lines to a chamber. They mix in the chamber via high-velocity impingement. The resulting mixture is then injected into a mold at low pressure, about 0.48 MPa, and at low temperature, about 66°C. This leads to a lower tooling cost and less lead time than for most other molding processes.

The process is similar in concept to an ordinary two-part epoxy mix. RIM reactants become a low-viscosity, exothermic, expanding material that readily flows into a mold. The mold-clamping machine typically is equipped with electronics that turn the mold so that material more easily flows into all cavities. When a successful series of movements is perfected, they are stored as a program to ensure uniform molding of the part.

As the mold fills, the reactants are still releasing heat while tiny bubbles swell the plastic into the finest details of the mold. Cooling coils built into the mold transfer heat to maintain an ideal working temperature. The reactants quickly harden, and parts can be removed in a few minutes.

RIM polyurethane has an integral, low-density, cellular core and a solid, high-density skin. RIM structural foam has a high strength-to-weight ratio and excellent chemical resistance. A variety of shapes can be molded this way without molded-in stresses.

The process accommodates rapid thickness variations, without sink marks. Substrates of wood, steel, or other materials can be encapsulated for fillers or for increased strength. Available RIM material systems are varied and range from flexible to rigid to structural composite. In general, RIM serves larger, thick-walled (6.4 mm or more) structural parts.

The process molds polyurethane, epoxy, and other liquid chemical systems. Mixing 2–4 components in the proper chemical ratio is accomplished by a high-pressure impingement-type mixing head, from which the mix material is delivered into the mold at low pressure, where it reacts (cures). See also *injection molding* (*plastics*).

Reactive Diluent

As used in epoxy formulations, a compound containing one or more epoxy groups that functions mainly to reduce the viscosity of the mixture.

Reactive Metal

A metal that readily combines with oxygen at elevated temperatures to form very stable oxides, for example, titanium, zirconium, and beryllium. Reactive metals may also become embrittled by the interstitial absorption of oxygen, hydrogen, and nitrogen.

Reactive Sputtering (of Interference Films)

A method of revealing the microstructure of metals with the aid of physically deposited interference layers (films) that is based on an optical-contrast mechanism without chemical or morphological alteration of the specimen surface. In interference layer microscopy, light that is incident on the deposited film is reflected at the air/layer and layer/specimen interfaces. Phases with different optical constants appear in various degrees of brightness and colors. The color of a phase is determined by its optical constants and by the thickness and optical constants of the interference layer. In this process, the film material is applied by cathode sputtering, transported through a gas chamber containing a reactive gas and deposited on an anodically connected specimen. Oxygen is frequently used as reaction gas, and interaction with the sputtered material leads to deposition of oxidic interference films. See also *sputtering* and *vacuum deposition* (of interference films).

Reactor

Any vessel or container in which a reaction occurs. Where the reaction is chemical, the reactor may be no more than a simple closed vessel, possibly heated and/or pressurized, in which the reaction progresses in isolation from the external environment. In a nuclear context, the term usually implies the containment vessel, everything within it and also directly associated external equipment such as shielding.

Reagent

A substance, chemical, or solution used in the laboratory to detect, measure, or react with other substances, chemicals, or solutions. See also *reagent chemicals*.

High-purity chemicals used for analytical reactions, for testing of new reactions where the effects of impurities are unknown, and for chemical work where impurities must either be absent or at a known concentration.

Real Area of Contact

In *tribology*, the sum of the local areas of contact between two solid surfaces, formed by contacting asperities, which transmit the interfacial force between the two surfaces. Contrast with *apparent area of contact*.

Real-Time Radiography

A method of nondestructive inspection in which a two-dimensional radiographic image can be immediately displayed on a viewing screen or television monitor. This technique does not involve the creation of a latent image; instead, the unabsorbed radiation is converted into an optical or electronic signal, which can be viewed immediately or can be processed in near real time with electronic and video equipment. The principal advantage of real-time radiography over film radiography is the opportunity to manipulate the test piece during radiographic inspection. This capability allows the inspection of internal mechanisms and enhances the detection of cracks and planar defects by manipulating the part to achieve the proper orientation for floor detection. See also *radiography*.

Reamed Extrusion Ingot

A cast hollow extrusion ingot that has been machined to remove the original inside surface.

Reamer

A rotary cutting tool with one or more cutting elements called teeth, used for enlarging a hole to desired size and contour. It is supported principally by the metal around the hole it cuts.

Reaming

An operation in which a previously formed hole is sized and contoured accurately by using a rotary cutting tool (*reamer*) with one or more cutting elements (teeth). The principal support for the reamer during the cutting action is obtained from the workpiece:

1. *Form reaming*: Reaming to a contour shape.
2. *Taper reaming*: Using a special reamer for taper pins.
3. *Hand reaming*: Using a long lead reamer that permits reaming by hand.
4. *Pressure coolant reaming* (*or gun reaming*): Using a multiple-lip, end cutting tools through which coolant is forced at high pressure to flush chips ahead of the tool or back through the flutes for finishing of deep holes.

Recalescence

(1) The increase in temperature that occurs after undercooling, because the rate of liberation of heat during transformation of a material exceeds the rate of dissipation of heat. (2) A phenomenon, associated with the transformation of gamma iron to alpha iron on cooling (supercooling) of iron or steel, that is revealed by the brightening (reglowing) of the metal surface owing to the sudden increase in temperature caused by the fast liberation of the latent heat of transformation. The effect is particularly pronounced in steels of eutectoid composition transforming at the eutectoid temperature. Contrast with *decalescence*.

Recarburize

(1) To increase the carbon content of molten cast iron or steel by adding carbonaceous material, high-carbon pig iron, or a high-carbon alloy. (2) To carburize a metal part to return surface carbon lost in processing; also known as *carbon restoration*.

Recess

A groove or depression in a surface.

Reciprocal Lattice

A lattice of points, each representing a set of planes in the crystal lattice, such that a vector from the original of the reciprocal lattice to any point is normal to the crystal planes represented by that point and has a length that is the reciprocal of the plane spacing.

Reciprocal Linear Dispersion

The derivative $d\lambda/dx$, where λ is the wavelength and x is the distance along the spectrum. The reciprocal linear dispersion usually is expressed in Å/mm. See also *linear dispersion*.

Reciprocating-Screw Injection Molding

A combination injection and plasticating unit in which an extrusion device with a reciprocating screw is used to plasticate the material. The injection of material into a mold can take place by direct extrusion into the mold, by reciprocating the screw as an injection plunger, or by a combination of the two. When the screw serves as an injection plunger, the unit acts as a holding, measuring, and injection chamber. See also *injection molding* (*plastics*).

Reclaim Rinse

A non-flowing rinse used to recover *dragout*.

Recoil Line

See *impact line*.

Reconstruction

The process by which raw digitized detector measurements are transformed into a cross-sectional computed tomography image. See also *computed tomography*.

Recovery

(1) The time-dependent portion of the decrease in strain following unloading of a specimen at the same constant temperature as the initial test. Recovery is equal to the total decrease in strain minus the instantaneous recovery. The recovery is expressed in the same units as *instantaneous recovery*. (2) Reduction or removal of work-hardening effects in metals without motion of large-angle grain boundaries. One example is the case of some severely work-hardened steels in which heating to a modest temperature, say 250°C–350°C, affects an improvement in toughness with little loss of strength. (3) The proportion of the desired component obtained by processing an ore, usually expressed as a percentage. (4) More generally than in the above usages, the return of properties or characteristics, usually

desirable, that have been lost or impaired as a result of some prior treatment including radiation as well as working. The recovery may occur naturally or be induced, for example, by heat treatment. In the case of fatigue, rest periods during testing may improve fatigue life, so such rests are said to provide some recovery.

Recrystallization

(1) The formation of a new, strain-free grain structure from that existing and cold-worked metal, usually accomplished by heating. (2) The change from one crystal structure to another, as occurs on heating or cooling through a critical temperature. (3) A process, usually physical, by which one crystal species is grown at the expense of another or at the expense of others of the same substance but smaller in size. (4) The process of forming in a solid material, a new grain (crystal) structure, usually during heat treatment of a work-hardened material but also when a metal undergoes a phase change in the solid state. Recrystallization commences at individual nuclei, each of which initiates a crystal that grows outward in three dimensions until it meets its neighbor or the surface. The results of grains are equiaxed, free from strain, and the material is normally in a softened state. See also *crystallization*.

Recrystallization Annealing

Annealing cold-worked metal to produce a new grain structure without phase change.

Recrystallization Temperature

(1) The lowest temperature at which the distorted grain structure of a cold-worked metal is replaced by a new, strain-free grain structure during prolonged heating. Time, purity of the metal, and prior deformation are important factors. (2) The approximate minimum temperature at which complete recrystallization of a cold-worked metal occurs within a specified time.

Recrystallized Grain Size

(1) The grain size developed by heating cold-worked metal. The time and temperature are selected so that, although recrystallization is complete, essentially no grain growth occurs. (2) In aluminum and magnesium alloys, the grain size after recrystallization, without regard to grain growth or the recrystallized conditions. See also *recrystallization*.

Recuperator

(1) Equipment for transferring heat from gaseous products of combustion to incoming air or fuel. The incoming material passes through pipes surrounded by a chamber through which the outgoing gases pass. (2) A continuous heat exchanger in which heat is conducted from the products of combustion to incoming air through flue walls.

Recyclate Plastics

Resins made at least partially from recycled thermoplastics. In 1996, about 17% of the 225×10^6 lb (102×10^6 kg) of plastic processes by a major U.S. automaker was recyclate. Polyurethane-bonded foam made from flexible polyurethane scrap is used for carpet underlays, gymnasium mats, auto sound damping and mud guards, particleboards, and, mixed with rubber chips, and adhesive pavements on athletic fields.

Caprolactam, a nylon feedstock, is recoverable from carpets and re-usable in carpets, textiles, and molded parts. Polystyrene foam with as much as 50% recycled polystyrene is used for extruded food trays and containers. Polyethylene film scrap and stretch and shrink wrap, bottles, and trash, dry-cleaning, and merchandizing bags are reprocessed into film products and used as fillers and molded products. *Petra 140*, of Allied Signal, is a 40% glass-reinforced polyethylene terephthalate (PET) made from recycled soda bottles.

Dow Plastics has a method for incorporating as much as 10% process scrap in urethane RIM autobody panels. Miles, Inc., has a process for compression-molding parts from 100% RIM scrap regrind. Polyvinyl chloride recycled from flexible PVC wire and cable is used to mold hidden auto parts that dampen noise and resist water.

Recyclates are widely used for *plastic lumber* or *synthetic wood*. Advantages over natural wood include weight reduction, especially by foaming, and rot resistance. They also resist splintering and color fade.

Polycarbonate (PC)—ABS blends, or alloys, using 25% recyclate are used for personal-computer housings.

Glass fibers used to reinforced plastics are also recyclable. **CSX hybrid fibers**, of Phoenix Fiberglass, are reclaimed from sheet molding compound waste and suitable for mixing with virgin fibers. Deja, Inc., makes footwear from recycled tire rubber, wetsuit trimmings, old file folders, coffee filters, seat cushions, soda bottles, and other materials.

Used Rubber USA makes while it's and belts from old inner tubes and scrap rubber. Michelin Americas can incorporate as much as 10% cured rubber from discarded tires in new ones without affecting their performance. Up to 60% recycled rubber from tires is used in *Symar-T*, a thermoplastic vulcanizate from NRI Industries, which provides performance and processing quality comparable to that of conventional thermoplastic elastomers. Discarded tires are also used to produce methanol, liquid petroleum gas, and gas oil. *EEKO recyclable synthetic leather* is used for handbags.

Red Brass

A series of copper alloys including one wrought alloy (C23000) and several cast alloys (C83300–C84800). Red brass C23000 contains 84%–86% copper and the balance zinc except for small amounts (0.05% maximum) of lead and iron. The cast alloys, which include leaded red brass, have a zinc content that can range from 1.0%–2.5% (C83500) to 13%–17% (C84800), and all but C83400, which is limited to 0.20% tin, contain substantial amounts of this element; 1.0%–6.5%, depending on the alloy. Lead content of the cast alloys can be as little as 0.50% maximum (C83400) or as much as 8.0% (C84400). Some of these alloys also contain about 1% nickel and most contain smaller amounts of other metals.

C23000 is available in most wrought mill forms and is used for condenser and heat-exchanger tubing, plumbing lines, electrical conduit, fasteners, and architectural and ornamental applications. It is quite ductile in the annealed condition and can be appreciably hardened by cold work.

Standard gilding metal (C21000), used for making cheap jewelry and small-arms ammunition, contains 94%–96% copper. It has a golden-red color, is stronger and harder than copper, but has only about half the electrical conductivity.

Red Hardness/Strength

Hardness and strength of a material at red heat. The term usually implies that the material in question has these characteristics to a useful extent.

Red Heat

An imprecise term since perceptions vary with the intensity and color of the background light. In natural dim light, a dark red is readily observed at about 540°C and above about 1000°C with the color becoming much more orange/yellow.

Red Mud

A residue, containing a high percentage of iron oxide, obtained in purifying bauxite in the production of alumina in the *Bayer process*.

Red Short

Susceptible to cracking during hot working because of poor ductility. The same as hot short except that it obviously only applies to materials that are worked at red heat.

Redox Potential

This *potential* of a reversible oxidation–reduction electrode measured with respect to a *reference electrode*, corrected to the hydrogen electrode, in a given *electrode*.

Redrawing

The second and successive deep-drawing operations in which cup-like shells are deepened and reduced in cross-sectional dimensions. See also *deep drawing*.

Reduced Powder

Generic term for any metal or nonmetal powder produced by the reduction of an oxide, hydroxide, carbonate, oxalate, or other compound without melting.

Reducing

(1) A broad term applied to many hot or cold working processes for reducing the cross section of a plate, bar, or tube. (2) An abbreviation for the cold reducing process, a cold forging technique in the final stages of tube making (of tube ends). (3) A reduction in the external and internal diameters without any deliberate change in wall thickness.

Reducing Agent

(1) A compound that causes *reduction*, thereby itself becoming oxidized. (2) A chemical that, at high temperatures, lowers the state of oxidation of other batch chemicals.

Reducing Atmosphere

(1) A furnace atmosphere that tends to remove oxygen from substances or materials placed in the furnace. (2) A chemically active protective atmosphere that at elevated temperature will reduce metal oxides to their metallic state. Reducing atmosphere is a relative term and such an atmosphere may be reducing to one oxide but not to another oxide. (3) An atmosphere or welding gas flame having an excess of fuel relative to oxygen, which can have a reducing or deoxidizing effect on the workpiece.

Reducing Flame

(1) A gas flame produced with excess fuel in the inner flame. (2) A gas flame resulting from combustion of a mixture containing too much fuel or too little air. See the term *neutral flame*.

Reduction

(1) In cupping and deep drawing, a measure of the percentage decrease from blank diameter to cup diameter, or of diameter reduction in redrawing. (2) In forging, rolling, and drawing, either the ratio of the original to final cross-sectional area or the percentage decrease in cross-sectional area. (3) A reaction in which there is a decrease in valence resulting from a gain in electrons. (4) In its chemical sense, the removal, from a metal ore, of oxygen and other elements combined with the metal. (5) In manufacturing, the relationship between the original dimensions and the final dimensions, usually stated as a ratio or a percentage. Contrast with *oxidation*.

Reduction Cell

A pot or tank in which either a water solution of a salt or a fused salt is reduced electrolytically to form free metals or other substances.

Reduction in Area (RA)

(1) The difference between the original cross-sectional area of a tensile specimen and the smallest area at or after fracture as specified for the material undergoing testing. Also known as reduction of area. (2) The amount, usually expressed as a percentage, by which the cross-sectional area of a tensile test specimen is reduced at the point of failure. See *tensile test*.

Reduction of Oxide

The process of converting a metal oxide to metal by applying sufficient heat in the presence of a solid or gaseous material, such as hydrogen, having a greater attraction for the oxygen than does the metal.

Redundancy

The availability of backup systems or additional load bearing features or material to accommodate failure of one component.

Redux Bonding

A proprietary adhesive system.

Redwood Viscosity

A commercial measure of viscosity expressed as the time in seconds required for 50 cm of a fluid to flow through a tube of 10 mm length and 1.5 mm diameter at a given temperature. It is recommended that fundamental viscosity units be used.

Reeding

The operation of forming serrations and corrugations in metals by coining or embossing.

Reel

(1) A spool or hub for coiling or feeding wire or strip. (2) To straighten and planish a round bar by passing it between contoured rolls.

Reel Breaks

Transverse breaks or ridges on successive inner laps of a coil that results from crimping of the lead end of the coil into a gripping segmented mandrel. Also called *reel kinks*.

Re-entrant

A surface feature that points inwards. A re-entrant angle acts as a notch: a re-entrant feature on a casting prevents its removal from the mold unless the mold has special provisions or is expendable.

Reference Electrodes

A nonpolarizable *electrode* with a known and highly reproducible *potential* used for potentiometric and voltammetric analyses. See also *calomel electrode*.

Reference Material

In materials characterization, a material of definite composition that closely resembles in chemical and physical nature the material with which an analyst expects to deal; used for calibration or standardization. See also *standard reference material*.

Reference Stress

A concept whereby the creep life of a component subject to a complex stress pattern is equal to the life of a simple creep test piece tested at the reference stress.

Refined

In a purified form.

Refined Iron

Originally, good-quality wrought iron, see *Puddling process*. The term is now used more vaguely referring either to a good-quality cast iron, particularly if it has a fine grain and flake size, or to high-purity iron.

Refinement (of Grain Size)

Techniques for producing a smaller, that is, finer, grain size.

Refining

The branch of *process metallurgy* dealing with the purification of crude or impure metals. Compare with *extractive metallurgy*.

Reflectance

The ratio of the radiant power or flux reflected by a medium to the radiant power or flux incident on it; generally expressed as a percentage.

Reflection (X-Ray)

See *diffraction*.

Reflection Grating

An optical component that employs reflection off a series of fine, equidistant ridges, rather than transmission through a pattern of slots, to diffract light into its component wavelengths. The gratings used in optical instrumentation are almost exclusively reflection gratings. See also *concave grating*, *diffraction grating*, *plane grating*, and *transmission grating*.

Reflection Method

The technique of producing a diffraction pattern by x-rays or electrons that have been reflected from a specimen surface.

Reflector Sheet

A clad product consisting of a facing layer of high-purity aluminum capable of taking a high polish, for reflecting heat or light, and a base of commercially pure aluminum or an aluminum–manganese alloy, for strength and formability.

Reflow Soldering

(1) A process for joining electronic components by tinning the mating surfaces, placing them together, heating until the solder fuses, and allowing to cool in the joined position. (2) A soldering process variation in which preplaced solder is melted to produce a soldered joint or coated surface. The use of this term is not recommended by the American Welding Society.

Reflowing

(1) The brief melting of a surface, often a dip coating, to improve flatness and brightness. (2) Melting of an electrodeposit followed by solidification. The surface has the appearance and physical characteristics of a hot-dipped surface (especially tin or tin alloy plates). Also called *flow brightening*.

Reflux

In materials characterization, heating a substance at the boiling temperature and returning the condensed vapors to the vessel to be reheated.

Refractive Index

The ratio of the phase velocity of monochromatic light in a vacuum to that in a specified medium. Refractive index is generally a function of wavelength and temperature. Also known as index of refraction.

Refractories, Specialty

These are ceramic materials used in high-temperature structures or equipment. The term *high temperatures* is somewhat indefinite but usually means above about 1000°C, or temperatures at which, because of melting or oxidation, the common metals cannot be used.

In some special high-temperature applications, the so-called refractory metals such as tungsten, molybdenum, niobium, and tantalum are used.

Materials, usually ceramics, are employed where resistance to very high temperature is required, as for furnace linings and metal-melting pots. Materials with a melting point above 1580°C are called refractory, and those with melting points above 1790°C are called highly refractory. However, in addition to the ability to resist softening and deformation at the operating temperatures, other factors are considered in the choice of a refractory, especially load-bearing capacity and resistance to slag attack and spalling. Heat transfer and electrical resistivity are sometimes also important. Many of the refractories are derived directly from natural minerals, but synthetic materials are much used. To manufacture refractory products, powders of the raw materials are mixed and usually dry-pressed to form the desired shape.

The greatest use of refractories is in the steel industry, where they are used for construction of linings of equipment such as blast furnaces, stoves, and open-hearth furnaces. Other important uses of refractories are for cement kilns, glass tanks, and nonferrous metallurgical furnaces, ceramic kilns, steam boilers, and paper plants. Special types of refractories are used in rockets, jets, and nuclear power plants. Many refractory materials, such as aluminum oxide and silicon carbide, are also very hard and are used in abrasives; some applications, for example, aircraft brake linings, use both characteristics.

Refractory materials are commonly grouped into (1) those containing mainly aluminosilicates; (2) those made predominantly of silica; (3) those made of magnesite, dolomite, or chrome ore, termed basic refractories (because of their chemical behavior); and (4) a miscellaneous category usually referred to as special refractories.

Aluminosilicate Refractories

Fireclay is the raw material from which the bulk (about 70%) of refractories is manufactured. Different grades are distinguished according to the softening temperature or the pyrometric cone equivalent (PCE), the number of the standard pyrometric cone that deforms under heat treatment in the same manner as the fireclay. Thus, the minimum PCEs for low, intermediate, high, and super duty fireclays are 19, 29, 31/32, and 33, respectively. Fireclays are also classified by their working properties into two other classes: plastic (those that form a moldable mass when mixed with water) and flint (a hard, rocklike clay that does not become plastic when mixed with water). In general, flint clays have higher PCEs than plastic clays and are mixed with them to form higher-grade fireclay brick.

High-alumina refractories are made from clays that contain, in addition to the alumina (Al_2O_3) in the clay minerals, hydrates of aluminum oxide. These raise the total Al_2O_3 content and make the material more refractory. Different grades are distinguished on the basis of the total Al_2O_3 content (50%, 60%, and 70% alumina refractories).

Sillimanite and kyanite are anhydrous aluminosilicate minerals used to make special refractory objects, such as crucibles, tubes, and muffles, or as an addition to fireclay to increase refractoriness and to control its shrinkage during firing.

Silica Refractories

These account for about 15% of total production. They are made from crushed and ground quartzite (ganister) to which about 2% lime has been added to assistant bonding, both before and after firing. The quality of silica refractories is to a great extent determined by the amount of Al_2O_3 impurity; even small amounts have a deleterious effect on refractoriness. This is just opposite to the case of alumina in fireclays, where a higher alumina content means greater refractoriness. High-grade silica brick contains less than 0.6% Al_2O_3, and even the standard grade contains less than 1%. During firing, the mineral quartz transforms to cristobalite and tridymite, the high-temperature forms of silica.

The outstanding characteristic of silica is its ability to withstand high loads at elevated temperatures, for example, as a sprung-arch roof 9–12 m wide over an open hearth. The hearth may be operated within (50°C) of the melting point of silica.

Basic Refractories

Magnesite refractories are so named because magnesium carbonate mineral was for many years the soul raw material. Since World War II, seawater has become a significant source of magnesium oxide refractory, and such material is often called seawater magnesite. In any case, the raw material was calcined to form a material largely magnesium oxide, MgO; about 5% iron oxide is usually added before calcining.

Chrome refractories are made from chrome ore, a complex mineral containing oxides of chromium, iron, magnesium, aluminum, and other oxides crystallized in the spinel structure. These crystals are usually embedded in a less refractory matrix called gangue.

In an attempt to combine the best properties of each, magnesite and chrome are often mixed to form chrome magnesite or magnesite–chrome refractories (the first-named is the dominant constituent).

Dolomite is a mixed calcium–magnesium carbonate, which, when calcined to a mixture of MgO and CaO, is used in granular form to patch the bottoms of open hearths and also to make bricks.

Miscellaneous Materials

Special refractories are made of a great many materials, and it is possible to mention here only a few of the more important ones.

Silicon carbide, SiC, is used for many refractory shapes; its outstanding properties are good thermal and electrical conductivity (it is used to make electric heating elements for furnaces), good heat-shock resistance, strength at high temperatures, and abrasion resistance. The first silica and carbide refractories were bonded with clay so that the refractory properties of the bond placed the ultimate limit on the material. A method of making self-bonded silicon carbide has been developed to remove this limitation. Although silicon carbide tends to oxidize to form SiO_2 and either CO or CO_2, the silica-oxidation product forms a glassy coating on the remaining material and to a certain extent protects it from further oxidation.

Insulating firebrick is made from refractory clays to which a combustible material (sawdust, cork, coal) has been added; when this burns during the firing operation, it leaves a brick of high porosity.

The low thermal conductivity of insulating brick reduces heat losses from furnaces, and the low bulk density and consequent low heat capacity reduce the amount of heat needed to bring the furnace itself up to temperature. The main disadvantage of such bricks is their low strength, but even this is useful in that they can be cut or ground to shape quite readily.

Pure oxides, of which alumina is the principal example, are used for many special refractories. Zircon and zirconia are finding increased and significant uses as refractory materials. Some, such as beryllia, thoria, and uranium oxide, are of particular interest for nuclear applications.

R

Carbides, nitrites, borides, silicides, and sulfides of various sorts have been considered refractory materials, and some study has been made of them; aside from a few carbides and nitrites, however, none has found much use.

Cermets are an intimate mixture of a metal and a nonmetal, for example, aluminum oxide and chromium.

Manufacture

Standard ceramic techniques are used. Hand molding, once widely used, is used only for special shapes and small orders. The extrusion or stiff mud process is used for plastic fireclays; very often, the extruded blanks are repressed or hydraulically rammed to form special shapes, for example, T-sections of refractory pipe. Power pressing of simple shapes is the most widely used forming method. Hot pressing and hydrostatic pressing are used for some special refractories. Slip casting is used for special refractory shapes. Fusion casting is commonly used for glass tank blocks; these are mainly either aluminum oxide or aluminum oxide with significant amounts of silica, zirconia, or both.

Refractories are generally fired in tunnel kilns, but some periodic kilns are still used, particularly for special shapes.

Some types of basic refractories, known as chemically bonded, are pressed with a chemical binder, such as magnesium oxychloride, and installed without firing. Some of these, the steel-clad refractories, are encased in a metal sheath at the time of pressing. When the refractory is heated after installation, the iron oxidizes and reacts with the refractory, forming a tight bond between the individual bricks.

In all refractory products and in unfired brick in particular, the maximum possible form density is desired. To this end, careful crushing and sizing of raw materials are carried out so that, as far as possible, the gaps between large pieces are filled with smaller particles, and the space between these with still smaller, and so on. In the case of clay refractories, it is customary to use prefired (calcined) clay or crushed, fired rejects (both are known as grog) to increase the density and to reduce the firing shrinkage.

Properties

A high melting point is, of course, necessary in a refractory, but many other properties must be considered in choosing a refractory for a specific application.

A definite melting point is characteristic of pure materials; however, actual minerals from which refractories are made, for example, clay, are far from pure and hence do not melt at a specific temperature. Rather, they form increasing amounts of liquid as the temperature is increased above a certain minimum temperature at which liquid first appears. This characteristic of gradual softening is indicated by the PCE of the material and the underload test.

High-temperature strength is important for refractories, but most materials become plastic and flow at elevated temperatures. Therefore, the rate of flow (creep rate) at a given temperature under a given load is a more important design criterion.

Thermal conductivity determines the amount of heat that will flow through a furnace wall under given conditions, and a knowledge of this property is essential to furnace design.

Thermal-shock resistance is the ability of a specimen to withstand, without cracking, a difference in temperature between one part and another. For example, if a red-hot brick is dropped into cold water, it is likely to shatter since the outside cools and contracts while the center is still hot. This cracking is often referred to as thermal spalling, the term *spalling* meaning any cracking off of large pieces of brick. Other causes of spalling are mechanical (hitting the brick and knocking off a piece) and structural (a reaction in the brick that changes the mineral structure and causes cracking). Thermal-shock resistance is enhanced by high-strength, low Young's modulus, low thermal expansion, and sometimes, depending on conditions, high thermal conductivity. Whether or not a given specimen cracks under heat shock depends not only on the material of which it is made, but also on its size and shape and on the test conditions, for example, whether it is dropped into water or into still air at the same temperature.

Various chemical properties are important in refractories. For example, the tendency of the magnesium oxide in basic brick to hydrate, that is, to react with water to form $Mg(OH)_2$, should be as low as possible. Turning to high-temperature chemistry, the rate of corrosion of refractories by molten slags and iron oxide fumes is vital to the length of service.

Carbon deposition is another chemical reaction that affects the life of refractories. The reaction is not with the refractory but is catalyzed by substances in it. When carbon monoxide, perhaps in the top of a blast furnace, comes in contact with certain iron compounds, which can occur in fireclays, its reduction to carbon is catalyzed. This carbon deposits at the site of the catalyst in the brick and causes the brick to shatter.

Refractoriness

In *refractories*, the capability of maintaining a desired degree of chemical and physical identities at high temperatures and in the environment and conditions of use.

Refractory

(1) A material (usually an inorganic, nonmetallic, ceramic material) a very high melting point with properties that make it suitable for such uses as furnace linings and kiln construction. (2) The quality of resisting heat. (3) Historically, the term implied mineral and ceramic materials used in applications such as furnace linings, but modern usage includes any material that does not react with its environment and has useful levels of strength for useful periods of time at high temperatures. See also *refractories*.

Refractory Alloy

(1) A heat-resistant alloy. (2) An alloy having at extremely high melting point. See also *refractory metals*. (3) An alloy difficult to work at elevated temperatures.

Refractory Brick

A refractory shape, the most widely used size being approximately $23 \times 11.3 \times 6.4$ (or 7.6) cm ($9 \times 4\ 7/16 \times 2.5$ (or 3) in.). These rectangular units are often referred to as straight brick to differentiate them from brick shapes having related overall dimensions, such as arch, key, and wedge brick, which have some interfacial angles that depart from 90°.

Refractory Cement

A large proportion of the commercial refractory cements used for furnace and oven linings and for fillers are fireclay–silica–ganister mixtures with a refractory range of 1427°C–1538°C.

Cheaper varieties may be mixtures of fireclay and crushed brick, fireclay and sodium silicate, or fireclay and silica sand. An important class of refractory cements is made of silicon carbide grains or silicon carbide-fire sand with clay bonds or synthetic mineral bonds. The temperature range of the ceramics is 1482°C–1871°C. Silicon carbide cements are acid resistant and have high thermal and electric conductivity. For crucible furnaces, the silicon carbide cements are widely used except for molten iron. Alumina and alumina–silicate cements are very refractory and have high thermal conductivity. Calcined kaolin, diaspore clay, mullite, sillimanite, and combinations of these make cements that are neutral to most slags and to metal attacks. They are electrical insulators. Chrome–ore cements are difficult to bond unless mixed with with magnesite. A chrome–magnesite cement is made of treated magnesite and high-grade chrome ore. It sets quickly and forms a hard, dense structure. The melting point is above 1982°C. It is used particularly for hot repairs in open-hearth furnaces.

Zircon–magnesite cement is made with 25% refined zircon sand, 10% milled zircon, 15% fused magnesia, and 50% low-iron deadburned magnesite bonded with sodium silicate. A wide range of refractory cements of varying compositions and characteristics are sold under trade names, and these are usually selected by their rated temperature resistance.

Carbofrax cement is silicon carbide with a small amount of binder in various grades for temperatures from 871°C to 1760°C, depending on the fineness and the bond. Firefrax cement is an aluminum silicate, sometimes used in mixtures with ganister for aligning furnaces. It is for temperatures up to 1649°C. Alfrax cement is fused silica also and various grades for temperatures from 899°C to 1816°C.

Refractory Hard Metals

Refractory hard metals (RHMs) are a ceramic-like class of materials made from metal-carbide particles bonded together by a metal matrix. Often classified as ceramics and sometimes called cemented or sintered carbides, these metals were developed for extreme hardness and wear resistance.

These are true chemical compounds of two or more metals in the form of crystals of very high melting point and high hardness. Because of their ceramic-like nature, they are often classified as ceramics, and they do not include the hard metallic carbides, some of which, with metal binders, have similar uses; nor do they include the hard cermets. The refractory hard metals may be single large crystals, or crystalline powder bonded to itself by recrystallization under heat and pressure. In general, parts made from them do not have binders or contain only a small percentage of stabilizing binder. The intermetallic compounds, or intermetals, are marketed regularly as powders of particle size from 150 to 325 mesh for pressing into mechanical parts or for plasma-arc deposition as refractory coatings, and the powders are referred to chemically, such as the borides, beryllides, and silicides. The oxides and carbides of the metals are also used for sintering and for coatings, and the oxides are called cermets.

The RHMs are more ductile and have better thermal shock resistance and impact resistance than ceramics, but they have lower compressive strength at high temperatures and lower operating temperatures than most ceramics. Generally, properties of RHMs are between those of conventional metals and ceramics. Parts are made by conventional powder-metallurgy compacting and sintering methods.

Many metal carbides such as SiC and BC are not RHMs but are true ceramics. The fine distinction is in particle binding: RHMs are always bonded together by a metal matrix, whereas ceramic particles are self-bonded. Some ceramics have a second metal phase, but the metal is not used primarily for bonding.

RHM System

Tungsten Carbide

Tungsten carbide with a 3%–20% matrix of cobalt is the most common structural RHM. The low-cobalt grades are used for applications requiring wear resistance; the high-cobalt grades serve where impact resistance is required.

Tantalum Carbide/Tungsten Carbide

Tantalum carbide and tungsten carbide combined in a matrix of nickel, cobalt, and/or chromium provide an RHM formulation especially suited for a combination of corrosion and wear resistance. Some grades are almost as corrosion resistant as platinum. Nozzles, orifice plates, and valve components are typical uses.

Titanium Carbide

Titanium carbide in a molybdenum and nickel matrix is formulated for high-temperature service. Tensile compressive strengths, hardness, and oxidation resistance are high at 1093°C. Critical parts for welding and thermal metalworking tools, valves, seals, and high-temperature gauging equipment are made from grades of this RHM.

Tungsten–Titanium Carbide

Tungsten–titanium carbide ($WTiC_2$) in cobalt is used primarily for metal-forming applications such as draw dies, tube-sizing mandrels, burnishing rolls, and flaring tools. The $WTiC_2$ is a gall-resistant phase in the RHM containing tungsten carbide as well as cobalt.

Borides

Zirconium boride is a microcrystalline gray powder of the composition ZrB_2. When compressed and sintered to a specific gravity of about 5.3, it has a Rockwell A hardness of 90, a melting point of 2980°C, and a tensile strength of 241–276 MPa. It is resistant to nitric and hydrochloric acids, to molten aluminum and silicon, and to oxidation. At 1204°C, it has a transverse rupture strengths of 379 MPa. It is used for crucibles and for rocket nozzles.

Chromium boride occurs as very hard crystalline powder in several phases: the CrB orthorhombic crystal, the hexagonal crystal Cr_2B, and the tetragonal crystal Cr_3B_2. Chromium boride parts produced by powder metallurgy have a specific gravity of 6.20–7.31, with a Rockwell A hardness of 77–88. They have good resistance to oxidation at high temperatures, are stable to strong acids, and have high heat-shock resistance up to 1316°C. The transverse rupture strength is from 552 to 931 MPa. CrB is used for oil-well drilling. A sintered material, used for gas-turbine blades, contains 85% CrB with 15% nickel binder. It has a Rockwell A hardness of 87 and a transverse rupture strength of 848 MPa.

Molybdenum boride, Mo_2B, has a specific gravity of 9.3, a Knoop hardness of 1660, and a melting point of about 1660°C. Tungsten boride, W_2B, has a specific gravity of 16.7 and a melting point of 2770°C. Titanium boride, TiB_2, is light in weight with a specific gravity of 4.5. It has a melting point of about 2593°C. Molded parts made from the powder have a Knoop hardness of 3300 and a flexural strength of 241 MPa, and they are resistant to oxidation up to 982°C with a very low oxidation rate above that point to about 1371°C. They are inert to molten aluminum. Intermetal powders of beryllium–tantalum, beryllium–zirconium, and beryllium–columbium are also marketed, and they are light weight and have high strength. Sintered parts resist oxidation to 1649°C.

Disilicides

Molybdenum disilicide, $MoSi_2$, has a crystalline structure in tetragonal prisms and has a Knoop hardness of 1240. The decomposition point is above 1870°C. It can be produced by sintering molybdenum and silicon powders or by growing single crystals from an arc melt. The specific gravity of the single crystal is 6.24. The tensile strength of sintered parts is 276 MPa, and compressive strength is 2296 MPa. The resistivity is 29 $\mu\Omega$ cm. It is used in rod form for heating elements in furnaces. The material is brittle but can be bent to shape at temperatures above 1093°C.

Kanthal Super is an $MoSi_2$ rod and an inert atmosphere, the operating temperature is 1600°C. Furnace gases containing active oxygen raise the operating temperature to about 1700°C, while gases containing active hydrogen lower it to about 1350°C.

Tungsten disilicide, WSi_2, is not as hard and not as resistant to oxidation at high temperatures, but has a higher melting point, 2050°C.

Other (Nitrides, Aluminides)

Titanium nitride, TiN, is a light brown powder with a cubic lattice crystal structure. Sintered bars are extremely hard and brittle, with a hardness above Mohs 9 and a melting point of 2950°C. It is not attacked by nitric, sulfuric, or hydrochloric acid and is resistant to oxidation at high temperatures. In recent years, titanium nitride coatings have been used to markedly extend the life of tool-steel cutters and forming tools. The coatings, golden in color, are deposited by chemical or physical vapor deposition. Aluminum nitride, AlN, when molded into shapes and sintered, forms a dense, nonporous structure with a hardness of Mohs 6. It resists the action of molten iron or silicon to 1704°C, and molten aluminum to 1427°C, but is attacked by oxygen and carbon dioxide at 760°C. At least 1% oxygen causes AlN properties to deteriorate rapidly. However, if oxygen is removed, AlN is highly thermally conductive, lightweight, and a good insulator. A newly developed process has removed oxygen concentration from AlN down to 0.5%, and subsequent nitriding and a self-purification reaction has produced commercial grade AlN.

Nickel aluminide is a chemical compound of the two metals, and when molded and sintered into shapes, has good oxidation resistance and heat resistance at high temperatures. It is a sintered nickel-aluminide material with a specific gravity of 5.9 and a transverse rupture strength of 1034 MPa at 1093°C, twice that of cobalt-bonded titanium carbide at the same temperature. The melting point is 1649°C. It resists oxidation at 1093°C. It is used for highly stressed parts in high-temperature equipment. This compound in wire form is used for welding, flame coating, and hard surfacing. An aluminum powder coated with nickel may be mixed with zirconia or alumina, which will increase the hardness and heat resistance of the nickel-aluminide coating. Columbium aluminide is used as a refractory coating as it is highly resistant for long periods at 1427°C. Tin aluminide is oxidation resistant to 1093°C, but a liquid phase forms at about this point.

Tribaloy intermetallic materials are composed of various combinations of nickel, cobalt, molybdenum, chromium, and silicon. For example, one composition is 50% nickel, 32% molybdenum, 15% chromium, and 3% silicon. Another is 52% cobalt, 28% molybdenum, 17% chromium, and 3% silicon. Supplied as alloy-metal powder, welding rod, or casting stock, they can be cast, deposited as a hard-facing surface plasma-sprayed, or consolidated by powder metallurgy. The materials have exceptional wear and corrosion resistance properties in corrosive media and in air up to 1093°C. Typical applications are in pumps, valves, bearings, seals, and other parts for chemical process equipment. Also, the materials are suited for marine and saltwater applications and for parts subject to wear in atomic energy plants.

Refractory Metals

Various new techniques are being utilized to produce refractory metals and their alloys. One of these processes is powder injection molding (PIM). Over 250 PIM operations serve a diverse range of applications worldwide; cemented carbides for wear and cutting applications, tungsten or molybdenum heat sinks, niobium rocket nozzles, heavy alloy projectiles and radiation shields, and titanium golf clubs. It is interesting to note the frequent PIM successes involving components with about 100 dimensions, such as a wristwatch case.

PIM

Currently, complex-shaped parts of tungsten–iron–nickel with 93% tungsten are also being produced by PIM. Special efforts are focused on the sintering process to develop high density and keep tight tolerances in the final components.

PAS

Plasma-activated sintering (PAS) as a means to consolidate tungsten powders is another effective process to produce dynamic compression properties.

VPS

Vacuum plasma spray (VPS) forming/sintering as a viable fabrication alternative to the near-net shape manufacturing a fully dense components of W-3.5% Ni-1% Fe, Ta-10% W, Mo-40% Re, and Nb-1% Zr alloys. Because the process allows direct shape-making in vacuum, it results in much lower oxygen and moisture pickup than conventional powder-metallurgy consolidation methods.

LPPS

Low-pressure plasma spray (LPPS) technique for depositing tungsten coatings on metallic, ceramic, and composite substrates is another novel technique.

SANS

Tungsten and molybdenum and their ongoing critical role have dominated lighting technology since the development of potassium-doped tungsten as a special dispersion-strengthened system, in which hardening centers are soft particles in the form of potassium-filled bubbles. Such bubbles act as a barrier against boundary migration and are essential for the mechanical stability of elongated grains in the doped (or non-sag) tungsten wire. A new method, based on small angle neutron scattering (SANS), has been developed to determine shaped distribution and the state of deformation of second-phase dispersoids. Correlation between the results of the SANS measurements and metal graphical examination has proved SANS to be suitable for characterizing the morphology of the bubbles and doped tungsten.

Refrigerants

Gases, or very-low-boiling-point liquids, used for the heat-absorbent cycle in refrigerating machines. The ideal refrigerant, besides having a low boiling point, should be noncorrosive, non-flammable, and nontoxic. It must be free of water since as little as 40 ppm of water may cause freezing in the system. Ammonia is a common refrigerant. Ethyl chloride, methyl ether, carbon dioxide, and various chlorinated and fluorinated hydrocarbons marketed under trade names such as the *Freons* of the Du Pont Co. and the *Genetrons* of Allied Chemical Corp. are also used. *Dichlorodifluoromethane*, a nonflammable, colorless, odorless gas is Freon 12. *Freon E3* is a fluorated hydrocarbon which boils at 306°F (152°C) and is stable to 570°F (299°C). It is pourable at 175°F (79°C). It is used for dielectric insulating and electronic equipment. *Trichloro-monofluoromethane* is used as a refrigerant in industrial systems employing centrifugal compressors, and in indirect expansion-type air-conditioning systems. *Genetron II* is this material. *Genetron 101* has a chlorinated constituent of chlorofluorocarbons that has been implicated in the depletion of stratospheric ozone, and alternatives have been developed that are less environmentally harmful.

Of the chlorofluorocarbons, CFC-11 and CFC-12 have the highest ozone-depletion potential (ODP), with CFC-11 having an ODP of 1.0. *Hydrochlorofluorocarbons* (HCFCs) are also ozone-depleting but much less so and, thus, are viewed as temporary replacements. HCFC-22 for example, has an ODP of 0.055, and HCFC-123 has an ODP of 0.02. *Hydrofluorocarbons* being chlorine-free, have no ozone-depletion potential. A large variety of HFC mixtures are viewed as possible replacements for HCFCs and CFCs.

Regenerator

(1) Same as *recuperator* except that the gaseous products or combustion heat brick checkerwork in a chamber connected to the exhaust side of the furnace while the incoming air and fuel are being heated by the brick checkerwork in a second chamber, connected to the entrance side. At intervals, the gas flow is reversed so that incoming air and fuel contact hot checkerwork while that in the second chamber is being reheated by exhaust gases. (2) A system of heat conservation in which the heat of the process exhaust gas is extracted and returned to the cool ingoing gas. The term usually refers to pairs of honeycomb brick structures, one of which is heated by hot exhaust gas, while at the same time the other has heat extracted from it by the ingoing cold gas stream, the gas flows being alternated from time to time. However, the term may also be applied to other devices such as those with a rotating element passing through the two gas streams.

Regrind

Waste plastic material (such as spruce, runners, excess parison material, and reject parts from injection molding, blow molding, or extrusion operations) that has been reclaimed by shredding or granulating. Regrind is usually mixed with virgin compound, at a predetermined percentage, for remolding.

Regular Reflection

See *specular reflection*.

Regulator

A device for controlling the delivery of welding or cutting gas at some substantially constant pressure.

Regulus

The impure button, globule, or mass of metal formed beneath the slag in the smelting and reduction of ores. The name was first applied by alchemists to metallic antimony because it readily alloyed with gold.

Regulus Metal/Regulus of Antimony

Pure antimony.

Rehbinder Effect

Modification of the mechanical properties at or near the surface of a solid, attributable to interaction with a *surfactant*.

Reheat Cracking

Generally any cracking that develops when a component is heated following cooling from casting or other heating cycle. More specifically, the term refers to cracking in welds when they are heated either during postweld heat treatment or when they enter service. This form of damage is most common in the affected zones of low alloy creep resisting steels such as 0.5% chromium, 0.5% molybdenum, and 0.25% vanadium. Reheating of such weldments causes precipitation of alloy carbides within the grains, which are consequently strong and rigid. The grain boundaries are relatively weak and suffer creep or intergranular hot tensile failure as a result of high levels of residual stress from the welding cycle. Externally imposed stresses exacerbate the problem.

Reheat Furnace

Any furnace in which a part formed item is heated for further processing.

Reheater

In a boiler, the system of tubing in which steam is reheated following its first passage through a steam turbine.

Reinforced Concrete

This is portland cement concrete containing higher-strength, solid materials to improve its structural properties. Generally, steel wires or bars are used for such reinforcement, but for some purposes, glass fibers or chop wires have provided desired results.

Unreinforced concrete cracks under relatively small loads or temperature changes because of low tensile strength. The cracks are unsightly and can cause structural failures. To prevent cracking or to control the size of crack openings, reinforcement is incorporated in the concrete. Reinforcement can also be used to help resist compressive forces or to improve dynamic properties.

Steel usually is used in concrete. It is elastic, yet has considerable reserve strength beyond its elastic limit. Under a specific axial load, it changes in length only about 1/10 as much as concrete. In compression, steel was more than 10 times stronger than concrete, and in tension, more than 100 times stronger.

As reinforcement for concrete, steel may be used in the form of bars or rods, wire, fibers, pipe, or structural shapes such as wide-flange beams.

Bars are the most commonly used form. The size of a bar usually is specified by a number that is about eight times the nominal diameter, sizes range from No. 2, a nominal 0.6 cm bar, to No. 18, a nominal 5.7 cm bar.

During construction, the bars are placed in a form and then concrete from a mixture is cast to embed them. After the concrete has hardened, deformation is resisted and stresses are transferred from concrete to reinforcement by friction and adhesion along the surface of the reinforcement. The bonding may be improved mechanically by giving reinforcing bars raised surfaces, such bars are known as deformed bars. They are generally manufactured to standard specifications such as those of the American Society for Testing and Materials (ASTM). Similarly, the form wires used as reinforcement provide greater bond than smooth wires. The form wires are designed by D followed by a number equal to 100 times the nominal area in square inches.

Individual wires or bars resist stretching and tensile stress and the concrete only in the direction in which such reinforcement extends. Tensile stresses and deformations, however, may occur simultaneously in other directions. Therefore, reinforcement must usually be placed in more than one direction. For this purpose, reinforcement sometimes is assembled as a rectangular grid, with clips or welded joints at the intersections of the wires or bars. For example, prefabricated wire grids, called welded-wire fabric, often are used for slab reinforcement in highway pavement and in buildings.

Under some conditions, fiber-reinforced concrete is an alternative to such arrangements. The fiber, made of fine steel wires, glass fibers, or plastic threads, is embedded in the concrete in short lengths, often only about 2.5 cm. They are added to the concrete mixer along with the other ingredients. To achieve crack control in all directions, the fiber should be uniformly spaced, at close intervals, and randomly oriented throughout the hardened concrete. Such reinforcement also improves concrete tensile strength, ductility, and dynamic properties.

Reinforced Containing Elements

They improve some properties. The term usually refers to materials such as reinforced concrete or reinforced plastics that contain, respectively, steel bars and various fibers to improve their tensile and bending strengths.

Reinforced Molding Compound

A plastic reinforced with special fillers or fibers (glass, synthetic fibers, minerals, and so forth) to meet specific requirements.

Reinforced Plastics

Molded, formed, filament-wound, tape-wrapped, or shaped plastic parts consisting of resins to which reinforcing fibers, mats, fabrics, and so forth, have been added before the forming operation to provide some strength properties greatly superior to those of the base resin. See also *resin–matrix composites*.

Reinforced Polypropylene

Polypropylene that is reinforced with mineral fillers, such as tile, mica, and calcium carbonate, as well as glass and carbon fibers. The maximum concentration usually uses 50 wt.%, although concentrates with higher levels of filler or reinforcement are available.

Reinforced Reaction Injection Molding (RRIM)

A reaction injection molding process with a reinforcement added. See also *reaction injection molding*.

Reinforcement

A strong material bonded into a matrix to improve its mechanical properties. Reinforcements are usually long fibers, chopped fibers, whiskers, particulates, and so forth. The term is not synonymous with *filler*. The most commonly used reinforcement materials are glass (E-glass and S-glass), aramid, silicon carbide, boron, alumina, fused silica, alumina–boria–silica, and carbon/graphite. See also *carbon–carbon composites*, *ceramic–matrix composites*, *metal–matrix composites*, and *resin–matrix composites*.

Reinforcement (of Weld)

Weld metal that stands proud of the straight line joining the toes at the weld to parent metal junctions. Some authorities prefer the alternative term, "Excess," but others consider this to have an undesirably pejorative implication bearing in mind that some overfilling is normal, usually fully acceptable and preferable to insufficient filling. Where the overfill is unacceptably large, it would be termed excessive reinforcement rather than excess excess! See also *face reinforcement* and *root reinforcement*.

Rejectable

See preferred term *nonconforming*.

Relative Atomic Mass

The strictly correct term for *atomic mass*.

Relative Dielectric Constant

See *permittivity*.

Relative Humidity

(1) The ratio of the molecular fraction of water vapor present in the air to the molecular fraction of water vapor present in saturated air at the same temperature and barometric pressure. Approximately, it equals the ratio of the partial pressure or density of the water vapor in the air to the saturation pressure or density, respectively, at the same temperature. (2) The ratio, expressed as a percentage, of the amount of water vapor present in a given volume of air at a given temperature to the amount required to saturate the air at that temperature.

Relative Rigidity

In dynamic mechanical measurements of plastics, the ratio of modulus at any temperature, frequency, or time to the modulus at a reference temperature, frequency, or time.

Relative Sintering Temperature

In powder metallurgy, the ratio of the sintering temperature to the melting temperature of the substance as expressed on the Kelvin scale.

Relative Standard Deviation (RSD)

The standard deviation expressed as a percentage of the mean value:

$$\mathrm{RSD} = 100\left[\frac{S}{X}\right]\frac{d^2}{n-1}$$

where
    S is the standard deviation
    d is the difference between the individual results and the average
    n is the number of individual results
    X is the average of individual results

Also known as coefficient of variation.

Relative Transmittance

The ratio of the transmittance of the object in question to that of a reference object. For a spectral line on a photographic emulsion, it is the ratio of the transmittance of the photographic image of the spectral line to the transmittance of a clear portion of the photographic emulsion. Relative transmittance may be total, specular, or diffuse. See also *transmittance*.

Relative Viscosity

For a polymer in solution, the ratio of the absolute viscosities of the solution (of stated concentration) and of the pure solvent at the same temperature.

Relaxation Curve

A plot of either the remaining or relaxed stress as a function of time. See also *relaxation rate*.

Relaxation Strictly Creep Relaxation

The reduction of imposed load as a result of deformation by creep. The mechanism involves the reduction in elastic strain with a corresponding increase in plastic, that is, creep, strain. The phenomenon occurs in components such as high-temperature bolts that are subject to a fixed strain rather than a fixed stress. See *tensile test* and *displacement-controlled loading*.

Relaxation Time

(1) The time taken for some characteristic or property to reduce to a predetermined value. The term is usually used in connection with diffusion control processes such as creep. (2) The time required for a stress under a sustained constant strain to diminish by a stated fraction of its initial value.

Relaxed Stress

The initial stress minus the remaining stress at a given time during a stress-relaxation test.

Relay

An electrically controlled device that opens and closes electrical contacts to affect the operation of other devices in the same or another electric circuit.

Release Agent

In forming of plastics, a material that is applied as a thin film to the surface of a mold to keep the resin from bonding to the mold. Also called parting agent. See also *mold release agent*.

Release Film

In forming of plastics, an impermeable layer of film but does not bond to the resin being cured. See also *separator*.

Release Paper

A sheet, serving as a protectant or carrier, or both, for an adhesive film or mass, which is easily removed from the film or mass prior to use.

Reliability

A quantitative measure of the ability of a product or service to fulfill its intended function for a specified period of time.

Relief

The result of the removal of tool material behind or adjacent to the cutting edge to provide clearance and prevent rubbing (heel drag). See also *relief angle*.

Relief Angle

The angle formed between a relieved surface and a given plane tangent to a cutting edge or to a point on a cutting edge. Also known as clearance angle. See also the term *single-point tool*.

Relieving

Buffing or other abrasive treatment of the high points of an embossed metal surface to produce highlights that contrast with the finish in the recesses.

Remaining Stress

The stress remaining at a given time during a stress-relaxation test. See also *stress relaxation*.

Remanence

The magnetic induction remaining in a magnetic circuit after removal of the applied magnetizing force. Sometimes called *remanent induction*.

Renormalizing

The normalizing of steel that has undergone some process, for example, welding, that has impaired some characteristic. The term obviously implies that the steel was in the normalized condition prior to the process.

Repeatability

A term used to refer to the test result variability associated with a limited set of specifically defined sources of variability within a single laboratory.

Repeated Impact

See *impingement*.

Replica

A reproduction of a surface in a material. It is usually accomplished by depositing a thin film of suitable material, such as a plastic, onto the specimen surface. The film is subsequently extracted and examined by optical microscopy, scanning electron microscopy, or transmission electron microscopy, the latter being the most common. Replication techniques can be classified as either surface replication or extraction replication. Surface replicas provide an image of the surface topography of the specimen, while extraction replica lift particles from the specimen. See also *atomic replica, cast replica, collodian replica, Formvar replica, gelatin replica, impression replica, negative replica, oxide film replica, plastic replica, positive replica, preshadowed replica, tape replica method (faxfilm)*, and *vapor-deposited replica.*

Replicate

In electron microscopy, to reproduce using a *replica.*

Repressing

The application of pressure to a previously pressed and sintered powder metallurgy compact, usually for the purpose of improving some physical or mechanical property or four-dimensional accuracy.

Reprocessed Plastic

A thermoplastic prepared from melt-processed scrap or reject parts by a plastics processor, or from nonstandard or nonuniform virgin material. The term scrap does not necessarily connote feedstock that is less desirable or usable than virgin material. Reprocessed plastic may or may not be reformulated by the addition of fillers, plasticizers, stabilizers, or pigments.

Reproducibility

A term used to describe test result variability associated with specifically defined components of variance obtained both from within a single laboratory and between laboratories.

Rerolling Quality

Rolled *billets* from which the surface defects have not been completely removed.

Resenes

The constituents of *rosin* that cannot be saponified with alcoholic alkali, but that contain carbon, hydrogen, and oxygen in the molecule. See also *saponification.*

Reset

The realigning or adjusting of metalforming dies or tools during a production run; not to be confused with the operation *setup* that occurs before a production run.

Residual Field

Same as *residual magnetic field.*

Residual Gas Analysis (RGA)

The study of residual gases in vacuum systems using mass spectrometry.

Residual Magnetic Field

The magnetic field that remains in a part after the magnetizing forces have been removed.

Residual Method

Method of magnetic-particle inspection in which the particles are applied after the magnetizing force has been removed.

Residual Strain

The strain associated with residual stress.

Residual Stress

(1) The stress existing in a body at rest, in equilibrium, at uniform temperature, and not subjected to external forces. Often caused by the forming or thermal processing curing process. (2) An internal stress not depending on external forces resulting from such factors as cold working, phase changes, or temperature gradients. (3) Stress present in a body that is free of external forces or thermal gradients. (4) Stress remaining in a structure or member as a result of thermal or mechanical treatment or both. Stress arises in fusion welding primarily because the weld metal contracts on cooling from the solidus to room temperature. (5) Such stresses exist as an internal system with compression stresses at one location balancing tensile stresses elsewhere. They can be induced in a component by many treatments such as solidification, deformation, machining, and uneven heating or cooling (particularly welding). A simple example can be envisaged by considering a tube that is slit longitudinally producing a wide gap. The gap can be closed inducing stress round the tube circumference, and if the gap edges are welded together, the tube is left with a system of residual hoop stresses, tensile on the exterior, and compression at the bore. In some cases, the level of stress may be very high, up to yield magnitude, and often it will cause, or contribute to, many damage mechanisms including creep, fatigue, brittle fracture, and stress corrosion. Generalizing, tensile residual stresses are potentially damaging, but compressive residual stresses can be beneficial and they may be deliberately induced by painting and similar techniques. See also *stress relief.*

Residuals, Residual Elements

Elements in a metal not added deliberately. They usually remain from the original ore or remelted scrap but may be introduced from other sources such as flux or the furnace environment.

Resilience

(1) The amount of energy per unit volume released on unloading. (2) The capacity of a material, by virtue of high yield strength and low elastic modulus, to exhibit considerable elastic recovery on release of load.

Resin

A solid, semisolid, or pseudosolid organic material that has an indefinite and often high molecular weight, exhibits a tendency to flow when subjected to stress, usually has a softening or melting range, and usually fracture conchoidally. (2) Liquid resin, that is, an organic polymeric liquid, which, when converted to its final state, becomes the resin. (3) An organic polymer that cross-links to form a thermosetting plastic when mixed with a curing agent. (4) In reinforced plastics, the material used to bind together the reinforcement material; the matrix. See also *polymer*, *reinforced plastics*, and *resin–matrix composites*.

Historically, **resins** is the term applied to an important group of substances obtained as gums from trees or manufactured synthetically. It is also frequently used interchangeably with the term **plastics**. The common resin of the pinetree is called **rosin**.

Originally a category of vegetable substances soluble in ethanol but insoluble in water, resin in modern technology is generally an organic polymer of indeterminate molecular weight. The class of flammable, amorphous secretions of conifers or legumes are considered true resins, and include kauri, copal, dammar, mastic, guaiacum, jalap, colophony, shellac, and numerous less well-known substances. Water-swellable secretions of various plants, especially the Burseraceae, are called gum resins and include myrrh and olibanum. The official resins are benzoin, guaiac, mastic, and resin. Other natural resins are copal, dammar, dragon's blood, elaterium, lac, and sandarac. The natural vegetable resins are largely polyterpenes and their acid derivatives, which find application in the manufacture of lacquers, adhesives, varnishes, and inks.

The synthetic resins, originally viewed as substitutes for copal, dammar, and elemi natural resins, have a large place of their own in industry and commerce. Phenol–formaldehyde, phenolurea, and phenyl-melamine resins have an important commercially for a long time. Any unplasticized organic polymer is considered a resin; thus, nearly any of the common plastics may be viewed as a synthetic resin. Water-soluble resins are marketed chiefly as substitutes for vegetable gums and in their own right for highly specialized applications. Carboxymethylcellulose, hydroxyalkylated cellulose derivatives, modified starches, polyvinyl alcohol, polyvinylpyrrolindone, and polyacrylamides are used as thickening agents for foods, paints, and drilling muds, as fiber sizings, in various kinds of protective coatings, and as encapsulating substances.

Oleoresins are natural resins containing essential oils of the plants. Gum resins are natural mixtures of true gums and resins and are not soluble in alcohol. They include rubber, gutta percha, gamboge, myrrh, and olibanum. Some of the more common natural resins are rosin, dammar, mastic, sandarac, lac, and animi. Fossil resins, such as amber and copal, are natural resins from ancient trees, which is then chemically altered by long exposure. The synthetic resins differ chemically from natural resins, and few of the natural resins have physical properties that make them suitable for mechanical parts.

Resin Content

The amount of *resin* in a *laminate*, expressed as either a percentage of total weight or total volume.

Resin Pocket

In plastics, an apparent accumulation of excess resin in a small, localized section visible on cut edges of molded surfaces, or internal to the structure and non-visible. See also *resin-rich area*.

Resin System

A mixture of *resin* and ingredients such as catalyst, initiator, diluents, and so forth, required for the intended processing method and final product.

Resin Transfer Molding (RTM)

A process by which catalyzed resin is transferred or injected into an enclosed mold in which reinforcement has been placed. The fiberglass reinforcement is usually woven, nonwoven, or knitted fabric.

Resin–Matrix Composites

Resin–matrix composites are advanced engineering materials that contain a reinforcement (such as fibers or particles) supported by an organic (plastic) binder (matrix). Resin–matrix composites were developed in response to demands of the aerospace community for stronger, more lightweight materials. Aluminum alloys, which provide high strength and fairly high stiffness at low weight, have provided good performance and have been the primary materials used in aircraft structures over the years. However, both corrosion and fatigue in aluminum alloys have produced problems that have been costly to remedy. World War II promoted a need for materials with improved structural properties. In response, fiber-reinforced composites were developed, and by the end of the war, fiberglass-reinforced plastics had been used successfully in filament-wound rocket motors and in various other structural applications. Today, resin–matrix composites are seeing significantly expanded levels of use and components where reduced weight is critical. This is primarily a result of their tailorability as well as their high-strength- and modulus-to-density ratios. In recent years, these materials have seen applications ranging from mass-produced tennis rackets to complex aerospace structures.

Resin–matrix composite materials can be divided into two basic categories, depending on the type of reinforcement: continuous fiber-reinforced and particulate/short fiber-reinforced composites. The first of these two materials typically uses a continuous array of oriented fibers, whereas the second category uses randomly dispersed particulates or chopped fibers. Continuous fiber composites, which are the most common type of composite, are typically made up of 3–30 μm diameter fibers that are oriented and surrounded in a supportive matrix material. Generally, the fibers used in these material systems are several orders of magnitude stiffer and stronger than the surrounding matrix. Glass fiber reinforced organic–matrix composites are the most familiar and widely used and have had extensive application in industrial, consumer, military, and aerospace markets. Carbon fiber-reinforced resin–matrix composites are the most commonly applied advanced (non-fiberglass) composites. They offer extremely high specific properties, high-quality materials that are readily available, reproducible material forms, increasingly favorable cost projections, and comparative ease of manufacture. Composites reinforced with aramid, other organics, boron fibers, and silicon carbide, alumina, and other ceramic fibers are also used. The high stiffness and strength of these fibers control the characteristic engineering properties of the composite.

The continuous matrix phase functions as a supportive element to the fibers. In this capacity, fiber orientation and alignment are maintained, load is transferred between fibers, and strength is provided in nonreinforced directions. A wide variety of matrix materials are available for use with each fiber type. By far, the most extensive matrices in current use are those employing either elevated-temperature curing epoxies or room-temperature curing vinyl esters.

Other matrix materials include bismaleimide resins, polyamide resins, and thermoplastic resins such as polyether ether ketone (PEEK), polyphenylene sulfide (PPS), and polyetherimide (PEI).

Although there are a wide variety of processing methods for these materials, continuous fiber-reinforced composites are usually fabricated by laminating together and curing multiple fiber plies impregnated with unreacted matrix resin. Such processing facilitates the arranging of plies of material at a variety of angles. This structuring produces a stacked, laminated construction. In the aerospace industry, the angle of plies used in the laminated stack are oriented at fixed angles to the direction of the major load. The most common fixed angles are 0°, 45°, −45°, and 90° to major load axis. See *quasi-isotropic laminate*. By selecting the proper number of plies at each of these angles, composite properties and their anisotropies can be designed for strength, modulus, or even the degree of thermal expansion along each principal direction.

Resinography

The science of morphology, structure, and related descriptive characteristics as correlated with composition or conditions and with properties or behavior of resins, polymers, plastics, and their products.

Resinoid

Any of the class of thermosetting synthetic resins, either in their initial temporarily fusible state or in their final infusible state. See also *novolac* and *thermosetting*.

Resinoid Bond

An organic bond usually of the phenol formaldehyde resin type but sometimes consisting of other synthetic resins.

Resinoid Wheel

A grinding wheel bonded with a synthetic resin. See the term *grinding wheels*.

Resin-Rich Area

A significant thickness of a nonreinforced resin layer of the same composition as that within the base material. In *reinforced plastics*, a localized area filled with resin and lacking reinforcing material. See also *resin pocket*.

Resin-Starved Area

In *reinforced plastics*, a localized area of insufficient resin, usually identified by low glass, dry spots, or fiber showing on the surface.

Resintering

(1) A second sintering operation on a powder compact. (2) Sintering a repressed compact. See also *repressing*.

Resist

(1) Coating material used to mask or protect selected areas of a substrate from the action of an etchant, solder, or plating. (2)

A material applied to prevent flow of brazing filler metal into unwanted areas. (3) Coating material, lacquer, etc., applied locally on a component, or generally to equipment to prevent deposition during electroplating or to prevent dissolution during etching. Also called *stop-off*, etc.

Resistance

The opposition that a device or material offers to the flow of direct current, equal to the voltage drop across the element divided by the current through the element. Also called electrical resistance.

Resistance Alloys

Electrical resistance alloys include both the types used in instruments and control equipment to measure and regulate electrical characteristics and those used in furnaces and appliances to generate heat. In the former applications, properties near ambient temperature are of primary interest; in the latter, elevated-temperature characteristics are of prime importance. In common commercial terminology, electrical resistance alloys used for control or regulation of electrical properties are called resistance alloys, and those used for generation of heat are referred to as resistance heating alloys.

The primary requirements for resistance alloys are uniform resistivity, stable resistance (no time-dependent aging effects), reproducible temperature coefficient of resistance, and low thermal electric potential versus copper. Properties of secondary importance are coefficient of expansion, mechanical strength, ductility, corrosion resistance, and ability to be joined to other metals by soldering, brazing, or welding. Alloys must be strong enough to withstand fabrication operations, and it must be easy to procure an alloy that has consistently reproducible properties and order to ensure resistor accuracy.

Resistors for electrical and electronic devices may be divided into two arbitrary classifications: those employed in precision instruments in which overall error is considerably less than 1%, and those employed where less precision is needed. The choice of alloys for a specific resistor application depends on the variation and properties that can be tolerated. Materials for resistors include copper–nickel (2%–22% Ni) alloys, generally referred to as radio alloys; copper–manganese–nickel alloys (10%–13% Mn and 4% Ni), generally referred to as manganins; constantin alloys, whose compositions vary from 50Cu–50Ni to 65Cu–35Ni; nickel–chromium–aluminum alloys that nominally contains 20% Cr, 3% Al, and 2%–5% of copper, iron, and/or manganese; 80Ni–20Cr alloys; and iron–chromium–aluminum alloys (nominally 73Fe–22Cr–5Al).

Resistance heating elements are used in many varied applications—from small household appliances to large industrial process heating systems and furnaces that may operate continuously at temperatures of 1300°C (2350°F) or higher. The primary requirements of materials used for heating elements are high melting point, high electrical resistivity, reproducible temperature coefficient of resistance, good oxidation resistance, absence of volatile components, and resistance to contamination. Other desirable properties are good creep strength, high emissivity, low thermal expansion, and low modulus (both of which help minimize thermal fatigue), good resistance to thermal shock, and good strength and ductility at fabricating temperature.

The most commonly used resistance heating alloys are nickel–chromium and nickel–chromium–iron alloys. Other materials include iron–chromium–aluminum alloys similar in composition to resistor alloys, high-melting temperature pure metals, and nonmetallic materials, which can be used effectively at temperatures as high as 1900°C (3450°F).

Resistance Brazing

A resistance joining process in which the workpieces are heated locally and filler metal that is preplaced between the workpieces is melted by the heat obtained from resistance to the flow of electric current through the electrodes and the work. In the usual application of resistance brazing, the heating current is passed through the joint itself.

Resistance Butt Welding

Resistance welding processes in which the components are butt joined, that is, face as edge to edge, as opposed to a lap joint. See preferred term *flash welding* and *upset welding*.

Resistance Heating

Resistance heating is the generation of heat by electric conductors carrying current. The degree of heating for a given current is proportional to the electrical resistance of the conductor. If resistance is high, a large amount of heat is generated, and the material is used as a resistor rather than as a conductor.

Resistor Materials

In addition to having high resistivity, heating elements must be able to withstand high temperatures without deteriorating or sagging. Other desirable characteristics are low-temperature coefficient of resistance, low cost, and deformability, and availability of materials. Most commercial resistance alloys containing chromium or aluminum or both, because a protective coating of chrome oxide or aluminum oxide forms on the surface upon heating and inhibits or retards further oxidation.

Heating Element Forms

Because heat is transmitted by radiation, convection, conduction, or combinations of these, the form of element is designed for the major mode of transmission. The simplest form is the helix, using a round wire resistor, with the pitch of the helix approximately three wire diameters. This form is adapted to radiation and convection and is generally used for room or air heating. It is also used in industrial furnaces, utilizing force convection up to about 650°C. Such helices are stretched over grooved high-aluminum refractory insulators and are otherwise open and unrestricted. These helices are suitable for mounting in air ducts or enclosed chambers, where there is no danger of human contact.

For such applications as water heating, electric range units, and die heating, where complete electrical isolation is necessary, the helix is embedded in magnesium oxide inside a metal tube, after which the tube is swaged to a small diameter to compact the oxide and increase thermal conductivity. Such units can then be formed and flattened into desired shapes. The metal tubing is usually copper for water heaters and stainless steel for radiant elements, such as range units. In some cases, the tubes may be cast into finned aluminum housings, or fins may be brazed directly to the tubing to increase surface area for convection heating.

Modification of the helix for high-temperature furnaces involves supporting each turn in a grooved refractory insulator, with the insulators strung on stainless alloy rods. Wire sizes for such elements are 5 mm in diameter or larger, or they may be edge-wound strap. Such elements may be used up to 980°C furnace temperature.

Another form of furnace heating element is the sinuous grid element, made of heavy wire or strap or casting and suspended from refractory or stainless supports built into the furnace walls, floor, and roof.

Silicon carbide elements are in rod form, with low-resistance integral terminals extending through the furnace walls.

Direct Heating

When heating metal strip or wire continuously, the supporting rolls can be used as electrodes, and the strip or wire can be used as the resistor.

Molten Salts

The electrical resistance of molten salts between immersed electrodes can be used to generate heat. Limiting temperatures are dependent on decomposition or evaporation temperatures of the salt. Parts to be heated are immersed in the salt. Heating is rapid and, since there is no exposure to wear, oxidation is largely prevented. Disadvantages are the personnel hazards and discomfort of working close to molten salts.

Major Applications

A major application of resistance heating is in electric home appliances, including electric ranges, clothes dryers, water heaters, coffee percolators, portable radiant heaters, and hair dryers. Resistance heating has also found application in home or space heating; some homes are designed with suitable thermal insulation to make electric heating practicable.

Resistance Seam Welding

A resistance welding process that produces coalescence at the faying surfaces by the heat obtained from resistance to electric current through workpieces that are held together under pressure by electrode wheels. The resulting weld is a series of overlapping resistance spot welds made progressively along a joint by rotating the electrodes. See the term *longitudinal resistance seam welding*.

Resistance Soldering

Soldering in which the joint is heated by electrical resistance. Filler metal is either face fed into the joint or preplaced in the joint.

Resistance Spot Welding

A process in which faying surfaces are joined in one or more spots by the heat generated by resistance to the flow of electric current through workpieces that are held together under force by electrodes. The contacting surfaces in the region of current concentration are heated by a short-time pulse of low-voltage, high-amperage current to form a fused nugget of weld metal. When the flow of current ceases, the electrode force is maintained while the weld metal rapidly cools and solidifies. The electrodes are retracted after each weld, which usually is completed in a fraction of a second.

Resistance Strain Gauge

A strip, wire, etc., of material where the electrical resistance of which varies progressively with strain. A calibrated piece of the material is bonded to the component to be monitored, and its change in resistance provides a measure of strain.

Resistance Welding

A group of welding processes that produce coalescence of metals with resistance heating and pressure. See also *flash welding*, *projection welding*, *resistance seam welding*, and *resistance spot welding*.

Resistance Welding Die

The part of a resistance welding machine, usually shaped to the work contour, in which the parts being welded are held and which conducts the welding current.

Resistance Welding Electrode

The part or parts of a resistance welding machine through which the welding current and, in most cases, pressure is applied directly to the workpiece. The electrode may be in the form of a rotating wheel, rotating roll, bar, cylinder, plate, clamp, chuck, or modification thereof.

Resistance Welding Gun

A manipulating device to transfer current and provide electrode force to the weld area (usually in reference to a portable gun).

Resistivity

Resistance to the passage of electricity. It is the reciprocal of conduction and in the S.I. system, is measured in micro ohm centimeters. See *electrical resistivity*.

Resite

Synonym for *C*-stage.

Resitol

Synonym for *B*-stage.

Resol

Synonym for *A*-stage.

Resole Resin

Linear phenolic resin produced by alkaline condensate of phenol and formaldehyde.

Resolution

The capacity of an optical or radiation system to separate closely spaced forms or entities; in addition, the degree to which such forms or entities can be discriminated. Resolution is usually specified as the minimum distance by which two lines or points in the object must be separated before they can be revealed as separate lines or points in the image. See also *resolving power* and *shape resolution*.

Resolving Power

(1) The ability of a given optical lens system to reveal fine detail and an object. Of a microscope or individual lens, the size of the smallest feature that can be distinguished. It is about 0.2 μm in a conventional light microscope. See also *resolution*.

Resonance

The excitation of a component or system at its natural frequency of vibration. Such excitation can induce a progressive increase in amplitude of vibration with associated increase in stress and consequent risk of fatigue failure.

Resonant Forced Vibration Technique

A technique for performing dynamic mechanical measurements on plastics, in which the sample is oscillated mechanically at the natural resonant frequency of the system. The amplitude of oscillation is maintained constant by the addition of makeup energy. Elastic modulus is calculated from the measured frequency. Damping is calculated from the additional energy required to maintain constant amplitude.

Resorcinol

A colorless crystalline material with a melting point of 230°F (110°C). It is very soluble in water and in alcohol. It is used in the production of plastics; in the manufacture of fluorescein; in the production of xanthene and azo dyes, particularly the fast *Alsace green*; in medicine as an antiseptic; and in making the explosive lead styphnate. Resorcinol polymerizes with formaldehyde to form the *resorcinol-formaldehyde plastics* that will cure at room temperature and with only slight pressure. They are used in strong adhesives for plywood and wood products and do not deteriorate from acid action as some other plastics do. *Resorcinol adhesives* remain water soluble during the working period for 2–4 h and are then insoluble and chemical resistant.

Response Curve for *N* Cycles

In fatigue data analysis, a curve fitted to observed values of percentage survival at N cycles for several stress levels, where N is a preassigned numbers such as 10^6 and 10^7. It is an estimate of the relationship between applied stress and the percentage of the population that would survive N cycles. See also *S–N curve*.

Rest Periods

With reference to fatigue, periods of significant length during which cyclic loading is suspended and some recovery may occur with consequent benefit to fatigue life.

Rest Potential

See *corrosion potential* and *open-circuit potential*.

Restrained Weld Test

A test in which the test piece being welded is firmly secured by some support system of bolts or substantial preliminary welds to minimize movement during the test welding cycle. The restraint imposes high tensile stresses as a result of weld metal contraction during cooling following welding. The test, therefore, indicates the likelihood of cracking being encountered when welds in a similar procedure are made during production.

Restraint

Any external mechanical force that prevents a part from moving to accommodate changes in dimension due to thermal expansion or contraction. Often applied to weldments made while clamped in a fixture. Compare with *constraint*.

Restrictor Rings

Rings, usually faced with *white metal*, placed outside a bearing to prevent fluid from being discharged.

Restrike

(1) A repeat of a forging or stamping operation to improve dimensions or quality of surface detail. (2) Additional compacting of a sintered powder metallurgy compact.

Restriking

(1) The striking of a trimmed but slightly misaligned or otherwise faulty forging with one or more blows to improve alignment, improve surface condition, maintain close tolerances, increase hardness, or affect other improvements. (2) A *sizing* operation in which coining or stretching is used to correct or alter profiles and to counteract distortion. (3) A salvage operation following a primary forging operation in which the parts involved are rehit in the same forging die in which the pieces were last forged.

Resultant Field

The magnetic field that is the result of two or more magnetizing forces impressed on the same area of a magnetizable object. Sometimes called a vector field.

Resultant Rake

The angle between the tooth face and an axial plane through which the tooth point measured in a plane perpendicular to the cutting edge. The resultant rake of a cutter is a function of three other angles: radial rate, axial rake, and corner angle.

Retained Austenite

See *steel*.

Retainer

See *cage*.

Retardation Plate

A plate placed in the path of a beam of polarized light in an *optical microscope* for the purpose of introducing a difference in phase. Usually, quarter-wave or half-wave plates are used, but if the light passes through them twice, the phase difference is doubled.

Retarder

Synonym for *inhibitor*.

Retention Time (t_R)

In *chromatography*, the amount of time a sample compound spends in the chromatographic column.

Retentivity

The capacity of a material to retain a portion of the magnetic field set up in it after the magnetizing force has been removed.

Retort

A vessel used for distillation of volatile materials, as in separation of some metals and in destructive distillation of coal.

Retrogradation

A change of starch pastes that are used in adhesive formulations from low to high consistency upon aging.

Retrogression

The phenomenon observed in the precipitation hardening of solution treated material whereby initial heat treatment at some low aging temperatures subsequently causes, at some higher temperature a more rapid hardening than normal and a consequent risk of over aging. See *precipitation hardening*.

Reverberatory Furnace

A furnace in which the flame used for melting the metal does not impinge on the metal surface itself but is reflected off the walls of the root of the furnace. The metal is actually melted by the generation of heat from the walls and the roof of the furnace. Although this furnace heating concept is still used in many foundries and smelters, more modern reverberatory furnaces use roof burners or side burners directed toward the metal surface. Such furnaces are designed in either the wet hearth or dry hearth configuration. In a wet hearth furnace, the products of combustion are in direct contact with the top of the molten bath, and the heat transfer is achieved by a combination of convection and radiation. In a dry hearth furnace, the charge of solid metal is positioned on a sloping hearth above the level of the molten metal so that the charge is completely enveloped by the hot gases. Heat is rapidly absorbed by the solid charge, which melts and subsequently drains from the sloping hearth into the wet holding basin or chamber.

Reverse Drawing

Redrawing of a sheet metal part in a direction opposite to that of the original drawing.

Reverse Engineering

The practice of copying a design by measuring and analyzing a sample, repairing drawings, specification and procedures, and producing the copy without infringing the original manufacturers rights. The legal aspects require great care.

Reverse Flange

A sheet metal flange made by shrinking, as opposed to one form by stretching.

Reverse Flow Forging

Processes in which a billet contained in a die is partially pierced by a RAM causing it to expand to fill the die and extrude back around and up the ram. Impact extrusion is similar but utilizes a high-velocity ram.

Reverse Helical Winding

In *filament winding*, as the fiber delivery arm traverses one circuit, a continuous helix is laid down, reversing direction at the polar ends, in contrast to biaxial, compact, or sequential winding. The fibers cross each other at definite equators, the number depending on the helix. The minimum region of crossover is three. See also the term *helical winding*.

Reverse Impact Test

A test in which one side of a sheet of plastic is struck by a pendulum or falling object, and the reverse side is inspected for damage.

Reverse Polarity

Direct-current arc welding circuit arrangement in which the electrode is connected to the positive terminal. A synonym for *direct current electrode positive (DCEP)*. Contrast with *straight polarity*.

Reverse Redrawing

A second drawing operation in a direction opposite to that of the original drawing.

Reverse(d) Polarity (Welding)

An ambiguous term as is its opposite, *straight polarity*. In some countries, including the United Kingdom, reversed polarity usually means direct current arc welding with the electrode connected to the negative pole of the supply and straight polarity has the electrode connected to the positive. Unfortunately, U.S. terminology is exactly opposite. Better terms are *electrode positive* and *electrode negative*.

Reverse-Current Cleaning

Electrolytic cleaning in which a current is passed between electrodes through a solution, and the part is set up as the anode. Also called *anodic cleaning*.

Reversed Stress

A cyclic stress where the compressive load and the tensile load are of the same value. See *fatigue*.

Reversed-Phase Chromatography (RPC)

Bonded-phase chromatography with a nonpolar stationary phase and a polar mobile phase. See also *bonded-phase chromatography*.

Reversible (Electrolytic) Cell

A cell in which the two electrodes are immersed in separate electrolytes and in which reversing the current flow results in a reversion of the reactions at the electrodes.

Reversing Mill

A rolling mill through which the product passes backward and forward a number of times, the direction of rotation of the rolls reversing on each pass.

Reversion

Generally the return to some previous condition. More specifically, the reversal of a martensite shear transformation in nonferrous alloys to restore precisely the structure of the original pre-quenched phase.

Reworked Plastic

A thermoplastic that is reground, pelletized, or solvated after having been previously processed by molding, extrusion, and so forth. In many specifications, the use of reworked material is limited to clean plastic that meets the requirements specified for virgin material and yields a product essentially equal in quality to one made from virgin material only.

Reyn

The former English unit of dynamic viscosity equal to approximately 14.9 poise.

Reynold's Equation

A basic equation of *hydrodynamic lubrication*.

R.F. Radio Frequency

As in R.F. welding that utilizes a R.F. power beam for radiation heating.

RGA

See *residual gas analysis*.

Rhenium and Alloys

Rhenium (symbol Re) is a rare, silvery-white element with a metallic luster. It is one of the hardest metals at 580 HV (30% cold work), is one of the heaviest with the density of 21 g/cm^3, and has one of the highest melting points at 5760°C. It also has no known ductile to brittle transition temperature, which enables rhenium parts to be thermally cycled without being degraded. It remains strong at high temperatures and does not become brittle after welding. Although it is a very ductile material, it has the third highest modulus of elasticity of any metal. The addition of rhenium to molybdenum and tungsten has enhanced their strength, ductility, and formability.

The crystal structure is closely packed hexagonal, making it more difficult to work than the cubic-structured tungsten, but the crystal grains are tiny, and small amounts of rhenium added to tungsten improve ductility and increase high-temperature strength of tungsten used in lamp filaments and wire.

During the past decade, interest has increased dramatically in rhenium as a construction material for parts with complex shapes in high-temperature service environments. A typical example of such an application is rocket thrusters, which have operating temperatures up to 2230°C. Other applications include welding rods, thermocouples, and cryogenic magnets. When rhenium is added to superalloys in amounts of 3%–6%, the turbine-inlet temperature in gas turbine engines is increased by 125°C.

Processing

The primary source of rhenium is the by-product molybdenite obtained from porphyry copper deposits.

Virtually, all of the wrought products of pure rhenium are produced from bars made by powder metallurgy. Although rhenium can be arc melted in an inert atmosphere, the resultant metal is not well suited for fabrication due to the coarse as-melted grain size and the possible segregation of small amounts of rhenium oxide at the grain boundaries. Even with vacuum melting, the workability has not equaled that of bars produced by powder metallurgy. The latter bars have a small grain size and are much more amenable to fabrication.

Rhenium powder produced by hydrogen reduction of ammonium perrhenate is consolidated by pressing and split rectangular dies at pressures of 25–30 tsi using nonlubricated powder. These press bars or vacuum pre-sintered at 1200°C and a pressure of 0.5–1 μ. They are subsequently resistance-sintered in dry hydrogen adding maximum temperature of approximately 90% of the melting point in a manner almost identical to that used for tungsten. The sintered bars are normally 90%–96% of the theoretical density, having undergone 15%–20% shrinkage in each dimension. These bars are then ready for fabrication.

Pure rhenium is fabricated by cold working with frequent intermediate recrystallizing anneals. Hot working of rhenium results in "hot short" failures, probably because of the presence of rhenium heptoxide at the grain boundaries. This volatile oxide has a melting point of 297°C and a boiling point of 363°C.

The fabrication of rhenium wire requires processing by swaging and drawing. Work hardening is so great in cold swaging that reductions are limited to one 10% reduction between anneals to avoid serious damage to the swaging dies. The reductions are somewhat greater in cold drawing beginning with 10% but increasing to about 40% between anneals. Diamond drawing dies are used for all pure rhenium wire drawing.

The production of rhenium strip is likewise accomplished by cold rolling. After several light initial reductions on the sintered bars, reductions up to 40% can be affected between anneals. Strip thinner than 0.13 mm is rolled using small-diameter tungsten carbide work roles in a "4 high" mill. Strip can be processed in this matter to thicknesses of 0.03 mm.

Because of the rapid rate of work hardening that accompanies the cold working of rhenium, annealing constitutes a very important part of the fabrication processes. Frequent anneals are necessary at temperatures of 1550°C–1700°C for times varying from 10 to 30 min in an atmosphere of dry hydrogen or a hydrogen–nitrogen mixture. Furnaces utilizing molybdenum heating elements and aluminum oxide refractories are used most extensively. Anneals performed as indicated result in complete recrystallization with the grain size usually between 0.010 and 0.040 mm and a hardness of about 250–275 VHN.

It is virtually impossible to apply conventional ingot casting and hot/warm forming techniques to rhenium because of its high melting point, poor oxidation resistance, and the difficulty of working an as-cast structure.

Rhenium and its alloys are generally manufactured in the United States and Europe through the powder-metallurgy route, with a tolerance on the rhenium addition in the range of ±0.5% or better. Various methods have been chosen to suppress second-phase formation. In Russia, vacuum melting is the preferred production process for alloys such as Mo-47% Re and W-27% Re, which have rhenium content tolerances of ±0.3%.

In alloys W-3%–5% Re for wire products, the tungsten-based material is intentionally doped with potassium, aluminum, and silicon to control recrystallization behavior. European and Russian companies also manufacture wire of these alloys in the undoped condition. In recent years, W-3% to 5% Re and up to 10% Re undoped compositions have been found to be ideal for fabricating x-ray targets, heat sinks, and other complex shapes.

Better performance can be achieved by additions of 11%–14% rhenium to molybdenum and additions of 10%, 15%, and 20% rhenium to tungsten, as so-called dilute or moderate alloys, Mo-41%–44% Re is the basis of another group of alloys for consideration.

Properties

Rhenium is a refractory metal with properties similar in many ways to tungsten and molybdenum. Pure rhenium has significantly greater room-temperature ductility combined with good high-temperature strength. When used as an alloy addition with tungsten or molybdenum forming body-centered cubic solid solutions, rhenium produces significant improvements in the ductility of those metals.

It is significant to note that the melting point of rhenium is the second highest of the metals, exceeded only by tungsten. Its density is surpassed only by osmium, iridium, and platinum and its modulus of elasticity only by osmium and iridium. Moreover, the hardness and strength of rhenium are very high, and its rate of work hardening is greater than that for any other metal.

The combination of high strength and ductility exhibited by rhenium and rhenium alloys is of great interest. Furthermore, these materials can be inert gas-arc-welded with a resultant weld that is ductile and possesses corrosion resistance as good as the parent metal. Note ductile-to-brittle transition is observed in pure rhenium or the rhenium–molybdenum alloys containing 30–50 wt.% Re as occurs in molybdenum at about room temperature and in tungsten at approximately 300°C. Although rhenium editions lower the transition temperature of tungsten, it still occurs above room temperature. However, even below the transition temperature for the recrystallized material, the ductility of the rhenium–tungsten alloys is higher than that of pure tungsten.

Fabrication

Rhenium cannot be machined by conventional methods because it work-hardens rapidly. However, electrical discharge machining is an effective manufacturing method, and parts can be brought to final finish through ceramic or diamond grinding. Rhenium can be plastically cold formed, although frequent annealings are required to restore formability after relatively small deformation.

Chemical vapor deposition is also a practical method for fabricating rhenium components with complex shapes. This method is particularly advantageous for products with thin metal structures, and for applications that require an oxidation-resistant protective layer on the rhenium surface. In addition, recently developed multiform powder-metallurgy products have the advantage of being isotropic, and they can be produced with a high relative density and consistent mechanical properties from lot to lot by well-established technologies.

Thermal spray is another alternative to the conventional powder-metallurgy route. Specifically, the cold gas dynamic spray method (CGSM) compacts rhenium powder as a coating onto a suitable substrate by ballistic impingement. Unfortunately, little success has been achieved because of the morphology and small particle size of rhenium metal powder.

Low-pressure plasma spray (LPPS) experiments have produced tantalum forms with thickness of 6 mm and a relative density exceeding 95%. These experiments indicate that LPPS may be a realistic way to produce near-net shape components of refractory metals, including rhenium. A preliminary attempt to make rhenium tubing resulted in a relative density of 83.5%, which was increased to over 95% after sintering. The prospects for manufacturing high-density isotropic products by plasma spraying of rhenium appear to be very good, although it is difficult to produce rhenium powder with good flow properties.

Among the newest technologies, the directed light fabrication (DLF) process is a multidimensional, laser-controlled deposition technique. Good results in casting complex near-net shape products have been achieved for metals with moderate melting points. The unidirectional solidification of a small molten pool makes this a single-step, almost waste-free process without risk of oxidation. Further studies of the DLF process are encouraged to develop the processing technique for refractory metals and to study the final properties of its products.

A long-standing effort has been made to apply the favorable combination of mechanical properties (high strength, outstanding plasticity, and formability) exhibited by rhenium at ambient temperature to pressure-shape the required profile from semifinished products such as sheet. Rhenium tube-spinning technology has recently emerged from being only a developmental effort.

Another example is the trend toward adopting explosive metalworking techniques for joining in fabricating rhenium parts. However, because rhenium work-hardens extraordinarily fast at low levels of cold deformation, frequent intermediate annealing constitutes a major and important part of such fabrication processes. As an interesting alternative, the appropriate shape design has been achieved by winding rhenium wire in layers, and filling the interwire void space with fine-grain particles by chemical vapor deposition.

Direct hot isostatic pressing (HIP) of near-net shape rhenium products is an intense plastic-deforming technique applied to powder that has not been explored in much detail. Limited data has been published on the effect of HIP on material properties of rhenium powder-metallurgy ingots and mill products. It is known that a combination of sintering and HIP procedures has demonstrated the possibility of developing both high density and good formability in powder-metallurgy rhenium parts obtained after further machining.

One of the challenges with direct HIP consolidation of near-net geometry is to reach near-full density of rhenium at temperatures far below typical sintering temperatures. During the development of the technique, careful evaluation of process parameters such as pressure, temperature, time, and heating/cooling rates may make it possible to control grain size, material properties, and dimensional variations. Because of the need to control so many parameters, success may depend on the nascent technology of intelligent HIP, in which the response of the material to processing parameters dictates the selection of optimal processing conditions.

Applications

The outstanding properties of rhenium and rhenium alloys suggest their use in many specialized applications.

Thermocouples

Rhenium versus tungsten thermocouples can be used for temperature measurement and control to approximately 2200°C, whereas previous thermocouple use was limited to temperatures below 1750°C. Indeed, 74% W–26% Re versus W thermocouples can be used at temperatures of at least 2750°C with a high emf output ensuring accurate precise temperature measurements with excellent reproducibility and reliability.

Electronic

Rhenium is now widely used for filaments for many spectrographs and for ion gauges for measuring high vacuum. Its ductility, chemical properties, including the fact that it does not react with carbon to form a carbide, and its emission characteristics make it superior to tungsten for these applications.

The elevated temperature properties of rhenium and rhenium alloys, their weldability, and electrical resistivity suggest their use for various components in electronic tubes.

Electrical

The superconducting properties of Re–Mo alloys have prompted their use for coils in compact-size electromagnets with high field strength. Although this application is still in the development stages, it looks most promising.

Rhenium has received considerable acclaim as an electrical contact material. It possesses excellent resistance to wear as well as arc erosion. Furthermore, the contact resistance of rhenium is extremely stable because of its good corrosion resistance in addition to the fact that possible formation of an oxide film on the contacts would not cause any appreciable change in the contact resistance since the resistivity of the oxide is almost the same as that of the metal. Extensive tests for some types of make-and-break switching contacts has shown rhenium to have 20 times the life of platinum–palladium contacts currently in use.

Heating elements for resistance heated vacuum or inert atmosphere furnaces seems to be a potential application for rhenium and especially rhenium–tungsten alloys. These materials would not undergo the embrittlement that occurs in tungsten upon heating, and they would allow for usage in vacuum, hydrogen, or inert atmospheres.

Welding Filler Rod

Rhenium and Re–Mo alloys can be readily welded. In fact, Re–Mo alloy can be used as a filler material for obtaining ductile welds in molybdenum. Although not all of the problems associated with the heat-affected zone are readily overcome, the properties of welds made with Re–Mo alloy filler wire are very much better.

Rheocasting

(1) Casting of a continuously stirred semisolid metal slurry. The process involves vigorous agitation of the melt during the early stages of solidification to break up solid dendrites into small spherulites. See also *semisolid metal forming*. (2) Diecasting or other processes utilizing the thixotropic characteristics of stirred, partially solidified alloys. When a partly solidified alloy is stirred, the

normal dendritic structure breaks down to form a slurry of rounded particles. In practice, the slurry contains little liquid, but it becomes increasingly fluid the more vigorously it is stirred. In the real casting process, the highly fluid slurry is fed into the shot sleeve of a pressure diecasting machine and injected into the die. *Thixocasting* is a related process in which rapidly cooled slugs of rheocast slurry are held at an intermediate temperature prior to reheating and injection. In **thixoforging,** the rapidly cooled rheocast slugs are forged in closed dies. Benefits claimed for rheocasting include lower injection temperature leading to reductions in die wear, energy costs, process cycle time, shrinkage and porosity, and a capability to maintain high quality in thin sections. See also *rheoforming*.

Rheodynamic Lubrication

A regime of lubrication in which the rheological (non-Newtonian) properties of the lubricant predominate. This term is especially applied to lubrication with grease.

Rheology

(1) The study of the interaction of time, temperature, and stress on the properties and characteristics of materials. The term is usually used in connection with noncrystalline materials such as glass, plastic, and rubbers but aspects are relevant to the creep of metals and also note **rheocasting**. (2) The science of deformation and the flow of matter.

Rheopectic Material

A material that shows an increase in viscosity with time under a constant shear stress. After removal of the shear stress, the viscosity slowly returns to its original level. Compare with *thixotropy*.

Rheotropic Brittleness

That portion of the brittleness characteristic of non-face-centered cubic metals, where tested in the presence of a stress concentration or at low temperatures or high strain rates, that may be eliminated by prestraining under milder conditions.

Rhodanizing

Electroplating with rhodium.

Rhodium

Metallic rhodium is the whitest of the platinum metals and does not tarnish under atmospheric conditions. This rare metal (symbol Rh), found in platinum ores, is very hard and is one of the most infusible of the metals. The melting point is 1963°C. It is insoluble in most acids, including aqua regia but is attacked by chlorine at elevated temperatures and by hot fuming sulfuric acid. Liquid rhodium dissolves oxygen, and ingots are made by argon-arc melting. At temperatures above 1200°C, rhodium reacts with oxygen to form rhodium oxide, Rh_2O_3. The specific gravity is 12.44. Rhodium is used to make the nibs of writing pens, to make resistance windings in high-temperature furnaces, for high-temperature thermocouples, as a catalyst, and for laboratory dishes. It is the hardest of the platinum group metals; the annealed metal has a Brinell hardness of 135. Rhodium also has considerable strength and rigidity, ultimate tensile strengths ranging from 952 to 2068 MPa, and tensile modulus from 28.9×10^4

to 37.9×10^4 MPa, depending on condition or hardness. Rhodium is also valued for electroplating jewelry, electric contacts, hospital and surgical instruments, and especially reflectors.

The most important alloys of rhodium are rhodium–platinum. They form solid solutions in any proportion, but alloys of more than 40% rhodium are rare. Rhodium is not a potent hardener of platinum but increases its high-temperature strength. It is easily workable and does not tarnish or oxidize at high temperatures. These alloys are used for thermocouples and in the glass industry.

Rhombohedral

Having three equal axes, with the included angles equal to each other, but not equal to 90°. See also the term *unit cell*.

Rib (Metals)

(1) A long V-shaped or radiused as indentation used to strengthen large sheet metal panels. (2) A long, usually thin protuberance is used to provide flexural strength to a forging (as in a rib-web forging).

Rib (Plastics)

A reinforcing member designed into a plastic part to provide lateral, horizontal, loop, or other structural support.

Rib Mark

A curved line on the crack surface of ceramics or glasses, usually convex in the general direction toward which the crack is running. The term is useful in referring to a mark of the shape until its specific nature is learned.

Riddle

A sieve used to separate foundry sand or other granular materials into various particle-size grades or to free such a material of undesirable foreign matter.

Ridging Wear

A deep form of scratching in parallel ridges usually caused by plastic flow of the subsurface layer.

Rifled Bore

Spiral ridges, grooves, or undulations deliberately introduced along the bore of tubes. In gun barrels, they spin and stabilize the shot. In tubes for heat exchangers, they cause the contents to swirl giving enhanced heat transfer capacity and reducing the possibility of hot spots and steam blanketing.

Rig Testing

Test of structures or components of assemblies intended to simulate service environments, usually on a large scale.

Rigging

The engineering design, layout, and fabrication of pattern equipment for producing castings, including a study of the casting solidification program, feeding and gating, rising, skimmers, and fitting flasks.

Right-Hand Cutting Tool

A cutter all of whose flutes twist away in a clockwise direction when viewed from either end.

Rightward Welding

See *backhand welding*.

Rigid Plastics

For purposes of general classification, a plastic that has a modulus of elasticity either in flexure or in tension greater than 690 MPa (100 ksi) at 23°C (73°F) and 50% relative humidity.

Rigid Resin

A resin having a modulus high enough to be of practical importance, for example, ≥690 MPa (100 ksi).

RIM

See *reaction injection molding*.

Rimmed Steel

A low-carbon steel that is only partially deoxidized before pouring. As a result, joy solidification some of the carbon in the steel reacts with the remaining oxygen to form carbon dioxide. This occurs fairly early during solidification, so the material forming the outer layers at the sides and bottom of the ingot tends to be lower in carbon and impurities than the interior. Also, a band of voids filled with carbon dioxide develops at about one-third the radius measured from the exterior. There is a concentration of impurities at the center line, but, because the volume of voids approximately balances the shrinkage, the central shrinkage pipe is small. During rolling, the voids collapse and fuse. The final product has a high-purity surface layer, which is particularly suitable for sheet of good surface quality and corrosion resistance. See the term *capped steel*.

Ring and Circle Shear

A cutting or shearing machine with two rotary-disk cutters driven in unison and equipped with a circle attachment for cutting inside circles or rings from sheet metal, where it is impossible to start the cut at the edge of the sheet. One cutter shaft is inclined to the other to provide cutting clearance so that the outside section remains flat and usable. See also *circle shear* and *rotary shear*.

Ring Riser

A *riser block* with openings matching those in the metal forming.

Ring Rolling

The process of shaping weldless rings from pierced disks or shaping thick-wall ring-shaped blanks between rolls that control wall thickness, ring diameter, height, and contour. See the term *axial rolls*.

Ring Seal

A piston ring-type seal that assumes its sealing position under the pressure of the fluid to be sealed.

Ring Tests

Various tests of welding characteristics in which a circular disc, hole, or groove are welded and checked for cracking.

Ringing

The audible or ultrasonic tone produced in a mechanical part by shock, and having the natural frequency or frequencies of the part. The quality, amplitude, or decay rate of the tone may sometimes be used to indicate the quality or soundness. A crude crack detection test. See also *sonic testing* and *ultrasonic testing*.

Rinsability

The relative ease with which a substance can be removed from a metal surface with a liquid such as water.

Ripple Formation

Formation of periodic ridges and valleys transverse to the direction of motion on a solid surface. Also referred to as *rippling*.

Ripple Mark

See *Wallner lines* (ceramics and glasses).

Rise Time

In urethane foam molding, the time between the pouring of the urethane mix and the completion of foaming.

Riser

(1) A reservoir of molten metal connected to a casting to provide additional metal to the casting, required as a result of shrinkage before and during solidification. (2) That section of pipeline extending from the ocean floor up the offshore oil-drilling platform. Also the vertical tube in a steam generator convection bank that circulates water and steam upward.

Riser Blocks

(1) Plates or pieces inserted between the top of a metal forming press bed or bolster and the die to decrease the height of the die space. (2) Spacers placed between bed and housings to increase *shut height* on a four-piece tie-rod straight-side press.

River Lines/Pattern

Steps on a cleavage fracture surface that radiate from the origin.

River Marks (Ceramics and Glasses)

Cleavage steps on individual grains of a polycrystalline material or on a single crystal. These markings spread out away from the point of origin. These are a special case of *twist hackle*.

River Pattern (Metals)

A term used in fractography to describe a characteristic pattern of cleavage steps running parallel to the local direction of crack propagation on the fracture surfaces of grains that have separated by cleavage.

Rivet

A rivet is a short rod with a head formed on one end. A rivet is inserted through aligned holes in two or more parts to be joined; then by pressing the protruding end, a second head is formed to hold the parts together permanently. The first is called the manufactured head and the second the point. In forming the point, a hold-on or dolly bar is used to back up the manufactured head and a rivet is driven, preferably by a machine riveter. For high-grade work such as boiler-joint riveting, the rivet holes are drilled and reamed to size, and a rivet is driven to fill the hole completely. Structural riveting uses punched holes.

Small rivets (11 mm and under) are for general-purpose work with head forms as follows: flat, countersunk, button, pan, and truss. These rivets are commonly made of rivet steel, although aluminum and copper are used for some applications. The fillet under the head may be up to 0.8 mm in radius.

Large rivets (13 mm and over) are used for structural work and in boiler and ship construction with heads as follows: roundtop countersunk, button (most common), high button or acorn, pan, cone (truncated), and flattop countersunk.

Boiler rivets have heads similar to large rivets with steeple (conical) added but have different proportions from large rivet heads in some cases.

Special-purpose rivets are tinner's rivets, which have flatheads for use in sheet metal work; cooper's rivets, which are used for riveting hoops for barrels, casks, and kegs; and belt rivets, use for joining belt ends.

Blind rivets are special rivets that can be set without access to the point. They are available in many designs but are of three general types: screw, mandrel, and explosive. In the mandrel type, the rivet is set as the mandrel is pulled through. In the explosive type, an explosive charge in the point is set off by a special hot iron; the explosion expands the point and sets the rivet.

Standard material for rivets in open-hearth steel (containing manganese, phosphorus, and sulfur) with tensile strengths of 310–380 MPa. Standards include acceptance tests for cold and hot ductility and hardness. Materials for some special-purpose rivets are aluminum and copper. In the past four decades, aluminum and titanium rivets have been used in the aerospace industry very effectively.

Riveting

Joining of two or more members of a structure by means of metal rivets, the unheaded end being upset after the rivet is in place.

Roasting

Heating an ore to affect more chemical change that will facilitate *smelting*.

Robber

An extra cathode or cathode extension that reduces the current density on what would otherwise be a high-current-density area on work being electroplated.

Robotics

Robotics is the field of engineering concerned with the development and application of robots, as well as computer systems for their control, sensory feedback, and information processing. There are many types of robotic devices, including robotic manipulators, robot hands, mobile robots, walking robots, aids for disabled persons, telerobots, and microelectromechanical systems (*MEMS*).

The term *robotics* has been very broadly interpreted. It includes research and engineering activities involving the design and development of robotic systems. Planning for the use of industrial robots in manufacturing or evaluation of the economic impact of robotic automation can also be viewed as robotics. This breath of usage arises from the interdisciplinary nature of robotics, a field involving mechanisms, computers, control systems, actuators, and software.

Robotic Mechanisms

Robots produce mechanical motion that, in most cases, results in manipulation or locomotion. For example, industrial robots manipulate parts or tools to perform manufacturing tasks such as material handling, welding, spray painting, or assembly; automated guided vehicles are used for transport of materials in factories and warehouses. Telerobotic mechanisms provide astronauts with a large manipulator for space applications. Walking machines have applications in hazardous environments. Mechanical characteristics for robotic mechanisms include degrees of freedom of movement, size and shape of the operating space, stiffness and strength of the structure, lifting capacity, velocity, and acceleration under load. Performance measures include repeatability and accuracy of positioning, speed, and freedom from vibration.

Manipulator Geometries

An important design for robotic manipulators is the articulated arm, a disfiguration with rotary joints that resembles a human arm. An extension of this idea is the selective compliance assembly robot arm (SCARA) configuration, which is been extensively applied to the assembly in manufacturing. Manipulators with this configuration have high stiffness during vertical motions. Cylindrical configurations with a Rotary joint at the base and a prismatic joint at the shoulder are useful for simple material transfer and assembly. Overhead gantry (Cartesian) robots are used for high-payload applications and those requiring rapid positioning over large workspaces.

Degrees of Freedom

Some manipulators have simple mechanical designs involving only two or three degrees of freedom of movement. Most robotic manipulators, however, have at least six degrees of freedom so that the device position a part or approach a part with any desired position or orientation. The wrist is positioned that any x, y, z position in the workspace. Then, the end effector is rotated to a desired orientation (roll, pitch, yaw). In effect, the wrist represents the origin of a three-axis coordinate fixed to the gripper. Moving the first three points of the arm translates this origin to any point in a three-axis coordinate system fixed to the work space; motion of the final three joints (in the wrist) orients the gripper coordinate system in rotation about an origin at the wrist point. Some robots have wrist joint axes that do not intersect at a single point. However, the mathematics of mechanical motion (kinematics) is considerably simplified if these axes do intersect at a point.

Types of Joints

Robotic mechanisms can have joints that are prismatic (linear motion), articulated (rotary motion), or a combination of both types. Although many robots use only articulated joints to imitate the human arm, limited actions can be produced by using prismatic joints alone. Robot locomotion can also be obtained by the use of wheels and treads. Walking robots that use articulated legs have been developed.

Joint Actuators

Actuators for moving joints of a robotic mechanism are typically electric or hydraulic motor. Although actuators can be placed directly at the joints, the weight and bulk of these motors and associated gear transmissions limit the performance of the robot, particularly at the wrist of industrial robots. Another design involves placement of the actuators in the base of the robot and transmission of motion to the joints through mechanical linkages such as shafts, belts, cables, or gears. This approach overcomes many of the problems associated with locating actuators at the joints but requires the design of backlash-free mechanical linkages that can transmit power effectively through the arm in all of its positions and orientations. Other devices include end effectors, robotic hands, and mobile robots.

Rochelle Copper

(1) A copper electrodeposit obtained from copper cyanide plating solution to which Rochelle salt (sodium potassium tartrate) has been added for grain refinement, better anode corrosion, and cathode efficiency. (2) The solution from which a Rochelle copper electrodeposit is obtained.

Rock Candy Fracture

A fracture that exhibits separated-grain facets; most often used to describe an *intergranular fracture* in a large-grained metal.

Rocking Curve

A method used in *x-ray topography* for determining the degree of imperfection in a crystal by using monochromatic, collimated x-rays reflecting off a "perfect" crystal to probe a second test crystal. A rocking curve is obtained by monitoring the x-ray intensity diffracted by the test cycle as it is slowly rocked or rotated, through the Bragg angle for the reflecting planes.

Rocking Shear

A type of guillotine shear that utilizes a curved blade to shear sheet metal progressively from side to side by a rocker motion.

Rockrite Tube-Reducing Process

See *tube reducing*.

Rockwell Hardness Number

A number derived from the net increase in the depth of impression as the load on an indentor is increased from a fixed minor load to a major load and then returned to the minor load. Various scales of Rockwall hardness numbers have been developed based on the hardness of the materials to be evaluated. The scales are designated by alphabetic suffixes to the hardness designation. For example, 64 HRC represents the Rockwall hardness number of 64 on the Rockwell C scale. See also *Rockwell superficial hardness number*.

Rockwell Hardness Test

An indentation hardness test using a calibrated machine that utilizes the depth of indentation, under constant load, as a measure of hardness. Either a 120° diamond cone with a slightly rounded point or a 1.6 or 3.2 mm (1/16 or 1/8 in.) diameter steel ball is used as the indenter. See also *Brale indenter*.

Rockwell Superficial Hardness Number

Like the *Rockwell hardness number*, the superficial Rockwell number is expressed by the symbol HR followed by a scale designation. For example, 81 HR 30N represents the Rockwell superficial hardness number of 81 on the Rockwell 30N scale.

Rockwell Superficial Hardness Test

The same test as used to determine the Rockwell hardness number except that smaller minor and major loads are used. In Rockwell testing, the minor load is 10 kgf, and the major load is 60, 100, or 150 kgf. In superficial Rockwell testing, the minor load is 3 kgf, and major loads are 15, 30, or 45 kgf. In both tests, the indenter may be either a diamond cone or a steel ball, depending principally on the characteristics of the material being tested.

Rod

A solid round metal section 9.5 mm (3/8 in.) or greater in diameter, whose length is great in relation to its diameter.

Rod Mill

(1) A *hot mill* for rolling rod. (2) A mill for fine grinding, somewhat similar to a *ball mill*, but employing long steel rods instead of balls to affect grinding.

Roke

A planar defect penetrating from a surface, due to oxide entrapment during casting.

Roll Bending

Curving sheets, bars, and sections by means of rolls. See the term *bending rolls*.

Roll Compacting

Progressive compacting of metal powders by use of a rolling mill.

Roll Flattening

The flattening of metal sheets that have been rolled in packs by passing them separately through a two-high cold mill with virtually no deformation. Not to be confused with *roller leveling*.

R

Roll Forging

A process of shaping stock between two driven rolls that rotate in opposite directions and have one or more matching sets of grooves in the rolls; used to produce finished parts or preforms for subsequent forging operations.

Roll Forming

(1) Metal forming through the use of power-driven rolls whose contour determines the shape of the product; sometimes used to denote power *spinning*. (2) A process in which a series of powered rolls progressively shape material usually strip, to some relatively complex configuration, for example, tubes of various shapes that may subsequently be seam welded.

Roll Resistance Spot Welding

Process for making separated resistance spot welds with one or more rotating circular electrodes. The rotation of the electrodes may or may not be stopped during the making of a weld.

Roll Scale

The scale formed on hot rolled material.

Roll Straightening

The straightening of metal stock of various shapes by passing it through a series of staggered rolls, the rolls usually being in horizontal and vertical planes or by reeling in two-roll straightening machines.

Roll Table

A conveyor table where rolls furnish the contact surface.

Roll Threading

See preferred term *thread rolling*.

Roll Welding

Solid-state welding in which metals are heated, then welded together by applying pressure, with rolls, sufficient to cause deformation at the faying surfaces. The rolls forced together the component faces (a lap joint) or edges (a butt joint) producing a forge weld or pressure weld at the interface. See also *forge welding*.

Rolled Compact

A compact made by passing metal powder between rollers so as to form a relatively long, sheetlike compact.

Rolled Gold

Same as *gold filled* except that the proportion of gold alloy to the weight of the entire article may be less than 1/20th. Fineness of the gold alloy may not be less than 10 k. See also *karat*.

Rolled Sections

Bars rolled to form a wide variety of cross sections including I, H, L, and X.

Rolled Steel Joist (RSJ)

A steel beam of I and H cross section for load-bearing structural applications.

Roller Air Analyzer

An air-elutriation apparatus suitable for the particles size determination of the metal powders, especially in the subsieve range. See also *elutriation*.

Roller Bearing

(1) A bearing in which the relatively moving parts are separated by rollers. See the term *rolling-element bearing*. (2) A rolling-element bearing comprising inner and outer races, the gap between them being occupied by rollers. The rollers may be parallel or tapered and are usually contained in a cage.

Roller Expansion

A technique for expanding the bore of tubes. A ring of hardened rollers supported on a tapered mandrel is inserted into and rotated around the tube bore. As the roles progress up the taper, they expand the bore. The technique is often used to secure tubes into tube plates or vessels and, with appropriate controls, strong and pressure tight joints can be achieved.

Roller Hearth Furnace

A modification of the pusher-type continuous furnace that provides for rollers in the hearth or muffle of the furnace whereby friction is greatly reduced and lightweight trays can be used repeatedly without risk of unacceptable distortion and damage to the work. See also *pusher furnace*.

Roller Leveler Breaks

Obvious transverse *breaks* usually about 3–6 mm (1/8–1/4 in.) apart caused by the sheet metal fluting during *roller leveling*. These will not be removed by stretching.

Roller Leveler Lines

Same as *leveler lines*.

Roller Leveling

(1) *Leveling* by passing flat sheet metal stock through a machine having a series of small-diameter staggered rolls that are adjusted to produce repeated reverse bending. (2) Processes for leveling sheet and straightening bar by passing the material through a series of staggered rolls. The first rollers induce a fairly severe bend and subsequent rollers apply bends of progressively reducing severity in alternate directions until the material emerges level/straight.

R

Roller Spot Welding

A resistance welding process in which a series of spot welds is produced in sheets traveling through a pair of rollers (discs would usually be a better description), which apply a driving and clamping force and deliver an intermittent electrical current of high amperage to cause welding. If the spot welds overlap, the process may be termed *seam welding*. If rotation of the rollers stops during the electrical supply, the processes may be termed "*step by step.*"

Roller Stamping Die

An engraved roller used for impressing designs and markings on sheet metal.

Roller Threading

A process for producing a thread form by forcing grooved roller dies against the bar being threaded. Compared to a cut thread, the roll threaded component is usually cheaper, slightly higher bulk strength, and contains compressive residual stresses at the surface, which can improve fatigue performance.

Rolling

The reduction of the cross-sectional area of metal stock, or the general shaping of metal products, through the use of rotating rolls. See also *rolling mills*.

Rolling (as in Surface Rolling)

A process in which a narrow roll bears heavily as it rolls against a component. The intention is to induce beneficial compressive residual stresses, which inhibits fatigue crack initiation. It is applied to features such as the radii of crankshafts.

Rolling (Pure Rolling with No Sliding and No Spin)

A motion of two relatively moving bodies, of opposite curvature, whose surface velocities in the common contact area are identical with regard to both magnitude and direction. See also *sliding* and *spin*. A process in which a bar, slab, etc. is reduced in cross section as it travels between a pair of rolls having a gap between them, which is smaller than the original thickness of the material. The rolls are usually driven, and tension may be applied to the material on the inlet or exit sides to assist or restrain its passage.

Rolling Angle

The angle between the line drawn through the working roll centers and the line drawn from one roll center to the point on the roll surface, which is first contacted by the material entering. Also termed *Angle of Bite*, *Nip (Angle)*, *contact angle*, and *similar variations*.

Rolling Direction (in Rolled Metals)

See *longitudinal direction*.

Rolling Element Bearing

A bearing in which the load between two components in relative motion is carried via rolling elements such as balls, rollers, or needles. These rolling elements may bear directly onto the shaft or the housing, or they may bear on inner and outer *races* carried by the respective component. Compare with *plain bearing*.

Rolling Mandrel

In *ring rolling*, a vertical roll of sufficient diameter to accept various sizes of ring blanks and to exert rolling force on an axis parallel to the main roll. See also the term *axial rolls*.

Rolling Mill Equipment (or the Building Containing the Equipment)

Rolling, that is, reducing the cross section of bar, slab, sheet, etc., as it travels through the gap between a pair of rolls. Mills are described by a wide variety of terms. The simplest mills comprise only two rolls, other rolls have more, the total number being indicated by terms such as two-high, four-high, etc. A *Cluster Mill* or *Sendzimer Mill* has six or more rolls. Some mills are reversing, that is the material is worked as it is passed through first in one direction and then in the other, the gap being reduced between passes. In a three high mill, all of the rolls are working rolls, that is, they reduce the material as it passes through one pair and returns through the other pair. In a four high and more complex mill only two of the rolls are working rolls, the other backing or supporting rolls minimize flexing of the working rolls. A *Universal Mill* has multiple rolls position to work both the vertical and horizontal axes of the material. Rolls are carried in *Stands* and material usually passes through a number of stands during a rolling cycle. In a *Tandem Mill*, two or more parallel stands are grouped close together so that the material enters the second and subsequent rolls before clearing the first. In a *Looping Mill*, two or more stands live side by side so that material, usually bar, emerging from one is directed in a loop to enter the second stand. In a *Cross Country Mill*, the material completely clears each stand before traveling to the next.

Rolling Mills

Machines used to decrease the cross-sectional area of metal stock and to produce certain desired shapes as the metal passes between rotating rolls mounted in a framework comprising a basic unit called *a stand*. Cylindrical rolls produce flat shapes; grooved rolls produce rounds, squares, and structural shapes. See also *four-high mill*, *Sendzimir mill*, and *two-high mill*.

Rolling Texture

Preferred orientation produced by rolling.

Rolling Velocity

See *sweep velocity*.

Rolling-Contact Fatigue

Repeated stressing of a solid surface due to rolling contact between it and another solid surface or surfaces. Continued rolling-contact fatigue of bearing or gear surfaces may result in rolling-contact damage in the form of subsurface fatigue cracks and/or material pitting and spallation.

Rolling-Contact Wear

Wear to a solid surface that results from rolling contact between that surface and another solid surface or surfaces.

Röntgen

The unit of radiation dose, $1R + 2.58 \times 10^{-4}$ C/kg.

Room Temperature

A temperature in the range of $-20°C$ to $30°C$ ($-68°F$ to $85°F$). The term room temperature is usually applied to an atmosphere of unspecified relative humidity.

Room-Temperature Setting Adhesive

An adhesive that sets (to handling strength) within an hour at temperatures from $20°C$ to $30°C$ ($-68°F$ to $85°F$) and later reaches full strength without heating. Compare with *cold-setting adhesive*, *hot-setting adhesive*, and *intermediate temperature setting adhesive*.

Room-Temperature Vulcanizing (RTV)

Vulcanization or curing at room temperature by chemical reaction, particularly of silicones and other rubbers.

Root

See preferred term *root of joint* and *root of weld*.

Root Concavity

A hollow in the root of a weld.

Root Crack

A crack in either the weld or heat-affected zone at the root of a weld. See the term *weld crack*.

Root Edge

A *root face* of zero width.

Root Face

The portion of a weld groove face adjacent to the root of the joint.

Root Gap

See preferred term *root opening*.

Root Head

A weld deposit that extends into or includes part or all of the root of the joint. See also *root of joint*.

Root Mean Square (rms)

(1) The square root of the arithmetic mean of the squares of the numbers in a given set of numbers. (2) The effective value of an alternating periodic voltage or current. (3) A term describing the surface roughness of a machined surface. (4) The area beneath a waveform. See also *surface roughness*.

Root of Joint

The portion of a weld joint where the members are closest to each other before welding. In cross section, this may be a point, a line, or an area. Same as *root of weld preparation*.

Root of Weld

(1) In the case of a weld to be made; it is the area of the parent materials where the first deposit is to be made. In the case of a completed weld, it is the face of the first weld run that is remote from the operator. (2) The points, at which the weld bead intersects the base-metal surfaces either nearest to or coincident with the *root of joint*.

Root of Weld Preparation

The area where two surfaces to be joined are in the closest approach in which the first run will be deposited.

Root Opening

In a weldment, the separation between the members at the *root of joint* prior to welding.

Root Pass

The first bead of a *multi-pass weld*, laid in the *root of joint*.

Root Penetration

The depth that a weld extends into the *root of joint*, measured on the centerline of the root cross section. See the term *joint penetration*.

Root Radius

See preferred term *groove radius*.

Root Rolling

The rolling of cut thread roots and similar features to induce favorable residual stresses to resist fatigue.

Root Run/Pass (of Weld)

The first run of a multiple run weld.

Rooting

See *oxide rooting*.

Root-Mean-Square End-to-End Distance

A measure of the average size of a coiled polymer molecule, usually determined by light scattering.

Rosebuds

Concentric rings of distorted coating, giving the effect of an opened rosebud. Noted only on *minimized spangle*.

Rosette

(1) Rounded configuration of microconstituents in metals arranged in whorls or radiating from a center. (2) Strain gages arranged to indicate at a single position strains in three different directions. (3) Typically, three gages are aligned at 120° to each other. See also *strain gage*.

Rosette (Star)Fracture

A tensile fracture that exhibits a central fibrous zone, and intermediate region of radial shear, and an outer circumferential shear-lip zone. Rosette fractures are often observed in temper-embrittled steels. See also *shear lip* and *temper embrittlement*.

Rosette Graphite

Arrangement of graphite flakes in which the flakes extend radially from the center of crystallized areas and gray cast iron. See the term *flake graphite*.

Rosin

The common resin of several varieties of pine tree, found widely distributed in North America and Europe. It is obtained by cutting a longitudinal slice in the tree and allowing the exudation to drip into containers. The liquid resin is then distilled to remove the turpentine, and the residue forms what is known as *gum rosin*, or *pine gum*. *Wood rosin* is obtained by distillation of old pine stumps. It is darker than gum rosin and is inferior for general use.

Rosin contains seven acids with very similar characteristics but consists chiefly of *abietic acid*. Normally, when gum rosin is heated, the natural pimaric acid isomerizes to form abietic acid; but in the production of turpentine and rosin from pine sap, the turpentine is removed by steam distillation, and various acids are then extracted. Pimaric acid is closely related to abietic acid. It reacts with maleic and anhydride, and *maleopimaric acid* is used in printing inks and coatings. Rosin has a specific gravity of about 1.08 and a melting point of about 180°F (82°C), and it is soluble in alcohol, turpentine, and alkalies. It is used in varnishes, paint driers, soluble oils, paper sizing, and belt dressings; and for compounding with rubber and other resins; and for producing many chemicals.

Rosin is generally graded commercially by letters according to color. The darkest grade is B, and the highest is W. Extra grades are A, nearly black, and WW, water-white. Thirteen color grades are designated under the Naval Stores Act. The dark grades of wood rosin are considered inferior. They have a high melting point and low acid number and are used for making rosin oil; for battery waxes, thermoplastics, and dark varnish; and for linoleum manufacture. Rosin is usually marketed in barrels of 280 lb (127 kg). **Naval stores** is an old name for rosin and turpentine. Palletized rosin consists of free-flowing, dustless pellets produced by coating droplets of molten rosin with inert powder.

Hardened rosin is a weak resonate made by adding 6%–8% high-calcium lime to melted rosin. It is used in some varnishes.

Hydrogenated rosin has greater resistance to oxidation than common rosin, has less odor and taste, and has a pale color that is more stable to light. It is used in protective coatings, in paper size, in adhesives, in soaps, and as a tackifier and plasticizer in rubber. Because of its saturated nature, it cannot be used for rosin-modified plastics.

Rosin is hardened by polymerization to form a dipolymer of abietic acid. The product is then pale in color and has a lower acid number and a higher melting point than rosin. *Poly-pale resin*, of Hercules, is a polymerized rosin with a melting point of 208°F–217°F (98°C–103°C), acid number 152–156, and saponification value 157–160. It can be substituted for natural copals in paints, and in gloss oil, it gives water resistance and high viscosity. In the making of metallic resinates, it gives higher melting points, higher viscosity, and better solubility than natural rosin. Compare with *resin*.

Rosin Flux

A soldering flux having a rosin base that becomes interactive after being subjected to the soldering temperature. See also *flux*.

Rotary Filling and Burring

Machining or smoothing surfaces with contour-fitting rotary tools where only a minimum amount of material is to be removed.

Rotary Forging

A process in which the workpiece is pressed between a flat anvil and a swiveling (rocking) die with a conical working face; the platens move toward each other during forging. Also called orbital forging. Compare with *radial forging*. See *Pilger mill* and *tube making*.

Rotary Furnace

A circular furnace constructed so that the hearth and workpieces rotate around the axis of the furnace during heating. The charge is inserted at one point and remains in the hearth for nearly a full circle before being withdrawn. Also called rotary hearth furnace.

Rotary Press

A machine for forming powder metallurgy parts that is fitted with a rotating table carrying multiple die assemblies in which powder is compacted.

Rotary Retort Furnace

A continuous-type furnace in which the work advances by means of an internal spiral, which gives good control of the retention time within the heated chamber.

Rotary Roughening

A method of surface roughening prior to thermal spraying wherein a revolving roughening tool is pressed against the surface being prepared, while either the work, or the tool, or both, move.

Rotary Seal

A mechanical seal that rotates with a shaft and is used with a stationary mating ring.

Rotary Shear

(1) A sheet metal cutting machine with two rotating-disk cutters mounted on parallel shafts driven in unison. (2) A device for cutting sheet and plate utilizing a pair of driven, overlapping disk cutters.

Rotary Swager

A swaging machine consisting of a power-driven ring that revolves at high speed, causing rollers to engage cam surfaces and force the dies to deliver hammer-like blows on the work at high frequency. Both straight and tapered sections can be produced.

Rotary Swaging

A *bulk forming* process for reducing the cross-sectional area or otherwise changing the shape of bars, tubes, or wires by repeated radial blows with one or more pairs of opposed dies.

Rotating Bending Fatigue Test

Various fatigue tests in which a cylindrical test piece is rotated while under a lateral load so that any point on its periphery undergoes one complete tensile/compression reverse cycle per revolution. The test piece may be supported as a cantilever, or in three-point or four-point bending depending on the particular machine.

Rotating Electrode Powder

An atomized powder consisting exclusively of solid spherical or near-spherical particles.

Rotating Electrode Process

A method for producing metal powders wherein a consumable metal or alloy electrode is rotated during arc melting at very high speeds, and the molten droplets are propelled by centrifugal force toward the wall of a large collecting chamber while solidifying in flight. If the consumable electrode is melted by an electric arc, the process is called the rotating electrode process, or REP; if it is melted by a plasma arc, the process is called the plasma rotating electrode process, or PREP; and if the bar is melted by a laser, the process is called the laser rotating electrode process (LREP). It is also possible to use electron beam melting. The melting technique used depends primarily on the need for cleanness of the powder and on the type of atmosphere required for the alloy.

Rotational Casting

A method used to make hollow articles from thermoplastic materials. The material is charged into a hollow mold capable of being rotated in one or two planes. The hot mold fuses the material into a gel after the rotation has caused it to cover all surfaces. The mold is then chilled, and the product is stripped out.

Rotational Molding

The preferred term for a variation of the *rotational casting* process that uses dry, finely divided (35 mesh, or 500 μ) plastic powders, such as polyethylene, rather than fluid materials. After the powders are heated, they are fused against the mold walls forming a hollow item with uniform wall thickness.

Rouge

A hydrated iron oxide used for polishing metals and in break-in lubricants for aluminum bronze bearings. It has a Mohs hardness of 5.5–6.5 and is made by calcining ferrous sulfate and driving off the sulfur. The color is varying shades of red; the darker the color, the harder the rouge. The grains are rounded, unlike the grains of crocus. The pale-red rouge is used for finishing operations; the other grades are used for various polishing of metal surfaces. *Stick rouge* is made of finely crushed powder. Although the word rouge means red, materials of other colors are used for buffing and are called rouge. *Black rouge*, also called *Glassite*, is magnetic iron oxide made by precipitating ferrous sulfate with caustic soda. It is used for buffing but is not popular because it stains the skin. *Green chrome rouge* is *chromium oxide*, CrO, made by the strong heating of chromic hydroxide. It is used for buffing stainless steels. When used as a paint pigment, it is called *Guignet's green*. *Satin rouge* is a name applied to lamp black when used as a polishing medium in the form of brick for polishing silverware. **Crocus** is a name applied to metal powders of the deep-yellow, brown, or red color made into cakes with grease for polishing. *Polishing crocus* is usually red ferric oxide used for a buffing glass and jewelry. Crocus cloth is a fabric coated with red iron oxide, marketed in sheets and used for polishing metals.

Rouge Finish

A highly reflective finish produced with rouge (finely divided, hydrated iron oxide) or other very fine abrasive, similar in appearance to the bright polish or mirror finish on sterling silver utensils.

Rough Blank

A *blank* for a metal forming or drawing operation, usually of a regular outline, with necessary stock allowance for process metal, which is trimmed after forming or drawing to the desired size.

Rough Grinding

Grinding without regard to finish, usually to be followed by a subsequent operation.

Roughing Stand

The first stand (or several stands) of rolls through which a reheated *billet* passes in front of the finishing stands. See also *rolling mills* and *stand*.

Rough Machining

(1) Machining without regard to finish, usually to be followed by a subsequent operation. (2) Machining to an approximate size, usually oversize so that subsequent final or finish machining achieves the required dimension and finish.

Roughness

(1) Relatively finely spaced surface irregularities, the heights, widths, and directions of which establish the predominant surface pattern. (2) The microscopic peak-to-valley distances of surface protuberances and depressions. See also *surface roughness* and *surface finish*.

Roughness-Width Cutoff

The maximum width of surface irregularities to be included in the measurement of roughness height.

Rough-Polishing Process

A polishing process having the primary objective of removing the layer of significant damage produced during earlier machining and abrasion stages of a medley graphic preparation sequence. A secondary objective is to produce a finish of such quality that a final polish can be produced easily.

Rough Threading

A method of surface roughening prior to thermal spraying, which consists of cutting threads with the sides and tops of the threads jagged and torn.

Routing

Cutting out and contouring edges of various shapes in a relatively thin material using a small diameter rotating cutter, which is operated at fairly high speeds.

Roving

A number of yarns, strands, tows, or ends collected into a parallel bundle with little or no twist. Rovings are used to produce fiber-reinforced plastics. See also *woven roving*.

Roving Cloth

A textile fabric, coarse in nature and woven from roving, used in reinforced plastics.

Row Nucleation

The mechanism by which stress-induced crystallization is initiated in reinforcing fibers for composites, usually during fiber spinning or hot drawing.

RRIM

See *reinforced reaction injection molding*.

RSJ

Rolled steel joist.

RTM

See *resin transfer molding*.

RTV

See *room-temperature vulcanizing*.

Rub Mark

See *abrasion*.

Rubber

This term originally referred to a natural or tree rubber, which is a hydrocarbon polymer of isoprene units. With the development of synthetic rubbers having some rubbery characteristics but differing in chemical structure as well as properties, a more general designation was needed to cover both natural and synthetic rubbers. The term *elastomer*, a contraction of the words elastic and polymer, was introduced and defined as a substance that can be stretched at room temperature to at least twice its original length and, after having been stretched and the stress removed, returns with force to approximately its original length in a short time.

Three requirements must be met for rubbery properties to be present in both natural and synthetic rubbers: long threadlike molecules, flexibility in the molecular chain to allow flexing and coiling, and some mechanical or chemical bonds between molecules.

A useful way of visualizing rubber structure is to consider as a model a bundle of wiggling snakes in constant motion. When the bundle is stretched and released, it tends to return to its original condition. If there were no entanglements, stretching the bundle would pull it apart. The more entanglements, the greater the tendency to recover, corresponding to cross-links in rubber. In rubbers, the characteristic property of reversible extensibility results from the randomly coil arrangement of the long polymer chains. Upon extension, the chains are elongated in a more or less orderly array. The tendency to revert to the original coiled disarray upon removal of the stress accounts for the elastic behavior. Vulcanizing rubber increases the number of cross-links and improves the properties.

Natural rubber and most synthetic rubbers are also commercially available in the form of latex, a colloidal suspension of polymers in an aqueous medium. Natural rubber comes from trees in this form and many synthetic rubbers are polymerized in this form; some other solid polymers can be dispersed in water.

Latexes are the basis for a technology and production methods completely different from the conventional methods used with solid rubbers.

Processing

In the crude state, natural and synthetic rubbers possess certain physical properties that must be modified to obtain useful end products. The raw or unmodified forms are weak and adhesive. They lose their elasticity with use, change markedly in physical properties with temperature, and are degraded by air and sunlight. Consequently, it is necessary to transform the crude rubbers by compounding and vulcanization procedures into products that can better fulfill a specific function.

Curatives and Vulcanization

After the addition of curing or vulcanizing agents to rubber, the application of heat causes cross-linking, yielding a durable product by binding the long chains together. Sulfur, the first successful curing ingredient, is the basis of vulcanization. However, various chemicals or combinations of chemicals are also capable of vulcanizing rubber. These include oxidizing agents, such as selenium or tellurium, organic peroxides, and nitro compounds, and also generators of free radicals, such as organic peroxides, azo compounds, and certain organic sulfur compounds such as the alkyl disulfides.

Sulfur alone results in very slow cure rates and develops less than optimum physical properties in the rubber. Several classes of compounds have been found that accelerate the rate of vulcanization, improve the efficiency of the vulcanization reaction, and reduce the sulfur requirements, in general enhancing the physical properties of the vulcanized rubber.

Accelerators are substances that act as catalysts in the vulcanization process, initiating free radicals. The first accelerators to be used in the rubber industry were inorganic chemicals, such as the basic carbonates and oxides of lead supplemented by magnesium or lime.

Some rubbers can be vulcanized by gamma radiation.

Pigments

Vulcanization improves the elasticity and aging properties of rubber, but in most cases it is necessary to enhance further such properties as tensile strength, abrasion resistance, and tear resistance by incorporation of fillers. Those that improve specific physical properties are known as reinforcing fillers; those that serve primarily as diluents are classed as inert fillers. The physical properties of the resulting vulcanization are affected by both the type and amount of filler.

The most universally used reinforcing filler in the rubber industry is carbon black. The types used commercially in the greatest bulk for this purpose include furnace and thermal blacks.

In addition to the carbon blacks, inorganic reinforcing agents, such as zinc oxide and the silicas, are used for reinforcement of light-colored end products.

Specific Types of Rubber

One rubber is obtained from natural sources in commercial quantities. In addition, a number of synthetic compounds are classified as rubbers.

Smoked sheet and pale crepe represent the forms in which the major portion of natural rubber is commercially available.

Natural Rubber (NR)

Although natural rubber may be obtained from hundreds of different plant species, the most important source is the rubber tree. Natural rubber is cis-1,4-polyisoprene, containing approximately 5000 isoprene units in the average polymer chain.

The economic competition from synthetic rubbers has stimulated research and development in natural rubber by increasing productivity in the field, improving uniformity and quality of the product and packaging, and developing modified natural rubbers with specific properties.

Increased productivity has been achieved by increasing the yield of the trees by cross-pollination of high-rubber-yielding clones, use of chemical stimulants, and better tapping and collection methods. These methods have resulted in large increases in yields on Malaysian estates, with much greater increases anticipated.

With improved productivity and better processing methods and controls, much better quality and uniformity have been achieved. This made possible the development of standard Malaysian rubbers (SMR) to meet specifications on a number of properties, including dirt and ash content, viscosity, and copper and manganese content.

In addition, experimental work has been done on improved collection of latex by microtapping and collecting in polyethylene bags, and trees having three parts: a high-yielding trunk system grafted onto a strong root system and an improved, more prolific leaf system grafted onto the trunk.

Despite competition from synthetic rubbers and a great reduction in percentage use worldwide of natural rubber, the tonnage of natural rubber continues a steady growth in parallel with the growth of the industry.

Styrene–Butadiene Rubbers (SBR)

The extensive development of the synthetic rubber industry originated with the World War II emergency, but continued expansion has been the result of the superiority of the various synthetic rubbers in certain properties and applications. The most important synthetic rubbers and the most widely used rubbers in the entire world are the styrene–butadiene rubbers (SBR).

Formally designated GR-S, SBRs are obtained by the emulsion polymerization of butadiene and styrene in varying ratios.

When SBR is used in light-duty tires, such as passenger car tires, the cold-rubber compounds have proved equal or superior to natural rubber treads. However, they are inferior to natural rubber for truck tires because of the greater heat buildup during flexing. The cold, oil-extended type, prepared by the replacement of a portion of the polymer by a heavy-traction oil, accounts for more than 50% of the cold-rubber production. Its advantage is primarily economic. Master batches, containing both oil and carbon black, have the advantage of process simplification.

In general, the compounding and processing methods for all the SBR types are similar to those of natural rubber. Although natural rubber is superior with respect to lower heat buildup, resilience, and hot tear strength, the SBR types are more resistant to abrasion and weathering. However, carbon black or some other reinforcing fillers must be added to the SBR to develop the best physical properties. Unlike natural rubber, SBR does not crystallize on stretching and thus has low tensile strength unless reinforced. The major use for SBR is in tires and tire products. Other uses include belting, hose, wire and cable coatings, flooring, shoe products, sponge, insulation, and molded goods.

cis-1,4-Polybutadiene (BR)

Work on the duplication of natural rubber stimulated interest in stereoregulated polymerization of butane, particularly of the high cis-1,4 structure.

Black-loaded vulcanizates of cis-1,4-polybutadiene rubbers exhibit good physical properties, such as lower heat generation, higher resilience, improved low-temperature properties, and greatly improved abrasion resistance. Processing properties are rather poor but can be greatly improved by employing blends with natural rubber or SBR. Tests on retreaded passenger tires gave outstanding abrasion resistance and increased resistance to cracking as compared with natural rubber. In passenger tires, cis-1,4-polybutadiene improves tread wear by about 1% for each percent of polybutadiene and the tread compound. In truck tires, a blend with natural rubber gives 14% more wear than natural rubber alone.

Butyl Rubber

Isobutylene and isoprene or butadiene obtained from cracked refinery gases are the primary raw materials required for the manufacture of butyl rubber.

Neither neoprene nor butyl rubber requires carbon black to increase its tensile strength, but the reinforcement of butyl rubber by carbon black or other fillers does improve the modulus and increases their resistance to tear and abrasion.

The excellent resistance of butyl rubbers to oxygen, ozone, and weathering can be attributed to the smaller amount of unsaturation present in the polymer molecule. In addition, these rubbers exhibit good electrical properties and high permeability to gases. The high permeability to gases results in use of butyl as an inner liner in tubeless tires. Other widespread uses are for wire and cable products, injection-molded and extruded products, hose, gaskets, and sealants, and where good damping characteristics are needed.

R

Ethylene–Propylene Polymers (EPM, EPDM)

Stereospecific catalysts are employed to make synthetic rubbers by the copolymerization of ethylene and propylene. Either monomer alone polymerizes to a hard, crystallizable plastic, but copolymers containing 35%–65% of either monomer are amorphous, rubbery solids. Special catalysts must be employed because ethylene polymerizes many times faster than propylene. The best results seem to be obtained with complex catalysts derived from an aluminum alkyl and a vanadium chloride or oxychloride.

Processing techniques and factory equipment used with other rubbers can also be applied to these copolymers. The mechanical properties of their vulcanizates are generally approximately equivalent to those of SBR.

Terpolymers containing ethylene, propylene, and a third monomer, such as dicyclopentadiene, have become more popular because they contain unsaturation and thus may be sulfur-cured by using more or less conventional curing systems.

Neoprene (CR)

One of the first synthetic rubbers used commercially in the rubber industry, neoprene is a polymer of chloroprene, 2-chloro butadiene-1,3. In the manufacturing process, acetylene, the basic raw material, is dimerized to vinyl-acetone and then hydrochlorinated to the chloroprene monomer.

Sulfur is used to vulcanize some types of neoprene, but most of the neoprenes are vulcanized by the addition of basic oxides such as magnesium oxide and zinc oxide. The cure proceeds through reaction of the metal oxide with the tertiary allylic chlorine that arises from the small amount of 1,2-polymerization that occurs. Other compounding and processing techniques follow similar procedures and use the same equipment as for natural rubber. One of the outstanding characteristics of neoprene is the good tensile strength without the addition of carbon black filler. However, carbon black and other fillers can be used when reinforcement is required for specific end-use applications that require increased tear and abrasion resistance.

The neoprenes have exceptional resistance to weather, sun, ozone, and abrasion. They are good in resilience, gas impermeability, and resistance to heat, oil, and flame. They are fairly good at low temperature and electrical properties. This versatility makes them useful in many applications requiring oil, weather, abrasion, or electrical resistance or combinations of these properties, such as wire and cable, hose, belts, molded and extruded goods, soles and heels, and adhesives.

Nitrile Rubber (NBR)

Much of the basic pioneering research on emulsion polymerization systems was with nitrile-type rubbers. These rubbers, first commercialized as the German Buna N types in 1930, are copolymers of acrylonitrile and a diene, usually butadiene.

Both sulfur and nonsulfur vulcanizing agents may be used to cure these rubbers. Carbon black or other reinforcing agents are necessary to obtain the optimum properties. If proper processing methods are followed, the nitrile rubbers can be blended with natural rubber, polysulfide rubbers, and various resins to provide characteristic such as increased tensile strength, better solvent resistance, and improved weathering resistance.

The nitrile rubbers have outstanding oil, grease, and solvent resistance. Consequently, the commercial usage of these rubbers is largely for items in which these properties are essential. Another major usage is the utilization of the latex form for adhesives and for the finishing of leather, impregnation of paper, and the manufacture of nonwoven fabrics.

cis-1,4-Polyisoprene (IR)

In 1954, synthetic cis-1,4-polyisoprene was made from isoprene with two different classes of catalysts. The first class includes lithium and the lithium alkyls. The second class uses a mixture of an aluminum alkyl and titanium tetrachloride, the system first used for the low-pressure polymerization of ethylene. Both catalyst system polymerizations are carried out in hydrocarbon solution and require highly purified monomer and solvent. Traces of air, moisture, and most polar compounds adversely affect reaction rates, polymer properties, and structure.

The cis-1,4 polymer structure obtained with these catalysts is also characteristic of natural Hevea rubber. The presence of a high cis content appears necessary for the desirable physical properties with Hevea in contrast to the inferior properties of emulsion-type polyisoprene, which contains mixed cis-, trans-1,2-, and 3,4-isoprene units.

This polymer and the corresponding butadiene polymer discussed above are called stereorubbers because of their preparation with stereospecific catalysts. The emulsion polymers are formed by a free-radical mechanism that does not permit control of the molecular structure. Stereorubbers are formed by anionic mechanisms that permit nearly complete control of the structure of the growing polymer chain in stereoregular fashion.

The type of catalyst employed influences the structure of the polymer. The aluminum–titanium-catalyzed rubbers exhibit, on stretching, a gradual crystallization somewhat slower than natural, and they are readily processed because the molecular weight range tends to be less than that of natural rubber. The lithium-catalyzed rubbers contain about 93% cis-1,4 structure; they exhibit very little tendency to crystallize and cannot be processed satisfactorily until they have undergone substantial mastication to reduce their very high molecular weight values to a level more nearly resembling that of well masticated natural rubber.

The differences in structure influence the properties of these rubbers. The alumina–titanium-catalyzed rubber, because it crystallizes more readily, exhibits better hot tensile strength. The lithium-catalyzed polymer, as it is of higher molecular weight exhibits higher resilience and less heat generation. Both types, when compounded and cured, produce physical properties that closely approach, or are equivalent to, natural rubber.

The tests with heavy-duty tires for trucks, buses, and airplanes have shown that with respect to wear and heat buildup the isoprene rubbers are comparable to natural rubber during tire operation. Polyisoprene rubber has passed qualification tests in high-speed jet aircraft tires to withstand landing speeds as high as 250 mi/h (112 m/s).

Other Rubbers

There are other specialty rubbers available that are important because of specific properties, but in the aggregate, they make up only approximately 2% of the world production of synthetic rubbers.

Silicone Rubbers

Silicone rubber is a linear condensation polymer based on dimethyl siloxane. In the preparation, dimethyl dichlorosilane is hydrolyzed to form dimethyl silanol, which is then condensed to dimethyl siloxane, and this, upon further condensation, yields dimethyl polysiloxane, the standard silicone rubber.

Various types of silicone rubbers are produced by substituting some of the methyl groups in the polymer with other groups such as phenol or vinyl. Advantages of this type of substitution are evidenced by improvements in specific properties. For example, the presence of phenol groups in the polymer chain gives further improvement in low-temperature properties. Fluorine-containing side groups improved chemical resistance. Many types are commercially available, ranging from fluid liquids to tough solids.

Because sulfur is not effective for the vulcanization of most silicone rubbers, a strong oxidizing agent, such as benzoyl peroxide, is used; the cross-linking produced is random.

Although the standard silicone rubbers are not reinforced by carbon black, the physical properties can be improved by the incorporation of various inorganic fillers such as titania, zinc oxide, and silica, which act as reinforcing and modifying agents. The physical, chemical, and electrical properties can be altered by varying the type and amount of these fillers. Carbon black can be used as a filler with vinyl-containing polymers.

In general, the silicone rubbers have relatively poor physical properties and are difficult to process. However, they are the most stable of rubbers that are capable of remaining flexible over a temperature range of −90°C to 316°C. They are unaffected by ozone, are resistant to hot oils, and have excellent electrical properties. Their most extensive uses are for wire and cable insulation, tubing, packings, and gaskets in aerospace and aircraft applications. In the form of dispersions and pastes, they are used for dip-coating, spraying, brushing, and spreading. Silicone rubbers are important in medical and surgical devices because of their property, unique among elastomers, of being compatible with body tissues. Fast, automatic, economical injection molding of liquid silicones has been developed.

Hypalon

Hypalon is the DuPont trade name for a family of chlorosulfonated polyethylenes prepared by treating polyethylenes with a mixture of chlorine and sulfur dioxide, whereby a few scattered chlorine and sulfonyl chloride groups are introduced into the polyethylene chain. By this treatment, polyethylene is converted to a rubberlike material in which the undesirable degree of crystallinity is destroyed but other desirable properties of polyethylene are retained. The outstanding chemical stability of Hypalon results from the complete absence of unsaturation in the polymer chain.

Vulcanization is accomplished by means of metallic oxides, such as litharge, magnesia, or red lead, and the presence of an accelerator as a stock in itself. Hypalon can be blended with other types of rubbers to provide a wide range of properties.

Hypalon has extreme resistance to ozone. Its chemical resistance to strong chemicals, such as nitric acid, hydrogen peroxide, and strong bleaching agents, is superior to any of the commonly used rubbers. These vulcanizates also have good heat resistance, mechanical properties, and unlimited color ability. Typical applications include white sidewall tires and a variety of sealing, waterproofing, insulating, and molded items.

Epichlorohydrin Elastomers

These elastomers are polymers of epichlorohydrin. The two main types of epichlorohydrin elastomers are the homopolymer (CO) and the copolymer (ECO) of epichlorohydrin and ethylene oxide. The copolymer has a lower brittle point, better resilience, and a lower specific gravity than the homopolymer but is not as good in high-temperature properties.

Since the backbone of the molecule is saturated, these elastomers cannot be vulcanized with sulfur. Cross-linking is achieved by reaction of the chloromethyl side group with diamines of thioureas. A metal oxide is also required.

They exhibit outstanding ease of processing; extreme impermeability to gases; moderate tensile strength and elongation; high modulus; good abrasion resistance; good heat aging; excellent resistance to hot oil, water, perchloroethylene, acids, bases, and ozone; good low-temperature properties; and electrical properties raging between those of a poor insulator and a good conductor, depending on compounding.

These properties make the epichlorohydrin elastomers useful in gaskets for oil field specialties, diaphragms, and pump and valve parts; hose for low-temperature flexibility; oil and fuel resistance, or gas impermeability applications, and mechanical goods such as belting, wire, and cable.

Thermoplastic Elastomers

Proper choice of catalyst and order of procedure in polymerization have led to development of thermoplastic elastomers. The leading commercial types are styrene block copolymers with a structure that, unlike the random distribution of monomer units in conventional polymers, consists of polystyrene segments or blocks connected by rubbery polymers such as polybutadiene, polyisoprene, or ethylene–butylene polymer. These types can be designated SBS, SIS, and SEBS, respectively. Other types of thermoplastic elastomers, such as the polyester type, have also been developed.

SBS, SIS, and SEBS polymers, when heated to a temperature above the softening point of the styrene blocks, can be processed and shaped like thermoplastics but, when cooled, act like vulcanized rubber, with the styrene block serving as cross-links. Thus, it is unnecessary to vulcanize. Also, scrap or excess material can be reused.

Thermoplastic elastomers are very useful in providing a fast and economical method of producing a variety of products; including molded goods, toys, sporting goods, and footwear. One of the disadvantages for many applications is the low softening point of the thermoplastic elastomers. Attempts to modify the composition and structure to raise the softening temperature have generally resulted in inferior rubbery properties.

Fluoroelastomers

These are basically copolymers of vinylidene fluoride and hexafluoropropylene.

Because of their fluorine content, they are the most chemically resistant of the elastomers and also have good properties under extremes of temperature conditions. They are useful in the aircraft, automotive, and industrial areas.

Polyurethane Elastomers

Polyurethane elastomers are of interest because of their versatility and variety of properties and uses. They can be used as liquids or solids in a number of manufacturing methods. The largest use has been for making foam for upholstery and bedding, but difficulties have been encountered in public transportation due to flammability.

Polysulfide Rubbers

These rubbers have a large amount of sulfur in the main polymer chain and are therefore very chemically resistant, particularly to oils and solvents. They are used in such applications as putties, caulks, and hose for paint spray, gasoline, and fuel.

Polyacrylate Rubbers

Polyacrylate rubbers are useful because of their resistance to oils at high temperatures, including sulfur-bearing extreme-pressure lubricants.

Products

A wide variety of products have a rubber, or elastomer, as an essential component. Most rubber products contain a significant amount of nonrubber materials, used to impart processing, performance, or cost advantages.

The automotive industry is the biggest consumer of rubber products. About 60% of all rubber used goes into the production of passenger, bus, truck, and off-the-road tires. In addition, a typical automobile contains about 68 kg of rubber products such as belts, hoses, and cushions. Outside the automotive area, a wide variety of rubber products are produced, including such familiar items as rubber bands, gloves, and shoe soles.

The formulation and manufacture of rubber products are as diverse as the wide variety of applications for which they are intended. This discussion is limited to the more common methods of manufacture and those that apply to the widest variety of products. In general, the manufacture of rubber products involves compounding, mixing, processing, building or assembly, and vulcanization.

Compounding

The technical process of determining what the formulation components of a rubber product should be is called compounding. This term is also applied to the actual process of weighing out the individual components in preparing for mixing.

Rubber is the backbone of any rubber product. Natural rubber, obtained from the latex of rubber trees, accounts for about 23% of all rubber consumed in the United States. The balance is synthetic rubber. Of the synthetic rubbers, SBR (styrene–butadiene rubber) and polybutadiene rubber are most important, accounting for 71% of synthetic rubber production. A variety of specialty synthetic rubbers, such as butyl, EPDM, polychloroprene, nitrile, and silicone, account for the balance of synthetic rubber production.

Choice of the rubber to be used depends on cost and performance requirements. The specialty rubbers often give superior performance properties but do so at higher product cost.

Many rubber products contain less than 50% by weight of rubber. The balance is a selection of fillers, extenders, and processing or protective coatings.

In products other than tires, clays may be added as extenders, silicas as reinforcing agents, and plasticizers for flex or fire-retardant properties; colors or brightening agents may also be used.

Mixing

This step accomplishes an intimate and homogeneous mix of the formulation components. Most large-volume stocks are mixed in internal mixers; these operate with two winged rotors with the compounding gradients forced into the rotors by an external ram.

There are thousands of different recipes, each designed for different purposes and each requiring a mixing procedure of its own. In addition, various manufacturers differ in their ideas as to precisely how the mixing operation should be conducted, and the equipment in various plants differs widely. The trend is toward more automation in this process, both in weighing and changing the ingredients into the mixer and in handling the stock after discharge.

Forming

Usually, forming operations involve either extrusion into the desired shape or calendering to sheet the material to some specified gauge or to apply a sheet of the material to a fabric.

Building

Building operations, varying from simple to complex, are required for products such as tires, shoes, fuel cells, press rolls, conveyor belts, and life rafts. These products may be built by combining stocks of different compositions or by combining rubber stocks with other construction materials such as textile cords, woven fabric, or metal. For a few products, no building operations are required. For example, many molded products are made by extruding the rubber, cutting it into lengths, and placing it in the mold cavities. Also, extruded products that are given their final shape by the die used in the extruder are ready for the final step of vulcanization, with no intermediate building operations.

The building operations for tires vary widely, depending on the kind of tire to be built at the equipment available in the particular plant. For example, the cord reinforcement may be of polyester, nylon, or steel wire; the number of plies required increases as the size and service requirements increase; the trip proper may be of a different composition than the tread base and side walls; one of the side walls may be made of a white stock; the construction may be of the tubeless variety, in which case an air-barrier ply of rubber is applied to the inside surface of the first ply; a puncture-sealing layer may be applied to the inside surface; or there may be several bead assemblies instead of only one.

Increasing popularity of radial tires has caused some basic changes in tire manufacturing operations. In a radial tire, the reinforcement cords in the body, or carcass, of the tire are placed in a radial direction from bead to bead. Radial tires require special building drums and result in a green tire of a different appearance from the barrel-shaped green bias-ply tire.

Vulcanization

The final process, vulcanization, follows the building operation or, if no building operations are involved, the forming operation.

Vulcanization is the process that converts the essentially plastic, raw rubber mixture to an elastic state. It is normally accomplished by applying heat for a specified time at the desired level. The most common methods for vulcanization are carried out in molds held closed by hydraulic presses and heated by contact with steam-heating platens, which are a part of the press in open steam in an autoclave; under water maintained at a pressure higher than that of saturated steam at the desired temperature; in air chambers in which hot air is circulated over the product; or by various combinations of these methods.

The vast majority of products are sulfur-cured; that is, sulfur cross-links join the rubber chains together. For special applications, vulcanization may take place without the use of sulfur, for example, in the resin- or peroxide-initiated cross-links or metal-oxide-cured polychloroprene.

The time and temperature required for vulcanization of a particular product may be varied over a wide range by proper selection of the vulcanizing system. The usual practice is to use as fast a system as can be tolerated by the processing steps through which the material will pass without "scorching," that is, without premature vulcanization caused by heat during these processing steps. Rapid vulcanization affects economies by producing the largest volume of goods possible from the available equipment. This is particularly the case for products made in molds, because molds are costly, and their output is determined by the number of heats that can be made per day.

The rate of vulcanization increases exponentially with an increase in temperature; hence, the tendency is to vulcanize at the highest temperature possible. In practice, this is limited by many factors, and the practical curing temperature range is 127°C–171°C.

There are numerous exceptions both below and above this range, but it probably covers 95% of the products made.

Finishing operations following vulcanization include removal of mold flash, sometimes cutting or punching to size, cleaning, inspection for defects, addition of fittings such as valves or couplings, painting or varnishing, and packing.

Latex Technology

In addition to the technology of solid rubber, there is a completely different, relatively small but important part of the rubber industry involving the manufacture of products directly from natural rubber latex from the tree, synthetic latex from emulsion polymerization, or aqueous dispersions made from solid rubbers.

In latex technology, materials to be added to the rubber are colloidally dispersed in water and mixed into the latex, a process involving the use of lighter equipment and less power than the mixing of solid rubber compounds. The latex compound can then be used in a variety of processes such as coating or impregnating of cords, fabrics, or paper; in paints or adhesives; molding (such as in toys); dipping (for thin articles like balloons, or household and surgeon's gloves); rubber thread (for garments); and production of foam. Latex technology is particularly important in producing articles for medical and surgical uses.

Rubber Blanket

A sheet of rubber or other resilient material used as an auxiliary tool in forming.

Rubber Forming

Forming a sheet metal wherein rubber or another resilient material is used as a functional die part. Processes in which rubber is employed only to contain the hydraulic fluid are not classified as rubber forming.

Rubber Wheel

A grinding wheel made with a rubber bond.

Rubber-Pad Forming

A sheet metal forming operation for shallow parts in which a confined, pliable rubber pad attached to the press slide (ram) is forced by hydraulic pressure to become a mating die for a punch or group of punches placed on the press bed or baseplate. Developed in the aircraft industry for the limited production of a large number of diversified parts, the process is limited to the forming of relatively shallow parts, normally not exceeding 40 mm (1.5 in.) deep. Also known as the Guerin process. Variations of the *Guerin process* include the *Marforming process*, the *fluid-cell process*, and *fluid forming*.

Rubbers

Cross-linked polymers having glass transition temperatures below room temperature that exhibit highly elastic deformation and have higher elongation.

Rubbing Bearing

A bearing in which the relatively moving parts slide without deliberate lubrication.

Rubbling

Stirring molten metal and or scraping dross from the surface.

Rubidium

A rare metallic element, symbol Rb, atomic weight 85.45, belonging to the group of alkali metals. The chief occurrence of rubidium is in the mineral lepidolite. There is no real rubidium ore, but the element is widely disseminated over the earth in tiny quantities. It is a necessary element in plant and animal life and is found in tea, coffee, tobacco, and other plants. It is silvery-white metal, with a specific gravity of 1.53, a melting point of 102°F (39°C), and boiling at 1270°F (688°C). It takes fire easily in air and decomposes water. Of all the alkali metals, it is next to cesium in the highest chemical activity. It can be obtained by electrolysis and has few industrial applications owing to its rarity. Its chief use is in electronics. For photoelectric cells, it is preferred to cesium, and a very thin film is effective. Like potassium, it has a weak radioactivity by the omission of beta particles, the beta emission being only about one-thousandth that of an equal weight of uranium.

Rubidium Titanyl Arsenate (RTA)

A derivative of potassium titanyl phosphate (KTP), is a nonlinear optical crystal material for improved optical parametric oscillators (OPOs), modulators, and Q-switches. OPOs are used to tune laser frequency; Q-switches are used to control release of laser energy. Because RTA crystals retain clarity with increasing laser power, they can extend a laser's operating transmission range 15% into the infrared. RTA will increase the tunable range of solid-state lasers in industrial, medical, and military applications.

Ruby

A red variety of the mineral **corundum** that ranks with the best grades of precious stones as a gemstone, while the off-color stones are used for watch and instrument bearings. Most of the best rubies, come from upper Myanmar, Thailand, and Cambodia. Before the advent of the synthetic ruby, the larger stones were more viable than diamond. The pink to deep-red colors of the ruby are due to varying percentages of chromic oxide. **Star rubies** contain also a small amount of titania which precipitates along crystallographic planes of the hexagonal crystal and shows as a movable six-ray star when the gem is cut with the axis normal to the base of the stone. **Spinel ruby** is not corundum but magnesium aluminate and is a spinel, often occurring in the same deposits.

Synthetic rubies are equal in all technical qualities to the natural, and synthetic star rubies surpass the natural stones in perfection and quality. Most of the ruby used for instrument bearings is synthetic corundum colored with chromic oxide, since the instrument makers prefer the red color, but the name *ruby* is often applied regardless of color. The U.S. practice in bearing manufacture is to start with a cylindrical rod of the diameter of the desired bearing and to slice to the required thickness. The single-crystal rods are flame polished. For industrial uses, the name *ruby* is often applied to the synthetic material even when it is not red. Synthetic rubies with 0.05% chromium are also used for **lasers**, or light amplifiers, to produce high-intensity light pulses in a narrow beam for communications. Lasers have a wide range of uses, and various other materials are used for specific purposes. Crystals of **potassium dihydrogen phosphate** are used to control the direction of narrow beams of light. The crystal is mounted on the face of the cathode-ray tube.

Rubies are also used in **masers** to detect radio signals for space rockets at great distances. The word *maser* means microwave amplification by stimulated emission of radiation. Ruby has the same physical and chemical properties as the sapphire and corundum. But the color inclusions do affect the electronic properties. The *ruby-sapphire crystals* produced by Linde for **optical masers** are grown with a core of ruby and an overlay sheath of sapphire.

The cutting of rubies, sapphires, and other hard crystals into tiny shape bearings was formally a specialized hand industry, and large stocks of cut stones were kept in the National Stockpile for wartime emergencies. But the slicing and shaping of hard crystals are now widely dispersed on a production basis to meet the needs of the electronic industries, and the equipment is regularly manufactured. For example, the *Accu-Cut wheels* of Aremco Products, for slicing and notching crystals and hard electronic ceramics, are metal disks 0.010–0.030 in. (0.025–0.076 cm) thick and 1–4 in. (2.54–10.2 cm) in diameter, with diamond grit metallurgically bonded to the periphery.

Rugosities

See *asperities*.

Ruling Section

The maximum section size of a steel bar that can be fully hardened through to the center by a particular quenching procedure. See *steel* and *hardenability*.

Rumbling

Treating components in a rotating barrel so that they rub against each other and, possibly, other abrasive materials, to remove surface flash or generally improve the surface finish. The same as tumbling except that larger components, and hence more noise may be implied.

Run-In (Noun)

(1) In tribology, an initial transition process occurring in newly established wearing contacts, often accompanied by transients in coefficient of friction, wear rate, or both, that are uncharacteristic of the given tribological system's long-term behavior. (2) In seals, the period of initial operation during which the seal-lip wear rate is greatest and the contact surface is developed.

Run-In (Verb)

In tribology, to apply a specified set of initial operating conditions to a tribological system in order to improve its long-term frictional or wear behavior, or both. The run-in may involve conditions either more severe or less severe than the normal operating conditions of the *tribosystem* and may also involve the use of special lubricants and/or surface chemical treatments. See also *break in* (verb).

Runner

(1) A channel through which molten metal flows from one receptacle to another. (2) The portion of the gate assembly of a casting that connects the sprue with the gate(s). (3) Parts of patterns and finished castings corresponding to the portion of the gate assembly described in (2). See the term *gating system*.

Runner Box

A distribution box that divides molten metal into several streams before it enters the casting mold cavity.

Runner System

All the sprues, runners, and gates through which a plastic material flows from the nozzle of an injection machine or the pot of a transfer mold to the mold cavity. See also *resin transfer molding* and *transfer molding*.

Running-In

The process by which machine parts improve in conformity, surface topography, and frictional compatibility during the initial stage of use. Chemical processes, including formation of an oxide skin, and metallurgical processes, such as strain hardening, may contribute.

Run-Off Plate (of Weld)

Metal attached to the finish position of a weld on which the weld terminates. Its purpose is to provide a site for terminal defects and thereby ensure that full-quality weld is maintained to the end of the main joint. It is normally removed following completion of welding.

Run-On Plate (of Weld)

Metal attached to the start position of a weld on which the weld is initiated. Its purpose is to provide a site for start defects and thereby ensure that a full-quality weld is established by the time the main joint is reached. It is normally removed following completion of welding.

Runout

(1) The unintentional escape of molten metal from a mold, crucible, or furnace. (2) An imperfection in a casting caused by the escape of metal from the mold. (3) In casting, overflow or other excessive leakage from a mold. (4) Of thread. The tapering of a thread to blend into the plain shank of a bolt. (5) Of a test specimen. A specimen that has survived the specified test program without failure. The program may be, for example, a specified number of fatigue cycles or a period of exposure to an environment capable of inducing stress corrosion. See also *axial runout* and *radial runout*.

Runout Table

A *roll table* used to receive a rolled or extruded section.

Rupture

The process of breaking or fracture often in a creep context. There are ill-defined subtleties to the use of this term in a metallurgical context. See *break*. Normally, when linked directly to a damage mechanism, "rupture" implies a break involving significant ductility (contrast with fracture). Hence, the term "creep rupture" is common, but "brittle rupture" would usually be regarded as clumsy. However, it is acceptable, when used in a less precise context, as a synonym for "failure" or "burst" even when a low-ductility failure mechanism

R

is involved, as in phrases such as "rupture of a pressure vessel by brittle fracture."

Rupture Data

Creep failure data. See *creep*, *stress rupture*, and *Larson–Miller parameter*.

Rupture Strength

Creep strength.

Rust

The visible corrosion product, ranging in color through red, brown, and yellow formed on iron and steel in oxygenated moist environments. It is largely ferric hydroxide, hydrated iron oxide with a variable composition including FeOH and $Fe_2O_3 \cdot H_2O$. It is bulky relative to the iron consumed, permeated to moisture, and is readily detached. It is not protective. Some scales on iron such as magnetite can be protective, but these are not formed in moist oxygenated environments. See also *white rust*.

Rust Preventives

Petroleum rust preventives have been used for the temporary preservation of corrodible metal surfaces ever since Colonel Drake discovered the first oil well over 100 years ago. From this first oil, production waxes were separated, refined, and made into "petroleum jelly," bringing into being the first petroleum rust preventives. The use of petroleum products as protective coatings for oil well equipment spread into all areas requiring low-cost, temporary rust protection.

The raw materials used today are produced in modern petroleum refineries in large volume and low cost. Petroleum has provided a wide range of physical properties required to meet the exacting and varied engineering specifications of the complex industrial and diversified military rust-preventive requirements. These physical properties are viscosity, melting point, pour point, consistency, adhesion, volatility, and density. Chemical additives are usually incorporated with petroleum rust preventives to improve lubricity, to modify viscosity–temperature relationships, to reduce pour point, to displace water from metal surfaces, to dissolve salts, to prevent foaming, to prevent oxidation of the compound, to impart antiwear properties, to improve detergency and dispersancy, and to prevent corrosion and rusting.

Atmospheric corrosion of ferrous metals accounts for a substantial part of why rust prevention is important and critical. The rust prevention of machined parts presents an overall cost saving because the cost of protection is much less than the cost of cleaning to remove rust, of reprocessing, or of scrapping parts that cannot be salvaged.

Petroleum rust preventatives provide temporary but complete protection against rusting, are easy to apply, easy to remove, and are low in cost. The results of rusting may result in dimensional change, structural weakening, and loss of decorative finish.

Rust-Preventive Types

Petroleum rust preventatives may be conveniently separated into four major types: (1) solid or semisolid waxes, (2) oils, (3) solvent cutbacks, and (4) emulsions.

Solid or Semisolid Waxes

These compounds are generally prepared from combinations of petrolatums and microwaxes with heavy lube oils added to obtain the desired consistency. Three consistencies are specified for military use: hard, medium, and soft.

1. The hard-film grade is applied in the molten state and is intended for long-term, temporary outdoor storage of heavy arms, jigs, and forms, for protecting small metal parts, either packaged or unpackaged, and for long-term indoor storage protection of brightly finished surfaces.
2. The medium-film grade is applied in the molten state or at about ambient temperature by dipping or brushing. It is intended for unshielded outdoor storage in relatively moderate climates at temperatures not exceeding the flow point temperature of the compound.
3. The soft-film grade is applied either by brushing or dipping at room temperature or from the molten state. It is intended primarily for the preservation of friction bearings and for use on machined surfaces indoors or for packaged parts.

Additives may or may not be present, depending on the type of service for which the compound is intended. Several advantages are claimed for the solid type compounds:

1. They form coatings of controllable thickness, they do not drain off during storage, and they provide a substantial physical barrier to both moisture and corrosive gases.
2. They form coatings resistant to mild abrasion, airborne contaminants, and ultraviolet light (sunlight) for extended periods.
3. The fast-setting wax coatings eliminate prolonged draining or drying periods during processing.

There are also several disadvantages:

1. Special thermostatically controlled dip tanks must be used to control thickness.
2. The coating, once removed accidentally during handling, is not self-healing and must be reapplied.
3. The compounds are difficult to remove by wiping or rinsing with solvents. Special facilities must be available for the efficient removal of these hard grades. They are not to be applied to intricate parts or inaccessible surfaces.

Preservative Oils

These are available in as many types as there are special uses for oil-type products. Examples are as follows:

1. Preservative lubricants may contain detergents, pour depressants, bearing corrosion inhibitors, and antioxidants, as well as antirust additives for gasoline or diesel engine use.
2. Special preservative oils for both reciprocating and turbojet aviation engines are available containing nonmetallic (nonashing), antirust additives.
3. Light lubricants for automatic rifles, machine guns, and aircraft, suitable for operation at temperatures as low as −55°C, are available in several viscosity grades.
4. General-purpose preservatives are produced specifically for marine use where salt spray as well as high humidity may rust machinery and shipboard cargo.

5. Preservative hydraulic fluids are compounded to preserve hydraulic systems during storage, shipment, and manufacture. These oils contain additives that will permit temporary operational use of the hydraulic systems. In addition to antirust additives, oxidation inhibitors, viscosity index improvers, antiwear agents, and pour depressants are present in hydraulic fluids.
6. Slushing oils are used extensively for the preservation of rolled steel and aluminum sheet rods and bar stock. These oils require careful selection of additives to prevent "black staining" of stacked or rolled sheet. This form of corrosion is a common problem in steel mills.

Metal-conditioning oils are used to loosen and soften scale and rust from heavily corroded surfaces such as marine bulkheads, decks, and structural supports of ballast tanks. These oils appreciably reduce the pitting and general corrosion resulting from the combination of humid atmospheres and frequent immersion in salt water.

Solvent Cutbacks

These are generally of three types and can be applied by spraying, dipping, or slushing. The volatile solvent used in these compounds normally has a flashpoint above 37.8°C and will evaporate completely in several hours leaving the residual protective coating. They include the following:

1. Asphalt cutbacks formulated to dry "to handle" in several hours leaving a smooth black coating approximately 2 mils thick. These coatings are recommended for severe outdoor exposure. They are resistant for several years to salt spray, sunlight, rain, and high humidity.
2. Petrolatum and waxlike coatings that can be either hand and "dry to touch" or soft and easily removed. The dry-to-touch or tack-free coating is usually transparent and provides an excellent preservative coating for spare parts that are stored for long periods indoors and are handled frequently. The soft coatings provide excellent protection against corrosive atmospheres, high humidity, and airborne contamination. These coatings are easily removed by wiping or solvent degreasing but must be protected by protective packaging or storage arrangements. These coatings are not fully self-healing if damaged.
3. Oil coatings are used largely for short-term, in-shop use where parts are still being processed. Humidity protection is provided, with additional emphasis placed on suppression of fingerprint rusting and on displacement of the water remaining from processing.

Emulsifiable Rust Preventives

The emulsifiable rust preventive is an oil or wax concentrate that on proper dilution and on mixing with water will provide a ready-to-use product in the form of an oil-in-water emulsion system. These emulsion systems are fire resistant, provide excellent protection, and are economical. They are intended for use whenever a low-cost, fire-resistant, soft-film, corrosion preventive can be utilized. The compound is formulated to be emulsified in distilled or deionized water, normally at a 1:4 ratio, reducing the volatile content of the corrosion preventive to a minimum. Its fire-resistant properties make it especially adaptable as a substitute for volatile solvent cutbacks where fire hazards increase the expense of application. It has proved efficient for use on small hardware or on parts of uniform contour. Caution should be exercised and applying it to parts of radical design with cracks, crevices, or depressions, which would prevent adequate drainage and cause subsequent corrosion by trapping excess amounts of water.

Selection of Rust Preventives

The selection of petroleum rust preventatives is based on their temporary nature, ease of application and removal, the economy of providing protection against corrosion, and the type of protection desired (e.g., indoor protection or outdoor exposure).

Indoor Protection

Preservatives of this group must provide a high degree of protection against high humidity, corrosive atmospheres, and moderate amounts of airborne contaminants. Resistance to abrasion is occasionally required where spare parts that are stored in bins or on uncovered racks. Normally, however, the more easily removed products are used, such as oils, soft petrolatums, or soft-film solvent cutbacks.

Outdoor Exposure

Preservatives of this group must withstand the effects of rain, snow, and sunlight as well as provide good resistance to abrasion. The products used for this service are the hard film, asphalt solvent cutbacks, and the hard grades of hot-dip waxes.

Short-Term Storage

These preservatives are commonly used to protect parts between machining operations or prior to final preservation with a heavy-duty product. These light oils or solvent cutbacks usually have fingerprint-suppressing properties to protect the parts during production handling.

Special-Purpose Lubricants

Lubricants for many purposes are formulated to provide primary or secondary rust-prevention properties. These special lubricants include nonashing lubricants for aircraft engine use, high-detergency oils for motor vehicles, low-temperature machine gun or rifle oils, and special salt-spray-resistant lubricating oils for shipboard use.

Functional Uses

Hydraulic oils possessing rust-preventive properties are used during the shipment or storage of equipment containing hydraulic power-transmission systems. This equipment is found in aerospace systems, submarines, aircraft carriers, and industrial machine tools. Various viscosity grades are available.

Rust Proofing

Any surface treatment applied to steel that is claimed to impede or delay rusting.

Ruthenium

A hard silvery-white metal (symbol Ru), ruthenium has a specific gravity of 12.4, a melting point of about 2310°C, and a Brinell hardness of 220 in the annealed state. The metal is obtained from the residue of platinum ores by heat reduction of ruthenium oxide, RuO_2, in hydrogen. Ruthenium is the most chemically resistant of the platinum metals and is not dissolved by aqua regia. It is used as a catalyst to combine nitrogen in chemicals. As ruthenium tetroxide, RuO_4, is a powerful catalyst for organic synthesis, oxidizing alcohols to acids, ethers to esters, and amides to imides. Ruthenium is a close-packed hexagonal crystal structure. It has a hardening effect on platinum; a 5% addition of ruthenium raises Brinell hardness from 30 to 130 and the electrical resistivity to double that of pure platinum.

Ruthenium is used commercially to harden alloys of palladium and platinum. The alloys are used as electrical contacts (12% ruthenium in platinum) and in jewelry and fountain pen tips. Application of ruthenium to industrial catalysis (hydrogenation of alkenes and ketones) and to automobile emission control (catalytic reduction of nitric oxide) and detection have been active areas of research.

Rutherford Backscattering Spectroscopy

A method of determining the concentration of various elements as a function of depth beneath the surface of a sample, by measuring the energy spectrum of ions that are backscattered out of a beam directed at the surface.

Rutherford Scattering

A general term for the classical elastic scattering of energetic ions by the nuclei of a target material.

Rutile

Rutile is the most frequent of the three polymorphs of titania, TiO_2, crystallizing in the tetragonal system is the rutile structure type.

The mineral occurs as striated tetragonal prisms and needles, commonly repeatedly twinned. The color is deep blood red, reddish brown, to black, rarely violet or yellow. Specific gravity is 4.2, and hardness 6.5 on the Mohs scale. Melting point is 3317°C. It is synthesized from $TiCl_4$ at red heat, and colorless crystals are grown by the flame-fusion method, which because of their adamantine luster are used as gem material.

In ceramic applications where titanium oxide is desirable (but where a pure white or certain shades are not required), the more economical rutile is frequently substituted for the pure chemical. Rutile is used to stain pottery bodies and glazes in colors ranging from ivory through yellows to dark tan, according to the amount introduced. Artificial teeth are among the ceramics so tinted.

Rutile is also used in the glass and porcelain enamel industries as a colorant, and to introduce TiO_2. The largest use of rutile is as a constituent in welding rod coatings.

Rutile Electrodes

See *electrode (welding)*.

r-Value

See *plastic strain ratio*.

RZ Powder

The term iron powder made in Germany from the scale of pig iron.

S

S Curve

See *isothermal transformation diagram.*

S/N Curve

A graph plotting stress range (S) against the number of cycles to failure (N) to predict the fatigue life of a material. See *fatigue.*

Sacrificial Protection

Reduction of corrosion of a metal in an *electrolyte* by galvanically coupling it to a more anodic metal; a form of *cathodic protection.*

Saddling

Forming a seamless metal ring by forging a pierced disk over a mandrel (or saddle).

Safe Working Load (S.W.L.)

The maximum load that a component should carry in a normal service. When used in its formal statutory sense, the term recognizes that there is an adequate margin between the safe working load and the *proof load* and a further large margin between the proof load and the breaking load. See *safety factor.*

Safety Critical

Identifies a component or system whose function is vital to the safety of plant operations or the population at large.

Safety Factor

An allowance made by a designer to accommodate unknowns. The factor can be applied to one or more of the properties on which the design is based, including yield, tensile, fatigue or creep strength, and corrosion or erosion rate; see respective entries. For example, a component could be designed on the basis that the expected maximum load will introduce a stress of only a third of the ultimate tensile stress. The safety factor of 3 allows for unpredictable factors such as excessive loading, variation in materials properties, minor manufacturing defects, or deterioration in service. Depending on the particular application, the safety factor may be calculated on the relationship between the expected breaking load at either the safe working load or the proof load. In general, considerations of economy encourage reduction in the safety factor toward unity but this requires increasingly tight quality assurance and accurate prediction and monitoring of service conditions, possibly supported by regular inspection of critical items.

Safety Glass

Glass so constructed, treated, or combined with other materials as to reduce, in comparison with ordinary sheet or plate glass, the likelihood of injury to persons by objects from exterior sources or by these safety glasses when they may be cracked or broken. Types of safety glass include (1) laminated safety glass—two or more pieces of glass held together by an intervening layer or layers of plastic materials. It will crack and break under sufficient impact, but the pieces of glass tend to adhere to the plastic and not to fly. If a hole is produced, the edges are likely to be less jagged than would be the case with ordinary glass; (2) tempered safety glass—a single piece of specially treated plate, sheet, or float glass. When broken at any point, the entire piece immediately breaks into innumerable small pieces, which may be described as granular, usually with no large jagged edges; (3) wire safety glass—a single piece of glass with a layer of meshed wire completely embedded in the glass, but not necessarily in the center of the sheet. It may or may not be ground and polished on both sides. When the glass is broken, the wire mesh holds the pieces together to a considerable extent.

Sag (Metals)

An increase or decrease in the section thickness of a casting caused by insufficient strength of the mold sand of the cope or of the core.

Sag (Plastics)

(1) A local extension (often near the die face) of the parison during extrusion by gravitational forces. This causes necking down of the parison. (2) The flow of a molten sheet in a thermoforming operation.

Sagging (Adhesives)

Run-off or flow-off of adhesive from an adherend surface due to application of excess or low-viscosity material.

Sagging (Ceramics and Glasses)

(1) A defect characterized by bending or slumping of a ceramic article fired at excessive temperature. (2) A process of forming glass by reheating until it conforms to the shape of the mold or form on which it rests. (3) A defect characterized by a wavy line or lines appearing on those surfaces of porcelain enamel that have been fired in a vertical position. (4) A defect characterized by irreversible downward bending in a ceramic article insufficiently supported during the firing cycle.

Sal-Ammoniac

Ammonium chloride, NH_4Cl, sometimes used as a soldering flux.

Salt

A salt is a compound formed when one or more of the hydrogen atoms of an acid are replaced by one or more cations of the base. The common example is sodium chloride in which the hydrogen ions of hydrochloric acid are replaced by the sodium ions (cations) of sodium hydroxide. There is a great variety of salts because of the large number of acids and bases that has become known.

Classification

Salts are classified in several ways. One method—normal, acid, and basic salts—depends upon whether all the hydrogen ions of the acid or all the hydroxide ions of the base have been replaced:

Class	Examples
Normal salts	$NaCl$, NH_4Cl, Na_2SO_4, Na_2CO_4, Na_3PO_4
Acid salts	$NaHCO_4$, NaH_2PO_4, Na_2HPO_4, $NaHSO_4$
Basic salts	$Pb(OH)Cl$, $Sn(OH)Cl$

The other method—simple salts, double salts (including alums), and complex salts—depends upon the character of completeness of the ionization.

Class	Examples
Simple salts	$NaCl$, $NaHCO_3$, $Pb(OH)Cl$
Double salts	KCl, $MgCl_2$
Alums	$KAl(SO_4)_2$, $NaFe(SO_4)_2$, $NH_4Cr(SO_4)_2$
Complex salts	$K_3Fe(CN)_6$, $Cu(NH_3)_4Cl_2$, $K_2Cr_2O_7$

In general, all salts in solution will give ions of each of the metal ions; an exception is a complex type of salts such as $K_3Fe(CN)_6$ and $K_2Cr_2O_7$.

Any aggregate of molecules, atoms, or ions joined together with a coordinate covalent bond can be correctly called salt; however, by common parlance, the term *salt* usually refers to an electrovalent compound, the classical example of which is sodium chloride.

Salt Bath

A vessel containing molten salts of various compositions. The salt offers rapid and uniform heat transfer with protection against oxidation. Some salts also are intended to carburizing the steel as the first stage of *case hardening*.

Salt Bath Heat Treatment

Heat treatment for metals carried out in a bath of molten salt. See the terms *immersed-electrode furnaces* and *submerged-electrode furnaces*.

Salt Fog Test

An *accelerated corrosion test* in which specimens are exposed to a fine mist of a solution usually containing sodium chloride, but sometimes modified with other chemicals. Also known as salt spray test.

Salting Out

Precipitating a substance in a solution by adding a second substance, usually a salt, without any chemical reaction, such as a double decomposition, taking place.

Saltpeter

Potassium nitrate, KNO_3, a constituent of gunpowder.

Salt Spray Test

Various tests of corrosion rate and stress corrosion susceptibility. Many such tests have been devised with the common factor being that a salt solution is sprayed onto the test specimens for a defined period at regular intervals. The solution is usually sodium chloride up to 20%, and the environment may be open air or an ambient cabinet with environmental control. The test specimens may be unstressed or stressed in tension or bending and may be welds or other joints or bimetallics. The test may be predetermined or continued until the test criterion is met. The possible criteria include failure, cracking, corrosion rate, or merely surface appearance. See *salt fog test*.

Samarium

A metallic element, one of the rare earth metals. It has little commercial application.

Sample

(1) One or more units of a product (or a relatively small quantity of a bulk material) withdrawn from a *lot* or process stream and then tested or inspected to provide information about the properties, dimensions, or other quality characteristics of the lot or process stream. (2) A portion of a material intended to be representative of the whole.

Sample Average

The sum of all the observed values in a sample divided by the sample size. It is a point estimate of the population mean. Also known as arithmetic mean.

Sample Median

The middle value when all observed values in a sample are arranged in order of magnitude. If an even number of samples are tested, the average of the two middlemost values is used. It is a point estimate of the population median, or 50% point.

Sample Percentage

The percentage of observed values between two stated values of the variable under consideration. It is a point estimate of the percentage of the population between the same two stated values.

Sample Splitter

A device to divide a metal powder pile for sampling.

Sample Standard Deviation (s)

The square root of the sample variance. It is a point estimate of the population standard deviation, a measure of the "spread" of the frequency distribution of a population. This value of s provides a statistic that is used in computing interval estimates and several test statistics. For small sample sizes, s underestimates the population standard deviation.

Sample Variance (s²)

The sum of the squares of the differences between each observed value on the sample average divided by the sample size minus 1. It is a point estimate of the population variance.

Sample Thief

A pointed, hollow, tubular device to withdraw a representative metal powder sample from a shipping drum or other packing unit.

Sampling

The process of obtaining a portion of a material that is adequate for making the required tests or analyses and that is representative of that portion of the material from which it is taken.

Sand

Sand is an unconsolidated granular material consisting of mineral or rock fragments between 6.5 μm and 2 mm in diameter. Finer material is referred to as silt and clay, coarser material as gravel. Sand is usually produced by the chemical or mechanical breakdown of older source rocks, but may also be formed by the direct chemical precipitation of grains or by biological processes. Accumulations of sand result from hydrodynamic sorting of sediment during transport and deposition.

Origin and Characteristics

Most sand originates from the chemical and mechanical breakdown, or weathering, of bedrock. Chemical weathering is most efficient to soils, and most sand grains originate within soils. Rocks may also be broken into sand-size fragments by mechanical processes, including diurnal temperature changes, freeze–thaw cycles, wedging by salt crystals or plant roots, and ice gouging beneath glaciers.

Transport

Sand can be transported by any medium with sufficient kinetic energy to keep the grains in movement. In nature, sand is commonly transported by rivers, ocean currents, wind, or ice. Rivers are responsible for transporting the greatest volume of sand over the greatest distances. When rivers deposit sand at the ocean edge, it is commonly remobilized by high-energy breaking waves or transported along the shore by powerful longshore currents. Sand, like coarser and finer sediments, may also be entrained by glacial ice and transported to the sea by glacial flow. Where sand is widely exposed to wind currents, such as in river bars, on beaches, and in glacial outwash plains, it may be picked up and transported by the wind. As it moves, whether subaqueously or subaerially, sand is organized into complex moving structures (bedforms) such as ripples and dunes.

Economic Importance

Sand and gravel production is second only to crushed-stone production among nonfuel minerals in the United States. Although sand and gravel have one of the lowest average per ton values of all mineral commodities, the vast demand makes it among the most economically important of all mineral resources. Sand and gravel are used primarily for construction purposes, mostly as concrete aggregate. Pure quartz sand is used in the production of glass, and some sand is enriched and rare commodities such as ilmenite (a source of titanium) and gold.

Types

An accumulation of grains of mineral matter derived from the disintegration of rocks. It is distinguished from gravel only by the size of the grains or particles but is distinct from clays that contain organic materials. Sands have been sorted out and separated from the organic material by the action of currents of water or by winds across arid lands and are generally quite uniform in size of grains. Usually commercial sand is obtained from riverbeds or from sand dunes originally formed by the action of winds. Much of the earth's surface is sandy, and these sands are usually quartz and other siliceous materials. The most useful commercially are *silica sands*, often above 98% pure. Silica sands for making glass must be free from iron. *Beach sands* usually have smooth, spherical, to ovaloid particles from the abrasive action of waves and tides and are free of organic matter. The white beach sands are largely silica but may also be of zircon, monazite, garnet, and other minerals, and are used for extracting various elements. *Monazite sand* is the chief source of thorium. The *black sands* of Oregon contain chromate, and those of Japan contain magnetite. Kyanite is found in the Florida sands.

Sand is used for making mortar and concrete and for polishing and sandblasting. Sands containing a little clay are used for making molds in foundries. Clear sands are employed for filtering water. Construction sand is not shipped great distances, and the quality of sands used for this purpose varies according to local supply. *Standard sand* is a silica sand used in making concrete and cement tests. The grains are free of organic matter and pass through a 20-mesh sieve, but are retained on a 30-mesh. *Engine sand*, or *traction sand*, is a high-silica sand of 20–80 mesh washed free of soft bond and fine particles, used to prevent the driving wheels of locomotives or cars from slipping on wet rails.

Molding sand, or *foundry sand*, is any sand employed for making molds for casting metals, but especially refers to sands that are refractory and also have binding qualities. Pure silica is ideal for heat resistance, but must contain enough alumina to make it bind together. Molding sands may contain 80% to 92% silica, up to about 15 alumina, about 2 iron oxide, and not more than a trace of lime. Some molding sand contains enough clay or loam to bond it when it is tamped into place. The amount of bond in *Grant sand* and in *Tuscarawa sand* is 17%–18%. About 33% of these natural sands pass through a No. 100 screen, and 20 through a No. 150 screen. The finer the grain, the smoother the casting, but fine-grain sand is not suitable for heavy work because of its impermeability to gases. Sands without natural bonds are more refractory and are used for steel molding. Sands for steel casting must be silica sands containing 90% silica, or preferably 98, and are mixed with 2–10 fireclay. For precision casting, finely ground aluminum silicate is used in the silica sand mixes, and it requires less of a bonding agent.

Zircon sand has high heat resistance and is used for alloy steel casting. *Zircon flour* is finely milled zirconite sand used as a mold wash. *Zirconite sand* for molding is 100–200 mesh in its natural state.

It is 70% heavier than silica sand and has a higher heat conductivity that gives more rapid chilling of castings. Common molding sands may contain 5% to 18% of clay materials and may be mixtures of sand, silt, and clay, but they must have the qualities of refractoriness, cohesiveness, fineness of grain, and permeability. To have refractory quality, they must be free of calcium carbonate, iron oxide, and hydrocarbons. *Core sands* also have these qualities, but they are of coarser grain and always require a bond that will bake solidly but will break down easily at the temperature of pouring. Sand with rounded grains is preferred, and the grains must be very uniform in size to prevent filling. When a molding sand is burned out, it is made suitable for reuse by adding bond; but when fireclay is used as a bond, it adheres to the sand grains and that makes it unsuitable for reuse. *Parting sand* is a round-grained sand without bond used on the joints of molds. Foundry parting is usually tripoli or bentonite. Cores are made with sand mixed with core oils. *Greensand cores* are unbaked cores made with molding sand.

Sandblast sand is sand employed in a blast of air for cleaning castings, removing paint, cleaning metal articles, giving a dull, rough finish to glass or metal goods, or renovating the walls of stone or brick buildings. Sandblast sand is not closely graded, and the grades vary with different producers.

Sharp grains cut faster, but rounded grains reduce smoother surfaces. The sand is usually employed over and over, screening out the dust. The dust and fine used sand may be blasted wet. This is known as *mud blasting* and produces a dull finish.

Sandblasting

(1) Abrasive blasting with sand. See also *blasting* or *blast cleaning* and compare with *shot blasting*. (2) Similar to shot blasting extent that sand or other sharp edge mineral grit is used.

Sand Blister/Buckle/Scab

A bulge on the surface of a sand casting. It contains a large proportion of sand mixed with the metal and is easily scraped or chipped off. It results from an area of soft sand being permeated by molten metal.

Sand Castings

Sand casting can be broken down into three general processes: green-sand molding, dry-sand molding, and pit molding. Green-sand molding, in all probability, produces the largest tonnage of castings but is limited in that long, thin projections are very difficult to cast. Dry-sand molds, on the other hand, can produce castings of any desired intricacy (within the normal dimensional limits of the process) because the sand is baked with a binder, increasing the strength of the thin mold sections. Pit molding is the sand process used to make very large, heavy castings.

The Process

Green-Sand Molding

The essential parts of a green-sand mold are pattern, gate, sprue, riser, pouring basin, party joint, cope, load, pin, drag, and bottom board. The pattern is placed in the flask (consisting of the cope and drag) and molding sand is tightly rammed around it. Usually, the mold is made in two halves, with the pattern lying at the parting line. Removing the pattern from the rammed sand leaves the desired cavity from which the casting is produced. The molder then produces a sprue for pouring metal into the cavity and an opening for a riser to permit air in the cavity to be expelled. Both the sprue and riser also act as a source of hot metal during solidification and help eliminate shrinkage cavities in the casting. Cores can be added to the mold cavity to shape internal casting surfaces.

Among the advantages of green-sand molding are the following: one pattern can be used to produce any number of castings; most non-ferrous alloys, gray irons, ductile irons, malleable irons, and steel can be cast; and finally, the fragility of the sand cores permits them to collapse after metal is cast around them, thereby eliminating or reducing casting stresses and tendency for hot tearing.

The limitation of the process is its inability to support long, thin projections in the sand.

Dry-Sand Molding

Core boxes, not patterns, are used to make the various parts of the mold. A mixture of sand and a binder is formed in a core box acid baked at 204°C–260°C to harden the sand. The various pieces are then assembled into a mold.

Dry-sand molds, because of the way which they are produced, can be used to make intricate castings. The baking operation strengthens the sand and permits thin projections to be cast without danger of collapsing the mold walls. Cores used in dry-sand molding are collapsible and help reduce hot tearing tendencies.

Pit Molding

If large, intricate parts must be produced, pit molding is considered the most economical production method. (Castings weighing up to 230,000 kg have been produced in pit molds.) Pit molding is a highly specialized operation and the equipment used—sand slingers, molding machines, etc.—precludes the use of much hand labor. The mold usually is dried, increasing sand strength and, consequently, the ability to resist mold erosion during pouring as well as the weight of the casting being poured.

Sand Control

Testing and regulation of the chemical, physical, and mechanical properties of foundry sand mixtures and their components.

Sand Grain Distribution

Variation or uniformity in particle size of a sand aggregate when properly screened by standard screen sizes. See also *grain fineness number*.

Sand Hole

A pit in the surface of a sand casting resulting from a deposit of loose sand on the surface of the mold.

Sandpaper

Originally a heavy paper coated with sand grains on one side, used as an abrasive, especially for finishing wood. Sharp grains obtained by crushing quartz later replaced sand, and the product was called *flint paper*. But most abrasive papers are now made with aluminum oxide or silicon carbide, although the term *sandpapering* is still employed in wood polishing. Quartz grains, however are still much used on papers for the wood industries. For this purpose the quartz grains are in grades from 20 mesh. Good *sandpaper quartz* will contain at least 98.9% silica. The paper used is heavy, tough, and flexible, usually 70 or 80 lb (32 or 36 kg) paper, and the grains are bonded with a strong glue. A process is also employed to deposit the grains on end by electrostatic attraction so that the sharp edges of the grains are presented to the work.

Sand Reclamation

Processing of used foundry sand by thermal, wet, or dry methods so that it can be used in place of new sand without substantially changing the foundry sand practice.

Sandstone

A consolidated sand rock, consisting of sand grains united with a natural cementing material. The size of the particles and the strength of the cement vary greatly in different natural sandstones. The most common sand in sandstone is quartz, with considerable feldspar, lime, mica, and clay matter. The cementing material is often fine chalcedony. *Silica sandstones* are hard and durable but difficult to work. *Calcareous sandstone*, in which the grains are cemented by calcium carbonate, is called *freestone* and is easily worked, but it disintegrates by weathering. Freestone is homogeneous and splits almost equally well in both directions. *Chert*, formally used as an abrasive and, when employed in building and paving, known under local names as *hearthstone*, *firestone*, and *malmstone*, is a siliceous stone of sedimentary origin. It has a radiating structure and splintery fracture and is closely allied to flint. In color it is light gray to black or banded. The colors of sandstones are due to impurities, pure siliceous and calcareous stones being white or cream-colored. The yellow to red colors usually come from iron oxides, black from manganese dioxide, and green from glauconite. *Crab Orchard stone* of Tennessee is high in silicate with practically no CaO and is often beautifully variegated with red and brown streaks.

Sandstones for building purposes are produced under innumerable names, usually referring to the locality. The *bluestone* of New York State is noted for its even grain and high crushing strength, up to 131 MPa. It contains about 70% silica sand with clay as the binder. *Amherst sandstone* from Ohio contains up to 95% silica with four aluminum oxides and is colored gray and buff with iron oxides. *Flexible sandstone*, which can be bent, comes from North Carolina. It is *itacolumite* and has symmetrically arranged quartz grains which interlock and rotate against one another in a binder of mica and talc.

Holystone is a block of close-grained sandstone, formally used for rubbing down the decks of ships and still used for rubbing down furniture in concrete work. *Briar Hill stone* and Macstone are trade names for building blocks consisting of lightweight concrete faced with a slab of sandstone. *Kemrock* is a sandstone impregnated with a black furfural resin and baked to a hard finish. It is used for table tops in chemical equipment to resist acids and alkalies. The term *reservoir rock* refers to friable, porous sandstone that contains oil or gas deposits.

Sand Tempering

Adding sufficient moisture to molding (foundry) sand to make it workable.

Sandwich

Any triple or even multilayer construction. Typically it might comprise outer faces of high-strength durable material and a thick interlayer of low density material with sufficient strength to maintain the bond and spacing between the outer and hence the rigidity of the assembly.

Sandwich Construction

A panel composed of a lightweight core material, such as honeycomb, cellular or foamed plastic, and so forth, to which to relatively thin, dense, high-strength or high-stiffness faces, or skins, are adhered. See the term *honeycomb*.

Sandwich Materials

These are a type of laminar composite composed of a relatively thick, low-density core between phases of comparatively higher density. Structural sandwiches can be compared to I beams. The facings correspond to the flanges; the objective is to place a high-density, high-strength material as far from the neutral axis as possible, thus increasing the section modulus. The bulk of a sandwich is the core. Therefore, it is usually lightweight for high strength-to-weight and stiffness-to-weight ratios. However, it must also be strong enough to withstand normal shear and compressive loadings, and it must be rigid enough to resist bending or flexure.

Core Materials

Core materials can be divided into three broad groups: cellular, solid, and foam. Paper, reinforced plastics, impregnated cotton fabrics, and metals are used in cellular form. Balsa wood, plywood, fiberboard, gypsum, cement-asbestos board, and calcium silicate are used as solid cores. Plastic foam cores—especially polystyrene, urethane, cellulose acetate, phenolic, epoxy, and silicone—are used for thermal-insulating and architectural applications. Foamed inorganic such as glass, ceramics, and concrete also find some use. Foam cores are particularly useful where the special properties of foams are desired, such as insulation. And the ability to foam in place is an added advantage in some applications, particularly in areas that are difficult to reach.

Cellular Cores

Of all the core types, however, the best for structural applications are the rigid cellular cores. The primary advantages of the cellular core are that (1) it provides the highest possible strength-to-weight ratio, and (2) nearly any material can be used, thereby satisfying virtually any service condition.

There are, essentially, three types of cellular cores: honeycomb, corrugated, and waffle. Other variations include small tubes or cones and mushroom shapes. All these configurations have certain advantages and limitations. Honeycomb sandwich materials, for example, can be isotropic, and they have a high strength-to-weight ratio, good thermal and acoustical properties, and excellent fatigue resistance. Corrugated-core sandwich is anisotropic and does not have as wide a range of application as honeycomb, but it is often more practical than honeycomb for high production and fabrication into panels.

Construction

Theoretically, any metal that can be made into a foil and then welded, brazed, or adhesive-bonded can be made into a cellular core. A number of materials are used, including aluminum, glass-reinforced plastics, and paper. In addition, stainless steel, titanium, ceramic, and some superalloy cores have been developed for special environments.

One of the advantages of sandwich construction is the wide choice of facings, as well as the opportunity to use thin sheet materials.

The facings carry the major applied loads and therefore determine the stiffness, stability, and to a large extent, the strength of the sandwich. Theoretically, any thin, bondable material with a high tensile or compressive strength-to-weight ratio is a potential facing material. The materials most commonly used are aluminum, stainless steel, glass-reinforced plastics, wood, paper, and vinyl and acrylic plastics, although magnesium, titanium, beryllium, molybdenum, and ceramics have also been used.

Theory

The theory of sandwich materials and functions of the individual components may best be described by making an analogy to an I-beam. The high-density facings of a sandwich correspond to the flanges of the I-beam; the objective is to place a high-density, high-strength material as far from the neutral axis as possible to increase the section modulus without adding much weight. Honeycomb in a sandwich is comparable to the I-beam web that supports the flanges and allows them to act as a unit. The web of the I-beam and honeycomb of the sandwich carry the beam shear stress. Honeycomb and a sandwich differs from the web of an I-beam in that it maintains a continuous area support for the facings, allowing them to carry stresses up to or above the yield strength without crippling or buckling. The adhesive that bonds honeycomb to its facings must be capable of transmitting shear loads between these two components, thus making the entire structure an integral unit.

When a sandwich panel is loaded as a beam, the honeycomb and the bond resist the shear loads while the facings resist the moments due to bending forces, and hence carry the beam bending as tensile and compressive load. When loaded as a column, the facings alone resist the column forces while the core stabilizes the thin facings to prevent buckling, wrinkling, or crippling.

Advantages

The largest single reason for the use of sandwich construction and its rapid growth to one of the standard structural approaches during the past 60 years is its high strength-to-weight or stiffness-to-weight ratio. As an example, consider a 0.6 m span beam with a width of 0.3 m and supporting a load of 1634.4 kg at the midspan. This beam, if constructed of solid steel, would have a deflection of 14.2 mm and weigh 31.15 kg. A honeycomb-sandwich beam using aluminum skins on aluminum cores and carrying the same total load at the same total deflection would weigh less than 3.6 kg. As an interesting further comparison, a magnesium plate to the same specifications would weigh 11.9 kg and an aluminum plate 15.7 kg. Although such clear and simple cases of comparative strength, weight, and stiffness are not normally found in actual designs, it has generally been found that equivalent structures of sandwich construction will weigh from 5% to 80% less than other minimum weight structures and frequently possess other significant advantages.

Other advantages of sandwich construction include extremely high resistance to vibration and sonic fatigue, relatively low noise transmission, either high or low heat transmission depending on the selection of core materials, electrical transparency (varying from almost completely transparent in the case of radome structures to completely opaque in the case of metal sandwich structures), relatively low-cost tooling when producing complex aircraft parts, ability to mass-produce complicated shapes, ability to absorb damage and absorb energy while retaining significant structural strength, and flexibility of design available.

Sandwich Rolling

Rolling two or more strips of metal in a pack, sometimes to form a roll-welded composite.

Sandwich Testing

A method of heating a thermoplastic sheet, prior to forming, that consists of heating both sides of the sheet simultaneously.

S.A.P.

Sintered aluminum powder, usually with reference to material that has been centered then extruded to provide a material with better strength at ambient and elevated temperatures compared with conventionally cast aluminum. The improved properties result from the *dispersion hardening* effect of particles of aluminum oxide from the powder surface.

Saponification

The alkaline hydrolysis of fats whereby a soap is formed; more generally, hydrolysis of an ester by an alkali with the formation of an alcohol and a salt of the acid portion.

Saponification Number

(1) A measure of the amount of constituents of petroleum that will easily *saponify* under test conditions, determined by the number of milligrams of sodium hydroxide that is consumed by 1 g of oil under test conditions. Saponification number is a measure of fatty materials compounded in an oil. (2) A number given to quenching oils that reflects the oils amount of compounding with fatty materials, which thereby helps evaluate the condition of these oils in service. See also *neutralization number* and the term *oil quenching*.

Saponify

(1) To convert into soap. (2) To subject to, or to undergo, saponification.

Sapphire

Sapphire is any gem variety of the mineral corundum (Al_2O_3) except those called ruby because of their medium- to dark-red color. Sapphire has a hardness of 9 (Mohs scale), a specific gravity near 4.00, and refractive indices of 1.76–1.77.

Although blue sapphires are most familiar to the public, sapphires occur in many other colors, including yellow, brown, green, pink, orange, and purple, as well as colorless, and black.

General Nature

Synthetic sapphire, chemically and physically, is the same as the natural mineral, which is now uneconomical to mine because of the low cost of the synthetic ($0.05 g^{-1}$). It is a single crystal of very high purity aluminum oxide and crystallizes in the hexagonal–rhombohedral crystal system. It has a hardness that is second only to diamond among naturally occurring minerals. It has unusual chemical stability, high thermal conductivity, excellent high-temperature strength, broadband optical transmission, and low electrical loss.

Chemical Properties

Sapphire is inert to most chemical agents. For example, it is not attacked by any acids, including hydrofluoric acid. It is resistant to alkalies and is not attacked by uranium hexafluoride up to 600°C.

Sapphire will dissolve in water at the rate of 1 mg/day at 240 MPa and 700°C. It is also attacked by anhydrous sodium borate–sodium carbonate at 1000°C, by silicon at 1500°C, by sodium hydroxide at 700°C, and by boiling phosphoric acid.

Forms Available and Uses

Synthetic sapphire is available in the form of rods, boule, and disks as well as optical lens shapes. Sapphires are fabricated with diamond grinding techniques into a variety of surface roughnesses, can be optically polished, and can be chemically or flame polished.

Most applications are for watch and instrument bearings, gemstones, phonograph needles, ball pointpen balls, infrared and ultraviolet optics, as well as electronic insulators and masers. Others include orifices, arc lamp tubes, radiation pyrometry, crucibles, film and textile guides, solar cell covers, optical flats, liquid nitrogen heat sinks, klystron windows, and optical filter substrates.

By deliberately adding impurities to the starting material, other modifications of sapphire crystals can be obtained. The most important today is ruby, which is alumina with 0.05% chromium oxide added and is the basic solid-state crystal in microwave and optical maser systems.

The submicron size high-purity alumina powder, which is used as a starting material for producing crystals, is also used as a polishing agent in dentifrice formulation and metallurgical polishing, as supports for high-purity chemical catalyst carriers, and as the raw material in specialty ceramic manufacture.

Satin Finish (Metals)

A diffusely reflecting surface finish on metals, lustrous but not mirrorlike. One type is a *butler finish*.

Satin Finish (Plastics)

A type of finish having a satin or velvety appearance, specified for plastics or composites.

Satin Weave

See *four-harness satin, eight-harness satin*, and *harness satin*.

Saturated

(1) A solution containing the maximum solute content at equilibrium.
(2) A magnet that is magnetized to the maximum level achievable.

Saturated Calomel Electrode

A *reference electrode* composed of mercury, mercurous chloride (calomel), and a saturated aqueous chloride solution.

Saturation Pressure

The pressure, for a pure substance at any given temperature, at which vapor and liquid, or vapor and solid, coexisted in a stable equilibrium.

Sauter Mean Diameter

The diameter of a liquid droplet that has the same ratio of volume to surface area as the ratio of total volume to total surface area in a distribution of drops, as shown in the following equation:

$$\text{SMD} = \frac{\sum_i n_i d_i^3}{\sum_i n_i d_i^2}$$

where
   i is a sampling size interval
   d_i is the drop diameter
   n_i is the number of drops in that interval

Sauveur's Diagram

A means for estimating the carbon content of annealed carbon steels (maximum of 0.5% manganese) by examining their microstructure to determine the amount of pearlite. The diagram is a straight line plot between 0% and 100% pearlite at eutectoid composition, usually taken as 0.87% carbon, followed by a further straight line falling to 85% pearlite at 1.5% carbon.

Saw Burn

Blackening or carbonization of a cut surface of a pultruded composite material or plastic section. Usually caused by cutting with a dull saw blade, cutting too slowly, or cutting a highly reinforced material with a diamond blade without water.

Saw Gumming

In saw manufacture, grinding away of punch marks or milling marks in the gullets (spaces between the teeth) and, in some cases, simultaneous sharpening of the teeth; in reconditioning of worn saws, restoration of the original gullet size and shape. See the term *sawing*.

Sawing

Using a toothed blade or disk to sever parts or cut contours.

1. *Circular sawing*: Using a circular saw fed into the work by motion of either the workpiece or the blade.
2. *Power band sawing*: Using a long, multiple-tooth continuous band resulting in a uniform cutting action as the workpiece is fed into the saw.
3. *Power hack sawing*: Sawing in which a reciprocating saw blade is fed into the workpiece.

Saybolt Universal Viscosity

A commercial measure of viscosity expressed as the time in seconds required for a 60 mL of fluid to flow through the orifice of the standard Saybolt Universal viscometer (calibrated tube) at a given temperature under specified conditions; used for lighter petroleum products and lubricating oils.

S-Basis

The S-basis property allowable is the minimum value specified by the appropriate federal, military, Society of Automotive Engineers,

American Society for Testing and Materials, or other recognized and approved specifications for the material.

SBR Rubber

SBR synthetic rubbers are versatile, general-purpose elastomers, developed as a substitute for natural rubber during World War II. Known as Buna-S (BU-tadiene, NA-trium, or sodium S-tyrene) in Germany, they were designated GR-S (G-overnment R-ubber S-tyrene) by the United States. The materials have been designated SBR (S-tyrene B-utadiene R-ubber) by the ASTM (American Society for Testing and Materials). They constitute about 80% of domestic synthetic-rubber consumption.

SBR Types

SBR polymers are generally available as dry rubbers (bale or crumb form) and its lattices. So-called hot polymers (polymerized at 50°C), it has been steadily replaced in recent years by the 20%–30% higher tensile, longer-wearing "cold" rubbers (polymerized at 5°C). Cold rubbers generally accept higher pigment loadings than the hot types. The remaining 12% usage of hot polymers is maintained by certain demands for the inherently easier processing characteristics of these materials.

Styrene content varies from as low as 9% and so-called Arctic or low-temperature-resistant rubbers up to 44% in a rubber designed for its excellent flow characteristics. Materials above an arbitrary 50% level are termed plastics, butadiene-styrene, or high-styrene resins. These resins are used as stiffening agents for SBR rubbers in applications such as shoe soles, and the lattices in this composition range are used in the paint industry and paper coating.

Lattices vary in styrene content from 0% to 46%; the higher levels are used where greater strength is needed. Solids vary from 28% to 60%. The higher solids content, large particle size, cold types provide the largest single outlet for SBR lattices as they are rapidly taking over the major share of the natural rubber foam market. Other applications for SBR lattices include adhesives, coatings, impregnating or spreading compounds for rug backing, textiles, and paper saturation.

Physical Properties

With natural and SBR compounds of the same hardness, it is found that the synthetic formulations not only extrude faster, more smoothly, and with a greater degree of dimensional stability, but also with less danger of sticking if the article is to be coiled during curing.

In molded items, SBR compounds are more nearly free of taste and odor than those made with natural rubber.

The inherently superior physical properties of SBR over natural rubber (or those superior characteristics obtainable with special polymers or through compounding of SBR) include the following:

1. Enhanced resistance to attack by organisms (e.g., soil bacteria)
2. Lower compression set
3. Less subject to scorch in curing operations
4. Superior flex resistance
5. Better resistance to animal and vegetable oils
6. Higher abrasion resistance
7. Superior water resistance
8. Lower freeze temperatures
9. Better dampening effect
10. Increased filler tolerance
11. Less high-temperature discoloration
12. Superior aging, including
    a. Heat resistance (steam or dry heat)
    b. Sunlight checking
    c. No reversion or tackiness
    d. Resistance to oxidation
    e. Ozone cracking
13. Enhanced electrical insulation properties

Natural rubber excels SBR in building tack, tensile strength, tear resistance, elongation, hysteresis (heating on flexing), and resilience properties.

Although SBR has a higher thermal conductivity than natural rubber, it is not sufficient to overcome its higher hysteresis value. Whereas the low conductivity of natural rubber is an engineering advantage in gaskets or seals used in low-temperature equipment, it becomes a distinct disadvantage in an application such as electrical insulation where overheating at the surface of the conductor is experienced during an overload condition.

Another drawback of the low thermal conductivity of natural rubber is encountered when thick slabs or sections are molded, as the time required to bring the stock up to temperature and reduce it again is excessive. This is particularly significant when the compound used is a hard rubber, containing large amounts of sulfur, which liberate internal heat as well. Although the internal heating factor is not important with soft compounds containing small amounts of sulfur, the polymer content of hard rubber may actually be destroyed by excessive internal temperatures.

Compounding

After selection of the proper polymer from the wide diversity of types available, the compounder is afforded a further latitude in building in desired properties through selection of vulcanizing agents, fillers, deodorants, age resistors, blends of other rubbers and resins, etc.

One of the outstanding differences between natural rubber (*Hevea*) and SBR is the need for reinforcement in the latter. Whereas the *Hevea* tensile strength might be increased 25% with the addition of carbon black from its uncompounded value of about 24.03–27.2 MPa, the tensile strength of SBR is increased to above 20.6 MPa from an initial value of less than 6.9 MPa.

Although not as effective as the carbon blacks, other inorganic fillers including clays, whiting, coated calcium carbonate, and precipitated silicas, barytes, zinc oxide, magnesium, lithopone, and mica are used. Titanium dioxide pigment is used where covering power is desired in white or very brightly colored stocks.

Various degrees of oil resistance may be imparted to SBR compounds (or, put conversely, lower-cost, oil-resistant rubbers can be produced) through blending SBR with the higher-cost neoprene or butadiene-acrylonitrile copolymer rubbers.

Scab

A defect on the surface of a casting that appears as a rough, slightly raised surface blemish, crusted over by a thin porous layer of metal, under which is a honeycomb or cavity that usually contains a layer of sand; defect common to thin-wall portions of the casting or around hot areas of the mold.

Scabbing

(1) In wear, a loosely used term referring to the formation of bulges in the surface. (2) In fracture mechanics, it is identified with *spalling*.

Scaffold

A fused mass of coke and other materials that build up in the shaft of a blast furnace and impede normal downflow of the charge.

Scale (Glass)

A small particle of foreign material embedded in the surface of molded glass articles.

Scale (Metals)

Surface oxidation, consisting of a partially adherent layers of corrosion products, left on metals by heating or casting in air or in other oxidizing atmospheres.

Scale (Plastics)

A condition in which resin plates or particles are on the surface of a pultrusion. Scales can often be readily removed, sometimes leaving surface voids or depressions.

Scale

(1) The material deposited on a surface exposed in pure water. Its primary constituent is mechanically deposited impurities but it may include a percentage of corrosion product. (2) Of a drawing, the ratio of the size as drawn to the size in reality.

Scale Pit

(1) A surface depression formed on a forging due to scale remaining in the dies during the forging operation. (2) A pit in the ground in which scale (such as that carried off by cooling water from rolling mills) is allowed to settle out as one step in the treatment of effluent waste water.

Scaling

(1) Forming a thick layer of oxidation products on metals at high temperature. Scaling should be distinguished from rusting, which involves the formation of hydrated oxides. See also *rust*. (2) Depositing water-insoluble constituents on a metal surface, as in cooling tubes and water boilers.

Scalped Extrusion Ingot

A cast, solid, or hollow extrusion *ingot* that has been machined on the outside surface.

Scalping

Removing surface layers from an *ingot*, *billet*, or *slab* by mechanical or flame cutting techniques of surface defects prior to working an *ingot*, *billet*, or *slab*. See also *die scalping*.

Scandium

Scandium (symbol Sc, element number 21) is as plentiful in the Earth's crust as tungsten and more than cadmium, and it is considered a rare element because it is so widely distributed that commercial quantities are difficult to obtain.

Chemically, scandium is somewhat similar to aluminum but is more nearly like yttrium and the rare earth metals. It is quite reactive, forming very stable compounds. The halides, however, can be reduced by calcium, sodium, and potassium, and it is in this manner that scandium metal is produced. For example, scandium fluoride is generally reduced with calcium metal in a tantalum or tungsten crucible in an inert atmosphere. As the scandium is purified by distillation at 1650°C–1700°C in a vacuum of 10^{-5} mm Hg. This produces a material with a metallic luster that has a slight yellow tinge but the general appearance of aluminum.

Chemical Properties

Scandium reacts very rapidly with dilute acids but reacts slowly in a mixture of concentrated nitric and hydrofluoric acids owing to the formation of an insoluble layer of scandium fluoride on the surface of the metal.

Scandium metal is quite stable in air at room temperature but oxidizes fairly rapidly at higher temperatures. Scandium oxidizes at a rate of 0.0187, 0.304, and 0.421 mg/cm^2/h at 400°C, 600°C, and 800°C, which are very dense, adherent films, while the coatings at 800°C or above are loose and nonprotective.

Fabrication

Scandium metal can be arc-melted in an inert atmosphere such as argon at pressures of 50 mmHg without loss by evaporation. It is extremely active, however, and will "getter" the furnace atmosphere, producing a dark film on the surface of the metal.

Pure scandium can be cold rolled or swaged into thin pieces without annealing if the oxygen content and is kept below 200 ppm. Excessive oxygen occurs as dispersed oxides in the grain boundaries and makes fabrication difficult by causing intergranular cracking.

Metal Preparation and Purification

Scandium is normally prepared by reduction of ScF_3 by heating with calcium or lithium in an inert atmosphere using tantalum crucibles. Unfortunately, molten scandium dissolves up to 5 wt% tantalum, and the metal must be further processed to remove this tantalum. Tungsten is more inert, but tungsten crucibles are seldom used because they are more difficult to fabricate and are more brittle than tantalum.

The as-reduced metal is purified by vacuum-induction melting to remove the more volatile impurities. Then the scandium is distilled away from the residual tantalum and tungsten crucible contamination and other less volatile impurities. The scandium vapor is condensed in a closed-end tantalum tube inverted over the crucible that holds the metal to be purified, and as long as the condensed scandium is not melted, it will not pick up any tantalum. Scandium can be purified by zone refining. But the purest metal is achieved by electrotransport purification of distilled scandium metal.

Uses

The amount of scandium produced and consequently used annually is less than a few hundred kilograms. This is due to a lack of a good source of a raw material containing scandium. If scandium were less costly, much more would be used, but in most applications, other,

S

less-expensive elements are substituted. The demand for scandium has begun to increase and may spur further development of supply.

Metallurgy

Scandium modifies the grain size and increases the strength of aluminum. If added to magnesium together with silver, cadmium, or yttrium, it also strengthens magnesium alloys. Scandium inhibits the oxidation of the light rare earths and, if added along with molybdenum, inhibits the corrosion of zirconium alloys in high-pressure steam.

Ceramics

The addition of 1 mol of ScC to 4 mol TiC has been reported to form the second-hardest material known at the time. Sc_2O_3 can be used in many other oxides and to improve electrical conductivity, resistant to thermal shock, stability, and density. In most of these applications, other rare earth oxides are used because of cost. Laboratory balance knife-edges made from Sc_2O_3 are reported to be better than those made from sapphire.

Electronics

Scandium is used in the preparation of the laser material $Gd_3ScGa_4O_{12}$, gadolinium–scandium–gallium–garnet (GSGG). This garnet when doped with both Cr^{3+} and Nd^{3+} ions is said to be 3½ times as efficient as the widely used Nd^{3+}-doped yttrium–aluminum–garnet (YAG:Nd^{3+}) laser. Ferrites and garnets containing scandium are used in switches in computers, in magnetically controlled switches that modulate light passing through the garnet, and in microwave equipment.

Lighting and Phosphors

The largest use of scandium outside of the research laboratory is probably in high-intensity lights. Scandium iodide is added because of its broad emission spectrum. Bulbs with mercury, NaI, and ScI produce a highly efficient light output of a color close to sunlight. This is especially important when televising presentations indoors or at night. When used with night displays, the bulbs give a natural daylight appearance.

Scandium compounds may be used as host for phosphors or as the activator ion, Sc_2O_3, $ScVO_4$, and its Sc_2O_2S are typical host materials, while $ZnCdS_2$ activated with a mixture of silver and scandium is a red, luminescent phosphor suitable for use in television. In most cases, other materials are used because of economics.

Compounds

Scandium Carbide

This is a gray powder with a density of 3.59 g/cm³ and a hexagonal structure that is soluble in mineral acids and has potential as a high-temperature semiconductor.

Scandium Nitride

This gray to red powder with a density of 3.6 g/cm³ has a face-centered (NaCl) structure. The material is of interest in space technology by virtue of its light weight and high melting point (2700°C). Also, scandium nitride crucibles are used for preparing single-crystal gallium arsenide or phosphide.

Scandium Oxide

This is a lightweight refractory oxide with a melting point of ~2300°C. The single-crystal density is 3.91 g/cm³ with a cubic structure.

Scandium oxide is now prepared in quantity from a variety of sources. It is now routinely separated from certain uranium tailings. Some recovery from beryllium ores can be expected plus potential recovery from some phosphate ores with up to 1%–2% scandium values.

It appears to provide better service than high-alumina compositions. The oxide can be flame-sprayed onto a variety of surfaces where it shows heat and thermal shock resistance superior to zirconia, alumina, and magnesia. Single crystals of the oxide are superior to sapphire for balance knife edges or laser crystal host matrices.

Scanning Acoustic Microscopy (SAM)

The use of a reflection-type acoustic microscope to generate very high resolution images of surface and near-surface features or defects in a material. The images are created by mechanically scanning a transducer with an acoustic lens in a raster pattern over the sample. Compared with conventional ultrasound imaging techniques, which operate in the 1–10 MHz range, SAM is carried out at 100–2000 MHz. See also *scanning laser acoustic microscopy (SLAM)*, *ultrasonic C-scan inspection*, and *ultrasonic testing*.

Scanning Auger Microscopy (SAM)

An analytical technique that measures the lateral distribution of elements on the surface of a material by recording the intensity of their Auger electrons versus the position of the electron beam.

Scanning Electron Microscope

A high-power magnifying and imaging instrument using an accelerated electron beam as an optical device and containing circuitry that causes the beam to traverse or scan an area of sample in the same manner as does an oscilloscope or TV tube. It may utilize reflective (*scanning electron microscopy*) or transmitted (*scanning transmission electron microscopy*) electron optics. This scanning electron microscope provides two outstanding improvements over the *optical microscope*; it extends the resolution limits so that picture magnifications can be increased from 1,000× to 2,000× up to 30,000× to 60,000×, and it improves the *depth-of-field* resolution more dramatically, by a factor of approximately 300, thus facilitating its use in fracture studies.

Scanning Electron Microscopy (SEM)

An analytical technique in which an image is formed on a cathode-ray tube whose raster is synchronized with the raster of a point beam of electrons scattered over an area of the sample surface. The brightness of the image at any point is proportional to the scattering by or secondary emissions from the point on the sample being struck by the electron beam.

Scanning Laser Acoustic Microscopy (SLAM)

A high-resolution, high-frequency (10–500 MHz) ultrasonic inspection technique that produces images of features in a sample throughout its entire thickness. In operation, ultrasound is introduced to the bottom surface of the sample by piezoelectric transducer, and the transmitted wave is detected on the topside via rapidly scanning laser beam. Compare with *scanning acoustic microscopy*.

Scanning Transmission Electron Microscopy (STEM)

An analytical technique in which an image is formed on a cathode-ray tube whose raster is synchronized with the raster of a point beam of electrons scanned over an area of the sample. The brightness of the image at any point is proportional to the number of electrons that are transmitted through the sample at the point where it is struck by the beam.

Scantlings

The dimensions of the elements of a structure.

Scarf

See preferred term *edge preparation*.

Scarfing

Cutting surface areas of metal objects, ordinarily by using an oxy-fuel gas torch. The operation permits surface imperfections to be cut from ingots, billets, or the edges of plate that are to be beveled for butt welding. See also *chipping*.

Scarf Joint (Adhesive Bonding)

A joint made by cutting away similar angular segments on two adherends and bonding the adherends with the cut areas fitted together. See also *lap joint*.

Scarf Joint (Welding)

A butt joint in which the plane of the joint is inclined with respect to the main axis of the members.

Scatterband

See *creep*.

Scattering (of Radiant Energy)

The deviations in the direction of propagation of *radiant energy*.

Scattering of x-Rays

In *radiology*, one of the two ways in which the x-ray beam interacts with matter and the transmitted intensity is diminished in passing through object. The x-rays are scattered in a direction different from the original beam direction. Such scattered x-rays falling on the detector do not add to the radiological image. Because they reduce the contrast in that image, steps are usually taken to eliminate the scattered radiation. Also known as *Compton scattering*.

Scavengers

Materials in a flux above a molten metal that combine with contaminants to form slag.

Scavenging

Processes in which some substance, particularly a gas, is added to or bubbled through molten metal to remove dissolved gases or other impurities. The gas may be inert and act purely physically by allowing the contaminating gas to diffuse into it or, alternatively, it may be active and react chemically with the contaminant.

S.C.F.

Stress concentration factor.

Schaeffler Diagram

A diagram predicting the microstructure of rapidly cooled high chromium nickel stainless steels, in particular welds. The various alloying elements are categorized as austenite formers or ferrite formers and ascribed a factor according to their relative potency compared with nickel or chromium, respectively. The total austenite forming effect is plotted on the vertical axis and the ferrite forming effect on a horizontal. Zones in the diagram then indicate whether the structure will be austenite, martensite, ferrite, or some mixture.

Scheelite

An ore of the metal tungsten, occurring usually with quartz in crystalline rocks associated with wolframite, fluorite, cassiterite, and some other minerals. Scheelite is *calcium tungstate*, $CaWO_4$, containing theoretically 80.6% tungsten trioxide and 19.4 lime. It is called *powellite* when it contains some molybdenum to replace a part of the tungsten. It occurs as massive granular or in crystals. The order is white, yellow, brown, or green, with a vitreous luster. *Tungstic acid* is a yellow powder of composition H_2WO_4 made from the ore by treating with HCl. It is not soluble in water but is soluble in alkaline and in hydrofluoric acid, and it is used as a mordant in dyeing, in plastics, and for making tungsten wire by reducing. *Phosphotungstic acid*, called *heavy acid*, is used as a catalyst in difficult synthesis operations on complex ring-compound chemicals. The molecular weight is 2, 879, of which three-fourths is tungsten. The three hydrogens produce a strong acid activity. The acid is soluble in water and in organic solvents.

Pure crystals of scheelite suitable for scintillation-counter phosphors for gamma-ray detection are found, but the natural crystal is rare. Calcium tungstate is grown synthetically as a clear, water-white crystal of tetragonal structure in rods and boules with the axis oriented perpendicular to the growth axis of the rod.

Schlieren

Regions of varying refraction in a transparent medium often caused by pressure or temperature differences and detectable especially by photographing the passage of a beam of light.

Schőniger Combustion

A method of decomposition of organic materials by combusting them in a sealed flask that contains a solution suitable for absorbing the combustion products. The flask is swept with oxygen before ignition.

Scintillator

A material that produces a rapid flash of visible light when an x-ray photon is absorbed.

Scleroscope Hardness Test

A dynamic indentation hardness test using a calibrated instrument that drops a diamond-tipped hammer from a fixed height onto the surface of the material being tested. The height of rebound of the hammer is a measure of the hardness of the material.

Scleroscope Hardness Number (HSc or HSd)

A number related to the height of rebound of a diamond-tipped hammer dropped on the material being tested. It is measured on a scale determined by dividing into 100 units the average rebound of the hammer from a quenched (to maximum hardness) and untempered AISI W-5 tool steel test block. See *shore scleroscope.*

Scorification

(1) Oxidation, in the presence of fluxes, of molten lead containing precious metals, to partly remove the lead in order to concentrate the precious metals. (2) Refining of metals, particularly gold and silver and other precious metals.

Scoring

(1) The formation of severe scratches in the direction of sliding. Scoring may be due to local solid-phase welding or to abrasion. In the United States, the term *scuffing* is sometimes used as a synonym for scoring. Minor damage should be called *scratching* rather than scoring. (2) In *tribology*, a severe form of wear characterized by the formation of extensive grooves and scratches in the direction of sliding. (3) The act of producing a scratch or narrow groove in a surface by causing a sharp instrument to move along that surface. (4) The marring or scratching of any form metal part by metal pickup on the punch or die. (5) The reduction in thickness of a material along a line to weaken it intentionally along that line.

Scouring

(1) A wet or dry cleaning process involving mechanical scrubbing. (2) A wet or dry mechanical finishing operation, using fine abrasive and low pressure, carried out by hand or with a cloth or wire wheel to produce *satin* or *butler*-type finishes.

Scouring Abrasive

Natural sand grains or pulverized quartz employed in scouring compounds and soaps, buffing compounds, and metal polishes. Federal specifications require that the abrasive grains used in grit cake soap and scouring compounds all pass a No. 100 screen; the grains for scouring compounds for marble floors must all pass a No. 100 screen, and 95% pass a No. 200 screen. For ceramic floors 90% must pass a No. 80, and 95% must pass a No. 60 screen. Very fine air-floated quartz is employed in metal polishes, and all grains pass a 325-mesh screen, but the extremely fine powders of metal oxides for polishes and fine finishes are generally called *soft abrasives* and are not classified as scouring materials. Same as *abrasion.*

Scragging

The deliberate pre-service straining of components, particularly springs, beyond yield to improve fatigue life. See *autofrettage,* which is a similar technique applied to hollow products.

Scrap

(1) Products that are discarded because they are defective or otherwise unsuitable for sale. (2) Previously used metal that is suitable for reprocessing, usually remounting, into a product of adequate quality. Material rejected or removed during production and return for remelting is often termed *process scrap.* (3) Material discarded or spoiled and not suitable for continued processing to a finished component.

Scraper

An exclusive seal that has metallic or other firm lips or scraping elements. It serves to remove foreign material from a reciprocating shaft.

Scratch

A groove produced in a solid surface by the cutting and/or plowing action of a sharp particle or protuberance moving along that surface.

Scratch Brush

Wire bristle brush.

Scratch Hardness Test

A form of hardness test in which a sharp-pointed stylus or corner of a mineral specimen is traversed along a surface so as to determine the resistance of that surface to cutting or abrasion. The *Mohs hardness* test is among the most widely used forms of scratch hardness tests but is mainly applied to many mineralogical specimens or abrasives. The file hardness test was one of the first scratch hardness tests used for evaluating the hardness of metallic materials. The file test, which involves pressing the flat face of a steel file heat treated to approximately 67–70 HRC against and slowly across the surface to be tested, is useful in estimating the hardness of steels in the high hardness ranges. It provides information on soft spots and decarburization quickly and easily, and is readily adaptable to odd shapes and sizes that are difficult to test by other methods. Other scratch hardness tests involve using diamond cones, pyramids, and spherical tips, but such scratch hardness tests have not been established and standardized to the extent that macro- and microindentation hardness tests have been.

Scratching

(1) In *tribology*, the mechanical removal and/or displacement of material from a surface by the action of abrasive particles or protuberances sliding across the surfaces. (2) The formation of fine scratches in the direction of sliding. Scratching may be due to asperities on the harder slider or to hard particles between the surface or embedded in one of them. Scratching is to be considered less damaging than *scoring.* See also *abrasion*, *plowing*, and *ridging wear.*

Scratching Abrasion

See *low-stress abrasion.*

Scratch-Resistant Coatings

Coating applied to glass surfaces to reduce the effects of frictive damage. Examples are SnO_2 or TiO_2 coatings applied to glass containers.

Scratch Test

Various fairly crude tests in which the component in question is scratched. The two concepts are either that the component is scratched by a series of calibrated implements (see *Mohs test* and *file test*) or it is scratched by a single implement to find variations in hardness.

Scratch Trace

In a metallographic specimen, a line of etch markings produced on a surface at the site of a preexisting scratch, the physical groove of the scratch having been removed. The scratch trace develops when the deformed material extending beneath the scratch has not been removed with the scratch groove and when the residual deformed material is attacked preferentially during etching.

Screen

(1) The woven wire or fabric cloth, having square openings, used in a sieve for retaining particles greater than the particular mesh size. U.S. standard, ISO, or Tyler screen sizes are commonly used. (2) One of a set of sieves, designated by the size of the openings, used to classify granular aggregates such as sand, ore, or coke by particle size. (3) A perforated sheet placed in the gating system of a mold to separate impurities from the molten metal.

Screen Analysis

See *sieve analysis.*

Screen Classification

See *sieve classification.*

Screening

Separation of a powder according to particle size by passing it through a screen having the desired mesh size.

Screw

(1) A screw is a cylindrical body with a helical groove cut into its surface. For practical purposes, a screw may be considered to be a wedge wound in the form of a helix so that the input motion is a rotation while the output remains translation. The screw is to the wedge much the same as the wheel and axle is to the lever in that it permits the exertion of force through a greatly increased distance.

The screw is by far the most useful form of inclined plane or wedge and finds application in the bolts and nuts used to fasten parts together; in lead and feed screws used to advance cutting tools or parts in machine tools, in screw jacks used to lift such objects as automobiles, houses, and heavy machinery; in screw-type conveyors used to move bulk materials; and in the propellers for airplanes and ships. (2) Some versions, wood screws or self-tapping screws, taper so that they can cut their way into unthreaded holes. A machine screw is similar to a bolt in having no taper and requiring a prethreaded hole to enter but it is threaded for the full length of the shank. A drive screw either has a driven member mounted on it or it runs in a fluid to which it imparts motion, termed an Archimedes screw.

Screw Dislocation

A dislocation that travels in a direction normal to the direction of formation of the slip step. See *dislocation.*

Screw-Machined Parts

Among the many screw-machine parts are such things as bushings, bearings, shafts, instrument parts, aircraft fittings, watch and clock parts, pins, bolts, studs, and nuts. Screw-machine parts can be produced at rates up to 4000 parts per hour, and tolerances of ± 0.03 mm are common. In addition, these parts can be made of practically any material that is machinable and that can be obtained in rod form. In comparison, die casting is restricted to a limited number of alloys of relatively low melting point, and cold heading requires materials that are ductile at room temperature.

However, screw-machine parts do have certain limitations: Because the material must be in the form of bar or rod, the cross-section of the part is generally limited to a circle, hexagon, square, or other readily available extruded cross-sections, although a special extruded cross-section can sometimes be ordered. The stock used must be as large as the greatest cross-section of the part, depending on the shape, this may or may not result in large scrap losses. In general, irregular and nonsymmetrical parts are not good screw-machine parts, although some parts of this type are produced.

Parts can become quite expensive unless certain restrictions are observed in selection of material and in design of parts.

Materials

Selection of the material is very important. Cost of producing a part is influenced greatly by ease of machining; poor machinability reduces tool life and causes frequent shutdown to replace tools. If service requirements permit, readily machinable materials, such as resulfurized steels or leaded brasses, should be given preference.

Some of the many metals used in screw-machine parts are carbon and low-alloy steels; stainless and heat-resisting steels; cast and malleable irons; copper alloys, particularly brass, bronze, and nickel–silver; aluminum and magnesium alloys; nickel and cobalt superalloys; and gold and silver. Among the nonmetallic materials used are vulcanized fiber, hard rubber, nylon, Teflon, methyl methacrylate, polyethylene, polystyrene, and phenolics.

Lot Size

Lot size should be considered in selecting the material, as it influences not only the cost of the part but the material that can be used.

Design

Screw-machine parts should be designed to be made from standard compositions, standard stock sizes, and standard shapes. Specifying nonstandard bar not only increases costs but may result in delays in obtaining the parts.

Closer dimensions can be held but should be specified only when necessary, because they increase the cost of producing and inspecting the part.

Holes

Standard drill sizes should be specified if possible. When holes require reaming, sizes that can be finished with standard reamers should be specified.

Threads

A major item in the design of a screw-machine product is the selection of a proper standard thread, with the aim of providing interchangeability. In new designs, use the Unified Screw Thread System, adopted by the United States, Canada, and Great Britain. For existing designs, however, continue to use the American National System. Indicate threads by size, pitch, series, and class, only; using a special pitch and specifying major and minor diameters is often unnecessary and always expensive.

Concentricity

When concentricity is required, specify it in terms of total indicator reading (TIR) rather than as a dimension. It is good engineering practice to indicate the diameters that must run true by an arrow leading to a note reading "diameters marked A must be concentric to within—TIR measured at points—." Specifying concentricity when it is not required increases cost of the part unnecessarily.

Burrs

Do not specify removal of burrs unless this operation is essential. Since there are some doubts about the definition of a burr, the screw-machine industry has established certain criteria:

1. Sharp corners are not considered burrs unless they have ragged edges and interfere with operation of the part.
2. Slight tears or roughness on the first two threads of a tapped hole or the male thread are not burrs unless they interfere with assembly.
3. A projection is not a burr if it must be found with a magnifying glass.

Burrs can be eliminated in most cases without additional cost by specifying chamfered or rounded corners on the parts.

Finish

In general, cost of finishing will depend on the degree of finish required; the finer the finish, the higher the cost. A finish for appearance only need not be as fine as a finish required on mating or bearing surfaces. In any case, finish should be specified in terms of microinch root-mean-square (rms) units. Microinch finish depends on material and operation and without secondary operations can range from the usual 125 rms to as low as 16 rms. The latter finish requires special care and is a very high cost operation on a production basis.

Dimensions

If a part requires heat treatment or plating, dimensions supplied to the part producer will depend on where these operations are to be done. If the supplier is to finish the part, give dimensions to apply after heat treatment or plating; if the buyer is to finish the part, give dimensions to apply before heat treating or plating.

Screw Plasticating Injection Molding

A technique in which the plastic is heated and converted from pellets to a viscous melt, and then forced into a mold by means of an extruder screw, which is an integral part of the molding machine.

Machines are either single stage, in which plastication and injection are done in the same cylinder, or double stage, in which the material is plasticated in one cylinder and then fed to a second for injection into a mold. See also *injection molding* (plastics).

Screw Press

A high-speed press in which the ram is activated by a large screw assembly powered by a drive mechanism.

Screws

See *extruder* (plastics).

Screw Stock

(1) Free-machining bar, rod, or wire. (2) A common term for steels and nonferrous metals, notably brass, having superior machinability and used to make screws and small turn parts on automatic screw machines. In steels, free-cutting, or free-machining, quality is imparted by the addition of small amounts of alloying elements, such as sulfur, phosphorus, lead, tellurium, bismuth, calcium, or selenium, sometimes in various combinations. In brass, lead is the usual free-machining additive.

Scrim

A low-cost reinforcing fabric for composites made from continuous filament yarn in an open-mesh construction. Used in the processing of tape or other *B-stage* material to facilitate handling.

Scruff

A mixture of tin oxide and iron-tin alloy formed as dross on a tin-coating bath.

Scuffing

(1) Localized damage caused by the occurrence of solid-phase welding between sliding surfaces, without local surface melting. In the United Kingdom, scuffing implies local solid-phase welding only. In the United States, scuffing may include abrasive effects and the term *scoring* is sometimes used as a synonym. (2) A mild degree of *galling* that results from the welding of asperities due to frictional heat. The welded asperities break, causing surface degradation. In general, the term *scuffing* has been used in so many different ways that its use should be avoided whenever possible, and instead replaced with a more precise description of the specific type of *surface damage* being considered. (3) Light surface damage by a mechanical rather than a corrosion process. The term is used for the early stage of both *adhesive wear* and *abrasive wear*.

Sea Coal

Finely ground coal, used as an ingredient in molding sands.

Seal

(1) In tribology, a device designed to prevent leakage between relatively moving parts. (2) A device designed to prevent the movement of fluid from one chamber to another, or to exclude contaminants.

Sealant

A material applied to a joint in paste or liquid form that hardens or cures in place, forming a seal against a gas or liquid entry. The primary application area for sealants is the building construction industry. Sealants for the commercial construction market are designed to act as a barrier to water, air, atmospheric pollution, vibration, insects, dirt, and noise, while compensating for movement that occurs in the structure being sealed. This movement can be caused by many factors, including changes in temperature and moisture content, permanent loading or load transfer, and chemical changes such as sulfate attack and wind loads. Other key sealant markets are the automotive industry, the aerospace market, and the electrical and electronics market. Major sealant types include oil-base caulks (25%–30% linseed or soy oil, 6%–12% fibrous filler, 40%–60% calcium carbonate filler, 0.05%–3% pigment, 0%–2% gelling agents, and 0%–1% catalyst), latex acrylic polymers, polyvinyl acetate caulks, solvent acrylics, butyl sealants, polysulfites, urethanes, and silicones.

Seal Coat (Thermal Spraying)

Material applied to infiltrate the pores of a thermal sprayed deposit.

Sealing

(1) Closing pores in anodic coatings to render them less absorbent. (2) Plugging leaks in a casting by introducing thermosetting plastics into porous areas and subsequently setting the plastic with heat.

Sealing Face

The lapped surface of a seal that comes in close proximity to the face of the mating ring of a face seal, thus forming the primary seal. With reference to lip seals, the preferred term is seal contact surface.

Seal Nose

The part of the primary seal ring of a face seal that comes in closest proximity to the mating surface and that, together with the mating surface, forms the primary seal.

Seal Oil

An oil resembling sperm oil obtained from the blubber of the *oil seal*, a sea mammal native to the Atlantic Ocean. The oil has a saponification value as high as 195 and an iodine value up to 150, and it was once valued for lubricating and cutting oils but is now scarce.

Some seal oil is obtained from Steller's sea lion, a large-eared seal found from southern California to the Bering Sea. The adult male weighs up to 2200 lb (998 kg). The blubber is about 75% oil, with an iodine value of 143 and saponification value of 190. From 40% to 50% of the carcass is a dense, dark-red, edible meat, but in the United States, *seal meat* is used only in animal foods. *Seal leather*, from the skin, is used for fancy specialty articles, but it has too many defects for general use. The product known as *sealskin* is a valuable fur skin from the fur seal, about 80% of which are caught off the Pribilof Islands where they return in June to breed. No killing is now permitted at sea. Each bull seal has as many as 50 females, and the killing is usually restricted to the surplus males. About 30% of the skins are black fur, which brings the highest price.

Sealing Processes

Various processes intended to improve the surface condition, particularly its resistance to fluid penetration but also its resistance to staining or tarnishing. Examples include coating with oil, grease, silicone, paraffin/lanoline mixtures, etc., immersing newly anodized coatings in boiling water to block the natural pores with the resultant aluminum hydroxide, pressure or vacuum impregnation of porous casings with resins and phosphate conversion coatings.

Sealing Run (of Weld)

The run laid on the remote side of a previously welded root. Compare with *seal weld*.

Seals (Ceramic-to-Metal Seals)

The resultant fabrication after brazing depends on the integrity of the seal. Application in electronics demands that the seal withstand high vacuum and that the materials used in its fabrication do not outgas to poison the vacuum. There are several types of ceramic-to-metal seals.

Butt Seals

A flat metal washer brazed to a flat metallized ceramic surface is typical of this class. The principal stress is a shear stress that may break the ceramic in brittle fracture. Ceramic backup rings are frequently used to establish a uniform stress distribution, which materially improves the ruggedness of the joint. Ductility of the braze and of the metal hardware is also significant.

External Seals

A flange, or collar, surrounding a cylindrical ceramic part is typical of this class. The metal is selected for its slightly higher coefficient of expansion and it produces the resultant seal.

Glass-to-Metal Seals

The coefficient of linear expansion of the glass is matched as closely as possible to that of the metal it is to join. Both materials are heated prior to forming the seal; the glass to flowing and the metal to a dull red heat. A thin oxide coating on the metal often is needed for good adhesion.

Internal Seals

The seals can be very reliable if a low-expansivity metal (e.g., tantalum) can be used for the lead. Low-expansion metals result in compressive stresses but their poor oxidation resistance frequently precludes their use. Kovar and nickel-iron alloys, plus the high expansivity braze metals, ordinarily result in tensile stresses at the inner wall of the ceramic. As a ceramic is weakest in tension, this design may be troublesome unless properly engineered.

Ram Seal (Crunch Seal)

The metal sleeve with a soft plating such as copper, silver, or nickel is distorted while moving over the sharp ceramic edge. Ceramic-metal contact is limited to a thin circumferential line.

Taper Seal (Telescoping Seal)

A thin metal sleeve is forced onto a thick ceramic cylinder whose outside diameter is tapered. Negligible stresses are introduced into the ceramic but the metal may be stressed beyond its yield point. During thermal cycling, the metal may fail through fatigue.

Seal Weld

A weld that forms a continuous impervious barrier to fluid but with no significant contribution to joint strength. Compare with *sealing run*.

Seam

(1) On a metal surface, an unwelded fold or lap that appears as a crack, usually resulting from a discontinuity. (2) A surface defect on a casting related to but of lesser degree than a *cold shut*. (3) A ridge on the surface of a casting caused by a crack in the mold face. (4) A defect in the form of an elongated, surface emergent fissure produced during working operations. (5) Any linear joint formed by soldering, welding, folding, etc.

Seamless Tubing

Tubing that does not contain a longitudinal or helical seam weld. It may be made by various processes including various processes such as *tube making* and *extrusion*.

Seam Weld

A continuous weld made between or upon overlapping members, in which coalescence may start and occur on the faying surfaces, or may have proceeded from the surface of one member. The continuous weld may consist of a single weld bead or a series of overlapping spot welds. Common seam weld types include (1) lap seam welds joining flat sheets, (2) flange-joint lap seam welds with at least one flange overlapping the mating piece, and (3) mash seam welds with work metal compressed at the joint to reduce joint thickness. See also *roller spot welding*.

Seam Welding

(1) Arc or resistance welding in which a series of overlapping spot welds is produced with rotating electrodes or rotating work, or both. (2) Making a longitudinal weld in sheet metal or tubing. See also *resistance seam welding*.

Season Cracking

An obsolete term for stress corrosion cracking in brass. The term is said to derive from the spontaneous cracking, during the Indian monsoon season, of brass cartridges containing significant residual stress and stored in the ammonia-rich environment of stables.

Seasoning

Various circumstances where a product is left to develop some characteristic over a period of time.

Seaweed

A plant growing in the sea, belonging to the extensive plant division known as *algae*. About 17,000 varieties of seaweed are listed, but only a few are exploited commercially. Algae are non-seed-bearing plants containing photosynthetic pigments. They have no vascular or food-conveying system and must remain submerged in the medium from which they acquire their food. They occur in both fresh and salt waters.

The *brown seaweeds*, which are the true kelps, grow in temperate and polar waters. They produce *algin*, *fucoidin*, and *laminarin*. The *red seaweeds* are the *carrageens*, which reduce *carrageenan*, and the *agarophytes*, which yield *agar* and *agaroid*. They grow in warm waters. But color is an indefinite classification; the chlorophyll in the green Irish moss is often so masked by other pigments that the weed may be purplish black. All the seaweed colloids, or phycocolloids, are polysaccharides, having galactose units linked in long chains of molecular weights from 100,000 to 500,000, varying in their chemical structure. They are anionic polyelectrolytes, with negative radicals on each repeating polymer unit. *Irish moss*, also called *chondrus*, *pearl moss*, and *carrageen*, is a dwarf variety of brown seaweed.

The seaweed of the north Atlantic is also used to produce agar and algin. It is bleached and treated to produce gelatin used in foodstuffs, as a clarifying agent, and as a sizing for textiles. It is a better suspending and gelatinizing medium than agar for foodstuffs and cosmetic emulsions. At least 25 mineral salts are known to be present in seaweed as well as several vitamins. In the utilization of the seaweed as gelatin or alginate, these are left in the *kelp meal*, which is marketed as poultry and stock feed. In Asia, the whole plant is cooked and eaten. *Seaweed flour*, made in Germany from *Iceland seaweed*, is the ground, dry seaweed containing all the minerals and vitamins. It is mixed with wheat and rye flours to make *algenbrot*, a bread with higher food value and better keeping qualities than ordinary wheat bread. But more than 8% gives a peculiar flavor to the bread. The Irish name *dulse* is applied to the dried or cooked seaweed, used in the Canadian Maritime Provinces for food. It is purple and rich in iodine and mineral salts. Other species, known as *laver* and *murlins*, are also used in Iceland, Ireland, and Scotland for food. When used for producing iodine in Scotland, the seaweed goes under the general name of *tangle*.

Dry seaweed contains up to 30% *alginic acid*; the water-soluble salts of this acid are called algin. It belongs to the group of complex, open-chain *uronic acids* that occur widely in plant and animal tissues and are related to the proteins and pectins. All the algins are edible, but they pass unchanged through the alimentary tract and add no food value. Carrageenan is much used as a stabilizer for chocolate in milk. Laminarin is used as *laminarin sulfate* as a blood-clotting agent. *Sodium alginate* is used as a stabilizer and ice-crystal retarder in ice cream, as an emulsifier in medicines, and to replace gum arabic, a colorless, water-soluble gum made by dissolving algin in sodium carbonate solution and neutralizing with hydrochloric acid. *Protan jelly*, used for coating fish for freezing, is algin in a dilute edible acid. When frozen, the jelly is impervious to air and prevents oxidation. It can be washed off with water.

Alginic fibers are silklike fibers made by forcing a sodium alginate solution through spinnerets into a calcium chloride bath and insolubilizing with beryllium acetate, but the fiber is soluble in sodium soaps and the fabrics must be dry-cleaned. Soluble alginic yarns are used for making fancy fabrics where uneven spacing of threads is desired without change in the loom. The alginate yarn is washed out of the fabric after weaving, leaving the desired spacing.

Secant Modulus

(1) The slope of the secant drawn from the origin to any specified point on the stress-strain curve; see also *modulus of elasticity*. (2) The ratio of stress to strain, analogous to the elastic modulus,

for cases where the stress strain relationship is not linear; see *tensile test*. The secant modulus ignores nonlinearity by calculating a specific value for each measured combination of stress and strain.

Secondary Alloy

Any alloy whose major constituent is obtained from recycled scrap metal. Compare with *primary alloy*.

Secondary Bonding

The joining together, by the process of adhesive bonding, of two or more already cured composite parts, during which the only chemical or thermal reaction occurring is the curing of the adhesive itself.

Secondary Bonds

The weaker interatomic bonds such as the van der Waals force.

Secondary Circuit

The portion of a welding machine that conducts the secondary current between the secondary terminals of the welding transformer and the electrodes, or electrode and work.

Secondary Creep

This stage of creep during which the rate of creep deformation is constant, assuming constant stress. See *creep*.

Secondary Crystallization

In processing of plastics, the slow crystallization process that occurs after the main solidification process is complete. Often associated with impure molecules.

Secondary Electron

A low-energy electron (0–50 eV) emitted from a surface that is struck by particles with higher energies.

Secondary Etching

Development of microstructures deviating from the primary structure of a metal or alloy through transformation and heat treatment in the solid state.

Secondary Extinction

In *x-ray diffraction*, a decrease in the intensity of a diffracted x-ray beam caused by parallelism or near-parallelism of mosaic blocks in a mosaic crystal; the lower blocks are partially screened from the incident radiation by the upper blocks, which have reflected some of it. See also *primary extinction*.

Secondary Gas

In thermal spraying, the gas constituting the minor constituent of the arc gas fed to the gun to produce the plasma. The primary arc gas, usually argon or nitrogen, is supplemented with secondary gases such as nitrogen, helium, and/or hydrogen, in order to increase the temperature of the plasma.

Secondary Hardening

The hardness increase in some alloy steels produced by tempering. It results from carbide precipitation effects plus, in some cases, the transformation to martensite of retained austenite following the reduction of its carbon content by the carbide precipitation. See *steel*.

Secondary Ion

An ion other than the probe ion that originates from and leaves the specimen surface as a result of bombardment with a beam of primary or probe ions; a sputtered ion. See also *sputtering*.

Secondary Ion Mass Spectroscopy (SIMS)

An analytical technique that measures the masses of ions emitted from the surface of a material when exposed to a beam of incident ions. The incident ions are usually monoenergetic and are all of the same species, for example, 5 keV Ne^+ ions. See also *secondary ion*.

Secondary Metal

Metal recovered from scrap by remelting and refining.

Secondary Nucleation

In processing of plastics, the mechanism by which crystals grow.

Secondary Operation

Any operation performed on a sintered powder metallurgy compact, such as sizing, coining, repressing, impregnation, infiltration, heat or steam treatment, machining, joining, plating, or other surface treatment.

Secondary Phases

Phases in alloy systems that do not include the pure metal in their range, that is, all phases in a system other than the two primary phases.

Secondary Recrystallization

The process by which a few large grains are nucleated and grow at the expense of a fine-grained, but essentially strain-free matrix. Also known as abnormal or discontinuous grain growth. See also *primary recrystallization*.

Secondary Seal

A device, such as bellows, piston ring, or O-ring, that allows axial movement of the primary seal of a mechanical face seal without undue leakage.

Secondary-Standard Dosimetry System

A system that measures energy deposition indirectly. It requires conversion factors to account for such considerations as geometry, dose rate, relative stopping power, incident energy spectrum, or other effects, in order to interpret the response of the system. Thus, it requires calibration against a primary dosimetry system or by means of a standard radiation source. See also *dosimeter*.

Secondary Stresses

(1) Bending stresses as opposed to direct tensile stresses. (2) Additional stresses superimposed on the main stress.

Secondary Structure

In aircraft and aerospace applications, a structure that is not critical to flight safety.

Secondary x-Rays

The x-rays emitted by a specimen following excitation by a primary x-ray beam or an electron beam.

Second-Degree Blocking

In adhesive bonding, an adherence of such degree that when surfaces under test are parted, one surface or the other will be found to be damaged.

Second-Phase Inhomogeneity

In ceramics and glasses, a microstructural irregularity related to the nonuniform distribution of a second phase, for example, an atypically large pocket of a second phase or a second-phase zone of composition or crystalline phase structure different from the matrix material.

Sectioning

The removal of a conveniently sized representative specimen from a larger sample for medley graphic inspection. Sectioning methods include shearing, *sawing* (using hack saws, band saws, and diamond wire saws), abrasive cutting, and *electrical discharge machining*.

Sedimentation

The settling of particles suspended and dispersed in a liquid through the influence of an external force, such as gravity or centrifugal force.

Seebeck Effect

The electromotive force (e.m.f.) due to the difference of temperature between two junctions of dissimilar conductors in a circuit. See also *thermoelectric effect*.

Seed

(1) A small, single crystal of a desired substance added to a solution to induce crystallization. (2) Small particles or agglomerates, crystals, or crystallites introduced in large numbers into a vessel to serve as nuclei or centers for further growth of material on their surfaces. (3) An extremely small gaseous inclusion in glass.

Seger Cone

Tall, narrow pyramids (about 8°–10° slope) of fireclay and refractory mixtures having compositions selected to cause the pyramid to collapse at some specific temperature. They are used usually in multiples covering a range of temperature to monitor the peak temperature in furnaces.

Segment

In *sampling*, a specifically demarked portion of a *lot*, either actual or hypothetical.

Segment Die

A die made of parts that can be separated for ready removal of the workpiece. Synonymous with *split die*.

Segregation (Metals)

(1) Nonuniform distribution of alloying elements, impurities, or microphases in metals and alloys. (2) A casting defect involving a concentration of alloying elements at specific regions, usually as a result of the primary crystallization of one phase with the subsequent concentration of other elements in the remaining liquid. Microsegregation refers to normal segregation on a microscopic scale in which material richer in an alloying element freezes in successive layers on the dendrites (*coring*) and in constituent network. Macrosegregation refers to gross differences in concentration (e.g., from one area of a casting to another). See also *inverse segregation* and *normal segregation*.

Segregation (Plastics)

A separation of components in a molded article usually denoted by wavy lines and color striations in thermoplastics. In thermoset plastics, usually a separation of resin and filler on the surface.

Segregation

Variations in impurity and alloy content across a grain or casting. The effect arises because the first metal to solidify is usually relatively pure so impurities and alloying elements tend to be rejected into the band of material immediately ahead of the advancing solidification front. Within a grain, significant alloy composition gradients can occur—termed *Coring*. On a larger scale, when advancing grains meet, impurities pushed ahead of the two grains concentrate at the grain boundaries—termed *minor segregation*. If solidification continues progressively from the periphery to the center of the ingot, impurities and the alloy constituents tend to become concentrated near the center line. This is called *normal segregation*, *major segregation*, or *V segregation*, the latter reflecting the characteristic distribution of the impurities revealed by sectioning the ingot. In large ingots, the impurity level in the band of molten metal ahead of the solidification front may depress its freezing point to such an extent that the purer metal at the center of the ingot solidifies first, leaving a band of impurities at about mid radius. This is termed *inverted V segregation*, again because of its appearance on sectioning. In extreme cases, material rich in low-melting-point alloy can be squeezed back along and solidify grain boundaries toward or onto the exterior surface, termed *inverse segregation*. Eruptions of this low-melting-point, alloy-rich material onto the surface are termed *blebs* or *blebbing* or *cauliflowers*.

Segregation Banding

Inhomogeneous distribution of alloying elements aligned in filaments or plates parallel to the direction of working. See also the term *banding*.

Segregation (Coring) Etching

Development of segregation (coring) mainly in macrostructures and microstructures of castings.

Seize

To prevent a metal part from being ejected from a die as a result of *galling*.

Seizing/Seizure/Seize-Up

(1) An extreme case of adhesive wear when the two contacting surfaces become permanently locked together. The engagement mechanism may be purely mechanical, resulting from mutual distortion and embedding of the two surfaces or there may be a measure of welding, perhaps considerable, resulting from the high local temperature generated by friction. (2) The stopping of relative motion as a result of interfacial friction. Seizure may be accompanied by gross surface welding. The term is sometimes used to denote *scuffing*. See also *solder lock*.

Séjournet Process

The use of glass as lubricant during extrusion at high temperature, particularly of steels. Immediately prior to entering the extrusion chamber, the heated billet is coated, typically by rolling it down a tray of powder, and a bonded pad of powder is placed in the die end of the chamber. See *Ugine-Séjournet process*.

Selected-Area Diffraction (SAD)

Electron diffraction from a portion of a sample selected by inserting an aperture into the magnification portion of the lens system of a transmission electron microscope. Areas as small as 0.5 μm in diameter can be examined in this way.

Selected Area Diffraction Pattern (SADP)

An electron diffraction pattern obtained from a restricted area of a sample. The sharp spots in the pattern correspond closely to points in the reciprocal lattice of the material being studied. Usually such patterns are taken from a single crystal or a small number of crystals. Any of the three types of electron diffraction patterns can be generated in an SADP: (1) ring patterns from fine-grained polycrystalline materials in which diffraction occurs simultaneously from many grains with different orientations relative to the incident beam; (2) spot patterns in which diffraction occurs from a single-crystal region of the specimen; and (3) Kikuchi line patterns in which diffraction occurs from a single-crystal region of the specimen, which is sufficiently thick that the diffracting electrons have undergone simultaneous elastic and inelastic scattering. See the term *diffraction pattern* and *Kikuchi lines*.

Selective Block Sequence

A block welding sequence in which successive blocks are completed in a certain order selected to create a predetermined stress pattern. See also the term *block sequence*.

Selective Freezing

When an alloy phase with a wide composition range commences to solidify the first crystals are considerably enriched with the higher melting point constituent. These may be selectively removed for use or to be discarded.

Selective Heating

Intentionally heating only certain portions of a workpiece.

Selective Leaching

(1) Corrosion in which one element is preferentially removed from an alloy, leaving a residue (often porous) of the elements that are more resistant to the particular environment. Also called *dealloying* or parting. (2) The dissolution and removal of one component from a mixture that may be an ore or other mix of powders or it may be a solid alloy. The term is sometimes used as a generic term for corrosion processes such as dezincification in which one element of an alloy is removed. However, purists may argue that dezincification and similar processes involve solution of both elements followed by redeposition of the survivor. See also *decarburization*, *decobaltification*, *denickelification*, *dezincification*, and *graphitic corrosion*.

Selective Quenching

Quenching only certain portions of an object.

Selective Transfer

A process involving the transfer and attachment of a specific species from one surface to the mating surface during sliding. This term is to be distinguished from the general tribological term *transfer*, which involves the same general process, but which, when used by itself, does not discriminate as to which of the species present as a multiconstituent surface is transferred during sliding.

Selectivity

In materials characterization, the ability of a method or instrument to respond to a desired substance or constituent and not to others.

Selenium

Selenium (symbol Se) is the third member of group VIB of the periodic arrangement of the elements; it is more metallic than sulfur but less metallic than tellurium, the two adjacent elements of this group.

It is rarely found in its native state but is usually associated with lead, copper, and nickel from which it is recovered as a by-product.

Chemical Properties

The chemical properties of selenium are very similar to other group VIB elements such as sulfur and tellurium. It reacts readily with oxygen and halogens to form their respective compounds, that is, oxides and halides. Selenous and selenic acids are easily prepared and in many respects resemble sulfurous and sulfuric acids. Selenium is not very soluble in aqueous alkali solutions, but selenites and selenates are readily prepared by fusion with oxidizing salts of the alkalies.

Selenium metal is odorless and tasteless, but the vapor has a putrid odor. The material was highly poisonous, and is used in insecticides and ship-hull paints. Foods grown on soils containing selenium may have toxic effects, and some weeds growing in the Western states have high concentrations of selenium and are poisonous to animals eating them. Selenium burns in air with a bright flame to form selenium dioxide, SeO_2, which is in white, four-sided, crystalline needles. The oxide dissolves in water to form selenous acid, H_2SeO_3, resembling sulfurous acid but very weak.

Forms

Commercial grades (99.5%) of selenium are usually sold as a powder of various mesh sizes packed in steel drums. High-purity selenium (99.99%), used primarily in the electronics industry, is available as small pellets.

Ferroselenium containing 50%–58% selenium and nickel-selenium containing 50%–60% selenium are available in the form of metallic chunks packed in wooden barrels or boxes.

Applications

The most important uses for selenium are in the electronics industry, and such components as current rectifiers, photoelectric cells, xerography plates, and as a component in intermetallic compounds for thermoelectric applications.

A selenium dry-plate rectifier consists of a base plate of aluminum or iron, either nickel plated or coated with a thin layer of bismuth; a layer of 0.05–0.08 mm of halogenated selenium; an artificial barrier layer; and a counter-electrode of cadmium or a low-melting alloy. The purpose of a controlled amount of halogen in the selenium is to accelerate the transformation from amorphous to the hexagonal form, which is done at temperatures slightly above 210°C during the manufacturing process.

The photoelectric properties of selenium make it useful for light-measuring instruments and for electric eyes. Amorphous or vitreous selenium is a poor conductor of electricity, but when heated, it takes the crystalline form, its electrical resistance is reduced, and it changes electrical resistance when exposed to light. The change of electric conductivity is instantaneous; even the light of small lamps has a marked effect since the resistance varies directly as the square of the illumination. The pure amorphous powder is also used for coating nickel-plated steel or aluminum plates in rectifiers for changing alternating current to pulsating direct current. The coated plates are subjected to heat and pressure to change the selenium to the metallic form, and the selenium coating is covered with a layer of cadmium bismuth alloy. Selenium rectifiers are smaller and more efficient than copper oxide rectifiers, but they require more space than silicon rectifiers and are limited to an ambient temperature of 85°C.

Selenium is also used in steels to make them free-machining, with up to 0.35% used. Up to 0.05% of selenium may also be used in forging steels. About 0.017 to 0.024 kg of selenium per 0.9 metric ton of glass may be used in glass to neutralize the green tint of iron compounds. Large amounts produce pink and ruby glass. Selenium gives the only pure red color for signal lenses. Pigment for glass may be in the form of the black powder, barium selenite, $BaSeO_3$, or as sodium selenite, Na_2SeO_3, and may be used with cadmium sulfide. Selenium is also used as an accelerator in rubber and to increase abrasion resistance.

In copper alloys, selenium improves machinability without hot shortness. Selenium copper is a free-cutting copper containing about 0.50% selenium. It machines easily, and the electric conductivity is nearly equal to that of pure copper. The tensile strength of the annealed selenium copper is about 207 MPa. Small amounts of selenium salts are added to lubricating oils to prevent oxidation and gumming.

Xerography is a dry printing method for the production of images by light or other rays. A thin layer of selenium is given a strong positive electrical charge and the design is projected on it, which discharges the selenium in proportion to the light intensity, producing a latent image. This image is developed by dusting with negatively charged particles that adhere to the plate where it was not struck by the projected light. Some further simple steps allow the image to be used as a master for many additional prints.

One of the oldest uses of selenium is in the glass and ceramic industries. The familiar ruby red that is commonly seen in glass is produced by cadmium selenide. The same cadmium-selenide reds find broad application in the field of ceramics and enamels.

Self-Absorption

In *optical emission spectroscopy*, reabsorption of a photon by the same species that emitted it. For example, light emitted by sodium atoms in the center of a flame may be reabsorbed by different sodium atoms near the outer portions of the flame.

Self-Acting Bearing

See *gas bearing* and *self-lubricating bearing*.

Self-Adjusting Arc Welding

This term usually refers to metal inert gas welding (MIG). It is so-called because the electrical characteristics of the equipment are such that the wire feed speed, the arc current, and the wire burn rate mutually interact to maintain a constant arc gap.

Self-Aligning Bearing

A *rolling-element bearing* with one spherical raceway that automatically provides compensation for shaft or housing deflection or misalignment. Compare with *aligning bearing*.

Self-Annealing Materials

Those materials that require no heat input to recrystallize when working at ambient temperatures, for example, tin and lead.

Self-Curing

See *self-vulcanizing*.

Self-Diffusion

Thermally activated movement of an atom to a new site in a crystal of its own species, as, for example, a copper atom within a crystal of copper.

Self-Extinguishing Resin

A resin formulation that will burn in the presence of a flame but will extinguish itself within a specified time after the flame is removed. The term is not universally accepted.

Self-Fluxing Alloys (Thermal Spraying)

Certain materials that "wet" the substrate and coalesce when heated to their melting point, without the addition of a fluxing agent.

Self-Hardening Steel

See preferred term *air-hardening steel*.

Self-Locking

Fastener bolts and similar threaded fasteners with antislackening features as described for self-locking nuts.

Self-Locking Nut

A nut with features that resist slackening resulting from in-service vibration and other inadvertent loads. Apart from the obvious risk of a joint completely parting if the nut is lost, even a limited reduction in preload raises the risk of fatigue failure. There are numerous locking features although they tend to fall into two categories. One general type, often termed *stiff nuts*, relies on some form of distortion of the nut thread to introduce a high level of friction against, and possibly indentation into, the mating thread. A second category has inserts of some resilient material, typically a hard plastic, set into the nut thread. With some types of locking features, it is good practice to discard the component after a single use as the locking effect can be much reduced on second tightening. Also see *lock nut*.

Self-Lubricating Bearing

(1) A bearing independent of external lubrication. These bearings may be sealed for life after packing with grease or may contain *self-lubricating material*. (2) A sintered product whose accessible pore volume is filled with a liquid lubricant that automatically produces a lubricating film on the bearing surface during running of the shaft. This is due to a pumping action of the shaft and frictional heat that lowers the viscosity of the oil. After completion of the running cycle, the oil is reabsorbed into the pore system of the bearing by capillary attraction. Porous self-lubricating bearings are divided into three groups: sintered bronze bearings, iron-based sintered bearings, and iron–bronze sintered bearings. The original of most widely used P/M bearing material is 90% Cu–10% Sn bronze, with or without the addition of graphite (1% fine natural graphite is often added to enhance fabrication, as well as to improve bearing properties). The 90% Cu–10% Sn bronze material is superior in bearing performance to the iron-based and iron–bronze compositions, which are lower in cost and used in less severe applications.

Self-Lubricating Material

Any solid material that shows low friction without application of a lubricant. Examples are graphite, molybdenum disulfide, and polytetrafluoroethylene. Taken in a broader context, the term can also refer to a composite (powder metallurgy) material into which a lubricous species has been incorporated. See also *self-lubricating bearing* and *solid lubricant*.

Self-Reversal

In optical emission spectroscopy, the extreme case of self-absorption. See also *self-absorption*.

Self-Skinning Foam

A urethane foam that produces a tough outer surface over a foam core upon curing.

Self-Tapping Screw

See *screw*.

Self-Vulcanizing

Pertaining to an adhesive that undergoes vulcanization without the application of heat. See also *vulcanization*.

Selvage

The woven-edge portion of a fabric used in reinforced plastics that is parallel to the warp.

Selvyt

A deep nap cloth used, with a suitable abrasive agent, for metallographic polishing.

S.E.M.

Scanning electron microscope. See *electron microscope*.

Semiautomatic Arc Welding

Arc welding with equipment that controls only the filler metal feed. The advance of the welding is manually controlled.

Semiautomatic Brazing

Brazing with equipment that controls only the brazing filler metal feed. The advance of the brazing is manually controlled.

Semiautomatic Plating

Plating in which prepared cathodes are mechanically conveyed through the plating baths, with intervening manual transfers.

Semiblind Joint

A weld joint in which one extremity of the joint is not visible.

Semiconductor

A solid crystalline material whose electrical conductivity is intermediate between that of a metal and an insulator, ranging from about 10^5 S to 10^{-7} S/m, and is usually strongly temperature dependent.

In a solid material, the electrons can exist only in certain bands of energy level that are separated by regions normally forbidden to electrons. Three general relationships between filled, unfilled, and forbidden bands can be considered. First, certain solids have some allowed bands completely filled with electrons and others empty. If the forbidden regions between the allowed bands are large, the electrons cannot move from one band to the next and the material is an *insulator*. Second, in materials where bands are not completely filled or their energy levels overlap, the electrons can move freely between bands. The electrons moving in this matter are termed *conductivity electrons* and the material is a *metallic conductor*. Finally, in cases where the energy gap between a filled and an empty gap is small, then a small input of energy can cause an electron to jump the gap. Silicon is one example. At very low temperatures, the four valence electrons of each atom is associated in a strong covalent bond with another four electrons of an adjacent atom. As the temperature increases, that is, energy is added, electrons can move to another band leaving a site for an incoming electron and conductivity is achieved. Pure metals with this characteristic are termed *intrinsic semiconductors*. The effect can also be induced by adding particular impurity elements that assist an electron to bridge the forbidden gaps. The impurity elements are termed *dopants* and the alloyed materials are *extrinsic semiconductors*. The dopants can work in two ways. Silicon is a group IV element with four valence electrons, which, in its pure state, is an intrinsic semiconductor, as described earlier. However, if an atom of a group V element with five valence electrons is added to form a substitutional solid solution, one of the electrons is in excess of the number required for bonding so it takes up a position in the normally forbidden gap from where it is easily dislodged by a small energy input to contribute to conduction. Dopant elements contributing an excess electron are termed *donors*. In a similar manner, a group III element atom in solid solution will have only three valence electrons required for bonding with the four required by the adjacent silicon. This effectively produces a hole in the valence band that can accept electrons jumping from a silicon to silicon bond again contributing to conduction. Such elements are termed *acceptor dopants*. Conventionally, conductivity in semiconductors is considered on the basis of the movement of holes into which electrons can jump. The movement of a whole is regarded as a "positive" movement and hence the materials with an acceptor dopant providing holes are referred to as producing "p"-type semiconductors. Similarly, the movement of an electron is regarded as a negative movement of a hole and hence materials with a donor dopant are "n"-type semiconductors.

Semiconducting Materials

Semiconductors may be defined as materials that conduct electricity better than insulators, but not as well as metals. An enormous range of conductivities can meet this requirement. At room temperature, the conductivities characteristic of metals are on the order of 10^4–10^6 Ω/cm, while those of insulators range from 10^{-25} to 10^{-9} Ω/cm. The materials classed as semiconductors have conductivities that range from 10^{-9} to 10^4 Ω/cm. The conductivity of metals normally decreases with an increase of temperature, but semiconductors have the distinctive feature that in some range of temperature their conductivity increases rather than decreases with an increase of temperature.

Another criterion for a semiconductor is that the conduction process be primarily by electrons and not ionic, since the latter involves the transfer of appreciable mass as well as charge.

Materials used are ones that are capable of being partly conductors of electricity and partly insulators, and are used in rectifiers for changing alternating current to pulsating direct current, and in transistors for amplifying currents. They can also be used for the conversion of heat energy to electric energy, as in the solar battery. In an electric conductor, the outer rings of electrons of the atoms are free

to move, and provide a means of conduction. In a semiconductor, the outer electrons, or valence electrons, are normally stable, but, when a doping element that serves to raise or lower energy is incorporated, the application of a weak electric current will cause displacement of valence electrons in the material. Silicon and germanium, each with a single stable valence of four outer electrons, are the most commonly used semiconductors. Elements such as boron, with a lower energy level, but with electrons available for bonding and thus accepting electrons into the valence ring, are called *hypoelectronic elements*. Elements such as arsenic, which have more valence electrons that are needed for bonding and may give up an electron, are called *hyperelectronic elements*. Another class of elements, like cobalt, can either accept or donate an electron, and these are called *buffer atoms*. All of these types of elements constitute the doping elements for semiconductors.

In a nonconducting material, used as an electrical insulator, the energy required to break the valence bond is very high, but there is always a limit at which an insulator will break the bond and become a conductor with high current energy. The resistivity of a conductor rises with increasing temperature, but in a semiconductor the resistivity decreases with temperature rise, and the semiconductor becomes useless beyond its temperature limit. Germanium can be used as a semiconductor to about 93°C, silicon can be used to about 204°C, and silicon carbide can be used to about 343°C.

Extrinsic and Intrinsic Semiconductors

Metals for use as semiconductors must be of great purity, since even minute quantities of impurities would cause erratic action. The highly purified material is called an *intrinsic metal*, and the desired electron movement must come only from the doping element, or *extrinsic conductor*, that is introduced. The semiconductors are usually made of single crystals, and the positive and negative elements need be applied only to the surfaces of the crystal, but methods are also used to incorporate the doping element uniformly throughout the crystal.

The process of electron movement, although varying for different uses and in different intrinsic materials, can be stated in general terms. In the silicon semiconductor, the atoms of silicon with four outer valence electrons bind themselves together in pairs surrounded by eight electrons. When a doping element with three outer electrons, such as boron or indium, is added to the crystal, it tends to take an electron from one of the pairs, leaving a hole and setting up an imbalance. This forms the p-type semiconductor. When an element with five outer electrons, such as antimony or bismuth, is added to the crystal, it gives off electrons, setting up a conductive band, which is the n-type semiconductor. Fusing together the two types forms a p–n junction, and a negative voltage applied to the p side attracts the electrons of the three-valence atoms away from the junction so that the crystal resists electronic flow. If the voltage is applied to the n side, it pushes electrons across the junction and the electrons flow. This is a diode, or rectifier, for rectifying alternating current into pulsating direct current. When the crystal wafers are assembled in three layers, p–n–p or n–p–n, a weak voltage applied to the middle wafer increases the flow of electrons across the whole unit. This is a transistor. Germanium and silicon are bipolar, but silicon carbide is unipolar and does not need a third voltage to accelerate the electrons.

Semiconductors can be used for rectifying or amplifying, or they can be used to modulate or limit the current. By the application of heat to ionize the atoms and cause movement, they can also be used to generate electric current; or in reverse, by the application of a current they can be used to generate heat or remove heat for heating or cooling purposes and air conditioning, heating, and refrigeration. But for uses other than rectifying or altering electric current, the materials are usually designated by other names and not called semiconductors. Varistors are materials, such as silicon carbide, whose resistance is a function of the applied voltage. They are used for such applications as frequency multiplication and voltage stabilization. Thermistors are thermally sensitive materials. Their resistance decreases as the temperature increases, which can be measured as close as 0.001°C, and they are used for controlling temperature or to control liquid level, flow, and other functions affected by rate of heat transfer. They are also used for the production or the removal of heat and air conditioning, and may then be called thermoelectric metals.

Hall Effect

Whether a given sample of semiconductor material is n- or p-type can be determined by observing the Hall effect. If an electric current is caused to flow through a sample of semiconductor material and a magnetic field is applied in a direction perpendicular to the current, the charge carriers are crowded to one side of the sample, giving rise to an electric field perpendicular to both the current and the magnetic field. This development of a transverse electric field is known as the Hall effect. The field is directed to one or the opposite direction depending on the sign of the charge of the carrier.

The magnitude of the Hall effect gives an estimate of the carrier concentration. The ratio of the transverse electric field strength to the product of the current and the magnetic field strength is called the Hall coefficient, and its magnitude is inversely proportional to the carrier concentration. The coefficient of proportionality involves a factor that depends on the energy distribution of the carriers and the way in which the carriers are scattered in their motion. However, the value of this factor normally does not differ from unity by more than a factor of 2. The situation is more complicated when more than one type of carrier is important for the conduction. The Hall coefficient then depends on the concentration of the various carriers and their relative mobilities.

The product of the Hall coefficient and conductivity is proportional to the mobility of the carriers when one type of carrier is dominant. The proportionality involves the same factor that is contained in the relationship between the Hall coefficient and the carrier concentration. The value obtained by taking this factor to be unity is referred to as the Hall mobility.

Types

Although some of the chemical elements are semiconductors, most semiconductors are chemical compounds. Compound semiconductors can be simple compounds, such as compounds formed from elements of the III–V, V–VI, II–IV, II–VI, I–VI, II–V, II–VI groups of the periodic table, as well as alloys of these compounds, ternary compounds, complicated oxides, and even organic complexes.

For organic semiconductors, it is generally true that the larger the molecule, the higher the conductivity. Inclusion of atoms other than carbon, hydrogen, and oxygen in the molecule often appears to increase the conductivity. The phthalocyanine molecule, which is much like the active center of several biologically active compounds, for example, chlorophyll, can accommodate a metallic atom in its center.

An organic semiconductor may be either n-type or p-type. For example, among the organic dyes, the cationic dyes appear to be n-type semiconductors and the anionic dyes p-type.

Magnetic ferrites are usually polycrystalline ceramics, composed of mixed semiconductor and oxides, and possess useful magnetic properties combined with high electrical resistivities.

The magnetically soft ferrites have the general form MFe_2O_4, where M represents one or more of the following divalent metals: copper, magnesium, manganese, nickel, iron, and zinc. The crystallites have the cubic structure of the spinel system.

Fabrication

Although some electronic devices require extremely small semiconductor elements, others require large masses of material. For example, some diodes and thermistors have volumes, exclusive of the lead wires, of less than 1 ten-millionth of a cubic inch, while some silicon domes require as much as a 1000 m^3 of material, and xerographic and luminescent panels cover many square centimeters. Semiconductors are prepared in the form of single crystals (as large as several millimeters in diameter and several centimeters in length), polycrystalline ingots, thin films, or sintered ceramic shapes. Semiconductors are generally brittle materials and some make excellent abrasives. However, at higher temperatures (near their melting points) they can be deformed elastically in a manner analogous to engineering metals at room temperature.

Single crystals and polycrystalline ingots can be cut into the desired shapes by using diamond or other appropriate abrasive wheels or by using ultrasonic cutting tools and a boron carbide abrasive. A continuous wire and an abrasive slurry are sometimes used for cutting semiconductors. Mapping and etching are commonly used for preparing small semiconductor elements of the desired shape and surface condition. Large-area selenium xerographic plates are prepared by vacuum evaporation, whereas ceramic molding and sintering techniques are used in the preparation of ferrites and some luminescent materials.

Single crystals are grown by the horizontal Bridgman, the vertical Bridgman, Czochralski, floating-zone, and vapor-phase deposition techniques. The impurities may be added to the starting materials or to the melt during growth. Controlled regions of impurities are created by diffusion, alloying, or vapor-phase deposition to form epitaxial layers on a single-crystal substrate.

Some semiconductor elements are formed by use of a semiconductor powder in an appropriate binder. In the manufacture of magnetic ferrites, the oxides are mixed in required proportions and milled to obtain a fine particle size. The powder is then partially sintered in a prefiring kiln, and subsequently crushed, milled, and granulated before being pressed or extruded into the required shape. The main firing cycle takes place at temperatures around 1250°C in a controlled atmosphere. As a result, the parts are sintered into a dense, homogeneous ceramic. During sintering, the parts shrink some 20% M linear dimensions.

Applications

The technology and practical importance of semiconductor devices have been growing steadily in the past decade. The major applications of semiconductors can be divided into 10 categories: diodes and transistors, luminescent devices, ferrites, special resistors, photovoltaic cells, infrared lenses and domes, thermoelectric devices, piezoelectric devices, xerography, and electron emission.

Indium antimonide, InSb, has a cubic crystal structure, and it is used for infrared detectors and for amplifiers in galvanomagnetic devices. Indium arsenide, InAs, also has a very high electron mobility, and is used in thermistors for heat-current conversion. It can be used to 816°C. Some materials can be used only for relatively low temperatures. Copper oxide and pure selenium have been much used in current rectifiers, but they are useful only at moderate temperatures, and they have the disadvantage of requiring much space. Indium phosphide, InP, has a mobility higher than that of germanium, and can be used in transistors above 316°C. Aluminum antimonide, AlSb, can be used at temperatures to 538°C. In lead selenide, PbSe, the mobility of the charge-carrying electrons decreases with rise in temperature, increasing resistivity. It is used in thermistors.

Bismuth telluride, Bi_2Te_3, maintains its operating properties between −46°C and 204°C, which is the most useful range for both heating and refrigeration. When doped as a p-type conductor, it has a temperature difference of 601°C and an efficiency of 5.8%. When doped as an n-type conductor, the temperature difference is lower, 232°C, but the efficiency within this range is more than doubled. Lead telluride, PbTe, has a higher efficiency, 13.5%, and temperature difference of 582°C, but it is not usable below 177°C, and is employed for conversion of the waste heat atomic reactors at about 371°C.

Gallium arsenide has high electron mobility, and can be used as a semiconductor. Cadmium sulfide, CdS, is thus deposited as a semiconductor film for photovoltaic cells, or solar batteries, with film thickness of about 2 μm. When radioactive isotopes, instead of solar rays, are added to provide the activating agent, the unit is called an atomic battery, and the large area of transparent backing for the semiconductor is not needed.

Manganese telluride, MnTe, with a temperature difference of 982°C, has also been used as a semiconductor. Many other materials can be used, as semiconductors with temperature differences at different gradients can be joined in series electrically to obtain a wider gradient, but the materials must have no diffusion at the junction.

Cesium sulfide, CeS, has good stability and thermoelectric properties at temperatures to 1093°C, and has a high temperature difference, 1110°C. It can thus be used as a high-stage unit in conversion devices. High conversion efficiency is necessary for transducers, while a high dielectric constant is desirable for capacitors. Low thermal conductivity makes it easier to maintain the temperature gradient, but for some uses high thermal conductivity is desirable. Silver–antimony–telluride, $AgSbTe_2$, has a high energy-conversion efficiency for converting heat to electric current, and it has a very low thermal conductivity, about 1% that of germanium.

The semiconductor-type intermetals are also used in magnetic devices, since the ferroelectric phenomenon of heat conversion is the electrical analog of ferromagnetism. Chromium–manganese–antimonide is nonmagnetic below about 250°C and magnetic above the temperature. Various compounds have different critical temperatures. Below the critical temperature, the distance between the atoms is less than that which determines the lineup of magnetic forces, but with increased temperature the atomic distance becomes greater and the forces swing into a magnetic pattern.

Organic semiconductors fall into two major classes: well-defined substances, such as molecular crystals and crystalline complexes, and isostatic and syndiotactic polymers; and disordered materials, such as atactic polymers and pyrolytic materials. Few of these materials have yet found commercial application.

Amorphous silicon containing hydrogen is promising for use in solar cells because of its low cost and suitable electrical and optical properties.

Semicrystalline

In plastics, materials that exhibit localized crystallinity. See also *crystalline plastic*.

Semifinisher

An impression in a series of forging dies that only approximates the finished dimension of the forging. Semifinishers are often used to extend die life or the finishing impression, to ensure proper control of grain flow during forging, and to assist in obtaining desired tolerances.

Semifinishing

Preliminary operations performed prior to finishing.

Semiguided Bend

The bend obtained by applying a force directly to the specimen than the portion that is to be bent. The specimen is either held at one end and forced around a pin or rounded edge, or is supported near the ends and then by a force applied on the side of the specimen opposite the supports and midway between them. In some instances, the bend is started in this manner and finished in the manner of a *free bend*.

Semikilled Steel

Steel that is incompletely deoxidized and contains sufficient dissolved oxygen to react with the carbon to form carbon monoxide and thus offset solidification shrinkage. The voids filled with carbon dioxide are closed during subsequent hot working. See the term *capped steel*.

Semimetal

An element having many but not all of the characteristics of a metal. Also termed a metalloid.

Semipermanent Mold

A *permanent mold* in which fresh sand cores are used repeatedly.

Semipermeable Membrane

A membrane that is permeable to the solvent but impervious to the solute.

Semipositive Mold

A mold for forming plastics that combines the capabilities of a *flash mold* and *positive mold*. As the two halves of a semipositive mold begin to close, the mold acts much like a flash mold, because the excess material is allowed to escape around the loose-fitting plunger and cavity. As the plunger telescopes further into the cavity, the mold becomes a positive mold with very little clearance, and full pressure is exerted on the material, producing a part of maximum density. This type of mold uses to advantage the free flow of material in a flash mold and the capability of producing dense parts and the positive mold.

Semirigid Plastic

For purposes of general classification, a plastic that has a modulus of elasticity and flexure or in tension between 70 and 690 MPa (10 and 100 ksi) at 23°C and 50% relative humidity.

Semisolid Metal Forming

A two-step casting/forging process in which a billet is cast in a mold equipped with a mixture that continuously stirs the thixotropic melt, thereby breaking up the dendritic structure of the casting into a fine-grained spherical structure. After cooling, the billet is stored for subsequent use. Later, a slug from the billet is cut, heated to the semisolid state, and forged in a die. Normally the cast billet is forged when 30%–40% is in the liquid state. See also *rheocasting*.

Semi Steel

Low carbon cast iron having graphite flakes small in size and quantity and well distributed and hence having relatively good ductility compared with normal gray cast iron. It is usually made by melting scrap steel with the pig iron.

Sendzimir Mill

A type of *cluster mill* with small-diameter work rolls and larger-diameter backup rolls, backed up by bearings on a shaft mounted eccentrically so that it can be rotated to increase the pressure between the bearing and the backup rolls. Used to roll precision and very thin sheet and strip.

Sensitive Tint Plate

A gypsum plate used in conjunction with polarizing filters in an *optical microscope* to provide very sensitive detection of birefringence and double refraction.

Sensitivity

(1) The capability of a method or instrument to discriminate between samples having different concentrations or amounts of the analyte. (2) The smallest difference in values that can be detected reliably with a given measuring instrument.

Sensitization

In austenitic stainless steels, the precipitation of chromium carbides, usually at grain boundaries, on exposure to temperatures of about 540°C–845°C, leaving the grain boundaries depleted of chromium and therefore susceptible to preferential attack by a corroding medium. Welding is the most common cause of sensitization. Weld decay (sensitization) caused by carbide precipitation in the weld heat-affected zone leads to *intergranular corrosion*.

Sensitizing Compounds

Supersensitizing compounds are metal salts in an aqueous or organic solution, which form an invisible film on the surface of glass and other ceramic surfaces. This film is not completely understood but is believed to be an ionizing or electronic effect that serves to initiate and hastened surface treatments such as silvering and plating. Supersensitizing refers to a second step involving the use of noble metal compounds, which further enhances the reduction properties of the metals about to be formed on the glass or ceramic surface.

Sensitizing compounds include aluminum compounds (basic aluminum acetate, aluminum chloride, aluminum formoacetate, aluminum nitrate), barium salts, boron trichloride, cadmium compounds, iron sulfate, tin chloride, titanium sulfate, and triethanolamine titanate.

Supersensitizing compounds include gold chloride, iridium salts, osmium compounds, palladium chloride, silver nitrate, and silver oxide.

Sensitizing Heat Treatment

A heat treatment, whether accidental, intentional, or incidental (as during welding), that causes precipitation of constituents at grain boundaries, often causing the alloy to become susceptible to *intergranular corrosion* or *intergranular stress-corrosion cracking*. See also *sensitization*.

Sensor

Any detection device, with or without a measuring capability.

Separate-Application Adhesive

An adhesive consisting of two parts, one part being applied to one adherend and the other part to the other adherend and the tube being brought together to form a joint.

Separator

In rolling-element bearings, the part of a cage that lies between the rolling elements. This term is sometimes used as a synonym for *cage*. See the term *rolling-element bearing*.

Separator (Composites)

In processing of composites, a permeable layer that allows volatiles and air to escape from the laminate and excess resin to be bled from the laminate into the bleeder plies during cure. Porous Teflon-coated fiberglass is an example. Often placed between lay-up and bleeder to facilitate bleeder systems removal from laminate after cure. See also *bleeder* and *lay-up*.

Sequence Timer

In resistance welding, a device used for controlling the sequence and duration of any or all of the elements of a complete welding cycle except *heat time* or *weld time*.

Sequence Weld Timer

Same as *sequence timer* except that either *weld time* or *heat time*, or both, are also controlled.

Sequestering Agent

A material that combines with metallic ions to form water-soluble complex compounds.

Serial Sectioning

A metallographic technique in which an identified area on a section surface is observed repeatedly after successive layers of known thickness have been removed from the surface. It is used to construct a 3D morphology of structural features. See also *sectioning*.

Series Submerged Arc Welding

A submerged arc welding process variation in which electric current is established between two (consumable) electrodes that meet just above the surface of the work. The work is not in the electrical circuit. See also *submerged arc welding*.

Series Welding

Resistance welding in which two or more spot, seam, or projection welds are made simultaneously by a single welding transformer with three or more electrodes forming a series circuit.

Serpentine

A mineral of theoretical formula $3MgO \cdot SiO_2 \cdot 2H_2O$, containing 43% magnesium oxide. It is used for building trim and for making ornaments and novelties. The chips are employed in terrazzo and for roofing granules. Actually, the stone rarely approaches the theoretical formula and usually contains 2%–8% iron oxide with much silica and aluminum. It has an asbestos-like structure. The attractively colored and veined serpentine of Vermont is marketed under the name of *verde antique marble*. The massive verde antique of Pennsylvania is used with dolomite in refractories. *Antigorite* is a form of serpentine found in California, which has a platy rather than a fibrous structure. The serpentine of Columbus County, Georgia, contains 36%–38% MgO and 2–5 chrome ore. It is used as a source of magnesia.

Sessile Dislocation

A dislocation formed by a plane of vacancies surrounded by normal lattice. This type of dislocation cannot move.

Set

(1) The shape of the solidifying surface of a metal, especially copper, with respect to concavity or convexity. May also be called pitch. (2) Deformation. (3) A substantial chisel for cutting metal.

Set (Polymerization)

To convert an adhesive into a fixed or hardened state by chemical or physical action, such as condensation, polymerization, oxidation, vulcanization, gelation, hydration, or evaporation of volatile constituents. See also *cure*.

Set Copper

An intermediate copper product containing about 3.5% cuprous oxide, obtained at the end of the oxidizing portion of the fire-refining cycle.

Set, Permanent

See *permanent set*.

Setting Temperature

The temperature to which an adhesive or an assembly is subjected to set the adhesive. The temperature attained by the adhesive in the process of setting (adhesive setting temperature) may differ from the temperature of the atmosphere surrounding the assembly (assembly setting temperature). See also *curing temperature* and *drying temperature*.

Setting Time

The period of time during which an adhesively bonded assembly is subjected to heat or pressure, or both, to set the adhesive. See also *drying time* and *joint-conditioning time*.

Settling

(1) Separation of solids from suspension in a fluid of lower density, solely by gravitational effects. (2) A process for removing iron from liquid magnesium alloys by holding the melt at a low temperature after manganese has been added to it.

Set Up

To harden, as in the curing of a polymer resin.

Severe Wear

A form of wear characterized by removal of material in relatively large fragments. Severe wear is an imprecise term, frequently used in research, and contrasted with *mild wear*. In fact, the phenomena studied usually involve the transition from mild to severe wear and the factors that influence that transition. With metals, the fragments are usually predominantly metallic rather than oxidic. Severe wear is frequently associated with heavy loads and/or adhesive contact.

Severity of Quench

Ability of quenching medium to extract heat from a hot steel workpiece; expressed in terms of the *Grossman number* (*H*).

S-Glass

A magnesium aluminosilicate composition that is especially designed to provide very high tensile strength glass filaments. S-glass and S-2 glass fibers have the same glass composition but different finishes (coatings), S-glass is made to more demanding specifications, and S-2 is considered a commercial grade. See the term *fiberglass*.

Shadow Angle

In shadowing of replicas, the angle between the line of motion of the evaporated atoms and the surface being shadowed. See also *replica* and *shadowing*.

Shadow Cast Replica

A replica that has been shadowed. See also *replica* and *shadowing*.

Shadowing

Directional deposition of carbon or a metallic film on a plastic replica so as to highlight features to be analyzed by transmission electron microscopy. Most often used to provide maximum detail and resolution of the features of fracture surfaces. The process is carried out in a vacuum chamber in which a suitable metal is evaporated and projected to an acute angle onto the replica surface to exaggerate topographical irregularities. See also *metal shadowing*, *oblique evaporation shadowing*, and *shadow angle*.

Shadow Mask (Thermal Spraying)

A thermal spraying process variation in which an area is partially shielded during the thermal spraying operation, thus permitting some overspray to produce a feathering at the coating edge.

Shadow Microscope

An electron microscope that forms a shadow image of an object using electrons emanating from a point source located close to the object.

Shaft Furnace

A vertical cylindrical furnace packed nearly to the top with the charge. The materials are continuously fed in the top to react as they progress down to be extracted at the base. See *blast furnace*.

Shaft Run-Out

Twice the distance that the center of a shaft is displaced from the axis of rotation; that is, twice the eccentricity.

Shakeout

Removal of castings from a sand mold. See also *knockout*.

Shaker-Hearth Furnace

A continuous type furnace that uses a reciprocating shaker motion to move the parts along the hearth.

Shale

A rock formed by deposition of colloidal particles of clay and mud, and consolidated by pressure. It is fine-grained and has a laminated structure, usually containing much sand colored by metal oxides. Unlike sandstones, shales are not usually porous; most are hard, slatelike rocks. Slate is a form of shale that has been subjected to intense pressure. Some shales are calcareous or dolomitic and are used with limestone in making portland cement. These are called *marlstone*. Oil shale is a hard shale with veins of greasy solid known as *kerogen*, which is oil mixed with organic matter. *Crude shale oil* is a black, viscous liquid containing up to 2% nitrogen and a high sulfur content. But when oil shale is heated above 399°C, the kerogen is cracked into gases condensable to oils, gases, and coke. Some shales also yield resins and waxes and others contain small amounts of uranium, vanadium, and molybdenum. The regular commercial by-products of *Swedish shale oil* recovery are sulfur, fuel gas, ammonium sulfate, tar, and lime. Shales contain abrasive particles of platelike shape. The term is applied particularly to diamond abrasives.

Shank

(1) The portion of a die or tool by which it is held in position in a forging unit or press. (2) The handle for carrying a small ladle or crucible. (3) The main body of a lathe tool. If the tool is an inserted type, the shank is the portion that supports the insert. (4) The plain shaft that carries the active end of a component, for example, the unthreaded central length of a bolt or the plain end of a drill bit.

Shank-Type Cutter

A cutter having a straight or tapered shank to fit into a machine-tool spindle or adapter.

Shape Accuracy

For an elastic slab loaded in compression, the ratio of the loaded area to the force-free area.

Shape Memory Alloys

These are a group of metallic materials that can return to some previously defined shape or size when subjected to the appropriate thermal procedure. That is, shape memory alloys (SMA) can be plastically deformed at some relatively low temperature and, upon exposure to some higher temperature, will return to their original shape. Materials that exhibit shape memory only upon heating are said to have a one-way shape memory, whereas those that also undergo a change in shape upon recooling have a two-way memory. Typical materials that exhibit the shape memory effect include a number of copper alloy systems and the alloys of gold–cadmium, nickel–aluminum, and iron–platinum.

A shape memory alloy may be further defined as one that yields a thermoelastic martensite, that is, a martensite phase that is crystallographically reversible. In this case, the alloy undergoes a martensitic transformation of a type that allows the alloy to be deformed by a twinning mechanism below the transformation temperature. A twinning mechanism is a herringbone structure exhibited by martensite during transformation. The deformation is then reversed when the twinned structure reverts upon heating to the parent phase.

Transformation Characteristics

The martensite transformation that occurs in shape memory alloys yields a thermoelastic martensite and develops from a high-temperature austenite phase with long-range order. The martensite typically occurs as alternately sheared platelets, which are seen as a herringbone structure when viewed metallographically. The transformation, although a first-order phase change, does not occur at a single temperature but over a range of temperatures that is characteristic for each alloy system.

There is a standard method of characterizing the transformation and naming each point in the cycle. Most of the transformation occurs over a relatively narrow temperature range, although the beginning and end of the transformation during heating and cooling actually extends over a much larger temperature range. The transformation also exhibits hysteresis in that the transformation on heating and cooling does not overlap. This transformation hysteresis varies with the alloy system.

Thermomechanical Behavior and Alloy Properties

The mechanical properties of shape memory alloys vary greatly over the temperature range spanning their transformation. The only two alloy systems that have achieved any level of commercial exploitation are the nickel–titanium alloys and the copper-base alloys. Properties of the two systems are quite different. The nickel–titanium alloys have greater shape memory strain, tend to be much more thermally stable, have excellent corrosion resistance compared to the copper-base alloys, and have higher ductility. The copper-base alloys are much less expensive, can be melted and extruded in air with ease, and have a wider range of potential transformation temperatures. The two alloy systems thus have advantages and disadvantages that must be considered in a particular application.

Materials

The most common shape memory alloy material is *nitinol*, an acronym for Ni-Ti-NOL (Naval Ordinance Laboratory). As suggested by the name, the material consists of approximately equal parts of nickel and titanium and was originally developed by the Naval Ordinance Laboratory.

Two different phases, martensite and austenite, are typically associated with the crystalline structure of shape memory alloys. The austenite phase is a highly ordered phase that occurs above a certain transition temperature. In this phase, the crystalline bonds must be at right angles with one another. The martensite phase occurs below the transition temperature, and right angle bonds are not required, where a twinned phase occurs. Strain is required to achieve a particular alignment, since the martensite phase is not in an ordered state.

The term *shape memory* is derived from the fact that a shape memory alloy can recover its original shape when heated above a certain transition temperature. Before a shape memory alloy such as nitinol can present shape memory behavior, it must first be trained. The shape memory alloy is annealed while it is constrained in the shape that it is to memorize. From a microscopic view, this corresponds to the austenite phase. For the nitinol wire, the annealing process corresponds to a contraction in the length of the wire. After annealing, the material is said to have a one-way shape memory.

The nitinol recovers its memorized shape when heated above the transition temperature and will remain in that shape upon cooling. For the shape memory alloy to return to its initial shape, an external force must be applied to deform it. The two-way shape memory corresponds to a martensite phase to austenite phase by heating (shape recovery) followed by a mechanically produced deformation of the material upon cooling.

Applications

The manufacturing and the technology associated with both of the two commercial classes of shape memory alloys are quite different as are the performance characteristics. Therefore, in applications where a highly reliable product with a long fatigue life is desired, the nickel–titanium alloys are the exclusive materials of choice. Typical applications of this kind include electric switches and actuators. However, if high-performance is not mandated and cost considerations are important, then the use of copper–zinc–aluminum shape memory alloys can be recommended. Typical applications of this kind include safety devices such as temperature fuses and fire alarms.

Heating a shape memory alloy product to a temperature above some critical temperature is not recommended. The critical temperature for nickel–titanium is approximately 250°C, and for copper–zinc–aluminum approximately 90°C. Extended exposure to thermal environments above these critical temperatures results in an impaired memory function regardless of the magnitude of the load.

Another technical consideration in the practical application of shape memory alloys pertains to fastening or joining these materials to conventional materials. This is a significant issue because shape memory alloys undergo expansions and contractions not encountered in traditional materials. Therefore, if shape memory alloys are welded or soldered to other materials, they can easily fail at the joint when subjected to repeated loading. Alloys of nickel–titanium and copper–zinc–aluminum can also be brazed by using silver filler metals; however, the brazed region can fail because of cyclic loading. It is therefore desirable to devise some other mechanism for joining shape memory alloys to traditional materials. Shape memory alloys cannot be plated or painted for similar reasons.

Free Recovery

In this case, a component fabricated from a shape memory alloy is deformed while martensitic, and the only function required of the shape memory is that the component return to its previous shape upon heating. A prime application is the blood-clot filter in which a nickel-titanium wire is shaped to anchor itself in a vein and catch passing clots. The part is chilled so it can be collapsed and inserted into the vein; then body heat is sufficient to return the part to its functional shape.

Constrained Recovery

The most successful example of this type of product is undoubtedly a hydraulic coupling. These studies are manufactured as cylindrical sleeves slightly smaller than the metal tubing that they are to join. Their diameters are then expanded while martensitic and, upon warming to austenite, they shrink in diameter and strongly hold the tube ends. The tubes prevent the coupling from fully recovering its manufactured shape, and the stresses created as the coupling attempts to do so are great enough to create a joint that, in many ways, is superior to a weld.

Force Actuators

In some applications, the shape memory component is designed to exert force over a considerable range of motion, often for many cycles. Such an application is the circuit-board edge connector. In this electrical connector system, the shape memory alloy component is used to force open a spring when the connector is heated. This allows force-free insertion or withdrawal of a circuit board in the connector. Upon cooling, the nickel–titanium actuator becomes weaker, and the spring easily deforms the actuator while it closes tightly on the circuit board and forms the connections.

An example based on the same principle is a fire safety valve, which incorporates a copper–zinc–aluminum actuator designed to shut off toxic or flammable gas flow when fire occurs.

Other Applications

A number of applications are based on the pseudoelastic (or superelastic) property of shape memory alloys. Some eyeglass frames use superelastic nickel–titanium alloy to absorb large deformations without damage. Guide wires for steering catheters into vessels in the body have been developed using wire fashioned of nickel–titanium alloy, which resists permanent deformation if bent severely. Arch wires for orthodontic correction also use this alloy.

Shape memory alloys have found application in the field of robotics. The two main types of actuators for robots using these alloys are biased and differential. Biasing uses a coil spring to generate the bias force that opposes the unidirectional force of the shape memory alloy. In the differential type, the spring is replaced with another shape memory alloy, and the opposing forces control the actuation. A microrobot was developed with five degrees of freedom corresponding to the capabilities of the human fingers, wrist, elbow, and shoulder. The variety of robotic maneuvers and operations are coordinated by activating the nickel–titanium coils in the fingers and wrist in addition to contraction and expansion of straight nickel–titanium wires in elbow and shoulders. Digital control techniques in which a current is modulated with pulse-width modulation are employed in all of the components to control their spatial positions and speeds of operation.

In medical applications, in addition to mechanical characteristics, highly reliable biological and chemical characteristics are very important. The material must not be vulnerable to degradation, decomposition, dissolution, or corrosion in the organism, and must be biocompatible.

Nickel–titanium shape memory alloys have also been employed in artificial hip joints. These alloys have also been used for bone plates, for marrow pins for healing bone fractures, and for connecting broken bones.

Shape Memory Effect (S.M.E.)

An effect, observed in some alloy systems, which is associated with a reversible and progressive transformation to martensite as temperature falls. The unusual characteristic is that if the material is deformed (within limits) while in one phase, it will, on changing temperature to the other phase, refer to the original shape.

Shaping

Producing flat surfaces using single-point tools. The work is held in a vise or fixture, or is clamped directly to the table. The ram supporting the tool is reciprocated in a linear motion past the work.

1. Form shaping: Shaping with a tool ground to provide a specified shape.
2. Contour shaping: Shaping of an irregular surface, usually with the aid of a tracing mechanism.
3. Internal shaping: Shaping of internal arms such as keyways and guides.

Shark Leather

A durable, nonscuffing leather used for bookbindings, handbags, and fancy shoes, made from the skin of sharks. The shark is the largest of the true fishes, but has a skin unlike fishskin. When tanned, the surface is hard, the epidermis thicker than cowhide, and the long fibers lie in a cross-weave. The shark is split on the back instead of the belly, as in cowhides, and the skins measure from 3 to 20 ft² (0.3–2 m²). The hard denticle, called the *shagreen*, is usually removed, after which the leather is pliable but firm, the exposed grain not pulling out. *Shagreen leather* is a hard, strong leather with the grain side covered with globular granules made to imitate the sharkskin. Eastern shark leather has a deep grain with beautiful markings. *Pearl sharkskin* is from the Japanese ray. It is used for trim on pocketbooks. *Morocco leather* is from a small shark of the Mediterranean, but the name is also applied to a vegetable-tanned Spanish goatskin on which a pebbly grain is worked up by hand boarding. It is now made from ordinary goatskin by embossing.

Most of the sharkskin is now a by-product of the catch for oil, which is used for medicinal purposes. The shark liver is about one-fourth the total weight of the animal, and *shark-liver oil* is 30 times higher in vitamin A than cod-liver oil. The oil is also used for soap, lubricating, and heat-treating oil, though normally it is too expensive for these purposes.

Shark's Teeth

In ceramics and glasses, a striation consisting of dagger-like step fractures starting at the scored edge and extending to or nearly to the compression edge.

Sharp-Notch Strength (σ_s)

The notch tensile strength measured using specimens with very small notch root radii (approaching the limit for machining capability); values of sharp-notch strength usually depend on notch root radius.

Shatter Cracking

Low ductility cracking, usually subsurface, by various mechanisms including brittle fracture and hydrogen embrittlement. It is also termed as fisheyes or flakes. See *flake* (metals).

Shaving

(1) As a finishing operation, the accurate removal of a thin layer of a work surface by straight line motion between a cutter and the surface. (2) Trimming parts such as stampings, forgings, and tubes to remove uneven sheared edges or to improve accuracy.

Shaw (Osborn-Shaw) Process

See *ceramic molding*.

Shear

(1) The type of force that causes or tends to cause two contiguous parts of the same body to slide relative to each other in a direction parallel to their plane of contact. (2) A machine or tool for cutting metal and other material by the closing motion of two sharp, closely adjoining edges; for example, *squaring shear* and *circular shear*. (3) An inclination between two cutting edges, such as between two straight knife blades or between the punch cutting edge and the die-cutting edge, so that a reduced area will be cut each time. This lessens the necessary force but increases the required length of the working stroke. This method is referred to as angular shear. (4) The act of cutting by shearing dies or blades, as in shearing lines. (5) The cutting action where a pair of blades or similar abutting faces slide across each other cutting the entrapped material as with scissors. Shear strength, in this context, refers to the capacity to resist this form of loading. (6) The deformation mechanism in which layers of atoms on the crystal lattice slide across one another like a pack of playing cards. The layers involved, termed the shear planes, are usually those at an angle of 45° to the principle stress. Shear strength in this context refers to the stress necessary to induce permanent deformation by such a mechanism.

Shear Angle

The angle that the *shear plane*, in metal cutting, makes with the work surface.

Shear Bands

(1) Bands of very high shear strain that are observed during rolling of sheet metal. During rolling, these form at approximately ±35°

to the rolling plane, parallel to the transverse direction. They are independent of grain orientation and at high strain rates traverse the entire thickness of the rolled sheet. (2) Highly localized deformation zones in metals that are observed at very high strain rates, such as those produced by high velocity (100–3,600 m/s, or 330–11,800 ft/s) projectile impacts or explosive rupture. During high-strain-rate shear, also known as adiabatic shear, the bulk of the plastic deformation is concentrated in narrow bands within the relatively undeformed matrix. The shear bands are believed to occur along slip planes and the local strain rate within the adiabatic shear bands can exceed 10^6 s^{-1}.

Shear Edge

The cut off edge of a mold.

Shear Fracture

A mode of fracture in crystalline materials resulting from translation along slip planes that are preferentially oriented in the direction of the shearing stress.

Shear Hackle

A *hackle* in ceramics and glasses generated by interaction of a shear component with the principal tension under which the crack is running.

Shear Ledges

Steps between areas of crystalline fracture on a brittle fracture surface. They are normally aligned along the line of crack propagation and hence pointed to the origin. See *radial marks*.

Shear Lip

(1) A narrow, slanting ridge along the edge of a fracture surface. The term sometimes also denotes a narrow, often crescent-shaped, fibrous region at the edge of a fracture that is otherwise of the cleavage type, even though this fibrous region is in the same plane as the rest of the fracture surface. (2) An area, at the edge of a flat fracture surface, where the plane of fracture is at about 45° to the direction of loading. It occurs by ductile shear at the final stage of crack propagation leaving, usually, a sharp fracture edge. See *shear* (5).

Shear Modulus (*G*)

The ratio of shear stress to the corresponding shear strain for shear stresses below the proportional limit of the material. Values of shear modulus are usually determined by torsion testing. Also known as modulus of rigidity. Same as *bulk modulus*.

Shear Plane

A confined zone along which shear takes place in metal cutting. It extends from the cutting edge to the work surface. See *shear* (6).

Shear Rate

With regard to viscous fluids, the relative rate of flow or movement.

Shear Stability

The ability of a lubricant to withstand shearing without degradation. See also *penetration* (of a grease).

Shear Strain

The tangent of the angular change, caused by a force between two lines originally perpendicular to each other through a point in a body. Also called *angular strain*.

Shear Strength

The maximum shear stress that a material is capable of sustaining. Shear strength is calculated from the maximum load during a shear or torsion test and is based on the original cross-sectional area of the specimen. See *shear, definitions* (5) *and* (6).

Shear Stress

(1) The stress component tangential to the plane on which the forces act. (2) A stress that exists when parallel planes in metal crystals slide across each other.

Shear Stress (Plastics)

The stress developing in a polymer melt when the layers in a cross section are gliding along each other or along the wall of the channel (in laminar flow). Shear stress is equal to force divided by the area sheared, yielding pounds per square inch.

Shearing

The parting of material that results when one blade forces the material past an opposing blade. See also *shear*.

Shear Thickening

An increase in viscosity of non-Newtonian fluids (e.g., polymers and their solutions, slurries, and suspensions) with an increase in shear stress or time. See also *shear thinning*.

Shear Thinning

A decrease in viscosity of non-Newtonian fluids with an increase in shear stress or time. The decrease in viscosity may be temporary or permanent. The latter happens when the shear stress is sufficiently large to rupture a chemical bond, so that the sheared liquid has a lower viscosity than it had prior to shearing. See also *shear thickening*.

Sheared-Off

This term is used loosely with respect to virtually any failure of a bolt or similar fastener. It is probably best limited to those cases where the bolt has been cut by a shearing action of one bolted surface sliding across the other.

Sheath

(1) The material, usually an extruded plastic or elastomer applied outermost to a wire or cable. Also called a jacket. (2) A sheet metal or

glass covering of a sintered billet to protect it from oxidation or other environmental contamination during hot working. See also *can*.

Shed

A feature intended to shed, that is, discard, unwanted deposits.

Sheepskin

The skin of numerous varieties of sheep, employed for fine leather for many uses. The best sheepskins come from the sheep yielding the poorest wool. When the hair is short, coarse, and sparse, the nourishment goes into the skin. The merino types having fine wool have the poorest pelts. Wild sheep and the low-wool crossbreeds of India, Brazil, and South Africa have close-fibered, firm pelts comparable in strength with some kidskin, and retain the softness of sheepskin. This type of sheepskin from the hair sheep is termed *cabretta* and is used almost entirely for making gloves and for shoe uppers. None is produced in the United States. The commercial difference between sheepskin and *lambskin* is one of weight only. Sheepskins usually run 3–3.5 lb (1.4–1.6 kg) per skin without wool, and lambskins are those below 3 lb (1.4 kg). Sheepskins are tanned with alum, chrome, or sumac. The large heavy skins from Argentina and Australia are often split, and the grain side tanned in sumac for bookbinding and other goods; the flesh side is tanned in oil or formaldehyde and marketed as chamois. The fine-grained sheepskin from Egypt, when skived and especially treated, are known as *mocha leather. Uda skins* and *white fulani skins*, from Nigerian sheep, are used for good-quality grain and suede glove leather. *Sheepskin shearlings* are skins taken from heavy-wooled sheep a few weeks after shearing. They are tanned with the wall on, and the leather is used for aviation flying suits and for coats.

Sheet

A flat-rolled metal product of some maximum thickness and minimum width arbitrarily dependent on the type of metal. It has a width-to-thickness ratio greater than about 50. Generally, such flat products under 6.5 mm (1/4 in.) thick are called sheets, and those 6.5 mm (1/4 in.) thick and over are called plates. Occasionally, the limiting thickness for steel to be designated as sheet steel is No. 10 Manufacturer's Standard Gauge for sheet steel, which is 3.42 mm (0.1345 in.) thick.

Sheet Forming

The plastic deformation of a piece of sheet metal by tensile loads into a three-dimensional shape, often without significant changes in sheet thickness or surface characteristics.

Sheeting

A form of plastic in which the thickness is very small in proportion to length and width and in which the plastic is present as a continuous phase throughout, with or without filler.

Sheet Metal Forming

This is the process of shaping thin sheets of metal (usually less than 6 mm) by applying pressure through male or female dies or both. Parts formed of sheet metal have such diverse geometries that it is difficult to classify them. In all sheet-forming processes, excluding shearing, the metal is subjected to primarily tensile or compressive stresses or both. Sheet forming is accomplished basically by processes such as stretching, bending, deep drawing, embossing, bulging, flanging, roll forming, and spinning. In most of these operations, there are no intentional major changes in the thickness of the sheet metal.

There are certain basic considerations that are common in all sheet forming. Grain size of the metal is important in that too large a grain produces a rough appearance when formed, a condition known as orange peel. For general forming, an American Society for Testing and Materials (ASTM) No. 7 grain size (average grain diameter 32 μm) is recommended. Another type of surface irregularity observed in materials such as low carbon steel is the phenomenon of yield-point elongation that results in stretcher strains or Lueder's bands, which are elongated depressions on the surface of the sheet. This is usually avoided by cold-rolling the original sheet with a reduction of only 1%–2% (temper rolling). Since yield-point elongation reappears after some time because of aging, the material should be formed within this time limit. Another defect is season cracking (stress cracking, stress corrosion cracking), which occurs when the form part is in a corrosive environment for some time. The susceptibility of metals to season cracking depends on factors such as type of metal, degree of deformation, magnitude of residual stresses in the formed part, and environment.

Anisotropy or directionality of the sheet metal is also important because the behavior of the material depends on the direction of deformation. Anisotropy is of two kinds: one in the direction of the sheet plane and the other in the thickness direction. These aspects are important, particularly in deep drawing.

Formability of sheet metals is of great interest, even though it is difficult to define this term because of the large number of variables involved. Failure in sheet forming usually occurs by localized necking or buckling or both, such as wrinkling or folding. For a simple tension-test specimen, the true (natural) necking strain is numerically equal to the strain-hardening exponent of the material; thus, for example, commercially pure annealed aluminum or common 304 stainless steel stretches more than cold-work steel before it begins to neck. However, because of the complex stress systems in most forming operations, the maximum strain before necking is difficult to determine, although some theoretical solutions are available for rather simple geometries.

Considerable effort has been expended to simulate sheet-forming operations by simple tests. In addition to bands or tear tests, cupping test have also been commonly used, such as the Swift, Olsen, and Erichsen tests. Although these tests are practical to perform and give some indication of the formability of the sheet metal, they generally cannot reproduce the exact conditions to be encountered in actual forming operations.

Stretch Forming

In this process, the sheet metal is clamped between jaws and stretched over a form block. The process is used in the aerospace industry to form large panels with varying curvatures. Stretch forming has the advantages of low die cost, small residual stresses, and virtual elimination of wrinkles in the formed part.

Bending

This is one of the most common processes in sheet forming. The part may be bent not only along a straight line but also along a curved path (stretching, flanging). The minimum bend radius, measured to the inside surface of the bend, is important and determines the limit at which the material cracks either on the outer surface of the bend or at the edges of the part. This radius, which is usually expressed in terms of multiples of the sheet thickness, depends on the ductility of the material, width of the part, and its edge conditions.

Springback in bending and other sheet-forming operations is due to the elastic recovery of the metal after it is deformed. Determination of springback is usually done in actual tests. Compensation for springback and practice is generally accomplished by overbending the part; adjustable tools are sometimes used for this purpose.

In addition to male and female dies used in most bending operations, the female die is replaced with a rubber pad. In this way, die cost is reduced and the bottom surface of the part is protected from scratches by a metal tool. The roll-forming process replaces the vertical motion of the dies by the rotary motion of rolls with various profiles. Each successive roll bends the strip a little further than the preceding roll. The process is economical for forming long sections in large quantities.

Rubber Forming

Although many sheet-forming processes are carried out in a press with male and female dies usually made of metal, there are four basic processes that utilize rubber to replace one of the dies. Rubber is a very effective material because of its flexibility and low compressibility. In addition, it is low in cost, is easy to fabricate into desired shapes, has a generally low wear rate, and also protects the workpiece surface from damage.

The simplest of these processes is the Guerin process. Auxiliary devices are also used in forming more complicated shapes. In the Verson–Wheelon process, hydraulic pressure is confined in a rubber bag, the pressure being about five times greater than that in the Guerin process. For deeper draws, the Marform process is used. This equipment is a package unit that can be installed easily into a hydraulic press. In deep drawing of critical parts, the hydroform process is quite suitable, where pressure in the dome is as high as 100 MPa. A particular advantage of this process is that the formed portions of the part travel with the punch, thus lowering tensile stresses, which can eventually cause failure.

Bulging of tubular components, such as coffee pots, is also carried out with the use of a rubber pad placed inside the workpiece; the part is then expanded into a split female die for easy removal.

Deep Drawing

A great variety of parts are formed by this process, the successful operation of which requires a careful control of factors such as blank-holder pressure, lubrication, clearance, material properties, and die geometry. Depending on many factors, the maximum ratio of blank diameter to punch diameter ranges from about 1.6 to 2.3.

This process has been extensively studied, and the results show two import material properties for deep drawability are the strain-hardening exponent and the stress ratio (anisotropy ratio) of the metal.

Spinning

The process forms parts with rotational symmetry over a mandrel with the use of a tool or roller. There are two basic types of spinning: conventional or manual spinning, and shear spinning. The conventional spinning process forms the material over a rotating mandrel with little or no change in the thickness of the original blank. Parts can be as large as 6 m in diameter. The operation may be carried out at room temperature or higher for materials with low ductility or greater thickness. Success in manual spinning depends largely on the skill of the operator. The process can be economically competitive with drawing; if a part can be made by both processes, spinning may be more economical than drawing for small quantities.

In shear spinning (hydrospinning, floturning), the deformation is carried out with a roller in such a manner that the diameter of the original blank does not change but the thickness of the part decreases by an amount dependent on the mandrel angle. The spinnability of a metal is related to its tensile reduction of area. For metals with a reduction of area of 50% or greater, it is possible to spin a flat blank to a cone of an included angle of 3° in one operation. The shear spinning produces parts with various shapes (conical, curvilinear, and also tubular by tube spinning on a cylindrical mandrel) with good surface finish, close tolerances, and improved mechanical properties.

Miscellaneous Processes

Many parts require one or more additional processes; some of these are described briefly here. Embossing consists of forming a pattern on the sheet by shallow drawing. Coining consists of putting impressions on the surface by a process that is essentially forging; the best example is the two phases of a coin.

Coining pressures are quite high, and control of lubrication is essential to bring out all the fine detail in a design. Shearing is separation of the material by the cutting action of a pair of sharp tools, similar to a pair of scissors. The clearance in shearing is important to obtaining a clean-cut. A variety of operations based on shearing are punching, blanking, perforating, slitting, notching, and trimming.

Die materials used are cast alloys, die steels, and cemented carbides for high-production work. Nonmetallic materials such as rubber, plastics, and hardwood are also used as die materials. The selection of the proper lubricant depends on many factors, such as die and workpiece materials, and severity of the operation. A great variety of lubricants are commercially available, such as drawing compounds, fatty acids, mineral oils, and soap solutions.

Pressures in sheet-metal forming generally range between 7 and 55 MPa (normal to the plane of the sheet); most parts require about 10 MPa.

Sheet Metal Parts

Stamping and pressing make up a large family of metal-forming processes. Included in this group are blanking, pressing, stamping, and drawing, all of which are used to cut or form metal plate, sheet, and strip. The steps common in all stamping and pressing operations are the preparation of a flat blank and shearing or stretching the metal into a die to attain the desired shape.

In drawing, the flat stock is either formed in a single operation, or progressive drawing steps may be needed to reach the final form. In spinning, flat disks are dished by a tool as they revolve on a lathe.

Stamping involves placing the flat stock in a die and then striking it with a movable die or punch. Beside shaping the part, the dies can perform perforating, blanking, bending, and shearing operations. Almost all metals can be stamped. In general, stampings are limited to metal thicknesses of 9.5 mm or less. Pressing and drawing operations can be performed on cold metals up to 19 mm thick and up to about 89 mm on hot metals.

In recent years, many new press-forming and drawing techniques have been developed. A number of them make use of rubber pads, bags, and diaphragms as part of the die or forming elements.

Some involve stretch forming over dies. Others combine forming and heat-treating operations. And still other methods, known as high-energy rate forming, employ explosive, electrical, or magnetic energy to produce shockwaves that form the material into the desired shape.

Stamping

Frequently, secondary operations, such as annealing in furnaces, trimming in lathes or rolls, brake bends, and tapping, make it difficult to define a part as a stamping. Such operations, as well as finishing operations, may cost more than comparatively fast and economical press operations.

Sizes and Materials

Presses and presslike machines are not necessarily limited to sheet-metal forming. They punch paper doilies and cut uppers for shoes. Multislide machines form round or flat wire; some presses impact-extrude aluminum, zinc, and steel into deep shells for toothpaste tubes or shell bodies, using slugs cut from bars, or cast for the purpose, or sometimes punched from sheets and plates, in bellmouth dies to burnish the edges. Presses forge from billets and they compress powdered metal and carbon into compacts.

Types of Work

Presses perform such operations as blanking (some blanks are made on shears) and cutting off; piercing (punching) holes, cutouts, or extruding holes; bending to almost any angle (this would include lance forming of tabs); embossing; or forming strengthening ribs or shallow pockets and hemming (bending edges up and then back flat on themselves). The edges of shells may be curled in or out. Coining changes thicknesses, for the raw material of stamping is almost always uniform.

Related to blanking are trimming operations to remove excess stock and shaving, a slight removal of stock. Examples are the first step of broaching to obtain close tolerances, small teeth, and straight edges where the breakout on stampings is objectionable, or to improve edge appearance to about 125 µin. on thinner materials, and about 150 µin. on stock 2.3 mm and over.

Sheet Molding Compound

Sheet molding compound (SMC) combine with matched-metal-die compression molding has been the most successful glass fiber composite in automotive exterior body panel applications. In the medium and heavy truck market, SMC is a clear winner. In the more aesthetically demanding car market, however, SMC remains a niche player relegated to medium-volume specialty vehicles. More recent successes point to a shift in focus to more functional and structural applications.

The most successful applications address the total system needs and offer unmatched value in terms of high quality, dependability, and cost-effectiveness. The radiator support assembly is such a case. This illustrates that SMC can win and retain large volume applications resulting in profitable growth. SMC solutions are perhaps the best known and most developed and thus offer valuable lessons for all composites.

A thermoset composite of fibers, usually an unsaturated polyester resin, and pigments, fillers, and other adhesives that have been compounded and processed into sheet form to facilitate handling and the subsequent compression molding operation (generally cross linked with styrene).

SMC Developments

Compression-molded SMC is perhaps the most talked about composite system for automotive applications. SMC is a versatile composite capable of satisfying structural and aesthetic needs at fast molding cycles, and it is suitable for large production volumes.

SMC is an accepted and commercially proven material for automotive part manufacture, serving the role of composite ambassador to the automotive world. SMC offers capital efficiency vs. steel. Capital efficiency makes SMC a viable option for both styling differentiation and new capacity expansion.

Because of longer cycles and more defects, however, the cost of low-density SMC remains high, relegating it to but a few niche applications. This illustrates that the tension resulting from simultaneous needs of the original equipment manufacturers (OEMs) for low-cost and weight reduction cannot be resolved with higher-cost technology.

SMC applications fall into three major categories: functional, structural, and appearance. Recent gains and functional applications, such as oil pans and heat shields, have been impressive. The main driver of this achievement has been resin development. Both elevated temperature and oil resistance needs are now being met with polyester resins, which are less expensive than the vinyl esters previously employed.

Structural applications offer deeper insights into the capabilities and limitations of the SMC. Two such applications, the radiator support assembly and cross car beam, illustrate the capability of SMC for parts consolidation. Yet the former leads to a low-cost product, whereas the latter remains a premium niche component. Some reasons behind this outcome can be explained from the total system perspective.

Appearance or Class A applications are of importance to both growth and survival of SMC. The opportunity for SMC in this arena is huge and significant inroads have been made. Yet SMC has penetrated only a small fraction of this potential and only, to a significant degree, in the United States. The battle with steel in this arena has been examined using cost models.

Structural Applications

Structural applications attempt to exploit the design freedom and parts consolidation potential of SMC leading to integration of multiple functional roles. Integrated front-end system design, for the Ford Taurus and Mercury Sable, incorporates an upper and lower radiator support molded in SMC. The lower radiator support consolidates 22 steel parts into two SMC parts, resulting in a 14% cost reduction vs. steel.

Another application for SMC is cross car beams, which are also known as supported instrument panels. Because of their use on light truck platforms, cross car beams consume significant amounts of SMC. The level of functional integration is as large as in the case of the radiator support but the end result is very different. The added complexity pushes the boundaries of SMC flow and moldability. To make matters worse, attempts have been made to use very low density SMC, resulting in even more molding difficulties. In the end, low-cost has not been achieved, making both OEMs and molders unhappy. Perhaps a simpler design, incorporating a steel structural insert, might be more successful.

These two examples illustrate that parts consolidation is a useful tool for achieving low-cost; however, there are limits to this

S

approach and often the increased complexity leads to higher cost than in an equivalent system of many specialized components.

Sheet Separation

In spot, seam, or projection welding, the gap that exists between faying surfaces surrounding the weld, after the joint has been welded.

Sheffield Plate

Decorative electrodeposited silver plating or copper or nickel silver.

Shelf Life

The length of time a material, substance, product, or reagent can be stored under specified environmental conditions and continue to meet all applicable specification requirements and/or remain suitable for its intended function.

Shelf Roughness

Roughness on upward-facing surfaces where undissolved solids have settled on parts during a plating operation.

Shell

(1) A hollow structure or vessel. (2) An article formed by deep drawing. (3) The metal sleeve remaining when a billet is extruded with a dummy block of somewhat smaller diameter. (4) In shell molding, a hard layer of sand and thermosetting plastic or resin formed over a pattern and used as the mold wall. (5) A tubular casting used in making seamless drawn tube. (6) A pierced forging.

Shellac

A product where an insect lives on various trees of southern Asia. The larvae of the lac insect settle on the branches, pierce the bark, and feed on the sap. The lac secretion produced by the insects forms a coating over their bodies and makes a thick incrustation over the twig. Eggs developed in the females are deposited in a space formed in the cell, and the hatched larvae emerge. This forming continues for 3 weeks and is repeated twice a year. The incrustation formed on the twigs is scraped off, dried in the shape, and is the commercial *stick lac*. It contains woody matter, lac resin, lac die, and bodies of insects. *Seed lac* is obtained by screening, grinding, and washing stick lac. The washing removes the lac die. *Lac die* was once an important dyestuff, giving about the same colors as cochineal but not as strong.

Shellac is prepared from seed lac by melting or by extraction with solvents. The molten material is spread over a hot cylinder and stretched, and the cooled sheet is broken into flakes of shellac. When pure, shellac varies from pale orange to lemon yellow, but the color of commercial shellac may be due to a high content of common resin. *White shellac* is made by bleaching with alkalies.

Hard lac has the soft constituents removed by solvent extraction. For electrical use, the wax content should be below 3.5%. By solvent extraction of the seed lac, the wax may be reduced to 1%. Shellac is graded by color and by its freedom from dirt. The first grade contains no resin, but other grades may contain up to 12%.

Most Indian exports of seed lac to the United States are of the special grade, which has a high bleach index. *Cut shellac* is shellac dissolved in alcohol but usually mixed with a high percentage of resin. Shellac has good adhesive properties and high dielectric strength, and is used in adhesives, varnishes, floor waxes, insulating compounds, and some molding plastics. Hard-face wax polishes contain a high percentage of shellac, up to 80%, to conserve carnauba and other waxes.

Shell Core

A shell-molded sand core.

Shell Hardening

A surface-hardening process in which a suitable steel workpiece, when heated through and quench hardened, develops a martensite layer or shell that closely follows the contour of the piece and surrounds a core of essentially pearlitic transformation product. This result is accomplished by a proper balance among section size, steel hardenability, and severity of quench.

Shelling

(1) A term used in railway engineering to describe an advanced phase of *spalling*. (2) A mechanism of deterioration of coated abrasive products in which the entire abrasive grains are removed from the cement coating that held the abrasive to the backing layer of the product.

Shell Mold Castings

Shell mold casting is a process that uses relatively thin-wall mold made by bonding silica or zircon sand with a thermosetting phenolic or urea resin. It has gained widespread use because it offers many advantages over conventional sand castings. Shell mold casting is a practical and economical way to meet the demand for weight reduction, thinner sections, and closer tolerances.

Advantages

There are five basic advantages:

1. Lower costs. High production rates and fewer finishing operations result in a lower unit cost for applicable parts.
2. Closer tolerances. Shell castings have closer tolerances than sand castings. Draft allowances are also reduced.
3. Smoother surface finish. Shell molding provides an improved surface compared with sand casting (250–1000 μin. rms for sand, 125–250 for shell).
4. Less machining. The precision of shell molding reduces, and in many cases illuminates, machining or grinding operations.
5. Uniformity. Insulating properties of the molds produce casting surfaces free of chill and with a more uniform grain structure.

Although a shell mold is more expensive than a green sand mold, the possibility of reducing weight, minimizing machining, and eliminating cores often result in savings sufficient to offset the mold cost.

The Process

The shell molding process can be broken down into five operations:

1. A match plate pattern is made from tool steel with dimensions calculated to allow for subsequent metal shrinkage.
2. The resin-sand mixture is applied to the metal pattern, which is then heated to 218°C–232°C. The hot pattern melts the resin, which flows between the grains of sand and binds them together. Thickness of the mold increases with time. After the desired thicknesses are reached, excess unbonded sand is poured from the pattern.
3. The pattern, with the soft shell adhering, is placed in an oven and heated to 566°C–649°C for 30 s to 1 min. This cures the shell and produces a hard, smooth mold that reproduces the pattern surface exactly. The shell is stripped from the pattern by ejector pins. The other half of the mold is produced in the same way.
4. Sprues and risers are opened and cores inserted to complete the cope and drag halves of the mold.
5. The two shell halves are glued together under pressure to form a tightly sealed mold that can be stacked and stored indefinitely.

Shell Molded Cores

Shell molded cores are an offshoot of the shell molding process. These cores have several advantages over the sand cores that they replace: the shell core usually costs less than the sand core; strength and rigidity permit handling without damage or distortion; sharp details, including threads, are accurately reproduced; core weight is reduced; cured cores are unaffected by moisture and can be stored for long periods of time; most shell cores are hollow and can function as vents during casting.

Shell Molding

A foundry process in which a mold is formed from thermosetting resin-bonded sand mixtures brought in contact with preheated (150°C–260°C) metal patterns, resulting in a firm shell with a cavity corresponding to the outline of the pattern. Also called *Croning process*.

Shell Tooling

A mold or bonding fixture for forming plastic parts that consists of a contoured surface shell supported by a substructure to provide dimensional stability.

Shielded Carbon Arc Welding

A carbon arc welding process variation that produces coalescence of metals by heating them with an electric arc between a carbon electrode and the work. Shielding is obtained from the combustion of a solid material fed into the arc or from a blanket of flux on the work, or both. Pressure may or may not be used, and filler metal may or may not be used.

Shielded Metal Arc Cutting

A metal arc cutting process in which metals are severed by melting them with the heat of an arc between a covered metal electrode and the base metal.

Shielded Metal Arc Welding (SMAW)

A manual arc welding process in which the heat for welding is generated by an arc established between a flux-covered consumable electrode in a workpiece. The electrode tip, molten metal pool, arc, and adjacent areas of the workpiece are protected from atmospheric contamination by a gaseous shield obtained from the combustion and decomposition of the electrode covering. Additional shielding is provided for the molten metal in the weld pool by a covering of molten flux or slag. Filler metal is supplied by the core of the consumable electrode and from metal powder mixed with the electrode covering of certain electrodes. Shielded metal arc welding is often referred to as arc welding with stick electrodes, manual metal arc welding, and stick welding.

Shielding

(1) A material barrier that prevents radiation or a flowing fluid from impinging on an object or a portion of an object. (2) In an electron-optical instrument, the protection of the electron beam from distortion due to extraneous electric and magnetic fields. Because the metallic column of the microscope is at ground potential, it provides electrostatic shielding. See also *magnetic shielding*. (3) Placing an object in an electrolytic bath so as to alter the current distribution on the cathode. A nonconductor is called a shield; a conductor is called a *robber*, a thief, or a guard.

Shielding Gas

(1) Protective gas used to prevent atmospheric contamination during welding. (2) A stream of an inert gas directed at the substrate during thermal spraying so as to envelop the plasma flame and substrate; intended to provide a barrier to the atmosphere in order to minimize oxidation.

Shift

A casting imperfection caused by mismatch of cope and drag or of cores and molds.

Shim

A thin piece of material used between two surfaces to obtain a proper fit, adjustment, or alignment.

Shimmy Die

See *flat edge trimmer*.

Shock Load

The sudden application of an external force that results in a very rapid buildup of stress—for example, piston loading in internal combustion engines.

Shoe

(1) A metal block used in a variety of bending operations to form or support the part being processed. (2) An anvil cap or *sow block*. (3) A device for gathering filaments into a strand, in glass fiber forming. See the term *glass filament bushing*.

Shore Hardness

A measure of the resistance of material to indentation by a spring-loaded indenter during scleroscope hardness testing. The higher the number, the greater the resistance. Normally used for rubber materials. See also *scleroscope hardness test*.

Short

An imperfection in a molded plastic part due to an incomplete fill. In reinforced plastics, this may be evident either from an absence of surface film in some areas or as lighter, unfused particles of material showing through a covering surface film, accompanied possibly by thin-skin blisters. In thermoplastics, also called *short-shot*.

Short Beam Shear

A flexural test of a plastic specimen having a low test span-to-thickness ratio (e.g., 4:1), such that failure is primarily in shear.

Short Circuiting Transfer

In consumable-electrode arc welding, a type of metal transfer similar to globular transfer, but in which the drops are so large that the arc is short-circuited momentarily during the transfer of each drop to the weld pool. See also *gas metal arc welding* and compare with *globular transfer* and *spray transfer*.

Shortness (Adhesives)

A qualitative term that describes an adhesive that does not string cotton, or otherwise form filaments or threads during application.

Shortness (Metals)

A form of brittleness in metal. It is designated as *cold shortness* or *hot shortness* to indicate the temperature range in which the brittleness occurs.

Shorts

The product that is retained on a specified screen in the screening of a crushed or ground material. See also *plus sieve*.

Short Shot

Insufficient injection of material into the mold during forming of plastics or composites.

Short-Term Etching

In metallographic preparation of specimens, etching times of seconds to a few minutes.

Short Transverse

See *transverse*.

Shot

(1) Small, spherical particles of metal. (2) The injection of molten metal into a die casting die. The metal is injected so quickly that it can be compared to the shooting of a gun.

Shotblasting

Blasting with metal *shot*; usually used to remove deposits or mill scale more rapidly or more effectively than can be done by *sandblasting*.

Shot Capacity

In forming plastic parts, the maximum weight of material an injection machine can provide from one forward motion of the ram, screw, or plunger.

Shot Peening

A method of cold working metals in which compressive stresses are induced in the exposed surface layers of parts by the impingement of a stream of *shot*, directed at the metal surface at high velocity under controlled conditions. It differs from blast cleaning in primary purpose and in the extent to which it is controlled to yield accurate and reproducible results. Although shot peening cleans the surface being peened, this function is incidental. The major purpose of shot peening is to increase fatigue strength. Shot for peening is made of iron, steel, or glass.

Shotting

The production of *shot* by pouring molten metal in finely divided streams. Solidified spherical particles are formed during descent in a tank of water.

Shoulder

See preferred term *root face*.

Shrinkage (Ceramics)

The fractional reduction in dimensions or volume of a material or object when subjected to drying, calcining, or firing (sintering).

Shrinkage (Metals)

(1) The contraction of metal during cooling after hot forging. Die impressions are made oversize according to precise shrinkage scales to allow the forgings to shrink to design dimensions and tolerances. (2) See *casting shrinkage*.

Shrinkage (Plastics)

The relative change in dimension from the length measured on the mold when it is cold to the length of the molded plastic object 24 h after it has been taken out of the mold.

Shrinkage Cavity

A void left in cast metal as a result of solidification shrinkage. Shrinkage cavities can appear as either isolated or interconnected irregularly shaped voids. See also *casting shrinkage*.

Shrinkage Rule

A measuring ruler with graduations expanded to compensate for the change in the dimensions of the solidified casting as it cools in the mold.

Shrinkage Stress

See preferred term *residual stress*.

Shrinkage Void

A cavity type discontinuity in weldments normally formed by shrinkage during solidification.

Shrink Fit

An *interference fit* produced by heating the outside member of mating parts to a practical temperature for easy assembly. Usually the inside member is kept at or near room temperature. Sometimes the inside member is cooled to increase ease of assembly.

Shrink Forming

Forming of metal wherein the inner fibers of a cross section undergo a reduction in a localized area by the application of heat, cold upset, or mechanically induced pressures.

Shroud

A protective, refractory-lined metal-delivery system to prevent reoxidation of molten steel when it is poured from ladle to tundish to mold during continuous casting.

Shunt

A device used to divert part of an electric current.

Shut Height

For a metal forming press, the distance from the top of the bed to the bottom of the slide with the stroke down and adjustment up. In general, it is the maximum die height that can be accommodated for normal operation, taking the *bolster plate* into consideration.

SI

See *silicones*.

Sialon

Sialon is a generic term for a family of compositions produced by reacting silicon nitride with aluminum oxide and aluminum nitride at high temperatures. In other words, a sialon is a silicon nitride-base ceramic in which some of the silicon has been replaced with aluminum and some of the nitrogen with oxygen, resulting in a substituted solid solution referred to as β'-sialon with a compositional range of $Si_{6-x}Al_xO_xN_{8-x}$ based on the β-Si_6N_8 unit cell. Two major types have been developed commercially, one consisting of β' grains in a semicontinuous intergranular crystalline phase of yttrium–aluminum–garnet (YAG). Parts made from β'-sialon are produced by cold isostatically pressing the powder, followed by sintering at a maximum temperature of 1800°C for approximately 1 h; yttrium oxide is used as a sintering aid.

A solid solution based on the alpha silicon nitride structure has received less attention, but α'-sialon shows promise commercially because it has greater hardness than β'-sialon. In addition, it offers an improved product with less grain-boundary glass as a result of the incorporation of the cation from the sintering additive (e.g., Y_2O_3) into a solid solution of $M_x(Si, Al)_{12}(O, N)_{16}$ based on the α-$Si_{12}N_{16}$ unit cell. Currently, single-phase α'-sialons are only available in hot pressed or hot isostatically pressed forms, but composites of $\beta' + \alpha'$ are available produced by pressureless sintering.

Properties of sialons, which are based on the sintering aids used in the fabrication route followed, are similar to silicon nitride. The advantages of sialons are their low coefficient of thermal expansion (2–$3 \times 10^{-6}/°C$) and good oxidation resistance. The array of applications for sialons is also similar to that of silicon nitride—namely, automotive applications in machine tool cutting inserts for machining difficult-to-cut materials such as cast irons in wrought nickel-base superalloys.

Side Corings

In forming of plastics, projections that are used to core a hole in a direction other than the line of closing of a mold, and that must be withdrawn before the part is ejected from the mold. Also called side draw pins.

Side Cutting-Edge Angle

See the term *single-point tool*.

Side Milling

Milling with cutters having peripheral and side teeth. They are usually profile sharpened but may be form relieved. See also *milling*.

Side Rake

In a single-point turning tool, the angle between the tool face and a reference plane, corresponding to radial rake in milling. It lies in a plane perpendicular to the tool base and parallel to the rotational axis of the work. See the term *single-point tool*.

Side Thrust

The lateral force exerted between the dies by reaction of a forged piece on the die impressions.

Siemens (S)

A unit of electrical conductivity. One siemens of conductance per cubic meter with a potential of 1 V allows the passage of 1 A/m². See also *conductance* (electrical).

Sieve

A standard wire mesh or screen used in graded sets to determine the mesh size or particle size distribution of particulate and granular solids. Sieves are stacked in order, with the largest mesh size at the top and a pan at the bottom. An appropriate sample weight of powder is spread on the top sieve and covered. The stack of sieves is agitated in a prescribed manner (shaking, rotating, or tapping) for a specified period of time. The powder fractions remaining on each sieve and contained in the bottom pan are weighed separately and reported as percentages retained or passed by each sieve. See also *sieve analysis*.

Sieve Analysis

A method of determining *particle size distribution*, usually expressed as the weight percentage retained upon each of a series of standard screens of decreasing mesh size. For example, a typical sieve analysis for titanium powder is as follows:

Mesh Size (U.S.)	Particle Size (μm)	Weight Percent Retained
+80	+177	0
−80 + 100	−177 + 149	0.1
−100 + 140	−149 + 105	11.2
−140 + 200	−105 + 74	32.9
−200 + 230	−74 + 64	5.0
−230 + 325	−64 + 45	23.3
−325	−45	27.5

Using the standard sieve designations and screen openings sizes, a user can determine that −140 + 299 powder is smaller than approximately 105 μm, yet larger than approximately 74 μm, that is, −33% is retained on the 200 mesh screen.

Sieve Classification

The separation of powder into particle size ranges by the use of a series of grated sieves. Also called screen analysis.

Sieve Fraction

That portion of a powder sample that passes through a sieve of specified number and is retained by some finer mesh sieve of specified number. See also *sieve analysis*.

Sieve Shaker

A device for shaking, knocking, or vibrating a single sieve or a stack of sieves. It consists of a frame, a motorized knocker, shaker or vibrator, and fasteners for the sieve(s).

Sieve Underside

The underside of the mesh to which loose powder often adheres. To ensure accuracy of a sieve analysis, this material must be removed and included in the weight determination.

Sigma Bonding

Covalent bonding between atoms in which *s* orbitals or hybrid orbitals between *s* and *p* electrons overlap in cylindrical symmetry along the axis joining the nuclei of the atoms. See also *pi bonding*.

Sigma Phase

A hard, brittle, nonmagnetic intermediate phase with a tetragonal crystal structure, containing 30 atoms per unit cell, space group, *P*4/*mnm*, occurring in many binary and ternary alloys of the transition elements. The composition of this phase in the various systems is not the same, and the phase usually exhibits a wide range in homogeneity. Alloying with a third transition element usually enlarges the field homogeneity and extends it deep into the ternary section.

Sigma-Phase Embrittlement

Embrittlement of iron–chromium alloys (most notably austenitic stainless steels) caused by precipitation at grain boundaries of the hard, brittle intermetallic *sigma phase* owing to long periods of exposure to temperatures between approximately 560°C and 980°C. Sigma-phase embrittlement results in severe loss in *toughness* and *ductility*, and can make the embrittled material susceptible to *intergranular corrosion*. See also *sensitization*.

Signal-to-Noise Ratio

(1) The ratio of the amplitude of a desired signal at any time to the amplitude of noise signals at the same time. (2) Ratio of the average response to the root-mean-square variation about the average response. Ratio of variances associated with the two parts of the performance measurement. See also *noise*.

Significance Level

The stated probability (risk) that a given test of significance will reject the hypothesis that a specified effect is absent when the hypothesis is true.

Significant

Statistically significant. An effect of difference between populations is said to be present if the value of a test statistic is significant, that is, lies outside the predetermined limits. See also *population*.

Silica (SiO$_2$)

The common oxide of silicon usually found naturally as quartz or in complex combination with other elements such as silicates. Various polymorphs and natural occurrences of silica include cristobalite, tridymite, cryptocrystalline chert, flint, chalcedony, and hydrated opal. Silica is the primary ingredient of sand, refractories, and glass.

Silica, as the dioxide of silicon is commonly termed, is the principal constituent of the solid crust of the Earth. Consequently, it is a major ingredient of most of the nonmetallic, inorganic materials used in industry. Silica occurs in a number of allotropic forms, which have different properties and uses.

The principal modifications of silica are quartz, silica glass, cristobalite, and tridymite. The last two are sometimes combined under the term of inverted silica. The valuable physical properties of these four principal modifications lead to a wide variety of applications for silica in industry and technology.

Quartz

The common low-temperature form of silica, quartz, is strong and insoluble in water. Consequently, sandstone, which is composed of grains of quartz held together by a siliceous cement, is an excellent building stone. Huge quantities of silica stone are used as aggregate for concrete. One curious variety has a structure that renders the rock flexible and is known as itacolumite or flexible sandstone. Quartzite, in which the quartz grains of an original sandstone have recrystallized and grown to a compact mass, and vein quartz are too hard and difficult to shape for use as building stones.

Rock crystal, as the euhedral quartz found in nature is termed, is often carved into ornaments of great beauty because of its perfect transparency and because of the high luster it produces by polishing.

The "crystal ball" of story and romance is a polished quartz sphere. Colored varieties of quartz are much valued as semiprecious stones. The purple amethyst, blue sapphire quartz, yellow citrine or false topaz, red or pink rose quartz, smoky quartz, and the dark brown morion are examples.

Sandstone is also useful as an abrasive. It is used for grindstones and pulpstones. "Berea" grit from northern Ohio and novaculite from Arkansas are examples of coarse and fine-grained natural stones that are still preferred for some purposes to the artificial abrasives made from silicon carbide or aluminum oxide.

The strength and hardness of various material forms of silica leads to use for crushing and for grinding. Flint pebbles are used in ball mills for grinding all sorts of materials. Agate mortars are insoluble in the chemical laboratory for pulverizing minerals before analysis. Sandstone and quartzite are used for millstones and buhrstones to crush and grind grain, paint pigments, fertilizers, and many other products. Quartz sand, driven by compressed air, constitutes the useful sandblast for cleaning metal, decorating glass, and refacing stone buildings. Coated on paper, quartz grains provide the carpenter and cabinet maker with the familiar "sandpaper," although much modern sandpaper is really coated with crushed glass containing only about 70% silica.

The abrasive properties of quartz grains are also employed by locomotives to gain traction on steel rails, in billiard cue chalk, and in the chicken gizzard, where quartz grains are used as a kind of natural ball mill.

The thermal properties of the various forms of silica also lead to important uses. Its high melting point (1710°C) makes it a good refractory, while its low cost and low thermal expansion have brought about a wide use in industrial furnaces, particularly for melting steel and glass.

Properties

In general, the thermal expansion of all of the common forms of silica is low at high temperatures. This makes silica refractories capable of withstanding sudden temperature changes and large thermal gradients very well in these high-temperature ranges. At lower temperatures, large volume changes, due to rapid inversions from a high-temperature crystalline forms of quartz, tridymite, cristobalite, to corresponding low-temperature forms, make these materials rather sensitive to sudden temperature changes. The low-temperature forms of quartz, tridymite and cristobalite, also have higher thermal expansions than the high-temperature forms and the amorphous form of silica. This form of silica, variously known as silica glass, vitrosil, and fused quartz, is thus the only form of silica that may be heated rapidly from room temperature without fear of breakage. A further limitation on the thermal behavior of silica refractories is occasioned by the large volume increase that occurs as quartz changes slowly to tridymite or cristobalite at temperatures above 870°C. This may lead to swelling and warping of silica bricks if they are not first converted to the forms stable at high temperatures by prolonged firing. Because the volume changes that occur when tridymite changes from its high-temperature form to the forms stable at lower temperatures are less than the corresponding change for cristobalite, an effort is made to convert the silica refractory as completely to tridymite as is feasible before using it.

Uses

Silica bricks are made from crushed quartzite rock, known as ganister, which is bonded with 1.5%–3.0% of lime. They are molded smaller than the dimensions desired in the finished brick to allow for an expansion of about 3/8 in./ft as the quartz inverts to tridymite during firing. A firing schedule of about 20 days at 1450°C (about cone 16) is required. Study of the phase diagram of the silica–lime system shows why considerable quantities of lime may be used to bond the quartzite in silica bricks without loss of refractoriness. The lime is taken up in an immiscible lime-rich glass phase of which only a small amount is formed because of its high lime content.

In use, silica bricks are characterized by retention of rigidity and load-bearing capacity to temperatures above 1600°C, without the slow yield characteristic of fireclay brick. If a cold silica brick is heated suddenly, it spalls and disintegrates owing to the sudden volume changes taking place at the high–low inversions of tridymite and cristobalite. If heated cautiously through this sensitive temperature region, silica brick is very resistant to temperature shock.

The refractory properties of quartz sand are also employed in molds for cast iron. Huge quantities of sand, with varying amounts of clay impurity to bond the grains, are used for this purpose in the foundries of the world.

Crystal quartz in piezoelectric oscillators have become indispensable for the control of radio transmitters, radar equipment, and other electronic "timing" gear.

This use and other electrical and optical uses of crystal quartz depend upon the symmetry of the material. Some experts state that low quartz is the most important and most familiar example of the trigonal enantiomorphous hemihedral or trigonal trapezohedral class of symmetry. This class is characterized by one axis of threefold symmetry with three axes of twofold symmetry perpendicular thereto and separated by angles of 120°. This class has no plane of symmetry and no center of symmetry.

Optical Properties

The transparency of crystal quartz, particularly to the short waves of the ultraviolet, makes it very useful in optical instruments. Quartz prisms are employed in spectrographs for analysis of light waves varying in length from almost 5 μm in the infrared to almost 0.2 μm in the ultraviolet. Lenses made of crystal quartz are also useful over this range for photography and microscopy. Crystal quartz is more transparent than fused quartz for these purposes, but its birefringence introduces complications and design.

Quartz also has the property of rotating the plane of vibration of polarized light traveling along its axis. This property is associated with its left- or right-handed character. Sugar solutions have the same optical rotating power and quartz planes or wedges are employed as standards and optical devices used for analyzing sugar solutions by means of this rotary power. Many other organic chemicals that have asymmetric right- or left-handed molecules can be studied and assayed by this optical means.

The variation of this rotary power of quartz with wavelength or color of light makes it possible to construct monochromators capable of selecting light of a desired color from a white source with very large optical apertures.

Amorphous Silica

The amorphous form of silica also has mechanical, optical, thermal, and electrical properties, which make it very useful in hundreds of technical applications. Vitreous silica is made by three different processes. The earliest process, which is still used, consists simply of melting quartz by application of high temperature. The oxyhydrogen flame was the first source of heat for this process and is still used to melt fragments of rock crystal that are combined to form rods and other shapes of silica glass. However, electrical heat from graphite resistors is more commonly employed at present since there is less

S

trouble from volatilization of the silica. Large masses can be made by this technique and shaped into various useful forms although the very high temperature required for working silica glass makes this method of manufacture difficult and expensive.

To obtain a clear transparent product, crystal must be melted. If sand is used, the product is white and opaque because of the numerous air bubbles trapped in the very viscous glass. It is impossible to heat silica glass hot enough to drive out the bubbles or, in glass parlance, to "fine" it. However, the opaque "vitrosil" made by fusing sand is very useful for chemical apparatus, when the low thermal expansion and insolubility of the silica glass play a role.

Recently, the Corning Glass Works developed a new process for making silica glass by hydrolysis of silicon tetrachloride in a flame. The resulting silica is deposited directly on a support in the form of transparent silica glass, which is more homogeneous and purer than the glass made by fusing quartz.

Large pieces can be made by this process and the unusual homogeneity of the product makes it especially useful for optical parts and for sonic delay lines where striations in the older fused quartz are detrimental. The sonic delay lines are polygons of the silica glass in which acoustic waves travel on a long path, being reflected at the numerous polygonal faces.

Finally, silica glass of 96% purity or better is made by an ingenious process invented by the Corning Glass Works. In this third process for making vitreous silica, an object is shaped first from a soft glass containing about 30% of borax and boric acid as fluxes. A suitable heat treatment causes a submicroscopic separation of two glass phases. One phase, which is composed chiefly of the fluxes, is then bleached from the glass in a hot, dilute nitric acid bath. If the composition and heat treatment are exactly right, the high silica phase is left as a porous "sponge" with the shape and size of the original soft glass article. This is carefully washed and dried and fired by heating to about 1200°C.

In the firing step, the porous silica sponge shrinks to a dense transparent object of glass containing 96% SiO$_2$ or more, which has the desirable properties of silica glass made by either of the other processes. The 4% of impurity in the glass made by this process consists chiefly of boric oxide. Because of its presence, this class is appreciably softer than pure silica glass. For corresponding viscosities, the "Vycor" 96% silica glass requires about 100°C lower temperature.

Silica glass made by any of these processes has many uses because of its unique combination of good properties.

Fibers of silica glass have very high tensile strength and almost perfect elasticity. This makes them useful in constructing microbalances, electrometers, and similar instruments. Silica fiber in diameters as small as 0.000076 cm comes in random matted form or in rovings.

Because of its small thermal expansion, fused silica is very resistant to sudden temperature changes. It is also a very hard glass so that it may be used in the laboratory for crucibles and combustion tubes to much better advantage than ordinary glass. Vitreous silica makes possible the construction of thermometers operating up to 1000°C. The small thermal expansion and durability of fused silica have led to its use in fabrication of standards of length. Fused silica plates, ground to optical flatness, are used in interferometers for measurement of thermal expansion.

The fact that fused silica is an excellent insulator with little or no tendency to condense surface films of moisture makes it valuable in the construction of electrical apparatus. It has a very high dielectric strength and low dielectric loss.

Silica glass also has very good transmission for visible and ultraviolet light. This makes it useful for the construction of mercury lamps and other optical equipment. In the mercury lamps, the strength and heat resistance of silica glass makes it possible to operate with high internal pressures, producing very high light intensities and great efficiency. The 96% silica glass made by the Vycor process may be fired in a reducing atmosphere or in vacuum to modify and improve its light-transmitting properties.

Silica glass is among the most chemically resistant of all glasses. This makes it particularly useful in the analytical laboratory where there is the added advantage that any contamination of contained solutions can only be by the one oxide, silica. Crucibles of fused silica glass may be used for pyrosulfate fusions. Condensers of fused silica are extremely useful for distilling acids (except hydrofluoric) and for preparation of an extremely pure water.

Another useful form of silica is obtained by dehydrating silicic acid. In this way, a porous "gel" is obtained that has an enormous surface area and is capable of adsorbing various gases and vapors, particularly water vapor. Silica gel is used in dehumidifiers to remove water vapor from the air. It is also used as a catalyst and as a support for other catalysts.

The porous 96% silica skeleton obtained in the Vycor process also has a very large surface and is useful as an adsorbent and drying agent. It has less capacity than silica gel, but better mechanical strength.

A very finely divided form of silica known as silica soot is obtained by hydrolizing silicon tetrachloride in a flame without heating the product hot enough to consolidate it in as a glass. This material is valuable as a thermal insulator and as a white filler for rubber and plastics.

Other Uses

The silica (actually ground silica) used in the pottery industry is called flint. The addition of flint affects warpage very little.

In ceramic bodies, potters' flint or pulverized quartz or sand is the constituent that reduces drying and burning shrinkage and assists promotion of refractoriness. Flint has an important bearing on the resistance of bodies to thermal and mechanical shock, because of the volume changes that accompany crystal transformation. In the unburned body, it lowers plasticity and workability, lowers shrinkage, and hastens drying. A coarse crystalline form of quartz, called macrocrystalline quartz, is more often used for potters' flint than the cryptocrystalline form.

Silica is used in all glazes as the chief, and often the only, acid radical (RO$_2$ group). It may be adjusted to regulate the melting temperature of the glaze. In common glazes, the ratio of silica to bases (RO group) is never less than 1:1 nor more than 3:1. By varying the relative proportions of the RO group and balancing the group against any desired silica content, the maturing temperature of a glaze may be quite closely controlled. In other words, the fusibility of the glazes used in the presence of equal proportions of fluxes depends on their relative silica contents.

In porcelain enamels, it may be taken as a general rule that, other things remaining constant, the higher the percentage of silica, the higher will be the melting point of the enamel and the greater its acid resistance. Silica has a low coefficient of expansion and increasing it in an enamel lowers the coefficient of expansion of that enamel. One method of regulating an enamel coating is to increase the silica content when the enamel is inclined to split off in cooling. Silica is a form of flint or quartz is used in both ground-coat and cover-coat enamels, and it has the same effect in either type.

The temperature required for melting an enamel is materially affected by the fineness of the silica. Cryolite, antimony, and tin oxide give their maximum value as pacifiers with minimum heat treatment in the smelter. The form and fineness of the silica should, therefore, be carefully watched and allowed for in compounding the batch. All forms of silica may be used with good results, but experience has shown that, in the same enamel, a smaller quantity of sand than of powdered quartz is necessary. Similarly, less sand than flint should be used, but the difference in this case is less than in the former. High SiO_2 tends to harden the enamel. The lower limit established in the usual run of enamels is 1:1 equivalents.

In the manufacture of semiconductors, monolithic circuits, and integral circuits for the electronics industry, the use of fused quartz is widespread for plumbing and diffusion furnace muffles. The need to prevent product contamination makes this choice mandatory.

Irish Refrasil is 98% silica and has a green color. It is used for ablative protective coatings. It resists temperatures to 1588°C. Silica flour, made by grinding sand, is used in paints, as a facing for sand molds, and for making flooring blocks. Silver bond silica is water-floated silica flour of 98.5% SiO_2, ground to 325 mesh. In zinc and lead paints, it gives a hard surface. Pulverized silica, made from crushed quartz, is used to replace tripoli as an abrasive. Ultrafine silica, a white powder having spherical particles of 4–25 µm, is made by burning silicon tetrachloride. It is used in rubber compounding, as a grease thickener, and as a flattening agent in paints. An example of this material is *Aerosil*.

A polymer-impregnated silica, *Polysil*, has twice the dielectric strength of porcelain as well as better strength. It is also cheaper to make, and its composition can be tailored to meet specific environmental and operating conditions.

Silica aerogel is a fine, white, semitransparent silica powder; its grains have a honeycomb structure, getting extreme lightness. It weighs 40 kg/m³ and is used as an insulating material in the walls of refrigerators, as a filler in molding plastics, as a flattening agent in paints, as a bodying agent in printing inks, and as a reinforcement for rubber. It is produced by treating sand with caustic soda to form sodium silicate, and then treating with sulfuric acid to form a jelly-like material called *silica gel*, which is washed and ground to a fine dry powder. It is also called *synthetic silica*.

Silicon monoxide, SiO, does not occur naturally but is made by reducing silica with carbon in the electric furnace and condensing the vapor out of contact with the air. It is lighter than silica, having a specific gravity of 2.24, and is less soluble in acid. It is brown powder valued as a pigment for oil painting, as it takes up a higher percentage of oil than ochres or red lead. It combines chemically with the oil. Fumed silica is a fine translucent powder of the simple amorphous silica formula made by calcining ethyl silicate. It is used instead of carbon black in rubber compounding to make light-colored products, and to coagulate oil slicks on water so that they can be burned off. It is often called white carbon, but the "white carbon black" called *Cab-O-Sil*, used for rubber, is a silica powder made from silica and tetrachloride. *Cab-O-Sil EH5*, a fumed colloidal form, is used as a thickener in resin coatings. The thermal expansion of amorphous fused silica is only about 1/8 that of alumina. Refractory ceramic parts made from it can be heated to 1093°C and cooled rapidly to subzero temperatures without fracture.

Silica Flour

A sand additive, containing about 99.5% silica, commonly produced by pulverizing quartz sand and large ball mills to a mesh size of 80–325.

Silica Gel

A precipitated colloidal mass or gel of indefinitely hydrated silica: also the dried or activated product of the same. Useful as a desiccant, scavenger, and catalyst substrate.

Silica Minerals

Silica occurs naturally in at least nine different varieties (polymorphs), which include tridymite (high-, middle-, and low-temperature forms), cristobalite (high- and low-temperature forms), coesite, and stishovite, in addition to high (β) and low (α) quartz. These forms have distinctive crystallography and optical characteristics.

The transformation between the various forms are of two types. Displacive transformations, such as inversions between-temperature (β) and low-temperature (α) forms, result in a displacement or change in bond direction but involve no breakage of existing bonds between silicon and oxygen atoms. These transformations take place rapidly over a small temperature interval and are reversible. Reconstructive transformations, in contrast, involve disruption of existing bonds and subsequent formation of new ones. These changes are sluggish, thereby permitting a species to exist metastably outside its defined pressure-temperature stability field. Two examples of reconstructive transformations are ≤ quartz and ≤ stishovite.

Silicate-Type Inclusions

Inclusions composed essentially of silicate glass, normally plastic at forging and hot-rolling temperatures, that appear in steel in the wrought condition as small elongated inclusions usually dark in color under reflected light as normally observed.

Silicides

Silicides are a group of substances, usually compounds, comprising silicon in combination with one or more metallic elements. These hard, crystalline materials are closely related to intermetallic compounds and have, therefore, many of the physical and chemical characteristics and some of the mechanical properties of metals.

Silicides are not natural products. They received but little attention prior to the development of the electrical furnace, which provided the first practical means of attaining and controlling the high temperatures generally required in their preparation.

Composition

Although a majority of the metals react with silicon, many of the resulting silicides do not have the properties usually required in engineering materials. The silicides that appear most promising for practical utilization in engineering and structural applications are, with few exceptions, limited to those of the refractory, or high melting metals of groups IV, V, and VI of the periodic table. Included in this category are the silicides of titanium, zirconium, hafnium, vanadium, columbium, tantalum, chromium, molybdenum, and tungsten.

Silicides can be prepared by direct synthesis from the elements, by reduction of silica or silica and halogenides and the appropriate metal oxide or halogenide with silicon, carbon, aluminum, magnesium, hydrogen, etc., and by electrolysis of molten compounds. They are also obtained as by-products in many metallurgical processes. High-purity silicides of stoichiometric composition are difficult to prepare.

The chemical composition of silicides cannot, in general, be predicted from a consideration of the customary valences of the elements. The zirconium–silicon system, for example, is reported to include the compounds Zr_4Si, Zr_2Si, Zr_3Si_2, Zr_6Si_5, $ZrSi$, and $ZrSi_2$. The disilicide composition (MSi_2) occurs in all of the refractory metal–silicon systems and probably will prove the most important, particularly in those applications requiring high-temperature stability.

General Properties

Silicides resemble silicon in their chemical properties with the degree of similarity roughly proportional to the silicon content. At normal temperatures, the refractory metal disilicides are inert to most ordinary chemical reagents. The compounds are not thermodynamically stable in the presence of oxygen and in the finely pulverized state they oxidize readily. However, in massive form, they are oxidation resistant because of the formation of a protective surface layer of silica. Bodies of $MoSi_2$ and WSi_2 are highly resistant to oxidation even at temperatures approaching their melting points.

The physical and mechanical characteristics of silicides are, to a large extent, determined by the properties of the component metal. The refractory metal silicides have highly crystalline structures and moderate densities. Their melting points are intermediate to relatively high. They have low electrical resistance, high thermal conductivity, and fair thermal shock resistance. They have high hardness, high compressive strength, and moderate tensile strength at both room and elevated temperatures. The elevated temperature stress-rupture and creep properties are good. Brittleness at low impact resistance are the most serious disadvantages of these materials.

Excellent oxidation resistance has prompted detailed studies of molybdenum disilicide. Quantitative information on the properties of most silicides has been developed within the past two decades.

Fabrication

The general methods used for consolidating powders can be applied to the silicides. High-density parts are obtained by cold pressing and sintering and also by hot pressing. Slip casting and extrusion are convenient methods for preparing certain shapes and sizes. Casting in the molten state is difficult due to the partial decomposition of silicides at their melting points. Silicide coatings can be prepared by vapor deposition techniques. Dense, fully sintered silicide parts are extremely difficult to work. They can be cut using silicon carbide or diamond wheels. Grinding has shown some promise but it is a slow process. "Green" or presintered compacts can, however, be shaped by conventional methods.

Availability

Although the availability of silicides is gradually increasing, the varieties, quantities, and shapes are limited. Molybdenum disulfide is commercially available in powder form and as furnace heating elements, and certain shapes and sizes have been produced on a custom basis. Some of the other compositions can be obtained on special order. The small market that has developed for $MoSi_2$ can be expected to stimulate general interest in the silicides and to foster their commercial development and production.

The practical utilization of silicides, with few exceptions, notably $MoSi_2$ furnace heating elements, has been implemented. Molybdenum disulfide is a structural material for gas turbine and missile components, which do not require high impact and thermal shock resistance. Igniter elements, thermocouple shields, gas probes, and nozzles are other potential applications. The high hardness of these materials has found usage in metalworking dies and tooling.

Silicide coatings, prepared by vapor-phase deposition, afford excellent oxidation protection to molybdenum and tungsten. This method of providing oxidation resistance is versatile because fabrication can be completed before the coating is applied. Similar coatings can be produced on other materials such as graphite by using silicide powders. Silicide coatings are commercially available and are a major item in the present market for silicide products.

Silicon

A metallic element (symbol Si), silicon is used chiefly in its combined state and is the most abundant solid element in the Earth's crust (28%).

The metal has been prepared by reducing the tetrachloride by hydrogen using a hot filament, or by aluminum, magnesium, or zinc. The fluoride or alkali fluorosilicates have been reduced with alkali metals or aluminum. Silica can be converted to metal by reduction in the electric furnace with carbon, silica carbide, aluminum, or magnesium. The element has also been produced by the fusion electrolysis of silica in molten alkali oxide–sodium chloride–aluminum chloride baths. Extremely pure metal has been made by the treatment of silanes with hydrogen.

It is a gray-white, brittle, metallic-appearing element, not readily attacked by acids except by a mixture of HF and HNO_3. It is soluble in hot NaOH or KOH and is prepared in the pure crystalline form by reduction of fractionally distilled $SiCl_4$.

Silica combines with many elements including boron, carbon, titanium, and zirconium in the electric furnace. It readily dissolves in molten magnesium, copper, iron, and nickel to form silicides. Most oxides are reduced by silicon at high temperatures.

Fabrication

Silicon can be cast by melting in a vacuum furnace and cooling in vacuum by withdrawing the element from the heated zone. Single-crystal ingots have been prepared by drawing from the melt and by the "floating zone" technique. It is claimed that silicon exhibits some workability above 1000°C, but at room temperature it is very brittle. Metals can be coated with a silicon-rich layer by reducing the tetrachloride with hydrogen on the hot metal surface.

Types

In very pure form, silicon is an intrinsic semiconductor, although the extent of its semiconduction is greatly increased by the introduction of minute amounts of impurities.

Pure silicon metal is used in transistors, rectifiers, and electronic devices. It is a semiconductor and is superior to germanium for transistors as it will withstand temperatures to 149°C and will carry more power. Rectifiers made with silicon instead of selenium can be smaller and will withstand higher temperatures. Its melting point when pure is about 1434°C, but it readily dissolves in molten metals. It is never found free in nature but, when combined with oxygen, it forms silica, SiO_2, one of the most common substances on Earth. Silicon can be obtained in three modifications.

Form

Amorphous silicon is a brown-colored powder with a specific gravity of 2.35. It is fusible and dissolves in molten metals. When heated in the air, it burns to form silica. Graphitoidal silicon consists of black glistening spangles and is not easily oxidized and not attacked by the common acids, but is soluble in alkalies. Crystalline silicon is obtained in dark, steel-gray globules of crystals or six-sided pyramids of specific gravity 2.4. It is less reactive than the amorphous form, but is attacked by boiling water. All these forms are obtainable by chemical reduction. Silicon is an important constituent of commercial metals. Molding sands are largely silica, and silicon carbides are used as abrasives. Commercial silicon is sold in the graphitoidal flake form, or as ferrosilicon, and silicon–copper. The latter forms are employed for adding silicon to iron and steels. Commercial refined silicon contains 97% pure silicon and less than 1% iron. It is used for adding silicon to aluminum alloys and for fluxing copper alloys. High-purity silicon metal, 99.95% pure, made in an arc furnace, is too expensive for common uses, which is employed for electronic devices and in making silicones. For electronic use, silicon must have extremely high purity, and the pure metal is a nonconductor with a resistivity of 300,000 Ω cm. For semiconductor use, it is "doped" with other atoms yielding electron activity for conducting current. Epitaxial silicon is a higher purified silicon doped with exact amounts of impurities added to the crystal to give the desired electronic properties.

Principal Compounds

Silicate is reported to form compounds with 64 of the 96 stable elements, and it probably forms silicides with 18 other elements. Besides the metal silicides, used in large quantities and metallurgy, silicon forms useful and important compounds with hydrogen, carbon, the halogen elements, nitrogen, oxygen, and sulfur. In addition, useful organosilicon derivatives have been prepared.

Hydrides

The hydrides of silica are named silanes; the compound SiH_4 is called monosilane, Si_2H_6 disilane, Si_3H_8 trisilane, and so on. Compounds in which oxygen atoms alternate with silicon atoms in the principal part of the structure are called siloxanes, and those with nitrogen between silicon atoms are called silazanes. All other covalent compounds of silicon are considered for the purpose of nomenclature to be derived from these silanes, and modified silanes are named according to substituent groups and their placement along the principal silicon-containing chain or ring.

Use

Because of its inherent brittleness, there are no engineering applications of silicon. The chief use of highly purified metal is as a semiconductor in transistors, rectifiers, and solar cells. It may be fired with ceramic materials to form heat-resistant articles. Silicon can serve as an autoxidation catalyst and as an element in photocells. Mirrors for dental use are formed with a reflecting surface of silicon. The metal is also employed to prepare silicides and alloys, and to coat various materials. In commercial quantities, it is also used as a starting material for the synthesis of silicones.

By far the largest use of silicon is as compounds in the ceramic industry. It is also employed as an alloying element in ferrous metals and is the basis of the family of chemicals known as silicones.

An important application of silicon is in the electronics industry where it has been widely employed in the manufacture of crystal rectifiers and integrated circuits. Sufficiently pure silicon has been produced by carefully controlled zone refining and crystal growth to make possible its use as transistors. Since the energy gap of silicon is 1.1 eV, compared with 0.75 eV for germanium, silicon transistors may be operated at higher temperatures and power levels than those made of germanium.

Silicon Bronze

Silicon bronze is a family of wrought copper-base alloys (C64700–C66100) and one cast copper alloy (C87200), the wrought alloys containing from 0.4% to 0.8% silicon (C64700) to 2.8% to 4.0% silicon (C65600), and the cast alloy 1.0% to 5.0%, along with other elements, usually lead, iron, and zinc. Other alloying elements may include manganese, aluminum, tin, nickel, chromium, and phosphorus. The most well-known alloys are probably silicon bronze C65100, or low-silicon bronze B, and silicon bronze C65500, or high-silicon bronze A, as they were formally called. As these names imply, they differ mainly in silicon content: 0.8%–2.0% and 2.8%–3.8%, respectively, although the latter alloy also may contain as much as 0.6% nickel. C87200 contains at least 89% copper, 1.5% silicon, and as much as 5% zinc, 2.5% iron, 1.5% aluminum, 1.5% manganese, 1% tin, and 0.5% lead. Regardless of alloying ingredients, copper content is typically 90% or greater.

Both of the common wrought alloys are quite ductile in the annealed condition, C65500 somewhat more ductile than C65100, and both can be appreciably strengthened by cold working. Annealed, tensile yield strengths are on the order of 103–172 MPa depending on mill form, with ultimate tensile strengths to about 414 MPa and elongations of 50%–60%. Cold working can increase yield strength to as much as 483 MPa. Electrical conductivity is 12% for C65100 and 7% for C65500 relative to copper, and thermal conductivity is 57 and 36 W/(m K), respectively. The alloys are used for hydraulic-fluid lines in aircraft, heat-exchanger tubing, marine hardware, bearing plates, and various fasteners.

Silicon bronze C87200 is suitable for centrifugal, investment, and sand, plaster, and permanent-mold castings. As sand cast, typical tensile properties are 379 MPa ultimate strength, 172 MPa yield strength, and 30% elongation. Hardness is Brinell 85, electrical conductivity 6%, and, relative to free-cutting brass, machinability is 40%. Uses include pump and valve parts, and marine fittings and bearings.

Silicon Carbide

The reduction of silica with excess carbon under appropriate conditions give silicon carbide (SiC), which crystallizes in a number of forms but is best known in the cubic-diamond form with spacing α_o of 0.435 nm (compared with 0.356 nm for diamond). In the pure form, silicon carbide is green (α-hexagonal) or yellow (β-cubic), but the commercial product is black and has a bluish or greenish iridescence. The carbide is not easily oxidized by air except above 1000°C, and retains its physical strength up to this temperature. For these reasons, it is a favorite structural refractory material for the ceramic arts. It also is extremely hard, with a Mohs hardness in excess of 9, and so has found wide application as an abrasive.

Silicon carbide does not melt without decomposition at atmospheric pressure, but does not melt at 2830°C at 3.5 MPa.

Forms

β-SiC (cubic) forms at 1400°C–1800°C and α-SiC (hexagonal) forms at temperatures > 1800°C.

SiC is used as an abrasive as loose powder, coated abrasive cloth and paper, wheels, and hones. It will withstand temperatures to its decomposing point of 2301°C and is valued as a refractory. It retains its strength at high temperatures, has a low thermal expansion, and its heat conductivity is 10 times that of fireclay. Silicon carbide is made by fusing sand and coke at a temperature above 2204°C.

Unlike aluminum oxide, the crystals of silicon carbide are large, and they are crushed to make the small grains used as abrasives. They are harder than aluminum oxide, and because they fracture less easily, they are more suited for grinding hard cast irons and ceramics. The standard grain sizes are usually from 100 to 1000 mesh. The crystalline powder in grain sizes from 60 to 240 mesh is also used in lightning arrestors. *Carborundum*, *Crystolon*, and *Carbolon* are trade names for silicon carbide.

Types

Three main types are produced commercially. Green SiC is an entirely new batch composition made from a sand and coke mixture, and is the highest purity of the three. Green is typically used for heating elements. Black SiC contains some free silicon and carbon and is less pure. A common use is as bonded SiC, and is not very pure. It is typically used as a steel additive.

Fabrication and Properties

Silicon carbide is manufactured in many complex bonded shapes, which are utilized for super-refractory purposes such as setter tile and kiln furniture, muffles, retorts and condensors, skid rails, hot cyclone liners, rocket nozzles and combustion chambers, and mechanical shaft seals. It is also used for erosion- and corrosion-resistant uses such as check valves, orifices, slag blocks, aluminum die casting machine parts, and sludge burner orifices. Electrical uses of SiC include lightning arrestors, heating elements, and nonlinear resistors.

Silicon carbide refractories are classified on the basis of the bonds used. Associated-type bonds are oxide or silica, clay, silicon oxynitride, and silicon nitride, as well as self-bonded.

A process for joining high-temperature-resistant silicon carbide structural parts that have customized thermomechanical properties has been developed, and the materials include SiC-based ceramics and composites reinforced by different fibers.

The method begins with the application of a carbonaceous mixture to the joints. The mixture is cured at a temperature between 90°C and 110°C. The joints are then locally infiltrated with molten silicon or with alloys of silicon and refractory metals. The molten metal reacts with the carbon in the joint to form silicon carbide and quantities of silicon and refractory disilicide phases that can be tailored by choosing the appropriate reactants.

In mechanical tests, the joints were found to retain their strength at temperatures from ambient to 1370°C. The technique can also be used in the repair of such parts.

Uses and Applications

SiC is used in the manufacture of grinding wheels and coated abrasives. Large tonnages are used in cutting granite with wire saws and as a metallurgical additive in the foundry and steel industries.

Other uses are in the refractory and structural ceramic industries. As an abrasive, silicon carbide is best used either on very hard materials, such as cemented carbide, granite, and glass, or on soft materials, such as wood, leather, plastics, and rubber.

Refrax Silicon Carbide and KT

The first material is bonded with silicon nitride. It is used for hot-spray nozzles, heat-resistant parts, and for lining electrolytic cells for smelting aluminum. Silicon carbide KT is molded without a binder. It has 96.5% SiC with about 2.5% silica. The specific gravity is about 3.1, and it is impermeable to gases. It is made in rods, tubes, and molded shapes.

Silicon Carbide Foam

This is a lightweight material made of self-bonded silicon carbide foamed into shapes. It is inert to hot chemicals and can be machined.

Silicon Carbide Crystals

These are used for semiconductors at temperatures above 343°C. As the cathode of electronic tubes instead of a hot-wire cathode, the crystals take less power and need no warm-up.

Silicon Carbide Fibers

SiC fiber is one of the most important fibers for high-temperature use. It has high strength and modulus and will withstand temperatures even under oxidizing conditions up to 1800°C, although the fibers show some deterioration in tensile strength and modulus properties at temperatures above 1200°C. It has advantages over carbon fibers for some uses, having greater resistance to oxidation at high temperatures, superior compressive strength, and greater electrical resistance.

There are two forms of SiC fibers. One consists of a pyrolytic deposit (chemical vapor deposition) of SiC on an electrically conductive usually carbon, continuous filament. Fiber diameter is about 140 μm. This technology has been used to make filaments with both graded and layered structures, including surface layers of carbon, which provide a toughness-enhancing parting layer in composites with a brittle matrix (e.g., silicon nitride).

The other forms of filamentary SiC are fibers that are extruded from sinterable SiC powder and allowed to sinter during free fall from the extruder. Fibers produced to date are 0.13–0.25 mm in diameter.

There are two commercial processes for making continuous silicon carbide fibers: (1) by coating silicon carbide on either a tungsten or a carbon filament by vapor deposition to produce a large filament (100–150 μm in diameter), or (2) by melt spinning an organic polymer containing silicon atoms as a precursor fiber followed by heating at an elevated temperature to produce a small filament (10–30 μm in diameter). Fibers from the two processes differ considerably from each other but both are used commercially.

Improved composites of SiC fibers in Si/SiC matrices have been invented for use in applications in which there are requirements for materials that can resist oxidation at high temperatures in the presence of air and steam. Such applications are likely to include advanced aircraft engines and gas turbines.

The need for improved composites arises: Although both the matrix and fiber components of older SiC/(Si/SiC) composites generally exhibit acceptably high resistance to oxidation, these composites become increasingly vulnerable to oxidation and consequent embrittlement whenever mechanical or thermomechanical loads become large enough to crack the matrices. Even the narrowest cracks become pathways for the diffusion of oxygen.

Typically, to impart toughness to an SiC/(Si/SiC) composite, the SiC fibers are coated with a material that yields a high stress to allow some slippage between the fibers and matrix. If oxygen infiltrates through the cracks to the fiber coatings, then, at high temperature, the oxygen reacts with the coatings (and eventually with the fibers), causing undesired local bonding between fibers in the matrix and consequent loss of toughness.

S

The improved composites incorporate matrix additives and fiber coatings that retard the infiltration of oxygen by reacting with oxygen in such a way as to seal cracks and fiber/matrix interfaces at high temperatures. These matrix additives and coating materials contain glass-forming elements—for example, boron and germanium.

Boron is particularly suitable for use in fiber coatings because it can react with oxygen to form boron oxide, which can, in turn, interact with the silica formed by oxidation of the matrix and fiber materials to produce borosilicate glasses. Boron and SiB_6 may prove to be the coating materials of choice because they do not introduce any elements beyond those needed to form borosilicate glasses.

One way to fabricate an SiC/(Si/SiC) composite object is to first make a preform of silicon carbide fibers interspersed with a mixture of silicon carbide and carbon particles, then infiltrate the preform with molten silicon. Boron can be incorporated by chemical vapor deposition onto the fibers prior to making the preform.

Boron and germanium can be incorporated into the matrix by adding these elements to either the matrix or the molten silicon infiltrant. Inasmuch as the solubility of boron and silicon is limited, it may be necessary to add the boron via the preform in a typical case.

Silicon Carbide Platelets

Single crystals of α-phase hexagonal crystal structure and four size ranges currently are produced: −100, + 200 mesh (100–300 μm in diameter, 5–15 μm thick); −200, + 325 mesh (50–150 μm in diameter, 1–10 μm thick); +325 mesh (5–70 mm in diameter, 0.5–5 mm thick); and −400 mesh (3–30 μm in diameter, 0.5–3 μm thick). The finest size is a research product, and additional development work has been conducted to produce an even smaller diameter platelet in the 0.5 μm range, which would be an ideal reinforcement material for ceramic-matrix composites.

In addition to reinforcing ceramics, silicon carbide platelets also are used to increase the strength, wear resistance, and thermal shock performance of aluminum matrices, and to enhance the properties of polymeric matrices. Because platelets are very free flowing, they can be processed in the same manner as particulates.

Silicon Carbide Whiskers

Whiskers as small as 7 μm in diameter can be made by a number of different processes. Although these whiskers have the disadvantage in some applications of not being in continuous filament form, it can be made with higher tensile strength and modulus values than continuous silicon carbide filaments.

Silicon carbide whiskers are single crystals of either α- or β-phase crystal structure. The SiC whiskers tend to exhibit a hexagonal, triangular, or rounded cross-section and may contain stacking faults.

SiC whiskers can be fabricated by the reaction of silicon and carbon to form a gaseous species that can be transported and reacted in the vapor phase. This type of formation is referred to as a vapor–solid reaction.

The reactions occur at temperatures greater than 1400°C and in an inert or nonoxidizing atmosphere. In addition, a catalyst is added to assure the formation of whiskers rather than particulate during the reaction.

Although SiC whiskers can be coated with several different materials, such as carbon, to enhance their performance, the as-produced SiC whiskers generally contain a 5–30 SiO_2 coating, which forms during synthesis.

SiC whiskers are added to a variety of matrices to increase the toughness and high-temperature strength of these materials. The elastic modulus for SiC whiskers is 400–500 GPa and the tensile strength ranges from 1 to 5 GPa. A variety of ceramic matrices, such as aluminum oxide, silicon nitride, molybdenum disilicide, aluminum nitride, mullite, cordierite, and glass ceramics, are combined with SiC whiskers to increase the overall mechanical properties of the resulting composite. For example, the wear resistance, toughness, and thermal shock of aluminum oxide is increased by the addition of SiC whiskers. The resulting composite has been used for such applications as high-performance cutting tool inserts. The addition of SiC whiskers to an alumina matrix can double the fracture toughness of the resulting composite, depending on whisker content and processing conditions.

SiC whiskers also can be combined with metals to increase the high-temperature strength of a material as well as provide a comparable substitution for heavier traditional materials, such as steel. Metal-matrix composites (MMCs) are being tested for such applications as piston ring grooves, cylinder block liners, brake calipers, and aerospace components. MMCs can be fabricated by infiltrating an SiC whisker preform with aluminum or by the addition of SiC whiskers to molten aluminum.

Polymer-matrix composites combine the strength and impact resistance of polymers with the thermal conductivity, fatigue, and wear resistance of the whiskers. Whisker-reinforced polymers have strong potential to replace traditional plastics in automotive, aerospace, and recreational applications.

Silicon Cast Iron

This is an acid-resistant cast iron containing a high percentage of silicon. When the amount of silicon and cast iron is above 10%, there is a notable increase in corrosion and acid resistance. The acid resistance is obtained from the compound Fe_3Si, which contains 14.5% silicon. The usual amount of silicon in acid-resistant castings is from 12% to 15%. The alloy casts well but is hard and cannot be machined. These castings usually contain 0.75%–0.85% carbon.

A 14%–14.5% silicon iron has a silvery-white structure and is resistant to hot sulfuric acid, nitric acid, and organic acids. Silicon irons are also very wear resistant and are valued for pump parts and for parts for chemical machinery.

Silicon Copper

An alloy of silicon and copper used for adding silicon to copper, brass, or bronze, silicon copper is also employed as a deoxidizer of copper and for making hard copper. Silicon alloys in almost any proportion with copper and is the best commercial partner of copper. A 50-50 alloy of silicon and copper is hard and extremely brittle and black in color. A 10% silicon and 90% copper alloy is as brittle as glass; in this proportion, silicon copper is used for making the addition to molten copper to produce hard, sound copper alloy castings of high strength. The resulting alloy is easy to cast in the foundry and does not dross. Silicate-copper grades in 5%, 10%, 15%, and 20% silicon are also marketed. A 10% silicon-copper melts at 816°C; a 20% alloy melts at 623°C.

Silicon Halides

Silicate tetrachloride, $SiCl_4$, is perhaps the best-known monomeric covalent compound of silicon. It is readily available commercially. It can be prepared by chlorinating elementary silicon, or by the action of chlorine on a mixture of silica with finely divided carbon, or by the chlorination of silicon carbide. It is a volatile liquid that fumes in moist air and hydrolyzes rapidly to silica and hydrochloric acid.

Pure silica with very high surface area, produced by this method, is used as a reinforcing filler (white carbon black) in silicone rubber and as a thickening agent in organic solutions. Silicate tetrachloride reacts readily with alcohols and glycols, for example, to form the corresponding ethers, which may also be considered to be esters of silicic acid.

Silicon Manganese

An alloy employed for adding manganese to steel, and also as a deoxidizer and scavenger of steel, silicon manganese usually contains 65%–70% manganese and 12%–25% silicon. It is graded according to the amount of carbon, generally 1%, 2%, and 2.5%. For making steels low in carbon and high in manganese, silicomanganese is more suitable than ferromanganese. A reverse alloy, called manganese–silicon, contains 73%–78% silicon and 20%–25% manganese, with 1.5% max iron and 0.25% max carbon. It is used for adding manganese and silicon to metals without the addition of iron. Still another alloy is called ferromanganese–silicon, containing 20%–25% manganese, about 50% silicon, and 25%–30% iron, with only about 0.50% or less carbon. This alloy has a low melting point, giving ready solubility in the metal.

Silicon Nitride

Silicon nitride (Si_3N_4) disassociates in air at 1800°C and at 1850°C under 1 atm N_2. There are two crystal structures: α (1400°C) and β (1400°C–1800°C), both hexagonal. Its hardness is approximately 2200 on the Knoop K100 scale, and it exhibits excellent corrosion and oxidation resistance over a wide temperature range. Typical applications are molten-metal-contacting parts, wear surfaces, special electrical insulator components, and metal forming dies. It is under evaluation for gas turbines and heat engine components as well as antifriction bearing members.

Processing

Pure silicon nitride powders are produced by several processes, including direct nitridation of silicon, carbothermal reduction—C + SiO_2 + N_2 yields Si_3N_4 (gas atmosphere)—and chemical vapor deposition—$3SiH_4 + 4NH_3$ yields $Si_3N_4 + 12H_2$. Reacting SiO_2 with ammonia or silanes with ammonia will also produce silicon nitride powders. It is found that the highest purity powders come from gas-phase reactions.

Types

Sinterable/Hot Pressed/Hot Isostatically Pressed Silicon Nitride

These types are SSN, HPSN, and HIPSN, respectively. They are used mainly in the higher performance applications. Powdered additives, known as sintering aids, are blended with the pure Si_3N_4 powder and allow densification to proceed via the liquid state. Pore-free bodies can be so produced by sintering or hot pressing. Of course, the properties of the material and dense pieces are dependent on the chemical nature of the sintering aids employed.

Sinterable silicon nitrides are a more recent innovation and allow more flexibility in shape fabrication than does HPSN. Highly complex shapes can be die-pressed or isostatically pressed. Densification

can be performed by either sintering or hot isostatic pressing (HIP). Properties of the dense piece are dependent on the additives, but in general the strength below 1400°C, as well as oxidation resistance of HPSN and SSN, far exceeds those properties for reaction-bonded silicon nitride (RBSN).

Reaction-Bonded Silicon Nitride

More common today is RBSN. Silicon powder is pressed, extruded, or cast into shape, then carefully nitrided in a N_2 atmosphere at 1100°C–1400°C, so as to prevent an exothermic reaction, which might melt the pure silicon.

The properties of RBSN are usually lower than those of HPSN or SSN, due mainly to the fact that bodies fabricated in this manner only reach 85% of the theoretical density of silicon nitride and no secondary phase between grains is present.

Silicon Nitride Fibers

Si_3N_4 fibers have been prepared by reaction between silicon oxide and nitrogen in the presence of a reducing agent in an electrical resistance furnace at 1400°C. Silicon nitride short fibers are used in composites for specialty electrical parts, and aircraft parts, and radomes (microwave windows). Silicon nitride whiskers have also been grown as a result of the chemical reaction between nitrogen and a mixture of silicon and silica.

Silicon Oxides

Silicon dioxide is perhaps best known as one of its crystalline modifications known as quartz, colorless crystals of which are also known as rhinestones and Glens Falls diamonds. Purple or lavender-colored quartz is called amethyst, the pink variety is rose quartz, and the yellow type is citrine.

Because rock crystal has been collected and admired for thousands of years, large and perfectly formed natural crystals of quartz are now very rare. With the growth of radio broadcasting and the electronics industry, piezoelectric crystals cut from perfect specimens of quartz have been used in increasing quantities, to the point of scarcity of natural crystals. As the supply diminished, considerable effort was devoted to the problem of growing crystals of quartz by artificial means. Some success has been achieved by growing the crystals hydrothermally from a solution of silica glass in water containing an alkali or a fluoride.

A number of natural noncrystalline varieties of silicon dioxide are also known, such as the hydrated silica known as opal and the dense unhydrated variety known as flint. Onyx and agate represent still other semiprecious forms.

Silicon Steel

All grades of steel contain some silicon and most of them contain 0.10% to 0.35% as a residual of the silicon used as a deoxidizer. But 3% to 5% silicon is sometimes added to increase the magnetic permeability, and larger amounts are added to obtain wear-resisting or acid-resisting properties. Silicon deoxidizes steel, and up to 1.75% increases the elastic limit and impact resistance without loss of ductility. Silicon steels within this range are used for structural purposes and for springs, giving a tensile strength of about 517 MPa and 25% elongation.

The structural silicon steels are ordinarily silicon–manganese steel, with the manganese above 0.50%. Low-carbon steels used as structural steels are made by careful control of carbon, manganese, and silicon and with special mill heat treatment.

The value of silicon steel as a transformer steel occurs where silicon increases the electrical resistivity and also decreases the hysteresis loss, making silicon steel valuable for magnetic circuits where alternating current is used.

Silicone (SI) Resins

Silicone resins are synthetic materials capable of cross-linking or polymerizing to form films, coatings, or molded shapes with outstanding resistance to high temperatures.

They are a group of resin-like materials in which silicon takes the place of the carbon of the organic synthetic resins. Silicon is quadrivalent like carbon. But, while the carbon also has a valence of 2, silicon has only one valence of 4, and the angles of molecular formation are different. The two elements also differ in electronegativity, and silicon is an amphoteric element, with both acid and basic properties. The molecular formation of the silicones varies from that of the common plastics, and they are designated as inorganic plastics as distinct from the organic plastics made with carbon.

In the long-chain organic synthetic resins, the carbon atoms repeat themselves, attaching on two sides to other carbon atoms, while in the silicones the silicon atom alternates with an oxygen atom so that the silicon atoms are not tied to each other. The simple silane formed by silicon and hydrogen corresponding to methane, CH_4, is also a gas, as is methane, and has the formula SiH_4. But, in general, the silicones do not have the SiH radicals, but contain CH radicals as in the organic plastics.

Composition

Silicones are made by first reducing quartz rock (SiO_2) to elemental silicon in an electric furnace, then preparing organochlorosilane monomers ($RSiCl_3$, R_2SiCl_2, R_3SiCl) from the silicon by one of several different methods. The monomers are then hydrolyzed into cross-linked polymers (resins) whose thermal stability is based on the same silicon–oxygen–silicon bonds found in quartz and glass. The properties of these resins will depend on the amount of cross-linking and on the type of organic groups (**R**) included in the original monomer. Methyl, vinyl, and phenyl groups are among those used in making silicone resins.

Properties

Silicone resins have, in general, more heat resistance than organic resins, have higher dielectric strength, and are highly water resistant. Like organic plastics, they can be compounded with plasticizers, fillers, and pigments. They are usually cured by heat. Because of the quartzlike structure, molded parts have exceptional thermal stability. Their maximum continuous-use service temperature is about 260°C. Special grades exceed this and go as high as 371°C–482°C. Their heat-deflection temperature for 1.8 MPa is 482°C. Their moisture absorption is low, and resistance to petroleum products and acids is good. Nonreinforced silicones have only moderate tensile and impact strength, but fillers and reinforcements provide substantial improvement. Because silicones are high in cost, they are premium plastics and are generally limited to critical or high-performance products such as high-temperature components in the aircraft, aerospace, and electronics fields.

Common characteristics shared by most silicone resins are outstanding thermal stability, water repellency, general inertness, and electrical insulating properties. These properties, among others, have resulted in the use of silicone resins in the following fields:

1. Laminating (reinforced plastics), molding, foaming, and potting resins
2. Impregnating cloth coating, and wire varnishes for Class II (high-performance) electric motors and generators
3. Protective coating resins
4. Water repellents for textiles, leather, and masonry
5. Release agents for baking pans

Laminating Resins

Laminates made from silicone resin and glass cloth are lightweight, strong, heat-resistant materials used for both mechanical and dielectric applications. Silicone-glass laminates have low moisture absorption and low dielectric losses, and retain most of their physical and electrical properties for long periods at 260°C.

Laminates may be separated into three groups according to the method of manufacture: high-pressure, low-pressure, and wet lay-up.

High Pressure

Silicone-glass laminates (industrial thermosetting laminates) have excellent dielectric strength and arc resistance, and are normally used as dielectric materials. To prepare high-pressure laminates, glass cloth is first impregnated by passing it through a solvent solution of silicone resin. The resin is dried of solvent and precured by passing the fabric through a curing tower. Laminates are prepared by laying up the proper number of plies of preimpregnated glass cloth and pressing them together at about 6.8 MPa and 177°C for about 1 h. They are then oven-cured at increasing temperatures, with the final cure at about 249°C. The resulting laminates can be drilled, sawed, punched, or ground into insulating components of almost any desired shape. Typical applications include transformer spacer bars and barrier sheets, slot sticks, panel boards, and coil bobbins.

Low Pressure

Silicone-glass laminates made by low-pressure reinforced plastics molding methods usually provide optimum flexural strength, for example, about 272 MPa even after heat aging. They are used for mechanical applications such as radomes, aircraft duct work, thermal barriers, covers for high-frequency equipment, and high-temperature missile parts. In making low-pressure laminates, glass cloth is first impregnated and laid up as described earlier. Since the required laminating pressure can be as low as 0.068 MPa, matched-metal-molding and bag-molding techniques can be used in laminating, making possible greater variety in laminated shapes. Laminates should be after-cured as already described.

Wet Lay-Up

Silicone-glass laminates can now be produced by wet lay-up techniques because of the solventless silicone resins recently developed. Such laminates can be cured without any pressure except that needed to hold the laminate together. This technique should prove especially useful in making prototype laminates, and in short production runs where expensive dies are not justified. Laminates are prepared by wrapping glass cloth around a form and spreading on catalyzed resin, repeating this process until the desired thickness is obtained. The laminate surface is then wrapped with a transparent film, and air bubbles are worked out. Laminates are cured at 149°C and, after the transparent film is removed, postcured at 204°C.

Molding Compounds

Silicone molding compounds consist of silicone resin, inorganic filler, and catalyst, which, when molded under heat and pressure, form thermosetting plastic parts. Molded parts retain exceptional physical and electrical properties at high temperatures, resist water and chemicals, and do not support combustion. Specification MIL-M-14E recognizes two distinctly different types of silicone molding compounds: type MSI-30 (glass-fiber filled) and type MSG (mineral filled). Requirements for both are as follows.

Type MSI-30

Glass-filled molded parts have high strengths, which become greater the longer the fiber length of the glass filler. Where simple parts can be compression-molded from a continuous fiber length compound, strengths will be approximately twice as great. Properly cured glass-fiber-filled molded parts can be exposed continuously to temperatures as high as 371°C and intermittently as high as 538°C. The heat-distortion temperature after postcure is 482°C. Because of the flow characteristics of these fiber-filled compounds, their use is generally limited to compression molding.

Type MSG

Mineral-filled compounds are free-flowing granular materials. They are suitable for transfer molding and can be used in automatic preforming at molding machines. They are used to make complex parts that retain their physical and electrical properties at temperatures above 260°C but that do not require impact strength. Silicon molding compounds are excellent materials for making Class H electrical insulating components such as coil forms, slot wedges, and connector plugs. They have many potential applications in the aircraft, missile, and electronic industries.

Foaming Powders

Silicone foaming powders are completely formulated, ready-to-use materials that produce heat-stable, nonflammable, low density silicone foam structures when heated. Densities vary from 162 to 288 kg/m³, and compressive strength from 0.68 to 2.23 MPa. Electrical properties are excellent, and water absorption after 24 h immersion is only 2.5%. The maximum continuous operating temperature of these forms is about 343°C.

Foams are prepared by heating the powders to between 149°C and 177°C for about 2 h. The powder can be foamed in place, or foamed into blocks and shaped into woodworking tools. Foams are normally after-cured to develop strength, but can often be cured in service.

Silicone foams are being used in the aircraft and missile industries to provide lightweight thermal insulation and to protect delicate electronic equipment from thermal shock. They can also be bonded to silicone-glass laminates or metals to form heat- and moisture-resistant sandwich structures.

Potting Resins

Solventless silicone resins can be used for impregnating, encapsulating, and potting of electrical and electronic units. Properly catalyzed, filled, and cured, they form tough materials with good physical and electrical properties, and will withstand continuous temperatures of 204°C and intermittent temperatures above 260°C.

Typical physical properties of cured resins include flexural strength of 48.6 MPa, compressive strength of 117 MPa, and water absorption of 0.04%.

Before use, resins are catalyzed with dicumyl peroxide or ditertiary butyl peroxide. Resins can be simply poured in place, although vacuum impregnation is suggested where fine voids must be filled. Fillers such as glass beads or silica flour are added to extend the resin; their use increases physical strength and thermal conductivity but decreases electrical properties. The resin is polymerized by heating it to about 149°C, and postcured, first at 204°C, then at the intended operating temperature if higher.

Electrical Varnishes

Silicone varnishes (solvent solutions of silicone resins) has made possible the new high-temperature classes of insulation for electrical motors and generators. Silicone insulating varnishes will withstand continuous operating temperatures at 177°C or higher.

Electrical equipment that operates at higher temperatures makes possible motors, generators, and transformers that are much smaller and lighter, or equipment that delivers 25%–50% more power from the same size and still has a much longer service life.

The resinous silicone materials used in Class H electrical insulating systems include the following:

1. Silicone bonding varnish for glass-fiber-covered magnet wire
2. Silicone varnishes for impregnating and bonding glass cloth, mica, asbestos paper. Sheet installations made of these heat-resistant materials are used as slot liners for electric motors and as phase insulation.
3. Silicone dipping varnish that impregnates, bonds, and seals all insulating components into an integrated system.

Other silicone materials used in electrical equipment include silicone rubber lead wire, silicone-adhesive-backed glass tape, and temperature-resistant silicone bearing greases. Silicone insulated motors, generators, and transformers are now being produced by the leading electrical equipment manufacturers.

Uses

The wide range of structural variations of silicone resins makes it possible to tailor compositions for many kinds of applications. Low-molecular-weight silanes containing amino or other functional groups are used as treating or coupling agents for glass fiber and other reinforcements to cause unsaturated polyesters and other resins to adhere better.

The liquids, generally dimethyl silicones of relatively low molecular weight, have a low surface tension, great wetting power and lubricity for metals, and very small change in viscosity with temperature. They are used as hydraulic fluids; as antifoaming agents; as treating and waterproofing agents for leather, textiles, and masonry; and in cosmetic preparations. The greases are particularly desired for applications requiring effective lubrication at very high and at very low temperatures.

Silicone resins are used for coating applications in which thermal stability in the range 300°C–500°C is required. The dielectric properties of the polymers make them suitable for many electrical applications, particularly in electrical insulation that is exposed to high temperatures and as encapsulating materials for electronic devices.

Silicone enamels and paints are more resistant to chemicals than most organic plastics, and when pigmented with mineral pigments will withstand temperatures up to 538°C. For lubricants, the liquid silicones are compounded with graphite or metallic soaps and will operate between −46°C and 260°C. The silicone liquids are stable at their boiling points, between 399°C and 427°C, and have low vapor pressures, so that they are also used for hydraulic fluids and heat-transfer media. Silicone oils, used for lubrication and as insulating and hydraulic fluids, are methyl silicone polymers. They retain a stable viscosity at both high and low temperatures. As hydraulic fluids, they permit smaller systems to operate at higher temperatures. In general, silicone oils are poor lubricants compared with petroleum oils, but they are used for high temperatures, 150°C–200°C, at low speeds and low loads.

Silicone resins are blended with alkyd resins for use in outside paints, usually modified with a drying oil. Silicone–alkyd resins are also used for baked finishes, combining the adhesiveness and flexibility of the alkyd with the heat resistance of the silicone. A phenyl ethyl silicone is used for impregnating glass-fiber cloth for electrical insulation and it has about double the insulating value of ordinary varnished cloth.

Silicone Rubber

Silicone rubbers are a group of synthetic elastomers noted for their (1) resilience over a very wide temperature range, (2) outstanding resistance to ozone and weathering, and (3) excellent electrical properties.

Composition

The basic silicone elastomer is a dimethyl polysiloxane. It consists of long chains of alternating silicon and oxygen atoms, with two methyl side chains attached to each silicon atom. By replacing a part of these methyl groups with other side chains, polymers with various desirable properties can be obtained. For example, where flexibility at temperatures lower than −57°C is desired, a polymer with about 10% of the metal side chains replaced by phenyl groups will provide compounds with brittle points below −101°C. Side chain modification can also be used to produce elastomers with lower compression set, to increase resistance to fuels, oils, or solvents, or to permit vulcanization at room temperature.

Curing, or vulcanization, is the process of introducing cross-links at intervals between the long chains of the polymer. Silicone rubbers are usually cross-linked by free radical-generating curing agents, such as benzoylperoxide, which are activated by heat, or the cross-linking can be accomplished by high-energy radiation beams. Room temperature vulcanized compounds are cross-linked by the condensation reaction resulting from the action of metal-organic salts, such as zinc and tin octoates. Pure polymers upon cross-linking change from viscous liquids into elastic gels with very low tensile strength. To attain satisfactory tensile strength, reinforcing agents are necessary. Synthetic and natural silicas and metallic oxides are commonly used for this purpose. In addition to the vulcanizing agents and reinforcing fillers described, other additives may be incorporated into silicone compounds to pigment the stock, to improve processing, or to reduce the compression set of certain types of silicone gum.

Ordinary silicone rubber has the molecular group $H \cdot CH_2 \cdot Si \cdot CH_2 \cdot H$ in a repeating chain connected with oxygen linkages, but in the nitrile–silicone rubber one of the end of hydrogens of every fourth group in the repeating chain is replaced by a C:N radical. These polar nitrile groups give a little affinity for oils, and the rubber does not swell with oils and solvents. They retain strength and flexibility at temperatures from −73°C to above 260°C, and is used for such

products as gaskets and chemical hose. As lubricants, silicones retain a nearly constant viscosity at varying temperatures. Fluorosilicones have fluoroalkyl groups substituted for some of the methyl groups attached to the siloxane polymer of dimethyl silicone. They are fluids, greases, and rubbers, incompatible with petroleum oils and insoluble in most solvents. The greases are the fluids thickened with lithium soap or with a mineral filler.

Types

Silicone-rubber compounds can be conveniently grouped into several major types according to characteristic properties. According to one such classification system, the types are (1) general purpose, (2) extremely low temperature, (3) extremely high temperature, (4) low compression set, (5) high strength, (6) fluid resistant, (7) electrical, and (8) room temperature vulcanizing rubbers.

1. General-purpose compounds are available in Shore A hardnesses from 30 to 90, tensile strengths of 4.8–8.24 MPa, and ultimate elongations of 100%–500%. The service temperature range extends from −55°C to 260°C, and they have good resistance to heat and oils, along with good electrical properties. Many of these compounds contain semireinforcing or extending fillers to lower their cost.
2. Extremely low temperature compounds have brittle points near −118°C and are quite flexible at −84°C to −90°C. Their physical properties are usually about the same as those of the general-purpose stocks, with some reduction in oil resistance.
3. Extremely high temperature compounds are considered serviceable for over 70 h at 343°C and will withstand brief exposures at higher temperatures; for example, 4–5 h at 371°C and 10–15 min at 399°C. In comparison, general-purpose compounds are limited to about 260°C for continuous service in 316°C for intermittent service.
4. Low-compression-set compounds provide typical values of 10%–20% compression set after 22 h at 149°C. They have improved resistance to petroleum oils and various hydraulic fluids, and are particularly suitable for use in O-rings and gaskets.
5. High-strength compounds, in Shore A hardnesses of 25–70, provide tensile strengths from 0.82 to over 136 MPa and elongations from 400% to 700%, with tear strengths from 26,790 to 58,045 g/cm. Compounds of this class may operate over a service temperature range from −90°C to 316°C.
6. Excellent resistance to a wide range of fuels, lubricants, and hydraulic fluids is offered by compounds based on a silicone polymer with 50% of its side methyl groups substituted by trifluoropropyl groups. Physical properties are similar to properties of other types of silicone compounds. However, service temperature range is somewhat limited. Its low-temperature properties are about the same as those of the dimethyl polymer, with the brittle point around −68°C and an upper service temperature of around 260°C.
7. In general, silicone-rubber compounds have excellent electrical properties, which, along with their resistance to high temperatures, make them suitable for many electrical applications. With proper compounding, dielectric constant can be easily vary from about 2.7 to 5.0 or higher, while the power factor can be varied from 0.0005 up to 0.1 or higher. The volume resistivity of a typical silicone

compound will be in the range of 10^{14} to 10^{16} Ω cm and its dielectric strength will be about 450–550 V/mil thickness (measured on a slab 2.0 mm). Resistance to corona is excellent and water absorption is low. In most cases, excellent electrical properties are retained over a wide temperature and frequency range. Compounds can also be prepared with very low resistivity, as low as about 10 Ω cm, for special applications. Insulated tapes for cable-wrapping applications can be prepared from electrical-grade compounds with a partial cure or from a completely cured self-adhering silicone compound.

8. Room-temperature vulcanizing silicone rubbers are available to provide most of the performance characteristics of silicone rubbers in compounds that cure at room temperature.

In addition to having excellent heat resistance, silicone rubbers retain their properties to a much greater extent at high temperatures than do most organic rubbers. For example, a silicone compound with a tensile strength at temperature of 1.36 MPa will have a tensile strength at 316°C of 0.48 MPa or over. Most organic rubbers, although their initial properties are much higher, will be virtually useless at 260°C (except for the fluoroelastomers). In applications where a silicone rubber part operates in low oxygen atmosphere, such as sealing on high-altitude aircraft, its heat resistance will be still further improved.

In using silicone rubber at high temperatures, care must be taken to prevent reversion or depolymerization, which may occur where a part is required to operate in an enclosed environment. Here again, where this problem cannot be eliminated by the design engineer, the silicone-rubber fabricator can produce compounds with relatively high resistance to reversion.

Fabrication and Uses

In general, silicone rubbers may be handled on standard rubber-processing equipment. Their fabrication differs from that of organic rubbers chiefly in that uncured silicone compounds are softer and tackier and have much lower green strength. Also, an oven postcure in a circulating air oven is often required after vulcanization to obtain optimum properties. Silicone rubbers can be extruded, molded, calendered, sponged, and foamed. Since compounding of the rubber stock determines to a great extent the processing characteristics of the material, the fabricator should be consulted before the material is specified, to determine whether compromises are necessary to obtain the best combination of physical properties and the most desirable shape. Silicone-rubber compounds can also be applied to fabrics by calender coating, knife spreading, or solvent dispersion techniques.

For certain applications, very soft or low durometer materials are required. Suitable for such applications are low durometer solid silicone rubbers, closed or open cell expanded silicone rubbers (designated here as sponge and foam, respectively), and fibrous silicone rubber.

Silicone-rubber sponge is available in molded sheets, extrusions, and simple molded shapes. As in the case of solid silicone rubber, improved resistance to fluids or abrasion can be obtained by bonding molded or extruded sponge to a fabric or plastic cover. Silicone foam rubber can be fabricated in heavy cross-sections and complex shapes, and can be foamed and vulcanized either at ambient or elevated temperature. It is suitable for use where an extremely soft, low-density silicone material is required. Like sponged, it can be bonded to fabrics on plastics.

For some applications a solid, low durometer material has advantages over sponge or foam. For example, it should probably be specified for gaskets or seals where the low compression set of foam, the compression–deflection characteristics of sponge, and the higher tensile and tear strength of solid silicone rubbers must be combined.

The last highly compressible material, fibrous silicone rubber, consists of hollow rubber fibers sprayed in a random manner and bonded into a low-density porous mat. Its properties include excellent compression set combined with good tear and tensile strength, and very high porosity. As manufactured at present, it is serviceable from −55°C to over 260°C, and is available in mats 3.2 mm thick and 228 mm wide.

Silicone rubbers are most widely used in the aircraft, electrical, and automotive industries, although their unique properties have created many other applications. Specific examples would include seals for aircraft canopies or access doors, insulation for wire and cable, dielectric encapsulation of electronic equipment, and gaskets or O-rings for use in aircraft automobile engines. An example of another field in which they are useful is the manufacture of stoppers for pharmaceutical vials, since silicone rubbers are tasteless, odorless, and nontoxic.

Siliconizing

(1) Diffusing silicon into solid metal, usually low carbon steels, at an elevated temperature in order to improve corrosion or wear resistance. (2) A process for developing a silicon rich layer on steel by packing a component in ferrosilicon and heating at about 900°C for a few hours. The layer has good wear and oxidation resistance.

Silk

Silk is the fibrous material in which the silkworm, or larva of the moth, envelops itself before passing into the chrysalis state. Silk is closely allied to cellulose and resembles wool in structure, but unlike wool it contains no sulfur. The natural silk is covered with a wax or silk glue, which is removed by scouring in manufacture, leaving the glossy fibroin, or raw-silk fiber. The fibroin consists largely of the amino acid alanine, which can be synthesized from pyruvic acid. Silk fabrics are used mostly for fine garments but are also valued for military powder bags because they burn without a sooty residue.

The fiber is unwound from the cocoon and spun into threads. The chief silk producing countries are China, Japan, India, Italy, and France. Floss silk is a soft silk yarn practically without twist, or is the loose waste silk produced by the worm when beginning to spin its cocoon.

Satin is a heavy silk fabric with a close twill weave in which the fine warp threads appear on the surface and the weft threads are covered by the peculiar twill. Common satin is of eight-leaf twill, the weft intersecting and binding down the warp at every eighth pick, but 16–20 twirls are also made. In the best satins, a fine quality of silk is used.

Silky Fracture

A metal fracture in which the broken metal surface has a fine texture, usually dull in appearance. Characteristic of tough and strong metals. Contrast with *crystalline fracture* and *granular fracture*.

Silver and Alloys

A white metal (symbol Ag), silver is very malleable and ductile, and is classed with the precious metals. It occurs in the native state, and also combined with sulfur and chlorine. Copper, lead, and zinc ores frequently contain silver; about 70% of the production of silver is a by-product of the refining of these metals.

Silver is the whitest of all the metals and takes a high polish, but easily tarnishes in the air because of the formation of a silver sulfide. It has the highest electrical and heat conductivity: 108% IACS relative to 100% for the copper standard and about 422 W/(m K), respectively. Cold work reduces conductivity slightly. The specific gravity is 10.7, and a melting point is 962°C. When heated above the boiling point (2163°C), it passes off as a green vapor. It is soluble in nitric acid and in hot sulfuric acid. The tensile strength of cast silver is 282 MPa, with Brinell hardness 59. The metal is marketed on a troy-ounce value.

Pure silver has the highest thermal and electrical conductivity of any metal, as well as the highest optical reflectivity. It is the most ductile and most malleable of any metal next to gold. Silver can be hammered into sheet 0.01 mm thick or drawn out in wire so fine 120 m would weigh only 1 g. Classified as one of the most corrosion-resistant metals, silver, under ordinary conditions, will not be affected by caustics or corrosive elements, unless hydrogen sulfide is present, causing silver sulfide to form. Silver will dissolve rapidly in nitric acid and more slowly in hot concentrated sulfuric acid. Unless oxidizing agents are present, the action of diluted or cold solutions of sulfuric acid is negligible. Organic acids generally do not attack the metal and caustic alkalies have but a slight effect on pure silver.

Although silver tarnishes quickly in the presence of sulfur and sulfur-bearing compounds, it oxidizes slowly in air and the oxide decomposes at a relatively low temperature.

Classification

Silver is classified by grades in parts per thousand based on the silver content (impurities are reported in parts per hundred). Commercial grades are fine silver and high fine silver. As ordinarily supplied, fine silver contains at least 999.0 parts silver per thousand, and they go as high as 999.3 parts per thousand. Any of the common base metals may be present, although copper is usually the major impurity. Any silver of higher purity than commercial fine contains its purity in its description, that is, 999.7 high fine silver. The purest silver obtainable in quantity is 999.9 plus; the impurities are less than 0.01 part per thousand. Fine silver may also contain small percentages of oxygen or hydrogen; deoxidized silver is available for applications where these elements may be a detriment.

Fabricability

Silver can be cold-worked, extruded, rolled, swaged, and drawn. It can be cold-rolled or cold-drawn drastically between anneals, and can be annealed at relatively low temperatures. To prevent oxidation when casting by conventional methods, silver should be protected by a layer of charcoal or by melting under neutral or reducing gas. Deoxidation by adding lithium or phosphorus can be obtained leaving a residual content of 0.01% max. The excellent ductility of silver makes it readily workable hot or cold.

Molten silver will absorb approximately 20 times its own volume of oxygen. Most of this oxygen is given up when the silver solidifies in cooling, but care should be taken in melting and casting because any oxygen left in the cast bars will cause cracking when they are fabricated and the castings may have blowholes.

Galling, seizing of the tool, and surface tearing are problems encountered when sheeting fine silver. This can be somewhat alleviated by using material cold-worked as much as possible.

Joining

Fine silver can be soldered without difficulty using tin–lead solders. Boron–silver filler metal can be used in brazing, and welding can be done by resistance methods and by atomic hydrogen or inert-gas shielded arc processes. A range of 204°C–427°C is recommended for annealing, with best strength and ductility achieved between 371°C and 427°C. Little additional softening occurs at higher temperatures, which may induce welding of adjacent surfaces. The lighter the gauge, the lower should be the annealing temperature.

Applications of Alloys and Compounds

Because silver is a very soft metal, it is not normally used industrially in a pure state, but is alloyed with a hardener usually copper. Sterling silver is a name given to a standard high-grade alloy containing a minimum of 925 parts in 1000 of silver. It is used for the best tableware, jewelry, and electrical contacts. This alloy of 7.5% copper work hardens and requires annealing between roll passes. Silver can also be hardened by alloying with other elements.

The standard types of commercial silver are fine silver, sterling silver, and coin silver. Fine silver is at least 99.9% pure and is used for plating, making chemicals, and for parts produced by powder metallurgy. Coin silver is usually an alloy of 90% silver and 10% copper, but when actually used for coins, the composition and weight of the coin are designated by law. Silver and gold are the only two metals that fulfill all the requirements for coinage. The so-called coins made from other metals are really official tokens, corresponding to paper money, and are not true coins. Coin silver has a Vickers hardness of 148 compared with a hardness of 76 for hard-rolled pure silver. It is also used for silverware, ornaments, plating, for alloying with gold, and for electric contacts. Silver is not an industrial metal in the ordinary sense. It derives its coinage value from its intrinsic aesthetic value for jewelry and plate, and in all civilized countries silver is a controlled metal.

Silver powder, 99.9% purity, for use in coatings, integrated circuits, and other electrical and electronic applications, is produced in several forms. Amorphous powder is made by chemical reduction and comes in particle sizes of 0.9–1.5 μm. Powder made electrolytically is in dendritic crystals with particle sizes from 10 to 200 μm. Atomized powder has spherical particles and may be as fine as 400 mesh. Silver-clad powder for electric contacts is a copper powder coated with silver to economize on silver. Silver flaky is in the form of laminar platelets and is particularly useful for conductive and reflective coatings and circuitry. The tiny flat plates are deposited in overlapping layers permitting a metal weight saving of as much as 30% without reduction of electrical properties.

Nickel-coated silver powder, for contacts and other parts made by powder metallurgy, comes in grades with 1/4%, 1/2%, 1%, and 2% nickel by weight.

The porous silver comes in sheets in standard porosity grades from 2 to 55 μm. It is used for chemical filtering.

Silver plating is sometimes done with a silver–tin alloy containing 20–40 parts silver and the remainder tin. It gives a plate having the appearance of silver but with better wear resistance. Silver plates have good reflectivity at high wavelengths, but reflectivity falls off at about 350 nm, and is zero at 3000, so that it is not used for heat reflectors.

Silver-clad sheet, made of a cheaper nonferrous sheet with a coating of silver rolled on, is used for food-processing equipment. It is resistant to organic acids but not to products containing sulfur. Silver-clad steel, used for machinery bearings, shims, and reflectors, is made with pure silver bonded to the billet of steel and then rolled.

For bearings, the silver is 0.025–0.889 cm thick, but for reflectors of silver is only 0.003–0.008 cm thick. Silver-clad stainless steel is stainless-steel sheet with a thin layer of silver rolled on one side for electrical conductivity.

Silver iodide is a pale-yellow powder of the composition AgI, best known for its use as a nucleating agent and for seeding rain clouds. Silver nitrate, formally known as lunar caustic, is a colorless, crystalline, poisonous, and corrosive material of the composition $AgNO_3$. It is used for silvering ring mirrors, for silver plating, in indelible inks, and medicine, and for making other silver chemicals. The high-purity material is made by dissolving silver and nitric acid, evaporating the solution, and crystallizing the nitrate, then redissolving the crystals in distilled water and recrystallizing. It is an active oxidizing agent. Silver chloride, AgCl, is a white granular powder used in silver-plating solutions. This salt of silver and other halogen compounds of silver, especially silver bromide, AgBr, are used for photographic plates and films. Silver chloride is used in the preparation of yellow glazes, purple of Cassius, and silver lusters. A yellowish silver luster is obtained by mixing silver chloride with three times its weight of clay and ochre and sufficient water to form a paste.

Silver chloride crystals in sizes up to 4.5 kg are grown synthetically. The crystals are cubic, and can be heated and pressed into sheets. The specific gravity is 5.56, index of refraction is 2.071, and melting point is 455°C. They are slightly soluble in water and soluble in alkalies. The crystals transmit more than 80% of the wavelengths from 50 to 200 μm.

Silver sulfide, Ag_2S, is a gray-black, heavy powder used for inlaying in metal work. It changes its crystal structure at about 179°C, with a drop in electrical resistivity, and is also used for self-resetting circuit breakers. Silver potassium cyanide, $KAg(CN)_2$, is a white, crystalline, poisonous solid used for silver-plating solutions. Silver tungstate, Ag_2WO_4, silver manganate, $AgMnO_4$, and other silver compounds are produced in purity grades for electronic and chemical uses.

Silver nitrate, $AgNO_3$, with a melting point of 212°C, and decomposes at 444°C and is soluble, corrosive, and poisonous. It is prepared by the action of nitric acid on metallic silver. Silver nitrate is the most convenient method of introducing silver into a glass; a solution of the compound is poured over the batch.

The photosensitive halides are used in photography, the cyanides are used in electroplating, and most of the minor silver salts are prepared from silver nitrate.

In advanced ceramic applications, silver is unsurpassed as a conductor of heat and electricity. Silver is used in conductive coatings for capacitors, printed wiring, and printed circuits on titanites, glass-bonded mica, steatite, alumina, porcelain, glass, and other ceramic bodies. These coatings are also used to metallize ceramic parts to serve as hermetically sealed enclosures, becoming integral sections of coils, transformers, semiconductors, and monolithic and integrated circuits.

Two types of conductive coatings can be used on ceramic parts; those that are fired on and those that are baked on or air dried. The fired-on type contains, in addition to silver powder, a finely divided low-melting glass powder, temporary organic binder, and liquid solvents in formulations with direct soldering properties and other suitable for electroplating, both possessing excellent adhesion and electrical conductivity. The baked-on and air-dry types contain, in addition to silver powder, a permanent organic binder and liquid solvents. These preparations have somewhat less adhesion, electrical conductivity, and solderability than the fired-on type, but can be electroplated if desired. The air-dry type is used when it is not desirable to subject the base material to elevated firing temperatures.

Any of the aforementioned silver compositions are available in a variety of vehicles suitable for application by squeegee, brushing, dipping, spraying, bonding wheel, roller coating, etc.

Firing temperatures for direct-solder silver preparations range from 677°C to 788°C. Silver compositions to be copper plated are fired at 1200°F–1250°F. The firing cycle used with these temperatures will vary from 10 min to 6 h, depending on the time required to equalize the temperature of the furnace charge.

A 62% Sn–36% Pb–2% Ag solder is generally used with the direct-solder silver compositions. It is recommended that this solder be used at a temperature of 213°C–219°C. Soldering to the plated silver coating is less critical and 50% Sn–50% Pb, as well as other soft solders, are being used with good results.

The air-dried silver compositions will, as the designation implies, air dry at room temperature in approximately 16 h. This drying time can be shortened by subjecting the coating to temperatures of 60°C–93°C for 10–30 min. The baked-on preparations must be cured at a minimum temperature of 149°C for 5–16 h. The time may be shortened to 1 h by raising the temperature to 301°C.

The same soft solders and techniques as recommended for the fired-on coatings may be used for the electroplated air-dried and baked-on preparations. It is extremely difficult to solder to air-dried or baked-on coatings without first electroplating.

The surface conductivity of the fired silver coating is far better than that of the air-dried or baked-on coating. Fired coatings have a surface electrical square resistance of approximately 0.01 Ω while the surface electrical square resistance of air-dried or baked-on is about 1 Ω.

Usually Alloyed

Because pure silver is so soft, it is usually alloyed with other metals for strength and durability. The most common alloying metal is copper, which imparts hardness and strength without appreciably changing the desirable characteristics of silver. Sterling silver has applications in manufacturing processes. For example, sterling silver plus lithium has been used in the aircraft industry for bracing honeycomb sections. Other silver–copper alloys are coin silver, 90.0% silver, 10.0% copper, and the silver–copper eutectic, 72% silver, 28% copper. This latter alloy has the highest combination of strength, hardness, and electrical properties of any of the silver alloys.

Silver braze filler metals are widely used for joining virtually all ferrous and nonferrous metals, with the exception of aluminum, magnesium, and some other lower-melting point metals. Whereas pure silver melts at 960°C, silver alloys, with compositions of 10%–85% silver (the alloying metals are copper, zinc, cadmium, and/or other base metals), have melting points of 618°C–960°C. These alloys have the ductility and malleability and can be rolled into sheet or drawn into wire of very small diameter. They may be employed in all brazing processes and are generally free-flowing when molten. Recommended joint clearances are 0.05–0.13 mm when used with flux. Whereas fluxes are usually required, zinc and cadmium-free alloys can be brazed in a vacuum or in reducing or inert atmospheres without flux. Joints made with silver brazing filler metals are strong, ductile, and highly resistant to shock and vibration. With proper design, there is no difficulty in obtaining joint strength equal to or greater than that of the metals joined. The strongest joints have but a few thousandths of an inch of the alloy as bonding material. Typical joints made with silver brazing filler metals, giving the greatest degree of safety, are scarf, lap, and butt joints.

For electrical contacts, silver is combined with a number of other metals, which increased hardness and reduce the tendency to sulfide tarnishing. Silver–cadmium, for example, is extensively used for contacts, with the cadmium ranging from 10% to 15%. The advantages of these alloys are resistance to sticking or welding, more uniform wear, and a decreased tendency for metal transfer.

Alloys used for contact and spring purposes are the silver–magnesium–nickel series (99.5% silver), which are used where electrical contacts are to be joined by brazing without loss of hardness, in miniature electron tubes for spring clips where high thermal conductivity is essential, and for instruments and relay springs requiring good electrical conductivity at high temperatures. These are unique, oxidation-hardening alloys. Before being hardened, the silver–magnesium–nickel alloy can be worked by standard procedures. After hardening in an oxidizing atmosphere, the room-temperature tensile properties are similar to those of hard rolled sterling silver or coin silver.

Gold and palladium are also combined with silver for contact use because they reduce welding and tarnishing and, to some extent, increase hardness.

When certain base metals do not combine with silver by conventional methods, powder metal processes are employed. This is particularly true of silver-iron, silver-nickel, silver-graphite, silver-tungsten, etc. these alloys are used in electrical contacts because of the desirable conductivity of silver and the mechanical properties of the base metals. They can be pressed, centered, and rolled into sheet and wire that as ductile unsuitable for forming into contacts by heading or stamping operations.

Other silver products are those produced chemically—powder, flake, oxide, nitrate, and paint. Silver powder and flake are composed of large amounts of silver with 0.03% or 0.04% copper and traces of lead, iron, and other volatiles.

Silvering

The application of a reflecting surface to glass. Techniques including chemical deposition and various solutions and sputtering or vapor deposition.

Silver Paints

These are used as conductive coatings that are pigmented with metallic silver flake or powder and bonding agents that are specially selected for the type of base material to which they are applied. These coatings are used to make conductive surfaces on such materials as ceramics, glass, quartz, mica, plastics, and paper, as well as on some metals. They are used for making printed circuits, resistor and capacitor terminals, and in miniature electrical instruments and equipment. Silver paints fall into two classifications: (1) fired-on types for base materials, they can withstand temperatures in the 399°C–927°C range, and (2) air-dried or baked-on types for organic base materials that are dried at temperatures ranging from 21.1°C to 427°C. The bonding agent in the fired-on type of coating is a powdered glass frit, whereas in the air-dried or baked-on type of coating organic resins are used. The viscosity and drying rate of each type varies, depending on the methods of application, such as spraying, dipping, brushing, roller coating, or screen stenciling.

Silver Solder

Brazing alloys, also termed hard solders are based mainly on silver, copper, zinc and cadmium, for example, 50% silver, 19% cadmium, 16% zinc and 15% copper. The various filler metals have melting ranges commencing at about 620°C (the eutectic for the alloy), i.e., intermediate between the tin lead solders and the high copper brass fillers.

Their strength is similar to the braze fillers i.e., better than the solders. Silver solders are not confined to joining silver components but have applications for steel and copper items where the requirements of strength and relatively low melting point justify the cost.

High-melting-point solder employed for soldering joints where more than ordinary strength and, sometimes, electrical conductivity are required. Most silver solders are copper–zinc brazing filler matals with the addition of silver. They may contain 9% to 80% silver, and the color varies from brass yellow to silver white. Cadmium may also be added to lower the melting point. Some solders do not necessarily contain zinc, and alloys of silver and copper in proportions are arranged to obtain the desired melting point and strength. A silver solder with their relatively low melting point contain 65% silver, 20% copper, and 15% zinc. It melts at 693°C, has a tensile strength of 447 MPa, and elongation 34%. The electrical conductivity is 21% that of pure copper. A solder melting at 760°C contains 20% silver, 45% copper, and 35% zinc. **ASTM silver solder No. 3** is this solder with 5% cadmium replacing an equal amount of the zinc. It is a general-purpose solder. **ASTM silver solder No. 5** contains 50% silver, 34% copper, and 16% zinc. It melts at 693°C and is used for soldering electrical work and refrigeration equipment.

Any tin present in silver solders makes them brittle; lead and iron make the solders difficult to work. Silver solders are malleable and ductile and have high strengths. They are also corrosion-resistant and are especially valuable for use in food machinery and apparatus where lead is objectionable. Small additions of lithium to silver solders increase fluidity and wetting properties, especially for brazing stainless steels or titanium. **Sil-Fos**, is a phosphor-silver brazing solder with a melting point of 704°C. It contains 15% silver, 80% copper, and 5% phosphorus. Lap joints brazed with Sil-Fos have a tensile strength of 207 MPa. The phosphorus in the alloy acts as a deoxidizer, and the solder requires little or no flux. It is used for brazing brass, bronze, and nickel alloys. The grade goes by the name of Easy solder contains 65% silver, melts at 718°C, and is a color match for sterling silver. **TL silver solder** has only 9% silver in melts at 871°C. It is brass yellow in color and is used for brazing nonferrous metals. **Sterling silver solder**, for brazing sterling silver, contains 92.8% silver, 7% copper, and 0.2% lithium. Flow temperature is 899°C.

A lead–silver solder recommended to replace tin solder contains 96% lead, 3% silver, and 1% indium. It melts at 310°C, spreads better than ordinary lead-silver solders, and gives a joint strength of 34 MPa. **Silver-palladium alloys** for high-temperature brazing contain 5% to 30% palladium. With 30%, the melting point is about 1232°C. These alloys have exceptional melting and flow qualities and are used in electronic and spacecraft applications.

Silver Soldering

Nonpreferred term used to denote brazing with a silver-based filler metal. See preferred terms *furnace brazing*, and *induction braising*, and *torch brazing*.

Silver Steel

A high carbon steel supplied as softened bright precision ground bar or strip for machining purposes. It does not contain silver. See *steel*.

Silver-Type Batteries

Because they are six times lighter and five times smaller than other batteries of similar capacity, silver–zinc batteries have found wide

S

use in guided missiles, Tele metering equipment, and guidance control circuits and mechanisms. Where longer life and ruggedness are more important than the weight, silver–cadmium rechargeable batteries are specified. Where seawater activation is required, silver chloride–magnesium couples are used. Another silver type of battery is the solid electrolyte type made with silver, silver iodide, and vanadium pentoxide. This battery, designed for low-current applications, weighs less than 1 oz and has almost unlimited shelf life.

Single-Action Press

A metal forming press that provides pressure from one side.

Single-Bevel Groove Weld

A groove weld in which the joint edge of one member is beveled from one side.

Single-Circuit Winding

In forming of plastics and composites, a winding in which the filament path makes a complete traverse of the chamber, after which the following traverse lies immediately adjacent to the previous one. See also *filament winding*.

Single Crystals

In crystalline solids the atoms or molecules are stacked in a regular manner, forming a three-dimensional pattern, which may be obtained by a three-dimensional repetition of a certain pattern unit called a unit cell. When the periodicity of the pattern extends throughout a certain piece of material, one speaks of a single crystal. A single crystal is formed by the growth of a crystal nucleus without secondary nucleation or impingement on other crystals.

Growth Techniques

Among the most common methods of growing single crystals are those of P. Bridgman and J. Czochralski. In the Bridgman method the material is melted in a vertical cylindrical vessel that tapers conically to a point at the bottom. The vessel then is lowered slowly into a cold zone. Crystallization begins in the tip and continues usually by growth from the first forward nucleus. In the Czochralski method, a small single crystal (seed) is introduced into the surface of the melt and then drawn slowly upward into a cold zone. Single crystals of ultrahigh purity have been grown by zone melting. Single crystals are also often grown by bathing a seed with a supersaturated solution; the super saturation is kept lower than is necessary for sensible nucleation.

When grown from a melt, single crystals usually take the form of their container. Crystals grown from solution (gas, liquid, or solid) often have a well-defined form that reflects the symmetry of the unit cell.

Physical Properties

Ideally, single crystals are free from internal boundaries. They give rise to a characteristic x-ray diffraction pattern. For example, the Laue pattern of a single crystal consists of a single characteristic set of sharp intensity maxima.

Many types of single crystals exhibit anisotropy, that is, a variation of some of their physical properties according to the direction along which they are measured. For example, the electrical resistivity of a randomly oriented aggregate of graphite crystallites is the same in all directions. The resistivity of a graphite single crystal is different, however, when measured along different crystal axes. This anisotropy exists for both structure-sensitive properties, which are not affected by imperfections (such as elastic coefficients).

Anisotropy of a structure-sensitive property is described by a characteristic set of coefficients that can be combined to give the macroscopic property along any particular direction in the crystal. The number of necessary coefficients can often be reduced substantially by consideration of the crystal symmetry; whether anisotropy, with respect to a given property, exists depends on crystal symmetry.

And the structure-sensitive properties of crystals (for example, strength and diffusion coefficients) seemed governed by internal defects, often on an atomic scale.

Industry and government are developing one of the first accurate computer-model predictions of molten metals and molding materials used in casting.

Currently, the computer information is being used to design and cast aircraft turbine blades. Similarly, another company is using the information from the computer models to improve the casting of automobile and light truck engine blocks.

Cast-metal parts are used in 90% of all durable goods such as washing machines, refrigerators, stoves, lawnmowers, cars, boats, and aircraft. The goal of the partnership is to produce accurate models for all alloys used by the casting industry. The information can then be used by manufacturers to standardize metal mixing recipes, allowing more effective competition in the marketplace.

Single-Crystal Superalloys

The nickel-base alloys that contain a single crystal, or more accurately, a single grain or primary dendrite. Because these materials have no grain boundaries, they exhibit improved high-temperature properties and corrosion resistance.

Single-Impulse Welding

Spot, projection, or upset welding by a single impulse of current. Where alternating current is used, an impulse may be any fraction or number of cycles.

Single-J Groove Weld

A groove weld in which the joint edge of one member is prepared in the form of a J, from one side. See also the term *single-bevel groove weld*.

Single-Lap Shear Specimen

In adhesive testing, a specimen made by bonding the overlapped edges of two sheets or strips of material. In testing, a single-lap specimen is usually loaded in tension at the ends, thereby creating shear stresses at the joint interface.

Single-Point Tool

See nomenclature—*side clearance, side relief, side rake angle, end clearance angle, end relief angle, end cutting edge angle, normal side relief angle, normal side clearance angle*, and *nose radius*.

Single-Port Nozzle (Plasma Arc Welding and Cutting)

A constricting nozzle containing one orifice, located below and concentric with the electrode. See the term *multiport nozzle*.

Single Relief Angle

See *single point tool*.

Single Spread

Application of adhesive to only one adherend of a joint.

Single-Stand Mill

A rolling mill designed such that the product contacts only two rolls at a given moment. Contrast with *tandem mill*.

Single-U Groove Weld

A groove weld in which each joint edge is prepared in the form of a J or half-U from one side.

Single-V Groove Weld

A groove weld in which each member is beveled from the same side.

Single Welded Joint

In arc and gas welding, any joint welded from one side only.

Sinkhead

Same as *riser*.

Sinking

(1) The operation of machining the impression of a desired forging into die blocks. (2) See *tube sinking*. (3) In the context of tube manufacture the term refers to drawing without the use of an internal mandrel or plug to control the bore size. See *drawing*. (4) Die sinking.

Sink Mark

A shallow depression on the surface of an injection-molded plastic part due to the collapsing of the surface following local internal shrinkage after the gate seals.

Sinter

To densified, crystallize, bond together, and/or stabilize a particulate material, agglomerate, or product by heating or firing close to but below the melting point. Often involves melting of minor components or constituents, and/or chemical reaction. Also, the product of such firing.

Sinter Hardening

Recognizing the great potential of sinter hardening, researchers continue to develop useful data that will help to attain benefits derived

from this heat treatment. Improved properties can be achieved by effective control of material composition, density, section size, sintering temperature, and cooling rate. By controlling these variables a variety of microstructures and resultant properties are achievable, enabling particular powder metallurgy parts to favorably perform under specific severe conditions. However, material options, process flexibility, and application requirements demand a better understanding of process, microstructure, and mechanical property relationships to capitalize fully on the opportunity of sinter hardening.

Sinter hardening refers to a process where the cooling rate experienced in the cooling zone of the sintering furnace is fast enough that a significant portion of the material matrix transforms to martensite. Interest in sinter hardening has grown because it offers good manufacturing economy by providing a one-step process and a unique combination of strength, toughness, and hardness.

A variety of microstructures and properties can be obtained by varying both the alloy type and content as well as the postsintering cooling rate. By controlling the cooling rate, the microstructure can be manipulated to produce the required proportion of martensite, which will lead to desired mechanical properties. By understanding how the sintering conditions affect the microstructure, materials can be modeled to produce the final properties that are desired.

Alloying

Alloying elements are used in cast, wrought, and P/M materials to promote hardenability and increase the mechanical strength of the parts. A graphical way of examining the effects of alloying elements on the final microstructure of the steel is by using the characteristic isothermal transformation (I-T) diagram. This indicates the time necessary for the isothermal transformation of phases in the material from start to finish, as well as the cooling time and temperature combinations needed to produce the final microstructure. A similar diagram, known as a continuous cooling (CCT) curve, also is frequently used to determine the variation in microstructure as a function of cooling rate.

Materials and Processing

As expected, increasing the cooling rate resulted in increased apparent hardness and strength values. On the whole, hardness values were increased between 2 and 10 HRC for a given material.

As expected, in all materials, the percent of martensite present increased significantly with the increase in cooling rate. The effect of the increased martensite levels is apparent in the hardness values for each of the materials. The effect of the higher levels of martensite on tensile properties is less obvious. In several cases, materials with significantly lower percentages of martensite and lower hardness values demonstrated higher tensile strengths.

Some results show the following trends:

1. Materials with 2 wt% Ni and 0.5 wt% Graphite Admixed
   a. Accelerated cooling resulted in increase strength and apparent hardness while decreasing elongation values only slightly. This result was a consequence of increased martensite content and finer pearlitic microstructures. In these materials, the martensite was the result of transformation of nickel-rich areas in the microstructure.
   b. The increase in prealloyed alloy content from 0.85 to 1.5 wt% Mo resulted in a larger increase in strength than the addition of 1.0 wt% admixed Cu.

c. Although the 0.85 wt% Mo materials exhibited higher percentages of martensite than identical chemistries based on the 1.5 wt% Mo prealloyed material, the higher molybdenum materials had higher apparent hardness and strength values. This surprising result was explained by the presence in the 1.5 wt% Mo-based material of significantly finer pearlite.

2. Materials with 2 wt% Cu and 0.9 wt% Graphite Admixed

a. As the cooling rate was increased for these materials, the apparent hardness increased. This was associated with higher martensite contents in the faster-cooled materials. Martensite contents of greater than 50% were found in all three base materials when accelerated cooling was utilized.

b. The materials with the highest apparent hardness values (0.5 wt% Ni, 1.5 wt% Mo pre-alloy) did not exhibit the highest tensile strength values. The highest UTS values were determined for the fast-cooled version of the 0.85 wt% Mo prealloyed material. Retained austenite may be one potential cause for the fall off in strength for the Mo–Ni material.

Sintered Density

The quotient of the mass (weight) over the volume of the sintered body expressed in grams per cubic centimeter.

Sintered Density Ratio

The ratio of the density of the sintered body to the solid, pore-free body of the same composition or theoretical density.

Sintering

A process, below the melting point but usually at elevated temperature, in which contacting particles mutually diffuse and bond. The process may be used to agglomerate fine mineral ore dust to make it more easily processed and handled or it may be used to make substantial objects of metal powders for further processing or to directly enter service. If the atmosphere is intended to react with the powder during the operation to promote bonding the process is termed *activated sintering*.

The bonding of adjacent surfaces of particles in a mass of powder or a compact by heating. Sintering strengthens a powder mass and normally produces densification and, in powdered metals, recrystallization. See also *liquid phase sintering* and *solid-state sintering*.

Sintering Atmospheres

See *protective atmosphere* (2).

Sintering Cycle

A predetermined and closely controlled time-temperature regime for sintering compacts, including the heating and cooling phases.

Sintering Temperature

The maximum temperature at which a powder compact is sintered. The temperature was either measured directly on the surface of the body by optical pyrometer, or indirectly by thermocouples installed in the furnace chamber.

Sintering Time

The time period during which a powder compact is at sintering temperature.

Sintrate

In powder metallurgy, controlled heating so that a compact is sintered before the melting point of an infiltrating material is reached. See also *infiltration*.

Sisal

The hard, strong, light-yellow to reddish fibers from the large leaves of the sisal plant, and the henequen plant. Sisal is employed for making rope, cordage, and sacking. About 80% of all binder twine is normally made from sisal, but sisal ropes have only 75% of the strength of Manila rope and are not as resistant to moisture. Sisal is a tropical plant, and grows best in semiarid regions. **Yucatan sisal**, or **henequen**, is from the henequen plant and is reddish, stiffer, and coarser, and is used for binder twine. The Indian word *henequen* means knife, from the knifelike leaves. The plant is more drought-resistant than sisal.

The fibers of sisal are not as long or as strong as those of Manila hemp, and they swell when wet, and they are soft and are preferred for binder twine either alone or mixed with Manila hemp. **Sisal fiber** is also used instead of hair in cement plasters for walls and in laminated plastics. **Corolite** is a molded plastic made with a mat of sisal fibers so as to give equal strength in all directions. **Tampico**, which yields a stiff, hard, but pliant fiber employed for circular power brushes, is valued for polishing wheels, as the fibers hold the grease buffing composition, and it is not brittle but abrades with flexibility.

Agava, of Agava Products, Inc., is a dark-brown, viscous liquid extracted from the leaves of agave plants, used as a water conditioner for boiler-water treatment. It is a complex mixture of sapogenines, enzymes, chlorophyllin, and polysaccharides.

A fine strong fiber is obtained from the long leaves of the **pineapple**, native to tropical America. The plant is grown chiefly for its fruit, known in South America under its Carib name **anana'** and marketed widely as canned fruit and juice, preserves, and confections. For fiber production the plants are spaced widely for leaf development and are harvested before the leaves are fully mature. The retted fibers are long, white, and of fine texture and may be woven into water-resistant fabrics. The very delicate and expensive **piña cloth** of the Philippines is made from **pineapple fiber**. The fabrics of Taiwan are usually coarser and harder.

Situ

See *In situ*.

Size

In composites manufacturing, a treatment consisting of starch, gelatin, oil, wax, or other suitable ingredients applied to yarn or fibers at the time of formation to protect the surface and aid the process of handling and fabrication or to control the fiber characteristics. The treatment contains ingredients that provide surface lubricity and binding action, but unlike a finish, contains no coupling agent. Before final fabrication into a composite, the size is usually removed by heat cleaning, and a finish is applied.

Size Effect

Effect of the dimensions of a piece of metal on its mechanical and other properties and on manufacturing variables such as forging reduction and heat treatment. In general, the mechanical properties are lower for a larger size.

Size-Exclusion Chromatography (SEC)

Liquid chromatography method that separates molecules on the basis of their physical size. This technique is most often used in the analysis of polymers. Also termed *gel permeation chromatography*.

Size Factor

The relationship between the size of atoms which determine, in part, how they form solutions and compounds. If atoms of different elements, or more strictly their positive ions, are similar in size then, if other factors are correct, they conform substitutional solid solutions. If they are widely different and other factors are appropriate they may form interstitial solid solutions or compounds. See *solution* and *size factor compounds*.

Size Factor Compounds

These are intermetallic phases formed when the size factor and other aspects are suitable and the elements are present in specific simple proportions. The laves phases are based on the relationship AB_2, hence $MgCu_2$ or $TiCr_2$, which form when the constituent atoms differ in size by about 22.5% allowing a particular form of close packing. The interstitial compounds form when certain metals and nonmetals having widely differing atomic sizes and so able to adopt an interstitial structure. Examples include carbides such as Fe_3C, WC and Mo_2C or nitrides such as TiN.

Size Fraction

A separated fraction of a powder whose particles lie between specified upper and lower size limits. See also *sieve analysis*.

Size of Weld

(1) The joint penetration in a groove weld. (2) The lengths of the nominal legs of a fillet weld. (3) The weld metal thickness measured at the root of a flange weld.

Sizing (Adhesives)

The process of applying a material on a surface in order to fill pores and thus reduce the absorption of the subsequently applied adhesive or coating or to otherwise modify the surface properties of the substrate to improve the adhesion, and also, the material used for this purpose. See also *primer*.

Sizing (Composites)

See *size*.

Sizing (Metals)

(1) Secondary forming or squeezing operations needed to square up, set down, flatten, or otherwise correct surfaces to produce specified dimensions and tolerances. See also *restriking*. (2) Burnishing, broaching, drawing, and shaving operations are also called sizing. (3) A finishing operation for correcting ovality in tubing. (4) Final pressing of a sintered powder metallurgy part to obtain a desired dimension.

Sizing Agents

These coatings are applied to glass textile fibers in the forming operation. The sizes use may contain one or more or any combination of binders, lubricants, and coupling agents.

Dextrin

A starch derivative commonly called starch in the fiberglass industry. It is of the low viscosity variety when used in fiberglass sizes. Practically all the fiberglass yarn that has been woven has been sized with this starch size. In industrial fiberglass fabrics, the size is removed in a process called heat cleaning, whereas in decorative woven material the size is burned off in the coronizing operation.

Gelatin

Used in small amounts as a constituent of a standard glass fiber textile size. Together with dextrin (starch), the gelatin acts as a binder for the other materials formulated into a size for the fiberglass textile yarns.

Polyvinyl Acetate

As a latex, it is used as a binder in a standard size formulation for continuous fiberglass yarn. Usually this continuous yarn is made into roving and a large package is usually made up of 60 ends of continuous yarn, although other end counts such as 30 ends, 20 ends, and 8 ends are also available. Roving is used primarily as reinforcement in reinforced plastics, and in preforming operations.

Sizing Content (Composites)

The percent of the total strand weight made up of sizing, usually determined by burning off or dissolving the organic sizing. Also known as loss on ignition. See also *size*.

Sizing Die

A die used for the *sizing* of a sintered compact.

Sizing Punch

A punch used for the pressing of a sintered compact during the *sizing operation*.

Sizing Knockout

An ejector punch used for rejecting a sintered compact from one *sizing die*.

Sizing Pass

The final light working pass in rolling or drawing to provide a high-quality surface with precise dimensions.

Sizing Stripper

A punch used during the *sizing* operation.

Skein

A continuous filament, strand, yarn, or roving for reinforced plastics, wound up to some measurable length and usually used to measure various physical properties.

Skeleton

An unsintered or sintered porous powder metallurgy compact with a large proportion of interconnected porosity that makes it suitable for *infiltration*.

Skelp

(1) The starting stock for making welded pipe or tubing, most often it is strip stock of suitable width, thickness, and edge configuration. (2) Slit strip with any edge preparation that is to be formed into a tube by longitudinally rolling and, or drawing through a die prior to seam welding.

Skidding

A form of nonuniform relative motion between solid surfaces due to rapid periodic changes in the traction between those surfaces.

Skid-Polishing Process

A mechanical polishing process in which the surface of the metallographic specimen to be polished is made to skid across a layer of paste, consisting of the abrasive and the polishing fluid, without contacting the fibers of the polishing cloth.

Skim Gate

In foundry practice, a gating arrangement designed to prevent the passage of slag and other undesirable materials into a casting.

Skimmer

A tool for measuring scum, slag, and dross from the surface of molten metal.

Skimming

Removing or holding back dirt or slag from the surface of the molten metal before or during pouring.

Skin (Metals)

A thin outside metal layer, not formed by bonding as in cladding or electroplating, that differs in composition, structure, or other characteristics from the main mass of metal.

Skin (Plastics)

The relatively dense material that sometimes forms on the surface of a cellular plastic or sandwich construction.

Skin Drying

Drying the surface of a foundry mold by direct application of heat.

Skin Effect

Concentration of material.

Skin Lamination

In flat-rolled metals, a surface rupture resulting from the exposure of a subsurface lamination by rolling.

Skin Pass

The confinement of high-frequency electrical currents to the skin, i.e., outer surface, of a conductor. See *temper rolling*.

Skip Sequence Weld

A welding technique in which a long run of weld is built up by a number of short deposits laid in a preplanned sequence such that the first series of welds are made with gaps between them and subsequent welds fill the gaps.

Skip Weld

See preferred term *intermittent weld*.

Skiving

(1) Removal of a material in layers or chips with a high degree of sheer or slippage, or both, of the cutting tool. (2) A machining operation in which the cut is made with a form tool with its face so angled that the cutting edge progresses from one end of the work to the other as the tool feeds tangentially past the rotating workpiece.

Skull

(1) A layer of solidified metal or dross on the walls of a pouring vessel after the metal has been poured. (2) The unmelted residue from a liquated weld filler metal.

Slab

(1) A flat-shaped semifinished rolled metal ingot with a width not less than 250 mm (10 in.) and a cross-sectional area not less than 105 cm^3 (16 in.^{2}). (2) An imprecise term referring to part rolled material with a thickness about half its width. It is implicit that it will subsequently be rolled to plate or sheet.

Slabbing

The hot working of an *ingot* to a flat rectangular shape.

Slabbing Mill

See preferred term *peripheral milling*.

Slack Quenching

The incomplete hardening of steel due to quenching from the austenitizing temperature at a rate slower than the critical cooling rate for the particular steel, resulting in the formation of one or more transformation products in addition to martensite. Cooling steel at a rate insufficient to form a fully martensitic structure but fast enough to form bainite. See *steel*.

Slag

(1) A nonmetallic product resulting from the mutual dissolution of flux and nonmetallic impurities in smelting, refining, and certain welding operations (see, for example, *electroslag welding*). In steelmaking operations, the slag serves to protect the molten metal from the air and to extract certain impurities. (2) The mixture of flux and impurities extracted from the metal, which warms during various smelting processes. It floats on, and remains separate from, the underlying metal and has the additional benefit in some cases that it protects the underlying metal from the environment.

The molten material that is drawn from the surface of iron in the blast furnace. Slag is formed from the earthy materials in the ore and from the flux. Slags are produced in the melting of other metals, but iron blast-furnace slag is usually meant by the term. Slag is used in cements and concrete, for roofing, and as a ballast for roads and railways. Finely crushed slag is used in agriculture for neutralizing acid soils. **Blast-furnace slag** is one of the lightest concrete aggregates available. It has a porous structure and, when crushed, is angular. It is also crushed and used for making pozzuolana and other cements. Slag contains about 32% silica, 14% alumina, 47% lime, 2% magnesia, and small amounts of other elements. It is crushed, screen, and graded for marketing. **Crushed slag** is about 30% lighter than gravel. **Honeycomb slag** is the finest grade of commercial slag is from 0.48 cm to dust; the run-of-crusher slag is from 10 cm to dust. **Basic phosphate slag**, a by-product in the manufacture of steel from phosphate ores, is finely ground and sold for fertilizer. It contains not less than 12% phosphoric oxide, P_2O_5, and is known in Europe as **Thomas slag**. **Foamed slag** is a name used in England for honeycomb slag used for making lightweight, heat-insulating blocks. A superphosphate cement is made in Belgium from a mixture of basic slag, slaked lime, and gypsum.

Slag Fiber/Wool

Fine stranded fibers formed from various slags and similar in texture to glass wool or fiber. It commonly substitutes for the glass product as an insulating material.

Slag Inclusion

(1) Slag or dross entrapped in a metal. (2) Nonmetallic solid material entrapped in weld metal or between weld metal and base metal.

Slag Trap (of Weld Joint)

Any joint geometry or feature which could retain molten slag and impede its subsequence removal.

Slant Fracture

A type of fracture in metals, typical of *plane-stress* fractures, in which the plane of separation is inclined at an angle (usually about 45°) to the axis of applied stress.

Slate

A shale having a straight cleavage. Most shales are of sedimentary origin, and their cleavage was the result of heavy or long-continued pressure. In some cases slates have been formed by the consolidation of volcanic ashes. The slaty cleavage does not usually coincide with the original stratification. Slate is of various colors: black, gray, green, and reddish. It is used for electric panels, chalkboards, slate pencils, tabletops, roofing shingles, floor tiles, and treads. The terms **flagstone** and **cleftstone** are given to large, flat sections of slate used for paving, but the names are also applied to blue sandstones cut for this purpose. Slate is quarried in large blocks, and then slabbed and split. The chief slate producing states are Pennsylvania, Vermont, Virginia, New York, and Maine. Roofing slates vary in size from 30 × 15 cm to 61 × 36 cm, and from 0.32 to 1.91 cm in thickness, and are usually of the harder varieties. The roofing slate from coal beds is black, fine-grained, and breaks into brittle thin sheets. It does not have the hardness or weather resistance of true slate. **Ribbon slate**, with streaks of hard material, is inferior for all purposes. Lime impurities can be detected by the application of dilute hydrochloric acid to the edges and noting if rapid effervescence occurs. Iron is a detriment to slates for electric purposes. **Slate granules** are small, graded chips used for surfacing prepared roofing. **Slate flour** is ground slate, largely a by-product of granule production. It is used in linoleum, caulking compounds, and asphalt surfacing mixtures. **Slate lime** is an intimate mixture of finely divided, calcined slate and lime, about 60% by weight lime to 40 slate. It is employed for making porous concrete for insulating partition walls. The process consists in adding a mixture of slate lime and powdered aluminum, zinc, or magnesium to the cement. The gas generated on the addition of water makes the cement porous.

Sleeve Bearing

A cylindrical plain bearing used to provide radial location for a shaft, which moves axially. Sleeve bearings usually consists of one or more layers of bearing alloy(s), or liner, bonded to a steel backing. Sleeve bearing is sometimes used to denote *journal bearing*. See also *sliding bearing*.

Slice

The cross-sectional plane through an object that is scanned to produce the image in *computed tomography*. See also *tomographic plane*.

Slices

Sections of ingots of single-crystal material that have been sawed from the ingot. Also known as wafers.

Slide

The main reciprocating member of a metal forming press, guided in the press frame, to which the punch or upper die is fastened; sometimes called the *ram*. The inner slide of a double-action press is called the plunger or punch-holder slide; the outer slide is called the blank-holder slide. The third slide of a triple-action press is called the lower slide, and the slide of a hydraulic press is often called the platen. See also the term *press forming* and *straight-side press*.

Slide Adjustment

The distance that a metal forming press slide position can be altered to change the shut height of the die space. The adjustment can be made by hand or by power mechanism.

Slide-Roll Ratio

See *slide-sweep ratio.*

Slide-Sweep Ratio

The ratio of sliding velocity to sweep velocity, for example, in a pair of gears. In rolling, the slide-sweep ratio is called the *slide-roll ratio.*

Sliding (Pure Sliding with no Rolling or Spin)

A motion of two relatively moving bodies, in which their surface velocities in the common contact area are different with regard to magnitude and/or direction. See also *rolling*, *spin*, and *specific sliding.*

Sliding Bearing

A bearing in which predominately sliding contact occurs between relatively moving surfaces. Sliding bearings may be either unlubricated, liquid lubricated, grease lubricated, or solid lubricated. See also *sleeve bearing.*

Sliding Fit

A fit with just sufficient slack to allow the mating components to move axially relative to each other without galling but without perceptible lateral play. This usually implies an interfacial gap just sufficient to accommodate a film of lubricant. Contrast with *push fit.* Similar to *slip fit.*

Sliding Velocity

The difference between the velocities of each of the two surfaces relative to the point of contact.

Slime

(1) A material extremely fine particle size encountered in ore treatment. (2) A mixture of metals and some insoluble compounds that forms on the anode in electrolysis.

Slip

(1) Deformation by planes of atoms in the crystal lattice sliding over each other. The sliding action is facilitated by the movement of dislocations. (2) The thick water-based paste used in slip casting.

Slip (Adhesives)

In an adhesively bonded components or specimens, the relative collinear displacement of the adherends on either side of the adhesive layer in the direction of the applied load.

Slip (Ceramics)

(1) A slurry or suspension of fine clay or other ceramic powders in water, having the consistency of cream, that is used in slip casting or as a cement or glass preparation. (2) A suspension of colloidal powder in an immiscible liquid (usually water).

Slip (Metals)

Plastic deformation by the irreversible shear displacement (translation) of one part of a crystal relative to another in a definite crystallographic direction and usually on specific crystallographic plane. Sometimes called glide.

Slip Angle

The angle at which a tensioned fiber will slide off a filament-wound dome. If the difference between the wind angle and the geodesic angle is less than the slip angle, the fiber will not slide off the dome. Slip angles for different fiber-resin systems vary and must be determined experimentally. See also *filament winding* and *wind angle.*

Slip Band

A group of parallel slip lines so closely spaced as to appear as a single line when observed under an optical microscope. See also *slip line.*

Slip Casting

(1) The ceramic forming process consisting of filling or coating a porous mold with a *slip*, allowing to dry, and removing for subsequent firing. (2) A technique in which a water and powder paste is poured into an absorbent plaster mold. The mold absorbs most of the water leaving a fragile replica of the mold interior.

Slip Coating

A ceramic material or mixture other than a glaze, applied to a ceramic body and fired to the maturity required to develop specified characteristics.

Slip Crack

See *pressing crack.*

Slip Direction

The crystallographic direction in which the translation of slip takes place. See the term *slip* (metals).

Slip Fit

A loosely defined *clearance fit* between parts assembled by hand without force, but implying slipping contact.

Slip Flask

A tapered *flask* that depends on a movable strip of metal to hold foundry sand in position. After closing the mold, the strip is refracted and the flask can be removed and reused. Molds thus made are usually supported by a *mold jacket* during pouring.

Slip Forming

A plastic sheet-forming technique in which some of the sheet material is allowed to slip through the mechanically operated clamping rings during a stretch-forming operation.

Slip-Interference Theory

Theory involving the resistance to deformation offered by a hard phase dispersed in a ductile matrix.

Slip Liner

(1) Visible traces of slip planes on metal surfaces; the traces are (usually) observable only if the surface has been polished before deformation. The usual observation on metal crystals (under a light microscope) is of a cluster of slip lines known as a *slip band*. (2) Markings produced on deformed surfaces that were polished prior to deformation. They arise because neighboring areas have deformed along differently aligned slip planes giving rise to surface irregularities.

Slippage

The movement of the adherends with respect to each other during the adhesive bonding process.

Slip Plane

The crystallographic plane in which *slip* occurs in a crystal.

Slitting

Cutting or shearing along single lines to cut strips from a metal sheet or to cut along the lines of a given length or contour in a sheet or workpiece.

Sliver (Composites)

A number of staple or continuous filament fibers are aligned in a continuous strand without twist. See also *continuous filament yarn*, *staple fiber*, and *strand*.

Sliver (Metals)

An imperfection consisting of a very thin elongated piece of metal attached by only one end to the parent metal into whose surface it has been worked.

Slope Control

Producing electronically a gradual increase or decrease in the welding current between definite limits and within a selected time interval.

Slot Extrusion

A method of extruding plastic film or sheet in which the molten thermoplastic compound is forced through a slot die (T-shaped or coat

hanger shape). Following extrusion, the film or sheet is cooled by passing it through a water bath or over water-cooled rolls.

Slot Furnace

A common batch furnace for heat treating metals where stock is charged and removed through a slot or opening.

Slotting

Cutting a narrow aperture or groove with a reciprocating tool in a vertical shaper or with a cutter, broach, or grinding wheel.

Slot Weld

A weld made in an elongated hole in one member of a lap or T-joint joining that member to that portion of the surface of the other member which is exposed through the hole. The hole may be open at one end and may be partially or completely filled with weld metal. A fillet welded slot should not be construed as conforming to this definition.

Slow Bend Test

A bend test to determine the ductility/brittleness characteristics of metals, usually steel. The term contrast with the fast loading rate of impact tests for determining such characteristics. The test specimen may be notched or plain and may be loaded in either three-point or four-point bending.

Slow Butt Welding

Same as *resistance butt welding*.

Slow Strain Rate Technique

An experimental technique for evaluating susceptibility to *stress-corrosion cracking*. It involves pulling the specimen to failure in uniaxial tension at a controlled slow strain rate while the specimen is in the test environment and examining the specimen for evidence of stress-corrosion cracking.

Sludge

A coagulated mass, often containing foreign matter, formed at low temperature in combustion engines from oil oxidation residues, carbon, and water.

Sluffing

An occurrence during protrusion of reinforced plastics in which scales peel off or become loose, either partially or entirely, from a pultrusion. Not to be confused with scraping, prying, or physically removing scale from a pultrusion. Sluffing is sometimes spelled sloughing. See also *pultrusion*.

Slug

(1) A short piece of metal to be placed in a die for forging or extrusion. (2) A small piece of material produced by piercing a hole in sheet material. See also *blank*.

Slug Test

Same as *peel test*.

Slugging

The unsound practice of adding a separate piece of material in a joint before or during welding, resulting in a welded joint in which the weld zone is not entirely built up by adding molten filler metal or by melting and recasting base metal, and which therefore does not comply with design, drawing, or specification requirements. Such material partly fills the joint and is hidden by subsequently deposited weld metal but is not fully fused and forms a serious weakness.

Slumpability

The flow of gravity of a grease in a container, allowing it to feed out into a pump or can. Slumpability also influences the leakage of grease from a bearing.

Slurry

(1) A thick mixture of liquid and solids, the solids being in suspension in the liquid. (2) Any pourable or pumpable suspension of a high content of insoluble particulate solids in a liquid medium, most often water. (3) The thick suspension of solids in water or other liquid is less viscous than a paste and easily poured. See also *slip*.

Slurry Abrasion Response (SAR) Number

A measure of the relative abrasion response of any material in any slurry, as related to the instantaneous rate of mass loss of a specimen at a specific time on the cumulative abrasion-corrosion time curve, converted to volume or thickness loss rate. See also *Miller number*.

Slurry Abrasivity

The relative tendency of a particular moving slurry to produce abrasive and corrosive wear compared with other slurries.

Slurry Erosion

Erosion produced by the movement of a slurry past a solid surface.

Slurry Preforming

Method of preparing a reinforced plastic preforms by wet processing technique similar to those used in the pulp molding industry. For example, glass fibers suspended in water are passed through a screen that passes the water but retains the fibers in the form of a mat.

Slush

Material in an intermediate stage between fully molten and fully solid.

Slush Casting/Molding

(1) A technique for producing hollow castings. Molten metal is poured into a mold, usually metal and often of complex shape. The molten metal is usually swirled round and, after a solid skin has formed over the mold interior, the remaining molten metal is poured out. In a variation on the process a measured amount of molten metal is introduced through an orifice which is then sealed prior to swirling. In this way a fully enclosed hollow is produced. (2) Processes in which the alloy is poured or injected in a pasty, parts solidified, state. See *rheocasting*.

Slushing Compound

An obsolete term for describing oil or grease coatings used to provide temporary protection against *atmospheric corrosion*.

Slushing Oil

A mineral oil containing additives that enable it to protect the parts of a machine against rusting.

Slush Molding

A method for casting thermoplastics, in which the resin in liquid form it is poured into a hot mold where a viscous skin forms. The excess slush is drained off, the mold is cooled, and the molding stripped out.

SMC

See *sheet molding compound*.

SME

Shape memory effect.

Smart Materials

Materials which alter some significant characteristic when subjected to some external stimulus, for example, piezoelectric crystals which generate a voltage when stressed.

Smearing

Mechanical removal of material from a surface, usually involving plastic shear deformation, and redeposition of the material as a thin layer on one or both surfaces. See also *transfer*.

Smelting

(1) Thermal processing wherein chemical reactions take place to produce liquid metal from a beneficiated ore. (2) Processes involving chemical reaction at high temperatures to reduce ore to molten metal or to some intermediate product. Where the final product is to be molten metal a flux is usually added to combine with the unwanted oxide and other impurities forming a slag.

Smith Forging

See *hand forge (smith forge)*.

Smoke

Airborne finely particulate materials usually products of combustion.

Smoke Agents

Chemicals used in warfare to produce an obscuring cloud of fog to high movements. Smokes may be harmless and are then called **screening smokes**, or **smoke screens**, or they may be toxic and called **blanketing clouds**. There are two types of smokes: those forming solid or liquid particles and those forming fogs or mists by chemical reaction. **White smokes**, which do not have light-absorbing particles, such as carbon, are formed by chemical reaction and have the best opacity or screening action. The first Naval smokescreens were made by limiting the admission of air to the fuel in the boilers, and the first Army smoke pots contained mixtures of pitch, tallow, saltpeter, and gunpowder. The British **smoke candles** contained 40% potassium nitrate, 29% pitch, 14% sulfur, 8% borax, and 9% coal dust. They gave a brown smoke, but one that lifted too easily.

Fog or military screening may be made by spraying an oil mixture into the air at high velocity. The microscopic droplets produce an impenetrable fog which remains for long period. **White phosphorus** is a dense, white smoke, called **WP smoke**, by burning to the pentoxide and changing to phosphoric acid in the moisture of the air. Its vapor is toxic. Smoke from **red phosphorus** is known as **RP smoke**. **Sulfuric trioxide**, SO_3, is an effective smoke producer in humid air. It is a mobile, colorless liquid vaporizing at 45°C to form dense, white clouds with an irritating effect. The French **opacite** is **tin tetrachloride** or **stannic chloride**, $SnCl_4$, a liquid that fumes in air. The smoke is not dense, but it is corrosive and it penetrates gas masks. **FS smoke** is made with a mixture of chlorosulfonic acid and sulfur trioxide. **Silicon tetrachloride**, $SiCl_4$, is a colorless liquid that boils at 60°C, and fumes in the air, forming a dense cloud. Mixed with ammonia vapor, it resembles a natural fog. The heavy mineral known as **amang**, separated from Malayan tin ore, containing ilmenite and zircon, is used for smoke screens. **Titanium tetrachloride**, $TiCl_4$, is a colorless to reddish liquid boiling at 136°C. It is used for smoke screens and for skywriting from airplanes. In most air it forms dense, white fumes of **titanic acid**, and hydrogen chloride. The commercial liquid contains about 25% titanium by weight.

A common smoke for airplanes is **oleum**. It is a mixture of sulfur trioxide in sulfuric acid, which forms fuming sulfuric acid, or **pyrosulfuric acid**. The dense liquid is squirted in the exhaust manifold. **Zinc smoke** is made with mixtures of zinc dust or zinc oxide with various chemicals to form clouds. **HC smoke** is zinc chloride with an oxidizing agent to burn up residual carbon so that the smoke will be gray and not black. **Signal smoke** is colored smoke used for ship distress signals and for radiation marking signals. They are mixtures of a fuel, an oxidizing agent, a dye, and sometimes a cooling agent to regulate the rate of burning and to prevent decomposition of the dye. Unmistakable colors are used so that the signals may be distinguished from fires, and the dyes are mainly anthraquinone derivatives, together with mixtures of a azo, azine, and diphenyl-methane compounds.

Smut

(1) A reaction product sometimes left on the surface of a metal after pickling, electroplating, or etching. (2) Finely particulate material released into the atmosphere usually as a result of combustion. (3) Surface blackening produced on some metals by pickling in caustic soda solution.

Snagging

(1) Heavy stock removal of superfluous material from a workpiece by using a portable or swing grinder mounted with a coarse grain abrasive wheel. (2) *Offhand grinding* on castings and forgings to remove surplus metal such as gate and riser pads, fins, and parting lines.

Snake

(1) The product formed by twisting and bending of hot metal rod prior to its next rolling process. (2) Any crooked surface imperfection in a plate, resembling a snake. (3) A flexible mandrel used in the inside of a shape to prevent flattening or collapse during a bending operation.

Snakeskins

The snakeskins employed for fancy leathers are in general the skins of large, tropical snakes which are notable for the beauty or oddity of their markings. Snakeskins for shoe-upper leathers, belts, and handbags are glazed like kid and calfskin after tanning. Small cuttings are used for inlaying on novelties. The leather is very thin, but is remarkably durable and is vegetable-tanned and finished in natural colors, or is dyed. **Python skins** are used for ladies' shoes. Diamond-backed rattlesnakes are raised on snake farms in the United States. The meat is canned as food, and the skins are tanned into leather. Only the back is used for leather, as the belly is colorless.

Snaky Edges

See *carbon edges*.

Snap Flask

A foundry flask hinged on one corner so that it can be opened and removed from the mold for reuse before the metal is poured.

Snap Temper

A precautionary interim stress-relieving treatment applied to high-hardenability steels immediately after quenching to prevent cracking because of delay and tempering them at the prescribed higher temperature.

S–N Curve

A plot of stress (S) against the number of cycles to failure (N). The stress can be the maximum stress (S_{max}) or the alternating stress amplitude (S_a). The stress values are usually nominal stress; i.e., there is no adjustment for stress concentration. The diagram indicates the *S–N* relationship for a specified value of the mean stress (S_m) or the stress ratio (A or R) and a specified probability of survival. For N a log scale is almost always used. For S a linear scale is used most often, but a log scale is sometimes used. Also known as *S–N* diagram.

S–N Curve for 50% Survival

A curve fitted to the median value of fatigue life at each of several stress levels. It is an estimate of the relationship between applied stress and the number of cycles-to-failure that 50% of the population would survive.

S–N Curve for *p*% Survival

A curve fitted to the fatigue life for *p*% survival values at each of several stress levels. It is an estimate of the relationship between applied stress and the number of cycles-to-failure that *p*% of the population would survive. *p* may be any number, such as 95, 90, etc.

Snowflakes

See *flakes*.

Snug Fit

A loosely defined fit implying the closest clearances that can be assembled manually for firm connection between parts. See also *clearance fit*.

Soak Cleaning

Immersion cleaning without electrolysis.

Soaking

In heat treating of metals, prolonged holding at a selected temperature to effect homogenization of structure or composition. See also *homogenizing*.

Soaking Pit

A chamber in which newly cast ingots, particularly steel, are stored to allow the temperature to equalize prior to the first hot rolling operation. There may also be some homogenization of the structure. The facility may be no more than an insulated but unheated hole in the ground usually with some form of copper but supplementary heating may be applied.

Soak Time

The length of time a ceramic material is peanut at the peak temperature of the firing cycle. See also *firing*.

Soap

In lubrication, a compound formed by the reaction of a fatty acid with a metal or metal compound. Metallic soaps formed by reaction *in situ* are an important group of boundary lubricants.

A cleansing compound produced by saponifying oils, fats, or grease with an alkali. When caustic soda is added to fat, glycerin separates out, leaving **sodium oleate**, which is soap. But since oils and fats are mixtures of various acid glycerides, the soaps made directly from vegetable and animal oils may be mixtures of oleates, palmitates, linoleates, and laurates. **Soap oils** in general, however, are those oils which have greater proportions of nearly saturated fatty acids, since the unsaturated fractions tend to oxidize to form aldehydes, ketones, or other acids, and turn rancid. If an excess of alkali is used, the soap will contain free alkali; and the greater the portion of the free alkali, the coarser is the action of the soap. ASTM standards for milled toilet soap permit only 0.17% free alkali. Soap makers now employ refined and bleached oils, which are then hydrolyzed into fatty acids and glycerol prior to saponification with caustic. This allows the fatty acids to be distilled, resulting in a more stable product. **Sodium soaps** are always harder than **potassium** soaps with the same fat or oil. Hard sodium soaps are used for chips, powders, and toilet soaps. Soft, caustic potash soaps are the liquid, soft, and semi soft pastes. Mixtures of the two are also used. Soaps are made by either the boiled process or the cold process. Chip soap is made by pouring the hot soap onto a cooled revolving cylinder from which the soap is scraped in the form of chips or ribbons which are then dried to reduce the moisture content from 30% to 10%. Soap flakes are made by passing chips through milling rollers to make thin, polished, easily soluble flakes.

Powdered soap is made from chips by further reducing the mixture and grinding. **Milled soaps** are made from chips by adding color and perfumes to the dry chips and then passing through milling rollers and finally pressing in molds. **Toilet soaps** are made in this way. Soap is used widely in industrial processing, and much of the production has consisted of chips, flakes, powdered, granulated, and scouring powders. Soaps have definitely limitations of use. They are unstable in acid solutions and may form insoluble salts. In hard waters they may form insoluble soaps of calcium or magnesium unless a phosphate is added. Many industrial cleansers, therefore, may be balanced combinations of soaps, synthetic detergents, phosphates, or alkalies, designed for particular purposes.

About half of all soap is made with tallow, 25% with coconut oil, and the remainder with palm oil, greases, fish oils, olive oil, soybean oil, or mixtures. A typical soap contains 80% mixed oils and 20 coconut oil, with not over 0.2 free alkali. Auxiliary ingredients are used in soap to improve the color, for perfuming, as an astringent, or for abrasive or harsh cleaning purposes. Phenol or cresylic acid compounds are used in **antiseptic soap**. The soft soaps and liquid soaps of USP grade have a therapeutic value and may be sold under trade names.

Solvents are added to industrial soaps for scouring textiles or when used in soluble oils in the metal industry. Zinc oxide, benzoic acid, and other materials are used in facial soaps with the idea of aiding complexion. Excessive alkalinity in soaps dries and irritates the skin, but **hand grit soap** usually has 2%–5% alkaline salts such as borax or soda ash and 10%–25% abrasive materials. Softer **hand soap** may contain marble flour. Silicate of soda, used as a filler, also irritates the skin. **Face soaps**, or **toilet soaps**, contain coloring agents, stabilizers, and perfuming agents. For special purposes, **cosmetic soaps** contain medications. **Deodorant soaps** contain antibacterial chemicals, such as **triclosan**, which inhibit the production of bacteria on the skin. Experts disagree on whether antibacterial ingredients are harmful to the skin. Some, such as **Dove**, are a blend of detergents and soap. **Castile soap** is a semitransparent soap made with olive oil. **Marseilles soap** and **Venetian soap** are names for castile soap with olive oil and soda. Ordinary soft soaps are used as bases for toilet soap are made with mixtures of linseed oil and olive oil. Linseed oil, however, gives a disagreeable odor. Soybean oil, corn oil, and peanut oil are also used, although peanut oil, unless the arachidic acid is removed, makes a hard soap. **Tall oil soaps** are sodium soaps made from the fatty acids of tall oil. They are inferior to sodium oleate in detergency, but superior to sodium rosinate. Many toilet soaps contain excess unsaponified oil, fatty acid, or lanolin and are known as **superfatted**.

Saddle soap is any soap used for cleaning leather goods which has the property of filling and smoothing the leather as well as cleaning. The original saddle soaps were made of palm oil, rosin, and lye, with glycerin and beeswax added. Oils for the best soaps are of the nondrying type. High-grade **soft soap** for industrial use is made with coconut or palm kernel oil with caustic potash. But soft soap in paste form is generally made of low-titer oils with caustic soda, usually linseed, soybean, or corn oil. The lauric acid of coconut oil gives the coconut-oil soaps their characteristic of profuse lathering, but lauric acid affects some skins by causing itching, and soaps when a high-coconut-oil content and low titer are also likely to break down in hot water and wash ineffectively. Palm-kernel oil develops free acids, and upon aging the soap acquires the odor of the oil. Palm oil produces a crumbly soap. It does not lather freely, but is mild to the skin. Olive oil is slow-lathering, but has good cleansing powers. It is often used in textile soaps. Cottonseed oil is used in some laundry soaps but develops yellow spots in the soap. Corn oil with potash makes a mild soft soap. Soybean oil also makes a soft soap. Rosin is used to make yellow laundry soaps. ASTM standards for **bar soap** permit up to 25% rosin. Sulfonated oils do not give as good cleansing action as straight oils, but are used in shampoos where it is desirable to have some oil or greasiness. Blending of various oils is necessary to obtain a balance of desired characteristics in a soap. Hand soaps may be made with trisodium phosphate or with disodium phosphate, or sodium perborate, known as **per-borin**, all of which are crystalline substances which are dissolved in water solution. **Soap powder** is granular soap made in a vacuum chamber or by other special processes. It usually contains 15%–20% soap and the balance sodium carbonate. **Scouring powder** is an intimate mixture of soap powder and in insoluble abrasives such as pumice. **Floating soaps** are made light by blowing air through them while in the vats. **Soapless shampoos** and tooth powders contain saponin or chemical detergents. **Liquid soaps** are made by saponification with potassium and ammonium hydroxide, or triethanolamine, to produce more-soluble products. The floating soaps, such as **Ivory** from Proctor & Gamble Co., are made by injecting air into the molten soap.

Soapstone

The bulk form of talc, also termed steatite, hydrated magnesium silicate. A very soft mineral, it is No. 1 on the Mohs scale of hardness. It has a high electrical resistance and is hence used for insulators. It is readily carved for ornamental applications. In powder form it is termed *French Chalk* or, if perfumed, *Talcum Powder.*

A massive variety of impure talc is employed for electric panels, gas-jet trips, stove linings, tank linings, and as an abrasive. It can be cut easily and becomes very hard when heated because of the loss of its combined water. The waste product from the cutting of soapstone is ground and used for the same purposes as talc powder. **Steatite** is a massive stone rich in talc that can be cut readily, while soapstone may be low in talc. When free of iron oxide and other impurities, blocks of steatite are used for making spacer insulators for electronic tubes and for special electrical insulators. **Block steatite** suitable for electrical insulation is mined in Montana, India, and Sardinia. Steatite is also ground and molded into insulators. It can be purified of iron and other metallic impurities by electrolytic osmosis. When fluxed with alkaline earths instead of feldspar, the molded steatite ceramics have a low loss factor at high frequencies, and have good electrical properties at high temperatures. **Talc crayons** for marking steel are sticks of soapstone.

Soda Ash

Sodium carbonate (Na_2CO_3) obtained from trona, a hydrated sodium carbonate sodium bicarbonate ore. Soda ash is used in petroleum refining and for soaps and detergents. Its primary use, however, is in glass manufacture. Soda ash is the third major constituent of soda-lime-silica glasses and is the main source of Na_2O in any glass that contains soda. It acts as a flux, reducing the temperature required to melt the silica. See also *Solvay process.*

The common name for anhydrous **sodium carbonate**, which is the most important industrial alkali. It is a grayish-white, lumpy material which loses any water of crystallization when heated. For household used in hydrous crystallize form it is called **washing soda**, **soda crystals**, or **sal soda**, as distinct from **baking soda**, which is **sodium hydrogen carbonate**, or **sodium bicarbonate**. Sal soda contains more than 60% water. Another grade, with one molecule of water, is the standard product for scouring solutions. Federal specifications call for this product to have a total alkalinity not less than 49.7% Na_2O. Commercial high-quality soda ash contains 99% minimum sodium carbonate, or 58 minimum Na_2O. It varies in size of particle and in bulk density, being marketed as extra-light, light, and dense. **Laundry soda** is soda ash mixed with sodium bicarbonate, with 39%–43% Na_2O. **Modified sodas**, used for cleansing where a mild detergent is required, are mixtures of sodium carbonate and sodium bicarbonate. They are used in both industrial and household cleaners. **Tanners' alkali**, used in processing fine leathers, and **textile soda**, used in fine wool and cotton textiles, are modified sodas. **Flour bland**, used by the milling industry in making free-flowing, self-raising food flours, is a mixture of sodium bicarbonate and tricalcium phosphate.

Soda ash is made by the Solvay process, which consists of treating a solution of common salt with ammonia and with carbon dioxide and calcining the resulting filter cake of sodium bicarbonate to make **light soda ash**. **Dense soda ash** is then made by adding water and recalcining. Soda ash is less expensive than caustic soda and is used for cleansing, for softening water, in glass as a flux and to prevent fogging, in the wood-pulp industry, for refining oils, and soapmaking, and for the treating of ores. **Caustic ash**, a strong cleaner for metal scouring and for paint removal, is a mixture of about 70% caustic soda and 30% soda ash. **Flake alkali**, of PPG Industries, contains 71% caustic soda and 29% soda ash. Soda ash is also used as a flux in melting iron to increase the fluxing action of the limestone, as it will carry off 11% sulfur in the slag. **Soda briquettes**, used for desulfurizing iron, are made of soda ash formed into pellets with a hydrocarbon bond. **Hennig purifier** is soda ash combined with other steel-purifying agents made into pellets.

The salt brine of Owens Lake, California, is an important source of soda ash. The brine, which contains 10.5% sodium carbonate and 2.5% **sodium borate decahydrate**, is concentrated and treated to precipitate the **trona** (Na_2CO_3). The Salt Lake area of Utah is a source of trona. Soda ash and sodium carbonate may be sold under trade names, **Purite** is a sodium carbonate. **Tronacarb** is an industrial grade, and **Tronalight**, as the name suggests, is a light soda ash.

Soderberg Diagram

See *fatigue.*

Sodium

A metallic element (symbol Na and atomic weight 23), sodium occurs naturally only in the form of its salts. The most important mineral containing sodium is the chloride, NaCl, which is common salt. It also occurs as the nitrate, Chile saltpeter, as a borate in borax, and as a fluoride and a sulfate. When pure, sodium is silvery white and ductile, and it melts at 97.8°C and boils at 882°C. The specific gravity is 0.97. It can be obtained in metallic form by the electrolysis of salt. When exposed to the air, it oxidizes rapidly, and must therefore be kept in airtight containers. It has a high affinity for oxygen, and it decomposes water violently. It also combines directly with the halogens, and is a good reducing agent for the metal chlorides. Sodium is one of the best conductors of electricity and heat.

The metal is a powerful desulfurizer of iron and steel even in combination. For this purpose it may be used in the form of soda ash pellets or in alloys. Desulfurizing alloys for brasses and bronzes are sodium–tin, with 95% tin and 5% sodium, or sodium–copper. Sodium–lead, used for adding sodium to alloys, contains 10% sodium, and is marketed as small spheroidal shot. It is also marketed as **sodium marbles**, which are spheres of pure sodium up to 2.54 cm in diameter coated with oil to reduce handling hazard.

Sodium bricks contain 50% sodium metal powder dispersed in a paraffin binder. They can be handled in the air, and are a source of active sodium. Sodium in combination with potassium is used as a heat-exchange fluid in reactors and high-temperature processing equipment. A sodium–potassium alloy, containing 56% sodium and 44% potassium, has a melting point of 19°C and a boiling point of 825°C. It is a silvery mobile liquid. High-surface sodium is sodium metal absorbed on common salt, alumina, or activated carbon to give a large surface area for use in the reduction of metals or in hydrocarbon refining. Common salt will adsorb up to 10% of its weight of sodium in a thin film on its surface, and this sodium is 100% available for chemical reaction. It is used in reducing titanium tetrachloride to titanium metal. Sodium vapor is used in electric lamps. When the vapor is used with a fused alumina tube it gives a golden-white color. A 400 W lamp produces 42,000 lumens and retains 85% of its efficiency after 6,000 h.

Inorganic Reactions

Sodium reacts rapidly with water, and even with snow and ice, to give sodium hydroxide and hydrogen. The reaction liberates sufficient heat to melt the sodium and ignite the hydrogen.

When exposed to air, freshly cut sodium metal loses its silvery appearance and becomes dull gray because of the formation of a coating of sodium oxide. Sodium probably oxidizes to the peroxide, Na_2O_2, which reacts with excess sodium present to give the monoxide, Na_2O. When sodium reacts with oxygen at elevated temperatures, sodium super oxide, NaO_2, is formed; this reacts with more sodium to form the peroxide.

Sodium does not react with nitrogen, even at very high temperatures. Sodium and hydrogen react above about 200°C to form sodium hydroxide. This compound decomposes at about 400°C and cannot be melted. Sodium hydride can be formed by the direct reaction of hydrogen and molten sodium or by hydrogenating dispersions of sodium metal in hydrocarbons. Sodium reacts with carbon with difficulty, if at all, and this reaction may be said to have been adequately studied.

At room temperature fluorine and sodium ignite, dry chlorine and sodium react slightly, bromine and sodium do not react, and iodine and sodium do not react. However, in the presence of moisture or at elevated temperatures all reactions take place at very high rates.

Sodium reacts with ammonia, forming sodium amide and liberating hydrogen. The reaction may be carried out between molten sodium and gaseous ammonia (–30°C) in the presence of catalysts of finely divided metals. Sodium reacts with ammonia in the presence of coke to form sodium cyanide.

Carbon monoxide reacts with sodium, but the resulting carbonyl, NaCO, is stable only at liquid ammonia temperatures. At high temperatures sodium carbide and sodium carbonate are formed from carbon monoxide and sodium.

The reactions of sodium with various metal halides to give the metal plus sodium chloride are very important. Thus, titanium tetrachloride is reduced to titanium metal. Similarly, the halides of zirconium, beryllium, and thorium can be reduced to the corresponding metals by sodium. The interaction between sodium and potassium chloride is used in the commercial production of potassium metal.

Sodium hydroxide, NaOH, is also commonly known as caustic soda, and also as sodium hydrate. Lye is an old name used in some industries and in household uses. It readily absorbs water from the atmosphere and must be protected in storage and handling. It is corrosive to the skin and must be handled with extreme care to avoid caustic burns.

Most sodium hydroxide is produced by the electrolysis of sodium chloride solutions in one of several types of electrolytic cells. An older processes the soda-lime process whereby soda ash is converted to caustic soda.

Organic Reactions

Sodium does not react with paraffin hydrocarbons but does form additional compounds with naphthalene and other polycyclic aromatic compounds and with arylated alkenes. It reacts with acetylene, replacing the acetylenic hydrogens to form sodium acetylides. Sodium adds to dienes, the reaction which forms the basis of the buna synthetic rubber process.

Principal Compounds

Sodium compounds are widely used in industry, particularly sodium chloride, sodium hydroxide, and soda ash. Sodium bichromate, $Na_2Cr_2O_7 \cdot 2H_2O$, a red crystalline powder, is used in leather tanning, textile dyeing, wood preservation, and in pigments. Sodium metavandate, $NaVO_3$, is used as a corrosion inhibitor to protect some chemical-processing piping. It dissolves in hot water, and a small amount in the water forms a tough impervious coating of magnetic iron oxide on the walls of the pipe. Sodium iodide crystals are used as scintillation probes for the detection and analysis of nuclear energies. Sodium oxalate is used as an antienzyme to retard tooth decay. In the drug industry sodium is used to compound with pharmaceuticals to make them water-soluble salts. Sodium is a plentiful element, easily available, and is one of the most widely used.

Sodium carbonate is best known under the name soda ash because sodium carbonate occurs in (and once was extracted from) plant ashes. Most sodium carbonate is produced by the Solvay or ammonia-soda process. In an initial reaction, salt is converted to sodium carbonate, which precipitates and is then separated.

Some soda ash is made synthetically by the Solvay process although an increasing amount is obtained from lake brines. Commercial grades of soda ash are available as 48% (Na_2O) light and dense and has 58% (Na_2O) light and dense; light and dense refers to apparent bulk density. Ordinarily 48%–58% grades are available in either light or dense but contain NaCl, which may affect certain ceramic uses. A 48% special grade is available in granular and extra light forms; it contains Na_2SO_4. The material derived from natural sources is almost NaCl-free.

About one half of the total American soda ash production is used as a fluxing ingredient by the glass industry. The quality of soda ash in glass batches varies with the type of glass being made.

Sodium sulfate, Na_2SO_4, is also known in the anhydrous form as salt cake. The decahydrate, $Na_2SO_4 \cdot 10\,H_2O$, is known as glauber salt.

Most sodium sulfate is produced synthetically as a byproduct or coproduct in various industries.

Sodium aluminate, $Na_2OAl_2O_3$, whose melting point is 1650°C, is soluble in water and sodium carbonate. Sodium aluminate has found use as a setting up agent for acid-resistant enamel. It is prepared by heating together bauxite and slips. When used in this capacity, it affords easier control of the slip than can be obtained by the use of alum or sulfuric acid, because of its tendency to stabilize the mobility and yield values. Sodium aluminate is also used as a substitute for sodium silicate and sodium carbonate and pottery slips.

Sodium antimonate (sodium meta-antimonate), $Na_2OSb_2O_5 \cdot 5H_2O$, is a white powder insoluble in water and fruit acids. Sodium antimonate is extremely stable at high temperatures and does not decompose below 1427°C. It is usually made from antimony oxide, caustic soda, and sodium nitrate. Sodium antimonate is used as the principal opacifier in dry-process enamel frits for cast iron sanitary-ware and in some of the acid-resistant enamel frits for sheet steel. It is used in cast iron enamels; sodium antimonate is generally recognized as being more desirable than antimony trioxide.

Sodium cyanide is a salt of hydrocyanic acid of the composition NaCN, used for carburizing steel for case hardening, for heat-treating baths, for electroplating, and for the extraction of gold and silver from their ores. For carburizing steel it is preferred to potassium cyanide because of its lower cost than its higher content of available carbon. It contains 53% CN, as compared with 40% of potassium cyanide. The nitrogen also aids in forming the hard case on the steel. The 30% grade of sodium cyanide, melting at 679°C, is used for heat-treating baths instead of lead, but it forms a slight case on the steel. Sodium cyanide is very unstable, and on exposure to moist air liberates the highly poisonous hydrocyanic acid gas, HCN. For gold and silver extraction it easily combines with the metals, forming soluble double salts, $NaAu(CN)_2$. Sodium cyanide is made by passing a stream of nitrogen gas over a hot mixture of sodium carbonate and carbon in the presence of a catalyst. It is a white crystalline powder, soluble in water. The white copper cyanide used in electroplating has the composition $Cu_2(CN)_2$, containing 70% copper. It melts at 474.5°C and is insoluble in water, but is soluble in sodium cyanide solution. Sodium ferrocyanide, or yellow prussiate of soda, is a lemon-yellow crystalline solid, used for carburizing steel for case hardening. It is also employed in paints, in printing inks, and for the purification of organic acids; in minute quantities, it is used in salt to make it free-flowing. It is soluble in water. Calcium cyanide in powder or granulated forms is used as an insecticide. It liberates 25% of hydrocyanic acid gas.

Sodium nitrate (soda niter), $NaNO_3$, with a melting point of 208°C, and decomposes at 380°C and is soluble. Sodium nitrate is used in enamel frits in quantities of 2%–8%. It is highly important that sufficient nitrate be present in enamels to prevent reduction of any easily reducible compounds in the batch, especially lead or antimony compounds. The function of sodium nitrate in glass is to oxidize organic matter that may contaminate batch materials, to prevent reduction of some of the batch constituents, to help maintain colors, and to speed the melt. It is the lowest melting of all glassmaking materials. Common applications of sodium nitrate are to ensure the pink color of manganese oxide and to prevent reductions of lead in potash lead glasses.

Sodium nitrite, $NaNO_2$, is soluble in water. It is prepared from sodium nitrate by reduction with lead. Sodium nitrite has been used for some years as a mill addition, or as an addition after milling, to enamel ground coats to prevent rust while drying, and also as a setting-up agent. More recently, sodium nitrite has been used rather generally in cover coats to correct for tearing.

Sodium phosphate, has a melting point of 346°C and is soluble in water. Sodium phosphate has been recently added to glass batches, producing an opal glass of unusual properties. Three other forms of the phosphate are available—monobasic, tribasic, and pyrophosphate. The last is most adaptable because it melts at 970°C in the anhydrous form. It is derived by the fusion of disodium phosphate.

Sodium silicate, is commonly made by melting sand and soda ash in a reverberatory furnace. Various proportions of the two ingredients are used and widely divergent characteristics result. The most alkaline liquid silicate made by this furnace process has a ratio of $1Na_2O:1.6SiO_2$ and the most siliceous liquid grade has a ratio of $1Na_2O:3.75SiO_2$.

Uses

The largest single use for sodium metal, accounting for about 60% of total production, is in the synthesis of tetraethyllead, and anti-knock agent for automotive gasolines.

A second major use is in the reduction of animal and vegetable oils to long-chain fatty alcohols; these alcohols are raw materials for detergent manufacture. This use has been decreasing in favor of production of such alcohols by high-pressure catalytic hydrogenation.

Sodium metal is also used in making sodium hydride, sodium amide, and sodium cyanide. It is also used in the synthesis of "isosebacic acid." The use of liquid sodium metal as a heat-transfer agent in nuclear reactors is also becoming increasingly important.

Sodium chloride is used in the manufacture of sodium hydroxide, sodium carbonate, sodium sulfate, and sodium metal. In sodium sulfate manufacture, hydrogen chloride is the coproduct; in metallic sodium manufacture, chlorine gas as is the coproduct.

Rock salt is used in curing fish, in meat packing, in curing hides, and in making freezing mixtures. Food preparation, including canning and preserving, consumes much salt. Table salt accounts for only a small percentage of sodium chloride consumption, most of it going into the industrial uses outlined above.

Sodium hydroxide is perhaps the most important industrial alkali. Its major use is in the manufacture of chemicals, about 30% attributed to this category. The next major use is the manufacture of cellulose film and rayon, both of which proceed through soda cellulose (the reaction product of sodium hydroxide and cellulose); this accounts for about 25% of the total caustic soda production. Soap manufacture, petroleum refining, and pulp and paper manufacture each account for a little less than 10% of total sodium hydroxide use.

Sodium carbonate finds its major use in the glass industry, which takes about one third of total production. Approximately another third goes into the manufacture of soap, detergent, and various cleansers. The manufacture of paper and textiles, nonferrous metals, and petroleum products accounts for much of the balance.

The major consumer of sodium sulfate (salt cake) is the kraft paper industry. Increasing quantities of sodium sulfate are used in the manufacture of flat glass. Other uses of salt cake are in detergents, ceramics, mineral stock feeds, and pharmaceuticals.

In the area of biological activity, the sodium ion (Na^+) is the main positive ions present in extracellular fluids and is essential for maintenance of the osmotic pressure and of the water and electrolytic balances of body fluids.

Soft

The opposite of hard in its variations.

Softening

Various processes of reducing hardness and tensile strength and, usually increasing ductility, malleability and toughness. See *annealing*, *tempering*, and *normalizing*.

Softening Point

That temperature at which a glass fiber of uniform diameter elongates at a specific rate under its own weight when measured by standard ASTM test methods. The viscosity at the softening point depends on the density and surface tension. For example, for a glass of density 2.5 g/cm^2 and surface tension 300 dynes/cm, the softening point temperature corresponds to a viscosity of 106.6 Pa s.

Softening Range

The range of temperatures within which a plastic changes from a rigid to a soft state. Actual values depend on the test method. Sometimes erroneously referred to as softening point.

Soft Magnetic Materials

Magnetic materials are broadly classified into two groups with either hard or soft magnetic characteristics. Hard magnetic materials are characterized by retaining a large amount of residual magnetism after exposure to a strong magnetic field. These materials typically have coercive force, H_c, values of several hundred to several thousand oersteds (Oe) and are considered to be permanent magnets. The coercive force is a measure of the magnetizing force required to reduce the magnetic induction to zero after the material has been magnetized. In contrast, soft magnetic materials become magnetized by relatively low strength magnetic fields, and when the applied field is removed, they return to a state of relatively low residual magnetism. Soft magnetic materials typically exhibit coercive force values of approximately 5 Oe to as low as 0.002 Oe. Soft magnetic behavior is essential in any application involving changing electromagnetic induction, such as solenoids, relays, motors, generators, transformers, magnetic shielding, and so on. Other important characteristics of magnetically soft materials include high permeability, high saturation induction, low hysteresis-energy loss, low eddy-current loss in alternating flux applications, and in specialized cases, constant permeability at low field strengths and/or a minimum or definite change in permeability with temperature.

Cost, availability, strength, corrosion resistance, and ease of processing are among the key factors that influence the final selection of a soft magnetic material. Magnetically soft materials manufactured in large quantities include high-purity irons, low-carbon (≤0.08% C) steels that contain additions of phosphorus (0.03%–0.15%) and manganese (0.25%–0.75%) to increase the electrical resistivity, silicon (electrical) steels containing 2%–3.5% Si, iron-nickel alloys with nickel contents ranging from 45% to 79%, iron–cobalt alloys (for example, 49Fe–49Co–2V and Fe–27Co–0.6Cr), ferritic stainless steels, and ferrites (manganese–zinc and nickel–zinc in particular). Soft magnetic amorphous materials are also being produced. See *metallic glass*.

Soft Solder

Alloys primarily of lead and tin. They meld at low temperatures to provide a fairly strong joint and seal between metals such as lead, steel (usually pre-tinned), copper or copper alloys. The term soft solder is in contradiction to hard solder, the latter term being applied to braze filler metals and silver solders which, apart from higher strength and hardness, have melting range commencing above 450°C, the usually accepted differentiation between solder and brazing temperatures. The soft solders offer a variety of compositions with a useful variation in freezing ranges. Plumbers solder,

70% lead, 30% tin, has a wide freezing range, 250°C–183°C, with a long pasty stage allowing the joint to be "wiped" to a smooth profile. Electricians solder, also termed Tinmans Solder is of approximately eutectic composition, 62% tin, 38% lead, freezing at 183°C. The soldering processes and techniques are largely similar to brazing apart from the lower temperatures and lower strength. However, because of the lower temperature, small solder joints can be made with the hard tip or "bit" of a soldering iron. Another common process is wave soldering in which the components to be soldered, for example, circuit boards, are suspended joint face down, just above a bath of solder. A wave is then introduced in the surface of the bath and the component passed over skimming the wave. In most cases a flux is applied prior to, or during, soldering to clean the surfaces of oxide and other contaminants which would inhibit wetting and bonding. See preferred term *solder*.

Soft Soldering

See preferred term *soldering*.

Soft Temper

Same as *dead soft* temper.

Soft Water

Water that is free of magnesium or calcium salts.

Soil

Undesirable material on a surface that is not an integral part of the surface. Oil, grease, and dirt can be soils; a decarburized skin and excess *hard chromium* are not soils. Loose scale is soil; hard scale may be an integral part of the surface and, hence, not soil.

Sol

A colloidal suspension comprised of discrete or separate solid particles suspended in a liquid. Differs from a solution, though one merges into the other. Compare with *gel*.

Solder Alloys

These are alloys of two or more metals used for joining other metals together by surface adhesion without melting the base metals as in welding and without requiring as high a temperature as in brazing. However, there is often no definite temperature line between soldering alloys and brazing filler metals. A requirement for a true solder is that it have a lower melting point than the metals being joined and an affinity for, or be capable of uniting with, the metals to be joined.

Types and Forms

The most common solder is called half-and-half, plumbers' solder, or ASTM solder class 50A, and is composed of equal parts of lead and tin. It melts at 183°C. The density of the solder is 8802 kg/m^3, the tensile strength is 39 MPa, and the electrical conductivity is 11% that of copper. SAE (Society of Automotive Engineers) solder No.1 has 49.5%–50.0% tin, 50% lead, 0.12% max antimony, and 0.08% max copper. It melts at 181°C. Much commercial half-and-half, however, usually contains larger proportions of lead and some antimony, with less tin. These mixtures have higher melting points, and solders with less than 50% tin have a wide melting range and do not solidify quickly. Sometimes a wide melting range is desired, in which case a wiping solder with 38%–45% of tin is used. A narrow-melting range solder, melting at 183°C–185°C, ASTM solder class 60A, contains 60% tin and 40% lead. A 42% tin and 58% lead solder has a melting range of 183°C–231°C. Slicker solder is the best quality of plumbers' solder, containing 63%–66% tin and the balance lead.

Solder alloys are available in a wide range of sizes and shapes, enabling users to select that one that best suits their application. Among these shapes are pig, slab, cake or ingot, bar, paste, ribbon or tape, segment or drop, powder, foil, sheet, solid wire, flux cored wire, and preforms. There are 11 major groups of solder alloys:

Tin-Antimony

Useful at moderately elevated operating temperatures, around 149°C, these solders have higher electrical conductivity than the tin-lead solders. They are recommended for use where lead contamination must be avoided. A 95% Sn–5% Sb alloy has a solidus of 235°C, a liquidus of 240°C, and a resulting pasty range of −11.1°C.

Tin–Lead

Constituting the largest group of all solders in use today, the tin–lead solders are used for joining a large variety of metals. Most are not satisfactory for use above 149°C under sustained load.

Tin-Antimony–Lead

These may normally be used for the same applications as tin-lead alloys with the following exceptions: aluminum, zinc, or galvanized iron. In the presence of zinc the solders form a brittle intermetallic compound of zinc and antimony.

Tin–Silver

These have advantages and limitations similar to those of tin-antimony solders. The tin silvers, however, are easier to apply with a rosin flux. Relatively high cost confines these solders to fine instrument work. Two standard compositions; 96.5% Sn–3.5% Ag, the eutectic; 95% Sn–5% Ag, with a solidus of 221°C is and liquidus of 245°C.

Tin–Zinc

These are principally for soldering of aluminum since they tend to minimize galvanic corrosion.

Lead–Silver

Tensile, creep, and shear strengths of the solders are usually satisfactory up to 177°C. Flow characteristics are rather poor and these solders are susceptible to humid atmospheric corrosion in storage. The use of a zinc chloride-base flux is recommended to produce a good joint on metals uncoated with solder.

Cadmium–Silver

The primary use of cadmium–silver solder is in applications where surface temperature will be higher than permissible with lower melting solder. Improper use may lead to health hazards. The solder has a composition of 95% Cd–5% Ag. Solidus is 338°C and liquidus is 393°C.

Indium–Lead Alloys

These are alkali-resistant solders. A solder with 50% lead and 50% indium melts at 182°C, and is very resistant to alkalies, but lead–tin solders with as little as 25% indium are resistant to alkaline

solutions, have better wetting characteristics, and are strong. Indium solders are expensive. Adding 0.85% silver to a 40% tin soft solder gives equivalent wetting on copper alloys to a 63% tin solder, but the addition is not effective on low-tin solders. A gold–copper solder used for making high-vacuum seals and for brazing difficult metals such as iron–cobalt alloys contain 37.5% gold and 62.5% copper.

Palladium–Nickel

A palladium–nickel alloy with 40% nickel has a melting point about 1237°C. The brazing filler metals containing palladium are useful for a wide range of metals and metal to ceramic joints.

The remaining four groups of Zn–Al, Cd–Zn, and solders containing bismuth and indium were covered earlier.

Silver Solder

Silver solder is high-melting-point solder employed for soldering joints where more than ordinary strength and, sometimes, electric conductivity are required. Most silver solders are copper–zinc brazing filler metals with the addition of silver. They may contain 9% to 80% silver, and the color varies from brass yellow to silver white. Cadmium may also be added to lower the melting point. Silver solders do not necessarily contain zinc, and brazing filler metals of silver and copper in proportions are arranged to obtain the desired melting point and strength. A silver braze filler metal with a relatively low melting point contains 65% silver, 20% copper, and 15% zinc. It melts at 693°C, has a tensile strength of 447 MPa, and elongation 34%. The electrical conductivity is 21% that of pure copper. A solder melting at 760°C contains 20% silver, 45% copper, and 35% zinc. ASTM silver solder No. 3 is this solder with 5% cadmium replacing an equal amount of the zinc. It is general-purpose solder. ASTM silver solder No. 5 contains 50% silver, 34% copper, and 16% zinc. It melts at 693°C, and is used for soldering electrical work and refrigeration equipment.

Any tin present in silver solders makes them brittle; lead and iron make the solders difficult to work. Silver solders are malleable and ductile and have high strength. They are also corrosion resistant and are especially valuable for use in food machinery and apparatus where lead is objectionable. Small additions of lithium to silver solders increase fluidity and wetting properties, especially for brazing stainless steels or titanium. Sil-Fos is a phosphor-silver brazing solder with a melting point of 704°C. It contains 15% silver, 80% copper, and 5% phosphorus. Lap joints brazed with Sil-Fos have a tensile strength of 206 MPa. The phosphorus in the alloy acts as a deoxidizer, and the solder requires little or no flux. It is used for brazing brass, bronze, and nickel alloys.

Another grade, Easy solder, contains 65% silver, melts at 718°C, and is a color match for sterling silver. TL silver solder has only 9% silver and melts at 871°C. It is brass yellow in color, and is used for brazing nonferrous metals. Sterling silver solder, for brazing sterling silver, contains 92.8% silver, 7% copper, and 0.2% lithium. Flow temperature is 899°C.

A lead–silver solder to replace tin solder contains 96% lead, 3% silver, and 1% indium. It melts at 310°C, spreads better than ordinary lead-silver solders, and gives a joint strength of 34 MPa. Silver–palladium alloys for high-temperature brazing contains from 5% to 30% palladium. With 30%, the melting point is about 1232°C. These alloys have exceptional melting and flow qualities and are used in electronic and spacecraft applications.

Cold Solders

Cold solder, use for filling cracks in metals, may be a mixture of a metal powder in a pyroxylin cement with or without a mineral filler, but the strong cold solders are made with synthetic resins, usually epoxies, cured with catalysts, and with no solvents to cause shrinkage. The metal content may be as high as 80%. Devcon F, for repairing holes in castings, has 80% aluminum powder and 20% epoxy resin. It is heat-cured at 66°C, giving high adhesion. Epoxyn solder is aluminum powder in an epoxy resin in the form of a putty for filling cracks or holes in sheet metal. It cures with a catalyst. The metal-epoxy mixtures give a shrinkage of less than 0.2%, and they can be machined and polished smooth.

Lead-Free Solder Replacements

In spite of the sustained efforts of researchers and technology leaders in packaging, to date there is no lead-free solder alloy that is a drop-in replacement for tin–lead solder in assembly processes. Because tin–lead solder has been used for so long and is so much a part of the typical process engineer's thinking, the quest for an affordable lead-free alloy replacement is facing mounting pessimism that the effort will be successful.

On the other hand, optimism that adhesive-type solders will prove to be a serious alternative to metallurgical materials continues to grow. These polymer-based conductive adhesives are being used in various applications previously "reserved" for tin–lead solders. Regular production equipment and traditional assembly processes are producing high-quality assemblies with demonstrated long-term reliability using the new solders. And for some products, polymers are considered an enabling technology. One major market is the polyester-based flexible circuit market, particularly those built using polymer thick film.

Polymer Solders

Polymer solders are also known as "conductive adhesives," or materials that provide the dual functions of electrical connection and mechanical bond. The adhesive components of a typical material are some form of polymer, i.e., long-chain molecules widely used to produce structural products, which are also known for their excellent dielectric properties. Already used extensively as electrical insulators, as solders their necessary conductivity is accomplished by adding highly conductive fillers to the polymer binders.

The most common polymer solders are silver-filled thermosetting epoxies supplied as one-part thixotropic pastes. Silver is used not only because it is usually cost-effective, but also for its unique conductive oxide. A blend of silver powder and flakes achieves high conductivity while maintaining good printability. Because the mechanical strength of the joint is provided by the polymer, the challenge in a formulation is to use the maximum metal loading without sacrificing the required strength. (Some polymer solders contain more than 80% metal filler by weight.)

Polymer solders do not typically form metallurgical interfaces in the usual sense. Electrical integrity requires that the metal filler particles be in close contact to form a conductive path between the circuit trace and the component lead. Ideally, the silver flakes will overlap and smaller particles will fill in the gaps to form a conductive chain.

In the past, polymer solders were successful only on circuits using precious-metal conductors. This was because junction resistance was seen to increase to unacceptable levels when ordinary printed circuit boards and components were joined with these materials. To solve the instability problem it was necessary to create stable, nonmetallurgical junctions with oxidizable surfaces. For example, one polymer-solder formulation, used on flexible circuits and recently optimized for rigid boards, provides junction stability between solder-coated and bare-copper surfaces via polymer shrinkage during curing to force irregular particles through the interface oxides.

Advantages in using polymer solders include compatibility with the range of surfaces including some nonsolderable substrates; low-temperature processing, resulting in lower thermal stress; no pre- or postclean requirements, thereby reducing equipment needs and cycle times, and lowering or eliminating the release of volatile organic compounds.

Solderability

The relative ease and speed with which a surface is wetted by molten solder.

Solder Embrittlement

Reduction in mechanical properties of a metal as a result of local penetration of solder along grain boundaries.

Solder Lock

The phenomenon whereby if a soft soldered joint between copper components is heated for too long a period, including remelting, the copper and solder mutually diffuse causing the melting point of the solder to increase to such an extent that it cannot be melted by the usual heating system. Plumbers sometimes use the term "Seized Joint" for the phenomenon.

Solder Materials

Fluxes

Fluxes range from very mild substances to those of extreme chemical activity. For centuries, rosin, a pine product, has been known as an effective and practically harmless flux. It is used widely for electrical connections in which utmost reliability, freedom from corrosion, and absence of electrical leakage are essential. When less stringent requirements exist and when less carefully prepared surfaces are to be soldered, rosin is mixed with chemically active agents that aid materially in soldering.

The rosin-type fluxes may be incorporated as the core of wire solders or dissolved in various solvents for direct application to joints prior to soldering.

Inorganic salts are widely used where stronger fluxes are needed. Zinc chloride and ammonium chloride, separately or in combination, are most common. They may also be obtained as so-called acid-core solder wire or in petroleum jelly as paste flux. All of the salt-type fluxes leave residues after soldering that may be a corrosion hazard. Washing with ample water accomplished by brushing is generally wise.

Solder Paint

A suspension of powdered soft solder and flux which can be painted onto the surface to be soldered. After heating to fuse the solder and form the joint the flux is cleaned off.

Soldering

A group of processes that join metals by heating them to a suitable temperature below the solidus of the base metals and applying a filler metal having a liquidus not exceeding 450°C. Molten filler metal is distributed between the closely fitted surfaces of the joint by capillary action. See also *solder.*

Soldering Flux

See *flux* (2).

Soldering Gun

An electrical soldering iron with a pistol grip and a quick heating, relatively small bit.

Soldering Iron

A soldering tool having an internally or externally heated metal bit usually made of copper.

Solder Short

See *bridging* (5).

Sol–Gel Process

This is a chemical synthesis technique for preparing gels, and glasses, and ceramic powders. The synthesis of materials by the sol–gel process generally involves the use of metal alkoxides, which undergo hydrolysis and condensation polymerization reactions to yield gels.

The production of glasses by the sol–gel method is an area that has important scientific and technological implications. For example, the sol–gel approach permits preparation of glasses at far lower temperatures than is possible by using conventional melting. It also makes possible synthesis of compositions that are difficult to obtain by conventional means because of problems associated with volatile authorization, high melting temperatures, or crystallization. In addition, the sol–gel approach is a high-purity process that leads to excellent homogeneity. Finally, the sol–gel approach is adaptable to producing films and fibers as well as bulk pieces, that is, monoliths (solid materials are macroscopic dimensions, at least a few millimeters on a side).

Glass Formation

Formation of silica-based materials is the most widely studied system. However, an anonymous range of multicomponent silicate glass compositions have also been prepared.

The sol–gel process can ordinarily be divided into the following steps: forming a solution, adulation, drying, and densification.

Hydrolysis and Condensation

In general, the processes of hydrolysis and condensation polymerization are difficult to separate. The hydrolysis of the alkoxide need not be complete before condensation starts; and in partially condensed silicate, hydrolysis can still occur at unhydrolyzed sites. Several parameters have been shown to influence the hydrolysis and condensation polymerization reactions: these include the temperature, solution pH, the particular alkoxide precursor, the solvent, and the relative concentration of each constituent. In addition, acids and bases catalyze the hydrolysis and condensation polymerization reactions; therefore, they are added to help control the rate and the extent of these reactions.

Microstructural Development

The conditions under which hydrolysis and condensation occur have a profound effect on gel growth and morphology. These structural conditions greatly influence the processing of sol-gel glasses into various forms. It is well established, for example, that acid-catalyzed solutions with low water content (that is conditions that produce linear polymers) offer the best type of solution for producing fibers.

Gelation and Aging

As the hydrolysis and condensation polymerization reactions continue, viscosity increases until the solution ceases to flow. The time required for gelation to occur is an important characteristic that is sensitive to the chemistry of the solution and the nature of the polymeric species. This sol-to-gel transition is irreversible, and there is little if any change in volume.

Drying

The drying process involves the removal of the liquid phase; the gel transforms from an alcogel to a xerogel. Low-temperature evaporation is frequently employed, and there is considerable weight loss and shrinkage. The drying stage is a critical part of the sol–gel process. As evaporation occurs, drying stresses arise that can cause catastrophic cracking of bulk materials.

Densification

The final stage of the sol–gel process is densification. At this point the gel-to-glass conversion occurs and the gel achieves the properties of the glass. As the temperature increases, several processes occur, including elimination of residual water and organic substances, relaxation of the gel structure, and, ultimately, densification.

Applications

The sol–gel process offers advantages for a broad spectrum of materials applications. The types of materials go well beyond silica and include inorganic compositions that possess specific properties such as ferroelectricity, electrochromism, or superconductivity. The most successful applications utilize the composition control, microstructure control, purity, and uniformity of the method combined with the ability to form various shapes at low temperatures. Films and coatings were the first commercial applications of the sol–gel process. The development of the sol–gel-based optical materials has also been quite successful, and applications include monoliths (lenses, prisms, lasers), fibers (waveguides), and a wide variety of optical films. Other important applications of sol–gel technology utilize controlled porosity and high surface area for catalyst supports, porous membranes, and thermal insulation.

Solid Cutters

Cutters made of a single piece of material rather than a composite of two or more materials.

Solid Density

See *density, absolute.*

Solid-Film Lubrication

Lubrication by application of a *solid lubricant.*

Solidification

The process whereby a metal changes, on cooling, from the fully liquid state to the fully solid-state. Solidification commences when a small group of atoms forms a nucleus from which a crystal (grain) grows by additional atoms attaching themselves at specific locations to build up the crystal lattice. Each crystal develops by forming, in three dimensions, a system of angular branches termed a dendrite. As a dendrite grows the gaps between main branches are filled by further branches. Eventually, in most practical cases, the dendrite will meet a neighboring dendrite growing to form another crystal. The line at which they meet is termed the crystal—or grain boundary. Provided sufficient molten metal is available in the immediate vicinity will be filled with solid material. If not, porosity will remain. This is termed *interdendritic porosity* when located within a grain at gaps in the dendritic branches or *inter-crystalline* or *intergranular porosity* if it is located at the grain boundaries. All pure metals solidify at a fixed temperature while alloys, except in special cases, solidify progressively over a temperature range. If an alloy is held at a temperature within the range it will comprise a pasty mixture, sometimes described as a *mush* or *slush,* of solid and liquid components in fixed proportions at specific compositions defined by the bulk alloy composition and the temperature. See also *segregation* and *lever rule.*

Solidification Cracking

Cracking occurring during solidification, usually as a result of stress and imposed by constraint of thermal contraction. Such cracking is normally located at grain boundaries as these are the last material to solidify and hence are relatively weak and unable to withstand the loads imposed by the surrounding material as it cools.

Solidification Range

The temperature between the liquidus and the solidus.

Solidification shrinkage

The reduction in volume of metal from beginning to end of solidification. See also *casting shrinkage.*

Solidification Shrinkage Crack

A crack that forms, usually at elevated temperature, because of the internal (shrinkage) stresses that develop during solidification of a metal casting. Also termed hot crack.

Solid Lubricant

Any solid used as a powder or thin film on a surface to provide protection from damage during relative movement and to reduce friction and wear. Examples include molybdenum disulfide, graphite, polytetrafluoroethylene (PTFE), and mica.

Solid-Metal Embrittlement

The occurrence of *embrittlement* in a material below the melting point of the embrittled link species. See also *liquid-metal embrittlement.*

Solid Particle Erosion

Erosion by solid particles entrained in a fluid. The term is usually used in the context of damage observed in the early stages of blades in a steam turbine. Such damage is caused by particles of oxide or other debris entrained in the steam from earlier parts of the system. Oxides may be released by spalling in the super heater and reheater stages and debris may be produced by poor practice during welding. This contrasts with the more common erosion observed in the final stages of blades that are exposed to steam cooled to a stage where large quantities of water droplets are produced. See *steam erosion.*

Solid-Phase Chemical Dosimeter

An apparatus that measures radioactivity by using plastic, dyed plastic, or glass with an optical density, usually in the visible range, that changes when exposed to ionizing radiation. Examples currently in use include dyed polymethyl methacrylate (red perspex), undyed polyvinyl chloride, dyed polyamide (blue dye in a nylon matrix), and dyed polychlorostyrene (green dye in a chlorostyrene matrix). Solid-phase chemical dosimetry is generally considered to be a secondary-standard dosimetry system.

Solid-Phase Diffusion Welding

See *diffusion welding.*

Solid-Phase Forming

The use of metalworking techniques to form thermoplastics in a solid phase. Procedure begins with a plastic blank that is heated and fabricated (that is, forged) by bulk deformation of the materials in constraining dies by the application of force. Also called solid-state stamping.

Solid Shrinkage

See *casting shrinkage.*

Solid Solubility

See *solution* and *phase.*

Solid Solution

A single, solid, homogeneous crystalline phase containing two or more chemical species. See *solution* and *phase.*

Solid Solution Hardening/Strengthening

See *solution hardening.*

Solid State

(1) Pertaining to circuits and components using semiconductors as substrates. (2) In the context of electronics this term implies electronic devices that comprise only solid materials such as transistors rather than devices containing gases such as thermionic valves.

Solids Content

The percentage by weight of the nonvolatile matter in an adhesive. The actual percentage of the nonvolatile matter in an adhesive will vary according to the analytical procedure that is used. A standard test method must be used to obtain consistent results.

Solid-State Sintering

A sintering procedure for compacts or loose powder aggregates during which no component melts. Contrast with *liquid phase sintering*.

Solid-State Welding

A group of welding processes that join metals at temperatures essentially below the melting points of the base materials, without the addition of a brazing or soldering filler metal. Pressure may or may not be applied to the joint. Welding that does not involve melting of either the parent materials or a filler. Examples include *cold welding*, *diffusion welding*, *forge welding*, *hot pressure welding*, and *roll welding*.

Solidus

(1) The highest temperature at which a metal or alloy is completely solid. (2) In a *phase diagram*, the locus of points representing the temperatures at which the various compositions stop freezing upon cooling or begin to melt upon heating. (3) The line defining the lower limit of the melting range of an alloy. A pure metal melts and solidifies at a specific temperature but, apart from special cases, alloys melt and solidify progressively over a range of temperature. Within the range the alloy will exist as a pasty mixture of solid plus liquid. When represented graphically the line defining the upper limit of the range is referred to as the *liquidus* and that defining the lower limit is the *solidus*. See also *liquidus*.

Soluble Anode Process

Various processes of refining metals by making the impure material the anode of a solute. The material taken into solution by the solvent.

Soluble Oil

A mineral oil containing additives that enable it to form a stable emulsion with water. Soluble oils are used as cutting or grinding fluids.

Solute

The component of either a liquid or solid solution that is present to a lesser or minor extent; the component that is dissolved in the solvent.

Solution

(1) In chemistry, a homogeneous dispersion of two or more types of molecular or ionic species. Solutions may be composed of any combination of liquids, solids, or gases, but they always consist of a single phase. (2) A phase containing more than one component. It is commonly recognized that a liquid can form a solution, i.e., water can dissolve salt to form a salt solution. Similarly, metals and the solid-state can form solutions. As with liquids the principal constituent will be termed the *solvent* and the secondary consistory is the *solute*. When the solvent contains the maximum possible amount to solute it is termed *saturated*. Extreme treatment, which precludes equilibrium, for example, fast cooling, can cause excess quantities of solute to be retained in the solvent. Such unstable solutions are termed *supersaturated*. A solid solution, in equilibrium, will have a uniform composition at all positions and, under the microscope will appear as a featureless single phase. In the cases of some pairs of metals, such as copper and nickel, the two are completely soluble in each other at all compositions but in most alloy systems solubility is limited and multiple phases occur. Where the composition range of a phase includes the pure metal it is termed a primary solid solution, otherwise it is termed a secondary solid solution or intermediate phase. Where the range of the intermediate phase is narrow and based on a simple ratio of the atoms of the two elements it may be termed an *intermetallic compound*. In a substitutional solid solution the solute atoms take positions on the crystal lattice normally occupied by a solvent atom. In an interstitial solid solution the relatively small solute atoms fit into the spaces between solvent metal atoms.

Solution Anneal

A heat treatment intended to affect annealing, i.e., softening and recrystallization of work hardened material, and to take into solution all (or most) precipitates. It is also implied that there will be no significant reprecipitation during subsequent cooling. The term is often used in the context of austenitic stainless steels and where such material is intended for high temperature service there may also be an application that's of grain growth is intended to enhance creep properties.

Solution Hardening

Hardening of an alloy as a result of one or more elements being in solid solution in another. The effect results from the different size of the solute atoms distorting the crystal lattice of the solvent and thereby impeding dislocation movement. The term is sometimes used casually to refer to the hardening process involving solution treatment plus precipitation hardening but this is usually regarded as erroneous.

Solution Heat Treatment

Heating an alloy to a suitable temperature, holding at that temperature long enough to cause one or more constituents to enter into *solid solution*, and then cooling rapidly enough to hold those constituents in solution.

Solution Potential

Electrode potential where half-cell reaction involves only the metal electrode and its ion.

Solvation

The process of swelling, gelling, or dissolving a resin by a solvent or *plasticizer*.

Solvay Process

A method for producing *soda ash* that involves the reaction of salt (NaCl) and limestone to form sodium carbonate (Na_2CO_3) with calcium chloride ($CaCl_2$) as a by-product.

Solvent

A material, usually a liquid, having the power of dissolving another material and forming a homogeneous mixture called a **solution**.

The mixture is physical, and no chemical action takes place. A solid solution is such a mixture of two metals, but the actual mixing occurs during the liquid or gaseous state. Some materials are soluble in certain other materials in all proportions, while others are soluble only up to a definite percentage and the residue is precipitated out of solution. Homogeneous mixtures of gases may technically be called solutions, but are generally referred to only as mixtures.

The usual industrial application of solvents are for putting solid materials into liquid solution for more convenient chemical processing, for thinning paints and coatings, and for dissolving away foreign matter as in dry-cleaning textiles. But they may have other uses, such as absorbing dust on roadways and killing weeds. They have an important use in separating materials, for example, in the extraction of oils from seeds. In such use a **clathrate** is a solid compound added to the solution containing a difficult-to-extract material, but which is trapped selectively by the clathrate. The solid clathrate is then filtered out and processed by heat or chemicals to separate the desired compound. **Antifoamers** are chemicals, such as the silicones, added to solvents to reduce foam so that processing equipment can be used to capacity without spillover.

The usual commercial solvents for organic substances are the alcohols, ether, benzene, and turpentine, the latter two being common solvents for paints and varnishes containing gums and resins. The so-called **coal-tar solvents** are light oils from coal tar. **Solvent oils**, from coal tar, are Amber to dark liquids with distillation ranges from about 150°C to 340°C, with specific gravity is 0.910–0.980. They are used as solvents for asphalt varnishes and bituminous paints. **Shingle stains** are amber to dark grades of solvent oils of specific gravity 0.910–0.930.

A valuable solvent for rubbers and many other products is **carbon bisulfide**, CS_2, also called, **carbon disulfide**, made by heating together carbon and sulfur. It is flammable and toxic. When pure, it is nearly odorless. **Ethyl acetate**, made from ethyl alcohol and acetic acid, is an important solvent for nitrocellulose and lacquers. It is liquid, boiling at 77°C. One of the best solvents for cellulose is **cuprammonium hydroxide**. Amyl and other alcohols, amyl acetate, and other volatile liquids are used for quick-drying lacquers, but many synthetic chemicals are available for such use. **Dioxan**, a water-white liquid is a good solvent for cellulose compounds, resins, and varnishes, and is used also in **paint removers**, which owe their action to their solvent power.

The chlorinated hydrocarbons have powerful solvent action on fats, waxes, and oils and are used in degreasing. Of major commercial significance are **perchloroethylene (PCE)**, **trichloroethylene (TCE)**, and **1,1,1-trichloroethane (1,1,1-TCA)**. The biggest industrial use of PCE, also known as **tetrachloroethylene** and **Perc**, is as a dry-cleaning solvent because of its nonflammability, and high solvency, vapor pressure, and stability. The largest applications of TCE and 1,1,1-TCA have been in metal cleaning, which also consumes significant quantities of PCE. Because 1,1,1-TCA has been implicated in ozone depletion of the stratosphere, its use is being discontinued. Hydrofluoroether-based solvents have similar boiling points to 1,1,1-TCA and CFC-113 and are possible alternatives to 1,1,1-TCA.

Dichloromethane, known also as **methylene chloride** and **carrene**, is a colorless, nonflammable liquid. It is soluble in alcohol and is used in paint removers, as a dewaxing solvents for oils, for degreasing textiles, and as a refrigerant. **Cyclohexane**, made by the hydrogenation of benzene, is a good solvent for rubbers, resins, fats, and waxes.

A **plasticizer** is a liquid or solid that dissolves in or is compatible with a resin, gum, or other material and renders it plastic, flexible or easy to work. A sufficient quantity of plasticizer will result in a viscous mixture which consists of a suspension of solid grains of the resin or gum in the liquid plasticizer. The plasticizer is in that sense a solvent, but unlike an ordinary solvent the plasticizer remains with the cured resin to give added properties to the materials, such as flexibility.

Solvent-Activated Adhesive

A dry-film adhesive that is rendered tacky by the application of a solvent just prior to use.

Solvent Adhesive

An adhesive having a volatile organic liquid as a vehicle. This term excludes water-base adhesives.

Solvent Molding

Process for forming thermoplastic articles by dipping a male mold in a solution or by dispersing the resin and drawing off the solvent, leaving a layer of plastic film adhering to the mold.

Solvus

In a phase or equilibrium diagram, the locus of points representing the temperature at which solid phases with various compositions coexist with other solid phases, that is, the limits of solid solubility.

Sommerfeld Number

A dimensionless number that is used to evaluate the performance of Journal bearings. It is numerically defined as follows:

$$\frac{P}{\eta U}\left(\frac{c}{r}\right)^2$$

where

P is the load per unit width
η is the dynamic viscosity
U is the surface velocity
c is the radial clearance
r is the bearing radius

At lower concentricities it is convenient to use the Sommerfeld number and the form given. Because it tends to infinity as the eccentricity approaches unity, the reciprocal form is frequently used in the case of heavily loaded bearings. The expression:

$$\frac{\eta N}{p}\left(\frac{r}{c}\right)^2$$

in which

N is the frequency of rotation
p is the pressure, is sometimes referred to as the Sommerfeld number, particularly in the United States

See also *Ocvirk number*.

Sonic Fatigue

Same as *acoustic fatigue*.

Sonic Testing

Any inspection method that uses sound waves (in the audible frequency range, about 20–20,000 Hz) to induce a response from a part or test specimen. Sometimes, but inadvisably, used as a synonym for *ultrasonic testing*. Usually, testing merely involves striking the component in question and listening to the "ring" emitted. A long pure tone implies freedom from gross cracks while a brief dull note indicates cracking. See also *ultrasonic testing*.

Sonotrode

The vibration emitting head of an ultrasonic welding unit.

Sound and Vibration Insulators

Materials used for reducing the transmission of noise. Insulators are used to impede the passage of sound waves, as distinct from isolators used under machines to absorb the vibrations that cause the sound. For factory use the walls, partitions, and ceilings offer the only media for the installation of sound insulators. All material substances offer resistance to the passage of sound waves, and even glass windows may be considered as insulators. But the term refers to the special materials placed in the walls for the specific purpose. Insulators may consist of mineral wool, hair felt, foamed plastics, fiber sheathing boards, or simple sheathing papers. Sound insulators are marketed under a variety of trade names, such as **Celotex**, made from the bagasse, and **Fibrofelt**, made from flax or rye fiber. Wheat straw is also used for making insulating board. Sound insulators are often also heat insulators. **Linofelt** is a sound- and heat-insulating material used for walls. It consists of a quilt of flax fiber between tough waterproof paper. It comes in sheets. **Fiber metal**, of Technetics Corp., comprises randomly interlocked similar metal fibers, with the fibers bonded by sintering at all contact points. Similar to nonwoven textile felts, its trade name is **Feltmetal**, and it is available in sheet form in various fibers, thickness, and porosity. 316 and 347 stainless steel and aluminum-alloy fibers are used mainly for noise reduction of aircraft turbines, turbine blowers, and high-speed fans. Noise reduction is by *resistive absorption*, by which the amplitude of sound waves is reduced by converting most of the acoustic energy into heat. Other applications include abradable seals, using Hastelloy X fibers, and high-temperature thermal insulation, using an iron, chromium, aluminum, and yttrium alloy.

Vibration insulators, or **isolators**, to reduce vibrations that produce noises, are usually felt or fiberboards placed between the machine base and the foundation, but for heavy pressures they may be metal wire helically wound or specially woven, deriving their effectiveness from the form rather than the material.

Source (x-Rays)

The area emitting primary x-rays in a diffraction experiment. The actual source is always the focal spot of the x-ray tube, but the virtual source may be a slit or pinhole, depending on the conditions of the experiment.

Sour Gas

A gaseous environment containing hydrogen sulfide and carbon dioxide in hydrocarbon reservoirs. Prolonged exposure to sour gas can lead to *hydrogen damage*, *sulfide-stress cracking*, and/or *stress-corrosion cracking* in ferrous alloys.

Sour Water

Waste waters containing fetid materials, usually sulfur compounds.

Sow Block

A block of heat-treated steel place between the anvil of the hammer and the forging die to prevent undue wear to the anvil. Sow blocks are occasionally used to hold insert dies. Also called anvil cap. See also the terms *drop hammer* and *gravity hammer*.

Soybean Oil

Biodiesel fuel, a combination of natural oil or fat with an alcohol such as methanol or ethanol, could help the U.S. reduce air pollution and its dependence on imported oil.

Soybean oil is the most commonly used feedstock in the U.S. Biodiesel works in all conventional diesel engines and can be distributed through the existing industry infrastructure.

Using 100% biodiesel fuel reduces carbon dioxide emissions by more than 75% compared to petroleum diesel. Using a blend of 20% biodiesel reduces them by 15%. The fuel also produces less particulate, carbon monoxide, and sulfur emissions, all targeted as public health risks. On the downside, biodiesel is more expensive than petroleum diesel and it produces slightly higher amounts of nitrogen oxide, a pollutant.

Space-Charge Aberration

In electron microscope, an aberration resulting from the mutual repulsion of the electrons in a beam. This aberration is most noticeable in low-voltage, high-current beams. This repulsion acts as a negative lens, causing rays, which were originally parallel, to diverge. See also *aberration*.

Space Frame

A load bearing structure of struts and stays.

Space Lattice

(1) Either an alternative term for crystal lattice or a notation system for defining the location of atoms on the crystal lattice. See *Bravais Lattice*. (2) A regular, periodic array of points (lattice points) in space that represents the locations of atoms of the same kind in a perfect crystal. The concept may be extended, where appropriate, to crystalline compounds and other substances, in which case the lattice points often represent locations of groups of atoms of identical composition, arrangement, and orientation. See also the terms *lattice* and *unit cell*.

Space Processing

The carrying out of various processes on materials aboard orbiting spacecraft is known as space processing. Until the space age, the Earth's gravity had always been considered a constant in the fluid-flow equations that govern heat and mass transport and materials processing. When the space shuttle became operational in the early 1980s, the potential benefits of suppressing the acceleration of gravity in certain processes began to be seriously considered. There have been numerous flight opportunities for microgravity experimentation.

Protein Crystal Growth

The importance of crystallography as a mechanism for determining three-dimensional structure of complex macromolecules has placed new demands on the ability to grow large (approximately 0.5 mm on a side), highly ordered crystals of a vast variety of biological macromolecules to obtain high-resolution x-ray diffraction data.

The growth of protein crystals in reduced gravity has the potential advantages of (1) the ability to suspend the growing crystals in the growth solution to provide a more uniform growth environment and (2) the ability to reduce the convective mass transport so that growth can take place to a diffusion-control led environment. The effect of convection on the growth of crystals is not well understood, but it is generally accepted that unsteady growth conditions that can result from convective flows are harmful to crystal growth.

Attempts to grow protein crystals in space produce mixed results. Sometimes no crystals or crystals that are inferior to those grown on Earth are produced. However, occasionally, the space-grown crystals are larger and better ordered than the best ever grown on Earth. In fact, the improvement in internal order obtained in protein growth in reduced gravity can be so dramatic as to allow structure to be solved or refined to higher resolution than is possible by using the diffraction data from the best available Earth-grown crystals.

There is still the question of why attempts to grow protein crystals in space produce superior results only part of the time. (It should be remembered that many unreported experiments on the ground are not successful either.) One possible explanation is that the growth process is developed and optimized on the ground before committing the experiment to flight. However, the conditions that are optimum under normal gravity may not take advantage of the microgravity environment. Therefore, it may be necessary to actually develop the optimal growth processes in space to improve the yield of protein crystal growth experiments there.

Electronic and Photonic Materials

Single-crystalline materials suitable for electronic and photonic applications have received much attention as candidates for microgravity processing for several reasons. These are critical, high-value materials whose applications demand extreme control of composition, purity, and defects. Despite the rapid advances made in electronic materials, progress on many fronts is still limited by available materials. Gravity-driven flows certainly influence mass transport in growth processes. Even though the primary purpose of most of the space processing with these materials has been to gain insight and understanding that can be used in Earth-based processing, limited production of certain specialty materials in space is a possibility if it turns out that there is no Earth-based alternative.

Vapor Growth

Several vapor crystal growth experiments on the shuttle produced some interesting results that are not at all understood. An example is the growth of unseeded germanium–selenium (GeSe) crystals by physical vapor transport using an inert noble gas as a buffer to a closed tube. When this is done on the ground, many small crystallites form a crust inside the growth ampoule at the cold end. Growth in space produces dramatically different results; The crystals apparently nucleate away from the walls and grow as thin platelets, which eventually become entwined with one another, forming a web that is loosely contained by the tube. Even more striking is the appearance of the surfaces of the space-grown crystals. The surfaces are mirrorlike and almost featureless, exhibiting only a few widely spaced growth terraces. By contrast, crystallites grown on the ground under identical thermal conditions have many pits and irregular, closely spaced growth terraces.

Another example is the growth of $Hg_{0.4}Cd_{0.5}Te$ by closed-tube chemical vapor deposition on mercury–cadmium–tellurium (HgCdTe) substrates using mercuric iodide (HgI_2) as a transport agent. Again, considerable improvements in the space-grown samples are observed, relative to those grown on the ground, in terms of surface morphology, chemical microhomogeneity, and crystalline perfection.

In the growth of thin films of copper phthalocyanine on copper substrates by physical vapor deposition, a dramatic difference is found in the appearance and morphology of the space-grown film as compared with the films produced on the ground. Scanning electron microscopy reveals a close-packed columnar structure for the space-grown films, roughly resembling a thick pile carpet. The ground-grown samples have a lower-density, randomly oriented structure that resembles a shag carpet.

Mercuric iodide crystals grown during orbital flight by physical vapor transport exhibit sharp, well-formed facets indicating good internal order. This is confirmed by gamma-ray rocking curves, which are approximately 1/3 the width of those taken on samples grown on the ground. Both electron and hole mobility are significantly enhanced in the flight crystals.

Solution Growth

Triglycine sulfate crystals can be grown from solution during orbital flight by using a novel cooled sting method. Supersaturation is maintained by extracting heat through the seed mounted on a small heat pipe, which in turn is attached to a thermoelectric device. Growth under diffusion-control transport conditions may avoid liquid and gas inclusions, the most common type of defect in solution-grown crystals, which are believed to be caused by unsteady growth conditions resulting from convective flows.

This growth technique has produced crystals of exceptional quality. The usual growth defects in the vicinity of the seed that forms during the transition from dissolution to growth (the so-called ghost of the seed) are notably absent. High-resolution x-ray topographs taken with synchrotron radiation indicated a high degree of perfection. In pyroelectric detection for far-infrared radiation, the detectivity of the space-grown crystal is significantly higher than the seed crystal and the Q (ratio of energy stored to energy loss per cycle) is more than doubled.

Metallic Alloys and Composites

Processing of metallic alloys and composites in space has been carried out to study dendrite growth, monotectic alloys, liquid-phase sintering, and electrodeposition.

Dendrite Growth

The microgravity environment provides an excellent opportunity to carry out critical tests of fundamental theories of solidification without the complicating effects introduced by buoyancy-driven flows. For example, one investigator carried out a series of experiments to elucidate dendrite growth kinetics under well-characterized diffusion-controlled conditions in pure succinonitrile. This constituted a rigorous test of various nonlinear dynamical pattern formation theories, which provide the basis for the prediction of the microstructure and physical properties achieved in a solidification process.

Comparison of dendrite tip velocities, measured as a function of undercooling over a range from 0.05°C to 1.5°C with ground-based measurements, shows that affects convection are more significant at the smaller undercoolings and are still important up to undercoolings as large as 1.3 K. Even in microgravity, there is a slight departure in the data at the smallest undercooling, which is attributed to the residual acceleration of the spacecraft. These data also allow the determination of the scaling constant important in the selection of the dynamic operating state, which the present theories have been unable to provide.

Monotectic Alloys

Some of the first microgravity experiments and metallurgy were attempts to form fine dispersions in monotectic alloys, that is, alloy systems that have liquid-phase immiscibilities. Attempts to solidify such alloys from the melt in normal gravity always result in macroscopic segregation because the densities of the two liquid phases are invariably different. It was thought that this phase separation could be avoided in microgravity and intimate mixtures of the two phases would result that might have interesting and unusual properties.

Liquid-Phase Sintering

Composites formed by liquid-phase sintering have many commercial applications, from cutting tools to electrical switch contacts. Fine particles of the more refractory phase are mixed with particles of a lower-melting material, which, when melted, forms a matrix to bind the nonmelting particles together. The system is stabilized during the sintering process by using a large volume fraction (80%–85%) of non-melting particles to support the structure while the molten phase interpenetrates the intergranular spaces. Fortunately, for many applications it is desirable to have a large volume fraction of the more refractory phase. However, there are some applications where there is a requirement to increase the volume fraction of the matrix material to amplify its properties. Space processing can be used to prepare composites (such as tungsten particles in a copper–nickel matrix, cobalt particles in a copper matrix, and iron particles in a copper matrix) with host-material volume fractions ranging from 30% to 50%, to increase the sintering time, and to provide valuable insight into evolution of pores and other defects that occur and sintered products produced on Earth.

Electrodeposition

Electrodeposition experiments in reduced gravity have produced some intriguing results. With higher current densities than can normally be used on Earth, nickel with a nanocrystalline structure can be deposited on gold substrates. Attempts to duplicate this result in normal gravity by the use of convectively stable geometries and porous media have not been successful. It is speculated that the morphology of the hydrogen bubbles that form on the cathode in microgravity somehow promotes the formation of nickel hydride, which produces a nanocrystalline structure.

Attempts have been made to codeposit diamond dust with copper, and small particles of Co_2C_3 with cobalt, to form cermets that would be extremely hard and wear resistant. A ground-based technique for depositing a bonelike hydroxyapatite coating on prosthetic implants, which is based on this work, has significantly better adhesion than currently available coatings. Another derivative of this work is a plating process using Cr(III), which poses significantly fewer environmental problems than the common Cr(VI) process.

Spacer Strip

A metal strip or bar inserted in the root of a joint prepared for groove welding to serve as a backing and to maintain root opening throughout the course of the welding operation.

Spacing (Lattice Planes)

See *interplanar distance*.

Space Drill

See preferred term *flat drill*.

Spalling (Ceramics)

(1) The cracking or rupturing of a refractory unit, which usually results in the detachment of a portion of the unit. (2) A defect characterized by separation of the porcelain enamel from the aluminum base metal without apparent external cause. Spalling can result from the use of improper alloys or enamel formulations, incorrect pretreatment of the base metal, or faulty application and firing procedures.

Spalling (Metals)

(1) Separation of particles from a surface in the form of flakes. The term spalling is commonly associated with rolling-element bearings and with gear teeth. Spalling is usually a result of subsurface fatigue and is more extensive than pitting. (2) In *tribology*, the separation of macroscopic particles from a surface in the form of flakes or chips, usually associated with rolling-element bearings and gear teeth, but also resulting from impact events. (3) The spontaneous chipping, fragmentation, or separation of a surface or surface coating. (4) A chipping or flaking of a surface due to any kind of improper heat treatment or material dissociation.

Spalls

The primary cause of premature failures of forged hardened steel rolls. Spalls are sections that have broken from the surface of the roll. In nearly all cases, they are observed in the outer hardened zone of the body surface, and they generally exhibit well-defined fatigue beach marks. The most common spalls are the circular spall and the line spall. Circular spalls exhibit subsurface fatigue marks and a circular, semicircular, or elliptical pattern. They are generally confined to a particular body area. A line spall has a narrow width of subsurface fatigue that extends circumferentially around the body of the roll. Most line spools originate at or beneath the surface in the outer hardened zone.

Spangle

The characteristic crystalline form in which a hot dipped zinc coating solidifies on steel strip, especially galvanized.

Spark

A series of electrical discharges, each of which is oscillatory and has a comparatively high maximum instantaneous current resulting from the breakdown of the analytical gap or the auxiliary gap, or both, by electrical energy stored in high-voltage in capacitors.

Each discharge is self-initiated and is extinguished when the voltage across the gap, or gaps, is no longer sufficient to maintain it.

Spark Erosion

See *electrical pitting*.

Spark Machining

Metal removal by repeatedly striking an electric arc against the component to remove material by melting and vaporization. Usually, the component is submerged in paraffin or a similar medium during machining and individual arc strikes are very small. The technique is useful for very hard materials or for holes which are of complex shape. The component is usually the anode and the tool the cathode.

Spark Sintering

In powder metallurgy, a pressure sintering or hot pressing method that provides for the surface activation of the powder particles by electric discharges generated by a high alternating current applied during the early stage of the consolidation process.

Spark Source Mass Spectrometry

An analytical technique in which a high-voltage spark in a vacuum is used to produce positive ions of a conductive sample material. The ions are injected into a mass spectrometer, and the resulting spectrum is recorded on a photographic plate or measured using an electronic detector. The position of a particular mass spectral signal determines the element and isotope, and the intensity of the signal is proportional to the concentration.

Spark Testing

A method used for the classification of ferrous alloys according to their chemical compositions, by visual examination of the spark pattern or stream that is thrown off when the alloys are held against a grinding wheel rotating at high speed.

Spatial Grain Size

The average size of the three-dimensional grains in polycrystalline materials as opposed to the more conventional grain size determined by a simple average of observations made on a cross section of the material.

Spatial Resolution

A measure of the ability of an imaging system to represent fine detail; the measure of the smallest separation between individually distinguishable structures. See also *resolution*.

Spatter

The metal particles expelled during arc or gas welding. They do not form part of the weld.

Spatter Loss

The metal lost due to *spatter*.

Specific Adhesion

Adhesion between surfaces that are held together by the valence forces of the same type as those that give rise to cohesion.

Specific Energy

In cutting or grinding, the energy expended or work done in removing a unit volume of material.

Specific Gravity (Gases)

The ratio of the density of a gas to the density of dry air at the same temperature and pressure.

Specific Gravity (Solids and Liquids)

The ratio of the density of a material to the density of substandard material, such as water, at a specified temperature. Also known as relative density.

Specific Heat

(1) The ratio of the amount of heat required to raise a mass of material 1° in temperature to the amount required to raise an equal mass of a reference substance, usually water, 1° in temperature; both measurements are made at a reference temperature, usually at constant pressure, or constant volume. (2) The quantity of heat required to raise a unit mass of a homogeneous material one degree in temperature in a specified way; it is assumed that during the process no phase or chemical change occurs.

Specific Humidity

In a mixture of water vapor and air, the mass of water vapor per unit mass of moist air.

Specific Power

Same as *unit power*.

Specific Pressure

In powder metallurgy, the pressure applied to a green or sintered compact per unit of area of punch cross section.

Specific Properties

Material properties divided by material density.

Specific Resistance Resistivity

The electrical resistance of unit length of unit cross-section of material. Measured in microhm centimeters.

Specific Sliding

The ratio of the algebraic difference between the surface velocities of two bodies in relative motion to their sum.

Specific Surface

The surface area of a powder expressed in square centimeters per gram of powder or square meters per kilogram of powder.

Specific Viscosity

The relative viscosity of a solution of known concentration of a polymer minus one. It is usually determined for a low concentration of the polymer.

Specific Volume

The volume of a substance per unit mass; the reciprocal of the density.

Specific Wear Rate

In journal bearings, the proportionality constant K in the equation:

$$h = Kpyt$$

where
    h is the radial wear in the bearing
    p is the apparent contact pressure
    y is the velocity of the journal
    t is the sliding time

The constant K has also been called the *wear factor*, but there are other definitions for the term wear factor that do not necessarily refer to journal bearings or derive their meanings from the above equation.

Specimen

A test object, often of standard dimensions and/or configuration, that is used for destructive on nondestructive testing. One or more specimens may be cut from each unit of a *sample*.

Specimen Chamber (Electron Optics)

The compartment located in the column of the electron microscope in which the specimen is placed for observation.

Specimen Charge (Electron Optics)

The electrical charge resulting from the impingement of electrons on a non-conducting specimen.

Specimen Contamination (Electron Optics)

The contamination of the specimen and caused by the condensation upon it of residual vapors in the microscope under the influence of electron bombardment.

Specimen Distortion (Electron Optics)

A physical change in the specimen caused by desiccation or heating by the electron beam.

Specimen Grid

See *specimen screen*.

Specimen Holder (Electron Optics)

A device that supports the specimen and specimen screen in the correct position in the specimen chamber of the microscope.

Specimen Screen (Electron Optics)

A disk of fine screen, usually 200-mesh stainless steel, copper, or nickel, that supports the replica or specimen support film for observation in the microscope.

Specimen Stage

The part of the microscope that supports the specimen holder and the specimen in the microscope and can be moved in a plane perpendicular to the optic axis from outside the column.

Specimen Strain

A distortion of the specimen resulting from stresses occurring during metallographic preparation or observation. In electron metallography, strain may be caused by stretching during removal of a replica or during subsequent washing or drying.

Spectral Background

In spectroscopy, a signal obtained when no analyte is being introduced into the instrument, or a signal from a species other than that of the analyte.

Spectral Distribution Curve

The curve showing the absolute or relative radiant power emitted or absorbed by a substance as a function of wavelength, frequency, or any other directly related variable.

Spectral Line

A wavelength of light with a narrow energy distribution or an image of a slit formed in the focal plane of a spectrometer or photographic plate that has a narrow energy distribution approximately equal to that formed by monochromatic radiation.

Spectral Order

The number of the intensity of a given line from the directly transmitted or specularly reflected light from a *diffraction grating*.

Spectrochemical (Spectrographic, Spectrometric, Spectroscopic) Analysis

The determination of the chemical elements or compounds in a sample qualitatively, semiquantitatively, or quantitatively by measurements of the wavelengths and intensities of spectral lines produced by suitable excitation procedures and dispersed by a suitable optical device.

Spectrogram

A photographic or graphic record of a spectrum.

Spectrograph

(1) An optical instrument with an entrance slit and dispersing device that uses photography to record a spectral range. The radiant power passing through the optical system is integrated over time, and the quantity recorded as a function of radiant energy. (2) An instrument for producing and displaying the spectrum of radiation emitted by a material when excited by, for example, an electric arc. It can be used for the analysis of metals. Where the instrument measures the intensity of the omission may be termed a *spectrometer.*

Spectrometer

An instrument with an entrance slit, a dispensing device, and one or more exit slits, with which measurements are made at selected wavelengths within the spectral range, or by scanning over the range. The quantity detected is a function of the radiant power.

Spectrophotometer

A spectrometer that measures the ratio (or a function of the ratio) of the intensity of two different wavelengths of light. These two beams may be separated in terms of time or space, or both.

Spectrophotometry

A method for identification of substances and determination of their concentration by measuring light transmittance in different parts of the spectrum.

Spectroscope

An instrument that disperses radiation into a spectrum for visual observation.

Spectroscopy

The branch of physical science treating the theory, measurement, and interpretation of spectra.

Spectrum

The ordered arrangement of electromagnetic radiation according to wavelength, wave number, or frequency.

Specular Reflection

The condition in which all the incident light is reflected at the same angle as the angle of the incident light relative to the normal at the point of incidence. The reflection surface then appears bright, or mirrorlike, when viewed with the naked eye. Sometimes termed regular reflection.

Specular Transmittance

The transmittance value obtained when the measured radiant energy in *emission spectroscopy* has passed from one source to the receiver without appreciable scattering.

Speculum Metal

An alloy formally used for mirrors and in optical instruments. It contains 65%–67% copper, the balance tin. It takes a beautiful polish and is hard and tough. An old Roman mirror contained about 64% copper, 19% tin, and 17% lead; and an Egyptian mirror contained 85% copper, 14% tin, and 1% iron. The old Greek mirrors were carefully worked out with 32% tin and 68% copper. They had 70% of the reflecting power of silver, with a slight red excess of reflection that gave a warm glow, without the blue of nickel or antimony. This alloy is now plated on metals for reflectors. A modern telescope mirror contained 70% copper and 30% tin. **Chinese speculum** contains about 8% antimony and 10% tin. Speculum plate, which has been advocated by the Tin Research Institute for electroplating, to give a hard, white, corrosion-resistant surface for food processing equipment and optical reflectors, has 55% copper and 45% tin. It is harder than nickel and retains its reflectivity better than silver.

Speed of Travel

In welding, the speed with which a weld is made along its longitudinal axis, usually measured in meters per second or inches per minute.

Speiss

Metallic arsenides and antimonides that result from smelting metal ores such as those of cobalt or lead.

Spelter

(1) Crude zinc obtained in smelting zinc ores. (2) Brass for brazing or, less commonly, as a cheap alternative to bronze for sculptures.

Spelter Solder

A brazing filler metal of approximately equal parts of copper and zinc.

SPF

See *superplastic forming.*

Spherical Aberration

A lens defect in an optical microscope in which image-forming rays passing through the outer zones of the lens focus at a distance from the principal plane different from that of the rays passing through the center of the lens. See also *aberration* and *chromatic aberration.*

Spherical Bearing

A bearing that is self-aligning by virtue of its partially spherical form.

Spherical Powder

A powder consisting of ball-shaped particles.

Spherical Roller Bearing

(1) A spherical bearing containing rollers. (2) A roller bearing containing barrel-shaped or hour glass-shaped rollers riding on spherical (concave or convex) races to provide self-aligning capability.

Spheroidal Graphite

(1) Graphite of spheroidal shape with a polycrystalline radial structure. This structure can be obtained, for example, by adding cerium or magnesium to the melt. See also *ductile iron* and *nodular graphite*. (2) Graphite, in cast iron, having a spherical rather than a flake shape. See *cast iron*.

Spheroidal Powder

A powder consisting of oval or rounded the particles.

Spheroidite

An aggregate of iron or alloy carbides of essentially spherical shape dispersed throughout a matrix of *ferrite*.

Spheroidized Carbide

Carbide in steel that has become spherical in shape as a result of the heating. See *steel*.

Spheroidized Structure

A microstructure consisting of a matrix containing spheroidal particles of another constituent.

Spheroidizing

Heating and cooling to produce a spheroidal or globular form of carbide in steel. Spheroidizing methods frequently used are:

1. Prolonged holding at a temperature just below A_{c1}.
2. Heating and cooling alternatively between temperatures that are just above and just below A_{c1}.
3. Heating to a temperature above A_{c1} or A_{c3} and then cooling very slowly in the furnace or holding at a temperature just below A_{c1}.
4. Cooling at a suitable rate from the minimum temperature at which all carbide is dissolved to prevent the re-formation of a carbide network, and then reheating in accordance with method one or two above. (Applicable to hypereutectoid steel containing a carbide network.) See *steel*.

Spherulite

A rounded aggregate of radiating laminar crystals with appearance of a pom-pom in plastics. Spherulites contain amorphous material between the crystals and usually impinge on one another, forming polyhedrons. Spherulites are present in most crystalline plastics and may range in diameter from a few tenths of a micron to several millimeters.

Spherulitic Graphite Cast Iron

Same as *ductile cast iron*.

Spider

In a plastic molding press, that part of an ejector mechanism that operates the ejector pins. In extrusion, the membranes supporting a mandrel within the head/die assembly.

Spider Die

Same as *porthole die*.

Spiegel, Spiegeleisen

Ferromanganese master alloys used in the final stages of steel production. Where they are differentiated the former has about 20%–30% manganese and 5% carbon, remainder iron, while the latter is similar but with about 5%–20% manganese. See *pig iron*.

Spiking

In electron beam welding and laser welding, a condition where the depth of penetration is nonuniform and changes abruptly over the length of the weld.

Spin

In bearings, rotation of a rolling element about an axis normal to the contact surfaces. See also *rolling* (pure rolling with no sliding and no spin) and *sliding*.

Spindle

(1) Shaft of a machine tool in which a cutter or grinding wheel may be mounted. (2) Metal shaft to which a mounted wheel is cemented.

Spindle Oil

An oil of low viscosity used to lubricate high-speed light spindles.

Spinel

Spinel is any of a family of important AB_2O_4 oxide minerals, where A and B represent cations with a simple cubic crystal lattice. Spinel minerals are widely distributed in the Earth, in meteorites, and in rocks from the moon. The ideal spinel formula is $MgOAl_2O_3$ or $MgAl_2O_4$. Spinel has a melting point of 2135°C and the mineral is found in small deposits. It is formed by solid-state reaction between MgO and Al_2O_3 and is an excellent refractory showing high resistance to attack by slags, glass, etc.

$MgOAl_2O_3$ or $MgAl_2O_4$ compound itself, is refractory and chemically near-neutral. Magnesium aluminate spinels are used as an addition to fired magnesia refractory bricks to improve thermal shock resistance. Lithium-based spinels are candidate materials for rechargeable lithium batteries.

High-purity spinel is a chemically derived spinel powder made by the co-precipitation of magnesium and aluminum complex sulfates, with subsequent calcination to form the oxide compound. Purities range from 99.98% to 99.995%. The ceramic powders prepared by this process can be hot-pressed into transparent window materials with exceptional infrared transmission range.

Applications

The major ceramic applications for spinels are the magnetic ferrospinels (ferrites), chromite brick, and spinel colors. Magnetic recording tape coated with α-Cr_2O_3 is a relatively recent development. It is also used as a porous protective coating in oxygen sensors for automotive emission controls.

The material is available as fused spinel in special refractory applications and also in a special particle shape and distribution for flame and plasma-arc spraying. The magnetic spinels are of special importance because of the wide-spread interest and application of the ceramic ferrospinels (ferrites). Two classes of ferrospinels occur: magnetic and nonmagnetic.

The magnetic are related to the inverse structure in the nonmagnetic to the normal structure.

Spin Glass

One of a wide variety of materials that contain interacting atomic magnetic moments and also possess some form of disorder, in which the temperature variation of the magnetic susceptibility undergoes an abrupt change in slope at a temperature generally referred to as the freezing temperature.

Spinneret

A type of extrusion die for plastics that consists of a metal plate with many tiny holes, through which a plastic mesh is forced, to make fine fibers and filaments. Filaments may be hardened by cooling in air, water, and so forth, or by chemical action.

Spinning

The forming of a seamless hollow metal part by forcing a rotating blank to conform to a shaped mandrel that rotates concentrically with a blank. In the typical application, a flat-rolled metal blank is forced against the mandrel by a blunt, rounded tool; however, other stock (notably, welded or seamless tubing) can be formed. A roller is sometimes used as the working end of the tool.

Spinning

A manufacturing process in which sheet or plate material is rapidly rotated and forced by a non-cutting tool against a shaped former. The variations range from the simplest where a plain shaping tool is handheld to high powered systems where a roller head tool is carried on a powered carriage; the latter may be termed *flow spinning*.

Spinodal Curve

A graph of the realizable limit of the supersaturation of a solution. See also *spinodal structure*.

Spinodal Hardening

See *aging*.

Spinodal Structure

A fine, homogeneous mixture of two phases that form by the growth of composition waves in a solid solution during suitable heat treatment. The phases of a spinodal structure differ in composition from each other and from the parent phase, but have the same crystal structure as the parent phase. Spinodal structures are resolvable only at high magnifications such as made possible by *transmission electron microscopy*.

Spin Wave

A sinusoidal variation, propagating through a crystal lattice, of that angular momentum associated with magnetism (mostly spin angular momentum of the electrons). See also *spin glass*.

Spin Welding

Same as friction welding particularly when applied to plastics.

Spiral-Flow Test

A method for determining the flow properties of a thermoplastic resin in which the resin flows along the path of a spiral cavity. The length of the material that flows into the cavity and its weight gives a relative indication of the flow properties of the resin.

Spiral Mold Cooling

A method of cooling injection molds or similar molds for forming plastics in which the cooling medium flows through a spiral cavity in the body of the mold. In injection molds, the cooling medium is introduced at the center of the spiral, near the sprue section, because more heat is localized in this section.

Spiral Welded Tube

Tube formed from strip rolled to a helix and welded on the spiral interface between the edges.

Spit

See preferred term *flash*.

Splash Lines

Damaging or unsightly material on the surface or in the interface of spot or seam welds, etc., caused by the ejection of material during welding.

Splash Lubrication

A system of lubrication in which the lubricant is splashed onto the moving parts.

Splat Casting

Dropping small quantities of molten metal onto a cold metal surface to achieve very rapid rates of cooling. A variation of the technique runs a thin stream of molten metal onto a cold metal wheel.

Splat Powder

A rapidly cooled or quenched powder whose particles have a flat shape and a small thickness compared to other dimensions. Similar to *flake powder*.

Splat Quenching

The process of producing splat powder.

Splay

The tendency of a rotating drill bit to drill off-center, out-of-round, non-perpendicular holes.

Splay (Plastics)

A fanlike surface defect near the gate on a plastic part.

Splay Lines

Lines found in a plastic part after molding, usually due to the flow of material in the mold. Sometimes called *silver streaking*.

Splice

The joining of two ends of glass fiber yarn or strand used for reinforcing plastics, usually by means of an air-drying adhesive.

Splicing

The joining of two multiply stranded ropes or cables by intertwining the individual strands of each into the other.

Spline

Any of a series of longitudinal, straight projections on a shaft that fit into slots on a mating part to transfer rotation to or from the shaft.

Split Die

A die made of part that can be separated for ready removal of the workpiece. Also known as segment die.

Split Pipe Backing

Backing in the form of a pipe segment used for welding round bars.

Split Punch

A segmented punch or a set of punches in a powder metallurgy forming press that allow(s) a separate positioning for different powder fill heights and compact levels in dual-step and multistep parts. See also *stepped compact*.

Split-Ring Mold

A mold for forming plastics in which a split-cavity block is assembled in a chase to permit the forming of undercuts in a molded plastic piece. These parts are ejected from the mold and then separated from the piece.

Split Seal

A seal that has its primary sealing elements split in a plane parallel to the axis of the shaft such that, instead of the rings being continuous, they are essentially two semicircles. Modified designs of lip seals feature units with a single lip separation and with one or more separations of the metallic stiffening mechanisms.

Spodumene

A mineral of composition $Li_2O \cdot Al_2O_3 \cdot 4SiO_2$, with some potassium and sodium oxides. It is the chief ore of the metal lithium, but it requires a higher temperature for sintering than lepidolite, and the sinter is more difficult to leach. **Lithospar** is a name for feldspar and spodumene from North Carolina. In Germany lithium is obtained from the lithium mica **zinnwaldite**, which is a mixture of potassium–aluminum orthosilicate and lithium orthosilicate with some iron, and contains less than 3% Li_2O.

Sponge

The cellular skeleton of a marine animal of the genus *Spongia*, of which there are about 3000 known species, only 13 of which are of commercial importance. It is employed chiefly for wiping and cleaning, as it will hold a great quantity of water in proportion to its weight, but it also has many industrial uses such as applying glaze to pottery. Sponges grow like plants, attached to rocks on the sea bottom. They are prepared for use by crushing to kill them, scraping off the rubbery skin, macerating in water to remove the gelatinous matter, and bleaching in the sun.

The prepared sponge is an elastic, fibrous structure chemically allied to silk. It has sievelike membranes with small pores leading into pear-shaped chambers. The best sponges are spheroidal, regular, and soft. Commercial sponges for the U.S. market must have a diameter of 11.4 cm or more. Most of the Florida sponges are the **sheepswool sponge**, use for cleaning and industrial sponging. The **Rock Island sponge**, from Florida, and the **Key wool sponge** are superior in texture and durability to the **Bahama wool sponge**, which is coarser, more open, and less absorbent. The **Key yellow sponge** is the finest grade. The **grass sponge**, is inferior in shape and texture. The fine **honeycomb sponge**, of the Mediterranean Sea, is of superior grade and has been preferred as a bath sponge. The **Turkey cup sponge** is rated as the finest, softest, and most elastic of the sponges, but the larger of the zimocca sponges are too hard for surgical use and are employed for industrial cleaning. Sponges for industrial and household uses have now been largely replaced by foamed rubbers and plastics.

Sponge

A form of metal characterized by a porous condition that is the result of the decomposition or reduction of a compound without fusion. The term is applied to forms of iron, titanium, zirconium, uranium, plutonium, and the platinum group metals.

Sponge Effect

See *squeeze effect*.

Sponge Iron

Iron made from ferrous sand and pressed into briquettes, which can be charged directly into steel furnaces instead of pig iron. It was originally made on a large scale in Japan where only low-grade sandy ores were available. Sponge iron is made by charging the sand continuously into a rotary furnace to drive off the light volatile products and reduce the iron oxide to metallic iron, which is passed through magnetic separators, and the finely divided iron briquetted. Unbriquetted sponge iron, with a specific gravity of 2, is difficult to melt because of the oxidation, but briquetted material, with a specific gravity of 6, can be melted in electric furnaces. Sponge iron, to replace scrap in steelmaking, is also made from low-grade ores by reducing the ore with coke-oven gas or natural gas. It is not melted, but the oxygen is driven off, leaving a spongy, granular product. As it is very low in carbon, it is also valuable for making high-grade alloy steels.

S

A form of sponge iron employed as a substitute for lead for coupling packings was made in Germany under the name of **sinterit**. The reduction is carried out in a reducing atmosphere at a temperature of 1200°C–1350°C, instead of heating the iron oxide with carbon. Since the porous iron corrodes easily, it is coated with asphalt for packing use. **Iron sponge**, employed as a purifier for removing sulfur and carbonic acid from illuminating gas, is a sesquioxide of iron obtained by heating together iron ore and carbon. It has a spongy texture and is filled with small cells.

Sponge Iron Powder

Ground and sized sponge iron that may have been purified or annealed or both.

Sponge Titanium Powder

Ground and sized titanium sponge. See also *Kroll process*.

Spongy

A porous condition in metal powder particles usually observed in reduced oxides.

Spool

A type of weld filler metal package consisting of a continuous length of electrode wound on a cylinder (called the barrel) which is flanged at both ends. The flange extends below the inside diameter of the barrel and contains a spindle hole.

Spot Drilling

Making an initial indentation in a work surface, with a drill, to serve as a sintering guide in a subsequent machining process.

Spotfacing

Using a rotary uneven staining of metal by entrapment, hole-piloted end-facing tool to produce a flat surface normal to the axis of rotation of the tool on or slightly below the workpiece surface.

Spot Test

A test in which spots of chemical reagents applied to the test surface produce reactions indicative of the composition.

Spotting Out

Delayed, uneven staining of metal by entrapment of chemicals during the finishing operation.

Spot Weld

A weld made between or upon overlapping members in which coalescence may start and occur on the faying surfaces or may proceed from the surface of one member. The weld cross section is approximately circular. See the term *resistance spot welding*.

Spot Welding

(1) Welding of the lapped parts in which fusion is confined to a relatively small circular area. It is generally resistance welding, but may also be gas tungsten-arc, gas metal-arc, or submerged-arc welding. (2) A resistance welding process in which the components, usually sheet, are clamped between two electrodes supplying heating current. The weld formed is approximately the size of the electrodes, or the smaller of them if they differ. (3) Any localized weld formed by any process. See the term *resistance spot welding*.

Spragging

Intermittent motion arising from design features that allow an increase in tangential force or displacement to produce an increase in normal force.

Spray Angle

In thermal spraying, the angle of particle approach, measured from the surface of the substrate to the axis of the spray nozzle.

Spray Deposit

A coating applied by any of the thermal spray methods. See also *thermal spraying*.

Spray Distance

In thermal spraying, the distance maintained between the gun nozzle and the substrate surface during spraying.

Spray Drier

A large vessel into which a slurry containing metal or ceramic powders is sprayed through orifices in a stationary or revolving head and thrown as droplets into a stream of heated air which dries them. The dried droplets are typically tiny agglomerates, often in hollow bead form, hence free-flowing.

Spray Drying

A powder-producing process in which a slurry of liquids and solids or a solution is atomized into droplets and a chamber through which heated gases, usually air, are passed. The liquids are evaporated from the droplets and the solids are collected continuously from the chamber. The resulting powder consists of free-flowing, spherical agglomerates.

Sprayed-Metal Molds

Molds for forming plastics made by spraying molten metal onto a master until a shell of predetermined thickness is achieved. The shell is then removed and backed with plaster, cement, casting resin, or other suitable material. Used primarily as a mold in the sheet-forming process.

Spraying Sequence (Thermal Spraying)

The order in which different layers of similar or different materials are applied in a planned relationship, such as overlapped, superimposed, or at certain angles.

Spray Lay-Up

A wet lay-up for processing of reinforced plastics in which a stream of chopped fibers (usually glass) is fed into a stream of liquid resin in a mold. The direction of the fibers is random, as opposed to the mats or woven fabrics that can be used in hand lay-up. See also *hand lay-up* and *wet lay-up*.

Spray Metal Forming

Spray metal forming is a rapid solidification technology for producing semifinished tubes, billets, plates, and simple forms in a single integrated operation. In contrast to other powder metallurgy processes, spray metal forming offers the distinct advantage of skipping the intermediate steps of atomization and consolidation by atomizing and collecting the spray in the form of a billet in a single operation. Also, the elimination of powder handling reduces oxide content and enhances ductility.

The Process

Spray forming involves converting a molten metal stream into a spray of droplets by high-pressure gas atomization. The droplets cool rapidly in flight and ideally arrive at a collector plate with just enough liquid content to spread and completely wet the surface. The metal then solidifies into one almost fully dense preform with a very fine, uniform microstructure. Steel, copper, nickel-based superalloys, and aluminum alloys have been successfully spray-formed. These billets are combined, in the downstream manufacturing process, with extrusion or forging, and then coupled with high-speed machining to produce components in final form.

Advantages

Rapid solidification processes such as spray metal forming offer some distinct advantages over conventional ingot metallurgy processing. Superior properties due to fine grain sizes; a fine, homogeneous distribution of second-phase precipitates; and the absence of macrosegregation result from cooling rates on the order of 10^3–10^5 K/s (gas atomization processes approach 10^6 K/s). The high cooling rate in spray forming is obtained by higher gas-to-metal ratios. In certain alloy systems, a high volume fraction of fine (0.05–0.2 μm) intermetallic dispersoids may be obtained with high gas-to-metal ratios.

Applications

Many different aluminum alloys and SiC-particulate-reinforced aluminum metal-matrix composites have been processed. They include conventional alloys in the 2XXX, 3XXX, 5XXX, 6XXX, and 7XXX series; high-temperature and high-strength alloys; and high-silicon-content alloys. The nonconventional alloys have been developed specifically for spray forming or adapted from alloys developed for rapid solidification rate (RSR) processes such as gas atomization or melt spinning. These alloys have shown superior properties such as wear resistance, room-and high-temperature strength, and creep resistance.

Alloys developed especially for spray forming, such as the ultrahigh zinc content alloys and 7050 aluminum with additional zinc, have been processed. Many of these alloys have been processed with the addition of SiC to improve the stiffness. These alloys have shown superior strength properties over conventionally processed material.

For principal alloy systems originally developed for RSR processing have been extensively spray-formed. These are Al–Fe–V–Si (FVS), Al–Fe–Ce–W (FCW), Al–Ce–Cr–Co (CCC), and Al–Ni–Co (NYC) alloys. To optimize mechanical properties, these alloys were spray-formed over a range of processing parameters. For ultrahigh-temperature aluminum alloys, the melt superheat, or pour temperature, and the gas-to-metal ratio, and the injection of SiC particulate, have the greatest effect on microstructure (e.g., droplet and dispersoid size and volume fraction) and resultant properties.

Uses

Two government-sponsored programs have been completed using spray-formed material to produce components for use in Department of Defense vehicles. The first program produced track pins for advanced tract ground combat vehicles using an ultrahigh-strength aluminum alloy. The second program developed the spray-forming processing parameters to produce an ultrahigh-temperature aluminum alloy for stator vanes in high-performance jet aircraft engines.

The goal of the track pin program was to develop processing that produces an ultrahigh-strength aluminum alloy that can replace the steel currently used in the manufactured pins for tract ground combat vehicles. The objective was to replace the hollow steel pin with a solid aluminum pin that has similar properties. Several alloys were spray-formed, extruded, heat-treated, and tested to determine the tensile strength, ductility, and modulus. The two alloys with the best combination of properties were selected for additional processing. To increase the stiffness, SiC was added during spray forming. The spray-formed ultrahigh-strength alloys have yield strengths in excess of 690 MPa and show good ductility even with the addition of SiC. Track pins have been produced and were successfully tested. The corrosion resistance of these materials is improved by spray forming. The use of the ultrahigh-strength aluminum would result in a weight reduction of over 204.3 kg.

The NYC alloy shows promise for use in static parts of jet engines. The material properties at high temperatures provide an opportunity to replace heavier titanium parts with a lighter-weight aluminum alloy resulting in substantial saving in life-cycle costs.

The high-temperature aluminum alloy has been produced as extrusions and forgings. Machining contractors have developed the most efficient and economical processes to deliver a finished part. The alloy has been extruded and machined into the final component configuration and the alloy has good machining characteristics. Forging trials have shown that the material can be readily deformed offering increased material yield.

Spray Metallizing

See *metallizing*.

Spray Nozzle

In atomizing of metal powders, an orifice through which a molten metal passes to form a stream that can be further disintegrated by a gas, a liquid, or by mechanical means.

Spray Quenching

A quenching process using spray nozzles to spray water or other liquids on a part. The quenched rate is controlled by the velocity and volume of liquid per unit area per unit of time of impingement.

Spray Rate

Same as *feed rate*.

Spray Transfer

In consumable-electrode arc welding, a type of metal transfer in which the molten filler metal is propelled across the arc as fine droplets. Compare with *globular transfer* and *short-circuiting transfer*. See also the term *short-circuiting transfer* and *metal transfer*.

Spray Welding

Various processes for producing coatings on metals. The basic characteristic is that the material is initially applied by some metal powder spraying technique and the component is then heated by a flame or other means to cause the deposited powder to fuse to the component.

Spread

The quantity of adhesive per unit joint area applied to an adherend, usually expressed in pounds of adhesive per thousand square feet of joint area.

Spreader

An axial groove in a plain bearing designed to spread oil along the bearing.

Spreader (Plastics)

A streamlined metal block placed in the path of flow of the plastic material in the heating cylinder of extruders and injection molding machines to spread it into thin layers, thus forcing it into intimate contact with the heating areas.

Spreader Pockets

Depressions in a sliding surface designed to distribute lubricant.

Sprengle Explosives

Chlorate compounds that have been rendered reasonably safe from violent explosion by separating the chlorate from the combustible matter. The potassium chlorate, made into porous cartridges and dipped, just before use, in a liquid combustible such as nitrobenzene or dead oil, was called **rack-a-rock**. It is a mixture of 79% chlorate and 21% nitrobenzene. **Rack-a-rock special** contains, in addition, 12%–16% picric acid. Sprengle explosives were formally used as military explosives, are very sensitive to friction and heat, and are now valued only for mining or when it is desired to economize on nitrates.

Sprigs

Pins or rods inserted to strengthen local weak areas of sand molds.

Springback

(1) The elastic recovery of metal after stressing. (2) The extent to which metal tends to return to its original shape or contour after undergoing a forming operation. This is compensated for by over-bending or by a secondary operation of *restriking*. (3) In flash, upset, or pressure welding, the deflection in the welding machine caused by the upset pressure.

Spring Brass

An imprecise term applied to various hard rolled brass alloys but particularly 70% copper, 30% zinc.

Spring Constant

The force required to compress a spring or specimen 25 mm (1 in.) in a prescribed test procedure.

Spring Steel

This is a term applied to any steel used for springs. The majority of springs are made of steel, but brass, bronze, nickel silver, and phosphor bronze are used where their corrosion resistance or electric conductivity is desired. Carbon steels, with from 0.50% to 1.0% carbon, are much used, but vanadium and chromium–vanadium steels are also employed, especially for heavy car and locomotive springs. Special requirements for springs are that the steel be low in sulfur and phosphorus, and that the analysis be uniform. For flat or spiral springs that are not heat-treated after manufacture, hard-drawn or rolled steels are used. These may be tempered in the mill shape. Music wire is widely employed for making small spiral springs. A much-used straight-carbon spring steel has 1% carbon and 0.30%–0.40% manganese, but becomes brittle when over-stressed. ASTM carbon steel for flat springs has 0.70%–0.80% carbon and 0.50%–0.80% manganese, with 0.04% max each of sulfur and phosphorus. Motor springs are made of this steel rolled hard to a tensile strength of 1723 MPa. Watch spring steel, for mainsprings, has 1.15% carbon, 0.15%–0.25% manganese, and in the hard-rolled condition, as an elastic limit above 2068 MPa.

Silicon Steels

These are used for springs and have high strength. These steels average about 0.4% carbon, 0.75% silicon, and 0.95% manganese, with or without copper, but the silicon may be as high as 2%. A steel, used for automobile leaf springs and recoil springs, contains 2% silicon, 0.75% manganese, and 0.60% carbon. The elastic limit is 689–2068 MPa, depending on drawing temperature, with a hardness 250–600 Brinell.

Manganese Steels

These steels for automotive springs contain about 1.25% manganese and 0.40% carbon, or about 2% manganese and 0.45% carbon. When heat-treated, the latter has a tensile strength of 1378 MPa and 10% elongation. Part of the manganese may be replaced by silicon and the silicon–manganese steels have tensile strengths as high as 1861 MPa. The addition of chromium or other elements increases ductility and improves physical properties. Manganese steels are deep-hardening but are sensitive to overheating. The addition of chromium, vanadium, or molybdenum widens the hardening range.

Forms

Wire for coil springs ranges in carbon from 0.50% to 1.20%, and in sulfur from 0.028% to 0.029%. Bessemer wire contains too much sulfur for spring use. Cold working is the method for hardening the wire and for raising the tensile strength.

The highest grades of wire are referred to as music wire. The second grade is called hard-drawn spring wire. The latter is a less expensive basic open-hearth steel with manganese content of 0.80%–1.10%, and an ultimate strength up to 2068 MPa.

Applications

For jet-engine springs and other applications wear resistance to high temperatures is required, stainless steel and high-alloy steels are used. But, while these may have the names and approximate compositions of standard stainless steels, for spring-wire use their manufacture is usually closely controlled. For example, when the carbon content is raised in high-chromium steels to obtain the needed spring qualities, the carbide tends to collect in the grain boundaries and cause intergranular corrosion unless small quantities of titanium, columbium, or other elements are added to immobilize the carbon. Types include Type 302 stainless steel of highly controlled analysis for coil springs. Alloy NS-355 is a stainless steel having a typical analysis of 15.64% chromium, 4.38% nickel, 2.68% molybdenum, 1% manganese, 0.32% silicon, 0.12% copper, with the carbon at 0.14%. The modulus of elasticity is 205,100 MPa at 27°C and 168,000 MPa at 427°C. 17-7 PH stainless steel has 17% chromium, 7% nickel, 1% aluminum, and 0.07% carbon. Wire has a tensile strength up to 2378 MPa. Spring wire for high-temperature coil springs may contain little or no iron. Alloy NS-25, for springs operating at 760°C, contains about 50% cobalt, 20% chromium, 15% tungsten, and 10% nickel, with not more than 0.15% carbon.

Spring Temper

A *temper* of nonferrous alloys and some ferrous alloys characterized by tensile strength and hardness about two-thirds of the way from *full hard* to *extra spring* temper.

Sprue (Metals)

(1) The mold channel that connects the *pouring basin* with the runner or, in the absence of a pouring basin, directly into which molten metal is poured. Sometimes referred to as down-sprue or downgate. (2) Sometimes used to mean all gates, risers, runners, and similar scrap that are removed from castings after shakeout.

Sprue (Plastics)

A single hole through which thermoset molding compounds are injected directly into the mold cavity.

Spun Parts

Metal spinning, essentially, involves forming flat sheet metal disks into seamless circular or cylindrical shapes. It is a useful processing technique when quantity does not warrant investment needed for draw dies.

The Process

The first step in the spinning process is to produce a form to the exact shape of the inside contours of the part to be made. The form can be of wood or metal. This form is secured to the headstock of a lathe and the metal blank is, in turn, secured to the form. Manual spinning techniques exist, and mechanical spinning lathes usually can be set up to force the blank against the form mechanically.

In addition to manual or power spinning, hot spinning is sometimes used either to anneal a spun part, eliminating the need to remove a partially formed blank from the lathe, or else to increase the plasticity of the metal being formed. In the latter category, some metals such as titanium or magnesium must be spun hot because their normal room-temperature crystal structure lacks ductility. Heavy parts (up to 122 mm in some cases) can also be spun with increased facility at elevated temperature.

Shapes and Tolerances

Basically, a component must be symmetrical about its axis to be adaptable to spinning. The three basic spinning shapes are the cone, hemisphere, and straight-sided cylinder. The shapes are listed in order of increasing difficulty to be formed by spinning.

Available spinning equipment is the limiting factor in determining the size of parts. Parts can be made ranging in diameter from 25.4 mm to almost 3.6 m. Thickness ranges from 0.010 to 122 mm. Most commonly, spun parts range in thickness from 0.059 to 4.7 mm.

Spun Roving

A heavy, low-cost glass or aramid fiber strand consisting of filaments that are continuous but doubled back on themselves. See also *roving*.

Sputtering

The bombardment of a solid surface with a flux of energetic particles (ions) that results in the ejection of atomic species. The ejected material may be used as a source for deposition. No heat is involved. The ejected atoms are deposited on surrounding surfaces. See also *physical vapor deposition* and the term *secondary ion*.

Sputter Texturing

Texturing of a metal surface improves the bonding of surfacing material to the substrate or the attachment of parts to the textured piece. One of many applications is the texturing of the metal surface of medical hip implants. These devices require an irregular surface to stimulate bone attachment.

Texturing of complex shapes can be improved with a sputter-etching method that uses temporarily attached ceramic particles. By controlling the size and distribution of the ceramic particles, the width and depth of the texture can be regulated.

The first stage of the process is the spraying or dripping of adhesive on the area to be textured. Microspheres of ceramic are forced into the adhesive and the area is heated. The part is then placed in a discharge chamber where it is bombarded with argon ions. This operation produces an etching on the surface of the part not covered by the ceramic spheres. The etch depth is controlled by voltage, current density, and sputtering duration. The adhesive, which is charred by the sputter-etch process, is removed with atomic oxygen in a plasma asher. The brushing away of the ceramic particles reveals a textured surface. See *Ind. Heating*, January, p. 45, 2000.

Square Drilling

Making square holes by means of a specially constructed drill made to rotate and also to oscillate so as to follow accurately the periphery of a square guide bushing or template.

Square Groove Weld

A groove weld in which the abutting surfaces are square.

Squaring Shear

A machining tool, used for cutting sheet metal or plate, consisting essentially of a fixed cutting knife (usually mounted on the rear of the bed) and another cutting knife mounted on the front of a reciprocally moving crosshead, which is guided vertically in side housings. Corner angles are usually 90°. See the term *shear* (2).

Squeeze Affect

(1) The production of lubricant from a porous retainer by application of pressure. Also known as sponge effect. (2) The persistence of a film of fluid between two surfaces that approach each other in the direction of their common normal plane.

Squeeze Casting

A hybrid liquid-metal forging process in which liquid metal is forced into a permanent mold by a hydraulic press.

Squeeze-Out

Adhesive pressed out of the bond line due to pressure applied on the adherends.

Squeeze Time

In resistance welding, the time between the initial applications of pressure and current.

Stable

Not liable to change physically or chemically.

Stabilization

In carbon fiber forming, the process used to render the carbon fiber precursor infusible prior to carbonization.

Stabilized Stainless Steel

Austenitic stainless steel containing sufficient quantities of certain elements such as titanium, or niobium (columbium). These combine preferentially with carbon which, in their absence would, upon heating in the 400°C–900°C range, combine with the chromium causing local chromium depletion, particularly at grain boundaries, leading to corrosion problems, for example, after welding. To achieve immunity the alloy addition needs to exceed some level related to the carbon content; typical figures being 10× carbon for niobium and 5× carbon for titanium. Immunity from this form of damage is also achieved by limiting the carbon to less than 0.03% or by quenching the steel from 1050°C or by adding an element, such as molybdenum

at about 3%, which induces a small percentage of ferrite which localizes the carbide precipitation. However, although these latter three treatments confer immunity the term 'stabilized' is usually taken to apply steel varieties with a strong carbide formers. See *steel, sensitization* and *weld decay*.

Stabilizers

Chemicals used in plastics formulation to help maintain physical and chemical properties during processing and service life. A specific type of stabilizer, known as an ultraviolet stabilizer, is designed to absorb ultraviolet rays and prevent them from attacking the plastic. Heat stabilizers are added to lessen the severity of thermal oxidation processes and their effect on properties.

Stabilizing Gas (Plasma Spraying)

The arc gas, which is ionized to form the plasma. Introduced into the arc chamber tangentially, the relatively cold gas chills the outer surface of the arc stream, tending to constrict the arc, raise its temperature, and force it out of the front anode nozzle in a steady, relatively un-fluctuating stream. See the terms *plasma spraying* and *transferred arc*.

Stabilizing Treatment

(1) Before finishing to final dimensions, repeatedly heating a ferrous or nonferrous part to or slightly above its normal operating temperature and then cooling to room temperature to ensure that there is dimensional stability in service. (2) Transforming retained austenite in quenched hardenable steels, usually by *cold treatment*. (3) Heating a solution-treated stabilized grade of austenitic stainless steel to 870°C–900°C to precipitate all carbon as TiC, NbC, or TaC so that *sensitization* is avoided on subsequent exposure to elevated temperature. (4) Any process intended to stabilize the microstructure, dimensions or other features of a component so that undesirable changes do not occur in subsequent service or treatment. For example, some complex component intended for high-temperature service may be subjected to a pre-service heat treatment at a slightly higher temperature.

Stack Cutting

Thermal cutting of stacked metal plates arranged so that all the plates are severed by a single cut.

Stacking Fault

A defect in the crystal lattice in which the normal stacking arrangement is disrupted with partial dislocations at the perimeter.

Stack Molding (Metals)

A foundry practice that makes use of both faces of a mold section, one face acting as the drag and the other as the cope. Sections, when assembled to other similar sections, form several tiers of mold cavities, all castings being poured together through a common screw.

Stacking Sequence

A description of a laminate that details the orientations of the plies and their sequence in the laminate. See also the terms *laminate* and *quasi-isotropic laminate*.

Stack Welding

Resistance spot welding of stacked plates, all being joined simultaneously.

Stage

A device for holding a specimen in the desired position in the optical path of a microscope.

Staggered-Intermittent Fillet Welding

Making a line of intermittent fillet welds on each side of a joint so that the increments on one side are not opposite those on the other. Contrast with *chain-intermittent fillet welding.*

Staggered-Tooth Cutters

Milling cutters with alternate flutes of oppositely directed helixes.

Staging

Heating a premixed resin system, such as in a *prepreg*, until the chemical reaction (curing) starts, but stopping the reaction before the gel point is reached. Staging is often used to reduce resin flow in subsequent press molding operations.

Staining

Precipitation etching that causes contrast by distinctive staining of microconstituents; different interference colors originate from surface layers of varying thickness. Also known as *color etching.*

Stainless Steel

Stainless steel comprises a large and widely used family of iron-chromium alloys known for their corrosion resistance—notably their "non-rusting" quality. This ability to resist corrosion is attributable to a chromium-oxide surface film that forms in the presence of oxygen. The film is essentially insoluble, self-healing, and nonporous. A minimum chromium content of 12% is required for the formation of the film, and 18% is sufficient to resist even severe atmospheric corrosion. Chromium content, however, may range to about 30% in several other alloying elements, such as manganese, silicon, nickel, or molybdenum, are usually present. Most stainless steels are also resistant to marine atmospheres, fresh water, oxidation at elevated temperatures, and mild and oxidizing chemicals. Some are also resistant to salt water and reducing media. They are also quite heat resistant, some retaining useful strength to 981°C. And some retain sufficient toughness at cryogenic temperatures. Thus, stainless steels are used in a wide range of applications requiring some degree of corrosion and/or heat resistance, including auto and truck trim, chemical and food-processing equipment, petroleum-refining equipment, furnace parts in heat treating hardware, marine components, architectural applications, cookware and housewares, pumps and valves, aircraft and aircraft-engine components, springs, instruments, and fasteners.

The 18-8 chromium–nickel steels were called super stainless steels in England to distinguish them from the plain chromium steels. Today, wrought stainless steels alone include some 70 standard compositions and many special compositions. They are categorized as austenitic, ferritic, martensitic, or precipitation-hardening (PH) stainless steels, depending on their microstructure or, in the case of the PH, their hardening and strengthening mechanism. There are also many cast stainless steels having these metallurgical structures. They are also known as cast corrosion-resistant steels, cast heat-resistant steels, and cast corrosion- and heat-resistant steels. Several compositions are also available in powder form for the manufacture of stainless steel powder-metal parts.

Fabrication

As with steels in general, the so-called wrought stainless steels come from the melting furnaces in the form of ingot or continuously cast slabs. Ingots require a roughing or primary hot working, which the other form commonly bypasses. All then go through fabricating and finishing operations such as welding, hot and cold forming, rolling, machining, spinning, and polishing. No stainless steel is excluded from any of the common industrial processes because of its special properties; yet all stainless steels require attention to certain modifications of technique.

Hot Working

Hot working is influenced by the fact that many of the stainless steels are heat-resisting alloys. They are stronger at elevated temperatures than ordinary steel. Therefore, they require greater roll and forge pressure, and perhaps less reductions per pass or per blow. The austenitic steels are particularly heat resistant.

Welding

Welding is influenced by another aspect of high-temperature resistance of these metals—the resistance to scaling. Oxidation during service at high temperatures does not become catastrophic with stainless steel because the steel immediately forms a hard and protective scale. But this, in turn, means that welding must be conducted under conditions that protect the metal from such reactions with the environment. This can be done with specially prepared coatings on electrodes, under cover of fluxes, or in vacuum; the first two techniques are particularly prominent. Inert-gas shielding also characterizes widely used processes among which at least a score are now numbered. As for weld cracking, care must be taken to prevent hydrogen absorption in the martensitic grades and martensite in the ferritic grades, whereas a small proportion of ferrite is almost a necessity in the austenitic grades. Metallurgical "phase balance" is an important aspect of welding the stainless steels because of these complications from a two-phase structure. Thus, a minor austenite fraction in ferritic stainless can cause martensitic cracking, while a minor ferrite fraction in austenite can prevent hot cracking. However, the most dangerous aspect of welding austenitic stainless steel is the potential "sensitization" affecting subsequent corrosion.

Machining and Forming

These processes adapt to all grades, and with these major precautions: First, the stainless steels are generally stronger and tougher than carbon steel, such that more power and rigidity are needed in tooling. Second, the powerful work-hardening effect gives the austenitic grades the property of being instantaneously strengthened upon the first touch of the tool or pass of the roll. Machine tools must therefore bite surely and securely, with care taken not to "ride" the piece. Difficult forming operations warrant careful attention to variations in grade that are available, also in heat treatment, for accomplishing end purposes without unnecessary work problems.

Finishing

These operations produce their best effects with stainless steels. No metal takes a more beautiful polish, and none holds it so long or so well. Stainlessness is not just skin-deep, but body through. And, of course, coatings are rendered entirely unnecessary.

Stainless Steel (Cast)

Cast stainless steels are divided into two classes: those intended primarily for uses requiring corrosion resistance and those intended mainly for uses requiring heat resistance. Both types are commonly known by the designations of the Alloy Casting Institute of the Steel Founders Society of America, and these designations generally begin with the letter C for those used mainly for corrosion resistance and with the letter H for those used primarily for heat resistance. All are basically iron–chromium or iron–chromium–nickel alloys, although they may also contain several other alloying ingredients, notably molybdenum in the heat-resistant type, and molybdenum, copper, and/or other elements in the corrosion-resistant type. The corrosion-resistant cast stainless steel type follows the general metallurgical classifications of the wrought stainless steels, that is, austenitic, ferritic, austenitic–ferritic, martensitic, and precipitation hardening. Specific alloys within each of these classifications are austenitic (CH-20, CK-20, CN-7M), ferritic (CB-30 and CC-50), austenitic-ferritic (CE-30, CF-3, CF-3A, CF-8, CF-8A, CF-20, CF-3M, CF-3MA, CF-8M, CF-8C, CF-16F, and CG-8M), martensitic (CA-15, CA-40, CA-15M, and CA-6NM), and precipitation hardening (CB-7Cu and CD-4MCu). The chromium content of these alloys may be as little as 11% or as much as 30%, depending on the alloy. The heat-resistant cast stainless steel types may contain as little as 9% chromium (Alloy HA), although most contain much greater amounts, such as 32% in HL. Although nickel content rarely exceeds chromium content in the corrosion-resistant type, it does in several heat-resistant types (HN, HP, HP, HU, HW, and HX). In fact, nickel is the major ingredient in HU, HW, and HX. Several of the heat-resistant types can be used at temperatures as high as 1149°C. C-series grades are used in valves, pumps, and fittings. H-series grades are used for furnace parts and turbine components.

Applications

Iron–chromium alloys containing from 11.5% to 30% chromium and iron–chromium–nickel alloys containing up to 30% chromium and 31% nickel are widely used in the cast form for industrial process equipment at temperatures from –257°C to 649°C. The largest area of use is in the temperature range from room temperature to the boiling points of the materials handled.

Typical stainless castings are pumps, valves, fittings, mixers, and similar equipment. Chemical industries employ them to resist nitric, sulfuric, phosphoric, and most organic acids, as well as many neutral and alkaline salt solutions. The pulp and paper industry is a large user of high alloy castings in digesters, filters, pumps, and other equipment for the manufacture of pulp. Fatty acids and other chemicals involved in soap-making processes are often handled by high alloy casting. Bleaching and dyeing operations in the textile industry require parts made from high alloys. These corrosion-resistant alloys are also widely used in making synthetic textile fibers. Pumps and

valves cast of various high alloy compositions find wide application in petroleum refining. Other fields of application are food and beverage processing and handling, plastics manufacturer, preparation of pharmaceuticals, atomic-energy processes, and explosives manufacture. Increasing use is being made of cast stainless alloys for handling liquid gases at cryogenic temperatures.

Stainless Steel (Wrought)

Except for the precipitation-hardening (PH) stainless steels, wrought stainless steels are commonly designated by a three-digit numbering system of the American Iron and Steel Institute. Wrought austenitic stainless steels constitute the 2XX and 3XX series and the wrought ferritic stainless steels are part of the 4XX series. Wrought martensitic stainless steels belong either in the 4XX or 5XX series. Suffix letters, such as I, for low carbon content or Se for selenium, are used to denote special compositional modifications. Cast stainless steels are commonly known by the designations of the Alloy Casting Institute of the Steel Founders Society of America, which begin with letters CA through CN and are followed by numbers or numbers and letters. Powder compositions are usually identified by the designations of the Metal Powder Industries Federation.

Of the austenitic, ferritic, and martensitic families of wrought stainless steels, each as a general-purpose alloy. All of the others in the family are derivatives of the basic alloy, with compositions tailored for special properties. The stainless steel 3XX series has the largest number of alloys and stainless steel 302, a stainless "18-8" alloy, is a general-purpose one. Besides its 17%–19% chromium and 8%–10% nickel, it contains a maximum of 0.15% carbon, 2% manganese, 1% silicon, 0.4% phosphorus, and 0.03% sulfur. 302B is similar except for greater silicon (2%–3%) to increase resistance to scaling. Stainless steels 303 and 303Se are also similar except for greater sulfur (0.15% minimum) and, optionally, 0.6% molybdenum and 303, and 0.06 maximum sulfur and 0.15 minimum selenium in 303Se. Both are more readily machinable than 302. 304 and 304L stainless steels are low-carbon (0.08% and 0.03% maximum, respectively) alternatives, intended to restrict carbide precipitation during welding and, thus, are preferred to 302 for applications requiring welding. They may also contain slightly more chromium and nickel. 304N is similar to 304 except for 0.10%–0.16% nitrogen. The nitrogen provides greater strength than 302 at just a small sacrifice in ductility and a minimal effect on corrosion resistance. 305 has 0.12% maximum carbon but greater nickel (10.5%–13%) to reduce the rate of work hardening for applications requiring severe forming operations. S30430, as designated by the Unified Numbering System, contains 0.08 maximum carbon, 17%–19% chromium, 8%–10% nickel, and 3%–4% copper. It features a still lower rate of work hardening and is used for severe cold-heading operations. 308 contains more chromium (19%–21%) and nickel (10%–12%) and, thus, is somewhat more corrosion and heat resistant. Although used for furnace parts and oil-refinery equipment, its principal uses for welding rods because its higher alloy content compensates for alloy content that may be reduced during welding.

Stainless steels 309, 3095, 310, 310S, and 314 have still greater chromium and nickel contents. 309S and 310S are low-carbon (0.08% maximum) versions of 309 and 310 for applications requiring welding. They are also noted for high creep strength. Stainless steel 314, which like 309 and 310 contains 0.25% maximum carbon, also has greater silicon (1.5 to 3%), thus providing greater oxidation resistance. Because of the high silicon content, however, it is prone to embrittlement during prolonged exposure at temperatures of 649°C–816°C. This embrittlement, however, is only evident at room temperature and is not considered harmful unless the alloy is subject to shock loads. These alloys are widely used for heaters and heat exchangers, radiant tubes, and chemical and oil-refinery equipment.

Stainless steels 316, 316L, 316F, 316N, 317, 317L, 321, and 329 are characterized by the addition of molybdenum, molybdenum and nitrogen (316N), or titanium (321). Stainless steel 360, with a 16%–18% chromium, 10%–14% nickel, and 2%–3% molybdenum, is more corrosion and creep resistant than 302- or 304-type alloys. Type 316L is the low-carbon version for welding applications; 316F, because of its greater phosphorus and sulfur, is the "free-machining" version; and 316N contains a small amount of nitrogen for greater strength. Stainless steels 317 and 317L are slightly richer and chromium, nickel, and molybdenum and, thus, somewhat more corrosion and heat resistant. Like 360, they are used for processing equipment in the oil, chemical, food, paper, and pharmaceutical industries. Type 321 is titanium-stabilized to inhibit carbide precipitation and provide greater resistance to intergranular corrosion in welds. Type 329, a high-chromium (25%–30%) low-nickel (3%–6%) alloy with 1%–2% molybdenum, is similar to 316 in general corrosion but more resistant to stress corrosion. Stainless steel 330, a high-nickel (34%–37%), normal chromium (17%–20%), 0.75%–1.5% silicon, molybdenum-free alloy, combines good resistance to carburization, heat, and thermal shock.

Stainless steels 347 and 348 are similar to 321 except for the use of columbium and tantalum instead of titanium for stabilization. Type 348 also contains a small amount (0.2%) of copper. Both have greater creep strength than 321 and they are used for welded components, radiant tubes, aircraft-engine exhaust manifolds, pressure vessels, and oil-refinery equipment. Stainless steel 384, with nominally 16% chromium and 18% nickel, is another low-work-hardening alloy used for severe cold-heading applications.

The stainless steel 2XX series of austenitics comprises 201, 202, and 205. They are normally chromium content (16%–19%), but low in nickel (1%–6%), high in manganese (5.5%–15.5%), and with 0.12%–0.25% carbon and some nitrogen. Types 201 and 202 have been called the low-nickel equivalents of 301 and 302, respectively. Type 202, with 17%–19% chromium, 7.5%–10% manganese, 4%–6% nickel, and a maximum of 1% silicon, 0.25% nitrogen, 0.15% carbon, 0.06% phosphorus, and 0.03% sulfur, is a general-purpose alloy. Type 201, which contains less nickel (3.5%–5.5%) and manganese (5.5%–7.5%), was prominent during the Korean War due to a nickel shortage. Type 205 has the least nickel (1%–1.75%), and the most manganese (14%–15.5%), carbon (0.12%–0.25%), and nitrogen (0.32%–0.40%) contents. It is said to be the low-nickel equivalent of 305 and has a low rate of work hardening that is useful for parts requiring severe forming operations.

Like stainless steels in general, austenitic stainless steels have a density of 7750–8027 kg/m³. Unlike some other stainless steels, they are essentially nonmagnetic, although most alloys will become slightly magnetic with cold work. Their melting range is 1371°C–1454°C, specific heat at 0°C–100°C is about 502 J/kg K, and electrical resistivity at room temperature ranges from 69 × 10⁻⁸ to 78 × 10⁻⁸ Ω m.

Types 309 and 310 have the highest resistivity, and 201 and 202 the lowest.

Most are available in many mill forms and are quite ductile in the annealed condition, tensile elongation's ranging from 35% to 70%, depending on the alloy. Although most cannot be strengthened by heat treatment, they can be strengthened appreciably by cold work. In the annealed condition, the tensile yield strength of all the austenitics falls in the range of 207–552 MPa, with ultimate strengths in the range of 517–827 MPa. But cold-working 201 or 301 sheet just to the half-hard temper increases yield strength to 758 MPa and ultimate strength to at least 1034 MPa. Tensile modulus is typically 193 × 10³

to 199 × 10³ MPa and decreases slightly with severe cold work. As to high-temperature strength, even in the annealed condition most alloys have tensile yield strengths of at least 83 MPa at 815°C, and some (308, 310) about 138 MPa. Types 310 and 347 have the highest creep strength, or stress-rupture strength, at 538°C–649°C. Annealing temperatures range from 954°C to 1149°C, initial-forging temperatures range from 1093°C to 1260°C, and their machinability index is typically 50 to 55, 65 for 303 and 303Se, relative to 100 for 1112 steel.

Among the many specialty wrought austenitic stainless steels are a number of nitrogen-strengthened stainless steels; Nitronic 20, 32, 33, 40, 50, and 60; 18–18 Plus and Marinaloy HN and 22; and SAF 2205 and 253MA. Nitrogen, unlike carbon, has the advantage of increasing strength without markedly reducing ductility. Some of these alloys are twice as strong as the standard austenitics and also provide better resistance to certain environments. All are normal or higher than normal in chromium content. Some are also normal or higher than normal in nickel content, whereas others are low in nickel and, in the case of 18-18 Plus, nickel-free. Nitronic 20, a 23% chromium, 8% nickel, 2.5% manganese alloy, combines high resistance to oxidation and sulfidation and was developed for engine exhaust valves. Unlike austenitics in general, it is hardenable by heat treatment. Solution treating at 1177°C, water quenching, and aging at 760°C provide tensile strengths of 579 MPa yield and 979 MPa ultimate. SAF 2205, and extra-low-carbon (0.03%), 22% chromium, 5.5% nickel, 3% molybdenum alloy, is a ferritic-austenitic alloy with high resistance to chloride-and hydrogen-sulfide-induced stress corrosion, pitting in chloride environments, and intergranular corrosion in welded applications.

The wrought ferritic stainless steels are magnetic and less ductile than the austenitics. Although some can be hardened slightly by heat treatment, they are generally not hardenable by heat treatment. All contain at least 10.5% chromium and, although the standard alloys are nickel-free, small amounts of nickel are common in the nonstandard ones. Among the standard alloys, stainless steel 430 is the general-purpose alloy. It contains 16%–18% chromium and a maximum of 0.12% carbon, 1% manganese, 1% silicon, 0.04% phosphorus, and 0.03% sulfur. Stainless steel 430F and 430FSe, the "free-machining" versions, contain more phosphorus (0.06% maximum) and sulfur (0.15% minimum in 430F, 0.06 maximum in 430FSe). 430FSe also contains 0.15% minimum selenium, and 0.6% molybdenum is an option for 430F. The other standard ferritics are stainless steels 405, 409, 429, 434, 436, 442, and 446. 405 and 409 are the lowest in carbon (0.08% maximum) and chromium (11.5%–14.5% and 10.5%–11.75%, respectively), the former containing 0.10%–0.30% aluminum to prevent hardening on cooling from elevated temperatures, and the latter containing 0.75% maximum titanium. Type 429 is identical to 430 except for less chromium (14%–16%) for better weldability. Types 434 and 436 are identical to 430 except for 0.75%–1.25% molybdenum in the former in this amount of molybdenum plus 0.70% maximum columbium and tantalum in the latter; these additives improve corrosion resistance in specific environments. Types 442 to 446 are the highest in chromium (18%–23% and 23%–27%, respectively) for superior corrosion and oxidation resistance, and in carbon (0.20% maximum). Type 446 also contains more silicon (1.50% maximum).

The standard alloys melt in the range of 1427°C–1532°C, thermal conductivities of 21–27 W/m K at 100°C, and electrical resistivities of 59–67 µΩ cm at 21°C. In the annealed condition, tensile yield strengths range from 241 to 276 MPa for 405 to as high as 414 MPa for 434, with ultimate strengths of 448–586 MPa and elongations of 20%–33%. For 1% creep in 10,000 h at 538°C, 430 has a stress-rupture strength of 59 MPa. Typical applications include automotive trim and exhaust components, chemical-processing equipment, furnace hardware and heat-treating fixtures, turbine blades, and molds for glass.

Wrought martensitic stainless steels are also magnetic and, as they are hardenable by heat treatment, provide high strength. Of those in the stainless steel 4XX series, 410, which contains 11.5%–13.0% chromium, is the general-purpose alloy. The others, 403, 414, 416, 416Se, 420, 420F, 422, 431, 440A, and 440C, have similar (403, 414) or more chromium (16%–18% in the 440s). Most are nickel-free or, as in the case of 414, 422, and 431, low in nickel. Most of the alloys also contain molybdenum, usually less than 1%, plus the usual 1% or so maximum of manganese and silicon. Carbon content ranges from 0.15% maximum in 403 through 416 and 416Se, to 0.60%–0.75% in 440A, and as much as 1.20% in 440C. Type 403 is the low-silicon (0.50% maximum) version of 410; 414 is a nickel (1.25%–2.50%)-modified version for better corrosion resistance. Types 416 and 416Se, which contain 12%–14% chromium, also contain more than the usual sulfur or sulfur, phosphorus, and selenium to enhance machinability. Type 420 is richer in carbon for greater strength, and 420F has more sulfur and phosphorus for better machinability. Type 422, which contains the greatest variety of alloying elements, has 0.22%–0.25% carbon, 11%–13% chromium, low silicon (0.75% maximum), low phosphorus, and sulfur (0.025% maximum), 0.5%–1.0% nickel, 0.75%–1.25% of both molybdenum and tungsten, and 0.15%–0.3% vanadium. This composition is intended to maximize toughness and strength at temperatures to 649°C. Type 431 is a higher-chromium (15%–17%) nickel (1.25%–2.50%) alloy for better corrosion resistance. The high-carbon, high-chromium 440 alloys combine considerable corrosion resistance with maximum hardness. The stainless steel 5XX series of wrought martensitic alloys—501, 501A, 501B, 502, 503, and 504—contain less chromium, ranging from 4% to 6% in 501 and 502, to 8%–10% in 501 B and 504. All contain some molybdenum, usually less than 1%, and are nickel-free.

Most of the 4XX alloys can provide yield strengths greater than 1034 MPa and some, such as the 440s, more than 1724 MPa. The martensitic stainless steels, however, are less machinable than the austenitic and ferritic alloys and they are also less weldable. Forging temperatures range from 1038°C to 1232°C. Most of the alloys are available in a wide range of mill forms and typical applications include turbine blades, springs, knife blades and cutlery, instruments, ball bearings, valves and pump parts, and heat exchangers.

The wrought PH stainless steels are also called age-hardenable stainless steels. Three basic types are now available: austenitic, semi-austenitic, and martensitic. Regardless of the type, the final hardening mechanism is precipitation hardening, brought about by small amounts of one or more alloying elements, such as aluminum, titanium, copper, and, sometimes, molybdenum. Their principal advantages are high strength, toughness, corrosion resistance, and relatively simple heat treatment.

Of the austenitic PH stainless steels, A-286 is the principal alloying. Also referred to as an iron-based superalloys, it contains about 15% chromium, 25% nickel, 2% titanium, 1.5% manganese, 1.3% molybdenum, 0.3% vanadium, 0.15% aluminum, 0.05% carbon, and 0.005% boron. It is widely used for aircraft turbine parts and high-strength fasteners. Heat treatment solution treating at 981°C, water or oil quenching, aging at 718°C–732°C for 16 to 18 h and air cooling) provides an ultimate tensile strength of about 1035 MPa and a tensile yield strength of about 690 MPa, with 25% elongation and a Charpy impact strength of 87 J. The alloy retains considerable strength at high temperatures. At 649°C, for example, tensile yield strength is 607 MPa. The alloy also has good weldability and its corrosion resistance in most environments is similar to that of 3XX stainless steels.

The semiaustenitic PH stainless steels are austenitic in the annealed or solution-treated condition and can be transformed to a martensitic structure by relatively simple thermal or thermomechanical treatments. They are available in all mill forms, although sheet and strip are the most common. True semi-austenitic PH 14-8 Mo, PH 15-7 Mo, and 17-7PH, AM-350 and AM-355 are also so classified, although they are said not truly to have a precipitation-hardening reaction. The above steels are lowest in carbon content (0.04% nominally in PH 14-8 Mo, 0.07% in the others). PH 14-8 Mo also nominally contains 15.1% chromium, 8.2% nickel, 2.2% molybdenum, 1.2% aluminum, 0.02% manganese, 0.02% silicon, and 0.005% nitrogen. PH 15-7 Mo contains 15.2% chromium, 7.1% nickel, 2.2% molybdenum, 1.2% aluminum, 0.50% manganese, 0.30% silicon, and 0.04% nitrogen. 17-7PH is similar to PH 15-7 Mo except for 17% chromium and being molybdenum-free. AM-350 contains 16.5% chromium, 4.25% nickel, 2.75% molybdenum, 0.75% manganese, 0.35% silicon, 0.10% nitrogen, and 0.10% carbon. AM-355 has 15.5% chromium, 4.25% nickel, 2.75% molybdenum, 0.85% manganese, 0.35% silicon, 0.12% nitrogen, and 0.13% carbon. In the solution-heat-treated condition in which these steels are supplied, they are readily deformable. They then can be strengthened to various strength levels by conditioning the austenite, transformation to martensite, and precipitation hardening. One such procedure, for 17-7PH, involves heating at 760°C, air cooling to 16°C, then heating to 565°C and air-cooling to room temperature. In their heat-treated conditions, these steels encompass tensile yield strengths ranging from about 1241 MPa for AM-355 to 1793 MPa for PH 15-7 Mo.

After solution treatment, the martensitic PH stainless steels always have a martensitic structure at room temperature. The steels include the progenitor of the PH stainless steels, Stainless W, PH 13-8 Mo, 15-5PH, 17-4PH, and Custom 455. Of these, PH 13-8 Mo and Custom 455, which contain 11%–13% chromium and about 8% nickel plus small amounts of other alloying elements, are the higher-strength alloys, providing tensile yield strengths of 1448 and 1620 MPa, respectively, in bar form after heat treatment. The other alloys range from 15% to 17% chromium and 4% to 6% nickel, and typically have tensile yield strengths of 1207–1276 MPa in heat-treated bar form. They are used mainly in bar form and forgings, and only to a small extent in sheet. Age hardening, following high-temperature solution treating, is performed at 427°C–677°C.

Stainless Steel Products

Metal fibers are used for weaving into fabrics for arctic heating clothing, heated draperies, chemical-resistant fabrics, and reinforcement in plastics and metals. Stainless-steel yarn made from the fibers is woven into stainless-steel fabric that has good creep resistance and retains its physical properties to 427°C. The fiber may be blended with cotton or wool for static control, particularly for carpeting.

Stainless (Steel/Iron)

Any iron or steel containing more than about 12% chromium can be termed stainless. With about 12% or more of chromium the material, when exposed to the normal atmosphere, rapidly forms a thin, impervious, chromium rich, oxide film which protects the underlying steel from further attack. There are many forms of stainless steel, see *steel*. The term Stainless Iron has also been used for many widely differing alloys ranging from 12% chromium low carbon steel to 30% chromium, high carbon cast-iron. It should be viewed with caution. See steel for further commentary on stainless steels.

Staking

Fastening two parts together permanently by recessing one part within the other and then causing plastic flow at the joint.

Stalagmometer

An apparatus for determining surface tension. The mass of a drop of liquid is measured by weighing a known number of drops or by counting the number of drops obtained from a given volume of the liquid.

Stamping

The general term used to denote all sheet metal pressworking. It includes blanking, shearing, hot or cold forming, drawing, bending, and coining. (1) Pressing and forging sheet and plate in closed dies.

Stand

(1) A piece of rolling mill equipment containing one set of work rolls. In the usual sense, any pass of a cold-or hot-rolling mill. (2) Two or more rolls in a housing. See *rolling mill*.

Standard

(1) A basis of comparison on which units are based. For example, units of measurement were traditionally based on a solid metal standard bar retained by some national or international body. (2) A specification covering all relevant aspects of a material, procedure, system, etc. Aspects defined include composition, properties, dimensions and performance.

Standard Addition

In chemical analysis, a method in which small increments of a substance under measurement are added to a sample under test to establish a response function or, by extrapolation, to determine the amount of a constituent originally present in the sample.

Standard Deviation

A measure of the dispersion of observed values or results from the average expressed as the positive square root of the variance.

Standard Electrode Potential

The reversible potential for an electrode process when all products and reactions are at unit activity on a scale in which the potential for the standard hydrogen half-cell is zero.

Standard Gold

(1) A gold alloy containing 10% copper; at one time used for legal coinage in the United States. (2) In the United Kingdom, 91.66% gold, remainder copper.

Standard Grain-Size Micrograph

A micrograph of a known grain size at a known magnification that is used to determine grain size of metals by direct comparison with another micrograph or with the image of a specimen.

Standardization

(1) The process of establishing, by common agreement, engineering criteria, terms, principles, practices, materials, items, processes, and equipment parts and components. (2) The adoption of generally accepted uniform procedures, dimensions, materials, or parts that directly affect the design of a product or a facility. (3) In analytical chemistry, the assignment of a compositional value to one standard on the basis of another standard.

Standard Reference Material

A reference material, the composition or properties of which are certified by a recognized standardizing agency or group.

Standard Silver

In the United Kingdom, 92.5% silver, remainder copper. In the United States, 90% gold, remainder copper.

Standard Wire Gauge

A series of standard diameters.

Standoff Distance

The distance between a nozzle on a welding or cutting torch and the base metal.

Staple Fibers

Fibers for reinforcing plastics that are of spinnable length manufactured directly or by cutting continuous filaments to short lengths (usually 13–50 mm, or 1/2–2 in., long, and 1–5 denier). See also *denier*.

Starch

A large group of natural carbohydrate compounds occurring in grains, tubers, and fruits. The common cereal grains contain 55% to 75% starch, and potatoes contain about 18%. Starches have a wide usage for foodstuffs, adhesives, textile and paper sizing, gelling agents, and fillers; in making explosives and many chemicals; and for making biodegradable detergents such as sodium tripolyphosphate. Starch is a basic need of all peoples and all industries. Much of it is employed in its natural form, but it is also easily converted to other forms, and more than 1000 different varieties of starch are usually on the U.S. market at any one time.

Most of the commercial starch comes from corn, potatoes, and mandioca. Starches from different plants have similar chemical reactions, but all have different granular structure, and the differences in size and shape of the grains have much to do with the physical properties. **Cornstarch** has a polygonal grain of simple structure. It is the chief food starch in the western world, although sweet-potato starch is used where high gelatinization is desired, and tapioca starch is used to give quick tack and high adhesion in glues. **Tapioca starch** has rounded grains truncated on one side and is of lamellar structure. It produces gels of clarity and flexibility, and because it has no serial flavor, it can be used directly for thickening foodstuffs. **Rice starch** is polygonal and lamellar, and has very small particles. It makes an opaque stiff gel and is also valued as a dusting starch for bakery products, although it is expensive for this purpose.

White-potato starch has conchoidal or ellipsoidal grains of lamellar structure. When cooked, it forms clear solutions easily controlled in viscosity, and gives tough, resilient films for coating paper and fabrics. Prolong grinding of grain starches reduces the molecular chain, and the lower weight then gives greater solubility in cold water. Green fruits, especially bananas often contain much starch, but the ripening process changes the starch to sugars.

In general, starch is a white, amorphous powder having a specific gravity from 0.499 to 0.513. It is insoluble in cold water but can be converted to **soluble starch** by treating with a dilute acid. When cooked in water, starch produces an adhesive paste. Starch is easily distinguished from dextrins as it gives a blue-color with iodine while dextrins give violet and red. The starch molecule is often described as a chain of glucose units, with the adhesive **waxy starches** as those with coiled chains. But starch is a complex member of the great group of natural plant compounds consisting of starches, sugars, and cellulose, and originally named carbohydrates because the molecular formula could be written as $C_n(H_2O)_x$; but not all now-known carbohydrates can be classified in this form, and many now-known acids and aldehydes can be indicated by this formula.

Star Craze

Multiple fine surface separation cracks in protruded reinforced plastics that appear to emanate from a central point in that exceed 6 mm (1/4 in.) in length, but do not penetrate deep equivalent depth of a full ply of reinforcement. This condition is often caused by impact damage. See also *crazing* (plastics).

Stardusting

An extremely fine form of roughness on the surface of a metal deposit.

Stark Effect

A shift in the energy of spectral lines due to an electrical field that is either externally applied or is an internal field caused by the presence of neighboring ions or atoms in a gas, solid, or liquid.

Starting Sheet

A thin sheet of metal used as the cathode in electrolyte refining.

Starting Torque

The torque that is required for initiating rotary motion.

Starved Area

An area in a reinforced plastic part that has an insufficient amount of resin to wet out the reinforcement completely. This condition may be due to improper wetting, impregnation, or resin flow; excessive molding pressure; or incorrect bleeder cloth thickness.

Starved Joint

An adhesive bonded joint that has an insufficient amount of adhesive to produce a satisfactory bond. This condition may result from too thin a spread to fill the gap between the adherends, excessive penetration of the adhesive into the adherend, too short an assembly time, or the use of excessive pressure.

State of Strain

A complete description of the deformation within a homogeneous deformed volume or at a point. The description requires, in general, the knowledge of the independent components of *strain*.

State of Stress

A complete description of the stresses within a homogeneously stressed volume or at a point. The description requires, in general, the knowledge of the independent components of *stress*.

Static

Stationary or very slow. Frequently used in connection with routine testing of metal specimens. Contrast with *dynamic*.

Static Coefficient of Friction

The *coefficient of friction* corresponding to the maximum friction force that must be overcome to initiate macroscopic motion between two bodies.

Static Electrode Force

The force between the electrodes in making spot, seam, or projection welds by resistance welding under welding conditions, but with no current flowing and no movement in the welding machine.

Static Equivalent Load ($P_{0\ metal}$)

In rolling-element bearings, the static load which, if applied, would give the same life as that which the bearing will attain under actual conditions of load and rotation. See also *rating life*.

Static Fatigue

A term sometimes used to identify a form of hydrogen embrittlement in which a metal appears to fracture spontaneously under a steady stress less than the yield stress. There almost always is a delay between the application of stress (or exposure of the stressed metal to hydrogen) and the onset of cracking. More properly referred to as *hydrogen-induced delayed cracking*. Contrast with *fatigue*.

Static Friction

See *limiting static friction*.

Static Hot Pressing

A method of applying a static load uniaxially during hot pressing of metal or ceramic powders. Contrast with *dynamic hot pressing* and *isostatic hot pressing*. See also *hot pressing*.

Static Load Rating (C_0)

In rolling-element bearings, the static load that corresponds to a permanent deformation of rolling element and race at the most heavily stressed contact of 0.00001 of the rolling-element diameter. See also *rating life*.

Static Modulus

The ratio of stress to strain under static conditions. It is calculated from static stress-strength tests, in shear, compression, or tension. Expressed in force per unit of area.

Static Plate

See *explosive welding*.

Static Stress

A stress in which the force is constant or slowly increasing with time, for example, test to failure without shock.

Static Viscosity

See *viscosity*.

Stationary Phase

In chromatography, a particulate material packed into the column or a coating on the inner walls of the column. A sample compound in the stationary phase is separated from compounds moving through the column as a result of being in the mobile phase. See also *mobile phase*.

Statistic

A summary value calculated from the observed values in a sample.

Statistical Process Control

The application of statistical techniques for measuring and analyzing the variation in processes.

Statistical Quality Control

The application of statistical techniques for measuring and improving the quality of processes and products (includes statistical process control, diagnostic tools, sampling plans, and other statistical techniques).

Statuary Bronze

Copper alloys used for casting statues, plaques, and ornamental objects that require fine detail and a smooth, reddish surface. Most of the famous large bronze statues of Europe contain 87% to 90% copper, with varying amounts of tin, zinc, and lead. Early Greek statues contain 9% to 11% tin with as much as 5% lead added apparently to give greater fluidity for crisp details. A general average bronze will contain 90% copper, 6% tin, 3% zinc, and 1% lead. Statuary bronze for cast plaques used in building construction contains 86% copper, 2% tin, 2% lead, 8% zinc, and 2% nickel. The nickel improves fluidity and hardness and strengthens the alloy, and the lead promotes an oxidized finish on exposure. The statuary bronze used for hardware has 83.5% copper, 4% lead, 2% tin, and 10% zinc.

Staving

A tube manipulation process for increasing the external diameter and maintaining the bore constant. Also termed *upsetting*.

Stave Bearing

A *sleeve bearing* consisting of several axially held slats or staves on the outer surface of which the bearing material is bonded.

Stay

A structural member of slender proportions, such as a bar, beam or wire, carry tensile loads.

Steadite

(1) A hard structural constituent of cast iron that consists of a binary eutectic of ferrite, containing some phosphorus in solution, and iron phosphide (Fe_3P). The eutectic consists of 10.2% P and 89.8% Fe. The melting temperature is 1050°C. (2) The eutectic between austenite and iron phosphide in cast iron.

Stead's Brittleness

A condition of brittleness that causes transcrystalline fracture in the coarse grain structure that results from prolonged annealing of thin sheets of low-carbon steel previously rolled at a temperature below about 705°C. The fracture usually occurs at about 45°C to the direction of rolling.

Steady Load

Loads that do not change in intensity, or change so slowly that they may be regarded as steady.

Steady-Rate Creep

See *creep*.

Steadyrest

In cutting or grinding, a stationary support for a long workpiece.

Steam Blanketing

Power plant boilers generate steam by circulating water through tubes fired, i.e., heated, on their exterior. Normally, the steam is generated as discrete bubbles termed (*nucleate boiling*) which rise by gravity or are swept by the pump circulation. However, where the heat flux is excessive steam can form a persistent layer along the bore towards the fire. This layer is termed a *steam blanket* and can lead to various adverse effects, including caustic attack, excessive thickness and fissuring of the normally protective magnetite film, overheating of the tube and distortion.

Steam Erosion

Erosion of high velocity steam impinging on a component. In practice, erosion by pure gaseous steam is rare and the term is often a misnomer (although see wire drawing). Usually, the damage is caused either by entrained solids, see *solid particle erosion* or by water droplets. The term is commonly used regarding the damage observed on the last stage of moving blades of a steam turbine where the steam is very "wet" containing a considerable amount of condensed water droplets. However, even in this case the steam velocity is not responsible for the metal loss since the damaging droplets are those that drip off the stationary blades to be impacted by the moving blades. Once droplets have been accelerated to steam velocity they follow the steamline flow and cause little damage. Strictly, also, this form of water droplet damage is not a form of abrasion by a cutting action, rather it is the result of repeated high velocity impact and hence is more akin to a surface fatigue process. The surface produced is not smooth or grooved but develops a "Cats Tongue" texture of sloping sharp spikes and pits which become progressively coarser as damage progresses.

Steam Hammer

A type of *drop hammer* in which the ram is raised for each stroke by a double-action steam cylinder and the energy delivered to the workpiece is supplied by the velocity and weight of the ram and attached upper die driven downward by steam pressure. The energy delivered during each stroke can be varied.

Steam Molding

A process used to mold plastic parts from pre expanded beads of polystyrene using steam as a source of heat to expand the blowing agent in the material. The steam in most cases is in direct, intimate contact with the beads. It may also be used indirectly, by heating mold surfaces that are in contact with the bead.

Steam Side

See *fire side*.

Steam Treatment

The treatment of a sintered ferrous part in steam at temperatures between 510°C and 595°C in order to produce a layer of black iron oxide (magnetite, or ferrous-ceric oxide, $FeO-Fe_2O_3$) on the exposed surface for the purpose of increasing hardness and wear resistance.

Steatite

A compact, massive rock composed principally of *talc* (magnesium silicate). Ground steatite is used in *porcelain enamels* and *ceramic whiteware*. See *soapstone*.

Steckel Mill

A cold reducing mill having two working rolls and two backup rolls, none of which is driven. The strip is drawn through the mill by a power reel in one direction as far as the strip will allow and then reversed by a second power reel, and so on until the desired thickness is attained.

Steel

Steel is iron alloyed with small amounts of carbon, 2.5% maximum, but usually much less. The two broad categories are carbon steels and alloy steels, but they are further classified in terms of composition, deoxidation method, mill-finishing practice, product form, and principal characteristics. Carbon is the principal influencing element in carbon steels, although manganese, phosphorus, and sulfur are also present in small amounts, and these steels are further classified as low-carbon steels and sometimes referred to as mild steel (up to 0.30% carbon), medium-carbon steels (0.30%–0.60%), and high-carbon steels (more than 0.60%). The greater the amount of carbon, the greater the strength and hardness, and the less ductility. Alloy steels are further classified as low-alloy steels, alloy steels, and high-alloy steels; those having as much as 5% alloy content are the most widely used. The most common designation systems for carbon and alloy steels are those of the American Iron and Steel Institute and the Society of Automotive Engineers, which follow a four- or five-digit numbering system based on the key element or elements, with the last two digits indicating carbon content in hundredths of a percent.

Plain carbon steels (with 1% maximum manganese) are designated 10XX; resulfurized carbon steels, 11XX; resulfurized and rephosphorized carbon steels, 12XX; and plain carbon steels with 1%–1.65% manganese, 15XX. Alloy steels include manganese steels (13XX), nickel steels (23XX and 25XX), nickel-chromium steels (31XX–34XX), molybdenum steels (40XX and 44XX), chromium–molybdenum steels (41XX), nickel–chromium–molybdenum steels (43XX, 47XX, and 81XX–98XX), nickel–molybdenum steels (46XX and 48XX), chromium steels (50XX–52XX), chromium–vanadium steels (61XX), and tungsten–chromium steels (72XX), and silicon–manganese steels (92XX). The letter B following the first two digits designates boron steels and the letter L leaded steels. The suffix H is used to indicate steels produced to specific hardenability requirements. High-strength, low-alloy steels are commonly identified by a 9XX designation of the SAE, where the last two digits indicating minimum tensile yield strength in 1000 psi (6.8 MPa).

In contrast to rimmed steels, which are not deoxidized, killed steels are deoxidized by the addition of the oxidizing elements, such as aluminum or silicon, in the ladle prior to ingot casting, thus, such terms as aluminum-killed steel. Deoxidation markedly improves the uniformity of the chemical composition and resulting mechanical properties of mill products. Semikilled steels are only partially deoxidized, those intermediate in uniformity to rimmed and killed steels. Capped steels have a low-carbon steel rim characteristic of rimmed-steel ingot and central uniformity more characteristic of killed-steel ingot, and are well suited for cold-forming operations.

Steels are also classified as air-melted, vacuum-melted, or vacuum-degassed. Air-melted steels are produced by conventional melting methods, such as open hearth, basic oxygen, and electric furnace. Vacuum-melted steels are produced by induction vacuum melting and consumable electrode vacuum melting. Vacuum-degassed steels are air-melted steels that are vacuum processed before solidification. Vacuum processing reduces gas content, nonmetallic inclusions, and center porosity, and segregation. Such steels are more costly, but have better ductility and impact and fatigue strengths.

Steel-mill products are reduced from ingot into such forms as blooms, billets, and slabs, which are then reduced to finished or semifinished shape by hot-working operations. If the final product is produced by hot working, the steel was known as hot-rolled steel. If the final product is shaped cold, the steel was known as cold-finished steel or, more specifically, cold-rolled steel, or cold-drawn steel. Hot-rolled mill products are usually limited to low- and medium-nonheat-treated carbon steels. They are the most economical steels, have good formability and weldability, and are widely used. Cold-finished steels, compared with hot-rolled products, have greater strength and hardness, better surface finish, and less ductility. Wrought steels are also classified in terms of mill-product form, such as bar steels, sheet steels, and plate steels. Cast steels refer to those used for castings, and P/M (powder metal) steels refer to powder compositions used for P/M parts. Steels are also known by their key characteristic from the standpoint of application, such as electrical steels, corrosion-resistant stainless steels, low-temperature steels, high-temperature steels, boiler steels, pressure-vessel steels, etc.

Steel Powder

Steel powder is used mainly for the production of steel powder metal parts made by consolidating the powder under pressure and then sintering, and, to a limited extent, for steel-mill products, principally tool-steel bar products. For powder metal parts, the powder may be admixed for the desired composition or prealloyed, that is, each powder particle is of the desired composition. For mill products, prealloyed powder is used primarily. Steel powder is widely used to make small- to moderate-size powder metal parts, with compositions closely matching those of wrought steels. Among the more common are carbon steels, copper steels, nickel steels, nickel-molybdenum steels, and stainless steels.

Steel Wool

Steel wool consists of long, fine fibers of steel used for abrading, chiefly for cleaning utensils and for polishing. It is made from low-carbon wire that has high tensile strength, usually having 0.10%–0.20% carbon and 0.50%–1% manganese. The wire is drawn over a track and shaved by stationary knife bearing down on it, and may be made in a continuous piece as long as 30,480 m. Steel wool usually has three edges but may have four or five, and strands of various types are mixed. There are nine standard grades of steel wool, the finest of which has no fibers greater than 0.0027 cm thick; the most commonly used grade has fibers that vary between 0.006 and 0.010 cm. Steel wool comes in batts, or in flat ribbon form on spools usually 10 cm wide. Stainless steel wool is also made, and copper wool is marketed for some cleaning operations.

Steel Types

Steel normally contain less than about 1.5% carbon and are categorized, rather imprecisely, on the basis of their carbon and alloy element content. Some of the more common terms are:

Plain carbon steels contain no significant quantity of alloying elements other than carbon and manganese and are subcategorized as follows:

Mild steel or, less commonly, low carbon steel, containing less than about 0.2% carbon. It has limited capability of hardened by heat treatment (as described below) and is usually used in the normalized or cold work condition.

Carbon steel containing more than about 0.25% carbon, often further subdivided into medium carbon steel containing 0.25% to about 0.5% carbon. High carbon steel containing 0.5% to about 1.2% or, rarely 2% carbon. Carbon manganese steel containing carbon in widely varying amounts plus manganese of about 1%–2%. Note that lesser amounts of manganese present in most steels do not merit this description.

Alloy steel has one or more elements, other than carbon and manganese below about 2%, deliberately added to approve some property. Such steels are often designated by their principal alloy content(s) e.g., "Nickel steel," "Chrome (ium) Moly(bdenum) steel," etc. Almost invariably, carbon is also present and unless the other alloying elements are present in large quantities, the steel responds to heat treatment broadly as summarized below. In a few specialized alloy steels, carbon may be deliberately excluded.

High alloy steel is another imprecise term but it usually implies a total alloy content above about 5%, excluding carbon and manganese.

Steel Physical Metallurgy and Heat Treatment

An important characteristic of iron is that it is a ferrite or alpha (α) phase, with a body centered cubic (BCC) crystal lattice structure. From 910°C to about 1400°C the stable phase is austenite or gamma (γ) with a face centered cubic (FCC) arrangement. Between 1400°C and a melting point at about 1530°C the structure returns to the BCC form but is referred to as delta (δ) phase or delta ferrite. Note, there is no beta phase in iron. It used to be thought that a beta phase existed between 910°C and the Curie point, that is the change from magnetic to nonmagnetic, at 770°C. In the case of pure iron, the change from one phase to another is not readily detectable outside the laboratory and has no great commercial significance. However, the addition of small amounts of carbon to form steel introduces a series of useful effects.

The solubility of carbon in iron varies considerably. In the low-temperature ferritic phase solubility is very low, about 0.03% maximum. However, the austenitic phase can hold in solid solution all of the carbon present in conventional steels. Furthermore, the addition of carbon extends the temperature range over which austenite can exist under equilibrium conditions. Thus, on slow cooling, the austenite does not transform fully to ferrite at 910°C which commences to transform in some lower temperature, termed the Upper Critical Temperature, which depends on the carbon content. Transformation then continues progressively as the temperature falls further but the ferrite which is formed contains very little carbon so the remaining austenite becomes progressively richer in carbon. Ultimately, at a specific temperature, all of the remaining carbon rich austenite will transform to a mixture of ferrite and iron carbide, Fe_3C also termed cementite. This temperature, 723°C in plain carbon steels, is termed the Eutectoid Temperature or the Lower Critical Temperature. The upper and lower critical temperatures are also referred to as, respectively, the A_3 and the A_1. The term A_2 is rarely used but refers to the change at 770°C at which iron ceases to be magnetic. Often, the "A" temperatures are further defined as in A_{r1} or A_{c1}. The "r" suffix denotes the temperature measured during cooling (from the French refroidissement) and 'c' the temperature measured during heating (chauffage). The range of temperature between A_1 and A_3 is variously described as the Critical Range, or Transition Range. The mixture of ferrite and cementite formed at 723°C is usually precipitated as the eutectoid form comprising alternate plates, or lamallae, of carbide and ferrite. This has an appearance under the microscope similar to mother of pearl, hence its name-pearlite. Microstructures formed in this way usually comprise grains of ferrite plus grades of pearlite, the amount of pearlite being roughly proportional to the quantity of carbon in the steel with about 0.8% carbon, termed the Eutectoid Composition, giving a fully pearlitic structure. The terms Hypo- and Hyper-eutectoid indicate, respectively, compositions of lower and higher carbon content. Hypereutectoid steels have a structure comprised predominantly of pearlite with a small amount of cementite as films or discrete areas. In isolation, cementite is hard and brittle but pearlite, with its alternate layers of cementite and relatively soft, tough ferrite, provides a good combination of hardness, strength and toughness. Generalizing, the hardness of the steel increases with a proportion of pearlite in the microstructure.

All of the treatments considered so far have involved slow cooling rates so that appreciable diffusion can occur and transformation effectively reaches the stable equilibrium state. The process of very slow cooling from above the transition range, often in a furnace, is termed annealing or full annealing. This produces a coarse pearlite and ferrite structure with the steel in a soft, low strength, condition. A slightly faster cooling, which is usually achieved by removing the steel from the furnace to cool in still air, is termed normalizing. This produces a finer pearlite with slightly higher strength and hardness but for practical purposes normalized steel is still usually regarded as being in a softened condition and a near equilibrium state. However, if a steel is cooled more rapidly, by quenching, into oil, water or brine etc., insufficient time is available for diffusion. This forces transformation to take place below the A_1 resulting in nonequilibrium transformation. With only a slight increase in cooling rate the effect is to precipitate a finer form of pearlite but as the rate increases the lamellar form is suppressed and the carbide is deposited as a dispersion of particles in an acicular i.e., needlelike, ferrite matrix. This structure is called bainite. Depending upon the precise cooling rate, bainite can form over a wide range of temperature below about 550°C. Transformation high in this range produces upper bainite with a micro-scopical appearance often described as "feathery." Faster cooling delays transformation to a lower temperature producing the more obviously acicular structure of lower bainite. Bainitic structures have a higher hardness than the ferritic/pearlitic structures.

At even higher cooling rates, i.e., when the Critical Cooling Rate is exceeded, Martensite is formed by a diffusionless shear process which commences at the M_s (martensite start) temperature and continues until the M_f (martensite finish) temperature. Transformation is not time-dependent but is a function of the temperature in the M_s–M_f range. In terms such as "M_{10} temperature" the subscript indicates the percentage transformation to martensite at the temperature in question. If the M_f temperature is below ambient, the transformation to martensite will remain incomplete and the structure will contain retained austenite. This may be transformed to martensite by further cooling and the steel will remain fully martensitic on return to ambient. Fully martensitic structures offer the maximum hardness that can be developed by heat treatment of plain carbon steels.

The process of heating a steel into the austenitic range and then cooling at a rate sufficient to cause transformation to bainite or martensite is usually referred to as hardening. The martensitic structures in particular are hard and of high tensile strength but they tend to be brittle. Consequently, quenched steels are usually reheated for an hour or so to temperatures in the range 100°C–600°C to produce progressively softer, but more tough steels. This treatment, termed tempering, causes the metastable bainite and martensite phases to transform progressively to ferrite and iron carbide. The carbide particles produced by low temperature tempering are extremely fine and difficult to resolve in a light microscope but become increasingly coarse the higher the tempering temperature. Modern practices have referred to the structures as tempered martensites or tempered bainites. Earlier practice used the term troostite, defined as a dark etching, unresolvable structure developed by light tempering, i.e., up to about 300°C and sorbite defined as a fine, but resolvable, iron carbide in a ferrite matrix produced by tempering above about 400°C. When the tempering temperature exceeds about 500°C the carbides become increasingly coarse and are usually described as spheroidized carbide although the terms sorbite or sorbitic carbide are still occasionally used. Cooling rate is not critical for tempering treatments but the time at temperature does affect the structure particularly at higher temperatures where longer times produce coarser carbides.

Up to this point interest has centered on plain carbon steels, that is ones that contain no deliberate alloying additions apart from carbon and up to about 2% manganese. The steels can develop very high strengths, but to do so, they need to be cooled very rapidly during the hardening process. This is readily achieved in the case of small components but with larger sizes the material, particularly at the center of this section, will not achieve the critical cooling rate to transform to martensite. A further problem associated with rapid cooling is that it can cause distortion or even cracking particularly of complex shaped items. This latter problem arises partly because of the thermal contraction that occurs as the temperature falls and partly because of the volume change associated with the various phase changes. These difficulties are overcome by adding to the steel further alloying elements that delay transformation so that slower cooling rates can develop the martensite or bainite structures. Many elements have this affect including manganese, nickel, chromium, molybdenum, vanadium and, of course increase carbon. They are said to confer hardenability. This term, therefore, refers to the ease with which a steel may be hardened; it does not refer to the level of hardness that can be achieved. Hardenability is measured in terms of the maximum section size, or ruling section, that can be fully hardened by a given cooling rate. Some of the elements that are added remain in solid solution but others can combine with the carbon to form alloy carbides which can confer further beneficial effects or, in other circumstances, present problems. When large amounts of some alloying elements, such as nickel and manganese, are added to the steel the transformation characteristics are changed to such an extent that the steel remains partially or fully austenitic down to ambient temperature even when slowly cooled. Such steels are termed austenitic steel, or, if they are partly austenitic, they may be termed duplex. Fully austenitic steels are nonmagnetic and, since they remain austenitic they cannot be hardened by rapid cooling. If, in addition to the austenite stabilizing elements (nickel and manganese), the steel contains more than about 12% chromium it can be termed austenitic stainless steel which is considered further below. It will be recognized, however, that not all austenitic steels are stainless.

Steels, Corrosion Resistant

An adverse characteristic of iron and steel that influences their usefulness is the lack of corrosion resistance. Iron reacts with oxygen in the atmosphere to form a range of iron oxides. The rate of attack is dramatically increased in the presence of moisture and the oxide formed in these circumstances is often combined with water to form rust. The unfortunate characteristics of rust are that it is very bulky compared with the iron that it replaces, it is readily detached and, most critically, it does not form a barrier to further contact with the environment. Consequently, rusting continues progressively in appropriate atmospheres until the iron is consumed. Most alloying elements have little effect on the corrosion resistance of iron. Elements such as copper, aluminum, and silicon can confer a small benefit in certain environments but only chromium has a major effect. Chromium reacts with atmospheric oxygen very strongly and rapidly but the oxide formed is thin, transparent, tightly adherent and acts as a barrier to further attack. This benefit is conferred on iron if the chromium content of the alloy exceeds about 12%. Thus, steels with more than this quantity of chromium are referred to as stainless steels. If chromium is the only significant element the alloy is sometimes referred to as stainless iron but this term has been applied to various steels and cast irons and it should be used and interpreted with caution. The term ferritic stainless steel usually indicates an iron chromium alloy containing in sufficient carbon to undergo a hardening heat treatment as described above, while higher carbon alloys that can be hard and are termed martensitic stainless steel. If more than about 8% nickel is added to a high chromium steel the austenite to ferrite transformation is suppressed to such an extent that the material remains austenitic at normal ambient temperatures and the alloy is referred to as austenitic stainless steel. A common alloy contains 18% chromium plus 8% nickel,—hence the popular designation '18/8' stainless. Stabilized stainless steels, contain elements, such as titanium and niobium, having a strong affinity for carbon which otherwise, in adverse circumstances could combine with some of the chromium at the grain boundaries with consequent susceptibility to intergranular corrosion. Austenitic steels cannot be hardened by conventional heat treatment as described above but certain more complex stainless steels can be hardened by precipitation hardening processes.

Steels-Alloy

So many specialized steels have been developed for particular applications that they cannot all be dealt with. However, some further limited information on a few of the more common special alloys is provided under headings such as Maraging, Silver and Tool Steels and on some specialized heat treatments under headings such as Austempering and Martempering.

Steels

For specific types, see *alloy steels, austenitic manganese steels, bearing steels, stainless steels, carbon steels, cast corrosion-resistant stainless steels, cast heat-resistant stainless steels, chromium-molybdenum heat-resistant steels, dual-phase steels, duplex stainless steels, high-strength low-alloy steels, low-alloy steels, stainless steels, tool steels,* and *ultrahigh-strength steels.*

Stellite

A range of proprietary alloys mostly based on cobalt and having excellent corrosion resistance, hardness and strength.

Step Aging

Aging of metals at two or more temperatures, by steps, without cooling to room temperature after each step. See also *aging,* and compare with *interrupted aging* and *progressive aging.*

Stepback Sequence

See preferred term *backstep sequence.*

Step Bearing

A plain surface bearing that supports the lower end of a vertical shaft. Other types of bearings may be thus described when they are mounted on a step or bracket. See also *Rayleigh step bearing* and *stepped bearing.*

Step Brazing

The brazing of successive joints on a given part with filler metals of successively lower brazing temperatures so as to accomplish the joining without disturbing the joint previously brazed. A similar result can be achieved at a single brazing temperature if the remelt temperature of prior joints is increased by metallurgical interaction.

Stepdown Test

A test involving the preparation of a series of machined steps progressing inward from the surface of a metal bar (usually steel) for the purpose of detecting by visual inspection the internal laminations caused by inclusion segregates.

Stepped Bearing

A thrust bearing in which the working face consists of one or more shallow steps. A distinction should be drawn between a stepped bearing and a *step bearing.*

Stepped Compact

A powder metallurgy compact with one (dual step) or more (multistep) abrupt cross-sectional changes, usually obtained by pressing with split punches, each section of which uses a different pressure and a different rate of compaction. See also *split punch.*

Stepped Extrusion

See *extrusion.*

Step Fracture (Glass)

See *striation* (glass).

Step Fracture (Metals)

(1) Cleavage fractures that initiate on many parallel cleavage planes. (2) Faceted cleavagelike fractures that occur during Stage I fatigue fractures (high-cycle, low-stress fractures).

Step Soldering

The soldering of successive joints on a given part with solders of successively lower soldering temperatures so as to accomplish the joining without disturbing the joints previously soldered.

Step (Fatigue) Test

A fatigue test program in which a specimen is subjected to a prescribed number of cycles at each of a series of progressively increasing stresses. The fatigue limit is then taken as the penultimate stress, i.e., the highest that did not cause failure or, alternatively, the average of the penultimate and the final stresses.

Stereo Angle

One half of the angle through which the specimen is tilted when taking a pair of *stereoscopic micrographs.* The axis of rotation lies in the plane of the specimen.

Stereographic Projection

A technique for presenting, in two dimensions, information on the three-dimensional orientation of crystallographic planes. The unit cell of a crystal comprises a small number of atoms arranged in a simple geometric pattern, for example, at the corners of a cube. The crystallographic planes are the planes on which the atoms lie, for example, the cube faces or the cube diagonals. The basic stereographic projection usually deals with a single crystal which, for the purpose of the exercise, is considered to be located at the center of a large sphere. For example, take the technique for a single plane, C. A line is projected perpendicularly from plane C to intersect the sphere surface at point P. A further line is then drawn from P to the pole of the other hemisphere, point S. The point D at which this line pierces the equatorial plane is the stereographic projection of the plane C. This procedure is repeated for all crystallographic planes of interest and the pattern of the series of D points forms the stereographic projection of the crystal. The technique can also be applied to multiple crystals such as a small piece of sheet. If the crystals of this sheet are aligned at random than the D points will be randomly distributed on the stereographic projection. However, a grouping of points will be evidence of preferred orientation. The term pole figure is often used interchangeably with the term stereographic projection. Where they are differentiated, pole figure usually refers to projections of multiple crystals.

Stereoisomer

An isomer in which atoms are linked in the same order but differ in their arrangement. See also *isomer* and *isotactic stereoisomerism.*

Stereolithography

Stereolithography uses a laser beam to convert a special photosensitive polymer from liquid to solid. The liquid polymer is held in a tank that also contains a platform that can be raised and lowered and onto which the part will be built.

The platform is first positioned just below the surface of the liquid polymer, and a laser beam is rastered across the surface of the polymer to solidify a two-dimensional image of the bottom layer of the part. The platform is then lowered a small distance to allow a thin layer of liquid polymer to cover the solidified layer, and the laser beam is again rastered to solidify the next layer on top of and bonded to the initial layer. The platform is lowered again to form the third layer, and so on.

When the final (top) layer of the part has been solidified, the platform is raised from the tank to drain away the liquid polymer from the finished part. Very complex shapes can be formed, including holes and even internal hollows (if a means is provided to drain out the unsolidified polymer).

Problems

Overhangs and undercuts are a challenge to form by stereolithography because the lower layers of such features will not be connected to the main body of the part, and temporary supporting structures (which can later be removed) must be fabricated as the layers are built up. By loading the polymer with ceramic or metal powder, stereolithography can be used to fabricate greenware that can be debinded and sintered to form a finished part.

Because this and the other rapid prototyping processes are rather slow, they would be appropriate production methods only for limited runs. By the same token, however, the avoidance by rapid prototyping of expensive tooling makes it all the more attractive for limited runs. Stereolithography is perhaps the most well-developed technique for rapid prototyping, and a number of vendors provide hardware, software, and special polymers to accomplish this technique. Some of the hardware is small enough that the term *desktop manufacturing* can legitimately be applied to it.

Stereophotogrammetry

A method of generating topographic maps of fracture surfaces by the use of a stereoscopic microscope interfaced to a microcomputer which calculates the three-dimensional coordinates of the fracture surface and produces the corresponding profile map, contour plot, or carpet plot.

Stereoradiography

A technique for producing paired radiographs that may be viewed with a stereoscope to exhibit a shadowgraph in three dimensions with various sections in perspective and spatial relation.

Stereoscopic Micrographs

A pair of micrographs (or fractographs) of the same area, but taken from different angles so that the two micrographs when properly mounted and viewed reveal the structures of the objects in their three-dimensional relationships.

Stereoscopic Specimen Holder

A specimen holder designed for the purpose of making stereoscopic micrographs that allows the tilting of the specimen through the *stereo angle.*

Stereospecific Plastics

Implies a specific or definite order of arrangement of molecules in space. This ordered regularity of the molecules in contrast to the branched or random arrangement found in other plastics permits close packing of the molecules and leads to like crystallinity (for example, in polypropylene).

Stereotype Metals

Tin antimony alloys similar to Type Metals. See *White Metals* and *Babbit Metals.*

Sterling Silver

A silver alloy containing at least 92.5% Ag, the remainder being unspecified but usually copper. Sterling silver is used for flat and hollow tableware and for various items of jewelry.

Stern-Tube Bearing

The final bearing through which a propeller shaft passes in a boat or ship.

Stick Electrode

A shop term for *covered electrode.* The coated rod electrode used for manual metal arc welding.

Stick Welding

See preferred term *shielded metal arc welding.*

Sticker Breaks

Arc-shaped *coil breaks,* usually located near the center of sheet or strip.

Stickout

See preferred term *electrode extension.*

Stick-Slip

A relaxation isolation usually associated with a decrease in the *coefficient of friction* as the relative velocity increases. Stick-slip was originally associated with the formation and destruction of interfacial junctions on a microscopic scale. This is often the basic cause. The period depends on the velocity and on the elastic characteristics of the system. Stick-slip will not occur if the static friction is equal to or less than the dynamic friction. The motion resulting from stick-slip is sometimes referred to as jerky motion. See also *spragging.*

Stiction

A term used with reference to the force necessary to overcome static friction.

Stiffness

(1) The rate of stress with respect to strain; the greater the stress required to produce a given strain, the stiffer the material is said to be. (2) The ability of a material or shape to resist elastic deflection. For identical shapes, the stiffness is proportional to the modulus of elasticity. For a given material, the stiffness increases with increasing moment of inertia, which is computed from cross-sectional dimensions.

Stiff Nut

See *self locking nut.*

Stippled Area

See *hackle.*

Stitch Weld

A line of overlapping spot welds. See preferred term *intermittent weld.*

Stitching

See *metal stitching.*

Stock

A general term used to refer to a supply of metal in any form or shape and also to an individual piece of metal that is formed, forged, or machined to make parts.

Stoichiometric

Having the precise weight relation of the elements in a chemical compound; or (quantities of reacting elements or compounds) being in the same weight relation as the theoretical combining weight of the elements involved.

Stoke (Centistoke)

The centimeter-gram-second (cgs) unit of kinematic viscosity.

Stokes Raman Line

A *Raman line* that has a frequency lower than that of the incident monochromatic beam. See also *Raman spectrum.*

Stoking (Obsolete)

See preferred term *continuous sintering.*

Stoneware

A vitreous or semivitreous ceramic ware of fine texture, made primarily from either non-refractory fireclay or some combination of clays, fluxes, and silica. Used for cookware, art wear, and tableware.

Stop

A device for positioning stock or parts in a die.

Stopoff

A material used on the surfaces adjacent to the joint to limit the spread of soldering or brazing filler metal. See also *resist.*

Stopper Rod

A device in a bottom-pour ladle for controlling the flow of metal through the nozzle into a mold. The stopper rod consists of a steel rod, protective refractory sleeves, and a graphite stopper head. See also the term *bottom-pour ladle.*

Stopping Off

(1) Applying a *resist.* (2) Depositing a metal (copper, for example) in localized areas to prevent carburization, decarburization, or nitriding in those areas. (3) Filling in a portion of a mold cavity to keep out molten metal. (4) Material applied to a surface to prevent, locally, some reaction or effect. For example, to prevent local plating or etching during such processes were to prevent adhesion of weld splatter.

Stops

Metal pieces inserted between die halves used to control the thickness of a press-molded plastic part. Not a recommended practice, because the resin will receive less pressure, which can result in voids.

Storage Life

The period of time during which a liquid resin, packaged adhesive, or prepreg can be stored under a specified temperature conditions and remain suitable for use. Also called *shelf life.*

Storage Modulus

A quantitative measure of the elastic properties in polymers. defined as the ratio of the stress, in phase with the strain, to the magnitude of the strain. The storage modulus may be measured in tension or flexure, compression, or shear.

Storage Stability

A measure of the ability of a lubricant to undergo prolonged periods of storage without showing any adverse conditions due to oxidation, oil separation, contamination, or any type of deterioration.

Stored-Energy Welding

Resistance welding with electrical energy accumulated electrostatically, electromagnetically, or electrochemically at a relatively low rate and made available at the higher rate required in welding.

Straddle Milling

Face milling a workpiece on both sides at once using two cutters spaced as required. See also *face milling* and *milling*.

Straightening

(1) Any bending, twisting, or stretching operation to correct any deviation from straightness and bars, tubes, or similar long parts or shapes. This deviation can be expressed as either camber (deviation from a straight line) or as total indicator reading (TIR) per unit of length. (2) A finishing operation for correcting misalignment in a forging or between various sections of a forging. See also *roll straightening*.

Straight Polarity

Direct-current arc welding circuit arrangement in which the electrode is connected to the negative terminal. A synonym for *direct current electrode negative* (*DCEN*). Contrast with *reverse polarity*.

Straight-Side Press

A sheet metal forming press which has a frame made up of a base, or bed; two columns; and a top member, or crown. In most straight-side presses, steel tie rods hold the base and crown against the columns. Straight-side presses have crankshaft, eccentric-shaft, or eccentric-gear drives. In a single-action straight-side press, the slide is equipped with air counterbalances to assist the drive in lifting the weight of the slide and the upper die to the top of the stroke. Counterbalance cylinders provide a smooth press operation and easy slide adjustment. Die cushions are used in the bed for blank-holding and for ejection of the work.

Strain

The unit of change in the size or shape of a body due to force. Also known as nominal strain. The term is also used in a broader sense to denote a dimensionless number that characterizes the change in dimensions of an object during a deformation or flow process. See also *engineering strain*, *linear strain*, and *true strain* and *tensile test*.

Strain-Age Embrittlement

A loss in *ductility* accompanied by an increase in hardness and strength that occurs when low-carbon steel (especially rimmed or capped steel) is aged following *plastic deformation*. The degree of *embrittlement* is a function of aging time and temperature, occurring in a matter of minutes at about 200°C, but requiring a few hours to a year at room temperature.

Strain Aging

(1) *Aging* following plastic deformation. (2) The changes in ductility, hardness, yield point, and tensile strength that occur when a metal or alloy that has been cold worked is stored for some time. In steel, strain aging is characterized by a loss of ductility and a corresponding increase in hardness, yield point, and tensile strength.

Strain Aging or Strain Age Hardening

A spontaneous increase in hardness occurring in some materials, particularly certain steels, after they have been deformed during manufacture or service. The hardness increase is usually small but it can cause serious embrittlement. It may occur over days or even years and can be accelerated by moderate heating as occurs, for example, during hot dip galvanizing. The effect results when dissolved interstitial elements such as carbon and particularly nitrogen diffuse through the crystal lattice until they become engaged with, and hence impede the movement of, dislocations. Contrast with *strain hardening*.

Strain Amplitude

The ratio of the maximum deformation, measured from the main deformation to the free length of the unrestrained test specimen.

Strain Annealing

A technique for developing a very large grain size by annealing a material after it has received a small but critical amount of cold working. See *critical strain*.

Strain Energy Release Rate

The elastic strain energy released in unit propagation of a crack.

Strain Energy

The potential energy stored in a body by virtue of elastic deformation, equal to the work that must be done to produce this deformation. See also *elastic energy*, *resilience*, and *toughness*.

Strainer Core

In foundry practice, a perforated core in the gating system for preventing slag and other extraneous material from entering the casting cavity.

Strain Etching

Metallographic etching that provides information on deformed and under forms areas if present side-by-side. In strained areas, more compounds are precipitated.

Strain Gauge

A device for measuring small amounts of strain produced during tensile and similar tests on metal. A coil of fine wire is mounted on a piece of paper, plastic, or similar carrier matrix (backing material), which is rectangular in shape and usually about 25 mm (1 in.) long. This is glued to a portion of metal under test. As the coil extends with the specimen, its electrical resistance increases in direct proportion. This is known as bonded resistance-strain gage. Other types of gauges measure the actual deformation. Mechanical, optical, or electronic devices are sometimes used to magnify the strain for easier reading. See also rosette.

Strain Hardening

An increase in hardness and strength of metals caused by plastic deformation at temperatures below the recrystallization range. Also known as *work hardening*.

Strain-Hardening Coefficient

See *strain-hardening exponent*.

Strain-Hardening Exponent

The value of n in the relationship:

$$\sigma = K\varepsilon^n$$

where
  σ is the *true stress*
  ε is the *true strain*
  K, which is called the strength coefficient, is equal to the true stress at a true strain of 1.0

The strain-hardening exponent, also called "n-value," is equal to the slope of the true stress/true strain curve up to maximum load, when plodded on a log-log coordinates. The n-value relates to the ability of a sheet material to be stretched in metal working operations. The higher the n-value, the better the formability (stretchability).

Strain Markings

Manifestations of prior plastic deformation visible after etching of a metallographic section. These markings may be referred to as slip strain markings, twin strain markings, and so on, to indicate the specific deformation mechanism of which they are a manifestation.

Strain Point

That temperature corresponding to a specific rate of elongation of a glass fiber or a specific rate of midpoint deflection of a glass beam. At the strain point of glass, internal stresses are substantially relieved in a matter of hours.

Strain Rate

The time rate of straining for the usual tensile test. Strain as measured directly on the specimen gauge length is used for determining strain rate. Because strain is dimensionless, the units of strain rate are reciprocal time.

Strain-Rate Sensitivity (*m*-Value)

The increase in stress (σ) needed to cause a certain increase in plastic strain rate ($\dot{\varepsilon}$) At a given level of plastic strain (ε) and a given temperature (T).

$$\text{Strain - rate sensitivity} = m = \left| \frac{\Delta \log \sigma}{\Delta \log \dot{\varepsilon}} \right|_{\varepsilon_T}$$

Strain Relaxation

Reduction in internal strain over time.

Strain Rods

(1) Rods sometimes used on gapframe metal forming presses to lessen the frame deflection. (2) Rods used to measure elastic strain, and thus stresses, in frames of metal forming presses.

Strain State

See *state of strain*.

Strand

Normally, an untwisted bundle or assembly of continuous filaments used as a unit to reinforced plastics, including slivers, tows, ends, yarn, and so forth. Sometimes a single fiber or filament is called a strand.

Strand Casting

A generic term describing *continuous casting* of one or more elongated shapes such as billets, blooms, or slabs; if two or more shapes are cast simultaneously, they are often of identical cross-section.

Strand Count

The number of *strands* in a plied *yarn* or *roving*.

Stranded Electrode

A composite filler metal electrode of stranded wires which may mechanically enclose materials to improve properties, stabilize the arc, or provide shielding.

Strand Integrity

The degree to which the individual filaments making up a *strand* or *end* are held together by the applied sizing. See also *size*.

Strand Tensile Test

A tensile test of a single resin-impregnated *strand* of any fiber.

Strap Joint

A joint formed by butting two plate edges together and laying a strap along the butt line. The strap is then joined to the individual plates by bolting, riveting, welding or brazing, etc.

S

Strada

In sampling, segments of a lot that may vary with respect to the property under study.

Strauss Test

A test for weld decay in which a specimen is boiled in a copper sulfate/sulfuric acid solution for 72 h and then bent to check for cracking.

Stray Arc/Flash

In arc welding, the unintentional striking of an arc against the component, including previously deposited weld metal, and the surface damage so caused.

Straight Current

(1) Current flowing through paths other than the intended circuit. (2) Current flowing in electrodeposition by way of an unplanned and undesired bipolar electrode that may be in the tank itself or a poorly connected electrode. (3) Unexpected electrical currents arising in processes or encountered in the environment. For example, earth return or other currents associated with tram ways can produce currents in other buried components.

Stray-Current Corrosion

(1) Corrosion resulting from direct current flow through paths other than the intended circuit. For example, by an extraneous current in the earth. (2) Electrolytic corrosion caused by stray currents in the environment. Electrical currents are found in soil as a result of its deliberate use as an earth return or as a result of accidental leakage. Such currents can cause severe corrosion of equipment at considerable distances from the location where the current enters the soil.

Strength

The maximum nominal stress a material can sustain. Always qualified by the type of stress (tensile, compressive, or shear).

Strength Coefficient

See *strain-hardening exponent*.

Stress

The intensity of the internally distributed forces or components of forces that resist a change in the volume or shape of a material that is or has been subjected to external forces. Stress is expressed in force per unit area. Stress can be normal (tension or compression) or shear. See also *compressive stress, engineering stress, mean stress, nominal stress, normal stress, residual stress, shear stress, tensile stress,* and *true stress*.

Stress (Glass)

Any condition of tension or compression existing within the glass, particularly due to incomplete annealing, temperature gradient, or inhomogeneity.

Stress Amplitude

One-half the algebraic difference between the maximum and minimum stresses in one cycle of a repetitively varying stress.

Stress Analysis

Any technique for determining the level and distribution of stress in complex components and structures. Examples include scale solid modeling, mathematical and computer modeling including finite element analysis and stress assisted/accelerated corrosion. Corrosion in which the rate of attack is significantly increased by stress. Where the attack causes cracking it would normally be termed *Stress Corrosion Cracking*.

Stress Concentration

(1) On a macromechanical level, the magnification of the level of an applied stress in the region of a notch, void, hole, or inclusion. (2) The local increase in stress at a crack, notch or other section change, also termed a stress riser. A component containing a crack is obviously weaker than one which is defect free but the reduction and strength is often more than would be expected simply from the reduction in cross-section. The cause can be visualized by considering a cracked bar in which the load is evenly distributed over many individual strands. Remote from the crack the strands and the load will be evenly spread across the section but at the crack the load in the severed strands will be transferred to the immediately adjacent, intact strands. The strands at the zone at the crack tip will, therefore, be subjected to their normal share of the load plus the load transferred from the severed strands. In ductile materials this zone will yield transferring load deeper into the surrounding material. However, non-ductile material will not yield and locally the stresses remain high leading to one extension of cracking at the crack tip. Stress concentrations can also be caused by holes, steps and section changes. Generalizing the sharper the step or the smaller the crack tip radius the more severe their stress concentration. See also *notch strengthening*.

Stress Concentration Factor (K_t)

A multiplying factor for applied stress that allows for the presence of a structural discontinuity such as a notch or hole; K_t equals the ratio of the greatest stress in the region of the discontinuity to the nominal stress for the entire section. Also called theoretical stress concentration factor (*S.C.F.*).

Stress Corrosion

Preferential attack of areas under stress in a corrosive environment, where such an environment alone would not have caused corrosion.

Stress-Corrosion Cracking (SCC)

(1) A cracking process that requires the simultaneous action of a corrodent and sustained tensile stress. This excludes corrosion-reduced sections that failed by fast fracture. It also excludes intercrystalline or transcrystalline corrosion, which can disintegrate an alloy without applied or residual stress. Stress-corrosion cracking may occur in combination with *hydrogen embrittlement*. (2) Cracking resulting from the combined and concurrent effect of corrosion and a stress. Some metals are highly susceptible to cracking when stressed while

exposed to specific corrodents. In the absence of the corrodent the stress has no adverse effect and, even when both stress and the corrodent are present, the volume of metal loss to the corrosion process is usually negligible. The attack normally takes the form of cracking commencing at the surface and penetrating along a multiple branching path. The stress, almost invariably tensile, may be externally imposed or residual and the path will be either intergranular or transgranular depending upon the particular material and environment. Examples include brass in ammonia solutions and austenitic stainless steel in chloride solutions. See *steels*.

Stress Crack

(1) External or internal cracks in a plastic caused by tensile stresses less than that of its short-time mechanical strength, frequently accelerated by the environment to which the plastic is exposed. The stresses that cause cracking may be present internally or externally or may be combinations of the stresses. (2) Cracking occurring as a result of the combined and concurrent effect of the hostile environment and stress. This term is commonly applied in the context of polymeric (plastic) materials in which case the responsible environmental factor may be liquid, gaseous or radiation including light. However, the term is occasionally, perhaps increasingly, used in the case of metals as an alternative to stress corrosion. See also *crazing*.

Stress-Cracking Failure

The failure of a plastic by cracking or crazing some time after it has been placed under load. Time-to-failure may range from minutes to years. Causes include molded-in stresses, post fabrication shrinkage or warpage, and hostile environment.

Stress Cycle

The smallest segment of the stress-time function that is repeated periodically.

Stress Cycles Endured (N)

The number of cycles of a specified character (that produce fluctuating stress and strain) that a specimen has endured at any time in its stress history.

Stress Equalizing

A low-temperature heat treatment used to balance stresses in cold-work material (metals) without an appreciable decrease in the mechanical strength produced by cold working.

Stress Force (Per Unit Area)

The three basic stresses are tension, compression and shear. See *tensile test*.

Stress Fracture

See *fracture stress*.

Stress-Induced Crystallization

The production of crystals in a polymer by the action of stress, usually in the form of an elongation. It occurs in fiber-spinning and rubber elongation and is responsible for enhanced mechanical properties.

Stress-Intensity Calibration

A mathematical expression, based on empirical or analytical results, that relates the *stress-intensity factor* to load and crack length for a specific specimen planar geometry. Also known as K calibration.

Stress-Intensity Factor

A scaling factor, usually denoted by the symbol K, used in *linear-elastic fracture mechanics* to describe the intensification of applied stress at the tip of a crack of known size and shape. At the onset of rapid crack propagation in any structure containing a crack, the factor is called the critical stress-intensity factor, or the *fracture toughness*. Various subscripts are used to denote different loading conditions or fracture toughnesses:

> K_c. Plane-stress fracture toughness. The value of stress intensity at which crack propagation becomes rapid in sections thinner than those in which plane-strain conditions prevail.
>
> K_1. Stress-intensity factor for a loading condition that displaces the crack faces in a direction normal to the crack plane (also known as the opening mode of deformation).
>
> K_{1c}. Plane-strain fracture toughness. The minimum value of K_c for any given material and condition, which is attained when rapid crack propagation in the opening mode is governed by plane-strain conditions.
>
> K_{1D}. Dynamic fracture toughness. The fracture toughness determined under dynamic loading conditions; it is used as an approximation of K_{1c} for very tough materials.
>
> K_{1scc}. Threshold stress intensity factor for stress-corrosion cracking. The critical plane-strain stress intensity at the onset of stress-corrosion cracking under specified conditions.
>
> K_Q. Provisional value for plane-strain fracture toughness.
>
> K_{th}. Threshold stress intensity for stress-corrosion cracking. The critical stress intensity at the onset of stress-corrosion cracking under specified conditions.
>
> ΔK. The range of the stress-intensity factor during a fatigue cycle. See also *fatigue crack growth rate*.

Stress-Intensity Factor Range (ΔK)

In fatigue, the variation in the *stress-intensity factor* in a cycle, that is, $K_{max} - K_{min}$. See the term *fatigue crack growth rate*.

Stress-Number Curve

The S/N curve depicting fatigue properties.

Stress Raisers

(1) Design features (such as sharp corners) or mechanical defects (such as notches) that act to intensify the stress at these locations. (2) Any feature such as a section change, crack or inclusion which causes an increase in the stress in its vicinity, i.e., a stress concentration.

Stress Relief

Any process for reducing residual stress induced by cold working, welding, etc. The term usually implies processes which involve heating the component to some moderate temperature for a short period, typically an hour or two. As examples, stress relief of welds in normalized plain carbon steel would be at about 550°C and creep resisting alloy steels at 700°C. Simplistically, at such temperature the yield strength is greatly reduced allowing plastic deformation to convert much of the elastic strain to plastic strain. Consequently, stress relief may cause some deformation of the component and, even after stress relief, some stresses will remain. The treatment may also produce some beneficial softening of hardened areas of the heat affected zones of welds but it is usually implicit that the tensile properties of the bulk component will not be adversely affected to any significant extent. In the case of cold work material there may be useful improvements in ductility and toughness. See also *vibratory stress relief.*

Stress Range

See *range of stress.*

Stress Ratio (*A* or *R*)

The algebraic ratio of two specified stress values in a stress cycle. Two commonly used stress ratios are: (1) the ratio of the alternating stress amplitude to the mean stress, $A = S_a/S_m$; and (2) the ratio of the minimum stress to the maximum stress. $R = S_{min}/S_{max}$. See *fatigue.*

Stress Relaxation

The time-dependent decrease in stress in a solid under constant constraint at constant temperature.

Stress-Relaxation Curve

A plot of the remaining or relaxed stress as a function of time. The relaxed stress equals the initial stress minus the remaining stress. Also known as *stress-time curve.*

Stress-Relief Cracking

Cracking in the *heat-affected zone* or weld metal that occurs during the exposure of weldments to elevated temperatures during postweld heat treatment, in order to reduce residual stresses and improve toughness, or high temperature service. Stress-relief cracking occurs only in metals that can precipitation-harden during such elevated-temperature exposure; it usually occurs as *stress raisers,* is intergranular in nature, and is generally observed in the coarse-grained region of the weld heat- affected zone. Also called postweld heat treatment cracking or stress relief embrittlement.

Stress Relieving

Heating to a suitable temperature, holding long enough to reduce residual stresses, and then cooling slowly enough to minimize the development of new residual stresses.

Stress Rupture (Properties) and Creep Failure (Properties)

There are various conventions for presenting graphically the interaction between stress, temperature and time to failure for a particular material. The term "Rupture Data" usually refers to a plot of stress on the vertical scale against temperature on the horizontal with lines on the graph identifying time to failure. This system is useful as it can also present proof stress data. See also *Larson Miller Parameter.*

Stress-Rupture Strength

See *creep-rupture strength.*

Stress-Rupture Test

See *creep-rupture test.*

Stress State

See *state of stress.*

Stress-Strain Curve

A graph in which corresponding values of stress and strain from a tension, compression, or torsion test are plotted against each other. The values of stress are usually plotted vertically (ordinates on *y*-axis) and values of strain horizontally (abscissas or *x*-axis). Also known as deformation curve and *stress-strain diagram.* See also *engineering strain* and *engineering stress.*

Stress-Strain Diagram

See *tensile test.*

Stretch-Bending Test

A simulative test for sheet metal formability in which a strip of sheet metal is clamped at its ends in lock beads and deformed in the center by a punch. Test conditions are chosen so that fracture occurs in the region of punch contact.

Stretcher Leveling

The leveling of a piece of sheet metal (that is, removing warp and distortion) by gripping it at both ends and subjecting it to a stress higher than its yield strength.

Stretcher Straightening

A process for straightening rod, tubing, and shapes by the application of tension at the ends of the stock. The products are elongated a definite amount to remove warpage.

Stretcher Strains

Elongated markings that appear on the surface of some sheet materials when deformed just past the yield point. These markings lie approximately parallel to the direction of maximum shear stress and are the result of localize yielding. See also the term *Lüders lines.*

Stretch Former

(1) A machine used to perform *stretch forming* operations. (2) A device adaptable to a conventional press for accomplishing stretch forming.

Stretch Forming

The shaping of a metal sheet or part, usually of uniform cross section, by first applying suitable tension or stretch and then wrapping it around a die of the desired shape. The four methods of stretch forming are stretch draw forming, stretch wrapping, compression forming, and radial draw forming. See the term *radial draw forming*.

Stretching

The extension of the surface of a metal sheet in all directions. In stretching, the flange of the flat blank is securely clamped. Deformation is restricted to the area initially within the die. The stretching limit is the onset of metal failure.

Striation (Glass)

A fracture surface marking consisting of a separation of the advancing crack front into separate fracture planes. Also known as coarse hackle, step fracture, or lance. Striations may also be called shark's teeth or whiskers.

Striation (Metals)

A fatigue fracture feature, often observed in electron micrographs, that indicates the position of the crack front after each succeeding cycle of stress. The distance between striations indicates the advance of the crack front across that crystal during one stress cycle, and a line normal to the striations indicates the direction of local crack propagation. See also *beach marks* and the term *fatigue striation* (metals).

Stribeck Curve

A graph showing the relationship between coefficient of friction and the dimensionless number ($\eta N/P$), where η is the dynamic viscosity, N is the speed (revolutions per minute for a journal), and P is the load per unit of projected area. The symbols Z and v (linear velocity) may be substituted for η and N, respectively.

Strike

(1) A thin electrodeposited film of metal to be overlaid with other plated coatings. (2) A plating solution of high covering power and low efficiency designed to electroplate a thin, adherent film of metal. (3) A defect produced during welding by accidental, brief contact of an arc welding electrode.

Striker/Striking Plate

An electric arc welding, a piece of metal, not part of the component to be welded, on which the arc is initiated immediately prior to its transfer to the joint being made. Its function is to avoid arc initiation defects on the joint.

Striking

Electrodepositing, other special conditions, a very thin film of metal that will facilitate further plating with another metal or with the same metal under different conditions.

Striking Surface

Those areas on the faces of a set of metal forming dies that are designed to meet when the upper die and lower die are brought together. The striking surface helps protect impressions from impact shock and aids in maintaining longer die life.

Stringer

In wrought materials, and elongated configuration of microconstituents or foreign material aligned in the direction of working. The term is commonly associated with elongated oxide or sulfide inclusions in steel.

Stringer Bead

A continuous weld bead made without appreciable transverse oscillation. Contrast with *weave bead*.

Stringers

Discontinuous lines of nonmetallic inclusions in the microstructure resulting from, and providing information on, deformation during manufacture.

Stringiness

The property of an adhesive that results in the formation of filaments or threads when adhesively bonded surfaces are separated.

Strip

(1) A flat-rolled metal product of some maximum thickness and width arbitrarily dependent on the type of metal; narrower than *sheet*. (2) A roll-compacted metal powder product. See also *roll compacting* and *rolled compact*. (3) Removal of a powder metallurgy compact from the die. An alternative to ejecting or knockout. See also *ejection*.

Strippable Coatings

Strippable coatings are those that are applied for temporary protection and that can be readily removed. They are composed of such resins as cellulosics, vinyl, acrylic, and polyethylene; they can be water based, solvent-based, or hot-melt. The choice of base depends on the surface to be protected. Water-base grades are neutral to plastic and painted surfaces, whereas solvent-base types affect those surfaces. Clear vinyl strippable coatings, perhaps the most widely used, are usually applied by spraying in thicknesses of 30–40 mils. Acrylic strippable coatings impart a clear, high-gloss, high-strength temporary film to metal parts. Polyethylene strippable coatings are relatively low-cost and can be used on almost all surfaces except glass. Cellulosic strippable coatings are designed for hot-dip application. Film thicknesses range widely and can go as high as 200 mils. The mineral oil often present in these coatings exudes and coats the metal surfaces to protect it from corrosion over long periods.

Types

Vinyl coatings like those described in MIL-C-3254 specification were first developed for ships. These are called cocooning systems and are applied over chicken wire or a similar frame over the object

S

to be protected. The interstices of the chicken wire are coated by a process known as webbing. This consists of spraying a specially designed vinyl coating in a web fashion so that it coats the interstices with a very thin, spiderweb-like, fragile covering. This in turn is coated with a material similar to an ordinary strip coat vinyl by spraying. A more rigid protective coat of asphalt is then applied, following which coats of vinyl or aluminum enamel are applied. The advantages of this coating system are long life, ability to cover irregular surfaces, and easy removal. Disadvantages include a cumbersome structure, which is expensive.

Strippable, Spray or Able, Vinyl Coatings

These materials are generally applied by spraying in thicknesses of 30–40 mils. Their tensile strength runs 3.4 MPa minimum with an elongation of 200%, minimum. These materials are designed to be strippable after years of protective service. They are suitable for bright steel, aluminum, painted surfaces, wood, etc. They can also be used to protect spray booths, for aircraft protection, and on tanks, trucks, ships, and similar equipment.

Coatings can be produced in a translucent, colored affect or in a clear form so as to show any defects in the substrate. These coatings are generally sprayed from 1 to 2 mils thick and are designed for protection and covered storage or in the transportation or fabrication of tools. They can be readily removed even though film thickness is low.

Ethyl Cellulose, Type 1, and Cellulose Acetobutyrate, Type 11

These 100% coatings are designed for dip applications in a hot-melt bath of 177°C. Thickness ranges from 100 to 200 mils, depending on the protection desired. The mineral oil generally present in these coatings exudes and coats the metal surface to keep it from corroding and in a strippable condition for long periods. Tensile strength of the coatings is about 2.06 MPa minimum, with an elongation of about 90% when originally made, and 70% when aged.

The coatings are generally used on tools, steel and aluminum parts, and many other parts that can withstand the temperature of dipping. Variations of specification types can be formulated that do not exude oil, which can be objectionable and handling, particularly with electrical equipment. Pourable variations are also commercially available.

Some special types of ethyl cellulose strippable materials can be used on painted surfaces and are formulated so that these surfaces are not affected. They can be applied by spraying. Other specially designed strippable materials are also practical; some of them can be used for packaging.

Stripped Die Method

A specific method of removal of a powder metallurgy compact after pressing, which keeps it in position between the punches while the die is retracted either upward or downward until the compact is fully exposed and freed by an upward withdrawal of the upper punch.

Stripper (Adhesives)

A chemical solvent or acid that can remove an adhesive bond.

Stripper (Metals)

A plate designed to remove, or strip, sheet metal stock from the punching members during the withdrawal cycle. Strippers are also used to guide small precision punches in close-tolerance dies, to guide scrap away from dies, and to assist in the cutting action. Strippers are made into types: fixed and movable.

Stripper Punch

A punch that serves as the top or bottom of a metal forming die cavity and later moves farther into the die to eject the part or compact. See also *ejector rod* and *knockout* (3).

Stripping

(1) Removing a coating from a metal surface. (2) Removing a foundry pattern from the mold or the core box from the core.

Strontium

A chemical element (symbol Sr), strontium is the least abundant of the alkaline-earth metals. It has a melting point about 770°C, and it decomposes in water. The metal is obtained by electrolysis of the fused chloride, and small amounts are used for doping semiconductors. Its compounds have been used for deoxidizing nonferrous alloys and for desulfurizing steel. But the chief uses have been in signal flares to give a red light, and in hard, heat-resistant greases. Strontium-90, produced atomically, is used in ship-deck signs as it emits no dangerous gamma rays. It gives a bright sign, and the color can be varied with the content of zinc, but it is short-lived. Strontium is very reactive and used only in compounds.

Compounds

Strontium nitrate is a yellowish-white crystalline powder, produced by roasting and leaching celestite and treating with nitric acid. The specific gravity is 2.96, the melting point is 645°C, and it is soluble in water. It gives a bright crimson flame, and is used in railway-signal lights and in military flares. It is also a source of oxygen, pyrotechnics, as well as a precursor for ceramic powders.

The strontium sulfate used as a brightening agent in paints is powdered celestite. Strontium sulfide, SrS, used in luminous paint, gives a blue-green glow, but it deteriorates rapidly unless sealed. Strontium carbonate, is used in pyrotechnics, ceramics, and ceramic permanent magnets for small motors.

The development of glazes for low-temperature vitreous bodies can be materially aided through the use of strontia. The added fluidity provided by strontia when replacing calcium and/or barium should promote interface reaction, improve glaze fit, while offsetting the slightly higher thermal expansion evidenced in some cases in the dinnerware glaze tested. Strontia additions to such glazes should materially increase glaze hardness and lower the solubility. Scratch resistance should be improved when replacements are made, especially at the expense of calcium and barium, which would be due, in part, to the earlier reaction of strontia enabling the glaze to clear with a minimum of pits.

Strontium hydrate loses its water of crystallization at 100°C and melts at 375°C. It is used in making lubricating greases and as a stabilizer in plastics. Strontium fluoride is produced in single crystals for use as a laser material. When doped with samarium it gives an output wavelength around 650 nm.

Strontium hexaboride, which is generally known as SrB_6, is stable to temperatures up to 2760°C, above which decomposition initiates. Possible uses for the material are energy sources when using the radioisotope, high-temperature insulation, nuclear reactor control rods, and control additives.

Strontium titanate has a melting point of 2080°C. Methods of compounding are (1) from mixed strontium carbonate and titanium dioxide, (2) from mixed strontium oxalate and titanium dioxide, and (3) from strontium titanyl oxalate. The strontium titanate is a high-dielectric constant material (225–250), which at lower temperatures has a temperature coefficient of dielectric constant somewhat higher than that of calcium titanate.

Strontium titanate can be used by itself or in combination with barium titanate in applications for capacitors and other parts.

The power factor of strontium titanate is unusually high at low frequencies with a great improvement in power factor in the neighborhood of 1 MHz. The thermal expansion of strontium titanate is linear over a wide temperature range (100°C–700°C).

Structural Adhesives

A structural adhesive is defined as a material used to transfer loads between adherends in service environments to which the assembly is typically exposed. Structural adhesives constitute about 35% of the total estimated sales of all adhesives and sealants. Their primary areas of application include automotive, aerospace, appliances, biomedical/dental construction, consumer electronics, fabrics, furniture, industrial machines, and marine and sports equipment.

The most common type of structural adhesive is classified as a chemically reactive adhesive. The most widely used materials included in this classification are of epoxies, polyurethanes, modified acrylics, cyanoacrylates, and anaerobics. Chemically reactive adhesives can be subdivided into two groups: one-component systems, which include moisture cure and heat-activated cure categories, and two-component systems, which are subdivided into mix-in and no-mix systems. One-component formulations that cure by moisture from the surrounding air or by adsorbed moisture from the surface of a substrate include polyurethanes, cyanoacrylates, and silicones. A one-component heat-activated system usually consists of two components that are premixed. Chemical families in this group include epoxies and epoxy-nylons, polyurethanes, polyimides, polybenzimidazoles, and phenolics.

Two-component mix-in systems consist of two separate components that must be metered in the proper ratio, mixed, and then dispersed. Chemical families in this group include epoxies, modified acrylics, polyurethanes, silicones, and phenolics. Two-component no-mix systems consist of two separate components that do not require careful metering because no mixing is involved. Adhesive is applied to one surface, while an accelerator is applied to a second surface. The surfaces are then joined. Modified acrylics are included in this group.

Other types of structural adhesives include evaporation or diffusion, hot-melt, delayed-tack, film, pressure-sensitive, and conductive adhesives. Evaporation or diffusion adhesives include materials based on organic solvents or water. In the solvent-base systems, which include rubbers, phenolics, and polyurethanes, the adhesive solution is coated on the porous substrates. Following solvent evaporation and/or absorption into the substrates, the surfaces are joined. Water-base adhesives comprise materials that are totally soluble or dispersive in water. Hot-melt adhesives are 100% solid thermoplastics that are very loosely classified as structural adhesives because most will not withstand elevated temperature loads without creep. Delayed-tack adhesives remain tacky following heat activation and cooling. Tack time ranges from minutes to days over a wide temperature range. Tack adhesives include styrene-butadiene copolymers, polyvinyl acetates, and polystyrene. Film adhesives are two-sided and one-sided tapes and films that are applied quickly and easily. Examples are nylon-epoxies, elastomer-epoxies, epoxy-phenolics, and high-temperature-resistant polyimides. Pressure-sensitive adhesives are capable of holding substrates together when they are brought into contact under brief pressure at room temperature. Conductive adhesives include both electrically and thermally conductive materials that are added as fillers to the adhesive (usually epoxies). The most commonly used electrically conductive filler is silver in powder or flake form. Gold, copper, and aluminum are also used. Thermally conductive fillers include alumina (the most common), beryllia, boron nitride, and silica. See *hot-melt adhesives*, *water-base adhesives*, *pressure-sensitive adhesives*.

Structural Bond

An adhesive bond that joins basic load-bearing parts of an assembly. The load may be either static or dynamic.

Structural Ceramics

Advanced structural ceramic materials are being used increasingly for load-bearing applications. In such applications require materials that have high strength at room temperature and/or retain high strength at elevated temperatures, resist deformation (slow crack growth or creep), are damage tolerant, and are resistant to corrosion and oxidation and/or to abrasion and friction. Ceramics appropriate for such use offer a significant weight savings over metals. Applications include heat exchangers, automotive engine components such as turbocharger rotors and roller cam followers, power generation components, cutting tools, biomedical and plants, and processing equipment used for fabricating a variety of polymer, metal, and ceramic parts. The materials can be either monolithics or composites. A major obstacle to be overcome before these materials see more widespread use is their cost. Many of the processes used for fabrication are labor-intensive or have a high rejection rate, resulting in unacceptably high costs for the final products.

The most important of the bulk monolithic materials for high-temperature structural applications are silicon nitride (Si_3N_4), silicon aluminum oxynitride (sialon), silicon carbide (SiC), partially stabilized or transformation-toughened zirconia (ZrO_2), and alumina (Al_2O_3). These materials can exhibit high strengths (>500 MPa, or 70 ksi), moderate to high fracture toughness (4–14 MPa$\sqrt{m}$, or 3.6–12.7 ksi$\sqrt{in}$.), and low creep rates (<10^{-9} s^{-1}) at 1300°C. Other candidates for structural applications are aluminum titanate (Al_2TiO_5), which has received much attention because of its good thermal shock resistance, and boron carbide (B_4C), the chief advantages of which are its high hardness (29 GPa, or 4.2×10^6 psi) and low density (2.50 g/cm^3).

Two-phase structural ceramics include siliconized silicon carbide (Si/SiC) and ceramic-matrix composites. The former is fabricated by first making a green body, or preform, of SiC, which can be infiltrated with liquid silicon to fill the open space in the preform. Silicon contents range from about 15% to 50%. Ceramic composites are designed to have improved damage tolerance or increased toughness through the addition of second-phase reinforcements in the form of particulates, whiskers, or fibers.

Particulate-reinforced ceramics include transformation-toughened zirconia, in which ZrO_2 particulates act to deflect cracks from the main propagation path. Several whiskers-reinforced compositions are available, the most successful to date being SiC$_w$–Al$_2$O$_3$ used as cutting tools. A number of continuous fibers are available for reinforcement of polymer, metal, glass, glass-ceramic, and ceramic matrices. These include compositions of carbon and graphite, silicon carbide, silicon carboxynitride, silicon nitride, alumina, mullite, and a variety of glasses. Single-crystal fibers of sapphire are also available.

TABLE S.1

Industry, Use, Properties, and Applications for Structural Ceramics

Industry	Use	Property	Application
Fluid handling	Transport and control of aggressive fluids	Resistance to corrosion, mechanical erosion, and abrasion	Mechanical seal faces, meter bearings, faucet valve plates, spray nozzles, micro-filtration membranes
Mineral processing power generation	Handling ores, slurries, pulverized coal, cement clinker, and flue gas neutralizing compounds	Hardness, corrosion resistance, and electrical insulation	Pipe linings, cyclone linings, grinding media, pump components, electrostatic precipitator insulators
Wire manufacturing	Wear applications and surface finish	Hardness, toughness	Capstans and draw blocks, pulleys and sheaves, guides, rolls, dies
Pulp and paper	High-speed paper manufacturing	Abrasion and corrosion resistance	Slitting and sizing knives, stock-preparation equipment
Machine tool and process tooling	Machine components and process tooling	Hardness, high stiffness-to-weight ratio, low inertial mass, and low thermal expansion	Bearings and bushings, close tolerance fittings, extrusion and forming dies, spindles, metal-forming rolls and tools, coordinate-measuring machine structures
Thermal processing	Heat recovery, hot-gas cleanup, general thermal processing	Thermal stress resistance, corrosion resistance, and dimensional stability at extreme temperatures	Compact heat exchangers, heat exchanger tubes, radiant tubes, furnace components, insulators, thermocouple protection tubes, kiln furniture
Internal combustion engine components	Engine components	High-temperature resistance, wear resistance, and corrosion resistance	Exhaust port liners, valve guides, head faceplates, wear surface inserts, piston caps, bearings, bushings, intake manifold liners
Medical and scientific products	Medical devices	Inertness in aggressive environments	Blood centrifuge, pacemaker components, surgical instruments, implant components, lab ware

The favorable thermal, chemical, and tribological properties of some of the structural ceramics can also be achieved by the use of ceramic coatings on other materials such as metals. A number of the ceramic coatings are in use, including ZrO_2, titanium nitride, titanium carbide, SiC, and diamond (see Table S.1).

Structural Foams

Expanded plastic materials having integral solid skins and porous cores that exhibit outstanding rigidity. Structural foams involve a variety of thermoplastic resins as well as urethanes.

Extending the size capabilities of molded parts beyond the limits of conventional injection molding is one of the main advantages of structural foam molding. Whereas injection moldings are usually referred to in terms of ounces and inches, foam moldings usually involve pounds and feet, despite the fact that foam density is much lower. Parts weighing 22.7 kg are not uncommon by low-pressure structural foam methods, and some molders can produce 45.4 kg parts in a single shot.

Foam Processing

Although structural foam parts are produced by several different methods, all systems disperse a gas into the polymer melt during processing, either by adding a chemical blowing agent to the compound or by inducing a gas directly into the melt. The gas creates the cellular core structure in the part. Regardless of the type or form a foaming agent used or when it is added to the melt, structural foam processes are classified as either low-pressure or high-pressure methods. Such a classification relates directly to size range, surface finish, economics, and properties of the molded part.

Low Pressure

Also known as short-shot, conventional structural foam processing methods are the most commonly used because they are the simplest and best suited for economical production of large, three-dimensional parts. In this process, a controlled mixture of resin and gas is injected into a mold creating a low cavity pressure—from 1.36 to 3.43 MPa. The mixture only partially fills the mold, and the bubbles of gas, having been at a higher pressure, expand immediately and fill the cavity. As the cells collapse against the mold surface, a solid skin of melt is formed over the rigid, foamed core.

Skin thickness is controlled by the amount of melt injected, mold temperature, type and amount of blowing agent, and temperature and pressure of the melt. With the use of multiple injection nozzles, extremely large parts can be molded; alternatively, several parts of varying sizes can be molded simultaneously in multi-cavity molds. Standard nominal wall thickness is 3.2 mm.

Another process variation, coinjection, involves the separate injection of two compatible resins. First, a solid resin is injected to form the solid, smooth skin against the mold surfaces. Then, the second material, a measured short shot containing a blowing agent, is injected to form the foamed part interior. The core material is usually a lower-cost resin than the skin material.

High Pressure

This is an expandable-mold, structural foam molding that is closer to conventional injection molding. The heating melt (with a blowing agent) is injected into the mold, creating cavity pressures of between 34 and 136 MPa. The mold is entirely filled, and the pressure prevents any foaming from occurring while the skin position solidifies against the mold surfaces.

At this point, the method departs from conventional injection molding. Mold pressure must be reduced and space provided to allow foaming to take place between the solid-skin surfaces. Depending on the type of equipment and size and configuration of the part, these two provisions are made either by withdrawing cores or by special press motions that partially open the mold halves.

Advantages

In addition to large-size capability and low-density structure, structural foam parts offer high stiffness-to-weight advantages. A 25% increase in wall thickness (over that of a solid section) can provide twice the rigidity—at equal weight—of a solid part. Strength-to-weight ratios of foam sections can be two to five times those of structural metals. Foamed parts made by any of the various methods are relatively stress-free because the foaming is done at a low pressure. For the same reason, sink marks do not occur in foamed parts behind ribs or at wall intersections.

Tooling for low-pressure structural foam molding is generally less expensive than that for injection molding because the low-pressure permits the use of lighter-weight mold materials. Tooling for high-pressure systems is more expensive than for conventional molding because of the special tooling motions involved to accommodate the foaming cycle.

Limitations

Surface finish is the most apparent difference and low-pressure structural foam parts, compared with conventional moldings. Part surfaces had a characteristic swirl pattern caused by the blowing agent, some of which becomes trapped between the mold surface and the skin of the part. The swirl pattern is both visual and tactile; surface roughness can be as much as 1000 µin. Parts that require smooth, finish surfaces require secondary operations, usually sanding, filling, and painting.

The principal process variables that control the swirl pattern are mold temperature, melt temperature, injection rate, and the nature and concentration of the blowing agent. Control of these variables can produce foamed parts with surfaces that replicate any mold surface, smooth or patterned. But this improvement is not achieved without trade-offs; changes involve slower injection rates, heating and cooling of the mold, and other alterations that increase mold costs.

Parts produced by gas counterpressure molding have a significantly reduced swirl pattern because the foaming gases are kept in solution until the solid skin is formed against mold surfaces. Surfaces of co-injected parts are comparable to those of solid, injection-molded parts. Surfaces of parts made by the high-pressure processes are comparable to those of injection-molded parts because the surface of the melt in contact with the mold solidifies while under pressure. But such parts have a "witness line," which has the appearance of a wide parting line at the edges—where the mold opening made provision for foaming. Such areas may require touching up.

Another limitation of high-pressure foam molding is part size and shape. The high pressures and the cost of tools limit these systems too much smaller parts that can be molded economically by low-pressure systems. Most high-pressure parts are relatively flat. See also polyurethanes.

Structural Glass

(1) Flat glass, usually colored or opaque, and frequently ground and polished, used for structural purposes. (2) Glass block, usually hollow, used for structural purposes.

Structural Materials

These are construction materials that, because of their ability to withstand external forces, are considered in the design of a structural framework. Materials used primarily for decoration, insulation, or other than structural purposes are not included in this group.

Clay Products

The principal products in this class are the solid masonry units such as brick and the hollow masonry units such as clay tile or terra-cotta.

Brick is the oldest of all artificial building materials. It is classified as face brick, common brick, and glazed brick. Face brick is used on the exterior of a wall and varies in color, texture, and mechanical perfection. Common brick consists of the kiln run of brick and is used principally as a backup masonry behind whatever facing material is employed. It provides the necessary wall thickness and additional structural strength. Glazed brick is employed largely for interiors where beauty, ease of cleaning, and sanitation are primary considerations.

Structural clay tiles are burned-clay masonry units having interior hollow spaces termed cells. Such tile is widely used because of its strength, lightweight, and insulating and fire protection qualities. It size varies with the intended use.

Load-bearing tile is used in walls that support, in addition to their own weight, loads that frame into them, for example, floors and the roof. Tiles manufactured for use as partition walls, for furring, and for fireproofing steel beams and columns are classed as non-load-bearing tile. Special units are manufactured for floor construction: some are used with reinforced-concrete joints, and others with the steel beams in flat-arch and segmental-arch construction.

Architectural terra-cotta is a burned-clay material used for decorative purposes. The shapes are molded either by hand in plaster-of-paris molds or by machine, using the stiff-mud process.

Building Stones

Building stones generally used are limestone, sandstone, granite, and marble. Until the advent of steel and concrete, stone was the most important building material. Its principal use now is as a decorative material because of its beauty, dignity, and durability.

Concrete

Concrete is a mixture of cement, mineral aggregate, and water, which, if combined in proper proportions, form a plastic mixture capable of being placed in forms and of hardening through the hydration of the cement.

Wood

The cellular structure of wood is largely responsible for its basic characteristics, unique among the common structural materials. The strength of wood depends on the thickness of the cell walls. Its tensile strength is generally greater than its compressive strength. The ratio of its strength to its stiffness is much higher than that of steel or concrete; therefore, it is important that deflection be carefully considered in the design of a wooden floor system.

Laminated structural lumber is formed by gluing together two or more layers of wood with the grain of all layers parallel to the length of the member. Both laminated lumber and plywood make use of modern gluing techniques to produce a greatly improved product.

The principal advantages derived from lamination are the ease with which large members are fabricated and the greater strength of the built-up members. Laminated lumber is used for beams, columns, arch ribs, cord members, and other structural members.

Plywood, while also laminated, is formed from three or more thin layers of wood that are cemented or bonded together, with the grain of the several layers alternately perpendicular and parallel to each other. Plywood is generally used as a replacement for sheathing or as form lumber for reinforced concrete structures. Both laminated structural lumber and plywood have the advantage of minimizing the effects of knots, shakes, and other lumber defects by preventing them from occurring in more than core lamination at a given cross-section.

Structural Metals

Of importance in this group are the structural steels, steel castings, aluminum alloys, magnesium alloys, and cast and wrought iron.

Steel castings are used for rocker bearings under the ends of large bridges. Shoes and bearing plates are usually cast in carbon steel, but rollers are often cast in stainless steel.

Aluminum alloys are strong, lightweight, and resistant to corrosion. The alloys most frequently used are comparable with the structural steels in strength. However, because aluminum alloys have a modulus of elasticity one third that of steel, the danger of local buckling is likely to determine the design of aluminum compression members.

Magnesium alloys are produced as extruded shapes, rolled plate, and forgings. The principal structural applications are in aircraft, truck bodies, and portable scaffolding.

Composite Materials

These are engineered materials synthesized with two distinct phases and comprising a load-bearing material housed in a relatively weak protective matrix. The combination of two or more constituent materials yields a composite material with engineering properties superior to those of the constituents. The associated materials are termed polymer-matrix composites (PMCs), ceramic-matrix composites (CMCs), and metal-matrix composites (MMCs).

The principal features of a fibrous composite material are the fibers, the matrix material, and the interface region between these two dissimilar materials. This class of structural material can be classified as metallic, ceramic, or polymeric, depending on the load-bearing or reinforcing material employed. The reinforcement may be particulates, whiskers, laminated fibers, or a woven fabric. These reinforcements are bonded together by the matrix, which distributes the loading between them. Generally, the reinforcement is a fibrous or particulate material, with the latter category permitting far superior structural properties to be achieved at the expense of more-challenging fabricating technologies and higher costs.

There are numerous examples in nature where this type of microstructure, comprising a load- bearing structural phase housed in a protective matrix, is present. An example is a tree, where the trunk and branches comprise flexible cellulose fibers in a rigid lignin matrix. Development of the class of synthetic materials with this type of structure, such as fiberglass composites and graphite-epoxy laminate, revolutionized the automotive, aerospace, and sporting goods industries. However, it should be noted that composite materials have been used for centuries; for example, bricks were manufactured in ancient Egypt that featured a composite material of clay and straw.

The latest generation of advanced composite materials includes some of the lightest, strongest, stiffest, and most corrosion-resistant materials available to the engineering community. For example, there is a striking contrast of the magnitudes of the stiffness-to-weight ratio or the strength-to-weight ratio of the commercial metals relative to those of the advanced composite materials. Whereas the specific stiffness of aluminum can be increased threefold by the addition of silicon carbide fibers to create a metal-matrix composite, the specific stiffness of graphite-epoxy, fiber-reinforced, polymeric materials can be over four times greater than the specific strength of steel. The ramifications of this comparison are immense, because of lightweight, high-strength, high-stiffness structures can be fabricated with these advanced polymeric composite materials with a weight savings of approximately 50%. This class of designs translates into superior performance for diverse products such as those in the aerospace, defense, automotive, biomedical, and sporting goods industries.

Structural Reaction Injection Molding (SRIM)

A molding process that is similar in practice to resin transfer molding. SRIM derives its name from the RIM process from which the resin chemistry and injection techniques have been adapted. The term structural is added to indicate the reinforced nature of the composite components manufactured by this process. In the SRIM process a preformed reinforcement is placed in a close mold, and a reactive resin mixture is mixed under high pressure in a specially designed mix head. Upon mixing, the reacting liquids flow at low-pressure through a runner system and fill the mold cavity, impregnating the reinforcement material in the process. Once the mold cavity has filled, the resin quickly completes its reaction. A completed component can often be removed from the mold in as little as 1 min.

Structural Shape

A piece of metal of any of several designs accepted as standard by the structural branch of the iron and steel industries.

Structure

As applied to a crystal, the shape and size of the unit cell and the location of all atoms within the unit cell. As applied to microstructure, the size, shape, and arrangement of phases. See also unit cell.

Structure Factor

A mathematically formulated term that relates the positions and identities of atoms in a crystalline material to the intensities of x-ray or electron beams diffracted from particular crystallographic planes.

Strut

A column or beam carrying a compressive load.

Stub-in

In welding, the entrapment of the electrode or filler wire in the solidifying weld metal.

Stub

The discarded end of some consumable such as a manual metallic arc welding electrode.

Stud

(1) Various components, which usually fit some larger assembly, which act as connectors, retainers, or locating devices. (2) A threaded fastener with both ends threaded, one for insertion into the main body of a component and the other end to carry a nut. For example, a set of studs would be screwed into a cylinder block prior to positioning the cylinder head and installing the nuts. Compare with studbolt.

Stud Welding

An arc welding process in which the contact surfaces of a stud, or similar fastener, and the workpiece are heated and melted by an arc drawn between them. The stud is then plunged rapidly onto the workpiece to form a weld. Partial shielding may be obtained by the use of a ceramic ferrule surrounding the stud. Shielding gas or flux may or may not be used. The two basic methods of stud welding are known as stud arc welding, which produces a large amount of weld metal around the stud base and a relatively deep penetration into the base metal, and capacitor discharge stud welding, which produces a very small amount of weld metal around the stud base and shallow penetration into the base metal. The heating processes include electric-arc, -resistance, capacitor discharge, friction, etc.

Studbolt

A threaded fastener with both ends threaded and carrying nuts at both ends. For example, a studbolt would pass through a pair of flanges to hold them together. Compare with *stud*.

Studding

Bar threaded over its full-length, often supplied in standard lengths to be cut to length as required. Also see multiple studs (1) and (2).

Styrene–Acrylonitrile (SAN)

A copolymer of about 70% styrene and 30% acrylonitrile, with higher strength, rigidity, and chemical resistance than can be attained with polystyrene alone. These copolymers may be blended with butadiene as a terpolymer or grafted onto the butadiene to make acrylonitrile–butadiene–styrene resins. They are transparent and have high heat-deflection properties, excellent gloss, chemical resistance, hardness, rigidity, dimensional stability and load-bearing capability.

Styrene–Maleic–Anhydride (SMA)

Copolymers made by the copolymerization of styrene and maleic anhydride, with higher heat resistance than the parent styrenic and acrylonitrile–butadiene–styrene families. For structurally demanding applications, SMAs are reinforced with glass fibers. Most reinforced grades contain 20 wt% reinforcement.

Styrene–Rubber Plastics

Plastics based on styrene polymers and rubbers, the styrene polymers being the greatest amount by mass.

Styrofoam Pattern

An expendable pattern of foamed plastic, especially expanded polystyrene, used in manufacturing castings by the lost foam process. See the term *lost foam casting*.

Subatomic

Particles forming an atom or processes occurring on a smaller scale than an atom.

Subboundary Structure (Subgrain Structure)

A network of low-angle boundaries, usually with misorientation less than 1° within the main grains of a microstructure.

Subcritical Annealing

Softening steels at a temperature just below the transition range, typically about 650°C for about an hour. Also termed *process annealing*. Such treatment is normally applied to heavily cold worked material so it causes the ferrite to recrystallize but does not take the carbide into solution. The resulting microstructure therefore comprises equiaxed grains of ferrite superimposed on elongated stringers or bands of spheroidized carbide. The benefits of the process compared with normalizing or full annealing include lower heating costs, less scaling and, in some cases, good machining characteristics. See *steel*, *annealing* and *transformation temperature*.

Subgrain

A portion of a *crystal* or *grain*, with an orientation slightly different from the orientation of neighboring portions of the same crystal.

Sublimation

Evaporation directly from the solid to vapor without an intermediate liquid phase.

Submerged Arc Welding

Arc welding in which the arc, between a bare metal electrode and the work, is shielded by a blanket of granular, fusible material overlying the joint. Pressure is not applied to the joint, and filler metal is obtained from the consumable electrode (and sometimes from a supplementary welding rod).

Submerged-Electrode Furnace

A furnace used for liquid carburizing of parts by heating molten salt baths with the use of electrodes submerged in the ceramic lining. See also *immersed-electrode furnace*.

Submicron Powder

Any powder whose particles are smaller than ~1 pm.

Submicroscopic

Below the resolution of a microscope.

Subsample

A portion taken from a *sample*. A laboratory sample may be a subsample of a gross sample; similarly, a test portion may be a subsample of a laboratory sample.

Subsieve Analysis

Size distribution of particles that will pass through a standard 325-mesh sieve having 44 pm openings. See the term *sieve analysis*.

Subsieve Fraction

Particles that will pass through a 44 pm (325-mesh) screen. See also sieve analysis.

Subsieve Size

See preferred term *subsieve fraction*.

Sub-Sow Block (Die Holder)

A block used as an adapter in order to permit the use of forging dies that otherwise would not have sufficient height to be used in the particular unit or to permit the use of dies in a unit with different *shank sizes*. See also *sow block*.

Substitutional Element

An alloying element with atomic size and other features similar to the solvent that can replace or substitute for the solvent atoms in the lattice and form a significant region of solid solution in the *phase diagram*. See *solution*.

Substitutional Solid Solution

A *solid solution* in which the solvent and solute atoms are located randomly at the atom sites in the crystal structure of the solution. See also *interstitial* solid *solution*.

Substrate

(1) The material, workpiece, or substance on which the coating is deposited. (2) A material upon the surface of which an adhesive-containing substance is spread for any purpose, such as bonding or coating. A broader term than *adherend*. (3) In electronic devices a body, board, or layer of material, on which some other active or useful material(s) or component(s) may be deposited or laid, for example, electronic circuitry laid on an alumina ceramic board. (4) In catalysts, the formed, porous, high-surface area carrier on which the catalytic agent is widely and thinly distributed for reasons of performance and economy. (5) The material underlying the surface film or outer layer.

Substrate Preparation

The set of operations, including cleaning, degreasing, and finishing applied to the base material prior to applying a coating; intended to ensure an adequate bond to the coating.

Substrate Temperature

In thermal spraying, the temperature attained by the base material as the coating is applied. Proper control of the substrate temperature by intermittent spraying or by the application of external cooling will minimize stresses caused by substrate and coating thermal expansion differences.

Substructure

Same as *sub-boundary structure*.

Subsurface Corrosion

Formation of isolated particles of corrosion products beneath a metal surface. The results from the preferential reactions of certain alloy constituents to inward diffusion of oxygen, nitrogen, or sulfur.

Subzero Machining

Using refrigerant or other means for cooling the workpiece during, or before, machining.

Subzero Treatment

Any treatment below 0°C or fahrenheit particularly cooling of steel to cause transformation of retained austenite. See *steel*.

Suck-Back

See preferred term *concave root surface*.

Suede

Also called **napped leather**. A soft-finished, chrome-tanned leather made from calf, kid, or cowhide splits, or from sheepskin. It is worked on a staking machine until it is soft and supple, and then buffed or polished on an abrasive wheel. It has a soft nap on the polished side and may be dyed any color. Suede is used for shoe uppers, coats, hats, and pocketbooks, but is now largely imitated with synthetic fabrics. **Artificial suede**, or **Izarine**, of Atlas Powder Co., has a base of rubber fabric. Fine cotton fibers dyed in colors are cemented to one side, and the underside of the sheet is beaten to make the fibers stand out until the cement hardens. The fabric looks and feels like fine suede. Some suede is also made by chemical treatment of sheepskins without staking. It has a delicate softness, but is not as wear-resistant as calfskin.

Sulfamic Acid

A white, crystalline, odorless solid, very soluble in water, but only slightly soluble in alcohol. The melting point is 178°C. The acid is stronger than other solid acids, approaching the strength of hydrochloric. It is used in bating and tanning leather, giving a silky, tight grain in the leather. An important use is for cleaning boiler and heat-exchanger tubes. It converts the calcium carbonate scale to the water-soluble calcium sulfamate, which can then be flushed off and combined with sodium chloride; it also converts the rust to ferric chloride and then to the water-soluble iron sulfamate. High purity grades are used for bleaching paper pulp and textiles, organic synthesis, gas-liberating compositions, and as a catalyst for urea-formaldehyde resins. **Ammonium sulfamate** is the ammonia salt of the acid, used as a cleanser and anodizer of metals, as a weedkiller, and for flame proofing paper and textiles.

Suds

Water or emulsion containing corrosion inhibitors and other additives used to cool and lubricate components and tooling during severe metal cutting operations.

Sulfating

The development of a layer of lead sulfate on discharged lead acid cells. It causes permanent damage as it is not reduced by subsequent charging.

Sulfidation

The reaction of a metal or alloy with a sulfur-containing species to produce a sulfur compound that forms on or beneath the surface on the metal or alloy.

Sulfide Spheroidization

A stage of overheating ferrous metals in which sulfide inclusions are partly or completely spheroidized.

Sulfide Stress Cracking (SSC)

Brittle fracture by cracking under the combined action of *tensile stress* and *corrosion* in the presence of water and hydrogen sulfide. See also *environmental cracking*.

Sulfide-Type Inclusions

In steels, nonmetallic inclusions composed essentially of manganese iron sulfide solid solutions (Fe, Mn)S. They are characterized by plasticity at hot-rolling and forging temperatures and, in the hot-worked product, appear as dove-gray elongated inclusions varying from a threadlike to oval outline. See also the term *stringer*.

Sulfinuz Process

A case hardening treatment for steel carried out in a salt bath which produces a case rich in nitrogen and sulfur and carbon having superior fatigue and fretting resistance.

Sulfochlorinated Lubricant

A lubricant containing chlorine and sulfur compounds, which react with a rubbing surface at elevated temperatures to form a protective film. There may be a synergistic effect, producing a faster reaction than with sulfur or chlorine additives alone.

Sulfonated Oil

A fatty oil that has been treated with sulfuric acid, the excess acid being washed out and only the chemically combined acid remaining. The oil is then neutralized with an alkali. Sulfonated oils are water-soluble and are used in cutting oils and in fat liquors for leather finishing. **Mahogany soap** is a name for oil-soluble petroleum sulfonates used as dispersing and wetting agents, corrosion inhibitors, emulsifiers, and to increase the oil absorption of the mineral pigments in paints. **Phosphorated oils**, or their sulfonates, may be used instead of the sulfonates as emulsifying agents or in treating textiles and leathers. They are more stable to alkalies.

Sulfone Polymers

Sulfones are amorphous engineering thermoplastics noted for high heat-deflection temperatures and outstanding dimensional stability. These strong, rigid polymers are the only thermoplastics that remain transparent at service temperatures as high as 204°C.

Three commercially important sulfone-based resins are: polysulfone (PSU), including **Udel** and **Ultrason S**; polyarylsulfone (PAS), including **Radel**; and polyethersulfone (PES), including **Ultrason E**. These materials are claimed to offer the highest performance profiles of any thermoplastics processable on conventional screw-injection and extrusion machinery. Processing temperatures, however, are higher than those of other thermoplastics; the sulfones are processed on equipment that can generate and monitor stock temperatures in the range of 343°C–382°C.

Properties

Heat resistance is the outstanding performance characteristic of the sulfones. Service temperature is limited by heat-deflection temperature, which ranges from 174°C to 204°C. A high percentage of physical, mechanical, and electrical properties is maintained at elevated temperatures, within limits defined by the heat-deflection temperatures. The strength and stiffness of PSU and PES are virtually unaffected up to their glass-transition temperature. For example, the flexural modulus of molded parts remains above 2040 MPa at service temperatures as high as 160°C. Even after prolonged exposure to such temperatures, the resins do not discolor or degrade. Thermal stability and oxidation resistance are excellent at service temperatures well above 149°C.

The continuous service temperature limit (CSTL) for PSU is 160°C, and 180°C for PES. With respect to flammability, PES is rated V-0 per UL 94, and PSU is rated at V-2.

Electrical insulating properties are generally in the midrange among those of other thermoplastics, and they change little after heat aging at the recommended service temperatures. Dissipation factor and dielectric constant—and thus, loss factor—are not affected significantly by increased temperature or frequency.

Creep of the sulfones compared with that of other thermoplastics is exceptionally low at elevated temperatures and under continuous load. For example, creep at 99°C is less than that of acetal or heat-treat resistant ABS (acrylonitrile–butadiene–styrene) at room temperature. This excellent dimensional stability qualifies these sulfone resins for precision-molded parts.

The hydrolytic stability of these resins makes them resistant to water absorption in aqueous acidic and alkaline environments. The combination of hydrolytic stability and heat resistance results in exceptional resistance to boiling water and steam, even under autoclave pressures and cyclic exposure of hot-to-cold and wet-to-dry. PES resins have excellent resistance to hot lubricants, engine fuels, and radiator fluids, and they are resistant to gasoline. The aromatic resins are also resistant to aqueous inorganic acids, organic acids, alkalies, aliphatic hydrocarbons, alcohols, and most cleaners and sterilizing agents.

The sulfones also share a common drawback: they absorb ultraviolet rays, giving them poor weather resistance. Thus, they are not recommended for outdoor service unless they are painted, plated, or UV-stabilized.

Sulfur

Sulfur (symbol S) is one of the most useful of the elements. Its occurrence in nature is little more than 1% that of aluminum, but it is easy to extract and is relatively plentiful. In economics, it belongs to the group of "S" materials—salt, sulfur, steel, sugars, starches—whose consumption is a measure of the industrialization and the rate of industrial growth of a nation.

Strict environmental laws are driving the production of sulfur recovered as a by-product of various industrial operations. It is also obtained by the distillation of iron pyrites, as a by-product of copper and other metal smelting, natural gas, and from gypsum. The sterri exported from Sicily for making sulfuric acid is broken rock rich in sulfur. Brimstone is an ancient name still in popular use for solid sulfur.

Sulfur forms a crystalline mass of a pale-yellow color, with a melting point of 111°C. It forms a ruby vapor at about 416°C. When melted and cast, it forms amorphous sulfur with a specific gravity of 1.955. The tensile strength is 1 MPa, and compressive strength 22 MPa. Since ancient times it has been used as a lute for setting metals into stone. Sulfur also condenses into light flakes known as flowers of sulfur, and the hydrogen sulfide gas, H_2S, separated from sour natural gas, yields a sulfur powder.

Elemental sulfur is widely used for the synthesis of sulfur compounds. It reacts directly with virtually all elements except the noble gases. In addition to its use as a chemical intermediate, sulfur is used increasingly as a construction material. For example, sulfur-impregnated concrete is much more resistant to acid corrosion that is conventional concrete. Highways have been paved with high-sulfur asphalts.

Properties

Sulfur is twice the atomic weight of oxygen but has many similar properties and has great affinity for most metals. The crystalline sulfur is orthorhombic, which converts to monoclinic crystals if cooled slowly from 120°C. This form remains stable below 120°C. When molten sulfur is cooled suddenly, it forms the amorphous sulfur, which has a ring molecular structure and is plastic, which converts gradually to the rhombic form. Sulfur has a wide variety of uses in all industries. The biggest outlet is for sulfuric acid, mainly for producing phosphate fertilizers.

Uses

Sulfur is used for making gunpowder and for vulcanizing rubber, but for most uses it is employed in compounds, especially as sulfuric acid or sulfur dioxide.

Sulfur is used in glass as a colorant to produce golden yellows and ambers, and also with cadmium sulfide and selenium ruby glass. In sulfur amber glasses, the element is introduced as flowers of sulfur, cadmium sulfide, or sodium sulfide. Its manufacture necessitates several precautions.

Compounds

Sulfur dioxide, or sulfurous acid anhydride, is a colorless gas of the composition SO_2, used as a refrigerant, as a preservative, in bleaching, and for making other chemicals. It liquefies at about –10°C. As a refrigerant it has a condensing pressure of 23.5 kg at 30°C. The gas is toxic and has a pungent, suffocating odor, so that leaks are detected easily. It is corrosive to organic materials but does not attack copper or brass. The gas is soluble in water, forming sulfurous acid, H_2SO_3, a colorless liquid with suffocating fumes. The acid form is the usual method of use of the gas for bleaching.

Sulfur Dome

An inverted container, holding a high concentration of sulfur dioxide gas, used in die casting to cover a pot of molten magnesium to prevent burning.

Sulfuric Acid

An oily, highly corrosive liquid of the composition H_2SO_4, sulfuric acid as a specific gravity of 1.841 and a boiling point of 330°C. It is miscible in water in all proportions, and the color is yellowish to brown according to the purity. It may be made by burning sulfur to the dioxide, oxidizing to the trioxide, and reacting with steam to form the acid. It is a strong acid, oxidizing organic materials and most metals. Sulfuric acid is used for pickling and cleaning metals, in electric batteries in plating baths, for making explosives and fertilizers, and for many other purposes. In the metal industries it is called dipping acid, and in the automotive trade it is called battery acid.

Uses

Sulfuric acid is used in the enameling industry for pickling purposes. The solutions vary in strength from 5% to 8%, although it is said that a 6% solution of sulfuric acid heated to 71°C–77°C will be the most effective in the pickling of sheet iron.

In making up H_2SO_4 solutions, always add the acid to the water, and never the water to the acid, as the latter method may cause a violent reaction. Sulfuric acid also has been used as a mill addition for acid-resisting enamels.

Compounds

Sulfur trioxide, or sulfuric anhydride, SO_3, is the acid minus water. It is a colorless liquid boiling at 46°C, and forms sulfuric acid when mixed with water.

Niter cake, which is sodium acid sulfate, $NaHSO_4$, or sodium bisulfate, contains 30%–35% available sulfuric acid and is used in hot solutions for pickling and cleaning metals. It comes in colorless crystals or white lumps, with a specific gravity of 2.435 and melting point 300°C. Sodium sulfate, or Glauber's salt, is a white crystalline material used in making kraft paper, rayon, and glass. **Salt cake**, Na_2SO_4, is impure sodium sulfate used in the cooking liquor in making paper pulp from wood. It is also used in freezing mixtures. Sodium sulfite is a white to tan crystalline powder very soluble in water but nonhygroscopic.

Sodium sulfide, Na_2S, is a pink flaky solid, used in tanneries for dehairing, and in the manufacture of dyes and pigments. The commercial product contains 60%–62% Na_2S, 3.5% NaCl, and other salts, and the balance water of crystallization.

Sulfurized Lubricant

A lubricant containing sulfur or a sulfur compound that reacts with a rubbing surface at elevated temperatures to form a protective film. The shear strength of the sulfide film formed on ferrous materials is lower than that of the metal but greater than that of the film formed by reaction with a chlorinated lubricant.

Sulfur Print

A print formed by placing a suitable photographic paper, previously soaked in 3% sulfuric acid, against a prepared steel surface to reveal the presence and distribution of sulfur compounds particularly manganese sulfide inclusions. These inclusions reveal the *flowlines*, i.e., the pattern of deformation, of the material during the various working operations since the material was cast.

Sum Peak

An artifact encountered in x-ray analysis during pulse pileup where two x-rays simultaneously entering the detector are counted as one x-ray, the energy of which is equal to the sum of both x-rays. See also *escape peak*.

Sun Hemp

The bast fiber is used for cordage and rope in place of jute, but is lighter in color and is more flexible, stronger, and more durable than jute. It resembles true hemp, but is not as strong. It is more properly called **sann hemp**. It is also known as **sunn fiber, Indian hemp,** and **Bombay hemp**. The plant, which is a shrub, is cultivated extensively in India. The method of extraction is the same as for true hemp. The best fibers are retained locally for making into cloth. It is used in the United States for making cigarette paper and for oakum. **Madras hemp** is from another species of the same plant.

Superabrasives

Superabrasives collectively refer to diamond and cubic boron nitride (CBN), both of which are produced by synthetic means. Diamond is the hardest material known, and CBN is the second hardest. Because of their high hardness, abrasion resistance, and other unique properties, these materials find extensive use in a wide variety of abrasive or cutting applications.

The primary objective in the synthesis of diamond and CBN is to transform a crystal structure from a soft hexagonal form to a hard cubic form. In the case of carbon, for example, hexagonal carbon (graphite) would be transformed into cubic carbon (diamond). The synthesis of CBN or diamond grit is normally achieved by subjecting hexagonal carbon or boron nitride to high temperatures and high pressures with large special-purpose presses. By the simultaneous application of heat and pressure, hexagonal carbon or boron nitride can be transformed into a hard cubic form. Synthetic CBN and diamond grains can be used as loose abrasives, as bonded abrasives in grinding wheels and hones, and as bonded abrasive and single-point cutting tools. When used for the latter application, diamond and CBN are referred to as *ultrahard tool materials*.

The synthesis of diamond was first demonstrated by the General Electric Company in the 1950s. This invention subsequently led to the rapid growth of diamond for industrial applications. In addition to its high hardness, diamond is an excellent heat conductor and has an extremely low coefficient of friction. However, at temperatures above 800°C, diamond tends to graphitize, thus losing its value as a wear-resistant abrasive. Diamond suffers rapid wear and chemical dissolution (erosion) when abraded against iron. Therefore, it is not normally used as an abrasive against ferrous materials. Diamond is used extensively to grind stone, concrete, carbides, glass, ceramics, plastics, and composites. Polycrystalline diamond tool (machining) applications include nonferrous materials (aluminum–silicon alloys, copper alloys, and tungsten carbides), fiberboard, composites (graphite-epoxy, carbon–carbon, and fiberglass-reinforced plastic), and ceramics.

Like diamond, CBN also has a cubic structure. In its hexagonal close-packed structure, hexagonal boron nitride is similar to graphite and is used as a solid lubricant. **Borazon**, the General Electric trade name for CBN, was first synthesized in 1959. Unlike diamond, CBN is not very reactive with iron and therefore is used for grinding and machining of ferrous materials such as hardened steels and cast irons. Cubic boron nitride is also used for precision grinding and machining of nickel- and cobalt-base superalloys and hardfacing materials.

Superalloys

The term *superalloy* is broadly applied to iron-base, nickel-base, and cobalt-base alloys, often quite complex, which combine high-temperature mechanical properties and oxidation resistance to an unusual degree. Alloy requirements for turbosuperchargers and, later, the jet engine, largely provided the incentive for superalloy development.

Because of their excellent high-temperature performance, they are also known as high-temperature, high-strength alloys. Their strength at high temperatures is usually measured in terms of stress-rupture strength or creep resistance. For high-stress applications, the iron-base alloys are generally limited to maximum service temperature of about 649°C, whereas the nickel-and cobalt-base alloys are used at temperatures to about 1093°C and higher. In general, the nickel alloys are stronger than the cobalt alloys at temperatures below 1093°C and the reverse is true at temperatures above 1093°C. Superalloys are probably best known for aircraft-turbine applications, although they are also used in steam and industrial turbines, nuclear-power systems, and chemical-and petroleum-processing equipment. A great variety of cast and wrought alloys are available and, in recent years, considerable attention has been focused on the use of powder-metallurgy techniques as a means of attaining greater compositional uniformity and finer grain size.

Strengthening Mechanisms

Superalloys of the nickel–chromium and iron–nickel–chromium types usually contain sufficient chromium to provide that needed oxidation resistance and are further strengthened by the addition of other elements. The strengthening mechanisms include solid solution, precipitation, and carbide hardening. Most of the superalloys combine at least two and frequently all three of the above mechanisms.

Solid Solution

This is accomplished by introducing elements having different atomic sizes than those of the matrix elements, to increase the lattice strain. In addition to chromium, needed also for oxidation resistance, the elements molybdenum, columbium, vanadium, cobalt, and tungsten are effective in varying degrees as solid-solution strengtheners when added in proper balance.

Superalloys of this type, such as 16–25–6, were used in gas turbines of older design for disks, employing "hot-cold work" to obtain the required yield strength in the hub area. However, hot-cold work is not effective as a means of getting high strength at temperatures much above about 538°C.

Superconductivity

Superconductivity is a phenomenon occurring in many electrical conductors, in which the electrons responsible for conduction undergo a collective transition into an ordered state with many unique and remarkable properties. These include the vanishing of resistance to the flow of electric current, the appearance of a large diamagnetism and other unusual magnetic effects, substantial alteration of many thermal properties, and the occurrence of quantum effects otherwise observable only at the atomic and subatomic level.

The ability of certain materials, when cooled to extremely low or cryogenic temperatures, to conduct electricity with essentially zero resistance to DC current, and to AC current below certain critical high-frequency ranges, was found in the scientific community.

Therefore, as gas turbines operating temperatures increased and turbine disk rim temperatures appreciably exceeded 538°C, other approaches were needed.

Cobalt-base alloys are strengthened principally by solid solution hardening, usually combined with a dispersion of stable carbides.

Precipitation Hardening

Precipitation hardening is the method now employed to impart high-strength at high temperatures to most of the superalloys used for critical components of aircraft gas turbines. This includes alloys ranging from 25% nickel A-286 wheel alloy to such recent and complex high-temperature nickel-base wrought and cast turbine blade alloys as "Udimet" 700, "Nimonic" alloy 115, and IN-100.

The presence of aluminum and titanium, usually jointly, in a nickel-chromium base, with or without iron, imparts unique age-hardening characteristics through the precipitation of gamma prime phase {$Ni_3(Al,Ti)$}.

The outstanding difference between the previously used age-hardening systems, typified by duralumin or beryllium-copper, and those utilizing gamma prime hardening lies in the fact that an alloy of the latter type can be heated in service appreciably above the optimum aging temperature without permanent loss of strength. In the conventional critical dispersion type age-hardening alloys, strength can only be restored by a complete cycle of heat treatment involving a high-temperature solution treatment and reaging.

The difference between the two types of age-hardening systems is that the aluminum-titanium hardened alloy retains most of its initial hardness (and high-temperature properties) when the overheat is removed. However, an alloy typical of the other type of system (such as beryllium copper or beryllium nickel) overages and does not regain its hardness when the overheat is removed. It is this "reversible" aging behavior, along with high resistance to overaging (agglomeration of gamma-prime hardening phase), that has led to the widespread use of aluminum-titanium age-hardened nickel-base alloys for first-stage turbine blades in commercial and advanced design jet engines. The somewhat less complex iron-nickel-chromium alloys, such as A-286 and "Incoloy" alloy 901, used for turbine disks, are similarly age-hardened, with Ni_3Ti comprising most of the age-hardening component in these two alloys. Sufficient aluminum is also present in the gamma-prime precipitate to improve structural stability.

Carbide Hardening

Carbides provide a major source of dispersed-phase strengthening and cobalt-base alloys. These alloys do not respond to age hardening with aluminum and titanium since the gamma-prime phase does not form unless substantial amounts of nickel are present. Although gamma prime {$Ni_3(Al, Ti)$} is the principal strengthener in the age-hardenable nickel-base alloys, important auxiliary strengthening can be obtained by precipitation of quite complex carbides. The nature of the carbides formed and their mode of distribution can usually be controlled by alloy formation in heat treatment.

Deoxidizers and Malleabilizers

In addition to the major alloying elements, small but effective amounts of malleabilizers such as boron and zirconium must be present to neutralize the effects of impurities that adversely affect hot ductility. As little as 0.005% boron is highly effective in nickel-base alloys, and zirconium, in a concentration about 10 times that of boron, is also useful. In air melting of nickel-base alloys, magnesium is also added to *fix* any sulfur picked up during melting.

The high-temperature properties of the age-hardenable alloys are governed to a large degree by the hardener (aluminum plus titanium) content. However, as the hardener increases, the temperature of incipient fusion decreases and a lower temperature limit of forgeability rises, because of increased high-temperature strength, until forging is no longer practical by conventional procedures. The use of cast turbine blade alloys to meet very high temperature requirements is a natural consequence.

While other co-present elements and the amount of hot and cold workability required are also factors, it may be considered that alloys with up to 4% total hardener (aluminum plus titanium) are generally available in all common mill forms and/or fabricable by conventional methods. As the hardener increases to about 8% or 9%, wrought alloys are still available, but in progressively fewer forms, ultimately being limited to small forgings such as turbine blades, at increasingly greater cost. The commercial cast turbine blade alloys contain from about 7% to 11% total hardener and are often chosen over forgings for reason of cost or necessity or both.

Phases

Superalloys consist of a face-centered cubic (fcc) austenitic gamma phase matrix plus a variety of secondary phases. The principal secondary phases are carbides (MC, $M_{23}C_4$, M_6C, and then M_7C_3) in all superalloy types and gamma prime (γ) fcc ordered $Ni_3(Al, Ti)$ intermetallic compound in Ni-and Fe-Ni-based alloys. In alloys containing niobium and tantalum, the primary strengthening phase is gamma double prime (γ), a body-centered tetragonal phase. Superalloys derive their strength from solid-solution hardeners and precipitating phases. Carbides may provide limited strengthening directly (e.g., through dispersion hardening) or, more commonly, indirectly (e.g., by stabilizing grain boundaries against excessive shear). In addition to those elements that produce solid-solution hardening and promote carbide and γ formation, and other elements (e.g., boron, zirconium, hafnium, and cerium) are added to enhance mechanical and/or chemical properties.

The three types of superalloys (Fe-Ni, Ni-, and Co-base) are further subdivided into wrought, cast, and powder metallurgy alloys. Cast alloys can be further broken down into polycrystalline, directionally solidified, and single-crystal superalloys. The most important class of Fe-Ni-base superalloys includes those alloys that are strengthened by intermetallic-compound precipitation in an fcc matrix. The most common precipitate is γ, typified by alloys A-286 and Incoloy 901, but some alloys precipitate γ, typified by Inconel 718. Another class of cast Fe-Ni-base superalloys is hardened by carbides, nitrides, and carbonitrides; some tungsten and molybdenum may be added to produce solid-solution hardening. Other Fe-Ni-base alloys are modified in stainless steels primarily strengthened by solid-solution hardening.

The most important class of Ni-base superalloys is strengthened by intermetallic-compound precipitation in an fcc matrix. The strengthening precipitate is γ, typified by Waspaloy and Udimet 700. Another class is represented by Hastelloy X, which is essentially solid-solution strengthened, but which also derives some strengthening from carbide precipitation produced through a working-plus-aging schedule. A third class includes oxide-dispersion-strengthened (ODS) alloys, which are strengthened by dispersions of inert particles such as yttria.

Cobalt-base superalloys are strengthened by solid-solution alloying and carbide precipitation. Unlike the Fe-Ni and Ni-base alloys, no intermetallic phase has been found that will strengthen

S

Co-base alloys to the same degree that γ or γ' strengthens the other superalloys.

Compositions and Properties

Although factors such as tensile properties, corrosion resistance, fatigue strength, expansion characteristics, etc., are important, the creep-rupture characteristics are usually the prime requisite in the selection of a superalloy.

Vacuum Melting

Vacuum induction-melted superalloys are generally more ductile than air-melted material, permitting the use of higher aluminum plus titanium levels while still retaining adequate ductility. Vacuum induction melting accomplishes refining and also permits closer control of composition than does air melting. The vacuum arc (consumable electrode) method is widely used for superalloys, especially those employed for turbine disks. Often the electrode for the vacuum arc-melting charge is obtained from a vacuum induction melt, the end product thus combining the benefits of refining occurring in vacuum induction melting with the controlled solidification and sound ingot structure associated with the vacuum arc process. Some refining is also accomplished in the vacuum arc process, but to a lesser degree than in vacuum induction melting.

Iron-Base Superalloys

The iron-base superalloys include solid-solution alloys and precipitation-hardening (PH), or precipitation-strengthened, alloys. Solid-solution types are alloyed primarily with nickel (20% to 36%) and chromium (16% to 21%) although other elements are also present in lesser amounts. Superalloy 16-25-6, for example, the alloy designation indicating its chromium, nickel, and molybdenum contents, respectively, also contains small amounts of manganese (1.35%), silicon (0.7%), nitrogen (0.15%), and carbon (0.06%). Superalloy 20Cb-3 contains 34% nickel, 20% chromium, 3.5% copper, 2.5% molybdenum, as much as 1% columbium, and 0.07% carbon. Incoloy 800, 801, and 802 contain slightly less nickel and slightly more chromium with small amounts of titanium, aluminum, and carbon. N-155, or Multimet, an early sheet alloy, contains about equal amounts of chromium, nickel, and cobalt (20% each), +3% molybdenum, 2.5% tungsten, 1% columbium, and small amounts of carbon, nitrogen, lanthanum, and zirconium. At 732°C, this alloy has a 1000-h stress-rupture strength of about 165 MPa.

PH iron-base super alloys provide greater strengthening by precipitation of a nickel-aluminum-titanium phase. One such alloy, which may be the most well known of all iron-base superalloys, is A-286. It contains 26% nickel, 15% chromium, 2% titanium, 1.25% molybdenum, 0.3% vanadium, 0.2% aluminum, 0.04% carbon, and 0.005% boron. At room temperature, it has a tensile yield strength of about 690 MPa and a tensile modulus of 145×10^3 MPa. At 649°C, tensile yield strength declines only slightly, to 607 MPa, and its modulus is about the same or slightly greater. It has a 1000-h stress-rupture strength of about 145 MPa at 732°C. Other PH iron-base superalloys are Discoloy, Haynes 556 (whose chromium, nickel, cobalt, molybdenum, and tungsten contents are similar to that N-155); Incoloy 903 and Pyromet CTX-1, which are virtually chromium-free but high in nickel (37% to 38%) and cobalt (15% to 16%); and V-57 and W-545, which contain about 14% chromium, 26% to 27% nickel, about 3% titanium, 1% to 1.5% molybdenum, plus aluminum, carbon, and boron. V-57 has a 1000-h stress rupture

strength of about 172 MPa at 732°C and greater tensile strength but similar ductility than A-286 at room and elevated temperatures.

Nickel-Base Superalloys

Nickel-base superalloys are solid-solution, precipitation, or oxide-dispersion strengthened. All contain substantial amounts of chromium, 9% to 25%, which, combined with the nickel, accounts for their excellent high-temperature oxidation resistance. Other common alloying elements include molybdenum, tungsten, cobalt, iron, columbium, aluminum, and titanium. Typical solid-solution alloys include Hastelloy X (22% to 23% chromium, 17% to 20% iron, 8% to 10% molybdenum, 0.5% to 2.5% cobalt, 2% aluminum, 0.2% to 1% tungsten, and 0.15% carbon), Inconel 600 (15.5% chromium, 8% iron, 0.25% copper maximum, 0.08% carbon), and Inconel 601, 604, 617, and 625, the latter containing 21.5% chromium, 9% molybdenum, 3.6% columbium, 2.5% iron, 0.2% titanium, 0.2% aluminum, and 0.05% carbon. At 732°C, wrought Hastelloy X (it is also available for castings) has a 1000-h stress-rupture strength of about 124 MPa and has high oxidation resistance at temperatures to 1204°C.

The precipitation-strengthened alloys, which are the most numerous, contain aluminum and titanium for the precipitation of a second strengthening phase, the intermetallic $Ni_3(Al, Ti)$ known as gamma prime (γ') or the intermetallic Ni_3Cb known as gamma double prime (γ"), during heat treatment. One such alloy, Inconel X-750 (15.5% chromium, 7% iron, 2.5% titanium, 1% columbium, 0.7% aluminum, 0.25% copper maximum, and 0.04% carbon), has more than twice the tensile yield strength Inconel 600 at room temperature and nearly 3 times as much at 760°C. Its 1000-h stress-rupture strength at 760°C is in the range of 138 to 207 MPa. Still great tensile yield strength at room and elevated temperatures and a 172-MPa stress-rupture strength at 760°C are provided by Inconel 718 (19% chromium, 18.5% iron, 5.1% columbium, 3% molybdenum, 0.9% titanium, 0.5% aluminum, 0.15% copper maximum, 0.08% carbon maximum), a wrought alloy originally that also has been used for castings. Among the strongest alloys in terms of stress-rupture strength is the wrought or cast IN-100 (10% chromium, 15% cobalt, 5.5% aluminum, 4.7% titanium, 3% molybdenum, 1% vanadium, less than 0.6% iron, 0.15% carbon, 0.06% zirconium, 0.015% boron). Investment cast, it provides a 1000-h stress-rupture strength of 517 MPa at 760°C, 255 MPa at 871°C, and 103 MPa at 982°C. Other precipitation-strengthened wrought alloys include Astroloy; D-979; IN 102; Inconel 706 and 751; M 252; Nimonic 80A, 90, 95, 100, 105, 115, and 263; René 41, 95, and 100; Udimet 500, 520, 630, 700, and 710; Unitemp AF2-IDA; and Waspaloy. Other cast alloys, mainly investment-cast, include B-1900; IN-738X; In-792; Inconel 713C; M 252; MAR-M 200, 246, 247, and 421; NX-188; René 77, 80, and 100; Udimet 500, 700, and 710; Waspaloy; and WAZ-20.

A few of the cast alloys, such as MAR-M 200, are used to produce directionally solidified castings, that is, investment castings in which the grain runs only unidirectionally, as long as the length of turbine blades. Eliminating transverse grains improves stress-rupture properties and fatigue resistance. Grain-free alloys, or single-crystal alloys, also have been cast, further improving high-temperature creep resistance. Regarding powder-metallurgy techniques, emphasis has been on the use of prealloy powder made by rapid solidification techniques (RST) and mechanical alloying (MA), a high-energy milling process using attrition mills or special ball mills. Dispersion-strengthened nickel alloys are alloys strengthened by a dispersed oxide phase, such as thoria, which markedly increases strength at very high temperatures but only moderately so at intermediate elevated temperatures, thus limiting applications. TD-nickel, or thoria-dispensed nickel, was the first of such superalloys, and it was subsequently modified with about 20% chromium, TD-NiCr, for greater oxidation resistance. MA 754 and MA 6000E alloys combine dispersion strengthening with yttria and gamma-prime strengthening.

Cobalt-Base Superalloys

Cobalt-base superalloys are for the most part solid-solution alloys, which, when aged, are strengthened by precipitation of carbide or intermetallic phases. Most contain 20% to 25% chromium, substantial nickel and tungsten and/or molybdenum, and other elements, such as iron, columbium, aluminum, or titanium. One of the most well-known, L-605, or Haynes 25, is mainly a wrought alloy, although it is also used for castings. In wrought form, it contains 20% chromium, 15% tungsten, 10% nickel, 3% iron, 1.5% manganese, and 0.1% carbon. At room temperature, it has a tensile yield strength of about 462 MPa, and at 871°C about 241 MPa. Its 1000-h stress-rupture strength at 815°C is 124 MPa. The more recent Haynes 188 (22% chromium, 22% nickel, 14.5% tungsten, 3% iron, 1.5% manganese, 0.9% lanthanum, 0.35% silicon, and 0.1% carbon), which was developed for aircraft-turbine sheet components, provides roughly similar strength and high oxidation resistance to above 1093°C. MP35N (35% nickel, 20 chromium, 10 molybdenum) is a work-hardening alloy used mainly for temperature corrosion-resistant fasteners. Another alloy, S-816, contains equal amounts of chromium and nickel (20% each), equal amounts of molybdenum, tungsten, columbium, and iron (4% each), and 0.38% carbon. Primarily a wrought alloy, although also used for castings, it has a 1000-h stress-rupture strength of 145 MPa at 815°C. Other casting alloys include AiResist 13, 213, and 215; Haynes 21 and 31, the latter also known as X-40; Haynes 151; J-1650; MAR-M 302, 322, 509, and 918; V-36; and WI-52. Their chromium content ranges from 19% (AiResist 215) to 27% (Haynes 21), and some are nickel free or low in nickel. Most contain substantial amounts of tungsten or tantalum, and various other alloying elements. Among the strongest in terms of 1000-h stress-rupture strength at 815°C are Haynes 21 and 31–290 MPa and 352 MPa, respectively.

Fabrication

Forging and Hot Working

Many of the commercial nickel-base age-hardenable superalloys can be forged or hot-worked with varying degrees of ease. As has been indicated, the top side of the forging temperature range is limited by such considerations as incipient fusion temperature, grain size requirements, tendency for "bursts," etc., while the lower side by the stiffness and ductility of the alloy. The recommendations of the metal producer should be sought for optimum forging practice for a given alloy.

The hot extrusion process increases the gamut of superalloys compositions that can be hot-worked. By the use of a suitable sheath on the extrusion billet, otherwise unworkable alloys can be reduced to bar by hot extrusion. In some instances, mild steel has been used for a sheath, while in others, nickel-chromium alloys such as Nimonic alloy 75 and Inconel alloy 600 have been employed. Some of the very high-speed hot extrusion processes may be of value in hot-working the more refractory superalloys.

Heat Treatment

The conventional heat-treating equipment and fixtures are generally suitable for nickel alloys, and austenitic stainless steels are also applicable to the nickel-base high-temperature alloys. Nickel-base alloys are more susceptible to sulfur and lead embrittlement than iron-base alloys. It is therefore essential that all foreign material, such as grease, oil, cutting lubricants, marking paints, etc., be removed by suitable solvents, vapor greasing, or other methods, before heat treatment.

When fabricated parts made from thin sheet or strip of age-hardening alloys such as Inconel alloy X-750 must be annealed during and after fabrication, it is desirable, especially in light gauges, to provide a protective atmosphere such as argon or dry hydrogen to lessen the possibility of surface depletion of the age-hardening elements. This precaution may not be as necessary in heavier sections, since the surface oxidation involves a much smaller proportion of the effective cross-section.

It is usually necessary after severe forming, or after welding, to apply a stress relief anneal (above 899°C) to assemblies fabricated from aluminum-titanium age-hardenable nickel-base alloys prior to aging. It is vitally important to heat the structure rapidly through the age-hardening temperature range of 649°C–760°C (which is also the low ductility range) so that stress relief can be achieved before any appreciable aging takes place. This is conveniently done by charging into a furnace at or above the desired annealing temperature. It has been found at times that the efficacy of this procedure has been vitiated in large welded structures by charging onto a cold car, resulting in a slower and nonuniform heating of the fabricated part when run into the hot furnace. Contrary to expectations, little difficulty has been encountered with distortion under the aforementioned rapid heating conditions. In fact, distortion of weldments of substantial size has been reported to be less than by conventional slow-heating methods.

Forming

All of the wrought nickel-base alloys available as sheet can be formed successfully into quite complex shapes and having much plastic flow. The lower-strength Inconel alloy 600 and Nimonic alloy 75 offer few problems. The high-strength age-hardening varieties, processed in the annealed condition, can be subjected to a surprising amount of cold work and deformation, provided sufficient power is available. Explosive forming has also been successfully employed on a number of nickel-base alloys.

Machining

All of the alloys discussed can be machined, the strongest and highest hardener content materials causing the most difficulty. The recommendations of the metal producer should be followed with respect to Altamont condition of heat treatment, type of tool, speed and feed, cutting lubricant, etc. Wrought alloys of quite high hardener content, such as Inconel alloy 700 and Udimet 500, although difficult to handle, can be machined with reasonable facility using high-speed-steel tools of the tungsten-cobalt type, and cemented carbide tools of the tungsten-cobalt and tungsten-tantalum-cobalt type.

Various electroerosion processes have been successfully used on a number of the age-hardened superalloys and, at hardener levels, may be necessary for some operations such as drilling.

Welding

Inconel alloy 600, Nimonic alloy 75, and other nickel-base alloys of the predominantly solid-solution strengthened type offer no serious problems in welding. All of the common resistance and fusion welding processes (except submerged arc) are regularly and successfully employed.

In handling the wrought superalloys age-hardened with gamma prime, it is necessary to observe certain precautions. Material should be welded in the annealed condition to minimize the hazard of cracking in weld or parent metal. If the components to be joined to have been severely worked or deformed, they should be stress-relief-annealed before welding by charging into a hot furnace to ensure rapid heating to this stress-relieving temperature. Similarly, weldments should be stress-relieved before attempting to apply the 704°C age-hardening treatment.

Where subassemblies must be joined in the age-hardened condition, the practice of "safe ending" with a compatible nonaging material prior to age hardening can be usefully employed. The final weldments joining the fully age-hardened components are then made on the "safe ends."

Such new welding processes as "short arc," electron, and laser beam have come into increasing use and have been helpful in joining some of the very high hardener content alloys.

Brazing

The solid solution type chromium-containing alloys, such as Inconel alloy 600, are quite readily brazed, using techniques and brazing filler metals applicable to the austenitic stainless steels. Generally speaking, it is desirable to braze annealed (stress-free) material to avoid embrittlement by the molten braze metal. Where brazing filler metals are employed that melt above the stress-relieving temperature, a prior anneal is usually not needed. As with the stainless steels, dry hydrogen, argon, and helium atmospheres are used successfully; and vacuum brazing is also very successfully employed.

The age-hardened nickel-base alloys containing titanium and aluminum are rather difficult to braze, unless some method of fluxing, solid or gaseous, is used. Alternatively, the common practice is to preplate the areas to be furnace brazed with 0.01 to 0.03 mm of nickel, which prevents the formation of aluminum or titanium oxide films and permits ready wetting by the brazing filler metal.

Silver brazing filler metals can be used for lower temperature applications. However, since the nickel-base superalloys are usually employed for high-temperature applications, the higher melting point and stronger and more oxidation-resistant brazing filler metals of the Ni-Cr-Si-B type are generally used. The silver-palladium-manganese and palladium-nickel filler metals also provide useful brazing materials for intermediate service temperatures.

Superbronze

A name applied to brasses containing both aluminumand manganese. They are ordinarily high brasses with 2%–3% manganese and 1–6 aluminum, with sometimes also some iron. They have greatly increased strength and hardness over the original brasses, but the ductility is reduced and they are difficult to work and machine. The early superbronze was known as **Heusler alloy**. Muntz metal is also frequently modified with manganese, iron, and aluminum. The alloys are used where high strength and corrosion resistance are required, and they are often marketed under trade names. The name *superbronze* is a shop term rather than a technical classification, and thus the name is often applied to any hard, highstrength, heat-treatable, copper-base alloy.

This current is referred to as supercurrent and is carried on the surface of the superconductor within a particular death characteristic of the material.

Until recently, superconductivity was only seen in materials at fantastically cold temperatures not exceeding 23 K (−250°C) in the intermetallic compound Nb_3Ge. This meant that all superconductors had to be cooled with liquid helium, which is expensive and cumbersome to handle. Applications for superconductors were quite limited. The main use was for nuclear magnetic resonance scanners used by hospitals to examine soft tissue without surgery. Nuclear magnetic resonance scanners tap the intensely powerful magnetic fields that superconductors can be made to generate.

However, in August 1986, researchers discovered a compound of the metals lanthanum, barium, and copper, along with oxygen, that would superconduct at 35 K (−238°C).

Following this discovery, researchers brought about a dramatic jump in T_c 93 K (with an onset temperature of 98 K) by substituting yttrium for lanthanum, while roughly switching the +2 and +3 ion ratios.

Superconductors

Superconductors are solid crystalline materials whose electrical resistance drops significantly as temperature decreases. Until recently, at temperatures approaching close to absolute zero (−272°C) were required for their resistivity to vanish. Some of the metals exhibiting superconductivity at near absolute zero include iridium, lead, mercury, columbium, tin, tantalum, vanadium, and many alloys and chemical compounds. Alloys considered among the best commercially available are lead–molybdenum–sulfur, columbium-10, and columbium–titanium.

In recent years, alloys and compounds have been developed that are superconductive at temperatures substantially above absolute zero. These include a compound of lanthanum, strontium, copper, and oxygen, which is superconductive at −240°C, and a barium–yttrium–copper oxide, which is superconductive at −183°C. A two-phase ceramic superconductor has been developed in which one of the phases is superconductive at −33°C. It is basically copper oxide containing barium and yttrium.

Properties and Processing

Because J_c is higher in single crystals than in polycrystalline materials, it can be increased by minimizing the presence of grain boundaries. This is done by single-crystal growth or techniques such as melt texturing, where grain boundaries become highly directionalized to allow for significantly less random disruption of current flow. Here, the grain boundaries become predominately parallel to the copper–oxygen chains (the direction of current flow) in all the crystals. Consequently, the microstructure consists of long, needle-like grains that are parallel and intermeshed.

Single crystals have been grown epitaxially by sputtering (as well as by pulsed Excimer laser deposition and electron beam epitaxy). Good deposition can be achieved when the material is reactively sputtered in an oxygen-containing environment from three separate metallic targets (yttrium, barium, copper) simultaneously. This is known as triode sputtering and allows for tight control of stoichiometry. Following this, as with bulk samples, the films typically must be annealed in pure oxygen at −500°C. Samples also must be slow cooled to 300°C, as gravimetric analysis shows that the capability of the compound to absorb more oxygen into its crystal structure increases with decreased temperature down to 300°C. Annealing can be successfully accomplished below 300°C by use of an oxygen-ion bombardment.

The greatest reproducibility of results in single-crystal formation has come when using single-crystal $SrTiO_3$ as the epitaxy substrate.

S

Some success has been found in depositing single crystals on single-crystal MgO, which has a much lower dielectric constant than $SrTiO_3$.

Applications

There are a number of practical applications of superconductivity. Powerful superconducting electromagnets guide elementary particles in particle accelerators, and they also provide the magnetic field needed for magnetic resonance imaging (MRI). Ultrasensitive superconducting circuits are used in medical studies of the human heart and brain and for a wide variety of physical science experiments. A completely superconducting prototype computer has even been built.

Most superconductive applications that have been considered to have reasonable possibility of being achieved in the next few years involve thin-film deposition of these materials. Thin films clearly have a higher J_c advantage over bulk superconductors.

Photovoltaic substances, for example, if interfaced with a superconductor, can act as signal detectors (e.g., infrared devices) because they will be sensitive to the most minute electrical fields.

One of the potential uses of the newer superconductors is for making more powerful and efficient electromagnets that could be used in trains to levitate them above their tracks, and thus make train speeds of hundreds of miles per hour possible.

Superconductors—Materials

Superconductors are materials that exhibit a complete disappearance of electrical resistance on lowering the temperature below a critical temperature (T_c). For all superconductors presently known, as is the critical temperatures are well below room temperature, and they are usually attained by cooling with liquefied gas (helium or nitrogen), either at or below atmospheric pressure. A superconducting material must also exhibit perfect diamagnetism, that is, complete exclusion of an applied magnetic field from the bulk of the superconductor. Superconductivity permits electric power generators and transmission lines to have capacities many times greater than recently possible. It also allows the development of levitated transit systems capable of high speeds and provides an economically feasible way of producing the large magnetic fields required for the confinement of ionized gases and controlled thermalnuclear fusion.

Superconductivity is observed in a broad range of materials. These include more than half of the metallic elements and a wide range of compounds and alloys. To date, however, the materials that have received the most attention are niobium–titanium superconductors (the most widely used superconductor), A15 compounds (in which class the important intermetallic Nb_3Sn lies), ternary molybdenum chalcogenides, and high-temperature ceramic superconductors. The chalcogenides and ceramics, however, are in the development stages.

Niobium–titanium superconductors are actually composite wires that consist of Nb–Ti for filaments (<10 μm in diameter) and that it is an oxygen-free, high-purity (99.99%) copper matrix. In commercially pure aluminum (alloy 1100, 99.0%) and copper–nickel alloys (typically in concentrations of 90:10 or 70:30), matrices have also been utilized. The filament alloy most widely used is Nb–46.5Ti. Binary Nb–Ti compositions in the range of 45%–50% Ti exhibit T_c values of 9.0–9.3 K. Composite conductors containing as few as one to as many as 25,000 filaments have been processed by advanced extrusion and wire-drawing techniques. The primary applications for Nb–Ti superconductors are magnets for use in magnetic

resonance imaging (MRI) devices used in hospitals and high-energy physics pulsed accelerator-magnet applications. An example of the latter application is the Superconducting Supercollider for studying the elementary particles of which all matter is composed and the forces through which matter interacts.

A15 superconductors are brittle superconductors and brittle intermetallic A_3B compounds with a body-centered cubic crystal structure. Of the 76 known A15 compounds, 46 are known to be superconducting. Because of its ease of fabrication, Nb_3Sn is also assembled into multifilamentary wires. Applications for Nb_3Sn-base superconductors include large commercial magnets, power generators and power transmission lines, and devices for magnetically confining high-energy plasma for thermonuclear fusion.

The ternary molybdenum chalcogenides represent a vast class of materials whose general formula is $M_xMo_6X_8$, where M is a cation and X a chalcogen (sulfur, selenium, or tellurium). Most of the research on these materials has been centered around $PbMo_6S_8$ and $SnMo_6S_8$, the former having a T_c of 14–15 K.

High-temperature superconductors (T_c values exceeding 90 K) are ceramic oxides in wire, tape, or thin-film form. The systems being studied include Y-Ba-Cu-O (most notably $YBa_2Cu_3O_7$), Bi-Sr-Ca-Cu-O, and Ti-Ba-Ca-Cu-O.

Type I superconductors lose this characteristic (perfect diamagnet in a weak magnetic field) at a specified field of strength; Type II loses it progressively over a range of field strength.

Supercooling

(1) Cooling of a substance below the temperature at which a change of state would ordinarily take place without such a change of state occurring, for example, the cooling of a liquid below its freezing point without freezing taking place; this results in a *metastable* state. (2) Cooling, without a phase transformation occurring, below a temperature at which transformation would occur under equilibrium conditions. The term is often used with particular reference to the rapid cooling of molten metal to below its normal freezing temperature or range. This promotes the formation of abnormally large numbers of the nuclei that start solidification and hence leads to a very fine grain size.

Supercritical Fluids

SCFs are fluids which, when compressed and heated above a critical pressure and temperature, have diffusion properties similar to those of gases and density similar to those of liquids and, thus, are efficient solvents. Critical parameters (pressure, temperature, and density, respectively) for **carbon dioxide** are 7.4 MPa, 31°C, and 471 kg/m^3; **water**, 22 MPa, 374°C, and 332 kg/m^3; **hydrogen**, 1.3 MPa, −240°C, and 33 kg/m^3; **ammonia**, 11.3 MPa, 132°C, and 235 kg/m^3; and **methyl alcohol**, or **methanol**, 8 MPa, 240°C, and 277 kg/m^3.

Current and potential applications for SCFs, especially carbon dioxide and water, stem largely from regulatory pressures on ecology and safety, and health trends. They were first used in the 1970s for decaffeinating coffee and tea, replacing trichloroethylene and methylene chloride, have replaced ethylene chloride in spice extraction, and have been recently introduced for metal cleaning, deasphalting, and spray painting without volatile organic compounds. Emerging or potential uses for carbon dioxide include ethanol purification: extraction of acetone from antibiotics, fat and cholesterol from egg yolks, and vitamin E from soybean oil and soil remediation. Carbon dioxide, being nonpolar, is effective in removing nonpolar contaminants from virtually any matrix and may prove economical for on-site removal of organic soil contaminants that do not easily volatilize.

Union Carbide's Unicarb process uses a compressed gas, usually carbon dioxide, in the supercritical state to replace most of the solvent in conventional and high-solids topcoatings as well as in primers to reduce volatile organic compounds by as much as 80% and increase transfer efficiency by reducing overspray. Concern about the known human carcinogen perchloroethylene in dry-cleaning processes may have driven some 10% of the 30,000 such establishments in North America to supercritical carbon dioxide systems by the year 2015.

Water has potential for treating various organic substances by supercritical oxidation. Organic liquids and gases mix with the water and are transformed into carbon dioxide and water while in organics dissolve only slightly, allowing them to concentrate and be recovered. Virtually all of the organics are destroyed. Potential abounds for waste treatment at chemical, pulp and paper, and weapons plants.

Water is also in contention for soil remediation, having proved effective in removing virtually all of the polycyclic aromatic hydrocarbons in soil. It is aimed at some 2000 sites in the United States and Canada where town gas has been made from coal.

Supercritical

Any action that takes a system or material past some critical point.

Supercritical Fluid

Water or other material that is pressurized to such an extent, that is, beyond the supercritical pressure, such that on heating the fluid does not boil in the conventional sense. It does not change from an identifiable liquid phase to an identifiable gas phase, there is no change in density, no latent heat is involved, and there is no physical interface such as exists normally between water and steam.

Supercritical Pressure

The pressure at which a substance becomes a supercritical fluid. For water, it is 3,208 lbs per sq in absolute.

Superficial Hardness Test

See *Rockwell superficial hardness test.*

Superfines

The portion of a metal powder that is composed of particles smaller than a specified size, usually 10 µm.

Superfinishing

An abrasive process utilizing either a curved bonded honing stick (stone) for a cylindrical workpiece or a cup wheel for flat and spherical work. A large contact area, 30% approximately, exists between workpiece and abrasive. The object of superfinishing is to remove surface fragmentation and to correct any inequalities in geometry, such as grinding feed marks and chatter marks. Also known as microhoning. See also *honing.*

Superheating

(1) Heating of a substance above the temperature at which a change of state would ordinarily take place without a change of state occurring, for example, the heating of a liquid above its boiling point without boiling taking place; this results in a metastable state. (2) Any increment of temperature above the melting point of a metal; sometimes construed to be any increment of temperature above normal casting temperatures introduced for the purpose of refining, alloying, or improving fluidity. (3) Heating, without a phase transformation occurring, above a temperature at which transformation could occur under equilibrium conditions. (4) Heating molten metal well above its melting point to improve fluidity, etc., during casting.

Superlattice

See *ordered structure.*

Superplastic Forming

This is a process for shaping superplastic materials, a unique class of crystalline materials that exhibit exceptionally high tensile ductility. Superplastic materials may be stretched in tension to elongation's typically in excess of 200% and more commonly in the range of 400%–2000%. There are rare reports of a higher tensile elongation's reaching as much as 8000%. The high ductility is obtained only for superplastic materials and requires both the temperature and rate of deformation (strain rate) to be within a limited range. The temperature and strain rate required depend on the specific material. A variety of forming processes can be used to shape these materials; most of the processes involve the use of gas pressure to induce the deformation under isothermal conditions at the suitable elevated temperature. The tools and dies used, as well as the superplastic material, are usually heated to the forming temperature. The forming capability and complexity of configurations producible by the processing methods of superplastic forming greatly exceed those possible with conventional sheet forming methods in which the materials typically exhibit 10%–50% tensile elongation.

Processes

Superplastic forming typically utilizes a gas pressure differential across the superplastic sheet to induce to superplastic deformation and cause forming. Two processes have been developed: blow forming and movable-tool forming.

Blow Forming

Where gas pressure alone is used, the process is term *blow forming.* Blow forming utilizes tooling heated to the superplastic temperature, and the gas pressure differential is usually applied according to a time-dependent schedule designed to maintain the average strain rate within the superplastic range. The tools and the superplastic sheet are heated to the same temperature, and the gas pressure is applied to cause a creep-like plastic stretching of the sheet that eventually contacts and takes the shape of the configuration die.

Movable-Tool Forming

For relatively deep shapes, forming methods involving movable tools combined with gas-pressure forming may permit greater thinning control and reduced forming times as compared with the blow-forming method. One method is essentially the same as the blow-forming process, except that the die may be moved during the forming process. Another method uses a more complex sequence. The bubbleplate holds the superplastic sheet in place and prevents gas breakage.

The plug-assisted forming method involves a movable die that is pushed into and stretches the superplastic sheet material, followed by the application of gas pressure on the same side of the sheet as the movable die. In snap-back forming, the sheet is first billowed by tree forming with gas pressure imposed on the movable-tool side of the sheet; the tool is then moved into the billowed sheet, and finally, the gas pressure is imposed on the opposite side of the sheet to form the superplastic material onto the tool.

Diffusion Bonding

Diffusion bonding, also known as diffusion welding, is sometimes used in conjunction with superplastic forming to produce parts of complexity not possible with a single-sheet forming process. Diffusion bonding is a solid-state joining process in which two or more materials are pressed together under sufficient pressure and at a sufficiently high temperature to result in joining. In diffusion bonding, there is usually little permanent deformation in the bulk of the parts being joined, although local deformation does occur at the interfaces on a microscopic scale. Because interfacial contamination, such as oxidation, will interfere with the bonding mechanisms, the process is usually conducted under an inert atmosphere, such as vacuum or inert gas.

Some superplastic materials are ideally suited for processing by diffusion bonding, because they deform easily at the superplastic temperature, and this temperature is consistent with that required for diffusion bonding. The most suitable alloys tend to have a high solubility for oxygen and nitrogen, so that these contaminants can be removed from the surface by diffusion into the base metal. For example, titanium alloys fall into this class and are readily diffusion bonded. Aluminum alloys form a very thin but tenacious oxide film and are therefore quite difficult to diffusion bond. For certain materials, and under conditions of proper processing, the diffusion-bond interfacial strength can be equal to that of the parent base material. It has been found that metals processed to fine grain size, as required for superplastic deformation, are the most suitable for diffusion bonding because they require lower bonding pressure than a coarse-grain metal of the same alloy composition.

Combined Methods

The processing conditions for superplastic forming and diffusion bonding are similar, both requiring an elevated temperature and benefiting from the fine grain size. Consequently, a combined process of superplastic forming with diffusion bonding has been developed that can produce parts of greater complexity than single-sheet forming alone. The combined process of superplastic forming and diffusion bonding can involve multiple-sheet forming after localized diffusion bonding, producing expanded structures and sandwich configurations of various types. It is also possible to form a sheet superplastically onto, and diffusion bonded to, a separate piece of material thereby producing structural configurations much more like forgings than sheet metal structures. There are other joining methods that have also been utilized as alternatives to diffusion bonding, such as spot welding, and have been combined with superplastic forming to produce complex structures.

Applications

There are a number of commercial applications of superplastic forming and combined superplastic forming and diffusion bonding, including aerospace, architectural, ground transportation, and numerous miscellaneous uses. Examples are wing axis panels in the Airbus A310 and A320, bathroom sinks in the Boeing 737, turbo-fan-engine-cooling duct components, external window frames in the space shuttle, front covers of slot machines, and architectural siding for buildings.

Superplastic Forming (SPF)

A strain rate sensitive sheet metal forming process that uses characteristics of materials exhibiting high tensile elongation. Superplastic forming methods include: blow molding, in which gas pressure is imposed on a superplastic diaphragm, causing the material to form into the die configuration; vacuum forming, a process similar to blow molding except that the forming pressure is limited to atmospheric pressure (100 kPa) versus the maximum pressure of 700–3400 kPa for blow molding; thermoforming methods adapted from plastics technology (see *thermoforming*), which involve a moving or adjustable die member in conjunction with gas pressure or vacuum; and superplastic forming/diffusion bonding (SPF/DB), which combines blow molding and solid-state bonding. See also *diffusion bonding* and *superplasticity*.

Superplasticity

(1) The ability of certain metals (the most notably aluminum- and titanium-base alloys) to develop extremely high tensile elongation's at elevated temperatures and under controlled rates of deformation. (2) The ability to accept very large amounts of deformation often greater than 2000% elongation, without cracking or necking. The phenomenon typically occurs in very fine grained material deformed at temperatures and strain rates such that progressive deformation is matched by continuous recrystallization and recovery at a rate sufficiently rapid to maintain the fine grain size. The temperature is typically about half the melting point on the absolute scale, and the strain rate is usually low compared with industrial manufacturing and testing practice. If the circumstances are correct than any neck which tends to form will become a site for an increase in strain rate with consequent increase in stress at work hardening. Deformation will therefore cease at this point and be transferred to material elsewhere.

Superpolymers

Many plastics developed in recent years can maintain their mechanical, electrical, and chemical resistance properties at temperatures over 213°C for extended periods of time. Among these materials are polyimide, polysulfone, polyphenylene sulfide, polyarylsulfone, novaloc epoxy, aromatic polyester, and polyamide-imide. In addition to high-temperature resistance, they have in common high strength and modulus of elasticity, and excellent resistance to solvents, oils, and corrosive environments. They are also high in cost. Their major disadvantage is processing difficulty. Molding temperatures and pressures are extremely high compared to conventional plastics. Some of them, including polyimide and aromatic polyester, are not molded conventionally. Because they do not melt, the molding process is more of a sintering operation. One indication of the high-temperature resistance of the superpolymers is their glass transition temperature of well over 260°C, as compared to less than 177°C for most conventional plastics. In the case of polyimides, the glass transition temperature is greater than 427°C and the material decomposes rather than softens when heated excessively. Aromatic polyester, a homopolymer also known as polyoxybenzoate, does not melt, but at 427°C can be made to flow in a nonviscous manner similar to metals. Thus, filled and unfilled forms and parts can be made by hot sintering, high-velocity forging, and plasma spraying.

Notable properties are high thermal stability, good strength at 316°C, high thermal conductivity, good wear resistance, and extra-high compressive strength. Aromatic polyesters have also been developed for injection and compression molding. They have long-term thermal stability and a strength of 20 MPa at 288°C. At room temperature, polyimide is the stiffest of the group with a top modulus of elasticity of 51,675 MPa, followed by polyphenylene sulfide with a modulus of 33,072 MPa. Polyarylsulfone has the best impact resistance of the superpolymers with impact strength of 0.27 kg·m/cm (notch).

Polyetherimide (PEI) is an amorphous thermoplastic that can be processed with conventional thermoplastic processing equipment. Its continuous-use temperature is 170°C, and its deflection temperature is 200°C at 2 MPa. The polymer also has inherent flame resistance without use of additives. This feature, along with its resistance to food stains and cleaning agents, makes it suitable for aircraft panels and seat component parts. Tensile strength ranges from 103 to 165 MPa. Flexural modulus at room temperature is 3300 MPa.

Polyimide (PI) foam is a spongy, lightweight, flame-resistant material that resists ignition up to 427°C and then only chars and decomposes. Some formulations result in harder materials that can be used as lightweight wallboard or floor panels while retaining fire resistance.

Aromatic polyketones are high-performance thermoplastics, which include polyetheretherketone (PEEK), glass transition temperature of 143°C and melting point of 335°C; polyetherketone (PEK), glass transition temperature of 154°C; polyetherketoneketone (PEKK), glass transition temperature of 154°C and melting point of 335°C; polyaryletherketone (PAEK), glass transition temperature of 170°C and melting point of 380°C. Glass fiber reinforcement improves the strength, stiffness, and dimensional stability of these materials. In addition, there are various ketone-based copolymers.

Supersaturated

A metastable solution in which the dissolved material exceeds the amount the solvent can hold and normal equilibrium at the temperature and other conditions that prevail. See *solution*.

Supersonic

Pertains to phenomenon in which the speed is higher than that of sound. Not synonymous with ultrasonic; see also *ultrasonic frequency*.

Supplemental Operation

See *secondary operation*.

Support Pins

Rods or pins of precise length used to support the overhang of irregularly shaped punches in metal forming presses.

Support Plate

A plate that supports a draw ring or draw plate in a sheet metal forming press. It also serves as a spacer. See also *draw plate* and *draw ring*.

Supporting Electrode

An electrode, other than a self-electrode, on which the sample is supported during spectro-chemical analysis.

Surface Alterations (Metals)

Irregularities or changes on the surface of a material due to machining or grinding operations. The types of surface alterations associated with metal removal practices include mechanical (e.g., plastic deformation, hardness variations, cracks, etc.), metallurgical (e.g., phase transformations, twinning, recrystallization, and untempered or overtempered martensite), chemical (e.g., intergranular attack, embrittlement, and pitting), thermal (heat-affected zone, recast, or redeposited metal, and resolidified material), and electrical surface alterations (conductivity change or resistive heating).

Surface Area

(1) The area, per unit weight of a granular or powdered solid, of all external and internal surfaces that are accessible to a penetrating gas or liquid. Surface area is given as square meters or kilogram (m²/kg) or square centimeters per gram (cm²/g). (2) The actual area of the surface of a casting or cavity. The surface area is always greater than the *projected area*.

Surface Checking

Same as *checks*.

Surface Contact Points

In powder metallurgy, the points at which abutting particles may contact during contacting and which grow into necks during sintering. See also *neck* and *neck formation*.

Surface Damage

In tribology, damage to a solid surface resulting from mechanical contact with another substance, surface, or surface moving relatively to it and involving the displacement or removal of material. In certain contexts, *wear* is a form of surface damage in which material is progressively removed. In another context, surface damage involves a deterioration of function of a solid surface even though there is no material loss from that surface. Surface damage may therefore precede wear.

Surface Diffusion

One of the primary diffusion mechanisms during sintering. It is predominant and for smaller particles and lower sintering temperatures as compared to other diffusion mechanisms, such as lattice or volume diffusion, which are prevalent for larger particles and higher temperatures. See also *volume diffusion*.

Surface Distress

In bearings and gears, damage to the contacting surfaces that occurs through intermittent solid contact involving some degree of sliding and/or surface fatigue. Surface distress can occur in numerous forms depending on the conditions under which the bearing or gear was operated and on the nature of the interaction between the contacting surfaces.

Surface Engineering

Matters concerning any aspect of surface condition or treatment encoding machining, peening, residual stress, polishing, plating, case hardening, diffusion treatment, conversion coatings, and key for painting.

Surface Effects

This term usually refers to the influence on fatigue properties of surface features such as polished, scratches, machining, pitting, local residual stresses, and fretting.

Surface Film

Any continuous contamination on the surface of a powder particle.

Surface Finish

(1) The geometric irregularities in the surface of a solid material. Measurement of surface finish shall not include inherent structural irregularities unless these are the characteristics being measured. (2) Condition of a surface as a result of a final treatment. (3) The form and quality of a surface. It may be defined in qualitative terms such as polished, or mill finished or, particularly in the case of machined surfaces, it may be measured quantitatively or semi-quantitatively. The semi-quantitative techniques compare the roughness of the surface in question with a series of standard machine surfaces, the comparison being made by eye and or by scratching with a fingernail. The quantitative techniques use instruments which measure the surface texture and profile by optical methods or by running a stylus over the surface and measuring the height and pitch of the undulations. The irregularities are rarely simple and the equipment integrates the various long and short range deviations to give a roughness figure in micro-meters, microns or micro-inches. See also *roughness*.

Surface Grinding

Producing a plain surface by grinding.

Surface Hardening

A generic term covering several processes applicable to a suitable ferrous alloy that produces, by quench hardening only, a surface layer that is harder or more wear resistant than the core. There is no significant alteration of the chemical composition of the surface layer. The processes commonly used are *carbonitriding, carburizing, induction hardening, flame hardening, nitriding,* and *nitrocarburizing.* See *case hardening.*

Surface Integrity (Metals)

A technology that involves the specification and manufacture of unimpaired or enhanced surfaces through the control of the many possible alterations produced in a surface layer during manufacture. Surface integrity is achieved by the proper selection and control of manufacturing processes and the ability to estimate their effects on the significant engineering properties of work materials. See also *surface alterations* (metals).

Surface Modification

The alteration of surface composition or structure by the use of energy or particle beams. Elements may be added to influence the surface characteristics of the substrate by the formation of alloys, metastable alloys or phases, or amorphous layers. Surface-modified layers are distinguished from conversion or coating layers by their greater similarity to metallurgical alloying versus chemically reacted, adhered, or physically bonded layers. However, surface structures are produced that differ significantly from those obtained by conventional metallurgical processes. This latter characteristic further distinguishes surface modification from other conventional processes, such as amalgamation or thermal diffusion. Two types of surface modification methods commonly employed are *ion implantation* and *laser surface processing.*

Surface Pigments (for Brick)

Brick has a long history of durability. Environmental issues such as energy consumption and waste disposal are increasing in importance, producing opportunities for brick usage to increase.

Facing brick need no rendering, painting, or regular maintenance. They look good year after year and often century after century with no energy or product input. The brick can be recycled again and again, and at the end of their lives can be used as aggregate. Brick are the material of the future, and as demand for brick increases, so will demand for choice. The range of choices can be expanded by using surface pigments.

Compositions

Surface pigments can be classified into several groups. Pigment mixtures, which on firing develop their intrinsic color, fall into the largest category of stains. The second class includes pigments that react with the brick surface sand or grog to produce a specific shade of color. Usually, the iron in the surface layers of the clay is the critical factor. There are also colors that have been stabilized by a prior treatment. These are normally used where it is necessary to have the maximum volume of brick one uniform color.

Ferrous and ferric oxides are the source of a large number of colors ranging from very bright yellow through reds and maroons to blacks. The tone of the red surface colors depends on the firing temperature, the compound from which the oxide was formed, and the kiln atmosphere.

The presence of zinc or titanium compounds modifies the iron color in pigments and produces a range of buffs and yellows.

The spectral range of brown pigments is extremely wide, ranging in shade from cream to dark farmhouse brown.

Black pigments are mixtures of compounds of iron and manganese. Occasionally, cobalt oxide is added to intensify the blue-black quality of some surface stains.

Normally, gray pigments are produced by diluting selected black stains with inert and reactive fillers. The dominant tone of any black pigment is apparent when it is diluted, and some satisfactory blacks show undesirable tones when used diluted as grays.

Dry and Wet Applications

Dry

There are several methods of applying pigments in dry foam. Surface pigments are trickle-fed through a sieve and then brushed or lightly rolled into the brick surface. Surface effects are applied by vibrating the granules onto the column and rolling them into the surface.

For surface pigments or frits added to aggregates, the aggregate (a sand, grog, or other suitable material) is mixed with the pigment or frit at a predetermined ratio.

Wet

Surface pigments can be added to water-based suspensions and applied as a spray, slurry, or engobe. For spraying, a ratio of one part by weight of pigment to four parts by weight of water is generally acceptable.

This thorough wetting of the finally powdered pigment ensures a smooth suspension.

Surface Preparation. The operations necessary to produce a desired or specified surface condition.

Surface Preparation (Adhesives). Physical and/or chemical preparation of an adherend to make it suitable for adhesive bonding.

Surface Rolling. See *rolling.*

Surface Roughness. Fine irregularities in the *surface texture* of a material, usually including those resulting from the inherent action of the production process. Surface roughness is usually reported as the arithmetic roughness average, R_a, and is given in micrometers or microinches. See also the term *roughness.*

Surface Tension. (1) The force acting on the surface of a liquid, tending to minimize the area of the surface. (2) The force existing in a liquid–vapor phase interface that tends to diminish the area of the interface. This force acts at each point on the interface in the plane tangent to that point. (3) The phenomenon by virtue of which a liquid appears to have a skin and form a meniscus or stand as globules on a non-wetted surface. It arises from the mutual attraction between molecules exposed at the surface and the consequent tendency of a surface to contract to the smallest possible area.

Surface Texture. The roughness, waviness, lay, and flaws associated with a surface as defined by the term *roughness*. See also *lay*.

Surface Treatment. In composites fabrication, a material (size or finish) applied to a fibrous material during the forming operation or in subsequent processes. For carbon fiber surface treatment, the process used to enhance bonding capability of fiber to resin. See also *size*.

Surface Void. A void which is located at the surface of a material and is a consequence of processing, that is, a surface reaction layer, as distinguished from a volume distributed flaw such as a *pore* or *inclusion*.

Surfacing. (1) The deposition of filler metal (material) on a base metal (substrate) to obtain desired properties or dimensions. (2) The deposition of material, particularly by welding or similar processes, to provide a surface coating having properties such as high hardness or corrosion resistance is superior to the substrate. Where the intention is merely to restore dimensions, the term *Building-Up* is often considered more appropriate. See also *buttering, cladding, coating,* and *hard facing*.

Surfacing Mat. A very thin mat, usually 180–510 µm (7–10 mils) thick, of highly filamentized fiberglass, used primarily to produce a smooth surface on a reinforced plastic laminate, or for precise machining or grinding.

Surfacing Weld. A type of weld composed of one or more stringer or weave beads deposited on an unbroken surface to obtain desired properties or dimensions. See also *stringer bead* and *weave bead*.

Surfactant. (1) A chemical substance characterized by a strong tendency to form adsorbed interfacial films when in solution, emulsion, or suspension, thus producing effects such as low surface tension, penetration, boundary lubrication, wetting, and dispersing. (2) A compound that affects interfacial tension between two liquids. It usually reduces surface tension. See also *Rehbinder effect*.

Suspension. A liquid containing a solid material in finely divided particulate form and evenly dispersed rather than settled as a deposit.

Sustained Load Failure. Failure in steels usually high tensile, at stress levels that would normally be acceptable and possibly after a considerable period of time. It results from excessive hydrogen content and is accompanied by negligible ductility. See *hydrogen damage*.

Swabbing. Wiping of the surface of a metallographic specimen with a cotton ball saturated with etchant to remove reaction products simultaneously.

Swage. (1) The operation of reducing or changing the cross-section area of stock by the fast impact of revolving dies. (2) The tapering of bar, rod, wire, or tubing by forging, hammering, or squeezing; reducing a section by progressively tapering lengthwise until the entire section attains the smaller dimension of the taper. See also the term *rotary swager*.

Swaging. (1) Reducing the diameter of bar, etc., by forging, hammering, or similar processes by hand or machine as in *rotary swaging*. (2) Tapering bar, rod, wire, or tubing by forging, hammering, or squeezing; reducing a section by progressively tapering lengthwise until the entire section attains the smaller dimension of the taper. See also *rotary swaging*.

Swarf. (1) The strands of chippings of scrap material cut from components during machining operations such as drilling and turning. Also the sludge formed by grinding and comprising metal and grit. (2) Intimate mixture of grinding chips and fine particles of abrasive and bond resulting from a grinding operation.

Sweat. (1) Exudation of a low-melting phase during solidification of metals. Also known as *sweat-back*. For tin bronzes, it is called tin sweat. (2) Low melting point material exuding from the surface of castings during the final stages of solidification. See *segregation*. (3) Moisture condensing on cold components heated with a gas flame. See also *exudation*.

Sweating. (1) Exudation of bearing material or lubricant due to high temperature. (2) A soldering technique in which two or more parts are precoated (tinned), then reheated and joined without adding more solder. Also called *sweat soldering*.

Sweating Out. Bringing small globules of one of the low-melting constituents of an alloy to the surface during heat treatment, such as lead out of bronze.

Sweat Soldering. See *sweating* (2).

Sweep. A type of foundry pattern that is a template cut to the profile of the desired mold shape that, when revolved around a stake or spindle, produces that shape in the mold.

Swedish Iron. An imprecise term usually implying high purity iron, particularly one made from high-grade ores smelted with charcoal and refined to minimize contamination.

Sweeps. Sweepings, dust, and other debris collected from the floors and other surfaces of premises handling precious metals. It is collected for extraction of the valuable metal.

Sweep Velocity. The mean of the surface velocities of two bodies at the area of contact. Occasionally, the sum of the velocities is quoted instead of the mean. In rolling, the sweep velocity is also called the rolling velocity.

Sweet Roast. Same as *dead roast*.

S.W.G. Standard Wire Gauge.

Swift Cup Test. A stimulative test for determining formability of sheet metal in which circular blanks of various diameters are clamped in a die ring and deep drawn into a cup by a flat-bottomed cylindrical punch. The ratio of the largest blank diameter that can be drawn successfully to the cup diameter is known as the *limiting drawing ratio (LDR)* or *deformation limit*.

Swing Forging Machine. Equipment for continuously hot reducing ingots, blooms, or billets to square flats, rounds, or rectangles by the crank-driven oscillating action of paired dies.

Swing Frame Grinder. A grinding machine suspended by a chain at the center point so that it may be turned and swung in any direction for grinding of billets, large castings, or other heavy work. Principal use is removing surface imperfections and roughness.

S.W.L. Safe Working Load.

Symmetrical Laminate. A *laminate* in which the stacking sequence of plies below its midplane is a mirror image of the stacking sequence above the midplane.

Synchronous Initiation. In resistance welding, the initiation and termination of each half-cycle of welding transformer primary current so that all half-cycles of such current are identical in making spot and seam welds or in making projection welds.

Synchronous Timing. See preferred term *synchronous initiation*.

Synchrotron. A device for accelerating charged particles by directing them along a roughly circular path in a magnetic guide field. As the particles pass through accelerating cavities placed along their orbits, their kinetic energy is increased repetitively, multiplying their initial energy by factors of hundreds or thousands. See also *synchrotron radiation*.

Synchrotron Radiation. Electromagnetic radiation emitted by charged particles in circular motion at relativistic energies.

Syndiotactic Stereoisomerism. A polymer molecule in which side atoms or side groups alternate regularly on opposite sides of the chain. See the term *isotactic stereoisomerism*.

Syneresis (of a Grease). See *bleeding* (2).

Syneresis. Spontaneous separation of a liquid from a gel due to contraction of the gel.

Syntactic Cellular Plastics. Reinforced plastics made by mixing hollow microspheres of glass, epoxy, phenolic, and so forth, into fluid resins (with additives and curing agents) to form a moldable, curable, lightweight, fluid mass; as opposed to foamed plastic, in which the cells are formed by gaseous bubbles released in the liquid plastic by either chemical or mechanical action. Also known as *syntactic foams*.

Syntectic. An isothermal reversible reaction in which a solid phase, on absorption of heat, is converted to two conjugate liquid phases.

Syntectic Equilibrium. A reversible univariant transformation in which a solid phase that is stable only at lower temperature decomposes into two conjugate liquid phases that remain stable at higher temperature.

Synthetic Cold Rolled Sheet. A hot rolled pickled sheet given a sufficient final temper pass to impart a surface approximating that of cold rolled steel.

Synthetic Natural Rubber (Isoprene)

Stereo regular polyisoprene and polybutadiene elastomers, high in *cis*-1,4 content, are of growing interest to the engineer, both because of engineering performance, and their competitive price. IR (*cis*-polyisoprene) has been called "synthetic natural rubber" because chemically and physically it is similar to *Hevea*.

General properties and examples of end-use performance show it to be a satisfactory supplement to natural rubber in a wide variety of products. Molecular weight can be controlled within quite wide limits and linearity can be maintained even with the longest chains. Higher-molecular-weight materials have been satisfactorily extended with oil to yield compositions with a desirable combination of low cost and attractive properties.

The development of IR latex is a noteworthy advance in latex technology. The low emulsifier level, stereoregularity of the polymer, large particle size, and low viscosity have not hitherto been available in a general-purpose synthetic latex. These properties, combined with high gum strength and elongation, offered advantages for many latex applications.

Vulcanizate Properties

IR can be processed in a manner similar to that used for natural rubber. Vulcanization can be carried out by means of curatives commonly used with natural or SBR. Properties of gum vulcanizates are quite similar to those of natural rubber, although IR has somewhat lower modulus and higher extensibility.

There are excellent hysteresis properties, low heat buildup and high resilience of both the IR and the polybutadiene tread vulcanizates, and IR is second only to natural rubber and tear strength.

Processing and compounding procedures for oil-extended IR are similar to those for an extended polymer, except that lower curative levels are recommended for maximum tensile strength and flat curing characteristics. Properties of extended vulcanizates approach those of the non-extended materials.

Applications

Similarity of performance between IR and natural rubber has permitted use of IR as a supplement for natural rubber in uses such as tire treads, carcasses, and white sidewalls.

In nontire uses the low ash content, light color, and good mold flow characteristics of IR are of particular advantage. Good electrical properties and low moisture absorption make it suitable for a number of electrical insulating uses. Parts molded in IR exhibit sharp definition and excellent color stability to light.

The low cost and good performance of oil-extended IR are promising for tire carcass compounds, molded mechanical goods, and footwear.

In latex form, IR is the first synthetic that possesses an average particle size as large and a particle size distribution as broad as that of natural rubber latex. It is highly promising for a number of coating and dipping applications, as well as foaming.

Synthetic Oil

Oil produced from chemical synthesis rather than from petroleum. Examples are esters, ethers, silicons, silanes, and halogenated hydrocarbons.

Systéme International d'Unités

The International System of Units adopted by the General Conference of Weights and Measures and endorsed by the International Organization for Standardization. *See S.I. System*.

T

T_g

See *glass transition temperature.*

T Joint

A joint between the end or age of one component and the face of another. Some usage of the term is confined to joints where the two components are at approximately right angles but other usage includes joints where the angle is much more acute and the weld is a fillet.

T.T.T. Curves

Time, temperature, transformation curves. See *isothermal transformation curves.*

Tabs

Extra lengths of reinforced plastic or other material at the ends of the tensile specimen to promote failure away from the grips. Also called doublers.

Tack

That property of an adhesive that enables it to form a bond of measurable strength immediately after adhesive and adherend are brought into contact under low pressure. See also *dry tack, tack range,* and *tacky dry.*

Tack Range

The period of time in which an adhesive will remain in the *tacky dry* condition after application to the adherend, under specified conditions of temperature and humidity.

Tack Welds

(1) Small, scattered welds made to hold parts of a weldment in proper alignment, while the final welds are being made. (2) Intermittent welds to secure weld backing bars. See also *backing* (2).

Tacking

Making *tack welds.*

Tacky Dry

The condition of an adhesive when the volatile constituents have evaporated or been absorbed sufficiently to leave it in a desired tacky state.

Taconite

A siliceous iron formation from which certain iron ores of the Lake Superior region are derived; consists chiefly of fine-grain silica mixed with magnetite and hematite.

Tafel Line, Tafel Slope, Tafel Diagram

When an electrode is polarized, it frequently will yield a current/potential relationship over a region that can be approximated by $\eta = \pm \beta \log(i/i_0)$, where η is the change in open-circuit potential, i is the current density, and β and i_0 are constants. The constant β is also known as the Tafel slope. If this behavior is observed, a plot on semilogarithmic coordinates is known as the Tafel line and the overall diagram is termed a Tafel diagram.

Tagging

Reducing the diameter of the end of a bar to allow its insertion through a drawing die or into a swaging machine.

Tailings

The material left after a low-grade ore has been concentrated. It will contain some of the desired metal but not in economically recoverable quantities.

Talc

Talc is a hydrous magnesium silicate with the composition 63.4% SiO_2, 31.9% MgO, and 4.7% H_2O when found in pure form. It is an extremely soft mineral with a Mohs hardness of 1.

Talc is a soft friable mineral of fine colloidal particles with a soapy feel. It has a composition of $4SiO_2 \cdot 3MgO \cdot H_2O$ and a specific gravity of 2.8. It is white whelf pure, but may be colored gray, green, brown, or red with impurities.

Talc is now used for cosmetics, for paper coatings, as a filler for paints and plastics, and for molding into electrical insulators, heater parts, and chemical ware. The massive block material, called steatite talc, is cut into electrical insulators. It is also called lava talc. The more impure block talcs are used for firebox linings and will withstand temperatures to 927°C. Gritty varieties contain carbonate minerals and are in the class of soapstones. Varieties containing lime are used for making porcelain.

Talc used in ceramics is usually mined, sorted, crushed, and milled to 95%–99%—200 mesh.

Talc is also termed soapstone or steatite and has a characteristic soapy or greasy feel. Used as an ingredient in ceramic whiteware and ceramic wall and floor tiles.

Cordierite is a talc-like mineral with a high percentage of magnesia used for refractory electronic parts. Extruded cordierite serves as a substrate for the active catalyst metals in auto catalytic converters.

Magnesium silicate, used as a filler in rubber and plastics and also as an alkaline bleaching agent for oils, waxes, and solvents, is a white, water-insoluble powder of composition $MgSiO_4$, having a pH of 7.5–8.5. In the cosmetic trade, it is known as talcum powder.

Applications

The major applications for talc are tile and hobbyware bodies, cordierite catalyst supports, kiln furniture, and electrical porcelains. There are minor applications in electronic packaging, sanitary ware, dinnerware, and glazes.

Talc is used as a flux for high alumina ceramics, sanitary ware, and dinnerware. It is a low-cost source of magnesium in these applications and helps to produce less porous bodies at lower firing temperatures. It is readily carved for ornamental applications. In powdered form, it is termed French chalk or, if perfumed, talcum powder.

Tall Oil

An orderly, resinous liquid obtained as a by-product of the sulfite paper-pulp mills. The alkali saponifies the acids, and the resulting soap is skimmed off and treated with sulfuric acid to produce tall oil. The name comes from the Swedish *talloel*, meaning pine oil. The crude oil is brown, and the refined oil is reddish yellow and nearly odorless. It has a specific gravity of 0.98. The oil from Florida paper mills contains 41%–45% rosin, 10%–15% pitch, and the balance chiefly fatty acids. The fatty acids can be obtained separately by fractionating the crude whole oil. The oil also contains up to 10% of the phytosterol *sitosterol*, used in making the drug cortisone.

Tall oil is used in scouring soaps, asphalt emulsions, cutting oils, insecticides, animal dips, in making factice, and in plastics and paint oils. It is marketed in processed and concentrated form.

Tallow

A general name for the heavy fats is obtained from all parts of the bodies of sheep and cattle. The best grades of internal fats, or *suet*, are used for edible purposes, but the external fats are employed for lubricants, for mixing with waxes and vegetable fats, for soaps and candles, and for producing chemicals. The tallows have the same general composition as lard but are higher in the harder saturated acids, with about 51% of palmitic and stearic acids, and lower in oleic acid. The edible grades known as *premier jus*, prime, and edible are white to pale yellow, almost tasteless, and free from disagreeable odor, but the nonedible or industrial tallows are yellow to brown unless bleached. The best grade of industrial tallow is Packers No. 1. *Peacock* is in acidless grade for metal working, lubricants and additives, soaps, mold release, animal feed supplements, inks, and pigments; other premium and custom grades of Peacock are also produced by George Pfau's Co.

Talysurf

See *surface finish*.

Tamp

To form or compact a ceramic mixture or a particulate solid by tamping or repeated impact, usually performed manually.

Tan Delta (Tan δ)

(1) The ratio of the loss modulus to the *storage modulus*, measured in compression, K; tension or flexure, E; or shear, G. See also *complex modulus* and *loss modulus*. (2) The ratio of the out-of-phase components of the dielectric constant (i.e., the loss) to the in-phase component of the dielectric constant (i.e., the permittivity). See also *loss factor*.

Tandem Die

Same as *follow die*.

Tandem Mill

A rolling mill consisting of two or more stands arranged so that the metal being processed travels in a straight line from stand to stand. In continuous rolling, the various stands are synchronized so that the strip can be rolled in all stands simultaneously. The rotational speeds of the individual stands are correlated to maintain a steady movement of the product without interstage kinking or excessive tension. In contrast with *single-stand mill*. See also *rolling mills*.

Tandem Seal

A multiple-seal arrangement consisting of two seals mounted one after the other, with the faces of the seal heads oriented in the same direction.

Tandem Welding

Arc welding in which two or more electrodes are in a plane parallel to the line of travel.

Tangent Bending

The forming of one or more identical bands having parallel axes by wiping sheet metal around one or more radius dies in a single operation. The sheet, which may have side flanges, is clamped against the radius die and then made to conform to the radius die by pressure from a rocker-plate die that moves along the periphery of the radius die. See also *wiper forming (wiping)*.

Tangent Modulus

(1) The elastic modulus measured over a small change in stress. It is relevant for those cases where stress is not proportional to strain so the stress–strain line in the elastic range is a curve rather than a straight line. As the term indicates, it is measured as the tangent to the line at the stress of interest. (2) The slope of the *stress–strain curve* at any specified stress or strain. See also *modulus of elasticity* and see *tensile test*.

Tangential Stress

See *shear stress*.

Tank Voltage

The total voltage between the anode and cathode of a plating bath or electrolytic cell during electrolysis. It is equal to the sum of (a) the equilibrium reaction potential, (b) the *IR* drop, and (c) the electrode potentials.

Tanning Agents

Materials, known as *tannins*, used for the treatment of skins and hides to preserve the hide substance and make it resistant to decay. The tanned leather is then treated with fats or greases to make it soft and pliable. Tannins may be natural or artificial. The natural tannins are chiefly vegetable, but some mineral tanning agents are used. The vegetable tannins are divided into two color classes: the *catechol* and the *pyrogallol*. The catechol tannins are cutch, quebracho, hemlock, larch, gambier, oak, and willow. The pyrogallol tannins are gallnuts, sumac, myrobalans, chestnut, valonia, divi-divi, and algarobilla.

In the ink industry, the catechol tannins are known as *iron greening*, and the pyrogallol tannins as *iron bluing*, and the latter are used for making writing inks. Catechol is also produced synthetically from coal tar. It is a water-soluble *dihydric phenol* and is used in some inks and for making dyestuffs, medicinals, and antioxidants.

Alum tanning is an ancient process but was introduced in Europe only about the year 1100, and the alum- and salt-tanned leather was called *Hungary leather*. *Formaldehyde* is also used as a tanning agent. Formaldehyde was patented as a tanning agent in 1898. A later patent covered a rapid process of tanning sheepskins with alcohol and formalin and then neutralizing in a solution of soda ash. Unlike vegetable agents, formaldehyde does not add weight to the skin. It is often used as a pretanning agent to lessen the astringency of the vegetable tannins and increase its rate of diffusion. *Melamine* resins are used for tanning to give a leather that is white throughout and does not yellow with age. Leather may also be tanned with chromic acid or chrome salts, which make the fibers insoluble and produce a soft, strong leather. Chrome alum, sodium or potassium dichromates, or products in which chromic acid has been used as an oxidizing agent may be used. *Chrome tanning* is rapid and is used chiefly for light leathers.

Tantalum and Alloys

Tantalum is a white lustrous metal (symbol Ta), resembling platinum. Tantalum is a high-density, ductile, refractory metal that exhibits exceptional corrosion resistance and good high-temperature strength over 1663°C. The annealed wrought metal in its pure form is easily worked and can be cold worked in much the same manner as fully annealed mild steel.

It is one of the most acid-resistant metals and is classed as a noble metal. Its specific gravity is 16.6, or about twice that of steel and, because of its high melting temperature (2996°C), it is called a refractory metal. In sheet form, it has a tensile yield strength of 345 MPa and is quite ductile. At very high temperatures, however, it absorbs oxygen, hydrogen, and nitrogen and becomes brittle. Its principal use is for electrolytic capacitors, but because of its resistance to many acids, including hydrochloric, nitric, and sulfuric, it is also widely used for chemical-processing equipment. It is attacked, however, by hydrofluoric acid, halogen gases at elevated temperatures, fuming sulfuric acid, and strong alkalies. Because of its heat resistance, tantalum is also used for heat shields, heating elements, vacuum-furnace parts, and special aerospace and nuclear applications. It is a common alloying element in superalloys. The metal is also used for prosthetic applications.

Tantalum metal is used in the manufacture of capacitors for electronic equipment, including citizen band radios, smoke detectors, heart pacemakers, and automobiles. An extremely stable film of tantalum oxide acts as an insulator in the capacitor. It is also used for heat-transfer surfaces in chemical production equipment, especially where extraordinarily corrosive conditions exist. Its chemical inertness has led to dental and surgical applications.

Tantalum forms alloys with a large number of metals. Of special importance is ferrotantalum, which is added to austenitic steels to reduce intergranular corrosion. Tantalum is used extensively in the chemical industry where its excellent fabrication and joining properties permit the application to acid-resistant heat exchangers, condensers, ductwork, chemical lines, and other chemical process equipment. Tantalum also finds use in the medical profession. Because of its nontoxic properties and immunity to body chemicals, tantalum is used for sutures, gauze, pins, and plates.

Fabricability

Hot Working

High-purity tantalum sintered bar, cast ingot, and annealed wrought forms can be worked at room temperature, although the working of large ingots and billets is sometimes performed at elevated temperature to permit working within equipment strength capacity. Cast ingots, protected by canning or coating materials, have been forged and extruded at temperatures up to 1316°C. Cold-work tantalum can be stress relieved or annealed at a variety of time–temperature schedules depending upon stress level and chemical purity of the material. A temperature of at least 1204°C is generally used for full annealing, while temperatures between 816°C and 927°C can be used for stress relieving. Annealing atmosphere must be high-purity argon, helium, or, preferably, a vacuum of 1/10 of a micron or less.

Cold Working

The excellent room-temperature ductility of stress-relieved and fully annealed tantalum makes the forming of tantalum comparatively simple. But the grain size of the material must be carefully considered for requirements where surface finish is of importance. The combination of a higher tensile strength and fair uniform elongation of stress-relieved tantalum sometimes makes it more satisfactory for drawing and forming than fully annealed material. It is very important to consider the temper properties of the material in the design of forming tools.

Joining

Tantalum can be joined by electron beam, tungsten-inert gas, and spot and resistance welding, but must be carefully protected from the effects of oxidation during welding. Uncontaminated welds are ductile and usually can be worked at room temperature.

Electron-beam-melted 90% tantalum–10% tungsten alloy can be formed and joined in a similar manner as pure tantalum provided that the higher strength and more rapid work-hardening characteristics of the alloy are considered.

Alloys

Tantalum alloys, including tungsten and tungsten–hafnium compositions, such as Ta-10W, T-111 (8% tungsten, 2% hafnium), and T-222 (9.6% tungsten, 2.4% hafnium, and 0.01% carbon), are used for rocket-engine parts and special aerospace applications. The tensile yield strength of Ta-10W is about 1089 MPa at room temperature and 621 MPa and 871°C.

Tantalum Beryllides and Carbides

Tantalum Beryllide

$TaBe_{12}$ and $TaBe_{17}$ are intermetallic compounds with good strength at elevated temperatures. $TaBe_{12}$ is tetragonal; density is 4.18 g/cm^3; melting point is 1849°C; and coefficient of thermal expansion is $8.42 \times 10^{-6}/°C$. $TaBe_{17}$ is hexagonal; density is 5.05 g/cm³; melting point is 2045°C; and coefficient of thermal expansion is $8.72 \times 10^{-6}/°C$. Both compounds can be formed by all of the known ceramic-forming methods plus flame and plasma-arc spraying. The materials are subject to safety requirements for all beryllium compounds.

Tantalum Carbide

TaC and Ta_2C are the two primary carbides. The ore tantalum carbide with a congruent melting point is TaC. It is dark to light brown in color with a metallic luster. Ta_2C melts in congruently and is gray with a metallic luster. Ta_2C melts at 3400°C; the melting point of TaC has been reported to be as high as 4820°C.

TaC burns in air with a bright flash and is only slightly soluble in acids. The tensile strength at room temperature is 13.6–27.2 MPa.

Tantalum carbide is used in cemented carbide-cutting tools.

Tap

A cylindrical or conical thread-cutting tool with one or more cutting elements having threads of a desired form on the periphery. By a combination of rotary and axial motions, the leading end cuts an internal thread, the tool deriving its principal support from the thread being produced. The tool is used for cutting threads on the surface of a whole.

Tap Density

The apparent density of a powder, obtained when the volume receptacle is tapped or vibrated during loading under specified conditions.

Tape

In composites manufacture, a unidirectional *prepreg* that consists of a thin sheet of fiber reinforced uncured resin usually wound on a cardboard core.

Tape Casting

Tape casting is a familiar technique to most ceramic engineers. It has been widely used as a method for fabricating improved capacitors. Conceptually, the process is simple. First, a slurry is prepared that contain ceramic powder (or powders) suspended in a solution of polymers. The slurry is read into a relatively thin liquid coating on a smooth surface. The solution is then removed, usually through evaporation, although absorption into a porous medium can also be used, and a dry film is formed that is a green ceramic powder compact, termed *green sheet*.

The tape usually has three qualitatively distinct phases: an inorganic powder, a continuous polymer–matrix, and a porosity phase. (The powder is dispersed within the polymer–matrix.) Each phase is important. In considering each of these phases, it is useful to recognize that the process to produce green sheet is usually best regarded as a primary process, and one or more secondary operations are necessary to complete the fabrication of a green part ready for firing.

Applications

The "classic" applications of tape casting are centered in the electronics industry and include the production of flat, smooth substrates for either thick film or integrated circuitry, capacitors, dielectrics, and piezoelectric elements.

Layered Manufacturing

Although the ability to mix materials and stack sheets has long been recognized in the electronics industry, a recent and important development has been the recognition that green sheet can play an important role as a feedstock for rapid prototyping (RP), also known as solid freeform fabrication. Although much that is useful can be derived from the work done for electronic applications, there are a number of distinct requirements for RP that produce additional constraints on green-sheet physical properties and, therefore, formulations.

The field of RP was developed in response to the high cost of tooling that is often an obstacle to implementing design changes or material substitution. All RP processes start with a computer-aided design, computationally approximated by a series of closely spaced 2D outlines, then use a process to construct each outline sequentially, assemble them, and suitably postprocess the assemblage to yield the final part.

Frequently, the term "layered manufacturing" is also applied to these processes. This name is particularly appropriate since each process involves the fabrication of a 3D object by automatic sequential stacking of appropriately contoured thin (pseudo-2D) sections. By controlling $x–y$ motions in each layer, an (in principle) arbitrary component can be built up.

Each RP technology is distinguished by the process and machinery used to define each thin section and create the final stack.

RP and Ceramics

The application of solid freeform fabrication to engineering ceramics is motivated by a desire to take advantage of the striking advances in these materials over the last 20 years (such as transformation-toughened oxides, high-toughness silicon nitrides, and ceramic-matrix composites) as a result of government-funded research programs throughout the world.

CAM-LEM

The computer-aided manufacturing of laminated engineering materials (CAM-LEM) process was developed specifically for production of ceramic parts.

Among the ceramic formulations that have been used with CAM-LEM are alumina, silicon nitride, zirconia-toughened oxides, and PZT.

The essence of the CAM-LEM process includes feeding the green sheet onto a movable platform, creating a cut outline through relative motion of the table and laser, automatic and selective extraction of the cut outline, and addition to the build start. Conversion to a dense ceramic article requires lamination of the green sheets followed by binder burnout and firing.

The total time to produce a ceramic part is the sum of that required to produce the green part plus the firing time, which includes binder burnout and sintering. The CAM-LEM process produces green parts that are compatible with conventional firing cycles for laminated tape-cast parts.

Tape Replica Method (Faxfilm)

A method of producing a *replica* by pressing the softened surface of tape or plastic sheet material onto the surface to be replicated.

Tape Wrapped

In composites fabrication, wrapping of heated fabric tape onto a rotating mandrel, which is subsequently cooled to form the surface for the next tape player application.

Taper Section

A section cut obliquely (acute angle) through a surface and prepared metallographically. The angle is often chosen to increase the vertical magnification of surface features by a factor of 5 or 10. Taper sectioning is usually carried out for microstructural examination of coated metal specimens.

Tapered Land Bearing

A *thrust bearing* containing pads of fixed taper.

Tapered Roller Bearing

A *rolling-element bearing* containing tapered rollers.

Tapping

(1) Producing internal threads with a cylindrical cutting tool having two or more peripheral cutting elements shaped to cut threads of the desired size and form. By a combination of rotary and axial motion, the leading end of the tap cuts the thread, while the tap is supported mainly by the thread it produces. See *tap*. (2) Opening the outlet of a melting furnace to remove molten metal. (3) Removing molten metal from a furnace.

Tar

A black, solid mass obtained in the destructive distillation of coal, peat, wood, petroleum, or other organic material. When coal is heated to redness in an enclosed oven, it yields volatile products and the residue coke. Upon cooling, the volatile matter, tar, and water are deposited, leaving the coal gases free. Various types of coal yield tars of different qualities and quantities, depending on the relative levels of more than 200 compounds, including *guaiacol, cresol, xylenol, crotonic acid, maltol,* and *ketones*. Anthracite gives little tar, and cannel coal yields large quantities of low-gravity tar. In the manufacture of gas, the tar produced from bituminous coal is a viscous, black liquid containing 20%–30% free carbon and is rich in benzene, toluene, naphthalene, and other aromatic compounds. In the dry state, this tar has a specific gravity of about 1.20. Tar is also produced as a by-product from coke ovens.

Coal tars are usually distilled to remove the light aromatics, which are used for making chemicals, and the residue tar, known as *treated tar*, or *pitch*, is employed for roofing, road making, and bituminous paints and waterproofing compounds. *Coal-tar pitch* is the most stable bituminous material for covering underground pipes, as a binder for electrodes used in aluminum smelting, and as an impregnant in refractories. *Coal-tar carbon* amounts to about 32% of the original tar. It is marketed in lump form for chemical use. The fixed carbon content is 92.5%–95.6%, sulfur about 0.30, and volatile matter 3–6. *Calcined carbon*, from coal tar, contains less than 0.5% sulfur and 0.5% volatile matter.

The lightest distillate of coal tar, benzene, is used as an automotive fuel. *Coal-tar oils* are used as solvents and plasticizers.

They consist of various distillates or fractions up to semisolids. *Tar oil* from brown coal tars was used for diesel fuel oil by extracting the phenols with methyl alcohol.

Pine tar is a by-product in the distillation of pinewood. It is a viscous, black mass and is much used for roofing. *Rosintene* is a light grade and is also sometimes called pitch, but pitch is the tar with the pine-tar oil removed, known as *pine pitch*. Navy pitch and ship pitch are names that refer to specification pine pitch for marine use. It is medium hard to solid; has a specific gravity of 1.08–1.10, as a melting point not less than 64°C; is completely soluble in benzol; and has uniform black color or red–brown in thin layers. *Wood tar* from the destructive distillation of other woods is a dark-brown, viscous liquid used as a preservative, deriving this property from its content of creosote.

Target

That part of an x-ray tube in an x-ray spectrometer, which the electrons strike and from which x-rays are emitted.

Tarnish

Surface discoloration of a metal caused by formation of a thin film of corrosion product.

Taylor Process

A process for making extremely fine metal wire by inserting a piece of larger-diameter wire into a glass tube and stretching the two together at high temperature. The wire may be coated when inserted into the glass tube or by being drawn through a small bath of molten glass.

Taylor Vortices

In a *journal bearing*, vortices formed in a liquid occupying the annular space between two concentric cylinders.

TD Nickel

Nickel that is dispersion strengthened by thoria (thorium oxide).

TEM

Transmission electron microscope. See *electron microscope*.

Teapot Ladle

A ladle in which, by means of an external spout, metal is removed from the bottom rather than the top of the ladle. See *ladle*.

Technetium

A metallic element having little commercial application.

Technical Ceramics

Same as *advanced ceramics*.

Technical Glass

A term that usually refers to glasses designed with some specific property essential for a mechanical, industrial, or scientific device.

Tee Joint

A joint in which the members are oriented in the form of a T.

Teeming

Pouring molten metal from a ladle into ingot molds. The term applies particularly to the specific operation of pouring either iron or steel into ingot molds.

Teeming Line/Height

The level in an ingot mold to which the molten metal was initially poured.

Teeth

The resultant surface irregularities or projections formed by the breaking of filaments or strings, which may form when adhesive-bonded substrates are separated. See also *legging*, *stringiness*, and *webbing*.

Telegraphing

In a laminate or other type of composite construction, a condition in which irregularities, imperfections, or patterns of an inner layer are visibly transmitted to the surface. Telegraphing is occasionally referred to as photographing.

Tellurium

An elementary metal (symbol Te) tellurium is obtained as a steel-gray powder of 99% purity by the reduction of tellurium oxide, or tellurite, TeO_2. The specific gravity is about 6.2 and the melting point is 450°C. The chief uses are in lead to harden and toughen the metal and in rubber as an accelerator and toughener. Less than 0.1% tellurium in lead makes the metal more resistant to corrosion and acids and gives a finer grain structure and higher endurance limit. Tellurium–lead pipe, with less than 0.1% tellurium, has a 75% greater resistance to hydraulic pressure than plain lead. Tellurium copper (C14500, C14510, and C14520) is a free-machining copper containing 0.3%–0.7% tellurium. It machines 25% more easily than free-cutting brass. The tensile strength, annealed, is 206 MPa, and the electric conductivity is 98% that of copper. A tellurium bronze containing 1% tellurium and 1.5% tin has a tensile strength, annealed, of 275 MPa and is free-machining. Tellurium is used in small amounts in some steels to make them free-machining without making the steel hot short as do increased amounts of sulfur. But tellurium is objectionable for this purpose because inhalation of dust or fumes by workers causes garlic breath for days after exposure, although the material is not toxic. As a secondary vulcanizing agent with sulfur in rubber, tellurium in very small proportions, 0.5%–1%, increases the tensile strength and aging qualities of the rubber. It is not as strong an accelerator as selenium, but it gives greater heat resistance to the rubber.

Tellurium is an important component of many thermoelectric devices, and such devices can be used for both power generation and cooling. The requirements of a good thermoelectric element are high thermoelectric power, low thermal conductivity, and low electrical resistivity. Lead telluride (PbTe), bismuth telluride (Bi_2Te_3), and silver antimony telluride meet these requirements better than any other currently known materials. By the addition of various other elements these compounds can be made either *p*-type or *n*-type semiconductors.

Oxides

The oxides of tellurium are tellurium monoxide, TeO; tellurium dioxide, TeO_2; and tellurium trioxide, TeO_3. The monoxide is reported as a black, amorphous powder that is stable in dry air in the cold but that is oxidized in moist air to the dioxide. On being heated in vacuum, it apparently disproportionates into the dioxide and elemental tellurium. It can be formed by heating the mixed oxide $TeSO_3$. The dioxide is the most stable oxide and is formed when tellurium is burned in air or oxygen or by oxidation of tellurium with cold nitric acid. It has two crystalline forms.

Telomer

A polymer composed of molecules having terminal groups incapable of reacting with additional monomers, under the conditions of the synthesis, to form larger polymer molecules of the same chemical type. See also *monomer* and *polymer*.

Temper

(1) In steels it refers to
  (1a) Reheating steel, previously hardened by heat treatment, and the range of 100°C–700°C. This alters the microstructure, reduces hardness and strength but increases ductility and toughness. This is the most widely accepted meaning among metallurgists.
  (1b) Carbon content.

Temper

(2) In many metals, but not usually steels, the various strength ranges are referred to as tempers. For example, aluminum has temper designations such as T4 or T6 with the higher number having higher strength. Also terms such as half hard, quarter hard, and full temper are termed tempers or temper grades.

Temper (Ceramics)

To moisten and mix clay, plaster, or mortar to proper consistency.

Temper (Glass)

(1) The degree of residual stress in annealed glass measured polarimetrically or by polariscopic comparison with a reference standard. (2) Term sometimes employed in referring to *tempered glass*.

Temper (Metals)

(1) In heat treatment, reheating hardened steel or hardened cast iron to some temperature below the eutectoid temperature for the purpose of decreasing hardness and increasing toughness. The process also is sometimes applied to normalized steel. (2) In tool steels, temper is sometimes used, but inadvisably, to denote the carbon content.

(3) In nonferrous alloys and in some ferrous alloys (steels that cannot be hardened by heat treatment), the hardness and strength produced by mechanical or thermal treatment, or both, and characterized by a certain structure, mechanical properties, or reduction in area during cold working. (4) To moisten *green sand* for casting molds with water.

Temper Brittleness

Reduced impact strength and low-fracture toughness of some martensitic or bainitic steels after heaving into or cooling through the 300°C–600°C range. Fracture is intergranular and probably arises as a result of segregation of impurity elements such as arsenic, antimony, phosphorus, and tin to the prior austenite grain boundaries. See *temper embrittlement*.

Temper Carbon

Clusters of finally divided graphite, such as that found in malleable iron, that are formed as a result of the composition of cementite, for example, by heating white cast iron above the ferrite–austenite transformation temperature and holding at these temperatures for a considerable period of time. Also known as annealing carbon. See also *malleable iron*.

Temper Color

A thin, tightly adhering oxide skin (only a few molecules thick) that forms when steel is tempered at a low temperature, or for a short time, in air or a mildly oxidizing atmosphere. The color, which ranges from straw to blue depending on the thickness of the oxide skin, varies with both tempering time and temperature and provides a guide to the metal temperature.

Temper Embrittlement

Embrittlement of low-alloy steels caused by holding within or cooling slowly through a temperature range (generally 300°C–600°C), just below the transformation range. Embrittlement is the result of the segregation at grain boundaries of impurities such as arsenic, antimony, phosphorus, and tin; it is usually manifested as an upward shift in ductile-to-brittle transition temperature. Temper embrittlement can be reversed by retempering above the critical temperature range, then cooling rapidly. Compare with *tempered martensite embrittlement*.

Temper Hardening

(1) Artificial aging as a part of a precipitation hardening process. (2) Secondary hardening.

Temper Rolling

Light cold rolling of sheet steel to improve flatness, to minimize the formation of *stretcher strains*, and to obtain a specified hardness or temper.

Temper Time

In resistance welding, that part of the postweld interval during which the current is suitable for tempering or heat treatment.

Temperature Coefficient of Resistance

The amount of resistance change of a material per degree of temperature rise.

Temperature Stress

The maximum stress that can be applied to a material at a given temperature without physical deformation.

Tempered Glass

Glass that has been subjected to a thermal treatment characterized by rapid cooling to produce a compressively stressed surface layer. See *safety glass*.

Tempered Layer

A surface or subsurface layer and a steel specimen that has been tempered by heating during some stage of the metallographic preparation sequence (usually grinding). When observed in a section after etching, the layer appears darker than the base material.

Tempered Martensite

The decomposition products that result from heating martensite before the ferrite–austenite transformation temperature. Under the optical microscope, darkening of the martensite needles is observed in the initial stages of tempering. Prolonged tempering at high temperatures produces spheroidized carbides in a matrix of ferrite. At the higher resolution of the electron microscope, the initial stage of tempering is observed to result in a structure containing a precipitate of fine iron carbide particles. At approximately 260°C, a transition occurs to a structure of larger and elongated cementite particles in a ferrite matrix. With further tempering at higher temperatures, the cementite particles become spheroidal, decreased in number, and increased in size.

Tempered Martensite Embrittlement

Embrittlement of high-strength alloy steels caused by tempering in the temperature range of 205°C–370°C; also called 350°C embrittlement. Tempered martensite embrittlement is thought to result from the combined effects of cementite precipitation on prior-austenite grain boundaries or interlath boundaries and the segregation of impurities at prior-austenite grain boundaries. It differs from *temper embrittlement* in the strength of the material and the temperature exposure range. In temper embrittlement, the steel is usually tempered at a relatively high temperature, producing lower strength and hardness, and embrittlement occurs upon slow cooling after tempering and during service at temperatures within the embrittlement range. In tempered martensite embrittlement, the steel is tempered within the embrittlement range, and service exposure is usually at room temperature.

Tempering

See *temper 1a and 1b*.

Tempering (Glass)

The process of rapidly cooling glass from near its softening point to induce compressive stresses on the surface balanced by interior tension, thereby imparting increased strength.

Tempering (Metals)

In heat treatment, reheating hardened steel to some temperature below the eutectoid temperature to decrease hardness and/or increase toughness.

Tempilstiks

A proprietary brand of crayon available in a wide range of melting points. The component is marked with one or more crayons and observed during, or examined after processing or testing, to determine the temperature reached.

Template (Templet)

(1) A guide or a pattern in manufacturing items. (2) A gauge or pattern made in a die department, usually from sheet steel; used to check dimensions on forgings and as an aid in sinking *die impressions* in order to correct dimensions. (3) A pattern used as a guide for cutting and laying plies of a *laminate*.

Temporary Weld

A weld made to attach a piece or pieces to a weldment for temporary use and handling, shipping, or working on the weldment.

Tenacity

The term generally used in yarn manufacture and textile engineering to denote the strength of a yarn or of a filament of a given size. Numerically, it is the grams of breaking force per denier unit of yarn or filament size. Grams per denier is expressed as gpd. See also *denier* and *tensile strength*.

Tensile Modulus

Stress is defined as the force per unit area and is calculated by dividing the applied load by the original cross section to the test piece. The SI unit of stress is the pascal (P), or more practical, megapascal (MPa); typical imperial units are tons per square inch (t/in.²) or, particularly in the United States, pounds per square inch (psi) or thousands of pounds per square inch (tpsi). Non-SI-metricated units include kilograms per square millimeter (kg/mm²) and newtons per square millimeter (N/mm²). The extension is the total amount by which the gauge length of the test piece is stretched. Strain is defined as the extension per unit length and is calculated by dividing the extension by the original gauge length. It is usually expressed as a percentage although terms such as inches per inch may be used.

For most tensile tests, it is conventional to plot a graph of stress against strain, often referred to as a stress/strain curve, to allow calculation of mechanical properties. As the load is increased progressively during a typical test, the strain increases in direct proportion to the increase in stress so that, on the graph, a straight-line relationship is maintained. During this stage, the specimen behaves in an elastic manner such that when the load is removed, all deformation disappears, and the specimen returns to its original dimensions (however, see *elastic after effect*). The slope of the straight-line stage, mathematically the stress divided by the strain, is variously termed the elastic modulus, the tensile modulus or Young's modulus. At some point, the straight-line relationship will cease to hold. This point and the associated stress are referred to as the limit of proportionality.

With many materials, a further small increase in load may introduce further elastic deformation, but the material soon reaches its elastic limit. Increasing the stress above this level then produces a combination of elastic deformation and plastic deformation, predominantly the latter. Plastic or permanent deformation/set is defined as that deformation not recovered by removal of the load and is measured as plastic strain. The position on the stress/strain curve at which plastic deformation commences is also loosely termed the "yield point," but the stricter use of this term refers to the phenomenon in some materials, particularly some steels, whereby at the yield point, an incremental increase in strain occurs without any increase in stress. The stress at the yield point is termed the "yield stress," which gives the yield strength. As further load is applied beyond the elastic limit and yield point, there will be a period during which deformation is uniform over the full gauge length of the specimen but, ultimately, at some maximum load, the specimen will commence to thin locally, termed "necking." This marks the onset of plastic instability following which the cross section of the neck reduces rapidly, the load registered by the test equipment falls rapidly, and the test piece breaks.

Engineering designs are often based on the stress at which plastic deformation will commence. In the case of many steels and some other metals, the yield point is conveniently identifiable as a distinct inflection, or kink, on the stress/strain graph; indeed, in some test circumstances, the line exhibits two inflections, termed the lower and upper yield points. However, for most metals, the elastic limit, that is, the transition from elastic to plastic deformation, is indistinct. It is normal practice to determine, from the stress/strain graph, the stress required to cause a specified small amount of plastic deformation, usually referred to as a percentage strain or offset. This is determined from the graph by identifying on the strain axis the required level of strain and, from it, drawing a line parallel with the straight-line portion of the stress–strain curve. The point at which this parallel line intersects the curve then identifies the required stress. The strain specified is often 0.2%, and the associated stresses are referred to as the 0.2% proof stress or 0.2% offset. Other percentages may be quoted but the differences, in practical terms, between the various levels of proof stress are small. The maximum load carried by the specimen during the test gives, by calculation based on the original cross section of the specimen, the most common measure of the material tensile strength, variously termed the maximum stress, tensile strength or ultimate tensile strength (UTS), etc. After testing, the broken specimen is reassembled for the gauge length to be remeasured. The increase in gauge length is calculated as a percentage of the original gauge life and reported by terms such as % elongation to failure or, simply, % elongation. Finally, the size of the area at the point of failure is measured to calculate the % reduction in area. The elongation and reduction in area are measures of ductility. In commercial activities, UTS, determined as described earlier, is the most commonly specified measure of material strength. However, references occasionally made to terms such as true stress–strain or true maximum stress. These terms recognize that in the conventional tensile test, all calculations are based on the original cross-section area, despite the progressive reduction in cross section following initial yielding. Calculations that take into account this progressive reduction will provide higher values of UTS than those conventionally calculated. These are termed the true values. See *elastic modulus* and *Young's modulus*.

Tensile Strength

In tensile testing, the ratio of maximum load to original cross-sectional area. Also called *ultimate strength*. Compare with *yield strength* and *tensile test*.

Tensile Stress

A stress that causes two parts of an elastic body, on either side of a typical stress plane, to pull apart. In contrast with *compressive stress.*

Tensile Testing

See *tension testing.*

Tension

The force or load that produces elongation.

Tension Set

The condition in which a plastic material shows permanent deformation caused by a stress, after the stress is removed.

Tension Testing

A method of determining the behavior of materials subjected to uniaxial loading, which tends to stretch the material. A longitudinal specimen of known length and diameter is gripped at both ends and stretched at a slow, controlled rate until rupture occurs. Also known as tensile testing.

Tensometer

A small testing machine capable of small scale tensile compression and then testing.

Tenth-Scale Vessel

A filament wound material test vessel based on a 1/10 subscale of the prototype. See also *filament winding.*

Terminal Erosion Rate

The final steady-state erosion rate that is reached (or appears to be approached asymptotically) after the erosion rate has declined from the maximum value. This occurs in some, but not all, cavitation and liquid impingement tests.

Terminal Period

In cavitation and liquid impingement erosion, a stage following the deceleration period, during which the erosion rate has leveled off and remains approximately constant (sometimes with superimposed fluctuations) at a value substantially lower than the maximum rate attained earlier.

Terminal Phase

A solid solution having a restricted range of compositions, one end of the range being a pure component of an alloy system. The same as *primary phase.*

Terminal Solid Solution

In a multicomponent system, any solid phase of limited composition range that includes the composition of one of the components of the system. See also *solid solution.*

Ternary

Containing three components, in particular alloy systems with three elements.

Ternary Alloy

An alloy that contains three principal elements.

Ternary System

The complete series of compositions produced by mixing three components in all proportions.

Terne

An alloy of lead containing 3%–15% Sn, used as a *hot-dip coating* for steel sheet or plate. The term "long terne" is used to describe terne-coated sheet, whereas a short terne is used for terne-coated plate. Terne coatings, which are smooth and dull in appearance (terne means dull or tarnished in French), give the steel better corrosion resistance and enhance its ability to be formed, soldered, or painted.

Terne Metal

An alloy of typically 80% lead, 20% tin, and 0.2% antimony.

Terneplate

Sheet or plate steel coated by hot-dip processes. The coating, intended to enhance corrosion resistance primarily, also improves formability and solderability. Because of the toxicity of red, however, special precautions are required and fabrication operations. Standard coating-thickness designations for sheet products range from LT10 (no minimum) to LT110 (336 g/m^2) total weight both sides based on a triple-spot test. Automobile gasoline–fuel tanks have been the major application, although terneplate also has been used for fuel tanks of lawn mowers and outboard marine motors as well as for roofing and building construction, caskets, and other applications.

Terpolymer

A polymeric system that contains three monomeric units.

Terra-Cotta

A general English term applied to fired, and glazed, yellow and red clay wares; in the United States, it refers particularly to the red-and-brown, square and hexagonal tiles made from common brick clay, always containing iron. Some special terra-cottas are nearly white, while for special architectural work, other shades are obtained. The clays are washed, and only very fine sands are mixed with them in order to secure a fine, open texture and smooth surface. Terra-cotta is used for roofing and for tile floors, for hollow building blocks, and in decorative construction work. Good, well-burned terra-cotta is less than 3.8 cm. thick. Terra-cotta is light, having a density of 1922 kg/m^3, and withstands fire and frost.

Tertiary Third

See *creep for tertiary creep.*

Tesselated Stress

The stress that develops within a material because of local variations from point to point in some physical property. Examples include differences in thermal expansion or elastic modulus between neighboring phrases.

Testing Machine (Load-Measuring Type)

A mechanical device for applying a load (force) to a specimen.

Testpiece

A sample of some larger item that has been prepared for some form of testing to determine the properties of the larger item.

Tetra

Four.

Tetrachloroethane

A colorless liquid employed as a solvent for organic compounds such as oils, resins, and tarry substances. It is an excellent solvent for sulfur, phosphorus, iodine, and various other elements. It is used as a paint remover and bleacher, as an insecticide, and in the production of other chlorine compounds. It is also called *acetylene tetrachloride* and is made by the combination of chlorine and acetylene. It is narcotic and toxic, and the breathing of the vapors is injurious. Mixed with dilute alkalies, it forms explosive compounds. In the presence of moisture, it is very corrosive to metals. Mixed with zinc dust and sawdust, it is employed as a smoke screen.

Tetragonal

At least three mutually perpendicular axes, two equal in length and unequal to the third.

Tetrahedron

A solid object with four faces, in particular, a pyramid with three sides and a base.

Tex

A units for expressing linear density equal to the mass of weight in grams of 1000 m of filament, fiber, yarn, or other textile strand.

Textile Fibers

Natural fibers constitute one of man's oldest sources of building materials. There is evidence to indicate that weaving and probably spinning were not unknown to our Stone Age ancestors. It is important to realize that there is no such thing as a natural textile fiber, although today there are human-made textile fibers. There are only natural fibers that have been diverted from their original function by mankind for use in textiles.

In man's search for fibers that can be used to further our own ends, literally dozens of naturally occurring fibers have been investigated. Only 23 are readily recognized by most textile authorities as being of commercial importance, and one fiber alone, cotton, accounts for approximately 70% of the total fibers consumed by the world population for textile purposes. If, however, this listing of natural fibers is carefully reviewed, it will be found that all fibers can be grouped into six different types of spinnable fibers, each differing fundamentally with respect to molecular and morphological structure. The distinctive characteristics possessed by the fibers of the six groups are such that the groups may be subjectively described as cotton-like, linen-like, sisal-like, wool-like, silk-like, and asbestos-like.

Textile

A textile is a material made mainly of natural or synthetic fibers. Modern textile products may be prepared from a number of combinations of fibers, yarns, plies, sheets, foams, furs, or leather. They are found in apparel, household and commercial furnishings, articles, and industrial products. Materials made directly from plastic sheet or film, leather, fur, or film are not usually considered to be textiles.

The term *fabric* may be defined as a thin, flexible material made of any combination of cloth, fiber, or polymer (film, sheet, or foams); *cloth* as a thin, flexible material made from yarns; *yarn* as a continuous strand of fibers: and *fiber* as a fine, rodlike object in which the length is greater than 100 times the diameter. The bulk of textile products is made from cloth.

The natural progression from raw material to finished product requires the cultivation or manufacture of fibers; the twisting of fibers into yarns (spinning); the interlacing (weaving) or interlooping (knitting) of yarns into cloth; and the finishing of cloth prior to sale.

Spinning Processes

The ease with which a fiber can be spun into yarn is dependent upon its flexibility, strength, surface friction, and length. Exceedingly stiff fibers or weak fibers break during spinning. Fibers that are very smooth and slick or fibers that are very short do not hold together. To varying degrees, the common natural fibers (wood, cotton, and linen) have the proper combinations of the aforementioned properties. The synthetic fibers are textured prior to use to improve their spinning properties by simulating the convolutions of the natural fibers. Natural and synthetic filament fibers, because of their great length, need not be twisted to make useful yarns.

The properties of a yam are influenced by the kind and quality of fiber, the amount of processing necessary to produce the required fineness, and the degree of twist. The purpose of the yarn determines the amount and kind of processing. The yarn number (yarn count) is an indication of the size of a yarn—the higher the number, the finer the yarn. The degree of twist is measured in turns per inch (tpi) and is varied from three to six times the square root of the yarn number for optimal performance.

The conversion of staple fiber into yarn requires the following steps: picking (sorting, clearing, and blending), carting and combing (separating and aligning), drawing (reblending), drafting (reblended fibers are drawn out into a long strand), and spinning (drafted fibers are further attenuated and twisted into yarn).

Textile Oil

(1) An oil used to lubricate thread or yarn to prevent breakage during spinning and weaving. (2) An oil acceptable for direct contact with fibers during textile production.

Textural Stress

Same as *tesselated stress.*

Texture

In a polycrystalline aggregate, the state of distribution of crystal orientations. In the usual sense, it is synonymous with *preferred orientation*, in which the distribution is not random. Not to be confused with *surface texture*. See also *fiber* (metals) and *fiber texture* (metals).

Texture

(1) The characteristics of a surface such as roughness, alignment, and sharpness of topographical features. (2) Preferred orientation. See *orientation*.

TGA

See *thermogravimetric analysis*.

Thallium

A soft bluish-white metal (symbol Tl) thallium resembles lead but is not as malleable. The specific gravity is 11.85 and melting point 302°C. At about 316°C, it ignites and burns with a green light. Electrical conductivity is low. It tarnishes in air, forming an oxide coating. It is attacked by nitric acid and by sulfuric acid. The metal has a tensile strength of 9 MPa and a Brinell hardness of 2. Thallium–mercury alloy, with 8.5% thallium, is liquid with a lower freezing point than mercury alone, −60°C, and is used in low-temperature switches. Thallium–lead alloys are corrosion resistant and are used for plates on some chemical-equipment parts.

Applications

The major use for thallium is as a rodenticide and insecticide. The sulfate compound is most commonly a ploy for this application; therefore, the largest commercial sale of the element is in the form of the sulfate. Thallium sulfate is a heavy white crystalline powder, odorless, tasteless, and soluble in water. The advantage of this compound over many other rodenticides is that it is not detected by the rodent.

Other commercially available thallium chemicals are thallous nitrate and thallic oxide. Further uses of thallium compounds are as follows: (1) thallium oxisulfide, employed in a photosensitive cell that has high sensitivity to wavelengths in the infrared range; (2) thallium bromide–iodide crystals, which have a good range of infrared transmission and are used in infrared optical instruments; and (3) alkaline earth phosphors, which are activated by the addition of thallium.

Other minor uses for thallium are in glasses with high indices of refraction, in the production of tungsten lamps as an oxygen getter, and in high-density liquids used for separating precious stones from ores by flotation.

Toxicity

Thallium and thallium compounds are toxic to humans as well as other forms of animal life. Therefore, special care must be taken that thallium is not touched by persons handling it. Rubber gloves should be used in handling both the metal and its compounds. Proper precautions should be taken for adequate ventilation of all working areas.

Theoretical Density

The density of a material calculated from the number of atoms per unit cell and measurement of the lattice parameters.

Theoretical Electrode Force

The force, neglecting friction and inertia, in making spot, seam, or projection welds by resistance welding, available at the electrodes of a resistance welding machine by virtue of the initial force application and the theoretical mechanical advantage of the system.

Theoretical Stress-Concentration Factor

See *stress-concentration factor*.

Theoretical Throat

See *throat of a fillet weld*.

Thermal Activation

The increase in temperature that provides the additional energy input necessary to initiate and maintain a process.

Thermal Aging

Exposure of a material or component to a given thermal condition or a programmed series of conditions for prescribed periods of time.

Thermal Alloying

The act of uniting two different metals to make one common metal by the use of heat.

Thermal Analysis

A method for determining transformations in a metal by noting the temperatures at which thermal arrests occur. These arrests are manifested by changes in slope of the plotted or mechanically traced heating and cooling curves. When such data are secured under nearly equilibrium conditions of heating and cooling, the method is commonly used for determining certain critical temperatures required for the construction of *phase diagrams*.

Thermal Conductivity

(1) Ability of a material to conduct heat across a temperature gradient. All materials conduct heat by the transmission of vibration between neighboring atoms and molecules. Metals also conduct energy via the cloud of free electrons that surround the atoms in the crystal lattice, a much more efficient process, hence the better thermal conductivity of metals. (2) The rate of heat flow, under steady conditions, through unit area, per unit temperature gradient in the direction perpendicular to the area. It is usually expressed in English units as Btu per square feet per degrees Fahrenheit ($Btu/ft^2/°F$). It is given in SI units as watts per meter kelvin (W/m K).

Thermal Cutting

(1) A group of cutting processes that melts the metal (material) to be cut. (2) Cutting by means of thermal processes such as electrical

arc and oxyacetylene, which heat and melt the material. The term includes processes in which excess oxygen (including air) is injected to assist cutting by reacting with the material and blasting away molten material and oxide. Powder coating is a variant in which a powder is injected to assist the process largely by a scouring action. See also *air carbon arc cutting, arc cutting, carbon arc cutting, electron beam cutting, laser beam cutting, metal powder coating, oxyfuel gas cutting, oxygen arc cutting, oxygen cutting,* and *plasma arc cutting.*

Thermal Cycling

The cyclic change in thermal environment.

Thermal Decomposition

(1) The decomposition of a compound into its elemental species at elevated temperatures. (2) A process whereby fine solid particles can be produced from a gaseous compound. See also *carbonyl powder.*

Thermal Electromotive Force

The *electromotive force* generated in a circuit containing two dissimilar metals when one junction is at a temperature different from that of the other. See also *thermocouple.*

Thermal Embrittlement

Intergranular fracture of maraging steels with decreased toughness resulting from improper processing after hot working. Thermal embrittlement occurs upon heating above 1095°C and then slow cooling through the temperature range of 980°C–815°C and has been attributed to precipitation of titanium carbides and titanium carbonitrides at austenite grain boundaries during cooling through the critical temperature range. See also *maraging steels.*

Thermal Endurance

The time required at a selected temperature for a material or system of materials to deteriorate to some predetermined level of electrical, mechanical, or chemical performance under prescribed test conditions.

Thermal Etching

(1) Heating a specimen (usually a ceramic) in air, vacuum, or inert gases in order to delineate the grain structure. Used primarily in high-temperature microscopy. (2) Usually the same as heat tinting but where the terms are differentiated. Thermal etching is carried out in a vacuum, oxidation is prevented, and some diffusion effects are subsequently observable.

Thermal Expansion

The change and waste of a material with change in temperature. See also *coefficient of thermal expansion.*

Thermal Expansion Molding

In forming of plastics, a process in which elastomeric tooling details are constrained within a rigid frame to generate consolidation pressure by thermal expansion during the curing cycle of the autoclave molding process.

Thermal Fatigue

Fracture resulting from the presence of temperature gradients that vary with time in such a manner as to produce cyclic stresses and a structure.

Damage, particularly cracking, resulting from thermal cycling. This form of damage is differentiated from repeated creep loading, which is seen as the simple loading and unloading of a component at a high temperature. The critical feature of thermal fatigue is that at some position, the normal thermal expansion (less commonly contraction) is constrained by neighboring material that has not expanded to the same extent because of a difference in temperature or, less common, a difference and coefficient of expansion. This induces localized stresses. In the simplest case, the stresses remain within the elastic range and disappear when the temperature gradient flattens. The result was a simple fatigue cycle leading to fatigue cracking; some damage may occur because of the high temperature, but it can be regarded as hot potential damage rather than creep in its usual sense. In more severe cases, the temperature differential is sufficient to induce, locally, stresses that exceed yield at the temperature reached. In many practical cases, the situation is one where a small volume of material, perhaps an edge or corner, rises rapidly to its high operating temperature but its expansion is constrained by the large surrounding volume rising in temperature much more slowly. As a result the high temperature, material yields in compression. When, subsequently, the temperature gradient flattens as the surrounding material leads to the operating temperature, the yielded material, being of reduced dimensions, reverts to a tensile stress. This tensile stress remains active at the operating temperature causing creep. The time spent under load and at temperature during each cycle is termed the dwell period. More severe temperature gradients or combinations of thermal and imposed service loads can cause more complex cycles of stress and plastic deformation. In some cases, material may experience both plastic compression and plastic extension during each thermal cycle; in other cases, the component may progressively extend or compress over a series of cycles, sometimes termed ratcheting or shortening. See *copper shortening.* In addition to the effects of the thermomechanical cycle, thermal fatigue is often accelerated by corrosion and oxidation effects. Such attack may help both crack initiation and propagation by, for example, preferential oxidation at grain boundaries or by forming stress raising, sharp tipped cracks in oxide films. Depending upon the relative contributions of the thermomechanical and the corrosion effects, the attack may become a form of high-temperature corrosion fatigue. See also *craze cracking.*

Thermal Inspection

A nondestructive test method in which heat-sensing devices are used to measure temperature variations in components, structures, systems, or physical processes. Thermal methods can be useful in the detection of subsurface flaws or voids, provided the depth of the flaw is not large compared to its diameter. Thermal inspection becomes less effective in the detection of subsurface flaws as the thickness of an object increases, because the possible depth of the defects increases.

Thermal Instability

The tendency to distort as the temperature is raised. The phenomenon results from various effects and can have various consequences. For example, small variations of thermal expansion coefficient within a component may induce distortion that disappears on cooling. This is difficult to eliminate. In other cases, a component will contain significant levels of residual stress, so that on heating, it experiences creep relaxation or even yield. The resultant dimensional change does not disappear on cooling, and full stability can usually be achieved by stress relief, sometimes termed a *thermal stability soak* or *cycle* at a temperature in excess of normal operating temperature.

Thermal Noise

The electrical noise produced in a resistor by thermally generated currents. These currents average zero but produce electrical power having a nonzero average, which can affect instrument response. Also known as *Johnson noise*.

Thermal Paint/Crayons/Strips

Materials that show some physical change, usually a color change, as they reach some predetermined temperature. Series of such materials, covering a wide range of temperature, are available for use as a form of temperature measurement.

Thermal Rating

The maximum or minimum temperature at which a material or component will perform its function without undue degradation.

Thermal Resistance

A measure of a body's ability to prevent heat from flowing through it, equal to the difference between the temperatures of opposite faces of the body divided by the rate of heat flow. Also known as heat resistance.

Thermal Severity Number

A means for quantifying the cooling capacity of a joint to be welded. Satisfactory welding of steel is critically dependent on avoiding excessive cooling rates otherwise hardening and cracking will occur, see *steel*. The cooling rate is directly dependent on the cross section of the material conducting the heat away, and the thermal severity number (TSN) is a measure of the total of the available conduction paths and units of ¼ in. For example, a butt joint between two pieces of plate ¼ in. thick would have a TSN of 2 (two conduction paths each one unit), and a tee joint between 1 in. plates would have a TSN of 12 (three conduction paths each four units). The TSN is correlated with other inputs to indicate any requirements for preheating the joint, see *weldability*. Also see *controlled thermal severity test*.

Thermal Shock

(1) The development of a steep temperature gradient and accompanying high stresses within a material or structure. (2) Effects, particularly high stress and resultant cracking, or less common spalling, resulting from severe temperature gradients with constraint of the associated thermal expansion or contraction. Compared with thermal fatigue, which results from broadly similar factors, the term thermal shock usually implies a small number of thermal cycles and the development of damage immediately when the temperature change is imposed.

Thermal Spray Powder

A metal, carbide, or ceramic powder mixture designed for use with *hardfacing* and *thermal spraying* operations.

Thermal Spraying

A group of coating or welding processes in which finely divided metallic or nonmetallic materials are deposited in a molten or semi-molten condition to form a coating. The coating material may be in the form of powder, ceramic rod, wire, or molten materials.

Thermal spray comprises a group of processes in which a heat source converts metallic or nonmetallic materials into a spray of molten or semimolten particles that are deposited onto a substrate. Any material that does not sublimate or decompose at temperatures close to its melting point can be applied by thermal spray, as long as it is available in wire or powder form.

Thermal spray coatings offer practical and economical solutions to a variety of industrial problems. They are most commonly applied to resist wear, heat, oxidation, and corrosion, provide electrical conductivity or resistance, and restore worn or undersized dimensions. Although the coating techniques have been around for some time, ongoing improvements are leading to lower application costs and a better understanding of how these coatings work. When properly selected and applied, thermally sprayed coatings can reduce downtime, lower production costs, and improve production yields.

Thermal spray is somewhat related to the welding process. In welding, the added material is actually fused to the base metal, forming a metallurgical bond, whereas a thermally sprayed coating generally adheres to the substrate through a mechanical bond. Nonetheless, some thermal spray processes are capable of achieving mechanical bond strengths that exceed 70 MPa.

The basic thermal spray technologies include plasma spray, wire arc spray, flame spray, detonation gun, and high-velocity oxygen fuel.

Plasma Spray

The plasma spray process requires a plasma gun or torch to generate an arc, which creates the plasma by ionizing a continuous flow of argon gas that is injected into the arc. The arc is struck between a water-cooled copper anode and a tungsten cathode. This type of process is also referred to as nontransferred arc spraying, because the arc is confined to the plasma gun. It is generally operated at energies in the neighborhood of 40–100 kW.

The high heat of plasma causes a large increase in the volume of inert gas introduced, and this produces a high-speed gas jet that accelerates the molten particles and propels them toward the substrate at high velocities. High particle velocities result in dense coatings and high bond strengths.

The plasma transferred arc process is somewhat of a hybrid between plasma spraying and welding. In this process, an arc is struck between the nonconsumable electrode of the plasma torch and the workpiece itself. The feedstock, in the form of wire or powder, is introduced into the resulting external plasma. The material is melted and puddled onto the substrate, producing a metallurgical bond similar to welding, but with a lot less dilution. This process is capable of producing dense and smooth coatings, but it is not capable of applying ceramics.

Wire Arc Spray

The wire arc spray process, like plasma spraying, requires an electrical heat source to melt materials. In this case, the feedstock consists of two conductive metal wires. These two wires act as electrodes that are continuously consumed as the tips are melted by heat from the electrical arc that is struck between them. An atomizing gas shears off the molten droplets and propels them toward the substrate.

T

The atomizing gas is usually compressed air, but it can also be an inert gas such as nitrogen or argon. Compressed air causes oxidation of metal particles, resulting in a large amount of metal oxide in the coating. Because of this, the coating is harder and more difficult to machine than the source material of the coating. This can be a disadvantage because some coatings have to be ground. However, the increased hardness can also enhance wear resistance.

In addition, the temperature of the arc far exceeds the melting point of the sprayed material, resulting in the formation of superheated particles. Consequently, localized metallurgical interactions or diffusion zones develop, which enable achievement of good cohesive and adhesive strengths.

The wire arc process also operates in higher spray rates than the other thermal spray processes. The spray rate, which is dependent on the applied current, makes this process relatively economical.

Flame Spray

In the flame spray process, powder or wire materials are melted through the release of chemical energy triggered by a combustion process. A fuel gas (or liquid) is burned in the presence of oxygen or compressed air. Acetylene fuel gas is most frequently selected, due to its high combustion temperature of 3100°C and low cost. Propane, hydrogen, MAPP, and natural gas are also common choices. The flame melts the feedstock and also accelerates and propels the molten particles. Compressed shop air is also used to assist and boost the particle velocities. However, a compressed inert gas such as argon or nitrogen is preferred if oxidation is a concern.

The setup of a flame spray system is relatively inexpensive and mobile. A basic setup requires only a flame spray torch, a supply of oxygen, and a fuel gas. To increase safety, the seller might have to be augmented with an enclosed spray booth and exhaust.

Because its particle velocities are lower than those of the other thermal spray processes, flame spray coatings are usually of lower quality; they have higher porosity and lower cohesive and adhesive strengths. However, coating quality can be improved by a "spray-and-fuse" process. After the coating is applied by flame spray, the combustion process is repeated to raise the substrate temperature to the point at which the previously applied coating starts to melt. Fusing temperatures exceed 1040°C. The final coating is extremely dense and well bonded by a metallurgical bond. A disadvantage of this technique is the high substrate temperature required and the possibility for deformation of the part.

Detonation Gun

The detonation gun process involves an intermittent series of explosions, which melt and propel the particles onto the substrate. Specifically, a spark plug ignites a mixture of powder and oxygen–acetylene gas in a barrel. After ignition, a detonation wave accelerates and heats the entrained powder particles. After each detonation, the barrel is purged with nitrogen gas, and the process is repeated several times per second.

Coatings produced by the detonation gun process are of excellent quality. The particle velocities are high, so the coatings are dense and exhibit high bond strengths. The drawback is that the process is relatively expensive to operate. It also produces noise levels that can exceed 140 dB and requires special sound- and explosion-proof chambers.

HVOF Spray

The high-velocity oxygen fuel (HVOF) thermal spray process is closely related to the flame spray process, except that combustion takes place in a small chamber rather than in ambient air. The HVOF combustion process generates a large volume of gas caused by the formation and thermal expansion of such exhaust gases as carbon dioxide and water vapor.

These gases must exit the chamber through a narrow barrel several inches long. Because of the extremely high pressure created in the combustion chamber, the gases exit the barrel at supersonic velocities, thereby accelerating the molten particles. Although the particles do not reach the speed at which the gases are traveling, they do reach very high velocities. Particle velocities of over 750 m/s have been measured. These high particle speeds, and subsequent high kinetic energy, translate into dense coatings with some of the highest bond strengths possible.

Coating Characteristics

The goal of all these thermal spray processes is to provide a functional coating that meets all of the necessary requirements. The quality of a coating depends on the final function of the coating and can be determined by evaluating a number of coating characteristics.

Characteristics that can be evaluated to determine coating quality include

- Microstructure (porosity, unmelts, oxidation level)
- Macrohardness (Rockwell B or C) and microhardness (Vickers or Knoop)
- Bond strength (adhesive and cohesive)
- Corrosion resistance
- Wear resistance
- Thermal shock resistance
- Dielectric strength

Coating Selection

Metal forming, paper and pulp, paper converting, printing (including offset and flexographic), chemical, petrochemical, textile, infrastructure, food processing, automotive, medical, power generation, and aerospace all take advantage of thermally sprayed coatings. For each application, the coating is selected to perform one or more functions. The five most encountered functions are wear resistance, heat and/or oxidation resistance, corrosion resistance, electrical conductivity or resistance, and the restoration of worn or undersized dimensions.

Basically, coatings fall into three categories: metal/alloys, ceramics, and cermets. Almost every metal and alloy available can be sprayed in some form. Frequent choices include copper, tungsten, molybdenum, tin, aluminum, and zinc. Frequently sprayed alloys include steels (carbon and stainless), nickel/chromium, cobalt-base alloys, nickel-base alloys, bronzes, brass, and babbitts. Ceramic materials are usually metal oxide ceramics such as chromium oxide, aluminum oxide, alumina–titania composites, and stabilized zirconias.

Cermets are coatings that combine a ceramic and a metal or alloy. Two examples include tungsten carbide (the ceramic constituent) in a cobalt matrix, and chromium carbide in a nickel–chromium matrix.

Overall, thousands of different products and components are coated with great success. See also *electric arc spraying*, *flame spraying*, *plasma spraying*, and *powder flame spraying*.

In addition, new applications are developed daily. Unfortunately, it is not possible to provide examples of every application and what coating is best for the specific function—see *Adv. Mater. Proc.*, 154(6), p. 32, 1998.

Thermal Spraying Gun

A device for heating, feeding, and directing the flow of a thermal spraying material.

Thermal Stability

See *thermal instability*.

Thermal Straightening

Same as *hot spot straightening*.

Thermal Stress

(1) Stress produced by restraint of thermal expansion or contraction. (2) Stresses in a material resulting from nonuniform temperature distribution.

Thermal Taper

See *thermal wedge*.

Thermal Wear

Removal of material due to softening, melting, or evaporation during sliding or rolling. Thermal shock and high-temperature erosion may be included in the general description of thermal wear. Wear by diffusion of separate atoms from one body to the other, at high temperatures, is also sometimes denoted as thermal wear.

Thermal Wedge

The increase in pressure due to the expansion of the lubricant, for example, in a parallel thrust bearing. Thermal distortion of the bearing surfaces may also form a wedge shape. This is referred to as *thermal taper*.

Thermally Induced Embrittlement

See *embrittlement*.

Thermal–Mechanical Treatment

See *thermomechanical working*.

Thermic Lance

A coating process in which an oxyacetylene (or other fuel gas) flame is formed at the tip of a steel tube and directed at the item to be cut. The tube is progressively burned away, and the heat from this reaction adds to the normal heat from the oxyacetylene flame producing very high temperatures. In some cases, additional heat may be developed by inserting steel rods into the tube.

Thermionic

Concerned with the emotion of electrons by material at high temperature.

Thermionic Cathode Gun

An *electron gun* that derives its electrons from a heated filament, which may also serve as the cathode. Also termed *hot cathode gun*.

Thermionic Emission

The ejection of a stream of electrons from a hot cathode, usually under the influence of an electrostatic field.

Thermistor

A contraction of thermal resistor, a semiconductor, the electrical resistance of which varies considerably and progressively with temperature.

Thermit Crucible

The vessel in which thermit reactions take place.

Thermit Mixture

A mixture of metal oxide and finally divided aluminum with the addition of alloying metals as required.

Thermit Mold

In *thermit welding*, a mold formed around the parts to be welded to receive the molten metal.

Thermit Reaction/Welding

The thermit process utilizes the powerful exothermic, self-propagating reaction resulting from ignition of finally divided aluminum powder mixed with the oxide of a metal such as iron, chromium, molybdenum, or manganese. The reaction oxidizes the aluminum and reduces the ore to its metal, and the heat released is sufficient to melt even high melting point metals. The process can be used to produce such metals but it is better known for its use in thermit welding. In a typical application, joining rail track, the reaction between the aluminum powder and the iron oxide is initiated in the upper chamber of a refractory mold and superheated molten iron flows down into a lower chamber to fill the gap between, and fuse to, the rail faces. See *thermit welding*.

Thermit Welding

A welding process that produces coalescence of metals by heating them with superheated liquid metal from a chemical reaction between a metal oxide and aluminum, with or without the application of pressure. Filler metal, when used, is obtained from the liquid metal. The process is used primarily for welding railroad track.

Thermochemical Machining

Removal of workpiece material—usually only burrs and fins—by exposure to hot fuel gases that are formed by igniting an explosive, combustible mixture of natural gas and oxygen. Also known as the *thermal energy method*.

Thermochemical Treatment

Heat treatment for steels carried out in a medium suitably chosen to produce a change in the chemical composition of the object by exchange with the medium.

Thermocompression Bonding

See preferred term *hot pressure welding.*

Thermocouple

Thermocouples are the most common type of temperature sensor used and nearly 16% of all process instrumentation measures, indicates, or controls temperature. Thomas Johann Seebeck is credited with inventing the thermocouple in 1821. His experiment consisted of two dissimilar metal wires joined at the ends to form a loop with each end held at a different temperature.

Seebeck detected the induced current by the displacement of a compass needle that was near one of the wires. Further study revealed that the temperature gradient induced an electric current and when this circuit was broken at the center, an open-circuit voltage was measured, that is, the Seebeck electromotive force. The thermocouple is based on the concept that for small changes in temperature (T_{hot} to T_{cold}), the voltage is proportional to the temperature difference.

Operating environment and temperature are important considerations for picking the correct thermocouple.

When using a thermocouple, it is very important to understand that the measured voltage is developed along the entire length of the thermocouple. Steep temperature gradients should be avoided because any defect in the wire within the gradient will contribute a large error. Steep gradients may also induce recrystallization and grain growth, thus changing the calibration. In this regard, feeding thermocouples through insulation is critically important because deformation of the wires may produce recrystallization during operation.

Thermocouple Materials

A device for measuring temperatures, consisting of lengths of two dissimilar metals or alloys that are electrically joined at one end and connected to a voltage-measuring instrument at the other end. When one junction is hotter than the other, a *thermal electromotive force* is produced that is roughly proportional to the difference in temperature between the hot and cold junctions. Standard thermocouple materials include

Fe, 44Ni-55Cu, 90Ni-9Cr, 94Ni-Al, Mn, Fe, Si, Co, 84Ni-14Cr-1.4Si, 95Ni-4.4Si-0.15Mg, OFHC Cu, 87Pt-13Rh, Pt, 90Pt-10Rh, 70Pt-30Rh, and 94Pt-6Rh. Nonstandard materials include nickel–molybdenum, nickel–cobalt, iridium–rhodium, platinum–molybdenum, gold–palladium, palladium–platinum, and tungsten–rhenium alloys.

Thermodynamics

The study of the interaction between heat and mechanical energy in any process.

Thermoelastic Instability

In sliding contact, sharp variations of local surface temperatures with the passing of asparities leading to stationary or slowly moving hot spots of significant magnitude.

Thermoelectric Cooling

The extraction of heat by the Peltier effect. See *thermoelectric effects.*

Thermoelectric Effect (s)

This term is used as a synonym for the Seebeck effect or as a collective term for all of the following effects. The Seebeck effect is the electromotive force (emf) due to the difference in temperature between the two junctions of dissimilar conductors in a circuit. The Thompson (alternatively the Kelvin) effect is the emf produced by a difference of temperature between two sites of a conductor or, alternatively, the release or absorption of heat associated with an electrical current flowing between two sites on the conductor, which are at different temperatures. The Peltier effect is the release or absorption of heat as an electric current crosses the junction from one conductor to a dissimilar one.

Thermoformed Plastic Sheet

Thermoplastic sheet forming consists of the following three steps; (1) a thermoplastic sheet or film is heated above its softening point; (2) the hot and pliable sheet is shaped along the contours of a mold, the necessary pressure being supplied by mechanical, hydraulic, or pneumatic force or by vacuum; (3) the formed sheet is removed from the mold after being cooled below its softening point.

Sheet Materials

The following five groups of thermoplastic materials account for the major share of the thermoforming business:

1. *Polystyrenes*: High-impact polystyrene sheet, ABS sheet, biaxially oriented polystyrene film, and polystyrene foam
2. *Acrylics*: Cast and extruded acrylic sheet and oriented acrylic film
3. *Vinyls*: Unplasticized rigid PVC, vinyl copolymers, and plasticized PVC sheeting
4. *Polyolefins*: Polyethylene, polypropylene, and their copolymer films
5. *Cellulosics*: Cellulose acetate, cellulose acetate butyrate, and ethyl cellulose sheet

A sixth group of increasing importance is the linear polycondensation products, such as polycarbonates, polycaprolactam (type 6 nylon), polyhexamethylene adipamide, oriented polyethylene terephthalate (polyester), and polyoxymethylene (acetal) films.

General Process Considerations

Thermoplastic sheets softened between 121°C and 232°C. It is important that the sheets be heated rapidly and uniformly to the optimum forming temperature. The fastest heating is brought about with infrared radiant heaters. Some thermoplastics cannot tolerate such intense heat and require convection heating and air-circulating ovens or conduction heating between platens. In a few instances, the sheets are formed "in line," making use of the heat of extrusion.

The following four basic forming methods and more than 20 modifications are known:

Matched Mold Forming

This process, in which the hot sheet is formed between a registering male and female mold section, employs mechanical or hydraulic pressure.

It is used for corrugating flat rigid sheeting either "in line" or in a separate operation. For continuous longitudinal corrugation, the hot sheet is pulled through a matched mold with registering top and bottom teeth. Transverse corrugation is accomplished with matched top and bottom rolls or molds mounted on an endless conveyor belt or chains.

For stationary molding operations, a rubber blanket, backed with a liquid or inflated by air, frequently replaces the male section. Another modification is the *plug and ring* technique in which the ring acts as a stationary clamping device and the moving plug resembles the top portion of the male mold.

Elastic and oriented thermoplastic sheets possess a plastic memory, that is, they tend to draw tight against the force that stretches them. This property occasionally permits forming against a single mold only.

Slip Forming

A loosely clamped sheet is allowed to slip between the clamps and is "wiped" around a male mold. This process has been used to avoid excessive thinning when forming articles with the draws.

Air Blowing

A hot sheet is blown with preheated compressed air into a female mold.

Variations of this process are *free blowing* without a mold into a bubble, *plug-assist blowing* in which a cored plug pushes the hot sheet ahead before blowing, and *trapped sheet forming* in which a clamping ring slides over the mold before applying the compressed air. The last process is employed in automatic roll-fed packaging machines with biaxially oriented films.

Vacuum Forming

This is the most common sheet-forming process with many modifications. *Modifications of vacuum forming*:

- In *straight vacuum forming*, the hot thermoplastic sheet is clamped tight to the top of a female mold or of a vacuum box that contains a male mold. Drawing sheets into a female mold results in excellent replica of fine details on the outer surface of the drawn articles. Straight vacuum forming has proved excellent for shallow draws; however, in articles with small radii or deeper dimensions, the corners and bottom are excessively thinned out. Employing a female mold with multiple cavities is more economical than forming over a number of the male molds, because it permits smaller spacing between cavities (without bridging), which in turn allows more pieces per sheet. Drawing sheets over a male mold produces articles with the thickest section on top. It is used for the production of 3D geographical maps, because it gives a greater accuracy of registration due to restricted shrinkage.
- *Free vacuum forming* into a hemisphere, without a mold, is similar to free air blowing and is employed with acrylic sheets where perfect optical clarity has to be maintained.
- *Vacuum snapback forming* makes use of the plastic memory of the sheet. The hot sheet is drawn by vacuum into an empty vacuum box, while a male mold on a plug is moved from the top into the box. The vacuum is released at the sheet, still hot and elastic, snaps back against the male mold and cools along the contours of the mold. This method is employed with ABS and plasticized vinyl sheets for the production of cases and luggage shells.
- *Drape forming* is a technique that allows deeper drawing. After clamping and heating, the sheet is mechanically stretched over a male mold, then formed by vacuum,

which picks up the detailed contours of the mold. Acrylic and polystyrene sheets slide easily over the mold, whereas polyethylene sheets tend to freeze on contact with the mold, causing differences in thickness.

- To overcome the problem of thinning, *vacuum plug-assist forming* into a female mold has been developed. Its principle is to force a heated and clamped sheet into a female mold using a plug-assist before applying the vacuum. This technique may be considered a reverse of the snapback method.
- *Vacuum air-slip forming* represents a modification of drape forming and is designed to reduce thinning on deep-drawn articles. It consists of prestretching the sheet pneumatically prior to vacuum forming over a male mold. Prestretching is accomplished either with entrapped air by moving the male mold like a piston in the vacuum chamber or by compressed air.
- There are at least three variations of the *reverse-draw technique*. All employ the principle of blowing a bubble of the hot plastic sheet and pushing a plug in reverse direction into the outside of the hot bubble. This accomplishes a folding operation permitting deeper draws than any other common practice.
- The technique of *reverse draw with plug assist* consists of heating the clamped sheet, raising a female mold so that a sealed cavity is formed, while a bubble is blown upward. The preheated plug assist is lowered and pushes the sheet into the cavity. The final shaping is accomplished with vacuum.
- A variation of this method is *reverse draw with air cushion*. The plug assist is furnished with holes through which hot air is blown downward and pushes the hot sheet ahead of the plug assist, minimizing mechanical contact. This technique is used in forming materials with sharp softening ranges have limited hot strength, such as polyethylene and polypropylene.
- *Reverse draw on a plug* uses a male mold on the plug to preserve the finish of the sheet.

Design and Construction of Vacuum Molds

Depth of draw is a prime factor controlling the wall thickness of the formed article. During straight vacuum forming into a female mold, the depth of draw should not exceed one-half of the cavity width. For drape forming over a male mold, the height-to-width ratio should be 1:1 or less. With plug assist, air slip, or one of the reverse-draw techniques, the ratio may exceed the 1:1 ratio.

Proper air evacuation assists material flow in the desired direction and in uniform wall thickness. In general, deep corners require intensified evacuation. The diameter of vacuum holes should be 0.25–0.6 mm for polyethylene sheets, 0.6–1 mm for other thin-gauge materials, and may increase to 1.5 mm for heavier rigid materials. Sharp bends and corners should be avoided, because they result in excessive stress concentration and in reduction of strength. The forming cycle can be accelerated and maintained by the use of mold temperature controls. Molds for permanent use are cast from aluminum or magnesium alloy.

Finishing

After the article has been formed, it must be cooled, removed from the forming machine, and separated from the remainder of the sheet. Trimming of thin-walled articles can be carried out hot or cold, in the forming machine or after removal. Heavy-gauge articles should be trimmed only after cooling. Clicker dies, high dies, and Walker dies are frequently used. Decorating formed articles is generally accomplished by printing the flat sheet before the forming separation. Formed articles may also be spray coated.

Thermoforming

The process of forming a thermoplastic sheet into a 3D shape after heating it to the point at which it is soft and flowable, and then applying differential pressure to make sheet conform to the shape of a mold or die positioned below the frame. When the thermoplastic material has been reinforced, it should be heated to the point that it is soft enough to be formed without cracking or breaking the reinforcing fibers. There are three basic mold types: female (concave), male (convex), and matched (a combination of the two). (1) In basic female forming, or straight vacuum forming, the heated sheet is positioned over the mold cavity and is pulled into the cavity by vacuum. (2) In basic male forming, also called drape forming, the heated plastic is drawn over the mold. As soon as a seal is created, the vacuum is activated, and the plastic is forced directly against the surface of the mold. See *drape forming*. (3) In matched-mold thermoforming, the stamping force of matched male and female molds is used. The male mold pushes the heated sheet into the female cavity. See (thermoforming variations) *air-slip thermoforming, plug-assist thermoforming, pressure bubble plug-assist thermoforming, trapped sheet, contact heat, pressure thermoforming*, and *vacuum snapback thermoforming*.

Thermogalvanic Corrosion

(1) Corrosion resulting from an *electrochemical cell* caused by a thermal gradient. (2) Electrochemical corrosion mechanisms in which a primary factor is the difference in temperature between the anode and the cathode. The corroded anode is usually the colder.

Thermogravimetric Analysis

The study of the change in mass of a material under various conditions of temperature and pressure.

Thermomechanical Working

A general term covering a variety of metal-forming processes combining controlled thermal and deformation treatments to obtain synergistic effects, such as improvement in strength without loss of toughness. Same as *thermal–mechanical treatment*.

Thermometal

An alternative term for bimetallic strips.

Thermometer

Any instrument for measuring temperature.

Thermonuclear

Usually a nuclear reaction in which a dominant factor is the impact of particles having a high kinetic energy such as in the hydrogen bomb.

Thermoplastic

Becoming plastic upon heating and returning to a rigid state on cooling. Capable of being repeatedly softened by an increase in temperature and hardened by a decrease in temperature. These polymeric materials that, when heated, undergo a substantially physical rather than chemical change and that in the softened stage can be shaped into articles by molding or extrusion. The transition usually develops at about 80°C–120°C.

One definition of thermoplastic covers the type of decorating glass enamels or overglaze applied through a hot screen. The media is a wax composition that is heated and added to the enamel powder. On cooling, it forms a solid case that is broken into cubes. The cubes are heated and flow onto a resistance-heated metal screen. When applied on ware, the thermoplastic freezes immediately, and other colors can be superimposed without a drying cycle. Thermoplastic permits wraparound decorations and reduces ware handling losses.

Other thermoplastic materials are used in conjunction with the injection molding of technical ceramics such as spark plugs.

Thermoplastic Continuous-Fiber-Reinforced Materials

These materials are tougher and withstand impacts better. They mold readily and can be recycled. They also have unlimited shelf life and admit no hazardous solvents during processing.

Processing

This continuous-fiber-reinforced thermoplastic (CFRTP) process weaves together strands of powder-resin-coated fibers to produce TowFlex fabrics. This is in contrast to other fabrics where raw reinforcement fibers are first woven then coated. Weaving individual coated strands makes the fiber highly drapable. This is because each strand within the fabric moves freely relative to adjacent strengths. In addition, the strands remain flexible because they are not fully wet out prior to molding. Complete wet out of the fabric comes during the compression-molding process.

CFRTP Basics

Fiber-reinforced thermoplastics are widely used in injection molding. The vast majority of these products use short or chopped fibers. These fibers generally measure less than 6.4 mm and will randomly orient themselves during molding. Typical injection molded parts contain only 20%–30% reinforcement fiber. Short, randomly oriented fibers in low percentages do not provide much reinforcement. And it is often difficult to mold such material into complex, large, or thick-walled parts without voids or knit lines.

In comparison, parts molded from new CFRTPs often contain more than 60% reinforcement. Reinforcement fibers run continuously throughout the entire part in specified directions to help optimize strength and stiffness. CFRTPs contain continuous-reinforcement fiber filaments that are powder coated with melt-fusible thermoplastic particles. The uniformly coated filaments are woven into fabrics or braid, formed into semirigid unidirectional tapes or ribbons, or laminated into panels. The resin particles wet out and consolidate quickly when compression molded.

The first step in compression molding CFRTP fabric is to cut and assemble fabric plies. The plies create a preform that approximates the flat pattern shape of the molded part. Automated cutting equipment such as reciprocating knives or ultrasonic gear may be an option for complicated patterns manufactured in high volume. Steel-rule dies, electric rotary shears, or hand shears/scissors might work best for lower-volume applications. It is often useful to machine a simple preform fixture with a cavity or recess in the shape of the flat pattern. The assembly fixture helps keep the precut fabric plies in proper order, location, and orientation. This includes any partial plies needed to build up additional thickness in specific areas.

Stack plies are tack welded ultrasonically or via a soldering iron before removal from the fixture. This helps keep them in the right orientation during handling and storage. The preforms have unlimited shelf life and can be produced in large quantities independent of the molding process or molds. They also need no additional layup labor before being compression molded into finished parts. The fabric drapes easily and makes for a straightforward molding process that forms and shapes the flat preform. In contrast, thermoset preimpregnated materials generally require cutting and splicing to mold complex shapes.

Molds

Match steel metal molds give the best surface finish and mold life when molding CFRTP parts. Nickel-plated aluminum molds are good for moderate volumes of material processing below 316°C. Steel molds are mandatory for higher-temperature molding of matrix resins such as polyphenylene sulfide or polyetheretherketone. Registration of the upper to lower mold halves takes place through either guide pins or the mold configuration itself.

Generally, molds are designed to fully "bottom out" on the CFRTP material rather than on thickness stops. This helps maintain pressure on the materials throughout the molding process. Thickness stops, however, are used to maintain flatness and help ensure that the mold does not "rock" during the process. They also establish a minimum parts thickness. Stops should typically be 0.25–0.50 mm lower than the desired nominal parts thickness. For relatively thin parts or plates of less than 3.81 mm, quantity of fabric plies loaded into the mold primarily controls thickness. Thicker parts or plates use a specified preform weight along with the number of plies to control parts thickness.

Compression Molding

CFRTP fabrics generally use three variations of the compression-molding process. All employ conventional equipment and do not need rapid closing speeds or excessive press tonnage.

Single-press/heated–cooled platens: The platens in a single press are heated and cooled to reach the right processing conditions. This approach applies best in situations requiring a variety of different parts and relatively low volumes that do not need a quick molding cycle. The molding process begins with loading of the preform into the lower mold half. Operators next install the upper mold half and load the complete mold into the press. It is then heated with pressures on the order of 68.6–343 kPa. Low pressure during heating helps ensure good heat transfer and initiates the forming of the preform. Full pressure, 0.68–5.44 MPa, comes during final compaction as the mold reaches its required processing temperature. Next, the press platen is cooled to bring the mold and CFRTP part to the removal temperature.

Hot/cold shuttle press: Hot/cold shuttle presses separate the basic segments of the compression-molding process—heating, cooling, and loading/unloading—for better efficiency. Separate heating and cooled platen presses apply pressure to the mold and CFRTP material. Cycle times are short because molds shuttle between preheated hot presses and precooled cooling presses. Cycle times are particularly fast for relatively thin parts at low-mass molds. Molds do not need individual heating and cooling

systems, which helps reduce costs. Molds for deep-draw parts may need to use heated or cooled press platens or bolsters. These approximate the part shape and mount to the hot and cold presses. This approach helps keep heating and cooling sources close to the material to increase processing speed. The hot/cold shuttle-press approach is useful for combinations of different parts in moderate to high volumes. Here, economics do not justify the expense of heating and cooling provisions in each mold. The hard-pressed station is first preheated to 10°C–37.8°C higher than the desired mold temperature. A preform goes in the lower mold half, the upper mold half lowers into place, and the mold shuttles into the preheated press. Low pressure, 68.6–343 kPa, is applied as the mold heats for good heat transfer and to start preform shaping. Full pressure of 0.68–3.4 MPa or more forces final compaction as the mold reaches the processing temperature range. Next, the hot press opens, pressure releases, and the mold shuttles into the cold fresh station. Full pressure again applies as the mold and material cool enough for part removal.

Single-press/heated–cooled molds: Here, integrally heated and cooled molds mount directly onto press platens. Processing cycles for complex-shaped parts can be fast because mold heating and cooling systems can be close to the CFRTP. Individual molds are more expensive, because each mold must contain integral heating and cooling. Press platens must also have sufficient travel to open wide enough for easy preform loading in part removal. Otherwise, molds would need to be removed from the press for part removal. The single-press/heated–cooled mold approach is especially appropriate for large production runs of a specific part or plate. Manufacturers can amortize molds over a large part count, and speedy processing helps drive cost down. It is also appropriate for large or deep molds where the shuttle process is impractical. The integrally heated and cooled mold halves are usually attached to the press platens to minimize handling. The mold temperature cycles between the processing and part-removal temperatures. The CFRTP preform loads into the mold in the press closes with low pressure, 68.6–343 kPa, for good heat transfer and to initiate preform shaping. The heated mold goes under full pressure, 0.68–3.4 MPa or more, for final compaction and forming. The mold and material then cool, and the part is demolded. Cycle times can be under 5 min depending on the mold mass, parts thickness, and part geometry.

Applications

There have been few commercial applications for CFRTP, however. One reason has been that early product forms of these materials were tougher to process and mold. The development of highly drapable, conformable CFRTP fabrics addresses such shortcomings. They are made from thermoplastic such as nylon, polypropylene, polyphenylene sulfide, polyetherimide, and polyetheretherketone with carbon, glass, or aramid reinforcement fibers. They can easily be molded into complex structural shapes and are ideally suited for production quantities of between 1,000 and 50,000 parts annually.

A CFRTP fabric called RF6 is used to make an eight-dihedral-faced kayak paddle. The paddle surface is said not only to grab the water more effectively than conventional blades but also to release surface pressure at eight precise locations along the outside edge of the blade. This produces a blade that has zero flutter while providing more bite per square inch of surface area. The thin blade design was made possible by the high stiffness and impact resistance of the CFRTP material.

Thermoplastic Elastomers

Thermoplastic elastomers (TPEs) are a group of polymeric materials having some characteristics of both plastics and elastomers. They are also called elastoplastics. Requiring no vulcanization or curing, they can be processed on standard plastics processing equipment. They are lightweight, resilient materials that perform well over a wide temperature range.

There are a half-dozen different types of elastoplastic.

Types

TPEs have been traditionally categorized into two classes: block copolymers, which include styrenics copolyesters, polyurethanes, and polyamides; and thermoplastic/elastomer blends and alloys, comprised of thermoplastic polyolefins and thermoplastic vulcanizates.

Conventional TPEs are considered two-phase materials composed of a hard thermoplastic phase that is mechanically or chemically mixed with a soft elastomer phase. The resulting material shares the characteristics of both.

Advantages and Drawbacks

TPEs give engineers several advantages over thermoset rubbers, including lower fabrication costs, faster processing times, little or no compounding, recyclable scrap, and processing by conventional thermoplastic equipment. In addition, the processing equipment consumes less energy and maintains tighter tolerances than that of rubber processing.

There are several drawbacks, however. TPEs have relatively low melting temperatures, making them unusable in high-temperature applications. They usually require drying before molding because they are hygroscopic or moisture absorbing. Manufacturers must use TPEs for high-volume applications for it to be economical. And molders accustomed to working with rubber have little experience using TPE materials and equipment.

Properties

As for material properties, TPEs come in durometers, or hardnesses, ranging from 3 Shore A to 70 Shore D, which roughly translates into going from very soft and flexible to semirigid. Although they have relatively low melt temperatures, TPEs resist continuous exposure to temperatures up to 135°C, with spikes up to 149°C. On the other end of the temperature spectrum, they remain flexible at temperatures down to −69°C. The polymers have outstanding dynamic fatigue resistance, good tear strength, and resist acids and alkalies, ultraviolet light, fuels and oils, and ozone, and maintain their grip in wet and dry conditions.

Advancements and Applications

The latest advancements in TPE formulations are resin blends that adhere directly to engineering thermoplastics without any special features and can be applied by insert molding (over molding) and two-shot injection molding. The new formulations let designers mold grips onto a wider range of thermoplastics, particularly high-strength materials such as high-impact polystyrene, glass-filled nylon, and polycarbonate, which are commonly used in power tools, sporting goods, and electronics.

New materials are also easier to mold in thin-wall sections, which is crucial for consumer electronics where weight is a concern. Consumer-electronics manufacturers can use TPEs for grip strips, for example, because new resins flow easily across long, thin channels in part molds. Engineers prefer to mold "soft-touch" elastomers onto rigid substrates instead of using adhesives and mechanical locks; yet most elastomers do not provide the necessary adhesion, compatibility, and durability.

To answer these needs, a material system was developed consisting of a rigid and a flexible thermoplastic polyurethane formulation. The two materials—Estaloc thermoplastic and Estane elastomer—have similar chemical makeup. This chemical compatibility helps the two materials form a strong bond without adhesives, giving automotive manufacturers, for example, a "one-stop shop" for interior applications such as ignition bezels.

Copolyether-ester thermoplastic elastomer applications include tubing and hose, V belts, couplings, oil-field parts, and jacketing for wire and cable. Their chief characteristic is toughness and impact resistance over a broad temperature range.

Thermoplastic Fluoropolymers

See *fluoroplastics*.

Thermoplastic Injection Molding

A process in which melted plastic is injected into a mold cavity, where it cools and takes the shape of the cavity. Bosses, screw threads, ribs, and other details can be integrated, which allows the molding operation to be accomplished in one step. The finished part usually does not require additional work before assembling. See also *reciprocating-screw injection molding*.

Thermoplastic Olefins

The olefinics, or thermoplastic olefins (TPOs), are produced in durometer hardnesses from 54A to 96A. Specialty flame-retardant and semiconductive grades are also available. The TPOs are used in autos for paintable body filler panels and air deflectors and as sound-deadening materials in diesel-powered vehicles.

The TPOs have room-temperature hardnesses ranging from 60 Shore A to 60 Shore D. These materials, as they are based on polyolefins, have the lowest specific gravities of all thermoplastic elastomers. They are uncured or have low levels of cross-linking. Material cost is midrange among the elastoplastics.

These elastomers remain flexible down to −51°C and are not brittle at −68°C. They are autoclavable and can be used at surface temperatures as high as 135°C in air. The TPOs have good resistance to some acids, most bases, many organic materials, butyl alcohol, ethyl acetate, formaldehyde, and nitrobenzene. They are attacked by chlorinated hydrocarbon solvents. Compounds rated V-Oby Underwriter Laboratories (UL) 94 methods are available.

Thermoplastic Polyesters

A class of thermoplastic polymers in which the repeating units are joined by ester groups. The two important types are *polyethylene terephthalate* (PET), which is widely used as film, fiber, and soda bottles; and *polybutylene terephthalate* (PBT), which is primarily a molding compound.

Known chemically as polybutylene terephthalate (PBT) and polyethylene terephthalate (PET), the thermoplastic polyester-molding compounds are crystalline, high-molecular-weight polymers. They have an excellent balance of properties and processing characteristics and, because they crystallize rapidly and flow readily, mold cycles are short.

In addition to several unreinforced molding resins, the polyesters are available in glass-reinforced grades. Unreinforced and glass-filled grades are available with UL flammability ratings of 94 HB and 5 V.

Properties

Thermoplastic polymers have excellent resistance to a broad range of chemicals at room temperature including aliphatic hydrocarbons, gasoline, carbon tetrachloride, perchloroethylene, oils, fats, alcohols, glycols, esters, ethers, and dilute acids and bases. They are attacked by strong acids and bases.

High creep resistance and low moisture absorption give the polyesters excellent dimensional stability. Equilibrium water absorption, after prolonged immersion at 22.75°C, ranges from 0.25% to 0.50% and, at 66°C, is 0.52% to 0.60%. Black-pigmented grades are recommended for maximum strength retention in outdoor uses.

Thermoplastic Polyimides

Fully imidized, linear polymers with exceptionally good thermo-mechanical performance characteristics. Aromatic thermoplastic polyimides are generally produced by the polycondensation reaction of aromatic dianhydrides with aromatic diamines or aromatic diisocyanates in a suitable reaction medium.

Thermoplastic Polypropylene

With long-fiber-reinforced materials, thermoplastic composite extrusions reportedly offer properties superior to those of PVC and polypropylene—as well as more costly engineered thermoplastics like polycarbonates. This development is referred to as very high modulus extrusion (VHME).

Parts reinforced by long-fiber materials have been injection molded for 15 years. But VHME thermoplastic is the first commercial application of long-fiber extrusions. The material consists of a long-fiber-based core for strength and impact resistance, sheathed by inner and outer layers of ABS, polypropylene, or weatherable PVC.

To date, the fibers extruded have been limited to glass and carbon, and the glass fibers are about 1.27 cm long. Quite a number of thermoplastics can be used as the matrix in VHME work. They include polyurethane, polypropylene, ABS, and polycarbonate.

The nylons or polyacetals have not initially been done because current equipment is not capable of processing at that high a temperature.

Applications

Recent applications for VHME include a commercial refrigeration system and the subfloor of a refrigerated semitruck trailer. In the commercial refrigeration application, VHME was used to make a corner support to hold glass in a glass assembly. Engineers chose the material because it offered low thermal conductivity and enough strength to retain the glass. In the trailer application, the material replaces the heavy wood that tended to rot. Aside from its resistance

to rot, VHME produced other benefits. It improved the thermal performance of the unit, and it also reduced the weight about 1 lb/lineal foot. Thus, on the semitrailer, it saved about 184 kg.

Thermoplastic Polyurethanes

Linear (segmented) block copolymers. Such a copolymer consists of repeating groups of diisocyanate and short-chain diol, or chain extender, for a rigid block, and repeating groups of diisocyanate and long-chain diol, or polyol, for a flexible block. Thermoplastic polyurethanes (TPUs) are commonly injection molded, blow molded, or extruded. See *polyurethanes*.

TPU is often the choice for critical tasks because it offers a broad range of high-performance properties. Moreover, it is known for reliability and has a long working life even in harsh end uses. TPUs routinely provide low-temperature flexibility, high abrasion and moisture resistance, and a long storage life.

TPU is a thermoplastic elastomer with many of the same physical and mechanical properties of vulcanized rubber, but with the wider range of processing options, common to other thermoplastic polymers. TPUs, like vulcanized rubber, have low hysteresis, high elongation, and good tensile strength. Processors can also tailor their durometer, or hardness, by varying their internal structure without adding special chemicals or plasticizers.

Urethane chemistry is built around four of the most common elements: carbon, hydrogen, nitrogen, and oxygen. And it uses the molecular urethane linkage ($NHCO_2$) to connect a series of block copolymers with alternating hard and soft segments. The ratio and molecular structure of these segments determine the specific properties of a TPU grade. The hard segments contain an isocyanate structure while the soft segments consist of different polyols. These polyol segments, either polyether or polyester, are used to distinguish different types of TPUs.

Types and Properties

Polyether TPU provides a softer "feel" or drape than polyester and is generally preferred where there is a skin contact. Compared with polyesters, it offers better moisture vapor transmission rates as superior low-temperature properties. It is inherently stable when exposed to high humidity and is naturally more resistant to fungus, mildew, and microbe attack.

However, polyester TPUs have better abrasion resistance with higher tensile and tear strengths for a given durometer. The polyester version also stands up better to fuels and oils, has superior barrier properties, and does not age as fast thanks to better oxidation resistance. However, polyester TPUs will eventually break down in high humidity.

TPUs are further identified by the chemical makeup of their isocyanate, or hard segment components. TPUs are classified into aliphatics (linear or branched chains) or aromatics (containing benzene rings). Aromatic TPUs are strong, general-purpose resins that resist attack by microbes, stand up well to chemicals, and are usually processed. An aesthetic drawback, however, has the tendency of aromatics to discolor, or yellow, when exposed to ultraviolet (UV) light or low-level gamma sterilization. The addition of UV stabilizers or absorbers can reduce this discoloration.

Aliphatic urethanes, on the other hand, are inherently light stable and resists discoloration from UV exposure or gamma sterilization. They are also optically clear, which makes them suitable laminates for encapsulating glass and security glazings.

From an environmental standpoint, use of TPU often makes sense.

Urethane burns more cleanly than polyvinyl chloride and other films that compete in the disposable medical-supply market. Using urethane helps avoid the toxic by-products resulting from incinerating disposables fabricated from PVC. In addition, TPU is also readily recycled.

Thermoplastic Styrenes

The styrenics are block copolymers, composed of polystyrene segments in a matrix of polybutadiene or polyisoprene. Lowest in cost of the elasticoplastics, they are available in crumb grades and molding grades and are produced in the durometer harnesses from 35A to 95A.

Thermoplastic Urethanes

Thermoplastic urethanes are of three types: polyester urethane, polyether urethane, and caproester urethane. All three are linear polymeric materials, and therefore do not have the heat resistance and compression set of the cross-linked urethanes. They are produced chiefly in three durometer hardness grades—55A, 80A, and 90A. The soft 80A grade is used where high flexibility is required, and the hard grade 70D is used for low-deflection load-bearing applications.

Thermoplastic Vulcanizates

Thermoplastic vulcanizates (TPVs), a type of thermoplastic elastomer, are more challenging to color because they consist of a blend of polypropylene and very fine ethylene propylene diene terpolymer (EPDM) rubber particles. The EPDM particles make it impossible to develop a clear resin. However, recent developments offer whiter and cleaner TPVs that can produce bright colors. This class of thermoplastic elastomers consists of mixtures of two or more polymers that have received a proprietary treatment to give them properties significantly superior to those of simple blends of the same constituents. The two types of commercial elastomeric alloys or melt-processable rubbers (MPRs) and TPVs. MPRs have a single phase; TPVs have two phases.

Properties

Thermoplastic vulcanizates are essentially a fine dispersion of highly vulcanized rubber in a continuous phase of a polyolefin. Critical to the properties of a TPV are the degree of vulcanization of the rubber and the fineness of its dispersion. The cross-linking and fine dispersion of the rubber phase gives a TPV high tensile strength (7.55–26.78 MPa), high elongation (375%–600%), resistance to compression and tension set, oil resistance, and resistance to flex fatigue. TPVs have excellent resistance to attack by polar fluids and fair-to-good resistance to hydrocarbon fluids. Maximum service temperature is 135°C.

Elastomeric alloys are available to the 55A to 50D hardness range, with ultimate tensile strength ranging from 5.44 to 27.2 MPa. Specific gravity of MPRs is 1.2–1.3; the TPV range is 0.9–1.0.

Uses

In 1981, a line of TPVs, called Santoprene, was commercialized based on EPDM rubber and polypropylene designed to compete with thermoset rubbers in the middle performance range. In 1985, a second TPV, Geolast, based on polypropylene and nitrile rubber was introduced. This TPV alloy was assigned to provide greater oil resistance than that of the EPDM-based material. The nitrile-based

TPV provides a thermoplastic replacement for thermoset nitrile and neoprene because oil resistance of the materials is comparable.

The MPR product line, called *Alcryn*, was introduced in 1985. It is a single-phase material, which gives it a stress–strain behavior similar to that of conventional thermoset rubbers. MPRs are plasticized alloys of partially cross-linked ethylene interpolymers and chlorinated polyolefins.

Thermoreactive Deposition/Diffusion Process

A method of coating steels with a hard, wear-resistant layer of carbides, nitrides, or carbonates. In the thermoreactive deposition/diffusion process, the carbon and nitrogen in the steel substrate diffuse into a deposited layer with a carbide-forming or nitride-forming element such as vanadium, niobium, tantalum, chromium, molybdenum, or tungsten. The diffused carbon or nitrogen reacts with the carbide- and nitride-forming elements in the deposited coating so as to form a dense and metallurgically bonded carbide or nitride coating at the substrate surface.

Thermoset

A resin that is cured, set, or hardened, usually by heating, into a permanent shape. The polymerization reaction is an irreversible reaction known as cross-linking. Once set, a thermosetting plastic cannot be remelted, although most soften with the application of heat.

Thermoset Injection Molding

A process in which thermoset material that has been heated to a liquid state is caused to flow into a cavity or several cavities and held at an elevated temperature for a specific time. After cross-linking is completed, the hardened part is removed from the open mold. The mold used is usually of through-hardened tool steel that has been chrome plated and highly polished.

Thermoset Plastics

Thermoset Composites

Thermoset matrix systems dominate the composites industry because of their reactive nature and ease of impregnation. They begin in a monomeric or oligomeric state, characterized by very low viscosity. This allows ready impregnation of fibers, complex shapes, and a means of achieving cross-linked networks in the cured part. The early high-performance thermoset-matrix materials were called advanced composites, differentiating them from the glass/polyester composites that were emerging commercially in the 1950s. The "advanced" term has come to denote, to most engineers, a resin-matrix material reinforced with high-strength, high-modulus fibers of glass, carbon, aramid, or even boron, and usually laid up in layers to form an engineered component. More specifically, the term has come to apply principally to epoxy-resin-matrix materials reinforced with oriented, continuous fibers of carbon or of a combination of carbon and glass fibers, laid up in multilayer fashion to form extremely rigid, strong structures.

Resin Systems

More than 95% of thermoset composite parts are based on polyester and epoxy resins; of the two, polyester systems predominate in volume by far. Other thermoset resins used in reinforced form are phenolics, silicones, and polyimides.

Polyesters

They can be molded by any process used for thermosetting resins. They can be cured at room temperature and atmospheric pressure, or at temperatures to 177°C and under higher pressure. These resins offer a balance of low cost and ease of handling, along with good mechanical, electrical, and chemical properties, and dimensional stability.

Polyesters can be compounded to be flexible and resilient, or hard and brittle, and to resist chemicals and weather. Halogenated (chlorinated or brominated) compounds are available for increased fire retardance. Low-profile (smoother surface) polyester compounds are made by adding thermoplastic resins to the compound.

Polyesters are also available in ready-to-mold resin/reinforcement forms—bulk-molding compound (BMC) and sheet-molding compound (SMC). BMC is a premixed material containing resin, filler, glass fibers, and various additives. It is supplied in a dough-like, bulk form and as extruded rope.

SMC consists of resin, glass fiber reinforcement, filler, and additives, processed in a continuous sheet form. Three types of SMC compounds are designated by Owens Corning Fiberglass Corp. as random (SMC-R), directional (SMC-D), and continuous fiber (SMC-C). SMC-R, the oldest and most versatile form, incorporates short glass fibers (usually about 25.4 mm long) in a random fashion. Complex parts with bosses and ribs are easily molded from SMC-R because it flows readily in a mold. SMC-C contains continuous glass fibers oriented in one direction, and SMC-D, long fibers (203–305 mm long), also oriented in one direction.

Moldings using SMC-C and SMC-D have significantly higher unidirectional strength but are limited to relatively simple shapes because the long glass fiber cannot stretch to conform to a shape. These two types of SMC are usually, but not always, used in combination with SMC-R. Various combinations are available that contain a total of as much as 65% glass by weight. These materials are used for structural, load-bearing components.

High-glass-content SMCs are also produced by PPG Industries, designated as XMC. These compounds contain up to 80% glass (or glass/carbon mixtures) as continuous fibers in an X pattern.

XMC is a high-glass content sheet molding compound. In one application which was designed by Motor Wheel Corp., and compression molded by Goodyear. A wheel composed of a three-ply hoop of Goodyear's XMC 65% long-glass and chopped fiber-reinforced resin on the periphery, and multiple plies of high-strength steel molding compound, reinforced with 50% random glass fiber, for the club. Forty percent lighter than a steel wheel and 10% to 20% lighter than an aluminum design, the composite wheel has undergone long-term and accelerated laboratory, proving ground, and field evaluations.

A class of resins produced by dissolving unsaturated, generally linear, alkyd resins in a vinyl-type active monomer such as styrene, methyl styrene, or diallyl phthalate. Cure is effected through vinyl polymerization using peroxide catalysts and promoters or heat to accelerate the reaction. One important commercial type is liquid resins that are cross-linked with styrene and used either as impregnants for glass or carbon fiber reinforcements in laminates, filament-wound structures, and other built-up constructions, or as binders for chopped-fiber reinforcements in molding compounds, such as *sheet molding compound* (SMC) and *bulk molding compound* (BMC). A second important type is liquid or solid resins that are cross-linked with other esters in chopped-fiber and mineral-filled molding compounds, such as alkyd and diallyl phthalate. See also *unsaturated polyesters*.

Epoxies

These are low-molecular-weight, syrup-like liquids that are cured with hardeners to cross-linked thermoset structures that are hard and tough. Because the hardeners or curing agents become part of the finished structure, they are chosen to provide desired properties in the molded part. (This is in contrast to polyester formulations wherein the function of the catalyst is primarily to initiate cure.) Epoxies can also be formulated for room-temperature curing, but heat-curing produces higher properties.

Epoxies have outstanding adhesive properties and are widely used in laminated structures. The cured resins have better resistance than polyesters to solvents and alkalies, with less resistance to acids. Electrical properties, thermal stability (to 288°C in some formulations), and wear resistance are excellent.

Phenolics

The oldest of the thermoset plastics have excellent insulating properties and resistance to moisture. Chemical resistance is good, except to strong acids and alkalies.

Reinforced phenolics are processed principally by high-pressure methods—compression molding and continuous laminating—because volatiles are condensed during the molding process. Recently developed injection-moldable grades, however, have made the processing of phenolics competitive with thermoplastic molding in some applications.

Silicones

These have outstanding thermal stability, even in the range of 260°C–371°C. Water absorption is low, and dielectric properties are excellent. Chemical resistance (except to strong alkalies) is very good.

Properties

Typical thermoset composites are brittle and have poor impact resistance. An impact may cause little visible surface damage but make the part dramatically weaker.

Thermoset resins need a chemical reaction, usually brought on by heat, to harden, or cure. The chemical reaction is irreversible; once cured, a thermoset material cannot be reprocessed or reformed. And molding cycle time for thermoset materials is largely determined by the curing time.

Thermosetting

(1) Having the property of undergoing a chemical reaction by the action of heat, catalysts, ultraviolet light, and so on, leading to a relatively infusible state. (2) Materials that, when first heated, can be molded and then "cured," that is, undergo a chemical change to become rigid. Subsequent heating does not cause softening.

Thermospray

A form of metal spraying in which powdered metal or other materials are injected into a high-temperature flame impinging on the surface to be coated.

Thermotropic Liquid Crystal

A liquid crystalline polymer that can be processed using thermoforming techniques.

Thickener

A solid material dispersed of a liquid lubricant to produce a grease. Silica, clays, and metallic soaps are widely used as thickeners.

Thick-Film Circuit

A circuit that is fabricated by the deposition of materials having between 5 and 20 μm (0.2 and 0.8 mils) thickness, such as screen-printed cermet pastes on a ceramic substrate, which are fired in a kiln to create permanent conductive patterns. Compare with *thin-film circuit*.

Thick-Film Lubrication

(1) A condition of lubrication in which the film thickness of the lubricant is appreciably greater than that required to cover the surface asperities when subjected to the operating load, so that the effect of the surface asperities is not noticeable. Also known as full-film lubrication. (2) The formation of a substantial film of lubricant between the faces of a plain bearing. This is the usual design intent, see *oil wedge*. See also *thin-film lubrication* and the term *lubrication regimes*.

Thief

A racking device or nonfunctional pattern area used in the electroplating process to provide a more uniform current density on plated parts. Thieves absorb the unevenly distributed current on irregularly shaped parts, thereby ensuring that the parts will receive an electroplated coating of uniform thickness. See also *robber*.

Thin-Film Circuit

A circuit fabricated by the deposition of material several thousand angstroms thick (such as a circuit fabricated by vapor deposition). Compare with *thick-film circuit*.

Thixotropic

A property of certain gels or adhesive systems to thin upon isothermal agitation (shearing) and to thicken upon subsequent test. The characteristic of a material whereby its viscosity is dependent on the shear rate. In practice, the more rapidly the material is stirred, the more fluid it becomes. The process is reversible so that when stirring ceases, the material becomes more viscous as in the case of nondrip paint. See also *rheocasting*.

Thixotropic Processing

Thixoforming

A method of producing aluminum castings with performance characteristics said to meet or exceed those of steel forgings with 65% less weight, and up to 25% lower cost, has been developed.

During the process, marketed under the name *thixoforming*, aluminum ingots are heated to the precise temperature at which the material begins to be transformed from the solid to the liquid state. In this form, it can be cast to make strong, lightweight parts that are cheaper and perform better than many forged steel parts.

The critical technology enabling this process is the temperature control system, which is capable of maintaining a uniform temperature throughout the ingot to within ±1°C. It is critical to the process that the entire ingot be precisely in the thixotropic state (borderline liquid/solid). If the center is harder than the exterior, the part will have less beneficial metallurgical properties. The process has reportedly had very good success in the automotive industries.

Thixotropic Casting

Ranging from setter tiles and saggers for use in the electronic ceramics industry, to furnace forehearth shapes for the container glass industry, to kiln-car furniture at structural ceramics, precast shapes offer significant benefits.

A new generation of precast shapes has been developed using a thixotropic casting process. This process provides a smoother surface, more precise tolerances, greater strength, improved thermal shock resistance, higher heat tolerances, and more resistance to chemical attack than traditional refractory castable shapes.

Process

Thixotropic casting uses a dispersion agent that allows the ceramic mix to flow when vibrated without requiring high water content. The mix, vibrated into a plaster of paris mold, deairs and consolidates without the use of a cement-bonding agent. The finely crafted plaster of paris mold provides greater dimensional uniformity than common wooden or metal molds and results in a much smoother surface finish.

The crafting process for plastic molds generally involves four separate steps. The first step is to make an exact model of the end product. A master mold, which is a negative of the model, is then made. Then a case mold is made. The case mold is a replica of the original model and is made to protect the original from damage. From the case mold comes the durable working mold, which is a copy of the master mold and a negative of the case mold.

The four-step process is in place to ensure consistency from mold to mold. Depending on the various idea factory processes, each working mold is used to produce roughly 10–50 pieces. The working mold is then replaced with another, again to ensure manufacturing uniformity.

All shapes produced by the thixotropic casting process are high temperature fired between 1412°C and 1524°C. This develops a ceramic bond that enables the products to withstand greater temperatures for a longer period of time.

The fact that these shapes are made without a cement bond—they use a high-fired ceramic bond, instead—accounts for their high heat tolerance. Above 1371°C, cement-bonded refractories begin to soften, the result of glass phase formation. But firing the shapes to up to 1524°C results in a completely formed ceramic bond that will not readily soften in temperatures above 1371°C.

Firing at high temperatures also has the advantage of adding stability to precast shapes. When you first fire a castable shape with a cement bond, a lot of mineralogical changes take place as the product heats up. It can change the overall size and shape of the product, which can lead to an inconsistent product for the customer. By firing the shapes to 1412°C and beyond, there will be no unexpected or unwanted mineralogical changes when first used. The changes that occur in firing are taken into consideration when the molds are designed. The result is that the product performs as it was designed.

The firing process for precast-fired shapes is a highly controlled operation. All shapes are fired in a 7-day cycle that includes a precise drying process. The periodic kilns that are used to dry and fire the shapes are temperature controlled to within 20°, which is close tolerance for this type of manufacturing. Even the kiln cars are loaded to facilitate a specific airflow pattern that enhances strength and consistency.

Applications

Typical applications for precast kiln furniture shapes include saggers, setter plates, and pusher tiles.

The most common use is with technical ceramic products that have to be fired at higher-temperature areas—above 1301°C.

Other applications include kiln furniture and structural ceramics. The process can handle very small pieces to the very large pieces like kiln furniture, kiln car shapes, posts, and beams—even tracks.

Other applications for precast shapes include glass contact parts in the container glass industry—stirrers, cover blocks, spouts, plungers, and tubes. These glass tank forehearth shapes enable glass manufacturers to maintain consistency in manufacturing their products.

Shapes are also available with a variety of chemical compositions, depending on the requirements of the application. Compositions range from 50% alumina products to shapes with ultrahigh alumina and alumina/zirconia/silica content.

Thixotropy

A property of certain gels or adhesive systems to thin upon isothermal agitation (shearing) and to thicken upon subsequent tests.

Thomas Converter

A Bessemer converter having a basic bottom and lining, usually dolomite, and employing a basic slag. Similar to Bessemer converter.

Thomson Effect (Alternatively the Kelvin Effect)

The emf produced by a difference of temperature between two sites of a conductor and, conversely, the release or absorption of heat associated with an electrical conductor flowing between two sites on the conductor which are at different temperatures. See also *thermoelectric effect*.

Thoria–Thorium Oxide, ThO₂

Small amounts are added to tungsten for lamp filaments to restrain grain growth during long-term high-temperature service.

Thorium

A soft ductile, silvery-white metal (symbol Th) thorium occurs in nature to about the same extent as lead but so widely disseminated in minute quantities difficult to extract that it is considered a rare metal.

It was once valued for use in incandescent gas mantles in the form thorium nitrate, $Th(NO_3)_4$, but is now used chiefly for nuclear electronic applications.

For many years, thorium oxide has been incorporated in tungsten metal, which is used for electric light filaments; small amounts of the oxide have also been found to be useful in other metals and alloys. The oxide is employed in catalysts for the promotion of certain organic chemical reactions. Thorium oxide has special uses as a high-temperature ceramic material.

Thoria–urania ceramics are used for reactor fuel elements. They are reinforced with columbium or zirconia fibers to increase thermal conductivity and shock resistance. The metal or its oxide is employed in some electronic tubes, photocells, and special welding electrodes. The metal can serve as a getter in vacuum systems and in gas purification, and it is also used as a scavenger in some metals.

Because of its high density, chemical reactivity, mediocre mechanical properties, and relatively high cost, thorium metal has no market value as a structural material. However, many alloys containing thorium metal have been studied in some detail, and thorium does have important applications as an alloying agent in some structural metals. Perhaps, the principal use for thorium metal, beyond its use in the nuclear field, is in magnesium technology. Approximately 3% thorium, added as an alloying ingredient, imparts to magnesium metal high-strength properties and creep resistance at elevated temperatures. The magnesium alloys containing thorium, because of their lightweight and desirable strength properties, are being used in aircraft engines and an airframe construction.

Thorium can be converted in a nuclear reactor to uranium-233, an atomic fuel. The system of thorium and uranium-233 gives promise of complete utilization of all thorium in the production of atomic power. The energy available from the world supply of thorium has been estimated as greater than the energy available from all of the world's uranium, coal, and oil combined.

Thorium is a radioactive metallic element, and about 1%–2% is added to tungsten to improve electron emission in applications such as tungsten inert gas (gas tungsten arc) welding electrodes and thermionic valve filaments.

Thread

The helical ridge and groove around a threaded fastener. See *fiber* (composites).

Thread Count

The number of yarns (threads) per inch in either the lengthwise (warp) or crosswise (fill or weft) direction of woven fabrics used for reinforcing plastics.

Thread Rolling

The production of threads by rolling the piece between two grooved die plates, one of which is in motion or between rotating grooved circular rolls. The result is less accurate than machining but may introduce favorable residual stresses. Also known as roll threading.

Threaded Fasteners

Devices such as bolts, studs, and screws and the associated internally threaded nuts or holes, which allow components to be clamped together. See *preload*.

Threading

Producing external threads on a cylindrical surface.

1. *Die threading*: A process for cutting external threads on cylindrical or tapered surfaces by the use of solid or self-opening dies.
2. *Single-point threading*: Turning threads on a lathe.
3. *Thread grinding*: See definition under *grinding*.
4. *Thread milling*: A method of cutting screw threads with a milling cutter.

Threading and Knurling (Thermal Spraying)

A method of surface roughening in which spiral threads are prepared, followed by upsetting with a knurling tool.

Three High Mill

See *rolling mill*.

Three-Point Bending

(1) The bending of a piece of metal or a structural member in which the object is placed across two supports, and force is applied between and in opposition to them. (2) A bend test in which the test piece is supported at two points and loaded at a point midway between them. The stress peaks at the central loading point unlike four-point bending. See also *V-bend die*.

Three-Quarters Hard

A *temper* of nonferrous alloys and some ferrous alloys characterized by tensile strength and hardness about midway between those of *half hard* and *full hard* tempers.

Threshold Stress

Threshold stress for *stress corrosion cracking*. The critical gross section stress at the onset of stress corrosion cracking under specified conditions.

Throat Depth (Resistance Welding)

The distance from the centerline of the electrodes or platens to the nearest point of interference for flat sheets in a resistance welding machine. In the case of a resistance seam welding machine with a universal head, the throat depth is measured with the machine arranged for transverse welding.

Throat Height (Resistance Welding)

The unobstructed dimension between arms throughout the throat depth in a resistance welding machine.

Throat of a Fillet Weld

A term that includes the theoretical throat, the actual throat, and the effective throat. (1) The theoretical throat is the distance from the beginning of the root of the joint perpendicular to the hypotenuse of the largest right triangle that can be inscribed within the fillet weld cross section. This dimension is based on the assumption that the root opening is equal to zero. (2) The actual throat is the shortest distance from the root of the weld to its face. (3) The effective throat is the minimum distance minus any reinforcement from the root of the weld to its face. See also *concave fillet weld* and *convex fillet weld*.

Throat of a Groove Weld

See preferred term *size of weld*.

Throat Opening

See preferred term *horn spacing*.

Through Weld

A nonpreferred term sometimes used to indicate a weld of substantial length made by melting through one member of a lap or tee joint and into the other member.

Throughput

Volume of charge passed in a time unit through a production sintering furnace.

Throw

The distance from the centerline of the crankshaft or main shaft to the centerline of the crankpin or eccentric in crank or eccentric presses. Equal to one-half of the stroke. See also *crank press* and the term *eccentric gear*.

Throwing Power

(1) The relationship between the *current density* at a point on a surface and its distance from the *counterelectrode*. The greater the ratio of the surface resistivity shown by the electrode reaction to the volume resistivity of the electrolyte, the better is the throwing power of the process. (2) The ability of a plating solution to produce a uniform metal distribution on an irregularly shaped *cathode*. Compare with *covering power*.

Thrust Bearing

A bearing in which the load acts in the direction of the axis of rotation.

Thulium

A metallic element, one of the rare earth groups, with little commercial application.

Thyristors

Thyristors (semiconductor controlled rectifiers) made from silicon carbide have been fabricated and tested as prototypes of power-switching devices capable of operating at temperatures up to 350°C. The highest-voltage-rated of these thyristors are capable of blocking current at forward or reverse bias as large as 900 V and can sustain forward current as large as 2 Å with a forward potential drop of

−3.9 V. the highest-power rated of these thyristors, which are also the highest-power-rated SiC thyristors, can block current at a forward or reverse bias of 700 V and can sustain an "on" current of 6 Å at a forward potential drop of −3.67 V. The highest-current rated of these thyristors can block current at a forward or reverse bias of 400 V and can sustain an "on" current of 10 Å.

These thyristors feature epitaxial *n*- and *p*-doped layers of 4H-SiC in the sequence *npnp* starting on the substrate; the structure stands in contrast to the *pnpn* structure of common silicon thyristors. The fabrication of the high-quality crystalline structures needed in these layers has been made possible by advances in growth of crystals, epitaxial growth of thin films, doping by both in situ and ion-implantation techniques, oxidation, formation of electrical contacts, and other techniques involved in the fabrication of electronic devices. The above *npnp* 4H-SiC thyristors have been found to exceed the speed of the fastest inverter-grade silicon thyristors.

Two other important parameters for a thyristor are (1) the maximum rate of increase of forward applied voltage that can be applied before the thyristor latches on and (2) the time taken to achieve a high forward current density. The 4H-SiC thyristors show no turn-on even when forward bias was ramped up at a rate of 900 V/μs. Measurements and pulsed operation showed that it took between 3 and 5 ns for these devices to start carrying currents at densities of 2800 Å/cm^2.

Tide Marks

See *beach marks*.

Tie Bar

A bar-shaped connection added to a casting to prevent distortion caused by uneven contraction between two separated members of the casting.

Tie Line

See *lever rule*.

Ties, Tie Rod Bars

Beams carrying significant tensile loads in a construction or assembly.

TIG Welding

Tungsten inert gas welding; see preferred term gas *tungsten-arc welding*.

Tiger Stripes

Continuous bright lines on sheet or strip in the rolling direction.

Tight Fit

A loosely defined fit of slight negative allowance the assembly of which requires a light press or driving force.

Tile

As a structural material, tile is a burned clay product in which the coring exceeds 25% of the gross volume; as a facing material, any thin, usually flat, square product. Structural tile used for load bearing may or may not be glazed; it may be cored horizontally or vertically. Two principal grades are manufactured: one for exposed masonry construction, and the other for unexposed construction. Among the forms of exposure is frost; tile for unexposed construction where temperatures drop below freezing is placed within the vapor barrier or otherwise projected by a facing, in contrast to roof tile.

Structural tile with a ceramic glaze is used for facing. The same clay material that is molded and fired into structural tile is also made into pipe, glazed for sewer lines, or unglazed for drain tile.

As a facing, clay products are formed into thin flat, curved, or embossed pieces, which are then glazed and burned. Commonly used on surfaces that are subject to water splash or that require frequent cleaning, such vitreous glazed wall tile is fireproof. Unglazed tile is laid as bathroom floor. By extension, any material formed into a size comparable to clay tile is called tile. Among the materials formed into tile are asphalt, cork, linoleum, vinyl, and porcelain.

Tile is a ceramic surfacing unit, usually relatively thin in relation to facial area, made from clay or a mixture of clay of other ceramic materials, having either a glazed or unglazed face and fired above red heat in the course of manufacture to a temperature sufficiently high to produce specific physical properties and characteristics.

Tilt

In electron microscopy, the angle of the specimen relative to the axis of the electron beam; at zero tilt, the specimen is perpendicular to the beam axis.

Tilt Boundary

An interface between two grains or the subgrains of the same grain where the angular misalignment is very small that the boundary is little more than a single plane of dislocations. Also consists of an array of edge *dislocations*.

Tilt Casting/Molding

Casting processes in which the mold is initially tilted to allow the molten metal to be poured gently onto a sloping side. As the mold fills, it is progressively rotated to its vertical position. The process is intended to eliminate turbulence which might entrain dross.

Tilt Furnace

A furnace for the infiltration of copper into porous sintered tungsten for heavy-duty contacts. The furnace is tilted to one position for the separate sintering of the tungsten and melting of the copper, and then tilted to the opposite position to let the melt run to and contact or infiltrate, respectively, the tungsten pieces.

Tilt Hammer

A primitive forge in which a centrally pivoted beam carries the hammer at one end and the other end is struck repeatedly by an eccentric on a rotating shaft. The eccentric raises the hammer, which then falls freely to strike the workpiece on the anvil.

The drive could be provided by various sources—often water wheel or steam or even animal power.

Tilt Mold

A casting mold, usually a book (permanent) mold, that rotates from a horizontal to a vertical position during pouring, which reduces agitation and thus the formation and entrapment of oxides. See also *permanent mold*.

Tilt Mold Ingot

An ingot made in a *tilt mold*.

Tilting Pad Bearing

A pad bearing in which the pads are free to take up a position at an angle to the opposing surface according to the hydrodynamic pressure distribution over its surface.

Time Delay (of Yield in Some Steels)

The phenomenon whereby rapid application of a load above the yield stress does not induce immediate yield. This is because a period, typically a few milliseconds, is required for dislocations to break away from the lattice features such as interstitial atoms, to which they are pinned.

Time Profile

A plot of the modulus, damping, or both, of a material versus time.

Time Quenching

A term used to describe a quench in which the cooling rate of the part being quenched must be changed abruptly at some time during the cooling cycle. Same as *martempering*.

Time–Temperature Curve

A curve produced by plotting time against temperature.

Time–Temperature–Transformation Diagram

See *isothermal transformation diagram*.

Timing Marks

Sharp lines produced by slightly changing the direction of the applied tension at regular time intervals. Also known as *arrest marks*.

Tin and Alloys

A silvery-white lustrous metal (symbol Sn), with a bluish tinge, tin is soft and malleable and can be rolled into foil as thin as 0.0051 cm. Tin melts at 232°C. Its specific gravity is 7.298, close to that of steel. Its tensile strength is 27 MPa. Its hardness is slightly greater than that of lead, and its electrical conductivity is about one-seventh that of silver. It is resistant to atmospheric corrosion but is dissolved in mineral acids. The cast metal has a crystalline structure, and the surface shows dendritic crystals when cast in a steel mold.

It alloys readily with nearly all metals.

Tin is a nontoxic, soft, and pliable metal adaptable to cold working such as rolling, extrusion, and spinning. It is highly fluid when molten and has a high boiling point, which facilitates its use as a coating for other metals. It can be electrodeposited readily on all common metals.

Properties

Tin reacts with strong acids and strong bases but is relatively inert to nearly neutral solutions. In indoor and outdoor exposure, it retains its white-silvery color because of its resistance to corrosion. A thin film of stannic oxide is formed in air, which provide surface protection.

Two allotropic forms exist: white tin (β) and gray tin (α). Although the transformation temperature is 13.2°C, the change does not take place unless the metal is of high purity, and only when the exposure temperature is well below 0°C. Commercial grades of tin (99.8%) resist transformation because of the inhibiting effect of the small amounts of bismuth, antimony, lead, and silver present as impurities.

Alloying elements such as copper, antimony, bismuth, cadmium, and silver increase its hardness. Tin tends rather easily to form hard, brittle intermetallic phases, which are often undesirable. It does not form wide solid solution ranges and other metals in general, and there are a few elements that have appreciable solid solubility in tin. Simple eutectic systems, however, occur with bismuth, gallium, lead, thallium, and zinc.

Tin is rarely used alone; rather, it is generally used as a coating for a base metal or as a constituent of an alloy. The range of useful alloys is extensive and extremely important.

Forms

Tin can be obtained in a number of forms; granulated, mossy, fine powder, sheet, foil, and wire. Tin-base alloys are available in many forms; solder can be obtained in 0.46 kg bars, solid and cored wire, powder, sheet, and foil; babbitt-type metal and casting alloys in bars and ingots; bronze in ingot; continuously cast bars at shapes, sheet, and foil.

Applications

Full use is made of the ductility, surface smoothness, corrosion resistance, and hygienic qualities of tin in the form of foil, pipe, wire, and collapsible tubes. Tin foil is devoid of springiness and is ideal for wrapping food products, as liners for bottle caps, and for electrical condensers. Heavy-walled tin pipe and tin-lined copper pipe are used by the food and beverage industries for conveying distilled water, beer, and soft drink syrups. Tin wire is used for electrical fuses and for packing glands in pumps of food machinery. Collapsible tubes, made by impact extrusion from disks of pure tin, are used for pharmaceutical and food products. Additionally, tin is used in brasses, bronzes, and the babbitts, and in soft solders.

The most important use of tin is for tin-coated steel containers (tin cans) used for preserving foods and beverages. Other important uses are solder alloys, bearing metals, bronzes, pewter, and miscellaneous industrial alloys. Tin chemicals, both inorganic and organic, find extensive use in the electroplating, ceramic, plastic, and agricultural industries.

Tinplate manufacture is now largely a continuous electrolytic process with only a small percentage of production in hot-tinning machines. The coating thickness may be less than 0.01 mm. Heavily coated tinned steel sheet is used in making gas meters and automotive parts such as filters and air coolers.

The electrical industry is a large user of tin-coated steel and copper in the form of connectors, capacitor and condenser cans, and tinned copper wire. Many kinds of food-handling machinery, including holding tanks, mixtures, separators, milk cans, pipes, and valves, are made of tinned steel, cast iron, copper, or brass. The tin coating may be applied by dipping in molten tin or by electrodeposition.

The plating industry utilizes tin as anodes for the electrodeposition of pure tin and tin alloy coatings. Plated tin, either as a matt or bright finish, provides easily solderable surfaces for steel, copper, or aluminum. Tin alloy coatings (tin–copper, tin–lead, tin–nickel, tin–zinc, tin–cadmium, tin–cobalt) have advantages over single-metal plates. They are denser and harder, more corrosion resistant, brighter or more easily buffed, and more protective to basis metals. Tin–copper (12% tin) has the appearance of 24-karat gold and, when lacquered, serves as an attractive finish for jewelry, trophies, wire goods, and hardwares. Tin–lead electroplates (40%–65% tin) have excellent corrosion resistance and solderability and are well adapted to the plating of printed circuits and electronic parts. Tin–zinc coatings (75% tin) are a good alternative coating for cadmium in particular applications, and they provide galvanic protection to steel in contact with aluminum. Tin–cadmium coatings (25% tin) are especially resistant to salt vapors and have a number of applications in the aircraft industry and as a coating for fasteners. A tin–nickel coating (65% tin) finds use as an etchant resist in the manufacture of printed circuit boards, as well as an ornamental and highly corrosion-resistant finish for watch parts, scientific instruments, and power connectors. The tin–cobalt alloy (80% tin) has an appearance similar to a chromium deposit and is used to plate fasteners, ancillary office equipment, hinges, kitchen utensils, hand tools, and tubular furniture.

Stannic Oxide (SnO₂)

SnO_2 has by far its largest commercial outlet in the ceramic industry, where it is used either as a white pigment (i.e., opacifier) or as a constituent of colored pigments in the glazes applied to, for example, crockery, lavatory ware, and decorative wall tile.

Tin oxide is an important constituent of ceramic stains for enamels, glazes, and bodies. Pink and maroon colors are obtained with tin oxide, chrotin oxide, and vanadium compounds. Tin oxide also is an important color stabilizer for some of the tin-bearing pink, gray, yellow, and blue coloring stains for glazes.

This oxide was formally an important opacifier for enamels on cast iron and sheet steel, although it has been replaced by substitute materials such as antimony, zirconium, titanium, and other compounds. The substitution has been largely for economic reasons, as tin oxide is still recognized as the superior opacifier from the standpoint of quality for both glazes and enamels.

In glass, stannic oxide is an important addition to cadmium–selenium and gold colors, especially reds. Stannous oxide, SnO, is a necessary ingredient in the development of copper ruby glass and is also used to produce black glasses. Stannous oxide is a black powder and is also a component of ruby-red and black glasses. Tin oxide, because of its resistance to solution and most glasses, especially those high in lead oxide, is being used for refractories for special applications, such as glass feeders, and conducting electrodes for electrical resistance melting of glass.

Alloys

Soft Solders

Soft solders constitute one of the most widely used and indispensable series of tin-containing alloys. Common solder is an alloy of tin and lead, usually containing 20%–70% tin. It is made easily by melting the two metals together. With 63% tin, a eutectic alloy melting sharply at 169°C is formed. This is much used in the electrical industry. A more general-purpose solder, containing equal parts of tin and lead, has a melting range of 31°C. With less tin, the melting range is increased further, and wiping joints such as plumbers make can be produced. Lead-free solders for special uses include tin containing up to 5% of either silver or antimony for use at temperatures somewhat higher than those for tin–lead solders and tin–zinc base solders often used in soldering aluminum.

Bronzes

Bronzes are among the most ancient of alloys and still form an important group of structural metals. Of the true copper–tin bronzes, up to 10% tin is used in wrought phosphor bronzes and from 5% to 10% tin in the most common cast bronzes. Many brasses, which are basically copper–zinc alloys, contain 0.75%–1.0% tin for additional corrosion resistance such wrought alloys as admiralty metal and naval brass, and up to 49% tin in cast leaded brasses. Among special cast, bronzes are bell metal, historically 20%–24% tin for best tonal quality, and speculum, a white bronze containing 33% tin that gained fame for high reflectivity before glass mirrors were invented. Although soft, conformable, and corrosion resistant, the low mechanical strength of bronze must be boosted by bonding to steel, cast iron, or bronze backing materials.

Pewters

Pewter is an easily formed tin-base alloy that originally contained considerable lead. Thus, because colonial pewter darkened and because of potential toxicity effects, its use was discouraged. Modern pewter is lead-free. The most favorable composition, *Britannia metal*, contains about 7% antimony and 2% copper. This has desired hardness and luster retention, yet it can be readily cast, spun, and hammered.

Alloys that contain from 1% to 8% antimony and 0.5% to 3% copper have excellent castability and workability. For spun pewter products, antimony content is usually below 7%, and pewter casting alloys contain 7.5% antimony and 0.5% copper. Because of the excellent drawing and spinning properties of tin, wrought parts are usually made from pewter that is first cast into slabs and then rolled into sheet.

Babbitt/bearing Metal

Babbitt or bearing metal for forming or lining a sleeve bearing is one of the most useful tin alloys. It is tin containing 4%–8% each of copper and antimony to give compressive strength and a structure desired for good bearing properties. An advantage of this alloy is the ease with which castings can be made or bearing shells relined with simple equipment and under emergency conditions. Aluminum–tin alloys are used in bearing applications that require higher loads than can be handled with conventional babbitt alloys.

Type Metals

Type metals are lead-base alloys containing 3%–15% tin and a somewhat larger proportion of antimony. As with most tin-bearing alloys, these are used and remelted repeatably with little loss of constituents. Tin adds fluidity, reduces brittleness, and gives a structure that reproduces fine detail.

Flake and nodular gray iron castings are improved by adding 0.1% tin to give a fully pearlitic matrix with attendant higher hardness, heat stability, and improved strength and machinability.

Tin is commonly an ingredient in costume jewelry, consisting of pewter-like alloys and bearing metal compositions often cast in rubber molds, in die castings hardened with antimony and copper for applications requiring close tolerances, thin walls, and bearing or nontoxic properties, and in low-melting alloys for safety appliances. The most common dental amalgam for filling teeth contains 12% tin.

Die-Cast Tin Base

Historically, these were the first materials to be die cast. Low melting point and extreme fluidity of these alloys produce sound, intricate castings inexpensively and with little wear on molds. Antimony, copper, and lead are the principal additions to tin in die-casting alloys. These alloys are mainly gravity or centrifugally die cast. Cast tin-alloy parts can be held to tolerances of 0.015 mm/mm, with wall thicknesses down to 0.79 mm. Shrinkage is negligible.

Fusible Tin

Melting temperatures for these alloys are usually below the solidus of eutectic-base tin–lead solders (184°C). Primary alloying elements include bismuth, lead, cadmium, and indium. Most of these alloys provide electrical or mechanical links in safety devices. Other applications include low-temperature solders, seals for glass and other heat-sensitive materials, foundry patterns, molds for low-volume production of plastic parts, internal support for tube bending, and localized thermal treatment of parts.

Tin Powders

Produced by atomization techniques, these powers are available in a number of mesh sizes. They are used in the manufacture of powder-metallurgy parts, and tinning and solder pastes, and in the spray metallization of surfaces.

In small amounts, tin is also combined with titanium, zirconium, and other metals to provide special properties. It is used as an alloy in nodular and gray irons to provide greater strength, increased and uniform hardness, and improve machinability. Tin–nickel and tin–zinc coatings are used in the braking systems of automobiles. The tin–nickel alloy is coated on disk-brake pistons because of its good resistance to wear and corrosion. Tin–zinc is used to plate master cylinders in automotive braking systems.

Noncritical parts such as costume jewelry and small decorative items such as figurines can be made by casting pewter and other low-melting tin-base alloys in rubber molds. Tin die-casting alloys are suitable for low-strength precision parts and bearings for household appliances, engines, motors and generators, and gas turbines. These bearings perform well even at start-up and run-down periods of operation, at which times they carry a heavy, unidirectional load without the benefit of a fully formed hydrodynamic film. Other applications for tin-base die castings include parts for food-handling equipment, instruments, gas meters, and speedometers.

Miscellaneous

Tin is allotropic, changing on cooling below about –20°C (–13.2°C is actual equilibrium temperature) from a fairly ductile body-centered tetragonal structure to a low ductility diamond structure. The damaging formation of the low-temperature phase during services referred to as *tin pest*. Tin offers protection only by excluding the environment. Unlike zinc, tin does not offer sacrificial protection so if a tin coating is scratched the exposed steel corrodes.

See *electrochemistry and galvanizing*. Tin is a major constituent in white metals and solders and is a minor alloying element in some brasses and bronzes.

Tin Can

A can made from tin-plated steel.

Tin Cry

The creaking noise produced by tin and its alloys during deformation.

Tin Foil

Very thin sheet of tin or high tin alloy. Incorrect term for aluminum foil, which is much more common for wrapping foodstuffs and similar applications.

Tin Sweat

The formation of tin-rich beads on the surface of cast bronze due to inverse segregation. It may also occur if bronze is heated close to its melting point. See *sweat*.

Tin Tossing

Oxidizing impurities of molten tin by pouring it from one vessel to one another in air, forming a dross that is mechanically separable.

Tinmans Solder

A soft solder of eutectic composition.

Tin–Nickel Plating

An electroplated coating of nickel and tin, usually on a brass substrate, giving a highly reflective, corrosion resistant, and low friction surface.

Tinning

Coating metal with a very thin layer of molten solder or brazing filler metal. Usually prior to making a soldered or brazed joint.

Tin-Pest

A polymorphic modification of tin that causes it to crumble into a powder known as gray tin. It is generally accepted that the maximum rate of transformation occurs at about –40°C, but transformation can occur at as high as about 13°C.

Tinplate

Tinplate is soft-steel plate containing a thin coating of pure tin on both sides. A large proportion of the tinplate use goes into the manufacture of food containers because of its resistance to the action of vegetable acids and its nonpoisonous character. It solders easily and also is easier to work in dies than terneplate, so that it also is preferred over terneplate for making toys and other cheap articles in spite of a higher cost. Tinplate is made by the hot-dip process using palm oil as a flux or by a continuous electroplating process.

Processing and Uses

Tinplate manufacture in the United States is largely a continuous, high-speed electrolytic process, with less than 1% of production from hot-tinning machines. Electrolytic tinplate can be produced from either alkaline or acid electrolytes. Tinplates have either tin on each steel surface or a differential tin-coating thickness.

Hot-dipped tinplate is used for special corrosive packs, kitchen utensils, hardware items, and automotive parts. For other industrial applications, hot-dip tin coatings are applied to copper wire and sheet, as well as steel and cast iron parts. Examples are tinned copper and copper alloy strip for manufacture of electrical connectors and tinned food-processing equipment. Hot-dip tin–lead (terne) coatings find service as coatings for gasoline tanks, roofing materials, electronic applications, radiator water tubes, and component leads.

Tinsel

A decorative lustrous material usually as thin strands of foil. It was originally an alloy of about 60% zinc and 40% lead but now often some plastic material.

Tint Etching

Immersing metallographic specimens is specially formulated chemical etchants in order to produce a stable film on the specimen surface. When viewed under an optical microscope, these surface films produce colors, which correspond to the various phases in the alloy. Also known as *color etching*.

Tinting

See *heat tinting*.

Tin–Zinc Plating

An electroplated coating of an alloy of tin and zinc, typically about 75% tin and 25% zinc. The coating on steel offers corrosion protection close to that of zinc alone (see *galvanizing*) but is more readily soft soldered.

Tip Skid

See preferred term *electrode skid*.

TIR

Abbreviation for *total indicator reading*.

Titanates

Titanates are compounds made by heating a mixture of an oxide or carbonate of a metal and titanium dioxide. High dielectric constants, high refractive indices, and ferroelectric properties contribute primarily to their commercial importance.

Ferroelectricity may be described as the electric analogue of ferromagnetism. As a field is applied to a ferroelectric material, a nonlinear relationship between polarization and field (similar to the magnetization curve for iron) is observed. This increase in polarization is a function of the orientation of ferroelectric domains within the crystal.

As these domains become aligned, a saturation point is reached. If the field is now removed, the domains tend to remain aligned, and a finite value of polarization (called remanent polarization) can be measured. Extrapolation of the polarization at high field strength to zero field gives a somewhat higher value (spontaneous polarization).

To eliminate the remanent polarization, the field must be applied in the opposite direction, and the field required to return to the original state is called the coercive field. On further increase in the electric field, polarization in the opposite direction is achieved. This behavior leads to a characteristic ferroelectric hysteresis loop as the field is alternated.

Type of Materials

Titanates, for the most part, are prepared by heating a mixture of the specific oxide or carbonate with titanium dioxide. Titanium dioxide is of exceptionally high refractive index for a white oxide (2.6–2.9 for the rutile form, 2.5 for anatase) and due to its high refractive index finds wide application as a white pigment of high reflectance for the opacification of paint, plastics, rubber, paper, and porcelain enamels.

For electronics, polycrystalline (ceramic) titanium dioxide with its moderately high dielectric constant (−95) has been used as a capacitor; it does not show ferroelectric behavior. Titanium dioxide is normally an insulator, but in the oxygen deficient state when some of the Ti^{4+} sites are occupied by Ti^{3+} ions, it becomes an "n-type" semiconductor with conductivities in the range $1–10/\Omega$ cm.

Barium Titanate

Barium titanate crystals, $BaTiO_3$, are made by die-pressing titanium dioxide and barium carbonate and sintering at high temperatures. This crystal belongs to the class perovskite in which the closely packed lattice of barium and oxygen ions has a barium ion in each corner and an oxygen ion in the center of each face of a cube with the titanium ion in the center of the oxygen octahedron. For piezoelectric use, the crystals are subjected to a high current, and they give a quick response to changes in pressure or electric current. They also store electric charges and are used for capacitors. *Glennite 103* is a piezoelectric ceramic molded from barium titanate modified with temperature stabilizers.

Ceramic barium titanate can be made piezoelectric by applying a polarizing field of about 30 kV/cm at room temperature. The remanent polarization after removal of the polarizing voltage is permanent unless the material is overheated or subjected to high reverse voltages.

The advantages of ceramic barium titanate as a transducer lie in its mechanical strength, chemical durability, and ease of fabrication into virtually any shape desired. Barium titanate transducers, as ultrasonic generators, are used in various applications (emulsification, mixing, cleaning, drilling); other applications include such things as phonograph pickups and accelerometers.

Calcium Titanate

Calcium titanate, $CaTiO_3$, occurs in nature as the mineral perovskite. As a ceramic, it has a room-temperature dielectric constant of about 160. It is frequently used as an addition to barium titanate or by itself as a temperature compensating capacitor.

Single crystals have been grown by the flame fusion technique; calcium titanate crystals show a strong tendency toward twinning and, although the material is not ferroelectric, the twinning in the crystal shows a marked resemblance to the domain structure observed in barium titanate crystals.

Strontium Titanate

Strontium titanate, $SrTiO_3$, has a cubic perovskite structure at room temperature. It has a dielectric constant of about 230 as a ceramic and is commonly used as an additive to barium titanate to decrease the Curie temperature. By itself, it is used as temperature-compensating material because of its negative temperature characteristics. Strontium titanate has been used as a brilliant diamond-like gemstone and is a strontium mesotrititanate. Stones are made up to 4 karat. The refractive index is 2.412. It has a cubic crystal similar to the diamond, but the crystal is opaque in the x-ray spectrum. Single crystals of strontium titanate have been grown by the flame fusion process and strontium titanate is essentially colorless.

Because of its cubic structure, it is somewhat more satisfactory as an optical material than rutile (tetragonal). Strontium titanate crystals show a room-temperature dielectric constant of about 300 with a loss tangent of 0.0003.

Magnesium Titanate

Magnesium titanate, $MgTiO_3$, crystallizes as an ilmenite rather than perovskite structure. It is not ferroelectric and is used with titanium dioxide to form temperature-compensating capacitors. It has also been used as an addition agent to barium titanate.

Lead Titanate

$PbTiO_3$ is used as a less costly substitute for titanium oxide. It is yellowish in color and has only 60% of the hiding power but is very durable and protects steel from rust. Good ceramic specimens of lead titanate are somewhat difficult to prepare owing to the volatility of lead oxide at the firing temperature. Lead titanate is commonly used as a solid solution additive to increase the Curie temperature of barium titanate.

By substitution of zirconium for titanium in lead titanate a solid solution, lead zirconium titanate may be produced. Lead zirconate–lead titanate is a piezoelectric ceramic that can be used at higher temperatures than barium titanate.

Miscellaneous Titanates

The metatitanates of cadmium, manganese, iron, nickel, and cobalt all have an ilmenite rather than perovskite structure. None of these, as far as is known, is ferroelectric, and they are not particularly important electrically other than perhaps as addition agents to barium titanate. Nickel titanate has been grown as a single crystal, and its use as a rectifier has been suggested after the addition of suitable impurities during growth.

Bismuth stannate, a crystalline powder that dehydrates at about 140°C, maybe used with barium titanate in capacitors to increase stability at high temperatures.

Butyl titanate is a yellow viscous liquid used in anticorrosion varnishes and for flame-proofing fabrics. It is a condensation product of the tetrabutyl ester of orthotitanic acid and contains about 36% titanium dioxide.

Titanate fibers can be used as reinforcement in thermoplastic moldings. The fibers, called *Fybex*, can also be used in plated plastics to reduce thermal expansion, warpage, and shrinkage. Titanate fibers in plastics also provide opacity.

Titania

A white water-insoluble powder of composition, TiO_2, that is produced commercially from the minerals ilmenite and rutile. Used in paints and cosmetics and as an ingredient in porcelain enamels, ceramic whiteware, and ophthalmic glasses. Pure TiO_2 is also used as thin- or thick-film semiconductors. It is also used as white pigment for paints and is a common constituent in the coating of arc welding electrodes.

The best quality of titanium dioxide, or titania, is produced from ilmenite and is higher in price than many white pigments but has great hiding power and durability. Off-color pigments, with a light buff tone, are made by grinding rutile ore. The pigments have fine physical qualities and may be used wherever the color is not important. Titania is also substituted for zinc oxide and lithopone in the manufacture of white rubber goods and for paper filler. The specific gravity is about 4.

Titania crystals are produced in the form of pale-yellow, single-crystal boules for making optical prisms and lenses for applications where the high refractive index is needed. The crystals are also used as semiconductors and for gemstones. They have a higher refractive index than the diamond, and the cut stones are more brilliant but are much softer. The hardness is about 925 Knoop, and the melting point is 1825°C. The refractive index of the rutile form is 2.7 and that of the anatase is 2.5; the synthetic crystals have a refractive index of 2.616 vertically and 2.903 horizontally.

Titanium oxide is a good refractory and electrical insulator. The finely ground material gets good plasticity without binders and is molded to make resistors for electronic use. A microsheet is titanium oxide in sheets as thin as 0.008 cm for use as a substitute for mica for electrical insulation where brittleness is not important. Titania–magnesia ceramics have been made in the form of extruded rods and plates and pressed parts.

Uses

Titanium dioxide is a most important ceramic finish coat for sheet metal products. The opacity of this enamel imparted by titanium dioxide has lowered film thickness of these finishes to the range of organic coatings while retaining the durability of porcelain. These enamels are self-opacified. That is, titanium dioxide is not dispersed as an insoluble suspension during smelting nor is it added to the mill. Rather, titanium dioxide is taken into solution during smelting of the batch and is held in supersaturated solution through fritting. Upon firing the enamel, titanium dioxide crystallizes or precipitates from the glassy matrix.

Trimmers or trimmer condensers employing TiB_2 bodies are used for minute adjustments of capacitance. Normally, the rotor consists of a TiO_2 body. Parts are made with extreme accuracy and are usually supplied in one of three temperature coefficient types. The base is a low-loss ceramic composition.

Mechanical and physical properties of TiO_2 include relatively low strength (MOR 123.5–150.9 MPa; tensile strength 40.8–54.4 MPa, and low thermal conductivity 0.14 cal/cm/s/°C).

Titanium and Alloys

A metallic element (symbol Ti), titanium occurs in a great variety of minerals. The chief commercial ores of titanium are rutile and ilmenite. In rutile it occurs as an oxide. It is an abundant element but is difficult to reduce from the oxide. High-purity titanium (99.9%) has a melting point of about 1668°C, a density of 4.507 g/cm^3, and tensile properties at room temperature of about 234 MPa ultimate strength, 138 MPa yield strength, and 54% elongation. It is paramagnetic and has low electrical conductivity and thermal expansion.

The commercial metal is produced from sponge titanium, which is made by converting the oxide to titanium tetrachloride followed by reduction with molten magnesium. The metal can also be produced in dendritic crystals of 99.6% purity by electrolytic deposition from titanium carbide. Despite its high melting point, titanium reacts readily in copper and other metals and is much used for alloying and for deoxidizing. It is a more powerful deoxidizer of steel than silicon or manganese.

Melting

Because titanium has a great affinity for oxygen, nitrogen, and hydrogen at elevated temperatures, particularly when molten, all melting operations must be conducted in a vacuum and/or an inert gas atmosphere. The last 10 years has seen a dramatic shift in titanium melting capacity. At least 100 million kg of cold hearth melting capacity has been added to an industry dominated historically by vacuum arc remelting (VAR). Cold hearth melting is a more complex melting process that brings advantages, such as scrap recycling, shape casting (rectangular slabs), and unique alloying capability. Titanium presents special problems during melting because of its high reactivity with oxygen, nitrogen, and carbon. Exposure to these elements during melting causes severe embrittlement, even at low concentrations. This means that all titanium melting must occur in a vacuum or inert atmosphere (argon or helium). It also means that ceramic or graphite containers or liners are not permissible, limiting the design of melting containers to water-cooled copper vessels.

VAR in water-cooled copper crucibles quickly became the standard melting method. When discussing titanium cold hearth melting, one must divide the subject into two major divisions: (1) the production of commercially pure (CP) titanium and (2) the production of titanium alloys. Cold hearth production of CP titanium is done principally using electron beam (EB) melting and occurs at very high melting rates, in some cases greater than 2300 kg/h.

Cold heart melting of CP titanium provides several advantages. First is the use of economical raw materials, such as machined turnings contaminated with tungsten carbide particles. The cold hearth allows with tungsten carbide particles to be removed efficiently by gravity separation. The second advantage, as mentioned earlier, is high production rates. A third advantage is the ability to produce rectangular slabs instead of round ingots.

Although titanium alloys are successfully produced in both plasma and EB cold hearth processes, advantages for cold hearth melting titanium alloys are similarly impressive. As with CP, cold hearth melting of titanium alloys allows the use of many low-cost raw material forms with excellent yield. If EB melting holds an advantage for producing CP, plasma is simpler to use for titanium alloy melting. Because no alloy changes occur during plasma melting, chemistry control is relatively easy. EB melting of alloys containing high vapor pressure components is more problematic. For a common alloy such as Ti-6% Al-4% V, as much as 30% of the aluminum content may evaporate during EB melting. Alloy control in this case requires accurate prediction and compensation for elemental losses, and then very precise process control to meet the predicted losses uniformly throughout the ingot.

A very important technical justification for the use of cold hearth melting for titanium is then the production of premium quality titanium alloys for aircraft turbine engine rotating parts. Inclusion defects contained in the titanium used to produce these parts can be detrimental, leading to low cycle fatigue cracking, engine failure, and even aircraft loss. In the past 20 years, the aircraft engine producers have increasingly specified the use of cold hearth melting for the most critical titanium parts.

Cold hearth melting provides important mechanisms for the removal of inclusion-forming contaminants from titanium raw materials. High-density inclusion sources, such as tungsten carbide, tungsten, tantalum, and molybdenum, are easily removed by gravity separation (settling) in the cold hearth. By contrast, VAR provides virtually no removal of these defects.

The titanium melting industry has undergone a revolution in melting technology with plasma EB cold hearth melting replacing traditional VAR melting. All major U.S. titanium producers now partly or wholly own large-scale cold hearth melting facilities.

Cold hearth melting provides both economic and technical advantages to the titanium melter. Inexpensive raw materials not usable in other melt methods are readily utilized with excellent yields at high production rates. Potentially damaging inclusion sources may be removed assuring high-quality titanium products for the most demanding and critical applications.

Structure and Properties

Titanium is one of the few allotropic metals (steel is another), that is, it can exist in two different crystallographic forms. At room temperature, it has a close-packed hexagonal structure, designated as the alpha phase. At around 884°C, the alpha phase transforms to a body-centered cubic structure, known as the beta phase, which is stable up to the melting point of titanium of about 1677°C. Alloying elements promote formation of one or the other of the two phases. Aluminum, for example, stabilizes the alpha phase, that is, it raises the alpha to the beta transformation temperature. Other alpha stabilizers are carbon, oxygen, and nitrogen. Beta stabilizers, such as copper, chromium, iron, molybdenum, and vanadium, lower the transformation temperature, therefore allowing the beta phase to remain stable at lower temperatures, and even at room temperature. The mechanical properties of titanium are closely related to these allotropic phases. For example, the beta phase is much stronger, but more brittle than the alpha phase. Titanium alloys therefore can be usefully classified into three groups on the basis of allotropic phases: alpha, beta, and alpha–beta alloys.

Titanium and its alloys have attractive engineering properties. They are about 40% lighter than steel and 60% heavier than aluminum. The combination of moderate weight and high strengths, up to 1378 MPa, gives titanium alloys the highest strength-to-weight ratios of any structural metal. Furthermore, this exceptional strength-to-weight ratio is maintained from −216°C up to 538°C. A second outstanding property of titanium materials is corrosion resistance. The presence of a thin, tough oxide surface film provides excellent resistance to atmospheric and sea environments as well as a wide range of chemicals, including chlorine and organics containing chlorides. As it is near the cathodic end of the galvanic series, titanium performs the function of a noble metal. Titanium and its alloys, however, can react pyrophorically in certain media. Explosive reactions can occur with fuming nitric acid containing less than 2% water or more than 6% nitrogen dioxide and, on impact, with liquid oxygen. Pyrophoric reactions also can occur in anhydrous liquid or gaseous chlorine, liquid bromine, hot gaseous fluorine, and oxygen-enriched atmospheres.

Fabrication

Fabrication is relatively difficult because of the susceptibility of titanium to hydrogen, oxygen, and nitrogen impurities, which cause embrittlement. Therefore, elevated-temperature processing, including welding, must be performed under special conditions that avoid diffusion of gases into the metal. Heat is usually required in postforming operations.

Generally, titanium is welded by gas-tungsten arc, EB, or plasma-arc techniques. Metal inert gas processes can be used under special conditions. Thorough cleaning and shielding are essential because molten titanium reacts with nitrogen, oxygen, and hydrogen and will dissolve large quantities of these gases, which embrittle the metal. In all other respects, gas-tungsten arc welding of titanium is similar to that of stainless steel. Normally, a sound weld appears bright silver with no discoloration on the surface or along the heat-affected zone.

Like stainless steel, titanium sheet and plate work harden significantly during forming. Minimum bend radius rules are nearly the same for both, although spring back is greater for titanium. Commercially pure grades of heavy plate are cold formed or, for more severe shapes, warm formed at temperatures to about 427°C. Alloy grades can be formed at temperatures as high as 760°C in inert gas atmospheres. Tubes can be cold bent to radii three times the tube outside diameter, provided that both inside and outside surfaces of the bend are in tension at the point of bending. In some cases, tighter bends can be made.

Despite their high strength, some alloys of titanium have superplastic characteristics in the range of 816°C–927°C. The alloy used for most superplastically formed parts in the standard Ti-6Al-4V alloy. Several aircraft manufacturers are producing components formed by this method. Some applications involve assembly by diffusion bonding.

Titanium plates or sheets can be sheared, punched, or perforated on standard equipment. Titanium and Ti–Pd alloy plates can be sheared subject to equipment limitations similar to those for stainless steel. The harder alloys are more difficult to shear, so fitness limitations are generally about two-thirds those for stainless steel.

Titanium and its alloys can be machined and abrasive ground; however, sharp tools and continuous feed are required to prevent work hardening. Tapping is difficult because of the metal galls. Coarse threads should be used where possible.

Forms

Commercially pure titanium and many of the titanium alloys are now available in most common wrought mill forms, such as plate, sheet, tubing, wire, extrusions, and forgings. Castings can also be produced in titanium and some of the alloys; investment casting and graphite-mold (rammed graphite) casting are the principal methods. Because of the highly reactive nature of titanium in the presence of such gases as oxygen, the casting must be done in a vacuum furnace. Because of their high strength-to-weight ratio primarily, titanium and titanium alloys are widely used for aircraft structures requiring greater heat resistance than aluminum alloys. Because of their exceptional corrosion resistance, however, they (unalloyed titanium primarily) are also used for chemical-processing, desalination, and power-generation equipment, marine hardware, valve and pump parts, and prosthetic devices.

Unalloyed and Alloy Types

There are five grades of commercially pure titanium, also called unalloyed titanium; ASTM Grade 1, Grade 2, Grade 3, Grade 4, and Grade 7. They are distinguished by their impurity content, that is, the maximum amount of carbon, nitrogen, hydrogen, iron,

and oxygen permitted. Regardless of grade, carbon and hydrogen contents are 0.10% and 0.015% maximum, respectively. Maximum nitrogen is 0.03%, except for 0.05% in Grades 3 and 4. Iron content ranges from as much of 0.20% in Grade 1, the most pure (99.5%) grade, to as much as 0.05% in Grade 4, the least pure (98.9%). Maximum oxygen ranges from 0.18% in Grade 1 to 20.40% in Grade 4. Grade 7, 99.1% pure based on maximum impurity content, is actually a series of alloys containing 0.12%–0.25% palladium for improved corrosion resistance in hydrochloric, phosphoric, and sulfuric acid solutions. Palladium content has little effect on tensile properties, but impurity content, especially oxygen and iron, has an appreciable effect. Minimum tensile yield strength range from 172 MPa for Grade 1 to 483 MPa for Grade 4.

There are three principal types of titanium alloys: alpha or near-alpha alloys, alpha–beta alloys, and beta alloys. All are available in wrought form and some of each type for castings as well. In recent years, some also have become available in powder compositions for processing by hot isostatic pressing and other powder-metallurgy techniques. Titanium alpha alloys typically contain aluminum and usually tin. Other alloying elements may include zirconium, molybdenum, and, less commonly, nitrogen, vanadium, columbium, tantalum, or silicon. Although they are generally not capable of being strengthened by heat treatment (some will respond slightly), they are more creep resistant at elevated temperature than the other two types, are also preferred for cryogenic applications, and are more weldable but less forgeable. Ti-5%Al-2%Sn, which is available in regular and ELI grades (extra-low inertial) in wrought and cast forms, is the most widely used. In wrought and cast form, minimum tensile yield strengths range from 621 to 793 MPa, and tensile modulus is on the order of 107,000–110,000 MPa. It has useful strength to about 482°C and is used for aircraft parts and chemical-processing equipment. The ELI grade is noted for its superior toughness and is preferred for containment of liquid gases at cryogenic temperatures. Other alpha or near-alpha alloys and their performance benefits include Ti-8% Al-1% Mo-1% V (high creep strength to 482 °C), Ti-6% Al-2% Sn-4% Zr-2% Mo (creep resistance and stress stability to 593°C), Ti-6% Al-2% Cb-1% Ta-0.8% Mo (toughness, strength weldability), and Ti-2.25% Al-11% Sn-5% Zr-1% Mo (high tensile strength 931 MPa yield, superior resistance to stress corrosion, and hot salt media at 482°C). Another alpha alloy, Ti-0.3% Mo-0.8% Ni, also known as TiCode 12, is noted for its greater strength than commercially pure grades and equivalent or superior corrosion resistance, especially to crevice corrosion in hot salt solutions.

Titanium alpha–beta alloys, which can be strengthened by solution heat treatment and aging, afford the opportunity of parts fabrication and the more ductile annealed condition and then can be heat treated for maximum strength. Ti-6% Al-4% V, which is available in regular and ELI grades, is the principal alloy, its production alone having accounted for about half of all titanium and titanium alloy production. In the annealed condition, tensile yield strength is about 896 MPa and 13% elongation. Solution treating and aging increase yield strength to about 1034 MPa. Yield strength decreases steadily with increasing temperature, to about 483 MPa at about 510°C for the aged alloy. At 454°C, aged bar has a 1000 h stress-rupture strength of about 345 MPa. Uses range from aircraft and aircraft-turbine parts to chemical-processing equipment, and marine hardware. The alloy is also the principal alloy used for superplastically formed, and superplastically formed and simultaneously diffusion-bonded parts. At 899°C–927°C and low strain rates, the alloy exhibits tensile elongations of 600%–1000%, a temperature range also amenable to diffusion bonding the alloy.

Following are other alpha–beta alloys and their noteworthy characteristics:

Ti-6% Al-6% V-2% Sn; high strength to about 315°C, but low toughness and fatigue resistance

Ti-8% Mn; limited use for flat mill products; not weldable

Ti-7% Al-4% Mo; a forging alloy mainly, but limited use; 1034 MPa yield strength in the aged condition

Ti-6% Al-2% Sn-4% Zr-6% Mo; high strength, 1172 MPa yield strength, decreasing to about 759 MPa at 427°C; for structural applications at 400°C–540°C

Ti-5% Al-2% Sn-2% Zr-4% Mo-4% Cr and Ti-6% Al-2% Sn-2% Zr-2% Mo-2% Cr; superior hardenability for thick-section forgings; high modulus—about 117,000–124,000 MPa, respectively; tensile yield strength of about 1138 MPa

Ti-10% V-2% Fe-3% Al; best of the alloys in toughness at a yield strength of 896 MPa; can also be aged to a yield strength of about 1186 MPa; intended for use at temperatures to about 315°C

Ti-3% Al-2.5% V; a tubing and fastener alloy primarily, moderate strength and ductility, weldable.

Beta titanium alloys, fewest in number, are noted for their hardenability, good cold conformability in the solution-treated condition, and high-strength after aging. On the other hand, they are heavier than titanium and the other alloy types, their density ranging from about 4.84 g/cm³ for Ti-13% V-11% Cr-3% Al, Ti-8% Mo-8% V-2% Fe-3% Al, and Ti-3% Al-8% V-6% Cr-4% Zr-4% Mo to 5.07 g/cm³ for Ti-11.5% Mo-6% Zr-4.5% Sn, which is also known as Beta III. They are also the least creep resistant of the alloys. Ti-13% V-11% Cr-3% Al, a weldable alloy, can be aged to tensile yield strengths as high as 1345 MPa and retains considerable strength at temperatures to 315°C but has limited stability at prolonged exposure to higher temperatures.

Applications

One of the chief uses of the metal has been in the form of titanium oxide as a white pigment. It is also valued as titanium carbide for hard facings and for cutting tools. Small percentages of titanium are added to steels and alloys to increase hardness and strength by the formation of carbides or oxides or, when nickel is present, by the formation of nickel titanide.

The major portion of the commercial applications of titanium has been in the chemical-processing industries in the form of reactors, vessels, and heat exchangers. The pulp and paper industry has used it in bleaching equipment primarily as chlorine dioxide mixers, while the electrochemical industry has utilized heating and cooling (tubing) coils and anodizing and plating racks. Pumps, valves, thermowells, and other miscellaneous items are additional examples of commercial applications of titanium.

Titanium Carbide

A hard crystalline powder of the composition TiC; titanium carbide is made by reacting titanium dioxide and carbon black at temperatures above 1800°C. It is compacted with cobalt or nickel for use in cutting tools and heat-resistant parts. It is lighter in weight and less costly than tungsten carbide, but in cutting tools, it is more brittle. When combined with tungsten carbide in sintered carbide tool materials, however, it reduces the tendency of cratering in the tool.

Properties

TiC theoretically contains 20.05% carbon and is light metallic gray in color. It is chemically stable and is almost inert to hydrochloric and sulfuric acids. In oxidizing chemicals, such as aqua regia and nitric or hydrofluoric acids, TiC is readily soluble. It also dissolves in alkaline oxidizing melts. When heated in atmospheres containing nitrogen, nitride formation occurs above –1500°C. TiC is attacked by chlorine gas and is readily oxidized in air at elevated temperatures.

The density of TiC is 4.94 g/cm³, Mohs hardness is 9+, microhardness is 3200 kg/mm², and modulus of elasticity is 309,706 MPa. The modulus of rupture at room temperature has been reported as 499.8–843.2 MPa for materials sintered at 2600°C–3000°C. Hot modulus of rupture values are given as 107.78–116.96 MPa at 982°C and 54.4–63.92 MPa at 2200°C.

The melting point of TiC is 3160°C, and electrical resistivity at room temperature is 180–250 μ cm. It can be used as a conductor at high temperatures. Coefficient of thermal expansion between room temperature and 593°C is 4.12×10^{-6}/°F. Thermal conductivity is 0.041 cal/cm/s/°C.

Grades and Uses

A general-purpose cutting tool of this type contains about 82% tungsten carbide, 8% titanium carbide, and 10% cobalt binder. Kentanium is titanium carbide and various grades with up to 40% of either cobalt or nickel as the binder, used for high-temperature, erosion-resistant parts. For highest oxidation resistance only, about 5% cobalt binder is used. Other grades with 20% cobalt are used for parts where higher strength and shock resistance are needed and where temperatures are below about 982°C. Oxidation resistance only about 5% cobalt binder is used. Other grades with 20% cobalt are used for parts where higher strength and shock resistance are needed and where temperatures are below about 982°C. This material has a tensile strength of 310 MPa, compressive strength of 3789 MPa, and Rockwell hardness A90. Another for resistance to molten glass or aluminum has a binder of 20% nickel. A titanium carbide alloy for tool bits has 80% titanium carbide dispersed in a binder of 10% nickel and 10% molybdenum. The material has a hardness of Rockwell A93 and a dense, fine-grained structure. Ferro-Tic has the titanium carbide bonded with stainless steel. It has a hardness of Rockwell C55. Machinable carbide is titanium carbide in a matrix of Ferro-Tic C tool steel. Titanium carbide tubing is produced in round or rectangular form 0.25–7.6 cm in diameter. It is made by vapor deposition of the carbide without a binder. The tubing has a hardness above 2000 Knoop and a melting point of 3249°C.

Grown single crystals of titanium carbide have the composition TiC$_{0.94}$, with 19% carbon. The melting point is 3250°C, density 4.93, and Vickers hardness 3230.

The material is finding new uses in cermet components such as jet engine blades and cemented carbide tool bits. Titanium carbide has a relatively low electrical resistivity (1×10^{-4}) and can be used as a conductor of electricity, especially at high temperatures.

Extreme hardness of titanium carbide makes it suitable for wear-resistant parts such as bearings, nozzles, and coating tools. It also serves for special refractories under either neutral or reducing conditions. See also cemented *carbides, cermets,* and the term *composite coating.*

Titanium Diboride

The material has a melting point of 2980°C, is stable in HCl and HF acid, but decomposes readily in alkali hydroxides, carbonates, and bisulfates. It reacts with hot sulfuric acid.

Processing

Sintered parts of titanium diboride, TiB$_2$, are usually produced by either hot pressing, pressureless sintering, or hot isostatic pressing. Hot pressing of titanium diboride parts is conducted at temperatures >1800°C in vacuum or 1900°C in an inert atmosphere. Hot-pressed parts generally have a final density of >99% of theoretical. Typical sintering aids used for hot-pressed parts include iron, nickel, cobalt, carbon, tungsten, and tungsten carbide.

Pressureless sintering of TiB$_2$ is a less expensive method for producing net shape parts. Because of the high melting point of titanium diboride, sintering temperatures in excess of 2000°C often are necessary to promote sintering. Several different sintering aids have been developed to produce dense pressureless sintered parts by liquid-phase sintering. A combination of carbon and chromium, iron, or chromium carbide can be used as a sintering aid to produce pressureless sintered parts with a final density >95% of the theoretical density. Boron carbide also is added to inhibit grain growth during sintering. These sintering aids as well as atmospheric conditions can be used to lower the sintering temperature necessary for full densification.

Properties

Typical mechanical properties for hot-pressed titanium diboride include a flexural strength of 350–575 MPa, a hardness of 1800–2700 kf/mm^2, and a fracture toughness of 5–7 MPa m$^{-1/2}$. The mechanical property values are dependent on the type of fabrication method used (pressureless sintering vs. hot pressing), the purity of the synthesized powder, and the amount of porosity remaining in the finished part.

The elastic modulus of titanium diboride can range from 510 to 575 GPa and the Poisson ratio is 0.18–0.20. Titanium diboride has a room-temperature electrical resistivity of 15×10^{-6} Ω cm and a thermal conductivity of 25 Ω/mK.

Uses

Titanium diboride is used for a variety of structural applications including lightweight ceramic armor, nozzles, seals, wear parts, and cutting tool composites. Titanium diboride also has shown exceptional resistance to attack by molten metals, including molten aluminum, which makes it a useful material for such applications as metalizing boats, molten metal crucibles, and Hall–Heroult cell cathodes because of its intrinsic electrical conductivity. TiB$_2$ can be combined with a variety of other nonoxide ceramic materials, such as silicon carbide and titanium carbide, and oxide materials, such as alumina, to increase the main strength and fracture toughness of the matrix material.

Titanium Nitride

A hard, high-melting-point ceramic (2950°C) of the composition TiN that is used in cermets and as a coating material for cemented carbide cutting tools. See also *cermets* and the term *carbide tools*.

Titanium nitride whiskers (TiN$_w$) are single-crystal, acicular-shaped particles of titanium nitride that typically range in size from 0.3 to 1.0 μm in diameter with aspect ratios ranging from 5:1 to 50:1. TiN$_w$ have been produced using several different approaches including carbothermic reduction, laser synthesis, plasma synthesis, solid/solid and solid/gas combustion synthesis, and vapor phase reactions. TiN$_w$ are reported to form from both vapor–solid and vapor–liquid–solid mechanisms.

TiN$_w$ purity can vary from near stoichiometric (nitrogen of 22.7 wt%) to solid solutions of TiCN and TiON where carbon or oxygen substitutes for nitrogen. TiN$_w$ typically form with smooth surfaces and are relatively free from internal defects, which limit the strength of other whiskers such as SiC$_w$. Unlike SiC$_w$, TiN$_w$ are electrically conductive and have a coefficient of thermal expansion that is closer to steel and intermetallic materials. TiN$_w$ also have been found to have increased ability with iron-containing metals, alloys, and intermetallic compounds.

Potential applications include reinforcements in various materials including alumina and tungsten carbide–base cutting tool inserts for machining ferrous alloys, iron, nickel, and titanium aluminides (intermetallics), iron, nickel, and titanium metal/alloys and polymers.

Improved chemical resistance of TiN$_w$ with iron compounds makes it an excellent candidate for use as tool inserts for the machining of cast iron and tool steels. The reactivity of SiC$_w$ with iron compounds has limited alumina/SiC$_w$ cutting tool insert use primarily to superalloys. Alumina/TiN$_w$ composites have been shown to be machine readable using electric discharge machining.

Other areas of interest include stable reinforcements for iron-, nickel-, and titanium-based intermetallic compounds and metal/alloys and electrically conducting reinforcements for polymers to promote electrical conduction and charge dissipation. TiN$_w$ provide chemical stability and current carrying characteristics desirable in many applications.

Titanium Ores

The most common titanium ores are ilmenite and rutile. Ilmenite is an iron-black mineral having a specific gravity of about 4.5 and containing about 52% titanic oxide, or titania. The ore of India is sold on the basis of titanium oxide content, and the high-grade ore averages about 60% titanium oxide, 22.5% iron. and 0.4% silica. Titanium ores are widely distributed and plentiful. Ilmenite is found in northern New York, Florida, North Carolina, and in Arkansas, but the most extensive, accessible resources are found in Canada. The Quebec ilmenite contains 30% iron. The concentrated ore has about 36% titanium oxide and 41% iron and is smelted to produce pig iron and a slag containing 70% titania, which is used to produce titanium oxide. The beach sands of Senegal are mixed ores, the ilmenite containing 55%–58% titanium oxide and the *zirconiferous quartz* containing 70%–90% zirconia. The beach sands of Brazil are washed to yield a product averaging 71.6% ilmenite, 13% zircon, and 6% monazite.

Rutile is a titanium dioxide, containing theoretically 60% titanium. Its usual occurrence is crystalline or compact massive, with a specific gravity of 4.18 and 4.25 and Mohs hardness 6–6.5. The color is red to brown, occasionally black. Rutile was found in granite, gneiss, limestone, or dolomite. It is obtained from beach sand of northern Florida and Espirito Santo, Brazil, and is also produced in Virginia, in Australia, and in India. Rutile is also produced in Arkansas and Massachusetts. Rutile is marketed in the form of concentrates on the basis of 79%–98.5% titanium oxide. It is used as an opacifier in ceramic glazes and to produce tan-colored glass. It is also employed for welding-rod coatings. On welding rods, it aids stabilization of the arc and freezes the metal of slag.

Titration

A method of determining the composition of a sample by adding known volumes of a solution of known concentration until a given reaction (color change, precipitation, or conductivity change) is produced. See also *volumetric analysis*.

TIV

Abbreviation for *total indicator variation*.

T-Joint

See *tee joint*.

TLC

See *thin-layer chromatography*.

Toe Crack

A crack in the base metal occurring at the toe of a weld. See *weld crack*.

Toe of Weld

(1) Usually, this term refers to the line on an exposed surface where the boundary of the deposited weld metal meets parent metal. (2) Alternatively, it can refer to any line where the surface boundary of a weld run meets parent metal or a previous weld run. It will be seen that (2) includes positions where the toe has been over run and used by a subsequent deposit. In most cases, references to "toes" without any further description will imply definition (1) but sometimes the terms primary toes and secondary toes are used to refer, respectively, to definitions (1) and (2). See *fillet weld*.

Toggle Press

A *mechanical press* in which the *slide* is actuated by one or more toggle links or mechanisms.

Tolerance

The specified permissible deviation from a specified nominal dimension or the permissible variation in size or other quality characteristic of a part.

Tolerance Limits

The extreme values (upper and lower) that define the range of permissible variation in size or other quality characteristic of a part. See also *quality characteristics*.

Tomographic Plane

A section of the part imaged by the tomographic process. Although in computed tomography the tomographic plane or slice is displayed as a 2D image, the measurements are of the materials within a defined slice thickness associated with the plane. See also *slice* and the term *computed tomography*.

Tomography

From the Greek "to write a slice or section." The process of imaging a particular plane or slice through an object. See also *computed tomography*.

Tonghold

The portion of a forging billet, usually on one end, that is gripped by the operators tongs. It is removed from the part at the end of the forging operation. Common to drop hammer and press-type forging.

Tool Set

See *die*.

Tool Side

The side of a plastic part that is cured against the tool (mold or mandrel).

Tool Steel

To develop their best properties, tool steels are always heat treated. Because the parts may distort during heat treatment, precision parts should be semifinished, heat treated, and then finished. Severe distortion is most likely to occur during liquid quenching, so an alloy should be selected that provides the needed mechanical properties with the least severe quench.

Steels are used primarily for cutters in machining, shearing, sawing, punching, and trimming operation, and for dies, punches, and molds in cold- and hot-forming operations. Some are also occasionally used for nontool applications. Tool steels are primarily ingot-cast wrought products, although some are now also powder-metal products. Regarding powder-metal products, there are two kinds: (1) mill products, mainly bar, produced by consolidating powder into "ingot" and reducing the ingot by conventional thermomechanical wrought techniques, and (2) end product tools, produced directly from powder by pressing and sintering techniques. There are seven major families of tool steels as classified by the American Iron and Steel Institute: (1) high-speed tool steels, (2) hot-work tool steels, (3) cold-work tool steels, (4) shock-resisting tool steels, (5) mold steels, (6) special-purpose tool steels, and (7) water-hardening tool steels.

High-Speed Tool Steels

These steels are subdivided into three principal groups or types: the molybdenum type, designated M1–M46, the tungsten type (T1–T15), and the intermediate molybdenum type (M50–M52). Virtually all M-types, which contained 3.75%–9.5% molybdenum, also contain 1.5%–6.75% tungsten, 3.75%–4.25% chromium, 1%–3.2% vanadium, and 0.85%–1.3% carbon. M33–M46 also contain 5%–8.25% cobalt, and M6, 12% cobalt. The T-types, which are molybdenum-free, contain 12%–18% tungsten, 4%–4.5% chromium, 1%–5% vanadium, and 0.75%–1.5% carbon. Except for T1, which is cobalt-free, they also contain 5%–12% cobalt.

Both M50 and M52 contain 4% molybdenum and 4% chromium; the former also contain 0.85% carbon and 1% vanadium, the latter 0.9% carbon, 1.25% tungsten, and 2% vanadium.

The molybdenum types are now by far the most widely used, and many of the T-types have M-type counterparts. All of the high-speed tool steels are similar in many respects. They all can be hardened to at least Rockwell C63, have fine-grain size, and deep-hardening characteristics. Their most important feature is hot hardness; they all can retain a hardness of Rockwell C52 or more at 538°C. The M-types, as a group, are somewhat tougher than the T-type at equivalent hardness but otherwise mechanical properties of the two types are similar. Cobalt improves hot hardness, but at the expense of toughness. Wear resistance increases with increasing carbon and vanadium contents. The M-types have a greater tendency to decarbonization and, thus, are more sensitive to heat treatment, especially austenitizing. Many of the T-types, however, are also sensitive in this respect, and they are hardened at somewhat higher temperatures. The single T-type that stands out is T-15, which is rated as the best of all high-speed tool steels from the standpoint of hot hardness and wear resistance. Typical applications for both the M-type and T-type include lathe tools, end mills, broaches, chasers, hobs, milling cutters, planar tools, punches, drills, reamers, routers, taps, and saws. The intermediate M-types are used for what somewhat similar cutting tools but, because of their lower alloy content, are limited to less-severe operating conditions.

Hot-Work Tool Steels

These steels are subdivided into three principal groups: (1) the chromium type (H10–H19), (2) the tungsten type (H21–H26), and (3) the molybdenum type (H42). All are medium-carbon (0.35%–0.60%) grades. The chromium types contain 3.25%–5.00% chromium and other carbide-forming elements, some of which, such as tungsten and molybdenum, also import hot strength, and vanadium, which increases high-temperature wear resistance. The tungsten types, with 9%–18% tungsten, also contain chromium, usually 2%–4%, although H23 contains 12% of each element. Tungsten hot-work tool steels with higher contents of alloying elements are more heat resistant at elevated temperatures than H11 and H13 chromium hot-work steels, but the higher percentage also tends to make them more brittle in heat treating.

The one molybdenum type, H42, contains slightly more tungsten (6%) than molybdenum (5%), and 4% chromium and 2% vanadium. These alloying elements (chromium, molybdenum, tungsten, and vanadium) make the steel more resistant to heat checking than tungsten hot-work steels. Also, their lower carbon content in relation to high-speed tool steels gives them a higher degree of toughness.

Typical applications include dies for forging, die casting, extrusion, heading, trim, piercing and punching, and shear blades.

Cold-Work Tool Steels

There are also three major groups of cold-work tool steels: (1) high carbon (1.5%–2.35%); high chromium (12), which are designated D2–D7; (2) medium alloy air-hardening (A2–A10), which may contain 0.5%–2.25% carbon, 0%–5.25% chromium, 1%–1.5% molybdenum, 0%–4.75% vanadium, 0%–1.25% tungsten, and, in some cases, nickel, manganese or silicon, or nickel and manganese; and (3) oil-hardening types (O1–O7). They are used mainly for cold-working operations, such as stamping dies, drawing dies, and other forming tools as well as for shear blades, burnishing tools, and coining tools.

Shock-Resistant Tool Steels

These steels (Slto S7) are, as a class, the toughest, although some chromium-type hot-work grades, such as H10–H13, are somewhat better in this respect. The S-types are medium-carbon (0.45%–0.55%) steels containing only 2.50% tungsten and 1.50% chromium (SI), only 3.25% chromium and 1.40% molybdenum (S7), or other combinations of elements, such as molybdenum and silicon, manganese and silicon, or molybdenum, manganese, and silicon. Typical uses include chisels, knockout pins, screwdriver blades, shear blades, punches, and riveting tools.

Mold Steels

There are three principal mold steels: (1) P6, containing 0.10% carbon, 3.5% nickel, and 1.5% chromium; P20, 0.35% carbon, 1.7% chromium, and 0.40% molybdenum; and P21, 0.20% carbon, 4% nickel, and 1.2% aluminum. P6 is basically a carburizing steel produced to tool-steel quality. It is intended for hubbing—producing die cavities by pressing with a male plug—then carburizing, hardening, and tempering. P20 and P21 are deep-hardening steels and may be supplied in hardened condition. P21 maybe carburized and hardened after machining. These steels are tough but low in wear resistance and moderate in hot hardness; P21 is best in this respect. All three are oil-hardening steels, and they are used mainly for injection and compression molds for forming plastics, but they also have been used for die-casting dies.

Special-Purpose Tool Steels

These steels include L2, containing 0.50%–1.10% carbon, 1.00% chromium, and 0.20% vanadium; and L6, having 0.70% carbon, 1.5% nickel, 0.75% chromium, and, sometimes, 0.25% molybdenum. L2 is usually hardened by water quenching and L6, which is deeper hardening, by quenching in oil. They are relatively tough and easy to machine and are used for brake-forming dies, arbors, punches, taps, wrenches, and drills.

Water-Hardening Tool Steels

Water-hardening tool steels include W1, which contains 0.60%–1.40% carbon and no alloying elements; W2, with the same carbon range and 0.25% vanadium; and W5, having 1.10% carbon and 0.50% chromium. All are shallow-hardening and the least qualified of tool steels in terms of hot hardness. However, they can be surface-hardened to high hardness and, those provide high resistance to surface wear. They are the most readily machined tool steels. Applications include blanking dies, cold-striking dies, files, drills, counter sinks, taps, reamers, and jewelry dies.

Coatings

To prolong tool life, tool steel end products, such as mills, hobs, drills, reamers, punches, and dies, can be nitrided or coated in several ways. Oxide coatings, imparted by heating to about 566°C in a steam atmosphere or by immersing in aqueous solutions of sodium hydroxide and sodium nitrite at 140°C, are not as effective as traditional nitriding but do reduce friction and adhesion between the workpiece and the tool. The thickness of the coating developed in the salt bath is typically less than 0.005 mm, and its nongalling tendency is especially useful for operations at which failure occurs this way. Hard-chromium plating to a thickness of 0.0025–0.0127 mm provides a hardness of DPH 950–1050 and is more effective than oxide coating, but the plate is brittle and, thus, not advisable for tools subject to shock loads. Its toughness may be improved somewhat without substantially reducing wear resistance by tempering at temperatures below 260°C, but higher tempering temperatures impair hardness, thus wear resistance, appreciably. An antiseize iron sulfide coating can be applied electrolytically at 191°C using sodium and potassium thiocyanate. Because of the low temperature, the tools can be coated in the fully hardened and tempered condition without affecting hardness. Tungsten carbide is another effective coating. One technique, called Rocklinizing, deposits 0.0025–0.0203 mm of carbide using a vibrating arcing electrode of the material in a handheld gun. Titanium carbide and titanium nitride are the latest coatings. Nitride, typically 0.008 mm thick, has stirred the greatest interest, although carbide may have advantages for press tools subject to high pressure. In just the past few years, all sorts of tools, primarily cutters but also dies, have been titanium nitride–coated, which imparts a gold- or brass-like look. The coating can be applied by chemical vapor deposition (CVD) at 954°C–1066°C or by physical vapor deposition (PVD) at 482°C or less. Thus, the PVD process has an advantage in that the temperature involved may be within or below the tempering temperature of the tool steels so that the coating can be applied to fully hardened and tempered tools. Also, the risk of distortion during coating is less.

Another method being used to prolong tool life is to subject the tools to a temperature of –196°C for about 30 h. The cryogenic treatment, which has been called Perm-O-Bond and Cryo-Tech, is said to rid the steel of any retained austenite—thus the improved tool life.

Properties

Toughness

Toughness in *tool steels* is best defined as the ability of a material to absorb energy without fracturing rather than the ability to deform plastically without breaking. Thus, a high elastic limit is required for best performance since large degrees of flow or deformation are rarely permissible in fine tools or dies. Hardness of a tool has considerable bearing on the toughness because the elastic limit increases with an increase in hardness. However, at very high hardness levels, increased notch sensitivity and brittleness are limiting factors.

In general, lower carbon tool steels are tougher than higher carbon tool steels. However, shallow hardening carbon (W-1) or carbon–vanadium (W-2) tool steels with a hard case and soft core will have good toughness regardless of carbon content. The higher alloy steels will range between good and poor toughness depending upon hardness and alloying content.

Abrasion Resistance

Some tool steels exhibit better resistance to abrasion than others. Attempts to measure absolute abrasion resistance are not always consistent, but in general, abrasion resistance increases as the carbon and alloy contents increase. Carbon is an influential factor. Additions of certain alloying elements (chromium, tungsten, molybdenum, and vanadium) balanced with carbon have a marked effect on increasing the abrasion resistance by forming extremely hard carbides.

Hardness

Maximum attainable hardness is primarily dependent upon the carbon content, except possibly in the more highly alloyed tool steels. Tool steels are generally used somewhat below maximum hardness except for deep-drawing dies, forming dies, coating tools, etc. Battering or impact tools are put in service at moderate hardness levels for improved toughness.

Hot Hardness

The ability to retain hardness with increasing temperature is defined as hot hardness or red hardness. This characteristic is important in steels used for hot-working dies. Generally, as the alloy content of the steel is increased (particularly in chromium, tungsten, cobalt, molybdenum, and vanadium, which form stable carbides), the resistance to softening at elevated temperatures is improved. High-alloy tool steels with a properly balanced composition will retain a high hardness up to 593°C. In the absence of other data, hardness after high-temperature tempering will indicate the hot hardness of a particular alloy.

Heat Treatment

Hardenability

Carbon tool steels are classified as shallow hardening, that is, when quenched in water from the hardening (austenitizing) temperature, they form a hardened case and a soft core. Increasing the alloy content increases the hardenability or depth of hardening of the case. A small increase in alloy content will result in a steel that will harden through the cross sections when quenched in oil. If the increase in alloy content is great enough, the steels will harden throughout when quenched in still air. For large tool or die sections, a high-alloy tool steel should be selected if strength is to be developed throughout the section in the finished part.

For carbon tool steels that are very shallow in hardening characteristics, the P/F test, Disc test, and PV test are methods for rating this characteristic. Oil-hardening tool steels of medium-alloy content are generally rated for hardenability by the Jominy End Quench test.

Dimensional Changes during Heat Treatment

Carbon tool steels are apt to distort because of the severity of the water quench required. In general, water-hardening steels distort more than oil hardening, and oil hardening distort more than air-hardening steels. Thus, if a tool or die is to be machined very close to the final size before heat treatment and little or no grinding is to be performed after treatment, an air-hardening tool steel would be the proper selection.

Resistance to Decarburization

During heat treatment, steels containing large amounts of silicon, molybdenum, and cobalt tend to lose carbon from the surface more rapidly than steels containing other alloying elements. Steels with extremely high carbon content are also susceptible to rapid decarburization. Extra precautions should be employed to provide a neutral atmosphere when heat treating these steels. Otherwise, danger of cracking during hardening will be present. Also, it would be necessary to allow a liberal grinding allowance for cleanup after heat treatment.

Machinability

Since most tool steels, even in the annealed state, contain wear-resistant carbides, they are generally more difficult to machine than the open-hearth grades or low-alloy steels. In general, the machinability tends to decrease with increasing alloying content. Microstructure also has a marked effect on machinability. For best machinability, a spheroidal microstructure is preferred over pearlitic.

The addition of small amounts of lead or sulfur to the steels to improve machinability has gained considerable acceptance in the tool steel industry. These free machining steels not only machine more easily but also give a better surface finish than the regular grades. However, some caution is advised in applications involving transverse loading since lead or sulfur additions actually add longitudinal inclusions in the steel.

Available Forms

Tool steels are available in billets, bars, rods, sheets, and coil. Special shapes can be furnished upon request. Generally, the material is furnished in the soft (or annealed) condition to facilitate machining. However, certain applications require that the steel be cold-drawn or prehardened to a specified hardness.

A word of caution: Mill decarburization is generally present on all steel except that guaranteed by the producer to be decarburization-free. It is important that all decarburized areas be removed prior to heat treating or the tool or die may crack during hardening.

Tooling

A generic term applied to die assemblies and related items used for forming and forging metals.

Tooling Marks

Indications imparted to the surface of the forged part from dies containing surface imperfections or dies on which some repair work has been done. These marks are usually slight rises or depressions in the metal.

Tooling Resin

Resins that have applications as tooling aids, coreboxes, prototypes, hammer forms, stretch forms, foundry patterns, and so forth. Epoxy and silicone are common examples.

Tooth

(1) A projection on a multipoint tool (such as on a sore, milling cutter, or file) designed to produce cutting. (2) A projection on the periphery of a wheel or segment thereof—as on a gear, spline, or sprocket, for example—designed to engage another mechanism and thereby transmit force or motion, or both. A similar projection on a flat member such as a rack.

Tooth Point

On a *face mill*, the chamfered cutting edge of the blade, in which a flat is sometimes added to produce a shaving effect and to improve finish.

Top-and-Bottom Process

A process for separating copper and nickel, in which their molten sulfides are separated into two liquid layers by the addition of sodium sulfide. The lower layer holds most of the nickel.

Top Hat Furnace

See *Bell Furnace*.

Top Pouring

The normal process of pouring molten metal into the mold from the top. This allows solidification to commence at the bottom and has the advantage of simplicity. However, it has the disadvantage that it may induce turbulence, splashing, erosion of the mold, and entrainment of dross.

Topaz

The naturally occurring mineral, aluminum fluorosilicate, $Al_2Fl_2SiO_4$, having various colors depending on the individual impurities. Its hard and is No. 8 on the Mohs scale.

Torch

See *blowpipe*, *cutting torch*, and *welding torch*.

Torch Brazing

A soldering process in which the heat required is furnished by a fuel gas flame.

Torch Tip (Brazing)

That part of an oxyfuel gas brazing torch from which the gases issue. See also *cutting tip* and *welding tip*.

Toroid

Doughnut-shaped piece of magnetic material, together with one or more coils of current-carrying wire wound about the doughnut, with the permeability of the magnetic material high enough so that the magnetic flux is almost completely confined within it. Also known as *toroidal coil* and *toroidal magnetic circuit*.

Torque

The force of rotation, the product of force and the perpendicular distance between the axis of rotation and the direction of the force.

Torque-Coil Magnetometer

A magnetometer that depends for its operation on the torque developed by a known current in a coil that can turn in the field to be measured. See also *magnetometer*.

Torr

A non SI unit of pressure used for measuring near vacuum conditions. 1 torr = 1 mm Hg = 133.322 Pa.

Torsion

(1) A twisting deformation of a solid or tubular body about an axis in which lines that were initially parallel to the axis become helices. (2) A twisting action resulting in shear stresses and strains.

Torsional Moment

In a body being twisted, the algebraic sum of the couples or the moments of the external forces about the axis of twist, or both.

Torsional Pendulum

A device for performing dynamic mechanical analysis of plastics, in which the sample is deformed torsionally and allowed to oscillate in free vibration. Modulus is determined by the frequency of the resultant oscillation, and damping is determined by the decreasing amplitude of the oscillation.

Torsional Stress

The shear stress on a transverse cross section caused by a twisting action.

Torsion Test

A test designed to provide data for the calculation of the *shear modulus, modulus of rupture in torsion*, and *yield strength* and shear yield.

Torsion Twisting

The state of strain induced by torque.

Total Carbon

The sum of the *free carbon* and *combined carbon* (including carbon in solution) in a ferrous alloy.

Total Cyanide

Cyanide content of an electroplating bath (including both simple and complex ions).

Total Elongation

The total amount of permanent extension of a test piece broken in a tensile test usually expressed as a percentage over a fixed gauge length. See also *elongation, percent*.

Total-Extension-Under-Load Yield Strength

See *yield strength*.

Total Indicator Reading

See *total indicator variation*.

Total Indicator Variation

The difference between the maximum and minimum indicator readings during a checking cycle.

Total Transmittance

The ratio of the radiant energy leaving one side of a region between two parallel planes to the radiant energy entering from the opposite side.

Touch Welding

Metal arc welding in which the tip of the coating on the consumable electrode remains in contact with the parent metal during the welding operation. The electrode has a coating formulated to form a cup standing around the metal core during welding to avoid core to parent contact.

Toughness

Ability of a material to absorb energy and deform plastically before fracturing. Toughness is proportional to the area under the *stress–strain curve* from the origin to the breaking point. In metals, toughness is usually measured by the energy absorbed in a notch impact test. See also *impact test*.

An imprecise term encompassing the capacity to absorb energy prior to and during the cracking process, to deform plastically without cracking and to resist crack growth. It implies resistance to impact and hence the results of impact test may be referred to as "toughness." A measure of toughness is also provided by the area under the curve plotted from a tensile test. Also see *fracture toughness*.

Tough Pitch Copper

Copper containing from 0.02% to 0.04% oxygen, obtained by refining copper in a reverberatory furnace. See *copper*.

Tow

An untwisted bundle of continuous filaments, usually referring to man-made fibers, particularly carbon and graphite, but also fiberglass and aramid. A tow designated as 140 K has 140,000 filaments.

T-Peel Strength

The average load per unit width of adhesive bond line required to produce progressive separation of two bonded, flexible adherends, under standard test conditions.

TPI

See *thermoplastic polyimides*.

TPUR

See *thermoplastic polyurethanes*.

Tracer

In composites fabrication, fiber, tow, or yarn added to a prepreg for verifying fiber alignment and, in the case of woven materials, for distinguishing warp fibers from fill fibers.

Tracer Milling

Duplication of a three-dimensional form by means of a cutter controlled by a tracer that is directed by a master form.

Tracer Techniques

The use of small quantities of radioactive materials to monitor processes such as corrosion, wear, diffusion, and pollution.

Track

The mark made by a seal on the surface with which it mates.

Tracking

The breakdown of electrical resistance via a narrow track across a surface of an insulator. It usually results from some external contamination.

Tracking Pattern

The path a seal ring makes when in rubbing contact with the mating ring or seal plate.

Traction

In rolling contacts, the tangential stress transmitted across the interface. The traction will in general vary from point to point over the contact area. More generally, traction may denote the force per unit area of contact.

Tractive Force

The integral of the tangential surface stress over the area of contact.

Traditional Ceramics

Traditional ceramics are generally classified as those ceramic products that use clay or have a significant clay component in the batch. A clay-based ceramic body usually consists of one or more clays or clay minerals mixed with nonclay mineral powders such as fluxes (for example, feldspar) and fillers (for example, silica and alumina). Each of these constituents contributes to the plastic forming and fired characteristics of the body, with the clay acting as a plasticizer and binder for the other constituents.

Commercial clays are grouped as being kaoline (China) clay or ball clay. Kaolines consist primarily of ordered kaolinite, $Al_2Si_2O_5(OH)_4$, with some mica and free quartz. Ball clays, which are dug out of the ground in "blocks" or "balls," tend to be very fine grained and are composed of ordered and disordered kaolinite and varying percentages of mica, illite, montmorillonite, free quartz, and organic matter.

Traditional ceramic bodies are formed into shapes using many different techniques. A general sequence of unit operations would include raw material preparation, batch preparation, forming, drawing, prefire operations (glazing, decorating, etc.), firing, and postfire operations (glazing, decorating, machining, and/or cleaning). Forming techniques include hand molding and pottery wheels, extrusion, die pressing, and *slip casting*.

The five principal product areas for traditional ceramics are whitewares, glazes and porcelain enamels, structural clay products, cement, and refractories. Whiteware is the name given to a group of ceramic products characterized by a white or light-colored body with a fine-grained structure. Most whiteware products are glazed or decorated with patterns or designs. Examples of whiteware products are sanitary ware, tableware, artware, stoneware, and floor and wall tile.

A glaze is defined as a continuous adherent layer of glass on the surface of a ceramic body that is hard, nonabsorbent, and easily cleaned. A glaze is usually applied as a suspension of glaze-forming ingredients in water. After the glaze layer dries on the surface of the piece, it is fired, whereupon the ingredients melt to form a thin layer of glass. Porcelain enamels are very durable alkali borosilicate glass coatings bonded by fusion to metal substrates at temperatures above 425°C. Porcelain enamels are applied primarily to steel sheet, cast iron, aluminum alloys (in sheet or cast form), and aluminum-coated steels. See also the term *porcelain enamel*.

Structural clay products are ceramic materials used in construction. The raw materials are naturally occurring clays or shales. A distinguishing manufacturing characteristic of structural clays is their exposure to elevated firing temperatures to develop a bond between the particulate constituents and to develop the desired pore structure (pore quality, pore size distribution, and pore connection) for the intended application. Structural clay products include facing materials for buildings, building brick, paving brick, roofing tile, sewer pipe, and drain tile.

Cements are inorganic powders consisting predominantly of calcium silicates, which, when mixed with water to form a paste, react slowly at ambient temperatures to produce a coherent, hardened mass with valuable engineering properties. The hardened powder product is porous and consists primarily of calcium silicate hydrate. Uses of cement include steel-reinforced pipe, panels, columns, and beams, highway pavements, foundations, canal linings, dams, bridges decks, and floor slabs. See also *cement*.

Refractories are construction materials that can withstand high temperatures and maintain their physical properties. They are used extensively in structures associated with iron and steel production, copper and aluminum smelting, and glass and ceramic manufacturing. The primary types of clay refractories are fireclay and high alumina. Each type is used to produce bricks, as well as insulating refractories. See also *refractories*.

Traffic Mark

See *abrasion*.

Tramp Alloys

Residual alloying elements that are introduced into steel when unidentified alloy steel is present in the scrap charge to a steelmaking furnace.

Tramp Element

Contaminant in the components of a furnace charge, or in the molten metal or castings, whose presence is thought to be either unimportant or undesirable to the quality of the casting. Also called trace element.

Tramp Materials

Any contaminating materials entrained in a process and undesirable because they might enter the product or damage the production equipment.

Transcrystalline/Granular

Running across the crystal/grain, for example, a crack running fully or partly across a grain rather than around the grain boundary. Compare with intragranular that refers to features such as precipitates within a grain. See *transgranular*.

Transducer

A material capable of converting electrical energy to mechanical energy or vice versa.

Transfer

In tribology, the process by which material from one sliding surface becomes attached to another surface, possibly as a result of interfacial adhesion. Transfer is usually associated with adhesion, but the possibility of mechanical interlocking adherence, without adhesive bonding, exists in certain occurrences. The material may also back transfer to the surface from which it came. See also *selective transfer*.

Transfer Ladle

A ladle that can be supported on a monorail or carried in a shank and used to transfer metal from the melting furnace to the holding furnace or from the furnace to the pouring ladles.

Transfer Molding

A method of molding thermosetting materials in which the plastic is first softened by heat and pressure in a transfer chamber and then forced by high pressure through suitable sprues, runners, and gates into a closed mold for final shaping and curing. See also *resin transfer molding*.

Transference

The movement of ions through the *electrolyte* associated with the passage of the electric current. Also called transport or migration.

Transference Number

The proportion of total electroplating current carried by ions of a given kind. Also called transport number.

Transferred Arc

A plasma arc established between the electrode and the workpiece during plasma arc welding, cutting, and thermal spraying. See also the term *nontransferred arc* and *plasma welding*.

Transformation

Change from one phase to another at the transformation temperature.

Transformation Hardening

Heat treatment of steels comprising austenitization followed by cooling under conditions such that the austenite transforms more or less completely into martensite and possibly into bainite.

Transformation-Induced Plasticity

A phenomenon, occurring chiefly in certain highly alloyed steels that have been heat treated to produce metastable austenite or metastable austenite plus martensite, whereby, on subsequent deformation, part of the austenite undergoes strain-induced transformation to martensite. Steels capable of transforming in this manner, commonly referred to as TRIP steels, are highly plastic after heat treatment, but exhibit a very high rate of strain hardening and thus have high tensile and yield strengths after plastic deformation at temperatures between about 20°C and 500°C. Cooling to −195°C may or may not be required to complete the transformation to martensite. Tempering usually is done following transformation.

Transformation Ranges

Those ranges of temperature without which austenite forms during heating and transforms during cooling. The two ranges are distinct, sometimes overlapping but never coinciding. The limiting temperatures of the ranges depend on the composition of the alloy and on the rate of change of temperature, particularly during cooling. See also *transformation temperature* and *steel*.

Transformation Temperature

The temperature at which a change in phase occurs. This term is sometimes used to denote the limiting temperature of a transformation range. The following symbols are used for irons and steels:

$\mathbf{Ac_{cm}}$: In a hypereutectoid steel, the temperature at which solution of cementite in austenite is completed during heating.

$\mathbf{Ac_1}$: The temperature at which austenite begins to form during heating.

$\mathbf{Ac_3}$: The temperature at which transformation of ferrite to austenite is completed during heating.

$\mathbf{Ac_4}$: The temperature at which austenite transforms to delta ferrite during heating.

$\mathbf{Ae_{cm}}$, $\mathbf{Ae_1}$, $\mathbf{Ae_3}$, $\mathbf{Ae_4}$: The temperatures of phase changes at equilibrium.

$\mathbf{Ar_{cm}}$: In hypereutectoid steel, the temperature at which precipitation of cementite starts during cooling.

$\mathbf{Ar_1}$: The temperature at which transformation of austenite to ferrite or to ferrite plus cementite is completed during cooling.

$\mathbf{Ar_3}$: The temperature at which austenite begins to transform to ferrite during cooling.

$\mathbf{Ar_4}$: The temperature at which delta ferrite transforms to austenite during cooling.

$\mathbf{A_{r'}}$: The temperature at which transformation of austenite to pearlite starts during cooling.

$\mathbf{M_f}$: The temperature at which transformation of austenite to martensite is completed during cooling.

$\mathbf{M_s}$ (or $\mathbf{Ar''}$): The temperature at which transformation of austenite to martensite starts during cooling.

Note: All these changes, except formation of martensite, occur at lower temperatures during cooling than during heating and depend on the rate of change of temperature.

Transformation-Toughened Zirconia

A generic term applied to stabilized zirconia systems in which the tetragonal symmetry is retained as the primary zirconia phase. The four most popular tetragonal phase stabilizers are ceria (CeO_2), calcia (CaO), magnesia (MgO), and yttria (Y_2O_3). See also *zirconia*.

Transformed Beta

A local or continuous structure in titanium alloys consisting of decomposition products arising by nucleation and growth processes during cooling from above the local or overall β transus. Primary and regrowth α may be present. Transformed β typically consists of α platelets that may or may not be separated by β phase.

Transformer Steel

Steel having low magnetic hysteresis, and hence used as sheet for the laminations of electrical transformers and generator stators. Such steels are typically very low carbon with 2%–5% silicon and are carefully cold rolled to maximize their magnetic characteristics.

Transgranular

Through or across crystals or grants. Also called intracrystalline or transcrystalline.

Transgranular Cracking

Cracking or fracturing that occurs through or across a crystal or grain. Also called transcrystalline cracking. Contrast with *intergranular cracking*. See the term step fracture (metals) (1).

Transgranular Fracture

Fracture through or across the crystals or grains of a material. Also called transcrystalline fracture or intracrystalline fracture. Contrast with *intergranular fracture*.

Transient Creep

See *creep* and *primary creep*.

Transistor

An active semiconductor device capable of providing power amplification and having three or more terminals.

Transitional Fit

A fit that may have either clearance or interference resulting from specified tolerances on hole and shaft.

Transition Diagram

In tribology, a plot of two or more experimental or operating variables that indicates the boundaries between various regimes of wear or surface damage. The *IRG transition diagram* is a plot of normal force (ordinate) versus sliding velocity (abscissa) and is used to identify three regions with differing lubrication effectiveness. Various plots have been called transition diagrams, and the context of usage must be established.

Transition Elements

The elements, all metals, having an incomplete inner electron shell.

Transition Joint (Weld)

A weld formed between two different metals.

Transition Lattice

An unstable crystallographic configuration that forms as an intermediate step in a solid-state reaction such as precipitation from solid solution or eutectoid decomposition. See *phase change*.

Transition Metal

(1) A metal in which the available electron energy levels are occupied in such a way that the *d*-band contains less than its maximum number of ten electrons per atom, for example, iron, cobalt, nickel, and tungsten. The distinctive properties of the transition metals result from the incompletely filled *d*-levels. (2) The metallic

elements usually shown as a block in the periodic table between Groups II and III. They arise because their atomic structure includes incompletely "filled" inner subshells. See *atomic structure* and *periodic table*.

Transition Phase

A nonequilibrium state that appears in a chemical system in the course of transformation between two equilibrium states.

Transition Point

At a stated pressure, the temperature (or at a stated temperature, the pressure) at which two solid phases exist in equilibrium—that is, an allotropic transformation temperature (or pressure).

Transition Scarp

A *rib mark* generated when a crack changes from one mode of growth to another, as when a wet crack accelerates abruptly from Region II (plateau) to Region III (dry) of a crack acceleration curve. See also *intersection scarp*.

Transition Structure

In precipitation from solid solution, a metastable precipitate that is coherent with the matrix.

Transition Temperature

(1) The temperature at which the properties of a material change. Depending on the material, the transition change may or may not be reversible. (2) Any temperature at which a transition occurs but in particular the temperature at which steel changes from ductile to brittle behavior. See also *fracture appearance transition temperature*.

Transition Temperature (Metals)

(1) An arbitrarily defined temperature that lies within the temperature range in which metal fracture characteristics (as usually determined by tests of notched specimens) change rapidly, such as the ductile-to-brittle transition temperature (DBTT). The DBTT can be assessed in several ways, the most common being the temperature for 50% ductile and 50% brittle fracture (50% fracture appearance transition temperature, or FATT), or the lowest temperature at which the fracture is 100% ductile (100% fibrous criterion). The DBTT is commonly associated with *temper embrittlement* and *radiation damage* (neutron irradiation) of low-alloy steels. (2) Sometimes used to denote an arbitrarily defined temperature within a range in which the ductility changes rapidly with temperature.

Translation

The relative movement as one block of atoms slides across another.

Transmission Electron Microscope

A microscope in which the image-forming rays pass through (are transmitted by) the specimen being observed. Using the transmission electron microscope, microstructural features can be imaged at 1000–450,000×. See *electron microscope*.

Transmission Electron Microscopy (TEM)

An analytical technique in which an image is formed on a cathode-ray tube whose raster is synchronized with the raster of an electron beam over an area of the sample surface. Image contrast is formed by the scattering electrons out of the beam. TEM is used for very high magnification characterization of metals, ceramics, metals, polymers, and biological materials.

Transmission Grating

In electron optics, a transparent diffraction grating through which light is transmitted. See also *concave grating, diffraction grating, plane grating*, and *reflection grating*.

Transmission Method

A method of x-ray or electron diffraction in which the recorded diffracted beams emerge on the same side of the specimen as the transmitted primary beam.

Transmission Oil

(1) Oil used for transmission of hydraulic power. (2) Oil used to lubricate automobile transmission systems.

Transmittance

The ratio of the light intensity transmitted by a material to the light intensity incident upon it. In emission spectrochemical analysis, the transmittance of a developed photographic emulsion, including its film or glass supporting base, is measured by a microphotometer. In absorption spectroscopy, the material is the sample. See also *diffuse transmittance, relative transmittance, specular transmittance*, and *total transmittance*.

Transpassive Region

The region of *anodic polarization curve*, noble to and above the passive *potential* range, in which there is a significant increase in current density (increased metal dissolution) as the potential becomes more positive (noble).

Transpassive State

(1) State of anodically passivated metal characterized by a considerable increase of the corrosion current, in the absence of pitting, when the *potential* is increased. (2) The noble region of potential where an electrode exhibits a higher than passive current density.

Transport

See *transference*.

Transport Number

Same as *transference* number.

Trans Stereoisomer

A stereoisomer in which atoms or groups of atoms are arranged on opposite sides of a chain of atoms. See also *isotactic stereoisomerism*.

Transuranic/Transuranium Elements

The man-made elements having an atomic number greater than 92, that of uranium.

Transverse Direction

Literally, "across," usually signifying a direction or plane perpendicular to the direction of working. In rolled plate or sheet, the direction across the width is often called long transverse; the direction through the thickness, short transverse.

Transverse Resistance Seam Welding

The making of a resistance seam weld in a direction essentially at right angles to the throat depth of a resistance seam welding machine. See also the term *resistance seam welding*. Contrast with *longitudinal resistance seam welding*.

Transverse Rolling Machine

Equipment for producing complex preforms or finished forgings from round billets inserted transversely between two or three rolls that rotate in the same direction and drive the billet. The rolls, carrying replaceable die segments with appropriate impressions, make several revolutions for each rotation of the workpiece.

Transverse Rupture Strength

The stress, as calculated from the flexure formula, required to break a sintered powder metallurgy specimen. The test for determining the transverse rupture strength involves applying the load at the center of a 31.8 by 12.7 by 6.4 mm beam, which is supported near its ends.

Transverse Strain

The linear strain in a plane perpendicular to the loading access of a specimen.

Transverse Test (Bend, Tension, etc.)

A test across the primary axis of the material or across the line of a weld.

Transversely Isotropic

(1) In reference to a material, exhibiting a special case of orthotropy in which properties are identical in two orthotropic dimensions but not the third. (2) Having identical properties in both transverse (short and long) but not in the longitudinal direction.

Trapped-Sheet, Contact Heat, Pressure Thermoforming

A *thermoforming process* for making plastic parts in which a hot, porous blow plate is used in both heating and forming processes.

The plastic sheet lies between the female mold cavity and the hot blow plate. Air forced through the plate and pressure from the female mold push the sheet onto the hot plate. When the sheet is sufficiently heated, air pressure forces it into the female mold.

Travel Angle

The angle that a welding electrode makes with a reference line perpendicular to the axis of the weld in the plane of the weld axis. This angle can be used to define the position of welding guns, welding torches, high-energy beams, welding rods, thermal cutting and thermal spraying torches, and thermal spraying guns. See also *drag angle* and *push angle* and the term *backhand welding* and *forehand welding*.

Travel Angle (Pipe)

The angle that a welding electrode makes with a reference line extending from the center of the pipe through the molten weld pool in the plane of the weld axis.

Traverse Speed

The lineal velocity at which the torch is passed across the substrate during the thermal spraying operation.

Treeing

Localized excessive deposition of electroplate due to high current density in the area.

Trees

Visible projections of electrodeposited metal formed at sites of high current density.

Trepanning

(1) A machining process for producing a circular hole or groove in solid stock, or for producing a disk, cylinder, or tube from solid stock, by the action of a tool containing one or more cutters (usually single-point) revolving around a center. (2) Boring a hole by cutting away only a narrow circumferential band rather than the full cross section. This releases a central core assuming that the process is completed to the far surface.

Triaxiality

In a *triaxial stress* state, all stresses being tensile.

Triaxial Stress

A state of stress in which none of the three principal stresses is zero. See also *principal stress* (*normal*).

Tribo

A prefix indicating a relationship to interacting surfaces in relative motion.

Tribochemistry

The part of chemistry dealing with interacting surfaces in relative motion. Tribochemistry broadly encompasses such areas as lubricant chemistry, changes in reactivity of surfaces due to mechanical contact, oxidative wear, and other phenomena.

Triboelement

A solid body that is bounded by one or more *tribosurfaces* and that resides within a *tribosystem*. For example, in a pin-on-disk tribosystem, the pin is one triboelement and the disk is another. See the term *pin-on-disk machine*.

Tribology

(1) The science and technology of interacting surfaces in relative motion and of the practices related thereto. (2) The science concerned with the design, friction, lubrication, and the wear of contacting surfaces that move relative to each other (as in bearings, cams, or gears, for example).

Tribometer

(1) An instrument or testing rig to measure normal and frictional forces of relatively moving surfaces. (2) Any device constructed for or capable of measuring the friction, lubrication, and wear behavior of materials or components.

Tribophysics

That part of physics dealing with interacting surfaces in relative motion.

Triboscience

The scientific discipline devoted to the systematic study of interacting surfaces in relative motion. Triboscience includes the scientific aspects of *tribochemistry*, *tribophysics*, contact mechanics, and materials and surface sciences as related to *tribology*.

Tribosurface

Any solid surface whose intermittent, repeated, or continuous contact with another surface or surfaces, in relative motion, results in friction, wear, and/or surface damage. The surface of a body subjected to a catastrophic collision would not generally be considered a tribosurface because significant damage to the entire body is involved.

Tribosystem

Any functional combination of *triboelements*, including thermal and chemical surroundings.

Tribotechnology

The aspect of *tribology* that involves the engineering application of *triboscience* and the design, development, analysis, and repair of components for tribological applications.

Trichloroethylene

A volatile, nonflammable hydrocarbon used as a solvent and degreasant.

Triclinic

Having three axes of any length, none of the included angles being equal to one another or equal to 90°.

Tridymite

An allotrope of quartz. See *quartz*.

Triggered Capacitor Discharge

A high-voltage electrical discharge used in *emission spectroscopy* for vaporization and excitation of a sample material. The energy for the discharge is obtained from capacitors that are charged from an ac or dc electrical supply. Each discharge may be either oscillatory, critically damped, or overdamped. It is initiated by separate means and is extinguished when the voltage across the analytical gap falls to a value that no longer is sufficient to maintain it.

Trimetal Bearing

A bearing consisting of three layers. Trimetal bearings are often made of bronze with a white metal facing and a steel backing. See the term *sleeve bearing*.

Trimmer

The dies used to remove the flash or excess stock from a forging.

Trimmer Blade

The portion of the trimmers through which a forging is pushed to shear off the flash.

Trimmer Die

The punch press die used for trimming flash from a forging.

Trimmer Punch

The upper portion of the trimmer that contacts the forging and pushes it through the trimmer blades; the lower end of the trimmer punch is generally shaped to fit the surface of the forging against which it pushes.

Trimmers

The combination of *trimmer punch, trimmer blades*, and perhaps *trimming shoe* used to remove the flash from the forging.

Trimming

(1) In forging, removing any parting-line flash or excess material from the part with a trimmer in a trim press; can be done hot or cold. (2) In drawing, shearing the irregular edge of the drawn part. (3) In casting, the removal of gates, risers, and fins.

Trimming Press

A power press suitable for trimming flash from forgings.

Trimming Shoe

The holder used to support *trimmers*. Sometimes called trimming chair.

Triple-Action Press

A mechanical or hydraulic press having three slides with three motions properly synchronized for triple-action drawing, redrawing, and forming. Usually, two slides—the blankholder slide and the plunger—are located above and a lower slide is located within the bed of the press. See also *hydraulic press, mechanical press*, and *slide*.

Triple Curve

In a *P-T diagram*, a line representing the sequence of pressure and temperature values along which two conjugate phases occur in univariant equilibrium.

Triple Point

(1) A point on a phase diagram where three phases of a substance coexist in equilibrium. (2) The intersection of the boundaries of three adjoining grains, as observed in a metallographic section.

Tripoli

A name given to finely granulated, white, porous, siliceous rock, used as an abrasive and as a filler. True tripoli is an infusorial, diatomaceous earth known as **tripolite**, and is a variety of opal, or **opaline silica**. In the abrasive industry, it is called **soft silica**.

The material marketed for oil-well drilling mud under the name of **Opalite**, is an amorphous silica. Tripoli is used in massive form for the manufacture of filter stones for filtering small supplies of water and is also used for the manufacture of foundry parting. Finely ground tripoli, free from iron oxide, is used as a paint roller and in rubber. Tripoli grains are soft, porous, and free from sharp cutting faces, and they give a fine polishing effect. It is the most commonly used polishing agent. The word **silex**, which is an old name for silica and is also used to designate the pulverized flint from Belgium, is sometimes applied to finely ground white tripoli employed as an inert filler for paints. It is used with oil on rag-wheel polishing. A 250-mesh powder is used as a filler in molding compounds.

TRIP Steel

A commercial steel product exhibiting *transformation-induced plasticity*.

Trisodium Phosphate

A white, crystalline substance also known as **phosphate cleaner**, used in soaps, cleaning compounds, plating, textile processing, and boiler compounds. The commercial grade is not less than 97% pure, with total alkalinity of 16%–19%. The anhydrous trisodium phosphate is 2.3 times as effective as the crystalline form but requires a longer time to dissolve. **Disodium phosphate** is a white, crystalline product used for weighting silk, boiler treatment, cheese making, and cattle feeds.

The medicinal, or USP, grade has only seven molecules of water and has a different crystal structure. The commercial grade is 99.4% pure and is readily soluble in water. **Trisodium phosphate hemihydrate** is a granular, crystalline grade for degreasing and water conditioning. Monosodium phosphate is made by reacting soda ash with phosphoric acid in molecular proportions; it is used in similar applications to the disodium variety. **Sodium tripolyphosphate** is a water-soluble, white powder used as a detergent, a water softener, and a deflocculating agent in portland cement to govern the viscosity of the shale slurry without excessive use of water. Large quantities of these phosphates are used in the processing of chemicals, textiles, and paper; and since they are toxic contaminants of ground and surface waters, mill waste must be deactivated before they are discharged. The use of phosphates in detergents and soap powders has been banned in many areas since they lead to rapid algae growth in surface waters.

Tritium

Hydrogen with three neutrons in the nucleus. See *atomic structure*.

Triton

The nucleus of tritium (^{3}H), the triton is the only known radioactive nuclide belonging to hydrogen and β-decays to ^{3}He with a half-life of 12.4 years.

Trommel

A revolving cylindrical screen used in grading coarsely crushed ore.

Tropenas Converter

A converter in which air, possibly oxygen enriched, is blown from the side across the charge of previously desulfurized molten pig iron and scrap. The process is used for steel castings of lower quality.

Troy Ounce

A unit of weight for *precious metals* that is equal to 31.1034768 g (1.0971699 avoirdupois).

True Current Density

See preferred term *local current density*.

True Rake

See preferred term *effective rake*.

True Strain

(1) The ratio of the change in dimension, resulting from a given load increment, to the magnitude of the dimension immediately prior to applying the load increment. (2) In a body subjected to axial force, the natural logarithm of the ratio of the gage length at the moment of observation to the original gage length. Also known as natural strain.

True Stress

The value obtained by dividing the load applied to a member at a given instant by the cross-sectional area over which it acts. See *tensile test*.

Truing

The removal of the outside layer of abrasive grains on a grinding wheel for the purpose of restoring its face.

Trumpet Alloy

A brass with about 20% zinc, 1% tin, remainder copper, used for musical instruments.

Trunnion Bearing

A bearing used as a pivot to swivel or turn an assembly.

Tryout

In metal forming or forging, a preparatory run to check or test equipment, lubricant, stock, tools, or methods prior to a production run. Production tryout is run with tools previously approved; new die tryout is run with new tools not previously approved.

T.S.N.

Thermal severity number. See *controlled thermal severity test*.

T.T.T.

See *time–temperature–transformation diagrams*.

Tube

A pipe or hollow section. The various terms are largely interchangeable but "tube" would usually indicate a fairly small diameter and a simple cross section.

Tube Furnace

A furnace used for continuous or batch sintering powder metallurgy parts that utilizes a dense ceramic tube or a metallic retort to contain the controlled sintering atmosphere.

Tube Making

The basic tube making sequences can be categorized as

- Piercing and subsequent processes
- Welding of rolled sheet or plate
- Extrusion

These are described briefly in the succeeding text.

Piercing sequences commence with hot piercing, followed by hot reducing and cold reducing. In the press piercing process, a heated cylindrical billet, usually set in a die, has a tool forced along the central axis to fully penetrate the billet or nearly so. Alternatively, the billet may be fully pierced in a barrel or Mannesmann rotary piercer. This comprises a pair of barrel-shaped rolls set not quite parallel and rotating in the same direction. The heated billet is introduced, end on, into the end gap of the rolls. The rotation of the rolls spins the billet, and their angled alignment draws it in. At the same time, the barrel taper compresses the spinning billet across the diameter, which tends to open up a zone of weakness at the billet center line. A piercing point carried on a mandrel is located part way along the roll gap to coincide with the zone of weakness and the forward motion then thrusts the weak end center over the point to form the tube. The disc or Stiefel piercer processes are similar in concept except that the rolls are mushroom shaped and set with their axes near parallel but displaced. The three-roll piercing process is again similar to the Mannesmann mill except that three rolls are used, the central weak zone is less pronounced, and the billet is effectively forced over the piercing point. Fully pierced billets are usually termed "blooms" or "hollows" and near pierced billets are termed "bottles."

After piercing, the bloom is hot reduced by processes that include *hot rolling*, the *push bench* and the *pilger mill* or *rotary forge*, and *Assel mill* process. In hot rolling, the blooms or bottles are passed through grooved, driven rollers usually with a mandrel set in the bore to provide support and to control its size. The push bench process is usually applied to bottles. Immediately after piercing and without further heating, the bottle is set on a mandrel to be pushed through a series of dies and or nondriven rolls of progressively reducing diameter. The pilger mill comprises a pair of rolls with deep tapered grooves. The rolls have parallel axes and run in opposite directions with their faces in contact so that as the rolls rotate, the pair of grooves form, at the contact line, an orifice of varying diameter. The rolls rotate continuously and, when the orifice diameter is at its maximum, the heated bloom, carried on a mandrel, is thrust forward into the rolls against the direction of rotation. The rolls then force the tube backward, and the taper on the rolls forges a small length of the tube reducing its exterior diameter. This sequence is repeated with the tube being rotated 90° on each forward movement, and the tube moves progressively through the mill. The Assel mill is similar in layout to the Mannesmann piercer except that three rolls are deployed and they usually have a step at midlength to forge the heated bloom, carried on a mandrel, with better dimensional control than can be achieved by the pilger mill.

Cold reducing is commonly achieved by a drawing process in which the tube is pulled through a die. Usually, the bore is controlled by a mandrel or by a plug located in the tube bore at a position coinciding with the die. Drawing without use of a mandrel or plug is termed "sinking." Multiple drawing passes may be made with, if necessary, interstage annealing. An alternative is the *cold-reducing process*, also termed the "rockrite process," which is similar in concept to the pilger described earlier in that it uses tapered grooved rolls to produce a forging action. However, in this case, the groove diameter is much smaller in proportion to the roll size, and the pair of rolls reciprocates along a track changing rotation direction at the extremities of each stroke. The tube, on a tapered mandrel, is fed in at the point of maximum opening at the end of a stroke. It is not pushed backward by the rolls but is rotated slightly at each forward step to produce an even product. Large reductions of external diameter, bore, and wall thickness can be achieved with good control of dimensions including eccentricity.

Welded tube is produced from rolled sheet or plate, cut to width, and rolled into the tubular form and then welded. The seam is usually longitudinal but may be spiral. Virtually, any forms of welding can be employed depending upon the tube size and application but for higher quality tubing and smaller size ranges electrical resistance welding is popular. See also *extrusion* as this is a common route for tube production. Often extruded tube receives no further working, but in some cases, one or more drawing passes may be required.

Tube Manipulation

Various processes applied to tube ends.

Tube Plate

The perforated plate into which the tubes of a heat exchanger are sealed by some process such as expanding, welding, or soldering.

Tube Reducing

Reducing both the diameter and wall thickness of tubing with a mandrel and a pair of rolls. See also *spinning*.

Tube Sinking

Drawing tubing through a die or passing it through rolls without the use of an interior tool (such as a mandrel or plug) to control inside diameter; sinking generally produces a tube of increased wall thickness and length.

Tube Stock

A semifinished tube suitable for subsequent reduction and finishing.

Tuberculation

The formation of *localized corrosion* products scattered over the surface in the form of knoblike mounds called tubercles. The formation of tubercles is usually associated with *biological corrosion*.

Tubular Products, Steel

The general term used to cover all hollow carbon and low-alloy steel products used as conveyors of fluids and as structural members. Although these products are usually produced in cylinder form, they are often subsequently altered by various processing methods to produce square, oval, rectangular, and other symmetrical shapes.

Tumble Grinding

Various surfacing operations ranging from deburring and polishing to honing and microfinishing metallic parts before and after plating.

Tumbling

A process in which components are enclosed in a rotating barrel with some other material. The other material may be an abrasive, polishing or plating agent or peening shot—hence terms such as *tumble polishing*, *tumble peening*, and *tumble plating*. If both plating and peening agents are included, the process may be termed "peen plating."

Tumbling (Metals)

Rotating workpieces, usually castings or forgings, in a barrel partly filled with metal slugs or abrasives, to remove sand, scale, or fins. It may be done dry or with an aqueous solution added to the contents of the barrel. See also *barrel finishing*.

Tumbling (Plastics)

Finishing operation for small plastic articles by which gates, flash, and fins are removed and/or surfaces are polished by rotating them in a barrel with wooden pegs, sawdust, and polishing compounds.

Tungsten and Alloys

In many respects, tungsten (symbol W) is similar to molybdenum. The two metals have about the same electrical conductivity and resistivity, coefficient of thermal expansion, and about the same resistance to corrosion by mineral acids. Both have high strength at temperatures above 1093°C, but because the melting point of tungsten is higher, it retains significant strength at higher temperatures than molybdenum does. The elastic modulus for tungsten is about 25% higher than that of molybdenum, and its density is almost twice that of molybdenum. All commercial unalloyed tungsten is produced by powder-metallurgy methods; it is available as rod, wire, plate, sheet, and some forged shapes. For some special applications, vacuum-arc-melted tungsten can be produced, but it is expensive and limited to relatively small sections.

Fabrication

Fabrication is a multistep process that converts tungsten metal from the original massive state (bars or ingots) to a more useful shape (sheet, tube, wire) and, at the same time, improves its physical properties. The exact details of fabrication depend on the method used for consolidating the metal and the type of product desired. Arc- or electron-beam-melted tungsten normally is extruded or forged to increase its ductility, whereas powder-processed material, because of its finer-grained structure and smaller tendency to crack, is less likely to require this initial step.

Tungsten is usually worked below its recrystallization temperature because the recrystallized metal tends to be brittle. Because increased working decreases the recrystallization temperature, successive lower temperatures are used in each fabrication step.

Full-density wrought tungsten can be hot forged, swaged, extruded, rolled, and drawn as secondary fabrication steps used to produce the final shape. Working temperature is usually 1100°C or above depending on the grain size and type of deformation.

Sintered billets are forged, swaged, or rolled initially at temperatures in excess of 1400°C. Working temperature can be progressively lowered as the amount of work increases, but consideration must be given to equipment capacity because of the high strength of tungsten.

Several tungsten alloys are produced by liquid-phase sintering of compacts of tungsten powder with binders of nickel–copper, iron–nickel, iron–copper, or nickel–cobalt–molybdenum combinations; tungsten usually comprises 85%–95% of the alloy by weight. These alloys are often identified as heavy metals or machinable tungsten alloys. In compact forms, the alloys can be machined by turning, drilling, boring, milling, and shaping; they are not available in mill products forms because they are unable to be wrought at any temperature.

Properties

Tungsten, element 74 on the periodic chart, has a melting point of approximately 3410°C, with values ranging between 3387°C and 3422°C reported in the literature. This value easily makes it the highest-melting-point metal. It has the lowest coefficient of thermal expansion of all metals, and with a density of 19.25 g/cm^3, it is one of the heaviest. It has the lowest vapor pressure of all metals, and high thermal and electrical conductivity.

Single crystals of tungsten are elastically isotropic and have very high tensile and bulk moduli, but mechanical properties are strongly temperature dependent, with the yield strength and ultimate tensile strength decreasing significantly with increasing temperature. At elevated temperatures, tungsten reacts rapidly with oxygen, forming a series of oxides that have stoichiometries ranging between WO_2 and WO_3.

The unique properties of tungsten make it the element of choice for such applications as filaments for incandescent lamps and x-ray tubes, electron sources for scanning and transmission electron microscopes, and connectors for circuit boards. Although these characteristics might suggest an even wider range of applications, several actually limit its utility. For example, the high density of tungsten makes it unsatisfactory for any weight-conserving application, and its aggressive reaction with oxygen limits its service at high temperatures. Welding is difficult because of the reactivity of tungsten with oxygen, and the presence of oxygen and other interstitials in the metal can make it very brittle at room temperature. Nonetheless, the special properties of tungsten are so beneficial that in many cases, it has been worth the cost and effort to engineer around the problems.

Tungsten retains a tensile strength of about 344 MPa at 1371°C, but because of its heavy weight is normally used in aircraft or missile parts only as coatings, usually sprayed on. It is also used for x-ray and gamma-ray shielding. Electroplates of tungsten or tungsten alloys give surface hardness to Vickers 700 or above.

Applications

Tungsten has a wide usage for alloy steels, magnets, heavy metals, electric contacts, rocket nozzles, and electronic applications. Tungsten resists oxidation at very high temperatures and is not attacked by nitric, hydrofluoric, or sulfuric acid solutions. Flame-sprayed coatings are used for nozzles and other parts subject to heat erosion.

Tungsten is usually added to iron and steel in the form of ferrotungsten, made by electric-furnace reduction of the oxide with iron or by reducing tungsten ores with carbon and silicon. Standard grades with 75%–85% tungsten have melting points from 1760°C to 1899°C. Tungsten powder is usually in sizes from 200 to 325 mesh and maybe had in a purity of 99.9%. Parts, rods, and sheet are made by powder metallurgy, and rolling and forging are done at high temperature.

The tungsten powder is used for spray coatings for radiation shielding and for powder-metal parts. Tungsten wire is used for spark plugs and electronic devices. Tungsten wire as fine as 0.00046 cm is used in electronic hardware. Tungsten whiskers, which are extremely fine fibers, are used in copper alloys to add strength. Copper wire, which normally has a tensile strength of 206 MPa, will have a strength of 827 MPa when 35% of the wire is tungsten whiskers. Tungsten yarns are made up of fine fibers of the metal. The yarns are flexible and can be woven into fabrics. Continuous tungsten filaments, usually 10–15 µm in diameter, are used for reinforcement in metal, ceramic, and plastic composites. Finer filaments of tungsten are used as cores, or substrates, for boron filaments.

The metal was also produced as arc-fused grown crystals, usually no larger than 0.952 cm in diameter and 25.4 cm long, and worked into rod, sheet, strip, and wire. Tungsten crystals, 99.9975% pure, are ductile even at very low temperatures, and wire as fine as 0.008 cm and strip as thin as 0.013 cm can be cold-drawn and cold-rolled from the crystal. The crystal metal has nearly zero porosity, and its electrical and heat conductivity are higher than ordinary tungsten.

One tungsten–aluminum alloy is a chemical compound made by reducing tungsten hexachloride with molten aluminum.

Tungsten wire is not used exclusively for lamp filaments. Because of its high melting temperature, tungsten can be heated to the point where it becomes a thermionic emitter of electrons, without losing its mechanical integrity. Consequently, tungsten filaments are often used as electron sources in scanning electron microscopes and transmission electron microscopes, and also as filaments in x-ray tubes.

In x-ray tubes, electrons produced from the tungsten filament are accelerated so that they strike a tungsten or tungsten–rhenium anode, which emits the x-rays. Again, this application takes advantage of the high melting point of tungsten, since the energy of the electron beam required to generate x-rays is very high, and the spot where the beam hits the surfaces becomes very hot. In most tubes, the anode is rotated to limit the peak temperature and to allow for cooling.

Finally, tungsten filaments of a much larger size are often selected as the heating elements in vacuum furnaces. Again, because of the high melting point of tungsten, these furnaces can achieve much higher temperatures than furnaces made with other heating elements. It is important to note that in vacuum furnaces, as well as all of the other applications, the tungsten is in a controlled environment that inhibits its oxidation.

For example, tungsten heavy alloys are materials in which tungsten powder is liquid-phase sintered, usually with the nickel–iron powders, to produce a composite material in which tungsten occupies about 95% of the volume. As the sintering process proceeds, the nickel–iron powder melts. Although the solubility of liquid nickel–iron in solid tungsten is small, solid tungsten readily dissolves in liquid nickel–iron. As the liquid wets the tungsten particles and dissolves part of the tungsten powder, the particles change shape, and internal pores are eliminated as the liquid flows into them. As processing continues, the particles coalesce and grow, producing a final product that is approximately 100% dense and has an optimized microstructure.

One of the main products made by this method is kinetic energy penetrators of military armored vehicles. This application takes advantage of the high density of tungsten, and it has been found that the liquid sintered materials have better impact properties than pure tungsten made by traditional powder processing.

Cutting tools and parts that must resist severe abrasion are often made of tungsten carbide. Tungsten carbide chips or inserts, with the cutting edges ground, are attached to the bodies of steel tools by brazing or by screws. The higher cutting speeds and longer tool life made feasible by the use of tungsten carbide tools are such that the inserts are discarded after one use.

Tungsten compounds (5% of tungsten consumption) have a number of industrial applications. Calcium and magnesium tungstates are used as phosphors in fluorescent lights and television tubes. Sodium tungstate is employed in the fireproofing of fabrics and in the preparation of tungsten-containing dyes and pigments used in paints and printing inks. Compounds such as WO_3 and WS_2 are catalysts for various chemical processes in the petroleum industry. Both WS_2 and WSe_2 are dry, high-temperature lubricants. Other applications of tungsten compounds have been made in the glass, ceramics, and tanning industries.

A completely new and different approach to produce bulk tungsten products from the powder-metallurgy process is through chemical vapor deposition (CVD), which provides a tungsten coating on a substrate.

Tungsten hexafluoride is the most common tungsten source for CVD processing. This compound is a liquid at room temperature, but its vapor pressure is high enough that the vapor can be continuously extracted and passed across the part that is to be coated.

$$WF_6 + 3H_2 \rightarrow W + 6HF$$

The reaction requires temperatures above approximately 300°C and a surface that causes the dissociation of molecular hydrogen into atomic hydrogen. Therefore, sections of a part may be selectively coated by having surfaces that either catalyze or prevent this reaction.

One of the most important applications of this process has been in the electronics industry, in which tungsten vias are placed in integrated circuits. The vias are small metal plugs that connect one level of wiring to another in the circuit board. They are generally about 0.4 mm in diameter, with an aspect ratio of about 2.5. In future applications, the diameter may shrink to less than 0.1 mm and have an aspect ratio greater than five. The metal for this application must have good electrical conductivity, must not react with the surrounding materials, must adhere to the wiring or silicon above or below the via, and must be deposited by a CVD reaction, as that is the only way to fill such small holes.

The most important method in the electronics industry is blanket CVD. In this technology, an adhesive layer is first put down to make certain that the CVD tungsten will stick to the surface. This adhesive layer is often titanium nitride, TiN. Tungsten is deposited on top of this layer, covering the surface and filling the vias. After the CVD is complete, the tungsten on the entire surface is removed by chemical–mechanical polishing. This procedure leaves the vias filled but cleans the surface of the unnecessary tungsten.

Alloys

A large number of tungsten-based alloys have been developed. Binary and ternary alloys of molybdenum, niobium, and tantalum with tungsten are used as substitutes for the pure metal because of their superior mechanical properties. Adding small amounts of other elements such as titanium, zirconium, hafnium, and carbon to these alloys improves their ductility. Tungsten–rhenium alloys possess excellent high-temperature strength and improved resistance to oxidation but are difficult to fabricate. This problem is ameliorated somewhat by the addition of molybdenum, a common composition being W (40 at%)–Re (30%)–Mo (30%). The strengths of tungsten or tungsten–rhenium systems can be increased by small amounts of a dispersed second phase such as an oxide (ThO_2, Ta_2O_5), carbide (HfC, TaC), or boride (HfB, ZrB). The so-called heavy alloys are three-component systems composed mainly of tungsten in combination with a nickel–copper or nickel–iron matrix. These materials are characterized by high density (17–19 g/cm^3), hardness, and good thermal conductivity.

Tungsten is used widely as a constituent in the alloys of other metals, since it generally enhances high-temperature strength. Several types of tool steels and some stainless steels contain tungsten. Heat-resistant alloys, also termed "superalloys," are nickel-, cobalt-, or iron-base systems containing varying amounts (typically 1.5–25 wt%) of tungsten. Wear-resistant alloys having the trade name Stellites are composed mainly of cobalt, chromium, and tungsten.

Cobalt–tungsten alloy, with 50% tungsten, gives a plate that retains a high hardness at red heat. Tungsten RhC is a tungsten–rhenium carbide alloy containing 4% rhenium carbide. It is used for parts requiring high strength and hardness at high temperatures. The alloy retains a tensile strength of 517 MPa at 1927°C.

Tungsten Carbide

Tungsten carbide is an iron-gray powder of minute cubical crystals with a Mohs hardness above 9.5 and a melting point of about 2982°C. It is produced by reacting a hydrocarbon vapor with tungsten at high temperature. The composition is WC, but at high heat, it may decompose into W_2C and carbon, and the carbide may be a mixture of the two forms. Other forms may also be produced, W_3C

and W_3C_4. Tungsten carbide is used chiefly for cutting tool bits and for heat- and erosion-resistant parts and coatings.

One of the earliest of the American bonded tungsten carbides was *Carboloy*, which was used for cutting tools, gauges, drawing dies, and wear parts. The carbides are now often mixed carbides. Carboloy 608 contains 83% chromium carbide, 2% tungsten carbide, and 15% nickel binder. It is lighter in weight than tungsten carbide, is non-magnetic, and has a hardness to Rockwell A93. It is used for wear-resistant parts and resists oxidation to 1092°C. Titanium carbide is more fragile but may be mixed with tungsten carbide to add hardness for dies. *Kennametal K601* is used for seal rings and wear parts and is a mixture of tantalum and tungsten carbides without a binder. It has a compressive strength of 4650 MPa, rupture strength of 689 MPa, and Rockwell hardness A94. *Kennametal K501* is tungsten carbide with a platinum binder for parts subject to severe heat erosion.

Tungsten carbide LW-1 is tungsten carbide with about 6% cobalt binder used for flame-coating metal parts to give high-temperature wear resistance. Deposited coatings have a Vickers hardness to 1450 and resists oxidation at 538°C. *Tungsten carbide LW-1N*, with 15% cobalt binder, as a much higher rupture strength, but the hardness is reduced to 1150.

Tungsten Electrode

A nonfiller metal electrode used in arc welding or cutting, made principally of tungsten. See *welding torch (arc)*.

Tungsten Filaments

Thin tungsten wire used principally in incandescent lamps and other filament applications that require resistance to creep at high temperatures.

Tungsten Inert Gas Welding

Welding in which the electric arc is struck between the component being welded and a nonconsumable tungsten electrode, and the weld zone is shielded by a gas that is inert during the welding operation. See *gas tungsten arc welding*.

Tungsten Steel

Tungsten steel is any steel containing tungsten as the alloying element imparting the chief characteristics to the steel. It is one of the oldest of the alloying elements in steel.

Tungsten increases the hardness of steel and gives it the property of red hardness, stabilizing the hard carbides at high temperatures. It also widens the hardening range of steel and gives deep hardening. Very small quantities serve to produce a fine grain and raise the yield point. The tungsten forms a very hard carbide and an iron tungstite, and the strength of the steel is also increased, but it is brittle when the tungsten content is high. When large percentages of tungsten are used in steel, they must be supplemented by other carbide-forming elements. Tungsten steels, except the low-tungsten chromium–tungsten steels, are not suitable for construction, but they are widely used for cutting tools, because the tungsten forms hard abrasion-resistant particles in high-carbon steels. Tungsten also increases the acid resistance and corrosion resistance of steels. The steels are difficult to forge and cannot be readily welded when tungsten exceeds 2%. Standard tungsten–chromium alloy steels 72XX contain 1.5%–2% tungsten and 0.50%–1% chromium. Many tool steels rely on tungsten as an alloying element, and it may range from 0.50% to 2.50% in cold-work and shock-resistant types to 9%–18% in the hot-work type, and 12%–20% in high-speed steels.

Tup

The moving head of a forge, in particular a drop forge.

Tup Impact Test

A falling-weight (tup) impact test developed specifically for plastic pipe and fittings.

Turbine Oil

An oil used to lubricate bearings in a steam or gas turbine.

Turk's Head Rolls

Four undriven working rolls, arranged in a square or rectangular pattern, through which metal strip, wire, or tubing is drawn to form square or rectangular sections.

Turning

(1) Removing material by forcing a single-point cutting tool against the surface of a rotating workpiece. The tool may or may not be moved toward or along the axis of rotation, while it cuts away a material. (2) A machining operation in which the workpiece is gripped in the chuck of a lathe and rotated against a tool bit, which cuts away a material.

Turnings

The swarf, that is, strands or chips of metal removed by turning.

Turns Per Inch

In composites, a measure of the amount of twist produced in a *yarn*, *toe*, or *roving* during its processing history. See also *twist*.

Turpentine

Also called in the paint industry *oil of turpentine*. An oil obtained by steam distillation of the oleoresin, which exudes when various conifer trees are cut. Longleaf pine and slash pine are the main sources.

Wood turpentine, called in the paint industry *spirits of turpentine*, is obtained from waste wood, chips, or sawdust by steam extraction or by destructive distillation. Wood turpentine forms more than 10% of all American commercial turpentines. Wood turpentine has a peculiar characteristic sawmill odor, and the residue of distillation has a camphor-like odor different from that of gum turpentine. It differs very little in competition, however, from the true turpentine. Steam-distilled wood turpentine contains about 90% terpenes, of which 80% is alpha pinene and 10% is a mixture of beta pinene and camphene. Some wood turpentine is produced as a by-product in the manufacture of cellulose.

Turpentine varies in composition according to the species of pine from which it is obtained. It is produced chiefly in the United States, France, and Spain.

American turpentine oil boils at 154°C, and the specific gravity is 0.860. It is a valuable drying oil for paints and varnishes, owing to its property of rapidly absorbing oxygen from the atmosphere and transferring it to the linseed or other drying oil, which leaves a tough and durable film of paint. Turpentine is also used in the manufacture of artificial camphor and rubber, and in linoleum, soap, and ink. *Gum thus*, used in artists' oil paints, is thickened turpentine, although gum thus was originally made from olibanum. Turpentine is often adulterated with other oils of the pine or with petroleum products, and the various states have laws regulating its adulteration for paint use.

Camphene is produced by isomerizing the alpha pinene of turpentine. Camphor is then produced by oxidation of camphene in acid. Camphene was also the name of a lamp oil of the early nineteenth century made from distilled turpentine and alcohol. It gave a bright white light but was explosive.

Turquoise

An opaque-blue gemstone with a waxy luster. It is a hydrous phosphate of aluminum and copper oxides. It is found in the western United States in streaks in volcanic rocks, but most of the turquoise has come from the Kuh-i-Firouzeh, or turquoise mountain, of Iran, which is a vast deposit of feldspar igneous rock. The valuable stones are the deep blue. The pale blue and green stones were called *Mecca stones* because they were sent to Mecca for sale to pilgrims. *Bone turquoise* or *odontolite*, used for jewelry, is fossil bone or tooth, colored by a phosphate of iron.

Tuyere

An opening in a cupola, blast furnace, or converter for the introduction of air or inert gas. See *blast furnace* and *cupola*.

Twill Weave

A basic fabric weave for reinforced plastics characterized by a diagonal rib or twill line. Each end floats over at least two consecutive picks, allowing a greater number of yarns per unit area than in a plain weave, while not losing a great deal of fabric stability. See also *plain weave*.

Twin

(1) Two portions of a crystal with a definite orientation relationship; one may be regarded as the parent, the other as the twin. The orientation of the twin is a mirror image of the orientation of the parent across a twinning plane or an orientation that can be derived by rotating the twin portion about a twinning axis. (2) An arrangement of the crystal lattice in which the planes of atoms on adjacent blocks in a crystal are aligned in a mirror image of each other. Under the microscope, the intersection between the blocks appears as a line. Annealing twins can occur during annealing of metals have a face-centered cubic structure. In this case, the individual twin lines in a pair are straight, run from grain boundary to grain boundary, and may be well separated. Mechanical twins can be formed during working of body-centered cubic and close-packed hexagonal structures. Microscopically, these twins appear within a grain as a pair of closely spaced, curved lines joined at their ends in a lenticular formation. Mechanical twins, formed in steel failing in a brittle manner, usually as a result of shock loading, are termed Neumann bands or Neumann lamellae. See also the *annealing twin* and *mechanical twin*.

Twin Bands

Bands across a crystal grain, observed on a polished and etched section, where crystallographic orientations have a mirror-image relationship to the orientation of the matrix grain across a composition plane that is usually parallel to the sides of the band.

Twin Carbon Arc Brazing

A brazing process that produces coalescence of metals by heating them with an electric arc between two carbon electrodes. The filler metal is distributed in the joint by capillary action.

Twin Carbon Arc Welding

A *carbon arc welding* process variation that produces coalescence of metals by heating them with an electric arc between two carbon electrodes. No shielding is used. Pressure and filler metal may or may not be used.

Twin-Sheet Thermoforming

A technique for *thermoforming* hollow plastic objects by introducing high-pressure air between two sheets and blowing the sheets into the mold halves (vacuum is also applied).

Twist

(1) In a yarn or other textile strand, the spiral turns about its axis per unit of length. Twist may be expressed as *turns per inch* (*tpi*). Twist provides additional integrity to yarn before it is subjected to the weaving process, a typical twist consisting of up to one turn per inch. In many instances, heavier yarns are needed for the weaving operation. This is normally accomplished by twisting together two or more single strands, followed by a plying operation. Plying essentially involves retwisting the twisted strands in the opposite direction from the original twist. The two types of twist normally used are known as S and Z, which indicate the direction in which the twisting is done. Usually, two or more strands twisted together with an S twist are plied with a Z twist in order to give a balanced yarn. Thus, the yarn properties, such as strength, bundle diameter, and yield, can be manipulated by the twisting and plying operations. (2) In pultruded parts, twist describes a condition of longitudinal, progressive rotation that can be easily detected for a noncircular cross-section by placing the pultrusion on a plane surface, holding one end flat with the surface, and observing whether one edge or side of the other end does not lie parallel with that surface.

Twist Boundary

The subgrain boundary consisting of an array of screw *dislocations*.

Twist Drill

A drill bit with a pair of cutting edges formed by the junction of the helical flutes and the conical tip. The function of the flutes is to lead away the swarf as the drill penetrates.

Twist Hackle (Ceramics and Glasses)

A *hackle* that separates portions of the crack surface, each of which has rotated from the original crack plane in response to a twist in the axis of principal tension. In a single crystal, a twist hackle separates portions of the crack surface, each of which follows the same cleavage plane, the normal to the cleavage plane being inclined to the principal tension. In a bicrystal or polycrystalline material, a hackle is initiated at a twist grain boundary.

Two-Component Adhesive

An adhesive supplied in two parts that are mixed before application. Such adhesives usually cure at room temperature.

Two-High Mill

A type of rolling mill in which only two rolls, the working rolls, are contained in a single housing. Compare with *four-high mill* and *cluster mill*.

T-X Diagram

A 2D graph of the isobaric phase relationships in a binary system; the coordinates of the graph are temperature and concentration.

Type Metal

Any of a series of alloys containing lead (58.5%–95%), antimony (2.5%–25%) used to make printing type. Small amounts of copper (1.5%–2.0%) are added to increase hardness in some applications.

Any metal used for making printing type, but the name generally refers to *lead–antimony–tin* alloys. Antimony has the property of expanding on cooling and thus fills the mold and produces sharp, accurate type. The properties required in a type metal are ability to make sharp, uniform castings strength and hardness, fairly low melting point, narrow freezing range to facilitate rapid manufacture in type-making machines, and resistance to dressing. A common type metal is composed of nine parts lead to one antimony, but many varieties of other mixtures are also used. The antimony content may be as high as 30%, 15%–20% being frequent. A common *monotype metal* has 72% lead, 18% antimony, and 10% tin. Larger and softer types are made of other alloys, sometimes containing bismuth; the hardest small type contains three parts lead to one antimony. A low melting point, soft-type metal contains 22% bismuth, 50% lead, and 28% antimony. It will melt at about 154°C. Copper, up to 2%, is sometimes added to type metal to increase the hardness but is not ordinarily used in metals employed in rapid-acting type machines. Some monotype metal has about 18% antimony, 8% tin, and 0.1% copper, but standard *linotype metal* for pressure casting has 79% lead, 16% antimony, and 5% tin. *Stereotype metal*, for sharp casting and hard-wearing qualities, is given as 80.0% lead, 13.5% antimony, 6% tin, and 0.5% copper. *Intertype metal* has 11%–14% antimony and 3%–5% tin. A typical formula for *electrotype metal* is 94% lead, 3% tin, and 3% antimony. The Brinell hardness of machine-molded type ranges from 17 to 23, and that of stereotype metal is up to 30. As constant remelting causes the separation of the tin and lead, and the loss of tin, or impoverishment of the metal, new metal must be constantly added to prevent deterioration of a standard metal into an inferior alloy. For many years, lead–antimony–tin alloys have been used as a weld seam filler in auto and truck bodies. In this application, they are commonly referred to as *body solder*. Because of advances in printing technology and auto manufacturing, use of these lead alloys is steadily declining. See *white metal* and *Babbitt metal*.

Typical Basis

An average property value. No statistical assurance is associated with this basis

U

U-Bend Die

A die, commonly used in pressure-break forming, that is machined horizontally with a square or rectangular cross-sectional opening that provides two edges over which metal is drawn into a channel shape.

Ugine-Séjournet Process

A direct extrusion process for metals that uses molten glass to insulate the hot billet and to act as a lubricant. Same as Séjournet process.

UHMWPE

See *ultrahigh-molecular-weight polyethylene.*

Ultimate Elongation

The elongation at rupture.

Ultimate Load (in Brinell Test)

The load that will just force the Brinell ball to a depth of half its diameter.

Ultimate Strength

The maximum strength (tensile, compressive, or shear) a material can sustain without fracture; determined by dividing maximum load by the original cross-sectional area of the specimen. Also known as nominal strength or maximum strength.

Ultimate Tensile Strength

The ultimate or final (highest) stress sustained by a specimen in a tension test. Same as *tensile strength.* See *tensile test.*

Ultrahard Tool Materials

Very hard, wear-resistant materials—specifically polycrystalline diamond and polycrystalline cubic boron nitride—that are fabricated into solid or layered cutting tool blanks for machining applications. See also *superabrasives.*

Ultrahigh-Molecular-Weight Polyethylene

Those polyethylene resins having weight-average molecular weights ranging from 3×10^6 to 6×10^6. These materials have both the highest abrasion resistance and the highest impact strength of any plastic. See also *high-density polyethylenes.*

Ultrahigh-Strength Steels

These are the highest-strength steels available. Arbitrarily, steels with tensile strengths of around 1378 MPa or higher are included in this category, and more than 100 alloy steels can be thus classified. They differ rather widely among themselves in composition or the way in which the ultrahigh strengths are achieved.

Medium-carbon low-alloy steels were the initial ultrahigh-strength steels, and within this group, a chromium–molybdenum steel (4130) grade and a chromium–nickel–molybdenum steel (4340) grade were the first developed. These steels have yield strengths as high as 1654 MPa and tensile strengths approaching 2068 MPa. They are particularly useful for thick sections because they are moderately priced and have deep hardenability. Several types of stainless steels are capable of strengths above 1378 MPa, including a number of martensitic, cold-rolled austenitic, and semiaustenitic grades. The typical martensitic grades are types 410, 420, and 431, as well as certain age-hardenable alloys. The cold-rolled austenitic stainless steels work-harden rapidly and can achieve 1241 MPa tensile yield strength and 1378 MPa ultimate strength. Semiaustenitic stainless steels can be heat treated for use at yield strengths as high as 1516 MPa and ultimate strengths of 1620 MPa.

Maraging steels contain 18%–25% nickel plus substantial amounts of cobalt and molybdenum. Some newer grades contain somewhat less than 10% nickel and between 10% and 14% chromium. Because of the low carbon (0.03% max) and nickel contents, maraging steels are martensitic in the annealed condition but are still readily formed, machined, and welded. By a simple aging treatment at about 482°C, yield strengths of as high as 2068 and 2413 MPa are attainable, depending on specific composition. In this condition, although ductility is fairly low, the material is still far from being brittle.

Among the strongest of plain carbon sheet steels are the low- and medium-carbon sheet grades called *MarTinsite.* Made by rapid water quenching after cold rolling, they provide tensile yield strengths of as high as 1517 MPa but are quite limited in ductility.

There are two types of ultrahigh-strength, low-carbon, hardenable steels. One, a chromium–nickel–molybdenum steel, named *Astralloy,* has 0.24% carbon and is air-hardened to a yield strength of 1241 MPa and heavy sections when it is normalized and tempered at 260°C. The other type is an iron–chromium–molybdenum–cobalt steel and is strengthened by a precipitation hardening and aging process to levels of up to 1654 MPa in yield strength. High-alloy quenched and tempered steels are another group that has extra-high strengths. They contain 9% nickel, 4% cobalt, and 0.20%–0.30% carbon and develop yield strengths close to 2068 MPa and ultimate strengths of 2413 MPa. Another group in this high-alloy category resembles high-speed tool steels but is modified to eliminate excess carbide, considerably improving ductility. These so-called matrix steels contain tungsten, molybdenum, chromium, vanadium, cobalt, and about 0.5% carbon. They can be heat treated to ultimate strengths of over 2757 MPa—one of the highest strengths achievable in steels, except for heavily cold-worked high-carbon steel strips used for razor blades and drawn wire for musical instruments, both of which have tensile strengths as high as 4136 MPa.

These steels have a high fracture toughness (K_{1c} of 100 MPa√m, or 91 ksi√in).

Ultra Light Metals

See *light metals.*

Ultramicroscopic

See *submicroscopic.*

Ultrasonic

Sound waves at frequencies above those audible to the human ear. They are readily transmitted through solids and be used to detect subsurface defects. See *nondestructive testing.* They are also utilized in ultrasonic cleaning to clean components immersed in a bath of liquid, and in ultrasonic soldering, they disrupt the normally tenacious oxide film, such as that on aluminum, which would otherwise impede bonding between the underlying metal and the solder.

Ultrasonic Beam

A beam of acoustical radiation with a frequency higher than the frequency range for audible sound—that is, above about 20 kHz.

Ultrasonic Bonding

A method of joining plastics using vibratory mechanical pressure at ultrasonic frequencies. Electrical energy is changed to ultrasonic vibrations by means of either a magnetostrictive or piezoelectric transducer. The ultrasonic vibrations generate frictional heat, melting the plastics and allowing them to join.

Ultrasonic Cleaning

Immersion cleaning aided by ultrasonic waves that cause microagitation.

Ultrasonic Coupler

In ultrasonic welding and soldering, the elements through which ultrasonic vibration is transmitted from the transducer to the tip.

Ultrasonic C-Scan Inspection

A method for displaying the relative attenuation of ultrasonic waves across the surface of a structural component. An ultrasonic transducer is used to scan the surface of a material mechanically in than x–y raster scan mode while generating and receiving waves. Either the material is immersed in a water bath or columns of water are provided between the transducer and the material as a medium for ultrasonic energy transmissions. The received wave signals are electronically conditioned and measured to determine relative energy losses of the wave as it progresses through the material at each particular location on the specimen. Ultrasonic C-scan has been used extensively to determine both the initial integrity of a manufactured part and the void content and to follow the initiation and progression of damage resulting from environmental loading.

Ultrasonic Frequency

A frequency, associated with elastic waves, that is greater than the highest audible frequency, generally regarded as being higher than 20 kHz.

Ultrasonic Gas Atomization

A variation of the gas atomization process that uses high-frequency gas pulses with velocities up to 4,600 m/s (15,100 ft/s) to break up the molten metal stream. This process can produce high yields of powder with particle diameters less than 20 μm.

Ultrasonic Impact Grinding

Material removal by means of an abrasive slurry and the ultrasonic vibration of a nonrotating tool. The abrasive slurry flows through a gap between the workpiece and the vibrating tool. Material removal occurs when the abrasive particles, suspended in the slurry, are struck on the downstroke of the vibrating tool. The velocity imparted to the abrasive particles causes microchipping and erosion as the particles impinge on the workpiece. See also *ultrasonic machining.*

Ultrasonic Inspection

A nondestructive method in which beams of high-frequency sound waves are introduced into materials for the detection of surface and subsurface flaws in the material. The sound waves travel through the material with some attendant loss of energy (attenuation) and are reflected at interfaces. The reflected beam is displayed and then analyzed to define the presence and location of flaws or discontinuities. Most ultrasonic inspection is done at frequencies between 0.1 and 25 MHz—well above the range of human hearing, which is almost 20 Hz–20 kHz.

Ultrasonic Machining

A process for machining hard, brittle, nonmetallic materials that involves the ultrasonic vibration of a rotating diamond core drill or milling tool. Rotary ultrasonic machining is similar to the conventional drilling of glass and ceramic with diamond core drills, except that the rotating core drill is vibrated at an ultrasonic frequency of 20 kHz. Rotary ultrasonic machining does not involve the flow of an abrasive slurry through a gap between the workpiece on the tool. Instead, the tool contacts and cuts the workpiece, and a liquid coolant, usually water, is forced through the bore of the tube to cool and flush away the removed material. See also *ultrasonic impact grinding.*

Ultrasonic Soldering

The soldering process variation in which high-frequency vibratory energy is transmitted through molten solder to remove undesirable surface films and thereby promote wetting of the base metal. This operation is usually accomplished without a flux.

Ultrasonic Testing

See *ultrasonic inspection.*

Ultrasonic Welding

A solid-state process in which materials are welded by locally applying high-frequency vibratory energy to a joint held together under pressure. Ultrasonic energy is produced through a transducer, which converts high-frequency electrical vibrations to mechanical vibrations at the same frequency, usually above 15 kHz (above the audible range). Mechanical vibrations are transmitted through a coupling system to the welding tip and into the workpieces. The tip vibrates laterally, essentially parallel to the weld interface, while static force is applied perpendicular to the interface.

The vibration, typically in the range of 20–40 kHz, is generated electrically with available power of 200–3000 W and is delivered to the weld zone by the machine head, sometimes termed a *sonotrode*. An additional static interfacial force is normally applied, and additional external heating is an option. See the term *ultrasonic coupler*.

Ultraspeed Welding

See preferred term *commutator-controlled welding*.

Ultraviolet

Pertaining to the region of the electromagnetic spectrum from approximately 10 to 380 nm. The term ultraviolet without further qualification usually refers to the region from 200 to 380 nm.

Ultraviolet-Curable Hot-Melt Adhesives

For years, ultraviolet (*UV*)-curable pressure-sensitive adhesives (PSAs) have been recognized as a fixture alternatively to solvent-borne products. The idea of achieving the solvent and heat resistance of an acrylic without facing the various safety and environmental ramifications has always been enticing to both PSA formulators and users. The promise of this technology has led to the development of a variety of adhesive technology platforms.

Photoinitiators can now be purchased that offer much better thermal stability for improved pot life and coatability. Polymers have been developed that are much more chemically active, dramatically reducing the amount of photoinitiator required to achieve proper cure (and consequently the total cost of the adhesive). Thanks to the response of their suppliers, adhesive manufacturers are making UV-curable products that are more versatile than ever.

Converting a pressure-sensitive hot-melt coating line over to UV curing no longer demands growing accustomed to radically different adhesives.

Properties and Applications

Conventional hot-melt PSAs are widely used for tape and label applications. Their room temperature performance is difficult to match with alternative chemistries. They possess an outstanding combination of a high tack, peel, and shear and adhere well to wet on low-energy surfaces.

In addition to the performance advantages, conventional hot melts possess some significant processing advantages because they require no solvent vehicle for application. The lack of a combustible solvent makes them safer and more environmentally friendly than any other adhesive.

Because they require no drying, they can be applied more easily at high depositions for use on slick or rough surfaces. Finally, since they require no dryers, hot-melt coaters are generally more compact and lower in cost than liquid coaters.

Unfortunately, users of conventional hot melts can only enjoy these benefits over a limited range of conditions. The products are hobbled by poor resistance to solvents, plasticizers, and heat. This precludes their use in some industrial applications where their high room temperature peel and shear could make them otherwise well suited.

Types and Forms

Traditional pressure-sensitive hot melts are formulated primarily with block copolymers and various tackifying resins. The cohesive strength of the product is largely determined by the block copolymer used. Some of the most common block copolymers used are styrenic triblock copolymers. These are long polymer chains with polystyrene molecules grouped together to form two end blocks surrounding one midblock made of an elastomeric material. Frequently used triblock copolymers are styrene–isoprene–styrene or styrene–butadiene–styrene.

Future

UV offers an outstanding combination of versatile performance and ease of use. The UV-curable hot melt is comparable to the solvent-borne acrylic. In addition, a significant performance advantage has been seen on low-energy surfaces. This makes the technology even more appealing because of the ever-increasing use of plastics.

With new tools at their disposal, the performance of UV curables is now up to any adhesive task required. Formulators have crafted newer and better adhesives capable of a variety of tasks. These products capture the traditional advantages of hot melts while meeting many of the standards of acrylics.

Ultraviolet Degradation

The degradation caused by long-term exposure of a material to sunlight or other ultraviolet rays containing radiation.

Ultraviolet/Electron Beam Cured Adhesives

Radiation curing involves the rapid conversion of specially formulated, 100% reactive liquids to solids. Potential energy sources include microwaves, visible infrared (IR), ultraviolet (UV), and electron beam (EB) sources, the latter two being the most commercially important. Radiation-cured materials are used as coatings, inks, adhesives, sealants, and potting compounds.

The UV curing process typically involves the exposure of a reactive liquid that contains a photoinitiator to UV radiation at a wavelength between 200 and 400 nm. The liquid is rapidly converted to a solid, usually in less than 60 s. In the EB curing process, electrons are artificially generated and accelerated to energies of less than 100 keV to greater than 1 billion keV. Generally, 50–350 keV electrons are used to cure adhesives that have bond lines with thicknesses of 25–38 μm. The reactive liquid in the EB process does not contain a photoinitiator.

Because the main advantage of UV-/EB-curable adhesives is rapid curing at room temperature, they can be used to bond the heat-sensitive substrates, such as polyvinyl chloride. UV-/EB-cured adhesives have been used to replace solvent-base adhesives because of the increasing cost of properly recovering and disposing of solvents. The cross-linked nature of UV-/EB-cured adhesives results in good chemical, heat, and abrasion resistance; toughness; dimensional stability; and adhesion to many substrates. Unlike thermal curing, EB curing can be selective, and the depth of penetration can be controlled.

Most UV/EB adhesives are based on an addition polymerization curing mechanism. Materials consist of acrylic acid esters of various forms or combinations of acrylates with aliphatic or aromatic epoxies, urethanes, polyesters, or polyethers. Although the epoxy-base systems have higher tensile strengths, their elongations are less than those of the urethane-base systems. In addition, the urethane-base systems have better abrasion resistance.

Typical UV-curable adhesive applications include the electronics, automotive, medical, optics, and packaging markets, as well as tapes and labels. EB-curable adhesives are used in magnetic tapes and floppy disks, where magnetic particles are bonded to films, as well as in packaging, tapes, and labels.

Ultraviolet Radiation

Electromagnetic radiation in the wavelength range of 10–380 nm. See also *electromagnetic radiation*.

Ultraviolet Stabilizer

Any chemical compound that, when admixed with a resin, selectively absorbs ultraviolet rays.

Ultraviolet/Visible Absorption Spectroscopy

An analytical technique that measures the wavelength-dependent attenuation of ultraviolet, visible, and near-infrared light by an atomic or molecular species; used in the detection, identification, and quantification of numerous atomic and molecular species.

Unary System

Composed of one component.

Unbond

An area within an adhesively bonded interface between two adherends in which the intended bonding action failed to take place, or an area in which two layers of prepreg in a cured component do not adhere. Also used to denote specific areas deliberately prevented from bonding in order to simulate a defective bond, such as in the generation of quality standard specimens.

Uncertainty

(1) An indication of the variability associated with a measured value that takes into account two major components of error: (a) *bias* and (b) the random error attributed to the imprecision of the measurement process. (2) The range of values within which the true value is estimated to lie. It is a best estimate of possible inaccuracy due to both random and systematic errors.

Unctuous

A general term expressing the slippery feel of a material, such as a lubricant, when rubbed with the fingers.

Underannealing

Generally, in a pejorative sense, annealing at a temperature or for a time insufficient to induce full softening. The term is also used in a more specific sense when referring to the deliberate annealing of steel in the transition range so that the pearlite areas and a proportion of the ferrite areas are recrystallized.

Underbead Crack

A crack in the heat-affected zone of a weld generally not extending to the surface of the base metal. See also the term *weld crack*.

Undercoat

A deposited coat of material that acts as a substrate for a subsequent thermal spray deposit. See also *bond coat*.

Undercooling

Same as *supercooling*.

Undercure

An undesirable condition of a molded plastic article resulting from the allowance of too little time and/or temperature or pressure for adequate hardening of the molding.

Undercut

(1) In weldments, a groove melted into the base metal adjacent to the toe or root of a weld and left unfilled by weld metal. (2) For castings or forgings, same as *back draft*. (3) A reentrant detail on a casting. (4) A localized reduction in some machined dimension, intentional or otherwise. If unintentional, the term may be used in the sense of a stress raiser.

Underdraft

A condition wherein a metal curves downward or leaving a set of rolls because of higher speed in the upper roll.

Underfill

(1) In weldments, a depression on the face of the weld or root surface extending below the surface of the adjacent base metal. (2) A portion of a forging that has insufficient metal to give it the true shape of the impression.

Underfilm Corrosion

Corrosion that occurs under organic films in the form of randomly distributed threadlike filaments or spots. In many cases, this is identical to *filiform corrosion*.

Underflushing (of Weld)

See *flushing*.

Undersize Powder

Powder particles smaller than the minimum permitted by a particle size specification.

Understressing

(1) Applying a cyclic stress lower than the *endurance limit*. This may improve fatigue life if the member is later cyclically stressed at levels above the endurance limit. (2) In fatigue, the deliberate application of one or more cycles at a stress below, but usually close to the *fatigue limit*. This can have a strain aging effect improving the fatigue properties half-life.

UNE, UNO, UNQ, etc.

It has been proposed that man-made elements, particularly those discovered in the future, be named in terms of their atomic number in pseudo-Latin. Hence, element 104 is un-nil-quadum, that is, "unnilquadum" shortened to UNQ, etc.

Uniaxial

Acting or aligned along one axis of a crystal or a component.

Uniaxial Compacting

Compacting of powder along one axis, either in one direction or in two opposing directions. Contrast with *isostatic pressing*.

Uniaxial Load

A condition in which a material or component is stressed in only one direction along its axis or center line.

Uniaxial Strain

See *axial strain*.

Uniaxial Stress

A state of stress in which two of the three principal stresses are zero. See also *principal stress (normal)*.

Unidirectional Compacting

Compacting of powder in one direction.

Unidirectional Laminate

A reinforced plastic laminate in which substantially all of the fibers are oriented in the same direction. See the terms *laminate* and *quasi-isotropic laminate*.

Uniform Corrosion

(1) A type of corrosion attack (deterioration) uniformly distributed over a metal surface. (2) Corrosion that proceeds at approximately the same rate over a metal surface. Also called general corrosion.

Uniform Elongation

The elongation at maximum load and immediately preceding the onset of necking in a tensile test.

Uniformly Distributed Impact Test

See *distributed impact test*.

Uniform Magnetic Heating

Uniform magnetic heating (UMH) is a system by which electrical energy is converted to heat within metallic materials in a very efficient and flexible manner. Although this system and conventional induction heating both require electrical coils to convert electricity to magnetic flux energy, the similarities stop there.

UMH versus Induction

The CoreFlux UMH system utilizes two coils that are permanently fixed around a C-shaped laminate core. Similar to induction systems, the coils convert electrical energy into magnetic flux energy. However, in conventional induction, the component to be heat treated is placed inside the coil. The energy is directly transferred to the part in the form of surface eddy currents, which are generated as the current flows around the component.

The CoreFlux system transfers energy in a different way. Energy is transferred into a laminate core (similar to a transformer core) and channeled directly to the part in a linear manner. In this way, the magnetic flux energy is distributed throughout the entire part. Key to the technology is that the flux direction in the core oscillates at a user-defined frequency from 20 to 400 Hz. As a result, the polarity of the flux field changes at this defined frequency. Each time the polarity changes, heat is released throughout the component via "hysteresis loss."

Simply put, hysteresis loss is the energy released throughout the material as the magnetic domains in the microstructure are forced to realign continually with the alternating magnetic field. The effect of this phenomenon is the uniform production of heat throughout the component.

Because of the inherent nature of this method, the core and surface temperatures show minimal thermal gradients throughout the entire heating process. With induction, localized overheating of components with holes and unusual characteristics is a common problem, but overheating is not typically a concern with UMH.

In addition to the benefits of uniform through-heating, UMH also provides several other key advantages:

- The coils are permanently fixed to the machine and require no maintenance or coil changeovers.
- The same coils can run a wide range of processes and bring the benefit of flexibility by running families of parts with no coil changes required.
- Metallurgical results are typically significantly better than conventional systems.
- The machine requires no water cooling, which results in lower facility costs, less maintenance, and no energy loss through heated water.
- Energy efficiency can be as much as twice that of conventional induction. The power supply is basically a standard ac variable motor drive. This eliminates costly custom power supplies and the problems associated with their maintenance. It also greatly reduces overall capital equipment costs.

Applications

Heating Press Dies

One simple application of the technology that offers significant benefits is the heating of press dies. A variety of die shapes may be placed in a common machine that can heat the dies to the required temperature in a matter of minutes. As an alternative to oven preheating, the UMH process offers much faster, cleaner, and more efficient heating and allows customers to change dies more quickly. Machines may also be mobile so that one machine can serve multiple press locations. Current size and shape capabilities range from very small to approximately $180 \times 45 \times 45$ cm. Larger components may be accommodated with custom designs.

Tempering Gears and Bearings

One of the most promising areas of publication for this technology is the tempering of gears and bearings. Parts do not need to be rotated, and taller parts do not require scanning. For example, if a component has a round shape and an inner bore, the CoreFlux UMH technology can be applied by utilizing a core extension. The component is placed around the core extension, which provides a secondary field but causes very rapid and uniform heating with no part contact. UMH generates a uniform field throughout the component, from the inner core extension. On the other hand, if the part were induction tempered, the coil would be placed around the outside diameter, and the field would be generated from the outside inward to the core.

Fortunately, the CoreFlux UMH process overcomes time at temperature and the skin effect phenomenon. Because of the uniformity of heating, the core of the gear comes to temperature at virtually the same rate as the teeth. Metallurgical results typically exceed expectations and return properties similar, and sometimes superior, to oven tempering. An additional benefit is the ability to run multiple parts around the same core extension. Depending on the actual geometry, parts can be stacked around the core extension with minimal effect on the total cycle time. Obviously, this can have a profound effect on production throughput and machine utilization.

Hardening Gears and Bearings

The same core extension approach described earlier may be utilized for high-temperature applications. Although the research and documentation related to hardening are less mature than the lower-temperature applications, the technology has again demonstrated substantial benefits over today's alternatives. Targeted as hardening applications only, the technology offers the same flexibility as tempering. In fact, UMH has potential for the design of an entire hardening and temporary line that could allow for the flexible running of components inside a large "family" grouping, with those setup changes.

Heating Aluminum Billets

The properties of aluminum make difficult to through-heat quickly and uniformly by conventional technology. With the CoreFlux UMH process, and aluminum billet may be placed directly on the insulated core cap, and the top core/coil assembly may then be lowered to make light contact with the billet. This clamping effect is utilized to create the most efficient transfer of energy into the billet and to facilitate holding the billet in place during heating. Clamping pressure is adjustable to eliminate any marking or deformation.

Capabilities have been documented and proven, and application development continues in the aluminum field, with preheating applications ranging from $370°C$ to semisolid temperatures. Although steel forging offers similar promise, development continues in this area to assure that the machine cores will endure long exposures to extreme forging temperatures.

Shrink-Fit Applications

Another simple through-heat application is preheating components for shrink fitting. Again, when compared with any other alternatives available for lower-temperature through-heating, the CoreFlux UMH process is an improvement.

Press Tempering

Although a relatively new development, press-temper applications have drawn considerable attention. In cases where thin parts must be stacked and held flat during tempering or stress relieving, the CoreFlux process is worth considering. The top coil and core assembly are typically lowered via pneumatic cylinders, and the clamping pressure is limited to as little as a pound. However, the pressure can be adjusted to provide more than a ton of damping force if necessary. If higher force is required, the machine can be customized. In addition, this system could allow for the pressure to be controlled as a function of temperature.

Uniform Strain

The strain occurring prior to the beginning of localization of strain (necking); the strain to maximum load in the tension test.

Unimeric

Pertaining to a single molecule that is not monomeric, oligomeric, or polymeric, such as saturated hydrocarbons.

Unipolarity Operation

A resistance welding process variation in which succeeding welds are made with pulses of the same polarity.

Unit Cell

A parallelepiped element of crystal structure, containing a certain number of atoms, the repetition of which through space will build up the complete crystal. See the term *lattice*.

Unit Power

The net amount of power required during machining or grinding to remove a unit volume of material in unit time.

Univariant Equilibrium

A stable state among several phases equal to one more than the number of components, that is, having one degree of freedom.

Universal Forging Mill

A combination of four hydraulic presses arranged in one plane equipped with the billet manipulators and automatic controls, used for radial or draw forging. See the term *radial forging*.

Universal Gas Constant

See *gas constant*.

Universal Mill

A rolling mill in which rolls with a vertical axis roll the edges of the metal stock between some of the passes through the horizontal rolls that are capable of working through all surfaces of the section.

Unlubricated Sliding

Sliding without lubricant but not necessarily under completely dry conditions. Unlubricated sliding is often used to mean "not intentionally lubricated," but surface species such as naturally formed surface oxides and other interfacial contaminants may act in a *lubricant manner* is nominally unlubricated sliding.

Unsaturated Compounds

Any chemical compound having more than one bond between two adjacent atoms, usually carbon atoms, and being capable of adding other atoms at that point to reduce it to a single bond, for example, *olefins.*

Unsaturated Polyester Resin

The use of unsaturated polyester resins in structural applications is well documented. There are, however, significant quantities of unsaturated polyester resins used in specialist compounded products, which are more likely to be unreinforced. The most well known of these technologies are formulated gel coats, a technology that has been changing rapidly in recent years with improvements in gloss retention, color retention, and volatile organic compound emissions.

The introduction of granite effect coatings and solid surface material is a further example of the versatility of unsaturated polyester resins. Although these materials have been predominantly used for interior applications, their potential for exterior use on buildings provides exciting possibilities for a new and varied range of composite building materials providing stone effects at a fraction of the weight of conventional building materials.

Other compounded resins that are especially important to the building and construction market are those with fire-resistant characteristics. In addition, the improvements in smoke reduction from unsaturated polyester resin systems make such materials attractive for cladding applications. Combining the advantages of these resins with decorative coatings and sandwich construction provides the basis for structural, insulating components.

Markets

The markets for reinforced plastics are frequently split into a number of generally accepted sectors, such as marine, land transport, building and construction, and chemical containment. There are, of course, subdivisions in each sector, for example, powered pleasure boats, powered work boats, sailboats, and offshore applications in the marine market, but most of the discussion in the literature is about the use of fiber-reinforced composites in these market sectors and market subgroups. In general, unfilled resins with good mechanical properties are preferred, but there are, very often, requirements for compounded products to provide special characteristics to meet specific performance requirements. Obviously, compounded fire-resistant materials fall into such a category and are used to impart resistance to ignition, resistance to surface spread of flame, and, increasingly, reduction in emissions of smoke and toxic fumes. Although such materials are often highly filled, they are used with fiber reinforcement for the manufacture of structural and semistructural components. The importance of these resins and their developments together with two other important compounded unsaturated polyester resin-based products has been disclosed. These latter materials are not used in conjunction with fiber reinforcement but are usually simply filled or pigmented; they are gel coats and are mainly used as "in-mold" coatings and solid surface materials for the manufacture of synthetic granite-type products.

Resin concrete and repair putties are also large consumers of unsaturated polyester resin in non-fiber-reinforced compounds.

Resistance to Fire

The use of glass-fiber-reinforced plastics in applications where fire resistance was particularly important was introduced into the building industry five decades ago. Generally, the structural performance of the material was not questioned for building applications because it had been well proven for the construction of boats. However, as with most plastic materials, its ability to perform under fire conditions was in question for use in buildings, even though it had been documented that fires in buildings originate from the contents and in a vast majority of circumstances the structure does not contribute to loss of life.

One of the most successful means to improve the resistance of plastics to fire is by the incorporation of fillers, which break down with heat to produce heavy vapors to prevent oxygen reaching the surface of the material and hence reduce the possibility of burning. The major problem associated with the high levels of filler required to render resins fire retardant is the increase in their viscosity, which results in handling difficulties when manufacturing structural components.

The use of halogenated additives, which work synergistically with some fire-retardant fillers, helps to overcome handling problems but results in the potential for toxic fume production under fire conditions.

The availability of improved viscosity modifiers is now enabling resins filled with high levels of nontoxic fillers, such as alumina trihydrate, to be used to manufacture laminates containing reasonable levels of reinforcement to produce, at least, semistructural components.

Such systems will meet the new International Maritime Organization requirements for use on passenger ships. Under the test conditions, the material has to exhibit low surface spread of flame characteristics, low smoke emissions, and low emissions of carbon monoxide.

Gel Coat Protection

In the early days of the GRP industry, the need for resin-rich surfaces was established to

- Improve the durability of components
- Protect the laminate from the environment
- Reduce fiber pattern
- Provide a smooth aesthetic finish
- Eliminate the need for painting

As a result of these requirements, a market for ready-formulated coatings was established and gel coat product ranges became established.

The availability of quality "in-mold" coatings, such as gel coats, to fabricators saves labor and wastage in the workshop and improves the quality of molded components. Gel coats are available in brush and spray versions with a variety of properties and performance characteristics to meet a range of needs. They must be applied carefully and correctly to avoid faults.

Gel Coat Developments

Over the years, the need for improved gloss and color retention in gel coats has been recognized as developments in ultraviolet resistance and color fastness have resulted in a range of gel coats that can be weathered under the severest tropical weather conditions without changes in appearance.

Solid Surfaces

Resins have often been used to bind together fillers and aggregates to produce materials such as resin concrete and synthetic cultured and onyx marble. For decorative surfaces, a clear (translucent) gel coat is used to improve the quality of the surface finish and remove the effects of surface porosity. Although the gel coat used is usually based on good-quality, water-resistant resins, the inferior quality of the backing systems often results in a material that is susceptible to crazing, cracking, poor water resistance, and poor thermal resistance. Because the gel coat surface is too thin for repairs to be effectively carried out, the problems cannot be easily rectified.

The monopoly of the acrylic-based solid surface material has been gradually eroded by the introduction of unsaturated polyester–based solid surface, which offers a much wider range of colors to provide improved customer choice. Raw materials and manufacturing processes have been designed to eliminate voids in polyester-based solid surfaces.

Traditionally, solid surface materials have been used for the manufacture of kitchen surfaces, sinks, and bathroom units. However, there is increasing interest in more diverse applications such as furniture, table tops, tiles, paneling, cutlery, and pens. It is also possible to use the material as a 2–3 mm thick coating for other materials, and the granite effect finish is reviving interest in GRP for cladding for buildings.

The unsaturated polyester resin–based material is comprised of three components:

1. *Chips*: The colored fillers or chips, used to provide the granite effect, can be based on thermoplastic or thermoset materials.
2. *Resins*: Resin must be clear and near "water white" to allow the depth of color of the chips to be appreciated. The resin must also be resistant to elevated temperature, water, staining, UV light, and cigarette burns. Hence, typical formulations giving an acceptable level of performance are based on isophthalic acid and neopentyl glycol.
3. *Fillers*: Only alumina trihydrate can be used in addition to the colored "chips" because it is translucent. It also offers fire-retardant characteristics.

Solid surface systems are nonreinforced that can be machined and cut with conventional woodworking equipment. Patterns can be routed in solid surface materials and cast resin "inlaid" to provide a variety of customized finishes.

It is important to ensure when manufacturing solid surface that the resin is formulated to accept high filler loading without air entrapment and will develop hardness rapidly. The final product must be resistant to chipping, cracking, hot–cold water cycling, "blushing," and UV light. It must also be easy to machine for shaping and finishing.

Future

Unsaturated polyester resin–based compounded products provide a range of materials with tailored performance characteristics for a variety of markets.

Gel coats are essential for most applications for GRP, providing aesthetic finishes in the marine, transport, building, and construction markets. They have well-proven durability, but improvements in gloss and color retention will ensure their position as a major coating for fiber-reinforced composite materials in the future.

Fire-retardant resins with exceptionally low smoke production under fire conditions are becoming a reality with unsaturated polyester resin–based systems. New standards are providing new challenges, which are being met successfully to ensure that materials meet new requirements for surface spread of flame for materials for use in construction applications.

Unsoundness

This term usually implies volumetric defects such as internal cavities, voids, and porosity as opposed to planar defects such as crack or upper yield. See *tensile test*.

Unsymmetric Laminate

A *laminate* having an arbitrary stacking sequence without midplane symmetry.

Upper Punch

The member of a die assembly or tool set for forming powder metallurgy parts that closes the die and forms the top of the part being produced.

Upper Ram

The part of a pneumatic or hydraulic press for forming powder metallurgy parts that is moving in an upper cylinder and transmits pressure to the upper punch or set of upper punches.

Upset

(1) The localized increase in cross-sectional area of a workpiece or weldment resulting from the application of pressure during mechanical fabrication or welding. (2) That portion of the welding cycle during which the cross-sectional area is increased by the application of pressure. (3) Bulk deformation resulting from the application of pressure in welding. The upset may be measured as a percent increase in interfacial area, a reduction in length, or a percent reduction in thickness (for lap joints). (4) A local increase in cross-sectional area resulting from a longitudinal force. The term is commonly used in operations where the deformation is delivered in a die that is used to shape the upset, for example, the formation of a bolt head. The process is beneficial as it produces a desirable flow, that is, pattern of deformation, and it reduces the amount of machining.

Upset Butt Welding

Same as *resistance butt welding*.

Upset Forging

A forging obtained by *upset* of a suitable length of bar, billet, or bloom.

Upset Pressing

The pressing of a powder compact in several stages, which results in an increase in the cross-sectional area of the part prior to its ejection.

Upsetter

A horizontal mechanical press used to make parts from bar stock or tubing by *upset forging*, piercing, bending, or otherwise forming in dies. Also known as a *header*.

Upsetting Force

In *upset welding*, the force exerted at the faying surfaces during upsetting. See also *upset* (3).

Upsetting Time

In *upset welding*, the time during upsetting. See also *upset* (3).

Upset Weld

A weld made by *upset welding*.

Upset Welding

A resistance welding process in which the weld is produced, simultaneously over the entire area of abutting surfaces or progressively along a joint, by applying mechanical force (pressure) to the joint and then causing electrical current to flow across the joint to heat the abutting surfaces. Pressure is maintained throughout the heating period. See also *open-gap upset welding*.

Upslope Time

In resistance welding, the time during which the welding current continuously increases from the beginning of welding current. See also *slope control*.

Uranium

An elementary metal (symbol U), uranium never occurs free in nature but is found chiefly as an oxide in the minerals pitchblende and carnotite where it is associated with radium. The metal has a specific gravity of 18.68 and atomic weight 238.2. The melting point is about 1133°C. It is hard but malleable, resembling nickel in color, but related to chromium, tungsten, and molybdenum. It is soluble in mineral acids.

Uranium has three forms. The alpha phase, or orthorhombic crystal, is stable at 660°C; the beta, or tetragonal, exists from 660°C to 760°C; and the gamma, or body-centered cubic, is from 760°C to the melting point. The cast metal has a hardness of 80–100 Rockwell B, making work-hardening easy. The metal is alloyed with iron to make ferrouranium, used to impart special properties to steel. It increases the elastic limit and the tensile strength of steels and is also a more powerful deoxidizer than vanadium. It will denitrogenize steel and has also carbide-forming qualities. It has been used in high-speed steels with 0.05%–5% to increase the strength and toughness, but because of its importance for atomic applications, its use in steel is now limited to the by-product nonradioactive isotope uranium-238.

Uses

Metallic uranium is used as a cathode in photoelectric tubes responsive to ultraviolet radiation. Uranium compounds, especially the uranium oxides, were used for making glazes in the ceramic industry and also for paint pigments. It produces a yellowish-green fluorescent glass, and a beautiful red with yellowish tinge is produced on pottery glazes. Uranium dioxide, UO_2, is used in sintered forms as fuel for power reactors. It is chemically stable and has a high melting point at about 2760°C, but a low thermal conductivity. For fuel use, the particles may be coated with about 0.003 cm of aluminum oxide. The coating is impervious to xenon and other radioactive isotopes so that only the useful power-providing rays can escape. These are not dangerous at a distance of about 15 cm, and thus less shielding is needed. For temperatures above 1260°C, a coating of pyrolytic graphite is used.

Uranium has isotopes from 234 to 239, and uranium-235, with 92 protons and 143 neutrons, is the one valued for atomic work.

Natural uranium does not normally undergo fission because of the high probability of the neutron being captured by the U^{238} that then merely ejects a gamma ray and becomes U^{239}. When natural uranium is not in concentrated form, but is embodied in a matrix of graphite or heavy water, it will sustain a slow chain reaction sufficient to produce heat. In the fission of U^{235}, neutrons are created that maintain the chain reaction and convert U^{238} to plutonium. About 40 elements of the central portion of the periodic table are also produced by the fission, and eventually these products build up to a point where the reaction is no longer self-sustaining. The slow, non-explosive disintegration of plutonium yields neptunium. *Uranium 233* is made by neutron bombardment of thorium. This isotope is fissionable and is used in thermonuclear reactors.

Uranium yellow, also called *yellow oxide*, is a *sodium diuranate*, obtained by reduction and treatment of the mineral pitchblende. It is used for yellow and greenish glazing enamels and for impacting an opalescent yellow to glass, which is green in reflected light. *Uranium oxide* is an olive-green powder of composition U_3O_8, used as a pigment. *Uranium trioxide*, UO_3, is an orange-yellow powder also used for ceramics and pigments. It is also called *uranic oxide*. As a pigment class, it produces a beautiful greenish-yellow *uranium glass*. *Uranium pentoxide*, U_2O_5, is a black powder, and *uranous oxide*, UO_2, is used in glass to give a fine black color. The uranium oxide colors give luster and iridescence, but because of the application of the metal-to-atom work, the uses in pigments and ceramics are now limited.

Uranium Ores

The chief source of radium and uranium is *uraninite*, or *pitchblende*, a black, massive, or granular mineral with pitch-like luster. The mineral is a combination of the oxides of uranium together with small amounts of lead, thorium, yttrium, serial, helium, argon, and radium. The process of separation of radium is chemically complicated. Numerous minor uranium ores occur in many areas. A low-grade ore of 0.1% uranium oxide can be upgraded to as high as 5% by ion exchange.

Uranyl

The chemical name designating the UO_2^{2+} group and compounds containing this group.

Urea

Also called *carbamide*, urea is a colorless to white crystalline powder best known for its use in plastics and fertilizers. The chemistry of urea and the carbamates is very complex, and a great variety of related products are produced. Urea is produced by combining ammonia and carbon dioxide, or from cyanamide. It is a normal waste product of animal protein metabolism and is the chief nitrogen constituent of urine. It was the first organic chemical ever synthesized commercially. It has a specific gravity of 1.323, with a melting point of 135°C.

Types

The formula for urea may be considered to be $O \cdot C(NH_2)_2$ and thus an amide substitution in carbonic acid, $O \cdot C(OH)_2$, an acid that really exists only in its compounds. The urea-type plastics are called amino resins. The carbamates can also be considered as deriving from carbamic acid, NH_2COOH, an aminoformic acid that likewise appears only in its compounds. The carbamates have the same structural formula as the bicarbamates so that sodium carbamate has an NH_2 group substituted for each OH group of the sodium bicarbonate. The urethanes used for plastics and rubber are alkyl carbamates made by reacting urea with an alcohol or by reacting isocyanates with alcohols or carboxyl compounds. They are white powders melting at 50°C.

Isocyanates are esters of isocyanic acid, which does not appear independently. That dibasic diisocyanate is made from a 36-carbon fatty acid. It reacts with compounds containing active hydrogen. With modified polyamines, it forms polyurea resins, and with other diisocyanates, it forms a wide range of urethanes. Tosyl isocyanate producing urethane resins without a catalyst is toluene sulfonyl isocyanate. The sulfonyl group increases the reactivity.

Methyl isocyanate is a colorless liquid with a specific gravity of 0.9599. It reacts with water. With a flashpoint of less than −6.6°C, it is flammable and creates risks for fire. It is a strong irritant and is highly toxic. One of its principal uses is as an intermediate in the production of pesticides.

Urea is used with acid phosphates in fertilizers. It contains about 45% nitrogen and is one of the most efficient sources of nitrogen. Urea reacted with malonic esters produces malonyl urea, which is the barbituric acid that forms the basis for the many soporific compounds such as *luminal*, *phenobarbital*, and *amytal*. The malonic esters are made from acetic acid, and malonic acid derived from the esters is a solid that decomposes at about 160°C to yield acetic acid and carbon dioxide.

For plastics manufacture, substitution on the sulfur atom in thiourea is easier than on the oxygen in urea. Thiourea, also called *thiocarbamide*, *sulfourea*, and *sulfocarbamide*, is a white, crystalline, water-soluble material of bitter taste, with a specific gravity of 1.405. It is used for making plastics and chemicals. On prolonged heating below its melting point, 182°C, it changes to ammonium thiocyanate, or ammonium sulfocyanide, a white, crystalline, water-soluble powder melting at 150°C. This material is also used in making plastics; used as a mordant in dyeing, to produce black nickel coatings; and used as a weed killer. *Permafresh*, used to control shrinkage and gives wash-and-wear properties to fabrics, is dimethylol urea, which gives clear solutions to warm water.

Urea–formaldehyde resins are made by condensing urea or thiourea with formaldehyde. They belong to the group known as amino-aldehyde resins made by the interaction of an amine and an aldehyde. An initial condensation product is obtained that is soluble in water and is used in coatings and adhesives. The final condensation product is insoluble in water and is highly chemical resistant. Molding is done with heat and pressure. The urea resins are noted for their transparency and ability to take translucent colors. Molded parts with cellulose filler have a specific gravity of about 1.50, tensile strength 41–89 MPa, elongation 15%, compressive strength 310 MPa, dielectric strength 16×10^6 V/m, and a heat distortion temperature of 138°C. Rockwell hardness is about M 118. Urea resins are marketed under a wide variety of trade names: the *Uformite* resins, which are water-soluble thermosetting resins for adhesives and sizing; the *Urac* resins; and the *Casco* resins and *Cascomite*, or urea–formaldehyde. They are used for plasterboard, plywood, and in wet-strength paper.

Urea–Formaldehyde Adhesive

(1) An aqueous colloidal dispersion of urea–formaldehyde polymer that may contain modifiers and secondary binders to provide specific adhesive properties. (2) A type of adhesive based on a dry urea–formaldehyde polymer and water. A curing agent is commonly used with this type of adhesive.

Urethane Hybrids

Urethane acrylic polymers that are formed by the reaction of two liquid components, an acrylesterol and a modified diphenylmethane-4,4′-diisocyanate (MDI). The acrylesterol is a hybrid of a urethane (monoalcohol) and an acrylic (unsaturated monoalcohol). The liquid-modified MDI contains two or more isocyanate groups that can react with the hydroxyl portion of the acrylesterol molecule. Acrylamate resin systems are reinforced with glass (30%–40%) and are used in automotive applications and recreational products. When reinforced with carbon mat or metallized glass cloth, these materials can be used in communication equipment, such as electromagnetic interference/radio-frequency interference devices.

Urethanes

Also termed polyurethanes, urethanes are a group of plastic materials based on polyether or polyester resin. The chemistry involved is the reaction of a diisocyanate with a hydroxyl-terminated polyester or polyether to form a higher-molecular-weight prepolymer, which in turn is chain extended by adding difunctional compounds containing active hydrogens, such as water, glycols, diamines, or amino alcohols. The urethanes are block polymers capable of being formed by a literally indeterminate number of combinations of these compounds. The urethanes have excellent tensile strength and elongation, good ozone resistance, and good abrasion resistance. Combinations of hardness and elasticity unobtainable with other systems are possible in urethanes, ranging from Shore hardnesses of 15–30 on the "A" scale (printing rolls, potting compounds) through the 60–90 A scale for most industrial or mechanical goods applications to the 70–85 Shore "D" scale. Urethanes are fairly resistant to many chemicals such as aliphatic solvents, alcohols, ether, certain fuels, and oils. They are attacked by hot water, polar solvents, and concentrated acids and bases.

Urethane Foams

Urethane foams are made by adding a compound that produces carbon dioxide or by reaction of a diisocyanate with a compound containing active hydrogen. Foams can be classified somewhat according to modulus as flexible, semiflexible or semirigid, and rigid. No sharp lines of demarcation have been set on these different classes as the gradation from the flexibles to the rigids is continuous. Densities of flexible foams range from about 16 kg/m³ at the lightest to 64–80 kg/m³ depending on the end use. Applications of flexible foams range from comfort cushioning of all types, for example, mattresses, pillows, sofa seats, backs, and arms, automobile topper pads, rug underlay, and clothing interliners for warmth at light weight.

Flexible Types

The techniques of manufacture of flexible urethane foam vary widely, from intermittent hand mixing to continuous machine operation, from prepolymer to one-shot techniques, from slab forming to molding, and from stuffing to foamed-in-place.

Future applications envision the flexible foam not as a substitute for latex rubber foam or cotton but as a new material of construction allowing for design of furniture, for example, which is essentially all foam with a simple cloth cover and a very simple metal-supporting framework.

Rigid Types

Densities from about 24 to 800 kg/m³ on the semirigid side have been produced with corresponding compression strengths again for particular end uses ranging from installation to fully supporting structural members. The usefulness of the urethane system has been in the foamed-in-place principle using a host of containing wall materials.

Applications to the more rigid foam field have been thermal insulation of all types (low-temperature refrigeration ranging from liquid nitrogen temperatures to the freezing point of water and high temperature insulation of steam pipes, oil lines, etc.); shock absorption such as packaging and crash pads, where the higher hysteresis values produce either a better one-time high-impact "crash" use or, more often, lower-amplitude but higher-frequency container end use; filtration (air, oil, etc., where a large surface-to-volume ratio is needed with a simple technique to produce a reusable filter to allow for its additionally higher cost factor); structural (building applications of all kinds, combining a good thermal as well as structural behavior, and filling of building voids; curtain walls are some basic applications); flotation (boats, buoys, and every other imaginable objects afloat represent some possible application of urethane foams); and general-purpose applications that include all other uses such as decorative applications.

Rigid foams can be produced using a simple spray technique, and a number of machines are sold on the market for this technique. Time-consuming layup of foam is eliminated using this method. Insulations of walls, tanks, etc., are applications in use today. With the use of low-vapor-pressure isocyanates such as 4,4'-diphenylmethane diisocyanate, the potential irritant hazard during spraying is greatly lowered. Self-adhesion of the sprayed foam is a valuable asset of this type of system.

Urethane foams offer advantages over many of the better-known foams such as latex foam rubber, polystyrene, and polyethylene, with the combination of excellent properties and lower installed costs. Depending on the application, a lower foam density can be used with similar load-bearing properties, also one having an extremely low thermal conductivity be fabricated. The oil resistance, high-temperature resistance, good high-tensile properties, good permanence properties, resistance to mildew, resistance to flammability, and so on are in general the types of properties that, combined with foamed-in-place technology, put urethane foam far ahead of competitive materials.

Other Urethanes

Thermoplastic polyurethanes (TPUs) include two basic types: esters and ethers. There are also TPUs based on polycaprolactone, which, while technically being esters, have better resistance to hydrolysis. TPUs are used when a combination of toughness, flex resistance, weatherability, and low-temperature properties is needed. These materials can be injection molded, blow molded, and extruded as profiles, sheet, and film. Further, TPUs are blended with other plastic resins, including polyvinyl chloride, ABS, acetyl, SAN, and polycarbonate.

Urethane elastomers are made with various isocyanates, the principal ones being tolylene diisocyanate and 4,4'-diphenylmethane diisocyanate, reacting with linear polyols of the polyester and polyether families. Various chain extenders, such as glycols, water, diamines, or aminoalcohols, are used in either a prepolymer or a one-shot type of system to form the long-chain polymer.

Flexible urethane fibers, used for flexible garments, are more durable than ordinary rubber fibers or filaments and are 30% lighter in weight. They are resistant to oils and to washing chemicals and also have the advantage because of their white color. Spandex fibers are stretchable fibers produced from a fiber-forming substance at which a long chain of synthetic molecules are composed of a segmented polyurethane. Stretch before break of these fibers is from 520% to 610%, compared to 760% for rubber. Recovery is not as good as in rubber. Spandex is white and dyeable. Resistance to chemicals is good, but it is degraded by hypochlorides.

There are six basic types of polyurethane coatings, or urethane coatings, as defined by the American Society for Testing and Materials, Specification D16. Types 1, 2, 3, and 6 have long storage life and are formulated to cure by oxidation, by reaction with atmospheric moisture, or by heat. Types 4 and 5 are catalyst cured and are used as coatings or leather and rubber and as fast-curing industrial product finishes. Urethane coatings have good weathering characteristics as well as high resistance to stains, water, and abrasion.

Fabrication

Urethane elastomers can be further characterized by the method of fabrication of the final article. Three principal types of revocation are possible: (1) casting technique where a liquid prepolymer or a liquid mixture of all initial components (one-shot) is cast into the final mold, allowed to "set" and harden, and is then removed for final cure; (2) millable gum technique where conventional rubber methods and equipment are used to build the gum, add fillers, color, etc., and/or banbury, extrude, calender, and compression mold the final shaped item; (3) thermoplastic processing techniques where the resin can be calendered, extruded, and injection- or blow molded on conventional plastic machinery in final form (an important benefit here is that scrap can be reground and reused in fabricating other parts).

The choice of the proper method of fabrication largely depends on the economics of the process, because the properties of the final product may be about the same regardless of the method of fabrication. If a few large-volume items are needed, casting these into a single mold is usually more economical. However, if many thousands of small, intricate pieces are needed, usually injection molding is the preferred, more economical method of fabrication.

Uses

Applications of urethane elastomers have been developed where high abrasion resistance, good oil resistance, and good load-bearing capacity are of value, as in solid tires and wheels, especially of industrial trucks, the shoe industry, drive and belting applications, printing rolls, gasketing in oil, etc. Other applications include vibration dampening, for example, in hammer heads, air hammer handles, and shock absorption underlays for heavy machinery; low coefficient of friction with the addition of molybdenum disulfide for self-lubricating uses as ball-and-socket joints, thrust bearings, leaf spring slide blocks in the electrical industry, cable jacketing, and potting compounds are developing as important uses. Various systems of urethane elastomers with specific fillers have been developed into an important class of caulks and sealants, which are just beginning to take hold in applications such as concrete road-expansion joints and building caulking, in direct competition with such older materials as the polysulfides but at a much lower price and superior properties.

A host of other applications varies from adhesive bonding of fibers of all kinds to rocket fuel binders of the more exotic variety, which are becoming so important in the U.S. National Defense picture. Therefore, it is imperative that design engineers understand fully the material they are using and how they intend to utilize it in the final piece of equipment. For example, one recommendation is to limit the use of urethanes to below 82°C in water for continuous exposures.

Dry uses can go somewhat higher, for example, to 107°C for certain systems. In oil, exposures can be up to 121°C. Disregard of such limitations can result in failures, but the design engineer can eliminate these by the proper choice of material. On the other hand, the design engineer should choose the urethanes for their virtues, such as hardness and elasticity, where other materials such as natural and other synthetic rubbers may fail.

Properties

The urethanes have excellent tensile strengths and elongation, good ozone resistance, and good abrasion resistance. Knowledge of these properties is mandatory for good engineering design.

The greater load-bearing capacity of urethanes as compared to other elastomers is noteworthy, for it leads to smaller, less costly, lower-weight parts in equivalent applications. Tear strength is extremely high, which may be important in particular applications along with the very high tensile strengths. The high abrasion resistance has made possible driving parts for which no other materials could compete. However, in every such dynamic application, the engineer must design the part to allow for the higher hysteresis losses in the urethane. Whereas in some applications such as dampening the higher hysteresis works to advantage, in others hysteresis will lead to part failure if the upper temperature limit is thereby exceeded. Redesign of the part (thinner walls) to allow for greater dissipation of the heat generated will permit the part to operate successfully. This has proved to be the case many times.

Urethane elastomers generally have good low-temperature properties. The same hysteresis effect works in reverse here so that a part in dynamic use at temperatures as low as −51°C, while stiff in static exposure, immediately generates enough heat in dynamic use to pass through its second-order transition and does not show any brittleness but becomes elastic and usable. By proper choice of the polyester or polyether molecular backbone, lower use temperatures (as low as −62°C) have been formulated in urethane elastomers.

In addition to good mechanical properties, urethanes have good electrical properties, which suggest a number of applications. Oxygen, ozone, and corona resistances of this system are generally excellent. See also *isocyanate plastics* and *polyurethanes*.

UTS

Ultimate tensile strength. See *tensile test*.

UV

See *ultraviolet*.

V

V Process

A molding (casting) process in which the sand is held in place in the mold by vacuum. The mold halves are covered with a thin sheet of plastic to retain the vacuum.

V Segregation

See *segregation.*

Vacancy

(1) A structural imperfection in which an individual atom site is temporarily unoccupied. (2) An unoccupied atomic site in a crystal lattice.

Vacancy Jump

The movement of atoms in substitutional solid solution into a vacant site. This can be regarded as a movement of the vacancy in the opposite direction.

Vacant Site

Same as *vacancy.*

Vacu-Blasting

A grit blasting process in which the grit or shot, after impacting the target component, is recovered by vacuuming for reuse.

Vacuum

A space that is devoid of matter. Loosely, a pressure less than atmospheric.

Vacuum Annealing

Annealing carried out at sub-atmospheric pressure.

Vacuum Arc Melting (VAM)

A purification process for metals such as steel and titanium and in which a DC electric arc is struck between an electrode of the impure metal and the water-cooled copper mold over which it is suspended. The impure electrode is progressively melted, and a new ingot solidifies in the mold. A high vacuum avoids oxidation and removes volatile elements. The process is deliberately slow with minimum turbulence allowing impurities to float to the surface of the molten metal.

Vacuum Arc Remelting (VAR)

A consumable-electrode remelting process in which heat is generated by an electrical arc between the electrode and the ingot. The process is performed inside a vacuum chamber. Exposure of the droplets of molten metal to the reduced pressure reduces the amount of dissolved gas in the middle. See the term *consumable-electrode remelting.*

Vacuum Atomization

A commercial batch powder product-ion process based on the principle that, when a molten metal supersaturated with gas under pressure is suddenly exposed to vacuum, the gas expands, comes out of solution, and causes the liquid metal to be atomized. Alloy powders based on nickel, copper, cobalt, iron, and aluminum can be vacuum atomized with hydrogen. Powders are spherical, clean, and of a high purity.

Vacuum Bag

A flexible bag in which pressure may be applied to an assembly (inside the bag) by means of evacuation of the bag. See also *vacuum bag molding.*

Vacuum Bag Molding

A process for manufacturing reinforced plastics in which a sheet of flexible, transparent material plus a bleeder cloth and release film are placed over the lay-up on the mold and sealed at the edges. A vacuum is applied between the sheet and the lay-up. The entrapped air is mechanically worked out of the lay-up and removed by the vacuum, and the part is cured with temperature, pressure, and time. Also called *bag molding* or pressure bag molding. See also *lay-up.*

Vacuum Brazing

And nonpreferred term used to denote furnace brazing that takes place in a chamber or retort below atmospheric pressure.

Vacuum Carburizing

A high-temperature carburizing process using furnace pressures between 13 and 67 kPa (0.1–0.5 torr) during the carburizing portion of the cycle. Steels undergoing this treatment are austenitized in a rough vacuum, carburized in a partial pressure of hydrocarbon gas, diffused in a rough vacuum, and then quenched in either oil or gas. Both batch and continuous furnaces are used. See the term *vacuum furnace.*

Vacuum Carburizing

Heat treatment with gas quenching has already been an established heat treatment process for two or three decades in the field of full hardening. At first, it was limited to the hardening of high-alloyed tool steels whose alloy structure enabled them to be hardened satisfactorily with a rather slow gas cooling rate.

The enhanced quenching action achieved with gas pressures above 10 bar has allowed successful extension of gas quenching to the field of low-alloyed tool steels, steels for hardening and tempering, antifriction-bearing steels, and case-hardening steels. The capability to carburize and gas-quench in vacuum furnace installations has provided the industry with a new, environmentally friendly case-hardening process.

The Process

Like plasma carburizing, vacuum carburizing can also be performed in a vacuum furnace system. Vacuum carburizing can be succinctly described by the following key points:

- Carburizing gas is propane.
- Pressure ranges up to 20 mbar (absolute).
- Temperature range is usually 900°C–1050°C, but higher temperatures are also possible.

Once the charge has been heated to the carburizing temperature under a neutral atmosphere (vacuum or nitrogen), propane is admitted into the evacuated heating chamber. Propane very rapidly undergoes 100% disassociation into more stable hydrocarbons and hydrogen. Carbon is also released and diffuses through the surface of the steel or component.

Vacuum carburizing is characterized by a high carbon mass flow rate, which carburizes the surface layer to near the carbon saturation limit within a short treatment time. In the subsequent diffusion phase, no more carburizing gas is fed in—rather, the existing carbon diffuses farther into the steel in accordance with the diffusion law until the desired carbon profile has been attained.

Process Comparisons

In contrast to protective-gas carburizing, vacuum carburizing can be performed with substantially higher case carbon contents. The case carbon percentage is already over 1.3% after a short period of carburization and then is held at 1.4%–1.5% at 930°C, which is about 0.2% higher than in protective-gas carburizing with a carbon level just below the sooting limit. The higher case carbon content in vacuum carburizing results in short treatment times, even at the same carburizing temperature. Raising the carburizing temperature results in a further considerable time savings.

Vacuum carburizing systems readily permit a carburizing temperature of over 1050°C, although the heat treatment racks made of heat-resistant cast steel (which are currently in use) are no longer usable at such high temperatures. With racks made of carbon-fiber-reinforced carbon (CFC), the limit is shifted to much higher temperatures. CFC material can only be used in an oxygen-free atmosphere such as that prevailing during vacuum carburizing. Of course, the carburizing action at the point of contact with the component must be taken into consideration.

The grain growth of case-hardening steels also does not permit such high temperatures over a long period of time. The vacuum furnace offers a pearlitizing treatment to refine the grain. Despite the time cost for pearlitizing, the result in comparison to the time required in a multipurpose protective-gas chamber furnace is a time savings of about 4 h for the case hardening of a 25%CrMo$_4$ steel to a case depth (550 HV) of 1.7 mm.

Distortion

Vacuum carburizing with gas quenching offers a potential for reduced parts distortion. A large number of experiments, conducted primarily on transmission parts, have shown that the scatter of the dimensional and shape changes after gas quenching is narrower than after oil quenching.

For example, a clutch body (O.D., 84 mm; I.D., 50 mm; height, 15 mm; mass, 0.2 kg each) made of 16%MnCr$_5$ with a case depth (550 HV) of 0.4–0.8 mm was tested. The study evaluated 50 clutch bodies after case hardening in the vacuum furnace (quenching with helium at 20 bar) and, for comparison, 50 others were evaluated after case hardening in the protective-gas furnace (oil quenching). The radial run-out of the clutch bodies was measured in the soft and hard states. The results are shown in *Ind. Heating*, January 54, 2000.

Advantages

Case hardening in vacuum heat treatment systems with gas quenching offers the user many advantages in comparison to conventional protective-gas carburizing with oil quenching. Parts are clean and dry after treatment requiring no washers or management or disposal of liquid waste. Leidenfrost phenomenon is avoided and with more uniform quenching, distortion is minimized. Vacuum carburizing also allows for carburizing at up to 1000°C.

As a protective atmosphere, vacuum can prevent case oxidation and eliminate toxic off-gases. Vacuum carburizing process also provides a high carbon mass flow rate with low consumption of carburizing gas. With regard to productivity, the vacuum process can be integrated into a production line without the burdening requirements for fire-protection and fire-extinguishing systems, excessive heat removal to the surroundings, or extensive exhaust gas handling.

In determining the carbon mass flow rate, it becomes clear that in the first few minutes of carburizing in this process (up to 30 min), there is a very high carbon mass flow rate of up to 100 g/m^2 h. The case-hardening steel can be carburized up to its limit of solubility in the surface layer without any sooting occurring. The system technology also makes it possible to carburize at temperatures above 1000°C. These two factors result in an enormously shortened process duration.

The carburization results are comparable with those of the protective-gas process with regard to case depth, case carbon content, and surface hardness. The advantages for component quality lie in reduced distortion. Investigations of various transmission parts have shown that the scatter of the dimensional and shape changes can be narrowed with gas quenching in comparison to oil quenching. The clean surface of the component and the absence of case oxidation after heat treatment are additional advantages of this technology.

Vacuum Casting

A casting process in which metal is melted and poured under very low atmospheric pressure; a form of permanent mold casting in which the mold is inserted into liquid metal, vacuum is applied, and metal is drawn out into the cavity.

Vacuum Coatings

The process of vacuum coating is used to modify a surface by evaporating a coating material under vacuum and condensing it on the surface. It is normally carried out under high-vacuum conditions (at approximately 1 millionth of an atmosphere pressure). The material to be evaporated is heated until its vapor pressure appreciably exceeds the residual pressure within the vacuum system.

Vacuum coating can be used for many applications. For example, optical lenses are coated with magnesium fluoride to a fraction of a wavelength to prevent glare and provide much better transmission of light and a more reliable optical system. The deposited film is extremely adherent and will withstand normal cleaning.

Silicon monoxide is frequently used as an abrasion-resistant coating material. As deposited, it is soft and requires postheat treatment in air to convert it to silicon dioxide, which is transparent and extremely hard. It is frequently used to protect front surface mirrors and increases abrasion resistance by a factor of 5,000–10,000, while maintaining equal or higher reflectivity. Similarly, titanium is sometimes used for coating and is subsequently oxidized to yield a titanium dioxide abrasion-resistant surface.

By far the most common type of vacuum coating is the process of vacuum vandalizing. In this process, metal is evaporated and used as deposited without further treatment as opposed to the evaporation of compounds or materials that require post-treatment.

Vacuum metallizing has generally been used as a decorative process whereby costume jewelry, toys, etc., are given a metallic sheen and are made highly reflective. The base material may be either plastic or metal. In either case, the part is frequently lacquered before metallizing to prevent the evolution of gas from the base and to provide a smooth surface without mechanical buffing. Because the metal deposit is only about 2 or 3 millionths of an inch thick, the smooth surface is necessary to give a specular reflection.

When the metal is on the outside of the coated part, it is referred to as front surface. However, in applications where it is used on the back of a transparent plastic (e.g., dashboards and taillight assemblies on automobiles), it is referred to as a second surface coating. The advantage of second surface coating is provided by using the plastic as the exposed surface. Front surface coatings must generally be protected with a transparent lacquer overcoat (applied after metallizing) because the thin decorative coatings are not wear resistant in themselves.

Aluminum is the most popular vacuum-metallizing coating material for most applications. However, other metals may be used, such as zinc, cadmium, copper, silver, gold, or chromium. Of all these metals, aluminum has the best general combination of reflectivity, conductivity, and stability in air. By adding color to the topcoat lacquer, the aluminum deposit may be made to appear like copper or gold as well as metallic sheens of blues, reds, yellow, etc. By using separate sources for each constituent, it is possible to deposit alloys as well as pure metals.

These applications are for parts produced by batch metallizing; that is, the individual parts are mounted on racks inserted in the vacuum system, and after the necessary vacuum and evaporation temperatures are obtained, the parts are rotated so that they are uniformly coated by the evaporating metal. In batch metallizing, the aluminum is evaporated from tungsten filaments, which are heated by direct resistance. Because of this, the amount of aluminum that can be charged is limited and only thin coatings can be produced. Similarly, only small surfaces (a few square centimeters), such as can be exposed within a matter of seconds, can be metallized.

When it is desirable to coat larger surfaces, for example, rolls of flexible material, a semi-continuous metallizing process must be employed. For semi-continuous metallizing, a roll of material is mounted in the vacuum chamber to coat either or both sides of the web, which is subsequently rewound in vacuum. This process is currently in use for coating rolls of plastic sheeting and paper. To coat continuously over a period of hours, it is necessary to have larger volumes of aluminum available for evaporation than can be held on resistance-heated tungsten filaments. Therefore, the aluminum is generally heated by induction in crucibles.

Coating of rolls of materials provided one of the first functional applications for coatings that used the electrical conductivity of the metal deposited. This conductive layer was deposited on thin insulating layers of either paper or plastic and could be used for winding miniature condensers. The electrical conductivity is also used in the metallizing process itself as a means of measuring the amount of metal deposited. Since the conductivity is a function of the thickness of the metal, continuously measuring conductivity provides a control for the amount of metal deposited. Other functional uses of the coating are based on its reflectivity (e.g., reflective insulation).

Vacuum metallizing has recently been extended to include thick films, that is, in the range of 1–3 mils. Such coatings serve as corrosion-resistant barriers, particularly on high-tensile-strength steel exposed to marine atmospheres. Where the temperature requirements of steel are less than 260°C, cadmium deposits can be used. For temperatures in excess of this, aluminum shows much better protection and does not react with the base steel as cadmium does.

Truly continuous operation is necessary for coating rolled steel. Here, the rolls are unwound and rewound in air with the strip passing through seals into the vacuum chamber where it is coated. The metallizing of rolled stock allows separate control on each side of the web and the composition of the coating, as well as thickness, may be changed from one side to the other. There are several typical advantages of vacuum metallizing:

1. Close control of coating thickness and composition
2. Uniform deposits without buildup at sharp discontinuities
3. High coating rate
4. Low coating costs in volume production
5. Long life of equipment since few moving parts

There are also disadvantages of the process:

1. Part must be extremely clean.
2. Surfaces to be metallized must not evolve gas under vacuum.
3. Parts must not be temperature sensitive, that is, must be stable to about 125°C.
4. Deposits form well only on a surface exposed to hot metal; reentrant angles are not well coated.

The cost for metallizing in production lots for corrosion-resistant coatings is comparable to electroplating. Decorative metallizing is generally much less expensive than electroplating.

Vacuum Degassing

The use of vacuum techniques to remove dissolved gases from molten alloys.

Vacuum Deposition

Deposition of a metal film onto a substrate in a vacuum by metal evaporation techniques.

Vacuum Deposition (of Interference Films)

A method of revealing the microstructures of metals and carbides with the aid of vacuum-deposited interference layers (films). The phase shift in light reflected at the interference film/substrate interface contributes to the colors produced by a transparent film on a metallographic specimen. This phase shift depends on the optical properties of the film and substrate. Materials that have been found to produce phase contrast and color when vacuum deposited include titanium dioxide (TiO_2), silicon dioxide (SiO_2), zirconium dioxide (ZrO_2), zinc sulfide (ZnS), tin oxide (SnO_2), and carbon. Of these, TiO_2 is the most commonly used. All these (except carbon) are supplied in powder or chips and must be contained in a tungsten wire basket, which is approximately 100 mm (4 in.) from the specimens. The vacuum chamber is evacuated to 10^{-4} torr or lower. See also *reactive sputtering* (of interference films).

Vacuum Forming

A method of sheet forming in which the plastic sheet is clamped in a stationary frame, heated, and drawn down by a vacuum into a mold. In a broad sense, the term is sometimes used to refer to all sheet-forming techniques, including *drape forming* involving the use of vacuum and stationary molds. See the terms *thermoforming* and *vacuum snapback thermoforming*.

Vacuum Furnace

(1) A furnace using low atmospheric pressures instead of a protective gas atmosphere like most heat-treating furnaces. Vacuum furnaces are categorized as hot wall or cold wall, depending on the location of the heating and insulating components. (2) Any furnace in which the charger is treated under vacuum. In melting furnaces, the vacuum extracts dissolved gases, as well as protecting from contamination. In heat treatment furnaces, it prevents oxidation, carburization, and contamination.

Vacuum Fusion

An analytic technique for determining the amount of gases in metals; ordinarily used for hydrogen and oxygen, and sometimes for nitrogen. Applicable to many metals, but not to alkali or alkaline earth metals.

Vacuum Hot Pressing

A method of processing materials (especially metal and ceramic powders) at elevated temperatures, consolidation pressures, and low atmospheric pressures.

Vacuum Induction Melting (VIM)

A process for remelting and refining metals in which the metal is melted inside a vacuum chamber by induction heating. The metal can be melted in a crucible and then poured into a mold.

Vacuum Injection Molding

A molding process for fabricating reinforced plastics that utilizes both a male and female mold in which reinforcements are placed, a vacuum is applied, and a room temperature curing liquid resin is introduced to saturate the reinforcement.

Vacuum Melting

Melting in a vacuum to prevent contamination from air and to remove gases already dissolved in the metal; the solidification can also be carried out in a vacuum or at low pressure.

Vacuum Metallizing

A process in which surfaces are simply coated by exposing them to a metal vapor under vacuum.

Vacuum Molding

See *V process*.

Vacuum Nitrocarburizing

A subatmospheric nitrocarburizing process using a basic atmosphere of 50% ammonia/50% methane, containing controlled oxygen additions of up to 2%.

Vacuum Plating

A form of vacuum deposition in which some reaction is involved such as between the surface and some vapor introduced into the vacuum chamber.

Vacuum Processing

Vacuum processing is used in many industrial applications. Some of these processes are shown as follows:

1. Annealing of metals
2. Degassing of metals
3. Electron beam melting
4. Electron beam welding
5. Evaporation
6. Sputtering of metals
7. Casting of resins and lacquers
8. Drying of plastics
9. Drying of insulating papers
10. Freeze-drying of bulk goods
11. Freeze-drying of pharmaceutical products

Through the use of vacuum, it is possible to create coatings with a high degree of uniform thickness ranging from several nanometers to more than 100 mm while still achieving very good reproducibility of the coating properties. Flat substrates, web and strip, as well as complex molded plastic parts, can be coated with virtually no restrictions as to the substrate material.

The variety of coating materials is also very large. In addition to metal and alloy coatings, layers may be produced from various chemical compounds or layers of different materials applied in sandwich form. A significant advantage of vacuum coating over other methods is that many special coating properties desired, such as structure, hardness, electrical conductivity, or refractive index, are obtained merely by selecting a specific coating method and the process parameters for a certain coating material.

Deposition of thin films is used to change the surface properties of a base material or substrate. For example, optical properties such as transmission or reflection of lenses and other glass products can be adjusted by applying suitable coating layer systems. Metal coatings on plastic web produce conductive coatings for film capacitors. Polymer layers on metals enhance the corrosion resistance of the substrate.

Coating Sources

In all vacuum coating methods, layers are formed by deposition of material from the gas phase. The coating material may be formed by physical processes such as evaporation and sputtering, or by chemical reaction. Therefore, a distinction is made between physical vapor deposition (PVD) and chemical vapor deposition (CVD).

Thermal Evaporators

In the evaporation process, the material to be deposited is heated to a temperature high enough to reach a sufficiently high vapor pressure and the desired evaporation or condensation rate is set. The simplest sources used in evaporation consist of wire filaments, boats of sheet metal, or electrically conductive ceramics that are heated by passing an electrical current through them. However, there are restrictions regarding the type of material to be heated. In some cases, it is not possible to achieve the necessary evaporator temperatures without significantly evaporating the source holder and thus contaminating the coating. Furthermore, chemical reactions between the holder and the material to be evaporated can occur resulting in either a reduction in the lifetime of the evaporator or contamination of the coating.

Electron Beam Evaporators (Electron Guns)

To evaporate coating material using an electron beam gun, the material, which is kept in a water-cooled crucible, is bombarded by a focused electron beam and thereby heated. Since the crucible remains cold, in principle, contamination of the coating by crucible material is avoided and a high degree of coating purity is achieved. With a focused electron beam, very high temperatures of the material to be evaporated can be obtained and thus very high evaporation rates. Consequently, high-melting-point compounds such as oxides can be evaporated in addition to metals and alloys. By changing the power of the electron beam, the evaporation rate is easily and rapidly controlled.

Catholic Sputtering

In the cathode sputtering process, the target, a solid, is bombarded with high-energy ions in a gas discharge. The impinging ions transfer their momentum to the atoms in the target material, knocking the atoms off. These displaced atoms—the sputtered particles—condense on the substrate facing the target. Compared to evaporated particles, sputtered particles have considerably higher kinetic energy. Therefore, the conditions for condensation and layer growth are very different in the two processes. Sputtered layers usually have higher adhesive strength and a denser coating structure than evaporated ones.

Sputter cathodes are available in many different geometric shapes and sizes as well as electrical circuit configurations. What all sputter cathodes have in common is a large particle source area compared to evaporators, and the capability to coat large substrates with a high degree of uniformity. In this type of process, metals and alloys of any composition, as well as oxides, can be used as coating materials.

Chemical Vapor Deposition

In contrast to physical vapor deposition methods, where the substance to be deposited is either solid or liquid, in chemical vapor deposition, the substance is already in the vapor phase when admitted to the vacuum system. To deposit it, the substance must be thermally excited, that is, by means of appropriate high temperatures or with plasma. Generally, in this type of process, a large number of chemical reactions take place, some of which are taken advantage of to control the desired composition and properties of the coating. For example, by using silicon–hydrogen monomers, soft silicon–hydrogen polymer coatings, hard silicon coatings, or—by the addition of oxygen—quartz coatings can be created by controlling process parameters.

Web Coating

Metal-coated plastic webs and papers play an important role in food packaging. Another important area of application of metal-coated web is the production of film capacitors for electrical and electronics applications.

Metal coating is carried out in vacuum web coating systems. The unit consists of two chambers, the winding chamber with the roll of web to be coated in the winding system and the coating chamber, where the evaporators are located. The two chambers are sealed from each other, except for two slits through which the web runs. This makes it possible to pump high gas loads from the web roll using a relatively small pumping set. The pressure in the winding chamber may be more than a factor of 100 higher than the pressure simultaneously established in the coating chamber.

During the coating process, the web, at a speed of more than 10 m/s, passes a group of evaporators consisting of ceramic boats from which aluminum is evaporated. To achieve the necessary aluminum coating thickness at these high web speeds, very high evaporation rates are required. The evaporators must be run at temperatures in excess of 1400°C. Thermal radiation of the evaporators, together with the heat of condensation of the growing layer, yields a considerable thermal load for the web. With the help of cooled rollers, the foil is cooled during and after coating so that it is not damaged during coating and has cooled significantly prior to winding.

During the entire coating process, the coating thickness is continuously monitored with an optical measuring system or by means of electrical resistance measurement devices. The measured values are compared with the coating thickness set points in the system, and the evaporator power is thus automatically controlled.

Optical Coatings

Vacuum coatings have a broad range of applications in production of ophthalmic optics, lenses for cameras, and other optical instruments, as well as a wide variety of optical filters and special mirrors. To obtain the desired transmission of reflection properties, at least 3, but sometimes up to 50, coatings are applied in the glass or plastic substrates. The coating properties, such as thickness and refractive index of the individual coatings, must be controlled very precisely and matched to each other.

Most of these coatings are produced using electron beam evaporators and single-chamber units. The evaporators are installed at the bottom of the chamber, usually with automatically operated crucibles, in which there are several different materials. The substrates are mounted on a rotating calotte above the evaporators. Application of suitable shielding, combined with relative movement between evaporators and substrates, results in a very high degree of coating uniformity. With the help of quartz coating thickness monitors and direct measurement of the attained optical properties of the coating system during coating, the coating process is fully controlled automatically.

One of the key requirements of coatings is that they retain their properties under usual ambient conditions over long periods of time. This requires the production of dense coatings, into which neither oxygen nor water can penetrate. Using glass lenses, this is achieved by keeping the substrates at temperatures up to 300°C during coating by means of radiation heaters. However, plastic lenses, as those used in eyeglass optics, are not allowed to be heated above 80°C.

To obtain dense, stable coatings, these substrates are bombarded with argon ions from an ion source during coating. Through ion bombardment, the right amount of energy is applied to the growing layer so that the coated particles are arranged on the energetically most favorable lattice sites, without the substrate temperature reaching unacceptably high values. At the same time, oxygen can be added to the argon. The resulting oxygen ions are very reactive and ensure that the oxygen is included in the growing layer as desired.

Glass Coating

Coated glass plays a major role in a number of applications such as heat-reflecting coating systems on windowpanes to lower heating costs; solar protection coatings to reduce air-conditioning costs in countries with high-intensity solar radiation; coated car windows to reduce the heating-up of the interior; and mirrors used both in the furniture and the automobile industries.

The individual glass panes are transported into an entrance chamber at atmospheric pressure. After the entrance valve is closed, the chamber is evacuated with a forepump set. As soon as the pressure is low enough, the valve to the evacuated transfer chamber can be opened. The glass pane is moved into the transfer chamber and from there at constant speed to the process chambers, where coating is carried out by means of splutter cathodes. On the exit side, there is, in analogy to the entrance side, a transfer chamber in which the pane is held until it can be transferred out through the exit chamber.

Most of the coatings consist of a stack of alternative layers of metal and oxide. Because metal layers may not be contaminated with oxygen, the individual process stations have to be vacuum isolated from each other and from the transfer stations. To avoid frequent and undesirable starting and stopping of the glass panes, the process chambers are vacuum separated through so-called "slit locks," that is, constantly open slits combined with an intermediate chamber with its own vacuum pump. The gaps in the slits are kept as small as technically possible to minimize clearance and therefore conductance as the glass panes are transported through them. The pumping speed at the intermediate chamber is kept as high as possible to achieve a considerably lower pressure in the intermediate chamber than in the process chambers. This lower pressure greatly reduces the gas flow from a process chamber via the intermediate chamber to the adjacent process chamber. For very stringent separation requirements, it may be necessary to place several intermediate chambers between two process chambers.

The glass coating process requires high gas flows for the sputter processes as well as low hydrocarbon concentration. Turbomolecular pumps are used almost exclusively because of their high pumping speed stability over time.

While the transfer and process chambers are constantly evacuated, the entrance and exit chambers must be periodically vented and then evacuated again. Because of the large volumes of these chambers and the short cycle times, a combination of rotary vane pumps and Roots pumps is typically used to provide the necessary pumping speed.

Data Storage Disks

Coatings for magnetic- or magneto-optic data storage media usually consist of several functional coatings that are applied to mechanically finished disks. Most disks must be coated on both sides, and there are substantially greater low-particle contamination requirements as compared to glass coating. The sputter cathodes in the process stations are mounted on both sides of the carrier so that the front and back of the disk can be coated simultaneously.

An entirely different concept is applied for coating of single disks. In this case, the different process stations are arranged in a circle in a vacuum chamber. The disks are transferred individually from a magazine to a star-shaped transport arm. The transport arm cycles one station farther after each process step and in this way transports to substrates from one process station to the next.

During cycling, all processes are switched off and the stations are vacuum linked to each other. As soon as the arm has reached the process position, the individual stations are separated from each other by closing seals. Each station is pumped by means of its own turbomolecular pump, and the individual processes are started. By sealing off the process stations, excellent vacuum separation of the individual processes can be achieved. However, since the slowest process step determines the cycle interval, two process stations may have to be dedicated for particularly time-consuming processes.

Vacuum Refining

Melting in a vacuum to remove gaseous contaminants from the metal.

Vacuum Residue

The residue from vacuum distillation of crude oil.

Vacuum Sintering

Sintering of ceramics or metals at subatmospheric pressure.

Vacuum Sintering Furnace

A furnace wherein sintering of ceramics or metals is conducted in a vacuum. The furnace may be of a design either for batch sintering or for continuous sintering. See also *vacuum sintering*.

Vacuum Snapback Thermoforming

A *thermoforming* process for production of plastic items with external deep draws, such as auto parts and luggage. First, the sheet is clamped over the female cavity. Air pressure is then introduced through the channel and the base plate, stretching the plastic. When the material has been sufficiently stretched, the pressure is turned off, and vacuum is turned on, pulling the plastic into the mold. There are many variations of this method, some of which employ plug assists. See also *plug-assist forming*.

Vacuum Treatment/Melting

Treatment of solid or molten metal under vacuum. Vacuum treatment of solid metals, usually at high temperature, assists removal of dissolved gases. Vacuum treatment of molten metal removes gases and, depending on the vacuum, other contaminants. Bismuth, calcium, magnesium, and zinc and lead are removed from molten steel at about 10 kPa; antimony at about 100 Pa; and phosphorus, sulfur, and arsenic at 1 Pa.

Vacuum-Assisted Mold Processing

The use of atmospheric pressure to hold closed molds together during injection was the early process of vacuum-assisted resin injection (VARI). Adding vacuum has enabled resin transfer molding (RTM) to challenge compression molding and autoclaving systems capable of making the best high-performance composites. Vacuum is used in two ways. First, it mixes resin and hardener under a 91 Pa vacuum just prior to injection. Using an impeller, it agitates the mixture to drive air to the surface where the vacuum removes it. Degassing, which takes about 2 h per tank, also removes volatiles at low-molecular-weight by-products.

Some companies place their RTM tools in a vacuum chamber rather than using a vacuum tool. The chamber creates a vacuum that does not vary, even as resin fills the tool. The hard vacuum pulls any air and water vapor off the preform and sucks resin into the mold.

This combination of degassed resin and vacuum keeps voids under 3% and often better. This ensures consistently high structural integrity because voids concentrate stresses that initiate fractures and cause premature failure. It takes only a 2% increase in voids to drain interlaminar shear strength 20% and flexural modulus 10%. The process matches equivalent compression molding and autoclaving fiber volumes and voids.

Compared with compression molding and autoclaving, VARI does not need to apply pressure over the entire skin surface to vanquish voids, simplifying cocuring. For example, the process can fabricate cores and reinforced skins in a single step rather than bonding them after fabrication.

More importantly, the process makes composites in fewer, more controllable steps. These resin injection and molding processes—RTM, VARI, vacuum resin-transfer molding (VRTM), and vacuum-assisted resin-transfer molding (VARTM)—are much simpler processes to use.

Each process step is also independent and controllable. In VRTM, air evacuation depends on only the vacuum and resin preparation depends on only the resin mixer. Preform production varies with automated fabric weaving and preform placement, whereas core manufacture depends on molding or a machining process. The cure depends on a programmable heat source.

VRTM composites also show excellent resistance to water, solvents, and chemicals. This is largely a function of resin type and surface finish. Rough surfaces pitted with micropores trap water and chemicals and act as tiny reaction chambers that set in motion their own destruction. VRTM yields parts with less than 20 µin. root mean square (rms) porosity. VRTM can achieve this fine finish repeatedly on all surfaces, depending on the finish of the tool. For high-quality finishes, compression and autoclaving processes depend on uniform resin flow under pressure, which they cannot always maintain.

The main attraction of VRTM, despite its competitive properties, remains cost, where it offers real advantages over compression molding and autoclaving.

Another production process is low-cost VARTM infusion technology.

VARTM is becoming a manufacturing method of choice because of its ability to produce fairly large structures out of the autoclave with the high quality usually associated with higher-price processes.

Valence

A positive number that characterizes the combining power of an element for other elements, as measured by the number of bonds to other atoms that one atom of the given element forms applied chemical combination; hydrogen is assigned valence 1, and the valence is the number of hydrogen atoms, or their equivalent, with which an atom of the given element combines.

Valence Electrons

The electrons in the atom's outer shell, the number of which establishes the valency.

Valency

A measure of the proportions in which atoms combine. The valency of any element is the number of atoms that will combine with or replace one atom of hydrogen. It is controlled by the number of electrons in the outer shell of the atom, the so-called valency electrons. These being the only electrons available for interactions with other atoms. See *interatomic bonding.*

Valve Alloys

Iron-, nickel-, and cobalt-base alloys are the principal materials for intake and exhaust valves and valve-seat inserts of reciprocating combustion engines. Requirements include resistance to adhesive wear, heat, corrosion, and fatigue. Intake valves for light-duty, lower-temperature service are made from plain carbon steels. Temperatures are generally less than 425°C in light-duty, spark-ignition engines and 500°C in heavy-duty ones. Low-alloy martensitic steels, high-alloy martensitic steels, and austenitic steels are used progressively as temperatures and pressures increase. Intake-valve seats are commonly hard faced with a seat-facing alloy for the most demanding applications. Exhaust valves require resistance to wear, seat-face burning or guttering, fatigue, and creep, the latest to present head doming or "tuliping." Operating temperatures are generally 700°C–760°C, with spikes as high as 850°C. Exhaust valves are typically made of austenitic stainless steels and, for the highest service temperatures, superalloys.

Valve alloys include 1541H as 1547 *carbon steels;* 3140, 4140H, 5150H, 8645, B16, and GM-8440 *low-alloy steels;* Sil 1, Sil XB, 422, and SUH 11M *martensitic stainless steels;* and 21-2N, 21-4N, 21-4N+Cb+W, 23-8N, Gaman H, and 302 HQ *austenitic stainless steels.* Among the *superalloys,* all nickel based, are Inconel 751, Nimonic 80A, Pyromet 31V, and Waspaloy. Titanium alloys Ti–6Al–4V and Ti–6Al–2Sn–4Zr–2Mo find limited specialty applications.

Iron-base alloys include M2 tool steel and vanadium-free M2, and Sil XB. The tool steels, which contained M_6C carbides, are more resistant to wear and heat than Sil XB, which contains iron and chromium carbides. Hardness, 38–52 Rockwell C at room temperature, falls only to 30–34.5 at 427°C and to 23.5–25 at 538°C. The tool steels are often used for exhaust applications and gasoline engines and intake applications in diesel engines.

Nickel alloys, most often used for diesel exhaust inserts, include GM 3550M, SAE J 610B, 13, and J 610B, 12. All of these nickel alloys are generally confined to exhaust applications, performing poorly as intake alloys because, perhaps, of the type of film formed at lower temperatures.

As a class, *cobalt alloys* are generally useful to somewhat higher temperatures (871°C) and also provide sulfidation resistance. They include Stellite 3, Alloy 21, Stellite 6, and Stellite 12. Tribaloy T400 has the advantage of the combined lubricity and hardness of the laves phase for greater wear resistance.

Van der Waals' Bond

A secondary bond arising from the fluctuating dipole nature of an atom with all occupied electron shells filled.

Vanadium and Alloys

An elementary metal (symbol V), vanadium is widely distributed and is a pale-gray metal with a silvery luster. Its specific gravity is 6.02, and it melts at 1780°C. It does not oxidize in the air and is not attacked by hydrochloric or dilute sulfuric acid. It dissolves with a blue color in solutions of nitric acid. It is marketed as 99.5% pure, and cast ingots, machined ingots, and buttons. The as-cast metal has a tensile strength of 372 MPa, yield strength of 10 MPa, and an elongation of 12%. Annealed sheet has a tensile strength of 537 MPa, a yield strength of 455 MPa, and an elongation of 20%, and the cold-rolled sheet has a tensile strength of 827 MPa with an elongation of 2%. Vanadium metal is expensive but is used for special purposes such as for springs of high flexural strength and corrosion resistance.

Commercially important as an oxidation catalyst, vanadium also is used in the production of ceramics and as a colorizing agent. Studies have demonstrated the biological occurrence of vanadium, especially in marine species; in mammals, vanadium has a pronounced effect on heart muscle contraction and renal function.

Fabrication

Hot Working

Since vanadium oxidizes rapidly at hot-working temperatures, forming a molten oxide, it must be protected during heating. This is most easily accomplished by heating in an inert gas atmosphere. Other common practices have been found less suitable.

Vanadium ingots up to 152 mm in size have been successfully hot worked, but the degree of contamination is a modifying factor. Generally, the procedures used in working alloy steels apply.

In view of the difficulties involved in heating the metal, reheating is generally avoided and the starting temperature is a function of the amount of hot work to be accomplished and of the desired finishing temperature. Starting temperatures can range as high as 1260°C, and the finishing temperatures are limited by the beginning of recrystallization. Straightening is performed between 171°C and 427°C but not at room temperature.

Cold Working

Vanadium has excellent cold-working properties, provided that its surfaces are uncontaminated. They are, therefore, machined clean by removing between 0.50 and 1 mm.

Strip can be readily made from hot-rolled sections of 31 × 152 mm in cross section, and 0.25 mm material has been produced without and 0.03 mm with intermediate annealing. Where incipient cracking is observed, vacuum annealing at 899°C becomes necessary.

Extrusion is one of the most suitable fabricating methods for vanadium since warm extrusion followed by cold rolling or drawing avoids hot working with the troublesome heating step. At temperatures below 538°C, tube blanks 50.8 mm outside diameter × 6.4 mm wall thickness have been produced from hot-rolled and turned bars as well as from ingots.

Wires can be drawn from 9.5 mm diameter stock down to 0.025 mm, especially after copper plating. Reductions are usually 10% per pass.

In machining, vanadium resembles the more difficult stainless steel. Low speeds with light to moderate feed are used, and very light finishing cuts at higher speeds are possible.

Welding is not difficult, but contamination of the metal must be avoided by shielding from air by means of an inert gas, that is, argon.

Uses and Applications

The greatest use of vanadium is for alloying. Ferrovanadium, for use in adding to steels, usually contains 30%–40% vanadium, 3%–6% carbon, and 8%–15% silicon, with the balance iron, but may also be had with very low carbon and silicon. Vanadium–boron, for alloying steels, is marketed as a master alloy containing 40%–45% vanadium, 8% boron, 5% titanium, 2.5% aluminum, and the balance iron, but the alloy may also be had with no titanium. *Van-Ad* alloy, for adding

vanadium to titanium alloys, contains 75% vanadium and the balance titanium. It comes as fine crystals. The vanadium–columbium alloys containing 20%–50% columbium have tensile strengths above 689 MPa at 700°C, 482 MPa at 1000°C, and 275 MPa at 1200°C.

Vanadium salts are used to color pottery and glass and as mordants in dyeing. Red cake, of crystalline vanadium oxide, is a reddish-brown material, containing about 85% vanadium pentoxide, V_2O_5, and 9% Na_2O, used as a catalyst and for making vanadium compounds. Vanadium oxide is also used to produce yellow glass; the pigment known as vanadium–tin yellow is a mixture of vanadium pentoxide and tin oxide.

Vanadium is used in the cladding of fuel elements in nuclear reactors because it does not alloy with uranium and has good thermal conductivity as well as satisfactory thermal neutron cross section.

Because the metal alloys with both titanium and steel, it has found application in providing a bond in the titanium cladding of steel. Also, the good corrosion resistance of vanadium offers interesting possibilities for the future; it has excellent resistance to hydrochloric and sulfuric acids and resists aerated salt water very well. But its stability in caustic solutions is only fair and, in nitric acid, inadequate.

Borides, Carbides, and Oxides

Vanadium boride, VB, has a melting point of 2100°C with oxidation at 1000°C–1100°C, density 5.1 g/cm^3, Mohs hardness 8–9, and electrical resistivity 16 Ω cm. It is also formed as VB_2.

Vanadium carbide, VC, has a density of 5.81 g/cm^3 and is silver gray in color. It is chemically very stable; among the cold acids, it is attacked only by nitric acid. Below 499°C, Cl_2 reacts with VC. It burns in oxygen or air but is stable up to 2500°C with nitrogen. VC is harder than corundum.

Vanadium pentoxide, V_2O_5, has a melting point of 690°C and is slightly soluble in water. V_2O_5 is used by the ceramic industry as coloring agents producing various tints of yellow and greenish yellow. Vanadium pentoxide is an excellent flux, and small amounts may be helpful in promoting vitrification of ceramic products. Vanadate glasses are relatively fusible when compared with other oxide types.

Vanadium Steel

Vanadium was originally used in steel as a cleanser but is now employed in small amounts, 0.15%–0.25%, especially with a small quantity of chromium, as an alloying element to make strong, tough, and hard low-alloy steels. It increases the tensile strength without lowering the ductility, reduces grain growth, and increases the fatigue-resisting qualities of steels. Larger amounts are used in high-speed steels and in special steels. Vanadium is a powerful deoxidizer in steels but is too expensive for this purpose alone. Steels with 0.45%–0.55% carbon and small amounts of vanadium are used for forgings, and cast steels for aircraft parts usually contain vanadium. In tool steels vanadium widens the hardening range, and by the formation of double carbides with chromium makes hard and keen-edge die and cutter steels. All these steels are classed as chromium–vanadium steels. The carbon–vanadium steels for forgings and castings, without chromium, have slightly higher manganese.

Vanadium steels require higher quenching temperatures than ordinary steels or nickel steels. SAE 6145 steel, with 0.18% vanadium and 1% chromium, has a fine grain structure and is used for gears. It has a tensile strength of 799–2013 MPa when heat treated, with Brinell hardness 248–566, depending on the temperature of drawing, and then an elongation of 7%–26%. In cast vanadium steels, it is usual to have from 0.18% to 0.25% vanadium with 0.35% to 0.45% carbon. Such castings have a tensile strength of about 551 MPa and an elongation of 22%. A nickel–vanadium cast steel has much higher strength, but high-alloy steels with only small amounts of vanadium are not usually classed as vanadium steels.

Vapor

The gaseous form of substances that are normally in the solid or liquid state and can be changed to these states either by increasing the pressure or decreasing the temperature.

Vapor Blasting

A cleaning process in which components are scoured by an aqueous suspension of abrasive entrained in a high-velocity air stream. Same as *liquid honing*.

Vapor Degreasing

Degreasing of work in the vapor over a boiling liquid solvent, the vapor being considerably heavier than air. At least one constituent of the soil must be soluble in the solvent. Modifications of this cleaning process include vapor-spray-vapor, warm liquid-vapor, boiling liquid-warm liquid-vapor, and ultrasonic degreasing.

Vapor Deposition

See *chemical vapor deposition*, *physical vapor deposition*, and *sputtering*.

Vapor Plating

Deposition of a metal or compound on a heated surface by reduction or decomposition of a volatile compound at a temperature below the melting points of the deposit and the base material. The reduction is usually accomplished by a gaseous reducing agent such as hydrogen. The decomposition process may involve thermal dissociation or reaction with the base material. Occasionally used to designate deposition on cold surfaces by vacuum evaporation. See also *vacuum deposition*.

Vapor Pressure

The measure of the tendency of a material to release molecules to its surroundings. In liquids, the phenomenon is readily recognized as evaporation, but similar effects occur more slowly in solid materials.

Vapor-Deposited Coatings

These are thin single or multilayer coatings applied to base surfaces by deposition of the coating metal from its vapor phase. Most metals and even some nonmetals, such as silicon oxide, can be vapor deposited. Vacuum-evaporated films or vacuum-metallized films were produced by vacuum evaporation. In addition to vacuum evaporation, vapor-deposited films can be produced by ion sputtering, chemical-vapor plating, and a glow-discharge process. The first two are discussed under vacuum processing.

In the glow-discharge process, applicable only to polymer films, a gas discharge deposits and polymerizes the plastic film on the base material.

Applications

Vapor plating is not considered to be competitive with electroplating. Its chief use (present and future) is to apply coating materials that cannot be electroplated or cannot be applied in a nonporous condition by other techniques. Such materials include titanium, zirconium, columbium, tantalum, molybdenum, and tungsten and refractory compounds such as the transition metal carbides, nitrides, borides, and silicides. Vapor plating will also continue to be useful in the preparation of ultrahigh-purity metals and compounds for use in electronic applications and in alloy development.

A few of the main commercial uses of vapor plating are as follows:

1. The application of high-chromium alloy coatings to iron and steel articles by the displacement-diffusion coating process (known as pack chromizing), for abrasion resistance and for protection from corrosion by food products, strong oxidizing acids, alkalies, salt solutions, and gaseous combustion products at temperature up to about 800°C.
2. The application of molybdenum disilicide coatings to molybdenum by gas-phase siliconizing, for protection against air oxidation at temperatures between 800°C and 1700°C.
3. The preparation of ultrahigh-purity titanium, zirconium, chromium, thorium, and silicon by iodide vapor decomposition processes.
4. The preparation of junction transistors by the controlled diffusion of boron from boron halide into the surface of silicon or germanium wafers.
5. The preparation of oriented graphite plates and shapes (pyrolytic graphite), by the high-temperature pyrolysis of hydrocarbon gases, for use in rocket and missile applications.

In addition, the following coatings have been developed:

1. Tantalum coatings on iron and steel for corrosion resistance
2. Vanadized, tungstenized, and molybdenized iron for wear resistance
3. Tungsten coatings on copper x-ray and cyclotron targets
4. Conductive metallic coatings on glass, porcelain, alundum, porous bodies, rubber, and plastics
5. Refractory metal coatings on copper wires
6. Oxidation-resistant carbide coatings on graphite tubes, nozzles, and vanes
7. Metallic coatings of all types on metallic and nonmetallic powders
8. Decorative, colored coatings on glass
9. High-purity boron, radium, vanadium, germanium, and aluminum

The displacement-diffusion plating processes, such as pack-chromizing, can plate uniformly somewhat larger pieces and more complex shapes with inaccessible areas. Sheets and rod up 0.6–0.9 m in dimension have been coated, and no technical obstacles are seen to scaling up the processes to coat even larger pieces. The pack coating processes have the advantage of minimizing the problems of specimen support and warpage during plating.

Plating uniformity varies somewhat with the particular process used, with the shape of the object being plated, and with the attention given to providing proper gas flow around the object. A variation in thickness of 10%–25% is usually obtained. However, some coating processes can be made self-limiting so that the variation in coating thickness is much less than this range.

Disadvantages

Vapor plating has the following disadvantages:

1. Relative instability and air and moisture sensitivity of most of the compounds used as plating agents
2. A tendency to produce nonuniform deposits due to unfavorable gas flow patterns around the work, or to uneven specimen temperature
3. Alteration of physical properties of the substrate due to the elevated processing temperatures
4. The possibility of poor coating quality arising from undesired side reactions in the plating process

In general, the materials used as plating compounds in vapor plating are relatively unstable and easily decomposed by air and moisture, thus rendering them more expensive and more difficult to store and handle than the compounds used in other plating techniques. Also, some of the metal carbonyls, hydrides, and organometallic compounds are highly toxic, and some of the hydrides of metal alkyls inflame spontaneously upon contact with air.

To develop optimal properties in deposits of many materials, the plating compounds, particularly the moisture-sensitive metal halides, must be purified and used without contamination from the atmosphere. This apparent disadvantages is sometimes put to good use, however, when intentional contamination of the coating atmosphere with nitrogen or moisture is used to produce harder deposits (e.g., of titanium or tantalum), or to reduce the codeposition of carbon (e.g., with molybdenum from the carbonyl).

Nonuniform plating may result in all vapor plating processes, except for the displacement diffusion process, if consideration is not given to the gas-flow pattern around, or through the article to be coated. The shape factor may also have to be taken into account in selecting the method of heating the article to avoid nonuniform deposition due to nonuniform heating of the part. These difficulties can be overcome in extreme cases by applying more than one coating and using a different direction of gas flow over the specimen for each application.

If necessary, the displacement-diffusion type of coating process can be carried out at very low gas flow rates (since solid-state diffusion is the rate-controlling factor) and still produce uniform coatings. For this reason, this type of coating process is ideally suited for coating large, or highly irregular objects, or large numbers of small objects.

The elevated processing temperatures required in vapor plating may produce undesired physical changes in the article, such as loss of temper, grain growth, warping, dimensional change, or precipitation or solution of alloying constituents. However, in many instances, a vapor-plating procedure can be selected that will avoid marked undesirable change.

Undesirable side reactions in vapor-plating processes must be watched for and avoided. A particularly troublesome one in plating from metal halide vapors is the interaction of the base material and the halide vapor to form lower-valent halides, either of the base material or of the coating vapor. If the substrate temperature is too low or the plating atmosphere too rich in plating vapor, these lower-valent halides will condense at the surface of the substrate, producing a plate underlaid or contaminated with halide salts.

Such deposits are always poorly adherent, porous, and sensitive to moisture. Incomplete reduction or decomposition of the plating vapor alone can produce the same result. Contamination of this type is less likely to occur when plating inert base materials such as graphite, glass, and some ceramics.

After having been plated in a hydrogen atmosphere, some metals such as tantalum, columbium, and titanium, with a strong affinity for hydrogen, must be vacuum annealed, or at least cooled in an inert-gas atmosphere to avoid excessive hydrogen absorption and embrittlement.

Also, carburization of the substrate may occur in processes employing the metal carbonyls to coat metals with a strong affinity for carbon, if the substrate temperature is too high. The metal of the deposit itself may be partially carburized in some cases, as when depositing molybdenum, chromium, and tungsten from their carbonyl vapors.

Vapor-Deposited Replica

A *replica* formed of a metal or a salt by the condensation of the vapors of the material onto the surface to be replicated.

Vapor–Liquid–Solid Process

A process that utilizes vapor feed gases and a liquid catalyst to produce solid crystalline whiskers, such as silicon carbide whiskers used in composite materials.

Vapor-Phase Inhibitors

Substances that release vapors that inhibit corrosion on neighboring metal surfaces in moist conditions. Various materials, often impregnated into paper, are used, but they tend to be material specific, protecting some metals but not protecting others or even promoting their attack.

Vapor-Phase Lubrication

A type of lubrication in which one or more gaseous reactants are supplied to the vicinity of the surface to be lubricated and that subsequently react to form a lubricious deposit on that surface.

Variability

The number of degrees of freedom of a heterogeneous phase equilibrium.

Variance

A measure of the squared dispersion of observed values or measurements expressed as a function of the sum of the squared deviations from the population mean or sample average.

Varistor

A material, such as zinc oxide (ZnO), having an electrical resistance that is sensitive to changes in applied voltage.

Varnish

(1) In lubrication, a deposit resulting from the oxidation and/or polymerization of fuels, lubricating oils, or organic constituents of

bearing materials. Harder deposits are described as *lacquers*, and softer deposits are described as *gums*. (2) A transparent surface coating that is applied as a liquid and then changes to a hard solid; all varnishes or solutions of resinous materials in a solvent.

V-Bend Die

A die commonly used in press-brake forming, usually machined with a triangular cross-sectional opening to provide two edges as fulcrums for accomplishing *three-point bending*. See also the term *press-break forming*.

V-Brain Seal

A seal consisting of a ring or nested rings that have a V-shaped cross section and that are commonly made from elastomeric material. Spring loading is sometimes used to maintain contact between the seal and its meeting surface. It is normally used to seal against axial motion.

V-Cone Blender

A machine for blending metal powders that has two cone-shaped containers arranged in a V and open to each other. See also *blending*.

Vector Field

Same as a *resultant field*.

Vegard's Law

The relationship that states that the lattice parameters of substitutional solid solutions vary linearly between the values for the components, with composition expressed in atomic percentage.

Veil

An ultrathin mat for reinforcing plastics similar to a *surfacing mat*, often composed of organic fibers as well as glass fibers. See also *mat*.

Veining

(1) A sub-boundary structure in a metal but can be delineated because of the presence of a greater than average concentration of precipitate or solute atoms. (2) The fine lines on a metallographic specimen formed at the boundaries of subgrains.

Vello Process

A process for continuously drawing glass tubing (or cane) in which glass is fed downward to the draw through an annular orifice.

Vent (Metals)

A small opening in a foundry mold for the escape of gases.

Vent (Plastics)

A small hole or shallow channel in a mold that allows air or gas to exit as the plastic molding material enters.

Vent Cloth

A layer or layers of open-weave cloth used to provide a path for vacuum to "reach" the area over a *laminate* being cured, such that volatiles and air can be removed. Also causes the pressure differential that results in the application of pressure to the part being cured. Also called breather cloth.

Vent Mark

A small protrusion resulting from the entrance of metal into die vent holes.

Venting

In autoclave curing of a composite part or assembly, turning off the vacuum source and venting the vacuum bag to the atmosphere. The pressure on the part is then the difference between pressure in the autoclave and atmospheric pressure. In injection molding, gases evolve from the melt and escape through vents machined in the barrel or mold.

Verdigris

The green corrosion product formed on copper exposed to the atmosphere. It is usually copper sulfate but may also contain chlorides and carbonates depending on local pollution. Same as *patina*. Historically, the term was used for green pigments, particularly basic copper acetate.

Verification

Checking or testing an instrument to ensure conformance with the specification.

Verified Loading Range

In the case of testing machines, the range of indicated loads for which the testing machine gives results within the permissible variation specified.

Vermicular Iron

Same as *compacted graphite cast-iron*.

Vermiculite

(1) A granular, clay mineral constituent that is used as a textual material in painting, as an aggregate in certain plaster formulations used in sculpture, or mixed with a resin to form a filler of relatively high compressive strength. (2) A mixture of hydrated silicates of aluminum, iron, and magnesium. On heating, it expands and fragments producing the low-density material commonly used as a thermal insulator and for other applications. See also *filler* (1).

Vermiculite is a foliated mineral employed in making plasters and board for heat, cold, and sound insulation, as a filler in caulking compounds, and for plastic mortars and refractory concrete. The mineral is an alteration product of **biotite** and other micas. It occurs in crystalline plates, specific gravity 2.3, and Mohs hardness 1.5. The color is yellowish to brown. Upon calcination at 954°C, vermiculite expands at right angles to the cleavage into threads with a vermicular motion like a mass of small worms; hence its name.

The corklike pellets of vermiculite used for insulating fill in house walls are called *mica pellets*. *Exfoliated mica* is a name for expanded vermiculite.

Vernier

A short auxiliary scale that slides along the main test instrument scale to permit more accurate fractional reading of the least main division of the main scale. See also *least count*.

Vertical Illumination

Light incident on an object from the objective side of an optical microscope so that smooth planes perpendicular to the optical axis of the objective appear bright. See also *objective* and *optical microscope*.

Vertical Position

The position of welding in which the axis of the weld is approximately vertical. See the term *welding position*.

Vertical Position (Pipe Welding)

The position of a pipe joint in which welding is performed in the horizontal position, and the pipe may or may not be rotated. See the terms *horizontal fixed position* (pipe welding) and *horizontal rolled position* (pipe welding).

Vertical Welding Position

At a position in which the weld run is approximately vertical. This is usually taken to mean a Vickers hardness test. A hardness test utilizing a diamond indentor that is thrust into the component by a known load and the size of the indentation provides a measure of the hardness. The standard indentor is a square-based pyramid with a 136° included angle and, on the standard machine, weights are hung on the machine to induce via a lever system, loads of up to 120 kg at the indentor. After the load has been applied for a few seconds, the indentation is measured by a suitable microscope usually attached to the machine. The hardness number could then be calculated as load per unit area of impression in kg/mm from-where P is load in kg and d is the diagonal of the impression in mm. The hardness is independent of the load; but in practice, to keep the impression size within sensible limits and to maximize accuracy, the load is matched to the material, typically 30 kg for steels and 10 kg for many nonferrous metals. Tables for standard loads relate diametral measurements to hardness. Vickers hardness results are commonly reported as Hardness-Vickers, abbreviated to HV, or as Hardness Diamond (HD) or Vickers Pyramid Number (VPN). For example, 212 HV_{30} (or HD30 or VPN30), that is, indicates a hardness on the Vickers scale of 212. The subscript, 30, merely indicates the load in kilograms applied during the test, and, in most cases, it can be ignored as the tables used for the calculation make due allowance for the load applied. Variations on the basic theme include microhardness devices where extremely small loads are used to measure microscopical features and the Knoop tester that utilizes an elongated base pyramid for testing thin sheet on edge. Also, various portable devices have been developed for site use that utilize spring or hydraulic systems to apply the load.

VI Improver

An additive, usually a polymer, that reduces the variation in viscosity with temperature, thereby increasing the viscosity index of an oil.

Vibration Density

The apparent density of a powder mass when the volume receptacle is vibrated under specified conditions while being loaded. Similar to *tap density*. See also apparent density (1).

Vibratory Cavitation

Cavitation caused by the pressure fluctuations within a liquid, adduced by the vibration of a solid surface immersed in the liquid.

Vibratory Compaction

A powder compacting process where vibration of the die assembly is used in addition to the usual pressure.

Vibratory Finishing

A process for deburring and surface finishing in which the product and an abrasive mixture are placed in a container and vibrated.

Vibratory Mill

A *bull mill* wherein the comminution is aided by subjecting the balls or rods to a vibratory force. See also *comminution*.

Vibratory Polishing

A mechanical polishing process in which a metallographic specimen is made to move around the polishing cloth by imparting a suitable vibratory motion to the polishing system. See also *polishing* (4).

Vicat Softening Point

The temperature at which a flat-ended needle of 1 mm² (0.0015 in.²) circular or square cross section will penetrate a thermoplastic specimen to a depth of 1 mm (0.040 in.) under a specified load, using a uniform rate of temperature rise.

Vickers Hardness Number (HV)

A number related to the applied load and the surface area of the permanent impression made by a square-based pyramidal diamond indenter having included face angles of 136°, computed from

$$HV = 2P\sin\frac{\alpha/2}{d^2} = \frac{1.8544P}{d^2}$$

where

P is the applied load (kgf)
d is the mean diagonal of the impression (mm)
α is the face angle of the indenter (136°)

Vickers Hardness Test

A microindentation hardness test employing a 136° diamond pyramid indenter (Vickers) and variable loads, enabling the use of one hardness scale for all ranges of hardness—from very soft lead to tungsten carbide. Also known as diamond pyramid hardness test. See also *microindentation* and *microindentation hardness number.*

Vidicon

A camera tube in which a charge-density pattern is formed by photoconduction and stored on a photoconductor surface that is scanned by an electron beam.

VIM

See *vacuum induction melting.*

VIM–VAR

A production route for very high-quality steel. See *vacuum induction melting* and *vacuum arc remelting.*

Vinyl Acetate Ethylene

Since their introduction, vinyl acetate ethylene (VAE) copolymer emulsions have been a staple base for adhesive manufacturers. As the performance requirements within the packaging and construction markets have increased and diversified, so too has the use of these emulsions.

First, VAE copolymer emulsions offer a tremendous balance between performance properties and ease of use. The internal plasticization of the vinyl acetate with ethylene gives these emulsions adhesion to many difficult-to-adhere substrates, while the polyvinyl alcohol (PVOH) stabilization system provides for high wet tack, good setting speeds, and excellent machinability.

Second, manufacturers of VAE emulsions have continued to advance the performance capabilities of these materials. Available today are functionalized VAE systems for adhesion to metallized surfaces, a range of glass transition temperatures for specific film properties, low volatile organic compound emulsions for sensitive food packaging applications, and higher solids technologies as an alternative to non-water-based systems. With these new VAE copolymer emulsions, adhesive compounders are better able to address the ever-changing needs of the adhesive industry.

At last, these types of the VAE emulsions are made even more versatile by their ability to be compounded with other raw materials and polymer systems. The additional formulations that can result from their compatibility with plasticizers, residence, fillers, humectants, surfactants, polyvinyl alcohol, etc., can offer various improvements in adhesion, tack, heat/cold resistance, flame retardancy, and range.

Processing and Applications

A most recent advance in VAE emulsion technology has been the introduction of a PVOH-stabilized, ultrahigh-solids copolymer emulsion that is polymerized at 72% solids and a 2000 cps viscosity. Its composition, structure, and colloidal properties provide faster setting speeds, higher wet tack, and improved adhesion to difficult-to-adhere substrates that was thought possible for VAE emulsions a few years ago. These performance features are allowing adhesive compounders to broaden greatly the applications utilizing waterborne technologies.

Polyurethanes have been available as adhesives for quite some time and are commonly found in vacuum-forming in plastics-bonding operations within the automotive and footwear industries. During the last 5 years, waterborne urethane chemistry has undergone a significant transformation from solvent-borne or high cosolvent-containing polymer systems to 100% waterborne systems.

These aqueous polyurethane dispersions, like their solvent-borne counterparts, have some unique performance characteristics. They offer low heat reactivation temperatures, good adhesion to difficult-to-bond substrates, rabid green strength development, and a high-temperature heat resistance. However, they also have some significant drawbacks. They are low in solids, low in wet tack, slow drying, and relatively high in cost.

The blending of polymers to improve adhesive properties is already widely done in the industry. With the commercially available aqueous polyurethane dispersions on the market today, the opportunity exists to enhance the performance of ultrahigh-solids VAE emulsions through blending because these technologies are so complementary. They are both 100% waterborne, and the characteristics of the ultrahigh-solids VAE emulsion can compensate for the disadvantages of urethane with its speed of set, wet tack, and minimal water content.

Future

Through the blending of an ultrahigh-solids VAE emulsion with many of the commercially available aqueous polyurethane dispersions on the market today, adhesive compounders can create a new class of stable high-performance waterborne adhesives.

Depending on the urethane grade selected and the level incorporated in the blend, the performance properties of the ultrahigh-solids VAE emulsion is dramatically enhanced in several areas:

- Cohesive strength
- Adhesion (vinyl)
- Heat sealability
- Cross-linker performance

Vinyl Acetate Plastics

Plastics based on polymers of vinyl chloride or copolymers of vinyl chloride with other monomers, the vinyl chloride having the greatest amount of mass.

Vinyl Esters

A class of thermosetting resins containing esters of acrylic and/or methacrylic acids, many of which have been made from epoxy resin. Cure is accomplished, as with unsaturated polyesters, by copolymerization with other vinyl monomers, such as styrene. Glass-reinforced vinyl esters are used in corrosion-resistant products, such as piping and storage tanks, used in the pulp and paper, chemical process, wastewater, and mining industries.

Vinyl Resins and Plastics

These are a group of products varying from liquids to hard solids, made by the polymerization of ethylene derivatives, employed for finishes, coatings, and molding resins, or it could be made directly by reacting acetic acid with ethylene and oxygen. In general, the term *vinyl* designates plastics made by polymerizing vinyl chloride, vinyl acetate, or vinylidene chloride, but may include plastics made from styrene and other chemicals. The term is generic for compounds of the basic formula RCH=CR′CR″. The simplest are the polyesters of vinyl alcohol, such as vinyl acetate. This resin is lightweight with a specific gravity of 1.18, and is transparent, but it has poor molding qualities and its strength is no more than 34 MPa. But the vinyl halides, CH$_2$=CHX, also polymerize readily to form vinylite resins, which mold well, have tensile strengths up to 62 MPa, high dielectric strength and high chemical resistance, and a widely useful range of resins are produced by copolymers of vinyl acetate and vinyl chloride.

The possibility of variation in the vinyl resins by change of the monomer, copolymerization, and difference in compounding is so great that the term *vinyl resin* is almost meaningless when used alone. The resins are marketed under a continuously increasing number of trade names. In general, each resin is designed for specific uses but is not limited to those uses.

Vinyl Alcohol

Vinyl alcohol, CH$_2$=CHOH, is a liquid boiling at 35.5°C. Polyvinyl alcohol is a white, odorless, tasteless powder which on drying from solutions forms a colorless and tough film. The material is used as a thickener for latex, in chewing gum, and for sizes and adhesives. It can be compounded with plasticizers and molded or extruded into tough and elastic products. Hydrolyzed polyvinyl alcohol has greater water resistance, higher adhesion, and its lower residual acetate gives lower foaming. Soluble film, for packaging detergents and other water-dispersible materials to eliminate the need of opening the package, is a clear polyvinyl alcohol film. Textile fibers are also made from polyvinyl alcohol, either water soluble or insolubilized with formaldehyde or another agent. Polyvinyl alcohol textile fiber is hot-drawn by a semimelt process and insolubilized after drawing. The fiber has a high degree of orientation and crystallinity, giving good strength and hot-water resistance.

Vinyl alcohol reacted with an aldehyde, and an acid catalyst produces a group of polymers known as vinyl acetal resins, and separately designated by type names, as polyvinyl butyral and polyvinyl formal. The polyvinyl alcohols are called **Solvars**, and the polyvinyl acetates are called **Gelvas**. The vinyl ethers range from vinyl methyl ether, to vinyl ethylhexyl ether, from soft compound so hard resins. Vinyl ether is a liquid that polymerizes, or that can be reacted with hydroxyl groups to form acetyl resins. Alkyl vinyl ethers are made by reacting acetylene with an alcohol under pressure, producing methyl vinyl ether, ethyl vinyl ether, or butyl vinyl ether. They have reactive double bonds that can be used to copolymerize with other vinyls to give a variety of physical properties. The polyvinyl formals, **Formvars**, are used in molding compounds, wire coatings, and impregnating compounds. They are one of the toughest of the thermoplastics.

Plastisol

A plastisol is a vinyl resin dissolved in a plasticizer to make a pourable liquid without a volatile solvent for casting. The poured liquid is solidified by heating. Plastigels are plastisols to which a gelling agent has been added to increase viscosity. The polyvinyl acetals, **Alvars**, are used in lacquers, adhesives, and phonograph records. The transparent polyvinyl butyrals, *Butvars*, are used as interlayers of laminated glass. They are made by reacting polyvinyl alcohol with butyraldehyde. Vinal is a general name for vinyl butyral resin used for laminated glass.

Vinyl Acetate

Vinyl acetate is a water-white mobile liquid with boiling point 70°C, usually shipped with a copper salt to prevent polymerization in transit. It may be polymerized in benzene and marketed in solution, or in water solution for use as an extender for rubber, and for adhesives and coatings. The higher the polymerization of the resin, the higher the softening point of the resin. Polyvinyl acetate resin is a colorless, odorless thermoplastic with density of 1.189, unaffected by water, gasoline, or oils, but soluble in the lower alcohols, benzene, and chlorinated hydrocarbons. Polyvinyl acetate resins are stable to light, transparent to ultraviolet light, and are valued for lacquers and coatings because of their high adhesion, durability, and ease of compounding with gums and resins. Residents of low molecular weight are used for coatings, and those of high molecular weight for molding. Vinyl acetate will copolymerize with maleic acrylonitrile or acrylic esters. With ethylene, it produces a copolymer latex of superior toughness and abrasion resistance for coatings.

Vinyl Benzoate

Vinyl benzoate is an oily liquid that can be polymerized to form resins with higher softening points than those of polyvinyl acetate, but that are more brittle at low temperatures. These resins, copolymerized with vinyl acetate, are used for water-repellent coatings. Vinyl crotonate is a liquid of specific gravity of 0.9434. Its copolymers are brittle resins, but it is used as a cross-linking agent for other resins to raise the softening point and to increase abrasion resistance. Vinyl formate is a colorless liquid that polymerizes to form clear polyvinyl formate resins that are harder and more resistant to solvents than polyvinyl acetate. The monomer is also copolymerized with ethylene monomers to form resins for mixing in specialty rubbers. Methyl vinyl pyridine is used in making resins, fibers, and oil-resistant rubbers. It is a colorless liquid boiling at 64.4°C. The active methyl groups give condensation reactions, and it will copolymerize with butadiene, styrene, or acrylonitrile. Polyvinyl carbazole, under the name of **Luvican**, is used as a mica substitute for high-frequency insulation. It is a brown resin, softening at 150°C.

Vinyl Chloride

Vinyl chloride, also called ethenyl chloride and chloroethylene, produced by reacting ethylene with oxygen from the air and ethylene dichloride, is the basic material for the polyvinyl chloride resins. It is a gas. The plastic was produced originally for cable installation and for tire tubes. The tensile strength of the plastic may vary from the flexible resins with about 20 MPa to the rigid resin with the tensile strength of 62 MPa and Shore hardness of 90. The dielectric strength is high, up to 52 × 10^6 V/m. It is resistant to acids and alkalies. Unplasticized polyvinyl chloride is used for rigid chemical-resistant pipe. Polyvinyl chloride sheet, unmodified, has a tensile strength of 57 MPa, flexural strength 86 MPa, and a light transmission of 78%.

Polyvinyl Chloride

Polyvinyl chloride (PVC) is a thermoplastic polymer formed by the polymerization of vinyl chloride. Resins of different properties can be made by variations in polymerization techniques. These resins can be compounded with plasticizers, color, mineral filler, etc., and processed into usable forms, varying widely in physical and electrical properties, chemical resistance, and processing versatility in coloring and design. Compared with other thermoplastics of comparable cost, articles produced from the vinyl chloride plastics have outstanding chemical, flame, and abrasion resistance, tensile properties, and resistance to heat distortion.

PVC homopolymer resins, the largest single type of vinyl chloride-containing plastics, are produced by several methods of polymerization:

1. *Suspension*: The largest-volume method that produces resins for general purpose use, processed by calendering, injection molding, extrusion, etc.
2. *Mass or solution*: Produces fine particle size resins used principally for calendering and solution coating.
3. *Emulsion*: Produces extremely fine particle size resins used for the preparation of liquid plastisols or organosols for use in slush molding, coatings, and foam.

The two largest-volume members of the family of vinyl chloride polymers are the pure polyvinyl chloride or homopolymer resins and the vinyl chloride-vinyl acetate copolymers containing approximately 5%–15% vinyl acetate.

Types of Vinyls/PVCs

Rigid Vinyls

Products made from rigid vinyls are perhaps of most interest in the engineering field. Rigid materials can be prepared by calendering, extrusion, injection molding, transfer molding, and solution casting processes. Rigid PVC products are available in sheets, films, rods, pipes, profiles, valves, nuts and bolts, etc. The products can be machined easily with wood and metal-working tools.

Sheets or other forms can be conventionally welded by hot-air guns, using extruded welding rods of essentially the same composition as that of the sheet. The welded joints have strength equal to that of the base material. Rigid sheets can be thermoformed into many intricate shapes by several different thermoforming techniques such as vacuum forming and ring and plug forming. Rigid pipe can be threaded and joined like steel pipe or sealed with adhesives in a manner similar to the sweating of copper pipe. Vinyl pipe is being used increasingly in waterworks, the petroleum industry, in natural gas distribution, irrigation, hazardous chemical application, and food processing.

Rigid vinyl made from vinyl acetate–vinyl chloride copolymers is prominent in sheeting used for thermoforming for such items as maps, packaging, advertising displays, and toys. Rigids made from homopolymer vinyl chloride resins are used in heavier structural designs, for example, pipe, pipe valve, heavy panels, electrical ducting, window and door framing parts, architectural moldings, gutters, downspouts, and automotive trim. In these fields, rigid vinyls compete with aluminum and other metals. Rigid vinyl products made from homopolymer resins are available in two types: type I, or unmodified PVC, is approximately 95% PVC and has outstanding chemical resistance but low impact strength. Type II PVC, containing 10%–20% of a resinous or rubbery polymeric modifier, has improved impact strength but reduced chemical resistance.

Flexible Vinyls

Flexible vinyl products are produced by the same general methods used for rigid vinyl products. Flexibility is achieved by the incorporation of plasticizers (mainly high-boiling organic esters) with the vinyl polymer. By proper choice of plasticizer type, flexible products can be obtained that excel in certain specific properties such as gasoline and oil resistance, low temperature flexibility, and flame resistance.

The flexible sheeting and film can be fabricated by heat sealing to itself or other substrates by induction or high-frequency methods, solvent sealing, sewing, etc. Flexible vinyl film and sheeting find applications in upholstery, packaging, agriculture, etc. The corrosion resistance of vinyl sheeting makes it ideal for a pipe wrap to prevent corrosion of underground installations. Flexible extensions in many different shapes and forms have applications as insulating and jacketing on electrical wire and cable, refrigerator gaskets, weather stripping, upholstery, and shoe welding. Injection-molded flexible vinyl products are used as shoes, electrical plugs, and installation of various sorts.

Abrasion and stain resistance, coupled with unlimited coloring and design possibilities, have made flexible vinyl flooring one of the largest items in the floor-covering field. The major revolution in vinyl flooring is the greater emphasis on the use of relatively low to very low molecular weight homopolymer resins in place of the more expensive vinyl chloride–vinyl acetate copolymers.

Coatings

Coatings based on PVC polymers and copolymers can be applied from solutions, latex, plastisols, or organosols. Plastisols are liquid dispersions of fine particle size emulsion PVC in plasticizers. Organosols are essentially the same as plastisols but contain a volatile liquid organic diluent to reduce viscosity and facilitate processing. In many coating operations, conventional paint-spraying equipment is used. In other techniques, articles can be dip-, knife-, or roller-coated. Plastisols and organosols are used extensively for dip coating of wire products such as household utensils and knife coating of fabrics and paper. Plastisol and organosols products can be varied from hard (rigidsols) to very soft (vinyl foam). Vinyl coatings are used for a variety of applications requiring corrosion and/or abrasion resistance.

Production coatings with vinyl plastics involve a fluidized-bed technique. A metallic object, heated to 204°C–260°C, is immersed in a bed of finely ground plastic which is "fluidized" by air entering the bottom of the container. Articles can be coated by this method to a thickness of 7–60 mils. Both rigid and plasticized vinyls can be applied by this method.

Vinyl-Metal Laminates

These products are made by direct lamination of free processed, embossed, and designed vinyl sheet to metal, or continuous plastisol coating of metal sheet followed by fusing or curing of the plastisol and subsequent embossing. In the former process, the vinyl can be laminated to both sides of the metal sheet. Steel, aluminum, magnesium, brass, and copper have been used. The hardness, elongation, general properties, and thickness of the vinyl can be modified within wide limits to meet particular needs.

These laminates are dimensionally stable below 100°C and combine the chemical and flame resistance, decorative and design possibilities of vinyl with the rigidity, strength, and fabricating attributes of metals. The vinyl-metal laminates can be worked without rupture by many of the metalworking techniques, such as deep-drawing, crimping, stamping, punching, shearing, and reverse-bending. Disadvantages are inability to spot-weld, and lack of covering of metal edges, which is necessary where severe exposure conditions are encountered. Vinyl-metal laminates find use in appliance cabinets, machined housings, lawn and office furniture, automotive parts, luggage, chemical tanks, etc. The cost of these laminates is comparable to some lacquered metal surfaces.

Vinylidene Chloride Plastics

Vinylidene chloride plastics are derived from ethylene and chlorine polymerized to produce a thermoplastic with softening point of 116°C–138°C. The resins are noted for their toughness and resistance to water and chemicals. The molded residents have a specific gravity of 1.68–1.75, tensile strength 27–48 MPa, and a flexural strength of 103–117 MPa. *Saran* is the name of vinylidene chloride plastic, extruded in the form of tubes for handling chemicals, brines, and solvents to temperatures as high as 135°C. It is also extruded into strands and woven into a box-weave material as a substitute for rattan for seating. Saran latex, a water dispersion of the plastic, is used for coating and impregnating fabrics. For coating food-packaging papers, it is waterproof and greaseproof, odorless, and tasteless, and gives the papers a high gloss. Saran is also produced as a strong transparent film for packaging. Saran bristles for brushes are made in diameters from 0.025 to 0.051 cm.

Applications

The two largest-volume applications are in upholstery made from monofilaments and film for food packaging. Other uses are in window screening (monofilaments), paper and other coatings, pipe and pipe linings, and staple fiber.

Saran pipe and Saran-lined metal pipe are of interest in the engineering field. Saran-lined pipe is prepared by swaging an oversize metal pipe on an extruded Saran tube. These products can be installed with ordinary piping tool. Fittings and valves lined with Saran and flange joints with Saran gaskets are available. Vinylidene chloride–acrylonitrile copolymer has applications as coatings for tank car and ship-hold linings. Lacquers of these polymers are used in cellophane coatings yielding a product with the low-moisture vapor permeability of vinylidene chloride polymers plus the handling ease of cellophane. The lacquers are also used for paper coatings, dip coatings, and sprayed packaging. Vinylidene chloride copolymers in latex form are used in paper coatings and specialty paints.

Vinylidene Fluoride

Vinylidene fluoride has a high molecular weight, about 500,000. It is a hard, white thermoplastic resin with a slippery surface and has a high resistance to chemicals. It resists temperatures up to 343°C and does not become brittle at low temperatures. It extrudes easily and has been used for wire insulation, gaskets, seals, molded parts, and piping.

Vinylidene Chloride Plastics

Plastics based on polymer resins made by the polymerization of vinylidene chloride or copolymerization of vinylidene chloride with other unsaturated compounds, the vinylidene chloride being the greatest amount by weight.

Virgin Filament

An individual *filament* that has not been in contact with any other fiber or any other hard material.

Virgin Material

A plastic material in the form of pellets, granules, powder, flock, or liquid that is not been subjected to use or processing other than that required for its initial manufacture.

Virgin Metal

Metal produced from ore and not contaminated by remelted scrap. Same as *primary metal*.

Viscoelastic

The phenomenon whereby the reversible relationship between stress and strain is time related.

Viscoelasticity

A property involving a combination of elastic and viscous behavior that makes deformation dependent upon both temperature and strain rate. A material having this property is considered to combine the features of a perfectly elastic solid and a perfect fluid.

Viscosity

The bulk property of a fluid, semifluid, or semisolid substance that causes it to resist flow. Viscosity is defined by the equation

$$\eta = \frac{\tau}{(dv/ds)}$$

where
    τ is the shear stress
    v is the velocity
    s is the thickness of an element measured perpendicular to the direction of flow
    (dv/ds) is known as the rate of shear

Newtonian viscosity is often called dynamic viscosity, or absolute viscosity. Kinematic viscosity, or static viscosity (v), is the ratio of dynamic viscosity (η) to density (p) at a specified temperature and pressure ($v = \eta/p$). Recommended units of measure for dynamic viscosity are the Pascal second (Pa s) in SI units and poise (P) in English units. Recommended units of measure for kinematic viscosity are square meters per second (m²/s) in SI units and the Stoke, or centistoke (cSt) in English units.

Viscosity Coefficient

The shearing stress tangentially applied that will induce a velocity gradient in a material.

Viscosity Index (VI)

A commonly used measure of the change in viscosity of a fluid with temperature. The higher the viscosity index, the smaller the relative change in viscosity with temperature. Two different indices are used; the earlier usage applies to oils having a VI from 0 to 100. Extended VI applies to oils having a VI of at least 100. It compares the oil with a reference oil of VI 100.

Viscous

Possessing viscosity. This term is frequently used to imply high viscosity.

Viscous Deformation

Any portion of the total deformation of a body that occurs as a function of time when load is applied but that remains permanently when the load is removed generally referred to as *elastic deformation*.

Viscous Flow

A mode of plastic deformation, particularly early stages of creep, in which the stress–strain relationship is time dependent, and in which tensile loading does not initiate necking.

Viscous Friction

See *fluid friction*.

Visible

Pertaining to radiant energy in the electromagnetic spectral range visible to the normal human eye (–380 to 780 nm).

Visible Radiation

Electromagnetic radiation in the spectral range visible to the human eye (–380 to 780 nm).

Visible-Light-Emitting Diode

An optoelectronic device containing a semiconductor junction that emits visible light went forward biased. The material is usually gallium phosphide or gallium arsenide phosphide. See also *gallium* and *gallium compounds*.

Visual Examination

The qualitative observation of physical characteristics, observed by using the unaided eye or perhaps aided by the use of a simple hand-held lens (up to 10×).

Vitreous

(1) Partially or completely comprised of a glass; often containing solid particles distributed therein. (2) Glasslike, particularly vitreous enamels applied by fusing a glass powder on the surface of a (usually) steel) component.

Vitreous Enamel

See *porcelain enamel*.

Vitrification

(1) The formation of a glassy or noncrystalline material. (2) The characteristic of a clay product resulting when the kiln temperature is sufficient to fuse grains and close the surface pores, forming an impervious mass. (3) The progressive reduction in porosity of a ceramic composition as a result of heat treatment, or the process involved.

Vitrify

To render vitreous, generally by heating; usually, achieving enough glassy phase to render impermeable.

Vitriol Sulfuric Acid

A_2SO_4.

V-Mixer

A machine for mixing metal powders that has two cylindrical containers arranged in the shape of a V and open to each other. See also *mixing*.

Void

A cavity within a solid or liquid.

Void (Composites)

Air or gas that has been trapped and cured into a laminate. Porosity is an aggregation of microvoids. Voids are essentially incapable of transmitting structural stresses or nonradiative energy fields.

Void (Metals)

(1) A *shrinkage cavity* produced in castings or weldments during solidification. (2) A term generally applied to paints to describe *holidays*, holes, and skips in a film.

Void Content

Volume percentage of voids, usually less than 1% in a properly cured composite. The experimental determination is indirect, that is, it is calculated from the measured density of a cured laminate and the "theoretical" density of the starting material.

Voidage

The measure, usually a volume percentage, of voids in a material.

Volatile Content

The percentage of *volatiles* that is driven off as a vapor from a plastic or an impregnated reinforcement.

Volatile Organic Compound

Many efforts have been underway regarding the biological control of volatile organic compound (VOC) emissions.

Water-based and hot-melt adhesives and coatings have been developed and evaluated extensively, but they are not satisfactory for all applications and may require solvent-based cleaners and primers. Various developments have shown how biological treatment can reduce solvent levels in air emissions and focuses in particular on footwear production. It is, however, applicable to all industries using organic solvents.

For example, the U.K. Environmental Protection Act 1990 sets a 5 metric tons/year adhesive solvent usage threshold, above which processes are subject to local authority air pollution control. This regulation apparently affects manufacturing plants producing as few as 5000 pairs of shoes per week.

By June 1998, shoemakers and material suppliers were set to meet an emission limit of 50 mg/m³ measured as carbon, or a stringent mass emission control regime of 20 g/pair for footwear; similar controls have been initiated throughout Europe.

Catalytic combustion has been shown to be technically effective, but involves high capital and running costs. For example, adsorption on activated carbon is an established technique, but apparently it is unsuitable for the mixed cocktail of solvents in footwear production. Biological treatment processes have shown to be less costly and more suitable for arresting emissions of moderate concentration at ambient temperature, as found in shoe factory exhausts.

Biological treatment has been shown suitable for halting VOC emissions from the manufacturing industry with an on-site biological treatment unit set up to demonstrate VOC abatement.

Process

The process biologically breaks down VOCs into biomass by mineralization and utilization of the carbon; the by-products of breakdown are carbon dioxide and water vapor. Within the system, microorganisms grow as a biofilm on selected media where they produce enzymes to break down the VOC contaminants. Liquid is continuously recirculated to solubilize the solvents from the contaminated inlet gas. Water is sprayed evenly where the microorganisms oxidize the solvents, and monitoring of the recirculation liquid for determinants, including pH, nitrogen, phosphorus, suspended solids, and so on apparently allows for tighter process control and greater removal efficiency.

In this system, there is a large interface between gas and liquid phases so that the bed can capture the compounds of poor solubility that are present in the off-gases. Inputs to the process include main water supply, electricity to run pumps, solenoid valves, and a programmable logic controller, as well as nutrients to maintain the biomass, including nitrogen, phosphate, sulfate, and trace elements that are added according to the carbon load to the reactor.

Although this process was used at a footwear manufacturing plant, biotechnology also has application within other industries producing low-concentration mixed solvent waste streams, where recovery is inappropriate and the capital and operating costs of thermal and catalytic systems can be prohibitively expensive. Examples include printing, painting, laminating, metal and leather finishing, and furniture coating.

In another example, the U.S. Navy studied a new environmentally safe paint coating for use on its fleet of helicopters. The three-shade flat, haze gray coats were sprayed on a new Navy fleet combat support helicopter. What makes the coating environmentally safe is the absence of VOCs that contribute to air pollution. The new coating eliminates the use of chemicals targeted by the federal government for reduction or elimination.

Normal Environmental Protection Agency (EPA)-compliant aircraft coatings currently use bT-fy that contain about 3.5 lb of VOC/gallon, whereas this new paint has zero VOC.

The zero-VOC paint was developed by Deft Coatings, Inc. This zero-VOC coating also offers a significant weight benefit compared to the current paint and is nonflammable.

The first zero-VOC-coated helicopter was tested and evaluated by the Navy to ensure that the new coating met stringent Military Standard requirements.

Volatiles

Materials, such as water and alcohol, in a sizing or resin formulation, that are capable of being driven off as a vapor at room temperature or slightly elevated temperature. See also *size*.

Volatilization

The conversion of a chemical substance from a liquid or solid state to a gaseous or vapor state by the application of heat, by reducing pressure, or by a combination of these processes. Also known as *vaporization*.

Volt

The unit of potential difference or electromotive force in the meter-kilogram-second system, equal to the potential difference between two points for which 1 C of electricity will do 1 J of work in going from one point to the other. Symbolized V.

Voltage Alignment

A condition of alignment of an electron microscope so that the image expands or contracts symmetrically about the center of the viewing screen when the accelerating voltage is changed. See also *alignment*.

Voltage Drop

The amount of voltage loss from original input in a conductor of given size and length.

Voltage Efficiency

The ratio, usually expressed as a percentage, of the equilibrium–reaction potential in a given electrochemical process to the bath voltage.

Voltage Regulator

An automatic electrical control device for maintaining a constant voltage supply to the primary of a welding transformer.

Voltage Stress

That stress found within a material when subjected to an electrical charge.

Voltaic Cell

An electrolytic cell in which an electric current flows between a pair of electrodes that are immersed in an electrolyte and electrically connected externally.

Voltammetry

An electrochemical technique in which the current between working (indicator) electrodes and counterelectrodes immersed in an electrolyte is measured as a function of the potential difference between the indicator electrode and a reference electrode.

Volume Diffusion

One of the primary diffusion mechanisms during sintering. It is predominant for larger particles at higher temperatures, and its diffusion coefficient for the same conditions is smaller than that for grain boundary diffusion, and much smaller than that for surface diffusion. See also *grain boundary diffusion* and *surface diffusion*.

Volume Filling

Filling the volume of a die cavity or receptacle with loose powder, and striking off any excess amount.

Volume Fraction

Fraction of the constituent material, such as fibers in a composite material, based on its volume.

Volume Ratio

The volume percentage of solid in the total volume of a sintered body.

Volume Resistance

The ratio of the direct voltage applied to two electrodes in contact with or embedded in a specimen to that portion of the current between them that is distributed through the volume of the specimen.

Volume Shrinkage

The volumetric size reduction a powder compact undergoes during sintering. Contrast with *linear shrinkage*.

Volumetric Analysis

(1) Quantitative analysis of solutions of known volume, but unknown strength by adding reagents of known concentration onto a reaction end point (color change or precipitation) is reached; the most common technique is by *titration*. (2) A form of chemical analysis in which the material in question is dissolved in known concentration and the solution reacted with other solutions of suitable reagents and standardized concentration.

Volumetric Modulus of Elasticity

See *bulk modulus of elasticity*.

Volumetric Strain

The algebraic sum of the strains induced by the three principal stresses.

Von Laue Technique

See *back reflection x-ray technique*.

Vortex (Shedding) Fatigue

Fatigue cracking due to the shedding of vortices arising from the rapid flow of fluid over a surface. The example usually quoted is of slender metal chimney stacks that in strong winds shed vortices on alternate sides causing the stack to vibrate across the airstream causing fatigue cracking. The effect may be countered by fitting spiral fins around the stack.

Voxel

Shortened term for volume element. In computed tomography, the volume within the object that corresponds to a single pixel element in the image. The box-shaped volume defined by the area of the pixel and the height of the slice thickness. See also *pixel*.

Vulcanization

A chemical reaction in which the physical properties of a rubber are changed in the direction of decreased plastic flow, less surface tackiness, and increased tensile strength by reacting it with sulfur or other suitable agents. See also *self-vulcanizing*.

Vulcanize

To subject to *vulcanization*.

Vulcanized Fiber

Vulcanized fiber is a pure, dense, cellulosic material with good electrical insulating properties and high mechanical strength. It is half the weight of aluminum, easily machined and formed, and is used for parts such as for barriers, abrasive-disk backing, high-strength bobbin heads, materials-handling equipment, railroad-track insulation, and athletic guards.

Forms

Most manufacturers provide vulcanized fiber in the form of sheets, coils, tubes, and rods. Sheets are made in a thickness of 0.06–50.8 mm, approximately 1.2 × 2 m in size, or in rolls and coils from 0.06 to 2.3 mm thick. Tubes are made in the outside diameter range of 4.7–111.5 mm, and rods are produced 2.3–50.4 mm in diameter.

Sheets can be machined and formed to produce a variety of useful shapes for insulating or shielding purposes. Sheets, tubes, and rods can be machined using standard practices for cutting, punching, tapping, milling, shaping, sanding, etc.

Properties

Vulcanized fiber possesses a versatile combination of properties, making it a useful material for practically all fields. It has outstanding arc resistance, high structural strength per unit area, and can be formed and machined. In thin sections, it possesses high tear strength, smoothness, and flexibility. In heavier thicknesses, it resists repeated impact and has high tensile, flexural, and compressive strength.

The material is unaffected by normal solvents, gasoline, and oils and therefore is recommended for applications where a structural support is required in the presence of these materials.

Moisture absorption is high, and dimensional stability is affected by conditions of humility when not protected by moisture-resistant coatings.

Vulcanized fiber is produced in 13 basic grades and numerous special grades to meet specific application requirements.

Applications

Vulcanized fiber serves as the insulating material in a signal block for railroad track insulations. At the end of a signal station, the rails are completely insulated from the next adjoining section to form what is termed a *block*. The two meeting rails and the coupling fixtures are insulated with formed parts made from vulcanized fiber. This junction is effective while absorbing the repeated impact from trains under all weather conditions.

Vulcanized fiber offers durability, ease of fabricating, excellent wear characteristics, and lightness of weight for materials-handling and luggage applications. The materials-handling equipment resists scuffing, battering, denting, rusting, and other general wearing conditions and provides protection by its hardness and resilience.

Formed pieces of vulcanized fiber offer outstanding service as arc barriers in circuit breakers. The arc-resistant properties of vulcanized fiber prevent a breakdown when the circuit breaker is subjected to an overload. The formed barrier is tested to take higher electrical loads than the maximum that can be produced by the circuit and, since repeated circuit breaks will not affect the performance of vulcanized fiber, the need for replacement is negligible.

Peerless control tape is used for programming data, processing equipment, or automatic machining equipment. The special properties that give this material outstanding service life are high tensile strength, high tear strength, low stretch, and good abrasive resistance.

Flame-resistant vulcanized fiber gives designers a structural material that can be used in those applications requiring a nonburning material and reduces fire hazard by containing a fire at its source. Flame-retardant parts serve as barriers in electrical equipment, materials-handling equipment, and wastebaskets.

Chemistry

Vulcanized fiber is produced by the chemical action of zinc chloride solution on a saturating grade of absorbent paper when processed under heat and pressure. The action of the zinc chloride converts the cellulosic fibers to a dense, homogeneous structure producing a laminated material that is refined to a chemically pure form. Final processing consists of drying to the proper moisture content and applying the proper calender to give smoothness and uniformity of thickness.

Grades

The 13 basic grades are as follows:

Electrical insulation grade: Primarily intended for electrical applications and others involving difficult bending or forming operations. It is sometimes referred to as "fishpaper."

Commercial grade: Considered to be the general-purpose grade, sometimes referred to as "mechanical and electrical grade." It possesses good physical and electrical properties and fabricates well.

Bone grade: Characterized by greater hardness and stiffness associated with higher specific gravity. It machines smoother with less tendency to separate the plies in the machining operations.

Trunk and case grade: Conforms to the mechanical requirements of "commercial grade," but has better bending qualities and smoother surface.

Flexible grade: Made sufficiently soft by incorporating a plasticizer, it is suitable for gaskets, packings, and similar applications. It is not recommended for electrical use.

Abrasive grade: Designed as the supporting base for abrasive grit for both disk and drum sanders. It has exceptional tear resistance, ply adhesion, resilience, and toughness.

White tag grade: It has smooth clean surfaces and can be printed or written on without danger of ink feathering.

Bobbin grade: Used for the manufacture of textile bobbin heads. It punches well under proper conditions, but is firm enough to resist denting in use. It machines to a very smooth surface.

Railroad grade: Used as railroad track joint, switch rods, and other insulating applications for track circuits.

Hermetic grade: Used as electric-motor insulation in hermetically sealed refrigeration units. High purity at low methanol extractables are essential because it is immersed in the refrigerant.

White grade: Recommended for applications where whiteness and cleanliness are essential requirements.

Shuttle grade: Design for gluing to wood shuttles to withstand the repeated pounding received in textile power looms.

Pattern grade: Made to provide maximum dimensional stability and minimum warpage for use as patterns in cutting cloth, leather, and similar materials.

Vulcanized Oils

Vegetable oils vulcanized with sulfur and used for compounding with rubber for rubber goods, or as a rubber substitute. Castor oil, corn oil, rapeseed oil, and soybean oil are used. Tokenized oil is a white to brown, spongy, odorless cake, or a sticky plastic, with a specific gravity of 1.04. The material was invented in France in 1847 and was known as **factice**.

Factice cake is solidified, vulcanized oils, cut in slab form. It is an oil modifier of rubber, to add softness and plasticity. It also has some elasticity. **White factice** is made from rapeseed oil, which is high in a characteristic acid, crucic acid, by slow addition of sulfur chloride up to 25% sulfur content. **Erasing rubbers** are rubber compound with white factice or the factice alone. **Black factice** has mineral bitumen added to brown factice. **Factice sheet** is specially processed factice made by treating warm oil with sulfur and then with sulfur chloride. The strength and elasticity are higher. **Mineral rubber** was a name applied to vulcanized oils mixed with bitumens, especially gilsonite.

V–X Diagram

A graph of the isothermal or isobaric phase relationships in a binary system, the coordinates of the graph being specific volume and concentration.

Vycor

See *glass*.

Wafer

A slice of a semiconductor crystalline ingot. See also the term *slices*.

Wake Hackle

A *hackle* line extending from a singularity at the crack front in the direction of cracking. Such markings are associated with inclusions in ceramics and glasses—pores, bubbles, and solid particles—and are useful in determining the direction of crack propagation.

Walking-Beam Furnace

A continuous-type heat treating or sintering furnace consisting of two sets of rails, one stationary and the other movable, that lift and advance parts inside the hearth. With this system, the moving rails lift the work from the stationary rails, move it forward, and then lower it back onto the stationary rails. The moving rails then return to the starting position and repeat the process to advance the parts again.

Wallner Line (Ceramics and Glasses)

A fracture surface marking, having a wavelike profile in the fracture surface. Such marks frequently appear as a series of curved lines, indicating the direction of propagation of the fracture from the concave to the convex side of a given Wallner line. Also known as *ripple mark*.

Wallner Lines (Metals)

A distinct pattern of intersecting sets of parallel lines, sometimes reducing a set of V-shaped lines, sometimes observed when viewing brittle fracture surfaces at high magnification in an electron microscope. Wallner lines are attributed to interaction between a shock wave and a brittle crack front propagating at high velocity. Sometimes, Wallner lines are misinterpreted as fatigue striations. See the term *fatigue striations* (metals).

Wandering Sequence

Same as *random sequence*.

Warm Hardening

The term is occasionally used as an alternative to warm working and more usually refers to precipitation hardening at elevated temperature.

Warm Work

Deformation at a temperature above ambient but below that at which recrystallization occurs. The purpose may be to avoid some brittle range, to benefit from the lower yield strength at higher temperature, or to minimize scaling. It induces work hardening similar to cold working and unlike hot working.

Warm Working

Deformation of metals at elevated temperature below the recrystallization temperature. The flow stress and rate of strain hardening are reduced within increasing temperature; therefore, lower forces are required than in cold working. See also *cold working* and *hot working*.

Warm-Setting Adhesive

Same as *intermediate-temperature-setting adhesive*.

Warp

A significant variation from the original true, or plane, surface or shape.

Warp (Composites)

The yarn running lengthwise in a woven fabric. A group of yarns in long lengths and approximately parallel. Also, in laminates, a change in dimension of a cured laminate from its original molded shape.

Warpage

Dimensional distortion in a plastic object.

Warpage (Metals)

(1) Deformation other than contraction that develops in a casting between solidification and room temperature. (2) The distortion that occurs during annealing, stress relieving, and high-temperature service.

Wash

(1) A coating applied to the face of a mold prior to casting. (2) An imperfection at a cast surface similar to a *cut*.

Wash Metal

Molten metal used to wash out a furnace, ladle, or other container.

Wash Primers

Wash primers are a special group of corrosion-inhibited coatings designed for use on clean metal surfaces. They are also known as "wash-coat primers," "metal conditioners," and "etch primers."

W

The most widely utilized primers consist of a two-part system that is prepared at the point of use by simple mixing of specified proportions. The base grind portion contains a corrosion-inhibiting pigment, basic zinc chromate (also known as zinc tetroxy chromate), and a small amount of talc extender ground in an alcohol solution of polyvinyl butyral resin. The reducer portion consists of phosphoric acid, alcohol, and water. When these are mixed, a slow chemical reaction ensues, resulting in partial reduction of the chromate pigment. The life of the mixed primer is usually 9–12 h. Single-package primers are now in use.

Wastage

Loss of section by general corrosion, erosion, or abrasion.

Watch

A watch is a portable timepiece. Its operation may be described as mechanical, electromechanical, or electronic.

Mechanical and Electromechanical Watches

In the mechanical watch, a mainspring in the barrel stores operating energy; the user retightens the spring daily by means of the winding stem. The wheel train advances at five increments per second under control by the escapement. From there, the dial train turns the minute and hour hands across the watch face. The momentarily engageable setting feature enables the user to position the hands in accordance with a primary clock. The wheel train of four pairs, with an overall turns ratio of 1:40,000, reduces the high torque from the barrel to a low value controllable by the escapement, yet sufficient to drive the dial train.

Among the variety of features incorporated into modern watches or self-winding mechanism, substitution of an electrochemical cell for the mechanical mainspring, several forms of electromechanical escapement in place of the balance-and-hairspring mechanism, and instead of hands over a dial, marked disks viewed through windows for readout, extended to days of the week in calendar watches.

A significant improvement in accuracy of the electromechanical watch is provided by relocation of the time base mechanism from the output of the wheel train to the input. Instead of deriving power from a mainspring, this arrangement obtains power from a dry cell. In place of an escapement at the end of the wheel train to control its incremental advance, a tuning fork as a resonant element in an electronic oscillator reciprocates an index finger that rapidly ratchets against a fine-toothed index wheel at the input to the wheel train to initiate its advance. Through these actions, the tuning fork with its drive circuit serves both as time base and as electrical-to-mechanical transducer, introducing power from the dry cell into the wheel train as an intermediate level of torque. In one style of this watch, the tuning fork vibrates at 360 Hz; in a smaller style, the fork resonates at 480 Hz.

Electronic Watches

When solid-state electronic integrated circuits became available in quantity from production stimulated by digital computers, the all-electronic watch became a commercial reality. And in it, a chain of binary dividers triggered from a crystal oscillator develops a train of second pulses. These pulses drive a digital counter or scaler, which develops minute and hour pulses to activate the digital display. An externally switched fast/slow capacitor in the crystal oscillator enables the user to set the readout in accordance with a standard time signal. The frequency of the crystal oscillator, typically in the tens of kilohertz, is chosen so that successive divisions by 2 produces desired 1 s pulse rate. For example, an oscillator with a 65.536 kHz crystal is followed by 16 binary dividers.

Power may be supplied by mercury or silver oxide cells, which are replaced annually, or lithium batteries, which operate for up to 5 years. To extend operating life, a solar cell may charge a nickel–cadmium power cell while the watch is illuminated.

One form of readout uses a light-emitting diode (LED). An assembly of these on a monolithic chip illuminates appropriate bars of a seven-segment display for each digit of the readout. The LED display itself illuminating and can be read in the dark. Because illumination of the readout consumes most of the power in an electronic watch, a liquid crystal display (LCD) is used where low-power consumption is a first consideration. The LCD readout depends for its indication on ambient illumination; the display is brighter in incident light. In it, glass plates confine a thin layer of liquid crystal. On the inside surface of the front plate, a transparent metallic coating in the seven-segment pattern receives signals from the readout counter. A highly reflective metal coating on the inside surface of the back plate operates at ground potential. When a bipolar high-frequency pulse train energizes a segment, the electric field established through the liquid causes that region to become turbulent and thereby scatter incident light so that the segment appears diffusely illuminated against a specularly illuminated background.

Water

Water is a chemical compound with two atoms of hydrogen and one atom of oxygen in each of its molecules. It is formed by the direct reaction:

$$2H_2 + O_2 \rightarrow 2H_2O \qquad (W.1)$$

of hydrogen with oxygen. The other compound of hydrogen and oxygen, hydrogen peroxide, really decomposes to form water, reaction:

$$2H_2O_2 \rightarrow 2H_2O + O_2 \qquad (W.2)$$

Water is also formed in the combustion of hydrogen-containing compounds, in the pyrolysis of hydrates, and an animal metabolism.

Gaseous State

Water vapor consists of water molecules that move nearly independently of each other. The atoms are held together in the molecule by chemical bonds, which are very polar—the hydrogen end of each bond is electrically positive relative to the oxygen. When two molecules near each other are suitably oriented, the positive hydrogen of one molecule attracts the negative oxygen of the other, and while in this orientation, the repulsion of the like charges is comparatively small. The net attraction is strong enough to hold molecules together in many circumstances and is called a hydrogen bond.

When heated above 1200°C, water vapor disassociates appreciably to form hydrogen atoms and hydroxyl free radicals, reaction:

$$H_2O \rightarrow H + OH \qquad (W.3)$$

These products recombine completely to form water when the temperature is lowered. Water vapor also undergoes most of the chemical reactions of liquid water and, at very high concentrations, even shows some of the unusual solvent properties of liquid water. At about 374°C, order vapor may be compressed to any density without liquefying, and at a density as high as 0.4 g/cm³, it can dissolve appreciable quantities of salt. These conditions of high temperature and pressure are found in efficient steam power plants.

Solid State

Ordinary ice consists of water molecules joined together by hydrogen bonds in a regular arrangement. This unusual feature is a result of the strong and directional hydrogen bonds taking precedence over all other intermolecular forces in determining the structure of the crystal. If the water molecules were rearranged to reduce the amount of empty space, their relative orientations would no longer be so well suited for hydrogen bonds. This rearrangement can be produced by compressing ice to pressures in excess of 14 MPa. Altogether, five different crystalline forms of solid water have been produced in this way, the form obtained depending upon the final pressure and temperature. They are all denser than water, and all revert to ordinary ice when the pressure is reduced.

Liquid State

The molecules in liquid water also are held together by hydrogen bonds. When ice melts, many of the hydrogen bonds are broken and those that remain are not numerous enough to keep the molecules in a regular arrangement. Many of the unusual properties of liquid water may be understood in terms of the hydrogen bonds that remain. As water is heated from 0°C, it contracts until 4°C is reached and then begins the expansion that is normally associated with increasing temperature. This phenomenon and the increase in density when ice melts both result from a breaking down of the open, hydrogen-bonded structure as the temperature is raised. The viscosity of water decreases 10-fold as the temperature is raised from 0°C to 100°C, and this is also associated with the decrease of icelike character of the water as hydrogen bonds are disrupted by increasing thermal agitation. Even at 100°C, the hydrogen bonds influence the properties of water strongly, for it has a high boiling point and a high heat of vaporization compared with other substances of similar molecular weight.

Properties

Pure water, either solid or liquid, is blue if viewed through a thickness of more than 2 m. The other colors often observed are due to impurities.

Water is an excellent solvent for many substances, but particularly for those that disassociate to form ions. Its principal scientific and industrial use as a solvent is to furnish a medium for purifying such substances and for carrying out reactions between them.

Among the substances that dissolve in water with little or no ionization and that are very soluble are ethanol and ammonia. These are examples of molecules that are able to form hydrogen bonds with water molecules, although, except for the hydrogen of the OH group in ethanol, it is the hydrogen of the water that makes the hydrogen bond. On the other hand, substances that cannot interact strongly with water, either by ionization or by hydrogen bonding, or only sparingly soluble in it. Examples of such substances are benzene, mercury, and phosphorus.

Water is not a strong oxidizing agent, although it may advance the oxidizing action of other oxidizing agents, notably oxygen. Examples are the oxidizing action of water itself or its reactions with the alkali and the alkaline earth metals, even in the cold.

Water is an even poorer reducing agent than oxidizing agent. One of the few substances that it reduces rapidly is fluorine.

Water reacts with a variety of substances to form solid compounds in which the water molecule is intact, but in which it becomes a part of the structure of the solid. Such compounds are called hydrates and are formed frequently with the evolution of considerable amounts of heat.

Water Absorption

The ratio of the weight of water absorbent by a material to the weight of the dry material.

Water Atomization

See *atomization* (powder metallurgy).

Water Break

The appearance of discontinuous film of water on a surface signifying nonuniform wetting and usually associated with a surface contamination.

Water Break Test

A test to determine whether a surface is chemically clean by the use of a drop of water, preferably distilled water. If the surfaces are clean, the water will break and spread; a contaminated surface will cause the water to bead.

Water Drip Test

See *drip test.*

Water Droplet Erosion

Erosion by water droplets usually entrained in a high-velocity stream of air or steam. Also see *steam erosion* as this term is also used in reference to a particular case of water droplet erosion.

Water Quenching

A quench in which water is the quenching medium. The major disadvantage of water quenching is its poor efficiency at the beginning or hot stage of the quenching process. See also *quenching.*

Water Repellants

Chemicals used for treating textiles, leather, and paper such as washable wallpaper, to make them resistant to wetting by water. They are different from waterproofing materials in that they are used where it is not desirable to make the material completely waterproof but to permit the leather or fabric to "breathe." Water repellents must not form acids that would destroy the material, and they must set the dyes rather than cause them to bleed on washing. There are two basic types: a durable type that resists cleaning and a renewable type that must be replaced after the fabric is dry-cleaned.

W

Zelan, a pyridinium-resin compound of DuPont, is representative of the first type. *Quilon*, of the same company, is used for paper, textiles, and glass fabric and forms a strong chemical bond with the surface of the material by an attachment of the chromium end of the molecule through the covalent bond to the negatively charged surface. It is a stearotochromic chloride. The second type is usually an emulsion of a mineral salt over which a wax emulsion is placed; the treatment may be a one-bath process or may be by two separate treatments. *Aluminum acetate* is one of the most common materials for this purpose. Basic aluminum acetate is a white, amorphous powder. It is only slightly soluble in water but is soluble in mineral acids. *Niaproof* is a concentrated aluminum acetate water-proofing textile, and *Romssit* and *Migasol* are similar materials. Zirconium acetate, a white, crystalline material and its sodium salt are used as water repellants. *Zirconyl acetate*, a light-yellow solution containing 13% ZrO_2, is used for both water repellancy and flame resistance of textile fibers. *Intumescent agents* are repellent coatings that swell and snuff out fire when they become hot. *Latex 744B*, a repellant of this type. It is the vinyl water emulsion compound that with pentaerythritol, dicyandiamide, and monosodium phosphate, is used in textiles, wallboard, and fiber tile.

Silicones have established their value as water-repellent finishes for a range of natural and synthetic textiles. The silicone polymers may be added as a solution, an emulsion, or by spraying a fine mist; alternatively, intermediates may be added that either polymerize in situ or attach themselves to the fibers. These techniques result in the pickup of 1%–3% of silicone resin on the cloth. Commercially, *di-chloromethylsilane* polymer is added as a solution or emulsion to a fabric; this is heated in the presence of a catalyst, such as zinc salt of an organic acid or an organotin compound, to condense the polymer to form a water-repellent sheath around each fiber. Fluorine-based polymers are also employed for treating fabrics. *Gore-Tex* is a polytetrafluoroethylene coating on nylon fabric; garments fashioned from this treated nylon are weatherproof and breathable. *Scotchgard*, from 3M, is a polymer containing fluoroalkyl groups that is effective for repelling both water and oil. *Scotchban*, from the same company, provides water, oil, and grease repellency to paper. Zepel B, from DuPont Co., is a fluoropolymer dispersion in water that does not promote yellowing or discoloration of coated outerwear. The *Quilon* series, also from DuPont, consists of greenish solutions of chrome complexes and iso-propanol that are water-repellant agents for packaging materials, nonwoven fabrics, and adhesive tapes. *Vinsol MM*, from Hercules, Inc., is a dark brown, free-flowing powder that is a sodium soap of a blend of Vinsol resin and a fatty acid. It was especially developed for use in masonry cements.

Water Side

The surface in contact with water, particularly a tube carrying water and heated on its exterior.

Water Side Corrosion

Severe corrosion on the water side of a tube. The term is commonly used in the context of externally fired steel tubes and steam-raising plants. Steel can develop a protective magnetite oxide on its surface when exposed to water of adequate purity but deviations from correct water conditions, for example, excessive oxygen, contamination with seawater or other chemicals can cause a severe attack, leading to large internal scabs, deep corrosion, and a form of hydrogen embrittlement.

Water Softeners

Chemical compounds used for converting soluble, scale-forming solids in water into insoluble forms. In the latter condition, they are then removed by setting or filtration. The hardness of water is due chiefly to the presence of carbonates, bicarbonates, and sulfates of calcium and magnesium; but many natural waters also contain other metal complexes that need special treatment for removal. Temporary *hard waters* are those that can be softened by boiling; permanent hard waters are those that require chemicals to change their condition. Sodium hydroxide is used to precipitate magnesium sulfate. Caustic lime is employed to precipitate bicarbonate of magnesium, and sodium aluminate is used as an accelerator. Barium carbonate may also be used. Prepared water softeners may consist of mixtures of lime, soda ash, and sodium aluminate, the three acting together. *Sodium aluminate* is a water-soluble, white powder melting at 1650°C, which is also used as a textile mordant, for sizing paper, and in making milky glass. Reynolds Metals Co. produces this material in flake form with iron content below 0.0056% for paint, water softeners, and paper coatings. Alum is used in settling tanks to precipitate mud, and zeolite is used extensively for filtering water. The liquids added to the washing water to produce fluffier textiles are *fabric softeners* and not water softeners. They are usually basic quaternary ammonium compounds such as distearyl dimethyl ammonium chloride with 16 and 18 carbon atoms, which are cationic, or positively charged. A thin coating is deposited on the negatively charged fabric, giving a lubricated cloth with a fluffy feel.

Water is also softened and purified with *ion-exchange agents*, which may be specially prepared synthetic resins. *Cation-exchange agents* substitute sodium for calcium and magnesium ions and produce soft waters. When the water is treated with a hydrogen derivative of a resin, the metal cations form acids from the salts. The carbonates are converted to carbonic acid that goes off in the air. When it is treated again with a basic resin derivative, or *anion-exchange agent*, the acids are removed. Water receiving this double treatment is equal to distilled water. Salt-cycle anion exchange substitutes chloride ions for other anions in the water, and when combined with cation exchange, it produces sodium chloride in the water in place of other ions.

BiQust, from Purolite Ltd., is an anion-exchange resin developed at Oak Ridge National Laboratories to remove radioactive pertechnetate from groundwater. It can also be used to treat perchlorate anion in industrial discharge waters. Ion-exchange resins are also being used to remove metals from metal-plating and electronic wastewaters.

In electrolytic ion exchangers for converting seawater to freshwater, the basic cell is divided into three compartments by two membranes, one permeable only to cations and the other only to anions. The sodium ions migrate toward the cathode, and the chlorine ions go toward the anode, leaving freshwater in the center compartment. *Ion-exchange membranes* for electrodialysis (salt splitting or separation), and also used in fuel cells, are theoretically the same as powdered exchange resins but with an inorganic binder. Such a membrane resin of the Armour Research Foundation is made by the reaction of zirconyl chloride and phosphoric acid, giving a chain molecule of zirconium–oxygen with side chains of dihydrogen phosphate. *Zeolites* are crystalline aluminosilicates that display cation-exchange properties. The most common zeolite for softening uses is zeolite 4A, a sodium aluminosilicate made by Union Carbide Corp.

Water Tube Boiler

See *boiler tube*.

Water Wall

See *boiler tube.*

Water Wash (Thermal Spraying)

The forcing of exhaust air and fumes from a spray booth through water so that the vented air is free of thermally sprayed particles or fumes.

Water-Base Adhesives

A water-base adhesive formulation can be classified as a solution in which the polymer is totally soluble in water or alkaline water, or as a latex, which consists of a stable dispersion of polymer in an essentially aqueous medium. Solids content in water-base dispersions can be as high as 50% by volume. Many of the materials used in water-base adhesives are also used in organic-solvent-base adhesives. In recent years, exposure to organic solvents has been increasingly controlled by federal regulations. Therefore, it is advantageous to use water-base rather than organic-solvent-base adhesives. The use of water as a solvent results in lower-cost, nonflammability, and lower toxicity.

Water-base adhesives have fairly good resistance to organic solvents. However, moisture resistance is usually poor, and adhesives are subject to freezing, which can affect properties. Although there are many advantages to using water-base adhesives, there is still a large and important market for solvent-base adhesives. Water-base adhesives are usually unsuitable for hydrophobic surfaces, such as plastics, because of poor wettability. In addition, water shrinks some substances, such as paper, textiles, and, cellulosics, and is corrosive to selected metals, such as copper. Organic-solvent-base adhesives are suitable for application on hydrophobic surfaces and are compatible with most metal surfaces.

Water-base adhesive solutions consist largely of natural adhesives. Materials that are soluble in water alone include animal glues, starch, dextran, methylcellulose, and polyvinyl alcohol. Materials that are soluble in alkaline water include casein, rosin, carboxymethylcellulose, shellac, vinyl acetate, and acrylate copolymers containing carboxyl groups. A water-base adhesive latex consists of a stable dispersion in an aqueous medium. Lattices can be classified as natural, synthetic, or artificial. A natural latex is formed from natural rubber. A synthetic latex is based on an aqueous dispersion of polymers obtained by emulsion polymerization. Adhesive families in this category include neoprene, styrene-butadiene rubber (SBR), nitrile rubber, polyvinyl acetate, polyacrylates, and polymethacrylates. An artificial latex is made simply by dispersing the solid polymer. Included in this category are natural rosin and its derivatives, synthetic butyl rubber, and reclaimed rubber.

Water-base adhesive families can be subdivided into materials used in making water-base adhesives only and materials used in making both water-base and organic-solvent-base adhesives. Included in the former classification are casein, dextran, starch, animal glues, polyvinyl alcohol, sodium carboxymethylcellulose, and sodium silicate. Included in the latter group are amino resins (urea and melamine formaldehydes), phenolic resins (phenol and resorcinol formaldehydes), polyacrylates, polyvinyl acetates, polyvinyl esters, neoprene, nitrile rubber, SBR, butyl rubber, natural rubber, and reclaimed rubber.

Water-base adhesive applications by demand include construction (15%), transportation (<1%), rigid bonding (<5%), packaging (60%), nonrigid bonding (<17%), consumer goods (<1%), and tapes (<3%). Rigid bonding applications for water-base adhesives include appliances, housewares, machinery, and electronics. Nonrigid bonding applications include fabrics, shoes, filters, and rugs.

Water-Extended Polyester

A casting formulation in which water is suspended in the polyester resin.

Waterjet/Abrasive Waterjet Machining

A hydrodynamic machining process that uses AII for velocity stream of water as a cutting tool. This process is limited to the cutting of nonmetallic materials when the jetstream consists solely of water. However, when fine abrasive particles are injected into the water stream, the process can be used to cut harder and denser materials. Abrasive waterjet machining has expanded the range of fluid jet machining to include the cutting of metals, glass, ceramics, and composite materials. Water pressures up to 410 MPa are used. The coherent jet of water is propelled at speeds up to approximately 850 m/s.

Waterline Corrosion

Corrosive attack at the water to air interface of a partially submerged structure. It usually results from the highly oxygenated conditions in the area and hence is a form of differential aeration corrosion, but additional factors like wave action or marine organisms may be involved.

Water-Soluble Plastics

Within the plastics industry, water-soluble materials offer a variety of desirable physical properties yet retain the advantages inherent in a water system. These advantages include ease of handling, negligible solvent costs, low toxicity, and low flammability.

There is no sharp dividing line between water-dispersible and water-soluble polymers. Many so-called water-soluble plastics form colloidal dispersions rather than true solutions. In this text, emulsions or dispersions of water-insoluble polymers are not discussed (such as acrylics, polyvinyl acetate, styrene butadiene, and polyvinyl butyral).

The water-soluble plastics can be roughly divided into two general classes: thermoplastic resins and thermosetting resins.

Thermoplastic Resins

These plastics are usually synthesized by addition polymerization techniques. That is, small units (or monomers) are joined together to develop the final molecular weight and polymer configuration. Rarely do these polymers develop into long straight chains; considerable branching often occurs. The molecular weight chemistry of side groups and extent of branching all determine the properties that are obtained. These plastics are available as white or light-colored powders or in solution. Films, moldings, and extrusions are also available based on some of the thermoplastic resins.

Alkali-Soluble Polyvinyl Acetate Copolymers

Polyvinyl acetate itself is water insoluble. However, copolymers are available in which final acetate is copolymerized with an acidic comonomer. Such products retain the organic solubility of polyvinyl acetate but are soluble in aqueous alkali. These polymers exhibit low viscosity in solution and deposit high gloss films that are water resistant, provided a volatile alkali such as ammonia is used. The use of a fixed alkali will result in a film with permanent water sensitivity. They generally possess good adhesion to cellulose and a wide variety of other surfaces.

Major uses include loom-finish warp sizes for dope dyed yarns, repulpable adhesives or sizes for paper and board, conditioning agents for masonry prior to painting, protective coatings for metals, and leveling agent and film former in self-polishing waxes.

Ethylene–Maleic Anhydride Copolymers

High-molecular-weight polymers have been prepared by copolymerizing ethylene and maleic anhydride. These resins are available either in "linear" form or cross-linked with either anhydride, free acid, or amide-ammonium salt side chains.

Major applications include general thickening and suspending in adhesives, agricultural chemicals, cleaning compounds, and ceramics. This resin is used as a thickener for latex and as a warp size for acetate filament.

Polyacrylates

Commercially important polymers are prepared by polymerizing either acrylic or methacrylic acid. Usually, these products are neutralized with bases to the salt form. Solution viscosity increases during neutralization. Cast films are hard, transparent, colorless, and somewhat brittle.

Polyacrylic acid itself is used as a warp size for nylon. The neutralized polymers (polyacrylates) are used in various coating and binding applications (ceramics, grinding wheels, etc.). Because of interesting solution properties, the polyacrylates are used as thickeners, flocculants, and sometimes as dispersants in applications such as ore processing, drilling muds, and oil recovery.

Polyethers

Two different polymer types are covered under this heading: polyoxyethylene (includes polyethylene glycols) and polyvinyl methyl ether and copolymers.

Polyoxyethylene (Polyethylene Glycol)

These resins are available over a wide molecular weight range. Low-molecular-weight members are slightly viscous liquids, whereas the medium-molecular-weight types (1,000–20,000) are waxy solids. Polymers up through this molecular weight level are known as polyethylene glycols. Extremely high-molecular-weight (several thousand to several million) homologues are also available. All types are soluble in water and in some organic solvents. Applications for the liquid and waxy solids include lubricants (rubber molds, textile fibers, and metal), bases for cosmetic and pharmaceutical preparations, and chemical intermediates for further reaction. The very high-molecular-weight types are useful principally as thickeners in many application areas.

Polyvinyl Methyl Ether

This unique family of vinyl polymers shows inverse solubility in that the resins precipitate out above 35°C. They do redissolve upon cooling, and the addition of low-molecular-weight alcohols increases the solubility in water and raises the precipitation temperature. Higher homologues are available that are water insoluble and quite tacky. The products exhibit pressure-sensitive adhesiveness coupled with good cohesive strength and high wet tack. Copolymers are available that contain maleic anhydride to modify physical properties, particularly solubility, or tolerance to water and organic solvents and ease of insolubilization. Major uses take advantage of properties such

as pressure-sensitive characteristics (adhesives), tackiness (various latex systems), thickening active, heat-sensitizing lattices for dip forming, and binding power (pigments).

Polyvinyl Alcohol

These resins are available commercially in a wide range of types, which vary in viscosity and chemical composition. Polyvinyl alcohol exhibits good water solubility, high resistance to organic solvents, oils, and greases, high tensile strength, adhesion, and flexibility. In addition, the polymer is resistant to oxidation and in film form is an excellent barrier for various gases. Certain types exhibit surface activity in solution and all types are soluble in both acid and alkaline media. The resins can be cross-linked by borax and numerous organic and inorganic agents to produce thickening or even insolubilization.

In adhesives, polyvinyl alcohol contributes machinability, viscosity control, specific adhesion, and in some cases remoistenability. Other major uses include paper coating and sizing (for increased strengths, ink hold-out, and grease resistance), textile sizing, wrinkle-resistant finishes (wash-and-wear fabrics in conjunction with thermosetting resins), polyvinyl acetate emulsion polymerization (protective colloid), binder (for nonwoven ribbons, filters, etc.), film (release agent in polyester and epoxy molding and water-soluble packaging), cement additive (for improved strength, toughness, and adhesion), and photosensitive coating (in the graphic arts industry).

Polyvinyl Pyrrolidone

Polyvinyl pyrrolidone (PVP) exhibits good solubility in both water and various organic solvents. A nontoxic material and tacky substance when wet, the polymer is a dispersant, suspending agent, and an adhesive component for bonding difficult surfaces.

Major uses include cosmetic preparations (hair sprays, etc.), tablet binding and coating, detoxifying of dyes, drug, and chemicals, beverage clarification, and specialty textile and paper applications involving sizing, dyeing, and printing.

Copolymers are also available (like PVP-vinyl acetate) that have some advantages over the homopolymers in heat sealability, pressure-sensitive adhesiveness, and other properties.

Polyacrylamide

These high-molecular-weight polymers are soluble in both cold and hot water and in selected organic solvents. The resin is an efficient thickener and by reaction can be changed in physical and chemical properties.

In addition to general uses for water-soluble resins, these resins have shown an outstanding ability to flocculate fines and increase the filtration rate of slurries. Consequently, polyacrylamide is used in ore processing and in other such systems where dispersed materials are encountered.

Styrene–Maleic Anhydride

Copolymers of these two monomers are soluble in some organic solvents and alkaline water. Styrene–maleic anhydride resins produce viscous and stable aqueous solutions. This resin is a strong polyelectrolyte. It is used as a textile warp size, paper coating, and static-electricity conductor. The polymer is also used in alkaline latex systems as a protective colloid, emulsifier, pigment dispersant, and filming aid.

Cellulosic Derivatives

Various commercial derivatives are prepared from alpha cellulose, which is obtained from several plant sources. One class of derivatives is the water-soluble ethers. These products produce viscous aqueous solutions. All have some resistance to organic solvents, are hygroscopic, and are difficult to insolubilize.

Major industries that use these polymers as well as the water-soluble synthetic resins include food, pharmaceutical, cosmetic, textile, paper, petroleum (drilling muds), ceramic, paint, emulsion polymerization, and leather.

Hydroxyethylcellulose

This polymer is manufactured by reacting alkali cellulose with ethylene oxide. It can be water-soluble or only alkali-soluble depending on the extent of reaction. The alkali-soluble types possess the advantage of increased water resistance in deposited films. This polymer is somewhat intermediate in properties between methylcellulose and sodium carboxymethylcellulose. It is a protective colloid and relatively insensitive to the inclusion of multivalent ions in solution. The polymer is soluble in both hot and cold water, nonionic, but depolymerized by strong acids.

Water-soluble hydroxyethylcellulose is used in polyvinyl acetate emulsions as a stabilizer and in latex paints as a thickener and leveling agent.

Methylcellulose

Methylcellulose exhibits inverse water solubility in that it is more soluble at low temperatures than at high temperatures. It is nonionic in solution and is a very efficient thickener.

A major uses of latex paints (both polyvinyl acetate and acrylic types). Methylcellulose thickens the paint and contributes to good brushing characteristics as well. Other uses include bulking in laxatives and binding and thickening in cosmetics and pharmaceuticals.

Sodium Carboxymethylcellulose

This polymer is soluble in both hot and cold water. It exhibits good thickening action and suspending ability for particulates. Since solutions are ionic in character, they are somewhat sensitive to pH shifts and salt additions. Major uses include soil suspension in synthetic detergents and viscosity control in oil-well drilling muds.

Thermosetting Resins

A number of thermosetting resins are available in water solutions or in water-soluble form. These are principally the addition reaction products of formaldehyde with urea, phenolic, or melamine. Resorcinol and thiourea may also be reacted with formaldehyde to form water-soluble precondensates, although these have not attained the volume of the three main classes defined earlier.

These residents develop high molecular weight by a condensation reaction. The properties may change as the reaction proceeds. A-stage resins are those in which the degree of polymerization is minor. In some cases, the degree of polymerization of such that only dimers, trimers, or similar small units are prepared. These products are water soluble or at least can tolerate the addition of significant amounts of water.

Should the reaction proceed further, the polymers enter an area roughly defined as B-stage, in which they will tolerate addition of only small amounts of water but are soluble in certain organic solvents.

Manufacturers of thermosetting resins carry the polymerization reaction to the A- or B-stage. Further reaction (to the C-stage) is carried out by consumers of these resins.

As the condensation reaction proceeds and the molecular weight builds to form a rigid, three-dimensional system, the polymers reach a point where they will not dissolve in organic solvents and are then termed *cured* (or C-staged). Further heating of the resin beyond this point may establish additional cross-links, but the physical properties do not change drastically. Once the resin has reached the C-stage, excessive heating leads to chemical breakdown of the material.

Thermosetting water-soluble polymers are treated with the and/or catalyst to advance the cure after deposition on a particular surface or within a particular structure. Thus, the water-solubility feature is important in that it allows easy manipulation without the cost and hazards of organic systems. However, water is an essential ingredient when cross-linking cellulose with low-molecular-weight thermosetting resins. The general characteristics of a fully cured, water-soluble, thermosetting resin are similar to those obtained by curing B-stage varnishes or molding compounds.

In many areas, the water-soluble thermosetting resins compete with one another. Certain ones may be preferred in an industry or in a certain particular application because of specific properties or cost. Generally, these resins offer high-temperature stability and hardness coupled with water and solvent resistance. Resistance to either acids or bases can also be obtained.

Cyclic Thermosetting Resins

In this category are cyclic ethylene urea–formaldehyde resins and triazones, which can be obtained by cyclization of dimethylolurea with a primary amine (usually ethylamine) and then adding 2 mol of formaldehyde. The cyclic thermosetting resins were developed principally for textile applications because they do not react or polymerize with themselves (as do other water-soluble thermosetting resins) but to react with the hydroxyls in cellulose through the methylol groups. These resins are called cellulose reactants and are by far the largest class of resins used in the wash-and-wear treatment of fabrics. Such resins impart crease resistance, wrinkle recovery, stiffness, tensile strength, water repellency, and good resistance to yellowing by chlorine-containing bleaches. In addition, there are increases in resiliency, dimensional stability, and permanent texturizing.

Cyclic ethylene urea–formaldehyde resins have the advantage over triazone in better color stability, absence of odor, and scorch resistance. Triazone resins have gained prominence particularly because they are more resistant to the effects of chlorine and are somewhat cheaper. Triazones are used principally to develop wash-and-wear properties on white cotton.

Melamine–Formaldehyde

In general, melamine–formaldehyde (MF) resins are the most expensive of the water-soluble thermosetting types. They possess good color, lack of odor, high abrasion resistance, and high resistance to alkali.

The major uses in decorating laminants for surfacing of wood, paper, and other products. Other uses include binding of rock wool and glass wool for thermal insulation, finishing of nylon for stiffness and resilience, and importing dimensional stability to wool and some cellulosics.

Phenol–Formaldehyde

This very popular class of water-soluble thermosetting resins is intermediate in cost between the MF and the urea–formaldehyde (UF) types. They possess good water resistance, toughness, and acid resistance, although they are somewhat poorer than the MF resins in color, odor, and flame resistance. In most other physical and chemical properties, they are superior to the UF types.

Applications include laminates (including plywood and fabrics), grinding wheels, thermal insulation, battery separators, brake linings, and foundry uses.

Urea–Formaldehyde

These resins are the lowest cost as a class of the three general types but still cure to hard and somewhat brittle resins that have many desirable properties. The UF resins have an added plus in what they can cure at room temperature with suitable catalysts, whereas both the melamine–formaldehyde and phenol–formaldehyde types normally require temperatures in the neighborhood of 149°C to develop their full properties. The UF resins suffer somewhat in comparison with the other two types in poorer water resistance, less toughness, and poorer resistance to cyclical changes in temperature or water exposure.

UF resins are used in plywood because of ease of handling and lower temperature of cure, on paper for increased wet strength in air filters, and on certain rayon fabrics for improved stabilization and water resistance. The UF resins are used as insolubilizers for hydroxyl-containing polymers and in many of the general application areas for water-soluble, thermosetting resins.

Waterline Corrosion

Corrosive attack at the water to air interface of a partially submerged structure. It usually results from the highly oxygenated conditions in the area and hence is a form of differential aeration corrosion but additional factors like wave action or marine organisms may be involved.

Watt

A unit of electrical power equal to 1 J/s. Symbolized W.

Wave Number

The number of waves per unit length. Commonly used in infrared and Raman spectroscopy, the wave number is expressed as the reciprocal of the wavelength. The usual unit of wave number is the reciprocal centimeter (cm^{-1}).

Wave Soldering

An automatic soldering process where work parts (usually printed circuit boards) are automatically passed through a wave of molten solder. See also *dip soldering* and *soft soldering*.

Wavelength-Dispersive Spectroscopy (WDS)

A method of x-ray analysis that employs a crystal spectrometer to discriminate characteristic x-ray wavelengths. Compare with *energy*-dispersive *spectroscopy*.

Waviness

A wavelike variation from a perfect surface, generally much larger and wider than the roughness caused by tool or grinding marks. See the term *roughness*.

Wax

(1) Any of a group of substances resembling beeswax in appearance and character, and in general distinguished by their composition of esters and higher alcohols, and by their freedom from fatty acids. (2) Preferred lubricant for pressing cemented carbide powder mixtures.

Wax

Wax is a general name for a variety of substances of animal and vegetable origin, which are fatty acids in combination with higher alcohols, instead of with glycerin as in fats and oils. They are usually harder than fats, less greasy, and more brittle, but when used alone to not mold as well. Chemically, the waxes differ from fats and oils in that they are composed of high-molecular-weight fatty acids with high-molecular-weight alcohols. The most familiar wax is beeswax from the honeybee, but commercial beeswax is usually greatly mixed or adulterated. Another animal wax is spermaceti from the sperm whale. Vegetable waxes include Japan wax, jojoba oil, candelilla, and carnauba wax.

Mineral waxes include paraffin wax from petroleum, ozokerite, ceresin, and montan wax. The mineral waxes differ from the true waxes and are mixtures of saturated hydrocarbons.

The animal and vegetable waxes are not plentiful materials and are often blended with or replaced by hydrocarbon waxes or waxy synthetic resins. However, waxes can be made from common oils and fats by splitting off the glycerin and re-esterifying selected mixtures of the fatty acids with higher alcohols.

Types and Uses

Some plastics have wax characteristics and may be used in polishes and coatings or for blending with waxes. Polyethylene waxes are light-colored, odorless solids of low modular weight, of to about 6000. Mixed in solid waxes to the extent of 50%, and in liquid waxes up to 20%, they add gloss and durability and increase toughness. In emulsions, they add stability.

Waxes are employed in polishes, coatings, leather dressings, sizings, waterproofing for paper, candles, carbon paper, insulation, and varnishes. They are softer and have lower melting points than resins; they are soluble in mineral spirits and in alcohol and insoluble in water.

Synthetic waxes are used in liquid floor waxes, temporary corrosion protection, release agents, and as a melting point booster. There is a micronized polyethylene wax that is a processing and performance additive for adhesives, coatings, color concentrates, cosmetics, inks, lubricants, paints, plastics, and rubber. It can also be constituted from low-molecular-weight homopolymer, oxidized homopolymer, or as a copolymer. Another type is a methylene polymer used to blend with vegetable or paraffin waxes to increase the melting point, strength, and hardness. This is a mixture of terphenyls. It is a light-buff, waxy solid, highly soluble in benzene, and with good resistance to heat, acids, and alkalies. It is used to blend with natural waxes in candles, coatings, and insulation. Waxes are not digestible, and the so-called edible waxes used as water-resistant coatings for cheese, meats, and dried fruits are not waxes, but modified glycerides. One is a white, odorless, tasteless waxy solid melting at 40 °C, and is an acetylated monoglyceride of fatty acids.

Microcrystalline waxes are used for the vacuum impregnation of inorganic-filled, organic-bonded electrical insulation and coatings for ceramic capacitors and other electronic components. The wax is chosen because of its low moisture permeability.

Waxy emulsions have been widely used as binders for dry-press mixes and glaze suspensions. A high melting point paraffin also will make an excellent binder for dry-press granules. The paraffin is melted, added to the body, and then thoroughly incorporated by means of a heated muller-type mixer.

Ordinary paraffin also can be used to bond ceramic parts to steel plates for attachment to magnetic chucks during grinding.

Wax Pattern

A precise duplicate, allowing for shrinkage, of the casting and required gates, usually formed by pouring or injecting molten wax into a die or mold. See also *investment casting.*

Wax Pattern (Thermit Welding)

Wax molded around the parts to be welded to the form desired for the completed weld. See also the term *thermit welding.*

Wax Vent

A narrow passage through the core for a casting that allows venting of any gases formed in the core during the casting process. It is formed by incorporating in the core a string thickly coated with wax. When the core is baked, the wax evaporates or melts into the core and the string either burns or is pulled out to leave the passage.

Wear. (1) Damage to a solid surface, generally involving progressive loss of material, due to a relative motion between that surface and a contacting surface or substance. (2) The removal of material from a surface by frictional or other mechanical effects as it moves relative to another contacting surface. Two broad categories are recognized (see *abrasive wear* and adhesive wear and also *fatigue* and *pitting.* Compare with *surface damage).*

Wear Debris

Particles that become detached in a wear process.

Wear Pad

In forming, an expendable pad of rubber or rubberlike material of nominal thickness that is placed against the diaphragm to lessen the wear on it. See also *diaphragm* (3).

Wear Rate

The rate of material removal or dimensional change due to *wear* per unit of exposure parameter – for example, quantity of material removed (mass, volume, thickness) at unit distance of sliding or unit time.

Wear Rate (of Seals)

The amount of seal-surface wear, stated in terms of mils, worn in some designated time period. One commonly used unit is mils per hundred hours.

Wear Resistance

The resistance of a body to removal of material by the wear processes, expressed as the reciprocal of the wear rate. Wear resistance is a function of the conditions under which the wear process takes place. These conditions should always be carefully specified.

Wear-Resistant Steel

Many types of steel have wear-resistant properties, but the term usually refers to high-carbon, high-alloy steels used for dies, tooling, and parts subject to abrasion and for wear-resistant castings. They are generally cast and ground to shape. They are mostly sold under trade names for specific purposes. The excess carbon of the steels is in spheroidal form rather than as graphite. One of the earlier materials of this kind for drawing and forming dies is **Adamite**. It is a chromium-nickel-iron alloy with up to 1.5% chromium, nickel equal to half that of the chromium, and from 1.5% to 3.5% carbon with silicon from 0.5% to 2%.

The Brinell hardness ranges from 185 to 475 as cast, with tensile strengths up to 861 MPa. The softer grades can be machined and then hardened, but the hard grades are finished by grinding. Others have about 13% chromium, 1.5% carbon, 1.1% molybdenum, 0.70% cobalt, 0.55% silicon, 0.50% manganese, and 0.40% nickel. They are used for blanking dies, forming dies, and cams. T15 tool steel, for extreme abrasion resistance in cutting tools, is classed as a super high-speed steel. It is 13.5% tungsten, 4.5% chromium, 5% cobalt, 4.75% vanadium, 0.50% molybdenum, and 1.5% carbon. Its great hardness comes from the hard vanadium carbide and the complex tungsten-chromium carbides, and it has full red-hardness. The property of abrasion or wear resistance in steels generally comes from the hard carbides, and is thus inherent with proper heat treatment in many types of steel.

Wear Scar

The portion of a solid surface that exhibits evidence that material has been removed from it due to the influence of one or more wear processes.

Wear Transition

Any change in the wear rate or in the dominant wear process occurring at a solid surface. Wear transitions can be produced by an external change in the applied conditions (e.g., load, velocity, temperature, or gaseous environment) or by time-dependent changes (aging) of the materials and restraining fixtures in the *tribosystem.*

Weathering

Exposure of materials to the outdoor environment.

Weathering Steels

Copper-bearing *high-strength low-alloy steels* that exhibit high resistance to atmospheric corrosion in the unpainted condition.

Weave

The particular manner in which a fabric is formed by interlacing yarns and usually assigned a style number. See also the terms *basketweave* (composites), *crowfoot satin, eight-harness satin, leno weave*, and *plain weave.*

Weave Bead

A weld bead made with oscillation transverse to the axis of the weld. Contrast with *stringer bead.*

Weaving (of Weld)

The deposition of weld metal with significant side to side deflection of the arc or flame, together with any filler wire, relative to the main line of progress.

Web (Metals)

(1) A relatively flat, thin portion of a forging that affects an interconnection between ribs and bosses; a panel or wall that is generally parallel to the forging plane. See also *rib.* (2) For twist drills and reamers, the central portion of the tool body that joins the lands. (3) A plate or thin portion between stiffening ribs or flanges, as in an I-beam, H-beam, or other similar section.

Web (Plastics)

A thin plastic sheet in process in a machine. The molten web is that which issues from the die. The substrate web is the substrate being coated.

Webbing

Filaments or threads that sometimes form when adhesively bonded surfaces are separated. See also *legging* and *stringiness*. Compare with *teeth*.

Wedge Effect

The establishment of a pressure wedge in a lubricant. See also *wedge formation* (2).

Wedge Formation

(1) In sliding metals, the formation of a widget or wedges of plastically sheared metal in local regions of interaction between sliding surfaces. This type of wedge is known as a *prow*. It is similar to a *built-up edge*. (2) In hydrodynamic lubrication, the establishment of a pressure gradient in a fluid flowing into a converging channel. This is also known as *wedge effect*.

Weepage

A minute amount of liquid leakage by a seal. It is commonly considered to be a leakage rate of less than one drop of liquid per minute.

Weft

The transverse threads or fibers in a woven fabric. Those fibers running perpendicular to the warp. Also called *fill*, *filling yarn*, or *woof*.

Weight Percent

Percentage composition by weight. Contrast with *atomic percent*.

Weld

(1) A localized coalescence of metals or nonmetals produced either by heating the materials to suitable temperatures, with or without the application of pressure, or by the application of pressure alone and with or without the use of filler material. (2) A union formed between two components as a result of local mutual fusion or diffusion resulting from heating and, or, pressure. Where additional molten material is added to form or fill a joint the term "weld" is still used provided there is fusion of the main components and provided the melting point of the filler material is approximately similar to the melting point of the parent materials. Where the filler melting point is significantly lower, the terms soldering and brazing are appropriate. In these cases, there is no melting of the parent metals although there may be diffusion between the various components of the joint. Many welding processes are in use, see friction-, fusion-, arc-, gas-, GTA (TIG)-, GMA (MIG)-, welding etc.

Weldability

A specific or relative measure of the ability of a material to be welded under a given set of conditions. Implicit in this definition is the ability of the completed weldment to fulfill all functions for which support was designed.

Weldability Index

See *weld zone*.

Weldability Test

Any test for establishing whether a satisfactory joint can be produced by a particular welding procedure.

Weld Bead

The filler material deposited in a single pass of the welding arc or torch across a joint. It may also be termed a *run*.

Weldbonding

A joining method that combines resistance spot welding or resistance seam welding with adhesive bonding. The adhesive may be applied to a faying surface before welding or may be applied to the areas of sheet separation after welding.

Weld Brazing

A joining method that combines resistance welding with brazing.

Weld Crack

A crack in weld metal. See also *crater crack, root crack, toe crack,* and *underbead crack*.

Weld Cracking

Cracking that occurs in the weld metal. See also *cold cracking, hot cracking, lamellar tcaring,* and *stress-relief cracking*.

Weld Decay

A form of intergranular corrosion and resultant cracking occurring in some austenitic stainless steels and certain nickel-based alloys that have been heated in the approximate temperature range 500–900 °C. This causes deposition of chromium carbides in the grain boundaries. The adjacent material is thereby depleted in chromium and consequently prone to corrosion in appropriate environments. Material in this condition is termed *sensitized*. The attack is a form of electrolytic corrosion (*see electrochemistry*) in which small anodic areas at the grain boundary are surrounded by large cathodic grains. Consequently, although the volume of material corroded may not be large, the rates of penetration and hence cracking can be very rapid. Most welding processes expose a narrowband of the heat-affected zone to the damaging temperature range so that weld decay can be observed as a line of cracking close to the edge of welds in susceptible steels. However, any heating in the appropriate temperature range can cause the same form of microstructural damage and the consequent attack may, perhaps confusingly, still be described as weld decay. Attack can be prevented by modification of the composition or other means, see *stabilized stainless steel*. No stress is necessary for corrosion of sensitized steel so the mechanism is not a form of stress corrosion. See the term *sensitization*.

Weld-Delay Time (Resistance Welding)

The amount of time the beginning of welding current is delayed with respect to the initiation of the forge-delay timer in order to synchronize the forging force with welding current flow.

Welder

A person making a weld. Colloquially, a machine for welding.

Weld Deposit

Fused weld metal comprising mainly filler but including some fused parent metal.

Weld Face

The exposed surface of the last metal to be deposited.

Weld Filler

Material added to a joint story welding. It is usually implicit that the filler is fused to become a load carrying component.

Weld Fusion Zone

Parent metal that has fused.

Weld Gage

A device for checking the shapes and sizes of welds.

Weld Heat Affected Zone

See *heat affected zone*.

Weld-Heat Time (Resistance Welding)

The time from the beginning of welding current to the beginning of post-heat time.

Welding

(1) Joining two or more pieces of material by applying heat or pressure, or both, with or without filler material, to produce a localized union through fusion or recrystallization across the interface. The thickness of the filler material is much greater than the capillary dimensions encountered in *brazing*. (2) May also be extended to include brazing and soldering. (3) In tribology, adhesion between solid surfaces in direct contact at any temperature.

Welding Alloys

Welding alloys are usually of the form of rod, wire, or powder used for either electric or gas welding, for building up surfaces, or for hardfacing surfaces. In the small sizes in continuous lengths, welding alloys are called welding wire. Nonferrous rods used for welding bronzes are usually referred to as brazing rods, as the metal to be welded is not fused when using them. Welding rods may be standard metals or special alloys, coated with a fluxing material or uncoated, and are normally in diameters from 0.239 to 0.635 cm. Compositions of standard welding rods follow the specifications of the American Welding Society. Molded carbon, in sizes from 0.318 to 2.54 cm in diameter, is also used for arc welding. Low-carbon steel rods for welding cast iron and steel contain less than 0.18% carbon. High-carbon rods produce a hard deposit that requires annealing, but these are used for producing a hard filler. High-carbon rods, with 0.85% to 1.10% carbon, will give deposits with an initial hardness of 575 Brinell, whereas high-manganese rod deposits will be below 200 Brinell but will work-harden to above 500 Brinell. For high-production automatic welding operations, carbon-steel wire may have a thin coating of copper to ease operation and prevent spattering. Stainless steel rods are marketed in various compositions. There are welding rods that comprise a range of stainless steels with either titania-lime or straight-lime coatings. Stainless C is an 18-8 type of stainless steel with 3.5% molybdenum. Aluminum-weld is a 5% silicon aluminum rod for welding silicon-aluminum alloys, and the Tungweld rods, for hard surfacing, are steel tubes containing fine particles of tungsten carbide. Kennametal KT-200 has a core of tungsten carbide and a sheathing of steel. It gives coatings with a hardness of Rockwell C63. Chromang, for welding high-alloy steels, is an 18-8 stainless steel modified with 2.5 to 4% manganese.

Welding rods with grades of high-manganese steel give hardnesses from 500 to 700 Brinell, and high-speed steel rods are used for facing worn cutting tools; others are used for facing surfaces requiring extreme hardness and have the alloy granules in a soft steel tube. The welded deposit has a composition of 30% chromium, 8% cobalt, 8% molybdenum, 5% tungsten, 0.05% boron, and 0.02% carbon.

There is a group of welding alloys made especially for welding machines. They are, in general, sintered tungsten or molybdenum carbides, combined with copper or silver, and are electrodes for spot welding rather than welding rods. Tungsten electrodes may be pure tungsten, thoriated tungsten, or zirconium tungsten, the latter two used for direct-current welding. Thoriated tungsten gives high arc stability, and the thoria increases the machinability of tungsten.

Zirconium tungsten provides adhesion between the solid electrode and the molten metal to give uniformity in the weld.

Thermit is a mixture of aluminum powder and iron oxide used for welding large sections of iron or steel or for filling large cavities. The process consists of the burning of the aluminum to react with the oxide, which frees the iron to molten form. To ignite the aluminum and start the reaction, a temperature of about 1538 °C is required, which is reached with the aid of a gas torch or ignition powder, and the exothermic temperature is about 2538 °C. Cast iron thermit, used for welding cast iron, is thermit with the addition of about 3% ferrosilicon and 20% steel punchings. Railroad thermit is thermit with additions of nickel, manganese, and steel.

The Stellite hardfacing rods are cobalt-based alloys that retain hardness at red heat and are very corrosion resistant. The grades have tensile strengths to 723 MPa and hardnesses to Rockwell C52. Inco-Weld Ais welding wire for stainless steels and for overlays. It contains 70% nickel, 16% chromium, 8% iron, 2% manganese, 3% titanium, and not more than 0.07% carbon. The annealed weld has a tensile strength of 551 MPa with elongation of 12%. Nickel welding rod is much used for cast iron, and the operation is brazing, with the base metal not melted. Nickel silver for brazing cast iron contains 46.5% copper, 43.4% zinc, 10% nickel, 0.10% silicon, and 0.02% phosphorous. The deposit matches the color of the iron. Colmonoy 23A is a nickel alloy welding powder for welding cast iron and filling blow holes in iron casting by torch application. It has a composition of 2.3% silicon, 1.25% boron, 0.10% carbon, not over 1.5% iron, and the balance nickel, with a melting point of 1066 °C. For welding on large structures where no heat treatment of the weldment is possible, the welding rods must have balanced compositions with no elements that form brittle compounds. Rockide rods are metal oxides for hard surfacing.

Welding Consumables

The materials used in producing a welded joint. The term includes fillers, electrodes, fluxes, etc., and possibly electricity and gases etc., but not parent material.

Welding Current (Automatic Arc Welding)

The current in the welding circuit during the making of a weld, but excluding upslope, downslope, start, and crater fill current.

Welding Current (Resistance Welding)

The current in the welding circuit during the making of a weld, but excluding preweld or postweld current.

Welding Cycle

The total series of activities and operations involved in a machine, or less commonly an operative, producing a single weld and returning to the start condition in the making of a weld.

Welding Electrode

See preferred term *electrode* (welding).

Welding Flux

Material applied to the weld zone primarily to chemically clean the surface, form a slag with impurities and prevent oxidation. Fluxes may contain components with other characteristics. See *electrode coating*.

Welding Force

See preferred terms *electrode force* and *platen force*.

Welding Generator

A generator used for supplying current for welding.

Welding Ground

Same as *work lead*.

Welding Head

The part of a welding machine or automatic welding equipment in which a welding gun or torch is incorporated.

Welding Leads

The electrical cables that serve as either *work lead* or *electrode lead* of an arc welding circuit.

Welding Machine

Equipment used to perform the welding operation. For example, spot welding machine, arc welding machine, seam welding machine, etc.

Weld (ing) Position

The orientation of components and weld deposits during the welding operation. A very wide range of terms and definitions have been used, often with some confusion. There are a limited number of positions by reference to the combination of weld slope and weld rotation. See horizontal-vertical, flat, vertical and overhead, these are the basic welding positions, and also inclined. Other authorities such as the American Welding Society Publications "Welding Terms and Symbols" offer alternative and more numerous definitions.

Welding Position

See also *flat position, horizontal position, horizontal fixed position, horizontal rolled position, inclined position, overhead position*, and *vertical position*.

Welding Pressure

The pressure exerted joins the welding operation on the parts being welded. See also *electrode force* and *platen force*.

Welding Primer

A painter or other coating applied to protect material until such time as it may be welded. It is implicit that the coating does not have any adverse effect on the weld and hence does not have to be removed prior to welding.

Welding Procedure

The detailed methods and practices, including joint preparation and welding procedures, involved in the production of a *weldment*.

Welding Process

A materials joining process that produces coalescence of materials by heating them to suitable temperatures, with or without the application of pressure or by the application of pressure alone, and with or without the use of filler metal.

Welding Rectifier

A device in a welding machine for converting alternating current to direct current.

Welding Rod

Material in the form of a rod that provides filler material. It may be bare solid rod, as used for most gas welding or coated rod or filled tube as used for most electric arc welding. See *electrode*. (2) It is used for welding or brazing which does not conduct the electrical current, and which may be either fed into the weld pool or preplaced in the joint.

Weld (ing) Sequence

The order in which a series of welding operations is performed. In particular, the order and direction in which a multi-run weld is deposited.

Welding Stress

Residual stress caused by localized heating and cooling during welding.

Welding Technique

The way in which the operative holds and moves the welding equipment, the electrode, or blowpipe, filler rod, etc., as they form the weld.

Welding Technology

Many terms are used in a specialized sense in welding. Those with a metallurgical aspect are included at their alphabetical location in this book. BS 499 "Welding Terms and Symbols" and American Welding Society Publication A.30-80 "Welding Terms and Symbols" detail many more terms of interest to the welding specialist.

Welding Tip

A welding torch tip designed for welding. See the term the *welding torch* (oxyfuel gas).

Welding Torch (Arc)

A device used in the gas tungsten and plasma arc welding processes to control the position of the electrode, to transfer current to the arc, and to direct the flow of shielding and plasma gas. See also *gas tungsten arc welding* and *plasma arc welding*.

Welding Torch (Oxyfuel Gas)

A device used in oxyfuel gas welding, torch brazing, and torch soldering for directing the heating flame produced by the controlled combustion of fuel gases. See also oxy*fuel gas welding*.

Welding Transformer

A transformer used for supplying current for welding.

Welding Voltage

See *arc voltage*.

Welding Wire

See preferred terms *electrode* and *welding rod*.

Weld Interface

The interface between weld metal and base metal in a fusion weld, between base metals in a solid-state weld without filler metal, or between filler metal and base metal in a solid-state weld with a filler metal and in a braze.

Weld Interval (Resistance Welding)

The total of all heat and cool time when making one multiple-impulse weld.

Weld-Interval Timer

A device used in resistance welding to control heat and cool times and weld interval when making multiple-impulse welds singly or simultaneously.

Weld Junction

The interface between weld metal and heat affected zone.

Weld Length

See preferred term *effective length of weld*.

Weld Line (Metals)

See preferred term *weld interface*.

Weld Line (Plastics)

The mark visible on a finished plastic part made by the meeting of two flow fronts of plastic material during molding. Also called weld mark, flow line, knit line, or stria.

Weldment

An assembly whose component parts are joined by welding.

Weld Metal

The portion of a weld that has been melted during welding or otherwise manipulated to form a joint.

Weld Metal Area

The area of the *weld metal* as measured on the cross section of a weld.

Weld Nugget

The weld metal in spot, seam, or projection welding. It is often termed the zone of fusion but little or none of the metal may have been above its melting point. See the terms *nugget* and *resistance spot welding*.

Weld Pass

A single progression of a welding or surfacing operation along a joint, weld deposit, or substrate. A single movement of the welding equipment moves across the component and which material is continuously fused. Also the material is fused or deposited in a single pass. The result of a pass is a weld bead, layer, or spray deposit.

Weld Penetration

See preferred terms *joint penetration* and *root penetration*.

Weld Prep (Aration)

Generally, any activity associated with preparing an item for welding. More specifically the formation of the faces of a component to receive the weld deposit. For example, the edges of two plates to be welded together have a Vee preparation in which a single chamfer is formed on each edge.

Weld Procedure

The total prescribed system for performing a weld. It will define all significant variables such as materials, joint geometry, equipment set up, electrical and gas settings, preheat, interpass, and post-heating temperatures, etc.

Weld Root

Of a weld to be made, it is the area of the parent materials where the first deposit is to be made. Of a completed weld, it is the face of the first weld run that is remote from the operator.

Weld Rotation

The angle, relative to the vertical, of a line bisecting the angle between the fusion faces. Compare with *weld slope*.

Weld Run

Same as *weld pass*.

Weld Size

See preferred term *size of weld*.

Weld Slope

The angle, relative to the horizontal, of a line along the weld root. Compare with *weld rotation*.

Weld Tab

Additional material on which the weld may be initiated or terminated.

Weld Time (Automatic Arc Welding)

The time interval from the end of start time or end of upslope to beginning of crater fill time or beginning of downslope.

Weld Time (Resistance Welding)

The time that welding current is applied to the work in making a weld by single-impulse welding or flash welding.

Weld Timer

A device used in resistance welding to the weld time only.

Weld Voltage

See *arc voltage*.

Weld Zone

The neighborhood of a weld comprising all material involved in the weld operation including the weld deposit, the tip of any electrode or filler, and the adjacent parent metals. More generally, the area that may be affected by any aspect of welding such as oxidation, tarnishing, or spatter but not normally the larger area affected by residual welding stresses.

Steels to determine whether the numerous factors involved in each case collectively introduce a risk of cracking and whether the risk can be eliminated by preheating. Various procedures have been developed for determining weldability but one well-established system recognizes that the basic factors, for a non-preheated joint, are (i) the Hardenability of the steel, which is a function of composition, (ii) the Joint Severity, which is a function of the cross section of the parent material available to conduct heat from the joint,(iii) the amount of damaging hydrogen introduced into the weld zone, which is characteristic of the electrode coating (assuming no contamination), (iv) the heat input in terms of electrode size and burn off rate, and (v) the reduction in cooling rate resulting from preheating. The joint severity is assessed as a thermal severity number (TSN), which totals the number of 1/4 inch units of thickness available to conduct heat away, for example, a weld between two 1 inch thick plates would have a TSN of 8. The hardenability is determined from the composition in terms of the carbon equivalent (CE), for example, a steel with 0.2% carbon, 1% manganese, and 1% chromium would have a CE of 0.35. The CE is then allocated a Weldability Index (WI) letter depending upon the type of electrode coating to be used, for example, the CE of 0.35 would merit a Weldability Index of "E" with a Rutile coated electrode but "C" with a Low Hydrogen (Basic) coating. The TSN and WI are then compared with tables of various electrode sizes at the given burn off rates (in terms of length of electrode consumed per unit of weld deposit). For example, assuming that 1.5 inches of 10 gauge (0.128 inch diameter) electrode will be depositing 1 inch of weld metal then, with a TSN of 8 and a WI of "E," a preheat of about 200 °C is needed to reduce the cooling rate sufficiently to avoid cracking. Using a low hydrogen electrode and hence a WI of "C," reduces the necessary preheat to 125 °C and if, in addition, the electrode size is increased to 4 gauge (0.232 inch diameter) preheat is not necessary.

Wenstrom Mill

A rolling mill similar to a universal mill but where the edges and sides of the rolled section are acted on simultaneously.

Wet, Wettability

The property of a liquid, such as molten metal, to spread on a solid surface due to a low contact angle. This angle is a measure of the degree of wetting obtained in the solid–liquid system.

Wet-Bag Tooling

A rubber or plastic sheet mold used in cold isostatic or hydrostatic pressing of powders. See also *cold isostatic pressing, hydrostatic mold*, and *hydrostatic pressing*.

Wet Blasting

A process for cleaning or finishing by means of a slurry of abrasive in water directed at high velocity against the workpieces. Many different kinds and sizes of abrasives can be used in wet blasting. Sizes range from 20-mesh (very coarse) to 5000-mesh (which is much finer than face powder). Among the types of abrasives used are: organic or agricultural materials such as walnut shells and peach pits; novaculite, which is a soft type (6 to 6.5 Mohs hardness) of silica (99.46% silica); silica, quartz, garnet, and aluminum oxide; other refractory abrasives; and glass beads.

Wet Etching

Development of microstructure in metals with liquids, such as acids, bases, neutral solutions, or mixtures of solutions.

Wet Installation

A bolted joint in which a sealant is applied to the head and shank of the fastener such that after assembly a seal is provided between the fastener and the elements being joined. See also *sealant*.

Wet Lay-Up

A method of making reinforced plastics by applying the resin system as a liquid when the reinforcement is put in place. Polyesters and vinyl esters are the most commonly used family of resins used for wet lay-up.

Wet-Out

In composites fabrication, the condition of an impregnated *roving* or *yarn* in which substantially all voids between the sized strands and filaments are filled with resin.

Wet Strength (Adhesives)

The strength of an adhesive joint determined immediately after removal from a liquid in which it has been immersed under specified conditions of time, temperature, and pressure. The term is commonly used alone to designate strength after immersion in water. In latex adhesive, the term is used to describe the joint strength when the adherends are brought together with the adhesives still in the wet state. Compare with *dry strength* (adhesives).

Wet Strength (Composites)

The strength of an organic matrix composite when the matrix resin is saturated with absorbed moisture, or is at a defined percentage of absorbed moisture less than saturation (saturation is an equilibrium condition in which the net rate of absorption under prescribed conditions falls essentially to zero).

Wetting

(1) The spreading and sometimes absorption of a fluid on or into a surface. (2) A condition in which the interface tension between a liquid and a solid is such that the contact angle is 0° to 90°. (3) The phenomenon whereby a liquid filler metal or flux spreads and adheres in a thin continuous layer on a solid base metal. (4) The formation of a relatively uniform, smooth, unbroken, and adherent film of solder to a basis metal. (5) This occurs when the attraction between the liquid and the surface exceeds the surface tension of the liquid. If the liquid tends to form discrete globules it is not wetting the surface and the angle formed between the two surfaces is a measure of wetting. An acute angle reflects poor wetting and an angle approaching 180 °C indicates good wetting.

Wetting Agent

(1) A substance that reduces the surface tension of a liquid, thereby causing it to spread more readily on a solid surface. (2) A surface-active agent that produces *wetting* by decreasing the *cohesion* within a liquid.

Wetting Agents

These are chemicals used in making solutions, emulsions, or compounded mixtures, such as paints, inks, cosmetics, starch pastes, oil emulsions, dentifrices, and detergents, to reduce the surface tension and give greater ease of mixing and stability to the solution. In the food industries, chemical wetting agents are added to the solutions for washing fruits and vegetables to produce a cleaner and bacteria-free product. Wetting agents are described in general as chemicals having a large hydrophilic group associated with a smaller hydrophilic group. Some liquids naturally wet pigments, oils, or waxes, but others require a proportion of a wetting agent to give mordant or wetting properties. Pine oil is a common wetting agent, but many are complex chemicals. They should be powerful enough not to be precipitated out of solutions in the form of salts, and they should be free of odor or any characteristic that would affect the solution. Aerosol wetting agents are in the form of liquids, waxy pellets, or free-flowing powders. There are other free-flowing powders, basically modified polyacrylates, that are soluble in water and less so in alcohol. There are sodium or ammonium dispersions of modified rosin, with 90% of the particles below 1 μm in size. Also there is a sodium lignosulfonate produced from lignin waste liquor. It is used for dye and pigment dispersion, oil-well drilling mud, ore flotation, and boiler feed-water treatment.

Increasingly used by the ceramic industry is a popular type of poly-oxyethylene alkylate ether with a very high resistance to water hardness. Sulfonated types and carboxylates have moderate wetting properties and strong detergent and solubilizing tendencies.

Wet Winding

The process of *filament winding* of composites in which strands are impregnated with resin before or during winding onto the mandrel. See also *dry winding*.

Whale Oil

An oil extracted by boiling and steaming the blubber of several species of whale that are found chiefly in the cold waters of the extreme north and south. Whales are mammals and are predaceous, living on animal food. The blubber blanket of fat protects the body, and the tissues and organs also contain deposits of fat. Most whale oil is true fat, namely, the glycerides of fatty acids, but the head contains a waxy fat. In the larger animals, the meat and bones yield more fat than the blubber. Both the whalebone whales and the toothed whales produce whale oil. The **bluehead whales** of the south are the largest and yield the most oil per weight. The whaling industry is under international control, and allocations are made on the basis of blue whale units averaging 20 tons (18 metric tons) of oil each. The **gray whale**, or **California whale**, of the northern Pacific, is a small 50-ft (15-m) species. The **Greenland whale** of the north, and the **finback whale** of the south, produce much oil. The **beluga,** or **white whale**, and the **narwhal**, of the north polar areas, produce **porpoise oil**. These species of porpoise measure up to 20 ft (6 m) in length.

Whale oil is sold according to grade, which depends upon its color and keeping qualities. The latter in turn depends largely upon proper cooking at extraction. Grades 0 and 1 are fine, pale-yellow oils, grade 2 is amber, grade 3 is a pale brown, and grade 4 has from 15% to 60% with a strong, fishy odor. The specific gravity is 0.920 to 0.927, saponification value is 180 to 197, and iodine value is 105 to 135. Whale oil contains oleic, stearic, palmitic, and other acids in varying amounts. But whales are now so scarce that the former uses of the oils and meat are restricted, particularly in the United States.

Whale oils of the lower grades have been used for quenching baths, heat-treating steels, and in lubricating oils. The best oils are used in soaps and candles, or for preparing textile fibers for spinning, or for treating leather. In Europe, whale oil is favored for making margarine because it requires less hydrogen than other oils for hardening, and the grouping of 16 to 22 carbon atom acids gives the hardened product greater plasticity over a wider temperature range. **Sod oil** is oil recovered from the treatment of leather in which whale or other marine mammal oil was used. It contains some of the tannins and nitrogenous matter, which makes it more emulsifiable and more penetrant than the original oil.

Whale meat was used for food in Japan and in dog food in the United States. When it is cured in air, the outside is hard and black, but the inside is soft. In young animals the flesh is pale, and in older animals it is dark red. It has a slight fishy flavor, but when cooked with vegetables is almost indistinguishable from beef. It contains 15 to 18% proteins. **Whale-meat extract** is used in bouillon cubes and dehydrated soups. It is 25% weaker than beef extract. **Whale liver oil** is used in medicine for its high vitamin A content. It also contains **kitol**, which has properties similar to vitamin A but is not absorbed in all animal metabolism. **Whalebones** are the elastic, hornlike strips in the upper jaw of the Greenland whale and some other species. Whalebone is lightweight, very flexible, elastic, tough, and durable. It consists of a conglomeration of hairy fibers covered with an enamel-like fibrous tissue. It is easily split and when softened in hot water is easily carved. Whalebone has a variety of uses in making whips, helmet frames, ribs, and brush fibers. **Balcen** is a trade name for strips of whalebone use for whips, and for products where great flexibility and elasticity are required.

Wheelabrator

A brand name of a machine for grit or shot blasting. In essence, it is a rotating wheel, which flings the shot from its periphery to impact the target component.

Wheelabrating

Deflashing molded plastic parts by bombarding them with small particles at high velocity. See also *flash* (plastics).

W

Whetstone

Stones of regular fine grains is composed largely of chalcedony silica, often with minute garnet and rutile crystals. They are used as fine abrasive stones for the final sharpening of edge tools. Whetstones are also sometimes selected, fine sandstones from the grindstone quarries. The finest Whetstones are called **oilstones**. A fine-grained **honestone**, known as **coticule**, comes from Belgium and is used for sharpening fine-edged tools. It is compact, yellow in color, and contains minute crystals of yellow manganese garnet, with also potash mica and tourmaline. Coticule is often cut double with blue-gray **phyllite** rock adhering to and supporting it. **Rubbing stones** are fine-grained Indiana sandstones.

Whirl (Oil)

(1) Instability of a rotating shaft associated with instability in the fluid film. (2) The deflection of a rotating shaft such that the axis of deflection works its way around the axis of rotation. Contrast with *catenary*.

Whisker

(1) A short single crystal fiber or filament used as a reinforcement in a matrix. Whisker diameters range from 1 to 25 μm, with *aspect ratios* generally between 50 and 150. (2) Single-crystal growths resembling fine wire, which may extend to 0.64 mm high. They most frequently occur on printed circuit boards or electronic components that have been electroplated with tin. (3) Metallic filamentary growths, often microscopic, sometimes formed during electrodeposition and sometimes spontaneously during storage or service, after finishing. (4) Fine, hair-like single crystal grown under controlled conditions and usually having only a single axial dislocation.

Whiskers are very fine single-crystal fibers that have length-to-diameter ratios of 50 to 10,000. Since they are single crystals, their strengths approach the calculated theoretical strengths of the materials. Alumina whiskers, which have received the most attention, have tensile strengths up to 0.2 million MPa and a modulus of elasticity of 0.5 million MPa. Other whisker materials are silicon carbide, silica nitride, magnesia, boron carbide, and beryllia.

Whiskers (Glass)

See striation (glass).

Whistlers

Narrow passages in a mold to allow the escape of pockets of entrapped air. The high velocity of the heated escaping air can cause a whistling noise.

White

Brass

White brass is a bearing metal that is actually outside of the range of the brasses, bronzes, or babbitt metals. It is used in various grades; the specification is tin, 65%; zinc, 28% to 30%; and copper, 3% to 6%. It is used for automobile bearings, and is close-grained, hard, and tough. It also casts well. A different alloy is known under the name of white brass in the cheap jewelry and novelty trade. It has no tin, small proportions of copper, and the remainder zinc. It is a

high-zinc brass, and varies in color from silvery white to yellow, depending on the copper content.

White nickel brass is a grade of nickel silver. The white brass used for castings where a white color is desired may contain up to 30% nickel. The 60: 20: 20 alloy is used for white plaque castings for buildings. The high-nickel brasses do not cast well unless they also contain lead. Those with 15% to 20% nickel and 2% lead are used for casting hardware and valves. White nickel alloy is a copper-nickel alloy containing some aluminum. White copper is a name sometimes used for copper-nickel alloy or nickel brass. Nickel brasses known as German silver are copper-nickel-zinc white alloys used as a base metal for plated silverware, for springs and contacts in electrical equipment, and for corrosion-resistant parts. The alloys are graded according to the nickel content. Extra-white metal, the highest grade, contains 50% copper, 30% nickel, and 20% zinc. The lower great, called fifths, for plated goods, has a yellowish color. It contains 57% copper, 7% nickel, and 36% zinc. All of the early German silvers contained up to 2% iron, which increased the strength, hardness, and whiteness, but is not desirable in the alloys used for electrical work. Some of the early English alloys also contained up to 2% tin, but tin embrittles alloys.

Cast Iron

White cast-iron solidifies with all its carbon in the combined state, mostly as iron carbide, F_3C (cementite). White iron contains no free graphite as does gray iron, malleable iron, and ductile iron. White iron derives its name from the fact that it shows a bright white fracture on a freshly broken surface.

The main use for white iron is as an intermediate product in the manufacture of malleable irons. In addition to this, white iron is made as an end product to serve specific applications that require a hard, abrasion-resistant material. White iron is very hard and resistant to wear, has a very high compressive strength, but has low resistance to impact, and is very difficult to machine.

By the proper balancing of chemical composition and section size, an iron casting can be made to solidify completely white throughout its entire section. By modifying the balance and adjusting the cooling rate, the casting can be made to solidify with a layer of white iron at the surface backed up by a core of gray iron. Castings with such a duplex structure are called "chilled iron" castings.

Castings of white iron and chilled iron find their main use in resistance to wear and abrasion. Typical applications include parts for crushers and grinders, grinding balls, coke and cinder chutes, shot-blasting nozzles and blades, parts for slurry pumps, car wheels, metalworking rolls, and grinding rolls.

White cast irons, which have alloy contents well above 4%, fall into three major groups:

1. Nickel-chromium white irons are low-chromium alloys containing 3% to 5% nickel and 1% to 4% chromium, with one alloy modification that contains 7% to 11% chromium. The nickel-chromium irons are also commonly identified as Ni-Hard types 1 to 4.
2. The high chromium irons contain 11% to 23% chromium, up to 3% molybdenum, and are often additionally alloyed with nickel or copper.
3. A third group comprises the 25% or 28% chromium white irons, which may contain other alloying additions of molybdenum and/or nickel up to 1.5%.

The high-alloy white irons are primarily used for abrasion-resistant applications and are readily cast into parts needed in machinery for crushing, grinding, and handling of abrasive materials. The chromium content of high-alloy white irons also enhances their corrosion-resistant properties.

By using a fairly low silicon content, cast iron can be made to solidify white without the use of any additional alloy. Carbon contents are kept high (about 3.6%) when high hardness (575 Bhn) is desired. Such irons have very low toughness and a strength of about 240 MPa. For somewhat higher toughness and strength, at some sacrifice of hardness, the carbon content is lowered to about 2.8%. Unalloyed white and chilled irons have a structure composed of particles of mass of iron carbide in a matrix of fine pearlite. For highest hardness, strength, and toughness, the white iron is alloyed to produce a martensitic matrix surrounding particles of massive carbide.

Gold

White gold is the name of a class of jewelers' white alloys used as substitutes for platinum. The name gives no idea of the relative value of the different grades, which vary widely. Gold and platinum may be alloyed together to make a white gold, but the usual alloys consist of 20% to 50% nickel, with the balance gold. Nickel and zinc with gold may also be used for white golds. The best commercial grades of white gold are made by melting the gold with a white alloy prepared for this purpose. This alloy contains nickel, silver, palladium, and zinc. The 14- karat white gold contains 14 parts pure gold and 10 parts white alloy. A superior class of white gold is made of 90% gold and 10% palladium. High-strength white gold contains copper, nickel, and zinc with the gold. Such an alloy, containing 37.5% gold, 28% copper, 17.5% nickel, and 17% zinc, when aged by heat treatment, has a tensile strength of about 689 MPa and an elongation of 35%. It is used for making jewelry, has a fine, white color, and is easily worked into intricate shapes. White-gold solder is made in many grades containing up to 12% nickel, up to 15% zinc, with usually also copper and silver, and from 30% to 80% gold. The melting points of eight grades range from 695 to 845 °C.

Metals

Although a great variety of combinations can be made with numerous metals to produce white or silvery alloys, the name usually refers to the lead-antimony-tin alloys employed for machine bearings, packings, and linings, to the low-melting-point alloys used for toys, ornaments, and fusible metals, and to the type metals. Slush castings, for ornamental articles and hollow parts, are made in a wide variety of soft white alloys, usually varying proportions of lead, tin, zinc, and antimony, depending on cost and the accuracy and finish desired. These castings are made by pouring the molten metal into a metal mold without a core, and immediately pouring the metal out, so that a thin shell of the alloy solidifies against the metal of the mold and forms a hollow product. A number of white metals are specified by the American Society for Testing and Materials for bearing use. These vary in a wide range from 2% to 91% tin, 4.5% to 15% antimony, up to 90% lead, and up to 8% copper. The alloy containing 75% tin, 12% antimony, 10% lead, and 3% copper melts at 184 °C, is poured at about 375 °C, and has an ultimate compressive strength of 111 MPa and a Brinell hardness of 24. The alloy containing 10% tin, 15% antimony, and 75% lead melts at 240 °C, and has a compressive strength of 108 MPa and a Brinell hardness of 22. The first of these two alloys contains copper-tin crystals and the second contains tin-antimony crystals.

Society of Automotive Engineers (SAE) Alloy 18 is a cadmium-nickel alloy with also small amounts of silver, copper, tin, and zinc.

A bismuth- lead alloy containing 58% bismuth and 42% lead melts at 123.5 °C. It casts to exact size without shrinkage or expansion, and is used for master patterns and for sealing.

Various high-tin or reverse bronzes have been used as corrosion-resistant metals, especially before the advent of the chromium, nickel, and aluminum alloys for this purpose.

A white metal sheet now much used for making stamped and formed parts for costume jewelry and electronic parts is zinc with up to 1.5% copper and up to 0.5% titanium. The titanium with the copper prevents coarse-grain formation, raising the recrystallization temperature. The alloy weighs 22% less than copper, and it plates and solders easily.

White-Etching Layer

A surface layer in a steel that, as viewed in a section after etching, appears whiter than the base metal. The presence of the layer may be due to a number of causes, including plastic deformation induced by machining, or surface rubbing, heating during a metallographic preparation stage to such an extent that the layer is austenitized and then hardened during cooling, and diffusion of extraneous elements into the surface.

Whiteheart Malleable

See *malleable cast iron*.

White Layer

(1) Compound layer that forms in steels as a result of the *nitriding* process. (2) In tribology, a *white-etching layer*, typically associated with ferrous alloys, that is visible in metallographic cross sections of bearing surfaces. See also *Beilby layer* and *highly deformed layer*.

White Liquor

Cooking liquor from the kraft pulping process produced by recausticizing *green liquor* with lime. See also *kraft process*.

White Radiation

See *continuum*.

White Rust

Zinc oxide; the powder product of corrosion of zinc or zinc-coated surfaces.

Whiteware Ceramics

Technical whitewares include clays, porcelains, china, white stoneware, and steatites. The modern oxide ceramics would also be in this group. For technical use, these whitewares are usually vitrified (nonporous) or very nearly so. Most commonly, the pieces are glazed. To produce white-bodied ceramics, the raw materials must be of superior quality and selection.

Composition and Types

Most whiteware clays are basically kaolins and chemically hydrous aluminum silicates. Ball clays are less pure than kaolins and contain free silica and small amounts of other contaminants. Kaolins are generally not highly plastic, whereas ball clays are very plastic. The plasticity derives from the physical form of the minute particles, which are colloidal in size. Ball clays are not as white-burning as kaolins and impart color to the fired body.

Feldspars are used as fluxes to provide the alkaline oxides for the glassy phase surrounding the mullite ($3Al_2O_3 \cdot 2SiO_2$) crystals forming the mass of the body

Porcelains, China, and stoneware are composed of clays, silicate, and feldspar. Steatites are composed of talc (magnesium silicate) and clay. Minor variations are made to enhance special properties.

The porcelains, China, and stoneware are composed of alumina (Al_2O_3), silica (SiO_2), sodium and potassium oxides (Na_2O and K_2O), as well as calcia (CaO), zinc oxide (ZnO), zirconia (ZrO_2), titania ($TiCh$), barium oxide (BaO), magnesia (MgO), and phosphoric oxide (P_2O_5). Some other oxides may be present as traces. Iron oxide is usually present in small amounts as an undesirable impurity.

The oxides are supplied in kaolin, ball clays, quartz, feldspar, whiting, magnesia, and talc. Oxide ceramics that do not use clays may use only mixtures of refined oxides.

Compounding and Forming

The raw materials are intimately mixed, usually by ball milling, and then prepared for forming into ware by use of the methods listed next. For jiggering and simple mechanical pressing or extruding, the ball-milled slip is dewatered, filter pressed, deaired, and extruded into convenient size billets.

For casting, the specific gravity is adjusted to give a good casting viscosity.

For dry pressing, the body is dried, shredded, and powdered or spray dried into minute granules.

The pieces are formed as close to size before firing as practical, allowing for the shrinkage during firing. This shrinkage may run as high as 25% and must be very closely controlled to avoid loss in the firing of off-dimension pieces.

The ware is formed in a number of ways, some of them unchanged for centuries or even thousands of years and others unknown 60 years ago. The following are the primary methods used:

1. Throwing or jiggering on mechanical potters wheel from plastic clay body
2. Casting in plaster of Paris mold from liquid slurry or slip
3. Pressing from plastic clay with simple mold
4. Pressing from dry powder in metal dies
5. Isostatic pressing from dry powder in rubber sack with hydraulic pressure
6. Hot pressing with heated clay blank and mold
7. Simple extrusion through die
8. Extrusion with thermoplastic resin to the body (injection molding)

Many pieces are formed by a combination of pressing or extrusion followed by mechanical shaping in lathes by special tools or dry grinding.

After forming, pieces are bisque fired, which drives out moisture and water of crystallization. Porcelain is not vitrified during the bisque firing. The porcelain bisque is dipped or sprayed with glaze and, in the second firing, the body and glaze mature or vitrify together.

China is vitrified in the first firing. In China manufacture, the glaze is applied to the fired body and the glaze matures in the second firing, which is at a lower temperature than the initial firing. In some cases, the glaze can be applied to an unfired piece and only one firing is needed.

Today, most kilns are fired with natural or manufactured gas, fuel oil, or electricity. The round beehive kiln fired periodically has been largely replaced with continuous tunnel kilns, which may be built with movable cars moving from one end to the other on a straight track or as a circular tunnel with a moving floor. Periodic kilns are often used in specialized work where the volume does not justify use of a tunnel kiln. They are also more versatile.

The firing of technical ceramics is performed under the most carefully controlled conditions possible. Both the temperature and the atmosphere must be known and controlled. The ware must be properly placed in the kilns or damage to the piece will result. Warpage, uneven firing, and cracking can easily occur.

Maximum firing temperatures for unglazed refractory porcelain are usually about 1760°C. Laboratory chemical porcelain that is glazed is fired to 1454°C. Hotel, sanitary china, and electrical porcelain are usually fired about 1260°C. Steatite bodies mature at around 1260°C–1316°C.

Wicking

(1) The flow of a liquid along a surface into a narrow space. This capillary action is caused by the attraction of the liquid molecules to each other and to the surface. (2) The flow of solder away from the desired area by *wetting* or *capillary action*.

Widmanstätten Structure

A structure characterized by a geometrical pattern resulting from the formation of a new phase along certain crystallographic planes of the parent solid solution. The orientation of the lattice in the new phase is related crystallographically to the orientation of the lattice in the parent phase. The structure was originally observed in meteorites but is readily produced in many alloys, such as titanium, by appropriate heat treatment. See the term *basketweave* (titanium).

Width

In the case of a beam, the shorter dimension perpendicular to the direction in which the load is applied.

Wildness

A condition that exists when molten metal, during cooling, evolves so much gas that it becomes violently agitated, forcibly ejecting metal from the mold or other container.

Williams Riser

An *atmospheric riser*.

Wind Angle

In filament winding of composites, the angular measure in degrees between the direction parallel to the filaments and an established reference. In filament-wound structures, it is the convention to measure the wind angle with reference to the centerline through the polar bosses, that is, the axis of rotation. See also *filament winding*.

Winding Pattern

In filament winding of composites, the total number of individual circuits required for a winding path to begin repeating by laying down immediately adjacent to the initial circuit. A regularly recurring pattern of the filament path after a certain number of mandrel revolutions, leading eventually to the complete coverage of the mandrel. See also *filament winding*.

Winding Tension

In filament winding or tape wrapping of composites, the amount of tension on the reinforcement as it makes contact with the mandrel. See also *filament winding*.

Window

A defect in thermoplastic film, sheet, or molding, caused by the incomplete plasticization of a piece of material during processing. It appears as a globule in an otherwise blended mass. See also *fisheye* (plastics).

Winning

Recovering a metal from an ore or chemical compound using any suitable hydro-metallurgical, pyrometallurgical, or electrometallurgical method.

Wipe Etching

See *swabbing*.

Wiped Coat

A hot-dipped galvanized coating from which virtually all free zinc is removed by wiping prior to solidification, leaving only a thin zinc–iron alloy layer.

Wiped Joint

A joint made with solder having a wide melting range and with the heat supplied by the molten solder poured onto the joint. The solder is manipulated with a hand-held cloth or paddle so as to obtain the required size and contour. Traditionally, plumbers form soldered joints by wiping with a "moleskin." See *solder*.

Wiper

A pad of felt or other material used to supply lubricant or to remove debris.

Wiper Forming, Wiping

Method of curving sheet metal sections or tubing over a form block or die in which this form block is moved relative to a wiper block or slide block.

Wire Flame Spraying

A thermal spraying process variation in which the material to be sprayed is in wire or rod form. See also *flame spraying*.

Wire Glass

A sheet glass used in building construction for windows, doors, floors, and skylights, having woven wire mesh embedded in the center of the plate. It does not splinter or fly apart as common glass when subjected to fire or shock, and it has higher strength than common glass. It is made in standard thicknesses from 0.125 to 0.375 in. (0.318–0.953 cm) and in plates 60 × 110 in. (1.5 × 2.8 m) and 61 × 140 in. (1.5 × 3.6 m). Wire glass is made from plain, rough, or polished surfaces, or with ribbed or cobweb surface on one side for diffusing the light and for decorative purposes. It is also obtainable from corrugated sheets. Plastic-coated wire mesh may be used to replace wire glass for hothouses or skylights where less weight and fuller penetration of light grays are desired. *Cel-O-Glass*, of DuPont, is a plastic-coated wire mesh in sheet form.

Wire Rod

Hot-rolled coiled stock that is to be cold drawn into wire.

Wire Straightener

A device used for controlling the cast of coiled weld filler-metal wire to enable it to be easily fed into the welding gun.

Wiring

Formation of a curl along the edge of a shell, tube, or sheet and insertion of a rod or wire within the curl for stiffening the edge. See also *curling*.

Witness Mark

A surface mark revealing or confirming some action or process.

Wohler Fatigue Test

A fatigue test in which the specimen is a rotating cantilever, driven at the supported end and loaded via a bearing at the other. A rotation produces a single reversed stress cycle.

Wolfram

An old name for tungsten now usually referring to tungsten ores.

Wollaston Wire

Any wire made by the Wollaston process of fine-wire drawing. It consists of inserting a length of bare drawn wire into a close-fitting tube of another metal, the tube and core then being treated as a single rod and drawn through dies down to the required size. The outside jacket of metal is then dissolved away by an acid that does not affect the core metal. **Platinum wire** as fine as 0.00005 in. (0.00013 cm) in diameter is made commercially by this method, and gold wire as fine as 0.00001 in. (0.00002 cm) in diameter is also drawn. Wires of this fineness are employed only in instruments. They are marketed as composite wires, the user dissolving off the jacket. **Taylor process wire** is a very fine wire made by the process of drawing in a glass tube. The process is used chiefly for obtaining fine wire from a material lacking ductility, such as antimony, or extremely fine wire from a ductile metal. The procedure is to melt the metal or alloy into a glass or quartz tube and then draw down this tube with its contained material. Wire as fine as 0.00004 in. (0.00012 cm) in diameter has been made, but only in short lengths.

Wood

For most purposes, wood may be defined as the dense fibrous substance that makes up the greater part of a tree. It is found beneath the bark and in the roots, stems, and branches of trees and shrubs. Of the three sources, the stem or trunk furnishes the bulk of raw material for lumber products.

Wood is a renewable resource. It is grown just about everywhere and can be produced in any reasonable quantity needed for future consumption. Wood products and the management of forested lands are changing to meet modern conditions; hence, trees were grown to meet modern production requirements for size, quality, and quantity. For example, with the developments in the modern technique of gluing, the former use of extremely wide, thick, and excessively long lumber is no longer necessary. Laminated lumber and plywood have generally taken the place of these large boards and timbers. Not only are the raw materials for such products easier to grow and more economical to obtain that are solid timbers of comparable size, but also the products are generally improved by the use of modern methods of fabricating.

Although there are many species of wood, the commercially important types can be grouped into two categories of about 25 for each group.

Anatomy

Wood is composed mostly of hollow, elongated, spindle-shaped cells that are arranged parallel to each other along the trunk of a tree. The characteristics of these fibrous cells and their arrangement affect strength properties, appearance, resistance to penetration by water and chemicals, resistance to decay, and many other properties.

The combined concentric bands of light and dark areas constitute annual growth rings. The age of a tree may be determined by counting these rings at the stump.

In temperate climates, trees often produce distinct growth layers. These increments are called growth rings or annual rings when associated with yearly growth; many tropical trees, however, lack growth rings. These rings vary in width according to environmental conditions. Where there is a visible contrast within a single growth ring, the first-form layer is called earlywood and the remainder latewood. The earlywood cells are usually larger and the cell walls thinner than the latewood cells. With the naked eye or a hand lens, earlywood is shown to be generally lighter in color than latewood.

Because of the extreme structural variations in wood, there are many possibilities for selecting a species for a specific purpose. Some species (e.g., spruce) combine light weight and relatively high stiffness and bending strength. Very heavy woods (e.g., lignum vitae) are extremely hard and resistant to abrasion. A very light wood (such as balsa) has high thermal insulation value; hickory has extremely high short resistance; mahogany has excellent dimensional stability.

Many mechanical properties of wood, such as bending strength, crushing strength, and hardness, depend on the density of wood; the heavier woods are generally stronger. Wood density is determined largely by the relative thickness of the cell wall and the proportions of thick- and thin-walled cells present.

Hardwoods and Softwoods

The terminology used in the classification of trees is confusing, but because it has become general in usage, it is important for those who make or purchase products of wood to understand it.

The terms hardwood and softwood have no direct application to hardness or softness of the materials. Basswood is a softer domestic species, yet the yellow pines, which are classed as softwood, are often much harder. Even balsa, a foreign species that everyone knows, the lightest and the softest wood used in commerce, is classed as hardwood. For practical purposes, the hardwoods have broad leaves, whereas the softwoods have needlelike leaves. Trees (hardwoods) with broad leaves usually shed them at some time during the year, while the conifers (softwoods) retain a covering of the needlelike foliage throughout the year. There are quite a few exceptions to these criteria; but as users gain familiarity with the various woods, they will soon learn by experience into which group a species falls.

Those who use wood must know something of the botanical classification because the lumber industry is also divided into two distinct groups. The methods of doing business and manufacturing and grading for quality differ from each other.

Generally, the hardwoods are used for the manufacture of factory-made products, such as tools, furniture, flooring, and instrument cases. The largest market for softwoods is in the home construction field or for other building purposes. But there is no line of demarcation that is reliable, for hardwoods and softwoods are often interchangeable in use.

The more important hardwoods are as follows:

Alder	Ash	Aspen	Basswood
Beech	Birch	Cherry	Chestnut
Cottonwood	Elm	Hackberry	Hickory and Pecan
Holly	Locust	Magnolia	Maple (hard and soft)
Oak	red Oak	White	Sweet or Red
Gum	Sycamore	Tupelo or Black Gum	Walnut black
Willow black	Yellow Popular		

The more important softwoods are the following:

Cedar (several species)	Cypress Douglas Fir (not a true fir)	Firs (eastern and western)
Hemlock (eastern and western)	Larch Pine	eastern white Pine
jack Pine	lodgepole Pine	pitch Pine
Ponderosa Pine	red Pine	southern yellow (several species) Pine
sugar Pine	Virginia Pine	western white Redwood Spruce
eastern Spruce	Englemann Spruce	Sitka Tamarack

Hardwood

The horizontal plane of a block of hardwood (e.g., oak or maple) corresponds to a minute portion of the top surface of a stump or end surface of a log. The vertical plane corresponds to a surface cut parallel to the radius and parallel to the wood rays. The vertical plane corresponds to a surface cut at right angles to the radius and the wood rays, or tangentially within the log. In hardwoods, these three major planes along which wood may be cut are known commonly as end-grain, quarter-sawed (edge-grain), and plain-sawed (flat-grain) surfaces.

Softwood

The rectangular units that make up the end grain of softwood are sections through long vertical cells called tracheids or fibers. Because softwoods do not contain vessel cells, the tracheids serve the dual function of transporting sap vertically and giving strength to the wood. Softwood fibers range from about 3 to 8 mm in length.

Cell Walls and Composition

The principal compound in mature wood cells is cellulose, a polysaccharide of repeating glucose molecules that may reach 4 μm in length. The cellulose molecules are arranged in an orderly manner into structures about 10–25 nm wide called microfibrils. This ordered arrangement in certain parts (micelles) gives the cell wall crystalline properties that can be observed in polarized light with a light microscope. The microfibrils wind together like strands in a cable to form macrofibrils that measure about 0.5 μm in width and may reach 4 μm in length. These cables are as strong as an equivalent thickness of steel.

Wood, regardless of the species, is composed of two principal materials: cellulose, which is about 70% of the volume, and lignin, nature's glue for holding the cells and fibers together, which is from 20% to 28%. Residues in the form of minerals, waxes, tannins, oils, etc., compose the remainder. The residues, although small in volume, often provide a species with unusual properties. The oils in cypress are responsible for it is renowned as a decay-resistant wood. Aromatic oils provide many of the cedars with distinctive odors that make them valuable for clothing storage chests. Other chemicals provide resistance to water absorption, which is useful for constructing light, high-speed boats that are relatively free from increase in weight due to water absorption.

Many of the chemical residues of wood can be removed by neutral solvents, such as water, alcohol, acetone, benzene, and ether. Some of them may be caused to migrate from one part of the wood to another.

Structure of Wood

The roots, stem, and branches of a tree increase in size by adding a new layer each year, just as the size of one's hand is increased by putting on a glove. The layer or growth ring will vary in thickness due to the age of the tree, growing conditions, amount of foliage, and other factors.

The growth ring is divided into two parts—spring wood and summer wood. The former is usually lighter in weight than the latter and is denser and stronger. Generally, a tree or portion of a tree with the most summer wood is stronger than one that has less.

The thickness of the growth ring and the relative amounts of spring and summer wood have a greater effect on the appearance of wood and are often a deciding factor in the choice of furniture materials. Excessively, thick growth rings usually provide a rather coarse-textured material, whereas narrow growth rings provide fine texture. However, in hardwoods, a thick ring usually does have more summer wood and is, therefore, the stronger of the two. The situation is somewhat different for the softwoods.

Grades of Lumber

Modern grading of lumber is the result of experience. Over a period of 90 years, there has been a gradual evolution to meet the changes in industry. This trend will continue as long as wood products are used.

The grading of hardwoods and softwoods is entirely different; the former are used almost entirely as raw materials for manufactured products. For these, basic grading is on the number of usable cuttings or pieces that can be cut from a board. For some grades, these cuttings need only be sound, but for others they must be clear at least on one face. The cuttings must also be of a certain size.

Hardwood Grading

The grades of hardwood lumber from the highest quality to the lowest are "Firsts," the top-quality grade; the next are known as "Seconds." These two grades are usually marketed as one and cold "Firsts and Seconds"; the designation for the grade is FAS. The next lowest grade is "Selects," followed by "No. 1 Common," "No. 2

Common," "No. 3A Common," and "No. 3B Common." Sometimes, a grading is further differentiated, such as "FAS One Face," which means that it is of a much higher grade on one face then on the other. A prefix "WHND" is also used sometimes. It means that wormholes are not to be considered as defects in evaluating cuttings nor as a reason for disqualifying them as might otherwise be done.

Softwood Grading

The theory of softwood lumber grading is probably somewhat more difficult for the layperson to understand than that used for the hardwoods. This is because many of the species are graded under separate association rulings, which are similar, but with some important differences.

Furthermore, softwoods are divided into three general classes of products, each of which is graded under a different set of rules.

1. *Yard lumber*, which is used for building construction and other ordinary uses, is unobtainable in "Finish Grades," which are called "A," "B," "C," "D"; "Common Boards," which are called "No. 1," "No. 2," "No. 3," "No. 4," and sometimes "No. 5"; and as "Common Dimension" in grades "No. 1," "No. 2," and "No. 3."
2. *Structural lumber* is a relatively modern concept in the field of lumber grading. It is an engineered product intended for use where definite strength requirements are specified. The allowable stresses designated for a piece of structural lumber depend upon the size, number, and placement of the defects. The relative position of a defect is of great importance; therefore, if the maximum strength of the piece is to be developed, it must be used in its entirety. It cannot be remanufactured for width, thickness, or length.
3. *Factory* and *shop lumber* is the third general category. These grades are similar to those for hardwoods in that the lumber is graded by the number of usable cuttings that can be taken from a board, but here the resemblance ceases, for both the grade descriptions and nomenclature, are different. The term "Factory and Shop" is descriptive of the uses for which the product is designed. Much of it is used for general millwork products, patterns, models, etc., and wherever it is necessary to cut up softwood lumber for the production of factory-made items.

Wood Chemicals

These are chemicals obtained from wood. The practice was carried out in the past and continues wherever technical utility and economic conditions have combined to make it feasible. Woody plants comprise the greatest part of the organic materials produced by photosynthesis on a renewable basis and were the precursors of the fossil coal deposits. Future shortages of the fossil hydrocarbons from which most organic chemicals are derived may be the result of the economic feasibility of the production of these chemicals from wood.

Wood is a mixture of three natural polymers—cellulose, hemicelluloses, and lignin—and an approximate abundance of 50:25:25. In addition to these polymeric cell wall components, which make up the major portion of the wood, different species contain varying amounts and kinds of extraneous materials called extractives. Cellulose is a long-chain polymer of glucose that is embedded in an amorphous matrix of the hemicelluloses and lignin. Hemicelluloses are shorter or branched polymer of five- and six-carbon sugars other than glucose. Lignin is a three-dimensional polymer formed

W

of phenol–propane units. Thus, the nature of the chemicals derived from wood depends on the wood component involved.

Modern Processes

Chemicals derived from wood at present include bark products, cellulose, cellulose esters, cellulose ethers, charcoal, dimethyl sulfoxide, ethyl alcohol, fatty acids, furfural, hemicellulose extracts, kraft lignin, lignin sulfonates, pine oil, rayons, rosin, sugars, tall oil, turpentine, and vanillin.

Most of these are either direct products or by-products of wood pulping, in which the lignin that cements the wood fibers together and stiffens them is dissolved away from the cellulose. High-purity chemical cellulose or dissolving pulp is the starting material for such polymeric cellulose derivatives as viscous rayon and cellophane (regenerated celluloses from the xanthate derivative in fiber or film form), cellulose esters such as the acetate and butyrate for fiber, film, and molding applications, and cellulose ethers such as carboxymethylcellulose, ethylcellulose, and hydroxy-ethylcellulose for use as gums.

Potential Chemicals

Considerable development effort has been devoted to the conversion of renewable biomass, of which wood is the major component, into the chemicals usually derived from petroleum. Processes for which technical feasibility has been demonstrated and economic feasibility is influenced by fossil hydrocarbon cost and availability.

Wood Degradation

This refers to decay of the components of wood. Despite its highly integrated matrix of cellulose, hemicellulose, and lignin, which gives the wood superior strength properties and a marked resistance to chemical and microbial attack, a variety of organisms and processes are capable of degrading wood. The decay process is a continuum, often involving a number of organisms over many years. Wood degrading agents are both biotic and abiotic, and include heat, strong acids and bases, organic chemicals, mechanical wear, and sunlight (ultraviolet degradation).

Engineering Design

This is the process of creating products, components, and structural systems with wood and wood-based materials. Wood engineering design applies concepts of engineering in the design of systems and products that must carry loads and perform in a safe and serviceable fashion. Common examples include structural systems such as buildings or electric power transmission structures, components such as trusses or prefabricated stressed-skin panels, and products such as furniture or pallets and containers. The design process considers the shape, size, physical and mechanical properties of the materials, type and size of the connections, and the type of system response needed to resist both stationary and moving (dynamic) loads, and function satisfactorily in the end-use environment.

Wood is used in both light frame and heavy timber structures. Light frame structures consist of many relatively small wood elements such as lumber covered with a sheathing material such as plywood. The lumber and sheathing are connected to act together as a system in resisting loads; an example is a residential house wood floor system where the plywood is nailed to lumber bending members or joists. In this system, no one joint is heavily loaded because the sheathing spreads the load out over many joists. Service factors such as deflection or vibration often govern the design of floor systems rather than strength.

Light frame systems are often designed as diaphragms or shear walls to resist lateral forces resulting from wind or earthquake.

In heavy timber construction, such as bridges or industrial buildings, there was less reliance on system action and, in general, large beams or columns carry more load transmitted through decking or panel assemblies. Strength, rather than deflection, often governs the selection of member size and connections. There are many variants of wood construction using poles, wood shells, folded plates, prefabricated panels, logs, and combinations with other materials.

Engineered Wood Composites

Wood composites are products composed of wood elements that have been glued together to make a different, more useful or more economical product than solid sawn wood. Plywood is a common example of a wood composite sheathing panel product where layers of veneer are glued together. Plywood is used as a sheathing material in light frame wood buildings, wood pallets, and containers to distribute the applied forces to beams of lumber or other materials. Shear strength, bending strength, and stiffness are the most important properties for these applications and may be engineered into the panel by adjusting the species and quality of veneer used in the manufacture.

Processing

Processing involves peeling, slicing, sawing, and chemically altering hardwoods and softwoods to form finished products such as boards or veneer; particles or chips for making paper, particle, or fiber products; and fuel.

Most logs are converted to boards in a sawmill that consists of a large circular or band saw, a carriage that holds the log and moves past the saw, and small circular saws that remove excess bark and defects from the edges and ends of the boards. One method is to saw the log to boards with a single pass through several saw blades mounted on a single shaft (a gang saw). Sometimes, the outside of the log is converted to boards or chips until a rectangular center or cant remains. The cant is then processed to boards with a gang saw.

Other steps taken may be drying or machining, including veneer cutting. Wood is also ground to fibers for hardboard, medium-density fiberboard, and paper products. It is sliced and flaked for particle-board products, including wafer boards and oriented strand boards. Whether made from waste products (sawdust, planer shavings, slabs, edgings), or roundwood, the individual particles generally exhibit the anisotropy and hygroscopicity of larger pieces of wood. The negative effects of these properties are minimized to the degree that the three wood directions (longitudinal, tangential, and radial) are distributed more or less randomly.

Wood Products

Wood products are those products, such as veneer, plywood, laminate of products, particle-board, waferboard, pulp and paper, hardboard, and fiberboard, made from the stems and branches of coniferous (softwood) and deciduous (hardwood) tree species. The living portion of the tree is the region closest to the bark as commonly referred to as sapwood; the dead portion of the tree is called heartwood. In many species, especially hardwoods, the heartwood changes color because of chemical changes in it. The heartwood of walnut, for example, is dark brown and the sapwood almost white.

Wood is one of the strongest natural materials for its weight. A microscopic view reveals thousands of hollow-tubed fibers held together with a chemical called lignin. These hollow-tubed fibers give wood its tremendous strength for its light weight. These fibers, after the lignin bonding material is removed, make paper. In addition to lignin, wood is composed of other chemicals, including cellulose and hemicellulose. This category includes lumber products, veneer, and structural plywood.

Lumber

Most small log sawmills try to maximize value and yield by automating as much of the manufacturing process as possible. The basic process of cutting lumber involves producing as many rectangular pieces of lumber as possible from a round tapered log. There are only a few sawing solutions, out of millions possible, that will yield the most lumber from any given log or larger timber.

Veneer and Structural Plywood

Structural plywood is constructed from individual sheets of veneer, often with the grain of the veneer in perpendicular directions in alternating plies. The most common construction is three-, four-, and five-ply panels. The alternating plies give superior strength and dimensional stability.

Composite Products

These include laminated products made from lumber, particleboard, waferboard, and oriented strand board.

Laminated Products

Laminated products are composite products, made from lumber, parallel laminated veneer, and sometimes plywood, particleboard, or other fiber product. The most common types are the laminated beam products composed of individual pieces of lumber glued together with a phenyl resorcinol-type adhesive. Laminated beams are constructed by placing high-quality straight-grained pieces of lumber on the top and bottom, where tension and compression stresses are the greatest, and lower-quality lumber in the center section, where the stresses are lower.

Another form of composite team is constructed from individual members made from parallel laminated veneer lumber. This type of lumber has the advantage that it can be made into any length, thus creating beams to span large sections. These beams can be made entirely of parallel laminated veneer lumber, or with the top and bottom flange made from parallel laminated veneer lumber, and the center web made from plywood or flakeboard. Some of these products use solid lumber for the flange material and resemble an I beam. All structural laminated products have the advantage over lumber in that much of the natural variation due to defects is removed, and wood structural members can be made much larger than the typical 51 × 305 mm lumber product.

Particleboard

Nonstructural particleboard is another type of composite product that is usually made from sawdust or planer shavings. It is sometimes made from flaked roundwood. This type of particleboard is one of the most widely used forms of wood product. Often hidden from view, it is used as a substrate under hardwood veneer or plastic laminates. It is commonly used in furniture, cabinets, shelving, and paneling. Particleboard is made by drying, screening, and sorting the sawdust and planer shavings into different size classifications.

Waferboard and Oriented Strand Board

Waferboard and oriented strand board are structural panels made from flakes or strands and are usually created from very small trees. Unlike nonstructural particleboard, waferboard is designed for use in applications similar to those of plywood. Waferboard and oriented strand board can have flakes or strands oriented in the same direction, thus giving the board greater strength in the long axis. The most common type of waferboard is made from randomly oriented flakes.

Fiber Products

The most common fiber products result from pulping processes that involve the chemical modification of wood chips, sawdust, and planer shavings. Such products include pulp and paper, hardboard, and fiberboard.

Pulp and Paper

Paper making begins with the pulping process. Pulp is made from wood chips created in the lumber manufacturing process, small roundwood that is chipped, and recycled paper. The fibers in the chips must be separated from each other by mechanically grinding the fibers or chemically dissolving the lignin from them. The most common chemical processes are sulfite and sulfate (kraft). Following the pulping process, the fibers are washed to remove pulping chemicals or impurities. In some processes (e.g., writing papers), the fibers are bleached.

Dry, finished paper emerges from the end of this section, and it is placed in rolls for further manufacture into paper products.

Hardboard

Hardboard is a medium- to high-density wood fiber product made in sheets from 1.6 to 12.7 mm. Hardboard is used in furniture, cabinets, garage door panels, vinyl overlaid wall panels, and pegboard. It is made by either a wet or dry process.

Medium-Density Fiberboard

Medium-density fiberboard is used in many of the same applications where particleboard is used. It can be used in siding and is especially well suited to cabinet and door panels where edges are exposed. Unlike particleboard, which has a rough edge, medium-density fiberboard has a very fine edge that can be molded very well. Medium-density fiberboard is produced in much the same way that dry-processed hardboard is produced in its early stages. The chips are thermomechanically pulped or refined prior to forming into a dry mat. Following refining, medium-density fiberboard is produced in a fashion similar to that of particleboard. The dried pulp is sprayed with adhesive (usually urea–formaldehyde or phenol–formaldehyde) and formed into a dry mat prior to pressing in a multiopening hot press. After the panels have been formed, they are cooled and cut to smaller final product sizes prior to shipment.

Other Engineered Wood Products

The list of processes and potential processes that are available for wood construction is almost endless. The laminating of wood can be extended to using species of one kind for the surfaces and another for the core. Also, other species having altogether different properties may be used in other places of the assembly. Decking for aircraft carriers is a good example; the surface of the decking must have high abrasion resistance, but great strength in bending is also required. The combination of several species in the same timber is, therefore, most effective.

Barrels and other cylindrical containers are made of laminated wood staves, plywood, or veneer. Hogsheads for tobacco export are of solid staves or plywood.

Wood can be treated to provide considerable fire resistance; it then can be destroyed by heat but will not support combustion.

Natural wood is not adversely affected by extremes of cold such as are encountered in high latitudes and is, therefore, much in use for shelters, tools, sporting equipment, and other products where these conditions exist.

Wood is widely used for structures and products that must be nonmagnetic; hence, it is used for minesweepers and similar craft. Laminated wood has no peer for this type of product.

Special treatments have been developed for use in power-line construction. These treatment materials must act as nonconductors as well as preservatives.

Wood has excellent thermal insulating properties. It is used for refrigerated spaces, refrigerated delivery trucks, and in a great many other places where light weight and thermal insulation, combined with strength, are necessary. For example, specially treated milk containers are in widespread use in the dairy industry.

Special forms of timbers are manufactured for boxcar decking. Decking of this type provides a medium for fastening cargo as well as furnishing other needed functions.

In the textile industry, wood, both treated and in the natural form, is used for shuttles and bobbins, pulleys, and many other types of machinery.

Wood, in a natural form, treated or laminated, is used for water and other liquid conduits, cooling towers, and chemical containers.

Preservation/Protection

There are several wood preservation methods. The system used depends upon the service requirements. For example, a railroad tie, once in place, must withstand continuous exposure to the elements and almost every conceivable condition that promotes decay, until it is worn out some 20–30 years later. It is, therefore, necessary to provide this and similar items with maximum protection.

The two general methods of treating are surface applications where the wood is exposed to the chemicals by dipping, soaking, or brushing, and the treatments wherein the chemicals are forced into the wood through pressure. These latter treatments are used for the most critical applications. Pressure-treating operations must be conducted in plants with considerable equipment. It is a specialized business, and a large industry has arisen to take care of the many needs for highly treated products.

The most effective way to prevent damage through exposure of untreated wood is to perform all major machining operations prior to treating. If this is impossible, the exposed parts should be given a surface treatment by brushing or dipping. Unfortunately, neither of these treatments is as effective as the original pressure treating. All domestic species can be pressure-treated, but some of them require it less than others. The inherently decay-resistant species, such as cypress, redwood, some of the cedars, and white oak, need no treatment for most uses. But this applies only to the heartwood of any specie.

Of the various chemicals used for wood preservative treatment, with the pressure system, creosote is one of the oldest and best. However, it has an odor that is sometimes offensive and is, therefore, not generally suitable for manufactured items that may be used in contact with the body, or adjacent to food or enclosed places where the fumes may become objectionable. It can also be somewhat of a fire hazard and is sometimes not used for this reason. A third objection to creosote is the difficulty of painting over it. However, creosote is very effective against decay fungi and insect damage.

There are several effective forms of copper salts. These can be painted over; they do not contribute to fire hazard and are odorless.

They are extensively used in the manufacture of boats and other marine appliances.

A relatively new chemical for decay and insect prevention is pentachlorophenol. This has most of the advantages of creosote, except that it is probably not as effective against termite damage and is more expensive, but it lacks most of the disadvantages and can be used in enclosed places.

For food containers, preservative treatments are not generally recommended unless the conditions are closely examined and there is assurance that the chemicals are approved under the existing laws and are in no way harmful to health.

The second method of treatment is by one of the surface systems. Generally, the chemicals used for this purpose are almost the same as for pressure treating. However, they are often specially prepared and their carriers may differ to obtain greater natural penetration.

Of the various methods of application, dipping is the most efficient, for in this way all of the surface is exposed to the chemicals and a maximum amount of usable chemical is deposited, which may not be the case with brush or spray treatments. Often, heating the chemicals and cooling the pieces during dipping will increase penetration. However, long periods of soaking are not usually of much advantage. It is better to dip the wood parts for several minutes (sufficient time to assure that surface exposure is complete) and then pile the pieces closely together soon after they are removed from the chemical bath. Several days of natural absorption under these conditions will often provide a surprising amount of penetration. The success of this method depends to a considerable extent on the type of chemical used and particularly the carrier.

In the millwork industry, the surface method of application is of tremendous importance and is widely used for nearly all its products. Windows, storm sash, exterior trim, and most other products exposed to weather are effectively treated. In addition, the preservative chemicals are often combined with waxes or oils to provide a reasonable amount of dimensional stability.

Properties

The physical and mechanical characteristics of wood are controlled by specific anatomy, moisture content, and, to a lesser extent, mineral and extractive content. The properties are also influenced by the directional nature of wood, which results in markedly different properties in the longitudinal, tangential, and radial directions or axes. Wood properties within a species vary greatly from tree to tree and within a single axis. The physical properties (other than appearance) are moisture content, shrinkage, density, permeability, and thermal and electrical properties.

The mechanical properties of wood include elastic, strength, and vibration characteristics. These properties are dependent upon species, grain orientation, moisture content, loading rate, and size and location of natural characteristics such as knots.

Because wood is an orthotropic material, it has unique and independent mechanical properties in each of three mutually perpendicular axes—longitudinal, radial, and tangential. This orthotropic nature of wood is interrupted by naturally occurring characteristics such as knots that, depending on size and location, can decrease the stiffness and strength of the wood.

Wood Failure

The rupturing of wood fibers in strength tests on adhesively bonded specimens, usually expressed as the percentage of the total area involved that shows such failure.

Wood Flour

A pulverized wood product used in the foundry to furnish a reducing atmosphere in the mold, help overcome sand expansion, increase flowability, improve casting finish, and provide easier shakeout.

Finely ground dried wood employed as a filler and as reinforcing material in molding plastics and in linoleum, and as an absorbent for nitroglycerin. It is made largely from white-colored softwoods, chiefly pine and spruce, but maple and ash flours are preferred where no resin content is desired. Woods containing essential oils, such as cedar, are not suitable. Wood flour is produced from sawdust and shavings by grinding in burr mills. It has the appearance of wheat flour. The sizes commonly used are 40, 60, and 80 mesh; the finest is 140 mesh. Grade 1, used as a filler in rubber and plastics, has a particle size of 60 mesh and a specific gravity of 1.25, but 80 and 100 mesh are also used for plastics filler. Since wood flour absorbs the resin or gums when mixed in molding plastics and sets hard, it is sometimes mixed with mineral powders to vary the hardness and toughness of the molded product.

These quantities of *sawdust* are obtained in the sawmill areas. Besides being used as a fuel, it is employed for packing, for finishing metal parts in tumbling machines, for making particleboard, and for distilling to obtain resins, alcohols, sugars, and other chemicals. Some sawdust is pulped, and as much as 20% of such pulp can be used in kraft paper without loss of strength. Hickory, walnut, and oak sawdusts are used for meat smoking, or for the making of *liquid smoke*, which is produced by burning the sawdust and absorbing the smoke into water. For the rapid production of bacon and other meats, immersion in liquid smoke imitates the flavor of smoked meat. Some sawdust is used for agricultural mulch and fertilizer by chemical treatment to accelerate decay. *Bark fuel* is shredded bark, flash-dried and palletized with powdered coal. *Particleboard*, made by compressing sawdust or wood particles with a resin binder into sheets, has uniform strength in all directions, and a smooth, grainless surface. When used as a core for veneer panels, it requires no cross-laminating. Mechanical pulp for newsprint can be made from sawdust, but the quantity available is usually not sufficient. The material known as *ground wood*, of fine-mesh fibers, is made from cord wood, about 1 ton of fibers being produced from one cord of pulpwood. *Plastic wood*, usually marketed as a paste in tubes for filling cavities or seams in wood products, is wood flour or wood cellulose compounded with a synthetic resin of high molecular weight that will give good adhesion but not penetrate the wood particles to destroy their nature. The solvent is kept low to reduce shrinkage. When cured in place, the material can be machined, polished, and painted.

Wood Laminate

See *built-up laminated wood*, *glue-laminated wood*, and *plywood*.

Wood Preservatives

These fall into two general classes: oils, such as *creosote* and petroleum solutions of *pentachlorophenol*; and *waterborne salts* that are applied as water solutions. **Coal tar** creosote, a black or brownish oil made by distilling coal tar, is the oldest and still one of the more important and useful wood preservatives. Because it has recently been classified as a carcinogen, its use is expected to decrease. Its advantages are high toxicity to wood-destroying organisms; relative insolubility in water and low volatility, which impart to it a great degree of permanence under the most varied use conditions; ease of application; ease with which its depth of penetration can be determined; general availability and relatively low cost; and long record of satisfactory use.

Creosotes distilled from tars other than coal tar are used to some extent for wood preservation. For many years, either *cold tar* or *petroleum oil* has been mixed with cold tar creosote in various proportions to lower preservative costs.

Water-repellent solutions containing chlorinate phenols, principally pentachlorophenol, in solvents of the mineral spirit type have been used in commercial treatment of wood by the millwork industry since about 1931.

Pentachlorophenol solutions for wood preservation generally contain 5% (by weight) of this chemical, although solutions with volatile solvents may contain lower or higher concentrations. Preservative systems containing water-repellent components are sold under various trade names, principally for the dip or equivalent treatment of window sash and other millwork. According to federal specifications, the preservative chemicals may not contain less than 5% pentachlorophenol.

Standard wood preservatives used in water solution include *acid copper chromate, ammoniacal copper arsenite, chromated copper arsenate, zinc naphthenate, chromated zinc chloride*, and *fluor chrome arsenate phenol*. These preservatives are often employed when cleanliness and paintability of the treated wood are required. The chromated zinc chloride and fluor chromate arsenate phenol formulations resist leaching less than preservative oils and are seldom used where a high degree of protection is required for wood in ground contact or for other wet installations. Several formulations involving combinations of copper, chromium, and arsenic have shown high resistance to leaching and very good performance in service. The ammoniacal copper arsenite and chromated copper arsenate are included in specifications for such items as building foundations, building poles, utility poles, marine piling, and piling for land and freshwater use. *Organic sulfones* are another class of wood preservatives offering high degrees of protection. One such product is *diiodomethyl p-tolyl sulfone*, with trade name *Amical*.

Wood Veneer

A thin sheet of wood generally within the thickness range from 0.3 to 6.3 mm (0.01–0.25 in.) to be used in a laminate.

Woods, Imported

There are over 100 different species of foreign woods that are imported into the United States. By far, the greatest number of these are used for decorative or nonengineering purposes—some are used for both.

Imported woods to be considered here are all defined as imported species, except those that are also native to the United States. Thus, all Canadian kinds are excluded because the United States has every species of wood that grows in Canada. Mexico, on the other hand, supplies some tropical woods not found in the United States. These are pines from the highlands of Mexico and Central America that are imported in fairly sizable volume at times, but the engineering aspects of their utilization coincide directly with the southern pine found in the United States; consequently, no attempt is made to describe these. No other softwoods (conifers) are imported for uses as engineering materials. Thus, all types discussed here are categorized as hardwood (broadleaf) species. Even balsa is a hardwood because the terms "hardwood" and "softwood" in lumber-industry parlance allude to the botanical classification rather than the actual hardness or softness of the wood.

There are over 13 kinds of tropical hardwoods known to be used for engineering products or purposes. In certain parts of the world, there are vast expanses of untouched tropical forests, which, no doubt, contain several other hardwoods of potential

W

engineering value. However, the woods and current use are typical imported woods and applications:

1. Refrigeration: balsa
2. Wharves and docks: Greenheart, ekki, jarrah, ironbark, apitong, angelique
3. Boat construction: Philippines mahogany, Central and South American mahogany, African mahogany, balsa, apitong, teak, iroko, ironbark, jarrah, lignum vitae
4. Tanks and vats: Philippine mahogany, apitong
5. Building construction: mahogany (all types), apitong, balsa, greenheart
6. Poles, piling: greenheart, ekki
7. Machinery: lignum vitae
8. Aircraft and missiles: balsa
9. Vehicles: apitong

Wood-Based Fiber and Particle Materials

Flat-formed board products can be classified into two groups: (1) those primarily from fiber interfelted during manufacture with a predominantly natural bond, although extraneous material may be added to improve some property such as bond (or other strength property) and water resistance and (2) those made from distinct fractions of wood with the primary bond produced by an added bonding material.

Moldings, based on a composition of wood-based fiber or particle and binding material similar to those used for boards, have found increasing use.

Production statistics for molded units of fiber and wood particle are difficult to obtain because the units are used in such widely different commodities as toilet seats, croquet balls, school desktops and seats, frames for luggage, and armrests, door panels, and other molded components for automobiles. The number of uses for these moldings is increasing because they combine shape with adequate structural strength and durability for many uses. See *wood*.

Wood–Metal Laminates

Wood–metal laminates constitute a composite panel construction in which the core material is made up of a wood or wood-derivative slab to which metal facing sheets are adhesively bonded.

The core material is usually plywood or a composition board of wood fibers or chips compressed and bonded together to form a flat core slab. Balsa wood and insulation boards are frequently used as core materials where thermal insulation is a requirement.

The metal facing sheets may be steel, aluminum, stainless steel, porcelain enameled metals, rigidized metals, or metals with decorative finishes. Light-gauge metals are usually used, ranging in thickness from 0.25 to 1.5 mm depending upon the specific strength, stiffness, and service requirements.

Adhesives used in bonding the metal to the core material are selected to meet the desired service requirements. For most standard laminates, the adhesive is water-resistant and fungus-proof so that the panels may be subjected to exterior as well as interior exposure. Continuous surface temperatures may range from a minimum of −51°C to a maximum of 77°C, although specially prepared laminates are available for continuous use as high as 177°C.

Greater Stiffness

Wood–metal laminates are designed to utilize the best properties of each of the component parts providing a panel that is not only light in weight but also has good structural strength and high flexural rigidity. Because the metal facing sheets are supported by a core material of substantial thickness, smooth flat panels free from waves and buckles are obtained with light-gauge metals.

Wood–metal laminates are limited in size only by the size of the press equipment and commercial sizes of the metal sheets and core material *available*.

The panels may be sawed to exact size from stock size sheets. Frequently, however, the facing sheets are fabricated prior to bonding to the core material to provide special edge details.

Applications

Wood–metal laminates are used extensively in the architectural building and transportation fields. Curtain wall panels, column enclosures, partition panels, facia, and soffit panels are typical applications. Truck and trailer bodies, shipping containers, and railroad car partition panels and doors are common uses of the laminate. In general, wherever light weight, combined with high rigidity and structural strength, is prerequisite, wood–metal laminates may be used to good advantage.

Woods Metal

An alloy of 50% bismuth, 27% lead, 13% tin, and 10% cadmium, which melts at 66°C.

Woody Fracture

(1) A fracture surface having a coarsely fibrous texture. It usually reflects an elongated grain structure or large quantities of strangers of inclusions. (2) A macrostructure, found particularly in wrought iron and in extruded rods of aluminum alloys, that shows elongated surfaces of separation when fractured.

Woof

See weft.

Wool Grease

A brownish, waxy fat of a faint, disagreeable odor, obtained as a by-product in the scouring of wool. The purified grease was formally known as **degras** and was used for leather dressing, and lubricating and slushing oils, and in soaps and ointments; but it is now largely employed for the production of lanolin and its derivatives, chiefly for cosmetics. Wool grease contains **lanoceric acid**; *lanopalmic acid*, and **lanosterol**, a high alcohol related to cholesterol. Although these can be broken down into derivatives.

Lanolin is a purified and hydrated grease, also known as *lanain*, and in pharmacy as *lanum* and *adeps lanae*. It has a melting point of about 40°C and is soluble in alcohol. Lanolin is basically a wax consisting of esters of sterol alcohols combined with straight-chain fatty acids, and with only a small proportion of free alcohols. It contains about 95% of fatty acid esters, but its direct use as an emollient depends on the 5% of free alcohols and acids. However, more than 30 derivatives are obtained from lanolin, and these are used in blends to give specific properties to cosmetics. They are often marketed under trade names, and some of the ingredients may be synthesized from raw materials other than wool grease, or chemically altered from wool-grease derivatives.

A variety of products used in cosmetics and pharmaceuticals are made by fractionation or chemical alteration of lanolin. They are also useful in compounding plastics and industrial coating but are generally too scarce and expensive for these purposes. *Ethoxylated lanolin* and ethoxylated *lanolin alcohols* are used in water-soluble emulsions and conditioners. *Solulan* is a general trade name for these materials. *Lanolin oil* and *lanolin wax* are made by solvent fractionation of lanolin. *Viscolan* and *Waxolan* are these products. *Barium lanolate*, made by saponification, is used as a anticorrosion agent. It is antiphobic and is also used as an anticaking agent. In a 25% barium concentration, it is used for hard lubricating grease.

Wootz Process

An ancient steelmaking process that utilized the natural wind, suitably channeled, as an air blast for crucible smelting. The iron was then subjected to a cementation process.

Work Angle

The angle that the electrode makes with the referenced plane or surface of the base metal in a plane perpendicular to the axis of the weld. See also *drag angle* and *push angle*.

Work Angle (Pipe)

The angle that the electrode makes with the referenced plane extending from the center of the pipe through the molten weld pool. See the term *travel angle* (pipe).

Work Coil

The inductor used when welding, brazing, or soldering with induction heating equipment. See also *induction work coil* and the term inductor.

Work Connection

In welding, the connection of the work lead to the work.

Work Factor

A measure of the stability of a lubricant when subjected to an endurance test. The work factor is expressed as the average value of the ratio of three characteristics (viscosity, carbon residue, and neutralization number) as measured before the test to those same characteristics as measured after the test.

Work Lead

The electrical conductor connecting the source of arc welding current to the work. Also called work connection, welding ground, or ground lead.

Work Hardening

Same as strain hardening.

Workability

See formability.

Working

Causing permanent deformation to material. The terms "work" and "plastically deform" are effectively synonymous; the former tends to be used in a manufacturing contest and the latter in an academic or mechanical testing context. Cold working refers to processes where no recrystallization occurs and the component work hardens, that is, becomes progressively harder, up to some limiting level, and less ductile. Hot working refers to a process, usually at elevated temperature, where deformation is accompanied by recrystallization, that is, no hardening occurs. Warm working is a special case of cold working involving deliberate heating, perhaps to assist deformation or minimize oxidation, but the temperature reached is insufficient to allow recrystallization.

Working Distance

The distance between the surface of the specimen being examined microscopically and the front surface of the objective lens.

Working Electrode

The test specimen electrode in an *electrochemical cell*.

Working Life

The period of time during which a liquid resin or adhesive, after mixing with catalyst, solvent, or other compounding ingredients, remains usable. See also *gelation* time and *pot* life.

Working Stress/Load

This usually means the safe working load or the associated stress.

Worm

An exudation (sweat) of molten metal forced through the top crust of solidifying metal by gas evolution. See also *zinc worms*.

Wormhole

Elongated forms of gas pores or blowholes, particularly in welds.

Woven Fabric

A material (usually a planar structure) constructed by interlacing yarns, fibers, or filaments to form such fabric patterns as plain, harness satin, and leno weaves. See the terms *basketweave* (composites), *crowfoot satin*, *eight-harness satin*, *leno weave*, and *plain weave*.

Woven Roving

A heavy glass fiber fabric made by weaving roving, or yarn bundles.

Wrap Forming

See *stretch forming*.

Wrap Seam

A depression or step in the surface finish of a reinforced plastic caused by the lap of the flexible mold or carrier strip after it is removed from the cured pultrusion. See also *pultrusion*.

W

Wrap-Around Bend

The bend obtained when a specimen is wrapped in a closed helix around a cylindrical mandrel.

Wrapped Bush (Bearing)

A thin-walled steel bush lined with a bearing alloy, or any other bearing bush made from strip.

Wrapping Test

A test in which wire or similar section is wound around a former or itself a specified number of times to assess its ductility.

Wringing Fit

A fit of nominally zero allowance.

Wrinkle

A surface imperfection in laminated plastics that has the appearance of a crease or fold in one or more outer sheets of the paper, fabric, or other base. Also occurs in vacuum bag molding when the bag is improperly placed, causing a crease.

Wrinkle Bend

A bend, usually in a tube, produced by forming one or, usually, more semi-circumferential corrugations along the side forming the bend inner surface. The process typically involves locally heating a narrow band on the intended inside of the bend. The tube is then bent a small amount. This is repeated a short distance along the tube until the required profile is achieved.

Wrinkle Depression

An undulation or series of undulations or waves on the surface of a pultruded composite part.

Wrinkling

A wavy condition obtained in deep drawing of sheet metal, in the area of the metal between the edge of the flange and the draw radius. Wrinkling may also occur in other forming operations when unbalanced compressive forces are set up.

Wrist Pin Bearing

The bearing at the crankshaft end of an articulated connecting rod in a "V" engine.

Wrought

Any material that has been shaped by a deformation process, as opposed to a casting.

Wrought and Cast Aluminum Alloy Designations

Systems for designating wrought and cast aluminum alloys that have been devised by the Aluminum Association in the United States and the American National Standards Institute. For wrought alloys, a four-digit system is used to produce a list of wrought composition families as follows:

Aluminum, ≥99.00%	1xxx
Aluminum alloys and grouped by major alloying element(s):	
Copper	2xxx
Manganese	3xxx
Silicon	4xxx
Magnesium	5xxx
Magnesium and silicon	6xxx
Zinc	7xxx
Other elements	8xxx
Unused series	9xxx

Wrought Iron

Wrought iron is commercially pure iron made by melting white cast iron and passing an oxidizing flame over it, leaving the iron in a porous condition, which is then rolled to unite it into one mass. As those made, it has a fibrous structure, with fibers of slag through the iron and the direction of rolling. It is also made by the Aston process of shooting Bessemer iron into a ladle of molten slag. Modern wrought iron has a fine dispersion of silicate inclusions that interrupt the granular pattern and gives it a fibrous nature.

Structurally, wrought iron is a composite material; the base metal and the slag are in physical association, in contrast to the chemical or alloy relationship that generally exists between the constituents of other metals.

Originally iron of fairly high purity made by various smelting techniques followed by repeated forging to produce a tough, fairly corrosion resistant material suitable for major structures as well as decorative purposes. The *Puddling process* was the most common process in the United Kingdom. The **Aston** process was more common in the United States. Originally, wrought iron was a major structural material. However, it was superseded by steel of lower cost, higher strength and is available in much larger sizes. Consequently, these older processes are now commercially obsolete and current usage of the term "Wrought Iron" usually refers to mild steel bar bent in an aesthetically pleasing manner for functional or decorative applications.

The form and distribution of the iron silicate particles may be stringerlike, ribbonlike, or platelets. Practically, the physical effects of the incorporated iron silicate slag must be taken into consideration in bending and forming wrought iron pipe, plate, bars, and shapes, but when properly handled—cold or hot—fabrication is accomplished without difficulty.

Mechanical Properties

The value of wrought iron is in its corrosion resistance and ductility. It is used chiefly for rivets, staybolts, water pipes, tank plates, and forged work. Minimum specifications for ASTM wrought iron call for a tensile strength of 275 MPa, yield strengths of 165 MPa, and elongation of 12%, with carbon not over 0.08%, but the physical properties are usually higher. Wrought iron 4D has only 0.02% carbon with 0.12% phosphorus, and the fine fibers are of a controlled composition of silicon, manganese, and phosphorus. This iron has a tensile strength of 330 MPa, elongation 14%, and Brinell hardness 105. Manganese wrought iron has 1% manganese for higher impact strength.

Ordinary wrought iron with slag may contain frequent slag cracks, and the quality grades are now made by controlled additions of silicate, and with controlled working to obtain uniformity. But for tanks and plate work, ingot iron is now usually substituted.

The Norway iron formally much used for bolts and rivets was a Swedish charcoal iron brought to America in Norwegian ships. This iron, with as low as 0.02% carbon, and extremely low silicon, sulfur, and phosphorus, was valued for its great ductility and toughness and also for its permeability qualities for transformer cores. Commercial wrought iron is now usually ingot iron or fibered low-carbon steel.

The tensile properties of wrought iron are largely those of ferrite plus the strengthening affect of any phosphorus content, which adds approximately 6.8 MPa for each 0.01% above 0.10% of contained phosphorus. Strength, elasticity, and ductility are affected to some degree by small variations in the metalloid content and in even greater degree by the amount of the incorporated slag and the character of its distribution.

Nickel, molybdenum, copper, and phosphorus are added to wrought iron to increase yield and ultimate strengths without materially detracting from toughness as measured by elongation and reduction of area.

Fabrication

Forging

Wrought iron is an easy material to forge using any of the common methods. The temperature at which the best results are obtained lies in the range of 1149°C–1316°C. Ordinarily "flat and edge" working is essential for good results. Limited upsetting must be accomplished at "sweating to welding" temperatures.

Bending

Wrought iron plates, bars, pipe, and structural may be bent either hot or cold, depending on the severity of the operation, keeping in mind that bending involves the directional ductility of the material. Hot bending ordinarily is accomplished at a dull red heat (704°C–760°C) below the critical "red-short" range of wrought iron (871°C–927°C). The ductility available for hot bending is about twice that available for cold bending. Forming of flanged and dished heads is accomplished hot from special forming, equal property plate.

Welding

Wrought iron can be welded easily by any of the commonly used processes, such as forge welding, electrical resistance welding, electric metallic arc welding, electric carbon arc welding, and gas or oxyacetylene welding. The iron silicate or slag included in wrought iron melts at a temperature below the fusion point of the iron-base metal so that the melting of the slag gives the metal surface a greasy appearance. This should not be mistaken for actual fusion of the base metal; heating should be continued until the iron reaches the state of fusion. The siliceous slag content provides a self-fluxing action to the material during the welding operation.

Threading

The machinability or free-cutting characteristics of most ferrous metals are adversely influenced by either excessive hardness or softness. Wrought iron displays almost ideal hardness for good machinability, and the entrained silicate produces chips that crumble and clear the dies. Standard threading equipment that incorporates minor variations in lip angle, lead, and clearance is usually satisfactory with wrought iron.

Protective Coatings

Wrought iron lends itself readily to such cleaning operations as pickling and sandblasting for the application of protective coatings. Where protective coatings such as paint or hot-dipped metallic coatings are to be applied, the coatings are found to adhere more firmly to wrought iron and a thicker coat will be attained compared with other wrought ferrous metals. This is because the natural surface of wrought iron is microscopically rougher than other metals after cleaning, thus providing a better anchorage for coatings.

Corrosion Resistance

The resistance of wrought iron to corrosion has been demonstrated by long years of service life and many applications. Some have attributed successful performance to the purity of the iron-based, the presence of a considerable quantity of phosphorus or copper, freedom from segregation, to the presence of the inert slag fibers disseminated throughout the metal, or to combinations of such attributes.

In actual service, the corrosion resistance of wrought iron has shown superior performance in such applications as radiant heating and snow-melting coils, skating-rink piping, condenser and heat-exchanger equipment, and other industrial and building piping services. Wrought iron has long been specified for steam condensate piping where dissolved oxygen and carbon dioxide present severe corrosion problems. Cooling water cycles of the once-through and open-recirculating variety are solved by the use of wrought-iron pipe.

Applications

Building Construction

Hot and cold potable water, soil, waste, vent, and downspout piping; radiant heating, snow melting, air-conditioning cooling, and chilled-water lines; gas, fire protection, and soap lines; condensate and steam returns, ice-rink and swimming-pool piping; underground service lines and electrical conduit.

Industrial

Unfired heat exchangers, brine coils, condenser tubes, caustic soda, concentrated sulfuric acid, ammonia, and miscellaneous process lines; sprinkler systems, boiler feed and blowoff lines, condenser water piping, runner buckets, skimmer bars, smokestacks and standpipes, and salt and water well pipe and casing.

Public Works/Infrastructure

Bridge railings, fenders, blast plates, drainage lines and troughs, traffic signal conduit, sludge digester heating coils, aeration tank piping, sewer outfall lines, large outside diameter intake and discharge lines, trash racks, wire plates, dam gates, pier-protection plates, sludge tanks and lines, and dredge pipe.

Railroad and Marine

Tie spacer bars, diesel exhaust and air-brake piping, ballast and brine protection plates, brine, cargo and washed down lines on ships, hull and deck plating, rudders, fire screens, breechings, tanker heating coils, car retarder and yard piping, spring bands, car charging lines, nipples, pontoons, and car and switch deicers.

Others

Gas collection hoods, staybolts, flue gas conductors, sulfur mining gut, air and transport lines, coal-handling equipment, chlorine, compressed air lines, distributor arms, cooling tower, and spray pond piping.

Wrought iron is available in the form of plates, sheets, bars, structural, forging blooms and billets, rivets, chain, and a wide range of tubular products, including pipe, tubing and casing, electrical conduit, cold-drawn tubing, and welded fittings.

Wrung Fit

See *interference fit*.

Wüstite

Iron oxide, FeO.

X

X-Axis

In reinforced plastic laminate, an axis in the plane of the laminate that is used as the 0° reference for designating the angle of a lamina. See also *laminate* and the term *quasi-isotropic laminate*.

Xenon

One of the noble or inert gases and a rare gas (symbol Xe; atomic number 54) that is used in photographic flash lamps, luminescent tubes, and lasers.

Xerography

A technique for producing photocopies.

X-Ray

(1) A penetrating electromagnetic radiation, usually generated by accelerating electrons to high velocity and suddenly stopping them by collision with a solid body. Wavelengths of x-rays range from about 10^{-1} to 10^2 Å, the average wavelength used in research being about 1 Å. Also known as roentgen ray or x-radiation. (2) Electromagnetic radiation of less than 500 Ångstom units emitted by decelerated electrons. See also *electromagnetic radiation*.

X-Ray Diffraction

(1) An analytical technique in which measurements are made of the angles at which x-rays are preferentially scattered from a sample (as well as of the intensities scattered at various angles) in order to deduce information on the crystalline nature of the sample—its crystalline structure, orientations, and so on. (2) Techniques utilizing the diffraction of x-rays, particularly to investigate crystal structures. The wavelength of x-rays is of the same order as the lattice spacing of metal crystals, so if an x-ray beam is directed at a crystal, the regular spacing of the planes of atoms causes diffraction effects that can be recorded on photographic film. The diffraction patterns can provide information on lattice spacings, extent of cold working and material composition, etc. See *Bragg* and *von Laue*.

X-Ray Diffraction Residual Stress Techniques

Diffraction method in which the strain in the crystal lattice is measured, and the residual stress producing the strain is calculated assuming a linear elastic distortion of the crystal lattice. See also *macroscopic stress*, *microscopic stress*, and *residual stress*.

X-Ray Emission Spectroscopy

Pertaining to *emission spectroscopy* in the x-ray wavelength region of the electromagnetic spectrum.

X-Ray Fluorescence

Emission by a substance of its characteristic x-ray line spectrum on exposure to x-rays.

X-Ray Fluorescence Spectroscopy

An analytical technique in which a beam of x-rays excites the characteristic radiation of elements in the item of interest. The radiation from the individual constituents is then measured in a spectrometer.

X-Ray Map

An intensity map (usually corresponding to an image) in which the intensity in any area is proportional to the concentration of a specific element in that area.

X-Ray Photoelectron Spectroscopy

An analytical technique that measures the energy spectra of electrons emitted from the surface of a material when exposed to monochromatic x-rays.

Carbon black–filled rubber parts may challenge many analytical methods because the compounded rubber may include up to 20 individual ingredients. Often, after mixing, it is difficult to extract ingredients from the polymer base because of their interactions with other ingredients such as carbon black.

Testing of bonded rubber-to-metal parts amplifies the problem twofold. The bonding adhesive is also a multicomponent polymeric blend, much like the rubber. Also involved are contaminants associated with the substrate. For example, a stamped steel insert may be contaminated with hydrocarbons. If the metal is porous, oil may bleed out of the metal grain over time and undermine a bond system. Or, if the oil is present initially, it may prevent adhesive wet-out, thus inhibiting adhesion to the substrate.

Disadvantages

Some techniques such as reflectants, Fourier transform infrared spectrometer, and energy dispersive x-ray (EDS) would not be adequate for testing rubber bonded to metal. EDS, using a windowless system, does not have sufficient resolution for lower-weight elements.

Process

To gain the most definitive picture of failed bonded surfaces, scanning electron microscopy and x-ray photoelectron spectroscopy (XPS or ESCA) methods have proven the most valuable.

XPS is an excellent method for determining chemical composition of a surface. This method penetrates the surface of the sample to a depth of only 50 Å. XPS measures the binding energies of the atoms present on the surface, yielding a very distinctive signature of the atomic species present. XPS is sensitive enough to generate signals that provide both a quantitative and qualitative picture of the surface composition over a wide elemental range. Computer analysis techniques convert these signals to atomic percentages to allow reconstruction of the various materials present.

Case Study

The following case study analyzes black rubber bonded to a steel part. The part was exposed to a hostile environment (hydrocarbon fluids at 150°C) for a relatively short period of time before failure occurred. Parts of this type are expected to last approximately 8 years; this part failed in under 6 months.

Initial visual analysis indicated failure primarily in an adhesive mode. However, no apparent cause of the failure was easily identifiable. Gaining a clear understanding of the cause of the adhesive failure at this point was difficult because the port was saturated with hydrocarbon oil.

The cross section of the remaining bonded area indicated that under normal conditions, the adhesive could not be distinguished from the rubber. This would indicate a high degree of commingling is necessary between rubber and adhesive to achieve a good bond.

At this point, the unbond was believed to be caused by the overbaking of the adhesive. In this situation, the poor bond is due to premature cross-linking of the adhesive prior to the application of rubber. The cross-linking of the adhesive causes mechanisms for adhesion to be compromised and inhibits physical interaction.

To confirm this, both a control part, with bond failure due to adhesive overbake, and the failed part were subjected to analysis by XPS. Only failed mating surfaces were examined by this method. Typical areas scanned by this method were 1 × 3 mm. It is not necessary to neutralize with the electron gun due to only minimal charging effects. Only a single sample was tested for each condition.

The XPS analysis of the mating surfaces of the defective part showed the surfaces to be very similar in elemental composition. Values from the failed part, when compared with the control part, indicated the hydrocarbon fluid did not affect the analysis. This is most likely because the high vacuum necessary in this testing technique removed the hydrocarbon fluid from the surface.

The trace elements present in analysis of both surfaces of both parts support the conclusion that the parts failed by the same mechanism. Test results support the idea that both samples failed in a boundary layer between the rubber and the adhesive. Trace elements, known not to be native to the rubber compound, indicate the source to be the adhesive.

From the presence of adhesive components on all mating surfaces tested, we can effectively eliminate contamination as the cause of the unbond. The presence of the adhesive and the pulled-out material present on the failed part would indicate that the part may have been weakly bonded. Finally, the evident lack of adhesive and rubber interspersion would indicate one or both components had achieved sufficient viscosity to resist flow. The combined evidence using XPS tends to support our initial observation that all adhesives had prematurely cross-linked. These data allow the cause to be correctly identified and will help produce a more robust part and processing in the future.

X-Ray Spectrograph

A photographic instrument for x-ray emission analysis. If the instrument for x-ray emission analysis does not employ photography, it is better described as an x-ray spectrometer.

X-Ray Spectrometry

Measurement of wavelengths of x-rays by observing their diffraction by crystals of known lattice spacing.

X-Ray Spectrum

The plot of the intensity or the number of x-ray photons versus energy (or wavelength).

X-Ray Technology

As with medical x-ray instruments, there are analytical x-ray measurements that can produce images of internal structures of objects that are opaque to visible light. There are instruments that can determine the chemical elemental composition of an object, which can identify the crystalline phases of a mixture of solids, and others that determine the complete atomic and molecular structures of a single crystal. The determination of particle size and structural information for fibers and polymers and the study of stress, texture, and thin films are x-ray applications that are growing in importance.

Characteristics and Generation of X-Rays

X-Ray Electromagnetic Spectrum

X-rays are a form of electromagnetic radiation and have a wavelength, λ, much shorter than visible light. The center of the visible light spectrum has a wavelength of about 0.56×10^{-6} m. The most commonly used methods for generating x-rays are the synchrotron and x-ray tubes.

Synchrotron Radiation

X-rays are produced when very energetic electrons traveling close to the speed of light are decelerated. In synchrotrons, electrons are accelerated with electromagnets while traveling along a linear path. Then, they are inserted into a nearly circular path, which is maintained by bending magnets.

X-Ray Tubes

X-ray tubes are the most widely used source for the generation of x-rays. In these tubes, electrons are accelerated by a high electric potential (20–120 kV). These electrons strike the target (anode) of the tube and decelerate as they pass through the electron clouds of the atoms. This phenomenon produces a continuous spectrum similar, but much less intense, to that of the synchrotron. In addition, some high-energy electrons knock electrons out of the atomic orbitals of the atoms of the target material. When these orbitals are refilled by electrons, x-ray photons are generated. The resulting x-ray spectrum of intensity versus wavelength has a series of peaks known as characteristic lines.

The materials that are used as targets in x-ray tubes depend on the application.

Applications

X-ray applications can be placed into three categories:

1. X-ray radiography permits the imaging of the internal structure of an object (e.g., bones of a hand). It is based on the comparative observance of photons as they travel through the different materials making up the object.

2. X-ray fluorescence spectrometry consists of the measurement of the incoherent scattering of x-rays. It is used primarily to determine the elemental composition of a sample.
3. X-ray diffraction consists of the measurement of the coherent scattering of x-rays. X-ray diffraction is used to determine the identity of crystalline phases in a multiphase powder sample and the atomic and molecular structures of single crystals. It can also be used to determine structural details of polymers, fibers, thin films, and amorphous solids and to study stress, texture, and particle size.

X-Ray Fluorescence Spectrometry

X-ray fluorescence spectrometry is a technique for measuring the elemental composition of samples. The basis of the technique is the relationship between the wavelength or energy of the emitted incoherently scattered x-ray photons and the atomic number of the element. When an atom is bombarded with x-ray photons of sufficient energy, an inner-orbital electron may be displaced, leaving the atom in an excited state. The atom can return to the ground state by transference of a higher orbital electron into the vacancy (the resulting higher-level vacancy is filled by an electron from a still higher level). In so doing, the difference in energy between the electron ousted from the lower shell and the energy of the higher orbital electron is emitted as an x-ray photon. Each element produces a fluorescence spectrum of intensity versus wavelength that is characteristic of that element.

X-Ray Radiography

X-ray imaging tests are widely used to examine interior regions of metal castings, fusion welds, composite structures, and brazed components. Radiographic tests are made on pipeline welds, pressure vessels, nuclear fuel rods, and other critical materials and components that may contain 3D voids, inclusions, gaps, or cracks. Since penetrating radiation tests depend upon the absorption properties of materials on x-ray photons, the tests can reveal changes in thickness and density and the presence of inclusions in the material.

X-ray fluoroscopy is used for direct online examination. A fluorescent screen is used to convert x-ray photons into visible light photons. A television camera receives the visible image and displays it on a television screen. This type of system is used for security screening of carry-on luggage at airports.

As in medical x-ray imaging, computerized tomography can reveal the details of the internal structure of complex objects. Many detectors are used to measure the transmittance of x-rays along many lines through the object. A computer uses this information to produce an image of the internal structure of a slice of the object.

X-Ray Topography

A technique that comprises topography and x-ray diffraction. The term topography refers to a detailed description and mapping of physical (surface) features in a region. In the context of the x-ray diffraction, topographic methods are used to survey the lattice structure and imperfections in crystalline materials.

X-Ray Tube

A device for the production of x-rays by the impact of high-speed electrons on a metal target.

Xylene

Xylenes and ethylbenzene and C_8 aromatic isomers with the molecular formula C_8H_{10}. Xylenes consist of three isomers: o-xylene (OX), m-xylene, and p-xylene (PX). These differ in the positions of the two methyl groups on the benzene ring.

Uses

The majority of xylenes that are mostly produced by catalytic reforming or petroleum functions are used in motor gasoline. The majority of the xylenes that are recovered for petrochemical use are used to produce PX and OX. PX is the most important commercial isomer.

Xylylene Polymers

In a process capable of producing pinhole-free coatings of outstanding conformality and thickness uniformity through the unique chemistry of PX, a substrate is exposed to a controlled atmosphere of pure gaseous monomer. The coating process is best described as a vapor deposition polymerization. The monomer molecule is thermally stable but kinetically very reactive toward polymerization with other molecules of its kind. Although it is stable as a rarefied gas, upon condensation, it polymerizes spontaneously to produce a coating of high-molecular-weight, linear poly (p-xylylene).

Applications

Because the parylenes are generally insoluble in most solvents, even at elevated temperatures, they cannot be used as solvent-based coatings; neither can they be cast as films nor spun as fibers from solution.

The most important application of parylenes is as a conformal coating for printed wiring assemblies. These coatings provide excellent chemical resistance and resistance to fungal attack. In addition, they exhibit stable dielectric properties over a wide range of temperatures.

The use of parylenes as a hybrid circuit coating is based on much the same rationale as its use in circuit boards. A significant distinction lies in obtaining adhesion to the ceramic substrate material, the success of which determines the eventual performance of the coated part. Adhesion to the substrate must be achieved using adhesion promoters, such as the organosilanes.

Parylenes are superior candidates for dielectrics and high-quality capacitors. Their dielectric constant and loss remain constant over a wide temperature range. The thermistor sensing probe of a disposable bathythermograph is coated with parylene. This instrument is used to chart the ocean water temperature as a function of depth.

Parylene is used in the manufacture of high-quality miniature stepping motors, such as those used in wristwatches, and as a coating for the ferrite cores of pulse transformers, magnetic tape recording heads, and miniature inductors.

Use of parylene in the medical field is linked to electronics, for example, as a protective conformal coating on pacemaker circuitry.

As books age, the paper of their pages becomes brittle. A relatively thick coating of parylene can make these embrittled pages stronger.

By separating the coating from the substrate after deposition, the unique coating features of parylenes, especially continuity and thickness control and uniformity, can be imparted to a freestanding film. Applications include optical beam splitters, a window for a micrometeoroid detector, a detector cathode 4*n* x-ray streak camera, and windows for x-ray proportional counters.

Parylenes can be used for contamination control, that is, securing small particles to prevent them from damaging a surface in a sealed unit, barrier coating, coating for corrosion control, and as dry lubricants.

XY Plane

In reinforced plastic laminates, the reference plane parallel to the plane of the laminate.

Y

Yarns

Yarns are assemblages or bundles of fibers twisted or laid together to form continuous strands. They are produced with either filaments or staple fibers. Single strands of yarns can be twisted together to form ply or plied yarns, and ply yarns in turn can be twisted together to form cabled yarn or cord. Important yarn characteristics related to behavior are fineness (diameter or linear density) and the number of twists per unit length. The measure of fineness is commonly referred to as yarn number. Yarn numbering systems are somewhat complex, and they are different for different types of fibers. Essentially, they provide a measure of fineness in terms of weight per unit or length per unit weight.

Cotton yarns are designated by numbers, or counts. The standard count of cotton is 840 yd/lb. Number 10 yarn is therefore 8,400 yd/lb. A No. 80 sewing cotton is 80×840, or 67,200 yd/lb.

Linen yarns are designated by the lea of 300 yd. A 10-count linen yarn is 10×300, or 3000 yd/lb.

The size or count of spun rayon yarns is on the same basis as cotton yarn. The size or count of rayon filament yarn is on the basis of the denier, the rayon denier being 450 m weighing 5 cg. If 450 m of yarn weighs 5 cg, it has a count of 1 denier. If it weighs 10 cg, it is No. 2 denier. Rayon yarns run from 15 denier, the finest, to 1200 denier, the coarsest.

Reeled silk yarn counts are designated in deniers. The international denier for reeled silk is 500 m of yarn weighing 0.05 g. If 500 m weighs 1 g, the denier is No. 20. Spun silk count under the English system is the same as the cotton count. Under the French system, a count is designated by the number of skeins weighing 1 kg. The skein of silk is 1000 m.

A ply yarn is one that has two or more yarns twisted together. A two-ply yarn has two separate yarns twisted together. The separate yarns may be of different materials, such as cotton and rayon. A six-ply yarn has six separate yarns. A ply yarn may have the different plies of different twists to give different effects. Ply yarns are stronger than single yarns of the same diameter. Tightly twisted yarns make strong, hard fabrics. Linen yarns are not twisted as tightly as cotton because the flux fiber is longer, stronger, and not so fuzzy as the cotton. Filament rayon yarn is made from long, continuous rayon fibers, and it requires only slight twist. Fabrics made from filament yarn are called twalle. Monofilament is fiber heavy enough to be used alone as yarn, usually more than 15 denier. Tow consists of multifilament reject strands suitable for cutting into staple lengths for spinning. Spun rayon yarn is made from staple fiber, which is rayon filament cut into standard short lengths.

Yarn Bundle

See *bundle*.

Y-Axis

In composite laminates, the axis in the plane of the laminate that is perpendicular to the *x*-axis. See also *x-axis*.

Y-Block

A single *keel block*.

Yellow Brass/Metal

Any brass particularly as castings. A name sometimes used in reference to the 65Cu–35Zn type of brass, and sometimes the term refers specifically to brass with 60% copper and 40% zinc.

Yield

(1) Evidence of plastic deformation in structural materials. Also known as plastic flow or creep. See also *flow*. (2) The ratio of the number of acceptable items produced in a production run to the total number that was attempted to be produced. (3) Comparison of casting weight to the total weight of metal poured into the mold. (4) To suffer permanent deformation. See *tensile test*.

Yield Extension

The amount of extension between the upper and lower yield points. See *tensile test*.

Yield Point

The first stress in a material, usually less than the maximum attainable stress, at which an increase in strain occurs without an increase in stress. Only certain materials—those that exhibit a localized, heterogeneous type of transition from elastic to plastic deformation—produce a yield point. If there is a decrease in stress after yielding, a distinction may be made between upper and lower yield points. The load at which a sudden drop in the flow curve occurs is called the upper yield point. The constant load in the flow curve is the lower yield point.

The level of stress and strain at which a material suffers permanent deformation. See *tensile test*.

Yield Point Elongation

In materials that exhibit a yield point, the difference between the elongation at the completion and at the start of discontinuous yield.

Yield Strength

The stress at which a material exhibits a specified deviation from proportionality of stress and strain. An offset of 0.2% is used for many materials, particularly metals. Compare with *tensile strength*.

Yield Stress

The stress level of highly ductile materials at which large strains take place without further increase in stress.

Yield Value

The stress (either normal or shear) at which a marked increase in deformation occurs without an increase in load.

Young's Modulus

(1) A term used synonymously with modulus of elasticity. The ratio of tensile or compressive stresses to the resulting strain. (2) The elastic modulus—see *tensile test*. See also *modulus of elasticity*.

Ytterbium

A metallic element, one of the rare earth groups.

Yttrium and Alloys

A chemical element, yttrium (symbol Y, atomic number 39, atomic weight 88.905) resembles the rare earth elements closely. The stable isotope yttrium-89 constitutes 100% of the natural element, which is always found associated with the rare earths and is frequently classified as one.

Yttrium forms a white oxide, Y_2O_3, which dissolves in acid to form trivalent yttrium salts. Yttrium has become commercially important since 1964. Yttrium forms the matrix for the europium-activated yttrium phosphors. These phosphors, when excited by electrons, emit a brilliant, clear-red light. The television industry uses these phosphors in manufacturing television screens. It is claimed that this phosphor gives better color reproduction and a much brighter screen than did the older non-rare-earth red phosphor. Also, the yttrium iron garnets, $Y_3Fe_5O_{12}$, and other garnets have found important uses in radar and communication devices. They transmit shortwave energy with very small losses.

Yttrium metal absorbs hydrogen, and in alloys up to a composition of YH_2, they resemble metals very closely. In fact, in certain composition ranges, the alloy is a better conductor of electricity than the purer metal. The density of hydrogen near the YH_2 composition is greater per cubic centimeter than it is in water or liquid hydrogen; therefore, such alloys make excellent potential moderators for nuclear reactors. Also, these alloys can be heated to a white heat (about 1260°C) before the vapor pressure of hydrogen exceeds 1 atm (10^2 kPa), and therefore the moderator in the reactor can be operated at very high temperatures. Yttrium metal has a low nuclear cross section, so it is also a potential structural material for reactors of the future.

Yttrium is used commercially in the metal industry for alloy purposes and as a "getter" to remove oxygen and nonmetallic impurities in other metals. Radioactive yttrium isotopes have been used in attempts at treating cancer.

Yttrium Aluminum Garnet

Yttrium aluminum garnet has the formula $Y_3Fe_5O_{12}$; its crystals are capable of sustaining laser activity when doped with neodymium.

Yttrium Oxide

This oxide (Y_2O_3) has a melting point of 2685°C, a density of 5.03 g/cm^3, and is soluble in acids, but only slightly soluble in water.

Yttrium is not a rare earth but always occurs with them in minerals because of similar general chemistry. Applications are in electrically conducting ceramics, refractories, insulators, phosphors, glass, special optical glasses, and other ceramics.

White powder has cubic crystal structure and small amounts of dysprosium oxide, gadolinium oxide, and terbium oxide as impurities.

Yttria can be compounded into polycrystalline as well as single-crystal garnets for use in microwave generation and detection devices. Such materials are important to microwave technology because they exhibit both good dielectric and magnetic properties, which can be controlled through compositional variations.

Yttria-stabilized zirconia can be used to produce a high-quality diamond substitute for jewelry or a rugged sensor for measuring oxygen in automotive exhaust, depending on the method of fabrication. Nd/yttrium aluminum garnet single crystal rods find many applications as lasers in industry and research.

Y_2O_3 can be used (with scandium, lanthanum, and cesium oxides) with TiO_2 bodies for better control of properties than with alkaline earths. In combination with europium oxide, yttria is used to make the red phosphor in color television picture tubes. Combined with ZrO_2, it makes good high-temperature refractories. It is also used in silicon nitride as a sintering aid.

Yttria is a rare earth oxide that is added in small amounts (0.5–1.3 wt.%) to nickel-base powder metallurgy superalloy compositions in order to improve their high-temperature creep resistance and oxidation resistance. Such alloys are referred to as oxide-dispersion-strengthened alloys. Yttria is also an additive to ceramic compositions, most notably zirconia. See also *dispersion-strengthened material*, *mechanical alloying*, and *zirconia*.

Y-TZP

Yttria tetragonal zirconia polycrystalline (Y-TZP) is a fine-grained ceramic used in special engineering applications that benefit from its high density, excellent wear resistance, and fine grain size, such as fiber-optic ferrules. High-purity fine reactive coprecipitated zirconia powders containing 3 mol% yttria are used to produce Y-TZP ceramics.

Z

ZAF Corrections

A quantitative x-ray program that corrects for atomic number (*Z*), absorption (*A*), and fluorescence (*F*) effects in a matrix.

Z-Axis

In laminates, the reference axis normal to the *xy* plane of the laminate.

Zeeman Effect

A splitting of the degenerate electron energy level into states of slightly different energies in the presence of an external magnetic field. This effect is useful for background correction in *atomic absorption spectrometry*.

Zero Bleed

A laminate fabrication procedure that does not allow loss of resin during cure. Also describes prepreg made with the amount of resin desired in the final part, such that no resin has to be removed during cure. See also *laminate* and *prepreg*.

Zero Time

The time at which the given loading or constraint conditions are initially obtained in creep and stress-relaxation tests, respectively.

Zeta Potential

See *electrokinetic potential*.

Ziegler–Natta Catalysts

Initially, a catalyst consisting of an alkylaluminum compound with a compound of the titanium group of the periodic table, a typical combination being triethylaluminum at either titanium tetrachloride or titanium trichloride. Subsequently, an enormous variety of such mixtures is used in polymerization to provide stereospecificity (isotactic or syndiotactic). See also *isotactic stereoisomerism* and *syndiotactic stereoisomerism*.

Zinc and Alloys

A bluish-white crystalline metal, zinc (symbol Zn) has a specific gravity of 7.13, melts at 420°C, and boils at 906°C. The commercially pure metal has a tensile strength, cast, of about 62 MPa with an elongation of 1%, and the rolled metal has a strength of 165 MPa with an elongation of 35%. But small amounts of alloying elements harden and strengthen the metal, and it is seldom used alone.

Zinc is seldom used alone except as a coating. In addition to its metal and alloy forms, zinc also extends the life of other materials such as steel (by hot dipping or electrogalvanizing), rubber and plastics (as an aging inhibitor), and wood (in paints). Zinc is also used to make brass, bronze, and nickel silver; die-casting alloys in plate, strip, and coil; foundry alloys; superplastic zinc; and activators and stabilizers for plastics. Additionally, zinc is used for electric batteries, for die castings, and in alloyed sheets for flashings, gutters, and stamped and formed parts. The metal is harder than tin, and an electrodeposited plate has a Vickers hardness of about 45. Zinc is also used for many chemicals.

Production

The metallurgy of zinc is dominated by the fact that its oxide is not reduced by carbon below the boiling point of the metal.

A large fraction of the world's zinc is still produced from relatively small horizontal retorts with one furnace (or bank) containing hundreds of such units.

Other large, continuously operated vertical retorts have operated with top charging of briquettes of zinc oxide and bituminous coal and metal tapping from an outside condenser.

Another continuous method involves electrothermic reduction, using a novel condenser in which the retort vapors are sucked through molten zinc.

Neither horizontal or vertical retorts nor electrothermic units nor blast furnaces normally produce zinc of the extreme high purity required by much of the total market for zinc. Since 1935, a redistillation process has been used as the thermal means of meeting this demand. The principles of fractional distillation are utilized, and zinc of 99.99+% purity is made.

There is also an electrolytic method of producing metallic zinc. Because the selective flotation process made additional quantities of zinc concentrates available in localities where electric power is cheap, the production of electrolytic zinc was increased.

In the electrolytic process, the zinc content of the roasted ore is leached out with dilute sulfuric acid. The zinc-bearing solution is filtered and purified and the zinc content recovered from the solution by electrolysis, using lead alloy anodes and sheet aluminum cathodes. Current passing through the electrolytic cell, from anode to cathode, deposits the metallic zinc on the cathodes from which it is stripped at regular intervals, melted, and cast into slabs. Zinc so produced is 99.9+ or 99.99+% pure depending on need and the process control exercised.

Compounds and Zinc Forms

Zinc is always divalent in its compounds, except for some of those with other metals, which are classed as zinc alloys. Most of the more important zinc compounds are inorganic, since they are much more widely used than the organic zinc compounds.

The old name "spelter," often applied to slab zinc, came from the name "spailter" used by Dutch traders for the zinc brought from China.

Sterling spelter was 99.5% pure. Special high-grade zinc is distilled, with a purity of 99.99%, containing no more than 0.006% lead and 0.004% cadmium. High-grade zinc, used in alloys for die casting, is 99.9% pure, with 0.07 max lead. Brass special zinc is 99.10% pure, with 0.6 max lead and 0.5 max cadmium. Prime western zinc, used for galvanizing, contains 1.60 max lead and 0.08 max iron. Zinc crystals produced for electronic uses are 99.999% pure metal.

On exposure to the air, zinc becomes coated with a film of carbonate and is then very corrosion resistant. Zinc foil comes in thicknesses from 0.003 to 0.015 cm. It is reduced by electrodeposition on an aluminum drum cathode and stripping off on a collecting reel. But most of the zinc sheet contains a small amount of alloying elements to increase the physical properties. Slight amounts of copper and titanium reduce grain size in sheet zinc. In cast zinc, the hexagonal columnar grain extends from the mold face to the surface or to other grains growing from another mold face, and even very slight additions of iron can control this grain growth. Aluminum is also much used in alloying zinc. In zink used for galvanizing, a small addition of aluminum prevents formation of brittle alloy layer, increases ductility of the coating, and gives a smoother surface. Small additions of tin give bright spangled coatings.

Zinc has 12 isotopes, but the natural material consists of 5 stable isotopes, of which nearly half is zinc-64. The stable isotope zinc-67, occurring to the extent of about 4% in natural zinc, is sensitive to tiny variations in transmitted energy, giving off electromagnetic radiations that permit high accuracy in measuring instruments. It measures gamma-ray vibrations with great sensitivity and is used in the nuclear clock.

Zinc powder, or zinc dust, is a fine gray powder of 97% minimum purity usually in 325-mesh particle size. It is used in pyrotechnics, in paints, and in rubbers, used as a reducing agent and catalyst and as a secondary dispersing agent, and used to increase flexing and to produce sherardized steel.

In paints, zinc powder is easily wetted by oils. It keeps zinc oxide in suspension and also hardens the film. Mossy zinc, used to obtain color effects on face brick, is a spangly zinc powder made by pouring the molten metal into water. Feathered zinc is a fine grade of mossy zinc. Photoengraving zinc for printing plates is made from pure zinc with only a small amount of iron to reduce grain size and is alloyed with not more than 0.2% each of cadmium, manganese, and magnesium. Cathodic zinc, used in the form of small bars or plates fastened to the hulls of ships or to underground pipelines to reduce electrolytic corrosion, is zinc of 99.99% purity with iron less than 0.0014% to prevent polarization.

Applications

For many years, the greatest use of zinc has been to protect iron and steel against atmospheric corrosion. Because of the relatively high electropotential of zinc, it is anodic to iron. It zinc and iron or steel are electrically connected and are jointly exposed in most corrosive media; the steel will be protected, while the zinc will be attacked preferentially and sacrificially. This, along with the fact that zinc corrodes far less rapidly than iron in most environments, forms the basis for one of the great fields of use of zinc and galvanizing (by hot dipping or electrolytically), metallizing, and sherardizing, in zinc-pigmented paints systems, as anodes in systems for cathodic protection. The six techniques are described in the succeeding text.

Hot-Dip Galvanizing

Zinc alloys are readily compatible with iron. Therefore, steel articles, suitably cleaned, will be wet by molten zinc and will acquire uniform coatings of zinc, the thickness of which will vary with time, temperature, and rate of withdrawal. Such coats are continuous and reasonably ductile. Ductility is improved considerably by the

restriction of immersion time and by the addition of small amounts of aluminum to the galvanizing bath.

Millions of tons of steel products are protected by zinc annually. The time before first rusting of the iron or steel base is proportional to the thickness of zinc coat, which in turn is subject to control—depending on product and processing—within a range from thin wiped coats on some products to as much as 0.20 mm on certain low-alloy steels allowed to acquire a full natural coat.

The usefulness of zinc as a coating material comes from its dual ability to protect, first as a long-lasting sheath, and then sacrificially when the sheath finally is perforated.

Electrogalvanizing

Zinc may be electrolytically deposited on essentially all iron and steel products. Wire and strip are commonly so treated as are many fabricated parts. Electrodeposited coats are ductile and uniform but normally are thinner and therefore find applications in less rigorous service.

Metallizing

Zinc wire or powder is melted and sprayed on suitably grit-blasted steel surfaces—a growing use. Its virtues are flexibility in application and substantial thicknesses that may be applied. The method is particularly useful for renewal of heavy coatings on areas exposed to particularly critical corrosive conditions and the coating of parts too large for hot dipping. Although metallized coats may be somewhat porous, the sacrificial nature of zinc nevertheless makes them protective. Suitable pore sealants may be used as a part of a metallizing system.

Sherardizing

Zinc powder is packed loosely around clean parts to be sherardized in an airtight container. When sealed, heated to temperature near but below the zinc melting point, and then slowly rotated, the zinc alloys with the steel form a thin, abrasion-resistant, and uniform protective coating (0.4–1.8 g/cm²).

Sherardizing is used commonly to coat small items such as nuts, bolts, and screws; an exception is tubular electrical conduit. Sherardized coats receive varnish, tanks, and lacquers particularly well.

Zinc-Pigmented Paints

Evidence has accumulated to demonstrate that paints heavily pigmented with zinc dust perform similarly to zinc coats otherwise applied. Electrical contact must exist between the steel and the zink-dust particles, consequently, special vehicles that must be used and the steel surface must be clean.

Zinc Anodes

High-purity zinc, normally alloyed with small additions of aluminum, with or without cadmium, is cast or rolled into anodes that, when electrically bonded to bare or painted steel, will protect large areas from the corrosive attack of such environments as seawater. The advantages of zinc in this application include self-regulation (no more current is generated than is required), a minimum generation of hydrogen, and long life. This is a growing application for the protection of ship hulls, cargo tanks in ballast, piers, pilings, etc.

General Comment

Reference has been made to the importance of coating thickness—the heavier the coat, the longer the time before first rusting.

All evidence at hand indicates that the amount of zinc in a coat is the controlling factor and the method of application is of secondary importance. Uniformity of coat and adhesion must be good. No data are known to demonstrate that common zinc impurities normally present in amounts to or slightly above specification limits have any significantly deleterious or beneficial influence on the ability of zinc to protect iron or steel against atmospheric corrosion. Although any greater zinc may be used for galvanizing, Prime Western is the one most commonly employed.

Die casting is a market for zinc that may soon become its largest market. These alloys melt readily, are highly fluid, and do not attack steel dies or equipment. When used under good temperature control and with good die design practices, casting surfaces are excellent and easily finished. Physical properties are good, and dimensional stability is excellent.

Alloy control within the specified limits ensures long life. Low aluminum results in decreased casting and mechanical properties and adversely affects the performance of plated coatings. High aluminum can lead to brittleness (an alloy eutectic forms at 5% aluminum). High copper content decreases dimensional stability. Iron is commonly encountered is not critical. Lead, tin, and cadmium, if present above specification limits, can lead to intercrystalline corrosion with objectionable growth and serious cracking or brittleness as a result. Magnesium minimizes the deleterious influence of lead, tin, or cadmium but at or near specification maximum decreases ductility and castability and can lead to objectionable hot shortness. Other impurities such as chromium and nickel, which may be encountered, are not critical.

Zink die castings are used by the automotive, truck, and bus industry for functional, decorative–functional, or decorative purposes. A majority is plated with copper–nickel–chromium in a variety of plating systems especially adapted to withstand severe service conditions.

Other major outlets for zinc die casting include household appliances, business machines, machine tools, air-brake systems, and communication equipment.

Even such nonstructural materials as cardboard can be zinc coated by low-temperature flame spraying. Other important uses of zinc are in brass and zinc die-casting alloys, in zinc sheet and strip, in electrical dry cells, in making certain zinc compounds, and in chemical preparations.

A so-called tumble-plating process coats small metal parts by applying zinc powder to them with an adhesive, then tumbling them with glass beads, to roll out the powder into a continuous coat of zinc. Rechargeable nickel–zinc batteries offer higher energy densities than conventional dry cells. Foamed zinc metal has been suggested for use in lightweight structures such as aircraft and spacecraft. Some other uses of zinc are in dry cells, roofing, lithographic plates, fuses, organ pipes, and wire coatings.

Zinc is believed to be needed for normal growth and development of all living species, including humans; actually, life without zinc would be impossible. Zinc is a common element that is present in virtually every type of human food, and zinc deficiency is therefore not considered to be a common problem in humans. Zinc is a trace element that is present in biological fluids at a concentration below 1 ppm, and only a small amount (normally < 25 mg) is required in the daily diet. (The recommended daily allowance for zinc is 15 mg/day for adults and 10 mg/day for growing children.) It is relatively nontoxic, without noticeable side effects at intake levels of up to 10 times that normal daily requirement.

The adult human body contains approximately 2 g of zinc distributed in all cells but especially in bones and muscle. It occurs almost exclusively in association with other molecules, typically proteins, where it exists as a divalent on Zn^{2+}. Many of the zinc-containing proteins are enzymes, known as metalloenzymes, in which the zinc is bound to three amino acid side chains and a molecule of water. Zinc metalloenzymes are involved in most of the key steps of the replication, transcription, and translation of genetic material; hence, they are critical for growth and development. They also help catalyze all major pathways of metabolism, as well as many specialized reactions.

Many biological processes seem to have a relationship to zinc metabolism, although they are undefined. The oxide has been used since the first century for treatment of skin rashes. Zinc has also been found to be helpful in other skin disorders such as burns, acne, and surgical wounds. It seems to modulate the immune system and consequently has been purported to have antiviral, antibacterial, and anticancer properties.

Zinc deficiency sometimes occurs due to an inadequate diet or one that includes a high content of phytic acid, fiber, phosphate, calcium, or copper, all of which diminish absorption.

Zinc telluride is a semiconductive material that has been found to become photorefractive when it is suitably doped with vanadium or with manganese and vanadium. The combination of photorefractivity and semiconductivity makes this material attractive for use in a variety of applications, including optical power limiting (for shielding eyes or delicate sensors against intense illumination), holographic interferometry, providing reconfigurable optical interconnections for optical computing and optical communication, and correcting for optical distortions and combining laser powers via phase conjugation. In comparison with other important photorefractive materials based on III–V and II–VI binary compounds, ZnTe/V offers superior photorefractive performance at wavelengths from 0.6 to 1.3 μm.

Undoped or doped ZnTe can be grown by physical vapor transport in a closed ampoule.

Experiments were performed to investigate the utility of ZnTe/V/Mn for real-time resonant holographic interferometry. These experiments involved, variously, two- or four-wave mixing, using pulsed dye or continuous-wave helium–neon or diode lasers. Holographic image transfer and two-wavelength resonant holographic interferometry were demonstrated; in particular, a ZnTe/V/Mn crystal was used in a demonstration of resonant holographic interferometric spectroscopy, which is a technique for obtaining chemical-species-specific interferograms by recording two holograms simultaneously at two slightly different wavelengths near an absorption spectral peak of the species in question.

A vehicle (AT-1) was fitted with zinc–air batteries and ran for 1043 miles at 20–25 mph. The batteries consisted of 180 zinc–air cells weighing 1.9 kg and outputting 1.1 V each. On the record run, they delivered 76 kWh of power with 10% remaining, compared to the total capacity of the Saturn EV-1 of 16.2 kWh from a lead–acid battery pack that weighs 50% more than that in the AT-1. A unique aspect of the batteries is that they are intended to be physically swapped from the vehicle in minutes instead of recharged onboard for hours.

Intended for emerging market countries, the AT-1 is a five-passenger utility vehicle that can be configured in a variety of ways. Weighing 635.6 kg, it can carry a 590.2 kg load. Depending on motor, it has a top speed of 40–70 mph.

Although zinc–air batteries offer about four times the energy density of lead–acid, their weakness is specific power. It should be noted that this could be addressed by combining zinc–air batteries with other batteries or an ultracapacitor. As vehicles based on crude oil lose their selling position, zinc–air fuel cells are expected to capture a substantial part of the market.

Alloys

Alloys of zinc are mostly used for die castings for decorative parts and for functional parts where the load-bearing and shock requirements are relatively low. Because the zinc alloys can be cast easily in high-speed machines, producing parts that weigh less than brass and have high accuracy and smooth surfaces that require minimum machining and finishing, they are widely used for such parts as handles and for gears, levers, pawls, and other small parts. Zinc alloys for sheet contain only small amounts of alloying elements, with 90%–98% zinc, and the sheet is generally referred to simply as zinc or by a trade name. The modified zinc sheet is used for stamped, drawn, or spun parts for costume jewelry and electronics and contains up to 1.5% copper and 0.5% titanium. The titanium raises the recrystallization temperature, permitting heat treatment without coarse grain formation.

Hartzink had 5% iron and 2%–3% lead, but iron forms various chemical compounds with zinc, and the alloy is hard and brittle. Copper reduces the brittleness. *Germania*-bearing bronze contained 1% iron, 10% tin, about 5% each of copper and lead, and the balance zinc. The *Fenton* alloy had 14% tin, 6% copper, and 80% zinc, and the *Ehrhard*-bearing metal contained 2.5% aluminum, 10% copper, 1% lead, and a small amount of tin to form copper–tin crystals. *Binding* metal, for wire-rope slings, has about 2.8% tin, 3.7% antimony, and the balance zinc. Pattern metal, for casting gates of small patterns, was almost any brass with more zinc and some lead added but is now standard die-casting metal.

Zinc also is commonly used in varying degrees as an alloy component with other base metals, such as copper, aluminum, and magnesium. A familiar example of the latter is the association of varying amounts of zinc (up to 45%) with copper to produce brass.

Zinc alloys are commonly used for die castings, and the zinc used is high-purity zinc known as special high-grade zinc. The ASTM AG40A and SAE 903 are the most widely used; others include AC41A (SAE 925), Alloy 7, and II.ZRO 16. All typically contain about 4% aluminum, small amounts of copper, and very small amounts of magnesium. AG40A has a density of 6.6 mg/m^3, an electrical conductivity of 27% that of copper, a thermal conductivity of 113 W/m · K, an ultimate tensile strength of 283 MPa, and a hardness of Brinell 82. AC41A is stronger (331 MPa) and harder (Brinell 91), a trifle less electrically and heat conductive, and similar in density. The alloys have much greater unnotched Charpy impact strength than either die-cast aluminum or magnesium alloys, but are not especially heat resistant, losing about one-third of their strength at temperatures above about 93°C. Both alloys have found wide use for auto and appliance parts, especially chromium-plated parts, as well as for office equipment parts, hardware, locks, toys, and novelties. Alloy 7 is noted primarily for its better castability and the smoother surface finish it provides. It is as strong as AG40A, although slightly less hard and more ductile. II.ZRO 16 is not nearly as strong (228 MPa), but more creep-resistant at room and elevated temperatures.

Casting Alloys

Zinc-casting alloys can be grouped into two general categories: standard zinc die-casting alloys and the newer zinc–aluminum casting alloys.

Standard Die-Casting Alloys

For pressure die casting, the established zinc alloys are the No. 3, 5, and 7 *Zamak* alloys. As die castings, they have good general-purpose tensile properties and can be cast in thin sections and with good dimensional accuracy. The alloys are often selected for plated or highly decorative applications because of their excellent finishing characteristics. Three major end-use areas for zinc die-cast components are automotive, building hardware, and electrical.

Zamak alloys contain approximately 4% aluminum with low percentages of magnesium, copper, and sometimes nickel. Impurities such as tin, lead, and cadmium are carefully controlled. These alloys are not recommended for gravity casting. They are cast by the hot-chamber die-casting process, which is different from, and more efficient than, the cold-chamber die-casting process commonly used for aluminum. In addition, a specialized process is used for efficient production of miniature die-cast components, using these alloys as well as ZA-8.

Typical tolerances of zinc die-cast parts are ±0.0015 in./in., for the first inch with an additional ± 0.002 in./in. for larger parts. New zinc-casting technology allows for thin walls down to 0.025 in. (0.6 mm), improved internal soundness, and surface finishes that range typically from 32 to 64 root mean square.

Part dimensions change slightly when zinc die castings are aged. Zamak alloys No. 3 and 7 can shrink about 0.0007 in./in. after several weeks at room temperature. Alloy No. 5 responds similarly, but total shrinkage can be 0.0009–0.0024 in./in., followed by expansion of 0.0020 in./in. over a period of years. When it is necessary to bring these changes to completion within a short time after casting, a stabilizing treatment of 4–6 h at 100°C is recommended.

ZA-Casting Alloys

Designated as ZA-8, ZA-12, and ZA-27 (the numerical suffix represents the approximate percent by weight of aluminum), the high-aluminum alloys differ radically from the standard Zamak alloys in composition, properties, and castability. Although the ZA alloys were first introduced for gravity casting (sand and permanent mold), they have expanded into pressure die castings as well as a new, precision graphite-mold process. *Important*: Alloys ZA-12 and ZA-27 must be cold-chamber die cast; alloy ZA-8 is hot-chamber castable.

Gravity casting into low-cost graphite permanent molds provides high-quality ZA castings with excellent precision, eliminating much machining. It is particularly competitive for production quantities of 500–15,000 parts/year, where die casting or plastic injection molding would be prohibited because of tooling costs.

ZA alloys combine high strength and hardness (up to 480 MPa and 120 Bhn), good machineability with good bearing properties, and wear resistance often superior to standard bronze alloys. ZA castings are now competing with cast iron, bronze, and aluminum because of various property and processing advantages.

Of the three alloys, ZA-12 is preferred for most applications, and particularly for gravity casting. However, ZA-27 offers the highest mechanical properties regardless of casting method. Both are excellent bearing materials. ZA-8, on the other hand, gives the best plating characteristics. Because of its hot-chamber die castability and high mechanical properties, ZA-8 is also used for high-performance applications where standard zinc alloys can be marginal. All ZA alloys offer superior creep resistance and performance at elevated temperatures compared with the Zamak alloys.

The most recent casting alloys discussed earlier are three high-aluminum zinc-casting alloys for sand and permanent-mold casting: ZA-8, ZA-12, and ZA-27; the numerals in the designations indicate approximate aluminum content. They also contain more copper than AG40A and AC41A, from 0.5% to 1.2% in ZA-12 to 2% to 2.5% in ZA-27, and a bit less magnesium. As sand cast, ultimate tensile strengths range from 248 to 276 MPa for ZA-8 and 400 to 441 MPa for ZA-27. Unlike the common die-casting alloys, the ZA alloys also exhibit clearly defined tensile yield strengths: from 193 MPa minimum for sand-cast ZA-8 to 365 MPa for sand-cast ZA-27.

Tensile modulus is roughly 83×10^3 MPa. Also, because of their greater aluminum content, they are lighter in weight than the die-casting alloys.

Wrought Alloys

Wrought zinc alloys are available in rolled sheet, strip, foil, and as drawn rod or wire. With controlled rolling, zinc alloys can be tailored to meet a wide range of hardness, luster, and ductility requirements. Rolled zinc can be worked by common fabricating methods, and then polished, lacquered, painted, or plated.

When zinc alloys are formed in progressive presses, as in battery shell manufacture, they are self-annealing. After successive forming operations in nonprogressive presses, however, the alloys work-harden and break. This can be overcome, in copper-free alloys, by intermediate annealing for 5 min in boiling water to which 20% glycerin has been added. Copper-bearing alloys should be heated 5–10 min at 177°C. The copper/titanium-containing alloy should be held at 199°C for about 15 min to bring about recrystallization. Excessive exposure to higher temperature should be avoided, however, to prevent grain growth, cleavage cracks, and property deterioration.

Highly workable and highly forgeable wrought-zinc alloys containing titanium, aluminum, lead, cadmium, copper, or iron in various quantities are easily machined. Forged or extruded parts are free from porosity and have good detail.

Other Alloys

Manganese–zinc alloys with up to 25% manganese, for high-strength extrusions and forgings, are really 60–40 brass with part of the copper replaced by an equal amount of manganese and are classed with manganese bronze. They have a bright white color and are corrosion resistant. Zam metal, for zinc-plating anodes, is zinc with small percentages of aluminum and mercury to stabilize against acid attack. Zinc solders are used for joining aluminum. The tin–zinc solders have 70%–80% tin, about 1.5% aluminum, and the balance zinc. The working range is 260°C–310°C. Zinc–cadmium solder has about 60% zinc and 40% cadmium. The pasty range is between 266°C and 315°C.

A group of wrought alloys, called superplastic zinc alloys, have elongations of up to 2500% in the annealed condition. These alloys contain about 22% aluminum. One grade can be annealed and air-cooled two strains of 489 MPa. Parts of these alloys have been produced by vacuum forming and by a compression molding technique similar to forging but requiring lower pressures.

Chemicals

With the exception of the oxide, the quantities of zinc compounds consumed are not large compared with many other metals, but zinc chemicals have a very wide range of use; they are essential in almost all industries and for the maintenance of animal and vegetable life. Zinc is a complex element and can provide some unusual conditions and alloys and chemicals.

Zinc Oxide

Zinc oxide, ZnO, is a white, water-insoluble, refractory powder melting at about 1975°C, with a specific gravity of 5.66. It is much used as a pigment and accelerator in paints and rubbers. Its high refractive index, about 2.01, absorption of ultraviolet light, and fine particle size give high hiding power in paints and make it also useful in such products as cosmetic creams to protect against sunburn. Commercial zinc oxide is always white and in the paint industry is called *zinc white* and *Chinese white*. But with a small excess of zinc atoms in the crystals, obtained by heat treatment, the color is brown to red.

In paints, zinc oxide is not as whitening as lithopone, but it resists the action of ultraviolet rays and is not affected by sulfur atmospheres, and is thus valued in outside paints. Leaded zinc oxide, consisting of zinc oxide and basic lead sulfate, is used in paints, but for use in rubber, the oxide must be free of lead. The lead-free variety is also called *French process zinc oxide*. In insulating compounds, zinc oxide improves electrical resistance. In paper coatings, it gives opacity and improves the finish. Zinc-white paste for paint mixing usually has 90% oxide and 10% oil. Zinc oxide stabilizers, composed of zinc oxides and other chemicals, can be added to plastic molding compounds to reduce the deteriorating effects of sunlight and other types of degrading atmospheres.

The rubber industry is the largest consumer of zinc oxide, accounting for more than 50% of the market. Zinc oxide is most effective as an activator of accelerators in the vulcanization process.

The chemical industry has been opening new markets for zinc oxide. Examples are lubricating oil additives, water treatment, and catalysts. For photocopying, photoconductivity is a unique electronic property of zinc oxide.

In the ceramic industry, zinc oxide is used in the manufacture of glasses, glazes, frits, porcelain enamels, and magnetic ferrites. Here, the largest consuming plants are in the tile industry.

One ceramic grade of zinc oxide has these properties: specific gravity 5.6, apparent density 1201 kg/m³, and weight 5595.5 kg/m³. Typical chemical analysis: 99.5% ZnO, 0.05% Pb, 0.02% Fe, 0.01% Cd, 0.02% S (total), 0.10% HCl (insoluble), and 5 ppm magnetic iron.

In glass, zinc oxide reduces the coefficient of thermal expansion, thus making possible the production of glass products of high resistance to thermal shock. It imparts high brilliance of luster and high stability against deformation under stress. As a replacement flux for the more soluble alkali constituents, it provides a viscosity curve of lower slope. Specific heat is decreased and conductivity increased by the substitution of zinc oxide for BaO and PbO.

A 1% addition of zinc oxide to tank window glass lowers the devitrification temperature and improves chemical resistivity while maintaining good workability for drawing. It is used consistently in high-grade fluoride opal glass in which it greatly increases opacity, whiteness, and luster by inducing precipitation of fluoride crystals of optimum number and size. Apparently, zinc oxide makes its contribution to opacity through reduction of the primary opacifiers. It is used in optical glasses of high barium content to reduce the tendency to crystallize on cooling. The resistance of phosphate glasses to chemical attack is improved by the presence of zinc oxide. About 10% zinc oxide assists in the development of the characteristic color of cadmium sulfoselenide ruby glass, although its exact function is obscure.

Zinc oxide is used in many types of glazes, its function varying according to the particular composition in which it is included. In general, it provides fluxing power, reduction of expansion, prevention of crazing, greater gloss and whiteness, the favorable effect on elasticity, increased maturing range, increased brilliance of colors, and correction of eggshell finish. It is useful in preventing volatilization of lead by partial substitution for CaO, as high CaO tends to satisfy SiO_2, leaving PbO in a more volatile form.

In *Bristol* glazes for earthenware products, zinc oxide in combination with alumina produces opacity and whiteness to a fair degree, provided that the lime content is low. The use of zinc oxide in wall tile glazes is very general; zinc oxide content of certain types is 10% or more. Small amounts are used in gloss or bright tile, where its higher percentages are used where it is desired to develop a highly pleasing matte finish.

Zinc oxide is commonly used in dry-process cast-iron porcelain enamels in amounts of 0.5%–1%–14%. In general, low lead content implies high zinc, and vice versa. Its specific functions are to increase fusibility, improve luster, contribute to opacity and whiteness, reduce expansion, and increase extensibility. It is probably a little stronger as a flux than is lime but does not produce the sudden fluidity characteristics of lime.

Of great benefit to producers of cast-iron enamels is the relative nontoxicity of zinc oxide.

A recent use for zinc oxide is its application to the manufacture of magnetic ferrites, which have been developed over the past 35–40 years. They usually are composed of ferric oxide in combination with zinc oxide (of high chemical purity) and any one or more of several other oxides of bivalent metals. The amount of zinc oxide used varies from 10% to 35%, depending on the characteristics desired in the finished magnetic ferrite. With their prime properties of high permeability and low hysteresis, they are used in the field of electronics for such devices as high-frequency transformer cores for television receivers.

Zinc oxide crystals are used for transducers and other piezoelectric devices. The crystals are hexagonal and are effective at elevated temperatures, as the crystal has no phase change up to its disassociation point. The resistivity range is 0.5–10 $\Omega \cdot cm$. Normally recognized as an n-type semiconductor, it has a resistivity less than 103 $\Omega \cdot cm$. When doped with lithium, resistivity rises to 1012 $\Omega \cdot cm$ and it exhibits piezoelectricity about four times that of quartz.

Zinc oxide has luminescent and light-sensitive properties that are utilized in phosphors and ferrites. But the oxygen-dominated zinc phosphors used for radar and television are modifications of zinc sulfide phosphors.

Zinc sulfide phosphors, which produce luminescence by exposure to light, are made with zinc sulfide mixed with about 2% sodium chloride and 0.005% copper, manganese, or other activator and fired in a nonoxidizing atmosphere. The cubic crystal structure of zinc sulfide changes to a stable hexagonal structure at 1020°C, but both forms have the phosphor properties. Thin films and crystals of zinc selenide with purities of 99.999% are used for photo and electroluminescent devices. Zinc sulfide, a white powder, is also used as a paint pigment and for whitening rubber and for paper coating. *Cryptone* is zinc sulfide for pigment use at various grades, some grades containing barium sulfate, calcium sulfide, or titanium dioxide.

Zinc Blend

Zinc sulfide, ZnS, ore.

Zinc Coating

See *galvanizing*.

Zinc Chromate

An orange-colored constituent of some primer paints. It is not simply a pigment but provides some sacrificial cathodic protection particularly for aluminum in marine environments.

Zinc (Based) Die-Casting Alloy

Various alloys, principally zinc but with additions of aluminum, typically about 4%, possibly with small amounts of copper, lead, cadmium, etc. These materials have relatively low melting points and provide castings with good finish and detail.

Zinc Ores

The metal zinc is obtained from a large number of ores, but the average zinc content of the ores in the United States is only about 3% so that they are concentrated to contain 35%–65% before treatment. The sulfide ores are marketed on the basis of 60% zinc content and the oxide ores on the basis of 40% zinc content. *Sphalerite*, or *zinc blende*, is the most important and has been found in quantities in Missouri and surrounding states and in Europe. Sphalerite is a zinc sulfide, ZnS, containing theoretically 67% zinc. It has a massive crystalline or granular structure and a Mohs hardness of about 4. When pure, its color is white; it colors yellow, brown, green, to black with impurities.

Calamine is found in New Jersey, Pennsylvania, Missouri, and Europe. It is the ore that was formally mixed directly with copper for making brass. The ore usually contains only about 3% zinc and is concentrated to 35%–45%, and then roasted and distilled. Calamine is zinc silicate. It is a mineral occurring in crystal groups of a vitreous luster, and it may be white, greenish, yellow, or brown. The specific gravity is 3.4 and Mohs hardness 4.5–5. It occurs in Arkansas with *smithsonite*, a zinc carbonate ore. *Franklinite* is an ore of both the metals zinc and manganese.

The ore *zincite* is used chiefly for the production of the zinc oxide known as zinc white employed as a pigment. Zincite has the composition ZnO, containing theoretically 80.3% zinc. The mineral has usually a massive granular structure with a deep-red to orange streaked color. It may be translucent or almost opaque. Deep-red specimens from the workings at Franklin, New Jersey, are cut into gemstones for costume jewelry. *Willemite* is an anhydrous silicate containing theoretically 58.5% zinc. When manganese replaces part of the zinc, the ore is called *troostite*. It is in hexagonal prisms of white, yellow, green, or blue; manganese makes it apple green, brown, or red. The specific gravity is about 4 and Mohs hardness 5.5. The crushed ore is used in making fluorescent glass. The ore is widely dispersed in the United States.

Zinc Tinsel

An alloy of about 60% zinc and 40% lead, which is bright and reasonably tarnish resistant for decorative applications.

Zinc White

Zinc oxide, ZnO.

Zinc Worms

Surface imperfections, characteristic of high-zinc brass castings, which occur when zinc vapor condenses at the mold/metal interface, where it is oxidized and then becomes entrapped in the solidifying metals.

Zircon

Zircon, a mineral with the idealized composition $ZrSiO_4$, is one of the chief sources of the element zirconium. Trace amounts of uranium and thorium are often present and in general may then be partly or entirely metamict. The name "cyrtolite" is applied to an altered type of zircon. Structurally, zircon is a nesosilicate, with isolated SiO_4 groups.

Zircon is tetragonal in crystallization. It often occurs as well-formed crystals, which commonly are square prisms terminated by a low pyramid. The color is variable, usually brown to reddish brown, but also colorless, pale yellowish, green, or blue. The transparent colorless or tinted varieties are popular gemstones. Hardness is 7 1/2 on Mohs scale; specific gravity is 4.7, decreasing in metamict types.

Because of its chemical and physical stability, zircon resists weathering and accumulates in residual deposits and in beach and river sands.

Other properties include specific heat of 0.55 J/g/°C. It is chemically inert and stable in very high temperatures (liquidus > 2205°C). Zircon has excellent thermal properties, and its thermal conductivity is 14.5 Btu/ft²/h/°F/in., and coefficient of thermal expansion is 1.4×10^{-6}.

The extremely high thermal conductivity and chilling action of zircon make it very useful in controlling directional solidification and shrinkage in heavy-metal sections.

Uses

Zircon sand is used as refractory bedding material for heat-treating metal parts. It is used as a sealing medium for prevention of atmospheric leaks around doors and parts of heat-treating furnaces. Also, it is a high-quality, uniform sandblasting medium for metal preparation prior to plating, enameling, or buffing. The heavy, rounded grains give consistent peening without stray digs or gouges to mar the finish. The tough, resilient grains resist breakdown and loss.

Zirconia (Zirconium Oxide)

Zirconium oxide, ZrO_2, is a white crystalline powder with a specific gravity of 5.7, hardness 6.5, and refractive index 2.2. When pure, its melting point is about 2760°C, and it is one of the most refractory of the ceramics. It is produced by reacting zircon sand and dolomite at 1371°C and leaching out the silicates. The material is used as fused or sintered ceramics and for crucibles and furnace bricks. From 4.5% to 6% of CaO or other oxide is added to convert the unstable monoclinic crystal to the stable cubic form with a lowered melting point.

Zirconia is produced from the zirconium ores known as zircon and *baddeleyite*. The latter is a natural zirconium oxide. It is also called *zirkite* and *brazilite*. Zircon is zirconium silicate and comes chiefly from the beach sands. The sands are also called *zirkelite* and *zirconite*, or merely zircon sand. The white zircon sand has a zirconia content of 62% and contains less than 1% iron.

Uses

Fused zirconia, used as a refractory ceramic, has a melting point of 2549°C and a usable temperature up to 2454°C. The *zinnorite* fused zirconia is a powder that contains less than 0.8% silica and has a melting point of 2704°C. A sintered zirconia can have a density of 5.4, a tensile strength of 82 MPa, a compressive strength of 1378 MPa, and a Knoop hardness of 1100. *Zircoa B* is stabilized cubic zirconia used for making ceramics. *Zircoa A* is the pure monoclinic zirconia used as a pigment, as a catalyst in glass, and as an opacifier in ceramic coatings.

As an opacifier, zirconium compounds are used in glazes and porcelain enamels. Zirconium dioxide is an important constituent of ceramic colors and an important component of lead–zirconate–titanate electronic ceramics.

Pure zirconia also is used as an additive to enhance the properties of other oxide refractories. It is particularly advantageous when added to high-fired magnesia bodies and alumina bodies. It promotes sinterability and, with alumina, contributes to abrasive characteristics.

Zirconia brick for lining electric furnaces has no more than 94% zirconia, with up to 5% CaO as a stabilizer, and some silica. It melts at about 2371°C but softens at about 1982°C. The IBC 4200 brick is zirconia with calcium and hafnium oxides for stabilizing. It withstands temperatures up to 2316°C in oxidizing atmospheres and up to 1849°C in reducing atmospheres. Zirconia foam is marketed in bricks and shapes for thermal insulation. With a porosity of 75%, it has a flexural strength above 3 MPa and a compressive strength above 0.7 MPa. For use in crucibles, zirconia is insoluble in most metals except the alkali metals and titanium. It is resistant to most oxides; but with silicate, it forms $ZrSiO_4$; and with titania, it forms $ZrTiO_4$. Because structural disintegration of zirconia refractories comes from crystal alteration, the phase changes are important considerations. The monoclinic material, with a specific gravity of 5.7, is stable up to 1010°C and then inverts to the tetragonal crystal with a specific gravity of 6.1 and a volume change of 7%. It reverts when the temperature again drops below 1010°C. The cubic material, with a specific gravity of 5.55, is stable at all temperatures to the melting point, which is not above 2649°C because of the contained stabilizers. A lime-stabilized zirconia refractory with the tensile strength of 138 MPa has a tensile strength of 68 MPa at 1299°C.

Stabilized zirconia has a very low coefficient of expansion, and white-hot parts can be plunged into cold water without breaking. The thermal conductivity is only about one-third that of magnesia. It is also resistant to acids and alkalies and is a good electrical insulator.

To prepare useful formed products from zirconium oxide, stabilizing agent such as lime, yttrium, or magnesia must be added to the zirconia, preferably during fusion, to convert the zirconia to the cubic form. Most commercial stabilized zirconia powders or products contain CaO as the stabilizing agent. The stabilized cubic form of zirconia undergoes no inversion during heating and cooling.

Stabilized zirconia refractories are used where extremely high temperatures are required. Above 1649°C, in contact with carbon, zirconia is converted to zirconium carbide.

Zirconia is of much interest as a construction material for nuclear energy applications because of its refractoriness, corrosion resistance, and low nuclear cross section. However, zirconia normally contains about 2% hafnia, which has a high nuclear cross section. The hafnia must be removed before the zirconia can be used in nuclear applications.

Forms

Zirconia is available in several distinct types. The most widely used form is stabilized in cubic crystal form by a small lime addition. This variety is essential to the fabrication of shapes because the so-called unstabilized monoclinic zirconia undergoes a crystalline inversion on heating, which is accompanied by a disruptive volume change.

Zirconium is not wetted by many metals and is therefore an excellent crucible material when slag is absent. It has been used very successfully for melting alloy steels and the noble metals. Zirconia refractories are rapidly finding application as setter plates for ferrite and titanate manufacture, and as matrix elements and wind tunnel liners for the aerospace industry.

Other Types

Toughening mechanisms, by which a crack in a ceramic can be arrested, complement processing techniques that seek to eliminate crack-initiating imperfections. Transformation toughening relies on a change in crystal structure (from tetragonal to monoclinic) that zirconia or zirconium dioxide (ZrO_2) grains undergo when they are subjected to stresses at a crack tip. Because the monoclinic grains have a slightly larger volume, they can "squeeze" a crack shut as of calcium and they expand in the course of transformation. Because of the transformation toughening abilities of ZrO_2, which impart higher fracture toughness, research interest in engine applications has been high. In order for ZrO_2 to be used in high-temperature, structural applications, it must be stabilized or partially stabilized to prevent a monoclinic–tetragonal phase change. Stabilization involves the addition of calcia, magnesia, or yttria followed by some form of heat treatment. Partially stabilized zirconia (PSZ) ceramic, the toughest known ceramic, has been investigated for diesel-engine applications.

PSZ is a transformation-toughened material consisting of a cubic zirconia matrix with 20–50 vol% free tetragonal zirconia added in the matrix. The material is converted into the stabilized cubic crystal structure using oxide stabilizers (magnesia, calcia, yttria). The conversion is accomplished by sintering the doped zirconia at 1700°C. Magnesia-stabilized zirconia exhibits serrated plastic flow during compression at room temperature. The flow stress is strain rate sensitive. Several different grades are available for commercial use, and the properties of the material can be tailored to fit many applications.

One typical PSZ used for applications requiring maximum thermal shock resistance has a four-point bend strength of 600 MPa; PSZ is used experimentally as heat engine components, such as cylinder liners, piston caps, and valve seats. Vanadium impurities from fuel oil can cause zirconia destabilization, and sodium, magnesium, and sulfur impurities can cause yttria to disassociate from yttria-stabilized zirconia. Another area of interest for PSZ is in bioceramics, where it has use in surgical implants.

A new zirconia ceramic being developed, tetragonal zirconia polycrystal (TZP) doped with Y_2O_3, has the most impressive room-temperature mechanical properties of any zirconia ceramic. The commercial applications of TZP zirconia include scissors with TZP blades suitable for industrial use for cutting tough fiber fabrics, for example, Kevlar, cables, and ceramic scalpels for surgical applications. One unique application is fish knives. The knife blades are Y-TZP and can be used when the delicate taste of raw fish would be tainted by slicing with knives with metal blades.

Another zirconia ceramic-developed material is zirconia-toughened alumina (ZTA). ZTA is a composite polycrystalline ceramic containing ZrO_2, as a dispersed phase (typically ~15 vol%). Close control of initial starting powder sizes and sintering schedules is thus necessary to attain the desired ZrO_2 particle dimensions in the finished ceramic. Hence, the mechanical properties of the composite ZTA ceramics limit current commercial applications to cutting tools and ceramic scissors.

PSZ is also finding application in the transformation toughening of metals used in the glass industry as orifices for glass fiber drawing. This material is termed zirconia grain-stabilized platinum.

Clear zircon crystals are valued as gemstones because the high refractive index gives great brilliance.

Zirconia fiber, used for high-temperature textiles, is produced from zirconia with about 5% lime for stabilization. The fiber is polycrystalline, has a melting point of 2593°C, and will withstand continuous temperatures above 1649°C. These fibers are as small as 3–10 μm and are made into fabrics for filter and fuel cell use.

Zirconia fabrics are woven, knitted, or felted of short-length fibers and are flexible. Ultratemp adhesive, for high-heat applications, is zirconia powder in solution. At 593°C, it adheres strongly to metals and will withstand temperatures up to 2427°C. *Zircar* is zirconia fiber compressed into sheets to a density of 320 kg/m³. It will withstand temperatures up to 2482°C and has low thermal conductivity. It is used for insulation and for high-temperature filtering.

Zirconium and Alloys

A silvery-white metal, zirconium (Zr), has a specific gravity of 6.5 and a melting point of about 1850°C. It is more abundant than nickel, but is difficult to reduce to metallic form as it combines easily with oxygen, nitrogen, carbon, and silicon. The metal is obtained from zircon sand by reacting with carbon and then converting to the tetrachloride, which is reduced to a sponge metal for the further production of shapes. The ordinary sponge zirconium contains about 2.5% hafnium, which is closely related and difficult to separate. The commercial metal usually contains hafnium, but reactor-grade zirconium, for use in atomic work, is hafnium-free.

Commercially pure zirconium is not a high-strength metal, with the tensile strength of about 220 MPa, elongation 40%, and Brinell hardness 30, or about the same physical properties as pure iron. But it is valued for atomic construction purposes because of its low neutron capture cross section, thermal stability, and corrosion resistance. It is employed mostly in the form of alloys but may be had in 99.99% pure single-crystal rods, sheets, foil, and wire for superconductors, surgical implants, and vacuum-tube parts. The neutron cross section of zirconium is 0.18 barn, compared with 2.4 for iron and 4.5 for nickel. The cold-worked metal, with 50% reduction, has a tensile strength of about 545 MPa, with an elongation of 18% and hardness of Brinell 95. The unalloyed metal is difficult to roll and is usually worked at temperatures up to 482°C. Although nontoxic, the metal is pyrophoric because of its heat-generating reaction with oxygen, necessitating special precautions in handling powder and fine chips resulting from machining operations.

The metal has a close-packed hexagonal crystal structure, which changes at 862°C to a body-centered cubic structure that is stable to the melting point. At 300°C–400°C, the metal absorbs hydrogen rapidly, and above 200°C, it picks up oxygen. At about 400°C, it takes up nitrogen, and at 800°C, the absorption is rapid, increasing the volume and embrittling the metal.

The metal is not attacked by nitric (except red fuming nitric), sulfuric, or hydrochloric acid, but is dissolved by hydrofluoric acid. Zirconium powder is very reactive, and for making sintered metals, it is usually marketed as zirconium hydride, ZrH_2, containing about 2% hydrogen, which is driven off when the powder is heated to 300°C. For making sintered parts, alloyed powders are also used. Zirconium copper (containing 35% zirconium), zirconium nickel (with 35%–50% zirconium), and zirconium cobalt (with 50% zirconium) are marketed as powders of 200–300 mesh.

Properties

In addition to resisting HCl at all concentrations and at temperatures above the boiling temperature, zirconium and its alloys also have excellent resistance in sulfuric acid at temperatures above boiling and concentrations to 70%. Corrosion rate in nitric acid is less than 1 mil/year at temperatures above boiling and concentrations to 90%. The metals also resist most organics such as acetic acid and acetic anhydride as well as citric, lactic, tartaric, oxalic, tannic, and chlorinated organic acids.

Relatively few metals beside zirconium can be used in chemical processes requiring alternate contact with strong acids and alkalies. However, zirconium has no resistance to hydrofluoric acid and is rapidly attacked, even at very low concentrations.

Uses

Small amounts of zirconium are used in many steels. It is a powerful deoxidizer, removes the nitrogen, and combines with the sulfur, reducing hot shortness and giving ductility. Zirconium steels with small amounts of residual zirconium have a fine grain and are shock resistant and fatigue resistant. In amounts above 0.15%, the zirconium form zirconium sulfide and improves the coating quality of the steel.

A noncrystalline metal that reportedly has twice the strength of steel and titanium has been developed. The material, known as *Vitreloy*, is an alloy composed of 61% zirconium, 12% titanium, 12% copper, 11% nickel, and 3% beryllium. Its yield strengths is 1900 MPa, compared with 800 MPa for titanium alloy, Ti-6%Al-4%V, and 850 MPa for cast stainless steel.

Fracture toughness is said to be 55 MPa $m^{1/2}$, the same as high-strength steel but half that of titanium. Its resistance to permanent deformation is said to be two to three times higher than that of conventional metals. The density of Vitreloy is 6.1 g/cm^3 between cast titanium at 4.5 g/cm^3 and cast stainless steel at 7.8 g/cm^3. The material is particularly recommended for aerospace applications because of its surface hardness of 50 HRC. Cast titanium and steel are both tested at 30 HRC.

The beneficial properties of the alloy are ascribed to its noncrystalline structure. Because there are no patterns or grains within the structure, weak areas caused by grain boundaries are eliminated.

An advanced machinable ceramic that may be used to produce thermal shock-resistant components for aerospace, automotive, electrical, heat treating, metallurgical, petrochemical, and plastics applications up to 1550°C has been introduced. The new material (Aremcolox™ 502-1550) is based on the zirconium phosphate system ($Ba_{1+x} Zr_4 P_{6-2x} Si_{2x} O_{24}$) and is especially unique because of its low coefficient of thermal expansion (CTE) of 0.5×10^{-6} in./in. °F. This characteristic sets the material apart from standard ceramic materials such as alumina and zirconia, which have CTEs of 4.0×10^{-6} and 2.5×10^{-6}, respectively.

A low CTE ensures that as a component is thermally cycled, the mechanical stress induced through expansion and contraction does not cause the part to crack. This feature enables engineers to adapt the material to high thermal shock applications, such as combustion and heater systems, that were not previously feasible.

Additional properties and applications of the machinable ceramic include their use as molds, optical stands, microwave housings, engine parts, and applications in which high mechanical strength, hardness, and low porosity are required. A low-density version of the material (502–1550 LD) is recommended for use as brazing fixtures, induction heating liners, rocket nozzles, and high-temperature gauges, tooling, and structures. The material is easily machined using carbide tooling, and no postfiring is required.

Alloys

Zirconium alloys generally have only small amounts of alloying elements to add strength and resist hydrogen pickup. Zircaloy 2, for reactor structural parts, has 1.5% tin, 0.12% iron, 0.10% chromium, 0.05% nickel, and the balance zirconium. Tensile strength is 468 MPa, elongation 37%, and hardness Rockwell B89; at 316°C, it retains its strength of 206 MPa.

Zirconium alloys can be machined by conventional methods, but they have a tendency to gall and work-harden during machining. Consequently, tools with higher than normal clearance angles are needed to penetrate previously work-hardened surfaces. Results can be satisfactory, however, with cemented carbide or high-speed steel tools. Carbide tools usually provide better finishes and higher productivity.

Mill products are available in four principal grades: 702, 704, 705, and 706. These metals can be formed, bent, and punched on standard shop equipment with a few modifications and special techniques. Grades 702 (unalloyed) and 704 (Zr-Sn-Cr-Fe alloy) sheet and strip can be bent on conventional press brake or roll-forming equipment to a 5*t* bend radius at room temperature and to 3*t* at 200°C. Grades 705 and 706 (Zr–Cb alloys) can be bent to 3*t* and 2.5*t* radius at room temperature and to about 1.5*t* at 200°C.

Small amounts of zirconium in copper give age hardening and increase the tensile strength. Copper alloys containing even small amounts of zirconium are called zirconium bronze. They pour more easily than bronzes with titanium, and they have good electric conductivity. Zirconium copper master alloy for adding zirconium to brasses and bronzes is marketed in grades with 12.5% and 35% zirconium. A nickel–zirconium master alloy has 40%–50% nickel, 25%–30% zirconium, 10% aluminum, and up to 10% silicon and 5% iron. Zirconium–ferrosilicon, for alloying with steel, contains 9%–12% zirconium, 40%–47% silicon, 40%–45% iron, and 0.20% max carbon, but other compositions are available for special uses. SMZ alloy (silicon, manganese, and zirconium), for making high-strength cast irons without leaving residual zirconium in the iron, has about 75% silicon, 7% manganese, 7% zirconium, and the balance iron. A typical zirconium copper for electrical use is *Amzirc*. It is oxygen-free copper with only 0.15% of zirconium added. At 400°C, it has a conductivity of 37% IACS, a tensile strength of 358 MPa, and an elongation of 9%. The softening temperature is 580°C.

Zirconium alloys with high zirconium content have few uses except for atomic applications. Zircaloy tubing is used to contain the uranium oxide fuel pellets in reactors because the zirconium does not have grain growth and deterioration from radiation. Zirconia ceramics are valued for electrical and high-temperature parts and refractory coatings. Zirconium oxide powder, for flame-sprayed coatings, comes in either hexagonal or cubic crystal forms. Zirconium silicate, $ZrSi_2$, comes as a tetragonal crystal powder. Its melting point is about 1649°C and hardness about 1000 Knoop.

Zirconium Beryllides

Intermetallic compounds, $ZrBe_{13}$ and Zr_2Be_{17}, have good strengths at elevated temperatures. $ZrBe_{13}$ is cubic, density 2.72 g/cm^3, and melting point 1925°C; Zr_2Be_{17} is hexagonal, density 3.08 g/cm^3, and melting point 1983°C; parts can be formed by all ceramic-forming methods plus flame and plasma-arc spraying. Materials are subject to safety requirements for all beryllium compounds.

These intermetallics, because of their greater densities (BeO = 1.85 g/cm^3), contain more beryllium atoms per unit volume than beryllia, a decided advantage for compact, beryllium-moderated nuclear reactors.

Zirconium Carbide

Zirconium carbide, ZrC_2, is produced by heating zirconia with carbon at about 2000°C. The cubic crystalline powder has a hardness of Knoop 2090 and a melting point of 3540°C. The powder is used as an abrasive and for hot pressing into heat-resistant and abrasion-resistant parts.

Z

Zirconium Diboride

Zirconium diboride (ZrB_2) has a density of 6.09 g/cm^3 and a hexagonal (AlB_2) crystal structure with a melting point of 3040°C.

Zirconium diboride is oxidation resistant at temperatures <1000°C and reacts slowly with nitric, hydrochloric, and hydrofluoric acids. It reacts with aqua regia and hot sulfuric acid, as well as with used alkalies, carbonates, and bisulfates. Zirconium diboride has a typical room temperature electrical resistivity of 9.2 × 10^{-6} $\Omega \cdot$ cm and is superconductive at temperatures less than 2 K. The consolidation of ZrB_2 powder into parts is accomplished by hot pressing or pressureless sintering.

Similar to titanium diboride, ZrB_2 is wet by molten metals but is not attacked by them, making it a useful material for molten metal crucibles, free-formed nozzles, electric discharge machining electrodes, Hall–Heroult cell cathodes, and Thermowell tubes for steel refining. The last use is one of the largest uses of zirconium diboride. Other uses for ZrB_2 include electrical devices, refractories, and applications where high oxidation resistance is required.

Others

Zirconium oxychloride is a cream-colored powder soluble in water that is used as a catalyst, in the manufacture of color lakes and in textile coatings. Zirconium-fused salt, used to refine aluminum and magnesium, is zirconium tetrachloride, a hygroscopic solid with 86% $ZrCl_4$. Zirconium sulfate comes in fine, white, water-soluble crystals. It is used in high-temperature lubricants, as a protein precipitant, and for tanning to produce white leathers. Soluble zirconium is sodium zirconium sulfate, used for the precipitation of proteins, as a stabilizer for pigments and as an opacifier in paper. Zirconium carbonate is used in ointments for poison ivy, as the zirconium combines with the hydroxy groups of the urushiol poison and neutralizes it. Zirconium hydride has been used as a neutron moderator, although the energy moderation may be chiefly from the hydrogen.

Zincrometal

A steel coil-coated product consisting of a mixed-oxide underlayer containing zinc particles and a zinc-rich organic (epoxy) topcoat. It is weldable, formable, paintable, and compatible with commonly used adhesives. Zincrometal is used to protect outer body door panels in automobiles from corrosion.

Zone

Any group of crystal planes that are all parallel to one line, which is called the zone axis.

Zone(d) Furnace

A furnace in which the charge progresses through a number of zones where the heating rate and/or intended temperature vary. For example, the first zone might be intended to slowly preheat, the second to quickly rise to the peak temperature, and the third to allow initial cooling at some controlled rate. The zones are not normally separated by any physical barrier but have individual heating input and control systems.

Zone Leveling

A technique for homogenizing a material. It is similar to zone refining except that the heated zone sweeps back and forth repeatedly. This has the effect of redistributing impurities evenly along the length except at the extremities.

Zone Melting

(1) Highly localized melting, usually by induction heating, of a small volume of an otherwise solid metal piece, usually a metal rod. By moving the induction coil along the rod, the melted zone can be transferred from one end to the other. In a binary mixture where there is a large difference in composition on the liquidus and solidus lines, high purity can be attained by concentrating one of the constituents in the liquid as it moves along the rod. (2) Melting a narrow zone and zone leveling and zone refining.

Zone Refining

A technique for refining metals that are already nearly pure. It utilizes a local heat source, which slowly and repeatedly sweeps a narrow molten zone along a bar of the metal. In the appropriate cases where the impurity element depresses the melting point of the solvent, the molten zone collects some of the impurity and deposits it at the bar ends.

Zone Sintering

Highly localized, progressive heating during sintering to produce a desired grain structure, such as grain orientation, and directional properties without subsequent working.